INORGANIC CHEMISTRY

INORGANIC CHEMISTRY

Gary Wulfsberg
MIDDLE TENNESSEE STATE UNIVERSITY

University Science Books
Sausalito, California

About the cover art:
An endohedral metallofullerene, $M@C_{60}$ (page 628).
Computations suggest that the metal sits off the center of the C_{60} icosahedron.

University Science Books
55D Gate Five Road
Sausalito, CA 94965
Fax: (415) 332-5393
www.uscibooks.com

Production manager: *Susanna Tadlock*
Manuscript editor: *Jeannette Stiefel*
Designer: *Robert Ishi*
Illustrations: *LineWorks*
Compositor: *Asco Typesetters*
Printer & Binder: *Maple-Vail Book Manufacturing Group*

This book is printed on acid-free paper.

Library of Congress Cataloging-in-Publication Data

Wulfsberg, Gary, 1944–
Inorganic chemistry / Gary Wulfsberg.
p. cm.
Includes bibliographical references and index.
ISBN 1-891389-01-7 (alk. Paper)
1. Chemistry, Inorganic. I. Title.

QD151.5.W84 2000
546—dc21 98-053513

Printed in the United States of America
10 9 8 7 6 5 4 3 2 1

To my family members who have given me
so much personal support—Marlys, Joanna, Paul—
and in memory of my mother, Blanche, and two
scientific mentors, Professor Alarich Weiss
and Professor Daiyu Nakamura

Contents†

List of Tables xii
Preface xv

PART I INORGANIC IONS AND SIMPLE MOLECULES IN CHEMISTRY AND IN OUR ENVIRONMENT 1

CHAPTER 1 Periodic Trends in Fundamental Properties of Atoms and Simple Ions 3

1.1 Why Study Inorganic Chemistry? 3
1.2 The Periodic Table 5
1.3 Quantum Mechanics: Waves/Particles 7
1.4 Electrons in Atoms: Angular Part of the Electronic Wave Function 10
1.5 Radial Part of the Wave Function 15
1.6 Shielding and Slater's Rules 18
1.7 Characteristic Valence Electron Configurations 21
1.8 Valence Electron Configurations, Charges, and Oxidation Numbers of Ions 27
1.9 Types of Atomic Radii, Their Periodic Trends, and Reasons for the Trends 30
1.10 Periodic Trends in Electronegativities of Atoms and Their Explanation 36
1.11 *Relativistic Effects on Orbitals 39
1.12 *Predicting the Chemistry of Superheavy Elements 41

CHAPTER 2 Monoatomic Ions and Their Acid–Base Reactivity 55

2.1 Hydration of Cations 55
2.2 Acidity of Cations 56
2.3 Classification of Cations by Acidity Category 61
2.4 Halides that Fail to Undergo Hydrolysis: Oxo Cations 67
2.5 Hydration and Hydrolysis of Monoatomic Nonmetal Anions 68
2.6 Predominance Diagrams for Anions: Partially Protonated Anions 73
2.7 Acidic Nonmetal Hydrides and Basic Metal Oxides 74
2.8 Nomenclature 79

CHAPTER 3 Polyatomic Ions and Their Acid–Base Properties WITH APPLICATIONS TO ENVIRONMENTAL CHEMISTRY AND ANALYTICAL CHEMISTRY 89

3.1 Covalent Bond Types and Periodicity of Bond Energies 89
3.2 Covalent Bonding to Ions: The Lewis Acid–Base Concept and Complex Ions 98
3.3 Lewis Structures, Coordination Numbers, and Fluoro Anions 101
3.4 The Shapes of *p*-Block Molecules and Complex Ions 105
3.5 Formulas of Oxo Anions 111
3.6 Basicity of Oxo Anions 116
3.7 Protonation of Oxo Anions: Hydroxo Anions and Oxo Acids 120
3.8 Most Common Forms of the Elements in Natural Waters 125
3.9 *Nomenclature of Fluoro and Oxo Anions and Acids 129
3.10 *Lewis Structures and Oxidation Numbers 132

† Starred sections can readily be omitted at instructor's discretion.

CHAPTER 4 Ionic Solids and Precipitation Reactions of Hydrated Ions WITH APPLICATIONS TO ANALYTICAL CHEMISTRY, ORGANIC CHEMISTRY, AND WATER CHEMISTRY 149

4.1 Solubility Rules for Salts of Oxo and Fluoro Anions 149
4.2 Thermodynamics and Solubility Rules for Salts 154
4.3 Entropy and Precipitation: The Structure of Hydrated Ions 157
4.4 Nonacidic Cations, Nonbasic Anions, and the Structure of Liquid Water 161
4.5 Enthalpy and Precipitation: The Importance of Lattice Energies 162
4.6 Ionic Solids: Coulombic Attractions and Lattice Energies 164
4.7 Radius Ratios and Lattice Types 167
4.8 Stability of Lattices and the Solubility Rules 171
4.9 *Writing and Interpreting Net Ionic Equations 173
4.10 *Nonaqueous Solvents for Ionic or Ionizable Compounds 175

CHAPTER 5 Trends in Coordination Equilibria WITH APPLICATIONS TO BIOCHEMISTRY, ENVIRONMENTAL CHEMISTRY, GEOCHEMISTRY, AND MEDICINAL CHEMISTRY 191

5.1 Classification of Ligands 191
5.2 Chelate and Macrocyclic Effects 197
5.3 The Hard and Soft Acid–Base Principle 198
5.4 Relative and Borderline Hardness and Softness 201
5.5 The HSAB Principle and Polar Covalent Bonding 204
5.6 The Need for Two Parameters: Softness and Strength 206
5.7 Applications of HSAB: Solubility of Halides and Chalcogenides 207
5.8 Applications of HSAB: The Qualitative Analysis Scheme for Metal Ions 209
5.9 Applications of HSAB: The Geochemical Classification and Differentiation of the Elements 212
5.10 *Metal Ions in Biochemistry 215
5.11 Applications of HSAB: Biological Functions and Toxicology of the Elements 220
5.12 *Applications of HSAB and Chelation: Medicinal Chemistry 225

CHAPTER 6 Principles of Oxidation–Reduction Reactivity WITH APPLICATIONS TO CHEMICAL SAFETY, ENVIRONMENTAL CHEMISTRY, AND INDUSTRIAL CHEMISTRY 243

6.1 Standard Reduction Potentials and Their Diagrammatic Representation 243
6.2 Explosives, Flammable Compounds, and the Handling of Strong Reducing Agents 254
6.3 *Redox Predominance Diagrams: Construction; Species that Disproportionate; Nonstandard Conditions 261
6.4 Charge-Transfer Spectra and Marginal Redox Stability 264
6.5 Periodicity in the Activity of Metals and the Reduction of Metal Ions 266
6.6 *Hard and Soft Acids and Bases and Redox Chemistry 270
6.7 Redox Chemistry of the Nonmetals 273
6.8 Industrial Processes for Extracting Elements from Their Ores 276
6.9 Effects of Concentration and pH on Redox Chemistry: Pourbaix Diagrams 287
6.10 *pH Effects on Redox Chemistry: The Synthesis of Oxo Anions and Acids and Corrosion of Metals 297

CHAPTER 7 Thermochemical Analyses of Reactivity Trends 315

7.1 Thermochemical Analysis of Redox Reactions: Born–Haber Cycle 315
7.2 Thermochemical Analysis of the Activity Series of the Elements 323
7.3 Significance of Atomization Energies in Redox Chemistry 325
7.4 Periodicity and Significance of Ionization Energies and Electron Affinities 327
7.5 Polar Covalent Bond Energies and Pauling Electronegativities 331
7.6 Effects of Changing Substituents and Oxidation Numbers on Electronegativities 337

7.7 *Definitions of Electronegativity and Hardness Related to Ionization Energies and Electron Affinities 339
7.8 *Thermodynamics of the Lewis Acid–Lewis Base Interaction: Drago's *E*, *C*, and *T* Parameters 341

CHAPTER 8 Introduction to Transition Metal Complexes WITH APPLICATIONS TO BIOCHEMISTRY 357

8.1 Introduction 357
8.2 The Crystal Field Theory 359
8.3 High- and Low-Spin Electron Configurations: Magnetic Properties 362
8.4 Electronic Absorption Spectra of Complex Ions 365
8.5 The Spectrochemical Series of Ligands and the Effects of Covalency 369
8.6 Thermodynamic and Structural Consequences of Crystal Field Effects 374
8.7 Crystal Field Splitting in Complexes of Other Geometries 380
8.8 Geometric Preferences: Rates and Mechanisms of Ligand Exchange 387
8.9 *Syntheses of *d*-Block Complexes 391
8.10 *Bioinorganic Chemistry and Crystal Field Theory: Heme and Hemoglobin 395
8.11 *Nomenclature of Coordination and Organometallic Compounds 398

PART II INORGANIC MOLECULES AND MATERIALS: THEORY AND APPLICATIONS 417

CHAPTER 9 Symmetry WITH APPLICATIONS TO ART, ENVIRONMENTAL CHEMISTRY AND ORGANIC CHEMISTRY 419

9.1 Symmetry Operations and Elements 419
9.2 Molecular Point Groups 425
9.3 Elementary Applications of Symmetry: Molecular Polarity, Chirality, and Fluxionality 430
9.4 Character Tables for Symmetry Point Groups 438
9.5 *Applications of Symmetry: Infrared and Raman Spectra 443

CHAPTER 10 Molecular Orbital Theory WITH APPLICATIONS TO ORGANIC CHEMISTRY AND MATERIALS SCIENCE 459

10.1 Molecular Orbital Theory for Homonuclear Diatomic Molecules 459
10.2 Bond Energies and Lengths, Spectroscopy, and Orbital Mixing in Homonuclear Diatomic Molecules 468
10.3 Heteroatomic Diatomic Molecules and Ions 476
10.4 Molecular Ionization Energies, Electron Affinities, Electronegativities, and Hardness 483
10.5 Molecular Orbital Theory for Linear Molecules 485
10.6 Molecular Orbital Theory for Cyclic Molecules: Generator Orbitals 492
10.7 *Orbital Symmetry Rules for Concerted Reactions 498
10.8 Molecular Orbitals for D_{nh} Molecules 501
10.9 Molecular Orbitals for Three-Dimensional Molecules in Higher Order Point Groups 505
10.10 *Molecular Orbitals for Cluster Compounds 510
10.11 *Molecular Orbitals for Metals 513

CHAPTER 11 Organometallic Chemistry of the *d*-Block Elements WITH APPLICATIONS TO INDUSTRIAL CHEMISTRY, ORGANIC CHEMISTRY, BIOCHEMISTRY AND ENVIRONMENTAL CHEMISTRY 531

11.1 Organometallic Compounds of the *d*-Block Metals: The 18-Electron Rule 531
11.2 Mononuclear Metal Carbonyls and Related Organometallics 534

11.3 Metal Complexes of π-Donor Ligands 539
11.4 Other Ligands Forming One σ Bond in Organometallics: Dinuclear and Polynuclear Organometallics 548
11.5 Bridging Ligands: Alkynes and Carbonyls 556
11.6 Ligands Forming Two or More Bonds to a Metal or Metals: The Isolobal Analogy 559
11.7 Reaction Types of *d*-Block Organometallic Compounds 563
11.8 Catalytic Cycles 573
11.9 *Organometallic Catalysis in Biochemistry and Environmental Chemistry 581

CHAPTER 12 The Elements and Their Physical Properties WITH APPLICATIONS TO MATERIALS SCIENCE 595

12.1 Interunit Forces and Physical Properties 595
12.2 Physical Properties, Allotropes, and Uses of the Nonmetallic Elements 599
12.3 Metals and Alloys 611
12.4 Magnetism, Electrical Conductivity, and Semiconduction 617
12.5 Homopolyatomic Anions and Cations of the Elements: Metal Complexes 624
12.6 Heterogeneous Catalysis by Metals 631
12.7 Metallic Chains, Sheets, and Clusters 635

CHAPTER 13 Oxides of the Elements WITH APPLICATIONS TO GEOCHEMISTRY, ENVIRONMENTAL CHEMISTRY, AND MATERIALS SCIENCE 657

13.1 Physical States and Structures of the Fluorides and Oxides of the Elements 657
13.2 Covalent Oxides: Periodic Trends in Structure and Physical State 667
13.3 Acidity, Solubility, Practical Uses, and Environmental Chemistry of Volatile Oxides and Oxo Acids 673
13.4 Close Packed Anions, Metal Oxides, and Electrical Conductivity of Solid Ionic Compounds 685
13.5 Spinels, Perovskites, High-Temperature Superconductors, and Cooperative Magnetic Properties in Mixed-Metal Oxides 691
13.6 Polysilicates: Basic Structural Types, Uses, and Chemistry 697
13.7 Aluminosilicates 705
13.8 *Oligomeric Polyoxometallate Anions of Other *p*- and *d*-Block Elements: Selected Structures and Uses 711

CHAPTER 14 The Halides, Nitrides, and Sulfides of the Elements WITH APPLICATIONS TO MATERIALS SCIENCE AND BIOCHEMISTRY 729

14.1 Structural Features of Halides 729
14.2 Physical States and Physical Properties of Halides 738
14.3 Methods of Synthesis of Halides 741
14.4 Reactions and Uses of Halides 744
14.5 *Catenated and Cluster Halides 748
14.6 More Complex Oxygen Derivatives: Oxo Halides, Silicones, Inorganic Esters, Alkoxides 754
14.7 Nitrides and Carbides of the Elements 761
14.8 Sulfides, Selenides, Tellurides, and Arsenides of the Elements 767
14.9 Relationships among Ionic, Covalent, and Metallic Bonding 775

CHAPTER 15 Hydrides, Alkyls, and Aryls of the Elements WITH APPLICATIONS TO MATERIALS SCIENCE AND ORGANIC CHEMISTRY 789

15.1 Classification of Hydrides: Hydrides of Metals 789
15.2 Electron-Precise and Electron-Rich Molecular Hydrides 792
15.3 Electron-Deficient Molecular Hydrides 799
15.4 Syntheses of the Hydrides and Hydride Anions of the Elements 804

15.5 Reactions and Uses of the Hydrides 807
15.6 Alkyls and Aryls of the Elements: Classification and Structures 810
15.7 Methods of Synthesis of Alkyls and Aryls 816
15.8 Reactions and Uses of the Alkyls and Aryls 820
15.9 *Catenated Alkyls and Aryls of the *p*-Block 823
15.10 *Multiply Bonded and Divalent Alkyls and Aryls of the Heavier *p*-Block Elements 828

CHAPTER 16 Inorganic Reaction Mechanisms WITH APPLICATIONS TO MATERIALS SCIENCE, ORGANIC CHEMISTRY, ENVIRONMENTAL CHEMISTRY, AND BIOCHEMISTRY 839

16.1 Rate Laws and Stoichiometric Reaction Mechanisms for Ligand Substitution Reactions 840
16.2 Intimate Mechanisms for Ligand Substitution Reactions 846
16.3 Entering Group, Leaving Group, and Metal Effects on Ligand Substitution Rates 851
16.4 Spectator Ligand Effects on Ligand Substitution Rates 853
16.5 Classification of Redox Reaction Mechanisms 857
16.6 Outer-Sphere Electron-Transfer Processes 858
16.7 One-Electron Inner-Sphere and Long-Range Electron-Transfer Processes 861
16.8 Two-Electron Inner-Sphere Electron-Transfer Processes 868

CHAPTER 17 Advanced Topics: Excited Electronic States, Photochemistry, and Activated Molecules WITH APPLICATIONS TO MATERIALS SCIENCE, ATMOSPHERIC CHEMISTRY, AND BIOCHEMISTRY 879

17.1 Electronic States and Term Symbols 879
17.2 Electronic States for Diatomic Molecules and Their Consequences 885
17.3 Electronic States for Octahedral Complexes 887
17.4 Tanabe–Sugano Diagrams and Spectra of *d*-Block Complexes 889
17.5 Luminescence, Lasers, and Spin–Orbit Coupling 898
17.6 Photochemically Assisted Substitution and Redox Processes 901
17.7 Chlorophyll and Photosynthesis 905
17.8 Biochemical Oxygen Transport and Activation 906
17.9 Biochemical Nitrogen Activation 913

APPENDIX A: Discovery Laboratory Experiments for Part I 927

A.1 Some Reactions of Cations 928
A.2 Nonaqueous Reactions of Metal Ions and Compounds 929
A.3 Some Reactions of Oxo Anions 931
A.4 Reaction of Anions with Cations 932
A.5 Competitive Precipitation and Complexation Reactions 933
A.6 Periodicity in the Activity (Electromotive) Series of Metals 935
A.7 The Widely Varying Colors of *d*-Block Metal Complexes 938

APPENDIX B: Character Tables 940

Answers to Selected Exercises 949

Index 969

List of Tables

Table A Pauling Electronegativities of the Elements *front endleaf*
B1 Most Common Positive Oxidation Numbers for Each Element *front endleaf*
B2 Highest Positive Oxidation Numbers for Each Element *front endleaf*
C Shannon–Prewitt Crystal Ionic Radii *back endleaf*
D Table of Atomic Weights *back endleaf*
1.1 The Long Form of the Periodic Table and Characteristic Valence Electron Configurations 7
1.2 Orbitals for Hydrogen-Like Atoms or Ions 12
1.3 Comparison of Effective Nuclear Charges Calculated by Slater's Rules and from SCF Quantum Mechanical Calculations 23
1.4 Different Types of Atomic Radii for Selected Atoms 31
1.5 Covalent- and Single-Bonded Metallic Radii of the Elements 32
1.6 Calculated Atomic Radii for Selected Atoms 35
1.7 Allred–Rochow Electronegativities of the Elements 37
1.8 Principal Isotopes of Transuranium Elements 42
2.1 Hydration Enthalpies of Metal Cations 57
2.2 Hydrolysis Constants for Metal Cations 59
2.3 Relationship between pK_a, Acidity, and Z^2/r Ratios of Metal Ions 66
2.4 Hydration Enthalpies of Some Anions 69
2.5 Aqueous Basicity of Monoatomic Anions and their Partially Protonated Forms 70
3.1 Element–Element Covalent Bond Dissociation Energies 93
3.2 Bond Angle Distortions Due to Unshared Electron Pairs in Molecules with a Total Coordination Number of 4 109
3.3 Names and Formulas of the Important Oxo Anions 113
3.4 Suggested Classification of Oxo Anions 117
3.5 Main Forms of the Elements in Moderately Aerated Water of pH 5.5–7 127
3.6 Relationship between Names of Oxo Anions and Oxidation Number of the Central Atom 131
4.1 Thermodynamic Data on Precipitation 156
4.2 Hydration Numbers and Hydrated Radii of Some Hydrated Ions 159
4.3 Born Exponents and Electron Configurations of Ions 165
4.4 Lattice Types and Madelung Constants for Different Stoichiometries and Radius Ratios of Cations and Anions 169
4.5 Effect of Coordination Number on the Radius of the Na^+ Ion 170
4.6 Thermochemical Radii of Polyatomic Ions 172
5.1 Hard and Soft Acids and Bases 199
5.2 Characteristic Properties of Hard and Soft Acids and Bases 200
5.3 The Complete Qualitative Analysis Scheme of the Cations 210
5.4 Geochemical Classification of Elements 212
5.5 Major Mineral Sources of Selected Elements 214
5.6 Functions and Preferred Ligand Binding Groups for Essential Metal Ions 216
5.7 The pK_a Values of Free and Coordinated Ligands 216
5.8 Occurrence and Significance of Trace Elements in Natural Waters and Wastewaters 222
6.1 Formulas and Colors of the Highest Halides of Some Elements 266

6.2 Standard Reduction Potentials of Hydrated Metal Ions 267
6.3 Methods of Extraction of the Elements; Redox Predominance Diagrams of the Reducing Agents Used 279
6.4 Estimated Crustal Abundances of the Elements 285
6.5 Costs of the Elements 286
6.6 Inorganic Chemicals among the Top 50 in U.S. Production in 1995 287
6.7 Standard Reduction Potentials of Some Useful Oxidizing Agents in Acidic and Basic Solution 298
7.1 Heats of Atomization of the Elements 317
7.2 First Ionization Energies of the Elements 318
7.3 Second Ionization Energies of the Elements 319
7.4 Third Ionization Energies of the Elements 320
7.5 Electron Affinities of the Elements 321
7.6 Element–Fluorine Bond Dissociation Energies 332
7.7 Element–Chlorine Bond Dissociation Energies 332
7.8 Element–Bromine Bond Dissociation Energies 333
7.9 Element–Iodine Bond Dissociation Energies 333
7.10 Effects of Substituents on the Pauling Electronegativity of an Atom 339
7.11 Drago Parameters for Lewis Acids and Bases 342
7.12 Parameters for Enthalpies of Dissociation for Catimers and Animers 344
8.1 Colors of Hydrated Metal Ions Having Partially Filled Sets of *d* or *f* Orbitals 358
8.2 Pairing Energies (*P*) for Some Gaseous 3*d* Metal Ions 362
8.3 Calculated and Typical Observed Magnetic Moments 364
8.4 Crystal Field Splittings of Hydrated Metal Ions and Metal-Ion *g* Factors 367
8.5 Spectrochemical Series of Ligands and Their *f* Factors 371
8.6 Energy Levels of *d* Orbitals in Complexes of Various Geometries 383
8.7 Radii of Selected *d*-Block Metal Ions as a Function of Geometry 385
9.1 Elements and Operations of Molecular Symmetry 421
9.2 Approximate Minimum Rates of Exchange Necessary to Affect Common Structural Techniques 435
9.3 Selected Character Tables 439
10.1 Valence Orbital Potential Energies 463
10.2 Bond Energies and Lengths of Some Homonuclear Diatomic Molecules and Ions 470
11.1 Typical Mononuclear Carbonyls of the *d*-Block Elements 535
11.2 Neutral Two-Electron Donor Ligands Found in Organometallics 537
11.3 Numbers of C≡O Stretching Bands for Substituted Carbonyls 537
11.4 Electron Counting for Polyene and Cyclopolyene Ligands 539
11.5 Simple Metallocenes and Metallicinium Ions 543
11.6 Electron Counting for Other Classes of Organometallic Ligands 549
11.7 Dinuclear and Polynuclear Carbonyls 554
11.8 Some Common Examples of Insertion and Elimination Reactions 569
11.9 Characteristic Changes in Metal Atom Counts During Major Classes of *d*-Block Organometallic Reactions 572
11.10 Some Homogeneous Catalytic Processes 573
11.11 Leading Organometallic Compounds and Metal Catalysts 574
12.1 Horizontal Periodic Trends in Numbers of Unpaired Electrons Available for Bonding 599
12.2 Crystal Structures of the Metals 613
12.3 Elements Exhibiting Ferromagnetism and Low-Temperature Superconductivity 618
12.4 Structures of Some Metal Borides 630
12.5 Numbers of Different Common Reactant Gases Chemisorbed by Different Metals 633

12.6 Electron Counts for Closed Deltahedral Clusters 640
12.7 Selected Other *d*-Block Cluster Geometries Predicted by MO Calculations 642
13.1 Melting Points of the Highest Fluoride of Each Element 663
13.2 Predictions of Physical Properties and Structures of Some Halides and Oxides 666
13.3 Higher Oxides of the Elements 668
13.4 Major Acidic Oxides of the *p*- and *d*-Block Elements 674
13.5 Soluble and Insoluble Acidic and Basic Oxides 676
13.6 Sets of Common Ions Suitable for Isomorphous Substitution 692
14.1 Structures of Hepta-, Hexa-, and Pentahalides of the Elements 732
14.2 Tetra- and Trihalides of the *p*-Block Elements 734
15.1 Hydrides of the Elements 790
15.2 Standard Enthalpies of Formation and Stability to Oxidation of Some Hydrides 793
15.3 Covalent- and Hydrogen-Bond Energies 794
15.4 Noncatenated Methyl Derivatives of the Elements 811
16.1 Methods for Studing Kinetics of Reactions, with their Time Scales in Seconds 839
16.2 Limiting Forms of Rate Laws Depending on Magnitude of Denominator Term in [Y] 844
16.3 Rate Constants for Acid Aquation of Some Octahedral Complexes of Co(III) at 25 °C 846
16.4 Rate Constants for Substitution Reactions of $[Co(NH_3)_5(H_2O)]^{3+}$ with Y^{n-} at 45 °C 846
16.5 Rates of Substitution of $[Ti(H_2O)_6]^{3+}$ by Y^{n-} at 13 °C 847
16.6 Main Intimate Mechanisms for Substitutions of Octahedral Complexes 848
16.7 Rate Constants for Ligand Displacement in Some Square Planar Pt(II) Complexes 851
16.8 Effect of Nonleaving Ligands on Acid Hydrolysis Rates of Some Co(III) Complexes at 25 °C 854
16.9 Self-Exchange Outer-Sphere Redox Rate Constants and Standard Reduction Potentials 859
16.10 Second-Order Rate Constants for Selected Inner-Sphere Reactions with Variable Bridging Ligands 861
16.11 Rate Parameters for Some Reductions by V^{2+} at 25 °C 863
16.12 Robin and Day Classification of Mixed-Valence Compounds 867
17.1 Letters Used in Atomic Term Symbols 882
17.2 Multiple Terms of Various Electron Configurations; The Splitting of Atomic Electronic States in Complexes of O_h Symmetry 883
17.3 Arrangement of Microstates for p^2 Electron Configuration According to Their Total Spins and Angular Momenta 884
17.4 Racah *B* and *C* Parameters for Free Gaseous *d*-Block Metal Ions 895
17.5 Ligand *h* and Metal *k* Nephelauxetic Parameters for Selected Ligands 897
17.6 Variations of O–O Stretching Frequencies and Bond Distances with Bond Order 908
17.7 Major Types of Reactions Catalyzed by Cytochrome P-450 911
Character Tables 940

Preface

This textbook is intended for the undergraduate or introductory graduate inorganic chemistry course or sequence of courses, such as the one-semester junior–senior level advanced inorganic chemistry course. It is intended to fill the gap between advanced inorganic textbooks that are encyclopedias of chemical reactions and those that are mainly textbooks of theoretical or physical inorganic chemistry. It covers virtually all topics suggested for the core inorganic requirement for a professional chemistry degree by the American Chemical Society Division of Inorganic Chemistry ad hoc Subcommittee.[1]

The text is divided into two parts. Part I, Inorganic Ions and Simple Molecules in Chemistry and in our Environment, corresponds approximately to the first third of most advanced inorganic texts, but differs in that the chemical concepts and reactions of Part I can be explained using valence bond theory, so that it is accessible to students who have had only general and organic chemistry courses. Part II, Inorganic Molecules and Materials: Theory and Applications, includes the remainder of the topics that are standard for an advanced inorganic course: symmetry, molecular orbital theory, materials chemistry, organometallic chemistry, bioinorganic chemistry, and inorganic reaction mechanisms. This part of the text is at a more advanced level and is appropriate for students who have had physical chemistry.

Use of The Text in One-Semester Advanced Inorganic Chemistry Courses

We anticipate that most junior–senior level advanced inorganic students will be able to go quickly and selectively through Part I. These students will spend most of their time in Part II, which is discussed first.[2] Part II begins with chapters introducing the theoretical tools (symmetry, group theory, and molecular orbital theory) needed for advanced study. Chapter 10 on molecular orbital theory is designed to help students visualize molecular orbitals in important classes of molecules and materials, including cluster compounds and even bulk metals. This chapter is then immediately followed by chapters in which molecular orbital theory can be applied and reinforced extensively: Chapter 11 covers transition metal organometallic chemistry; Chapter 12 covers the elements and their clusters.

Chapter 12 begins a sequence of chapters (Chapter 12 is on the elements; Chapter 13 is on oxides; Chapter 14 is on halides, nitrides, and sulfides; Chapter 15 is on hydrides) that deal with the important area of inorganic materials. These chapters are introduced by key sections (Sections 12.1, 12.2, 13.1, and 14.1) that help a student understand why (and even to predict whether) some compounds (e.g., SiO_2, TiO_2, and ReO_3) have extended structures that make them of potential interest as materials or in geochemistry, while others (e.g., SO_2 and OsO_4) have molecular structures and are therefore important in other areas of chemistry (e.g., environmental chemistry). In each chapter, different sections compare and contrast physical and chemical properties and uses of the molecular compounds with those of the macromolecular materials, giving students a good picture of the importance of overall structure on properties and uses.

In order to stimulate student interest early in the course, other interesting applications of the principles of inorganic chemistry are integrated throughout the textbook, rather than leaving them confined to the back of the book. The subheadings of the chapters suggest the locations of these applications: environmental, geological, and aquatic inorganic chemistry appear in Chapters 3–6, 13, and 16; bioinorganic and medicinal inorganic chemistry and chemical safety are found in Chapters 5, 6, 8, 11, 16, and 17; physical chemistry is found so often that it is not separately indicated. Mechanisms of reactions are likewise discussed to varying degrees throughout the text, most prominently in Chapters 8, 9, 11, 12, 16, and 17. It is likely that students have had enough background in mechanisms in their organic course, and in kinetics in their physical chemistry course, to follow the mechanisms in the earlier chapters, but if not, the main chapter on kinetics and mechanisms (Chapter 16) can easily be moved to an early position in Part II.

Part I

In this textbook, Part I contains topics that can be understood without using molecular orbital theory; a one-semester advanced inorganic course may include at least some of these topics, particularly those in Chapters 1 (periodicity of fundamental atomic properties) and 8 (transition metal complexes). Chapter 8 can be placed after Chapter 10 if the instructor wishes to make use of symmetry and molecular orbital theory in developing ligand field theory for transition metal complexes.

Chapters 2–7 include reviews and extensions of topics covered briefly in general chemistry and then developed further in the early parts of most advanced inorganic texts. The fact that these chapters are written to be understood without a prior mastery of physical chemistry or molecular orbital theory means that they can be covered relatively quickly. Students who have had physical chemistry but nonetheless need further study of certain topics likely can do this on their own outside of class time; we believe that the many applications in environmental and medical fields found in Part I will interest these students enough so that they will do this. Students have commented that Part I helps them actually understand procedures that were learned by rote in general chemistry courses; therefore, Part I is especially valuable for beginning graduate students who aspire to teach general chemistry.

Much of the inorganic chemistry needed by analytical chemists, biochemists, industrial chemists, and environmental chemists is *reaction chemistry of simple ions in aqueous solution*, which can be treated quite well without the use of advanced bonding concepts. Aqueous reaction chemistry and its periodicity have traditionally been given very inadequate treatment both in freshman and in advanced inorganic chemistry courses, and is often being omitted for lack of time. A major goal of Part I is to give this important area its due, while improving the presentation over the traditional deadly dull march through the groups of the periodic table. Instead, inorganic reactivity is organized by familiar broad categories: acid–base reactivity in Chapters 2 and 3, precipitation processes in Chapter 4, complexation in Chapter 5, and redox chemistry in Chapter 6. In each chapter, general principles are developed that allow the student to *classify* reactive ions (and simple molecules) and then *predict* and *explain* how reactive they will be in the type of reaction under discussion. This organization was widely acclaimed when first presented in the author's earlier *Principles of Descriptive Inorganic Chemistry* (University Science Books, 1991), to which Part I of this text is closely related. The student who can apply and extend the principles of aqueous reactivity from Part I of the text will also gain several new insights into the more advanced chemistry covered in Part II, such as the chemistry of inorganic materials.

A number of new references and pedagogical improvements have been incorporated in the other partially reorganized chapters of Part I; perhaps the most notable of these is the introduction of redox predominance diagrams in Chapter 6 as a visual representation of standard reduction potentials, which also helps the student build up to the use of the more complex $E°$/pH or Pourbaix diagrams for redox chemistry.

Other Uses of This Text

This book is structured with an ascending level of difficulty as it progresses from Part I to Part II. This organization should allow it to be especially useful in international settings, which have the tradition of teaching some inorganic chemistry each year of the undergraduate curriculum, and is ideal for those American schools having both a "sophomore" and a senior-level inorganic course.

Many teachers with introductory inorganic chemistry courses are finding them to be populated more and more by students from a variety of fields—geology, environmental science, industrial chemistry, materials and polymer science, metallurgy, chemical safety, medicinal chemistry, molecular biology—who are finding that their discipline involves inorganic materials and reactivity. Part I emphasizes the *fundamentals* of inorganic chemistry that these students can use, for example, to anticipate and prevent new environmental problems, not just react to known ones. Such a course can develop a substantial enrollment as a valuable part of a chemistry minor. Optional sections at the ends of chapters (indicated by stars) may readily be omitted for these students.

If the inorganic chemists at a university want to tap into the market for an introductory inorganic chemistry course serving chemistry minors from allied sciences, we might suggest that this course, using Part I, also be available as an elective for majors; the applications of inorganic chemistry are also subjects that chemistry majors find fascinating and rewarding to study, and such study will prepare these students for the growing opportunities for interdisciplinary collaboration with the allied sciences. Having one text with a consistent approach for both courses would serve as a financial and intellectual incentive to students from each course to consider taking the other.

Another possible usage is at universities that have a rigorous freshman-level course for chemistry majors, which is intended to include significant coverage of descriptive inorganic chemistry. In such a course, Part I could be begun at the freshman level,[3] then the textbook can be kept for later use in the advanced inorganic course.

Other Aids to Learning

The text has an unusually large selection of challenging practice exercises (~40–70 per chapter); each chapter also has several worked-out examples and a list of concepts or study objectives. Exercises that are starred (*) have answers provided in the Answers section at the back of the book; the answers to the other exercises are provided in the

Solutions Manual, which is available from the publisher. Sections that are starred can be omitted at the instructor's discretion with little or no effect on subsequent chapters.

The most basic level of understanding of phenomena of inorganic reactivity is the observation of them. In Appendices A.1–A.7, we present several *optional* simple inorganic experiments appropriate for use in Part I in which students not only observe trends in inorganic reactivity, but are challenged to reason inductively from them to discover many of the main principles of the text. These technically simple experiments include asking the student to *design* parts of the procedure, in which they must understand how to control multiple variables—these experiments are practice in the scientific method of a sort not normally encountered until students do research.[4] In courses that do not have laboratories, these experiments can be done as *demonstrations* followed by a *discussion* among small groups of students to try to design the procedure and reason inductively to the principles that they discover. To make these more easily performed and more widely available, videos of two of these demonstration–discussions have been prepared.[5] Other useful videodisk and other computer simulations of inorganic reactivity and its trends are also becoming readily available.[6] If the course does have a laboratory, or for advanced inorganic laboratories, the instructor will want to add traditional synthesis and characterization experiments, such as those found in the laboratory manual by Girolami et al.[7]

Acknowledgments

I am much indebted to my editor, Jane Ellis, and to Jeannette Stiefel, Susanna Tadlock, and Bob Ishi, and to the following reviewers who provided useful and insightful comments on the manuscript: Robert Angelici, Roger DeKock, Art Ellis, Walther Ellis, Harry Gray, James Penner-Hahn, William Robinson, Alan Stolzenberg, William Trogler, David Tyler, and David Westmoreland.

I also wish to thank William Ilsley for class testing and error checking the manuscript, thus providing an invaluable supplement to my own class testing of almost all of the chapters of this manuscript. Users of this text are encouraged to contact me by e-mail (wulfsberg@mtsu.edu) about errors they find or with suggestions; I wish to thank such helpful readers in advance. I also thank Pekka Pyykkö, John Verkade, Clark Landis, Preston MacDougall, Martin Stewart, and Judith Iriarte-Gross for helpful suggestions related to their areas of specialization. I am also indebted to Barry Farris, Michael Kearney, Paul H. Wulfsberg, Joanna Wulfsberg, Candace Stacey, Lisa Campbell and Jennifer Watkins for assistance in proofreading and clerical chores, and Middle Tennessee State University for providing released time to work on the research for the manuscript.

Notes

1. These are listed by J. G. Verkade, *J. Chem. Educ.*, **68**, 911 (1991), Table 2.
2. We have had many students begin their use of this text with Part II; this has caused them no difficulties.
3. In this case, some of the more theoretical parts of Chapter 1 should be postponed until the advanced inorganic course.

4. See also G. Wulfsberg, *J. Chem. Educ.,* **60**, 725 (1983). We do not recommend doing all of these experiments, but doing only enough for the students to develop the ability to reason inductively and design a sound scientific experiment themselves. We have found that our students, even the seniors, usually do not understand how to design an experiment in which *two* variables must be controlled.
5. Lyubov V. Hoffman, "Design and Evaluation of Effectiveness of Audio-Visual Discovery Job Experiments in Teaching Descriptive Inorganic Chemistry," M.S. Thesis, Middle Tennessee State University, 1996. Faculty who adopt this text may obtain copies of the video forms of Experiments A.1 and A.6 from the author by writing him on letterhead stationery (Box 405, Chemistry Department, Middle Tennessee State University, Murfreesboro, TN 37132) and sending him two blank videocassette tapes.
6. Many of the principles of Part I have been incorporated into the program PIREXs (Predicting Inorganic Reactivity Expert System) [J. P. Birk, *J. Chem. Educ.: Software,* 3B, No. 1, Disks 1-3 (1990); see also J. P. Birk, *ACS Symp. Ser. 408* (Expert System Applications in Chemistry), 20 (1988); J. P. Birk, *J. Chem. Educ.,* **69**, 294 (1992)]. Birk has also developed an adjustable computer version of the Pourbaix diagrams of Chapter 6, which allows the student to superimpose the diagrams of two potentially reacting species; this is also expected to be published in *J. Chem. Educ.: Software*.
7. G. Girolami, T. Rauchfuss, and R. Angelici, *Synthesis and Technique in Inorganic Chemistry*, 3rd ed., University Science Books, Sausalito, CA, 1999.

INORGANIC CHEMISTRY

PART I

INORGANIC IONS AND SIMPLE MOLECULES IN CHEMISTRY AND IN OUR ENVIRONMENT

The first one-half of this text emphasizes basic reactivity trends in simple inorganic species, particularly ions and some simple molecules. Such species are found everywhere, so their reactivities are important to chemists and other scientists in a large variety of fields. Since the traditional specialities within chemistry and even the natural sciences, including inorganic chemistry, are becoming more and more interrelated in practice, we anticipate that the type of students taking a first-semester inorganic chemistry course will become more and more diverse. Therefore, we have tried to present principles that have useful consequences in a variety of scientific subdisciplines (in the subtitle of Chapters 3–6 and 8 in this part, we have indicated what these are). Fortunately, most of the reactivity trends of ions of the elements can be understood and predicted without the use of sophisticated bonding approaches and calculations; for this half of the text, only general chemistry is a necessary prerequisite, although since a fair number of organic ligands are involved in Chapters 5–8, it would also be advisable for the student to begin his or her study of organic chemistry.

CHAPTER

1

Periodic Trends in Fundamental Properties of Atoms and Simple Ions

1.1 Why Study Inorganic Chemistry?

About 1 in 10 professional chemists is an inorganic chemist, but *all* chemists and many other scientists must work with inorganic compounds: in the laboratory, in the field, or in theory. This book is intended to provide the fundamental facts and principles of inorganic chemistry needed, not only by practicing inorganic chemists, but by chemists and other scientists who work with inorganic materials in the different fields of chemistry and in several allied sciences, and who may profit by collaborating with inorganic chemists in research or industrial development.

Among chemists, organic chemists have always relied on inorganic reagents to carry out syntheses; this trend is increasing as organic synthesis turns more and more to the use of specific transition metal catalysts and organometallic compounds (Chapter 11). Analytical chemists are often concerned with the detection and quantification of elements other than carbon, and often use chelating ligands of appropriate hardness or softness (Chapter 5) to concentrate and detect metallic elements. Physical and theoretical chemists are concerned with measuring or calculating the fundamental properties of inorganic and organic substances. Modern biochemists are becoming increasingly aware of the critical role played in living systems by metal ions (Chapter 5). In addition there are many times when any chemist must make up a solution of a new type of inorganic reagent, modify a synthesis, detect an element in a new form, or study the properties of a different type of inorganic compound. At this point, the chemist needs to have some ability to anticipate the properties of inorganic compounds he or she has not dealt with in the past. This book is designed principally to enable any chemist to develop an understanding of and an ability to anticipate the chemistry of the elements and their most common compounds.

Academic chemists are not the only ones who deal in a nonroutine way with the compounds of elements other than carbon. Most of the largest volume industrial chemicals are inorganic compounds. Manufacturing processes are discussed in several chapters: nonmetals and metals in Chapter 12, oxo acids and glass in Chapter 13, and inorganic polymers in Chapters 14 and 15. We need to know something about the processes for manufacturing these chemicals to understand their pollution aspects, for example, or their dependence on a raw material produced only in some politically unstable or hostile country. We need to know inorganic chemistry to be able to invent manufacturing processes that use other raw materials or that produce fewer pollutants (such as the air pollutants discussed in Chapter 12 and the water pollutants discussed in Chapter 5). Industrial organic chemistry has invested heavily in the study of organometallic transition metal catalysts (Chapter 11), which produce industrial organic chemicals in high yield and purity, thus eliminating unwanted and often toxic byproducts that often had been dumped into the environment. All chemists need to be concerned with the fire or explosion hazards presented by some inorganic compounds (Chapter 6).

Many scientists and engineers who do not even consider themselves to be chemists must also deal with inorganic chemicals in a safe and insightful way. Biologists may have to make up a solution of an inorganic reagent, and may need to anticipate whether the reagent will be toxic to their organism (Chapter 5). Certainly, they need to know whether the material will explode on contact with water, or will fail to dissolve when they try to make up the solution! Environmental and aquatic chemistry (Chapters 2–6), toxicology and medicinal chemistry (Chapter 5), industrial chemistry (Chapters 11–15), chemical safety (Chapter 6), geochemistry (Chapters 5 and 13), and materials science and solid state physics and chemistry (Chapters 4, 12, and 13) all deal with a wide variety of inorganic compounds. Scientists in these fields need ways to anticipate the properties of new inorganic compounds. Chemists who develop a good fundamental understanding of the facts and principles of inorganic chemistry not only make better chemists, but they also may find opportunities to contribute to fields allied with traditional chemistry.

In recent years, the trend for inorganic chemists to interact with scientists from these other fields has paid some spectacular dividends. The mid-1980s discovery of *superconductivity*, a property that persists to unusually high temperatures in an (to most scientists) unexpected class of materials known as the *metal oxides* (e.g., $YBa_2Cu_3O_7$, see Chapter 13) resulted both in a Nobel prize for the discoverers and in an increased study of inorganic materials. Since this discovery, the rate of these new vista-opening discoveries has grown to a point where each December the very eminent journal *Science* began choosing a "Molecule of the Year."

Of the first four Molecules (or, more broadly, chemical discoveries) of the Year, three have been inorganic materials. The first (1989) was a biochemical material, DNA polymerase; but in 1990 the Molecule of the Year was *diamond*, which, of course, was no new substance, but which was being produced by new methods and in new forms (doped as a semiconductor; in thin films, Chapters 13 and 15). The 1991 Molecule of the Year was another form of the element carbon, the famous molecule C_{60}, named "buckminsterfullerene" because its architecture and symmetry (Chapters 9, 10, and 12) resembled those of the strong "geodesic domes" previously proposed by the environmental architect Buckminster Fuller; the chain of events leading to the discovery of this molecule began in *outer space*, where astronomers were seeking to explain certain unassigned spectroscopic frequencies that they thought were originating with interstellar carbon-containing molecules. The 1992 Molecule of the Year was another old-time inorganic molecule, one of the 10 simplest in the universe, *nitric oxide* (NO). This molecule was discovered by a medical researcher, Solomon Snyder, to be an important chemical messenger in the body, functioning as a neurotransmitter alongside of more familiar biochemical substances such as acetylcholine, and being involved intimately in the chemistry of the brain and of human sexual performance (Chapter 8).

No one can predict what next year's discovery of the year will be; only a few things seem reasonably predictable about these future revolutionary discoveries. More and more, they are *interdisciplinary* discoveries, in which inorganic chemistry plays a crucial role but by no means the only role. These molecules may originate in the research laboratories of inorganic chemistry, in the environmental field, or in outer space; they may involve exotic new molecules of types no one has ever imagined, or they may involve familiar old inorganic molecules that may have disappeared from inorganic texts that strive to emphasize only the current "hot" areas of research. Although no one can predict these discoveries, it does seem likely that they will be made, and more quickly exploited, by scientists who have a broad understanding of all aspects of inorganic chemistry, from the "pedestrian" realms of aqueous reaction chemistry (important in medicine and environmental chemistry), to modern sol–gel chemistry and to "hot" current areas such as materials science and catalysis, where this discovery might evolve from your professor's own area of research.

Hopefully, this textbook, while providing a broad, firm groundwork of principles that will enable you to understand and be able to predict a broad range of inorganic reactions and phenomena, will at the same time be short enough for you and your professor to have time to explore: (1) what the pleasures of inorganic research really are like, using as examples what really excites your professor in his or her field of research; and (2) what discoveries have been made since this text went to press, by leaving time at the end of the year for you to go to the library to read the most recent literature and review articles.

1.2 The Periodic Table

Inorganic chemists no longer attempt to memorize the physical properties and chemical reactions of the 112 known elements and their countless compounds. Instead, they try to understand these properties so as to be able to anticipate them. For this to be accomplished both quickly and reliably, their understanding must proceed on several levels.

First, it is difficult to understand physical explanations of chemical phenomena that a person has never observed; actual observation of chemical reactivity contributes a distinct and crucial level of understanding. The appendixes contain simple laboratory experiments that your instructor may have you try (or demonstrate) and discuss before you begin many of the chapters; these experiments give you an opportunity to observe the type of chemical reactivity discussed in that chapter. These experiments were written in such a way that you may be able to discover periodic trends in reactivity before these trends have been discussed in a lecture or in this text. Rest assured that there are often many ways of interpreting periodic trends, and organizing and explaining them; if your discovery does not match the interpretation given in the text it might just mean that you are more creative than the author. (The test would be, of course, in how many other physical and chemical properties you could predict using your idea.) Although these experiments are as simple to perform as experiments in general chemistry courses, they often involve chemicals that are too reactive for use by freshmen, and certainly involve much more sophisticated reasoning than was required earlier. If your course has a laboratory component, your instructor will very likely add other experiments that have different purposes: the syntheses and/or characterization of specific inorganic compounds. (A useful collection of this type of experiment is that of Girolami et al.[1])

Second, many of the most important properties of the elements and their compounds can be adequately understood and often predicted by understanding how they are related to some atomic properties that are already familiar to you from general chemistry: *electronegativity*, *ionic charge*, *atomic radius*, and *electron configuration*. One objective of Chapter 1 is to review these properties and how they vary systematically throughout the periodic table, since this level of understanding of chemical periodicity is very useful to all types of chemists as well as scientists in allied fields, and will be used extensively in the first one-half of this book.

Third, we believe that all of the familiar chemical properties of atoms and molecules, as well as the trends in electronegativities, atomic radii, electron configurations, and even the form of the periodic table, should ultimately be explainable in terms of our quantum mechanical model of the atom. This objective will be developed for atoms in this chapter, and for molecules in the second half of this text. Such understanding can also proceed at two levels—a qualitative understanding of the results of quantum mechanical calculations and of their causes, which is what we attempt in this text—and the development of the ability to follow the mathematical derivation of the quantum mechanical model of the atom, which some of you will encounter in courses in physical chemistry and quantum mechanics.

The invention of the periodic table, however, preceded any knowledge of subatomic structure or quantum mechanics, and had as its objective the systematization of the known physical properties and chemical reactions of the elements. The periodic table was organized in a form such that, if the elements were listed in order of an increasing fundamental property (originally atomic weight, now atomic number; i.e., the number of protons in the atom), very similar chemical properties would recur periodically through the list. When the properties first recurred, the horizontal listing of the elements would be disrupted, and a new **period** of elements would be begun, with the chemically similar elements listed below each other in a **group**.

To do this with the known elements, it is necessary to set up seven horizontal periods of elements. As it turns out, however, these periods become considerably longer as atomic numbers increase, so that the periodic table becomes inconveniently wide (32 groups at the bottom); hence, one usually sees a shorter form of the periodic table, with the last groups to appear (those of the "lanthanides" and "actinides") separately listed at the bottom. We will usually resort to this shorter form, but in Table 1.1 we present a fully expanded version of the periodic table.[2] As we can see by its shape, the periodic table has four regions; some of these regions are named differently by different authors. Both the two groups at the far left and the six groups at the far right are often called the representative elements. To distinguish them, we will call the two groups at the far left the ***s*-block elements**. According to the latest recommendations of the International Union for Pure and Applied Chemistry (IUPAC), these groups should be numbered "1" and "2". The six groups at the far right we shall call the ***p*-block elements**. The IUPAC recommends that (for clarity in literature searching) these be numbered 13–18, but traditionally they have been numbered either IIIA–VIIIA or IIIB–VIIIB. We will attempt to compromise these three numbering traditions in the *p* block: that is, the boron group will be numbered 13(III).

The low, long block of elements next to the *s*-block elements has a variety of names: "rare earth elements," "inner-transition elements," or "lanthanides and actinides". For simplicity, we will call these the ***f*-block elements**. They are not usually numbered, but for figuring valence electron configurations it is useful to do so; for this purpose, we will number them from 3*F* to 16*F*. Although the third block of elements is universally recognized as the transition elements, for consistency we will call them the

Table 1.1
The Long Form of the Periodic Table and Characteristic Valence Electron Configurations[a]

Group	1	2	3F	4F	5F	6F	7F	8F	9F	10F	11F	12F	13F	14F	15F	16F	3	4	5	6	7	8	9	10	11	12	13 (III)	14 (IV)	15 (V)	16 (VI)	17 (VII)	18 (VIII)
Period																																
1	$1s^1$																															$1s^2$
	1																															2
	H																															He
2	$2s^n$																										$2s^22p^{n-2}$					
	3	4																									5	6	7	8	9	10
	Li	Be																									B	C	N	O	F	Ne
3	$3s^n$																										$3s^23p^{n-2}$					
	11	12																									13	14	15	16	17	18
	Na	Mg																									Al	Si	P	S	Cl	Ar
4	$4s^n$																$4s^23d^{n-2}$										$4s^24p^{n-2}$					
	19	20															21	22	23	24	25	26	27	28	29	30	31	32	33	34	35	36
	K	Ca															Sc	Ti	V	Cr	Mn	Fe	Co	Ni	Cu	Zn	Ga	Ge	As	Se	Br	Kr
5	$5s^n$																$5s^24d^{n-2}$										$5s^25p^{n-2}$					
	37	38															39	40	41	42	43	44	45	46	47	48	49	50	51	52	53	54
	Rb	Sr															Y	Zr	Nb	Mo	Tc	Ru	Rh	Pd	Ag	Cd	In	Sn	Sb	Te	I	Xe
6	$6s^n$		$6s^24f^{n-2}$														$6s^25d^{n-2}$										$6s^26p^{n-2}$					
	55	56	57	58	59	60	61	62	63	64	65	66	67	68	69	70	71	72	73	74	75	76	77	78	79	80	81	82	83	84	85	86
	Cs	Ba	La	Ce	Pr	Nd	Pm	Sm	Eu	Gd	Tb	Dy	Ho	Er	Tm	Yb	Lu	Hf	Ta	W	Re	Os	Ir	Pt	Au	Hg	Ti	Pb	Bi	Po	At	Rn
7	$7s^n$		$7s^25f^{n-2}$														$7s^26d^{n-2}$															
	87	88	89	90	91	92	93	94	95	96	97	98	99	100	101	102	103	104	105	106	107	108	109	110	111	112						
	Fr	Ra	Ac	Th	Pa	U	Np	Pu	Am	Cm	Bk	Cf	Es	Fm	Md	No	Lr	Rf	Db	Sg	Bh	Hs	Mt									

[a] To obtain the characteristic valence electron configuration of an element, use the formulas given, in which *n* refers to the numerical part of the group designation. Thus for Np, in Group 7*F*, the characteristic valence electron configuration $7s^25f^{n-2}$ is $7s^25f^5$.

***d*-block elements**, and shall number their groups from 3–12, as is recommended by the IUPAC.

Note the general shape of the periodic table: every *two* periods a new block of elements is introduced, just after the *s* block. Note also that each new block of elements is wider than the ones introduced before: the *s* block of elements is 2 elements wide, the *p* block is 6 elements wide, the *d* block is 10 elements wide, and the *f* block is 14 elements wide.

1.3 Quantum Mechanics: Waves/Particles

Since the late 1920s and early 1930s we have realized that the atomic basis for the organization of the periodic table lies in the electrons of the atoms and in the types of paths those electrons follow as they move around the nuclei of atoms (i.e., in the types of **orbitals** occupied by electrons). Chemical change occurs among the electrons of atoms; the theoretical study of chemical change focuses on the study of electrons in orbitals. Just before this period of time, Bohr had developed a model of the atom in which the electrons were particles that followed *orbits* around the nucleus, in close analogy to the orbits of planets, asteroids, and comets around the sun. However, by the known laws of physics, negatively charged particles should not have remained in orbit about the positively charged nucleus, but should have spiraled inward, leading to collapse of the atom. Bohr avoided this catastrophe by introducing the assumption

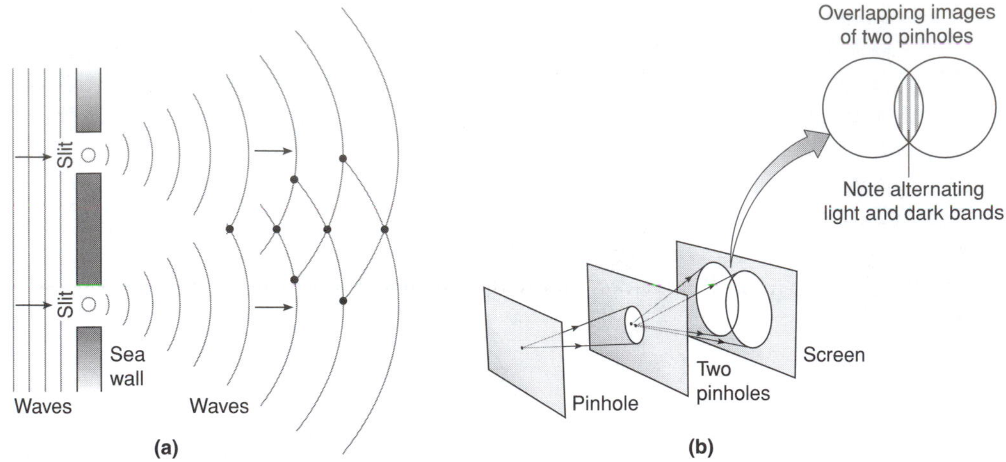

Figure 1.1
(*a*) Water waves (crests shown as lines, troughs between lines) passing through two gaps in a sea wall, being diffracted, and engaging in interference on the other side (constructive interference of crests with crests shown with heavy dots); (*b*) corresponding experiment of passing light through two pinholes, giving an interference pattern on the other side. [Adapted from W. J. Lehmann, *Atomic and Molecular Structure: the Development of Our Concepts*, Wiley, New York, 1972; pp. 178, 180.]

(modeled on earlier work of Max Planck) that only certain well-defined, "quantized" orbits were allowed for electrons, and not any intermediate ones, but he was unable to explain why this should be so.

The rationale for this assumption came with the development of **quantum mechanics**, the formulation of which began with the discarding by de Broglie of the earlier view that things in the universe were either waves or particles. De Broglie suggested that this is a false dichotomy: All things have properties of waves, such as characteristic **wavelengths** λ, in circumstances in which such properties can show up, and have properties of particles, such as **momentum** [mass m in kilograms (kg) times velocity v in meters per second ($\mathrm{m\,s^{-1}}$)], in circumstances in which these properties can be exhibited. The **de Broglie relationship** connects these two properties of a wave/particle:

$$mv = h/\lambda \tag{1.1}$$

The constant h, **Planck's constant**, has the value 6.626×10^{-34} J s.

A typical wave property is that of **diffraction**, in which a front of waves strikes a barrier or grating that has openings or slits spaced by distances comparable to the wavelengths of the waves: for example, as with waves of the ocean passing through a seawall [Fig. 1.1(*a*)]. As the waves pass through the openings, new wave fronts develop at each opening; these then spread out to strike other wave fronts, creating higher amplitude crests and troughs at some points via **constructive interference** of the wave fronts, and annihilating each other at other points in which the crest of one wave front coincides with the trough of another (**destructive interference**).

Electromagnetic radiation, which had always been thought of as waves, shows this property: For example, X-rays have very short wavelengths (3 pm and below), which are comparable in magnitude to the spacings between atoms and ions in crystals ($\sim$100 pm). The experiment of **X-ray diffraction** involves observing the diffraction pattern of X-rays passing through the gaps between the nuclei of atoms or ions in a crystal and creating characteristic diffraction patterns that can be analyzed to determine the arrangement and spacing of the atoms or ions in a crystal: its **crystal structure**. **Electron diffraction** can also be done using the wave property of electrons, if they are accelerated by a potential difference of about 20,000 V, so that they have a wavelength of about 10 pm. This method is used with gaseous molecules. We are thus justified in thinking of an electron as a wave located in a particular wave pattern around the nucleus, which has its crests and troughs (points of maximum amplitude) and its level or calm points (**nodes**, at which the wave has zero amplitude). But macroscopic objects of much larger mass (say 1 mg) have such short wavelengths that there are no physically real gratings with slits spaced closely enough to produce diffraction or other observable wave properties; we are justified in neglecting the wave properties of macroscopic objects. (However, physicists have suggested that objects as large as an amoeba could exhibit diffraction properties if they could be cooled sufficiently and caused to move at a uniform slow rate of a few centimeters per year.[3])

Another important wave property of particles, such as electrons that are confined within atoms or other regions of space, is that they cannot take any path whatsoever; their paths are described by equations called **wave functions**. Wave functions contain constants called **quantum numbers** that cannot have just any value, but must take values that are integers (or half-integers in some cases). Simpler than the case of an electron in a three-dimensional atom is the case of an electron confined to a one-dimensional path of length L by two barriers at either end (Fig. 1.2). The wave function for this electron is a sine function:

$$\psi = \sin(2\pi x)/\lambda \tag{1.2}$$

For a wave pattern of arbitrary wavelength λ, it is likely that the wave will reflect off of one barrier out of phase with itself, so that crests and troughs do not match up with each other after reflection; destructive interference then results. A stable standing wave pattern can only result if the wave has a wavelength that is integrally related to the distance L between the walls:

$$\lambda = 2L/n \qquad n = 1, 2, 3, \ldots \tag{1.3}$$

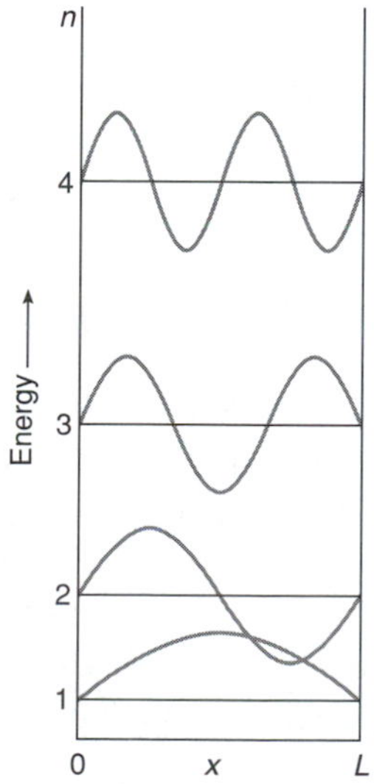

Figure 1.2
The first four wave functions of a particle in a box. [Adapted from D. F. Shriver, P. W. Atkins, and C. H. Langford, *Inorganic Chemistry*, Freeman, New York, 1990; p. 11.]

The number n is a quantum number; for each whole number that is assigned to it there is a corresponding standing wave that can persist inside the walls of this one-dimensional box. Combining these two equations, we obtain the wave functions for possible standing waves:

$$\psi = \sin n\pi x/L \qquad n = 1, 2, 3, \ldots \tag{1.4}$$

The standing waves drawn in Figure 1.2 may be characterized by the number of **nodes** present: For example, the standing wave drawn at the bottom of Figure 1.2 has no nodes (points of zero amplitude) within the box itself; each successive wave has one more node. These wave patterns are also related to those obtained by plucking a stringed musical instrument: Then, the zero-node wave is known as the fundamental, the one-node mode is called the first overtone, and so on.

The energy of the particle in this one-dimensional box can also be calculated, and depends strongly on the value chosen for n. The particle has only kinetic energy:

$$E = \tfrac{1}{2}mv^2 \tag{1.5}$$

Substitution in this equation of the de Broglie relationship (1.1) gives us:

$$E = \tfrac{1}{2}m(h^2/\lambda^2) \tag{1.6}$$

We now introduce the wavelength of the electron in the box, from Eq. (1.3):

$$E = \tfrac{1}{2}(h^2/m)(n/2L)^2 = n^2h^2/(8mL^2) \tag{1.7}$$

The kinetic energy of the electron in the box depends strongly on the value chosen for the quantum number n, as is also suggested by Figure 1.2. Note that its kinetic energy also increases as the number of nodes in the standing wave increases.

1.4 Electrons in Atoms: Angular Part of the Electronic Wave Function

An electron in an atom is in a more complex environment: It is confined in *three* dimensions by a potential derived from its attraction to the oppositely charged nucleus, and modified by the repulsions between it and any other electrons present. The equation that summarizes this principle is known as the **Schrödinger wave equation**:

$$\frac{-h^2}{8\pi^2 m}\left(\frac{\partial^2\psi}{\partial x^2} + \frac{\partial^2\psi}{\partial y^2} + \frac{\partial^2\psi}{\partial z^2}\right) + V\psi = E\psi \tag{1.8}$$

The first term of this partial differential equation is essentially the kinetic energy of the electron. The potential V in the second term includes the attractive potential of the nucleus for the electron, any electron–electron repulsions with other electrons, and (in molecules, to be considered in the second half of the book) the attractions to electrons in any other nuclei in a molecule. The parameter E then gives the total energy of the electron in the atom (or molecule). This equation is so formidable that it can be solved in an algebraic form only if just two particles are present: the nucleus and one electron.

In this case, the potential function V is just the Coulombic force of attraction between the nucleus and the electron:

$$V = (\text{charge of nucleus})(\text{charge of electron})/(\text{distance between}) = -Ze^2/r \quad (1.9)$$

In this equation, e is the charge of the electron, $-Ze$ is the charge of the nucleus, and r is the distance between them.

It is easier to solve this equation, and hence to describe the wave function ψ (path of the electron) in terms of spherical polar coordinates (r, θ, ϕ) than it is in terms of Cartesian coordinates (x, y, z). In spherical polar coordinates, r is the distance from the center of the atom (the nucleus), while θ and ϕ give the angles of the point in question away from the polar axis through the nucleus and a reference vertical plane, respectively (Fig. 1.3). The great advantage of this system of coordinates is that it is possible

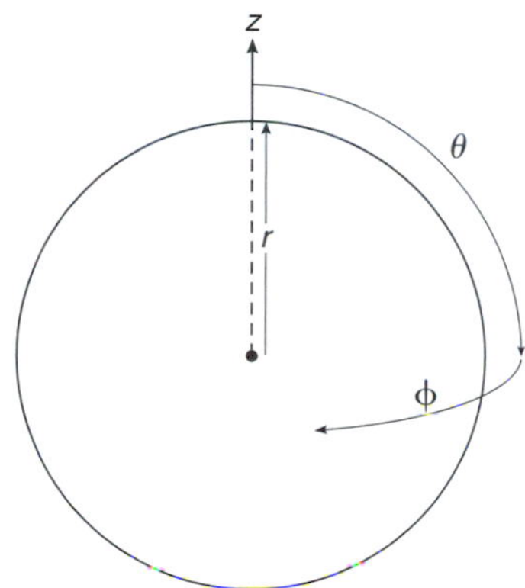

Figure 1.3
Spherical polar coordinates: r is the distance from the nucleus; θ is the angular distance from the z axis (north pole); ϕ is the azimuthal angular distance from the reference vertical plane (Greenwich meridian).

to factor the wave function ψ into a radial wave function R, a description of the distance of the electron (or of its crest, trough, etc.) from the nucleus, and an angular wave function Y, a description of the angular distance of the electron from the "north pole" and the reference vertical plane of the atom:

$$\psi = R_{n,\ell}(r) \cdot Y_{l,m_\ell}(\theta, \phi) \quad (1.10)$$

Let us first illustrate this distinction using the orbits of the planets as illustrations. The *angular part* of any planet's orbit is substantially the same. All move roughly in the equatorial plane of the solar system, so that $\theta = 90°$ ($\pi/2$ rad); the orbit is substantially unaltered as the planet sweeps the angle ϕ around the sun. These angular parts of the planetary orbits define the *shape* and the *orientation* of the planetary orbits: All are more or less circular and in the equatorial plane of the solar system. (Of course, comets move in orbits of different shapes and orientations.) The planetary orbits differ predominantly in their *radial parts*: Each planet moves in an orbit of a different radius.

The general form of the solution of the Schrödinger wave equation is too complex to make much sense on casual inspection. We do note here that, since the atom is a three-dimensional "box," the general (nonrelativistic) form of this solution for ψ contains three quantum numbers, n, ℓ, and m_ℓ, to be discussed below. When specific values of these three quantum numbers are substituted in the general form of the equation, the specific solutions obtained are much simpler; it is these that are normally tabulated for inspection (Table 1.2). We note that the solutions have been separated into radial

Table 1.2
Orbitals for Hydrogen-Like Atoms or Ions

A. Radial Wave Functions

$$R_{n,l}(r) = f(r)\left(\frac{Z}{a_0}\right)^{3/2} e^{-\rho/2}$$

with a_0 the Bohr Radius (0.53 Å) and $\rho = 2Zr/na_0$

n	l	$f(r)$	Orbital Designation
1	0	2	1s
2	0	$(\frac{1}{2}\sqrt{2})(2-\rho)$	2s
2	1	$(\frac{1}{2}\sqrt{6})\rho$	2p
3	0	$(\frac{1}{9}\sqrt{3})(6-6\rho+\rho^2)$	3s
3	1	$(\frac{1}{9}\sqrt{6})(4-\rho)\rho$	3p
3	2	$(\frac{1}{9}\sqrt{30})\rho^2$	3d

B. Angular Wave Functions

$$Y_{l,m_l}(\theta,\phi) = \left(\frac{1}{4\pi}\right)^{1/2} y(\theta,\phi)$$

l	m_l	$y(\theta,\phi)$	Orbital Designation
0	0	1	s
1	0	$3^{1/2}\cos\theta$	p_z
1	± 1	$\mp(\frac{3}{2})^{1/2}\sin\theta\, e^{\pm i\phi}$	p_x, p_y
2	0	$(\frac{5}{4})^{1/2}(3\cos^2\theta - 1)$	d_{z^2}
2	± 1	$\mp(\frac{15}{4})^{1/2}\cos\theta\sin\theta\, e^{\pm i\phi}$	d_{xz}, d_{yz}
2	± 2	$(\frac{15}{8})^{1/2}\sin^2\theta\, e^{\pm 2i\phi}$	$d_{xy}, d_{x^2-y^2}$

SOURCE: Adapted from D. F. Shriver, P. W. Atkins, and C. H. Langford, *Inorganic Chemistry*, Freeman, New York; 1990, p. 11.

and angular parts; the overall ψ is obtained by multiplying the two parts together.

The **principal quantum number**, n, occurs only in the radial part of the wave function and therefore is important in determining the *size* of the orbital; it can take any integral value from 1 upward. It is the first number shown in the usual representation of orbitals: thus, the 1s orbital is smaller than the 2s orbital.

The **secondary** or **orbital angular momentum quantum number**, ℓ, occurs in both parts of the wave function; it determines the *shape* of an orbital, which will be described in more detail below. This quantum number can take any integral value from zero up to one less than the principal quantum number. In designating orbitals, it is common to replace the numerical value of this quantum number with a corresponding letter designation:

Secondary quantum number ℓ	0	1	2	3	4	5
Orbital shape designation	s	p	d	f	g	h

The **magnetic quantum number**, m_ℓ, occurs only in the angular part of the wave function, and determines the *orientation* of the orbital of a given (nonspherical) shape;

this quantum number can take any positive or negative whole-number value, including zero, provided it falls between ℓ and $-\ell$ inclusive. For a given orbital size and shape (a *subshell* of orbitals, such as the 2*p* orbitals) there are $2l+1$ possible orientations; this is the total number of orbitals of this size and shape in a given atom.

Although not part of the orbital wave function, it also turns out that the electron, as it occupies a given orbital with a unique set of values of (n, l, m_l), can either go in with its spin *up* or *down*, depending on the value of its **spin angular momentum quantum number**, m_s, which can be either $+\frac{1}{2}$ or $-\frac{1}{2}$. A filled orbital has two electrons, one with each value of m_s. Consequently, an *s* subshell is filled by 2 electrons, a *p* subshell by 6, a *d* subshell by 10 electrons, and an *f* subshell by 14 electrons. On building up the periodic table atom by atom across a period, it takes this many electrons, hence this many atoms, to complete the corresponding block of the periodic table.

Our main concern in this course is not with the mathematical representations of the orbitals and the quantum numbers, but with their *visual* representation, and what consequences the natures of these orbitals have on periodicity in chemical reactivity. In this section, we describe the *angular* parts of the possible wave functions for electrons in a (hydrogen-like) atom. (The wave functions for orbitals normally tabulated are those for *hydrogen-like* atoms or ions with just one electron; it is found that orbitals in polyelectronic atoms move in basically similar paths.) These angular wave functions result (after some mathematical manipulation to remove imaginary numbers) in the familiar shapes of orbitals.

The *s* orbitals are spherical. In terms of a particle model of the electron, it is as likely to be at one angle from the north pole or from the reference vertical plane as another. In terms of the wave model, the wave path has only a crest; at no angle does the amplitude of the wave drop back to zero.

The *p* orbitals have the familiar dumbbell shape (Fig. 1.4), with two **lobes**, one on each side of the nucleus at the center. The mathematical angular wave function for a *p* orbital involves some constants times $\cos\theta$ or $\sin\theta$. The angular wave function for the p_z orbital contains $\cos\theta$, which goes from $+1$ (a crest) to zero (a node) to -1 (a trough) as we move from the north pole to the equator to the south pole. The *amplitude* of the wave is proportional to the *square* of the wave function, so it is equally positive in both lobes. This amplitude corresponds to the *probability of finding the electron at that spot*. The shaded plane in Figure 1.4 through the nucleus and the "equator" of the p_z orbital is the **nodal plane** of the p_z orbital, the plane in which the wave amplitude is zero and the electron cannot be found.[4]

A *d* orbital generally has the double-dumbbell shape indicated in Figure 1.4, with four lobes and *two* nodal planes. The four lobes alternately correspond to crests and troughs, but all, of course, have positive wave amplitudes and correspond to regions in which there are positive probabilities of finding the electron.

A typical *f* orbital has the quadruple-dumbbell shape shown in Figure 1.4. Other *f* orbitals in a given atom have a somewhat different appearance due to the different mathematical operations used to eliminate the imaginary numbers from the wave functions, but all of them share the characteristic of having *three* nodal surfaces passing through the nucleus. Note that *the number of nodal surfaces passing through the nucleus is equal to the secondary quantum number.*

The *p*, *d*, and *f* orbitals not only have characteristic shapes but also are oriented in certain directions. The *p* orbital we described earlier has its crest at its north pole and its trough at its south pole and is thus oriented along this polar axis, which we label the *z* axis. A given atom also has two otherwise-identical *p* orbitals aligned at right angles

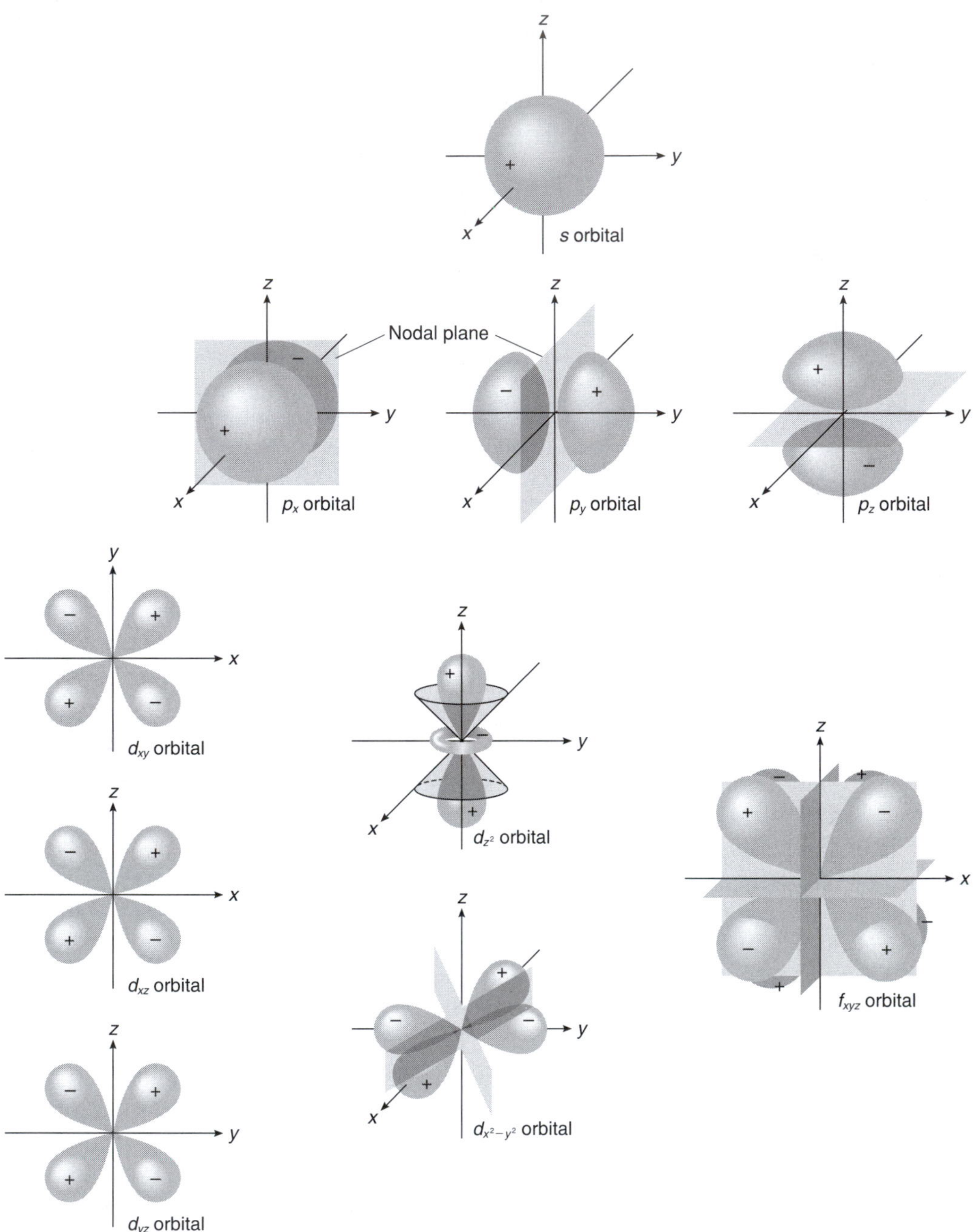

Figure 1.4

Angular wave functions of *s*, *p*, *d*, and an *f* orbital showing signs of the wave functions in different lobes and indicating nodal planes with dark planes or solid lines. [From *Chemical Structure and Bonding* by R. L. DeKock and H. B. Gray. Copyright © 1989 by University Science Books, Sausalito, CA.]

to the first one and to each other along an x and a y axis; we conventionally label these three orbitals as the p_x, p_y, and p_z orbitals (Fig. 1.4). For the p_z orbital $m_\ell = 0$. The p_x and p_y orbitals are associated with different combinations of $m_\ell = +1$ and $m_\ell = -1$. Since m occurs in the wave function associated with the imaginary number i (the square root of -1), combinations are needed to produce equations that contain only real numbers.

In a given atom, there are five otherwise-identical d orbitals, seven otherwise-identical f orbitals, and nine otherwise-identical g orbitals that differ in their orientations. These orientations are shown in Figure 1.4 for the d orbitals only. The difference in appearance of the last of these, the d_{z^2}, is an artifact of mathematical operations; like the others it has two nodal surfaces through the nuclei, but these surfaces happen to be planes wrapped around to form conical surfaces. When discussing bonding in d-block complexes (Chapter 8) and organometallic compounds (Chapter 11), you must know the designations, shapes, and orientations of these five orbitals.

1.5 Radial Part of the Wave Function

The relative size of an orbital (i.e., the most probable single radius at which the electron will be found) is a function of its *principal quantum number, n*. In addition to this major feature, orbitals also have structure (lobes and nodal surfaces) outward from the nucleus. This structure can be seen in the radial portion of the wave function for an electron, which passes from crests through zero levels to troughs and back again for most types of orbitals. There are many ways that we can plot the radial wave function: In Figure 1.5, we plot what is known as the radial probability function, which involves squaring the original wave function, so that negative parts of the wave function (troughs) become positive wave amplitudes or probabilities of finding the electrons. The radial probability function also involves multiplying the squared function by $4\pi r^2$, so that the result is not just the probability of finding the electron at *one* point at a distance r from the nucleus but is also the probability of finding the electron *at any point* in the spherical shell that is at a distance r from the nucleus.

Figure 1.5 clearly shows that, in a given atom, the most probable distance from the nucleus for the electron increases with increasing values of the principal quantum number, n. But it also shows that there are some differences for different shapes of orbitals. The $1s, 2p, 3d, 4f, \ldots,$ orbitals have the simplest radial probability functions, with one maximum and no radius at which there is zero probability of finding the electron (other than the nucleus for $2p, 3d, 4f$, etc.). But $2s, 3p, 4d, \ldots,$ orbitals all have one such distance, which corresponds to a *nodal sphere*: a nodal surface that does *not* pass through the nucleus. On one side of this surface, the wave function has a positive sign (crest); on the other side, it has a negative sign (trough). The $3s$ and other orbitals have two nodal spheres; the $4s$ and others have three nodal spheres. An orbital with a given principal quantum number n and a secondary quantum number ℓ has a total of $n - \ell - 1$ nodal spheres in its radial wave function. We have previously noted that it has ℓ nodal planes through the nucleus in its angular wave function; this gives an orbital a total of $n - 1$ nodal surfaces, which is the same number found earlier for the electron in the one-dimensional box.

It is possible and useful to combine the radial and angular wave function probability plots to give a total picture of where in the atom the electron in a given orbital is likely to be found. Without a laser holograph it is difficult to do this in three dimen-

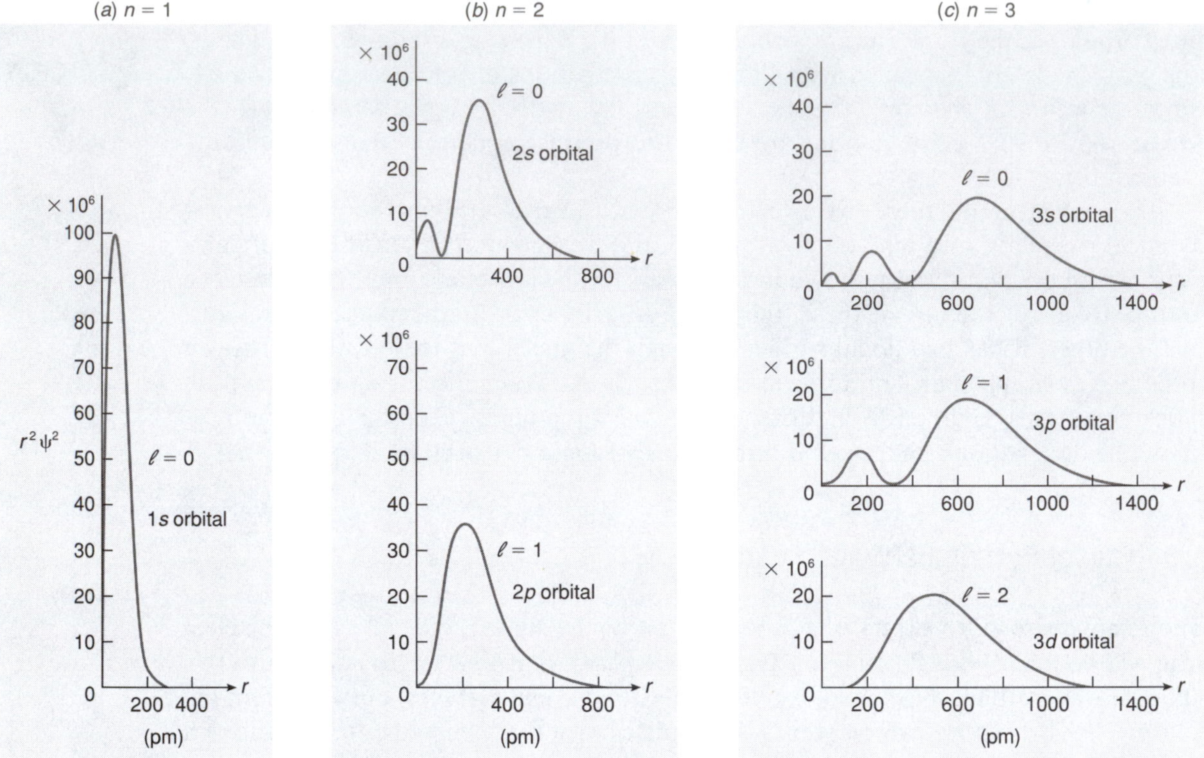

Figure 1.5
Radial probability functions plotted against distance from the nucleus for (*a*) $n = 1$, (*b*) $n = 2$, and (*c*) $n = 3$ for the hydrogen atom.

sions, but we can draw cross sections of orbitals, which is done for selected orbitals in Figure 1.6. These cross sections also contain (as a result of detailed calculations) contours of equal probability of finding the electron, much like geological topographic maps. Note that nodal spheres are indicated by dashed circles, while nodal planes are indicated by dashed lines.

Example 1.1

Sketch cross sections in the x–y plane (without density contours) of the following orbitals: (a) $4d_{x^2-y^2}$, (b) $5d_{xy}$, and (c) $5s$. Indicate nodal planes and spheres by dashed lines; show positive and negative signs of the wave function in the different lobes.

Solution:

It is best to start by counting up the number of nodal spheres (equal to $n - \ell - 1$) and nodal surfaces through the nucleus (equal to ℓ), then drawing these before attempting to fill in the actual lobes of the orbital. For the $4d$ orbital, there will be $4 - 2 - 1 = 1$ nodal sphere; draw a dashed circle out a distance from the nucleus. For the $5d$ orbital, there will be two concentric nodal spheres. The $5s$ orbital will have $5 - 0 - 1 = 4$ nodal spheres to be

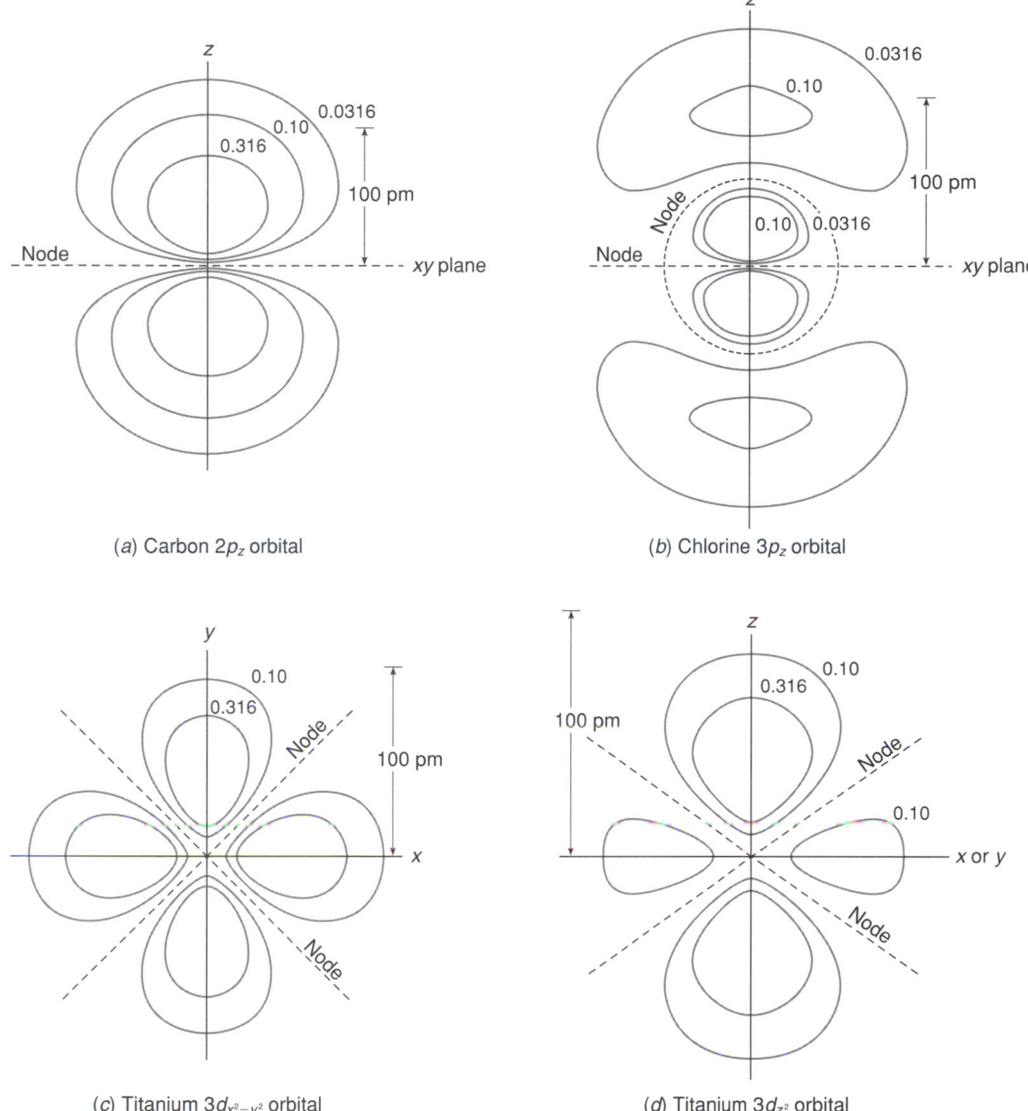

Figure 1.6
Electron-density contour maps for various orbitals. [Adapted from E. A. Ogryzlo and G. B. Porter, *J. Chem. Educ.*, **40**, 256 (1963).]

drawn concentrically about the nucleus. There will be two nodal surfaces through the nucleus for the two *d* orbitals; draw these as dashed lines through the nucleus. For the $d_{x^2-y^2}$ orbital, the lobes fall along the x and y axes, so draw the nodal planes (lines) midway between the axes; the reverse is true for the d_{xy} orbital.

Now shade in the lobes, which will fall in between the nodal lines and circles; the last lobe(s) will fall outside the outer nodal sphere. The inner lobes in a cross section will look approximately like circles; the outer ones can be drawn as shields or curved disks, or concentric circles.

Finally, fill in plus and minus signs to indicate the signs of the wave function. Start anywhere with either $a+$ or $-$ sign (the first choice is completely arbitrary), but thereafter any time you cross a nodal surface you must change signs. If this procedure results in any contradictions, a mistake has been made. The result is shown in Figure 1.7.

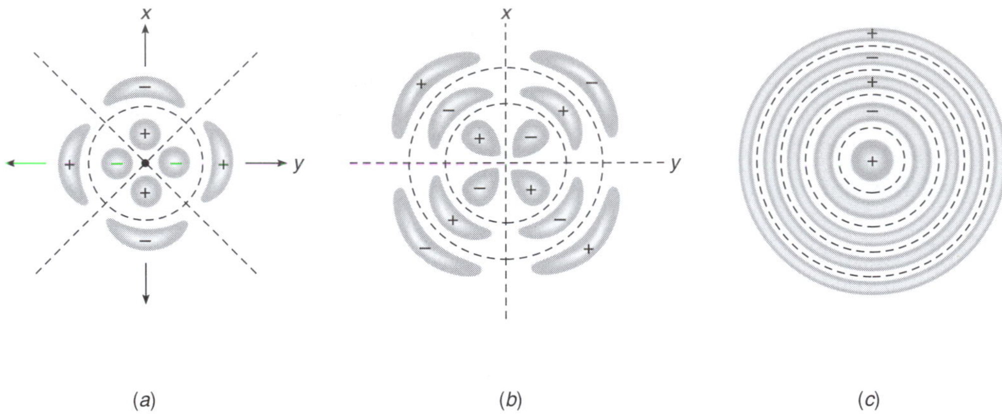

Figure 1.7
Cross-section sketches in the x–y plane of (a) the $4d_{x^2-y^2}$ orbital; (b) the $5d_{xy}$ orbital; (c) the $5s$ orbital.

1.6 Shielding and Slater's Rules

In the hydrogen atom (or other one-electron species), the energy with which the one electron in the orbital is held[5] depends only on the principal quantum number:

$$E = -\frac{2\pi^2 m Z^2 e^4}{n^2 h^2} \tag{1.11}$$

Although the one electron of such an atom or ion can be excited to any other orbital, in the normal *ground state* it is in the $1s$ orbital.

Equation (1.11) indicates that the energies of electrons in orbitals quickly become more negative with increasing nuclear charge (E is proportional to Z^2). This finding would appear to predict, for example, that the Li atom should be much harder to ionize than the H atom, whereas the converse is true. Evidently, the outermost ($2s$) electron of the Li atom does not experience the full attractive power of all three Li nuclear protons. Equation (1.11) does not consider the repulsion of the inner ($1s^2$) electrons on the outer ($2s^1$) Li electron, which make that electron much easier to ionize.

According to classical electrostatics, if the $2s$ electron of Li is completely outside of both the $+3$ charged nucleus and the two $1s$ electrons (-2 charge), it is the same as if it were outside a nucleus of $+3 - 2 = +1$ charge. Thus we may say that the $2s$ Li electron feels an **effective nuclear charge** that is less than the true nuclear charge due to the **shielding** ability of the core $1s$ electrons. Therefore, in Eq. (1.11), we might not want to use Z, the true nuclear charge, but an effective nuclear charge, Z^*, that is less than Z by the sum of the *screening constants* of the inner orbitals:

$$Z^* = Z - \Sigma S \tag{1.12}$$

But the 2*s* electron of Li does not behave as if it were experiencing exactly a +1 charge either: Its properties are more consistent with those of an electron feeling an effective nuclear charge of +1.279. This difference can be explained by the fact that the 2*s* orbital has *two* maxima in its radial probability function (Fig. 1.5) and by the fact that the lesser of these **penetrates** within the maximum of the inner 1*s* orbital; hence, although most of the time the 2*s* electron is in the outer lobe of that orbital, feeling a Z^* of +1, for some of the time it is inside of the 1*s* orbital, experiencing the full nuclear charge of +3. This penetration of inner orbitals is obviously going to affect the chemical properties of that valence Li electron: It will affect the ease of ionization of that electron, the size of its orbital, the electronegativity of the atom, and so on.

Equation (1.11) contains only the *principal* quantum number as a variable, and thus implies that the third electron of Li is equally stable in the 2*s* and the 2*p* orbitals. But, in fact, when two or more electrons are present, the electrons repel each other, and they repel each other differently, depending on the structures of the orbitals in which they are located. The 2*p* orbital of Li lacks the penetrating inner lobe that the 2*s* orbital has. Consequently, the valence electron of Li (and other atoms) prefers the 2*s* orbital, and fills it first. Likewise, we can see [from Fig. 1.5(*c*)] the superior penetrating power of the 3*s* orbital as compared to the 3*p* orbital, which is better than the 3*d*. Thus electrons in orbitals of common n but increasing ℓ have progressively poorer penetration of core electrons, so the orbitals of increasing ℓ are filled later. In multielectron atoms, the energies of electrons in orbitals increase with increases in either the principal or the secondary quantum numbers of the orbitals.

Once the 3*s*, 3*p*, and 3*d* orbitals are filled, for example, they have differing abilities to shield outer electrons (such as 4*s*) from the charge of the nucleus. It is easier for these outer electrons to penetrate the electrons of the 3*d* orbitals, which have little likelihood of being near the nucleus, than to penetrate the 3*p* or especially the 3*s* orbital. Consequently, electrons in orbitals of common n but increasing ℓ exhibit progressively poorer *shielding* of outer electrons and should have smaller shielding constants for use in Eq. (1.12).

It is useful to have some simple rules for estimating the degree to which electrons in the various types of orbitals shield other electrons from the nucleus, and hence for estimating the effective nuclear charges experienced by other electrons. Based on calculations done in 1930 and earlier, Slater[6] proposed some simplified rules for making such estimations. Although more modern calculations, such as those of Clementi and Raimondi,[7] give more accurate values of Z^*, Slater's rules are adequate for predicting most periodic *trends* in Z^*, which then allow us to predict periodic *trends* in the properties we will use most to predict reaction chemistry in this book: electron configurations, radii, electronegativities, and so on. And the fact that Slater's calculations can be stated as rules allows us to see *why* these trends exist, in terms of the different penetrating and shielding properties of different kinds of orbitals.

Slater's rules give different results for electrons in different types of orbitals. Therefore, we need to note which orbital type (subshell) is specified, as well as the atom's nuclear charge Z. Rules 2–5 compute the shielding constants of the other electrons for the particular electron in question; Rule 6 simply applies Eq. (1.12) to compute the effective nuclear charge for the specified electron. Rule 7 is a completely separate calculation of an *effective principal quantum number* n^* for the electron in question, which is needed for some applications.

1. After writing out the full electron configuration in the way you learned in general chemistry (i.e., in the order in which the orbitals are filled), rearrange the electron configuration of the element *in the following groupings and order*. Group *ns* and *np* orbitals together (their shielding properties are roughly equal); group all other types of orbitals separately (their shielding properties are dissimilar). The order is numerical:

$$(1s)(2s, 2p)(3s, 3p)(3d)(4s, 4p)(4d)(4f)(5s, 5p)\cdots.$$

2. Electrons to the *right* of the group of electrons in question contribute *nothing* to the shielding of that group of electrons. (This approximation cannot always be true, since outer orbitals generally have inner lobes and therefore have some penetrating ability.)

3. All *other* electrons in the same group (enclosed in the same parentheses) as the electron in question shield that electron to an extent of 0.35 unit of nuclear charge each. (Some of the time the other 3*p* electrons will be between the nucleus and the given 3*p* electron, for example.)

4. *If the electron in question is an s or p electron*: (a) All electrons with principal quantum number one less than the electron in question shield it to an extent of 0.85 unit of nuclear charge each. (b) All electrons with principal quantum number two or more less than the electron in question shield it completely (i.e., to an extent of 1.00 unit of nuclear charge each).

5. *If the electron in question is a d or f electron*: *All* electrons to the *left* of the group of the electron in question shield the *d* or *f* electron completely (i.e., to an extent of 1.00 unit of nuclear charge each). This is a manifestation of the poor penetrating power of the *d* or *f* electron.

6. Sum the shielding constants from Steps 2–5, and subtract them from the true nuclear charge Z of the atom in question to obtain the effective nuclear charge Z^* felt by the electron in question.

7. Values of n^* are not the same as the true principal quantum number for higher values of n:

n	1	2	3	4	5	6
n^*	1	2	3	3.7	4.0	4.2

Example 1.2

Calculate Z^* for an electron in each type of occupied orbital in Sc ($Z = 21$).

SOLUTION:

In Step 1, we rewrite the full electron configuration of Sc in the order $(1s)^2(2s, 2p)^8(3s, 3p)^8(3d)^1(4s)^2$. There are five different groups of occupied orbitals, hence the question calls for five different values of Z^* to be calculated.

(a) For any 1*s* electron, the *only* shielding is from the one other 1*s* electron, which by Rule 3 shields out 0.35 unit of nuclear charge. Hence, by Rule 6 (Eq. 1.12) $Z^*_{1s} = 21 - 0.35 = +20.65$.

(b) For any 2*s* and 2*p* electron (treated the same by Slater's rules), we have two sources of shielding. The seven other 2*s* and 2*p* orbitals contribute shielding of

$7 \times 0.35 = 2.45$ units of nuclear charge (Rule 3); the two 1*s* electrons contribute shielding of $2 \times 0.85 = 1.70$ units of nuclear charge (Rule 4a). By Rule 6, $Z^*_{2sp} = 21 - (2.45 + 1.70) = +16.85$.

(c) For each 3*s* and 3*p* electron, there are three sources of shielding: the seven other 3*s* and 3*p* electrons, which contribute shielding of 2.45 units (Rule 3); the eight 2*s* and 2*p* electrons, which contribute $8 \times 0.85 = 6.80$ units (Rule 4a); and the two 1*s* electrons, which shield completely, that is, 2.00 units of nuclear charge (Rule 4b). By Rule 6, $Z^*_{3sp} = 21 - (2.45 + 6.80 + 2.00) = +9.75$.

(d) For the 3*d* electron (since it has no other 3*d* companions to which Rule 3 would apply), only Rule 5 need be applied: all 18 electrons inside of it shield by 1.00 unit each, so $Z^*_{3d} = 21 - 18.00 = +3.00$.

(e) For each 4*s* electron, Rule 3 shows one companion 4*s* electron shielding 0.35 unit of nuclear charge, Rule 4a shows nine 9 of principal quantum number 3 $(3s, 3p, 3d)$ shielding $9 \times 0.85 = 7.65$ units, and Rule 4b shows 10 electrons of principal quantum number 2 or below $(1s, 2s, 2p)$ shielding 10.00 units of nuclear charge. Hence, $Z^*_{4s} = 21 - (0.35 + 7.65 + 10.00) = +3.00$.

1.7 Characteristic Valence Electron Configurations

The calculations of Example 1.2 indicate that two types of orbital in Sc, the 4*s* and 3*d*, can be distinguished from all of the other types by the relatively low effective nuclear charge that they feel. When appropriate calculations of the type suggested by Eq. (1.11) are made,[8] it is confirmed that the 4*s* and 3*d* orbitals are nowhere near as low in energy as the 1*s* through 3*p*. This justifies our dividing the electrons and orbitals into two classes: (1) The low-energy, strongly bound **core electrons** (and orbitals) of an atom—electrons in these orbitals are too strongly bound to be affected by the kinds of energy released in chemical reactions; (2) the higher energy, loosely bound (low Z^*) **valence electrons** (and orbitals), which are chemically active.

It is instructive to repeat these calculations for the element with 10 more units of nuclear charge than Sc (e.g., Ga). The electron configuration written in the style of Rule 1 is $(1s)^2(2s, 2p)^8(3s, 3p)^8(3d)^{10}(4s, 4p)^3$. Repetition of the calculations for each type of electron gives the following results. For 1*s*, the shielding is unchanged from Sc, so Z^* increases by the 10 additional protons ($Z^*_{1s} = +30.65$); the shielding is also unchanged for $(2s, 2p)$ and $(3s, 3p)$ electrons, so $Z^*_{2sp} = +26.85$ and $Z^*_{3sp} = +19.75$. However, the shielding *does* change for the remaining electrons. For 3*d* electrons, the shielding has increased somewhat from Sc to Ga, since other 3*d* electrons have been added. Therefore, Z^*_{3d} has increased moderately in Ga to +9.85. For $(4s, 4p)$ electrons, the shielding has increased quite a bit, since the $(4s, 4p)$ electrons are rather well shielded by all of the 3*d* electrons that have been added: $Z^*_{4sp} = +5.00$, only two units higher than in Ga. *The 4s and 4p electrons now stand alone at relatively low Z* values; these are the only valence electrons in Ga. The 3d electrons are now part of the core electrons.* The gradual plunge of the 3*d* orbitals into the core can be seen from the results of calculations across the *d* block (Fig. 1.8).

Many applications of these calculations specifically require the Z^* value for the valence electron of highest principal quantum number, that is, for the *ns* or *np* valence

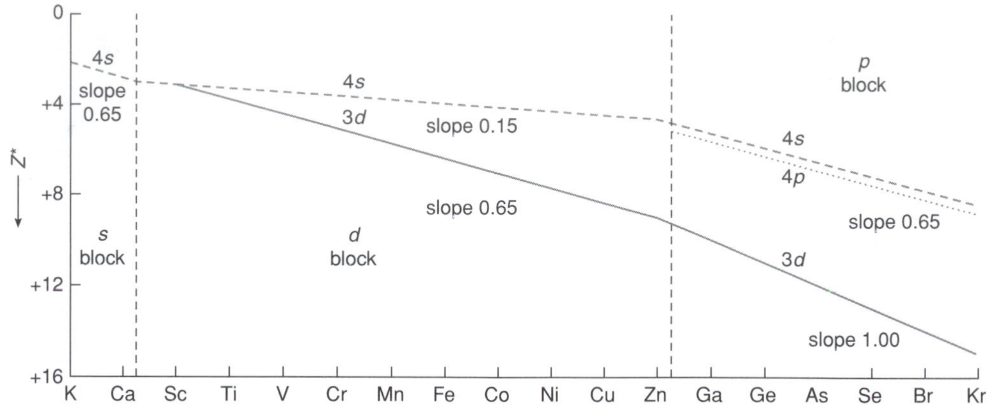

Figure 1.8
The calculated effective nuclear charges (Z^*) for the $3d$ (–), $4s$ (- -), and $4p$ (· · · ·) electrons of the elements of the fourth period. As Z^* values increase in magnitude, orbital energies drop, although orbital energies are not a function of Z^* alone.

electron. Since we will be interested in periodic trends in this property, it is useful to calculate variations in Z^* across the different blocks and down the different groups of the periodic table.

Example 1.3

Calculate Z^* for the last valence electron in each atom of the second period.

SOLUTION:

In Step 1, we can write the electron configurations of these atoms in the general form $(1s)^2(2s, 2p)^n$, in which n is the number of valence electrons (and also the group number). Step 2 does not apply. In Step 3, we obtain a total shielding constant of $(n - 1) \times 0.35 = -0.35 + 0.35n$ for the $2s$ and $2p$ electrons. In Step 4, we obtain $2 \times 0.85 = 1.70$ for shielding by the two $1s$ electrons. In Step 6, we add the shielding constants to obtain $1.35 + 0.35n$, which we subtract from the atomic number of $(2 + n)$ to obtain $Z^*_{2sp} = 0.65 + 0.65n$.

The results from this example are compared in Table 1.3 with the values obtained for lighter atoms by self-consistent field (SCF) quantum mechanical calculations:[6] As we can see, the agreement is quite good, although it is not quite so good at the left side of the third or in the fourth period. What is more important for our purposes is how well these functions work in predicting and thus explaining the periodic trends in the sizes of atoms, ionization energies, and electronegativities. We can see clearly that Z^* *increases* rather rapidly *on going to the right across the s and p blocks*, by +0.65 unit of charge per element.

A similar calculation of the variation of Z^* across the d block shows (Fig. 1.8) that Z^*_{4s} for the elements having the characteristic electron configuration $1s^22s^22p^63s^23p^63d^n4s^2$ equals $2.85 + 0.15n$. Thus Z^* values for valence s electrons

Table 1.3
Comparison of Effective Nuclear Charges Calculated by Slater's Rules and from SCF Quantum Mechanical Calculations[a]

Atom	Li	Be	B	C	N	O	F	Ne
Z* (Slater)	1.30	1.95	2.60	3.25	3.90	4.55	5.20	5.85
Z* (SCF)	1.279	1.912	2.421	3.136	3.834	4.453	5.100	5.758
Atom	Na	Mg	Al	Si	P	S	Cl	Ar
Z* (Slater)	2.20	2.85	3.50	4.15	4.80	5.45	6.10	6.75
Z* (SCF)	2.507	3.308	4.066	4.285	4.886	5.482	6.116	6.764
Atom	K	Ca	Ga	Ge	As	Se	Br	Kr
Z* (Slater)	2.20	2.85	5.00	5.65	6.30	6.95	7.60	8.25
Z* (SCF)	3.495	4.398	6.222	6.780	7.449	8.287	9.028	9.769

SOURCE: The SCF values are from J. E. Huheey, E. A. Keiter, and R. L. Keiter, *Inorganic Chemistry: Principles of Structure and Reactivity*, 4th ed., Harper-Collins, New York, 1993; p. 34.

[a] Calculations are for the outermost valence electron; the SCF values for the outermost *s* electron of *p*-block elements are somewhat higher than the values listed, which are for the *p* electron.

increase more slowly on crossing the *d* block (0.15 units of nuclear charge per element). This occurs because the $(n-1)d$ electrons being added across the *d* block are basically inside the *ns* electrons, and hence do a better job of shielding the outer *ns* electrons than can the *np* electrons in the *p* block.

The computation of Z^* for the outermost *s* orbital of the *f*-block elements produces a very interesting result: The value of Z^* is 2.85 regardless of how many *f* electrons are present in the element! This result is a consequence of the fact that the $(n-2)f$ orbitals are *two* shells below the *s* electrons. Hence, according to Rule 4b, the deeply buried $(n-2)f$ electrons completely shield the *ns* electrons.

Vertical Trends in Z^*. The calculation of Z^* for the outermost *s* electrons in a vertical group of atoms by Slater's rules also gives interesting results. For Group 2, we obtain $Z^*(\text{Be}) = 1.95$, $Z^*(\text{Mg}) = 2.85$, $Z^*(\text{Ca}) = 2.85$, $Z^*(\text{Sr}) = 2.85$, and $Z^*(\text{Ba}) = 2.85$. (Unfortunately, SCF calculations are not available for the heavier atoms for comparison.) We see that, in the *s* block, the Z^* values are unchanged from the third period down, which contributes to the similarity in chemical properties of the members of a group of elements.

If we carry out the same calculations for Group 13(III), the trends are somewhat different, however. Here we obtain $Z^*(\text{B}) = 2.50$, $Z^*(\text{Al}) = 3.50$, $Z^*(\text{Ga}) = 5.00$, $Z^*(\text{In}) = 5.00$, and $Z^*(\text{Tl}) = 5.00$. The values for B and Al have increased by $+0.65$ from those of Be and Mg in Group 2, the normal trend in the *p* block. However, the values for Ga, In, and Tl have increased by $+2.15$ from those of their Group 2 analogues. The difference is that, in the meanwhile, *d* orbitals have been filling. Even though Z^* for outer *s* electrons increases only slowly across the *d* block, it does increase 10 times. Consequently, the Z^* values of the elements beyond the *d* block—Ga, In, Tl—are 0.65 *plus 1.50* higher than the corresponding elements in Group 2. In terms of Group 13(III) alone, this produces an extra jump in Z^* between Al and Ga, which was not present in the corresponding part of Group 2. This increase in effective nuclear charge occurs among the elements Ga, Ge, As, Se, Br, and Kr, which immediately follow the first filling of the *d* orbitals.

Likewise, the fact that Z^* of Mg is higher than that of Be is due to Mg following the first filling of the *p* orbitals. However, we fail to notice any sign of an increase in Z^* in Tl, which follows the first filling of the *f* orbitals. We shall return to that point in subsequent sections

Order of Energies of Orbitals. The fact that the energies of the orbitals increase with increases in both the principal and the secondary quantum numbers results in the familiar but seemingly irregular order of filling of the orbitals in neutral atoms:

Orbital Filled
$1s < 2s < 2p < 3s < 3p < 4s \leq 3d < 4p < 5s \leq 4d < 5p < 6s \leq 4f \leq 5d < 6p$

1 2 2 3 3 4 4 4 5 5 5 6 6 6 6
Period when filled

This is the order in which the orbitals are encountered in the periodic table as one scans it from low to high atomic number (Table 1.1). Figure 1.9 shows the apparent relative energies of orbitals.[9] In each period the orbitals being encountered and filled are highlighted in the shaded regions of Figure 1.9; within these shaded regions the above orders hold.

However, calculations (even those as simple as our Z^* calculations of Example 1.2) show that this order does not persist at all times, but only applies *at the time of filling*. We saw in Example 1.2 and Figure 1.8 that 3*d* and 4*s* electrons have equal Z^* in Sc, but that Z^* is clearly greater for 3*d* than for 4*s* electrons once we have left the *d* block, in Ga. More detailed calculations (Fig. 1.9) show that the order of energy of orbitals is generally quite different at the times when orbitals are *not* filling. In hydrogen itself, with only one electron, all orbitals of the same principal quantum number are of equal energy, as implied by Eq. (1.11). In addition, once the orbitals have become core orbitals and have long since been filled, the same hydrogen-like order once again very nearly applies (as for most electrons in Fm, $Z = 100$, right side of Fig. 1.9.) At the time of filling in neutral atoms, however, the very non-hydrogen-like order given above prevails.

This means, as we saw earlier, that the designation of a particular orbital as core or valence depends not only on the period of the element under discussion, but also on the block of the periodic table. As our calculations in Example 1.2 and the text that followed suggested, 3*d* is a valence orbital in the *d* block of the fourth period, but it is a core orbital in the *p* block of the fourth period. Our calculations, and Figures 1.8 and 1.9, show how its energy drops rapidly across the fourth period. The 3*d* orbitals that fill in the *d* block do not do as good a job of shielding each other (Rule 3) as they do of shielding outer 4*s* or 4*p* electrons (Rule 4a); hence, only the 4*s* and 4*p* can remain valence electrons.

In general, we find that *only the last-filled s orbital remains a valence orbital outside of its block*, the *s* block; it remains a valence orbital across the entire period before becoming a core orbital in the next period. All other less penetrating types of orbitals are shielded out by the end of the block in which they fill, so they remain valence orbitals only in their characteristic block of the periodic table.

Hence, a given element has two types of valence electrons: those in the last-filled *ns* orbital and those in the last-filled orbital characteristic of the block, which consist of *np* in the *p* block; $(n-1)d$ in the *d* block; $(n-2)f$ in the *f* block (Table 1.1). The **characteristic valence electron configuration** shows the expected occupancy of just these two types of orbital (one in the *s* block). Other occupied orbitals are called **core** orbitals; as-yet-unoccupied orbitals will be designated **postvalence** or **virtual** orbitals.

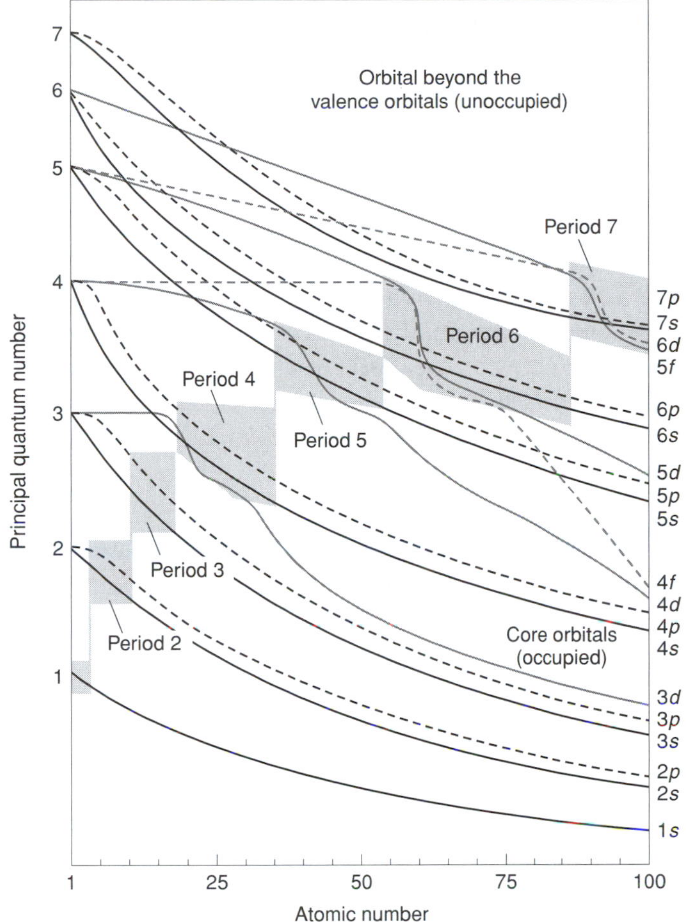

Figure 1.9
The apparent variation in energies of electrons in atomic orbitals with increasing atomic number in neutral atoms (energies not strictly to scale). The shaded boxes enclose the valence orbitals of the atom with the given atomic number. [Adapted from *Basic Inorganic Chemistry* by F. A. Cotton, G. Wilkinson and P. L. Gaus. Copyright © 1995 by John Wiley & Sons, Inc.] Caution is needed in interpreting fine details from such a diagram. For example, although 3*d* orbitals are always lower in energy than 4*s* orbitals, the total energies of all electrons in some atoms is lower when 4*s* is filled in preference to 3*d*, due to the effects of repulsions among different electrons. This figure is drawn to reflect total energies, which determines the order of filling of orbitals.

Electrons in core orbitals are attracted too strongly to the nucleus as a result of too-high effective atomic numbers to be much influenced by chemical reaction energies; postvalence orbitals are normally too high in energy to hold electrons. Chemical change is concentrated in the valence electrons occupying the valence orbitals.

Number of Valence Electrons of an Atom. One of the justifications for grouping elements as we do in the periodic table is that *elements in a group of the periodic table have the same number of valence electrons in the same shapes of orbitals*. These **valence electrons** are the ones symbolized by dots in Lewis (electron-dot) symbol used in general and organic chemistry (e.g., for the Group 13(III) atoms, ·B: ·Al: ·Ga: ·In: and

·Tl:). In general, *the number of valence electrons of an atom equals its group number*. In the new IUPAC system for numbering groups, this holds *except in the p block*. (This exception is the principal objection to the new numbering system.) Thus atoms in the boron group, Group 13(III), all have three valence electrons; the carbon group atoms, Group 14(IV), have four valence electrons, and so on.[10] *Throughout the text,* when a statement uses the numerical value of a group number (as for the number of valence electrons, or for an oxidation number), for a *p* block element, this number should be based on the *old* designation of the *p*-block group (or on its *new* designation *minus ten*). Thus the numerical value for the halogen Group 17(VII) will be taken to be 7, while that for the noble gases will be taken as 8.

In older periodic tables, the number of valence electrons equals the group number as long as the groups in a given part of the periodic table are consecutively numbered. Toward the end of the transition metals (*d*-block elements), the numbering is *not* consecutive, so that we must continue numbering upward. The groups of *f*-block elements are not numbered at all in a conventional periodic table, so that we must count over from La and Ac as "3" or Ce and Th as "4." Then, Lu and Lr should be numbered as having three valence electrons.

Example 1.4

How many valence electrons do the following atoms have: Se, Fe, Co, Pr?

SOLUTION:

Selenium is located in Group 16(VI) of the periodic table, so it has six valence electrons. Iron is located in Group 8, so it has eight valence electrons. Cobalt is located in Group 9 of our table but also in Group VIII or VIIIB of an older one. Clearly, though, it has to have one more electron than Fe; as this is a valence electron, we assign nine valence electrons to Co. Praseodymium is located in Group *5F* of our Table 1.1, so it has five valence electrons. On a conventional table we count over from La in Group IIIA or IIIB, then follow atomic numbers down to Ce with four valence electrons, finally arriving at Pr with five valence electrons.

When we draw Lewis (or electron-dot) symbols for atoms, we represent each valence electron with one dot. Due to the large number of dots that may result with the *d*- and *f*-block elements, Lewis symbols are used mainly with *s*- and *p*-block elements.

Valence Electron Configurations of Atoms. Having determined the number of valence electrons and the types of orbitals in which they go, to obtain the valence electron configuration we simply assign the valence electrons to the valence orbitals, starting with the *s* orbital.

Example 1.5

Write the characteristic valence electron configurations of the following atoms: Ba, As, Zn, and Tm.

SOLUTION:

First, we note the blocks and periods in which the elements are located. Barium is in the *s* block in the sixth period. Arsenic is in the *p* block in the fourth period. Zinc is in the *d* block in the fourth period. Thulium is in the *f* block in the sixth period. All of the elements will use the *ns* orbital as a valence orbital, where *n* is the period number of the element. In addition, As will use the *np* orbital, Zn will use the $(n-1)d$ orbital, and Tm will use the $(n-2)f$ orbital. Thus the valence orbitals are Ba, only the 6*s* orbital; As, the 4*s* and 4*p* orbitals; Zn, the 4*s* and 3*d* orbitals; and Tm, the 6*s* and 4*f* orbitals.

Second, we find the group numbers. Barium is in Group 2; As is in Group 15(V); Zn is in Group 12; and Tm is in Group 15*F*. Hence, these are the number of valence electrons (for As, there are five). By filling in these numbers of valence electrons into the *ns* orbitals first, we obtain Ba, $6s^2$; As, $4s^24p^3$; Zn, $4s^23d^{10}$; and Tm, $6s^24f^{13}$.

Alternately, since the elements of a group have the same number of valence electrons in the same shapes of orbitals, we can obtain the valence electron configuration of a later element in a group from the configuration of the first member of the group. The only difference is that the principal quantum number of each orbital increases by one for each descent of one period in the periodic table. For example, if we recall that the valence electron configuration of N is $2s^22p^3$, we can count down two periods to As, which thus has the configuration $4s^24p^3$.

Anomalous Electron Configurations. With all of the orbital crossing going on in the (shaded) valence regions of Figure 1.9, it is not surprising that there are many atoms that have anomalous orders of filling. There are 21 of these in the *d* and *f* blocks, in which one or more electrons shift from the expected valence orbital to the other expected valence orbital, or even to a nonexpected orbital—many of the neutral atoms of the *f* block have one or two electrons in the subsequent *d* orbital. There is little predictability to these anomalies; fortunately, they apply only to *isolated gaseous atoms*. Since very little of the chemistry we will discuss in this book is of isolated gaseous atoms, we recommend *not* memorizing or worrying about the 21 anomalous electron configurations.

There is a similar shift in energy of orbitals when an atom ceases being neutral and becomes positively charged: the late-filling $(n-1)d$ or $(n-2)f$ orbitals, under the influence of the extra positive charge, drop in energy relative to the *ns* or *np* orbital. Consequently, the order of filling in *positive ions* of the *d* and *f* blocks is also more hydrogen-like, in that the $(n-1)d$ or $(n-2)f$ orbitals are occupied in preference to the *ns* orbital. Consequently, there are no anomalous electron configurations among the chemically important positive ions.

1.8 Valence Electron Configurations, Charges, and Oxidation Numbers of Ions

We seldom deal with elements as isolated atoms. Nonmetals are often encountered as monoatomic **anions** (negative ions). The valence electron configurations of these anions are simply obtained by adding *np* electrons (and negative charges) until the *ns* and *np* orbitals are filled with a total of eight valence electrons. The Lewis symbols of these monoatomic anions will show an *octet* of (eight) valence electrons. The *charges* on these ions will be the negative of the number of extra electrons they have acquired, which is *the (old style) group number minus eight.* This charge is also the *oxidation number* that we assign to the nonmetal element in these anions. Thus, the Lewis symbol of the anion formed by sulfur will have eight dots representing eight valence electrons. Since this is two more electrons than the neutral sulfur atom has, the anion has a −2 charge; the oxidation number of the sulfur in this S^{2-} ion is −2.

The metallic elements are usually encountered in the form of positive ions (**cations**). For cations, the valence electron configuration is obtained by *deleting* enough electrons from the one or two types of valence orbital of the neutral atom to achieve the specified positive charge. These electrons are deleted *first* from the orbital of *highest principal quantum number*. Thus for *d*- and *f*-block elements, the electrons are deleted from the *ns* orbital first, then from the remaining $(n-1)d$ or $(n-2)f$ orbitals if necessary. For *p*-block elements, the electrons are deleted from the *np* orbital first, then from the *ns* orbital if necessary.

Example 1.6

Write the valence electron configurations of the following ions: Ba^{2+}, As^{3+}, Zn^{2+}, and Tm^{3+}.

SOLUTION:

First, we write the valence electron configurations of the neutral atoms (we have already done this in Example 1.5). Next, we remove one electron for each unit of positive charge. For Ba^{2+}, this means removing two *s* electrons to give a final configuration of $6s^0$. Removing three *p* electrons gives As^{3+} the configuration of $4s^2$ (i.e., $4s^2 4p^0$). Loss of two *s* electrons leaves Zn^{2+} with a $3d^{10}$ configuration. Thulium must lose two *s* electrons followed by one *f* electron, which results in a valence electron configuration of $4f^{12}$ for Tm^{3+}.

Periodic Trends in Oxidation Numbers. Among cations of the elements, the oxidation number (often expressed as a Roman numeral in parentheses) is also equal to the ionic charge. A necessary trend is that the *highest oxidation number of an element cannot exceed its group number*, since it cannot, by the definition of valence electrons, lose or share anything beyond its valence electrons. In ions that have lost all of their valence electrons, the Lewis symbol will, of course, be without dots; the valence electron configuration shows the valence orbitals holding zero electrons. Table B2 (inside the front cover) shows that the highest or *group oxidation number* is achievable at the *left side* of each block, where it not that high in an absolute sense, but the group oxidation number is generally not achievable at the right side of most blocks, where it would be +8 (*p* block), +12 (*f* block), or +16 (*f* block). In fact, no oxidation number over +8 is known.

In the *p* block, the second likely cationic electron configuration is that in which only the *np* electrons have been lost; the valence electron configurations of these ions are ns^2. The oxidation number of such a *p*-block cation is then *two less than the (old-style) group number*. Table B1 (inside the front cover) shows the *most common* positive oxidation number of each element. The oxidation state that is two less than the group number becomes the common one *at the right of the p block* and also more often so *at the very top and the bottom* of many of the other groups of the *p* block.

In the *d* block, the group oxidation number is common for less than one-half of the elements, those being at the left of the *d* block. Otherwise, there are no uniquely favored electron configurations—instead, we often find oxidation numbers and ionic charges equal to *+2, +3, or +4*. These lower oxidation numbers are also preferred more frequently at the *top* of the *d* block.

The group oxidation number is found only at the far left of the *f* block, but more extensively at the bottom of the block. The most common oxidation number for most of the *f*-block elements is +3.

Numerical Rules for Calculating Oxidation Numbers. When the formula of a compound containing ions is written, the charges on the ions normally are suppressed. Hence, the oxidation numbers in the formula cannot be obtained by inspection, and need to be calculated. The following set of rules work for calculating oxidation numbers as long as the specified elements do not have unusual bonding features and provided the species does not have more than three elements in it, or if it has more than three, that it can be separated into ions of three or fewer elements each.

1. Identify the *least* electronegative element in the species; it is probably really present as an ion of predictable charge. If it is in Group 1, assign it an oxidation number of +1; if it is in Group 2, +2; if it is in Group 3*F* or 3, an oxidation number of +3 (Al, Ga, and In can also normally be assigned oxidation numbers of +3).

2. Identify the *most* electronegative element. If it is the *last* remaining element, assign it an unknown oxidation number, x. More commonly, two elements remain to be assigned. If the most electronegative element is in Group 17(VII), assign it an oxidation number of −1; if it is in Group 16(VI), −2; if it is nitrogen from Group 15(V), −3. Only one element should now remain unassigned; assign it an unknown oxidation number of x.

3. Multiply each element's oxidation number by the number of atoms of that type present in the molecule or ion. Add all of these terms up; the sum *must* equal the *charge on the ion*, or zero if this is a neutal molecule. Solve this equation for the unknown oxidation number x.

This procedure gives the average oxidation number of all atoms of a given element in a given species, but will not differentiate atoms of an element in different bonding environments. Consequently, it can give fractional oxidation numbers. Also, especially in the realm of coordination and organometallic chemistry, these numerical rules are inadequate for assigning oxidation numbers. Hence, in Chapter 3 we will develop a more versatile approach to the assignment of oxidation numbers.

Example 1.7

Compute the oxidation numbers of each element in the ions (a) $H_2IO_6^{3-}$; (b) $S_3O_6^{2-}$; and (c) Na_2O_2.

SOLUTION:

(a) The sum of all oxidation numbers in this ion must equal the ion charge, −3. First, we assign the oxidation number of the least electronegative atom, H, as +1. Then, we assign oxygen the oxidation number −2; we assign I the unknown oxidation number x. Taking into account the number of atoms of each type, we have $2(+1) + x + 6(-2) = -3$. We solve the equation for x: The oxidation number of I must be +7.

(b) The sum of all oxidation numbers in this ion must be −2. Rule 2 allows us to assign −2 as the oxidation number of oxygen; sulfur is assigned the unknown oxidation number x. Since there are three S and six O atoms, $3x + 6(-2) = -2$. We solve this equation to find $x = 3\frac{1}{3}$.

(c) The sum of all oxidation numbers in this neutral compound must be zero. Rule 1 allows us to assign the oxidation number of +1 to each Na atom. By Rule 2, the oxidation number of the last element, oxygen, must be x. Since there are two of each type of atom, we have $2(+1) + 2x = 0$; so x, the oxidation number of oxygen, must be −1.

1.9 Types of Atomic Radii, Their Periodic Trends, and Reasons for the Trends

Experimental Radii. Before discussing periodic trends in the sizes of atoms and ions, it is necessary to note that there are different types of radii of atoms and ions that are measured and used in different bonding situations. We commonly distinguish five different types of atomic radii: covalent radii, metallic radii, ionic radii of cations, ionic radii of anions, and van der Waals radii. (There is no practical way of measuring the radii of isolated gaseous atoms.) Table 1.4 compares the values of these different types of radii for selected atoms.

To illustrate the differences in these types of radii, we have drawn a diatomic species in Figure 1.10(a). The inner circle for each atom represents the outermost extent

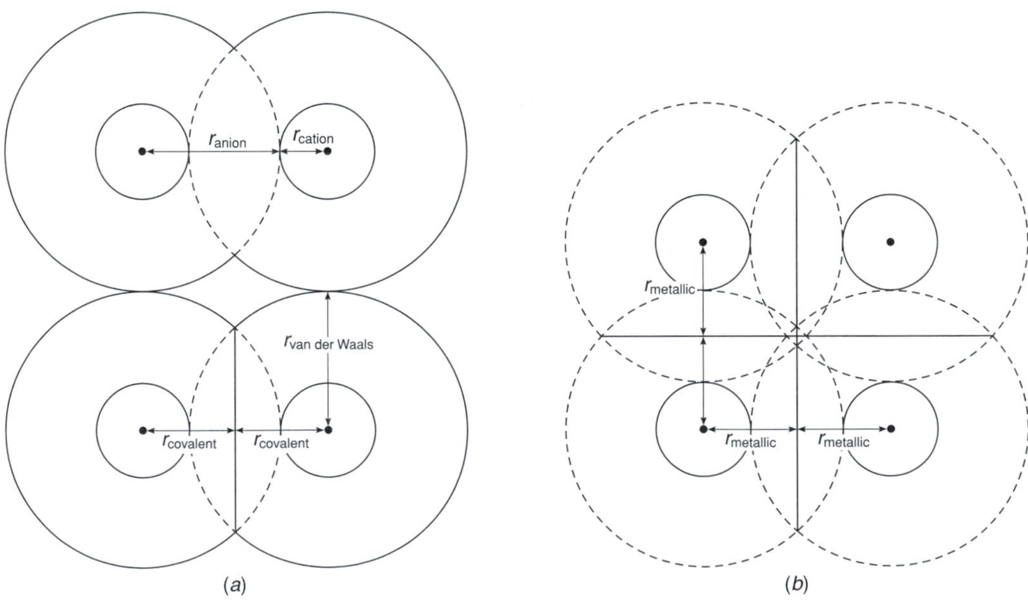

Figure 1.10
(a) Drawing of two adjacent molecules of a diatomic species; (b) drawing of four adjacent atoms in a metallic solid. The inner circles represent the outer extent of the core electrons of each atom; the outer circles represent the extent of the outer valence electrons. For explanations of the different lines and radii see the text.

Table 1.4
Different Types of Atomic Radii[a] for Selected Atoms

Element	Cationic	Covalent	Metallic	Anionic	van der Waals	Cov–Cat[b]	An–Cov[b]
Li	90	134	155		180	44	
Na	116	154	190		230	38	
K	152	196	235		280	44	
Be	59	125	112			66	
Mg	86	145	160		170	59	
B	41	90	98			49	
Al	67	130	143			63	
Ga	76	120	140		190	44	
In	94	144	158		190	50	
Tl	102	147	159			45	
N		70		132	150		62
P		110	128		185		
As	60	122	148		200	62	
Sb	74	143	166		220	69	
O		73		126	150		53
S	43	102	127	170	180	59	68
Se	56	117	140	184	190	61	67
Te	70	135	160	207	210	65	72
F		71 (54)[c]		119	155		48 (65)[c]
Cl	41	99		167	180	58	68
Br	53	114		182	190	61	68
I	67	133		206	204	66	73

SOURCES: Ionic radii are from Table C of this text; covalent and van der Waals radii are from J. E. Huheey, E. A. Keiter, and R. L. Keiter, *Inorganic Chemistry: Principles of Structure and Reactivity*, 4th ed., Harper-Collins, New York, 1993, p. 292; metallic and some covalent radii are from M. C. Ball and A. H. Norbury, *Physical Data for Inorganic Chemists*, Longman, London, 1974. The metallic radii are Pauling's metallic radii adjusted to correspond to a coordination number of 12 for the metal atoms.

[a] Units are in picometers (pm).

[b] The column headed "Cov–Cat" gives the difference of the covalent and cationic radii of the element; the column headed "An–Cov" gives the difference of the anionic and covalent radii.

[c] Values suggested by R. J. Gillespie and E. A. Robinson, *Inorg. Chem.*, **31**, 1960 (1992); see also Section 3.1 of this text.

of its *core* electrons; the outer circle for each atom represents the limit of its outer *valence* electrons. (Although we cannot measure such radii directly, it is useful to relate observed radii to the radii we anticipate for core and valence electrons.)

Let us first suppose that the species of Figure 1.10(*a*) are two *covalent molecules* of an element, such as Cl_2. In a covalent molecule, the valence orbitals overlap so that a valence electron of each atom can be shared with the other atom. Overlap is improved by bringing the two atoms closer, but ultimately we would expect the overlapping valence electrons of one atom to "bump into" and be repelled by the core electrons of the other atom. In a molecule such as Cl_2 in which both atoms are identical, we define the **covalent radius** to be *one-half of the internuclear distance,* that is, the distance from either nucleus to the straight line in Figure 1.10. Clearly, this distance should be intermediate between the radius of the core electrons and the radius of the valence electrons of either atom.

The two Cl_2 molecules of Figure 1.10 are supposed to be in the solid state, in a crystal in which each one just touches other molecules above and below it. We take *one-half of the internuclear distance* to the nearest chlorine nucleus in the next molecule to be the **nonbonded** or **van der Waals radius** of the chlorine atom. Since this distance should roughly correspond to the radius of the valence electrons, we expect the van der Waals radius of an atom to exceed its covalent radius. Unfortunately, for most elements, it is difficult to find cases of molecules in which two atoms of that element touch but in which we are certain that they do not bond. Hence, quoted van der Waals radii differ considerably and are not known accurately.

In Figure 1.10(*b*) we represent four atoms of a solid metallic element. In solid metals each atom characteristically has 8 or 12 nearest-neighbor atoms, all at the same distance. Hence, we have pushed the two molecules of Figure 1.10(*a*) together until they merged, with all distances becoming equal. This merger results in a **metallic radius** for each atom that somewhat exceeds the covalent radius of the "unmerged" atoms of Figure 1.10(*a*), due to the effect of the large number of atoms being brought into close proximity. Pauling[11] carried out calculations designed to compensate for this effect; the resulting *single-bond metallic radii* are quite comparable to covalent radii found in covalent compounds of the metals (Table 1.5). Since pure covalent compounds of the

Table 1.5
Covalent- and Single-Bonded Metallic Radii of the Elements

Group	1	2	3	4	5	6	7	8	9	10	11	12	13 (III)	14 (IV)	15 (V)	16 (VI)	17 (VII)	18 (VIII)
Period																		
1	H 37																	He 32
2	Li 134 122	Be 125 89											B 90 80	C 77	N 75	O 73	F 71 (54)[a]	Ne 69
3	Na 154 157	Mg 145 136											Al 130 125	Si 118 117	P 110 110	S 102 104	Cl 99	Ar 97
4	K 196 202	Ca 174	Sc 144	Ti 136 132	V 122	Cr 119	Mn 139 118	Fe 125 117	Co 126 116	Ni 121 115	Cu 135 118	Zn 120 121	Ga 120 125	Ge 122 124	As 122 121	Se 117 117	Br 114	Kr 110
5	Rb 216	Sr 191	Y 162	Zr 148 145	Nb 134	Mo 130	Tc 127	Ru 133 125	Rh 132 125	Pd 131 128	Ag 152 134	Cd 148 138	In 144 142	Sn 140 142	Sb 143 139	Te 135 137	I 133	Xe 130
6	Cs 235	Ba 198	Lu 156	Hf 144	Ta 134	W 130	Re 128	Os 133 126	Ir 132 126	Pt 131 129	Au 140 134	Hg 148 139	Tl 147 144	Pb 146 150	Bi 146 151	Po	At	Rn 145
			3F	*4F*	*5F*	*6F*	*7F*	*8F*	*9F*	*10F*	*11F*	*12F*	*13F*	*14F*	*15F*	*16F*		
6			La 169	Ce 165	Pr 164	Nd 164	Pm 163	Sm 162	Eu 185	Gd 162	Tb 161	Dy 160	Ho 158	Er 158	Tm 158	Yb 170		
7			Ac	Th 165	Pa	U 143	Np	Pu	Am	Cm	Bk	Cf	Es	Fm	Md	No		

SOURCES: Covalent radii are listed in the first row under the symbol for each element; these are taken from J. E. Huheey, E. A. Keiter, and R. C. Keiter, *Inorganic Chemistry: Principles of Structure and Reactivity*, 4th ed., Harper-Collins, New York; 1993, p. 292, or, if not available there, from M. C. Ball and A. H. Norbury, *Physical Data for Inorganic Chemists*, Longman, London, 1974, pp. 139–144. Single-bonded metallic radii are listed in the second row under the symbol for each element; these are taken from Ball and Norbury, op. cit.

[a] Value in parentheses was suggested by R. J. Gillespie and E. A. Robinson, *Inorg. Chem.*, **31**, 1960 (1992); see also Section 3.1 of this text.

metals are often hard to find, these single-bond metallic radii are often useful substitutes in looking at periodic trends in covalent radii.

Finally, let us suppose that the diatomic species of Figure 1.10(*a*) is an *ion pair*: A cation touching an anion. For many elements, the characteristic cation has lost all of its valence electrons; hence, we may suppose that the cationic radius equals the radii of the core electrons, and should be smaller than any of our other radii. Monoatomic anions characteristically fill their valence orbitals, so we may suppose that the anionic radius equals the radius of the valence electrons.

In looking at periodic trends in atomic radii, we must be sure to compare radii of the same type. But since a given element does not generally engage in all of these types of interactions, we often do not know a given type of radius for a particular atom, or we may not know it with much certainty. In such cases, we can deduce trends approximately by noting the *usual* relationships among the radii, as illustrated in Table 1.4: (1) the smallest radius is the *cationic radius*; (2) roughly 45–65 pm larger than this are the *covalent radius* and *metallic radius*; (3) roughly 55–75 pm larger than the latter two radii are the *anionic radius* and the *van der Waals* radius.

Trends in Covalent Radii. We will first examine periodic trends in covalent and metallic radii, since charge is not a variable in this series, and since one or the other of these radii is available for almost all elements of the periodic table (Table 1.5). Note that neutral atoms (as found in the elements, metals, or in covalent compounds) tend to become *smaller* from *left to right* in the periodic table. This trend is greatest in the *p* block, diminishes in the *d* block (with some anomalies at the end), and is almost but not quite absent in the *f* block (with anomalies at Eu and Yb).

In general, atoms get *larger* from *top* to *bottom* within a group, and this trend is well established in the *s* block. In the early part of the *p* block, however, we notice two interruptions of this trend: the fourth-period elements Ga and Ge are not significantly larger than the third-period elements Al and Si; the sixth-period elements Tl and Pb are almost not significantly larger than the fifth-period elements In and Sn. This latter anomaly also applies to the sixth-period *d*-block elements as compared to the fifth-period *d*-block elements.

Trends in Ionic Radii. The best available measurements of ionic radii, the Shannon–Prewitt values, are listed in Table C (inside the back cover). Horizontal trends in ionic radii may be examined in either of two ways. The simple horizontal trend easily observed in atoms is more difficult to see in ions, which are constantly changing charge across the *s* and *p* blocks. The most extensive horizontal series of cations of common charge is in the *f* block of elements, among the +3 ions of the lanthanides (elements of the sixth period); quite analogously to the neutral atoms, their +3 cations *decrease slightly in radii from left to right.* The same trend may be seen in a given period among the *d*-block elements of constant charge (i.e., +2, +3, or +4), although some anomalies may be noted.

We may also select a horizontal series of ions that are **isoelectronic**—that is, a series of ions of differing atomic number that have the same electron configuration. To examine such a trend, we may start with a Group 1 cation (with one more proton than its total number of electrons), then add protons, producing cations in later groups with increased charge due to increased numbers of protons but with the same number of electrons. For example, we may compare the cationic radii of the following isoelectronic ions in the fifth period: Rb^{+}, Sr^{2+}, Y^{3+}, Zr^{4+}, Nb^{5+}, Mo^{6+}, and Tc^{7+}. We note that as the positive charge on the nucleus increases, the attraction for the 36 electrons becomes stronger, resulting in a strong *contraction of the ion with increasing positive charge*. We may also extend this series to the left of Group 1 back to the preceding isoelectronic anions Br^- and Se^{2-}; these ions are larger with their increasing negative charges.

Among metals that form more than one cation, we find that with increasing positive charge, the radius of the cation decreases [see, e.g., the data for V, Cr, or Mn in Table C]. By removing electrons, we increase the surplus of protons over electrons and reduce the tendency of electron clouds to expand due to repulsion of their like charges.

Finally, taking some group such as Group 1, we may examine the vertical trends in atomic radii. Since the principal quantum number of the valence orbital steadily increases down a group, we expect to and do find that the radius of the ion increases in size. This trend holds quite well among the *s*-, *p*-, and (to a lesser extent) *f*-block elements. Just as there were with neutral atoms, there are diminishments or interruptions of the normal trend in the *p* block between the third and the fourth periods, and between the fifth and the sixth periods. In the *d* block, a corresponding reversal of trends occurs from the fifth to the sixth period.

Calculated Radii. The old Bohr model of the atom, which puts electrons in orbits, allows an easy calculation of the radius of the orbit of an electron:

$$r = a_0(n^*)^2/Z^* \tag{1.13}$$

This equation includes the radius of the Bohr hydrogen atom, $a_0 = 52.9$ pm, and indicates, as we might expect, that atoms should get larger as the effective principal quantum number n^* for the outermost electron increases but that they should get smaller as the effective nuclear charge for that electron increases. Of course, the Bohr model is no longer accepted, but the equation can still be used to give us a rough calculation of $\langle r_{max} \rangle$, the expected distance from the nucleus to the maximum in the radial probability function for the atom. As an extreme and interesting example, radio astronomers have detected, in interstellar gas regions, hydrogen atoms in which the electron is located in the 252*s* orbital! Application of Eq. (1.13) indicates that such an atom should have a radius of 0.00339 mm, and be visible under an ordinary microscope.[12]

Returning to earthly, ground-state atoms, the calculation for Li, for example, gives $\langle r_{max} \rangle = (52.9)2^2/1.30 = 162$ pm. Since $\langle r_{max} \rangle$ is not exactly the same as any of the types of experimental radii we have examined, and since the Bohr atom is not the last word in atomic theories, it is not surprising that our result differs from the covalent radius of Li, 134 pm (Table 1.5). The usefulness of $\langle r_{max} \rangle$ lies in looking at its periodic trends and seeing whether we can duplicate the periodic trends in other types of radii (we will use covalent and metallic radii). Table 1.6 presents calculated values across some periods and down some groups and compares the results with the observed covalent (or metallic) radii from Table 1.5.

First, we note that the atoms of the second period are calculated to contract across the *s* and *p* blocks, which they indeed do. According to Eq. (1.13), the radius decreases because of the increase in Z^*, which suggests that atoms decrease in size across the *s* and *p* blocks because additional *s* and *p* electrons of the same principal quantum number do a poor job of shielding each other; hence the additional protons in each nucleus across the period pull in *all* the valence electrons more strongly.

Table 1.6
Calculated Atomic Radii for Selected Atoms

Atom	Li	Be	B	C	N	O	F	Ne		
$\langle r_{max} \rangle$	162	108	81	63	54	46	41	36		
r_{cov}	134	125	90	77	75	73	71 (54)[a]			
Atom	Be	Mg	Ca	Sr	Ba					
$\langle r_{max} \rangle$	108	167	254	297	327					
r_{met}	89	136	174	191	198					
Atom	Sc	Ti	V	Cr	Mn	Fe	Co	Ni	Cu	Zn
$\langle r_{max} \rangle$	241	230	219	210	201	193	186	179	172	166
r_{met}	144	132	122	119	118	117	116	115	118	121
Atom	C	Si	Ge	Sn	Pb					
$\langle r_{max} \rangle$	65	115	128	150	165					
r_{cov}	77	118	122	140	146					
Atom	Zn	Cd	Hg							
$\langle r_{max} \rangle$	166	195	214							
r_{cov}	121	138	139							

[a] Value suggested by R. J. Gillespie and E. A. Robinson, *Inorg. Chem.*, **31**, 1960 (1992); see also Section 3.1 of this text.

Second, we note that atoms of Group 2 are calculated to increase in size down the group, which is in fact observed. (The calculated values seriously exaggerate the trend, however.) From the calculations, this trend must simply be a consequence of increasing the (effective) principal quantum number of the valence orbital: higher n or n^* means a larger orbital.

The third row of Table 1.6 shows a calculation across the *d* block. Atoms are calculated to decrease in size here too due to the increases in Z^*_{4s} (Z^*_{3d} is not relevant here since the size of the atom is determined by the size of the largest valence orbital, 4*s*). Since the increases in Z^* are smaller in the *d* than the *p* block, we calculate smaller changes in size per element. By and large this is found, although we have not accounted for the increases in size at Cu and Zn.

The fourth row shows a calculation down the *p* block [Group 14(IV)]. As expected, the atoms increase in size, but there is a smaller than expected increase between Si and Ge [or, in Group 13(III), between Al and Ga]. This smaller than expected increase is found even more emphatically in actual covalent radii. Between these two periods, n^* is increasing as usual, but Z^* partly counteracts this effect by increasing 1.50 more units than expected due to the first insertion of the *d*-block elements, as we discussed in Section 1.7. This effect is sometimes known as the **scandide contraction** of the elements Ga through Kr, since it is caused by the appearance of a new orbital type, the 3*d*, first found in scandium. Ironically, it is not really a contraction, since the expanding effect of the increase in n^* is still larger than the effect of the increase in Z^* [n^* is squared in Eq. (1.13), whereas Z^* is not]; but it is a contraction relative to normal trends. There is no contraction between Ge and Sn because both have *d* orbitals and equally increased Z^* values. In general, for properties depending on size, we may anticipate unusual resemblance of the corresponding *third- and fourth-period elements beyond the d block* because of the scandide nonexpansion.

A calculation not shown in Table 1.6 is that for crossing the *f* block, for the simple reason that $\langle r_{max} \rangle$ is 327 pm for the whole series of elements La through Yb. (Neither n^* nor Z^* changes in this series of elements.) This trend is almost, but not quite, observed in the metallic radii of these elements (Table 1.5), which decrease slowly from 169 pm for La to 158 pm at Tm (Eu and Yb are anomalies). This finding suggests that there really is some increase in Z^* across this series, contrary to Slater's rules, and that the 6*s* orbital really does penetrate the 4*f* orbital to some extent. Since we know that *s* orbitals are good penetrators and *f* orbitals are poor shielders, this finding is not very surprising. Slater's rules, designed for simplicity, are just not subtle enough to catch this effect.

The last calculation shown in Table 1.6 is for the Group 12 elements. These are calculated to expand normally on going down from Zn to Cd to Hg, but they do not in fact expand on going from Cd to Hg. Evidently, there is a **lanthanide contraction** (or nonexpansion) occurring in the elements after the first filling of *f* orbitals (which starts at La). This contraction prevents the post-*f*-block sixth-period elements from being larger (or much larger) than the corresponding fifth-period elements, which is confirmed in the metallic radii of Table 1.5. For properties of the elements that depend on size, we expect an unusual resemblance of the corresponding fifth- and sixth-period elements (beyond the *f* block).

1.10 Periodic Trends in Electronegativities of Atoms and Their Explanation

When atoms of different kinds form a covalent bond, we expect that the attraction of the two nuclei for the shared electron pair will not in general be the same and that the electron pair will, on the average, be closer to one of the nuclei. Pauling sought a measure of the relative abilities of the different atoms to attract such bond electrons to themselves, which he called the atom's **electronegativity**. It is not clear how such a property of atoms is to be measured experimentally; different people have proposed different measures of electronegativity. Pauling himself examined bond energies of molecules (Section 7.5) and obtained a table of what we now call **Pauling electronegativities**. An updated version of these values is included inside the front cover as Table A. (The method of measuring these values has an uncertainty of ± 0.05 units, so differences of less than this may not be meaningful.)

The overall trends in any scale of electronegativity are (1) electronegativities of atoms generally *increase* from *left to right* across the periodic table; (2) electronegativities generally *decrease* as we go from top to bottom of the periodic table. But there are parts of the periodic table in which these trends disappear or even reverse. If Pauling electronegativities strongly influence chemical properties, we would expect to find some anomalous chemical trends in these areas. We note that the general horizontal trend holds well across the second and third periods, with large increases in electronegativity from one element to the next one on its right. Electronegativities rise more slowly and sometimes decrease across the *d*-block elements in the fourth to sixth periods; electronegativities scarcely vary at all across the *f*-block elements.

Going *down groups* of the *s*-block elements, we see that all goes well; but going down groups of the *f*- and *d*-block elements, we often note significant *increases* in electronegativities. Going down Groups 13(III) and 14(IV), we notice that electronegativity values first decrease, then increase, then decrease again. This anomalous trend seems to dampen out and disappear in the later *p* groups. To the extent that electronegativity is an important predictor of chemical properties in the remaining chapters, we should also note unusual trends in chemical reactivity in these groups.

Allred and Rochow[13] attempted to produce a more theoretically based electronegativity scale. To do this, they interpreted Pauling's concept of electronegativity as being the force exerted by an atom on its (s and p) valence electrons:

$$F = e^2 Z^* / r^2 \tag{1.14}$$

For the charge, they used the effective nuclear charge Z^* from Slater's rules, and for the radius they used the covalent radius ($\langle r_{max} \rangle$ is clearly too inaccurate to be used here). They added certain parameters so that the range of their numbers would correspond to the Pauling scale of electronegativities and obtained the following equation:

$$\chi_{AR} = 3590 \frac{(Z^* - 0.35)}{r^2_{cov}} + 0.74 \tag{1.15}$$

The resulting Allred–Rochow electronegativities, χ_{AR}, of the elements (obtained using covalent radii available at the time) are shown in Table 1.7.

The computed Allred–Rochow and experimental Pauling electronegativities of the elements are in excellent agreement in the second, third, and fourth periods, and below these in the s blocks; hence the Allred–Rochow electronegativity calculations provide us with explanations of the main periodic trends in Pauling electronegativities.

Table 1.7
Allred–Rochow Electronegativities of the Elements

Group / **Period**	1	2	3	4	5	6	7	8	9	10	11	12	13 (III)	14 (IV)	15 (V)	16 (VI)	17 (VII)	18 (VIII)
1	H 2.20																	He 5.50
2	Li 0.97	Be 1.47											B 2.01	C 2.50	N 3.07	O 3.50	F 4.10	Ne 4.84
3	Na 1.01	Mg 1.23											Al 1.47	Si 1.74	P 2.06	S 2.44	Cl 2.83	Ar 3.20
4	K 0.91	Ca 1.04	Sc 1.20	Ti 1.32	V 1.45	Cr 1.56	Mn 1.60	Fe 1.64	Co 1.70	Ni 1.75	Cu 1.75	Zn 1.66	Ga 1.82	Ge 2.02	As 2.20	Se 2.48	Br 2.74	Kr 2.94
5	Rb 0.89	Sr 0.99	Y 1.11	Zr 1.22	Nb 1.23	Mo 1.30	Tc 1.36	Ru 1.42	Rh 1.45	Pd 1.35	Ag 1.42	Cd 1.46	In 1.49	Sn 1.72	Sb 1.82	Te 2.01	I 2.21	Xe 2.40
6	Cs 0.86	Ba 0.97	Lu 1.14	Hf 1.23	Ta 1.33	W 1.40	Re 1.46	Os 1.52	Ir 1.55	Pt 1.44	Au 1.42	Hg 1.44	Tl 1.44	Pb 1.55	Bi 1.67	Po 1.76	At 1.90	Rn 2.06
7	Fr 0.86	Ra 0.97																
			3*F*	4*F*	5*F*	6*F*	7*F*	8*F*	9*F*	10*F*	11*F*	12*F*	13*F*	14*F*	15*F*	16*F*		
6			La 1.08	Ce 1.08	Pr 1.07	Nd 1.07	Pm 1.07	Sm 1.07	Eu 1.01	Gd 1.11	Tb 1.10	Dy 1.10	Ho 1.10	Er 1.11	Tm 1.11	Yb 1.06		
7			Ac 1.00	Th 1.11	Pa 1.14	U 1.22	Np 1.22	Pu 1.22	Am	Cm	Bk	Cf	Es	Fm	Md	No		

SOURCES: Values from A. L. Allred and E. G. Rochow, *J. Inorg. Nucl. Chem.*, **5**, 264 (1958); E. J. Little and M. M. Jones, *J. Chem. Educ.*, **37**, 231 (1960); and L. C. Allen and J. E. Huheey, *J. Inorg. Nucl. Chem.*, **42**, 1523 (1980).

1. Electronegativities increase from left to right because Z^* increases, according to Rule 3 of Slater's rules: As we add additional electrons on crossing a period, we are also adding additional protons, which are only 35% shielded by the additional electrons; hence, all valence electrons (including those in the bond) are more strongly attracted to the nucleus.

2. Down Groups 1 and 2, Z^* often remains constant, so the small decline in electronegativity is due to the larger size: The bond electrons are farther from the nucleus, and hence somewhat more weakly attracted to it.

The Allred–Rochow electronegativities also duplicate the anomalously high electronegativities of Ga (>Al) and Ge (>Si). This trend is due to the increase in Z^*, which results from the first filling of the poorly shielding *d* orbitals—the same factor responsible for the scandide contraction discussed in Section 1.9. Even though Slater's rules (and hence the Allred–Rochow electronegativities) do not predict the lanthanide contraction, we know it is there and expect that it too should cause the Pauling electronegativities of the post-*f*-block sixth-period elements to be higher than those of the corresponding elements of the fifth period. This is, of course, often the case: We find the electronegativity relationships Au > Ag, Hg > Cd, Tl > In, Pb > Sn in the Pauling but not the Allred–Rochow scales.

But not every anomaly is explained: In the Pauling scale, the fifth-period elements of the *d* block are strikingly more electronegative than the corresponding fourth-period elements and not so much less electronegative than the sixth-period elements, except at gold. The lanthanide contraction cannot explain that, nor can it explain the fact that the deviation between the Pauling and the Allred–Rochow scales reaches its maximum at gold, in the *middle* of the sixth period and yet far beyond the *f* block.

Example 1.8

The greatest discrepancy between the Pauling and Allred–Rochow electronegativity scales occurs at gold. (a) Use Slater's rules to calculate (in the normal manner) Z^*_{6s} for the outer valence *s* electron of gold. (b) Rearrange the equation for Allred–Rochow electronegativities so that Z^* is the unknown. Assuming that 2.54 is what the Allred-Rochow electronegativity of gold should be, calculate what Z^* of gold should be.

SOLUTION:

(a) The characteristic electron configuration of Au, written in Slater's order, is $1s^2(2s,2p)^8(3s,3p)^8 3d^{10}(4s,4p)^8 4d^{10}4f^{14}(5s,5p)^8 5d^9 6s^2$. By Rule 3, the other 6*s* orbital contributes +0.35 to the shielding. By Rule 4a, the 17 electrons of principal quantum number 5 contribute 14.45, and the other 60 electrons contribute 60, to the shielding of the gold 6*s* electron, so $Z^*_{6s} = +4.20$.[14]

(b) Equation (1.15) can be rearranged to read

$$Z^* = 0.35 + \frac{r_{\text{cov}}^2(\chi_{\text{AR}} - 0.74)}{3590}$$

Substituting the covalent radius of gold, 140 pm, and the Pauling electronegativity of 2.54 in place of χ_{AR}, we calculate $Z^*_{6s} = +10.18$, which is considerably larger than the value calculated by Slater's rules.

1.11 *Relativistic Effects on Orbitals

The orbital wave functions we have been discussing up to this point do not incorporate any effects of relativity theory. But solutions of the Schrödinger wave equation that do incorporate relativity theory show that both the radial and the angular parts of the wave functions for the heavier atoms in the periodic table are appreciably altered by relativistic effects.[15,16]

The radial effect is known as **relativistic contraction**. We can think of the electron as a particle that is accelerated to a certain radial velocity by the attraction of the nucleus; as the nuclear charge builds up, this radial velocity builds up too: The radial velocity approaches the speed of light as the nuclear charge approaches $Z = 137.036$. The average radial velocity of the 1*s* electrons of an atom is proportional to the ratio of the nuclear charge of that atom to 137; the 1*s* electrons of a mercury atom move at $80/137 = 0.58$ times the speed of light. According to Einstein's theory of relativity, the mass, m, of a particle increases over its rest mass, m_0, when its velocity, v, approaches the speed of light, c:

$$m = m_0 \Big/ \sqrt{1 - (v/c)^2} \tag{1.16}$$

Hence, the mass of a 1*s* electron in mercury is about 1.2 times its rest mass. But the radius of the Bohr orbit of an electron is inversely proportional to the mass of the electron, so we expect the radius of the 1*s* orbital to be about 20% less than otherwise expected. This value is the relativistic contraction. It also affects the 2*s* and higher *s* orbitals roughly as much because of their inner lobes, which are close to the highly charged nucleus.

The effect is present to a lesser extent among *p* orbitals and is nearly absent among *d* and *f* orbitals, which have fewer lobes near the nucleus. Indeed, the *d* and *f* orbitals are more effectively screened out by the contracted *s* (and *p*) orbitals, so they undergo **relativistic expansion**. This expansion of the 5*d* orbitals increases their $\langle r_{max} \rangle$ to become comparable with the $\langle r_{max} \rangle$ of the 6*s* electrons, which increases the ability of the 5*d* orbitals to form additional and stronger covalent bonds in the sixth period. Along with their relativistic expansion, the energies of the 5*d* orbitals are raised, allowing electrons to be removed more readily, and hence the ability of the sixth-period *d*-block elements to form compounds in high oxidation states. For similar reasons affecting the 4*f* orbitals, uranium characteristically is found in higher oxidation states than neodymium.

While the 5*d* orbitals are filling and expanding, they are doing their usual poor job of shielding the 6*s* orbital from the nucleus. Once the 5*d* orbitals are filled, they are no longer valence orbitals, and the Pauling electronegativity must be determined by the 6*s* orbitals in Group 11 and beyond. These orbitals are contracted strongly in gold and mercury (Fig. 1.11)[17]. Beyond mercury, the less strongly affected and better shielding *p* orbitals are valence orbitals, so the effect diminishes somewhat. Eventually, however, the increase in nuclear charge again causes atomic contraction.

Since relativistic contraction has its maximum effect in the sixth period at the 6*s* orbital of gold and mercury, anomalous effects appear for these elements. One effect is the enhanced electronegativity of gold and mercury as compared with the elements above them, silver and cadmium. (Although the lanthanide contraction is partially responsible, it cannot account for the fact that enhancement is at a maximum in this part of the period.) Since the 6*s* orbital is contracted and bound more tightly, its separation in energy from the 6*p* orbitals is anomalously large, so much so that the 6*p* orbitals are almost not valence orbitals, but rather are nearly postvalence orbitals. Mercury, with a

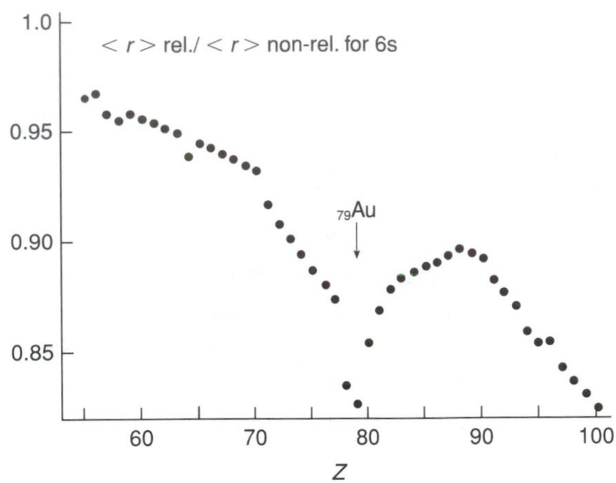

Figure 1.11
The relativistic contraction of the 6*s* orbital in the elements Cs ($Z = 55$) to Fm ($Z = 100$). [Adapted from P. Pyykko and J. P. Deselaux, *Accounts Chem. Res.*, **12**, 276 (1979).]

$5d^{10}6s^2$ valence electron configuration, has essentially filled all of its valence orbitals; this makes it very near to being a noble gas (or rather a noble liquid). Gold may then be thought of as one electron short of a noble liquid electron configuration and thus might be expected to show some resemblance to a halogen such as iodine. Indeed, gold has the highest electron affinity of any atom other than the halogens and forms a stable Au^- ion, similar in size to the Br^- ion, in the compounds RbAu, CsAu, and $(Cs^+)_3(Au^-)(O^{2-})$,[18] and in solution in liquid ammonia.

Beyond gold and mercury, the 6*s* electrons are still difficult to ionize, which accounts for the low stability of the group oxidation state in the subsequent elements. These elements prefer the oxidation state equal to the group number minus two; this is sometimes known as the **inert pair** (of 6*s* electrons) **effect**. Thus Tl normally occurs as Tl^+, Pb as Pb^{2+}, and Bi as Bi^{3+}. Another consequence is that it is advisable to use separate electronegativity values for Tl^+ and Tl^{3+}, and for Pb^{2+} and Pb^{4+}, since the 6*s* orbital participates directly in the covalent bond in the one case but not in the other.

In addition to the relativistic contraction of *s* orbitals and the relativistic expansion of *d* and *f* orbitals, there is another relativistic effect in heavy atoms known as **spin–orbit coupling**. The relativistic treatment of the Schrödinger wave equation was the first to reveal that electrons in atoms also have spin, and need a fourth quantum number, m_s, which can have the value of either $+\frac{1}{2}$ or $-\frac{1}{2}$. This number corresponds to angular rotation of the electron in either a clockwise or a counterclockwise direction. This angular rotation, of course, occurs even in the lightest atoms. But in heavy atoms the angular motion due to the spin of the electron about its own axis begins to mix with the angular motion of the electron among the lobes of the nonspherical orbitals. These two types of motion may either reinforce or cancel each other out in part. (We say that the electron spin angular momentum and the electron orbital angular momentum are coupled with each other.)

Consequently, in atoms heavier than bromine, we observe that the *p, d,* or *f* orbital electrons ionize at either of two energies, depending on whether the two types of angular motion are in harmony or are opposed. In the heaviest atoms, the difference between these two energies may be hundreds of kilojoules per mole, which is as large or larger than their covalent bond energies. Consequently, their bonding abilities are affected, and it becomes necessary to speak not just of a 6*p* electron but to distinguish which type of 6*p* electron is meant. The distinction is made by specifying an additional quantum number *j*, which can have either of two values: $j = l + \frac{1}{2}$ or $j = l - \frac{1}{2}$. This number is attached as a subscript on the right of the orbital designation: Thus we find two types of 6*p* orbitals, designated the $6p_{3/2}$ and the $6p_{1/2}$ orbitals. (The *nd* orbitals are also split into $nd_{5/2}$ and $nd_{3/2}$ orbitals.) Each set of orbitals can hold $2j + 1$ electrons; thus there can be four $6p_{3/2}$ electrons and two $6p_{1/2}$ electrons. The $6p_{1/2}$ orbitals penetrate the core better and experience relativistic contraction; hence, they are filled first and are more stable than the $6p_{3/2}$ orbitals by many kilojoules per mole (kJ mol^{-1}).

Consequently, in the sixth period of the *p* block the $6p_{1/2}$ orbital is a valence orbital, while the $6p_{3/2}$ orbitals are higher enough in energy that they are not necessarily used in stable compounds. This tendency is not very pronounced in the sixth period, but the ion Bi^{+} (valence electron configuration $6s^2 6p_{1/2}{}^2$) is known in a solid compound. This tendency is expected to be much more pronounced in the seventh period.

1.12 *Predicting the Chemistry of Superheavy Elements

At the time of this writing, 112 elements have been reported, with element 110 having been synthesized in November 1994,[19] 111 in December 1994,[20] and 112 in February 1996.[21] Although chemical properties are known only out to dubnium (element 105),[22] it is presumed that these elements are filling in the seventh period of the periodic table (the *d* block through element 112, then the *p* block through element 118). These elements should show enhanced relativistic properties over those of the sixth period, hence clarifying the nature of these effects. In element 112, the element below mercury, the stability of the $7s^2$ electron configuration might be great enough to cause it to be the next noble gas (actually liquid, since the van der Waals forces of 112 electrons will be considerable). Possibly element 114, the element below lead with the favored $7s^2 7p_{1/2}{}^2$ valence electron configuration, will also be a noble liquid.

Even without relativistic considerations, the yet-undiscovered elements of the eighth period should be very interesting to inorganic chemists. As can be seen in the longest form of the periodic table (Table 1.1), in every other period a new shape of orbital is first filled. The eighth period is expected to be one of those, in which a new type of orbital, the *g* orbital, is first occupied. Thus a new *g* block, 18 elements wide, would be expected between the *s* and *f* block. The 5*g* orbitals will surely be deeply buried inside the 8*s* orbitals and should, by Slater's rules and perhaps in fact, have shielding constants of 1.00. Consequently, these 18 elements may have identical radii, electronegativities, and chemistry, and could be a nightmare for the analytical chemist to separate. Following the filling of the 8*s* and the 5*g* orbitals, we would expect to fill the 6*f*, 7*d*, and 8*p* orbitals, reaching the next noble liquid at element 168, two periods below radon.

But, if we add in relativistic effects, the periodic table begins to look different from Table 1.1. In particular, calculations predict that the order of filling may not be the expected $8s < 5g < 6f < 7d < 8p$. Instead, spin–orbit coupling effects may be so great in the 8*p* orbitals that the $8p_{1/2}$ orbital may fill after 8*s*, while the $8p_{3/2}$ orbital may not fill at all in this period.[23] Consequently, the eighth period may end with the filling of the 7*d* orbitals, in element 164, which we may place two periods below mercury. The

Table 1.8
Principal Isotopes of Transuranium Elements

Isotope	Half-Life	Quantities Available
^{237}Np	2,200,000 years	Many kilograms
^{239}Pu	24,360 years	Many kilograms
^{244}Pu	82,800,000 years	>1 mg
^{243}Am	7650 years	>100 g
^{244}Cm	18.12 years	>100 g
^{247}Cm	16,000,000 years	Traces
^{247}Bk	1400 years	Traces
^{249}Bk	314 days	>1 mg
^{251}Cf	800 years	Traces
^{252}Cf	2.57 years	>1 mg
^{254}Es	276 days	>1 mg
^{257}Fm	94 days	>0.001 mg
^{258}Md	53 days	Traces
^{255}No	3 min	Traces
^{256}Lr	45 s	
^{261}Rf	70 s	

SOURCES: Data from F. A. Cotton and G. Wilkinson, *Advanced Inorganic Chemistry: A Comprehensive Text*, 5th ed., Wiley–Interscience, New York, 1988, p. 982; and the *Handbook of Chemistry and Physics*, 50th ed., Chemical Rubber Co., Cleveland, 1969, pp. B-267 to B-561.

properties of these elements may be quite different: The first element, number 119 below francium, is predicted to have a stable oxidation state of +4 due to the ease of ionization of its no-longer-core $7p_{3/2}$ electrons.[24]

Naturally, scientists are interested in the possibility of synthesizing superheavy atoms in order to test these predictions. Unfortunately, all naturally occurring elements of atomic number greater than 83 are radioactive; their half-lives are much less than the estimated age of the earth, 4.5 billion years. As they get heavier, their half-lives get shorter (Table 1.8), and the available world supply (from nuclear reactions) gets smaller. At the extreme, element 112 was produced in 1996 in the quantity of one atom, which lasted for 0.00028 s! (It was identified from its decay products.)[21]

Neutrons are important to the stability of nuclei: They act as a "glue" to bind protons together in the nucleus despite their electrostatic repulsion for each other. Nuclei with 20 or fewer protons are effectively bound together by an approximately equal number of neutrons, but as the number of protons grows to 83, it takes a proportionately larger number of neutrons (ultimately ~5 neutrons for each 3 protons) to hold the nucleus together. Figure 1.12 indicates the combinations of neutrons and protons that give rise to stable nuclei.

Nuclei that have insufficient numbers of neutrons to bind together their protons are called "proton rich," and undergo either of two types of radioactive decay that convert protons to neutrons: capture of a core electron to combine with the proton, or emission of a positron (positive electron). Nuclei that have too many neutrons are called "neutron rich," but seldom can emit the excess neutrons. Instead, the neutron in effect converts to a proton and an electron, the latter of which is emitted as a *beta* particle.

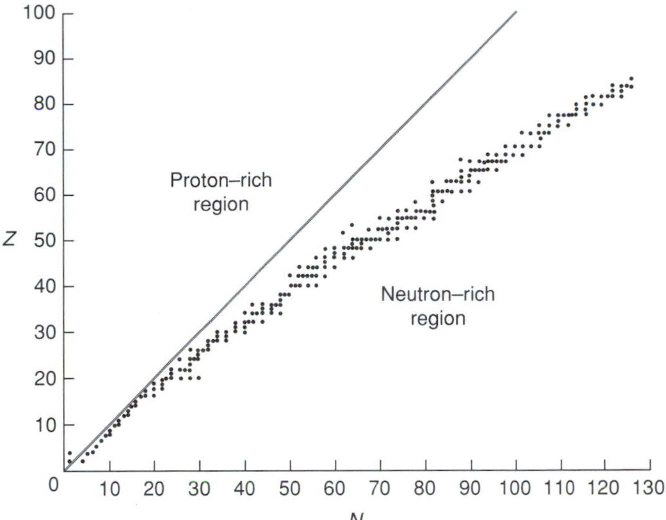

Figure 1.12
Combinations of neutrons and protons that form naturally occurring stable nuclei. [Adapted from *Nuclear and Radiochemistry*, 2nd ed., by G. Friedlander, J. W. Kennedy, and J. M. Miller. Copyright © 1964 by John Wiley & Sons, Inc.]

There is apparently no number of neutrons sufficient to bind nuclei with more than 83 protons together in a stable nucleus. These "overweight" atoms often emit *alpha* particles, which are helium-4 nuclei, that is, two protons and two neutrons. Many especially heavy atoms decay by *nuclear* fission: They break apart into two much smaller nuclei plus some spare neutrons.

Given these trends, it would seem impossible to verify the interesting chemical predictions made for superheavy elements in Section 1.11. However, the study of the occurrence of stable isotopes and the binding energies within them reveals that there are nuclear energy levels somewhat comparable to those of electrons, and that especial nuclear stability is reached with certain "magic numbers" of neutrons and protons, corresponding to noble gas electron configurations among electrons. Even numbers of neutrons and of protons are favored. Beyond that, filled nuclear energy levels (hence extra stability) seem to result with 2, 8, 20, 28, 50, and 82 protons (i.e., with helium, oxygen, calcium, tin, and lead); calculations suggest that 114 and 164 protons might fill the next two proton energy levels. Similarly, filled nuclear energy levels result from 2, 8, 20, 28, 50, 82, and 126 neutrons; calculations suggest that 184 and 196 neutrons might fill the next two energy levels. Hence, the hope exists that, although the elements from 93 to 109 protons lie far from the regions of nuclear stability, there might be one or more "islands of stability" for superheavy elements.[25]

Combining the proton magic number of 114 with the neutron magic number of 184, one might hope that extraordinary stability would be found for the isotope of mass number 298 of the element of atomic number 114. Calculations indicate that this isotope ought not to decay by beta particle emission, and that its half-life with respect to spontaneous nuclear fission ought to be very long: 10^{16} years! Unfortunately, the conditions for stability against alpha particle emission, beta particle emission, and nuclear fission differ somewhat; the half-life of this isotope with regard to alpha particle emission is expected to be only 1–1000 years. The longest half-lives have been calculated for the isotope of superheavy element 110 with 184 neutrons; this might have a half-life of 10^5 years. This element could offer many chemical possibilities and practical uses could perhaps be found.

Synthesizing such nuclei provides a formidable challenge, however. The early transuranic elements were synthesized by bombarding uranium (and later, heavier) atoms with neutrons or alpha particles, but this can only raise the atomic numbers by at most two; the necessary target nuclei are not available for element 114 or 164. Instead, it is proposed to use heavier bombarding particles, such as nickel-64 nuclei, on reasonably heavy targets such as bismuth-209 (this reaction actually led to element 111). However, in order to overcome electron–electron repulsions, *all* of the electrons must be stripped from the bombarding particle. Removal of all of the electrons from nickel to produce a Ni^{28+} ion requires the use of an extremely high-energy particle accelerator. The ion must also be given enough kinetic energy to overcome the repulsion of the Ni^{28+} nucleus and the Bi^{83+} nuclei, but not so much energy as to blast the target to pieces! Finally, there is the serious problem that most feasible projectiles do not have the 5:3 ratio of neutrons to protons, which will be needed in a stable superheavy nucleus. At the time of this writing, the theoretical "island of stability" of long-lived superheavy elements had not yet been reached, although efforts continue.[26]

Study Objectives

1. Draw the angular part of the wave function for *s*, *p*, *d*, and *f* orbitals. Include the radial part of the wave function in drawing specified *s*, *p*, or *d* orbitals; identify nodal planes and surfaces and the signs of the wave functions. Exercises 1–5.
2. Explain the concept of penetration or shielding of two different orbitals on the same atom, and how this gives rise to an effective nuclear charge that differs from the actual nuclear charge. Exercises 6–8.
3. Use Slater's rules to calculate Z^* for a given atom. Exercises 9–14.
4. Know the long form of the periodic table, so that, given a short-form table and the symbol (or atomic number) of an element (and the charge of a monoatomic cation or anion), you can write its characteristic valence electron configuration. Exercises 15–21.
5. Using numerical rules, calculate the oxidation numbers of atoms in specified molecules or ions. Exercises 22–26.
6. Without referring to Tables B1 or B2, describe the main trends in common and maximum oxidation numbers in different blocks of the periodic table. Exercises 27–33.
7. Distinguish the following types of atomic radii and compare them in magnitude for a given atom: covalent radius, metallic radius, cationic radius, anionic radius, and van der Waals radius. Exercises 34–39, and 69.
8. Without referring to Table C, describe the two main horizontal and the main vertical periodic trends in radii of atoms or ions. Exercises 40–43.
9. Use Z^* to calculate the radius $\langle r_{max} \rangle$ for a given atom; explain periodic trends in atomic radii. Explain the lanthanide and scandide contractions. Exercises 44–45.
10. Without referring to Table A, describe the characteristic horizontal and vertical periodic trends in Pauling electronegativities. Exercises 46–50.
11. Given the covalent radius of an element, use Z^* to calculate the Allred–Rochow electronegativity of an element. Explain periodic trends in electronegativities in terms of periodic trends in Z^*. Explain discrepancies between Allred–Rochow and Pauling electronegativities. Exercises 51–57.
12. Know that relativistic and spin–orbit coupling effects alter the shielding of outer electrons in very heavy atoms; know how this affects their sizes, electronegativities, and likely oxidation numbers. Exercises 58–61.
13. Extend the periodic table to include additional elements; extrapolate plausible valence electron configurations, electronegativities, ionic radii, and common positive oxidation numbers for these elements. Exercises 62–64.
14. Know the relationship of the periodic table to the quantum numbers of the atoms. Exercises 65–68.

Exercises

An asterisk (*) beside the exercise number indicates that the answers are in the back of the book.

1. *How many nodal planes do each of the following orbitals have? How many nodal spheres? Does the orbital have any inner lobes? (a) 13*s*; (b) 11*d*; (c) 5*g*; (d) 6*h*; (e) 9*g*; and (f) an orbital for which $n = 232$ and $\ell = 117$.

2. How would we designate (i.e., as 6*f*) an orbital having the following numbers of nodal planes and nodal spheres, respectively: (a) 2 and 1; (b) 0 and 4; (c) 2 and 5; (d) 4 and 0; and (e) 5 and 3.

3. Sketch *xy*-plane cross-section diagrams for the orbitals listed below. Indicate nodal planes and spheres by dashed lines, and regions of high electron probability by shading. Show the *x* and *y* axes. Indicate the positive and the negative parts of the wave function with plus and minus signs. (a) $4p_x$; (b) 3*s*; (c) $4d_{xy}$; (d) $3d_{x^2-y^2}$; and (e) 4*s*.

4. Some orbitals are drawn in Figure 1.13(*a*), along with their axis systems. (a) Tell how many nodal planes are present in each orbital; sketch these in. (b) Tell how many nodal spheres are present in each orbital; sketch these in. (c) Fill in the signs of the wave function on each lobe of each orbital. (d) Identify each orbital (e.g., $4f_{xyz}$).

5. (a) Tell how many nodal planes are present in each orbital sketched in Figure 1.13(*b*). (b) Tell how many nodal spheres are present. (c) Identify each orbital as specifically as possible (i.e., $4f_{xyz}$).

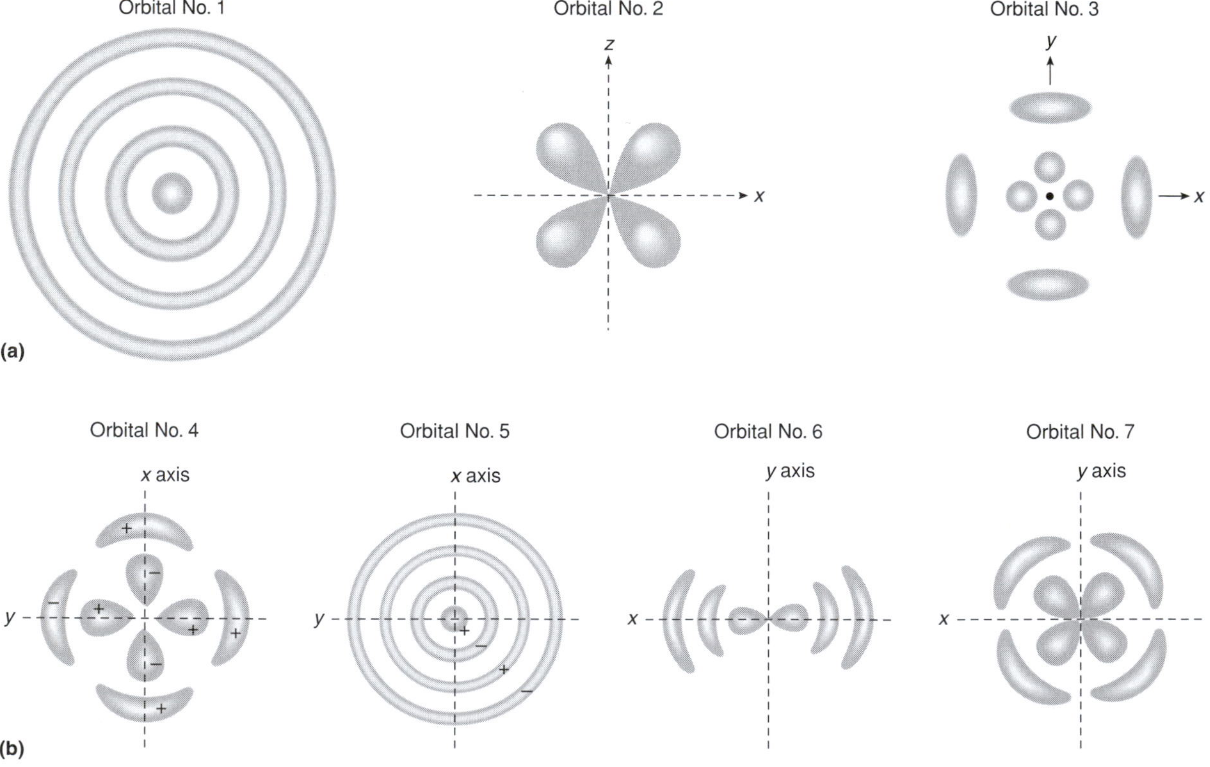

Figure 1.13
(*a*) Drawings of three orbitals for use in Exercises 4 and 7. (*b*) Drawings of four orbitals for use in Exercises 5 and 8.

6. *Rank the 5*s*, 5*p*, 5*f*, 5*d*, and 5*g* orbitals in order of increasing (a) penetrating ability; (b) shielding ability; and (c) degree to which they are shielded by inner electrons.

7. Of the orbitals drawn in Figure 1.13(*a*): (a) Which one will best penetrate the nucleus? (b) Which one will best shield other outer orbitals from the nucleus?

8. Of the orbitals drawn in Figure 1.13(*b*): (a) Which two will be poorest at penetrating the nucleus? (b) Which one will best shield other outer orbitals from the nucleus?

9. Write the full electron configurations of the following atoms, then rearrange them in the manner required for the application of Slater's rules: (a) Mg; (b) Si; (c) V; (d) Pm; (e) I; (f) Ir; and (g) No.

10. *Calculate Z^* for the 6*s* and the 4*f* valence electrons of Ce.

11. Tungsten has an atomic number of 74 and a characteristic electron configuration of $1s^2 2s^2 2p^6 3s^2 3p^6 4s^2 3d^{10} 4p^6 5s^2 4d^{10} 5p^6 6s^2 4f^{14} 5d^4$. (a) Rewrite this electron configuration in the order and groupings required for use with Slater's rules. (b) Compute Z^* for a 6*s* electron in this atom. (c) Compute Z^* for a 5*d* electron in this atom.

12. You have discovered a new atom, which has an atomic number of 125 and an electron configuration of

$$1s^2 2s^2 2p^6 3s^2 3p^6 4s^2 3d^{10} 4p^6 5s^2 4d^{10} 5p^6 6s^2 4f^{14} 5d^{10} 6p^6 7s^2 5f^{14} 6d^{10} 7p^6 8s^2 5g^5.$$

(a) Rewrite this electron configuration in the order and groupings required for use with Slater's rules. (b) Compute Z^* for a 5*s* electron in this atom. (c) Compute Z^* for a 5*f* electron in this atom.

13. (a) Calculate Z^* for the outermost *s* electron of the following sixth-period elements: Ba, Yb, Hg, Rn. (b) Across which block of the periodic table does Z^* of the outermost valence *s* electron vary most rapidly? least rapidly? (c) Explain why these answers are so, in terms of shielding.

14. (a) Calculate Z^* for electrons in the following types of orbital found in the Rn atom: 3*p*, 3*d*; 4*p*, 4*d*; 5*p*, 5*d*. (b) Show by calculation for which principal quantum number ($n = 3$, $n = 4$, or $n = 5$) the *np* and *nd* orbitals have the most nearly equal Z^* values. Would it be true to say that the deeper in the core one of these sets of orbitals is, the more hydrogen-like is the ordering of their energies?

15. *Write the characteristic valence electron configurations of the following atoms: (a) Fr; (b) As; (c) Pt; (d) Dy; (e) Ge; (f) Sr; (g) At; (h) W; (i) Bk; (j) Mn; and (k) Pa.

16. Write the characteristic valence electron configurations of the following atoms: (a) Pd; (b) Ce; (c) Co; (d) Au; (e) Am; (f) Bi; (g) Nd; (h) F; (i) Os; (j) U; (k) Mo; (l) Pu; and (m) Se.

17. Write the characteristic valence electron configurations of the following ions: (a) Bi^{3-}; (b) Bi^{3+}; (c) Ba^{2+}; (d) Pt^{2+}; (e) Dy^{3+}; (f) Ge^{2+}; (g) Ge^{4-}; (h) Pd^{2+}; (i) Sb^{3-}; (j) Nd^{3+}; (k) Am^{4+}; (l) Pb^{2+}; and (m) Co^{3+}.

18. Write the characteristic valence electron configurations of the following ions: (a) Cl^-; (b) At^{3+}; (c) Os^{2+}; (d) Mo^{4+}; (e) Se^{4+}; (f) Se^{2-}; (g) S^{2-}; (h) Tl^+; (i) Cs^+; (j) W^{4+}; (k) Bk^{3+}; and (l) Mn^{4+}.

19. *The characteristic valence electron configurations listed first are found in which of the species listed afterward? (a) $5s^25p^6$ is found in Xe, Te^{2+}, Te^{2-}, or none of these. (b) $4s^23d^5$ is found in V, Mn, Co^{2+}, or none of these. (c) $4f^2$ is found in Ce, Pr, Pr^{3+}, Nd^{3+}, Pm^{3+}, or none of these. (d) $6s^25d^{10}$ is found in Hg, Pt, Au^+, Au^-, Pb^{2+}, or none of these.

20. List all elements in the periodic table that (a) have five valence electrons; (b) have a Lewis symbol :X: (include elements for which we do not normally draw Lewis symbols); (c) have eight valence electrons; (d) can form ions with −2 charges.

21. Several chemical educators, most recently E. R. Scerri [*J. Chem. Educ.*, **68**, 122 (1991)] suggested an alternate form of the periodic table in which the *s* block is placed at the right, rather than the left, of the table. Discuss the advantages and disadvantages of this type of table in presenting the periodic trends of this chapter.

22. *Assign oxidation numbers to each atom in the molecules and ions of (a) CO_2; (b) NO_2^-; (c) NO^+; (d) H_2O_2; (e) CH_4; (f) IF_5; and (g) OsO_4.

23. Assign oxidation numbers to each atom in the molecules and ions of (a) HCN; (b) SO_3^{2-}; (c) NO_2^+; (d) XeF_4; and (e) ICl_4^-.

24. *Assuming the following species to consist of ions, write the valence electron configurations and (if appropriate) the Lewis symbols of the cations in (a) $FeCl_2$; (b) SnO_2; (c) PCl_3; and (d) $SbCl_6^{3-}$.

25. Assuming the following species to consist of ions, write the valence electron configurations and (if appropriate) the Lewis symbols of the cations in (a) UF_4; (b) WF_8^{3-}; (c) SCl_3^+; and (d) ClO_2^-.

26. Assign oxidation numbers to each kind of atom in each of the following: (a) BF_4^-; (b) NaH; (c) $H_2S_5O_6$; (d) As_7^{3-}; and (e) O_2F_2.

27. Without referring to Table B1, describe the most common positive oxidation numbers to be expected in the following regions of the periodic table: (a) *s* block; (b) left side of the *p* block; (c) right side of the *p* block; (d) left side of the *d* block; (e) right side of the *d* block; (f) left side of the *f* block; (g) right side of the *f* block.

28. *List all of the known chemical elements that have six valence electrons (and no more). Then, without referring to Tables B or C, tell which of these commonly achieve an oxidation number of 6. Give the atomic numbers of the next five elements that can be expected to have six valence electrons.

29. *Considering the location of each of the following elements within the *s, p, d,* or *f* blocks of elements, predict a likely common (positive or negative) oxidation number for each element: (a) Th; (b) Cl; (c) Zn; (d) W; (e) Ra; and (f) Si.

30. (a) Which of the following elements in fact show maximum positive oxidation numbers equal to their group numbers (take group numbers to be 3–8 in the *p* block): S, Ne, Ba, Ni, Ta, Er, U. (b) For each of the elements that do not show such high oxidation numbers, give a likely positive oxidation number.

31. The following elements form cations or at least exhibit positive oxidation numbers. Without reference to Tables B1 or B2, predict one or (if possible) two likely positive charges or oxidation numbers to be expected for each: (a) Be; (b) Ce; (c) V; (d) Sb; (e) N; (f) Se; and (g) I.

32. Which element is more likely to show the indicated oxidation number? (a) (Oxidation number +7) F or Cl; (b) (oxidation number +8) Fe or Os; (c) (oxidation number +6) Nd or U; (d) (oxidation number +6) S or Si; (e) (oxidation number −3) S or P.

33. For each element listed below, which is the *more* likely oxidation number? (Note that neither oxidation number may be common.) (a) Si +2 or −2; (b) Mt −3 or +5; (c) P +1 or +3; (d) P −1 or −3; (e) P +5 or −5; (f) Hs −2 or +8; (g) Po +4 or +2; and (h) Po −2 or +2.

34. *The single-bonded metallic radius of Na is 157 pm. Give a reasonable estimate of the following radii, assuming that the increment between radii of different magnitudes is 60 pm: (a) the covalent radius of Na; (b) the cationic radius of Na; (c) the anionic radius of Na; (d) the van der Waals radius of Na.

35. The cationic radius of Au (in Au^{3+}) is 99 pm. Give a reasonable estimate of the following radii for Au (assume the increment between radii of different magnitudes is 60 pm): (a) its covalent radius; (b) its anionic radius; (c) its single-bonded metallic radius; (d) its van der Waals radius.

36. The anionic radius of I^- is 206 pm. Give a reasonable estimate of the following radii, assuming that the increment between radii of different magnitudes is 60 pm: (a) covalent radius of I; (b) cationic radius of I^{7+}; (c) van der Waals radius of I.

37. The metallic radius of Pt is 129 pm. Give a reasonable estimate of the following radii, assuming that the increment between radii of different magnitudes is 60 pm: (a) the covalent radius of Pt; (b) the cationic radius of Pt^{6+}; (c) the anionic radius of Pt^{2-}.

38. Arrange the following types of atomic radii in order of increasing size for a given element: anionic radius, cationic radius, covalent radius, metallic radius, van der Waals radius. Be sure to indicate when two types of radius are approximately equal.

39. Suppose that you have a metallic atom bonded to a nonmetallic atom, and you are uncertain whether to classify the bonding as ionic or covalent. Using appropriate tabulated radii, calculate the internuclear distances you would expect to find for each type of bonding for the following cases: (a) Al and Cl; (b) Tl and I; and (c) Na and F. Which type of bonding appears to give longer bonds?

40. *Without the use of tables, arrange each of the following sets in order of increasing size: (a) Li, C, F, Ne; (b) Be, Ca, Ba, Ra; (c) B, Al, Ga, In, Tl; (d) V, Nb, Ta, Db (element 105).

41. Without the use of tables, arrange each of the following sets of atoms in order of increasing size, noting any cases in which two atoms should be of very similar size. (a) As, Br, Ca, Ga; (b) B, In, Al, Ga; (c) C, Si, Ge, Sn; and (d) Ge, Se, K, Zn.

42. Without referring to Table C, arrange the following sets of ions in order of increasing radii: (a) Cr^{6+}, Cr^{4+}, Cr^{2+}, Cr^{3+}; (b) Ra^{2+}, Mg^{2+}, Be^{2+}, Sr^{2+}; (c) the ions of valence electron configuration $6s^0$ between Cs and Re; and (d) the *f* block +3 ions of the seventh period.

43. Below are listed some series of cations; put each series in order of increasing size. (a) La^{3+}, Lu^{3+}, Eu^{3+}, Gd^{3+}; (b) Tl^{3+}, Ga^{3+}, B^{3+}; (c) Mn^{7+}, K^+, Sc^{3+}, V^{5+}; (d) V^{5+}, V^{2+}, V^{4+}, V^{3+}; (e) No^{3+}, No^{2+}, Md^{2+}, Fm^{2+}; (f) Cl^+, At^+, I^+, F^+ Br^+; (g) Au^+, Au^{7+}, Au^{3+}, Au^{5+}; (h) Pb^{2+}, Au^-, Tl^+, Bi^{3+}; (i) H^+, Fr^+, Rb^+, Li^+; and (j) Te^{4+}, In^+, Sb^{3+}, Sn^{2+}.

44. *(a) Using Slater's rules, calculate Z^* for the 4*s*, 3*d*, and 3*p* electrons in the Fe atom. (b) Calculate $\langle r_{max} \rangle$ for the 4*s*, 3*d*, and 3*p* electrons in the Fe atom.

45. In Exercise 13, you calculated Z^* for the outermost *s* electron of the following sixth-period elements: Ba, Yb, Hg, Rn. Use these numbers to (a) Calculate $\langle r_{max} \rangle$ for these four elements. (b) Explain what happens to these radii of atoms as one crosses the *s*, *p*-block of elements and why (in terms of Z^* and of shielding or penetration); (c) Explain what happens to $\langle r_{max} \rangle$ on crossing the *d* block of elements and how the magnitude of this change compares with that in the *s*, *p* block and why; (d) Explain what happens to $\langle r_{max} \rangle$ on crossing the *f* block of elements. (e) If your calculations fail to capture the observed trend in any of these cases, tell what the true trend is, and (if possible) why your calculations do not catch it.

46. *Fill in each blank with one of the following: increase; decrease; stay the same; more rapidly; less rapidly.
(a) Electronegativities of atoms ____ from left to right across a period. (b) Electronegativities of atoms ____ from top to bottom down a group. (c) Electronegativities of atoms change ____ when crossing the *d* block of elements than when crossing the *p* block. (d) Theoretical radii of atoms $\langle r_{max} \rangle$ ____ from left to right across the *f* block in the sixth period, but real radii ____ across the *f* block ____ than they do across the *p* block. (e) Radii of atoms ____ from top to bottom going down group 2; radii ____ ____ from top to bottom going down Group 13(III). Where in Group 13(III) do anomalies in this trend show up? Name the effects that are responsible.

47. Describe the main periodic trends of the following from left to right in the periodic table: (a) electronegativity; (b) radius, and (c) common oxidation numbers. Tell in what blocks of the periodic table (if any) one finds anomalous countertrends.

48. Describe the main periodic trends of the following from top to bottom in the periodic table: (a) electronegativity; (b) radius, and (c) common oxidation numbers. Tell in what blocks of the periodic table (if any) one finds anomalous countertrends.

49. Without the use of tables, arrange each of the following sets in order of increasing Pauling electronegativity: (a) Ba, Be, Ca, Mg, Sr; (b) Al, Cl, Na, P; (c) C, Si, Ge; and (d) Ag, Au, element below Au.

50. In what regions of the periodic table are there anomalous vertical trends in Pauling electronegativities? Are there corresponding anomalous vertical trends in these regions in ionic radii? Common positive oxidation numbers?

51. The Allred–Rochow electronegativity of Rn is not known; assuming a covalent radius for Rn of 130 pm, calculate it.

52. *(a) Using Slater's rules, calculate Z^*_{4s} for Zn. (b) Using Slater's rules, calculate Z^*_{3d} for Zn. (c) Calculate $\langle r_{max} \rangle$ for these two orbitals of zinc. (d) The covalent radius of zinc is 121 pm; calculate the Allred–Rochow electronegativity of Zn. How do you think this value compares with the Pauling value for Zn? If you think that it is either substantially (>0.5 units) higher or substantially lower, explain why this is the case.

53. In a certain atom, an electron in orbital A experiences an effective nuclear charge Z^* of +4.00, while an electron in orbital B experiences a Z^* of +21.85. Suppose also that the outermost *s* orbital of this atom is a 6*s* orbital ($n = 6$, $n^* = 4.2$), which experiences a Z^* of +2.60, and suppose that the covalent radius of this atom is 170 pm. (a) Is orbital A or orbital B more likely a *valence* orbital? (b) Compute $\langle r_{max} \rangle$ for this atom. (c) Compute χ_{AR} for this atom.

54. In a certain atom, an electron in orbital C experiences an effective nuclear charge Z^* of $+14.00$, while an electron in orbital D experiences a Z^* of $+4.85$. Suppose also that the outermost *s* orbital of this atom is a 6*s* orbital ($n = 6$, $n^* = 4.2$), which experiences a Z^* of $+3.60$, and suppose that the covalent radius of this atom is 190 pm. (a) Is orbital C or orbital D more likely a *valence* orbital? (b) Compute $\langle r_{max} \rangle$ for this atom. (c) Compute χ_{AR} for this atom.

55. Identify all elements in the periodic table for which the Pauling and Allred–Rochow electronegativities differ by more than 0.5. In what parts of the table do these occur?

56. Derive an equation for χ_{AR} in terms of Z^* and n^* as the only variables. (*Hint*: Make use of the equation for $\langle r_{max} \rangle$. Look for some simplifications to make the equation less messy.) Why is this equation not used?

57. *Compute and compare Z^* for the valence *s* orbital, $\langle r_{max} \rangle$, and χ_{AR}: (a) for Na and Cu; (b) for W, Nd, and U. Before World War II periodic tables had U listed below W, not Nd. From these numbers, does U appear to be more closely related to Nd or to W?

58. Explain why relativistic effects can cause gold to have both (a) a lower oxidation number than either Ag or Cu (-1 in CsAu) and (b) a higher oxidation number than either Ag or Cu ($+5$ in AuF_5 and $+7$ in AuF_7).

59. In Example 1.8, there are two effects that are responsible for the discrepancy between the Z^* calculated from Slater's rules and the Z^* calculated from the Pauling electronegativity of Au. (a) Name these two effects and describe them in terms of penetration or shielding. (b) For each of the two causes, tell whether the discrepancy should be the same, worse, or less for the element below gold. Overall, would you expect the electronegativity of this "eka-gold" to be greater than, less than, or equal to 2.54? (c) Name an element in an entirely different group of the periodic table to which the chemistry of eka-gold might be more related (than it would be to the chemistry of copper in Group 11).

60. *Keeping in mind the effects of relativistic contraction and expansion of orbitals and spin–orbit coupling of *p* orbitals, list some plausible likely common oxidation numbers for the elements of the seventh period beyond number 102. What element in this period has the best possibility of showing the hitherto-unknown oxidation number of $+9$, and what factor might make this possible?

61. Choose one of the following causes as the best explanation of each of the effects listed later. CAUSES: (1) Scandide contraction. (2) Lanthanide contraction. (3) Relativistic increase in mass of a valence *s* electron penetrating a heavy nucleus. (4) Relativistic shielding of a valence *d* or *f* electron by massive *s* electrons. (5) Spin–orbit coupling separating a subshell of valence *p* orbitals into sets with $j = \frac{3}{2}$ and $j = \frac{1}{2}$. (6) No special cause needed; this is a normal periodic trend. EFFECTS: (a) Gold has the highest electron affinity of any element other than a halogen. (b) Gold has a $+7$ oxidation state, much higher than is known for Cu or Ag. (c) The $8s^2 8p^2$ electron configuration of the element below lead is expected to be that of a noble gas or liquid. (d) It took 100 years to discover that the element hafnium is always present in samples of the element zirconium. (e) Gallium is more electronegative than aluminum. (f) Indium is less electronegative than gallium.

62. *Make reasonable predictions of the following: (a) the valence electron configurations of the elements of atomic number 126, 144, and 162; (b) the Pauling electronegativities of the three elements *directly below* U, Au, and At; (c) the common positive oxidation number for Rn, for element number 121, and for the elements *directly below* Np, No, Os, and Pb; (d) using Table C, the ionic radii for the cations of the elements with the common positive oxidation numbers in part (c) of this question.

63. (a) Iridium has a Pauling electronegativity of 2.20, and a common positive oxidation number of +4 with an ionic radius of 76 pm. Predict these properties for meitnerium, the element below iridium. (b) Bismuth has a Pauling electronegativity of 2.02, and a common positive oxidation number of +3 with an ionic radius of 117 pm. Predict these properties for the element below bismuth.

64. *Let us suppose that you have discovered a new element of atomic number 162, which you choose to name khalidium (Kh). Assume that the normal order of filling of orbitals still prevails in the eighth period. (a) Write the full electron configuration of this element. (b) Write the valence electron configuration of this element and of its +2 ion. (c) Note that a new type of orbital is used in Kh that is not used in any known atom. How many nodal planes will there be in a sketch of this type of orbital? How many nodal spheres will there be? Will the electrons in this type of orbital be good or poor at shielding the valence electrons of Kh? Will electrons in this type of orbital be good or poor at penetrating the inner orbitals of Kh? (d) Calculate Z^* for the outermost valence electron of Kh. (e) Using an estimated covalent radius of 150 pm, calculate the Allred–Rochow electronegativity of Kh. Do you think that the Pauling electronegativity of Kh will be higher, lower, or about the same as this value?

65. Suppose that some all-powerful being suddenly decided to change the Schrödinger wave equation. After this change, the energies of electrons in all atoms depends on the principal quantum number *only*. Everything else would remain the same. (a) Using the new rules, list the order of filling of the *s*, *p*, and so on orbitals for the first four values of the principal quantum number. (b) Write the electron configuration of nickel under the new rules. (c) What ions, if any, would nickel now tend to form? Could it achieve a noble gas electron configuration by covalent bonding? What would be the nearest noble gas configuration? Classify nickel now as a metal or a nonmetal. (d) What would happen to the nickels in your pocket when this change was made? What effect would this have on the economy?

66. Suppose that the solution to the Schrödinger wave equation were to be changed in this one way only: The secondary quantum number ℓ would be allowed to go to as high a value as the principal quantum number. (a) Write the order of filling of the eight lowest types of orbitals in an atom. (b) Draw a new periodic table showing the arrangement of the first 36 elements. (c) Assuming that ns^2np^6 remains a noble gas electron configuration, list the noble gases in your new periodic table. (d) If the changeover were made at the stroke of 12 midnight, discuss what would then happen to the water in the world.

67. Suppose that the inner lobes of orbitals were to be abolished, so that the 2*s* orbitals, for example, would lie completely outside the 1*s* orbitals of the same atom. (a) Explain how this would alter the shielding of orbitals. (b) Revise Slater's rules to accommodate this change. (c) Use the equations for $\langle r_{max} \rangle$ and Allred–Rochow electronegativities to reevaluate these quantities for the second and third-period elements. (d) Discuss any changes in periodic trends that would result.

68. Answer the two questions involving hypothetical sets of quantum numbers that were posed by C. H. Yoder and C. S. Yoder, *J. Chem. Educ.,* **67**, 759 (1990).

69. Recently, solids containing anions of the Group 1 metals (Na^-, K^-, Rb^-, Cs^-) have been isolated. (a) Compare and contrast the valence electron configurations of these anions with those of the *p*-block anions and of the *d*-block anion Au^-. (b) Predict the radii of these ions and compare them with the reported radii: R. H. Huang, D. L. Ward, and J. L. Dye, *J. Am. Chem. Soc.,* **111**, 5707 (1989); R. H. Huang, D. L. Ward, M. E. Kuchenmeister, and J. L. Dye, *J. Am. Chem. Soc.*, **109**, 5561 (1987).

70. Explain the fact that the element tin has an unusually high number of stable isotopes (10).

71. The stable isotope ^{48}Ca is unusual among light-element isotopes in its neutron/proton ratio. (a) Compute this ratio and explain why it is unusual. (b) Explain why this isotope is stable anyway. (c) Explain why this isotope is an attractive choice for use as a projectile in trying to create stable superheavy nuclei.

Notes

1. G. S. Girolami, T. B. Rauchfuss, and R. J. Angelici, *Synthesis & Technique in Inorganic Chemistry, A Laboratory Manual* 3rd ed., University Science Books, Mill Valley, CA, 1999.
2. There are a few elements that have such similar chemical and electronic properties that their placement into groups of the periodic table is somewhat arbitrary. Such a case is that of the pairs of elements lanthanum (La) and lutetium (Lu), and the pair actinium (Ac) and lawrencium (Lr), each of which has a claim to resemble the early elements in Group 3, scandium (Sc) and yttrium (Y). As pointed out by W. B. Jensen [*J. Chem. Educ.*, **59**, 634 (1982)], the metallurgical resemblance is much stronger for lutetium than for lanthanum, so we have adopted the metallurgist's convention of listing Lu (and by extension Lr) below Sc and Y. An important additional advantage of this is that the periodic table becomes more symmetrical, and it becomes easier to predict electron configurations. E. R. Scerri [*J. Chem. Educ.*, **68**, 122 (1991)] points out that recent determinations of the electron configurations of most of the *f*-block elements now are more compatible with this placement of Lu and Lr.
3. M. W. Browne, *New York Times*, Aug. 15, 1995, p. B5.
4. In relativistic quantum mechanical calculations, the probability here is not precisely zero, but is at a very low minimum.
5. This expression is somewhat more complex in Système International d'Unites (SI) units.
6. J. S. Slater, *Phys. Rev.*, **36**, 57 (1930).
7. E. Clementi and D. L. Raimondi, *J. Chem. Phys.*, **38**, 2868 (1963).
8. The calculations required for complete rationalization of core versus valence electron properties are actually those of ionization energies, which require applying Eq. (1.11) to each electron in the atom, then to each electron in the product ion, then taking the difference. We will not do these somewhat tedious calculations, which are outlined by K. F. Purcell and J. C. Kotz, *Inorganic Chemistry*, Saunders, Philadelphia, 1977, p. 45.
9. The electron configurations of atoms involve more than just the energies of orbitals, but also depend on interelectronic repulsion. It may be the case that, for example, the 4*s* orbital is always higher in energy than 3*d*, without altering the above order of filling; contributions to sorting this out include M. P. Melrose and E. R. Scerri, *J. Chem. Educ.*, **73**, 498 (1996); L. G. Vanquickenborne, K. Pierloot, and D. Devoghel, *Inorg. Chem.*, **28**, 1805 (1989).
10. Perhaps it is best, in the *p* block, just to think of the "1" in 13, 14, and so on, as serving the same function as the A or B in the old group numbers.
11. L. Pauling, *The Nature of the Chemical Bond*, 3rd ed., Cornell University Press, Ithaca, NY, 1960; Chapter 11.
12. D. B. Clark, *J. Chem. Educ.*, **68**, 454 (1991); **69**, 946 (1992).
13. A. L. Allred and E. G. Rochow, *J. Inorg. Nucl. Chem.*, **5**, 264 (1958).

14. Gold actually has an anomalous valence electron configuration of $5d^{10}6s^1$, for which $Z^*_{6s} = +3.70$.

15. K. S. Pitzer, *Acc. Chem. Res.*, **12**, 271 (1979).
16. P. Pyykkö and J.-P. Desclaux, *Acc. Chem. Res.*, **12**, 276 (1979); P. Pyykkö, *Chem. Rev.*, **88**, 563 (1988).
17. M. S. Liao and W. H. E. Schwarz [*Acta Crystallogr*, *Sect. B*, **50**, 9 (1994)] and A. Bayler, A. Schier, G. A. Bowmaker, and H. Schmidbaur [*J. Am. Chem. Soc.*, **118**, 7006 (1996)] recommend a radius for Au^+, which is 8–10 pm *smaller* than that of Ag^+, and which is considerably smaller than that given in Table C (inside the front cover). Data also exists to suggest a radius for Hg^{2+}, which is slightly less than that of Cd^{2+}: K. W. Klinkhammer and J. Weidlein, *Z. Anorg. Allg. Chem.*, **622**, 1209 (1996).
18. A. Pantelouris, G. Küper, J. Hormes, C. Feldmann, and M. Jansen, *J. Am. Chem. Soc.*, **117**, 11749 (1995); C. Feldmann and M. Jansen, *Z. Anorg. Allgem. Chem.*, **621**, 1907 (1995).
19. M. Freemantle, *Chem. Eng. News*, Nov. 28, 1994, p. 5.
20. M. Freemantle, *Chem. Eng. News*, Jan. 2, 1995, p. 7.
21. M. Rouhi, *Chem. Eng. News*, Feb. 26, 1996, p. 6.
22. D. C. Hoffman, *Chem. Eng. News*, May 2, 1994, p. 24, V. G. Pershina, *Chem. Rev.*, **96**, 1977 (1996).
23. B. Fricke and J. McMinn, *Naturwissenschaft*, **63**, 162 (1976).
24. "Extrapolation of Periodic Table Could Err," *Chem. Eng. News*, Sept. 3, 1973, p. 27.
25. G. T. Seaborg and W. D. Loveland, *The Elements Beyond Uranium*, Wiley-Interscience: New York, 1990; Chapter 7.
26. M. Freemantle, *Chem. Eng. News*, March 13, 1995, p. 35.

CHAPTER

2

Monoatomic Ions and Their Acid–Base Reactivity

Most elements are found in nature not as free elements but in the form of ions. Hence, we begin our study of the chemistry of the elements with some chemical properties of their ions. In this chapter, we will investigate the interaction of some common cations and monoatomic anions of the elements with water and see how the periodic trends in these reaction tendencies can be related to the atomic properties reviewed in Chapter 1. Your instructor may choose to have you begin your study of the chemistry of these ions with a laboratory investigation (or classroom demonstration or video and discussion) of the process of dissolving the cations of the elements (in the form of their chlorides) in water. This process may sound trivial, but you will find some unexpected drama in the process. You will find, upon analyzing the results, that even a simple reaction involves some important chemistry.

You instructor may now assign Experiment A.1. Then, we will construct a physical model of what is happening during these reactions. You will find that the principles you derive, and the classification scheme that is developed, apply to far more than just the reaction of a cation or an anion with the humble water molecule; ions react similarly with many other chemical species. The ways in which you begin looking at positively and negatively charged species in this chapter will be useful in subsequent chapters.

2.1 Hydration of Cations

When writing chemical equations for reactions of ions in solution we often write ions as if they were simple particles in solution; for example, we may write the sodium ion as Na^+ or perhaps as $Na^+(aq)$. But there are definite reactions between ions and molecules of the polar solvent, water. Since the oxygen atom of the water molecule is much more electronegative than the hydrogen atoms, each H–O bond is a polar covalent bond in which the bond electrons are (on the average) closer to the oxygen atom than the hydrogen atom, giving rise to a partial negative charge on oxygen and a partial

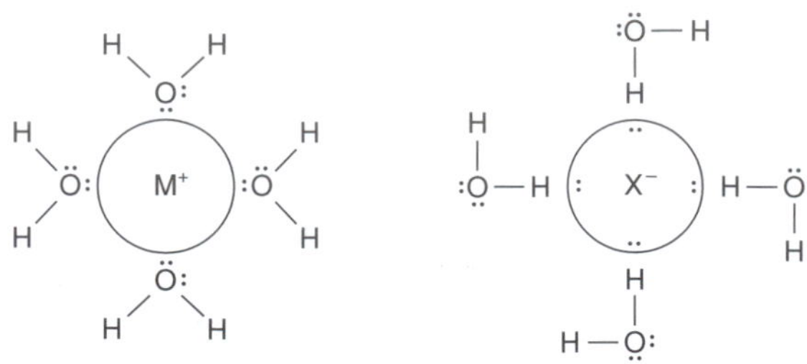

Figure 2.1
A hydrated cation and a hydrated anion.

positive charge on hydrogen. Since H_2O is not a linear molecule, it has a partial negatively charged end (its oxygen end) and a partial positively charged end (the hydrogen end). Since opposite charges attract, a positive ion (cation) placed in water surrounds itself with water molecules, with the oxygen ends inward toward the ion (Fig. 2.1). Conversely, a negative ion surrounds itself with water molecules, with the hydrogen ends inward. The products of these reactions are called **hydrated ions**.

The attraction of opposite charges is really quite a strong force. If we were to plunge 1 mol of gaseous cations into water, they would form hydrated ions and release a large amount of energy, which we call the **hydration energy** of the cation.[1] Hydration energies of a number of cations are listed in Table 2.1; by any normal chemical standard these are large energies. The data in Table 2.1 show that the hydration energy of a cation depends on the charge and the radius of the cation, as expected qualitatively from Coulomb's law, and also depends on the electronegativity of the element. Latimer[2] observed that if the electronegativity of the metal is not too great, the hydration energies of metal ions are given approximately by the equation

$$\Delta H_{\text{hyd}} = -60{,}900Z^2/(r + 50) \text{ kJ mol}^{-1} \tag{2.1}$$

where Z is the charge on the cation and r is the cationic radius (in picometers). (We can loosely equate the constant added to the radius of the cation with the radius of the oxygen in the water.)

No attempt is made in Latimer's equation to include the effects of electronegativity, but examination of the data for metals of Pauling electronegativities (χ_P) greater than 1.5 (on the right side of Table 2.1) shows that their hydration energies are substantially higher than those of ions of comparable radius and charge on the left side of the table. Such metals have electronegativities within about two units of that of oxygen, which suggests that for these metals there is not just an electrostatic attraction between the metal ion and the negative end of the water molecule, but that there also may be some degree of covalent bond formation, in which an unshared electron pair on water is shared with the metal ion.

2.2 Acidity of Cations

If the attraction of the metal ion for the negative end of the water dipole is strong enough, the water molecule itself is affected (Fig. 2.2). As the unshared electron pairs of the water molecule are pulled closer to (or even shared with) the metal ion, the

Table 2.1
Hydration Enthalpies of Metal Cations[a]

Electronegativity < 1.5			Electronegativity > 1.5		
Ion	Radius	ΔH_{hyd}	Ion	Radius	ΔH_{hyd}
		+1 Ions			
Cs	181	−263			
Rb	166	−296	Tl	164	−326
K	152	−321			
Na	116	−405	Ag	129	−475
Li	90	−515	Cu	91	−594
H		−1091			
		+2 Ions			
Ra		−1259			
Ba	149	−1304			
Eu	131	−1458			
Sr	132	−1445	Pb	133	−1480
No	124	−1485	Sn		−1554
Yb	116	−1594	Hg	116	−1824
Ca	114	−1592	Cd	109	−1806
			Ag	108	−1931
			Ti	100	−1862
			V	93	−1918
			Cr	94	−1850
			Mn	97	−1845
			Fe	92	−1920
			Co	88	−2054
			Ni	83	−2106
			Cu	91	−2100
Mg	86	−1922	Zn	88	−2044
			Be	59	−2487
		+3 Ions			
Pu	114	−3441			
La	117	−3283			
Lu	100	−3758	Tl	102	−4184
Y	104	−3620	In	94	−4109
Sc	88	−3960	Ga	76	−4685
			Ti	81	−4154
			V	78	−4375
			Cr	75	−4402
			Mn	78	−4544
			Fe	78	−4376
			Co	75	−4651
			Al	67	−4660
		+4 Ions			
Th	108	−6136			
U	103	−6470			
Ce	101	−6489			
Zr	86	−6593			
Hf	85	−7120			

SOURCES: Ionic radii are from Table C (inside the back cover); hydration enthalpies are taken from J. Burgess, *Metal Ions in Solution*, Ellis Horwood, Chichester, UK, 1978; pp. 182–183, and D. W. Smith, *Inorganic Substances: A Prelude to the Study of Descriptive Inorganic Chemistry*, Cambridge University Press, Cambridge, UK, 1990; p. 160. Additional enthalpies may be found in Y. Marcus, *Ion Solvation*, Wiley-Interscience, Chichester, UK, 1985; pp. 107–109; there are discrepancies of up to 2 or 3% among these sources.

[a] Units are in kilojoules per mole (kJ mol^{-1}).

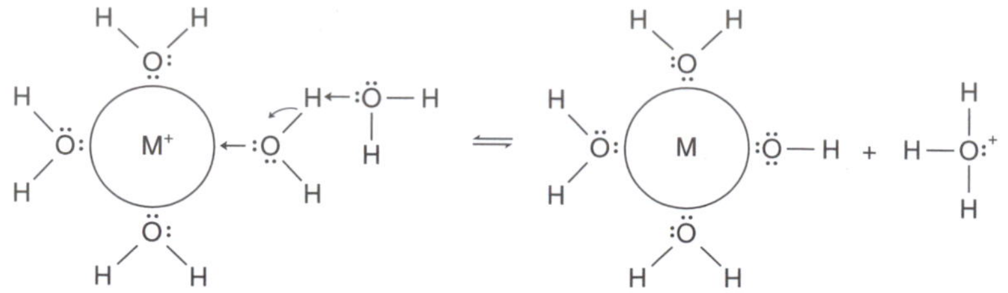

Figure 2.2
Hydrolysis of a hydrated cation.

electrons in the H–O bonds move closer to the oxygen to compensate some of its loss of electron density. Consequently, the hydrogen ends up with an increased positive charge, which makes it more closely resemble a hydrogen ion. Eventually, the H^+ may dissociate completely, leaving a hydroxide group attached to the metal and attaching itself to solvent water molecules to make a hydronium ion: the solution becomes acidic. We may represent this equilibrium by Eq. (2.2):

$$[M(H_2O)_6]^{z+} + H_2O \rightleftharpoons [M(H_2O)_5(OH)]^{(z-1)^+} + H_3O^+ \tag{2.2}$$

Often, this equation is simplified by representing the intact water molecules as (aq):

$$M^{z+}(aq) + H_2O \rightleftharpoons M(OH)^{(z-1)^+}(aq) + H^+(aq) \tag{2.3}$$

The chromium ion, for example, readily undergoes this reaction:

$$[Cr(H_2O)_6]^{3+} + H_2O \rightleftharpoons [Cr(H_2O)_5(OH)]^{2+} + H_3O^+ \tag{2.4}$$

or

$$Cr^{3+}(aq) + H_2O \rightleftharpoons Cr(OH)^{2+}(aq) + H^+(aq) \tag{2.5}$$

Note the similarity of these equations to the equation for the equilibrium process of ionization of a weak acid such as acetic acid:

$$HC_2H_3O_2 + H_2O \rightleftharpoons C_2H_3O_2{}^- + H_3O^+ \tag{2.6}$$

The equilibrium constant expression for this acid-producing reaction of metal ions, commonly called **hydrolysis**, is simply obtained

$$K_a = [Cr(OH)^{2+}][H^+]/[Cr^{3+}] \tag{2.7}$$

The equilibrium constant may be designated as K_h, or, since the reaction produces acid, as K_a, and may be measured.[3] One commonly finds tabulated[4] the negative logarithm of the equilibrium constant, pK_a:

$$pK_a = -\log K_a = -\log[Cr(OH)^{2+}]/[Cr^{3+}] + pH \tag{2.8}$$

The simple numbers in pK_a values are easier to compare at a glance than are the exponential numbers in K_a values, but an important consequence of taking *negative* logarithms, as in pH values, is that *lower* values of pK_a and pH correspond to *higher* degrees of hydrolysis, and *higher* acidities of solutions. Table 2.2 lists pK_a values for metal ions.

Equation (2.8) shows us that the proportion of the chromium in the +3 oxidation state, Cr(III), that has hydrolyzed to $Cr(OH)^{2+}$ depends on the pH; equal concen-

Table 2.2
Hydrolysis Constants for Metal Cations

Electronegativity < 1.5				**Electronegativity > 1.5**				
Ion	**Radius**	**(*a*)**	**pK_a**	**Ion**	**Radius**	**(*a*)**	**(*b*)**	**pK_a**
				+1 Ions				
K	152	0.007	14.5	Tl	164	0.006	0.016	13.2
Na	116	0.009	14.2	Ag	129	0.008	0.049	12.0
Li	90	0.011	13.6					
				+2 Ions				
Ba	149	0.027	13.5					
Sr	132	0.030	13.3	Pb	133	0.030	0.066	7.7
				Sn				3.4
Ca	114	0.035	12.8	Hg	116	0.034	0.082	3.4
				Cd	109	0.037	0.055	10.1
				Cr	94	0.043	0.043	10.0
				Mn	97	0.041	0.046	10.6
				Fe	92	0.043	0.075	9.5
				Co	88	0.045	0.082	9.6
				Ni	83	0.048	0.088	9.9
Mg	86	0.047	11.4	Zn	88	0.045	0.060	9.0
				Be	59	0.068	0.074	6.2
				+3 Ions				
Pu	114	0.079	7.0					
La	117	0.077	8.5	Bi	117	0.077	0.127	1.1
Lu	100	0.090	7.6	Tl	102	0.088	0.140	0.6
Y	104	0.086	7.7	Au	99	0.091	0.191	−1.5
Sc	88	0.102	4.3	In	94	0.096	0.123	4.0
				Ti	81	0.111	0.115	2.2
				Ga	76	0.118	0.148	2.6
				Fe	78	0.115	0.147	2.2
				Cr	75	0.120	0.135	4.0
				Al	67	0.134	0.145	5.0
				+4 Ions				
Th	108	0.148	3.2					
Pa	104	0.154	−0.8					
U	103	0.155	0.6					
Np	101	0.158	1.5					
Pu	100	0.160	0.5					
Ce	101	0.158	−1.1					
Hf	85	0.188	0.2	Sn	83	0.193	0.222	−0.6
Zr	86	0.186	−0.3	Ti	74	0.216	0.220	−4.0

SOURCES: Values of hydrolysis constants (pK_a) taken from C. F. Baes and R. E. Mesmer, *The Hydrolysis of Cations*, Wiley-Interscience, New York, 1976; and from J. Burgess, *Metal Ions in Solution*, Ellis Horwood, Chichester, UK, 1978; pp. 264–267.

[a] The Z^2/r ratio for the cation.

[b] The $Z^2/r + 0.096\ (\chi_P - 1.50)$ for the cation, as in Eq. (2.12).

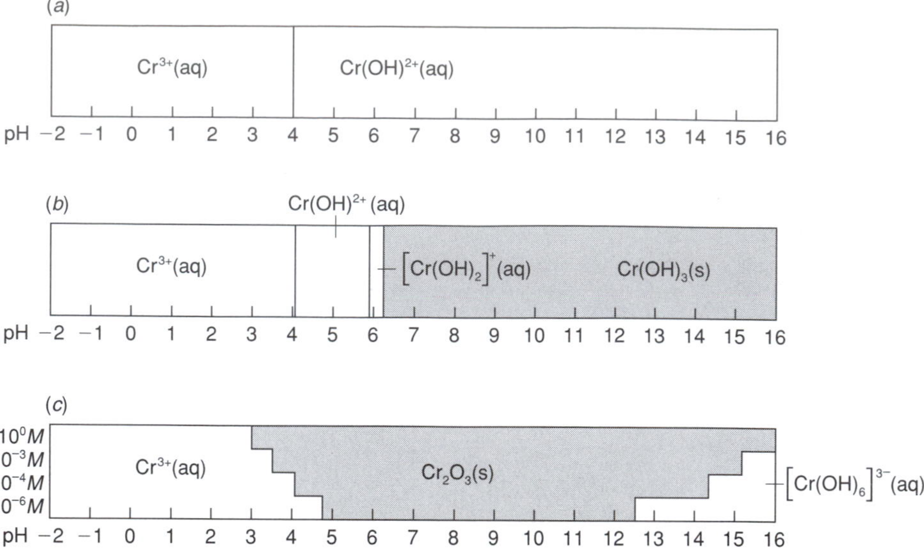

Figure 2.3
Simplified acid–base predominance diagrams for chromium(III). (*a*) Predominance diagram showing only Cr^{3+}(aq) and $Cr(OH)^{2+}$(aq). (*b*) Predominance diagram also showing $[Cr(OH)_2]^+$(aq) and the precipitate $Cr(OH)_3$(s). (*c*) Predominance diagram considering only Cr^{3+}(aq) and Cr_2O_3(s) but showing the effects of changing the total concentration of chromium(III) from 10^0 *M* (top) down to 10^{-6} *M* (bottom). Shaded regions of pH are those in which precipitates (insoluble substances) predominate.

trations of the two will be present when the pH equals pK_a (if no other equilibria are affecting either concentration). At a pH below the pK_a (4.0 for Cr^{3+}), the unhydrolyzed M^{z+} cation will be the *predominant* form of the metal in the solution; above that pH the *hydroxy cation* will be predominant. This relationship can easily be expressed visually by an **(acid–base) predominance diagram**, Figure 2.3(*a*), which shows the predominant form of an element (in a specified oxidation state) as a function of pH.

Reaction (2.2) is only the first of several reactions that may occur to a metal cation. A second, then a third water molecule in the hydrated ion may hydrolyze, producing more hydronium ions and giving rise to hydrated hydroxy cations containing more than one hydroxy group [$Cr(OH)_2{}^+$ in the case of chromium(III)]. When the process has continued to the point at which the original charge of the metal ion is completely neutralized, a *metal hydroxide* results. This species is no longer an ion, does not hydrate as readily, and is usually insoluble; hence, a precipitate appears in the solution. Often the insoluble metal hydroxides will subsequently lose molecules of water to give insoluble *metal oxides*, as represented by Eq. (2.9):

$$M(OH)_z(s) \rightarrow MO_{z/2}(s) + \frac{z}{2}\,H_2O \qquad (2.9)$$

Since it is difficult for us to tell when this has happened, we will not attempt to distinguish metal hydroxide and metal oxide precipitates.

Each of these reactions has its own K_a (numbered K_{a2}, K_{a3}, etc.), and determines the pH values at which the various intermediate hydroxy cations will be predominant [Fig. 2.3(*b*)]. Fortunately, for most metals the predominance ranges of the intermediate hydroxy cations are narrow, as they are for chromium(III) in Figure 2.3(*b*). Therefore, for simplicity, we can ignore the intermediate hydroxo cations and focus on only the original metal cation in solution and its hydroxide or oxide as a precipitate, as shown for chromium(III) in Figure 2.3(*c*).

It can be shown[5] that a metal hydroxide or oxide will precipitate at a pH that is fairly close to the pK_a value of the metal ion:

$$pH = pK_a - (1/z)\log[M^{z+}] - 5.6/z \tag{2.10}$$

The exact pH at which the precipitated oxide or hydroxide becomes the predominant form of the metal does depend somewhat on the concentration, however. Figure 2.3(*c*) shows the effects on the predominance ranges of the different chromium(III) species as total concentration of all chromium(III) species is changed in steps from 1 to 10^{-2} *M* to 10^{-4} to 10^{-6} *M*. In general, the predominance regions of soluble species expand with increasing dilution, while those of insoluble species contract. (In the dilution process, we add more water, which dissolves more of the precipitate, producing soluble species such as the hydroxy cations.) Since the standard concentration (activity) is 1.0 *M*, we will use that total concentration in our predominance diagrams. Geochemists and environmental chemists generally use much lower total concentrations, but the same principles and approximately the same conclusions apply in either case.

2.3 Classification of Cations by Acidity Categories

There are important practical consequences of the acidic properties and hydrolysis reactions of cations. The suggested laboratory experiment hints at some of these: A compound with an innocent-looking formula such as $TiCl_4$ may react quite violently with water. It may react with the water vapor in the air and fill the laboratory with choking fumes of the acid HCl. You may be trying to prepare a solution of a metal ion for some experiment in biology, only to find a precipitate forming. You may need to know the form that a metal ion pollutant takes in a lake, to know whether it will end up as an insoluble sludge of oxide at the bottom of the lake or whether it will remain in solution as a cation or as an oxo anion (Chapter 3). (Cations and anions may be taken up or rejected by quite different mechanisms by living organisms.) Thus it is important for us to be able to gauge the approximate acidity of a given cation, so that we can anticipate (even in the laboratory without a calculator) how violently a given compound will react with water or atmospheric humidity and whether its hydroxide or oxide will precipitate or an oxo anion will be produced.

There are over 150 cations listed in Table C; hence, it is very useful to devise **categories of acidity** into which these cations can easily be grouped, so that we can easily anticipate their reactivity. Experimentally, we find that the ions in a given category of acidity share several important chemical properties. These properties include not only the degree to which the ions react with water but also (as we shall see in subsequent chapters) the solubility or insolubility of salts formed by these ions and the properties of compounds formed by them (such as their oxides and halides) in the absence of water.

Cations with *pK_a values of 14 or greater* display such negligible hydrolysis reactions that their acidity does not manifest itself in any important way; we call these cations **nonacidic cations**. Included are such ions as Cs^+ and Rb^+, whose hydrolysis constants (pK_a values) are absent from Table 2.2 because the hydrolysis of these ions is too slight to be measured. The predominance diagrams of these cations [Fig. 2.4(*a*)] are very simple: The cation is predominant at all practical pH values; the hydroxide will not precipitate from a 1 *M* solution at any pH.

Cations with *pK_a values between 11.5 and 14* (such values can be measured in sensitive experiments) have acidities that are also not directly detectable, but that occasionally have significant consequences; we will refer to these ions as **feebly acidic cations**. As shown in Figure 2.4(*b*), the hydroxides of feebly acidic cations (such as Li^+, Ba^{2+}, Sr^{2+}, and Ca^{2+}) precipitate at pH values over their pK_a values (i.e., in strongly basic solutions).

Cations with *pK_a values between 6 and 11.5* we term **weakly acidic cations**. Acidity is an important part of their chemistry, although this acidity is not apparent in simple pH measurements because the solution normally also contains dissolved carbon dioxide, which is at least as strong an acid as these ions. Included in this group are many important +2 ions such as Mg^{2+} and the +2 charged *d*-block ions, whose biological functions, as we will see later, revolve around their acidity. These ions show enough acidity to precipitate insoluble metal hydroxides at pH values over the pK_a (i.e., in neutral or just slightly basic solutions) but if the solution is not made basic, the hydrated metal ion will be the predominant form [Fig. 2.4(*c*)].

Moderately acidic cations such as Al^{3+} and the +3 charged *d*-block ions have *pK_a values of 1–6* and are unmistakably acidic, since they detectably lower the pH values even of water containing dissolved CO_2. (The pK_a values of these ions are comparable to those of typical organic acids such as acetic acid.) These ions can be predominant in solution only at quite low pH values [Fig. 2.4(*d*)]; if the solutions of these ions are not kept highly acidic, metal hydroxides will precipitate (giving at least a cloudy appearance to the solution).

Strongly acidic cations have *pK_a values between −4 and 1*. These ions (e.g., Ti^{4+}) react violently and nearly completely with water, giving strongly acidic solutions and copious amounts of precipitate of insoluble metal oxide or hydroxide; the latter predominate except in very concentrated acids (in which the $[H^+]$ is >1 *M* so that the pH is <0).

Very strongly acidic cations presumably have *pK_a values below −4* (these values are not readily measured in practice). These ions may be expected to react irreversibly with water, generating an oxide or hydroxide and a high concentration of hydrogen ion. We say that these very strongly acidic cations are **leveled** by reaction with water to generate the characteristic acid of water solutions, H_3O^+. Thus no acid stronger than H_3O^+ can persist in water, and there is no pH at which these "cations" can predominate or even exist in water, so we have shown no example of a predominance diagram for any of these "cations." The "cations" of the purely nonmetallic elements of the periodic table have pK_a ratios in this range, which is to say that they cannot exist in water as cations. (This is one of the characteristics by which we differentiate metals and nonmetals.) But some metals in high oxidation states (V and VI) also fit in this category.

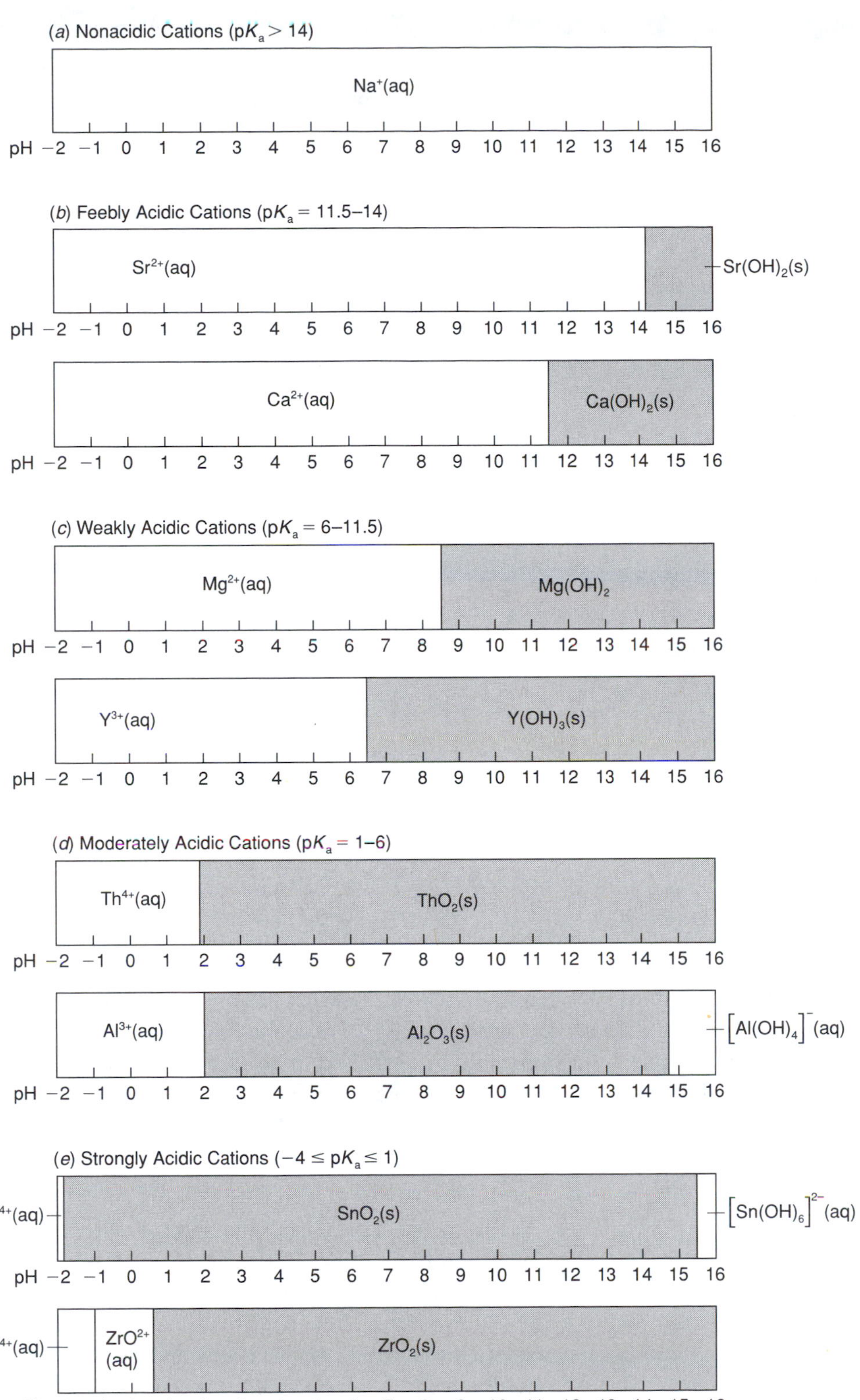

Figure 2.4
Acid–base predominance diagrams for (*a*) a typical nonacidic cation, (*b*) two typical feebly acidic cations, (*c*) two typical weakly acidic cations, (*d*) two typical moderately acidic cations, (*e*) two typical strongly acidic cations. [Data from M. Pourbaix, *Atlas of Electrochemical Equilibria in Aqueous Solutions*, National Association of Corrosion Engineers, Houston, TX, 1974. This data differs from that found in Table 2.2, but is used for consistency with the Pourbaix diagrams in Chapter 6.]

Example 2.1

Figure 2.5 shows some predominance diagrams for some unknown cations, which you have established by titration experiments (to determine pK_a values) and conductance

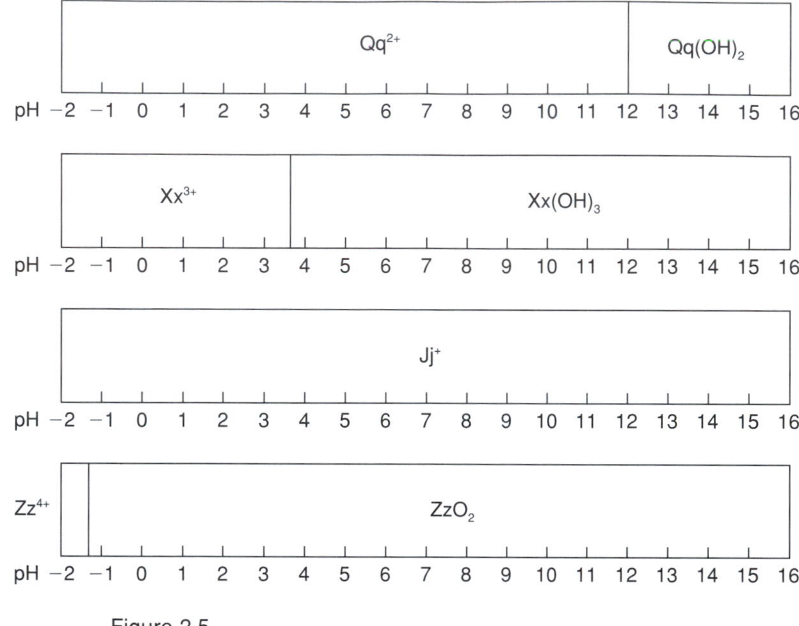

Figure 2.5
Imaginary predominance diagrams for use in Example 2.1.

experiments (to determine ionic charges). (a) Which of the cations shown is most acidic? (b) Roughly speaking, what is the pK_a of each cation? (c) Give the acidity classification of each cation. (d) Which of the above cations will, if dissolved in normal tap water, render the solution appreciably more acidic? (e) On each predominance diagram, shade in the pH region in which you expect to see a precipitate.

SOLUTION:

(a) The most acidic cation is the one most confined to the left side of its predominance diagram, Zz^{4+}. (b) Very approximately according to Eq. (2.10), the pK_a of each cation will be the pH at which it predominantly converts to the hydroxide or oxide: 12 for Qq^{2+}; 4 for Xx^{3+}; >16 for Jj^{+}; -1 for Zz^{4+}. (c) Based on these estimated pK_a values, we would classify Qq^{2+} as feebly acidic, Xx^{3+} as moderately acidic; Jj^{+} as nonacidic; and Zz^{4+} as strongly acidic. (d) Given that ordinary water is already weakly acidic from dissolved carbon dioxide, only those cations that are moderately acidic or more can add noticeably to the acidity: Xx^{3+} and Zz^{4+}. (e) Precipitates of hydroxides or oxides predominate to the right of (on the high pH or basic side of) the bars in these predominance diagrams.

Classifying Cations into Acidity Categories: Effects of Charge and Radius. If you performed Experiment A.1, you discovered that increasing charge of a metal cation increases its extent of hydrolysis, as does a decreasing radius; these are the same factors that influence the hydration energy of cations as expressed in the Latimer equation (2.1). Rather accurate equations have been presented for predicting the extent of hydrolysis of metal ions,[6,7] but these are not simple enough to be used for our purposes of quick estimation of approximate acidity. To obtain such a relationship, we have graphed (in Fig. 2.6, using closed circles) the pK_a values of metal cations of χ_P of 1.5 or below versus the Z^2/r ratios of these ions. Equation (2.11) expresses the relationship we find empirically.

$$pK_a = 15.14 - 88.16\,Z^2/r \tag{2.11}$$

Normally, we will find it quicker not to solve Eq. (2.11) to obtain values of pK_a for each of the 150 or so cations, but instead we will evaluate Z^2/r in order to place it in one of the six ranges or categories of acidity enclosed by boxes in Figure 2.6. (Since we

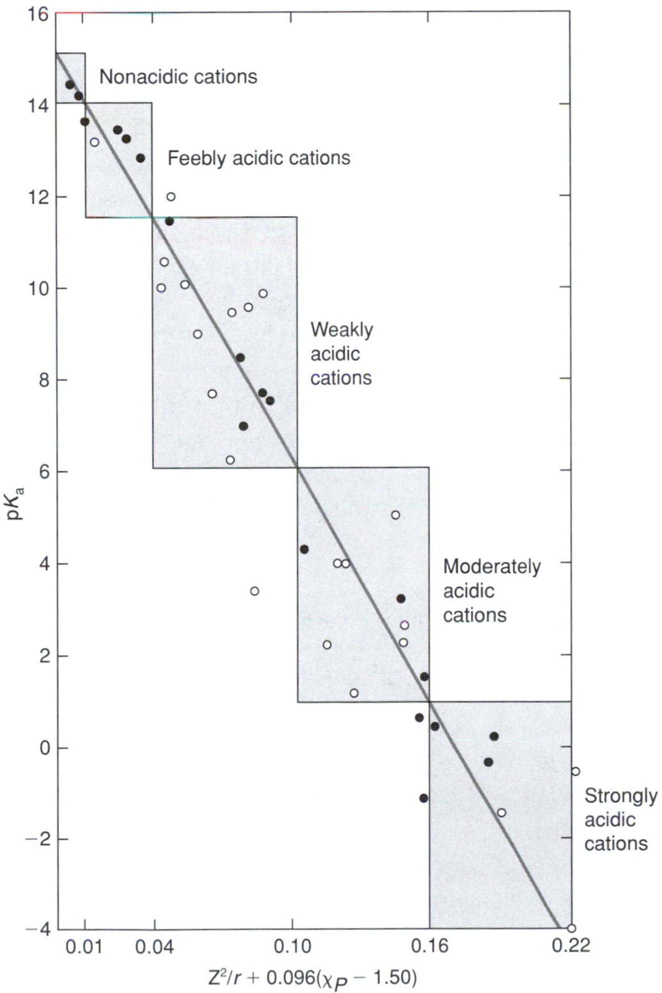

Figure 2.6
The pK_a values of cations as a function of charge, size, and electronegativity. Closed circles equal metals of electronegativity (χ_P) less than 1.5, plotted as a function of Z^2/r. Open circles equal metals of χ_P greater than 1.5, plotted as a function of $Z^2/r + 0.096(\chi_P - 1.50)$.

are not yet considering the effects of electronegativity, we will assume that the Pauling electronegativity of the element in the cation is <1.50.)

If the Z^2/r *ratio of the cation is less than 0.01*, it is *nonacidic.*

If Z^2/r *is between 0.01 and 0.04*, the cation is *feebly acidic.*

If Z^2/r *is between 0.04 and 0.10*, the cation is *weakly acidic.*

If Z^2/r *is between 0.10 and 0.16*, the cation is *moderately acidic.*

If Z^2/r is *between 0.16 and 0.22*, the cation is *strongly acidic.*

If Z^2/r is *over 0.22*, the cation is *very strongly acidic*.

Effects of Electronegativity. So far, we have not attempted to take into account the effects of high electronegativity of the cation, which you may have discovered if you performed Experiment A.1. From Table 2.2, we can see that the cations of metals with χ_P over 1.5 are more acidic than other metal ions of similar charge and size. We have derived a rough relationship between the "excess" Pauling electronegativity of a cation and its "excess" acidity, which allows us to modify Eq. (2.11) to include the effect of electronegativity:

$$pK_a = 15.14 - 88.16[Z^2/r + 0.096(\chi_P - 1.50)] \tag{2.12}$$

Note that this equation should be used *if and only if* the Pauling electronegativity of the metal exceeds 1.50.

In Figure 2.6, we have used open circles to show the relationship of the pK_a of cations to their "modified" Z^2/r ratio, $Z^2/r + 0.096(\chi_P - 1.50)$. There is more scatter in the data for these more electronegative cations than there was for the "simple" less electronegative cations, which suggests that we have oversimplified the relationship to electronegativity. But our main purpose is to categorize the relative acidity of these cations quickly; the scatter is not so serious as to prevent us from suggesting an even simpler rule of thumb: *If the electronegativity of the metal ion is 1.8 or greater, move it up one category in acidity*. The suggested categories of acidity, their corresponding pK_a values, and their relationships to Z^2/r and χ_P are summarized in Table 2.3.

Table 2.3
Relationship Between pK_a, Acidity, and Z^2/r Ratios of Metal Ions[a]

pK_a Range	Category	Z^2/r Ratio	X_P	Examples
14–15	Nonacidic	0.00–0.01	<1.8	Most +1 ions of the *s* block
11.5–14	Feebly acidic	0.00–0.01	>1.8	Tl^+
11.5–14	Feebly acidic	0.01–0.04	<1.8	Most +2 ions of *s*, *f* blocks
6–11.5	Weakly acidic	0.01–0.04	>1.8	Most +2 ions of the *d* block
6–11.5	Weakly acidic	0.04–0.10	<1.8	All +3 ions of the *f* block
1–6	Moderately acidic	0.04–0.10	>1.8	Most +3 ions of the *d* block
1–6	Moderately acidic	0.10–0.16	<1.8	Most +4 ions of the *f* block
(−4)–1	Strongly acidic	0.10–0.16	>1.8	Most +4 ions of the *d* block
(−4)–1	Strongly acidic	0.16–0.22	<1.8	
<(−4)	Very strongly acidic	0.16 and up	>1.8	
<(−4)	Very strongly acidic	0.22 and up	<1.8	

[a] The electronegativities of the *p*-block elements vary too greatly to allow their inclusion in one category of "cation".

Example 2.2

Classify each of the following cations and describe their reactions with water: Eu^{2+}, B^{3+}, W^{6+}.

SOLUTION:

We begin by finding ionic radii in Table C; then we compute Z^2/r. For Eu^{2+}, this ratio works out to be $2^2/131 = 0.031$; for B^{3+} it is $3^2/41 = 0.220$; and for W^{6+} it is $6^2/74 = 0.487$. Without considering electronegativities, we would classify Eu^{2+} as a feebly acidic cation, B^{3+} as strongly or very strongly acidic, and W^{6+} as very strongly acidic. Checking the table of electronegativities, we find that the latter two elements have electronegativities in excess of 1.8, so both are definitely in the class of very strongly acidic cations. The Eu^{2+} ion remains classified as feebly acidic, and should be present largely unchanged (as a hydrated ion) in solutions of normal pH. The latter two "cations" will not actually be present at all in water. Halides of these two cations will react violently with water to generate hydrohalic acid and the oxides or hydroxides (or, as we will see in Chapter 3, oxo acids or oxo anions) of these elements.

The relative acidity of different metal ions may be assessed by inspection if acid–base predominance diagrams are available for them. *The more acidic the species* (such as a metal ion), *the more it is confined to predominance in small ranges of extremely low pH at the left of a diagram.* Very strongly acidic species will not even be on the diagram, but would be found off the left side, so they represent the ultimate in small predominance ranges.

2.4 Halides that Fail to Undergo Hydrolysis; Oxo Cations

According to the rules we have developed, carbon tetrachloride should be one of the most reactive of all the chlorides of the elements, since Z^2/r for C^{4+} is quite impressive (0.533). The predicted explosive reaction of CCl_4 and water fails to materialize, however. Indeed, CCl_4 is used as an inert solvent that is insoluble in and unreactive with water. The key to this seeming violation of our concepts lies in the *very* small size of the carbon atom, which is so crowded in by the four large chlorine atoms that the water molecule cannot get in to start the reaction.

This nonreaction (at low temperatures) illustrates an aspect of the radius in the central atom in a molecule: There is a limit to the number of atoms that may be bonded to it. This number is called the **maximum coordination number** of the element, and although it depends on the size of the atoms around the central atom, it obviously *increases down the periodic table*. Elements of the second period almost never have more than four atoms bonded to them in compounds, which is an important basis of the **octet rule**. Hence, we say that the maximum coordination number for an element in the second period is 4. But (see Table C) the ions of the third and fourth periods, which have similar sizes to each other because of the scandide contraction, are substantially larger than those of the second period; hence, they can bond to as many as six other atoms. In the fifth and sixth periods, the ions, which are similar in size to each other due to the lanthanide contraction, are larger still, and cases of atoms bonded to more than six neighboring atoms are known.

Another case of "unexpected" nonreactivity of a halide is that of sulfur hexafluoride,[8] SF_6, which resists reacting with steam at 500 °C and resists molten KOH. The atoms get larger as we go down Group 16(VI), however; thus SeF_6 is unreactive with water at 25 °C but is hydrolyzed in the respiratory tract (with quite undesirable consequences); TeF_6 reacts slowly with cold water.

Although the rate of reaction of water and carbon tetrachloride is negligible at room temperature, it is still true that the end products of their very slow reaction (CO_2 and 4 HCl) are more stable than the reactants (CCl_4 and 2 H_2O) by 52 kJ. At high temperatures, the reaction does indeed occur. An intermediate product that forms in this reaction, which may happen when a carbon tetrachloride fire extinguisher is used along with water on a very hot fire, is carbonyl chloride or phosgene, $COCl_2$, a deadly poison. It has been suggested that a main reason that life can exist based on carbon chemistry rather than silicon chemistry is that the small size of carbon causes the reactions of carbon compounds to be much slower, so that many more carbon compounds that are not inherently thermodynamically stable can exist in the presence of reactive compounds such as water and oxygen.[9]

Oddly enough, for some ions near the center of the fifth and sixth periods (e.g., Ag^+, Sn^{2+}, Hg^{2+}, and Au^+) the preferred coordination number is often quite small—as low as 2. As you might expect, there is some anomalous chemistry in this part of the periodic table (relativistic effects are often involved here[10]). Thus the acidity of the hydrated Hg^{2+} and Sn^{2+} ions are much higher than expected from Eq. (2.12), which may be attributed to the fact that there are unexpectedly few water molecules coordinated to the metal ion. Thus the electrostatic force pulling on the electrons of each water molecule is stronger than in more normal hydrated ions. Similarly, the smallest hydrated ion, Be^{2+}, is also more acidic than one would predict.[6]

Oxo Cations. Although we have generally ignored the complex chemistry of the intermediate species that result after a metal ion hydrolyzes but before it precipitates as a hydroxide, in some cases this species is persistent and important enough to merit mention. Sometimes, when two or four hydroxide groups are attached to a (highly charged) cation, they may lose molecules of water and form **oxo cations** [see the predominance diagram of Zr(IV), Fig. 2.4(*e*), for an example]. Such species are particularly persistent for the +5 and +6 oxidation states of the *f*-block elements U, Np, Pu, and Am; the most common form in which you are likely to encounter uranium is in salts of the yellow uranyl ion, $UO_2{}^{2+}$. Similar species are relatively common in the chemistry of molybdenum(V) and (VI), and vanadium(IV) and (V). Many nonmetal halides, on partial hydrolysis, form species that resembles these compounds in formula (e.g., $POCl_3$). These compounds generally hydrolyze further with more water, however. (In the suggested experiment, the precipitates formed by antimony and bismuth are not the oxides, but antimonyl chloride, SbOCl, and bismuthyl chloride, BiOCl.)

2.5 Hydration and Hydrolysis of Monoatomic Nonmetallic Anions

Just as metal cations do not exist "bare" in aqueous solutions, neither do anions. They also attract water molecules to form hydrated ions[11], although in this case we would expect to (and do) find that the *positive* end of the dipolar water molecule (i.e., one of the hydrogen atoms) is attracted to each of the unshared electron pairs on the anion,

Table 2.4
Hydration Enthalpies[a] of Some Anions

Anion	Radius	Hydration Energy	Anion	Radius	Hydration Energy
F^-	119	−513	OH^-	119	−460
Cl^-	167	−370	NO_3^-	165	−314
			CN^-	177	−342
Br^-	182	−339	N_3^-	181	−298
I^-	206	−294	SH^-	193	−336
			BF_4^-	215	−223
			ClO_4^-	226	−235
S^{2-}	170	−1372	CO_3^{2-}	164	−1314
			SO_4^{2-}	244	−1059

SOURCES: Radii for monoatomic anions are from Table C; "radii" for polyatomic anions, which of course are not truly spherical, are thermochemical radii, taken from J. E. Huheey, E. A. Keiter, and R. L. Keiter, *Inorganic Chemistry: Principles of Structure and Reactivity*, 4th ed., Harper-Collins, New York, 1993, p. 118. Hydration enthalpies are from A. G. Sharpe, *J. Chem. Educ.*, **67**, 309 (1990); D. W. Smith, *Inorganic Substances*, Cambridge University Press, Cambridge, UK, 1990; p. 160, and M. C. Ball and A. H. Norbury, *Physical Data for Inorganic Chemists*, Longman, London, 1974.

[a]Units are in kilojoules per mole (kJ mol^{-1}).

forming a hydrogen bond to the anion (Fig. 2.1). Substantial hydration energies also result from this interaction (Table 2.4). As we would expect, these energies increase with increasing charge and decreasing size of the anion, and appear[12] (Fig. 2.7) to fit an equation similar to the Latimer equation:

$$\Delta H_{\text{hyd}} \approx -57{,}000 Z^2/r \text{ kJ mol}^{-1} \tag{2.13}$$

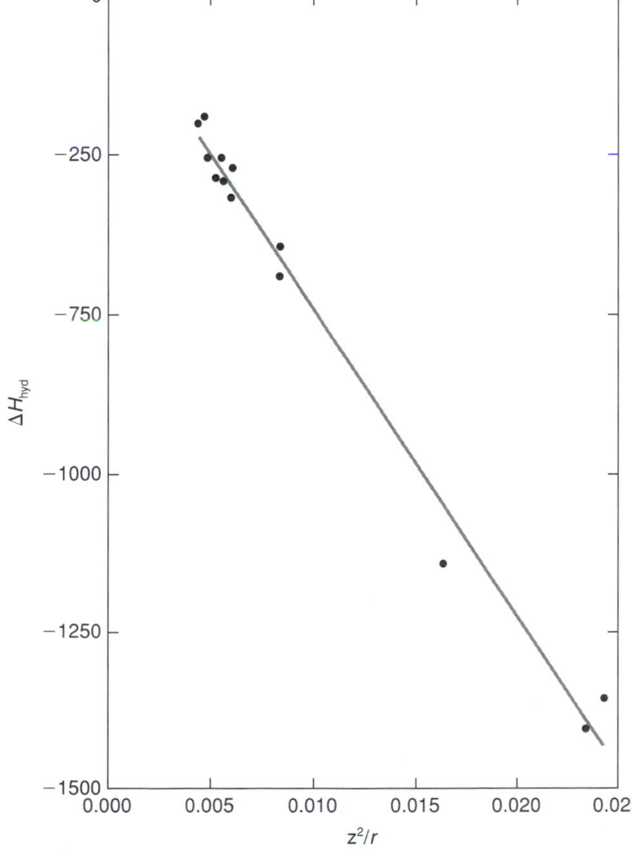

Figure 2.7
Hydration enthalpies of anions versus their Z^2/r ratios. Due to the paucity of the data for monoatomic anions, data for the polyatomic anions from Table 2.4 are also included in this plot; this does not affect the slope or intercept of the correlation.

Once again the interaction between an ion (in this case, an anion) and its waters of hydration frequently is great enough that one or more of the water molecules may be pulled apart in the process of hydrolysis. In this case, the partial positively charged hydrogen atom of the water bonds to the anion, releasing the remainder of the water molecule—a hydroxide ion—and producing a basic solution:

$$X^{y-} + H_2O \rightleftharpoons XH^{(y-1)-} + OH^- \qquad (2.14)$$

$$S^{2-} + H_2O \rightleftharpoons SH^- + OH^- \qquad (2.15)$$

An equilibrium constant can be written for the above reaction; this is often called a hydrolysis constant, symbolized by K_b in recognition of the fact that the anion acts as a base. As before, in Table 2.5A, we list the negative logarithms of these equilibrium constants, which are symbolized by pK_b or pK_{b1}. A large value of pK_{b1} corresponds to a case in which Reaction (2.14) goes very little to the right, so that the anion is acting as a weak base, producing a solution with a pH not much above that of distilled water. A small or negative value of pK_{b1} identifies a very reactive, very basic anion, which causes a strong change in the pH of the solution, this time to high values. Some pK_b values in this table are clearly too extreme to be measurable in water, and are estimated from values in other types of solutions. Others (the values for S^{2-}, Se^{2-}. and Te^{2-}) are subject to uncertainty and even controversy[13], so most of these values should be treated as illustrative, rather than quantitatively reliable.

Table 2.5
Aqueous Basicity of Monoatomic Anions and their Partially Protonated Forms

	A. Monoatomic Anions		
C^{4-} very strong	N^{3-} very strong	O^{2-} very strong $pK_1 = -22$	F^- weak $pK_1 = 10.85$
Si^{4-} very strong	P^{3-} very strong	S^{2-} strong $pK_1 = -4$	Cl^- nonbasic $pK_1 = 20.3$
Ge^{4-} very strong	As^{3-} very strong	Se^{2-} strong $pK_1 = -1$	Br^- nonbasic $pK_1 = 22.7$
		Te^{2-} moderate $pK_1 = 3.0$	I^- nonbasic $pK_1 = 23.3$
	B. Partially Protonated Anions		
CH_3^- very strong $pK_4 = -30$	NH_2^- very strong $pK_3 = -25$	OH^- strong $pK_2 = -1.74$	
SiH_3^- very strong $pK_4 = -21$	PH_2^- very strong $pK_3 = -13$	SH^- moderate $pK_2 = 7.11$	
GeH_3^- very strong $pK_4 = -11$	AsH_2^- very strong $pK_3 = -9$	SeH^- weak $pK_2 = 10.3$	
		TeH^- feeble $pK_2 = 11.4$	

SOURCES: The pK_b values are calculated from pK_a values given in R. J. Myers, *J. Chem. Educ.*, **63**, 687 (1986); S. Licht, F. Forouzan, and K. Longo, *Anal. Chem.*, **62**, 1356 (1990), W. L. Jolly, *Modern Inorganic Chemistry*, McGraw-Hill, New York, 1984; p. 177; R. V. Dilts, *Analytical Chemistry*, Van Nostrand, New York, 1974; p. 553; and W. H. Nebergall, H. H. Holtzclaw, Jr., and W. R. Robinson, *General Chemistry*, 6th ed., Heath, Lexington, MA, 1980.

It seems useful to categorize *anions* in categories of *basicity* that match those used for acidity of cations: We can think of these as categories of *acid–base reactivity*. Thus we classify the simple anions that have no detectable basicity, hence *pK_b values over 14* (Cl^-, Br^-, I^-) as *nonbasic anions.* These anions have no tendency to combine with hydrogen ions *in aqueous solution,* which means that their basicity is sufficiently less than that of water molecules, so they cannot compete with water for hydrogen ions. Later, we will see other circumstances in which basic properties of the halide ions are evident.

Anions that have *pK_b values between 11.5 and 14* we will call *feebly basic anions*; although there are no examples immediately at hand we will soon encounter some. The first anions with some commonly noticeable basicity, such as fluoride ion, have *pK_b values between 6 and 11.5*; we label these *weakly basic anions. Moderately basic anions* will be those with *pK_b values between 1 and 6* (e.g., Te^{2-}); *strongly basic anions* have *pK_b values between −4 and 1* (e.g., Se^{2-} and S^{2-}).

Very strongly basic anions with *pK_b values below −4* are those that react irreversibly with water, such as the oxide ion:

$$O^{2-} + H_2O \rightarrow 2\,OH^- \tag{2.16}$$

This reaction is an illustration of the *leveling* properties of very strong bases: Any base that is stronger than the characteristic base of the solvent will react with the solvent (in this case, water) to generate its characteristic base (i.e., hydroxide ion). No base stronger than OH^- can persist in water.

As we would expect from the corresponding trends with cations, the basicities of these ions increase strongly with increasing negative charge (i.e., from right to left in a period); the anions with charges of −3 or −4 are very strong bases that cannot persist in water. We also note that the basicities of these anions decrease with increasing size (down a group). Both of these periodic trends match those found with cations: Acid–base *reactivity increases with increasing charge and decreasing radius.*

A plot of the pK_b values in Table 2.5A versus their Z^2/r ratios appears (Fig. 2.8)[6] to show a reasonable fit to the equation

$$\mathrm{p}K_{b1} \approx 29 - 1200Z^2/r \tag{2.17}$$

Comparison of this equation with Eq. (2.11) shows that the basicity of anions increases much more rapidly with increasing Z^2/r than does the acidity of cations. Hence, the Z^2/r ranges used to classify acidity of cations (e.g., <0.01 = nonacidic) cannot be carried over to classify the basicity of anions without modification. We could select new Z^2/r ranges for use with anions, but there are really too few monoatomic anions to justify this. We still can compare Z^2/r ratios of different monoatomic anions to rank their relative basicities, however.

There are corresponding periodic trends in acidity of cations and basicity of anions. In Figure 2.9, we indicate these reactivity classifications for a few anions and cations (the latter with charges equal to their group numbers). It can be seen that the most *nonreactive* (nonacidic, nonbasic) ions are found at the *outside bottom* of the periodic table; the further up and into the periodic table one goes, the more reactive the ion becomes, as long as the charges on the ion increase with movement into the table.

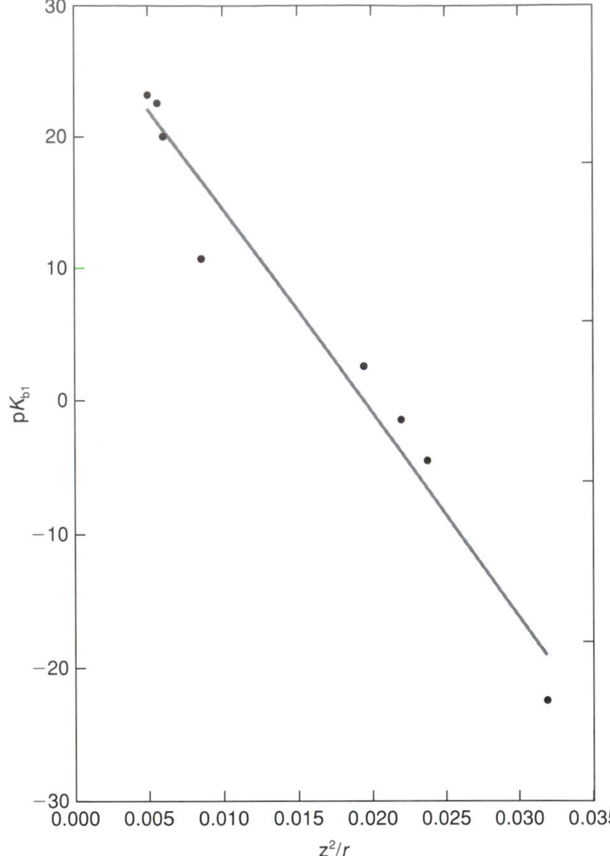

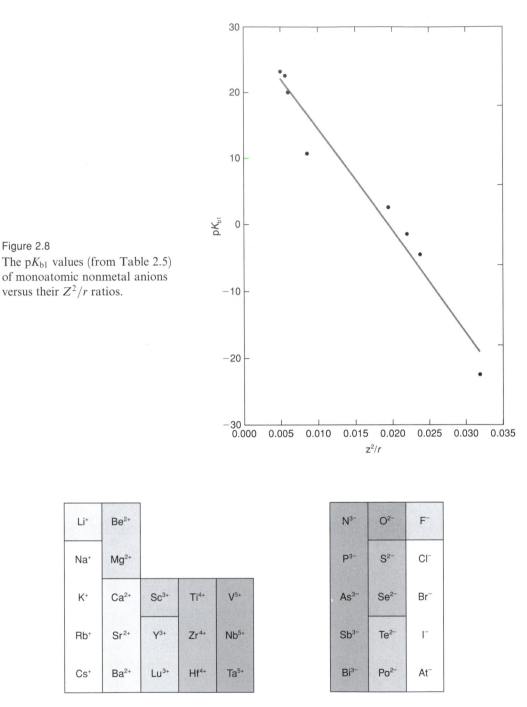

Figure 2.8
The pK_{b1} values (from Table 2.5) of monoatomic nonmetal anions versus their Z^2/r ratios.

Li^+	Be^{2+}			
Na^+	Mg^{2+}			
K^+	Ca^{2+}	Sc^{3+}	Ti^{4+}	V^{5+}
Rb^+	Sr^{2+}	Y^{3+}	Zr^{4+}	Nb^{5+}
Cs^+	Ba^{2+}	Lu^{3+}	Hf^{4+}	Ta^{5+}

N^{3-}	O^{2-}	F^-
P^{3-}	S^{2-}	Cl^-
As^{3-}	Se^{2-}	Br^-
Sb^{3-}	Te^{2-}	I^-
Bi^{3-}	Po^{2-}	At^-

Figure 2.9
Reactivity (acidity, basicity) classifications of selected ions as a function of periodic table position. Unshaded is nonacidic or nonbasic; feebly shaded is feebly acidic or basic; weakly shaded is weakly acidic or basic; moderately shaded is moderately acidic or basic; strongly shaded is strongly acidic or basic; very strongly shaded is very strongly acidic or basic.

Example 2.3

Rank the following anions in order of increasing basicity, and discuss periodic trends in their basicity: C^{4-}; Sn^{4-}; O^{2-}; and Te^{2-}.

SOLUTION:

The Z^2/r ratios for O^{2-} is 0.031; the ratio for Te^{2-} is 0.019. Table C does not give the anionic radii of C^{4-} or Sn^{4-}, but from periodic trends we might extrapolate radii that are a little larger than those of O^{2-} and Te^{2-}, respectively. With $r = 138$ pm, Z^2/r for C^{4-} becomes 0.116; with $r = 188$ pm, Z^2/r for Sn^{4-} becomes 0.085. The predicted order of basicity is $Te^{2-} < O^{2-} \ll Sn^{4-} < C^{4-}$.

This order makes sense in terms of periodic trends. We expect dramatically greater basicity with greater charge, which puts the −4 charged anions at the head of the list. Of lesser significance is the effect of decreasing size to give increasing basicity, which puts C^{4-} ahead of Sn^{4-} and O^{2-} ahead of Te^{2-}. This reasoning could have allowed us to rank these anions without extrapolating the unknown radii.

2.6 Predominance Diagrams for Anions: Partially Protonated Anions

In Figure 2.10, predominance diagrams for several of the simple anions are shown. As

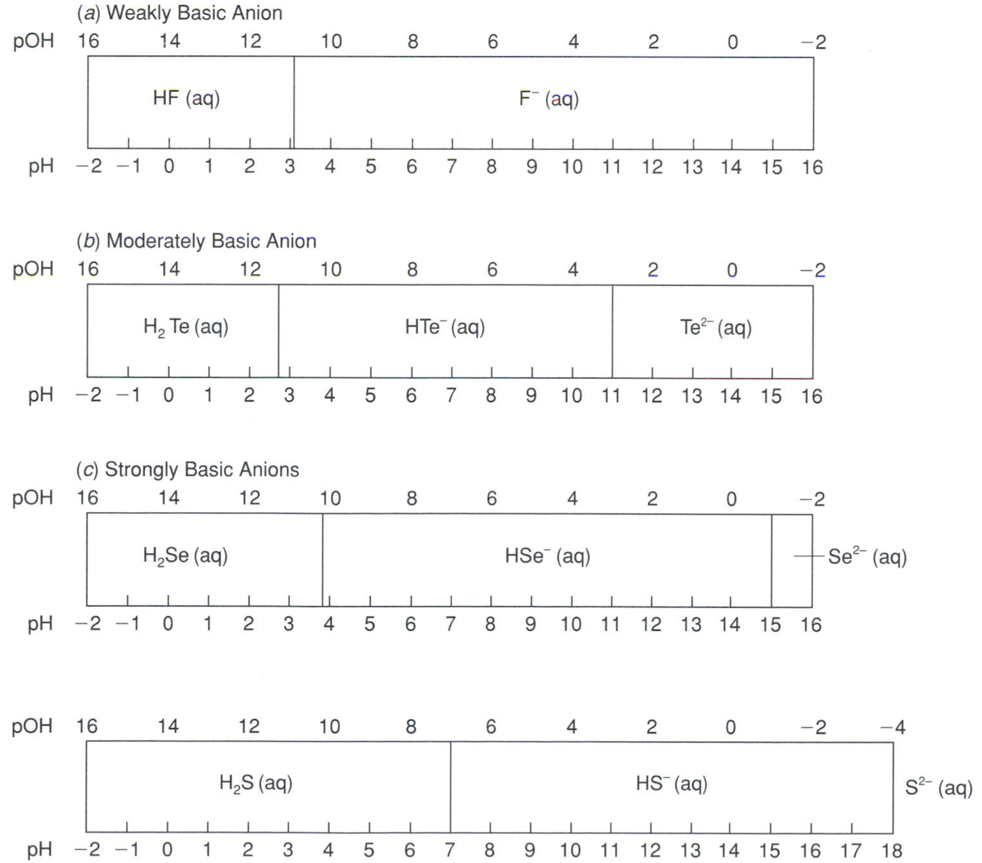

Figure 2.10
Acid–base predominance diagrams for (1 M total concentrations of) nonmetal monoatomic anions and their partially protonated and fully protonated forms.

illustrated by these anions, *the more strongly basic the species, the more it is confined to an extreme right position* in a predominance diagram. Since the nonbasic (neutral) anions Cl^-, Br^-, and I^- predominate at all pH values, they are not shown in Figure 2.10; neither are the very strongly basic anions, which do not predominate at any pH. The left border to the predominance region is given by the relationship pOH $\approx$ pK_b. Since we do not use pOH as the variable in setting up predominance diagrams, however, we note that pH = 14 – pOH, which finally sets the boundary as equal to 14 – pK_b.[14]

The product of hydrolysis of the simple anions [Eq. (2.12)] are *partially protonated anions.* In contrast with the cases of most hydroxy cations, the partially protonated anions *do* predominate over a significant pH range in aqueous solutions, so they are shown explicitly in Figure 2.10. These partially protonated anions can in turn undergo subsequent steps of hydrolysis [Eqs. (2.18) and (2.19)] and are distinguished by the symbols pK_{b2} or pK_2, pK_{b3} or pK_3, and so on. Some measured or estimated values of these equilibrium constants are listed in Table 2.5B.

$$SH^- + H_2O \rightleftharpoons H_2S + OH^- \qquad pK_2 = 7.11 \tag{2.18}$$

$$NH_2^- + H_2O \rightleftharpoons NH_3 + OH^- \qquad pK_3 = -25 \text{ (estimated)} \tag{2.19}$$

The values listed in Table 2.5B are all for the last stage of hydrolysis of the –1 charged partially protonated anions. These, of course, are less basic than the parent monoatomic ions with multiple negative charges; even so these anions cannot persist in water for the elements of Groups 14(IV) and 15(V). Even with charges held constant, the same periodic trends mentioned above still hold. Basicity increases from right to left, which suggests that the reduced electronegativities of the nonmetal atoms to the left allows them to form stronger bonds to the hydrogen of the water, leading to more hydrolysis.[15] The vertical tendency for basicity to decrease as size increases down a group holds as well with the partially protonated anions.

Some of the partially protonated anions have predominance regions that are confined to the *middle* of their predominance diagrams. These anions are *both* acidic and basic, but are not strong in either sense. The basic reaction of HS^-, for example, is given by Eq. (2.18); the acidic reaction is given by Eq. (2.20):

$$HS^- + H_2O \rightleftharpoons H_3O^+ + S^{2-} \tag{2.20}$$

The relative positions of the predominance regions suggests which property is more significant in aqueous solution. For example, dissolving the salt Na^+SH^-, which contains a nonacidic cation that will not affect the pH, will produce a solution that is *basic*, because the predominance region of SH^- is mainly on the basic (right) side of the diagram.

2.7 Acidic Nonmetal Hydrides and Basic Metal Oxides

Acidic Nonmetal Hydrides. Continuing to the left in the predominance diagrams of Figure 2.10, we eventually come to the *hydrides* of the nonmetallic elements. The hydrides that are confined to the left are *acidic hydrides*, in full analogy to metal ions occupying similar left-side positions. Because a predominance diagram has a fixed width, it follows that (assuming the predominance regions of the intermediate partially protonated anions are comparable) the larger the predominance region of a nonmetal anion, the smaller the region of its hydride must be. In general, *the more basic the anion, the less acidic the corresponding hydride will be.*

Periodic trends in the acidities of the hydrides follow, then, from the periodic trends in the basicities of their anions. The nonbasic and weakly basic halide ions of Group 17(VII) give rise to the *acidic hydrides* HF, HCl, HBr, and HI, with acidity increasing in the stated order. The lower Group 16(VI) anions, which have intermediate basicity, still give rise to hydrides confined to the left side of the predominance diagrams of Figure 2.10: H_2S, H_2Se, and H_2Te. These compounds are also acidic hydrides. But the very strongly basic anions of Groups 15(V) and 14(IV) give rise to *nonacidic* hydrides: CH_4, PH_3, and so on. Note that it is even customary to write the formulas of the acidic hydrides of Groups 17(VII) and 16(VI) with hydrogen first, as is common for acids, while the hydrides in Groups 14(IV) and 15(V) are written with hydrogen last to emphasize that they are *not* predominantly acidic.

Example 2.4

Some anions of new nonmetallic elements have been discovered. Sketch a likely predominance diagram for (a) Mm^-, which is nonbasic; (b) Nn^-, which is moderately basic; (c) Pp^{2-}, which is moderately basic; and (d) QqH^-, which is weakly basic and also weakly acidic.

SOLUTION:

(a) A nonbasic species such as Mm^- will not be protonated at any pH, so it will predominate over the entire pH range [Fig. 2.11(*a*)].

(*a*)

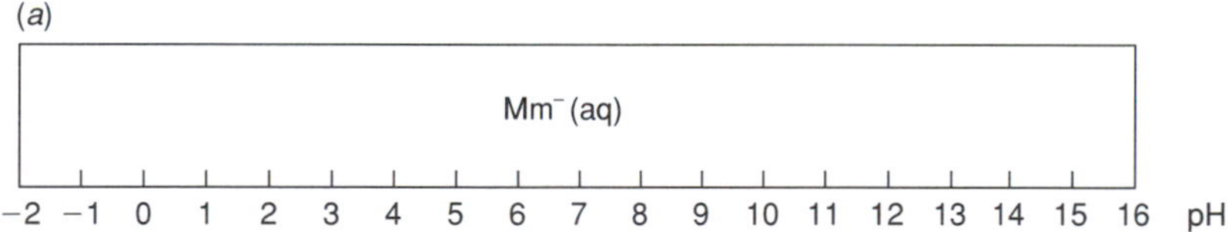

(*b*)

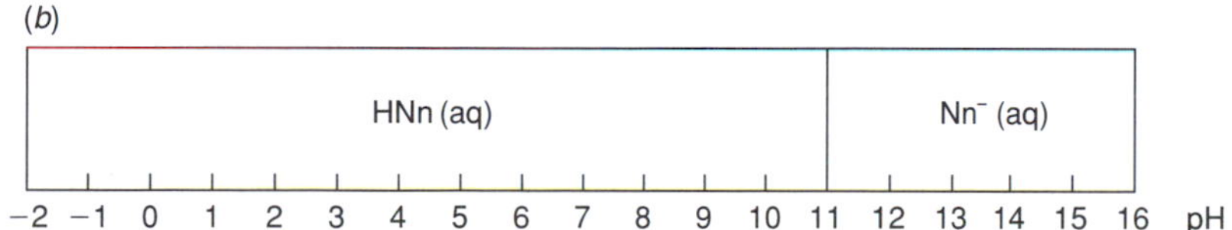

(*c*)

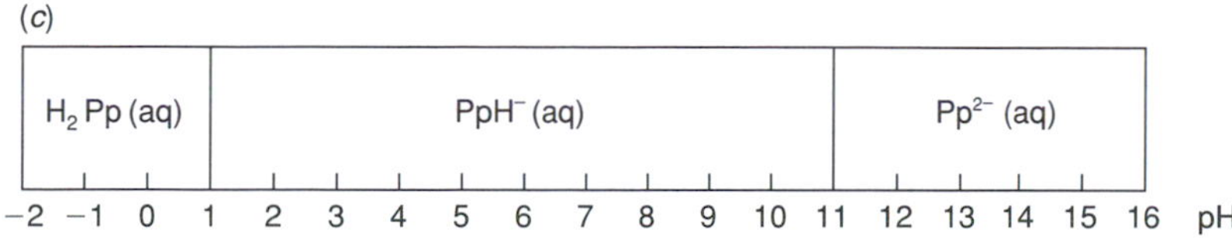

(*d*)

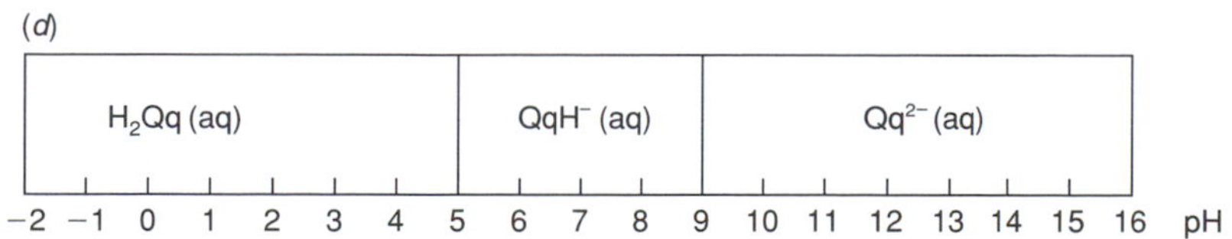

Figure 2.11
Answers for Example 2.4.

(b) Since Nn^- is moderately basic, its pK_b will be in the range of 1–6: let us say 3. The pOH at which it will stop predominating is therefore about 3: this is a pH of 11. Below pH 11 we will find the once-protonated species, HNn [Fig. 2.11(*b*)].

(c) Since Pp^{2-} is also moderately basic, we can apply the same reasoning as for Nn^- and set up the left edge of its predominance region at a pH of about 11. To the left of this will be the once-protonated species, PpH^-. Since PpH^- is still an anion, somewhere to the left of it (we have not developed principles to say just where) there will be another predominance region for H_2Pp [Fig. 2.11(*c*)].

(d) Since QqH^- is a partially protonated anion, there must be a predominance region for Qq^{2-} somewhere to the right of its predominance region and a predominance region for H_2Qq somewhere to the left. Since QqH^- is a weak base, its pK_b is between 6 and 11.5: say 9. Thus its left boundary is at pOH 9, which is pH 5. Because QqH^- is also a weak acid, its pK_a is also between 6 and 11.5: say 9. This pK_a value is the approximate right pH boundary for its predominance region [Fig. 2.11(*d*)].

Example 2.5

Rank these nonmetal hydrides in order of increasing acidic strength: CH_4; H_2O; H_2Se; and GeH_4.

SOLUTION:

More than one approach to this question is possible. For example, we may begin by ranking the basicity of the −1 charged partially protonated anions of these hydrides: $SeH^- < OH^-$ and $GeH_3^- < CH_3^-$ from vertical (radius) trends, and $SeH^- < GeH_3^-$ and $OH^- < CH_3^-$ from horizontal (electronegativity) trends. Putting these together, $SeH^- < OH^- \leq GeH_3^- < CH_3^-$. (To order OH^- below GeH_3^-, we would have to note that the horizontal trends in Table 2.5B are more pronounced than the vertical trends.) Since higher basicity means lower acidity, we deduce the acidity trend $CH_4 < GeH_4 \leq H_2O < H_2Se$.

Basic Metal Oxides. Returning now to the predominance diagrams of Figure 2.4, we can find a corresponding relationship between the acidity of metal ions and the basicity of their metal hydroxides—*the less acidic the metal ion* is, the more the predominance region of its hydroxide or oxide is strongly confined to the right, so the *more basic the metal oxide or hydroxide* is.

Clearly, the most basic oxides or hydroxides will be those of the nonacidic and feebly acidic metal ions. These oxides and hydroxides contain two kinds of ions: Cations with no or only feeble acidity, and anions with very strong basicity. Unsurprisingly, the basicity of the anion prevails; these anions give basic solutions in water.

The oxides and hydroxides of the nonacidic cations, in fact, are very water soluble. The oxides of the nonacidic cations react so strongly with water according to Eq. (2.16) that they are seldom seen; they cannot be prepared by heating the hydroxides to drive off water. The solid hydroxides of the nonacidic cations are so soluble that they **deliquesce** in humid air, removing water from the air to form a solution, and release heat on dissolving. By far, the most important strongly basic compound of the nonacidic cations is NaOH, which is manufactured on a very large scale by electrolysis of NaCl solutions (an oxidation–reduction process that will be discussed later).

The hydroxides of the feebly acidic cations are somewhat soluble in water, again giving basic solutions. The oxides of the feebly acidic cations are commonly encountered. Calcium oxide, CaO, commonly called lime, is the cheapest basic oxide and is easily manufactured by heating the abundant $CaCO_3$ (limestone):

$$CaCO_3(s) + \Delta \rightarrow CaO(s) + CO_2(g) \quad (2.21)$$

where Δ = heat.

Oxides of the feebly acidic cations react exothermically with water to give the hydroxide (with CaO this reaction is called *slaking*):

$$CaO + H_2O \rightarrow Ca(OH)_2 \quad (2.22)$$

This reaction can be reversed by heating the hydroxide to a high temperature (853 K) to drive off the water. These oxides and hydroxides are not notably deliquescent.

The hydroxides and oxides of the weakly and moderately acidic cations are insoluble in neutral water, so they do not raise the pH of water when added to it. (The subject of solubility will be explored further in Chapter 4.) But these oxides have a characteristic property of a base, which is reacting with and dissolving readily in acids:

$$MgO + 2\,H_3O^+(aq) \rightarrow Mg^{2+}(aq) + 3\,H_2O \quad (2.23)$$

Hence, they are still called *basic oxides* and hydroxides: The basicity (reactivity) of the oxide ion still outweighs the acidity of the metal ion.

Acid–Base Reactivity from Predominance Diagrams. At this point, we are ready to predict a limited amount of reactivity of different acids and bases with each other; predominance diagrams will help make these predictions possible by inspection.

Let us consider whether a solution of a strong acid such as HCl should react with a solution of a salt of the basic anion F^-. If HCl is to transfer a proton, it will become the nonbasic anion Cl^-, which predominates over the entire pH range of water; so the predominance region of the strong acid HCl is off-range to the left. The predominance region of the basic anion F^- is shown in Figure 2.10(*a*), and lies above pH 3: It does not overlap the predominance region of HCl. *If two related reactants* (e.g., HCl and F^-) *have nonoverlapping predominance regions,* they cannot coexist as predominant forms at the pH that will result after mixing, so *reaction tends to occur to give products* that *have overlapping predominance diagrams* and thus can coexist at the same pH. In this reaction mixture, the two products with overlapping predominance regions are Cl^- (pH -2 to $+16$) and HF (pH -2 to $+3$). *Stronger acids* (such as HCl) *and stronger bases* (such as F^-) *tend to react to give weaker acids* (such as HF) *and weaker bases* (such as Cl^-).

The farther apart the predominance regions lie, the stronger the reaction tendency will be. The reactants HCl and F^- have distant predominance regions, so the above reaction is quite complete, but a reaction between 1 *M* H_2Te [Fig. 2.10(*b*)] and 1 *M* $F^-(aq)$ [Fig. 2.10(*a*)], with almost-overlapping predominance regions, would result in an equilibrium mixture containing only a slight preponderance of the products HTe^- and HF. If the two related reactants have overlapping predominance regions, no reaction is expected.

The same principle can be used to predict reactions involving different metal ions and their conjugate bases, the hydroxides or oxides. For example, we may want to predict whether we can precipitate $Mg(OH)_2$ from a Mg^{2+} solution using $Sr(OH)_2$ as the precipitating agent:

$$Mg^{2+}(aq) + Sr(OH)_2(s) \rightleftharpoons Mg(OH)_2(s) + Sr^{2+}(aq) \qquad (2.24)$$

The predominance diagrams for Mg(II) and Sr(II) are shown in Figure 2.4(*c* and *b*), respectively. From the relatively small sizes of their predominance regions, it may be seen that the strongest acid is Mg^{2+} and the strongest base is $Sr(OH)_2$. Hence, they will indeed react to give the weaker acid Sr^{2+} and the weaker base $Mg(OH)_2$. We suggest that you not yet try to use this idea to predict reactions between different classes of substances (e.g., nonmetal hydrides and metal cations), since more precise principles for reactivity among diverse substances will be encountered in future chapters.

Example 2.6

What are plausible products when each of the following solid (at least somewhat) soluble salts are stirred with water? (a) K_2O and (b) Ti_3N_4. Write balanced equations for the reactions.

SOLUTION:

In these cases the two reacting species are provided by the same salt. Let us assume that the final pH will be mainly governed by the more reactive of the two ions, so we will analyze that ion first.

(a) The salt K_2O contains the very strongly basic anion O^{2-} ($pK_1 = -22$, Table 2.5), which reacts with water to give the partially protonated species OH^- ($pK_2 = -1.74$, therefore this species predominates above pH 15.74 and can exist in water). The net ionic equation for this hydrolysis is $O^{2-} + H_2O = 2\,OH^-(aq)$; the solution is now quite basic. The salt K_2O also contains the nonacidic cation K^+, which is nonreactive with water or hydroxide ion at any pH (its predominance region overlaps that of water, -2 to $+16$). Thus K^+ does not react, but can be used as a spectator ion to balance the overall (non-net ionic) equation:

$$H_2O + K_2O(s)\ (= 2K^+ + O^{2-}) \rightarrow 2K^+(aq) + 2\ OH^-(aq)$$

(b) The salt Ti_3N_4 contains the N^{3-} ion, which is very strongly basic, and Ti^{4+}, which is "only" strongly basic, therefore we consider the fate of the 4 mol of N^{3-} first. Not only do they react completely with water, but so do their partially protonated forms, NH^{2-} and NH_2^-:

$$4\ N^{3-} + 12\ H_2O \rightarrow 4\ NH_3(aq) + 12\ OH^-(aq)$$

The strongly basic solution now reacts with the 3 mol of strongly acidic Ti^{4+} from the Ti_3N_4 to give the very predominant $TiO_2(s)$:

$$3\ Ti^{4+} + 12\ OH^- \rightarrow 3\ TiO_2(s) + 6\ H_2O$$

Adding these two reactions together and canceling out the OH^- ions and half the H_2O molecules gives us:

$$Ti_3N_4(s) + 6\ H_2O \rightarrow 4\ NH_3(aq) + 3\ TiO_2(s)$$

2.8 Nomenclature

Cations. Since the elements of the *s* block (or others that follow the noble gases by three or fewer positions in the periodic table) form only one cation, they have very simple names: Ca^{2+} is named calcium ion; Al^{3+} is named aluminum ion. Since most of the elements outside of the *s* block form more than one cation (give more than one positive oxidation state), when we name ions or compounds of such an element we designate the oxidation state by using the corresponding Roman numeral (in parentheses) after the name of the element: Bi^{3+} is called the bismuth(III) ion. Oxo cations are commonly named by replacing the *-ium* ending of the metal by *-yl*: BiO^{+} is commonly called the bismuthyl ion. Oxo cations may be more systematically named using *oxo* with numerical prefixes [e.g., $UO_2{}^{2+}$ may be designated the dioxouranium(VI) cation].

Anions. The monoatomic anions of the nonmetallic elements are named using the suffix *-ide*. Thus I^- is named iodide; S^{2-}, sulfide; N^{3-}, nitride; and C^{4-}, carbide. Collectively, the anions of Group 17(VII) are often called the halide ions, those of Group 16(VI), the chalcogenide ions, and those of Group 15(V) the pnicogenide ions.

The partially protonated anions sometimes have special names. Although SH^- and SeH^- are just called the hydrogen sulfide ion and the hydrogen selenide ion, respectively, OH^- is, of course, called the hydroxide ion, $NH_2{}^-$ is called the amide ion, and $CH_3{}^-$ may be called the methide ion or the methyl anion.

The nomenclature of the fully protonated hydrides of the nonmetals also has some peculiarities. The acidic hydrides of Groups 16(VI) and 17(VII) are named differently when dissolved in water and showing acidic properties—using a prefix *hydro-* and a suffix *-ic acid*—than when present as pure, largely unionized compounds. Thus HF by itself is named hydrogen fluoride, but its aqueous solution is known as hydrofluoric acid; H_2S by itself is named hydrogen sulfide, and its aqueous solution is named hydrosulfuric acid. The same pattern applies to the other Group 16(VI) and Group 17(VII) hydrides, except for H_2O, which is always just plain old water.

The nonacidic hydrides of the Group 14(IV) and Group 15(V) elements have a different pattern of naming, in which the hydrogen is not directly referred to but is implied by a special suffix, *-ane* in Group 14(IV) and (usually) *-ine* in Group 15(V). The element is also named using its Latin form [Table D (inside the back cover)] if such exists. The hydrides of Group 14(IV) are thus plumbane, stannane, germane, silane, and (exceptionally) methane; those of Group 15(V) are bismuthine, stibine, arsine, phosphine, and (exceptionally) ammonia.

Naming Salts (including oxides and hydroxides)

1. From the formula of the salt identify the anion and its charge, and write its name.

2. Using the fact that the salt must be electrically neutral, derive the charge of the cation. (For example, the salt Cr_3N_2 contains two N^{3-} ions, named nitride, with a total of six negative units of charge. The three Cr ions must total six positive units of charge, so each is a Cr^{2+} ion.) Name the cation, using its oxidation number in Roman numerals if needed. [In Cr_3N_2, this is chromium(II) ion.]

3. List the name of the cation before the name of the anion. [Cr_3N_2 therefore is chromium(II) nitride.]

Writing Formulas of Salts

1. Write the formula of the cation.

2. Write the formula of the anion.

3. Find the electrically neutral combination. For example, calcium arsenide contains the Ca^{2+} cation and the As^{3-} anion. For the salt to be electrically neutral, the total charges of all cations and anions must add up to zero. The simplest formula results when the smallest possible number of each is used (three Ca^{2+} ions and two As^{3-} anions). Dropping charges gives us a formula of Ca_3As_2.

Study Objectives

1. From its pK_a value or its predominance diagram, classify the acidity of a cation. Draw the predominance diagram of a cation given the pK_a value or acidity classification. Exercises 1–6.

2. Using its charge, radius, and electronegativity, classify a metal ion as nonacidic, feebly acidic, weakly acidic, moderately acidic, strongly acidic, or very strongly acidic. Explain the relationship of hydration energy and acidity of a cation to its charge, radius, and electronegativity. Exercises 7–13.

3. Predict the extent of hydrolysis of a metal ion or metal salt in water, and tell whether it is possible to reverse this hydrolysis and how. Exercises 14–19.

4. Describe the effects of changing maximum coordination numbers on the hydrolysis of halides. Exercises 20–22.

5. Know periodic trends in hydration energies and basicities of monoatomic and partially protonated anions. Explain the relationships of hydration energies and basicities of these anions to charges, radii, and electronegativities. Exercises 23–29.

6. Select a likely predominance diagram for a simple monoatomic anion of a given identity or basic strength; locate the hydride and (if pertinent) partially protonated anions on this diagram. Use predominance diagrams to tell whether a given acid and base will tend to react. Exercises 30–36.

7. Rank nonmetal hydrides and metal oxides as to relative strength as acids or bases. Predict at what pH a metal oxide or hydroxide will precipitate. Exercises 37–40.

8. Name the monoatomic anions of the nonmetals, their partially protonated forms, and their fully protonated (hydride) forms. Name salts of these anions, or write their formulas. Exercises 41–49.

9. Understand how to control variables in designing a scientific experiment. Exercise 50.

Exercises

1. Which category of metal (or nonmetal) cation: (a) gives a neutral solution in water; (b) gives a faintly acidic solution in water, but the acidity is masked by that due to dissolved carbon dioxide; (c) gives a weakly acidic solution (comparable in acidity to vinegar); (d) hydrolyzes reversibly to give a strongly acidic solution; (e) hydrolyzes irreversibly in water.

2. The pK_{a1} values of some new metal cations have been measured. Classify the acidity of each cation, and tell if it will predominantly be in solution at pH 4, and then at pH 11. (a) Heltonium ion, $pK_a = 18.0$; (b) azharium, $pK_a = -18.0$; (c) lawhornium, $pK_a = 8.0$; (d) oyegbium, $pK_a = -3.0$; (e) robertsium, $pK_a = 3.0$.

3. For each predominance diagram shown below, tell to which acidity category the cation belongs, and give its approximate pK_{a1} value.

(a) pH→ 12.5

Cation(aq)	Oxide(s)

(b) 3.0 pH→

Cation(aq)	Hydroxide(s)

(c) −4 pH→ 16

Cation(aq)

(d) −4 pH→ 16

Oxide(s)

4. Consider the following cations, which have approximate pK_a values as listed: Bi^{3+}, 1.1; Bi^{5+}, −13.8; K^+, 14.5; Eu^{2+}, 12.5.

(a) Classify the acidity of these cations.

(b) Below are shown some unlabeled predominance diagrams. Assign each predominance diagram to one of the above ions.

0 pH 2 4 6 8 10 12 14

Cation	(Oxide ppt)	(Oxo anion)

(Cation)	(Oxide ppt)

(Cation)

(Oxide ppt)	(Oxo anion)

(c) Which two of these will have the most nearly neutral solutions?

(d) Which cation does not exist in this form in water under any conditions of pH?

5. Consider the following cations, which have approximate pK_a values as listed: Pb^{4+}, −7.4; Pb^{2+}, 7.7; Yb^{2+}, 12.1.

(a) Classify the acidity of these cations.

(b) Below are shown some unlabeled predominance diagrams. Assign each predominance diagram to one of the above ions.

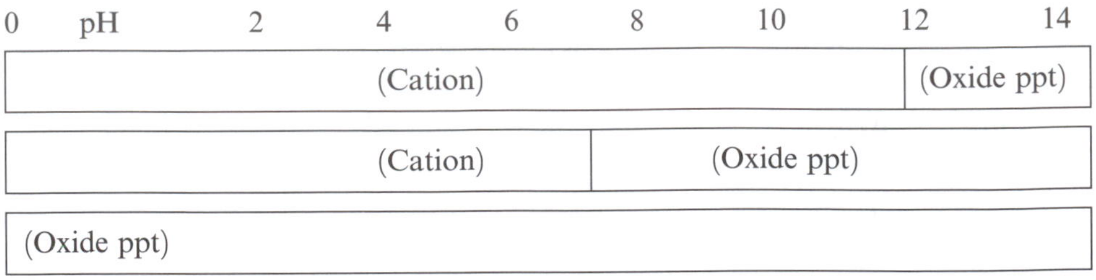

pH: 0–12 (pH scale 0, 2, 4, 6, 8, 10, 12)	pH: 12–14
(Cation)	(Oxide ppt)

pH: 0–~8	pH: ~8–14
(Cation)	(Oxide ppt)

pH: 0–14
(Oxide ppt)

(c) Which cation does not exist as a cation in water under any conditions of pH?

6. Use the pK_a data in Table 2.2 for the ions Au^{3+}, Sc^{3+}, Tl^{+}, and Co^{2+} to (a) Draw predominance diagrams including these cations; (b) classify the acidity of these cations; (c) estimate the pH above which their hydroxides would precipitate; (d) select the most acidic of these cations.

7. What are the three main factors that influence the acidity and the hydration energy of a cation? Explain (in terms of Coulombic and/or covalent interactions) why each factor influences the acidity of a cation the way that it does.

8. *Consider the following cations: U^{3+}, Ag^{+}, Pa^{5+}, C^{4+}, As^{3+}, Tl^{+}, and Th^{4+}. (a) Classify the acidity of each of these cations and describe the reactions of their chlorides with water. (b) Which of these would give "cloudiness" or precipitation upon dissolving in water? What could you do to rectify this if it occurred? (c) If the solutions of these were adjusted to final pH values of 5.5–7, in what chemical form will each element be present?

9. Use Eq. (2.11) and (2.12) to calculate pK_a values for each of the cations in the previous question, and determine whether any of the ions need to be shifted to another category of acidity.

10. Use charge/size ratios to classify the acidity of each of the cations listed in Table C for (a) chromium; (b) neptunium. Tell which of these cations would be predominantly in solution in mildly acidic water (pH of ~5.5–7) for (c) chromium and (d) neptunium.

11. For each of these four categories of metal ion—feebly acidic, moderately acidic, weakly acidic, and strongly acidic—give (a) the typical range of Z^2/r ratios found if the χ_P of the metal is less than 1.8; (b) the typical range of Z^2/r ratios found if the χ_P is greater than 1.8; (c) the approximate pH range at or above which the hydroxide of the metal will precipitate; (d) two real metal ions from the *d* block of metals that fall in that category.

12. *(a) Referring to Table 2.3, compute the smallest and the largest radius that will allow a +1 charged ion to fit into each category of cation acidity. Then, referring to Table C and the electronegativity table, list all +1 ions from Table C that fit into each of the categories. Describe the location in the periodic table of the ions of each category. (b) Do the same for +2 charged cations.

13. Referring to Tables A and C, and using approximate rules of classification rather than Eq. (2.13), derive a list of (a) all nonacidic cations and (b) all feebly acidic cations.

14. *(a) Calculate Z^2/r for each of the cations B^{3+}, Ba^{2+}, Ag^{+}, Th^{4+}, and As^{5+}. (b) Without considering electronegativity, classify the acidity (feebly acidic, etc.) of each of the cations. (c) Now reclassify each, including the effects of electronegativity.

(d) Which two ions will have the most nearly neutral solutions? Which ion(s) do not exist in this form in water under any conditions of pH?

15. (a) Calculate Z^2/r for each of the following cations: As^{3+}; Pb^{2+}; Rb^+; and N^{3+} (assume a 40-pm radius for N^{3+}). (b) Without considering electronegativity, classify the acidity of these cations. (c) Now reclassify each, including the effects of electronegativity. (d) Which one of these will have the most nearly neutral solution? (e) Which cation does not exist in this form in water under any conditions of pH?

16. *Describe briefly what will happen when you try to dissolve each of the following compounds in water: (a) KCl; (b) $NbCl_5$; (c) $AlBr_3$; (d) CBr_4; (e) BaI_2; and (f) IF_7. Which of these would be likely to fume in air and why?

17. (a) Classify the acidity (feebly acidic, etc.) of each of the following cations. Show the calculations you use to carry out the classifications: U^{4+}; Pd^{4+}; K^+; and Ag^+. (b) Which one of these ions would most likely react with water to give a solution of pH about 0? (c) Which one of these ions would most likely have a pK_a of about 14? (d) The hydroxides of which (one or more) of these ions would be insoluble in a solution of pH 7? (e) If you ordered and received sealed bottles of the chlorides of these cations, which should you be most reluctant to open in humid air?

18. (a) Classify the acidity of each of the cations Rb^+, La^{3+}, and P^{5+}. (b) The pK_a values of these ions are −33.1, 8.5, and 14.6, not necessarily in that order. Without actually calculating pK_a, show that you understand pK_a by assigning each pK_a value to the metal ion most likely to show it. (c) The pH values of some solutions of these ions are 0, 5.5, and 6.5. Without actually calculating pH, show that you understand pH by assigning each pH value to the appropriate metal ion. (d) Describe briefly what will happen when you try to dissolve soluble salts of each of these cations in water.

19. Below are shown some unlabeled predominance diagrams. Assign each predominance diagram to one of the following elements in the specified oxidation state: As(III), Pb(II), Rb(I), and N(III).

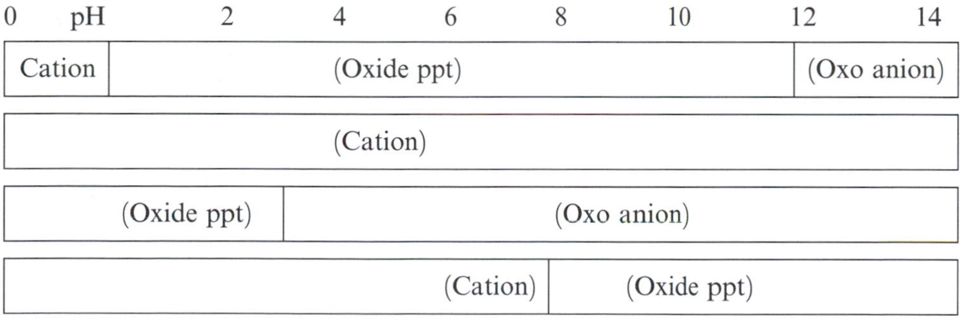

0 pH 2 4 6 8 10 12 14

Cation	(Oxide ppt)	(Oxo anion)

(Cation)

(Oxide ppt)	(Oxo anion)

(Cation)	(Oxide ppt)

20. *In each of the following pairs of compounds, choose the one that would react with water much more rapidly (or tell if both would react at similar rates). (a) UF_6 or SF_6; (b) $COCl_2$ or CCl_4; and (c) SO_2F_2 or SF_6. Give the name of the concept we use to predict these results; use it to explain your answers.

21. (a) Explain briefly what the concept of the maximum coordination number of an element is. Tell what this number is for each period of the periodic table (starting with the second period). (b) Briefly explain why BCl_3 and $TiCl_4$ are more likely to fume in tropical air than is CCl_4.

22. Explain briefly why WCl_6 fumes in humid air, but SF_6 does not.

23. (a) Which of the following anions are very strongly basic? S^{2-}, NH_2^-, SH^-, N^{3-}, or I^-. (b) Which are nonbasic?

24. Consider the following six anions: C^{4-}, Ge^{4-}, O^{2-}, Se^{2-}, F^-, and Br^-. (a) Which one is the weakest base? (b) Which of these are too strongly basic to exist in water? (c) Two of these anions can give partially protonated anions that can exist in water. Write the formulas (including charges) of the two water-stable partially protonated anions.

25. *(a) Which one of the following four anions is the least basic: CH_3^-; C^{4-}; F^-; or I^-. (b) Which is the most strongly basic?

26. *Rank the following imaginary nonmetal anions in order of increasing basicity: Aa^{3-} (radius = 140 pm); Bb^- (radius = 330 pm); Cc^{2-} (radius = 220 pm); Dd^- (radius = 170 pm); Ee^{3-} (radius = 355 pm).

27. The following elements form anions the basicity of which has not been studied. Predict the basicity classification of each of the following anions: (a) the Au^- ion (radius = 188 pm); (b) the anion of Po; (c) the anion of Bi. For the last two anions, use periodic trends to estimate expected charges and radii.

28. What do a *feebly acidic* cation and a *feebly basic* anion have in common?

29. Fill in each blank with one of the following words: INCREASES, DECREASES, POSITIVE, NEGATIVE, OXYGEN, HYDROGEN, METAL. (a) The acidity of a cation ____ as its charge increases because the ____ charge of the cation attracts the ____ end of the water molecule. (b) The basicity of an anion ____ as its charge increases because the ____ charge of the anion attracts the ____ end of the water molecule. (c) The acidity of a cation ____ as its radius increases because the charge of the cation is then further away from the ____ atom of the water molecule. (d) The acidity of a cation ____ as its electronegativity increases. When this happens, the covalency of the cation–oxygen bond ____, which allows more electrons to flow from the ____ atom to the ____ atom.

30. In the quantitative analysis laboratory, you are titrating four unidentified anions in order to find their pK_{b1} values. The ion names and their pK_{b1} values are beckwithide ion = −6; campbellide ion = +10; gallianide ion = 0; udezide ion = +18. (a) Which of these four anions is the most basic? Which is the least basic? (b) Assign each anion to a basicity category. (c) Which of these four (gaseous) anions will release the most energy when it forms a hydrated ion? Which will release the least? (d) If it later turns out that these four ions are in the same group of the periodic table, which one would you expect to find at the top of the group? Which would you expect at the bottom? (e) Match each predominance diagram in the set below with the corresponding anion from the above list.

−2 pH 0 2 4 6 8 10 12 14 16

(Hydride)	(Protonated Anion)	(Anion)

(Monoatomic anion)

(Protonated anion)	(Anion)

(Hydride)	(Protonated anion)

31. Below are shown some unlabeled predominance diagrams. Assign each predominance diagram to one of the following elements in the specified oxidation state: As(–III), Te(–II), F(–I), I(–I).

0 pH 2 4 6 8 10 12 14

(Hydride)	(Protonated Anion)	(Anion)

(Monoatomic anion)

(Hydride)	(Monoatomic anion)

(Hydride)

32. *Consider the range of metal oxides and hydroxides—from those containing nonacidic cations to those containing very strongly acidic cations. Which classes of cations should have oxides that will react with $Mg^{2+}(aq)$ to precipitate $Mg(OH)_2$?

33. Some anions of new nonmetallic elements have been discovered. For each one, sketch a likely predominance diagram. (a) Ff^{2-}, which is weakly basic. (b) GgH^-, which is feebly basic and also feebly acidic. (c) Jj^{3-}, which is very strongly basic. (d) Kk^-, which is moderately basic.

34. Use the predominance diagrams in the text to predict whether the following will react with each other. If so, predict the predominant products when the mixing results in the final pH mentioned. (a) H_2Te and HS^-, pH of 6.0; (b) Al^{3+} and S^{2-}, pH of 9.0; (c) HTe^- and HS^-, pH of 10.0; and (d) $Ca(OH)_2$ and Th^{4+}, pH of 8.0.

35. What are plausible products when each of the following salts are stirred with water? (a) K_2S; (b) $NaNH_2$; (c) Mg_3N_2; (d) Al_4C_3; and (e) BaO. If a reaction is predicted with the water, write a balanced chemical equation for it.

36. *List a plausible product when 1 mol of solids containing each of the following ions or substances are stirred together in 1 L of water. (a) Na^+ and Fe_2O_3; (b) Fe^{3+} and LiOH; (c) H_2S and Te^{2-}; (d) HCl and S^{2-}; (e) HTe^- and Se^{2-}; (f) $Al(OH)_3$ and Ca^{2+}; and (g) $Ca(OH)_2$ and Al^{3+}.

37. Rank these nonmetal hydrides in order of increasing acidic strength: HI, HF, BiH_3, and NH_3.

38. Rank these metal oxides in order of increasing basic strength: Al_2O_3; ZrO_2; CaO; K_2O; and WO_3.

39. *Chromium forms the following oxides: CrO; CrO_2; CrO_3; and Cr_2O_3. Rank these in order of increasing basic strength.

40. For each of the categories of metal ion acidity, tell whether the hydroxides or oxides of metal cations of that category would be soluble or insoluble in solutions of pH: (a) 0; (b) 7; and (c) 14.

41. (a) Name each of the following anions: C^{4-}, Ge^{4-}, O^{2-}, Se^{2-}, F^-, and Br^-. (b) Write the formulas of and name the partially protonated forms of these anions that have –1 charges (if such forms exist).

42. (a) Give the name of the following hydrides: SiH_4; PH_3. (b) Which of these hydrides is most acidic?

43. (a) Name these anions: S^{2-}; NH_2^-; SH^-; N^{3-}; and I^-. (b) Write the formulas of the salts of these anions with the La^{3+} cation.

44. Name the following ions and salts, indicating oxidation numbers if appropriate: (a) Cr^{2+}; (b) Cr^{3+}; (c) Se^{2-}; (d) HSe^-; (e) CrSe; (f) $CrCl_3$; (g) CrN; and (h) $Cr(NH_2)_3$.

45. (a) Write the formulas of the Ca^{2+} salts of the following anions: NH_2^-; SeH^-; P^{3-}; and I^-. (b) Write the names of the Fe^{3+} salts of the same anions.

46. *Divide each set of ions and salts into smaller sets, each containing the metal ion in the same oxidation state. Name the ions and salts, indicating oxidation numbers. (a) Fe^{2+}, Fe^{3+}, FeS, $FeCl_3$, FeN; (b) Sn^{2+}, Sn^{4+}, $SnCl_2$, SnO_2, Sn_3N_2, Sn_3N_4; and (c) Tl^+, Tl^{3+}, TlCl, TlN, Tl_2O, Tl_2O_3.

47. *Write the formulas of the following ions or salts: (a) copper(I) ion; copper(II) ion; copper(II) oxide; copper(I) chloride; copper(II) nitride; (b) chromium(III) ion; chromium(III) oxide; chromium(III) fluoride; chromium(VI) ion; chromium(VI) oxide; and chromium(VI) fluoride.

48. Write the formulas of the following ions or salts: (a) selenide ion; (b) amide ion; (c) titanium(IV) selenide; (d) titanium(III) hydroxide; (e) titanium(II) nitride; and (f) titanium(IV) carbide.

49. Write the formulas of the following ions or salts: (a) gold(I) ion; (b) gold(III) ion; (c) arsenide ion; (d) sulfide ion; (e) gold(III) oxide; (f) gold(I) iodide; (g) gold(III) arsenide; and (h) gold(I) sulfide.

50. Measuring the pH values of which of the following sets of ions represents the best-designed test of the effects of electronegativity on acidity? (a) Ca^{2+} versus Fe^{2+} versus Mn^{2+} versus Zn^{2+}; (b) Pr^{3+} versus Bi^{3+}; (c) Tl^+ versus Ba^{2+}; and (d) Pr^{3+} versus P^{3+}.

Notes

1. This experiment is quite impossible to perform, but the energy released can be determined indirectly. Note that in the experiment that you just did, you added cations to water from the solid state and also added anions. This makes a large difference, as we will see in Chapter 4, and consequently you did not usually detect a lot of energy being released.
2. Latimer, W. M., Pitzer, K. S., and Slansky, C. M., *J. Chem. Phys.*, **7**, 108 (1939). The original equation given by Latimer has been updated by Roger W. Todd to fit the more modern thermodynamic data and radii.
3. There are many experimental difficulties—in some cases different investigators differ by a factor of 10 in their determinations of a particular equilibrium constant—so we should not attempt to interpret small differences between different ionization constants.
4. Burgess, J., *Metal Ions in Solution*, Wiley, Chichester, UK, 1978; S. J. Hawkes, *J. Chem. Educ.*, **73**, 516 (1996).
5. Wulfsberg, G., *Principles of Descriptive Inorganic Chemistry*, University Science Books, Mill Valley, CA, 1991; p. 57.
6. Baes, C. F., Jr., and R. E. Mesmer, *The Hydrolysis of Cations*, Wiley-Interscience, New York, 1976; Chapter 18.
7. Barnum, D.W., *Inorg. Chem.*, **22**, 2297 (1983).
8. D. M. Sullivan, *J. Chem. Educ.*, **70**, 100 (1993).
9. M. J. S. Dewar, Organometallics, **1**, 1705 (1982).

10. M. Kaupp and H. G. von Schnering, *Inorg. Chem.*, **33**, 2555 (1994).
11. A. G. Sharpe, *J. Chem. Educ.*, **67**, 309 (1990).
12. Equations for hydration and hydrolysis of simple anions depend strongly on the ΔH_{hyd} and pK_b values for the sulfide, selenide, and telluride ions, which are very difficult to measure because these anions are easily oxidized and exist in low concentrations. Hence, these equations only show trends and cannot give reliable absolute values of these constants.
13. R. J. Myers, *J. Chem. Educ.*, **63**, 687 (1986); S. Licht, F. Forouzan, and K. Longo, *Anal. Chem.*, **62**, 1356 (1990).
14. This condition holds regardless of concentrations if the species on both sides of the boundaries are soluble at the level in question (1 *M* total concentration for our diagram). Many of the hydrides are gases of limited solubility, however, so that the boundary with a hydride on the left may vary depending on the partial pressure of the gaseous hydride.
15. Note that increasing electronegativity *increases* the acidity of a metal cation but *decreases* the basicity of a nonmetal anion. This makes sense if one thinks carefully about the processes illustrated in Figures 2.2 and (implicitly) in Figure 2.7—hydrolysis is favored by good covalent bonding between the cation or anion and the nearby part of the water molecule. This bonding is improved for either cations or anions if there is a good *match of electronegativities* between the cation and anion and the nearby part of the water molecule. Another way of putting this is that lower electronegativity of the cation or anion allows electrons to go onto or remain on the water molecule, hence favoring basicity of anions but reducing acidity of cations.

CHAPTER

3

Polyatomic Ions and Their Acid–Base Properties

With Applications to Environmental and Analytical Chemistry

In the early part of this book, we focus mainly on the chemistry of ions. Among the most important of these ions are the **polyatomic anions**, in which several atoms are held together by covalent or coordinate covalent bonds. Thus we first consider the types and periodic trends in the strengths of covalent bonds.

3.1 Covalent Bond Types and Periodicity of Bond Energies

For a covalent bond to form by sharing electrons between two orbitals on two different atoms, the shapes and wave function signs (Section 1.4) must be such that the *two atomic orbitals have positive overlap.* Positive overlap means that the *signs* of the two orbitals *match* in the region of overlap, as in Figure 3.1(*c*). Under these conditions, where the crest of one atomic orbital is in phase with the crest of the other atomic orbital (or the two troughs are in phase), the two electrons that are allowed to be in such an orbital have an enhanced probability of being present in the region between the two nuclei. In this region, the two electrons are attracted to both nuclei and serve to bond the two (otherwise mutually repulsive) nuclei to each other.

No bonding interaction or molecular orbital can result if equal amounts of positive and negative overlap occur. Suppose that we were to bring two atoms together along their mutual z axes, and try to overlap the p_x orbital of one atom with the p_y orbital of the other [Fig. 3.1(*a*)]. No bond would result, since the positive lobe of the p_x orbital on the one atom would have equal overlap with the positive and the negative lobes of the p_y orbital on the other atoms, and the same would hold for the negative lobe of the p_x orbital on the first atom.

If the signs of two overlapping orbitals *oppose* each other in the region of overlap, the crest of one atomic orbital overlaps with the trough of the other, resulting in destructive interference and loss of wave amplitude in the desirable region between the nuclei. The electrons must then spend most of their time beyond the nuclei and do not serve to attract the nuclei together. The mutual repulsions of the like-charged nuclei are unabated, and a repulsive or *antibonding* interaction results, as in Figure 3.1(*b*).[1]

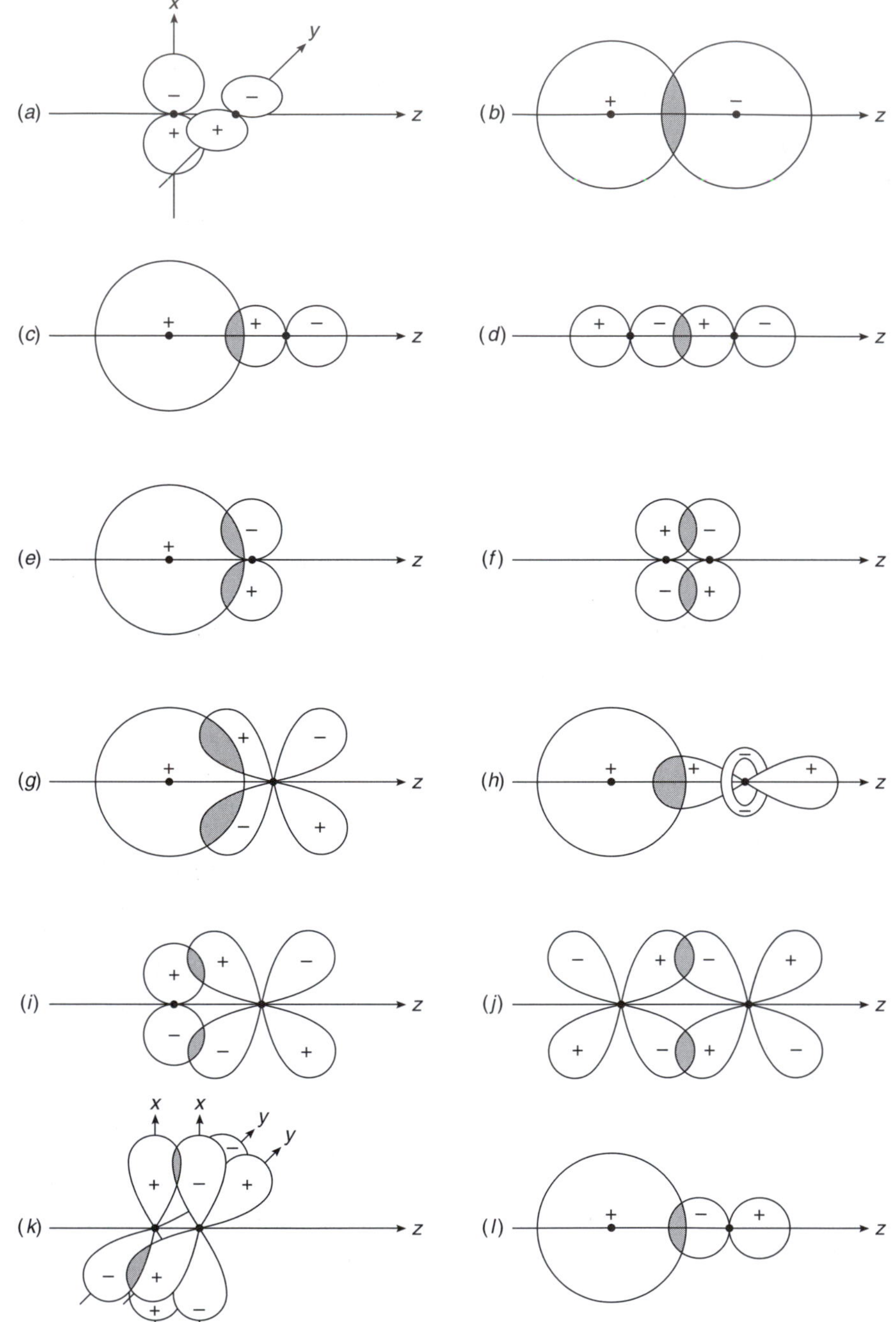

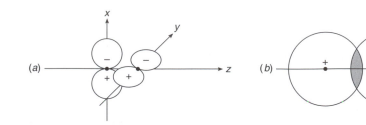

Figure 3.1
Combinations of atomic orbitals to be tested for positive, negative, and zero overlap (see the exercises). In every case, x is the vertical and z is the horizontal axis.

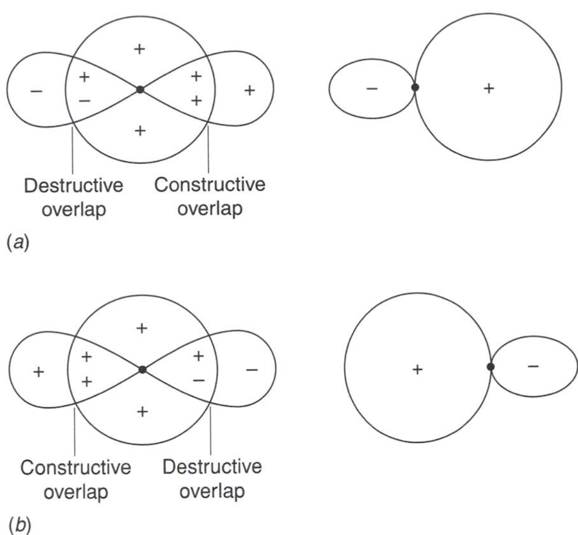

Figure 3.2
Hybridization of an *s* and a *p* orbital to produce two *sp* hybrid orbitals.

Hybrid Orbitals. It is possible to imagine and to construct mathematical orbital types that show better overlap properties than do the pure *s*, *p*, and *d* orbitals, which have lobes (or hemispheres for *s* orbitals) that are aimed away from the region of overlap, and hence are "wasted". For example, an *s* and a *p* orbital *on the same atom* can be mixed or *hybridized* with each other to produce two new *sp hybrid orbitals* (Fig. 3.2). Combining the crest (+ lobe) of an *s* orbital with a *p* orbital with its crest to the right produces an *sp* hybrid orbital with most of its amplitude at the right [Fig. 3.2(*a*)], where it can overlap well with an orbital of a neighboring atom. (Of the original trough part of the *p* orbital, only a vestigial tail remains.) The other possible combination (which must also be made) is identical in shape but is oriented so as to give maximal overlap with an orbital of an atom on the left [Fig. 3.2(*b*)]. Although it requires energy to mix different orbitals in this manner, the improved overlap that results afterward often make this feasible.

Hybrid orbitals are part of the valence bond theory description of chemical bonding; the main alternative approach, the molecular orbital theory, will be introduced in Chapter 10. Both approaches are valid descriptions of most molecules in which all electrons are paired, and since the valence bond theory with hybrid orbitals is simpler to follow and apply in some situations, we will use it first. However, this theory is not readily adapted to use in many molecules with unpaired electrons, such as those resulting from spectroscopic processes,[2] and the hybridization aspect of it has low predictive power (it is better for rationalization after the fact). Hence, hybrid orbitals are no longer as frequently used in bonding descriptions in inorganic chemistry as they used to be, or as they still are in organic chemistry. Therefore, we leave details on hybridization schemes to the organic chemistry course.

Sigma Bonds. Bonding molecular orbitals that result from positive overlap of two atomic orbitals are commonly classified as **sigma** (σ) **bonds**, **pi** (π) **bonds**, and **delta** (δ) **bonds**. (Additional types are possible—in theory, at least.) If a bonding molecular orbital has *no* nodal plane through *both* atomic nuclei, it is called a σ bonding orbital.

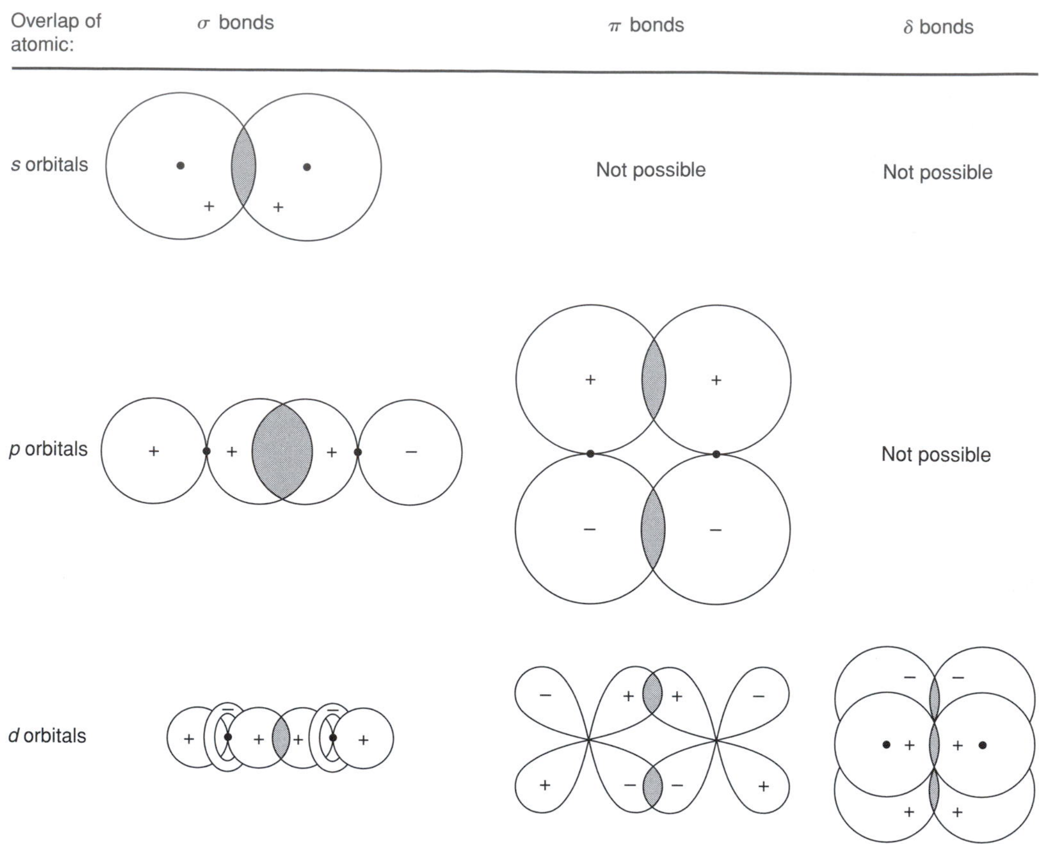

Figure 3.3
Overlap of atomic orbitals of two atoms, the nuclei of which are represented by dots, to give covalent bonds. Regions of positive orbital overlap are indicated by shading; the z axis for each atom is taken as the internuclear axis.

As shown in the first column on the left of Figure 3.3, σ bonds can be formed from two atomic s, p_z, or d_{z^2} orbitals; each combination overlaps in one shaded region to give a σ bond. (Although not pictured, a σ bond can also arise from the overlap of an s orbital on one atom with a p_z orbital on the other atom, and so on, as well as from appropriate hybridized orbitals.) If two atoms have just one covalent bonding orbital linking them (i.e., are joined by a *single bond*), this bonding orbital is nearly always a σ-bonding orbital.

Periodic trends in the strengths of σ bonds show up most clearly among the **bond (dissociation) energies** of homoatomic single bonds of the elements (Table 3.1). (These are the energies required to break single covalent bonds.) We see that *single-bond energies tend to increase as we proceed to the right in a given period.* We may also note the general vertical trend manifested in the s and p blocks—*bond dissociation energies increase as we go up a group.*

These trends may be connected with the trends in covalent and metallic radii (Section 1.9): *Small atoms generally form stronger covalent bonds* (Fig. 3.4). In bonds between smaller atoms, the shared electrons are closer to (and therefore more strongly attracted to) both atomic nuclei. An additional factor in this trend is that valence

Table 3.1
Element–Element Covalent Bond Dissociation Energies[a]

A. Single (Sigma) Bond Energies

H–H 432						
Li–Li 105	Be–Be (208)	B–B 293	C–C 346	N–N 167	O–O 142	F–F 155
Na–Na 72	Mg–Mg (129)	Al–Al	Si–Si 222	P–P 201	S–S 226	Cl–Cl 240
K–K 49	Ca–Ca (105)	Ga–Ga 113	Ge–Ge 188	As–As 146	Se–Se 172	Br–Br 190
Rb–Rb 45	Sr–Sr (84)	In–In 100	Sn–Sn 146	Sb–Sb 121	Te–Te 126	I–I 149
Cs–Cs 43						At–At 116

B. Pi Bond Energies[b]

C–C 256	N–N 387[c]	O–O 352
Si–Si 105	P–P 140	S–S 199
Ge–Ge 84	As–As 117	Se–Se 100
	Sb–Sb 87	Te–Te 92

SOURCE: Data from J. E. Huheey, E. A. Keiter, and R. L. Keiter, *Inorganic Chemistry: Principles of Structure and Reactivity*, 4th ed., Harper-Collins, New York; 1993, Table E-1 pp. A-25–A-33, and M. W. Schmidt, P. N. Truong, and M. S. Gordon, *J. Am. Chem. Soc.*, **109**, 5217 (1987).

[a] Units are in kilojoules per mole (kJ mol^{-1}).

[b] Pi bond energies are calculated as double-bond energies minus single-bond energies for the same element, or (triple-bond energies minus single-bond energies) times one-half.

[c] Schmidt et al. give a value for the pi bond energy of the N–N bond, 251 kJ mol^{-1}, which is not calculated as described above, and which therefore does not also include the effect of reducing unshared electron-pair repulsion on going from the single to the double bond. The 387-kJ mol^{-1} value should be used in calculating double- or triple-bond energies, but the 251-kJ mol^{-1} value is a better representation of the intrinsic strength of the pi bond itself, and so is better to compare with neighboring values.

orbitals at the bottom of the table have more nodal spheres, and hence inner lobes, than the same shape of orbital at the top. The valence orbitals of the heavier atoms are said to be more **diffuse**: The electrons spend more time in inner lobes that cannot overlap with orbitals of other atoms; hence, the covalent bonds of these atoms are weaker.

An anomaly occurs in the upper right of the *p* block. The N–N, O–O, and F–F single bonds are substantially weaker than P–P, S–S, and Cl–Cl bonds, even though the second-period atoms are much smaller. This example is another case in which too small a size leads to diminishing returns. Molecules containing these single bonds (such as H_2N–NH_2, HO–OH, and F_2) also have unshared pairs of electrons. These electrons are brought so close to each other that substantial electron pair–electron pair repulsion[3] results, weakening the net bonding (Fig. 3.5). Consequently, fluorine shows a much larger covalent radius (71 pm, Table 1.5) in F_2 than it does in most of its com-

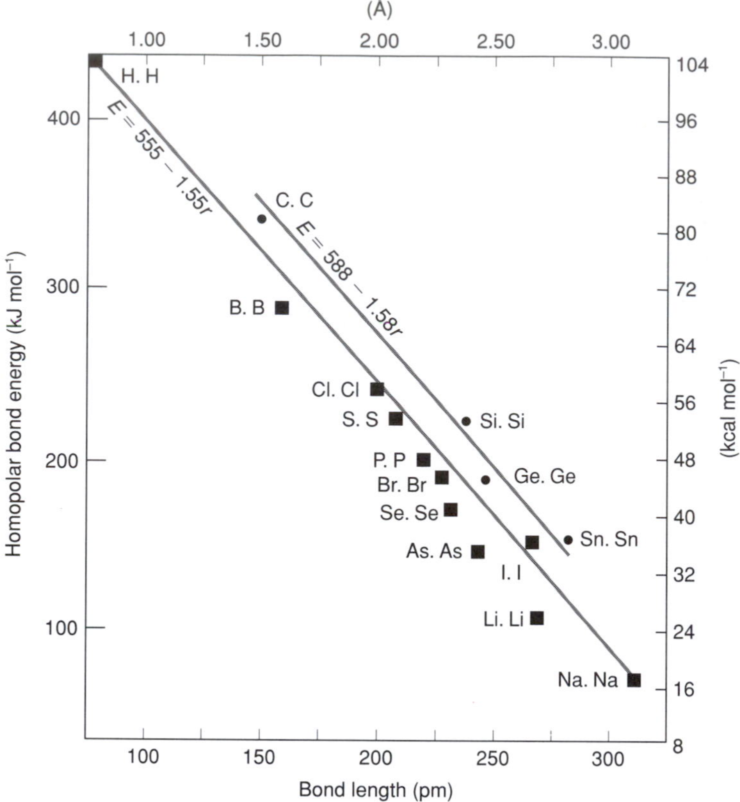

Figure 3.4
Relationship between bond length and bond energy for bonds between like atoms. The upper line is for Group 14(IV) elements only; the lower line is for other *s*- and *p*-block elements. [Adapted from J. E. Huheey and R. S. Evans, *J. Inorg. Nucl. Chem.*, **32**, 383 (1970). Copyright © 1970 by Pergamon Press.]

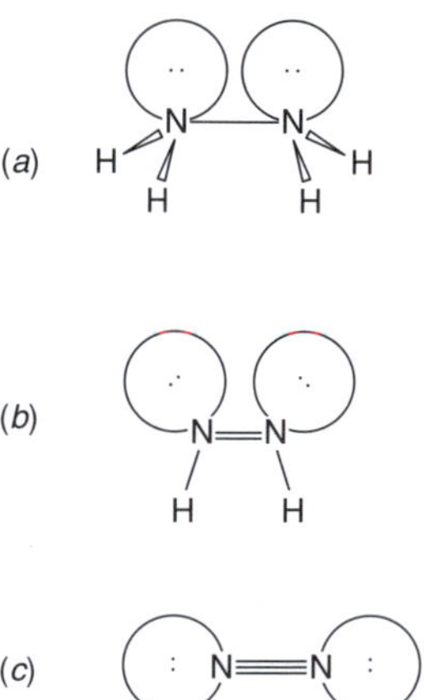

Figure 3.5
(*a*) Electron pair–electron pair repulsion between two small second-period atoms, illustrated for one conformation of $H_2N{-}NH_2$; (*b*) reduction of the repulsion when a double bond is present, illustrated for HN=NH; (*c*) absence of the repulsion when a triple bond is present, illustrated for N≡N.

pounds: Gillespie and Robinson[4] suggested a much smaller covalent radius (54 pm) for use in compounds in which this interaction is absent. (Logically, adjustments should also be made in the radii of O_2 and N_2.)

Pi Bonds. If the molecular orbital formed has *one* nodal plane passing through both atomic nuclei, it is called a π bond (center of Fig. 3.3). Two *p*-block atoms can form two such π bonds, one using their valence p_x orbitals, and one using their valence p_y orbitals. Two *d*-block atoms can also form two π bonds, one with their d_{xz} orbitals and one with their d_{yz} orbitals. Mixed π bonds involving a valence d_{xz} or d_{yz} orbital on one atom and a valence p_x or p_y orbital on the other are also formed between atoms of these two blocks of the periodic table. *Double bonds* nearly always consist of one σ plus one π bond; *triple bonds* consist of one σ plus two π bonds.

As double and triple covalent bond energies are available for many of the *p*-block elements, we can subtract σ- (single-) bond energies from these to obtain estimates[5] for *π-bond dissociation energies*, shown at the bottom of Table 3.1. (In the case of triple-bond energies, after subtraction the result is divided by 2, since a triple bond contains two π bonds.) The π-bond energies show the same basic periodic trends as σ-bond energies in the *p* block: They strengthen going to the right or up the periodic table. The data also show that *π bonds are generally weaker than σ bonds between the same elements*. Although Figure 3.3 does not give a computationally precise representation of the overlap of atomic orbitals, it suggests that this trend arises from the smaller total degree of overlap involved in π bonding than in σ bonding between orbitals of like shape and size.

The data of Table 3.1 also suggest that *π-bond energies tend to drop off with increasing distance even faster than σ bonds*. Consequently, the tendency of *p*-block elements to form double or triple bonds *drops off in the order* O > N=C ≫ S > P > Si.[6] Figure 3.3 suggests that what overlap there is in π bonds may be more sensitive to a small lengthening of the bond (as on going to a larger atom down a group of the periodic table) than the overlap in the corresponding σ bond. As Figure 3.5 also suggests, the repulsion of electron pairs on neighboring N, O, or F atoms is greatly diminished or absent in double or triple bonds, so the π (hence, double- and triple-) bond energies of N and O are not anomalously weak.

Because the π-bond energies are added to the already-existing σ-bond energies, *double bonds are stronger* than single bonds (although not twice as strong) and triple bonds are stronger than double bonds. As a result of the additional orbital overlaps and electron-pair sharing in multiple bonds, the *greater the bond order* (multiple nature) of the bond between a given pair of atoms, *the shorter* it is. Figure 3.6 shows a molecule that features a single, a double, and a triple bond between tungsten and carbon, all in the same molecule. The shortness of the bond (as compared to known

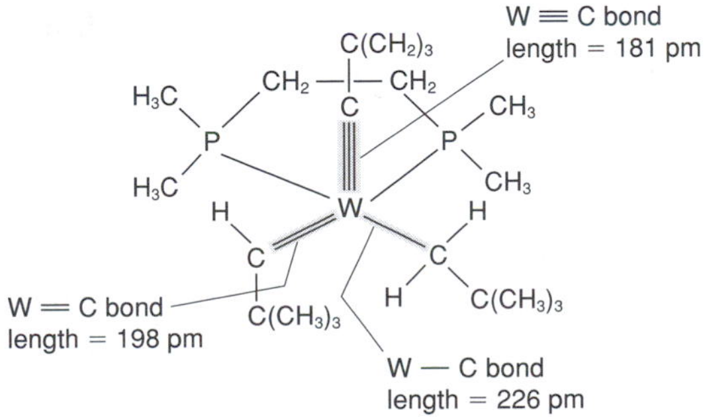

Figure 3.6
Structure of a molecule containing W–C, W=C, and W≡C bonds. [Data from M. R. Churchill and W. J. Youngs, *Inorg. Chem.*, **18**, 2454 (1979). Adapted from D. E. Ebbing, *General Chemistry*, 3rd ed., Houghton Mifflin, Boston, 1990; p. 338.]

bonds of lower order between the same pair of elements) is the feature most frequently used to identify multiple bonds in X-ray structural determinations on crystalline substances.

The periodic trends in the *d* and *f* blocks are more subtle. In the fourth period, the 3*d* orbitals have their maxima ($\langle r_{max} \rangle$) inside the maxima of the core 3*s* and 3*p* and the valence 4*s* orbitals (Fig. 1.5) . Since the electrons in the filled core 3*s* and 3*p* orbitals repel other atoms, it is possible for the 3*d* orbitals to be too deeply buried inside other electrons to be able to overlap the orbitals of the other atom to form covalent bonds. In the middle and later part of the *d* block in the fourth period, the increased effective nuclear charge (Z^*) for the *d* orbitals in fact shrinks them to the point that they show reduced possibility for overlapping orbitals of other atoms. This contributes especially to a reduction of the likelihood of π bonding, where the overlap is weaker to begin with. It also affects the σ overlap, which contributes to the loss of importance and then disappearance of the group oxidation state in the middle and right of the fourth period (Tables B1 and B2).

In the next two periods (fifth and sixth), however, the $\langle r_{max} \rangle$ of the valence *d* orbitals fall closer to the $\langle r_{max} \rangle$ of the relevant *s* and *p* orbitals, so that π bonding and high oxidation states are both found more frequently in the *d* block in the fifth and sixth periods.

In the sixth period, the 4*f* orbitals are very deeply buried indeed under other core and valence orbitals; consequently, there is no double bonding and very little tendency to show the group oxidation state among the *f*-block elements of the sixth period. Again, this problem is less pronounced in the seventh period, so occasional hints of π bonding and frequent occurrences of high oxidation states are found in the seventh period. Relativistic expansion of the *d* and *f* orbitals at the bottom of the *d* and *f* blocks also helps enlarge their radii sufficiently to allow overlap with orbitals of external atoms.

These effects result in the previously observed (Section 1.7) increased tendency to show the group oxidation number at the bottom of the *d* and *f* blocks (Table B2). And, although bond dissociation energy data are not available to quantify the trend, chemical experience indicates that *d- and f-block atoms of the lower periods have increased ability to form multiple bonds.* This tendency seems to reach a maximum at Mo and W, but is also strong in Tc and Re. Not only do these elements (and to a lesser extent their neighbors) form multiple bonds to other like atoms, but also to *p*-block elements such as N and O. These are the elements most likely to form oxo cations (Section 2.4); we will see later in this chapter that these elements also have altered abilities to form oxo anions.

Delta bonds. If the molecular orbital formed by overlap of two atomic orbitals has *two* nodal planes passing through both atomic nuclei, it is a δ bond (third column of Fig. 3.3; one region of orbital overlap is obscured in this view). Only *d*- (or *f*-) block elements can form δ bonds. Two δ bonds can be formed per pair of atoms, one using two d_{xy} and the other using two $d_{x^2-y^2}$ orbitals that overlap in four regions (one hidden from view). A *p*-block element cannot form a δ bond, and hence cannot participate in a quadruple or quintuple bond, but these are possible in the *d* block, and are well known in a series of acetates and other derivatives of quadruply bonded diatomic ions of early *d*-block metals: Cr_2^{4+}, Mo_2^{4+}, W_2^{4+}, Re_2^{6+}, and so on.

From chemical experience, the periodic trends among δ-bond energies are much like those of π bonds in the *d* block. As suggested qualitatively by the drawings of Figure 3.3, overlap is even weaker in the δ-bonding situation than in the π-bonding situation, so that *δ bonds are weaker than π bonds*. But the tendency to form δ bonds increases down the *d* block: The quadruply bonded species Mo_2^{4+} and W_2^{4+} have proportionately shorter and more robust bonds than Cr_2^{4+},[7] with the δ bonds of W_2^{4+} being slightly stronger than those of Mo_2^{4+}.[8]

Covalent Bonding Involving Postvalence Orbitals. In Section 2.4, we noted that the maximum coordination number of a cation increases down a group. Thus, in the second period of the *p* block, no more than four atoms may be bonded to the "cation", while in the third and fourth periods up to six may bond to it. So long as the bonding is truly ionic, no orbitals of the central atom need be involved in the bonding; but if the bonding is covalent, the central atom needs to have the appropriate number of orbitals to form this number of bonds. Among the nonmetallic elements of the *p* block, we know that the "cations" are too strongly acidic and too electronegative to exist; the bonding to other atoms does indeed involve covalent bonding. In the second period, the four available valence orbitals ($2s, 2p_x, 2p_y$, and $2p_z$) match the maximum coordination number in number, so each orbital (or some hybridized version of them) can overlap one orbital of another atom to form one covalent bond.

In the third and fourth periods, there are still only four available valence orbitals, but we know that as many as six (at least partially) covalent bonds can be formed; in such a **hypervalent**[9] molecule, the central atom has more than an octet of electrons. Two possibilities have been suggested to explain hypervalent bonding. In the valence bond picture of bonding (used in organic chemistry and in the valence shell electron-pair repulsion model for predicting geometry, Section 3.4), six equivalent bonds require the hybridization of six atomic orbitals; in addition to the valence *ns* and three *np* orbitals, the d^2sp^3 hybrid orbitals incorporate two *postvalence nd orbitals*. (Since there is no such thing as a 2*d* orbital, this is not possible in the second period). In the molecular orbital theory of bonding (now predominant in inorganic chemistry, to be discussed in detail in Chapter 10), four atomic orbitals on a central atom may suffice to form six bonds because one orbital can be used to overlap the atomic orbitals of *more than one* other atom. In general, the postvalence *nd* orbitals are unoccupied in free *p*-block atoms and are of high energy in part because they are *too large*: They are much more diffuse than the occupied valence orbitals. The postvalence *nd* orbitals are often incorporated into the molecular orbital description of the bonding in a hypervalent compound such as SF_6, but it is not clear how important these *nd* orbitals are to that bonding.

In the *d* block, there is no doubt that unoccupied orbitals beyond the valence *ns* and $(n-1)d$ orbitals—namely, the *np* orbitals—do become involved in the bonding, particularly later in the period. This movement gives a total of *nine* available orbitals to form conventional covalent bonds to other atoms, so that the high numbers of covalent bonds present in compounds such as those shown in Figure 3.6 are not surprising. Later, we will see the "18-electron rule" for bonding in *d*-electron organometallic compounds that is related to the number of available orbitals. Similarly, in the *f* block there is some involvement of the postvalence *d* orbitals, which give rise, for instance, to the frequent anomalous valence electron configurations ($6s^24f^x5d^1$, and so on) found in gaseous atoms of these elements.

3.2 Covalent Bonding to Ions: The Lewis Acid–Base Concept and Complex Ions

Acid and *base* are collective terms defined by chemists in such a way as to include all compounds having similar chemical properties. Thus a statement that "acids do thus and so" economically describes hundreds or thousands of reactions of the compounds included in the definition. Several times in the history of chemistry, chemists have noted additional groups of compounds that show many properties or reactions analogous to those of recognized acids and bases; hence, several new, broader definitions of acids and bases have appeared. In order to allow you to discover one of these broadened concepts of acidity and basicity, and to give you more concrete experience with this type of chemistry, your instructor may now assign Experiment A.2.

G. N. Lewis is considered to be one of the greatest American chemists. Among his many contributions to chemistry, he noted that many reactions and properties that resembled acid–base reactions and properties, occurred in compounds in the complete absence of water and even occurred in compounds that did not contain the elements hydrogen and oxygen. Organic amines such as pyridine [Fig. 3.7(*a*)] act like bases in that they react with (neutralize) acids, yielding solid salts such as pyridinium chloride, $[C_5H_5NH]^+Cl^-$. But Lewis thought it significant that bases such as pyridine also react

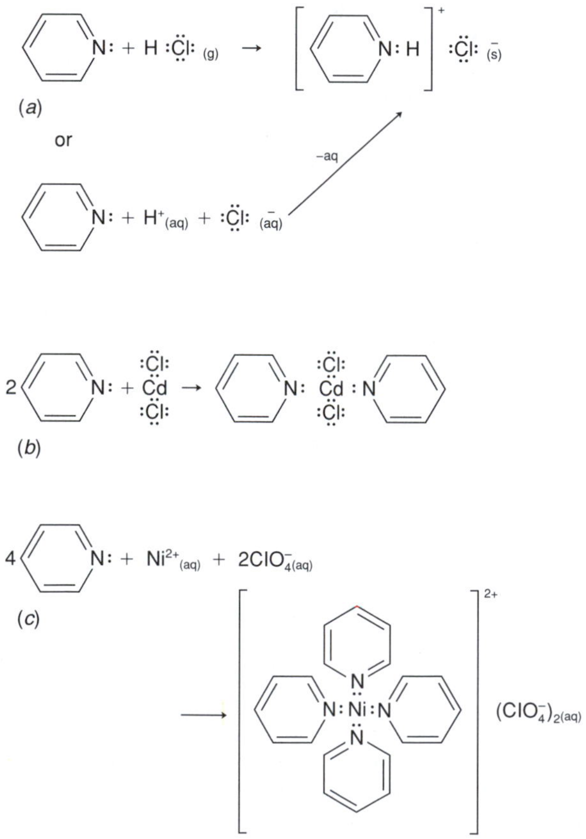

Figure 3.7
Representation of different Lewis acid–base neutralization reactions, all involving the same Lewis base, pyridine. (*a*) Reaction with gaseous or aqueous HCl; (*b*) reaction with $CdCl_2$ to give a neutral, nonionic coordination compound; (*c*) reaction with $Ni(ClO_4)_2$ to give a product containing a complex ion and unattached perchlorate ions.

with numerous metal ions and metal compounds to give neutralized solids and solid salts such as $[Cd(C_5H_5N)_2Cl_2]$ and $[Ni(C_5H_5N)_4]^{2+}(ClO_4^-)_2$. In these reactions, the metal ions and salts seem to be reacting very much like familiar, proton-containing acids. Lewis applied his work on the octet rule and on the drawing of dot structures to these reactions, which he felt should also be classified as acid–base reactions. In such reactions, the product was found to possess a new covalent bond (two shared electrons) not present in either reactant. The two electrons in this bond both came from the *same* reactant, which was in all cases the species that had been accepted as the *base* in the reaction (Fig. 3.7).

Thus Lewis proposed broadening the chemists' definition of bases to include *any species that can donate a pair of electrons*. Such species are now commonly called **Lewis bases**, **donors**, or **ligands**; organic chemists commonly call these **nucleophiles**. Correspondingly, Lewis proposed broadening the definition of acids to include not only hydrogen ion producing materials, but such species as metal ions, since they are also *species that can accept a pair of electrons*. These species are now often called **Lewis acids** or **acceptors**; organic chemists commonly call these **electrophiles** (electron lovers). In the product (usually a solid), the two atoms are held together by sharing of the pair of electrons that came from the same atom in the Lewis base, the **donor atom** of the base. The resulting bond is called a **coordinate covalent** or **dative bond**.[10] The species in which the Lewis acid and base are joined together in chemical matrimony is known by various names: It may be called an **acid–base adduct**, a **coordination complex** or **coordination compound**, or (if it is charged) a **complex ion**.

Any atom in the Lewis base that possesses unshared electron pairs capable of being donated is a potential donor atom. There are compounds and ions in which any nonmetal atom other than hydrogen (except in the hydride ion, $:H^-$) and the noble gases[11] may have unshared pairs of electrons and thus be a potential donor atom; even metal atoms occasionally can be donors. When carbon atoms (such as the C atom in the methide ion, $:CH_3^-$) act as donor atoms, metal–carbon bonds are formed; such compounds are known as **organometallic compounds**.

Not only inorganic anions, and many inorganic molecules, but all classes of organic compounds containing oxygen, nitrogen, sulfur, or halogen atoms[12] with unshared electron pairs can function as ligands. Thus we may see that the number of potential Lewis acid–base reactions in the world, or even in our bodies, is enormous. (From an inorganic chemist's point of view, cell fluids are "soups full of ligands" waiting to attach to metal ion Lewis acids.) If we can develop the acid–base concepts we already have to be used with this greatly expanded list of acids and bases, our power to predict chemical reactions and properties will certainly be enhanced.

A complex ion, when isolated from solution, must be accompanied by ion(s) of opposite charge **(counterions)**. To show that there is no sharing of electrons between these ions and the Lewis acid or base in the complex ion, the complex ion itself is traditionally enclosed in brackets. For example, we earlier showed formulas and structures (Fig. 3.7) of two coordination complexes, $[Cd(C_5H_5N)_2Cl_2]$ and $[Ni(C_5H_5N)_4]^{2+}(ClO_4^-)_2$. The cadmium compound is one uncharged molecule, enclosed by the brackets, with no charge. But the nickel compound contains a complex ion enclosed by brackets, the +2 charge of which is balanced by two perchlorate ions not joined to the nickel ion by any coordinate covalent bond. This distinction is real: Only the latter compound, when dissolved in an appropriate solvent, will conduct an electrical current.

The distinction did not come easily to chemists, however: Alfred Werner, a Swiss chemist, won the first Nobel prize awarded in inorganic chemistry (1913) for his study of the chemistry of coordination complexes. He prepared numerous such complexes, among them a series of complexes of the Co^{3+} ion with the ligand NH_3 (ammonia), which differed in composition, color, and reactivity. He found that aqueous solutions of the yellow compound $CoCl_3 \cdot 6NH_3$ conducted electricity as well in solution as, say, $LaCl_3$, and reacted with 3 mol $AgNO_3$ rapidly to precipitate 3 mol AgCl. The purple compound $CoCl_3 \cdot 5NH_3$ had a lesser conductivity, comparable to $BaCl_2$, and yielded only 2 mol AgCl immediately on treatment with excess $AgNO_3$. He also found two isomers (green and violet, respectively) of $CoCl_3 \cdot 4NH_3$, each of which had a conductivity comparable to that of NaCl, and each of which released 1 mol AgCl. Later, he prepared the compound $CoCl_3 \cdot 3NH_3$, the solution of which did not conduct electricity and did not react immediately with $AgNO_3$. Werner saw that this data could be explained by assuming that there were two kinds of chloride in these compounds. The chloride not present as free chloride ions could not conduct an electric current or react with Ag^+ because it was attached to the Co^{3+} (i.e., was a ligand). He rewrote the above formulas to make this distinction: $[Co(NH_3)_6]Cl_3$, $[Co(NH_3)_5Cl]Cl_2$, $[Co(NH_3)_4Cl_2]Cl$, and $[Co(NH_3)_3Cl_3]$, respectively.

The term *coordination number*, which we have been using all along to refer to the number of nearest-neighbor atoms or ions to a given atom or ion, is especially useful here. The coordination number of the Lewis acid or acceptor atom or ion in coordination compounds is an important property and is in most circumstances equal to the number of donor atoms attached to the acceptor atom. Provided that there is no multiple bonding to the acceptor atom, it is also equal to the number of "ordinary" and coordinate covalent bonds to the acceptor atom.

Chemists of Werner's time thought only in terms of the number of ordinary covalent bonds characteristically formed by an element—its **valence**. Thus they drew some strange-looking (to modern eyes) structures to account for the complexes of $CoCl_3$ with up to six ammonia molecules, since they assumed that, with a valence of 3, cobalt could only form three bonds. With our modern appreciation of the relative sizes of cations, anions, and donor atoms, we can see that cobalt would be expected to have a coordination number of greater than 3. Werner noticed that in each of the cobalt complexes mentioned above and in many others he studied, the experimental data could best be explained by assuming a constant coordination number of 6 for the Co^{3+} ion.[13]

Acid–Base Strength and Complex Ions. In Chapter 2, we observed that metal cations can distort the electron distribution in water to the point that hydrogen (hydronium) ions are released, so we classified metal cations as acids. Similarly, we classified anions as bases due to their ability to accept protons from the solvent, water, thus releasing hydroxide ions. But in a sense the presence of water is not essential to the acidic and basic functions of these ions. Indeed, too much water actually *diminishes* the acidity of metal ions or basicity of anions. In very concentrated solutions, such species (as well as their basic counterparts) behave as if they are even more concentrated than they actually are. For example, a concentrated (19 *M*) solution of KOH behaves in many respects in a manner that we would expect from a 10,000 *M* solution of hydrated hydroxide ions! A characteristic of such concentrated solutions is that there are not enough water molecules present to leave many water molecules free to act as solvent or even to complete the hydration of the ions.

Analytical and physical chemists are often concerned with this phenomenon. Concentrated solutions of ions do not show ideal colligative properties (freezing-point depression, etc.) or ideal compliance to the **law of chemical equilibrium** unless account is taken of the enhanced activity such ions have. This activity is often measured in terms of the **activity coefficient** of a dissolved ion, which is the ratio of its apparent or effective concentration to its actual concentration. In sufficiently dilute solutions, in which nearly all the ions are completely hydrated, the activity coefficients of ions approach 1.00. In solutions of moderate concentrations, activity coefficients are less than 1 due to the greater tendency for pairing of cations and anions to occur. But in highly concentrated solutions, activity coefficients become much greater than 1.

This finding suggests that the Lewis acid–base concept identifies the ultimate source of the acidity and basicity of the type of complex ions known as hydrated ions. The acidity or basicity inheres not in the water, but in the cation or the anion; the water actually serves to dampen out the activity of the metal ion or the anion. Consequently, inorganic chemists nowadays do many of their reactions in the absence of water.

Cobalt complex cations containing coordinated ammonia are not all that different from hydrated cobalt ions, but there is one significant difference: It is much more difficult for the central Co^{3+} cation to polarize the bonds to hydrogen in the ammonia molecule than in the water molecule, so hydrogen ions are not as readily transferred to the solution. (If the molecule attached to the Co^{3+} has no protons on its donor atom, e.g., pyridine, this is impossible.) Thus complex cations such as these provide important additional members of the class of *nonacidic cations.* Complex cations are, strictly speaking, not spherical, so it is impossible to rigorously define a radius for them, but we can readily imagine some kind of "effective" radius that is likely to be substantially larger than that of Cs^+ or Fr^+, so that a Z^2/r calculation, if possible, would usually support this classification. It is possible to prepare a hydroxide of the nonacidic $[Co(NH_3)_6]^{3+}$ cation; in accord with this classification, $[Co(NH_3)_6](OH)_3$ is a soluble strong base.

The simplest "complex" cation is that of ammonia with the smallest cation, H^+: the **ammonium** ion, NH_4^+. In size and charge, it closely resembles the K^+ ion, so in many respects it behaves as a nonacidic cation. However, its solution is definitely acidic due to the very strong acidity of the extremely tiny H^+ ion acting on the coordinated ammonia molecule. If we replace the hydrogens on this ion by organic R groups [where R = methyl (CH_3), ethyl (C_2H_5), etc.] the resulting cations are unambiguously nonacidic and, being even larger than Cs^+, show the properties of nonacidic cations even more clearly. Hence, we will frequently use the **tetramethylammonium ion**, $(CH_3)_4N^+$, and the **tetraethylammonium ion**, $(C_2H_5)_4N^+$, to illustrate the characteristic properties of nonacidic cations.

3.3 Lewis Structures, Coordination Numbers, and Fluoro Anions

As previously mentioned, many bonds may either be formed as normal covalent bonds, by sharing one electron from each atom, or as coordinate covalent bonds, with both electrons coming from a donor atom. The method of drawing Lewis structures taught in general chemistry works well in both situations. For the case of $AB_n^{z\pm}$ ions or molecules of charge $\pm z$, an alternate procedure for assembling Lewis structures, based on the concept of coordinate covalent bond formation between A^{x+} cations (Lewis acids) and B^{y-} anions, offers some advantages: It not only gives a satisfactory

Lewis structure (as the usual method also does), but automatically locates unshared electron pairs on the central atom. In many cases, it can also predict the number n of B groups in, and the overall charge of, the $AB_n^{z\pm}$ product.

Example 3.1

Draw the Lewis structures of (a) PCl_3, (b) SO_3, (c) SO_3^{2-}, (d) XeO_4^{2-}, (e) $SbCl_5$, and (f) $SbCl_5^{2-}$, treating each as being formed from the appropriate Lewis acids and bases.

SOLUTION:

1. *Calculate the oxidation numbers of each element* according to the simple method in Section 1.7. In each of the molecules and ions in question, you should assign the oxidation number of −1 to chlorine and the oxidation number of −2 to oxygen. The oxidation numbers of the other elements are (a) +3 for P, (b) +6 for S, (c) +4 for S, (d) +6 for Xe, (e) +5 for Sb, and (f) +3 for Sb.

2. *Treat each element with a negative oxidation number as an anion; write its Lewis symbol* (as in Section 1.7). Both Cl^- and O^{2-} are each drawn with an octet of electrons. *Treat each element with a positive oxidation number as a cation; write its Lewis symbol. Any unshared electron pairs found on the cation will still be on the same atom in the Lewis structure of the final product.* (a) The P^{3+} ion has an oxidation number two less than its group number. Hence, it retains two *s* electrons and its Lewis symbol is $:P^{3+}$. Similarly, for 1.(c) we obtain $:S^{4+}$, for 1.(d) we obtain $:Xe^{6+}$, and for 1.(f) we obtain $:Sb^{3+}$. In the other two cases, the elements are in the group oxidation states, so their Lewis symbols show no valence electrons: 1.(b) S^{6+} and 1.(e) Sb^{5+}.

3A. *Attach each Lewis base (anion) to the Lewis acid (cation) with a coordinate covalent bond,* normally represented by a line. This completes all of the Lewis structures except that of SO_3 (Fig. 3.8). **3B.** *If the central atom remains short of an octet of electrons* (i.e., <4 Lewis bases are attached) *and the central and Lewis-base donor atoms are capable of forming π bonds,* convert unshared electron pair(s) on the donor atom to π bond(s). Forming one of these bonds from O to S completes the Lewis structure of SO_3 (Fig. 3.8.) Oxygen is good at forming π bonds. Hence, it can act as a *π-donor ligand,* while chlorine is not likely to function in this manner.

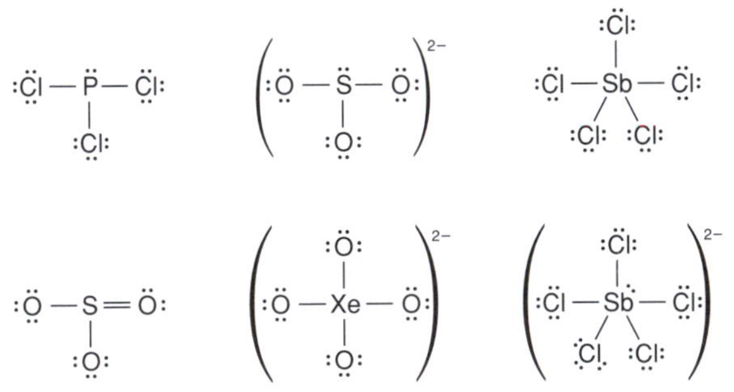

Figure 3.8
Completed Lewis structures for Example 3.1.

By using the Lewis acid–base concept, it is even possible to predict the formulas and the charges of these $AB_n{}^{z\pm}$ species, if the cation has a predictable coordination number, as it often does, at least in the second through fourth periods. Most cations reach their *maximum coordination number* (Section 2.4) when the donor atom is small (hence, sterically undemanding) and not too negatively charged. Adding this number of ligands gives us the "n" in the formula of the $AB_n{}^{z\pm}$; adding the charges of these ligands to the charge of the cation gives us the charge of the $AB_n{}^{z\pm}$.

Fluoride ion is probably the ideal ligand in this respect. Second period cations of moderate, strong, or very strong acidity characteristically form coordinate covalent bonds to *four*[14] fluoride ions; this results in either a neutral covalent fluoride, a *fluoro cation,* or a *fluoro anion.* Larger acidic third and fourth period cations characteristically show a maximum coordination number of *six* fluoride ions. The cations from the fifth and sixth periods are less predictable, but can exceed a coordination number of 6: coordination numbers of 6, 7, and 8 are commonly observed in fluoro anions of these periods, so that the *maximum* coordination number can be taken as 8 for fifth and sixth period fluoro anions.

Example 3.2

Given as many fluoride ions as they can accommodate, what complex species will the second period "cations" from Be^{2+} onward (each in their group oxidation state) form? What will be the Lewis structures of these species?

SOLUTION:

The formulas can be obtained by adding four fluoride ions to each "cation," being certain also to add one negative charge for each fluoride ion to the charge of the cation. The resulting formulas are $BeF_4{}^{2-}$, $BF_4{}^-$, CF_4, and $NF_4{}^+$. Note that this particular series includes one fluoro cation, one fluoride, and two fluoro anions. All of these are known species; since the group oxidation number is not known for O, F, or Ne, the series terminates at N. Feebly acidic Li^+ does not form a complex ion with F^-.

The Lewis structures of these species can easily be obtained by drawing four fluoride ions, each with its octet of electrons, then donating one electron pair from each fluoride ion to the central "cation". (The resulting covalent or coordinate covalent bond is normally represented by a line rather than two dots.) In each case, the result looks the same, since these are **isoelectronic** species:

$$M^{z+} + 4\,:\ddot{\underset{..}{F}}:^- \longrightarrow \left(\begin{array}{ccccc} & & :\ddot{F}: & & \\ & & | & & \\ :\ddot{\underset{..}{F}} & - & M & - & \ddot{\underset{..}{F}}: \\ & & | & & \\ & & :\underset{..}{F}: & & \end{array} \right)^{(4-z)^-}$$

Here M represents the second period element and z is its group number. Only the charge on the complex species varies across the period.

Now, let us consider the situation when the Lewis acid is a *p*-block "cation" with one (or more) electron pairs already on it. If only one electron pair is present, we know that pair is located in a valence *ns* orbital in a true cation. But that orbital, being full, cannot be used to accept electrons from a donor atom of a Lewis base. Furthermore, since its radius ($\langle r_{max} \rangle$) is comparable to that of the empty valence *np* orbitals, that spherical *ns* electron pair would repel all the donor atoms in all directions from the Lewis acid, preventing the formation of any very good coordinate covalent (or covalent) bonds. Rather than give up this advantage, the Lewis acid will usually confine the electron pair to some type of *sp* hybrid orbital, as in Figure 3.2.[15] Then the repulsive effect is confined to one direction, so that the Lewis bases can approach from all of the directions except that one. So *the total number of fluoride ions* (ideal Lewis bases) expected to coordinate to the Lewis acid is *one less than the maximum coordination number.*

Note that this situation would not arise in the *d* or *f* blocks, because any valence electrons present on the cation are *not* in spherical *ns* orbitals, but in the more directional $(n-1)d$ or $(n-2)f$ orbitals. Frequently, these orbitals are also *inside the valence* or postvalence orbitals used to accept the donated electrons. Therefore, they have no effect on the number of donor atoms coordinated to the Lewis acid. Even in most of the other cases, the narrow lobes of the $(n-1)d$ or $(n-2)f$ leave enough openings for all of the ligands to coordinate. [The more detailed analysis of the crystal field theory (Chapter 8) will identify cases in which problems arise.] Hence, in the case of *d*- or *f*-block Lewis acids, we will *ignore any remaining valence electrons in predicting the formula of the species.*

To render the above discussion more precise, it is useful to introduce the concept of the **total coordination number**,[16] which is the sum of the *number of other atoms attached* to the central atom (Lewis acid) in question plus the *number of unshared p-block valence electron pairs* (*sp* hybrids, etc.) about it. Then we may state that there is a **maximum total coordination number** for cations in each period: *four* in the second period; *six* in the third and fourth periods; *in excess of six* in the fifth and sixth periods.

Example 3.3

Predict the formulas and draw the Lewis structures (if appropriate) of the fluoride, fluoro anion, or fluoro cation formed by the following Lewis acids when an excess of F^- is available: (a) Cl^{5+}; (b) I^{5+}; (c) Cl^{3+}; and (d) Fe^{3+}.

SOLUTION:

(a) The Lewis formula of the Lewis acid is $:Cl^{5+}$, which leaves room for one less than the maximum total coordination number 6, of fluoride ions to coordinate: We expect the highest fluoro species to be neutral ClF_5.[17] (b) The isoelectronic $:I^{5+}$ cation is from the fifth period, so it has a maximum total coordination number *over 6*. The indefiniteness of this number makes our predictions less concrete, but if this number is 7, then the species $:IF_6^-$ should be (and is)[18] formed. (c) The Lewis formula of the Lewis acid is $:Cl:^{3+}$, with *two* unshared pairs of electrons; these should block the approach of two fluoride ions, allowing four to enter, giving ClF_4^-. (d) The valence electron configuration of the fourth period Lewis acid Fe^{3+} is $3d^5$. But since these $(n-1)d$ electrons tend not to block access of donor atoms, we ignore them and predict the formula FeF_6^{3-}. Lewis structures are customarily not drawn for *d*-block complexes (they would be crowded with dots on the *d*-block atom); the Lewis structures of the other examples are drawn in Figure 3.9.

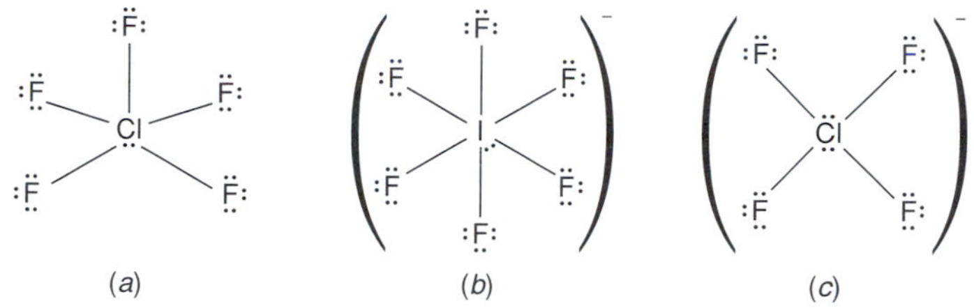

(*a*) (*b*) (*c*)

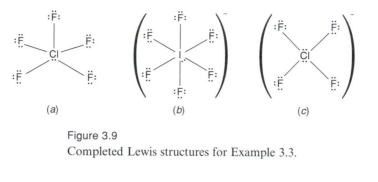

Figure 3.9
Completed Lewis structures for Example 3.3.

3.4 The Shapes of *p*-Block Molecules and Complex Ions

The geometric shapes of molecules, ions, and coordination compounds of the *p*-block elements, and the approximate angles between their covalent bonds, may be predicted remarkably well using a simple scheme known as the **Valence Shell Electron-Pair Repulsion (VSEPR)** model. This model assumes that the shapes of molecules and ions are determined primarily by the electrostatic repulsions of the negatively charged covalent bond electron pairs and unshared pairs of hybridized valence electrons about the central atom. The geometry and bond angles adopted are those that allow these pairs of electrons to move as far apart as possible. This model was originally based on the valence bond model of chemical bonding, but it has also been justified using more sophisticated types of theoretical calculations.[19]

We begin the determination of molecular geometry with a correct Lewis dot structure of the species. If it is at all large there will be more than one central atom, and each central atom may have its own geometry and bond angles that must be determined separately. For each central atom in this structure, we determine the total coordination number. Whether the atoms are attached by single, double, or triple bonds does not matter in this count, since all the electrons in double or triple bonds must occupy the same general region of space—between the two atoms in the double or triple bond. The only kind of unshared valence electron pairs we will consider are *sp*-hybridized electron pairs. (As discussed earlier, frequently *d* electrons in *d*-block compounds are in orbitals of shapes and orientations that do not block incoming ligands.) For each total coordination number, there is ideally one bond angle (or set of angles) that minimizes the repulsions of the electron pairs. There is then a term for the geometry that describes the locations of the atoms in a molecule having these bond angles.

Compounds with No Unshared *sp*-Hybridized Electron Pairs on the Central Atom. We begin with this, the simpler case. For these compounds, the *total* coordination number of each central atom equals its coordination number. The least repulsive arrangements of electron pairs for total coordination numbers between 2 and 9 are shown and named in Figure 3.10.

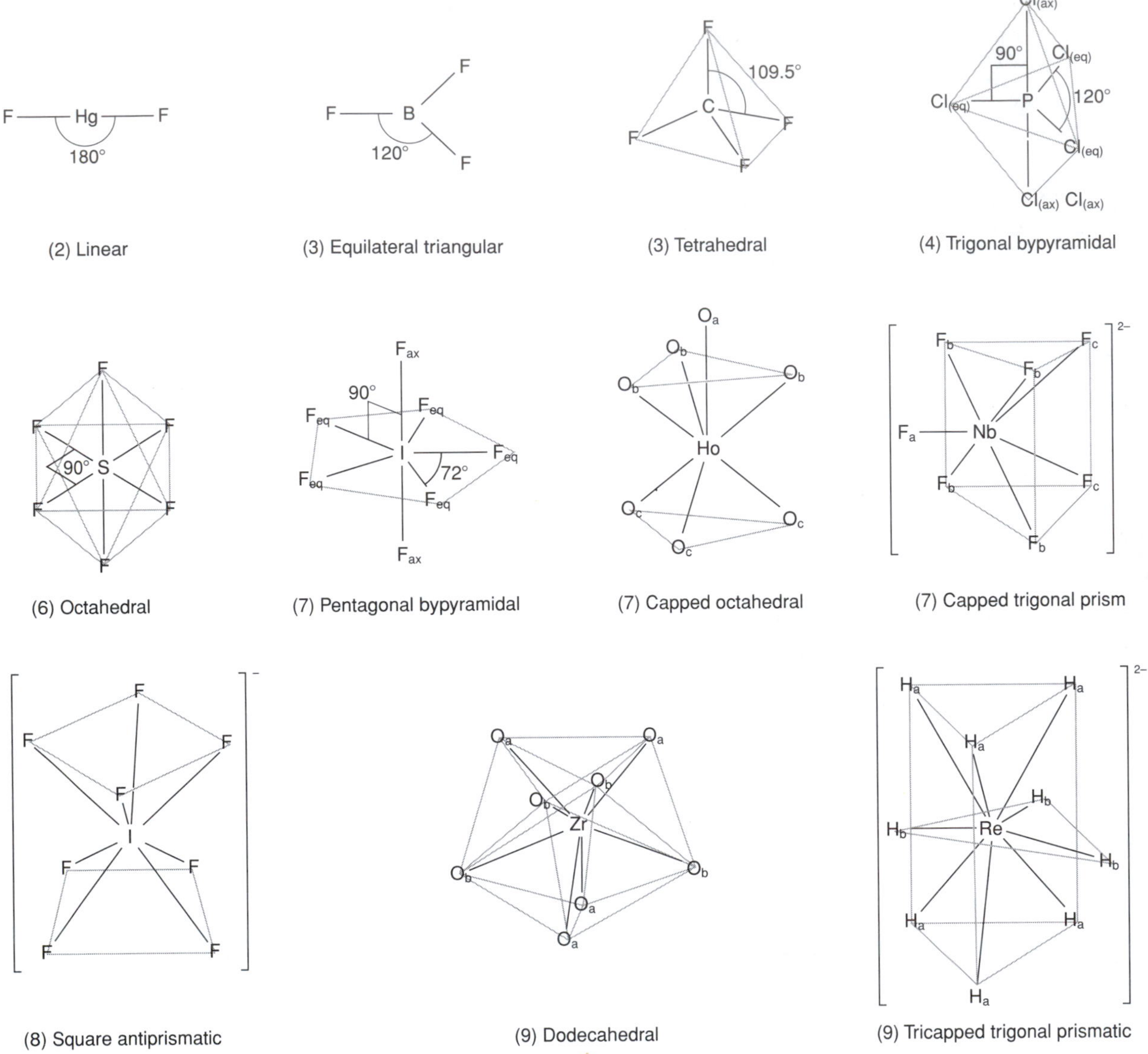

Figure 3.10
Idealized arrangements of electron pairs that minimize repulsions among two to nine electron pairs (the total coordination number, shown in parentheses) about a central atom. The examples shown are complete molecules or complex ions except for (7) the capped octahedron, part of the structure of $[Ho(H_2O)(C_{15}H_{11}O_2)_3]$; (8) the dodecahedron, part of the complex $[Zr\ (C_5H_7O_2)_4]$. Equivalent sets of outer atoms are indicated by the same subscripts: either a, b, or c or (for pyramidal structures) "ax" for axial and "eq" for equatorial.

A total coordination number of 2 about a given atom (as in HgF_2 or O=C=O) means that there are two independent pairs of electrons repelling each other. Repulsion is minimized with the two electron pairs at a bond angle of 180°, giving a *linear* geometry about that central atom. The hybridization that is associated with the linear geometry is *sp* (Fig. 3.2).

The total coordination number of 3 gives rise to bond angles of 120° between electron pairs and a equilateral triangular arrangement of bonds; the hybridization required to rationalize this geometry is sp^2, which means that one valence s and two valence p orbitals are mixed to produce the same total number (three) of equivalent hybrid orbitals oriented toward the three donor atoms.

For a total coordination number of four or higher, three-dimensional structures give less repulsion than structures that confine all the atoms and bond electrons to a single plane. Thus a square array of four electron pairs about a central atom separates the pairs by 90°, but a tetrahedral array gives a greater separation of 109.5°. Chemists had become so used to drawing molecules on planar sheets of paper and blackboards that this concept was surprisingly hard to accept at first. The central-atom hybridizations that produces tetrahedrally oriented hybrid orbitals are sp^3 and sd^3. (In the d block, both hybridizations may be involved to varying degrees.)

For a total coordination number of 5, it is impossible for all of the electron pairs to be equivalently situated about the central atom, even if all of the outer atoms are identical. Instead, there is a three-layered *trigonal bipyramidal* arrangement of outer atoms or unshared electron pairs into two different types. Those above and below the central atom are said to be in **axial** positions; those in the same plane as the central atom are said to be in **equatorial** positions. The angles between equatorial electron pairs are 120°; the angles between any of these pairs and any axial electron pair is 90°. [Note that one axial pair and the three equatorial pairs together form a pyramid with a triangular (trigonal) base; the name of the geometry arises from the presence of two such pyramids sharing a base.] The hybridization that corresponds to this geometry is dsp^3, which can be split approximately into sp^2 for the equatorial bonds and dp for the axial bonds.

Although it may not be apparent from the figure (it is a good idea to examine as many of these arrangements as possible using molecular models), the preferred arrangement of six electron pairs gives all 90° bond angles. This geometric shape, the *octahedron*, is highly symmetrical: No two pairs are any more axial than any other two pairs. The d^2sp^3 hybridization scheme corresponds to this geometry.

For a total coordination number of 7, there are three geometries of nearly equal efficiency in separating electron pairs: pentagonal bipyramidal, capped octahedral, and capped trigonal prismatic. Except in the case of the pentagonal bipyramid, the bond angles are variable. Seppelt and co-workers[20] suggested that, among these, the pentagonal bipyramid is preferred. A total coordination number of 8 gives rise to two arrangements of minimal repulsion, the square antiprism and the dodecahedron. For total coordination number 9, a unique geometry, the tricapped trigonal prism, is predicted to minimize the electron-pair repulsions.

Although total coordination numbers up to 14 are known for some very large central atoms with tiny outer atoms, real examples are not common and predictions of the favored geometry are not very clear.

Compounds with Unshared p-Electron Pairs on the Central Atom. Until recently, our methods of determining the geometry of molecules and ions (such as X-ray crystallography) could not locate unshared electron pairs; this is still usually the case. Hence, we describe what we can observe, the geometry of the *central* and *outer* or *donor* atoms only, which results in some new geometries (Fig. 3.11).[21] Although the approximate bond angles are still determined by the *total coordination number*, the choice of the term for the molecular geometry is based on the *coordination number*, which excludes the unshared central-atom electron pairs.

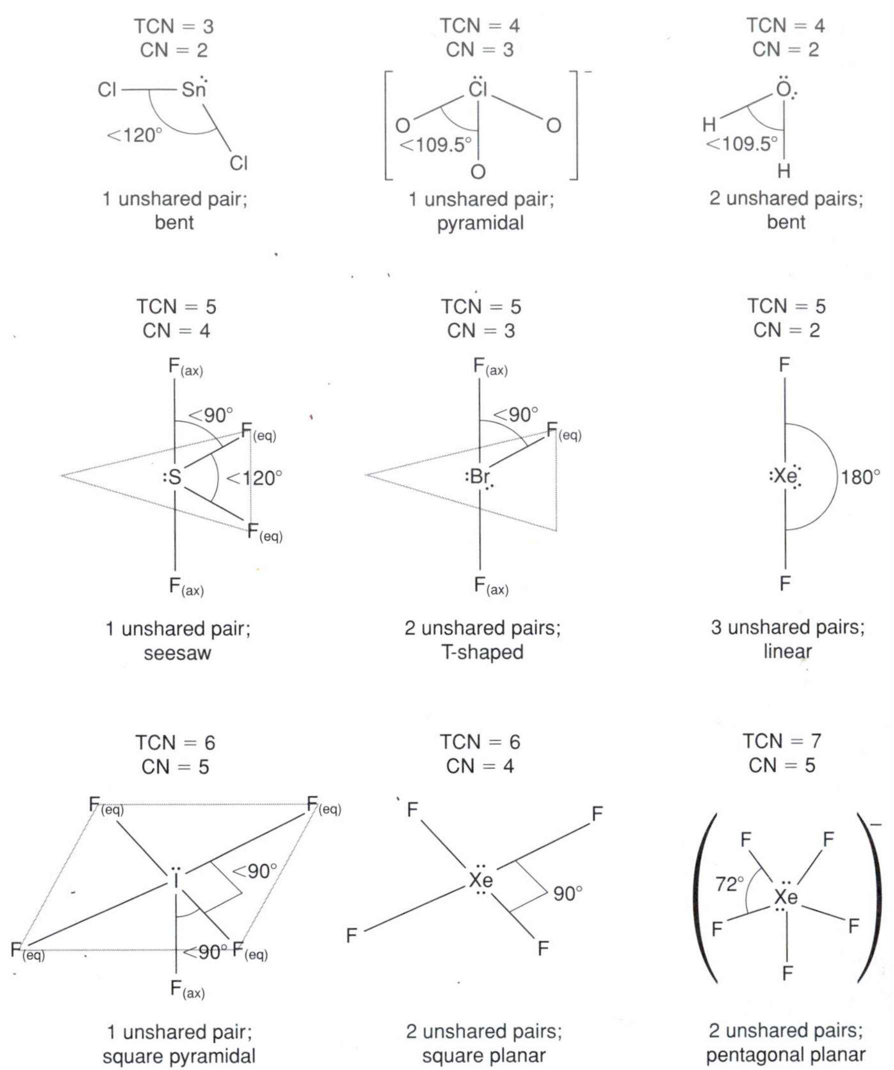

Figure 3.11
Geometries described by the central and outer atoms in molecules and ions also having unshared p electron pairs about the central atom. Total coordination numbers TCN, and coordination numbers CN, of central atoms are shown. For pyramidal structures, axial atoms are indicated by "ax" and equatorial by "eq".

When unshared electron pairs are present, VSEPR does predict slight distortions of the ideal bond angles of Figure 3.10. It is assumed that an outer atom "pulls out" an electron pair and makes it require less angular space about a central atom than does an unshared pair of electrons (Fig. 3.12).[22] We expect this assumption to have two kinds of consequences: (1) the bond angles will tend to distort to give more room to unshared electron pairs and (2) in geometries with positions that are not equivalent (such as the trigonal bipyramidal) the unshared electron pairs will occupy the more spacious positions.

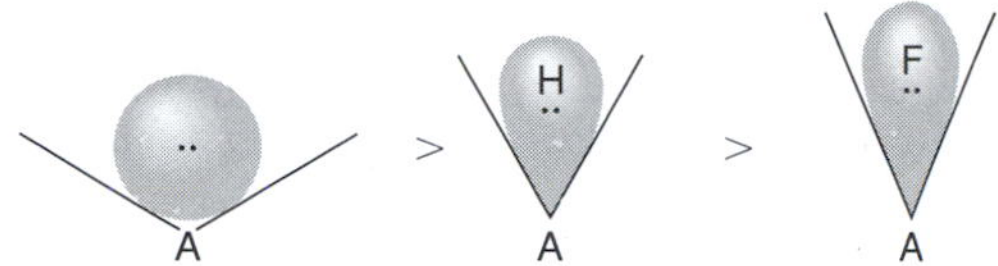

Figure 3.12
A schematic illustration of two assumptions of VSEPR. The first and second drawings illustrate an unshared electron pair occupying more space (taking a greater *cone angle*) about a central atom than an electron pair shared with an outer atom. The second and third drawings illustrate a bond pair to an atom of low electronegativity occupying more space about the central atom than a bond pair to a highly electronegative atom.

Since unshared electron pairs are expected to take more space around the central atoms,[23] the angles we can measure, those between bonds to different outer atoms, should generally be less than those shown in Figure 3.10. For example, the ideal tetrahedral angle of 109.5° in CF_4 is expected to be reduced in other molecules with a total coordination number of 4 but which include unshared *p*-block electron pairs. Bond angle data given in Table 3.2 confirm the predictions of reduced bond angles. Only when the unshared-pair repulsions are balanced on all sides of a given pair of bonds (as in XeF_2 in Fig. 3.11) is an undistorted bond angle expected.

In the trigonal bipyramid, the second consequence comes into play. Electron pairs in the axial positions have three near neighbors—the three equatorial pairs, only 90° away—so they instead adopt the equatorial position in order to have only two near neighbors (the two axial pairs at 90°). (The two equatorial pairs at 120° are far enough away not to matter as much.) This preference for the equatorial position results in the unusual geometries seen in Figure 3.11, such as the "seesaw" or "teeter-totter" geometry that is reminiscent of the children's balancing toy. Additionally, the remaining bond angles are reduced from 90 or 120° due to the spatial demands of the unshared electron pairs, except in the case of three equatorial pairs, the repulsions of which cancel each other out.

With a total coordination number of 6 and two lone pairs, the second principle leads us to expect the two lone pairs to go to opposite sides of the octahedron (as far from each other as possible), giving rise to the square planar geometry. Again bond angles will not distort, as the repulsions cancel each other out. Similarly, the pentagonal

Table 3.2
Bond Angle Distortions Due to Unshared Electron Pairs in Molecules with a Total Coordination Number of 4

Molecule	Angle (°)	Molecule	Angle (°)	Molecule	Angle (°)	Molecule	Angle (°)
NH_3	107.3			NF_3	102.1	OH_2	104.5
PH_3	93.3	PCl_3	100.3	PF_3	97.8	SH_2	92.2
AsH_3	91.8	$AsCl_3$	98.7	AsF_3	96.2	SeH_2	91
SbH_3	91.3	$SbCl_3$	99.5	SbF_3	88	TeH_2	89.5

SOURCE: Data from F. A. Cotton and G. Wilkinson, *Advanced Inorganic Chemistry: A Comprehensive Treatise*, 5th ed., Wiley-Interscience, New York, 1988, p. 23.

planar geometry occurs in the anion XeF_5^-, which has a total coordination number of 7 and two lone pairs.[24]

A most interesting case is that of total coordination number 7 with one unshared electron pair. Since we cannot predict the geometry even in the absence of unshared electron pairs, it is not easy to do so if an unshared electron pair is present. Various distorted geometries often result, but often a completely unexpected result turns up: an undistorted octahedron, in which the unshared electron pair occupies no angular space at all; finally this electron pair apparently has decided to remain in the valence *s* orbital in which it was originally located! Similar results are found for total coordination numbers 8 and 9 with one unshared electron pair: the XeF_7^- ion has an undistorted capped trigonal prismatic geometry,[25] and XeF_8^{2-} has an undistorted square antiprismatic geometry. The more sophisticated molecular orbital theory is usually used to explain these geometries, although it has been suggested that steric crowding, which may accompany such high total coordination numbers, leaves no room for the unshared electron pair to occupy a *sp*-hybrid orbital. Instead, the electron is confined to the *s* orbital, and all the ligands are repelled by it, which results in lengthened metal–ligand bonds.[26]

One final consequence of our assumptions, which are illustrated in Figure 3.12, is that a bond pair to a more electronegative outer atom should occupy less space than a bond pair to a less electronegative outer atom. At least one example of this theory generally works out in what is known as **Bent's rule**[27]: In trigonal bipyramidal molecules of the *p*-block elements having more than one kind of outer atom, it is found that the more electronegative substituent preferentially occupies axial positions. Thus in the mixed halide PF_2Cl_3 the fluorine atoms are axial and the less electronegative chlorine atoms are equatorial.

Example 3.4

Predict the geometric shapes and the bond angles of the following species: (a) NO_2^-; (b) ClO_2^-; (c) CO_3^{2-}; (d) SO_3^{2-}; (e) PCl_2F_3; and (f) TeF_7^-.

SOLUTION:

(1) Draw a correct Lewis structure of the species. **(2)** Determine the total coordination number about each central atom. **(3)** Count the number of unshared *sp*-hybridized electron pairs on each central atom; locate these in the most uncrowded positions. **(4)** Describe the resulting geometry (Fig. 3.10 or 3.11) and identify the undistorted bond angles (Fig. 3.10). **(5)** Predict any bond angle distortions due to repulsions from unshared electron pairs. **(6)** If two different types of outer atoms are involved in a trigonal bipyramidal structure, locate the most electronegative atoms in axial positions if possible.

Applying Steps **(2)** and **(3)**, you should obtain the total coordination number and number of unshared electron pairs for the oxo anions as follows: (a) For NO_2^- it is 3 and 1, respectively; (b) for ClO_2^- it is 4 and 2; (c) for CO_3^{2-} it is 3 and 0; (d) for SO_3^{2-} it is 4 and 1. From these, the application of Steps **(4)** and **(5)** gives the following geometries and bond angles: (a) bent and $<120°$ for NO_2^-; (b) bent and $<109.5°$ for ClO_2^-; (c) equilateral triangular and 120° for CO_3^{2-}; (d) pyramidal and $<109.5°$ for SO_3^{2-}.

(e) The compound PCl_2F_3 has a total coordination number of 5 and a trigonal bipyramidal structure. The more electronegative fluorines will prefer the axial sites, but as there are three fluorines and only two axial sites, the third fluorine will have to go equatorial.

(f) The compound TeF_7^- has a total coordination number of 7 and no unshared electron pairs. Of the three possible geometries, this ion happens to take the pentagonal bipyramidal geometry, with 90 and 72° bond angles.

VSEPR in the *d*-Block: An Aside

The VSEPR model also works reasonably well for predicting geometries of *d*-block complexes[28]; the major exception is the failure to predict the *d*-block complexes that adapt the square planar geometry, which must be predicted using the crystal field theory (Chapter 8). There are also a handful of cases at the beginning of the *d* and *f* blocks and the end of the *s* block in which the geometries are less symmetric than those predicted by VSEPR: that is, bent molecules with total coordination numbers of 2,[29] trigonal prismatic molecules with total coordination numbers of 6,[30] and reversals of Bent's rule in tetrahedral molecules.[31] These reversals, which need not concern us now, are explained in various ways as resulting from the participation in the bonding, in place of central-atom *p* orbitals of energy higher than the central-atom *s* orbitals, of central-atom *d* orbitals of different shapes and of energy lower than the central-atom *s* orbitals. Thus it is certainly sometimes necessary to go beyond the simple but remarkably useful premises of VSEPR theory.

3.5 The Formulas of Oxo Anions

Fluoro anions are useful for illustrating the assembly of a complex anion from a Lewis acid plus some Lewis bases, and they are useful in the laboratory, but fluorine is a relatively rare element in nature, so fluoro anions have comparatively few practical applications. **Oxo anions**, MO_x^{y-}, which can be assembled from a Lewis acid plus some oxide ions, O^{2-}, as Lewis bases, are far more important in ordinary life; most of the crust of the earth on which we stand consists of the salts of oxo anions. (These anions are somewhat more irregular in composition than the fluoro anions, which is why we introduced the fluoro anions first.)

Oxo anions are among the several classes of inorganic compounds in which the central Lewis acid does not normally achieve the maximum total coordination number, but instead achieves the *next lower* common coordination number, which we will call the **penultimate total coordination number**. In the second period, this number is 3; in the similarly sized third and fourth periods, this number is 4; in the fifth, sixth, and seventh periods this number is 6.

Example 3.5

Predict the formulas of the oxo anions formed by the following elements, each with the specified oxidation number: N(III), P(V), As(III), Se(IV), I(VII), I(V), and Mn(VI).

SOLUTION:

(The three steps of this process are represented pictorially in Fig. 3.13.)

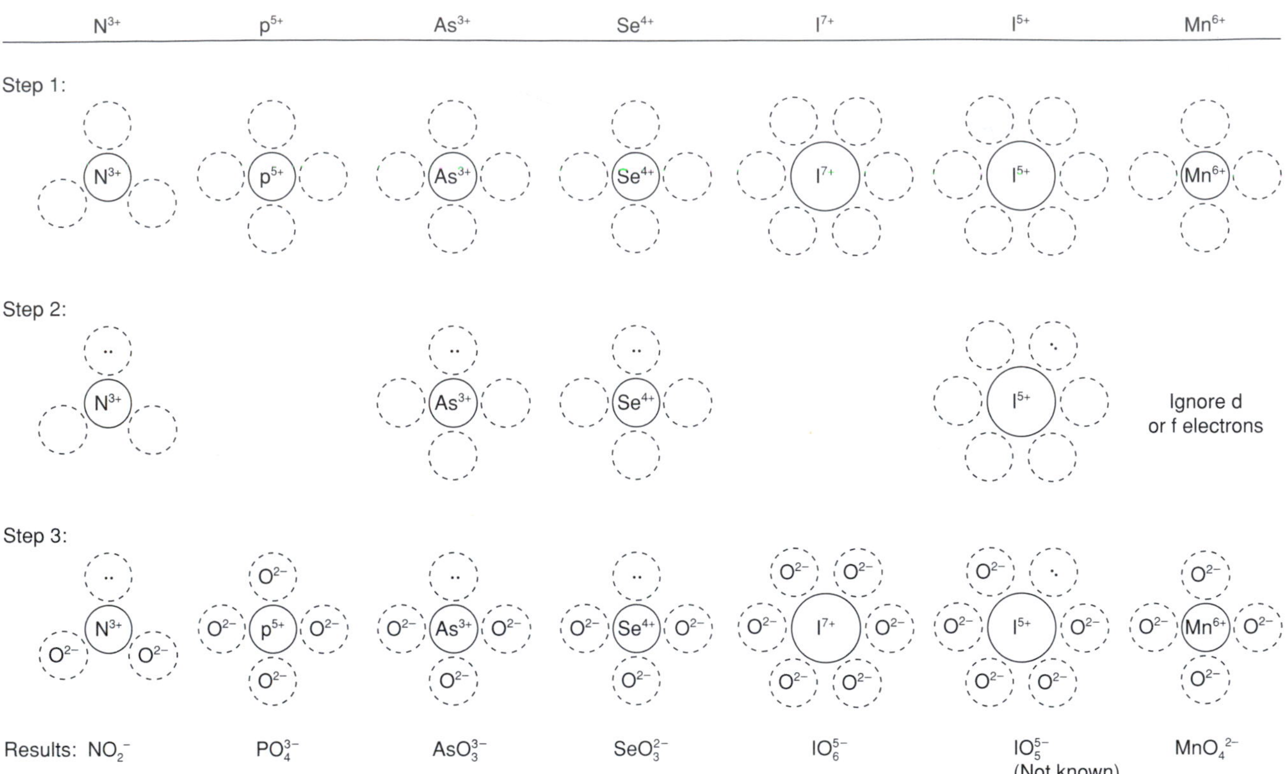

Figure 3.13
Pictorial assembly of oxo anions from cations, unshared *p*-electron pairs, and oxide ions. For steps and examples see the text.

1. The penultimate total coordination numbers expected for these elements in oxo anions are, based on their period numbers: 3 for N(III); 4 for P(V), As(III), Se(IV), and Mn(VI); and 6 for I(VII) and I(V). (In Fig. 3.13, each predicted total coordination site is represented by a broken circle.)

2. Next, we compare the oxidation numbers with group numbers to see whether any of these elements retain unshared *sp*-hybridized electron pairs that will block an incoming oxo anion. These elements can be represented as "cations" with the indicated unshared *sp* electron pairs if appropriate: $:N^{3+}$; P^{5+}; $:As^{3+}$; $:Se^{4+}$; I^{7+}; $:I^{5+}$; and Mn^{6+}, which has one *d* electron, but recall that *d* electrons are not expected to bar incoming ligands.

3. Now we subtract the number of unshared *sp* electron pairs from the penultimate total coordination number to give us the number of vacant coordination sites around the Lewis acid: 2 for $:N^{3+}$; 4 for P^{5+}; 3 for $:As^{3+}$ and $:Se^{4+}$; 4–6 for I^{7+}; 3–5 for $:I^{5+}$, and 4 for Mn^{6+}. We assemble the oxo anions by adding this number of oxide ions; we determine the *charge* of the oxo anion by adding charges of the central Lewis acid and the oxide ions

$:N^{3+} + 2\,O^{2-} \rightarrow NO_2^-$ $\qquad$ $P^{5+} + 4\,O^{2-} \rightarrow PO_4^{3-}$

$:As^{3+} + 3\,O^{2-} \rightarrow AsO_3^{3-}$ $\qquad$ $:Se^{4+} + 3\,O^{2-} \rightarrow SeO_3^{2-}$

$I^{7+} + 6\,O^{2-} \rightarrow IO_6^{5-}$ $\qquad$ $:I^{5+} + 5\,O^{2-} \rightarrow IO_5^{5-}$

$Mn^{6+} + 4\,O^{2-} \rightarrow MnO_4^{2-}$.

Table 3.3 shows the major oxo anions of the elements. As can be seen, our first five predictions are all confirmed; the relatively unimportant MnO_4^{2-} also has the

Table 3.3
Names and Formulas of the Important Oxo Anions

Period No.	Oxo Anions in Which the Central Atom Oxidation Number Equals the Group Number						Total Coordination No.
2*	BO_3^{3-} Borate	CO_3^{2-} Carbonate	NO_3^- Nitrate				3
3	AlO_4^{5-} Aluminate	SiO_4^{4-} Silicate	PO_4^{3-} Phosphate	SO_4^{2-} Sulfate	ClO_4^- Perchlorate		4
4			VO_4^{3-} Vanadate	CrO_4^{2-} Chromate	MnO_4^- Permanganate		
4	GaO_4^{5-} Gallate	GeO_4^{4-} Germanate	AsO_4^{3-} Arsenate	SeO_4^{2-} Selenate	BrO_4^- Perbromate		
5				MoO_4^{2-} Molybdate	TcO_4^- Pertechnetate		4–6
5		SnO_6^{8-} Stannate	SbO_6^{7-} Antimonate	TeO_6^{6-} Tellurate	IO_6^{5-} Periodate	XeO_6^{4-} Perxenate	
6				WO_4^{2-} Tungstate	ReO_4^- Perrhenate	OsO_6^{4-} Perosmate	
6		PbO_6^{8-} Plumbate			NpO_6^{5-} Perneptunate		
	Oxo Anions in Which the Central Atom Oxidation Number Is Two Less Than the Group Number						
2			$:NO_2^-$ Nitrite				3
3				$:SO_3^{2-}$ Sulfite	$:ClO_3^-$ Chlorate		4
4						$:FeO_4^{2-}$ ferrate	
4			$:AsO_3^{3-}$ arsenite	$:SeO_3^{2-}$ selenite	$:BrO_3^-$ bromate		
5						$:RuO_4^{2-}$ ruthenate	4–6
5		$:SnO_3^{4-}$ stannite	$:SbO_3^{3-}$ antimonite	$:TeO_3^{2-}$ tellurite	$:IO_3^-$ iodate	$:XeO_4^{2-}$ xenate	
6						$:OsO_4^{2-}$ osmate	

predicted formula. Deviations are found in the fifth and sixth periods, which we will discuss shortly: The iodine(V) oxo anion is in fact IO_3^-.

There are several reasons why the penultimate rather than the maximum total coordination number may be realized in a given class of compounds. The simplest reason occurs when the ligand being attached is relatively large, so we expect that a lower number of donor atoms can be coordinated. This trend is easily illustrated with the halide ions that are larger than fluoride: although the fluoro anion of Al is AlF_6^{3-}, the chloro anion is $AlCl_4^-$. On descending Group 13(III), however, the larger Tl^{3+} ion can accommodate six chloride ions, so that the chloro anion is $TlCl_6^{3-}$; however, the iodo anion is TlI_4^-.

This explanation is inapplicable for oxo anions, since the oxide ion is insignificantly larger than the fluoride ion (Table C). One obvious difference of the oxide ion is that it has a -2 charge, double that of the fluoride ion, so that attaching a large number of these ions builds up oxo anions with extremely high negative charges. Those negative charges would be expected to repel each other, perhaps enough to destabilize the anion. There may be some truth in this supposition, but it is not likely controlling since there are also *hydroxo anions*, formed by coordinating -1 charged hydroxide ions instead of -2 charged oxo anions, which also usually show the penultimate total coordination number.

Instead it seems likely that a main reason for the low total coordination number is one advanced by Mayer[32]: the tendency of the oxide ion to act as a *π-donor ligand*. Unlike fluorine, oxygen readily forms π bonds. Unlike F^-, O^{2-} is a very strong base, which is likely to continue donating electrons through π bonds if it can.

For this to happen, there must be empty acceptor orbitals on the Lewis acid. In a hypothetical second period oxo anion showing the *maximum* total coordination number, such as CO_4^{4-}, there would be no empty valence orbital. For carbon to accept a second pair of electrons from oxygen the oxide ion must be expelled. (Fig. 3.14). Once

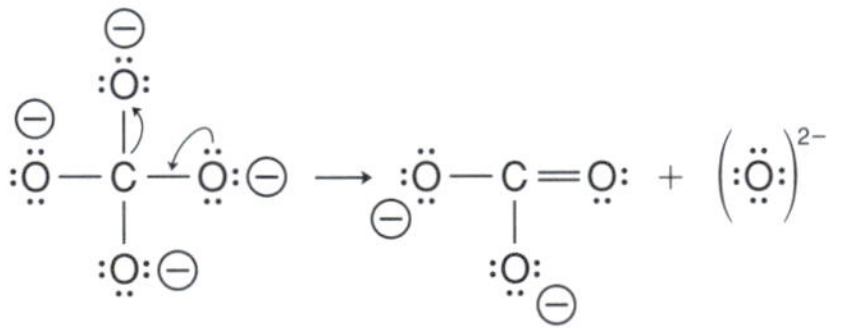

Figure 3.14
Expulsion of an oxide ion from a hypothetical CO_4^{4-} ion to give a stable π-bonded CO_3^{2-} ion.

this is done, a strong second period π bond is formed, excessive negative charge is dispelled, and some relief of overcrowding may be achieved. Thus π-bonded oxo anions are preferred for carbon and nitrogen. (The larger boron atom is somewhat ambivalent, and may go either to three or four coordination in different circumstances.)

Fifth and Sixth Periods. In these periods, oxo anions are sometimes found in which the total coordination number is even less than the penultimate coordination number of 6. There are two reasons for this behavior: the presence of strong π bonding, and the excessive spatial demands of an unshared electron pair.

In *d*-block oxo anions, the π-acceptor *d* orbitals are valence orbitals and are available if overlap is possible and if the orbitals are empty. These conditions are met with high oxidation states in the fifth and sixth periods. Hence, although the higher penultimate coordination number of 6 is most common in the fifth and sixth periods of most of the periodic table, for the Mo^{6+}, W^{6+}, Tc^{7+}, and Re^{7+} cations[33] the conditions are so favorable for π bonding that the very low coordination number of 4 is preferred.[34] The prevalent oxo anions are MoO_4^{2-}, WO_4^{2-}, TcO_4^-, and ReO_4^-. The alternate choices are not unknown, however: under some conditions ReO_6^{5-} can be made, and the molybdate and tungstate species often polymerize (Section 13.8) to give large anions in which Mo and W have total coordination numbers of 6.

In the *p* block, a total coordination number less than the penultimate can occur if an unshared *sp*-electron pair is present on the central atom, since such an electron pair occupies *more* space than does a bond pair, especially in these periods.[23] For example, the actual I^V oxo anion (Example 3.5) is IO_3^-, not IO_5^{5-}.

Resonance Structures and Formal Charge. There is some evidence (this point is controversial)[35] that the oxo anions in the later periods of the *p* block involve π bonding above and beyond what we have shown in our Lewis structures. The central-atom acceptor orbital in such a π-bonded oxo anion would have to be a postvalence *d* (or d^nsp^3 hybrid) orbital. Since postvalence *d* orbitals are less available than *p* orbitals, these bonds are not full-strength π bonds, and the element–oxygen bonds seem to be somewhere between single and double bonds. The bonding is more elegantly treated with molecular orbital theory, but it can be represented fairly adequately in valence bond theory by showing *resonance structures* that contribute to the bonding (Fig. 3.15). In general, we defer to the organic chemistry course for a discussion of resonance structures.[36]

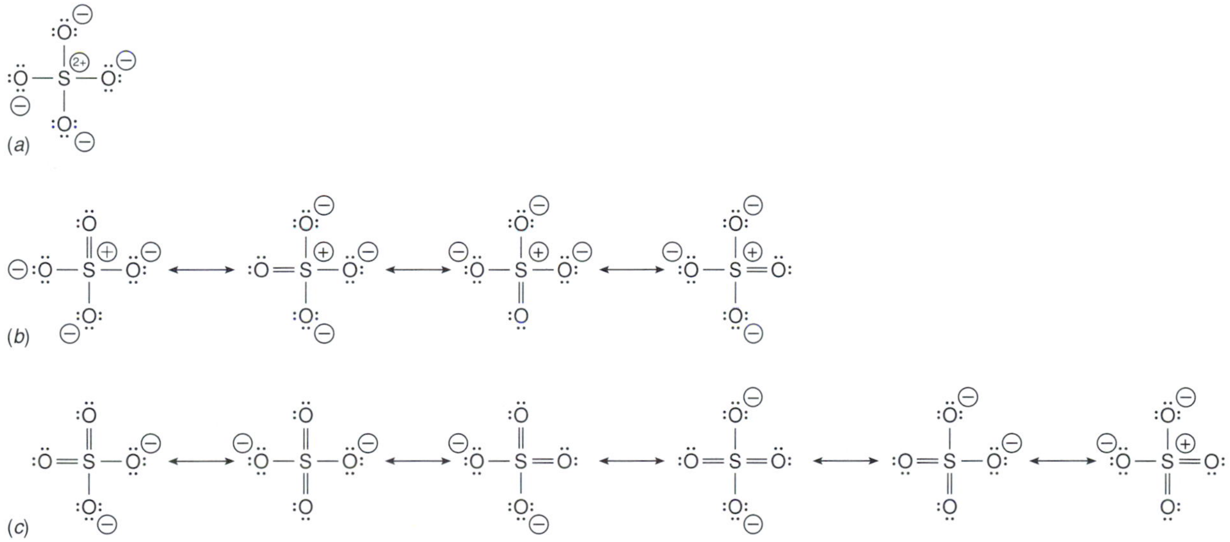

(*a*)

(*b*)

(*c*)

Figure 3.15
Resonance structures for the sulfate ion, SO_4^{2-}. The circled numbers are formal charges located on the indicated atoms of each structure. (*a*) Structure with only single bonds; this has the advantage of not using postvalence *d* orbitals or their hybrids, but has the disadvantage of requiring the separation of many formal charges. (*b*) Structures with one S=O double bond. (*c*) Structures with two S=O double bonds, which minimize formal charge buildup but require maximal utilization of postvalence orbitals on S. See a textbook of advanced organic chemistry (e.g., Ref. 36) for a discussion of choosing preferred resonance structures.

Formal Charge: An Aside

Creating a double bond in a resonance structure results in the reduction of *formal charge* (shown in circles) on both the nonmetal and oxygen atoms:

$$\begin{aligned}\text{Formal charge of atom} \\ &= \text{(its group number)} \\ &\quad - \text{(its No. of unshared electrons)} \\ &\quad - 0.5\text{(its No. of bonding electrons)}\end{aligned} \quad (3.1)$$

Counting its own and one-half of its shared electrons, a single-bonded oxygen has seven valence electrons of its own and acts as if it has a -1 formal charge, while a double-bonded oxygen has six electrons of its own and 0 formal charge. At the same time, those sulfur atoms in resonance structures with six bonds [Fig. 3.15(*c*)] also have zero charges of their own; less electrical work has to be done in separating opposite charges while assembling such resonance structures, so structures such as these and the structures in Figure 3.15(*b*) tend to be preferred over that in Figure 3.15(*a*). Another preference is for structures in which negative formal charges are placed on more electronegative atoms while positive formal charges are placed on more electropositive charges; unfortunately, both these principles must be violated severely in some very stable coordination compounds, including simple hydrated ions such as $[Al(H_2O)_6]^{3+}$, in which Al has a formal charge of -3, while each electronegative oxygen has a formal charge of $+1$. In view of difficulties such as these, formal charges have not been used as frequently in inorganic chemistry as they have in organic chemistry.[37]

3.6 The Basicity of Oxo Anions

Your instructor may now assign Experiment A.3.

Oxo anions are hydrated in the same manner as monoatomic nonmetal ions (Section 2.5); a selection of their hydration energies was included in Table 2.4.[38] Hydrated oxo anions also undergo hydrolysis reactions to give basic solutions containing OH^- and partially protonated anions:

$$MO_x^{y-} + H_2O \rightleftharpoons [MO_{(x-1)}OH]^{(y-1)^-} + OH^- \quad (3.2)$$

A familiar example of this process is the hydrolysis of carbonate ion to give bicarbonate:

$$CO_3^{2-} + H_2O \rightleftharpoons HCO_3^- + OH^- \quad (3.3)$$

Simple oxo anions span the whole range of basicity found in monoatomic anions. Again we will use the same terms to classify the degrees of basicity. As it happens, with just a slight degree of "rounding off" of pK_b values, only four categories will be needed to include all simple oxo anions. As before, **nonbasic** oxo anions show no detectable tendencies to undergo hydrolysis and do not alter the pH of their aqueous solution. **Feebly basic** oxo anions only occasionally manifest any basic properties. The category

Table 3.4
Suggested Classification of Oxo Anions

Classification	Type	Calculated pK_{b1}	Examples with Known pK_{b1} Values
Nonbasic anions	MO_4^-	22.6	M = Cl, Br, Mn, Tc, Re
	MO_3^-	16.9	M = N, Cl, Br, I(13.2)
Feebly basic anions	MO_4^{2-}	12.4	M = S(12.1), Se(12.0), Xe, Cr(7.5), Mo(9.9), W(9.4), Fe(6.2), Ru, Os
	MO_2^-	11.2	M = N(10.7), Cl(12.1)
Moderately basic anions	MO_6^{4-}	3.4	M = Xe, Os
	MO_4^{3-}	2.2	M = P(2.0), As(1.5), V(1.0)
	MO_3^{2-}	6.7	M = C, S(6.8), Se(7.4), Te(6.3)
	MO^-	5.5	M = Cl(6.5), Br(5.3), I(3.4)
Very strongly basic anions	MO_6^{5-}	−6.8	M = I, Np
	MO_4^{4-}	−8.0	M = Si, Ge
(Exist as hydroxo anions in solution)	MO_3^{3-}	−3.5	M = As, Sb
	MO_6^{6-}	−17.0	M = Te
	MO_4^{5-}	−18.2	M = B, Al, Ga
	MO_3^{4-}	−14.7	M = Sn

SOURCES: Known pK_{b1} values are calculated from the appropriate pK_a values given in F. A. Cotton and G. Wilkinson, *Advanced Inorganic Chemistry: A Comprehensive Text*, 5th ed., Wiley-Interscience, New York; 1988, p. 105; R. C. Weast, Ed., *Handbook of Physics and Chemistry*, 50th ed., Chemical Rubber Publishing Co., Cleveland, 1969; J. A. Dean, Ed., *Lange's Handbook of Chemistry*, 13th ed., New York, McGraw-Hill, 1985; and B. H. J. Bielski, *Free Radical Res. Commun.*, **12–13**, 469 (1991).

of **moderately basic** anions includes the ones on which equilibrium calculations were done in general chemistry, probably under the heading of "salts of weak acids"; their solutions are distinctly basic. The **very strongly basic** anions, however, are normally found only in anhydrous solids, since they hydrolyze completely or nearly completely in water according to Eq. (3.2). Table 3.4 gives examples of these categories of oxo anions, with their measured pK_b values when these are available.

Predictions of Basicity. Again we will try to develop simple rules that will enable us to predict just how basic the solution of an oxo anion should be, and that will enable us to predict some practical consequences. (For example, any solution that is highly basic is also corrosive to many materials, including human tissue!) With cations we looked at three variables: charge, size, and electronegativity of the metal ion. The size of oxo anions is determined not so much by the size of the central atom as by the number of **oxo groups** (oxygen atoms bonded to M and to no other atom) present. (This indirect approach to size is also useful because oxo anions are not spheres and do not have precisely definable radii.) Thus in an oxo anion of formula MO_x^{y-} we will be particularly interested in the effects on basicity of charge of the oxo anion, $-y$, its number of oxo groups, x, and (secondarily) the electronegativity of the central atom M.

Effect of Charge. As we by now would expect, increasing negative charge on an anion increases its tendency to hydrolyze and give rise to basic solutions. Referring to the (partial) list of oxo anions in Table 3.3, we see that the negative charges of oxo anions can be quite substantial indeed. Therefore, it is not surprising that many of them are very strongly basic. Comparison of the pK_{b1} values of otherwise-similar oxo anions with increasing negative charges leads to the conclusion that *the pK_{b1} of an oxo anion decreases by 10.2 units for each (additional) negative charge on it*[39]. This *increase* in basicity is substantial with increasing negative charge: It is enough, in favorable cases, to change the basicity by two categories with a change of only one unit of charge. (Thus, −3 charged PO_4^{3-}, with $pK_b = 2.0$, is moderately basic, while −2 charged SO_4^{2-}, with $pK_b = 12.1$, is not weakly, but rather feebly, basic.)

Effect of Number of Oxo Groups. Most of the nonmetals show more than one oxidation number and can form oxo anions that differ in the number of oxo groups that are attached to the nonmetal atom. These different oxo anions differ substantially in basicity. The most complete series of oxo anions is that of chlorine, which forms four different oxo anions with the following pK_{b1} values: ClO^-, 6.5; ClO_2^-, 12.1; ClO_3^-, and ClO_4^-, unmeasurable (these are such weak bases that their hydrolysis cannot be detected). Examination of the pK_{b1} values for a number of such sets of oxo anions shows that, on the average, *each additional oxo group in an oxo anion increases its pK_{b1} by 5.7 units.*

Why should the number of oxo groups have an effect on the basicity of an oxo anion? Students contemplating this question in the past have devised a number of explanations that fit the facts and are chemically reasonable.

1. Those who have already studied organic chemistry recognize that the more oxo groups there are in an oxo anion, the more resonance structures can be drawn by which the negative charge of the anion is delocalized (cf. the three plausible double-bonded resonance structures possible for SO_3^{2-} with the 10 for SO_4^{2-} drawn in Fig. 3.5). There is greater resonance stabilization of anions with more oxo groups.

2. Another explanation looks at the Lewis dot structures of oxo anions and asks the question, on which atoms of this oxo anion are the negative charges formally located? Of course, charges are not really on any one atom, since oxo anions are (internally) covalently bonded species, but knowing that oxygen is more electronegative than the other atom in the oxo anion, we can assume that the charges formally are on oxygen. But in an oxo anion such as ClO_4^-, there is no reason to assume that the negative charge is on one particular oxygen any more than on any of the others, so we may suppose that, overall, each oxygen atom bears a *partial charge* of (formally) $-\frac{1}{4}$. Likewise in ClO_3^-, we suppose a greater partial charge of $-\frac{1}{3}$; while the oxygen atom in ClO^- is stuck with the whole charge of −1. The oxygen atom in ClO^- therefore should be more basic than any other oxygen atom in any of the other oxo anions of chlorine. Although these numbers are fictitious, there should indeed be partial negative charges on oxygen, which should decrease as the number of oxo groups increases, resulting in reduced basicity. From a slightly different perspective, the more oxo groups are present, the larger the oxo anion is, so the more the charge is dispersed over space.

3. The conversion of ClO^- to ClO_2^- is actually an oxidation process (Chapter 6), which can be achieved by attaching an O atom to an unshared pair of electrons that was on the Cl atom of ClO^-. Oxygen, of course, is a very electronegative atom; when we add an oxo group to an anion such as ClO^-, the new oxygen atom strongly withdraws these electrons from the Cl. (In effect, the Cl now acts as a "cation" with a positive "charge" increased by 2.) The Cl now compensates its electron loss by attracting electrons more strongly from the other oxygen atom(s); hence, there is less effective negative charge at each oxo group for attracting the hydrogen atom of the water of hydration.

It is not infrequent in science to have more than one explanation of a phenomenon such as this. If the explanations are all satisfactory, we may continue to use more than one, depending on the situation or on personal preference. Of course, if you think about the preceding explanations, you will see that they are far from being completely different from one another—to a large degree they are just different ways of saying the same thing.

Effect of Electronegativity. The electronegativity of the nonmetal atom also influences the basicity of an oxo anion. Comparing the pK_{b1} values of ClO^- (6.50), BrO^- (5.3), and IO^- (3.4), we find, as expected, that reducing the electronegativity of the halogen atom increases the basicity of the oxo anion. For *d*-block metals of much lower electronegativity than *p*-block elements, this effect is greater: pK_{b1} for $SO_4{}^{2-} = 12.1$ and for $SeO_4{}^{2-} = 12.0$; but pK_{b1} for $CrO_4{}^{2-} = 7.5$ and for $FeO_4{}^{2-} = 6.2$.[40] Insufficient data, however, are available to allow a quantitative estimate of the effect of electronegativity; qualitatively it appears to be a somewhat smaller effect than that of the charge or the number of oxo groups in the oxo anion. Therefore for purposes of developing approximate rules we will ignore the influence of electronegativity.

We may incorporate the effects of the number of oxo groups and of the negative charge of an oxo anion into one equation that allows a reasonably accurate calculation of the constant for the first hydrolysis of an oxo anion:

$$pK_{b1} = 10.0 + 5.7x - 10.2y \pm 1.0 \tag{3.4}$$

Here x is the number of oxo groups and y is the number of units of negative charge in the oxo anion of formula $MO_x{}^{y-}$. Keep in mind the physical reasoning behind this equation: (1) higher pK_{b1} values correspond to weaker basicity (lower pH); (2) additional oxo groups weaken basicity, hence add to pK_{b1}; (3) additional negative charges increases basicity, hence subtract from pK_{b1}. In Table 3.4, we have also tabulated the calculated values of pK_{b1} for the important simple oxo anions of the elements.

Quick Estimates of Basicities. A quicker but rougher estimate of the basicity of an oxo anion may also be made by direct inspection of its chemical formula. We may note that in Eq. (3.4) the effect of one additional charge is roughly twice as great (and opposite in sign to) the effect of one additional oxo group; hence, it is possible to estimate relative basicities by matching single negative charges and pairs of oxo groups and deleting them together from the formula of the oxo anion. Thus for a nonbasic anion such as $MO_4{}^-$, we can cancel out the effects of the unit of negative charge and the effects of two of the oxo groups, to be left with "MO_2". If all the negative charge cancels out but oxo groups remain, we might expect to (and do) find that the original oxo anion was *nonbasic*.

For the feebly basic anions $MO_4{}^{2-}$ and $MO_2{}^-$, we find that this cancellation process removes *both* the negative charge and the oxo groups, leaving us only with "M". Oxo anions that reduce to an oxygen- and charge-free formula are expected to be *feebly basic.*

Species that end up with negative charge but no oxo groups we might expect to be appreciably basic. We find that oxo anions with -0.5 or -1 charges after cancellation are *moderately basic* (e.g., MO_3^{2-} becomes $M^{-0.5}$, and MO_6^{4-} becomes M^-.) Species that end up with charges that are *more* negative than -1 (e.g., MO_3^{4-}, which becomes $M^{-2.5}$) are *very strongly basic.*

Since fluorine and oxygen share similar sizes and similarly high electronegativities, it appears that these rules of thumb might work with fluoro anions as well; although there is a paucity of quantitative data on the basicity of fluoro anions, it appears that most are *nonbasic.* Most fluoro anions have charges from -1 to -3, and have at least six fluoro groups, so a process of deleting two fluoro groups per negative charge usually leaves only excess fluoro groups (MF_2, etc.). The main use of fluoro anions in inorganic chemistry is as valuable nonbasic anions.

Example 3.6

(a) Qualitatively classify the basicity of each of the following oxo or fluoro anions (these include some that are hypothetical or quite unstable): BrO_3^{3-}; MnO_4^{3-}; UO_4^{2-}; SbF_5^{2-}; and AlF_6^{3-}. (b) Compute the pK_{b1} values of the oxo anions in this set.

SOLUTION:

(a) Deleting two oxo (or fluoro) groups per unit of negative charge leaves the following residues: $Br^{-1.5}$; Mn^-; U; SbF; and Al. Those with oxo or fluoro groups remaining will be nonbasic: SbF_5^{2-}. Those with neither oxo nor fluoro groups nor negative charges remaining will be feebly basic: UO_4^{2-} and AlF_6^{3-}. The basic anions will have negative charges remaining after deletion: MnO_4^{3-}, with a -1 charge remaining, will be moderately basic, while BrO_3^{3-}, with a -1.5 charge remaining, will be very strongly basic.

(b) Equation (3.4) can be applied only to the oxo anions. For BrO_3^{3-}, this equation gives $pK_{b1} = 10.0 + 5.7(3) - 10.2(3) = -3.5$, which with an error of ± 1.0 may fit in the category of very strongly basic, as we have suggested. For MnO_4^{3-} $pK_{b1} = 10.0 + 5.7(4) - 10.2(3) = 2.2$, which fits the defined range for a moderately basic anion. For UO_4^{2-}, $pK_{b1} = 10.0 + 5.7(4) - 10.2(2) = 12.4$, which falls in the range defined for a feebly basic anion. (However, the quite low electronegativity of uranium would probably make this anion more basic than this.)

3.7 Protonation of Oxo Anions: Hydroxo Anions and Oxo Acids

The product of Reaction (3.2), the once-protonated oxo anion $[MO_{(x-1)}OH]^{(y-1)^-}$, is still a base, but it now has one less negative charge and one less oxo group, since one oxo group has been converted by protonation to a *hydroxo group.* If we calculate the basicity of this monoprotonated oxo anion using Eq. (3.4), we find that its pK_{b2} is 4.5 *units more than its* pK_{b1}. Likewise, the resulting diprotonated oxo anion is still a base, but of further diminished basicity due to a pK_{b3} that is yet another 4.5 units higher.

These calculations give us the information needed to construct predominance diagrams, in the manner that was done for monoatomic anions in Section 2.6. The process will now be easier because we can see from the above discussion that *the width of the predominance region of each partially protonated oxo anion* should be *4.5* pH (or pOH) units. (In Section 2.6, we could not predict the width of the corresponding predominance regions, undoubtedly because the proton was going on many types of donor atoms.)

Example 3.7

Sketch the expected predominance diagram for PO_4^{3-}.

SOLUTION:

Applying Eq. (3.4), we calculate the pK_b of PO_4^{3-} to be 2.2. This number equals the pOH of the left boundary of this moderately basic oxo anion. We suggest that the actual sketch be set up originally with pOH as the variable. To the left of this boundary, the once-protonated anion, $[(HO)PO_3]^{2-}$ is predominant; its pK_b is calculated to be 6.7, that is, this pK_{b2} is 4.5 units more than the previously calculated pK_{b1}. This pOH value is at the left boundary of $[(HO)PO_3]^{2-}$. To the left of this, $[(HO)_2PO_2]^-$ predominates, with a calculated pK_{b3} of 11.2. To the left of this pOH value, phosphoric acid, $(HO)_3PO$ or (more familiarly) H_3PO_4, predominates.

The predominance diagram is sketched in Figure 3.16(*c*), which is finally completed by adding the pH scale at the bottom, using pH = 14 – pOH. Then it can be seen that the moderately acidic H_3PO_4 predominates up to a pH of 2.8; $H_2PO_4^-$ predominates from there up to pH 7.3; HPO_4^{2-} predominates from there up to a pH of 11.8; PO_4^{3-} is predominant above that pH.

Example 3.8

Sketch the expected predominance diagram for AlO_4^{5-}.

SOLUTION:

From Eq. (3.4), we calculate the pK_b of AlO_4^{5-} to be −18.2. The predominance region of this (very strongly basic) anion will have only a left boundary, at a pOH equal to its pK_b, −18.2. Beyond this predominance region (to its left) will be the predominance region of the once-protonated from $[(HO)AlO_3]^{4-}$; the left boundary of this region will be at its pK_b, which is shifted to a pOH 4.5 units higher, −13.7. This process can be continued in steps of 4.5 pOH units until the negative charges of the original oxo anion are completely neutralized. Finally, the pH scale is added (Figure 3.17).

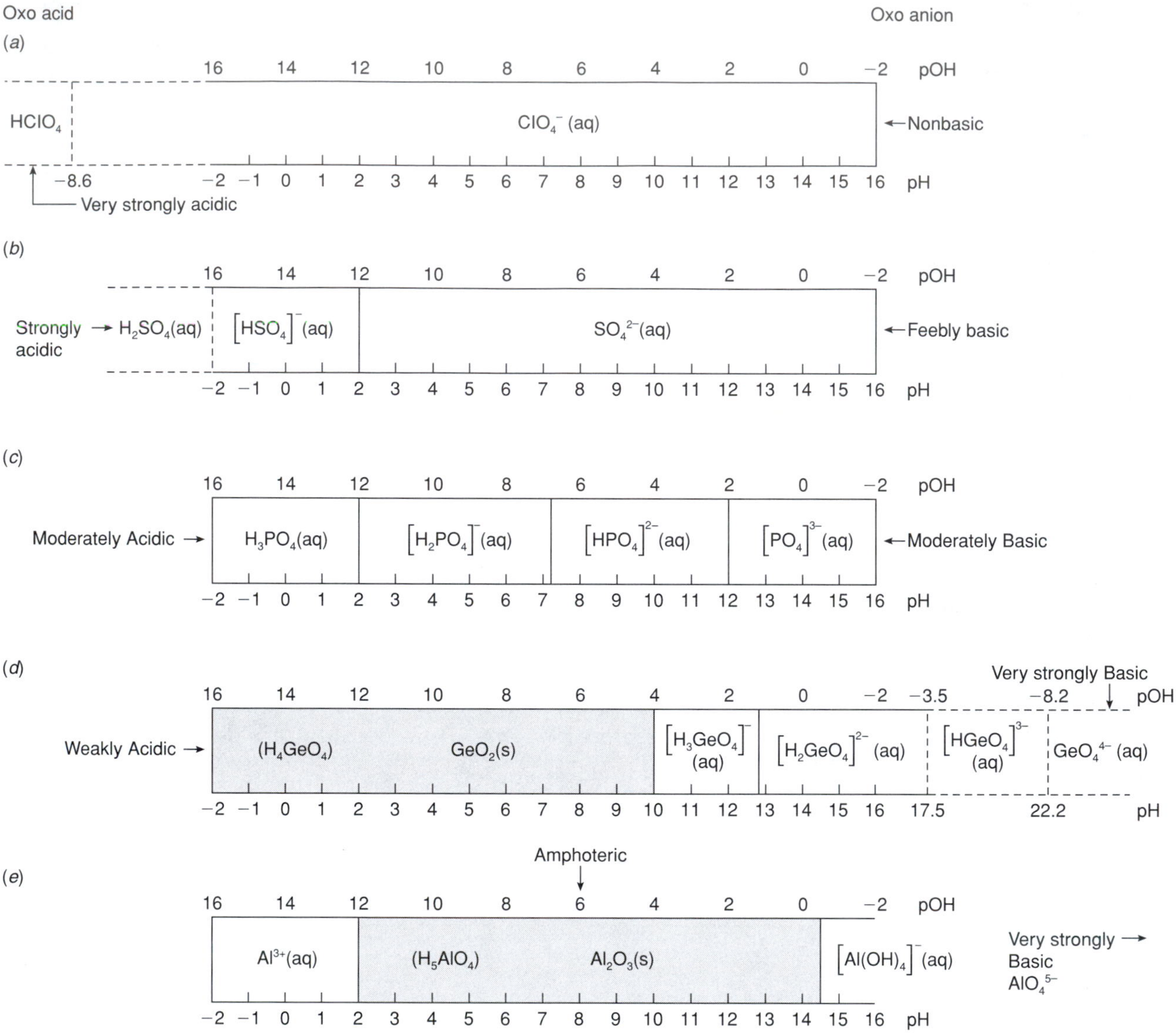

Oxo acid

Oxo anion

(*a*)

16 14 12 10 8 6 4 2 0 −2 pOH

$HClO_4$ | ClO_4^- (aq) ←Nonbasic

−8.6 −2 −1 0 1 2 3 4 5 6 7 8 9 10 11 12 13 14 15 16 pH

Very strongly acidic

(*b*)

16 14 12 10 8 6 4 2 0 −2 pOH

Strongly acidic → H_2SO_4(aq) | $[HSO_4]^-$(aq) | SO_4^{2-}(aq) ←Feebly basic

−2 −1 0 1 2 3 4 5 6 7 8 9 10 11 12 13 14 15 16 pH

(*c*)

16 14 12 10 8 6 4 2 0 −2 pOH

Moderately Acidic → H_3PO_4(aq) | $[H_2PO_4]^-$(aq) | $[HPO_4]^{2-}$(aq) | $[PO_4]^{3-}$(aq) ←Moderately Basic

−2 −1 0 1 2 3 4 5 6 7 8 9 10 11 12 13 14 15 16 pH

(*d*)

Very strongly Basic

16 14 12 10 8 6 4 2 0 −2 −3.5 −8.2 pOH

Weakly Acidic → (H_4GeO_4) GeO_2(s) | $[H_3GeO_4]^-$ (aq) | $[H_2GeO_4]^{2-}$ (aq) | $[HGeO_4]^{3-}$ (aq) | GeO_4^{4-} (aq)

−2 −1 0 1 2 3 4 5 6 7 8 9 10 11 12 13 14 15 16 17.5 22.2 pH

(*e*)

Amphoteric

16 14 12 10 8 6 4 2 0 −2 pOH

Al^{3+}(aq) | (H_5AlO_4) Al_2O_3(s) | $[Al(OH)_4]^-$ (aq) Very strongly Basic → AlO_4^{5-}

−2 −1 0 1 2 3 4 5 6 7 8 9 10 11 12 13 14 15 16 pH

Figure 3.16

Predominance diagrams for several oxo anions having four oxo groups, MO_4^{y-}, and their partially protonated forms and fully protonated oxo acids. The dashed lines represent theoretical boundaries computed for pH ranges which are not normally achievable in water.

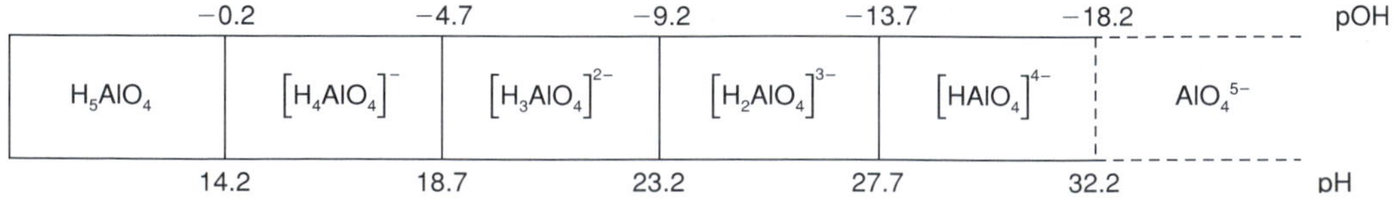

Figure 3.17

Predominance diagram for AlO_4^{5-}.

Much of Figure 3.17, of course, is meaningless in aqueous solution. In Section 3.2, we mentioned that a concentrated KOH solution behaves as if it were 10,000 *M* (because not enough water molecules are present to fully hydrate the hydroxide ions). Such a concentration corresponds to a pOH of −4 or a pH of 18. Hence, the only two species in the above diagram that can exist in water are H_5AlO_4 and $H_4AlO_4^-$.

When the original oxo anion has more negative charges than oxo groups ($y > x$, as in the above example) it is possible to transfer protons to each of the oxo groups and still have an anion. The resulting species, which no longer has any oxo groups, is perhaps better called a *hydroxo anion*, and may usefully be written $Al(OH)_4^-$ as an alternative to $H_4AlO_4^-$. Continued protonation to neutralize the last charge then puts protons on the hydroxyl groups, converting them to water molecules of hydration attached to the insoluble *hydroxide* of the element. As mentioned earlier, these are easily lost, and water loss may continue until the *oxide* of the element remains. Thus the formula H_5AlO_4 can be rewritten as $Al(OH)_3 \cdot H_2O$, which may dehydrate to $Al(OH)_3$. This compound may further dehydrate to AlO(OH), or ultimately (as on heating) to the oxide Al_2O_3.

Figure 3.16(*e*) shows the final form of the predominance diagram that includes (at impossible pH values) the very strongly basic oxo anion AlO_4^{5-}. For comparison, the predominance diagrams of the progressively less-charged (and therefore less basic) oxo anions MO_4^{y-} are also shown. The predominance regions of the oxo anions expand as they become less basic (less reactive).

In water, the protonation of very strongly basic anions continues until there results an equilibrium mixture of partially protonated oxo anions that have reasonable pK_b values (above −4 to exist in concentrated KOH; above 0 to predominate at the more usual pH of 14). Thus although the silicate anion SiO_4^{4-} is found in several minerals, it is not really present in a solution of sodium silicate due to its very strong basicity (calculated $pK_{b1} = -8.0$). Instead, the main species in solution is apparently the diprotonated ion $[SiO_2(OH)_2]^{2-}$, with a calculated pK_{b3} of 1.0. (The formula of this ion is often written as if it had lost a molecule of water and become the so-called metasilicate ion, SiO_3^{2-}, but there is no evidence for the existence of such a species, which has too low a total coordination number.)

In general, calculations on the oxo anions listed in Table 3.4 show that, even at pH 14, the protonation must proceed far enough to reduce the charge on the anions to −3, −2, or −1. (The one exception is MO_6^{4-}, which is expected to be only moderately basic due to its large number of oxo groups.) The other oxo or partially protonated oxo anions of −4 or greater charge are still too basic to exist in water, although they can be prepared by nonaqueous reactions or persist in the form of very insoluble salts (we will pursue this topic in Chapter 4).

For example, the impossibly basic antimonate ion, SbO_6^{7-}, with a calculated pK_{b1} of −27.2, is actually found in aqueous solution as $[Sb(OH)_6]^-$, with a much more moderate calculated pK_b of −0.2. Likewise, the anionic species in solution for Sn(IV) and Pb(IV) are actually $[Sn(OH)_6]^{2-}$ and $[Pb(OH)_6]^{2-}$; those for B, Al, and Ga have the formula $[M(OH)_4]^-$ (M = B, Al, or Ga). The formulas of these species are often inaccurately written as if they were oxo anions, for example, AlO_2^- instead of $[Al(OH)_4]^-$, but notice that such a formula, although it includes the correct charge, implies an absurdly low coordination number for the central atom.

Since the hydroxo anions are much less basic than the corresponding oxo anions, they can form with some metal ions that have too low a Z^2/r ratio (and too low an electronegativity) to form oxo anions. Thus in strongly basic solutions, hydroxo anions are known for several metals that have only moderately acidic cations: examples include $[Be(OH)_4]^{2-}$, $[Cr(OH)_6]^{3-}$, $[Fe(OH)_6]^{4-}$, $[Fe(OH)_6]^{3-}$, $[Ag(OH)_2]^-$, $[Au(OH)_4]^-$, $[Zn(OH)_4]^{2-}$, and $[Cd(OH)_4]^{2-}$. These hydroxo anions are often formed by dissolving the corresponding metal oxide or hydroxide in strongly basic (NaOH) solutions. Since these same metal oxides or hydroxides also dissolve in strongly acidic solutions to give hydrated cations, such oxides or hydroxides are capable of acting either as acids (reacting with NaOH) or as bases (reacting with an acid). Oxides or hydroxides (such as that of Al) that can react both ways are termed **amphoteric**.

Oxo Acids. As we continue to add protons to an oxo anion MO_x^{y-} (either by adding acid and lowering the pH of the solution or by letting hydrolysis occur), we ultimately arrive at an **oxo acid**[41] of formula H_yMO_x, which is expected to have a structure corresponding to $MO_{x-y}(OH)_y$, still possessing $(x - y)$ oxo groups. Similarly, protonation of a hydroxo anion, followed by the loss of some water molecules, leads to a metal hydroxide $M(OH)_y$, which can be considered to be an oxo acid with no oxo groups. These oxo acids have various tendencies to ionize:

$$H_yMO_x + H_2O = H_3O^+ + H_{y-1}MO_x^- \tag{3.5}$$

We can characterize the strength of these oxo acids by their pK_a values. If necessary, these can be calculated from the pK_b of the final protonated oxo anion (conjugate base) on the right side of Eq. (3.5), using the relationship:

$$pK_a = 14 - pK_b \tag{3.6}$$

From the calculations, it emerges[42] that *the strength of the oxo acid depends only on the number of oxo groups present*:

$$pK_a = 8.5 - 5.7(x - y) \tag{3.7}$$

(Writing the alternate formula for the oxo acid, $(HO)_yMO_{x-y}$, shows the number of oxo groups directly.) Since existing oxo acids of this type have from zero to three oxo groups, this allows us to categorize the oxo acids very readily:

1. Oxo acids with three oxo groups (in practice, HMO_4) are expected to have pK_a values of about -8.6 (i.e., to be *very strong acids*). In practice, the concentrations of un-ionized HMO_4 left in equilibrium in solutions of such acids are too small to be measured at all accurately, so pK_a cannot be measured in water solution; but there is no doubt that acids such as $HClO_4$ do ionize nearly 100%, and are thus very strong acids. Note that the very strong acids are the conjugate acids of nonbasic anions.

2. Oxo acids with two oxo groups (HMO_3, H_2MO_4, and H_4MO_6) are expected to have pK_a values of about -2.9 (i.e., to be *strong acids*). Again, exact measurements cannot be made, but these acids (such as HNO_3 and H_2SO_4) are well known to ionize extensively in solution. (The second ionization of an acid such as H_2SO_4 has a pK_{a2} that is $\sim$4.5 higher than the first ionization, and thus is a positive number. This indicates that, in its *second* ionization, sulfuric acid does not act as a strong acid.)

3. Oxo acids with one oxo group (HMO_2, H_2MO_3, H_3MO_4, and H_5MO_6) are expected to have pK_a values of about 2.8 (i.e., to be *moderately acidic*). We can measure pK_a values such as this, and have been for a number of these acids: They are found to be within a standard deviation of ± 0.9 of this value.[28]

4. "Oxo acids" without oxo groups (hydroxides of the nonmetals) are expected to have pK_a values of about 8.5 (i.e., to act as *weak acids*). The measured values for these acids fall within ± 1.0 of this value. Note that the ultimate conjugate bases of most of these weak acids are classified as very strongly basic oxo anions. Generally speaking, these hydroxides differ from the true oxo acids (that have oxo groups) by being *insoluble in water*. (Exceptions include boric acid, H_3BO_3, and HOCl.)

There are some oxo acids that have anomalous acidities. Notable among these are two oxo acids of phosphorus, H_3PO_3 and H_3PO_2, which have about the same pK_a (~ 2) as H_3PO_4, suggesting that all have one oxo group. In fact, there is independent evidence that this is the case, and that the structures of H_3PO_3 and H_3PO_2 have P–H instead of only O–H bonds (Fig. 3.18). Thus on neutralizing H_3PO_3 with strong base,

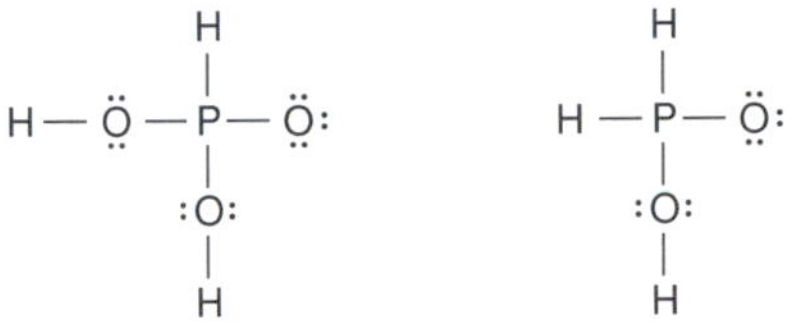

Figure 3.18
Structures of H_3PO_3 and H_3PO_2.

only two of the three protons can be removed; with H_3PO_2 only one can be removed.

The measured pK_a for carbonic acid (6.38) also seems to be out of line. But it is found that, in a solution of carbonic acid, most of the carbon is in the form of hydrated CO_2, not H_2CO_3. If a correction is made for this, it is found that the pK_a for the H_2CO_3 in the solution is 3.58, which is in the expected range.

5. Fluoro acids (such as $HAsF_6$) are *very strong acids*; some of these are components of "superacids" (Section 4.10).

3.8 Most Common Forms of the Elements in Natural Waters

In this section, we wish to sum up the concepts in Chapters 2 and 3 by using them to predict the forms in which the different chemical elements are likely to be found in water—say in a natural water such as a lake. If an element is found in a soluble form, such as a cation or oxo anion, it will be much more *available* for biological functions than if it is found in an insoluble form such as an insoluble oxide or hydroxide. Availability, of course, is a good thing if the element is a nutrient, but it is bad if it is toxic; we will discuss the biological effects of the elements in Sections 5.10–5.12.

The chemistry of the elements in natural waters is actually quite complex and involves types of chemistry that we have not yet dealt with, such as precipitation of salts, complexation by ligands, and oxidation by air or reduction by pollutants. At this time, we will have to confine ourselves to discussing a rather unnatural body of natural water, one that contains no oxidizing or reducing agents or complexing agents, and in which no salts (except hydroxides) are allowed to precipitate. (Distilled water fits this description better than any natural bodies of water, of course.) Since no oxidation–reduction reactions are allowed to occur, we assume that each element is in the common positive oxidation state specified in Table B1. We allow only those reactions that we have discussed in this chapter: the hydrolysis of hydrated cations to yield hydroxides or oxides of the elements, oxo acids, hydroxo anions, and oxo anions of the elements. We allow the pH of natural water to vary—as it may due to natural buffering on one hand or acid rain deposition on the other—to see how this will affect the availability of the different elements as either nutrients or pollutants.

Natural waters usually have pH values between 6 and 9. Thus we can see from predominance diagrams such as those in Figure 2.5 that the elements that we might expect to find in such waters, predominantly *in solution as hydrated cations*, are the nonacidic, feebly acidic, and *some* weakly acidic metal ions. Since charge is the most important single variable determining acidity, these would be mostly +1 and +2 charged cations (plus large +3 ions, but excluding very small +2 ions). The elements we might expect to find predominantly *as insoluble oxides or hydroxides in the bottom sediments* (or perhaps in colloidal form) are those that give the moderately and strongly acidic cations—small +3 cations, most elements with +4 oxidation numbers, and metals with +5 oxidation numbers.

Many of the elements that have very strongly acidic "cations" might also be in solution in the lake, but as *oxo anions*, perhaps partially hydrolyzed (protonated). Nonbasic and feebly basic anions would be expected to persist in solution in non-protonated form. Partially protonated forms would be expected in natural waters for moderately basic anions (e.g., HCO_3^- for the carbonate ion). Very strongly basic oxo anions are likely to be protonated even beyond the hydroxo-anion state: In such a case the insoluble hydroxide or oxide is the likely species.

If we combine these predictions with our information on the most common oxidation number for each element and exclude the possibility of oxidation–reduction reactions, we can finally predict the probable species and location (hydrated cation in solution, oxide or hydroxide as a precipitate, or oxo anion in solution) for each of the elements. Such a series of predictions is embodied in Table 3.5. Please note, however, that these are not actual results obtained by environmental chemists; they are only rough predictions made under constraints that do not much resemble a real body of water!

Example 3.9

The following radioisotopes (as fallout from an atomic bomb blast) are being deposited in a lake of pH 5.5–7: ^{99}Tc; ^{90}Sr; and ^{244}Pu. Without reference to Table 3.5 or predominance diagrams, predict the chemical forms these elements will take in this lake.

SOLUTION:

First, we need to predict the oxidation states each element will take. Table B1 allows us to postulate the presence of the following "cations": Tc^{7+}; Sr^{2+}; and Pu^{4+}.

Next, we need to evaluate the acidity of each cation to determine whether it can remain as a hydrated ion in neutral water: The Tc^{7+} ion is very strongly acidic with a $pK_a < -4$; Sr^{2+} is feebly acidic with a pK_a between 11.5 and 14; Pu^{4+} is on the line between moderately and strongly acidic with a pK_a of about 1. Recalling that the pH above which precipitation of a hydroxide occurs is roughly equal to its pK_a, we expect to find hydroxides precipitating if the cations have pK_a values below about 6 (i.e., for moderately, strongly, or very strongly acidic cations). Of the ions in question, only feebly acidic Sr^{2+} would be expected to stay in solution as a hydrated cation.

For the very strongly acidic cations (Tc^{7+} in this example), we need to evaluate the possibility of forming an oxo anion. Technetium is in the fifth period and in an oxo anion would be expected to show a penultimate total coordination number of 4–6. Since it is in the region of good π-bonding elements, we predict its oxo anion to have the formula TcO_4^-. We calculate the pK_{b1} of TcO_4^- to be 22.6. This ion is nonbasic and would thus be expected to persist in neutral water. However, for seventh period Pu^{IV} we predict an oxo anion, if formed, would have the formula PuO_6^{8-}, which would be very strongly basic and could not persist in water. (Detailed calculations, if carried out, would also show that none of the partially protonated forms could persist at neutral pH values either.) By default, the predominant form of Pu(IV) is expected to be the oxide, PuO_2, or the hydroxide, $Pu(OH)_4$.

The elements with moderately and strongly acidic cations are the ones most likely to be insoluble (as oxides or hydroxides) in unpolluted natural waters. Consequently, these elements are seldom available for biological activity, either in natural waters or in the digestive system or in body fluids, unless chemical processes other than acid–base reactions can somehow make them available. These *unavailable* elements thus are seldom either essential for life (despite the abundance of titanium and silicon, most

Table 3.5
Main Forms of the Elements in Moderately Aerated Water of pH 5.5–7[a]

Groups 3–5: Stable Oxidation Number Equals the Group Number. Groups 6–12: Lower Oxidation Number More Stable. Groups 13–17 (upper part): Stable Oxidation Number Equals Group Number. Lower part of groups 13–16: Stable Oxidation Number is 2 less than the Group Number.

1	2	3	4	5	6	7	8	9	10	11	12	13 (III)	14 (IVa)	15 (Va)	16 (VIa)	17 (VIIa)
H_2O																
Li^+	$Be(OH)_2$											$B(OH)_3$	CO_2 / HCO_3^-	NO_3^-	H_2O	F^-
Na^+	Mg^{2+}											$Al(OH)_3$	SiO_2	$H_2PO_4^-$ / HPO_4^{2-}	SO_4^{2-}	Cl^-
K^+	Ca^{2+}	$Sc(OH)_3$	TiO_2	$H_3V_2O_7^-$ / $H_2VO_4^-$	$Cr(OH)_3$	Mn^{2+} / MnO_2	$Fe(OH)_3$	Co^{2+}	Ni^{2+}	Cu^{2+}	Zn^{2+}	$Ga(OH)_3$	GeO_2	$H_2AsO_4^-$ / $HAsO_4^{2-}$	SeO_4^{2-}	Br^-
Rb^+	Sr^{2+}	Y^{3+} / $Y(OH)_3$	ZrO_2	Nb_2O_5	MoO_4^{2-}	TcO_4^-	$Ru(OH)_3$	Rh_2O_3	$Pd(OH)_2$	Ag^+	Cd^{2+}	$In(OH)_3$	SnO_2	Sb_2O_3	$HTeO_3^-$	IO_3^-
Cs^+	Ba^{2+}	Lu^{3+}	HfO_2	Ta_2O_5	WO_3 / WO_4^{2-}	ReO_4	OsO_2	IrO_2	PtO_2	Au metal	HgO	Tl^+	Pb^{2+}	Bi_2O_3	$HPoO_3^-$?	
		La^{3+}	Ce^{3+}	Pr^{3+}	Nd^{3+}	Pm^{3+}	Sm^{3+}	Eu^{3+}	Gd^{3+}	Tb^{3+}	Dy^{3+}	Ho^{3+}	Er^{3+}	Tm^{3+}		
		Ac^{3+}	ThO_2	Pa_2O_5	UO_2^{2+}	NpO_2^+	PuO_2	Am^{3+} / $Am(OH)_4$								

SOURCE: Forms chosen based on data given in M. Pourbaix, *Atlas of Electrochemical Equilibria in Aqueous Solutions*, NACE, Houston, TX, 1974, with preference being given to positive oxidation states when possible.

[a] Shaded areas represent insoluble compounds.

forms of life do not make use of them) or toxic to life. (Note, however, that if an element is *exceedingly toxic*, as is Pu^{4+}, the fact that most of it may end up in the sludge of a lake may be of little consolation, since even this small amount of the element that equilibria allow to remain in solution may be harmful.)

The biochemically more significant elements are usually those that can be present in solution (as cations, simple anions, or oxo anions). One exception is the element *iron*, which is biochemically very important even though its most common form, Fe^{3+}, is moderately acidic and therefore unavailable in natural waters at pH values above about 2.0. Organisms have very elaborate strategies for obtaining and conserving this common yet inaccessible nutrient (iron-deficiency anemia may be the most prevalent dietary deficiency disease among humans). In the Antarctic ocean this nutrient is not even common, since iron is often brought to ocean waters in the form of dust blowing from dry soils or deserts, which are scarce in Antarctica. The Antarctic ocean is incredibly fertile in other nutrients, however, and is quite productive of plant and animal life. But it has been calculated that the productivity of the Antarctic is limited by the low availability of iron, and could be multiplied *10-fold* if enough iron were supplied. It has been estimated that enough photosynthesis would result to remove a significant fraction of the excess CO_2 that humans have put into the atmosphere; this CO_2 is a major contributor to the greenhouse effect (see Section 13.3).[43] It has been confirmed experimentally that adding more iron to the ocean results in a spectacular increase in photosynthesis and a major reduction in CO_2 levels in the nearby waters.[44]

Polluted Waters. If a body of water becomes highly polluted with acidic pollutants, it is possible for moderately acidic cations to go into solution. The adverse consequences of acid rain for fish seem to involve just this process. Lakes that are poorly buffered can undergo quite substantial changes in pH when acid rainfall runs off into them. The pH of such a lake can drop to the point that normally insoluble aluminum is converted to $Al^{3+}(aq)$. The toxic effects on fish seem to be due not to the enhanced concentration of hydrogen ion but to the presence of aluminum ion, which is normally not encountered in natural waters. Ingested Al^{3+} may have toxic effects on humans as well.[45]

A rise in the pH of an acidic body of water by dilution or neutralization can cause the weakly and moderately acidic ions to precipitate as hydroxides. Such a pH change could be a way of removing aluminum ion from a lake "killed" by acid rain, but caution is called for! This difference in pH between the acid water and the more neutral environment of the gills of fish apparently causes gelatinous aluminum hydroxide to precipitate there, coating the gills. The fish seem to sneeze themselves to death in an attempt to get rid of this gelatinous precipitate. Similar chemical reactions are involved in the *acid mine drainage* problem of old coal mines, in which the mineral pyrite, FeS_2, is slowly oxidized by air to iron(III) sulfate, $Fe_2(SO_4)_3$. This compound dissolves in the water draining from the mine and, of course, undergoes extensive hydrolysis to give sulfuric acid. As the pH of the acid water is raised again by dilution with uncontaminated streams of water, the hydroxy cation of iron converts to yellow insoluble iron(III) hydroxide, which precipitates as unsightly *yellow boy* along the stream banks.

Some elements that are desired nutrients for plants and animals are also weakly or moderately acidic cations and so may be unavailable in neutral or slightly basic solutions (unless some other type of chemistry can be used to keep them in solution). Many essential micronutrients are weakly acidic +2 charged cations: Zn^{2+}, Cu^{2+}, Co^{2+}, Fe^{2+}, and Mn^{2+}. These precipitate from dilute solutions at pH values above 5.3–8.5, so liming (adding CaO to) an acidic soil or lake may remove essential micronutrients.

3.9 *Nomenclature of Fluoro and Oxo Anions and Acids

Fluoro Anions. We begin by covering the nomenclature of fluoro anions, which is both simpler than that of oxo anions and more characteristic of coordination compounds in general (a topic that will be covered in more detail in Section 8.11). Three elements are involved in naming a fluoro cation or anion.

1. The number of fluoro groups is indicated by a prefix (di- for two, tri- for three, tetra- for four, penta- for five, hexa- for six, hepta- for seven, octa- for eight), placed in front of fluoro-, which is placed in front of the name of the central atom.

2. In a fluoro *anion*, the ending of *-ate* is added after the name of the central atom, and the Latin root name of the element is used if it has one (Table D). (Fluoro cations use the ordinary name of the element with no *-ate* suffix.)

3. If the central atom can vary its oxidation number, this is indicated in either of two ways within parentheses at the end of the name of the species. In the *Stock* convention, the *oxidation number* of the central atom is indicated in Roman numerals in the parentheses; in the *Ewens–Basset* convention, the *charge on the ion* is indicated in Arabic numerals in the parentheses. (The Stock convention is probably more common, but the Ewens–Bassett convention is used in *Chemical Abstracts.*)

As examples, the ion SiF_6^{2-} may be named hexafluorosilicate(IV) ion (Stock convention) or hexafluorosilicate(2−) ion (Ewens–Bassett convention); FeF_6^{3-} may be named hexafluoroferrate(III) ion or hexafluoroferrate(3−) ion.

Naming Oxo Anions. If oxo anions were also named by the above system, which is sanctioned by the International Union of Pure and Applied Chemistry (IUPAC), their names would be easier to derive but would be longer and (above all) unfamiliar: NO_2^- would be named dioxonitrate(III) ion or dioxonitrate(−1) ion. This naming system has not caught on at all for oxo anions, which have been extremely important in many fields for so many centuries that an older, more difficult naming system is entrenched in usage.

This older but far more common naming system uses suffixes (and sometimes prefixes) to do the job of all three of the above steps, by indicating *relative* oxidation numbers. All names replace the final suffix of the name of the element (e.g., *-ium*) with either the suffix *-ate* or the suffix *-ite*. In addition, a prefix (*per-* or *hypo-*) may be added. The oxo anion in the most common oxidation state is given the name ending in *-ate* (e.g., arsenate for AsO_4^{3-}). The oxo anion in which the central atom has the next lower oxidation number is given the name ending in *-ite* (e.g., arsenite for AsO_3^{3-}). Any oxo anion in which the central atom has a still lower oxidation number is given the prefix *hypo-* and the suffix *-ite*. Finally, if there is a central-atom oxidation number higher than the most common one, the prefix per- and the suffix *-ate* are used.

For this system of nomenclature the key question is, Which oxo anion is in the most common oxidation state? The answer to this is not scientifically precise, and can vary substantially throughout the periodic table, as indicated in Table B1. First, we note that very few elements commonly show an oxidation number of +7; for purposes of nomenclature, +6 is the highest "common" oxidation number. Second, although the prevalence of the group oxidation state depends on the block and period of the periodic table (Section 1.7), for purposes of nomenclature all periods and blocks are treated the same. The general pattern is that *the most common oxo anion* (the one that is named with *-ate*) is the one in which the *central-atom oxidation number equals the group number, provided that this does not exceed 6*. In these groups numbered 1–6, the oxo anion in the (group number −2) oxidation state are named with *-ite*; any still lower oxidation states (such as occur in $H_2PO_2^-$) are named with the *hypo-* prefix and *-ite* suffix (hypophosphite).

Groups numbered 7 and 8 [Groups 17(VII), 7, *7F*, 18(VIII), and 8] are treated differently, since the group oxidation state is treated as uncommonly high. Oxo anions exhibiting the group oxidation number here are designated "per-ate," as in perchlorate. The next oxidation number is two below this one, and is the oxidation state considered common in these groups, hence it is designated by the suffix *-ate*.[46] These names are illustrated in Table 3.3.

One other set of prefixes is sometimes encountered. In the fifth and sixth periods, in which the expected total coordination number is variable, either type of oxo anion may sometimes be encountered, and may be present in equilibrium:

$$H_4IO_6^- \rightleftharpoons IO_4^- + 2\,H_2O \qquad K = 29 \tag{3.8}$$

In such cases, the oxo anion with the lower total coordination number is given a pre-prefix of *meta-* (IO_4^- is metaperiodate); the one with the higher total coordination number has the pre-prefix of *ortho-* [$H_4IO_6^-$ is the (tetrahydrogen) orthoperiodate ion].

Example 3.10

Give the names of the following oxo anions: NO_2^-, PO_4^{3-}, AsO_3^{3-}, SeO_3^{2-}, FeO_4^{2-}, and IO_3^-.

SOLUTION:

It is necessary to check two things: how the oxidation number of the central atom compares with the group number and the group number itself. If the latter is Groups 17(VII), 18(VIII), 7, or 8, the naming system slips into "high gear". For NO_2^-, PO_4^{3-}, AsO_3^{3-}, and SeO_3^{2-}, the element is in a group numbered 6 or less, so if the central-atom oxidation number equals the group number, the suffix *-ate* is used: PO_4^{3-} is phosphate. For the other three, the oxidation number equals the group number minus two, so NO_2^- is nitrite, AsO_3^{3-} is arsenite, and SeO_3^{2-} is selenite.

For the iodine and iron oxo anions, the names move up one category: Those showing the group oxidation numbers use per- and *-ate* [IO_6^{5-} is (ortho)*per*iodate]. Since the oxidation numbers in the iodine and iron oxidation states are two less than the group numbers, IO_3^- is iodate and FeO_4^{2-} is named ferrate (Latin roots are still used when available).

The nomenclature of oxo acids is based on that of their ultimate conjugate bases, the oxo anions. Put quite simply, if the name of the conjugate oxo anion ends in *-ate*, the ending of the oxo acid is changed to *-ic acid*. If the name of the conjugate oxo anion ends in *-ite*, the ending of the oxo acid becomes *-ous acid*. Thus the acid obtained by protonating the perchlor*ate* ion is called perchlor*ic acid*; that obtained from the phosph*ate* ion is called phosphor*ic acid*; that obtained from the sulf*ite* ion is called sulfur*ous acid*; and that obtained from the hypochlor*ite* ion is hypochlor*ous acid*.

Formulas of Oxo Anions from their Names. To go from the *name* of an oxo anion to its *formula*, somewhat the reverse process is suggested. First, determine whether the group number of the element is a "7" or an "8" and note the prefixes and suffixes in the name of the oxo anion. This determines how the oxidation number of the central atom relates to the group number (Table 3.6). Once the oxidation number of the central atom has been determined, the problem is just like Example 3.5, and the same three steps are used.

Table 3.6
Relationship between Names of Oxo Anions and the Oxidation Number of the Central Atom

Prefixes	Suffixes	In Most Groups	In Groups 17(VII), 18 (VIII), 7, 8
pre-	-ate	Does not occur	Oxidation No. = group No.
	-ate	Oxidation No. = group No.	Oxidation No. = group No.−2
	-ite	Oxidation No. = group No. −2	Oxidation No. = group No. −4
hypo-	-ite	Oxidation No. = group No. −4	Oxidation No. = group No. −6

Example 3.11
Give the formulas of the following ions: perxenate, vanadate, and chlorite.

SOLUTION:

First, vanadium is in a group numbered 5, which is named normally. The suffix *-ate* tells us that the oxidation number of vanadium equals its group number, so it is +5. Both Xe and Cl are in Groups 17(VII) and 18(VII); their names are "geared up". The *per-* and *-ate* with xenon tells us, in this case, that the oxidation number equals the group number, +8. The *-ite* suffix for chlorine is two steps below the top modifiers (as in perchlorate), so the oxidation number is two steps (four oxidation numbers) below the group number of Cl, that is, the oxidation number is +3.

Second, the penultimate total coordination numbers for third period chlorine and fourth period vanadium are each 4, while that of fifth period xenon is from 4 to 6.

Third, we examine oxidation numbers to see whether *p*-block elements have unshared pairs. These occur only for Cl(III) in chlorite, which (since the oxidation number is 4 below the group number) has two unshared pairs of electrons.

Finally, we mentally merge cations with enough oxide ions to generate oxo anions with the expected penultimate total coordination numbers. Combining V^{5+} with four O^{2-}, we obtain $VO_4{}^{3-}$. Combining $:Cl:^{3+}$ with two O^{2-}, we obtain $ClO_2{}^{-}$, which has a total coordination number of 4. Combining Xe^{8+} with four O^{2-} does not even give an oxo anion, so we combine it with six O^{2-} to obtain $XeO_6{}^{4-}$.

3.10 *Lewis Structures and Oxidation Numbers

The procedure for drawing Lewis structures given in Section 3.3 is quick, but there are situations in which it is difficult or impossible to use: for the occasional molecule in which an electron pair must be shared between more than two atoms, and (more commonly) in large molecules with many central atoms or molecules in which it is not easy to envision the bonding as being coordinate covalent. Any molecule in which a given type of central atom bonds to other atoms of the same type (a *catenated* compound) is an example.

For more complex molecules such as these, the procedure generally taught in general chemistry is more appropriate; we review it here, using as examples (a) *cyclo*-$(HBNH)_3$; (b) $[O_3S\text{–}S\text{–}S\text{–}SO_3]^{2-}$, and (c) NSF_3.

1. Count the total number of valence electrons in the molecule or ion. To do this, add up the group numbers for each atom in the molecule; add one extra electron for each unit of negative charge, and deduct one electron for each unit of positive charge. Thus in (a) there are three times $(1 + 3 + 5 + 1) = 30$ valence electrons; in (b) there are 10 Group 16(VI) atoms and two negative charges for a total of 62 valence electrons; in (c) there are $5 + 6 + 3(7) = 32$ valence electrons.

2. Draw a skeleton structure of the molecule or ion, joining atoms with lines to represent shared electron pairs, that is, single bonds. In more complex molecules, some indications must be given or deduced concerning the atoms that are connected. For example, in (a) hydrogens normally form only one bond, so the connections between the three HBNH units to form the cyclic ring implied by *cyclo* must be between B and N atoms [Fig. 3.19(*a*), top]. In (b), the unique atom in each SO_3 group, S, is taken as the center of that group; the catenated S–S connections are given, so that the skeleton structure at the top of Figure 3.19(*b*) results. In (c), the atom in the middle of the formula is taken as the central atom, and is connected to the others as in the top of Figure 3.19(*c*).

3. Now deduct two electrons for each bond drawn from the total number of valence electrons (Step 1) to obtain the number of electrons still available to complete the Lewis structure. We calculate for (a) $30 - 24 = 6$ electrons still needed to complete the Lewis structure, for (b) $62 - 18 = 44$ electrons still available, and for (c) $32 - 8 = 24$ electrons available.

4. Draw a circle around each non-hydrogen atom, including all bonds to it. See how many electrons each atom still needs to achieve an octet of electrons; total these numbers. (We will say that an atom such as P in PCl_5 that starts with more than an octet of electrons needs *zero* additional electrons.) Thus, as shown in Figure 3.19, the total number of electrons *needed* by the atoms of each structure are (a) 2 for each B and 2 for each N for a total of 12; (b) 6 for each O and 4 for each of the two central S atoms for a total of 44; (c) 6 for N and 6 for each F for a total of 24.

5. If the number of valence electrons needed (Step 4) matches the number of electrons available (Step 3), only unshared electron pairs are needed to complete the octets of each atom; fill these in. This is true in examples (b) and (c). In each of these the outer O, N, and F atoms get three additional unshared electron pairs, and the two needy S atoms in (b) each get their two electron pairs, to give the Lewis structures shown in the middle of Figure 3.19 (*b* and *c*).

6. If more electrons are needed (Step 4) than are available (Step 3), make multiple

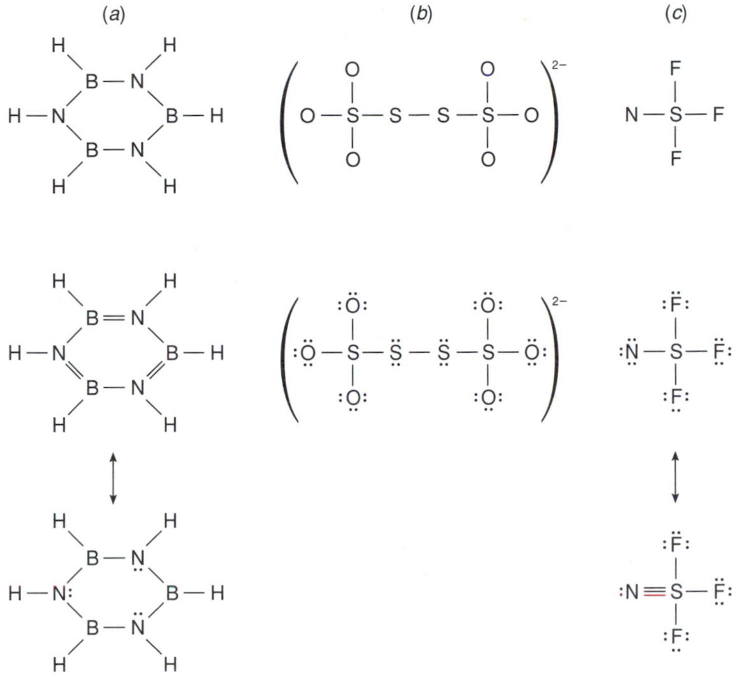

Figure 3.19
Drawing Lewis structures for examples (*a*)–(*c*). (Top) Skeleton structures from Step 2; (middle and bottom) alternate resonance Lewis structures.

bonds by adding one extra line between pairs of atoms for each two electrons you are short. If there is a choice of different outer atoms to which the multiple bonds may be drawn, it is best to draw them to O, N, C, or S; they cannot be to H. Then complete the placement of the remaining electrons as unshared electron pairs. In (a), three more pairs are needed than are available; these can be shared between adjacent B and N atoms as shown in the middle of Figure 3.19(*a*). However, since boron is not a good double-bonding atom, one can also draw a resonance structure (Section 3.5) without double bonds, in which the electron pairs are only on the more electronegative N atoms [bottom of Fig. 3.19(*a*)].

7. If there are too many electrons to locate, put the extra electrons on the central atom, which must be from the third period of the periodic table or below and will then violate the octet rule. Violation of the octet rule is not required for any of these examples, but can be employed to generate additional resonance structures for some of these species. In the structure that we have drawn for (c), the S atom has a formal charge of +2 and the N atom has a formal charge of −2; these excessive formal charges can be relieved by converting unshared electron pairs on N to double bonds to S, which then exceeds an octet [bottom of Fig. 3.19(*c*)]. Although we have not shown it in Figure 3.19(*b*), the same can be done if desired to relieve the +2 formal charges on the outer S and the −1 formal charges on the outer O atoms.

8. Finally, *check your answer*! Be sure that (a) the total number of electrons shown agrees with the number calculated in Step 1 and (b) each atom has an octet of electrons (except hydrogens with two, and, if necessary, central atoms from the third period or below, which may have more than eight).

Oxidation Numbers from Lewis Structures. The rules given in Section 1.7 similarly work only in comparatively simple situations; a more general procedure can be applied once the Lewis structure of a molecule or ion has been successfully drawn. The degree to which a bond electron pair is unequally shared between two atoms of differing electronegativity is difficult to measure. If the two atoms are of extremely different electronegativities, the electron pair may reside completely on the more electronegative atom, giving a positive and a negative ion (ionic bonding). If the two atoms are identical, the electrons should be equally shared in a pure covalent bond. Most cases fall in between, however, with a partial negative charge being assignable to the more electronegative atom and a partial positive charge to the less electronegative atom. In such cases, rather than attempt to measure the actual magnitude of the partial charges in the molecule, for bookkeeping purposes we may decide to assign the bond pair of electrons completely to the more electronegative atom, that is, we pretend that the bonding is ionic. After doing this to all the bonds in the molecule, we have produced a number of (sometimes imaginary) positive and negative ions. The imaginary charge that we have produced on each of the atoms of the molecule may be assigned as the oxidation number of that atom.

To determine oxidation numbers in this manner, we begin with the Lewis dot structure of the molecule; as examples we use the molecules for which we drew the Lewis structures in Figure 3.19(*a–c*).

1. Move the electrons of each bond onto the more electronegative atom of that bond (if the two atoms are equal in electronegativity, move one electron onto each atom). In (a), the order of electronegativities is B < H < N; electrons in B–N single or double bonds are all placed on N; electrons in B–H bonds are placed on H; electrons in N–H bonds are placed on N. It does not matter which resonance structure is used—the results are the same [Fig. 3.20(*a*)]. In (b), O is more electronegative than S, but the S–S bonds must be divided equally [Fig. 3.20(*b*)]. In (c), the order of electronegativities is S < N < F, so the redistribution shown in Figure 3.20(*c*) results.

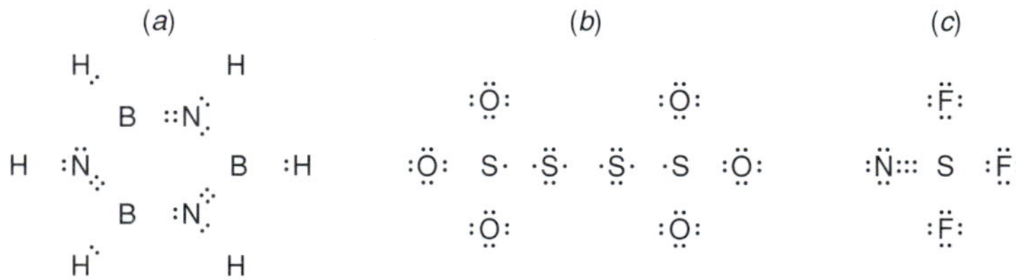

Figure 3.20
Disassembly of structures for examples (*a–c*) in the process of assigning oxidation numbers.

2. Compare the remaining number of valence electrons on each atom with the number of valence electrons found on a free atom of that element (i.e., its group number). Each electron that has been lost in forming the molecule or ion adds one to the oxidation number of the element; each electron that has been gained contributes −1 to its oxidation number. In (a), the borons have no valence electrons, which is three less than their group number, so they have a +3 oxidation number (for bookkeeping purposes, they are treated as B^{3+} ions). Similarly, the N atoms have acquired three new electrons in this compound, so show a −3 oxidation number; the H atoms *on B* have a −1 oxidation number and the H atoms *on N* have a +1 oxidation number. In (b), the O atoms show their usual −2 oxidation state; the S atoms *in the SO_3 groups* have a +5 oxidation state; the central S atoms have a 0 oxidation state. (If it is desired to give only one oxidation number for all S atoms, for example, as a step in balancing a redox reaction, these two numbers can be appropriately averaged.) In (c), the F atoms show their usual -1 oxidation number, N shows the -3 oxidation number, and S has a $+6$ oxidation number.

This procedure is especially useful for assigning oxidation numbers to metal atoms in complex ions; oxidation numbers are used later to name complex compounds (Chapter 8).

Example 3.12.

Draw structures of the following complex ions or coordination compounds, and in each one deduce the oxidation number of the central metal or nonmetal ion: (a) $[Al(H_2O)_6]^{3+}$, (b) $[Co(NH_3)_5Cl]SO_4$, and (c) $K_3[FeF_4(OH)_2]$.

SOLUTION:

The rules given above are applied within the complex ion itself; the counterions shown for (b) and (c) are unwanted complications that should be deleted. The brackets in these formulas separate the complex ion from the counterions, which we recognize as (b) the sulfate ion, SO_4^{2-}, and (c) three potassium ions, K^+. When the charges of these ions are subtracted out, we obtain the charges on the complex ions: $+2$ in (b) and -3 in (c).

Within the brackets, we should look for the Lewis acid, which is commonly a metal atom or ion but can be a nonmetal cation; more often than not the Lewis acid is written just inside the left bracket. The other species within the brackets are ligands. It is customary to enclose each polyatomic ligand inside a set of parentheses; monoatomic ligands such as Cl^- are not enclosed. With this information, we can draw structures (Fig. 3.21) in which each ligand is connected to the metal ion through a likely donor atom by a coordinate covalent bond, represented by a line in Figure 3.21(*a*). Octets should be completed on each donor atom.

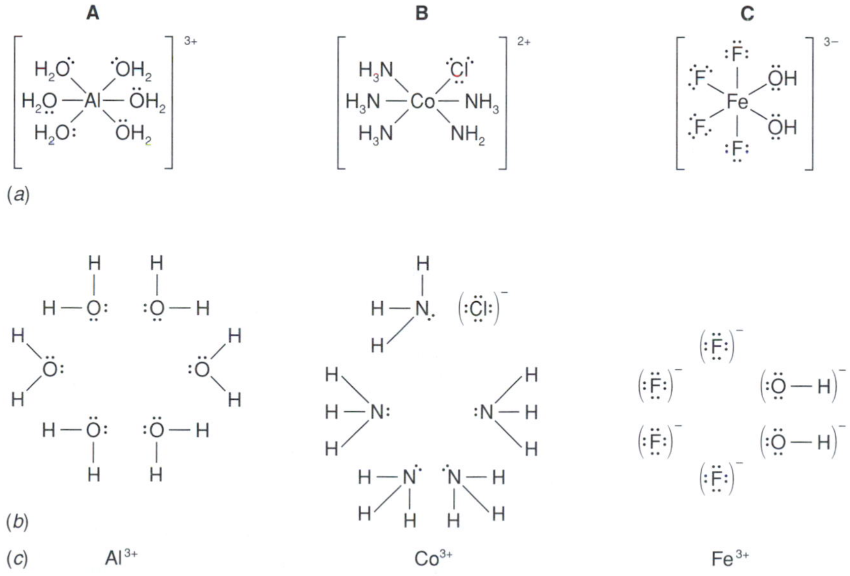

(*c*) Al^{3+} Co^{3+} Fe^{3+}

Figure 3.21
(*a*) Structures of the coordination complexes or complex ions in the example; (*b*) the ligands removed from each complex; (*c*) the Lewis acids that remain after removal of the ligands.

Next, we apply Step (1): We dissect each complex ion into its parent ligands and metal ion, as shown in Figure 3.21(*b*), by pushing the two electrons in each coordinate covalent bond back onto the more-electronegative donor atoms,[47] producing Lewis structures of the parent ligands as shown.

In Step 2, the Lewis structure of each ligand is compared with the sums of group numbers to determine whether each ligand is neutral or negatively charged. (Positive ions are almost never ligands.) Thus we see that the numbers of electrons represented in the Lewis structures of H_2O and NH_3 equals the sum of group numbers; the numbers we have drawn for Cl, F, and OH exceed the sums of group numbers by 1, so these are −1 charged anions: Cl^-, F^-, and OH^-.

Next, we can deduce the charge on the central metal ion without drawing its Lewis structure: This is simply equal to the charge on the whole complex ion minus any charges on anionic ligands that we have removed. Thus for $[Al(H_2O)_6]^{3+}$, removal of six neutral water molecules still leaves a +3 charge; the Lewis acid is Al^{3+}. Removal of some neutral ammonia molecules and a Cl^- from $[Co(NH_3)_5Cl]^{2+}$ leaves a Co^{3+} ion as the Lewis acid. Removal of six −1 charged ligands from the −3 charged iron complex ion leaves a Fe^{3+} ion as the Lewis acid. These metal–atom charges are the oxidation numbers of the metals.

Study Objectives

1. Identify cases of positive, negative, and zero overlap between specified *s*, *p*, *d*, or *f* orbitals on two different atoms. Identify cases of σ, π, and δ bonds resulting from such overlap. Compare the overlapping ability of hybridized and unhybridized orbitals; of $(n-2)f$, $(n-1)d$, ns, and np orbitals as a function of periodic table position. Exercises 1–7.

2. Without the use of the tables in the earlier chapters, describe and explain the main and the anomalous periodic trends in covalent single and multiple bond energies. Rationalize periodic trends in σ, π, and δ bond energies. Exercises 8–12.

3. Know the use of the common terms for Lewis acid–base (coordination) chemistry; identify examples for these terms in the formulas or structures of given coordination compounds. Predict possible products of Lewis acid–base reactions like those of Experiment A.2. Exercises 13–19.

4. Draw the Lewis dot structure of a specified molecule or ion. Using the concept of the maximum total coordination number of a cation, predict the formulas of its highest fluoride, fluoro cation, or fluoro anion. Exercises 20–21.

5. Predict the geometries of molecules and ions, including bond angles and distortions. Exercises 22–28.

6. Predict the formulas of oxo anions (and acids) of elements in specified oxidation states. Exercises 29–34.

7. Explain why some oxo anions are more basic than others. From its charge and number of oxo groups, calculate the pK_b of an oxo anion. Then classify it as nonbasic, feebly basic, moderately basic, or strongly basic. Exercises 35–46.

8. Construct predominance diagrams involving oxo anions; identify hydroxo anions and amphoteric oxides that occur in such diagrams, and the form that can be predominant at a given pH. Exercises 47–50.

9. Classify an oxo acid as weakly, moderately, strongly, or very strongly acidic. Exercises 51–54.

10. Using the most likely oxidation state of a given element, predict the form in which it would likely be found (hydrated cation, insoluble oxide or hydroxide, or oxo anion) in nonpolluted or (acid–base) polluted natural waters. Exercises 55–63.

11. Name oxo acids, oxo anions, and their salts; write the formulas of oxo acids, oxo anions, and their salts given their names. Exercises 64–72.

12. Draw Lewis structures of more complicated molecules or ions. Exercises 74–76.

13. Calculate the oxidation numbers of atoms in specified molecules or ions working from Lewis dot structures. Exercises 77–80.

Exercises

1. In Figure 3.1 are drawn different pairs of atoms with orbitals overlapping. (a) Identify the specific atomic orbital (e.g., d_{xy}) used by each atom in each case. (b) In each case, tell whether there is positive, negative, or zero overlap between the orbitals of the two atoms. (c) For each case in which there is positive overlap, identify the bond that is formed as σ, δ, or π.

2. *In Figure 3.22 are drawn different pairs of atoms (A on the left and B on the right) with orbitals overlapping. (a) Identify the specific atomic orbital used by each atom

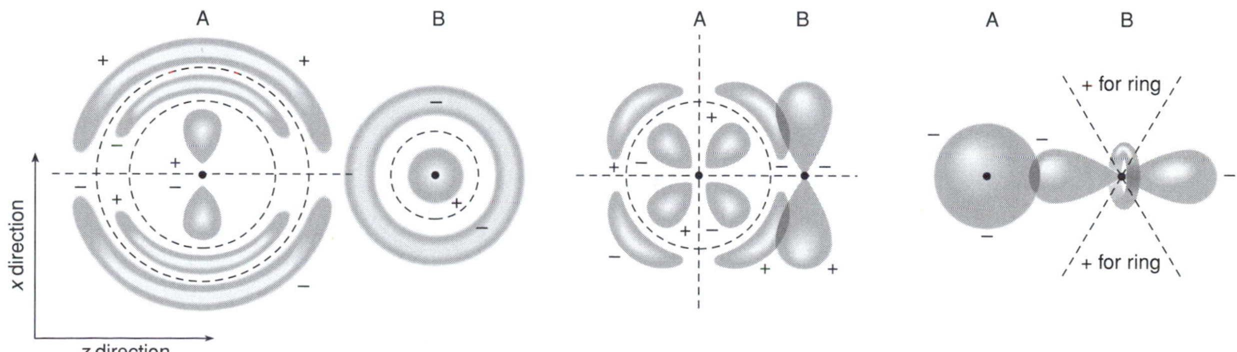

Figure 3.22
Three different cases of overlapping atomic orbitals on atoms A and B.

in each case. (b) In each case, tell whether there is positive, negative, or zero overlap between the orbitals of A and of B. (c) For each case in which there is positive overlap, identify the bond that is formed as σ, δ, or π.

3. In Figure 3.23 are drawn different pairs of atoms with orbitals overlapping. (a) Identify the specific atomic orbital used by each atom in each case. (b) In each case, tell

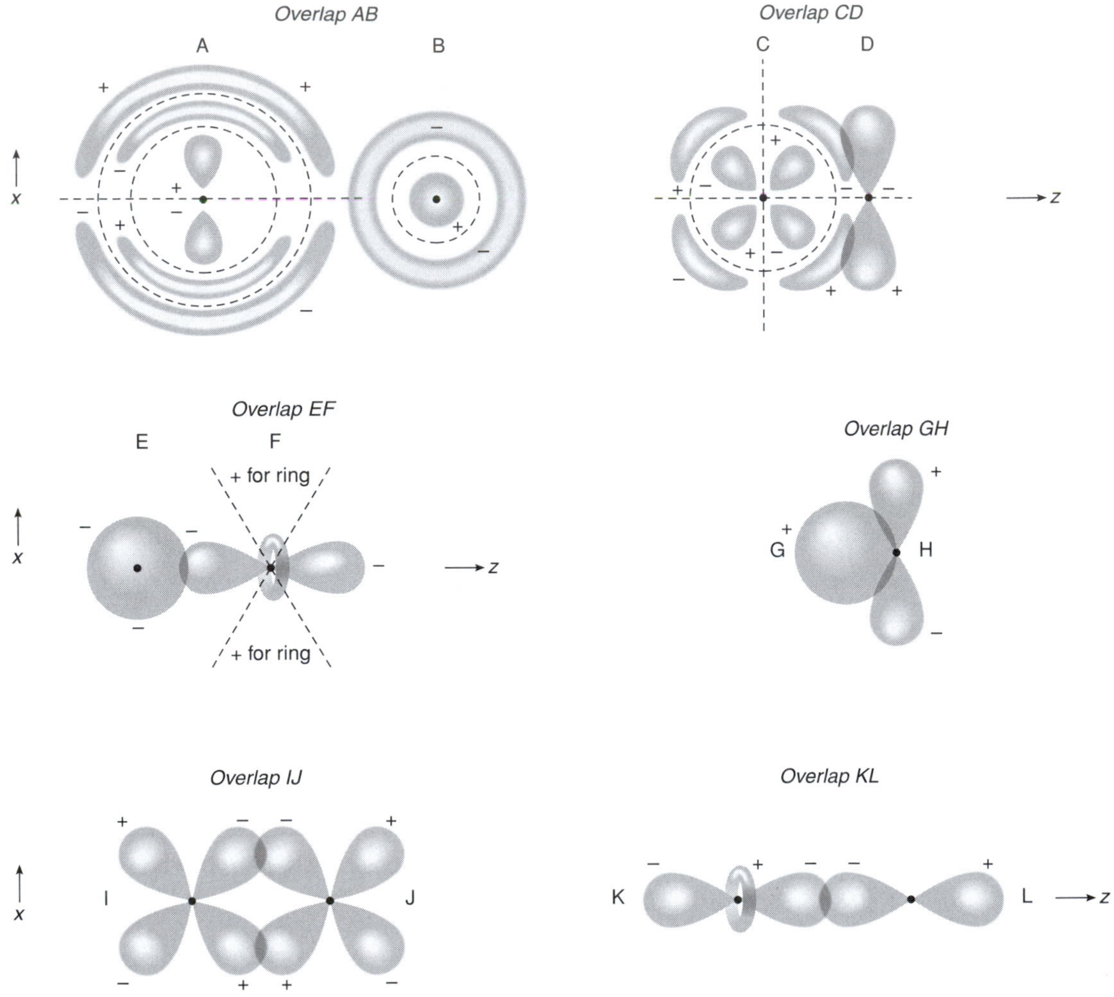

Figure 3.23

whether there is positive, negative, or zero overlap between the orbitals of the two atoms. (c) For each case in which there is positive overlap, identify the bond that is formed as σ, δ, or π. (d) Among the cases in which there is positive overlap, identify the two cases that probably give the strongest bonds; the two cases that give the weakest bonds.

4. (a) Using Slater's rules, calculate the following Z^* values for the Fe atom: Z^*_{4s}; Z^*_{3d}. (b) Calculate the radii $\langle r_{\max} \rangle$ for the 4s and the 3d electrons in the Fe atom. (c) Which orbital of Fe (4s or 3d) is likely to give better overlap with the orbital of another atom bonded to Fe? Briefly explain why, using your calculations.

5. Application of Slater's rules leads one to calculate the following Z^* values for Ti and Zn: 3.15 for the 4s orbital of Ti, 3.65 for the 3d orbital of Ti, and 4.35 for the 4s orbital of Zn. (a) Use Slater's rules to calculate Z^* for the 3d orbital of Zn. (b) Now calculate $\langle r_{\max} \rangle$ for the 3d and for the 4s orbitals in Ti; for the 3d and for the 4s

orbitals in Zn. (c) In the text, it was noted that 3*d* orbitals are not as good at covalent overlap with other orbitals as 4*s* orbitals. Is this equally true for Zn and for Ti? If not, for which element (Ti or Zn) is the statement more strongly true? (d) Explain your calculated trend in terms of penetration or shielding of orbitals. Explain how your results are relevant to the fact that Ti and its neighbors form more stable oxo cations than do Zn and its neighbors.

6. *Using normally predicted electron configurations, compute Z^* and $\langle r_{max} \rangle$ for electrons in the valence *d* and *s* orbitals of (a) Cr, (b) Mo, and (c) W. Explain whether or not these calculations show the lesser covalent-bond overlapping ability of the Cr 3*d* orbitals as compared to the Mo 4*d* and the W 5*d* orbitals. What factor in Slater's rules is responsible?

7. Two of the elements in Exercise 6, Cr and Mo, actually have the "anomalous" electron configurations $3d^5 4s^1$ and $4d^5 5s^1$, respectively, while W has the predicted electron configuration. Repeat the calculations of Exercise 6 using the actual electron configurations, and determine whether the conclusions of Exercise 6 are altered.

8. *Without the use of tables, arrange each of the following sets in order of increasing covalent-bond energies (increasing stability of covalent bonds): (a) the following single bonds: C–C, Si–Si, Ge–Ge, Sn–Sn, Pb–Pb; (b) the following single bonds: Li–Li, Be–Be, B–B, C–C; (c) the following Mo–Mo bonds: σ, δ, π.

9. Which element shows the greatest tendency to form double or triple bonds? C, Si, Ge, Sn, or Pb

10. *Give the name (σ, etc.) of each of the kinds of bonds drawn in Figure 3.24, and

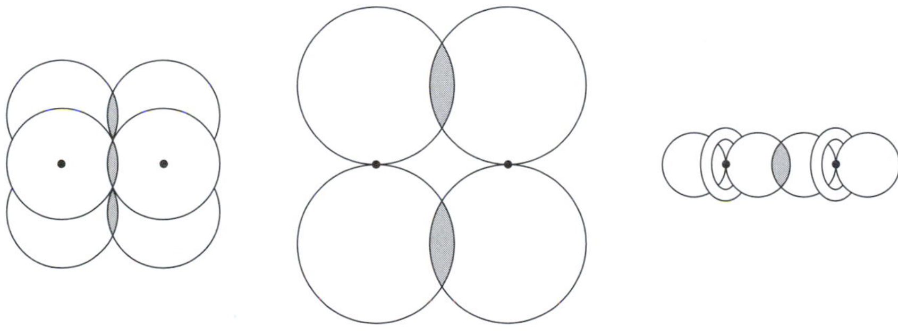

Figure 3.24

arrange them in order of increasing bond dissociation energy. Fill in the signs of the wave functions necessary to give bonding interactions in each case.

11. Which is a better conclusion? (a) Pi bonds are especially stable among the second period elements; or (b) σ bonds between certain second-period elements are especially unstable. Explain your conclusion, using data from Table 3.1.

12. Consider the compound with the structure shown in Figure 3.6. (a) How many σ and how many π bonds are present in each W–C bond? (b) Which one is the strongest of these three W–C bonds? Is it three times as strong as the weakest one? Why or why not? (c) Suppose that your friend developed the ambition to prepare an similar but better compound that not only contains a W–C single bond, a W–C double bond, and a W–C triple bond, but also a W–C quadruple bond. Can your friend succeed? Why or why not?

13. *Draw the Lewis dot structures of the following molecules or ions, and identify the potential donor atoms in each: (a) NH_3; (b) $NH_2–CH_2–COOH$ (an amino acid); (c) SO_4^{2-}; and (d) SO_3^{2-}.

14. Consider the following complexes: $[Au(CH_3)_2(NH_3)_2]^+$; SnI_6^{2-}; $HgCl_2$; and $[Hg(CH_3)_2]$. Treat all the bonds to the metal as coordinate covalent bonds. (a) Give the coordination number of the metal atoms in each of these complexes. (b) Identify the donor atoms in the ligands found in each complex.

15. Consider the compound $[AlF_2(NH_3)_4]Cl$. Write the formula (including charge, if any) of (a) the counterion; (b) the Lewis acid; and (c) the two different types of ligands. (d) What is the coordination number of the Lewis acid? (e) What is the donor atom of the larger of the two ligands?

16. Pretend you are Alfred Werner. You have, by painstaking work, isolated the following set of compounds:

$PdCl_2$:4 $(CH_3)_3P$	Contains 3 ions; 2 Cl^- ions precipitated by Ag^+
$PdCl_2$:3 $(CH_3)_3P$	Contains 2 ions; 1 Cl^- ion precipitated by Ag^+
$PdCl_2$:2 $(CH_3)_3P$	Contains 0 ions; 0 Cl^- ions precipitated by Ag^+
$PdCl_2$:$(CH_3)_3P$:KCl	Contains 2 ions; 0 Cl^- ions precipitated by Ag^+
$PdCl_2$:2 KCl	Contains 3 ions; 0 Cl^- ions precipitated by Ag^+

(a) Write the formulas of each of these compounds in the modern form, with complex ions enclosed in brackets. (b) Identify the counterions in each compound. (c) What is the coordination number of palladium in this series of compounds? (It is constant.)

17. Some metalloidal elements such as As can act either as Lewis acids or bases in different situations. Write a likely reaction forming the indicated bonds to As from likely starting materials for each of the following compounds, and tell in each case whether the As acted as a Lewis acid or base. (a) As–(-F)$_5$; (b) $(CH_3)_2O–AsCl_3$; (c) $[Cl–AsCl_3]^-$; and (d) Ni–(-AsF_3)$_4^{2+}$.

18. *Give plausible products (or just one product) for each of the following reactions, or tell if no reaction is expected. (a) $CsCl + AlCl_3 \rightarrow$; (b) $SiCl_4 + 4\ C_5H_5N \rightarrow$ (A conductor of electricity in solution); and (c) $PbCl_4 + 2\ C_5H_5N \rightarrow$ (A nonconductor of electricity in solution).

19. For each set of reactants, label the Lewis acid and the Lewis base. Then write a likely formula for the product formed. If it is a complex ion, enclose it in brackets, and label the counterion that accompanies the complex ion. (a) $H^+(aq) + Cl^-(aq) + C_5H_5N \rightarrow$; (b) $2\ C_5H_5N + SnCl_4 \rightarrow$; (c) $2\ Cs^+ + 2\ Cl^- + SnCl_4 \rightarrow$.

20. Predict the formulas of the highest fluoro anions of the following elements in the specified oxidation states: (a) Si^{4+}; (b) B^{3+}; (c) As^{5+}. (d) Cr^{3+}; (e) Mo^{5+}; (f) Se^{4+}; (g) Xe^{6+}; and (h) Xe^{8+}.

21. Assuming that an excess of F^- is available and referring to Tables B1, B2, and C, predict the formulas of the fluoro species expected for (a) each third period element from Al to Cl in each of its positive oxidation states; (b) the fourth period *d*-block elements in their highest oxidation states; (c) the fifth period *p*-block elements in their highest oxidation states; (d) the Group 15(V) elements in their +5 oxidation states; (e) the Group 15(V) elements in their +3 oxidation states.

22. *Write the Lewis dot structures of the following molecules or ions (treat hydrogen as the hydride anion, H^-): (a) CO_2; (b) NO_2^-; (c) NO^+; (d) CH_4; (e) H_2SiBr_2; (f) IF_5; (g) OsO_4; (h) NO_2^+; (i) XeF_4; (j) SiH_4; and (k) ICl_4^-.

23. Draw Lewis dot structures of the following molecules and ions. (a) SO_2^{2+}; (b) XeF_6; (c) SiS; (d) IF_4^+; (e) $TeCl_6^{2-}$; (f) IO_3^-; (g) IF_4^-; (h) XeF_5^-; (i) BCl_4^-; and (j) TeF_4.

24. *Draw the Lewis dot structures of each of the following molecules or ions. Predict the bond angles (including distortions) about the central atom in each species. Give the name of the resulting geometry (atoms only, not unshared electron pairs): (a) NO_2^+; (b) XeF_4; (c) $(CH_3)_2TeCl_2$; (d) $SbCl_5$; (e) IF_5; (f) ClNO; (g) XeF_3^+; (h) $SbCl_5^{2-}$; (i) XeO_4; (j) XeF_5^+; (k) $(CH_3)_2ICl$; and (l) NCN^{2-}.

25. Draw the Lewis dot structures of each of the following molecules or ions. Give the geometric shapes of and the bond angles (including distortions) around the central atoms of the following: (a) $[Te(CH_3)_2Cl_3]^-$ (central atom is Te); (b) $(CH_3)_2S{=}O$ (central atom is S); (c) BrF_4^-; (d) NO_2^-; (e) ICl_4^-; (f) BO_3^{3-}; (g) $XeOF_4$; (h) ClO_2^-; (i) CN_2^{2-}; and (j) SCl_4.

26. Draw the structure of the following, showing their geometric arrangement of groups (cis or trans, axial or equatorial, etc.): (a) $(CH_3)_2TeCl_2$; (b) $SbBr_2F_2^-$; and (c) $(CH_3)_2SbCl_3$.

27. Give the geometric shapes of and the bond angles (including distortions) around the central atoms of the fluoro species of Exercise 20.

28. Give the geometric shapes of and the bond angles (including distortions) around the central atoms of the fluoro species of Exercise 21.

29. *Write the formula (including charge) of the oxo anion of (a) boron with oxidation number +3; (b) bromine with oxidation number +3; (c) arsenic with oxidation number +5; (d) sulfur with oxidation number +4;

30. Write the formulas of the oxo anions that would be formed by each of the Group 14(IV) elements (carbon group) in the +2 oxidation state (do not worry that some of these do not actually exist). Then explain briefly why the formulas change as you go down the group.

31. *Write the formulas of the oxo anions that the following elements would have if they had the oxidation number and the (hypothetical) total coordination number listed: (a) boron with oxidation number +3 and total coordination number 4; (b) boron with oxidation number +3 and total coordination number 3; (c) arsenic with oxidation number +3 and total coordination number 5; (d) ruthenium with oxidation number +8 and total coordination number 5; (e) iodine with oxidation number +5 and total coordination number 5.

32. Predict the formulas of both the highest fluoro anion (or cation, or neutral compound) and of the highest oxo anion of each of the following elements in the specified oxidation state: (a) C^{4+}; (b) Si^{4+}; (c) Cl^{3+}; (d) Cl^{7+}; and (e) I^{7+}.

33. Draw the Lewis structures of both the highest fluoro anion (or cation, or neutral compound) and of the highest oxo anion of each of the following elements in the specified oxidation state: (a) Xe^{4+}; (b) S^{4+}; and (c) N^{5+}.

34. In addition to the bromate ion formed by bromine with the common oxidation number +5, bromine also forms another oxo anion with an oxidation number higher than +5, and two other oxo anions with oxidation numbers lower than +5. Predict the formulas of each of these three oxo anions.

35. *Explain how and why the basicity of an oxo anion depends on its charge, the number of oxo groups present, and the electronegativity. Use your reasoning to try to predict the approximate basicity of the following more complicated oxo anions: $(O_3P{-}O{-}PO_3)^{4-}$ (pyrophosphate ion); $(O_3P{-}O{-}PO_2{-}O{-}PO_3)^{5-}$ (triphosphate ion).

36. (a) Explain briefly why a given cation might have a *maximum* total coordination number. Tell what this number is for each period of the periodic table (starting with the second period). (b) Give one possible reason why, in another compound, the same cation might only achieve its *penultimate* total coordination number. Tell what this number is for each period of the periodic table (starting with the second period). (c) Which total coordination number (maximum or penultimate) is likely to apply in determining the formulas of *fluoro anions?* Apply the appropriate number to the case of Si^{4+} and predict the formula of the fluoro anion of Si^{4+}. (d) Which total coordination number (maximum or penultimate) is likely to apply in determining the formulas of *oxo anions?* Apply the appropriate number to the case of Si^{4+} and predict the formula of the oxo anion of Si^{4+}. Apply the appropriate number to the case of S^{4+} and predict the formula of the oxo anion of S^{4+}. (e) Explain the difference between the *coordination number* and the *total coordination number* of an element. Illustrate it by giving the actual values of these two numbers for the sulfur atom in the oxo anion of S^{4+}.

37. (a) Draw a reasonable structure of a hydrated oxo anion, and use it to explain what effect increasing the negative charge of the oxo anion has on the basicity of the oxo anion. (b) Although the $[O_2C{-}CH_2{-}CH_2{-}CO_2]^{2-}$ ion has twice the negative charge of the $CH_3CO_2^-$ ion, its basicity is not significantly different. Explain why not.

38. The elements N, Se, and I each exhibit two common positive oxidation numbers; each occurs in an oxo anion. Predict the formulas (including charges) of those oxo anions. List all of the above six oxo anions that would be classified as (a) nonbasic; (b) feebly basic; (c) moderately basic; and (d) very strongly basic.

39. *(a) Write the formula of the oxo anion likely to be formed by the following elements with the oxidation states shown in parentheses: Re(VII); As(III); S(II) (imaginary); Cl(V); Sb(V); P(V). (b) Calculate the pK_b for the formula you have written. (c) Classify the basicity of this anion (weakly basic, etc.)

40. (a) Write the formula of the oxo anion likely to be formed by the following elements with the oxidation states shown in parentheses: P(III); Br(V); Br(III); Si(IV); C(IV); S(IV). (b) Calculate the pK_b for the formula you have written. (c) Classify the basicity of this anion (weakly basic, etc.).

41. *Without using Eq. (3.3), cancel oxo groups and negative charges to select the appropriate category of basicity for each of the following hypothetical oxo anions: (a) MO_5^{2-}; (b) MO_5^{5-}; and (c) MO_7^{6-}. List them in order of increasing basicity.

42. When the use of triphosphates in detergents was under attack, sodium carbonate and sodium silicate were tried as replacements for use in high concentration in detergents. However, concern was expressed that such detergents would be very dangerous if infants should swallow them; the detergents might also be quite corrosive to the washing machines. Explain why.

43. Consider the three oxo anions ReO_4^-, AlO_4^{5-}, and AsO_4^{3-}. (a) Classify the basicity of each oxo anion. (b) The pK_{b1} values of these three anions are -18.2, 2.2 and 22.6 (not necessarily in order). Without actually calculating pK_b values, associate each value with the oxo anion most likely to show it. (c) Solutions of the three anions are found to have pH values of 6.5, 12.0, and 14.0 (not necessarily in order). List each pH with the ion most likely to have a solution of that pH. (d) Can any of these oxo anions *not* exist in water?

44. Arsenic in the +3 oxidation state forms the AsO_3^{3-} oxo anion. (a) Calculate the pK_{b1} for this oxo anion. (b) Classify the strength of this oxo anion. (c) Name this oxo anion. (d) Write the formula of the calcium salt of this oxo anion. (e) Draw the Lewis structure of this anion.

45. Arsenic in the +5 oxidation state forms the AsO_4^{3-} oxo anion. (a) Calculate the pK_{b1} for this oxo anion. (b) Classify the strength of this oxo anion. (c) Name this oxo anion. (d) Write the formula of the thorium(IV) salt of this oxo anion. (e) Draw the Lewis structure of this anion. (f) Name the corresponding fluoro anion, AsF_6^-.

46. *Assuming that the basicity of fluoro anions can be qualitatively evaluated by the same procedure as can that of oxo anions, rank the following known fluoro anions in order of increasing basicity, and identify all which may *not* be nonbasic: ClF_2^-; BeF_4^{2-}; IF_4^-; GaF_6^{3-}; SbF_6^{3-}; PbF_6^{2-}; SbF_6^-; TeF_7^-; TaF_7^{2-}; ZrF_7^{3-}; TaF_8^{3-}; and WF_8^{2-}.

47. Which oxides shown in the predominance diagrams of Figure 2.4 are *amphoteric?*

48. *(a) Sketch predominance diagrams over the pH range 0–14 for each of the following oxo anions and their protonated forms: MO_3^-; MO_3^{2-}; MO_3^{3-}; and MO_3^{4-}. (b) Which of these anions gives rise to hydroxy anion(s)? (c) Which partially protonated form of each anion is expected to predominate at a pH of 14? (d) Classify the acidity of the oxo acids derived from these four anions.

49. (a) Sketch predominance diagrams over the pH range 0–14 for each of the following oxo anions and their protonated forms: MO_6^{4-}; MO_6^{5-}; MO_6^{6-}; and MO_6^{7-}. (b) Which of these anions gives rise to hydroxy anion(s)? (c) Which partially protonated form of each anion is expected to predominate at a pH of 14? (d) Classify the acidity of the oxo acids derived from these four anions.

50. The partially protonated anion $H_4IO_6^-$ is in equilibrium in solution with the anion IO_4^- (plus two water molecules). (a) Classify the basicity of these two ions. (b) Would there be much of a change in pH as this equilibrium was reached? (c) Classify the basicities of the partially protonated anion $H_4TeO_6^{2-}$ and of the (hypothetical) anion TeO_4^{2-}. (d) Would there be much of a change in pH as these two ions interconverted?

51. The sodium salt of the oxo acid *triphosphoric acid* is a major component of detergents; the acid itself has the formula $H_5P_3O_{10}$ and can be written out as $(HO)_2P(=O)–O–P(=O)(OH)–O–P(=O)(OH)_2$. (a) How many oxo groups are present in this acid? Given the number of oxo groups, how would you classify its acidity using the steps in Section 3.7? (b) Actually, its acid strength is very similar to that of H_3PO_4. How can you rationalize this, and what caution would you recommend in applying the steps in Section 3.7 to the case of very large oxo acids?

52. In saturated KOH solution, the acid H_3AsO_3 can be fully neutralized to give the AsO_3^{3-} ion, but the third hydrogen of the acid H_3PO_3 and the second and third hydrogens of the acid H_3PO_2 cannot. Explain.

53. *Classify the acidity of each of the following oxo acids: (a) H_4XeO_6; (b) H_6TeO_6; and (c) a hypothetical oxo acid formed by Te(IV) in which Te has a total coordination number of 6; (d) the oxo acid that may someday be formed by meitnerium in the +9 oxidation state.

54. Molybdic acid is a substantially weaker acid than chromic acid, sulfuric acid, or selenic acid. How might this be explained, given that Mo is in the fifth period, in which the penultimate total coordination number is not always achieved? [See J. J. Cruywagen and J. B. B. Heyns, *J. Chem. Educ.*, **66**, 861 (1989).]

55. Assume that the following elements in the specified oxidation states will form oxo anions: N(III); Mn(VII); Xe(IV). (a) Predict the formula of the oxo anion formed by each element in that oxidation state. (b) Classify the basicity of each oxo anion. (c) Predict whether each element would be found in solution as an oxo anion in a lake of normal pH. (d) Tell whether any oxo anion could never be found in water at any pH, but would only exist in water in a partially protonated form.

56. *For each of the following elements, give its expected positive oxidation number, and give the formula of the form (species) in which you would expect to find it in water of pH 5.5–7. (a) Li; (b) Al; (c) W. Referring to your answers to Exercises 62c and 62d in Chapter 1, do the same for the following: (d) element number 121; (e) Rn; and the elements directly below (f) Np, (g) No, (h) Os, and (i) Pb.

57. A lake near the Oak Ridge National Laboratory in Tennessee has become contaminated with plutonium from the reprocessing of spent fuel from nuclear reactors. Predict whether most of the plutonium in this lake is likely to be dissolved in the water or would be found in the sediments at the bottom of the lake. Also predict whether this might be altered if the lake were strongly subject to the effects of acid rain.

58. Two radioactive elements, technetium (Tc, atomic No. 43) and promethium (Pm, atomic No. 61) do not occur naturally on earth but are found in the fallout from atomic bomb explosions. Predict the formulas of the forms (cation, anion, or oxide) in which each would likely be found in a lake of pH 5.5–7.

59. *The recent concern about the presence of radon is actually mainly concerned with its radioactive decay product, polonium. Assume that polonium in aqueous solution would be present with an oxidation state of +4. (a) Predict whether Po would more likely be present in a lake of pH 5.5–7 as a cation or as an oxide. (b) Predict two possible formulas for the oxo anion of polonium. Tell whether each could be present in a lake of pH 5.5–7.

60. Consider the following cations: Gd^{3+}; Au^{3+}; B^{3+}; Pa^{5+}; Ac^{3+}; and Tl^{+}. (a) Classify the acidity of each of the cations. (b) Describe briefly what will happen when you try to dissolve each of these cations in water. (c) Write the formula of the predominant form that each cation would take in a lake highly polluted with acid rain (pH 3).

61. Among the elements near uranium in the periodic table are neptunium (Np), plutonium (Pu), and protactinium(Pa). (a) Carry out the calculations necessary to predict whether the +3 cations of these elements could appear in natural waters of normal pH as hydrated cations. (b) Which one oxidation state listed for each one of these elements in Table C would give rise to the least basic oxo anion of that element? (c) Assuming that these three oxo anions can exist, predict their formulas (including charges) and give their names. (d) Now determine whether these oxo anions could in fact exist in natural waters. Show your calculations or reasoning. (e) Write the formulas of the calcium salts of each of these three oxo anions.

62. *Consider the oxo anions of the following elements in their +5 oxidation states: nitrogen (N); bromine (Br); phosphorus (P); antimony (Sb); neptunium (Np).

(a) Write the total coordination number (or numbers) you expect each element to show in its oxo anions. (b) Write the expected formula, including charge, of each oxo anion. (If more than one total coordination number is possible, use the most likely one.) (c) Write the name of each oxo anion. (d) List the basicity classification of each oxo anion. (e) Tell whether that oxo anion is likely to be *dissolved* in the water of a lake of normal pH, or whether it will be found in the *sediments* of the lake as an oxide or hydroxide instead. (f) Which oxo anion that you mentioned in part (e) as being soluble in natural waters is actually present in natural waters in partially protonated form (i.e., as a type of hydroxo anion)? Write the formula (including charge) of a partially protonated form of this oxo anion that you would actually expect to find in natural waters.

63. This question concerns the following cations and oxo anions: Rb^+; P^{5+}; La^{3+}; ReO_4^-; AlO_4^{5-}; and AsO_4^{3-}. (a) Draw reasonable structures for the hydrated ions that would be formed by the above cations and anions upon their being plunged into water. Use the drawings to explain why some of these cations or anions could not actually exist in water; tell what products would be produced by their reaction with water. (b) The three cations in the above list have pK_a values of $-33.1, 8.5$, and 14.6. Make the most likely match of each cation with its pK_a. (c) Solutions of the three cations (of similar and reasonable concentrations) have pH values of 0, 5.5, and 6.5. Make the most likely match of each cation with the pH of its solution. (d) The three anions in the above list have pK_b values of $-18.2, 2.2$, and 22.6. Make the most likely match of each anion with its pK_b. (e) Solutions of the three anions (of similar and reasonable concentrations) have pH values of 14.0, 12.0, and 6.5. Make the most likely match of each cation with the pH of its solution. (f) Which of the above cations and anions would you expect to find biochemically available in solution in a lake of normal pH? Which (if any) would only be rendered soluble in a lake highly contaminated by acid rain? by "basic rain"? (g) Suppose that the three anions in the above list were transported to a moon of a planet on which the relative abundance of the elements oxygen and fluorine were reversed. On this moon the lakes would consist of liquid HF, and the oxo anions would be converted to fluoro anions. Assume that this happens unaccompanied by any redox chemistry. Write likely formulas for the fluoro anions of Re, Al, and As. Which one of these is actually found in a real mineral on earth? Compare the basicity of each fluoro anion (in aqueous solution) to the oxo anion of the same element (in aqueous solution).

64. Name each of the following oxo anions, calculate its approximate pK_{b1}, and tell whether its solution in water will be neutral, feebly basic, moderately basic, or so strongly basic that it will react with the water to form a hydroxo anion: (a) CO_3^{2-}; (b) BrO_4^-; (c) IO_6^{5-}; (d) XeO_6^{4-}; (e) AsO_3^{3-}; and (f) IO^-.

65. *Write the formulas of each of the following oxo anions and classify each as nonbasic, feebly basic, moderately basic, or very strongly basic: (a) silicate; (b) tellurate; (c) perbromate; (d) sulfite; (e) hypochlorite; (f) perneptunate; (g) nitrite; (h) ferrate.

66. Write the formula of each of the following oxo acids and tell whether its solution in water will be very strongly acidic, strongly acidic, moderately acidic, or weakly acidic: (a) permanganic acid; (b) selenic acid; (c) arsenious acid; (d) selenous acid; (e) telluric acid; (f) molybdic acid; (g) arsenic acid; (h) ferric acid; (i) hypobromous acid; and (j) hyposelenous acid (hypothetical).

67. *Name each of the following salts: (a) UO_2SO_4; (b) $TiCl_3$; (c) SO_2Cl_2; (d) $FePO_4$; (e) Ag_5IO_6; and (f) Hg_3TeO_6.

68. *Give the formulas of the following salts: (a) bismuthyl nitrate; (b) strontium perchlorate; (c) europium(II) sulfite; (d) iron(II) phosphate; (e) chromium(II) carbonate; (f) cesium phosphate; (g) zinc(II) perbromate; (h) potassium perbromate; (i) calcium bromite; and (j) calcium borate.

69. Name each of the following salts: (a) $Sr(BrO_3)_2$; (b) $Au_2(SeO_4)_3$; (c) $UO_2(NO_3)_2$; (d) $Hg_5(IO_6)_2$; (e) $TlIO_3$; (f) $Zn_5(NpO_6)_2$; (g) $TlClO$; (h) Sn_3N_2; (i) $Al(RuO_4)_3$; (j) $HgSO_3$; (k) K_8SnO_6; (l) $Sr(ClO)_2$; and (m) K_4SnO_3.

70. Write the formulas of the following salts: (a) thorium(IV) phosphate; (b) barium perxenate; (c) aluminum(III) sulfite; (d) magnesium sulfite; (e) cerium(III) pertechnetate; (f) molybdenum(IV) sulfide; (g) calcium phosphate; (h) calcium nitrite; (i) silver(I) perxenate; (j) thorium(IV) chlorite; (k) iron(III) silicate; and (l) chromium(III) silicate.

71. (a) Name each of the following anions: MnO_4^-; NpO_6^{5-}; WF_8^{2-}; and IO^-. (b) Classify each of the above anions as nonbasic, feebly basic, moderately basic, or very strongly basic. (c) Calculate pK_{b1} for each oxo anion. (d) Calculate the pK_b for the partially protonated oxo anion $H_4NpO_6^-$. (e) Which of these five ions could be present in a lake of pH 3–10?

72. Consider an oxo anion with the formula MO_3^{2-}. (a) Classify the basicity of this ion (as weakly basic, etc.) (b) This oxo anion will also form a partially protonated (hydroxo) anion of formula HMO_3^-. Classify its basicity. (c) Which (if any) Group 14(IV) elements would give an oxo anion with this formula? Give the name of this oxo anion or these oxo anions (if any). (d) Which Group 16(VI) elements would give an oxo anion with this formula? Give the name(s) of this oxo anion or these oxo anions. (e) If the element Xe gave such an oxo anion (i.e., XeO_3^{2-}), what would this ion be named?

73. If you did not do Experiments A.2 and A.3, go back and try to apply the principles of this chapter to predict what would have happened in each experiment. Also answer the questions included in each experiment.

74. Draw Lewis structures of the following molecules or ions: (a) H_2O_2; (b) HCN; (c) $H_2N–NH_2$; (d) $H_2N–OH$; (e) $H_3C–SiH_3$; and (f) $[H_3C–C(=O)(–O–O)]^-$.

75. *Draw Lewis structures of the following molecules or ions: (a) $HCCl_3$; (b) NSF; (c) C_2H_6; (d) C_3H_7Cl; (e) $Fe(CO)_5$; and (f) $Cl–CH_2–SiH_2–SiH_2–CH_2Cl$.

76. Draw Lewis structures of the following molecules or ions: (a) HOF; (b) Si_2H_6; (c) Si_3H_8; (d) CSe_2; (e) $[V(CO)_6]^-$; and (f) $ClCH_2–CH_2Cl$.

77. Assign oxidation numbers to each atom in the molecules and ions of Exercise 74.

78. Assign oxidation numbers to each atom in the molecules and ions of Exercise 75.

79. Assign oxidation numbers to each atom in the molecules and ions of Exercise 76.

80. Assign oxidation numbers to the central metal (or nonmetal) atom in each of the following complex ions or compounds: (a) $[Ru(NH_3)_5I](ClO_4)_3$; (b) $[SbCl_5(OH_2)]$; (c) UO_2^{2+}; and (d) $Ba_3[AlF_6]_2$.

Notes

1. When dealing with atomic orbitals that have inner lobes, only the sign of the outermost part of the wave function is important, since overlap between orbitals from different atoms never extends farther into any orbital than the outermost lobe.
2. G. A. Gallup, *J. Chem. Educ.*, **65**, 671 (1988).
3. Conventionally, this is thought to be repulsion between the unshared electron pairs on neighboring atoms, but many authors (e.g., R. T. Sanderson, *Polar Covalence*, Academic, New York, 1983, Chapter 5) contend that it is due to repulsion between an unshared pair and the bond pair of electrons.
4. R. J. Gillespie and E. A. Robinson, *Inorg. Chem.*, **31**, 1960 (1992).
5. This process measures more than just the strength of the π bond, since when it is broken, the remaining σ bond is excessively short, and must lengthen, altering the energy balance.
6. M. W. Schmidt, P. N. Truong, and M. S. Gordon, *J. Am. Chem. Soc.*, **109**, 5217 (1987).
7. The quadruple bonding in the latter is actually so weak that it can be disrupted merely by changing the coordination environment of another ion in its crystal lattice: S. Hao, S. Gambarotta, and C. Bensimon, *J. Am. Chem. Soc.*, **114**, 3556 (1992).
8. J. P. Collman and H. J. Arnold, *J. Cluster Sci.*, **5**, 37 (1994).
9. J. I. Musher, *Angew. Chem. Intl. Ed. Engl.*, **8**, 54 (1969).
10. There are many cases in which covalent bonds can be formed either using two electrons from one atom and none from the other, or using one electron from each atom, and the bonds have the same properties regardless of which way they form. A. Haaland [*Angew. Chem. Intl. Ed. Engl.*, **28**, 992 (1989)] has proposed that bonds can, however, be unambiguously classified as covalent or coordinate covalent based on how they *break apart* in the gas phase. With this distinction, a number of differences in the two types of bond emerge, which are beyond the scope of the present discussion.
11. Krypton and xenon atoms have been reported as ligands in short-lived coordination complexes such as $[Cr[CO)_5Xe]$: M. B. Simpson, M. Poliakoff, J. J. Turner, W. B. Maier III, and J. G. McLaughlin, *J. Chem. Soc., Chem. Commun.*, 1355 (1983); B. H. Weiller, *J. Am. Chem. Soc.*, **114**, 10910 (1992).
12. R. J. Kulawiec and R. H. Crabtree, *Coord. Chem. Rev.*, **99**, 89 (1990); T. D. Newbound, M. R. Colsman, M. M. Miller, G. P. Wulfsberg, O. P. Anderson, and S. H. Strauss, *J. Am. Chem. Soc.*, **111**, 3762 (1989).
13. A software simulation of the historic controversy between Werner and Jørgensen is available: D. M. Whisnant, *J. Chem. Ed. Software*, Vol. 1D, No. 2 (1993); see also *J. Chem. Edue.*, **70**, 902 (1993).
14. Theoreticians have predicted the possibility of fluoro anions of C^{4+} and N^{5+} in which these "cations" have a coordination number of 6: C. S. Ewig and J. R. Van Wazer, *J. Am. Chem. Soc.*, **112**, 109 (1990).
15. A. Vogler and H. Nikol, *Comments Inorg. Chem.*, **14**, 245 (1993).
16. W. L. Jolly, *Modern Inorganic Chemistry*, McGraw-Hill, New York, 1984; Chapter 3.
17. Unexpectedly, very unstable ClF_6^- and more stable BrF_6^- ions have been prepared. K. O. Christe, W. W. Wilson, R. V. Chirakal, J. C. P. Sanders, and G. J. Schrobilgen, *Inorg. Chem.*, **29**, 3506 (1990); K. O. Christe and W. W. Wilson, *Inorg. Chem.*, **28**, 3275 (1989).
18. A.-R. Mahjoub and K. Seppelt, *Angew. Chem. Int. Ed. Engl.*, **30**, 323 (1991).
19. R. J. Gillespie and I. Hargittai, *The VSEPR Model of Molecular Geometry*, Allyn and Bacon, Boston, 1991; R. F. W. Bader, R. J. Gillespie, and P. J. MacDougall, *J. Am. Chem. Soc.*, **110**, 7329 (1988).
20. A.-R. Mahjoub, T. Drews, and K. Seppelt, *Angew. Chem. Int. Ed. Engl.*, **31**, 1036 (1992); K. Seppelt, *Comments Inorg. Chem.*, **12**, 199 (1991); K. O. Christe, J. C. P. Sanders, G. J. Schrobilgen, and W. W. Wilson, *J. Chem. Soc. Chem. Commun.*, 837 (1991).
21. Sometimes authors describe the geometry *including* the (unseen) electron pairs: For example, if a molecule has a total coordination number of 6, which includes some unshared electron pairs, these authors may describe the compound as *pseudo*-octahedral, abbreviated as Ψ-octahedral.
22. Although theoretical calculations indicate that this assumption is incorrect [M. B. Hall, *J. Am. Chem. Soc.*, **100**, 6333 (1978); *Inorg. Chem.*, **17**, 2261 (1978)], it is so much easier to use than more "correct" models in predicting observed results that chemists continue to use it.

23. Gillespie and Robinson[4] suggested that lone pairs take *twice* as much space in the third and fourth periods, and *three times* as much space in the fifth period. If this proves to be generally applicable, the definition of total coordination number given earlier in this chapter would need a more complicated redefinition.
24. K. O. Christe, E. C. Curtis, D. A. Dixon, H. P. Mercier, J. C. P. Sanders, and G. J. Schrobilgen, *J. Am. Chem. Soc.*, **113**, 3351 (1991).
25. A. Ellern, A.-R. Mahjoub and K. Seppelt, *Angew. Chem. Int. Ed. Engl.*, **35**, 1123 (1996).
26. A.-R. Mahjoub and K. Seppelt, *Angew. Chem. Int. Ed. Engl.*, **30**, 876 (1991); for other explanations see O. Knop, A. Linden, B. R. Vincent, S. C. Choi, T. S. Cameron, and R. J. Boyd, *Can. J. Chem.*, **67**, 1984 (1989) and M. Kaupp, Ch. van Wüllen, R. Franke, F. Schmitz, and W. Kutzelnigg, *J. Am. Chem. Soc.*, **118**, 11939 (1996).
27. H. A. Bent, *Chem. Rev.*, **61**, 275 (1961).
28. M. Gerloch, *Coord. Chem. Rev.*, **99**, 117 (1990); D. L. Kepert, *Inorganic Stereochemistry*, Springer, Berlin, 1982.
29. T. P. Hanusa, *Polyhedron*, **9**, 1345 (1990); R. L. DeKock, M. A. Peterson, L. K. Timmer, E. J. Baerends, and P. Vernooijs, *Polyhedron*, **9**, 1919 (1990).
30. R. J. Gillespie, I. Bytheway, T.-H. Tang, and R. F. W. Bader, *Inorg. Chem.*, **35**, 3954 (1996); Z. Lin and I. Bytheway, *Inorg. Chem.*, **35**, 594 (1996).
31. V. Jonas, C. Boehme, and G. Frenking, *Inorg. Chem.*, **35**, 2097 (1996).
32. J. M. Mayer, *Comments Inorg. Chem.*, **8**, 125 (1988).
33. In the *p* block, this may happen to a very limited extent: Iodine(VII) occasionally forms the IO_4^- ion instead of the more common IO_6^{5-} ion.
34. There does seem to be a reluctance to adopt the total coordination number 5.
35. Contrast the points of view, each rooted in a different computational model, in L. Suidan, J. K. Badenhoop, E. D. Glendening, and F. Weinhold, *J. Chem. Educ.*, **72**, 583 (1995) and that in D. K. Straub, *J. Chem. Educ.*, **72**, 889 (1995).
36. See a textbook of organic chemistry for more details on principles used to choose the most appropriate resonance structures: for example, T. H. Lowry and K. S. Richardson, *Mechanism and Theory in Organic Chemistry*, 3rd ed., Harper & Row, New York, 1987; pp. 1–10.
37. For a contrary point of view, see D. G. DeWit, *J. Chem. Educ.*, **71**, 750 (1994).
38. Additional hydration energies of oxo anions may be found in Y. Marcus, *Ion Solvation*, 5th ed. Wiley-Interscience, Chichester, UK, 1985; pp. 107–109.
39. This and the following conclusions concerning pK_b are derived from generalizations given for pK_a values of oxo acids: F. A. Cotton and G. Wilkinson, *Advanced Inorganic Chemistry: A Comprehensive Text*, 5th ed., Wiley-Interscience, New York, 1988; p. 104.
40. B. H. J. Bielski, *Free Rad. Res. Comms.*, 12–13, 469 (1991).
41. A useful procedure for the direct derivation of the formulas of the oxo acids is found in N. K. Kildahl, *J. Chem. Educ.*, **68**, 1001 (1991).
42. This is a variation on the Ricci equation: J. E. Ricci, *J. Am. Chem. Soc.*, **70**, 109 (1948).
43. R. Baum, "Adding Iron to Ocean Makes Waves As Way to cut Greenhouse CO_2," *Chem. Eng. News*, July 2, 1990. p. 21.
44. R. Rawls, "Ironing the Ocean," *Chem. Eng. News*, Nov. 4, 1996, p. 40.
45. R. B. Martin, *Acc. Chem. Res.*, **27**, 204 (1994).
46. A complication occurs in the *d* block, where the oxidation number can be one less than the group number. Here one often sees MnO_4^{2-} called "manganate" and RuO_4^- called "perruthenate;" for clarity in such cases it may be best to use the IUPAC system.
47. In the rare case in which the donor atom is *less* electronegative than the central metal atom (as in some hydrides), the assignment of oxidation numbers is ambiguous: The usual procedure seems to be to give the electron pair to the most electronegative atom if the central atom is from the *p* block, but to give the unshared electron pair to the hydride ligand if the metal ion is from the *d* block.

CHAPTER

4

Ionic Solids and Precipitation Reactions of Hydrated Ions

With Applications to Analytical, Organic, and Water Chemistry

4.1 Solubility Rules for Salts of Oxo and Fluoro Anions

In Chapters 2 and 3, we examined the acid–base properties cations and anions in aqueous solution. In general, we did not consider possible interactions between cations and anions present in the same solution. In Experiment A.1, the solutions contained a nonbasic anion, Cl^-, that did not affect the chemistry of the cations being studied; in Experiment A.3, the nonacidic cation present, Na^+, did not appreciably alter the reactions of the anions. But, as you may see in Experiment A.4, many combinations of cations and anions react with each other to give insoluble precipitates.

It is important to be able to anticipate when a precipitation reaction will occur in a solution containing several cations and anions. This reaction happens, for example, in the body when the products of the concentrations of certain cations and anions come to exceed the solubility products of their salts, as when calcium ions and oxalate ions precipitate as calcium oxalate, one form of kidney stones. A laboratory worker preparing a buffered solution of a certain metal ion for biological or medical studies may be upset to find the metal ion giving a precipitate with the anion of the buffering agent. Conversely, a worker preparing a solution of a metal ion or of an oxo anion nearly always does this by dissolving an ionic salt in water. In preparing a standard solution of the barium ion for use in analytical chemistry, he or she may go to the stockroom to find a barium salt, and find not one, but many different salts. They certainly do not want to select a salt that is insoluble in water! If a table containing the relevant solubility products (K_{sp} values) is at hand, there is no problem; but often this is not the case. Qualitative knowledge of the solubility (precipitation) properties of inorganic salts is then quite useful.

Unfortunately, the solubility properties of ionic salts in water present certain complexities not found in the solubilities of other classes of materials. There is a general saying, often encountered in general chemistry, that Like dissolves like. Nonpolar covalent materials quite generally dissolve in nonpolar covalent solvents; polar hydrogen-bonding molecules generally dissolve in hydrogen-bonding solvents. Ionic salts are usually not used as solvents, however, because they are not generally liquids below inconveniently high temperatures. Unless some specific bonding interaction is present, ionic salts will not dissolve in nonpolar covalent solvents. But with respect to the most important polar covalent solvent, water, no simple blanket statement can be made—many ionic salts are soluble in water, and many are not.

The need to be able to anticipate the solubility properties of salts leads many general chemistry textbooks to present a series of solubility rules, arranged by an anion: All sulfates are soluble except those of Ca^{2+}, Sr^{2+}, Ba^{2+}, Hg^{2+}, Pb^{2+}, and Ag^{+}, and so on. Due to the limited appeal of memorizing rules such as these, the rules do not cover all the cations, and cover only a few anions, so even if the student remembers these rules after the exam, they may not cover the situation at hand. A more reliable source of information is a chemical handbook such as the *Handbook of Chemistry and Physics*,[1] which has an extensive tabulation of "Physical Constants of Inorganic Compounds". Even this tabulation can cover only a fraction of all the possible or known inorganic salts, however. Thus there is a need for *generalized solubility rules* that can be easily applied to predict the solubility of a wide variety of inorganic salts.

At this point your instructor may want you to carry out (or observe and discuss) Experiment A.4, which gives observations from which you may be able to devise generalized solubility rules for salts of oxo and fluoro anions. Although the principles are relevant for salts of anions with donor atoms other than O and F, there is another principle (Chapter 5) that also comes into play with other donor atoms. Thus the principles in this chapter are not sufficient by themselves for complete prediction of the solubilities of salts of anions other than oxo and fluoro anions.

The results of experiments such as Experiment A.4 are best organized (Fig. 4.1) by the acidity classifications of the cations and the basicity classifications of the anions being combined to form a given (soluble or insoluble) salt. When this is done, patterns emerge: Precipitation of insoluble salts occurs with cation–anion combinations that fall in the shaded regions of Figure 4.1; these are mainly in the lower right part of the diagram but also occur along the diagonal line representing matching of the reactivity classifications of the cation and anion. No precipitation occurs (salts are soluble) in the clear regions at the lower left and upper right. Less consistent results (some precipitation, some cases of salts remaining in solution) occur in the remaining borderline regions (partially shaded). A generalized solubility rule or tendency can be written for each region, based on the reactivity classifications of the cations and anions involved.[2]

Solubility Rule I applies in Region I of Figure 4.1 to combinations of acidic cations plus basic anions. Since most metal cations found in aqueous solution are at least weakly acidic, and most oxo anions are at least moderately basic, the reaction of *acidic cations and basic anions to give insoluble salts* is an important general property. By *acidic cations* we mean any weakly, moderately, strongly, or very strongly acidic cations. The term *basic anions* refers to any anions that are weakly, moderately, strongly, or very strongly basic. (Not all of these categories are found among simple oxo anions, of course.)

This generalized solubility rule approximately sums up two commonly given *solubility rules*. (a) All carbonates are insoluble except those of the Group 1 elements and NH_4^+; and (b) all hydroxides are insoluble except those of the Group 1 elements, Sr^{2+}, and Ba^{2+}. The generalized solubility rule gives up some exactness in terms of the slightly different behavior of hydroxides and carbonates, but it is clearly applicable to the salts of a number of other important oxo anions, such as phosphate, arsenate, silicate, and borate.

ANIONS: / CATIONS:	Nonbasic	Feebly basic		Moderately basic		Very strongly basic
Nonacidic	III	II A				
Feebly acidic	II B	IV C	IV A			
Weakly acidic		IV B	I			
Moderately acidic						
Strongly acidic						
Very strongly acidic						

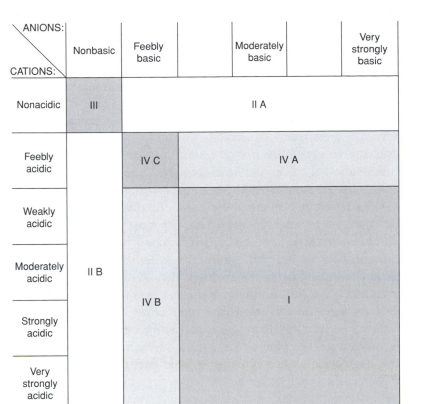

Figure 4.1
Characteristic pattern of precipitation on mixing 1 N (0.33–1 M) solutions of cations with 1 N solutions of oxo anions. Shaded regions represent combinations that give precipitates; in unshaded regions no precipitation occurs; in partially shaded regions some combinations give precipitates and others do not.

Solubility Rule II applies in regions IIA and IIB of Figure 4.1 to *cross-combinations* of ions: nonacidic cations plus all other basic anions (including even feebly basic anions) in region IIA; acidic cations (including even feebly acidic cations) plus nonbasic anions in region IIB. These *cross-combinations give soluble salts*. The most common solubility rule related to this rule of thumb is that for nitrate: All nitrates are soluble. Since most cations are acidic, their soluble oxo salts will be of nonbasic anions such as nitrate, perchlorate, and the fluoro anions. Provided that the principles of Chapter 5 do not interfere, this rule also covers the solubility of most chlorides, bromides, and iodides. Salts of these anions are the ones you would choose in order to make up solutions of most of the metal cations for use in experiments. Conversely, if you need a solution of a given oxo or fluoro anion, you would dissolve its salt with a nonacidic cation (generally Na^+ or K^+ for economic reasons).

Solubility Tendency III applies in region III of Figure 4.1 to combinations of nonacidic cations plus nonbasic anions. The generalized solubility tendency in this region is that *large nonacidic cations and large nonbasic anions give insoluble salts*. This is a tendency rather than a rule, since the driving force here (as we shall see in Section 4.2) is not strong enough to give *highly* insoluble salts. (Most combinations in this region have solubilities less than that required to give standard concentrations of 1 *M*; but our usual qualitative perception of the limits of solubility are closer to 0.1 *M*, and many of these combinations give solubilities between these numbers.) If you performed Experiment A.4, you may have also noticed this qualitative difference: In region I, due to their highly insoluble nature, the precipitates form immediately upon mixing, which does not allow time for visibly crystalline material to grow. (Only very tiny crystals are formed.) But in region III, the precipitation often does not immediately begin, which allows time for the growth of large crystals.

This tendency to crystallize out of solution or precipitate is almost completely absent for the smallest of the nonacidic cations, Na^+ (hence, the specific solubility rule that all sodium salts are soluble), and the smallest of the nonbasic anions, NO_3^-, Cl^-, Br^-, and I^-. This tendency becomes more pronounced with Cs^+ salts and perchlorates ($CsClO_4$ itself precipitates readily), and it becomes most pronounced (and most useful) with the salts of *complex cations* and *complex anions*, which are generally very large and nonacidic and very large and nonbasic, respectively. When inorganic chemists synthesize a new complex cation in solution, they generally crystallize it out of the solution by adding a solution of a large nonbasic anion such as ClO_4^- **(WARNING: See Chapter 6 for a reason why this choice often is very hazardous)** or BF_4^- or PF_6^- (much safer choices); even nitrate or iodide may work well if used in excess (so that the solubility is reduced by the common ion effect). The same principle is used by organic natural products chemists, when isolating organic nitrogen compounds (amines, alkaloids, etc.): They extract their organic nitrogen compound into an aqueous strong acid, which converts the compound into a large organic *ammonium* cation. This cation is then precipitated by a suitable large nonbasic anion. Conversely, new complex or large organic anions are best crystallized out of solution by adding a large nonacidic cation such as Cs^+ or $[(CH_3)_4N]^+$, $[(C_2H_5)_4N]^+$, and so on. Other things being equal, it helps minimize the solubility (and maximize the yield) to match, not only the size of the cation and anion, but also to match the charges: For example, to precipitate a complex -3 charged anion with a complex $+3$ charged cation.[3]

Example 4.1

The most highly symmetrical of all nonlinear polyatomic ions is the icosahedral ion $B_{12}H_{12}^{2-}$. Estimate the basicity of this ion using the principles you would for an ion of formula $B_{12}F_{12}^{2-}$. If you prepared this ion in high concentration in solution, which one of the following ions would be the best to add to form a crystalline precipitate with this anion? Li^+, Be^{2+}, Al^{3+}, $[(CH_3)_4N]^+$, $[C_2H_4N_2(CH_3)_6]^{2+}$, $[Co(NH_3)_6]^{3+}$, ClO_4^-, SiF_6^{2-}, AlF_6^{3-}.

SOLUTION:

A large fluoro anion of formula $B_{12}F_{12}^{2-}$ would certainly be quite nonbasic; even though H is not as electronegative as F, we might suppose that $B_{12}H_{12}^{2-}$ might also be nonbasic. To precipitate this nonbasic anion we need a nonacidic cation, of which we have three choices: $[(CH_3)_4N]^+$, $[C_2H_4N_2(CH_3)_6]^{2+}$, or $[Co(NH_3)_6]^{3+}$. Of these, the best would likely be the one with the charge that matches that of our nonbasic anion, $[C_2H_4N_2(CH_3)_6]^{2+}$.

Solubility Tendency IV applies to salts containing feebly acidic cations (in region IVA of Fig. 4.1) and/or feebly basic anions (in region IVB). The driving forces (to be discussed in the remainder of the chapter) favoring insolubility in region I and solubility in region II overlap with each other and come into conflict in region IV. The solubility of salts of these ions is particularly complex and difficult to predict. In region *IVA*, *precipitation* usually prevails, but oxides and hydroxides of feebly acidic cations are (at least somewhat) soluble, as we saw in Chapter 2. In region *IVB*, *solubility* usually prevails, but, for example, sulfates of some large, weakly acidic cations also show low solubility. (When dealing with salts in these regions, you are well advised to check a chemistry handbook for solubilities.) The clearest result occurs at the intersection of these two regions, region IVC: *large, feebly acidic cations* (such as Ba^{2+}) *give precipitates with large, feebly basic anions* (such as $SO_4{}^{2-}$ and $SeO_4{}^{2-}$).[4] (Thus, in gravimetric determinations in the quantitative analysis course, barium and sulfate ions are always precipitated with each other.) This tendency applies somewhat to the fluoride ion itself, although with $pK_b = 10.85$ it is actually just barely *weakly* basic: LiF is insoluble, and the most insoluble[5] fluoride is CaF_2, widely found in nature as the mineral fluorite.

Example 4.2

Without reference to a table of solubility products, classify each of the following salts as soluble or insoluble in water: Ag_2SeO_4, K_3PO_4, $ZnSO_4$, and $Th_3(PO_4)_4$.

SOLUTION:

The charges of the cations and anions in each salt must be identified so that their acidity and basicity can be classified. The cations are Ag^+, K^+, Zn^{2+}, and Th^{4+}. Referring to Table C for radii, we compute Z^2/r ratios for these four cations: 0.008, 0.007, 0.046, and 0.145, respectively, which allows the first two to be categorized as nonacidic, the third as weakly acidic, and the fourth as moderately acidic. The table of electronegativities, however, shows that Ag has an electronegativity exceeding 1.8, so Ag^+ should be reclassified as feebly acidic.

The anions are $SeO_4{}^{2-}$, $PO_4{}^{3-}$, and $SO_4{}^{2-}$. Any oxo anion with a -3 or higher charge (such as phosphate) is automatically at least moderately basic. A calculation using Eq. (3.4) (or reference to Table 3.4) classifies sulfate and selenate as feebly basic; hence, the salt Ag_2SeO_4 is a combination of a feebly acidic cation and a feebly basic anion and should be insoluble; K_3PO_4 is a combination of a nonacidic cation and a basic anion and should be soluble; $ZnSO_4$ is a combination of an (weakly) acidic cation and a feebly basic anion (region IVB) and should probably be soluble; $Th_3(PO_4)_4$ is a combination of an (moderately) acidic cation and a (moderately) basic anion and should be insoluble. Check in a *Handbook of Chemistry and Physics* to confirm all of these predictions.

In addition, it is perhaps useful here to point out that the *"solubilities" of insoluble salts of basic anions will be enhanced in solutions of strong acids.* This solubility enhancement is not a true case of solubility, perhaps, but of reaction: A basic anion tends to react more completely with the strong acid H_3O^+ than with a less acidic metal ion, thus generating a new salt of that metal ion and a weak acid in solution. The equilibrium that results from the competition between H_3O^+ and the metal ion will be especially enhanced if the weak oxo acid decomposes; thus carbonates and sulfites readily "dissolve" in strong acids since the carbonic and sulfurous acids produced decompose to carbon dioxide and sulfur dioxide, gases that escape from the solution and allow the equilibrium to shift further toward dissolution.

4.2 Thermodynamics and Solubility Rules for Salts

Although it is useful to identify reactivity trends such as the generalized solubility rules and tendencies of Section 4.1, we soon want to push our level of understanding further and ask questions such as, Why is the tendency to react (i.e., to precipitate) greater in some cases than in others? Questions such as this can be answered in several ways; often valuable clues as to what is happening may be obtained by analyzing thermodynamic data on the equilibrium involved, which in this case is the precipitation reaction (4.1) (or of its reverse, the process of dissolution of an ionic salt):

$$y\,M^{m+}(aq) + m\,X^{y-}(aq) \rightleftharpoons M_yX_m(s) + p\,H_2O \tag{4.1}$$

The thermodynamic measure of the tendency of a reaction to proceed to products (to be spontaneous) or not to proceed to products (to be nonspontaneous) is the **Gibbs free energy change**, ΔG. If the reaction system tends to go to the right [to give products, i.e., a precipitate in Reaction (4.1)], it will show a *negative* value of ΔG (it will also be characterized by an equilibrium constant, K_{eq}, that is >1). A chemical reaction that does *not* tend to proceed to products shows the opposite characteristics, that is, $\Delta G > 0$ (and $K_{eq} < 1$). Such a reaction tends to go in the reverse direction [i.e., in Reaction (4.1), the salt listed on the right will dissolve].

On the level of chemical thermodynamics, two factors are connected with the tendency of a reaction to proceed to products (to be spontaneous) or not to proceed to products (to be nonspontaneous). The first of these factors is that *products are more likely to be favored if they are lower in energy than the reactants.* For example, if the products of a reaction contain stronger covalent bonds than the reactants, the reaction is more likely to favor products. If the reaction is carried out in an open container (at constant pressure), the energy change can be measured as an evolution of heat by the reacting system and is known as the **enthalpy change** of the reaction, ΔH (or $\Delta H°$, if measured with standard conditions and concentrations). A system that evolves heat (and is thus more likely to proceed to products) is said by convention to have a *negative enthalpy change* ($\Delta H < 0$).

However, not all spontaneous reactions involve the liberation of heat by the reacting system. For example, ice tends to melt at room temperature, but a melting ice cube certainly does not give off heat! There must be another reason for reactions to proceed, one that sometimes can override the tendency to go to lower energy products. This tendency of reactions to proceed in the direction of products that are more *disordered* or *random* is typical. It is measured thermodynamically as the **entropy change**, ΔS, of the system. A system that proceeds toward greater disorder or a more random state is said by convention to exhibit a *positive entropy change* ($\Delta S > 0$).

We find that the amount of heat evolved by a reaction (its enthalpy change) and the degree of disorder produced (its entropy change) are not very much affected by changes in temperature, but reaction tendencies are affected. In particular, at low temperatures the tendency toward heat evolution usually determines whether the reaction will proceed, while at higher temperatures the tendency (or lack of it) toward increasing disorder becomes more and more significant. Thus the relationship between the two tendencies involves not only ΔH and ΔS, but also the temperature T (in the kelvin scale). In particular, the free energy change, which indicates whether the reaction tends to proceed or not, is related to these three parameters by the equation

$$\Delta G = \Delta H - T\Delta S \tag{4.2}$$

It is often useful to determine whether the sign of ΔG is governed predominantly by the sign of the ΔH term or by the sign of the $-T\Delta S$ term. At high enough temperatures, the $-T\Delta S$ term must predominate; all complex molecules are unstable at very high temperatures and tend to decompose to large numbers of smaller gaseous molecules or atoms, which spread out over a larger volume of space (at constant pressure). But at room temperature we usually find that the ΔH term is larger in magnitude than the $-T\Delta S$ term, so we usually concentrate our attention on factors such as the strengths of the covalent bonds involved. However, if we find the $-T\Delta S$ term to be larger in magnitude at room temperature, we are alerted to the fact that something is going on in the reaction that causes an unusual degree of disorder (such as a production of a large number of gas molecules).

Thermodynamics of Precipitation Reactions. Such is often the case in precipitation reactions. If you performed Experiment A.4 and felt the test tubes after mixing the solutions, you normally found that there is little heat evolution or absorption involved in the precipitation (or dissolution) of an ionic salt; hence, in analyzing the thermodynamics of this reaction, it is necessary to consider both the enthalpy and entropy change. Thermodynamic data for the precipitation of some salts is shown in Table 4.1.

We may note the following thermodynamic tendencies from the data in Table 4.1.

1. Insoluble salts in region I (and sometimes in region IV) are insoluble for reasons that manifest themselves in the entropy term (ΔG is negative because $-T\Delta S$ is negative); the *precipitation* of these salts is accompanied by *increasing disorder*.

2. Soluble salts (in region II and sometimes in region IV) are soluble for reasons that manifest themselves in the enthalpy term [ΔG is positive for Reaction (4.1) because ΔH is positive]; the *dissolution* of these salts is accompanied by the *evolution of heat energy*.

3. Insoluble salts in region III are insoluble for reasons that manifest themselves in the enthalpy term; the *precipitation* of these salts is accompanied by the *evolution of heat energy*.

4. In regions III and IV, the effects of the ΔH and the $-T\Delta S$ terms oppose each other but are similar in magnitude. The opposed effects sometimes cause the salts to be only somewhat insoluble, and allow exceptions to occur, which is why we spoke of tendencies rather than rules in these regions.

These are tantalizing hints to a variety of processes that must be going on in such a "simple" reaction as precipitation (or dissolution) of an ionic solid. Thermodynamics cannot, however, determine *why* disorder increases with one set of cations and anions in one region while heat energy is released in another; other physical methods must be

Table 4.1
Thermodynamic Data on Precipitation[a]

Salt	ΔG°	ΔH°	$-T\Delta S^\circ$ (298 K)	Solubility (mol kg^{-1} H_2O)
		Rule I. Acidic cations + basic anions → precipitates		
$Be(OH)_2$	−121	−31	−90	0.000008
$Mg(OH)_2$	−63	−3	−61	0.0002
$Ca(OH)_2$	−28	16	−44	0.025
Li_2CO_3	−17	18	−34	0.18
$MgCO_3$	−45	28	−74	0.0093
$CaCO_3$	−48	10	−57	0.0002
$SrCO_3$	−52	3	−56	0.00007
$BaCO_3$	−47	−4	−43	0.00011
$FePO_4$	−102	78	−180	Slight
		Rule II. Cross combinations → soluble salts		
		IIA. Nonacidic cations + basic anions		
KOH	62	55	7	19.1
RbOH	74	63	11	17.6
CsOH	83	71	12	26.4
K_2CO_3	36	35	1	8.12
Rb_2CO_3	50	41	9	19.5
Cs_2CO_3	73	62	10	8.0
		IIB. Acidic cations + nonbasic anions		
$Mg(NO_3)_2$	89	85	4	1.65
$Ca(NO_3)_2$	32	20	13	2.08
$Sr(NO_3)_2$	−3	−18	14	1.89
$Ba(NO_3)_2$	−13	−40	27	0.33
$Mg(ClO_4)_2$	144	141	4	2.24
$Ba(ClO_4)_2$	46	12	34	5.91
		Tendency III. Nonacidic cations + nonbasic anions → precipitates		
$KClO_4$	−12	−51	39	0.054
$RbClO_4$	−14	−57	43	0.027
$CsClO_4$	−12	−55	44	0.034
$NaNO_3$	6	−21	27	8.59
		Tendency IV. Feebly basic anions (sulfates)		
		IVB. Sulfates of acidic cations → usually soluble salts		
$Al_2(SO_4)_3$	96	338	−241	0.92
$BeSO_4$	59	123	−64	2.40
$MgSO_4$	30	91	−61	2.88
		IVC. Sulfates of feebly acidic cations → precipitates		
$CaSO_4$	−27	18	−45	0.014
$SrSO_4$	−34	9	−43	0.0006
$BaSO_4$	−50	−19	−31	0.00001

SOURCES: C. S. G. Phillips and R. J. P. Williams, *Inorganic Chemistry*, Oxford University Press, Oxford, UK, 1965; p. 254; D. A. Johnson, *Some Thermodynamic Aspects of Inorganic Chemistry*, Cambridge University Press; Cambridge, UK, 1968; p. 107; B. G. Cox and A. J. Parker, *J. Am. Chem. Soc.*, **95**, 6879 (1973); *Handbook of Chemistry and Physics*, 36th ed., Cleveland, Chemical Rubber Publishing, 1954, p. 1682. Solubilities listed are often for the hydrated salts, are for temperatures between 0 and 30 °C, and are calculated from data in the *Handbook of Chemistry and Physics*.

[a] Numbers given in the table are the standard free energies, enthalpies, and entropies of precipitation [i.e., for Reaction (4.1)] of the salts listed, in kilojoules per mole (kJ mol^{-1}).

employed for this. The results of these investigations are given in the following sections, with emphasis on the reasons in the "well-behaved" regions I and II, and the synthetically important region III.

4.3 Entropy and Precipitation: The Structure of Hydrated Ions

It may seem surprising at first that precipitation reactions of acidic cations and basic anions should be due to a negative entropy term, $-T\Delta S$, because this term corresponds to increasing randomness or disorder in the system, which is not what we expect in a reaction that produces a crystalline solid precipitate. But the reacting cations and anions exist as hydrated ions, and upon formation of the precipitate, water molecules are released. If enough water molecules are released (and if these molecules are sufficiently disordered as part of the water solvent), then the resulting positive entropy change (negative $-T\Delta S$) may exceed the ordering effect of producing a crystalline precipitate.

From Table 4.1, we can see that large negative $-T\Delta S$ terms occur characteristically only upon the reaction of acidic cations with basic anions. It has been suggested by Latimer[6] that the entropy change in precipitation reactions is independent of the type of crystalline product formed. Thus it has become possible, with some further assumptions, to assign separate entropies of precipitation for the hydrated cations and for the hydrated anions. [The actual practice has been to consider the reverse of Eq. (4.1)—the dissolution of an ionic solid—and to tabulate entropies of solution for the cations and anions.[7]] Theoretically, these entropies of solution of cations or anions should also be related to the familiar ratio Z^2/r,[8] as suggested by Figure 4.2.

It is readily apparent from Figure 4.2 that *all but the nonacidic cations* and *nonbasic anions* make negative contributions to the entropy term for Reaction (4.1): Most ions release enough water molecules to give a net disordering effect upon precipitation or a net ordering effect upon dissolution of a salt. Such ions are referred to as **electrostatic structure makers**.

We also note that the more acidic or basic the ion (the higher its charge or the smaller its radius), the greater is the magnitude of the ion's structure-making property. The inverse relationship to size may seem surprising: Since larger cations have greater maximum coordination numbers, we might expect them to attach more water molecules. But, evidently, the smaller ion attaches more water molecules or at least orders them more effectively. Similarly, although increasing the charge of a cation does not increase its maximum coordination number, it does increase the ordering of water molecules in the hydrated ion.

These seeming contradictions can be explained by postulating that hydrated cations and anions have more complex structures than we have assumed up to now. Certainly, all hydrated metal ions and nonmetal or oxo anions have a layer of water molecules surrounding the ion, as shown in Figure 2.1. This layer is referred to as the **primary hydration sphere** or **inner sphere** of the hydrated ion, and the larger the "bare" cation or anion, the more water molecules can be accommodated in this primary hydration sphere. But ions that are at all acidic or basic exert such a strong attraction for the water molecules in their primary hydration sphere, pulling electron pairs toward a cation or hydrogens toward an anion, that these water molecules exert an enhanced hydrogen-bonding attraction for other water molecules and organize them

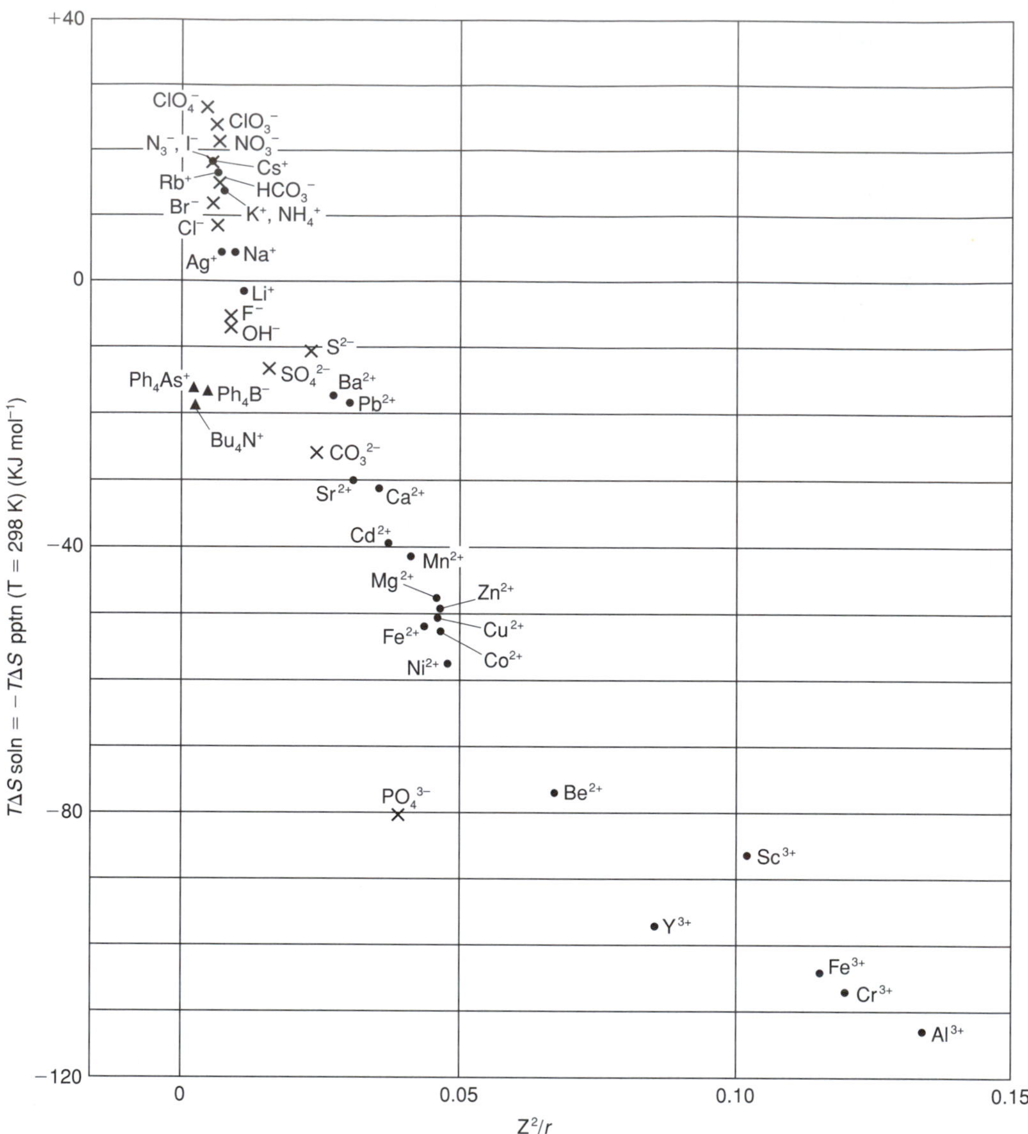

Figure 4.2
The 298 K entropy terms ($T\Delta S_{\text{soln}} = -T\Delta S_{\text{pptn}}$) for cations and anions as a function of Z^2/r. Crosses equal anions; dots equal metal cations; triangles equal organic ions. Thermodynamic data from Cox and Parker[7]; thermochemical radii of oxo anions from E. A. Keiter, R. L. Keiter, and J. E. Huheey, *Inorganic Chemistry*, Harper-Collins, New York, 1993; p. 118.

into a **secondary hydration sphere** or **outer sphere** around the first layer (Fig. 4.3); without the moderating effect of this secondary hydration sphere most metal cations would be strongly or very strongly acidic.[9] Note that this phenomenon has the same cause as does the phenomenon of hydrolysis that we discussed in Chapter 2. (In this case, however, the attraction does not have to be strong enough to pull the water molecule of the second layer apart—it only needs to attach it by strengthened hydrogen bonds.) Thus we can see why this phenomenon should be directly related to the acidity or basicity of the cation or anion.

This secondary hydration sphere may consist of more than one layer of water

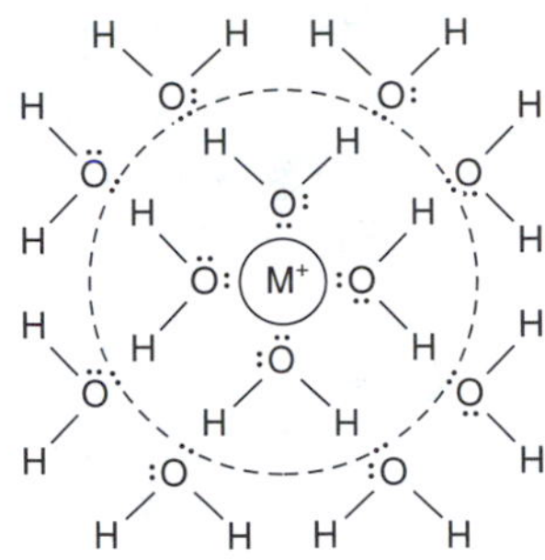

Figure 4.3
Primary and secondary hydration spheres of a hydrated cation.

molecules—the stronger the attraction of the bare ion for water molecules (i.e., the greater its acidity or basicity), the more water molecules that may be attached in additional layers. Each subsequent layer would be expected to involve weaker attractions, however, so water molecules would stay immobilized by hydrogen bonds for shorter and shorter periods of time. Many methods have been used to study this phenomenon of multiple hydration layers, but these diverse methods of measuring the number of waters attached or the radius of the hydrated ion[10] give inconsistent answers, since each method requires a different amount of time for the water molecules to remain attached in order to be counted. But nearly all measurements give consistent *trends* in hydration numbers among series of ions, such as the results for the ions in Table 4.2, obtained from transference measurements by way of the *Stokes radii* or *hydrated radii* of the hydrated ions, also in Table 4.2. In summary, the reason that most acidic cations and basic anions react to give precipitates is due to the disorder resulting from the release of numerous water molecules from the multiple hydration spheres of the cations and of the anions. This finding is apparent when we write the seemingly simple precipitation reaction of, say, magnesium and carbonate ions in full, including a reasonable number of water molecules:

$$[Mg(H_2O)_{36}]^{2+} + [CO_3(H_2O)_{28}]^{2-} \rightarrow MgCO_3(s) + 64\,H_2O \qquad (4.3)$$

Table 4.2
Hydration Numbers and Hydrated Radii of Some Hydrated Ions

Ion	Z^2/r^a	Hydration Number	Hydrated Radius (pm)
Cs^+	0.0055	6	228
K^+	0.0066	7	232
Na^+	0.0088	13	276
Li^+	0.0111	22	340
Ba^{2+}	0.0268	28	
Sr^{2+}	0.0303	29	
Ca^{2+}	0.0351	29	
Mg^{2+}	0.0465	36	
Cd^{2+}	0.0549	39	
Zn^{2+}	0.0599	44	

SOURCES: Hydration numbers from A. T. Rutgers and Y. Hendrikx, *Trans. Faraday Soc.*, **58**, 2184 (1962). Hydrated radii from R. P. Hanzlik, *Inorganic Aspects of Biological and Organic Chemistry*, Academic, New York, 1976; p. 31.

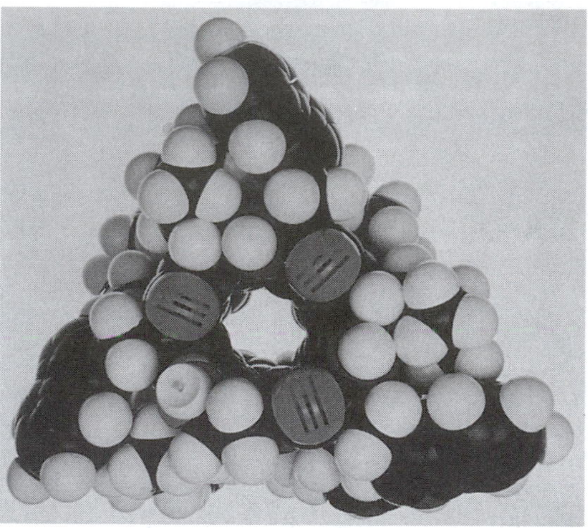

Figure 4.4
A model of the structure of gramicidin A, showing the pore at the center. [Reprinted with permission from D. Urry, *Proc. Natl. Acad. Sci. USA*, **69**, 1610 (1972).]

Since the *hydrated* ions of smaller bare ions may be larger than those of larger bare ions, we sometimes see unexpected chemical results. For example, an important question in biochemistry has to do with how metal ions can migrate across cell membranes, which consist largely of lipids in which metal ions are quite insoluble. One model of this process, called the *pore model*,[11] envisions the cell membrane as containing pores through which hydrated ions can pass without having to dissolve in the lipids. The antibiotic gramicidin A in fact contains a hollow center of about 400-pm diameter that, when in biological membranes, causes cells to pass ions readily (Fig. 4.4). The relative rate of passage of ions through this pore is $Cs^+ > Rb^+ > K^+ > Na^+ > Li^+$, which makes sense if the hydrated cesium ion is indeed the smallest hydrated ion in this series.

Example 4.3

How does the partial protonation of a multiply charged oxo anion alter its $T\Delta S_{soln}$? Describe what happens to the structure-making properties of such an ion upon partial protonation. Does this entropy effect favor the partial protonation of such ions? (The hydronium ion itself is only a weakly structure-making ion.)

SOLUTION:

Adding a proton to a multiply charged oxo anion reduces its negative charge, and hence reduces its ability to organize multiple hydration spheres around it, so that its $T\Delta S_{soln}$ is made more negative. (Compare the $+25.5\,kJ\,mol^{-1}$ value for CO_3^{2-} in Fig. 4.2 with the $-13.8\,kJ\,mol^{-1}$ value for HCO_3^-.) Since the disappearance of the hydrated proton itself has little effect on the overall order in the solution, this change has reduced the order in the solution. Hence, protonation of the anion is favored by this entropy effect.

4.4 Nonacidic Cations, Nonbasic Anions, and the Structure of Liquid Water

Since all metal ions form hydrated ions in which water molecules are presumably more ordered than they are in liquid water, it may seem surprising that not all metal ions are classified as electrostatic structure makers. As can be seen from Figure 4.2, the nonacidic cations (Na^+ through Cs^+) and the nonbasic anions have positive entropies of solution and thus have been termed **electrostatic structure breakers**. To understand how this can be, we must realize that liquid water itself has a structure that can be broken by large ions dissolved in it.

Due to the high polarity of the O–H bond, each water molecule (like many other kinds of molecule) can form hydrogen bonds to other molecules of the same kind. Unlike any other small molecule, however, the H_2O molecule can form *two* hydrogen bonds using its two hydrogen atoms, and it can complete *two other* hydrogen bonds using its two unshared electron pairs. Thus in ice, a very extensive network of hydrogen bonds allows each water molecule to be attached to four other water molecules (Fig. 4.5). This arrangement of hydrogen bonds between water molecules results in large cavities within the network of water molecules, which causes ice to be of rather low density. Molecules are normally more tightly packed in the solid state than they are in the liquid state. Thus the solid form of a substance is usually more dense than its liquid form. But this is not true of ice—ice floats in water!

When ice melts to form liquid water, enough hydrogen bonds must be broken to allow molecules to flow around each other: Thus ice has an unusually high heat of fusion. But there is good evidence that not all the hydrogen bonds are broken, and not all the extensive organization and structure of ice is lost. The structure of liquid water is a controversial topic, but one school of scientists[12] describes it as containing short-lived "icebergs" of structured water amid other non-hydrogen-bonded water molecules ($\sim$80% of the water molecules are supposed to be present in icebergs). Some of the free

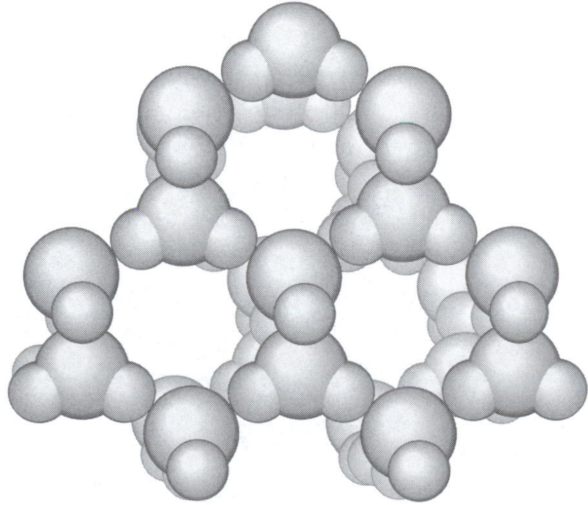

Figure 4.5
The structure of ice. The large spheres represent oxygen atoms, and the small ones hydrogen. [Figure from *Chemical Principles*, Fifth Edition, by William L. Masterson and Emil J. Slowinski, copyright © 1987 by Saunders College Publishing, reproduced by permission of the publisher.]

water molecules then can fill up the cavities present in the icebergs, giving liquid water a density higher than that of ice. (X-ray studies indicate that each water molecule has an average of 4.4 other water molecules near to it, rather than 4.0 as in ice.)

The large size of the nonacidic cations and nonbasic anions is thought to cause disruption of the iceberg structure in the nearby liquid water, hence giving these ions their positive entropies of solution and earning the title *electrostatic structure breakers*. When salts of nonacidic cations with nonbasic anions precipitate, the icebergs are able to restore their structure, creating order. Since nature does not prefer order, this effect resists the precipitation of these salts. Recalled from Section 4.2 that these salts precipitate because of the associated favorable *enthalpy* change; the unfavorable *entropy* change that accompanies precipitation is what prevents these salts from being highly insoluble, and allows the exceptions to Solubility Tendency III.

Nonacidic cations and anions having organic chains attached {such as the tetrabutylammonium cation, $[(C_4H_9)_4N]^+$, or the tetrabutylborate anion, $[(C_4H_9)_4B]^-$} have quite different effects. Water cannot penetrate to the charged N or B atoms, so these ions are generally not even hydrated in solution. Their long hydrocarbon chains (containing nonpolar C–H and C–C bonds) do not hydrogen bond to water, but instead slip inside the cavities of the icebergs, forcing out the loose water, which results in the production of more iceberg structure. Thus such ions are known as **hydrophobic structure makers**. On precipitation, the icebergs break apart and refill the cavities, so for these *organic* ions entropy effects do indeed contribute to the insolubility tendency noted in Solubility Tendency III.

4.5 Enthalpy and Precipitation: The Importance of Lattice Energies

We noted in Section 4.2 that the *enthalpy* change upon precipitation is involved both with the solubility of cross-combination salts (Solubility Rule II) and with the insolubility of salts of nonacidic cations with nonbasic anions (Solubility Tendency III). The enthalpy of precipitation of an ionic salt [i.e., ΔH for Reaction (4.1)] can be related to the properties of the cation and anion involved if we break down the relatively complex Reaction (4.1) into three simpler steps: (1) the hydrated cation is dehydrated and converted to a gaseous bare cation; (2) the same is done to the anion; (3) the gaseous cations and anions are then allowed to come together to form the ionic solid. (This process is shown schematically in Fig. 4.6.) By Hess' law, the enthalpy change for the overall reaction (4.1) must equal the sum of the enthalpy changes for each of the component steps.

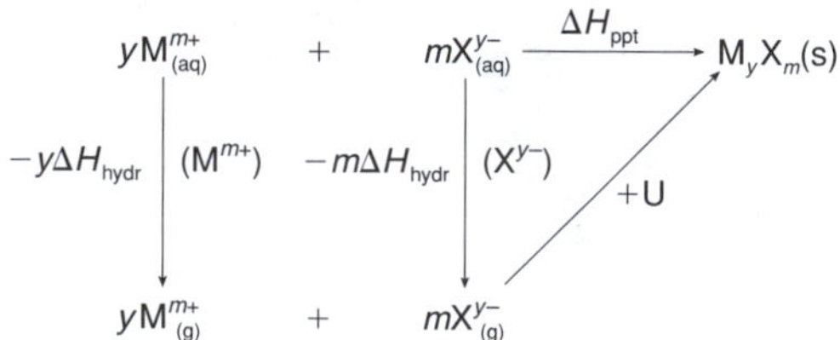

Figure 4.6
Thermodynamic cycle for analysis of the precipitation of a solid.

Steps 1 and 2 are simply the reverse of the processes of formation of the hydrated cations and hydrated anions, so their enthalpy changes are readily obtained by reversing the signs of the hydration enthalpies given in Tables 2.1 and 2.4. The enthalpy change for Step 3, called the **lattice energy**, is due to the attraction of oppositely charged ions for each other, which results in their condensing to form a crystalline solid. These three energy terms add up to give the overall enthalpy of precipitation, given in Table 4.1. Thus we can evaluate the lattice energies of several salts. For example, for the reaction of nonacidic K^+ with nonbasic ClO_4^-, we have the overall relationship for the enthalpy of precipitation

$$\Delta H_{\text{pptn}}(\text{KClO}_4) = -\Delta H_{\text{hyd}}(\text{K}^+) - \Delta H_{\text{hyd}}(\text{ClO}_4^-) + U(\text{KClO}_4) \tag{4.4}$$

where U represents the lattice energy. Taking data from Tables 2.1, 2.4, and 4.1, we find numerically

$$-51\,\text{kJ} = -(-321)\,\text{kJ} - (-235)\,\text{kJ} + U(\text{KClO}_4) \tag{4.5}$$

from which we find $U(\text{KClO}_4)$ to be -607 kJ, a very large energy term indeed.[13]

Example 4.4
Calculate the lattice energies of $Ba(ClO_4)_2$ and of $Mg(ClO_4)_2$.

SOLUTION:

In doing this summation, it is necessary to multiply each energy (in kJ mol^{-1}) from the tables by the number of moles of that species involved. The hydration enthalpies are given in kilojoules per mole of ion; for each of these salts, we are using 2 mol of perchlorate ion. The enthalpies of precipitation are per mole of product; the question implies that 1 mol of each salt is produced. Thus for $Ba(ClO_4)_2$ we have $(+12\,\text{kJ mol}^{-1})\,(1\,\text{mol}) = -(-1304\,\text{kJ mol}^{-1})\,(1\,\text{mol}) - (-235\,\text{kJ mol}^{-1})\,(2\,\text{mol}) + U$; solving for U gives us $-1762\,\text{kJ mol}^{-1}$. Similarly, the data for $Mg(ClO_4)_2$ gives us a lattice energy of $-2251\,\text{kJ mol}^{-1}$.

Clearly, lattice energies are very large (negative) energy terms. Hydration enthalpies are also very large (negative) energies; when we take the difference of these, only a (usually) very small enthalpy of precipitation remains; what matters is whether the lattice energy is larger than the combined hydration enthalpies of the component ions. In the following sections, we will examine the nature of the lattice energy in much greater depth, and we will see that it has much in common with the hydration enthalpies. We will try to see why, in the case of nonacidic cations and nonbasic anions, the lattice energies are larger (more negative) than the sums of the hydration energies, thus producing insoluble salts, while in the case of acidic cations plus nonbasic anions (or nonacidic cations plus basic anions) the reverse is true, producing soluble salts.

4.6 Ionic Solids: Coulombic Attractions and Lattice Energies

The large lattice energies of ionic solids result from the strong attractions of the oppositely charged cations and anions. The energy of such an attraction can be derived from Coulomb's law in physics and can be expressed as

$$E_{\text{attraction}} = \frac{Z_+ Z_- e^2 N}{4\pi e_0 r} \tag{4.6}$$

In this equation, the variables are the charge on the cation, Z_+, the charge on the anion, Z_-, and the distance between the two, r. Thus it has dimensions of charge squared over distance and is another example of this familiar type of ratio in chemistry. This equation includes certain constants needed to give the answers in SI units: e is the charge on the electron, 1.602×10^{-19} C (coulombs); N is Avogadro's number (so that the result will be expressed per mole of ionic compound); and ε_0 is the dielectric constant (permittivity) of a vacuum, $8.854 \times 10^{-12}\,\text{C}^2\text{m}^{-1}\text{J}^{-1}$. We prefer, however, to use picometers instead of Systéme International (SI) units of length (m), and to have our answer in kilojoules per mole (kJ mol^{-1}), not in joules per mole (J mol^{-1}). Incorporating the appropriate conversion factors and collecting all constant terms into one term, we simplify this equation to

$$E_{\text{attraction}} = \frac{138{,}900 Z_+ Z_-}{r} \tag{4.7}$$

This interaction is exothermic (E is negative), since the charge of the anion is negative.

In developing the relationship between this attractive energy and the lattice energy, we find it desirable to use spherical cations and anions (i.e., not oxo anions). Hence, we will consider the attraction of 1 mol of sodium ions and 1 mol of chloride ions; experimentally the lattice energy of NaCl is found to be -774 kJ mol^{-1}. Let us start with cations and anions at opposite ends of the universe (r = infinity), so that the energy of attraction is zero. We now allow the cations and anions to approach until each cation just touches one other anion, but no other interactions occur; now we have a mole of gaseous **ion pairs** of NaCl. Experimentally, the NaCl distance in such an ion pair is 236 pm. Substituting this distance in Eq. (4.7), we find an attractive energy of -589 kJ mol^{-1}.

Although bringing the ions closer would bring the answer closer to -774 kJ mol^{-1}, this cannot happen because the like-charged electrons of the oppositely charged ions cause repulsion between the ions. The energy of the repulsion is given by

$$E_{\text{repulsion}} = NB/r^n \tag{4.8}$$

where B is a constant and n is known as the Born exponent; its value depends on the principal quantum numbers of the electrons involved in the repulsion (Table 4.3) and can be determined by measuring the compressibility of an ionic compound.

The total energy of the ion pairs will be the sum of the attraction represented by Eq. (4.7) and the repulsion represented by Eq. (4.8). It is a minimum (at its most negative) when the ions are at their equilibrium separation r_0 (in the case of NaCl, 236 pm apart). In the general case, the minimum in energy is found (using calculus)

Table 4.3
Born Exponents and Electron Configurations of Ions

Born Exponent	Principal Quantum Number of Outermost Electrons of Ion
5	1 ($1s^2$: H^-, Li^+)
7	2 ($2s^22p^6$: F^-, Na^+)
9	3 ($3s^23p^6$, perhaps $3d^{10}$: Cl^-, K^+, Zn^{2+}, Ga^{3+})
10	4 ($4s^24p^6$, perhaps $4d^{10}$: Br^-, Rb^+, Cd^{2+}, In^{3+})
12	5 ($5s^25p^6$, perhaps $5d^{10}$: I^-, Cs^+, Au^+, Tl^{3+})

by taking the derivative of the total energy with respect to distance and setting that derivative equal to zero:

$$0 = \frac{dE_{\text{total}}}{dr} = \frac{-138{,}900Z_+Z_-}{{r_0}^2} - \frac{nNB}{{r_0}^{n+1}} \tag{4.9}$$

This equation can be solved for the constant B, and the new value for B can be substituted back into the total energy $(4.7 + 4.8)$ to give the final form for the total energy of 1 mol of ion pairs:

$$E_{\text{total}} = \frac{138{,}900Z_+Z_-}{r_0}\left(1 - \frac{1}{n}\right) \tag{4.10}$$

To evaluate this expression for the NaCl ion pair, we note that the Born exponent for Na^+ is 7 and that for Cl^- is 9; hence, with an average Born exponent of 8, the last term of Eq. (4.10) becomes $(1 - \frac{1}{8})$, and the *total* energy of formation of the NaCl ion pair from the infinitely separated ions is $\frac{7}{8}$ of $-589\,\text{kJ mol}^{-1}$, or $-515\,\text{kJ mol}^{-1}$.

This value is still quite a bit short of the experimental lattice energy, but of course we know that solid NaCl does not consist of isolated ion pairs. The maximum coordination numbers of Na^+ and of Cl^- are by no means met in the ion pair; additional chloride ions from other ion pairs can still be attracted to the sodium ion in one ion pair, and vice versa. This attraction leads ultimately to the structure of solid NaCl (Fig. 4.7), in which each Na^+ surrounds itself with as many Cl^- ions as can fit (6), and each Cl^- likewise is surrounded by six Na^+ ions. The equilibrium Na–Cl distance r_0 in this structure is 283 pm.

This crystal lattice has many more Coulombic attractions between oppositely charged ions than does the ion pair. In addition, there are many Coulombic repulsions between like-charged ions. Each one can be evaluated using Eq. (4.10); they must then be counted up, the distance involved in each interaction noted, and all the energy terms added over the entire crystal!

To begin, we note from Figure 4.7 that each Na^+ ion has 6 nearest-neighbor Cl^- ions at a distance of r_0; these contribute six times the energy term of Eq. (4.10) to the total lattice energy, U. However, the next nearest neighbors to each Na^+ ion are 12 other Na^+ ions, each across the diagonal of a square from the original Na^+ and thus at a distance from it of $\sqrt{2}$ times r_0; these *subtract* a total of $12/\sqrt{2}$ times the energy term of Eq. (4.10) from the lattice energy U. Likewise the next, next nearest neighbors to each Na^+ ion are the eight Cl^- ions across the body diagonal of a cube (at a distance of $\sqrt{3}r_0$) from the Na^+; this adds $8/\sqrt{3}$ energy terms to the lattice energy. And so the

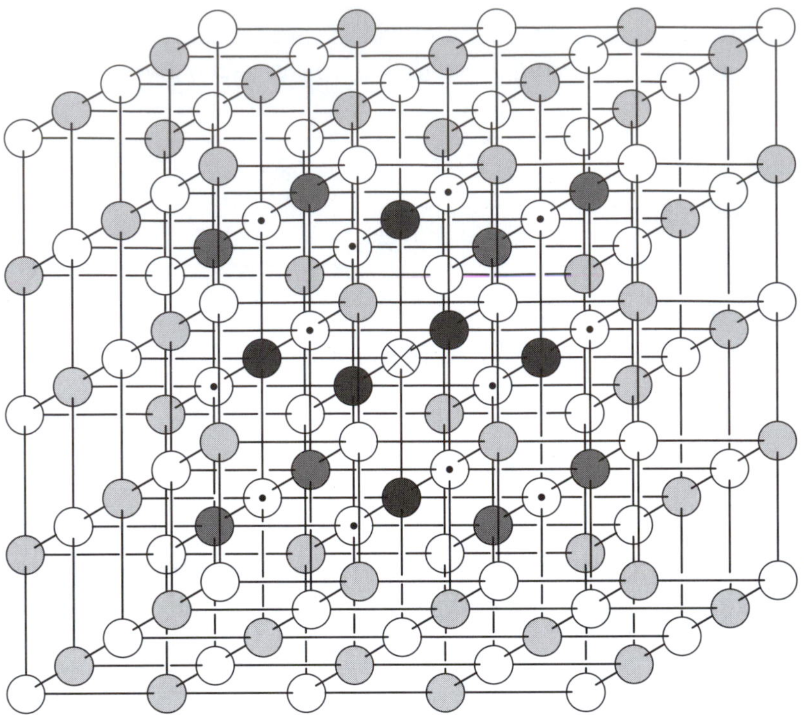

Figure 4.7
Crystal structure of NaCl. Starting with the Na^+ ion marked ⊗, there are 6 nearest neighbors (•), 12 next nearest neighbors (⊙), 8 next, next nearest neighbors (darkly shaded), etc. [From *Inorganic Chemistry*, 4th Edition, by J. E. Huheey, E. A. Keiter and R. L. Keiter; Copyright © 1993, by Harper Collins Publishers. Reprinted by permission of Addison-Wesley Educational Publishers.]

process continues, giving rise to the infinite series shown as the last term in Eq. (4.11).

$$U = \frac{138{,}900 Z_+ Z_-}{r_0}\left(1 - \frac{1}{n}\right)\left(\frac{6}{1} - \frac{12}{\sqrt{2}} + \frac{8}{\sqrt{3}} - \frac{6}{\sqrt{4}} + \frac{24}{\sqrt{5}} - \cdots\right) \tag{4.11}$$

The infinite series in Eq. (4.11) converges rather slowly to a limiting number that is characteristic of the NaCl crystal lattice geometry but that is independent of the charges on the ions and the distance r_0; this number is called the **Madelung constant**, M; for the NaCl lattice type, it has the value 1.74756, and applies to any of the dozens of ionic compounds in which the ions are arranged in this manner. Thus Eq. (4.11) can be rewritten as

$$U = \frac{138{,}900\,\mathrm{M}\,Z_+ Z_-}{r_0}\left(1 - \frac{1}{n}\right) \tag{4.12}$$

Finally, we can substitute into Eq. (4.12) the values of the Madelung constant, ionic charges, and interionic distance to obtain the theoretical lattice energy of NaCl, $U = -751\,\mathrm{kJ\,mol^{-1}}$. This value is quite close to the experimental lattice energy of $-774\,\mathrm{kJ\,mol^{-1}}$.

Small corrections for other minor energy terms can be added, but this treatment shows us the major factor responsible for the stability of an ionic lattice: the strong Coulombic attraction of the oppositely charged ions, which extends in all directions from a given ion and throughout space. This attraction is strong enough to overcome the Coulombic repulsions that must exist between like-charged ions and the repulsions that exist between outer electrons of adjacent ions.

4.7 Radius Ratios and Lattice Types

The NaCl lattice type shown in Figure 4.7 is only one of many known arrangements of ions (lattice types) in ionic solids. Some other lattice types that are important for binary ionic substances are shown in Figure 4.8.

Three major reasons why most ionic solids must adopt a lattice type other than the NaCl type are

1. The NaCl lattice incorporates an equal number of cations and anions. If the *stoichiometry* of the compound is other than 1:1, another lattice type is required.

2. If there is significant covalent bonding involved between the cation and the anion, the orbitals of the two touching atoms must overlap well; consequently, certain approximate bond angles are needed in the lattice type. The NaCl (or some other) lattice type may not incorporate these angles. Thus for a lattice involving saturated

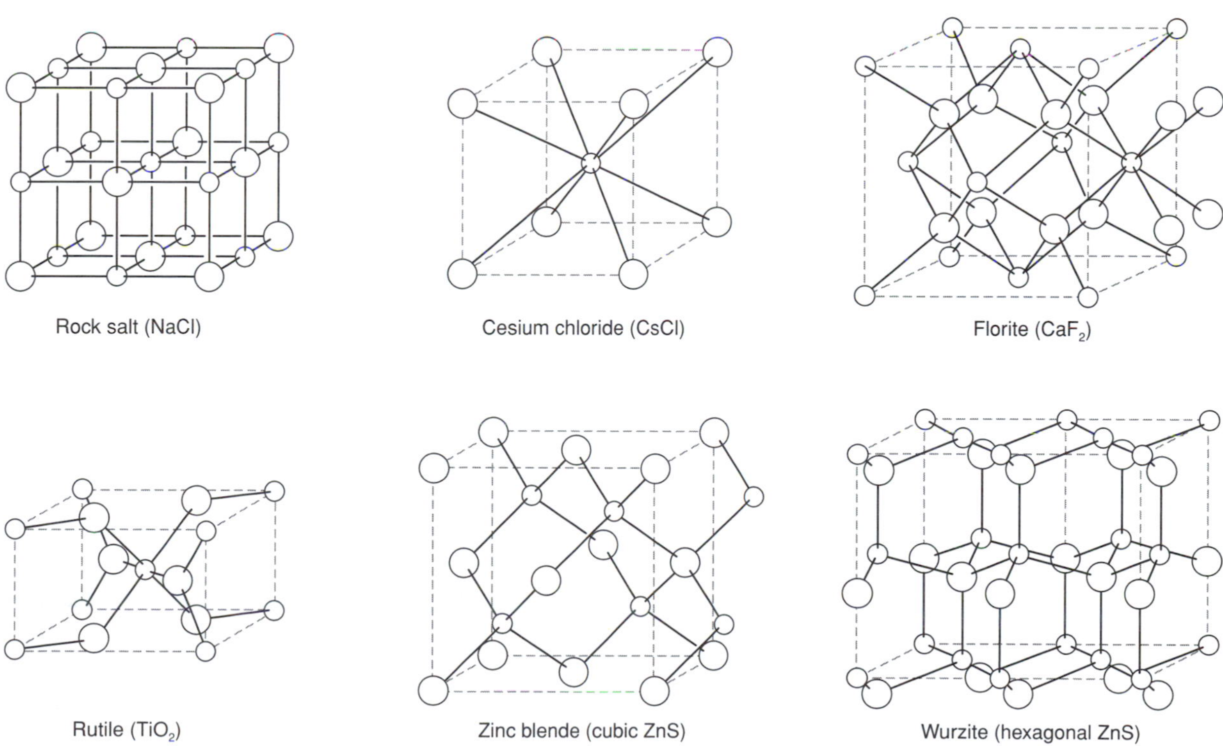

Figure 4.8
Important lattice types for binary ionic compounds. Small circles denote metal cations; large circles denote anions. [Adapted from F. A. Cotton and G. Wilkinson, *Advanced Inorganic Chemistry*, 5th ed., Wiley-Interscience, New York, p. 5. Copyright © 1988 by John Wiley & Sons, Inc.]

carbon, bond angles somewhere near 109° are needed, whereas the "bond" angles Cl–Na–Cl and Na–Cl–Na in the NaCl structure are all 90 or 180°. Some unusual lattices can result, as will be seen, for example, in Chapter 14 for some partially covalent halides. We will leave this consideration for that chapter, however, and (after checking to see that our electronegativity differences between touching atoms are at least close to 2.0) consider only highly ionic compounds in this chapter.

3. The NaCl lattice incorporates coordination numbers of 6 for the cation and 6 for the anion. This coordination number is not optimal for many combinations of cations and anions. (a) If the cation is too small relative to the anion, it may not touch the anions, while the large anions still touch each other. This reduces the attractive forces while keeping the repulsive forces strong—a situation to be avoided, if possible, by going to a lattice in which the coordination number of the cation is lower so that the smaller number of anions can touch the cation without touching each other. (b) If the cation is too large relative to the anion, it may be possible to fit additional anions around the cation without introducing excessive anion–anion repulsion; hence, the compound will be expected to adopt a different lattice, one in which the cation has a higher coordination number.

Thus, the relative size of the cation and anion (the **radius ratio** of the cation to the anion) is an important element in determining the type of crystal lattice that an ionic compound will adopt. We can use solid geometry to work out the ideal radius ratio for "perfect packing" (cations and anions just touching each other; anions and anions just touching each other) of many lattices. For example, for the NaCl lattice type (Fig. 4.7 or 4.8) we can consider a square of two Na^+ ions at opposite corners and two Cl^- ions at the other corners. The Na^+ and Cl^- ions touch along the edges, and the larger Cl^- ions touch each other across the diagonal of the square. Consequently, we have the following two equations:

$$r_{\text{cation}} + r_{\text{anion}} = a \text{ (the edge length)} \tag{4.13}$$

$$2\,r_{\text{anion}} = \sqrt{2}a \text{ (the diagonal length)} \tag{4.14}$$

By eliminating a between these two equations and rearranging, we can solve for the "ideal" radius ratio $r_{\text{cation}}/r_{\text{anion}} = \sqrt{2} - 1 = 0.414$. Likewise, starting with the cube of the CsCl lattice (Fig. 4.8) and using the fact that the cube diagonal is equal to $\sqrt{3}$ times the edge, we can calculate the ideal radius ratio for this lattice to be 0.732. If a cation and an anion have a radius ratio between 0.414 and 0.732, the NaCl lattice will be expected to be chosen, since enlarging the cation in a NaCl lattice spreads the touching anions apart, reducing repulsions, while shrinking the cation in the CsCl structure reduces cation–anion contact, reducing attractions. Hence, the range of radius ratios between 0.414 and 0.732 should be ideal for the NaCl crystal structure to be adapted. Similar ranges for the other common binary lattice types for different stoichiometries are shown in Table 4.4.

It is possible to predict coordination numbers for salts not corresponding to any of the types specified in Table 4.4.

1. The radius ratio should predict the coordination number for the less abundant ion in any lattice type and for any stoichiometry. (The less abundant ions have more neighbors of the opposite charge and hence the most critical problems of crowding.)

Table 4.4
Lattice Types and Madelung Constants for Different Stoichiometries and Radius Ratios of Cations and Anions

Radius Ratio (Cation/Anion)	Lattice Type	Coordination Number of Cation	Coordination Number of Anion	Madelung Constant	Reduced[a] Madelung Constant
	A. 1 : 1 Stoichiometry of Salt (MX)				
0.225–0.414	Wurtzite (ZnS)	4	4	1.63805	1.63805
	Zinc blende (ZnS)	4	4	1.64132	1.64132
0.414–0.732	Rock salt (NaCl)	6	6	1.74756	1.74756
0.732–1.000	CsCl	8	8	1.76267	1.76267
	B. 1 : 2 Stoichiometry of Salt (MX_2)				
0.225–0.414	Beta-quartz (SiO_2)	4	2	2.201	1.467
0.414–0.732	Rutile (TiO_2)	6	3	2.408[a]	1.605
0.732–1.000	Fluorite (CaF_2)	8	4	2.51939	1.6796
	C. 2 : 3 Stoichiometry of Salt (M_2X_3)				
0.414–0.732	Corundum (Al_2O_3)	6	4	4.1719[b]	1.6688
	D. Other Stoichiometries and Lattice Types				
Never favored	Ion pair	1	1	1.00000	1.0000
0.000–0.155		2			
0.155–0.225		3			
0.225–0.414		4			
0.414–0.732		6			
0.732–1.000		8			
1.000		12			

[a]Reduced Madelung constant = Madelung constant $\times 2/p$, where p = number of ions in the simplest formula of the salt.
[b]Exact value dependent on details of the structure.

Except in some compounds such as K_2O, the less abundant ion is the cation; hence, from the radius ratios given at the bottom of Table 4.4, the expected coordination number of the cation (more properly, the less abundant ion) can be predicted. (Intermediate coordination numbers such as 5, 7, and 9 do occur with some stoichiometries, but these are difficult to predict.)

2. The average coordination number of the anion (the more abundant ion) can then be calculated from the stoichiometry of the salt of M and X:

$$\begin{aligned}&(\text{Coordination No. of M})(\text{No. of M in formula})\\&\quad = (\text{Coordination No. of X})(\text{No. of X in formula})\end{aligned} \tag{4.15}$$

Normally, anions are larger than cations, so the radius ratio is less than 1.00; but occasionally the cation is the larger. In this case, we often find that one of the known lattice types is adopted, but the positions of the cations and the anions are reversed—the cation, which is now the larger ion, takes the position held by the usually larger anion. Thus if the radius ratio is greater than 1.00, we can calculate the inverse radius ratio $r_{\text{anion}}/r_{\text{cation}}$ to pick the lattice type in which the cations and anions will reverse roles. The lattice type is then named with the prefix *anti-*.

Example 4.5

Predict the lattice types that are adopted by the following oxides: Tl_2O, BeO, and ThO_2.

SOLUTION:

First, we obtain the appropriate radii from Table C, and then we calculate the radius ratios r_{cation}/r_{anion}: $164/126 = 1.30$, $59/126 = 0.47$, and $108/126 = 0.86$, respectively. (Since the first radius ratio is > 1.00, we calculate the inverse radius ratio $126/164 = 0.77$.)

Second, we refer to Table 4.4D to obtain the expected coordination number of the cation (6 for Be^{2+} and 8 for Th^{4+}) or the anion if it is less abundant in the chemical formula. The O^{2-} is less abundant in Tl_2O, so we predict its coordination number to be 8.

Third, we calculate the coordination number of the other ion using Eq. (4.15). We compute that each Tl^+ should be surrounded by four O^{2-} ions, each O^{2-} in BeO ought to be surrounded by six Be^{2+} ions, and each O^{2-} in ThO_2 should be surrounded by four Th^{4+} ions.

Finally, if these stoichiometries and sets of coordination numbers match any of the entries in Table 4.4A–C, we can read the lattice type from the table. We thus predict that BeO ought to adopt the NaCl structure, ThO_2 ought to adopt the fluorite structure, and Tl_2O should adopt the antifluorite structure, in which each Tl^+ is surrounded by four O^{2-} ions, and each O^{2-} ion is surrounded by eight Tl^+ ions. According to Wells,[14] ThO_2 does indeed adopt the fluorite structure, but BeO actually adopts the wurzite structure, and Tl_2O adopts a structure with a coordination number of 6 for oxygen.

The radius ratio rule succeeds in predicting lattice types about two-thirds of the time.[15] The trends are generally right—for a given anion, larger cations adopt lattices in which the cations have larger coordination numbers—but the radius ratio at which the transition from one lattice type to another occurs is often substantially different than predicted. Certainly, one reason for this in many cases is the second one given above—covalency in the bonding. Another source of error may arise from the fact that the measured radii, particularly of cations, change when their coordination numbers change: As more anions crowd around the cation, they repel each other more and are less able to compress the cation. Thus, for example, the radius of Na^+ varies with its coordination number, as shown in Table 4.5. Despite the middling success of quantitative predictions with the radius ratio rule, it gives us further insight into the consequences of Coulombic attractions and repulsions in ionic compounds.

Table 4.5
Effect of Coordination Number on the Radius of the Na^+ Ion

Coordination Number of Na^+	4	5	6	7	8	12
Radius of Na^+ (pm)	113	114	116	126	132	138

Example 4.6
Compute the lattice energy of ThO_2.

SOLUTION:

Equation (4.12) requires us to provide the cation and anion charges (+4 and −2), the Madelung constant, the interionic distance, and the average Born exponent for the two ions. In the previous example, we determined that ThO_2 should and does adopt the fluorite lattice, which from Table 4.4 has a Madelung constant of 2.51939. The interionic distance is presumed to be the sum of the ionic radii of Th^{4+} and O^{2-}, $108 + 126 = 234$ pm. From Table 4.3, we find the Born exponent for O^{2-} (electron configuration $2s^2 2p^6$) to be 7. The Th^{4+} ion has the valence electron configuration $7s^0$ or, considering its outermost electrons, $6s^2 6p^6$. The Born exponent for this configuration is not listed, so let us extrapolate a value of 14. Averaging the two Born exponents 7 and 14 gives us $n = 10.5$ to use in Eq. (4.12). Substituting these numbers gives us

$$\frac{138{,}900(2.51939)(4)(-2)}{234}\left(1 - \frac{1}{10.5}\right)$$

which is a lattice energy of $-10{,}824\,\text{kJ mol}^{-1}$.

If the lattice type is unknown or unusual, the Madelung constant for it is not available. Nonetheless a reasonable estimate of the lattice energy can be made using the Kapustinskii[16] equation

$$U = \frac{120{,}000 v Z^+ Z^-}{r_0}\left(1 - \frac{34.5}{r_0}\right) \tag{4.16}$$

where v is the total number of ions in the formula unit of the ionic compound. And, although polyatomic ions are not strictly spherical and do not have true radii, *thermochemical radii* have been derived[17] (Table 4.6) that will allow reasonable calculations of lattice energies of the salts of polyatomic ions.

4.8 Stability of Lattices and the Solubility Rules

Suppose that we wish to compare the lattice energies of three salts of the same 1:1 stoichiometry, A^+X^-, B^+Y^-, and C^+Z^-, chosen such that (1) the sum of the cationic and anionic radii, r_0, is the same in all three cases; (2) the average Born exponent is the same in each case; but (3) the radius ratio differs such that the first salt takes one of the ZnS structures, the second salt takes the NaCl structure, and the third salt takes the CsCl structure. From Eq. (4.12) we can see that the lattice energies of the three salts will differ due to the different Madelung constants, and from Table 4.4 we can see that the higher the radius ratio, the greater the Madelung constant. Thus C^+Z^- should have the most negative lattice energy. We conclude that *more stable lattices are formed by cations and anions that are relatively close in size.*

Table 4.6
Thermochemical Radii of Polyatomic Ions

Ion	(pm)	Ion	(pm)	Ion	(pm)
Cations		Anions		Anions	
NH_4^+	151	$GeCl_6^{2-}$	314	$PdCl_6^{2-}$	305
$(CH_3)_4N^+$	215	GeF_6^{2-}	252	$PtBr_6^{2-}$	328
PH_4^+	171	HCl_2^-	187	$PtCl_4^{2-}$	279
Anions		HCO_2^-	155	$PtCl_6^{2-}$	299
		HCO_3^-	142	PtF_6^{2-}	282
$AlCl_4^-$	281	HF_2^-	158	PtI_6^{2-}	328
BCl_4^-	296	HS^-	193	$SbCl_6^-$	337
BF_4^-	218	HSe^-	191	SeO_3^{2-}	225
BH_4^-	179	IO_3^-	108	SeO_4^{2-}	235
BrO_3^-	140	$IO_2F_2^-$	163	SiF_6^{2-}	245
$CH_3CO_2^-$	148	$IrCl_6^{2-}$	221	$SnBr_6^{2-}$	349
ClO_3^-	157	$MnCl_6^{2-}$	308	$SnCl_6^{2-}$	335
ClO_4^-	226	MnF_6^{2-}	242	SnI_6^{2-}	382
CN^-	177	MnO_4^-	215	SO_4^{2-}	244
CNS^-	199	N_3^-	181	$TiBr_6^{2-}$	338
CO_3^{2-}	164	NCO^-	189	$TiCl_6^{2-}$	317
$CoCl_4^{2-}$	305	$NH_2CH_2CO_2^-$	176	TiF_6^{2-}	275
CoF_6^{2-}	230	NO_2^-	178	VO_4^{3-}	246
CrF_6^{2-}	238	NO_3^-	165	$ZnBr_4^{2-}$	285
CrO_4^{2-}	242	O_2^-	144	$ZnCl_4^{2-}$	272
$CuCl_4^{2-}$	307	O_2^{2-}	159	ZnI_4^{2-}	309
$FeCl_4^-$	344	OH^-	119		
$GaCl_4^-$	275	$PbCl_6^{2-}$	334		

SOURCE: Data from H. D. B. Jenkins and K. P. Thakur, *J. Chem. Educ.*, **56**, 576 (1979), adjusted to be compatible with Shannon-Prewitt crystal radii. [From J. E. Huheey, E. A. Keiter, and R. L. Keiter, *Inorganic Chemistry: Principles of Structure and Reactivity*, 4th ed., Harper-Collins, New York, 1993; p. 118.]

Now let us compare the total hydration energies of the cations and anions in each of these three salts. In the manner of the Latimer equation (2.1), we expect the sum of these energies to be proportional to Z^2/r for the cations plus Z^2/r for the anions (ignoring the correction for the radius of water). By assuming some arbitrary values for these radii, it may be verified that the sum of hydration energies should be most negative for the salt with the most disparate radii, A^+X^-. Hence, according to the thermochemical cycle for solubility (Fig. 4.6), the enthalpy of precipitation should be most positive for A^+X^- and most negative for C^+Z^-. As noted by Morris,[18] the most insoluble salts are expected to be those in which the hydration energies of the cations and anions are most nearly matched. These are also expected to be the cations and anions that are most nearly matched in the strength of their acidity and basicity, respectively.

Solubility Tendency III. Looking back at our solubility rules (Section 4.2), we can now more fully explain our rules for those classes of salts for which solubilities are determined principally by the enthalpy (rather than the entropy) of precipitation. The ultimate reason that nonacidic cations and nonbasic anions give rise to insoluble salts is that such cations and anions are both large (i.e., similar in size). Therefore they form an especially stable crystal lattice, but do not give especially good hydration energies (in comparison to the lattice energy).

We noted earlier that (other things being equal) matching the charges of the two ions also seems to help in precipitating or crystallizing a given ion. Comparing Madelung constants of lattices involving different charges on one of the ions cannot be done directly, since different stoichiometries must result. Kapustinskii[16] found it useful to convert the Madelung constants for stoichiometries other than 1:1 to a 1:1 basis (e.g., for 2 mol of ions) by dividing by one-half the number of ions in the formula. Such *reduced Madelung constants* (Table 4.4) are close to 1.6 for most lattice types, but are the largest for 1:1 salts, suggesting that the best packing of ions occurs with equal numbers of cations and anions. Hence, to precipitate a large -1 charged (i.e., nonbasic) anion we are likely to do best if we add a large $+1$ (nonacidic) cation.

The favorable lattice energies resulting from good matching of cation and anion charges and radii not only render such products more insoluble, aiding their isolation, but also may stabilize the salts against thermal decomposition. For example, many oxo anions such as carbonate and sulfate decompose upon heating to give gases such as carbon dioxide and sulfur trioxide, plus the oxide ion, which is much smaller than the original oxo anion and thus forms more stable lattices with small (more acidic) metal ions. Since the larger carbonate and sulfate ions form better lattices with large cations, carbonates and sulfates of large (nonacidic) cations are stable to higher temperatures than are the carbonates and sulfates of small, acidic cations. (The carbonate ion is more susceptible to this effect, and in general, carbonates of even moderately acidic cations are not stable at all at room temperature.)

Solubility Rule II. Likewise, the reason that acidic cations and nonbasic anions give rise to soluble salts is that such ions are quite different in size and give poorer lattice energies than hydration energies. In extreme cases, the cation may be so much smaller than the anion that even in one of the ZnS structures there may not be good contact between cations and anions. In solution, however, good contact may be achieved between cations and small water molecules, and there is "bonus" energy resulting from the formation of multiple spheres of hydration. Hence, not only are such salts quite soluble in water, but if the solution is evaporated until the salt does indeed crystallize, it normally crystallizes as a **hydrated salt**, in which the primary hydration sphere of the cation (usually) is retained. (Of course, the radius ratio of the *hydrated* cation to the anion is much larger than that of the anhydrous cation to the anion.) For example, most *d*-block metals form hydrated sulfates, nitrates, chlorides, and so on, which are often written $CuSO_4 \cdot 5\,H_2O$, and so on, but which could more accurately be written as including hydrated cations, such as $[Cu(H_2O)_4]^{2+}$. When such salts are obtained anhydrous, they are often very good **desiccants** or drying agents, capable of removing water from solvents or from the air: $Mg(ClO_4)_2$, $CaCl_2$, and $CaSO_4$ are used in this manner.

Note that the solubility rules and tendencies implicitly depend on the ions involved forming relatively normal hydrated ions and lattices. Hence, they may not apply well to salts of ions that are very irregularly shaped, as, for example, long-chain organic carboxylate anions, $C_nH_{2n+1}CO_2^-$, which can only hydrate on one end and would likely form atypical crystal lattices as well.

4.9 *Writing and Interpreting Net Ionic Equations

Writing Net Ionic Equations. Equations for chemical equilibria are made easier to write if the dissolved ionic compounds involved in them are written as separate ions [e.g., as $Na^+(aq)$ and $NO_3^-(aq)$] rather than as complete formula units [$NaNO_3(aq)$].

Such equations are known as **ionic equations**. It is often found that in equilibria involving such ionic compounds, one of the ions appears both as reactant and as product (i.e., does not react at all). We say that such an ion is a **spectator ion**. Deleting spectator ions from an ionic equation simplifies the equation and gives a **net ionic equation**. It is easier to write an equilibrium constant expression starting from a net ionic equation. Such an equation is also more general—we can see that it applies as well for other ionic salts with different spectator ions.

In writing ionic equations, (1) *soluble* ionic salts are written as separate ions, while (2) *insoluble* ionic salts are written as complete formula units [often with (s) for solid or an arrow after them]. Similarly, (3) *very strong* and *strong* oxo acids exist in solution predominantly as separate ions, and are written as such, while (4) *moderate* and *weak* oxo acids exist in solution mostly as molecules and are written as complete formula units.

For example, one of the precipitation reactions in Experiment A.4 involves combining solutions of $Mg(NO_3)_2$ and Na_3PO_4; a precipitate of $Mg_3(PO_4)_2$ results. Writing full formula units for this reaction gives us the long "molecular" form of the reaction:

$$3\,Mg(NO_3)_2 + 2\,Na_3PO_4 \rightarrow Mg_3(PO_4)_2 + 6\,NaNO_3 \tag{4.17}$$

We obtain the ionic form of this reaction by writing the three soluble salts as separate ions:

$$\begin{aligned} &3\,Mg^{2+}(aq) + 6\,NO_3^{-}(aq) + 6\,Na^{+}(aq) + 2\,PO_4^{3-}(aq) \\ &\quad \rightarrow Mg_3(PO_4)_2(s) + 6\,Na^{+}(aq) + 6\,NO_3^{-}(aq) \end{aligned} \tag{4.18}$$

Although Reaction (4.18) does not yet save any space, we note that $6\,Na^+$ and $6\,NO_3^-$ occur on both sides of the equation. These spectator ions can be deleted to give the net ionic equation:

$$3\,Mg^{2+}(aq) + 2\,PO_4^{3-}(aq) \rightarrow Mg_3(PO_4)_2(s) \tag{4.19}$$

Not only is Eq. (4.19) more compact, but it also suggests that *any* soluble Mg^{2+} salt and *any* soluble NO_3^- salt would give the same reaction. Hence, dozens of reactions are summarized by this one equation.

Interpreting Net Ionic Equations; Using Inorganic Handbooks. Due to the efficiency of net ionic equations in presenting information, they are commonly encountered in textbooks. To carry out these reactions, however, some interpretation is needed, since we must use salts that also contain the (unspecified) spectator ions.

Let us suppose that we want to prepare 0.100 mol of cerium(III) phosphate, $CePO_4$. Knowing that this salt is the combination of an acidic cation and a basic anion, we suspect that it can be prepared by precipitation:

$$Ce^{3+}(aq) + PO_4^{3-}(aq) \rightarrow CePO_4(s) \tag{4.20}$$

To carry out this reaction, we must decide what cerium salt and what phosphate salt to use. To get satisfactory reaction rates, we normally choose *soluble* salts, which we can choose with the aid of our solubility rules. For Ce^{3+}, we may select a salt of a nonbasic anion such as ClO_4^-, NO_3^-, or Cl^-, or a feebly basic ion such as SO_4^{2-}. For PO_4^{3-}, we select the salt of a nonacidic cation such as Na^+ or K^+.

But we must also guard against the possibility that the spectator ions we choose will combine to give a second precipitate, which would contaminate our product. In theory, Solubility Tendency III states that the nonacidic cation we have chosen should react with the nonbasic anion we have chosen to give a second precipitate! Fortunately, this tendency has exceptions: The choice of Na^+ as a cation seldom results in a precipitate with a nonbasic anion, which can usually safely be chosen as Cl^- or NO_3^-. Since the feebly basic sulfate ion is also acceptable, we actually have several choices available to us, depending on what is available in the stockroom. However, we would not want to take the perchlorate of one ion and the potassium salt of the other.

When we go to weigh out the reactants, we must include the weight of any water of hydration that the salt includes. Sometimes the label on the bottle mentions the presence of waters of hydration only in the fine print, if at all! To learn the likely number of waters of hydration in the reactants or products and to check that our predictions of solubility are correct, it is useful to look up the reactants and products in the inorganic section of a chemistry handbook such as the *Handbook of Chemistry and Physics*. Let us suppose that we have decided to use cerium(III) nitrate and sodium phosphate to synthesize the cerium(III) phosphate. Under "cerium" in the handbook we find the nitrate hexahydrate, $Ce(NO_3)_3 \cdot 6\,H_2O$. Looking under the columns labeled "Solubility ... in cold water" we find either abbreviations such as i., d., sl.s., s., or v.s., or we find numerical data such as 0.053°. The abbreviations mean, respectively, *insoluble*, *decomposed*, *slightly soluble*, *soluble*, or *very soluble*; the number means that 0.053 g of salt dissolves in 100 mL of water at the temperature of the superscript. (An infinity sign indicates complete solubility.) If we find i. or d. for a reactant, we had better choose again; if we find sl. s. or a number under (say) 5, we will need a lot of water to dissolve the reactant. Beside $Ce(NO_3)_3 \cdot 6\,H_2O$ we find v.s., which makes this a good choice. To make 0.100 mol of $CePO_4$ we need 0.100 mol of the hydrated nitrate, which (from its listed formula weight) is 43.4 g.

To find the data for the phosphate, we must look under the counterion we have chosen. Under "sodium" we find many "phosphate" entries. Some of these are partially protonated forms such as $NaH_2PO_4 \cdot H_2O$. These might be suitable, but it is better to choose a hydrate of Na_3PO_4. As expected, these are soluble. Both a 10 and a 12 hydrate are listed. We should see which is in the stockroom and use its formula weight to calculate the quantity to be used in the reaction. After checking to be sure that $CePO_4$ is insoluble and that $NaNO_3$ is soluble (they are), we are ready to go to work in the laboratory.[19]

4.10 *Nonaqueous Solvents for Ionic or Ionizable Compounds

Many ionic compounds are insoluble in water, and many of those that are soluble undergo hydrolysis in water. Therefore it is often very useful to find **nonaqueous solvents** for ionic (as well as polar and nonpolar covalent) compounds. The Lewis acid–base properties of many liquid substances allow them to be used as solvents for selected ionic and other compounds. There is a wide variety of such solvents; here we can only indicate categories and give a few examples.

Nonaqueous solvents are usually divided into two broad categories: **protic solvents**, which have ionizable hydrogen atoms, and **aprotic solvents**, which have either no hydrogen atoms or only very nonacidic hydrogen atoms.

Aprotic Solvents. The aprotic solvent category is usually subdivided into three groups. The first of these is the group of **nonpolar, weakly solvating** solvents. These include the saturated hydrocarbons such as pentane, hexane, heptane, "petroleum ether" (a mixture of saturated hydrocarbons), and cyclohexane. Also included in this group are aromatic hydrocarbons such as benzene and toluene, which are usually weakly solvating, the very slightly polar and weakly Lewis-basic chlorocarbons such as CH_2Cl_2 and $CHCl_3$, and the fats and lipids in our bodies. Their advantage as solvents is that they are unlikely to interfere in reactions of the solute with Lewis acids or bases.

Such solvents cannot generally interact effectively either with the cation or the anion of an ionic salt, and thus there is no energy of solvation to offset the lattice energy; hence, most ionic salts are insoluble in such solvents. But these can be good solvents for nonpolar *hydrophobic* solutes such as nonpolar organic compounds and *uncharged* complexes and organometallic compounds that have carbon- and hydrogen-rich groups on their outer surfaces; the organic groups may be (among others) *alkyl groups* C_nH_{2n+1} (collectively symbolized by R) or *aryl groups* (collectively symbolized by Ar or a hexagon with an inscribed circle, and based on substitution within the cyclic *phenyl group* C_6H_5). The hydrophobic compounds may have these groups directly attached to the metal (in which case they are organometallic compounds), or the compounds may be neutral complexes of ligands derived from partially protonated anions by replacement of the hydrogen(s) with alkyl or aryl groups. Examples of such organic-substituted anions include the methoxide ion, CH_3O^-, the methylthiolate ion, CH_3S^-, the dimethylamide ion, $(CH_3)_2N^-$, or collectively the alkoxide ions, RO^-, the alkylthiolate or mercaptide ions, RS^-, and the dialkylamide ions, R_2N^-. These uncharged complexes and organometallic compounds are generally *insoluble in water*, since they cannot form hydrated ions; but they also have very small lattice energies since they are uncharged. These complexes are hydrophobic structure makers, and hence are repelled by water (hydrophobic) but dissolve in nonpolar solvents.

This property is frequently used in analytical chemistry. A metal ion present in very low concentration in a natural water is complexed with an organic ligand to give an uncharged complex. This complex is then extracted into a much smaller volume of a nonpolar solvent and is thus separated from many impurities and concentrated at the same time. Complexes of this type tend to accumulate in fatty tissue and lipids in the body and are not easily excreted in the watery urine; hence, toxic effects of metals in these complexes persist longer. When an animal dies and is eaten by another animal higher up the food chain, the uncharged complex is passed on to it. Since such animals eat many times their own weight of the lower chain animal, the concentration of the uncharged metal complex builds up; this process is known as **bioamplification** of the toxin.

A related property of such uncharged complexes, which also depends on the absence of large lattice energies to attract complex molecules to each other, is that they can be quite *volatile*. This property will be explored in more detail in Part II, but it does play a role in toxicity properties to be discussed in Chapter 5, because it allows a toxic metal ion to be introduced into the body through the lungs.

Even charged complexes can be rendered soluble in nonpolar solvents if either the cation or the anion can be converted to the form of large complex ions (with small

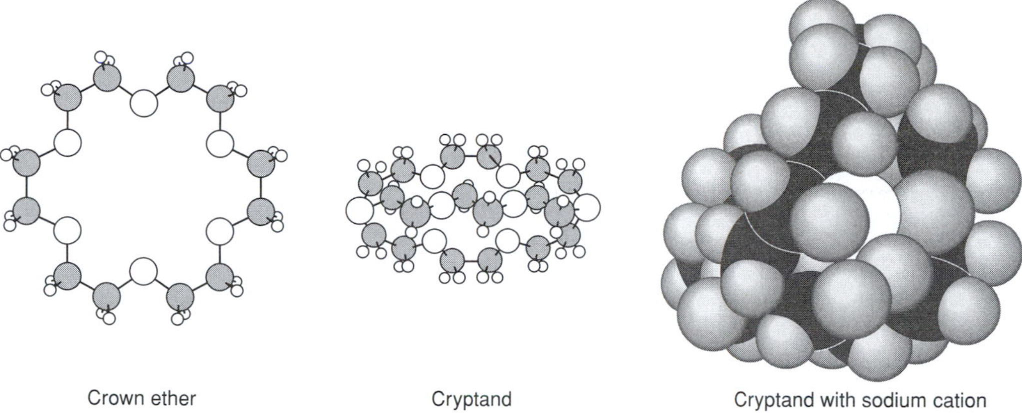

Figure 4.9
Molecular models of a crown ether, a *cryptand* or crypt polyether, which also contains two nitrogen atoms, and a Group 1 cation (in white) encapsulated by a crypt polyether ligand. [Adapted with permission from "Anions of the Alkali Metals," by James L. Dye. In *Sci. Am.*, July 1977. Copyright © 1977 by Scientific American, Inc.]

lattice energies resisting dissolution) with hydrophobic organic groups on the outside. The *tetraalkylammonium* cations {such as $[(CH_3)_4N]^+$ and $[(C_2H_5)_4N]^+$ but especially those with even longer organic groups} already fit this description, being hydrophobic structure makers, and can often carry a polar counterion from water into a nonpolar organic solvents. For example, potassium permanganate, $KMnO_4$, is a very useful oxidizing agent in organic chemistry, but it has some disadvantages: It is not soluble in the nonpolar organic solvents that best dissolve the organic compound to be oxidized; it is not even very soluble in water (Solubility Tendency III); when it does dissolve the MnO_4^- ion hydrates, which reduces its reactivity. These problems can be overcome by adding a small, catalytic quantity of a tetraalkylammonium cation, which will form some ion pairs with MnO_4^-. These ion pairs can then dissolve in unhydrated, highly reactive form in the organic solvent where the organic reactant is very soluble. Such tetraalkylammonium salts function as **phase-transfer catalysts**.

Alternately, the existing metal cation can be complexed with a large hydrophobic ligand to form a very nonacidic, organic-soluble complex ion. The most widely used ligands for this purpose are long-chain cyclic ethers known as *crown ethers* and *cryptates* or crypt ethers (Fig. 4.9). These have cavities in their centers in which appropriately sized cations will fit well; in the resulting complexes the oxygen-donor atoms all face inward to coordinate to the cation, and the hydrophobic, nonpolar C–H bonds all face outward to give solubility in nonpolar solvents. Small concentrations of crown ethers also function effectively as phase-transfer catalysts, and render the cation more nonacidic, stabilizing large anions that might otherwise be unstable.

If *both* the cation and the anion are complex ions with organic groups on the outside, we expect perhaps even larger solubilities in nonpolar solvents, but now *reduced* solubilities in water, since neither the cation nor the anion would be expected to hydrate well.

Example 4.7

Tell whether the following salts or complexes are likely to be soluble in water or in nonpolar organic solvents: (a) $[PtCl_2(NH_3)_2]$; (b) PtO; (c) $[PtCl(NH_2C_{10}H_{21})_3]Cl$; and (d) $[PtCl(NH_2C_{10}H_{21})_3]TcO_4$.

SOLUTION:

Complex (a) is uncharged, so is not likely to be water soluble (some solubility might result from hydrogen bonding of the coordinated NH_3 molecules to the water). The ligands do not have C–H or other nonpolar bonds, so the complex will also not be soluble in nonpolar organic solvents. Salt (b) is a salt of an (moderately) acidic cation and a (very strongly) basic anion, so it is insoluble in water; there is nothing organic in its structure to make it soluble in nonpolar organic solvents. Complex (c) contains a large nonacidic cation with a long organic chain and a small nonbasic anion, Cl^-. This salt would *probably* be soluble in water (as an exception to Solubility Tendency III) and should be soluble in nonpolar organic solvents; it should be a good (but expensive) phase-transfer catalyst. Complex (d) differs only in that its anion is a larger nonbasic anion. It probably will still be (at least somewhat) soluble in the nonpolar organic solvent, but will be insoluble in water.

The second group of aprotic solvents is that of **polar, strongly solvating** but **non-ionized** solvents. These solvents may dissolve a number of ionic solutes by virtue of their properties either as Lewis acids or bases. It is more important in overcoming the lattice energy of an ionic compound to solvate the small cation rather than the large anion, since the cation will generally provide the larger solvation energy; hence, most of these solvents are good *Lewis bases*. Numerous organic compounds fit into this group: For example, the ethers (R_2O), the ketones ($R_2C{=}O$) such as acetone, the esters (RCO_2R'), various sorts of organic nitrogen compounds (amines, nitriles, etc.), amides ($RCONR'_2$), and organic nitro compounds (RNO_2).

A few useful solvents are predominantly Lewis acids. Sulfur dioxide can be liquefied by mild cooling and is often used as a solvent. Some nonmetal halides such as SbF_5 and $SbCl_5$ are very strong Lewis acids and can be used to promote ionization of very weak Lewis bases:

$$(C_6H_5)_3CCl + SbCl_5 \rightarrow [(C_6H_5)_3C]^+ + SbCl_6^- \tag{4.21}$$

The third group of aprotic solvents have both Lewis acid and Lewis base properties (i.e., are *amphoteric*). These solvents, such as bromine trifluoride, can dissolve both types of solutes:

$$KF + BrF_3 \rightarrow K^+ + BrF_4^- \tag{4.22}$$

$$SbF_5 + BrF_3 \rightarrow BrF_2^+ + SbF_6^- \tag{4.23}$$

In a manner analogous to the autoionization of the amphoteric solvent water (into H^+ and OH^-), these are also **autoionizing** *solvents*:

$$2\,BrF_3 \rightleftharpoons BrF_2^+ + BrF_4^- \tag{4.24}$$

In this type of solvent, it is possible to use another definition of acids and bases, the **solvent system** definition. In this system, an acid is defined as *a species that increases the concentration of the cation characteristically produced by autoionization of the solvent* (BrF_2^+ in the case of BrF_3 or H_3O^+ in the case of water); a base is *a species that increases the concentration of the corresponding anion* (BrF_4^- and OH^-, respectively). Then the reverse of Eq. (4.24) can be called an acid–base neutralization reaction (analogous to the reaction of H^+ and OH^-): Titrations can be carried out using solutions of SbF_5 as the Lewis acid and KBr as the Lewis base. Due to their combined Lewis acid–base properties, the autoionizing solvents are quite reactive, for example, with the water of the air, and are rather difficult to handle without contamination.

Molten ionic salts may be considered to be a fourth class of aprotic solvents. A few of these that incorporate large nonacidic cations and large nonbasic anions of irregular shape have such low lattice energies that they are liquid at or just above room temperature, and can be useful for studying nonaqueous reactions involving dissolved ions. An example of such a solvent is *N*-butylpyridinium heptachlorodialuminate, $[C_5H_5NC_4H_9]^+[Al_2Cl_7]^-$.

Protic Solvents. Protic solvents are fundamentally like the aprotic solvents, but because of the presence of acidic hydrogen atoms, they possess two additional features that we may utilize. Because some form of hydrogen ion is released, it is possible to use the concept of pH. In addition, reactions analogous to hydrolysis can occur: Metal ions may enhance the acidity of such solutions to the point that protons are lost, and nonmetal anions may enhance the basicity to the point that protonation of the anion may occur. Such reactions, known as **solvolyses**, often give rise to new and useful chemical products; these will be discussed in Chapter 14.

The protic equilibria may be illustrated with a very important **basic protic solvent**, liquid ammonia (bp −33 °C). Like water, these solvents are amphoteric and undergo autoionization:

$$2\,NH_3 \rightleftharpoons NH_4^+ + NH_2^- \qquad (4.25)$$

In such a solution, the hydrogen ion is attached (via a coordinate covalent bond) to an ammonia molecule to give the ammonium ion, which is the analogue (in ammonia) to the hydronium ion H_3O^+ (in water). The equilibrium constant for this autoionization is about 10^{-27}, so it would be possible to define a pH scale for liquid ammonia in which $pH = -\log\,[NH_4^+]$. With this definition, a neutral solution (one in which the concentrations of NH_4^+ and NH_2^- were equal) would have a pH 13.5 rather than 7 as in water. It would be difficult, however, to meaningfully compare a pH measured in liquid ammonia with one measured in water.

The concept of "effective pH" has been devised to render such comparisons possible. The **effective pH** of a nonaqueous solution is the pH of a hypothetical aqueous solution in which a dissolved test acid would have the same equilibrium cation–anion ratio as found in the nonaqueous solution.[20] An equilibrium mixture of NH_4^+ and NH_2^- could not be achieved in water due to the complete reaction (leveling) of NH_2^- with water to give OH^- and NH_3, but we can compute that this would happen if a pH of 23.5 could be achieved in aqueous solution; hence, the "neutral" solution in liquid ammonia has an effective pH of 23.5. Since by aqueous standards this is very basic, ammonia is categorized as a basic solvent.

Figure 4.10 shows the effective pH ranges of various protic solvents. In each case, the neutral effective pH of that solvent is in the middle of the range shown; its acidic

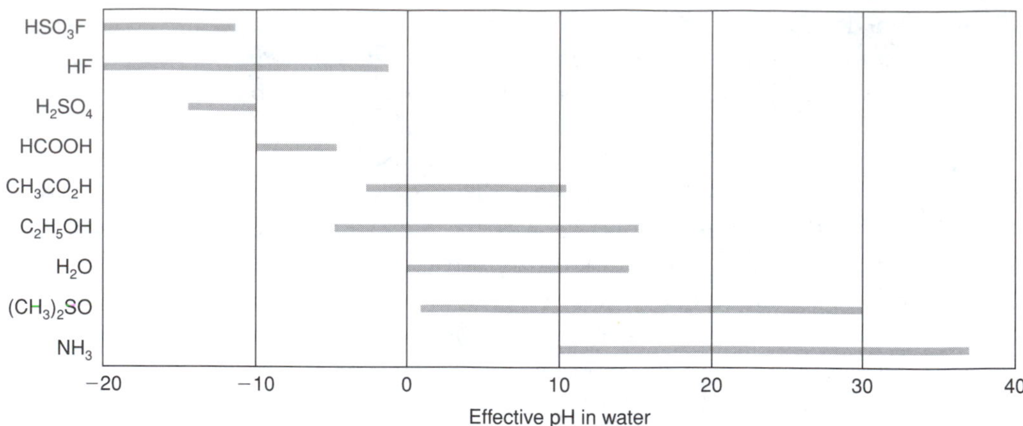

Figure 4.10
Effective pH ranges of various protic solvents. [Adapted from W. L. Jolly, *Modern Inorganic Chemistry*. Copyright 1984 by McGraw-Hill Book Company.]

pH values are to the left, and its basic pH values are to the right. If the middle of the effective pH range of a solvent falls to the right of the middle of the pH range of water (pH 7), we classify the solvent as a basic solvent [e.g., NH_3 and dimethyl sulfoxide, DMSO or $(CH_3)_2SO$]. Basic solvents such as NH_3 are useful for investigating the properties of very strongly basic anions, because they may not be very strongly basic relative to the solvent and will undergo less solvolysis than they would in water.

The solvents above ethanol (C_2H_5OH) in Figure 4.10 are among those in the group of **acidic protic solvents**; these are useful for studying the properties of very acidic cations. The most acidic solvent in Figure 4.10, fluorosulfonic acid (HSO_3F), may be further enhanced in acidity by the addition of the aprotic Lewis acid SbF_5. This acid combines with the fluorosulfonate anion (a *very* weak Lewis base indeed), shifting the ionization equilibrium farther to the right:

$$HSO_3F + SbF_5 = FSO_3SbF_5^- + H^+ \qquad \text{(solvated to give } H_2SO_3F^+\text{)} \qquad (4.26)$$

Such a solution is called a **superacid**, since it is capable of protonating any known Lewis base, no matter how feeble (e.g., Xe is converted, in part, to XeH^+). It is even capable of protonating some species that lack unshared electron pairs. For example, $(CH_3)_4C$ is protonated to form an intermediate cation $(CH_3)_4CH^+$, in which a *shared sigma-bond pair* is protonated. Such an electron pair is then shared among *three* atoms, which is a concept foreign to the valence bond theory used in Part I, but which is basic to the molecular orbital theory that we will cover in Part II.

Study Objectives

1. Predict whether a given salt of an oxo or fluoro anion is soluble or insoluble in water. Select a suitable counterion to precipitate a given ion. Exercises 1–13.
2. Know what entropy and enthalpy represent, and how they influence the tendency of a reaction to proceed. Exercises 14–17.
3. Compare different ions in terms of their primary and secondary hydration spheres, their hydration numbers, and their classifications as electrostatic structure makers, electrostatic structure breakers, or hydrophobic structure makers. Describe the structure of ice, liquid water, and of hydrated ions. Exercises 18–23.
4. Calculate energy terms from thermochemical cycles for the precipitation of ionic salts from water. Exercises 24–26.
5. Calculate the Madelung constant for a simple geometry. Exercise 27.
6. Calculate the lattice energy of a salt from its ionic radii, given a table of Madelung constants and Born exponents. Exercises 28–30.
7. Use the stoichiometry of a salt and the ratio of its ionic radii to predict the coordination numbers of its anions and cations and (if possible) the lattice type it adopts. Exercises 31–41.
8. In terms of entropy and enthalpy changes and the above classifications, explain *why* a given class of oxo salt is soluble or insoluble. Exercises 42–49.
9. Pick out examples of salts that are likely to be good desiccants and that are likely to crystallize from water as hydrates. Exercises 50–52.
10. Convert the molecular form of an equation to the net ionic form. Convert the net ionic form to a useful molecular form, and check in a handbook to determine whether the salts involved likely are hydrates. Exercises 53–56.
11. Predict the solubility of a given coordination compound or complex ion in (a) water; (b) nonpolar solvents. Exercises 57–60.
12. Classify different solvents as protic or aprotic, then also within the subgroups of these classifications; choose an appropriate solvent to study the properties of a strongly acidic cation or a strongly basic anion. Exercises 61–64.

Exercises

1. *Which of the following salts will be insoluble in water? $Co(NO_3)_2$, $CsBrO_4$, $CePO_4$, Cs_3AsO_4, $BaSeO_4$, $Hg_5(IO_6)_2$, or TiO_2.
2. Which of the following will be insoluble in water? $CuMoO_4$, Cs_2MoO_4, Cr_2TeO_6, $Cr(NO_3)_3$, HgO, NaOH, $RbTcO_4$, KIO_4, or K_5IO_6.
3. Which of the following will be insoluble in water? $Co(ReO_4)_2$, $CsReO_4$, K_2FeO_4, Al_2TeO_6, EuO; or Ga_2O_3.

4. *Write general chemistry type specific solubility rules for the salts of the following oxo anions: (a) chromate; (b) ferrate; (c) pertechnetate; (d) silicate; (e) tellurate. These rules should state whether most salts of this anion are soluble or insoluble, then list some ions that are exceptions.

5. Liming a lake to neutralize acid rain in it could seriously affect the availability of nutrients such as phosphate and molybdate, but not nitrate. Explain.

6. If you did not do the experiment at the beginning of this chapter, go back and try to apply the principles of this chapter to predict what would have happened at each step. Also answer the questions included.

7. Consider each of the following cations: U^{3+}; Rb^+; Pb^{4+}; and Al^{3+}. (a) Classify the acidity of each of these cations. (b) Which of these cations give a precipitate with the ClO_4^- ion? (c) Which of these cations give a precipitate with the PO_4^{3-} ion?

8. Assume that the following elements will form oxo anions when they are in the indicated oxidation states: Tc(VII); Pu(VIII); and As(III). (a) Predict the formula of the oxo anion formed by each element in that oxidation state. (b) Name each oxo anion. (c) Classify the basicity of each oxo anion. (d) Predict whether the K^+ salt of each will be insoluble. (e) Predict whether the Fe^{3+} salt of each will be insoluble.

9. When you mix an aqueous solution of the Hg^{2+} ion with an aqueous solution of the SiO_4^{4-} ion, you obtain two precipitates, one yellow and one white. (a) What are the logical formulas for those two precipitates? (b) Design a simple mixing experiment using Hg^{2+} and another anion to determine which precipitate is which compound. (c) When you mix Hg^{2+} with PO_4^{3-} instead of with SiO_4^{4-}, you obtain only one precipitate. Why would PO_4^{3-} be less likely to give the second precipitate?

10. *In your organic lab, you have just synthesized the listed new ions in solution. Tell how you would isolate each ion from solution to get a stable crystalline solid. (a) Tropylium ion, $C_7H_7^+$. (b) Squarate ion, $C_4O_4^{2-}$. (c) Cyclopentadienide ion, $C_5H_5^-$.

11. What are the possible insoluble salts that can arise on mixing solutions of the following cations—Pu^{4+}, $[(CH_3)_4N]^+$, Eu^{2+}—with solutions of the following anions—PF_6^-, PO_4^{3-}, SO_4^{2-}?

12. In order to get tenure and a promotion at your school, you have to publish some new laboratory experiment. You decide to devise a qualitative analysis scheme of the anions that use three cations to separate the following anions by precipitation: ReO_4^-; SeO_4^{2-}; SeO_3^{2-}; PO_4^{3-}; SiF_6^{2-}; and OH^-. (a) Your anion group I is precipitated with the $[(CH_3)_4N]^+$ ion. Which anions will precipitate in this group? Write the formulas of the salts that will precipitate. (b) Your anion group II is precipitated with the Sr^{2+} ion. Write the formulas of the salts that will precipitate in this group. (c) Your anion group III is precipitated with the Ga^{3+} ion. Write the formulas of the salts that will precipitate in this group.

13. The generalized solubility rules and tendencies implicitly treat the phenomenon of precipitation as a type of acid–base reactivity. However, uncharacteristically one of the solubility rules and tendencies shows that there is (precipitation) reactivity between two types of acid–base species that have completely *overlapping* predominance areas. Which two types of species are these, and which solubility rule or tendency is involved?

14. *Which of the following represent reactions that have favorable entropy terms ($-T\Delta S$ at $T > 0$ K)? What are the signs of the $-T\Delta S$ terms? (a) Egg → broken egg, contents spilled; (b) unassembled jigsaw puzzle → assembled jigsaw puzzle; (c) crystalline solid → gaseous cations + gaseous anions; (d) liquid water → steam; (e) liquid water → ice; (f) 50 coins in roll, heads all up → 50 coins in roll but some heads up, some down.

15. *Which of the changes in Exercise 14 have favorable enthalpy terms: Which unfavorable? Which are zero? What is the sign of each ΔH term?

16. What will most favor the changes in Exercise 14: high temperatures or low? (If necessary, interpret higher temperature as increased agitation of the items described.)

17. Tell which of each of the following changes favor products and which favor reactants (left side). In each case, tell whether your answer is due mainly to a favorable entropy term [in Eq. (4.2)], an unfavorable entropy term, a favorable enthalpy term, or an unfavorable enthalpy term (in some cases more than one answer is possible). Briefly explain why each answer is correct. (a) Broken glass beaker → reassembled intact glass beaker; (b) coal (C) + O_2(g) → CO_2(g); (c) unexploded nitroglycerine → exploded gaseous products; (d) Humpty Dumpty (an egg) on top of the wall → Humpty Dumpty after the fall; (e) an unassembled jigsaw puzzle on the table → an assembled jigsaw puzzle on the same table.

18. Describe the structure of ice and of liquid water, and explain why ice floats in water.

19. *Select one of the following ions {Al^{3+}, Li^+, Rb^+, or $[(C_4H_9)_4N]^+$} as an example of each of the following: (a) the ion that attaches the fewest water molecules when it dissolves in water; (b) the ion that forms the largest hydrated ion; (c) a hydrophobic structure maker; (d) an electrostatic structure maker; (e) an electrostatic structure breaker.

20. Select one of the following ions (Fr^+, Fe^{3+}, Li^+, or Ac^{3+}) as an example of each of the following: (a) the ion that attaches the greatest number of water molecules when it dissolves in water; (b) the ion that forms the smallest hydrated ion; (c) an electrostatic structure maker; (d) an electrostatic structure breaker.

21. *Consider the following set of ions: K^+; Zn^{2+}; Al^{3+}; $SiO_4{}^{4-}$; $ClO_4{}^-$; and $SeO_4{}^{2-}$. (a) Which cation will form the largest hydrated ion? (b) Which cation(s) will be electrostatic structure makers? (c) Which cation(s) will disrupt the icebergs in liquid water? (d) Which anion is most likely to be an electrostatic structure breaker? (e) Which cation(s) will give insoluble salt(s) with the $ClO_4{}^-$ ion? (f) Which cation(s) will give insoluble salt(s) with the $SiO_4{}^{4-}$ ion? (g) Write the formula of any salt containing just these ions that will precipitate for reasons connected with an entropy change. (h) Write the formula of any soluble salt containing just these ions and explain (in terms of lattices, etc.) why it is soluble.

22. Consider the following set of ions: Cs^+; Li^+; Ga^{3+}; $CO_3{}^{2-}$; $MnO_4{}^-$; and $XeO_6{}^{4-}$. (a) Which positive ion (cation) will form the largest hydrated ion? (b) Which cation(s) will be electrostatic structure makers? (c) Which cation(s) will disrupt the icebergs in the liquid water? (d) Which anion is most likely to be an electrostatic structure breaker? (e) Which cation(s) will give insoluble salt(s) with the $MnO_4{}^-$ ion? (f) Which cation(s) will give insoluble salt(s) with the $CO_3{}^{2-}$ ion?

23. Fill in the blanks with the appropriate term from the following list (soluble salts, precipitates, makers, breakers, ΔH, $-T\Delta S$), then complete the explanation in your own words. (a) Acidic cations and basic anions tend to give ____. Thermodynamically this is favored because of the ____ term in ΔG. Acidic cations and basic anions are classified as electrostatic structure ____ because: (b) Nonacidic metal cations are classified as electrostatic structure ____ because: (c) Cross-combinations (such as nonacidic cations plus basic anions) tend to give ____. Thermodynamically, this is favored because of the ____ term in ΔG. The reason that this happens is ____.

24. *Enthalpies of precipitation are known for the following halides: LiF, $-5\,\text{kJ mol}^{-1}$; NaCl, $-4\,\text{kJ mol}^{-1}$; AgCl, $-65\,\text{kJ mol}^{-1}$; $CaCl_2$, $+83\,\text{kJ mol}^{-1}$. Use data on hydration enthalpies (Tables 2.1 and 2.4) to calculate the lattice energies of these compounds.

25. (a) Use the known lattice energies for $CaSO_4$, $-2653\,\text{kJ mol}^{-1}$, $SrSO_4$, $-2603\,\text{kJ mol}^{-1}$, and $BaSO_4$, $-2423\,\text{kJ mol}^{-1}$, along with data from Tables 2.1 and 4.1, to compute the hydration energy of the sulfate ion. How well do your results agree for each of these three salts? How well do they agree with the value quoted in Table 2.4? (b) Compare your result with the hydration energy of the perchlorate ion, which has a similar size, and explain any differences between the two.

26. Usually, ΔH° for the precipitation of an ionic salt is about equal to zero. Use this approximate value and the hydration energies for Al^{3+} and for ClO_4^- (Tables 2.1 and 2.4) to estimate the lattice energy of aluminum perchlorate.

27. Calculate the Madelung constant for (a) a simple cube consisting of four cations alternating corners with four anions; (b) a gaseous "ion triplet" consisting of a cation enclosed by two anions 180° apart from each other; (c) a similar "ion triplet" in which the anions are 90° apart; (d) an infinite linear arrangement of alternating cations and anions. This case involves an infinite series: Do you know the sum of this series from your calculus course? If not, add up the first 10 terms.

28. (a) Calculate the theoretical lattice energy for the salt CaF_2. (b) The hydration enthalpy of Ca^{2+} is $-1592\,\text{kJ mol}^{-1}$; the hydration enthalpy of F^- is $-497\,\text{kJ mol}^{-1}$. Calculate the enthalpy of precipitation of CaF_2. (c) Ignoring entropy effects, use the results of your calculations to predict whether CaF_2 is soluble or insoluble. Can entropy effects generally safely be ignored in questions like this?

29. Consider the ionic salt ThS_2, which adopts the TiO_2 lattice type. (a) Estimating a suitable Born exponent for the thorium ion, calculate the lattice energy of ThS_2. (b) If the hydration energies of the thorium and sulfide ions are -6136 and $-1372\,\text{kJ mol}^{-1}$, respectively, calculate ΔH_{pptn} for ThS_2.

30. In honor of the 50th anniversary of the United Nations, let us consider the hypothetical ionic salt uranium(III) nitride. (a) Write the formula of uranium(III) nitride. (b) Let us assume that it adopts the NaCl lattice type. Compute the expected lattice energy for uranium(III) nitride, extrapolating a reasonable value for the Born exponent for uranium.

31. Select the most likely lattice types for each of the following salts: (a) BeF_2; (b) CaO; (c) BeI_2; (d) BeTe; and (e) CaF_2.

32. *Using Tables C and 4.4: (a) calculate the radius ratio for CeO_2 and predict the coordination numbers of the cerium and oxide ions and the type of lattice that CeO_2 would adopt and (b) calculate the lattice energy of CeO_2.

33. Cesium and gold form an ionic compound, Cs^+Au^-, with a cesium–gold distance of 369 pm. (a) What type of lattice will CsAu adopt? [See U. Zachwieja, *Z. Anorg. Allg. Chem.*, **619**, 1095 (1993) for the actual lattice type.] (b) Estimating a suitable Born exponent for the Au^- ion, calculate the lattice energy of CsAu. (c) Assuming that the enthalpy of precipitation of CsAu is zero, calculate the enthalpy of hydration of the Au^- ion.

34. *Assume that the rare, radioactive elements astatine and polonium form anions At^- and Po^{2-}, each of which has a radius of about 226 pm. Which of the following lattice types (ZnS, NaCl, CsCl, SiO_2, TiO_2, or CaF_2) would be expected to be adopted by (a) the astatide of each Group 1 metal (Na through Cs); (b) the astatide of each Group 2 metal (Mg through Ba); (c) the polonide of each Group 2 metal (Mg through Ba).

35. Using your results from Exercise 34 and Table 4.4, calculate the lattice energy of barium astatide. If we assume that the enthalpy of solution of barium astatide in water is zero, calculate the hydration energy of the astatide ion. Draw the hydrated astatide ion, showing how the water molecules would orient themselves.

36. The elements radium and astatine both arise during the radioactive decay of uranium. Assume that they get together in the form of the salt $RaAt_2$; let us take the radius of the Ra^{2+} ion as 162 pm and the radius of the At^- ion as 226 pm. (a) Predict the coordination numbers of Ra and At and the type of lattice that $RaAt_2$ will adopt. (b) Calculate the lattice energy of $RaAt_2$. (c) The hydration enthalpy of the Ra^{2+} ion is $-1259\ \text{kJ mol}^{-1}$; we may extrapolate an estimated hydration enthalpy for the At^- ion of $-260\ \text{kJ mol}^{-1}$. Calculate the enthalpy of precipitation of $RaAt_2$.

37. The elements polonium and thorium both arise during the radioactive decay of uranium. Assume that they get together in the form of the salt $ThPo_2$; let us take the radius of the Th^{4+} ion as 108 pm and the radius of the Po^{2-} ion as 226 pm. (a) Predict the coordination numbers of Th and Po and the type of lattice that $ThPo_2$ will adopt. (b) Calculate the lattice energy of $ThPo_2$. (c) If we assume that the enthalpy of solution of $ThPo_2$ is zero, calculate the hydration enthalpy of the Po^{2-} ion.

38. *The separation of the *f*-block elements of the sixth period (the "lanthanides" or "rare earths") and the associated Group 3 ions Sc^{3+} and Y^{3+} is a very tricky business due to the identical +3 charges they normally show in their cations, and the close similarity of their ionic radii. In the early days, the separation was attempted by fractional crystallization, which would work if the ions would crystallize in different lattice types. Assume for simplicity that you have only the largest and the smallest of these ions, La^{3+} and Sc^{3+}, and that you want to crystallize their nitrides. (The ionic radius of N^{3-} is 132 pm.) (a) Predict the coordination numbers that will be found in LaN. Which lattice type is expected for LaN? (b) Predict the coordination numbers that will be found in ScN. Which lattice type is expected for ScN? (c) Calculate the lattice energy of lanthanum nitride, LaN.

39. Calculate radius ratios to determine the most likely lattice type (or if this cannot be identified, the cation and anion coordination numbers) for (a) UN; (b) UN_2; and (c) U_3N_4.

40. All but one of the halides in Exercise 24 (which one?) in fact have the NaCl structure. Calculate the lattice energies of these compounds from their ionic radii, Madelung constants, and so on. (Use the radius ratio to choose the lattice type for the halide with the non-NaCl structure.) For which halide is there the greatest discrepancy between your answers to Exercise 24 and to this exercise? How does the cation in this halide differ from the other three cations, and how might this explain the discrepancy?

41. Life is built on the chemistry of carbon, whereas the solid earth that supports life is built on the chemistry of the element below it, silicon. Although the "cations" of these elements, C^{4+} and Si^{4+}, have no real independent existence, we can understand some of the profound differences in the chemistry of these two elements on earth starting from the ionic radii of these "cations" (Table C). (a) Combination of these cations with an appropriate number of oxide ions gives the characteristic oxo anion of each element. Use radius ratios to predict the formulas of the oxo anions of carbon and silicon. Should the two oxo anions contain the same number of oxygens? (b) What are the actual formulas of the carbonate and simple silicate ions? Contrast the Lewis structures of these in terms of types of bonding available to silicon and carbon. (c) Which anion, carbonate, or silicate should be the more basic? Which (if either) might actually exist in water at some appropriate pH?

42. *Give an example of a salt composed of a cation and an anion, each of which is an electrostatic structure maker. Is such a salt likely to be soluble or insoluble? Give the physical reason for this (in terms of the structure of the hydrated ions, or the suitability of its lattice, etc.). Is this solubility/insolubility associated with entropy or enthalpy effects?

43. Aluminum phosphate is quite insoluble. Which one of the following effects is responsible for its insolubility? (a) An enthalpy effect resulting from the very stable lattice formed; (b) An entropy effect resulting from the very stable lattice formed; (c) An enthalpy effect resulting from the release of numerous waters of hydration when forming aluminum phosphate from its ions; (d) An entropy effect resulting from the release of numerous waters of hydration when forming aluminum phosphate from its ions; (e) an enthalpy effect resulting from the large ions disrupting the structure of liquid water; (f) an entropy effect resulting from the large ions disrupting the structure of liquid water.

44. *Explain, with reference to the structure of water, of crystal lattices, and/or of hydrated ions, why: (a) iron(III) perchlorate is soluble in water; (b) iron(III) phosphate is insoluble in water; (c) rubidium permanganate is fairly insoluble in water; (d) tetrabutylammonium tetrabutylborate is insoluble in water.

45. Generalized Solubility Tendency IV in Section 4.2 says that many feebly acidic cations and feebly basic anions give insoluble salts. Both Li^+ and Ba^{2+} are feebly acidic cations, and NO_2^-, ClO_2^-, SO_4^{2-}, and SeO_4^{2-} are all feebly basic anions. Although $BaSO_4$ and $BaSeO_4$ are indeed insoluble, $Ba(NO_2)_2$, $Ba(ClO_2)_2$, Li_2SO_4, and Li_2SeO_4 are all soluble. Can you explain this?

46. *(a) Assign each of the cations Sr^{2+}, Ga^{3+}, and $[(CH_3)_4N]^+$ to one of the following acidity categories: nonacidic; feebly acidic; acidic. (b) Which one of these cations will form the largest hydrated ion? (c) Which cation(s) will be electrostatic structure makers? (d) Which cation will give the least soluble perchlorate? Briefly explain why that perchlorate is insoluble. (e) Briefly explain why the perchlorates of the other two cations will be soluble.

47. Consider the following set of ions: Cs^+; Li^+; Ga^{3+}; O^{2-}; ReO_4^-; and TeO_6^{6-}. (a) Which positive ion (cation) will form the largest hydrated ion? (b) Which cation(s) will be electrostatic structure makers? (c) Which cation(s) will disrupt the icebergs in the liquid water? (d) Which anion is most likely to be an electrostatic structure breaker? (e) Write the formula of a salt containing a cation and anion from this set that will be insoluble because there would be a favorable entropy change upon precipitation. (f) Write the formula of a salt containing a cation and anion from this set that will be soluble because there would be an unfavorable enthalpy change upon its precipitation.

48. Explain, with reference to the structure of water, of crystal lattices, and of hydrated ions, why: (a) aluminum perchlorate is soluble in water; (b) cesium hexafluorophosphate is insoluble in water; (c) yttrium phosphate is insoluble in water.

49. Consider the following three cations: In^{3+}; Ca^{2+}; and $[(C_4H_9)_4N]^+$. (a) Which one of these cations will form the largest hydrated ion? (b) Which cation(s) will be hydrophobic structure makers? (c) Which cation will give the least soluble perchlorate? Briefly explain why that perchlorate is insoluble. (d) Briefly explain why the perchlorates of the other two cations will be soluble. (e) The perchlorate of which cation would be most likely to be a useful desiccant? When crystallized from water, would this perchlorate most likely be a hydrate?

50. Will $BeSO_4$ or $BaSO_4$ be more likely to be (a) a good desiccant, or (b) crystallize as a hydrate?

51. Consider the following ions: Be^{2+}; Cs^+; Ba^{2+}; [Ba(crown ether)]$^{2+}$; Ga^{3+}; ReO_4^-; FeO_4^{2-}; O^{2-}; TeO_6^{6-}; SiF_6^{2-}. (a) Assign each of the ions to one of the following six categories: nonacidic, feebly acidic, acidic, nonbasic, feebly basic, basic. (b) For each cation pick one of the following anions—ReO_4^-; FeO_4^{2-}; O^{2-}; TeO_6^{6-}; SiF_6^{2-}—that should give a water-insoluble salt with the cation; then write the formula of the resulting insoluble salt. (c) From the original list of anions and cations in this question, pick any combination that will produce a salt that is likely to be a good desiccant. Write its formula. Why will this salt likely form a hydrate?

52. *Among the possible salts that can arise on mixing solutions of the following cations—Pu^{4+}, $[(CH_3)_4N]^+$, or Eu^{2+}—with solutions of the following anions—PF_6^-, PO_4^{3-}, or SO_4^{2-}: (a) which one salt is most likely, when its aqueous solution is evaporated, to crystallize as a hydrate? (b) which one salt, if synthesized by some method not involving water, is most likely to be a good desiccant?

53. *Rewrite each of the following molecular equations as a net ionic equation:

(a) $Ba(ClO_4)_2(aq) + H_2SO_4(aq) \rightarrow BaSO_4(s) + 2\,HClO_4(aq)$

(b) $Na_2SeO_3(aq) + H_2SO_4(aq) \rightarrow H_2SeO_3(aq) + Na_2SO_4(aq)$

(c) $2\,H_2SO_4(aq) + Na_4SiO_4(aq) \rightarrow Si(OH)_4(s) + 2\,Na_2SO_4(aq)$

(d) $3\,Th(NO_3)_4(aq) + 4\,K_2NaPO_4(aq) \rightarrow$
$Th_3(PO_4)_4(s) + 8\,KNO_3(aq) + 4\,NaNO_3(aq)$

(e) $2\,NaHSO_3(aq) + H_2SO_4(aq) \rightarrow Na_2SO_4(aq) + 2\,H_2O(l) + 2\,SO_2(g)$

54. Convert the following to net ionic equations.

(a) $Sr(ClO_4)_2(aq) + H_2SeO_4(aq) \rightarrow SrSeO_4(s) + 2\,HClO_4(aq)$

(b) $Na_2SeO_3(aq) + H_2SeO_4(aq) \rightarrow H_2SeO_3(aq) + Na_2SeO_4(aq)$

(c) $Sr(ClO_2)_2(aq) + H_2SO_4(aq) \rightarrow SrSO_4(s) + 2\,HClO_2(aq)$

(d) $2\,La(NO_3)_3(aq) + CuSO_4(aq) + 8\,NaOH(aq) \rightarrow$
$La_2CuO_4(s) + Na_2SO_4(aq) + 6\,NaNO_3(aq) + 4\,H_2O$

(e) $Sr(ClO_4)_2(aq) + Li_2SeO_4(aq) \rightarrow SrSeO_4(s) + 2\,LiClO_4(aq)$

(f) $Al_2(SeO_3)_3(s) + 3\,H_2SeO_4(aq) \rightarrow 3\,H_2SeO_3(aq) + Al_2(SeO_4)_3(aq)$

55. *Using a chemistry handbook, write specific instructions (compounds used and weight of each compound needed) to carry out the following net ionic equations and produce 0.20 mol of product: (a) $3\,Co^{2+}(aq) + 2\,VO_4{}^{3-}(aq) \rightarrow Co_3(VO_4)_2(s)$; (b) $2\,Ag^+(aq) + SO_4{}^{2-}(aq) \rightarrow Ag_2SO_4(s)$.

56. Using a chemistry handbook, write specific instructions to carry out the following net ionic equations and produce 0.20 mol of product: (a) $3\,Mg^{2+}(aq) + 2\,AsO_4{}^{3-}(aq) \rightarrow Mg_3(AsO_4)_2(s)$; (b) $3\,UO_2{}^{2+}(aq) + 2\,VO_4{}^{3-}(aq) \rightarrow (UO_2)_3(VO_4)_2(s)$; (c) $3\,Sr^{2+}(aq) + 2\,AsO_4{}^{3-}(aq) \rightarrow Sr_3(AsO_4)_2(s)$, with 0.10 mol of product desired.

57. *Consider the following forms of mercury: $Hg(CH_3)_2(\ell)$, $HgCl_2(aq)$, HgS(s, insoluble). Which form or forms will (a) pass unabsorbed through the digestive system (absorption requires solubility in lipids or water); (b) be most easily eliminated in urine through the kidneys; (c) most readily cross the (nonpolar) blood–brain barrier; or (d) most readily undergo bioamplification?

58. Consider the following forms of platinum: $[Pt(CH_2CH_3)_4]$; $Li_2[PtCl_6]$; and insoluble PtS. Which form will most likely (a) be soluble in water but not nonpolar organic solvents; (b) be soluble in nonpolar organic solvents but not water; (c) pass unabsorbed through the digestive system; (d) be eliminated from the body through the kidneys; or (e) cross the (nonpolar) blood–brain barrier.

59. Tell whether the following salts or complexes are likely to be soluble in water, in nonpolar solvents, or in neither: (a) FeO; (b) $KClO_4$; (c) $[(C_3H_7)_4N]Cl$; (d) $[(C_3H_7)_4N]PF_6$; (e) $Ba(NO_3)_2$; (f) [Ba(crown ether)]$(NO_3)_2$; (g) $[Cr(NH_3)_3Cl_3]$; or (h) $[Cr(NH_2C_8H_{17})_3Cl_3]$.

60. Of the compounds $Fe(BrO_4)_3$, $Zn_5(IO_6)_2$, tetraethylammonium perchlorate, and $BaFeO_4$: (a) Which one is the best desiccant and is obtained as a hydrate upon evaporating its aqueous solution? (b) Which one is most likely to be soluble in body fat?

61. Classify the following nonaqueous solvents, first as acidic, basic, or nonpolar, and second, as protic or aprotic: (a) hexane, (b) ammonia; (c) HSO_3F; (d) pyridine; (e) sulfur dioxide; (f) benzene; (g) HF; (h) diethyl ether; or (i) $SbCl_5$.

62. *With the use of Figure 4.10, select a suitable nonaqueous solvent in which to do an acid–base titration of (a) the nonmetal anions listed in Table 2.5; (b) the +4 charged ions in Table 2.2; (c) the first ionizable proton of telluric acid; (d) the first ionizable proton of selenic acid; (e) the third ionizable proton of telluric acid; and (f) the ions XeH^+ and $[(CH_3)_4CH]^+$.

63. What solvents would you choose to enhance (a) the acidic properties of acetone, $(CH_3)_2C{=}O$; (b) the basic properties of HNO_3; (c) the basic properties of SiH_4; and (d) the acidic properties of SiH_4? Write an equation showing a reasonable reaction of each solute with its solvent.

64. When ionic compounds are dissolved in water, not only primary but also secondary hydration spheres are formed around the ions. Would you expect primary and/or secondary solvation spheres to form around ions dissolved in (a) nonpolar aprotic solvents; (b) polar aprotic solvents; or (c) protic solvents?

65. In the Smoking Hills of the Northwest Territories of Canada, on the shores of the Arctic ocean, are exposed seams of peat and coal that were ignited by lightning strikes (probably) thousands of years ago, and have been burning ever since, emitting SO_2, which has resulted in acid rain in the vicinity. Some of the nearby arctic lakes have pH values as low as 1.0. (a) The acid rain has dramatically enhanced the concentrations of some metal ions in these lakes. List some metal ions that would have dramatically increased concentrations and some that would be much less affected, explaining your choices. (b) Which ions in these lakes would form the largest hydrated ions? Which ions would most affect the structure of the liquid water in the lakes? Describe the structure of liquid water and how it affects the density of water; tell how this density factor changes the stratification of lakes as the seasons pass. (c) Adding what kinds of oxides to these lakes would most likely remove the metal ions that are in elevated concentrations? Write an equation for such a process.

Notes

1. Annual editions, Chemical Rubber Publishing Company, Cleveland, OH.
2. D. A. Johnson (*Some Thermodynamic Aspects of Inorganic Chemistry*, Cambridge University Press, Cambridge, UK, 1968; Chapter 5) similarly analyzed solubility tendencies mathematically in terms of the radii and of the charges of the cations and anions involved, and R. Rich (*Periodic Correlations*, Benjamin, New York, 1965, Chapter 8) and M. Monroe and K. Abrams [*J. Chem. Educ.*, **61**, 885 (1984)] analyzed them in terms of the basicity of the anions involved. Obviously, the approaches are related since acidity and basicity of cations and anions are functions of their charges and radii.
3. F. Basolo, *Coord. Chem. Rev.*, **3**, 213 (1968).
4. Another example is SiF_6^{2-}. Although we classified it as nonbasic, and it does precipitate nonacidic cations, due to its −2 charge it must be nearly feebly basic: It also precipitates feebly acidic cations.
5. In region IVB of Figure 4.1, we find F^- manifesting its weak basicity more clearly: The fluorides of most acidic cations are relatively insoluble.
6. W. M. Latimer, *The Oxidation States of the Elements and their Potentials in Aqueous Solution*, 2nd ed., Prentice-Hall, Englewood Cliffs, NJ, 1952.
7. B. G. Cox and A. J. Parker, *J. Am. Chem. Soc.*, **95**, 6879 (1973).
8. C. S. G. Phillips and R. J. P. Williams, *Inorganic Chemistry*, Oxford University Press, New York, 1965; pp. 258–261.
9. J. A. Duffy and M. D. Ingram, *Inorg. Chem.*, **17**, 2798 (1978).
10. Y. Marcus, *Chem. Rev.*, **88**, 1475 (1988); H. Ohtaki and T. Radnai, *Chem. Rev.*, **93**, 1157 (1993).
11. R. P. Hanzlik, *Inorganic Aspects of Biological and Organic Chemistry*, Academic, New York, 1976; p. 61.
12. J. Burgess, *Metal Ions in Solution*, Ellis Horwood, Chichester, UK, 1978; pp. 16–17; H. S. Frank and W.-Y. Wen, *Discuss. Faraday Soc.*, **24**, 133 (1957).
13. Since the data on enthalpies of precipitation (Table 4.1) are rather limited, lattice energies are normally calculated via another thermochemical cycle based on oxidation–reduction reactions: This *Born–Haber cycle* is covered in Chapter 7 along with other thermochemical analyses of oxidation–reduction reactions.
14. A. F. Wells, *Structural Inorganic Chemistry*, 4th ed., Oxford University Press, Oxford, UK, 1975.
15. L. C. Nathan, *J. Chem. Educ.*, **62**, 215 (1985).

16. A. F. Kapustinskii, *Q. Rev. Chem. Soc.*, **10**, 283 (1956); see also T. C. Waddington, *Adv. Inorg. Chem. Radiochem.*, **1**, 157 (1959). An extension of this equation to deal with lattice energies of mixed metal oxides (Chapter 13) has been proposed: L. Glasser, *Inorg. Chem.*, **34**, 4935 (1995).
17. K. B. Yatsimirskii, *Izv. Akad. Nauk SSSR, Otdel. Khim. Nauk*, 453 (1947); 398 (1948).
18. D. F. C. Morris, *Struct. Bonding* (*Berlin*), **6**, 157 (1969).
19. If the cation is strongly acidic and/or the anion is strongly basic, hydrogen ions may be more abundant than the desired cation, or hydroxide ions may be more abundant than the desired anion. The hydroxide of the cation, or the oxo acid form of the anion, may then precipitate along with or instead of the desired salt. Sometimes this may be avoided by the proper order of addition or by techniques such as homogeneous precipitation—the literature should be consulted for techniques appropriate for the desired salt.
20. See, for example, W. L. Jolly, *Modern Inorganic Chemistry*, McGraw-Hill, New York, 1984; p. 202.

CHAPTER
5

Trends in Coordination Equilibria

With Applications to Biochemistry, Environmental Chemistry, Geochemistry, and Medicinal Chemistry

The majority of organic compounds and inorganic anions possess donor atoms and can act as ligands. Since there are so many different metal ions, there are a limitless number of coordination compounds that may be formed in natural waters, in biological fluids, or in industrial process waters or wastewaters. Important properties of metal ions, such as their solubilities in fat or water and their toxicities, will depend on which of the countless number of possible ligands ends up attached to the metal ion. It is no longer sufficient to determine analytically the concentration of a given element in a biological or environmental sample, it is also important to know the chemical forms in which the element is present, which is referred to as the **speciation** of the element.

From the mathematical point of view, the question, Which metal ion will combine with which ligand?[1] is rather formidable. In principle, we must measure the equilibrium constants relating to the stepwise or overall formation of each possible combination of metal ion and ligand. We must know the initial concentration of each metal ion and each ligand in a given natural water, biological fluid, or analytical sample, and the pH and the reducing potential of the sample. All this information, and a large computer, are needed to calculate the equilibrium concentration of each possible complex. Fortunately, there are two very important qualitative generalizations that we will emphasize in this chapter that will enable us to predict some of the most likely complexes in a complicated mixture of metal ions and ligands.

5.1 Classification of Ligands

Since so many organic and inorganic molecules have so many unshared pairs of electrons in so many geometric arrangements, it is quite useful to develop some general classes of ligands. There are several ways of doing this. The first way we will examine is structural, based on the number of unshared electron pairs geometrically available for donation to a Lewis acid or to different Lewis acids.

Monodentate Ligands. The simplest Lewis bases have only one donor atom and only one unshared pair of electrons. Clearly, such ligands can form only one bond to one metal ion or other Lewis acid. Such ligands are called **monodentate** ligands, as are monoatomic anions that have more than one electron pair, such as the halide and oxide ions, in situations in which they donate to only one metal ion.

Bridging Ligands. Since the halide and other monoatomic anions (and many of their partially protonated or organic-substituted derivatives) have more than one unshared electron pair, the second (and third) electron pair may be donated to a second (and a third) metal ion. A ligand that donates more than one electron pair to more than one metal ion or other Lewis acid is known as a **bridging ligand**. This tendency to act as a bridging ligand is often manifested in place of donating the second or third pairs to the original metal ion, since in the latter cases double or triple bonds are formed, which involve π bonds that are normally weaker than σ bonds.

Many important bridging ligands are linear, with donor atoms at the two ends of the ligand. Among the most important of these ligands are some of the **pseudohalide ions**. These ligands got their name because of certain resemblances in their acid–base and oxidation–reduction chemistry to the halide ions; these are shown in Figure 5.1. Like the halide ions, the pseudohalide ions have not only the capability of functioning as bridging ligands, but also have relatively weak basicity. In redox chemistry, the **cyanide**, **thiocyanate**, and **thiosulfate** ions resemble the halide ions in that oxidation produces dimeric species (known as cyanogen, thiocyanogen, and tetrathionate ions, respectively):

$$2\,Cl^- \rightarrow Cl_2(g) + 2\,e^- \tag{5.1}$$

$$2\,CN^- \rightleftharpoons (CN)_2(g) + 2\,e^- \tag{5.2}$$

$$2\,SCN^- \rightleftharpoons (SCN)_2(\ell) + 2\,e^- \tag{5.3}$$

$$2\,S_2O_3{}^{2-} \rightleftharpoons S_4O_6{}^{2-}(aq) + 2\,e^- \tag{5.4}$$

The **azide** ion is a very powerful reducing agent that reacts to give, not a dimer, but 3 mol of (very stable) nitrogen molecules:

$$2\,N_3{}^- \rightarrow 3\,N_2 + 2\,e^- \tag{5.5}$$

The resemblance of the **cyanate** ion, NCO^-, to the halide ions lies more in the area of acid–base than of redox chemistry.

Ligands such as these anions are capable of donating electron pairs to different atoms and thus are *potential* bridging ligands. In many complexes, however, this potential is unfulfilled, and only one donor atom is used to coordinate one metal ion. If a ligand has two *different types* of donor atoms, but uses only one, a choice as to which donor atom is used must be made. Ligands that can use either of two (or more) different donor atoms are called **ambidentate ligands.**[2] In some cases, different complexes are known in which the donor atom is chosen differently; such complexes are known as **linkage isomers** of each other.

Chelating Ligands. Many ligands with more than one donor atom have geometries such that they can form more than one coordinate covalent bond to the same metal

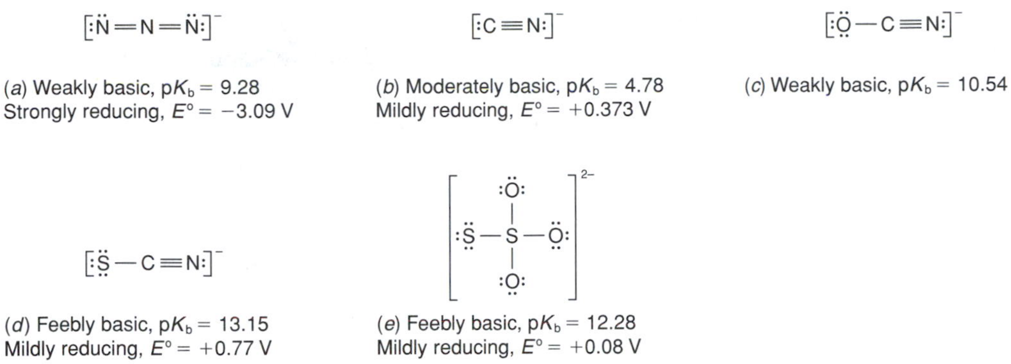

$[:\ddot{N}=N=\ddot{N}:]^-$

(*a*) Weakly basic, pK_b = 9.28
Strongly reducing, $E°$ = −3.09 V

$[:C\equiv N:]^-$

(*b*) Moderately basic, pK_b = 4.78
Mildly reducing, $E°$ = +0.373 V

$[:\ddot{O}-C\equiv N:]^-$

(*c*) Weakly basic, pK_b = 10.54

$[:\ddot{S}-C\equiv N:]^-$

(*d*) Feebly basic, pK_b = 13.15
Mildly reducing, $E°$ = +0.77 V

$[S_2O_3]^{2-}$

(*e*) Feebly basic, pK_b = 12.28
Mildly reducing, $E°$ = +0.08 V

Figure 5.1
Lewis dot structures and basic properties of some pseudohalide ions. (*a*) The azide ion, N_3^-; (*b*) the cyanide ion, CN^-; (*c*) the cyanate ion, CNO^-; (*d*) the thiocyanate ion CNS^-; (*e*) the thiosulfate ion $S_2O_3^{2-}$. [The pK_b values are from R. V. Dilts, *Analytical Chemistry*, Van Nostrand, New York, 1974, p. 533. Standard reduction potentials, for later use in Chapter 6, are from J. E. Huheey, *Inorganic Chemistry: Principles of Structure and Reactivity*, 3rd ed., Harper & Row, New York, 1983; pp. A-45–A-56.]

ion. Such ligands are known as potential **chelating ligands**. Such ligands, of course, must be nonlinear, and they must form bonds with reasonable angles (i.e., somewhere near 90 or 109.5°) at the metal atom. This bond angle requirement normally means that the different donor atoms must be far enough apart in the ligand to form *five-* or *six-membered rings*, counting the metal ion. Thus there should be two or three atoms intervening between the two donor atoms (Fig. 5.2). Note that neither the pseudohalide ions nor the simple oxo anions meet these requirements: They normally act either as monodentate or as bridging ligands.

Chelating ligands are classified according to the number of donor atoms they possess that are suitably positioned to be simultaneously donated to the same metal ion. (Often they have additional donor atoms that are "out of position" and that could bridge to another metal ion, but usually do not.) Ligands with two suitably

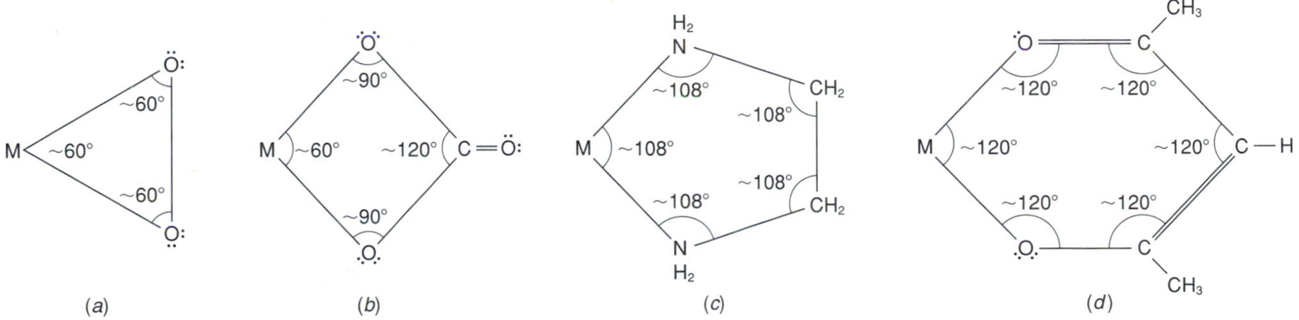

Figure 5.2
Geometric requirements for chelation by a ligand. (*a*, *b*) Bond angles at the metal are unfavorable if chelation is attempted with O_2^{2-} or CO_3^{2-}, giving three- or four-membered rings. (*c*, *d*) Bond angles are favorable upon chelation with ethylenediamine or the acetylacetonate anion, giving five- and six-membered rings.

positioned donor atoms (such as ethylenediamine, $NH_2CH_2CH_2NH_2$) are called (potential) **bidentate chelating ligands**; those with three (such as diethylenetriamine, $NH_2CH_2CH_2NHCH_2CH_2NH_2$) are called **tridentate chelating ligands**, and so on. In general, such ligands are called **polydentate ligands**. As we will see in Section 5.2, ligands that *can* form chelate rings with a metal ion usually *do* so.

Most chelating ligands are organic molecules or ions; the identification of their chelating potential is more challenging when (as is usually the case) you are given only a condensed structural formula of the organic species, with no indication of the presence of unshared electron pairs. The following procedure will help identify potentially chelating ligands.

1. Expand the condensed structural formula into a full structural formula that shows all bonds in the organic species. Add unshared electron pairs to complete the octets of the non-hydrogen atoms; identify the potential donor atoms. (Organic species nearly always obey the octet rule; we are assuming that the condensed structural formula has shown all double or triple bonds.)

2. To function as a chelating ligand, the donor atoms must be adequately far apart. Many organic functional groups contain two or more donor atoms, but in almost all cases the donor atoms in a given functional group are too close together to lead to chelating abilities. Hence, identify all the functional groups containing donor atoms; if there is only one, the organic group can only be a monodentate ligand, not a chelating ligand.

3. Pick one donor atom in one functional group, and label it No. 1. Proceed along the backbone of the organic molecule (mostly along carbon atoms, and not along side atoms such as hydrogen) toward the next functional group, numbering upward until another donor atom is reached. If the number of this donor atom is No. 4 or 5, the two donor atoms have the proper number of "spacer atoms" between them (two or three) to chelate.

4. It is also necessary that donor atoms Nos. 1 and 4 or 5 be free to form a closed ring along with the metal ion. Circumstances that can prevent this include (a) the two donor atoms being tied back in a small (under $\sim$12 atom) ring compound, or (b) a bond angle of 180° rather than the usual 109.5 or 120° occurring along the backbone. (The 180° bond angle will promote bridging rather than chelating properties.)

5. If these geometric conditions allow chelation, the ligand is a potential bidentate chelating ligand. To determine whether the ligand could be tridentate, repeat Steps **3** and **4** starting from either of the identified donor atoms toward some other functional group in the molecule. Repeat until all functional groups have been checked.

Example 5.1

Which of the following are potential chelating ligands: (a) malonate ion, $^{-}OC({=}O)CH_2C({=}O)O^{-}$; (b) 2,5,8-nonanetrione, $CH_3C({=}O)CH_2CH_2C({=}O)CH_2CH_2C({=}O)CH_3$?

SOLUTION:

First, draw the full structural formulas of these species, completing octets; the atoms with unshared electron pairs are potential donor atoms [shaded in the full structural formulas shown in Fig. 5.3(*a*)].

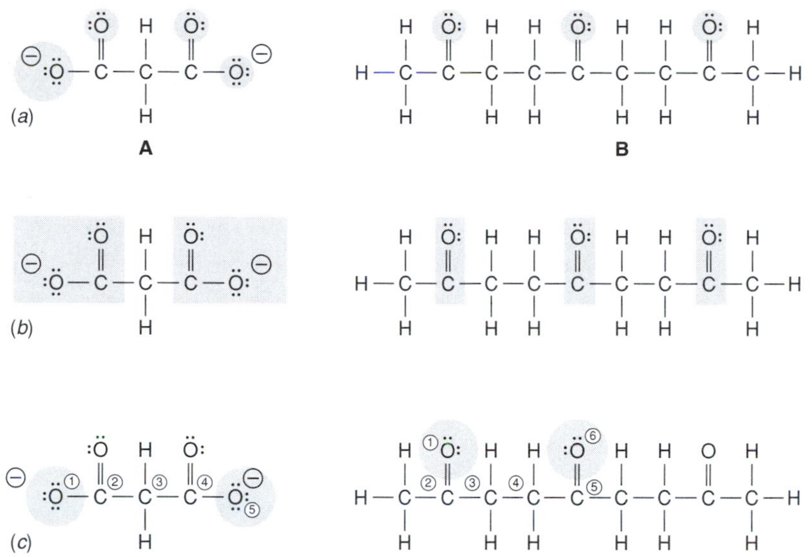

Figure 5.3
Analyses of two organic structures for Example 5.1.

Second, ion **A** contains two carboxylate functional groups and molecule **B** contains three carbonyl (ketone) functional groups [shaded in Fig. 5.3(*b*)].

Third, beginning at the leftmost donor atom of each species, the backbone atoms are numbered as shown in Figure 5.3(*c*) until a donor atom in the next functional group is reached. In ion **A**, either O in the second carboxylate functional group could be No. 5, but in molecule **B** the O in the second carbonyl group is No. 6.

Fourth, neither geometric factor that could prevent chelation is present: In neither case are the two donor atoms already part of a ring; applying valence shell election-pair repulsion (VSEPR) theory, we find that no backbone carbon atoms have 180° bond angles (all are 109 or 120°).

Fifth, we conclude that **A** has the proper spacing and bond angles to be a potential bidentate chelating ligand. Since there are no further functional groups, that is its final classification. Molecule **B** has its donor atoms spaced too far apart, and fails this test. But since there is an additional carbonyl functional group, the third and fourth steps should be repeated, by renumbering the oxygen in the middle functional group as atom No. 1, and counting out to the oxygen in the right functional group, which again is atom No. 6. Hence, **B** is *not* a potential chelating ligand.

Macrocyclic Ligands. A special class of chelating ligand includes those that are large ring compounds even without a metal atom present and that can position several donor atoms inside their ring to donate to a metal ion. Such ligands are called **macrocyclic** ligands. Chlorophyll, heme, and vitamin B_{12} all contain tetradentate macrocyclic ligands (Fig. 5.4): The organic ligands of chlorophyll and heme contain a large 16-membered *porphyrin* ring that includes four nitrogen donor atoms; vitamin B_{12} contains a slightly smaller *corrin* ring that also includes four nitrogen atoms. The crypt-

Figure 5.4

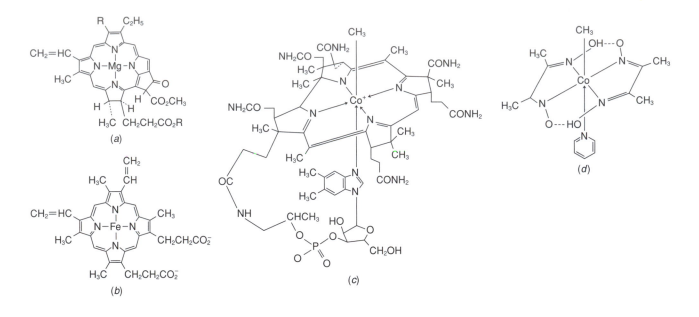

Structures of some biologically significant coordination compounds: (*a*) chlorophyll, containing a Mg^{2+} ion coordinated to four nitrogen donor atoms of an organic (porphyrin) ligand; (*b*) heme, containing a Fe^{2+} ion coordinated to four nitrogen donor atoms of a porphyrin ligand and a nitrogen donor atom from a protein; (*c*) methylcobalamin, one derivative of vitamin B_{12}; (*d*) methyl-cobaloxime, a synthetic mimic of methylcobalamin. [Structures adapted from F. A. Cotton and G. Wilkinson, *Basic Inorganic Chemistry*. Copyright © 1976 by John Wiley & Sons, Inc.]

ands and crown ethers (Fig. 4.9) are also commonly used macrocyclic ligands; the crown ether shown in Figure 4.9, *cyclo*-$(CH_2CH_2O)_6$, contains an 18-membered ring with 6 oxygen donor atoms and is commonly called 18-crown-6.

Lewis acids have been synthesized that have more than one acceptor atom arranged in an appropriate geometry to coordinate a single Lewis base, and hence function as *chelating* or *macrocyclic Lewis acids* (Fig. 5.5).[3]

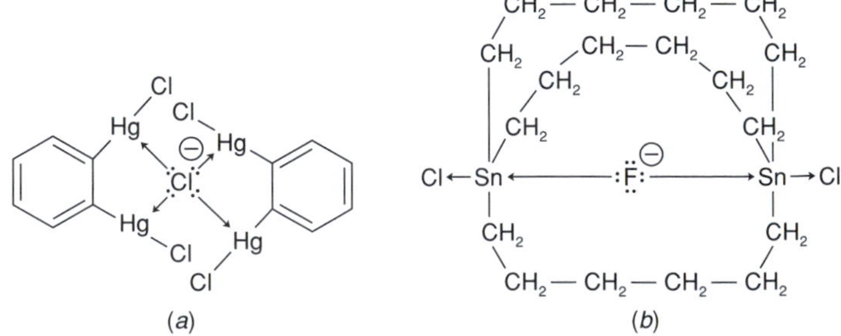

Figure 5.5

Lewis bases complexed by chelating and macrocyclic Lewis acids. (*a*) Chloride ion complexed in the complex anion of $[(C_6H_5)_4P]^+[Cl{:}(C_6H_4Hg_2Cl_2)_2]^-$ (Beauchamp et al.[3a]); (b) Fluoride ion complexed in the complex anion of $[(C_4H_9)_4N]^+[F{:}SnCl(C_6H_{12})_3ClSn]^-$ (Newcomb et al.[3b]).

Pi-Donor Ligands. There are certain compounds (mostly organic) containing double or triple bonds, and hence containing π bonds that can actually donate the electrons in a π bond to a metal ion, even though these are already shared. The existence of such *nonclassical* or π **donor ligands** requires an extension of our definitions: Not only unshared but also loosely shared electrons can be donated. In such cases, it is hard to define the number of donor atoms, and hence the coordination number of the metal and the geometry of the complex. Since such compounds have carbon donor atoms, they are classified as organometallic compounds, and will be discussed further later in the book.

5.2 The Chelate and Macrocyclic Effects

The first generalization that is used to predict which ligands will be complexed with which metal ions is the following: *Chelate ligands form more stable complexes than analogous monodentate ligands* (generally chelates of higher denticity will give more stable complexes than chelates of lower denticity); *macrocyclic ligands of appropriate size form more stable complexes than chelate ligands.*

To see the **chelate effect**, we need to set up a competition (for a metal ion) between 1 mol of an *n*-dentate chelating ligand and *n* mol of a very similar monodentate ligand. For example, if we mix 1 mol Ni^{2+} ion with 3 mol ethylenediamine and 6 mol ammonia, the equilibrium will favor the formation of the complex with ethylenediamine:

$$[Ni(NH_3)_6]^{2+} + 3\,NH_2CH_2CH_2NH_2 \rightleftharpoons [Ni(NH_2CH_2CH_2NH_2)_3]^{2+} + 6\,NH_3 \quad (5.6)$$

The free energy change driving this reaction to the right is $-67\,\text{kJ mol}^{-1}$, which breaks down into a very small enthalpy change of $-13\,\text{kJ mol}^{-1}$ and a larger entropy change term ($-T\Delta S = -54\,\text{kJ mol}^{-1}$). The entropy effect is responsible for the shift of this and similar equilibria in favor of the complex with the chelated ligand attached to the metal ion. The enthalpy change is small since each complex ion in Eq. (5.6) involves six very similar Ni–N coordinate covalent bonds. The reason for the favorable entropy change is more complex and involves several factors,[4,5] the largest of which arises from the fact that there are 4 mol of solute particles on the left side of Eq. (5.6) and 7 mol on the right; the increasing degrees of freedom (disorder) among the products favor complex formation by the chelate ligand.

Similarly, a competition between a noncyclic chelating ligand and a macrocyclic (chelating) ligand having the same number and type of donor atoms will generally lead to complex formation predominantly by the macrocyclic ligand. This is known as the **macrocyclic effect**.

$$[K\{CH_3O(CH_2CH_2O)_5CH_3\}]^+ + cyclo\text{-}(CH_2CH_2O)_6$$
$$\rightleftharpoons [K(CH_2CH_2O)_6]^+ + CH_3O(CH_2CH_2O)_5CH_3 \qquad K_{eq} = 10^4 \quad (5.7)$$

Even though the same number of moles and reactants are involved in this reaction, the entropy term[5] also drives this type of reaction to the right. The reason is that a long noncyclic ligand such as $CH_3O(CH_2CH_2O)_5CH_3$ is more flexible than the corresponding macrocyclic ligand $(CH_2CH_2O)_6$ and can adopt many more conformations than the macrocycle when it is *not* coordinated.

5.3 The Hard and Soft Acid–Base Principle

Your instructor may assign Experiment A.5 at this point; if you did the qualitative analysis scheme in your general chemistry course, he or she may assign only the alternate version of this experiment. In this experiment, you can discover the principle that enables us to make qualitative predictions of the predominant products that result when Lewis acids and Lewis bases (of the same denticity, e.g., all monodentate or all bidentate) compete with each other to form precipitates or complex ions. After we develop this principle, we will use it to organize the chemistry of such diverse phenomena as (1) the qualitative analyses scheme that you may have done in general chemistry, (2) the activity series experiment that you also may have done, (3) additional solubility rules for inorganic compounds, (4) the geochemistry of the elements and why they occur in the ores in which they are found, (5) the nutritional and toxic effects of many metals and nonmetals in the body (e.g., micronutrients and toxic effects of heavy metals), and (6) the ways in which medicinal chemists devise drugs to counteract the effects of heavy metal poisoning.

In Experiment A.5, you may have made the observation that in mixtures of several Lewis acids and Lewis bases, a certain set of metal ions tend to combine with the iodide ion in preference to the fluoride ion. This same set of metal ions prefers the sulfide ion to the hydroxide ion, thiourea to urea, and the sulfide ion to the silicate ion. This last experiment is, of course, related to the **differentiation** of the elements in nature. Berzelius first noted in 1796 that certain metal ions tend to occur in Nature as sulfides, while others tend to occur as oxides, carbonates, sulfates, or silicates. Only about 30 years ago was it realized just how broadly these observations could be generalized and how many thousands of chemical reactions and phenomena could be (reasonably well) predicted using the generalized observation. The general answer to the question, Which metal ion will tend preferentially to form a complex ion with which ligand? was best summarized several years ago by Pearson[6a]: *Hard (Lewis) acids tend to combine with hard (Lewis) bases; soft acids prefer soft bases.*

This statement is now known as the **hard–soft acid–base**, or **HSAB**, **principle**. It is simply a summary of the observed results of thousands of chemical reactions in inorganic and organic chemistry, biochemistry, aquatic chemistry, medicinal chemistry, geochemistry, and so on. Put into equation form, for the following chemical reaction among complex ions, the equilibrium will tend to favor the products on the right side of Eq. (5.8):

$$[\text{Hard acid(:soft base)}_n] + [\text{soft acid(:hard base)}_n] \rightarrow [\text{hard acid(:hard base)}_n] + [\text{soft acid(:soft base)}_n] \quad (5.8)$$

To be able to use the HSAB principle, we must find a way to divide metal ions into the two classes *hard* and *soft* (other authors use the terms *class a* and *class b* instead). Next, we must divide all ligands into the two classes called hard and soft (or class a and class b, respectively). We can do this reasonably well, although some acids and bases will not fall clearly into either class; we will call them *borderline*. The classification scheme that is usually devised is shown in Table 5.1, superimposed upon a table of Pauling electronegativities.

Table 5.1
Hard and Soft Acids and Bases[a]

H 2.2																	He
													Most often seen as Lewis bases	Borderline bases		Hard bases	
Li 0.98	Be 1.57			Hard acids						Hard acids		B 2.04	C 2.55	N 3.04	O 3.44	F 3.98	Ne
Na 0.93	Mg 1.31								Borderline acids			Al 1.61	Si 1.90	P 2.19	S 2.58	Cl 3.16	Ar
										(+2)					Soft bases		
K 0.82	Ca 1.00	Sc 1.36	Ti 1.54	V 1.63	Cr 1.66	Mn 1.55	Fe(+3) 1.83 (+2)	Co(+3) 1.88 (+2)	Ni 1.91	Cu(+1) 2.0	Zn 1.65	Ga 1.81	Ge 2.01	As 2.18	Se 2.55	Br 2.96	Kr 3.0
															Softest bases		
Rb 0.82	Sr 0.95	Y 1.22	Zr 1.33	Nb 1.6	Mo 2.16?	Tc 1.9?	Ru 2.2	Rh (+3) 2.28 (+1)	Pd 2.20	Ag 1.93	Cd 1.69	In (+3) 1.78 (+1)	Sn (+4) 1.96 (+2)	Sb 2.05	Te 2.1	I 2.66	Xe 2.6
Cs 0.79	Ba 0.89	Lu 1.27	Hf 1.3	Ta 1.5	W 2.36?	Re 1.9?	Os 2.2	Ir (+3) 2.2 (+1)	Pt 2.28	Au 2.54	Hg 2.0	Tl (+1)1.60 (+3)2.04	Pb (+2)1.87 (+4)2.33	Bi 2.02			
										Soft acids							
Fr 0.7	Ra 0.9						Borderline acids										
		La 1.10	Ce 1.12	Pr 1.13	Nd 1.14	Pm	Sm 1.17	Eu	Gd 1.20	Tb	Dy 1.22	Ho 1.25	Er 1.24	Tm 1.25	Yb		
		Ac 1.1	Th 1.3	Pa 1.5	U 1.38	Np 1.36	Pu 1.28	Am 1.3	Cm 1.3	Bk 1.3	Cf 1.3	Es 1.3	Fm 1.3	Md 1.3	No 1.3		
									Hard acids								

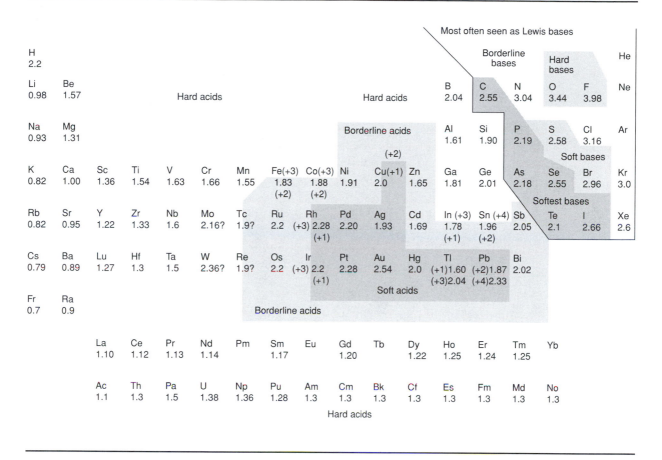

[a]Numbers in parentheses are oxidation numbers. The number below each atomic symbol is the Pauling electronegativity of that element.

Soft Acids or Class B Metal Ions. Despite their presence in different groups of the periodic table, these metal ions, popularly known as the *heavy metal* ions, have a number of chemical similarities. They have insoluble chlorides, so they tend to occur in Group I of the qualitative analysis scheme. They are chemically rather inert and thus are low on the activity series of metals. They occur in nature as the free elements or in sulfide and related minerals. In terms of fundamental atomic properties (as you may have noted in the lab experiment), these metals are characterized by quite *high* Pauling electronegativities for metals, generally in the range of 1.9–2.54. Other characteristics that they share to a lesser extent are large size (ionic radii in excess of 90 pm) and low charge (usually +1 or +2). Simply put, they form a triangle of metals around the element gold, which is the most electronegative of all metals.

Soft Bases or Class B Ligands. According to the HSAB principle, the group of ligands with which these metal ions tend to associate will be called *soft bases* or *class b* ligands. The soft-acid metals are found in nature either in the form of the native (free) elements or as chlorides, bromides, iodides, sulfides, selenides, tellurides, arsenides, and so on. They are not found as oxides or fluorides or as the salts of various oxo anions such as sulfate, silicate, carbonate, and so on. If we have the misfortune to swallow the ions of the soft-acid metals and live through the experience, we will find the ions bonded to certain donor atoms in the proteins and enzymes of our body: to sulfur and to selenium, but *not* to oxygen, fluorine, or (probably) nitrogen. Table 5.1 indicates that the following donor atoms are characteristic of soft bases: C, P, As, S, Se, Te, Br, and I. (We may also add H when it occurs as a donor atom in the hydride ion, H^-.) Fundamentally, these are nonmetals with moderate Pauling electronegativities (χ_P) of 2.1–2.96. These are also the largest nonmetal atoms, with anionic radii in excess of 170 pm.

Hard Bases or Class A Ligands. Contrasting most strongly with the soft nonmetal donor atoms are *oxygen* and *fluorine*, which are seldom found associated in nature with the soft acids. These donor atoms are of very high electronegativity (3.44 and 3.98) and are the smallest of the nonmetal atoms (anionic radii of ~120 pm). Although this seems like a small group of atoms, they (oxygen in particular) are the donor atoms in countless ligands—the oxo anions such as sulfate, carbonate, and silicate; the anions of organic acids such as acetate; and several classes of organic molecules such as alcohols and ketones.

Hard Acids or Type A Metal Ions. Although there are only two unambiguously hard donor atoms, in nature a large number of metal ions occur associated with ligands containing these two. These *hard acids* occupy a good share of the periodic table (Table 5.1); they have in common *low* electronegativities of (usually) 0.7–1.6. They frequently (but not invariably) have relatively small cationic radii ($<$ perhaps 90 pm) and may often have high charges (+3 or higher)[7].

As you may have noticed in the lab experiment, the most important atomic variable in classifying hard and soft acids and bases is the Pauling electronegativity of the donor or acceptor atom (Table 5.2). Sometimes the other variables (charge and size) override considerations of electronegativity. Thus H^+ is classified as a hard acid despite its high electronegativity on the basis of its extremely small size (0 pm). Both B^{3+} and C^{4+} are considered hard acids despite their electronegativities both because of their very small size and their high charges.

Once we have classified the two Lewis acids and the two Lewis bases, we apply the HSAB principle to determine whether the equilibrium in question favors products or reactants. Note that if there are less than two Lewis acids or less than two Lewis bases, the principle does not apply.

Table 5.2
Characteristic Properties of Hard and Soft Acids and Bases

Property	Hard Acids	Soft Acids	Soft Bases	Hard Bases
Electronegativity	0.7–1.6	1.9–2.5	2.1–3.0	3.4–4.0
Ionic radius (pm)	<90	>90	>170	~120
Ionic charge	$\geq +3$	$\leq +2$		

Example 5.2

Predict whether reactants or products are favored in each of the following equilibria:

(a) $Nb_2S_5 + 5\,HgO \rightleftharpoons Nb_2O_5 + 5\,HgS$

(b) $La_2(CO_3)_3 + Tl_2S_3 \rightleftharpoons La_2S_3 + Tl_2(CO_3)_3$

(c) $2\,CH_3MgF + HgF_2 \rightleftharpoons (CH_3)_2Hg + 2\,MgF_2$

SOLUTION:

(a) The two cations are Nb^{5+} and Hg^{2+}. On the basis of its electronegativity and charge, the former is clearly a hard acid, whereas the latter is a soft acid. The two anions are O^{2-} and S^{2-}, which are hard and soft bases, respectively. It is convenient to write "HA", "HB", "SA", and "SB" under the species as they occur in the equation:

$$\underset{\text{HA SB}}{Nb_2S_5} + \underset{\text{SAHB}}{5\,HgO} \rightleftharpoons \underset{\text{HA HB}}{Nb_2O_5} + \underset{\text{SASB}}{5\,HgS}$$

The side of the equation under which the combinations HAHB and SASB occur will be favored.

(b) Based on electronegativity considerations, we classify La^{3+} as a hard acid and Tl^{3+} as a soft acid. We classify S^{2-} as a soft base. In $CO_3{}^{2-}$, it is important to identify the donor atom: It is oxygen, hence this is a hard base. The combinations HAHB and SASB occur under the reactants in this case.

(c) Clearly, Mg^{2+} is a hard acid and Hg^{2+} is a soft acid; F^- is a hard base and $CH_3{}^-$, with a carbon donor atom, is a soft base. Confusingly, the species CH_3MgF will have written under it SB–HA–HB. But you will notice that on the right side, the F^- is still coordinated to the Mg^{2+}. It has not participated in the reaction, so that the relevant exchange is of the soft base $CH_3{}^-$ away from the hard acid Mg^{2+} to the soft acid Hg^{2+}. This equilibrium favors products.

5.4 Relative and Borderline Hardness and Softness

Suppose that we are considering a reaction in which both Lewis acids are soft and both Lewis bases are soft, for example,

$$CdSe + HgS \rightleftharpoons CdS + HgSe \tag{5.9}$$

The HSAB principle can even be used to predict the result here (the reaction goes to the right), because it is possible to assign *relative* softnesses to different soft acids and bases. Among soft bases the following orders of softness have been observed:

Group 17(VII)	$I > Br > Cl > F$
Group 16(VI)	$Te = Se > S \gg O$
Group 15(V)	$Sb < As = P > N$

As we see in Table 5.1, the softest donor atoms are those of the lowest electronegativity, and they form a "ridge" along the metal–nonmetal boundary. Thus we find, in a competitive precipitation experiment involving all the halide ions, that AgI will precipitate preferentially (is least soluble). Silver bromide (AgBr) is not quite as insoluble as AgI; AgCl is not quite as insoluble as AgBr; AgF, which involves a hard base, is much more soluble than any of the others.

Evaluating the relative softness of the metal ions has proved somewhat harder. There are some minor discrepancies among different competitive equilibria involving different soft metal ions with different soft bases. For our purposes, we can say that the further the metal is from *gold* (which all seem to agree is the softest acid), the harder it is. (Gold is, of course, the metal with the *highest* Pauling electronegativity.)

Example 5.3

Students in an inorganic lab have just discovered some new elements, which have the following symbols and properties:

SYMBOL	Fl	Mq	Mr	Az	Za	Mm
Electronegativity	2.22	1.33	2.08	2.60	3.33	1.91
Ion formed	Fl^{2-}	Mq^{3+}	Mr^{+}	Az^{2-}	Za^{2-}	Mm^{2+}
Ion radius (pm)	188	62	125	155	117	103

(a) Based on the information given, classify each of the ions as a hard acid, soft acid, hard base, or soft base. (b) Rank the soft acids and the soft bases in order of increasing softness.

SOLUTION:

The anions are expected to be bases; those with electronegativities below 2.96 (Fl^{2-} and Az^{2-}) likely will be soft bases; the radius of Fl^{2-} is consistent with this while the radius of Az^{2-} is a little small. Since its electronegativity is also higher, Az^{2-} will be less soft than Fl^{2-}. The Za^{2-} ion is small and of high electronegativity, so is a hard base.

The cations are expected to be Lewis acids; those with electronegativities above 1.9 (Mr^{+} and Mm^{2+}) likely will be soft acids; their low charges and radii in excess of 90 pm are consistent with this. Since Mr^{+} has a lower charge, larger radius, and a higher electronegativity than Mm^{2+}, it will be the softer of the two. Based on its high charge, low radius, and low electronegativity, Mq^{3+} is expected to be a hard acid.

For use with competitive equilibria in which softness is present in more than one Lewis acid and more than one Lewis base, we may restate the HSAB principle as follows: **Less soft acids tend to combine with less soft bases; softer acids prefer softer bases.** Thus in Reaction (5.9), mercury, being closer to gold in the periodic table, should be softer, and should prefer the softer base, the selenide ion.

So far, it does not seem to be profitable to attempt to assign relative hardnesses to different hard acids and bases, although Pearson is currently working on this problem.[8] If a competitive equilibrium involves only one species with any degree of softness, it is best to use, not the HSAB principle, but the concepts in the first few chapters of this book. For example, the aqueous equilibrium

$$Hg(ClO_4)_2 + 2\,KOH \rightleftharpoons Hg(OH)_2 + 2\,KClO_4 \qquad (5.10)$$

involves only one soft acid and no soft bases. We can predict that products will be favored and $Hg(OH)_2$ and $KClO_4$ will precipitate, by evaluating the *strengths* of the Lewis acids and Lewis bases. The K^+ ion is a nonacidic cation and should give a precipitate with the nonbasic ClO_4^- anion; Hg^{2+} is a weakly to moderately acidic cation and should give a precipitate with the strongly basic hydroxide anion.

Borderline Acids and Bases. Given that there are degrees of softness, it is not surprising that there are some metal and nonmetal atoms that show such a small degree of softness that softness will not consistently govern the results of their competitive equilibria. As Table 5.1 shows, there are several such metal ions, which are labeled borderline acids. (Many of these have insoluble sulfides but soluble halides, for example.) Likewise, two of the nonmetal atoms, chlorine and nitrogen, are classified as borderline bases. In these cases, the electronegativities of the atoms or their sizes may fall between the ranges that are typical for soft and for hard species. Although the chemistry of borderline acids and bases is a little harder to predict, we can still use the HSAB principle: A borderline base (such as Cl) will be *softer* than F but *not so soft* as I.

Modifying the Softness of an Atom. Since high oxidation numbers are one characteristic of a hard acid, and low oxidation numbers are a characteristic of a soft acid, it is possible to alter the softness of a metal ion by changing its oxidation number. This is a particularly important characteristic of the later *d*-block metals of the fourth period, which are often borderline acids to begin with. Among the metals Fe to Zn the +1 oxidation state, which occurs only for Cu, is definitely soft, whereas the +2 oxidation state is borderline, and the +3 ion (such as for Fe) is characteristically hard. In Figure 5.6, we see this effect in a chelate complex of iron in which three strands of the ligand offer a total of six hard oxygen donor atoms low in the strands and six borderline nitrogen donor atoms higher in the strands. When the iron is oxidized to the harder

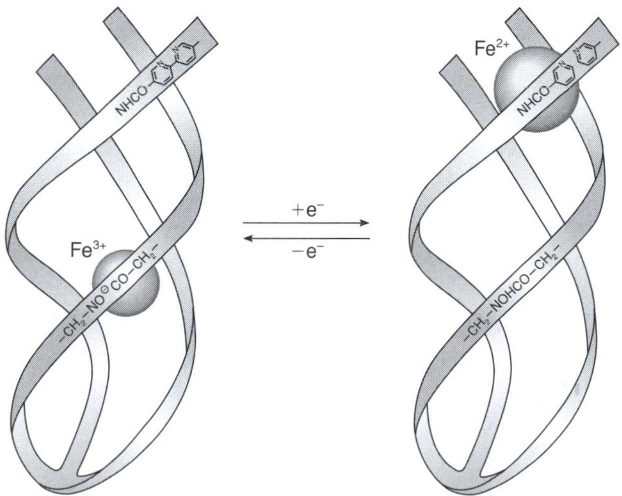

Figure 5.6
Schematic representation of the iron complex of a triple-stranded chelate ligand offering a choice of six hard oxygen or six borderline nitrogen donor atoms. [Adapted from L. Zeloikovich, J. Libman, and A. Shanzer, *Nature* (*London*), **374**, 790 (1995).]

Fe^{3+}, it prefers the six hard donor atoms; when it is reduced to Fe^{2+}, it jumps along the strands in order to be chelated by the six borderline donor atoms.[9]

Several other examples of changes in categories of a metal with change in its oxidation state can also be seen in Table 5.1. We also note that, in terms of their catalytic effects, metallic forms of elements (oxidation number = 0) are generally soft acids.

The softness of a donor or acceptor atom can also be altered by changing the substituents on the atom. Attaching soft bases to a Lewis acid generally softens it as a Lewis acid in any further reactions with additional Lewis bases; the converse holds true for the attachment of hard bases. For example, in the form of hydrated ions (water is a hard base), all of the Group 13(III) +3 metal ions except Tl^{3+} are hard acids; in the form of their methyl compounds $M(CH_3)_3$, all are soft acids except $B(CH_3)_3$ and $Al(CH_3)_3$.

Example 5.4

Predict whether reactants or products are favored in the following equilibria:

(a) $ZnI_2 + HgCl_2 \rightleftharpoons ZnCl_2 + HgI_2$

(b) $(CH_3)_2O{:}BF_3 + (CH_3)_2S{:}BH_3 \rightleftharpoons (CH_3)_2S{:}BF_3 + (CH_3)_2O{:}BH_3$

(c) $CuI_2 + Cu_2O \rightleftharpoons 2\,CuI + CuO$

SOLUTION:

(a) The I^- ion is a soft base and Hg^{2+} is a soft acid. Although Zn^{2+} is a borderline acid (Table 5.1) and Cl^- is a borderline base, relative to Hg^{2+} and I^-, they are each the harder species. Thus the SA–SB combination is HgI_2 and the harder acid–harder base combination is $ZnCl_2$; products are favored.

(b) In the four coordination compounds chosen, the change that occurs is a switch of B–O and B–S coordinate covalent bonds. Breaking apart the compounds at these bonds, we identify the hard base $(CH_3)_2O$, and the soft base, $(CH_3)_2S$. The two Lewis acids BH_3 and BF_3 differ in their substitutents; BF_3 should be the harder. Hence, the reactants should be preferred.

(c) The bases are easily classified; I^- is soft and O^{2-} is hard. The copper occurs in two oxidation states: Cu^+ in Cu_2O and CuI, and Cu^{2+} in CuO and CuI_2. Since Cu^+ is softer, the SA–SB combination is CuI, and products are favored.

5.5 The HSAB Principle and Polar Covalent Bonding

The HSAB principle is what we call *empirical*: It is based on observation. Your conclusions upon doing the lab experiments have also been empirical. Although empirical relationships are very valuable, science also seeks to find theoretical reasons for the relationships. (We have generally done this in the chapters following the lab experiments.) For the HSAB principle, this has been more difficult than usual, in part because it is a qualitative relationship: We do not have good numerical values for softness of species (cf. Parr and Pearson[8]). Sophisticated methods and arguments (such as perturbation molecular orbital theory[10]) have been applied to the question of why the principle works; some of these will be examined later in the text. But for now let us analyze the principle at a very simple level, looking at the fundamental properties of hard and soft acids and bases (Table 5.2).

Of these properties, the most important in determining hardness or softness is the Pauling electronegativity.[6b] We note that hard acids and hard bases have Pauling electronegativities that differ by a large amount (at least 1.8 units). We recall from general chemistry that a large electronegativity difference between bonded atoms favors a highly ionic (electrostatic) bond. Hard acids and hard bases also tend to be small, and hard acids tend to be highly charged. All these factors contribute to high lattice energies, and hence stability for ionic hard acid–hard base reactants or products.

On the other hand, soft acids and soft bases have very similar Pauling electronegativities; the small difference between them favors the formation of nonpolar covalent bonds. The large size and low charges of the atoms, although not favorable for covalent bonding, are even less favorable for ionic bonding.

On examining the combination of a hard base and a soft acid (or a hard acid and a soft base), we find that the relative electronegativities and sizes are not optimal for either ionic or covalent bonding. The electronegativity differences of about 1 are characteristic of **polar covalent bonds**. So the HSAB principle, in a sense, is restating the observation (e.g., common in organic chemistry) that polar covalent compounds tend to be rather reactive if they can react to give an ionic product and a nonpolar covalent product. Many of the more reactive species in organic chemistry (such as Grignard reagents and alkyl halides) have polar covalent bonds (C–Mg and C–halogen, respectively), and they often react according to the HSAB principle to generate new covalent and new ionic bonds:

$$\begin{array}{ll} R_2C{=}O & \text{(polar covalent)} \\ + & \\ H_3C{-}MgBr & \text{(polar covalent)} \end{array} \longrightarrow \text{(covalent)} \begin{array}{l} R_2C{-}O^{-}\,{}^{+}MgBr \text{ (ionic)} \\ | \\ H_3C \end{array} \tag{5.11}$$

We also observe a thermodynamic difference between the hard acid–hard base and the soft acid–soft base interactions. We noted in Chapter 4 that the most common hard acid–hard base precipitation reaction, that of an acidic cation and a basic anion, is driven by its *entropy change*, as many solvent water molecules are released. For example, the typical hard acid–hard base reaction (5.12) involves a small, unfavorable ΔH of $+51\ \text{kJ mol}^{-1}$, and is driven by its $-T\Delta S$ of $-121\ \text{kJ mol}^{-1}$.

$$H^+(aq) + F^-(aq) \rightleftharpoons HF(aq) \tag{5.12}$$

Reactions of the large, low-charged soft acids and bases generally do not involve much of an entropy change but are associated with a favorable *enthalpy change*. Reaction (5.13), a typical soft acid–soft base reaction, involves a negligible $-T\Delta S$ of $+10\ \text{kJ mol}^{-1}$ and is driven by the magnitude of its ΔH, $-315\ \text{kJ mol}^{-1}$.

$$Hg^{2+}(aq) + I^-(aq) \rightleftharpoons HgI^+(aq) + (aq)(aq) \tag{5.13}$$

This enthalpy change accompanies the formation of *stronger bonds: a better covalent bond* (Hg–I), and a better electrostatic interaction (aq)(aq), and will be analyzed in Chapter 7.

Another factor that contributes to hardness is a matter of size: Soft bases with large donor atoms have difficulty replacing small hard–base water molecules around small hard acids.[11]

5.6 The Need for Two Parameters: Softness and Strength

Despite its many successful predictions (many of which will be illustrated in the remainder of the chapter), we are by no means ready to replace all of the principles from earlier in the book with the HSAB principle. *In general, it is impossible to characterize the Lewis acid–Lewis base interaction by a single parameter.* Softness is just one parameter, which must be considered in addition to another parameter or concept: the *strength* of the acid or base. When we evaluate the Z^2/r ratio for an acid or base, or count its oxo groups, and so on, we are evaluating its strength. (We may, if we wish, set up a counterpart statement dealing with strength: Stronger acids tend to react with stronger bases; weaker acids prefer weaker bases. This statement summarizes our results from comparing predominance diagrams of potentially reacting ions in Chapters 2 and 3, and approximately summarizes the solubility rules and tendencies in Chapter 4.)

Since a competitive Lewis acid–base reaction is governed by at least two independent principles[12], we may expect to find cases in which the two principles come into conflict and in which the HSAB principle by itself fails to give a correct prediction. Reaction (5.14) is an example of such a case:

$$\underset{\substack{\textbf{SASB} \\ \textbf{Strong A–Weak B}}}{[CH_3HgSO_3]^-} + \underset{\substack{\textbf{HAHB} \\ \textbf{Weak A–Strong B}}}{[OH_2{:}OH]^-} \rightleftharpoons \underset{\substack{\textbf{SAHB} \\ \textbf{Strong A–Strong B}}}{CH_3HgOH} + \underset{\substack{\textbf{HASB} \\ \textbf{Weak A–Weak B}}}{[OH_2{:}SO_3]^{2-}} \tag{5.14}$$

(Note that we have included the solvent, water, as a hard acid—it has a hard acceptor atom, H, engaged in hydrogen bonding to the anion.) The analyses beneath the equation show that the HSAB principle favors reactants, while considerations of acid and base strengths favor products. The observed equilibrium constant of only 10 indicates that neither is strongly preferred over the other.

Fortunately, in everyday experience there are relatively few cases in which the principle of strength (or the chelate principle) overrides or even counterbalances the HSAB principle. Probably, this is due to the fact that the typical enthalpy changes of covalent-bond formation are much larger than the typical entropy changes driving chelation or the aqueous electrostatic reactions of strong acids and strong bases.

As our discussion of Reaction (5.14) indicates, it is often important to include the solvent explicitly in the reaction. Most solvents are Lewis acids and bases, and can be classified as hard, borderline, or soft; their contribution to the direction of the reaction may be crucial. For the reaction

$$KCl + AgNO_3 \rightleftharpoons AgCl(s) + KNO_3 \tag{5.15}$$

we are, in aqueous solution, actually dealing with hydrated ions. Water is a hard Lewis base (oxygen donor atom) and a hard Lewis acid (hydrogen acceptor atom). In net ionic form in aqueous solution, the above reaction may be represented as

$$Cl^-{:}H_2O + Ag^+{:}OH_2 \rightleftharpoons AgCl(s) + H_2O{:}H_2O \qquad (5.16)$$

Clearly, the HSAB principle favors products. But if the solvent is liquid ammonia, which has a borderline nitrogen donor atom, the prediction is not so clear, since the borderline base NH_3 is competing with the borderline base Cl^-. Actually, Reaction (5.15) goes in the opposite direction in liquid ammonia, and KCl precipitates. It should be noted that the HSAB principle is most commonly used for competitive equilibria in aqueous solutions.

Although soft bases generally prefer to combine with soft acids in competitive acid–base equilibria, we shall note in Chapter 8 that there are some soft bases that distinctly prefer *d*-block soft acids and will not combine with *p*-block soft acids. These bases, such as carbon monoxide, contain empty π-bonding orbitals that can form a π bond only with filled π-bonding *d* orbitals. Such a specific interaction is of course not included either in the general concept of softness or of strength of CO as a base.

Although the HSAB principle clearly is not infallible, in real life (especially in aqueous solutions) it is highly useful, and can help us understand and predict a number of phenomena in chemistry and related fields; examples of these will be given in the Sections 5.7–5.12.

5.7 Applications of HSAB: Solubility of Halides and Chalcogenides

In Chapter 4, we developed solubility rules, but these were only for oxo and fluoro anions: hard bases. Using the HSAB principle, we can now supplement these rules with solubility rules for salts of soft bases.

At first glance, it may seem that a precipitation equilibrium reaction such as $Ag^+ + Cl^- \rightleftharpoons AgCl(s)$ could not be treated by the HSAB principle, since the equation appears to contain only one acid and one base. But when we write a metal ion such as Ag^+ and are speaking of an aqueous solution, we really mean a *hydrated* ion of the type $[Ag(H_2O)_n]^+$, in which the soft silver ion is coordinated by the hard oxygen donor atoms of several water molecules. Likewise, by Cl^- we really mean $[Cl(H_2O)_m]^-$, in which the borderline base, the chloride ion, is hydrogen bonded to the hard-acid hydrogen atoms of several water molecules. If we rewrite this precipitation reaction in the form of Eq. (5.16), we have an equation that is indeed of the form of Eq. (5.8). Thus we can predict that *the chlorides, bromides, and iodides of the soft acids are insoluble*. Indeed, this turns out generally to be true for the +1 and +2 ions of the soft-acid metals: the insoluble chlorides are CuCl (but not $CuCl_2$), AgCl, AuCl (but not $AuCl_3$), TlCl, Hg_2Cl_2 (but not $HgCl_2$), $OsCl_2$, $IrCl_2$, $PdCl_2$, $PtCl_2$ (but not $PtCl_4$), and $PbCl_2$. The lists of insoluble bromides and of insoluble iodides are similar but slightly larger.

We may make similar predictions of the solubilities of *pseudohalide* salts (Section 5.1). The cyanide ion (using its carbon donor atom) and the thiocyanate ion (using its sulfur donor atom) qualify as soft bases, while the azide ion can be identified as a borderline base. We would predict, and generally do find, that the cyanides, thiocyanates, and azides of the soft acids are insoluble.

We may anticipate a similar solubility rule for sulfides, selenides, and tellurides. As it turns out, the sulfides, selenides, and tellurides of the soft *and borderline* acids are insoluble. We may rationalize this solubility rule by noting that the S^{2-}, Se^{2-}, and

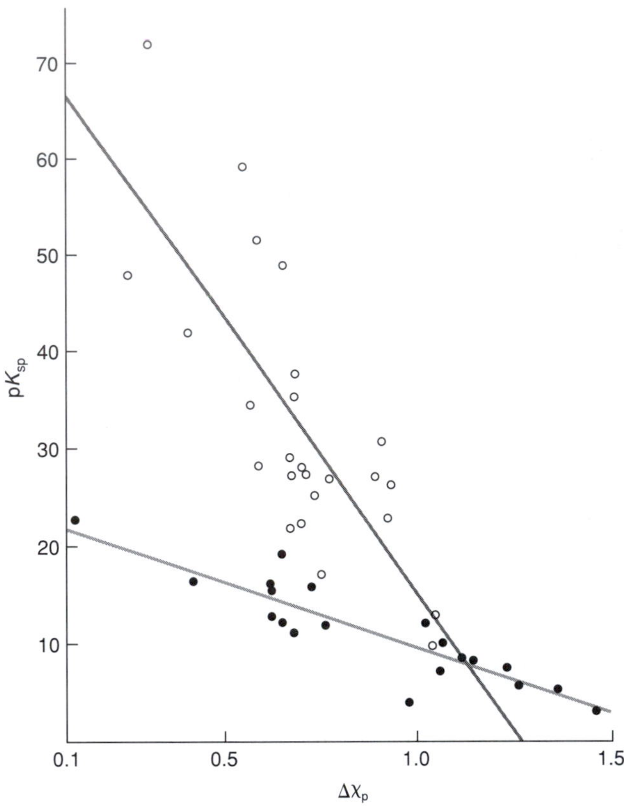

Figure 5.7
Negative logarithms of solubility products (pK_{sp}) of MS, MSe, and MTe (open circles), and of MCl, MBr, MI, MN_3, MCN, and MSCN (closed circles), versus differences in Pauling electronegativities of the Lewis acids and bases. [The pK_{sp} data are from A. F. Clifford, *Inorganic Chemistry of Qualitative Analysis*, Prentice-Hall, Englewood Cliffs, NJ, 1961; Appendix 7.]

Te^{2-} ions are *stronger* bases than the Cl^-, Br^-, and I^- ions. This combination of softness and some strength enables them to combine with borderline acids, which although not as soft are generally smaller and/or more highly charged than the purely soft acids. Thus they are *stronger* acids.

We can also make use of relative softness here: We find that, for a given soft acid, the iodide is less soluble than the bromide, which is less soluble than the chloride; similarly, the telluride and selenide are less soluble than the sulfide. In fact, a relationship has been noted[13] between the solubility products of sulfides, selenides, and tellurides of 1:1 stoichiometry (i.e., of MS, MSe, and MTe) and the electronegativity difference between the cation and the anion (Fig. 5.7).

In contrast, the HSAB principle is not relevant to the prediction of the water solubilities of the salts of the hard bases F^-, OH^-, and O^{2-}, since there is no soft base involved in their precipitation equilibria. Instead, we must note that the fluoride ion is a weak base, while the hydroxide and oxide ions are very strong bases. Consequently, *the fluorides, oxides, and hydroxides of acidic cations are insoluble.* Since there is a substantial difference in strength between the fluoride and the other two anions (fluoride ion is almost feebly basic), it is not too surprising that there are some differences in detail between fluorides and oxides or hydroxides. Thus feebly acidic cations have soluble oxides and hydroxides but insoluble fluorides, while some of the soft-acid cations (Ag^+ and Hg^{2+}) have soluble fluorides but insoluble oxides.

Technically, the HSAB principle cannot be used to predict the solubility of the sulfides, selenides, and tellurides of the hard acids, since no soft acid is involved in their solubility equilibria. The concept of strength can be used, however. These anions are basic anions, and thus would be expected to give insoluble salts with hard acidic cations, such as La^{3+}, Ti^{4+}, or Al^{3+}. To some extent, this is the case. But as we may expect intuitively from HSAB, the oxides of such metal ions are much less soluble than the sulfides, selenides, and tellurides. Consequently, the sulfides, selenides, and tellurides of the hard acids undergo *hydrolysis* to give the oxides or hydroxides of the cation, which may precipitate, while the basic anions hydrolyze to give HS^- or H_2S (or the selenium or tellurium analogues).

5.8 Applications of HSAB: The Qualitative Analysis Scheme for Metal Ions

In your general chemistry course, you may have spent many weeks in laboratory learning to separate and identify the metal ions present in a solution, via the **qualitative analysis scheme** (qual scheme) for the cations. You may recall that all the metal ions are first separated into five groups:

Group I, precipitated with 0.3 *M* HCl

Group II, precipitated with H_2S from acidic solution

Group III, precipitated with $(NH_4)_2S$ from basic solution

Group IV, precipitated with $(NH_4)_2CO_3$

Group V, which remains in solution throughout

Perhaps at the time you wondered how the principle of periodicity applied to the results of this series of experiments. (Obviously, the members of qual scheme group I (Ag^+, $Hg_2{}^{2+}$, Pb^{2+}) do not belong to Group I of the periodic table, or to any other group; neither are they in the same period). It helps, in trying to see the periodic pattern, to know the pattern that results when we do a separation and analysis of *all* of the metal ions. The results of this Herculean task have been described[14] and are summarized in Table 5.3.

As you may have foreseen, the metals of qual scheme group I are the soft acids. These metal ions combine strongly with the borderline base, the chloride ion, to give insoluble chlorides. But in the qual scheme a large excess of chloride ion is present, and many of these ions combine with the excess chloride present to give chloro anions, $[MCl_n]^{x-}$, which remain in solution. The Ag^+, $Hg_2{}^{2+}$, Tl^+, and Pb^{2+} ions precipitate, while the ions of the metals Cu, Pd, Os, Ir, Pt, and Au form soluble chloro anions.

The group precipitating reagent for qual scheme group II, the sulfide ion, is very soft and is a stronger base than the chloride ion, so it combines with additional metal ions to give more insoluble sulfides. Had we neglected to put the chloride ion in, the sulfide ion would now precipitate all of the soft acids of qual scheme group I, plus the

Table 5.3
The Complete Qualitative Analysis Scheme of the Cations

H 2.2														Nonmetals present as *anions* (not included in scheme)			He
Li 0.98	Be 1.57											B 2.04	C 2.55	N 3.04	O 3.44	F 3.98	Ne
Na 0.93	Mg 1.31				Group III (as hydroxides except * as sulfides)							Al 1.61	Si 1.90	P 2.19	S 2.58	Cl 3.16	Ar
K 0.82	Ca 1.00	Sc 1.36	Ti 1.54	V 1.63	Cr 1.66	Mn* 1.55	Fe* 1.83	Co* 1.88	Ni* 1.91	Cu 2.0	Zn* 1.65	Ga* 1.81	Ge 2.01	As 2.18	Se 2.55	Br 2.96	Kr 3.0
Rb 0.82	Sr 0.95	Y 1.22	Zr 1.33	Nb 1.6	Mo 2.16	Tc 1.9	Ru 2.2	Rh 2.28	Pd 2.20	Ag 1.93	Cd 1.69	In* 1.78	Sn 1.80	Sb 2.05	Te 2.1	I 2.66	Xe 2.6
Cs 0.79	Ba 0.89	Lu 1.27	Hf 1.3	Ta 1.5	W 2.36	Re 1.9	Os 2.2	Ir 2.2	Pt 2.28	Au 2.54	Hg 2.0	Tl 1.60	Pb 1.87	Bi 2.02	Po 2.0	At 2.2	Rn
Fr 0.7	Ra 0.9				Insoluble oxide			Group II				Group I		Group II			
Group V	Group IV							Group III									
		La 1.10	Ce 1.12	Pr 1.13	Nd 1.14	Pm	Sm 1.17	Eu	Gd 1.20	Tb	Dy 1.22	Ho 1.25	Er 1.24	Tm 1.25	Yb		
		Ac 1.1	Th 1.3	Pa 1.5	U 1.38	Np 1.36	Pu 1.28	Am 1.3	Cm 1.3	Bk 1.3	Cf 1.3	Es 1.3	Fm 1.3	Md 1.3	No 1.3		

largest number of the borderline acids. If we do the experiment properly and first delete the metals of qual scheme group I, we are left with the metals not quite so near to gold—approximately the borderline acids, minus those of the first row of the transition metals.

The group precipitation of qual scheme group III involves a basic solution of the sulfide ion. The sulfide ion is now present in much higher concentration, but it is also competing with another strong base, the much harder hydroxide ion. In solution, we now have only hard-acid metal ions and a small group of borderline acids from the +2 ions of the first row of the *d*-block metals, Mn^{2+}, Fe^{2+}, Co^{2+}, Ni^{2+}, and Zn^{2+}. These precipitate as *sulfides*, while the numerous *strong* hard acids present (Table 5.3, including also all of the *f*-block elements) combine with the *strong* hard base, OH^-, to precipitate the hydroxides or oxides of these metals.

At this point, only the *nonacidic* and *feebly acidic* hard acids of the first two groups of the periodic table are present. In qual scheme group IV, the moderately basic carbonate ion is added, since it precipitates only the feebly acidic cations. (Of course, had we not carried out the earlier separations, it would precipitate almost all of the preceding metals too.) Remaining in solution to the end are the nonacidic cations of qual scheme group V, which prefer association with the (nonbasic) hard base water, in hydrated-ion form. (Some of the confirmatory tests involve adding nonbasic anions to get precipitates of these metals.)

Example 5.5

Returning to the hypothetical elements classified in Example 5.3: (a) Tell whether their bromides, thiocyanates, cyanates, and sulfides will be soluble. (b) Tell in which group of the qual scheme each of the cations will fall, and write the formula of the compound in which it will precipitate.

SOLUTION:

(a) The two soft acids were Mr^+ and Mm^{2+}. Since neither showed any reason to be classified as borderline, it is likely that both will have insoluble bromides, thiocyanates, and sulfides. Since cyanate has one borderline N and one hard O donor atom, it is probably a hard base, and its solubilities would be based on its strength as a base. It is a weak base (Fig. 5.1), so it might give precipitates with acidic cations; based on their Z^2/r ratios and electronegativities Mr^+ is feebly acidic, Mm^{2+} is weakly acidic, and the hard-acid Mq^{3+} is moderately acidic. We predict that $Mm(OCN)_2$ and $Mq(OCN)_3$ would probably be insoluble.

(b) As soft acids, Mr^+ and Mm^{2+} would precipitate as the chlorides MrCl and $MmCl_2$ in group I unless they formed chloro anions with the excess chloride ion, in which case they would precipitate as the sulfides Mr_2S and MmS in group II. (We have no way to predict this.) As a hard acid, Mq^{3+} would precipitate in one of the last three groups. Since it is moderately acidic, it would precipitate with the basic hydroxide ion in Group III as $Mq(OH)_3$ or Mq_2O_3.

Example 5.6

You are carrying out the qual scheme separation of the cations, but forget to add the HCl for the group I separation or the acidic H_2S for the group II separation; for the group III separation you make the solution strongly basic, but no sulfide ion is present. Under these conditions tell in which of the remaining groups of the qual scheme each of these cations will fall: (a) Pb^{2+}; (b) Ag^+; and (c) Tl^+.

SOLUTION:

In this mismanaged experiment no soft base is present, so the separations will depend entirely on the strengths of the cations. Based on their Z^2/r ratios and electronegativities, Pb^{2+} is weakly acidic, so will precipitate in group III [as $Pb(OH)_2$]; Ag^+ is feebly acidic, so may partially precipitate in group III as Ag_2O (oxides of feebly acidic cations are somewhat soluble) and partially in group IV as Ag_2CO_3; Tl^+ is nonacidic and will remain in solution in group V.

5.9 Applications of HSAB: The Geochemical Classification and Differentiation of the Elements

In studying the types of minerals in which different elements are found, Berzelius long ago noted that certain metals tend to occur as sulfides and others as carbonates or oxides. Nowadays, geochemists classify the elements into four classes, according to their predominant geological pattern of behavior (Table 5.4):

Table 5.4
Geochemical Classification of the Elements[a]

								Lithophiles							Atmophiles		
																	He
Li	Be											B	C	N	O	F	Ne
Na	Mg											Al	Si	P	S	Cl	Ar
K	Ca	Sc	Ti	V	Cr	Mn	Fe	Co	Ni	Cu	Zn	Ga	Ge	As	Se	Br	Kr
Rb	Sr	Y	Zr	Nb	Mo		Ru	Rh	Pd	Ag	Cd	In	Sn	Sb	Te	I	Xe
Cs	Ba	Lu	Hf	Ta	W	Re	Os	Ir	Pt	Au	Hg	Tl	Pb	Bi			
								Siderophiles				Chalcophiles					

[a]This classification emphasizes behavior under conditions at the surface of the earth; many elements can display other classes of behavior as well.

1. The **lithophiles**, the metals and nonmetals that tend to occur as cations in oxides, silicates, sulfates, or carbonates. Since these anions all possess oxygen as a donor atom, the metal ions involved are the *hard acids*. The nonmetals that are classified as lithophiles either are inherently hard bases or have been oxidized by atmospheric oxygen to oxo anions, which are *hard bases*.

2. The **chalcophiles**, which occur in nature as cations in sulfides (less commonly with other soft bases such as telluride, arsenide, etc., since sulfide is much the most common soft base). The chalcophiles are mostly the *borderline* and some *soft acids*. Logically, many of the *soft bases* are also listed as chalcophiles, since they are found in such minerals.

We will pay less attention in this chapter to the other two categories, which are based on behavior in oxidation–reduction chemistry (Chapter 6).

3. The **atmophiles**, the chemically unreactive nonmetals that occur in the atmosphere in elemental form (N_2 and the noble gases).

4. The **siderophiles**, the metals that tend to occur native (in the elemental form); these are also soft acids (Section 6.6), so in this chapter the siderophiles will be treated as a subdivision of the chalcophiles.

The process by which the elements separated into these classes is known as the **primary differentiation of the elements**. It is believed that early in the history of the earth it was hot, so that many materials were molten. Overall, metals are in excess over nonmetals, so only the more active metals were present as salts (silicon is assumed also to have been present as silica and silicates). The abundant, only moderately active metal iron is presumed to have been present in molten form, dissolving nickel and the less-active siderophile metals as it sank (due to its high density) to form the *core* of the earth. Hard-acid metal ions tended to associate with the silicate ions and formed the *mantle* and the *crust* of the earth; the crust also contains borderline- and soft-acid metal sulfides.

These original mineral deposits have not remain unchanged, however. A variety of chemical processes have continued to alter them (the **secondary differentiation of the elements**), giving rise to new minerals. Sometimes particular minerals were concentrated in deposits that are rich enough to mine; such minerals are known as **ores.**[15] Table 5.5 lists the commercially most important ores of some of the elements that are utilized in the greatest quantities. The hard-acid metals occur only as hydrated ions or as salts of hard bases, while the soft acids occur very predominantly either as salts of soft bases or as the elements themselves. The sulfides of the borderline acids, not being as insoluble as the sulfides of the soft acids, are more susceptible to oxidation by O_2 to give soluble sulfates or other secondary minerals that involve hard bases as anions. Hence, the minerals bearing these elements are more diverse than those of the hard and soft acids.

Oceanographic research has revealed that these secondary processes that generate minerals are still going on beneath the ocean, at the midoceanic ridges.[16] Here magma from the mantle at a temperature of 1200 °C continuously rises and is cooled to form the oceanic crust. This rock cracks as it moves off the midoceanic ridges, allowing seawater to seep in close to the hot magma. In this superheated water (350 °C but still liquid, due to the very high pressures at the bottom of the ocean), chemical processes are speeded up. In particular, Mg^{2+} and $SO_4{}^{2-}$ are removed from the ocean water. The Mg^{2+} reacts with the basaltic rock, which acts as a source of SiO_2, and water to give an insoluble magnesium hydroxy-silicate and hydrogen ion. The sulfate is reduced by iron(II) silicate found in the mantle rock to give sulfide (as pyrite):

$$2\,SO_4{}^{2-} + 4\,H^+ + 11\,Fe_2SiO_4 \rightarrow FeS_2 + 7\,Fe_3O_4 + 11\,SiO_2 + 2\,H_2O \qquad (5.17)$$

The superheated water finally emerges from vents at the top of the midoceanic ridge at 350 °C, containing no Mg^{2+} or $SO_4{}^{2-}$, but containing a great deal of dissolved H_2S (from the hydrogen and sulfide ions produced) and dissolved *d*-block metal ions (released from the basalt by the action of the hydrogen ions). As this superheated solution spews into the cold ocean water, the qualitative analysis scheme of the anions is replayed: The softer *d*-block metal ions (Fe^{2+}, Ni^{2+}, Cu^{2+}, and Zn^{2+}) give a precipitate of their sulfides as a black *smoker* that gives rise to a black *chimney*. The sulfide of the harder Mn^{2+} ion, however, tends to be carried away from the smoker, where it is oxidized by bacteria using dissolved O_2 to give MnO_2 in the form of manganese-rich nodules found on the ocean floor.[17]

As we see from Table 5.5, the soft-acid metals and many of the borderline-acid metals are generally used in much lower quantities than are the hard-acid metals because their ores are much less abundant, both in the universe and in the crust of the earth. They are, however, both scarce and extremely useful, and consequently, in general, they command very high prices. A black smoker has been found to be a significant ore deposit, containing on the order of 30,000 tons of Cu at a concentration of 1–2%.[18] Such situations have focused a lot of attention on the resources (such as manganese nodules) that lie at the bottom of the sea and have made quite important the question of ownership of such resources.[19]

Table 5.5
Major Mineral Sources of Selected Elements

Element	Mineral Names, Formulas	Tonnes Used/Year	Uses of Elements
	A. Hard-Acid Metals		
Sodium	Halite, $NaCl$	90,700,000	To make $NaOH$
	Natron, $Na_2CO_3 \cdot 10\,H_2O$		
Calcium	Limestone, $CaCO_3$	54,400,000	To make CaO, cement
	Gypsum, $CaSO_4 \cdot 2\,H_2O$		
Potassium	Sylvite, KCl	9,060,000	Fertilizer
Magnesium	Seawater, $Mg^{2+}(aq)$	8,160,000	Light metal
	Magnesite, $MgCO_3$		
Aluminum	Bauxite, Al_2O_3	5,500,000	Light metal
Chromium	Chromite, $FeCr_2O_4$	1,270,000	Alloys with Fe
Titanium	Ilmenite, $FeTiO_3$	907,000	Light metal
	Rutile, TiO_2		
	B. Borderline-Acid Metals		
Iron	Hematite, Fe_2O_3	281,000,000	Structural metal
	Magnetite, Fe_3O_4		
	Siderite, $FeCO_3$		
Manganese	Pyrolusite, MnO_2	5,440,000	Alloys with Fe
	Psilomelane, $BaMn_5O_{11}$		
Zinc	Sphalerite, ZnS	3,400,000	Coating metals
Nickel	Pentlandite, $(Ni,Fe)_9S_8$	362,000	Alloys with Fe
	Garnierite, $Ni_3Si_2O_5(OH)_4$		
Tin	Cassiterite, SnO_2	172,000	Coating metals
Antimony	Stibnite, Sb_2S_3	54,500	Alloys with Pb
Molybdenum	Molybdenite, MoS_2	40,800	Alloys with Fe
	C. Soft-Acid Metals		
Copper	Chalcopyrite, $CuFeS_2$	4,800,000	Electrical
	Chalcocite, Cu_2S		
Lead	Galena, PbS	2,500,000	Metal in batteries
Mercury	Cinnabar, HgS	8,160	Electrical, to make $NaOH$
Silver	Argentite, Ag_2S	7,250	Photography
Gold	Gold, Au	1,450	Coinage, jewelry
	Calaverite, $AuTe_2$		
Platinum	Platinum, Pt	27	Catalysts
	Sperrylite, $PtAs_2$		
	D. Soft-Base Nonmetals		
Carbon	Coal	1,600,000,000	Reducing agent, fuel
Sulfur	Sulfur, S	18,100,000	To make H_2SO_4
	Pyrite, FeS_2		
Phosphorus	Apatite, $Ca_5(PO_4)_3OH$	6,300,000	Fertilizer
Bromine	Seawater, brines, $Br^-(aq)$	99,700	Organobromine compounds
Arsenic	Arsenopyrite, $FeAsS$	36,300	Alloys with Pb
	Enargite, Cu_3AsS_4		
Iodine	Brines, $I^-(aq)$	3,630	Photography, medicine
	Caliche, $NaIO_3$		
	E. Borderline-Base Nonmetals		
Nitrogen	Air, as N_2	15,000,000	Fertilizer
	Soda niter, $NaNO_3$		
Chlorine	Halite, $NaCl$	4,500,000	Organochlorine compounds, HCl
	F. Hard-Base Nonmetals		
Oxygen	Air, as O_2	18,100,000	Steel manufacture
Fluorine	Fluorite, CaF_2	997,000	To make fluorides

SOURCES: Data from J. W. Moore and E. A. Moore, *Environmental Chemistry*, Academic, New York, 1976, and F. H. Day, *The Chemical Elements in Nature*, Reinhold, New York, 1964.

5.10 *Metal Ions in Biochemistry

In the last four decades, the study of coordination compounds has grown tremendously in scope. Chemists began to realize that many important biochemicals are coordination compounds involving one or more metal ions coordinated to organic groups of great size and complexity. In many cases, the metal ions are so outnumbered by the atoms of C, H, N, O, S, and P that their presence was not even realized until modern methods of analytical chemistry came along that were capable of detecting very low concentrations of metals. Two very important and famous compounds (viz, chlorophyll and hemoglobin) have relatively small organic portions, so their metal content has long been known (Fig. 5.4). Chlorophyll is the key to photosynthesis in plants. The heme unit in hemoglobin transports the oxygen molecule in the body: The O_2 molecule acts as a ligand and is coordinated to the iron atom of heme during its transport. One fairly complex vitamin, vitamin B_{12} or cobalamin [Fig. 5.4(c)], incorporates a cobalt atom that, in the various forms of this vitamin, is identifiable as a Co^{3+}, a Co^{2+}, or a Co^{+} ion. The Co^{3+} ion functions by coordinating ligands with carbon donor atoms (such as CH_3^-), which it then transfers to other compounds. (This vitamin is one of the few naturally occurring organometallic compounds.)

Enzymes are nature's catalysts for speeding up chemical reactions in the body to rates as much as 100,000 times as fast as they would occur without catalysis. Many enzymes contain from one to several metal ions firmly incorporated into the protein structure of the enzyme; these are known as **metalloenzymes**. Other enzymes, although not incorporating the metals irreversibly into their structures, do require the reversible coordination of metal ions in order to become active; these are called **metal-activated enzymes**. Most of the metal ions of the third and fourth periods of the periodic table function in one or more metalloenzymes or in activating numerous enzymes; these metals, essential for animal life, are listed in Table 5.6 along with a summary of their functions.

Metal ions have several overall functions in biochemical and other catalytic systems, which have been summarized[20] as listed below. Note that the first three of these clearly involve the function of the metal ions as Lewis acids.

1. If two molecules that are to react with each other can each act as a ligand and be coordinated to the same metal ion, the statistical odds of their finding each other and colliding may be greatly enhanced. Often the geometries of the ligands and of the coordination complex may be such that the reactive atoms of the two molecules may be brought close together. This tendency is called *the template effect.*

2. Metal ions can change the whole conformation of an enzyme or biomolecule containing negatively charged functional groups that are close together (such as triphosphate ion groups in DNA and RNA). In the absence of +2 charged metal ions, DNA and RNA double helices tend to unwind due to the repulsion of the triphosphate groups. The mutual attraction of the negatively charged triphosphate groups to (especially) the weakly acidic Mg^{2+} ion helps retain the double helix.

3. Just as the acidity of a metal ion polarizes the water molecule in the complex ion $[M(H_2O)_x]^{y+}$ and causes it to release hydrogen ions, an acidic metal ion will enhance the acidity of other coordinated molecules (Table 5.7).[21] Then, for example, bonds other than O–H may also be attacked by water (hydrolyzed), producing products more readily than would otherwise be the case.

Table 5.6
Functions and Preferred Ligand Binding Groups for Essential Metal Ions

Metal ion	Function	Ligand Groups, with Donor Atoms in Parentheses
		A. Hard Acids
Na^+	As charged ion	Hydrated ions (O)
K^+	As charged ion	Singly charged oxygen donor atoms or neutral oxygen ligands (O)
Mg^{2+}	As charged ion, structural	Carboxylate (O), phosphate (O), nitrogen donors (N)
Ca^{2+}	As charged ion, structural	Like Mg^{2+} but less affinity for nitrogen donors, phosphate, and other multidentate anions
Fe^{3+}	Redox reactions	Carboxylate (O), tyrosine (O), $-NH_2$ (N), porphyrin ("hard" N)
Co^{3+}	Redox reactions	Similar to Fe^{3+}
V^{n+}	Essential to sea squirts; mimics insulin.	
Cr^{3+}	Glucose tolerance factor	
Al^{3+}	May activate two enzymes	
		B. Borderline Acids
Mn^{2+}	Lewis acid	Similar to Mg^{2+}
Fe^{2+}	Redox reactions	$-SH$ (S), $-NH_2$ (N) > carboxylates (O)
Zn^{2+}	Lewis acid	Imidazole (N), cysteine (S)
Cu^{2+}	Redox reactions	Amines (N) $\gg$ carboxylates (O)
Ni^{2+}	In urease; stabilizes coiled ribosomes	
Mo^{2+}	Redox reactions	$-SH$ (S)
W^{n+}	Essential for high-temperature microbes near black smokers	
		C. Soft Acids
Cu^+	Redox reactions	Cysteine (S)

SOURCES: Data from M. N. Hughes, *Inorganic Chemistry of Biological Processes*, 2d ed., Wiley & Sons, Inc., Chichester, UK, 1981; J. E. Huheey, E. A. Keiter, and R. L. Keiter, *Inorganic Chemistry: Principles of Structure and Reactivity*, 4th ed., Harper-Collins, New York, 1993; pp. 943–948, *Chem. Eng. News*, Apr. 12, 1993, p. 39, Feb. 21, 1994, p. 35, and May 22, 1995, p. 23.

Table 5.7
The pK_a Values of Free and Coordinated Ligands[a]

	No Metal	$M = Ca^{2+}$	$M = Ni^{2+}$	$M = Cu^{2+}$
$[M(NH_3)]^{2+} \rightleftharpoons H^+ + [M(NH_2)]^+$	35.0		32.2	30.7
$[M(H_2O)]^{2+} \rightleftharpoons H^+ + [M(OH)]^+$	14.0	12.8	9.9	7.3
$[M(HImid)]^{3+} \rightleftharpoons H^+ + [M(Imid)]^{2+}$	7.0		4.0	3.8
$[M(CH_3CO_2H)]^{2+} \rightleftharpoons H^+ + [M(CH_3CO_2)]^+$	4.7	4.2	4.0	3.0

SOURCE: Data from S. J. Lippard and J. M. Berg, *Principles of Bioinorganic Chemistry*, University Science Books, Sausalito, CA, 1994, p. 25, and from Table 2.2 and references cited therein.

[a] Imid = Imidazole = :N(ring)N–H

4. The ions K^+, Na^+, Ca^{2+}, and Mg^{2+} trigger and control certain biochemical mechanisms; the passage of Na^+ ions across nerve cell walls constitutes an electrical current involved in nerve impulse transmission.

5. The *d*-block metals with several oxidation states are particularly useful in catalyzing biological oxidation–reduction reactions: In biochemistry, copper, iron, cobalt, and molybdenum are frequently involved.

Although a detailed study of the biological chemistry of metal ions is beyond the scope of this chapter, we may briefly illustrate some of these functions with two examples.

The trigger, control, and nerve transmission functions of the Na^+ and K^+ ions involve their sudden passage across nonpolar cell membranes; such passage of charged particles constitutes an electric current. In order to be ready for such a discharge, they must be pumped across cell membranes to a state of disequilibrium in which the concentration of K^+ within cells ($\sim 0.1\ M$) is about 10 times that in extracellular fluids; the concentration of Na^+ must be kept about 10 times as high in the extracellular fluids as in the cells. This requires both an active pumping mechanism and some fairly sophisticated ion selectivity to distinguish these two rather similar nonacidic cations.

In early work on the mechanism of this selectivity and transport, two models were developed: the *pore* or *channel* model and the *carrier* model. Both models are based on the mechanisms of action of certain antibiotics that kill cells by causing a sudden leaking of ions across the cell membranes. The pore or channel model was briefly mentioned in Section 4.3; it is modeled on the behavior of the antibiotic gramicidin A (Fig. 4.4), which inserts in the cell wall and has a channel of 400-pm diameter through which hydrated ions may pass. The carrier model is based on the structures and behavior of antibiotics such as valinomycin and nonactin (Fig. 5.8), which resemble the crown ethers in structure and ion selectivity (nonactin selects for K^+ over Na^+ and Cs^+).

Through the use of microelectrodes attached to the outsides of cell membranes, it has been determined that there are in fact ion channels through cell membranes.[22]

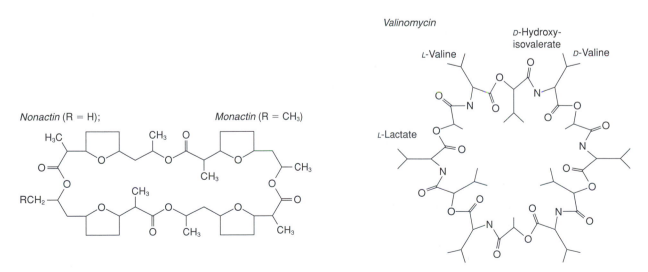

Nonactin (R = H); *Monactin* ($R = CH_3$)

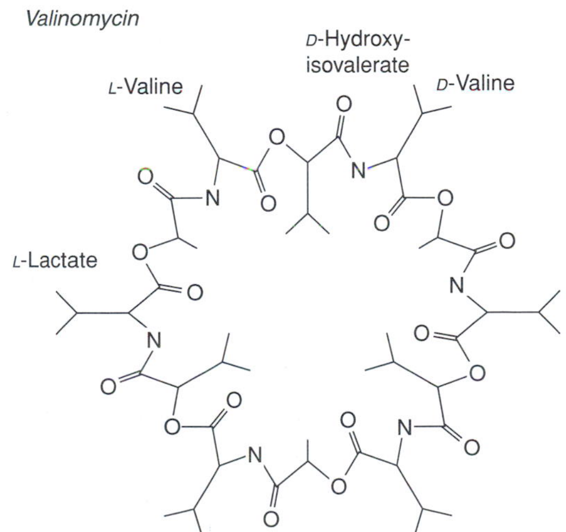

Figure 5.8
Some macrocyclic antibiotics. [Adapted from M. N. Hughes, *The Inorganic Chemistry of Biological Processes*, 2nd ed., p. 267. Copyright © 1981 by John Wiley & Sons, Inc., Chichester, UK.]

The malfunctions of the channels for the passage of Na^+ and K^+ are involved in the disease epilepsy; the absence of the channel for Cl^- is implicated in cystic fibrosis.[23] The abnormal amyloid β protein found in the plaque in the brains of patients with Alzheimer's disease forms cation-selective channels that probably lead to undesirably high levels of Ca^{2+} in brain cells,[24] or result in a defect in the normal passage of K^+ into and out of cells.[25] Neither pore nor carrier antibiotics can actually *pump* ions against a concentration gradient, of course; the mechanism by which the cell does this is thought to involve some kind of "revolving door" protein that picks up Na^+ (and phosphate ion) inside the cell, rotates to the outside of the cell, discharging the Na^+ while picking up K^+, then rotating back to the inside of the cell to discharge the K^+.

One of the best-understood metalloenzymes is *carboxypeptidase A*, shown in Figure 5.9, which helps cleave protein molecules by hydrolyzing the amide group linking the carboxyl-terminal amino acid to the polypeptide or protein:

$$-C(=O)-NH-CHR-C(=O)-O^- + H_2O \rightarrow -C(=O)-O^- + {}^+NH_3-CHR-C(=O)-O^- \quad (5.18)$$

In the center of the 307 amino acids linked together to form this enzyme is one Zn^{2+} ion, which is intimately involved in the mechanism of action of this enzyme.[26] There are several changes involved in the structure of this enzyme before [Fig. 5.9(*a*)] and after [Fig. 5.9(*b*)] the polypeptide enters the "pocket" of the enzyme; we mention only those involving the zinc. Before the polypeptide enters, the Zn^{2+} is coordinated to a H_2O molecule and to the enzyme at three sites. Upon entry of the polypeptide the Zn^{2+} coordinated water, which has enhanced acidity, is deprotonated to a hydroxide ion, which is more basic than water and better able to attack the peptide N–C bond. The oxygen of the carbonyl group (C=O) next to the N–C bond is also attached to zinc, which enhances its acidity (ability to react with the OH^-); a transient five-coordinate Zn^{2+} ion is present.

The template effect is perhaps best illustrated in the syntheses of the crown ether 18-crown-6 itself, from the reaction of $\lbrace CH_2OCH_2CH_2OH)_2$, $\lbrace CH_2OCH_2CH_2Br)_2$, and 2 mol of 2 MOR. If MOR is $(C_4H_9)_4N^+OH^-$, the reaction is slow and the main product is a long polymer $(CH_2CH_2O)_x$: It is improbable that the ends can get together and form the large ring present in 18-crown-6. But if MOR is $K^{+-}OC_4H_9$, the two reactants presumably first coordinate to the K^+, which brings their ends close together, and allows the formation of the product, 18-crown-6 coordinated to K^+, in good yield:

$$\lbrace CH_2OCH_2CH_2OH)_2 + \lbrace CH_2OCH_2CH_2Br)_2 + 2\,KOC_4H_9 \rightarrow [K(CH_2CH_2O)_6]Br + KBr + 2\,C_4H_9OH \quad (5.19)$$

The search for simpler coordination complexes that mimic the behavior of complicated biochemical complexes is one of the activities of the field of research called **bioinorganic chemistry**. With simpler ligands present, there is more hope of being able to sort out or even calculate the reasons for the functions of the catalyst. For example, possible mechanisms for the activity of the Zn^{2+} in carboxypeptidase A in promoting the hydrolysis of polypeptides has been modeled in much simpler Co^{3+} complexes of the polypeptide.[27] Many of the functions of vitamin B_{12} can be mimicked using the cobalt complex of a much simpler organic ligand, shown in Figure 5.4(*d*)[28].

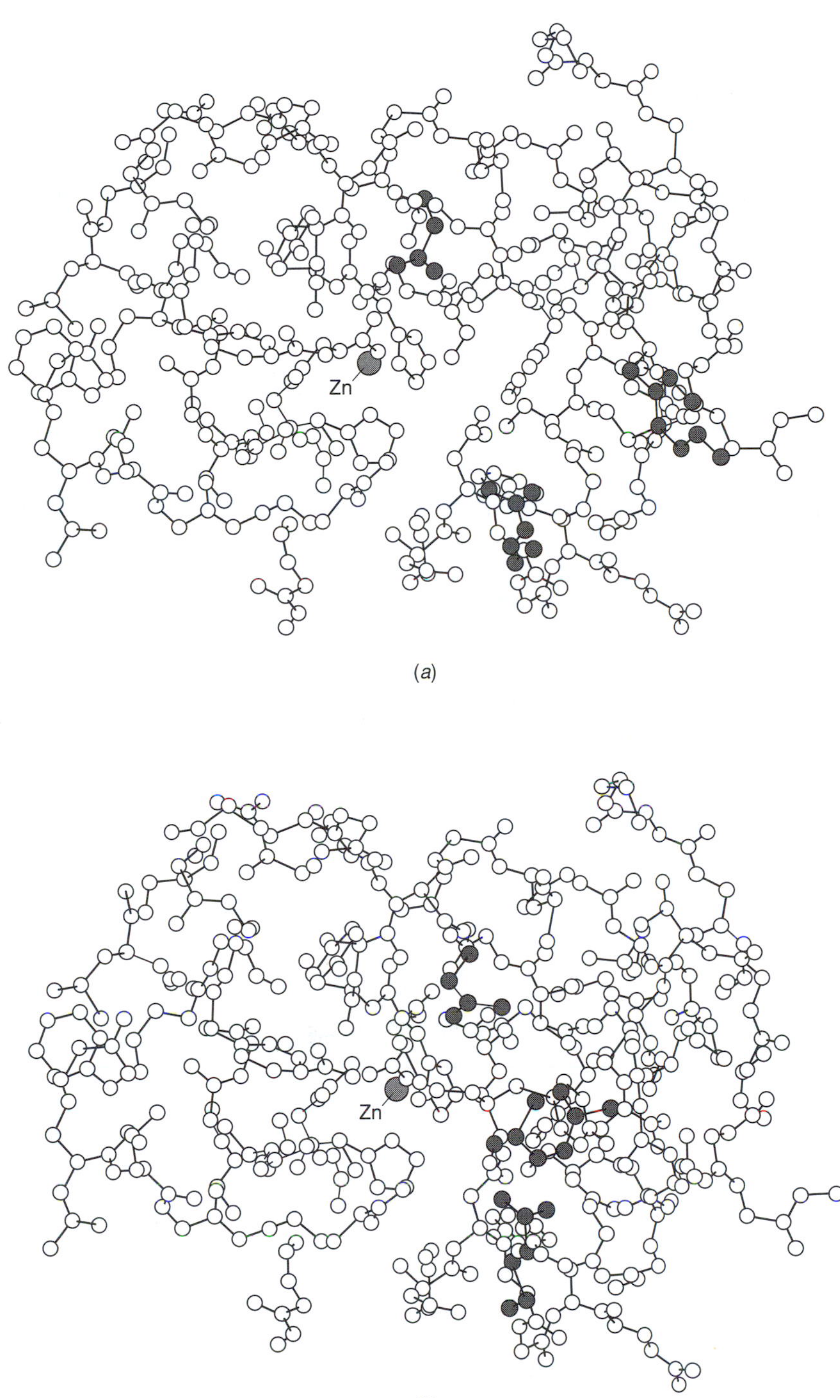

Figure 5.9
About one-fourth of the structure of the zinc-containing metalloenzyme carboxypeptidase A (*a*) in its "resting" state, and (*b*) after incorporating glycyl-L-tyrosine. [Adapted with permission from W. N. Lipscomb, *Chem. Soc. Rev.*, **1**, 319 (1972).]

5.11 Applications of HSAB: Biological Functions and Toxicology of the Elements

The HSAB principle is very useful in organizing the chemistry of the metal ions in ecology, biochemistry, and medicine.[18] Table 5.6 also indicates the donor atoms of biochemical ligands to which the essential metal ions prefer to bind; these follow the general pattern we would expect from the HSAB principle. Other metal and nonmetal ions are toxic components of natural waters and wastewaters; they are listed in Table 5.8 along with their sources and effects. We note immediately that soft acids are seldom essential for life and are often toxic. The HSAB principle also tells us the kind of biochemical sites at which toxic metal ions are likely to bind.

Hard Acids. As we see from Table 5.6, several hard acids are essential for life. The metal ions Na^+, K^+, Mg^{2+}, and Ca^{2+} often function in the form of their hydrated ions, which the body allows to pass selectively across various barriers. This passage of charged ions constitutes an electric current, which can transmit a nerve impulse or trigger some response. Both Mg^{2+} and Ca^{2+} also serve structural purposes: the former in preserving the double helical structure of DNA; the latter in the form of calcium phosphate or carbonate as bones, teeth, and shells.

Most other hard-acid metal ions are small and/or highly charged. Thus they are unavailable in natural waters (Chapter 2), because they precipitate as hydroxides. Hence, most hard-acid metal ions are neither essential nor toxic. Beryllium is one exception—it is very dangerously poisonous. Apparently, it substitutes for Mg^{2+} in certain enzymes in the body, but being much more acidic, it alters the functions of those enzymes so that they can no longer function. Barium is also toxic if soluble, but in the form of insoluble $BaSO_4$ it is safely given as "barium milkshakes" to make the stomach opaque to X-rays. Some of the *f*-block elements (such as plutonium) of the seventh period are exceedingly toxic due to their radioactivity. Plutonium tends to concentrate in the bones, where even a few millionths of a gram emit enough dangerous alpha particles to irradiate the bone marrow and cause leukemia.

Borderline Acids. Many of the lighter borderline-acid metal ions, although needed only in tiny quantities, have critical biochemical functions. Their appreciable Lewis acidity enables them to bind certain ligands (e.g., O_2) or to enhance the acidity and reactivity of those ligands as a step in synthesizing or metabolizing them. Many of these metal ions also have two or more accessible oxidation states differing by only one electron, which allows them (in appropriate enzymes) to catalyze important redox reactions. In excessive quantities, these metal ions can be toxic: For example, excess iron causes siderosis and may result in an increased risk of heart attacks.[29]

Soft Acids. In order to function properly, enzymes must maintain their correct three-dimensional conformational structure, which is held in position by comparatively weak bonds: hydrogen bonds and the sulfur–sulfur single bonds found in the component amino acid cystine. The latter bond, which is formed by oxidation of the S–H bonds of two cysteine amino acid components, is easily ruptured by the action of soft-acid metal ions (giving S-metal bonds instead). (Compounds such as cysteine that contain the SH functional group are often called *mercaptans*. This term is short for "mercury capturer.") Critical enzymes, being very efficient at catalyzing biochemical reactions, are often present in quite low concentrations and are quite susceptible to deactivation by soft-metal ions; hence, all soft-metal ions are toxic when present in the body at concentrations of tens of milligrams per kilogram of body weight (tens of parts per million, ppm); none is essential (other than Cu^+ while present during redox reactions of Cu^{2+}).

Perhaps the best known of these is *mercury*.[30] The toxic effects of this element, as well as many others, depends on the chemical form in which it is found. As metallic mercury it is insoluble and passes through the digestive system unchanged. (It was once used as a laxative!) However, it is volatile, and the vapors are exceedingly toxic. (In general, the lungs lack the kinds of defense mechanisms against toxins that the digestive system possesses.) Spills of metallic mercury need to be cleaned up carefully, since mercury spilled in a laboratory can poison the scientists working there over many years.[31]

Soluble inorganic mercury salts (such as $HgCl_2$) are absorbed by the gastrointestinal tract, and are toxic. The expression "mad as a hatter" refers to the long-term effects of inorganic mercury ingestion by persons who engaged in the old practice of using $HgCl_2$ in treating beaver pelts to make hats. If a single dose of a soluble inorganic mercury salt is ingested and does not cause death, the body can excrete the mercury over a period of months, since it is water soluble. In contrast, very insoluble mercury salts such as Hg_2Cl_2 (*calomel*) and HgS are relatively nontoxic to begin with, since they are nonvolatile and insoluble and cannot be absorbed by the body.

Of particular danger are the derivatives of the *methylmercury* cation, CH_3Hg^+, which is perhaps the softest of all cations. Its derivatives, such as CH_3HgCl and $(CH_3)_2Hg$, are both volatile and absorbable through the digestive system. They are much more soluble in fats and lipids than in water, however, so that they cannot readily be excreted. Due to this lipid solubility, they can penetrate the blood–brain and the placental barriers, causing brain damage and birth defects. Cases of methylmercury poisoning have occurred where fish were eaten that came from waters into which only inorganic mercury was being dumped. It seems that certain microorganisms are able to methylate inorganic mercury via the natural methylating agents vitamin B_{12} (Fig. 5.4) and *S*-adenosylmethionine;[32] bacteria confronted with mercury can also respond by activating a mercury-detoxification enzyme that reduces the metal ion to elemental mercury.[33]

Lead is also a notorious soft-acid poison, occurring in the modern environment as a result of the former practice of using a lead hydroxy-carbonate as a white paint pigment and more recently from the use of tetraethyllead, $(CH_3CH_2)_4Pb$, as an antiknock agent in gasoline. As expected, lead reacts with enzyme SH groups; its most serious effects arise from the consequent deactivation of two enzymes needed for the biosynthesis of heme and from adverse affects upon the function of the brain. Biochemically, *cadmium* behaves as a soft acid and is also quite toxic. In excess, it causes itai-itai disease, in which the bones become brittle and break easily and painfully. Its levels also correlate with increased levels in the population of diseases related to high blood pressure. *Thallium* is used principally as a rat poison, which speaks for its toxicity.[34]

Gold and *platinum* are much too valuable to allow them to escape into wastewaters, but they are toxic. Both have found medicinal uses. Gold is used in antiarthritic drugs such as chloro(triethylphosphine)gold(I), $(C_2H_5)_3P{:}AuCl$, or Myocrisin, the gold(I) thiomalate $[Au{-}S{-}CH(COO^-Na^+)(CH_2COO^-Na^+)]_n$. Although quite effective, there are often toxic side effects due to the long-term accumulation of gold in the body. Cisplatin, an important anticancer drug, is the complex $[PtCl_2(NH_3)_2]$. Somewhat surprisingly from the HSAB point of view, the platinum bonds to the nitrogen donor atoms of adjacent guanine groups in DNA, bending the double helix. As we may expect, toxic side effects accompany such chemotherapy. The structure of

Table 5.8
Occurrence and Significance of Trace Elements in Natural Waters and Wastewaters

Element	Sources	Effects and Significance	USPHS Limit (mg L^{-1})[a]	Occurrence: % of Samples, High, Mean (μg L^{-1})[b]
		A. Hard Acids		
Beryllium	Coal, nuclear power and space industries	Acute and chronic toxicity, possibly carcinogenic	Not given	Not given
Boron	Coal, detergent formulations, industrial wastes	Toxic to some plants	1.0	98% (above 1 μg L^{-1}), 5000,101
		B. Borderline Acids		
Chromium	Metal plating, cooling-tower water additive (chromate), normally found as Cr(VI) in polluted water	Essential trace element (glucose tolerance factor), possibly carcinogenic as Cr(VI)	0.05	24.5%, 112, 9.7
Manganese	Mining, industrial waste, acid mine drainage, microbial action on manganese minerals at low pE	Relatively nontoxic to animals, toxic to plants at higher levels, stains materials (bathroom fixtures and clothing)	0.05	51.4% (>0.3 μg L^{-1}), 3230, 58
Iron	Corroded metal, industrial wastes, acid mine drainage, low pE water in contact with iron minerals	Essential nutrient (component of hemoglobin), not very toxic, damages materials (bathroom fixtures and clothing)	0.05	75.6%, 4600, 52
Copper	Metal plating, industrial and domestic wastes, mining, mineral leaching	Essential trace element, not very toxic to animals, toxic to plants and algae at moderate levels	1.0	74.4%, 280, 15
Molybdenum	Industrial waste, natural sources	Possibly toxic to animals, essential for plants	Not given	32.7 (>2 μg L^{-1}), 5400, 120
Zinc	Industrial waste, metal plating, plumbing	Essential element in many metalloenzymes, aids wound healing, toxic to plants at higher levels, major component of sewage sludge limiting land disposal of sludge	5.0	76.5% (>2 μg L^{-1}), 1180, 64

Table 5.8 (*continued*)

Element	Sources	Effects and Significance	USPHS Limit (mg L^{-1})[a]	Occurrence: % of Samples, High, Mean (μg L^{-1})[b]
		C. Soft Acids and the Softer Borderline Acids (Heavy Metals)		
Cadmium	Industrial discharge, mining waste, metal plating, water pipes	Replaces zinc biochemically, causes high blood pressure, kidney damage, destruction of testicular tissue and red blood cells, toxic to aquatic biota	0.01	2.5%, not given, 9.5
Lead	Industry, mining, plumbing, coal, gasoline	Toxic (anemia, kidney disease, nervous system), wildlife destruction	0.05	19.3% (>2 μg L^{-1}), 140, 23
Mercury	Industrial waste, mining, pesticides, coal	Acute and chronic toxicity	Not given	Not given
Silver	Natural geological sources, mining, electroplating, film-processing wastes, disinfection of water	Causes blue-gray discoloration of skin, mucous membranes, eyes	0.05	6.6% (>0.1 μg L^{-1}), 38, 2.6
		D. Soft Bases		
Iodine (iodide)	Industrial waste, natural brines, seawater intrusion	Prevents goiter	Not given	Rare in fresh water
Arsenic	Mining byproduct, pesticides, chemical waste	Toxic, possibly carcinogenic	0.05	5.5% (>5 μg L^{-1}) 336, 64
Selenium	Natural geological sources, sulfur, coal	Essential at low levels, toxic at higher levels, causes "alkali disease" and "blind staggers" in cattle, possibly carcinogenic	0.01	Not given
		E. Hard Base		
Fluorine (fluoride ion)	Natural geological sources, industrial waste, water additive	Prevents tooth decay at about 1 mg L^{-1}, causes mottled teeth and bone damage at around 5 mg/L in water	0.8–1.7 depending on temperature	Not given

SOURCE: Adapted From S. E. Manahan, *Environmental Chemistry*, 3d ed., p. 147. Copyright © 1979 by Willard Grant Press, Boston.

[a] *Public Health Service Drinking Water Standards*, U.S. Public Health Service (USPHS), 1962.

[b] John F. Kopp and Robert C. Kroner, *Trace Metals in Waters of the United States*, United States Environmental Protection Agency, 1969. The first figure is the percentage of samples showing the element; the second is the highest value found; the third is the mean value in positive samples.

cisplatin has also been modified by incorporating a bridging long diamine molecule, $H_2N(CH_2)_nNH_2$, which joins two platinum metal ions; the two platinum ions are anticipated to be able to link two DNA strands and to bond to other groups in DNA.[35]

Soft Bases. Many of the soft-base nonmetals are beneficial in small doses and may even be essential. *Sulfur* and *phosphorus* are, of course, major components of amino acids, nucleic acids, and so on. *Chlorine* is essential as Cl^-; *iodine* is essential in the form of the hormone thyroxin. Of particular interest are the very soft *arsenic* and *selenium*. Both are essential, but only the functions of selenium are known. It is a component of glutathione peroxidase, and as a result of its reducing properties it protects against free radicals, which may be implicated in aging and causing cancer; selenocysteine is incorporated in peptides synthesized in cells along with the 20 amino acids normally listed in textbooks.[36] Since selenium is softer than sulfur, it attracts soft metal ions preferentially and thus protects against their toxic effects (such as carcinogenesis). Thus despite the fact that tuna have been found with elevated mercury levels, no one has been found to suffer mercury poisoning from high consumption of tuna. Tuna is also high in selenium, which may tie up the mercury in the form of very stable complexes.

On the other hand, both of these elements are *very toxic*. Arsenic poisoning, of course, has been well known for millennia. Selenium occurs in nature at high levels as selenite and selenate ions in certain soils in New Zealand and the western United States, where it is concentrated in certain plants such as locoweed. Cattle eating this weed develop a disease known as the *blind staggers* due to the selenium. Using HSAB, one may speculate that accumulation of Se much in excess of that needed to tie up soft-acid metals may result in binding to borderline-acid metals that are essential in enzyme functions. Whether or not this happens, it is not a contradiction to say that selenium (or other substances) can be both *essential* and (in slightly higher doses) also *highly toxic*. With medication, it is certainly true that more is not necessarily better.

Other soft bases are also well-known toxins. The soft carbon donor atoms of carbon monoxide, :CO, and the cyanide ion, $:CN^-$, bind strongly to borderline-acid metal ions such as Fe^{2+} in heme. Thus the hard base O_2 is unable to compete equitably with :CO for the iron ion in hemoglobin, and cyanide prevents the heme-based enzyme cytochrome *c* oxidase from binding O_2. Many nonmetal hydrogen or methyl compounds (such as phosphine, arsine, hydrogen sulfide, hydrogen selenide, and their methylated forms) have soft donor atoms and are very poisonous.

Example 5.7

(a) Which of the hypothetical ions of Example 5.3 are most likely to be *toxic*? (b) Suppose two of the toxic ions occurred together in one (binary) compound. Would this compound likely be *more* or *less* toxic than the separate ions ingested at different times?

SOLUTION:

(a) The soft acids Mr^+ and Mm^{2+} will be toxic; probably the two soft bases, Fl^{2-} and Az^{2-} will be too. (b) Binary compounds involving these ions would have one soft acid and one soft base. The resulting compounds (Mr_2Az, Mr_2Fl, MmAz, or MmFl) would be insoluble, so they likely would be less toxic than the separate hydrated ions.

5.12 *Applications of HSAB and Chelation: Medicinal Chemistry

Given the potency and variety of metal ions in biochemical systems, it can easily be appreciated that the removal of excess or unwanted metal ions selectively from the body (or even from analytical samples or from wastewaters) is no simple task. Any number of ligands can be imagined that may form a complex with a given metal ion, but a number of additional constraints are usually imposed. Concentrations of the offending metal ion and the reagent to be used may both need to be low, which does not favor complex formation unless a very stable complex results. At the same time, the ligand should not form too stable complexes with other metal ions, since these may need to be undisturbed. In analytical chemistry, it may be desirable for the complex to be soluble in nonpolar solvents so that it can be extracted and concentrated. To some extent, this is necessary in medicinal chemistry too, so that the ligand and metal complex can cross nonpolar cell membranes, but if the complex is too water insoluble, it cannot be excreted in the urine.[37]

In designing reagents for these tasks, the medicinal, analytical, or environmental chemist has two principles to guide him or her: the HSAB principle and the principle that chelating ligands form more stable complexes than do monodentate ones (Section 5.2). Figure 5.10(*a*) shows some drugs used to remove excess or toxic metal ions from the body.

The chelate principle is best illustrated by the drug EDTA (ethylenediaminetetraacetic acid), which when deprotonated at the four COOH groups is a hexadentate ligand, and is a very powerful chelating ligand indeed (it is used extensively in analytical chemistry to titrate metal ions). The problem lies in its nonselectivity—it prefers highly charged metal ions but will form very stable complexes with almost any metal ion but a +1 charged one. It is the treatment of choice for plutonium poisoning, being one of the few ligands capable of dislodging the Pu^{4+} ion from its precipitated form in the bones, but excesses will then remove the calcium from the bone. This problem is overcome by administering EDTA with calcium ions already chelated in it. Ethylenediaminetetraacetic acid is also used as a food preservative, since it chelates metal ion impurities in foods that would otherwise catalyze air oxidation (spoiling) of the food.

In other ligands that are less overwhelmingly chelating, greater selectivity is possible using the HSAB principle. For example, the drug British Anti-Lewisite (BAL), when deprotonated at the –SH groups, is chelating with soft-base sulfur donor atoms, and thus selectively chelates soft acids in the body. Originally developed during World War I as an antidote for the war gas Lewisite, $ClCH{=}CH_2AsCl_2$, it is now used in treating poisoning by other soft acids such as mercury and thallium.

As we previously mentioned, some borderline acids, though essential, are toxic at higher levels. Wilson's disease, for example, is a metabolic disorder that results in the inability to excrete excess Cu^{2+} ion, which accumulates to toxic levels. It is treated with the ligand penicillamine, containing a soft and a borderline donor atom in a chelate ring. In human history, iron has often been a deficiency problem (anemia), since Fe^{2+} is normally unstable to oxidation and Fe^{3+} is too acidic to be available at the pH of the intestine, where absorption must take place; hence in vitamins, simple iron salts are not used. Instead, salts with chelating organic anions are commonly included. But the body has evolved *no* mechanism (other than bleeding) for excreting excess iron, since historically this situation hardly ever arose. But iron poisoning does now occur and is treated with the hard-base chelating ligand desferrioxamine. Another hard-base

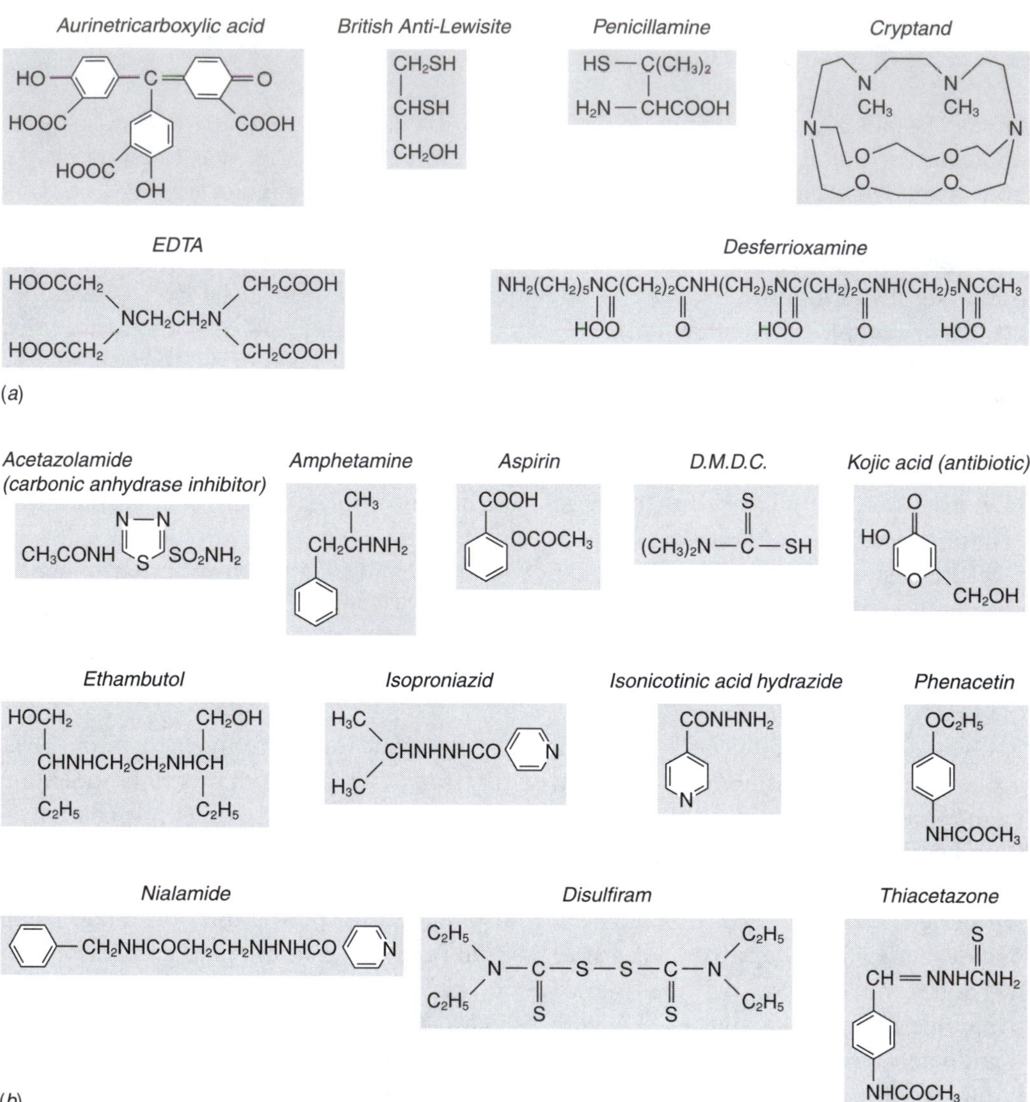

Figure 5.10
(*a*) Some chelating ligands used as drugs to remove metal ions from the body; (*b*) other pharmaceuticals, the actions of which *may* involve metal ions. [Adapted from M. N. Hughes, *The Inorganic Chemistry of Biological Processes*, 2d ed. Copyright © 1981 by John Wiley and Sons Ltd., Chichester, UK.]

chelating ligand, aurinetricarboxylic acid, is used in analytical chemistry as a reagent for aluminum; medicinally it is used to treat poisoning by the Be^{2+} ion (similar size and acidity).

A number of drugs used to treat various conditions not caused by metal poisoning per se are nonetheless good candidates to act as ligands, and their mechanisms of action in the body may involve metal ions (e.g., aspirin action may involve copper). Some of these are shown in Figure 5.10(*b*).

The value of macrocyclic ligands in chelation therapy (and in extraction of specific metal ions in analytical chemistry) is enhanced by the possibility of designing them so as to be *size selective.* For example, in the conformation shown in Figure 4.9, the crown ether 18-crown-6 has a cavity of a size that is governed by the van der Waals radius of the oxygen atoms, 150 pm (Table 1.3)—the six oxygen donor atoms touch each other at this radius, forming a hexagon with sides of 300 pm that resists being pulled in any further by cations that are too small for the hole in the center of the hexagon. The distance from any oxygen atom of the hexagon to the center of the metal ion in the hole can be calculated by trigonometry, after dividing the polygon into small metal–oxygen–oxygen triangles. In the case of 18-crown-6, the optimal metal–oxygen distance is calculated to be 300 pm, which results (after deducting the van der Waals radius of oxygen) in an optimal metal–ion radius of 150 pm. Consistent with this calculation, 18-crown-6 forms a more stable complex with K^+ (radius = 152 pm) than with other Group 1 ions, which are smaller or larger than K^+.

Many other features can be built into the structures of crown ethers and related macrocycles to alter this simple type of selectivity; these features are beyond the scope of this text. Even on going to divalent cations, a different result obtains: Pb^{2+} (radius = 133 pm) is selected over Ba^{2+} (radius = 149 pm). Keep in mind that atoms and ions such as ether oxygen atoms are compressible (as measured by the Born exponent, Section 4.5) and so do not have strictly invariant radii. Crown ethers are also flexible molecules that can adopt nonplanar conformations using less than their full complement of oxygen donor atoms if need be; the choice of conformations will be affected also by the resulting repulsions within CH_2CH_2 groups. Although these repulsions are too complex for simple qualitative evaluation, there are now accessible computer programs for doing molecular mechanics calculations to evaluate the repulsions involved in the chemistry of metal-ion selection by chelate and macrocyclic ligands.[38] Macrocyclic ligands containing alternating double bonds (such as the porphyrins, Fig. 5.4) have much less flexibility in structure, and have much more invariant hole sizes.

Macrocyclic cryptand ligands form more stable complexes more selectively even than the crown ethers; the cryptand shown in Figure 4.9 removes radioactive strontium and radium, as well as lead, from the body without disturbing calcium. The cryptand shown in Figure 5.10(*a*) selects Cd^{2+} over Ca^{2+} and Zn^{2+} by a factor of 1 million and thus can be used for the selective removal of cadmium.

For some purposes, it is actually desirable to introduce radioactive metal ions such as ^{99}Tc into the body when complexed in the form of radiopharmaceuticals.[39] Various tissues process organics in a different manner depending on whether they are healthy or diseased; the radiation from the ^{99}Tc then allows imaging of these tissues, giving different images for healthy or for diseased tissue.

Many biological organisms have developed their own methods (not always known to man) to deal with excesses of toxic elements or deficiencies of essential elements. In the body, cadmium tends to concentrate in the kidneys, where it is bound to sulfur donor atoms in the protein metallothionein, the amino acid concentration of which is about 30%. Metallothionein also binds other soft acids, and has no other known biological function.[40]

Hyperaccumulator plants,[41] such as Alpine pennycress, have evolved in areas with naturally high concentrations of certain normally toxic metal or nonmetal ions such as Zn^{2+}, Cd^{2+}, Ni^{2+}, Pb^{2+}, and selenium, and have developed ways of storing them in startlingly high concentrations without harm to themselves. These plants are being investigated for use in a natural process of cleaning up hazardous waste dumps; some are even suitable for beautifying the dump (the National Endowment for the Arts once funded an artistic planting of these plants on a hazardous waste dump). The ash from

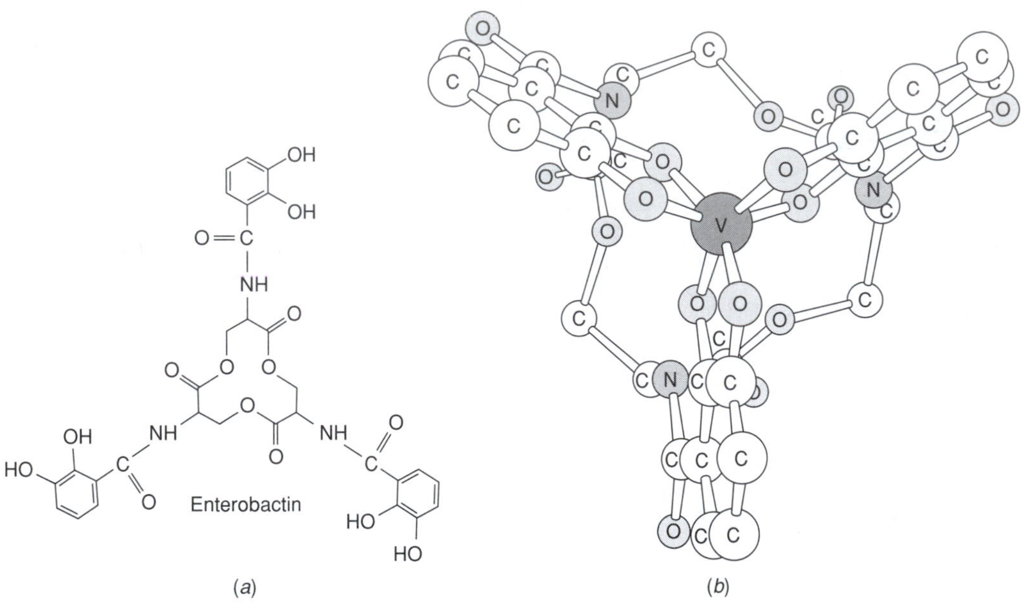

Figure 5.11
(*a*) The structure of enterobactin and (*b*) enterobactin chelating a V^{4+} cation. [The latter figure is reproduced from S. J. Lippard and J. M. Berg, *Principles of Bioinorganic Chemistry*, University Science Books, Sausalito, CA, 1994; p. 109.]

such plants can have a very high concentration of the specific metal accumulated, hence it can be a useful ore.

On the other hand, iron ions are an essential nutrient for life, and the common iron ion, Fe^{3+}, is so acidic that it is unavailable at normal pH values. Hence, organisms have had to evolve careful mechanisms for obtaining scarce Fe^{3+} from natural waters (e.g., the Antarctic ocean), and for storing it internally. Bacteria achieve this by secreting hard-base chelating ligands known as **siderophores**.[42] Enterobactin (Fig. 5.11) is a hexadentate chelating ligand (binding with six deprotonated phenolic oxygen donor atoms), which has a formation constant for complexation of one Fe^{3+} ion to one enterobactin hexaanion of 10^{49}! Desferrioxamine (Fig. 5.10) is used medicinally to eliminate excess Fe^{3+} from the body; it contains three of the unusual *hydroxamic acid* functional group, any one of which [when in the deprotonated form $-C(=O)-N(-O^-)-$] is already able to chelate iron. The bacterial siderophore alcaligin[43] also contains three hydroxamic acid groups on a more preorganized (and therefore more efficient) organic backbone.

Study Objectives

1. Draw the structures of the pseudohalide ions; name them or give their formulas alone or in salts. Exercises 1–4.

2. Classify ligands. Exercises 5–13.

3. Predict the positions of equilibria involving monodentate, chelating, and macrocyclic ligands; give possible explanations of the chelate and macrocyclic effects. Exercises 14–21.

4. List a group of Lewis acids or of Lewis bases in order of increasing softness, taking into account the effects of (a) changing the oxidation number of the donor or acceptor atom; (b) changing the hardness or softness of other groups attached to it. Exercises 22–28.

5. Use HSAB to predict whether a given reaction will go to the left or to the right. Exercises 29–33.

6. Discuss explanations (in terms of thermodynamics and bonding types) of why the HSAB principle works. Know that the HSAB principle may not work well if there is a great difference in strength in the acids or bases involved. Exercises 34–39.

7. Use HSAB and acidity classifications to predict whether a given halide, sulfide, selenide, or telluride will be soluble or insoluble in water. Exercises 40, 41.

8. Use HSAB and acidity classifications to predict in which group of the qualitative analysis scheme a given metal ion will fall. Exercises 42–47.

9. Use HSAB to classify an element as an atmophile, a lithophile, or a chalcophile or siderophile. Identify likely mineral sources of given elements, and forms in which they would occur in natural waters. Exercises 48–54.

10. Use HSAB to predict the binding sites of different metal ions to biological molecules. Identify and discuss the toxicities of elements using the HSAB principle. Exercises 55–62.

11. Use HSAB and the chelate principle to select likely ligands for medicinal use. Exercises 63–69.

12. Know and identify examples of the functions of metal ions in biochemistry. Exercises 70–71.

Exercises

1. Draw the Lewis structures (and indicate the charges) of the following pseudohalide or other polyatomic ions: (a) the azide ion; (b) the thiocyanate ion; (c) the cyanate ion; and (d) the thiosulfate ion.

2. *Write the formulas of the following: (a) sodium azide; (b) chromium(II) cyanide; (c) silver(I) thiosulfate; (d) mercury(II) azide; and (e) barium dimethylamide.

3. Name the following: (a) $Al(N_3)_3$; (b) SCN^-; (c) $Fe(CN)_2$; (d) $Fe(CN)_3$; (e) $Al(OCN)_3$; (f) $Fe(SCN)_3$; and (g) $Hg(SCH_3)_2$.

4. Name the following: (a) $S_2O_3^{2-}$; (b) $Pb(N_3)_2$; (c) $Pb(OCN)_4$; and (d) $[N(CH_3)_4](CN)$.

5. Consider the following molecules: (1) $H_2N–CH_2–(C{=}O)–OH$; (2) $CH_3–O–CH_2–O–CH_3$; (3) $H_3N–CH_2–C({=}O)O$; (4) $[HO(O{=})C–CH_2]_3N$; and (5) $H–C({=}O)–OH$. (a) Complete the Lewis structures of these molecules by drawing in their unshared electron pairs. Underline each potential donor atom. (b) Circle the functional groups in the molecules. Which of the molecules are *polyfunctional* (have more than one functional group)? (c) Which of these molecules have donor atoms appropriately spaced for them to be *chelating ligands*? (d) Classify each chelating ligand as bidentate, tridentate, tetradentate, and so on.

6. *Which of the following are *chelating* ligands? (a) $PO_4{}^{3-}$; (b) (cyclic) $(O–CH_2CH_2)_5$; (c) $CH_3–O–CH_2–O–CH_3$; (d) $(CH_3)_3N$; (e) $N(CH_2–CO_2{}^-)_3$; and (f) $[NH_3CH_2CH_2NH_3]^{2+}$.

7. Draw the Lewis dot structure (including unshared electron pairs) of any example of your choice of the following: (a) a bidentate chelating ligand; (b) a tetradentate macrocyclic ligand; and (c) an ambidentate ligand that cannot chelate.

8. (a) Identify donor atoms in the molecules $(CH_3)_2N–N(CH_3)_2$, *cyclo*-$(SCH_2CH_2)_3$, and $CH_3SCH_2CH_2SCH_2CH_2SCH_3$ by filling in the proper number of unshared electron pairs on the atoms that should have them in complete Lewis structures of each molecule. (b) Which of these molecules are chelating ligands? (c) Which are macrocyclic ligands? (d) Classify each chelating or macrocyclic ligand as bidentate, tridentate, and so on.

9. (a) Which of the ligands drawn below are chelating ligands? (b) The Pd^{2+} ion normally has a coordination number of 4. Predict the formula (including overall charge) of the complex ion formed by the Pd^{2+} ion, showing a coordination number of 4, with (one or more moles of) each ligand, and allowing each chelating ligand to function as a chelating ligand.

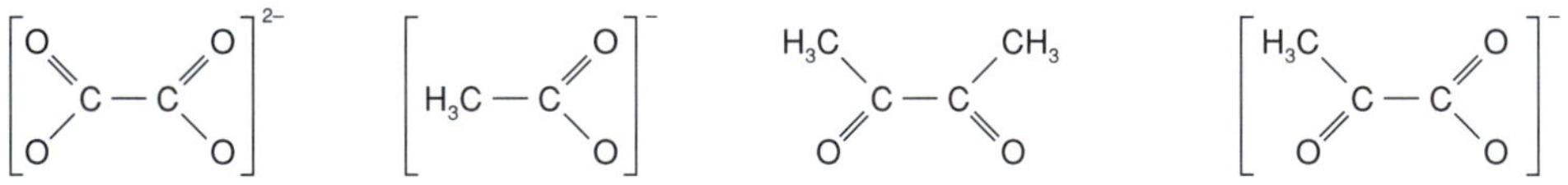

$[O_2C–CO_2]^{2-}$ $[H_3C–CO_2]^-$ $H_3C–C(=O)–C(=O)–CH_3$ $[H_3C–C(=O)–CO_2]^-$

10. (a) Complete the Lewis dot structures of the following potential ligands (all obey the octet rule), showing unshared electron pairs with dots, and circling the potential donor atoms in each:

Ligand "**A**": H–N(H)–C(H)(H)–C(=O)–O–H

Ligand "**B**": $[AsO_3]^{3-}$ (As bonded to three O)

Ligand "**C**": $[O–P(O)(O)–O–P(O)(O)–O–P(O)(O)–O]^{5-}$

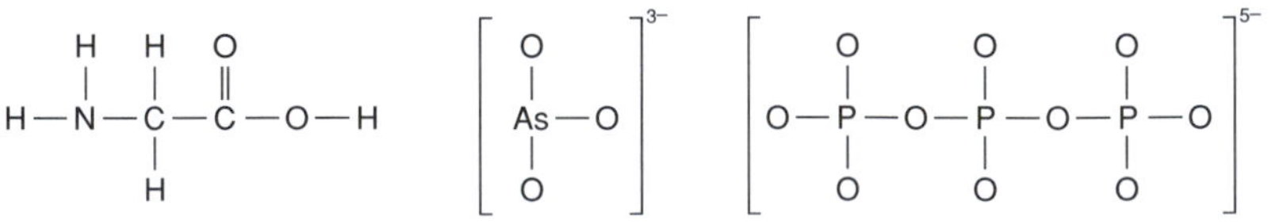

(b) Which one of the above three ligands is least likely to be a chelate ligand? (c) Which one is most likely to be a tridentate chelating ligand? (d) Which one is an ambidentate ligand? (e) Which one is more likely to be a bridging (rather than a chelate) ligand? (f) Which one (if any) is a macrocyclic ligand? (g) The $:Pb^{2+}$ ion tends to have a maximum total coordination number of a little over 6. Write the formula (including charge) of the complex ion it would form, having such a maximum total coordination number, with one or more moles of the highest order chelating ligand from the above set.

11. *Below is drawn the structure of a (hypothetical) truly complex complex ion.

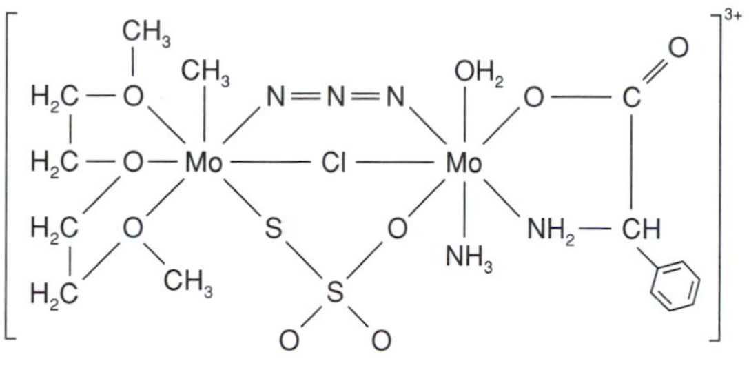

(a) Write the formulas, including charges if any, of the ligands found in this complex. Underline the donor atoms on each ligand. (b) List all chelating ligands in the above complex, if any. Identify each as bidentate, and so on. (c) List all bridging ligands in the above complex, if any. (d) Give the coordination number of each Mo ion in the above complex. (They may or may not be the same.)

12. Shown below are some molecules and ions that *may* be capable of acting as ligands. (a) Pick out all of these that are capable of acting as chelating ligands. (b) Classify each potentially chelating ligand as (potentially) *bidentate*, *tridentate*, and so on. (Choose the highest possible classification for each.)

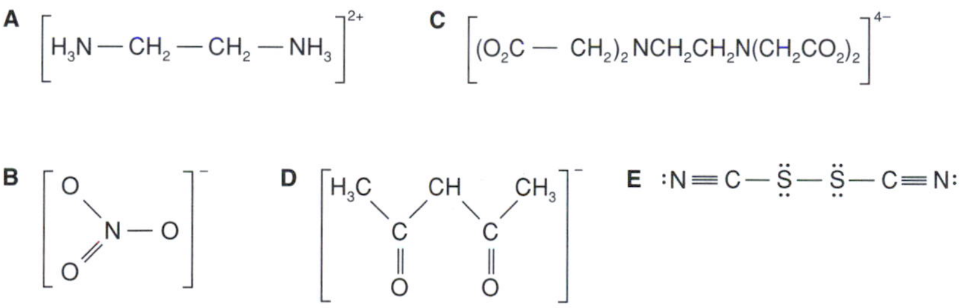

A $[H_3N-CH_2-CH_2-NH_3]^{2+}$

C $[(O_2C-CH_2)_2NCH_2CH_2N(CH_2CO_2)_2]^{4-}$

B $[O_2N-O]^-$

D $[H_3C-C(=O)-CH-C(=O)-CH_3]^-$

E $:N{\equiv}C-\ddot{S}-\ddot{S}-C{\equiv}N:$

13. Below is drawn a neutral coordination compound of the metal niobium, Nb. (a) What is the coordination number of Nb in this complex? (b) What is the likely geometric shape about the Nb atom? (c) What are the likely approximate bond angles about the Nb atom? (d) List all chelating ligands present in this complex. (e) List all ambidentate ligands present in this complex.

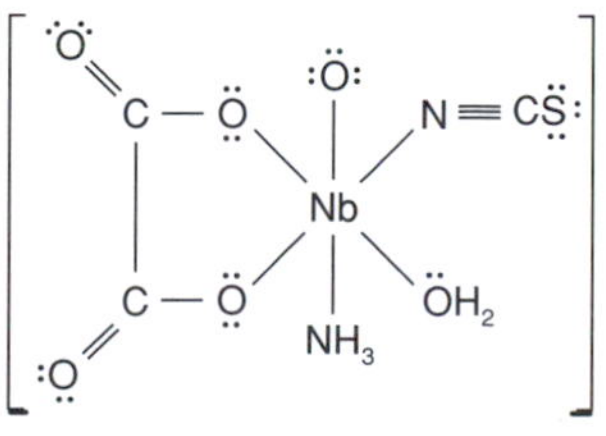

14. Below are drawn the Lewis structures of some chelated complexes. (a) For each atom that is part of the actual chelate ring (i.e., the metal atom, the two donor atoms, and the atoms that connect the donor atoms) use VSEPR theory to predict the ideal bond angles within the chelate ring at that atom. (b) If the chelate ring is to be planar, these angles must add up to $(n - 2)$ times 180°, where n is the number of atoms in the chelate ring. Add up the predicted angles for each chelated complex. Which complexes, with which values of n, come closest to the VSEPR ideal? (c) For some structures that show strain due to nonideal angles in the planar structure, a nonplanar structure can be found that will allow nearly ideal angles to be achieved. Using molecular models, determine for which nonideal ring sizes (values of n) this is feasible.

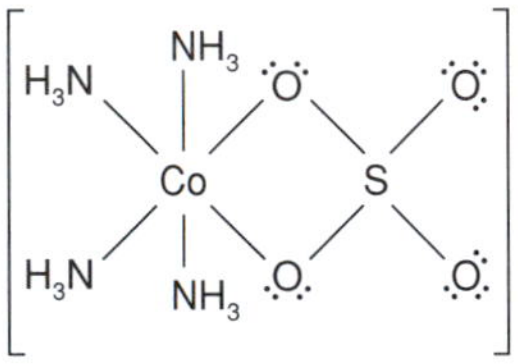

A 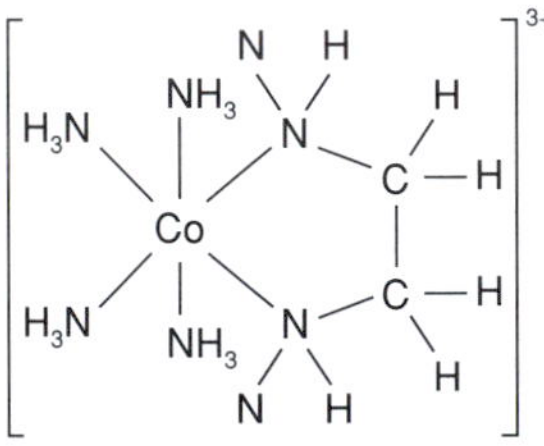 **B** 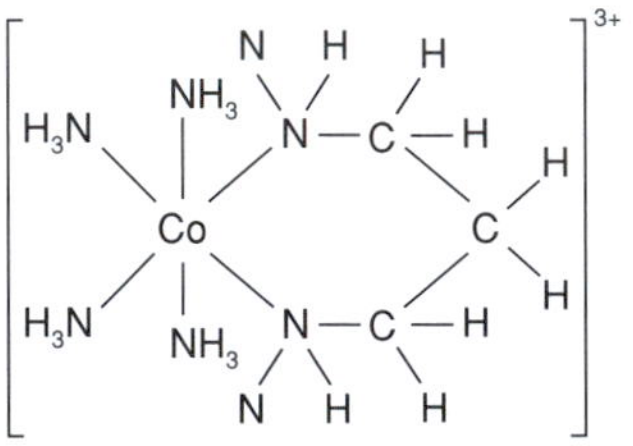 **C** 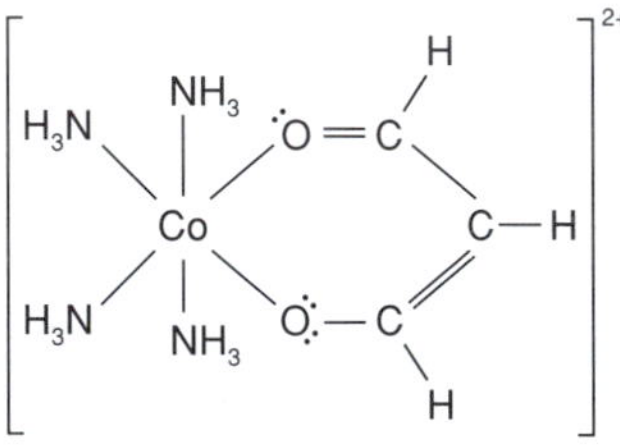 **D**

15. *Some ligands are drawn below. Which of these is/are (a) tridentate ligands; (b) macrocyclic ligands; (c) ligands that can chelate; (d) ligands that can bridge metal ions but not chelate them; (e) the ligand that would form the least stable complex with a metal ion?

$NH_2—CH_2—CH_2—NH_2$ **A**

$H_2N—NH_2$ **B**

$(NH_2CH_2CH_2)_2NH$ **C**

$(NH_2CH_2CH_2)_3N$ **D**

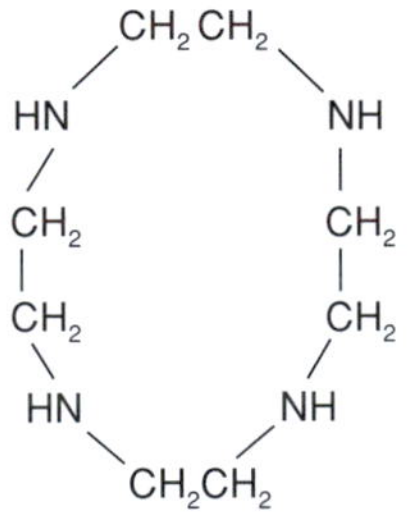

E

16. Three correct Lewis structures (**A**, **B**, and **C**) are drawn below.

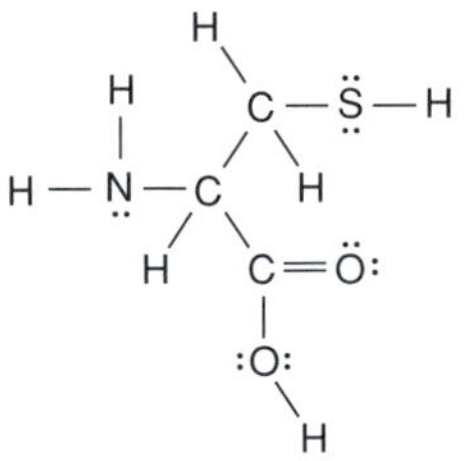

A 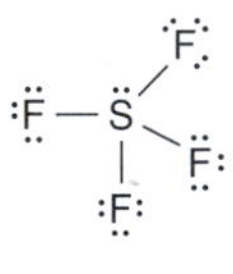 **B** 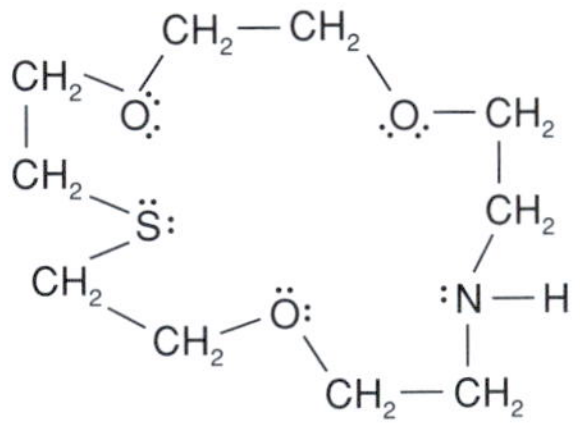 **C**

(a) Identify all possible donor atoms in molecules **A**, **B**, and **C**. (b) Which (if any) of the molecules **A**, **B**, **C** can be chelating ligands; which can be macrocyclic ligands? (c) Which one of these molecules would probably form the most stable complex with a metal ion? Would this extra stability be connected with an enthalpy effect, or an entropy effect?

17. Explain why chelating ligands give more stable complexes than monodentate ligands; use a balanced chemical equation in your explanation. Explain why macrocyclic ligands give more stable complexes than chelating ligands.

18. Write a balanced chemical equation, including formulas of complex ions, for any equilibrium reaction involving the donor atoms specified that gives a good illustration of the macrocyclic effect. Indicate which side of the equilibrium is favored. Donor atoms: (a) S; (b) N; (c) O.

19. *Consider the three complexes: (**A**) $[Pt(CN)_2(NH_2CH_2CH_2NH_2)_2](ClO_4)_2$; (**B**) $[Pt(CN)_2(NH_3)_4]SO_4$; (**C**) which is drawn below. (a) List these three complexes in order of increasing stability. (b) Which complex (**A**, **B**, **C**, or none) contains a chelating ligand? Which complex contains a macrocyclic ligand? Which complex contains a ligand that is bridging?

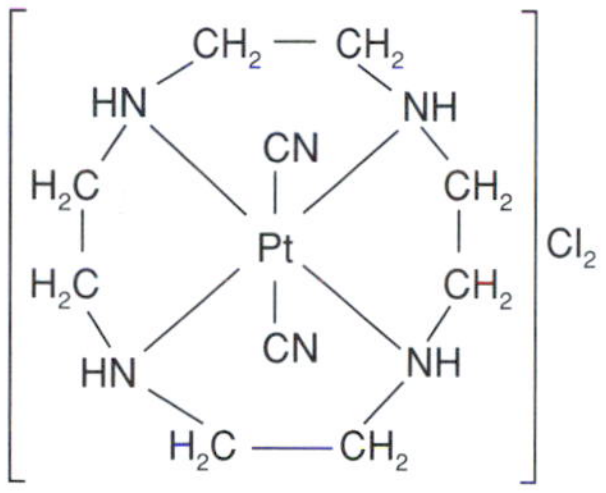

C

20. Predict whether reactants (left side) or products (right side) will be favored in each of the following equilibria:

(a) $Cd(SCH_3)_2 + HSCH_2CH_2SH \rightleftharpoons Cd(SCH_2CH_2S) + 2\,HSCH_3$

(b) $[Zn(OCH_2CH_2O)_2]^{2-} + 4\,HOCH_3 \rightleftharpoons [Zn(OCH_3)_4]^{2-} + 2\,HOCH_2CH_2OH$

(c) $[HgCl_2(\text{18-crown-6})] + CH_3O(CH_2CH_2O)_5CH_3 \rightleftharpoons$
$[HgCl_2\{CH_3O(CH_2CH_2O)_5CH_3\}] + \text{18-crown-6}$

(d) $[MgCl_2\{CH_3O(CH_2CH_2O)_4CH_3\}] + \text{15-crown-5} \rightleftharpoons$
$[MgCl_2(\text{15-crown-5})] + CH_3O(CH_2CH_2O)_4CH_3$.

21. Two different ligands are involved in each of the following complex–ion equilibria. (a) Classify each ligand in each equilibrium reaction as monodentate, chelating, or macrocyclic. (b) Classify each chelating or macrocyclic ligand as bidentate, tridentate, and so on. (c) Predict whether reactants (left side) or products (right side) will be favored in each of the reactions. (d) The position of the equilibrium in each of these reactions is determined by which: the ΔH term, or the $-T\Delta S$ term?

Reaction 1: $Pb(SCH_3)_2 + HSCH_2CH_2SCH_2CH_2SH \rightleftharpoons$
$Pb(SCH_2CH_2SCH_2CH_2S) + 2\,HSCH_3$.

Reaction 2: $[ZnCl_2(\text{15-crown-5})] + CH_3O(CH_2CH_2O)_4CH_3 \rightleftharpoons$
$[ZnCl_2\{CH_3O(CH_2CH_2O)_4CH_3\}] + \text{15-crown-5}$.

22. In each of the following complexes, classify both the metal ions and each ligand as *hard*, *soft*, or *borderline*: (a) $(NH_4)_2[Pd(-SCN)_4]$; (b) $[Co(NH_3)_4(SeO_4)]ClO_4$.

23. *Identify the potential donor atoms in each of the following molecules or ions, and classify the donor atoms as hard, soft, or borderline bases: (a) NH_3; (b) $NH_2{-}CH_2{-}COOH$ (an amino acid); (c) SO_4^{2-}; and (d) SO_3^{2-}.

24. Classify each Lewis acid and each Lewis base in the following complexes as hard, soft, or borderline. (a) $[Na(CH_3{-}O{-}CH_2{-}CH_2{-}O{-}CH_3)_3]^+$; (b) $[SnI_6]^{2-}$; (c) $[HgCl_2]$; and (d) $[Hg(CH_3)_2]$.

25. *Rewrite each of these lists of Lewis acids or bases so that it is in increasing order of softness: (a) Cu^+, Au^+, Ag^+, K^+; (b) Br^-, I^-, F^-, Cl^-; (c) Mo^{2+}, Mo^{6+}, Mo^{4+}; (d) BF_3, $B(OCH_3)_3$, $B(CH_3)_3$; (e) FHg^+, CH_3Hg^+, Fe^{2+}, Fe^{3+}; and (f) $(CH_3)_2S$, $(CH_3)_2Se$, $(CH_3)_2O$.

26. Most oxo anions are hard bases, but some are ambidentate and might also act as soft bases. List some of these.

27. Rewrite each of these lists of Lewis acids or bases so that it is in increasing order of softness: (a) Cu^+, Cu^{2+}, Cu^{3+}; (b) $(CH_3)_2SO$, $(CH_3)_2S$, Cl_2SO (all using S as donor atom); (c) $(CH_3)_2O$, $(CH_3)_2Se$, $(CH_3)_2S$, $(CF_3)_2O$; (d) BF_3, BCl_3, $B(CH_3)_3$; (e) Sr^{2+}, Hg^{2+}, Pb^{2+}; (f) I^-, Cl^-, Br^-, F^-; (g) Ir^+, Ir^{3+}, Ir^{4+}; (h) Po^{2+}, Pt^{2+}, Pd^{2+}, Po^{4+}, Pa^{5+}; (i) Ag^{2+}, As^{3+}, Ac^{3+}, Au^+, Ag^+; (j) InF_3, BF_3, InI_3, TlI_3; and (k) $[Co(NH_3)_5]^{3+}$, $[Co(H_2O)_5]^{3+}$, $[Co(NH_3)_5]^{2+}$, $[Co(PH_3)_5]^+$.

28. Students in inorganic lab have just discovered some new elements, which have the following names, symbols, and properties:

ELEMENT:	DeVersine (Dv)	Malium (Ml)	Derricium (Dm)	Carterogen (Ct)
Electronegativity	2.22	1.43	2.33	3.78
Ion Formed	Dv^{3-}	Ml^{4+}	Dm^+	Ct^{2-}
Ion Radius	198	52	144	125

Based on the information given, classify these ions as hard acids, soft acids, hard bases, or soft bases.

29. *Predict whether reactants (left side) or products (right side) will be favored in each of the following equilibria:

(a) $As_2S_5 + 5\,HgO \rightleftharpoons As_2O_5 + 5\,HgS$

(b) $ZnI_2 + HgCl_2 \rightleftharpoons ZnCl_2 + HgI_2$

(c) $La_2(CO_3)_3 + Bi_2S_3 \rightleftharpoons La_2S_3 + Bi_2(CO_3)_3$

(d) $2\,CH_3MgF + HgF_2 \rightleftharpoons (CH_3)_2Hg + 2\,MgF_2$

(e) $AgF + LiI \rightleftharpoons AgI + LiF$

30. Predict whether reactants (left side) or products (right side) will be favored in each of the following equilibria:

(a) $[Cu(thiourea)_4]^+ + [Cu(urea)_4]^{2+} \rightleftharpoons [Cu(thiourea)_4]^{2+} + [Cu(urea)_4]^+$

(b) $PbSe + HgS \rightleftharpoons HgSe + PbS$

(c) $3\,FeO + Fe_2S_3 \rightleftharpoons Fe_2O_3 + 3\,FeS$

(d) $CdSO_4 + CaS \rightleftharpoons CdS + CaSO_4$

(e) $Zn(SCH_3)_2 + Hg(SeCH_3)_2 \rightleftharpoons Hg(SCH_3)_2 + Zn(SeCH_3)_2$

(f) $2\,Fe(SCN)_3 + 3\,Fe(OCN)_2 \rightleftharpoons 2\,Fe(OCN)_3 + 3\,Fe(SCN)_2$

(g) $CaS + ZnSeO_4 \rightleftharpoons ZnS + CaSeO_4$

31. Give plausible products (or just one product) for each of the following reactions, or tell if no reaction is expected.

(a) $CdSO_4 + MgS \rightarrow$

(b) $(CH_3)_2Hg + CaF_2 \rightarrow$.

32. Predict any products that will precipitate in each of the following mixtures in water:

(a) $HgCl_2 + KI + KF \rightarrow$

(b) $PrCl_3 + Na_2S \rightarrow$

(c) $TlNO_3 + KI + KF \rightarrow$.

33. Predict whether reactants (left side) or products (right side) will be favored in each of the following equilibria:

(a) $Hg(CN)_2 + Ca(ClO_4)_2 \rightleftharpoons Hg(ClO_4)_2 + Ca(CN)_2$

(b) $CH_3HgI + ClHgCl \rightleftharpoons CH_3HgCl + ClHgI$

(c) $HgSeO_4 + SrS \rightleftharpoons HgS + SrSeO_4$

(d) $CuF + LiI \rightleftharpoons CuI + LiF$

(e) $PbSeO_4 + HgS \rightleftharpoons PbS + HgSeO_4$

(f) $3\,FeBr_2 + 2\,FeF_3 \rightleftharpoons 3\,FeF_2 + 2\,FeBr_3$

(g) $2\,CuF + CaBr_2 \rightleftharpoons 2\,CuBr + CaF_2$

(h) $PtSeO_4 + PbSe \rightleftharpoons PtSe + PbSeO_4$

34. What do relative electronegativities suggest about the nature of the bond between a hard acid and a hard base? A soft acid and a soft base?

35. Are the ionic radii and charges of hard acids and bases also favorable to the kind of bonding you cited in the previous question? Explain.

36. *Hydrogen and boron do not have electronegativities characteristic of hard acids, yet the H^+ and B^{3+} ions are classified as hard acids. Which of their properties other than electronegativity might justify such a classification?

37. Tell whether the position of equilibrium of each of these reactions is governed by the enthalpy or the entropy change of the reaction: (a) a reaction that goes to products due to the chelate effect; (b) a reaction that goes to products due to the HSAB principle; and (c) a reaction that goes to products due to the macrocyclic effect.

38. What other characteristics of acids and of bases—besides their hardness and softness—determines whether a reaction will go to the left or to the right? Find thermodynamically characterized equations for reactions involving these characteristics, and compare the magnitudes of the entropy or enthalpy changes with the magnitude involved in a typical HSAB reaction.

39. *The equilibrium constant for the reaction

$$HgI_2 + 2\,KOH \rightleftharpoons Hg(OH)_2 + 2\,KI$$

is relatively close to 1. Briefly explain why this reaction does not strongly tend to go to one side or the other.

40. *Identify all *insoluble* compounds: CdTe, AgI, AgF, KI, EuSe, TiO_2, $TiTe_2$, and $PtAs_2$.

41. Identify all insoluble compounds in each series: (a) BaTe, $TiTe_2$, TlI, Tl_2S, $CoAs_2$ (contains Co^{2+} ions), CoI_2, CaTe, CaI_2, Li_3As; (b) BaO, $CoSO_4$, TiO_2, CaF_2. (c) Which of these compounds are hydrolyzed?

42. Give the symbols for two metal ions that would fall into each of the groups of the qualitative analysis scheme.

43. In which group of the qualitative analysis scheme will each of the following ions fall? (a) Tl^+; (b) Cu^+; (c) Rb^+; (d) Cr^{3+}; and (e) Sn^{2+}.

44. *Predict in which group of the qualitative analysis scheme each of the following metal ions would occur. If the metal ion gives a precipitate, write the formula of the precipitate: (a) Sn^{2+}; (b) Pr^{3+}; (c) Cu^+; (d) Sr^{2+}; (e) Sb^{3+}; (f) Eu^{3+}; (g) Au^+; (h) Ra^{2+}; and (i) Rb^+.

45. *Which of the following anions could be used to precipitate the cations of qual scheme group V: OH^-, ClO_4^-, ClO_2^-, ReO_4^-, PF_6^-, ICl_4^-.

46. (a) If a qual scheme unknown were inadvertently *not* treated with HCl to precipitate group I or H_2S to precipitate group II, in which of the remaining groups would the following cations then precipitate? Ag^+, Hg^{2+}, Bi^{3+}. (b) If, in addition, the group III precipitation used a basic solution from which sulfide were again omitted, in which of the remaining groups would the same cations precipitate?

47. You are directing a freshman student in a research project to devise a new qualitative analysis scheme that should separate the cations into approximately the existing five groups, but by using novel precipitating anions. Among the choices listed, which is the best substitute for (a) the HCl precipitant in qual scheme Group I: HBr, HF, $HClO_4$, or HClO; (b) the H_2S precipitant in Group II: H_2SO_3, H_2O, H_2Te, or H_2SO_4; and (c) the $(NH_4)_2CO_3$ precipitant in group IV: $(NH_4)_2SO_4$, $(NH_4)ClO_4$, $(NH_4)_4C$, or NH_4OH; (d) If you wanted to precipitate group V, which compound would do this: Li_2SO_4, $LiClO_4$, Li_4C, or LiOH?

48. *Answer the questions about the following metal ions: Zr^{4+}; Ag^+; and Sb^{3+}. (a) Is its bromide soluble or insoluble? (b) Is its selenide soluble or insoluble? (c) Is its oxide soluble or insoluble? (d) In which group of the qualitative analysis scheme does this metal ion occur? (e) Is this metal a lithophile? (f) Which is the more likely mineral source: a silicate, a sulfide, or in seawater?

49. *Classify each element as either (A) a lithophile or (B) either a chalcophile or a siderophile. Then choose its most likely mineral source.

(a) Element: La; source: $LaPO_4$, LaAs, LaI_3, or $LaCl_3$.

(b) Element: Pt; source: $PtAs_2$, PtN_2, $PtSiO_4$, or PtF_2.

(c) Element: Zr; source: $ZrSiO_4$, $ZrPbS_4$, or $ZrCl_4$.

(d) Element: Sb; source: $Sb_2(SiO_3)_3$, Sb_2S_3, SbF_3, or $SbCl_3$.

(e) Element: Te; source: TeF_4, Na_2Te, PbTe, or TeI_4.

50. Classify each element as either (A) a lithophile or (B) either a chalcophile or a siderophile. Then choose its most likely mineral source.

(a) Element: Be; source: $Be^{2+}(aq)$, $Be_3Al_2Si_6O_{18}$, or $BeSeS_2$.

(b) Element: F; source: F^-(aq), CaF_2, or HgF_2.

(c) Element: Co; source: $CoCO_3$, $CoAs_2$, or Co^{2+}(aq).

(d) Element: Th; source: ThS_2, ThO_2, or Th^{4+}(aq).

51. Answer the following questions about each of these four ions: Co^{2+}, Tl^+, Pr^{3+}, and Cd^{2+}. (a) Is its bromide soluble or insoluble? (b) Is its selenide soluble, insoluble, or hydrolyzed? (c) In which group of the qualitative analysis scheme (I, II, III as sulfide, III as hydroxide, IV, or V) does this metal ion occur? (d) Is this metal more likely a lithophile or a chalcophile? (e) Which is the more likely mineral source: a silicate or a sulfide? (f) Which is the more likely mineral source: an arsenate or an arsenide?

52. Answer the following questions about each of these four ions: Pd^{2+}, Bk^{3+}, Sb^{3+}, and Ra^{2+}. (a) Is its bromide soluble or insoluble? (b) Is its sulfide soluble, insoluble, or hydrolyzed? (c) In which group of the qualitative analysis scheme (I, II, III as sulfide, III as hydroxide, IV, or V) does this metal ion occur? (d) Is this metal more likely a lithophile or a chalcophile? (e) Which is the more likely mineral source: a silicate or a telluride?

53. Answer the following questions about each of these five ions: Sr^{2+}, Tl^+, Bi^{3+}, Eu^{2+}, and Co^{2+}. (a) Is its chloride soluble or insoluble? (b) Is its telluride soluble or insoluble? (c) Is its oxide soluble or insoluble? (d) In which group of the qualitative analysis scheme does this metal ion occur? (e) Is this metal a lithophile or a chalcophile? (f) Which is the more likely mineral source: a silicate or a sulfide?

54. Answer the following questions about each of these four ions: Ag^+, Eu^{3+}, Eu^{2+}, and Bi^{3+}. (a) Classify each as either (A) a lithophile or (B) a chalcophile or siderophile. (b) Give the group of the qualitative analysis scheme in which each would fall (I, II, III, IV, or V). (c) Write the formula of a plausible mineral source of that ion.

55. *Which are more generally toxic, soft-acid metal ions or hard-acid metal ions? To which type of biochemical ligand would soft-acid metal ions bind most strongly? (a) phosphate groups; (b) porphyrin groups (nitrogen donor); (c) cysteine groups (sulfur donor).

56. Briefly explain why the element selenium is necessary for life and protects against mercury poisoning, but is also highly poisonous.

57. Which of the following manners of disrupting protein structure would you expect to be most uniquely associated with the heavy metal ions? (a) Hydrolysis of the CO–NH (peptide) linkages; (b) Disruption of hydrogen bonding between different CO–NH groups; (c) Disruption of S–S bond formation between different cysteine amino acid S–H groups; (d) Disruption of the electrostatic attraction of oppositely charged amino acid side chains.

58. Which two of the following classes of metal ions and ligands are most likely to be both essential to life and toxic to life: hard acids; borderline acids; soft acids; hard bases; borderline bases; soft bases. Explain briefly for one of these classes (or for a specific element from one of these classes) how this seeming paradox can be true.

59. *You are the health officer for a company that has just had a spill of some radioactive ions: francium ion (Fr^+); actinium ion (Ac^{3+}); polonide ion (Po^{2-}); perastatate ion ($AtO_6{}^{5-}$); meitnerium ion (Mt^{2+}). Ignore the effects of the radioactivity of these ions and think only of their possible chemical toxicity. (a) Which ion is the one most likely to bind to triphosphate groups ($P_3O_{10}{}^{5-}$) that are part of DNA? (b) Which ion is most likely to disrupt the quaternary structural elements of enzymes that depend on S–S bonds? (c) Which ion is most likely to bind to Fe^{2+} ions in essential enzymes? (d) Which ion is most likely to be methylated by vitamin B_{12} in the system and rendered even more toxic? (e) Which ion is most likely to remain as an (unhydrolyzed) hydrated ion without binding to anything, and hence be easily excreted in the urine?

60. You are the health officer for a company that has just had a spill of some ions: Rb^+; Pr^{3+}; HTe^-; $H_4TeO_6{}^{2-}$; and Pt^{2+}. Think of their possible chemical toxicity. (a) Which ion is the one most likely to bind to triphosphate groups ($P_3O_{10}{}^{5-}$) that are part of DNA? (b) Which ion is most likely to disrupt the quaternary structural elements of enzymes that depend on S–S bonds? (c) Which ion is most likely to remain as an (unhydrolyzed) hydrated ion without binding to anything, and hence be easily excreted in the urine? (d) Which ion is most likely to bind to Fe^{2+} ions in essential enzymes? (e) Which ion is most likely to be methylated by vitamin B_{12} in the system and rendered even more toxic?

61. You are the county health officer. Three new industries are moving to your area; each will discharge a different inorganic pollutant. In each case, briefly discuss the likelihood and nature of health effects resulting from that industry. Finally, tell which one industry should concern you the most. (a) The Shahrokhi Semiconductor Co., which will discharge some hydrogen telluride vapors. (b) The Kinningham Ceramic Engines Corp. of America, which will discharge some zirconium(IV) oxide in its wastewaters. (c) The Bingham Brewery, Inc., which will discharge some cobalt(II) chloride in its wastewaters. (d) Which industry is of most concern?

62. Tell whether each of the following classes of metal ions and ligands is *essential to life*, or *toxic to life*, or *neither*, or *both essential and toxic* to life: (a) typical hard acids; (b) typical borderline acids; (c) typical soft acids; (d) selenium.

63. *Which of the following medicinal chemicals would be most effective in combating poisoning by a soft-acid metal ion:

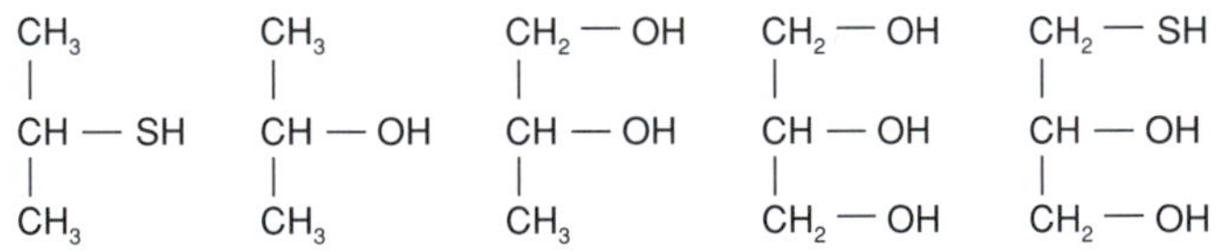

CH_3–CH(SH)–CH_3 CH_3–CH(OH)–CH_3 CH_2(OH)–CH(OH)–CH_3 CH_2(OH)–CH(OH)–CH_2(OH) CH_2(SH)–CH(OH)–CH_2(OH)

64. Consider the following six ligands. (a) List them in order of increasing softness (note that some will be equally soft). (b) Which one of these might make the best medicine to combat poisoning by the Pt^{2+} ion? What principles did you use to choose this ligand?

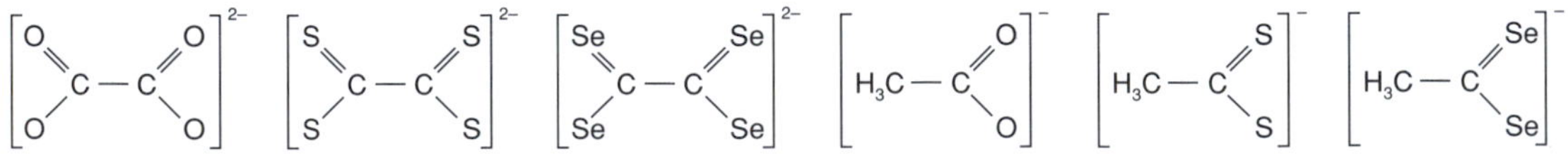

$[C_2O_4]^{2-}$ (O,O / C–C / O,O) $[C_2S_4]^{2-}$ (S,S / C–C / S,S) $[C_2Se_4]^{2-}$ (Se,Se / C–C / Se,Se) $[H_3C–CO_2]^-$ $[H_3C–CS_2]^-$ $[H_3C–CSe_2]^-$

65. You are trying to design drugs to treat poisoning by the Hg^{2+} and Be^{2+} ions by forming very stable, excretable complexes of these metals. Choose the mercury-containing complex below that would give the best excretion; likewise choose the best beryllium-containing complex. Explain your choice in a few words.

(a) Hg complexes:

HgF_4^{2-} $CH_3—Hg—CH_3$ $[Hg(SCH_2CH(S)CH_2OH)_2]^{2-}$ $[Hg(S(CH_3)CH_2CH(S(CH_3))CH_2OH)_2]$ $[Hg(O_2C\text{-}C_6H_3(O)\text{-}C(=C_6H_3(=O)CO_2H)\text{-}C_6H_3(OH)CO_2H)_2]^{2-}$

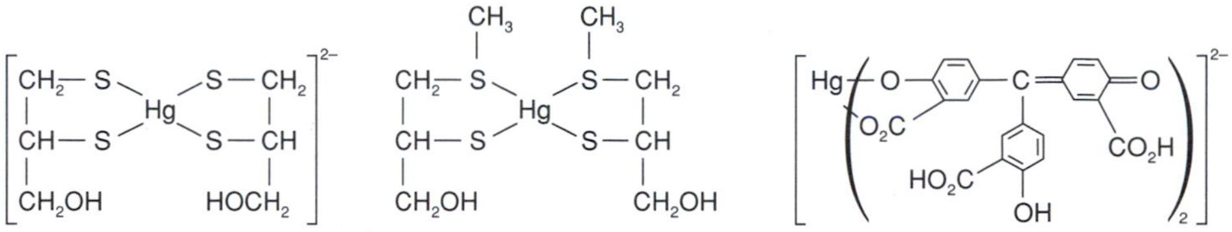

(b) Be complexes:

BeF_4^{2-} $CH_3—Be—CH_3$ $[Be(SCH_2CH(S)CH_2OH)_2]^{2-}$ $[Be(O(CH_3)CH_2CH(O(CH_3))CH_2OH)_2]$ $[Be(O_2C\text{-}C_6H_3(O)\text{-}C(=C_6H_3(=O)CO_2H)\text{-}C_6H_3(OH)CO_2H)_2]^{2-}$

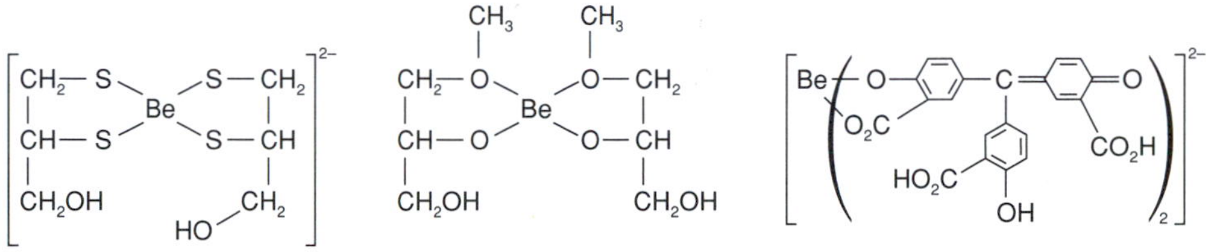

66. Answer each of the following questions for these four metal ions: Co^{2+}, Tl^+, Pm^{3+}, and Cd^{2+}. (a) Which *two* of these ions are most likely to have toxic chemical properties? (b) Which *one* of these ions is most likely to have nutritional value? (c) Which *one* of these ions would most likely be eliminated from the body by the use of the medicinal $EDTA^{4-}$ (ethylenediaminetetraacetate ion, $[COO(CH_2)_2NCH_2 \cdot CH_2N(CH_2COO)_2]^{4-}$) [Fig. 5.10(*a*)]? (d) Which *one* of these ions would most likely be eliminated from the body by the use of the medicinal penicillamine [Fig. 5.10(*a*)]?

67. Three of the six compounds listed below are used to treat poisoning by the Hg^{2+}, Be^{2+}, and Cu^{2+} ions; the other three are not useful for any of these. Identify the useless compounds; for each medicinally useful ligand identify the one metal ion of the above three that it would be used to treat.

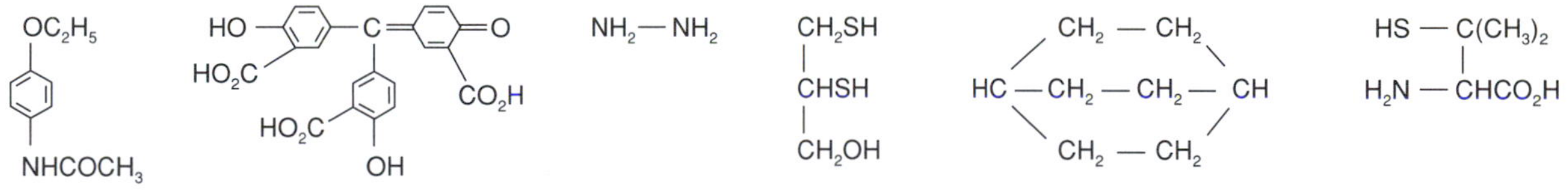

$OC_2H_5\text{-}C_6H_4\text{-}NHCOCH_3$ $HO\text{-}C_6H_3(CO_2H)\text{-}C(=C_6H_3(=O)CO_2H)\text{-}C_6H_3(OH)CO_2H$ $NH_2—NH_2$ $CH_2SH\text{-}CHSH\text{-}CH_2OH$ $HC(CH_2CH_2)_3CH$ $HS—C(CH_3)_2\text{-}CH(NH_2)CO_2H$

68. You are a physician dealing with victims of a spill of some radioactive ions: francium ion (Fr^+); actinium ion (Ac^{3+}); polonide ion (Po^{2-}); perastatate ion (AtO_6^{5-}); and meitnerium ion ($Z = 109$) Mt^{2+}. (a) Which ion is the one most likely to removed from the body upon administration of the drug $EDTA^{4-}$? [Fig. 5.10(*a*)]? (b) Which ion is most likely to be removed from the body upon administration of the drug British anti-Lewisite [Fig. 5.10(*a*)].

69. Below are the structures of seven ligands. From among them find: (a) the ligand that would be best suited for reacting with K^+ in the body; (b) the ligand that would be best suited for removing Hg^{2+} from the body.

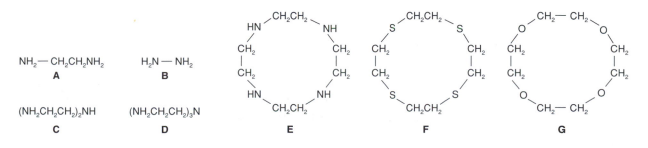

A $NH_2—CH_2CH_2NH_2$

B $H_2N—NH_2$

C $(NH_2CH_2CH_2)_2NH$

D $(NH_2CH_2CH_2)_3N$

E cyclic $(NHCH_2CH_2)_4$

F cyclic $(SCH_2CH_2)_4$

G cyclic $(OCH_2CH_2)_4$

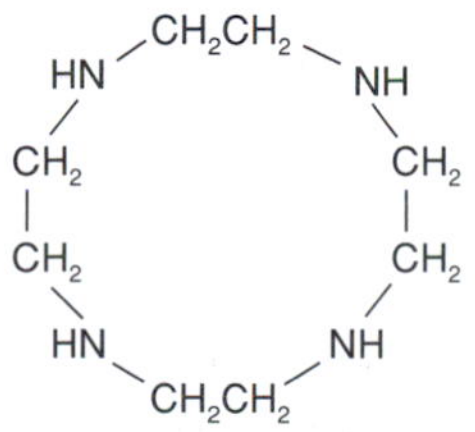

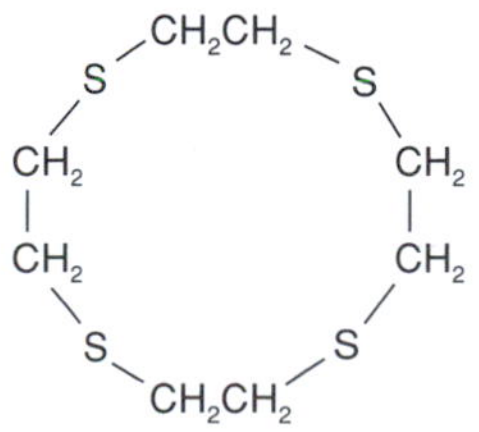

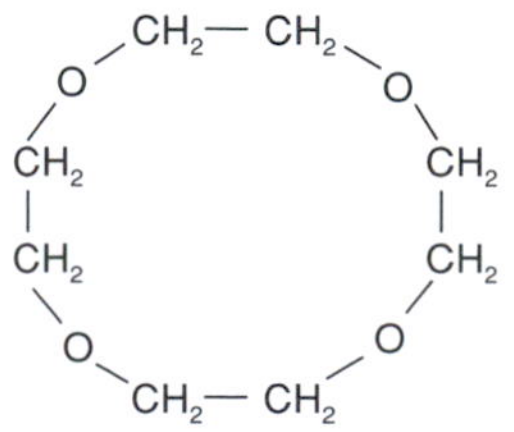

70. List the functions of metal ions in enzymes.

71. *Complexes containing coordinated NH_3 and H_2O are common, as are complexes containing coordinated R_3N, R_2O, R_3P, R_2S, and so on. But complexes containing coordinated PH_3 and H_2S (or RPH_2 and RSH) are very seldom seen. Suggest a possible reason why, based on the functions of metal ions in enzymes.

72. Using trigonometry, calculate the radius of the hole in the following crown ethers, and select the best-fitting +1 charged metal ion for each: (a) 12-crown-4; (b) 15-crown-5; (c) 21-crown-7.

73. Deep under the ocean at the midoceanic ridges where plates are separating and molten magma at 1200 °C approaches the ocean floor, oceanic sulfate ion is reduced to an acidic H_2S solution, and transition metal ions are leached from the rock. The superheated solution emerges from "black smokers" into the ocean floor and is quickly cooled from 350 to 4 °C, whereupon a chemistry resembling that of the qualitative analysis scheme of the cations occurs. (a) In the traditional qualitative analysis scheme of the cations (qual scheme), the group I cations Ag^+, Hg_2^{2+}, and Pb^{2+} are traditionally precipitated as chlorides by the addition of excess HCl. If a stoichiometric amount of HCl were used instead, a number of other metal cations would also precipitate as chlorides. List four such additional cations that might precipitate in black smokers as chlorides (from oceanic chloride ion). In the traditional qual scheme, these ions are kept in solution by the excess HCl present. Write a net ionic equation showing what soluble product is formed by each of your four additional ions with four moles of HCl. (Although the true number may be other than four in some cases, use four in your equation.) (b) In group II of the traditional qual scheme, H_2S in acidic solution is used to precipitate the following cations: Hg^{2+}, Pb^{2+}, Bi^{3+}, Cu^{2+}, Cd^{2+}, As^{3+}. Sb^{3+}, and Sn^{2+}. (Note that this resembles conditions in the black smoker fume as it first cools.) List three additional metals that could precipitate in this group if they were included in the unknown (or the leachate from the crustal rock by the smoker). List two additional compounds other than H_2S that could be used to precipitate about the same group of metals. (c) In group III of the traditional qual scheme, alkaline $(NH_4)_2S$ is used to precipitate seven additional cations: Co^{2+}, Ni^{2+}, Fe^{2+}, Mn^{2+}, Al^{3+}, Cr^{3+}, and Zn^{2+}. Note that this would resemble conditions as the acidic smoker solution is neutralized by the buffering capacity of the ocean; also playing a role are dissolved O_2 and manganese-metabolizing bacteria. As what compound does each cations precipitate under these conditions? Suggest six additional cations that would come down in group III if included in the unknown, and tell the form in which they would precipitate. (d) In the primary and secondary differentiation of the elements in the earlier geochemical history of the earth, separations somewhat related to those of the qual scheme occur. Geochemists traditionally divide the metals into three classes: the siderophiles, which tend to occur in nature in the elemental form; the chalcophiles, that occur in nature as sulfides or related minerals; and the lithophiles, the minerals that occur in nature as silicates or related minerals. Tell how the five qual scheme groups are related to the two geochemical classifications other than siderophiles. Clarify the term "related minerals": identify some mineral types that are related to the sulfides; to the silicates.

74. Any amino acid R–CH(NH_2)–COOH is capable of chelating a metal ion using the N donor atom of the amino group and an O donor atom of the carboxylic acid group. (a) Consult a standard textbook list of the common amino acids, and identify those amino acids that can function as **tridentate** chelating ligands. (b) Which donor atoms of these ligands are most likely to be deprotonated on forming complexes and why? How much negative charge will each tridentate chelating amino acid likely bear when coordinated? (c) As representative metal ions, let us take Be^{2+} and Cr^{3+}. What are likely coordination numbers for these metal ions? Using the example of Cr^{3+}, write equations showing the formation of complex ions from Cr^{3+} and the expected number of each tridentate chelating amino acids. Be sure to show the net complex-ion charges of the chelated products. (d) Suppose the person in whose body this chemistry is going on has arthritis, and is taking anti-arthritis drugs containing gold present as Au^{3+}. From which of the above complex ions would the gold most likely displace the chromium ion?

Notes

1. R. J. P. Williams, *The Metals of Life*, Van Nostrand-Reinhold, New York, 1971; Chapter 4.
2. J. L. Burmeister, *Coord. Chem. Rev.*, **105**, 77 (1990).
3. (a) A. L. Beauchamp, M. J. Olivier, J. D. Wuest, and B. Zacharie, *J. Am. Chem. Soc.*, **108**, 73 (1986). (b) M. Newcomb, J. H. Horner, M. T. Blanda, and P. J. Squattrito, *J. Am. Chem. Soc.*, **111**, 6294 (1989).
4. C.-S. Chung, *J. Chem. Educ.*, **61**, 1062 (1984); C.-S. Chung, *Inorg. Chem.*, **18**, 1321 (1979); R. T. Myers, *Inorg. Chem.*, **17**, 952 (1978).
5. R. D. Hancock (*J. Chem. Educ.*, **69**, *615 (1992)*) argues that enthalpy contributions determine whether six- or five-membered chelates are favored, and that this in turn is behind the macrocyclic effect; see also A. E. Martell and R. D. Hancock, in *Coordination Chemistry: A Century of Progress, ACS Symp. Ser.*, **565**, 240 (1994).
6. (a) R. G. Pearson, *Hard and Soft Acids and Bases*, Dowden, Hutchinson & Ross, Stroudsburg, PA, 1973; (b) op. cit., p. 6.
7. Note that many of the nonmetals may function as "cations" or Lewis acidic sites in some compounds and as donor atoms in others. For example, arsenic as As^{3+} is a hard acid, but as a donor (As^{3-}, etc.) it is a soft base.
8. R. G. Parr and R. G. Pearson, *J. Am. Chem. Soc.*, **105**, 7512 (1983); R. G. Pearson, *J. Chem. Educ.*, **64**, 561 (1987); R. G. Pearson, *Coord. Chem. Rev.*, **100**, 403 (1990); see also Chapter 10.
9. L. Zeloikovich, J. Libman, and A. Shanzer, *Nature* (*London*), **374**, 790 (1995).
10. G. Klopman, *J. Am. Chem. Soc.*, **90**, 223 (1968).
11. R. D. Hancock and A. E. Martell, *J. Chem. Educ.*, **73**, 654 (1996).
12. If the ligands have differing numbers of usable donor atoms, the principle of the chelate and macrocyclic effects would also come into play.
13. A. F. Clifford, *J. Am. Chem. Soc.*, **79**, 5404 (1957).
14. C. S. G. Phillips and R. J. P. Williams, *Inorganic Chemistry*, Oxford University Press, New York, 1965, Volume 2, Chapter 34.
15. See, for example, G. Brimhall, *Sci. Am.*, **265** (5), 84 (1991).
16. J. M. Edmond, K. L. Von Damm, R. E. McDuff, and C. I. Measures, *Nature* (*London*), **297**, 187 (1982); J. M. Edmond and K. Von Damm, *Sci. Am.*, **248** (4), 78 (1983); P. A. Rona, *Sci. Am.*, **254** (1), 84 (1986).
17. S. Manahan, *Environmental Chemistry*, 5th ed., Lewis Publishers, Chelsea, MI, 1991, p. 138.
18. A. M. Rouhi, *Chem. Eng. News*, Dec. 11, 1995, p. 37.
19. J. Haggin, *Chem. Eng. News*, Nov. 7, 1988, p. 18.
20. See, for instance, M. N. Hughes, *The Inorganic Chemistry of Biological Processes*, 2nd ed., Wiley, Chichester, UK, 1981.
21. R. W. Hay, "Lewis Acid Catalysis and the Reactions of Coordinated Ligands," in *Comprehensive Coordination Chemsitry*, G. Wilkinson, Ed., Pergamon, Oxford, 1987; Vol. 6, p. 411.
22. "Ion Channels: Discoverers win physiology Nobel," *Chem. Eng. News*, Oct. 14, 1991, p. 4.
23. M. J. Welsh and A. E. Smith, *Sci. Am.*, **273** (6), 52 (1995).
24. References cited in R. Baum, *Chem. Eng. News*, Jan. 25, 1993, p. 5.
25. R. Etcheberrigaray, E. Ito, K. Oka, B. Tofel-Grehl, G. E. Gibson, and D. L. Alkon, *Proc. Natl. Acad. Sci.*, **90**, 8209 (1993); *Chem. Eng. News*, Sept. 6, 1993, p. 7.

26. D. W. Christianson and W. N. Lipscomb, *Acc. Chem. Res.*, **22**, 62 (1989).
27. P. A. Sutton and D. A. Buckingham, *Acc. Chem. Res.*, **20**, 357 (1987).
28. G. N. Schrauzer, *Acc. Chem. Res.*, **1**, 97 (1968).
29. L. K. Altman, *New York Times*, Sept. 8, 1992.
30. Of interest is B. Roueché, "Annals of Medicine: Cinnabar," *New Yorker*, Dec. 8, 1986 p. 94.
31. Although it seems logical from the point of view of the HSAB principle, the traditional practice of sprinkling powdered sulfur over spillled mercury is not a good clean-up procedure since the reaction of elemental mercury and sulfur is *very* slow.
32. S. Krishnamurthy, *J. Chem. Educ.*, **69**, 347 (1992), and references cited therein.
33. *New York Times*, April 27, 1993. Bacteria have also developed mechanisms to resist 11 other toxic metal ions.
34. See D. Labianca, *J. Chem. Educ.*, **67**, 1019 (1990).
35. N. P. Farrell, T. G. Appleton, Y. Qu, J. D. Roberts, A. P. Soares Fontes, K. A. Skov, P. Wu, and Y. Zou, *Biochemistry*, **34**, 15480 (1995).
36. W. Leinfelder, E. Zehelein, and A. Böck, *Nature* (*London*), **331**, 723 (1987).
37. As an example of this type of research, see R. C. Hider and A. D. Hall, *Prog. Med. Chem.*, **28**, 40 (1991).
38. R. D. Hancock, *Acc. Chem. Res.*, **23**, 253 (1990).
39. S. Jurisson, D. Berning, W. Jia, and D. Ma, *Chem. Rev.*, **93**, 1137 (1993).
40. R. Baum, *Chem. Eng. News*, Jan. 8, 1990, p. 20.
41. A. M. Rouhi, *Chem. Eng. News*, Jan. 13, 1997, p. 21.
42. S. J. Lippard and J. M. Berg, *Principles of Bioinorganic Chemistry*, University Science Books, Sausalito, CA, 1994; pp. 103–111.
43. Z. Hou, C. J. Sunderland, T. Nishio, and K. N. Raymond, *J. Am. Chem. Soc.*, **118**, 5148 (1996).

CHAPTER

6

Principles of Oxidation–Reduction Reactivity

With Applications to Chemical Safety, Environmental Chemistry, and Industrial Chemistry

6.1 Standard Reduction Potentials and Their Diagrammatic Representation

In the previous chapters, we examined the acid–base, complexation, and precipitation reactions of compounds of the elements in fixed oxidation states. We saw, though, that most elements have more than one positive oxidation state; all can be prepared as the free elements with oxidation state 0; some have negative oxidation states. Reactions in which elements change their oxidation numbers are known as **oxidation–reduction** or **redox** reactions and will be the focus of this chapter.

Redox reactions normally require the presence of two reactants: In one reactant, the oxidation number of an element is reduced, and in the other, an oxidation number of an element increases. These two processes can be treated as separate **half-reactions**, which can often even be carried out in separate beakers of an **electrochemical cell** if these two beakers are connected by a wire and a salt bridge, to conduct the electrons (and ions) involved in these half-reactions.

For example, the redox reaction

$$Pb(s) + 2\,AgNO_3(aq) \rightarrow Pb(NO_3)_2(aq) + 2\,Ag(s) \tag{6.1}$$

can be separated into one half-reaction involving lead and one involving silver. The oxidation number of lead is increased from 0 in the lead metal to +2 in the lead(II) ion (found in its nitrate), and electrons come from the lead; we say that the lead is being **oxidized**, and that this is an **oxidation half-reaction:**

$$Pb(s) \rightarrow Pb^{2+}(aq) + 2\,e^- \tag{6.2}$$

These electrons find their way to the silver ions, where each electron reduces the oxidation number of one silver species (the silver is being **reduced**), and this is a **reduction half-reaction:**

$$Ag^+(aq) + e^- \rightarrow Ag(s) \tag{6.3}$$

Half-reactions often involve more complex species than these, such as oxo anions. In such cases, the balancing of the half-reactions is a little less obvious, and is reviewed here. As an example, let us consider the reduction of $Cr_2O_7{}^{2-}$ (the dichromate ion) to Cr^{3+}.

1. Assign oxidation numbers (Sections 1.8 and 3.10), and identify the element being oxidized or reduced: Its oxidation number changes. In the example, the oxidation number of each Cr atom changes from +6 in $Cr_2O_7{}^{2-}$ to +3 in Cr^{3+}. We note that this is a reduction of three for each Cr atom.

2. Balance the number of atoms of the elements being oxidized or reduced. Thus: $Cr_2O_7{}^{2-} \rightarrow 2\,Cr^{3+}$.

3. Balance the total change in oxidation number of all atoms of this element by adding the needed number of electrons, which can be considered, in effect, to have oxidation numbers of −1. Reduction half-reactions involve electrons as reactants; oxidation half-reactions produce electrons. In the example, the two Cr atoms are reduced by a total of six oxidation numbers, so they consume six electrons as reactants: $Cr_2O_7{}^{2-} + 6\,e^- \rightarrow 2\,Cr^{3+}$.

4. Balance the charges in the half-reaction by adding the needed number of H^+ ions (if the reaction is being done in acid solution, which will be standard in this chapter: If we were in basic solution, we would use hydroxide ions). Since we have eight negative charges on the left side of our example reaction and six positive charges on the right, we must add 14 hydrogen ions to the left side, so that it also will have six positive charges: $Cr_2O_7{}^{2-} + 6\,e^- + 14\,H^+ \rightarrow 2\,Cr^{3+}$.

5. Balance the hydrogen and oxygen atoms in the half-reaction by adding the appropriate number of water molecules: $Cr_2O_7{}^{2-} + 6\,e^- + 14\,H^+ \rightarrow 2\,Cr^{3+} + 7\,H_2O$.

In Reaction (6.1), the oxidation of the lead is accomplished by the silver(I) nitrate, which we therefore call the **oxidizing agent** in this redox reaction. The reduction of the silver from the +1 to the zero oxidation state is accomplished by the lead, which we call the **reducing** agent.

Thermodynamically, Reaction (6.1) and other redox reactions will go if the free energy change, ΔG, for the reaction is negative. Experimentally, instead of measuring ΔG we measure the **cell electromotive force** (**emf**, symbolized as E and measured in volts) generated when the reaction occurs spontaneously in a voltaic (galvanic) electrochemical cell. (Or we measure the emf required to force it to go in an electrolytic cell.) A spontaneous reaction occurring in a galvanic cell is considered to generate a positive emf, E. If the reaction occurs under reversible conditions, the emf can be related to the free energy change by the equation

$$\Delta G = -nFE \qquad (6.4)$$

where n is the number of electrons exchanged in the reaction as written, and F is the conversion between electrochemical and thermodynamic units, $96.5\,\mathrm{kJ\,V^{-1}(mol\,e^-)^{-1}}$.

In an electrochemical cell, the overall cell emf, E_{cell}, is presumed to be the sum of the potentials of the two half-reactions. We can measure only this *sum*, so we do not know the actual potential generated by any half-reaction. By convention, we assign a potential of 0.000 V to the half-reaction in which hydrogen ion is being reduced to hydrogen gas under standard conditions:

$$2\,H^+(aq,\ \text{activity} = 1) + 2\,e^- = H_2(\text{pressure} = 1\ \text{atm}) \qquad E^\circ = 0.000\ \text{V} \qquad (6.5)$$

The potential generated by this and other half-cells depends on the conditions of the reactants and products, so we define conditions such that each substance is in its standard state. For gases, the standard state means a pressure of 1.000 atm for an ideal gas (or its equivalent if the gas is not ideal); for pure liquids, solvents, and solids, this is the pure liquid or solid at 1.000-atm pressure. The standard state for a solute is a concentration that gives an activity or thermodynamic concentration of 1.000, in units of either molalities or molarities.[1] When H^+ is a product or reactant, its standard concentration will also be 1.000 *M*—hence the standard pH will be 0.00. If all reactants and products are in their standard states, we place a superscript (°) on the E (and the ΔG) to designate this fact.

We may now set up electrochemical cells in which all reactants and products are present in their standard state. In one half-cell, hydrogen gas is being oxidized to hydrogen ion, and in the other half-cell a chemical species is being reduced (reversibly) to a product in which an element is in a lower oxidation state. The emf generated in this cell (or required as an input if the reaction is not spontaneous) is called the **standard reduction potential**, E°, for reduction of that species. The more positive the E° of a species is, the more easily it is reduced (and the more reactive it is as an oxidizing agent). If the species has a negative standard reduction potential, it cannot be reduced by hydrogen (under standard conditions), and it is thus a poor oxidizing agent.

Extensive tabulations have been made of standard reduction potentials of various chemical species.[2] However, it is sometimes difficult to keep the arbitrary sign conventions in mind, or to get a "feel" for chemical reactivity and its periodic variation by looking at tables, so several diagrammatic ways of representing half-reaction potentials have been devised; these include Latimer and Frost diagrams.[3] Most areas of applied chemistry (environmental, geochemistry, and corrosion chemistry) use **Pourbaix diagrams**. These are a bit complex to learn directly, since they include both potential and pH as variables. Hence, we will first introduce redox predominance diagrams, in which the pH is fixed at zero by the standard hydrogen ion concentration of 1.000 *M*.

Redox predominance diagrams are modeled on the acid–base predominance diagrams of Chapter 2, so a brief review of the latter is in order. In the acid–base predominance diagrams (as, e.g., those of Fig. 2.4) the predominant acid–base form of each element (in a fixed oxidation state) is shown as a function of pH. Species that are nonacidic and nonbasic persist at all pH values and are therefore shown as predominant over the entire diagram. The more *acidic* a species, the more it is confined to the *left* side of its acid–base predominance diagram. The more *basic* a species, the more it is confined to the *right* side of its acid–base predominance diagram. If two different species can both persist at some common pH, they will not tend to react with each other; but if they have *nonoverlapping* predominance regions, they will tend to react via an acid–base reaction to give products that can coexist at a common pH (i.e. have overlapping predominance regions).

A redox predominance diagram shows the predominant (thermodynamically most stable) oxidation state and chemical form of an element at any given potential. In this type of diagram, elements with no redox chemistry (such as Ar) will predominate at all potentials, and will cover the entire diagram. Among those with more than one thermodynamically stable oxidation state, *more strongly oxidizing stable chemical*

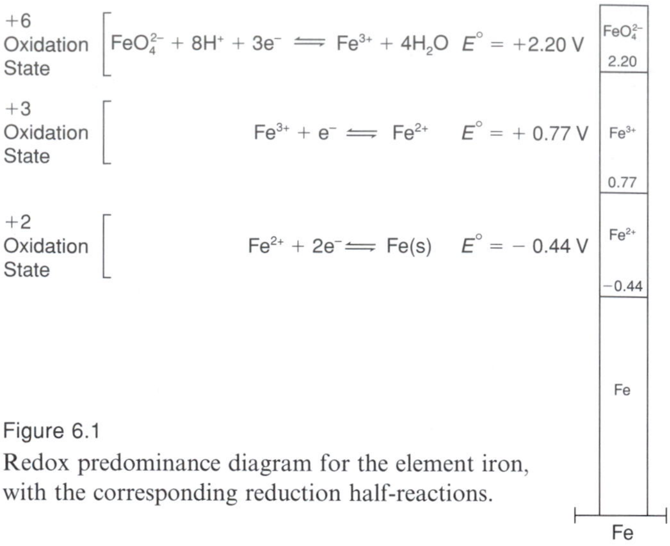

+6 Oxidation State: $FeO_4^{2-} + 8H^+ + 3e^- \rightleftharpoons Fe^{3+} + 4H_2O$ $E^\circ = +2.20$ V

+3 Oxidation State: $Fe^{3+} + e^- \rightleftharpoons Fe^{2+}$ $E^\circ = +0.77$ V

+2 Oxidation State: $Fe^{2+} + 2e^- \rightleftharpoons Fe(s)$ $E^\circ = -0.44$ V

FeO_4^{2-}
2.20
Fe^{3+}
0.77
Fe^{2+}
−0.44
Fe

Fe

Figure 6.1
Redox predominance diagram for the element iron, with the corresponding reduction half-reactions.

forms are confined to higher regions of redox predominance diagrams. Correspondingly, strongly reducing stable chemical forms will be low in the predominance diagrams. In Figure 6.1, we show a typical redox predominance diagram, that of iron. Of the chemical forms of iron shown, the most strongly oxidizing is the ferrate ion, $FeO_4{}^{2-}$; the most strongly reducing is iron metal, Fe.

Each numbered horizontal bar in a redox predominance diagram separates two chemical forms of the element in question. Above the E° shown at the bar, the (more strongly oxidizing) chemical form above the bar is the predominant form; below that E°, the (less oxidizing) chemical form below the bar is the predominant form. Thus, in a body of water that is kept well aerated, so that oxidizing conditions prevail (E° moderately above +0.77 V) more iron will be present as Fe^{3+}(aq) than as Fe^{2+}(aq); in poorly aerated parts of a body of water containing reducing agents such as decaying plant matter (E° moderately below +0.77 V) more iron will be present in the reducing form, Fe^{2+}(aq).

Note that this arrangement in the redox predominance diagrams corresponds to the way real bodies of water (lakes) stratify: the upper layer (epilimnion) tends to become well aerated, hence it contain oxidized forms of elements; the lower layer (hypolimnion) tends to accumulate reducing impurities, hence it contains reduced forms of elements.

Note also that the thermodynamically stable forms of the elements are arranged in the diagrams so that *higher oxidation states of elements occur higher in redox predominance diagrams.* This finding can be confirmed by assigning oxidation states to the species shown in the iron redox predominance diagram. Thus, iron(VI), in the chemical form $FeO_4{}^{2-}$, is the most stable form in a solution of pH 0 and a total iron concentration of 1.000 *M* above a potential of +2.20 V. As the potential is lowered, the stable oxidation state drops, first to +3, then to +2, then to 0, in elemental iron.

The reduction half-reactions that accompany Figure 6.1 have E° values equal to the numerical boundaries shown in the redox predominance diagrams. The reactant form of the element is predominant above the boundary; the product form is predominant below the boundary. Thus, reduction half-reaction (6.6) corresponds to the +2.20-V boundary in the iron predominance diagram:

$$FeO_4{}^{2-} + 8\,H^+ + 3\,e^- \rightarrow Fe^{3+} + 4\,H_2O \qquad E^\circ = +2.20\text{ V} \tag{6.6}$$

Note that the essence of the half-reaction can be read from the redox predominance diagram, but the balancing is incomplete: electrons, water, and H^+ must be added and balanced. Redox predominance diagrams of all of the elements are presented in Figures 6.2–6.5.[4]

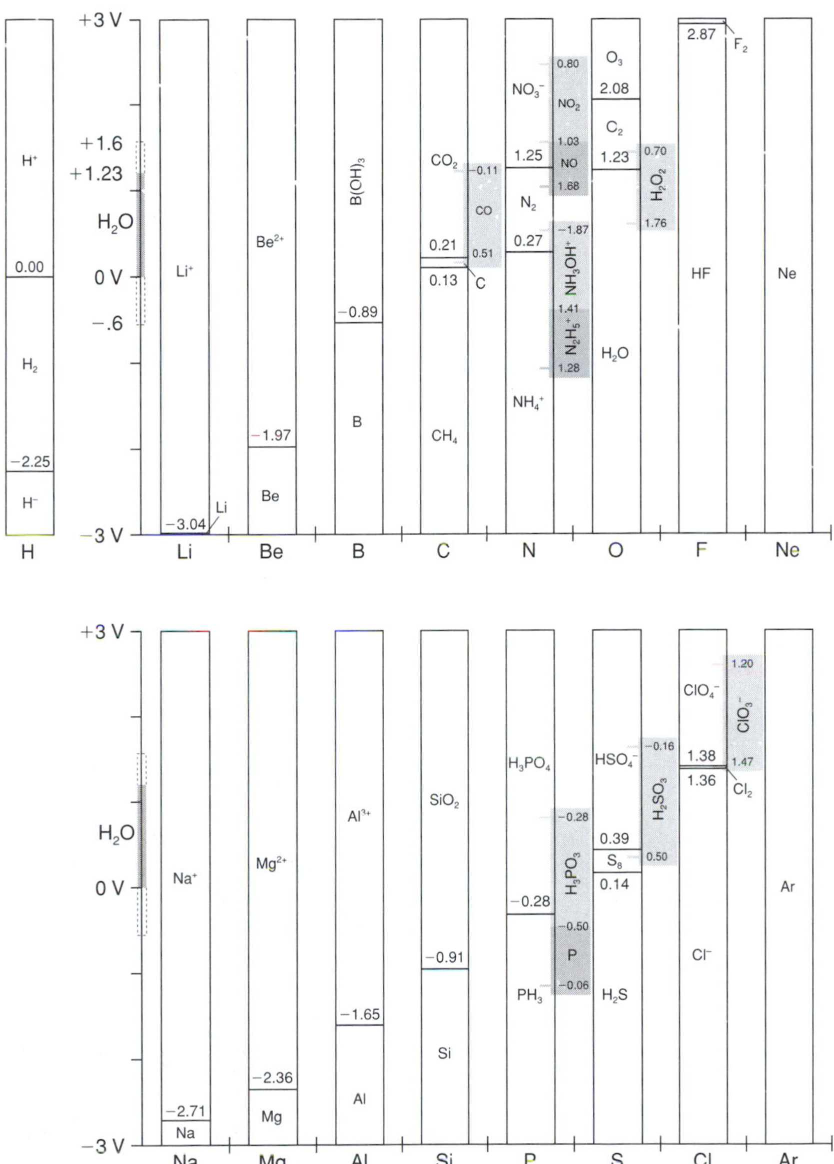

Figure 6.2
Redox predominance diagrams of the lighter *s*- and *p*-block elements, showing the thermodynamically stable form of each element at the potential indicated on the left. The numbers shown within each bar graph are the standard reduction potentials between the two adjacent forms of the element. Selected standard reduction potentials involving thermodynamically unstable (metastable) species and the stable or unstable species directly adjacent to them are indicated in shaded boxes for certain elements. [Sources: B. Douglas, D. McDaniel, and J. J. Alexander, *Concepts and Models of Inorganic Chemistry*, 2d ed., Wiley, New York, 1983; pp. 772–782; D. F. Shriver, P. W. Atkins, and C. H. Langford, *Inorganic Chemistry*, Freeman, New York, 1990; pp. 642–663; Birk, J. P., *Predicting Inorganic Reactivity: Expert System*, Version 2.30, Project SERAPHIM, 1989 (also in *J. Chem. Educ. Software*). Some significant inconsistencies were noted for Ge and No.]

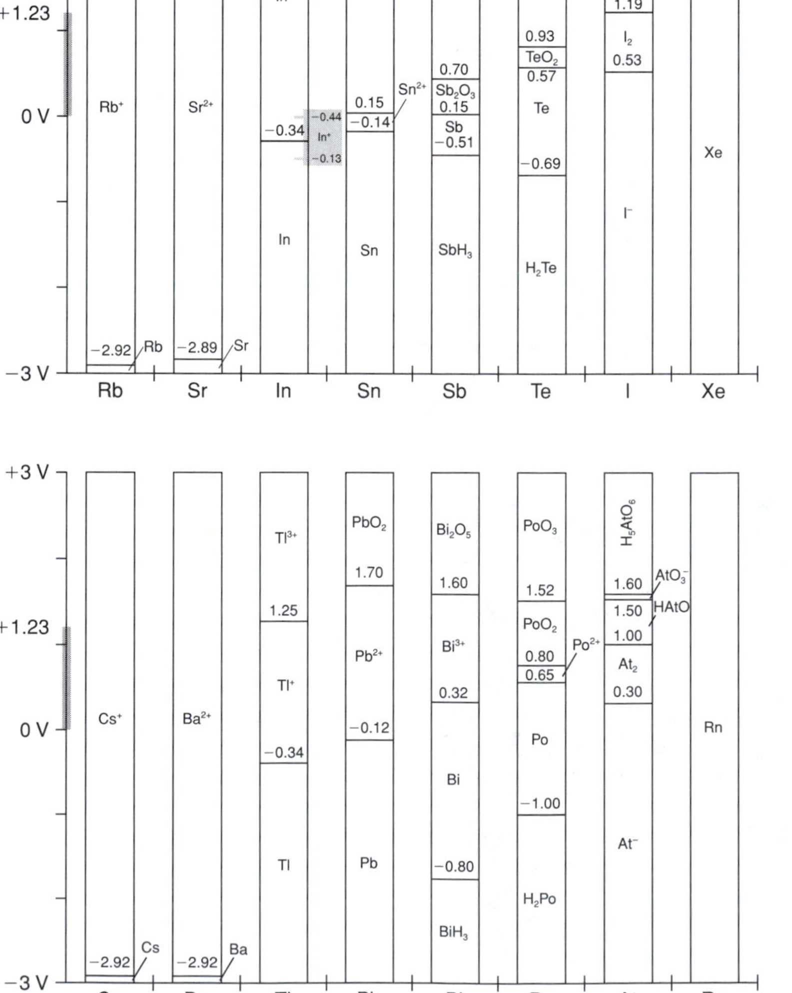

Figure 6.3

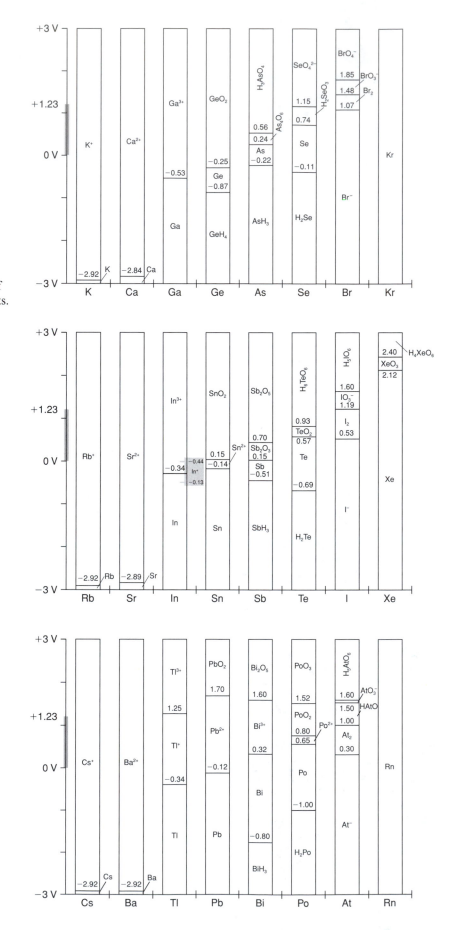

Redox predominance diagrams of the heavier *s*- and *p*-block elements. [Sources and notes same as for Figure 6.2.]

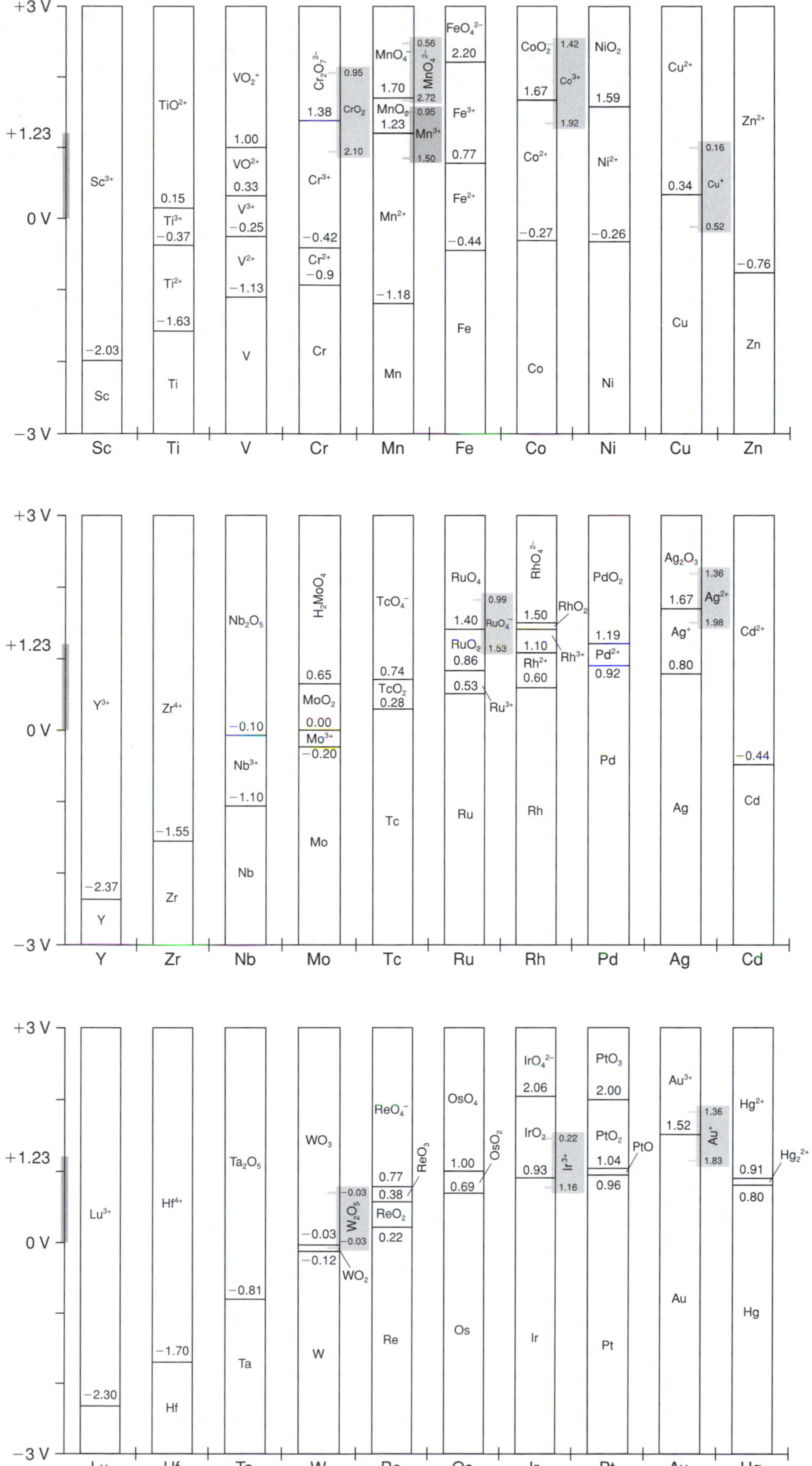

Figure 6.4
Redox predominance diagrams of the *d*-block elements.

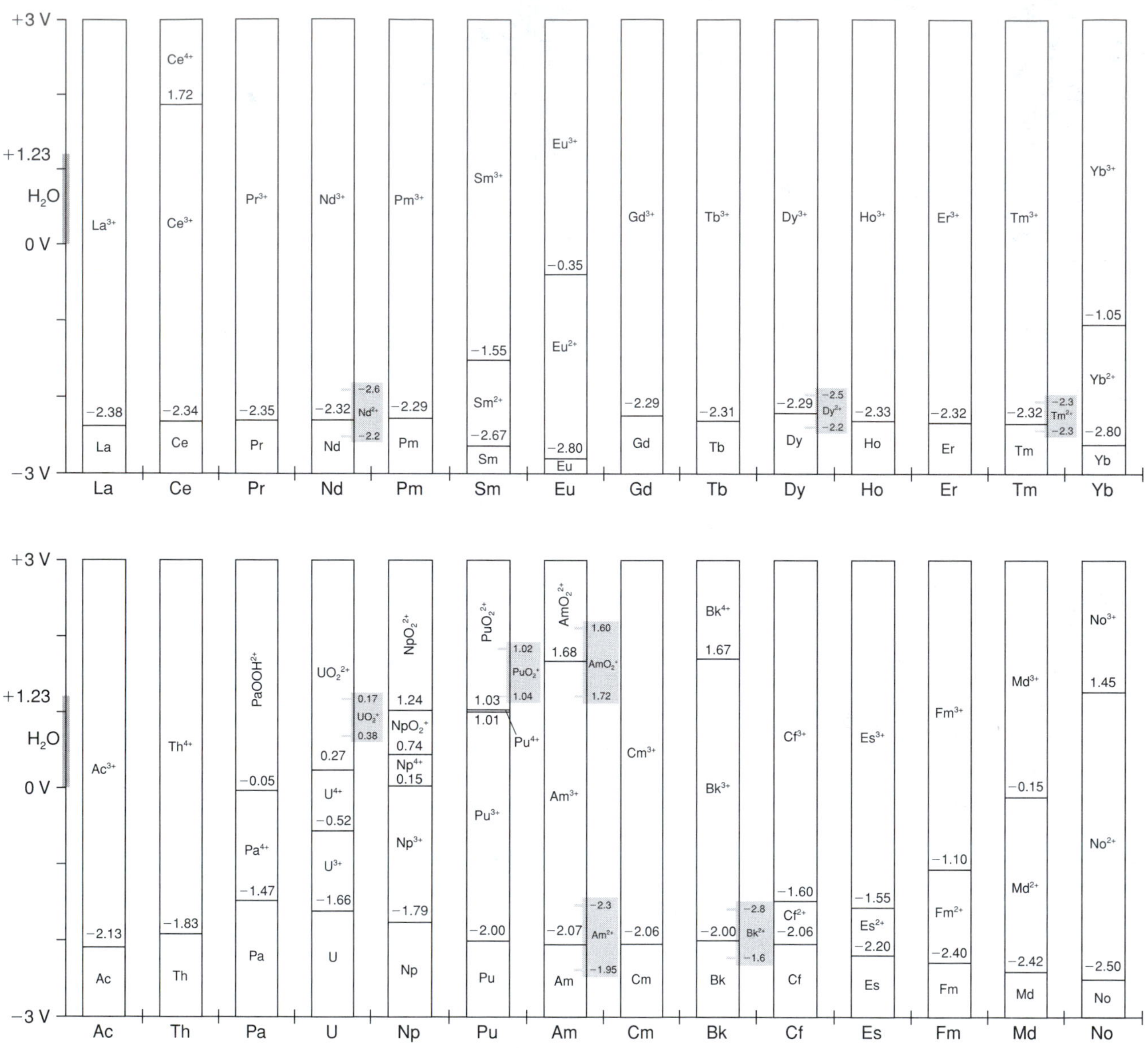

Figure 6.5
Redox predominance diagrams of the *f*-block elements.

Example 6.1
From the redox predominance diagrams (Figs. 6.2–6.5), find the following: (a) the strongest stable oxidizing agent in the *p* block; (b) the strongest oxidizing agent among stable oxo acids or anions in the *p* block; (c) the strongest stable reducing agent in the *d* block. For each of these, write the balanced half-reaction by which it is oxidized or reduced, and tell the standard reduction potential of that half-reaction.

SOLUTION:

(a) The strongest oxidizing agent in the *p* block will be the species in Figures 6.2 or 6.3 that is most severely confined to the top of its diagram; this is F_2, which is reduced to F^- at a standard reduction potential of +2.87 V. The reduction of F_2 produces 2 F^-. Each of the two fluorines is reduced from oxidation state 0 to oxidation state −1; two electrons are required to give the balanced half-reaction:

$$F_2 + 2\,e^- \rightarrow 2\,F^- \qquad E^\circ = +2.87\ \text{V}$$

(b) The strongest oxidizing agent among oxo acids or anions in the *p* block similarly is the one most confined to the top: H_4XeO_6. It is reduced to XeO_3 at an E° of +2.40 V. The half-reaction for this reduction requires $2\,e^-$ to reduce the Xe atom from an oxidation state of +8 in H_4XeO_6 to +6 in XeO_3. The charge of the $2\,e^-$ must be balanced with $2\,H^+$ as a reactant to give an equal charge (zero) on both sides of the half-reaction. Finally, water is added as a product to give the balanced half-reaction:

$$H_4XeO_6 + 2\,e^- + 2\,H^+ \rightarrow XeO_3 + 3\,H_2O \qquad E^\circ = +2.40\ \text{V}$$

(c) The strongest stable reducing agent in the *d* block is the species most severely confined to the *bottom* of its predominance diagram; this is Y, which has a standard reduction potential of −2.37 V. The negative sign means that a potential will not be *generated*, but must be *applied* to a solution of Y^{3+} to reduce it to Y. Three electrons are required to carry out this reduction:

$$Y^{3+} + 3\,e^- \rightarrow Y \qquad E^\circ = -2.37\ \text{V}$$

Predicting Redox Reactivity. Redox reactivity may be predicted from the redox predominance diagrams in much the same way as acid–base reactivity may be predicted from the acid–base diagrams. *Two species that have touching or overlapping redox predominance regions* have a potential at which they can coexist and *will not tend to react with each other.*[5] Thus, FeO_4^{2-} and Fe^{3+} have predominance areas that touch at +2.20 V; this means that they are at equilibrium with each other at a total iron concentration of 1 *M* at pH 0.

Two species that have nonoverlapping predominance areas are expected to react with each other to give products that are stable in each other's presence, that is, have overlapping predominance areas. This redox reaction will generate a standard cell emf E° that equals the numerical gap between their predominance regions. Thus, FeO_4^{2-} will oxidize either Fe^{2+} or Fe metal. The cell emf generated by the reaction with Fe^{2+} will be (2.20) − (0.77) = +1.43 V. In this case, the only possible product of each species will be the form of iron between these two, Fe^{3+}.

The balanced redox reaction is obtained by combining the reduction half-reaction (6.6) with the oxidation half-reaction (6.7).

$$Fe^{2+} \rightarrow Fe^{3+} + e^- \qquad (6.7)$$

The balanced overall redox reaction *must contain no electrons:* All of the electrons produced by the oxidation half-reaction are consumed by the reduction half-reaction.

To balance the overall redox reaction, we must multiply Reaction (6.7) by 3 to produce the three electrons used up in half-reaction (6.6); hence the balanced overall three-electron redox reaction is Eq. (6.8):

$$FeO_4{}^{2-} + 8\,H^+ + 3\,Fe^{2+} \rightarrow 4\,Fe^{3+} + 4\,H_2O \qquad E^\circ = +1.43\text{ V} \tag{6.8}$$

Example 6.2

(1) Predict the products (if any) of the following aqueous reactions under standard conditions: (a) $Ca^{2+} + H_2$; (b) $Ag_2O_3 + H_2$; (c) $Mn + H^+$; (d) $Mn^{2+} + Pb$; (e) $Pb^{2+} + Mn$; and (f) $MnO_4{}^- + Pb$. (2) Write complete balanced equations for those reactions that do occur.

SOLUTION:

(a) The predominance areas of Ca^{2+} and H_2 overlap [Fig. 6.6(*a*)]; no reaction will occur.

(b) The predominance areas of Ag_2O_3 and H_2 do *not* overlap [Fig. 6.6(*b*)], so they will react to give species with overlapping predominance areas: H^+ and either Ag^+ or Ag(s). Which will form depends on how much reducing agent is present. Each possible balanced equation is obtained by combining the corresponding silver half-reaction with the oxidation half-reaction $H_2 \rightarrow 2\,H^+ + 2\,e^-$. Using the half-reaction for the reduction to Ag^+

$$Ag_2O_3(s) + 4\,e^- + 6\,H^+ \rightarrow 2\,Ag^+ + 3\,H_2O$$

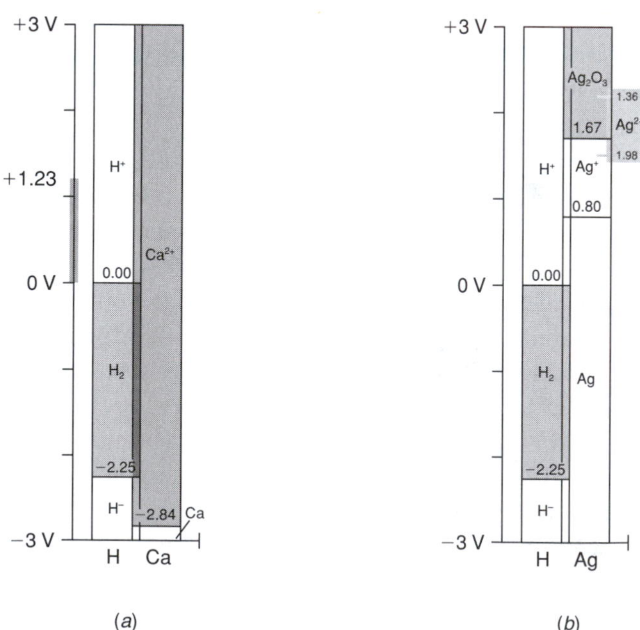

Figure 6.6
(*a*) Superposition of the redox predominance diagrams for H and Ca, showing that the predominance regions of Ca^{2+} and H_2 *do* overlap (darkened area). (b) The redox predominance regions of Ag_2O_3 (shaded) and H_2 (shaded) do *not* overlap.

we obtain:

$$Ag_2O_3 + 2\,H_2 + 2\,H^+ \rightarrow 2\,Ag^+ + 3\,H_2O$$

The reduction to Ag similarly uses the half-reaction:

$$Ag_2O_3 + 6\,e^- + 6\,H^+ \rightarrow 2\,Ag(s) + 3\,H_2O$$

to give

$$Ag_2O_3 + 3\,H_2 \rightarrow 2\,Ag + 3\,H_2O$$

(c) Areas for Mn and H^+ do not overlap, so we expect them to react to give Mn^{2+} and H_2, which have overlapping predominance areas:

$$Mn + 2\,H^+ \rightarrow Mn^{2+} + H_2$$

(d) There will be no reaction, since the predominance areas of Mn^{2+} and Pb overlap.

(e) There will be reaction to give species with overlapping predominance areas:

$$Pb^{2+} + Mn \rightarrow Pb + Mn^{2+}$$

(f) There will be reaction to give species with overlapping predominance areas, Pb^{2+} and either MnO_2 or Mn^{2+}, depending on quantities of reactants present:

$$2\,MnO_4{}^- + 3\,Pb + 8\,H^+ \rightarrow 3\,Pb^{2+} + 2\,MnO_2 + 4\,H_2O$$

or

$$2\,MnO_4{}^- + 5\,Pb + 16\,H^+ \rightarrow 2\,Mn^{2+} + 5\,Pb^{2+} + 8H_2O$$

Redox Predominance Diagrams for Water at pH 0. If one is carrying out a synthesis or reaction in aqueous solution, water is a potential reactant! Above an $E°$ of +1.229 V, water is oxidized to oxygen:

$$2\,H_2O \rightarrow 4\,H^+(aq) + O_2 + 4\,e^- \qquad E^\circ_{redn} = +1.229\ V \tag{6.9}$$

As discussed in Section 6.2, the rate of this reaction is slow unless an overpotential of about 0.5–0.6 V is also present. So, oxidizing agents with predominance regions confined to above +1.8 V placed in water (at pH 0) will decompose rapidly, releasing O_2, which is in fact the fate of the iron(VI) species $FeO_4{}^{2-}$ under standard conditions.

If a reducing agent is introduced that has a predominance area entirely below 0.00 V (standard conditions), the hydrogen ion in water will be reduced to H_2 by half-reaction (6.5). This reaction may also be slow unless an additional overpotential is provided, so rapid reaction is expected if $E°$ is below −0.6 V. Consequently, the species of an element under consideration will react rapidly with water and decompose if its predominance range does not overlap the short-term predominance (stability) range of water, −0.6 to +1.8 V.

For long-term work, as in geochemistry, the narrower predominance range for water of 0.0 to +1.229 V is more appropriate. These ranges are indicated at the left edges of Figures 6.2–6.5. Compounds or ions with predominance regions not overlapping these ranges at the appropriate pH will not persist in aqueous reactions, but may be important in nonaqueous or solid-state reactions. Hence, the most important oxidation states of the elements (at pH 0) will be those that fall within this stability range of the most common solvent, water.[6]

6.2 Explosives, Inflammable Compounds, and the Handling of Strong Reducing Agents

Oxidation–reduction reactions can be extremely exothermic, giving off much more energy than is usually observed for acid–base reactions. [In the case of the combustion of the hydrides of carbon (hydrocarbons) as fuels, we carry out such an oxidation–reduction reaction for the express purpose of generating energy.] But many reactions of oxidizing and reducing species with nonoverlapping predominance ranges, although they are quite exothermic, may not occur immediately upon contact of the reactants because there are difficult steps in the mechanism. For example, reactions that oxidize monoatomic ions (such as H^+) to diatomic gases (H_2) must first produce very high energy monoatomic species (H); an extra **activation energy** is needed to start this reaction, which translates into an **overpotential**[7] of approximately an extra 0.5–0.6 V that is needed to begin producing the high-energy intermediate. (Otherwise the intermediate does not build up to a high concentration, so that subsequent steps remain slow.)

This reaction may also occur if the two half-reactions are **noncomplementary**, that is, they involve different numbers of electrons. For example, the reduction of O_2 to stable products such as $2\,H_2O$ requires that the reducing agent supply four electrons; few common standard reductions produce four electrons, since few elements have stable forms in which the oxidation states differ by four. In such cases, a one-step mechanism is impossible, and other mechanisms are needed, which may involve unstable intermediates such as the high-energy, toxic superoxide ion, O_2^- (the removal of which from cells is necessary and which is accomplished by the enzyme superoxide dismutase). However, if a **catalyst** is added, the mechanism may change, which allows the reaction to proceed by another mechanism that does not involve unstable intermediates. Since *d*-block metals often have several fairly stable cations differing in charge from each other by only +1, they are especially useful as catalysts for redox reactions, such as those involving O_2.

Hence, strong oxidizing and strong reducing agents may often be mixed or placed in contact with each other with no immediate result (i.e., gasoline in contact with air), especially if the potential difference (nonoverlap of predominance ranges) is less than 0.6 V. However, once a catalyst or a source of the activation energy (i.e., a spark or flame) is provided, the exothermic reaction may begin rapidly. Often, if the potential difference is greater than about 1.0 V, so much energy is then liberated that it then cannot be dissipated but instead goes to provide the activation energy for several other molecules or ions to react. In that case, a fire or **explosion** may occur.

Example 6.3

Which of the following appear the most likely to show explosive properties: (a) $Eu(ClO_4)_2$, (b) $AgClO_4$, or (c) $TlClO_4$.

SOLUTION:

The standard reduction potentials of the cations and anions of these salts are obtained from Figures 6.2–6.5. The Eu^{2+} ion is (mainly) a reducing agent, confined below an $E°$ of −0.35 V; Ag^+ is (mainly) an oxidizing agent, predominant above an $E°$ of +0.80 V; ClO_4^- is exclusively an oxidizing agent, found above an $E°$ of +1.38 V.

(a) As shown in Figure 6.7(*a*), there is a gap of +1.73 V between the predominance regions of Eu^{2+} and ClO_4^-. Hence, these two ions thermodynamically should react with each other. If this reaction is slow enough to allow the salt to be isolated, the salt should be treated as a potential explosive.

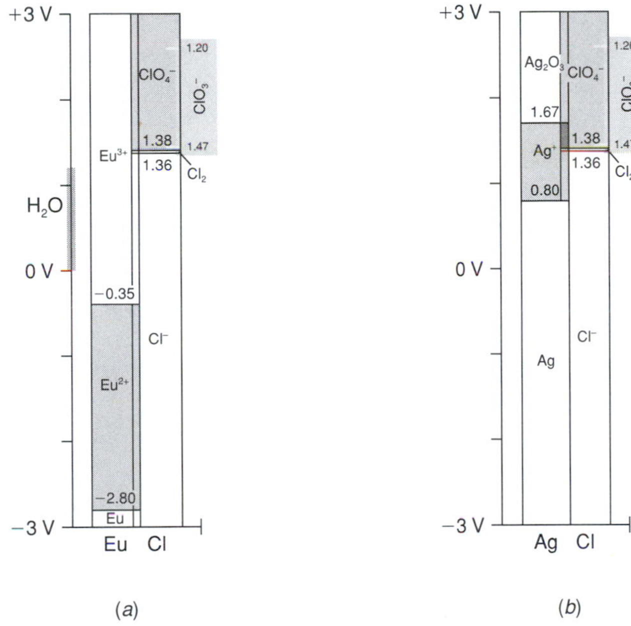

Figure 6.7
(*a*) Nonoverlap of (shaded) predominance regions of ClO_4^- and Eu^{2+}, with a gap exceeding 1.0 V. (*b*) Overlap of predominance regions of Ag^+ and ClO_4^- (in darkened area).

(b) The salt $AgClO_4$ contains two oxidizing ions, which have predominance regions that overlap between +1.38 and +1.67 V (Fig. 6.7). Hence, the salt should not decompose by a redox reaction—at least in the absence of dust or other external reducing agents.

(c) The predominance region for ClO_4^-, with a lower boundary at +1.38 V, does not overlap the region for Tl^+, which has an upper boundary at 1.25 V. The gap, however, is much less than 1.0 V; so we *cautiously* conclude that, althought this salt should not be theremodynamically stable, if it does decompose, it will *probably* not release enough energy to cause an explosion. Do not bet your life on this, however, especially if you are handling large quantities of this salt!

One of the most important reasons for seeing periodic trends in redox chemistry is to become aware of species that, when mixed, have the potential to inflame or explode. If you do well in learning the skills involved in this chapter, your life expectancy may increase! As a first general lesson in **chemical safety**, always take precautions when you are carrying out a reaction of a strong oxidizing and a strong reducing agent. Unless you are doing this reaction in a dilute aqueous solution (in which reactions are slowed by dilution and by the high heat capacity of the water), you should take such precautions as providing external cooling, allowing a vent for any gases produced, and recalling the locations of the appropriate types of fire extinguishers. If an explosion should occur, the glass of the reaction vessel may be shattered into shards, which may be hurled into your skin and eyes, so **wear safety goggles** and **do the reaction behind a safety shield**, such as the window of a hood or fume cupboard. (The hood also serves to confine flames and toxic gases that might be given off.) Be especially careful if the expected exothermic reaction does **not** start upon initial cautious addition of the first small amount of the second reagent. This means that an activation energy or catalyst is needed but is (temporarily) absent. Under these conditions heating or disturbing the system may suddenly provide the activation energy for part of the molecules; the energy released may provide it for the rest, and result in a violent explosion. **DO NOT ADD MORE REACTANT TO GET THE REACTION GOING**—instead, get the safety shield down. If you have foolishly already added a lot of the second reactant, **EVACUATE THE LABORATORY IMMEDIATELY!**

Unfortunately, mechanisms are so diverse that the 0.6- and 1.0-V guidelines suggested above are by no means reliable.[8] Neither is it possible to provide a complete list of chemicals, or combinations of chemicals, that give rise to explosions—chemists are discovering new ones every month, usually unintentionally! (Such new discoveries are usually promptly reported in the Letters to the Editor section of *Chemical & Engineering News*.) Of course, many combinations of strong oxidizing and reducing agents react smoothly at room temperature, since only a small activation energy is needed (e.g., the reaction may be complementary), or the reactions are only mildly exothermic. But an incomplete list of potentially hazardous combinations can be suggested here.[9]

1. A salt that is the combination of a strongly reducing cation and a strongly oxidizing anion may persist for kinetic reasons, but is thermodynamically unstable and is an explosion hazard. From the predominance diagram for nitrogen (Fig. 6.2) it can be seen that the **ammonium** ion, NH_4^+, has a limited predominance range; its salts with strongly oxidizing oxo anions are often explosive. Thus ammonium nitrate, a common fertilizer, when ignited is a powerful explosive (predominance gap = 0.98 V); the worst industrial accident in U.S. history occurred in 1947 in Texas City, TX when two shiploads detonated, killing 581 people and destroying two-thirds of the city.[10] Ammonium perchlorate (predominance gap = 1.11 V) is used as an oxidant for solid-fueled rockets; in 1988 one of two American plants manufacturing this salt, at Henderson, NV, was destroyed by a series of explosions.[11] There are also many organic ammonium ions containing one-to-four organic R groups in place of the ammonium hydrogen ions, and many complex cations involving ammonia or organic amines as ligands; these are large cations that form nicely insoluble perchlorates, but these perchlorates are treacherous explosives! If one wishes to precipitate these complex ions or organic ammonium ions, it is much safer to use nonbasic fluoro anions such as BF_4^- or PF_6^- (in which B or P show little oxidizing ability).

2. The same considerations apply with the combination of a strongly oxidizing cation and a strongly reducing anion. The cations of the electronegative metals are strongly oxidizing. Two important strongly reducing anions are the **azide** ion, N_3^- ($E° = -3.09$ V) and the **acetylide** ion, C_2^{2-}. Thus, lead(II) azide (predominance gap = +2.91 V) is very shock sensitive and is used to provide the activation energy needed to detonate more stable and powerful explosives. Sodium azide mixed with potassium nitrate is used in the automobile air bag,[12] since the reaction of this mixture is readily initiated, and the air bag is then inflated by the large volume of N_2 released:

$$10\,NaN_3(s) + 2\,KNO_3(s) \rightarrow 5\,Na_2O(s) + K_2O(s) + 16\,N_2(g) \qquad (6.10)$$

Similarly, acetylides such as Ag_2C_2 and Cu_2C_2 are very explosive. Several explosive silver salts can be produced in common laboratory situations.[13]

3. Naturally, mixtures of strongly oxidizing and reducing substances present the possibility of a fire or an explosion. Such substances should be stored in separate areas of the stockroom or laboratory. Mixture is particularly easy if both substances are gases; the physical chemistry that determines whether a fire or an explosion will result has been well studied.[14] Likely gaseous reducing agents include hydrogen and many of the hydrides of the nonmetals (such as the hydrocarbons); likely gaseous oxidizing agents include F_2, Cl_2, Br_2, O_2, and the volatile oxides of the elements (Chapter 12). The F_2 molecule is so vigorous an oxidizing agents that most substances will catch fire or explode in it; its usefulness is correspondingly limited. Although a powerful oxidizing agent, O_2 usually requires a significant activation energy—but as classroom demonstrations of igniting a mixture of H_2 and O_2 demonstrate, the activation energy is readily available. Explosions or fires may also occur, however, even if one or neither reactant is a gas. Active metals can be ignited in air; finely powdered combustible solids (such as dust in grain elevators) can form explosive mixtures with air; mixtures of solids such as gunpowder (S, C, KNO_3) also can be explosive.

4. It is possible to combine an oxidizing group of atoms (functional group) and a reducing group in the same molecule; such substances are often explosive. Familiar explosives such as nitroglycerine, $C_3H_5(NO_3)_3$, trinitrotoluene (TNT), $C_7H_5(NO_2)_3$, and picric acid, $C_6H_3O(NO_2)_3$ contain oxidizing nitrogen groups and reducing hydrocarbon groups. Upon reaction, these compounds generate not only a great deal of energy but large volumes of gaseous CO, CO_2, H_2O, and N_2. Similarly hazardous organic perchlorates can be produced by combining perchloric acid and alcohols—this should never be attempted.

5. Finally, recall from Chapter 3 that single covalent bonds between the very electronegative nonmetals (especially the second-period elements N, O, and F) are intrinsically weak. Compounds containing these bonds are therefore not only good oxidizing agents, but are often thermodynamically metastable and prone to quite exothermic disproportionation. Hence, compounds containing these bonds should be treated as potentially explosive: O–O (e.g., hydrogen peroxide, H_2O_2, or organic peroxides, R_2O_2); N–N (e.g., azides and diazonium salts); O–Cl (perchloric acid, chlorates, chlorites, and the oxides of chlorine); O–N (the nitro organics mentioned in the previous paragraph and compounds such as silver fulminate, AgONC); and even N–Cl, N–Br, and N–I ($NI_3{\cdot}NH_3$ is detonated by the force of a housefly stomping its foot).

Handling Strong Reducing Agents. Since O_2 is a strong oxidizing agent whose reactions are not always as slow as we would like, it is often important to handle strong reducing agents under air-free (inert-atmosphere) conditions, which also insure that

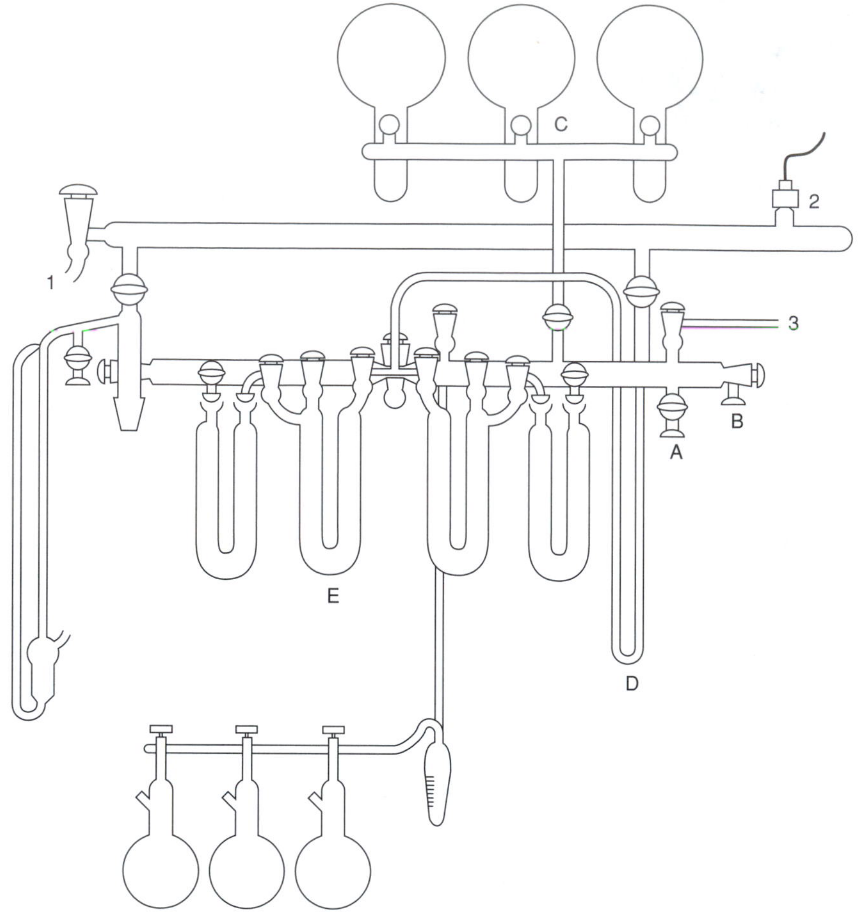

Figure 6.8
Multipurpose high-vacuum line. [Adapted from D. F. Shriver, and M. A. Drezdzon, *The Manipulation of Air-Sensitive Compounds*, 2nd ed., Wiley, New York, 1986; p. 101.]

acidic cations and basic anions have no opportunity to react with the moisture in the air. Much of the progress that inorganic chemistry has made in recent decades has involved strongly reducing compounds of the elements (e.g., organometallic compounds), and would not have been possible without the development of these inert atmosphere techniques. Alfred Stock pioneered these techniques in his study of the silanes, Si_nH_{2n+2}. There are four principal inert-atmosphere techniques.[15]

The best technique for the handling of reducing gaseous reactants and products, which pose especially high hazards of explosive oxidation, involve the use of the **vacuum line** (Fig. 6.8). In this technique, vacuum pumps are used to pump the air out of the vacuum line, which consists of glass tubing and containers separated by stopcocks. The containers are cooled in baths of differing low temperatures such as to condense certain gaseous reagents and products as liquids, thus retaining them, while allowing more volatile ones to pass on to other, colder, baths, where they are condensed.

For nonvolatile solids, either of two types of apparatus are used: the **inert-atmosphere drybox** (Fig. 6.9), or the **glovebag** (Fig. 6.10); the drybox may also be used to handle infectious microorganisms or radioactive materials. In the case of the inert-

Figure 6.9
Commercial drybox (glovebox, inert-atmosphere box). Photograph shows a Vacuum Atmospheres Nexus Dri-Lab System with video microscope, courtesy of Vacuum Atmospheres.

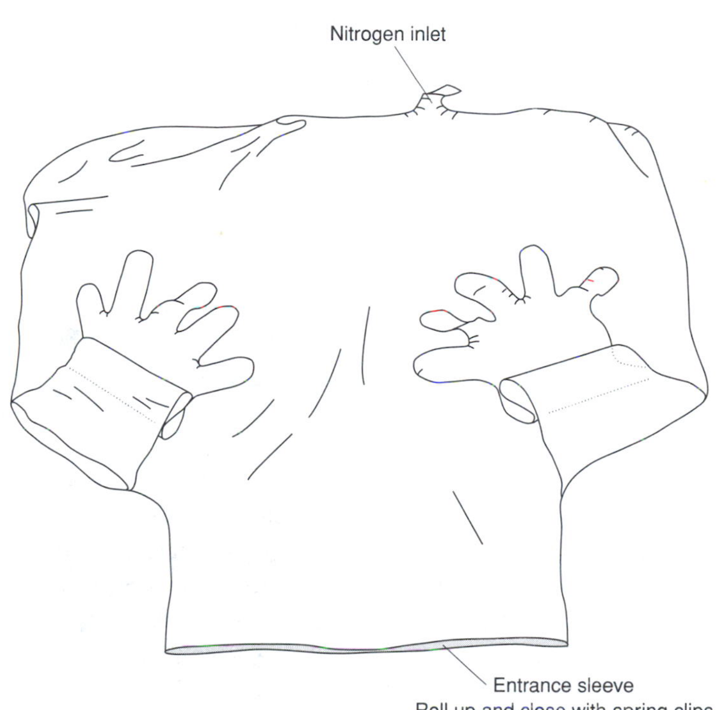

Figure 6.10
Commercial (IRI) polyethylene glovebag. [Adapted from D. F. Shriver, and M. A. Drezdzon, *The Manipulation of Air-Sensitive Compounds*, 2nd ed., John Wiley & Sons, Inc., New York, 1986; p. 46.]

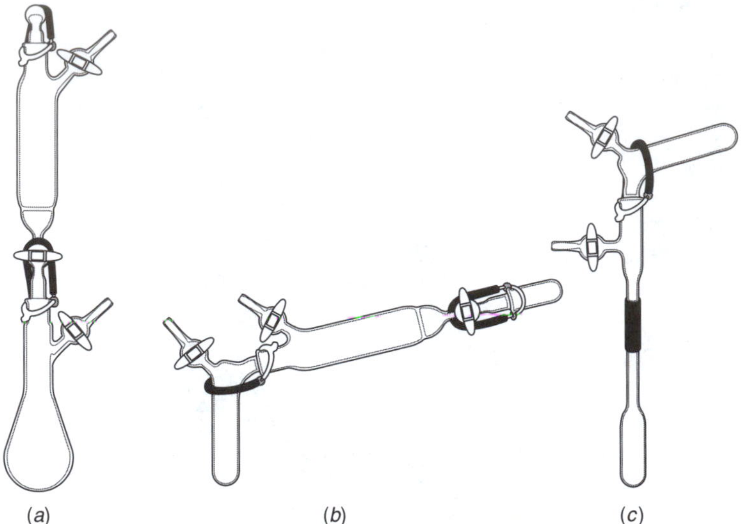

Figure 6.11
Typical assemblies of commercial (Kontes) Schlenkware for (*a*) filtering a solution; (*b*) pouring a solid from a filter into a storage tube; and (*c*) pouring a solid from a storage tube into a glass ampoule, which can be sealed off. [Adapted from D. F. Shriver, *Manipulation of Air-Sensitive Compounds*, 2nd ed., John Wiley & Sons, Inc., 1986; p. 32.]

atmosphere box, a nonoxidizing gas such as N_2 or Ar is continually pumped through the box, then through canisters containing desiccants and strong reducing agents. Reactants and pieces of apparatus are introduced through the port at the right of the side chamber of the box, which is pumped out and purged several times with the inert atmosphere before the port leading from the right chamber to the box itself is opened, and the materials are transferred in. The more quickly operated glovebag is suitable for less-sensitive reactions, and relies on a cylinder of prepurified inert gas to be used to purge the bag repeatedly of air, before the reaction is carried out inside the plastic bag. Reactants and pieces of apparatus are admitted at the bottom through an opening while inert gas is flowing rapidly through; this is followed by several cycles of (1) inflating the bag with the inert gas, followed by (2) a quick opening of the bottom entrance while the bag is manually compressed so as to sweep out its residual content of air.

Finally, solutions can be handled either in inert-atmosphere boxes or glovebags, but are more conveniently handled in a modified type of standard-taper glassware called **Schlenkware** (Fig. 6.11), which feature sidearms that can be connected to a cylinder of a prepurified inert gas (in this case Ar is preferred to N_2). Reagents are introduced normally into separate parts of the apparatus, which is then put under vacuum to remove air, followed by the admission of argon. Repetition of several cycles of vacuum removal of air followed by admission of argon gives a sufficiently air-free atmosphere to allow the reactants to be mixed (by tipping the whole apparatus, etc.), and the products to be filtered, and so on. If a solid reactant must be added, or a different piece of Schlenkware must be connected, this can be done by quickly removing a top stopper while the system is under slight argon pressure—since argon is heavier than air and is streaming out of the joint, almost no air will enter during these few seconds.

6.3 *Redox Predominance Diagrams: Construction; Species that Disproportionate; Nonstandard Conditions

Construction of Redox Predominance Diagrams. The redox predominance diagrams of Figures 6.2–6.5 can be constructed from tables of half-reactions and their standard reduction potentials according to the following procedure, which is illustrated for iron.

Step 1. Collect all of the reduction half-reactions with standard reduction potentials for a given element. List them in *decreasing* order of oxidation number of the element in the reactant. For iron, this list is headed by Eq. (6.6), followed by

$$Fe^{3+} + e^- \rightarrow Fe^{2+} \qquad E^\circ = 0.77\,V \tag{6.11}$$

$$Fe^{2+} + 2\,e^- \rightarrow Fe(s) \qquad E^\circ = -0.44\,V \tag{6.12}$$

Step 2. Set up a potential scale from +3.0 V at the top to −3.0 V at the bottom. (This includes all chemically realistic species.) Draw in boundaries at the E° values given. Above each boundary, put the oxidized form of the element involved with that E°; below it put the reduced form. This results in the iron diagram shown in Figure 6.1.

Metastable Species that Should Disproportionate. Next, let us consider the case of Ag. We list the available redox half-reactions in decreasing order of silver oxidation state:

$$Ag_2O_3(s) + 4\,e^- + 6\,H^+ \rightarrow 2\,Ag^+ + 3\,H_2O \qquad E^\circ = +1.67\,V \tag{6.13}$$

$$Ag_2O_3(s) + 2\,e^- + 6\,H^+ \rightarrow 2\,Ag^{2+} + 3\,H_2O \qquad E^\circ = +1.36\,V \tag{6.14}$$

$$Ag^{2+}(aq, 1\,M) + e^- \rightarrow Ag^+ \qquad E^\circ = +1.98\,V \tag{6.15}$$

$$Ag^+(aq, 1\,M) + e^- \rightarrow Ag(s) \qquad E^\circ = +0.80\,V \tag{6.16}$$

We note, on reading down this list, that the common pattern for E° values to decrease as oxidation states decrease is interrupted when we come to the reduction of Ag^{2+}; its E° is the highest of all! If we attempt to draw partial predominance diagrams involving Ag^{2+} and its two half-reactions, we get a peculiar result (Fig. 6.12): One half-reaction indicates that the predominance region of Ag^{2+} must be below +1.36 V, while the other half-reaction indicates that it must be above +1.98 V! The peculiarity is

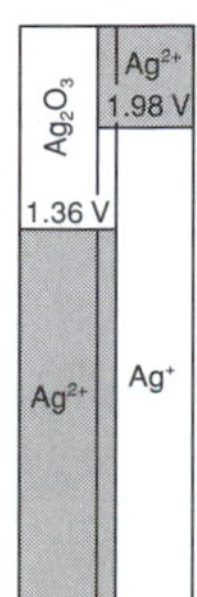

Figure 6.12
Attempts to construct a partial redox predominance diagram including Ag^{2+}, using (left side) the standard reduction potential for the formation of Ag^{2+}, Eq. (6.14), and (right side) the standard potential for the reduction of Ag^{2+} to Ag^+, Eq. (6.15).

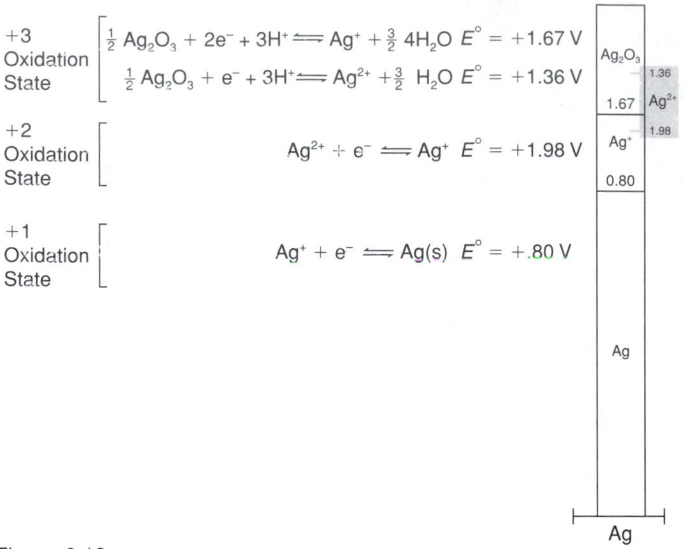

+3 Oxidation State

$\frac{1}{2}\,Ag_2O_3 + 2e^- + 3H^+ \rightleftharpoons Ag^+ + \frac{3}{2}\,4H_2O \quad E^\circ = +1.67\ V$

$\frac{1}{2}\,Ag_2O_3 + e^- + 3H^+ \rightleftharpoons Ag^{2+} + \frac{3}{2}\,H_2O \quad E^\circ = +1.36\ V$

+2 Oxidation State

$Ag^{2+} + e^- \rightleftharpoons Ag^+ \quad E^\circ = +1.98\ V$

+1 Oxidation State

$Ag^+ + e^- \rightleftharpoons Ag(s) \quad E^\circ = +.80\ V$

Ag_2O_3
1.36
1.67
Ag^{2+}
1.98
Ag^+
0.80
Ag
Ag

Figure 6.13
Construction of redox predominance diagram for Ag, excluding the thermodynamically metastable Ag^{2+} to the right of the diagram.

that there is a gap of 0.62 V between the two incomplete (and therefore invalid) "predominance regions" for Ag^{2+}. This gap correctly suggests that the species Ag^{2+} is thermodynamically unstable and *reacts with itself*; such reactions are known as **disproportionation** reactions.

$$3\,H_2O + 2\,Ag^{2+} + 2\,Ag^{2+} \rightarrow Ag_2O_3 + 2\,Ag^+ + 6\,H^+ \tag{6.17}$$

Although the positive E° for this reaction means that Ag^{2+} will eventually react with itself, the kinetics may be slow, in which case Ag^{2+} may be described as **metastable**. If so, Ag^{2+} may be a useful reactant for inorganic or other chemists, since its high standard reduction potential does make it a powerful oxidizing agent.

Hence, in constructing the predominance diagram of Ag (Fig. 6.13), we must bypass Ag^{2+}; it does not predominate at any E°. Of the two half-reactions listed for the reduction of Ag(III) (as Ag_2O_3), we skip the one that produces Ag^{2+}, and use the one that produces Ag^+. Note that the lower boundary for the predominance region for Ag_2O_3 is *the first potential that is encountered on reducing the potential down from* +3 V (the top of the diagram). This half-reaction takes us into the predominance region of Ag^+, for which we must next find a lower boundary, at which Ag^+ is reduced to a still lower oxidation state. Thus the next half-reaction that is incorporated into the redox predominance diagram is Eq. (6.16), which gives the lower boundary for Ag^+, 0.80 V.

Although metastable species such as Ag^{2+} cannot be incorporated in the predominance diagram itself, and are unlikely to be found in natural waters, soil, or as corrosion products, we can append their E° data on the side of the diagram, as shown for Ag^{2+} in the shaded part of Figure 6.13. At the lower boundary, Ag^{2+} is reduced to the adjacent species (at its left) in the next lower oxidation stage, Ag^+, at an E° of +1.98 V. At the upper boundary Ag^{2+} is oxidized to the adjacent species, Ag_2O_3, at an

$E°$ of +1.36 V. For metastable species such as Ag^+ the *lower* $E°$ characteristically has a higher value than the *upper* $E°$, so the shaded area is an *imaginary* predominance area. Nonetheless, the numerical $E°$'s can be used in essentially the standard way to compute whether a reaction will proceed, as shown in the following example.[16]

Example 6.4

Will the following redox reactions tend to occur? If so, what are the cell emfs? (a) $Br_2 + 2\,NH_3OH^+ \rightarrow 2\,Br^- + N_2 + 2\,H_2O + 4\,H^+$; and (b) $2\,MnO_4{}^{2-} + 6\,Br^- + 8\,H^+ \rightarrow 3\,Br_2 + 2\,MnO_2 + 4\,H_2O$.

SOLUTION:

(a) Its relatively high predominance region means that Br_2 is probably acting as an oxidizing agent. (Its oxidizing ability is confirmed by the listing of Br^- as a product.) The relevant predominance region boundary for Br_2 is thus the lower one, +1.07 V. The boundary applying to NH_3OH^+ as a reducing agent is its upper one, −1.87 V, at which it is converted to the adjacent species in the next higher oxidation state, N_2. There is a large gap between the two regions, (+1.07) − (−1.87) = +2.94 V, so this reaction will proceed, possibly explosively.

(b) The Br^- ion is clearly a reducing agent with an upper boundary of +1.07 V. Hence, $MnO_4{}^{2-}$, if it is to react, would have to act as an oxidizing agent, so its lower boundary of +2.72 V is relevant. There is a gap of +1.65 V between this lower boundary for $MnO_4{}^{2-}$ and the upper boundary for Br^-, so the reaction will be expected to proceed, perhaps explosively.

Among the thermodynamically metastable species that do not appear in redox predominance diagrams are the soft-base pseudohalide ions cyanide, thiocyanate, and thiosulfate, which resemble the halide ions in that oxidation produces dimeric species:

$$2\,I^- \rightarrow I_2(g) + 2\,e^- \qquad E° = +0.54\ \text{V} \tag{6.18}$$

$$2\,CN^- \rightarrow (CN)_2(g) + 2\,e^- \qquad E° = +0.37\ \text{V} \tag{6.19}$$

$$2\,SCN^- \rightarrow (SCN)_2(\ell) + 2\,e^- \qquad E° = +0.77\ \text{V} \tag{6.20}$$

$$2\,S_2O_3{}^{2-} \rightarrow S_4O_6{}^{2-}(aq) + 2\,e^- \qquad E° = +0.08\ \text{V} \tag{6.21}$$

This reaction of the thiosulfate ion is used extensively in analytical chemistry in the quantitative analysis of oxidizing agents, since many oxidizing agents react with iodide ion to give (intensely colored) iodine, which reacts completely with easily prepared solutions of the thiosulfate ion to give iodide ion and the tetrathionate ion, $S_4O_6{}^{2-}$.

The azide ion is a very powerful reducing agent that reacts to give, not a dimer, but 3 mol of (very stable) nitrogen molecules:

$$2\,N_3{}^- \rightarrow 3\,N_2 + 2\,e^- \qquad E° = -3.09\ \text{V} \tag{6.22}$$

The azide ion will even reduce the sodium ion: Heating solid sodium azide produces sodium metal. Organic anions such as the phenyl anion, the methyl anion, and the **acetylide** ion, C_2^{2-}, as well as the **hydride** ion, H^-, are not only very strongly basic but also strongly reducing.

Effects of Concentration on Predominance Regions. The predominance regions of Ag^+ and I^- do not overlap under standard conditions, which could lead us to expect Ag^+ to oxidize I^-; this does not happen. The problem is that the product, AgI, is an insoluble soft acid–soft base combination, and the solution concentrations of Ag^+ and I^- can never reach the standard 1.0 M concentration assumed in the predominance diagram. The effect of nonstandard concentrations and activities on potential at room temperature is given by the well-known Nernst equation:

$$E = E^\circ - 0.059/n \log Q \tag{6.23}$$

In the Nernst equation, n represents the number of electrons transferred and Q is the reaction quotient, which has the same form as the equilibrium expression but includes concentrations that apply when the system is not at equilibrium. For example, application of the Nernst equation to Reaction (6.16) gives

$$E = 0.80\,\text{V} - 0.059/1 \log 1/[Ag^+] = 0.80\,\text{V} + 0.059 \log[Ag^+] \tag{6.24}$$

If the concentration of Ag^+ is reduced below 1 M, as, for example, if Ag^+ is precipitated as AgCl, in a saturated solution in which $[Ag^+] \approx 10^{-5}\,M$, the lower predominance-region boundary for Ag^+ is lowered. *At concentrations below the standard, the predominance region of a soluble species is enlarged.* When adjusted for the lower concentrations present in a saturated solution of AgI, the redox predominance regions of Ag^+ and I^- overlap.

Conversely, *at concentrations above the standard, the predominance region of a soluble species contracts.* For example, the predominance region for sulfuric acid (as HSO_4^-) has a lower boundary at 0.39 V, while that for iodide ion has an upper boundary at 0.53 V. The two regions overlap; 1 M H_2SO_4 does not oxidize I^-. However, *concentrated* sulfuric acid is a stronger oxidizing agent because of its contracted predominance region; sulfuric acid oxidizes iodide ion during attempts to prepare hydrogen iodide from NaI and concentrated H_2SO_4:

$$2\,NaI(s) + 3\,H_2SO_4(\ell) \rightarrow I_2 + SO_2 + 2\,H_2O + 2\,NaHSO_4 \tag{6.25}$$

The corresponding experiment with the less easily oxidized bromide ion (as NaBr) gives some HBr and some Br_2; NaCl gives exclusively HCl.

$$NaCl + H_2SO_4 \rightarrow HCl + NaHSO_4 \tag{6.26}$$

6.4 Charge-Transfer Spectra and Marginal Redox Stability

Sometimes when we attempt to carry out a Lewis acid–base reaction, a redox reaction results instead. For example, suppose that we are attempting the preparation of all of the iron(III) halides. The Fe^{3+} ion is an oxidizing cation ($E° = +0.77$ V), while the halide ions have quite varying reducing properties ($F^- < Cl^- < Br^- < I^-$). The predominance region of Fe^{3+} overlaps the predominance regions of F^- (by 2.10 V), Cl^- (by 0.59 V), and Br^- (by 0.30 V), but not that of I^- (there is a gap of 0.24 V). We can indeed prepare FeF_3, $FeCl_3$, and $FeBr_3$, but not FeI_3; iodide is reducing enough to reduce Fe^{3+} under normal circumstances:

$$2\,Fe^{3+} + 6\,I^- \rightarrow 2\,FeI_2 + I_2 \qquad (6.27)$$

Once we have isolated our products, we will be struck by another, seemingly unrelated property that they have: their colors change dramatically as the halide ion is changed. Iron(III) fluoride is a white solid, while $FeCl_3$ and $FeBr_3$ are intensely colored (red-brown if anhydrous); the very difficult-to-prepare FeI_3 is intensely black.[17] It turns out that, if the oxidizing power of an ion and the reducing power of its counterion are just insufficient to produce a redox reaction, the compound can exist (although it will have a low heat of formation), and it will likely be *intensely colored.* This color results from an electronic **charge-transfer absorption**: Light energy is capable of transferring the electron from the valence orbital of the reducing ion (e.g., Cl^- or Br^-) to the empty valence orbital of the oxidizing ion (e.g., Fe^{3+}). Because it can be moving in the same direction as the electron being transferred, the light wave has an unusually high probability of being absorbed in the charge-transfer process. Therefore the color depletion of the reflected light is strong, and we perceive intense color. In contrast, the process of light absorption in *d*- and *f*-block metal complex ions (Chapter 8) does not involve transfer of the electron—it stays on the same atom in the same type of orbital—and a given light wave is much less likely to be absorbed; hence, the colors of these complex ions are relatively faint. Compounds exhibiting visible charge-transfer spectra are often of marginal stability; this is the source of a commonplace observation that intensely colored compounds are sometimes quite reactive (in a redox sense) and should be treated with respect.

Normally, the transferred electron subsequently returns to the original atom, but sometimes the compound is permanently decomposed, so that it is **photosensitive**. Silver bromide is an example of a photosensitive halide; its decomposition after exposure to light is the basis of the process of photography.

Even in cases for which relevant predominance diagrams are not available, similar trends among halides can be noted. In Section 6.3 we noted that compounds containing nitrogen–halogen bonds tend to be unstable and may be explosive. This tendency is particularly strong, among the nitrogen trihalides, for nitrogen triiodide, which is not only a very easily detonated explosive, but is also intensely colored. Nitrogen trichloride and tribromide are not only not quite as easily detonated, but they show less color: NCl_3 is yellow. In contrast, nitrogen trifluoride is a stable, nonexplosive compound, which is also colorless.

The highest halides of the *p* block (Table 6.1) show some interesting resulting periodic trends. (1) Quite a number of iodides of the elements are intensely colored, while few fluorides of the elements are. (2) The halides of the elements at the left of the *s* and *p* blocks, in low oxidation states, are colorless. The chlorides, bromides, and iodides of elements in more oxidizing (higher) oxidation states toward the right of the *p* block are intensely colored; the halides at the far right, where oxidation numbers and oxidizing ability are very high, are nonexistent. (4) On moving to the right among group oxidation states, both intense colors and nonexistence are first encountered among the iodides and last among the fluorides.

Table 6.1
Formulas and Colors of the Highest Halides of Some Elements[a]

				A. Fluorides					
LiF	BeF_2			BF_3	CF_4	NF_3	(OF_2)	(F_2) yellow	
NaF	MgF_2			AlF_3	SiF_4	PF_5	SF_6	ClF_5	
KF	CaF_2	ScF_3	ZnF_2	GaF_3	GeF_4	AsF_5	SeF_6	BrF_5	(KrF_2)
RbF	SrF_2	YF_3	CdF_2	InF_3	SnF_4	SbF_5	TeF_6	IF_7	XeF_6
CsF	BaF_2	LuF_3	HgF_2	TlF_3	PbF_4	BiF_5			
				B. Chlorides					
LiCl	$BeCl_2$			BCl_3	CCl_4	NCl_3 yellow			
NaCl	$MgCl_2$			$AlCl_3$	$SiCl_4$	PCl_5	SCl_4 yellow	(Cl_2) yellow-green	
KCl	$CaCl_2$	$ScCl_3$	$ZnCl_2$	$GaCl_3$	$GeCl_4$	$AsCl_5$ yellow	$SeCl_4$ yellow	$(BrCl)$ red	
RbCl	$SrCl_2$	YCl_3	$CdCl_2$	$InCl_3$	$SnCl_4$	$SbCl_5$	$TeCl_4$ yellow	(ICl_3) orange	$(XeCl_2)$?
CsCl	$BaCl_2$	$LuCl_3$	$HgCl_2$	$TlCl_3$	$PbCl_4$ yellow	$BiCl_3$			
				C. Bromides					
LiBr	$BeBr_2$			BBr_3	CBr_4	NBr_3 ?			
NaBr	$MgBr_2$			$AlBr_3$	$SiBr_4$	PBr_5 yellow	(S_2Br_2) red		
KBr	$CaBr_2$	$ScBr_3$	$ZnBr_2$	$GaBr_3$	$GeBr_4$	$AsBr_3$ yellow	$SeBr_4$ orange	(Br_2) red	
RbBr	$SrBr_2$	YBr_3	$CdBr_2$	$InBr_3$ yellow	$SnBr_4$	$SbBr_3$	$TeBr_4$ orange	(IBr) red	
CsBr	$BaBr_2$	$LuBr_3$	$HgBr_2$	$TlBr_3$ yellow	$PbBr_2$	$BiBr_3$ yellow			
				D. Iodides					
LiI	BeI_2			BI_3	CI_4 red	NI_3 red			
NaI	MgI_2			AlI_3	SiI_4	PI_5 black			
KI	CaI_2	ScI_3	ZnI_2	GaI_3 yellow	GeI_4 orange	AsI_3 red			
RbI	SrI_2	YI_3	CdI_2	InI_3 yellow, red	SnI_4 yellow	SbI_3 yellow, red	TeI_4 gray	(I_2) purple	
CsI	BaI_2	LuI_3	HgI_2 yellow, red	TlI yellow	PbI_2 yellow	BiI_3 black			

[a] If no color is listed, the compound is colorless or white. Compounds to the left of the bar are in the group oxidation state; those to the right are in the (group number −2) oxidation state; those in parentheses are in still lower oxidation states.

6.5 Periodicity in the Activity of Metals and the Reduction of Metal Ions

At this point, your instructor may ask you to discover this periodicity experimentally for yourself by performing or observing and discussing Experiment A.6 (in the Appendix). This experiment involves interpretation of the **activity series of metals**, which you may have investigated in general chemistry. In this case, your instructor may assign only part F of the experiment.

Among the most common and simplest redox reactions are those in which metals are oxidized to hydrated metal ions in common oxidation states, or in which the hydrated metal ions are reduced back to metals. In the simplest version of this reaction, $H^+(aq)$ (perhaps from the self-ionization of water) is used to oxidize the metal, or $H_2(g)$ is used to reduce the metal ion; the simple predominance diagram of hydrogen is shown in Figure 6.2. We may expect that metals with predominance areas entirely below that of $H^+(aq)$ (metals with negative standard reduction potentials) will react with $H^+(aq)$ to give a hydrated metal ion and $H_2(g)$; the further apart the predominance areas, the more vigorous the reaction is expected to be. This vigor of the reaction is the basis of the activity series of metals experiment, in which reactivity is judged qualitatively. It can be expressed quantitatively by the E° values; hence, the activity series is sometimes also called the electromotive series. Table 6.2 summarizes the standard reduction potentials (in acid solution) of the most common metal ion of each metallic element.[18]

Table 6.2
Standard Reduction Potentials of Hydrated Metal Ions[a]

Group	1	2	3	4	5	6	7	8	9	10	11	12	13 (III)	14 (IV)	15 (V)	16 (VI)
Period																
1	H +1 0.00															
2	Li +1 −3.04	Be +2 −1.85														
3	Na +1 −2.71	Mg +2 −2.37											Al +3 −1.66			
4	K +1 −2.93	Ca +2 −2.87	Sc +3 −2.02	Ti +3 −1.21	V +3 −0.87	Cr +3 −0.74	Mn +2 −1.19	Fe +2 −0.47	Co +2 −0.23	Ni +2 −0.23	Cu +2 +0.34	Zn +2 −0.76	Ga +3 −0.56	Ge +2 0.00		
5	Rb +1 −2.92	Sr +2 −2.89	Y +3 −2.37	Zr +4 −1.54	Nb +3 −1.10	Mo +3 −0.20	Tc +2 −0.50	Ru +3 +0.38	Rh +3 +0.80	Pd +2 +0.92	Ag +1 −0.80	Cd +2 −0.40	In +3 −0.34	Sn +2 −0.14		
6	Cs +1 −3.08	Ba +2 −2.91	Lu +3 −2.30	Hf +4 −1.70	Ta	W +3 −0.11	Re +3 +0.3	Os +2 +0.85	Ir +3 +1.16	Pt +2 +1.2	Au +3 +1.52	Hg +2 +0.85	Tl +1 −0.34	Pb +2 −0.13	Bi +3 +0.29	Po +2 +0.65
7		Ra +2 −2.92	Lr +3 −2.06													
Group			3*F*	4*F*	5*F*	6*F*	7*F*	8*F*	9*F*	10*F*	11*F*	12*F*	13*F*	14*F*	15*F*	16*F*
6			La +3 −2.36	Ce +3 −2.34	Pr +3 −2.35	Nd +3 −2.32	Pm +3 −2.29	Sm +3 −2.30	Eu +3 −1.99	Gd +3 −2.29	Tb +3 −2.30	Dy +3 −2.99	Ho +3 −2.33	Er +3 −2.31	Tm +3 −2.31	Yb +3 −2.22
7			Ac +3 −2.13	Th +4 −1.80	Pa +4 −1.62	U +4 −1.37	Np +4 −1.75	Pu +4 −1.25	Am +3 −2.07	Cm +3 −2.06	Bk +2 −1.97	Cf +3 −2.01	Es +3 −1.98	Fm +3 −1.95	Md +3 −1.66	No +3 −1.18

SOURCES: Data are taken from B. Douglas, D. H. McDaniel, and J. J. Alexander, *Concepts and Models of Inorganic Chemistry*, Wiley, New York, 1983, and from M. C. Ball and A. H. Norbury, *Physical Data for Inorganic Chemists*, Longman, London, 1974.

[a]The first figure under the symbol of the element represents the charge of the cation being reduced to the element. (For ions with more than one cation, the more prevalent cation was chosen.) The second figure is the standard reduction potential for the reduction of that cation to the elemental form.

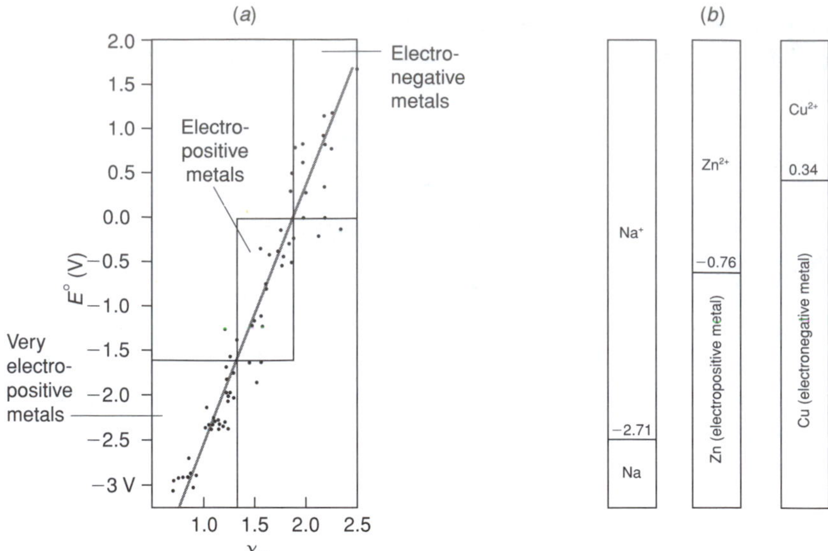

Figure 6.14
(*a*) The standard reduction potential ($E°$) of the most common cation of each metallic element versus its Pauling electronegativity, with the three categories of metals indicated. (*b*) The redox predominance diagram of one representative metal of each category.

As you may have discovered in analyzing Experiment A.6, the activity of a metal correlates fairly well with its Pauling electronegativity, as shown in Figure 6.14. Although this correlation is not exact enough for precise ordering of elements of very similar activity (recall that Pauling electronegativities have uncertainties of ± 0.05), it does enable us to classify the redox activity of elements, based on their Pauling electronegativities.

1. The most active metals have low electronegativities. We may include in a group of *very electropositive metals* all metals with Pauling electronegativities below 1.4. The cations of these metals generally have standard reduction potentials of −1.6 V or below, so the predominance areas of the metals are small and low: These metals are *very good reducing agents*, and are so reactive that they react with the low concentration of H^+ present in neutral water to release hydrogen.[19] (For some of the least active of these, the water must be boiling or in the form of steam.) This group includes the metals at the far left of the periodic table (Groups 1–3, and the groups of the *f* block). Conversely, their metal ions are not good oxidizing agents at all: They have very large predominance areas and cannot be reduced to the metal in aqueous solution, since water is more easily reduced than these cations.

2. We may identify a group of *electropositive metals* with Pauling electronegativities between 1.4 and 1.9; their cations generally have standard reduction potentials between 0.0 and −1.6 V. This group includes the *d*-block elements of the fourth period, and the *p*-group metals of the fourth and fifth periods. Although these metals do not react very readily with neutral water, they do react with hydrogen ion (acids).

3. There is a group of *electronegative metals* with Pauling electronegativities between 1.9 and 2.54; the metals have large predominance areas while their cations

generally have positive standard reduction potentials, and hence fairly small predominance areas. This group includes most of the *d*-block elements of the fifth and sixth periods and the *p*-group metals of the sixth period. This group of metals is not oxidized by hydrogen ion; the metal ions of this group are good oxidizing agents that are easily reduced by hydrogen gas.

This relative order of reactivity of metals applies not only to oxidation by water or the hydrogen ion, but (approximately) to oxidation by other oxidizing agents as well. Although H^+ cannot attack the electronegative metals, oxidizing acids such as concentrated nitric acid contain anions that can attack and dissolve most of them. The lower boundary of the predominance area of NO_3^- is much higher (Fig. 6.2) than that of H^+, so it is a stronger oxidizing agent, especially when concentrated.

$$Ag(s) + 2\,H^+ + NO_3^- \rightarrow Ag^+ + NO_2(g) + H_2O \tag{6.28}$$

Other examples of oxidizing oxo acids will be discussed later in this chapter.

As another example, in a **displacement** reaction a more-reactive (more electropositive) metal reacts with the cation of a less-active (more electronegative) metal, to give the (positive) cation of the more electropositive metal, while the more electronegative element ends up in the zero oxidation state in the elemental form, after combining with (negative) electrons. Displacement is also readily predicted by examining the redox predominance (or Pourbaix) diagrams of the two elements.

Example 6.5

Which of the following displacement reactions are expected to go to the products indicated? (a) $Fe^{2+} + Cu \rightarrow Fe + Cu^{2+}$; (b) $Mn + 2\,Ag^+ \rightarrow Mn^{2+} + 2\,Ag$; (c) $2\,Na + Cu^{2+} \rightarrow 2\,Na^+ + Cu$; and (d) $Bh + Hs^{2+} \rightarrow Bh^{2+} + Hs$.

SLOUTION:

(a) As seen from the predominance diagram of Figure 6.4, the predominance regions of Fe^{2+} and Cu overlap, so no reaction is expected. This failure to react can also be predicted from electronegativities: The most electronegative metal, Cu, is expected to remain in elemental form.

(b) The predominance diagrams of Mn and Ag^+ do not overlap, so reaction is predicted. The most electronegative metal, Ag, is expected to be produced in elemental form, which is done by Reaction (b).

(c) As in (b), this reaction would be expected to go, and would in fact go readily in suitable nonaqueous solvents. In water, however, a complication would arise: Na, a very electropositive metal, it also reacts with water to give NaOH(aq) and $H_2(g)$, and the concentration of water in the solution is much higher than the concentration of Cu^{2+}. Furthermore, the NaOH byproduct of the reaction with water would tend to precipitate the Cu^{2+} as $Cu(OH)_2$. This reaction would not proceed as cleanly in water as expected from a simple examination of electronegativities or predominance diagrams.

(d) Neither predominance diagrams nor electronegativities are known for the newly discovered elements Bh and Hs. However, from periodic trends we expect Hs to be more electronegative than Bh, so elemental Hs is the expected product of this reaction, as shown.

The classifications of metals by redox activity is also useful for predicting non-aqueous chemistry, such as the oxidation of metals by oxygen itself. Thus the very electropositive metals readily ignite in air, burning to produce oxides.[20] Fires that result from the combustion of these metals cannot be fought with water, since water also reacts with these metals to produce H_2, which burns explosively. Neither can such fires be extinguished with CCl_4, which is actually a rather strong oxidizing agent, or with CO_2: Such fires are generally best extinguished with sand (SiO_2). In contrast, many of the electronegative metals are not oxidized even by oxygen. This resistance to oxidation (plus their scarcity) has contributed to their being prized by humans for millennia as the *noble metals*, and to their use in coinage and jewelry.

6.6 *Hard and Soft Acids and Bases and Redox Chemistry

You may have noticed that the three redox groupings of metal activity are defined by Pauling electronegativity ranges that are very similar to those used to classify hard and soft acids (Chapter 5). The most active metals, the *very electropositive metals*, are largely the metals that give *hard acid* metal ions in solution. The metals of intermediate activity, the *electropositive metals*, are, in general, those that give borderline-acid metal ions in solution; while the inactive *electronegative metals* correspond well to the metals that have cations that are *soft acids*.

The correspondence works rather well in both directions: *Soft-acid cations* are usually *good oxidizing agents*. Likewise, *soft-base anions* are usually fairly *good reducing agents*. As discussed in Section 6.3, what prevents many soft-acid–soft-base salts from being thermodynamically unstable or even explosive is their *insolubility*, which may cause sufficient expansion of the soft-acid and soft-base predominance regions to give overlap; nevertheless some such combinations (e.g., silver azide and acetylide) are notorious explosives (Section 6.2.).

But salts of soft acids and soft bases, even if stable, often have low enthalpies of formation and are of low thermodynamic stability; it is thus not surprising that they are often *intensely colored* due to the presence of charge-transfer absorptions (Section 6.4). You may recall, for example, the bright yellow, orange, red, and black colors of the sulfides that precipitate in groups II and III of the qualitative analysis scheme, which contrast with the pale or nonexistent colors of the hard-acid–hard base products found later in the scheme. Likewise, chalcophile minerals often have intense colors (even with a semimetallic sheen in some cases), in contrast with lithophile minerals, which may be somewhat colored due to impurities of *d*-block metal ions, or which may be colorless.

The formation of complex ions is often used to increase the activity of a metal in redox reactions. For example, nitric acid is a good oxidizing agent, but it coats many electropositive metals with tough films of insoluble oxides. If the metal is a hard acid, this may often be overcome by the use of a mixture of nitric and hydrofluoric acids; the surface oxide then dissolves with the aid of the hard base F^- to give a soluble fluoro anion:

$$Nb + 5\,HNO_3 + 6\,HF \rightarrow H^+ + NbF_6^- + 5\,NO_2 + 5\,H_2O \qquad (6.29)$$

Even nitric acid is not a strong enough oxidizing agent to oxidize the softest acid metal, gold. However, a mixture of hydrochloric and nitric acid, known as *aqua regia*, is able to dissolve gold, since the gold ion is complexed by the chloride ion:

$$2\,Au + 11\,HCl + 3\,HNO_3 \rightarrow 2\,H^+ + 2\,AuCl_4^- + 3\,NOCl + 6\,H_2O \tag{6.30}$$

The even softer base, the cyanide ion, forms such stable complexes with the soft acids that the parent metals may be oxidized with atmospheric oxygen. This reaction is exploited in the extraction of these precious metals from deposits having very low percentages of metal:

$$4\,Au + 8\,CN^- + O_2 + 2\,H_2O \rightarrow 4\,[Au(CN)_2]^- + 4\,OH^- \tag{6.31}$$

Related to this is the well-known observation that, although silver is quite unreactive to the strong oxidizing agent in the air, O_2, it is readily tarnished by sulfur or by H_2S and air, to give black silver sulfide.

Because of the increased softness associated with low oxidation numbers, we generally find that low oxidation states of metals are stabilized by the presence of soft-base ligands. Oxidation states for metals of zero or below are mainly found in their organometallic compounds, in which the soft donor atom carbon is attached to the metal atom. High oxidation states of metals are usually stabilized by the presence of hard-base ligands. Thus the maximum oxidation states of metals are usually found in their fluorides, oxides, fluoro anions, or oxo anions.

As we have seen, the versatile Lewis acid–base concept encompasses much of chemistry: It includes not only the earlier, more restricted types of acid–base reactions, but also precipitation reactions and what we usually call coordination chemistry. Only oxidation–reduction (redox) chemistry is excluded. A reducing agent does "donate" (or transfer) an electron to an electron acceptor (oxidizing agent), but of course this donation or transfer does not fit the Lewis definition because an electron *pair* is not involved.

However, acids and bases are what we define them to be, and a Russian chemist, M. Usanovich, has proposed (in effect) removing the word "pair" from the Lewis definition, and thus including redox chemistry within the scope of acid–base chemistry.[21] This definition has not caught on, but this is not due to any intrinsic flaw in such a definition. In some reactions in which a soft base, which is also a good reducing agent, reacts with a soft acid, which is also a good oxidizing agent, the mechanism turns out to be, not one of electron-pair donation, but of *single electron transfer* (i.e., the reaction is a redox reaction).

Within the Usanovich definition the electron itself could be thought of as a (Usanovich) base. And, since soft-acid cations most readily combine with electrons to give metals, the electron can then be classified as a *soft base*—some say the "ultimate soft base". And the results of displacement reactions can be easily rationalized: The soft-acid metal cation ends up combined with the ultimate soft base, the electron:

$$\underset{\text{SA:HB}}{Hg^{2+}(OH_2)_n(aq)} + \underset{\text{HA:SB}}{Mn^{2+}(e^-)_2(s)} \rightarrow \underset{\text{SA:SB}}{Hg^{2+}(e^-)_2(\ell)} + \underset{\text{HA:HB}}{Mn^{2+}(OH_2)_n(aq)} \tag{6.32}$$

In attempting to study a very strong reducing agent such as the electron itself, nonaqueous solvents can be useful. Liquid ammonia is more difficult to reduce (to H_2 and the NH_2^- ion) than water (to H_2 and the OH^- ion). Thus the very electropositive metals (such as the Group 1 metals and Ca, Sr, Ba, Eu, and Yb) react violently with water (Experiment A.6), releasing H_2. Unless a catalyst is present, these metals do not react with liquid ammonia, but instead dissolve to give intensely blue solutions of remarkably low density. These solutions contain the ammonia equivalents of hydrated cations, $[M(NH_3)_n]^{2+ \text{ or } 1+}$; the nature of the anions present was the fascinating question.

If additional Group 1 metal is added to one of the above blue solutions, a second, even less dense liquid of bronze color forms, which is insoluble in the first and floats on it. These properties, which have been known since 1864, have fascinated chemists ever since, but the nature of the solutions proved difficult to determine, in part because evaporating either the blue or bronze solution to dryness generally gave back the metal. But this problem was overcome by the application of the principles of Section 4.1: The unknown but presumably large anions in the blue and the bronze solutions should best be stabilized in the solid state by using larger cations than simple Group 1 cations. Relatively stable crystals were produced[22] by adding macrocyclic ligands (the *crown* or *crypt* polyethers), which form more stable and larger complex cations with the Group 1 metal than does ammonia. By this method, crystalline salts containing the blue and bronze anions were isolated.

The bronze solutions gave golden crystals of a material containing one molecule of crown or crypt polyether and *two* atoms of Group 1 metal ion. X-ray crystallography of the product with Na showed that one Na atom sits at the center of the polyether as a typical complex cation (Fig. 4.9). The other sodium atom sits outside by itself, and hence must be present as the Na^- ion. The combination of a sodium cation and a sodide anion is normally less stable than metallic sodium, but evidently either the extra energy of coordinate covalent bond formation with the polyether or its physical barrier to electron transfer (or both) gives marginal stability to these remarkable compounds.

The blue solution gave dark blue crystals containing solvated *electrons*, $e^-(NH_3)_m$. The deep blue color results from excitement of the electron by light; this has led to the facetious proposal that electrons are blue! The low density results from the fact that the ammonia molecules solvating the electron are remarkably far apart: The solvated electrons act as if they are "electride" ions with a surprisingly large "ionic radius" of about 150–170 pm, which is a characteristic of soft bases.

In Chapter 5, we noted that the soft and borderline acids occur in nature as sulfides and related minerals and can be called chalcophiles. The softest acids show an even greater preference for combination with electrons, and hence occur in nature in the reduced, uncombined form (**native**), as the free metals. Geochemists call these metals **siderophiles**. Since metals are in excess over nonmetals in the earth, not all of the metals can be oxidized. Much of the very abundant metal, iron, did not get oxidized in the early history of the molten earth and sank to form the **core** of the earth, taking with it much of the supply of the metals less active than iron. These metals are depleted by a factor of 1000 as compared to their abundance in asteroids, which never were melted and did not develop cores. Sediments deposited around the earth about 65,000,000 years ago are a thousand times enriched in (at least) two of these elements, osmium and iridium, as compared to normal sediments. This enrichment has been taken as evidence that an extraterrestrial body (perhaps an asteroid) may have collided with the earth at that time, providing a shower of dust over the earth that was (incidentally) iridium rich and (more importantly) may have shut out a good deal of sunlight for many years. This was the period in which the dinosaurs and all other large animals of the time became extinct.[23] Volcanic eruptions and mid-oceanic "black smokers" also produced enriched amounts of such metals; a volcano in Columbia has been found to be emitting over 500 g of gold per day into the atmosphere.[24]

Some (and perhaps many) mineral deposits may be the result of the action of microorganisms, which obtain their energy by mediating redox reactions.[25] The more familiar **heterotrophic bacteria** oxidize organic compounds, but there are also **autotrophic bacteria** that can thrive in a completely inorganic environment, even deep within the earth's crust,[26] oxidizing inorganic materials and using the energy to reduce CO_2 to organic compounds. *Gallionella* lives by oxidizing iron(II), and excretes copious quantities of Fe_2O_3; *Thiobacillus thiooxidans* oxidizes H_2S, S_8, or $S_2O_3{}^{2-}$ and excretes (and survives in) 0.5 M sulfuric acid! In Section 5.9, the MnO_2-rich nodules produced by manganese-eating bacteria were mentioned.

Bacteria and some other organisms can also use oxidized inorganic materials as oxidizing agents when O_2 is not available; nitrate and sulfate are commonly reduced (thus producing the vast deposits of very pure elemental sulfur found in Louisiana). Some bacteria (including those in the human digestive tract) have a mercury-reducing *merA* gene that enables them to reduce toxic Hg^{2+} to volatile elemental mercury, which then escapes from them, reducing their exposure. Evidence is also accumulating that other microbes can produce crystals of elemental gold as a crust on their surfaces; this process may be responsible for a gold field found in the Amazon jungle, far from volcanic activity.[27]

6.7 Redox Chemistry of the Nonmetals

Redox Classifications. The redox chemistry, Latimer diagrams, and Pourbaix and redox predominance diagrams of the nonmetals are complicated by the presence of positive, negative, and zero oxidation states for most of these elements; in addition, most of them have two positive oxidation states and some have more. Nonetheless, it will still be useful to categorize the nonmetals into two groups, based on their Pauling electronegativities. The first group we may call the *very electronegative nonmetals.* These nonmetals (F_2, Cl_2, Br_2, O_2, and N_2) are those with Pauling electronegativities over 2.8; mainly these are the nonmetals that, when reduced to ions or compounds in which they function as donor atoms, we classify as *hard or borderline bases.* Their redox predominance diagrams are shown together in Figure 6.15(*a*), with the predominance regions of the elemental forms being shaded. The remaining nonmetals, with Pauling electronegativities between 1.9 and 2.8, we may term the *electronegative nonmetals*; their predominance diagrams are shown in Figure 6.15(*b*). When reduced to ions or compounds, these nonmetals generally function as *soft-base* donor atoms.

Positive Oxidation States. Examination of the redox predominance or Pourbaix diagrams of the elements allow us to be more exact about the periodic trends in stability of positive oxidation states that are suggested in Tables B1 and B2. In comparing the diagrams for the electronegative and the very electronegative groups of nonmetals, first note that *positive oxidation states are either nonexistent (F, O) or very oxidizing* (have small, high predominance areas) for the very electronegative nonmetals, whereas the positive oxidation states are generally more predominant for the electronegative nonmetals (except Xe and I). *Semiquantitatively*, we can extend the relationship of activity and electronegativity to the nonmetals: The more electronegative a nonmetal atom is, the harder it is to oxidize.[28]

Since the electronegativities of nonmetals increase from left to right across a given period, this means that the predominance areas of the very high oxidation states con-

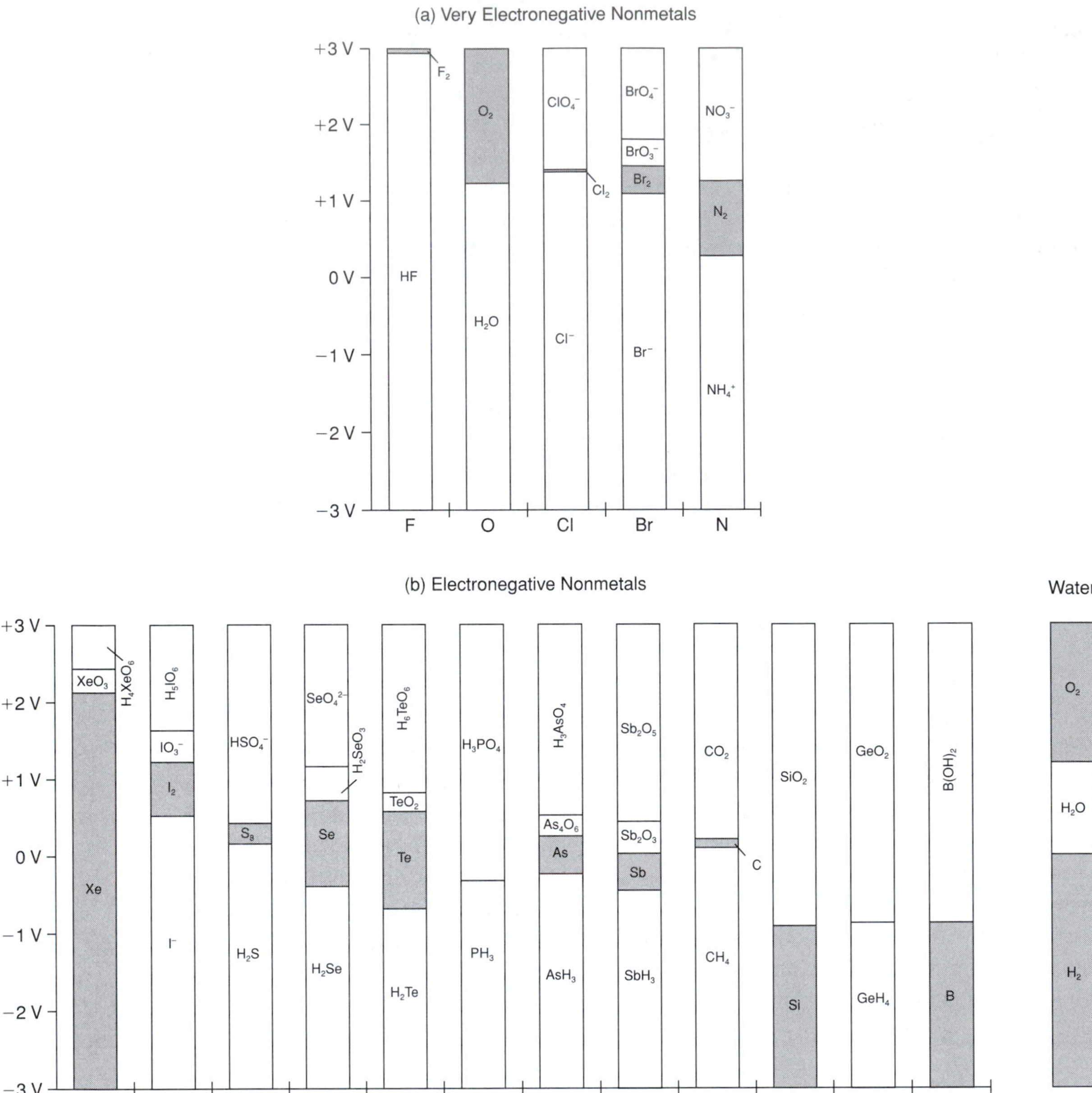

Figure 6.15
Redox predominance diagrams under standard conditions for (*a*) the very electronegative nonmetals; (*b*) the electronegative nonmetals.

strict: The nonmetal oxo acids and anions (or oxides) become better oxidizing agents as one goes to the right in a given period (until eventually, for F, O, and the lighter noble gases, they will oxidize their own oxo groups and cannot exist). Thus, the last element of the second period that is able to achieve the group oxidation number is nitrogen; nitric acid and nitrates are commonly used as strong oxidizing agents. At the end of the third period, perchloric acid and perchlorates are similarly used; chloric acid and chlorates are also good oxidizing agents. Further left in these periods, among the electronegative nonmetals, oxo acids and anions or oxides such as those of B, C, and Si have such large predominance areas that they are of no practical use as oxidizing agents.

Conversely, then, the electronegative nonmetals themselves have lower predominance areas than do the very electronegative nonmetals, and will be more susceptible to oxidation. One form of phosphorus inflames spontaneously in air when dry; at the other extreme, fluorine and the earlier noble gases cannot be oxidized.

The vertical trends in stability among the high oxidation states of the nonmetals are perhaps a bit more surprising. Comparing the predominance diagrams for the Group 15(V) elements (Fig. 6.15), one notices that the largest predominance area for the group oxidation state belongs to the third-period element, phosphorus (as H_3PO_4); stability drops off going up to nitrogen (NO_3^-); it also drops off going down to arsenic (AsO_4^{3-}), antimony (Sb_2O_5), and bismuth (Bi_2O_5). In the *p* block, the group oxidation number is least stable at the top (second period) and at the bottom (sixth period). Thus nitrate ion in acid solution (i.e., nitric acid) is a useful oxidizing agent; acid solutions of "bismuthate" ion are very powerful oxidizing agents. The oxidizing ability of nitric acid is related to the high electronegativity of nitrogen, but the oxidizing ability of bismuthate must have another cause, the relativistic destabilization of the group oxidation state in the sixth period (Section 1.11).

A more subtle vertical periodicity occurs in Groups 16(VI) and 17(VII) [and almost in Group 15(V)] on going from the third through the fourth to the fifth periods: the stability of the high oxidation state follows the sequence: Per. 3 element $\gg$ Per. 4 element $<$ Per. 5 element. Although small, this reversal is enough to have had some interesting chemical consequences: The perbromate ion was not synthesized until 1968, long after perchlorates and periodates were made and came into routine use. This subtle periodic effect results from the scandide contraction discussed in Section 1.9.

By examination of Figure 6.3, we can see the corresponding trends in the stability of the group oxidation state among the *d*-block elements. The horizontal periodic trends are the same as for the *p*-block elements: As one goes to the right, the increasingly high oxidation states lose stability, disappearing after Group 7 or 8. But the vertical periodicity is different: This time the fourth-period elements show the greatest reluctance to adopt the group oxidation state, which is more stable in the fifth and especially the sixth periods. Again relativistic effects are at least partly involved in this trend.

As the group oxidation state approaches nonexistence on going to higher group numbers, we sometimes find the appearance of intense colors due to charge-transfer absorptions. For example, we find that TiO_2 and VO_4^{3-} are colorless and of low redox activity, while the good oxidizing agent CrO_4^{2-} is intensely yellow, the very good oxidizing agent MnO_4^- is intensely purple, and the corresponding color-producing absorption band of the powerfully oxidizing FeO_4 is shifted out of the visible spectrum into the near-infrared (IR) region.[29]

Negative oxidation states. The nonmetallic elements can also act as oxidizing agents. Upon oxidizing an element, the nonmetals may be reduced to their monoatomic anions or the protonated forms of these (Chapter 2). As the electronegativity of a nonmetal increases, its activity as an *oxidizing* agent increases, and the anion or hydride acquires a larger predominance area, so it is more stable (Fig. 6.15). Anions and hydrides are very much predominant in the chemistry of the very electronegative nonmetals, whereas the anions or hydrides of the electronegative nonmetals mainly fall below the predominance area of water, so they are easily oxidized (are good reducing agents).

Given the periodic trends of electronegativities, it follows that the activity of a nonmetal as an oxidizing agent increases on going to the right across a period, and on going up a group. Fluorine is such an active oxidizing agent that it will oxidize all metals and most other substances. (Even water, in the form of steam, will catch fire in an atmosphere of fluorine!) This poses a problem: In what container can one store fluorine gas? Fortunately, the metals nickel and copper can be used, since they are **passivated** by fluorine: Impervious films of nickel or copper fluorides are formed on the surfaces of these metals.

The other halogens are successively less oxidizing. A significant reaction of the halogens is the displacement reaction: A more electronegative (more active) halogen displaces a less active halogen from its anion. Bromine is made in this way from bromide ion in seawater:

$$2\,Br^- + Cl_2 \rightarrow Br_2 + 2\,Cl^- \qquad (6.33)$$

Oxygen is actually a strong oxidizing agent, but its oxidizing reactions are generally quite slow (Section 6.2). Hence, very flammable materials can last indefinitely in the presence of air, at least until a flame or spark comes along to begin the reaction. Nitrogen is also not really as inert as we commonly think, but its very stable triple bond means that high temperatures or strong reducing agents are required to reduce it.

In contrast, the electronegative nonmetals find little laboratory use as oxidizing agents, but sulfur, for example, reacts slowly with some of the electronegative metals (such as Ag in silverware) to tarnish them. In the early history of the earth, reducing conditions are thought to have prevailed, so many of these nonmetals formed minerals in which they take negative oxidation numbers. The earth's atmosphere is now oxidizing, so many of these elements are now more often found oxidized (as oxo anions or oxides) than reduced.

6.8 Industrial Processes for Extracting Elements from Their Ores

Although humans have been mining and using metals for thousands of years, and continue to discover new deposits of ores, by and large we have used up many of the ores that have the highest percent composition of the element being sought. Consequently, we now have to use many ores containing very low percentages of the elements, which requires more and more elaborate chemical processes to produce metals and nonmetals from their ores.[30,31] Many of these processes can be divided into a few general steps, however.

Often the first step is one called **beneficiation** of the ore. The particular mineral containing the desired element may be associated with large amounts of other materials such as rock and gravel. Beneficiation concentrates the ore so that energy and chemicals do not have to be wasted treating rock and gravel. To achieve beneficiation, chemical or physical differences between the ore and the other material are exploited. For example, the principal aluminum ore, bauxite, contains Al_2O_3 in association with Fe_2O_3 and TiO_2. In the Bayer process, differences in the acidic properties of the three metal ions are exploited. The bauxite is heated with 30% NaOH solution at 190°C and

8-atm pressure to cause the oxide Al_2O_3 to dissolve by forming $[Al(OH)_4]^-$. The oxides Fe_2O_3 and TiO_2 do not form hydroxo anions, so they do not dissolve and are filtered off. Then the solution is cooled, which reduces the tendency of the aluminum to form the $[Al(OH)_4]^-$ ion, so purified $Al(OH)_3$ crystallizes out and is filtered off.

Copper is now extracted from ores having less than 1% Cu content, so beneficiation is quite important. Copper ores are often sulfides, which have different physical properties than silicates. Thus, detergents made from organic soft-base anions will wet the copper ore but not the silicate rock. Vigorous agitation produces a copper sulfide-containing froth that floats to the surface of the water and can be skimmed away from the water and rock.

Once the element has been concentrated, it may need to be **chemically converted** to a different salt of the metal for further processing. For example, sulfide ores are much more difficult to process than oxide ores, so they are *roasted* in air to convert the sulfide to sulfur dioxide and the oxide of the metal:

$$2\,MS(s) + 3\,O_2(g) = 2\,MO(s) + 2\,SO_2(g) \tag{6.34}$$

The sulfur dioxide released at smelters producing such metals as copper and nickel has been a serious air-pollution problem. In the case of titanium, the chloride is more easily processed than the oxide, so that rutile is heated with chlorine and carbon:

$$TiO_2(s) + C(s) + 2\,Cl_2(g) \rightarrow TiCl_4(\ell) + CO_2(g) \tag{6.35}$$

In some modern processes of **hydrometallurgy**, the beneficiation and the conversion are carried out in one step. Thus, action of air and acid on copper sulfide ore can both separate copper from rock and convert it to a more usable form:

$$2\,Cu_2S(s) + 5\,O_2(g) + 4\,H^+(aq) \rightarrow 4\,Cu^{2+}(aq) + 2\,SO_4{}^{2-}(aq) + 2\,H_2O \tag{6.36}$$

The third step in the production of metals is **reduction**, in which the preferred salt of the metal ion is reduced to the elemental form. This process is closely related to the activity series of the metals: A reducing agent is needed that is more active than the metal being produced. The reducing agent chosen is the cheapest one that is active enough to do the job. Correspondingly, for *nonmetals* occurring in negative oxidation states in nature, the third step is **oxidation**; the cheapest effective oxidizing agent is chosen.

In the case of the very inactive metals (the electronegative metals), no reducing agent at all may be needed, since these siderophiles often occur **native** (uncombined) in nature. When these metals do occur combined (often as sulfides), the heat (Δ) generated by the roasting process is often sufficient to decompose the oxide to the metal, a reaction that is promoted by the favorable entropy change accompanying the formation of gaseous O_2.

$$Cu_2S(s) + 2\,O_2(g) \rightarrow 2\,CuO(s) + SO_2(g) + \Delta \tag{6.37}$$

$$2\,CuO(s) + \Delta \rightarrow 2\,Cu(s) + O_2(g) \tag{6.38}$$

The very inactive nonmetals (the noble gases and N_2) also occur uncombined in the atmosphere and do not need to be oxidized; they (and O_2) are obtained by fractional distillation of liquid air. Oxygen, sulfur, and carbon also occur uncombined in nature, even though they are not particularly inactive: these deposits are due to past biological processes. Since oxidation or reduction is not necessary, carbon and sulfur are very cheap nonmetallic elements.

Most of the moderately active metals and nonmetals (the electropositive metals and the electronegative nonmetals) are reduced from their oxides using the inexpensive but mild reducing agent carbon (coke from coal) and heat. The metals and nonmetals that can be produced by reduction with carbon are shown in Table 6.3. Note that they fall in a relatively narrow range of electronegativities, going only as far left in the periodic table as Fe.

The activity of carbon as a reducing agent increases at higher temperatures: The *formation*, not the decomposition, of carbon monoxide is increasingly favored at higher temperatures, since the number of moles of gas increases during the process:

$$C(s) + 0.5\,O_2(g) \rightarrow CO(g) \tag{6.39}$$

Consequently, at high enough temperatures carbon can reduce a more "active" metal:

$$FeO(s) + C(s) \rightarrow Fe(\ell) + CO(g) \tag{6.40}$$

Ellingham diagrams (Fig. 6.16) show the free energy change of formation for metal and carbon oxides as a function of temperature. (Note that when ΔG becomes negative

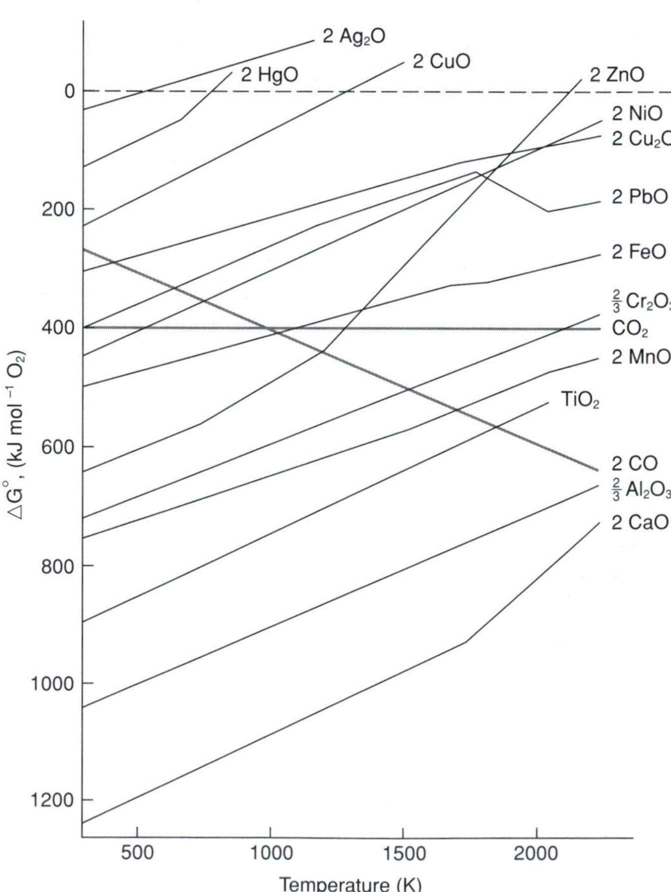

Figure 6.16
An Ellingham diagram showing the free energies of decomposition of metal oxides as a function of temperature. (Discontinuities result from phase changes of either the metal or the oxide.) [Adapted from D. J. G. Ives, *Monographs for Teachers, Number 3*, The Chemical Society, London, 1969.]

Table 6.3A
The Method of Extraction (Oxidation or Reduction) of the Elements as a Function of Periodic Table Position and Electronegativity

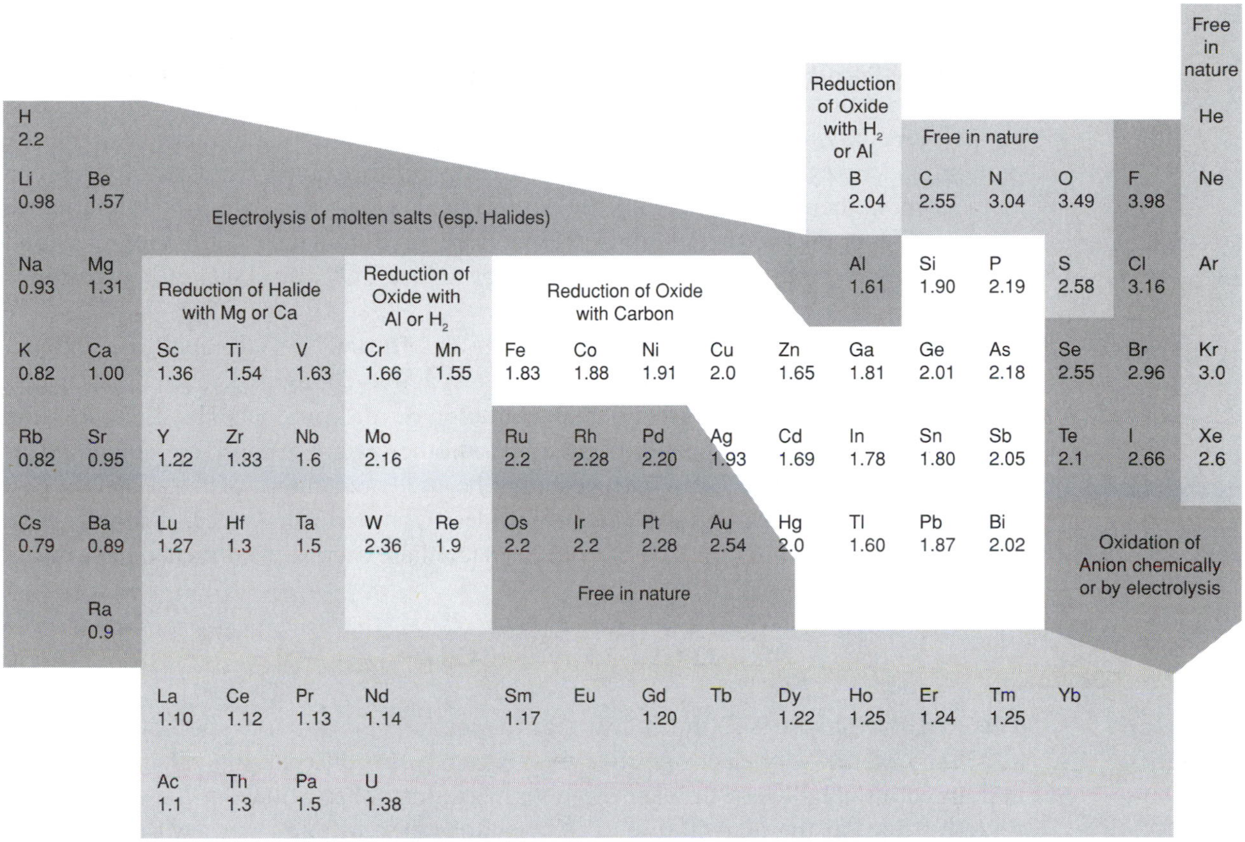

H 2.2																	He
Li 0.98	Be 1.57											B 2.04	C 2.55	N 3.04	O 3.49	F 3.98	Ne
Na 0.93	Mg 1.31											Al 1.61	Si 1.90	P 2.19	S 2.58	Cl 3.16	Ar
K 0.82	Ca 1.00	Sc 1.36	Ti 1.54	V 1.63	Cr 1.66	Mn 1.55	Fe 1.83	Co 1.88	Ni 1.91	Cu 2.0	Zn 1.65	Ga 1.81	Ge 2.01	As 2.18	Se 2.55	Br 2.96	Kr 3.0
Rb 0.82	Sr 0.95	Y 1.22	Zr 1.33	Nb 1.6	Mo 2.16		Ru 2.2	Rh 2.28	Pd 2.20	Ag 1.93	Cd 1.69	In 1.78	Sn 1.80	Sb 2.05	Te 2.1	I 2.66	Xe 2.6
Cs 0.79	Ba 0.89	Lu 1.27	Hf 1.3	Ta 1.5	W 2.36	Re 1.9	Os 2.2	Ir 2.2	Pt 2.28	Au 2.54	Hg 2.0	Tl 1.60	Pb 1.87	Bi 2.02			
	Ra 0.9																
		La 1.10	Ce 1.12	Pr 1.13	Nd 1.14		Sm 1.17	Eu	Gd 1.20	Tb	Dy 1.22	Ho 1.25	Er 1.24	Tm 1.25	Yb		
		Ac 1.1	Th 1.3	Pa 1.5	U 1.38												

Region labels: Electrolysis of molten salts (esp. Halides); Reduction of Halide with Mg or Ca; Reduction of Oxide with Al or H_2; Reduction of Oxide with Carbon; Reduction of Oxide with H_2 or Al; Free in nature; Free in nature; Free in nature; Oxidation of Anion chemically or by electrolysis

Table 6.3B
The Redox Predominance Diagrams of the Reducing Agents Used in the Production of Elements

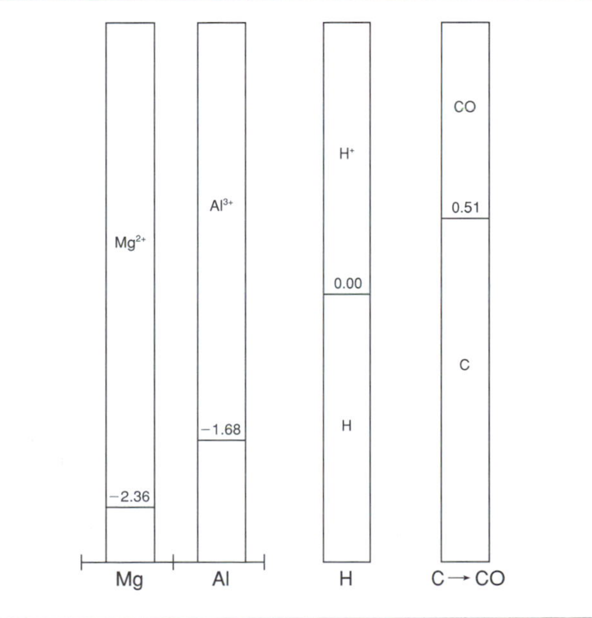

in Fig. 6.16, by Eq. 6.4 E° becomes positive, and the reaction of decomposition to metal and O_2 is favored.) The diagram shows that increasing temperature favors the *decomposition* of metal oxides at higher temperatures but also favors the *formation* of carbon monoxide. Consequently, at any temperature above which the heavy line for CO in Figure 6.16 crosses the line for the decomposition of the metal oxide in question, carbon can reduce the metal oxide. In terms of redox predominance diagrams, high temperatures favor the *decomposition* of metal oxides to metals, and hence expand the predominance regions of metals; but high temperatures favor the *formation* of CO from carbon, hence push the C/CO boundary *down* in the carbon predominance diagram. Thus, although the room temperature predominance regions of C(s) and Fe^{2+} (as in FeO) overlap, so that C(s) does not reduce Fe^{2+}, by 2000 K the situation is reversed.[32]

Some of the electronegative metals (such as Mn and Ti), although they can in principle be produced by carbon reduction, are active enough to react with carbon itself to produce metallic carbides that contaminate the product. Hence, unless the carbide is a desirable component of an alloy, another, more expensive reducing agent must be chosen (Table 6.3). Hydrogen may be used, but many of these metals also combine with hydrogen to give metal hydrides (Chapter 15). Hence, a more active metal such as aluminum may be used to displace these metals from their oxides in the spectacular *thermite reaction*:

$$Cr_2O_3(s) + 2\,Al(s) \rightarrow 2\,Cr(\ell) + Al_2O_3(s) \tag{6.41}$$

This reaction generates so much heat that the metal is produced as red-hot liquid.

Many of the very electropositive metals can be produced using calcium or magnesium with the halide of the element in an inert atmosphere. Some of these metals are so active that, at the temperature of the reaction, they will also react with N_2 to give nitrides. Hence, the production of titanium metal is carried out at 900 K in an argon atmosphere by the Kroll process:

$$TiCl_4(g) + 2\,Mg(\ell) \rightarrow Ti(s) + 2\,MgCl_2(\ell) \tag{6.42}$$

Titanium is an abundant, very desirable metal that is as strong as steel but is considerably lighter, so that it would no doubt be very widely used were it not for the very high expense of this process. (It must also be welded in an argon atmosphere, which is difficult). In the time of Napoleon III, aluminum was made by a similar, very expensive process. It is said that Napoleon III once entertained at a state dinner at which the lesser guests had to content themselves with eating from gold or silver plates, while Napoleon III and the most important guests ate from aluminum plates!

For the most active metals, of course, there exists no more active metal, which can be used to displace them from their salts. These metals are produced by **electrolysis** of their fused (molten) halides (these metals are too active to be liberated by electrolysis of an aqueous solution). The most important of these metals are aluminum and sodium.

Aluminum is made by electrolysis of Al_2O_3 obtained by beneficiation and dehydration of bauxite. An oxide such as Al_2O_3 has too high a melting point to be used in its fused form, however, so a nonaqueous solvent (not itself decomposed by electrolysis) had to be found for the Al_2O_3. This problem was difficult, but in 1886 Hall in the United States and Heroult in France found a suitable mineral, cryolite (Na_3AlF_6, found only in Ivigtut, Greenland, but now made synthetically). The electrolysis (Fig.

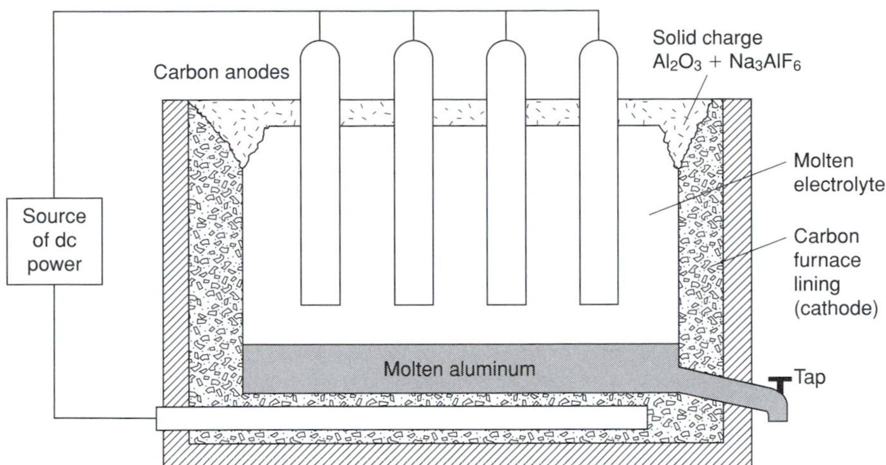

Figure 6.17
Schematic diagram of a Hall–Heroult aluminum reduction cell. [Adapted from *Environmental Chemistry*, by J. W. Moore and E. A. Moore. Copyright © 1976 by Academic Press.]

6.17) operates at a temperature of nearly 1000 °C, a voltage of 4.5 V, and a current of 100,000 A. The furnace lining of graphite serves as the cathode, at which molten aluminum forms and is drawn off. Oxygen forms at carbon anodes, where it reacts with them to generate CO. This helps reduce the demand on electrical energy; nonetheless the production of aluminum requires an enormous inputs of electricity. Consequently, the recycling of scrap aluminum (as in cans) is profitable (even in the absence of any environmental considerations) since this step is avoided.

For the most active (very electronegative) nonmetals, electrolytic *oxidation* must also be used. Fluorine gas is produced by electrolysis of a molten adduct of KF and HF, using a stainless steel cathode and a carbon anode (the byproduct H_2 must be kept rigorously separate from the F_2 or violent explosions will result). Chlorine gas is produced on a very large scale by electrolysis of NaCl, either fused (in which case sodium metal is also produced), or in solution (in which case H_2 and NaOH are the byproducts).

The less-active nonmetals Br_2 and I_2 can be produced by chemical oxidation of their aqueous anions by the relatively inexpensive Cl_2. Elemental Br_2 is obtained directly from bromide ion in sea-water, despite its low concentration there, by passing Cl_2 through the sea-water and blowing the volatile liquid Br_2 out with air. Elemental I_2 is similarly obtained from certain brines having a high concentration of I^-.

The final step in the production of most elements is **purification** or **refinement** of the metal (or nonmetal). One of the main contaminants for metals such as iron is a residue of SiO_2. The charge for a blast furnace (Fig. 6.18) includes not only iron ore and coke, but also limestone ($CaCO_3$), which at the temperature of the blast furnace breaks down into CaO. The CaO combines with SiO_2 to form calcium silicate or **slag**, which melts along with the iron and falls to the bottom of the blast furnace. The slag is less dense than iron, so it floats on the molten iron and is drawn off separately. (The CaO and coke together also help remove sulfur, a harmful impurity as FeS, by the reaction

$$FeS + CaO + C \rightarrow Fe + CaS + CO \qquad (6.43)$$

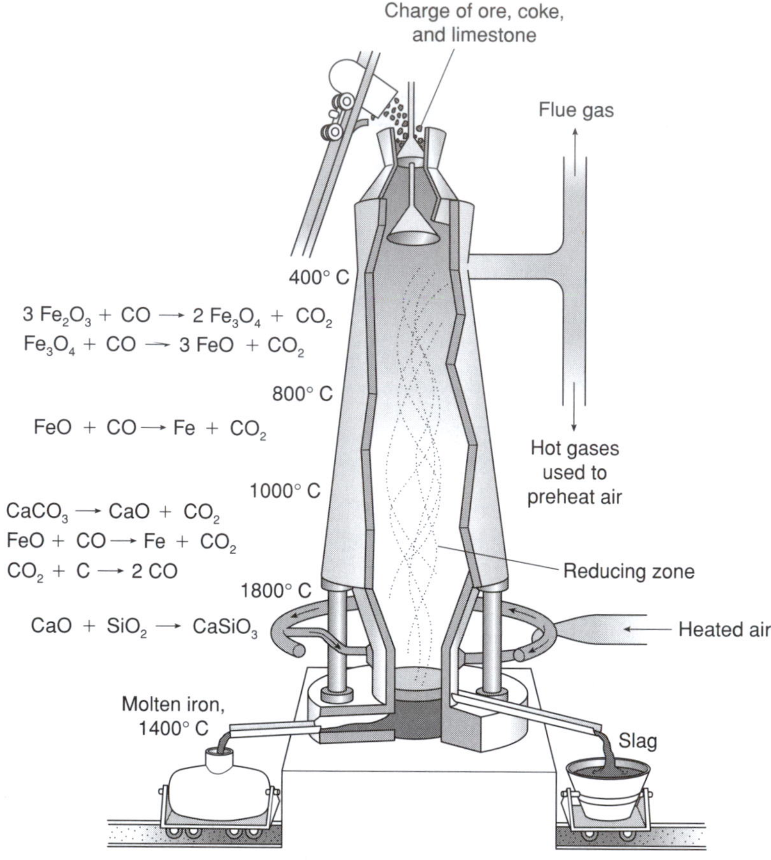

Figure 6.18
Schematic diagram of a blast furnace, showing temperatures and chemical reactions occurring in different parts of the furnace. [Adapted from *Modern Descriptive Chemistry* by E. G. Rochow. Copyright © 1977 by W. B. Saunders Company.]

The CaS is also a component of the slag that is removed.)

The iron that is produced in the blast furnace, called *pig iron*, still contains more than 2% each of several impurities that make it brittle: excess carbon (as iron carbide), phosphorus (as iron phosphide), silicon, and manganese. To produce **steel**, these impurities are removed by several processes, the most modern of which is known as the *basic oxygen process*. In this process, O_2 is blown over or through the molten iron in the presence of CaO; the more active impurities (carbide, phosphide, Si, and Mn) have a greater tendency to be oxidized to their oxides or calcium oxo anion salts than does iron, although some iron oxides are also produced. More molten slag is formed, which melts and is drawn off the molten steel.[33]

The purification or refining of copper is also important, since it must be quite pure for electrical purposes, and since its impurities include metals such as silver and gold, which are even more valuable than the copper. Impure copper is cast into plates that are used as the anode in an electrolytic cell. A controlled voltage is applied, which is sufficient to oxidize copper to Cu^{2+} and other more active metals to their cations, but which does not oxidize the less active metals such as Ag, Au, and the platinum metals. These fall to the bottom of the electrolytic cell as *anode mud* or *slime* and are recovered. The cathode is a sheet of highly purified copper; at the controlled voltage only Cu^{2+} is active enough to be reduced back to the metal. (The Fe^{2+}, Ni^{2+}, etc., remain in the solution as cations.)

Perhaps one ought to add another stage to the process of extraction of elements from their ores: that of **pollution control**. As you can appreciate on rereading the descriptions of the processes of beneficiation, conversion, reduction, and purification, many kinds of impurities form that can be solid, water, and air pollutants. Increasingly, the design of emission controls and waste disposal processes[30,34] play an important part in the economics of the production of the elements.

Example 6.6

In Exercise 30, you are given standard reduction potentials for four imaginary elements found on Venus: (a) Nk; (b) Ss; (c) Hm; and (d) Pz. Using the answers to Exercise 30 (provided at the back of the book), predict which two of these elements could most likely be produced by reduction of their oxides or oxo acid with carbon; tell how the other two might be produced.

SOLUTION 1:

(a) The boundary between elemental Nk and its reduced form, Nk^-, is very high, at +2.50 V. Elemental Nk is thus a strongly oxidizing, very electronegative nonmetal (much like fluorine), and would have to be produced from Nk^- by electrochemical oxidation of this anion.

(b) The predominance region of the element Ss is sandwiched between that of its oxo acid, H_6SsO_6, above +1.65 V, and that of its oxo anion, Ss^{2-}, at +0.10 V. Elemental Nk is thus an electronegative nonmetal (somewhat like Te) with a highly oxidizing oxo acid that ought to be easily reduced by carbon. But if its anion is found on Venus instead, it could be oxidized chemically or electrochemically.

(c) The predominance region of Hm metal reaches up to that of its cation at +0.85 V, which means the cation is a fairly good oxidizing agent. Likely Hm^{2+} is a soft acid; its oxide surely could be reduced by carbon, but it might also occur native on Venus (depending on how oxidizing the atmosphere of Venus is).

(d) The predominance region of Pz is very low, only extending up to −2.30 V. The element Pz is thus a very active metal. Hence, it is likely a very electropositive metal. Carbon would not suffice for its reduction; likely one would need to use a more active metal (Ca or Mg) with its halide.

SOLUTION 2:

Alternatively, we can use Figure 6.15 to predict the electronegativities of the two metals from their standard reduction potentials. From its $E°$ of +0.85 V we would estimate the χ_P of Hm to be about 2.1; from its $E°$ of −2.30 we would estimate the χ_P of P_z to be about 1.2. Then, from Table 6.3 we can see that metals of $\chi_P = 2.1$ are found (on earth) free in nature, while those of $\chi_P = 1.2$ are produced by the reduction of the halide with Mg or Ca. The nonmetals could also be classified as very electronegative or electronegative by analogy with predominance diagrams in Figure 6.15.

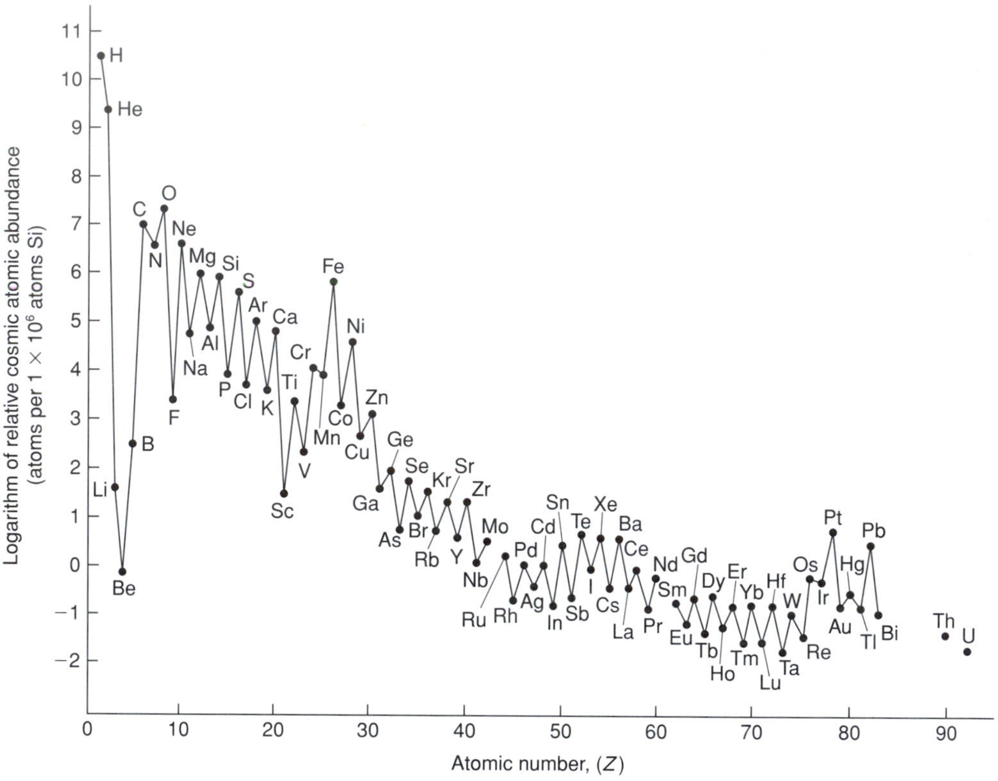

Figure 6.19
Logarithms of the relative cosmic abundance of the elements (per each 10^6 atoms of silicon) versus atomic numbers. [Adapted from *Inorganic Chemistry: A Modern Introduction*, by T. Moeller. Copyright © 1982 by Wiley-Interscience, New York. John Wiley & Sons, Inc.]

Cost and Abundance of the Elements. One final property of an element that profoundly affects its practical uses is its abundance, which in part depends on the stability of the nuclei of its isotopes. Figure 6.19 gives the estimated relative cosmic abundance of the elements (per each 10^6 atoms of silicon). Some trends are apparent: (1) Hydrogen and helium, the primordial nuclear fuel and its fusion product, constitute over 99% of the atoms in the universe. (2) Abundances generally drop off from them, although the next three, Li, Be, and B, which are readily transmuted by nuclear bombardment, are of exceptionally low abundance. Iron is of unexpectedly high abundance, since the nuclei of iron have the greatest stability per nuclear particle. (3) Elements with even atomic numbers are more abundant, and have more stable isotopes, than those with odd atomic numbers.

The abundances of the elements in the crust of the earth are not always similar to those in the cosmos, as they are affected by the primary and the secondary differentiation of the elements. For example, elements present as light gaseous molecules or noble gases tend, over time, to escape from the earth's atmosphere when their velocities exceed the escape velocity from the earth. Metals less active than iron have tended to concentrate in the core of the earth, dissolved in the iron, and thus are inaccessible to us. On the other hand, some rare hard-acid elements with unusual sizes and charges

Table 6.4
Estimated Crustal Abundances of the Elements

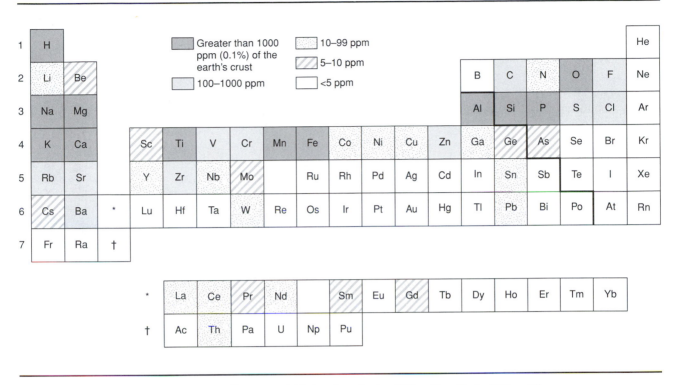

Greater than 1000 ppm (0.1%) of the earth's crust
100–1000 ppm
10–99 ppm
5–10 ppm
<5 ppm

1	H																	He
2	Li	Be											B	C	N	O	F	Ne
3	Na	Mg											Al	Si	P	S	Cl	Ar
4	K	Ca		Sc	Ti	V	Cr	Mn	Fe	Co	Ni	Cu	Zn	Ga	Ge	As	Se	Br	Kr
5	Rb	Sr		Y	Zr	Nb	Mo		Ru	Rh	Pd	Ag	Cd	In	Sn	Sb	Te	I	Xe
6	Cs	Ba	*	Lu	Hf	Ta	W	Re	Os	Ir	Pt	Au	Hg	Tl	Pb	Bi	Po	At	Rn
7	Fr	Ra	†																

*	La	Ce	Pr	Nd		Sm	Eu	Gd	Tb	Dy	Ho	Er	Tm	Yb
†	Ac	Th	Pa	U	Np	Pu								

SOURCE: Data from T. Moeller, *Inorganic Chemistry: A Modern Introduction*, Wiley, New York, 1982; p. 24.

did not fit into the lattices of the minerals that crystallized out during the solidification of the mantle, and are especially abundant in the crust of the earth. The estimated crustal abundance of the more common elements are indicated in Table 6.4.

Most important for the overall usefulness of an element is its cost. Although related to its abundance, the cost is also related to how many other demands there are for the element; whether it is obtained as a byproduct of the production of other more abundant elements; whether it is widely dispersed in many minerals (due to having common charges, ionic radius, and electronegativity) or is concentrated by secondary differentiation in convenient ore deposits; and the political situation in the countries in which it is located, if it is found in but a few places. In Table 6.5, we give costs of the elements (in elemental form). These costs are for small quantities of high purity research grade samples, and would be much lower on an industrial scale, but the relative trends among elements ought to be reasonably similar. The costs, of course, are also affected by the relative activity of the different elements: more active ones are more difficult to produce (sodium is much cheaper as sodium ion in rock salt than as the metal).

Many of the highly industrialized countries of the world depend almost completely on imports for their supply of many strategic metals, and have become accustomed to their ready availability on the marketplace. Hence, many technologies have been developed based on the elements with the optimal combination of cost and physical and/or chemical properties; the telephone handset, for example, contains over 40 dif-

Table 6.5
Costs of the Elements[a]

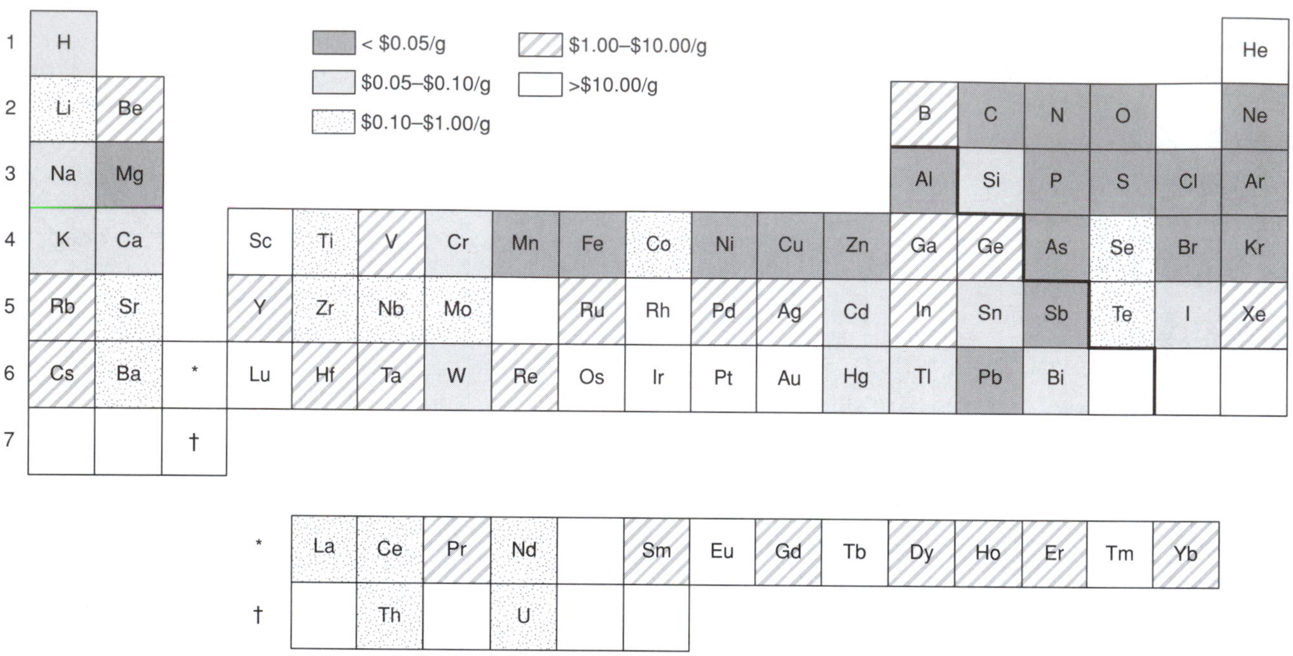

< \$0.05/g
\$0.05–\$0.10/g
\$0.10–\$1.00/g
\$1.00–\$10.00/g
>\$10.00/g

1	H																		He
2	Li	Be												B	C	N	O		Ne
3	Na	Mg												Al	Si	P	S	Cl	Ar
4	K	Ca		Sc	Ti	V	Cr	Mn	Fe	Co	Ni	Cu	Zn	Ga	Ge	As	Se	Br	Kr
5	Rb	Sr		Y	Zr	Nb	Mo		Ru	Rh	Pd	Ag	Cd	In	Sn	Sb	Te	I	Xe
6	Cs	Ba	*	Lu	Hf	Ta	W	Re	Os	Ir	Pt	Au	Hg	Tl	Pb	Bi			
7			†																

*	La	Ce	Pr	Nd		Sm	Eu	Gd	Tb	Dy	Ho	Er	Tm	Yb
†		Th		U										

[a]Costs are for research grade materials in small quantitites and are taken from the *Alfa Catalog, 1983–1984*, Alfa Products Division of Morton Thiokol, Inc., Danvers, MA; 1983.

ferent elements. Some of these elements are mined in only a few countries that may be politically unstable; for example, most of the world's cobalt comes from the Katanga province of Zaire, which has been shut off from the rest of the world more than once by secessionist wars or invasions. Consequently, the price of cobalt fluctuated wildly during these times. Even in the absence of such sudden interruptions, many people expect producer countries to attempt to put together cartels in the future to attempt to increase the prices for these critical materials.[35] Then the problem of substitution of materials in industrial processes may suddenly become critical ones for inorganic chemists and materials scientists. Finding substitutes becomes a problem too when a material such as mercury or asbestos is found to cause such severe environmental problems that it must be replaced.

It is worth noting also that the problem of cost is much more critical for an element and its compounds if it is used in very large quantities in the production of basic industrial inorganic chemicals (i.e., those in the table included in the exercises); there are often not many economically viable possibilities for substitution. But the chemical industry is also involved in the production of *specialty* chemicals, which are produced in small quantities at high prices for uses the importance of which cannot be measured by tons of output. (The solid-state semiconductor industry is much more important than would be guessed by the number of tons of microchips produced.) So the chemistry of rare elements can indeed be very important. As an admittedly extreme example, the unstable radioactive element technetium, which virtually does not even exist naturally on the earth, is nonetheless important in the radiopharmaceutical industry.

Finally, we should note that, most frequently, the inorganic chemical industry needs, not elements, but their compounds, so that the expensive step of reduction (or oxidation) can be avoided. Table 6.6 lists the inorganic chemicals found in the *Chemical & Engineering News* annual listing of the top 50 in terms of total tonnage.

Table 6.6
Inorganic Chemicals among the Top 50 in U.S. Production in 1995[a]

Rank	Substance	Production (billions of pounds)
1	Sulfuric acid	95.36
2	Nitrogen	68.04
3	Oxygen	53.48
5	Calcium oxide (lime)	41.23
6	Ammonia	35.60
7	Phosphoric acid	26.19
8	Sodium hydroxide	26.19
10	Chlorine	25.09
11	Sodium carbonate	22.28
14	Nitric acid	17.24
15	Ammonium nitrate	15.99
17	Urea	15.59
22	Carbon dioxide	10.89
27	Hydrochloric acid	7.33
31	Ammonium sulfate	5.24
37	Carbon black	3.32
39	Potash	3.22
42	Titanium dioxide	2.77
45	Aluminum sulfate	2.41
46	Sodium silicate	2.25
	Total inorganics	464.10
	Total organics	285.89
	Grand total	749.99

SOURCE: Data from *Chem. Eng. News*, April 8, 1996, p. 17.
[a] Does not include the production of metals.

6.9 Effects of Concentration and pH on Redox Chemistry: Pourbaix Diagrams

In Chapter 2, we visually represented acid–base equilibria by horizontal predominance diagrams; in this chapter, we have represented redox equilibria by vertical predominance diagrams. As was implicitly suggested by Usanovich in his acid–base definition, however, these two broad types of chemical equilibria are not as cleanly distinct from each other as we have made them out to be (acid–base chemistry in Chapters 2–4, redox chemistry now). The predominant form of an element in the environment (in soil or natural waters) will be influenced by *both* types of chemical equilibria simultaneously. Hence, in applied work it is necessary to have a visual representation of the effects of both redox potential and pH on the predominant form of an element. Such a diagram, having these two variables as its two axes, is known variously as a **Pourbaix diagram**,[36] a predominance-area diagram, a $E°$–pH diagram, or a pE–pH diagram.[37]

In a sense, a Pourbaix diagram is a merger of the acid–base and redox predominance diagrams that we have been using so far. Figure 6.20(*a*) shows the redox predominance diagram for iron; this matches exactly the Pourbaix diagram if the pH is held at zero. In both types of diagram, redox reactions involve vertical motion (from high oxidation states and potentials at the top to low oxidation states and potentials at the bottom); crossing a horizontal (double-line) boundary such as the one at 0.77 V in either diagram means that most of the iron has undergone a redox reaction:

$$Fe^{3+} + e^- \rightarrow Fe^{2+} \qquad (6.44)$$

The acid–base predominance diagrams for Fe(III) and Fe(II), respectively, are reproduced at the bottom of Figure 6.20. These *approximately* duplicate the part of the Pourbaix diagram in which pH is a variable but the potential is held constant at +0.78 V for Fe(III) and −0.43 V for Fe(II), respectively. In both types of diagram, motion across a vertical (single-line) boundary [as at pH 2 for Fe(III)] corresponds to an acid–base reaction of the iron:

$$Fe^{3+} + 3\,OH^- \rightarrow Fe(OH)_3(s) \qquad (6.45)$$

Pourbaix diagrams are more than the sum of the separate types of predominance diagrams, however. Many of the boundary lines in Figure 6.20 are *diagonal.* During a change of pH and/or potential that crosses one of these lines, a reaction occurs that is *both an acid–base and a redox reaction.* For example, the reaction by which FeO_4^{2-} is converted to Fe^{3+} (top of Pourbaix diagram, between pH 0 and 2) involves *both electrons and hydrogen ions:*

$$FeO_4^{2-} + 3\,e^- + 8\,H^+ = Fe^{3+} + 4\,H_2O \qquad (6.46)$$

It follows from LeChatelier's principle that the position of this equilibrium, hence its *redox potential, depends on the pH.* The appearance of H^+ (or OH^-) as a reactant or product is characteristic of most half-reactions involving hydrides, oxo anions, oxo acids, and oxides and hydroxides of the elements. The potentials for these species can be *strongly* influenced by the pH.

The effect of nonstandard concentrations and activities (as of H^+) on potential at room temperature is given by the well-known Nernst equation (6.23). In the case of half-reaction (6.46), the potential is

$$\begin{aligned} E &= 2.20 - \frac{0.059}{3}\log\frac{[Fe^{3+}]}{[FeO_4^{2-}][H^+]^8} \\ &= 2.20 - 0.157\,\text{pH} - 0.020\log\frac{[Fe^{3+}]}{[FeO_4^{2-}]} \end{aligned} \qquad (6.47)$$

The very high half-cell potential for ferrate ion means that this ion is quite unstable under acidic conditions. Removing hydrogen ion from the equilibrium (6.46) by making the solution strongly basic strongly shifts the equilibrium to the left and increases the stability of ferrate ion. At pH 14 the potential is 0.00 V minus the last

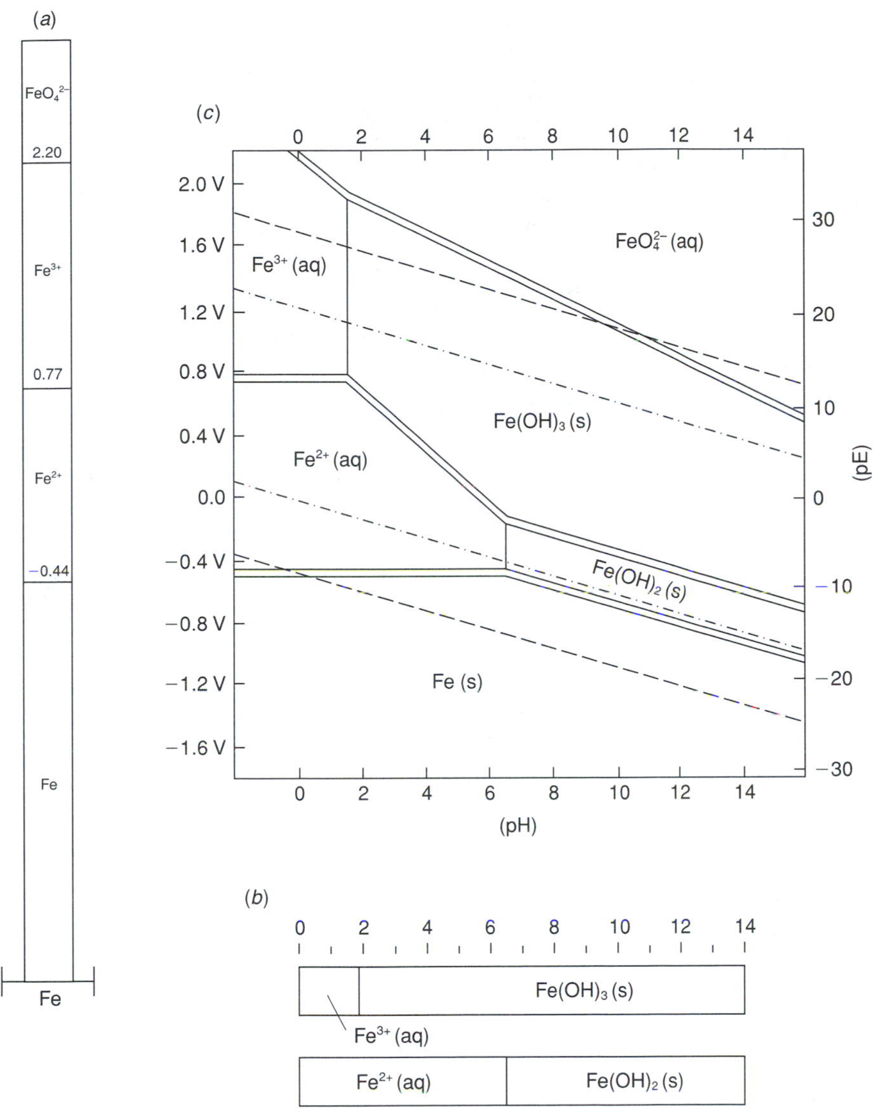

Figure 6.20
(*a*) At left, the redox predominance diagram for iron; (*b*) below, the acid–base predominance diagrams for Fe(III) and Fe(II); and (c) above right, the Pourbaix diagram showing the predominant form of iron at any given combination of redox potential and pH for 1 *M* iron solutions. Solid single lines (——) separate species related by acid–base equilibria; solid double lines (══) separate species in different oxidation states. Dash–dotted lines (—·—·—) enclose the theoretical and dashed lines (- - - -) enclose the practical region of stability of the water to oxidation or to reduction.

term of Eq. (6.47)—but as there can be very little Fe^{3+} ion in solution at pH 14, the potential has dropped all the way to 0.72 V for 1 *M* ferrate ion.

The half-reactions involving the oxidation and the reduction of water [Eq. (6.9) and (6.5), Section 6.1] are both pH dependent. After applying the Nernst equation and using the definition of pH, it is found that the potentials for these two half-reactions are (for all conditions standard except the concentration of H^+):

$$E^\circ \text{ for half-reaction (6.9)} = +1.229\,\text{V} - 0.059\,\text{pH} \tag{6.48}$$

$$E^\circ \text{ for half-reaction (6.5)} = -0.059\,\text{pH} \tag{6.49}$$

The potential for half-reaction (6.9) gives the theoretical and long-term upper boundary for the predominance area of water, shown in Figure 6.20(*c*) as the upper diagonal dash–dotted line. The potential for half-reaction (6.5) gives the lower theoretical and long-term boundary for the predominance area of water, represented by the lower diagonal dash–dotted line. The region between these two lines is sometimes called the **window of** (redox) **stability of water**.

Since both of these half-reactions involve forming diatomic gases (O_2 and H_2, respectively), overpotentials are significant for both boundaries of water's window of stability. Adding in 0.6 V to either boundary gives the diagonal dashed boundary lines of Figure 6.20. It may be seen that the predominance area of $FeO_4{}^{2-}$ falls within this expanded window of stability of water above a pH of 10. This means that a 1 *M* solution of ferrate ion may be kinetically stable in water in strongly basic solutions for short periods of time. In fact, it can be prepared and used under these conditions if it is used quickly, but since it does not fall within the narrower window of stability of water, the ferrate ion will eventually oxidize the water to oxygen:

$$4\,FeO_4{}^{2-} + 10\,H_2O \rightarrow 4\,Fe(OH)_3 + 3\,O_2 + 8\,OH^- \tag{6.50}$$

Changes in pH can thus alter the results of redox reactions very significantly, and change the stability of species in either direction. Since most practical chemistry occurs at pH value that are not close to zero, there is a definite need for Pourbaix diagrams.

Interpretation of Pourbaix Diagrams. In a Pourbaix diagram, the separate conventions of the acid–base and the redox predominance diagrams both hold:

Strong *oxidizing agents* are confined to the *top* of the diagram.

Strong *reducing agents* are confined to the *bottom* of the diagram.

Strong *acids* are confined to the *left* of the diagram.

Strong *bases* are confined to the *right* of the diagram.

Strong oxidizing agents and reducing agents tend to react: so species that do *not* overlap in a vertical sense will tend to undergo redox reactions with each other. If the predominance areas fail to overlap by more than 1 V, the reaction will be very exothermic, and may even produce an explosion.

Species that predominate over wide areas have feeble or no redox or acid–base properties.

Any given (well-mixed) solution can be characterized by a particular *E* and a particular pH. Finding this point on the Pourbaix diagram of an element will give us the thermodynamically most stable (in principle, most abundant) form of that element at that *E* and pH. For example, we can see from Figure 6.20(*c*) that, at a reduction potential of +0.8 V and a pH of 14, the predominant form of iron in a solution with a total of 1 mol of iron present per liter is FeO_4^{2-}.

Example 6.7

On the Pourbaix diagram for iron find (a) the chemical form of iron that is the strongest oxidizing agent; (b) the form of iron that is the strongest reducing agent; (c) the form of iron that would predominate in a neutral solution at a potential of 0.00 V; (d) the standard reduction potential for FeO_4^{2-} being reduced to Fe^{3+}; (e) the standard reduction potential for Fe^{2+} being reduced to Fe metal, and (f) the acid–base conditions under which Fe metal will most vigorously reduce water.

SOLUTION:

(a) The most strongly oxidizing form of iron shown in Figure 6.20(*c*) is the form with the highest lower boundary to its predominance area, the ferrate ion FeO_4^{2-}. This form contains iron with its highest oxidation number, +6.

(b,f) Iron metal has the lowest upper boundary on the Pourbaix diagram and is the most strongly reducing form of iron shown in the Pourbaix diagram. In strongly acidic solution, its predominance area does not overlap even the long-term predominance area of water, so iron metal is expected to react most vigorously with water at low pH value.

(c) A line drawn straight up at pH 7 intersects a line drawn across at a potential of 0.00 V in the area labeled "$Fe(OH)_3(s)$". Under these conditions, this is the most abundant form of iron present. Since we are close to the regions labeled "$Fe^{2+}(aq)$" and $Fe(OH)_2(s)$, at equilibrium lesser amouts of these will also be present.

(d), (e) Standard reduction potentials are found at pH 0, since 1 *M* is the standard concentration of H^+. The standard reduction potential for the reduction of ferrate ion to iron(III) ion is found at the intersection of the pH 0 vertical line with the boundary separating FeO_4^{2-} and Fe^{3+}. This is approximately at 2.2 V on the diagram. Similarly, the standard reduction potential for $Fe^{2+}(aq)$ being reduced to metallic iron is found at about −0.5 V.

Example 6.8

Without consulting the textbook diagrams:

(a) fill in the blanks in the Pourbaix diagram for plutonium (Fig. 6.21) with these species: PuO_2, PuO_3, Pu_2O_3, PuO_2^{2+}, Pu^{4+}, Pu^{3+}, and Pu.

(b) Which species is the strongest oxidizing agent? Realistically, is this species a strong enough oxidizing agent to oxidize the water it is dissolved in at a reasonable rate? (The boundaries for water shown in the diagram are the theoretical ones).

(c) Which species is the strongest reducing agent? Realistically, is this species a strong enough reducing agent to reduce the water it is dissolved in at a reasonable rate?

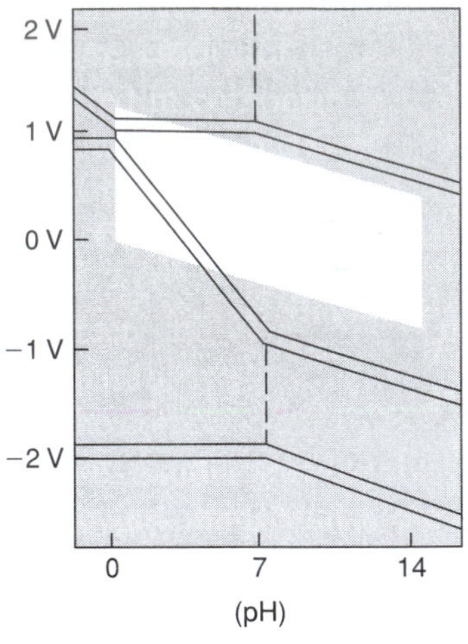

Figure 6.21
Blank Pourbaix diagram for use in Example 6.8.

SOLUTION:

(a) Plutonium in its highest oxidation state (VI, in PuO_2^{2+} and PuO_3) will go in the top predominance areas, above the top double line. Of these two forms of Pu(VI), the more acidic one, PuO_2^{2+}, will predominate at the left, and the less acidic or more basic one, PuO_3, will predominate at the right. The diagram is then filled in downward with lower and lower oxidation states of Pu, ending with Pu metal at the bottom.

(b) The strongest oxidizing agent would probably be taken with reference to the stability window of water, so PuO_3 would be chosen, although in an absolute sense PuO_2^{2+} has a higher $E°$ at most pH values. However, neither appears to be far enough outside the stability window of water to cause the water to be oxidized rapidly.

(c) The strongest reducing agent is Pu metal. It is below even an extended window of stability of water, so it should reduce water at a rapid rate.

Pourbaix diagrams for 1 *M* total concentrations of the elements in the periodic table are given in Figures 6.22–6.24. In these diagrams, only the long-term (thermodynamic) stability area for water is represented, as white windows.[38]

Note that the predominance area for iron(II) species [Fe^{2+} and $Fe(OH)_2$] narrows considerable at higher pH values. In some cases [e.g., chlorine(0), which appears as Cl_2 in Fig. 6.22] the predominance area for a given oxidation state may disappear completely above or below a given pH. When the predominance area of a given oxidation state disappears, that oxidation state loses its thermodynamic stability and is likely to undergo *disproportionation*. Thus Cl_2 is stable in water at pH < 0, but at higher pH values it undergoes disproportionation to chlorine in negative and positive oxidation states.

The positions of acid–base and redox equilibria are also dependent on the total concentration of iron. The 1 *M* total concentrations of the standard Pourbaix diagrams are inappropriate for use by geochemists, soil chemists, and water chemists. Figure 6.25 is a Pourbaix diagram for a 10^{-6} *M* total iron concentration. Comparison with Figure 6.20(*c*) shows that, for example, the right boundary for Fe^{3+} moves from a pH of about 1.7 to about 3.4. In general, increasing dilution enlarges the predominance area of *soluble* species. (This expansion can be thought of as an application of LeChatelier's principle.)

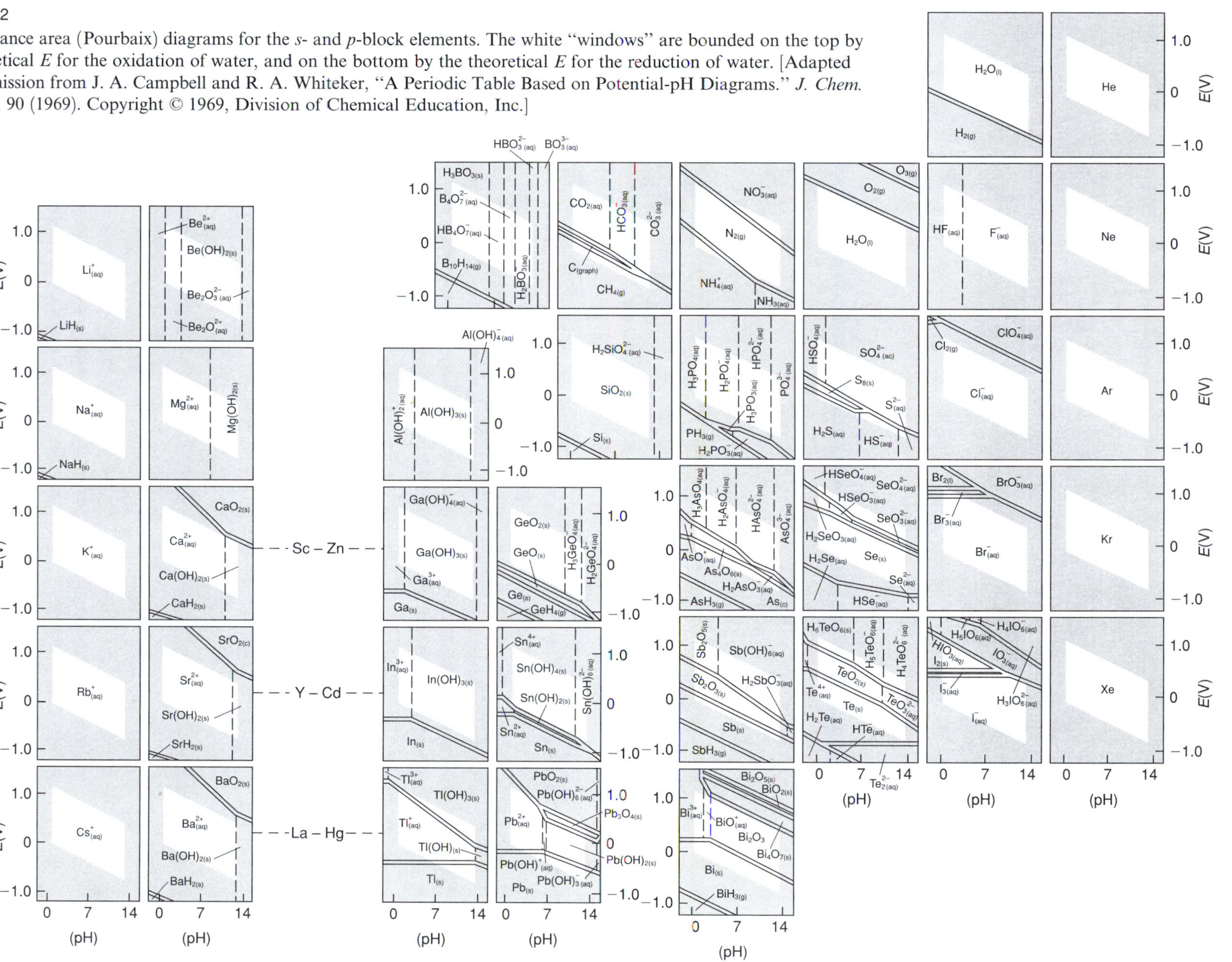

Figure 6.22

Predominance area (Pourbaix) diagrams for the *s*- and *p*-block elements. The white "windows" are bounded on the top by the theoretical *E* for the oxidation of water, and on the bottom by the theoretical *E* for the reduction of water. [Adapted with permission from J. A. Campbell and R. A. Whiteker, "A Periodic Table Based on Potential-pH Diagrams." *J. Chem. Educ.*, **46**, 90 (1969). Copyright © 1969, Division of Chemical Education, Inc.]

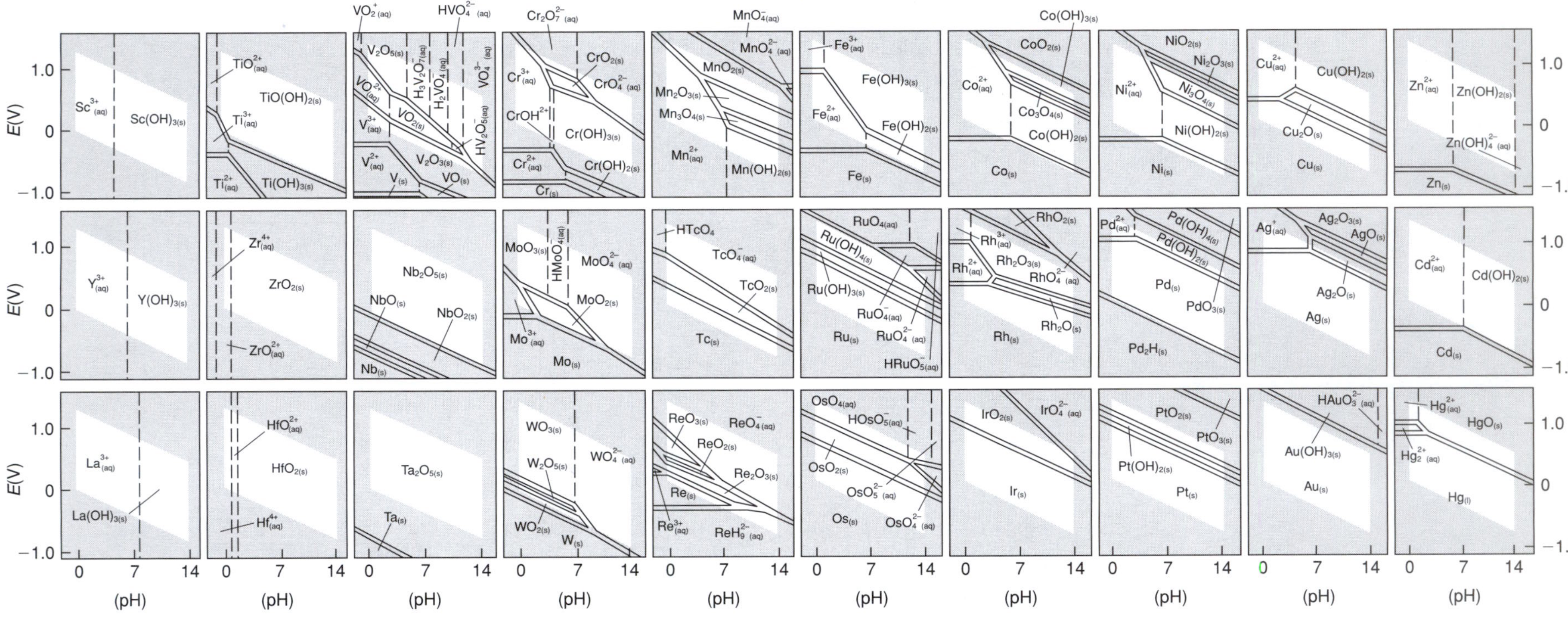

Figure 6.23
Predominance area (Pourbaix) diagrams for the *d*-block elements. The "window" boundaries are the same as those given in Figure 6.22 [Adapted with permission from J. A. Campbell and R. A. Whiteker, "A Periodic Table Based on Potential-pH Diagrams," *J. Chem. Educ.*, **46**, 90 (1969). Copyright © 1969, Division of Chemical Education, Inc.]

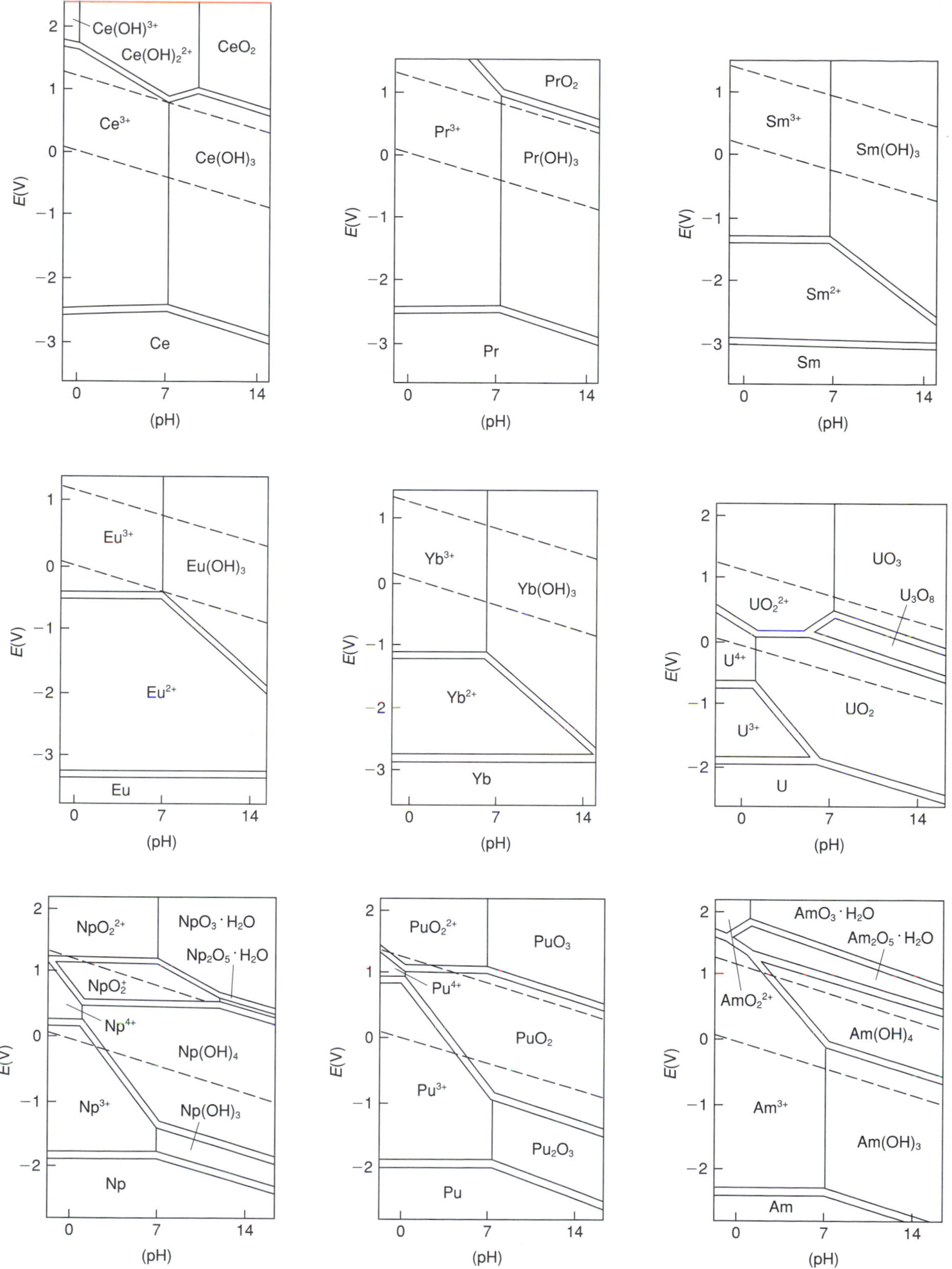

Figure 6.24

Predominance area (Pourbaix) diagrams for the *f*-block elements showing redox chemistry. [Adapted from M. Pourbaix, *Atlas of Electrochemical Equilibria in Aqueous Solutions*, National Association of Corrosion Engineers, Houston, 1974.]

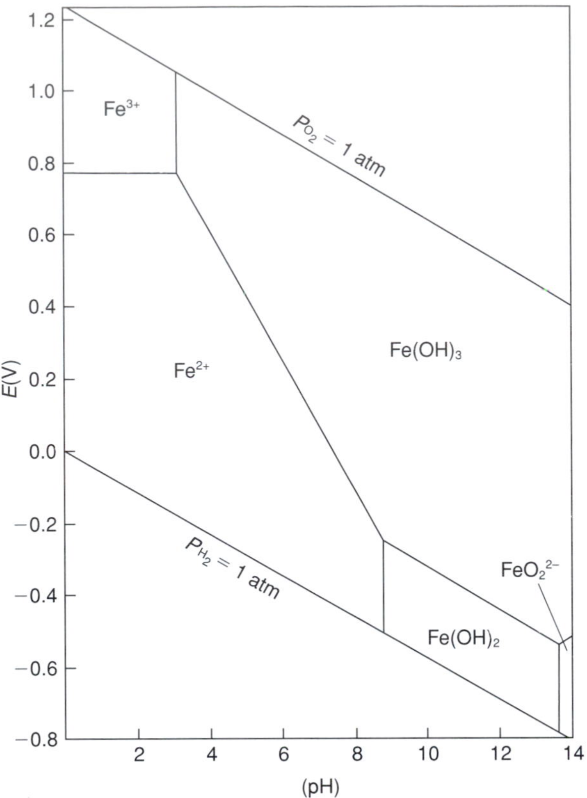

Figure 6.25
Pourbaix diagram for iron at a total activity of 10^{-6} *M* (only region within window of stability of water is shown.) [Adapted from D. G. Brookins, *Eh–pH Diagrams for Geochemistry*, Springer-Verlag, New York, 1988; p. 75.]

Pourbaix diagrams with appropriate total concentrations allow us to make more realistic predictions of the forms that the different elements will take in natural waters (Section 3.8). The surface waters of a clean lake are well aerated, and have dissolved oxygen concentrations that are high enough that their potentials are reasonably close to the oxygen standard reduction potential (Figure 6.26). In a lake that is highly polluted with organic reducing agents, in the bottom layer of a thermally stratified lake, or in a swamp, conditions may be not only quite anaerobic, but actively reducing, so that the lower boundary of reduction to H_2 may be approached. Under these conditions, many of the most common nutrient elements are converted to unfamiliar forms: By reference to Figure 6.22 we see that carbon may be converted to methane, CH_4, sometimes referred to as *swamp gas*. Nitrogen is reduced to the ammonium ion, NH_4^+, or to ammonia, NH_3. This latter compound does no good for the odor of the water, and is quite toxic. Sulfur is reduced to hydrogen sulfide, H_2S; emissions of this gas to the atmosphere are a major source of acid rain after the sulfur has been reoxidized. Given overpotentials, it is even possible for phosphorus to be reduced to low concentrations of phosphine, PH_3, a toxic, foul-smelling gas that commonly ignites upon exposure to air, giving rise to eerie light emission over swamps at night.

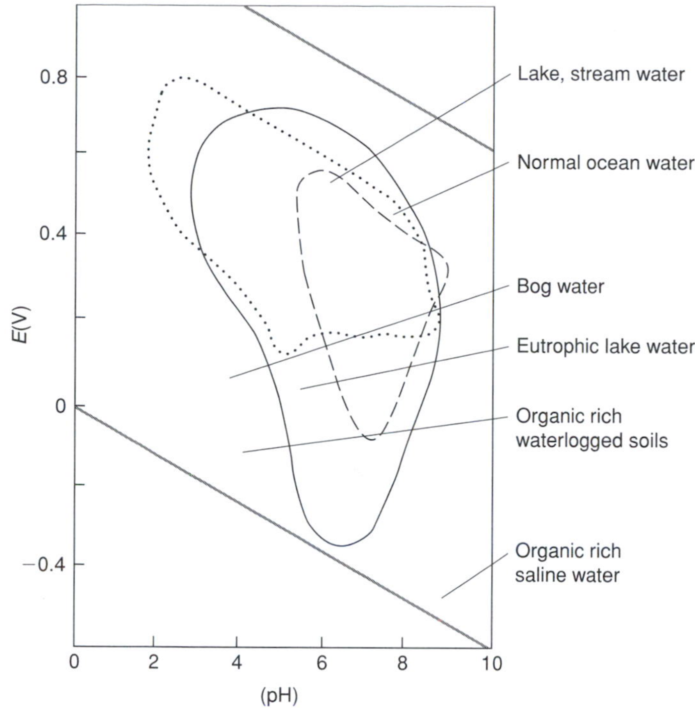

Figure 6.26
The approximate potential and pH values for some environmental waters, indicated by the lines shown or by the regions enclosed as follows: (——) soil water, (- - - - -) subsurface water, (· · · · ·) acid mine-drainage water. [Adapted from J. E. Fergusson, *Inorganic Chemistry and the Earth*, Copyright © 1982 by Pergamon Press.]

6.10 *pH Effects on Redox Chemistry: The Synthesis of Oxo Anions and Acids and Corrosion of Metals

In Section 6.9, we determined that the ferrate ion can have only short-term stability in water, and then only at high pH values. It will be noted that ferrate, *and most oxidizing oxo anions and oxo acids in Figures 6.22–6.24, have diagonal lower boundaries that slope downward at higher pH values* with a slope that is greater than that of the stability window of water. This means that, in general, the stability of oxidizing oxo anions and acids is enhanced in basic solutions. On the other hand, their oxidizing ability is enhanced in acidic solution.

Consequently, it is advantageous or even necessary to carry out the synthesis of a salt of an oxo anion or acid with a redox predominance region confined to *above +1.2 V* [39] (i.e., with a very highly oxidized central atom[40]) in a *basic* solution. To carry out the oxidation, an oxidizing agent that not only has a high redox predominance region but also involves few or no hydrogen ions in its reduction half-reaction is preferred; kinetic and economic factors also play a role. Table 6.7 lists some of the common strong oxidizing agents that may be chosen, along with their E° values under standard and standard basic (pH 14) conditions.

Thus a very oxidizing species such as the ferrate ion, which at pH 0 will oxidize the water it is dissolved in, is normally prepared in very alkaline solution, using chlorine,

Table 6.7
Standard Reduction Potentials of Some Useful Oxidizing Agents in Acidic and Basic Solution[a]

Oxidizing Agent and Half-Reactions	E°_{acid} (V)	E°_{base} (V)
Fluorine		
$F_2 + 2e^- = 2F^-$	+2.87	+2.87
Chlorine		
$Cl_2 + 2e^- = 2Cl^-$	+1.36	+1.36
Ozone		
$O_3 + 2e^- + 2H^+ = O_2 + H_2O$	+2.076	
$O_3 + 2e^- + H_2O = O_2 + 2OH^-$		+1.24
Hydrogen peroxide		
$H_2O_2 + 2e^- + 2H^+ = 2H_2O$	+1.776	
$HO_2^- + 2e^- + H_2O = 3OH^-$		+0.88
Oxygen		
$O_2 + 4e^- + 4H^+ = 2H_2O$	+1.229	
$O_2 + 4e^- + 2H_2O = 4OH^-$		+0.40
Nitric acid; nitrates		
$NO_3^- + e^- + 2H^+ = NO_2 + H_2O$	+0.79	
$NO_3^- + 3e^- + 2H_2O = NO + 4OH^-$		+0.427

SOURCES: Standard reduction potentials are taken or derived from data given in B. Douglas, D. H. McDaniel, and J. J. Alexander, *Concepts and Models of Inorganic Chemistry*, 2d ed., Wiley, New York, 1983.
[a] The pH is 0 and 14.

which is converted in part to hypochlorite ion in such solutions, as the oxidizing agent:

$$2\,Fe(OH)_3 + 3\,OCl^- + 4\,OH^- \rightarrow 2\,FeO_4{}^{2-} + 5\,H_2O + 3\,Cl^- \qquad (6.51)$$

To further assist in the synthesis, the ferrate ion can be precipitated by the use of an appropriate cation (Ba^{2+} in this case, since ferrate is feebly basic). This reduces the concentration of ferrate ion in solution, which [from Eq. (6.47)] further improves the cell potential. Periodate is similarly prepared by alkaline oxidation of iodate ion, and tellurate from tellurite.

Another useful oxidizing agent in such syntheses is hydrogen peroxide, H_2O_2. Although the reduction of H_2O_2 does involve hydrogen ion, it involves less than most oxo anions do, so basic conditions are still useful. Thus chromates may be synthesized as follows:

$$2\,Cr(OH)_3 + 3\,H_2O_2 + 4\,OH^- \rightarrow 2\,CrO_4{}^{2-} + 8\,H_2O \qquad (6.52)$$

Another powerful oxidizing agent, ozone (O_3) in basic solution, is similarly used to oxidize xenon trioxide to perxenate; fluorine gas is used to oxidize bromate ion to perbromate ion.

Difficult oxidations can often be carried out by heating solid reactants (oxide of the element, basic oxide, and an oxidizing agent such as sodium peroxide) until they melt and react:

$$Na_2O + 2\,Na_2O_2(s) + Bi_2O_3(s) \rightarrow 2\,Na_3BiO_4 \qquad (6.53)$$

Often KNO_3 may be heated with a reactant—the KNO_3 decomposes at high temperatures to liberate (in effect) O_2 and oxide ion, giving basic conditions that favor oxidation:

$$2\,KNO_3 + Se \rightarrow K_2SeO_4 + 2\,NO \qquad (6.54)$$

This reaction is similar to that involved in the detonation of gunpowder, in which saltpeter, KNO_3, is used to oxidize carbon and sulfur to their oxides. Similar reactions are used to prepare potassium manganate, K_2MnO_4, from MnO_2 and potassium ruthenate, K_2RuO_4, from ruthenium metal.

Often the oxygen of the air will suffice as the oxidizing agent:

$$10\,Li_2O + 4\,NpO_2 + 3\,O_2 \rightarrow 4\,Li_5NpO_6 \qquad (6.55)$$

Many oxo anions can also be prepared by electrolysis of aqueous solutions containing the elements in low oxidation states. Reduction of the water at the cathode evolves hydrogen and produces OH^- to make the solution basic; the oxo anion is produced at the anode. Hypochlorite ion, ClO^-, can be produced by electrolysis of cold dilute solutions of NaCl; further reaction yields chlorate ion, ClO_3^-; electrolysis of solutions of the chlorate ion yields the perchlorate ion. Permanganate ion is similarly produced by electrolysis of the manganate ion, MnO_4^{2-}.

If the oxo acid or anion has a standard reduction potential of about *+0.3 to +1.2 V* (i.e., the oxo anion does not involve too high an oxidation state), basic conditions may not be necessary. Strongly oxidizing oxo acids such as *nitric acid* ($E^\circ = +0.79\,V$) may then be used. In this manner, iodine is oxidized to iodic acid, selenium to selenous acid, As_2O_3 to arsenic acid, and technetium and rhenium compounds to $HTcO_4$ and $HReO_4$. Or it may be possible to oxidize the element in air (without base) to give its oxide, then react the oxide with water to give the oxo acid; nitric and sulfuric acids are prepared commercially in this manner.

If the desired species has a reduction potential *below* about *+0.2 V*, simpler syntheses (such as acid–base ones) may be possible, since the element may well occur in nature in that oxidation state. However, redox chemistry may be employed if one wishes. Very vigorous reactions of the common metals zinc and aluminum with base produce their hydroxo anions; the heat and hydrogen gas generated by this reaction are used to clean stopped-up drains. For this reason, household lye (NaOH) should never be used in aluminum pans!

If a desired oxo anion involves the element in an unusually *low oxidation state*, one might carry out an electrochemical *reduction* of a higher oxo anion, or use a chemical reducing agent [e.g., the synthesis of nitrite ion, Eq. (6.1)]. Dissolution of the element itself in a strongly basic solution may be appropriate, as in the synthesis of hypophosphorous acid:

$$P_4 + 4\,OH^- + 4\,H_2O \rightarrow 4\,[H_2PO_2]^- + 2\,H_2 \qquad (6.56)$$

Similar to this reaction is the preparation of sodium hypochlorite (household bleach) by disproportionation of chlorine gas:

$$Cl_2 + 2\,OH^- = ClO^- + Cl^- + H_2O \qquad (6.57)$$

In this case, chloride ion is an inevitable byproduct; in acidic solution the above reaction is not favored, as the chloride and hypochlorite ions react to give chlorine gas back. For this reason, household bleach should never be combined with any acidic cleansing agents.

As we have seen, the synthetic inorganic or industrial chemist has a variety of methods at his or her disposal to prepare oxo acids and anions. Although many of these can be purchased, some cannot due to their instability. Even some of those that can be purchased may deteriorate during their storage or transportation; hence, synthesis may be necessary. Useful sources of synthetic details include the serial *Inorganic Syntheses,* which tends to focus on more exotic chemicals; for simpler compounds such as oxo acids and their salts the work by Brauer[41] is probably more useful.

Positions of equilibria such as that of Eq. (6.46) are shifted to the left at low pH, which enhances the oxidizing abilities of oxidizing oxo anions or acids. Hence, *oxidizing reactions of oxo anions or acids are generally carried out in acidic solution.* Many oxidizing oxo anions find use in organic chemistry to carry out different sorts of oxidations, but these are beyond the scope of this book. In analytical chemistry, dichromate and permanganate are often used to determine the concentration of various reducing agents by titration in acidic solutions; other oxidizing agents such as bismuthate and perchlorate are used in preparing samples for titration. Overall, nitric acid is undoubtedly most frequently used (among the oxo acids) for oxidative purposes.

Corrosion. The surfaces of many metals generally tarnish and become covered with a surface layer of oxide, which is often tough enough and insoluble enough to protect them against further oxidation, although thermodynamically the production of oxide is favored. [This is the reason why the aluminum metal used in Experiment A.6 had to be cleaned before its activity could be determined.] In contact with air and water the process of **corrosion**[42] may occur, in which the metal acts as an anode and is oxidized

$$Fe(s) = Fe^{2+} + 2\,e^- \tag{6.58}$$

Some other point on the surface of the metal, or the junction of the metal with a more electronegative metal, acts as an cathode, at which (typically) oxygen is reduced to hydroxide ion. The electrochemical circuit is completed by migration of hydroxide ions (or other ions—salt water assists corrosion strongly) back to the Fe^{2+} ions to precipitate $Fe(OH)_2$. The $Fe(OH)_2$ further oxidizes and dehydrates to give hydrated iron(III) oxide, the familiar rust.

The prevention of corrosion is economically a very important activity; it has been estimated that as much as one-half of all the metal that has been produced by humans has corroded away. Since the mechanism of corrosion involves oxidation of iron to Fe^{2+}, corrosion of iron occurs principally at pH and E values within the predominance area of Fe^{2+} (Fig. 6.20). Corrosion does not occur in adequately basic solutions due to the coating of the iron with FeO [or $Fe(OH)_2$]. If the E is below the standard reduction potential of iron, the iron is stable and will not rust. This reduction of the potential may be accomplished by coating or attaching a more active metal (such as Zn or Mg) to the iron; the more active metal will then corrode instead. Or the metal may actually be **passivated** by the use of a strong oxidizing agent (such as dichromate or pertechnetate or sometimes even nitric acid), since that coats the iron with an impervious layer of a very insoluble metal oxide (Fe_2O_3 for iron). (In some of these cases, we would have predicted very vigorous reactions indeed with these strong oxidizing agents!) The great corrosion resistance of stainless steel is due to its high (11% or greater) chromium content; this oxidizes on the surface to give an exceptionally insoluble oxide, $FeCr_2O_4$. The HSAB principle (Section 5.3) is also useful in designing coatings to prevent corrosion[43]; its use in Section 6.4 to choose conditions to promote activity of metals could be reviewed now, since these conditions are those which *promote* corrosion.

Study Objectives

1. Write balanced oxidation–reduction half-reactions for species in redox predominance diagrams. From their locations in redox predominance diagrams, identify oxidizing and reducing agents and rank them in order of strength. Exercises 1–11.

2. From their locations in redox predominance diagrams, tell whether two species (or one species plus solvent water) should undergo a redox reaction; if so, identify possible products, and compute the standard cell voltage for the expected reaction. Balance the equation for the expected redox reaction. Exercises 12–17.

3. Recognize compounds likely to be inflammable, oxidizing, or explosive; know appropriate safety precautions to take in handling these compounds. Exercises 18–27.

4. From tables of standard reduction potentials for an element, (a) construct a redox predominance diagram for that element; (b) tell whether a given species is thermodynamically unstable or metastable, and what may become of it (i.e., disproportionate). Exercises 27–30.

5. Select halides or sulfides, and so on, which are most likely to be nonexistent or intensely colored; explain the source of the color. Exercises 31, 32.

6. Rank a group of metals into an activity series based on their Pauling electronegativities; predict the results of the reaction of any metal with H_2O, HCl, or HNO_3. Exercises 33–38.

7. Know how redox reactions can be influenced by formation of complex ions involving a metal ion with hard or soft ligands. Exercises 39–43.

8. Know the general periodic trends in the stability of high oxidation states of elements; classify high-oxidation-state oxo acids and anions as to relative strength as oxidizing agents. Exercises 44–45.

9. Classify nonmetallic elements, their anions and hydrides, and pseudohalide ions as to relative strength as oxidizing and reducing agents. Exercises 46–52.

10. Give a feasible and economically preferable method of preparing any element from the form in which it occurs in nature. Exercises 53–63.

11. Construct or interpret Pourbaix diagrams for thermodynamically stable species; use them to classify and rank these species as oxidizing or reducing agents. Use Pourbaix diagrams to identify forms of an element that will oxidize or reduce water at high or low pH values. Suggest the forms in which given elements will occur in natural waters. Exercises 64–71.

12. Suggest pH conditions and methods of synthesizing given oxo acids or oxo anions; of corroding or passivating a metal. Exercises 72–81.

Exercises

1. The following several examples are reductions of an element. In each case: (a) identify the oxidation numbers of the element in the reactant and then in the product; (b) tell in which block in the redox predominance diagram below each compound should be located; (c) complete and balance the reduction half-reactions (using H^+, in acid solution).

Diagram	$WO_3 \rightarrow WO_2$		$H_2MoO_4 \rightarrow MoO_2$		$Cr_2O_7^{2-} \rightarrow Cr^{3+}$	
	$E° = -0.03$ V	A	$E° = +0.65$ V	C	$E° = +1.38$ V	E
		B		D		F

2. The following several examples are reductions of an element. In each case, first identify the oxidation states of the element in the reactant and then in the product; second, complete and balance the reduction half-reaction (using H^+, in acid solution); an finally find the standard reduction potential for the half-reaction from the redox predominance diagrams (Figs. 6.2–6.5):

(a) $MnO_4^- \rightarrow MnO_2$
(b) $TcO_4^- \rightarrow TcO_2$
(c) $ReO_4^- \rightarrow ReO_3$
(d) $GeO_2 \rightarrow Ge$
(e) $H_3AsO_4 \rightarrow As_4O_6$
(f) $SeO_4^{2-} \rightarrow H_2SeO_3$
(g) $BrO_4^- \rightarrow BrO_3^-$

3. *According to redox predominance or Pourbaix diagrams: (a) What is the most oxidizing chemical form of cobalt? (b) What is the most strongly reducing form of selenium in acidic solutions? (c) Which +2 ion of the *f*-block metals from the sixth period is least strongly reducing? (d) What is the most strongly oxidizing form of iridium? (e) Which MO_2^{2+} ion of the *f*-block metals from the seventh period is most strongly oxidizing? (f) Which elements from the periodic table have only one stable oxidation state in aqueous solution?

4. *According to redox predominance or Pourbaix diagrams, for which of the following fourth period elements is the group oxidation number most strongly oxidizing (least stable): Zn, Ga, Ge, As, Se, on Br?

5. *According to redox predominance or Pourbaix diagrams, for which Group 5, 15(V), or 5*F* element is the +5 oxidation state: (a) least oxidizing (most stable) and (b) most oxidizing (least stable)?

6. According to redox predominance or Pourbaix diagrams, for which Group 7, 17(VII), or 7*F* element is the +7 oxidation state: (a) least oxidizing (most stable) and (b) most oxidizing (least stable)?

7. Using redox predominance or Pourbaix diagrams, choose the best oxidizing agent in each set: (a) MnO_4^-, TcO_4^-, ReO_4^-; (b) GeO_2, H_3AsO_4, SeO_4^{2-}, BrO_4^-; (c) SO_4^{2-}, SeO_4^{2-}, H_6TeO_6, PoO_3; and (d) $Cr_2O_7^{2-}$, H_2MoO_4, WO_3.

8. For which Group 15(V) element X should the following reaction proceed most readily: $4\,X + 5O_2 + 12\,OH^- \rightarrow 4\,XO_4^{3-} + 6\,H_2O$

9. Which species in each set is least likely to oxidize something (i.e., is most stable): (a) FeO_4 (recently discovered—see Chapter 13), RuO_4, OsO_4; (b) NO_3^-, H_3PO_4, H_3AsO_4, H_7SbO_6, H_7BiO_6; (c) H_7SbO_6, H_6TeO_6, H_5IO_6, H_4XeO_6; (d) H_4GeO_4, H_3AsO_4, SeO_4^{2-}, BrO_4^-; and (e) SO_4^{2-}, SeO_4^{2-}, H_6TeO_6, H_6PoO_6.

10. (a) Identify the most oxidizing chemical form of manganese. (b) Identify the most reducing form of manganese. (c) Write the balanced half-reaction for the reduction process that occurs in the manganese redox predominance diagram at -1.18 V. (d) Write the balanced half-reaction for the reduction process that occurs in the manganese redox predominance diagram at $+1.70$ V.

11. (a) Write the balanced half-reaction for the reduction process that occurs in the nitrogen redox predominance diagram at $+1.25$ V. (b) Identify the most oxidizing chemical form of nitrogen. (c) Identify the most reducing form of nitrogen.

12. Using your results from Exercise 2, compute the cell potentials $E°$ for the following redox reactions, and tell whether they will go to give products (support your answer either by computation or using the redox predominance diagrams):
(a) $MnO_4^- + TcO_2 \rightarrow MnO_2 + TcO_4^-$
(b) $MnO_4^- + ReO_3 \rightarrow MnO_2 + ReO_4^-$
(c) $As_4O_6 + GeO_2 \rightarrow Ge + H_3AsO_4$
(d) $H_2SeO_3 + BrO_4^- \rightarrow BrO_3^- + H_2SeO_4$

13. Using your results and the predominance diagrams from Exercise 1, (1) balance the following redox reactions, (2) compute their cell potentials $E°$, and (3) tell whether they will go to give products:
(a) $H_2MoO_4 + WO_2 \rightarrow MoO_2 + WO_3$
(b) $Cr_2O_7^{2-} + WO_2 \rightarrow WO_3 + Cr^{3+}$.

14. *Which of the following elements—Zn, Cl_2, At_2, Au, Bi, Sc, Hg, Tm, Os—will react (a) with a solution of I^- to generate I_2; (b) with a solution of Ag^+ to generate Ag; and (c) with a solution of Na^+ to generate Na?

15. Which of the following elements—O_2, Rb, Sb, Pt, Y, Sm, Mn, Rh, Sn, Li—will react (a) with a solution of I^- to generate I_2; (b) with a solution of Ag^+ to generate Ag; and (c) with a solution of Na^+ to generate Na?

16. (a) What are the likely thermodynamically stable products of the redox reaction between NH_4^+ and IO_3^- in aqueous acid? (b) Write a balanced redox equation for this predicted reaction. (c) Compute the emf for this predicted reaction.

17. *(a) List two chemical forms of nitrogen that theoretically should oxidize Mn to Mn^{2+}. (b) Balance the equation for the redox reaction that should occur in acid solution between one of these forms of nitrogen (your choice) and Mn, yielding Mn^{2+} and a likely nitrogen-containing product. (c) List two chemical forms of manganese that should oxidize NH_4^+ to N_2. (d) Balance the equations for the two

redox reactions that should occur between each of these two chemical forms of Mn and NH_4^+ to give N_2 and a Mn-containing product.

18. *For each of the following salts, tell whether (1) it will be thermodynamically stable in the sense that its cations and anions will not undergo redox reactions with each other; (2) it will be so unstable that, if it can be made at all, it will be potentially explosive, or (3) its ions may react with each other, but probably not explosively. (a) $CrFeO_4$; (b) $CsMnO_4$; (c) NH_4MnO_4; (d) $EuMnO_4$; (e) Ca_2C_2; (f) Ag_2C_2; (g)$Cr(BrO_3)_2$; and (h) $Tl(N_3)_3$.

19. Using redox predominance (or Pourbaix) diagrams for the appropriate elements, tell which of the following compounds or mixtures are likely to be thermodynamically stable. If they are not stable compounds or combinations, which are likely to be potentially explosive? (a)$Ni^{2+} + NO_3^-$; (b) $CH_4 + F_2$; (c) $PH_3 + ClO_4^-$; (d) $NH_4^+ + Br^-$; (e) $Mn^{2+} + H_4XeO_6$; (f) $NH_4^+ + H_3PO_4$; (g) $Ag^+ + N_3^-$; (h) $CH_4 + H_4XeO_6$; (i) $Se + O_2$; and (j) $Hg^{2+} + ClO_4^-$.

20. Refer to the redox predominance diagrams for N, I, Co, and Cr. (a) Which of the following salts are expected to be nonexplosive and thermodynamically stable to redox reactions: $Co(NO_3)_3$; NH_4I; $(NH_4)_2Cr_2O_7$; and $Cr(IO_3)_2$? (b) Which of the following mixtures are potentially explosive: HNO_3 and I_2; Cr and I_2; $K_2Cr_2O_7$ and Cr; and NH_4I and CrI_2?

21. Use redox predominance or Pourbaix diagrams to answer the following questions: (a) Which of the following are expected to be nonexplosive and thermodynamically stable to redox reactions: $AgIO_3$, NH_4Br, NH_4Br_3, $TlReO_4$, and $FeBr_3$? (b) Which of the following mixtures are potentially explosive: HNO_3 and BiH_3, NH_3 and H_2S, $HClO_4$ and SiO_2, Pu^{4+} and H_2Te; Yb^{2+} and H_2Te? (c) Write the formula of the salt containing Se and Fe that is most likely to be unstable (explosive or nonexistent).

22. *The cations of silver show the following standard reduction potentials:

$$Ag^+ + e^- = Ag(s) \qquad E^\circ = +0.799\ V$$

$$Ag^{2+} + e^- = Ag^+ \qquad E^\circ = +1.98\ V$$

(a) Is the Ag^+ ion more likely to act as a good oxidizing agent or a good reducing agent? (b) Which one of the following is most likely to be capable of converting Ag^+ to Ag^{2+}: F_2, F^-, I_2, or I^-? (c) Given that the BrO_4^- ion has a standard reduction potential of + 1.743 V (when being reduced to BrO_3^-), is $AgBrO_4$, if it exists, likely to be explosive? (d) Given that the standard reduction potential for the reaction $H_2 + 2\,e^- = 2\,H^-$ is -2.25 V, is AgH, if it exists, likely to be explosive?

23. Give one example of an explosive (or potentially explosive) compound in each of the following classes: (a) a compound containing both strongly oxidizing and strongly reducing functional groups; (b) a salt containing a strongly oxidizing cation and a strongly reducing anion; (c) a compound containing unstable covalent bonds; (d) a salt containing a strongly reducing cation and a strongly oxidizing anion.

24. Which would most likely be an explosive compound: (a) methyl cyanate, CH_3OCN; (b) methyl isocyanate, CH_3NCO; or (c) methyl fulminate, CH_3ONC.

25. *Use the redox predominance or Pourbaix diagrams for manganese and nitrogen to answer these questions. (a) Which two chemical forms of nitrogen should oxidize Mn to Mn^{2+}? (b) Which chemical forms of manganese should oxidize NH_4^+ to N_2? (c) Ammonium permanganate, NH_4MnO_4, is a known salt. Comment on its expected solubility or insolubility, and on its expected stability upon being struck with a hammer. (d) We can write a hypothetical reaction for the decomposition of this salt: $NH_4MnO_4 \rightarrow 2\,H_2O + MnO_2 + 0.5\,N_2$. Would these three products likely coexist with each other, or would you expect them (if formed at all) to react further with each other? (e) Ammonium manganate, $(NH_4)_2MnO_4$, is not a well-known salt. Comment on its expected solubility or insolubility, and on its expected stability upon being struck with a hammer. (f) We can write a hypothetical reaction for the decomposition of this salt: $(NH_4)_2MnO_4 \rightarrow 4\,H_2O + Mn + N_2$. According to the Pourbaix diagrams, Would these three products likely coexist with each other, or would you expect them (if formed at all) to react further with each other? (g) Comment on the likely safety aspects of grinding together, in a mortar and pestle, finely divided: (g1) potassium nitrate with manganese metal; (g2) potassium nitrate with manganese dioxide.

26. Use the redox predominance or Pourbaix diagrams of the three nonmetals Br, N, and Sb to answer the following questions. (a) List the anions and hydrides of these elements (NH_4^+, Br^-, SbH_3) in order of increasing activity as reducing agents. (b) List the highest oxo acids or anions of these elements (NO_3^-, BrO_3^-, $H_7Sb(OH)_6$) in order of increasing activity as oxidizing agents. (c) List the elements themselves (Br_2, N_2, and Sb) in order of increasing activity as oxidizing agents. (d) Of the nine species mentioned in the part (c) of this question, the redox reaction of which two species would likely be most exothermic, and therefore might present the greatest risk of an explosion? (e) Which one of these elements occurs native (uncombined) in nature? Which one of these elements would likely be prepared by reduction of an oxide with carbon? Which would likely be prepared by chemical or electrochemical oxidation?

27. (a) Draw a redox predominance diagram for the element berkelium (Bk), given the following standard reduction potentials:

$$Bk^{3+} + 3\,e^- \rightleftharpoons Bk(s) \qquad E^\circ = -1.97\,V$$

$$Bk^{4+} + e^- \rightleftharpoons Bk^{3+} \qquad E^\circ = +1.64V$$

(b) Briefly explain why the species Bk^{2+} cannot be placed successfully on the predominance diagram, given the following standard reduction potentials:

$$Bk^{4+} + 2\,e^- \rightleftharpoons Bk(s) \qquad E^\circ = -1.6V$$

$$Bk^{3+} + e^- \rightleftharpoons Bk^{2+} \qquad E^\circ = -2.8V$$

(c) Will berkelium metal react with water? (d) Suppose berkelium becomes a water pollution problem. In what oxidation state would you expect to find it in contact with natural waters?

28. The first manned mission to Pluto returns to Earth with a sample of a hitherto-unknown element, which is named waltonium (Wa). Studies of its electrochemistry revealed the following standard reduction potentials:

$$WaO_4^- + e^- \rightleftharpoons WaO_4^{2-} \qquad E° = +2.17\,V$$

$$WaO_4^{2-} + 2\,e^- + 8\,H^+ \rightleftharpoons Wa^{4+} + 4\,H_2O \qquad E° = +1.67\,V$$

$$Wa^{4+} + e^- \rightleftharpoons Wa^{3+} \qquad E° = +0.17\,V$$

$$Wa^{3+} + e^- \rightleftharpoons Wa^{2+} \qquad E° = -1.98\,V$$

$$Wa^{3+} + 3\,e^- \rightleftharpoons Wa(s) \qquad E° = -1.50\,V$$

$$Wa^{2+} + 2\,e^- \rightleftharpoons Wa(s) \qquad E° = -0.50V$$

(a) Sketch this information into a redox predominance diagram, identifying the position of each chemical form of waltonium. (b) One of the chemical forms of waltonium is thermodynamically unstable, and hence has only an imaginary predominance area in this diagram. Which chemical form is this? (c) Will waltonium metal react with 1 *M* HCl to liberate hydrogen gas? (d) Which chemical form of waltonium (if any) will react with water to liberate oxygen gas?

29. The first manned mission to Mars returns to Earth with samples of four hitherto-unknown elements, which are named kruegerine (Kg), bailogen (Bl), birdwellium (Bw), and sanchezium (Sz), respectively. In early experiments with the nonmetals, Kg is found to form an anion Kg^-; the standard reduction potential of Kg to give its anion is found to be +2.10 V. The nonmetal Bl also forms an anion Bl^{2-}, showing a standard reduction potential of +0.20 V. The metals Bw and Sz form cations with standard reduction potentials of +0.55 V for Bw^{2+} and −2.50 V for Sz^{2+}. (a) Sketch this information into redox predominance diagrams, identifying the position of each element and its cation or anion. (b) Which of these elements (if any) will react with water to liberate hydrogen gas? (c) Which of these elements (if any) will react with water to liberate oxygen gas? (d) The cation of which of these elements (if any) will react with the anion of Kg to produce elemental Kg? (e) Will elemental Sz react with the Bw^{2+} cation to liberate Bw metal?

30. *The first manned mission to Venus returns to Earth with samples of four hitherto-unknown elements, which are named nekoorine (Nk), sengsavanogen (Ss), herrium (Hm), and przybylskium (Pz), respectively. In early experiments with the nonmetals, Nk is found to form an anion Nk^-; the standard reduction potential of Nk to give its anion is found to be +2.50 V. The nonmetal Ss also forms an anion Ss^{2-}, showing a standard reduction potential of +0.10 V, and an oxo acid H_6SsO_6, with a standard reduction potential of +1.65 V. The metals Hm and Pz form cations with standard reduction potentials of +0.85 V for Hm^{2+} and −2.30 V for Pz^{2+}. (a) Sketch this information into redox predominance diagrams, identifying the position of each element and its cation, anion, or oxo acid. (b) Complete and balance the half-reaction that shows the reduction of H_6SsO_6 to elemental Ss in acid solution. (c) Which of these elements (if any) will react with water to liberate hydrogen gas? (d) Which of these elements (if any) will react with water to liberate oxygen gas? (e) The cation of which of these elements (if any) will react with the anion Nk^- to produce elemental Nk? (f) Will elemental Pz react with the Hm^{2+} cation to liberate Hm metal?

31. *In each of the following sets of compounds, identify the one compound that is most probably either nonexistent or very intensely colored, and the one compound that not only exists, but is least intensely colored or even colorless: (a) BiF_3, $BiCl_3$, $BiBr_3$, BiI_3; (b) InI_3, SnI_4, SbI_5; (c) PF_5, PBr_5, PI_5, PAt_5; and (d) FO_4^-, MnO_4^-, ReO_4^-. Give the name for the type of light absorption responsible for intense color among some of the above compounds. Explain it briefly.

32. In each set of compounds, all of which exist, select the compound that likely shows the most intense color: (a) CF_4, CCl_4, CBr_4, CI_4; (b) CrF_3, $CrCl_3$, $CrBr_3$, CrI_3; (c) TlF_6^{3-}, $TlCl_4^-$, $TlBr_4^-$, TlI_4^-; (d) CaS, HgS, CdS, ZnS; and (e) PI_5, SiI_4, AlI_3, MgI_2, NaI.

33. *Based on their electronegativities, arrange in an activity series all the metals the names of which begin (a) with the letter *T* and (b) with the letter *P*.

34. (a) Based on their electronegativities, arrange the following metals in an activity series: K, Au, Cu, Mg, Mn, Zn, Fe, and Ni. (b) Similarly list the following metals in order of decreasing activity: Ca, Cd, Ce, Cf, Co, Cr, Cs, Cu. (c) In each series, which of the metals should visibly react with (cold or hot) water to liberate hydrogen? (d) In each series, which of the metals should fail even to react with hot HCl to liberate hydrogen?

35. *Describe the activity of each of the following elements with (1) cold water; (2) hot HCl solution; (3) concentrated HNO_3: (a) La; (b) Pt; (c) Co; (d) Sc; (e) Os; (f) Sr; (g) Cr; (h) In; (i) Ir; (j) Nd; (k) Cu; (l) Zn; (m) Sn; and (n) Ba.

36. From the following lists of metals, select those that would (1) react with water; (2) react with HCl but not with water; (3) not react either with HCl or with water. (a) Au, Be, Bi, Ce, Fe, Ga, Ir, Pu. (b) Rb, Sb, Pt, Y, Sm, Mn, Rh, Sn; and (c) Li, Bi, Sc, Hg, Tm, Os, Zn, Au.

37. Consider the following metals: Ba, Lu, Os, Ti, Bi, Sn, Cu, Au, and Pu. (a) Which of these will react with water? (b) Which of these will react with HCl but not with water? (c) Which of these will dissolve only in HNO_3, or will not dissolve in any of the above?

38. If you did not perform Experiment A.6, go back to it now: Describe what would have happened during the experiment, and answer the questions in part E.

39. You are a metallurgist employed in industry, and need to dissolve some samples of metal in acid to get solutions for analysis by atomic absorption spectroscopy. You are having difficulty in dissolving samples of certain metals in any single acid, even the good oxidizing agent nitric acid. Which acid, HF or HCl, would better help dissolve the indicated metals in nitric acid? How would you explain how it helped? (Include, if possible, a plausible formula for the product.) Metals: (a) tantalum (Ta) and (b) platinum (Pt).

40. Name the Russian chemist who extended the Lewis acid and base definitions to include redox reactions. Under the Russian definition how is the electron classified in the HSAB scheme?

41. (a) List the following elements in an activity series for their reaction with water or aqueous acids: Na, Pt, Co, Bi. (b) Might the position of Pt in this activity series be altered if the reaction were run in liquid hydrogen cyanide, HCN, as a solvent, instead of water? If so, how would its position be altered and why?

42. *Which ligand would best stabilize the −1 oxidation state of Mn? F^-; H_2O; NH_3; or CO.

43. Which ligand would best stabilize the +6 oxidation state of Pt? F^-; H_2O; NH_3; or CO.

44. Answer Exercises 4–9 without the use of redox predominance or Pourbaix diagram, using your knowledge of periodic trends.

45. *Gold will not dissolve in concentrated sulfuric acid but will dissolve in concentrated selenic acid. Explain why.

46. Based on periodic trends, which element in each set is least likely to oxidize something: (a) F_2, I_2, Br_2, Cl_2 and (b) F_2, C, O_2.

47. Based on the electronegativities of Cl and Sb, predict: (a) Which element will be the stronger oxidizing agent, chlorine or antimony? (b) Which oxo acid will be the stronger oxidizing agent, perchloric acid or antimonic acid? (c) Which hydride will be the better reducing agent, HCl or SbH_3?

48. (a) Write the name and formula of the *least* explosive salt you can make by combining an acid from Exercise 47, part (b) with a basic hydride from Exercise 47, part (c). (b) The mixing of the vapors of which element from Exercise 47, part (a) with which hydride from part (c) gives the *greatest* danger of an explosion?

49. *Based on periodic trends, which *two* of the following six anions would be the weakest reducing agents? CH_3^-, Ge^{4-}, O^{2-}, Se^{2-}, F^-, and H^-.

50. Consider the following four anions: C^{4-}, Sn^{4-}, F^-, and I^-. Based on periodic trends: (a) Which one is the strongest reducing agent? (b) Which one is the weakest reducing agent?

51. Consider the following six anions: C^{4-}, Ge^{4-}, O^{2-}, Se^{2-}, F^-, and Br^-. Based on periodic trends: (a) Which one is the strongest reducing agent? (b) Which one is the weakest reducing agent?

52. *According to electrochemical data, redox predominance diagrams, or Pourbaix diagrams, which of these oxidizing agents—NO_3^-, $Cr_2O_7^{2-}$, H_2MoO_4, ClO_4^-, $Br_2(l)$, HSO_4^-—are capable of oxidizing the reducing agent: (a) SCN^- and (b) N_3^-.

53. Give the symbols of three elements for which each of the following is the preferred commercial method of preparation: (a) occur native (in the elemental form); (b) prepared by electrolytic or chemical oxidation of their anions; (c) prepared by reduction of their oxides or oxo salts with carbon; (d) prepared by reduction of their halides with a more active metal; and (e) by electrolytic reduction of its cation in a fused-salt medium.

54. From each of the following sets of metals, select two that are prepared by the reduction of their oxides with carbon, and two that are best prepared by the reduction of their halides with the metals Mg or Ca. (a) Au, Be, Bi, Ce, Fe, Ga, Ir, Pu; (b) Rb, Sb, Pt, Y, Sm, Mn, Rh, Sn; and (c) Li, Bi, Sc, Hg, Tm, Os, Zn, Au.

55. *Select the best reducing agent or method—carbon; none needed (occurs native); a more active metal; electrolysis—for the industrial production of each of the following elements. (a) P; (b) Pt; (c) Na; (d) Os; (e) Co; (f) Li; and (g) Zr.

56. Using balanced chemical equations, explain: (a) why the element carbon (and no other) is a so much better reducing agent at higher temperatures than it is at room temperature and (b) how copper is separated both from more active and from less active metals by electrolytic refining.

57. Explain why aluminum metal is produced in Norway and Canada rather than in Guyana, where bauxite is mined.

58. Explain the function of each of the following processes in the production of a metal, and give an example for each: (a) beneficiation of the ore; (b) chemical conversion of an ore; (c) reduction of the ore; and (d) purification of the metal.

59. *What is the purpose (beneficiation, chemical conversion, reduction, or purification) of each of the following chemical reactions in producing the metal specified?

(a) Titanium: $TiO_2 + 2\,Cl_2 + 2\,C \rightarrow TiCl_4 + 2\,CO$

(b) Aluminum: $2\,Al_2O_3 + 6\,C +$ electrical energy $\rightarrow 4\,Al + 6\,CO$

(c) Aluminum: $OH^- + (Fe_2O_3, Al_2O_3, TiO_2)(s) \rightarrow Fe_2O_3(s) + TiO_2(s) + Al(OH)_4{}^-(aq)$; filter; cool $\rightarrow Al_2O_3(s)$

(d) Nickel (Mond process): $(Fe, Co, Ni) + CO \rightarrow Fe(s) + Co(s) + Ni(CO)_4(g)$; then heat gas $\rightarrow 4\,CO(g) + Ni(s)$

(e) Copper: (Fe, Cu, Ag, Au) + electrolytic oxidation $\rightarrow Fe^{2+}(aq) + Cu^{2+}(aq) + Ag(s) + Au(s)$; solution + electrolytic reduction $\rightarrow Cu(s) + Fe^{2+}(aq)$

60. Choose one of the following purposes—purification, reduction, oxidation, conversion, beneficiation—for each of the following chemical reactions; then identify the element M to which the reaction would apply.

(a) $(M, Ag, Au, Fe) \rightarrow M^{2+} + Fe^{2+} + Au(s) + Ag(s) + 4\,e^-$; followed by $M^{2+} + Fe^{2+} + 2\,e^- \rightarrow M(s) + Fe^{2+}$ (electrolysis)

Which is M likely to be: Te, Pt, Cu, or Na?

(b) $2\,MCl_5 + 5\,Mg \rightarrow 2\,M + 5\,MgCl_2$

Which is M likely to be: Bi, V, Be, or Ir?

(c) $2\,KM(\ell)+$ electrical energy $\rightarrow 2\,K(\ell) + M_2(g)$.

Which is M likely to be: F, Na, H, S, or C?

61. *With the help of other sources such as Fergusson[30] and Swaddle[31], describe and write chemical equations for as many of the four processes in the production of a metal as apply to the production of the following: (a) steel; (b) aluminum; (c) copper; and (d) titanium.

62. Refer to the redox predominance diagrams for N, I, Co, and Cr. (a) You want to prepare these elements by reduction of their ores in higher oxidation states by the most economical method possible, reduction with carbon (C). Which reduction is *least* likely to succeed? (Equations are not balanced.) $Cr_2O_3 + C \rightarrow Cr + CO$; $Co_2O_3 + C \rightarrow Co + CO$; $HIO_3 + C \rightarrow H_2O + I_2 + CO$; and $HNO_3 + C \rightarrow N_2 + CO$.

(b) If one of the other above reactions does not proceed to products at room temperature, what would you do to increase the activity of the carbon as a reducing agent? Explain briefly why this would help.

(c) You decide to produce cobalt metal (Co) by the carbon reduction process, after discovering a deposit of the mineral linneite, Co_3S_4, deposited in a low percentage in rock deposits. Name and describe the steps you would have to carry out to obtain pure cobalt metal.

63. (a) Assuming that coal can be found on Pluto, is it likely that waltonium metal (Exercise 28) could be produced there by reduction of its oxide or oxo anion with carbon? (b) Assuming that coal can be found on Mars, which two of the elements in Exercise 29 could most likely be produced there by reduction of their oxides or oxo anions with carbon? (c) On the trip back from Mars, which two of the elements in Exercise 29 should be kept as far apart as possible on the space ship because they might produce an explosion if they came in contact?

64. A new element, wattsium (Wa), has just been discovered. Careful electrochemical measurements have established the Pourbaix diagram (Fig. 6.27). The following

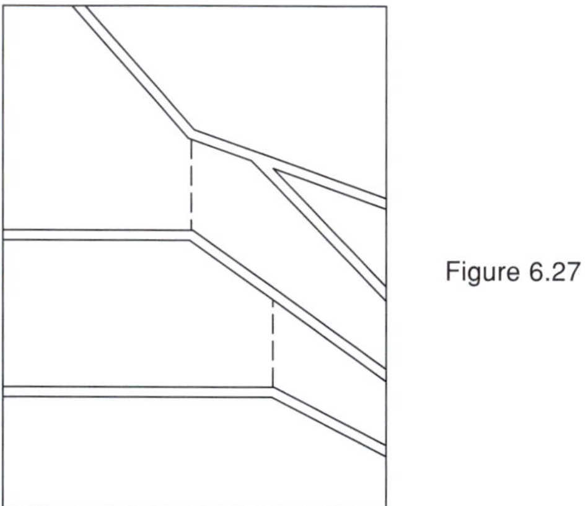

Figure 6.27

thermodynamically stable species are involved: Wa^{3+}, Wa^{2+}, Wa(s), $Wa(OH)_3$, $Wa(OH)_2$, $WaO_4{}^{3-}$, and $WaO_4{}^{2-}$. (a) Locate each species in the appropriate predominance region in the diagram. (b) Which species is the best reducing agent? (c) Which species is the best oxidizing agent?

65. A blank Pourbaix diagram for the element ruthenium, Ru, is shown (Fig. 6.28).

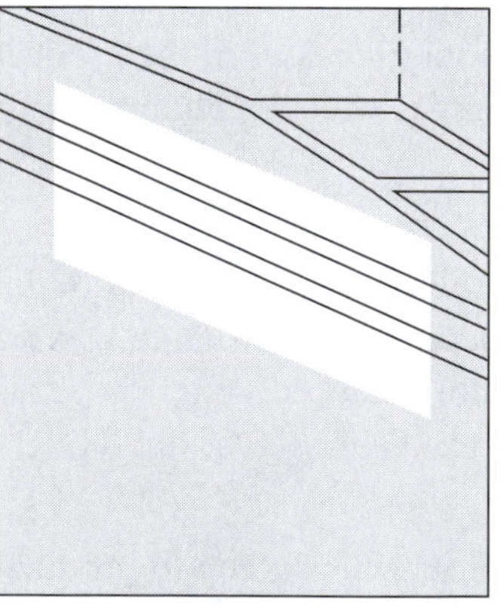

Figure 6.28

Without reference to the text, fill in the following chemical forms of Ru in the predominance areas of this diagram: Ru, RuO_4, $RuO_4{}^-$, $RuO_4{}^{2-}$, $Ru(OH)_3$, $Ru(OH)_4$, and $RuO_5{}^{2-}$.

66. Figure 6.24 shows the $E°$/pH (Pourbaix) diagram for uranium. You are attempting to study the fate of waste uranium (from nuclear fuel reprocessing) in a natural lake.

(a) If the lake is well aerated and not polluted, in what form will the uranium be found? Will it be present in solution or found in the sludge at the bottom of the lake? (b) If the lake is well aerated but is highly contaminated with acid rain (pH 3), in what form will the uranium be found? Will it be present in solution or found in the sludge at the bottom of the lake? (c) Suppose some metallic uranium were dropped into this lake (with pH 3), but it fell to the bottom of the lake, where oxygen was absent and reducing impurities (from decaying organic matter) were present. Would it remain as elemental uranium? If not, write a balanced chemical reaction for what would happen, and describe its vigor.

67. Figure 6.23 shows the Pourbaix diagram for manganese. People often find that clear well water that they draw from wells will deposit a black manganese-containing solid on standing in their toilet bowls. Explain what that solid is and why it forms in the toilet bowl and not underground.

68. *Figure 6.24 shows the Pourbaix diagram for plutonium. You are attempting to study the fate of waste plutonium (from atomic bomb assembly) in a lake. (a) If the lake is well aerated and of normal pH, in what chemical form will the plutonium be found? Will it be present in solution or found in the sludge at the bottom of the lake? (b) Will metallic plutonium dumped into the lake remain in the metallic form, or will it react with the water? If the latter, write a balanced chemical equation showing the reaction with the water. (c) Would any corrosion of the metallic plutonium by the water be most severe in acidic or in basic solution? (d) If the lake is anaerobic and highly polluted with acid rain, in what chemical form will the plutonium be found? Will it be present in solution or found in the sludge at the bottom of the lake?

69. Identify the chemical form of each of the following elements that is predominant in well-aerated lakes of normal pH; in highly anaerobic lakes that are strongly contaminated with acid rain. (a) Mn or (b) N.

70. *Examine the Pourbaix diagrams for the elements Am, Cr, Mn, Bi, and I. If one exists, find a chemical form of each of these elements that, at some pH, would (a) disproportionate; (b) oxidize water; and (c) reduce water.

71. Consult the Pourbaix diagram for ruthenium, Ru, to answer these questions. (a) Which chemical form of Ru listed is the best oxidizing agent at pH 14? (b) Which chemical form of Ru listed is the best reducing agent at pH 0? (c) Will any chemical form of Ru listed release H_2 in contact with water? If so, which form? (d) Which chemical form of Ru is most likely to be found in well-aerated natural waters of pH 7? (e) Which chemical form of Ru is most likely to be found in the anaerobic waters of a swamp at pH 7?

72. (a) Choose the best oxidizing agent in each set of ions: NO_3^- in acid solution; NO_3^- in basic solution; NO_3^- in neutral solution. (b) Which species is least likely to oxidize something (i.e., is most stable): NO_3^- in acid solution; NO_3^- in neutral solution; and NO_3^- in basic solution.

73. *The most suitable pH to carry out the synthesis of the FeO_4^{2-} ion is: 0, 7, or 14?

74. The most suitable pH to use the FeO_4^{2-} ion as a powerful oxidizing agent is: 0, 7, or 14?

75. The standard reduction potentials for the oxo anions FeO_4^{2-}, RuO_4^{2-}, and OsO_6^{6-} are +0.96 V, +1.40 V, and +2.20 V, but not necessarily in that order. (a) Which standard reduction potential belongs to which ion? (b) Which of these ions would be the most powerful oxidizing agent? (c) Write balanced chemical equations showing good methods of synthesis of each of these oxo anions or acids from the oxide of each element in its most common positive oxidation state (or from the element itself).

76. *Referring to predominance or Pourbaix diagrams, for each of the oxo anions listed below, choose the best of the following three synthetic approaches: (1) the anion must be prepared in strongly basic solution by electrolysis or by use of a powerful oxidizing agent such as Cl_2, F_2, O_2, O_3, or H_2O_2; (2) the anion can be prepared in acidic solution (as an oxo acid) by the action of an oxidizing agent such as HNO_3; (3) the anion is likely to be prepared by reduction of a higher oxo anion or by the action of base on the element itself; (4) the oxo anion does not exist. OXO ANIONS: (a) KrO_4^{2-}; (b) MnO_4^{2-}; (c) CrO_4^{2-}; (d) $[Zn(OH)_4]^{2-}$; and (e) IO_3^-.

77. Using the oxidizing agent of your choice, write a balanced chemical equation for the synthesis of each real oxo anion from the previous exercise from an oxide of the element in a common oxidation state, or from the element itself.

78. It is desired to prepare a bismuthate salt. Potassium bismuthate is an insoluble red salt of empirical formula $KBiO_3 \cdot 0.33\ H_2O$; sodium bismuthate is an insoluble yellow salt of empirical formula $NaBiO_3 \cdot 3.5\ H_2O$. (a) This synthesis involves the reaction of Bi_2O_3 with Br_2 in 50% KOH solution, which has an effective pH equal to 19! Explain why such strongly basic conditions are used. (b) Predict what would happen if the mixture of products of this reaction were acidified (to pH 3). (c) *Review*: What is the formula of the corresponding antimony oxo anion in aqueous solution? How likely is it that $KBiO_3 \cdot 0.33\ H_2O$ and $NaBiO_3 \cdot 3.5\ H_2O$ contain a simple oxo anion of formula BiO_3^-? Explain. Given their empirical formulas, which of these could have a structure analogous to that of the antimony analogue? Which of these could not?

79. Consult Pourbaix diagrams for Mn, W, and P in order to answer the following questions. (a) Which of the following oxo anions should be capable of oxidizing the Mn^{2+} ion: MnO_4^-; WO_4^{2-}; $[H_2PO_3]^-$; and $[H_2PO_4]^-$. (b) Suppose that this reaction (or one of these reactions) that should occur does not in fact take place when you begin to mix the solutions; nothing happens. Would it then be advisable to add a lot more of the second solution to get the reaction going? Why or why not? (c) Identify one chemical species that undergoes disproportionation at some pH. (d) From the possibilities given at the end of this exercise, select the best method of synthesis of the following ions: MnO_4^-; WO_4^{2-}; and $[H_2PO_3]^-$. METHODS OF SYNTHESIS: (1) redox not needed: the element will occur in nature in this oxidation state; (2) react the element in its common oxidation state with Cl_2 or H_2O_2 in basic solution; (3) react the element in its common oxidation state with Cl_2 or H_2O_2 in acidic solution; (4) react the element in its common oxidation state with HNO_3 in acidic solution; (5) reduce an element from a higher oxidation state, or react base with the element itself.

80. *Choose from each set of oxo anions the one that requires the most strongly oxidizing conditions for synthesis: (a) FeO_4^{2-}, RuO_4^{2-}, OsO_6^{6-}; (b) GeO_4^{4-}, AsO_4^{3-}, SeO_4^{2-}, BrO_4^-; and (c) SeO_4^{2-}, TeO_6^{6-}, PoO_6^{6-}.

81. Chromium dissolves readily in HCl but is passivated by HNO_3; explain with the help of the Pourbaix diagram of Cr.

Notes

1. In dilute aqueous solutions, molalities and molarities converge to the same value, as do activities and concentrations. Unfortunately, a concentration of 1.000 mol L^{-1} is not very dilute, but we will not concern ourselves here with the distinction of activities and concentrations, which result from disruptions of the primary and secondary hydration spheres of the ions in nondilute solutions.
2. For example, A. J. Bard, R. Parsons, and J. Jordan, *Standard Potentials in Aqueous Solution*, Dekker, New York, 1985.
3. See, for example, D. F. Shriver, P. W. Atkins, and C. H. Langford, *Inorganic Chemistry*, Freeman, New York, 1990; pp. 246–250. The use of different types of diagrams is not just an illustration of the perversity of chemists; each has advantages in the fields in which it is used.
4. Many chemical species cannot be placed directly on a redox predominance diagram, because if they are arranged in order of decreasing oxidation states they are then out of order of decreasing E° values. Such species will be discussed further in Section 6.3; these species are indicated by locating them (along with their reduction potentials) *alongside* the predominance diagram.
5. Some reaction may occur to reduce their concentrations somewhat from 1.000 M, as can be predicted from the Nernst equation, but the reactants will still be predominant over the products.
6. If your instructor wishes to proceed directly to the use of Pourbaix diagrams, you will now go to Sections 6.9 and 6.10 before returning to Sections 6.2 and beyond, for which either predominance or Pourbaix diagrams can be used.
7. T. W. Swaddle, *Applied Inorganic Chemistry*, University of Calgary Press, Calgary, 1990; pp. 190–196.
8. A more commonly used criterion is that the enthalpy change for the decomposition be more negative than about -2.9 kJ g^{-1}, but other factors, such as the number of moles of gas produced, are also involved: see, for example, M. Laing, *J. Chem. Educ.* **70**, 392 (1993); R. Meyer, *Explosives*, 2nd ed., Verlag Chemie, Weinheim, 1981; pp. 18–20.
9. See also L. Bretherick, "Chemical Reactivity: Instability and Incompatible Combinations," in *Improving Safety in the Chemical Laboratory: A Practical Guide*, J. A. Young, Ed., Wiley, New York, 1987; Chapter 7.
10. Commemorations 50 years later are in the *New York Times*, April 16, 1997, p. A10, and in *Chem. Eng. News*, April 21, 1997, p. 11.
11. R. J. Seltzer, *Chem. Eng. News*, August 8, 1988, p. 7.
12. A. Madlung, *J. Chem. Educ.*, **73**, 347 (1996).
13. J. L. Ennis, and E. S. Shanley, *J. Chem. Educ.*, **68**, A6 (1991).
14. W. L. Jolly, *Modern Inorganic Chemistry*, McGraw-Hill, New York, 1984; p. 155.
15. D. F. Shriver, and M. A. Drezdzon, *The Manipulation of Air-Sensitive Compounds*, 2nd ed., Wiley, New York, 1986.
16. Alternately, as suggested by William Ilsley, we can include the metastable species in a partial redox diagram along with the adjacent stable species if we use one boundary only: the upper boundary if we wish to examine the oxidation of the metastable species, and the lower boundary if we wish to study the reduction of the metastable species.
17. K. B. Yoon, and J. K. Kochi, *Z, Anorg. Allg. Chem.*, **561**, 174 (1988); K. B. Yoon, and J. K. Kochi, *Inorg. Chem.*, **29**, 869 (1990).
18. For elements with lower oxidation states that are oxidized by water (Cr^{2+} and Eu^{2+}) the standard reduction potential is that of the stable ion in water (Cr^{3+} and Eu^{3+}). This potential is not shown in the redox predominance diagrams, but can be computed from the values shown by the equation $nE^\circ(M^{3+} \rightarrow M) = nE^\circ(M^{3+} \rightarrow M^{2+}) + nE^\circ(M^{2+} \rightarrow M)$. Otherwise, the data in Table 6.2 should match (within experimental errors) that given in the redox predominance diagrams.
19. The predominance diagram for *neutral* water has an upper boundary at +0.81 V and a lower boundary at -0.41 V, to which overpotential effects must be added or subtracted.
20. The heavier Group 1 metals produce peroxides (salts of the $O_2{}^{2-}$ ion) and superoxides (salts of the $O_2{}^{-}$ ion), since these larger and/or less-charged anions give better lattice energies with these large cations.
21. M. Usanovich, *Zhur. Obschei. Khim.*, **9**, 182 (1939).

22. D. Issa and J. L. Dye, *J. Am. Chem.* Soc., **104**, 3781 (1982); J. L. Dye, *Sci. Am.*, **257** (3), 66 (1987); J. L. Dye, *Chemtracts: Inorg. Chem.*, **5**, 243 (1993).
23. For a summary of the evidence of the cause of the extinction, see: W. Alvarez and F. Asaro, *Sci. Am.*, **263** (4), 78 (1990); V. E. Courtillot, *Sci. Am.*, **263** (4), 85 (1990).
24. S. Blakeslee, *New York Times*, Oct. 28, 1994, p. A1.
25. S. E. Manahan, *Environmental Chemistry*, 5th ed., Lewis Publishers, Chelsea, MI, 1991. See Chapter 6 and particularly Section 6.11, p. 137.
26. J. K. Frederickson and T. C. Onstott, *Sci. Am.*, **275** (4), 68 (1996).
27. W. J. Broad, *New York Times*, Oct. 15, 1996, p. B5.
28. The standard reduction potentials of nonmetals do not fit the correlation of Figure 6.14 as well, however, since they oxidize to give diverse oxo acids, anions, or oxides rather than hydrated cations; Figure 7.5 shows the result of including nonmetals.
29. Yu. M. Kiselev, N. S. Kopelev, and A. P. Bobylev, *Russ. J. Inorg. Chem. (Engl. trans.)*, **34**, 1517 (1989); Yu. M. Kiselev, G. V. Ionova, A. A. Kiseleva, N. S. Kopelev, A. P. Bobylev, A. B. Yatskevich, and V. I. Spitsyn, *Dokl. Chem. (Engl. Trans.)*, **293**, 403 (1987).
30. J. E. Fergusson, *Inorganic Chemistry and the Earth: Chemical Resources, Their Extraction, Use, and Environmental Impact*, Pergamon, Oxford, UK, 1982; pp. 82–106.
31. T. W. Swaddle, *Applied Inorganic Chemistry*, University of Calgary Press, Calgary, 1990; Chapter 15.
32. In practice, furnace temperatures in excess of 2000 K are impractical due to energy costs and/or damage to the furnace lining.
33. Traditional African methods of iron production, in use from 600 B.C. into the twentieth century, produce high-phosphorus iron: P. R. Schmidt and S. T. Childs, *Am. Sci.*, **83**, 524 (1995).
34. J. W. Moore, and E. A. Moore, *Environmental Chemistry*, Academic, New York, 1976; Chapter 11.
35. W. Lepkowski, *Chem. Eng. News*, June 4, 1979, p. 14.
36. M. Pourbaix, *Atlas of Electrochemical Equilibria in Aqueous Solutions*, National Association of Corrosion Engineers, Houston, 1974.
37. The p*E* scale is intended to represent the concentration of the standard reducing agent, the electron, as the pH scale represents the concentration of the standard acid, the hydrogen ion. p*E* values are obtained from reduction potentials by dividing by 0.059.
38. Since these diagrams come from different sources than the redox predominance diagrams, there are some slight discrepancies, especially in the formulas indicated for precipitates that may be hydroxides or may be (hydrated) oxides.
39. Standard reduction potentials reflect not only the stability of the higher oxidation state but also of the reduced state, so their values vary depending on the reduced state chosen. Thus a boundary such as +1.2 V can only be a rough guideline.
40. Necessary if the central-atom oxidation number is over 6, or somewhat lower if the element is in the second or sixth period of the *p*-block or the fourth period of the *d*-block elements.
41. G. Brauer, Ed., *Handbook of Preparative Inorganic Chemistry*, 2 Vols., 2nd ed., Academic, New York, 1963.
42. T. W. Swaddle, *Applied Inorganic Chemistry*, University of Calgary Press, Calgary, 1990; pp. 190–196, Chapter 14.
43. F. H. Walters, *J. Chem. Educ.*, **68**, 29 (1991).

CHAPTER

7

Thermochemical Analyses of Reactivity Trends

7.1 Thermochemical Analysis of Redox Reactions: The Born–Haber Cycle

In developing an understanding of the reasons for the solubility rules in Chapter 3, we found it useful to do a thermochemical analysis of the process of precipitation of an ionic salt. In this chapter, we will carry out additional thermochemical analyses of some of the other reactivity trends we have found since then, beginning with some of the redox reactions of the elements that we discussed in Chapter 6.

Even such a simple redox reaction as the burning of sodium in chlorine [Eq. (7.1)] is complex enough that it is best broken down into a sum of simpler reactions that can be easily analyzed thermochemically. Then, by Hess's law, the enthalpy change for the reaction—ΔH_f°, the enthalpy of formation of NaCl—is equal to the sum of the enthalpy changes for the different simpler reactions that add up to Reaction (7.1).

$$Na(s) + 0.5\,Cl_2(g) \rightarrow NaCl(s) \qquad \Delta H_f^\circ = -411\ \mathrm{kJ\,mol^{-1}} \tag{7.1}$$

The thermochemical cycle of simpler reactions we construct should relate this complex redox reaction to simpler gas-phase reactions; hence, we use the following general strategy, involving three steps:

1. Take all reactants to the gaseous state; identify and assign all energy changes involved.

2. Let the reactants react in the gaseous state to give gaseous products; identify and assign all energy changes involved.

3. Take the products to their final state; identify and assign all energy changes involved.

Then, by Hess's law, the energy change for the overall reaction is equal to the sum of the energy changes for each of the steps.

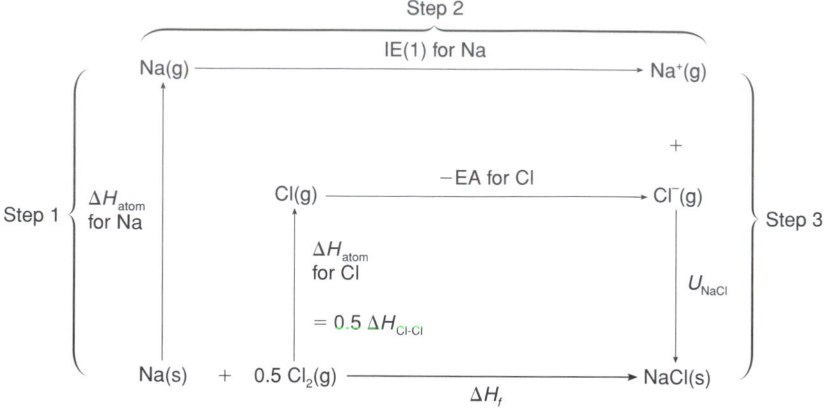

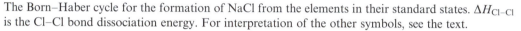

Figure 7.1
The Born–Haber cycle for the formation of NaCl from the elements in their standard states. ΔH_{Cl-Cl} is the Cl–Cl bond dissociation energy. For interpretation of the other symbols, see the text.

Analysis of the reactions by which ionic compounds are formed from their elements in the standard state was pioneered by Born[1a] and Haber,[1b] so such an analysis is now known as a **Born–Haber cycle**. The classical Born–Haber cycle for the formation of NaCl from Na metal and Cl_2 gas is shown in Figure 7.1.

In Step 1 of this thermochemical cycle, we convert sodium metal (a solid) and chlorine gas, which consists of diatomic molecules, to gaseous atoms. The energy required to produce 1 mol of gaseous atoms of an element from the form of the element that is usual at room temperature and 1-atm pressure is known as the **heat**[2] **of atomization** of that element; these are tabulated in Table 7.1. For Na, this is $+107\,\text{kJ mol}^{-1}$; for Cl, this is $+122\,\text{kJ mol}^{-3}$.

In Step 2, we allow the gaseous atoms to react to give the products in gaseous form. In this case, we remove an electron from Na to form the gaseous Na^+ ion, and we give the electron to the Cl atom to form the gaseous Cl^- ion. The energy required to remove the outermost (most loosely bound) electron from a gaseous atom to produce a gaseous cation of +1 charge is known as its **first ionization energy** (often called ionization potential, and abbreviated IE or IP).

$$\text{Na(g)} \rightarrow \text{Na}^+\text{(g)} + \text{e}^- \qquad \Delta H = \text{IE}(1) = +495.8\,\text{kJ mol}^{-1} \tag{7.2}$$

Such a process is inherently endothermic, as separating positive and negative charges (cations and electrons) in the gas phase requires an input of energy. The first ionization energies of the elements are tabulated in Table 7.2.

We may then remove a second electron to produce a gaseous cation of +2 charge. This removal requires a **second ionization energy** [IE(2), Table 7.3], which is substantially larger than the first ionization energy, since an electron must be pulled away from an already positively charged ion. Likewise, third ionization energies (Table 7.4) are larger yet. These energy terms would need to be included in the Born–Haber cycle for the formation of a metal di- or trihalide.

Table 7.1

Heats of Atomization of the Elements[a,b]

Period \ Group	1	2	3	4	5	6	7	8	9	10	11	12	13 (III)	14 (IV)	15 (V)	16 (VI)	17 (VII)	18 (VIII)
1	H 218																	He 0
2	Li 159	Be 324											B 563	C 717	N 473	O 249	F 79	Ne 0
3	Na 107	Mg 146											Al 326	Si 456	P 315	S 279	Cl 122	Ar 0
4	K 89	Ca 178	Sc 378	Ti 471	V 515	Cr 397	Mn 283	Fe 415	Co 426	Ni 431	Cu 338	Zn 131	Ga 277	Ge 377	As 303	Se 227	Br 112	Kr 0
5	Rb 81	Sr 165	Y 423	Zr 605	Nb 733	Mo 659	Tc 661	Ru 652	Rh 556	Pd 377	Ag 285	Cd 112	In 244	Sn 302	Sb 262	Te 197	I 107	Xe 0
6	Cs 76	Ba 182	Lu [b]414	Hf 621	Ta 782	W 860	Re 776	Os 789	Ir 671	Pt 564	Au 368	Hg 64	Tl 182	Pb 195	Bi 207	Po 142	At	Rn 0

Period	*3F*	*4F*	*5F*	*6F*	*7F*	*8F*	*9F*	*10F*	*11F*	*12F*	*13F*	*14F*	*15F*	*16F*
6	La 423	Ce 419	Pr 356	Nd 328	Pm 301	Sm 207	Eu 178	Gd 398	Tb 389	Dy 291	Ho 301	Er 317	Tm 232	Yb 152
7	Ac [b]293	Th 575	Pa [b]481	U 482	Np [b]337	Pu 352	Am [b]239	Cm	Bk	Cf	Es	Fm	Md	No

SOURCES: Heats (enthalpies) of atomization of the *s*- and *d*-block elements were taken from W. L. Jolly, *Modern Inorganic Chemistry*, McGraw-Hill, New York, 1984; p. 292; those of the *d*-block elements were taken from W. W. Porterfield, *Inorganic Chemistry: A Unified Approach*, Addison-Wesley, Reading, MA, 1984, p. 84; those of the *f*-block elements are from N. N. Greenwood and A. Earnshaw, *Chemistry of the Elements*, Pergamon, Oxford, UK, 1984.

[a] Units are in kilojoules per mole (kJ mol^{-1}).

[b] Values preceded by [b] are enthalpies of *vaporization*, which are normally slightly less than true heats of atomization, since the metals vaporize in part as diatomic or polyatomic molecules.

Table 7.2
First Ionization Energies of the Elements[a,b]

Period	s^1	s^2	s^2d^1	s^2d^2	s^2d^3	s^2d^4	s^2d^5	s^2d^6	s^2d^7	s^2d^8	s^2d^9	s^2d^{10}	s^2p^1	s^2p^2	s^2p^3	s^2p^4	s^2p^5	s^2p^6
1	H 1312																	He 2372
2	Li 520	Be 899											B 801	C 1086	N 1402	O 1314	F 1681	Ne 2081
3	Na 496	Mg 738											Al 578	Si 786	P 1012	S 1000	Cl 1251	Ar 1520
4	K 419	Ca 590	Sc 631	Ti 658	V 650	Cr 653	Mn 717	Fe 759	Co 758	Ni 737	Cu 746	Zn 906	Ga 579	Ge 762	As 944	Se 941	Br 1140	Kr 1351
5	Rb 403	Sr 550	Y 616	Zr 660	Nb 664	Mo 685	Tc 702	Ru 711	Rh 720	Pd 805	Ag 731	Cd 868	In 558	Sn 709	Sb 832	Te 869	I 1008	Xe 1170
6	Cs 376	Ba 503	Lu 524	Hf 654	Ta 761	W 770	Re 760	Os 840	Ir 880	Pt 870	Au 890	Hg 1007	Tl 589	Pb 716	Bi 703	Po 812	At 930	Rn 1037
7	Fr 400	Ra 509	Lr 444															

	s^2f^2	s^2f^2	s^2f^3	s^2f^4	s^2f^5	s^2f^6	s^2f^7	s^2f^8	s^2f^9	s^2f^{10}	s^2f^{11}	s^2f^{12}	s^2f^{13}	s^2f^{14}
6	La 538	Ce 528	Pr 523	Nd 530	Pm 536	Sm 543	Eu 547	Gd 592	Tb 564	Dy 572	Ho 581	Er 589	Tm 597	Yb 603
7	Ac 490	Th 590	Pa 570	U 590	Np 600	Pu 585	Am 578	Cm 581	Bk 601	Cf 608	Es 619	Fm 627	Md 635	No 642

SOURCE: First ionization energies, IE(1), of the elements listed are taken from J. E. Huheey, E. A. Keiter, and R. L. Keiter, *Inorganic Chemistry: Principles of Structure and Reactivity*, 4th ed., Harper-Collins; New York, 1993; pp. 36, 37.

[a] Units are in kilojoules per mole (kJ mol^{-1}).

[b] Group headings indicate the characteristic valence-electron configurations of the atoms being ionized.

Table 7.3

Second Ionization Energies of (+1 Ions of) the Elements[a,b]

	s^1											s^2	s^2p^1	s^2p^2	s^2p^3	s^2p^4	s^2p^5	s^2p^6
Period																		
1																	He 5250	Li 7298
2	Be 1757											B 2427	C 2353	N 2856	O 3388	F 3374	Ne 3952	Na 4562
3	Mg 1451											Al 1817	Si 1577	P 1903	S 2251	Cl 2297	Ar 2666	K 3051
4	Ca 1145	Sc 1235	Ti 1310	V 1414	Cr 1496	Mn 1509	Fe 1561	Co 1646	Ni 1753	Cu 1958	Zn 1733	Ga 1979	Ge 1537	As 1798	Se 2045	Br 2100	Kr 2350	Rb 2633
5	Sr 1064	Y 1181	Zr 1267	Nb 1382	Mo 1558	Tc 1472	Ru 1617	Rh 1744	Pd 1875	Ag 2074	Cd 1631	In 1821	Sn 1412	Sb 1595	Te 1790	I 1846	Xe 2046	Cs 2230
6	Ba 965	Lu 1340	Hf 1440	Ta 1500	W 1700	Re 1260	Os 1600	Ir 1680	Pt 1791	Au 1980	Hg 1810	Tl 1971	Pb 1450	Bi 1610	Po 1800	At 1600	Rn	Fr 2100
7	Ra 979	Lr 1428																
6		La 1067	Ce 1047	Pr 1018	Nd 1034	Pm 1052	Sm 1068	Eu 1085	Gd 1170	Tb 1112	Dy 1126	Ho 1139	Er 1151	Tm 1163	Yb 1175			
7		Ac 1145	Th 1147	Pa 1129	U 1438	Np 1129	Pu 1129	Am 1158	Cm 1196	Bk 1187	Cf 1206	Es 1216	Fm 1225	Md 1235	No 1254			

SOURCE: Second ionization energies, IE(2), of the +1 ions of the elements listed are taken from J. E. Huheey, E. A. Keiter, and R. L. Keiter, *Inorganic Chemistry: Principles of Structure and Reactivity*, 4th ed., Harper-Collins, New York, 1993; pp. 36, 37.

[a] Units are in kilojoules per mole (kJ mol^{-1}).

[b] Group headings indicate the valence electron configurations of the ions being ionized a second time.

Table 7.4
Third Ionization Energies of (+2 Ions of) the Elements[a,b]

Period	d^1	d^2	d^3	d^4	d^5	d^6	d^7	d^8	d^9	d^{10}	s^1	s^2	s^2p^1	s^2p^2	s^2p^3	s^2p^4	s^2p^5	s^2p^6
1																	Li 11815	Be 14849
2											B 3660	C 4621	N 4578	O 5300	F 6050	Ne 6122	Na 6912	Mg 7733
3											Al 2745	Si 3232	P 2912	S 3361	Cl 3822	Ar 3931	K 4411	Ca 4912
4	Sc 2389	Ti 2652	V 2828	Cr 2987	Mn 3248	Fe 2957	Co 3232	Ni 3393	Cu 3554	Zn 3833	Ga 2963	Ge 3302	As 2736	Se 2973	Br 3500	Kr 3565	Rb 3900	Sr 4210
5	Y 1980	Zr 2218	Nb 2416	Mo 2621	Tc 2850	Ru 2747	Rh 2997	Pd 3177	Ag 3361	Cd 3616	In 2705	Sn 2943	Sb 2440	Te 2698	I 3200	Xe 3100	Cs 3400	Ba 3600
6	Lu 2022	Hf 2250	Ta 2100	W 2300	Re 2510	Os 2400	Ir 2600	Pt 2800	Au 2899	Hg 3300	Tl 2878	Pb 2082	Bi 2466	Po 2700	At 2900	Rn	Fr 3100	Ra 3300

Period	f^1	f^2	f^3	f^4	f^5	f^6	f^7	f^8	f^9	f^{10}	f^{11}	f^{12}	f^{13}	f^{14}
6	La 1850	Ce 1949	Pr 2086	Nd 2130	Pm 2150	Sm 2260	Eu 2400	Gd 1990	Tb 2110	Dy 2200	Ho 2200	Er 2190	Tm 2284	Yb 2415
7	Ac 1900	Th 1978	Pa 1814	U 1843	Np 1872	Pu 2103	Am 2161	Cm 2045	Bk 2152	Cf 2277	Es 2325	Fm 2354	Md 2451	No 2605

SOURCE: Third ionization energies, IE(3), of the +2 ions of the elements listed are taken from J. E. Huheey, E. A. Keiter, and R. L. Keiter, *Inorganic Chemistry: Principles of Structure and Reactivity*, 4th ed., Harper-Collins, New York, 1993; pp. 36, 37, and D. F. Shriver, P. W. Atkins, and C. H. Langford, *Inorganic Chemistry*, Freeman, New York, 1993; pp. 638–639.

[a] Units are in kilojoules per mole (kJ mol^{-1}).

[b] Group headings indicate the valence electron configurations of the ions being ionized a third time.

Table 7.5
Electron Affinities of the Elements[a]

First Electron Affinities (Ionization Energies of the −1 Ions of the Elements)

Period	s^1	s^2	s^2p^1											s^2p^2	s^2p^3	s^2p^4	s^2p^5	s^2p^6
1		H 73																
2	He −48	Li 60	Be −48											B 27	C 122	N −7	O 141	F 328
3	Ne −116	Na 53	Mg −39											Al 43	Si 134	P 72	S 200	Cl 349
4	Ar −96	K 48	Ca −29	Sc 18	Ti 8	V 51	Cr 64	Mn 0	Fe 16	Co 64	Ni 112	Cu 118	Zn 0	Ga 29	Ge 130	As 78	Se 195	Br 325
5	Kr −96	Rb 47	Sr −29	Y 30	Zr 41	Nb 86	Mo 72	Tc 53	Ru 101	Rh 110	Pd 54	Ag 126	Cd 0	In 29	Sn 116	Sb 103	Te 190	I 295
6	Xe −77	Cs 46	Ba 0	Lu 48	Hf 0	Ta 31	W 79	Re 14	Os 106	Ir 151	Pt 205	Au 223	Hg 0	Tl 19	Pb 35	Bi 91	Po 183	At 270

Second and Third Electron Affinities (Ionization Energies of the −2 and −3 Ions of the Elements)

3rd EAs	*2nd EAs*	
N −1070	N −673	O −744
P −886	P −468	S −456
As −802	As −435	Se −410

SOURCE: Electron affinities (EA) of the elements listed are taken from R. T. Myers, *J. Chem. Educ.*, **67**, 307 (1990); group headings indicate the characteristic valence electron configurations of the anions being ionized. Second and third electron affinities are taken from R. G. Pearson, *Inorg. Chem.*, **30**, 2856 (1991).

[a] Units are in kilojoules per mole (kJ mol^{-1}).

The energy change involved in *adding* an electron to an atom to form an anion is known as the **electron affinity** of that element (Table 7.5). Due to an unfortunate tradition regarding the signs of electron affinities, they are better regarded as the energies required to *remove* the electron of a gaseous anion of −1 charge to produce a gaseous atom of that element[4]:

$$Cl^-(g) \rightarrow Cl(g) + e^- \qquad \Delta H = \text{EA} = +348.8 \text{ kJ mol}^{-1} \tag{7.3}$$

As we see from Table 7.5, not all gaseous anions hold on to their electrons well; many have zero or negative electron affinities. No gaseous anion holds onto a second extra electron well: The second electron affinity even of oxygen (as the O^{2-} ion) is $-744\,\text{kJ mol}^{-1}$.

Thus in Step 2, we must add the first IE of sodium, $+496\,\text{kJ mol}^{-1}$, but *subtract* the EA of chlorine, since the chloride ion is not being stripped of its electron but is instead being formed. Note that adding all the energy terms listed for the first two steps gives $+376\,\text{kJ mol}^{-1}$; the reaction of two of the most active elements would be hopelessly endothermic were it not for the final step.

Finally, in Step 3, we allow the gaseous ions to come together to form the final solid product. As we recall from Chapter 4, a very large amount of energy, the *lattice energy, U,* is released at this step. Adding all of these energy terms should then give us the observed enthalpy of formation of a metal halide such as NaCl:

$$\Delta H_\text{f}(\text{MX}_n) = \Delta H_\text{atom}(\text{M}) + n\,\Delta H_\text{atom}(\text{X}) + \sum \text{IE(n)} - n\,\text{EA(X)} + U \tag{7.4}$$

In fact, one of the practical uses of the Born–Haber cycle is to obtain experimental values of lattice energies, to which calculated values such as the $-751\,\text{kJ mol}^{-1}$ for NaCl can be compared. Subtracting the energy terms obtained for Steps 1 and 2 ($+376\,\text{kJ mol}^{-1}$) from the enthalpy of formation of NaCl ($-411\,\text{kJ mol}^{-1}$) gives us an experimental lattice energy of $-787\,\text{kJ mol}^{-1}$. The calculated value of the lattice energy for NaCl ($-751\,\text{kJ mol}^{-1}$, Section 4.6) is close enough to this to give us confidence in the essential correctness of our model for the Coulombic attractive forces in ionic compounds (Chapter 4).

Not all such calculations agree so well with experimental lattice energies, however. For example, the enthalpy of formation of silver bromide is $-96\,\text{kJ mol}^{-1}$.[5] Subtracting from this the other component energy terms analogous to those in Figure 7.1, we obtain an experimental lattice energy of $-899\,\text{kJ mol}^{-1}$. Silver bromide crystallizes in the NaCl lattice; substituting the appropriate ionic radii, Madelung constant, Born exponents, and charge into Eq. (4.12) gives us a calculated lattice energy of $-702\,\text{kJ mol}^{-1}$. This is seriously short of the experimental value. In fact, had we taken this calculated value as valid and added it to the other terms in Figure 7.1, we would have predicted an enthalpy of formation for AgBr of $+101\,\text{kJ mol}^{-1}$. This value suggests that AgBr is thermodynamically unstable, and that if it exists it could be explosive! Since cameras using photographic processes based on AgBr seldom if ever explode, we may conclude that the Coulombic model for the attraction of Ag^+ and Br^- fails to account completely for the lattice energy of this compound; the lattice energy is more exothermic than calculated. As we saw in Chapter 5, a contribution to ΔH due to covalent bond formation is a characteristic of the reactions of soft acids and soft bases.

Another use of Born–Haber calculations is in predicting whether unknown compounds will be thermodynamically stable or not. When Bartlett prepared the surprising ionic compound $O_2{}^+PtF_6{}^-$, he noticed that the first ionization energy of the "noble gas" xenon was lower than the first ionization energy of the O_2 molecule. He made a rough Born–Haber calculation to predict that there might be a stable compound $Xe^+PtF_6{}^-$. So (in a vacuum system) he mixed xenon gas and PtF_6 gas and obtained a solid ionic product[6] that created quite a sensation in inorganic chemistry, since at the time it was thought that noble gases could form no compounds.

Example 7.1

Extrapolate the appropriate properties of the newly discovered element 112 just below mercury, "eka-mercury", and use these to predict whether this element will form a fluoride analogous to HgF_2.

SOLUTION:

Extrapolation of numerical values down a short and irregular group such as 12 is of course risky business, but let us try anyway just for fun. We extrapolate the following values for eka-Hg: $\Delta H_{atom} = +20\,kJ\,mol^{-1}$; $IE(1) = +1130\,kJ\,mol^{-1}$; $IE(2) = +2000\,kJ\,mol^{-1}$. We also need the known terms for fluorine: $2\ (\Delta H_{atom}) = +158\,kJ\,mol^{-1}$; $-2(EA) = -656\,kJ\,mol^{-1}$. Thus the enthalpy of formation of eka-HgF_2 should be roughly $+2652\,kJ\,mol^{-1}$ plus the lattice energy of this compound.

Since F^- is a hard base while eka-Hg^{2+} is expected to be a soft acid, we will assume that Eq. (4.12) will also give us a reasonably accurate lattice energy for eka-HgF_2. We extrapolate an ionic radius of 123 pm for eka-Hg^{2+}, which is virtually identical with that of F^-(119 pm), so the fluorite lattice type should be adopted (Madelung constant 2.51939). We will take the Born exponent of eka-Hg^{2+} to be 14; averaging this with the Born exponent for F^-, 7, gives us 10.5. Substituting these numbers into Eq. (4.12), we predict a lattice energy for eka-HgF_2 of about $-2617\,kJ\,mol^{-1}$.

If we believe these numbers (and in view of their uncertainties we should *not*), ΔH_f for the fluoride of eka-Hg will be roughly $+35\,kJ\,mol^{-1}$. Unless covalent bonding stabilizes this compound, it will be thermodynamically unstable. Since fluorine is the strongest of oxidizing agents and gives some of the greatest lattice energies, this suggests the possibility that the element below mercury may not exhibit the +2 group oxidation state: It may be a noble gas or liquid, which we had anticipated in Section 1.12 as a consequence of relativistic effects. On the other hand, some calculations[7] suggests that, due to relativistic effects, mercury might form HgF_4, so that eka-HgF_4 would be an even stronger possibility.

7.2 Thermochemical Analysis of the Activity Series of the Elements

In Experiment A.6, we examined the activity of the metals as reducing agents (for water or hydrogen ion), and found a better correlation of their activity with their Pauling electronegativities than with their (first) ionization energies. This finding may well have surprised you, since the reaction of the metals with water or hydrogen ion produces metal ions and obviously does involve ionization of the metals.

One problem in correlating activity and first ionization energies is that most of the metals form, not +1, but +2 or +3 (or even +4) ions. Hence, the second and third (and perhaps fourth) ionization energies should have been involved in the correlation. But there is still another problem: Ionization energies apply to *gaseous* metal *atoms* giving *gaseous* metal ions, whereas the reactions of metals with water or hydrogen ion involve *solid* (bulk) metals giving *hydrated* metal ions. To find the relationship of activity to ionization energies, we need to construct a thermochemical cycle for the relatively complicated process of dissolving a metal in standard (1 *M*) acid:

$$M(s) + z\,H^+(aq) \rightarrow M^{+z}(aq) + z/2\,H_2(g) \tag{7.5}$$

We can generalize the thermochemical cycle for the reaction of any metal with 1 *M* hydrogen ion; the generalized cycle is shown schematically in Figure 7.2.

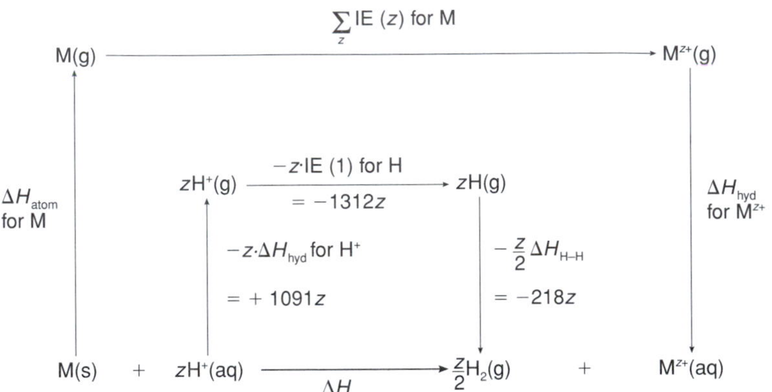

Figure 7.2
Thermochemical cycle for the reaction of a metal in its standard state with 1 *M* aqueous hydrogen ion to give an aqueous metal ion of charge $+z$ and gaseous hydrogen. $\Delta H_{\text{H–H}}$ is the H–H bond dissociation energy. For interpretation of the symbols for the other enthalpy changes, see the text.

Let us apply this process to the (dangerously exothermic) reaction of sodium metal with 1 *M* aqueous hydrogen ion. In Step 1, we take the two reactants to the gaseous state. For the hydrogen ion, we must remove its water of hydration; to do this we must *supply* the energy of hydration of H^+, $-\Delta H_{\text{hyd}} = +1091\,\text{kJ mol}^{-1}$ (Table 2.1). The sodium metal must be atomized; its heat of atomization ΔH_{atom} is $+107\,\text{kJ mol}^{-1}$.

In Step 2, we allow the gaseous sodium atom and hydrogen ion to react—an electron is ionized from sodium (giving gaseous Na^+) and is acquired by the hydrogen ion (giving a gaseous H atom). To do this, we must supply the first ionization energy of Na, $\Delta H_{\text{IE(1)}} = +496\,\text{kJ mol}^{-1}$ from Table 7.2, but we allow the *reverse* of the first ionization of hydrogen to occur, $-\Delta H_{\text{IE(1)}} = -1312\,\text{kJ mol}^{-1}$.

In Step 3, we allow the gaseous sodium ion to hydrate itself, $\Delta H_{\text{hyd}} = -405\,\text{kJ mol}^{-1}$. We allow the individual hydrogen atoms to come to their final state, as gaseous diatomic molecules. Energy is released when 0.5 mol of covalent H–H bonds is formed per mole of H atoms. This energy is the negative of the heat of atomization of H (or one-half the reverse of the conventional H–H bond dissociation energy, $-0.5\Delta H_{\text{H–H}}$), $-218\,\text{kJ mol}^{-1}$.

Finally, we apply Hess's law and add the energy changes in all these steps. The terms involving the hydrogen add up to $-439\,\text{kJ mol}^{-1}$, while those involving sodium give $+198\,\text{kJ mol}^{-1}$; overall the theoretically calculated enthalpy change for the reaction of sodium metal with 1 *M* acid is $-241\,\text{kJ mol}^{-1}$. Experimentally, we know that the standard reduction potential of sodium ion is -2.71 V (Table 6.1). This value can be converted to thermodynamic units using the relationship

$$\Delta G^\circ = -n\mathscr{F}E^\circ \tag{7.6}$$

where n is the number of electrons involved in the redox reaction (one in this case) and $\mathscr{F}$, the Faraday, is $96.5\,\mathrm{kJ\,eV^{-1}}$. (In the thermochemical cycle of Fig. 7.2, the metal ions are not being reduced but are being formed by oxidation; hence, we also need to reverse the sign of the standard reduction potential.) If we neglect entropy changes (i.e., formation of gaseous H_2, hydration and dehydration of hydrated ions), we can equate ΔG and ΔH and obtain ΔG for the oxidation of sodium, $-262\,\mathrm{kJ\,mol^{-1}}$. In view of our neglect of entropy changes, the agreement with the calculated enthalpy change, $-241\,\mathrm{kJ\,mol^{-1}}$, is good.

The calculations for the activity of sodium can be applied to the activity of any metal by substituting the appropriate atomization and ionization energies. We may note that the overall theoretical enthalpy changes involving hydrogen are constant at $-439\,\mathrm{kJ\,mol^{-1}}$ of H^+ involved. Thus the generalized ΔH for the reaction is

$$\Delta H = \Delta H_{\text{atom}} + \sum \mathrm{IE}(z) + \Delta H_{\text{hyd}} - 439z \tag{7.7}$$

where z is the number of electrons and hydrogen ions involved and is the charge on the metal cation formed. Finally, assuming the equality of ΔH and ΔG, we apply Eq. (7.6) to replace ΔG by $-z\mathscr{F}E^\circ$, and, noting that E° for this oxidation reaction is the reverse of the standard reduction potential, we obtain in general:

$$96.5\,E^\circ = \frac{\Delta H_{\text{atom}}}{z} + \frac{\sum \mathrm{IE}(z)}{z} + \frac{\Delta H_{\text{hyd}}}{z} - 439\,\mathrm{kJ\,mol^{-1}} \tag{7.8}$$

7.3 Significance of Atomization Energies in Redox Chemistry

In Table 7.1, we note some periodic trends in heats of atomization that are quite different from any we have seen before. In crossing the s and p blocks in the second and third periods, we find the highest atomization energies in the *middle* of the period: Carbon and silicon are the hardest elements to atomize or vaporize in their periods. Likewise, in crossing the s and d blocks in the fifth and sixth periods, the highest atomization energies are also in the middle, at or near Group 6. As we will see in Chapter 12, it is in the middle of these blocks of the periodic table that the greatest number of covalent bonds must be broken in order to produce atoms of elements.

In a pattern that we have seen before, the vertical trends are opposite in the p and in the d blocks. Atomization energies are highest for elements at the *top* of the p block [i.e., carbon in Group 14(IV)], while they are highest at the *bottom* of the d block (tungsten in Group 6). These trends are due to the corresponding trends in bond energies that were discussed in Chapter 3.

To see the chemical significance of the atomization energies of the metals, let us calculate standard reduction potentials from Eq. (7.8) with the ΔH_{atom} term deleted (i.e., for the reaction of *gaseous* metal atoms with acids). This calculation amounts to subtracting a factor of $\Delta H_{\text{atom}}/96.5z$ from the recorded standard reduction potentials (Table 6.2); in the case of sodium this changes its standard reduction potential from -2.71 to $-3.82\,\mathrm{V}$. More interestingly, it changes the reduction potential of the least active metal, gold, from $+1.68$ to $-2.13\,\mathrm{V}$ (i.e., makes it almost as active as magnesium actually is).[8]

In fact, it *is* now possible to start reactions with gaseous metal atoms, by providing the atomization energy separately with a high-temperature resistance heater in a high-vacuum apparatus known as a *metal atom reactor*. The metal atoms produced by this means are in fact much more reactive than bulk metals as we know them. They must be handled in high vacuum at very low temperatures to prevent them from reacting with each other, with other gaseous substances, or with the walls of the reaction vessel. With the enhanced activity possessed by gaseous metal atoms, it is possible to carry out many reactions that are completely impossible with solid or liquid metals.[9]

Example 7.2

Compare the activity of the metals Hg, Pt, Pd, and Ba (all of which form +2 ions) in the solid state; as gaseous atoms.

SOLUTION:

We may use the standard reduction potentials of these metals as indicators of their solid-state activities. Referring to Table 6.2, we can set up an activity series: activity of Ba (−2.91 V) ≫ Pb (−0.13 V) ≫ Hg (+0.85 V) > Pt (+1.2 V). This order is what we would have expected based on the electronegativities of the elements.

To calculate the gas-phase activities we can either use Eq. (7.8), omitting the atomization term,[10] or we can subtract the atomization energy (in volts) from the standard reduction potentials. Either way of calculating may introduce some errors resulting from entropy changes, but we will disregard these. If we take the latter approach, we find that we must subtract the following amounts from the standard reduction potentials: Ba, 0.94 V; Pb, 1.01 V; Hg, 0.33 V; and Pt, 2.92 V. This causes platinum to jump up dramatically in the activity series: Ba (−3.85 V) ≫ Pt (−1.72 V) > Pb (−1.14 V) ≫ Hg (+0.52 V).

As another illustration of the significance of atomization energies, let us consider further some of the redox chemistry of the elements with especially high atomization energies (e.g., carbon and the metals in the vicinity of tungsten). Certainly, we can understand how the high atomization energies of these elements reduce their activities. But although these elements are relatively reluctant to assume positive oxidation states, when they do so they prefer to assume *very high* positive oxidation states. Thus although Mo and W are much less active metals than Cr, they show a strong preference for the +6 oxidation state when they are oxidized, while Cr prefers +3 (Table B1).

To analyze this, let us suppose that we have 2 mol of Mo^{3+} or W^{3+} ions. Let us consider the possibility of a *disproportionation* reaction:

$$2\,M^{3+}(aq) \rightleftharpoons M^{6+}(aq) + M(s) \tag{7.9}$$

This disproportionation reaction will be more favorable for Mo and W than for Cr, since producing the solid metal will recover the heat of atomization of one-half the metal atoms, which is more significant for Mo and W than for Cr. Looking at this from another angle, we note that the 3*d* orbitals of Cr are not so readily available for bonding; in the Cr^{3+}(aq) ion these hold three electrons that are not used in bonding to the waters of hydration. In the products M^{6+}(aq) [or, more realistically, MO_3(s)], bonding to oxygen does occur; in M(s) bonding of metal atoms to each other does occur. The 4*d* orbitals of Mo and the 5*d* orbitals of W are more suitable for such extensive bonding; it is energetically unfavorable to let these orbitals be "wasted" as in Cr^{3+}(aq).

7.4 Periodicity and Significance of Ionization Energies and Electron Affinities

Successive Ionization Energies. As we see by paging through the tables of electron affinities (zeroth ionization energies), followed by the first, second, and third ionization energies, for a given element successive ionizations generally become more difficult in a relatively regular pattern. But if the *n*th ionization is one that begins removing core electrons from an atom, that ionization will naturally be much more difficult than would otherwise have been the case. In Figure 7.3 the square roots of the successive ionization energies of the first 39 elements are plotted[11]; large gaps are noted between the energies of ionizations of core and of valence electrons of an atom of given atomic number Z.

Figure 7.3 also shows that ionization energies fall into bands or groups of $2, 6, 10, \ldots,$ energies, which clearly correspond to ionizations from sets of $s, p, d, \ldots,$ orbitals. However, not all electrons from a set of orbitals ionize at the same energy: The energy of ionization depends on the number of other electrons from the same set that has been ionized, since this alters the number of electron–electron repulsions present. When we drew an energy level diagram such as that of Figure 1.9, we were applying the *orbital approximation* that a given orbital holds any (possible) number of electrons, each at the same energy. This is not actually the case; however, calculating

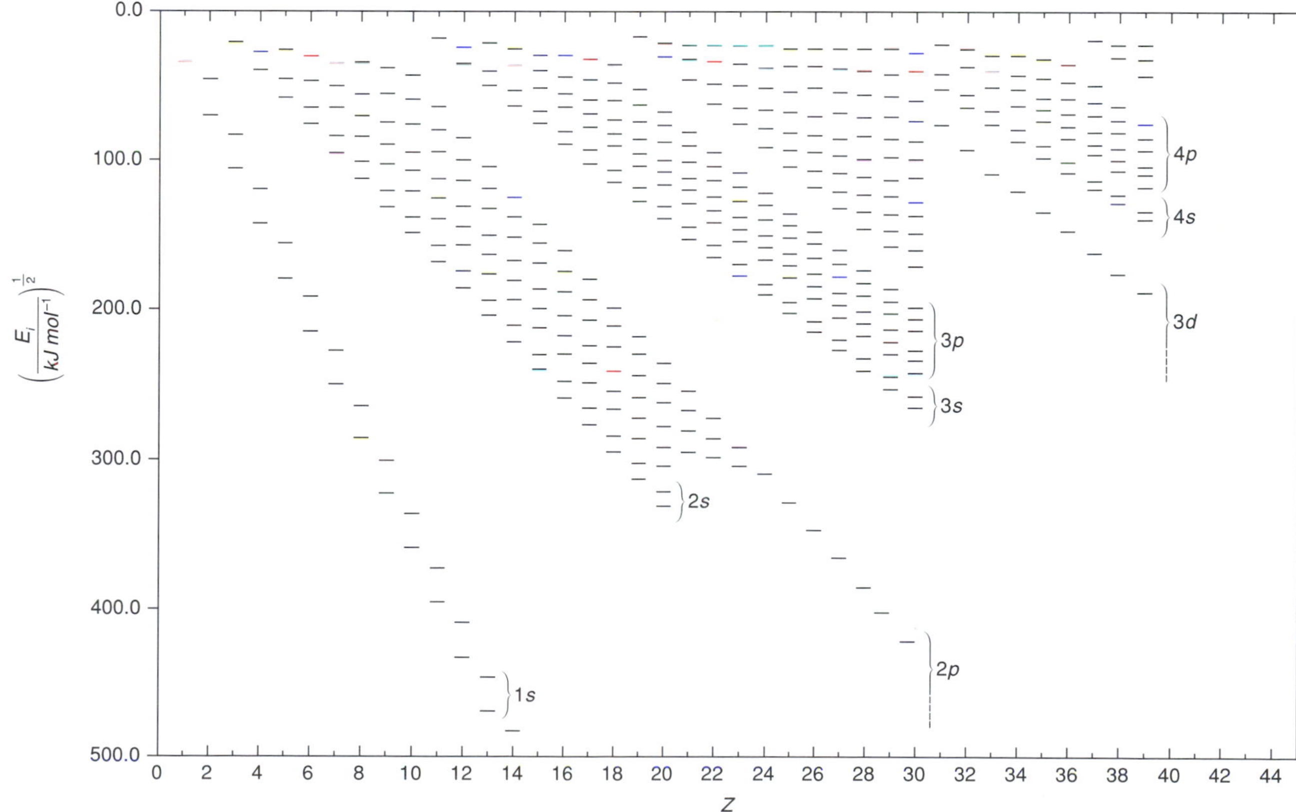

Figure 7.3
Square roots of the known ionization energies of the first 39 elements. [Adapted from P. Mirone, *J. Chem. Educ.*, **68**, 132 (1991).]

the effects of the changing numbers of electron–electron repulsions presents so many difficulties that we normally make this approximation.

Note that there is a relationship between Figures 7.3 and 1.9: the order of orbital energies plotted in Figure 1.9 is derived qualitatively from the mean ionization energies for a given orbital type as shown in Figure 7.3. Note also that as the bands develop clearly to the right in Figure 7.3, they show the ordering $3s < 3p < 3d < 4s$, which is the ordering found in Figure 1.9 after these orbitals have been filled for some time. Extrapolating back to the left in Figure 7.3 shows, for example, that the energies of 3*d* orbitals may well cross over the energies of 4*s* orbitals.

Horizontal Periodic Trends. We generally find that a given ionization (including the zeroth) becomes more difficult across a period. Among atoms using 2*s* and 2*p* valence electrons, for example, ionization is easiest for the $2s^1$ electron configuration [IE(1) of Li, IE(2) of Be^+, IE(3) of B^{2+}, EA of He^-; this trend is discernible in Fig. 7.3 and is amplified in Fig. 7.4]. Ionization becomes more difficult farther to the right for atoms or ions of the same charge. The highest ionization energy is required for atoms or ions of the $2s^22p^6$ electron configuration [IE(1) of Ne, IE(2) of Na^+, IE(3) of Mg^{2+}, EA of F^-]. When the electron being lost changes from being an *s* valence electron to a *p* valence electron, however, ionization becomes easier (e.g., compare the first IE values of B and Be). When the electron being removed first comes from a more than half-filled subshell, ionization is also somewhat easier, since an electron in a filled orbital is repelled by the other electron occupying the same orbital. Thus the first IE of O ($2s^22p^4$, with one paired-up *p* electron) is somewhat closer to zero than that of N ($2s^22p^3$, with each *p* electron in its own orbital).

As an illustration of the significance of the horizontal variation of ionization energies, let us consider the question of the relative stabilities of the +2 and +3 ions of the *d*-block metals of the fourth period. As we see from Table B1, there is a rather

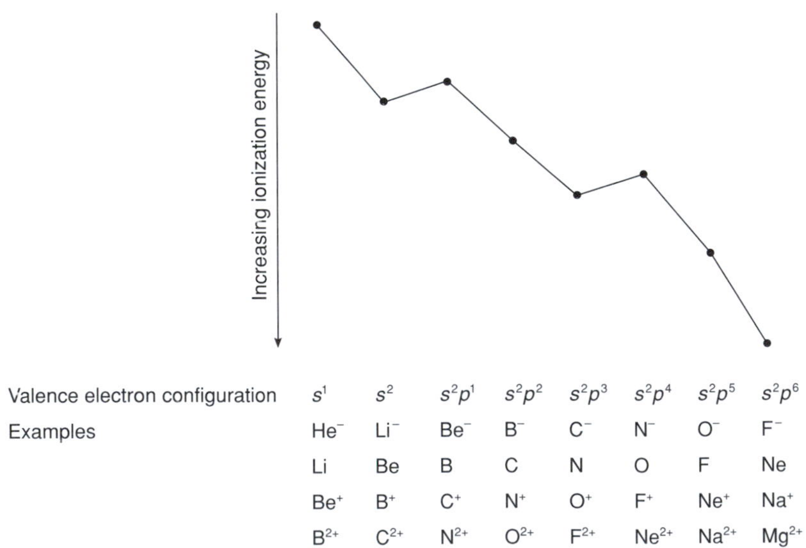

Figure 7.4
Horizontal periodic trends in ionization energies and electron affinities.

irregular alternation in stability of these two oxidation states as we cross the fourth period. In the production of a +3 ion by dissolving a *d*-block metal in acid, we may consider the final stage to be the following:

$$M^{2+}(aq) + H^{+}(aq) \rightarrow M^{3+}(aq) + 0.5\,H_2(g) \tag{7.10}$$

A thermochemical cycle may be set up for this reaction for which the energy terms add up as follows:

$$96.5\,E^{\circ} = \mathrm{IE}(3) + \Delta H_{\mathrm{hyd}}(+3\,\mathrm{ion}) - \Delta H_{\mathrm{hyd}}(+2\,\mathrm{ion}) - 439 \tag{7.11}$$

By using the Latimer equation (2.1), we can estimate the hydration energies of the +3 and +2 ions for which radii are given in Table C. Taking the difference between these hydration energies gives us a contribution ranging from $-2560\,\mathrm{kJ\,mol^{-1}}$ for Ti to $-2693\,\mathrm{kJ\,mol^{-1}}$ for Cr. (The hydration energies are always more favorable for the +3 ions, of course). This variation is, however, much smaller than the variation in IE(3), which ranges from $+2389\,\mathrm{kJ\,mol^{-1}}$ for Sc to $+3833\,\mathrm{kJ\,mol^{-1}}$ for Zn.

Thus the variation in the third ionization energies of the *d*-block elements of the fourth period is largely responsible for the varying stability of the +3 ions as compared to the +2 ions. In general, IE(3) increases across the fourth period in the *d* block; hence, +3 ions generally are energetically unfavorable after Cr. However, the +3 ion of Fe is more easily formed, since it involves the relatively easy ionization of the sixth *d* electron of Fe^{2+}, which is one electron above a half-filled subshell.

Similarly, the third ionization energy increases across the sixth period in the *f* block, resulting in increasing stability of the +2 oxidation state—until Gd^{2+} is reached. The Gd^{2+} ion has one electron above a half-filled subshell and is therefore very readily oxidized to Gd^{3+}. But the third ionization energy then increases again to the end of the *f* block. Hence, although the +3 oxidation state is the most stable for all the *f*-block elements in the sixth period, the +2 oxidation state is least unfavorable for the element before Gd (Eu^{2+}, with a half-filled subshell) and at the end of the *f* block (Yb^{2+}, with a filled subshell).

It has been suggested[12] that, if the *n*th ionization energy of an element is more than 2400 kJ greater than the $(n-1)$th ionization energy, that the other factors such as enhanced hydration energies cannot compensate, and the $+n$ oxidation state of that element is not achievable by chemical means. For a few cases in which the *n*th ionization energy is only 1200–2400 kJ greater, the higher oxidation state remains as a challenge to be discovered: Hg(III), reported once in a short-lived complex[13] (see also Example 7.1); Cd(III) and Zn(III); and Rb(II) and Cs(II) (see Exercise 17).

Vertical Periodic Trends. Generally speaking, we find that the *n*th ionization energy or electron affinity *decreases* down a group: The electron being ionized is in a larger orbital of higher principal quantum number, and hence is not attracted as strongly to the nucleus. A couple of irregularities are noteworthy, however. (1) The electron affinities of the *p*-block elements of the second period are unexpectedly less than those of the third (or later) periods because of the great electron–electron repulsions found in such small anions as F^- and O^-. (2) The ionization energies of some of the later *d*-block elements of the sixth period and early *f*-block elements of the seventh period are higher than expected, because of the relativistic effects discussed in Section 1.11: Gold has the highest electron affinity of any element other than a halogen.

Calculations of Ionization Energies. There are two attractive features about relating chemical reactivity trends to ionization energies: (1) they can be measured with fair accuracy; (2) they can be calculated from effective nuclear charges Z^*, such as those developed from Slater's rules in Section 1.6, or more accurately from the Z^* values of Clementi and Raimondi.[14] The basic equation for the energy of the nth electron in an atom or ion is derived from Eq. (1.11):

$$E_n = -R_{\mathrm{H}}(Z^*/n^*)^2 \tag{7.12}$$

in which the constants of Eq. (1.11) are gathered into the new constant R_{H}, which has the value 1312.0 kJ mol^{-1}. Technically, to obtain an ionization energy it is necessary to apply Eq. (7.12) to *each* electron in the atom being ionized, then to *each* electron in the ion produced, then take the difference[15]:

$$\mathrm{IE} = \sum E_n \text{ (ionized species)} - \sum E_n \text{ (un-ionized species)} \tag{7.13}$$

In practice, Z^* values are the same in the atom and the ion for electrons deep enough in the core, which simplifies the calculation somewhat.

Example 7.3

Calculate the first ionization energies of (a) Li; (b) F; and (c) Cu with electron configuration $4s^2 3d^9$ ionizing to Cu^+ with electron configuration $3d^{10}$.

SOLUTION:

(a) The electron configuration of Li is $1s^2 2s^1$; that of the product of ionization, Li^+, is $1s^2$. The Z^* value for the 2s electron of Li is +1.30 (Table 1.3; Section 1.7). The Z^* value for each 1s electron in either Li or Li^+ is $3 - (0.35) = +2.65$; since it is unaltered upon ionization it makes no contribution to the ionization energy, which is therefore equal to $1312(1.30/2)^2 = 554$ kJ mol^{-1}, as compared to 520 kJ mol^{-1} in Table 7.2.

(b) The electron configuration of F is $1s^2(2s, p)^7$; that of its ionization product, F^+, is $1s^2(2s, p)^6$. Again ionization results in no change in the effective nuclear charge felt by the core 1s electron, so the ionization energy depends only on the changes felt by the valence electrons. The Z^* parameter for each of the seven $(2s, 2p)$ electrons in F is +5.20, while Z^* for each of the six $(2s, 2p)$ electrons in F^+ is +5.55. Applying Eq. (7.12) to these electrons and subtracting the energies for F from the energies for F^+, we obtain: $\mathrm{IE}_{(1)} = -1312[6(5.55/2)^2 - 7(5.2/2)^2] = -60{,}600 + 62{,}100 = 1500$ kJ mol^{-1}, as compared to the experimental 1681 kJ mol^{-1}. (Note that since these calculations involve the differences of two large numbers, there are not many significant figures in the answers.)

(c) For the Cu atom, $Z^*_{4s} = +4.20$, while $Z^*_{3d} = +8.20$; for the Cu^+ ion with the rearranged electron configuration, $Z^*_{3d} = +7.85$. The Z^* parameter is unchanged for all of the core electrons upon ionization, so we can ignore them in the calculations: $\mathrm{IE} = -1312[10(7.85/3.00)^2 - 2(4.20/3.70)^2 - 9(8.20/3.00)^2] = -1312(68.5 - 2.58 - 67.2) = 1700$ kJ mol^{-1}. This ionization energy differs from the experimental value of 746 kJ mol^{-1}; however, the gaseous Cu atom actually has an anomalous electron configuration of $3d^{10}4s^1$ rather than the configuration assumed in this problem.

Isoelectronic Periodic Trends. For an ns^1 isoelectronic series (such as the $3s^1$ series Na, Mg^+, Al^{2+}, . . .), the ionization energy is simply given by Eq. (7.12); this allows us to see why the square root of the ionization energy of a given electron in a given orbital increases linearly as Z^* increases across a period, as can be seen by following the ionization of the first $3s^1$ electron in Figure 7.3. In attempting to prepare some of the superheavy nuclei discussed in Section 1.12, it is necessary to produce (in a particle accelerator) electron-free ions such as Fe^{26+} to be used to bombard atoms such as uranium (the projectiles must be free of electrons so that electron–electron repulsions between the iron and uranium species do not prevent the iron nucleus from reaching the uranium nucleus in order to undergo nuclear fusion with it.) Use of Eq. (7.12) to calculate the ionization energy of just the 1*s* electron of Fe will show why this cannot be done by chemical means!

7.5 Polar Covalent Bond Energies and Pauling Electronegativities

Although thermochemical analyses based on ionization, hydration, and atomization energies and electron affinities are satisfying theoretically because good values of these quantities are available and neat thermochemical cycles can be constructed containing them, these analyses suffer from a defect: Reactivity trends seldom depend on only one of these quantities, so that they are only indirectly related to reactivity. Of far more direct use in predicting reactivity is the Pauling electronegativity of the reactive element or ion. In the following sections, we will investigate the use of electronegativities in another type of thermochemical cycle.

Pauling electronegativities are derived from measurements of bond dissociation energies of bonds between unlike atoms. Bond dissociation energies for gaseous elements (i.e., bonds involving like atoms) were considered in Chapter 3 and were tabulated in Table 3.1. Bond dissociation energies can also be evaluated for gaseous single-bonded polar covalent compounds provided that there is only one type of bond present. (If more than one type of bond is present, all of them may dissociate; there is then no unambiguous way to divide the enthalpy change among the different bond types.) The bond energies of the gaseous halides of many of the elements are tabulated in Tables 7.6–7.9.

The chemist Linus Pauling noted some interesting periodic trends in this data.[16] First, the polar bonds of elements to halogens are also almost always stronger than the nonpolar bonds of the elements to themselves or the bonds of the halogens to themselves (Table 3.1). Second, the strongest bonds in the gas state are for the halides of the metals at the far left of the periodic table. These gaseous halides, of course, consist of ion pairs, triplets, and so on. Thus it seems that the bond energies of gaseous compounds are directly related to their degree of polarity: Ion pairs show higher bond energies than polar covalent bonds, which show higher bond energies than nonpolar covalent bonds.

Pauling thus suggested that bond energies can be analyzed as a sum of contributions from covalent bonding and ionic bonding. Pauling chose to take the covalent contribution to the bond energy as the average of the halogen–halogen and element–element covalent bond energies (Table 3.1). We may suppose that the ionic contribution should depend on the magnitude of the partial positive charge of the element times the magnitude of the partial negative charge of the halide divided by the bond distance

Table 7.6
Element–Fluorine Bond Dissociation Energies[a,b]

HF	565														
LiF	573	BeF_2	632	BF_3	613	CF_4	485	NF_3	283	OF_2	189	F_2	155		
NaF	477	MgF_2	513	AlF_3	583	SiF_4	565	PF_3	490	SF_6	284	ClF_5	142		
												ClF_3	172		
												ClF	249		
KF	490	CaF_2	550	GaF_3	469	GeF_4	452	AsF_5	406	SeF_6	285	BrF_5	187	KrF_2	50
						GeF_2	481	AsF_3	484	SeF_4	310	BrF_3	201		
										SeF_2	351	BrF	249		
RbF	490	SrF_2	553	InF_3	444	SnF_4	414	SbF_5	402	TeF_6	330	IF_7	231	XeF_6	126
				InF	523	SnF_2	481	SbF_3	440	TeF_4	335	IF_5	268	XeF_4	130
										TeF_2	393	IF_3	272	XeF_2	131
CsF	502	BaF_2	578	TlF	439	PbF_4	331	BiF_5	297						
						PbF_2	394	BiF_3	393						

SOURCE: Data are from J. E. Huheey, E. A. Keiter, and R. L. Keiter, *Inorganic Chemistry: Principles of Structure and Reactivity*, 4th ed., Harper-Collins, New York, 1993; pp. A-25–A-33.
[a]Units are in kilojoules per mole (kJ mol^{-1}).
[b]Compounds are shown for which each bond dissociation energy applies.

Table 7.7
Element–Chlorine Bond Dissociation Energies[a,b]

HCl	428														
LiCl	464	$BeCl_2$	461	BCl_3	456	CCl_4	327	NCl_3	313	HOCl	218	FCl	249		
NaCl	408	$MgCl_2$	406	$AlCl_3$	421	$SiCl_4$	381	PCl_3	326	S_2Cl_2	255	Cl_2	240		
KCl	423	$CaCl_2$	429	$GaCl_3$	354	$GeCl_4$	349	$AsCl_3$	322	$SeCl_4$	192	BrCl	216		
						$GeCl_2$	385			$SeCl_2$	243				
RbCl	444	$SrCl_2$	469	$InCl_3$	328	$SnCl_4$	323	$SbCl_5$	248	$TeCl_4$	311	ICl	208		
				InCl	435	$SnCl_2$	386	$SbCl_3$	315	$TeCl_2$	284				
CsCl	435	$BaCl_2$	475	TlCl	364	$PbCl_4$	243	$BiCl_5$	274						
						$PbCl_2$	304								

SOURCE: Data are from J. E. Huheey, E. A. Keiter, and R. L. Keiter, *Inorganic Chemistry: Principles of Structure and Reactivity*, 4th ed., Harper-Collins, New York, 1993; pp. A-25–A-33.
[a]Units are in kilojoules per mole (kJ mol^{-1}).
[b]Compounds are shown for which each bond dissociation energy applies.

Table 7.8
Element–Bromine Bond Dissociation Energies[a,b]

HBr	362												
LiBr	418	$BeBr_2$	372	BBr_3	377	CBr_4	285			HOBr	201	FBr	249
NaBr	363	$MgBr_2$	339			$SiBr_4$	310	PBr_3	264	S_2Br_2	217	ClBr	216
KBr	379	$CaBr_2$	402	$GaBr_3$	302	$GeBr_4$	276	$AsBr_3$	258	$SeBr_4$	151	Br_2	190
						$GeBr_2$	325			$SeBr_2$	201		
RbBr	385	$SrBr_2$	405	$InBr_3$	279	$SnBr_4$	273	$SbBr_5$	184	$TeBr_4$	176	IBr	175
				InBr	406	$SnBr_2$	329	$SbBr_3$	260	$TeBr_2$	243		
CsBr	416	$BaBr_2$	427	TlBr	326	$PbBr_4$	201	$BiBr_3$	232				
						$PbBr_2$	260						

SOURCE: Data are from J. E. Huheey, E. A. Keiter, and R. L. Keiter, *Inorganic Chemistry: Principles of Structure and Reactivity*, 4th ed., Harper-Collins, New York, 1993; pp. A-25–A-33.
[a]Units are in kilojoules per mole (kJ mol^{-1}).
[b]Compounds are shown for which each bond dissociation energy applies.

Table 7.9
Element–Iodine Bond Dissociation Energies[a,b]

HI	295												
LiI	347	BeI_2	289			CI_4	213			HOI	201	FI	278
NaI	304	MgI_2	264			SiI_4	234	PI_3	184			ClI	208
KI	326	CaI_2	326	GaI_3	237	GeI_4	212	AsI_3	200	SeI_2	151	BrI	175
						GeI_2	264						
RbI	331	SrI_2	335	InI_3	225	SnI_4	205	SbI_3	195	TeI_4	121	I_2	149
						SnI_2	261			TeI_2	192		
CsI	335	BaI_2	360	TlI	280	PbI_2	205	BiI_3	168				

SOURCE: Data are from J. E. Huheey, E. A. Keiter, and R. L. Keiter, *Inorganic Chemistry: Principles of Structure and Reactivity*, 4th ed., Harper-Collins, New York, 1993; pp. A-25–A-33.
[a]Units are in kilojoules per mole (kJ mol^{-1}).
[b]Compounds are shown for which each bond dissociation energy applies.

(i.e., Z^2/r). If we neglect the variability of the bond distance, an equation for the element(E)–halogen(X) bond energy, $\Delta H(\text{E–X})$, begins to emerge:

$$\Delta H(\text{E–X}) = 0.5[\Delta H(\text{X–X}) + \Delta H(\text{E–E})] + k(\text{partial positive charge})(\text{partial negative charge}) \quad (7.14)$$

The magnitudes of the partial positive (+) and negative (−) charges in polar covalent bond are even now difficult to determine unambiguously. But logically each should depend on the *difference* between the relative electron-attracting abilities of the two elements: The more unequal this ability, the more charge will build up at each end of the bond. This electron-attracting ability Pauling called the **electronegativity** χ_P of that element; he defined it as the attraction of an atom in a molecule (or polyatomic ion) for the electrons in its covalent bonds. Thus each of the partial charges would be expected to build up in proportion to the difference between the electronegativities of the two atoms, until the charge separation became as complete as it could be in the ion pairs:

$$\text{Partial positive charge} = -\text{partial negative charge} = k'[\chi_P(\text{X}) - \chi_P(\text{E})] \quad (7.15)$$

Substituting this expression into Eq. (7.14), collecting the constants $-kk'^2$ together, and setting them equal to 96.5 kJ eV^{-1} gives us the relationship between bond energies and differences of Pauling electronegativities:

$$\Delta H(\text{E–X}) = 0.5[\Delta H(\text{X–X}) + \Delta H(\text{E–E})] + 96.5[\chi_P(\text{X}) - \chi_P(\text{E})]^2 \quad (7.16)$$

Provided that we arbitrarily set one electronegativity value (Pauling chose 4 for F; we now use 3.98), Eq. (7.16) can be used, along with modern experimental gaseous bond dissociation energies, to determine the values of the Pauling electronegativities given in Table A (see inside front cover).[17] Values have been obtained separately for each kind of gaseous halide and for other single-bonded compounds of the elements, such as hydrides. The values in Table A represent mean values for these different compounds, and have standard deviations of ± 0.05.[18]

Equation (7.16) is unreliable for any electronegativity differences greater than about 1.8. As noted from Tables 7.6–7.9, at the left of the periodic table the bond energies are approximately constant despite changes in Pauling electronegativities. This observation is one source of the suggestion that the bonding in a halide or other compound becomes as ionic as it can get with electronegativity differences much over 1.8.

For the *d*- and *f*-block metals the element–element single-bond energies are generally unknown. The electronegativity difference $\chi_P(\text{E}) - \chi_P(\text{X})$ must therefore be calculated by another route, which results from the analysis of the formation reaction of the metal halides:

$$0.5\,\text{M–M(s)} + n/2\,\text{X–X(g)} \rightarrow \text{MX}_n\text{(g)} \quad (7.17)$$

ΔH for this reaction is ΔH_f of the gaseous halide.

The total bond energies of the products are given by the Expression (7.18):

$$-n\{0.5(\Delta H_{\text{M–M}} + \Delta H_{\text{X–X}}) + 96.5[\chi_P(\text{M}) - \chi_P(\text{X})]^2\} \quad (7.18)$$

For the $n/2$ mol of X–X(g) reactant we simply add $n/2\Delta H_{X-X}$, since there is no electronegativity difference between the X atoms in the X–X molecules; this cancels the corresponding term in the Expression (7.18). However, unless the solid metal also contains $n/2$ mol of M–M bonds, its bond-energy term will not cancel the corresponding term in the Expression (7.18), and we will not obtain a usable result. (The Pauling equation is most conveniently applied to the study of reactivity trends in which there are the same number of bonds among reactants as among products.) As we shall see in Chapter 12, this treatment of the number of M–M bonds is fairly defensible. Then we finally obtain the simple equation:

$$\Delta H_f = -96.5\, n[\chi_P(M) - \chi_P(X)]^2 \tag{7.19}$$

This equation is in fact used, along with the experimental enthalpies of formation of *d*- and *f*-block metal halides and other salts, to obtain the Pauling electronegativities of these metals.

Analysis of Reactivity trends. We can now apply the Pauling electronegativity Eq. (7.16) to the analysis of a simple HSAB equilibrium (Section 5.3) occurring in the gas phase:

$$\text{HA:SB(g)} + \text{SA:HB(g)} \rightleftharpoons \text{HA:HB(g)} + \text{SA:SB(g)} \tag{7.20}$$

We evaluate the bond energies of each of the four reactants and products using Eq. (7.16). We then take the sum of the bond dissociation energies of the *reactants* minus the sum of the bond dissociation energies of the *products* (in which bonds are formed, not dissociated). When we do this, all the terms involving element–element bond energies [ΔH(HA–HA), etc.] drop out, leaving four terms:

$$\Delta H = 96.5\{[\chi_P(\text{HA}) - \chi_P(\text{SB})]^2 + [\chi_P(\text{SA}) - \chi_P(\text{HB})]^2 - [\chi_P(\text{HA}) - \chi_P(\text{HB})]^2 - [\chi_P(\text{SA}) - \chi_P(\text{SB})]^2\} \tag{7.21}$$

Let us now take some typical electronegativity values for the donor and acceptor atoms of hard and soft acids and bases: $\chi_P(\text{HA}) = 1.6$; $\chi_P(\text{HB}) = 3.4$; $\chi_P(\text{SA}) = \chi_P(\text{SB}) = 2.5$. Substituting these values in the previous equation gives us

$$\Delta H = 96.5(0.81 + 0.81 - 3.24 - 0) = -156\,\text{kJ mol}^{-1} \tag{7.22}$$

This calculation indicates, as expected from the evidence presented in Section 5.5, that the reaction to give the products HA–HB and SA–SB is accompanied by a favorable ΔH, but it also indicates why: *The driving force of the HSAB reaction lies in the formation of the very stable ionic hard acid–hard base product.*

Qualitatively, we can also apply this conclusion to Reaction (7.5) involved in the activity series of metals. The solid metal put into acid and the gaseous H_2 produced each contain only nonpolar covalent bonds with no electronegativity differences at all. The only electronegativity differences occur in the M–OH_2 and H–OH_2 coordinate covalent bonds found in the hydrated metal ion produced and the hydronium ion reacted. We expect greater stability for whichever of these two has the greater electronegativity difference between oxygen and the species coordinated (the metal ion or hydrogen ion). This is qualitatively in agreement with our conclusions in Chapter 6;

if the metal is less electronegative than hydrogen,[19] the M–O bond should be more ionic than the H–O bond. Hence, the hydrated metal ion will be favored instead of the hydrated hydrogen ion (hydronium ion), and the metal will be more active than hydrogen.

A Semiquantitative Relationship of Metal Activity and Pauling Electronegativity: An Aside

The activity series is difficult to analyze quantitatively since the number or M–M bonds present per M atom in the solid metal reactant and the number of M–O bonds present in the hydrated-ion product are not so clearly determined. We can come closest to a satisfactory theoretical analysis if we consider first a *displacement* reaction of a more active metal M with the cation M′ of a less active metal,

$$\text{M–M(s)} + 2\,\text{M}'(\text{–OH}_2)_n{}^{z+} \rightarrow \text{M}'\text{–M}'\text{(s)} + 2\text{M}(\text{–OH}_2)_n{}^{z+} \quad (7.23)$$

in which M and M′ both have the same hydration numbers n (hence, the same number of M–O bonds) and the same number of M–M covalent bonds. Under these conditions, all terms in $\Delta H_{\text{M–M}}$, $\Delta H_{\text{M}'\text{–M}'}$, $\Delta H_{\text{M–O}}$, and $\Delta H_{\text{M}'\text{–O}}$ cancel out, leaving only two terms in electronegativity differences:

$$\Delta H = -2n \cdot 96.5\{[3.44 - \chi_\text{P}(\text{M})]^2 - [3.44 - \chi_\text{P}(\text{M}')]^2\} \quad (7.24)$$

It can be seen that ΔH will become more negative and (neglecting entropy) $E°$ for this reaction will increase as M becomes less electronegative than M′, so that the more ionic hydrated ion (that of M^{z+}) is formed.

To proceed to the analysis of the activity series, let M′ be H^+ in Eq. (7.23); then $z = 1$. In the hydronium ion product, we let $n = 1$; hence, this equation can be rewritten as:

$$\text{M–M} + 2\,\text{H–OH}_2{}^+ \rightarrow \text{H–H} + 2\,\text{M–OH}_2{}^+ \quad (7.25)$$

We now have greater problems in counting the total number of bonds, but if we make some semi-reasonable assumptions about these numbers,[20] and about the electronegativity of H *in the hydronium ion*, we obtain:

$$E° = -[(3.44 - \chi_\text{M})^2 - (3.44 - 1.9)^2] = -\chi_{\text{M}^2} + 6.88\chi_\text{M} - 9.46 \quad (7.26)$$

The fit of this equation to the standard reduction potentials of the metal ions and nonmetal compounds (in common oxidation states) to the elements is shown in Figure 7.5. Equation (7.26) reflects the fact that the activity of a metal or a nonmetal as a reducing agent increases as its electronegativity decreases, because there is a greater electronegativity difference and a greater ionic character in the $\text{M}^+\text{–OH}_2$ bond than in the $\text{H}^+\text{–OH}_2$ bond. Again the driving force for the reaction results from maximizing the ionic bonding in one product.

The mediocre fit of this equation to the experimental data clearly shows the effects of the drastic assumptions used in deriving it and the presence of only one variable: the Pauling electronegativity; much better fits are found with the three variables present in the thermochemical cycle (atomization energies, ionization energies, and hydration energies). The advantage to the single-variable relationship is that it makes it possible for a practicing chemist to determine *quickly* what safety hazards might be present when handling an unfamiliar element.

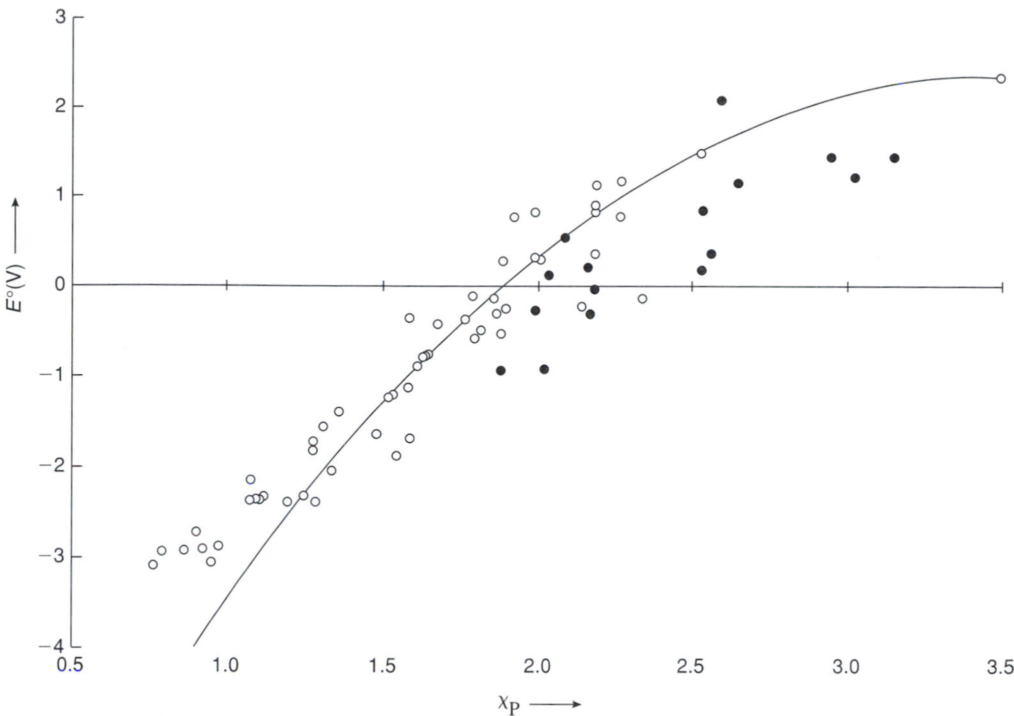

Figure 7.5
Standard reduction potentials of the metal ions (open circles) and common nonmetal oxides, oxo acids, and oxo anions (closed circles, in common oxidation states) to the elements, versus the Pauling electronegativities of the elements. The curve is a plot of Eq. (7.26).

7.6 Effects of Changing Substituents and Oxidation Numbers on Electronegativities

Just as we have seen that changing substituents and oxidation numbers can change the softness of the atom, we also find that changing these factors can change the Pauling electronegativity of an atom.

Example 7.4

The I–F bond energy differs in the different iodine fluorides; it is 231 kJ mol^{-1} in IF_7, 268 kJ mol^{-1} in IF_5, 272 kJ mol^{-1} in IF_3, and 278 kJ mol^{-1} in IF. Compute the Pauling electronegativity of iodine in each of these compounds.

Solution:

We assume the electronegativity of fluorine to be 3.98 and take the I–I and F–F bond energies from Table 3.1; these are 149 and 155 kJ mol^{-1}, respectively, for an average of 152 kJ mol^{-1}. Referring to the Pauling electronegativity equation, if we deduct this average from the bond energies, then divide by 96.5, we get $[\chi_P(F) - \chi_P(I)]^2$. For I(VII) in IF_7, we obtain 0.82; for I(V) in IF_5 we obtain 1.20; for I(III) we obtain 1.24; for I(I) we obtain 1.30. Taking square roots and deducting the square root from the electronegativity of F, 3.98, we obtain the following Pauling electronegativities: 3.07 for I(VII), 2.88 for I(V), 2.87 for I(III), and 2.84 for I(I).

The example shows that the Pauling electronegativity of an element increases with increasing oxidation number, which makes sense, since a more electron-poor atom should have a greater attraction for shared electrons in its covalent bonds. The example also shows that the change in electronegativity with changing oxidation state is usually not great enough to justify tabulating separate values, especially since the values have inherent uncertainties of ± 0.05. In the few cases where such data are available, the variation is not much greater than 0.20, even over a range of several oxidation numbers. There are two exceptions, however; the electronegativities of thallium and lead are very sensitive to their oxidation number (changes of over 0.40 for a change of only 2 in their oxidation states). Hence, Table A contains separate values for the two oxidation states indicated for these two elements. (The oxidation numbers implied for the other elements in Table A are $\pm$ the Roman numerals indicated at the head of each group—but these values are normally used for any oxidation state.)

The effects of changing substituents on Pauling electronegativities cannot be investigated by looking at bond dissociation energies; the enthalpy change for the dissociation of a compound containing two different substituents cannot cleanly be divided into enthalpy changes for the two different kinds of bonds; however, other methods can be used to estimate Pauling electronegativities on such compounds. Results obtained by one such method on different compounds of C, Si, Ge, B, and N are shown in Table 7.10. They indicate that the electronegativity of an atom can vary by up to 0.4 Pauling units depending on its substituents; even greater variability is suggested by other methods.[21]

Although such differences are perhaps worth tabulating, the size of the table that would be needed to include all such *group electronegativities* is impractical. Instead, both inorganic and organic chemists, working from the "base" Pauling electronegativities in Table A, then rationalize variations in terms of the **inductive effects** of the substituents. From the examples in Table 7.10, we see that *electron-donating substituents* (those of low electronegativities themselves) *lower the electronegativity of an atom; electron-withdrawing substituents on an atom increase its electronegativity*. This is eminently logical: An atom losing bond electrons to some of its substituents may be expected to try to make up its losses by withdrawing bond electrons from other substituents.

There is a general agreement between Pauling's and Pearson's (HSAB) approaches to the results of competitive Lewis acid–base equilibria. But it has been noted[22] that, under certain circumstances, an apparent paradox results (the *Pearson–Pauling paradox*).

Let the acceptor atom be the cation of an electronegative metal. Let us now either increase the oxidation number of the cation or put more electron-withdrawing ligands on it. Either step ought to increase the electronegativity of the cation. According to the Pauling electronegativity equation (7.16), this increase will diminish the cation's relative affinity for very electronegative donor atoms. Thus we may say that this metal atom has become a softer acid. However, according to Section 5.4, increasing oxida-

Table 7.10
Effects of Substituents on the Pauling Electronegativity of an Atom[a]

Compound	Electronegativity of Element	Tabulated χ_P
	A. Carbon Compounds	2.55
$(CH_3)_3C-Cl$	2.29	
$(CH_3)_2$ CH–Cl	2.33	
CH_3CH_2-Cl	2.36	
CH_3-Cl	2.40	
$ClCH_2-Cl$	2.47	
Cl_2CH-Cl	2.56	
Cl_3C-Cl	2.64	
	B. Extreme Values for Atoms Other Than Carbon	
$(CH_3)Si-Cl$	1.76	1.90
Cl_3Si-Cl	1.92	
$(CH_3)_3$ Ge–Cl	1.82	2.01
Cl_3 Ge–Cl	2.11	
$(HNBCl)_3$	1.90	2.04
$HCB_{10}Cl_{10}CH$	2.13	
CH_2CH_2N-Cl	2.88	3.04
$(CH_3)_2C(CO)_2(NCl)_2$	3.28	

SOURCES: The source of details on these, and of the NQR frequencies themselves, is G. K. Semin, T. A. Babushkina, and G. G. Yakobson, *Nuclear Quadrupole Resonance in Chemistry*, Israel Program for Scientific Translations (Jerusalem), Wiley, New York, 1975. For a review of methods of determining group electronegativities, see P. R. Wells, *Prog. Phys. Org. Chem.*, **6**, 111 (1968).

[a] Electronegativities were computed by the author from the ^{35}Cl nuclear quadrupole resonance (NQR) frequencies of the chlorines in the compounds indicated, using the relationship found among simple singly bonded chlorides of the elements, NQR frequency = 26.0 (χ_P − 1.13) (author's unpublished correlation), which may also be derived from the Townes–Dailey theory for NQR frequencies using the Gordy relationship of ionic character to electronegativity differences.

tion numbers and attaching very electronegative (hard) substituents make an acceptor atom a *harder* acid. Pearson[22] notes that when these contradictory predictions arise, the HSAB principle almost invariably gives the correct prediction. This may be true partly because the influence of oxidation state on electronegativity is really quite small. In the two cases in which it is *not* small, Tl and Pb, it is generally felt that the unusual trend prevails: Tl(III) seems to be softer than Tl(I) and Pb(IV) seems softer than Pb(II). Smith[23] attempted to resolve this conflict by including an additional parameter in the Pauling equations; the ratio of this second parameter to electronegativity is a function of the softness of the metal ion.

7.7 *Definitions of Electronegativity and Hardness Related to Ionization Energies and Electron Affinities

Although much inorganic reactivity can be predicted using Pauling electronegativities, it is clear that much more satisfactory thermochemical analyses of reactivity are possible using electron affinities, ionization energies, and so on. Given this situation, it is not surprising that chemists have tried to find theoretical connections between the two sets of concepts.

The best known of these attempts is that of Mulliken, who produced a set of **Mulliken** electronegativities that is obtained as the *average of the electron affinity and first ionization energy* of each element:

$$\chi_M = 0.5[\mathrm{IE}(1)_v + \mathrm{EA}_v] \tag{7.27}$$

Of course, *electronegativity* refers to the attraction of an atom *in a compound* for its *shared electron pairs*, while *ionization energy* and *electron affinity* refer to attractions of *isolated gaseous* atoms or ions for their (unshared) electrons. Mulliken reasoned that the degree of sharing in the bond would be an average of the tendency of the isolated atom to lose electrons (ionization energy) and to gain electrons (electron affinity). The electron affinities and ionization energies used, however, are not those from Tables 7.2–7.5, but are for electrons coming from *hybrid* orbitals. For the *p*-block elements, in which *s* and *p* valence orbitals are used, the Mulliken electronegativities do indeed agree satisfactorily with the Pauling electronegativities that we have been using, but different hybridizations may be used by a given element in different compounds, resulting in different electronegativities. Mulliken electronegativities are not generally available for the *d*-block elements, since it is very difficult to determine to what extent each of the three types of available orbitals (*s*, *p*, and *d*) are being hybridized in the compounds of *d*-block elements.

A related attempt is that of Allen,[24] who avoided the difficulties of predicting hybridizations by defining a *spectroscopic electronegativity* as a weighted average of the (one-electron) ionization energies of all of the valence-shell electrons (these energies can be computed from spectroscopic data). Allen's values do not include electron affinities, but these are substantially smaller than ionization energies; his values are also in good agreement with Pauling's in the *s* and *p* blocks. Spectroscopic electronegativities can also be obtained for early *d*-block atoms and show good agreement with Pauling values, but late in the *d* block it is also impossible to determine precisely how many *d* (and *p*) electrons are being used in the valence shell. Calculations for the sixth period, which involve relativistic effects, are very difficult, but also seem to give good agreement with the trends in Pauling values.

Parr and Pearson,[25a] and Pearson[25b] introduced a property called the electronic chemical potential, μ, which is derived from the molecular orbital theory (and will be discussed again in Chapter 10), but which has approximately the same definition [Eq. (7.27)] as the Mulliken electronegativity. Their work also allows a related definition of the *hardness* η (in the HSAB sense) of an atom, ion, or molecule as being related to the *difference* of the ionization energy and electron affinities of the atom, ion, or molecule:

$$\eta = 0.5[\mathrm{IE}(1)_v - \mathrm{EA}_v] \tag{7.28}$$

The *softness*, σ, of the species is then simply the inverse of its hardness η.

Given the importance of electronegativities and the imprecise definition of them, it is not surprising that other scales, each with their merits, have been proposed. We have, of course, already introduced the Allred–Rochow electronegativity scale in Chapter 1; two other scales, those of Sanderson[26] and of Boyd and Edgecombe,[27] are based on electron densities.

7.8 *Thermodynamics of the Lewis Acid–Base Interaction: Drago's *E*, *C*, and *T* Parameters

Pauling electronegativities are based on the strengths of bonds measured in the gas phase in binary compounds. Coordination compounds are not, in general, binary compounds, but they may in many cases be formed in the gas phase from volatile Lewis acids and Lewis bases; enthalpy changes can readily be measured for such reactions in which coordinate covalent bonds are formed. Since many Lewis acids and Lewis bases are nonvolatile, it is often necessary to carry out these reactions instead in very nonpolar (nonacidic and nonbasic) solvents, and assume that none of the measured enthalpy changes are affected by solute–solvent interactions.

Hundreds of such thermodynamic measurements have been made. Drago et al.[28] found that these measurements can be quite adequately predicted using the *Drago equation*:

$$\Delta H \text{ (in kJ mol}^{-1}) = -4.184(E_A E_B + C_A C_B + R_A T_B) \qquad (7.29)$$

In this equation, there are three parameters for each acid, C_A, E_A, and R_A, and three parameters for each base, which must be found by fitting experimental enthalpy-change data for reactions involving that acid or that base.[29] The *E* parameters are intended to measure relative[30] strength in *electrostatic bonding* (strength at bonding through dipole–dipole forces), the *C* parameters to measure strength in *covalent bonding* (strength as a soft base); and the *R* and *T* parameters to measure the amount of *charge transfer* from the base (*T* stands for transmittance) to the acid (*R* stands for receptance) *if either the acid (A) or the base (B)* (or both) *is ionic.* (The *R* and *T* factors are quite small for neutral acids and bases.) Acids that bond well covalently therefore tend to bond to bases with the corresponding property, since the product $C_A C_B$ then is large; the corresponding statement can be made about electrostatic (e.g., hard) acids and bases.

Once these parameters have been obtained, they can be used to predict the enthalpy changes for thousands of acid–base combinations that have not been tested. If such predictions are tested later, the predictions are usually found to be quite accurate unless that particular acid–base combination involves steric hindrance (spatial interference of groups attached to the donor atom with other groups attached to the acceptor atom). Table 7.11 contains Drago parameters for a number of Lewis acids and Lewis bases.[31]

Example 7.5

Calculate the enthaply changes of the following Lewis acid–base reactions: (a) I_2 + ethyl acetate; (b) HF + tetrahydrofuran (THF).

SOLUTION:

Taking the appropriate *E*, *C*, and *T* parameters from Table 7.11, we find that
(a) $\Delta H = -4.184(0.50 \times 1.62 + 2.00 \times 0.98 + 0 \times 0.89) = -11.6 \text{ kJ mol}^{-1}$;
(b) $\Delta H = -4.184(2.03 \times 1.64 + 0.30 \times 2.18 + 0.47 \times 0.75) = -18.1 \text{ kJ mol}^{-1}$.

Table 7.11
Drago Parameters for Lewis Acids and Bases

Acid	E_A	C_A	R_A	Acid	E_A	C_A	R_A
I_2	0.50	2.00		H^+	45.00	13.03	130.21
H_2O	1.54	0.13	0.20	CH_3^+	19.70	12.61	55.09
SO_2	0.56	1.52	0.85	Li^+	11.72	1.45	24.21
HF[a]	2.03	0.30	0.47	K^+[a]	3.78	0.10	20.79
HCN[a]	1.77	0.50	0.54	NO^+[a]	0.1	6.86	45.99
CH_3OH	1.25	0.75	0.39	NH_4^+[a]	4.31	4.31	18.52
H_2S[a]	0.77	1.46	0.56	$(CH_3)_2NH_2^+$[a]	3.21	0.70	20.72
HCl[a]	3.69	0.74	0.55	$(CH_3)_4N^+$[a]	1.96	2.36	8.33
C_6H_5OH	2.27	1.07	0.39	$C_5H_5NH^+$[a]	1.81	1.33	21.72
$(CH_3)_3COH$	1.36	0.51	0.48	$(C_2H_5)_3NH^+$[a]	2.43	2.05	11.81
$HCCl_3$	1.49	0.46	0.45	$(CH_3)_3NH^+$[a]	2.60	1.33	15.95
CH_3CO_2H[a]	1.72	0.86	0.63	H_3O^+	13.27	7.89	20.01
CF_3CH_2OH	2.07	1.06	0.38	$(H_2O)_2H^+$	11.39	6.03	7.36
C_2H_5OH	1.34	0.69	0.41	$(H_2O)_3H^+$	11.21	4.66	2.34
i-C_3H_7OH	1.14	0.90	0.46	$(H_2O)_4H^+$[a]	10.68	4.11	3.25
PF_3[a]	0.61	0.36	0.87	$(CH_3)_3Sn^+$	7.05	3.15	26.93
$B(OCH_3)_3$[a]	0.54	1.22	0.84	$(C_5H_5)Ni^+$	11.88	3.49	32.64
AsF_3[a]	1.48	1.14	0.78	$(CH_3)NH_3^+$[a]	2.18	2.38	20.68
$Fe(CO)_5$[a]	0.10	0.27	1.00				
CHF_3[a]	1.32	0.91	0.27				
$B(C_2H_5)_3$[a]	1.70	2.71	0.61				

Base[b]	E_B	C_B	T_B	Base[b]	E_B	C_B	T_B
NH_3	2.31	2.04	0.56	C_5H_5NO	2.29	2.33	0.67
CH_3NH_2	2.16	3.12	0.59	$(CH_3)_3P$	1.46	3.44	0.90
$(CH_3)_2NH$	1.80	4.21	0.64	$(CH_3)_2O$	1.68	1.50	0.73
$(CH_3)_3N$	1.21	5.61	0.75	$(CH_3)_2S$	0.25	3.75	1.07
$C_2H_5NH_2$	2.35	3.30	0.54	CH_3OH	1.80	0.65	0.70
$(C_2H_5)_3N$	1.32	5.73	0.76	C_2H_5OH	1.85	1.09	0.70
$HC(C_2H_4)_3N$	0.80	6.72	0.83[c]	C_6H_6	0.70	0.45	0.81
C_5H_5N	1.78	3.54	0.73	H_2S[a]	0.04	1.56	1.13
4-$CH_3C_5H_4N$	1.74	3.93	0.73[c]	HCN[a]	1.19	0.10	0.90
3-$CH_3C_5H_4N$	1.76	3.72	0.74[c]	H_2CO[a]	1.56	0.10	0.76
3-ClC_5H_4N	1.78	2.81	0.75[c]	CH_3Cl[a]	2.54	0.10	0.23
CH_3CN	1.64	0.71	0.83	CH_3CHO[a]	1.76	0.81	0.74
$CH_3C(O)CH_3$	1.74	1.26	0.80	H_2O[a]	2.28	0.10	0.43
$CH_3C(O)OCH_3$	1.63	0.95	0.86	$(CH_3)_3COH$[a]	1.92	1.22	0.71
$CH_3C(O)OC_2H_5$	1.62	0.98	0.89	C_6H_5CN[a]	1.75	0.62	0.85
$HC(O)N(CH_3)_2$	2.19	1.31	0.74[c]	F^-	9.73	4.28	37.40
$(C_2H_5)_2O$	1.80	1.63	0.76	Cl^-[a]	7.50	3.76	12.30
$O(CH_2CH_2)_2O$	1.86	1.29	0.71	Br^-[a]	6.74	3.21	5.86
$(CH_2)_4O$	1.64	2.18	0.75	I^-	5.48	2.97	6.26
$(CH_2)_5O$	1.70	2.02	0.74[c]	CN^-	7.23	6.52	9.20
$(C_2H_5)_2S$	0.24	3.92	1.10[c]	OH^-[a]	10.43	4.60	50.73
$(CH_3)_2SO$	2.40	1.47	0.65	CH_3O^-[a]	10.03	4.42	33.77

SOURCE: R. S. Drago, D. C. Ferris, and N. Wong, *J. Am. Chem. Soc.* **112**, 8953–8961 (1990). R. S. Drago, N. Wong, and D. C. Ferris, *J. Am. Chem. Soc.* **113**, 1970–1977 (1991).

[a] Tentative parameters from limited enthalpy data.

[b] If not indicated otherwise, the bases in this table have E_B and C_B determined from the fit of neutral acid–neutral base adducts.

[c] The E_B and C_B for these bases are well determined. The T_B values are tentative for they have been determined from limited data. See the original papers for methods and accuracy of the values.

Covalent Bond Cleavage. The bond between a given pair of atoms can be disrupted in the gas phase either to leave one electron on each atom (to dissociate to atoms or free radicals, as in Eqs. (7.30) and (7.31)), or they may dissociate with both electrons going onto the Lewis base, as in Eqs. (7.32)–(7.34).

$$H_3C\!:\!CH_3(g) + \text{energy} \rightarrow H_3C\cdot(g) + \cdot CH_3(g) \tag{7.30}$$

$$Na\!:\!Cl(g) + \text{energy} \rightarrow Na\cdot(g) + \cdot Cl(g) \tag{7.31}$$

$$H_3N\!:\!BH_3(g) + \text{energy} \rightarrow H_3B(g) + :NH_3(g) \tag{7.32}$$

$$H_3N\!:\!Ag^+(g) + \text{energy} \rightarrow Ag^+(g) + :NH_3(g) \tag{7.33}$$

$$Na\!:\!Cl(g) + \text{energy} \rightarrow Na^+(g) + :Cl^-(g) \tag{7.34}$$

Since Reactions (7.32)–(7.34) are the reverse of coordinate covalent bond formation, the Drago equation can be used to calculate ΔH for these reactions. The first two reactions are the reverse of covalent bond formation, and are the types of dissociations for which bond dissociation energies are measured (and then used to calculate Pauling electronegativities). Drago et al.[32] extended their approach to cover these Pauling-type bond dissociation energies as well, and fit them to another equation:

$$\Delta H(\text{catimer-animer}) = -4.184(C_A C_B + E_{an}E_{cat} + R_{an}T_{cat}) \tag{7.35}$$

Values of these constants for the catimer (the less electronegative atom in the covalent bond) and the animer (the more electronegative atom in the covalent bond) are shown in Table 7.12. This equation gives a better fit to the known bond energies than does the Pauling electronegativity equation (7.16), which of course is to be expected when the equation contains three times as many adjustable parameters.

Equations (7.31) and (7.34) illustrate that, in principle, any given bond could dissociate either into two radicals, or into a Lewis acid and a Lewis base. The following example shows that there is generally a strong energy preference for one pathway of dissociation.

Example 7.6

We can conceive of the gaseous HF molecule dissociating in either of two ways: (a) with the shared bond pair returning to the F atom, generating $H^+(g)$ and $F^-(g)$ in the reverse of a Lewis acid–base reaction; (b) with one electron from the shared bond pair returning to each atom, generating $H\cdot(g)$ and $F\cdot(g)$ in a bond dissociation reaction. Use the appropriate Drago parameters to estimate ΔH for each process; which requires the least energy?

SOLUTION:

For process (a) we use the Drago equation (7.29) and the data from Table 7.11: $\Delta H = +4.184[(13.03)(4.28) + (45.00)(9.73) + (130.21)(37.40)] = +4.184(55.77 + 437.85 + 4869.85) = +22{,}441$ kJ mol! (This extraordinarily large result does not at all agree with the corresponding experimental value, 1554 kJ mol^{-1}, derived from the proton affinity of gaseous F^-; since both H^+ and F^- are among the strongest gaseous acids and

Table 7.12
Parameters for Enthalpies of Dissociation for Catimers and Animers

Catimer[a]	E_{cat}	C_A	T_{cat}
H	8.05	13.03	0.06
Li	6.52	1.45	31.46
K	5.54	0.30	30.67
CH_3	4.90	12.61	1.97
Mg	10.33	0.88	2.29
Al	9.41	4.20	21.04
Mn	3.59	2.70	25.16
$(^5\eta\text{-}C_5H_4CH_3)IrP(CH_3)_3$	3.83	7.37	16.51
Ag*	1.61	3.21	27.45
Hg*	2.12	0.63	0.21
Tl	6.01	1.84	20.39
Ba*	10.85	0.15	14.92
Na*	3.83	1.03	30.01
Rb*	5.93	0.14	29.58
Cu	3.62	6.52	18.90
CH_3CH_2	5.55	11.42	0.55
$C_6H_5CH_2$	3.56	11.55	0.20
CH_3CO*	7.13	8.00	0.20
C_6H_5	8.11	7.59	5.72
Ni	5.02	5.01	18.54
$MoC_5H_5(CO)_3$	2.09	7.10	14.08
Cr*	5.89	1.72	20.20
Zn*	4.26	0.10	10.73
I	3.48	6.18	1.93
Br*	1.76	10.01	1.76
Cl*	0.74	13.69	1.04
Animer[a]	E_{an}	C_B	R_{an}
F	10.00	4.28	2.03
Cl	6.72	3.76	2.10
Br	5.67	3.21	1.87
I	4.03	2.97	1.76
CN*	4.82	6.52	8.02
OH	7.32	4.60	1.62
H	2.23	6.60	0.97
NH_2*	5.29	4.82	0.39
CH_3	4.26	5.42	0.03
CF_3	3.76	5.84	3.69
CCl_3*	3.64	5.10	2.70

[a] Tentative values that were constrained to fit because of a limited data set or large errors in the reported data are indicated with an asterisk. [Adapted from R. S. Drago, N. Wong, and D. C. Ferris *J. Am. Chem. Soc.* **113**, 1970 (1991). Copyright © 1991 American Chemical Society.]

bases, they have very large *E*, *C*, and *T* parameters, which become unrealistically high when multiplied together.)

For process (b), we use the Drago equation (7.35) and the data from Table 7.12: $\Delta H = +4.184[(13.03)(4.28) + (8.05)(10.00) + (0.06)(2.03)] = +4.184(55.77 + 80.5 + 0.12) = +571\ \text{kJ mol}^{-1}$. The latter is close to the observed H–F bond energy of $+565\ \text{kJ mol}^{-1}$ (Table 7.8) and is the process that requires the least expenditure of energy.

Clearly, gaseous neutral molecules do not dissociate into oppositely charged gaseous ions as shown in Eq. (7.34) but rather into neutral gaseous atoms as shown in Eq. (7.31): Dissociation into gaseous ions requires an enormous amount of work to be done to separate the charges to give gaseous cations and anions. (Recall the discussion in Section 4.6 of the enormously exothermic attraction of oppositely charged gaseous ions to give gaseous ion pairs.)

The other four bonds shown above do dissociate in the gas phase as written, but the energy requirements of Reactions (7.30) and (7.31) are much greater than those of Reactions (7.32) and (7.33). Haaland[33] pointed out that the latter two bonds can unambiguously be classified as coordinate covalent or *dative* bonds based on their gas-phase dissociation behavior, and that such bonds are considerably weaker than those that dissociate as ordinary covalent bonds, with one electron going to each atom. Not only are dative bonds weaker, they are longer, and they take less space in the VSEPR model than do true polar covalent bonds. This difference in bond types is reflected in the necessity for two Drago-type equations to account for polar bonding in general.

In general (as previously indicated in Section 5.6), *it is impossible to characterize the Lewis acid–base interaction by a single parameter* such as strength or softness. Our characterizations of relative acidities of metal ions in Chapter 2 worked relatively well because we always had the same Lewis base present, H_2O. (The bonding of water is mainly of electrostatic and charge-transfer nature to cations of metals of low electronegativity.) But, in general, the Lewis acid–base interaction involves varying mixes of electrostatic, charge-transfer, and covalent bonding and requires more than one parameter to characterize each species.

Given this, it is possible to get contradictory predictions using the *E*, *C*, and *T* parameters and the HSAB theory. In general, however, the types of chemical reactions to which each has been applied successfully are quite different. It is best to use the *E*, *C*, and *T* parameters in predicting the positions of equilibria in gas-phase reactions or reactions in nonpolar solvents; in polar solvents such as water, in which solvation energies are important, the HSAB principle seems to be the more applicable principle. For example, despite the above example, we know that, due to the powerful effects of hydration, *in solution* HF and NaCl both dissociate into ions (ionize) rather than dissociating into atoms. Solvents have powerful influences on reactions when ions are involved!

Study Objectives

1. Given thermodynamic and electrochemical data, set up a Born–Haber cycle and an equation for the oxidation of a metal, and calculate an energy term from it. Exercises 1–3.

2. Set up thermochemical cycles for other redox reactions (such as that of the activity series of elements) and calculate energy terms or explain periodic trends in activity using it. Exercises 4–12.

3. Know the main periodic trends in heats of atomization, ionization energies, and electron affinities. Exercises 13–17.

4. Calculate selected ionization energies or electron affinities using Slater's rules. Exercises 18–22.

5. Calculate bond energies or heats of formation from Pauling electronegativities, or vice versa; write thermochemical cycles involving bond energies. Exercises 23–31.

6. Know how the Pauling electronegativity of an element is affected by changing substituents on it, or by changing its oxidation number. Exercises 32–41.

7. Ignoring questions of hybridization, calculate the Mullikin electronegativities and the softness of selected atoms. Exercises 42, 43.

8. Calculate the enthalpy change of formation of a coordinate covalent bond from E and C parameters, or vice versa. Exercises 44–48.

Exercises

1. *Set up a Born–Haber cycle for each of the following reactions, and calculate the experimental lattice energies of the products.

(a) $Ca(s) + F_2(g) \rightarrow CaF_2(s)$; $\Delta H_f = -1215\,kJ\,mol^{-1}$

(b) $Ca(s) + 0.5\,O_2(g) \rightarrow CaO(s)$; $\Delta H_f = -636\,kJ\,mol^{-1}$

(c) Calculate the theoretical lattice energies for these two compounds and compare them to your experimental values.

2. The standard enthalpies of formation (in kilojoules per mole) of the metal(II) oxides of the fourth-period elements of the d block are as follows (M. C. Ball and A. H. Norbury, *Physical Data for Inorganic Chemists*, Longman, London, 1974; pp. 62–64):

TiO	VO	MnO	FeO	CoO	NiO	CuO	ZnO
−518	−431	−385	−264	−239	−241	−155	−348

(a) Write a general equation suitable for calculating the lattice energy of any of these oxides, based on thermochemical data. (b) Find the lattice energies of TiO, CuO, and ZnO. (c) Overall the most dissimilar ΔH_f values are those of TiO and CuO; the most dissimilar ΔH_f values of neighboring oxides are those of CuO and ZnO. Which thermodynamic factor is apparently responsible for each of these dissimilarities? (d) What factor in the theoretical lattice-energy calculations (as done in Section 4.6) is most responsible for the substantial change in lattice energies between TiO and CuO?

3. The formula of the superconductor $YBa_2Cu_3O_7$ suggests that one-third of its copper ions are in the unusual +3 oxidation state, but some research suggested the surprising alternative that, instead of a structural subunit involving Cu^{3+} and O^{2-} ions, the superconductor may involve Cu^{2+} and singly charged O^- ions. Sketch a thermochemical cycle to estimate the energetics of the transformation $(Cu^{3+})_2(O^{2-})_3(s) \rightarrow (Cu^{2+})_2(O^-)_2(O^{2-})(s)$. Supply as many numbers as you can to complete the thermochemical analysis, but you will not be able to compute the lattice energies of the reactant and the product. Which of the two lattice energies should be larger? Which energy terms in this thermochemical cycle favor the $Cu^{3+}O^{2-}$ structure? Which favor the $Cu^{2+}O^-$ structure?

4. *Consider the possibility of a reaction between tin metal and aqueous strontium ion: $Sr^{2+}(aq) + Sn(s) \rightarrow Sn^{2+}(aq) + Sr(s)$. (a) Draw a diagram of the thermochemical cycle for this reaction, and (using appropriate data) calculate ΔH for this reaction. (b) Does this reaction go as written to products? Neglecting entropy effects, calculate the cell potential $E°$ for the reaction as written.

5. (a) Set up and evaluate a thermochemical cycle to evaluate the enthalpy change for the reaction: $2\,Cr^{3+}(aq) + H_2 \rightarrow 2\,Cr^{2+}(aq) + 2\,H^+(aq)$. (b) Could H_2 be used as a reducing agent to prepare $Cr^{2+}(aq)$? (c) Ignoring entropy effects, calculate $E°$ for this reaction.

6. The new transplutonium element qinjium (Qj) has just been discovered; like Pu, in water it forms a +3 ion of radius 114 pm and hydration energy ΔH_{hyd} of -3441 kJ mol^{-1}. Other properties are compared in the following table:

Atomization energy DH_{atom}	Qj 440 kJ mol^{-1}	Pu 352 kJ mol^{-1}
First ionization energy IE(1)	Qj 499 kJ mol^{-1}	Pu 585 kJ mol^{-1}
Second ionization energy IE(2)	Qj 1480 kJ mol^{-1}	Pu 1129 kJ mol^{-1}
Third ionization energy IE(3)	Qj 2114 kJ mol^{-1}	Pu 2103 kJ mol^{-1}

You want to determine which metal will be higher in the activity series: Qj or Pu. (a) Draw a thermochemical cycle for the reaction: $Qj(s) + Pu^{3+}(aq) \rightarrow Pu(s) + Qj^{3+}(aq)$. Label the energy terms involved. (b) Calculate ΔH for the above reaction. Which metal is more active? (c) Estimate the voltage generated by the above reaction.

7. The first and second ionization energies of Mg and of Pb do not differ greatly, but their activities (in reacting with acid to form hydrated +2 ions) do. (a) Calculate ΔH for the reaction of Mg metal with 1 M hydrogen ion to produce hydrated Mg^{2+} ions and H_2. (b) Make the same energy calculation for Pb metal. (c) Which metal is more active? Calculate the approximate standard reduction potential $E°$ for each metal from your ΔH value. (d) Fundamentally, what is the cause of the difference in activity between Mg and Pb?

8. Consider the possibility of a reaction between magnesium metal and aqueous lead ion: $Pb^{2+}(aq) + Mg(s) \rightarrow Mg^{2+}(aq) + Pb(s)$. (a) Draw a diagram of the thermochemical cycle for this reaction, and (using appropriate data from the tables in the text) calculate ΔH for this reaction. (b) Does this reaction go as written to products? Which metal is more active? Neglecting entropy effects, calculate the cell potential $E°$ for the reaction as written.

9. *Consider the possibility of a reaction between *gaseous* lead metal (from a metal-atom reactor apparatus) and aqueous magnesium ion: $Pb(g) + Mg^{2+}(aq) \rightarrow Pb^{2+}(aq) +$

Mg(s). (a) Draw a diagram of the thermochemical cycle for this reaction, and calculate ΔH for this reaction. (b) Does this reaction go as written to products? Neglecting entropy effects, calculate the cell potential E° for the reaction as written.

10. Consider the possibility of a (balanced) reaction between Pb metal and aqueous Ag^+ to give Ag metal and Pb^{2+} ion. (a) Draw a diagram of the thermochemical cycle for this (balanced) reaction. (b) Calculate ΔH for this reaction. (c) Does this reaction go as written to products? Which metal is more active? Neglecting entropy effects, calculate the cell potential E° for the reaction as correctly balanced.

11. Suggest a better way to attempt to correlate activities of metals (Experiment A.6) with ionization energy than simply correlating with the first ionization energies of metals. (*Hint*: Look at Eq. 7.8.) Try this correlation (on graph paper) with the E° values from Table 6.2. Do you get a more satisfactory correlation with activity using your ionization-energy function than the correlation of activity with electronegativity (Fig. 6.14)? If not, why not?

12. Sketch thermochemical cycles for the following reactions, showing the products formed after each step. Use standard abbreviations to indicate the energy change expected for each step, and show how these should be added or subtracted to obtain ΔH for the reaction.

(a) $U(s) + 2F_2(g) \rightarrow UF_4(s)$

(b) $2Cr^{3+}(aq) + Zn(s) \rightarrow Zn^{2+}(aq) + 2Cr^{2+}(aq)$

(c) $Co^{3+}(aq) + Cr^{2+}(aq) \rightarrow Co^{2+}(aq) + Cr^{3+}(aq)$

Although you are not doing actual calculations, from the principles in the text identify which one energy term is likely responsible for the fact that: (d) the formation of UF_4 from the elements is exothermic; (e) the reaction of Co^{3+}(aq) with Cr^{2+}(aq) is exothermic.

13. *Without referring to Table 7.1, choose the element from each set that should have the highest atomization energy and the element that should be easiest to atomize, and explain why you made the choice you did: (a) C, Si, Ge, Sn, Pb; (b) V, Nb, Ta; (c) Li, Be, B, C, N, O, F, Ne; and (d) Sr, Zr, Mo, Ru, Pd, Cd, Sn, Te, Xe.

14. (a) Which element of Group 13(III) is the hardest to atomize? Explain why. (b) Which element of Group 7 is the easiest to atomize? Explain why.

15. (a) Which atom of the sixth period has the highest electron affinity? Why? (b) A secondary maximum value is reached at which atom in the middle of the period? Why?

16. *Below are listed some important periodic trends in the properties of elements. For each property, a selection of elements from a given group or period are listed. Choose the element from the set at which that property reaches a maximum, or indicate if this property is the same for all elements of the set.

Atomization energy	Rb	Y	Mo	Rh	Ag	Cd
Electron affinity	Na		Al	P	Cl	Ar
First ionization energy	Na	Mg	Al	P	Cl	Ar
Second ionization energy	Na	Mg	Al	P	Cl	Ar

17. It has recently been reported [Moock and Seppelt, *Angew. Chem. Intl. Ed. Engl.* **28**, 1676 (1989)] that the common cesium ion Cs^+ can be oxidized in acetonitrile solution to give the Cs^{3+} ion, at a potential of +3.0 V ($E°$). This finding was subsequently disputed [C. Jehoulet and A. J. Bard, *Angew. Chem. Int. Ed. Engl.*, **30**, 836 (1991)]. Perhaps thermochemical computations could illuminate this disagreement. Let us assume for this problem that the reaction can be done in aqueous solution. (a) Explain why there is such a huge jump between IE(1) and IE(2) for Cs as compared to the smaller jump between IE(2) and IE(3). If this report is confirmed, what would be unique in all of chemistry about the product? (b) Draw a thermochemical cycle showing the steps involved in the oxidation: $Cs^+(aq) \rightarrow Cs^{3+}(aq) + 2e^-$. Describe briefly what happens in each step, and show what energy terms are required in a thermochemical analysis of this process. (c) One energy term you need for your thermochemical cycle is not known or listed either, but if you have identified it correctly you should be able to identify about how many times larger it should be than another term. (d) Do IE(2) and IE(3) appear to be prohibitively high for Cs for this reaction to occur? In discussing this, compare them with IE(2) and IE(3) for other elements in the periodic table that do form compounds in the +3 oxidation state. Also, complete your calculation of the thermochemical cycle using your estimate from part (c).

18. *Calculate (using Slater's rules) the zeroth ionization energies of (a) Li^- and (b) F^-. (That is to say, calculate the electron affinities of Li and F.)

19. Using the preliminary calculations in Example 1.2, calculate the first ionization energy of Sc: (a) assuming that a 4s electron is ionized and (b) assuming that the 3*d* electron is ionized.

20. Application of Slater's rules in Chapter 1 allowed us to calculate Z^* for the valence (3*s*) orbital in the Mg atom as +2.85. (a) Calculate Z^* for the valence (3*s*) orbital in the Mg ion, Mg^+. (b) From this data, calculate the ionization energy of the Mg atom.

21. Calculate the second ionization energies of (a) Li; (b) F; (c) Sc^+ with electron configuration $3d^14s^1$ going to Sc^{2+} with electron configuration $3d^1$; and (d) Sc^+ with electron configuration $4s^2$ going to Sc^{2+} with electron configuration $4s^1$.

22. Calculate the ionization energy for the 26th electron of Fe; express this both in kilojoules per mole and volts. Is any chemical oxidizing agent strong enough to remove this electron (and, by extension, the other 25 electrons of Fe that would precede it in ionization)?

23. *The electronegativity of carbon in CCl_4 is estimated as 2.64; the electronegativity of carbon in $(CH_3)_3CCl$ is estimated as 2.29. Compute the expected C–Cl bond energy in each compound.

24. Consider an element, udezium (Ud), in the middle of the periodic table. The Ud–Ud bond energy is 193 kJ mol^{-1}; the Ud–F bond energy is 543 kJ mol^{-1}; the F–F bond energy is 155 kJ mol^{-1}. From this information calculate the Pauling electronegativity of Ud.

25. Consider an element, taylorium (Ty), in the middle of the periodic table. The Ty–Ty bond energy is 114 kJ mol^{-1}; the Ty–F bond energy is 503 kJ mol^{-1}; the F–F bond energy is 155 kJ mol^{-1}. From this information, calculate the Pauling electronegativity of Ty.

26. *In a course in qualitative organic analysis you are warned that carrying out a sodium (metal) fusion test with a chlorocarbon such as CCl_4 could cause an explosion. Given that the bond dissociation energy of the C–Cl bond is 327 kJ mol^{-1}, set up a thermochemical cycle for the reaction

$$4\,Na(s) + CCl_4(g) \rightarrow 4\,NaCl(s) + C(s, \text{graphite})$$

Evaluate ΔH and $E°$ for this reaction. Could it be explosive?

27. Set up and evaluate a thermochemical cycle to evaluate the enthalpy change for the reaction

$$2\,Na(s) \rightarrow Na_2(g).$$

28. (a) Set up thermochemical cycles for and, if possible, evaluate ΔH for the following reactions (bond dissociation energies can be found in Table 7.7):

$$CCl_4(g) + C(s) \rightarrow 2\,CCl_2(g)$$

$$SnCl_4(g) + Sn(s) \rightarrow 2\,SnCl_2(g)$$

(b) What explanation do these calculations suggest for the greater stability of the +2 oxidation state at the bottom of Group 14(IV)? (c) How would the principles discussed in Section 7.6 modify your discussion in (b)?

29. Some new elements have just been discovered in the author's laboratory. The following atomic parameters have been obtained:

Element	Sheltonium	Bennerine	Coatsium	Kamelogen
Electronegativity	1.06	3.82	1.97	2.34
Radius (pm)	(cation) 107	(anion) 123	(cation) 105	(anion) 189
Element–element single-bond energy (kJ mol^{-1})	78	166	88	183

(a) Will products or reactants be favored in the equilibrium: sheltonium benneride + coatsium kamelide ⇌ sheltonium kamelide + coatsium benneride? (b) Calculate the covalent bond energies expected for the coatsium–kamelogen bond; for the sheltonium–bennerine bond.

30. Consider an element, przybylskium (Pz) in the middle of the periodic table. The Pz–Pz bond energy is 293 kJ mol^{-1}; the Pz–F bond energy is 613 kJ mol^{-1}; the F–F bond energy is 155 kJ mol; the I–I bond energy is 149 kJ mol^{-1}. (a) From this information, calculate the Pauling electronegativity of Pz. (b) From your Pauling electronegativity of Pz calculate the expected bond energy of the Pz–I bond.

31. Write the equation analogous to Eq. (7.23) for the reaction of a metal with acid to give $M^{2+}(aq)$ ions, and derive the dependence of $E°$ on χ_P for this metal. Is the result different from Eq. (7.26)?

32. Consider the four carbon compounds: F_3CCl; H_3CCl; FCCl (fluorochlorocarbene); HCCl (chlorocarbene). (a) In which one is the Pauling electronegativity of carbon highest? (b) In which one is the Pauling electronegativity of carbon lowest? (c) In which one is the C–Cl bond strongest? (d) In which one is the C–Cl bond weakest?

33. *You are given the following bond energies: Ge–Ge = 188 kJ mol^{-1}; Cl–Cl = 240 kJ mol^{-1}; Ge–Cl (in $GeCl_2$) = 385 kJ mol^{-1}. Calculate the electronegativity of Ge in $GeCl_2$. Does your answer differ from the Pauling electronegativity of Ge in $GeCl_4$ (2.01) in the expected manner? Explain.

34. Compare the compounds $(CH_3)_3SnCl$ and F_3SnCl. (a) Which should show the highest electronegativity for tin? (b) Which should show the strongest Sn–Cl bond?

35. *For which elements (other than iodine, used in an example) are there appropriate data in Tables 3.1 and 7.6–7.9 for the calculation of the effects of changing oxidation numbers on the Pauling electronegativities of elements? Carry out such calculations for two of those elements.

36. The enthalpies of formation of the iodine fluorides are as follows: (a) IF, −94.7 kJ mol^{-1}; (b) IF_5, −821.5 kJ mol^{-1}; and (c) IF_7, −958.6 kJ mol^{-1}. Calculate the Pauling electronegativity of iodine in each of these three oxidation states, and compare your results with those computed from bond energies in Example 7.4.

37. In Figure 6.14, two of the points most deviant from the correlation shown are those for molybdenum and tungsten. Calculate the electronegativities of Mo and W from the following enthalpies of formation (in kJ mol^{-1}): (a) $MoCl_2$, −184; (b) $MoCl_3$, −272; (c) $MoCl_4$, −330; (d) $MoCl_5$, −379; (e) $MoCl_6$, −377; (f) WCl_2, −151; (g) WCl_4, −289; (h) WCl_5, −343; and (i) WCl_6, −405. Are the deviations likely to be due to choosing data for halides of Mo and W in the "wrong" oxidation state?

38. *From the following enthalpies of formation (in kilojoules per mole), compute and compare the electronegativities of the Group 11 metals as their oxidation numbers change. Are any of the changes abnormal, or abnormally large? (a) CuBr, −105, and $CuBr_2$, −132; (b) AuBr, −38, and $AuBr_3$, −84.

39. Some new elements have just been discovered in the author's laboratory. The following atomic parameters have been obtained:

Element	Wanium	Robbinsine	Cupellium	Lukogen
Electronegativity	2.02	3.50	1.33	2.08
Single-bond energy to another atom of same (kJ)	77	133	146	171
Cation radius (pm)	125 (+2 ion)		62 (+3 ion)	
Anion radius (pm)		120 (−1 ion)		188 (−2 ion)

(a) Wanium is found to exhibit two positive oxidation states, +2 and +4, in its compounds with robbinsine. In which positive oxidation state will wanium be more electronegative? In which positive oxidation state will the wanium–robbinsine bond be stronger? (b) In the +4 oxidation state wanium forms various derivatives such as $(H_3Wa)F$ and $(F_3Wa)F$. Suppose you could measure the Wa–F bond energy in each compound. Would the two bond energy measurements come out the same? If not, in which compound would the bond energy come out higher? Again if not, in which compound would the calculated Pauling electronegativity come out higher? (c) Based on the information given, classify each of the following as a hard acid, soft acid, hard base, or soft base: the wanium cation, the robbinside anion, the cupellium cation, the lukogide anion. (d) What type of bonding (polar covalent, etc.) will predominate in the wanium–lukogen bond? Will the formation of this bond from the aqueous ions result mainly in a favorable *entropy* or *enthalpy* change? (e) The ions of which two of these elements are most likely to be toxic? (f) In which Group (I–V) of the qualitative analysis scheme of the cations will the wanium cation fall? the cupellium cation?

40. *Some new elements have just been discovered in the author's laboratory. The following atomic parameters have been obtained:

Element	Williamsium	Manningine	Galicium	Norrellogen
Electronegativity	1.06	3.82	2.37	2.34
Single-bond energy to another atom of same (kJ)	87	146	88	138

(a) Calculate the bond energy of the galicium–manningine bond. (b) Which one of the following compounds is least likely to be stable? Williamsium–manningide, galicium–norrellide, or norrellogen–manningide. Explain why. (c) Norrellogen is found to exhibit two positive oxidation states, +1 and +3, in its compounds with manningine. In which positive oxidation state will norrellogen be more electronegative? In which oxidation state will the norrellogen–manningine bond be stronger?

41. Some new elements have just been discovered in the author's laboratory. The following atomic parameters have been obtained:

Element	Wattsium	Davisine	Jonesium	Welchogen
Electronegativity	2.42	3.50	1.33	2.30
Single-bond energy to another atom of same (kJ)	83	140	133	171
Cation Radius (pm)	130 (+1 ion)		60 (+3 ion)	
Anion Radius (pm)		120 (−1 ion)		190 (−2 ion)

(a) Calculate the bond energy of the jonesium–welchogen bond. (b) Wattsium is found to exhibit two positive oxidation states, +1 and +3, in its compounds with davisine. In which positive oxidation state will wattsium be more electronegative? In which positive oxidation state will the wattsium–davisine bond be stronger? (c) Based on the information given, classify each of the following as a hard or soft acid or base: the wattsium cation; the daviside anion; the jonesium cation; the welchide anion. (d) What type of bonding (polar covalent, etc.) will predominate in the wattsium–welchogen bond? Will the formation of this bond from the aqueous ions result mainly in a favorable *entropy* or *enthalpy* change? (e) The ions of which two of these elements are most likely to be toxic? (f) In which Group (I–V) of the qualitative analysis scheme of the cations will the wattsium cation fall? the jonesium cation?

42. Ignoring any questions about hybridization, calculate (in units of kilojoules per mole): (a) the Mullikin electronegativities (χ_M) of Mg and of Pb. Which is more electronegative according to Mullikin? Does Pauling agree? (b) The softnesses (σ) of Mg and of Pb. Which is softer? Does the HSAB principle agree?

43. Assuming no hybridization, calculate Mullikin electronegativities and softnesses of the Group 11 elements. Are the trends in these calculations within the group in agreement with the trends in Pauling electronegativities and the trend indicated by the HSAB principle?

44. *Calculate and compare the enthalpy change for the gas-phase *dissociation* versus *ionization* of the following: (a) KF and (b) KI. Relatively speaking, which is easier to ionize and why?

45. Evaluate ΔH for the: (a) ionization of H_2O:HF into the ions H_3O^+ and F^-; how does this compare to ΔH for the dissociation of HF into ions (see Example 7.6); (b) dissociation of H_3O^+ into H_2O and H^+; (c) dissociation of the adduct H_2O:HF into separate molecules H_2O and HF. (d) Determine whether or not these calculations plus that of Example 7.6 obey Hess' Law.

46. Calculate the dissociation enthalpies of the following reactions: (a) $H_2O{:}I_2(g) \rightarrow H_2O(g) + I_2(g)$ and (b) $HOI(g) \rightarrow HO(g) + I(g)$. The first O–I bond dissociates in a Lewis acid–base pattern, with both electrons from the bond returning to oxygen; the second O–I bond dissociates into free radicals, with one electron going to O and one to I. Do your calculations support Haaland's conclusion that the coordinate covalent bond is weaker than the covalent bond?

47. The HSAB principle and the Drago equation each give predictions of directions of reactions in which coordinate covalent bonds are formed. Assuming that the necessary data are available, which of these approaches is likely to be more appropriate for predicting the positions of each of the following equilibria, and why?

(a) $Cu^{2+}(aq) + S^{2-}(aq) \rightleftharpoons CuS(s) + aq$

(b) $BH_3(g) + CO(g) \rightleftharpoons BH_3CO(g)$

(c) $HgCl_2 + 2\,NaI \rightleftharpoons HgI_2 + 2\,NaCl$ (all in methanol)

(d) $(C_2H_5)_2O{:}BF_3 + (CH_3)_2S{:}InI_3 \rightleftharpoons (CH_3)_2S{:}BF_3 + (C_2H_5)_2O{:}InI_3$ (all in hexane)

(e) Equilibrium (d) run in ethanol as a solvent.

48. Use the appropriate Drago equation to evaluate the energies involved in the formation or dissociation of the bonds given in the following reactions, then combine the results to evaluate ΔH for the complete reactions. Do these *gas-phase* reactions obey the HSAB principle, which is normally applied in the solution phase?

(a) $TlI(g) + KF(g) \rightarrow TlF(g) + KI(g)$. (Also check this reaction using experimental bond energies.)

(b) $H_3O^+{:}I^-(g) + I_2{:}F^-(g) \rightarrow I_2{:}I^-(g) + H_3O^+{:}F^-(g)$

(c) $KCF_3(g) + TlCCl_3(g) \rightarrow KCCl_3(g) + TlCF_3(g)$

Notes

1. a. M. Born, *Verhandl. Deut. Physik. Ges.*, **21**, 13 (1919); b. F. Haber, *Verhandl. Deut. Physik. Ges.*, **21**, 750 (1919).
2. If we do this under constant pressure, this heat is actually an enthalpy change.
3. The atomization of a nonmetal may be regarded as the sum of two other steps: the vaporization of the nonmetallic molecule X_n, requiring its *heat of vaporization*, and the breaking of the bonds in the gaseous X_n molecule, requiring the input of the *bond dissociation energy*. Thus to obtain 1 mol of Cl atoms, we would not need a heat of vaporization, since Cl_2 is already gaseous, but we would need to invest one-half the Cl–Cl bond dissociation energy in order to break 0.5 mol of Cl–Cl bonds.

4. Since this process logically precedes the first ionization of the gaseous atom, it has been suggested that its enthalpy change be called the *zeroth* ionization energy of the element.
5. Useful compilations of enthalpies of formation and of lattice energies are in M. C. Ball and A. H. Norbury, *Physical Data for Inorganic Chemists*, Longman, London, 1974, pp. 59–96.
6. N. Bartlett, *Proc. Chem. Soc. (London)*, 1962, 218; N. Bartlett, *Endeavor*, **88**, 3 (1963).
7. M. Kaupp, M. Dolg, M. Stoll, and H. G. von Schnering, *Inorg. Chem.*, **33**, 2122 (1994).
8. This neglects the fact that converting a hot gaseous metal to cold liquid-phase hydrated ions will have a new unfavorable entropy term, but this is presumably a relatively minor correction.
9. K. J. Klabunde, *Chemistry of Free Atoms and Particles*, Academic, New York, 1980.
10. There is one problem in trying to calculate the gas-phase activities of these metals from Eq. (7.8) without the atomization term: The hydration energy of Pt^{2+} is not known. On the other hand, we could use Eq. (7.8) in its complete form to calculate those hydration energies, although the resulting values would necessarily assume a zero entropy change in the activity reaction, which is not likely the case. Doing so anyway gives an estimated hydration energy of $-2115\,\text{kJ mol}^{-1}$ for Pt^{2+}, which is qualitatively in line with those given in Table 2.1. Alternately, the hydration energies could be estimated using the Latimer equation (2.1). But the Latimer equation does not include a term for the effects of the high electronegativities of Hg and Pt, and it cannot correct for the fact that Pt^{2+} has an anomalously small coordination number.
11. P. Mirone, *J. Chem. Educ.*, **68**, 132 (1991).
12. L. H. Ahrens, *J. Inorg. Nucl. Chem.*, 2, 290 (1956); L. H. Ahrens, *Geochim. Cosmochim. Acta*, **2**, 155 (1952); **3**, 1 (1953).
13. R. L. Deming, A. L. Allred, A. R. Dahl, A. W. Herlinger, and M. O. Kestner, *J. Am. Chem. Soc.*, **98**, 4132 (1976).
14. E. Clementi and D. L. Raimondi, *J. Chem. Phys.*, **38**, 2686 (1963).
15. K. F. Purcell and J. C. Kotz, *Inorganic Chemistry*, Saunders, Philadelphia, 1977; p. 45.
16. L. Pauling, *The Nature of the Chemical Bond*, 3rd ed., Cornell University Press, Ithaca, NY, 1960; Chapter 3. For a historical treatment of the idea of electronegativity before Pauling, see W. B. Jensen, *J. Chem. Educ.*, **73**, 11 (1996).
17. A. A. Allred, *J. Inorg. Nucl. Chem.*, **17**, 215 (1961).
18. Faulty thermodynamic data will, of course, lead to faulty electronegativities; this is thought to be the case, for example, for lead(IV).
19. In practice, if the metal has an electronegativity less than 1.9.
20. Letting M' be H^+ results in two new problems: (1) the product $M{-}OH_2^+$ has the same hydration number as H^+, which is an unrealistically low number; (2) in Figure 6.14, it may be seen that the point for H^+ (at $E^\circ = 0\,\text{V}$ and $\chi_P = 2.2$) deviates considerably from the best-fit line (perhaps for the reason just mentioned, and because its reduction product, H_2, is a gas and has a considerably different entropy than a metal). The H^+ ion fits the correlation better if we give it an empirical electronegativity of 1.9 (in H_3O^+); making this adjustment and the appropriate substitution in Eq. (7.24), we obtain Eq. (7.26).
21. P. R. Wells, *Prog. Phys. Org. Chem.*, **6**, 111 (1968).
22. R. G. Pearson, *J. Chem. Soc., Chem. Commun.*, 1968, 65.
23. D. W. Smith, *J. Chem. Educ.*, **67**, 911 (1990).
24. L. C. Allen, *J. Am. Chem. Soc.*, **111**, 9003 (1989); summarized by S. A. Borman, *Chem. Eng. News*, 18, Jan. 1, 1990.
25. a. R. G. Parr and R. G. Pearson, *J. Am. Chem. Soc.*, **105**, 7512 (1983); b. R. G. Pearson, *Inorg. Chem.*, **27**, 734 (1988).
26. R. T. Sanderson, *J. Chem. Educ.*, **29**, 539 (1952); **31**, 2 (1954); 238 (1954).
27. R. J. Boyd and K. E. Edgecombe, *J. Am. Chem. Soc.*, **110**, 4182 (1988).
28. R. S. Drago, D. C. Ferris, and N. Wong, *J. Am. Chem. Soc.*, **112**, 8953 (1990); R. S. Drago, N. Wong, and D. C. Ferris, *J. Am. Chem. Soc.*, **113**, 1970 (1991).
29. The parameter 4.184 converts the energy units from the original kilocalories per mole to kilojoules per mole.
30. In the derivation of the Drago parameters, there are always two more parameters than there are measurable enthalpy changes, so that two of these parameters must be arbitrarily set. These are the E_A and the C_A terms for I_2, which were arbitrarily set at $E_A = 0.50$ and $C_A = 2.00$. However, we cannot therefore assert that I_2 is four times as good at covalent bonding as it is at electrostatic (dipole–dipole) bonding. And we cannot meaningfully compare the E and C values to each other—we can only compare E_A values to other E_A values, and so on.

31. Drago and workers continue to add to their databases and recalculate *E*, *C*, and related values. Newer values for more extensive sets of donors and acceptors are found in R. S. Drago, A. P. Dadmun, and G. C. Vogel, *Inorg. Chem.*, **32**, 2473 (1993) and also G. C. Vogel and R. S. Drago, *J. Chem. Educ.*, **73**, 701 (1996). One should not mix values from newer and older sets of parameters.
32. R. S. Drago, N. Wong, and D. C. Ferris, *J. Am. Chem. Soc.*, **113**, 1970 (1991). A more extensive (and somewhat altered) set of values than are found in Table 7.12 is in R. S. Drago and N. M. Wong, *Inorg. Chem.*, **34**, 4004 (1995).
33. A. Haaland, *Angew. Chem. Intl. Ed. Engl.*, **28**, 992 (1989).

CHAPTER

8

Introduction to Transition Metal Complexes

With Applications to Biochemistry

8.1 Introduction

We discussed the chemistry of complex ions in Chapters 3–5; in this chapter, we emphasize some special features present in the complex ions of the *d*-block metals. First, we should stress that in most respects, the complex ions of the *d*-block metals are like those found in the rest of the periodic table: their stability results mainly from the donation of electrons from the Lewis base ligands to (empty orbitals in) the Lewis acid metal ion; their complexes with chelating ligands are more stable than with monodentate ligands; the hard and soft acid–base (HSAB) principle applies; and so on.

But there are certain features about complexes formed from metal ions having incomplete filling of the *d* (and *f*) orbitals that are not observed in other complexes. The first and most obvious of these features is the great variety of colors observed in these complexes—even in the simplest complex ions, the hydrated ions (Table 8.1). A second unique feature is the magnetic properties of these ions—they may have unpaired electrons, which results in the magnetic property called *paramagnetism* (Section 8.3). A third important feature is the large number of possible oxidation states, which (in contrast with elements in the *p* block) may differ by only one electron; even zero or negative oxidation states are possible. A fourth feature is that the *d*-block metal ions may bond well with certain types of ligands such as carbon monoxide (the *carbonyl* ligand) that very seldom bond to *p*-block elements.

Some of these special properties are found and will be illustrated in all parts of the *d* and *f* blocks among complexes having incompletely filled *d* and *f* orbitals, but they *all* are commonly found together only in the *d*-block complexes of *the fourth period* in which the *d* orbitals are incompletely filled: that is, the complexes of the eight metals Ti through Cu. It is the complexes of these metal ions that will be emphasized.

In this chapter, we develop two special theories used specifically to interpret the special properties of *d*-block complexes. The oldest of these is the **crystal field theory**, which was developed by physicists around 1930[1] to explain similar special properties that result when ionic crystals (such as NaCl) are irradiated; its existence and relevance

Table 8.1
Colors of Hydrated Metal Ions Having Partially Filled Sets of *d* or *f* Orbitals

	Period 4, *d* Block		Period 6, *f* Block		Period 7, *f* Block	
n^a	Ion	Color	Ion	Color	Ion	Color
1	Ti^{3+}	Violet	Ce^{3+}	Colorless	Pa^{4+}	Colorless
2	V^{3+}	Blue	Pr^{3+}	Green		
3	V^{2+}	Violet	Nd^{3+}	Lilac	U^{3+}	Red-brown
	Cr^{3+}	Violet			Np^{4+}	Yellow-green
4	Cr^{2+}	Blue	Pm^{3+}	Pink	Np^{3+}	Purple
					Pu^{4+}	Tan
5	Mn^{2+}	Pale pink	Sm^{3+}	Yellow	Pu^{3+}	Blue-violet
	Fe^{3+}	Pale purple			Am^{4+}	Red
6	Fe^{2+}	Pale green	Eu^{3+}	Pink	Am^{3+}	Pink
			Sm^{2+}	Red	Cm^{4+}	Yellow
7	Co^{2+}	Pink	Gd^{3+}	Colorless		
			Eu^{2+}	Colorless		
8	Ni^{2+}	Green	Tb^{3+}	Pink		
9	Cu^{2+}	Blue	Dy^{3+}	Yellow		
10			Ho^{3+}	Yellow		
11			Er^{3+}	Lilac		
12			Tm^{3+}	Green		
13			Yb^{3+}	Colorless		

SOURCE: F. A. Cotton and G. Wilkinson, *Advanced Inorganic Chemistry: A Comprehensive Text*, 5th ed., Wiley-Interscience, New York, 1988; Chapter 18 959 and 989.
[a]Number of electrons in valence *d* or *f* orbitals = *n*.

to *d*-block complex ion chemistry was unknown to chemists until the 1950s. As befits its origin, it treats bonding from an electrostatic point of view: Ligands are modeled as point negative charges. This obvious oversimplification works surprisingly well. Many of the terms and quantitative parameters used even today with the more sophisticated bonding theories have their origins in the crystal field theory. For some purposes, however, it becomes imperative to consider the fact that the metal–ligand interaction involves covalent overlap; when this adjustment is made to the crystal field theory we have the **ligand field theory**, which will also be introduced in this chapter.

The two main theories applied in the *p* block, *valence bond theory* and *molecular orbital theory*, have their applications in the *d* block as well. The valence bond theory can account for the main stabilization factor in complexes, the metal–ligand Lewis acid–base interaction, just as well in the *d* block as elsewhere; but it has more difficulties in cleanly and unambiguously predicting some of the special properties of *d*-block complexes, so it is not much emphasized anymore by inorganic chemists for this type of application. Far more powerful is the molecular orbital theory, which is introduced in Chapter 10 and also applied to *d*-block complexes there. Instructors whose classes consist entirely of chemistry majors may elect to do this chapter after Chapters 9 and 10, so as to be able to utilize the molecular orbital approach and show the symmetry origins of some of the labels. However, if the class contains biochemistry or biology majors or others who do not normally have much use for molecular orbital theory, it will be desirable to do this chapter now: The *d*-block complexes are too important in life not to be discussed with these students!

8.2 The Crystal Field Theory

Many of the special properties of the *d*- (and *f*-) block complexes result from the fact that the central metal ion in the complex is generally *nonspherical* due to partial occupancy of the valence (*d* or *f*) orbitals. Table 8.1 shows examples among hydrated ions of every possible partially complete electron configuration from d^1 to d^9 and from f^1 through f^{13}. Metal ions having these electronic configurations will have an irregular surface that will interact differently with incoming ligands, depending on the direction from which they approach the metal. This is true for all electron configurations except those in which the *d* and *f* orbitals are *equally* occupied. The d^0 and f^0 configurations are obviously spherical; it is less obvious, but can be shown from the wave functions, that the fully occupied d^{10} and f^{14} configurations are also spherical, as are the half-filled electron configurations d^5 and f^7 provided that each of the *d* or *f* orbitals is occupied by one electron. Many of the special properties that we will discuss are absent in these cases (e.g., in Table 8.1, the hydrated ions are either colorless or have only pale colors).

Since the maximum total coordination number among the *d*-block metal ions of the fourth period is 6, let us begin by discussing octahedral complexes of these ions, in which the ligands come in from the $+x$, $-x$, $+y$, $-y$, $+z$ and $-z$ directions. Each of these ligands has a pair of electrons to be donated to the metal ion; in the crystal field theory model these are treated as point charges. If the metal ion has a d^0 electron configuration, a complex is formed with no special properties. But if the metal ion has a d^1 electron configuration (e.g., Ti^{3+}), the properties will vary depending on which *d* orbital holds the electron. In Figure 8.1, we take the drawings of the shapes of the *d* orbitals from Figure 1.4, and locate them inside an octahedron of ligands (point negative charges). It can be seen that two of the *d* orbitals, the d_{z^2} and $d_{x^2-y^2}$, have lobes that are oriented along the *x*, *y*, and *z* axes. If the electron is in one of these orbitals, it will repel the incoming point charge, producing a less stable complex. On the other hand, the other three orbitals, the d_{xy}, d_{xz}, and d_{yz} orbitals, have lobes that fall *between* the *x*, *y*, and *z* axes, so that there is much less repulsion. Clearly, it will be preferable, in an octahedral complex, for the electron to occupy the d_{xy}, d_{xz}, or d_{yz} orbital (labeled the t_{2g} set of orbitals; for the origin of the label see Chapter 9) rather than the d_{z^2} or $d_{x^2-y^2}$ orbital (the e_g set of orbitals).

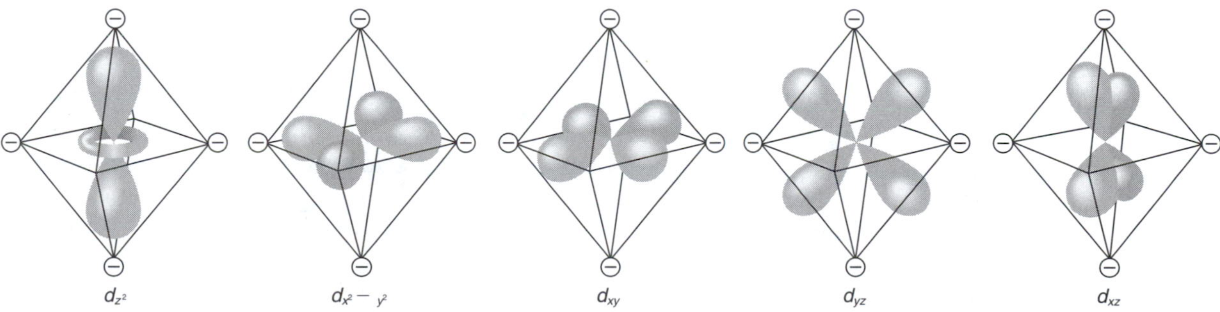

Figure 8.1
Orientations of the *d* orbitals inside an octahedron of ligands or point negative charges. [From Zumdahl, Steven, *Chemistry*, Third Edition. Copyright © 1993 by D. C. Heath and Company. Used by permission of Houghton Mifflin Company.]

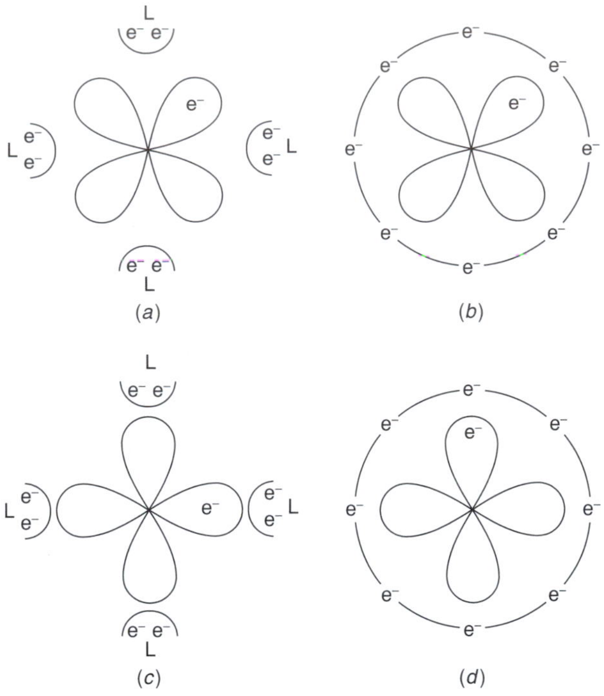

Figure 8.2
The xy plane representation of the interaction of two metal d orbitals with the electrons of donor atoms in a complex. (a, c): the real octahedral set of donor atoms, with eight electrons being donated. (b, d): the hypothetical spherical complex, with eight electrons being donated equally from all positions on the sphere at the same distance. (a, b): One metal d electron in the d_{xy} orbital. (c, d): one metal d electron in the $d_{x^2-y^2}$ orbital.

We want to compare this actual situation with the hypothetical one in which the consequences of the nonspherical electron distribution is removed. For a d^1 electron configuration, this could be achieved, for example, by putting 0.2 electrons in each of the five d orbitals. Since this is somewhat hard to envision, the equivalent hypothetical complex ion is conventionally taken in which the nonsperical shape is removed from the set of ligands instead, by distributing the electrons of the six octahedral donor atoms evenly over a sphere at a common distance, as sketched in Figure 8.2 (b and d).

The complex ion with the one electron in the t_{2g} orbital (top of Figure 8.2) experiences *less* repulsion of the ligand in the real octahedral complex [Fig. 8.2(a)] than in the hypothetical spherical complex ion [Fig. 8.2(b)]. The ligands are able to move *closer* to the metal ion along the d-electron free x, y, and z axes, thereby donating their electrons more effectively to the metal ion and produce a more stable complex. The extra energy released as a result of this nonspherical interaction is called the **crystal field stabilization energy (CFSE)** of the complex ion. Since this energy term also arises when considerations of covalency in the ligand–d orbital interaction are introduced as part of the ligand field theory, this energy term is also called the ligand field stabilization energy (LFSE).

On the other hand, if the one electron of the metal ion were in the e_g set of orbitals (bottom of Fig. 8.2), it would be experiencing *more* repulsion in the real octahedral

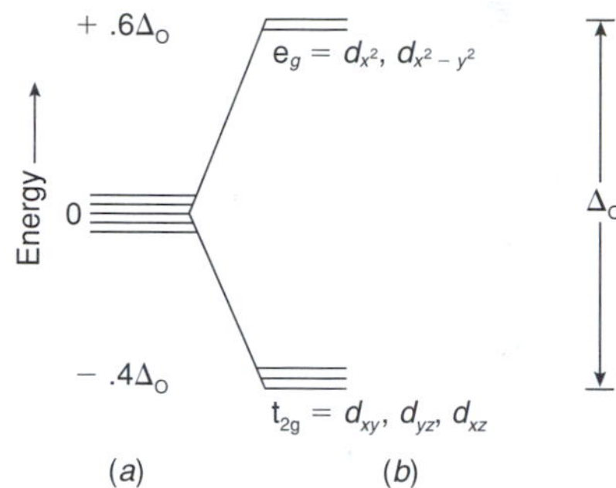

Figure 8.3
Relative energies of the metal *d* orbitals in (*a*) a hypothetical spherical set of donor atoms (left side), taken as the zero of energy and (*b*) a real octahedral set of donor atoms (right side.) The Δ_0 parameter is the crystal field splitting (also represented as 10 *Dq*).

complex ion [Fig. 8.2(*c*)] than in the hypothetical spherical complex ion [Fig. 8.2(*d*)]. In Figure 8.3, we show an **energy level diagram** comparing the relative energies of orbitals that the *d* electron could occupy in these two geometries of complexes: Any *d* orbital in the spherical complex (on the left) is taken as the zero of energy, while in the octahedral complex (on the right) the $t_{2g}{}^1$ set of orbitals is lower in energy and the $e_g{}^1$ set of orbitals is higher. The difference in energy between the two sets of orbitals (t_{2g} and e_g) is called the (octahedral) **crystal field splitting** (or ligand field splitting) and is symbolized as **10*Dq*** or Δ_o (in which the subscript "o" stands for octahedral).

To determine just where the energy level of the hypothetical spherical ion is located with respect to the two real ions, we can briefly move ahead to the case of the evenly distributed d^5 electron configuration $t_{2g}{}^3e_g{}^2$. This metal ion is spherical, so it has zero crystal field stabilization energy. Therefore, the stabilizations resulting from the three t_{2g} orbitals must just balance the repulsions from the two e_g electrons. Consequently, the repulsion per e_g electron must be $\frac{3}{2}$ times the stabilization per t_{2g} electron. Taking the energy of the hypothetical spherical ion as zero, the energy of the t_{2g} orbital must then be $-0.4\,\Delta_o$ (or $-4D_q$); the energy of the e_g orbital must be $+0.6\,\Delta_o$ (or $+6D_q$). Returning to the case of the d^1 metal ion, this ion will have a crystal field stabilization energy of $-0.4\,\Delta_o$ in the preferred $t_{2g}{}^1$ configuration.

The behavior of a complexed d^2 metal ion such as V^{3+} will be quite similar: In such an ion the electrons preferentially occupy two of the three t_{2g} orbitals, with both spins aligned the same way (e.g., both spins up—Hund's rule). This gives twice the crystal field stabilization energy of the d^1 case, $-0.8\,\Delta_o$. In a complexed d^3 metal ion such as Cr^{3+}, the three electrons will each occupy one of the t_{2g} orbitals, each with its spin in the same direction, to give a crystal field stabilization energy of $-1.2\,\Delta_o$.

It is also true that the energies of the seven *f* orbitals of a given *f*-block metal ion are split in energy in an octahedral complex ion. However, since $\langle r_{max}\rangle$ for the valence $(n-2)f$ orbitals is so much less than $\langle r_{max}\rangle$ for the valence *ns* [or even the core $(n-1)p$] orbitals, the ligands cannot in fact get close enough to the *f* orbitals to cause much crystal field splitting at all, particularly for metals with 4*f* valence electrons. With regard to the *d*-block metals, the crystal field splitting is large among metals with 4*d* and 5*d* valence electrons and intermediate in magnitude among metals with 3*d* valence electrons.

8.3 High- and Low-Spin Electron Configurations: Magnetic Properties

The fourth electron of a d^4 metal ion such as Cr^{2+} must make a choice between two evils. If it goes into one of the three t_{2g} orbitals, thereby avoiding the repulsion of the ligand electrons, it is repelled by the other electron already in that orbital. It must enter this orbital with its spin aligned opposite to that of the other electron already in the orbital; we say that the two spins are *paired.* The energy required to force these two electrons into the same orbital is called the **pairing energy**, ***P*** (Table 8.2).[2] From Table 8.2, it may be seen that pairing is more difficult in +3 ions (which are smaller) than in the larger +2 ions.

Table 8.2
Pairing Energies (P) for Some Gaseous $3d$ Metal Ions[a]

Configuration	M^{2+}	P (kJ mol^{-1})	P (cm^{-1})[b]	M^{3+}	P (kJ mol^{-1})	P (cm^{-1})
d^4	Cr^{2+}	281	23,500	Mn^{3+}	335	28,000
d^5	Mn^{2+}	305	25,500	Fe^{3+}	359	30,000
d^6	Fe^{2+}	211	17,600	Co^{3+}	251	21,000
d^7	Co^{2+}	269	22,500	Ni^{3+}	323	27,000

SOURCE: D. S. McClure, "The Effects of Inner Orbitals on Thermodynamic Properties," in *Some Aspects of Crystal Field Theory*, T. M. Dunn, D. S. McClure, and R. G. Pearson, Eds., Harper & Row, New York, 1965; p. 82.

[a] The pairing energies are likely to be 15–30% lower in complex ions due to the effects of covalency in the metal–ligand bonding.

[b] An energy unit in spectroscopy (see Section 8.4).

If the fourth electron is to avoid paying the price of this pairing energy, it must instead go into one of the e_g orbitals, which are higher in energy by the crystal field splitting, Δ_o. So two possible electron configurations are possible for a d^4 metal ion. If P for the metal ion in question is smaller than Δ_o (e.g., the electrons are relatively easy to pair together in one orbital, if that orbital is large), the first electron configuration, $t_{2g}{}^4$, is selected, as shown in Figure 8.4(a). Such a complex has a CFSE of $-1.6\,\Delta_o + P$.[3] If the electric field of the ligands is relatively weak, resulting in P being larger than Δ_o, the electron configuration $t_{2g}{}^3e_g{}^1$ results, as shown in Figure 8.4(b). Such a complex has a CFSE of $3\,(-0.4\,\Delta_o) + 1(0.6\,\Delta_o) = -0.6\,\Delta_o$.

Complexes in which the metal d electrons would rather pair than fight the strong electric field of the ligand donor atoms are known as **strong-field complexes**; an example for Cr^{2+} is the cyanide complex $[Cr(CN)_6]^{4-}$. Complexes in which the metal d electrons can more readily overcome the electric field of the ligands are known as **weak-field complexes**; an example for Cr^{2+} is the hydrated ion $[Cr(H_2O)_6]^{2+}$. Both types of complexes are common for metals with $3d$ valence electrons (top period of the d block). Nearly all complexes of metals with $4d$ and $5d$ valence orbitals are strong-field complexes, since $4d$ and $5d$ valence orbitals are better able to reach out to and interact with (be repelled by) ligand orbitals.

The same dichotomy exists among the complexes formed by metal ions having five, six, and seven valence d electrons. In strong ligand fields, which are generally present in cyanide complexes, a d^5 metal ion such as Mn^{2+} or Fe^{3+} adopts the $t_{2g}{}^5$

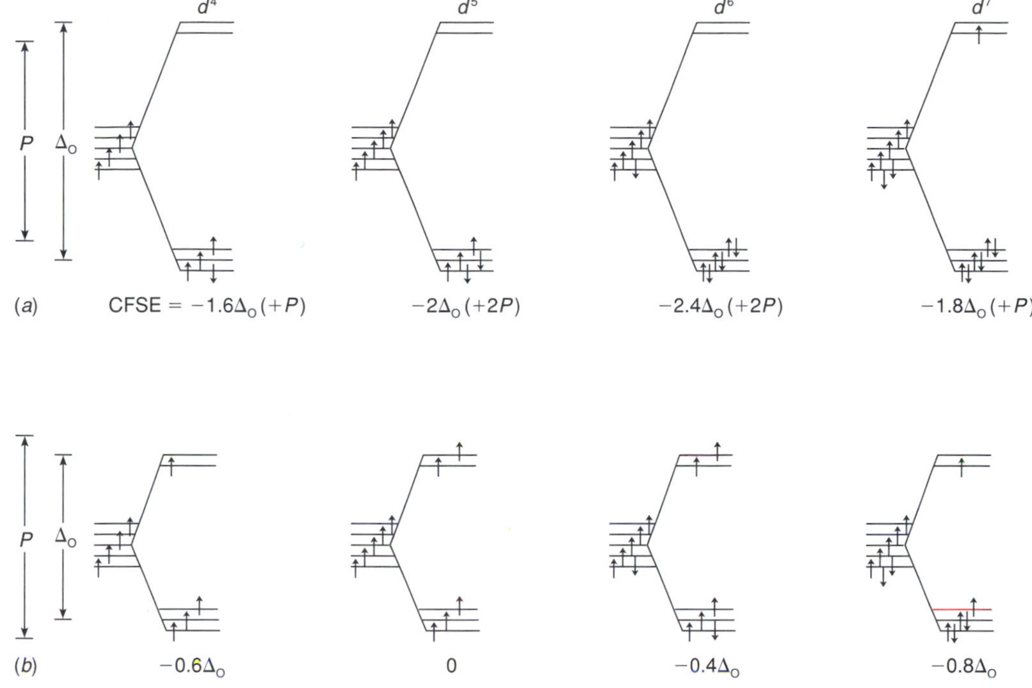

Figure 8.4
Energy level diagrams showing the (*a*) strong-field and (b) weak-field electron configurations and resulting CFSEs for d^4, d^5, d^6 and d^7 metal ions. (The contributions of pairing energies P to the CFSE *in excess of those found in the uncomplexed gaseous or the hypothetical spherically complexed ion* are shown in parentheses.)

electron configuration with a CFSE of $-2\,\Delta_o$ ($+2P$). In weak ligand fields (generally present in hydrated ions in the fourth period), d^5 metal ions adopt the $t_{2g}{}^3e_g{}^2$ electron configuration with a CFSE of zero. (*Remember*: This electron configuration is spherical.) The process continues for d^6 and d^7 metal ions (e.g., Fe^{2+} and Co^{2+}, respectively) as illustrated in Figure 8.4.

When we add one more electron to either the weak- or strong-field d^7 electron configuration to arrive at the d^8 electron configuration (as found in Ni^{2+}), we arrive at only one possible electron configuration: $t_{2g}{}^6e_g{}^2$, with a CFSE of $-1.2\,\Delta_o$. All octahedral d^8 complexes have the same electron configuration, so there is no longer any need of the weak-field/strong-field distinction. Similarly, d^9 ions (e.g., Cu^{2+}) adopt the $t_{2g}{}^6e_g{}^3$ electron configuration, with a CFSE of $-0.6\,\Delta_o$. The d^{10} ions, of course, are spherical, with zero CFSE and the $t_{2g}{}^6e_g{}^4$ electron configuration.

Unpaired Electrons and Magnetic Properties of Ions. Any ion, molecule, or atom in which all of the electrons are paired (e.g., most species in the p block except those known as *free radicals*) exhibits a weak magnetic property known as *diamagnetism*: The substance is weakly repelled by a magnetic field. (This property is unusually strong in the materials known as superconductors, to be discussed in Chapter 13.) Of more significance, any ion, molecule, or atom that has unpaired electrons exhibits the stronger magnetic property known as *paramagnetism*: Because the magnetic effects of all the electrons with spin "up" are not canceled by the opposed effects of an equal

Table 8.3
Calculated and Typical Observed Magnetic Moments[a]

n[b]	μ_{calc}	μ_{obs}
1	1.73	1.7–2.2 (in Cu^{2+} complexes)
2	2.83	2.8–3.3 (in Ni^{2+} complexes)
3	3.87	4.5–5.2 (in Co^{2+} complexes)
4	4.90	5.1–5.7 (in Fe^{2+} complexes)
5	5.92	5.6–6.1 (in Mn^{2+} complexes)

[a] In units of Bohr magnetons (BM). The deviation of the observed from the calculated values is due to a contribution from the *orbital motion* (not just the spin) of the unpaired electrons. Observed values are for octahedral complexes.
[b] Number of unpaired electrons.

number of electrons with spin "down", the species is attracted to a magnetic field. The magnitude of this effect for the species is known as its *magnetic moment*, μ; its theoretical value is most simply related to the number of unpaired electrons n by the equation

$$\mu = \sqrt{n(n+2)} \qquad \text{Bohr magnetons (BM)} \tag{8.1}$$

Calculated and commonly observed values of μ are given in Table 8.3.

The experimental magnetic moment is calculated from the observed force with which 1 mol of sample is pulled into a magnetic field.[4] This force, after corrections for the diamagnetism of the paired electrons in the metal ion, ligands, and counterions, is known as the corrected molar *magnetic susceptibility* of the substance, $\chi_m{}^{corr}$ [in units conventionally labeled as centimeter–gram–second (cgs)]. By *Curie's law*, the magnetic susceptibility of a substance is also inversely proportional to the absolute temperature of the substance.[5] The magnetic moment can be obtained from the absolute temperature, T, and the magnetic susceptibility:

$$\mu = 2.83\sqrt{\chi_m{}^{corr}\,T} \tag{8.2}$$

Measuring the magnetic susceptibility of a complex is therefore the most direct way of determining whether a given complex is a weak or strong field. Because of the importance of the electron spin in this measurement, weak-field complexes are also commonly called **high-spin** complexes and strong-field complexes are also called **low-spin** complexes.[6]

Example 8.1
Predict the number of unpaired electrons, the magnetic moment, and the molar magnetic susceptibility at 25 °C for each of the following complex ions. If the ion has the possibility of being either high spin or low spin, make predictions for both cases, but indicate the more likely possibility: (a) $[Fe(CN)_6]^{4-}$; (b) $[Ru(NH_3)_6]^{3+}$; (c) $[Cr(NH_3)_6]^{2+}$; and (d) $[EuCl_6]^{4-}$.

Solution:

First, in order to determine the d^n electron configuration in each complex ion, it is necessary to determine the oxidation number of each central metal atom; if this causes any difficulty, you may want to look ahead to Section 8.11 now. In (a), we have a complex of CN^- with the d^6 ion Fe^{2+}, which can be either high spin with four unpaired electrons, or low spin with 0 unpaired electrons; with CN^- as a ligand the latter is more probable. In (b), we have a complex of NH_3 with the d^5 Ru^{3+} ion, which can be either high spin with five unpaired electrons or low spin with one unpaired electron; in the fifth period, the latter is more likely. In (c), we have a complex of NH_3 with the d^4 Cr^{2+} ion; in the fourth period, either the high-spin configuration with four unpaired electrons or the low-spin configuration with two unpaired electrons is possible. In (d), we have a complex of Cl^- ions with the Eu^{2+} ion, which has a f^7 electronic configuration that will certainly be high spin with seven unpaired electrons, since f orbitals have so little overlap and interaction with ligand orbitals.

Using Eq. (8.1), the theoretical values of the magnetic moment μ are found: (a) maybe 4.90 but probably 0 BM; (b) maybe 5.92 but probably 1.73 BM; (c) either 4.90 or 2.83 BM; (d) 7.94 BM.

The predicted molar magnetic susceptibilities, χ_m^{corr} at 298 K, are obtained from a rearranged form of Eq. (8.2), $\chi_m^{corr} = 1/T(\mu/2.83)^2$: (a) may be 0.0101 cgs units but probably 0 cgs units; (b) may be 0.0147 but probably 0.0013 cgs units; (c) either 0.0101 or 0.0034 cgs units; and (d) 0.0264 cgs units per mole.

Occasionally, a complex is discovered in which the crystal field splitting Δ_o is almost exactly equal to the pairing energy P. In such cases, a change in temperature can cause a partial **spin crossover** from high- to low-spin behavior, with unusual magnetic behavior resulting. For example, some monothiocarbamate complexes of iron(III) were discovered[7] in which the magnetic moment dropped from 5.7 to 5.8 BM at 300 K, to 4.7 to 5.0 BM at 150 K, to 3.6 to 4 BM at 78 K, as the proportion of molecules in the high- and low-spin states changed with temperature. It has been suggested that this phenomenon could be the basis for a new technology of electronic information storage.[8]

8.4 Electronic Absorption Spectra of Complex Ions

The magnitude of the crystal field splitting, Δ_o, of a given complex is conveniently measured by recording its absorption spectrum in the visible (vis), ultraviolet (UV), and/or near-infrared (near-IR) regions of the electromagnetic spectrum (Fig. 8.5). It is the portion of the absorption spectrum that occurs in the visible (middle) region that is responsible for the beautiful and varied colors such as those of the hydrated d- and f-block ions (Table 8.1). Taking the first spectrum in this figure, that of the octahedral d^1 ion $[Ti(H_2O)_6]^{3+}$, as an example, the one d electron occupies a t_{2g} orbital. If the ion captures a photon (particle of light) of the appropriate energy, that d electron can be promoted to the higher energy e_g orbital. The energy required to do this is just Δ_o. We can, of course, express this energy in kilojoules per mole, as shown at the bottom of Figure 8.5, but spectroscopists customarily express the energy in the same units as the frequency of the light absorbed, reciprocal centimeters ($1000\,cm^{-1} = 12.0\,kJ\,mol^{-1}$;

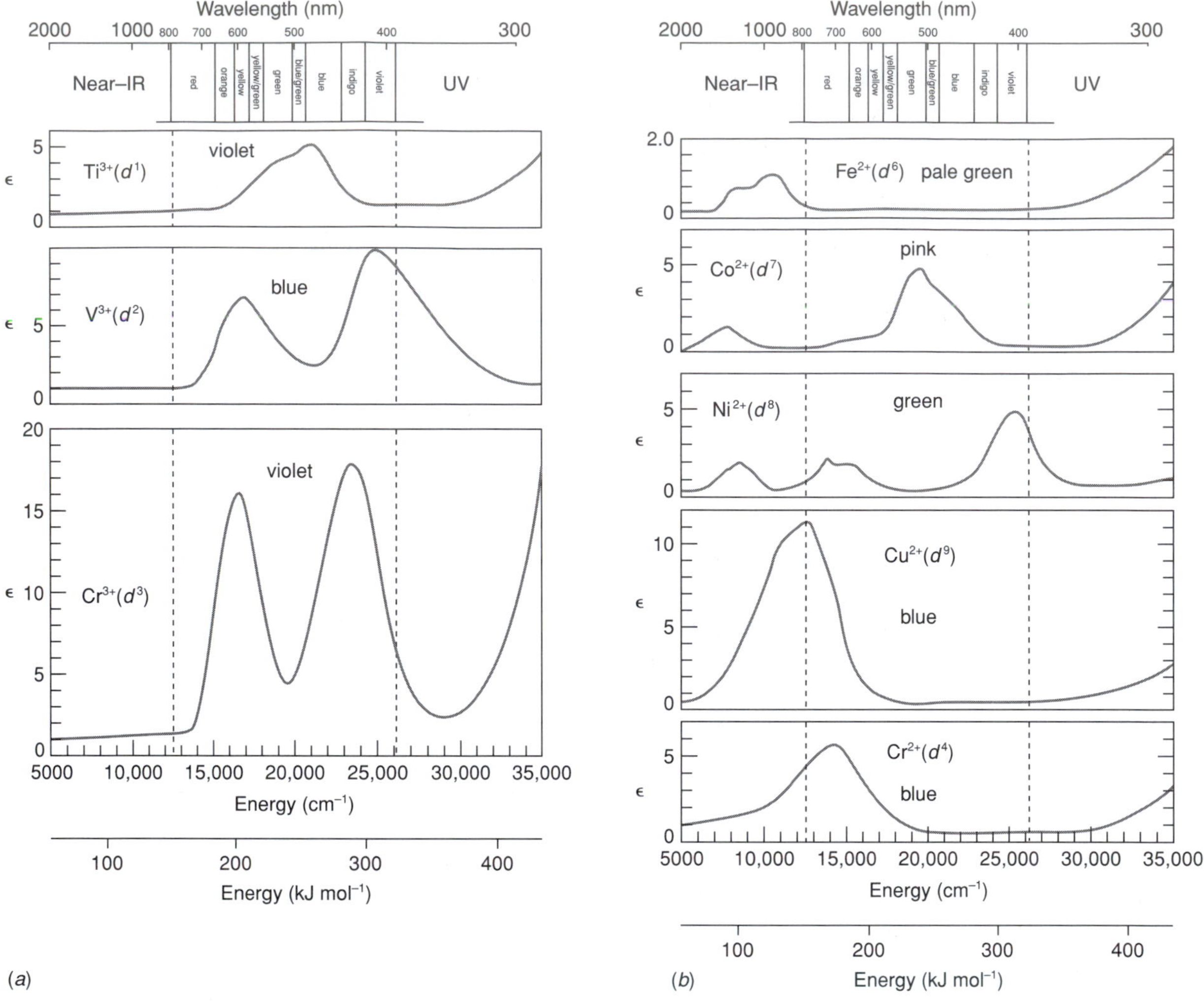

Figure 8.5
Electronic absorption spectra of some $3d^n$ hydrated metal ions in the ultraviolet, visible, and near-IR regions. (*a*) Spectra of some +3 charged ions; (*b*) spectra of some +2-charged ions. [Adapted from B. N. Figgis, *Introduction to Ligand Fields*, Interscience, New York, 1966.]

the instruments themselves are often calibrated in terms of the wavelength of the light in nm = 10^7 divided by the frequency in cm^{-1}.) The absorption peak for $[Ti(H_2O)_6]^{3+}$ occurs at 493 nm = 20,300 cm^{-1} = 243 kJ mol^{-1}, right in the middle of the visible region. Referring to the top of Figure 8.5, this means that blue and green light are absorbed out of the visible light; the other colors of light (particularly red and violet) pass through the solution of this ion and give it the color we observe, red-violet.

In the case of ions having more than one *d* electron in the t_{2g} orbitals (and more than one vacancy in the e_g orbitals), such as the $t_{2g}{}^2$ V^{3+} ion shown next in Figure 8.5, the situation is more complicated because of the presence of electron–electron repulsion effects. After the photon is absorbed, the excited V^{3+} ion will have an electron configuration of $t_{2g}{}^1 e_g{}^1$, but this could represent several pairs of orbitals. If the configuration is actually $d_{xy}{}^1 (d_{x^2-y^2})^1$, there will be a greater degree of repulsion between the two d electrons than if the configuration is actually $d_{xy}{}^1(d_{z^2})^1$, and so on. This topic will be explored in more depth in Chapter 17. The bottom line is that there will be more than one absorption band, because there will be different changes in the inter-electron repulsion energy for different combinations of orbitals involved (different electronic states). Fortunately, the lowest energy absorption band (the one closest to the near-IR) observed in the typical octahedral complex does not involve this complication, but occurs at the energy Δ_o.[9] In the case of the hydrated V^{3+} ion, the absorption is at 18,000 cm^{-1}, which is its Δ_o; in the case of the hydrated Cr^{3+} ion it is at 17,400 cm^{-1}, and so on. These and other values for Δ_o are listed in Table 8.4.

Trends in Crystal Field Splittings. Two trends in Δ_o are apparent from Table 8.4. The first trend can also be observed in the spectra of Figure 8.5: whereas the lowest energy absorption (of frequency $= \Delta_o$) of the +3 charged hydrated ions (left side of the figure) fall in the visible region, the same absorption of the +2 charged hydrated ions (left side) generally fall outside of the visible region, in the low-energy near-IR region.[10] As clarified in Table 8.4, within a given period, the crystal field splitting increases with increasing charge on the metal ion. We know that more positively charged metal ions are smaller and have much greater hydration energies; the ligands will be drawn in more closely and will repel the e_g electrons more strongly.

The second trend is that, for a given charge, the crystal field splitting increases on going down the periodic table. We noted earlier that $4d$ and $5d$ orbitals have better overlap with external orbitals than $3d$, so that there will be more repulsion between ligand electrons and electrons in $4d$ and $5d$ orbitals. Putting these trends together, we

Table 8.4
Crystal Field Splittings[a] of Hydrated Metal Ions and Metal-Ion g Factors[b]

(Valence Orbital Type) 3*d*						**4*d***		**5*d***	
				+2 Charged Ions					
V^{2+}	12,000; **11,800**	Cr^{2+}	**14,000**	Mn^{2+}	8,000; **7,500**	Ru^{2+}	20,000; **19,800**		
Fe^{2+}	**10,000**	Co^{2+}	9,000; **9,200**	Ni^{2+}	8,700; **8,600**				
Cu^{2+}	**13,000**								
				+3 Charged Ions					
Ti^{3+}	**20,300**	V^{3+}	**18,000**	Cr^{3+}	17,400; **17,400**	Mo^{3+}	24,600; **26,000**	Ir^{3+}	32,000
Mn^{3+}	**21,000**	Fe^{3+}	14,000; **14,000**	Co^{3+}	18,200; **20,760**	Rh^{3+}	27,000; **27,200**		
						Ru^{3+}	**28,600**		
				+4 Charged Ions					
Mn^{4+}	23,000					Tc^{4+}	30,000	Pt^{4+}	36,000

SOURCE: Data from J. E. Huheey, *Inorganic Chemistry: Principles of Structure and Reactivity*, 3rd ed., Harper & Row, New York; pp. 382, 384, 386; J. E. Huheey, E. A. Keiter, and R. L. Keiter, *Inorganic Chemistry: Principles of Structure and Reactivity*, 4th ed., Harper-Collins, New York; p. 406.

[a] Units for the crystal field splitting parameter Δ_o are reciprocal centimeters (cm^{-1}).

[b] Metal ion g factors, derived from crystal field splittings of complexes of several ligands, are given first; these are defined and used in Eq. (8.3). Measured Δ_o values of hydrated metal ions are given in boldface; ideally the two should be the same.

may note that the more highly charged metal ions of the last two periods (Ir^{3+}, Tc^{4+}, and Pt^{4+}) would have absorption bands mainly or entirely in the UV region of the spectrum, which would mean that they would be yellow or colorless; the most colorful part of the *d* block is definitely the fourth period! (Note that the hydrated ions of these particular metal ions would be combinations of hard bases with soft acids that are good enough oxidizing agents to oxidize water; these hydrated ions are seldom if ever seen. Nonetheless, the trends do apply to real complexes with borderline and soft base ligands.)

Intensity of Spectra. The intensities of absorption spectra are measured by their molar absorptivities, ε, as shown in Figure 8.5. If you have ever measured the UV spectra of organic compounds such as dyes, or of many analytically detected species, you will have been dealing with molar absorptivities of perhaps 10 thousand times the typical values indicated in Figure 8.5 (1–20). In theory, for a wave of light to be absorbed as it passes through a chemical species, there must be certain relationships between the orbital from which the electron is excited and the orbital into which it is excited. The electronic transition is, in principle, *forbidden* if the two orbitals are on the same atom and have the same value of the secondary quantum number, ℓ. Since both the t_{2g} and the e_g sets are of *d* orbitals, this "*d–d*" transition is theoretically forbidden; hence, thousands of photons will pass through the complex ion unabsorbed before one is finally absorbed. In contrast, charge-transfer spectra (Section 6.4) involve electronic transitions between different atoms, so they feature much more intense absorptions.

A number of mechanisms can account for the occasional photon being absorbed, without which all of these ions would be colorless. These mechanisms involve various sorts of covalent interactions between the *d* orbital and the ligand, which are weak. The problem is much more severe for the *f*-block ions in the sixth period, since this covalent overlap is virtually nonexistent: these hydrated ions (Table 8.1) have molar absorptivities well below 1 and have very pale, "pastel" colors.

The two *d*-block hydrated metal ions that also have pale colors, Mn^{2+} and Fe^{3+}, have the high-spin d^5 electron configuration, with each orbital having one electron aligned in the same direction. An electronic *d–d* transition is impossible unless the t_{2g} electron being excited simultaneously flips its spin so that it can pair with the electron already in the e_g orbital; this event is extremely improbable. The absorption spectrum of $[Mn(H_2O)_6]^{2+}$ ion is not shown in Figure 8.5 because on this scale it would appear to be a flat line; ε for this ion is at best 0.05.

When a change in spin state occurs, the color of the complex is also likely to undergo a significant change; if this occurs merely as a result of a change in temperature, the phenomenon is known as **thermochromism**. For example, the iron(III) monothiocarbamate complexes mentioned in Section 8.3 change in color from red to orange as the temperature is changed. High-spin iron(III) is a d^5 ion with no allowed *d–d* electronic transition,[11] while low-spin iron(III) has a ${t_{2g}}^5$ configuration before and a ${t_{2g}}^4{e_g}^1$ configuration after the absorption of light; with the many possible arrangements of these five electrons among the different *d* orbitals, the low-spin iron(III) complex has more than one *d–d* absorption band and thus has a different color than the high–spin iron(III) complex.

The change in spin state can also result from a change of ligands, which may then also produce a dramatic change in color of the complex, even with a constant metal ion. For example, high-spin (d^6) Co^{3+} in $[CoF_6]^{3-}$ is blue; low-spin Co^{3+} is characteristically yellow to orange in complex ions such as $[Co(NH_3)_6]^{3+}$.

Example 8.2
Give a qualitative explanation of the dramatic difference in color of high- and low-spin Co^{3+} complexes.

SOLUTION:

Part of the explanation lies simply in the change of ligands; each different ligand causes a different crystal field splitting, as will be amplified in Section 8.5. However, a discontinuous change from a series of blue to a series of yellow to orange complexes must involve more than this.

The spectrum of high-spin d^6 Co^{3+} would be expected to be similar to that of the high-spin d^6 hydrated Fe^{2+} (Fig. 8.5): No electronic transition is possible in the spherical $t_{2g}{}^3e_g{}^2$ set of five electrons with the same spin, unless (very improbably) the electron being promoted simultaneously flips its spin. Hence, the spectrum is mainly due to the one electron of opposite spin being promoted from the t_{2g} to the e_g orbital. Since the other five electrons in this ion are spherical, there is no question of different electron–electron repulsion energies depending on which e_g orbital is used after excitation. Hence, only one absorption band is found; in slight contrast to the Fe^{2+} case, this is shifted into the visible region at 13,000 cm^{-1} by the +3 charge of the cobalt ion, and the change of ligand.

The low-spin Co^{3+} ion has the $t_{2g}{}^6$ electronic configuration before light is absorbed, and has the $t_{2g}{}^5e_g{}^1$ configuration after absorption of the photon. In the $t_{2g}{}^5e_g{}^1$ configuration, there is more than one possible energy of *d*-electron–*d*-electron repulsion, and consequently there is more than one absorption band involving the visible region (at 21,200 and 29,550 cm^{-1} in $[Co(NH_3)_6]^{3+}$). Hence, both the number and the position of the absorption bands changes from high to low spin, which greatly alters the colors of light that pass through the solutions.

8.5 The Spectrochemical Series of Ligands and the Effects of Covalency

At this point, your instructor may ask you to perform Experiment A.7.

As if the colors of the $3d^n$ hydrated ions were not attractive enough, it turns out that addition of different ligands to the hydrated ions produces yet other colors, with each ligand seeming to produce a subtly or dramatically different color. (The beauty of some of the complexes, especially in the form of well-formed crystals, is what seduced me into inorganic chemistry in the first place!) With Ni^{2+} in particular, practically every color of the rainbow can be produced, although some colors are only produced in complexes that are not octahedral. As you many have discovered if you performed Experiment A.7, it is possible to arrange the complexes of Ni^{2+} in a "rainbow" order of colors depending on the ligand added. The same can be done with the complexes of Cu^{2+} and of Co^{2+}; when you examine your "rainbow" orders of ligands you find that they are approximately the same in each case. (If you actually measure the lowest energy absorption of each complex, you can improve the matching of the three lists of ligands, since the spectrometer is more sensitive to small changes in color than is the eye.) This ordering of ligands is known as the **spectrochemical series of ligands**; it

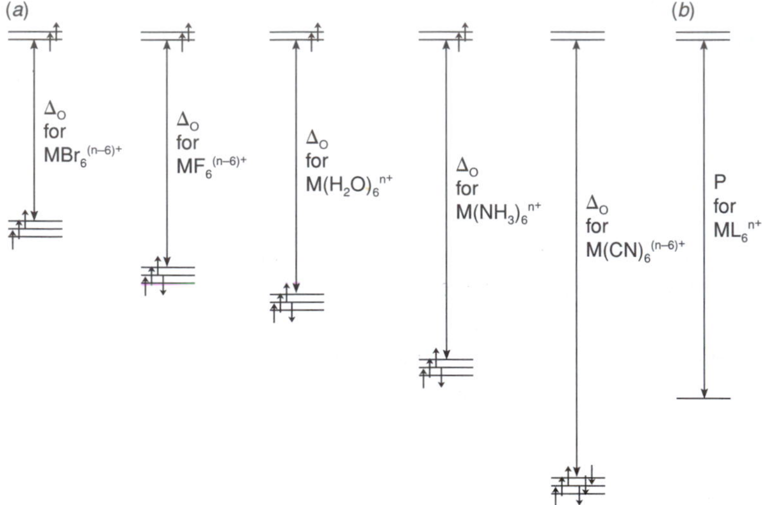

Figure 8.6
Some consequences of changing the ligand L in an octahedral complex ML_6 as L moves further along the spectrochemical series (to increasing *f* factors). (*a*) The arrows represent the relative magnitudes of the crystal field splitting Δ_0 in the complexes of Br^- ($f = 0.72$), F^- ($f = 0.90$), H_2O ($f = 1.00$), NH_3 ($f = 1.25$), and CN^- ($f = 1.70$). (*b*) The pairing energy *P* of the hypothetical d^6 metal ion M^{n+} in the complex; when Δ_0 finally exceeds this, the complex changes from high spin (weak field) to low spin (strong field).

should also be an ordering of ligands that produce complexes, for a given metal ion, arranged from the smallest value of Δ_o to the largest value, as suggested in Figure 8.6. Table 8.5 gives a more extensive spectrochemical series of ligands, together with the relative value of Δ_o for each ligand (relative to $H_2O = 1$).

Since not only the position but also (more or less) the relative value of Δ_o for each ligand commonly holds from one metal ion to another, it is often possible to estimate the absolute value of Δ_o by multiplying Δ_o for the hydrated form of a given metal ion (the ion *g* factor, Table 8.4) by the relative value for the ligand (the ligand *f* factor, Table 8.5):

$$\Delta_o = f_{\text{ligand}} \times g_{\text{ion}} \tag{8.3}$$

As suggested by Figure 8.6, somewhere along the spectrochemical series for most metal ions, Δ_o will finally become large enough to exceed the pairing energy *P* for the metal ion in its complexes, and the complexes from there on will no longer be weak field and high spin, but will be strong field and low spin. In principle, the value of Δ_o predicted using Eq. (8.3) can be used, for example, in predicting whether a complex will be strong or weak field. Thus, if $f_{\text{ligand}} \times g_{ion} > P$, we predict a strong-field complex, although the accuracy of the prediction is limited by the fact that *P* values are measured in ligand-free gaseous metal ions, not in actual complex ions where pairing energies are often rather different (see footnote *a* in Table 8.2).

Next, we look for periodic trends in the spectrochemical series of ligands. Given the crystal field theory's model of a ligand as a point source of charge, we might expect that the more negatively charged (or the more basic) a ligand is, the stronger the crystal field splitting it should produce (the higher it should fall in the spectrochemical series).

Table 8.5
Spectrochemical Series of Ligands and Their f Factors[a]

Ligands Below H_2O	f Factors	Ligands Above H_2O	f Factors
$\mathbf{I}^-$		$\mathbf{N}CS^-$	1.02
$\mathbf{Br}^-$	0.72	p-$CH_3C_6H_4\mathbf{N}H_2$	1.15
$\mathbf{S}^{2-}$		$\mathbf{N}C^-$	1.15
$\mathbf{S}CN^-$	0.73	$CH_3\mathbf{N}H_2$	1.17
$\mathbf{Cl}^-$	0.78	$\mathbf{N}H_2CH_2C\mathbf{O}_2{}^-$	1.18
$(C_2H_5O)_2P\mathbf{Se}_2{}^-$	0.8	$CH_3C\mathbf{N}$	1.22
$\mathbf{N}_3{}^-$	0.83	$C_5H_5\mathbf{N}$	1.23
$(C_2H_5O)_2P\mathbf{S}_2{}^-$	0.83	$\mathbf{N}H_3$	1.25
$N\mathbf{O}_3{}^-$	[b]	$\mathbf{N}H_2CH_2CH_2\mathbf{N}H_2$	1.28
$\mathbf{F}^-$	0.9	$\mathbf{N}H(CH_2CH_2\mathbf{N}H_2)_2$	1.30
$(C_2H_5)_2NC\mathbf{S}_2{}^-$	0.90	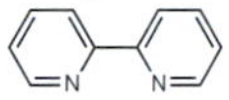 (bpy)[d]	1.33
$(CH_3)_2S\mathbf{O}$	0.91		
$(NH_2)_2C\mathbf{O}$	0.92	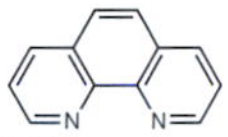 (phen)[e]	1.34
$CH_3C\mathbf{O}OH$	0.94		
$C_2H_5\mathbf{O}H$	0.97	$\mathbf{N}O_2{}^-$	
$(CH_3)_2NCH\mathbf{O}$	0.98	$CH_3C(CH_2CH_2O)_3\mathbf{P}$	
$\mathbf{O}H^-$	[c]	$\mathbf{C}N^-$	~1.7
$C_2\mathbf{O}_4{}^{2-}$	0.99	$\mathbf{C}O$	
$H_2\mathbf{O}$	1.00		

[a] The donor atoms of the ligands are shown in boldface. Data is from J. E. Huheey, E. A. Keiter, and R. L. Keiter, *Inorganic Chemistry: Principles of Structure and Reactivity*, 4th ed., Harper-Collins, New York; 1993, p. 408.
[b] Falls between Cl^- and F^-.
[c] Falls between F^- and $C_2O_4{}^{2-}$.
[d] Ligand is 2,2′-bipyridyl (abbreviated bpy).
[e] Ligand is 1,10-phenanthroline (abbreviated phen).

Clearly, this is not the case: The two −2 charged ligands, sulfide and oxalate ($C_2O_4{}^{2-}$), fall in the lower half of the series; charged OH^- falls below uncharged H_2O.

Thinking back to Chapter 5, we might look to see if the softness of the donor atom (in **boldface** in Table 8.5) is the determining factor. Indeed, at the bottom of the spectrochemical series there are many soft donor atoms: I, Br, (Cl), S, and Se. Next, we find ligands with *hard* donor atoms, F and O, followed with those with a *borderline* donor atom, N, followed by ligands with more *soft* donor atoms, P and C!

Instead, the spectrochemical series arranges ligands approximately by the group of the periodic table in which the donor atom is found

$$\text{Group 17(VII) donors} \leq \text{Group 16(VI) donors} < \text{Group 15(V) donors}$$
$$< \text{Group 14(IV) donors.}$$

This is an order of electronic structure of the donor atom, since Group 17(VII) and 16(VI) donor atoms generally have two or more unshared electron pairs, while Groups 15(V) and 14(IV) donor atoms generally have one unshared electron pair.

The importance of having two or more unshared electron pairs on the donor atom is that it enables the ligand to act as a *π-donor ligand* and to attempt to form a π bond by donating the second unshared electron pair to the t_{2g} set of metal d orbitals [Fig.

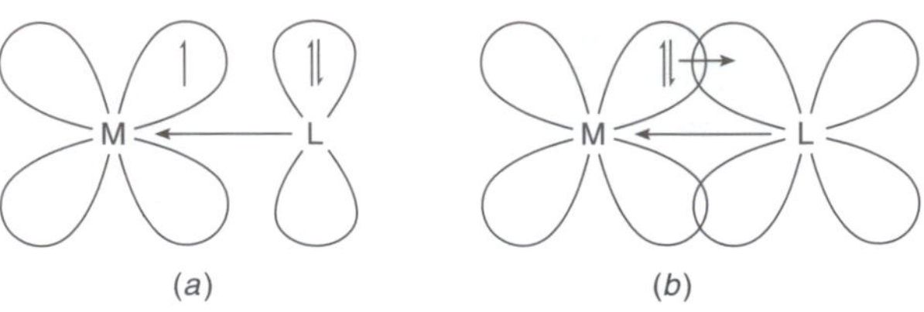

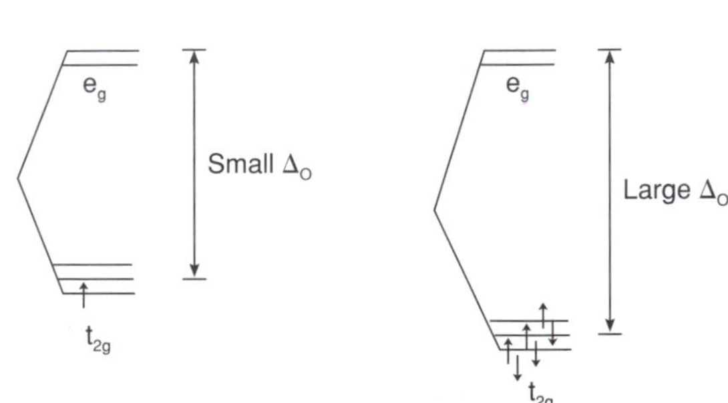

Figure 8.7
The π-type overlap of ligands with metal d orbitals in octahedral complexes. (*a*) The π-donor ligand has two electrons in a π-type orbital, but the metal has (at least) one electron in a t_{2g} orbital, resulting in repulsion that raises the energy of the t_{2g} orbitals relative to the e_g orbitals; (*b*) the π-acceptor ligand has an empty π-type orbital, and the metal has (at least one, normally two) electrons in a t_{2g} orbital, resulting in π bonding that lowers the energy of the t_{2g} orbitals relative to the e_g orbitals.

8.7(*a*)]. However, if the metal ion already has an electron in that d orbital, the interaction must be one of repulsion or *antibonding*, since a bond can involve only two electrons. In this case, the repulsion raises the energies of the electrons in the t_{2g} orbitals, resulting in smaller crystal field splittings than are found in similar complexes in which either the metal has no t_{2g} electrons or the ligand has no π-donor electrons. Consequently, π-donor ligands usually fall at the low end of the spectrochemical series.

Nitrogen-donor ligands with only one unshared electron pair are normally just *σ-donor ligands*, as is :CH_3^-, which has spectrochemical properties very similar to those of :NH_3.[12] Phosphorus- and carbon-donor ligands often have empty orbitals of π-type symmetry such as the beyond-valence 3d orbitals on phosphorus. Such ligands act as *π-acceptor* ligands and form π bonds by *accepting* electrons from the π-symmetry t_{2g} set of metal d orbitals [Fig. 8.7(*b*)]. The formation of this additional π bond stabilizes the complex further and results in complexes of such ligands falling at the *top* of the spectrochemical series.[13]

The π-acceptor ligands can be difficult to recognize unless one does quantitative molecular orbital calculations (these will be considered qualitatively in Chapter 10). Often, however, they can be recognized[14] by a characteristic property: They often can form an *oxide* by the addition of an oxygen atom, as in the following examples:

$$:C{\equiv}N:^- + :\underset{..}{O}: \rightarrow :\overset{..}{\underset{..}{O}}{-}C{\equiv}N:^- \leftrightarrow :\underset{..}{O}{=}C{=}\underset{..}{N}:^- \quad (8.4)$$

Cyanate ion

$$:C{\equiv}O: + :\underset{..}{O}: \rightarrow :\underset{..}{O}{=}C{=}\underset{..}{O}: \quad (8.5)$$

$$:PR_3 + :\underset{..}{O}: \rightarrow :\overset{..}{\underset{..}{O}}{-}PR_3 \leftrightarrow :\underset{..}{O}{=}PR_3 \quad (8.6)$$

$$:\dot{N}{=}\underset{..}{O}: + :\underset{..}{O}: \rightarrow :\overset{..}{\underset{..}{O}}{-}\dot{N}{=}\underset{..}{O}: \leftrightarrow :\underset{..}{O}{=}\dot{N}{-}\overset{..}{\underset{..}{O}}: \quad (8.7)$$

Nitrogen dioxide

These chemical tests will be easier to apply as you study the chemistry of these classes of compounds in Part II. This test, although not infallible, makes some sense in that the ligand must (1) donate a pair of electrons to the oxygen atom to form a σ bond and (2) at least in some resonance structures, the oxygen atom, a π donor, donates one of its unshared electron pairs back to the π-acceptor atom of the ligand to form a double bond to oxygen.

Hence, we are seeing a major conceptual defect of the crystal field theory: The important spectroscopic series of ligands very clearly depends on the covalent-bonding abilities of the ligands, not their point charges! It really is quite amazing that the simple crystal field model can predict as well as it does; the improvement in it, the ligand field theory, takes the σ- and π-covalent bonding into account. In doing so, however, some of the clarity and simplicity of the crystal field theory are lost.

Example 8.3

Using the Cr^{3+} ion, which has a g factor of 17,400 cm^{-1}, and the chelating glycinate ion $NH_2CH_2CO_2^-$, which has an f factor of 1.18, you prepare the octahedral complex $[Cr(NH_2CH_2CO_2)_3]$. (a) Compute the crystal field splitting Δ_o you expect to find in this complex. (b) Compute the crystal field stabilization energy you expect to find in this complex. (c) If Cr^{3+} were replaced by Cr^{2+} in the above complex, is it likely that Δ_o would be larger or smaller than in the above complex? (d) If Cr^{3+} were replaced by Mo^{3+} in the above complex, is it likely that Δ_o would be larger or smaller than in the above complex? (e) If, instead of using glycinate ion as a ligand, you used the dipeptide ion glycylglycinate, $[NH_2CH_2C(=O)NHCH_2CO_2]^-$, would the ligand field splitting likely increase or decrease. (Analyze the spectrochemical series for an additional trend to predict this.) (f) If, instead of using glycinate ion, you used the hypothetical phosphinoglycinate ion, $PH_2CH_2CO_2^-$, would the ligand field splitting increase or decrease? (g) Which one of the five spectra shown below (in Fig. 8.8) would you expect to observe for the original octahedral Cr^{3+} complex $[Cr(NH_2CH_2CO_2)_3]$?

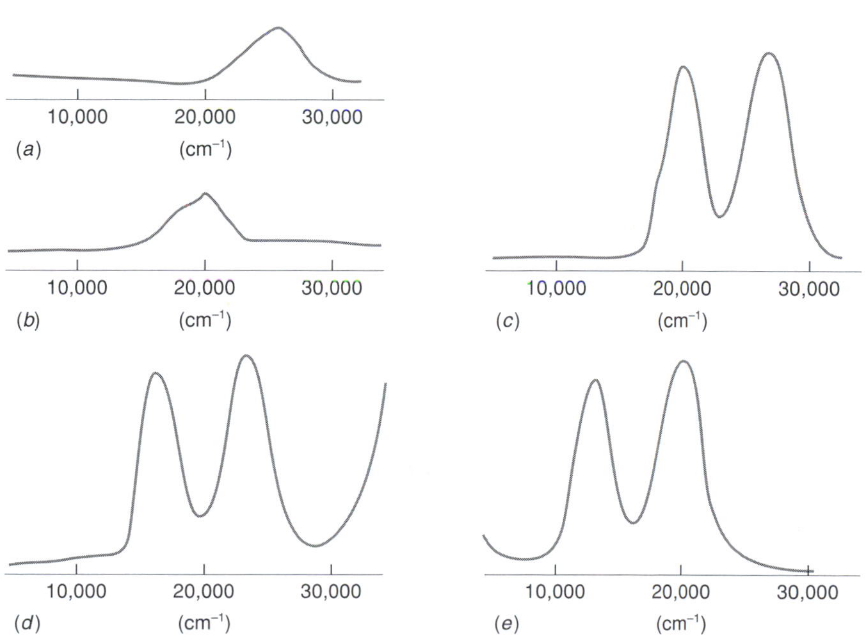

Figure 8.8
Spectra that may apply to $[Cr(NH_2CH_2CO_2)_3]$ in Example 8.3.

SOLUTION:

(a) $\Delta_o = f \times g = 17{,}400 \times 1.18 = 20{,}500\ \text{cm}^{-1}$. (b) For a d^3 ion, the CFSE $= -1.2\,\Delta_o = -24{,}600\ \text{cm}^{-1}$. (c) Smaller: Lower charged ions have smaller crystal field splittings and smaller g factors. (d) Larger: The 4d metal ions have larger crystal field splittings and larger g factors. (e) Glycylglycinate is a tridentate chelating ligand, whereas glycinate is only a bidentate chelating ligand. Increasing the number of chelate rings formable by a ligand increases its f factor slightly, hence raising its position in the spectrochemical series [cf. $NH(CH_2CH_2NH_2)_2$ with $NH_2CH_2CH_2NH_2$, or the latter with CH_3NH_2]. Hence, the crystal field splitting with glycylglycine likely would increase slightly. (f) Substituting a P donor atom, a π acceptor, for a N donor atom, a σ donor, is likely to promote the ligand substantially in the spectrochemical series, and give it a substantially larger Δ_o. (g) A Cr^{3+} ion, after electronic excitation, will have the $t_{2g}{}^2e_g{}^1$ electronic configuration, in which different electron–electron repulsions are possible, since both the $t_{2g}{}^2$ and the $e_g{}^1$ parts of the configuration can be occupied as different combinations of specific d orbitals. Hence, one of the spectra with more than one band will be chosen. In (a), we calculated that Δ_o should be 20,500 cm^{-1}; the spectrum in which the lowest energy absorption band is nearest to this value is the one drawn in Figure 8.8(c).

8.6 Thermodynamic and Structural Consequences of Crystal Field Effects

The CFSE computed in Section 8.2 adds to the stability of complex ions in a number of ways, and in characteristic periodic patterns. Let us begin by examining the hydration energies of the +2 ions of the 3d set of metals, beginning with the $3d^0$ ion Ca^{2+} (hydration energy $= -1592\ \text{kJ mol}^{-1}$, Table 2.1), and ending with the $3d^{10}$ ion Zn^{2+} (hydration energy $= -2044\ \text{kJ mol}^{-1}$). If 3$d$ metal ions were our hypothetical spherical ions, we would expect each successive ion between Ca^{2+} and Zn^{2+} to become smaller and therefore to have a greater (more negative) hydration energy. In fact, however, a plot of hydration energies across the $3d^n$ metal ions of +2 charge results in a characteristic *double-humped curve* [solid line in Fig. 8.9(a)].

The two humps in such a curve are due to addition of the contribution of the weak-field, high-spin crystal (ligand) field stabilization energies (CFSE or LFSE, Section 8.2) to the hydration energies expected for hypothetical spherical metal ions of steadily decreasing radii. Given also the energies of the crystal field splittings in the hydrated ions (Table 8.4), it is possible to compute the CFSE in kilojoules per mole [Fig. 8.9(b)]. When these are subtracted from the measured lattice energies a reasonably straight line remains [dashed line in Fig. 8.9(a)], which represents the hydration energies of the hypothetical spherical hydrated ions. The CFSEs are of the order of 5–10% of the hydration energies; this again serves to remind us that the main source of hydration energies (and of the other energies we will discuss in this section) are those discussed in Chapters 2 and 3. To some extent, the CFSEs amount to "fine tuning" the calculations, but this tuning has the dramatic consequences in magnetism, color, and the other ways described in this chapter.

Similarly, CFSEs contribute to the stability of solid compounds of the d-block elements, as measured by the lattice energies of these compounds. As seen in Figure

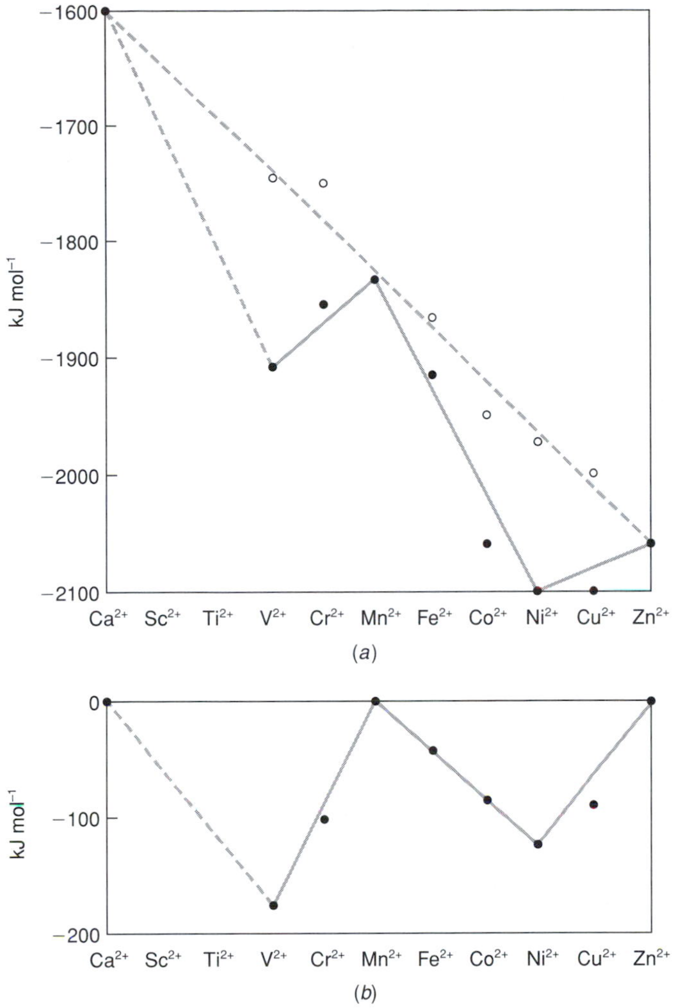

Figure 8.9
(*a*) Hydration energies of the +2-charged $3d^n$ ions (solid circles); (*b*) Crystal field stabilization energies of these ions; (*a*, open circles) hydration energies with the CFSEs subtracted.

8.10, plotting lattice energies of the $3d^n$-metal dihalides also produces double-humped curves; again CFSEs are responsible for about 5–10% of the lattice energies of these compounds.

The CFSEs resulting from the nonspherical nature of the *d*-block metal ions also affect their relative tendencies to form complexes with a given ligand, for example, NH_3 or $NH_2CH_2CH_2NH_2$:

$$[M(H_2O)_6]^{2+} + NH_3 \rightleftharpoons [M(H_2O)_5(NH_3)]^{2+} + H_2O \tag{8.8}$$

$$[M(H_2O)_6]^{2+} + NH_2CH_2CH_2NH_2 \rightleftharpoons [M(H_2O)_4(NH_2CH_2CH_2NH_2)]^{2+} + 2\,H_2O \tag{8.9}$$

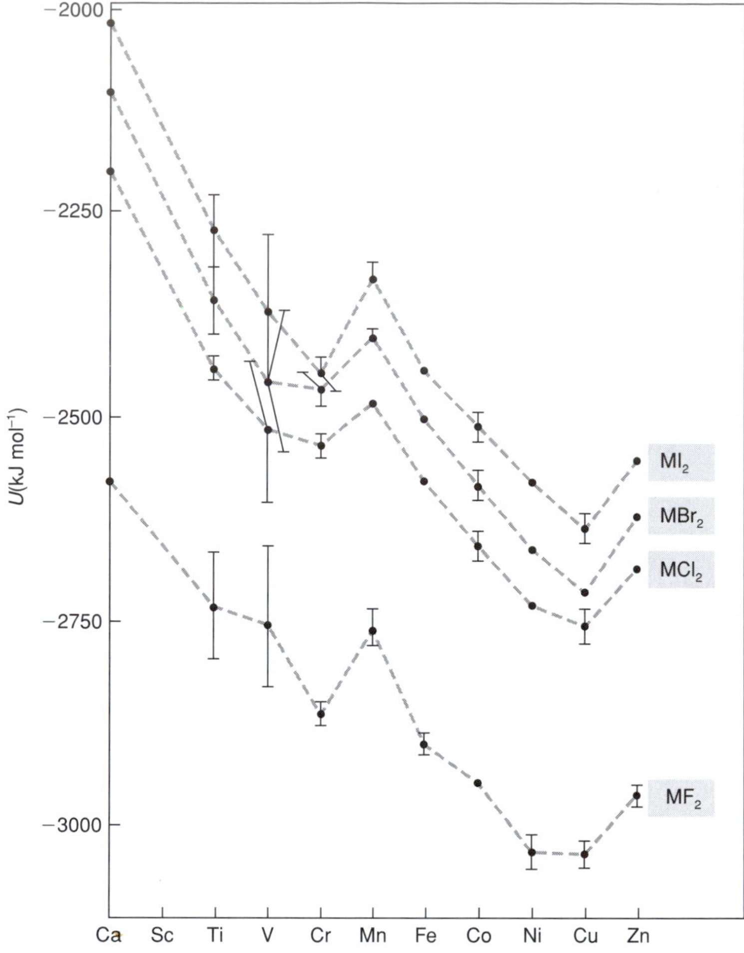

Figure 8.10
Lattice energies of the $3d^n$ halides MX_2. Vertical lines indicate uncertainties in experimental values. [Adapted from J. E. Huheey, E. A. Keiter, and R. L. Keiter, *Inorganic Chemistry: Principles of Structure and Reactivity*, 4th ed., Harper-Collins, New York, 1993; p. 409.]

These reactions represent the first steps in forming the fully substituted nitrogen–ligand complexes; their first *stepwise formation constants* K_1 should also follow a double-humped curve. Figure 8.11 shows data only for the latter half of the d block, however: Since these nitrogen-donor ligands are borderline bases, the formation of the complexes is very unfavorable for the hard acid +2 charged $3d^n$ metal ions early in the d block, and the stepwise formation constants are unmeasurable. For any weak-field ligand, the formation constants follow the order $Mn^{2+} < Fe^{2+} < Co^{2+} < Ni^{2+} < Cu^{2+} > Zn^{2+}$. This order is sometimes called the *Irving–Williams series*, and is often used in discussing metalloenzyme stabilities (e.g., in bioinorganic chemistry). It may, however, be modified if a particular ligand has strong size selectivity (e.g., macrocyclic ligands such as the crown ethers in Section 5.12).

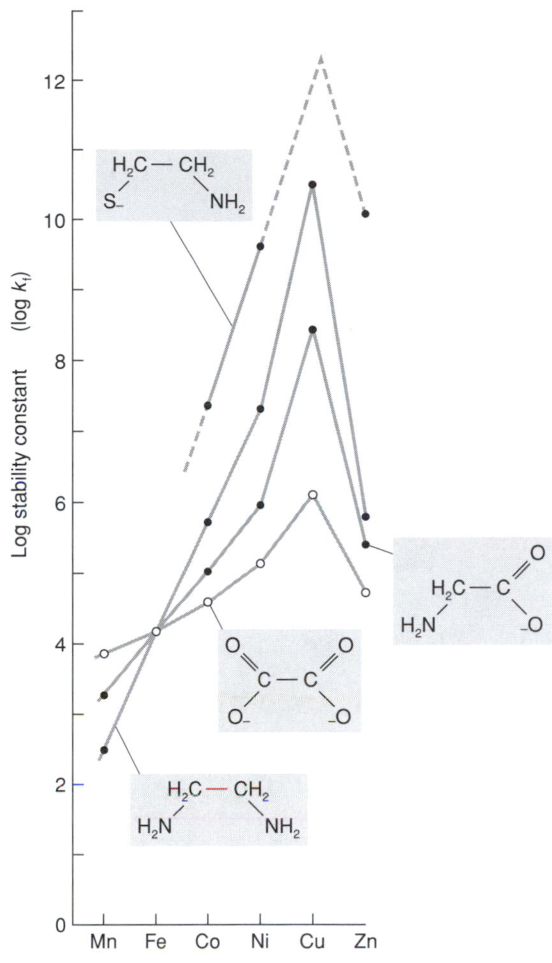

Figure 8.11
Logarithms of stability constants for a given ligand (shown next to each curve) versus the +2-charged metal ion forming the complex. [Adapted from H. Sigen and D. B. McCormick, *Acc. Chem. Res.*, **3**, 201 (1970).]

Effects on Fundamental Properties of Ions. The properties of *d*-block complex ions are not as different from those found in the rest of the periodic table as we often suppose[15]; these properties are still mainly determined by the same fundamental atomic properties such as size, electronegativity, and charge or oxidation state. In Chapter 1, we noted the overall periodic trends and certain anomalies in the horizontal periodic trends in these atomic and ionic properties. Now, we will see that many of the anomalies across the *d* block are related to the presence of crystal field stabilization of certain d^n electron configurations.

Radii. In Figure 8.12, we show the horizontal periodic variation in the radii of octahedrally coordinated $3d^n$ ions of +2 charge (*a*) and of +3 charge (*b*). At first, the ions decrease in size to the right in the period, which is the normal trend. The $3d^4$ through $3d^7$ ions can take either high- or low-spin electron configurations; the radii observed for the ions depend on the choice made. (In cases of spin crossover, a given metal ion can change its radius by up to ~15 pm; a vital biochemical consequence of such a change will be discussed in Section 8.10.) Electrons in e_g orbitals repel the ligands, resulting in *an increase in radius each time an electron is added to an e_g orbital.*

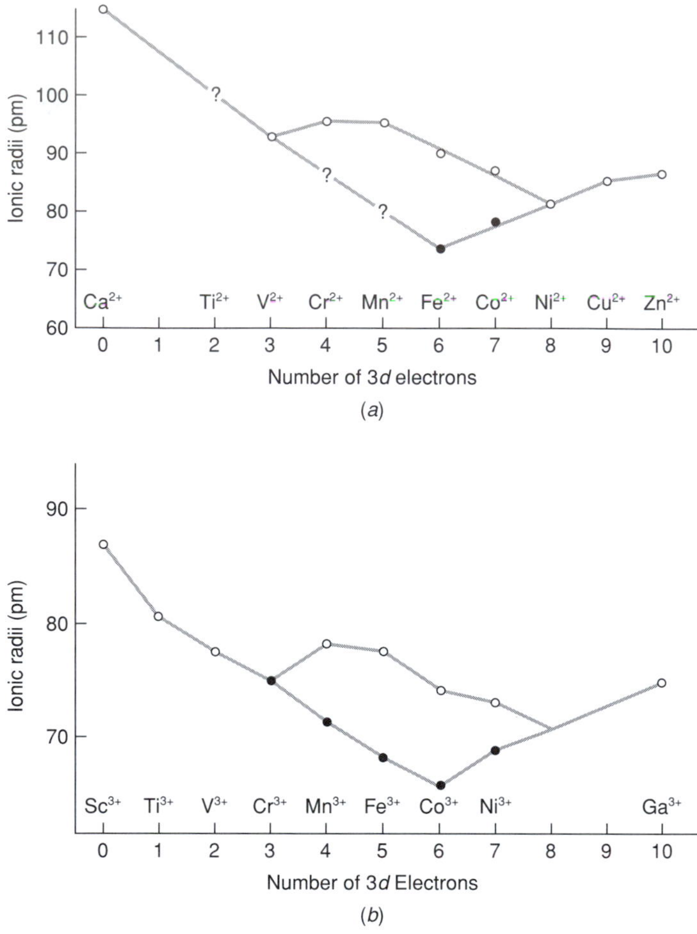

Figure 8.12
Radii of the +2 ions (*a*) and the +3 ions (*b*) as a function of the $3d^n$ electron configuration. Open circles represent high-spin (weak-field) ions; closed circles represent low-spin (strong-field) ions. [Adapted from R. D. Shannon and C. T. Prewitt, *Acta Crystallogr. Sect. B*, **26**, 1076 (1970).]

Since the high-spin ions occupy the e_g set of orbitals first at d^4, this is the point at which radii first begin to increase among the high-spin ions (open circles in Fig. 8.12); the familiar double-humped curve results. The radii of low-spin ions (closed circles) decrease across the period until the e_g orbitals are first occupied at the d^7 electron configuration; the "anomalous" increases in radius continue until the e_g orbitals are filled at Zn^{2+}. Consequently, plots of periodic properties of low-spin complex ions should show a single-humped curve, with the extreme anomaly at d^6.

When we take into account these true radii of the 3*d* metal ions, the value of the fundamental periodic trends we have studied earlier is increased. The double hump in the hydration-energy trend (Fig. 8.9) are accounted for by the double hump in the radius trend (Fig. 8.12), so that when hydration energies are plotted as a function of the radius of the ion, no humps remain, and the 3*d* ions fit the trend represented by the Latimer equation (2.1).

Pauling electronegativities are also affected by crystal field splittings; a plot of the

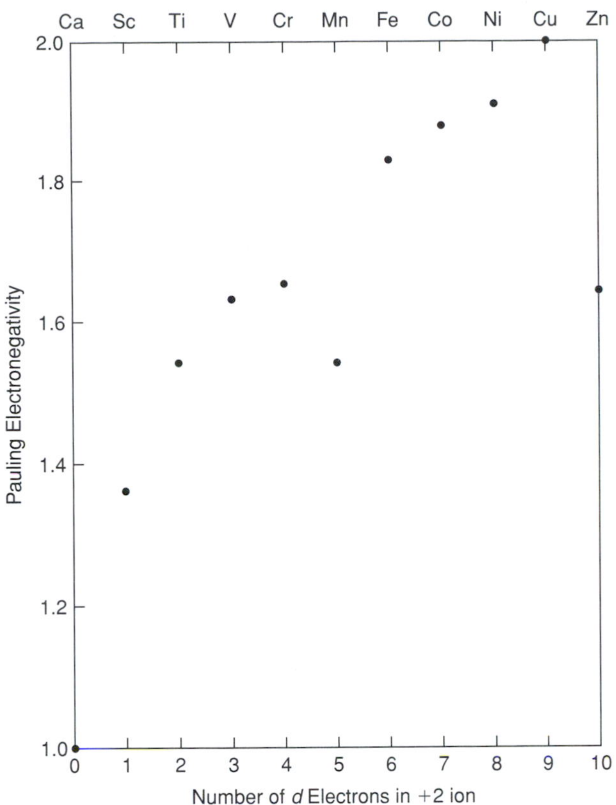

Figure 8.13
Pauling electronegativities of the $3d$ metals as a function of the $3d^n$ electron configurations. The electronegativities are computed from the enthalpies of formation of binary compounds of the +2 charged ions, except for Sc ($3d^1$), which is for the +3 charged ion.

Pauling electronegativities of the $3d$ metals is also a double-humped plot (Fig. 8.13). In Chapter 1, we noted the anomalous decreases in electronegativity at Mn and Zn; these elements have zero CFSE in their weak-field +2 charged ions. In Chapter 7, we discussed how Pauling electronegativities are generally obtained from bond energies of gaseous molecules; in the d-block another approach is necessary, and the data is obtained from the enthalpies of formation of salts of the metal (especially oxides and halides) in the +2 oxidation states. The enthalpies of formation also include contributions from the CFSE; these halides and oxides have lattices in which the metal ion has octahedral coordination, surrounded by weak-field halide and oxide ligands.

It is likely that a different series of Pauling electronegativities would be more appropriate among strong-field octahedral complexes, however. Since the CFSE is at a maximum at the d^6 electron configuration, the "effective" Pauling electronegativity should presumably peak at Fe^{2+} among +2 charged ions and Co^{3+} among +3 charged ions. Although Co^{3+} would normally be ranked as a hard acid, in low-spin complexes its electronegativity might be high enough to soften it; indeed Co^{3+} forms more numerous complexes with nitrogen-donor ligands than with oxygen-donor ligands. In summary, one of the advantages of using appropriate ionic radii and Pauling electronegativities is that they build in some of the features of crystal field effects.

8.7 Crystal Field Splitting in Complexes of Other Geometries

Since octahedral complexes have the maximum number of Lewis base–acid interactions feasible in the fourth period, the octahedral geometry is the most common, but by no means all complexes in this or the later periods are octahedral. First, we will consider two other complex ion geometries that originate in the octahedral geometry itself for particular d^n electron configurations as a result of additional crystal field effects.

Let us begin by considering a d^9 ion in an octahedral complex (e.g., Cu^{2+}). In such an ion, there is an *unequal population of the two* e_g *orbitals* $d_{x^2-y^2}$ and d_{z^2}. The electrons in the $d_{x^2-y^2}$ orbital repel (and are repelled by) the electrons donated by the four ligands located along the x and y axes; the electrons in the d_{z^2} orbital repel the electrons donated by the two ligands located along the z axis. If the two orbitals have equal populations, there are equal repulsions along all axes, but if the two orbitals are unequally populated, intuituively the repulsions cannot be the same along the z as along the x and y axes. It is to be expected, and indeed is found, that the ligands along the most-populated axis will move further away, while the ligands along the less-populated axis will move closer; this motion produces a *distorted octahedral* geometry. Experimentally, we find the lengthening occurs along the d_{z^2} orbital, from which we deduce that the greatest repulsion is along the z axis; hence, the electron configuration of Cu^{2+} is $t_{2g}{}^6(d_{z^2})^2(d_{x^2-y^2})^1$.

In Figure 8.14, we summarize the effects on the geometry and the d-orbital energy levels of this **Jahn–Teller distortion.** In 1937, Jahn and Teller proved a theorem stating that a nonlinear molecule cannot be stable in a degenerate electronic state, but must become distorted in such a way as to break down the degeneracy. Although we will not consider electronic states in detail until Chapter 17, the term "degenerate electronic state" refers to the situation in which the electrons can be arranged in either of two energy-equivalent ways: for the octahedrally coordinated Cu^{2+} ion, the $e_g{}^3$ electrons may be either $(d_{z^2})^2(d_{x^2-y^2})^1$ or $(d_{z^2})^1(d_{x^2-y^2})^2$; the distortion removes the degeneracy by making the former electronic configuration preferable.

The other important situation in which the e_g orbitals are unequally populated, and therefore Jahn–Teller distortion is observed, is in the electronic configuration $e_g{}^1$, as in the weak-field, high-spin case for d^4 ions such as Cr^{2+}. (Although ions with even population of the e_g orbitals but unequal populations of the t_{2g} orbitals also have degenerate electronic states, there is sufficiently less interaction between the t_{2g} electrons and the ligands so that little difference in bond lengths along the three axes are produced. But when a Jahn–Teller distortion is produced as a result of an unequal population of the e_g orbitals, it does split the energies of the t_{2g} set of orbitals as well, as seen in Fig. 8.14.)

The Jahn–Teller distortion has a definite effect on the stability and the spectra of ions having $e_g{}^1$ and $e_g{}^3$ electron configurations.[16] As diagrammed at the bottom of Figure 8.14, one effect of the Jahn–Teller distortion is to increase the energy difference between the t_{2g} electrons and the unoccupied "hole" in the e_g electrons; consequently, higher energy light must be absorbed, and the measured crystal field splitting is increased. Referring to Figure 8.5(*b*), we see that among +2 ions, only in the hydrated Cr^{2+} and Cu^{2+} ions is the lowest energy absorption shifted into the visible region of the spectrum. This shift accounts for the blue color of these ions and the larger crystal field splittings of these ions (Table 8.4). The effect has real thermodynamic consequences, since during the distortion *two* electrons are lowered in energy while only

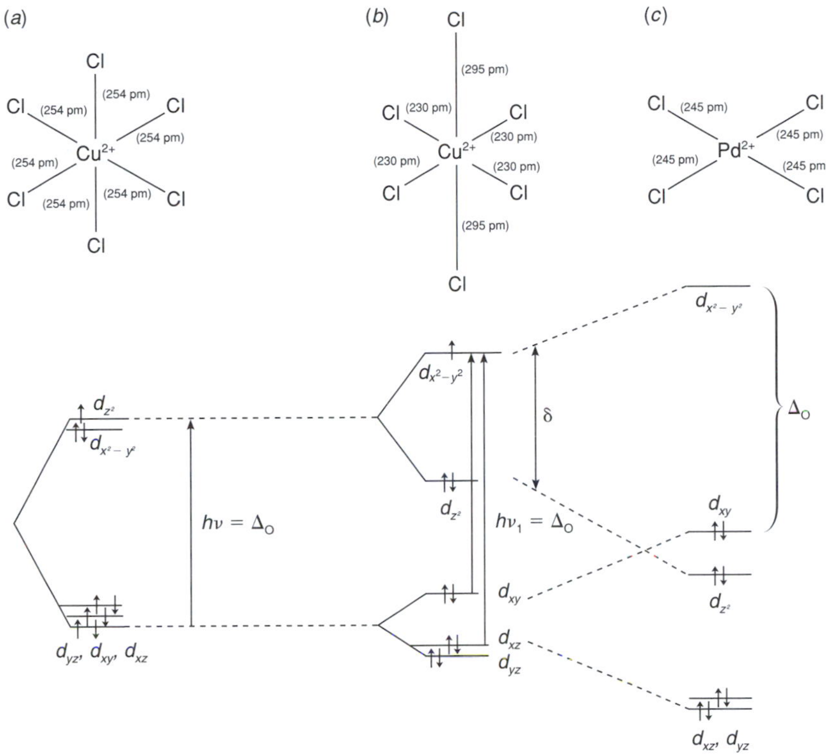

Figure 8.14
Effects of Jahn–Teller distortion of an octahedral complex of a d^9 ion. (*a*) The Cu^{2+} ion octahedrally coordinated by six Cl^- ligands in a hypothetical undistorted complex, showing the normal octahedral energy level diagram and (hypothetical) bond lengths in pm. (*b*) The same ion after the distorted octahedral geometry is adopted. (*c*) In a strong-field d^8 complex, the axial ligands are completely removed, producing a square-planar complex.

one is raised an equal amount in energy. This accounts for the fact that Cu^{2+} outranks Ni^{2+} in the Irving–Williams series and in electronegativity (Fig. 8.13).

Square Planar Geometry. Although it is *not* driven by the Jahn–Teller effect, we can imagine artificially compressing the distorted octahedral complex further along the *x* and *y* axes and allowing it to elongate further along the *z* axis until finally the (axial) ligands dissociate completely along the *z* axis. The resulting geometry is that of a *square planar* complex. This process is especially favorable (can release energy) for a d^8 complex ion [having one fewer d_{z^2} electron than those shown in Fig. 8.14 (*a* and *b*)] provided that the splitting that results [Δ_o in Fig. 8.14(*c*)] exceeds the pairing energy P for the ion, so that the unpaired electron in the $d_{x^2-y^2}$ orbital can drop down and pair with the other electron already in the d_{z^2} orbital. Ligands with strong enough fields to do this to the Ni^{2+} ion are infrequent: CN^- produces diamagnetic, square planar $[Ni(CN)_4]^{2-}$, and the chelate ligand dimethylglyoxime produces the characteristic red test species for Ni^{2+}, [Ni(dimethylglyoximate)$_2$]. As a result of the larger crystal field splitting for fifth- and sixth-period metal ions, the square planar geometry is normal for the d^8 Pd^{2+} and Pt^{2+} ions.

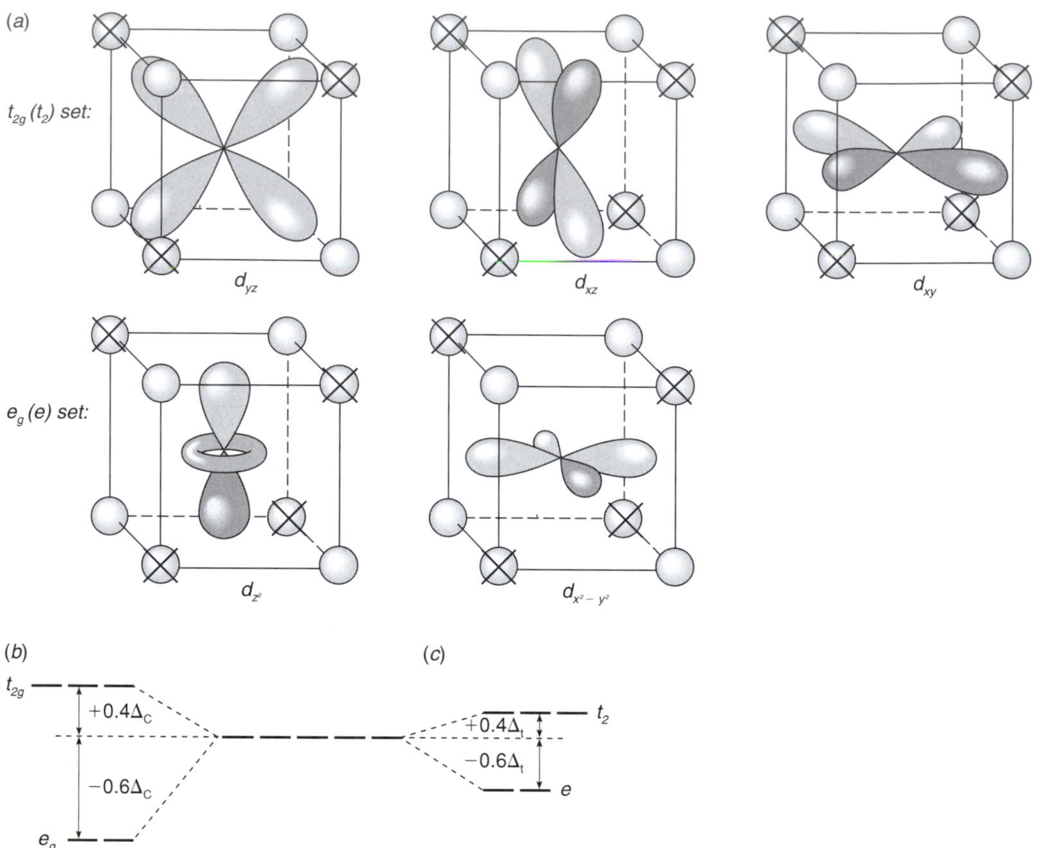

Figure 8.15
(*a*) Interaction of the *d* orbitals with a cubic set of ligands or, when ligands indicated by an X are omitted, a tetrahedral set of ligands. (*b*) Energy-level diagram for the cubic set of ligands; (*c*) Energy-level diagram for the tetrahedral set of ligands. [Adapted from I. S. Butler and J. F. Harrod, *Inorganic Chemistry: Principles and Applications*, 1989. Benjamin/Cummings, Redwood Park, CA; p. 394.]

Cubic and Tetrahedral Complexes. Tetrahedral coordination results when a fourth-period metal ion adopts its penultimate coordination number, 4, as a consequence of the unusual steric bulk of the ligands, or their possession of good π-donor abilities. To see how the energies of the *d* orbitals split in a tetrahedral crystal field, it is useful first to look at a related but far more uncommon geometry, that of an eight-ligand *cubic* coordination geometry (Fig. 8.15).

As suggested by Figure 8.15(*a*), the outstanding characteristic of cubic coordination is that there are *no* strong repulsions between the *d* orbitals and the ligands. The ligands still fall into the same two sets as in octahedral geometry—t_{2g} and e_g—but now it is the t_{2g} set that experiences the greater repulsion. The reason is that the lobes of this set of orbitals fall in the edge centers, closer to the ligands than the lobes of the e_g set, which fall in the face centers. Consequently, the energy level diagram [Fig. 8.15(*b*)] is

Table 8.6
Energy Levels of *d* Orbitals in Complexes of Various Geometries[a]

CN[b]	Geometry	d_{z^2}	$d_{x^2-y^2}$	d_{xy}	d_{xz}	d_{yz}	Largest Splitting
2	Linear[c]	1.028	−0.628	−0.628	0.114	0.114	0.914
3	Trigonal[d]	−0.321	0.546	0.546	−0.386	−0.386	0.867
4	Tetrahedral	−0.267	−0.267	0.178	0.178	0.178	0.445
4	Square planar[d]	−0.428	1.228	0.228	−0.514	−0.514	1.000
5	Trigonal bipyramid[e]	0.707	−0.082	−0.082	−0.272	−0.272	0.789
5	Square pyramid[e]	0.086	0.914	−0.086	−0.457	−0.457	0.828
6	Octahedron	0.600	0.600	−0.400	−0.400	−0.400	1.000
6	Trigonal prism	0.096	−0.584	−0.584	0.536	0.536	0.680
7	Pentagonal bipyramid[e]	0.493	0.282	0.282	−0.528	−0.528	0.810
8	Cube	−0.534	−0.534	0.356	0.356	0.356	0.890
8	Square antiprism	−0.534	−0.089	−0.089	0.356	0.356	0.445
9	Tricapped trigonal prism	−0.225	−0.038	−0.038	0.151	0.151	0.189
12	Icosahedron	0.000	0.000	0.000	0.000	0.000	0.000

[a] Units of Δ_o, the octahedral crystal field splitting, assuming the same overall charge density and distance. [Adapted from J. E. Huheey, E. A. Keiter, and R. L. Keiter, *Inorganic Chemistry: Principles of Structure and Reactivity*, 4th ed., Harper-Collins, New York, 1993, p. 405.]
[b] Coordination number = CN.
[c] Ligands lie along *z* axis.
[d] Ligands lie in *xy* plane.
[e] Pyramid base in *xy* plane.

reversed from that for octahedral complexes; the total crystal field splitting, Δ_c (c for cubic) can be shown to be approximately eight-ninths of that present in an octahedral complex, Δ_o.

The more important case of *tetrahedral* complexation is simply generated by deleting the one-half of the cubic set of ligands crossed over with an *X* in Figure 8.15(*a*). The pattern of interactions [Fig. 8.15(*c*)] remains the same,[17] but the total splitting is now halved; the total splitting Δ_t (t for tetrahedral) is now only four-ninths of Δ_o. Consequently, Δ_t is never greater than the pairing energy, *P*, and all tetrahedral complexes are weak field (high spin).

Other Geometries. Other geometries have occasional importance in *d*-block coordination chemistry; the crystal field splitting patterns for these can also be worked out and are tabulated in Table 8.6 in common units of Δ_o, the crystal field splitting in an octahedral field. (In many of these geometries, if too many of the ligands are different, the spectra become too complex to be fit completely using this table.)

Example 8.4
Calculate the CFSE of the d^1 through d^5 ions in the (a) tetrahedral and (b) square pyramidal geometries. In the tetrahedral geometry, express your answers both in terms of Δ_t and Δ_o; in the square pyramidal geometry, in terms of Δ_o.

SOLUTION:

(a) The energies in terms of Δ_t can be derived from the energy level diagram in Figure 8.15(c); they can then be expressed in terms of Δ_o by multiplying by four-ninths. The electron configurations and energies are as follows:

Free-ion configuration	d^1	d^2	d^3	d^4	d^5
Tetrahedral configuration	e^1	e^2	$e^2t_2{}^1$	$e^2t_2{}^2$	$e^2t_2{}^3$
CFSE (units of Δ_t)	−0.6	−1.2	−0.8	−0.4	0
CFSE (units of Δ_o)	−0.267	−0.533	−0.356	−0.178	0

(b) Table 8.6 gives the energies for the square pyramidal geometry in terms of Δ_o. The electron configurations are built up by filling the lowest orbital first; there is a fairly large energy gap between $d_{x^2-y^2}$ and d_{z^2}, which could perhaps result in spin pairing (a low-spin complex). Since this gap is less than Δ_o, let us assume that pairing does not result, and that we have a high-spin complex after this gap is encountered. The electron configurtions and energies are as follows:

	d^1	d^2	d^3	d^4	d^5
	$d_{xz,yz}{}^1$	$d_{xz,yz}{}^2$	$d_{xz,yz}{}^2d_{xy}{}^1$	$d_{xz,yz}{}^2d_{xy}{}^1d_{z^2}{}^1$	$d_{xz,yz}{}^2d_{xy}{}^1d_{z^2}{}^1d_{x^2-y^2}{}^1$
CFSE	−0.457	−0.914	−1.000	−0.914	0

Radii in Other Geometries. Recall from Section 4.7 that the radius of a cation is not an invariant property, but depends on its coordination number; Table C assumes (when possible) a coordination number of 6 and the most common spin state. As fewer ligands are placed around a cation, their electrons repel each other less, and the measured cationic radius decreases. In Table 8.7, the data for selected d-block metal ions shows that radii for four-coordinate (tetrahedral or square-planar) ions are smaller than for six-coordinate (high- or low-spin octahedral) ions; the data also illustrate the characteristically larger radii of high-spin over low-spin octahedral ions.

Spectra of Complexes in Other Geometries. The electronic absorption spectrum of a d- (or f) ion is very sensitive to and carries important information about the geometry of the coordination about that ion. We may mention three features about the spectra that carry useful information:

1. In a geometry such as tetrahedral, where the total crystal field splitting is unusually small (or in square planar, in which it is unusually large), the frequency of the first absorption band will fall at an unusually low (or an unusually high) energy.

Example 8.5

Estimate the position of the absorption band and the color of the tetrahedral $CuCl_4{}^{2-}$ anion.

SOLUTION:

We expect the crystal field splitting and the energy of the absorption band of the tetrahedral chloro complex of Cu^{2+} to be approximately four-ninths of that of a corresponding octahedral chloro complex. The spectrum of the octahedral chloro complex can be predicted approximately using Eq. (8.3), the f factor for Cl^- from Table 8.5, and the g factor for Cu^{2+} from Table 8.4:

$$\Delta_t = \frac{4}{9} \times f(Cl^-) \times g(Cu^{2+}) = 0.445 \times 0.78 \times 13{,}000 = 4500\,cm^{-1}$$

In fact, this absoption band occurs at 4800 cm^{-1}; by referring to Figure 8.5, we can see that this absorption falls off the left side of the spectral range shown, well into the near-IR region, so that this complex ion would be predicted to be colorless. In fact, $[CuCl_4]^{2-}$ is yellow (and undergoes some Jahn–Teller distortion from pure tetrahedral geometry); the color is due to a charge-transfer absorption.

Table 8.7
Radii of Selected d-Block Metal Ions as a Function of Geometry

		CN = 4	CN = 4	CN = 6 (Octahedral)	
Ion	**Electron Configuration**	**Square Planar**	**Tetrahedral**	**Low Spin**	**High Spin**
		+2 Charged Ions of Period 4			
Cr^{2+}	d^4			87	94
Mn^{2+}	d^5		80	81	97
Fe^{2+}	d^6	78	77	75	92
Co^{2+}	d^7		72	79	89
Ni^{2+}	d^8	63	69	83	
Cu^{2+}	d^9	71	71	87	
Zn^{2+}	d^{10}		74	88	
		+3 Charged Ions of Period 4			
Mn^{3+}	d^4			72	79
Fe^{3+}	d^5		63	69	79
Co^{3+}	d^6			69	75
Ni^{3+}	d^7			70	74
		+4 Charged Ions of Period 4			
Ti^{4+}	d^0		56	75	
Cr^{4+}	d^2		55	69	
Mn^{4+}	d^3		53	67	
Co^{4+}	d^5		54		67
		Selected Ions of Periods 5 and 6			
Pd^{2+}	d^8	78		100	
Pt^{2+}	d^8	74		94	
Ag^{3+}	d^8	81		89	
Au^{3+}	d^8	82		99	
Ag^{2+}	d^9	93		108	
Cd^{2+}	d^{10}		92	109	
Hg^{2+}	d^{10}		110	116	
Ag^+	d^{10}	116	114	129	

SOURCE: Data from R. D. Shannon, *Acta Crystallogr. Sect. A*, **32**, 751 (1976).

2. In many of the geometries tabulated in Table 8.6, the five *d* orbitals are split into *more than two* energy levels. This means that often even in the absence of electron–electron repulsion effects (e.g., a d^1 or d^9 ion) there may be more than one absorption band, since the electron can be excited to or from more than one energy level. When electron–electron repulsion effects are added in, the spectrum can become characteristically even more complex than those shown in Figure 8.5.

3. Although it is beyond the scope of this chapter to explain, the absorption coefficients in many other geometries such as the tetrahedral geometry are often larger than those found for octahedral complexes, so that the colors are more intense in many nonoctahedral complexes. Often a change in color intensity signals a change in coordination geometry. Indicating Drierite™ is mainly anhydrous $CaSO_4$ impregnated with anhydrous $CoCl_2$. Once the Drierite has absorbed its capacity of water from the solvent, the following changes in geometry and color occur about the Co^{2+} ion:

$$\underset{\textbf{Deep blue}}{[CoCl_4]^{2-}\ (\text{tetrahedral})} + 6\,H_2O \rightleftharpoons \underset{\textbf{Pink, not intense color}}{[Co(H_2O)_6]^{2+}\ (\text{octahedral})} + 4\,Cl^- \tag{8.10}$$

We may note that the blue color cannot be due to the lowest energy absorption of $CoCl_4{}^{2-}$, which must also be well into the near-IR spectral region, but must be due to one of the other absorption bands that result from electron–electron repulsion. The intensity of the color results from the nonoctahedral geometry of the chloro complex.

Sometimes a geometric change and consequent dramatic color change can be the result of a temperature change, in which case we have another type of thermochromism.[18] Many thermochromic phase transitions occur in salts of the $[CuCl_4]^{2-}$ ion with alkylammonium cations, $[R_nNH_{4-n}]^+$. At low temperatures, hydrogen bonding of the cations to the chlorines of the anion stabilize the square planar geometry; in this geometry the anion is yellow. Higher temperatures cause motion in the cation that breaks up the hydrogen bonding, allowing the $[CuCl_4]^{2-}$ anion to adopt a (Jahn–Teller distorted) tetrahedral geometry, in which it has a deep green color.

The useful spectroscopic properties of *d*- (and to some extent of *f*-) block elements are often exploited in biochemistry. Many metalloenzymes contain *silent* metal ions such as Zn^{2+} and Ca^{2+} that have no accessible electronic absorption spectra, or any other useful properties such as those being appropriate for nuclear magnetic resonance (NMR) studies. If these metal ions can be replaced by *very* similar ions from the *d* or *f* blocks that are colored or are suitable for NMR or other studies, the metalloenzyme will hopefully retain its structure and perhaps even its activity. Measurement of the spectrum of the reporter metal ion will then tell us about the coordination geometry about the metal ion.

This approach worked very successfully in the study of the zinc-containing enzyme carboxypeptidase A (Section 5.10). When the Zn^{2+} is removed and replaced by Co^{2+} (with identical radius and charge and not too different an electronegativity), the enzyme not only retains its activity, but the activity is *enhanced!* The spectrum of Co^{2+} substituted carboxypeptidase A was recorded[19] (peaks at 18,800 cm^{-1}, $\varepsilon = 210$; 17,500 cm^{-1}, $\varepsilon = 190$; 10,600 cm^{-1}, $\varepsilon = 20$; 6,400 cm^{-1}, $\varepsilon = 17$) and compared with the spectra of model Co^{2+} complexes of octahedral, tetrahedral, and trigonal bipyramidal geometry, shown in Figure 8.16. The spectrum of the metalloenzyme was not like that of the octahedral model compound, but bore resemblance both to that of the tetrahedral and the trigonal bipyramidal complexes. When the crystal structure of the zinc-

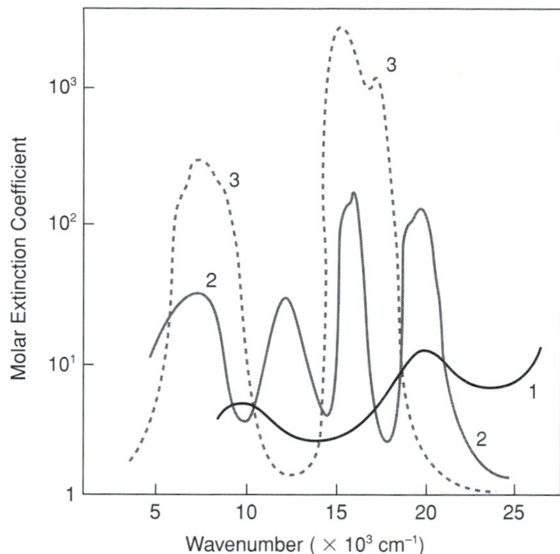

Figure 8.16
Electronic absorption spectra of model complexes of Co^{2+}. 1. (——) Octahedral Co^{2+} in $[Co(NH_2CH_2CH_2NH_2)_3]^{2+}$; 2. (——) trigonal bipyramidal Co^{2+} in [CoBr(hexamethyldiethylenetriamine)]$^+$; 3. (- - - -) Tetrahedral Co^{2+} in $Co(NCS)_4{}^{2-}$. [Adapted from E.-I. Ochiai, *Bioinorganic Chemistry: An Introduction*, Allyn and Bacon, Boston, 1977; p. 374.]

containing enzyme was determined, it was found that the coordination at Zn^{2+} is distorted from tetrahedral: Three bond angles at Zn are roughly near 90° and three are roughly near 120°, so that the structure resembles a trigonal bipyramidal structure with one ligand missing.

An appropriate substitution for silent Ca^{2+} in its enzymes is more difficult to find, since all *d*-block +2 charged ions are too small, and all *f* block +2 charged ions are very sensitive to air. Often it has proved possible to use *f*-block +3 charged ions, which are only slightly more acidic than Ca^{2+} and do have useful colors (Table 8.1). Of course, crystal field splittings in *f*-block ions are very small, but they have proved sufficient for this substitution to be useful (usually other optical properties such as fluorescence are monitored rather than absorption).

8.8 Geometric Preferences: Rates and Mechanisms of Ligand Exchange

The actual geometry chosen by the complex of a given d^n metal ion with a given set of ligands is the result of a balance of several factors. (1) The crystal field stabilization energies for a given d^n ion can vary quite substantially from one geometry to another, as suggested by Example 8.4. (2) More energy is released as a result of ligand–metal ion electron-pair donation when the maximum coordination number is reached. (3) Contrariwise, larger destabilizing ligand–ligand repulsive energies can result from the crowding of bulky ligands that occurs with high coordination numbers. The latter two energies are important, and cannot be calculated using the crystal field theory. But they

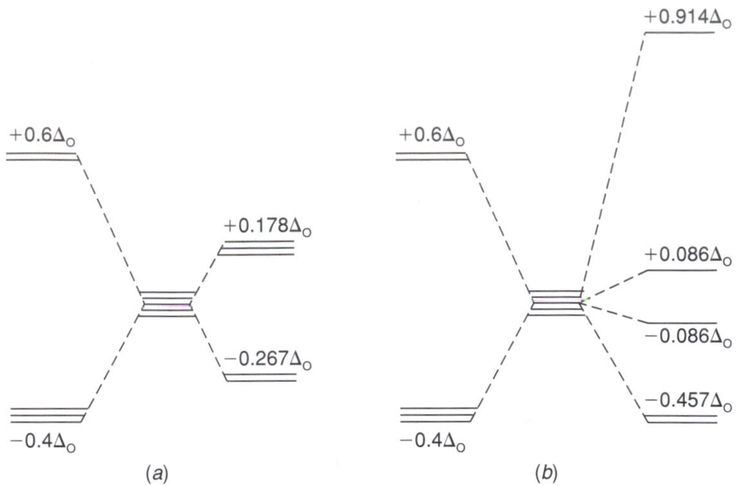

Figure 8.17
Comparative energy level diagrams for different pairs of geometries: (*a*) octahedral ML_6 (on the left) versus tetrahedral ML_4 (on the right) and (*b*) octahedral ML_6 (on the left) versus square pyramidal ML_5 (on the right).

should not vary much across a series of $3d^n$ metal ions, while the CFSEs vary dramatically and predictably. Hence, it is possible, using the crystal field theory, to predict which $3d^n$ electron configurations show the *greatest* preference for a given geometry (such as the usually preferred octahedral geometry) over some other geometry.

The two most common geometries for *d*-block ions are tetrahedral and octahedral, their energy level diagrams are compared in Figure 8.17(*a*). We can predict the relative preferences of different high-spin d^n ions for these two geometries by computing crystal field stabilization energies for each value of *n* in the two geometries, then taking the energy difference CFSE(octahedral) – CFSE(tetrahedral). This difference is known as the octahedral site preference (or stabilization) energy (OSSE) and varies as *n* varies. We computed CFSE(octahedral) in Section 8.3 for these ions and CFSE(tetrahedral) in Example 8.4 for the first five d^n ions; in order to make the subtraction we must use CFSE(tetrahedral) in units of Δ_o, as shown in the figure. Taking these differences we obtain

Free-ion configuration	d^1	d^2	d^3	d^4	d^5
OSSE (units of Δ_o)	−0.133	−0.267	−0.845	−0.422	0

It may be seen that ions with the d^3 electron configuration have the greatest preference for octahedral coordination; an extension of this calculation would also add ions with the d^8 electron configuration. The d^3 ion Cr^{3+} virtually never forms tetrahedral complex ions, and the d^8 ion Ni^{2+} seldom does, even with fairly bulky ligands, while the d^5, d^6, and d^7 ions Mn^{2+}, Fe^{2+}, and Co^{2+} frequently form tetrahedral complexes, especially with bulky ligands or ligands that fall fairly low in the spectrochemical series.

Another useful choice of geometries is between octahedral and square pyramidal geometries [Fig. 8.17(*b*)]. Formation of a square pyramidal five-coordinate complex is presumed to be a necessary step in the *substitution* reactions of metal ions (such as $3d^n$

ions, including hydrated ions) that show their maximum coordination numbers of 6. Such ions would be expected to show mechanisms in which, before a new ligand can come in, one of the existing ligands must first break its coordinate covalent bond to the metal, leaving a square pyramidal intermediate.[20] The simplest possible ligand-exchange reaction is that of solvent water for coordinated water in hydrated ions. We would expect the relative rates of this reaction to be related to at least two factors: (1) the hydration energy of the metal ion, which is a function of its acidity classification; and (2) for *d*-block metal ions, the octahedral versus square pyramidal preference energies.

Figure 8.18 shows some measurements of rate constants for exchange of water molecules by hydrated cations. It may be seen that there is a rough relationship of

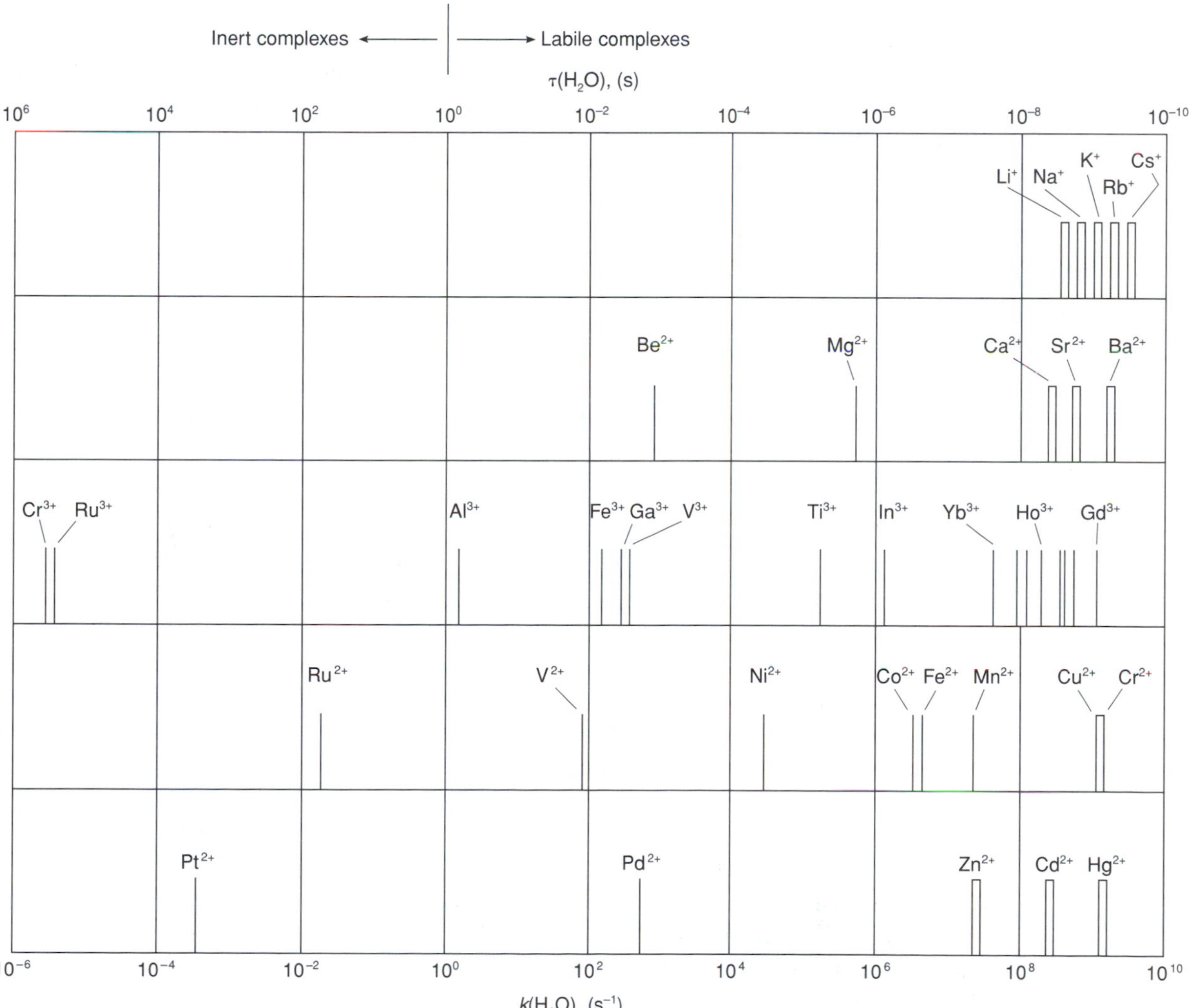

Figure 8.18
Mean lifetimes of a water molecule in the primary coordination sphere, $\tau(H_2O)$, and rate constants for exchange of water, $k(H_2O)$, for solvated cations as measured by NMR (lines) or as derived from complex formation reactions (open bars). [Adapted from Y. Docommun and A. E. Merbach, *Inorganic High Pressure Chemistry*: R. van Eldik, Ed.; Elsevier, Amsterdam, The Netherlands, 1986; p. 20.]

these rates to the acidity classifications of the cations:

1. The *nonacidic* and *feebly acidic cations* exchange solvent and coordinated water molecules with rate constants of 10^8–$10^9\ s^{-1}$; that is, a given water molecule will stay in the hydration sphere of such a cation for a period measured in nanoseconds. This value is close to the limiting factor, the rate of diffusion of molecules in solution at room temperature.

2. Most *weakly acidic* cations have smaller rate constants of 10^4–$10^7\ s^{-1}$; most *moderately acidic* cations have still smaller rate constants of 10^0–$10^3\ s^{-1}$. Water molecules in these ions are much more strongly bound to the cation, and it takes them periods ranging from microseconds to seconds to break away and be replaced by solvent water molecules.

However, some *d*-block weakly and moderately acidic cations deviate widely from the tendencies for non-*d*-block ions of similar charge and size. These deviations can be correlated with octahedral versus square pyramidal preference energies, which we can visualize using Figure 8.17(*b*), and calculate from the CFSEs given in Section 8.3 and Example 8.4:

Free-ion configuration:	d^1	d^2	d^3	d^4	d^5
OSSE (units of Δ_o)	+0.057	+0.114	−0.2	+0.314	0

In this calculation for weak-field (high-spin) ions, the d^3 (and also the d^8) ions stand out for their preference for octahedral coordination, which causes them to resist the dissociation of coordinated water. Their rates of exchange of coordinated and free water are in fact strikingly low (Fig. 8.19); the rates of ligand exchange for the d^3 ions V^{2+} and Cr^{3+} are on the order of 1 million times slower than for analogous +2 and +3 charged ions, respectively. A similar calculation for strong-field ions shows that there is a strong maximum in preference for octahedral coordination at the d^6 electron configuration; since most Co^{3+} complexes and fifth- and sixth-period *d*-block complexes are strong field, they also strongly resist ligand exchange.

Octahedral site preference energies can strongly affect the rates at which *d*-block metal ions react. Most metal complexes, provided that they are not immobilized by excessive chelation or macrocylic-ligand coordination, are *labile*: Their ligands exchange with other ligands or with solvent water in less than a minute. This limits the number of types of complexes (such as isomers, Chapter 9) that can successfully be isolated from solution. The complexes of Cr^{3+}, Co^{3+}, and the d^6 complexes of the fifth and sixth periods last for well over 1 min before reacting with other ligands and the solvent. Hence, they are more easily isolated and termed *inert*. In the extreme cases, these ions can be very slow to react indeed: The mean residence time of a given water molecule in the primary hydration sphere of a $[Cr(H_2O)_6]^{3+}$ ion at 25 °C is over 100 h; for fifth-period d^6 $[Rh(H_2O)_6]^{3+}$ the time is nearly 1.5 years; for sixth-period d^6 $[Ir(H_2O)_6]^{3+}$ it is nearly 300 years[21]! In striking contrast, the computed time for exchange of a water molecule in the *secondary* hydration sphere of a $[Cr(H_2O)_6]^{3+}$ ion is only 1.3×10^{-10} s.[22]

Werner's work in developing the original theory of coordination chemistry (Section 3.2) depended on being able to isolate and count all possible isomers of various octahedral Co^{3+}, Cr^{3+}, and Pt^{4+} complexes; this would have been impossible had

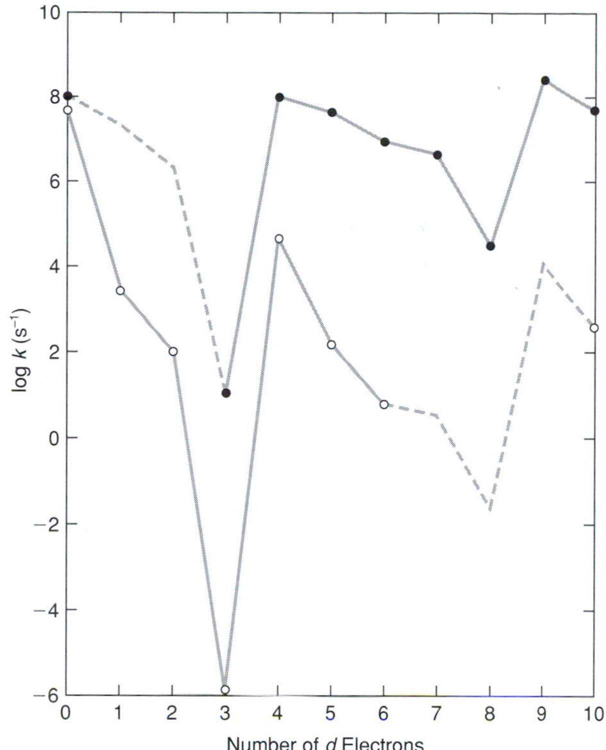

Figure 8.19
Rates of exchange of free and coordinated water for weak-field (high-spin) hydrated $3d^n$ metal ions. Solid circles are for +2-charged ions; open circles are for +3-charged ions. [Data from R. P. Hanzlik, *Inorganic Aspects of Biological and Organic Chemistry*, Academic, New York, 1976; p. 107, and J. Burgess, *Metal Ions in Solution*, Ellis Horwood, Chichester, UK, 1978; pp. 316–31.]

these been labile complexes. The methods of synthesis of complex ions discussed in Section 8.9 also hinge on this; for rapid syntheses it is desirable to avoid inert reactants.

Square planar complexes, which are typically of d^7, d^8, and d^9 metal ions with strong ligand fields, are more likely to exchange ligands by a different mechanism, prior *association* of the incoming ligand to form a square pyramidal transition state, before the outgoing ligand is lost. Given the strong crystal field splitting in the square planar geometry, it is not surprising that square planar d^8 complexes of such ions as Pd^{2+} and Pt^{2+} are also generally inert (resistant to *rapid* substitution reactions). In Chapter 16, we will go into much more detail on the rates of reactions of complex ions.

8.9 *Syntheses of *d*-Block Complexes

The syntheses of *d*-block complexes is an art that is both aesthetically satisfying and of great practical importance. For example, complexes of radioactive technetium are used in medicine as imaging agents, and the complex *cis*-$[PtCl_2(NH_3)_2]$, cisplatin, is the most widely used chemotherapeutic agent. A variety of synthetic methods[23] have been devised that employ the principles we have used; we summarize some of these here.

Regardless of how they are synthesized, the complexes must be separated from by-products and isolated, preferably in crystalline form. Recall (Chapter 4) that if the complexes are *uncharged*, they are unlikely to be water soluble, which may aid in their precipitation. If they are *charged*, they likely can be crystallized using Solubility Tendency III, by adding a large counterion of opposite charge; the five-coordinate $[Ni(CN)_5]^{3-}$ ion resisted isolation from solution until the matched counteranion $[Cr(NH_2CH_2CH_2NH_2)_3]^{3+}$ was added, whereupon $[Cr(NH_2CH_2CH_2NH_2)_3][Ni(CN)_5]\cdot 1.5\,H_2O$ crystallized out. Evaporating the solvent or adding another solvent that mixes with the first but that does not dissolve the product (e.g., adding ethanol to the aqueous solution) can help crystallize a salt of the product.

1. Substitution Reactions in Aqueous Solutions. This is the most common technique of synthesis, and usually employs an excess of the ligand to insure complete substitution and the absence of contamination by any intermediate mixed-ligand complexes. If the starting complex is labile, the reactions are quick, and are unlikely to stop at a mixed-ligand complex:

$$\underset{\text{Blue}}{[Cu(H_2O)_4]^{2+}} + 4\,NH_3 \xrightleftharpoons{-4\,H_2O} \underset{\text{Deep blue}}{[Cu(NH_3)_4]^{2+}} \xrightarrow[\text{Ethanol cosolvent}]{SO_4^{2-}\text{ Counterion}} [Cu(NH_3)_4](SO_4)(s) \tag{8.11}$$

With inert starting complexes, the reaction may require boiling for some hours:

$$\underset{\text{Wine red}}{[RhCl_6]^{3-}} + 3\,C_2O_4^{2-} \xrightarrow{100\,^\circ C,\,2\,h} \underset{\text{Yellow}}{[Rh(C_2O_4)_3]^{3-}} \xrightarrow[\text{Evaporation}]{K^+\text{ Counterion}} K_3[Rh(C_2O_4)_3]\ (s) \tag{8.12}$$

The reaction can stop at an intermediate mixed-ligand complex, especially if the latter is uncharged and precipitates:

$$[PtCl_4]^{2-} + NH_2CH_2CH_2NH_2 \xrightarrow{H_2O,\,\Delta} [PtCl_2(NH_2CH_2CH_2NH_2)](s) + 2\,Cl^- \tag{8.13}$$

2. Substitution Reactions in Nonaqueous Solvents. Water may be inadequate as the solvent in two situations. If the ligand is water insoluble, as many organic ligands are, the ligand may need to be dissolved in an organic solvent:

$$\underset{\text{Pale violet}}{[Fe(H_2O)_6]^{2+}} + \underset{\text{(in ethanol)}}{3\,C_{10}H_8N_2} \longrightarrow \underset{\text{Intense red}}{[Fe\,(C_{10}H_8N_2)_3]^{2+}} + 6\,H_2O \tag{8.14}$$

If the ligand is basic in water, and the metal ion is appreciably acidic (i.e., of +3 or higher charge) as well as being a hard acid, the metal ion may prefer to react with the hydroxide ions generated by the ligand in water, rather than with the ligand itself:

$$\underset{\substack{\text{Moderately acidic}\\ \text{hard acid}}}{[Cr(H_2O)_6]^{3+}} + \underset{\substack{\text{Moderately basic}\\ \text{borderline base}}}{3\,NH_2CH_2CH_2NH_2} \xrightarrow{\otimes} [Cr(NH_2CH_2CH_2NH_2)_3]^{3+} \tag{8.15}$$

$$[Cr(H_2O)_6]^{3+} + 3\,NH_2CH_2CH_2NH_2 \longrightarrow [Cr(OH)_3(H_2O)_3](s) + 3\,NH_2CH_2CH_2NH_3^+ \tag{8.16}$$

In this particular case, the reaction is also hindered by the fact that the d^3 Cr^{3+} ion is inert, and the water molecules do not dissociate as rapidly as the basic ligand deprotonates the acidic coordinated water molecules. Hence, a form of chromium(III) hydroxide is the main product.

This problem can be avoided by using an *anhydrous* chromium(III) salt (e.g., purple $CrCl_3$) in a dry nonaqueous solvent such as ether, to give the expected yellow product, $[Cr(NH_2CH_2CH_2NH_2)_3]Cl_3$. The reaction will still be slow, but it will at least give the desired product.

If the ligand is volatile (e.g., NH_3, as liquid ammonia), it may be advantageous to use the ligand itself as the solvent, then evaporate the excess ligand (liquid ammonia boils at $-33\,°C$):

$$\underset{\text{Purple}}{CrCl_3} + 6\,NH_3(\ell) \longrightarrow \underset{\text{Pink}}{[CrCl(NH_3)_5]Cl_2} + \underset{\text{Yellow}}{[Cr(NH_3)_6]Cl_3} \tag{8.17}$$

3. Substitution Reactions on Metal Ions in More Labile Oxidation States. Reaction (8.17) gives two products because of the kinetic inertness of the d^3 Cr^{3+} ion, so that replacement of the final chloride ion is incomplete. However, Cr^{2+} has a labile electronic configuration d^4. Hence, substitution reactions of Cr^{3+} can often be speeded enormously by adding a small amount of a reducing agent to convert some of the chromium reactant to chromium(II), which reacts very rapidly, then is reoxidized back to chromium(III).

A variation on this strategy is usually used to synthesize low-spin complexes of the Co^{3+} ion, which is an inert d^6 ion. The Co^{2+} ion is generally high spin and labile, and is also more commonly available as a starting material such as $[Co(H_2O)_6]Cl_2$. Therefore, the labile cobalt(II) analogue of the desired complex is generated in the presence of an oxidizing agent (air, O_2, H_2O_2) capable of oxidizing it to the corresponding cobalt(III) complex:

$$\underset{\text{Pink}}{[Co(H_2O)_6]Cl_2} + 6\,NH_3 \longrightarrow \underset{\text{Rose}}{[Co(NH_3)_6]Cl_2} + 6\,H_2O \tag{8.18}$$

$$4\,[Co(NH_3)_6]Cl_2 + 4\,NH_4Cl + O_2 \longrightarrow 4\,\underset{\text{Orange}}{[Co(NH_3)_6]Cl_3} + 4\,NH_3 + 2\,H_2O \tag{8.19}$$

4. Substitution of Weakly Bound or Volatile Ligands. It can be advantageous in a synthesis to prepare an intermediate complex containing a weakly bound or a volatile ligand that can easily be displaced by a desired ligand. Volatile ligands such as carbon monoxide, water, or ammonia are driven off by heating and then are replaced by other ligands in solution or by the counterions of the complex:

$$\underset{\text{White}}{[Pt(NH_3)_4]Cl_2(s)} \xrightarrow{250\,°C} \underset{\text{Yellow}}{\textit{trans}\text{-}[PtCl_2(NH_3)_2]} + 2\,NH_3(g) \tag{8.20}$$

Heating or irradiation of solutions of complexes of CO often leads to dissociation of this gaseous ligand; the intermediate unsaturated complexes are very reactive and readily add another ligand:

$$[W(CO)_6] \xrightarrow{h\nu} CO(g) + W(CO)_5 + P(C_6H_5)_3 \longrightarrow [W(CO)_5P(C_6H_5)_3] \tag{8.21}$$

Some common organic solvents such as C_4H_8O (tetrahydrofuran, THF) and CH_3CN (acetonitrile) are quite weak bases and dissociate more rapidly than most ligands, even from complexes of metals with inert electronic configurations. Weakly bonded ligands can often be recognized as those that are nonbasic (do not protonate readily in water), and thus lack good σ-donating ability. They also should lack good π-acceptor abilities (i.e., do not form oxides, see Section 8.5.) It can be advantageous to put weak ligands on a metal, then later to displace them with the desired ligands. Some chromium(III) complexes are conveniently made indirectly from anhydrous chromium(III) chloride by first preparing the THF complex:

$$CrCl_3(s) + 3\,C_4H_8O \xrightarrow{Zn,\Delta} [CrCl_3(C_4H_8O)_3] \tag{8.22}$$

In this case, the function of the Zn is to reduce a small amount of the inert $CrCl_3(s)$ to labile $CrCl_2(s)$, which reacts quickly with the THF solvent to generate a complex that is easily oxidized to the chromium(III)–THF complex. The THF can subsequently be displaced from this complex by most ligands.

Finally, it may be possible, in a noncoordinating solvent, to prepare a salt of the metal ion with a very nonbasic anion.[24] There is probably no anion so nonbasic that it will not coordinate to a metal ion if no other ligand is present, but some very nonbasic fluoro anions[25] such as SbF_6^- and $[OTeF_5]^-$ are easily displaced (especially from soft-acid cations) by almost any other ligand.

5. Ligand Construction or Destruction by Reactions of Coordinated Ligands. In the case of inert complexes or complexes of chelating ligands as reactants, the reactions of the metal–ligand bond may be slower than those of the covalent bonds in the ligands themselves. In such a case, it may be more practical to carry out a reaction on the coordinated ligand itself to generate a new ligand without breaking the metal–donor atom bond. For example, the carbonate ion is readily incorporated into inert cobalt(III) complexes as a monodentate ligand. The O–C bond in the coordinated carbonate ion is readily broken by acids, liberating $CO_2(g)$ and generating the H_2O complex:

$$\underset{\textbf{Pink}}{[(NH_3)_5Co\text{–}O\text{–}|\text{–}CO_2]^+} + 2\,H^+ \longrightarrow \underset{\textbf{Pink}}{[(NH_3)_5Co(OH_2)]^{3+}} + CO_2(g) \tag{8.23}$$

It can be shown by the use of isotopically labeled reactants that the oxygen of the coordinated water came from the original carbonate ion, not from the solvent water. Similarly, the NO_2 group of an inert Pt(IV) complex can be reduced to a coordinated NH_3 molecule in a manner reminiscent of that used in organic chemistry to reduce aromatic nitro compounds to aromatic amines:

$$[(NH_3)_3Pt\text{–}NO_2]^+ \xrightarrow{Zn,\,HCl} [(NH_3)_3Pt\text{–}NH_3]^{2+} \tag{8.24}$$

Irradiation by visible or UV light may be used to promote a reaction within a ligand. The C–C bond of chelated oxalate ion, $[C_2O_4]^{2-}$, is readily broken upon irradiation. This reaction produces 2 mol of $CO_2(g)$ and 2 mol of electrons. Hence, a metal ion is obtained with a lower oxidation number and two vacant coordination sites. The new metal ion can readily add additional ligands to form new complexes:

$$[Pt(C_2O_4)(PR)_3]_2 \xrightarrow{h\nu} [Pt(PR_3)_2] + 2\,CO_2(g) \xrightarrow{+2\,L} [PtL_2(PR_3)_2] \tag{8.25}$$

In some cases, the relative inertness of a complex may not be due to the d^n electronic configuration of the metal, but rather to the presence of chelating ligands that are slow to dissociate, since more than one coordinate covalent bond must first be broken. Such complexes may persist long enough to allow reactions to occur on the coordinated ligands to generate macrocyclic ligands that are otherwise difficult to prepare. These reactions may be facilitated by the enhanced acidity of the coordinated ligand and by the template effect (Section 5.10). For example, the chelate complex $[Co(NH_2CH_2CH_2NH_2)_3]^{3+}$ undergoes a reaction with ammonia and formaldehyde (CH_2O) similar to the Mannich condensation in organic chemistry:

$$[Co(NH_2CH_2CH_2NH_2)_3]^{3+} + 6\,CH_2O + 2\,NH_3 \longrightarrow 6\,H_2O + [\text{Co(sepulchrate)}]^{3+} \quad (8.26)$$

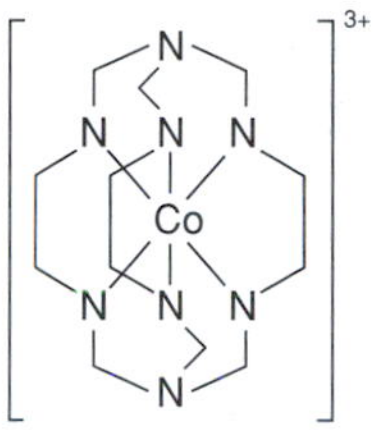

This type of ligand is known as a *sepulchrate* ligand; it is analogous in structure and function to the cryptand ligand of Figure 4.9.

Phthalocyanines are an important type of macrocyclic–ligand complex, often used as dyes but also as models for the isoelectronic metal–porphyrin complexes discussed in Section 8.10. These are readily assembled in a template reaction:

$$4\,C_6H_4(CN)_2 + Ni^{2+} + C_4H_9OH \xrightarrow{\Delta} C_3H_7CH{=}O + 2\,H^+ + [\text{Ni(phthalocyanine)}] \quad (8.27)$$

As a final example of this type of reaction, we might recall the template synthesis of the crown ethers themselves [Eq. (5.19)].

8.10 *Bioinorganic Chemistry and Crystal Field Theory: Heme and Hemoglobin

Isoelectronic to the phthalcyanine ligand shown in Eq. (8.27), and of considerably greater biological importance, are the macrocyclic ligands known as the porphyrins [Fig. 8.20(*a*); the metal-free ligands have −2 charges]. A type of porphyrin ligand complexes Mg^{2+} in *chlorophyll,* the importance of which needs no further emphasis, but in this section we wish to focus on the other major type of porphyrin–metal complex, *heme* [Fig. 8.20(*a*)].[26] Since this macrocyclic ligand has a planar, conjugated system of π bonds around its perimeter, it is a much more rigid macrocyclic ligand than are the crown ethers. Consequently, the ligand is even more selective for a certain metal-donor atom distance (of close to 200 pm) than are the crown ethers; its enforced square planar geometry gives it a stronger preference for the d^8 Ni^{2+} ion than predicted by the Irving–Williams series. But other metal ions may add a fifth and maybe a sixth ligand above or below the square plane; the function of heme in hemoglobin, of course, is to bond to the weak ligand O_2 and carry it from the lungs to the cells, where it is released to another heme-containing protein, myoglobin.

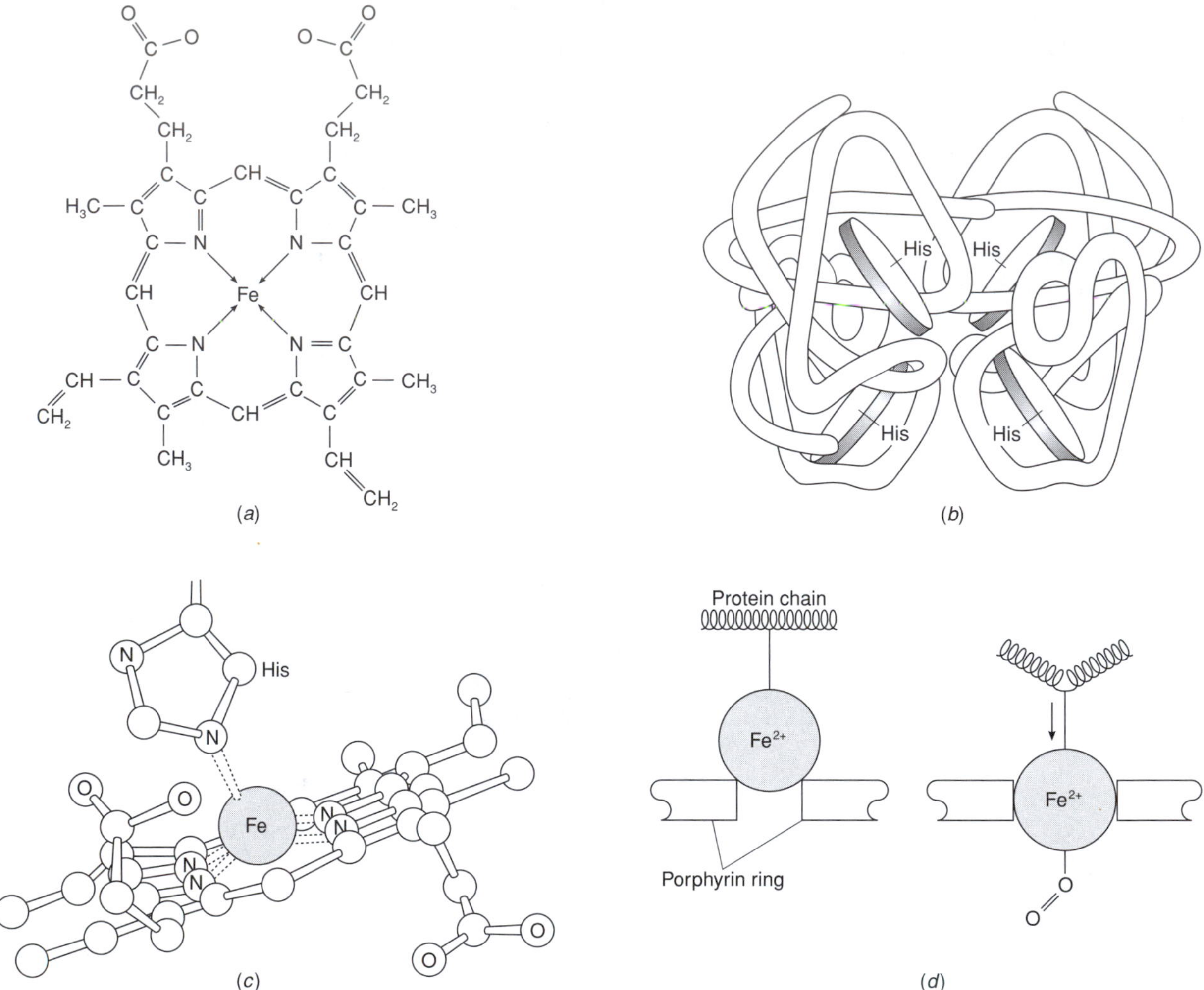

Figure 8.20
(*a*) The structure of the heme group, containing a Fe^{2+} ion and a (macrocyclic) porphyrin(2−) ligand. (*b*) The tetrameric hemoglobin molecule, containing four heme groups (represented by disks) bound to the protein by the amino acid histidine. (*c*) Details of the bonding of the heme group to histidine in myoglobin and hemoglobin, in the absence of bound O_2. Note that the Fe^{2+} is above the plane of the porphyrin ligand. (*d*) Sketch of the effects of coordination of O_2: The Fe^{2+} changes spin state, shrinks in size, falls into the plane of the porphyrin ligand, and pulls on the protein chain via the coordinated histidine. [Adapted from J. E. Huheey, *Inorganic Chemistry: Principles of Structure and Reactivity*, 3rd ed., Harper & Row, New York, 1983; pp. 892–899, 901.]

Hemoglobin itself [Fig. 8.20(*b*)] is composed of a protein, globin, which contains (among many other amino acids) four units of histidine, which contains a five-membered imidazole ring with nitrogen-donor atoms. These nitrogen donor atoms bond four separate heme molecules to the globin, causing the Fe^{2+} to be five coordinate [Fig. 8.20(*c*)]. The vacant sixth site of the potential octahedron about Fe^{2+} is reserved for O_2 as a ligand. In the absence of the O_2, the ligand field is (just barely) weak, so the Fe^{2+} is high-spin, with the $t_{2g}{}^4e_g{}^2$ electron configuration; in this spin state Fe^{2+} has a substantially larger radius than in the low-spin state (cf. data for octahedral Fe^{2+} in Table 8.7); its radius is *too large* for it to fit in the hole of the porphyrin ligand, so the iron sits *above* the plane of the porphyrin ring.

The binding ability of hemoglobin for O_2 must be delicately balanced with that of the intracellular O_2-binding agent, myoglobin (Mb), which contains only *one* heme unit and reversibly attaches O_2 [27]:

$$\text{Mb} + O_2 \rightleftharpoons \text{Mb:}O_2; \qquad K_{\text{Mb}} = \frac{[\text{Mb:}O_2]}{[\text{Mb}][O_2]} \tag{8.28}$$

Since hemoglobin (Hb) contains four heme groups, it often bonds four O_2 groups; nonetheless the equilibrium constant for this bonding (in the range of O_2 pressures significant for tissues) has a peculiar form:

$$K_{\text{Hb}} = \frac{[\text{HbO}_2]}{[\text{Hb}][O_2]^{2.8}} \tag{8.29}$$

This is an indication of the biochemical phenomenon known as *cooperative binding*, which means that the ability of a particular heme group to bind O_2 is strongly influenced by whether or not the other three heme groups in hemoglobin already possess O_2 ligands (i.e., information is passed from one heme group to the others). Consequently, in the lungs, where the pressure of O_2 is high and much O_2 is bound, the affinity of heme for O_2 becomes abnormally high, so that it is very efficiently loads up with as much O_2 as possible. But when the hemoglobin reaches the cells, where the pressure of O_2 is low, and O_2 begins to dissociate from the complex, this information is passed on so that the remaining heme groups are caused to unload their O_2 groups as well. This loss of O_2 allows the myoglobin to pick up all of the O_2—myoglobin, having only one heme group, has no cooperative binding, so it does not lose its affinity for O_2. Were it not for this cooperative binding, the O_2 could not efficiently be transferred from the blood to the cells.

The mechanism by which this cooperative binding is achieved is a beautiful example of design employing both the principles of biochemistry and of inorganic chemistry (specifically, ligand or crystal field theory). Although O_2 is not expected to be high on the spectrochemical series, upon bonding to heme in hemoglobin it adds just enough to the ligand field strength to cause the Fe^{2+} to go from the *high*-spin state to the *low*-spin state.[28] As indicated in Table 8.7, this causes the radius of the Fe^{2+} to decrease by a substantial 17 pm, enough that the Fe^{2+} can now fit in the hole of the porphyrin ring, into which it now falls [Fig. 8.20(*d*)]. Since the iron is attached to the globin protein by the histidine amino acid, this causes various sorts of motion in the globin itself, which are passed along to the regions of the other three heme groups, thus transmitting the necessary message that an O_2 has been attached, and that the other heme groups should go and do likewise.

There are some other necessary features of the chemistry of heme in hemoglobin. The O_2 molecule is the classical oxidizing agent, and Fe^{2+} is a good reducing agent, but if the Fe^{2+} in heme is oxidized to Fe^{3+} (*hematin*), it will not carry O_2! This oxidation to hematin happens readily in free heme in the presence of O_2 *and a water molecule*. Part of the function of the globin protein is to prevent access of H_2O to the heme while the O_2 is coming in. Also, if O_2 can act as a bridging ligand between two heme groups, the iron will also be oxidized: The globin separates the two heme groups so much that this is impossible.

Since O_2 is neither a strong field nor a soft ligand, it is not easy to get it to coordinate to the fairly soft iron ion if soft-base ligands such as CO and CN^- are present. These ligands form much more stable complexes with iron in heme groups in hemoglobin, myoglobin, or other species containing heme, such as the cytochromes (Section 17.8); this complexation is the basis of the toxicity of these soft bases. Recent astonishing studies indicate that the π-acceptor ligands nitric oxide (NO)[29] and carbon monoxide (CO)[30] are generated deliberately in the body, where they bond to metal-bearing proteins containing heme units; this paradoxically either activates or poisons the protein or enzyme. This bonding is done for a variety of purposes: (a) for both NO and CO, as an apparent new type of neurotransmitter, to carry nerve messages from certain types of neurons to others, at which it again coordinates to a heme group in the enzyme guanylyl cyclase, causing a change of spin state and consequent conformational changes in this enzyme; (b) for NO and CO, in long-term learning and in storing long-term memories in the brain; (c) for NO only, to affect the contractions and relaxations of smooth muscles, as in penile erections or blood vessels (NO apparently prevents blood clotting but also, in response to septic bacterial infection, lowers blood pressure so dramatically as to cause deadly septic shock); (d) to suppress pathogens in the body; and (e) unfortunately, during strokes massive release of NO causes the death of nerve and brain cells. Unlike most biotransmitters, the potent activity of NO is associated, not primarily with its molecular shape, but rather with its chemical bonding ability.

8.11 *Nomenclature of Coordination and Organometallic Compounds

Given that several complex organic ligands may be present in a coordination compound and that geometric isomers may be present, the name of a coordination compound can be quite formidable, sometimes taking more than a line of type. We will cover the naming of only some of the simpler coordination and organometallic compounds, omitting complex organic ligands and consideration of geometric and stereoisomers, and so on; the rules are those already used in Section 3.9 to name fluoro anions.

Naming of Ligands

1. Neutral ligands take the names they normally use as neutral molecules. There are a few specific exceptions: Water (H_2O) as a ligand is known as *aqua* (formerly *aquo*), NH_3 is named *ammine*, CO is named *carbonyl*, and NO is named *nitrosyl* (CS is also named *thiocarbonyl* and NS is named *thionitrosyl*).

2. The names of anionic ligands have *-o* in place of the final *-e*. (Thus *carbonate* as a ligand is known as *carbonato*.) Halide ions and some pseudohalide or similar anions ending in *-ide* drop the whole *-ide* ending: Thus F^- is *fluoro*, Cl^- is *chloro*, Br^- is *bromo*, I^- is *iodo*, O^{2-} is *oxo*, S^{2-} is *thio* (an exception), OH^- is *hydroxo*, CN^- is *cyano*, and $O_2{}^{2-}$ is *peroxo*. (But $N_3{}^-$ is *azido* and H^- is usually *hydrido* to avoid confusion with other usages of the terms azo and hydro.) The *-o* ending is *not* used, however, with carbon anions, which keep their *-yl* names: thus $CH_3{}^-$ as a ligand is just named *methyl*, and $C_6H_5{}^-$ is still named *phenyl*.

Naming the Central (Metal) Atom

3. In a complex cation or a neutral coordination compound, the name of the central (metal) atom is used unchanged. If the ion is a complex *anion*, the name of the central atom is converted to its Latin form (if there is one; see Table D for Latin roots), and the suffix *-ate* is added. This is analogous to the naming of oxo anions; thus an iron-containing complex anion is named *-ferrate*.

4. If the central (metal) atom is capable of having more than one oxidation number, the oxidation number of the central atom is shown last, in Roman numerals enclosed in parentheses (the *Stock convention*). (You may want to review the procedure given earlier in Section 3.10 for deriving these oxidation numbers.) Because of the occasional ambiguities that arise in assigning oxidation numbers, there is an alternate convention, the *Ewens–Bassett convention*, by which the net charge on the *whole* complex ion, in Arabic numerals, is enclosed in the parentheses (instead of the oxidation number of the central atom.)[31] Thus, FeO_4^{2-} can be called *ferrate(VI)* (Stock convention) or *ferrate(2−)* (Ewens–Bassett). Note that in coordination and especially in *d*-block organometallic chemistry, that metal atom oxidation numbers can be unusual (i.e., they may not appear in either Table B or C, and can, in some cases, be zero or even negative).

5. An alternate practice is commonplace with *p-block organometallic compounds*: If *neutral*, the central atom takes the name of its *hydride* (Section 2.8); if *cationic*, the *-ane* or *-ine* ending of the hydride becomes *-onium*. (If anionic, Rule 3 still holds.) Thus the arsenic in R_3As is named *arsine,* while in R_4As^+ it is named *arsonium*, and in R_6As^- it is named *arsenate(1−)* (by the Ewens–Bassett convention).

Order of Naming

6. The *cation* of a complex compound is named first, then the *anion*—as with ordinary salts, and regardless of which one is a complex ion, or whether both are.

7. Within the name of the coordination complex, the *ligands* are named first, then the *metal* or *central* atom or ion.

8. The ligands within a given complex ion are *listed in alphabetical order*. (Before 1971, anionic ligands were listed before neutral ligands, but now their listing is integrated.)

Use of Parentheses and Indication of Numbers of Ligands Present

9. The names of all organic ligands (except ethylene, $CH_2{=}CH_2$, and common radicals such as methyl and ethyl) are enclosed in parentheses. The rules for inorganic ligands are not as specific: Parentheses should be used if the ligand contains numerical prefixes (as with triphenylphosphine) or include the names of more than one type of atom (as with thiosulfato), since otherwise these could be confused with alternate complexes containing three C_6H_5 and a PH_3, or an S^{2-} and an SO_4^{2-}, ligands. In usual practice, it seems to be a good idea to use parentheses around *long* ligand names (say, four syllables or more). Shorter names, such as those of the specially named ligands H_2O, NH_3, CO, and NO, and those of fairly simple anions such as chloro and sulfato, are not seen enclosed in parentheses.

10. The *number of each type of ligand* present is indicated by prefixes. (a) If the ligand name is not enclosed in parentheses, *di-*, *tri-*, *tetra-*, *penta-*, and *hexa-* are added as prefixes to the name of the ligand to indicate the presence of 2, 3, 4, 5, 6, and so on, ligands. (b) If the ligand name is enclosed in parentheses, the prefixes *bis-*, *tris-*, *tetrakis-*, *pentakis-*, *hexakis-*, and so on, are used before the parentheses enclosing the name of the ligand. If no prefix is used, the presence of one ligand is implied; but if it is necessary to emphasize this, *mono-* may be used.

Example 8.6

Name the following compounds: (a) $B(CH_3)_3$; (b) $K[B(C_6H_5)_4]$; (c) $(CH_3)_3P$; and (d) $[(CH_3)_4P]Cl$.

SOLUTION:

The compounds listed all have carbon donor atoms in their methyl and phenyl ligands and are thus *p*-block organometallic compounds. Hence, the special provisions of Rules 2 and 5 come into play.

(a, c) The ligand CH_3^- is named *methyl* (without parentheses) by Rule 2. By the common practice of Rule 5, the boron is named *borane* and the phosphorus *phosphine* (without oxidation numbers). Rules 7 and 10 tell us to assemble these names as follows: (a) trimethylborane; (c) trimethylphosphine.

(b) By Rule 2, the ligand is *phenyl*; by Rules 5, 3, and 4 the central atom is named *borate*(III) (Stock convention) or *borate*(*1*−) (Ewens–Bassett convention). Rules 6, 7, and 10 lead to the final name: potassium tetraphenylborate(III) (or 1−). (The oxidation state could be omitted since boron has no other common oxidation number.)

(d) This compound is named similarly except that Rule 5 tells us to name the phosphorus as *phosphonium*. The compound thus is named tetramethylphosphonium chloride.

We have first illustrated the naming of organometallic compounds (Example 8.6) because the organic derivatives of the Groups 15(V), 16(VI), and 17(VII) elements show unshared electron pairs on the Group 15(V), 16(VI), and 17(VII) atoms; many of these compounds are found as Lewis bases or ligands in complex compounds.

Example 8.7

Name the following coordination complexes: (a) $K_2[SiF_6]$; (b) $[Co(NH_3)_4Cl_2]Cl$; (c) $[Pt(S_2O_3)(SCN)_2(H_2O)]$; and (d) $[ReH_3(P(C_6H_5)_3)_5]$.

SOLUTION:

(a) By Rule 2, the ligand of this complex anion is *fluoro*; by Rules 3 and 4, the central atom is *silicate(IV)* (oxidation number IV is so common for Si that this may be omitted) or *silicate(2−)*. Rule 6 tells us to name the cation, K^+, first: potassium hexafluorosilicate (with nothing or IV or 2− in parentheses).

(b) Rule 1 tells us to name NH_3 *ammine*; by Rule 2, Cl^- is *chloro*. By Rules 3 and 4, the Co is *cobalt(III)* or *cobalt(1+)*. By Rules 6–8, we name the components in this order: ammine, then chloro, then the cobalt, then the counteranion, chloride. Adding the numbering of the ligands as in Rule 10, we finally obtain tetramminedichorocobalt(III) chloride (using the Stock convention).

(c) The anionic ligands $S_2O_3{}^{2-}$ and SCN^- are named (*thiosulfato*) and (*thiocyanato*) by Rule 2, with the parentheses courtesy of Rule 9. Water is *aqua* by Rule 1. The Pt is *platinum(IV)* in the Stock convention (Rule 4). Applying Rules 7 and 8, we name the ligands first, in alphabetical order, and indicate the presence of two SCN^- ligands with *bis*. The compound is named aquabis(thiocyanato)(thiosulfato)platinum(IV).

(d) This compound contains a common organophosphorus compound, triphenylphosphine, as a ligand. Once that is named, the complex is named trihydridopentakis-(triphenylphosphine)rhenium(III).

Additional rules are also used to designate bridging ligands, to tell which donor atom is used by an ambidentate ligand, and so on. For these rules, refer to the latest rules for inorganic nomenclature of the International Union of Pure and Applied Chemistry (IUPAC).[32]

Searching Chemical Abstracts. The procedures for the synthesis of complex compounds given in Section 8.9 are not detailed enough to be carried out as is in the laboratory; the chemist needing to make a given complex is well advised to search the literature to find a known synthesis if there is one. The most general starting place for such a search is, of course, *Chemical Abstracts.* Even if you have had experience in searching *Chemical Abstracts* for organic compounds, some additional guidance is needed to search for coordination compounds. In addition, although detailed procedures have been published for carrying out an on-line search for coordination compounds,[33] these are not simple, and on-line searches can generate lots of expensive garbage if the search is not carefully constructed. Hence, it is useful to begin by learning to search printed *Chemical Abstracts.*

Often, it seems easiest to begin by searching the cumulated formula indexes. The search for coordination compounds in the formula indexes differs from that for organic compounds in that *counterions (and waters of hydration, etc.) are omitted from the formula*. (This procedure is reasonable, since the searcher generally wants to obtain information on the complex ion itself, and does not care much what the counterion is.) Within the formula of the complex ion, elements are listed in alphabetical order, except for carbon and hydrogen; carbon is always first, followed by hydrogen, based on the practice for organic compounds. (If carbon is absent, hydrogen is in alphabetical position.) The compounds in Examples 8.6 and 8.7 would be looked up under the formulas: (8.6a) C_3H_9B; (8.6b) $C_{24}H_{20}B$; (8.6d) $C_4H_{12}P$; and (8.7a) F_6Si; (8.7b) $Cl_2CoH_{12}N_4$; (8.7c) $C_2H_2N_2O_4PtS_4$; (8.7d) $C_{90}H_{78}P_5Re$.

Much more information will be found in the chemical substances index than in the formula index, however, so that the search is more profitably carried out using the name of the compound, especially if there are many references to be sorted through.[34] In the chemical substances index, both inorganic and organic compounds are not listed simply alphabetically, but rather under the name of the base compound, followed by modifiers. In the case of coordination compounds, the base name is that of the central atom, as produced by Rules 4 or 5 *using Ewens–Bassett nomenclature.* The compounds in Examples 8.6 and 8.7 would be looked up under the names: (8.6a) Borane, trimethyl; (8.6b) borate(1−), tetraphenyl[35]; (8.6d) phosphonium, tetramethyl; and (8.7a) silicate(2−), hexafluoro; (8.7b) cobalt(1+), tetraamminedichloro; (8.7c) platinum, aquabis(thiocyanato)(thiosulfato); (8.7d) rhenium, trihydridopentakis(triphenyl phosphine).

Study Objectives

1. Be able to identify and draw the *d* orbitals, and classify them as t_{2g} or e_g in an octahedral complex. Tell why *f* orbitals experience little crystal field splitting. Exercises 1–3.

2. Write and draw the electron configuration and calculate the ligand field stabilization energy for an octahedral complex in which the metal ion has 1, 2, 3, 8, 9, or 10 *d* electrons. Exercises 4, 5.

3. Write and draw the electron configuration, calculate the ligand field stabilization energy, and compute the number of unpaired electrons for the octahedral complex of a transition metal ion for which both high-spin and low-spin electron configurations are possible. Know what energy considerations determine which possibility will be selected in a given complex. Exercises 6–9.

4. Know the terms describing magnetic properties of transition metals and their complexes, and know how to predict magnetic moments and magnetic susceptibilities from the number of unpaired electrons. Exercises 10–12.

5. Know why the spectra of many complex ions show more than one absorption band; know which band in such spectra should be used to obtain the value of Δ_0. With the aid of Figure 8.5, from the color of light absorbed by a complex or from a drawing of its spectrum, be able to give the approximate value of Δ_0 for that complex. Exercises 13, 14.

6. Estimate Δ_0 from ligand *f* and metal *g* factors, or vice versa, and compare Δ_0 values with pairing energies to predict whether a given complex will be high or low spin. Know and explain how Δ_0 varies with changes in the period of the metal ion; the charge of the metal ion, and (in the spectrochemical series) with the π-donor, σ-donor, or π-acceptor nature of the ligand. Exercises 15–26.

7. Explain why there are irregularities in the hydration energies, lattice energies, and complex-ion formation constants across the *d* block. Know and explain the Irving–Williams series. Exercises 27–31.

8. Explain why ionic radii and electronegativities vary irregularly across the *d* block, and why some metal ions have two appropriate ionic radii. Exercises 32–34.

9. Be able to reason the crystal field splitting pattern of the *d* orbitals in a given (reasonable) geometry of ligands, such as tetrahedral. Calculate the ligand field stabilization energy for a given d^n electron configuration in a tetrahedral complex. Estimate the crystal field splitting and (roughly) the position of the first electronic absorption of a nonoctahedral complex. Exercises 35–42.

10. Explain the Jahn–Teller distortion, and predict the electron configurations in which it will occur. Exercises 43–48.

11. Calculate or predict the relative preference of different *d*-block metal ions for one geometry of complex or another; relate this preference to the relative rates (for ions of different d^n configurations) undergoing a specified mechanism of reaction. Exercises 49–54.

12. Select an appropriate method of synthesis of a given complex ion; predict the products of a given synthesis. Exercises 55–60.

13. Choose the best metal ions to be complexed by a porphyrin, and describe the geometry of the resulting complexes. Understand the significance of high spin–low spin and Fe^{2+}/Fe^{3+} distinctions in the Perutz model of the function of hemoglobin. Exercises 61, 62.

14. Dissect the structures of coordination compounds to identify the charges on ions and the central metal ions (hence, the oxidation number of the central metal). Given names of coordination compounds, write their formulas in the standard way; in the way they will be found in the formula index of *Chemical Abstracts*. Exercises 63–69.

15. Name coordination compounds in the standard way; in the way they will be found in the subject index of *Chemical Abstracts*. Exercises 70–76.

Exercises

1. Below are drawn some *d* orbitals in their *xyz* axes systems.

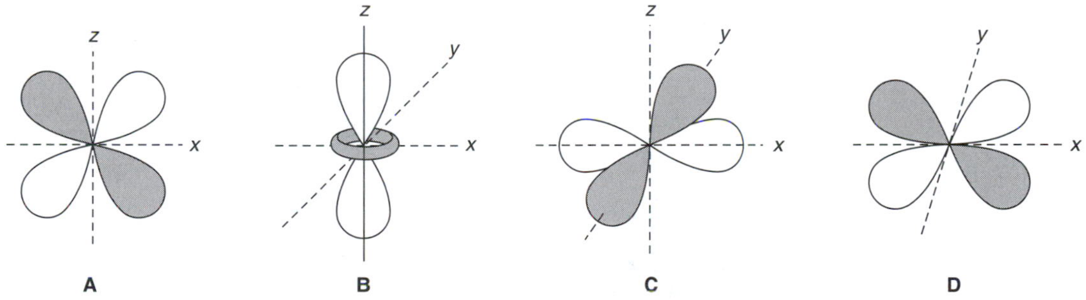

(a) List the names of these orbitals (d_{xy}, etc.). (b) Tell whether each belongs to the t_{2g} or the e_g set in an octahedral complex.

2. Below are drawn some *d* orbitals in their *xyz* axes systems. One of these is *not* a *d* orbital: do not answer the questions about it!

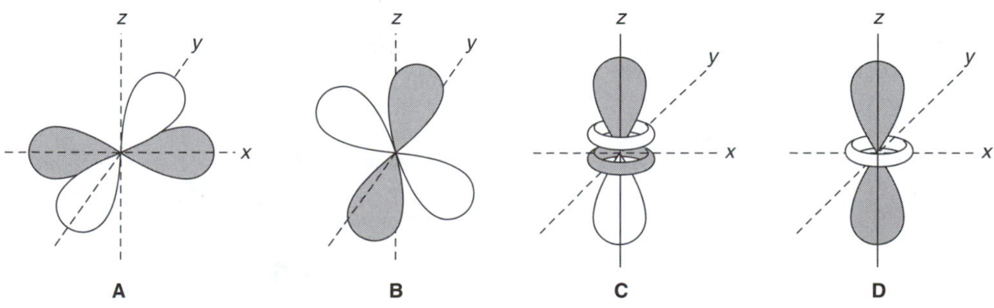

(a) List the names of these orbitals (d_{xy}, etc.). (b) Tell whether each belongs to the t_{2g} or the e_g set in an octahedral complex.

3. *The f orbitals are drawn and labeled in Figure 8.21 in forms that are appropriate for complexes of octahedral symmetry. (a) Group these into sets that will show equivalent interactions with external octahedral point charges or ligands. (b) Which of these sets will experience the greatest degree of repulsion? Which will experience the greatest stabilization? (c) Explain why these stabilizations and repulsions are relatively small.

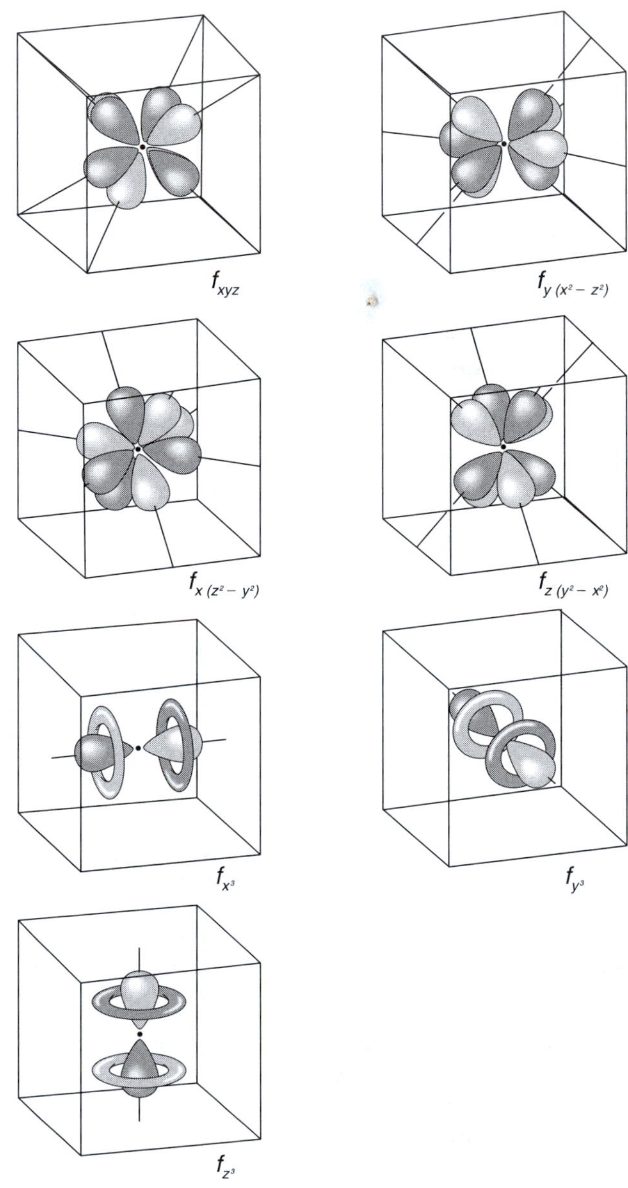

Figure 8.21
The f orbitals drawn in forms appropriate for complexes of octahedral symmetry. [From *Inorganic Chemistry* by D. F. Shriver, et al. © 1990, 1994 by Durward F. Shriver, Peter Atkins, and Cooper H. Langford. Used with permission of W. H. Freeman and Company.]

4. (a) Draw an energy level diagram that indicates qualitatively how the energy levels of the *d* orbitals are altered as a result of varying degrees of interaction of *d* and ligand electrons in an octahedral complex $[ML_6]^{n+}$. Show which *d* orbitals occur at which energy levels. (b) Show how the orbitals will be occupied by electrons for each of the following electron configurations: d^1, d^2, d^3. (c) Write the expected electron configuration in terms of t_{2g} and e_g. (d) For each electron configuration, compute the CFSE in units of Δ_o. (e) From Table C, select and list all *d*-block ions having each of these electron configurations.

5. Repeat the preceding exercise for the following electron configurations: d^8, d^9, d^{10}.

6. *For each of the following metal *d*-electron configurations and ligand field splitting patterns in octahedral complexes, (1) write the expected electron configuration in terms of t_{2g} and e_g, (2) compute the ligand field stabilization energy as a multiple of Δ_o (and *P*), (3) list the number of unpaired electrons expected, and (4) choose a likely example of this metal *d*-electron configuration and ligand field splitting from the following choices: $[Co(CN)_6]^{3-}$; $[Fe(H_2O)_6]^{3+}$; $[Fe(CN)_6]^{3-}$; and $[Co(H_2O)_6]^{3+}$: (a) d^6 strong field; (b) d^5 weak field; (c) d^6 weak field; and (d) d^5 strong field.

7. For each of the following metal *d*-electron configurations and ligand field splitting patterns in octahedral complexes, (1) write the expected electron configuration in terms of t_{2g} and e_g, (2) compute the ligand field stabilization energy as a multiple of Δ_o (and *P*), (3) list the number of unpaired electrons expected, and (4) choose a likely example of this metal *d*-electron configuration and ligand field splitting from the following choices: $[Co(CN)_6]^{4-}$; $[Mn(H_2O)_6]^{3+}$; $[Mn(CN)_6]^{3-}$; and $[Co(H_2O)_6]^{2+}$: (a) d^4 strong field; (b) d^7 weak field; (c) d^4 weak field; and (d) d^7 strong field.

8. Would the *f*-block metal ions be expected to form strong-field complexes? Why or why not?

9. In the fifth and sixth periods, strong-field complexes are very predominant in the *d*-block. Why is this? Find all examples listed in Table C of fifth- and sixth-period *d*-block metal ions having each of these electron configurations: (a) d^4; (b) d^5; (c) d^6; and (d) d^7. (e) Given that Ru^{2+} and Os^{2+} are also well known in complexes, which of these four electron configurations seem to have the most examples in the fifth and sixth periods, and why?

10. A Ni^{2+} complex is found to have an experimental magnetic moment of 2.83 BM (i.e., the square root of 8). (a) How many unpaired electrons are present on the Ni^{2+} ion? (b) Is the complex paramagnetic or diamagnetic? (c) Compute the corrected (molar) magnetic susceptibility χ_m^{corr} that you would expect to measure for this complex at 298 K; at 100 K.

11. *Predict the magnetic moments and the corrected molar susceptibilities at 273 K of the following species: (a) $[Fe(H_2O)_6]^{3+}$; (b) $[Fe(CN)_6]^{4-}$; (c) $[Fe(H_2O)_6]^{2+}$; (d) $[Fe(CN)_6]^{3-}$; and (e) $[HoF_6]^{3-}$.

12. Predict the magnetic moments and the corrected molar susceptibilities at 273 K of the following species: (a) $[Ni(H_2O)_6]^{2+}$; (b) $[Co(CN)_6]^{3-}$; (c) $[Co(CN)_6]^{4-}$; and (d) $[Gd(H_2O)_6]^{3+}$.

13. *Below are shown two spectra: the first is of a Cr^{3+} octahedral complex ion, while the second is of a Ti^{3+} octahedral complex ion.

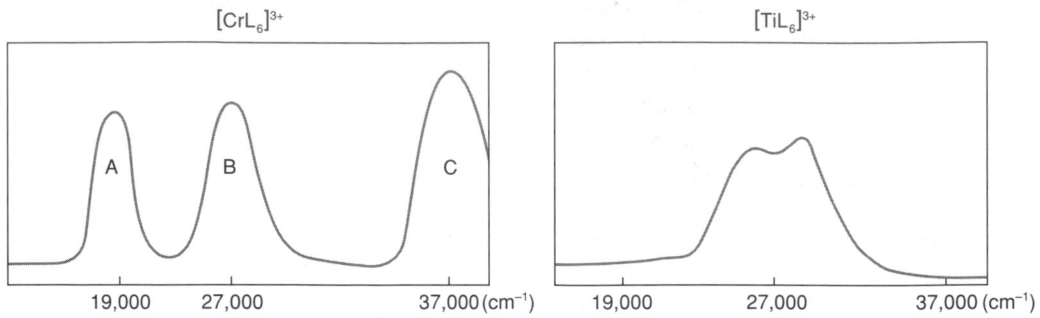

(a) Briefly explain why there is more than one absorption band in the spectrum of the chromium ion but not the titanium ion. (b) Which absorption band (A, B, or C) would you use to determine the value of Δ_o in the chromium complex ion? What is the value of Δ_o in this complex ion? (c) With the help of Figure 8.5, describe approximately the colors of these two complex ions.

14. Below are shown three spectra of three octahedral complex ions of the same metal. (a) Which of the following metal ions are almost certainly *not* the metal in these complex ions: Ti^{3+}, Ni^{2+}, Pt^{4+}, and Cu^{2+}. Briefly explain why you can rule out those metal ions. (b) Which absorption band (ν_α, ν_β, or ν_γ) would you use to determine the value of Δ_o in these complex ions? (c) What are the approximate values of Δ_o in each of these complex ions? (d) With the help of Figure 8.5, approximately describe the colors of these complex ions.

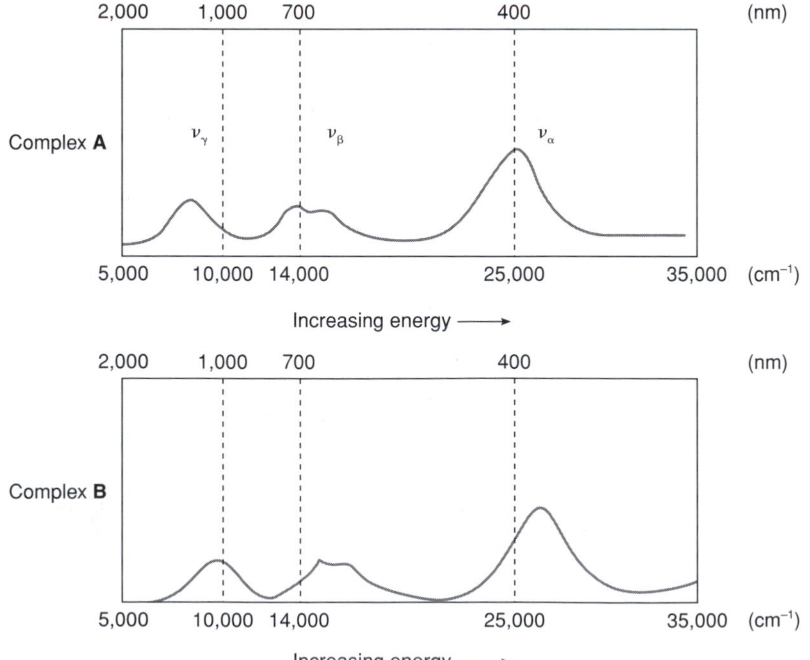

Figure 8.22
Spectra of three octahedral complexions of the same metal.

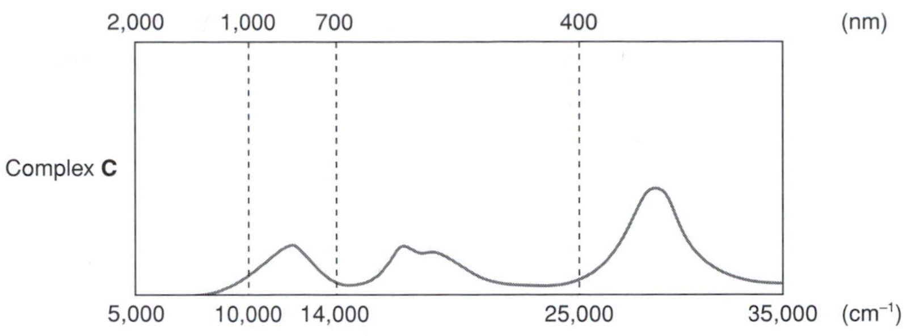

Figure 8.22 (*continued*)

15. Below are given the electronic spectra of some octahedral Ni^{2+} complexes.

Complex Ion	ν_1 (cm^{-1})	ν_2 (cm^{-1})	ν_3 (cm^{-1})
$[Ni(H_2O)_6]^{2+}$	8,130	13,500	25,100
$[NiA_6]^{2+}$	11,500	18,500	30,000
$[NiB_6]^{2+}$	12,700	19,300	30,000
$[NiC_6]^{4-}$	6,460	9,860	17,000

(a) What is Δ_o for the hydrated nickel ion? (b) What is the g factor for the nickel ion? (c) What is the f factor for ligand A; for ligand B; for ligand C? (d) If the Ni^{2+} ions in the above complexes were replaced by Ni^{3+} ions, would the values of Δ_o increase or decrease? (e) If the Ni^{2+} ions in the above complexes were replaced by Pt^{2+} ions, would the values of Δ_o increase or decrease? (f) The three ligands A, B, and C include the following: iodide ion, ammonia, and carbon monoxide. Identify each ligand and explain your reasoning.

16. Below are shown two spectra of two octahedral complex ions, MA_6 and NB_6.

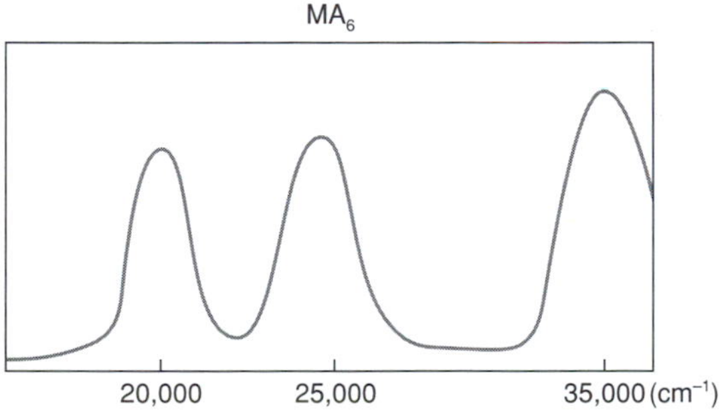

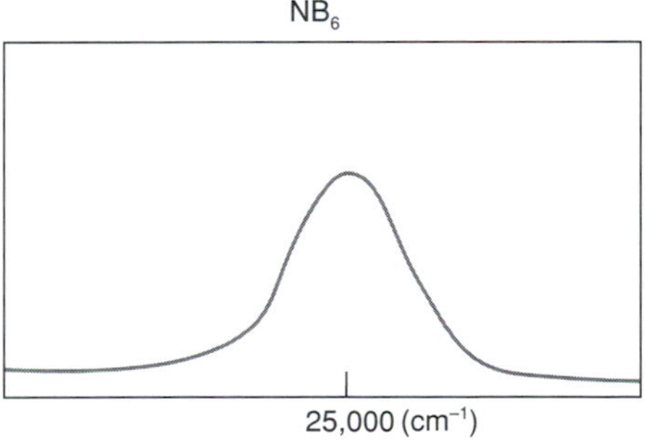

(a) What is the value of Δ_o (in cm^{-1}) for the complex MA_6? What is the value of Δ_o (in cm^{-1}) for the complex NB_6? Which octahedral complex is more likely that of a d^1 metal ion? Why do you choose that answer? (b) If the ligands in MA_6 are six identical molecules or ions with an f factor of 0.80, what is the g factor of the metal ion M^{n+}? (c) If the ligands in NB_6 are six identical molecules or ions with an f factor of 1.80, what is the g factor of the metal ion N^{n+}? (d) Which ligand (A in MA_6 or B in NB_6) is higher in the spectrochemical series? Which ligand (A or B) is more likely a π-donor ligand? Which ligand (A or B) is more likely a π-acceptor ligand?

17. Referring to Tables 8.2, 8.4, and 8.5 and assuming that the pairing energies in Table 8.2 are 22% too high to apply properly to complex ions, predict whether each of the following complexes will be strong field, weak field, or will be "too close to call": (a) $[CrBr_6]^{4-}$; (b) $[Cr(CN)_6]^{4-}$; (c) $[Mn(NH_2CH_2CH_2NH_2)_3]^{3+}$; (d) $[Co(NCS)_6]^{3-}$; (e) $[Co(SCN)_6]^{3-}$; and (f) $[CoF_6]^{3-}$.

18. *Assuming that the pairing energies in Table 8.2 are 22% too high to apply properly to complex ions, estimate the minimum ligand *f* factor required to produce a low-spin complex ion for each of the following metal ions; then identify the ligand that is closest to this value: (a) Co^{3+}; (b) Co^{2+}; (c) Fe^{2+}; (d) Mn^{2+}; (e) Mn^{3+}; and (f) Cr^{2+}.

19. The g factor for the Cr^{3+} complex ion in Exercise 13 is 17,400 cm^{-1}. Calculate the *f* factor for the ligand in this complex ion. Does this ligand fall above or below H_2O in the spectrochemical series?

20. *Assuming that the *g* factor for the metal ion in Exercise 14 is 8500 cm^{-1}, calculate the *f* factors for the three unidentified ligands in complexes **A**, **B**, and **C** (Fig. 8.22), and give plausible identities of those ligands.

21. The three different classes of ligands have different consequences of overlap with the t_{2g} set of orbitals in octahedral complexes. Answer the following about each class of ligands by filling in one of the choices in each of the blanks: lower in energy, unchanged in energy; higher in energy; strong or large; moderate, weak or small; Cl^-, NH_3, or CO.

(a) Pi-donor ligands make the t_{2g} electrons ____ and result in a ____ ligand field splitting Δ_o; an example of a π-donor ligand is ____; (b) Pi-acceptor ligands make the t_{2g} orbitals ____ and result in a ____ ligand field splitting Δ_o; an example of a π-acceptor ligand is ____; (c) Sigma-donor ligands make the t_{2g} electrons ____ and result in a ____ ligand field splitting Δ_o; an example of a σ-donor ligand is ____.

22. If you attempt Experiment A.7 using some beautifully colored *f*-block ion such as green Pr^{3+}(aq), you will be disappointed in the outcome, because all of your products will be the same color. Why?

23. *Based on their Lewis structures, predict whether each of the following ligands would be in the lower part, the middle, or the upper part of the spectrochemical series, and explain why: (a) CCl_2 (dichlorocarbene, with a C-donor atom); (b) NH_2^- (amide ion); and (c) the hypothetical O donor ligand

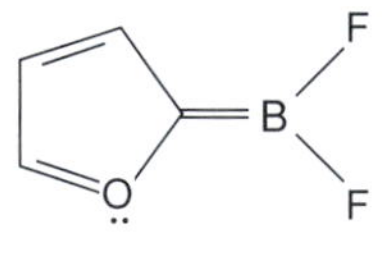

24. In Table 8.5, the *f* factor for the ligand CN^- is given only an approximate value, and no *f* factors are listed at all for the other very strong-field ligands. Why can chemists not determine reliable *f* factors for these ligands?

25. Another strong-field ligand not listed in Table 8.5 is the hydride ion, H^-; its position in the spectrochemical series is similar to that of $CH_3C(CH_2CH_2O)_3P$. Is H^- a strong-field ligand for the same reason that the phosphine ligand is? If not, how can you rationalize its strong ligand field?

26. *Consider the ambidentate ligands fulminate, $C{\equiv}N{-}O^-$, and cyanate, $N{\equiv}C{-}O^-$, each of which can bond to a metal ion either via the oxygen donor atom or via the donor atom at the other end. List these four possible ligands in the order you would expect to find them in the spectrochemical series, starting at the weak-field end. Underline the donor atom you are using in each case. Which two of these cases should have approximately equivalent positions in the spectrochemical series?

27. *Compute the actual crystal field stabilization energies (in kilojoules per mole) for the fourth-period +2 charged hydrated ions, and express each CFSE as a percentage of the total hydration energy for that ion.

28. Describe the order of stabilities of metal complexes of a given ligand as the metal is varied in the second half of the *d* block in the fourth period among ions (a) of +2 charge in strong-field complexes and (b) of +3 charge in weak-field complexes.

29. Explain the overall trends and the reversals in trends that occur in the following series of oxide lattice energies (in kilojoules per mole), all of which have the NaCl lattice structure: CaO(−3461), TiO(−3879), VO(−3912), MnO(−3808), FeO(−3921), CoO(−3988), and NiO(−4071). [Data from D. F. Shriver, P. W. Atkins, and C. H. Langford, *Inorganic Chemistry*, Freeman, New York, 1990; p. 226.]

30. Sketch plots of lattice energies versus d^n electron configuration for octahedrally coordinated fourth-period +2 charged ions in the following lattices: (For simplicity, assume all compounds use the same lattice type.) (a) Lattice with a weak-field anion [e.g., $MCl_2(s)$]; and (b) Lattice with a strong-field anion [e.g., $M(CN)_2(s)$].

31. Sketch plots of how you expect ΔH to vary as a function of the d^n electron configuration for fourth-period +2 charged ions in the following equilibria:

(a) $[M(H_2O)_6]^{2+} + 6\,CH_3NH_2 \rightleftharpoons [M(CH_3NH_2)_6]^{2+} + 6\,H_2O$

(b) $[M(H_2O)_6]^{2+} + 6\,CN^- \rightleftharpoons [M(CN)_6]^{4-} + 6\,H_2O$

(c) Explain why the equilibrium constants for these reactions would be smaller, on the average, on the *left* side of the *d* block than on the right side.

32. Sketch a plot of radii of fourth-period *d*-block metal ions in octahedral complexes, showing one plot for weak-field and one for strong-field ligands, showing clearly where the "humps" occur. Explain why there are "humps" in these plots and why they occur at the specific d^n electron configurations at which they do occur.

33. Plot radii from Table C for a series of like-charged ions (your choice of charges, but there should be several examples) in the fifth (or sixth) period as a function of the d^n electron configuration of the ion. Tell whether this plot resembles one expected for weak- or for strong-field complexes; indicate where you expect the maxima and the minima in this graph to fall.

34. Plot the trend you would expect for the Pauling electronegativities of the fourth-period elements when they have the +2 oxidation state and are in strong-field (low-spin) complexes. For which elements would new electronegativities result? Would these be higher or lower than those given in Table A?

35. Recall the energy level diagram for an octahedral complex ion $[ML_6]^{n+}$, and recall the sets of *d* orbitals that are equal in energy in such a complex ion. Now imagine that one ligand is removed from along the *z* axis, leaving a square pyramidal complex $[ML_5]^{n+}$. Do any of the *d* orbitals that were of equal energy to each other in $[ML_6]^{n+}$ remain equal in energy to each other in $[ML_5]^{n+}$? If so, which orbitals?

36. *Without reference to Table 8.6, reason whether each of the five d orbitals will be *stabilized* or *repelled* or *at about the same energy* (as compared to the hypothetical spherical reference case) if a metal ion containing one *d* electron is placed in a two-coordinate linear complex ion. (Note that this complex, by convention, is considered to lie along the *z* axis.)

37. A real example of a linear two-coordinate complex ion is NiO_2^{3-} [A. Möller, M. A. Hitchman, E. Krausz, and R. Hoppe, *Inorg. Chem.*, **34**, 2684 (1995)]. (a) Look up the spectrum of this ion, and determine whether the authors agree with the order of *d* orbitals deduced in Exercise 36. How do you explain the number of bands observed? (b) Estimate the value of Δ_o for this ion. (c) Compare this value with those found for hydrated ions in Table 8.4. Comment on the reasonableness of this value.

38. Let us consider the possibility of Cr^{3+} forming a tetrahedral complex with the ligand glycine, $[Cr(NH_2CH_2CO_2)_2]^+$. (a) From the appropriate *f* and *g* factors, compute the expected tetrahedral ligand field splitting Δ_t in this complex ion. (b) Compute the expected tetrahedral crystal field stabilization energy in reciprocal centimeters (cm^{-1}). (c) Compare this result with the octahedral crystal field stabilization energy computed in Example 8.3 to determine which geometry of complex is favored, octahedral or tetrahedral. By how many reciprocal centimeters is it preferred?

39. What factors tend to favor the loss of two ligands by an octahedral complex to form a tetrahedral complex? What factors tend to oppose this?

40. The *d* orbitals in a particular geometry of complex are found to be split by the crystal field of that geometry of ligands into the following energy levels: $d_{xz} = d_{yz} = -0.457\,\Delta_o$; $d_{xy} = -0.086\,\Delta_o$; $d_{z^2} = +0.086\,\Delta_o$; $d_{x^2-y^2} = +0.914\,\Delta_o$. (a) Which one of the following geometries could produce this particular splitting: octahedral, tetrahedral, square pyramidal (pyramid base in *xy* plane), cube, linear (ligands along *z* axis). (b) Assuming that your ligands were strong-field ligands, what metal ion electron configuration (d^3, etc.) would be most likely to favor this geometry? Draw and fill in the energy level diagram with arrows representing electrons to support your answer. (c) Give the symbol of a common transition metal ion that actually shows this electron configuration. (d) Calculate the CFSE for a Fe^{3+} ion in this geometry, with strong-field ligands being present.

41. *In Experiment A.7, the color changes developed (and the range of ν_1 absorptions) in the spectrochemical series of ligands are more spectacular for the complexes of nickel(II) ion than for any other metal ion. (a) The chloro complex of Ni^{2+} is yellow, with $\Delta = 3850\,cm^{-1}$. Compute the Δ_o expected for octahedral $[NiCl_6]^{4-}$, and give a plausible interpretation of the discrepancy. (b) The cyano complex $[Ni(CN)_4]^{2-}$ is red, with ν_1 at $31{,}100\,cm^{-1}$. Compute the Δ_t expected for tetrahedral $[Ni(CN)_4]^{2-}$, and give a plausible interpretation of the discrepancy. (c) The four-coordinate dimethylglyoxime complex of Ni^{2+} is bright pink and is used in lipstick and in the qualitative detection of Ni^{2+}; it has four nitrogen donor atoms attached to Ni and has no absorption in the near-IR region. What is its geometry? (d) Which of these geometries of nickel(II) complexes could you identify on the basis of magnetic susceptibility measurements alone?

42. When we were studying the formation of oxo anions in Chapter 3, we emphasized the role of *sp*-hybrid orbitals in occupying space that would otherwise be used to bind ligands, but deemphasized the roles of *d* electrons, stating that these tend to fall *between* the ligands. (a) For which of the following types of *d* electrons is this presumption basically true? basically false? The t_{2g} electrons in octahedral complexes; e_g electrons in octahedral complexes; t_2 electrons in tetrahedral complexes; e electrons in tetrahedral complexes; d_{z^2} electrons in square-planar d^8 complexes; $d_{x^2-y^2}$ electrons in square-planar d^8 complexes. (b) What geometry does valence shell electron-pair repulsion (VSEPR) predict for four-coordinate d^8 complexes if the d electrons are presumed to be stereochemically inactive (i.e., not to block incoming ligands)? What geometry does VSEPR predict if allowance *is* made for the type of d electrons that *do* block electrons? Is either prediction correct?

43. Would either the Cr^{3+} or the weak-field Cr^{2+} complex in Example 8.3 be subject to the Jahn–Teller (tetragonal) distortion? If so, which one? Would either the Cr^{3+} or the Cr^{2+} complex likely become square planar? If so, which one?

44. What strong-field d^n octahedral complexes are strongly subject to the Jahn–Teller distortion?

45. *(a) Which of the following octahedral complex metal ion electron configurations are expected to undergo a strong Jahn–Teller distortion: ${t_{2g}}^1$, ${t_{2g}}^3$, ${t_{2g}}^3{e_g}^1$, ${t_{2g}}^3{e_g}^2$, ${t_{2g}}^6$, ${t_{2g}}^6{e_g}^1$, ${t_{2g}}^6{e_g}^3$, and ${t_{2g}}^6{e_g}^4$. (b) Briefly describe how the shape of an octahedral complex changes when a Jahn–Teller distortion occurs.

46. Which of the following octahedral-complex metal ion electron configurations are expected to undergo a strong Jahn–Teller distortion? (Some of these are excited-state configurations that could only be found after absorption of light or other energy, but ignore this factor and answer for these as well.) (a) ${t_{2g}}^2$; (b) ${t_{2g}}^4$; (c) ${t_{2g}}^6$; (d) ${t_{2g}}^4{e_g}^2$; (e) ${e_g}^1$; (f) ${e_g}^2$; (g) ${t_{2g}}^3{e_g}^3$; and (h) ${e_g}^3$.

47. The formation of an octahedral complex of the bidentate chelating ligand ethylenediamine (en, $NH_2CH_2CH_2NH_2$) proceeds in three steps; for each step an equilibrium constant (the *stepwise formation constant*) can be measured. When en coordinates to Ni^{2+}, the three stepwise formation constants are each about 10^{10}. When en coordinates to Cu^{2+}, the first two steps also proceed with similar stepwise formation constants, but great difficulty is experienced in attempting to chelate a third en ligand (the third stepwise formation constant is only 0.1). What difficulty might there be in the third chelation of Cu^{2+} that is not present for Ni^{2+}?

48. Predict the two crystal lattice types likely to be found for solid fluorides of the fourth-period +2 charged *d*-block metal ions. Which two of these fluorides would be expected to show substantial differences from either possible predicted lattice type; what would be the nature of these differences? Would you expect the Madelung constants for these two lattice types to apply accurately to these two fluorides?

49. *Platinum has two common positive ions, Pt^{2+} and Pt^{4+}. One of these strongly prefers to form octahedral complexes, and the other strongly prefers to form square planar complexes. Which ion prefers octahedral? Which ion prefers square planar? Briefly explain your choices.

50. Six-coordinate complexes are normally octahedral, but there is an alternate geometry, *trigonal prismatic*. From Table 8.6, the energy levels of the d orbitals in a trigonal prismatic complex are $d_{z^2} + 0.096\Delta_o$; $d_{x^2-y^2} - 0.584\Delta_o$; $d_{xy} - 0.584\Delta_o$; $d_{xz} + 0.536\Delta_o$; and $d_{yz} + 0.536\Delta_o$. (a) Draw an energy level diagram comparing (on the left side) octahedral-complex energy levels with (on the right side) energy levels for a trigonal prismatic complex. (b) For a d^2 metal ion, the crystal field stabilization energy is of greater magnitude for which geometry? By how much energy (in units of Δ_o)? (c) For a d^3 metal ion, the crystal field stabilization energy is of greater magnitude for which geometry? By how much energy (in units of Δ_o)? (d) Which geometry is favored by the VSEPR theory?

51. Although the low position of the heavier halide ions (Cl^-, Br^-, I^-) in the spectrochemical series (and their fairly large size) allows most halo ions of the fourth-period *d*-block metal ions to be tetrahedral, this tendency is not pronounced for Ni^{2+} and is completely absent for Cr^{3+}, which only forms octahedral halo anions. Explain. Also explain why the tendency is more completely absent for Cr^{3+}.

52. If you were to react the Cr^{3+} complex of Example 8.3 with another ligand to displace one of the glycinate ions, how fast would you expect the reaction to proceed at room temperature: (Choose the most plausible answer.) in nanoseconds (10^{-9} s); in 10^{-8} s; in 10^{-4} s; in minutes or even years. The [Cr(glycinate)$_3$] complex is best described as which: inert, or labile.

53. The substitution reactions of octahedral d^3 and d^8 metal complex ions are very slow. (a) Why are the reactions of $[NiL_6]^{2+}$ ions not nearly so slow as the reactions of $[CrL_6]^{3+}$ ions, which may require days or weeks at room temperature? (b) The reactions of octahedral $[CrL_6]^{3+}$ can be speeded up enormously by adding some reducing agent (such as Zn) that will reduce Cr^{3+} to Cr^{2+}, then letting the Cr^{2+} complex be reoxidized to Cr^{3+}. Explain why this works.

54. *Using Figure 8.15, explain what would have to be done to the ligands in an eight-coordinate *cubic* complex to convert their positions to those found in an eight-coordinate *square antiprismatic* complex. Would this raise or lower the *ligand–ligand* electronic repulsions described by VSEPR theory? Would this raise or lower the *ligand–d electron* repulsions described by crystal field theory for the following *d* orbitals: d_{xy}; $d_{x^2-y^2}$; and d_{z^2}?

55. Why is it possible to separate and isolate the isomers of square planar *cis*-$[Pt(NH_3)_2Cl_2]$ and *trans*-$[Pt(NH_3)_2Cl_2]$? Why is it likely to be impossible to separate and isolate the isomers of square planar *cis*-$[Cu(NH_3)_2Cl_2]$ and *trans*-$[Cu(NH_3)_2Cl_2]$?

56. *In the synthesis of which of the following complexes would it be *unsatisfactory* to employ a plain substitution reaction on a hydrated metal ion in aqueous solution? (a) $[Co(NH_3)_6]Cl_3$; (b) $[Fe(C_5H_5N)_6]Cl_3$; (c) $[Ni(C_5H_5N)_6]Cl_2$; and (d) $[Tc(NH_3)_6](PF_6)_4$.

57. In the synthesis of which of the following complexes would there be a *significant* advantage in adding some reducing agent, or in using a starting metal compound in which the metal ion had an oxidation state one lower than in the product? (a) $[Fe(NH_2CH_2CH_2NH_2)_3](NO_3)_3$; (b) $[Ir(NH_2CH_2CH_2NH_2)_3](NO_3)_3$; (c) $K_3[Co(C_2O_4)_3]$ (contains low-spin cobalt); and (d) $[Cr(NH_3)_6]Cl_2$.

58. Which *three* of the following complexes would be the most suitable starting materials for substitution reactions based on their containing *weakly bound* or *volatile* ligands? (a) $[Fe(CN)_6]^{3-}$; (b) $[Cr(H_2O)_6]Cl_3$; (c) $[Cr(CO)_5Xe]$; (d) $[Pt(PF_6)_4]^{2-}$; (e) $[Zn\{(C_2H_5)_2O\}_4]^{2+}$; and (f) $[Ir(SCH_3)_6]^{3-}$.

59. *Some syntheses carried out in an inorganic laboratory course are listed below. Classify each syntheses as an example of one of the five synthetic methods enumerated in Section 8.9.

(a) $[Cr(H_2O)_6]^{3+} + 3\,C_5H_7O_2^-$ (in hot water) $\rightarrow [Cr(C_5H_7O_2)_3] + 6\,H_2O(\ell)$

(b) $6\,KSCN$(molten in flame) + $[Cr(H_2O)_6]^{3+} \rightarrow [Cr(SCN)_6]^{3-} + 6\,H_2O(g)$

(c) $Cr_2O_7{}^{2-} + 14H^+ + 9\,C_2O_4{}^{2-}$ (in hot water) $\rightarrow$
$2[Cr(C_2O_4)_3]^{3-} + 6\,CO_2(g) + 7\,H_2O(\ell)$

(d) $[Co(H_2O)_6]^{2+} + 2\,C_4H_8N_2O_2 + 0.5\,O_2 + Cl^-$ (in hot alcohol) $\rightarrow$
$[Co(C_4H_7N_2O_2)_2Cl] + 7\,H_2O + H^+$

60. The commercial drying agent Drierite is $CaSO_4$, and is regenerated by heating in an oven at 200–250 °C for 2 h. You have some indicating Drierite, which also contains $CoCl_2$; this has been in contact with $NH_3(g)$, and has been converted to $[Co(NH_3)_6]Cl_3$. Outline a plan to convert this contaminated material to Drierite.

61. Explain how it could be that the act of coordinating an O_2 molecule to the Fe^{2+} ion in hemoglobin could cause the Fe^{2+} ion to move a fair distance from its original position out of the plane of the porphyrin ligand to a new position in the plane of the porphyrin ligand. With what unusual aspect of the O_2 binding ability of hemoglobin is this connected?

62. Give some reasons why biochemists, in their research, might want to replace the Zn^{2+} ion in carboxypeptidase A with another *d*-block metal ion, and why they would end up replacing the Zn^{2+} ion with Co^{2+} rather than Ni^{2+} or Cu^{2+}.

63. (a) Draw Lewis dot structures of the following complex ions: $[Au(CH_3)_2(NH_3)_2]^+$; $[Na(CH_3{-}O{-}CH_2{-}CH_2{-}O{-}CH_3)_3]^+$; $[SnI_6]^{2-}$; $[HgCl_2]$; and $[Hg(CH_3)_2]$. Treat all the bonds to the metal as coordinate covalent bonds. (b) Write the formulas (including charges, if any) of the ligands found in each complex ion. (c) Write the formulas of the Lewis acids (including charges, if any) found in these complex ions.

64. *Below are drawn Lewis structures of two coordination compounds. Treat all bonds to the central metal atom in each as coordinate covalent bonds. For each complex, (a) write the formulas (including charges) of each ligand and (b) write the formula (including charge) of the central metal ion.

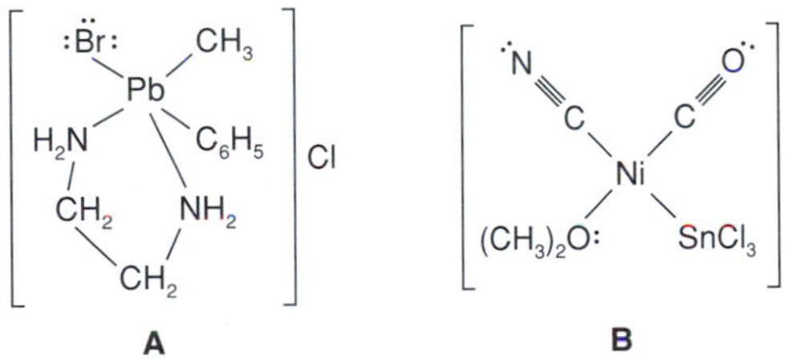

A

B

65. For the complexes $(NH_4)_2[Pd(SCN)_4]$ and $[Co(NH_3)_5(SeO_4)]ClO_4$: (a) Write the electron-dot structure for each kind of ligand, showing unshared electron pairs and underlining its donor atom(s). (b) Give the charge on each free ligand. (c) Give the oxidation number of the central (metal) atom.

66. Write the formulas of the following coordination compounds: (a) sodium tris(ethanedioato)chromate(III); (b) dicyanobis(1,2-ethanediamine)platinum(IV) perchlorate; (c) tetramminedicyanoplatinum(IV) sulfate; (d) dibromobis(1,2-ethanediamine)-platinum(2+) chloride; (e) barium diamminetriaqua(carbonato)dioxouranate(1−).

67. *Write the formulas of the coordination compounds of Exercise 66 in the form in which you would look them up in the formula index of *Chemical Abstracts*.

68. Write the formulas of the following: (a) diaquabis(1,2-ethanediol)iron(III) sulfate and (b) barium bis(bromomethyl)(carbonato)dicarbonylchromate(1−).

69. Rewrite the names in Exercise 68 in the way that they would be found in *Chemical Abstracts*.

70. Below is drawn a neutral coordination compound of the metal niobium, Nb. Assume each bond to Nb to be a coordinate covalent bond. (a) What charge is present on each of the ligands before they coordinate to the Nb? (b) What charge is present on the Nb atom before it coordinates all of these ligands? (c) Give the name of each *ligand* as it will be named in the nomenclature for coordination compounds.

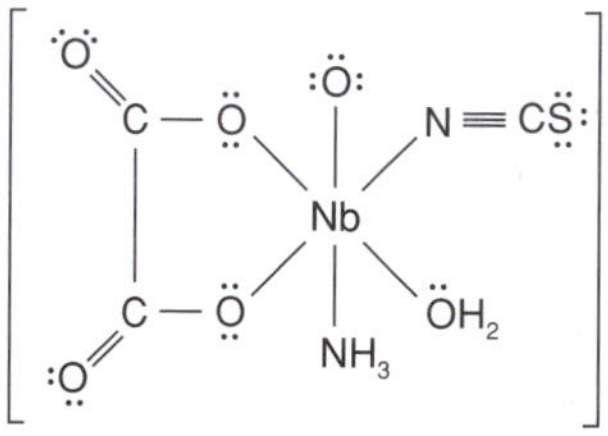

71. Give the name of each complex given in Exercises 64 and 65.

72. Consider the following two coordination compounds: (**I**) $K[PtCl_3(CH_2{=}CH_2)]$ and (**II**) $[Pt(CH_3)_3(NH_3)_3]Cl$. (a) Write the formula (including charges, if any) of one of each type of ligand found in each complex. (b) Write the formula (including charges, if any) of the platinum Lewis acid found in each complex. (c) Name the above two coordination compounds, using either the Stock convention or the Ewens–Bassett convention.

73. Name each of the following complex compounds: (a) $Ca_3[VCl_6]_2$; (b) $[Co(H_2O)_4(OH)_2][AlCl_4]$; (c) $Cs[V(CO)_3(P(C_6H_5)_3)_3]$; (d) $K_2[ReH_9]$; (e) $[Co(CO)_5][BF_4]$; (f) $K[Te(CH_3)_2Cl_3]$; (g) $[(CH_3)_4P][BrF_4]$; and (h) $[Ag(C_6H_6)_2][Nb(CO)_6]$.

74. *Name the following coordination compounds, using either the Stock convention or the Ewens–Bassett convention. (a) $[PtCl_2(NH_3)_2]$ (b) $K_2[Fe(CN)_5(CO)]$ (c) $[(C_6H_5)_4As]_2[Fe(CO)_4]$ (d) $[Fe(C_6H_6)\{(C_6H_5)_3P\}_2]$ (e) $[(C_2H_5)_4N]_3[CrCl_6]$; (f) $[PtIBrClF(H_2O)(NH_3)]$; and (g) $[Cr(C_6H_6)_2]I$.

75. Write the names of the coordination compounds in Exercise 74 in the way you would look them up in the chemical substance index of *chemical abstracts*.

76. Look up one of the following complex compounds in the latest three cumulated indexes of *Chemical Abstracts*. (Note that *Chemical Abstracts* uses Ewens–Bassett nomenclature.) (1) Make a photocopy of the 1 or 2 pages of each chemical substances index that contain all the references to the complex ion and all of its salts (or the neutral complex). Mark where these citations begin and where they end. (2) Circle the part of these pages that contains the citations of the specific complexes listed below (or their hydrates). (3) Find an abstract for an article from a journal in your school's library that covers the synthesis of your compound. (4) Look up the same complex ion in *Inorganic Syntheses*, in the indexes at the ends of volumes 10, 15, 20, 25, . . . , and the latest issue. Is a synthesis of your complex ion present? Complex compounds: (a) $[Cr(CH_3COCHCOCH_3)_3]$; (b) $K_3[Cr(SCN)_6]$; (c) $K_3[Cr(C_2O_4)_3]$; (d) $[Cr(NH_2CH_2CO_2)_3]$; (e) $[Cr(NH_2CH_2CH_2NH_2)_3]Cl_3$; and (f) $Na_3[CrCl_6]$.

77. *Iron forms a number of well-known complexes with the CN^- ion. The complexes $K_4[Fe(CN)_6]$, $K_3[Fe(CN)_6]$, $Fe[Fe(CN)_6]$, and $Fe_2[Fe(CN)_6]$ are not as intensely colored as $KFe[Fe(CN)_6]$, commonly known as Prussian blue. (a) Insofar as possible, name each of these five complexes. (b) In each complex in which an iron atom is written outside of the brackets, it is also acting as a Lewis acid. What is the Lewis base, and what donor atom(s) is it using? What is the classification of this kind of ligand? (c) Explain why Prussian blue is more intensely colored than the other complex compounds.

Notes

1. H. Bethe, *Ann. Phys.* (*Leipzig*), [5], **3**, 135 (1929); J. H. Van Vleck, *Phys. Rev.*, **41**, 208 (1932).
2. The tabulated pairing energy includes the sacrifice that must be made of the ability of the electron to undergo quantum mechanical exchange with other electrons of the same spin; this *exchange energy* is part of the basis of Hund's rule and contributes to the stability of half-filled electron shells such as d^5.
3. Often the P term in the crystal field stabilization energy is expressed as the number of units of pairing energy *in excess of those found in the hypothetical spherical ion* (e.g., $1P$ for d^6, $2P$ for d^7), as shown at the left side of each energy level diagram in Figure 8.4.
4. This measurement is made in a *Guoy balance*, with which a sample is weighed in the absence of and in the presence of a magnetic field; the magnetic field pulls on the suspended sample and adds to its measured weight. In the newer and more compact *Johnson–Matthey magnetic balance*, the mobile magnets are pulled around the immobile sample.
5. For most substances, it is necessary to add a constant to the absolute temperature in Eq. (8.2). This constant, θ, is known as the Weiss constant; the substance is then said to obey the Curie–Weiss law.
6. Other forms of magnetic properties (ferromagnetism and antiferromagnetism) arise when paramagnetic atoms or ions are in close proximity to each other, in bulk metal and salts (such as oxides), respectively; these properties will be covered in Chapter 12.
7. K. R. Kunze, D. L. Perry, and L. J. Wilson, *Inorg. Chem.*, **16**, 594 (1977).
8. J. Zarembowitch and O. Kahn, *New J. Chem.*, **15**, 181 (1991).
9. The energy term may be more complicated for low-spin complexes or nonoctahedral complexes. Even if these complications do not occur, a better value of Δ_o can be obtained by also taking into account the frequencies of the other absorption bands, as discussed in Chapter 17.
10. Many laboratories lack spectrometers capable of covering the near-IR region; fortunately Δ_o can also be determined from the energies of the next two absorptions, as indicated in Chapter 17.
11. However, charge-transfer adsorptions involving the reducing soft base, monothiocarbamate, and the oxidizing cation, Fe(III), are certainly possible.
12. P. Kofod, *Inorg. Chem.*, **34**, 2768 (1995).
13. Provided that the metal has t_{2g} electrons; these ligands often do not form very stable complexes at all with metal ions lacking such electrons.
14. R. J. Kulaweic and R. H. Crabtree, *Coord. Chem. Rev.*, **99**, 89 (1990).
15. M. Gerloch, *Coord. Chem. Rev.*, **99**, 117 (1990).
16. The alert reader may have noticed that the absorption spectrum of hydrated Cu^{2+} in Figure 8.5 has a shoulder. As suggested by Figure 8.14(*b*), this second absorption arises from the Jahn–Teller splitting of the t_{2g} set of orbitals. Splitting of spectral absorptions is also observed in those ions that have $e_g{}^1$ and $e_g{}^3$ electron configurations after absorption of light (e.g., Ti^{3+}).
17. For reasons that will be seen in Chapter 9, it is now necessary to remove the g from the symmetry labels for the sets of orbitals.
18. D. R. Bloomquist and R. D. Willett, *Coord. Chem. Rev.*, **47**, 125 (1982).
19. S. A. Latt and B. L. Vallee, *Biochemistry*, **10**, 4263 (1971); E.-I. Ochiai, *Bioinorganic Chemistry: An Introduction*, Allyn and Bacon, Boston, 1977, p. 374.
20. Given that hydrated ions have secondary spheres of hydration, more complex mechanisms can be imagined and are in fact found in which the incoming ligand might enter the secondary sphere

before the water leaves the primary sphere, and so on; these will be considered in more detail in Chapter 16.

21. A. Cusanelli, U. Frey, D. T. Richens, and A. E. Merbach, *J. Am. Chem. Soc.*, **118**, 5265 (1996).
22. A. Bleuzen, F. Foglia, E. Furet, L. Helm, A. E. Merbach, and J. Weber, *J. Am. Chem. Soc.*, **118**, 12777 (1996).
23. F. Basolo and R. C. Johnson, *Coordination Chemistry: the Chemistry of Metal Complexes*, 2nd ed., Benjamin, New York, 1964, Chapter 4.
24. W. Beck and K. Sünkel, *Chem. Rev.*, **88**, 1405 (1988); W. Beck, *Inorg. Syn.*, **28**, Chapter 3 (1990).
25. F. Sladky, *Inorg. Syn.*, **24**, 33 (1986).
26. J. J. R. Fraústo da Silva and R. J. P. Williams, *The Biological Chemistry of the Elements: The Inorganic Chemistry of Life*, Clarendon Press, Oxford, 1991, Chapter 13.
27. This reaction is actually more complex than shown, and involves more than the one step indicated.
28. Perhaps as Fe^{3+} (Section 17.8).
29. S. H. Snyder and D. S. Bredt, *Sci. Am.*, **266**(5), 68 (1992); A. L. Burnett, C. J. Lowenstein, D. S. Bredt, T. S. K. Chang, and S. H. Snyder, *Science*, **257**, 401 (1992); J. R. Lancaster, Jr., *Am. Sci.*, **80**, 248 (1992); P. L. Feldman, O. W. C. Griffith, and D. J. Stuehr, *Chem. Eng. News*, Dec. 20, 1993, p. 26; C. J. Lowenstein, J. L. Dinerman, and S. H. Snyder, *Ann. Int. Med.*, **120**, 227 (1994); E. W. Ainscough and A. W. Brodie, *J. Chem. Educ.*, **72**, 686 (1995).
30. A. Verma, D. J. Hirsch, C. E. Glatt, G. V. Ronnett, and S. H. Snyder, *Science*, **259**, 381 (1993).
31. If the oxidation number or the charge is zero, the parentheses and zero are omitted.
32. International Union of Pure and Applied Chemistry, *Nomenclature of Inorganic Chemistry: Recommendations 1990*, G. J. Leigh, Ed., Blackwell, Oxford, UK, 1990; B. P. Block, W. H. Powell, and W. C. Fernelius, *Inorganic Chemical Nomenclature: Principles and Practice*, American Chemical Society, Washington DC, 1990.
33. A. Kozlowski, *Searching Coordination Compounds in CAS ONLINE*, Chemical Abstracts Service, Columbus, OH, 1986.
34. As you carry a search back in time, however, you cross the years in which the nomenclature for complex compounds changes, so that if your searches in the chemical substances index suddenly go away, it is advisable to go back to the formula index to see what the old nomenclature was.
35. A specific entry under the counterion, potassium salt, might then follow the general headings.

PART

II

INORGANIC MOLECULES AND MATERIALS: THEORY AND APPLICATIONS

We have now completed our survey of the basic reaction types and reactivity trends of inorganic ions (including complex ions) and some simple inorganic molecules. In Part II, we wish (1) to look in more depth at molecular substances such as main group and transition metal organometallic compounds, and (2) to study the reaction types and reactivity trends of materials with more extended (polymeric) structures: the elements themselves and their ionic and polymeric covalent solid derivatives, including those having importance to the growing area of materials science. Many of these materials are important not only for their chemical reactivity (or lack of it), but also for their physical properties; so we will also look for periodic trends in important physical properties such as structural strength, bulk magnetic properties, electrical conductivity, ans so on. To study these, it is often necessary to use more advanced structural and bonding concepts than we have used in Part I, so we begin with a study of symmetry and the molecular orbital bonding theory.

CHAPTER

9

Symmetry

With Applications to Art, Environmental Chemistry, and Organic Chemistry

9.1 Symmetry Operations and Elements

We begin our study of more advanced bonding concepts with a consideration of a topic that would seem to belong to the realm of aesthetics. One of the properties that goes into making a work of art or nature (Fig. 9.1),[1] or a person, beautiful, is the presence of elements of **symmetry**. Molecules, too, may have greater or lesser degrees of symmetry; one of the things that surely has contributed to the interest in the discovery of the new form of the element carbon, buckminsterfullerene [C_{60}, Fig. 9.1(*h*)], is the fact that it is more symmetrical than almost any other nonlinear polyatomic molecule that has ever been discovered. Although buckminsterfullerene has so beautiful a structure that we feel compelled to humble it by calling it "buckyballs", such beauty (rooted in symmetry) has some real chemical consequences that we shall see in this chapter and in chapter 10: for example, its infrared (IR) spectrum is exceptionally simple; the shapes of its *molecular orbitals* are exceptionally predictable; its capability or inability to undergo certain types of chemical reactions will be more sharply defined.

Symmetry is defined in Webster's dictionary as "similarity of form or arrangement on either side of a dividing line or plane," with "correspondence of opposite parts in size, shape, and position." Operationally, we can state that if a molecule or ion has two or more orientations that are indistinguishable, that molecule or ion possesses symmetry.

As it turns out, there are only a few ways in which we can reorient molecules to see whether or not all parts coincide after the operation. These reorientations take place about the five kinds of points, lines (axes), or planes listed in Table 9.1, which are the possible **symmetry elements** of the molecule or ion being tested (other types of symmetry elements and operations are possible for extended solids). The operations that we attempt are known as **symmetry operations.**

The effects of these operations are best appreciated by working with physical models of the molecules being discussed. Some of the operations are difficult to envision with two-dimensional representations of molecules on paper. You will want to

Figure 9.1
Symmetry in Art and Nature. (*a*) The Eiffel Tower. Photograph by Michael Howell. Reproduced with permission. (*b*) Photograph of "Star of Bethlehem," *Campanula isophylla 'Mayi,'* by Tovah Martin. Reproduced with permission. (*c*) Three-blade propeller. (*d*) Children's toy pinwheel. (*e, f*) Radiolarians. Adapted from K. Wolf and R. Wolff, *Symmetrie*, Boelan-Verlag, Münster/Köln, Germany, 1956, and E. Haeckel, *Kunstformen der Natur*, Verlag der Bibliographischen Instituts, Leipzig, 1904. (*g*) Japanese crest. Adapted with permission from D. Hornung, *Traditional Japanese Crest Designs*, Dover, New York, 1986, p. 5. (*h*) Buckminsterfullerene, C_{60}, Adapted with permission of Springer-Verlag New York from the cover of *Chemtracts: Inorganic Chemistry*.

Table 9.1
Elements and Operations of Molecular Symmetry

Symmetry Element	Symbol	Operation	Example
Identity	E	Do nothing: leave all parts in place	
n-Fold proper axis	C_n	Rotation by $360°/n$ about this axis	Figure 9.2(a)
Mirror plane	σ	Reflection through this plane	Figure 9.2(b)
Inversion center	i	Inversion through this point	Figure 9.3
n-Fold improper axis	S_n	Rotation by $360°/n$ followed by reflection through a plane $\perp$ this axis	Figure 9.4

attempt to carry out or envision the following operations on molecular models while closing (or pretending to close) your eyes; if on opening your eyes you cannot tell from what you see whether you have carried out that operation, then the molecule or ion is symmetrical with respect to that operation. (Only the first two types of symmetry operations can actually be carried out with molecular models without detaching bonds, but it is easier to envision the others on three-dimensional models.)

Of these elements and operations, the most trivial seeming is that of the **identity** operation, which consists of leaving the molecule or ion alone while your eyes are closed; of course, all molecules remain unchanged upon carrying out this operation. Its listing is required by the mathematical properties of *group theory*[2]; the symbol for the identity element and operation is $\boldsymbol{E}$.

The nature of the other elements and operations are best illustrated with concrete examples; we begin with H_2O. The most prominent symmetry element in most molecules is that of the ***n*-fold proper axis** (symbol $\boldsymbol{C_n}$); the corresponding operation is one of rotation of the molecule by an angle of $360°/n$ about this axis. In water, this axis passes through the oxygen atom and midway between the hydrogen atoms. The two hydrogen atoms are interchanged by rotating 180° about this axis [Fig. 9.2(a); one hydrogen atom is starred so that you can see that the operation has indeed been carried out], which is a C_2 (*twofold* rotation) axis. One rotation of 180° about this axis generates a new but indistinguishable orientation of the molecule; carrying out the rotation operation twice, however, puts each H atom back to its original position, so it is the same as the identity operation. (We may express this by the identity equation $C_2{}^2 \equiv E$; in which the superscript ^{2} indicates that the operation has been carried out twice.) Molecules with more symmetry than H_2O may have more than one C_n axis, and may have axes in which n exceeds 2. The rotation axis with the highest value of n is designated the *principal rotation axis*; its direction is taken as the z (vertical) direction in the molecule for purposes of labeling in symmetry.

The second most prominent type of symmetry element is the **mirror plane(s)** (symbol σ) that may be present in a molecule or ion. As shown in Figure 9.2(b), H_2O has two mirror planes. Both of these include the vertical C_2 axis, so both are **vertical** mirror planes, and are designated by the symbol σ_v. The mirror plane including all three atoms of this planar molecule is finally labeled the $\sigma_v(xz)$ mirror plane; the plane bisecting the molecule is labeled the $\sigma_v'(yz)$ plane. There is one independent operation—reflection one time—that can be carried out about each of these planes; two reflections about either of them is identical to the identity operation. Although they are not present in the water molecule, there are two other types of mirror planes. **Hori-**

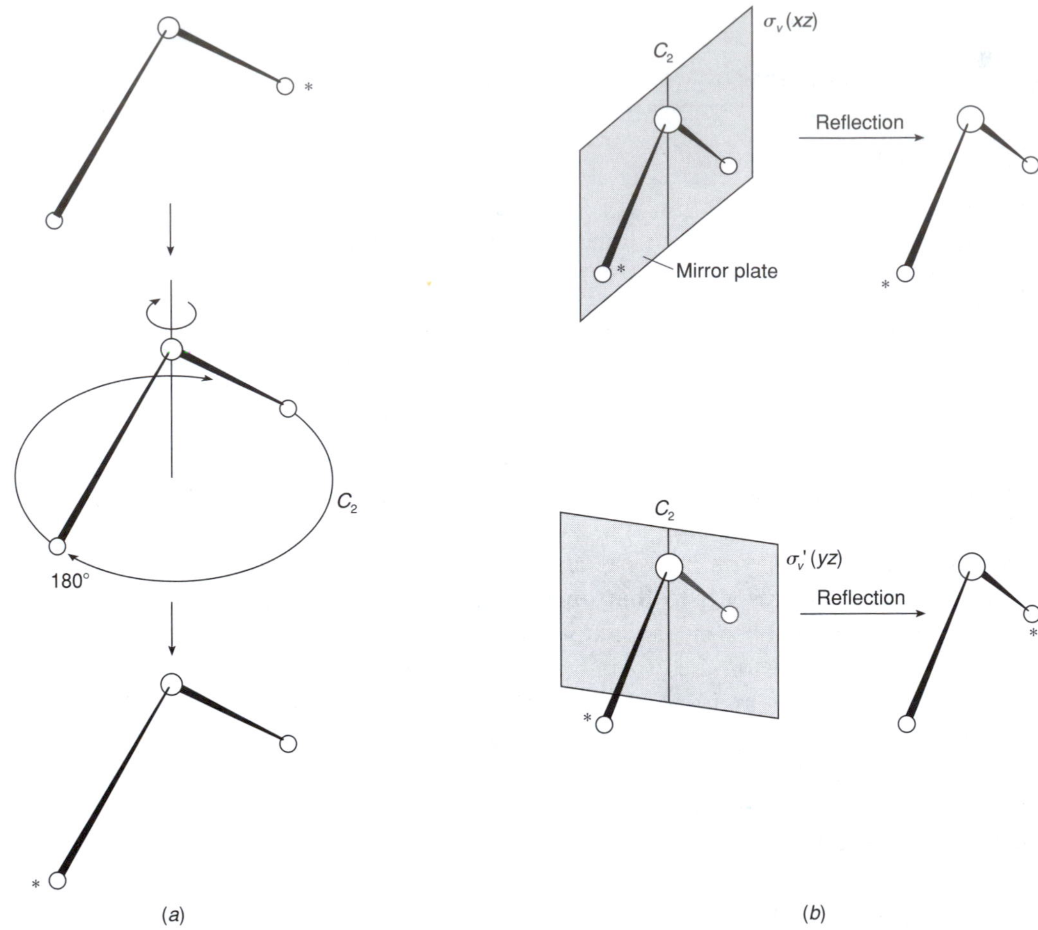

Figure 9.2
(*a*) The 180° rotation of a water molecule about its C_{2v} axis, an operation which leaves it apparently unchanged but actually interchanges the starred and unstarred hydrogen atoms; (*b*) The two vertical mirror planes in the water molecule, $\sigma_v(xz)$ and $\sigma_v'(yz)$, and the corresponding reflection operations. [Adapted from D. F. Shriver, P. W. Atkins, and C. H. Langford, *Inorganic Chemistry*, W. H. Freeman and Co., New York, 1990; pp. 50 and 51.]

zontal mirror planes, σ_h, are perpendicular to the (vertically oriented) principal rotation axis. **Dihedral** mirror planes, σ_d, are similar to vertical mirror planes in that they also include the principal rotation axis, but they are found to fall between adjacent bonds in the molecule, thus bisecting these bond angles.

Since there are no other elements of symmetry present in the water molecule, there are four (independent) symmetry operations possible: E, C_2, $\sigma_v(xz)$, and $\sigma_v(yz)$ (an operation carried out once has no superscript applied to it). These labels also identify the four symmetry elements of the water molecule.

To illustrate the other two types of symmetry elements, we select examples in which each type is especially prominent, beginning with the staggered conformation of *meso*-1,2-dibromo-1,2-dichloroethane, BrClHC–CHClBr, shown in Figure 9.3. The only symmetry operation and element (other than identity) present in this conformation is **inversion** about an **inversion center** (symbol ***i***). This operation consists of taking every atom with atomic coordinates (x, y, z) through the center of the molecule and out

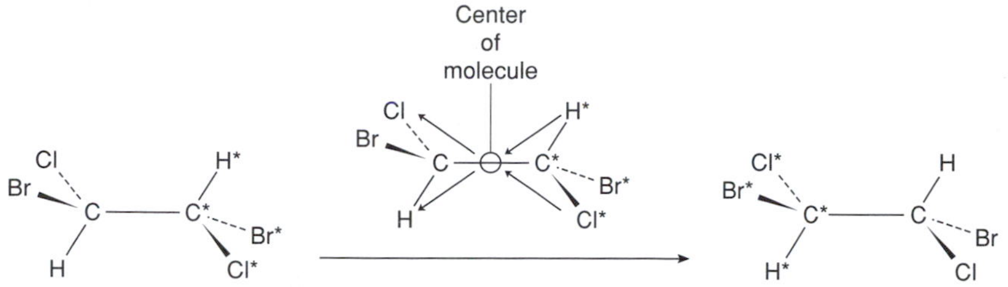

Figure 9.3
One conformation of *meso*-CHBrCl–CHBrCl, in which the inversion operation of the pairs of identical substituents (starred and unstarred) is partially illustrated by showing the motion of Cl* and H* through the center of the molecule and out in the same direction the same distance to the positions of Cl and H, respectively. Not shown are the simultaneous motions of Cl and H to the positions of Cl* and H*, and the simultaneous exchanges of Br with Br* and C with C*.

the other side an equal distance until the position $(-x, -y, -z)$ is reached. In this process, each Br passes through the center of the molecule (the midpoint of the C–C bond) and is exchanged with the other Br; the same occurs for the H, Cl, and C atoms.

The ***n*-fold improper rotation** (or **rotation–reflection**) axis is especially prominent in the organic molecule allene (propadiene) (Fig. 9.4). In this molecule, the two π bonds are cumulated (they both involve the same carbon atom), so one must be in the plane of the paper while the other is above and below the paper. This forces the CH_2 group at one end of the paper to be perpendicular to the plane of the paper while the other CH_2 group is in the plane of the paper. In this molecule, the long molecular axis is a fourfold improper rotation axis, S_4. An S_n improper rotation consists of rotation by $360°/n$ about the axis, followed by a reflection through a plane perpendicular to the rotation axis. In the case of allene, note that neither the rotation nor the reflection by itself is a symmetry element: The fourfold rotation takes hydrogen atom A 90° into the plane of the paper, where no H atom previously was located; the reflection then takes it across the central mirror plane to the position where H_D was previously located, meanwhile placing H_D where H_A had been; simultaneously, H_B and H_C are interchanged.

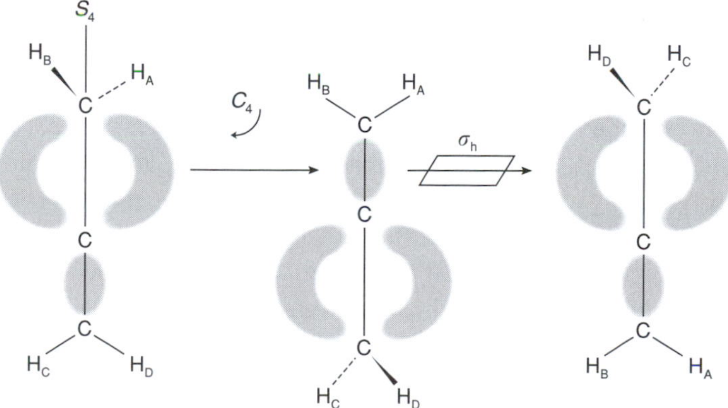

Figure 9.4
Improper rotation about the S_4 axis in allene, C_3H_4.

Example 9.1

Identify and count all symmetry elements and operations in the trigonal bipyramidal molecule PCl_5 [Fig. 9.5(*a*)].

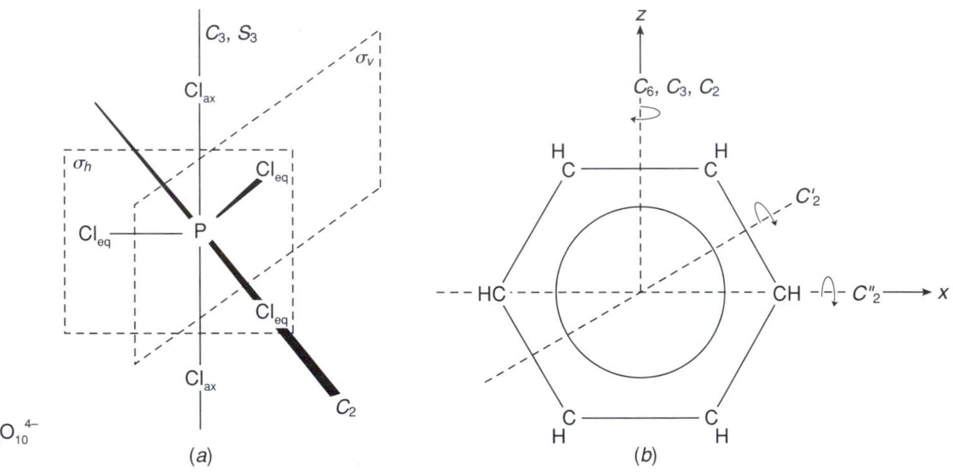

Figure 9.5
(*a*) Phosphorus pentachloride, showing its C_3 axis, its σ_h mirror plane, one of three σ_v mirror planes, and one of three C_2 axes. (*b*) Benzene, showing its C_6, C_3, and C_2 axes.

SOLUTION:

This molecule, like all others, possesses the identity element, *E*. Its most obvious element of symmetry is the threefold rotation axis, C_3, which passes through P and the two axial Cl atoms. Two independent operations can be carried out about this axis: rotation one time, designated C_3, and rotation twice, designated $C_3{}^2$; each generates a new (although indistinguishable) orientation of the molecule.

The C_3 axis is the principal rotation axis of this molecule, but it is not its only rotation axis. Phosphorus pentachloride can also be rotated by 180° about any of the three equatorial P–Cl bonds; so the molecule also has three C_2 axes and three C_2 operations.

Among mirror planes, the most prominent is the equatorial plane of the molecule. With the principal C_3 axis being vertical, this plane is placed horizontally ($\perp C_3$), so it is designated as a horizontal mirror plane, σ_h. But this is not the only mirror plane: There are three vertical mirror planes, σ_v, each including the vertical C_3 axis and one of the three equatorial P–Cl bonds.

The PCl_5 has a threefold improper rotation axis, S_3; it coincides with the C_3 axis. The rotation part of this operation does nothing to the axial chlorines but does interchange the equatorial ones; the reflection part of the operation through the (horizontal, in this case) reflection (mirror) plane does nothing to the equatorial chlorine atoms in this plane, but does interchange the axial chlorines; overall every chlorine undergoes an interchange upon carrying out the S_3 operation. There is no inversion center in this molecule, since inversion carries an equatorial chlorine into a position that was vacant before.

In summary, the PCl_5 molecule has the following symmetry elements: *E*, C_3, three different C_2 axes, σ_h, three different σ_v planes, and S_3, for a total of 10. Each of these elements has one (independent) operation associated with it, except that each threefold axis—C_3 and S_3—has two operations associated with each C_3, C_3^2; S_3, S_3^2; hence, there are 12 symmetry operations.

Example 9.2

Identify and count all symmetry elements and operations in the hexagonal planar molecule benzene, C_6H_6 [Fig. 9.5(*b*)].

SOLUTION:

Benzene, of course, has the E element. The highest rotation axis in this molecule is the sixfold one, C_6. The operations involving this axis include rotation by 60° (C_6), 120° ($C_6^2 \equiv C_3$), 180° ($C_6^3 \equiv C_2$), 240° ($C_6^4 \equiv C_3^2$), 300° (C_6^5), and 360° ($C_6^6 \equiv E$). This sixfold axis is also a threefold (we see the operations C_3 and C_3^2 that result from rotation about a C_3 axis) and a twofold axis. There are two other types of C_2 axes, however, running across the plane of the benzene ring: three of them cut between carbon atoms and three of them include pairs of C–H bonds; these are labeled C_2' and C_2'' axes to distinguish them from the C_2 axis that coincides with the C_6 axis.

Benzene is also rich in mirror planes. The plane of the molecule is perpendicular to the (vertical) principal (C_6) axis, so it is a σ_h (horizontal mirror plane). There are also six vertical mirror planes: three of them include pairs of C–H bonds and are labeled σ_v (vertical mirror planes); three of them cut between the C atoms and C–H bonds and are labeled σ_d (dihedral mirror planes).

The principal rotation axis of benzene coincides with an improper rotation axis: It is an S_6 axis (with two operations) and an S_3 axis (with two operations). The principal rotation axis is also an S_2 axis; however, this symmetry element is identically equal to the remaining type of symmetry element, the inversion center.

The complete list of symmetry elements in as symmetric a molecule as benzene is substantial: E, C_6, C_3, C_2, three C_2', three C_2'', σ_h, three σ_v, three σ_d, S_6, S_3, and i, for a total of 20 elements of symmetry of 12 types. Since the C_6, C_3, S_6, and S_3 elements each have two operations, the total number of symmetry operations is 24.

9.2 Molecular Point Groups

It is fortunate that there is no need to go through as detailed an analysis and as careful a count for every molecule as we did for benzene! First of all, it is apparent that many molecules and ions (e.g., C_6F_6, C_6Cl_6, C_6Me_6 if the methyl groups are treated as freely rotating symmetric tops) have exactly the same symmetry operations and elements as benzene itself does. Such molecules or ions with the same symmetry operations and elements are said to belong to the same **molecular point group.** There are symbols for each point group: The point group including benzene and its relatives is labeled the D_{6h} point group. Similarly, the point group of water is labeled C_{2v}, and includes not only very similar molecules such as H_2S, but also H_2CCl_2 and $COCl_2$ (phosgene);

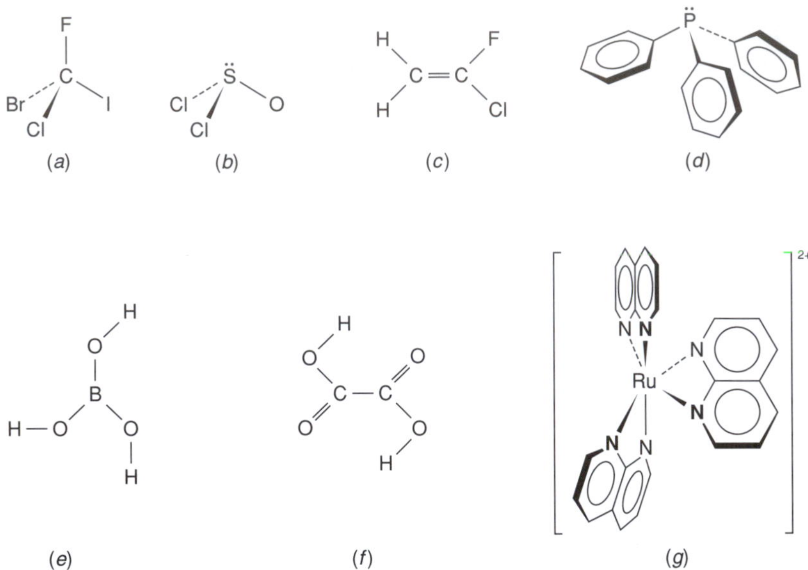

Figure 9.6
Examples of molecules or conformations of molecules for the nonaxial, C_n, C_{nh}, and D_n point groups.

the point group of PCl_5 is labeled D_{3h}, and also includes such diverse molecules as the trigonal planar BF_3, the planar benzene derivative 1,3,5-tribromobenzene, and the nine-coordinate tricapped trigonal planar anion ReH_9^{2-}.

The chemically most important point groups can be grouped into several sets; we list them roughly in order of the increasing number and complexity of symmetry elements and operations present.

A. The **nonaxial** groups have no rotational axes and can also be characterized as **low-symmetry** groups. These include three point groups: C_1, which has only the identity element, E [*Example*: CFBrClI, Fig. 9.6(*a*)]; C_s, which has only E and a mirror plane, σ (*Examples*: H_2C=CFCl [Fig. 9.6(*c*)], $SOCl_2$ [Fig. 9.6(*b*)], and the external human body); and C_i, which has only E and an inversion center, i (*Example*: BrClHC–CHClBr [Fig. 9.3]).

B. The C_n groups have a C_n axis (and E, which we will assume henceforth). An example of a molecule in the C_3 point group is triphenylphosphine, :PPh_3, in which the three phenyl groups have a propeller-like arrangement about the trigonal pyramidal P atom [Fig. 9.6(*d*)].

C. The C_{nv} groups have a C_n axis and n vertical and/or dihedral mirror planes. We have seen H_2O and other examples of the C_{2v} point group; pyramidal NH_3 is an example of the C_{3v} point group.

D. The C_{nh} groups characteristically have a C_n axis and σ_h (other elements and operations are present in particular point groups). An example in the C_{3h} point group is $B(OH)_3$ [Fig. 9.6(*e*)], which has a propeller-like arrangement of OH groups around the trigonal planar B atom; $H_2C_2O_4$ [oxalic acid; Fig. 9.6(*f*)] is similarly an example of the C_{2h} point group.

E. The D_n groups have a C_n axis and n C_2 axes perpendicular to this principal axis. An example of an ion in the D_3 point group is the tris-chelated complex ion $[Ru(1,10\text{-phenanthroline})_3]^{2+}$ [Fig. 9.6(*g*)].

F. The D_{nd} point groups have the C_n axis and n perpendicular C_2 axes characteristic of all the D groups, and characteristically add n σ_d planes. Allene, $CH_2{=}C{=}CH_2$ [Fig. 9.4], falls in the D_{2d} point group.

G. The D_{nh} point groups characteristically add σ_h. A square planar complex such as $[PtCl_4]^{2-}$ falls in the D_{4h} point group; we have already mentioned PCl_5 and other examples of the D_{3h} point group, and benzene as an example of the D_{6h} point group.

H. The S_n groups (n = an even number ≥ 4) have E plus S_n symmetry elements, and no other elements except the $C_{n/2}$ elements that arise from such identities as $C_2 \equiv S_4{}^2$, and so on.

I. The **linear** point groups $D_{\infty h}$ and $C_{\infty v}$ apply to homonuclear diatomic molecules, A–A, and heteronuclear diatomic molecules, A–B, respectively. Their symmetry is preserved upon rotation about an infinitesimally small angle; this operation can be carried out an infinite number of times.

J. The **higher order (high symmetry)** groups have more than one three- or higher fold rotation axis; these groups include the tetrahedral point group T_d, which has four C_3 axes and includes CH_4; the octahedral point group O_h, which has three C_4 axes, four C_3 axes, and i, and includes SF_6; and the icosahedral point group I_h, which includes six C_5 axes and a total of 120 symmetry operations; buckminsterfullerene (Fig. 9.1) is included in this group.

The use of the applications of symmetry and group theory depend on the assignment of molecules and ions to their appropriate symmetry point groups, which in turn must follow the determination of their shape using valence shell electron-pair repulsion (VSEPR) theory, Section 3.4. We suggest the following procedure.

1. Determine whether the molecule belongs to one of the low-symmetry groups (**A** in the above list) or one of the linear (**I** above) or high-symmetry groups (**J** above). These symmetry groups are readily identified by inspection.

2. If the molecule has not been identified in Step 1, find the highest order rotation axis (C_n with highest n) in the molecule.

3. Does the molecule have any C_2 axes perpendicular to the C_n axis? If the answer is *Yes*, the molecule falls in a D point group; if the answer is *No*, the molecule falls in a C or S point group.

4. Does the molecule have a horizontal mirror plane (σ_h) perpendicular to the highest order rotation axis? If the answer is *Yes, a* D-group molecule is finally classified in a D_{nh} point group (**G** above); a C-group molecule is finally classified in a C_{nh} point group (**D** above). If the answer is *No*, proceed to the next step.

5. Does the molecule have any vertical or dihedral mirror planes (these will contain the highest order axis)? If the answer is *Yes, a* D-group molecule is finally classified in a D_{nd} point group (**F** above); a C-group molecule is finally classified in a C_{nv} point group (**C** above). If the answer is *No* for a D-group molecule, it is finally classified in a D_n point group (**E** above); if the answer is *No* for a C- or S-group molecule, proceed to the next step.

6. Is there an S_{2n} axis colinear with the C_n axis? If the answer is *Yes*, the molecule is finally classified in an S_{2n} point group (**H** above); if the answer is *No*, the molecule is finally classified in a C_n point group (**B** above).

Figure 9.7 illustrates the steps of this method. Alternately, you can classify a molecule or ion by identifying its elements and operations of symmetry, then matching your list with those given for the chemically important point groups in the **character tables** for these point groups, which are found in Appendix B. We will have more to say about these character tables later; for now we point out that the symmetry operations for each point group are listed across the top (heading) row of each character table, following the symbol for the point group itself. (Operations of the same type are listed together; e.g., in the D_{6h} character table, the heading "$2C_6$" refers to the two operations C_6 and $C_6{}^5$, both of which rotate an atom by 60°: but one of which rotates in a clockwise and one in a counterclockwise direction.)

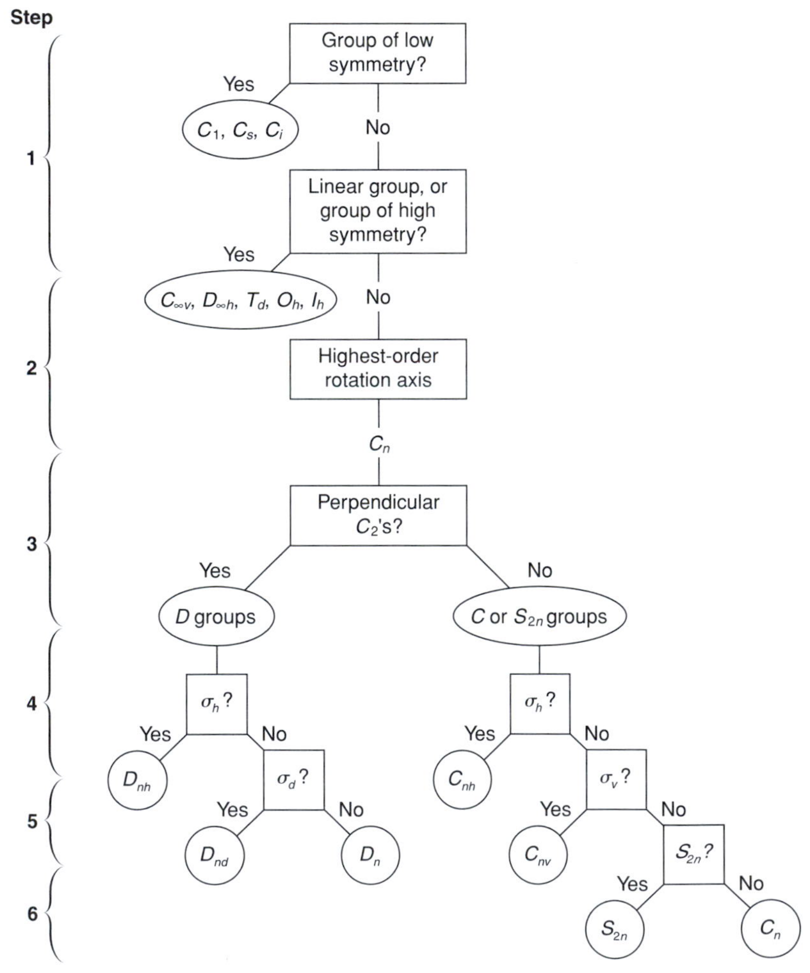

Figure 9.7
Diagram of procedure for assigning point groups. [Adapted from G. L. Miessler and D. A. Tarr, *Inorganic Chemistry*, Prentice Hall, Englewood Cliffs, NJ, 1991; p. 97.]

Example 9.3

Assign the following complexes, ions, and molecules to the appropriate symmetry point groups: (a) octahedral *cis*-$[CrL_2(CO)_4]$; (b) octahedral *trans*-$[CrL_2(CO)_4]$; (c) a free SO_4^{2-} anion; (d) a SO_4^{2-} ion coordinated to a metal in a linear fashion through one O donor atom; (e) a SO_4^{2-} ion bridging two metal ions; (f) S_2Cl_2; (g) $C_8H_4Cl_4$; (h) planar $C_8H_8^{2-}$; and (i) $B_{12}H_{12}^{2-}$. The structures of these species are indicated in Figure 9.8.

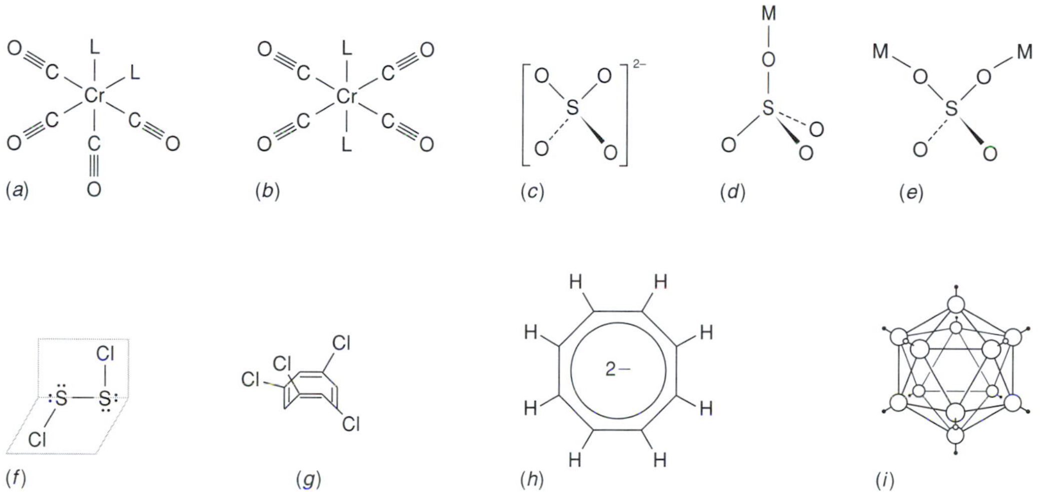

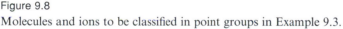

Figure 9.8
Molecules and ions to be classified in point groups in Example 9.3.

SOLUTION:

Applying Step **1** intuitively, we note the presence of more than one of the highest order C_3 axis in the tetrahedral SO_4^{2-} ion, which therefore likely belongs to the high-symmetry T_d point group, and the presence of more than one of the highest order C_5 axes in the $B_{12}H_{12}^{2-}$ ion, which therefore likely belongs to the high-symmetry I_h (icosahedral) point group.

2. The highest order C_n axis in each of the remaining molecules is as follows: (a) a C_2 axis bisecting the angle between the two Cr–L bonds; (b) a C_4 axis including both Cr–L bonds; (d) a C_3 axis along the M–O–S axis; (e) a C_2 axis bisecting the angle between the two donor O–S bonds; (f) a C_2 axis located halfway between the two S–Cl bond planes and passing through the middle of the S–S bond; (g) a C_2 axis running top to bottom through the center of the molecule; and (h) a C_8 axis running top to bottom through the center of the ion. These axes are shown in Figure 9.9.

3. The following cases have C_2 axes $\perp$ the C_n axis: (b) along the opposite OC–Cr–CO bonds; (h) along the opposite pairs of C–H bonds. Species (b) and (h) therefore belong to *D* point groups; (a), (d), (e), (f), and (g) belong to *C* or *S* point groups.

4. Horizontal mirror planes (σ_h, $\perp$ the highest order C_n axis) are found in (b), which therefore belongs to point group D_{4h}, and (h), which belongs to point group D_{8h}.

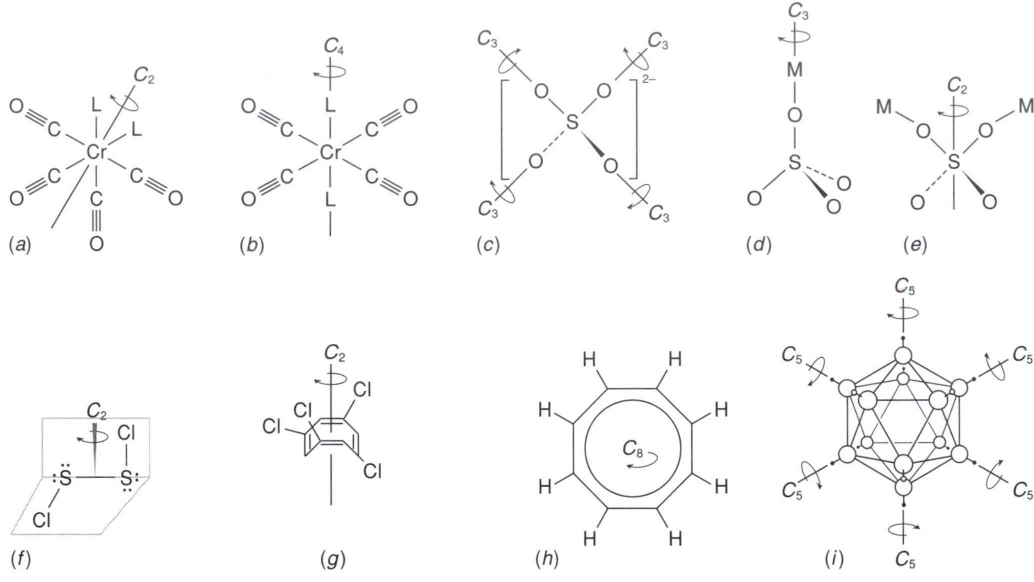

Figure 9.9
Highest order axes for molecules and ions in Example 9.3.

5. Vertical or dihedral mirror planes (those containing the highest order axis) are found in (a) and (e), both of which belong to point group C_{2v}, and (d), which belongs to point group C_{3v}.

6. An S_4 axis colinear with the C_2 axis is found in (g), which therefore falls in point group S_4. No such axis is found in (f), which therefore falls in point group C_2.

9.3 Elementary Applications of Symmetry: Molecular Polarity, Chirality, and Fluxionality

Polarity of Molecules. One of the properties of molecules determined by their symmetry is the **polarity** of molecules. Polar bonds, of course, result when two atoms of differing electronegativity are bonded together.[3] A polar molecule is a molecule with a permanent electric dipole moment: The *molecule* has positive and negative ends. This dipole moment results mainly from an unbalanced arrangement of the polar bonds in the molecule, but has other contributions, such as from the polarity of any unshared electron pairs. To be polar, the molecule cannot have a symmetry axis interchanging the positive and negative ends of the molecule. If a molecule has a principal C_n axis, this interchanges the x with the $-x$ direction and the y with the $-y$ direction, so there can be no polarity in these two directions; any polarity must lie along the z direction. If the molecule additionally has either a C_2 axis $\perp$ to the principal axis, or a σ_h plane, these also interchange z with $-z$; so a molecule having either of these two combinations of symmetry elements cannot be polar. Hence, a molecule belonging to a C_{nh}, any D, or any of the normal higher order point groups cannot be polar.

Chiral Molecules. **Chiral molecules** are those that are not superimposable on their mirror images in the same way that our right and left hands cannot be superimposed on each other; such molecules rotate the plane of polarized light, and hence are optically active. A chiral molecule and its mirror-image isomer are called **enantiomers**; you are no doubt familiar with this topic from your organic chemistry course. Technically, the symmetry requirement for a molecule to be chiral is that it *lack any kind of improper-rotation* (S_n) *axis*; but since $S_2 \equiv i$ and $S_1 \equiv \sigma$, the practical statement of the requirement is that a chiral molecule *cannot possess an improper rotational axis, an inversion center, or a mirror plane*. Hence, the molecule must belong to one of the C_n or D_n point groups (including the low-symmetry C_1 group; there are some additional chemically insignificant groups that could be listed as well).

The greatest interest in chirality among inorganic chemists is in its presence in certain isomers of coordination or organometallic complexes. The complexes mentioned in Example 9.3a–e all fall in point groups that preclude optical activity; but the tris-chelate complex $[Ru(1,10\text{-phenanthroline})_3]^{2+}$ [Fig. 9.6(*g*)] falls in point group D_3, which allows chirality; chirality occurs more frequently in complexes of chelating ligands, due to their reduced symmetry. Note that the enantiomer of $[Ru(1,10\text{-phenanthroline})_3]^{2+}$ in Figure 9.6 is drawn along a C_3 axis, with the three N atoms shown in boldface being above the other three. In this enantiomer, the lower N atom of each chelate ligand is located in a *counterclockwise* direction from the upper **N** atom of the same chelate ligand. Counterclockwise rotation on a clock is rotation to the left; the enantiomer shown is designated the Λ enantiomer, Λ-$[Ru(1,10\text{-phenanthroline})_3]^{2+}$. In the mirror-image enantiomer, the lower donor atoms would be located in a clockwise or right-handed direction from the upper donor atoms; this enantiomer is designated as Δ-$[Ru(1,10\text{-phenanthroline})_3]^{2+}$. These symbols indicate the absolute configuration, not the direction in which the plane-polarized light is rotated: We do not know from the absolute configuration which enantiomer would rotate plane-polarized light to the right and be designated the *d* or (+) stereoisomer.

The following example will also serve to introduce two common types of **geometric isomers** among octahedral complexes. In disubstituted octahedral complexes, MA_4B_2, the two B groups may either be cis (adjacent) to each other [Fig. 9.10(*a*)], or trans to (across from) each other [Figure 9.10(*b*)]; these two geometric isomers fall in different point groups. In trisubstituted octahedral complexes, MA_3B_3, there are also two geometric isomers possible: The three ligands may all be on the same triangular face of the octahedron [the *facial* or fac isomer, Fig. 9.10(*c*)], or they may lie in the same equatorial plane of the complex (the *meridional* or mer isomer, Fig. 9.10(*d*)].

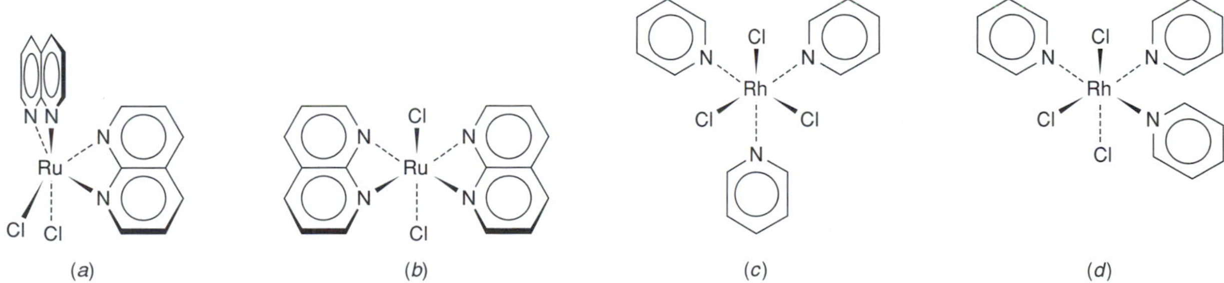

Figure 9.10
Examples of geometric isomers: (*a*) *cis*- and (*b*) *trans*-$[Ru(1,10\text{-phenanthroline})_2Cl_2]$; (*c*) *fac*- and (*d*) *mer*-$[RhCl_3(\text{pyridine})_3]$.

Example 9.4

Determine whether the following isomers are polar, and whether they possess optical activity: (a) *cis*-$[RuCl_2(1,10\text{-phenanthroline})_2]$; (b) *trans*-$[RuCl_2(1,10\text{-phenanthroline})_2]$; (c) *fac*-$[RhCl_3(\text{pyridine})_3]$; and (d) *mer*-$[RhCl_3(\text{pyridine})_3]$. The structures of these complexes are drawn in Figure 9.10.

SOLUTION:

We may begin by assigning the molecules to point groups, from which the identification of the two properties follows readily. Isomer (a) possesses, as its only nontrivial element of symmetry, a C_2 axis that bisects the Cl–Ru–Cl angle. It thus falls into the C_2 point group and, with this low symmetry, can be (and is) polar and chiral; the two enantiomers are the *cis*-Λ and the *cis*-Δ complexes. In isomer (b), the two phenanthroline rings are coplanar; the molecule possesses a principal C_2 axis (the Cl–Ru–Cl bonds), two other C_2 axes ⊥ to the principal axis, a σ_h plane, and so on: It falls in the C_{2h} point group, and cannot be either polar or chiral.

Isomer (c) possesses an obvious C_3 axis as viewed in Figure 9.10; there are also three mirror planes including the C_3 axis; the molecule falls in point group C_{3v}. It can be polar, but cannot be chiral. In isomer (d), the three N atoms form a coplanar arc, as do the three Cl atoms; the central N–Rh–central Cl axis is a C_2 axis. There are two vertical mirror planes: one including the three Cl and the central N donor atoms; the other including the three N and the central Cl donor atoms; the molecule is a member of the C_{2v} point group. It can be polar, but cannot be chiral.

The problem of separating (*resolving*) enantiomers is like the problem in organic chemistry. The individual enantiomers have no useful differences in physical properties unless they are somehow combined with an enantiomer of another chiral species to produce diastereomers. This is most easily done if the enantiomers are complex *ions*; then an optically active (say of Δ absolute configuration) counterion can be added that gives a salt of low solubility; the salt of the Δ counterion with the Δ complex ion will have a different solubility than the salt of the Δ counterion with the Λ complex ion; the two can be separated by fractional crystallization. If the complexes are uncharged (as in the above example) the resolution is not so simple.

The founder of the systematic study of coordination compounds, Alfred Werner, relied heavily on the isolation and counting of isomers—both geometric (i.e., cis or trans, meridional or facial) and optical—to establish the characteristic geometry about a given metal ion (this was done about 100 years ago, when X-ray crystallography was unavailable for structure determination). For example, if ruthenium(II) complexes were based on a trigonal prismatic (D_{3h}) rather than an octahedral geometry, there would still be two geometric isomers of $[Ru(1,10\text{-phenanthroline})_3]^{2+}$, but neither of these would be optically active. The detection of optical activity in an inorganic complex such as *cis*-$[Ru(1,10\text{-phenanthroline})_3]^{2+}$ was one more step in destroying the idea that the characteristic properties of bioorganic compounds such as optical activity could be produced only by living organisms and hence found only in organic compounds. Some skeptics still pointed out that the chelate ligands were organic; perhaps the optical activity somehow resided in these (optically inactive) ligands. So as a final proof, Werner prepared and resolved a carbon-free coordination compound [Fig. 9.11(*a*)].

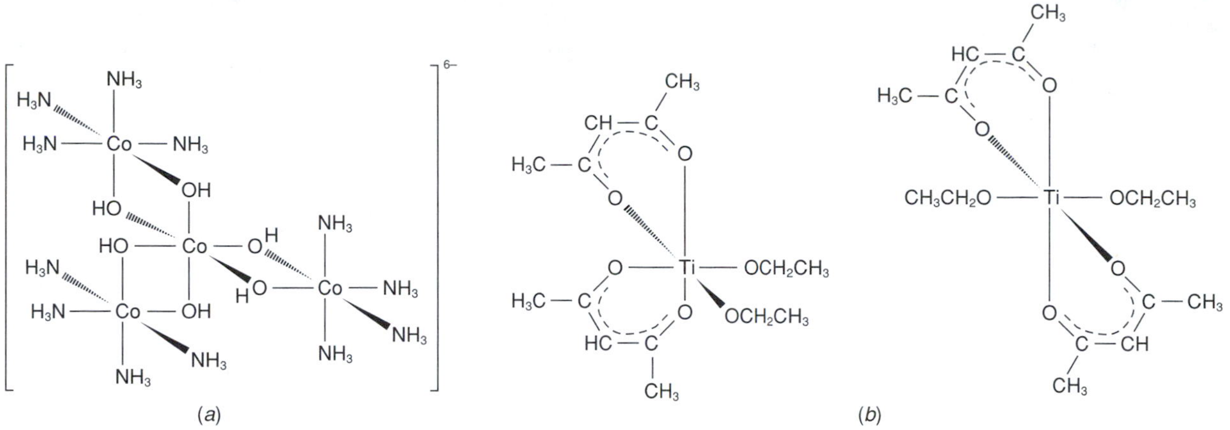

Figure 9.11
(*a*) The optically active but carbon-free complex cation resolved by Werner. [Adapted from I. S. Butler and J. F. Harrod, *Inorganic Chemistry: Principles and Applications*, Redwood City, Benjamin-Cummings, CA, 1989, p. 364.] (*b*) The cis and trans isomers of $[Ti(CH_3COCHCOCH_3)_2(OCH_2CH_3)_2]$.

The detection of enantiomers nowadays is enhanced by the use of nuclear magnetic resonance (NMR) spectroscopy, the basics of which we presume you are familiar with from your organic chemistry course. For example, in proton (^1H) NMR spectra, the groups of a CH_2 group are equivalent provided that they sit equivalently across a mirror plane from each other. The NMR spectra can be used to detect the two isomers of $[Ti(CH_3COCHCOCH_3)_2(OCH_2CH_3)_2]$ [Fig. 9.11(*b*)]: In the trans isomer, the CH_2 groups of the ethoxide ligands sit astride a mirror plane, so they give only one NMR signal; the same is true of the (rapidly rotating) CH_3 groups of the $CH_3COCHCOCH_3$ ligand. In the cis isomer, there is no mirror plane, so the two CH_2 protons are not equivalent by symmetry, and give separate NMR signals, as do the protons of the two CH_3 groups.

Fluxionality. The detection or resolution of optical isomers depends on their being kinetically inert (Sections 8.8 and 8.9), so that the isomers persist long enough to be crystallized separately, or detected separately in an NMR experiment. As we saw in those earlier sections, the complexes of many metal ions exchange ligands at a very rapid rate; the rate is often too rapid to allow the detection of particular isomers or structures.

Furthermore, it is not always necessary for a ligand to dissociate in order to exchange with another ligand: If vibrations of the molecule can take it from the geometry that is preferred to an alternate geometry that is not much higher in energy, then back again, the distinctions of particular coordination positions may be lost. For example, among five-coordinate species, the trigonal bipyramidal geometry, which has three equatorial ligands and two axial ligands, is preferred, but the square pyramidal geometry, which has four basal ligands and one axial ligand, is not too much different in its bond angles nor too much worse in energy. The conversion merely requires a bending of four bonds by less than 30° each [Fig. 9.12(*a*)]. The base of the square pyramid [Fig. 9.12(*b*)] includes the two original axial ligands, but also two of the original equatorial ligands; all four are now equivalent by the C_{4v} symmetry. The intermediate square planar compound, being less stable, quickly reverts to the trigonal bipyramidal

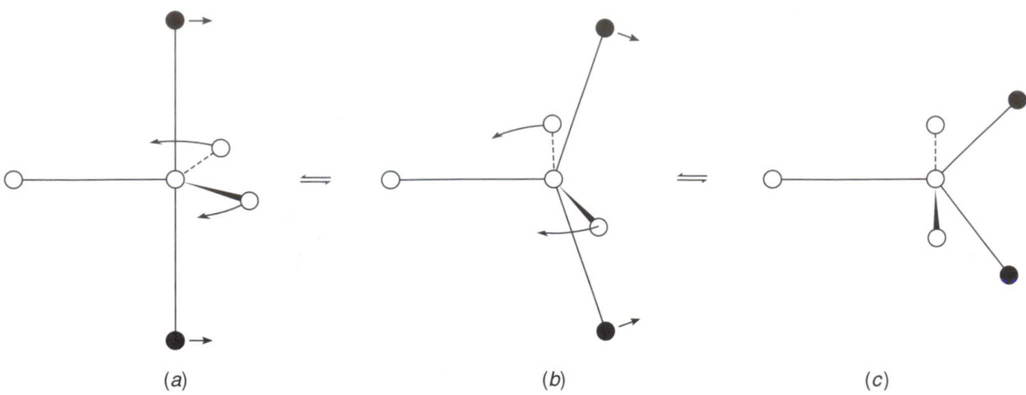

Figure 9.12
The Berry pseudorotation mechanism for exchanging axial and equatorial ligands in a trigonal bipyramidal complex. (*a*) Starting complex, with axial ligands shaded, and ligands moving in directions indicated as part of vibrational motion. (*b*) Square pyramidal transition state, with basal square plane to the right. Motion may now reverse to go back to (*a*), or continue as shown to give (*c*), a new conformation of the trigonal bipyramidal complex, with the new axial ligands (above and below the plane of the paper) being former equatorial ligands, and vice versa.

geometry by another small bending of bonds, but there is a 50% probability that the two shaded (originally axial) atoms now go into equatorial positions, while the two originally equatorial atoms (unshaded) now become axial [Fig. 9.12(*c*)]. Given enough time, this mechanism, called a *Berry pseudorotation*, can involve all of the ligands, interchanging them many times with each other. (Similar mechanisms are common for coordination numbers for which VSEPR allows two or more geometric choices: for example, seven coordination and eight coordination, but are less common for coordination numbers with strong geometric preferences, such as six coordination.) Molecules that have more than one equivalent minimum-energy conformation, and that are readily able to change from one conformation to another, are known as **fluxional molecules**.

If the method of detection is not quick enough, the different environments implied by a low-symmetry structural conformation may not be detectable by that method. For example, PF_5 has a trigonal bipyramidal structure, but in the ^{19}F NMR spectrum at room temperature or above only one type of fluorine can be detected. In this case, the rate of exchange of fluorine atoms between axial and equatorial positions is too rapid to allow separate radio wave absorption by distinct axial and equatorial fluorines; the molecule appears to have higher symmetry than it really does instantaneously. However, if the solution of such a molecule is cooled sufficiently (or perhaps if the molecule is placed in the solid state), the rate of ligand exchange may slow sufficiently that each ligand remains in position during the absorption of the radio wave; in such a case the axial and equatorial ligands show distinct NMR peaks, and the spectrum will be that expected from the symmetry of the static molecule.

So the appearance of the spectrum of such a molecule changes dramatically over certain temperature ranges, above which a significant fraction of the molecules acquire the thermal activation energy needed to change conformations; the rates of exchange of nuclei among different positions therefore increase. At low temperatures, each symmetry distinct bond or nucleus has its own absorption at different characteristic frequencies $v_1, v_2, \ldots$; there is a frequency difference Δv separating two of these frequencies. When the temperature is high enough that the exchange rate is much larger than this frequency difference, each nucleus has an absorption frequency that is the weighted average of the characteristic frequencies $v_1, v_2, \ldots,$ of the positions through which it is rapidly passing. Between these temperatures, the spectra of the nuclei broaden, at some temperatures perhaps to such an extent that they cannot be seen at all.

Each spectroscopic method uses its own characteristic frequency of electromagnetic radiation to excite nuclei, bonds, and so on. In general, the higher the frequency, the greater the differences in frequencies that can exist between the different exchanging bonds, nuclei, and so on. In the ^{1}H NMR experiment, the radio waves used are of low frequency (say, 60 MHz); at this radio frequency a typical 1 ppm chemical shift difference between two exchanging protons works out to 60 Hz = 60 s^{-1}. The protons need only jump between the equivalent positions significantly faster than 60 times a second in order to show the seemingly high symmetry spectrum. In ^{13}C NMR spectra, the chemical shifts are larger, and the exchanges do not become fast enough until a somewhat higher temperature. When we go to a completely different technique such as IR spectroscopy, the results are more dramatic. We usually express IR data in reciprocal centimeters (cm^{-1}); two inequivalent but exchangeable functional groups might differ in frequency by, say, 100 cm^{-1}. To convert this value to reciprocal seconds (s^{-1}), we multiply by the speed of light, 3×10^{10} cm s^{-1}; this gives us a frequency difference of 3×10^{12} s^{-1}. Achieving jumping rates in excess of 10^{12} s^{-1} is impossible, so IR spectra characteristically reveal the instantaneous structure of fluxional molecules.[4] Table 9.2 shows the characteristic (orders of magnitude of the) rates of exchange necessary with various spectroscopic methods to detect fluxional behavior in normal temperature ranges.

Table 9.2
Approximate Minimum Rates of Exchange Necessary to Affect Common Structural Techniques

Technique	Rate of Exchange (s^{-1})[a]
X-ray Diffraction[b]	10^{-3}
Chemical separation of isomers	10^{-2}
Stop-flow kinetics	10^{-2}–10^{3}
Nuclear magnetic resonance[c]	10^{1}–10^{9}
Nuclear quadrupole resonance[c]	10^{1}–10^{8}
Electron spin resonance[c]	10^{4}–10^{8}
Mössbauer spectroscopy (iron)	10^{7}
Infrared or raman spectroscopy	10^{13}
Visible spectroscopy	10^{14}
Ultraviolet spectroscopy	10^{15}

[a] Taken as the inverse of the time scale of the technique, as given by E. L. Muetterties, *Inorg. Chem.*, **4**, 769 (1965).

[b] Based on the data collection time (typically hours) rather than the characteristics of the absorbed electromagnetic radiation, which would give a figure of 10^{18} s^{-1}.

[c] Necessary rate is very dependent on the chemical system under investigation.

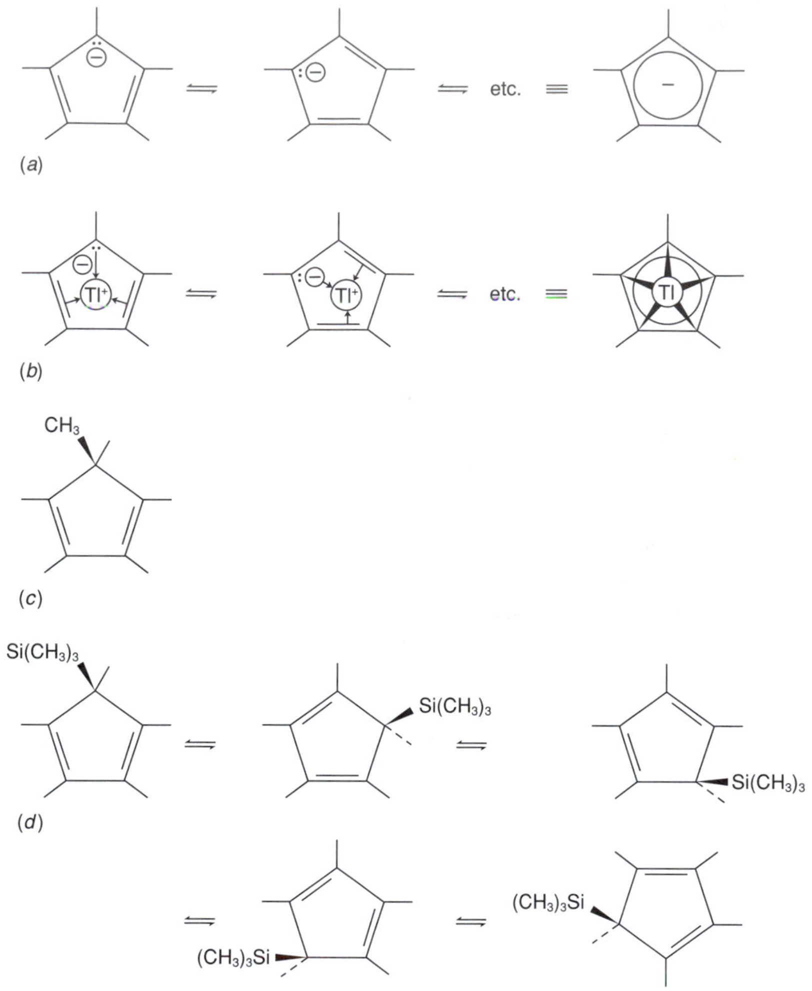

Figure 9.13
Different modes of bonding of a cyclopentadienide ligand. (*a*) Free anion, in which the negative charge is delocalized equally to all five carbon atoms via five resonance structures; (*b*) *Pentahapto*-cyclopentadienide mode, in which the ligand is bonded to a metal atom on one side equally by all five carbon atoms, as in gaseous η^5-C_5H_5Tl; (*c*) *Monohapto*-cyclopentadienide mode, in which the ligand is bonded through only one carbon atom, as in η^1-methylcyclopentadiene; (d) Fluxional *monohapto*-cyclopentadienyltrimethylsilane, in which the one carbon–silicon bond relocates itself over time so as to involve all carbon atoms equally.

Of even greater interest are molecules in which different pairs of atoms become connected in the different conformations that a fluxional molecule passes through. These are particularly important in organometallic derivatives of certain π-bonded organometallic species such as the cyclopentadienide ion, $C_5H_5^-$ [Fig. 9.13(*a*)]. This six π-electron aromatic compound is anionic, since the six π electrons are located on only five carbon atoms. The free anion has D_{5h} symmetry, and gives only one ^1H and one ^{13}C NMR signal at any temperature. It is capable of acting as a π-donor ligand, bonding to metal atoms in two common manners: It often bonds to *d*-block metal ions

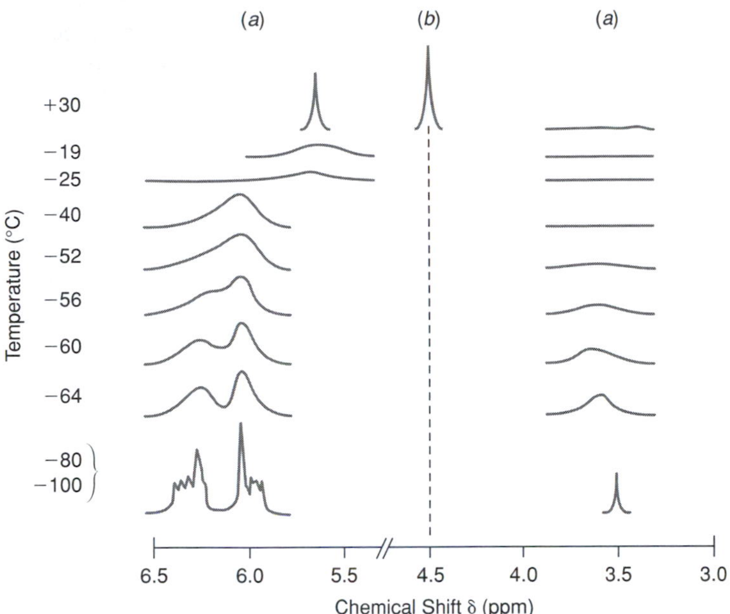

Figure 9.14
Variable temperature ^{1}H NMR spectrum of fluxional $[Fe(\eta^1\text{-}C_5H_5)(\eta^5\text{-}C_5H_5)(CO)_2]$. (*a*) Assigned to $\eta^1\text{-}C_5H_5$; (*b*) assigned to $\eta^5\text{-}C_5H_5$. [NMR spectrum adapted from M. J. Bennett, Jr., F. A. Cotton, A. Davison, J. W. Faller, S. J. Lippard, and S. M. Morehouse, *J. Am. Chem. Soc.*, **88**, 4371 (1966).]

using *all five* carbon atoms [Fig. 9.13(*b*)], donating three electron pairs. In this mode of bonding, the local symmetry of the ring is C_{5v}; in this bonding mode it also gives one ^{1}H and one ^{13}C NMR signal at any temperature, although the chemical shifts are of course different from those observed in the free ion. The $C_5H_5^-$ anion is also capable of bonding through *only one* carbon atom [Fig. 9.13(*c*)], donating only one electron pair, as in its simple organic derivatives such as $C_5H_5CH_3$; here the local symmetry falls drastically to C_s, and (not considering the CH_3 group) the compound gives *three* ^{1}H and *three* ^{13}C NMR signals (in intensity ratio 2:2:1) at any temperature. The mode of bonding can be distinguished in nomenclature by using the prefix *pentahapto-* to identify the ring bonded through five carbon atoms [e.g., *pentahapto*cyclopentadienylthallium(I)], and *monohapto-* to indicate bonding through one carbon atom. In chemical formulas, the prefixes used are η^5 and η^1, respectively (e.g., $\eta^5\text{-}C_5H_5Tl$ and $\eta^1\text{-}C_5H_5CH_3$).

The $C_5H_5^-$ anion is also capable of bonding through one carbon atom to metal (or metalloid) atoms, for example, in the compound $\eta^1\text{-}C_5H_5Si(CH_3)_3$; although the symmetry is the same, the NMR spectra typically are not: At *low* temperatures such compounds give *three* ^{1}H and *three* ^{13}C NMR signals; at *high* temperatures there is but one of each kind of NMR signal (Fig. 9.14), at a chemical shift that is the weighted average of the chemical shifts in the low-temperature spectrum:

$$\delta_{\text{high temp}} \text{ for } \eta^1\text{-}C_5H_5 = 0.2(2\delta_1 + 2\delta_2 + \delta_3)_{\text{low temp}} \quad (9.1)$$

A careful analysis of the collapse of the NMR spectrum in the intermediate temperature range is capable of determining the mechanism of exchange: It is found that the proton–carbon coupling constants $^1J_{C-H}$ collapse at a slightly lower temperature for the proton and unsaturated carbon next to the original bond, as compared to the proton and unsaturated carbon atoms most remote from the original bond; this can be used to show that the jump of the metal atom is a *1,2-shift* (to the adjacent carbon atom) rather than a 1,3-shift (to the most distant carbon atom), or a random shift (through a *pentahapto* intermediate). This finding indicates that the bonds of the ring to the metal are not completely broken during the fluxional process, but are rearranged with a rather low activation energy. The reason that this fluxional rearrangement or "ring whizzing" does not occur in organic derivatives of the C_5H_5 ring is that bonds do have to be weakened during the process; the bonds of carbon to metals are much weaker to start with than bonds of carbon to carbon.

The jumping rate of the metal around the ring is typically on the order of 10^2–10^3 jumps per second at room temperature, and lower when the solution is cooled. This value matches well the rate of exchange needed for NMR spectroscopic consequences to show up. Consequently, nearly all detection of fluxional behavior has been done by solution-phase (or solid-state) NMR spectroscopy; but we have been able to use the next-most suitable method, ^{35}Cl nuclear quadrupole resonance (Table 9.2), to detect fluxional behavior (and its necessary activation energy) in solid-state samples of $PhHgC_5Cl_5$ and some analogues.[5] Incidentally, this establishes that the motion is not the jumping of the metal around the ring, but rather the rotation of the ring about its lost (*pseudo*) fivefold axis, hindered by a weak chemical bond that must be stretched and partially broken, then re-formed to a different carbon atom, hundreds or thousands of times a second.

9.4 Character Tables for Symmetry Point Groups

Many of the other applications of symmetry take advantages of some of the consequences of the fact that symmetric molecules are covered by the mathematical theory known as **group theory**. The mathematics of this theory are beyond the scope of this book, but we shall make use of the fact that, for each symmetry point group, there is a corresponding **character table**. Character tables for the major chemically important point groups are found in Appendix B; for convenience, we reproduce some examples in Table 9.3, beginning with the character table for the C_{2v} point group of our first molecular example, water.

The meanings of the headings of the different columns of the character table are already familiar: in the upper left corner the point group is listed, and over the main body of the table the symmetry operations of the point group are listed (grouped into classes of closely related operations such as $2C_6 = C_6$ and $C_6{}^5$).

On the left side of the table are listed the **irreducible representations** in that point group, which we will try to explain in a nonmathematical way. Many parts of a molecule, such as most of the individual atomic orbitals, do not have the full symmetry of the whole molecule. For example, consider the p_x orbital of the oxygen atom of the water molecule: referring to Figure 9.15(a), this orbital lies in the xz plane (the plane of the H_2O molecule), along the x axis, so it is not directed exactly at either H atom. Let us consider what each operation does in turn to this orbital. The identity operation E leaves the orbital unchanged; the **character** of an operation leaving something

Table 9.3
Selected Character Tables

C_{2v}	E	C_2	$\sigma_v(xz)$	$\sigma_v'(yz)$		
A_1	1	1	1	1	z	x^2, y^2, z^2
A_2	1	1	−1	−1	R_z	xy
B_1	1	−1	1	−1	x, R_y	xz
B_2	1	−1	−1	1	y, R_z	yz

C_{3v}	E	$2C_3$	$3\sigma_v$		
A_1	1	1	1	z	$x^2 + y^2, z^2$
A_2	1	1	−1	R_z	
E	2	−1	0	$(x, y), (R_x, R_y)$	$(x^2 - y^2, xy), (xz, yz)$

D_{4h}	E	$2C_4$	C_2	$2C_2'$	$2C_2''$	i	$2S_4$	σ_h	$2\sigma_v$	$2\sigma_d$	(x axis coincident with C_2')	
A_{1g}	1	1	1	1	1	1	1	1	1	1		$x^2 + y^2, z^2$
A_{2g}	1	1	1	−1	−1	1	1	1	−1	−1	R_z	
B_{1g}	1	−1	1	1	−1	1	−1	1	1	−1		$x^2 - y^2$
B_{2g}	1	−1	1	−1	1	1	−1	1	−1	1		xy
E_g	2	0	−2	0	0	2	0	−2	0	0	(R_x, R_y)	(xz, yz)
A_{1u}	1	1	1	1	1	−1	−1	−1	−1	−1		
A_{2u}	1	1	1	−1	−1	−1	−1	−1	1	1	z	
B_{1u}	1	−1	1	1	−1	−1	1	−1	−1	1		
B_{2u}	1	−1	1	−1	1	−1	1	−1	1	−1		
E_u	2	0	−2	0	0	−2	0	2	0	0	(x, y)	

$D_{\infty h}$	E	$2C_\infty^\phi$	$\infty\sigma_v$	i	$2S_\infty^\phi$	∞C_2		
Σ_g^+	1	1	1	1	1	1		$x^2 + y^2, z^2$
Σ_g^-	1	1	−1	1	1	−1	R_z	
Π_g	2	$2\cos\phi$	0	2	$-2\cos\phi$	0	(R_x, R_y)	(xz, yz)
Δ_g	2	$2\cos 2\phi$	0	2	$2\cos 2\phi$	0		$(x^2 - y^2, xy)$
Σ_u^+	1	1	1	−1	−1	−1	z	
Σ_u^-	1	1	−1	−1	−1	1		
Π_u	2	$2\cos\phi$	0	−2	$2\cos\phi$	0	(x, y)	
Δ_u	2	$2\cos 2\phi$	0	−2	$-2\cos 2\phi$	0		

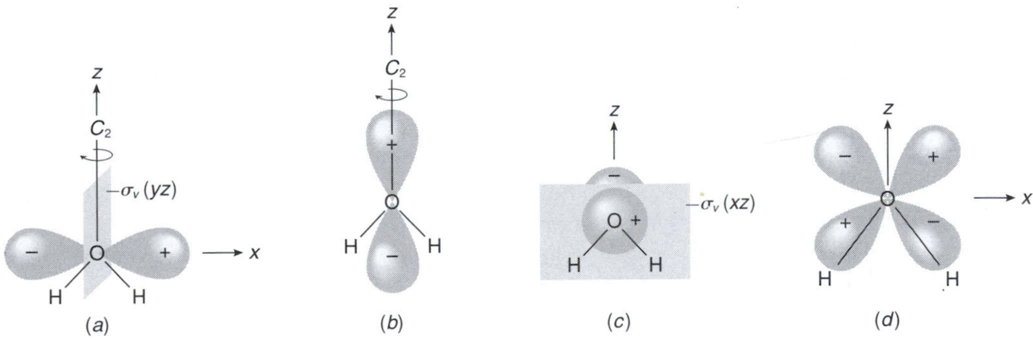

Figure 9.15
Drawings of selected oxygen atomic orbitals in the water molecule, to be used in determining the irreducible representations of these orbitals: (*a*) p_x orbital; (*b*) p_z orbital; (*c*) p_y orbital; and (*d*) d_{xz} orbital.

unchanged is +1. The C_2 operation, however, does interchange the positive and the negative lobes of the p_x orbital; the character for an operation interchanging positive and negative parts is −1. The $\sigma_v(xz)$ reflection leaves the p_x orbital unchanged, so the character of this operation is +1. Finally, the $\sigma_v(yz)$ operation once again interchanges the positive and negative lobes of the p_x orbital, so it has a character of −1. In the order in which the operations are listed in the character table, the characters of the operations are 1, −1, 1, −1, which corresponds to the third row of characters found in the character table, following the irreducible representation label B_1. We say that the p_x orbital transforms under the symmetry operations of the C_{2v} point group according to the irreducible representation B_1.

Not only orbitals, but many other things associated with a molecule transform according to one or another of the irreducible representations of that group. For example, a vector along the x axis, with its head on one side of the oxygen atom and its tail on the other side, transforms identically to the p_x orbital. For convenient reference, an x is listed at the right side of the row; this indicates that anything with symmetry like a vector in the x direction, such as a p_x orbital, transforms in this irreducible representation.

Example 9.5

Determine the irreducible representations of the following orbitals on the oxygen atom of water: s, p_z, p_y, d_{xz}.

SOLUTION:

The s orbital of oxygen is spherical and located at the center of the molecule; its one lobe has the same sign everywhere, so *all* operations leave it unchanged in sign and position. The characters of all of the operations are 1, so the s orbital transforms in the A_1 irreducible representation, which is the first representation listed. In any character table the first irreducible representation listed has all +1 characters; anything that transforms this way *does* have the full symmetry of the molecule.

The positive and negative lobes of the p_z orbital [Fig. 9.15(*b*)] are never interchanged by any of the four symmetry operations of this point group, so the p_z orbital also transforms in the totally symmetric A_1 irreducible representation of this point group. So does a vector centered on oxygen and pointing in the z direction; the z shown at the right of the A_1 row indicates both that p_z and a z vector transform in this manner.

The p_y orbital is affected differently by the four symmetry operations; the identity operation of course leaves it unchanged and has a character of 1; the C_2 operation interchanges positive and negative lobes (character of −1); the $\sigma_v(xz)$ operation does the same and also has a −1 character; but the $\sigma_v(yz)$ leaves the orbital unchanged, so it has a character of +1. This set of characters matches that of the B_2 irreducible representation; y is listed on the right side of this representation for a convenient reminder of this.

The d_{xz} orbital is unchanged by E (automatic character of +1); all its positive lobes and negative lobes are interchanged by the C_2 and the $\sigma_v(yz)$ operations (characters of −1 under each); its lobes are unchanged by the $\sigma_v(xz)$ operation (character of +1). This set of characters matches that of the B_1 irreducible representation, which is also indicated by the xz entry at the far right of the row.

The irreducible representations of the other d orbitals are also indicated by the entries at the far right of the table. (Sometimes the entries show this indirectly; since both z^2 and y^2 are listed with the A_1 irreducible representation, this implies that $d_{x^2-y^2}$ transforms in this irreducible representation as well.)

In general, the symbols used for the irreducible representations also carry some of this information. The letter A indicates that representation is symmetric to the principal rotation operation (has a character of $+1$), while B indicates that it is antisymmetric to that operation (has a character of -1). The subscript 1 indicates that the representation is symmetric either to the perpendicular C_2 axis (if present), or to a vertical mirror plane; the subscript 2 indicates antisymmetry. Of particular importance in many groups are the subscripts $\boldsymbol{g}$ (from the German *gerade*, meaning even), meaning that the representation is symmetric to inversion, and $\boldsymbol{u}$ (from the German *ungerade*, meaning odd), meaning that the representation changes sign upon inversion.

Additional symbols appear in groups of higher order: those with C_3 or higher axes. Let us consider as an example the XeF_4 molecule, which has a C_4 axis and falls in the D_{4h} point group (see Table 9.3).

Clearly, the sign of the wave function of the s orbital of the Xe atom is unchanged by any symmetry operation, so that the s orbital falls in the totally symmetric A_{1g} irreducible representation. The Xe p_z orbital has the signs of the wave function at its two lobes interchanged when the operation of inversion is carried out, so it must fall in one of the *ungerade* irreducible representations found in the lower half of the character table. The p_z orbital is unchanged in sign on carrying out rotation about the principal (C_4) axis, so it falls in an A irreducible representation. Reference to the character table for D_{4h} shows an entry "z" at the right side of the row for the A_{2u} irreducible representation, so the p_z orbital is classified here (the subscript 2 is used because this orbital is reversed in sign upon rotation about a C_2 axis perpendicular to the principal axis).

When we carry out a C_4 rotation on the p_x axis, however, we find that it neither keeps nor changes its sign, but rather is superimposed on the original position of the p_y orbital; simultaneously, the p_y orbital is rotated so as to coincide with the p_x orbital (but with a reversal in the sign of the wave function). Clearly, neither the p_x nor the p_y orbital considered alone show even the reduced symmetry necessary to transform according to any of the irreducible representations of this point group, but *together, considered as a pair*, they do interchange lobes and signs *with each other.* (In fact, in a molecule as symmetric as XeF_4, there is no way to determine which two fluorines fall on the x axis and which two fall on the y axis; the two directions are clearly equivalent in symmetry to each other.) The p_x and p_y orbitals thus must be considered together as a **doubly degenerate** set of orbitals. Doubly degenerate pairs of orbitals fall in irreducible representations beginning with the letter E. Since all p orbitals change sign upon inversion, this pair of orbitals falls in the doubly degenerate irreducible representation E_u, which is indicated by the appearance of the pair (x, y) on the right side of the character table. The character "2" appears under the identity operation (also coincidentally labeled E), since *two* orbitals are unchanged in sign upon this operation. The character "0" appears under the C_4 operation, since this operation retains the sign of the p_x orbital (partial character of $+1$) while at the same time inverting the sign of the p_y orbital (partial character of -1; the sum of the two contributions to the character is zero).

Example 9.6

Assign the s and p orbitals of the NH_3 molecule (Fig. 9.16) to the proper irreducible representations (NH_3 is in point group C_{3v}; its character table is also in Table 9.3).

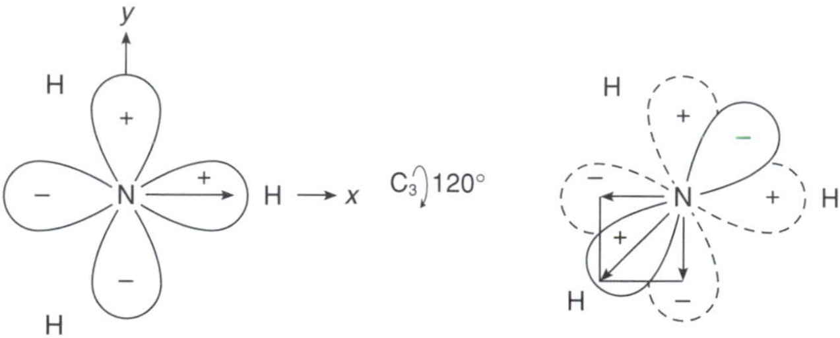

Figure 9.16

View down the C_3 axis of the NH_3 molecule: (a) before rotation, with the positive lobe of the p_x orbital also represented by a vector; (b) after rotation, with the former orbital positions shown by dotted lines; the vector representing the positive lobe of the p_x orbital is also shown as a vector sum of the negative lobes of the former p_x and p_y orbitals.

SOLUTION:

Ammonia, NH_3, is drawn in Figure 9.16 in a view down its threefold (z) axis, and with its valence p_x and p_y orbitals shown explicitly. Without further thought we can assign the s orbital of the central atom to the totally symmetric A_1 irreducible representation. With a little thought, it will be seen that the p_z orbital in ammonia is very similar in symmetry to the p_z orbital in water: Neither is altered by any symmetry operation, so p_z also transforms in the A_1 irreducible representation. The situation with regard to the p_x and p_y orbitals is more complex, however. A threefold rotation does not take the positive lobe of p_x (shown with a vector inside it) all the way to its negative lobe, but instead leaves it in between the negative p_x and the negative p_y lobes. By vector analysis, we see that it can be expressed as a resultant of the two. A somewhat similar situation applies to the p_y orbital upon threefold rotation. These two orbitals must therefore fall into the doubly degenerate irreducible representation in the C_{3v} point group, E. The effects of the C_3 operation are difficult to envision, since they intermix the two orbitals. The positive lobe of the p_x orbital is partially converted to the negative lobe; vector analysis shows that the partial character of this part of the operation is -0.5. Since the same result holds for the lobe of the p_y orbital, the overall character of this operation is $(-\frac{1}{2}) + (-\frac{1}{2}) = -1$. Fortunately for us, the character table spares us the necessity of doing this vector analysis, since it shows (p_x, p_y) at the right side of the E irreducible representation. The calculations of the characters of the other operations also take into account the fact that the operations act on more than one orbital. Thus, the identity operation E leaves *both* the p_x and the p_y components of the set unchanged; so the character of this operation is 2. Let us next consider any one of the σ_v operations, say the one shown in Figure 9.16: This mirror reflection interchanges positive and negative lobes of the p_x orbital for a *partial* character of -1, but also leaves the p_y orbital unchanged, for a partial character of $+1$. The overall character for the operation on both parts of the set is therefore 0. Finally, we see that the three characters, in order, are $2, -1, 0$; this set of characters is found under the irreducible representation labeled **E**, which is confirmed by the appearance of the (x, y) set at the right of this row.

In the higher order symmetry point groups, even greater degeneracy of orbitals can occur. For example, the p_x, p_y, and p_z orbitals are completely equivalent to each other in strictly tetrahedral (point group T_d) and octahedral (point group O_h) molecules and are interchanged with each other during symmetry operations; such **triply degenerate** irreducible representations use the letter **T** (or **F**) in their labels. Only in the I_h (icosahedral) point group (*Examples*: buckminsterfullerene, C_{60}, and dodecahedrane, $C_{20}H_{20}$) are higher degeneracies possible: Quadruply degenerate representations include the letter **G**, while quintuply degenerate representations include the letter **H.**

In classifying the central-atom orbitals of molecules of higher order, it is probably best to consider the complicated rotation operations last. (Of course, the results for *p* and *d* orbitals are listed at the right side of the table, but we may have to classify some *f* orbitals or things other than central-atom orbitals.) We can narrow the possibilities by first noting that *x* and *y* components of the items being classified are interchanged in groups with C_3 or higher axes, while *x*, *y*, and *z* components are only interchanged with each other in the higher order groups. Second, look at the effects of the inversion operation, if present, to determine whether the irreducible representation is *gerade* or *ungerade*; note that *s*, *d*, and *g* orbitals are gerade, while *p* and *f* orbitals are ungerade. Third, look at the effects of mirror planes. Hopefully, the classification can be reached without having to consider the effects of rotations, particularly of the C_3, C_5, and so on, types: some of these operations produce imaginary or trigonometric characters that we will not consider here.

The labels of irreducible representations have a number of uses in inorganic chemistry. We have already seen one use in Chapter 8, in which the *d* orbitals were grouped in symmetry-equivalent sets (d_{xy}, d_{xz}, d_{yz}) and (d_{z^2}, $d_{x^2-y^2}$); we then labeled these sets as t_{2g} and e_g, respectively. Consulting the character table for the O_h point group shows that these are the (uncapitalized) labels of the irreducible representations of these sets of orbitals.

9.5 *Applications of Symmetry: Infrared and Raman Spectra

Starred sections such as this at the end of a chapter can be omitted at the instructor's discretion with relatively minor effects on subsequent chapters.

The symmetry of a molecule or ion often has a major impact on the complexity of its IR spectrum. This impact can be used on the observed spectrum of a known ion or functional group to deduce whether or not, in a particular derivative, the ion or functional group is in a relatively high-order symmetry point group, which in turn may tell us something about what geometric isomer is present, or whether the species is a free ion, is a monodentate ligand, or is a polydentate ligand. We assume that you have used IR spectra in your organic chemistry course to identify functional groups in a compound, and thus are familiar with some of the factors that determine how high or how low the IR frequencies of a given group are.

The symmetry of a ion, molecule, or the local symmetry of an isolated[6] functional group determines *how many* IR absorptions should be present in the spectrum. [It does not determine the frequencies or energies of those absorptions; this calculation requires molecular orbital theory (Chapter 10)]. In a polyatomic species, each of the *N* atoms has three directions in which it can move (*x*, *y*, and *z*); so the total number of types of motion available to the polyatomic species is 3*N*. The molecule as a whole can still move in the three directions; these are called the three *translational modes* of the molecule and do not contribute to its IR spectrum. In addition, a molecule can rotate about its independent axes; for a *nonlinear* molecule there are thus *three rotational modes* that also do not contribute[7]; in a *linear* molecule rotations about the x and y axes are indistinguishable, so there are *two* noncontributing rotational modes. All the remaining types of motion (degrees of freedom) of the molecule are *internal vibrational modes* of the molecule and are what we are concerned with now:

Nonlinear molecules have $3N - 6$ internal vibrations.

Linear molecules have $3N - 5$ internal vibrations.

Thus we can see that a large molecule such as buckminsterfullerene, C_{60}, has a very large number of internal vibrations, 174! If all of these gave rise to IR absorptions, the IR spectrum of buckyballs would be very complex indeed; in fact, however, it has only *four* IR absorptions. Its high symmetry is responsible for its very simple IR spectrum; conversely, the observation of a simple IR spectrum for a molecule of that known size shows that it has high symmetry.

All vibrational modes of a species must be classifiable into one or another of the irreducible representations of the point group of the species. The simplest type of species is a *homonuclear diatomic molecule* such as H_2, N_2, or O_2; let us begin with molecules that belong to the $D_{\infty h}$ point group. For these simple linear molecules, $3N - 5 = 1$: there is only one possible vibrational mode, which is the one in which the two atoms stretch apart and then return, and which we can simply represent as H $\longleftrightarrow$ H. Referring to the character table for this point group (Table 9.3) and applying the operations to this vibrational mode, we see that no operation alters it, since the two atoms stretch together and return together. This one vibrational mode belongs to the first-listed, totally symmetric $\Sigma_g{}^+$ irreducible representation.

Let us next consider a nonlinear triatomic molecule, H_2O; this molecule has $3(3) - 6 = 3$ vibrational modes; we have already classified it in the C_{2v} point group. Its vibrational modes must fall in one of the four irreducible representations of this point group. However, if we try to allow one O–H bond at a time to vibrate, we find that the C_2 operation interchanges the vibrating and the nonvibrating O–H bonds; consequently, the two O–H bonds must be considered together in any vibrational mode of this molecule. Thus individual bonds do not vibrate separately; vibrations are of symmetry-adapted combinations of individual bonds that can be classified in one of the irreducible representations of the point group.

Through methods that, although not complex, are somewhat beyond the scope of this text, it is possible to derive pictures of the three collective vibrational modes of the water molecule. These methods are shown in Figure 9.17. By applying the operations of the C_{2v} point group to these vibrational modes, we can see that the first and the third of these are unchanged on applying any of the four operations, so they fall in the A_1 irreducible representation; the second one has heads and tails of the arrows interchanged during the C_2 and the $\sigma_v(yz)$ operations, so it falls in the B_1 irreducible representation. (By also applying what you may have learned in your organic or physical chemistry course, we see that the first two modes are O–H *stretching* modes, since bonds are stretched. These modes fall at higher frequencies than the third, which is a bond *bending* mode. In liquid water, the first two IR absorptions are at 3450 and 3615 cm^{-1}, while the bending absorption is at 1640 cm^{-1}.[8])

Now, let us briefly consider the IR spectrum of C_{60}, which falls in the high-

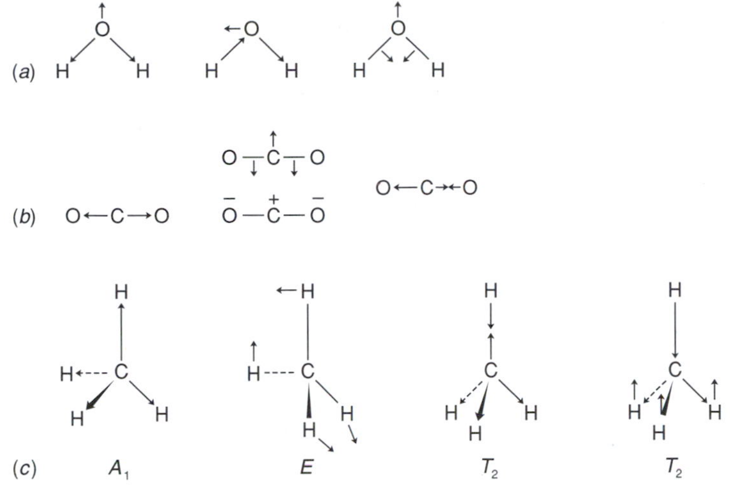

Figure 9.17
Vibrational modes of some simple molecules: (*a*) H_2O; (*b*) CO_2; (*c*) CH_4; (*b*) and (*c*) are for use in selected exercises.

symmetry I_h point group. Referring to the character table of this point group (Appendix B), we see that most of the irreducible representations are labeled *T*, *G*, or *H*. These representations are degenerate. This means that, if one vibrational mode falls in one of these degenerate irreducible representations, others must too. Vibrational modes must come in sets of three in a *T* representation, in sets of four in a *G* representation, and in sets of five in a *H* representation. But all vibrational modes occurring together in one of these degenerate sets of modes need the same energy to vibrate; each set of five modes in an *H* representation gives rise to only one IR absorption band. Degeneracy is one part of the story of the simplicity of the IR spectrum of buckyballs: Many vibrational modes are degenerate in energy with each other, so they cannot give separate IR frequencies.

To see the other part of the story, we must consider what happens when a molecule absorbs IR light. The wave of light can be considered to be a vector passing through the molecule in either the *x*, the *y*, or the *z* direction. In order for the light to be absorbed, the vibrational mode must involve a net motion of charge along one of these directions as well. That is, there must be *polar bonds* being excited in the vibrational mode, and the (inward or outward) stretch or bend must be asymmetric in one of these directions, so that the centers of positive and/or negative partial charges move; the net effect must be that there is a *change in dipole moment* during the excitation if the light wave is to be absorbed. Fortunately, this is easily determined by looking at a character table; irreducible representations of vibrational modes that involve a change in dipole moment have the entries *x*, *y*, or *z* on the right side of the table.

First, let us begin by considering the one vibrational mode of a homonuclear diatomic molecule such as H_2, N_2, or O_2. We see immediately that the covalent bond is nonpolar, so there can be no change in polarity upon its excitation. Second, it is in the totally symmetric $\Sigma_g{}^+$ irreducible representation, which includes neither *x*, *y*, nor *z* in its character table entry; even if there were polarity in this bond, the polarity would not change during this excitation since the left atom moves as far left as the right atom moves right; no change in center of the (hypothetical) charge occurs. Consequently, these molecules *have no IR absorptions.* This IR inactivity is environmentally very important, because the earth cools off at night by emitting IR (heat) radiation; if this were absorbed in the atmosphere, the cooling would be diminished and the earth would heat up. Fortunately, most of the gases in the atmosphere are either monoatomic (Ar), which have no vibrational modes at all, or are homonuclear diatomic (N_2 and O_2). The greenhouse effect in the atmosphere is due to gases that are either heteronuclear diatomic or are polyatomic (such as H_2O and CO_2); as we shall now see in the case of H_2O, these molecules have IR absorptions.

The character table for the water molecule (C_{2v}) is simple, and has only four irreducible representations; only the A_2 representation has neither x nor y nor z listed next to it, therefore it implies IR inactivity. However, none of the three vibrational modes of water (Fig. 9.17) falls in the A_2 irreducible representation; all are IR active. (You should examine these modes and satisfy yourself that each one involves net motion of the partially positive hydrogen end of the molecule and the partially negative oxygen end away from each other in some direction.)

The character table for C_{60} (I_h) has 10 irreducible representations, and (x, y, z) occur together in just one of them, T_{1u}—this means that vibrational modes in nine-tenths of the irreducible representations are IR inactive. Although the vibrational modes are not evenly distributed among the irreducible representations, this is qualitatively the second reason why the IR spectrum of C_{60} is so simple, with only four IR frequencies[9] allowed by the symmetry of the molecule.

A frequent use of these ideas by inorganic chemists has to do with how these IR spectra *change* upon complexation. Complexation of very symmetrical ligands is normally accompanied by a drastic loss of symmetry: The coordinated ligand then falls in a point group with a smaller total number of irreducible representations, so fewer of them fail to include x, y, and z; furthermore, few if any of the remaining irreducible representations are degenerate and include more than one vibrational mode at the same energy; consequently, the ligand develops additional IR frequencies.

Of course, on complexation new bonds are formed from the ligand to the metal, which adds IR bands independently to the above, but generally the metal–ligand bond is much weaker than the bonds within the ligands, and the metal atom is much heavier than the atoms within the ligand. Both of these factors mean that there is a great difference in the IR frequencies of the two types of bond, which results in little interaction of the vibrational modes within the ligand with those on the metal or beyond.

Soon after the discovery of a practical synthesis of C_{60}, inorganic chemists speculated that it might act as a π-donor ligand, donating electrons either from the five-membered rings on its surface (by analogy with cyclopentadienide ion) or from the six-membered rings (by analogy with benzene complexes). In fact, the early complexes isolated were *dihapto* complexes, such as $[(\eta^2\text{-}C_{60})Pt(PPh_3)_2]$. A complex of this type must belong to a drastically lower symmetry point group such as (probably) C_{2v}; in this point group there are no degenerate irreducible representations, and only the vibrational modes falling in the A_2 irreducible representation are IR inactive by virtue of symmetry. The ligand in this complex should almost approach having 174 IR bands! (In practice, in such a complex situation some vibrational modes would accidentally fall at the same energy as others (*accidental degeneracy*), and many of the IR bands would be so close to others that they could not be resolved—but the difference between this spectrum and that of pure C_{60} is quite dramatic nonetheless.[10])

Example 9.7

Would IR be useful to determine the bonding mode, if any, of the pentachlorocyclopentadienide ion $(C_5Cl_5)^-$ to metals? (The chlorinated analogue is chosen here because the Cl atoms are substantially heavier than the C atoms, and the C–Cl bonds are weaker than the C–C and C=C bonds, therefore one can analyze the C_5 ring by itself with little interference from involvement of the C–Cl vibrations in the C–C vibrational modes.)

SOLUTION:

The C_5 ring by itself has $3(5) - 6 = 9$ vibrational modes. The free anion falls in the D_{5h} point group; the character table for this point group (Appendix B) shows that one-half of the irreducible representations are degenerate, and only two of the eight irreducible representations allow IR activity. Hence, a very simple IR spectrum in the C–C stretching and bending regions is expected.

The *pentahapto*-C_5Cl_5 ring has a local symmetry of C_{5v}. Again, one-half of the irreducible representations are degenerate, so the spectrum will still be simplified by degeneracy of many of the vibrational modes; now two of the four irreducible representations allow IR activity. This is a higher fraction allowing IR activity, so a somewhat more complex spectrum is to be expected—in the absence of a more quantitative analysis, and given that combination and overtone bands may occur. The difference may not be strong enough to be decisive, however.

The *monohapto*-C_5Cl_5 but nonfluxional ring has a local symmetry of C_s; in this low-order point group *all* nine vibrational modes are IR active. This situation should be clearly distinguishable from the preceding cases.

The *monohapto*-C_5Cl_5 but fluxional ring also has an instantaneous local symmetry of C_s. The rate of exchange needed to alter IR spectra is prohibitively large (Table 9.2), so this IR spectrum will have the same number of bands as in the nonfluxional case. But the IR measurement in connection with the very much altered NMR spectrum will show the presence of a fluxional η^1-C_5Cl_5 ring instead of, say, a η^5-C_5Cl_5 ring (the NMR spectra of both of these would show but one peak at high temperatures).

In fact, the $C_5Cl_5^-$ ion shows one C–C IR absorption, a stretching frequency at about $1415\,cm^{-1}$; $(\eta^5\text{-}C_5Cl_5)_2Fe$ shows two C–C stretching frequencies at 1350 and $1307\,cm^{-1}$; and $(\eta^1\text{-}C_5Cl_5)R$ (R = organic group or mercury group) show about nine C–C stretching and bending frequencies ranging between 1604 and $339\,cm^{-1}$; from NMR and nuclear quadruple resonance (NQR) criteria, some of the Hg compounds are fluxional, while the organic derivatives are not.[11]

This type of criterion is of no use for detecting coordination modes of ligands that are already of low symmetry, but many oxo anions, for example, are of high symmetry when free but of substantially lower symmetry when coordinated. In addition, in many cases there is enough coupling of ligand vibrational modes through coordinated metal atoms to allow the identification of geometric isomers of complexes. Identification of isomers is especially practical in complexes of triply bonded π-donor ligands such as carbon monoxide, since the C≡O IR frequency is far removed from almost any other IR frequency (at $\sim 2170\,cm^{-1}$) and is easily identified (and interacts little with other vibrational modes). In a complex *cis*- or *trans*- $[ML_2(CO)_4]$ the four C≡O stretching vibrations couple with each other to give four new vibrational modes that fall in the irreducible representations of the C_{2v} point group (cis isomer) and the point group D_{4h} (trans isomer), respectively. There is a substantial difference in the character tables of these two point groups: D_{4h} has 10 irreducible representations, including two degenerate ones, eight of which confer IR inactivity; unsurprisingly, only one of the four C≡O stretching vibrations is IR active [at 1890 cm^{-1} for L = Ph_3P in the trans isomer]; with the much smaller C_{2v} character table, it turns out that all four are IR active (at 2016, 1915, 1900, and 1890 cm^{-1} in the cis isomer).[12] Section 11.2 and Table 11.3 give more details on the use of C≡O stretching vibrations in studying various metal carbonyls.

Although it is not as commonly available, there is another technique for obtaining vibrational spectra, **Raman spectroscopy**.[13] In this technique, visible laser light is reflected by a molecule, changing directions; some of the reflected light changes in energy when part of its energy is used to excite molecular vibrations. Given that a change in directions is involved, it is perhaps not surprising that the criterion for Raman activity is different: If a given irreducible representation allows Raman activity, on the right it will list xy, xz, yz, x^2, y^2, z^2, and/or combinations such as $x^2 - y^2$ or $x^2 + y^2 + z^2$. In higher order symmetry groups these often do not coincide with the IR active bands, so independent verification of the symmetry is possible. In addition, if the molecule possesses an inversion center, it works out that there are no lines in common between the IR and Raman spectra. Thus, whereas C_{60} has 4 IR active absorptions, all in the T_{1u} irreducible representation, it has 10 Raman active absorptions, in the A_g and H_g irreducible representations; in the lower symmetry $C_{60}M(PPh_3)_2$ (M = Pd or Pt) the quintuply degenerate H_g absorptions split into up to 5 other absorptions, while the nondegenerate A_g modes do not split.[14] Furthermore, for efficient Raman absorption, the requirement is not a change in dipole moment, but rather a change in polarizability. So Raman absorptions are strong for nonpolar bonds that are undetectable in the IR spectra.

Study Objectives

1. Identify symmetry elements in a molecule, ion, or object. Exercises 1–5.
2. Use the symmetry elements of a molecule or ion to identify the symmetry point group of a molecule. Exercises 6–16.
3. Know and apply the symmetry criteria for determining whether a molecule is polar or chiral. Exercises 17–21.
4. Describe the NMR spectra to be expected for given chemical species given their hapto nomenclature and/or the information that they are fluxional; distinguish different derivatives of appropriate ligands using the hapto nomenclature and symbolism. From the chemical shifts in the low-temperature NMR spectrum of a fluxional molecule, compute the expected high-temperature chemical shift. Exercises 22–27.
5. Determine the irreducible representation of an object (such as an orbital) located at the center of a molecule. Exercises 28–37.
6. From the formula of a molecule and the knowledge of whether or not it is linear, determine the total number of vibrational modes of that molecule. Identify the irreducible representation (symmetry species) in the character table of a given point group (a) that will result in IR absorptions; (b) that will result in Raman absorptions. Given drawings of the normal vibrational modes of a molecule, tell which vibrational modes will be IR active and which Raman active, and which vibrational modes correspond to stretching modes (which will tend to be higher in frequency). Exercises 38–46.

Exercises

An asterisk (*) beside the exercise number indicates that the answers are in the back of the book.

1. Figure 9.18 shows a number of two-dimensional representations of natural organisms, objects, or works of art. For each object: identify the highest order rotation axis present; tell whether an inversion center is present; and tell whether any mirror planes (other than the plane of the object itself) are present.

2. *Figure 9.1 shows a number of three-dimensional natural and artificial objects. For each object: (a) identify the highest order rotation axis present; (b) tell whether an inversion center is present; (c) tell whether any mirror planes are present; and (d) tell whether any improper rotation axes are present.

3. What kinds of symmetry elements (aside from E) are present in the molecules shown in Figure 3.11 as examples for VSEPR predictions?

4. What kinds of symmetry elements (aside from E) are present in the molecules shown in Figure 3.10 as examples for VSEPR predictions?

5. Below are shown several molecules or ligands, along with their basic geometric shapes as predicted by VSEPR theory. For each molecule or ligand: identify the highest

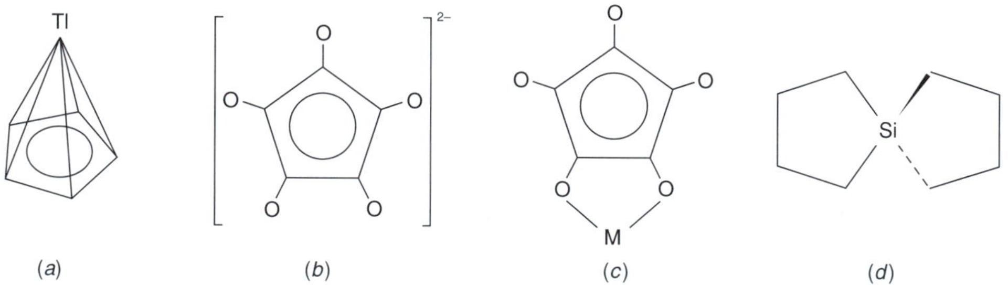

order rotation axis present; tell whether there is a C_2 rotation axis perpendicular to this highest order axis; tell whether an inversion center is present; tell whether any mirror planes are present: (a) pentagonal pyramidal TlC_5H_5; (b) pentagonal planar $[C_5O_5]^{2-}$; (c) bidentate chelated C_5O_5M; (d) chelated tetrahedral $Si(CH_2CH_2CH_2\text{-}CH_2)_2$.

Figure 9.18 (*facing page*)
Two-dimensional artistic or natural patterns for use in the Exercises: (*a*) American Indian decoration. Adapted with permission from L. H. R. Appleton, *American Indian Design and Decoration*, Mineola, NY, Dover Publications, 1971. (*b*) Quilt pattern with oak leaf wreath. Adapted with permission from M. Ickis, *Standard Book of Quilt Making and Collecting*, Mineola, NY, Dover Publications, 1959. (*c–h*) Pueblo Indian pottery designs. Reprinted with permission from Dorothy Koster Washburn, *A Symmetry Analysis of Upper Gila Area Ceramic Design*. Papers of the Peabody Museum of Archaeology and Ethnology, vol. 68, figures 66, 84, 90, 110a, 134, and 225. Copyright 1977 by the President and Fellows of Harvard College. (*i, j*) M. C. Escher "Whirlpools" and "Circle Limit IV" respectively. © 1998 Cordon Art B. V., Baarn, Holland. All rights reserved. Reprinted with permission. (*k, m*) Stalked jellyfish, and starfish and other organisms, respectively. Adapted from E. Haeckel, *Kunstformen der Natur*, Verlag der Bibliographischen Instituts, Leipzig, 1904. (*l*) Cross section of protein disk of tobacco mosaic virus, a cover diagram from J. M. Champness, A. C. Bloomer, C. Bricagne, P. J. C. Butler, and Aaron Klug, *Nature*, Vol. 259, 1976. Reprinted with permission of Macmillan Journals Ltd. and Dr. Aaron Klug. (*n*) Photograph of color-stained snowflake crystal by Emmitt Pollard. Reproduced with permission.

(a) (b) (c) (d) (e) (f) (g) (h) (i) (j) (k) (l) (n)

6. *Classify the objects you analyzed in Exercise 2 into symmetry point groups.

7. Classify the following isomers of coordination compounds into symmetry point groups: *cis*-$[PtCl_2(NH_3)_2]$ and *trans*-$[PtCl_2(NH_3)_2]$. For this purpose, assume that the NH_3 groups rotate freely so that the hydrogens do not lower the symmetry of these complexes.

8. Classify the following isomers of coordination compounds into symmetry point groups: *mer*-$[CoCl_3(NH_3)_3]$ and *fac*-$[CoCl_3(NH_3)_3]$. [The fac and mer isomers are illustrated in Fig. 9.10(*c* and *d*).] Assume that the NH_3 groups rotate freely.

9. *Classify the following derivatives of the perxenate ion into symmetry point groups: (a) the free XeO_6^{4-} ion itself; (b) the free oxo acid H_4XeO_6, in which the two oxo groups lie in trans positions and the four OH groups are bent at oxygen in a conformation resembling a swastika; (c) the free oxo acid H_4XeO_6, in which the two oxo groups lie in cis positions and the OH groups are all bent away from the oxo groups; and (d) a hypothetical organic derivative (*ortho*-$C_6H_4O_2)_3Xe^{2+}$, in which three benzenedioxy groups each chelate the Xe^{8+} "cation".

10. The rhodizonate ion, $C_6O_6^{2-}$, is a planar hexagonal organic anion in which (due to resonance) all oxygen atoms are equivalent. Give the symmetry point groups of the following rhodizonate derivatives: (a) the free rhodizonate anion; (b) a bidentate chelate complex $[C_6O_6M]$; and (c) a complex in which the rhodizonate anion bridges three different metal atoms of the same type, chelating each one of them: $[C_6O_6M_3]$.

11. *Classify into a symmetry point group the crown ether 18-crown-6 (Fig. 4.9, at the left): (a) assuming that the crown ether is completely planar; (b) assuming (more realistically) that the CH_2CH_2 groups alternately lie above and below the plane of the six O atoms.

12. Classify into symmetry point groups the examples of the basic VSEPR geometry types shown in Figure 3.10.

13. Classify the examples of the basic VSEPR geometry types having unshared electron pairs shown in Figure 3.11 into symmetry point groups.

14. Consider the following molecules or ions: ICl_4^-, SbFClBr, TeO_6^{6-}, and XeO_4^{2-}. (a) What is the highest order rotation axis present in each molecule or ion? (b) Which of the molecules or ions have centers of inversion? (c) Which of the molecules or ions have mirror planes? (d) Give the symmetry point groups of each of the above molecules or ions.

15. For each of the following species, list the highest order rotation axis present, and give the symmetry point group to which the molecule belongs: (a) $CHCl_3$; (b) the aromatic hydrocarbon $C_5H_5^-$, as a free ion; (c) the aromatic hydrocarbon $C_5H_5^-$, coordinated to a metal ion on one face at the center of the five carbon atoms; (d) the anion $Cl_3CSO_3^-$ with an eclipsed conformation (Cl atoms directly above O atoms); and (e) $Cl_3CSO_3^-$ with a staggered conformation (Cl atoms falling midway between the O atoms).

16. Which of the following molecules fall in the point group C_{3v}? (a) $:NH_3$; (b) CH_4; (c) 1,3,5-trichlorobenzene; (d) *fac*-$[Cr(NH_3)_3Cl_3)]$; and (e) *mer*-$[Cr(NH_3)_3Cl_3)]$. [The fac and mer isomers are illustrated in Fig. 9.10(*c* and *d*).]

17. *Which of the objects shown in Figure 9.1 are chiral, and which would be polar molecules (if they were molecules containing polar covalent bonds)?

18. (a) Molecules in which of the following point groups *cannot* be *polar (i.e., must* be *nonpolar)*? D_{3d}, T_d, C_s, O_h, $D_{\infty h}$, $C_{\infty v}$, C_{17}, D_2. (b) Molecules in which of the following point groups are *chiral*? D_{3d}, T_d, C_s, O_h, $D_{\infty h}$, $C_{\infty v}$, C_{17}, D_2. (c) Which are more likely to be chiral? Complexes of monodentate ligands, or complexes of chelating ligands, or both types are equally likely.

19. The dianion of 1,2-benzenediol (catechol), *ortho*-$[O_2C_6H_4]^{2-}$, readily forms chelated complexes with many metal and nonmetal cations. Draw structures of the chelate complexes that this dianion would form if it displaced all of the fluoro, oxo, and so on, ligands already present in the selected structures drawn out in Figures 3.10 and 3.11. (Where isomeric choices are possible, use the benzenediolato dianion to span neighboring axial and equatorial positions, or top and bottom planes, etc.) Then tell whether the resulting complex is polar; whether it is chiral. (a) $[C(O_2C_6H_4)_2]$, formed from CF_4; (b) $[S(O_2C_6H_4)_3]$, formed from SF_6; (c) $[I(O_2C_6H_4)_4]^-$, formed from IF_8^-; and (d) $[S(O_2C_6H_4)_2]$, formed from SF_4.

20. Which (if any) of the isomers of $[PtCl_2(NH_3)_2]$ (Exercise 7) or $[CoCl_3(NH_3)_3]$ (Exercise 8) are (a) polar? (b) chiral?

21. Each of the following pairs of trisubstituted octahedral complexes of Cr can exist as two geometric isomers: facial and meridional. Assume that monodentate ligands freely rotate around their donor atoms, so only the donor atoms need be considered in assigning the point group, but recall that chelated ligands are not free to rotate. (a) Is *fac*-$[Cr(NH_3)_3(CH_3CO_2)_3]$ chiral? (b) Is *mer*-$[Cr(NH_3)_3(CH_3CO_2)_3]$ chiral? (c) Draw all four possible isomers of the octahedral complex $[Cr(NH_2CH_2CO_2)_3]$, in which $[NH_2CH_2CO_2]^-$ is the anion of the simplest amino acid, glycine. (*Hint*: Two will be mer and two will be fac isomers.) How many pairs of enantiomers are included among the isomers?

22. Estimate the low-temperature chemical shifts of the protons of the fluxional C_5H_5 ring in Figure 9.14, and use them to compute the expected chemical shift observed for these protons in the high-temperature spectrum. Is this high-temperature chemical shift the same as that observed for the η^5-C_5H_5 ring in the same compound? Should it be?

23. *The ring compound cyclooctatetraene, C_8H_8, is a π-donor ligand. (a) What numbers of π electrons can it donate to a metal atom above one of its faces? For each answer, give the symmetry point group for the product C_8H_8M; tell the number of distinct ^1H NMR peaks to be expected, and name the ligand using the hapto nomenclature. (b) The metal derivative $[Fe(C_8H_8)(CO)_3]$ shows four ^1H NMR peaks at low temperatures, with approximate chemical shifts (δ) as follows: 4.7, 4.9, 6.1, and 6.5 ppm. It is fluxional; at high temperatures how many NMR peaks should be present, at what approximate chemical shift(s)? (c) Given the fact that the low-temperature chemical shifts are grouped into two sets of two, what is the most plausible hapto number for $[Fe(C_8H_8)(CO)_3]$?

24. The ring compound $C_7H_7^+$, the cycloheptatrienylium or tropylium cation, is a π-donor ligand. (a) What numbers of π electrons can it donate to a metal atom above one of its faces? (Consider the formal positively charged carbon atom of this ligand as coordinating to the metal ion in all cases.) For each answer, give the symmetry point group for the product C_7H_7M; tell the number of distinct ^1H NMR peaks to be expected, and name the ligand using the hapto nomenclature. (b) Let us presume that the metal derivative $[Fe(C_7H_7)(CO)_3]$ shows four ^1H NMR peaks at low temperatures, with approximate chemical shifts (δ) as follows: 4.7, 4.9, 6.1, and 6.5 ppm (with the 6.5 ppm peak being of one-half the intensity of the others). It is fluxional; at high temperatures how many NMR peaks should be present, at what approximate chemical shift(s)? (c) Given the fact that the low-temperature chemical shifts are grouped into two sets of two, what is the most plausible hapto number for $[Fe(C_7H_7)(CO)_3]$?

25. The ring compound benzene, C_6H_6, is a π-donor ligand. (a) What numbers of π electrons can it donate to a metal atom above one of its faces? For each answer, give the symmetry point group for the product C_6H_6M; tell the number of distinct ^1H NMR peaks to be expected, and name the ligand using the hapto nomenclature. (b) Let us presume that the metal derivative $[Fe(C_6H_6)(CO)_3]$ shows three ^1H NMR peaks at low temperatures, with approximate chemical shifts (δ) as follows: 4.7, 6.1, and 6.5 ppm. It is fluxional; at high temperatures how many NMR peaks should be present, at what approximate chemical shift(s)? (c) Given the way that the low-temperature chemical shifts are grouped, what are the two most plausible hapto numbers for $[Fe(C_6H_6)(CO)_3]$?

26. The aromatic ring cyclononatetraenide anion, $C_9H_9^-$, is a π-donor ligand; let us assume that it is a planar regular nonagon (nine-sided figure). (a) What numbers of π electrons can it donate to a metal atom? (b) The metal derivative $[Co(C_9H_9)(CO)_3]$ has a $Co(CO)_3$ group attached to one side of the ring. This compound shows five ^1H NMR peaks at low temperatures, with approximate chemical shifts (δ) as follows: 4.5, 4.7, 4.9, 6.1, and 6.5 ppm, with the 4.5 ppm peak having one-half the intensity of the others. Given this particular grouping of chemical shifts, what are the two most plausible hapto number for $[Co(C_9H_9)(CO)_3]$? These correspond to how many π electrons being donated to the metal atom? This evidence suggests which one of the following point groups for the molecule: C_{9v} or C_{3v} or C_s or O_h or T_d. (c) At high temperatures the ring in this cobalt compound is completely fluxional, so that only one NMR peak is observed. At what chemical shift should this peak appear? This evidence by itself would seem (falsely) to suggest which one of the following point groups for the molecule: C_{9v} or C_{3v} or C_s or O_h or T_d.

27. The ring compound cyclodecapentaene, $C_{10}H_{10}$, is a π-donor ligand; let us assume that it is planar. (a) What numbers of π electrons can it donate to a metal atom? (b) The metal derivative $[Fe(C_{10}H_{10})(CO)_3]$ shows five ^1H NMR peaks at low temperatures, with approximate chemical shifts (δ) as follows: 4.5, 4.7, 4.9, 6.1, and 6.5 ppm. Given this particular grouping of chemical shifts, what are the two most plausible hapto numbers for $[Fe(C_{10}H_{10})(CO)_3]$? (c) It is fluxional; at high temperatures how many NMR peaks should be present? At what approximate chemical shift(s)?

28. *Figure 8.21 shows the appearances and names of the *f* orbitals in forms appropriate for a central atom in an octahedral (O_h) complex ion. (a) Are the *x*, *y*, and *z* directions equivalent in such a complex? Using your answer, group the seven *f* orbitals into three smaller sets of orbitals that are likely interchanged with each other during the symmetry operations of the O_h point group. (b) The irreducible representation labels for each set of *f* orbitals would begin with which of the following letters: A, B, E, T, G, or H? (c) The labels for each set would include which subscript: *g* or *u*?

29. In the endohedral fullerene TiC_{60} (Section 12.5), the Ti atom and its *d* orbitals sit at the center of a complex of icosahedral symmetry. Table 8.6 indicates that all five *d* orbitals of the Ti atom will experience the same crystal field splitting, namely, none. In the I_h point group, will the irreducible representation for the five *d* orbitals be labeled with a *g* subscript or a *u* subscript? Which of the following letters—A, B, E, G, H, or T—will begin the label of this irreducible representation?

30. In the endohedral fullerene LaC_{60} (Section 12.5) the *f* orbitals of the La atom sit at the center of a complex of icosahedral symmetry. In this symmetry, the *f* orbitals are grouped into two energy levels, a set of three *f* orbitals and a different set of four *f* orbitals. (a) In the I_h point group, will the irreducible representations for the *f* orbitals be labeled with *g* subscripts or *u* subscripts, or neither? (I_h contains the inversion center *i*.) (b) Which two of the following letters—*A, B, E, G, H*, or *T*—will begin the labels used for the irreducible representations to which the *f* orbitals belong?

31. For a trigonal prismatic complex, $[TiL_6]$, Table 8.6 indicates that the d_{z^2} orbital is at an energy level of 0.096 Δ_o, $d_{x^2-y^2}$ and d_{xy} are at an energy level of $-0.584\Delta_o$, and d_{xz} and d_{yz} fall at an energy level of $+0.536\,\Delta_o$. Which of these *d* orbitals will be classified in irreducible representations beginning with the capital letter *A*? Which beginning with capital letter *E*? Which beginning with capital letter *T*?

32. The UF_6 molecule is octahedral and falls in the O_h point group; you may consult the right side of the O_h character table to answer these questions. (a) In which irreducible representation does the uranium f_{xyz} orbital fall? Explain what the capital letter in the label for this irreducible representation means. Explain what the lower case letter in the label for this irreducible representation means. (b) In which irreducible representation does the uranium f_{z^3} orbital fall? Explain what the capital letter in the label for this irreducible representation means. Symmetry operations of this point group interchange the f_{z^3} orbital with which other *f* orbital(s)?

33. *There are suggestions that plutonium may show the +8 oxidation state. Let us suppose that it does, in the compound PuO_4, which would be expected to fall in the T_d point group. (a) Give the symmetry labels (irreducible representations) for each of the *p* orbitals and for each of the *d* orbitals of the plutonium ion at the center of the molecule. (b) Referring to the drawing of the plutonium f_{xyz} orbital in Figure 8.21, mentally orient this orbital inside the PuO_4 tetrahedron such that each negative lobe coincides with a Pu–O bond, and decide in which irreducible representation the f_{xyz} orbital belongs.

34. Table 8.6 shows the energy levels of the central-atom *d* orbitals in complexes of various geometries; except for the rare *accidental degeneracy*, all orbitals at the same energy level are interconverted by symmetry operations of the point group, and therefore belong to the same irreducible representation. Assume that all ligands for the complexes in question are identical, so that these geometries lose none of their symmetry. List the irreducible-representation labels that would be used for the sets of *d* orbitals in complexes of the following geometries: (a) square planar; (b) trigonal prismatic; and (c) icosahedral.

35. Let us assume that the gaseous mercury(II) hydroxide molecule has the planar

```
           H
            \
structure   O–Hg–O   (linear next to Hg but bent at both ends), so that it
                  \
                   H
```

falls in the C_{2h} point group. Using the symmetry operations from the character table for C_{2h} but without consulting the right side of the table, determine in which irreducible representation each of the following orbitals on the Hg atom fall: (a) *s*; (b) the *p* orbital that lies along the O–Hg–O direction; (c) p_z; (d) p_x; and (e) $d_{x^2-y^2}$.

36. The commonly used labels for the sets of metal *d* electrons in octahedral complexes, t_{2g} and e_g, are labels of irreducible representations if the complex belongs to the O_h point group {e.g., $[Co(NH_3)_6]^{3+}$ with free rotation of NH_3 groups}; but they are not strictly applicable if the complex belongs to a lower symmetry point group. Referring to the character tables, pick the proper irreducible-representation labels for each metal *d* orbital in each of the following complex ions: (a) *cis*-$[CoCl_2(NH_3)_4]^+$; (b) *trans*-$[CoCl_2(NH_3)_4]^+$. (c) Excluding the possibility of accidental degeneracy of different *d*-orbital energy levels (Exercise 34), would you expect the visible spectra of these complex ions (see Chapter 8) to be more or less complex than that of $[Co(NH_3)_6]^{3+}$?

37. *The crown ether complex $[K(C_{12}H_{24}O_6)]^+$ was classified in an appropriate point group in Exercise 11b. Referring to the character table for this point group, give the symmetry labels (irreducible representations) for each of the *p* orbitals and for each of the *d* orbitals of the potassium ion at the center of the complex ion.

38. (a) What is the total number of vibrational modes expected for the sulfate ion, SO_4^{2-}? (b) In which irreducible representations of the appropriate point group will vibrational modes be IR active in the free sulfate ion? (c) In which irreducible representations will vibrational modes of the free SO_4^{2-} ion be Raman active? (d) What symmetry operation is necessary in a point group for there to be no common frequencies (*mutual exclusion*) between the IR and Raman spectrum of a species? Would the sulfate ion show this mutual exclusion? (e) Give the symmetry point group of a SO_4^{2-} ion chelated through two O atoms to a metal ion. Should IR spectroscopy be able to distinguish a free sulfate ion clearly from a sulfate ligand chelated to a metal ion? Explain.

39. *(a) What is the total number of vibrational modes expected for the perxenate ion, XeO_6^{4-}? (b) In which irreducible representations of the O_h point group will all vibrational modes of the perxenate ion be IR active? (c) In which irreducible representations of the O_h point group will all vibrational modes of the perxenate ion be Raman active? (d) Give the symmetry point group of a perxenate ion chelated to a metal ion through three O atoms. Should IR spectroscopy be able clearly to distinguish such a chelated perxenate ion from a free perxenate ion? Explain briefly.

40. (a) If the fullerene LaC_{60} (Section 12.5) were to have no symmetry whatsoever (i.e., belonged to the C_1 point group), how many vibrational modes should it exhibit in its IR spectrum? (b) If LaC_{60} were to belong to the I_h point group, list the irreducible representations in which the vibrational modes of C_{60} would be IR active; list those in which the vibrations would be Raman active. (c) A more likely structure of LaC_{60} is shown on the front cover. In what point group would this structure fall?

41. (a) Compute from its molecular formula how many vibrational modes a linear triatomic molecule such as CO_2 must have. (b) The vibrational modes of CO_2 are shown in Figure 9.17(*b*). Which two of these modes belong to a degenerate irreducible representation? (c) Which of these modes does not contribute to the IR spectrum of CO_2? Which is/are responsible for the fact that CO_2 is a greenhouse gas?

42. Carbon suboxide, C_3O_2, is a reactive molecule whose structure you are trying to determine by IR spectroscopy. You are considering three hypothetical structures: **A**, which is O=C=C=C=O; **B**, which is a cyclic structure drawn below, and **C**, which is a bicyclic spiro structure also drawn below. (a) Construct molecular models of

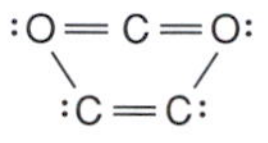

:O=C=O:
:C=C:
B

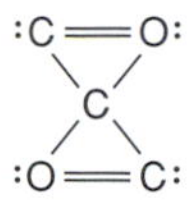

:C=O:
C
:O=C:
C

each hypothetical structure and assign each to a point group. (b) Give the irreducible representations of all vibrational modes that would be IR active in each point group. (c) Give the irreducible representations of all vibrational modes that would be Raman active in each point group. (d) Explain qualitatively how you could use the IR and Raman spectra to help determine the structure of carbon suboxide. (e) Which structures of carbon suboxide could be chiral? Which could be polar?

43. The simplest organic compound, methane, CH_4, has two IR absorptions, at 3019 and $1306\,cm^{-1}$. (a) What is the total number of vibrational modes that this molecule should have? (b) Figure 9.17(*c*) is considered a complete enough drawing of the vibrational modes (with irreducible representations) of a tetrahedral molecule such as CH_4; but you should have calculated a larger number of modes than this in part (a). Explain the apparent discrepancy. (c) Tell which of the four vibrational modes in Figure 9.17(*c*) is the one responsible for the IR absorption at $3019\,cm^{-1}$, and which one is responsible for the IR absorption at $1306\,cm^{-1}$; explain how you knew.

44. *Which will have the greater number of bands in its IR spectrum: SO_3 or $SO_3{}^{2-}$? Explain.

45. The Raman spectrum of C_{60} has been measured (see Chase and Fagan[14]). (a) In what irreducible representations of the I_h point group will the Raman-active vibrational modes appear? (b) Will any vibrational modes that are Raman active also be IR active? (c) Which spectrum of C_{60} would you expect to be more intense, the IR spectrum or the Raman spectrum? Why? (d) If the burning of forests were to spew large quantities of C_{60} instead of soot into the atmosphere, would the C_{60} be likely to act as an important greenhouse gas?

46. The main failures of VSEPR theory occur with molecules having a coordination number of 6 and a total coordination number of 7: Among such molecules and ions, some show an unexpected octahedral geometry while others show lower symmetry. K. O. Christe and W. W. Wilson [*Inorg. Chem.*, **28**, 3275 (1989)] studied the ions $BrF_6{}^-$ and $IF_6{}^-$; their NMR spectra suggested that both are fluxional. (a) They also studied the Raman spectra of these ions in solution. Which geometry (octahedral or distorted octahedral) should give the more complex Raman spectrum? What did they find in practice? (b) Did the authors cite any a priori reason for expecting these two ions to have different geometries? What is this reason?

Notes

1. Many examples in art and nature are found in I. Hargittai and M. Hargittai, *Symmetry through the Eyes of a Chemist*, VCH, New York, 1986. Even automobile tires have interesting elements of symmetry: W. G. Jackson, *J. Chem. Educ.*, **69**, 624 (1992). In addition, there may be evolutionary reasons why animals prefer symmetrical (beautiful) mates: N. Angier, *New York Times*, Feb. 8, 1994, p. B5.
2. The combination of any two operations in the group must be another operation in the group; One operation (i.e., E) must commute with all other operations in the group and leave them unchanged; The associative law of combination must hold: $\mathbf{A} \times (\mathbf{B} \times \mathbf{C}) = (\mathbf{A} \times \mathbf{B}) \times \mathbf{C}$; The reciprocal of each element is also in the group.
3. It is very useful in organic chemistry to treat C–H bonds, in which the bonded atoms differ in electronegativity by 0.35, as nonpolar; hence, your instructor may prefer to define polar bonds as those between atoms that differ in electronegativity by 0.4 or more.
4. A controversy currently exists over whether the unusual IR spectrum of tricarbonyl-(norbornadiene)iron is due to exchange of carbonyl groups that is rapid on the IR time scale: See J. J. Turner, F.-W. Grevels, S. M. Howdle, J. Jacke, M. T. Haward, and W. E. Klotzbücher, *J. Am. Chem. Soc.*, **113**, 8347 (1991); A. Vlcek, Jr., *Chemtracts: Inorg. Chem.*, **4**, 189 (1992); H. L. Strauss, *J. Am. Chem. Soc.*, **114**, 905 (1992).
5. G. P. Wulfsberg, J. S. Frye, A. C. Buchanan III, A. Weiss, C. C.-C. Jui, D. A. Davis, K. Bass, and R. W. Todd, *Organometallics*, **6**, 2363 (1987); G. P. Wulfsberg, A. C. Buchanan III, I. Rubin, A. Weiss, D. A. Davis, K. Bass, R. W. Todd, and C. C.-C. Jui, *Z. Naturforsch.*, **41a**, 175 (1986).
6. You may recall that IR frequencies depend on the masses of the vibrating atoms; functional groups that are connected through atoms of greatly different masses do not influence each other much as far as IR spectra go, so act as isolated functional groups.
7. These are indicated by the symbols R_x, R_y, and R_z in the character tables.
8. K. Nakamoto, *Infrared and Raman Spectra of Inorganic and Coordination Compounds*, 3rd ed., Wiley-Interscience, New York, 1977.
9. In practice, some absorptions can be overtones or combinations of other absorptions, so the numbers of predicted bands are not always precisely realized. The mathematical methods for predicting how many bands fall in each irreducible representation are not unduly complicated and are given in any text on chemical applications of group theory or physical methods in inorganic chemistry, but these methods are beyond the scope of this text.
10. B. Chase and P. J. Fagan, *J. Am. Chem. Soc.*, **114**, 2252 (1992); see also F. Chung and S. Sternberg, *Am. Sci.*, **81**, 56 (1993).
11. G. Wulfsberg, R. West, and V. N. M. Rao, *J. Am. Chem. Soc.*, **95**, 8658 (1973).
12. F. A. Cotton and G. Wilkinson, *Advanced Inorganic Chemistry*, 5th ed., Wiley, New York, 1988, p. 1035.
13. A fascinating biography of the Indian scientist C. V. Raman was given by F. A. Miller and G. B. Kauffman, *J. Chem. Educ.*, **66**, 795 (1989).
14. B. Chase and P. J. Fagan, *J. Am. Chem. Soc.*, **114**, 2252 (1992).

CHAPTER

10

Molecular Orbital Theory

With Applications to Organic Chemistry and Materials Science

10.1 Molecular Orbital Theory for Homonuclear Diatomic Molecules

In Part I, the bonding theory we used to describe molecules and ions was the *valence bond theory*, which is indeed an adequate bonding theory to describe the vast majority of molecules and ions in common experience, and which is relied upon in general chemistry and most of organic chemistry. However, from time to time we indicated that there are known substances in which the bonding is quite difficult to describe satisfactorily with valence bond theory. These difficulties occur especially in the realm of *electron-deficient molecules*, those molecules that do not have enough electrons in them to write a Lewis structure, but which exist and may be relatively stable nonetheless. (Indeed, it has been pointed out that these molecules are not deficient in electrons, since they are stable; they are really "theory-deficient molecules," i.e., the valence bonding theory was inadequate to describe them.) Occasionally, there are discrepancies that arise even when a "good" Lewis structure can be drawn; the classic example of this is the simple O_2 molecule, for which the Lewis structure shows an O=O double bond and two sets of (paired) unshared electrons on each O atom. However, the O_2 molecule in fact turns out to be *paramagnetic* (Section 8.3), with two unpaired electrons, and in liquid form it can be seen to be attracted to the poles of a strong magnet. When we discuss many extended or polymeric substances such as the metals themselves, the valence bond theory often proves very awkward to use.

The inadequacies of the valence bond theory are more serious when we attempt to interpret the various types of *electronic spectroscopy* of polyatomic ions and compounds, in which electrons often are promoted from orbitals that are readily described using valence bond theory into other orbitals that are much more difficult to describe in this manner. Even if the electrons are completely ionized away from the molecule, as in the method of *photoelectron spectroscopy* to be described in this chapter, the properties of the resulting ion are also difficult to describe classically. Therefore, the spectra are far more easily interpreted using molecular orbital theory.

Hence, we now need to introduce the **molecular orbital** (**MO**) **theory**. In this theory, molecules, by analogy with atoms, have orbitals called **molecular orbitals**; in analogy with atomic orbitals (AOs), which spread over (more or less) the entire atom, *MOs involve several atoms or the entire molecule*. Again, like atomic orbitals, each MO has a definite energy; there can be ionization energies and electron affinities for MOs. The electron configuration of a molecule is obtained by filling electrons into MOs starting with those at the lowest energy and in accord with the Pauli exclusion principle, and so on.

However, the computational problems become formidable when we try to apply the Schrödinger wave equation (Section 1.4) to electrons moving around several nuclei that attract them. Therefore, it is common to assume that, when the electrons are close to a given atomic orbital, their wave function closely resembles an atomic orbital of that atom. Consequently, MOs are commonly generated as a *linear combination of atomic orbitals* on the nuclei in question, which are allowed to overlap with each other. (So far, this is similar to the approach of valence bond theory).

In Section 3.1, we saw that the overlap of two atomic orbitals can be of three types: positive, negative, and zero. Valence bond theory emphasizes only the result of positive overlap: When two atomic orbitals (e.g., two 1*s* orbitals on two H atoms) have positive overlap, the (one or two) electrons that can reside in that orbital can spend a lot of time in the region of positive overlap, which is between the two nuclei. In this region, the electrons are attracted to the positive charges of the two nuclei; this attraction is a major source of the bond energy that makes H_2 a more stable species than two separate H atoms. Molecular orbital theory also takes this view of positive overlap, but differs in that the region of overlap is not restricted to being between *two* nuclei—it can fall between three, four, or any number of nuclei.

In contrast to valence bond theory, MO theory also stresses the importance of the case of *negative overlap* of two atomic orbitals (as in Fig. 3.1 parts *b*, *d*, *f*, *j–l*.) When two atomic orbitals have negative overlap, they act as waves that are out of phase with one another: The crest of one coincides with the trough of the other in the region of overlap, canceling out any wave amplitude. The electrons then must spend their time where the wave still has amplitude, which is in the nonoverlapping parts *outside* of the two nuclei. This leaves the two positive nuclei repelling each other with little or no electron density in between to help pull them together. Putting electrons into a MO built upon negative overlap not only does not contribute to bonding, but allows internuclear repulsion to proceed unchecked: If only such an orbital is involved, the atoms will fly apart; this is an **antibonding** interaction. Alternatively, if a molecule has two electron pairs, and one goes into a bonding orbital composed of (say) 1*s* orbitals on each atom showing positive overlap, while the other goes into an antibonding overlap composed of 1*s* orbitals showing negative overlap, the two effects cancel out: There is no bonding, and the electron pairs behave as unshared, nonbonding electron pairs do in Lewis structures. Actually, when the calculations are done, it turns out that the antibonding effect is somewhat larger than the bonding, so the two atoms (if using only these two electron pairs) not only do not bond, they fly apart: Two He atoms form no stable bond at all.[1]

Antibonding orbitals are important in MO theory in part because, in assembling MOs as linear combinations of atomic orbitals, there is a type of conservation of orbitals. If we combine *N* atomic orbitals to make molecular orbitals, we obtain *N* molecular orbitals; no atomic orbitals are "lost." Thus, if 1*s* orbitals are combined in a diatomic molecule to give a MO featuring positive overlap, there must also exist a corresponding MO featuring negative overlap. The existance of a MO with negative overlap does not mean, however, that we must put electrons into such an orbital: Commonly observed molecules have some or all of their antibonding orbitals unoccupied.

An important property of MOs is that they must have enough symmetry so that they belong to one of the irreducible representations of the point group of the molecule in which they are found; their labeling is based on the labels of these irreducible representations. In this section, we discuss the simplest type of molecules, **homonuclear diatomics** (those composed of just two identical atoms), which belong to the point group $D_{\infty h}$. The linear combination of two 1s orbitals with positive overlap has the full symmetry of the point group, so it falls in the $\Sigma_g{}^+$ irreducible representation. It is customary when labeling orbitals to convert the capital letters of irreducible representations to lower case letters, even when they are in Greek, so this MO is commonly called a σ_g molecular orbital; it matches the term "sigma bond" commonly used in valence bond theory for this type of bond. Since many types of positive overlap are possible in a given molecule, we may wish to show the type of atomic orbitals being combined on the two atoms A and B by designating this as a $\sigma_g(1s)$ molecular orbital. We can express it as a linear combination of the two 1s atomic orbital as follows:

$$\sigma_g(1s) = c_{\mathrm{A}}1s(\mathrm{A}) + c_{\mathrm{B}}1s(\mathrm{B}) \tag{10.1}$$

In a homonuclear diatomic molecule (but not a heteronuclear one), the two atoms are equally involved in the overlap, so the coefficients c_{A} and c_{B} are equal; we set them equal to one and ignore them, although they are affected by a mathematical requirement of *normalization* that is beyond the scope of our discussion.[2]

Since this MO was obtained by combining two atomic orbitals, there must be a distinct second MO, the one based on negative overlap, that is produced by taking one of the atomic 1s orbitals (say on atom B) as a trough rather than a crest, with a negative sign to its contribution to the wave function. This combination changes sign on inversion through the center of the molecule, so it falls in a different irreducible representation, $\Sigma_u{}^+$; it is labeled as a $\sigma_u{}^*$ orbital; the asterisk (*) is added to indicate that this is an antibonding orbital:

$$\sigma_u{}^*(1s) = c_{\mathrm{A}}1s(\mathrm{A}) - c_{\mathrm{B}}1s(\mathrm{B}) \tag{10.2}$$

We write electron configurations of molecules just as we do those of atoms. Molecular and atomic orbitals can take, at most, two electrons each, and the lowest energy one (in this case, the one with positive overlap) is filled first. We can write electron configurations for molecules or ions using the above two orbitals to hold from one to four electrons:

One electron (e.g., $H_2{}^+$)	$\sigma_g(1s)^1$ or just $\sigma_g{}^1$
Two electrons (e.g., H_2)	$\sigma_g{}^2$
Three electrons (e.g., $He_2{}^+$)	$\sigma_g{}^2\sigma_u{}^{*1}$
Four electrons (e.g., He_2)	$\sigma_g{}^2\sigma_u{}^{*2}$ (10.3)

In valence bond theory, an important classification of bonds is by their **bond order**: single, double, triple, or quadruple. In MO theory these bonds can occur along with nonintegral bond orders, so a formal definition of bond order is needed: It is *one-half*

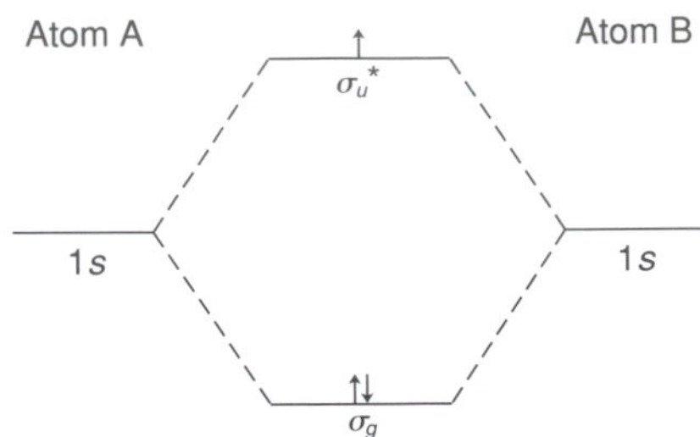

Figure 10.1
Energy level diagram for a homonuclear diatomic molecule or ion using $1s$ atomic orbitals (at the relative energies indicated on the left or the right) to form the MOs at the relative energies indicated at the center. Illustrated is the electron configuration of the ion He_2^+.

the difference between the number of electrons in bonding orbitals, n_b, and the number in antibonding orbitals, n^*

$$\text{Bond order} = 0.5\ (n_b - n^*) \tag{10.4}$$

Thus the bond order in H_2 is 1; in agreement with its Lewis structure from valence bond theory, it has a single bond. The bond order in He_2 is $0.5\ (2-2) = 0$, in agreement with our inability to draw a Lewis structure for such a molecule, it has a bond order of 0, and cannot exist because the atoms fly apart. Although one cannot draw likely looking Lewis structures for them, the other two ions, which each have nonintegral bond orders of $\frac{1}{2}$, can exist in high vacuum; green Xe_2^+ has been synthesized and is fairly stable in SbF_5 solution.[3]

It is worthwhile to put the notion of "stability" in its proper context here. We shall see that H_2^+ and He_2^+ have substantial bond energies and are stable *compared to the separate atoms and ions* (e.g., H and H^+). However, as we proceed through the rest of the text, we will not further mention these species, or most of the diatomic molecules that we may treat by MO theory in this chapter. The reason is that most of them are not stable to *polymerization* or to reaction with other molecules. We study them now because they are the simplest species in which we can investigate bonding; knowledge of them will also be of value to those who become radio astronomers, spectroscopists, or who study the chemistry of the upper atmosphere.

In Figure 1.9, we indicated (qualitatively) the relative energies of different atomic orbitals in an **energy level diagram**; we drew more of these in Chapter 8 to illustrate the effects of the crystal field theory. Similar diagrams can be drawn for molecules (Fig. 10.1). These diagrams are in fact even more valuable for molecules because they can express, not only the fact that the $\sigma_u^*(1s)$ is higher in energy (less stable) than the $\sigma_g(1s)$ molecular orbital, but it can also indicate how each stands in relationship to its parent $(1s)$ atomic orbitals.

To draw MO energy level diagrams, we first draw lines on either side of the diagram at the energy levels of the parent atoms (in polyatomics, central atoms are placed on one side and outer atoms on the other). These energy levels may be taken as the atomic *valence orbital (potential) energies* tabulated in Table 10.1.[4]

Next, we draw in bars in the center of the diagram to represent the energies of the MOs. Although computations are necessary to locate these energy levels quantitatively, we know that the energy level of a *bonding* MO is located *below* the levels of its

Table 10.1
Valence Orbital Potential Energies[a]

Atomic Number	Element	1s	2s	2p	3s	3p	3d	4s	4p
1	H	−13.6							
2	He	−24.6							
3	Li		−5.4						
4	Be		−9.3						
5	B		−14.0	−8.3					
6	C		−19.4	−10.6					
7	N		−25.6	−13.2					
8	O		−32.3	−15.8					
9	F		−40.2	−18.6					
10	Ne		−48.5	−21.6					
11	Na				−5.1				
12	Mg				−7.6				
13	Al				−11.3	−5.9			
14	Si				−14.9	−7.7			
15	P				−18.8	−10.1			
16	S				−20.7	−11.6			
17	Cl				−25.3	−13.7			
18	Ar				−29.2	−15.8			
19	K							−4.3	
20	Ca							−6.1	
21	Sc						−4.7	−5.7	−3.2
22	Ti						−5.6	−6.1	−3.3
23	V						−6.3	−6.3	−3.5
24	Cr						−7.2	−6.6	−3.5
25	Mn						−7.9	−6.8	−3.6
26	Fe						−8.7	−7.1	−3.7
27	Co						−9.4	−7.3	−3.8
28	Ni						−10.0	−7.6	−3.8
29	Cu						−10.7	−7.7	−4.0
30	Zn							−9.4	
31	Ga							−12.6	−6.0
32	Ge							−15.6	−7.6
33	As							−17.6	−9.1
34	Se							−20.8	−10.8
35	Br							−24.1	−12.5
36	Kr							−27.5	−14.3

SOURCE: From R. L. DeKock and H. B. Gray, *Chemical Structure and Bonding*, University Science Books, Sausalito, CA, 1989; pp. 227 and 330.

[a] All energies are the negatives of ionizations energies, for convenience of use in energy level diagrams, and represent averages for all terms (Chapter 17) of the specified orbitals. The *d*-block energies in the last three columns represent the ionization $3d^{n-1}4s \rightarrow 3d^{n-2}4s$, $3d^{n-1}4s \rightarrow 3d^{n-1}$, and $3d^{n-1}4p \rightarrow 3d^{n-1}$, respectively.

parent atomic orbitals, while the energy level of an *antibonding* MO is located (approximately) the corresponding distance *above* that of its parent atomic orbitals.

Finally, we draw in lighter or dashed lines to connect the energy levels of the MOs to those of their parent atomic orbitals; the MOs are labeled by symmetry type, and arrows can be filled in to represent occupancy of the orbitals (in Fig. 10.1 we show the electrons of He_2^+).

When we go beyond four electrons, we must occupy a third MO, which requires the use of additional parent atomic orbitals. The next type of atomic orbital available is the 2*s*, which can be used by either atom A or B or both. It would seem possible to have positive overlap between 2*s* orbitals on both atoms or between a 1*s* orbital on one atom and a 2*s* orbital on the other atom, but there are two criteria that rule out exclusive reliance on the latter, mixed combination of atomic orbitals. The first criterion is that *strong covalent bonding interactions are possible only between atomic orbitals that are fairly close in energy.* In an atom in which 2*s* orbitals are valence orbitals, the 1*s* orbitals are core electrons with much higher ionization energies (Section 7.4); the dissimilarities in energy levels make the overlap ineffective. The second criterion is that *molecular orbitals must have the symmetry of one of the irreducible representations* of the point group; a combination of $1s(\text{A}) + 2s(\text{B})$ is, for example, neither *gerade* nor *ungerade.* The combinations we need to consider next are

$$\sigma_g(2s) = 2s(\text{A}) + 2s(\text{B}), \tag{10.5}$$

and

$$\sigma_u{}^*(2s) = 2s(\text{A}) - 2s(\text{B}) \tag{10.6}$$

These orbitals have little or no involvement of 1*s* orbitals on atoms A and B because the 1*s* orbitals are much lower in energy than the 2*s* orbitals.

Now we can write electron configurations for molecules or ions with from five to eight electrons:

Five electrons (e.g., $Li_2{}^+$)	$\sigma_g{}^2\sigma_u{}^{*2}\sigma_g(2s)^1$ or (showing only valence electrons) $\sigma_g{}^1$
Six electrons (e.g., Li_2)	$\sigma_g{}^2\sigma_u{}^{*2}\sigma_g{}^2$ or $\sigma_g{}^2$
Seven electrons (e.g., $Be_2{}^+$)	$\sigma_g{}^2\sigma_u{}^{*2}\sigma_g{}^2\sigma_u{}^{*1}$ or $\sigma_g{}^2\sigma_u{}^{*1}$
Eight electrons (e.g., Be_2)	$\sigma_g{}^2\sigma_u{}^{*2}\sigma_g{}^2\sigma_u{}^{*2}$ or $\sigma_g{}^2\sigma_u{}^{*2}$ (10.7)

The latter versions of the electron configurations are really *valence* electron configurations of the molecules; since the two lower energy MOs are constructed from core atomic orbitals, they are chemically inactive as well. In fact, these core MOs are virtually indistinguishable from atomic 1*s* orbitals on the two atoms, since a third criterion is that in order to form a MO, there must be *significant overlap of the parent atomic orbitals*. Core orbitals, of course, generally have substantially smaller radii than valence orbitals and do not overlap well with orbitals on adjacent atoms.

When we go to molecules or ions possessing beyond eight electrons, we need to incorporate atomic 2*p* orbitals among the parent atomic orbitals; at this point, the possibilities become more extensive, since the $2p_z$ orbital has different symmetry properties than the doubly degenerate ($2p_x$, $2p_y$) set in the $D_{\infty h}$ point group. Taking linear combinations of the $2p_z$ orbitals as shown in Figure 10.2(*c*,*d*) on the two atoms gives two new MOs in the same irreducible representations as previously seen

$$\sigma_g(2p_z) = 2p_z(\text{A}) - 2p_z(\text{B}) \tag{10.8}$$

$$\sigma_u{}^*(2p_z) = 2p_z(\text{A}) + 2p_z(\text{B}) \tag{10.9}$$

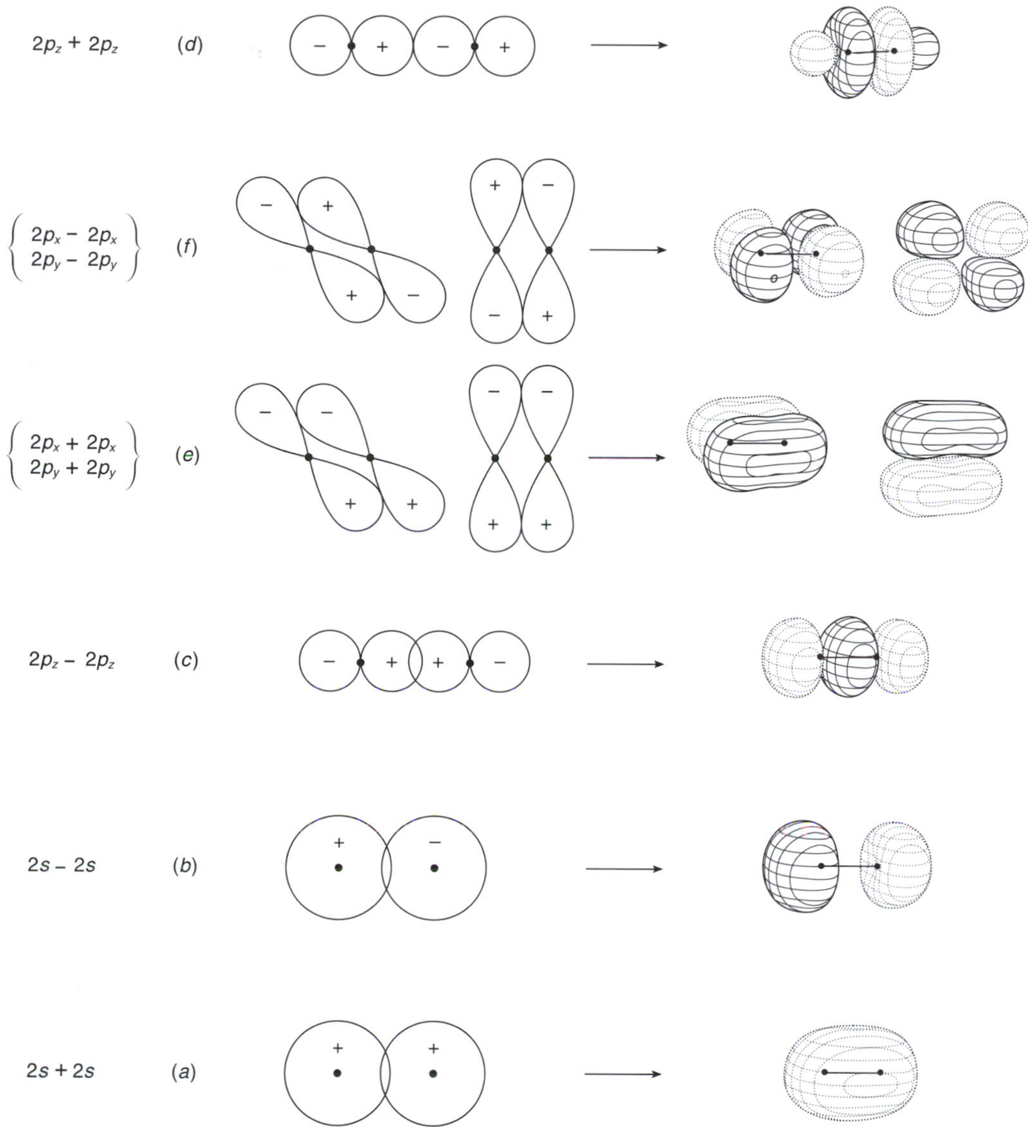

Figure 10.2
Overlap of valence orbitals to give MOs in the second period (no s–p mixing). The contour diagrams are the results of calculations for N_2. [Adapted from W. L. Jorgensen and L. Salem, *The Organic Chemists's Book of Orbitals*, Academic, New York, 1973; p. 79.]

The doubly degenerate $(2p_x, 2p_y)$ set of orbitals give positive overlap as seen in Figure 10.2(e) to produce the familiar π bonding orbitals (i.e., the degenerate pair of π_x and π_y bonds). These orbitals have the symmetry of the Π_u irreducible representation, so they are properly labeled as π_u bonding MOs. There is also a corresponding degenerate antibonding set of orbitals resulting from negative overlap; these have the symmetry of the Π_g irreducible representation, so they are properly labeled as $\pi_g{}^*$ antibonding MOs.

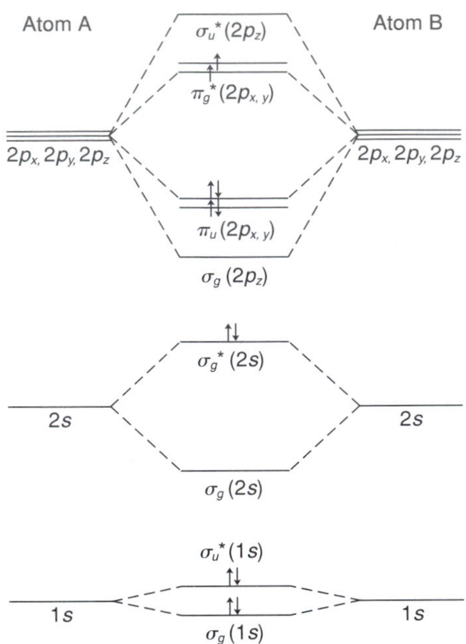

Figure 10.3
Energy level diagram for a homonuclear diatomic molecule or ion of the second period, with valence $2s$ and $2p$ atomic orbitals; no interaction is shown of the σ_g or of the $\sigma_u{}^*$ orbitals arising from valence $2s$ and $2p$ atomic orbitals. The electron configuration of O_2 is illustrated.

An energy level diagram including these MOs having parentage in p orbitals is shown in Figure 10.3. Note that in the isolated parent atoms, electrons in all p orbitals are at the same energy level. However, as we previously noted in Section 3.1, σ-type overlap is more effective than, and gives more bonding than, π-type overlap; σ-type negative overlap is also more antibonding than π-type negative overlap. Consequently, the bonding MOs resulting from σ-type overlap drop in energy from the parent atomic orbitals more than the bonding MOs resulting from π-type overlap, and the antibonding MOs resulting from σ-type overlap rise in energy from the parent atomic orbitals more than the bonding MOs resulting from π-type overlap. The order of filling of these MOs is $\sigma_g(2p_z) < \pi_u(2p_x, 2p_y) < \pi_g{}^*(2p_x, 2p_y) < \sigma_u{}^*(2p_z)$. The valence electron configuration of the 12-valence electron molecule O_2 is therefore $\sigma_g(2s)^2\sigma_u{}^*(2s)^2\sigma_g(2p_z)^2 \cdot \pi_u(2p_x, 2p_y)^4\pi_g{}^*(2p_x, 2p_y)^2$. The bond order can always be calculated from valence electron configurations only. In this case, it is $\frac{1}{2}(2 - 2 + 2 + 4 - 2) = 2$ and the O_2 molecule has the expected double bond. However, by Hund's rule the two electrons in the degenerate $\pi_g{}^*$ pair of orbitals must each occupy one of the orbitals with parallel spins; this accounts for the presence of two unpaired electrons and paramagnetism in O_2.

Frontier Orbitals. As we shall discuss in more detail in Sections 10.4 and 10.7, much important chemistry of the Lewis-base and redox types is concentrated in the *highest occupied molecular orbital*,[5] which is commonly abbreviated as the *HOMO*. Much Lewis acid and redox chemistry focuses on the *lowest unoccupied molecular*

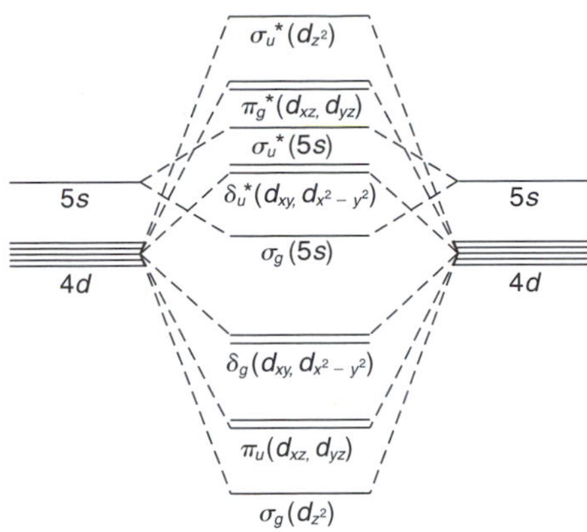

Figure 10.4
Energy level diagram for a homonuclear diatomic molecule or ion of the fifth period of the *d* block, with valence 5*s* and 4*d* orbitals as parent atomic orbitals.

orbital, which is commonly abbreviated as the *LUMO*. Redox chemistry of the *free radical* type is likely to involve any *singly occupied molecular orbital(s)* or *SOMOs*, such as the degenerate $\pi_g{}^*$ pair of orbitals in O_2. Collectively, these orbitals are the focus of much of chemical reactivity and are known as the *frontier orbitals* of the molecule or ion.

Homonuclear Diatomic Molecules Involving *d* Orbitals. At high temperatures, the *d*-block elements vaporize; although they are mostly atomized (Section 7.3), it is possible to generate a sufficient quantity of gaseous diatomic molecules of these elements to allow their study. The bonding in the diatomic molecules of the fourth-period elements is complicated by the fact that 3*d* orbitals give poor overlap with 3*d* orbitals on adjacent atoms (Section 3.1), so we will discuss only examples from the fifth and sixth periods, in which *d* orbitals on adjacent atoms can better overlap each other.

As indicated in the bottom row of Figure 3.3, there are three ways in which five *d* orbitals on each of two adjacent atoms can have positive overlap with each other. These types of positive overlap give rise to a σ_g bonding MO, two degenerate π_u bonding orbitals, and two degenerate δ_g bonding orbitals. The effectiveness of the covalent overlap, and hence the degree to which the bonding MO is more stable than the *d* orbitals of the parent atoms, falls in the order $\sigma_g > \pi_u > \delta_g$; in Section 3.1 we mentioned the consequence that the bonding strength falls in this order. This order results in an energy level diagram such as that shown in Figure 10.4, which also includes the obligatory antibonding orbitals resulting from the possible cases of negative overlap between the parent *d* orbitals. Again, the effectiveness of negative overlap is greatest for the $\sigma_u{}^*$ antibonding combination, so this is higher in energy than the other two antibonding combinations.

One complication that comes up in the *d* block is that the parent atoms have *two* types of valence orbitals. The *s* orbitals are also valence orbitals, and give rise to two additional MOs: a σ_g and a $\sigma_u{}^*$. Where these fall in the energy level diagram depends on how much higher in energy the parent *s* orbitals are than the parent *d* orbitals, and on the relative magnitudes of the two types of overlap. These factors cannot be predicted qualitatively and vary across a period. For the purpose of the following example, we assume the energy level diagram has the order of orbitals shown in Figure 10.4.

Example 10.1

Write the expected electron configuration and predict the bond order and the number of unpaired electrons for the homonuclear diatomic molecule Mo_2; identify the frontier orbitals in this molecule.

SOLUTION:

This diatomic molecule would have 12 valence electrons, which is enough to fill the six lowest orbitals in Figure 10.4 (since there are two sets of doubly degenerate energy levels, this means the lowest four distinct energy levels). Consequently, the electron configuration is expected to be $\sigma_g{}^2\pi_u{}^4\delta_g{}^4\sigma_g(5s)^2$. Since all of these orbitals are bonding, the predicted bond order is 6; this molecule should have a *sextuple* bond! (Actual calculations have supported this prediction.[6]) It should have no unpaired electrons, and hence be diamagnetic. The frontier orbitals include $\sigma_g(5s)$ as the HOMO and $\delta_u{}^*(d_{x^2-y^2}, d_{xy})$ as the LUMO.

Homonuclear Diatomic Molecules Involving *f* Orbitals. Predictions for these high-temperature molecules are fundamentally similar, and are most readily made for the seventh rather than the sixth period, since there is such poor overlap of *f* orbitals for sixth-period elements. Matched combinations of the seven *f* orbitals on two atoms in a homonuclear diatomic molecule give rise to four types of positive overlap, as illustrated in Figure 10.5(*a*): one σ_g, two degenerate π_u, two degenerate δ_g, and, representing a new type illustrated in Figure 10.5(*b*), two degenerate ϕ_u bonding MOs (these fall in the Φ_u irreducible representation implied but not actually listed in the character table for the $D_{\infty h}$ point group). Nonrelativistic calculations have predicted the electron configuration $\sigma_g{}^2\pi_u{}^4\delta_g{}^4\phi_u{}^4$ and a bond order of 7 for the molecule Np_2; relativistic corrections alter this slightly.[7]

You may note that the sequence of bonding types σ, π, δ, ϕ is simply the Greek letter equivalent of the sequence of atomic orbital types *s*, *p*, *d*, *f*. In addition, the end views of each type of orbital are the same: *s* and σ orbitals have no nodal planes including the nucleus or the internuclear axis; *p* and π orbitals have one nodal plane including the nucleus or the internuclear axis; *d* and δ orbitals have two such nodal planes; and *f* and ϕ orbitals have three.

10.2 Bond Energies and Lengths, Spectroscopy, and Orbital Mixing in Homonuclear Diatomic Molecules

Bond Energies and Lengths. As the bond order in a diatomic molecule or ion increases, there are two important practical consequences: (1) the bond energy increases (the bond becomes stronger) and (2) the bond becomes shorter. These phenomena are readily illustrated in the first period by the data in Table 10.2 for the species with 1–3 electrons, $H_2{}^+$ through $He_2{}^+$ (no data is available for He_2 because it does not exist). The maximum bond energy and the minimum bond lengths are found for H_2, the species with the maximum bond order. Species such as $H_2{}^+$ and $He_2{}^+$, which have bond orders of 0.5, have bond energies roughly one-half that of H_2, and bond lengths substantially longer than H_2.

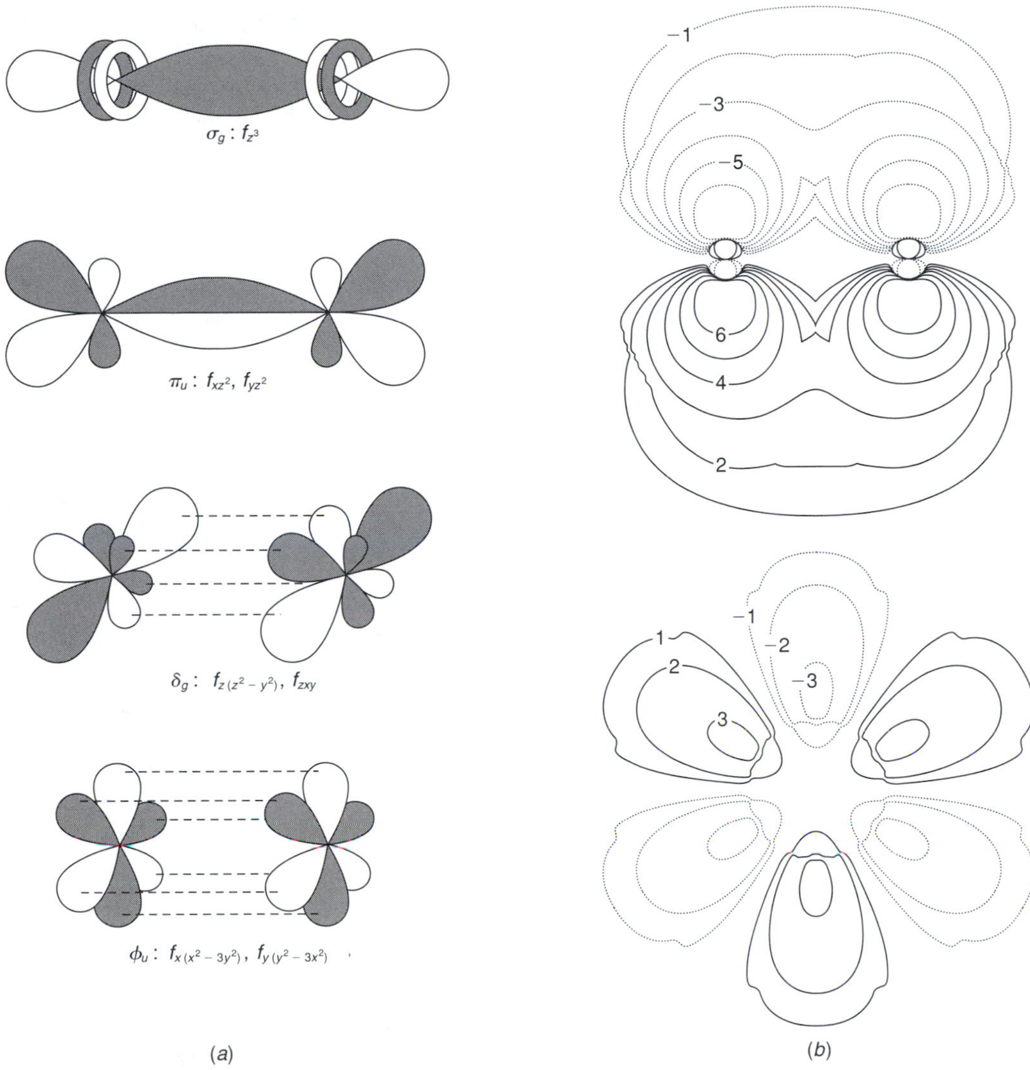

Figure 10.5
(*a*) Qualitative drawing of types of positive overlap possible between *f* orbitals on two adjacent atoms; positive and negative lobes are shown in black and white. (The parent *f* orbitals are drawn and named differently than in Fig. 8.21, since the symmetry of the molecule is different: *f* orbitals are normally drawn in different manners for molecules in higher order point groups than in others). (*b*) Contour diagrams of the ϕ_u bonding orbitals from a side view (top) and from an end view (bottom). [Adapted from B. E. Bursten and G. A. Ozin, *Inorg. Chem.*, **23**, 2910 (1984).]

Example 10.2
Partial bond strength and bond length data for the series of diatomic oxygen ions and the molecule are given in Table 10.2. Explain this data, and fill in the gaps, including a prediction of the data for O_2^{2+}.

SOLUTION:

The electron configurations of these diatomics can be predicted as in Example 10.1. These electron configurations lead to predictions of bond orders as follows: O_2^{2+}, 3.0; O_2^{+}, 2.5; O_2, 2.0; O_2^{-}, 1.5; and O_2^{2-}, 1.0. Bond energies should decrease and bond lengths should increase along this sequence. The strongest and shortest bond should therefore be that of O_2^{2+} (bond energy about 846 kJ mol^{-1} from data in Table 3.1 for one σ plus two π bonds, bond length < 112 pm for O_2^{2+}); the unknown bond energy for O_2^{+} should be about 670 kJ mol^{-1} and for O_2^{2-} about 142 kJ mol^{-1} for a single σ bond.

Ultraviolet and Visible Spectra. Electrons can be promoted from full (or partially full) orbitals to vacant (or partially vacant) orbitals by photons of light, as previously discussed in Section 8.4. The lowest energy absorption usually occurs in the visible or ultraviolet (UV) region of the spectrum, and results in the promotion of an electron from the HOMO to the LUMO of the molecule or ion. The diatomic halogen molecules F_2 through I_2 have a $\pi_g{}^{*4}$ HOMO and a $\sigma_u{}^{*0}$ LUMO. The intense color of these molecules is due to the absorption of light while an electron is being promoted from the HOMO to the LUMO. Since overlap and bond strength fall down a group of the periodic table as the orbitals become more diffuse (Section 3.1; Table 10.2), the separation of the HOMO and the LUMO decrease from F_2 through I_2. As we shall see in Chapter 12, F_2 is virtually colorless since this absorption occurs in the UV region of the spectrum; Cl_2 is yellow since the absorption band spans some of the UV and some of the adjacent visible region; Br_2 is deep red since the band is clearly in the visible; I_2 is purple since the band is at even lower energy in the visible region.

Table 10.2
Bond Energies and Lengths of Some Homonuclear Diatomic Molecules and Ions[a]

Molecule	Bond Energy (kJ mol^{-1})	Bond Length (pm)	Molecule	Bond Energy (kJ mol^{-1})	Bond Length (pm)
H_2^+	255	106	N_2^+	842	112
H_2	432	74	N_2	942	110
He_2^+	230	108	P_2	477	189
O_2^+		112	As_2	382	229
O_2	494	121	Sb_2	298	221
O_2^-	393	126	O_2	494	121
O_2^{2-}		149	S_2	421	189
B_2	274	159	Se_2	325	217
C_2	602	124	Te_2	261	256
Si_2	314	225	Cu_2	198	222
Ge_2	272		Ag_2	162	
Sn_2	192		Au_2	226	247
Pb_2	96				

SOURCE: Data from R. L. DeKock and H. B. Gray, *Chemical Structure and Bonding*, University Science Books, Sausalito, CA, 1989; p. 229.

[a] For data for diatomic Groups 1 and 17(VII) molecules, see Table 3.1.

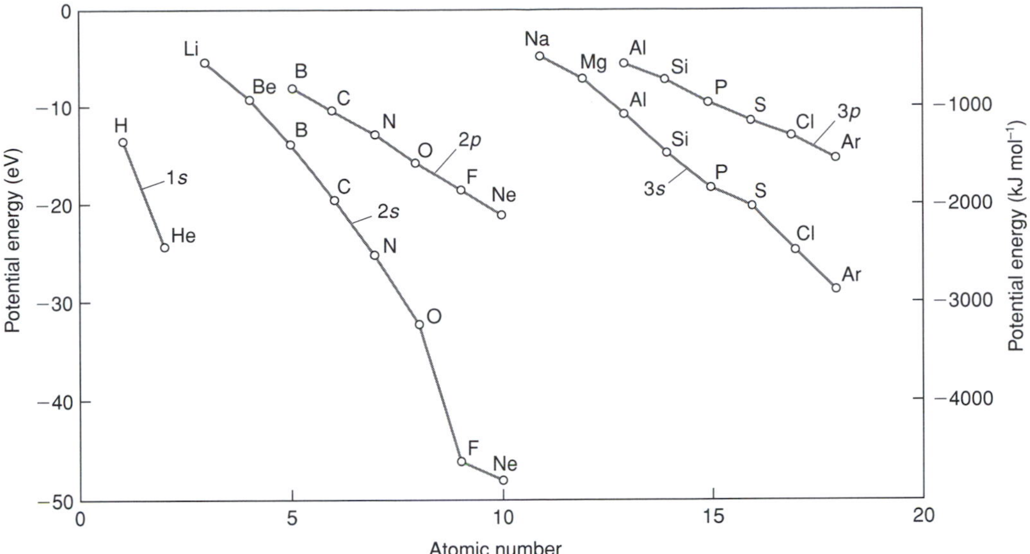

Figure 10.6
Valence orbital potential energies. [Adapted from G. L. Miessler and D. A. Tarr, *Inorganic Chemistry*, Prentice-Hall, Englewood Cliffs, NJ, 1991; p. 137.]

The intensity of the light absorption (and color of the compound) is restricted by the symmetries of the MO being vacated in the *ground state* and the one being occupied by the promoted electron in the *excited state* of the molecule. In order for the wave of light (which can be thought of as similar in symmetry to a **x**, **y**, or **z** vector, and therefore *ungerade*) to be absorbed, *either the HOMO or the LUMO* (but not both) *must be ungerade* if the molecule possesses an inversion center. (You will recall that the weak *d–d* absorptions of transition metal complexes discussed in Chapter 8 for octahedral complexes were from the set of *d* orbitals labeled t_{2g} to the set labeled e_g. These absorptions are weak because both orbital sets are *gerade*.) In the case of the diatomic halogen molecules, the absorptions (and the colors) are intense because the HOMO is *gerade* while the LUMO is *ungerade*.

Mixing of Molecular Orbitals of the Same Symmetry Type. In Section 10.1 we selected our linear combinations of atomic orbitals for construction of MOs based on three criteria, which included that the atomic orbitals be of comparable energy, and that the resulting MOs have the symmetry of one of the irreducible representations. These criteria do not actually exclude some possibilities that we have ignored so far. Atoms from the *p*, *d*, and *f* blocks have more than one type of valence orbital, and the energies of these different types of valence orbitals are often rather close, particularly at the beginning of a block, as shown by the *valence orbital (potential) energies* tabulated in Table 10.1 and graphed in Figure 10.6. Consequently, when we formed the $\sigma_g(2s)$ orbital for a second-period diatomic (Fig. 10.3), we did not consider the possibility that not only the 2*s* but also the $2p_z$ orbitals of the two atoms would contribute to this σ_g molecular orbital.

A corollary of these criteria is that *if two MOs falling in the same irreducible representation have energies close to each other* (i.e., are formed from atomic orbitals with

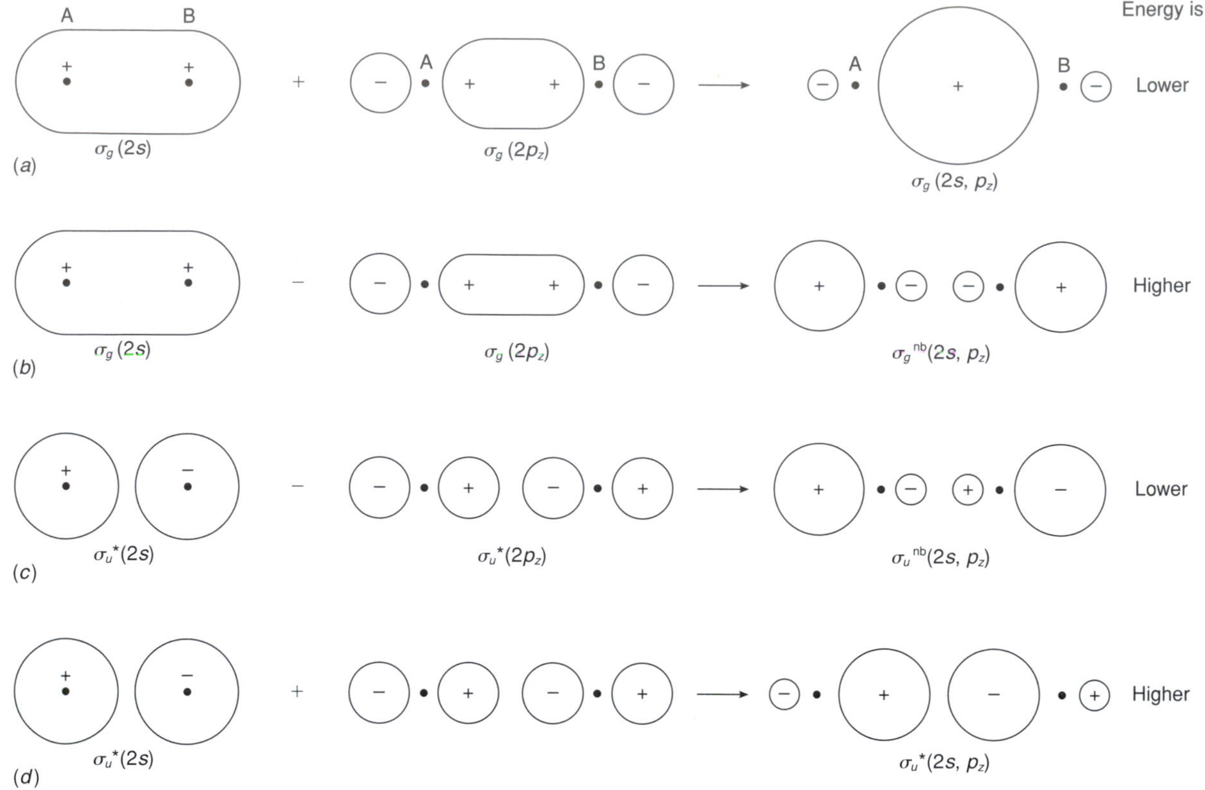

Figure 10.7

Mixing or hybridization of MOs of the same symmetry type and similar energies. (*a*) Positive and (*b*) negative combinations of $\sigma_g(2s)$ with $\sigma_g(2p_z)$ molecular orbitals; (*c*, *d*) positive and negative combinations of $\sigma_u{}^*(2s)$ with $\sigma_u{}^*(2p_z)$ molecular orbitals.

energies close to each other), *they will interact with each other* (i.e., undergo a type of electron configuration interaction, orbital mixing, or "hybridization"). The 2*s* orbitals are the parents of the $\sigma_g(2s)$ molecular orbital, and the 2*p* orbitals are the parents of the $\sigma_g(2p)$ molecular orbital. But these MO's are close in energy to each other, and they both belong to the same symmetry type, σ_g. Consequently the actual MOs are each partly derived from the 2*s* and partly derived from the 2*p* atomic orbitals:

$$\sigma_g(2s, p_z) = c_A 2s(\mathrm{A}) + c_B 2s(\mathrm{B}) + d_A 2p_z(\mathrm{A}) + d_B 2p_z(\mathrm{B}) \qquad (10.10)$$

In this "*s–p* hybridized" wave function, although $c_A = c_B$ and $d_A = d_B$ by the requirements of symmetry, $c_A \neq d_A$; the *s* orbitals are somewhat emphasized in the lower energy MOs and the *p* orbitals, being higher in energy to start with, are somewhat more highly emphasized in the higher energy MOs.

It is perhaps easiest to view this "hybridization" process as one of mixing the two σ_g molecular orbitals with each other. In combining two MOs, which *must* be of the same irreducible representation, we can make two new hybridized MOs (Fig. 10.7). The lower energy σ_g orbitals result from positive overlap of the two original MOs and have more positive overlap than either one, hence they are *lower in energy* than either starting MO [Fig. 10.7(*a*)]. The other, higher energy σ_g orbital [Fig. 10.7(*b*)] is made

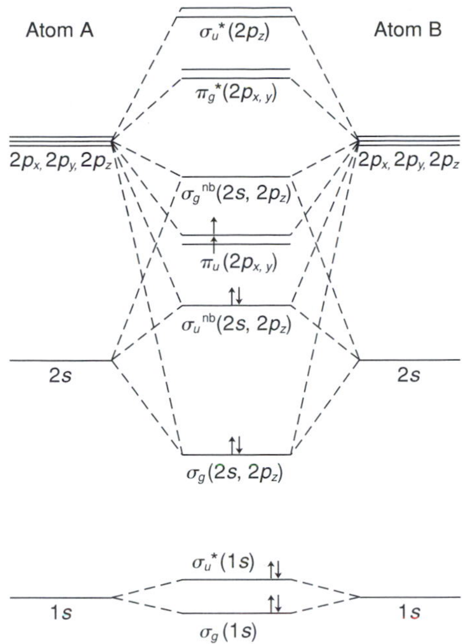

Figure 10.8
Energy level diagram for a homonuclear diatomic molecule with *s–p* hybridization or mixing.

with negative overlap of the two starting orbitals. The new negative overlap just about cancels out the original positive overlap present in the original MOs, so this σ_g orbital has *close to zero* overlap (it is only slightly bonding). It is *higher in energy* than either of the two starting MOs, and because of its low overlap of atomic orbitals, is an approximately *nonbonding* MO, σ_g^{nb}.

If we examine the energy level diagram for second-row diatomic molecules (Fig. 10.3) carefully, we see that there is only one other pair of MOs that share the same irreducible representation, and that can therefore hybridize if they are close in energy: $\sigma_u^*(2s)$ and $\sigma_u^*(2p_z)$. Upon combining these two antibonding orbitals, they give rise to one stabilized, approximately nonbonding (actually slightly antibonding) MO, σ_u^{nb} [Fig. 10.7(*c*)], and one *very* antibonding MO, σ_u^* [Fig. 10.7(*d*)]. When we incorporate the changes in energies of these four orbitals into the energy level diagram for second-row diatomics, we get a new diagram that may have one significant change in order: σ_g^{nb} may rise above π_u (Fig. 10.8).

The key condition that may make this hybridized energy level diagram applicable rather than the simpler diagram of Figure 10.3 is for the parent atoms' valence *s* and *p* orbitals to be close in energy. As indicated by the graph of valence orbital energies (Fig. 10.6), this condition is best met by the early elements of the *p* block, B and C; consequently, these elements form some unusual diatomic molecules and ions.

Example 10.3

Using the energy level diagram appropriate for *s–p* mixing (hybridization), write the electron configuration and describe the bonding in the B_2 molecule.

SOLUTION:

The B_2 molecule has six valence electrons. Using the energy level diagram of Figure 10.8, we obtain the following valence electron configuration: $\sigma_g{}^2(\sigma_u{}^{nb})^2\pi_u{}^2$. The bond order is $0.5\,(2 - \text{ca. } 0 + 2) =$ somewhat less than 2. The bond consists of one π bond, containing two unpaired electrons, plus part of a σ bond (since $\sigma_u{}^{nb}$ is actually slightly antibonding): B_2 is a strange species! But it does exist and has a bond energy of 274 kJ mol^{-1}; its bond length of 159 pm is distinctly shorter than twice the 90-pm covalent radius of B (Table 1.5).

When MOs involve the mixing of different parent orbitals, or when they involve several atoms, the clear distinction of bonding, nonbonding, and antibonding can be lost; some orbitals may be "approximately nonbonding" or may be more bonding than other bonding MOs. Consequently, the literature often omits the suffixes "nb" and "*" due to these ambiguities, and also does not show the parent atomic orbitals, which may be of several types. To distinguish the several σ molecular orbitals that may be present in a molecule, they are often listed numerically: "1σ" is the lowest energy σ molecular orbital in the molecule; "2σ" is the second lowest (regardless of whether it is bonding or antibonding), and so on. We have elected not to enumerate the MOs in this text, however, since this requires keeping track of all core MOs, which is otherwise fairly pointless. We also assign approximate nonbonding status since it allows us to compute approximate bond orders, hence it helps us to visualize the overall bonding.

Since students are generally more comfortable with valence bond theory at this point from their organic chemistry courses, it may be useful to translate some of the MO bonding pictures into their nearest valence bond equivalents (Lewis dot structures or resonance structures). For example, we know that F_2 has the MO valence electron configuration $\sigma_g(2s)^2\sigma_u{}^*(2s)^2\sigma_g(2p_z)^2\pi_u(2p_x, 2p_y)^4\pi_g{}^*(2p_x, 2p_y)^4$. Does this or does this not disagree with the Lewis dot structure of F_2, which shows a single bond and three unshared electron pairs per F atom? The "guidelines of translation" are as follows:

1. Electron pairs in bonding MOs in diatomic molecules correspond to individual bonds of the indicated type (σ, π, and δ).

2. When both a bonding MO and an antibonding MO of the same symmetry type (e.g., σ and σ^*) and atomic parentage are occupied by two electrons, their combined effect is to cancel each other's bonding contributions, and act as *two unshared electron pairs*, one on each atom.

3. When an electron pair occupies a *nonbonding* MO in a homonuclear diatomic molecule, it corresponds to *one* unshared electron pair located equally on *each* atom, a situation that can only be represented in valence bond theory by drawing resonance hybrids.

For example, the F_2 molecule has four electron pairs in bonding MOs and three in antibonding MOs, which of course results in a bond order of one: a single bond. The three pairs in antibonding MOs cancel out three pairs found in the corresponding type of bonding MO, so these together account for the six unshared electron pairs in the Lewis structure of F_2. There is no discrepancy between the MO and valence bond pictures of the bonding in this molecule.

There is no way that we can draw a Lewis structure for B_2 that obeys the octet rule. What Lewis structure is suggested by its electron configuration from MO theory? In Example 10.3, we obtained an electron configuration (with *s*–*p* mixing) of $\sigma_g{}^2(\sigma_u{}^{nb})^2\pi_u{}^2$, which includes one σ bond, one π bond, and one unshared electron pair located equally on *both* boron atoms. This latter pair can only be represented in valence bond theory by using resonance structures: :B=B ↔ B=B:.

It is also possible to carry out the reverse process, starting with a Lewis structure and adding electrons to get an indication as to whether the LUMO is bonding (it is if the bond order increases) or antibonding (it is if the bond order decreases).[8] For example, we can see that the LUMO of O_2 is antibonding (π^*) by adding two electrons to the Lewis structure of :Ö=Ö:, to give [:Ö–Ö:]$^{2-}$, which has a lower bond order (one less π bond) than O_2. By subtracting electrons from the Lewis structure and then rearranging to give the best possible structure of the electron-depleted substance, one can also surmise whether the HOMO is bonding or antibonding. The HOMO of O_2 is π^*, since removing its two electrons generates [:O≡O:]$^{2+}$, with one more π bond than O_2. (The degeneracy of the HOMO and LUMO, so that both are SOMOs, is not revealed by this process.)

Photoelectron Spectroscopy. Photoelectron spectroscopy is a technique that uses photons of high but constant energy to strike gaseous molecules of a substance and eject *photoelectrons* from the different occupied MOs in the substance. The kinetic energy of the photoelectrons is then measured; it is equal to the energy of the original photons minus the energy needed to ionize the electrons. Measuring energies of photoelectrons allows us to determine the ionization energies of *molecules*. For example, we know that the ionization energy of the H *atom*, expressed in electron volts, is 13.60 eV (data converted from Table 7.2 using the fact that 96.5 kJ mol^{-1} = 1 eV). From its photoelectron spectrum, we find that the ionization energy of the H_2 *molecule* is 15.45 eV, which reflects the fact that the σ_g electrons in the H molecule are more firmly bonded to the H nuclei than is the 1*s* electron in the single H atom.

Koopman's theorem states that the ionization energy we observe is equal to the negative of the orbital energy of the electron in the molecule, provided that, on ionizing that electron, the other electrons do not reorganize themselves and thereby change energy. Although the latter does in fact sometimes happen, usually Koopman's theorem holds pretty well, so that we have an experimental method of determining the energies of orbitals, that is, the levels at which we should draw the lines in diagrams such as those of Figures 10.1, 10.3, 10.4, and 10.8.

In practice, it is often observed that photoelectron spectra mainly contain, not single absorptions at precise energies such as 15.45 eV, but multiplets with even spacings between several lines; in H_2 the absorption at 15.45 eV is followed by several other evenly spaced lines at higher energies. The extra energies of the higher absorptions is used to excite the vibrations of the molecule; the energy spacing corresponds to the H–H stretching frequency of the molecule. This phenomenon occurs when the electron being ejected comes from a *bonding* or *antibonding* MO; in such cases the product ion (e.g., $H_2{}^+$) has a substantially shorter or longer bond at equilibrium. It is not formed with this equilibrium bond length, but rather with a nonequilibrium one (in a *vibrationally excited state*), and may use some of the energy of the photon to vibrate toward the equilibrium bond length. However, if the electron is ejected from a *nonbonding* MO, the bond order and the bond length in the product ion is unchanged from that in the molecule, so no vibrations need be excited to reach an equilibrium bond length.

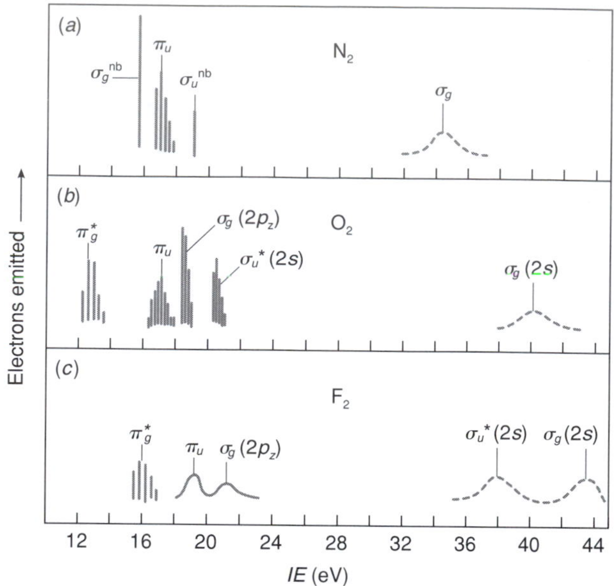

Figure 10.9
Schematic photoelectron spectra of (*a*) N_2, (*b*) O_2, (*c*) F_2 molecules. [Adapted from W. C. Price in *Electron Spectroscopy: Theory, Techniques, and Applications,* Vol. 1, C. R. Brundle and A. D. Baker, Eds., Academic, New York, 1977; p. 151.]

Hence, *if a photoelectron band has vibrational fine structure, the ionization producing it is ionization from a bonding or antibonding MO; if it has no fine structure, the ionization is from a nonbonding molecular orbital.*

In Figure 10.9, we show the photoelectron spectra of the three most stable diatomic molecules of the second period. We note that sharp, single absorption bands that indicate nonbonding MOs are found only in the spectrum of N_2; approximately nonbonding orbitals occur in homonuclear diatomic molecules only with *s–p* mixing, which therefore applies to N_2. By referring to Figure 10.8 and adding enough electrons to give the 14-electron configuration of N_2, we see that the HOMO predicted by this energy level diagram is σ_g^{nb}, which is approximately nonbonding; the only other nonbonding level is the third one down, σ_u^{nb}. In accord with this, we observe in the photoelectron spectrum that the first and third bands from the top lack vibrational structure. So the photoelectron spectrum not only confirms these assignments,[9] it suggests quantitatively at what energies the different levels in the energy level diagram should go. If we have done quantitative MO calculations, the photoelectron spectrum can support or contradict our calculations.

10.3 Heteroatomic Diatomic Molecules and Ions

The construction of MOs for heteronuclear diatomic molecules and ions (those composed of atoms of two different elements), while fundamentally no different than above, does involve two changes. First, heteroatomic diatomic molecules fall in a different point group, $C_{\infty v}$, which does not have an inversion center, so that MOs are no longer *gerade* or *ungerade*. Second, the orbitals of the two parent atoms are no longer at the same energies. It then becomes more important to know their actual energies, since the further apart in energy atomic orbitals are on two parent atoms, the less effectively they can overlap.

Let us begin by considering the diatomic molecule HF. Hydrogen has only one valence orbital, $1s(\text{H})$, while F has one core orbital, $1s(\text{F})$, and four valence orbitals, $2s(\text{F})$, $2p_z(\text{F})$, $2p_x(\text{F})$, and $2p_y(\text{F})$. From considerations of the signs of the atomic wave functions (as in Fig. 3.1) we quickly determine that the only hydrogen orbital, $1s(\text{H})$, must have *zero* overlap with two fluorine orbitals, $2p_x(\text{F})$ and $2p_y(\text{F})$; consequently, the $2p_x(\text{F})$ and $2p_y(\text{F})$ orbitals must be *strictly nonbonding*. Since the p_x and p_y orbitals are degenerate in the $C_{\infty v}$ point group, they are at one energy level, which we label π^{nb}; there is no direct change[10] in the energy level of this MO from that of the parent $2p_x(\text{F})$ and $2p_y(\text{F})$ atomic orbitals.

Symmetry alone does not exclude overlap between $1s(\text{H})$ and any of the remaining three orbitals on fluorine, so we need to see how close in energy these orbitals are. From Table 10.1, we see that the orbital energy of $1s(\text{H})$ is $-13.6\,\text{eV}$; the orbital energies for the three fluorine orbitals are $1s(\text{F}) \ll -40.2\,\text{eV}$ (not listed); $2s(\text{F}) = -40.2\,\text{eV}$; $2p_z(\text{F}) = -18.6\,\text{eV}$; these are the levels at which we draw the atomic orbitals on the sides of the energy level diagram for HF. Clearly, there must be an extreme mismatch of energies between the core $1s(\text{F})$ orbital and the valence $1s(\text{H})$ orbital, so the core $1s(\text{F})$ orbital is also a nonbonding orbital of unaltered energy. The main (positive and negative) overlap is between the $1s(\text{H})$ and the $2p_z(\text{F})$ orbitals; the positive combination of these orbitals gives a σ bonding MO, while the negative combination gives a σ^* antibonding combination. Some small degree of hybridization of the $2s(\text{F})$ orbital into these MOs may be possible, but the character of the $2s(\text{F})$ orbital is likely to be mainly nonbonding due to the large energy gap between it and the $\text{H}(1s)$ orbital.

The energy level diagram and the MOs that result are shown in Figure 10.10. We can write the wave function of the σ bonding orbital as a combination of the two parent orbitals:

$$\sigma = c_{\text{H}} 1s(\text{H}) + c_{\text{F}} 2p_z(\text{F}) \tag{10.11}$$

However, because the two parent orbitals are at dissimilar energies, we no longer have the case that $c_{\text{H}} = c_{\text{F}}$. Instead, *a given MO has a larger contribution (coefficient) from the atomic orbital that is closest to it in energy.* In the bonding MO represented by Eq. (10.11), c_{F} is substantially greater than c_{H} for this reason. Consequently, electrons in this orbital spend the greatest part of their time in the vicinity of the fluorine atom, which we have attempted to represent pictorially on the left side of Figure 10.10; occupation of this bonding orbital results in a *polar bond* polarized toward fluorine.

The opposite weighting holds for the antibonding σ^* orbital, which is closer in energy to its hydrogen parent, $1s(\text{H})$. Electrons in this orbital would spend more time closer to the hydrogen atom. However, this orbital is unoccupied in HF, which has the valence electron configuration $(\sigma^{\text{nb}})^2\sigma^2(\pi^{\text{nb}})^4$ with a bond order of 1.0 (as expected from the Lewis structure of the molecule). The photoelectron spectrum of HF shows a sharp band without vibrational structure at 16.0 eV assignable to the HOMO, π^{nb}, a broadened band at about 20 eV assignable to σ, and a band assignable to σ^{nb} at 40 eV.[10]

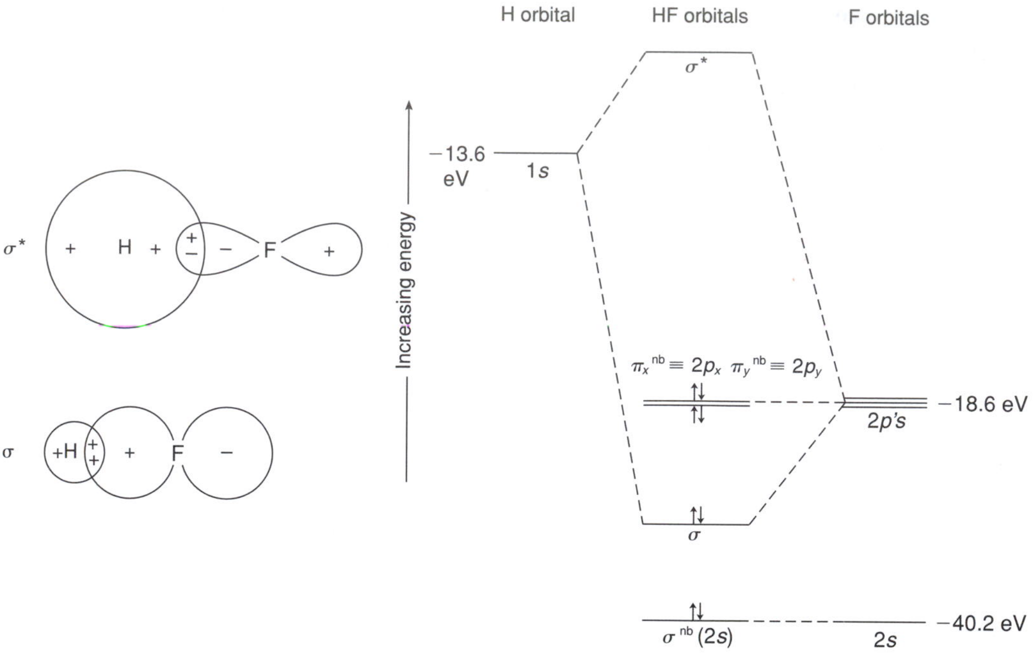

Figure 10.10
Energy level diagram for HF. The diagram is not to scale in that the 2*s* atomic orbital and $\sigma^{nb}(2s)$ molecular orbital are much lower in energy than shown. [Adapted from R. L. DeKock and H. B. Gray, *Chemical Structure and Bonding*, University Science Books, Sausalito, CA, 1989; p. 253.]

The energy level diagrams of the heavier Group 17(VII) hydrides (HX) are qualitatively similar; we shall shortly need to compare the energies of their frontier orbitals quantitatively, however. The energies of the HOMOs, which in all cases are π^{nb} molecular orbitals, are available from photoelectron spectra: −16.0 eV for HF; −12.7 eV for HCl; −10.5 eV for HI; these parallel the valence orbital energies of the *p* orbitals of the parent atoms (and, of course, their electronegativities).[11] The energies of LUMOs are never available from photoelectron spectra, since LUMOs do not contain electrons; computations have given the following estimates for the σ^* LUMOs +6.0 eV for HF; +3.3 eV for HCl; +0.0 eV for HI.[12] The LUMO energies drop through this series because there is less overlap between the hydrogen 1*s* orbital and the halogen *np* orbitals as *n* increases and the orbital becomes more diffuse. Consequently, the antibonding LUMOs are less antibonding in an absolute sense.

Carbon Monoxide. This important ligand molecule is isoelectronic with N_2, so it has the same number of valence electrons and an energy level diagram that is related to that of N_2 (Fig. 10.8), but with the significant difference that the parent atoms are not identical. Therefore they have different valence orbital energies. The same sequence of MOs does occur, however, as shown in Figure 10.11, without the *gerade* and *ungerade* labels, of course. Once again the possibility of *s*–*p* hybridization of the four σ molecular orbitals derived from the four atomic orbitals 2*s*(C), $2p_z$(C), 2*s*(O), and $2p_z$(O) exists and actually occurs. By far, the lowest in energy of these four parent atomic orbitals is 2*s*(O); the lowest σ hybridized MO, although hybridized, emphasizes this

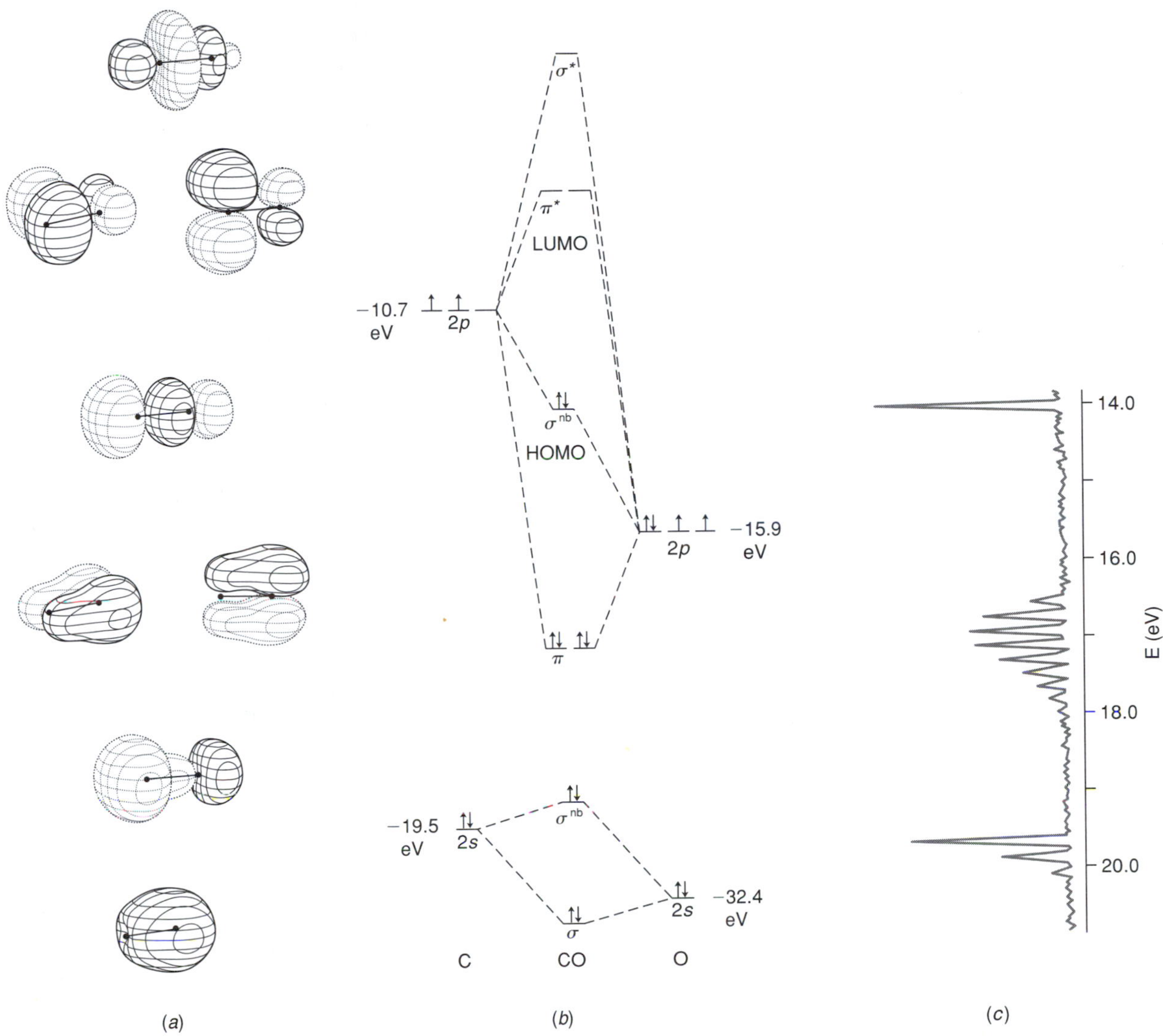

Figure 10.11
Carbon monoxide (CO): (a) contours of MOs; (b) energy-level diagram; (c) photoelectron spectrum. The diagram is not to scale in that the 2*s* atomic orbital of oxygen and the $\sigma^{nb}(2s)$ molecular orbitals are much lower in energy than shown. (a) Reprinted with permission from W. L. Jorgensen and L. Salem, *The Organic Chemists's Book of Orbitals*, Academic Press, San Diego, 1973, p. 78; (c) adapted with permission from L. Gardner and J. A. R. Samson, *Journal of Chemical Physics*, **62**, 1447 (1975). Copyright 1975 American Institute of Physics.

parent orbital in its composition, as indicated in the contour drawing at the left of Figure 10.11. The highest energy σ^* molecular orbital is strongly antibonding and emphasizes the highest energy parent atomic orbital, $2p_z$(C). The intermediate-energy σ^* and the intermediate-energy σ molecular orbitals both have reduced overlap between the C and O atoms and therefore tend in the direction of being nonbonding. Recall that the same thing happened in the case of N_2. The HOMO is the intermediate-energy

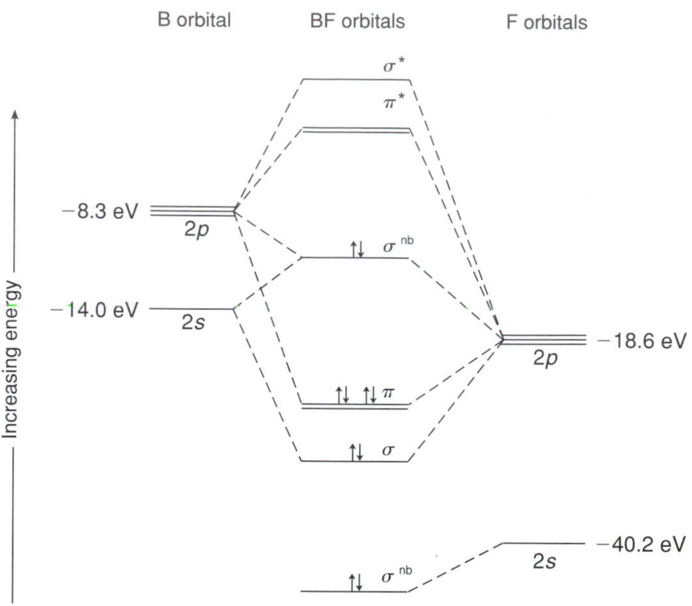

Figure 10.12
Energy level diagram for BF. The diagram is not to scale in that the 2*s* atomic orbital on F and the $\sigma^{nb}(2s)$ molecular orbital are much lower in energy than shown. [Adapted with permission from R. L. DeKock and H. B. Gray, *Chemical Structure and Bonding*, University Science Books, Sausalito, CA, 1989; p. 258.]

σ molecular orbital, which has a greater amplitude on carbon than on oxygen. Therefore it can roughly be equated with the lone pair of electrons on carbon in the Lewis structure of CO; the donor atom of CO is carbon rather than oxygen, which also has a lone pair, because the lone pair in the most reactive filled MO (the HOMO) is mainly on carbon.

The π and the π^* molecular orbitals of CO do not undergo any hybridization, so they are relatively uncomplex. The π orbitals are closer in energy to the parent oxygen $2p_x$ and $2p_y$ atomic orbitals, so they have greater amplitude on oxygen. The π^* orbitals are closer in energy to the carbon parent atomic orbitals, so they have greater amplitude on carbon. These orbitals are the LUMOs of the molecule.

Boron Monofluoride, BF. Although this molecule is isoelectronic with CO and N_2, the valence orbital energies of its two parent atoms are quite different from each other, since fluorine is much more electronegative than boron. The only good matchup of valence orbital potential energies between B and F is of the boron 2*s* (−14.0 eV) with the fluorine 2*p* (−18.7 eV). The fluorine 2*s* orbital lies far below the energy of any boron orbital, and the matchup of the fluorine 2*p* with the boron 2*p* (−8.3 eV) is much poorer in BF than was the corresponding matchup in CO or N_2.

Considering symmetry only, the energy level diagram (Fig. 10.12) of BF has the same sequence of MOs as CO, so the electron configuration comes out the same. Since effective overlap also requires that the parent atomic orbitals be similar in energy, however, some of the MOs that are bonding in N_2 and CO may be more or less nonbonding in BF. Only computation can determine this for certain, but qualitatively we can see that there is such an energy level mismatch between the parent orbitals of the lowest energy σ molecular orbital that this is likely to be nonbonding in BF even though it was not in N_2. Similarly, the 10.4-eV energy gap between the parent orbitals of the π molecular orbital, which inherently has poorer overlap than a σ molecular orbital, certainly makes this orbital less bonding in BF than it is in N_2 or CO.

If we assume that there is indeed π overlap in BF, this molecule has a valence electron configuration of $(\sigma^{nb})^2\sigma^2\pi^4(\sigma^{nb})^2$, for a bond order of 3; if we assume that there is no π overlap, the valence electron configuration is $(\sigma^{nb})^2\sigma^2(\pi^{nb})^4(\sigma^{nb})^2$, for a bond order of 1. Experimentally, the bond energy of BF, 548 kJ mol^{-1}, is considerably less than that of CO, 1070 kJ mol^{-1}, and in fact is close to the B–F single bond energy, 613 kJ mol^{-1} (Table 7.6). And in terms of the corresponding valence bond resonance structures, the triple-bonded Lewis dot structure, :B≡F:, shows a formal charge (Section 3.5) of −2 on boron and +2 on fluorine and is clearly not a good resonance structure, especially since the multiple positive charge is on the more electronegative atom. So, even though it does not satisfy the octet rule, the single-bonded resonance structure :B–Ḟ̤: could be more important for BF; computations would be necessary to determine this for certain.[13]

Finally, we may return to the question of stability. The B–F bond in BF is in no sense of the term a *weak* bond; the molecule would likely be very stable so long as it did not have a chance to react with other molecules (as in a high vacuum). However, we can imagine a thermochemical cycle (Chapter 7) for its disproportionation if it has a chance to encounter other BF molecules:

$$3\,BF(g) \rightarrow 2\,B(s) + BF_3(g) \qquad (10.12)$$

The products of this reaction would be favored powerfully by the recovery of the high atomization energy of elemental boron (563 kJ mol^{-1}; Table 7.1); note that there is no change in the number (three) of boron–fluorine bonds in the reactants and in the products. So we will never find BF in a bottle.

Example 10.4

The series of diatomic molecules C_2, BN, BeO, and LiF are isoelectronic, but because the sets of parent atoms have increasingly diverse valence orbital potential energies (Table 10.1), their energy level diagrams progressively diverge in character. (a) Set up the left and right sides of the energy level diagrams for each of these four molecules. How do the two sides change through this series? (b) Qualitatively, what effects does this have on the shapes of the MOs and their energy level through this series? (c) The bond lengths in the gaseous diatomic molecules in this series are 124 pm in C_2, 128 pm in BN, 133 pm in BeO, 156 pm in LiF; the bond energies (in kJ mol^{-1}) are 602 in C_2, 385 in BN, 444 in BeO, and 568 in LiF.[14] Discuss these trends.

SOLUTION:

(a) The left and right sides of the energy level diagrams are constructed from the data in Table 10.1 and are shown in Figure 10.13. As we progress through the series there is less and less matching of energy levels between the atoms on either side of the diagram.

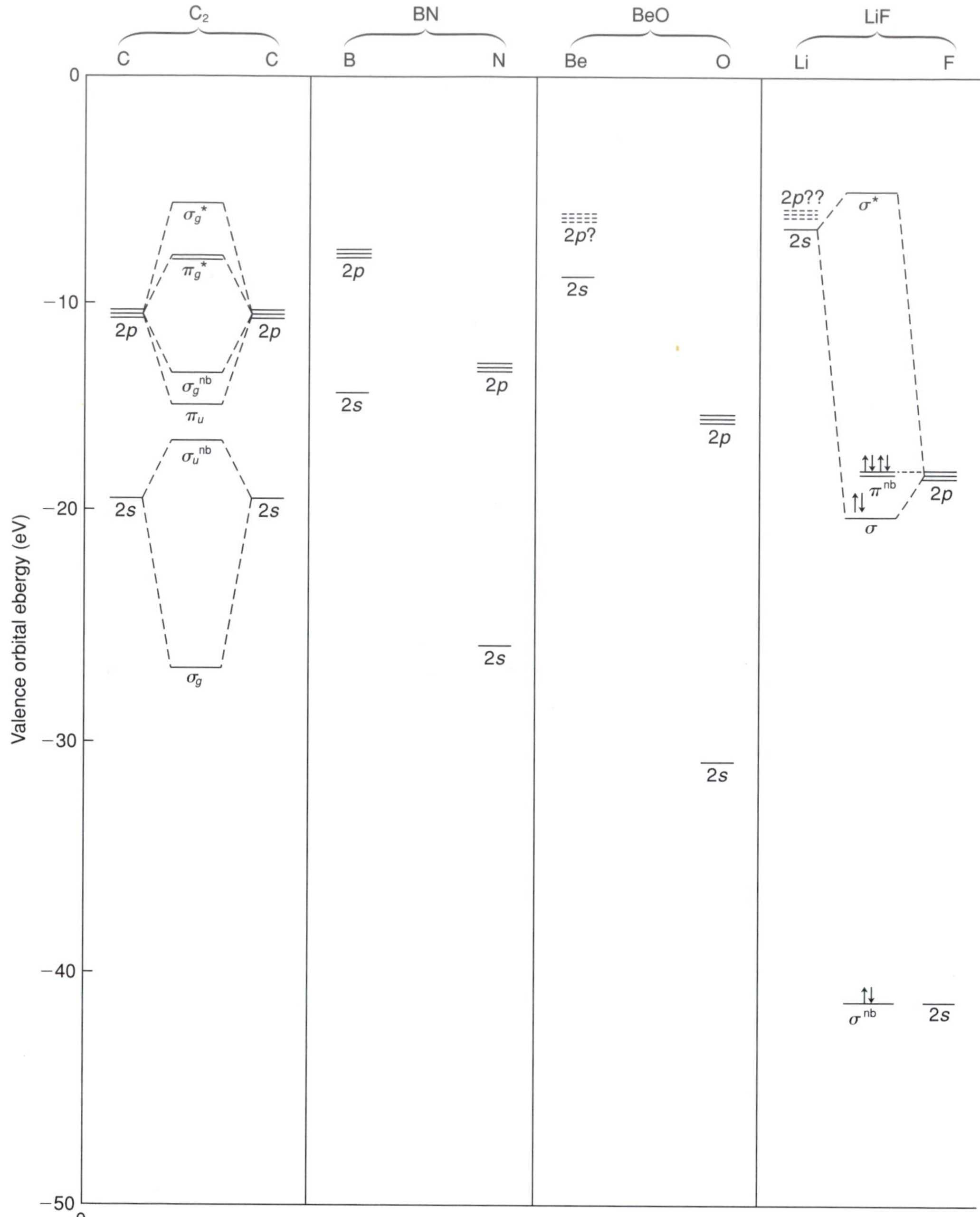

Figure 10.13
Partial energy level diagrams, showing orbital energies of parent orbitals only, for the series of isoelectronic molecules C_2, BN, BeO, and LiF; the approximate MO energy levels for LiF are also included to complete the last energy level diagram.

(b) This means that the degree of overlap of parent atomic orbitals decreases from C_2 to LiF, which results in less splitting of corresponding bonding and antibonding MOs (it also reduces the magnitude of *s*–*p* hybridization). Gradually, the low-energy occupied MOs will more and more resemble the atomic orbitals of the more electronegative element on the right. Consequently, the polarity of the bonds increases, so that the bonding in LiF comes about as close as possible to being completely ionic (we might call it an ion-pair rather than a molecule). An approximate energy level diagram for LiF is also shown in Figure 10.13. Note that all electrons are in orbitals that are much like fluorine orbitals in energy (and appearance).

(c) The bond lengths grow somewhat since the double bonding of C_2 is lost later in the series, then the covalent single bonding; ionic bond lengths tend to be somewhat longer than comparable covalent bond lengths (Exercise 39 in Chapter 1). However, the bond energies show no clear trends: The ionic (electrostatic) attractions of Li^+ and F^- in the gaseous ion pair give similar stabilizations to the covalent interactions in a molecule such as C_2.

10.4 Molecular Ionization Energies, Electron Affinities, Electronegativities, and Hardness

In Section 7.7, we introduced the concept of the Mulliken electronegativity of atoms, which is defined by Eq. (7.27) as one-half the average of the electron affinity and the first ionization energy of the atom with appropriately hybridized atomic orbitals, and mentioned that Parr and Pearson,[15a] and Pearson[15b] extended this definition to include the electronegativities χ of molecules:

$$\chi = 0.5\,(\mathrm{IE} + \mathrm{EA}) \qquad (10.13)$$

Ionization energies (IE) of molecules are available from photoelectron spectra; electron affinities (EA) should be (according to Koopman's theorem) equal to the negative of the orbital energies of the LUMO; these electron affinities can sometimes be measured and sometimes must be calculated. Hence, this equation may be restated as

$$\chi = -\frac{1}{2}(\text{orbital energy of HOMO} + \text{orbital energy of LUMO}) \qquad (10.14)$$

Since these two orbitals are the centers of chemical reactivity, electronegativities then measure the average attraction of molecules for electrons going into or out of their frontier orbitals (relative to the zero of energy being complete ionization; this electronegativity scale does not match the Pauling range of 0.7–4.0). Applying this equation to the data for the hydrogen halides from Section 10.3, we obtain the following electronegativities: for HF, $\chi = 0.5(-16.0 + 6.0) = 5.0$; for HCl, $\chi = 4.7$; for HI, $\chi = 5.3$. Note, however, that this is a different kind of electronegativity than the Pauling type, since (unlike χ_P) it is the same for all atoms within the molecule.[16]

Also recall that the same theoretical treatment led to a mathematical definition of the *hardness* η of a molecule [in the hard–soft acid–base (HSAB) sense, Chapter 5]:

$$\eta = \frac{1}{2}(\mathrm{IE} - \mathrm{EA}) = \frac{1}{2}(\text{orbital energy of LUMO} - \text{orbital energy of HOMO}) \quad (10.15)$$

Hard acids and hard bases therefore are those in which the frontier orbitals are far apart. The frontier orbitals of water are very well separated in energy: The HOMO of H_2O is at -12.5 eV while the LUMO is at $+6.4$ eV; the calculated hardness of water, 9.5 eV, considerably exceeds that of H_2S, 6.2 eV. Likewise, the order of hardness of the hydrogen halides is as expected: 11.0 eV for HF; 8.0 eV for HCl; and 5.3 eV for HI. Large energy separations of the HOMO and LUMO tend to make electronic rearrangements more difficult. Therefore it has been noted that harder molecules tend to be more stable[17]: H_2O and HF have stronger bonds and are more resistant to oxidation than are H_2S and HI.[18]

In this scheme, the softness of a molecule is then simply the inverse of its hardness:

$$\sigma = 1/\eta \tag{10.16}$$

Soft acids and soft bases are those in which the frontier orbitals are close in energy; such species are therefore generally more reactive. The insight into the source of softness in the closeness in energy of the frontier orbitals is valuable and was pointed out much earlier on the basis of MO calculations carried out by Klopman[19]; this closeness in energy means that the valence electrons in a soft acid or base can easily be promoted into another orbital, the LUMO. Hence, these species are described as "polarizable" or "soft". Thus, the relative softnesses of the hydrogen halides are $\sigma(\text{HF}) = 1/11.0 < \sigma(\text{HCl}) = 1/8.0 < \sigma(\text{HI}) = 1/5.3$, which is consistent with the periodic trends discussed in Chapter 5. Again, however, this calculation of softness applies to the whole molecule, regardless of how many donor atoms of what type are present, so it is not the same as we used in Chapter 5.[20] Nonetheless, it usually successfully calculates the results of HSAB exchange reactions.[21]

It is interesting to discuss the position of carbon monoxide in this system of ranking. With its orbital energies of -14.0 eV for the HOMO and $+1.8$ eV for the LUMO, its hardness parameter is 7.9 eV, which is close to that of HCl. We think of chlorides as borderline bases, and have classified CO as a soft base. However, as we shall see more clearly in subsequent chapters on organometallic chemistry, CO does not bond at all well to *p*-block elements, some of which are fairly soft, while it does bond well to *d*-block metals in low oxidation states, which are, of course, soft acids. Its difficulty in bonding to *p*-block Lewis acids lies in the fact that the orbital energy of its HOMO, -14.0 eV, is really quite low; CO is a very poor *σ donor ligand.* In Chapter 8, we pointed out that CO is a representative of the class of *π acceptor ligands*, which have low-energy orbitals capable of giving positive overlap with filled metal *d* orbitals. Carbon monoxide is customarily classified as soft because of the importance of this interaction, which is due in part to its π^* LUMO being of low energy. Also helpful is the fact that the LUMO has a high amplitude on its carbon donor atom, which enhances its ability to overlap filled metal *d* orbitals in the manner pictured in Figure 8.7(*b*).

With this model, we can further illuminate the nature of the soft acid–base interaction. The key element for a soft base is its HOMO, which contains its most reactive (most easily donated) electron pair; the key element for a soft acid is its LUMO. Soft bases tend to have high-energy HOMOs while soft acids tend to have low-energy LUMOs. Such a combination means that the HOMO of the acid has good overlap with the LUMO of the base, which favors the formation of a fairly stable chemical bond. However, in most cases it also helps that there can be *back-bonding*: Electron pair(s) from the high-energy HOMO of the soft acid may be donated to the relatively low-energy LUMO of the soft base, giving a slight character of a double bond to the soft acid–soft base bond, in which one electron pair is donated by the Lewis acid and the other by the Lewis base. This arrangement avoids building up formal charge on either the donor or the acceptor atoms.

It is also fairly characteristic of soft acid–soft base combinations to be intensely colored due to a charge-transfer interaction (Section 6.4) in which the energy of visible light completes the partial transfer of the shared electron from the high-energy HOMO of the Lewis base to the low-energy LUMO of the Lewis acid (or in some cases from the HOMO of the Lewis acid to the LUMO of the Lewis base).

10.5 Molecular Orbital Theory for Linear Molecules

We begin our extension of MO theory to larger molecules by considering linear polyatomic molecules and ions, which also belong to the point group $D_{\infty h}$, so that their MOs are labeled with the same irreducible-representation labels. The number of MOs that will be found in a species is equal to the number of atomic orbitals that are used to construct them. The simplest possible polyatomic molecules or ions are those constructed from hydrogen atoms, $(H_n)^{z\pm}$. In these cases, there are n valence atomic orbitals (1s on each atom), so we can construct n molecular orbitals.

The MOs for linear molecules are wave functions and show one-dimensional standing wave patterns characteristic of vibrating stringed musical instruments or of particles in one-dimensional boxes (Section 1.3, Fig. 1.2): Such wave functions are sine waves (Eq. 1.2) that can be differentiated by the numbers of nodal planes present that cut across the z direction of the string. (The first wave pattern has no nodes, the second has one, and the nth has $(n-1)$, much as in an atom.)

As for an atom, the wave functions are most conveniently generated by first locating the nodal planes.

1. Draw strings of n empty circles to represent the (z axis) chain of hydrogen-atom 1s orbitals in the H_n species. A total of n such chains should be drawn, one for each MO.

2. No nodal planes should cut the z axis of the first chain; one nodal plane should cut through the center of the second chain; two nodal planes should be located one-third of the way and two-thirds of the way along the second chain, and so on.

3. Now draw in sine waves that start at zero to the left of the leftmost atom and return to zero every time a nodal plane is reached, as well as to the right of the rightmost atom. The first four waves should look like those illustrated in Figure 1.2.

4. The signs of all atoms located beneath *crests* of the sine wave are now filled in as positive; the signs of all atoms located above *troughs* of the sine wave are now filled in as negative. Any atom that is cut in half by a nodal plane is *omitted* from the wave function (is neither + nor −), since an s orbital cannot be positive on one side and negative on the other. (A refinement is possible but not obligatory here. Orbitals that are partially cut by the sine wave can be diminished in size until the sine wave no longer cuts through them, and those that do not fill the space up to the sine wave can be enlarged. This refinement is a consequence of the fact that not all atomic orbitals contribute equally to each MO.)

5. The irreducible-representation labels can now be assigned to each MO. Since they are constructed from s orbitals, all are necessarily σ orbitals. Since the first orbital is positive everywhere, it is *gerade*. Since the next orbital has one nodal plane, it changes sign once and is consequently *ungerade*. Since each successive orbital has one more nodal plane, each changes sign one more time than the previous one. Consequently, the orbitals alternate between *gerade* and *ungerade*.

6. The bonding, antibonding, or nonbonding nature of each orbital is now assessed. (a) Count the number of positive overlaps between adjacent orbitals. (b) Count the number of negative overlaps between *adjacent* orbitals. Note that if an atomic orbital is deleted from a particular MO, its neighbors on either side do not overlap each other spatially, and the situation is neither positive nor negative overlap. (c) Compute the net number of positive overlaps, which equals the number of positive overlaps minus the number of negative overlaps. If this number comes out *positive*, the MO is a *bonding* MO. If it comes out *zero*, the MO is a *nonbonding* MO and can be labeled "nb"; if it is negative, the MO is *antibonding* and can be labeled with a "*".

7. Construct an energy level diagram for the species between the levels of the hydrogen 1*s* orbitals. The relative positions of the different energy levels are approximately those of the net number of positive overlaps, with the most positive (most bonding) located at the bottom. Since each successive MO of a homoatomic linear molecule has two fewer positive overlaps, the MOs will be spaced evenly up the diagram, with all bonding MOs located below the level of the H atomic orbitals, all antibonding MOs located above it, and any nonbonding MO located at the same level. There will be no degeneracy of orbitals.

In Figure 10.14, we draw the three MOs of a triatomic H_3 species: superimposed on each qualitative drawing of a MO we show the corresponding standing wave pattern of that MO. Mathematically, we can express the three MOs of the H_3 species as linear combinations of atomic orbitals:

$$\sigma_g{}^* = 1s(\mathrm{H_A}) - 1s(\mathrm{H_B}) + 1s(\mathrm{H_C}) \tag{10.17}$$

$$\sigma_u{}^{\mathrm{nb}} = 1s(\mathrm{H_A}) - 1s(\mathrm{H_C}) \tag{10.18}$$

$$\sigma_g = 1s(\mathrm{H_A}) + 1s(\mathrm{H_B}) + 1s(\mathrm{H_C}) \tag{10.19}$$

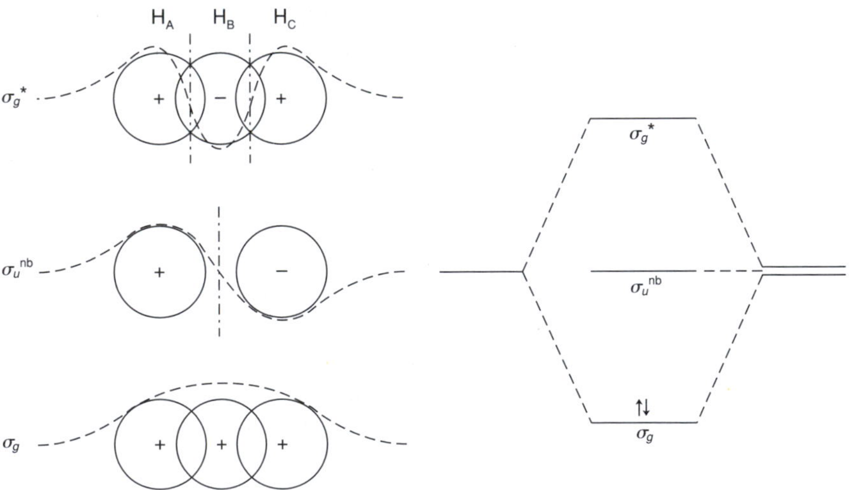

Figure 10.14
Molecular orbitals and energy level diagram for a homonuclear linear triatomic molecule or ion using 1*s* atomic orbitals. The wave pattern of each MO is sketched as a dashed line (- - - -); the nodal planes for each MO are indicated by vertical dash–dotted lines (—·—·—). Illustrated is the electron configuration of the ion $H_3{}^+$.

We also show the energy level diagram of an H_3 species. A species consisting of atoms of moderate electronegativity (such as H_n or hydrocarbons) would normally be expected to fill all of its bonding MOs, which are of low energy, and none of its antibonding MOs, which are of high energy. The nonbonding MOs may or may not be filled since they are of intermediate energy and contribute nothing to the stabilization of the molecule. (If highly electronegative atoms such as halogens are present in the molecule, all orbitals are of lower energy, so the nonbonding and even some antibonding MOs would be filled, as we shall see later in the case of XeF_2. Stable species usually fill up all orbitals below about -7 to -10 eV in energy.)

We can conceive of three linear H_3 species that might be reasonably stable: H_3^+, with a σ_g^2 electron configuration; H_3, with a $\sigma_g^2(\sigma_u^{nb})^1$ electron configuration, and H_3^-, with a $\sigma_g^2(\sigma_u^{nb})^2$ electron configuration. Each of these populates only one bonding MO, so according to Eq. (10.4) each has a bond order of one. However, this one bonding MO uses two electrons to bond together *three* positively charged nuclei. This type of bonding is known as **three-center, two-electron bonding** (3c–2e).

Clearly, the bonding in H_3 cannot be as strong an interaction as the bonding in H_2, where two electrons bond two nuclei. Indeed, the neutral species H_3 is very unstable and, if ever synthesized, decomposes quite exothermically:

$$2\,H_3 \rightarrow 3\,H_2 \tag{10.20}$$

This is because the products of this reaction have a total bond order of 3 in all molecules, whereas the reactants have a total bond order of 2. Instability of this sort is more readily predicted if we regard the number calculated from Eq. (10.4) as a *total bond order* that must be divided between pairs of connected nuclei, so that the *bond order per link*, taken as the total bond order divided by the number of interatomic links (two in the case of H–H–H), is a better indication of the relative instability of such molecules:

$$\text{Bond order per link} = \frac{\text{Total bond order from Eq. (10.4)}}{\text{Number of interatomic links}} \tag{10.21}$$

Thus, in H_3 each H–H link has a bond order of 0.5, which accounts for the instability of this species relative to H_2, in which the link has a bond order of 1.

Although H_3 is an unstable molecule, H_3^+ is thought to be important in the chemistry of interstellar clouds, where it has been detected.[22] Some molecules that employ three-center, two-electron bonding are remarkably stable. As an example, we mention the molecule diborane, B_2H_6, which has two BH_2 units linked by two *bridging hydrogens*. Therefore it has eight links:

```
H     H     H
 \   / \   /
  B     B
 /   \ /   \
H     H     H
```

This molecule, however, has only 12 valence electrons, not enough to put an electron pair into each link; it is sometimes called an **electron-deficient molecule**. In diborane, each boron-terminal hydrogen bond is a "normal" 2c–2e bond; each three-centered B–H–B link then involves two electrons in a bonding MO very much like that in H_3 (Fig. 10.14), with the corresponding nonbonding and antibonding MOs being empty. As we shall see in Chapters 12 and 15, this type of bonding is very common in boron and its compounds, since boron does not have one electron for each atomic orbital that participates in its covalent bonding; many such compounds are very stable thermally.

The H_3 molecule, however, is a species of limited interest. The MOs that we have derived in Figure 10.14 are of more practical use if we view them as being a *top view of a conjugated π-bonding system*, as in alkenes in organic chemistry. Figure 10.14 gives a top view of the π-bonding orbitals of allyl, C_3H_5; the bottom view of the π molecular orbitals shows lobes of reversed signs. This entails a change of inversion symmetry, so that the MOs labeled σ_g in Figure 10.14 are relabeled π_u, while the orbital labeled σ_u is relabeled π_g.[23] Just as with H_3, the allyl system can exist as a cation $C_3H_5^+$, a free radical $C_3H_5\cdot$, and as an anion $C_3H_5^-$. All of them involve only one π bond between the three carbon atoms, a situation that is represented in valence bond theory by drawing resonance hybrid structures (e.g., $^+CH_2\text{–}CH\text{=}CH_2 \leftrightarrow CH_2\text{=}CH\text{–}CH_2^+$ for $C_3H_5^+$). In the allyl cation, free radical, or anion, the π bond order between two linked carbon atoms is $\frac{1}{2}$; the complete $(\sigma + \pi)$ bond order per C–C link is $\frac{3}{2}$.

Although the structure of the allyl species can be accommodated by valence bond theory by the use of resonance hybrid structures, the MO theory is much more useful for predicting the nature of the electronic (UV) spectra of the allyl species. For example, in the allyl cation the σ_g orbital is occupied but the other two are not; *two* electronic transitions are thus possible, $\sigma_g \rightarrow \sigma_u{}^{nb}$ and $\sigma_g \rightarrow \sigma_g{}^*$.

Example 10.5

Draw and label the MOs of a chain of 10 hydrogen atoms, H_{10}. Draw the energy level diagram of this species, and predict what charge (if any) it would have.

SOLUTION:

The strings of atomic orbitals and nodal planes are shown in Figure 10.15. (Each atom was originally drawn as dots 10 mm apart on graph paper for convenience, and the nodal planes were drawn 10×10 mm apart, $10 \times 10/2$ mm apart, $10 \times 10/3$ mm apart, and so on. Then the sine waves were drawn in as shown, and the circles were filled in, with positive signs if they are in a crest and negative signs if they are in a trough; they were omitted if a nodal plane passed through or very close to their center.) The numbers of positive and negative overlaps were counted up as shown to give the tabulated net numbers of positive overlaps, which identified the top five orbitals as antibonding and the bottom five as bonding. Since the net numbers of positive overlaps increase by two for each MO, the energy level diagram is drawn with equal spacing of energy levels as shown at the right. Filling up the five bonding MOs requires 10 electrons, which is just the number brought in by 10 hydrogen atoms, so the H_{10} species is predicted to be neutral.

From Eq. (10.4), we compute a total bond order of 5 for this molecule and, applying Eq. (10.21), a bond order per H–H link of $\frac{5}{9}$. The H_{10} molecule is not known, and is less stable than five H_2 molecules.

Again, the more practical application of the MOs and energy level diagram of Figure 10.15 is to conjugated organic polyenes, in this case decapentaene, $C_{10}H_{12}$; the

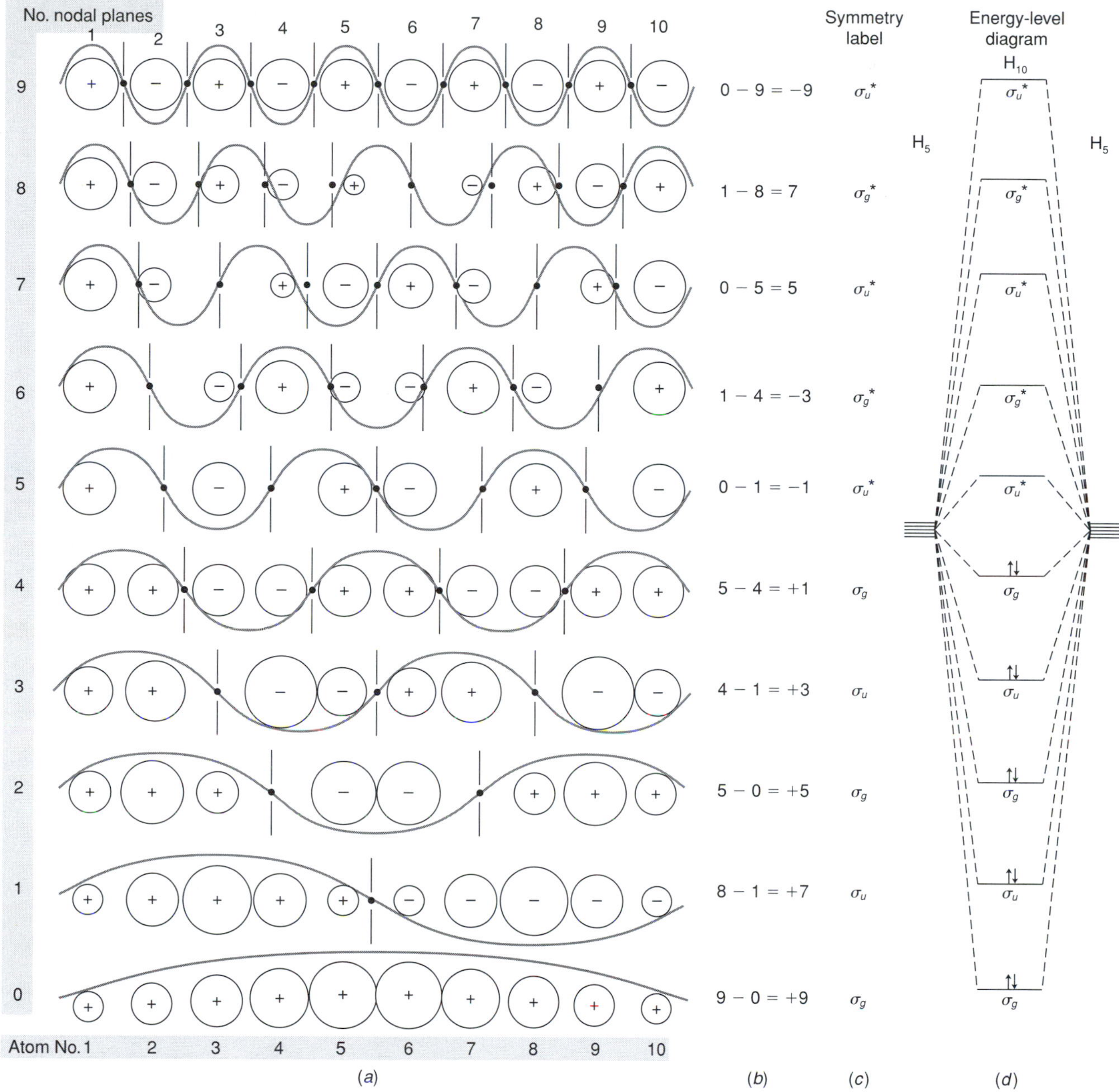

Figure 10.15
(*a*) Drawings of the 10 MOs of H_{10}, listed by the number of nodal planes present in each; (*b*) computations of net positive overlaps in each MO; (*c*) symmetry labels for each MO; (*d*) energy level diagram for H_{10}.

MOs add lobes of opposite sign below the σ_h mirror plane of the molecule and are relabeled as π rather than σ orbitals. The total π bond order of 5 matches that in the Lewis structure for $C_{10}H_{12}$, CH_2=CH–CH=CH–CH=CH–CH=CH–CH=CH_2. The energy level diagram can be used to show that there are nine different energies of electronic absorption that could show up in the UV spectrum of decapentaene; however, several are transitions from *gerade* to *gerade* or from *ungerade* to *ungerade* MOs, which are of lower intensity and may be hidden in the spectrum.[24]

Not only do the number of MOs and the number of UV absorption bands increase as the length of a conjugated π system increases, but the separation of the individual MOs, and hence the energy of light needed to promote electrons between successive MOs decreases as well. Ethylene, C_2H_4, has one absorption, $\pi \rightarrow \pi^*$, which is difficult to detect since it is in the vacuum UV region with a wavelength of less than 200 nm; butadiene has its lowest energy absorption, also $\pi \rightarrow \pi^*$, at 220 nm; 1,3,5-hexatriene, C_6H_{10}, has its lowest energy absorption at 258 nm. Since the lowest energy absorptions involve the electron moving from a *gerade* MO to an *ungerade* MO or vice versa, they are allowed by symmetry and give rise to quite intense absorptions. Eventually, when the carbon chain lengthens sufficiently, the lowest energy absorption moves into the visible region, and the organic compound becomes intensely colored. Human vision involves a long conjugated colored polyene known as 11-*cis*-retinal, $C_{17}H_{28}O$. When this molecule absorbs light, an electron is promoted from a MO that has positive π overlap between the eleventh and twelfth carbon atoms to an antibonding MO that has negative π overlap at that position. In effect, this converts the bond between these two carbon atoms from a double bond, which is not free to rotate, to a single bond, which is free to rotate, allowing the polyene to isomerize to 11-*trans*-retinal; this change initiates the process of detection of the light by the eye.

Heteronuclear Linear Molecules and Ions. We will consider only one example of this large class of substances, XeF_2. The Lewis dot structure of this molecule shows, in addition to two Xe–F bonds, three unshared electron pairs around Xe. In valence bond theory, it is possible to accommodate the fifth pair of electrons on the Xe atom by resorting to dsp^3 hybridization. Molecular orbital theory has the advantage that it is easily capable of explaining the bonding in a species such as XeF_2 without necessarily requiring the use of some of the high-energy postvalence *d* orbitals.

By leaving out *d* orbitals, we find that the Xe and each F atom each has four valence atomic orbitals, for a total of 12: This result means that we can construct 12 MOs. A complete description of these 12 orbitals and of the corresponding energy level diagram, although possible now, would be more cluttered than we wish to attempt at this point. To simplify matters, we note that XeF_2 has a total of 22 valence electrons, which is enough to fill *all but one* of the MOs, which will be an antibonding MO and the LUMO of the molecule. We know, of course, that σ overlap is stronger than π overlap, so that the usual order of MOs is $\sigma < \pi < \pi^* < \sigma^*$; hence, the LUMO is expected to be a σ^* orbital. All six π, π^{nb}, and π^* molecular orbitals (based on p_x and p_y orbitals on Xe and each F) are occupied, so that π bonding contributes nothing to the total bond order.

Hence, for simplicity's sake we examine only the σ bonding in this molecule. In the $D_{\infty h}$ point group, the only atomic orbitals that fall in Σ type irreducible representations are the *s* and p_z orbitals on Xe and F. Of these, the *s* orbitals of Xe and F fall at very low energy (by analogy with Ne or Ar in Table 10.1), so they have little contribution to the highest energy orbital, σ^*. Hence, we (over)simplify our bonding scheme to consider the interactions of only three orbitals, the p_z orbitals on the central Xe atom and the two terminal F atoms. The three resulting MOs are shown in Figure 10.16; just as in H_3 or allyl, one of these is bonding, one is nonbonding, and one is antibonding. Only the antibonding MO is unoccupied. Since the π orbitals contribute nothing to the total bond order, bonding is entirely due to the occupation of the σ_g bonding MO. The total bond order is just one, and the bond order per Xe–F link is therefore just $\frac{1}{2}$. The

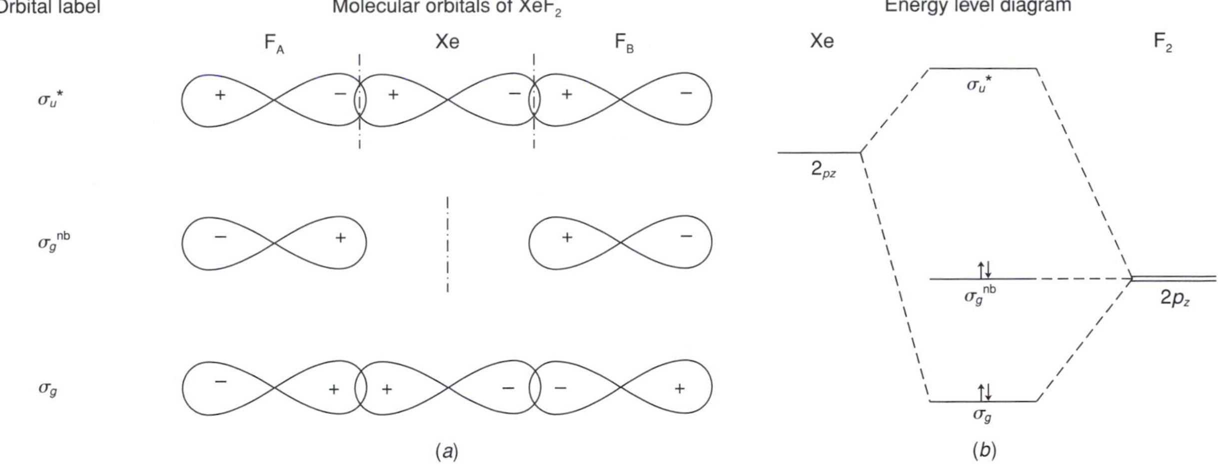

Figure 10.16
(*a*) Simplified σ MOs and (*b*) energy level diagram for XeF_2 to illustrate 3c–4e bonding. Nodal planes are indicated by vertical dash–dotted lines (—·—·—).

molecule can exist, but it is expected to have weak Xe–F bonds.[25] Although four electrons are available for the σ bonding in this molecule, two fill a nonbonding MO, and hence do not contribute to the bond order. This type of bonding interaction is therefore often called **three-center, four-electron (3e–4e) bonding.**

The XeF_2 molecule is a good illustration of the class of **hypervalent molecules**[26]: The *p*-block molecules, which in their Lewis structures or at least in their major resonance structures, exceed octets of electrons on their central atoms. Hypervalent molecules are often labeled with the **N-X-L** notation (10-Xe-2 for XeF_2) in which the first number is the number of electrons assigned (>8 in a hypervalent species) around the central atom whose atomic symbol is then given; the second number is the coordination number of the central atom. It has been suggested that hypervalence is not a useful concept, since central atoms from the third periods and below, with maximum total coordination numbers of 6 or higher, quite commonly have a "duodecet" of electrons.[27] And, like its parent term "valence," which has fallen out of favor for lack of usefulness and difficulty of simple definition, the above definition of hypervalence is awkward: SO_4^{2-} is hypervalent if resonance structures involving double bonds are emphasized, but is not if they are not used. Another definition[28] instead characterizes hypervalent bonding as 3e–4e electron bonding.

Three-center, four-electron or hypervalent bonding is stabilized when the outer atoms in the molecule or ion have either or both of two characteristics: They are part of a small chelate ring that stabilizes a small (90°) bond angle at the central atom (thus making space for additional outer atoms); and/or they are very electronegative. This latter characteristic can be justified by noting that the nonbonding MO in a 3c–4e electron bond puts excess electron density on the outer atoms. This characteristic can be justified from the valence bond point of view by noting that it puts excess positive charge on the central atom, which would contract and lower the energy of the post-valence *d* orbitals needed for the hybridized bonding. Thus, hypervalence is known among hydrides (Chapter 15) only[29] in SiH_5^-; among hydrides the octet "rule" reigns supreme.

10.6 Molecular Orbital Theory for Cyclic Molecules: Generator Orbitals

We begin the study of cyclic homoatomic molecules with simple cyclic H_n molecules, and then proceed to the much more important *conjugated cyclic polyenes* $C_nH_n{}^{z\pm}$ (including the *aromatic hydrocarbons*), in which carbon p_z atoms overlap to form a ring structure. These compounds include many important organometallic ligands. If homoatomic, these molecules fall in the D_{nh} point groups, in which the x and y directions are indistinguishable. This indistinguishability results in many of the MOs falling in doubly degenerate pairs of equal energy. Because of the great nonlinearity of the molecules or ions, we will use irreducible-representation labels appropriate to the D_{nh} point group rather than σ or π.

We begin by recognizing that the MOs of horizontal planar rings will be characterized by increasing numbers of vertical nodal planes. But where should we locate these nodal planes? Once they are located, we can identify crests and troughs and put signs on the atomic orbitals participating in the MOs.

A quick way of generating the MOs of cyclic (and other compact) molecules is known as the *Generator Orbital (GO)* approach.[30] In this approach, it is recognized that the *vertical nodal planes of the MOs of a molecule*, which must pass through the center of the molecule, *match the vertical nodal planes of appropriate atomic orbitals of a* real or imaginary *atom situated at the center of the molecule*. The orbitals of the real or imaginary atom at the center can be used to generate the proper positions of the nodal planes of the MOs of the real outer part of the molecule, and are hence known as *generator orbitals*.

For a two-dimensional ring of n atoms, there will be n MOs. Successive MOs, and the GOs used to generate them, will have increasing numbers of *vertical* nodal planes: $0, 1, 1, 2, 2, \ldots$, as drawn in Figure 10.17. (The numbers are doubled after 0 because, in these D_{nh} point groups, an xz nodal plane generated by a p_y generator orbital is doubly degenerate with a corresponding nodal plane in the second dimension, the yz plane generated by the p_x GO.) For example, the three MOs of *cyclo*-H_3 have 0, 1, and 1 nodal planes, respectively, and match the first three patterns shown in Figure 10.17, which are the patterns generated by the nodal planes of the first three generator orbitals in the xy (ring) plane , s, p_x, and p_y.

In addition, if we are dealing with the conjugated cyclic polyenes, we are interested in π-type MOs that are characterized by the presence of *one horizontal nodal plane* that passes through all of the nuclei of the ring. This nodal plane is the same one that separates the positive and negative lobes of the carbon $2p_z$ orbitals that go into forming the π molecular orbitals. The generator orbitals for these MOs have the corresponding z component as well: they add oppositely signed lobes behind the lobes of GOs for *cyclo*-H_n systems, which transforms the s generator orbital to a p_z GO for π systems, and transforms the p_x and p_y GOs to d_{xz} and d_{yz} GOs, and so on. As can be seen, the π-system GOs not only change s to p and change p to d as a consequence of the extra (horizontal) nodal plane, but also add the letter "z" to their designations.

Therefore the three π molecular orbitals of *cyclo*-C_3H_3 have the same number of vertical nodal planes as those of *cyclo*-H_3, but also have one horizontal nodal plane each. Therefore they are generated by central-atom p_z(with no vertical nodal planes) and d_{xz} and d_{yz} GOs (with one vertical nodal plane each). To construct the actual MOs, an n-sided polygon is inscribed symmetrically in n of the first nodal plane patterns from Figure 10.17. (For cases in which n is evenly divisible by 4, it is important to

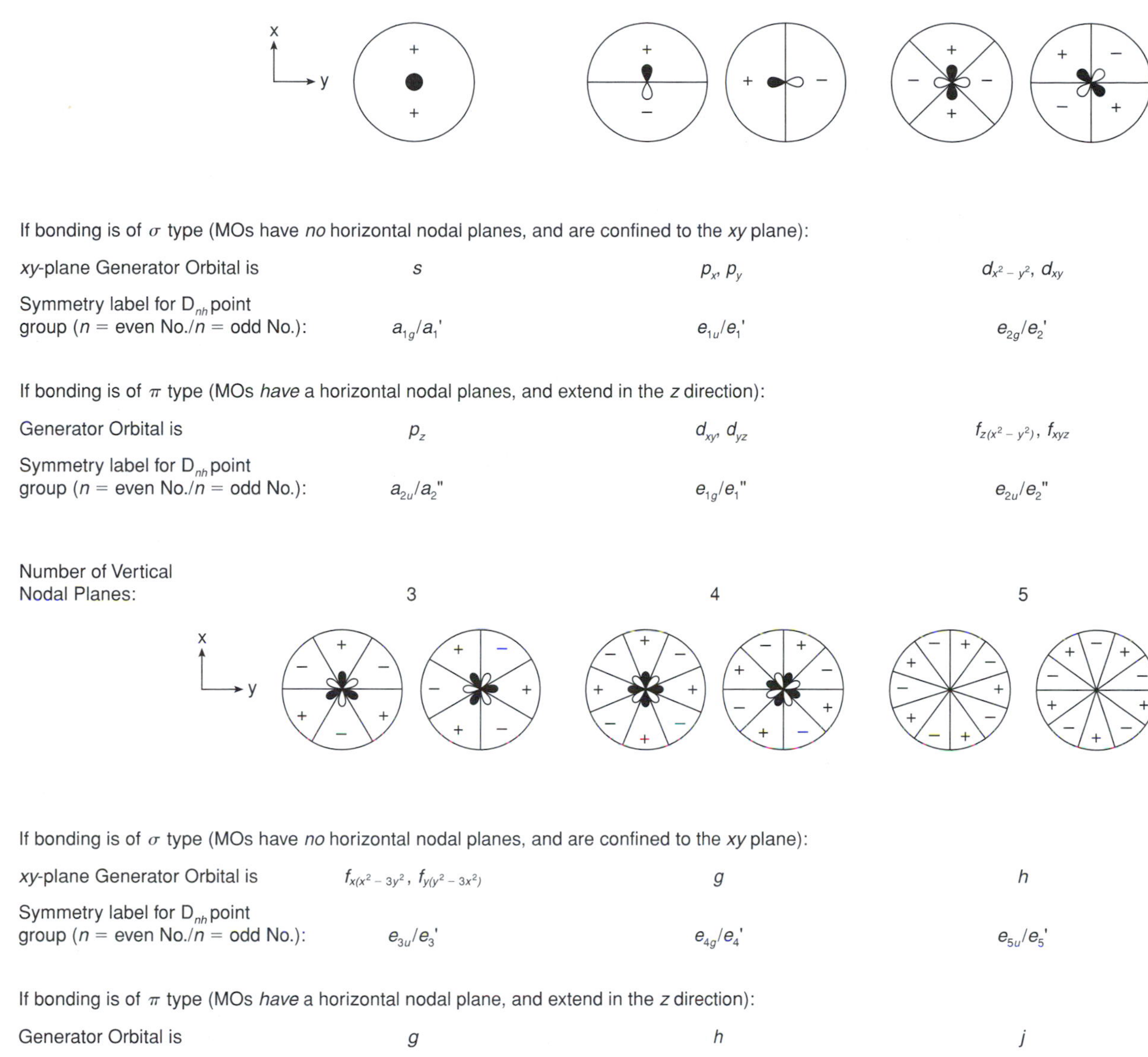

Figure 10.17
Nodal planes, generator orbitals and symmetry labels for cyclic molecules.

orient the polygon so that the points do *not* directly fall on the x and y axes.) Then circles are drawn at the corners of the polygon to represent the hydrogen 1s orbitals (for *cyclo*-H_3) or the top lobes of the carbon $2p_z$ orbitals (for cyclo-C_3H_3). If a nodal plane cuts through a corner of the polygon in a particular MO, that atomic orbital is omitted from the corresponding MO. (One can optionally draw circles of different sizes as for the linear molecules: The further the corner is from the nodal plane, the higher the amplitude of the wave function is at that atomic orbital.) Figure 10.18(*a*) shows the three MOs of *cyclo*-H_3. To obtain the MOs of *cyclo*-C_3H_3, we can simply take the MOs of *cyclo*-H_3 and add lobes of opposite sign behind each lobe of *cyclo*-H_3, as shown in Figure 10.18(*b*).

The symmetry label (irreducible representation) of the generator orbital also applies to the MO with which it shares a set of common nodal planes; this allows us to find the irreducible representation of the MOs in the character table of the D_{nh} cyclic molecule easily. Generally, very similar labels are used in the different point groups for the same nodal patterns and generator orbitals, and are summarized in Figure 10.17. However, the character table should be consulted for slight variations: Thus, the three MOs of *cyclo*-H_3 are actually labeled in the D_{3h} point group as a_1' and e' rather than a_1' and e_1', since there is no e_2' to be differentiated from the first e'.

The last MO to be filled for even-membered rings requires some special treatment. Figure 10.17 offers a doubly degenerate pair of nodal planes for this one orbital, but both cannot be used for one MO. It will be found that one of the two choices puts nodal planes through *all* of the atoms, which makes an MO impossible: The other nodal pattern is the one to be used. Since this MO is not degenerate, the *e* label given in the figure cannot be used; instead a label beginning with the letter *b* is used. The label is most easily obtained by locating the corresponding generator orbital in the appropriate character table.[31]

The energy level diagram for *cyclo*-H_3 or the π electrons of *cyclo*-C_3H_3 is shown in Figure 10.18(*c*). The a_1' or a_2'' MO has three positive overlaps and is bonding. One of the two e' or e'' orbitals has one positive and two negative overlaps, while the other has one negative overlap: Each has a net of one negative overlap, so the two are degenerate and mildly antibonding. The hypothetical H_3 or the real C_3H_3 species accommodates only two bonding electrons, and hence are expected to be found as the cations H_3^+ and $C_3H_3^+$. The H_3^+ cation has a low total bond order (1) and bond order per H–H link ($\frac{1}{3}$), so it is likely to lose a proton to some more basic molecule and become the much more stable H_2 molecule. However, the $C_3H_3^+$ cation is also held together by σ bonds, which we have ignored in this treatment, so it is actually found in the laboratories of organic chemists.

There is an alternate, simple way of drawing the energy level diagrams for conjugated cyclic polyenes, with which many organic chemistry students are familiar: inscribe the polygon with the proper number of carbon atoms inside a circle, with one apex of the polygon pointed straight down. The energy level diagram for that polyene has an energy level at the energy corresponding to each apex of the polygon. This automatically generates a unique bonding MO of lowest energy, followed by as many doubly degenerate pairs of MOs as possible. Those energy levels falling in the lower half of the circle are bonding MOs; those in the upper half are antibonding; those at the halfway point are nonbonding.[32] Filling the bonding energy levels only requires 2, 6, 10, 14, . . . , electrons, depending on the size of the polygon; these are the number of π electrons required to impart aromatic stability to a conjugated cyclic polyene.

Example 10.6

Draw and label the MOs and the energy level diagram of the conjugated cyclic polyene, C_4H_4, cyclobutadiene. Predict its number of unpaired electrons.

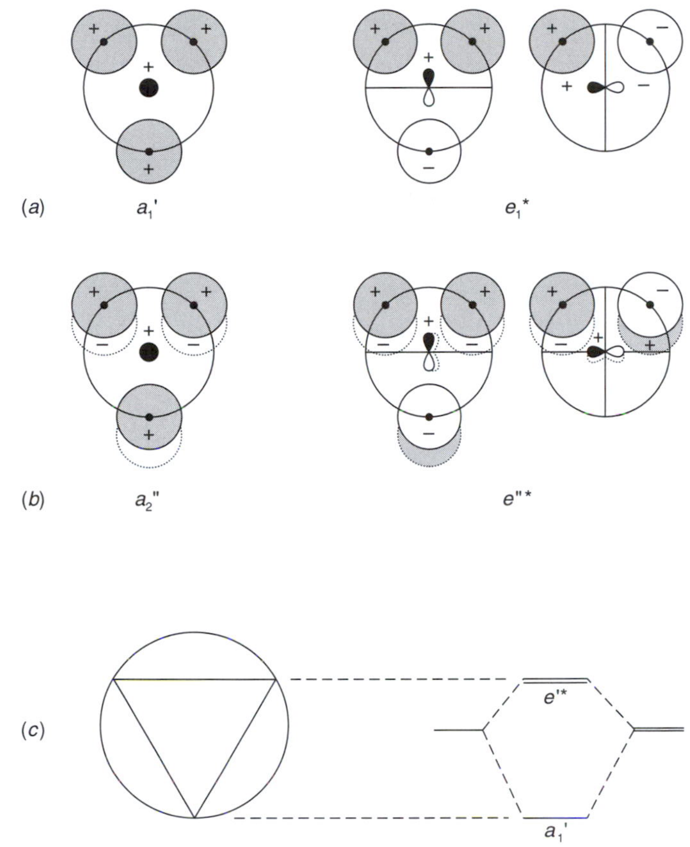

Figure 10.18
(*a*) Molecular orbitals of *cyclo*-H_3. (*b*) Pi MOs of *cyclo*-C_3H_3, with dashed lines representing lobes below the plane of the paper. (*c*) Energy level diagram for either species, shown with symmetry labels appropriate for *cyclo*-H_3.

SOLUTION:

This four-carbon-atom polyene has four π molecular orbitals, which utilize the zero-nodal-plane pattern, the two one-nodal-plane patterns, and *one* of the two two-nodal-plane patterns of Figure 10.17. We must be careful to orient this molecule so that we do not have its four corners falling on the x and y axes, so we draw it as a square rather than as a diamond.

The first MO is labeled a_{2u} and has four positive overlaps, so it is bonding and is lowest in the energy level diagram (Fig. 10.19). (The energy level diagram can also be obtained by inscribing a square in a circle, but *for this purpose* the square *must* be inscribed with its corners on the x and y axes.) The second and third MOs are a degenerate pair with two positive and two negative overlaps, and hence are nonbonding and are labeled e_g^{nb}. The top MO has four negative overlaps and is antibonding; it is not degenerate and cannot be labeled e_{2u}; we consult the D_{4h} character table to locate its f_{xyz} generator orbital [consult

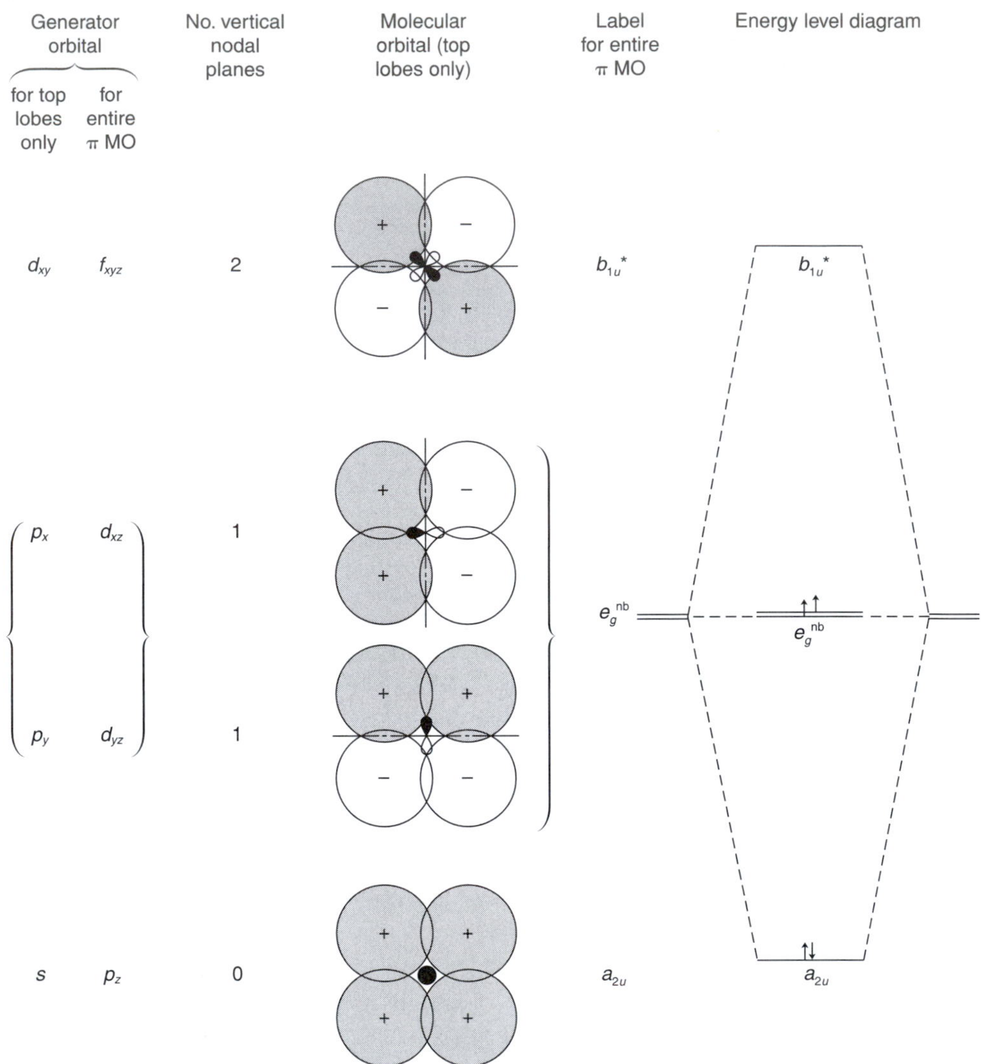

Figure 10.19
Generator orbitals, MOs, symmetry labels, and energy level diagram for the π electrons of *cyclo*-C_4H_4. The drawings of the MOs show tiny generator orbitals at the centers, but do not show the lobes of opposite sign for each MO, which fall below the plane of the paper.

Fig. 8.21 to see that this is used rather than $f_{z(y^2-x^2)}$] in the B_{1u} irreducible representation, and label this orbital b_{1u}^*.

The cyclobutadiene molecule is predicted to have two unpaired electrons in a degenerate pair of MOs, so its π bond order is only 1, which is 0.25 per C–C link. It is held together, not only by π, but also by σ bonds that we have not discussed, so it does not fly apart into two C_2H_2 molecules, but it retains some other peculiarities. By a variation of the Jahn–Teller theorem (Section 8.7), it undergoes a distortion to relieve the double orbital degeneracy of its partially filled e_g^{nb} orbitals: It elongates to become a *rectangular* molecule, which resembles two ethylene molecules loosely attached by very long σ bonds; with a nonsquare geometry the energy level diagram of Figure 10.19 no longer applies and two MOs are bonding and two are antibonding; all π electrons are paired in the bonding MOs, so the molecule is now diamagnetic. But with such a strained geometry it is not at all stable: It can be isolated at $-78\,^{\circ}$C, but rapidly decomposes at room temperature. In contrast to the familiar case of the cyclic aromatic hydrocarbon benzene, C_6H_6, cyclobutadiene has been *destabilized* by the conjugation of its π electrons in a ring; it is an example of an *antiaromatic* molecule.

Example 10.7

Draw and label the six π molecular orbitals of benzene. Compute the net positive overlaps of each, and draw the energy level diagram.

Solution:

With six carbon atoms utilizing six p_z orbitals, this molecule will have six π molecular orbitals, which will need six of the first seven nodal patterns shown in Figure 10.17. The first MO has no nodal planes, it has six positive overlaps, and is a a_{2u} bonding orbital. The next two MOs each have one vertical nodal plane. As drawn in Figure 10.20, the two MOs have slightly different appearances, since in one two atoms do not participate—but the end result is the same, two net positive overlaps each; these are labeled e_{1g}. These are followed by two slightly antibonding MOs with two vertical nodal planes each and a net of two

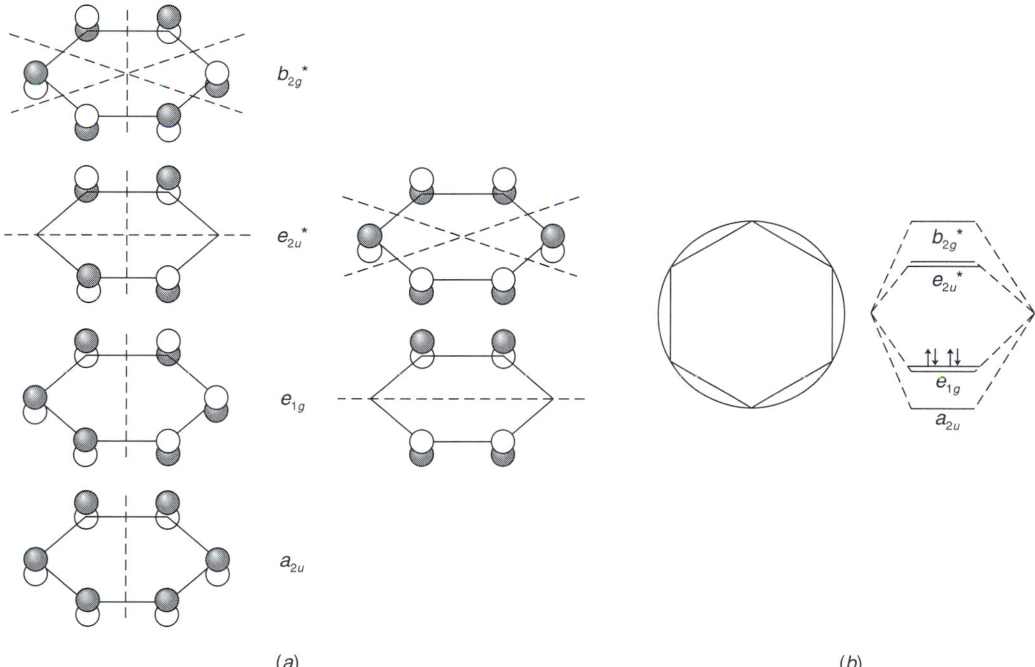

Figure 10.20
(*a*) Pi molecular orbitals of benzene, *cyclo*-C_6H_6; (*b*) energy level diagram of benzene. [Adapted with permission from R. L. DeKock and H. B. Gray, *Chemical Structure and Bonding*, University Science Books, Sausalito, CA, 1989; p. 314.]

negative overlaps; these are labeled e^*_{2u}. The sixth and highest energy MO uses just one of the two patterns of three nodal planes shown in Figure 10.17, the one that does *not* cut through and eliminate all six carbon atoms. Since this one MO is not degenerate, it is not labeled e^*_{3g}. Its generator orbital is a type of g orbital, so we cannot look up the irreducible representation in the D_{6h} character table; let us content ourselves with labeling this orbital as b^*_{xg} ($x = 1$ or 2.) (Detailed examination of the effects of the σ_v and σ_d operations on this MO show that it is actually $b_{2g}{}^*$.)

The energy level diagram of C_6H_6 can be generated from a hexagon inscribed point down in a circle [as shown in Fig. 10.20(*b*)]. The most stable form of C_6H_6 will fill the three bonding MOs with six carbon valence p_z electrons, hence C_6H_6 is an uncharged molecule.

It has been noted[33] that one way of characterizing the aromaticity (degree of extra stability) of conjugated cyclic polyenes is by their hardness, as computed by Eq. (10.15) from the energy separation of the HOMO and LUMO. The least aromatic is cyclobutadiene, in which the HOMO is degenerate with the LUMO (so that each is a SOMO) and the hardness is 0 eV. The most aromatic is, as expected, benzene, for which the hardness is 5.3 eV.

10.7 *Orbital Symmetry Rules for Concerted Reactions

In organic chemistry, we often draw mechanisms for reactions in which we "push arrows" to show a smooth flow of electron pairs around a series of bonds as two species react. It is easy to draw electron pairs flowing, but can such a Lewis acid–base reaction really happen as shown (e.g., in one single step in a **concerted** mechanism of electron-pair flow)? The frontier orbital concept has been extended to allow us to predict whether such a concerted reaction between two molecules or polyatomic ions can occur readily, or whether it will likely have a high activation energy. If the activation energy is high an alternate mechanism may apply, which involves several steps and intermediates and perhaps free radicals instead of acids and bases. These orbital symmetry rules are[34]

1. As the reactants approach each other, electron density must flow from the HOMO of the donor to the LUMO of the acceptor.

2. The HOMO of the donor and the LUMO of the acceptor must approach each other so that they have a net positive overlap, and they must be relatively close in energy (within ~6 eV).

3. The net effect of HOMO → LUMO electron transfer must be to make the bonds to be the formed and weaken the bonds to be broken.

A classic application of these rules is to the mechanism for the reaction of H_2 with I_2. For many years, this mechanism was thought to be a concerted mechanism in which the H_2 and I_2 molecules formed a square species involving some kind of electron-deficient bonding. This bonding finally broke down to give the observed products, 2 mol of HI [Figure 10.21(*a*)].

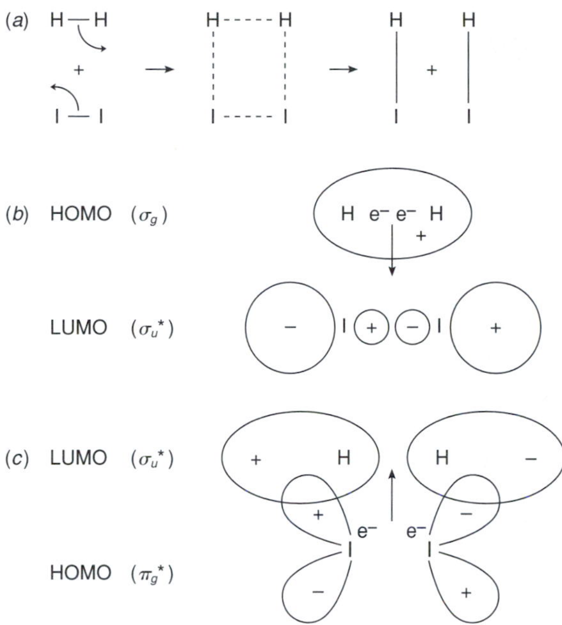

Figure 10.21
Orbital symmetry rules applied to the (*a*) proposed bimolecular reaction of H_2 with I_2 to give 2 HI. (*b*) Zero overlap between the HOMO of $H_2(\sigma_g)$ as the source of the electron pair being donated to the LUMO of I_2 (σ_u*). (*c*) Positive overlap between the HOMO of $I_2(\pi_g{}^*)$ as theo source of the electron pair being donated to the LUMO of H_2 (σ_u*).

First, let us consider the possibility of I_2 acting as the acceptor: It contains the more electronegative atoms, and often acts as a Lewis acid (Section 7.8). In the square transition state, there must then be positive overlap between the HOMO of H_2, σ_g, and the LUMO of I_2, $\sigma_u{}^*$, or this complex could not bond together. However, as shown in Figure 10.21(*b*), there is zero overlap between the *gerade* and the *ungerade* orbital.

Perhaps I_2 acts as the donor. Its HOMO is $\pi_g{}^*$; the one component of this degenerate pair that lies in the plane of the square transition state does indeed give positive overlap with the LUMO of H_2, $\sigma_u{}^*$ [Fig. 10.21(*c*)]. We cannot, of course, judge from orbital pictures whether these two MOs lie within 6 eV of each other, but we can test the last rule: The net effect of this electron transfer should be to weaken the H–H and I–I bonding, which is to be broken, and strengthen the H–I bonds that are to be formed. Transferring electrons to the H_2 $\sigma_u{}^*$ orbital weakens the H–H bond, since this orbital is antibonding. However, transferring electrons *from* the I–I $\pi_g{}^*$ depopulates an antibonding MO, and therefore *strengthens* the I–I bond, taking it closer to a double bond, rather than weakening it.

For many years, textbooks stated that this simple reaction occurred by this concerted mechanism; people were astonished in 1967 when it was shown that this mechanism is incorrect. Instead, the reaction is multistep, in which I_2 molecules must first dissociate into I atoms, which then react with H_2 molecules to give the product. (Reactions of free atoms and simple free radicals are not normally restricted by orbital symmetry rules.)

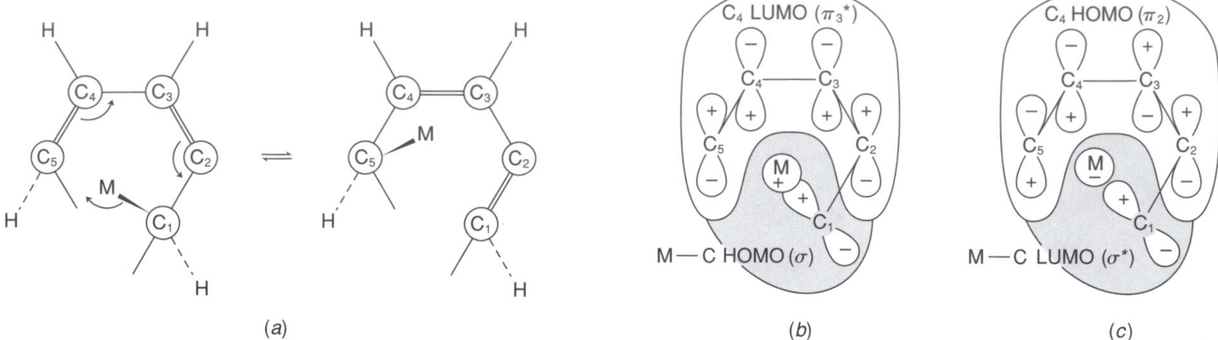

Figure 10.22

Mechanism for the 1,5-shift in η^1-MC_5H_7 or η^1-MC_5H_5. (*a*) The flow of electrons in the valence bond picture. (The dangling pair of σ bonds at the bottom of the molecule connect to two H atoms in η^1-MC_5H_7, but connect to each other in η^1-MC_5H_5.) (*b*) The overlap of the M–C HOMO (σ) for the donation of its electron pair and M to the C_4 LUMO ($\pi_3{}^*$). (*c*) The overlap of the C_4 HOMO (π_2) for the donation of its electron pair to the M–C LUMO (σ^*).

As another example, the mechanisms of fluxional behavior are governed by orbital symmetry rules, since fluxional behavior is defined so as to require concerted rather than free-radical mechanisms. Let us consider the fluxional motion of a metal atom in a pentadienyl derivative, C_5H_7M, or a *monohapto*-cyclopentadienyl derivative, η^1-C_5H_5M. Since the metal atom is of low electronegativity, we will first consider donation from the M–C σ bond as HOMO to the LUMO of the π system of the four remaining carbon atoms. First of all, we note that even though the HOMO is a σ bond and the LUMO is a π^* molecular orbital, due to the geometry of these systems there is spatial overlap between them (Fig. 10.22): This is known as σ–π conjugation or *hyperconjugation*.[35]

The LUMO of the C_4 system has two vertical nodal planes, with positive signs on the two end carbon atoms [Fig. 10.22(*b*)]. Hence, the M–C bond has positive overlap, not with the two interior carbon p_z orbitals (C3 and C4 if the carbon bearing the metal is numbered C1), but with those on the two ends (C2 and C5). Hence, a mechanism of a "1,3-shift" or a "1,4-shift" is not favored by orbital symmetry rules.

For an open (pentadienyl) system, the 1,2-shift does not form the bonds to be formed: It interrupts the π system by saturating C2, creating a high-energy $-CH_2^+$ site at C1. But a shift of the metal atom to C5 and the transfer of its electrons to the LUMO, π_3, allows new π overlap to result between the pairs of carbon atoms (C1, C2) and (C3, C4), where the new π bonds are supposed to be in the fluxionally rearranged product [Fig. 10.22(*b*)]; it also introduces negative overlap between C2 and C3, helping to break this π bond, which is to disappear in the rearranged product. Hence, the mechanism predicted in the open-chain pentadienyl derivative is that of a *1,5-shift.*

With the cyclopentadienyl derivative, C2 and C5 are equivalent by symmetry, and a 1,2-shift does not leave a dangling, unconjugated carbon atom, so the shift can be called either a 1,2- or a 1,5-shift. However, it can be confirmed (by detailed NMR measurements) that the fluxional behavior does *not* occur by a 1,3-shift in cyclopentadienylmetal compounds.

Example 10.8

Suppose that the electron and metal–atom transfer in η^1-C_5H_5M occurs in the other direction: from the HOMO of the C_4 π system to the LUMO of the M–C σ system. Does a 1,5-shift still obey the orbital symmetry rules?

SOLUTION:

The LUMO of the M–C σ system is σ^*, with a negative lobe at the top; the HOMO of the C_4 π system is the π_2 molecular orbital with one vertical nodal plane, as drawn in Fig. 10.22(*c*). Transfer of electrons to σ^* helps break the M–C bond. Loss of electrons from π_2 reduces the π bond order between the pairs of atoms (C2, C3) and (C4, C5), helping to break these bonds, which are not present in the rearranged product. This loss of electrons finally helps eliminate negative overlap between (C3, C4), which helps build a π bond between these two atoms.

In reality, both types of electron transfer (from C–H HOMO to C_4 LUMO, and back again from the C_4 HOMO to the C–H LUMO) occur in this concerted mechanism, completing the pushing of electrons so that neither part of the molecule ends up with any more charge than they had before the fluxional rearrangement.

There are additional examples of applications of orbital symmetry rules to organic cycloaddition reactions, which are beyond the scope of this text.[36] It would now appear that many concerted reactions are indeed not allowed by these rules, but proceed nonetheless by, for example, free radical mechanisms, while some others that are allowed may proceed with a more complex mechanism than suggested by the orbital symmetry rules.[37] We might also note that one of the advantages of *d*-block metal ions in catalysis of reactions is that they have so many types of valence orbitals of such diverse symmetry: If overlap of one type of valence orbital does not satisfy the orbital symmetry rules, the odds are good that another type of valence orbital can do the job.

10.8 Molecular Orbitals for D_{nh} Complexes

Borane, BH_3, and Boron Trifluoride, BF_3. Next, we consider the cases of polygonal (and later polygonal bipyramidal) molecules in which the central generator atom is in fact a real atom, and in which the outer atoms around the polygon use their valence orbitals to bond, not to each other, but to the real central atom. As our first example, we generate the MOs of BH_3 from those of *cyclo*-H_3. The MOs of *cyclo*-H_3 [Fig. 10.18(*a*)] now become a set of **terminal atom symmetry adapted orbitals (TASOs)** that can have positive overlap with, and bond to, real boron central-atom orbitals *only if both belong to the same irreducible representation*. Hence, a bonding a_1' molecular orbital [Fig. 10.23(*a*)] is formed by positive overlap between the a_1' TASO of the *cyclo*-H_3 group and the only valence boron orbital in the A_1' irreducible representation, the 2*s* orbital. An $a_1'^*$ orbital is also formed [Fig. 10.23(*b*)] but remains unoccupied. Likewise, a bonding pair of e' molecular orbitals [Fig. 10.23(*a*)] are formed by positive overlap between the e' TASO of the *cyclo*-H_3 group and the valence boron

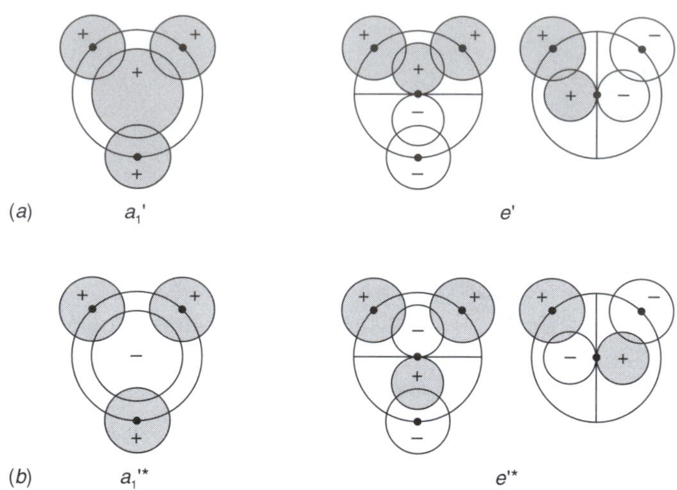

Figure 10.23
Sigma MOs of BH_3 (and of BF_3 and of the equatorial part of PF_5, but with only the inner lobes of the fluorine p orbitals shown). (a) Bonding MOs; (b) antibonding MOs. Not shown is (c) the nonbonding boron $2p_z$ orbital.

orbitals in the E' irreducible representation: the $2p_x$ and $2p_y$ orbitals. These three bonding MOs hold the six valence electrons of BH_3. The electron configuration is $(a_1')^2(e')^4$. One boron valence orbital, the $2p_z$, is then nonbonding and unoccupied $(a_2''^{\text{nb}})$.

Borane is mainly of academic interest, since it dimerizes to give diborane, B_2H_6, which was mentioned in Section 10.5. Of more real interest is boron trifluoride, BF_3, which does not dimerize. The σ bonding in this molecule is quite like that in BH_3, except that the lobes shown for the terminal atoms are not H 1s orbitals, but are the inner lobes of F $2p_x$ orbitals [Fig. 10.23(a)]. Each terminal F ligand also has three formally unshared electron pairs, one in 2s, which is too low in energy for effective overlap with anything; one in $2p_y$, which finds no mate in boron; and one in $2p_z$ extending along the z axis (threefold axis) of the molecule. The three $2p_z$ orbitals of the three terminal F atoms generate three TASOs that look just like the π orbitals in C_3H_3 [Fig. 10.18(b)], and are similarly labeled a_2'' and e''. There is no valence orbital on boron in the E'' irreducible representation, so the e'' TASO indeed corresponds to nonbonding (lone-pair) behavior. However, the a_2'' TASO has positive overlap with the a_2'' valence orbital on boron, $2p_z$, and gives rise to some π bonding in this molecule. (The degree of π bonding is limited, however, by the poor energy match between the boron and fluorine orbitals.) The electron configuration of BF_3 is then $(a_1')^2(e')^4(a_2'')^2$ (various nonbonding F orbitals).

Phosphorus Pentafluoride, PF_5. Phosphorus pentafluoride is a trigonal bipyramidal molecule with two types of fluorines, axial and equatorial. The σ bonding of the three equatorial fluorines to phosphorus is substantially like that of the three equatorial fluorines in BF_3 [Fig. 10.23(a)]. Since there must be bonding of the two axial fluorines to P in PF_5, we will withdraw the central-atom p_z orbital from use in π bonding and use it mainly to bond the two axial fluorines. The bonding in the axial PF_2 unit is then substantially like the 3c–4e bonding in XeF_2 (Section 10.5; Fig. 10.16), with the occupied σ-bonding MO being relabeled a_2'', and the occupied nonbonding MO being relabeled a_1'. Note that the bonding in PF_5 is almost completely separable into axial and equatorial subsets: Only the A_1' irreducible representation occurs in both subsets, so in fact the boron $2s$ orbital is involved in the bonding to both subsets of fluorines. The electron configuration of PF_5 is then $(a_1')^2(e')^4(a_2'')^2(a_1'^{nb})^2$(various nonbonding F orbitals).

Metal Complexes of Cyclic Polyenes. As was mentioned earlier (Chapters 3 and 8), organic compounds such as polyenes possessing shared π electron pairs can donate these electron to metal ions to act as nonclassical or *π-donor* ligands. Having derived the MOs of some polyenes, we can now see that the filled π molecular orbitals are the source of the donated electrons; we now seek to match these MOs with metal orbitals of appropriate symmetry to interact with them. We consider as examples only the cases in which *all* of the polyene unsaturated carbon atoms are coordinated to the metal atom, that is, the polyene is showing its highest category of hapticity (Section 9.3).

We expect the metal ion to bond above one face of the polyene, with overlap occurring between the lobes of the π bonds on that face of the polyene and the lobes of the metal orbitals on the side of the metal ion adjacent to the polyene. (Hence, it is not necessary to worry about the signs on the orbitals on the other half of the polyene.) The main criterion for overlap is that *the metal ion and the ligand MO share the same number of vertical nodal planes* (as listed in Fig. 10.17). This is equivalent to saying that the metal orbitals and polyene orbitals must fall in the same irreducible representation, if we take into account the fact that we have now eliminated the inversion center i and lowered the symmetry from D_{nh} to C_{nv}.

Let us take the common π ligand, benzene, C_6H_6, as an example; its MOs are shown in Figure 10.20. To find the metal ion orbitals of appropriate symmetry to interact with these TASOs, we can either recall those having the same number of vertical nodal planes, or else consult the C_{6v} character table.

The π molecular orbitals of benzene, when lowered to C_{6v} symmetry, have the g or u removed from their labels, and may have other changes: the MOs in Figure 10.20, from bottom to top, in the C_{6v} point group have symmetry labels a_1, e_1, e_2, and b_1, with 0, 1, 2, and 3 vertical nodal planes, respectively. If the metal atom sitting above the ligand face is a d-block metal ion, the metal orbitals eligible for bonding with these ligand MOs are s, p_z, and d_{z^2} with zero vertical nodal planes (a_1 irreducible representation); (p_x, p_y) and (d_{xz}, d_{yz}) with one vertical nodal plane (e_1 irreducible representation); $(d_{x^2-y^2}, d_{xy})$ with two vertical nodal planes (e_2 irreducible representation); and no d orbitals with three vertical nodal planes (b_1 irreducible representation).

Now suppose that we place another benzene ring on the opposite side of the metal ion to give a "sandwich molecule" $M(C_6H_6)_2$. Organometallic molecules of this type are discussed in Chapter 11; important specific examples are $[Cr(C_6H_6)_2]$ and $[Cr(C_6H_6)_2]^+$. This addition restores the molecule to D_{6h} symmetry and returns the inversion center i and the u and g labels. As the second benzene ring comes in, it can either come in with the same lobe signs pointing to the metal ion as the first ring had, which generates a *gerade* MO, or it can come in with the reverse-sign face toward the metal, which generates an *ungerade* MO. Thus a total of 12 TASOs will be generated for the two benzene rings. Some of these, however, will find no d-block metal-ion orbitals of matching symmetry, since d-block metal ions have only nine valence orbitals (including their np orbitals); these TASOs must be nonbonding to the metal ion. In

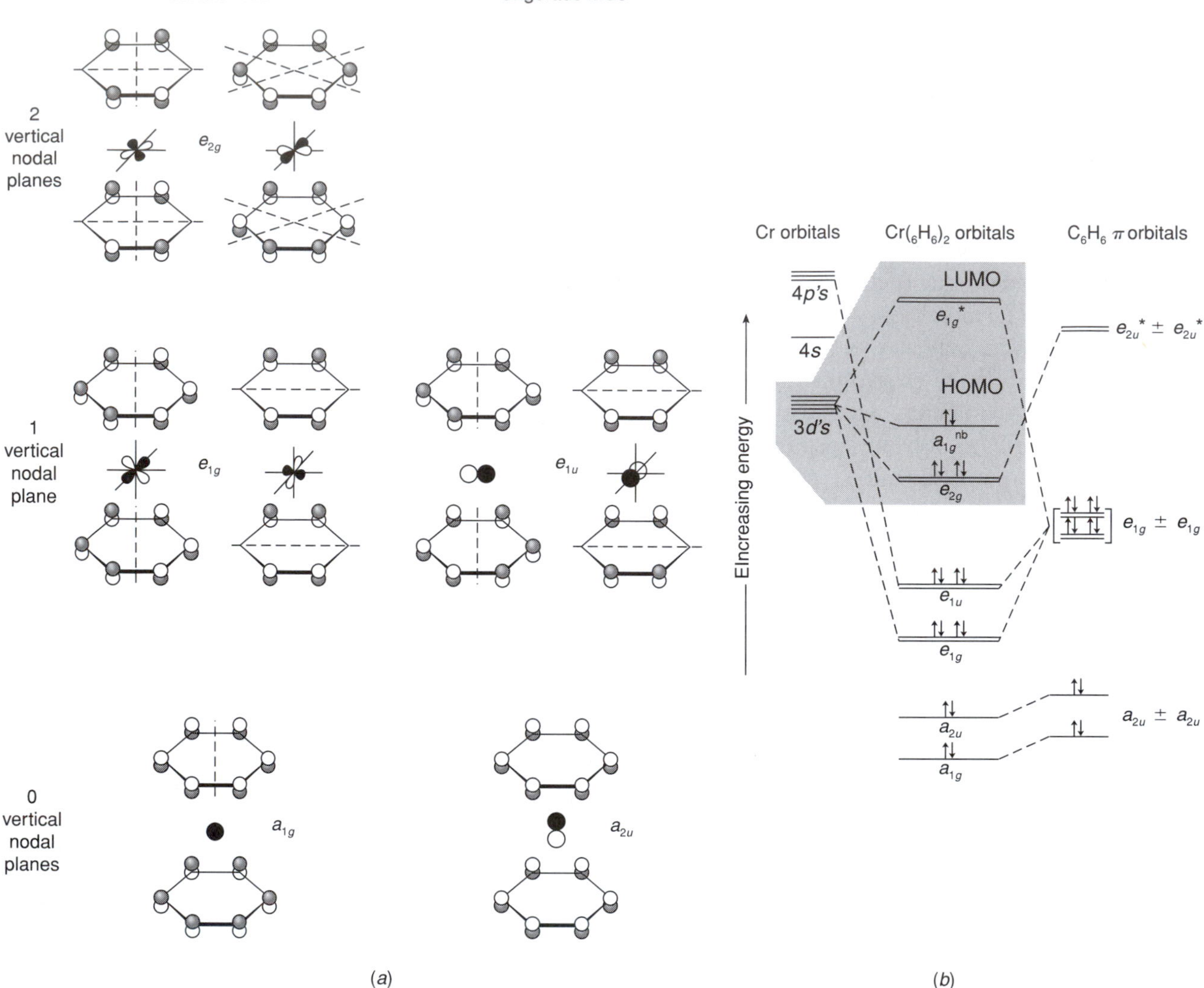

Figure 10.24

Selected MOs for bis(benzene)chromium. (*a*) Bonding MOs only, arranged by numbers of vertical nodal planes, with *gerade* MOs on the left and *ungerade* MOs on the right; (b) Energy level diagram, showing only one antibonding MO, the LUMO, and enclosing in the box the region emphasized in crystal field theory for the corresponding geometry, the trigonal prism (Table 8.6). (In order to simplify the diagram, not all of the dashed correlation lines have been drawn.) [Adapted from R. L. DeKock and H. B. Gray, *Chemical Structure and Bonding*, University Science Books, Sausalito, CA, 1989; pp. 389, 390.]

Figure 10.24(*a*), we draw the eight benzene MOs that do find metal-ion orbital partners; to save space we do not show the corresponding antibonding MOs. The *ungerade* *p* orbitals on the metal ions end up paired with the *ungerade* benzene TASOs, while the *gerade* *s* and *d* orbitals overlap the *gerade* TASOs.

In Figure 10.24(*b*), we draw the lower part only of the MO diagram for bis(benzene)chromium, which shows the energy levels of the eight bonding MOs shown in Figure 10.24(*a*); it shows the level of only the lowest of the antibonding MOs (the one that will turn out to be the LUMO). One metal-ion orbital remains approximately nonbonding in bis(benzene)chromium: the d_{z^2}. (The reason is that both d_{z^2} and s overlap with the ligand a_{1g} MO; consequently, the metal uses *sd* hybridization for this MO, with one metal orbital becoming approximately nonbonding, as in Section 10.2; also, the ring orbitals tend to fall in the conic nodes of d_{z^2}, further reducing overlap.)

This molecule, composed of atoms of intermediate electronegativity, is expected to fill its bonding MOs and may or may not fill its nonbonding MO. The eight bonding MOs require 16 electrons, 12 of which come from the π systems of the two benzene rings, and 4 of which come from Cr. If the nonbonding MO is also filled (to give a total of 18 valence electrons), Cr must provide the 2 extra electrons: Cr with 6 valence electrons is neutral Cr^0, so the resulting complex is neutral bis(benzene)chromium, $[Cr(C_6H_6)_2]$. If the nonbonding MO takes only 1 electron, the other known derivative results: $[Cr(C_6H_6)_2]^+$ [bis(benzene)chromium(I) ion].

Note that the bonding e_{2u} molecular orbitals in these organochromium species originate from e_{2u} benzene MOs that were originally unoccupied; the electrons in this pair of MOs must have been accepted from the metal d_{xy} and $d_{x^2-y^2}$ orbitals. This example is of the *back-donation* of electrons from metal atom to ligands, which is characteristic of ligands such as the polyenes that are high on the spectrochemical series and that are soft bases interacting with soft acids [chromium(0) and chromium(I)]. Polyenes are not only good π-donor ligands, they also act as π-*acceptor* ligands.

10.9 Molecular Orbitals for Three-Dimensional Molecules in Higher Order Point Groups

Next, we generate appropriate MOs for some three-dimensional molecules, beginning with AB_n molecules that have real central A atoms at the center to generate the TASOs of the terminal atoms B_n. For these three-dimensional molecules, we may consider orbitals in all three dimensions as potential generator orbitals. Since, for example, (p_x, p_y, p_z) are triply degenerate in the higher order point groups, they will generate a triply degenerate MO. We may note that, when f orbitals are used as real or generator orbitals in molecules in the higher order tetrahedral, octahedral, and icosahedral point groups, the forms of f orbitals illustrated in Figure 8.21 are to be used in preference to those illustrated in Figure 10.5.

Methane, CH_4 (T_d Point Group). The central carbon atom has four valence atomic orbitals, 2*s* (A_1 irreducible representation) and the three degenerate 2*p* orbitals (T_2 irreducible representation), oriented with respect to the tetrahedron as shown in Figure 10.25. The four H atoms can also give rise to four TASOs, the irreducible representations of which match those of the central atom. Positive overlap of the carbon valence orbitals and the H_4 TASOs of the same irreducible representation produce the two levels of bonding MOs that are fully occupied in the electron configuration of methane, $(a_1)^2(t_2)^6$. Negative overlap of the carbon valence orbitals and the H_4 TASOs of the same irreducible representation produce the two levels of antibonding MOs that are vacant (and not shown in the figure): $a_1{}^*$ and $t_2{}^*$. There are no orbitals left over to be nonbonding.

There are apparent discrepancies between this description of the bonding in CH_4 and that of valence bond theory, which describes methane as having sp^3-hybridized carbon, so that all H–C bonding orbitals are equivalent. In qualified[38] support of the

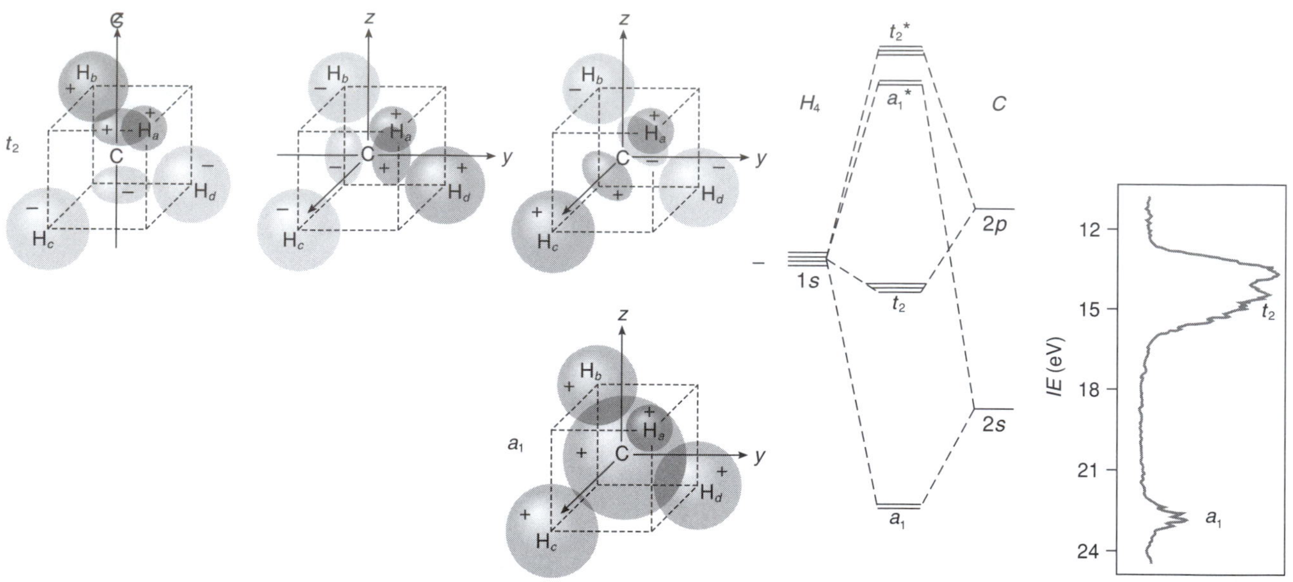

Figure 10.25
Molecular orbitals, energy level diagram, and photoelectron spectrum of the CH_4 molecule. [Adapted from R. L. DeKock and H. B. Gray, *Chemical Structure and Bonding*, University Science Books, Sausalito, CA, 1989; p. 283.]

MO description of methane, we may cite its photoelectron spectrum (Fig. 10.25), which shows *two* valence absorption bands: one from ionization from the bonding t_2 HOMO, at about 14 eV; and one from ionization from bonding a_1 MO, at about 23 eV. However, each set of MO bonding orbitals involves all hydrogens equally, so that MO theory, just like valence bond theory, results in all H–C links being equivalent.

Sulfur Hexafluoride, SF_6 (O_h Point Group). The hypervalent central sulfur atom has four valence orbitals, 3*s* (A_{1g} irreducible representation) and the three 3*p* orbitals (T_{1u} irreducible representation); it may also be able to use the postvalence 3*d* orbitals, especially because the oxidation state of sulfur is high in SF_6 and these *d* orbitals are likely to be contracted; nonetheless, let us show how the bonding can be accommodated without using the 3*d* orbitals. (We will use the *d* orbitals as hypothetical generator orbitals, however.)

The six terminal F atoms have 24 atomic orbitals. To make the treatment of these manageable, we will treat the six 2*s* orbitals as nonbonding, since they are so much lower in energy than any sulfur orbitals. We divide the eighteen 2*p* orbitals of the six fluorines into two sets: the **radial** set, which is the set of six 2*p* orbitals (one per atom) directed toward the sulfur at the center of the octahedron; and the **tangential** set, which is the set of twelve 2*p* orbitals that lie more or less along the outer surface of the octahedron. The radial set of orbitals is oriented so as to be able to form σ bonds with sulfur, while the tangential set is destined to hold nonbonding fluorine lone electron pairs, unless overlap with an appropriate central-atom set of orbitals can result in π bonding.

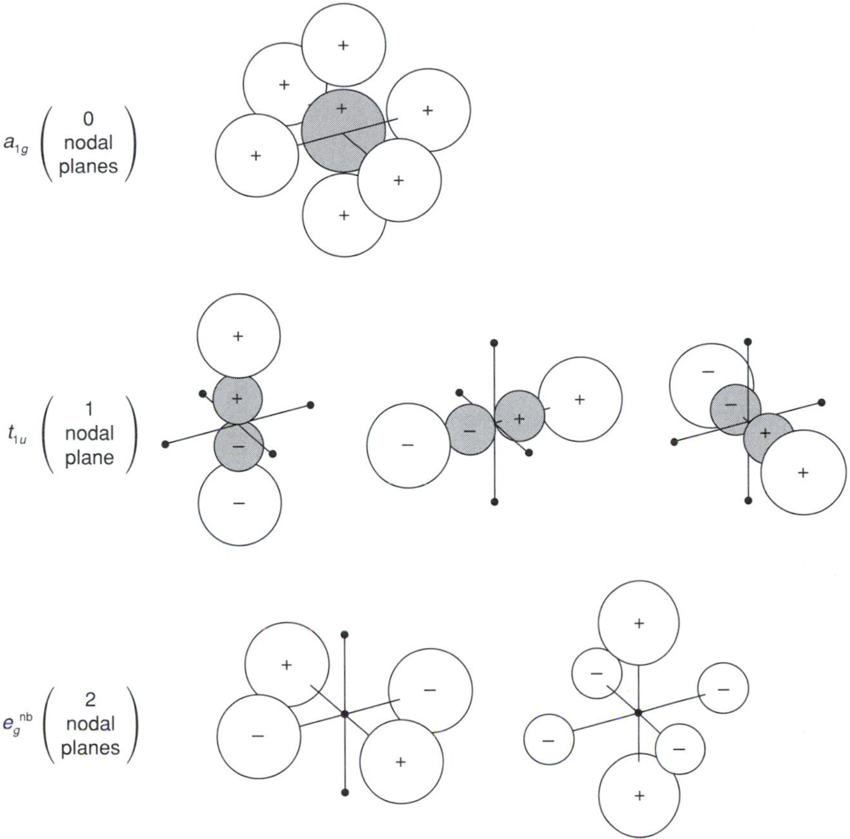

Figure 10.26
Sigma bonding and nonbonding MOs for an octahedral (O_h) complex in which the central atom cannot use *d* orbitals. (Orbitals of the central atom are shaded.) [Adapted from D. F. Shriver, P. W. Atkins, and C. H. Langford, *Inorganic Chemistry*, Freeman, New York, 1990; p. 679.]

Let us first consider the MOs that result from overlap of the radial set of six F_6 TASOs with the four sulfur valence orbitals. Since there are two more F_6 TASOs than there are sulfur valence orbitals, two F_6 TASOs are destined to be nonbonding. Otherwise, we can anticipate an equal number of bonding and antibonding MOs (four each).

Figure 10.26 shows the four bonding and two nonbonding MOs that result for SF_6 when no *d* orbitals are used in bonding. The first (a_{1g}, no nodal planes) involves the first valence orbital of the central S atom, its *s* orbital, both as a generator orbital and as a real partner in bonding. The second (t_{1u}, one nodal plane) involves the second S set of degenerate valence orbitals, (p_x, p_y, p_z), to give an occupied set of three bonding MOs (and, of course, the corresponding unoccupied set of antibonding MOs, which are not shown). Two nonbonding TASOs remain to be determined; the next type of S orbitals to use (as generator orbitals only) are *d* orbitals (two nodal planes), which fall in the irreducible representations t_{2g} and e_g. Only the latter is doubly degenerate, and e_g is indeed the symmetry label for the remaining TASOs. If the sulfur *d* orbitals are indeed not used in bonding , this e_g pair of TASOs is nonbonding and should give a sharp band in the photoelectron spectrum; if the sulfur *d* orbitals indeed do participate in bonding, they overlap with this e_g pair of TASOs to give a bonding MO with a broad or structured band in the photoelectron spectrum.[39]

Although it may well be the case that sulfur uses its 3*d* orbitals in the bonding in SF_6, this analysis indicates that similar hypervalent species might exist in which *d* orbitals do not participate. For example, computations have suggested that the hypervalent species NF_6^- and CF_6^{2-}, in which the use of *d* orbitals is out of the question, might possibly be stable enough to be isolated.[40]

Octahedral (O_h) Transition Metal Complex Ions. In *d*-block complexes, the central-atom *d* orbitals are not only involved in the bonding, they are lower in energy than the valence *s* and *p* orbitals. Consequently, not only the a_{1g} and the t_{1u} TASOs but also the e_g pair of TASOs give rise to bonding and antibonding molecular orbitals. If the ligands in these complexes are strictly σ-donor ligands (Section 8.5), there are no TASOs to match the t_{2g} set of central-atom *d* orbitals, which therefore is nonbonding. The six σ-donor ligands contribute six pairs of electrons, enough to fill the six bonding MOs (Fig. 10.27). If the central metal ion has a d^1 through a d^6 electron configuration, this electron density occupies the $t_{2g}{}^{nb}$ orbitals. Since these orbitals are nonbonding, their occupancy or nonoccupancy does not alter the bond order of the complex: Hence, there is no strong preference for d^0 or d^6 over the intermediate electron configurations, all of which are found in *d*-block complexes. For these ions, the HOMO is therefore $t_{2g}{}^{nb}$ and the LUMO is the first antibonding orbital, $e_g{}^*$.

Note that it is these frontier $t_{2g}{}^{nb}$ and $e_g{}^*$ orbitals that are emphasized in crystal and ligand field theory (Chapter 8); these frontier orbitals are enclosed in the box of Figure 10.27. We may note that the orbitals described as "nonbonding" in MO theory were previously described as holding electrons that avoid repulsion by the ligands in crystal field theory; the orbitals described as "antibonding" in MO theory are those that were described as holding electrons that experience maximum repulsion with the ligands in crystal field theory. The crystal field splitting, Δ_o, is simply the separation of the HOMO and the LUMO in MO theory.

Complexes with π Bonding. If the octahedral transition metal complex involves π-donor or π-acceptor ligands such as halide ions or carbon monoxide, respectively, then we must consider how the *tangential* p_x and p_y or the π^* orbitals of these respective ligands interact with the metal *d* orbitals. Six of these ligands have a total of 12 such orbitals; these must be combined into a total of 12 tangential TASOs, as shown in Figure 10.28. These are probably most reliably generated by more mathematical methods of group theory that are beyond the scope of this text, but we can point out that they do match the reasonable generator orbitals listed in the figure. They occur in triply degenerate sets, which is reasonable since the *x*, *y*, and *z* directions cannot be distinguished in octahedral complexes.

One of these sets, t_{1u} (generated by *p* atomic orbitals), has the minimum number of nodal planes (one) for a π-type MO, and could overlap with the metal-atom orbitals with one nodal plane, (p_x, p_y, p_z); however, these are already in use for forming stronger σ bonds. The third and fourth sets of TASOs have three and four nodal planes, respectively, and only overlap energetically inaccessible central-atom *f* and *g* orbitals, respectively, so are nonbonding in *d*-block complexes.

Consequently, we focus on the second of these sets, t_{2g}, which has two nodal planes and is generated by and overlaps with metal d_{xy}, d_{xz}, and d_{yz} atomic orbitals. Hence, *in a metal complex with π-donor or π-acceptor ligands, the metal-ion t_{2g} orbitals are no*

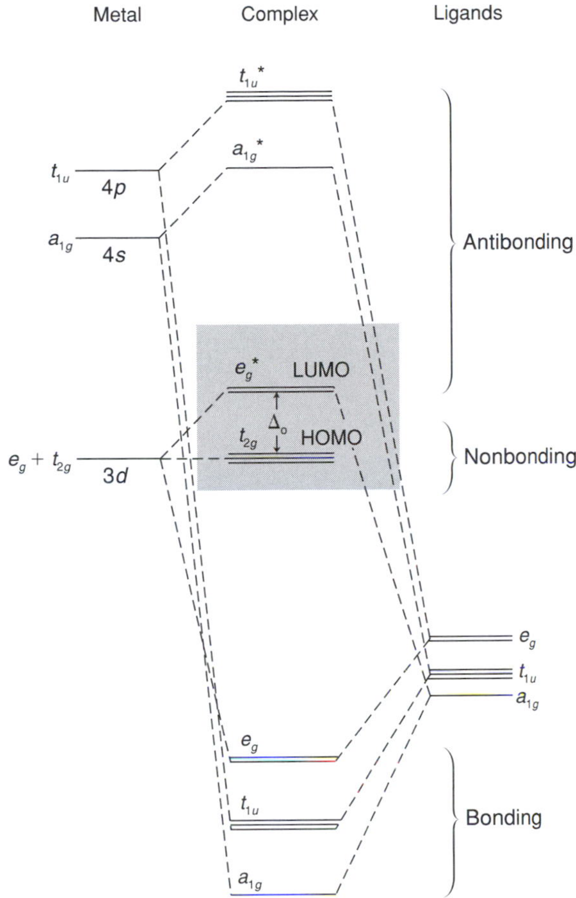

Figure 10.27
Energy level diagram for a typical octahedral *d*-block complex ion (σ-donor ligands). The frontier orbitals (emphasized in crystal field theory) are enclosed inside the box. [Adapted from D. F. Shriver, P. W. Atkins, and C. H. Langford, *Inorganic Chemistry*, Freeman, New York, 1990; p. 207.]

longer nonbonding, but become either bonding or antibonding MOs. With π-acceptor ligands such as carbon monoxide having empty tangential orbitals, and with metals having electrons in the t_{2g} set, positive overlap occurs with donation of metal t_{2g} electrons to the ligand tangential t_{2g} TASOs. This donation is the so-called "back-bonding" from the metal ion as Lewis base to the ligand as Lewis acid. This bonding interaction lowers the energy of the metal t_{2g} set of orbitals in Figure 10.27, which increases the magnitude of Δ_o for such π-acceptor ligands; note that t_{2g} is no longer nonbonding, but is now bonding.

With electronegative π-donor ligands such as the halide ions, the tangential t_{2g} set of TASOs is filled and is relatively low in energy. Its interaction with filled metal t_{2g} orbitals is of an antibonding sort, which causes the metal orbitals to be labeled t_{2g}^* and raised in energy above that shown in Figure 10.27. This elevation of t_{2g}^* then reduces the magnitude of Δ_o in such π-donor complexes.

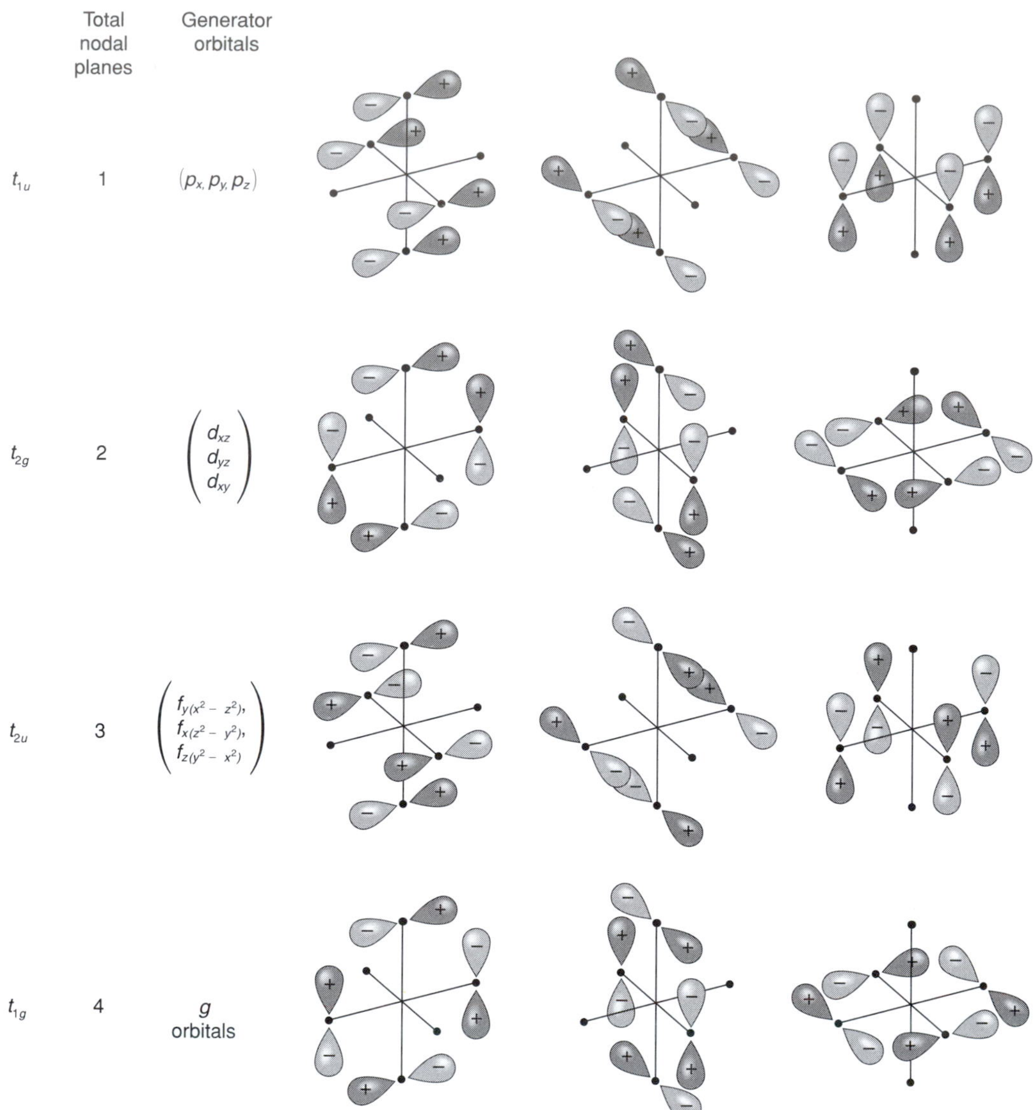

Figure 10.28
Terminal-atom symmetry-adapted orbitals (TASOs) for the π bonding in an octahedral (O_h) complex of π-donor ligands. [From *Inorganic Chemistry* by D. F. Shriver, et al. © 1990, 1994 by Durward F. Shriver, Peter Atkins, and Cooper H. Langford. Used with permission of W. H. Freeman and Company.]

10.10 *Molecular Orbitals for Cluster Compounds

Cluster compounds, B_n, can be thought of as polyhedral molecules or ions derived from polyhedral compounds, AB_n, by (1) the removal of the central atom, A, and (2) by the subsequent movement of the outer B atoms toward the center to the point that their orbitals overlap each other's. Thus they (generally) have no central atom, and their TASOs directly become their MOs by virtue of the overlap now present within the TASOs. To illustrate the bonding properties in clusters, we consider two *closo*-borane anions (Chapter 15), $B_6H_6{}^{2-}$ and $B_{12}H_{12}{}^{2-}$.

The $B_6H_6{}^{2-}$ anion (O_h point group) is composed of an octahedron of boron atoms, with each boron atom bonded to one terminal hydrogen atom by a normal single bond. Each of the six boron atoms has two tangential (p_x and p_y) valence orbitals and two radial (let us assume sp_z hybrid) orbitals. Let us assume that one radial orbital is involved exclusively in the bond to the terminal hydrogen atom, leaving another radial sp_z hybrid orbital on each boron that is oriented toward the center of the octahedral cluster.

The six radial orbitals need to be combined into six TASOs, which have the same appearance in this cluster as they did in the preceding octahedral complexes (Fig. 10.26); now, however, each contracts inward so that overlap develops within each TASO, and the central-atom atomic orbital is no longer present (except as a hypothetical generator orbital). By shrinking the a_{1g} TASO, we find that only positive overlaps occur, so a_{1g} becomes a *bonding MO*. Shrinking the t_{1u} TASOs until internal overlap occurs results in only negative overlap, so t_{1u} becomes a set of *antibonding MOs*. In looking at the e_g TASOs, it is easiest to see the result by considering the TASO that looks like an expanded $d_{x^2-y^2}$ orbital: mainly negative overlap occurs after contraction, so this orbital is also an antibonding orbital. Since the TASO that looks like an expanded $d_{x^2-y^2}$ orbital is degenerate in energy with the d_{z^2}-like orbital, the latter must be equally antibonding. The net effect is that the radial boron atomic orbitals give rise to *only one* bonding MO and *five* antibonding MOs, a sharp contrast to the usual approximately equal distribution of bonding and antibonding MOs.

The 12 tangential boron atomic orbitals give rise to 12 TASOs with the same appearance as the π-type TASOs found in octahedral complexes (Fig. 10.28). These TASOs are contracted inward until overlap occurs. It may be seen from Figure 10.28 that *four positive interatomic overlaps* then occur in the contracted t_{1u} and t_{2g} TASOs, which therefore become *bonding molecular orbitals*. As the other two TASOs (t_{1g} and t_{2u}) are contracted, four negative overlaps occur; these become antibonding MOs. So the 12 tangential MOs split in the conventional pattern: two sets of three bonding MOs and two sets of three antibonding MOs.

The overall bonding scheme for the 12 boron–boron links in $B_6H_6^{2-}$ thus includes 7 bonding MOs and 11 antibonding MOs; most of the bonding is in fact due to the tangential MOs. Each boron atom contributes three valence electrons, of which one is reserved for the normal covalent bond to the terminal hydrogen, so the six boron atoms contribute 12 valence electrons to the cluster. This is two short of being enough to fill the seven bonding MOs, which explains why B_6H_6 is not a neutral molecule but instead acquires two additional electrons to form a −2 charged anion. The anion is, in many respects, quite stable, since it fills all of its bonding MOs and leaves empty all of its antibonding MOs, even though the bond order per boron–boron link is only $\frac{7}{12}$. Once again, electron-deficient bonding is not necessarily connected with extraordinary instability in compounds.

Example 10.9

Consider the icosahedral anion $B_{12}H_{12}^{2-}$ (point group I_h). (a) Considering only radial boron sp_z hybrid orbitals and the TASOs formed from them, how many radial MOs are formed and what symmetry labels do these have? (b) Draw one radial MO for each symmetry label and determine whether it is bonding or antibonding. (c) Given that the tangential MOs distribute equally between bonding and antibonding MOs in the normal manner, explain why this very stable anion has a −2 charge.

SOLUTION:

(a) There are 12 boron atoms, each with one radial sp_z hybrid orbital, so there are 12 radial TASOs, hence, 12 radial MOs. We can use an imaginary generator atom at the center of this cluster to generate these MOs. The first MO has no nodal planes, matching the s generator orbital; consulting the I_h character table, we see that this MO should be labeled a_g. The next MOs have one nodal plane each, matching the (p_x, p_y, p_z) set of generator orbitals, and therefore are labeled t_{1u}. The next MOs have two nodal planes, matching the set of five d generator orbitals and having the irreducible-representation label h_g. Of the 12 radial MOs; three remain that have three nodal planes and match three f generator orbitals. In the I_h point group, the f orbitals are split into two sets, a degenerate set of three labeled t_{2u} and a degenerate set of four labeled g_u; since we are dealing with a degenerate set of three, it must be labeled t_{2u}.

(b) We can draw one each of the a_g, t_{1u}, h_g, and t_{2u} molecular orbitals by matching their symmetry patterns to that of suitable generator orbitals: s and (say) p_z, d_{z^2}, and f_{z^3} (Fig. 8.21). These MOs are sketched in Figure 10.29. Clearly, the a_g molecular orbital has

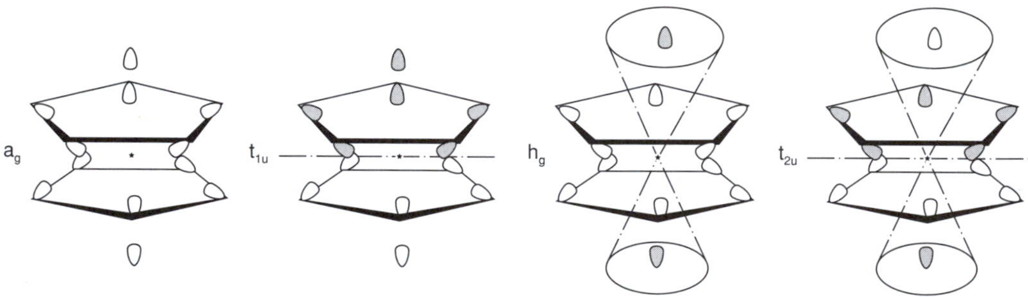

Figure 10.29
Typical molecular orbitals (qualitative sketches) for each of the four energy levels among radial molecular orbitals of the $B_{12}H_{12}{}^{2-}$ ion. Boron atoms with attached hydrogens are implied to be located at the top, bottom, and at each apex of the two pentagons. A shortened version of each boron's radial *sp*-hybrid orbital is shown (black = positive; white = negative sign of the wave function). These twelve orbitals would actually overlap at the center of the ion (indicated by the star, ⋆). Dash-dotted lines (—·—·) represent nodal planes and nodal conical surfaces in the MO's.

only positive overlaps, and is strongly bonding; the h_g and t_{2u} molecular orbitals are riddled with negative overlaps each time a nodal plane is crossed, and are strongly antibonding. The t_{1u} molecular orbital is more difficult to assess, so we consider it in more detail. The top boron atom clearly has positive overlaps with each of its five neighbors in the next plane below it; the same is true for the bottom boron atom and its neighbors; this gives a total of 10 positive overlaps. Negative overlaps occur between each of the five boron atoms in one pentagonal plane and its two neighbors in the pentagonal plane just across the nodal plane, which gives 10 negative overlaps. In addition, there are negative overlaps of boron atoms in the bottom half with the top boron atom in the top half, and so on: clearly, negative overlaps predominate, and this is an antibonding MO.[41]

(c) The tangential MOs are fomed from the p_x and p_y orbitals on each of 12 boron atoms, so there are a total of 24 MOs. It is given that one-half of these are bonding, so there are a total of 13 bonding MOs in $B_{12}H_{12}{}^{2-}$. Since each boron atom contributes two valence electrons for the cluster bonding (one being reserved for bonding to H), and there are two electrons for the −2 charge, this is just enough to fill the 13 bonding MOs. (Since an icosahedron has 30 edges (30 B–B links), the bond order per B–B link is 13/30. Although $B_{12}H_{12}^{2-}$ is very electron deficient, it is very stable.)

10.11 *Molecular Orbitals for Metals

We will discuss the properties of metals in more detail in Chapter 12. Here, we simply wish to consider the nature of the MOs used in bonding these stable species, which can have very low fractional bond orders per metal–metal link. Metals are *extended solids* in which the molecule is much larger than those discussed so far: The molecule is the entire crystal, and may include on the order of Avogadro's number of atoms. The bonding in the metal is pure covalent bonding based on MOs involving most or all of the sextillions of atoms in the crystal–molecule; there are therefore sextillions of MOs for a crystal of a metal. In order to be able to draw MOs of metals, we must consider a very tiny crystallite, then extend our results qualitatively.

As shall be seen in Chapter 12, the atoms in metals characteristically have very high coordination numbers, either 8 or 12. Let us consider the case of coordination number 6, in which the metal atoms sit in a cubic array (as in α-polonium). We will take a very small crystallite containing a square prismatic array of 16 metal atoms, 4 atoms high by 2 atoms wide by 2 atoms deep; we will generate the MOs and the energy level diagram for this small crystallite, based on the approach of Verkade.[42] For simplicity, we consider a Group 1 or 2 metal in which the only valence orbital is an *s* orbital; the 16-atom crystallite then has 16 MOs.

These 16 MOs can be generated by combining, side by side, the MOs of strings of four Group 1 linear tetraatomic molecules, that is, the 4 MOs of the H_4 molecule. First, we can build up the 8 MOs of a rectangle of eight Group 1 metal atoms, M_8, by taking all possible distinct permutations of the 4 MOs of H_4; these eight combinations are shown in Figure 10.30.

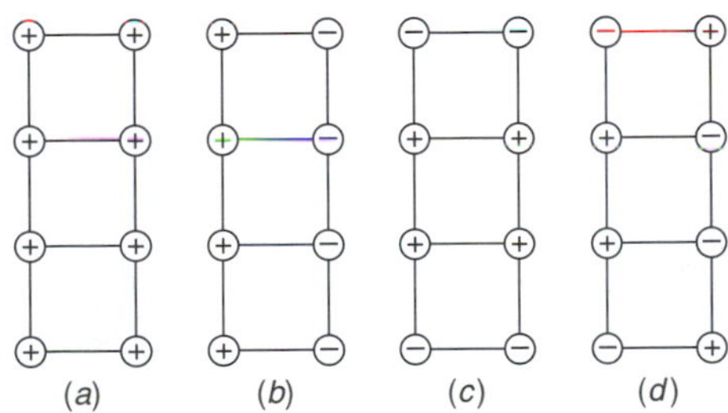

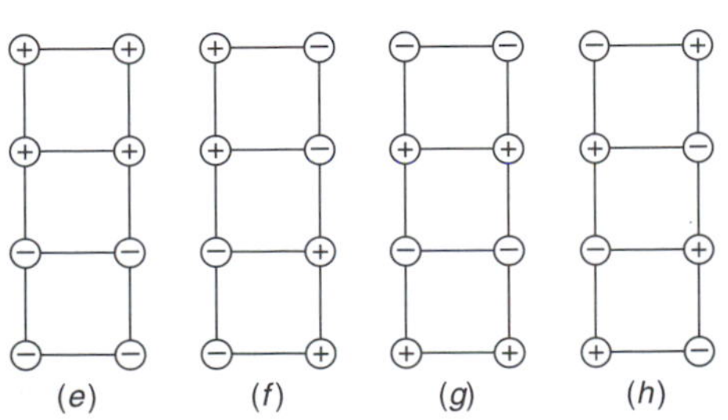

Figure 10.30
Distinct linear combinations of the MOs of H_4 chains overlapping to give an M_8 rectangular crystallite. [Adapted from J. G. Verkade, *J. Chem. Educ.*, **68**, 739 (1991).]

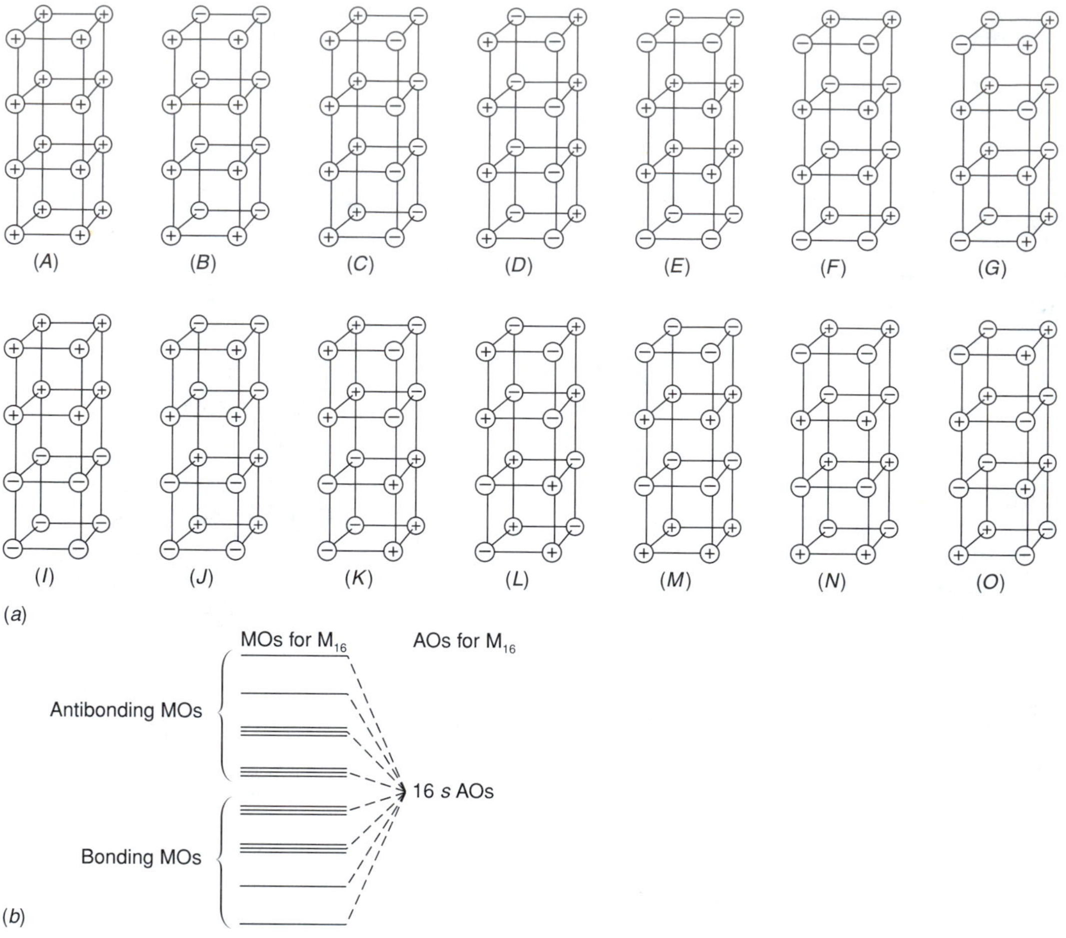

Figure 10.31
(*a*) Distinct linear combinations of the MOs of a pair of M_8 rectangles overlapping to give an M_{16} rectangular prismatic crystallite; (*b*) energy level diagram of M_{16}. [Adapted from J. G. Verkade, *J. Chem. Educ.*, **68**, 739 (1991).]

Next, we combine pairs of M_8 rectangles face to face, to generate the 16 distinct MOs of the M_{16} square prism. These 16 MOs are shown in Figure 10.31(*a*), and are labeled (A–P).

Each of the MOs has 28 M–M links, so there are 28 places in which either positive or negative overlap of adjacent atoms can occur. To develop the MO diagram of our M_{16}, we must count the net number of positive overlaps in each MO: The greater the number, the more strongly the MO is bound, and the MO is lower in energy. Molecular orbital (A) clearly has 28 positive overlaps, and is the most strongly bonding MO possible. Molecular orbital (I) has 24 positive and 4 negative overlaps, for a net of +20 overlaps, and is next in energy; it is followed by a triply degenerate set of MOs (B),

(C), and (E), each with 20 positive and 8 negative overlaps (net +12 overlaps). Other combinations can be found that give net totals of +4, −4, −12, −20, and −28 overlaps. Consequently, the energy level diagram [Fig. 10.31(*b*)] is easily produced; the 28 MOs are spaced across a **band** of energies centered at the energy of the original atomic *s* orbitals.

It is useful to compare this energy level diagram with that of H_{10} (Fig. 10.15), H_3 (Fig. 10.14), and H_2 (Fig. 10.1). It can clearly be seen that, as the array of atoms grows in space, the density of energy levels increases; not so easily determined by inspection is the fact that, although the separation between the highest and lowest MO (the width of the band of MOs) does increase, it increases slowly, so that successive MOs come closer and closer together in energy. If we now polymerize our M_{16} crystallite to produce a $M_{602,300,000,000,000,000,000,000}$ crystal of a metal, the 602,300,000,000,000,000,000,000 energy levels are now so closely crowded together that there is an infinitesimally small gap between each energy level in this band of MOs, while the total width of the band of MOs is still quite finite.[43]

If this metal is a crystal of sodium metal, the 602,300,000,000,000,000,000,000 valence electrons (1 per Na atom) are enough to half-fill the band of MOs. We may also compute the bond order per link of a Group 1 metal in the cubic lattice we have been discussing. In this lattice, each atom is linked to six other atoms; to avoid double counting of links when they occur at the other atom, we find a total of three links per atom. So a crystal of a Group 1 metal with 602,300,000,000,000,000,000,000 metal atoms has 3 × 602,300,000,000,000,000,000,000 links but has a total bond order of only $\frac{1}{2}$ × 602,300,000,000,000,000,000,000 (since it takes two electrons to make a bond). Hence, the bond order per link in such a metal is a very low $\frac{1}{6}$; if the coordination number were 12, the bond order per link would be an even lower $\frac{1}{12}$.

A similar but more intricate band of MOs can be expected to originate from the valence *p* orbitals of elements; this band, of course, would contain three times as many orbitals. If the valence *s* and *p* atomic orbitals of the element in question are well separated (from Table 10.2, we can see that this holds at the right of the *p* block), then the bands based on *s* and *p* electrons do not overlap in energy; there is a **band gap** between energy levels that electrons can no longer cross so readily. However, at the left of the *p* block the *s* and *p* atomic orbitals are separated by only a few electron volts, so that the bands do overlap. This overlap has important consequences for a metal such as magnesium. If its electrons could only occupy the *s* band of MOs, they would completely fill the band; since the top half of the band of MOs must consist of antibonding MOs, the net bond order would be zero, and the Mg atoms would fly apart; Mg could not be bonded as a solid metal. In fact, however, the band of *p* molecular orbitals overlaps the band of *s* molecular orbitals in Mg and other Group 2 metals, so that some of its electrons move into bonding MOs based on *p* orbitals; this restores the nonzero bond order of Mg.

Because the valence *d* orbitals on adjacent atoms have poorer overlap than valence *s* and *p* orbitals, the MO bands developed from them are narrower, and may often not overlap the bands of *s,p* orbitals found in the *d*-block metals. But the total number of (at least somewhat) bonding MOs grows as the number of possible valence orbitals grows: for a crystal of N atoms, from $\frac{1}{2}N$ in the *s* block (idealized, with no participation of *p* orbitals), to $\frac{4}{2}N$ in the *p* block, to $\frac{9}{2}N$ in the *d* block. Further consequences of this structure of MOs for the properties of metallic solids will be discussed in Chapter 12.

Study Objectives

1. Classify MOs in linear diatomic molecules as σ, π, δ, or ϕ. Label these orbitals according to their irreducible representation. Construct an energy level diagram for linear diatomic molecules assuming no *s–p* mixing or "hybridization." Exercises 1–3.

2. Write the electron configuration of a given molecule, using the proper symmetry labels, and give its bond order, its total number of unpaired electrons, and its paramagnetism or diamagnetism. Identify the highest occupied molecular orbital (HOMO), lowest unoccupied molecular orbital (LUMO), and any singly occupied molecular orbitals (SOMOs) for a given molecule. Exercises 4–9.

3. Compare the bond energies and lengths of closely related species. Exercises 10–13.

4. On a energy level diagram of a molecule or ion, sketch lines that would correspond to bands that might be observed in the UV or visible spectrum of the molecule or ion. Note whether the spectrum should contain intense absorption bands. Exercises 14, 15.

5. Tell which given MOs in a homonuclear diatomic molecule can undergo mixing or "hybridization". Interpret the resulting electron configurations in terms of an equivalent valence bond structure or set of resonance hybrid structures. Exercises 16–19.

6. Describe how the bonding and antibonding orbitals are normally influenced if the molecule is changed to heteronuclear, that is, CO instead of N_2. Exercise 20.

7. Explain or assign the photoelectron spectrum of a given molecule or ion. Interpret a photoelectron spectrum to determine whether a given MO is nonbonding, or bonding, or antibonding. Exercises 21–24.

8. From its ionization energy and electron affinity (or orbital energies) calculate the electronegativity and the hardness of a molecule. Exercises 25–31.

9. For linear polyatomic molecules (H_n, polyenes, etc.): construct qualitative pictures of the MOs, showing nodal planes; give them appropriate symmetry labels; give the number of positive and negative overlaps in each MO, and classify each as being overall bonding, nonbonding, or antibonding; draw the energy level diagram for such a molecule; write the electron configuration of a molecule, and give its total bond order, bond order per atom–atom link, and its total number of unpaired electrons; identify the HOMO and LUMO (or SOMO). Exercises 32–36.

10. Know how to carry out the same operations for cyclic polyatomic molecules. Exercises 37–40.

11. Tell which orbitals of a cyclic molecule are suitable for overlap with which orbitals of an atom M in forming an "open-face sandwich" molecule C_nH_nM (i.e., one-half of a sandwich molecule). Exercises 41–45.

12. Use the orbital symmetry rules to evaluate the plausibility of a proposed concerted mechanism for a reaction. Exercises 46, 47.

13. Know how to carry out the same operations (as in Study Objective 9) for planar or three-dimensional polyhedral complexes or molecules, AB_n, where A is a *p*- or a *d*-block atom or ion. Identify, in the energy level diagram of an octahedral complex, the part that is emphasized in crystal field theory for octahedral complexes. Explain how, in an octahedral complex, the presence of filled or empty π-type orbitals on the ligands changes the nonbonding status of the t_{2g} metal orbitals. Explain how this alters the position of the ligand in the spectrochemical series. Exercises 48–53.

14. Draw the radial MOs for a cluster complex such as $B_6H_6^{2-}$ and classify the MOs as bonding or antibonding. Exercises 54, 55.

15. Draw the MOs for a small crystallite of a metallic element, and from the overlaps in the MOs classify the MOs as bonding, nonbonding, or antibonding, and arrange them in an energy level diagram. Exercises 56–58.

Exercises

1. Drawn in Figure 10.32 are four combinations of orbitals located on the two adjacent atoms of a homonuclear diatomic molecule. (a) Tell whether each of the combina-

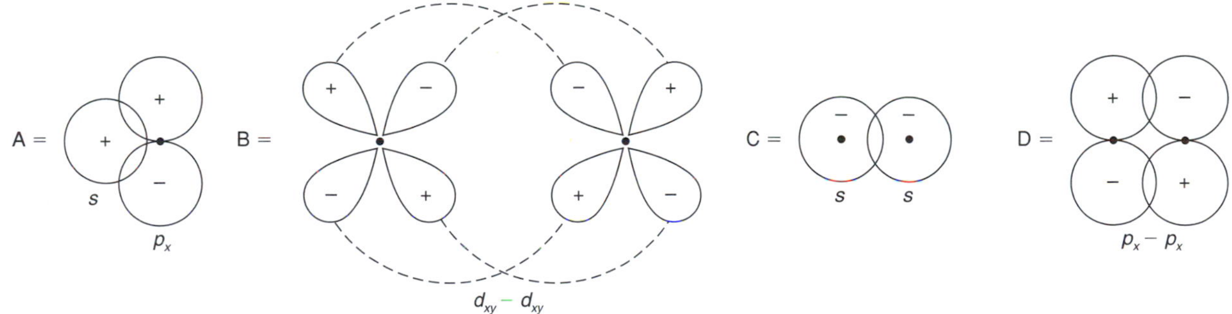

Figure 10.32

tions of orbitals has positive, negative, or zero overlap. (b) Describe each of the combinations as bonding, antibonding, or as an improper combination (for a homonuclear diatomic molecule). (c) Classify each combination as gerade, ungerade, or neither. (d) Classify each bonding or antibonding combination using the symmetry labels $\sigma_g{}^*$, and so on.

2. A hypothetical and incompletely labeled energy level diagram is shown in Figure 10.33 for a diatomic molecule M_2, which is composed of two transition metal atoms bonding to each other with their valence *s* and *d* orbitals. (a) Label each of the eight

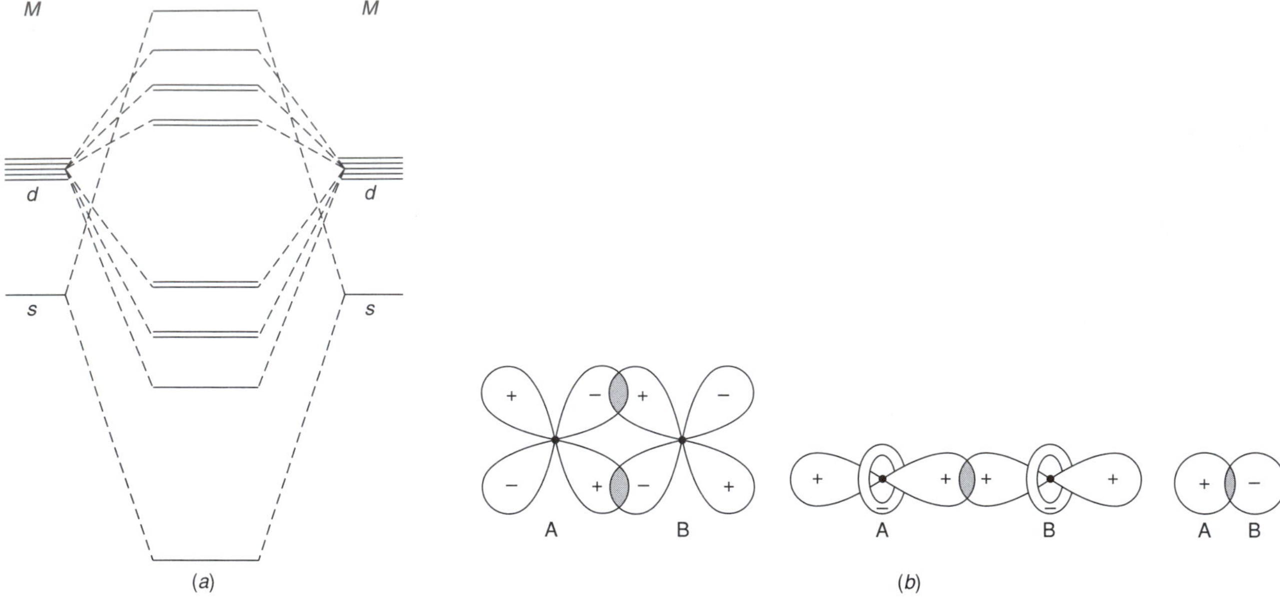

(a)
(b)

Figure 10.33

MO energy levels in Figure 10.33(*a*) using the appropriate combinations of the following symbols: σ, g, u, δ, π, nb, and $*$. (b) Some of these MOs are drawn in Figure 10.33(*b*). Put the appropriate labels below each drawing.

3. Figure 10.34 is an energy level diagram for a diatomic molecule M_2 composed of two transition metal atoms bonding to each other with their valence *s* and *d* orbitals. The MOs and their energy levels have been labeled *alphabetically* starting from the lowest energy levels. (a) Orbitals *A* and *E* are connected by dashed lines to single lines at the

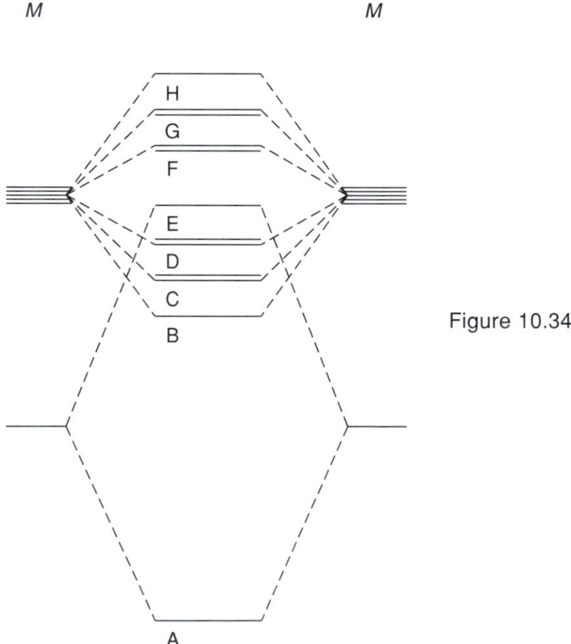

Figure 10.34

left and right of the diagram. These single lines represent the energies of which type of orbitals of the individual M atoms (*s* or *d*)? (b) Orbital *A* is which type of MO (bonding, antibonding, or nonbonding)? (c) Orbital *E* is which type of MO? (d) Draw a picture of the appearance of orbital *E*, showing signs of the wave function. (e) Which lettered MO(s), if any, should really be labeled $\sigma_u{}^*$? Which should really be labeled δ_g? Which should really be labeled $\phi_g{}^*$? (f) The $\delta_u{}^*$ MO results from which type (positive or negative) of overlap of which atomic orbitals of the two M atoms? Is the energy level of the $\delta_u{}^*$ molecular orbital singly or doubly degenerate? What does this mean?

4. *Assuming no *s–p* mixing, write the (valence) electron configurations of the following diatomic molecules or ions. For each molecule give the bond order, and the number of unpaired electrons. Also, tell which orbitals are SOMOs, if any; otherwise, identify the HOMO and the LUMO. (a) $B_2{}^{2-}$; (b) $N_2{}^+$; (c) $O_2{}^-$; (d) F_2; and (e) H_2.

5. Assuming no *s–p* mixing, write the (valence) electron configurations of the following diatomic molecules or ions. For each molecule give the bond order, and the number of unpaired electrons. Also, tell which orbitals are SOMOs, if any; otherwise, identify the HOMO and the LUMO. (a) B_2; (b) $C_2{}^{2-}$; (c) $N_2{}^{2+}$; (d) Be_2; (e) C_2.

6. *Using Figure 10.4, write the valence electron configurations of, compute the bond orders of, and tell the number of unpaired electrons in each of the following molecules. Also, tell which orbitals are SOMOs, if any; otherwise, identify the HOMO and the LUMO. For *f*-block elements, use the order of filling given in Section 10.1, and assume that these orbitals are immediately followed by other MOs based on atomic *f* orbitals. (a) Ti_2; (b) V_2; (c) Pd_2; (d) Am_2; (e) Th_2; and (f) Bk_2.

7. Using the same assumptions as in Exercise 6, write the expected valence electron configurations of, compute the bond orders of, and tell the number of unpaired electrons in each of the following ions: (a) $Mo_2{}^{4+}$; (b) $Np_2{}^{2+}$; (c) $Hg_2{}^{2+}$ (treat *d* electrons as core electrons); and (d) $Pt_2{}^{2-}$.

8. Using the energy level diagram and lettering scheme in Figure 10.34, and assuming normal orders of filling, for each of the following diatomic molecules or ions: Hf_2, $Mo_2{}^{2-}$, and $Pd_2{}^{4+}$. (a) Write valence electron configurations, adding stars (*) to indicate antibonding MOs (i.e., A^3B^{*7}); (b) indicate the number of unpaired electrons in the molecule or ion, (c) indicate its bond order, and (d) identify the HOMO or SOMO.

9. The fourth-period diatomic molecule V_2 is calculated to have the following electron configuration: $\sigma_g{}^2\pi_u{}^2\delta_g{}^2\sigma_g(s)^1\delta_u{}^{*2}\pi_g{}^{*1}$. (a) What is the bond order for this electron configuration? (b) Why is this so dissimilar from the electron configuration calculated for Nb_2, which resembles that of Mo_2 as given in the text example? (c) How can you explain the fact that the lowest energy orbitals are not always filled before the highest energy orbitals? (Think back to Chapter 8 on crystal field theory for some other cases in which this also happens.) (d) How many unpaired electrons is this configuration likely to represent? (Apply Hund's rule to the maximum extent.)

10. *Explain the following trends in bond energies and bond lengths, based on data from Tables 3.1 and 10.2: (a) The bond in $N_2{}^+$ is longer and weaker than the bond in N_2, but the bond in $Cl_2{}^+$ is shorter (189 pm) and stronger (415 kJ mol^{-1}) than the bond in Cl_2. (b) Bond energies fall among diatomic molecules going down Groups 14(IV),

15(V), and 16(VI), but increase going down Group 11. (c) Bond energies decrease much more rapidly going down Group 14(IV) than going down Group 17(VII).

11. Predict trends in bond energies and bond lengths in the following series: (a) ${C_2}^{2+}$, C_2, ${C_2}^{2-}$; (b) Mo_2; ${Mo_2}^{2+}$; ${Mo_2}^{4+}$; (c) Cr_2, Mo_2; and (d) Nb_2, Mo_2, Tc_2, Ru_2.

12. For each of the following species, write the electron configuration (assuming no *s–p* hybridization) and compute the bond order. Then tell: (a) Which should have the longer bond, O_2 or ${O_2}^{2-}$? (b) Which should have the stronger bond, B_2 or ${B_2}^{2-}$? (c) Which should have the weaker bond, C_2 or ${C_2}^{2-}$? (d) Which should have the shorter bond, ${O_2}^+$ or ${O_2}^{2-}$?

13. Referring to the electron configurations in Exercise 8 and the energy level diagram in Figure 10.34, which should have (a) the longer bond, Mo_2 or Mo_2^{2-}? (b) The stronger bond, Hf_2 or ${Hf_2}^{2-}$? (c) The weaker bond, ${Pd_2}^{2+}$ or ${Pd_2}^{4+}$?

14. Although the acetylide ion ${C_2}^{2-}$ is colorless, the C_2 molecule absorbs visible light of frequency 19,300 cm^{-1}. (a) Referring to Section 8.4, convert this frequency to energy units of kilojoules per mole; of electron volts. What color is C_2 likely to be? (b) Is C_2 likely to show *sp* hybridization or to be unhybridized? Explain, with the aid of its most likely electron configuration, why C_2 absorbs relatively low-energy visible light (${C_2}^{2-}$, O_2, and N_2 absorb mainly or exclusively in the UV or vacuum UV regions). (c) Is the color of C_2 likely to be intense? Why or why not?

15. Choose the (vis, UV, or near-IR) electronic transition(s) from the following list that should be intense: $\sigma_g \rightarrow {\sigma_u}^*$, $\sigma_g \rightarrow \delta_g$, $\pi_u \rightarrow \delta_g$, ${\delta_u}^* \rightarrow {\sigma_u}^*$.

16. (a) Draw an energy level diagram for a second-row homonuclear diatomic molecule with no *s–p* mixing or hybridization. Label each energy level with the appropriate symmetry label from the $D_{\infty h}$ point group, and use the superscripts *b*, *nb*, and $*$ to indicate whether the orbital is bonding, nonbonding, or antibonding. (b) Give the symmetry labels of the orbitals in your energy level diagram that are altered as a result of *s–p* mixing or hybridization.

17. *(a) Give the symmetry labels of the orbitals in Figure 10.4 that may be altered in energy and appearance as a result of *s–d* mixing or hybridization. What condition is necessary for this to occur? On which side of the *d* block are these conditions most likely to be met? (b) In the *f* block, hybridization or mixing is possible involving the $(n-2)f$, the $(n-1)d$, and the *ns* orbitals. Molecular orbitals based on which *f* orbital types (Fig. 10.5) might be affected by hybridization involving MOs based on which other atomic orbital types if the parent atomic orbitals are of appropriate relative energies?

18. Which is the best Lewis dot structure for (a) B_2 with no *s–p* hybridization (choose one) :B–B: or :B≡B: or B≡B or the resonance hybrids :B=B ↔ B=B:; (b) C_2 with no *s–p* hybridization :C=C: or C≡C or :C–C: or the resonance hybrids :C≡C ↔ C≡C:; (c) C_2 with *s–p* hybridization? :C=C: or C≡C or :C–C: or the resonance hybrids :C≡C ↔ C≡C:.

19. By adding or subtracting electron pairs from Lewis structures of the following, predict whether the HOMO and the LUMO of each is bonding, antibonding, or nonbonding: (a) H_2; (b) XeF_2; and (c) CH_4.

20. Suppose that you changed a homonuclear diatomic transition metal molecule M_2 to a *heteroatomic* molecule MM′, composed of two different transition metals of differing electronegativities. Briefly describe (a) one type of change that you would make on the left and right of the energy level diagram of M_2 to make it appropriate for MM', and (b) one type of change that you would make in the irreducible-representation symmetry labels.

21. Sketch the expected photoelectron spectrum of (a) B_2 and (b) of C_2. Will any of the bands be lacking fine structure?

22. *(a) Using Figure 10.11 as a starting point, write the electron configuration of the molecule NO, and calculate its bond order. (b) Is the bonding in NO weaker or stronger than that in CO? (c) Is the N–O bond longer or shorter than the CO bond? (d) Suggest how the photoelectron spectrum of NO should compare with that of CO.

23. Diatomic molecules having hybridization have a type of MO that is not present in unhybridized diatomic molecules. Describe briefly how the photoelectron spectrum should consequently differ in appearance from that of an otherwise-identical but unhybridized molecule.

24. (a) Which of the following three should have the highest bond order: NO, NO^+, or NO^-? (b) Which of these three should have the weakest bond? (c) Which of these three should have the shortest bond? (d) Which of these should have the photoelectron spectrum with the fewest bands? (e) Which (if any) of these should be paramagnetic, with how many unpaired electrons?

25. *The ionization energy of the N_2 molecule is 15.58 eV; its electron affinity is −2.2 eV. Based only on this information and the corresponding data for CO given in the text: (a) Which is the softer ligand, N_2 or CO? (b) Which should be the better σ-donor ligand? (c) Which should be the better π-acceptor ligand? (d) Which of the two ligands should have MOs with a shape or contour better suited to act as a σ donor? As a π acceptor?

26. In the *E&C* treatment of the Lewis acid–Lewis base interaction, I_2 is taken as the reference Lewis acid. (a) Which MO of I_2 is being used when it functions as a Lewis acid? (b) Show by calculation which halogen molecule is the softest acid. The ionization energies are 11.6 eV for Cl_2, 10.56 eV for Br_2, and 9.4 eV for I_2; the electron affinities are 2.4 eV for Cl_2, 2.6 eV for Br_2, and 2.6 eV for I_2.

27. Based on qualitative principles, would you expect the diatomic molecules of the lower part of the *d* block (such as Mo_2) to be harder or softer than the diatomic molecules of the top of the *p* block (such as N_2)? Why?

28. Sketch lines in the energy level diagram of LiF (Fig. 10.13) to represent the following energies. (For example, a line sketched from the level of the HOMO to the level of 0 eV represents the ionization energy.) (a) The electron affinity of LiF; (b) The electronegativity of LiF; and (c) The hardness of LiF.

29. *Explain why the electron affinity of C_2 is much greater (341 kJ mol^{-1}; 3.54 eV) than that of O_2 (~42 kJ mol^{-1}; 0.40 eV) or N_2 (about −212 kJ mol^{-1}; −2.2 eV).

30. Figure 10.35 shows the energy level diagrams of two hypothetical atoms, sengsavangium (Ss) and herrium (Hr). The numbers beside the levels represent the actual energies of the orbitals in electron volts, relative to zero for an electron at an infinite distance from the nucleus. (a) Draw an arrow on each diagram to indicate the process responsible for the first ionization energy of each atom. (b) Draw an arrow on each diagram to illustrate the process responsible for the electron affinity of each atom.

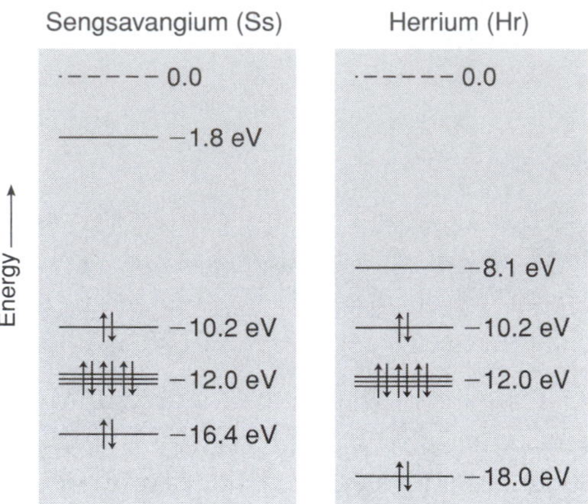

Figure 10.35

(c) Which atom is more electronegative, Ss or Hr? Calculate the Mullikin electronegativities (in electron volts) of each atom. (d) Which atom is harder, Ss or Hr? Calculate the hardness (in electron volts) of each atom.

31. For each of the following sets of molecules, ionization energies (IE) and electron affinities (EA) taken from R. G. Pearson, *Inorg. Chem.*, **27**, 734 (1988) are given in electron volts. Compute the electronegativity, hardness, and softness of each molecule, then identify the most electronegative molecule, the softest molecule, and the most stable molecule in each set. (a) CH_4 (IE = 12.7, EA = −7.8), CMe_4 (IE = 10.4, EA = −6.1), MeF (IE = 12.5, EA = −6.2), MeCl (IE = 11.2, EA = −3.7), MeI (IE = 9.5, EA = 0.2). (b) BF_3 (IE = 15.8, EA = −3.5), BCl_3 (IE = 11.60, EA = 0.33), BBr_3 (IE = 10.51, EA = 0.82). (c) PH_3 (IE = 10.0, EA = −1.9), PMe_3 (IE = 8.6, EA = −3.1), PF_3 (IE = 12.3, EA = −1.0), PCl_3 (IE = 10.2, EA = 0.8), PBr_3 (IE = 9.9, EA = 1.6).

32. *Consider a linear string of five hydrogen atoms, H_5. Draw the five MOs of this molecule. Indicate nodes (nodal lines or planes) with dashed lines; give the total number of nodal planes. Indicate positive and negative signs of the wave function within the MOs. Give the number of positive overlaps, the number of negative overlaps, and the net number of positive overlaps for each MO. Give the symmetry label for each MO. Draw the energy level diagram for linear H_5.

33. Consider a linear string of six hydrogen atoms, H_6. (a) Draw the six MOs of this molecule. Indicate nodes (nodal lines or planes) with dashed lines; give the total number of nodal planes. Indicate positive and negative signs of the wave function within the MOs. Give the number of positive overlaps, the number of negative overlaps, and the net number of positive overlaps for each MO. Give the symmetry label for each MO ($\sigma_g{}^b$, etc.). (b) Draw the energy level diagram for H_6 and fill in arrows for the electrons occupying those orbitals.

34. Which (if any) of the photoelectron spectra sketched in Figure 10.36 most nearly fits with the expected bonding in the H_3 molecule?

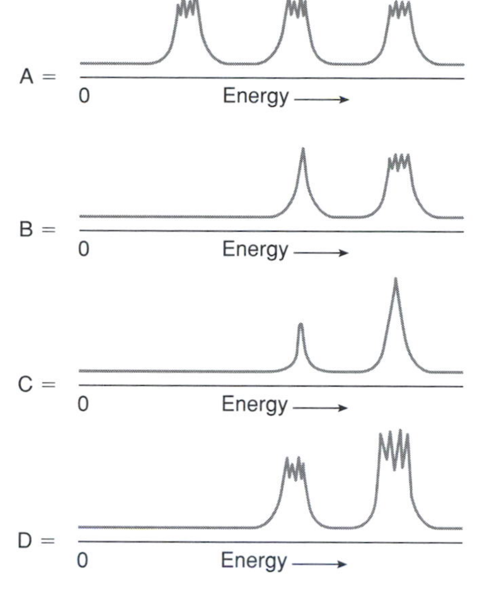

Figure 10.36

35. Draw and label all π-type MOs of the XeF_2 molecule (assume that Xe cannot use its postvalence *d* orbitals).

36. The six π molecular orbitals of (linear) hexatriene, C_6H_8, can easily be generated from the σ molecular orbitals of H_6 (Exercise 33). Describe the changes that must be made to the MOs of H_6 and their symmetry labels in order for these to apply to hexatriene.

37. *Consider a ring of six hydrogen atoms, *cyclo*-H_6. Draw the six MOs of this molecule. Indicate nodes (nodal lines or planes) with dashed lines. Indicate positive and negative signs of the wave function within the MOs. Give the number of positive overlaps, the number of negative overlaps, and the net number of positive overlaps for each MO. Classify each as bonding, antibonding, or nonbonding. Draw the energy level diagram for *cyclo*-H_6.

38. Consider the planar ring compound, cyclooctatetraene (*cyclo*-C_8H_8). (a) Draw the eight π molecular orbitals of this molecule (from a top view). Indicate nodes (nodal lines or planes) with dashed lines. Indicate positive and negative signs of the wave function within the MOs. Give the number of positive overlaps, the number of negative overlaps, and the net number of positive overlaps for each MO. Give the symmetry label for each MO. (b) Draw the energy level diagram for *cyclo*-C_8H_8 and use the appropriate number of arrows to show electrons occupying the orbitals. How many unpaired electrons would the planar molecule cyclooctatetraene have? How many UV/vis absorptions would it have? Would this be classified as an aromatic or an antiaromatic molecule? Would it be expected to adapt a distorted geometry?

39. Draw and label the MOs and the energy level diagram, and determine the expected charges, for the following cyclic species: (a) *cyclo*-H_5; (b) *cyclo*-C_5H_5 (π electrons only); (c) *cyclo*-H_7; (d) *cyclo*-C_7H_7 (π electrons only).

40. Saturated organosilicon compounds such as *cyclo*-Si_6Me_{12} show absorptions in their UV spectra that resemble those found in cyclic polyenes but which are not observed in the *UV or visible* regions of the spectra of saturated organic compounds such as *cyclo*-C_6Me_{12}; these absorptions involve the electrons in the Si–Si links. Describe the nature of these absorptions, and explain why they are not observed for the carbon analogues of these compounds.

41. Suppose that cyclobutadiene were to form a "half-sandwich" molecule by coordinating a transition metal ion above its top face (i.e., forming a complex $C_4H_4M^{n+}$.) Identify which *s*, *p*, and *d* orbitals of the transition metal would have the appropriate symmetry to bond with each of the four π molecular orbitals of cyclobutadiene.

42. *Consider a cyclopentadienide ion ($C_5H_5^-$) ring coordinated in a *pentahapto* manner to a *d*-block or an *f*-block metal ion. Identify which *s*, *p*, *d*, and *f* orbitals of the transition metal ion would have the appropriate symmetry to bond with each of the π molecular orbitals of $C_5H_5^-$; and tell in which resulting MOs the $C_5H_5^-$ ring is acting as a π-donor ligand and in which it can be accepting electrons from the metal ion.

43. Consider the planar η^8-$C_8H_8^{2-}$ (cyclooctatetraenide dianion) ligand. In Exercise 38, you drew facial views of each of the π molecular orbitals of this ligand. Now identify the metal-ion orbitals (if any) that can overlap with each of these ligand orbitals if the metal ion is a (a) *d*-block metal ion; (b) *f*-block metal ion.

44. Drawn in Figure 10.37 are one student's idea of the top views of the five π-type MOs to be found in $C_5H_5^-$ (D_{5h} point group). For purposes of answering some of the questions, each MO has been given a personal "name," such as Harry. (a) Draw in

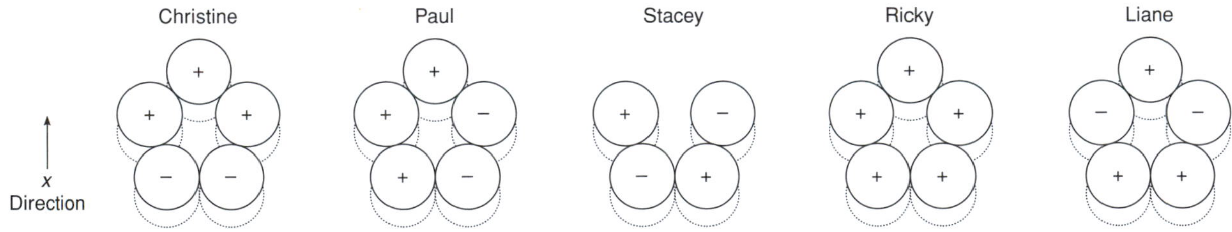

Figure 10.37

the vertical nodal planes of these five MOs. In doing this, you should find that one of these MOs is improper and cannot be a true MO of this molecule. Give the irreducible representation label that should be used for each proper orbital; include the superscripts nb and $*$ to indicate any orbitals that are nonbonding or antibonding. (b) Suppose that a *d*-block metal ion were to approach the $C_5H_5^-$ ring from above and to coordinate to it in a *pentahapto* manner, just above the center of the ring. Identify the specific *s*, *p*, and *d* orbitals of the metal ion that can bond to each proper named $C_5H_5^-$ molecular orbital. (c) Draw the expected energy level diagram for the π orbitals of $C_5H_5^-$, and place the names of the four proper orbitals at the appropriate energy levels.

45. Below are drawn one student's idea of the π-type MOs to be found in H_4 (D_{4h} point group). For purposes of answering some of the questions, each MO has been given a personal "name". (a) Draw in the nodal planes of these five MOs. In doing this, you

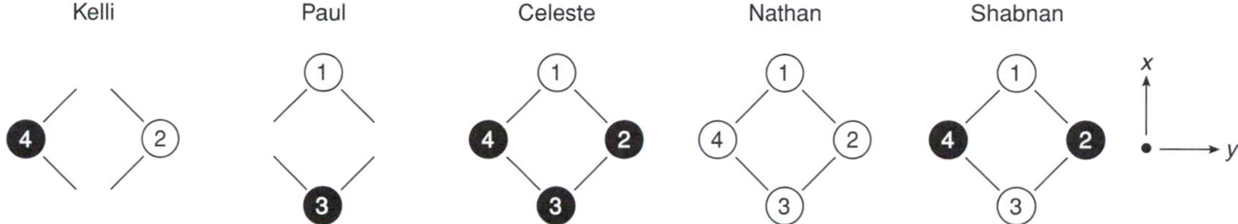

should find that one of these MOs is improper and cannot be a true MO of this molecule. Give the irreducible representation of each proper MO; include the superscripts nb and $*$ to indicate any orbitals that are nonbonding or antibonding. (b) Suppose that a *d*-block metal ion were to approach the H_4 ring from above and to coordinate to it just above the center of the ring. Identify the specific *s, p,* and *d* orbitals of the metal ion that can bond to each named proper H_4 molecular orbital. (c) Draw the expected energy level diagram for the π orbitals of H_4, and place the names of the four proper orbitals at the appropriate energy levels.

46. Consider the ozone molecule, O_3. Let us take its HOMO as π^{nb} and its LUMO as π^*, as drawn in Figure 10.38. Use the symmetry rules to determine whether the ozonolysis of ethylene can occur through the mechanism shown.

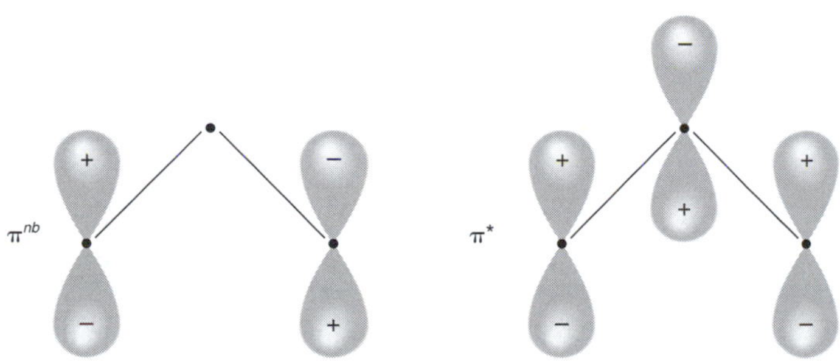

Figure 10.38

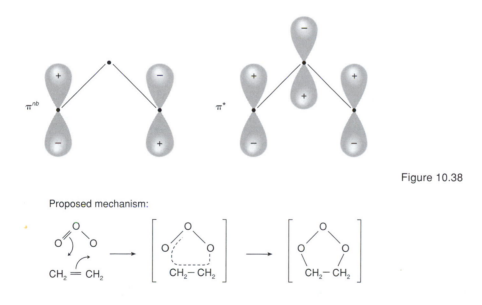

Proposed mechanism:

O_3 + $CH_2{=}CH_2$ → [$O_3 \cdots CH_2{-}CH_2$] → [$O_3CH_2{-}CH_2$ ring]

47. The cycloheptatrienyl derivative $C_7H_7SnMe_3$ has been found to undergo fluxional rearrangement of the $SnMe_3$ group, albeit with a rather high activation energy. (a) The π system of this molecule can be treated as hexatriene, a C_6 chain. Draw the HOMO and the LUMO of the π system, starting always with the positive lobe up at C_2. (b) Draw the HOMO and the LUMO of the σ system (C–Sn bond), starting in the HOMO with the positive lobe up. (c) Use the symmetry rules to predict to which carbon of the hexatriene system the $SnMe_3$ group will migrate during fluxional rearrangement. Explain your reasoning.

48. *Consider the hypothetical square planar molecule XeH_4. (a) Draw the H_4 TASOs. (b) Which (if any) Xe valence atomic orbitals are nonbonding? Which (if any) symmetry-adapted sets of H_4 orbitals are nonbonding? What is the maximum degeneracy found in any MO of XeH_4? (c) Draw in the Xe valence orbitals in the middle of the above sets of H_4 TASOs so as to produce the $a_{1g}{}^*$ molecular orbital of XeH_4; the e_u set of bonding MOs of XeH_4. (d) The energy level diagram for XeH_4 is shown in Figure 10.39. Fill in the appropriate symmetry labels in the diagram, and show the proper number of valence electrons as arrows on the diagram. (e) Compute the total bond order of XeH_4; the bond order per Xe–H link.

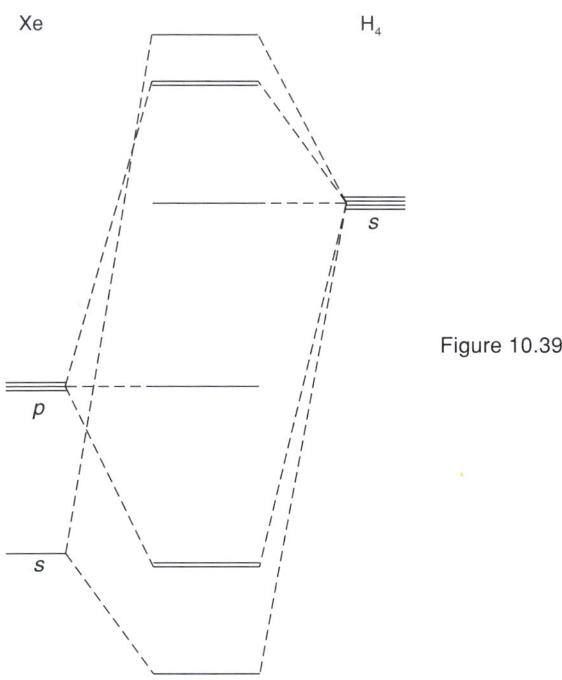

Figure 10.39

49. (a) Draw and label (using proper symmetry labels) the possible σ-type TASOs in the pentagonal planar $XeF_5{}^-$ ion. (b) How many unshared electron pairs are there on Xe in this ion, and where are they located? If appropriately hybridized, Xe atomic orbitals are used to hold these unshared electron pairs, and if Xe cannot use *d* orbitals in this bonding, what Xe atomic orbitals remain to bond with these TASOs? Which TASOs must remain nonbonding? (c) With these assumptions, what is the bond order per Xe–F link in this ion?

50. Consider a tetrahedral (T_d) transition metal complex $MH_4{}^{n-}$. (a) Draw the four TASOs of the H_4 ligand set, giving each its proper symmetry label. (b) Find the valence orbitals of the transition metal that match each of the TASOs of the four hydride ions in symmetry. (c) Draw the energy level diagram for this tetrahedral complex, and label the orbitals with their irreducible representations and with nb, or * if appropriate. Include only metal valence *s* and *d* orbitals. (In many cases, the exact ordering of energy levels is uncertain without calculations, so put down a reasonable order.) Circle the part of the energy level diagram that is emphasized by the crystal field theory (CFT), and show where the crystal field splitting ($10Dq$ or Δ_{tet}) is located.

51. Consider the tetrahedral molecule CH_4 (T_d point group); its energy level diagram is shown in Figure 10.40. (a) Place the appropriate irreducible-representation symmetry

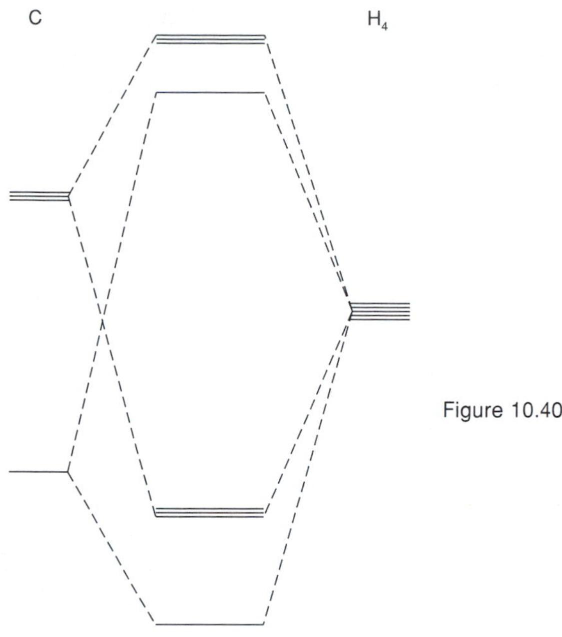

Figure 10.40

labels next to each MO energy level. Also label each carbon atomic orbital energy level at the left (as being s, p_x, etc.). (b) Why would someone say that CH_4 is not sp^3 hybridized? What type of experimental evidence could be cited to support this view? (c) How could you argue the opposing view, that methane *is* sp^3 hybridized? (d) Suppose the molecule were now changed to tetrahedral TiH_4. What new orbitals would have to be added to the left side of the energy level diagram? In what irreducible representations do these orbitals fall? Is there now any possibility of hybridization in the TiH_4 molecule? If so, which atomic orbitals on Ti would be involved?

52. Consider the hypothetical octahedral molecule SH_6 (O_h point group); its energy level diagram is shown in Figure 10.41. (a) Place the appropriate irreducible-

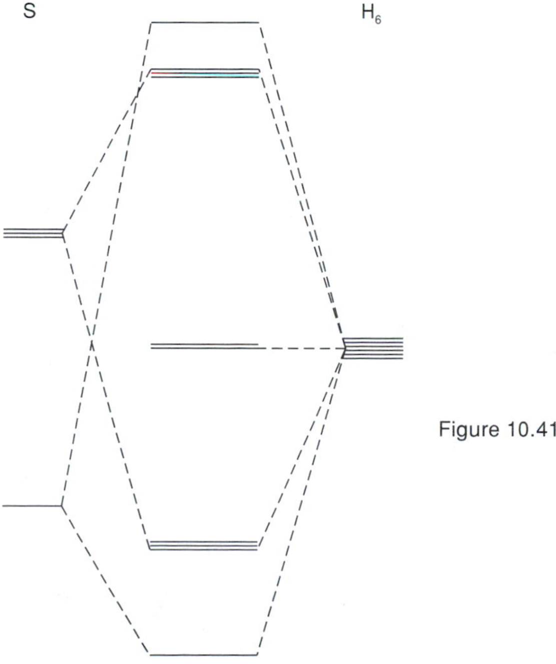

Figure 10.41

representation symmetry labels next to each MO energy level. Also label each sulfur valence atomic orbital energy level at the left (as being *s*, p_x, etc.) (b) Suppose the molecule were now changed to octahedral WH_6. What new valence orbitals would have to be added to the left side of the energy level diagram? In what irreducible representations do these valence atomic orbitals fall? (c) After adding these new valence AOs, are there now any nonbonding MOs in the WH_6 molecule? If so, give their symmetry label. (d) If the H_6 set of ligands were replaced by a set of six π-acceptor ligands, would there now be any nonbonding MOs?

53. (a) What are the appropriate generator orbitals to use to draw the TASOs for the ligands in the tricapped trigonal bipyramidal (D_{3h}) complex, ReH_9^{2-}? (b) Draw these TASOs, labeling them appropriately.

54. Draw the radial MOs for a tetrahedral cluster compound $B_4H_4^{2-}$, and classify each as bonding, nonbonding, or antibonding.

55. (a) Draw the radial MOs for an octahedral cluster compound $B_6H_6^{2-}$, and classify each as bonding, nonbonding, or antibonding. (b) Draw an energy level diagram for the radial MOs only of $B_6H_6^{2-}$, show which orbital(s) are occupied, and draw arrows to indicate all possible electronic (e.g., UV) absorptions in this ion. (c) Which UV absorption should be most intense, and why?

56. *Count the net positive overlaps in each of the MOs for the M_8 rectangular crystallite in Figure 10.30, and draw the energy level diagram for this crystallite.

57. Label each bonding and antibonding molecular orbital energy level in Figure 10.31(*b*) with the letter of the corresponding MO in Figure 10.31(*a*) [e.g., use label the energy level diagram with the appropriate letters from (*A*)–(*P*)].

58. Work out the corresponding MOs and energy level diagram for a shorter rectangular prismatic metal crystallite M_{12} (3 atoms high by 2 atoms wide by 2 atoms deep). Label each energy level as belonging to a bonding, an antibonding, or a nonbonding MO.

Notes

1. The He_2 "molecule" has been reported at 0.001 K [F. Luo, G. Kim, G. C. McBane, C. F. Giese, and W. R. Gentry, *J. Chem. Phys.*, **98**, 3564 (1993)], but is bound only by the extremely weak van der Waals attractions discussed in Section 12.1, not by a covalent bond.
2. A good presentation of MO theory that is somewhat, but not excessively, more in depth than the present one is found in R. L. DeKock and H. B. Gray, *Chemical Structure and Bonding*, University Science Books, Mill Valley, CA, 1989.
3. D. R. Brown, M. J. Clegg, A. J. Downs, R. C. Fowler, A. R. Minihan, J. R. Norris, and L. Stein, *Inorg. Chem.*, **31**, 5041 (1992).
4. These potential energies are given in the standard units of electron volts ($1\,eV = 96.5\,kJ\,mol^{-1}$) and are negatives of ionization energies that are the average energies for all electrons at that level. Although Table 10.1 does not include data to show the closeness of the valence state potential energies in the early *d* block, the discussion of trends in Z^* in Chapter 1 can be used to illustrate this.
5. Or set of orbitals, if they are degenerate in energy.
6. B. E. Bursten, F. A. Cotton, M. B. Hall, *J. Am. Chem. Soc.*, **102**, 6348 (1980).
7. B. E. Bursten and G. A. Ozin, *Inorg. Chem.*, **23**, 2910 (1984).
8. J. L. Reed, *J. Chem. Educ.*, **71**, 98 (1994).

9. Somewhat falsely in this case: There is an exception to Koopman's theorem—in the neutral N_2 molecule the HOMO is actually π_u, while in the product N_2^+ molecule ion the HOMO is σ_g^{nb}; the latter produces the observed ordering in the spectrum.
10. Some shift of the bands from the nonbonding orbitals from those of the parent atomic orbitals results from the polarity of the single bond in HF: Since electrons drift toward fluorine, the ionization energies are more like those expected from a fluoride ion and not so much like those from a neutral F atom.
11. The photoelectron peaks for ionization from the HOMOs of HBr and HI are split into two equal peaks by the phenomenon of *spin–orbit coupling* mentioned as a consequence of relativistic effects in Section 1.11. This coupling, it will be recalled, splits the $5p$ orbitals of iodine, for example, into two sets: $5p_{3/2}$ and $5p_{1/2}$; these differ in energy by nearly 1 eV in HI.
12. R. G. Pearson, *Inorg. Chem.*, **27**, 734 (1988).
13. Recall that, in Section 3.1, we omitted boron from the list of elements that commonly engage in multiple bonds; we see that there is some justification for this, since its multiple bonding partners are likely to be much more electronegative than boron. However, there is also evidence, in molecules such as the boron trihalides, for some π bonding.
14. R. L. DeKock and H. B. Gray, *Chemical Structure and Bonding*, University Science Books, Mill Valley, CA, 1989; pp. 229 and 259.
15. a. R. G. Parr and R. G. Pearson, *J. Am. Chem. Soc.*, **105**, 7512 (1983); b. R. G. Pearson, *Inorg. Chem.* **27**, 734 (1988).
16. L. C. Allen, *Acc. Chem. Res.*, **23**, 175 (1990). It may be mentioned that a procedure known as *electronegativity equalization* of all species in a reacting chemical system is often employed: R. T. Sanderson, *J. Chem. Educ.*, **31**, 2 (1954).
17. R. G. Pearson, *J. Am. Chem. Soc.*, **110**, 2092 (1988); R. G. Pearson, *Acc. Chem. Res.*, **26**, 250 (1993).
18. One deficiency of the numerical values of hardness obtained by these calculations is that they do not appear to have completely separated out the factor of *acid–base strength* (Section 5.6); the numerical values obtained for hardness for +1 charged cations, for example, are not readily compared in any practical way with those of +2-charged cations or of neutral molecules or anions.
19. G. Klopman, *J. Am. Chem. Soc.*, **90**, 223 (1968).
20. Hardness and softness localized at particular donor or acceptor atoms can be calculated with some difficulty: R. G. Pearson, *Chemtracts Inorg. Chem.*, **3**, 317 (1991).
21. D. Datta, *Inorg. Chem.*, **31**, 2797 (1992); P. K. Chattaraj and P. v. R. Schleyer, *J. Am. Chem. Soc.*, **116**, 1067 (1994).
22. T. R. Geballe and T. Oka, *Nature* (*London*), **384**, 334 (1996).
23. Furthermore, in reality the allyl system is not linear. It does not belong to the $D_{\infty h}$ point group, so that other labels are more proper—but the labels π_g and π_u are commonly used anyway.
24. Assuming that the molecule is linear, or at least is in a conformation that has an inversion center.
25. Adding the d_{z^2} orbital (Σ_g^+ irreducible representation) to the list of available Xe orbitals would allow positive overlap within the Σ_g^{+nb} MO, thereby increasing the bond order; but since there must be a considerable difference in energy between the F valence and the Xe postvalence orbitals, the likelihood of effective overlap is open to question.
26. J. I. Musher, *Angew. Chem. Intl. Ed. Engl.*, **8**, 54 (1969).
27. R. J. Gillespie and E. A. Robinson, *Inorg. Chem.*, **34**, 978 (1995).
28. J. C. Martin, *Science*, **221**, 509 (1983); M. Lattman, *Encyclopedia of Inorganic Chemistry*: R. B. King, Ed., Wiley, Chichester, UK, 1994, p. 1496.
29. D. J. Hajdasz and R. R. Squires, *J. Am. Chem. Soc.*, **108**, 3139 (1986).
30. D. K. Hoffman, K. Ruedenberg, and J. G. Verkade, *J. Chem. Educ.*, **54**, 590 (1977); D. K. Hoffman, K. Ruedenberg, and J. G. Verkade, *Struct. Bonding* (*Berlin*), **33**, 57 (1977); J. G. Verkade, *A Pictorial Approach to Molecular Bonding*, 2nd ed., Springer-Verlag, New York, 1997, with accompanying software available on the World Wide Web at http://www.public.iastate.edu/~jverkade/nodegame.hmtl.
31. This label gives the most difficulty, and it is usually for an antibonding orbital higher in energy than the LUMO: a relatively unimportant orbital. At the instructor's discretion, you may elect to label this orbital b_{xg} or b_{xu}, without deciding on the value of x.
32. The energy levels produced by this method are numerically correct in their relative energies.

33. Z. Zhou and R. G. Parr, *J. Am. Chem. Soc.*, **111**, 7371 (1989).
34. R. G. Pearson, *Chem. Eng. News*, 66, Sept. 28, 1970; R. L. DeKock and H. B. Gray, *Chemical Structure and Bonding*, University Science Books, Mill Valley, CA, 1989; p. 324.
35. This type of conjugation assures that the M–C σ bond has positive overlap with the lowest energy π orbital of the C_4 system, that is, if we put a positive sign on the one, we put a positive sign on the other.
36. R. G. Woodward and R. Hoffmann, *The Conservation of Orbital Symmetry*, Academic, New York, 1969.
37. M. J. S. Dewar and C. Jie, *Acc. Chem. Res.*, **25**, 537 (1992).
38. For a discussion of the conditions under which the simple valence bond and the simple molecular orbital pictures of the bonding in molecules such as methane and the related molecule water are valid, see G. Gallup, *J. Chem. Educ.*, **65**, 671 (1988), R. B. Martin, *J. Chem. Educ.*, **65**, 668 (1988), and M. Laing, *J. Chem. Educ.*, **64**, 124 (1987).
39. This band appears to be broad, although much of it is obscured by overlap with bands due to the nonbonding orbitals of fluorine: R. E. Ballard, *Photoelectron Spectroscopy and Molecular Orbital Theory*, Wiley, New York, 1978; p. 166.
40. C. S. Ewig and J. R. Van Wazer, *J. Am. Chem. Soc.*, **112**, 109 (1990).
41. Actually, one must do more than count overlaps, since MOs in many cases such as these have greater wave amplitudes at some atoms than at others, so that some overlaps are more positive or more negative than others. The approximate conclusion given above remains valid, however.
42. J. G. Verkade, *J. Chem. Educ.*, **68**, 739 (1991).
43. The properties of metallic solids are usually discussed by solid-state physicists in terms of these bands in the *band theory*, which incorporates language customary for physicists but not for chemists; for a useful translation guide, see R. Hoffman, *Solids and Surfaces: A Chemist's View of Bonding in Extended Structures*, VCH, New York, 1988.

CHAPTER

11

Organometallic Chemistry of the *d*-Block Elements

With Applications to Industrial Chemistry, Organic Chemistry, Biochemistry, and Environmental Chemistry

11.1 Organometallic Compounds of the *d*-Block Metals: The 18-Electron Rule

Organometallic compounds are those containing direct metal–carbon bonds (excluding a few cases in which the carbon compound or species is considered to be inorganic, e.g., carbides and cyanides). Before 1950, almost all known organometallic chemistry was of the *s* and *p* elements, and included compounds in which organic groups (R) [such as the methyl (Me) and phenyl (Ph) groups] were σ bonded to the metal atom; such compounds will be discussed in Chapter 15. Attempts to prepare corresponding methyls and phenyls of the *d*-block elements met almost universal failure, however. A few *d*-block metal derivatives of carbon monoxide, the **carbonyls**, had been discovered, starting with $Ni(CO)_4$ in 1890 (which was odd, since CO seldom bonds to *s*- or *p*-block elements). And Zeise, way back in 1827, had prepared a singular anion of platinum and ethylene, $[PtCl_3(C_2H_4)]^-$, which was odder still. But science always has to live with a few odd unexplained facts.

In 1951, Kealy and Pauson[1] were attempting a synthesis of the elusive organic compound, fulvalene, which contains two cyclopentadienyl (Cp) rings joined by a double bond [Fig. 11.1(*a*)], by the oxidation of the σ-bonded Grignard reagent C_5H_5MgX with Fe^{3+}:

$$2\,C_5H_5MgX + 4\,Fe^{3+} \not\longrightarrow C_5H_4{=}C_5H_4 + 2\,HX + 4\,Fe^{2+} + 2\,Mg^{2+} \qquad (11.1)$$

To their surprise, they obtained instead an orange solid, stable to 400 °C, which analyzed as $C_{10}H_{10}Fe$ and which they reported[1] with the suggestion that the Cp ligands were each σ bonded to iron [in the manner we now call *monohapto*, giving the iron a coordination number of 2, Fig. 11.1(*b*)]. Almost simultaneously, Miller et al.[2] reported another failed attempt at an organic synthesis (of an amine by heating cyclopentadiene and N_2 over an iron catalyst), which also gave Cp_2Fe and for which they proposed the same structure.

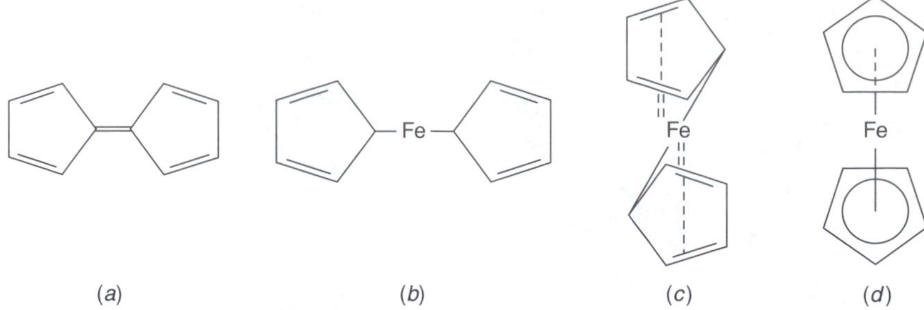

Figure 11.1
(*a*) Structure of the organic compound fulvalene, which Kealy and Pauson were attempting to synthesize when they obtained Cp_2Fe. (*b*) Structure proposed by Kealy and Pauson[1] for their product. (*c*) Structure that originally occurred to Wilkinson,[3] containing two σ bonds and one π bond from each ring to Fe. (*d*) Structure eventually conceived by Wilkinson and Woodward, in which all bonds from each ring to Fe are equivalent.

At Harvard, Wilkinson[3] read the first of these two papers, saw the proposed structure of this unprecedentedly stable *d*-block organometallic compound, said to himself, "[Expletive deleted] it can't be that!", and, having read about Zeise's salt, thought that the π bonds of the Cp groups ought also to be coordinated [Fig. 11.1(*c*)]. Then, recalling lectures on the aromaticity of 6 π-electron systems such as Cp^-, he conceived of the structure in Figure 11.1(*d*), with all carbon and hydrogen atoms being equivalent, and thought, "It's a sandwich." (To this day, such derivatives of two polyene ligands above and below a metal atom are known as **sandwich molecules** or **metallocenes.)** The next day Wilkinson and Woodward conceived of the proof of this equivalence: The infrared (IR) spectrum of the compound ought to be especially simple (due to its high symmetry), which was confirmed in another day or so.

Thus the prototype ligands for *d*-block metals in organometallic compounds turned out to be the aromatic polyenes and carbon monoxide. Although *d*-block metals do form compounds containing methyl and phenyl groups, these are most commonly present in compounds containing one of the prototype ligands, or some other related strong-field ligand (Chapter 8) such as organophosphines. As we saw in Chapter 10, these types of ligands are π-acceptor's, which promote low-spin complex formation; the other immediate confirmation of the structure of ferrocene was the prediction of and observation of the property of diamagnetism. Low-spin electron configurations are promoted because the normally nonbonding *d* highest occupied molecular orbitals (HOMOs) in complexes (t_{2g} in octahedral complexes) become bonding, and hence are well separated in energy from the lowest unoccupied molecular orbitals (LUMOs) ($e_g{}^*$ in octahedral complexes).

Soon, it was realized that for most *d*-block complex geometries, filling all of the bonding orbitals and leaving the antibonding orbitals vacant, led to a total of nine MOs being filled, one per valence *s*, *p*, and *d* orbital of the metal ion. Hence, we often predict the electron configurations of *d*-block organometallics using the **18-electron rule.** Corresponding to the frequent occurrence of octets in compounds of *p*-block elements, *stable organometallic compounds of the d-block elements tend to have 18 valence electrons about the metal atom.*

Before we begin applying this rule, let us see how it emerges from energy level diagrams of typical geometries found in the *d* block. Figure 10.27 shows the diagram for an octahedral complex. We point out three regions of orbitals in this diagram: (1) Below the boxed region are six bonding MOs. These orbitals are closer in energy to the original ligand orbitals (right side of the diagram) than the metal atomic orbitals (left side), so they are still mainly ligand-like orbitals; there are one per electron pair donated by the set of ligands. (2) Within the boxed region are the five orbitals emphasized by the crystal field theory, which are either equal in energy or not too far removed from the energies of the parent metal *d* atomic orbitals; the HOMO and LUMO are in this region. (3) Of less importance, since they are neither occupied nor are they the LUMOs, are the four antibonding orbitals above the boxed region; they are closest in energy to, and more or less correspond to, the metal ion's valence *s* and *p* orbitals.[4]

The stable electron configuration for a strong-field octahedral complex contains 18 valence electrons because (1) each of the six ligands contributes a pair of electrons to the six orbitals below the boxed region, for a total of 12 electrons; (2) in octahedral complexes of π-acceptor ligands the t_{2g} set of orbitals in the boxed region is also bonding and can and does prefer to hold an additional six electrons, to give a total of 18 valence electrons around the *d*-block atom.

Similarly, the energy level diagram for the metallocene bis(benzene)chromium (Fig. 10.24) begins with six ligand-based bonding MOs containing the six pairs of π electrons that the two benzene molecules contribute. The remaining five orbitals are most like the *d* orbitals on the metal atom; in this energy level diagram two are bonding (e_{1g} and e_{1u}), one is nonbonding ($a_{1g}{}^{\text{nb}}$), and two are antibonding. Filling the two bonding and the nonbonding orbital gives a total of 18 valence electrons around Cr, as is found in $(C_6H_6)_2Cr$.

Bis(benzene)chromium was discovered within months of ferrocene by E. O. Fischer as the product of a reduction of $CrCl_3$ in the presence of benzene as a π-donor ligand and $AlCl_3$ as a Lewis acid (chloride-ion acceptor):

$$3\,CrCl_3 + 6\,C_6H_6 + AlCl_3 + 2\,Al \rightarrow 3\,[Cr(C_6H_6)_2][AlCl_4] \qquad (11.2)$$

This cation was then reduced in aqueous solution to give the product, a black solid. Fischer shared the 1973 Nobel prize with Wilkinson.

The energy level diagram of $(C_6H_6)_2Cr$ is very similar to that of ferrocene, Cp_2Fe; both have a total of 18 valence electrons around the metal atom. Since the HOMO of each molecule is an approximately nonbonding d_{z^2} orbital, however, there is no strong reason to have exactly 18 valence electrons: Both compounds are readily oxidized to 17-electron cations [the product of Reaction (11.2), and the blue *ferricenium cation*, $FeCp_2{}^+$].

For other geometries, we may wish to refer back to the crystal field splittings given in Table 8.6, keeping in mind that these are presumed[5] to account for the middle part of the MO energy level diagrams. Thus, for a trigonal bipyramidal five-coordinate organometallic, first we need to count the 10 electrons donated by the five ligands. Next, we consider the five *d* orbitals whose energies are given in Table 8.6. Four of the five have negative crystal field splittings (low energies); in the presence of π-acceptor ligands, we presume that MO calculations would find these to be bonding (or perhaps nonbonding) MOs. These should logically be filled to give a total of 18 valence electrons; the high-energy d_{z^2} orbital likely would turn out to be antibonding, and would remain empty.

For a tetrahedral complex, this procedure does not appear to work. Four ligands donate eight electrons, and only two *d* orbitals $(e = d_{z^2}, d_{x^2-y^2})$ have negative crystal field splittings. The 18-electron rule is, however, obeyed in tetrahedral complexes of π-acceptor ligands such as CO, since (a) the splitting of *e* and t_2 in tetrahedral complexes is small, and (b) both the *e* and the t_2 sets of *d* orbitals donate electrons to the CO π-accepting orbitals of the CO ligands, so that both interactions become bonding; MO calculations show that these π-bonding interactions are more important than the σ-bonding interactions that give rise to the e–t_2 splitting.[4]

One major geometry definitely does not obey the 18-electron rule, however: the square planar geometry. In this four-coordinate geometry, the ligands donate eight electrons, and four of the five metal *d* orbitals either have negative or only small positive crystal field splittings. The $d_{x^2-y^2}$ orbital has a very large positive crystal field splitting, and normally therefore remains unoccupied; in MO terms it is definitely antibonding. When we add the number of electrons donated to the metal atom by the ligands (i.e., from below the "boxed region") to the number of *d* electrons contributed by the metal (i.e., from within the "boxed region") we obtain what we shall call the metal's **total valence electron count (VEC)**; in this case, this is $8 + 8 = 16$. Hence, *square planar organometallic compounds tend to have a total of 16 valence electrons.*

There is no uniform tendency to obey the 16- and 18-electron rule across the *d* block, however. At the beginning of the *d* block, some of the orbitals are too high in energy (Table 10.1) for effective usage in bonding. In addition, the number of ligands required to supply a complete set of 18 electrons may exceed the maximum coordination number of the element. At the end of the *d* block, the valence *d* orbitals may be too low in energy to give effective overlap with the π-accepting ligands, and at the lower right of the *d* block, relativistic effects (Chapter 1) make some of the *p* orbitals unavailable for effective bonding. Compliance with the 16- and 18-electron rules is greatest with the majority of the elements in the center of the *d* block.

11.2 Mononuclear Metal Carbonyls and Related Organometallics

The mononuclear (containing just one metal atom) carbonyls of the *d*-block elements make a useful first application of the 18-electron rule, since this rule is obeyed by the large majority of these carbonyls, and each CO ligand unambiguously donates just one electron pair to its metal atom or ion. As a soft base, CO characteristically forms organometallic compounds with soft-acid *d*-block metal atoms, which include not only our traditional soft acids from Chapter 5 but also many of the other *d*-block elements in *very low oxidation states*, such as $+1, 0, -1$, or even lower.[6] Table 11.1 shows representatives of the mononuclear carbonyls of the *d*-block elements. The carbonyls with an overall negative charge (lower left of the table) are called *carbonylate anions*; those with an overall positive charge (upper right of the table) are called *carbonyl cations*.

Example 11.1

Use the 18-electron rule to predict the formulas of the metal carbonyls with $0, +1$, and -1 overall charges.

Table 11.1
Typical Mononuclear Carbonyls of the *d*-Block Elements[a]

Octahedral d^3	$[Ti(CO)_6]^{2-}$	$[V(CO)_6]^-$	$Cr(CO)_6$	$[Mn(CO)_6]^+$	$[Ru(CO)_6]^{2+}$				
Trigonal bipyramedal d^8		$[V(CO)_5]^{3-}$	$[Cr(CO)_5]^{2-}$	$[Mn(CO)_5]^-$	$Fe(CO)_5$	$[Co(CO)_5]^+$			
Tetrahedral d^{10}			$[Cr(CO)_4]^{4-}$	$[Mn(CO)_4]^{3-}$	$[Fe(CO)_4]^{2-}$	$[Co(CO)_4]^-$	$Ni(CO)_4$		
Other		$V(CO)_6$				$[Co(CO)_3]^{3-}$	$[Pt(CO)_4]^{2+}$	$[Cu(CO)_3]^+$	
								$[Cu(CO)_2]^+$	$[Hg(CO)_2]^{2+}$

[a] Neutral carbonyls are enclosed in shaded boxes; carbonyl cations are above and to the right of the dividing lines; carbonylate anions are below and to the left of the dividing lines. In most cases the remaining elements in each group also form the corresponding carbonyl, although among neutral carbonyls $Pd(CO)_4$ and $Pt(CO)_4$ are only stable at low temperatures.

SOLUTION:

We will not consider any coordination numbers above 6, since this is the maximum total coordination number for the fourth period. Octahedral, trigonal bipyramidal, and tetrahedral carbonyls should give a total count of 18 valence electrons. The six CO atoms of a hexacarbonyl contribute 12 valence electrons, so the parent metal ion or atom should therefore have 6 valence electrons. Therefore, only Group 6 metals are expected to form neutral $M(CO)_6$; in fact Group 6 metals as well as vanadium do. The possible M^+ species in a carbonyl cation $M(CO)_6^+$ must be Group 7 metals in order to have 6 valence electrons; this is confirmed in Table 11.1. The possible M^- species in a $M(CO)_6^-$ carbonyl anion should be and are Group 5 metals.

For trigonal bipyramidal pentacarbonyl species, the ligands donate 10 electrons, so it is up to M, M^+, or M^- to provide 8 valence electrons. Hence, M in $M(CO)_5$ should be and is a Group 8 metal; M^+ should be and is a Group 9 metal ion, and M^- should be and is a Group 7 metal.

For tetrahedral tetracarbonyl species, in which the ligands donate 8 electrons, M or M^+ should donate 10 valence electrons. Hence, neutral M should be a Group 10 metal atom; in practice it can be Ni. Anionic M^- should be a Group 9 metal; in practice it is Co. Cationic M^+ should be a Group 11 metal, but no $[M(CO)_4]^+$ salt has been isolated, although Cu^+ may form such a species in solution.

Recalling the procedure for assigning central-atom oxidation numbers given in Section 3.10, it is a trivial matter to assign metal oxidation numbers in these simple carbonyl species, since all of the ligands are neutral. From this, the compounds and ions can be named as previously described in Section 8.11. However, organometallic chemists have developed a parallel set of informal ways of naming, which for these compounds puts the metal first: for example, $Fe(CO)_5$ as iron pentacarbonyl (or even just iron carbonyl). If we have the formula of a carbonyl species, we may recall from Section 3.4 that valence shell electron-pair repulsion (VSEPR) principles may also be used to predict any of these geometries *except the square planar*, provided that we *ignore* the unshared pairs of *d* electrons in assigning geometries. {Two square planar tetracarbonyl cations, $[Pd(CO)_4]^{2+}$ and $[Pt(CO)_4]^{2+}$, have been synthesized.[7]}

The mononuclear metal carbonyls are colorless or lightly colored volatile liquids or solids. The M–C bonds in these compounds are not especially strong; the bond strengths are around 160 $kJ\,mol^{-1}$, and are highest in the carbonylate anions and lowest in the infrequently seen carbonyl cations.[8] Metal carbonyls readily lose CO on heating or upon irradiation with ultraviolet (UV) light. Thus these compounds should be handled in hoods to avoid CO poisoning. Some of them are even more toxic than CO, however. Nickel carbonyl, for example, is 100 times as toxic as CO; it apparently decomposes in the lungs, in effect nickel-plating them.

Some metal carbonyls can be prepared by the **direct reaction** of the metal with CO; Mond discovered $Ni(CO)_4$ while investigating the corrosion of nickel valves by CO(g). This reaction is generally not as easy as one may hope, however, since the atomization energies of many of the metals are so high. Except in the case of nickel, high temperatures, long reaction times, and high pressures are required. Nickel reacts directly with CO at 30 °C and 1-atm pressure to give the carbonyl, which decomposes back to nickel metal and CO at somewhat higher temperatures; this reaction was the basis of the old Mond process for purifying nickel. Synthesizing other carbonyls such as those of iron, cobalt, molybdenum, and ruthenium requires temperatures of hundreds of degrees and pressures of hundreds of atmospheres for hours or days. However, metal atom reactors (Section 7.3) have been used to produce $Pd(CO)_4$, which is only stable at the very low temperatures at which it is produced.

Alternatively, many carbonyls, carbonylate anions, and carbonyl cations can be made by **reductive ligation** from CO, a metal halide, and a reducing agent (an active metal, an organometallic compound, or H_2, or even CO gas itself):

$$VCl_3 + 6\,CO + 4\,Na + 2\,MeO(CH_2CH_2O)_2Me$$
$$\rightarrow 3\,NaCl + \{Na[MeO(CH_2CH_2O)_2Me]_2\}^+[V(CO)_6]^- \qquad (11.3)$$

Equation (11.2) gives another example of this type of synthesis. Note in both cases the presence of a large counterion to stabilize the large carbonyl ion produced. Carbonylate anions can also be made by Group 1 metal reduction of a neutral carbonyl, or a less negatively charged carbonylate anion:

$$M(CO)_6 + 2\,Na \rightarrow 2\,Na^+ + [M(CO)_5]^{2-} + CO \qquad (11.4)$$

$$[V(CO)_6]^- + 2\,Cs \rightarrow [V(CO)_5]^{3-} + 2\,Cs^+ + CO \qquad (11.5)$$

Infrared spectroscopy is extensively used to characterize metal carbonyls, since the C≡O stretching frequency occurs in a very uncrowded region of the spectrum: 2143 cm^{-1} in CO itself. It is also very sensitive to the degree of backbonding to the CO (π acceptance by CO), since this reduces the bond order of CO toward two, and hence lowers the stretching frequency toward the C=O region of 1700–1750 cm^{-1} (as in organic ketones). Thus in the adduct $H_3B{:}C{\equiv}O$, in which no π back-bonding is possible, ν_{CO} is actually raised (to 2165 cm^{-1}). Among carbonyls, we expect back-bonding to be least in the cations and most in the anions, and this finding is confirmed by the trends in ν_{CO}: a 2251 cm^{-1} average in $[Pt(CO)_4]^{2+}$, 2094 cm^{-1} in $[Mn(CO)_6]^+$, 1984 cm^{-1} in $Cr(CO)_6$, 1858 cm^{-1} in $[V(CO)_6]^-$, and 1748 cm^{-1} in $[Ti(CO)_6]^{2-}$.[9] With further increases in negative charge, the stretching frequency is lowered below the C=O region

Table 11.2
Neutral Two-Electron Donor Ligands (besides CO) Found in Organometallics

A. Π-Acceptor Ligands		B. Mainly σ-Donor Ligands	
:C≡N–R	Isonitriles	$:OH_2$	Aqua
:C≡S	Thiocarbonyl	$:OR_2$	Ethers
:C≡Se	Selenocarbonyl	$:SR_2$	Sulfides
:N≡N:	Dinitrogen	$:SeR_2$	Selenides
$:PR_3$	Phosphines	$:NR_3$	Amines
$:PF_3$, $:PCl_3$	Phosphorus Trihalides	$:NC_5H_5$, etc.	Heterocyclic Amines
$:P(OR)_3$	Phosphites	:N≡C–R	Nitriles
$:AsR_3$	Arsines		
$:SbR_3$	Stibines		

into regions more characteristic of a single C–O bond: 1670 cm^{-1} in $[Mn(CO)_4]^{3-}$ and 1460 cm^{-1} in $[Cr(CO)_4]^{4-}$.[10]

The carbonyls are useful as starting materials for the production of other organometallic compounds, since CO(g) is relatively readily lost from them to give reactive 16-electron species that generally initially capture solvent molecules [in the case of $Cr(CO)_5$, even Xe atoms] to give $M(CO)_{n-1}$ (solvent); then the solvent is replaced by the final ligand. Among the simplest derivatives are those that result from the entry of another σ-donor ligand to replace the lost CO ligand. Table 11.2 lists a number of ligands that readily substitute for CO on a 1:1 basis because they also donate one pair of electrons; those in Table 11.2A do so most readily and extensively because they are also good π-acceptor ligands. The ligands in Table 11.2B are mainly or exclusively σ-donor ligands and can only replace CO a very limited number of times without causing the complex to go high spin.

The progress of a ligand substitution reaction can readily be followed in metal carbonyls: as the ligands are replaced, the symmetry point group of the molecule drops, resulting in an increase in the number of carbonyl stretching bands (to a limit of one per CO ligand in very low-symmetry molecules). Table 11.3 gives the number of C≡O stretching frequencies expected as a function of the degree of substitution.

Table 11.3
Numbers of C≡O Stretching Bands for Substituted Carbonyls

Complex	Isomer	Point Group	No. of Bands
$M(CO)_6$		O_h	1
$M(CO)_5L$		C_{4v}	2 or 3
$M(CO)_4L_2$	trans	D_{4h}	1
	cis	C_{2v}	3 or 4
$M(CO)_3L_3$	fac	C_{3v}	2
	mer	C_{2v}	3
$M(CO)_5$		D_{3h}	2
$M(CO)_4L$	axial	C_{3v}	2 or 3
	equatorial	C_{2v}	4
$M(CO)_3L_2$	ax–ax	D_{3h}	1
	ax–eq	C_s	3
$M(CO)_4$		T_d	1

Note that different geometric isomers generally should also give different numbers of absorptions, so they can also be distinguished by IR spectroscopy.

Full substitution of all CO groups by incoming ligands is usually not practical, but fully substituted **homoleptic** (having only one type of ligand) products are known in several cases (e.g., $[V(CNR)_6]^+$, $Ni(PF_3)_4$, $\{Co[P(OMe)_3]_5\}^+$, (and $[Pt(PPh_3)_4]$). These can be made by the same two methods used to make metal carbonyls: direct reaction of metal and excess ligand (often in a metal atom reactor), or reductive ligation of a metal salt in the presence of a ligand.

$$Ni(OH)_2 + 5\,CNR \rightarrow Ni(CNR)_4 + RNCO + H_2O \qquad (11.6)$$

Steric effects become important in complexes of phosphines and related ligands from Table 11.2A. Molecular modeling programs can now assess steric effects quite quantitatively; an earlier approximate evaluation of steric effects in ligands is obtained by measuring their *cone angle*[11] θ, the angle subtended by a cone around the ligand with its point located at the metal atom (for P donor atoms, the metal is assumed to be at a distance of 228 pm). Carbon monoxide has a modest θ of 95°; but θ for phosphines range from 118° for $:PMe_3$ through 145° for $:PPh_3$ to 182° for $:P(t\text{-Bu})_3$, where *t*-Bu is *tert*-C_4H_9. Large cone angles (of 140°–160° or higher) would not only affect the choice of geometric isomers from a reaction, but also favor dissociation of ligands, even to the point of violating the 18-electron rule: the complex $Ni[P(t\text{-Bu})_3]_4$ dissociates to such an extent that the 14-electron complex $Ni[P(t\text{-Bu})_3]_2$ can be isolated.

Carbonyl IR stretching frequencies are also useful in categorizing the relative σ-donor and π-acceptor properties of the incoming ligands when substitution is not complete. As the electron density supplied by the incoming ligand increases, the flow of electrons into the π^* orbital of CO increases, and the bond order and the ν_{CO} decrease. On the other hand, if the incoming ligand is a better π acceptor than CO, it will compete successfully for electron density, reducing the flow of electrons into the π^* orbital of CO: hence, the bond order and the ν_{CO} increase.

This and other data allow the ranking of related ligands from Table 11.2. For example, the phosphine and related ligands show the following order of basicity (π-donor ability)[12]:

$$PR_3 > PAr_3 > P(OR)_3 > P(OAr)_3 \gg PF_3 \qquad (11.7)$$

where Ar = aryl and R = alkyl; their order of π-accepting ability is thought to be the reverse:

$$PR_3 < PAr_3 < P(OR)_3 < P(OAr)_3 < PF_3 \qquad (11.8)$$

And triple-bonded ligands can also be rated in order of π-acceptor abilities[13]:

$$:C{\equiv}N^- < :N{\equiv}N < :C{\equiv}NR < :C{\equiv}O < :C{\equiv}S < :N{\equiv}O^+ \qquad (11.9)$$

The two charged end members of this series differ substantially from the other members: electron-rich cyanide ion is not only a poor π acceptor, but is also the best π donor of the series. It is compatible with higher oxidation states in complexes than the other ligands, and its complexes are sufficiently lower on the spectrochemical series to no longer be considered organometallic compounds, despite the presence of M–C bonds. At the other end of the series, NO^+ is strongly electron withdrawing.

Some of the substituted complexes are of importance for what they do to the stability and reactivity of the coordinated ligand. Both CS and CSe are stable in complexes, but not as free ligands; coordinated N_2, in which the extraordinarily stable N≡N triple bond (Raman stretching frequency of 2360 cm^{-1}) has been reduced closer to a double bond {1763 cm^{-1} in $[V(N_2)_2(Me_2PCH_2CH_2PMe_2)_2]^-$}, is sometimes thereby activated toward protonation[14] and further reduction by H_2, which could be an important step in improved methods of manufacture of ammonia fertilizers. Further details on this process are found in Section 17.9.

11.3 Metal Complexes of π-Donor Ligands

A property that is almost confined to the *d*-block metals is the ability to form complexes with organic compounds having *no* unshared pair of electrons but having instead one or more π bonds. As discussed in Section 10.8, in complexes of such *π-donor* ligands, the π-bond electron pairs are donated to *d* orbitals of the metal; there is also some degree of donation of electrons back from other *d* orbitals to the π^* orbitals of the organic ligand. Numerous π-donor ligands are known, starting with the simplest, ethylene ($CH_2{=}CH_2$), but in this section we will concentrate on the conjugated cyclic polyenes $C_nH_n{}^{\pm x}$ (Section 10.6), which are stabilized aromatic species when $n = 2, 6, 10, \ldots$ and are antiaromatic and destabilized as planar conjugated systems when $n = 4, 8, \ldots$. Although C_6H_6 and $C_5H_5{}^-$ are the most important of these, it is also possible to form sandwich or other organometallic compounds from other polyene species, or from mixtures of two of these ring compounds. Thus complexes are known containing the 2-electron donor $C_3H_3{}^+$ (cyclopropenium ion), the 4-electron donor C_4H_4 (cyclobutadiene, stable only in complexes), the 6-electron donor $C_7H_7{}^+$ (tropylium ion),[15] and the 10-electron donor $C_8H_8{}^{2-}$ (cyclooctatetraenide ion) (Table 11.4).

Table 11.4
Electron Counting for Polyene and Cyclopolyene Ligands

Ligands as Lewis Bases	Electron Count	Ligands as Neutral Species	Electron Count
	A. Cyclopolyenes with Full Hapticity		
η^3-$C_3H_3{}^+$ (Cyclopropenium ion)	2	η^3-$C_3H_3\cdot$ (Cyclopropenyl radical)	3
η^4-C_4H_4 (Cyclobutadiene)	4	η^4-C_4H_4 (Cyclobutadiene)	4
η^5-$C_5H_5{}^-$ (Cyclopentadienide ion)	6	η^5-$C_5H_5\cdot$ (Cyclopentadienyl radical)	5
η^6-C_6H_6 (Benzene)	6	η^6-C_6H_6 (Benzene)	6
η^7-$C_7H_7{}^+$ (Tropylium ion)	6	η^7-$C_7H_7\cdot$ (Cycloheptatrienyl radical)	7
η^8-$C_8H_8{}^{2-}$ (Cyclooctatetraenide ion)	10	η^8-C_8H_8 (Cyclooctatetraene)	8
	B. Polyenes		
η^2-C_2H_4 (Ethylene)	2	η^2-C_2H_4 (Ethylene)	2
η^4-C_4H_6 (Butadiene)	4	η^4-C_4H_6 (Butadiene)	4
η^5-$C_5H_7{}^-$ (Pentadienide anion)	6	η^5-C_5H_7 (Pentadienyl radical)	5
η^3-$C_3H_5{}^-$ (Allyl anion)	4	η^3-C_3H_5 (Allyl radical)	3
η^1-$C_3H_5{}^-$ (Allyl anion)	2	η^1-C_3H_5 (Allyl radical)	1
	C. Cyclopolyenes with Reduced Hapticity		
η^1-$C_5H_5{}^-$	2	η^1-C_5H_5	1
η^2-C_6H_6	2	η^2-C_6H_6	2
η^2-C_8H_8	2	η^2-C_8H_8	2
η^4-C_8H_8	4	η^4-C_8H_8	4
η^6-C_8H_8	6	η^6-C_8H_8	6

Collectively, such complexes are known as **metallocenes** (and metallicenium ions). Although the largest polyene, $C_8H_8^{2-}$, is too large to form a good sandwich with most *d*-block metal ions, it does form the most important *f*-block organometallic compounds,[16] such as $U(C_8H_8)_2$, commonly called *uranocene*.

The bis(cyclopentadienyl) metallocenes are usually made by **Lewis acid–base (LAB) reactions** of metal halides in donor solvents with sodium or thallium(I) cyclopentadienide:

$$FeCl_2 + 2\,Tl^+Cp^- \rightarrow 2\,TlCl(s) + FeCp_2 \tag{11.10}$$

Kealy and Pauson's original synthesis of ferrocene was a variation on this [in which the reducing power of $C_5H_5^-$ also came into play, since Fe^{3+} ions were reduced to iron(II) in the product], as is the alternate synthesis from cyclopentadiene, an amine, and a metal halide:

$$MCl_2 + 2\,C_5H_6 + 2\,Et_2NH \rightarrow MCp_2 + 2\,[Et_2NH_2]Cl \tag{11.11}$$

where $C_2H_5 = Et$.

However, Miller's original synthesis is not likely to have involved ions acting as Lewis acids and Lewis bases, but rather to proceed via the cyclopentadienyl free radical and iron atoms in a type of "redox-radical" reaction:

$$2\,C_5H_6 + Fe \xrightarrow{400\,°C} FeCp_2 + H_2 \tag{11.12}$$

The bis(benzene) sandwich compounds are made by Lewis acid–base reactions. Sandwich compounds such as $W(C_6H_6)_2$ of metals in the zero oxidation state are often best made from gaseous metal atoms and gaseous ligands using metal atom reactors (Section 7.3). Derivatives such as $[Cr(C_6H_6)_2]^+$ in which the metal is in a reduced oxidation state are made from ligand, metal halide, Al as reducing agent, and $AlCl_3$ as chloride ion acceptor [Eq. (11.2)]. If the metal is in the normal oxidation state, as in $[Ni(C_6Me_6)_2]^{2+}$, ligand, metal halide, and $AlCl_3$ may be used.[17]

Among the other polyenes, only the eight-membered ring has to date given sandwich compounds involving two rings of the same size; these are made by the Lewis acid–base reactions of $C_8H_8^{2-}$ (generated from neutral C_8H_8 and 2 mol K metal) and *f*-block metal halides:

$$UCl_4 + 2\,K_2(C_8H_8) \xrightarrow{THF,\,0\,°C} [U(C_8H_8)_2] + 4\,KCl \tag{11.13}$$

where tetrahydrofuran = THF (solvent).

Since uranium is not a *d*-block metal, the 18-electron rule does not apply; the donor- and π-acceptor orbitals of C_8H_8 interact both with valence $(n-2)f$ orbitals of the metal and with postvalence $(n-1)d$ orbitals. These and the related zerovalent *f-block* derivatives of sterically hindered benzenes, $[M(C_6H_3(t\text{-Bu})_3)_2]$, are among the few *f*-block metal compounds with a pronounced degree of covalent bonding.[18]

There are two ways in which the VECs in metallocenes can be computed to check compliance with the 18-electron rule; these two approaches are based on the two methods of synthesis of these compounds, and can be called the **Lewis acid–base** (LAB) and the **redox-radical** (RR) approaches. We illustrate each by checking the electron count in our two prototypic metallocenes, $FeCp_2$ and $Cr(C_6H_6)_2$.

In the LAB approach, the polyene ligands are identified in their Lewis base forms: the aromatic species donating 2, 6, or 10 π electrons (Table 11.4, first columns).[19] Both C_6H_6 and Cp^- can be counted as 6-electron donors. The neutral Cr atom has 6 electrons, so bis(benzene)chromium obeys the 18-electron rule. Since Cp^- is being counted as an anion, to conserve the electroneutrality of $FeCp_2$ the iron must be counted as the cation Fe^{2+} with 6 valence electrons. Thus ferrocene is also an 18-electron species. This characterization of the bonding in metallocenes as a LAB interaction between two stable ionic species has the advantages that it gives the correct number of *d* electrons to locate in the boxed (middle) region of a MO diagram (or in a crystal field splitting diagram), and it gives the oxidation number normally calculated for the metal ion.

The alternative RR counting system is also advocated by many; it is easier to use at first. In this system, each ligand is treated as a *neutral species* donating 1 electron per coordinated polyene carbon atom: η^n-C_nH_n donates *n* electrons. The metal, for purposes of calculation, is assumed to bear whatever charge is calculated from the formula of the species (0 for neutral metallocenes, +1 for +1 charged metallicenium ions, and so on) and to have the number of *d* electrons appropriate to its assumed charge. In this approach, the electron count in $FeCp_2$ is 2×5 electrons (from two η^5-Cp radicals) + 8 electrons (from an Fe atom) = 18 electrons.

Either system works equivalently in computing the total number of valence electrons on the organometallic species, but we suggest that you avoid confusion by beginning with one system or the other, and staying with it for awhile before diversifying. We illustrate both systems in Example 11.2.

Example 11.2

Calculate the metal VEC in each of the following metallocenes or metallocenium ions: (a) $NiCp_2$; (b) $Ni(C_4H_4)_2$; (c) $TiCp(C_7H_7)$; (d) $U(C_8H_8)_2$; (e) $[CoCp(C_3H_3)]^+$.

SOLUTION (LAB METHOD):

1. Each ligand is recognized as the corresponding aromatic species with the indicated charge and donating the cited number of electrons: (a) two −1 ions each donating 6 electrons; (b) two neutral molecules each donating 4 electrons; (c) a +1 ion and a −1 ion each donating 6 electrons; (d) two −2 ions each donating 10 electrons; (e) a +1 ion donating 2 electrons and a −1 ion donating 6 electrons.

2. The charge on the metal ion is then deduced, and its number of *d* (or *f*) electrons is noted: (a) Ni^{2+}, d^8; (b) Ni^0, d^{10}; (c) Ti^0, d^4; (d) U^{4+}, f^2; and (e) including the metallicenium ion charge, Co^+, d^8.

3. The ligand and metal-ion valence electrons are then totaled to give the VEC: (a) 20; (b) 18; (c) 16; (d) 22; and (e) 16 electrons.

SOLUTION (RR METHOD):

1. Each ligand is assumed to donate or share the same number of electrons as it has ring carbon atoms: (a) two neutral radicals donating 5 electrons each; (b) two neutral molecules donating 4 electrons each; (c) one 7-electron and one 5-electron donor; (d) two 8-electron donors; (e) one 3-electron radical and one 5-electron radical.

2. The charge on the metal ion is then taken as the charge on the species (0, except +1 in the last case), and its contribution of valence electrons is noted: (a) Ni^0, d^{10}; (b) Ni^0, d^{10}; (c) Ti^0, d^4; (d) U^0, f^6; (e) Co^+, d^8.

3. The ligand and metal-ion valence electrons are then totaled to give the VEC: (a) 20; (b) 18; (c) 16; (d) 22; and (e) 16 electrons.

Incidentally, the counting of coordination numbers and the description of geometries becomes more complex with π-donor ligands, since there is no equivalence between the number of donor atoms present and the number of electron pairs being donated. In bis(benzene)chromium there are 12 carbon atoms equidistant from the chromium atom, but one is reluctant to assign a coordination number of 12 to the chromium atom, since this grossly violates the expected maximum coordination number of a fourth-period atom. It is perhaps more usual in such a case to take the contribution to the coordination number from π-donor ligands as equal to the number of π-electron pairs donated. For both $Cr(C_6H_6)_2$ and $FeCp_2$, this number is two times three pairs from each ring, for a reasonable coordination number of 6. For uranocene, one then calculates a coordination number of 10, which is plausible in the seventh period. Thus C_5H_5 and C_8H_8 can be considered as somewhat similar to chelating tridentate and pentadentate ligands, respectively, but because their donor atoms are directly attached to each other, the cone angles of these ligands are much smaller than would be typical for tridentate or pentadentate ligands.

By far, the largest number of sandwich compounds are formed by the Cp^- anion. It forms sandwich molecules not only with Fe, Rh, and Os, but also with the *d*-block metals V, Cr, Mn, Co, and Ni; metallocenium cations are formed with Fe, Ru, V, Co, Ru, Ir, Cr, Mn, and Ni (Table 11.5). Remarkably, the fourth-period neutral metallocenes (e.g., vanadocene, chromocene, manganocene, ferrocene, cobaltocene, and nickelocene) all have melting points very close to 173 °C: the slightly differing metal atoms are evidently so buried within the sandwiches that they have no effect on the intermolecular forces that determine physical properties such as melting points (Chapter 12).

As a consequence of the presence of closely spaced (including nonbonding) MOs, non-18-electron metallocenes can exist and show a variety of magnetic properties. It is clear from Table 11.5 that manganocene has an anomalously large number of unpaired electrons; this anomaly disappears on substituting the C_5 ring with Me groups. Evidently, the small electronic change is sufficient to cause Δ_o to exceed P, the pairing energy (Section 8.3), in $Mn(C_5Me_5)_2$ and $Mn(C_5H_4Me)_2$, which are low-spin complexes (Chapter 8), but not in $MnCp_2$, which is a high-spin complex. (The C_5Me_5 ligand has assumed quite an importance of its own: This ligand confers enhanced hydrocarbon solubility on metallocenes and their derivatives, and its enhanced cone angle provides steric protection against some kinds of reactions. In many chemical formulas, the C_5Me_5 ligand is abbreviated as Cp^*.)

The non-18-electron metallocenes are much more reactive than ferrocene, however: Thus the 19-electron species cobaltocene is about as good a reducing agent as sodium metal, and is oxidized readily to the stable 18-electron *cobaltocenium ion*, $(CoCp_2)^+$; the 20-electron nickelocene is readily oxidized to the 19-electron $NiCp_2{}^+$ ($E^\circ = +0.18$ V), and even to the 18-electron dication $NiCp_2{}^{2+}$ ($E^\circ = +1.11$ V). This formally contains nickel in the unusually high +4 oxidation state; in contrast to CO, Cp^- as a ligand is compatible with relatively high metal oxidation states.

Table 11.5
Simple Metallocenes and Metallocenium Ions[a]

Neutral Metallocene Temperature (°c) Color	VCp_2 167 °C Purple 3 unpaired electrons M—C 228	$CrCp_2$ 172 °C Scarlet 2 unpaired electrons M—C 217	$Mn(C_5Me_5)_2$ Orange 1 unpaired electron M—C 211	$FeCp_2$ 173 °C Orange 0 unpaired electrons M—C 206	$CoCp_2$ 173 °C Purple 1 unpaired electron M—C 212	$NiCp_2$ 173 °C Green 2 unpaired electrons M—C 220
Neutral Metallocene Temperature (°c) Color			$MnCp_2$ 172 °C Pink 5 unpaired electrons M—C 238	$RuCp_2$ 199 °C Yellow 0 unpaired electrons		
Neutral Metallocene Temperature (°c) Color			$Re(C_5Me_5)_2$ Purple 1 unpaired electron	$OsCp_2$ 230 °C Colorless 0 unpaired electrons		
Metallocenium Ions Color		$[CrCp_2]^+$ Black 3 unpaired electrons	$[Re(C_5Me_5)_2]^-$ Orange 0 unpaired electrons	$[FeCp_2]^+$ Blue 1 unpaired electron	$[CoCp_2]^+$ Yellow 0 unpaired electrons	$[NiCp_2]^+$ Yellow 1 unpaired electron
Metallocenium Ions Color		$[CrCp_2]^-$		$[RuCp_2]^+$ Yellow 1 unpaired electron	$[RhCp_2]^+$ Colorless 0 unpaired electrons	$[NiCp_2]^+$ Yellow 0 unpaired electrons
Metallocenium Ions Color					$[IrCp_2]^+$ Yellow 0 unpaired electrons	$[Pd(C_5Me_5)_2]^{2+}$
Metallocenium Ions						$[Pt(C_5Me_5)_2]^{2+}$

[a]Entries in table include melting point (°C), color, numbers of unpaired electrons, and M–C bond distances (pm).

Polyene Carbonyl Derivatives. Mixed organometallic compounds containing both carbonyl ligands and π-donor ligands are readily available, such as the important $[Cr(C_6H_6)(CO)_3]$ (Fig. 11.2) and $[MnCp(CO)_3]$; both are 18-electron species (containing Cr in oxidation state 0 and Mn in oxidation state +1). The closely related

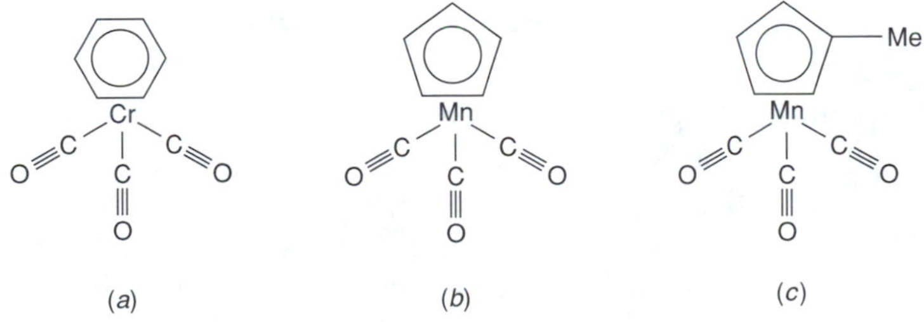

Figure 11.2
Important organometallic derivatives of *cyclo*polyenes and CO. (*a*) Benzenechromium tricarbonyl, $[Cr(C_6H_6)(CO)_3]$; (*b*) cyclopentadienylmanganese tricarbonyl, $[MnCp(CO)_3]$; and (*c*) methylcyclopentadienylmanganese tricarbonyl, $[Mn(C_5H_4Me)(CO)_3]$.

$[Mn(C_5H_4CH_3)(CO)_3]$ has been investigated as a possible replacement for organolead compounds in high-octane gasolines. Many of the less common polyenes, which do not form sandwich compounds by themselves, do form mixed compounds with the appropriate number of carbonyl ligands. Because of the resulting geometry, these are sometimes known as "piano-stool" molecules.

Example 11.3

Predict what fourth-period *d*-block metals and how many carbonyl groups would be present in the neutral piano-stool molecules formed by each of the polyenes in Table 11.4.

SOLUTION:

By using the LAB approach, $C_3H_3^+$ donates 2 electrons. Therefore in order to make 18 electrons, $M^- + n$ carbonyls must donate 16 electrons; so (Group No. of M + 1) $+2n = 16$. By using the RR approach, C_3H_3 donates 3 electrons. Therefore $M + n$ carbonyls must donate 15 electrons: Group No. of $M + 2n = 15$, which is equivalent to the equation from the other method. Only odd-numbered groups give whole number values of n: for Co, $n = 3$, so one solution is $[Co(C_3H_3)(CO)_3]$; for Mn, $n = 4$, therefore another solution is $[Mn(C_3H_3)(CO)_4]$; for V another solution is $[V(C_3H_3)(CO)_5]$; no further solution is likely since the maximum coordination number of 6 would thereby be exceeded. A phenyl-substituted version of the cobalt compound is known, $[Co(C_3Ph_3)(CO)_3]$. We leave the solutions for the higher ring sizes to the student as an exercise.

A variety of methods is used to prepare these compounds, such as (1) by reaction of a reactive metallocene with a metal carbonyl or with CO; and (2) by reaction of a metal carbonyl with the polyene, or with its parent hydrocarbon with elimination of hydrogen:

$$Cr(CO)_6 + C_7H_8 \xrightarrow{\text{reflux 16 hr}} 3\,CO + Cr(\eta^6\text{-}C_7H_8)(CO)_3$$

$$Cr(\eta^6\text{-}C_7H_8)(CO)_3 + [Ph_3C][BF_4] \rightarrow Ph_3CH + [Cr(\eta^7\text{-}C_7H_7)(CO)_3][BF_4] \qquad (11.14)$$

or (3) by reductive ligation of a metal halide in the presence of the polyene and CO. In the case of the C_4H_4 complexes, the parent ligand does not exist, so it must be both synthesized and captured as part of the same reaction:

$$cyclo\text{-}C_4H_4Cl_2 + Mo(CO)_6 + 2\,Na \rightarrow Mo(C_4H_4)(CO)_4 + 2\,CO + 2\,NaCl \qquad (11.15)$$

These compounds are properly named by the nomenclature given in Section 8.11, but an alternate nomenclature is also common, in which the name of the metal is sandwiched in the middle, after the polyene but before the carbonyls: Thus $[Cr(C_6H_6)(CO)_3]$ is commonly called benzenechromium tricarbonyl (contracted informally to "benchrotrene"); $[MnCp(CO)_3]$ is often called cyclopentadienylmanganese tricarbonyl or "cymantrene", and $[Mn(C_5H_4CH_3)(CO_3)]$, of interest as a possible replacement for tetraethyllead as an antiknock ingredient in gasoline, is called methylcyclopentadienylmanganese tricarbonyl or MMC.

The first stable *f*-block derivative of this type, $[U(C_5Me_4H)_3CO]$, has a quite short U–C bond distance as well as a low C–O stretching frequency of $1900\,cm^{-1}$. Both of these factors suggest that the CO is acting as a π-acceptor ligand, giving a U–C bond with some of the multiple-bond character found in most *d*-block carbonyls.[20]

Effects of Complexation on the Reactivity of Coordinated Organic Ligands. Previously, we noted (Sections 2.2 and 5.10) that complexation increases the acidity of a coordinated ligand; examples of this effect will be seen in Section 11.7. For now, we note that coordinated polyenes tend to act as aromatic systems: even those such as C_4H_4 that are antiaromatically destabilized as free molecules. For example, metallocenes and polyene carbonyls undergo two typical reactions of aromatic organic compounds, Friedel–Crafts acylation [Eq. (11.16)] and Mannich condensation with formaldehyde and an amine [Eq. (11.17)]:

$$FeCp_2 + (MeCO)_2O \xrightarrow{H_3PO_4} CpFe(C_5H_4COMe) + MeCO_2H \qquad (11.16)$$

$$(C_4H_4)Fe(CO)_3 + CH_2O + Me_2NH \xrightarrow{MeCO_2H} (C_4H_3\text{–}CH_2NMe_2)Fe(CO)_3 + H_2O \qquad (11.17)$$

Often, advantage can be taken of the electron-withdrawing (Lewis acidic) effect of the coordinated metal atom, which is greater in a polyene carbonyl than in a pure metallocene. Nucleophilic aromatic substitution on unactivated benzene derivatives is a notoriously difficult reaction that is inhibited by repulsion between π electrons in the aromatic ring and the electron pair of the entering nucleophile. However, in an arene-chromium tricarbonyl, the π electrons of the aromatic ring have been donated to a good extent to the metal atom, so that the nucleophile is free to approach the ring and carry out a substitution:

$$(PhCl)Cr(CO)_3 + {}^{-}{:}OMe \rightarrow (PhOMe)Cr(CO)_3 + Cl^- \qquad (11.18)$$

The metal atom is often in a low oxidation state in these compounds, and so can readily be oxidized; if it is oxidized sufficiently it loses its back-bonding ability, and releases the ligand as a free organic molecule. Thus such reactions become powerful tools in the hands of the synthetic organic chemist. Even cyclobutadiene can be released to enjoy a very fleeting existence:

$$(C_4H_4)Fe(CO)_3 + 3\,Ce^{4+} \rightarrow C_4H_4 + Fe^{3+} + 3\,Ce^{3+} + 3\,CO \qquad (11.19)$$

Cobaltocene and nickelocene, which are 19- and 20-electrons species having 1 and 2 unpaired electrons, respectively, can transfer some of that free radical reactivity to the organic ligands, allowing new organic syntheses:

$$Cp_2Ni + CCl_4 + PPh_3 \rightarrow CpNiCl(PPh_3) + Cp\text{–}CCl_3 \qquad (11.20)$$

The presence of the metal atom does impart some limitations on the reactivity of the organic ligand, however, since the metal atom may be a competing source of electrons. For example, attempts to carry out the Friedel–Crafts chlorination of ferrocene to give chloroferrocenes fail because the chlorine attacks the iron instead of the Cp

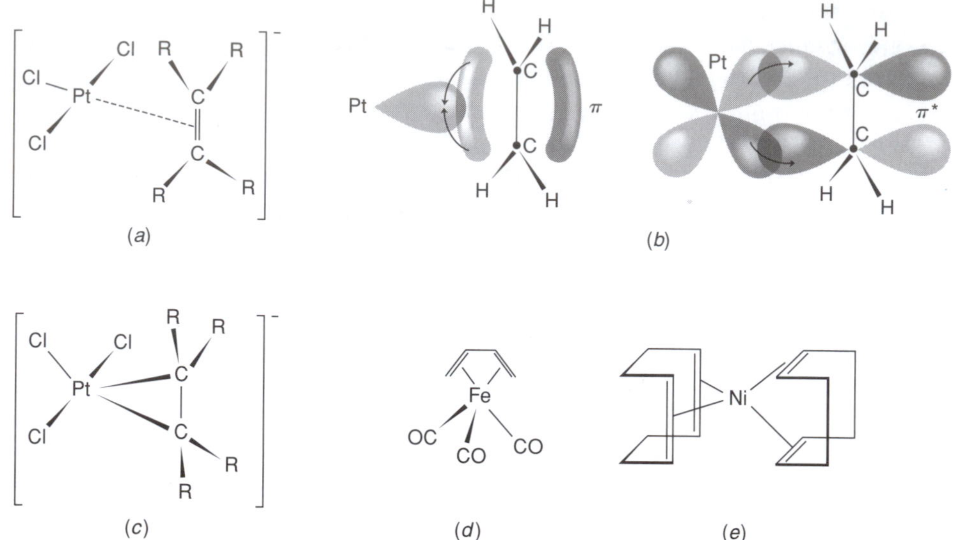

Figure 11.3
Alkene complexes. (*a*) The anion of Zeise's salt, $K[PtCl_3(CH_2{=}CH_2)$, emphasizing its role as a donor; (*b*) (left) electron donation from the π HOMO of an alkene to an empty orbital on Pt; (right) back donation from an occupied Pt *d* orbital to the π^* LUMO of the alkene; (*c*) an analogue of Zeise's salt with strong back-donation, producing a metallacyclopropane structure; (*d*) (butadiene)iron tricarbonyl; (*e*) bis(1,5-cyclooctadiene)nickel(0). [Adapted from D. F. Shriver, P. Atkins, and C. H. Langford, *Inorganic Chemistry*, 2nd ed., Freeman, New York, 1994; pp. 685–687.]

ring, producing ferricenium ion:

$$2\,Cp_2Fe + Cl_2 + 2\,AlCl_3 \rightarrow 2\,[Cp_2Fe][AlCl_4] \tag{11.21}$$

Complexes of Alkenes and Cyclic Polyenes Exhibiting Reduced Hapticity. Organic π systems need not be cyclic to coordinate to metals; linear alkenes, or cyclic alkenes not using all unsaturated carbon atoms, also are important ligands. The earliest organometallic compound of all, Zeise's salt, was prepared in 1827 by the reaction of $K_2[PtCl_4]$ with ethylene, C_2H_4:

$$K_2[PtCl_4] + CH_2{=}CH_2 \xrightarrow{SnCl_2} K[PtCl_3(CH_2{=}CH_2)] + KCl \tag{11.22}$$

As shown in Figure 11.3, this simple organometallic compound also involves donation from the π HOMO of $CH_2{=}CH_2$ to a vacant orbital of Pt, and back-donation from the Pt d_{xy} orbital to the π^* LUMO of $CH_2{=}CH_2$, so ethylene is also a π-acceptor ligand. Ethylenes substituted with electron-withdrawing substituents, such as tetracyanoethylene, $(NC)_2C{=}C(CN)_2$, are strongly π accepting, to the point that the bond order of the C=C bond is reduced toward 1, lengthening it significantly, while the bonds from each C to Pt are strengthened and bond angles are changed, producing a saturated three-membered *metallacyclopropane* ring [Fig. 11.3(*c*)].

Dienes and (linear or cyclic, nonconjugated) polyenes such as butadiene and 1,5-cyclooctadiene form similar but more stable π complexes [Fig. 11.3(*d*)], since these are stabilized by the chelate effect.

Odd-membered linear π chains also form similar π complexes; the most common of these are derived from the C_3H_5 (*allyl*) ligand, although the η^5-C_5H_7 (pentadienyl) chain has also been extensively investigated for its ability to form acyclic analogues of metallocenes. With the odd-membered chains, electron counting differs in the LAB and the RR systems (Table 11.4B). Complexes of the odd-membered chains are usually made from *s*-block metal derivatives of their anions:

$$NiBr_2 + 2\,C_3H_5MgCl \xrightarrow{Et_2O,\,-10\,^\circ C} Ni(\eta^3\text{-}C_3H_5)_2 \qquad (11.23)$$

although they can also be made from allyl chloride as the Lewis acid with a metal carbonylate anion as the Lewis base:

$$C_3H_5Cl + Mn(CO)_5^- \xrightarrow{18\,hr} (\eta^1\text{-}C_3H_5)Mn(CO)_5 \xrightarrow{86\,^\circ C} (\eta^3\text{-}C_3H_5)Mn(CO)_4 \qquad (11.24)$$

As shown in Eq. (11.24), the allyl ligand readily changes hapticity as well, becoming chelating in the η^3 form. Changes in hapticity add to the types of reactivity available to allyl and other polyene complexes; these may be either accompanied by the entry or loss of an additional ligand (as above) or may result in a change in the electron count of the compound.

Lower hapticity is also a possibility for the conjugated cyclic polyenes as well (Table 11.4C), although it is not as common as full hapticity since both chelate and aromatic stabilization may be lost in the process. For example, the compound $Cp_2Fe(CO)_2$ is known; this formula would appear to be that of a 22-electron molecule. But its structure shows that one ring is bonded flat over the iron with all five carbon atoms essentially equidistant from the iron, while the other ring is attached to the iron at only one carbon atom (two double bonds of the second ring are far from the iron atom, and are not bonding to it.) The two rings have quite different chemical and spectroscopic properties: One is *pentahapto* and one is a fluxional (Section 9.3) *monohapto* ring. The compound is best formulated as $[Fe(\eta^5\text{-}Cp)(\eta^1\text{-}Cp)(CO)_2]$. Reduced hapticity is the rule rather than the exception for the C_8H_8 ring, which is too large and often donates too many electrons for *d*-block complexes. It is frequently found as a nonplanar *tetrahapto* chelating ligand (Fig. 11.4(*a*)). This ligand is also fluxional, as

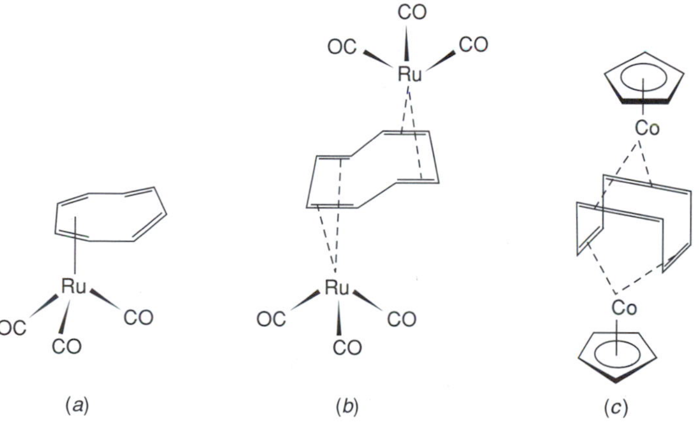

Figure 11.4
Different modes of bonding of cyclooctatetraene. (*a*) With reduced hapticity to one metal, in $(\eta^4\text{-}C_8H_8)Ru(CO)_3$; (*b*) With full hapticity achieved by bridging two metals while in its chair conformation, in $(CO)_3Ru(\mu^2\text{-}\eta^4\text{:}\eta^4\text{-}C_8H_8)Ru(CO)_3$; (*c*) With full hapticity achieved by bridging two metals while in its boat conformation, in $CpCo(\mu^2\text{-}\eta^4\text{:}\eta^4\text{-}C_8H_8)CoCp$. [Adapted from D. F. Shriver, P. Atkins, and C. H. Langford, *Inorganic Chemistry*, 2nd ed., Freeman, New York, 1994; p. 694.]

the ring rotates the four carbon atoms involved in bonding to the metal. Or the ligand can simultaneously use the other four π electrons to bond a second metal atom from either its boat or its chair conformation [Fig. 11.4(*b* and *c*)]; it then functions as both a chelating and a bridging ligand.

Reduced hapticity is also often seen in the tris- and tetrakis(cyclopentadienyls) of the early *d* and the *f* blocks. Among the tetrakis(cyclopentadienyls), those of Ti and Hf are found to be $M(\eta^5\text{-Cp})_2(\eta^1\text{-Cp})_2$, while that of the Hf atom's identical twin (Zr) is $M(\eta^5\text{-Cp})_3(\eta^1\text{-Cp})_1$, and those of Np, U, and Th are $M(\eta^5\text{-Cp})_4$. The coordination numbers one calculates for these formulas are unusually high, but all of these atoms are large for their periods. Undoubtedly, the cone angle of a η^5-Cp ligand would be substantially smaller than the combined angles of three normal monodentate ligands. In addition, the sizes of the central atoms do increase approximately in the order given, allowing for yet larger coordination numbers; for Np, U, and Th there is also no need to be constrained by the 18-electron rule.

Reduced hapticity is also generally found in the bis(cyclopentadienyls) at the end of the *d* block, which would greatly exceed 18 valence electrons if they were to adopt sandwich structures. As examples, we may cite $Hg(\eta^1\text{-Cp})_2$, $[Zn(\eta^5\text{-C}_5Me_5)\cdot(\eta^1\text{-C}_5Me_5)]$, and $[Cu(\eta^2\text{-Cp})_2]^-$.[21]

11.4 Other Ligands Forming One σ Bond in Organometallics; Dinuclear and Polynuclear Organometallics

Many organometallics contain various other types of ligands in addition to carbonyl and its π-acceptor companions and benzene and its π-donor companions; Table 11.6 lists a number of these other ligands. With a couple of exceptions, these are not the only or even the main ligands found in organometallics; nonetheless, they can be very important functional groups in carbonyls or polyene complexes, for example. We will use the 16- and 18-electron rule for these compounds, of course, but some of them do introduce some complications, in some cases, for the use of the LAB approach and in one case for the use of the RR approach.

Simple Ligands Forming One σ Bond to the Metal. We begin, however, with a quite simple and straightforward, but important, group of ancillary ligands: The *halides*, in which, for example, one or a few carbonyl ligands are replaced by Cl, Br, or I (less often by F, since organometallic groups as cations are generally soft acids). The organometallic halides, like their organic analogues, are useful compounds in that the halogen is often then easily replaced by other functional groups in substitution reactions:

$$CpFe(CO)_2Cl + Na^+Cp^- \xrightarrow{\text{THF}} (\eta^5\text{-Cp})Fe(CO)_2(\eta^1\text{-Cp}) + NaCl \qquad (11.25)$$

Halogen (and pseudohalogen, e.g., cyano) substitution products are not only known corresponding to many of the carbonyls listed in Table 11.1, but also for some metals not forming simple carbonyls: Pd, Pt, Cu, Ag, and Au form carbonyl halides {e.g., $[Pt(CO)_2Cl_2]$; $[Cu(CO)Cl]$, formed when a solution of CuCl in concentrated HCl is exposed to gaseous CO; Zeise's salt}. Several of the catalytically important organometallic compounds that we will discuss shortly are 16-electron square planar carbonyl halides: Wilkinson's catalyst, $[RhCl(PPh_3)_3]$; Vaska's compound, $[Ir(CO)(PPh_3)_2Cl]$; and the catalyst in the Monsanto acetic acid process, $[Rh(CO)_2I_2]^-$.

Table 11.6
Electron Counting for Other Classes of Organometallic Ligands

Ligands as Lewis Bases	Electron Count	Ligands as Neutral Species	Electron Count
		A. Ligands Forming One σ Bond to the Metal	
$:Cl^-$, $:Br^-$, $:I^-$, $:C{\equiv}N^-$,	2	$\cdot Cl$, $\cdot Br$, $\cdot I$, $\cdot OH$, $\cdot C{\equiv}N$,	1
$:H^-$, $:CH_3^-$,		$\cdot H$, $\cdot CH_3$,	
$:Mn(CO)_5^-$, $:Co(CO)_4^-$		$\cdot Mn(CO)_5$, $\cdot Co(CO)_4$	
$:MoCp(CO)_3^-$, $:FeCp(CO)_2^-$		$\cdot MoCp(CO)_3$, $\cdot FeCp(CO)_2$,	
$:NiCpCO^-$		$\cdot NiCpCO$	
$:N{\equiv}O^+$ (nitrosyl)	2	$\dot{:}N{\equiv}O$ (nitrosyl)	3
η^2-H_2	2	η^2-H_2	2
		B. Ligands Forming Two or Three Bonds to Two or Three Metals	
μ_2-$:\dot{C}{=}O^{2-}$	2 + 2	μ_2-$\cdot\dot{C}{=}O$(bridging carbonyl)	1 + 1
μ_3-$:\underset{\cdot\cdot}{C}{=}O^{4-}$	2 + 2 + 2	μ_3-$\cdot\dot{C}{=}O$ (apical carbonyl)	1 + 1 + 0
μ_2-$RC{\equiv}CR$ (alkynes)	2 + 2	μ_2-$RC{\equiv}CR$ (alkynes)	2 + 2
		C. Ligands Forming Two Bonds to the Metal (or Bridging Two Metals)	
$:\ddot{O}^{2-}$, $:\ddot{S}^{2-}$; $:\ddot{S}e^{2-}$,	4	$\cdot\dot{O}$, $\cdot\dot{S}$; $\cdot\dot{S}e$,	2
$:\ddot{N}R^{2-}$; $:\ddot{C}R_2^{2-}$,		$\cdot\dot{M}nCp(CO)_2$, $\cdot\dot{C}oCp(CO)$	
$:\ddot{W}(CO)_5^{2-}$, $:\ddot{F}e(CO)_4^{2-}$, $:\ddot{N}i(CO)_3^{2-}$		$\cdot\dot{N}R$; $\cdot\dot{C}R_2$ (carbene, alkylidene),	
$:\ddot{M}nCp(CO)_2^{2-}$, $:\ddot{C}oCp(CO)^{2-}$		$\cdot\dot{W}(CO)_5$, $\cdot\dot{F}e(CO)_4$, $\cdot\dot{N}i(CO)_3$	
		D. Ligands Forming Three Bonds to the Metal (or Bridging Three Metals)	
$:\underset{\cdot\cdot}{\ddot{N}}^{3-}$, $:\underset{\cdot\cdot}{\ddot{P}}^{3-}$, $:\underset{\cdot\cdot}{\ddot{C}}R^{3-}$,	6	$\cdot\underset{\cdot}{\dot{N}}$, $\cdot\underset{\cdot}{\dot{P}}$, $\cdot\underset{\cdot}{\dot{C}}R$ (alkylidyne),	3
$:\underset{\cdot\cdot}{\ddot{M}}n(CO)_4^{3-}$, $:\underset{\cdot\cdot}{\ddot{C}}o(CO)_3^{3-}$,		$\cdot\underset{\cdot}{\dot{M}}n(CO)_4$, $\cdot\underset{\cdot}{\dot{C}}o(CO)_3$,	
$:\underset{\cdot\cdot}{\ddot{M}}oCp(CO)_2^{3-}$, $:\underset{\cdot\cdot}{\ddot{F}}eCp(CO)^{3-}$,		$\cdot\underset{\cdot}{\dot{M}}oCp(CO)_2$, $\cdot\underset{\cdot}{\dot{F}}eCp(CO)$, $\cdot\underset{\cdot}{\dot{N}}iCp$	
$:\underset{\cdot\cdot}{\ddot{N}}iCp^{3-}$			

In the LAB approach, these ions contribute as many electrons as the CO but differ in that they contribute a unit of negative charge, as illustrated by the following synthesis:

$$Cr(CO)_6 + Et_4N^+Br^- \xrightarrow{120\,^\circ C} Et_4N^+[Cr(CO)_5Br]^- + CO \qquad (11.26)$$

The organometallic halides can also be generated by redox reactions:

$$Fe(CO)_5 + Br_2 \rightarrow Fe(CO)_4Br_2 + CO \qquad (11.27)$$

The RR approach reflects this by identifying the incoming species as halogen atoms, which are 1-electron donors.

Since organometallic compounds are often made from inorganic or organic halides, another source of halogens in the organometallic halides is the halogen present in the reactant:

$$[Ph_3C_3]^+Cl^- + Ni(CO)_4 \xrightarrow{\Delta,\,MeOH} [(C_3Ph_3)Ni(CO)_2Cl] + 2\,CO(g) \qquad (11.28)$$

$$TiCl_4 + 2\,Na^+Cp^- \longrightarrow Cp_2TiCl_2 + 2\,NaCl \qquad (11.29)$$

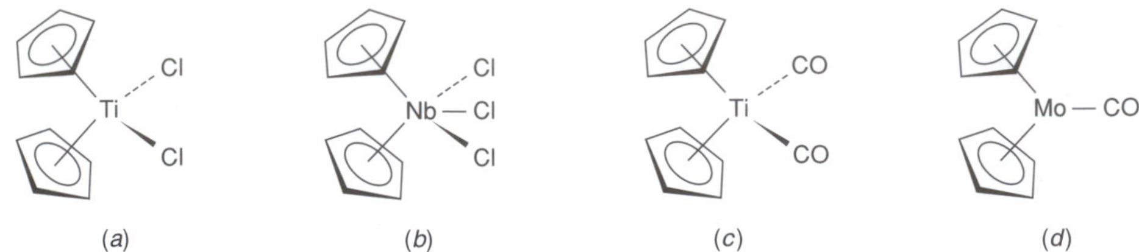

Figure 11.5
Some "bent sandwich" molecules. (*a*) Titanocene dichloride (Cp_2TiCl_2); (*b*) niobocene trichloride (Cp_2NbCl_3); (*c*) titanocene dicarbonyl [$Cp_2Ti(CO)_2$]; and (*d*) molybdocene carbonyl [$Cp_2Mo(CO)$].

This last product is often called "titanocene dichloride", and it is not only useful in preparing titanocene derivatives but is of interest in its own right, especially since it (and related metallocene dihalides) have been found to have anticancer properties.[22] Titanocene dichloride is an example of a "bent sandwich" molecule, since the two Cp rings are no longer parallel to each other, but are tilted to make room for the chlorines (Fig. 11.5).

The *alkyl* (R, such as CH_3) and *aryl* (Ar, such as C_6H_5) derivatives of the *d*-block organometallics are similar in that they are treated as 2-electron donors in the LAB approach and 1-electron donors in the RR approach, although in practice they would normally be available only by reactions that seem[23] to be LAB reactions:

$$Mn(CO)_5Cl + MeMgBr \rightarrow MeMn(CO)_5 + MgClBr \quad (11.30)$$

Alkyl and aryl derivatives are less common and less stable than might be expected, however, since some of the common reaction types (Section 11.7) of *d*-block organometallics transform alkyl and aryl groups into other chemical groups. Nonetheless, there are some homoleptic alkyls and aryls for resistant types of organic groups (e.g., WMe_6). Most discussion of these compounds is postponed until Chapter 15, where their *p*-block analogues are discussed, since (a) these fully alkylated or arylated compounds are much less common in the *d* block than in the *p* block, and (b) since simple alkyl and aryl ligands are σ-donor ligands only with no π-acceptor properties, the homoleptic alkyls and aryls show no tendency to obey 16- and 18-electron rules. Historically, organometallic chemists began looking for σ-bonded homoleptic alkyls in the first half of the twentieth century, and were rewarded for their efforts when looking in the *p* block but were endlessly frustrated in the *d* block; they had just about given up when the discovery of ferrocene was announced.

The elements below carbon also give ligands analogous to the simple alkyls and aryls: One of the earliest such functional groups known was the *trichlorostannyl* group ($SnCl_3^-$ in the LAB approach; $\cdot SnCl_3$ in the RR approach). Such derivatives are sometimes called "inorganometallic compounds".[24]

More Complex Ligands Forming One σ Bond to a Metal: Nitrosyl and Hydride. These two ligands are somewhat difficult to fit comfortably into one or the other of the electron-counting schemes. Nitric oxide (of new importance for its physiological functions, Chapter 8) is a powerful π-acceptor ligand that is related to CO [as in Eq. (11.9)], but it has one more valence electron, which must be accounted for somehow. In the LAB scheme, this electron is ionized off, giving the triple-bonded *nitrosonium ion*, $:N\equiv O:^+$, which is isoelectronic to CO and also is counted as a 2-electron donor. The only complication is that it introduces a *positive* charge into a complex ion. One homoleptic nitrosyl is known, $Cr(NO)_4$, a brown-black solid that is isoelectronic to the considerably less stable carbonyl tetraanion, $[Cr(CO)_4]^{4-}$, and tallies as an 18-electron species (8 electrons from four NO^+ plus 10 electrons from Cr^{4-}). In the RR scheme, things are more counterintuitive: For correct bookkeeping, neutral NO must donate *three* electrons (the one ionized in the other scheme, plus the σ-donated pair) to the metal. From this point of view, $Cr(NO)_4$ also tallies as having 18 valence electrons, 6 from Cr^0 and 12 from the four 3-electron donating neutral NO groups; it is then usually compared to $Cr(CO)_6$, which also has 6 electrons from Cr^0 and 12 electrons from its ligands.[25]

Example 11.4

Predict the formulas of the neutral fourth period: (a) cyclopentadienyl metal carbonyl monochlorides and (b) four-coordinate metal carbonyl nitrosyls.

SOLUTION (LAB METHOD):

(a) Here Cp^- donates 6 electrons and Cl^- donates 2 electrons, which leaves 10 electrons to be supplied by the M^{2+} ion and the carbonyls. Since each CO supplies 2 electrons, M must be from the even-numbered groups of the periodic table. Possible solutions are $CpTi(CO)_4Cl$, $CpCr(CO)_3Cl$, $CpFe(CO)_2Cl$, and $CpNi(CO)Cl$, but the first two exceed the likely maximum coordination number of 6. (b) The four ligands supply 8 electrons; 10 electrons must come from the metal, which must also have one negative charge per nitrosyl ligand, so the metal must be Co^- in the mononitrosyl, Fe^{2-} in the dinitrosyl, and so on: $Co(CO)_3(NO)$, $Fe(CO)_2(NO)_2$, and $Mn(CO)(NO)_3$.

SOLUTION (RR METHOD):

(a) Here Cp donates 5 electrons and Cl donates 1 electron, which leaves 12 electrons to be supplied by M and the carbonyls. Since each CO supplies 2 electrons, M must be from the even-numbered groups of the periodic table. Possible solutions are $CpTi(CO)_4Cl$, $CpCr(CO)_3Cl$, $CpFe(CO)_2Cl$, and $CpNi(CO)Cl$, but the first two exceed the likely maximum coordination number of 6. (b) The four ligands supply $8 + n$ electrons, where n is the number of nitrosyls; $(10 - n)$ electrons must come from the metal, so the metal must be Co in the mononitrosyl, Fe in the dinitrosyl, and so on: $Co(CO)_3(NO)$, $Fe(CO)_2(NO)_2$, and $Mn(CO)(NO)_3$.

An interesting property of the organometallic anions (such as the carbonylate and polyene carbonylate anions) and even neutral organometallics (such as the sandwich compounds) is that they act as Lewis bases, using the *d*-block metal atoms as *donor* atoms to form hydrogen–metal bonded species upon reactions with strong acids. The site of basicity may be the normal HOMO, one of the *d* electron pairs on the metal atom:

$$(\eta^5\text{-Cp})_2Fe + H^+ \rightarrow (\eta^5\text{-Cp})_2FeH^+ \tag{11.31}$$

Protons directly attached to metals have very unusual proton nuclear magnetic resonance (^{1}H NMR) shifts: They can be as far upfield as $\delta - 50$ pm, way out of the range of organic protons (0 ppm $< \delta <$ 15 ppm); this seems to suggest a very high electron density on the protons, which tends to justify classifying them as organometallic *hydrides*, even though they are more often formed from reaction with H^+ than with H^-, and can also be prepared by oxidation–reduction reactions involving H_2 as the hydrogen source. Since the electronegativities of hydrogen and of some *d*-block metals are very close, assignment of oxidation numbers to the metal and to hydrogen in these compounds is quite arbitrary: For purposes of standard nomenclature and the assignment of oxidation numbers, hydrogen in *d*-block organometallic compounds should be assigned the oxidation number -1 and called "hydrido".

Treating the hydrogen as hydrides is consistent with the assumptions of the LAB counting scheme and their NMR spectra, but not with their reactivity in water. The parent organometallic anions are clearly basic, showing the same range of basicity as monoatomic or oxo anions: in water octahedral $[V(CO)_6]^-$ and tetrahedral $[Co(CO)_4]^-$ are nonbasic, while trigonal bipyramidal $[Mn(CO)_5]^-$ and $[HFe(CO)_4]^-$ are weakly basic; the dianion $[Fe(CO)_4]^{2-}$ is strongly basic, and is hydrolyzed by water:

$$[Fe(CO)_4]^{2-} + H_2O \rightarrow [HFe(CO)_4]^- + OH^- \qquad (11.32)$$

The corresponding hydrides, $HV(CO)_6$ and $HCo(CO)_4$, are strong acids, while $HMn(CO)_5$ (pK_a = 7.1 in water) and the diprotic $H_2Fe(CO)_4$ (pK_{a1} = 6.8, pK_{a2} = 15) are weakly acidic; $HRe(CO)_5$ and $H_2Os(CO)_4$ are very weakly or feebly acidic (pK_1 of the latter = 12.8).[26] It is interesting to note that, except for the highly charged $[Fe(CO)_4]^{2-}$, the carbonyl acids and bases with even coordination numbers (4 or 6) are more stable (less acidic or basic) than those with odd coordination numbers (5 or 7).

Sometimes, the site of protonation is on the organic ring, as in the case of nickelocene, where ring protonation reduces the hapticity of one ring and brings the product into compliance with the 18-electron rule:

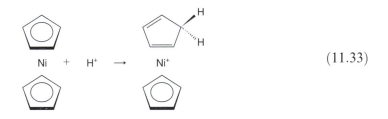

Ni + H⁺ → Ni⁺ (11.33)

Since *d*-block metal atoms act as very soft bases (their electronegativities are very low as compared to other *donor* atoms), they react even more readily with soft acids than with H^+ to form complexes with metal ions such as Hg^{2+} that have mercury–metal bonds:

$$2[Co(CO)_4]^- + Hg^{2+} \rightarrow Hg[Co(CO)_4]_2 \qquad (11.34)$$

On other occasions, the soft-base polyene carbon atoms may prove more attractive than the soft-base metal atom, and the soft acid may attack the carbon ring:

$$(C_5H_5)Mn(CO)_3 + 5\,Hg^{2+} + 5\,Cl^- \rightarrow [C_5(HgCl)_5]Mn(CO)_3 + 5\,H^+ \qquad (11.35)$$

Thus the chemistry of organometallic compounds involves transformations both at the metal atoms, which tend to fascinate inorganic chemists, and at the organic ligands, which seems most important to organic chemists, especially since these transformations are often difficult or impossible on the parent organic compounds.

Dinuclear Organometallics. Carbonylate and related organometallic anions not only react with inorganic soft acids such as Hg^{2+}, but also with organometallic soft acids such as carbonylate cations or halides:

$$[Re(CO)_5]^- + [Mn(CO)_5Br] \rightarrow [(CO)_5Re\text{–}Mn(CO)_5] + Br^- \qquad (11.36)$$

The *d*-block organometallic compounds such as this, containing two *d*-block metal atoms directly bonded to each other, are known as *dinuclear* organometallics. Such compounds can be prepared not only by LAB reactions such as the above, but also by reaction of two unstable free radicals (e.g., 17-electron species):[27]

$$2\,Cp_2Re\cdot \rightarrow Cp_2Re\text{–}ReCp_2 \qquad (11.37)$$

$$2\,CpCr(CO)_3{}^{\cdot} \rightleftharpoons Cp(CO)_3Cr\text{–}Cr(CO_3)Cp \qquad (11.38)$$

The unstable free radicals might in turn be intermediates when the normal syntheses for metal carbonyls are applied to metals from odd-numbered groups of the *d* block; since these radicals would then dimerize, many of the methods of syntheses already given can be used to prepare dinuclear carbonyls of these metals. Two examples of synthesis by reductive ligation follow:

$$2\,MnCl_2 + 4\,Na + 10\,CO \xrightarrow{\text{diglyme, }Ph_2CO} Mn_2(CO)_{10} + 4\,NaCl \qquad (11.39)$$

$$Re_2O_7 + 17\,CO \rightarrow Re_2(CO)_{10} + 7\,CO_2 \qquad (11.40)$$

Dinuclear carbonyls may also be made by heating or photolysis of mononuclear carbonyls, which drives off CO(g):

$$2\,Fe(CO)_5 \xrightarrow{h\nu} Fe_2(CO)_9 + CO \qquad (11.41)$$

Table 11.7 shows the formulas of a number of di- (and poly-) nuclear carbonyls and polyene carbonyls. In a certain sense, it is possible to think of many of these as pseudohalogens: reduction of 1 mol of one of these with 2 mol of a Group 1 metal (in the presence of a chelating solvent) generates 2 mol of a carbonyl or polyene carbonyl anion (as given in Table 11.6A):

$$Co_2(CO)_8 + 2\,Na \xrightarrow{\text{diglyme}} 2\,[Na(\text{diglyme})_x][Co(CO)_4]^- \qquad (11.42)$$

For homonuclear dinuclear organometallics, the M–M bond is clearly a nonpolar covalent bond, in which *each M–M single bond contributes one electron to the VEC of the other metal atom.*[28] (Dinuclear organometallics with multiple metal–metal bonds such as V≡V are also known[29] and are counted accordingly.)

Numerous polynuclear metal carbonyls are also known (Table 11.7). These carbonyls are found especially for the *d*-block metals of the fifth and sixth periods. They

Table 11.7
Dinuclear and Polynuclear Carbonyls

	$Mn_2(CO)_{10}$	$Fe_2(CO)_9$	$Co_2(CO)_8$	
	$Tc_2(CO)_{10}$	$Ru_2(CO)_9$[a]	$Rh_2(CO)_8$[a]	
	$Re_2(CO)_{10}$	$Os_2(CO)_9$[a]	$Ir_2(CO)_8$[a]	
$[Cr_2(CO)_{10}]^{2-}$		$[Fe_2(CO)_8]^{2-}$		
$Cp_2Cr_2(CO)_6$		$Cp_2Fe_2(CO)_4$		$Cp_2Ni_2(CO)_2$
$Cp_2Mo_2(CO)_6$		$Cp_2Ru_2(CO)_4$		
$Cp_2W_2(CO)_6$		$Cp_2Os_2(CO)_4$		
		$Fe_3(CO)_{12}$	$Co_4(CO)_{12}$	
		$Ru_3(CO)_{12}$	$Rh_4(CO)_{12}$	
		$Os_3(CO)_{12}$	$Ir_4(CO)_{12}$	

[a] Unstable at room temperature and standard pressure.

contain from three to dozens of metal atoms, are generally not linear chains or even rings, but contain the metal atoms in clusters so as not to sacrifice the substantial metal–metal atomization energies found among such metals (Section 7.3). These are typically generated by heating carbonyls of lower nuclearity so as to drive off CO:

$$3\,Fe_2(CO)_9 + \Delta \rightarrow 2\,Fe_3(CO)_{12} + 3\,CO \tag{11.43}$$

Polynuclear carbonyls containing less than six metal atoms generally also obey the 18-electron rule per metal atom if account is taken of the number of metal–metal bands; those with six or more metals often do not obey the 18-electron rules; as the clusters become more and more like small chunks of the metal itself, different electron-counting systems are called for (Section 12.7).

Example 11.5

Predict the formulas of the possible neutral trinuclear carbonyls in which the three metal atoms are single bonded in a metal triangle.

SOLUTION:

In a metal triangle, each metal atom has two metal–metal bonds to two neighbors, which at 1 electron each provide it with 2 electrons; the remaining 16 come from its own valence supply and the carbonyls. Let us assume a maximum coordination number of 6, of which two are the adjacent metal atoms in the triangle, so there can be no more than four carbonyls per metal atom. Four CO groups donate 8 electrons, so 8 electrons come from the Group 8 metal: The carbonyls with four CO groups per M are $Fe_3(CO)_{12}$, $Ru_3(CO)_{12}$, and $Os_3(CO)_{12}$. Three carbonyls per metal would give 6 electrons, leaving the metal to supply 10 electrons; the possible compounds $Ni_3(CO)_9$, $Pd_3(CO)_9$, and $Pt_3(CO)_9$ are not known, however.

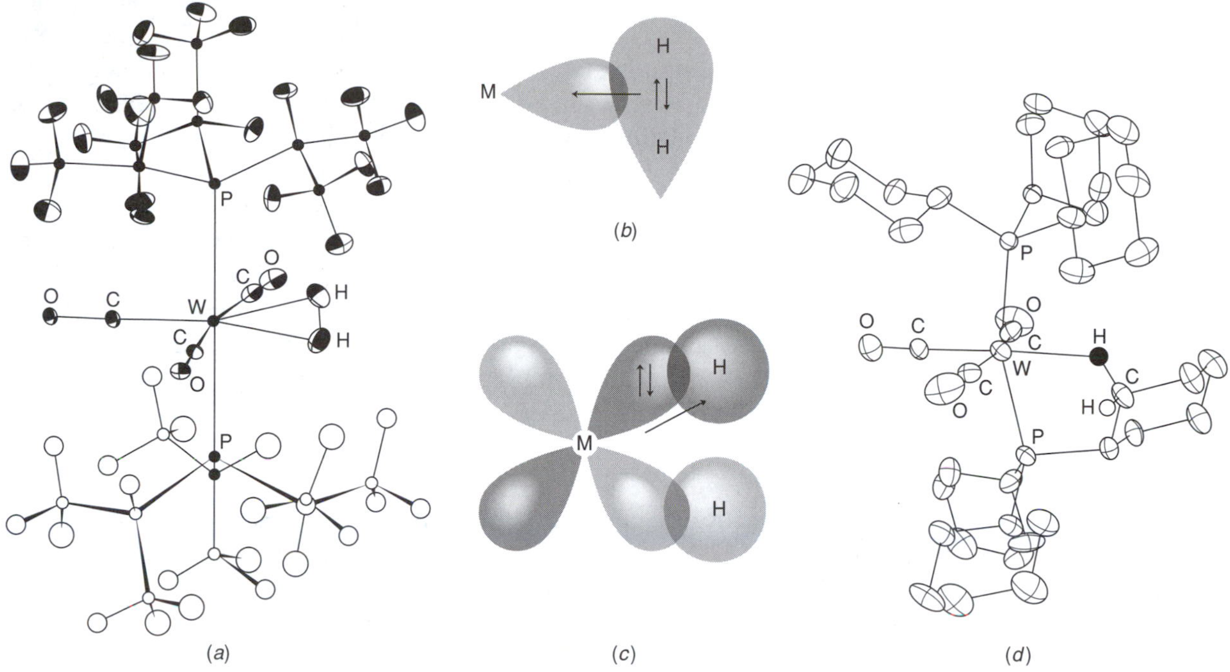

Figure 11.6
(*a*) Structure of $[W(CO)_3[P\text{-}i\text{-}Pr_3]_2(\eta^2\text{-}H_2)]$. [Adapted from G. J. Kubas, R. R. Ryan, B. I. Swanson, P. J. Vergamini, and H. J. Wasserman, *J. Am. Chem. Soc.*, **106**, 451 (1984).] (*b*) Electron donation from the σ HOMO of H_2 to an empty orbital on M; (*c*) back-donation from an occupied M *d* orbital to the σ^* LUMO of H_2; (*d*) structure of $W(CO)_3[P(C_6H_{11})_3]_2$ showing agostic interaction of a C–H bond with the metal atom. [Adapted from H. J. Wasserman, G. J. Kubas, and R. R. Ryan, *J. Am. Chem. Soc.*, **108**, 2294 (1986).]

Two-Electron Donation from a σ Bond: Hydrogen and Alkanes as Ligands. Perhaps the most unusual ligands forming σ bonds to *d*-block elements are those such as H_2,[30] in which the electron pair being donated is already present in a σ bond; the first such characterized complex was $[W(CO)_3[P\text{-}i\text{-}Pr_3]_2(\eta^2\text{-}H_2)]$ [Fig. 11.6(*a*)]. When the H_2 σ-bonding electron pair is donated to a vacant metal *d* orbital of a quite acidic cation [Fig. 11.6(*b*)], the result is a 3c–2e bond such as was discussed in Section 10.5; this interaction is supplemented in greater or lesser degree by back-bonding from a filled metal *d* orbital to the σ^* LUMO of H_2 [Fig. 11.6(*c*)]. Detection of coordination of H_2 is more than normally challenging, since it is difficult, using X-ray diffraction, to detect very light hydrogen atoms in the vicinity of relatively massive metal atoms. The acidity of coordinated H_2 shows a dramatic enhancement: pK_a rises from about 35 in free H_2 to -2 in $[(C_5Me_5)Re(CO)(NO)(H_2)]$.

As back-bonding by the metal becomes greater, the results are analogous to those discussed with ethylene complexes (Section 11.3): (1) the increased overlap between M and H_2 results in the formation of an additional M–H σ bond; (2) the increased population of the σ^* orbital results in the rupture of the H–H bond. Consequently, the η^2-H_2 ligand is transformed into two hydride ligands, while the loss of electrons by the metal results in an increase in its oxidation state by 2. Once it was realized that there can be σ-donor ligands such as H_2, and that they are related to more normal ligands such as H^- by this type of orbital interaction, old literature reports of hydrides showing unusually high metal oxidation states were reexamined, and some were shown to be incorrect. The compound $FeH_4(PR_3)_3$ does not contain iron in the oxidation state of +4, since the proper formulation of this compound is $[Fe(\eta^2\text{-}H_2)(H)_2(PR_3)_3]$.

Ligand attachment via 3c–2e bonding is not limited to the case of H_2; the same type of bonding results when superacids interact with saturated hydrocarbons (Section 4.10). To date, only one metal complex of an alkane has been isolated,[31] but there is spectroscopic evidence for the low-temperature existence of several complexes of the type $[M(CO)_5(\text{alkane})]$ (M = Cr, Mo, W; alkane = pentane, heptane, cyclohexane, etc.).[32] Also, in several cases the close approach of C–H bonds of (already coordinated) metal alkyl groups to vacant metal coordination sites [Fig. 11.6(*d*)] suggests that the C–H bond is engaged in a type of 3c–2e bonding known as an *agostic interaction*. Isolation of a complex with CH_4 as a ligand, and subsequent back-donation to rupture coordinated CH_4 into CH_3^- and H^-, would be of great interest, as this type of *alkane activation* reaction would allow the conversion of relatively available but unreactive CH_4 from natural or synthesis gas to be converted into much more reactive CH_3^- and subsequently into many types of organic compounds. An iridium complex ion containing coordinated dichloromethane, $[Ir(C_5Me_5)(PMe_3)(\eta^1\text{-}ClCH_2Cl)Me]^+$, has been found to activate and react with C–H bonds in benzene at −30 °C and in CH_4 at 10 °C[33]; the presence of the extremely weak and easily displaced dichloromethane ligand is thought to be important to the ease of these reactions.

11.5 Bridging Ligands: Alkynes and Carbonyls

Alkynes. Alkynes or substituted acetylenes, RC≡CR, have two mutually perpendicular π orbitals, which allows a variety of bonding interactions with metal ions.[34] In many cases, the alkyne may donate only one of the two π electron pairs to a metal ion in a manner analogous to that shown by alkenes such as ethylene (Fig. 11.3). The other π electron pair is, of course, also available for donation, but its orientation is such that it is most readily donated to a different metal atom, so that the alkyne ligand acts as a ligand *bridging* two different metals, as in the complex $[Co_2(CO)_6(\mu_2\text{-}t\text{-BuC}{\equiv}\text{C-}t\text{-Bu})]$ [Fig. 11.7(*a*)]. (In this formula μ is used to indicate a bridging ligand, and the suffix 2 indicates that the bridging involves two metals; we shall shortly see that other ligands may bridge three metals.)

Bridging and Apical Carbonyls. That alkynes should function as bridging ligands is not surprising in view of the arrangement of their π electron pairs; more surprising is the fact that the "simple" carbonyl ligand may also do so in some of the dinuclear and polynuclear carbonyls listed in Table 11.7. Evidence for this may be obtained from comparison of their carbonyl-region IR spectra: For example, $Ir_4(CO)_{12}$ has a simpler spectrum than does either $Rh_4(CO)_{12}$ or $Co_4(CO)_{12}$, consistent with a higher symmetry for the former tetranuclear carbonyl (T_d) than the latter two (C_{3v}). The carbonyl stretching frequencies of neutral carbonyls normally occur in the 2050–1900-cm^{-1} range, but carbonyls such as $Rh_4(CO)_{12}$ or $Co_4(CO)_{12}$ show in addition bands in the 1900–1750-cm^{-1} range, which are more nearly suggestive of a C=O double bond, which could be characteristic of a bridging carbonyl group (in acetone the carbonyl group "bridges" between two methyl groups).

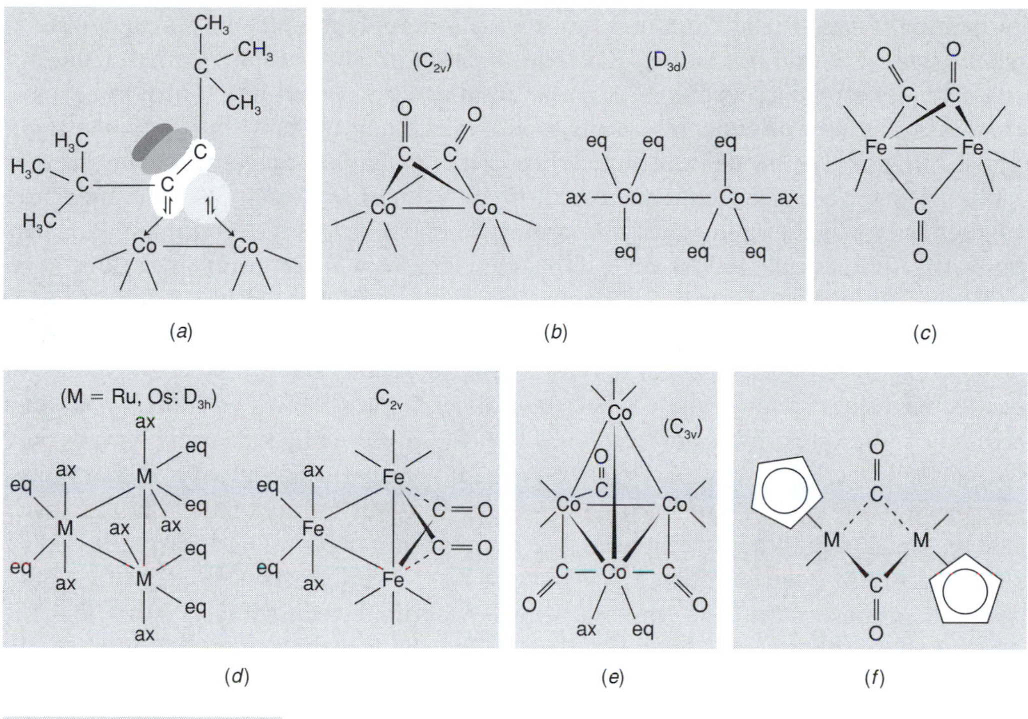

(a) (b) (c)

(d) (e) (f)

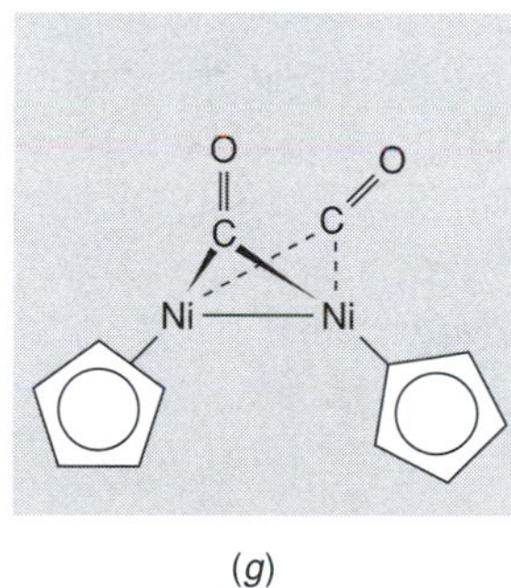

(g)

Figure 11.7
Structures of carbonyl derivatives having bridging ligands. For clarity, terminal C≡O groups are represented by dashes (–) drawn outward from metal atoms; *ax* = ligand is axial in the metal coordination geometry; *eq* = ligand is equatorial in the metal coordination geometry. (*a*) $[Co_2(CO)_6(\mu_2\text{-}t\text{-BuC}{\equiv}\text{C-}t\text{-Bu})]$, showing donation from two perpendicular π orbitals to two different Co atoms; (*b*) alternate structures of $Co_2(CO)_8$; (*c*) structure of $Fe_2(CO)_9$; (d) structures of $M_3(CO)_{12}$ (M=Ru, Os) and of $Fe_3(CO)_{10}(\mu_2\text{-}CO)_2$; (*e*) structure of $Co_4(CO)_9(\mu_2\text{-}CO)_3$; (*f*) structure of the trans isomer of $[CpM(CO)(\mu_2\text{-}CO)]_2$ (M = Fe, Ru); and (*g*) structure of $[CpNi(\mu_2\text{-}CO)]_2$.

Crystal structure determinations have subsequently verified that polynuclear carbonyls of lighter elements in a group often have different, less symmetric structures than the corresponding carbonyls of the heavier elements: These less symmetric structures may contain *bridging carbonyl groups*, as shown in Figure 11.7.

Pairs of bridging carbonyls are formed from pairs of terminal carbonyls in a process that is reminiscent of back-bonding, but in which electrons are supplied to the LUMOs of each terminal CO from the *d* orbitals of the *adjacent* metal atom:

$$\begin{array}{c} :M{-}C{\equiv}O: \\ | \\ :O{\equiv}C{-}M: \end{array} \rightleftharpoons :\ddot{O}{=}C\begin{array}{c} M \\ | \\ M \end{array}C{=}\ddot{O}: \qquad (11.44)$$

This process is readily reversible as is shown by the case of $Co_2(CO)_8$. In the solid state, this compound has the C_{2v} structure shown in Figure 11.7(*b*) with two bridging carbonyls, but its IR spectrum in solution is considerably more complicated, and indicates the presence of both the C_{2v} and D_{3d} isomers, along with a third form of unknown structure. The ready coexistence of these forms seems not to be due to any kinetic barrier to their interchange, but rather to all forms being of very nearly equal energy.

In addition, in some substituted trinuclear or higher carbonyls, some carbon monoxide IR stretching frequencies occur in a still lower frequency range, more nearly suggestive of a C–O bond order as low as 1. These *apical* or triply bridging (μ_3) carbonyls are presumably formed by back-donation from *d* electrons of yet a third adjacent metal atom, in a second process analogous to that illustrated in Eq. (11.44).

As more crystal structure determinations have appeared and been analyzed, it has become more evident that bridging and apical carbonyls are more characteristic of the lighter elements in a group, while the heavier elements form carbonyls with only terminal CO groups. Johnson[35] observed that the carbonyl ligands in the polynuclear carbonyls of the lighter elements tend to pack into regular geometric forms: For example, the 12 carbonyls in either $M_3(CO)_{12}$ or $M_4(CO)_{12}$ lie approximately at the apices of a regular icosahedron (I_h symmetry), seemingly independently of the arrangement of the metal atoms beneath them (a D_{3h} triangle or a T_d tetrahedron, respectively). Thus Johnson proposes that steric packing factors determine the arrangement of the carbonyl groups for the smaller metal atoms only; these carbonyl groups bond to the metal atoms either as terminal, or bridging, or apical carbonyls depending on how many metal atoms are nearby. Clearly, this could be the case only if the three bonding arrangements are of nearly equal energy, because normally steric forces of repulsions between groups are not nearly as significant as bond energies. Supporting an interpretation of this type is the observation that the bonding in these bridging polynuclear carbonyls tends to be fluxional: For example, the tetrahedron of cobalt atoms in $Co_4(CO)_{12}$ rotates almost freely beneath the icosahedron of carbon atoms. And these processes also illustrate just how versatile the *d*-block metal atoms can be when they have nine valence orbitals of such diverse shapes and patterns of nodal planes.

Electron Counting for Bridging and Apical Carbonyls. Since bridging carbonyls are so easily interconverted with terminal carbonyls, there is much to be said for the electron-counting scheme which treats the two the same: the RR scheme. In this scheme, a carbonyl is treated as a two-electron donor in either case: If it is terminal, it donates two electrons to one metal; if it is bridging, it donates one electron to each of two metals. (Apical bonding is a little trickier, however; it may then be counted as donating one electron to each of two metals but none to the third, which must supply the electrons for that bond.)

However, since the mechanism of bridge formation [Eq. (11.44)] shows electrons being accepted by CO, and since its bond order is reduced according to the IR and other evidence, the LAB scheme can also be defended, in which the bridging CO group is treated as a dianion, $C{=}O^{2-}$, which is isoelectronic to O_2 and has two unshared electron pairs on carbon. This dianion then donates two electrons to each of the two metal atoms. (Awkwardly, however, the metal atoms in neutral carbonyls then must be assigned positive oxidation states.) Apical carbonyls then are treated as complexes of the tetraanion, $C{-}O^{4-}$, which is isoelectronic with O_2^{2-} or F_2 and donates two electrons to each of the three adjacent metal ions.

Example 11.6

Compute the VEC per metal atom in each of the two types of $M_3(CO)_{12}$ structure shown in Figure 11.7(*d*). Which of the two types should give the simpler IR spectra? How many ^{13}C NMR peaks should each give?

SOLUTION (LAB METHOD):

For the D_{3h} structure each metal atom is identical, so it has the same electron count, 8 from four terminal CO groups and 2 from the two M–M bonds each is involved in; each Group 8 metal atom then brings in 8 electrons of its own for a total of 18. The same count applies to the left Fe atom shown in the C_{2v} drawing. The other two Fe ions are associated with two CO^{2-} bridging ligands, so each are counted as Fe^{2+} species with 6 valence electrons, to which we add 2 electrons each from M–M bonds, 6 electrons each from three terminal CO groups, and 2×2 electrons from the two bridging CO^{2-} ligands, for a total of 18 electrons.

SOLUTION (RR METHOD):

For the D_{3h} structure each metal atom is identical, so it has the same electron count, 8 from four terminal CO groups and 2 from the two M–M bonds each is involved in; each Group 8 metal atom then brings in 8 electrons of its own for a total of 18. The same count applies to the left Fe atom shown in the C_{2v} drawing. The other two Fe irons each have 8 valence electrons of their own, to which we add 2 electrons each from M–M bonds, 6 electrons each from three terminal CO groups, and 2×1 electrons from the two bridging CO ligands, for a total of 18 electrons.

SPECTRA:

The D_{3h} structure, being of higher symmetry, should give the simpler IR spectrum, especially since only its spectrum will be devoid of bridging IR frequencies below $1900\,cm^{-1}$. The ^{13}C NMR spectrum of the D_{3h} structure should show two peaks, one due to the equatorial CO groups and one due to the axial ones. Since the C_{2v} structure is likely to be fluxional, we expect a spectrum that varies with temperature: If we assume that the iron triangle moves under immobile CO ligands, then at high temperatures the axial terminal CO groups are interchanged with the bridging CO groups, while the equatorial CO groups remain distinct, to give two NMR peaks. At low temperatures, a more complex spectrum is to be expected: the C_2 axis running through the left Fe and between the two bridging CO groups interchanges pairs of CO groups, so that six NMR peaks are to be expected.

11.6 Ligands Forming Two or More Bonds to a Metal or Metals: The Isolobal Analogy

Analogy is a powerful tool in chemistry, especially in the creative aspects of imagining possible new classes of compounds. For example, we might ask whether it is possible to bridge two metal atoms using some ligand other than μ^2-CO. We might hypothesize that it should be possible to do this with other ligands that also have (from the RR approach) two orbitals with one electron in each. Such a fragment of a molecule could also serve to link the two metals in a manner analogous to μ^2-CO, since it has the same number of electrons prepared to bond to the metal, in the same number of orbitals, which have similarly shaped lobes. Both μ^2-CO and the new ligands could be said to be **isolobal fragments**.

Some sets of isolobal and isoelectronic fragments are drawn (using the RR approach) in Figure 11.8. Each row of the figure contains a different isolobal family. For example, Fig. 11.8(*a*) illustrates, and Table 11.6A lists, a number of fragments in which all but one valence orbital are occupied by either bonding or lone pairs, and one frontier orbital remains, in which is located one electron. These fragments are suitable to act as terminal groups (not as bridging links) in molecules. This isolobal relationship can exist, not only between molecular fragments or functional groups from the same block of the periodic table, but also between a fragment from the *p* block and one from the *d* block. The synthetic chemist has a reasonable hope that the combination of any two of these fragments would give a di(fragment) molecule: for example, H_2, HCl, H–Me, $HMn(CO)_5$, $HCo(CO)_4$, $CpFe(CO)_2H$. This possibility is enhanced if the energy levels of the frontier orbitals are similar: for example, if the fragments being exchanged are of similar hardness or softness. Two ligands or functional groups are said to be isolobal and isoelectronic "*if the number, symmetry properties, approximate energy, and shape of the frontier orbitals and the number of electrons in them are similar.*[36]"

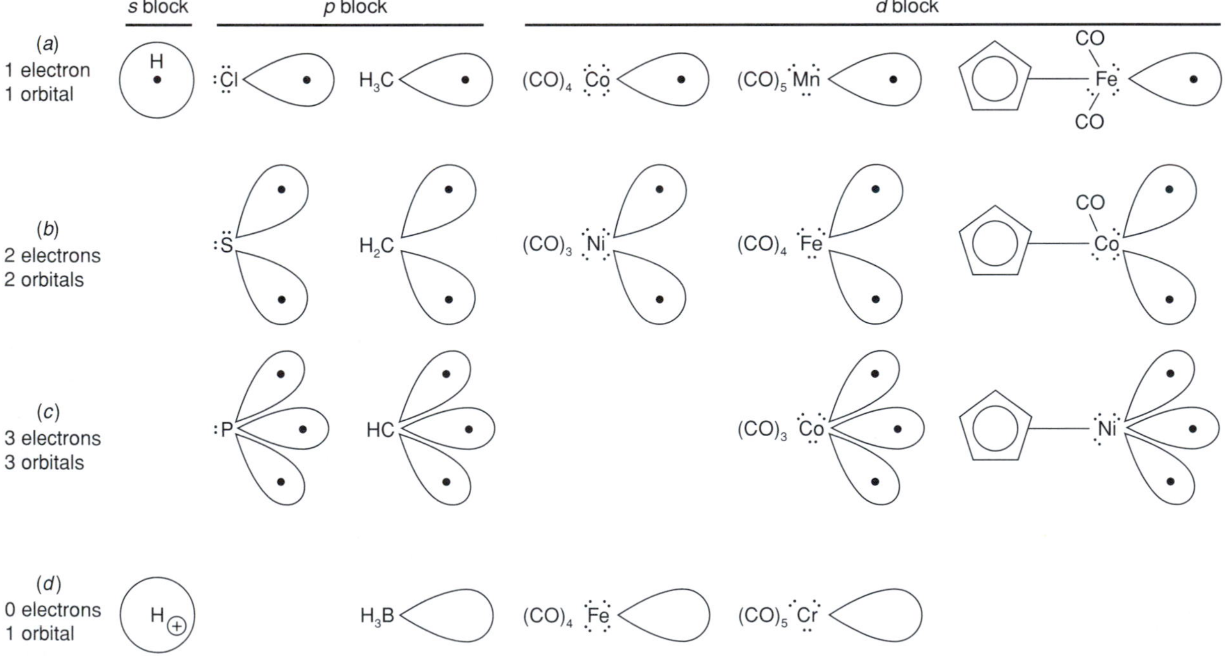

Figure 11.8

Isolobal and isoelectronic molecular fragments. (*a*) Species in which all valence orbitals are occupied, except one that holds one electron; (*b*) Similar except that two valence orbitals hold two electrons. (*c*) Similar except that three valence orbitals hold three electrons. (*d*) Similar except that one valence orbital is empty. (*Left columns*) Species from the *s* and *p* blocks; (*right columns*) species from the *d* block; the first of the *d*-block columns contains species in which the metal atom will (after compound formation) have a coordination number of 5; in the species in the other columns the metal atom will have a coordination number of 6.

Example 11.7

Justify the isolobality of the species in Figure 11.8(*a*).

SOLUTION:

We need to show that the correct number of valence electrons are present, and that all but one valence orbital is doubly occupied in each, while the last is singly occupied. Here H· has only one valence orbital, which is singly occupied. The Cl· and $CH_3^{\bullet}$ group each have four valence orbitals; in Cl· three are doubly occupied by lone pairs while in $CH_3^{\bullet}$ three are doubly occupied by bond pairs; a singly occupied orbital remains on each. (The energy level requirement can be crudely assessed based on the similarity of the electronegativities of H and C; Cl may be not as close a match.)

The *d*-block species each have nine valence orbitals; in $Mn(CO)_5^{\bullet}$ five are occupied by electron pairs donated by the CO groups, three are doubly occupied by six of the seven valence electrons of neutral Mn, and the last electron singly occupies the last orbital; the accounting for $Co(CO)_4^{\bullet}$ is similar. In $CpFe(CO)_2^{\bullet}$ (RR approach), two orbitals are doubly occupied by electrons from the two CO groups; three orbitals are occupied by the five electrons from neutral $C_5H_5^{\bullet}$ plus one from iron; three orbitals are occupied by three unshared electron pairs of the neutral Fe, and the eighth electron of neutral Fe singly occupies the ninth (last) valence orbital. (The electronegativities of all of these metal ions are broadly comparable to those of C and H as well.)

Oxo, Carbene, and Alkylidene Complexes. Figure 11.8(*b*), as well as Table 11.6C, give a number of ligands with an unpaired electron in each of *two* orbitals. Those ligands that have good π bonding capability can form dimeric double-bonded species such as O_2, RN=NR (a diazene), and $R_2C{=}CR_2$; any of them can form two bonds to two other groups, as in OH_2, RNH_2, R_2CH_2, SCl_2, $RNCl_2$, and $Fe(CO)_4Cl_2$. These are characteristic reactions of Group 16(VI) elements, which could allow us to classify these ligands or molecular fragments as "pseudochalcogens".[37]

At this time, we wish to emphasize the bonding of these fragments to *d*-block metals in organometallics, of course: of most importance are some of these ligands that are good π bonders: O or O^{2-}, NR or NR^{2-}, and CR_2 (also known as transient free molecules called *carbenes* or *alkylidenes*) or $CR_2{}^{2-}$. With *d*-block metals that are also good π bonders (especially Mo, W, Tc, and Re, but also nearby metals), multiply bonded organometallics can be formed:

$$CpRe({=}O)_3,\ MeRe({=}O)_3,\ (CO)_5W{=}CR_2,\ Cp_2TaMe({=}CH_2).$$

Since carbenes or alkylidenes are not stable as free molecules, they must be constructed as ligands from other groups. Schrock[38] alkylidene derivatives such as $(Me_3CCH_2)_3Ta{=}CHCMe_3$ are apparently constructed by the decomposition of coordinated alkyl groups:

$$(Me_3CCH_2)_3TaCl_2 + 2\,Me_3CCH_2Li \rightarrow (Me_3CCH_2)_5Ta \rightarrow (Me_3CCH_2)_3Ta{=}CHCMe_3 + CMe_4 \quad (11.45)$$

In contrast, E. O. Fischer's carbene species contain an electron-withdrawing R group because they are constructed from coordinated CO in a sequence of two reactions:

$$W(CO)_6 + PhLi \rightarrow Li^+[(CO)_5W\overset{\overset{\displaystyle O}{\|}}{C}Ph]^- \qquad (11.46)$$

$$[(CO)_5W\overset{\overset{\displaystyle O}{\|}}{C}Ph]^- + [Et_3O]^+ \rightarrow [(CO)_5W{=}\overset{\overset{\displaystyle OEt}{|}}{C}Ph] + Et_2O \qquad (11.47)$$

Although the polarity and reactivity of these two types of CR_2 derivatives differ, they share the feature expected of double bonding: the M–C distances are shorter than single M–C bond distances. In contrast, carbene adducts of the d^{10} metal ions Cu^+, Ag^+, and Au^+, which are not good π bonders, show M–C distances that fall in the (broad) range of normal single bonds.[39]

Carbyne and Alkylidyne Complexes. Similarly, the ligands from Figure 11.8(*c*), or from Table 11.6D, that have good π bonding capability can form dimeric triple-bonded species such as N_2 and RC≡CR, or they can form three bonds to three other groups, as in NH_3 and RMe. These are characteristic reactions of Group 15(V) elements, which could allow us to classify these ligand groups as "pseudopnicogens".[37] These may be in the form of a triple bond to one metal: Thus, further extension of the Schrock reaction conditions can remove another hydrogen atom from a coordinated =CRH group, giving an alkylidyne complex such as that shown earlier in Figure 3.6 and containing single, double, and triple W–C bonds of lengths 226, 198, and 181 pm, respectively. In addition, the electronegative group can be removed from a Fischer carbene by the use of a strong hard Lewis acid such as BBr_3:

$$[(CO)_5W{=}\overset{\overset{\displaystyle OEt}{|}}{C}Ph] + BBr_3 \rightarrow EtOBBr_2 + [(CO)_5W{\equiv}CPh]^+Br^-$$
$$\rightarrow trans\text{-}[Br(CO)_4W{\equiv}CPh] + CO \qquad (11.48A)$$

The groups from Table 11.6D that are not good π bonders are restricted to forming three bonds to three other groups, in effect acting as apical ligands between them, as, for example, in PCl_3 and $Co(CNR)_3Cl_3$. The three bonds formed by CR (a *carbyne* or *alkylidyne*) can also be used, by analogy to apical μ^3-CO in polynuclear carbonyls, to bridge apically three different metal atoms. Thus, not only is the tetrahedral cluster $Co_4(CO)_{12}$ known, but clusters can be made by replacing one $Co(CO)_3$ fragment by (a) an isolobal CR fragment, to give the tetrahedral cluster $RCCo_3(CO)_9$:

$$9\,Co_2(CO)_8 + 4\,MeCCl_3 \rightarrow 4\,MeC[Co(CO)_3]_3 + 6\,CoCl_2 + 36\,CO \qquad (11.49)$$

(b) an Sb atom, to give the known cluster $SbCo_3(CO)_9$; (c) a CpNi fragment, to give the cluster $CpNiCo_3(CO)_9$. In addition, substitution by Sb atoms could be done twice, three times, or four times to give, respectively, $Sb_2Co_2(CO)_6$, $Sb_3Co(CO)_3$, and the yellow allotrope of Sb, Sb_4.

11.7 Reaction Types of *d*-Block Organometallic Compounds

In Section 5.10, we discussed five functions of metal ions in biochemical and other catalytic systems. The *d*-block metal ions and complex ions are particularly good at catalyzing reactions; among these, *d*-block organometallic compounds are especially good. In addition to the five functions mentioned earlier, we have by now seen some additional advantages that *d*-block metals may possess. Because of the variety of nodal properties that their nine different valence orbitals possess, reactions involving *d*-block metals are less likely to be constrained by orbital symmetry restrictions[40]; *d*-block metal atoms in organometallic compounds show not only Lewis-acidic but also Lewis-basic properties.

In a text at this level, we can only hope to suggest some of the many reactions of organometallic compounds. We present some types of reactions, beginning with (A) those types of reactions that only involve the gain or loss of single groups from the metal atoms, followed by (B) reactions that involve only the transformations of the ligands, and concluded by (C) more complex types of reactions that change both the ligands and the nature of their bonding to the metal. As we give examples of these types of reactions, we will tabulate and track changes in various counts of metal-atom properties, which will help us in identifying the reaction types and their consequences. We will tabulate the following metal-atom counts before and after each reaction type: (1) the VEC at the metal; (2) the coordination number of the metal; (3) the oxidation number of the metal (computed via the LAB approach); and (4) the geometry of the complex.

Reaction Type A1. Lewis Base Association or Dissociation (Gain or Loss of a Ligand)

Most organometallic compounds act as strong-field complexes with large crystal field stabilization energies for certain electron configurations (e.g., 18 valence electrons for d^6 octahedral and d^8 trigonal bipyramidal or square pyramidal; 16 valence electrons for d^8 square planar). Hence, the gain or loss of an additional ligand is generally nowhere near as rapid as for simple hydrated metal ions (Section 8.8): You may have noted how many reactions given in this chapter required heating times of many hours. However, the fact that the 18-electron rule has a "16-electron loophole" does facilitate fairly ready *association of Lewis bases* with 16-electron square planar species to give 18-electron five-coordinate products, along with the reverse process, the *dissociation of Lewis bases* from the 18-electron five-coordinate product:

Dissociation $\longrightarrow$	$Fe(CO)_5$ $\rightleftharpoons$	$Fe(CO)_4 + CO$	$\longleftarrow$ Association
Metal VEC	**18**	**16**	**Change of 2**
Metal coordination No.	**5**	**4**	**Change of 1**
Metal oxidation No.	**0**	**0**	**Change of 0**
Metal geometry	**Trigonal bipyramidal**	**Square planar**	

(11.50)

Square planar complexes are valuable in catalysis because they are *coordinately unsaturated*: They have a vacant coordination site at which a molecule (undergoing polymerization, etc.) can be associated to a metal.

An 18-electron octahedral species such as $Cr(CO)_6$ is unlikely to undergo ligand association, since a seven-coordinate, 20-electron species would result; dissociation is preferred as a first step, to give a five-coordinate (presumably square pyramidal) intermediate. Since this intermediate does not have its preferred number of valence electrons, this step is a slow, high-energy one, generally requiring photolysis with UV light; on the other hand, the intermediate, not being very stable, reacts quite rapidly to associate another ligand and return to the preferred 18-electron configuration. Although the reason is not entirely clear, experimentally it is found that the dissociation is least difficult for the fifth-period element of a given group, which accounts for the fact that the homogeneous catalysts in use (Section 11.8) are generally compounds of the fifth-period metals.[41]

Dissociation $\longrightarrow$	$Cr(CO)_6$	$\rightleftharpoons$	$Cr(CO)_5 + CO$	$\longleftarrow$ Association
Metal VEC	**18**		**16**	**Change of 2**
Metal coordination No.	**6**		**5**	**Change of 1**
Metal oxidation No.	**0**		**0**	**Change of 0**
Metal geometry	**Octahedral**		**Square pyramidal**	

(11.51)

Variant 1. In some cases in which the 18-electron species includes a cyclopolyene ligand, the substitution can nonetheless involve association of a ligand; the mechanism does not, however, involve a 20-electron intermediate, since a change of hapticity of the ligand (a "ring slippage") can occur during the association[42]:

Slippage + association	$\eta^5\text{-CpCo(CO)}_2 + L \rightarrow$	$\eta^3\text{-CpCo(CO)}_2L$	
Metal VEC	**18**	**18**	**Change of 0**
Slippage + dissociation	$\eta^3\text{-CpCo(CO)}_2L \rightarrow$	$CO + \eta^5\text{-CpCo(CO)}L$	
Metal VEC	**18**	**18**	

(11.52)

Variant 2. Although *stable*, isolatable organometallic compounds tend to follow the 18- and 16-electron rules, the most *reactive* species are those that do *not* follow these rules. In recent years, it has been found that many association and dissociation reactions proceed mainly via pathways that involve 17- and 19-electron species.[43] Although these species have less crystal field stabilization energy and are therefore less stable and found in only minute concentrations as intermediates in many reactions, their low stabilization energies allow them to react (often) millions of times faster than their 16- and 18-electron brethren. Therefore they can account for most or all of the substitution going on. Such a reaction can then be initiated by, for example, the prior *gain or loss of an electron* by the predominant 16- or 18-electron species as in the following simplified mechanism:

(Slow) Oxidation	$CpMn(CO)_3 \rightarrow$	$CpMn(CO)_3^+ + e^-$	
Metal VEC	**18**	**17**	**Change of 1**
Metal oxidation No.	**+1**	**+2**	**Change of 1**
(Rapid) Association	$CpMn(CO)_3^+ + L \rightarrow$	$CpMn(CO)_3L^+$	
Metal VEC	**17**	**19**	**Change of 2**
Metal oxidation No.	**+2**	**+2**	**Change of 0**

(Rapid) Dissociation	$CpMn(CO)_3L^+ \rightarrow$	$CpMn(CO)_2L^+ + CO$	
Metal VEC	**19**	**17**	**Change of 2**
Metal oxidation No.	+2	+2	**Change of 0**

(Rapid) Reduction	$CpMn(CO)_2L^+ + e^- \rightarrow$	$CpMn(CO)_2L$	
Metal electron count	**17**	**18**	**Change of 1**
Metal oxidation No.	+2	+1	

(11.53)

Reaction Type A2. Lewis Acid Association or Dissociation (Gain or Loss of a Lewis Acid)

In Section 11.4, we gave several examples in which a carbonylate anion associates with a Lewis acid [Eqs. (11.32) and (11.34)]. Related to these but of more importance in catalysis of organic reactions are cases [Eqs. (11.36) and (11.54)] in which the organometallic anion reacts with an organometallic or organic halide as Lewis acid to displace the halide ion and associate with the Lewis-acidic organic or organometallic part of the molecule: Organic chemists refer to these as *nucleophilic displacement reactions.*

Lewis Acid Association	$[Fe(CO)_4]^{2-} + MeBr \rightarrow$	$[MeFe(CO)_4]^- + Br^-$	
Metal VEC:	**18**	**18**	**Change of 0**
Metal coordination No.	**4**	**5**	**Change of 1**
Metal oxidation No.	**−2**	**0**	**Change of 2**
Metal geometry	**Tetrahedral**	**Trigonal bipyramidal**	

(11.54)

Nucleophilic displacement reactions can also be carried out on acyl halides:

$$[Fe(CO)_4]^{2-} + MeC({=}O)Br \rightarrow [MeC({=}O)Fe(CO)_4]^- + Br^- \qquad (11.55)$$

Organometallic anions very often give extremely rapid rates of nucleophilic displacement, since such rates generally are faster for softer bases.

For the reaction types involving only the gain or loss of simple groups at the metal atom (*not* the variants), the following counting characteristics may be noted: (1) both the Lewis-acid and the Lewis-base reactions are characterized by changes of 1 in the metal coordination number; (2) Lewis-base association and dissociation are characterized by a change of 2 in the metal VEC, while Lewis-acid association (and dissociation) involves no change in the metal VEC, but rather a change of 2 in the metal atom oxidation number.

Reaction Type B. Lewis Base or Lewis Acid Attack on Coordinated Ligands

The Lewis-acid properties of metal cations (including those in organometallic compounds) extend to the coordinated ligands, from the simplest ones such as H_2O (Chapter 2) to CO and coordinated organic compounds, rendering them more susceptible to reaction with Lewis bases. The carbon atoms of coordinated CO in many carbonyls act as Lewis-acidic sites, and can be attacked by bases such as Ph^- in reactions such as Eq. (11.46), which is the first step in generating Fischer carbenes. Neutral metal carbonyls are also susceptible to attack at the carbonyl carbon atom by OH^- to produce a coordinated carboxylic acid group:

$$(CO)_nM{-}C{\equiv}O + OH^- \rightarrow [(CO)_nM{-}C({=}O)OH]^- \qquad (11.56)$$

The role of metal atoms in accelerating the nucleophilic displacement attacks of Lewis bases on coordinated aryl halides was already mentioned in Section 11.3. In these reactions, the additional electrons provided by the attacking base flow out to other atoms on the ligand (the carbonyl oxygen or Cl^- on the benzene ring), so there is no change in any count that we make at the metal atom.

The converse reaction, the abstraction of Lewis bases from coordinated ligands, is not a spontaneous type of reaction, since coordinated ligands generally have enhanced acidity. This abstraction can be achieved, however, by reaction with an even stronger Lewis acid. The strong organic Lewis acid Ph_3C^+ (triphenylcarbonium ion) may be used to remove a hydride ion from the β position of a coordinated ethyl group, producing a coordinated alkene:

$$[CpFe(CO)_2CH_2CH_3] + Ph_3C^+ \rightarrow [CpFe(CO)_2(CH_2{=}CH_2)]^+ + Ph_3CH \qquad (11.57)$$

Again, no change occurs in any count made at the metal atom, although there is a subtle change at the metal atom: To replace the lost electron pair at the β carbon atom, the electron pair donated to the metal atom is converted from a σ bond pair to a π bond pair.

Variants: Lewis Base or Lewis Acid Attack on Coordinated Ligands at a Saturated α Carbon Atom. When Lewis bases attack saturated α carbon atoms of coordinated ligands, the extra electron density must flow instead to the metal atom, creating a multiple bond and causing some changes in its electron count: An example of this process is the possible attack of CH_3^- (from CH_3Li) on a coordinated CH_3 group in the following hypothesized step in the process of generating a Schrock carbene [cf. with Eq. (11.45)]:

$$Cp_2Ta(CH_3)_2Cl + CH_3^-Li^+ \rightarrow Cp_2Ta(CH_3)({=}CH_2) + CH_4 + LiCl \qquad (11.58)$$

Most counts at the metal do not change, but in this case the coordination number of Ta is reduced.

When Lewis base abstraction occurs at the α carbon atom [as in forming a carbyne complex from a Fischer carbene, Eq. (11.48A) and the β carbon is unable to make up the deficit, then the metal atom must do so, so that a change in oxidation state does occur.

$$BBr_3 + [(CO)_5W{=}C(OEt)Ph] \rightarrow [(CO)_5W{\equiv}CPh]^+Br^- + EtOBBr_2$$

(OEt is drawn attached above the carbene carbon.)

	Reactant	Product	
Metal VEC	**18**	**18**	**Change of 0**
Metal coordination No.	**6**	**6**	**Change of 0**
Metal oxidation No.	**+2**	**+4**	**Change of 2**
Metal geometry	**Octahedral**	**Octahedral**	

(11.48B)

Reaction Type C1. Oxidative Addition and Reductive Elimination Reactions

Since the *d*-block metal atoms in organometallic compounds are usually in low oxidation states, they are susceptible to oxidation (many of them must routinely be handled under inert atmospheres using the techniques outlined in Section 6.2). Square planar complexes are also coordinately unsaturated; *oxidative addition* is a process in which a (at least diatomic) ligand acquires two electrons from a metal atom, which go into an antibonding orbital, dissociating the ligand into two (perhaps) monoatomic ligands, which both coordinate to the metal. Since the reactant is a stable 16-electron square complex and the product is a stable 18-electron octahedral complex, both this reaction and its reverse, *reductive elimination,* proceed fairly readily. A simple inorganic example would be the oxidation of the $PtCl_4^{2-}$ ion by elemental chlorine:

$$\text{Oxidative addition} \rightarrow Cl_2 + PtCl_4^{2-} \rightleftharpoons PtCl_6^{2-} \leftarrow \text{Reductive elimination}$$

Metal VEC	**16**	**18**	**Change of 2**
Metal coordination No.	**4**	**6**	**Change of 2**
Metal oxidation No.	**+2**	**+4**	**Change of 2**
Metal geometry	**Square planar**	**Octahedral**	

(11.59)

Many simple diatomic molecules are capable of undergoing oxidative addition reactions, which includes not only obvious good oxidizing agents such as the halogens (to give dihalides), but also the hydrogen halides (to give halo, hydrido complexes), alkyl and aryl halides (to give alkyl or aryl, halo complexes), hydrogen itself (to give dihydrido complexes), and hydrocarbons (to give alkyl, hydrido complexes).

This relatively complicated reaction can occur via more than one type of mechanism. For example, the oxidative addition of H_2 to a square planar complex may begin with the coordination of H_2 in the vacant coordination site as (η^2-H_2). Then, it is followed by electron transfer from a metal *d* orbital to the H_2 σ^*, as illustrated in Figure 11.6(*c*), whereupon the H–H bond order is reduced to 0 and the η^2-H_2 ligand becomes two hydrido ligands. Oxidative addition to five-coordinate and to 17-electron organometallics also occurs, although with different mechanisms and different electron counts. But all mechanisms of oxidative addition by definition involve increases in the metal oxidation number and metal coordination number.

Agostic interactions (Section 11.4) may also put C–H bonds in position to undergo oxidative addition to metal atoms. Oxidative addition is particularly likely to happen to the ortho hydrogens of coordinated PPh_3 or $P(OPh)_3$, since a (four- or five-membered) chelate ring is formed in the process [Fig. 11.9(*a*)]; such reactions are also known as *cyclometalation* reactions.

Oxidative addition can also involve species that are multiply bonded, in which case only the π bond is nominally broken, so that the two new donor atoms remain attached to each other to form a three-membered chelate ring. Thus, the platinum(II) complex of the strongly electron-withdrawing (oxidizing) alkene tetracyanoethylene [Fig. 11.3(*a*)] withdraws enough electron density from Pt(II) [Fig. 11.3(*b*), right] to produce a metallacyclopropane [Fig. 11.3(*c*)] in which Pt has a formal oxidation state of +4.[44] Other multiply bonded species that produce metallacyles after oxidative additions include alkynes, tetrachloro-*o*-benzoquinone [Fig. 11.9(*b*)], and O_2 [Fig. 11.9(*c*)] to produce, with sufficient back-bonding, cyclic complexes of the peroxide ion, O_2^{2-}.

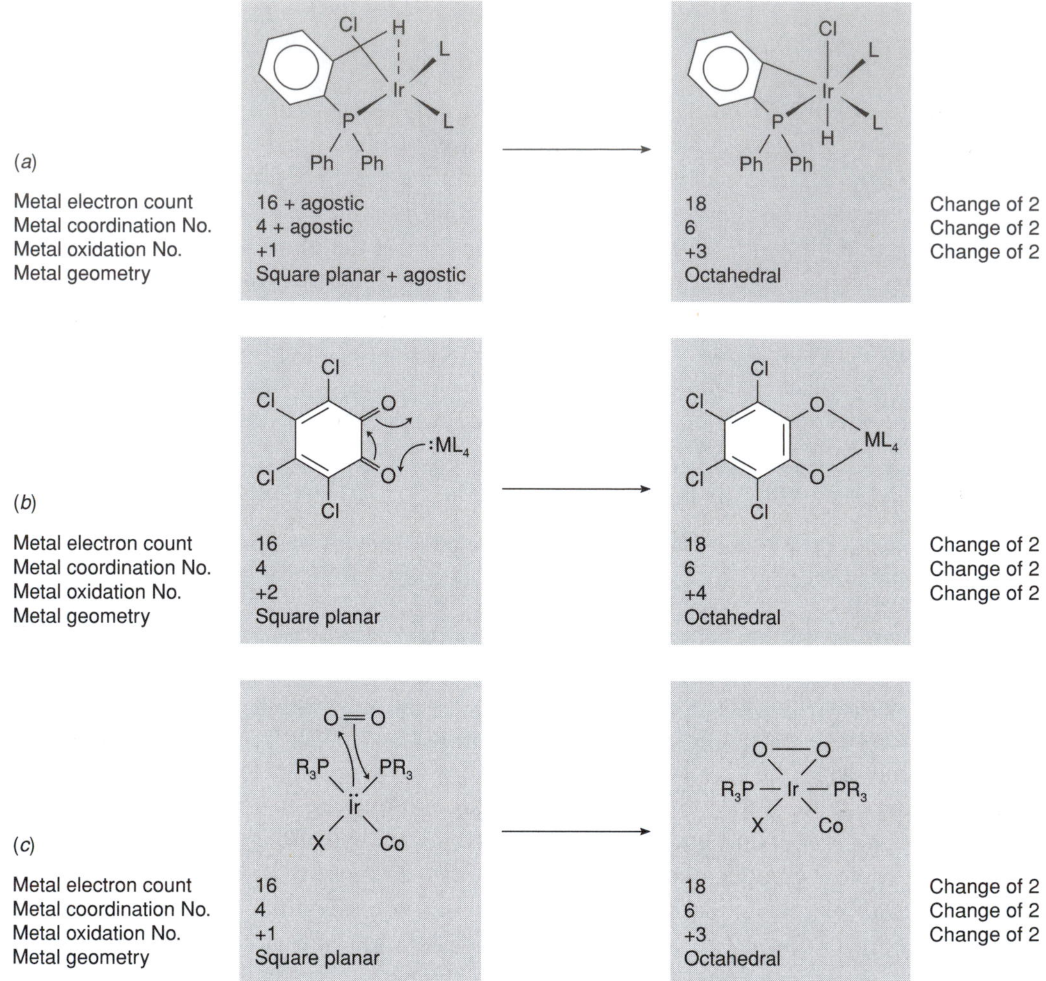

Figure 11.9

Oxidative addition reactions. (*a*) The ortho-metalation of a coordinated triphenylphosphine ligand. (*b*) Oxidative addition of tetrachloro-*o*-benzoquinone (3,4,5,6.-tetrachlorocyclohexadiene-1,2-dione). (*c*) Oxidative addition of O_2 to Vaska's compound to give a complex of the μ_2-$O_2{}^{2-}$ ligand.

Oxidative addition reactions are most readily experienced by metal atoms that are in low oxidation states, with high-energy HOMOs: that is, by the softest acids. The order of ease of oxidative addition by d^8 metal ions in square planar complexes follows the order: Os(0) > Ru(0) > Fe(0) ≫ Ir(I) > Rh(I) > Co(I) ≫ Pt(II) > Pd(II) ≫ Ni(II), Au(III) (note that for the last two, the product oxidation states of Ni^{IV} and Au^{V} are unusually high, hence disfavored).

Reductive elimination reactions would of course tend to be favored by the reverse sequence of metals. These reactions are most significant when an alkyl and a hydrido ligand are eliminated as an alkane, or especially when two alkyl ligands can be eliminated as an alkane with an elongated carbon chain, since this is an important element of organic synthesis.

Reaction Type C2. Insertion and Elimination (Deinsertion) Reactions

Another large class of complex reactions of organometallics is known as *insertion* reactions, because they result in the *insertion of one ligand into the bond of another ligand to the metal*: M–X + L → M–L–X. This reaction results in a larger ligand, which is often of use in organic synthesis. The converse of this process, the *elimination* reaction, results in an internal fragment of a ligand being removed and expelled from the metal's coordination sphere: M–L–X → M–X + L. Table 11.8 gives an abbreviated list of some of the most important types of insertion and elimination reactions.

Table 11.8
Some Common Examples of Insertion and Elimination Reactions

Insertion Direction →				**← Elimination Direction**
Bond	+	*Molecule*	⇌	Product
M–CH_3	+	CO	⇌	M–C(=O)–CH_3
M–CH_3	+	SO_2	⇌	M–S(=O)$_2$–CH_3
M–H	+	CH_2=CH_2	⇌	M–CH_2–CH_3
M–H	+	RC≡CR	⇌	M–C(R)=CHR
M–(RC≡CR)	+	RC≡CR	⇌	cyclo-(M–RC=CR–CR=CR)[a]

[a]Sometimes classified as an "oxidative coupling ⇌ reductive decoupling" pair since a change in formal metal oxidation state is involved.

This complex class of reactions also involves a variety of mechanisms, each with more than one step, so that the ligand is not directly "inserted" from outside the coordination sphere into a preexisting metal–ligand bond; instead it is often the case that there is *migration* of one ligand from being bonded to the metal to being bonded to the other ligand, or vice versa: L–M–X ⇌ X–L–M.

Thus, the accepted mechanism for the *carbonyl insertion reaction*, which converts a methyl complex into an acyl complex after "insertion" of CO, instead involves migration of a coordinated methyl group to attack (nucleophilically) the carbon atom of an adjacent, precoordinated carbonyl. This reaction leaves a vacant coordination site that is then filled by an incoming CO molecule:

CH_3 Migration → [(CO)$_5$MnMe] ⇌ [(CO)$_4$MnC(=O)Me]

	[(CO)$_5$MnMe]	[(CO)$_4$MnC(=O)Me]	
Metal VEC	**18**	**16**	**Change of 2**
Metal coordination No.	**6**	**5**	**Change of 1**
Metal oxidation No.	**+1**	**+1**	**Change of 0**
Metal geometry	**Octahedral**	**Square pyramidal**	

(11.60)

Base association → ← Base dissociation

CO + [(CO)$_4$MnC(=O)Me] ⇌ [(CO)$_5$MnC(=O)Me]

	[(CO)$_4$MnC(=O)Me]	[(CO)$_5$MnC(=O)Me]	
Metal VEC	**16**	**18**	**Change of 2**
Metal coordination No.	**5**	**6**	**Change of 1**
Metal oxidation No.	**+1**	**+1**	**Change of 0**
Metal geometry	**Square pyramidal**	**Octahedral**	

(11.61)

That the "inserted" CO does not in fact end up as part of the larger acyl ligand is shown by using labeled CO (the isotopic label does not end up in the acyl group). Earlier we saw an independent synthesis of acyl complexes [Eq. (11.55)]; since this insertion reaction is reversible, alkyl complexes can also be made from acyl complexes.

The *sulfur dioxide insertion reaction* is not reversible and begins with the bonding of the SO_2, not to the metal, but to the outside of the coordinated methyl group. This reaction is followed by an ionization and then recoordination of the soft sulfur donor atom to the soft-acid metal atom:

$$\mathrm{M{-}CR_3 + SO_2 \rightarrow {}^{\delta+}M{-}CR_3{-}SO_2{}^{\delta-} \rightarrow M^+[O_2SCR_3]^- \rightarrow M{-}\underset{\underset{\displaystyle O}{\|}}{\overset{\overset{\displaystyle O}{\|}}{S}}{-}CR_3} \qquad (11.62)$$

The *alkene* insertion reaction is readily reversible; its reverse is known as the *β-hydride elimination* reaction and is responsible for the ready decomposition of σ-bonded alkyls (such as ethyl derivatives or others having β hydrogens) of the *d*-block metals:

$$\text{Insertion} \rightarrow [\mathrm{L_4M(-H)(CH_2{=}CH_2)}] \rightleftharpoons [\mathrm{L_4M(CH_2CH_3)}] \leftarrow \text{Elimination}$$

Metal VEC	**18**	**16**	**Change of 2**
Metal coordination No.	**6**	**5**	**Change of 1**
Metal oxidation No.	**+1**	**+1**	**Change of 0**
Metal geometry	**Octahedral**	**Square pyramidal**	

(11.63)

In some cases, coordinated $CH_2{=}CH_2$ can insert into a $M{-}CH_2CH_3$ or other M–alkyl bond to give a $M{-}CH_2CH_2CH_2CH_3$ group; if this step can then be repeated many times it provides a mechanism for the polymerization of alkenes (Section 11.8).

Variant. The final example of an "insertion" reaction listed in Table 11.8 is the *cyclooligomerization* or "oxidative coupling" of an alkyne to give its cyclodimer, a substituted cyclobutadiene; or its cyclotrimer, a substituted benzene; or its cyclotetramer, a cyclooctatetraene. We might think of a key early step of such a reaction as the insertion of one coordinated alkyne in the bond of another alkyne to a metal such as cobalt, but with the twist that the other end of the building chain completes a chelate ring to form a *metallacyclopentadiene* ring (Fig. 11.10, Step 5), so that no change in coordination number occurs (instead, a change in formal oxidation number results).

Depending on the metal involved, another, more normal insertion into the metallacyclopentadiene ring can occur (Step 7), to give a metallacycloheptatriene ring. Finally, a reductive elimination of the organic cyclopolyene is shown as Step 8, to give a hexasubstituted benzene; if this step occurs earlier (as with iron) a coordinated cyclobutadiene ring is formed; if the step occurs later a cyclooctatetraene ring is formed. Alternatively, one CO ligand may also insert, to give a coordinated cyclopentadienone ring. In each of the key "insertion" steps of this possible mechanism except the first, the electron count of the metal decreases by two, while its coordination number decreases by one, and its oxidation number does not change, as is typical of insertion reactions.

Analysis of Reaction Types. The type of reaction an organometallic compound is undergoing is best determined by looking at what is happening to the metal atoms and ligands: oxidative additions add electrons to antibonding orbitals of ligands, reducing

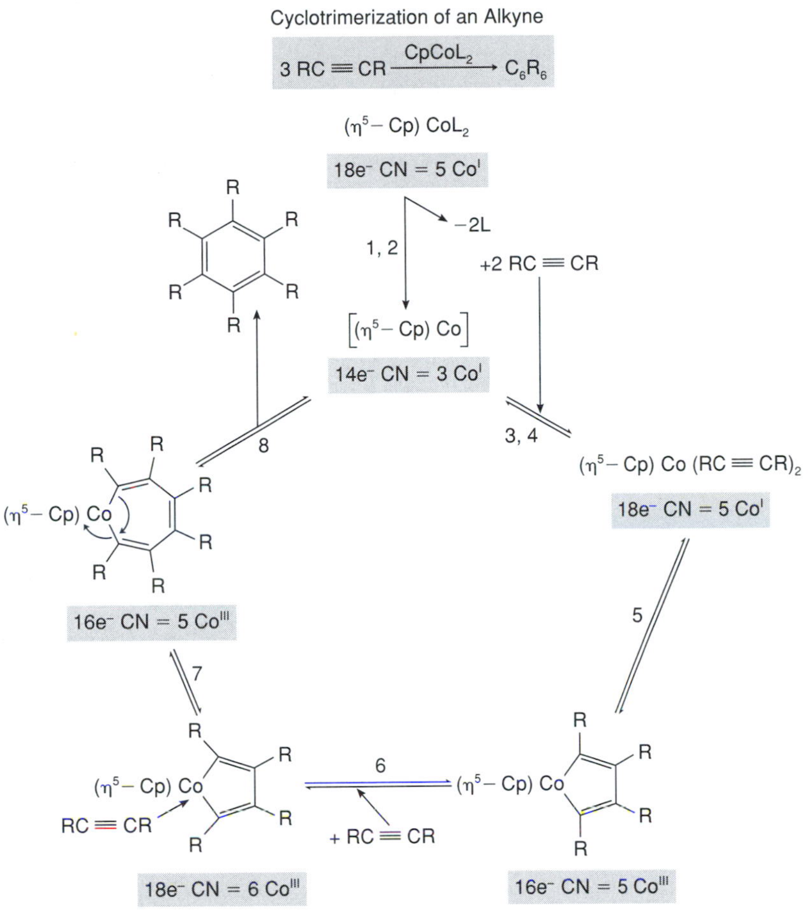

Cyclotrimerization of an Alkyne

$$3\ RC{\equiv}CR \xrightarrow{CpCoL_2} C_6R_6$$

Figure 11.10
Some steps typically found in various simplified mechanisms for the cyclooligomerization of an alkyne; actual mechanisms are quite complex. In the box below each organometallic species are given: the metal VEC, its coordination number (CN), and its oxidation state. Reactants are shown entering and products leaving the cycle as it is read in the clockwise direction, with steps being indicated by numbers. These steps are simplified from a partial mechanism given in C. M. Lukehart, *Fundamental Transition Metal Inorganic Chemistry*, Brooks/Cole, Monterey, CA, 1985; p. 163.

their bond order, while insertions build larger ligands, and so on. But in addition, most of the major types of organometallic reactions that we have described result in distinctive patterns of changes in the key counts for metal atoms. First, we note that all of the B types of reactions, which occur at the coordinated ligands, result in few if any changes in counts at the metal atom. Second, the main types of association or dissociation of groups (the A types) logically change the coordination numbers by one, and in addition change either the oxidation number or the number of valence electrons. Third, the more complex reaction types (C) generally involve a change of two in the number of valence electrons. The ideal numerical changes for each reaction type are tabulated in Table 11.9 for ease in classifying reaction types; the actual numerical changes may differ if there are complications occurring with a given reaction, such as loss of a ligand while oxidative addition is occurring, or a change in hapticity while a

Table 11.9
Characteristic Changes in Metal Atom Counts During Major Classes of *d*-Block Organometallic Reactions

	Lewis Base Association or Dissociation	Lewis Acid Association or Dissociation	Lewis Acid or Base Attack on Coordinated Ligand[a]	Oxidative Addition or Reductive Elimination[b]	Insertion or Elimination[c]
VEC	± 2	0	0	± 2	$\pm 2(0)$
Metal coordination No.	± 1	± 1	0	± 2	$\pm 1(0)$
Metal oxidation No.	0	± 2	0	± 2	0

[a] For attack at a saturated α carbon, one number must be nonzero.
[b] These numbers differ slightly for five-coordinate or 17-electron organometallics: But there is always a nonzero change in the coordination number and in the oxidation number.
[c] The first number given is for the actual insertion or elimination step; the zeros include, in addition, the effect of the Lewis base association or dissociation step that normally accompanies insertion or elimination.

ligand is being added. Even the idealized numerical changes in metal counts are identical for two of the categories in the table: Watching the ligand is necessary in order to distinguish insertions–eliminations from Lewis base associations–dissociations.

Example 11.8
Classify the following organometallic reactions: (a) the synthesis of tropyliumchromium carbonyl cation [Eq. (11.14)]; (b) Step 6 of Figure 11.10; (c) the two steps in the formation of $(\eta^3\text{-}C_3H_5)Mn(CO)_4$ [Eq. (11.24)].

SOLUTION:

(a) $Cr(\eta^6\text{-}C_7H_8)(CO)_3$ is an 18-electron, six-coordinate Cr^0 complex that becomes $[Cr(\eta^7\text{-}C_7H_7)(CO)_3]^+$, which is also an 18-electron, six-coordinate Cr^0 complex; the lack of any change in the counts at the metal suggest Lewis base attack or abstraction at the coordinated ligand. Examination of Eq. 11.14 shows that the Lewis base H^- has in fact been abstracted by the attacking Lewis acid, $[Ph_3C]^+$, to give Ph_3CH. (b) In Step 6 of Figure 11.10, the 16-electron, five-coordinate Co(III) cobaltacyclopentadiene adds a Lewis base, RC≡CR, to give a 18-electron, six-coordinate Co(III) complex; that it is a Lewis base association is confirmed by the changes of 2 in the metal electron count, 1 in the coordination number, and 0 in the oxidation state. (c) In Step 1 of Eq. (11.24), the 18-electron, five-coordinate, Mn^{-I} $Mn(CO)_5^-$ is transformed into the 18-electron, six-coordinate, $Mn^I(\eta^1\text{-}C_3H_5)Mn(CO)_5$, which confirms the appearance of the equation as a Lewis acid association reaction (allyl chloride being the Lewis acid). In step 2, the 18-electron, six-coordinate, $Mn^I(\eta^1\text{-}C_3H_5)Mn(CO)_5$ is transformed into the 18-electron, six-coordinate, $Mn^I(\eta^3\text{-}C_3H_5)Mn(CO)_4$. However, examination of the reaction shows that this cannot be Lewis base attack on the coordinated ligand, as suggested by the absence of change in all metal counts. The equation shows that, in fact, a Lewis base, CO, has dissociated; the expected changes in electron counts and coordination numbers have not materialized because the coordinated ligand has undergone a hapticity change so as to fill the vacated coordination site and replace the missing two electrons left after the dissociation of CO.

11.8 Catalytic Cycles

A *catalyst* is a substance that increases the rate of a reaction without itself being consumed; it does this by participating repeatedly in a cyclic series of reaction steps that both produce the product and ultimately regenerate the catalyst. Many *d*-block organometallic compounds are important catalysts in organic synthesis, both in the laboratory and in industrial processes (Table 11.10); either homogeneous or heteroge-

Table 11.10
Some Homogeneous Catalytic Processes

Hydroformylation of alkenes (Oxo process)

$$RCH{=}CH_2 + CO + H_2 \xrightarrow{\text{Co (I) or Rh (I)}} RCH_2CH_2\overset{\overset{\displaystyle O}{\|}}{C}H$$

Oxidation of alkenes (Wacker process)

$$H_2C{=}CH_2 + O_2 \xrightarrow{\text{Pd (II) or Cu (II)}} H_3C\overset{\overset{\displaystyle O}{\|}}{C}H$$

Carbonylation of methanol to acetic acid (Monsanto process)

$$CH_3OH + CO \xrightarrow{[RhI_2(CO)_2]^-} H_3C\overset{\overset{\displaystyle O}{\|}}{C}OH$$

Hydrocyanation of butadiene to adioponitrile

$$H_2C{=}CHCH{=}CH_2 + 2HCN \xrightarrow{Ni[P(OR)_3]_4} NCCH_2CH_2CH_2CH_2CN$$

Oligomerization of ethylene

$$nH_2C{=}CH_2 \xrightarrow{NiHL} H_2C{=}CH(CH_2CH_2)_{n-2}CH_2CH_3$$

Alkene dismutation (alkene metathesis)

$$2H_2C{=}CHCH_3 \xrightarrow{WOCl_4/AlCl_2Et} H_2C{=}CH_2 + H_3CCH{=}CHCH_3$$

Asymmetric hydrogenation of prochial alkenes

$$(H)_2C{=}C(COOR)(NHCOR) + H_2 \xrightarrow{[Rh)(diPAMP)_2]^+} RCH_2C^*H(COOR)(NHCOR)$$

Cyclotrimerization of acetylene

$$3\,CH{\equiv}CH \xrightarrow{Ni(acac)_2} C_6H_6 \text{ (benzene)}$$

SOURCE: Adapted from J. Halpern, *Inorg. Chim. Acta*, **50**, 11 (1981) and from D. F. Shriver, P. Atkins, and C. H. Langford, *Inorganic Chemistry*, 2nd ed., Freeman, New York, 1994; p. 720.

[a] The ligand diPAMP is a chiral diphosphine, $(MeOC_6H_4)PhPCH_2CH_2CH_2PPh(C_6H_4OMe)$.

[b] The abbreviation acac = acetylacetonate or 2,4-pentanedionate $[Me\overset{\overset{O}{\|}}{C}CH\overset{\overset{O}{\|}}{C}Me]^-$ [see also Fig. 14.11(*e*)].

Table 11.11
Leading Organic Compounds and Metal Catalysts

Rank	Compound	U.S. Production, 1988 ($\times 10^9$ kg)	Metal-Containing Catalysts Used
1	Ethylene	16.58	
2	Propylene	9.06	$TiCl_3$ or $TiCl_4 + AlR_3$ (R = alkyl) (Ziegler–Natta)
3	Urea	7.15	
4	Ethylene dichloride	6.19	$FeCl_3$, $AlCl_3$
5	Benzene	5.37	Pt on Al_2O_3 support
6	Xylene (all isomers)	5.23	Pt + Re
7	Ethylbenzene	4.51	$AlCl_3$
8	Terephthalic acid	4.35	Co, Mn compounds
9	Vinyl chloride	4.11	$CuCl_2$ on KCl support
10	Styrene	3.90	ZnO, Cr_2O_3
11	Methanol	3.33	ZnO + other metal oxides
12	Formaldehyde	3.05	Cu, Ag
13	Toluene	2.93	Pt on Al_2O_3 support
14	Ethylene oxide	2.44	Ag
15	Ethylene glycol	2.22	Rh complexes
16	Cumene	2.18	
17	Methyl-*tert*-butyl ether	2.12	
18	Phenol	1.60	
19	Butadiene	1.45	Fe_2O_3, other metal oxides
20	Acetic acid	1.43	Mn acetate, Rh organometallic complexes

SOURCE: *Chem. Eng. News*, June 19, 1989, p. 39; G. L. Miessler and D. A. Tarr, *Inorganic Chemistry*, Prentice-Hall, Englewood Cliffs, NJ, 1991; p. 494.

neous catalysis are involved in the production of most of the leading organic chemicals (Table 11.11).

Heterogeneous catalysts are present in a different phase (normally, as insoluble solids) than the reactants; these are the most important economically since they are easily recovered (e.g., by filtration), but the mechanisms of their action are more difficult to study than are those of *homogeneous* catalysts, including most organometallic catalysts, since the latter are present in solution and can have their concentrations varied. In this section, we discuss organometallic catalysts, hence mainly homogeneous ones; heterogeneous catalysts will be discussed in subsequent chapters that deal with the properties of solids.

Hydroformylation of Alkenes (Oxo Process). In this process, an alkene, CO, and H_2 react to give an aldehyde having one additional carbon atom, which is normally subsequently reduced to an alcohol; the reaction is catalyzed by either of the 16-electron, four-coordinate M^I species $[HCo(CO)_3]$ (Fig. 11.11) or $[HRh(CO)(PPh_3)_2]$. Although these are the species that are repeatedly regenerated in catalytic cycles, they are not readily available; instead *catalytic precursors* are introduced that form these species under reaction conditions: for the cobalt-catalyzed system, $Co_2(CO)_8$ and H_2 are introduced, which react to give 2 $HCo(CO)_4$, which then dissociates CO. In the drawing of a catalytic cycle such as that shown in Figure 11.11, catalytic precursors such as these species are drawn in a branch above the main catalytic cycle. The mechanism drawn here was proposed in 1961, but still has not been rigorously tested with respect

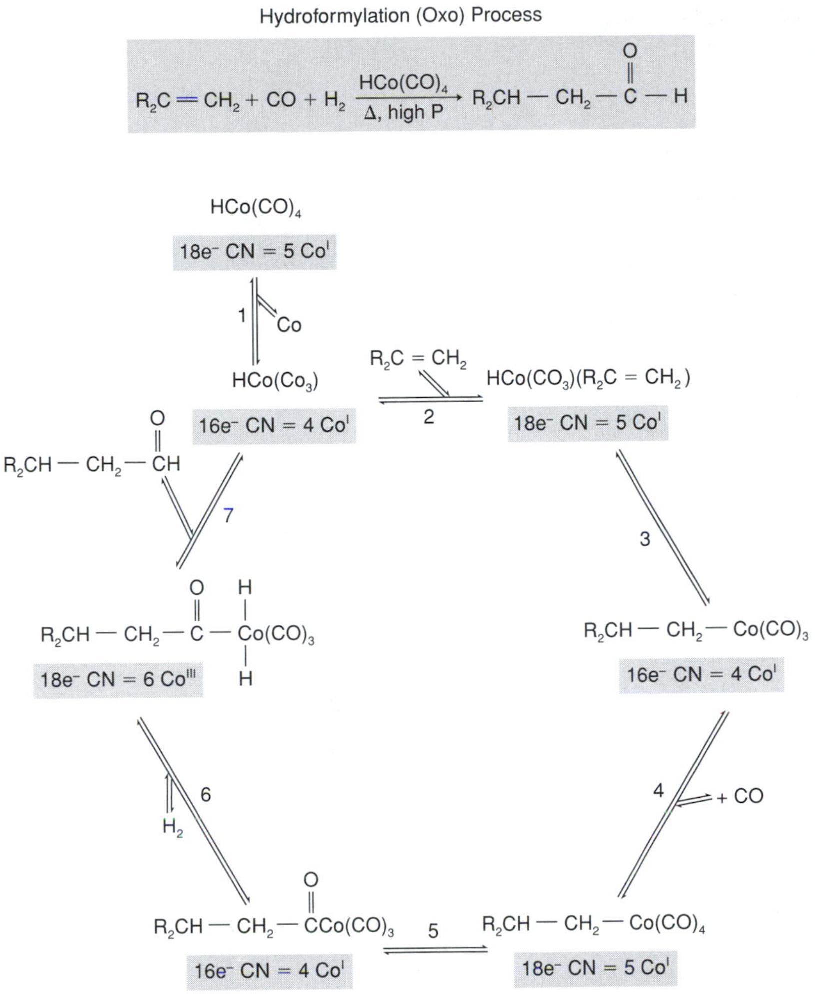

Hydroformylation (Oxo) Process

$$R_2C{=}CH_2 + CO + H_2 \xrightarrow[\Delta,\ \text{high P}]{HCo(CO)_4} R_2CH{-}CH_2{-}\overset{O}{\overset{\|}{C}}{-}H$$

$HCo(CO)_4$
$18e^-$ CN = 5 Co^I
1 Co
$HCo(Co_3)$
$16e^-$ CN = 4 Co^I
$R_2C = CH_2$
2
$HCo(CO_3)(R_2C = CH_2)$
$18e^-$ CN = 5 Co^I
3
$R_2CH-CH_2-Co(CO)_3$
$16e^-$ CN = 4 Co^I
4 + CO
$R_2CH-CH_2-Co(CO)_4$
$18e^-$ CN = 5 Co^I
5
$R_2CH-CH_2-\overset{O}{\overset{\|}{C}}Co(CO)_3$
$16e^-$ CN = 4 Co^I
6 H_2
$R_2CH-CH_2-\overset{O}{\overset{\|}{C}}-Co(CO)_3(H)_2$
$18e^-$ CN = 6 Co^{III}
7
$R_2CH-CH_2-\overset{O}{\overset{\|}{C}}H$

Figure 11.11
Proposed mechanism for the hydroformylation of alkenes (oxo process) using a cobalt carbonyl catalyst. The forward direction of the reaction is read from the top and then clockwise around the cycle, with reactants ($R_2C{=}CH_2$ at Step 2, CO at Step 4, and H_2 at Step 6 (shown entering from the rear on the cycle) and products shown leaving forward on the cycle ($R_2CH{-}CH_2{-}CH{=}O$ at Step 7).

to all steps shown; this formidable task has been completed for relatively few organometallic catalytic reactions.

The steps on the main catalytic cycle (2–7) may be described in the manner that we have done in Example 11.8. In Step 2, the alkene to be hydroformylated associates as a Lewis base (ligand) with the coordinately unsaturated $HCo(CO)_3$. The product of this reaction then undergoes an insertion reaction in Step 3. In Steps 4 and 5, another ligand, CO, is associated, followed by the insertion of a CO in the alkyl–cobalt bond to give the acyl–cobalt complex, $R_2CH{-}CH_2{-}C({=}O){-}Co(CO)_3$. In Step 6, the reactant H_2 is added oxidatively to the coordinately unsaturated acyl–cobalt complex to give a Co(III) complex, which finally undergoes reductive elimination of the product, $R_2CHCH{=}O$, and re-forms the original catalytic species, $HCo(CO)_3$.

Of the two catalytic species, the rhodium one promotes reactions under milder conditions (moderate temperature at 1-atm pressure) than does the cobalt one (150 °C and 250-atm pressure), which shows the typical faster reactivity of 4*d*-metal complexes mentioned earlier in this chapter. Its bulky phosphine ligands also suppress the formation of secondary aldehydes (secondary alkyls attached to a metal require a larger cone angle); hence, the rhodium catalyst competes economically despite its higher cost.

Hydrogenation of Alkenes Using Wilkinson's Catalyst. Wilkinson's catalyst, $ClRh(PPh_3)_3$, can be used for very selective hydrogenation of double bonds of alkenes at hydrogen pressures close to 1 atm. Some of its features illustrate the "fine tuning" and the delicate balances that go into making one compound a very selective catalyst while an analogue is inactive. The replacement of triphenylphosphine with alkylphosphines causes the loss of catalytic activity, presumably because the more strongly basic alkylphosphines do not dissociate as readily. Among alkenes, neither the smallest alkene, $CH_2{=}CH_2$, nor very sterically hindered alkenes, are hydrogenated; presumably ethylene forms too strong a complex, while the sterically hindered alkenes are blocked from coordination by the bulky triphenylphosphine ligands. If the alkene has several double bonds, only the least hindered of these is hydrogenated.

When chiral chelating phosphine ligands are used in place of triphenylphosphine, the catalyst can be used to synthesize optically active organic products such as pharmaceuticals. Most potentially chiral pharmaceuticals produced in the past were produced as racemic mixtures of the two chiral enantiomers, and were so administered. However, only one form is normally pharmaceutically active, which makes for a 50% waste; the wrong form is furthermore capable of causing undesired side effects in the patient. If the alkene becomes chiral upon coordination to the catalyst that already contains a chiral phosphine ligand, two diastereomeric complexes are produced; in favorable cases the desirable one of these is hydrogenated more rapidly to give the biologically active form of the pharmaceutical. Since the catalyst is recycled many times, a small investment in a chiral phosphine ligand pays off in the production of many moles of such medications as L-dopa, the chiral amino acid used to treat Parkinson's disease, and L-phenylalanine, another chiral amino acid used to produce the artificial sweetener, aspartame.

The mechanism of action of Wilkinson's catalyst is one of the few that has been quite thoroughly studied, and is now known to be more complex than the one shown in Figure 11.12; however, for pedagogic simplicity we wish to illustrate only the main part of the mechanism, which involves Lewis base dissociation of one Ph_3P in Step 1, oxidative addition of H_2 to Wilkinson's catalyst in Step 2, Lewis base association of the alkene in Step 3, insertion of the coordinated alkene into the Rh–H bond in Step 4, and reductive elimination of the product in Step 5.

The main point of controversy in this mechanism has been the involvement of a 14-electron species, $ClRh(PPh_3)_2$, that is not expected to be very stable by the 16- and 18-electron rule, and that has not been detected spectroscopically in the solution. But the kinetic evidence suggests that this species reacts 10^7 times as fast in oxidative addition of H_2 as does its precursor, $ClRh(PPh_3)_3$. Therefore most of the reactivity proceeds through the steps shown [some does proceed through an alternate first step, oxidative addition of H_2 to $ClRh(PPh_3)_3$]. As in the earlier cases of the 17- and 19-electron ligand substitution steps, a less stable and less abundant species may be sufficiently more reactive than a 16- or 18-electron species that it may be the one featured in the predominant branch of the mechanism.

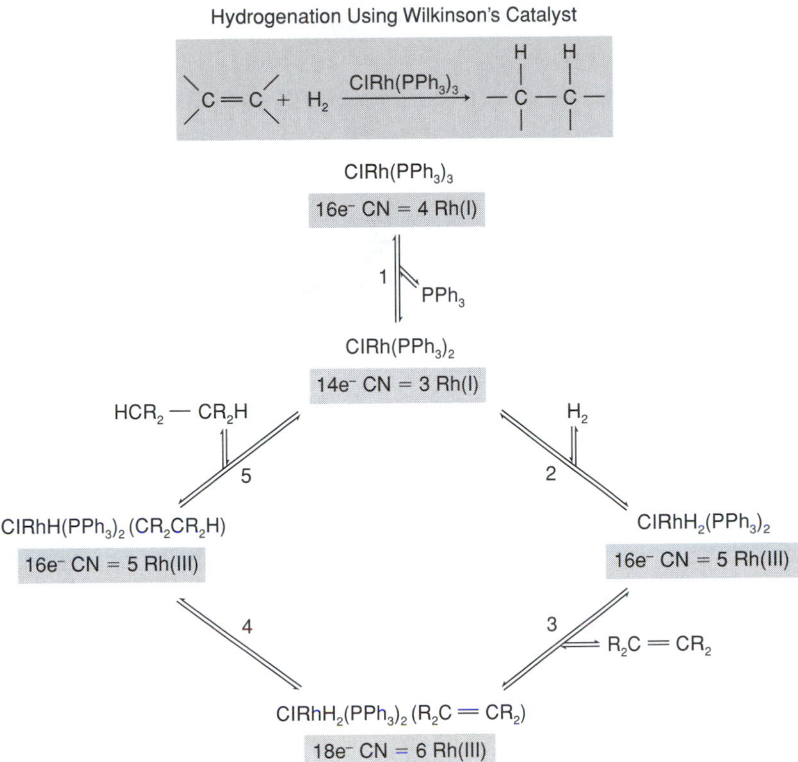

Figure 11.12
Main cycle of the mechanism for the hydrogenation of alkenes by Wilkinson's catalyst, $ClRh(PPh_3)_3$.

Monsanto Acetic Acid Process. Since 1971 this process has been used for producing concentrated (glacial) acetic acid from methanol and carbon monoxide. Although cobalt and iridium catalysts can be used, and chloride and bromide ligands can also be incorporated, the best results are obtained by adding the fifth-period element, rhodium, as $RhCl_3(H_2O)_4$, along with iodide ion and the carbon monoxide, which in the pre-catalytic parts of the cycle reduce the rhodium to Rh(I) in the form $[RhI_2(CO)_2]^-$. Figure 11.13 shows the principal catalytic cycle involved; the electron counts of the various organometallic species and the types of reactions involved are left to the student to fill out in Exercise 11.59.

Under acidic reaction conditions, the iodide ion and H^+ act as the strong acid HI, which readily reacts with methanol to generate water and iodomethane, MeI. This compound then undergoes reaction with the catalyst in Step 1; bromomethane or chloromethane would be less reactive here. At the end of the cycle (Step 4), the product actually emerges from the catalytic cycle as acetyl iodide, MeC(=O)I, which is very reactive with the water produced at Step 1 to give the final product, MeC(=O)OH, along with HI, which makes the system acidic and is then used up the next time around the cycle at Step 1.[45]

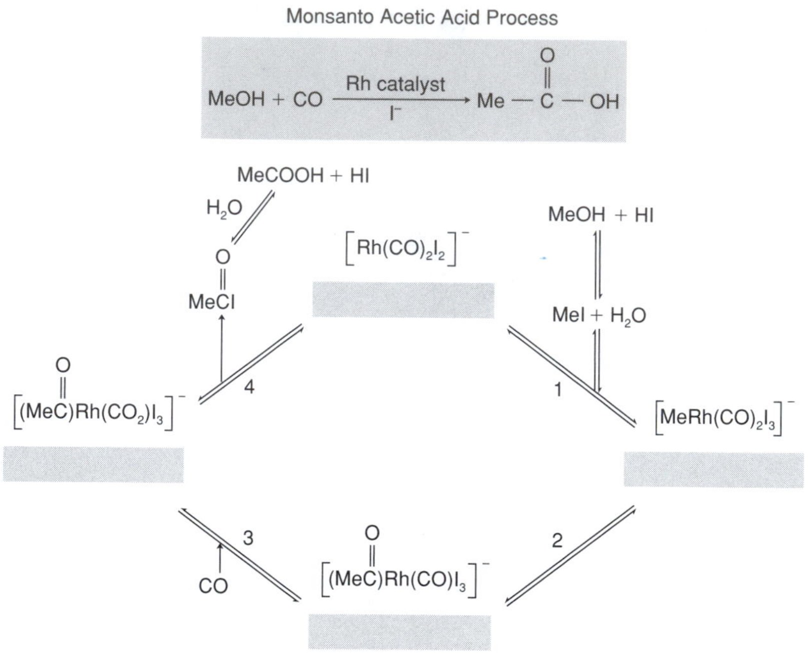

Monsanto Acetic Acid Process

$$\text{MeOH} + \text{CO} \xrightarrow[\text{I}^-]{\text{Rh catalyst}} \text{Me}-\overset{\overset{\large\text{O}}{\|}}{\text{C}}-\text{OH}$$

$[Rh(CO)_2I_2]^-$

$MeOH + HI \rightleftharpoons MeI + H_2O$

1 $[MeRh(CO)_2I_3]^-$

2 $[(MeC(=O))Rh(CO)I_3]^-$

3 CO

$[(MeC(=O))Rh(CO_2)I_3]^-$

4 $MeC(=O)I$ $\xrightarrow{H_2O}$ $MeCOOH + HI$

Figure 11.13

Principal catalytic cycle involved in the Monsanto process for producing glacial acetic acid from MeOH and CO. Electron counts are left for the student to fill in in Exercise 11.59.

Alkene Metathesis. This process is a sequence of reactions by which the two carbon atoms (with their substituents) at the two ends of a double bond are (in effect) snipped apart and then reattached in a random way:

$$CH_2{=}CH_2 + RCH{=}CHR \rightarrow 2\,RCH{=}CH_2 \tag{11.64}$$

This reaction is catalyzed by a variety of *d*-block metal catalysts through incompletely understood mechanisms that are generally thought to involve steps such as those sketched in Figure 11.14. These steps include the sequential coordination of each of the alkenes (Steps 1 and 5) to a carbene complex, their insertion into the M=C bond to generate metallacyclobutanes (Steps 2 and 6), the elimination from the other one-half of the metallacyclobutane ring of a different alkene (Steps 3 and 7), and the Lewis base dissociation of the altered alkene (Steps 4 and 8).

Although simple alkenes are not produced this way industrially, a number of more complex operations can be carried out conveniently, including the metathesis of C_{20} alkenes with small (C_2) alkenes to generate more saleable intermediate-sized (C_{11}) alkenes,

$$C_9H_{19}CH{=}CHC_9H_{19} + CH_2 = CH_2 \rightarrow 2\,C_9H_{19}CH{=}CH_2 \tag{11.65}$$

the opening of cyclic alkenes at their double-bonded carbon atoms and their reattach-

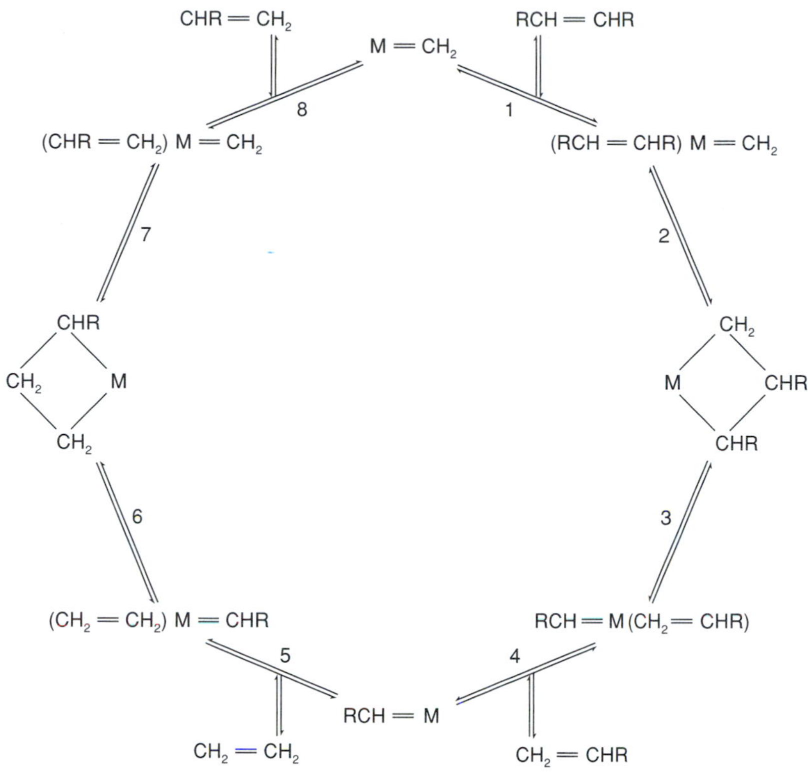

Figure 11.14
Key steps thought to be involved in the metathesis of alkenes,
$CH_2{=}CH_2 + RCH{=}CHR \rightarrow 2\,RCH{=}CH_2$.

ment in noncyclic (linear or polymeric) form to give polymeric alkenes, and the metatheses of alkynes.

Ziegler–Natta Polymerization of Alkenes. As our final example, this extremely important process, developed by Karl Ziegler in Germany and G. Natta in Italy during the 1950s, is used to polymerize alkenes such as ethylene and propylene stereospecifically to give polymers such as polyethylene and polypropylene, the production of which meant a quantum leap forward into the Age of Plastics. Ziegler discovered that, in hydrocarbon solution, the reactive halide $TiCl_4$ (Chapter 2) reacts with a *p*-block organometallic (Chapter 15), $AlEt_3$, to give $TiCl_3$ and then a hydrocarbon-insoluble titanium(III) alkyl, which functions as the heterogeneous catalyst. Heterogeneous catalysts such as these are advantageous to industry because they can be easily recovered by filtration. Unfortunately, it is difficult to determine their mechanisms of action.

Catalytic activity is also present in organic-soluble homogeneous catalysts based on titanocene dichloride (Cp_2TiCl_2), hafnocene dichloride, and especially zirconocene dichloride, when these are treated with a Lewis acidic organoaluminum polymer, methylaluminoxane, $(MeAlO)_n$, which both alkylates the Group 4 metal and abstracts

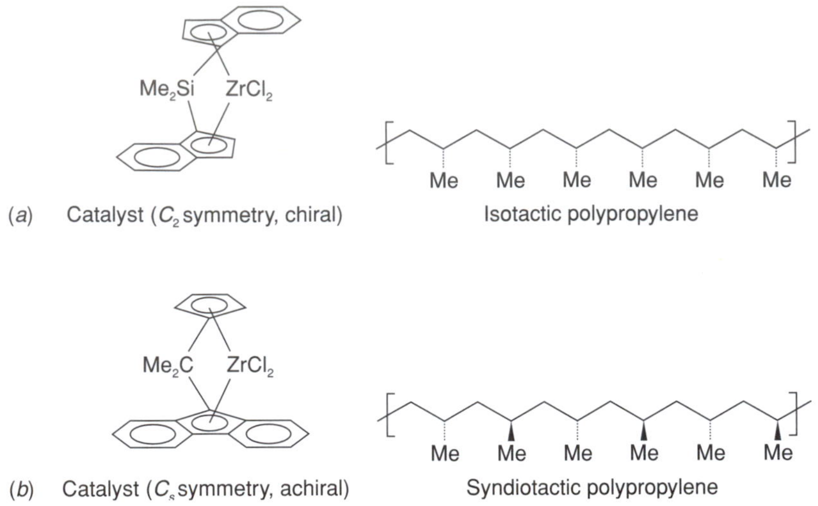

(a) Catalyst (C_2 symmetry, chiral) Isotactic polypropylene

(b) Catalyst (C_s symmetry, achiral) Syndiotactic polypropylene

Figure 11.15
(*a*) An *ansa*-*b*is(indenyl)zirconium dichloride, a catalyst precursor in the C_2 point group that gives isotactic polypropylene. (*b*) An *ansa*-(cyclopentadienyl)(fluorenyl)zirconium, a catalyst precursor in the C_s point group that gives syndiotactic polypropylene. [Adapted from *Chem. Eng. News*, Sept. 11, 1995, p. 15.]

an anion to generate the active catalyst species, for example, $(Cp_2ZrMe)^+$. This coordinately unsaturated cation coordinates an alkene next to the Zr–Me bond, into which it then inserts to give a Zr-Pr group. When the alkene is ethylene, the polymerization continues satisfactorily to give polyethylene, but with propylene ($CH_3CH{=}CH_2$), the chain terminates at much too short a length.

A major problem with propylene polymerization with either homogeneous or heterogeneous polymerization has to do with the stereoregularity of the process: If the CH_3 groups in the polypropylene polymer have random orientations along the chain, the plastic is weak. Two orderly arrangements are possible, which give two distinct polypropylenes, isotactic and syndiotactic (Fig. 11.15). Each has valuable but different properties; heterogeneous catalysis gives only the isotactic form.

A breakthrough improvement in the Ziegler–Natta process was achieved[46] with the recognition that both the stereoregularity of the polymer formed and the size of the alkene that could coordinate to the Zr atom could be controlled by changing the symmetry of the zirconocene and the bulkiness of the substituents on it. In an *ansa*-metallocene (Fig. 11.15), a short alkyl or silyl link between the rings pulls the two cyclopentadienyl rings toward each other, opening more space for coordination of bulky alkenes; such *ansa*-metallocenes can polymerize propylene to give long chains. A chiral (C_2 point group) *ansa*-metallocene catalyst involving two substituted cyclopentadienyl rings [Fig. 11.15(*a*)] produces isotactic polypropylene; a non-chiral (C_s point group) *ansa*-metallocene catalyst involving one doubly substituted cyclopentadienyl ring [Fig. 11.15(*b*)] produces the previously unavailable pure syndiotactic polypropylene. The new metallocene-based catalysts are technically superior to the older heterogeneous ones, but they are also more expensive; perhaps as much as $3 billion has been spent to develop this process for industrial use.

11.9 *Organometallic Catalysis in Biochemistry and Environmental Chemistry

Some industrial organometallic catalytic processes may have geochemical implications. The low-temperature activation of C–C bonds in hydrocarbons for reaction with H_2 to generate smaller hydrocarbon molecules would be helpful in cracking less marketable high molecular weight petroleum hydrocarbons to give the gasoline fraction. This type of chemistry has been achieved using a zirconium hydride catalyst on the surface of SiO_2, generated by the reaction of an organometallic zirconium compound with SiO_2.[47] This catalyzes the reaction of CMe_4 with H_2 to give ethane and methane; the hypothesized mechanism requires that a Me_3CCH_2Zr group undergo β elimination, not of hydride as mentioned in Section 11.8 (the β carbon bears no hydrogen) but rather of a methyl group, giving $Me_2C{=}CH_2$ and a MeZr group, which is then protonated by H_2 to give methane. The discovery of this reaction supports the suggestion[48] that the production of natural gas (CH_4) deposits in sedimentary deposits in the Earth's crust may be the result, not of high-temperature cracking reactions such as those now used in the petroleum industry, but of this type of ambient-temperature catalytic cleavage of higher hydrocarbons, involving *d*-block metals in the sediment and H_2 found abundantly in natural gas.

An extremely important advantage of organometallic catalysts, with their versatile variety of valence orbitals, their choice of many metallic atoms of varying electronegativities, and the wide variety of ligands of different donating and accepting types, is that they often can be fine tuned so as to give product yields of close to 100% with no byproducts. Older catalysts (such as strong acids) often produced many byproducts that had to be discarded, which created a serious hazardous-waste disposal problem, as well as wasting valuable reactants. From the chemical industry's point of view, the best solution to the problem of disposing of hazardous byproduct wastes is not to generate them in the first place; hence, research into catalysis of major industrial organic reactions by transition metal metallic and organometallic catalysts received top priority in funding among inorganic chemistry research projects in the United States for many years. This research was very successful, and transformed the way that industrial organic chemistry is carried out in many cases.

For many industrial processes, the main source of pollution at the present time is the solvent itself, which is volatile and is readily lost. Currently, a new wave of "green organic chemistry" research[49] is directed at adapting the homogeneous catalysts to function in more "environmentally benign" solvents such as supercritical carbon dioxide (which can exist only at high pressures, and which immediately evaporates to gaseous CO_2 when the pressure is released), or even water itself. First efforts in this area are focused on modifying the ligands in organometallic catalysts to impart solubility in the "green" solvent, rather than in the traditional organic solvents. In the case of water solubility,[50] this means adding additional ionic or polar functional groups to the phosphine ligands commonly employed in the catalysts to give modified phosphine ligands such as $(^{-}O_3SC_6H_4)_3P{:}$.

Producing 100% yields of desired compounds with the proper chirality and with the absence of toxic byproducts is also desirable in biochemical processes. But the known biochemical use of the catalytic ability of *d*-block organometallics (as opposed to the widely utilized coordination compounds) was, for a long time, confined to one example, that of coenzyme B_{12} or cobalamin (Fig. 11.16(*b*)). This compound is involved as a coenzyme in several enzymatic reactions that exchange **X** and **H** sub-

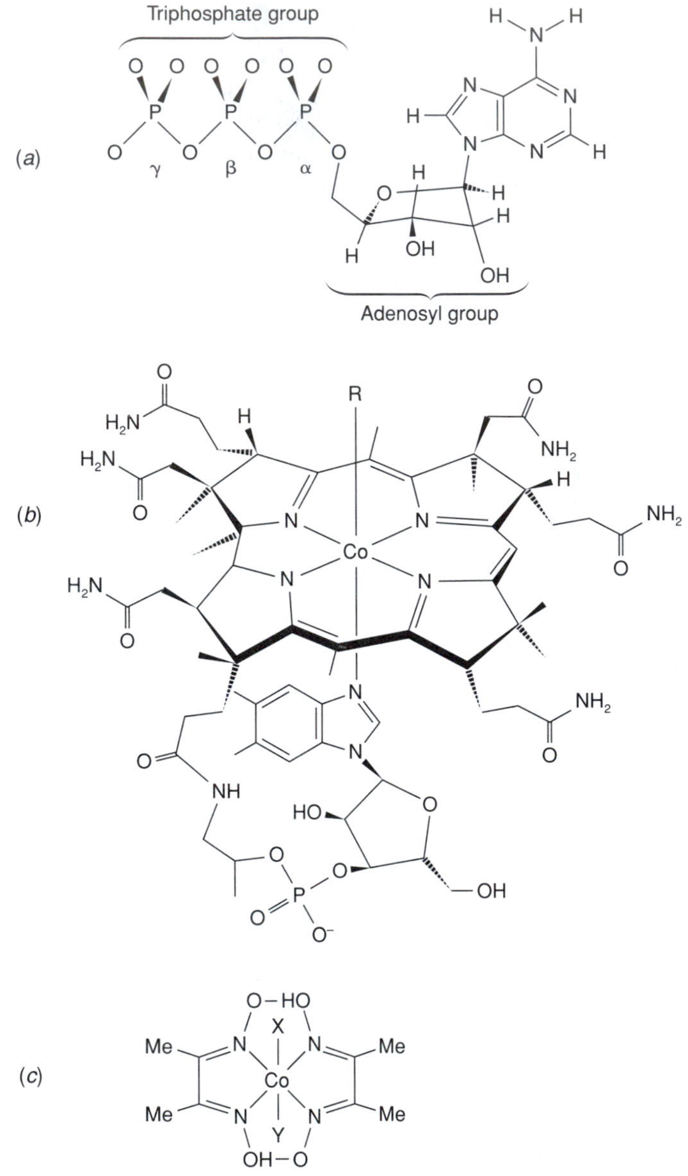

Figure 11.16
Structures related to coenzyme B_{12} (cobalamin). (*a*) The adenosyl group (when the triphosphate group shown is attached, this is ATP). (*b*) Cobalamins. For coenzyme B_{12}, Co = Co(III) and R = adenosyl; for Vitamin B_{12}, Co = Co(III) and R = CN; for methylcobalamin, Co = Co(III) and R = CH_3. For B_{12r}, Co = Co(II) and R = an electron, or an electron and H_2O (aquacobalamin). For B_{12s}, Co = Co(I) and R = an electron pair. (In all forms except B_{12s}, there is an overall positive charge on the complex.) (*c*) The cobaloxime [bis(dimethylglyoximato)cobalt] model system. [Adapted from S. J. Lippard and J. M. Berg, *Principles of Bioinorganic Chemistry*, University Science Books, Sausalito, CA, 1994; pp. 13, 133.

stituents on neighboring carbon atoms[51]:

$$\text{R–CH}\mathbf{X}\text{–CH}\mathbf{H}\text{–R}' \rightleftharpoons \text{R–CH}\mathbf{H}\text{–CH}\mathbf{X}\text{–R}' \qquad (11.66)$$

The X substituent being exchanged for a neighboring H atom can be CO_2H (in the action of methylmalonyl-coenzyme A mutase and glutamine mutase), NH_2 (ethanolamine ammonia lyase and L-β-lysine mutase), and OH (diol dehydrase).

The structure of cobalamin in its various forms is based on a cobalt atom having oxidation states of either +3 (in B_{12} itself), +2 (in B_{12r}), or +1 (in B_{12s}), coordinated by a tetradentate macrocyclic corrin ligand. The corrin ligand is broadly comparable to the porphyrin ligand found in chlorophyll and heme [Fig. 5.4(*a* and *b*), 8.20(*a*)], but differs in two significant ways: The macrocyclic ring is one atom smaller, and the ligand is saturated at many points on its outer periphery, which means that it is not as rigidly planar as the porphyrin ring. The fifth coordination site is occupied by a nitrogen donor atom from a benzimidazole ring attached by a long chain to the corrin ring; the chemical changes occur at the cobalt atom and its sixth coordination site.

The cobalt atom manifests typical *d*-block versatility as it is shuttled between three oxidation states, +3, +2, and +1, while the macrocyclic ring maintains a strong-field, low-spin environment. To generate coenzyme B_{12} itself, the cobalt is first reduced to the +1 oxidation state (d^8 electron configuration) in the form of square pyramidal B_{12s}, in which the cobalt atom possesses an active unshared electron pair in the d_{z^2} orbital, and can function as a very soft donor atom, as it does in the $[Co(CO)_4]^-$ anion. The B_{12s} then reacts with adenosine triphosphate [ATP, Fig. 11.16(*a*)] to displace the triphosphate ion, giving coenzyme B_{12}, with a cobalt–carbon bond to the adenosyl group, in which the oxidation state of cobalt is now +3, so that the reaction is one of oxidative addition.

It is interesting that much B_{12} chemistry can be carried out in a simpler model set of compounds, the cobaloximes [Fig. 11.16(*c*)], which are easily synthesized from dimethylglyoxime, the ligand commonly used to detect nickel(II) ion. (Note that in this structure, there is only a *quasi*-macrocyclic ring, completed by two strong hydrogen bonds; this *quasi*-macrocylic ring is of course more flexible than a truly conjugated, covalently bonded ring.) These also are most stable in d^6 strong-field cobalt(III) forms, but can be reduced either to cobalt(II) or to anionic cobalt(I) forms, which are also very soft bases, and react extremely rapidly in nucleophilic displacement reactions with even normally unreactive species such as CH_2Br_2, to give cobalt(III) again, now in an organometallic form containing a Co–CH_2Br group.

The actual action of coenzyme B_{12} in its characteristic reactions is not that of acid–base chemistry, however. Instead, the Co–C bond is cleaved homolytically, generating the reduced cobalt(II)-containing B_{12r} (d^7 square pyramidal) and an adenosyl radical. The adenosyl radical then abstracts the hydrogen atom to be exchanged, giving an organic radical, in which the X group migrates to the carbon atom having the unpaired electron, leaving the electron on the other carbon atom. This radical then takes the hydrogen atom back from the adenosyl hydrocarbon, giving the adenosyl radical back again.

These reactions of coenzyme B_{12} are vital to life, and humans suffer the deficiency disease pernicious anemia if they do not obtain enough B_{12} vitamin, which is normally made by bacteria in our small intestines. However, bacteria also make coenzyme B_{12} in natural waters, often in the form of methylcobalamin. In this form, the soft Me^- readily attaches to environmental soft-acid pollutants such as Hg^{2+}, Pb^{2+}, and Sn^{2+}

(and/or attaches Me radicals to the metallic forms of such elements), forming even softer and more toxic forms of these elements such as $MeHg^+$ and Me_2Hg (Section 5.11).[52]

Two other enzymatic systems have also been found to involve metal–carbon bonds. Anaerobic bacteria alkylate carbon monoxide or carbon dioxide to give acetyl groups using the enzyme carbon monoxide dehydrogenase; this enzyme involves a methyl–nickel bond.[53] A side reaction of xanthine oxidase produces a form of this enzyme now shown to involve a molybdenum–acyl bond with bonding both through C and O; it is suggested that the main reaction of this enzyme may produce a similar form.[54] Whether these and cobaloxime are rare examples of biological and environmentally active organometallic compounds, or are the first of many, remains to be discovered.

Study Objectives

1. Use a MO energy level diagram (or crystal field splittings) to justify the 18- (or 16-) electron rule for a given geometry. Exercises 1–3.

2. Compute the total valence electron counts (VECs) of *d*-block carbonyl or related organometallics to check for compliance with the 18-electron rule; use this rule to predict formulas of these compounds or ions. Assign metal atom oxidation numbers, name, and predict geometries of the compounds or ions. Exercises 4–14.

3. Know the syntheses and reactions of carbonyl organometallics. Exercises 15–19.

4. Understand the kind of information that can be obtained (e.g., about the presence of bridging carbonyl groups) from the carbonyl stretching frequencies in carbonyl and substituted carbonyl organometallics. Exercises 20–23.

5. Check the formulas of *d*-block metallocene or polyene carbonyl organometallics for compliance with the 18-electron rule; use this rule to predict formulas of these compounds or ions. Assign metal atom oxidation numbers and name the compounds or ions. Exercises 24–32.

6. Suggest methods of synthesis of metallocene or polyene carbonyl organometallics. Exercises 33–36.

7. Analyze the formulas of *d*-block organometallics containing ligands forming one σ bond for plausible coordination numbers, oxidation numbers, names, and compliance with the 18-electron rule; use this rule to predict formulas of these compounds or ions. Exercises 37–43.

8. Give possible syntheses and Lewis acid–base reactions (if any) of specified *d*-block organometallics containing ligands forming one σ bond, including dinuclear and polynuclear carbonyls. Identify the spectroscopic consequences of fluxional behavior in the latter. Exercises 44–48.

9. Use the isolobal analogy to suggest structures of possible dimeric molecules or multiply bonded organometallics, or bridged polynuclear organometallics. Exercises 49–53.

10. Classify organometallic reactions (such as those found in catalytic cycles) by type; or, given a reaction type and reactants, predict products or vice versa. Exercises 54–63.

Exercises

1. *(a) How many 2-electron donor ligands and what valence electron configuration of metal ion are necessary in order to have a square planar 16-electron *d*-block organometallic compound? (b) List all of the qualifying *d*-block neutral metal atoms; +1 ions; +2 ions; and +3 ions.

2. Referring to Table 8.6, decide whether (and if so, how) the 18-electron rule can be justified for the following geometries. In each case, if 18 electrons is not the preferred VEC, what number would be? (a) Trigonal prism; (b) pentagonal bipyramid; (c) cube; (d) square antiprism; (e) trigonal; and (f) tricapped trigonal prism.

3. Suppose a new geometry of organometallic complex produces crystal field splittings of the five *d* orbitals such that two *d* orbitals go *very strongly down* in energy, while the other three *d* orbitals go *very strongly up* in energy. (Let us say that this is so even after π-accepting effects have been taken into account.) The MO energy level diagram of this species would include not only these electrons, but also the electrons donated by the ligands. Would the 18-electron rule be expected to hold for this geometry if the complex were (a) five-coordinate? (b) seven-coordinate? In each case, if the 18-electron rule would not hold, tell how many valence electrons should be found for the metal.

4. *Compute the VEC in each of the following organometallic species, and tell which are and which are not 18-electron species: (a) $Pd(CO)_4$; (b) $Ni(PPh_3)_4$; (c) $Ni(PPh_3)_3$; (d) $[Fe(CO)_4]^{2-}$; (e) $[Cr(CO)_4]^{4-}$; and (f) $[Cr(CO)_5Xe]$.

5. Predict the formulas (including charges) of the following mononuclear organometallics: (a) the neutral carbonyl of Pd; (b) the least charged carbonyl anion of Cr; (c) the least charged carbonyl anion of Zr; (d) the cationic cobalt complex of MeNC; and (e) a square planar cationic complex of triphenylarsine with iridium.

6. Predict the formulas (including charges) of the following mononuclear organometallics. (a) The neutral carbonyl of Os; (b) the least charged carbonyl anion of Fe; and (c) the least charged anionic tantalum complex of PF_3.

7. Which *d*-block elements would be expected to form species of the following stoichiometries? (a) $[M(CO)_5]^{2-}$; (b) $[M(CO)_6]^{2-}$; (c) $[M(CO)_4]^{2-}$; (d) $[M(CO)_5]^{3-}$; (e) $[M(CO)_6]^{3-}$; and (f) $[M(CO)_4]^{3-}$.

8. *Assign oxidation numbers to the metal atom in the following compounds, name each species, and predict its geometry (including geometric isomer most likely by VSEPR theory and cone angles, if applicable): (a) $[Co(CO)_5]^+$; (b) $[Co(CO)_4]^-$; (c) $[Co(CNCH_3)_5]^+$; and (d) $Fe(CO)_4(PPh_3)$.

9. Assign oxidation numbers to the metal atom in the following species, name each species, and predict its geometry (including geometric isomer most likely by VSEPR theory and cone angles, if applicable): (a) $Cr(CO)_4(PPh_3)_2$; (b) $Mo(CO)_5(PF_3)$; (c) $[Re(CO)_5(CS)]^+$; and (d) $[Ir(PPh_3)_3(N_2)_2]^+$.

10. Compute the VEC in each of the following organometallic species. Then write the most likely geometry for each: (a) $V(CO)_6$; (b) $[Ir(CO)_5]^+$; and (c) $[Ni(PPh_3)_4]^{2+}$.

11. Compute the VEC, the metal-ion oxidation number, and the metal-ion *d* electron configuration (e.g., d^7) in each of the following species: (a) $[Co(C_5H_5)_2]^+$; (b) $[Fe(C_6H_6)(C_5H_5)]^+$; and (c) $[Pd(CO)_2(Me)Cl]$.

12. Write the formulas of the following compounds: (a) tetraethylammonium tetracarbonylferrate(−II); (b) cobalt pentacarbonyl cation; (c) [(18-crown-6)barium] pentacarbonyl(triethylstibine)niobate(1−); and (d) tetrakis(phosphorus trifluoride)-palladium.

13. Ellis has recently reported the synthesis of the diphosphine-substituted carbonyl $[Ti(CO)_5(Me_2PCH_2CH_2PMe_2)]$. (a) Classify the diphosphine ligand into one of the categories discussed in Section 5.1. (b) With this information, what is the likely coordination number of Ti in this complex? Discuss this in view of the maximum coordination number expected in the fourth period, and the maximum observed in neutral carbonyls. (c) Does this compound obey the 18-electron rule?

14. *Select the species from Table 11.1 that do not obey the 18-electron rule; identify the VEC and d^n electron configuration for the metal ion in each. If possible, discuss the reasons they might exist.

15. *Give reasonable syntheses of the following organometallic species; if they are ions, suggest a counterion to use in their final isolation and recrystallization from aqueous solution: (a) $[Nb(CO)_6]^-$ (b) $Fe(CO)_5$; (c) $Fe(CO)_4(PPh_3)$; (d) $\{Co[P(OPh_3)]_5\}^+$; and (e) $Pt[P(t\text{-}Bu)_3]_2$.

16. The conditions for the synthesis of $Cr(CO)_6$:

$$CrCl_3 + Al + 6\,CO \xrightarrow{AlCl_3,\,C_6H_6} Cr(CO)_6 + AlCl_3$$

are similar to those used to obtain bis(benzene)chromium [Eq. (11.2)], but the reaction does not stop at an analogous cation $[Cr(CO)_6]^+$. Explain this difference in the two reactions.

17. Why is 1 mol of CO given off in the synthesis of reduced carbonylate anions [Eqs. (11.4) and (11.5)]?

18. *What logical connection is there between the colorlessness of carbonyls, the position of CO in the spectrochemical series, and the fact that it is necessary to use UV rather than visible light to promote the dissociation of CO ligands in substitution reactions?

19. Write thermochemical cycles for the decomposition of the gaseous neutral metal carbonyls to solid metals and gaseous CO. Assuming an average M–CO bond dissociation energy of $160\,kJ\,mol^{-1}$, compute ΔH for this decomposition for known carbonyls. Does this explain why, for example, $Pt(CO)_4$ cannot be isolated at room temperature?

20. Explain why ν_{CO} decreases in each of these series: (a) $Ni(CO)_4$ ($2046\,cm^{-1}$) > $[Co(CO)_4]^-$ ($1883\,cm^{-1}$) > $[Fe(CO)_4]^{2-}$ ($1788\,cm^{-1}$) > $Mn(CO)_4]^{3-}$ ($1670\,cm^{-1}$). (b) In *fac*-$L_3Mo(CO)_3$, $L = PF_3$ (2090, $2055\,cm^{-1}$) > $L = PPh_3$ (1934, $1835\,cm^{-1}$) > $L = C_5H_5N$ (1888, $1746\,cm^{-1}$). (c) For the A_1 symmetry bands in $Ni(CO)_4$ ($2131\,cm^{-1}$) > $Ni(CO)_3\,[P(OMe)_3]$ ($2082\,cm^{-1}$) > $Ni(CO)_2[P(OMe)_3]_2$ ($2025\,cm^{-1}$) > $Ni(CO)[P(OMe)_3]_3$ ($1964\,cm^{-1}$). (d) For the A_1 bands in $Ni(CO)_3(PF_3)$ ($2111\,cm^{-1}$) > $Ni(CO)_3[P(OPh)_3]$ ($2085\,cm^{-1}$) > $Ni(CO)_3\,(PMe_3)$ ($2064\,cm^{-1}$).

21. Tell how you would use IR spectroscopy to tell which of each of the following sets of geometric isomers you had isolated: (a) *axial* or *equatorial* $Fe(CO)_4L$; (b) *axial–axial* or *axial–equatorial* or *equatorial–equatorial* $Fe(CO)_3L_2$; (c) *cis* or *trans* $Mo(CO)_4L_2$; (d) *fac* or *mer* $Mo(CO)_3L_3$; and (e) tetrahedral or square planar $M(CO)_4$.

22. *Draw resonance structures for the M–CO fragment of a carbonyl to account for the CO stretching frequency of (a) 1748 cm^{-1} in $[Ti(CO)_6]^{2-}$; (b) 1460 cm^{-1} in $[Cr(CO)_4]^{4-}$; and (c) 2251 cm^{-1} in $[Pt(CO)_4]^{2+}$.

23. A neutral polynuclear metal carbonyl has a rhodium atom single bonded to a technetium atom; there are also two bridging carbonyls joining the technetium and the rhodium atoms. Use the 18-electron rule to complete the Lewis structure drawing of the structure of this polynuclear carbonyl, showing (a) how many terminal carbonyls should be attached to each metal atom, and (b) the difference of the bond orders within the terminal and bridging carbonyls.

24. *Compute the VEC in each of the following species: (a) $NiCp_2$; (b) $Fe(C_6H_6)_2$; (c) $[Np(\eta^5\text{-}Cp)_3Me]$; and (d) $[CpFe(CO)_3]^+$.

25. Compute the metal oxidation numbers and VECs in, and then name, the following compounds or ions: (a) $CpMn(CO)_3$; (b) $C_6H_6Cr(CO)_3$; (c) $[Cr(C_6H_6)_2]^+$; (d) $[FeCp_2]^+$; (e) $[V(C_6H_6)_2]^-$; (f) $K_2[Ti(C_6H_6)_2]$; and (g) $[Nd(\eta^5\text{-}Cp)_3(OMe_2)]$.

26. Write the formulas of the following compounds and compute their VECs and coordination numbers: (a) ruthenocene; (b) rhodicenium ion; (c) benzene(cyclopentadienyl)iron(II) tetrachloroaluminate; (d) (η^6-benzene)(η^4-benzene)iron; and (e) bis(η^3-allyl)nickel.

27. Predict the formulas of the (a) cyclopentadienyl carbonyl of Co; (b) most stable sandwich species (Cp rings) for Ir; (c) most stable sandwich species (benzene rings) for Mn; and (d) a plausible mixed-(cyclic polyene) neutral sandwich species for Mn.

28. *In principle, what metals should form the most stable neutral sandwich molecules employing (a) two benzene rings; (b) two cyclopentadienyl rings; (c) one benzene and one cyclopentadienyl ring; (d) two cyclobutadiene rings; (e) one tropylium and one benzene ring; and (f) one cyclopropenium and one cyclopentadienide ring?

29. Predict what fourth-period *d*-block metals and how many carbonyl groups would be present in the neutral piano-stool molecules formed by each of the following polyenes. Also compute plausible coordination numbers for the metal atom in each: (a) C_4H_4; (b) C_5H_5; (c) C_6H_6; (d) C_7H_7; and (e) C_8H_8.

30. Show calculations to determine what fourth-period *d*-block metal should form the most stable −1 charged anionic sandwich or piano-stool molecules or ions employing: (a) one benzene and one cyclopentadienyl ring; (b) one tropylium and one cyclopentadienyl ring; and (c) one cyclobutadiene ring and three CO's.

31. The energy level diagram of metallocenes in the *d*-orbital region is similar to the corresponding region of bis(benzene)chromium (Fig. 10.24): electrons may occupy a_2'', e_1'', and e_2'' MOs. The spacing between these three orbitals can vary so as to give low-spin and high-spin compounds; in some cases intermediate-spin compounds may also be possible. (a) Write two possible valence *d* electron configurations (in terms of the a_2'', e_1'', and e_2'' MOs) for each of the metallocenes listed in the first row of Table 11.5. Use these to explain (b) the numbers of unpaired electrons given in the table for these metallocenes and also $MnCp_2$; (c) the M–C bond distances given for these metallocenes.

32. *Compute the coordination numbers and electron counts for the tetrakis(cyclopentadienyls) discussed at the end of Section 11.3.

33. Give an example of a water-soluble and a water-insoluble salt for each of the following cations and anions, and tell which one you would aim for in a synthesis; explain your choice. (a) $[CoCp_2]^+$; (b) $[Cr(C_6H_6)_2]^+$; (c) $[Co(CO)_4]^-$; (d) $[Co(CO)_5]^+$; and (e) $[Fe(CO)_4]^{2-}$.

34. Give reasonable syntheses of the following organometallic compounds: (a) $OsCp_2$; (b) $Mo(C_6H_6)_2$; (c) $MnCp(CO)_3$; (d) $Cr(C_6H_6)(CO)_3$; and (e) $[\eta^4$-(1,3-butadiene)$]Fe(CO)_3$.

35. What are the intended organometallics products of the following syntheses?

(a) $CoCp_2 + CO \xrightarrow{100\,\text{atm},\,130\,°\text{C}}$

(b) $C_4H_4Cl_2 + Cr(CO)_6 \xrightarrow{h\nu}$

(c) $3\,CoCl_2 + 6\,C_6Me_6 + 2\,AlCl_3 + Al \rightarrow$

(d) $Ti(g) + \text{excess}\ C_6H_6(g) \xrightarrow{77\,\text{K}}$

(e) $W(CO)_6 + C_6Me_6 \xrightarrow{h\nu}$

(f) $CpCo(CO)_2 + \text{cyclooctadiene} \xrightarrow{\Delta,\,16\,\text{hr}}$

36. Draw structures of the intended products of the following syntheses.

(a) $C_4H_4Cl_2 + Fe(CO)_5 \xrightarrow{h\nu}$

(b) $3\,RhCl_3 + 6\,C_6Me_6 + AlCl_3 + 2\,Al \rightarrow$

(c) $W(CO)_6 + C_6Me_6 + \text{1,5-cyclooctadiene} \xrightarrow{h\nu}$; then $\xrightarrow{\Delta,\,16\,\text{hr}}$

(d) $3\,MoCl_4 + 6\,C_6Me_6 + 3Al \xrightarrow{AlCl_3}$

(e) $Cr(CO)_6 + C_7Me_7{}^+ \xrightarrow{h\nu}$

37. *Compute the coordination number, metal oxidation number, VEC, metal *d*-electron count, and give the name of: (a) $MnCl(CO)_5$; (b) $Fe_2(CO)_9$; (c) D_{3d} $Co_2(CO)_8$; (d) $Ni(PPh_3)_2Cl_2$; and (e) $PdCl_2(CO)_2$.

38. For each of the following compounds, compute the VEC at the metal atom, and assign the oxidation number of the metal atom: (a) $Os(NO)_2Cl_2$; (b) $[Fe(NO)(CN)_5]^{2-}$; and (c) $Cr(NO)_4$.

39. In Exercise 1, the *d*-block metal atoms, +1 ions, +2 ions, and +3 ions giving 16-electron square planar complexes were identified. Write the formulas of the possible square planar carbonyl chlorides of these elements and ions.

40. Compute the coordination number, metal oxidation number, VEC, and the metal *d*- (or *f*-) electron count of (a) $Ti(\eta^5\text{-}Cp)_2(\eta^1\text{-}Cp)_2$; (b) $Zr(\eta^5\text{-}Cp)_3(\eta^1\text{-}Cp)_1$; (c) $Np(\eta^5\text{-}Cp)_4$; (d) $Ti(\eta^5\text{-}Cp)_2Cl_2$; (e) $Nb(\eta^5\text{-}Cp)_2Br_3$. (f) Explain why the coordination numbers can be so high in many of these compounds. (g) Explain why the VEC's can be less than 18 in many of these compounds.

41. Propose hapto numbers for the *cyclo*polyene ligands in the following organometallics that will result in compliance with the 18-electron rule: (a) $Cp_3Mo(NO)$; (b) $Cp^*RhC_6Me_6$; and (c) $Cp_2Cr(CO)_2$.

42. *Assign oxidation numbers in the following series of compounds:

(a) $Cr(NO)_4$, $Mn(NO)_3(CO)$, $Fe(NO)_2(CO)_2$, $Co(NO)(CO)_3$, $Ni(CO)_4$.

(b) $[Cr(NO)(CN)_5]^{4-}$, $[Mn(NO)(CN)_5]^{3-}$, and $[Fe(NO)(CN)_5]^{2-}$.

43. Compute the VEC's, metal oxidation numbers, and metal coordination numbers in the following chemical compounds: (a) Cp^*ReH_6; (b) $Cp^*Re(H_2)_3$; (c) $[Fe(H_2)H_2(PR_3)_3]$; and (d) $[Fe(H_2)_2(PR_3)_3]$.

44. Assuming metal coordination numbers of 6 and the applicability of the 18-electron rule, predict the formulas of (including identities of metals in) the following types of metal carbonyls: (a) tetranuclear, with the metal atoms forming a tetrahedron; (b) hexanuclear, with the metal atoms forming an octahedron; and (c) pentanuclear, with the metal atoms forming a trigonal bipyramid.

45. Considering inductive and other effects, arrange the following organometallic anions or molecules into a likely order of increasing basicity:

(a) $[Fe(CO)_4]^{2-}$, $[V(CO)_6]^-$, $[Mn(CO)_5]^-$, $[Co(CO)_4]^-$.

(b) $[Co(CO)_3L]^-$, where L = $P(OPh)_3$, PPh_3, PMe_3, PF_3.

(c) $Fe(CO)_3L_2$, where L = PPh_3, PMe_3, PMe_2Ph, $PMePh_2$ [see J. R. Sowa, V. Zanotti, G. Facchin, and R. J. Angelici, *J. Am. Chem. Soc.*, **113**, 9185 (1991)].

(d) $[(C_5X_5)Cr(CO)_3]^-$, where X = Cl, Me, H.

(e) FeL_5, where L = CN^-, NO^+, N_2, CO.

46. Identify plausible products of photolysis or pyrolysis (heating) of the following carbonyls: (a) $Fe(CO)_5$; (b) $Os_2(CO)_9$; and (c) $Rh_2(CO)_8$.

47. *Identify plausible products when the following dinuclear or polynuclear carbonyls are reacted with Na metal in the solvent diglyme, $MeOCH_2CH_2OCH_2CH_2OMe$: (a) $Mn_2(CO)_{10}$; (b) $Fe_3(CO)_{12}$; (c) $Co_2(CO)_8$; and (d) $Cp_2Fe_2(CO)_4$.

48. Predict the numbers of carbonyl ^{13}C NMR peaks you would observe in each of the following as a function of temperature, assuming complete fluxional exchange of bridging and terminal carbonyls at high temperatures. (It helps to make molecular models of these, or to consider their symmetry point groups.) (a) $Co_4(CO)_{12}$; (b) $Fe_2(CO)_9$; (c) $Fe_3(CO)_{12}$; and (d) C_{2v} $Co_2(CO)_8$.

49. *Which organometallic "pseudohalide" ion(s) listed in Tables 11.1 and 11.6 do not have a corresponding dinuclear "pseudohalogen" molecule listed in Table 11.7? Why might the pseudohalogen molecule not be formed?

50. Devise one fragment that falls in each of the three isolobal and isoelectronic series in Figure 11.8(*a*)–(*c*), using the following ligands attached to the central atom: (a) a *d*-block fragment including benzene and (some or no) carbonyls; (b) a *d*-block fragment including cyclobutadiene and (some or no) carbonyls; and (c) a *p*-block fragment including hydrogen, with B^- replacing C.

51. Propose a molecule of analogous structure to each of the following, but constructed from isolobal structural units from the *d* block of the periodic table: (a) *cyclo*-C_3H_6; (b) Br_2; (c) tetrahedral P_4; (d) SCl_2; and (e) the T_d cluster adamantane, $(CH_2)_6(CH)_4$.

52. Draw the structure of a simple organic compound that is isolobal to each of the following (some of the organics will not be stable compounds): (a) $[CpCr(CO)_3]_2$; (b) $Ir_4(CO)_{12}$; (c) $[CpMo(CO)_2]_2$; (d) $Os_3(CO)_{12}$; and (e) $[Co(CO)_3]_2(\mu_2\text{-}CO)_2$.

53. What would be the characteristic type of chemical reactivity of the isolobal and isoelectronic species listed in Figure 11.8(*d*)?

54. Classify the following organometallic reactions by type. (Note that some reactions may consist of two steps of different types.)

(a) $[(\eta^6\text{-}C_6H_6)Mn(CO)_3]^+ + BH_4^- \rightarrow [(\eta^5\text{-}C_6H_7)Mn(CO)_3] + 0.5\ B_2H_6$

(b) $[(\eta^4\text{-}CH_2{=}CH{-}CH{=}CH{-}CH_2OH)Fe(CO)_3] + H^+ \rightarrow$
$[(\eta^5\text{-}CH_2{=}CH{-}CH{=}CH{-}CH_2)Fe(CO)_3]^+ + H_2O$

(c) $[PhC({=}O)Mn(CO)_5] \rightarrow [PhMn(CO)_5] + CO$

(d) $[(ClCH_2CH_2)Re(CO)_5] \rightarrow [ClRe(CO)_5] + CH_2{=}CH_2$

(e) $[(CO)_nM{-}C({=}O)OH]^-$ [product of Reaction (11.56)] $\rightarrow [(CO)_nMH]^- + CO_2$

(f) $[Ni(CNR)_4] + Cl_2 \rightarrow [Ni(CNR)_4Cl_2]$

(g) $[Fe(CO)_5] + I_2 \rightarrow [Fe(CO)_4I_2] + CO$

55. Classify the following organometallic reactions by type; support your classification by computing the changes in the appropriate metal counts.

(a) $[Ni(PPh_3)_4] \rightarrow [Ni(PPh_3)_3] + PPh_3$

(b) $[(\eta^7\text{-}C_7H_7)Mn(CO)_3]^{2+} + BH_4^- \rightarrow [(\eta^6\text{-}C_7H_8)Mn(CO)_3]^+ + 0.5\ B_2H_6$

(c) $[Ni(CO)_4I_2] \rightarrow [Ni(CO)_4] + I_2$

56. Suggest likely products of the following reactions:

(a) Reductive elimination with $[Me_3Au(PPh_3)]$.

(b) Insertion of four $SnCl_2$ in $[PtCl_4]^{2-}$.

(c) Oxidative addition of the C–H bond of one ligand phenyl group to the metal in $[(Ph_3P)_3IrCl]$.

(d) Lewis base association of Ph_3P with $(\eta^5\text{-}Cp_2)_2Ni$.

(e) Lewis acid attack of $[Me_3O]^+$ on metal atom of $[(\eta^5\text{-}Cp)Fe(CO)_2]^-$.

(f) Insertion of SO_2 in $MeCo(CO)_4$.

57. Suggest likely organometallic products of the following reactions: (a) Lewis acid dissociation from $HMn(CO)_4(PPh_3)$; (b) Lewis base dissociation from $HMn(CO)_4(PPh_3)$; (c) Reductive elimination from $HPtCl_3(PPh_3)_2$; (d) Lewis acid attack of $[Me_3O]^+$ on the coordinated ligand of $(\eta^5\text{-}Cp)_2Ni$.

58. *What are likely reactants to use to obtain the following products by the reaction type indicated?

(a) $Pt[P(t\text{-}Bu)_3]_2$ by Lewis base dissociation.

(b) $[IrCl_3(PPh_3)_3]$ by oxidative addition.

(c) $[MeFe(CO)_4]^-$ by ligand elimination.

(d) $[MeFe(CO)_4]^-$ by Lewis acid association.

(e) A cyclic compound, $[HgFe(CO)_4]_4$, by Lewis acid association.

(f) $[(\eta^4\text{-}C_4H_4)FeCl_2(CO)_2]$ by oxidative addition and Lewis base dissociation.

(g) $[Cp_2Ti({=}CH_2)]$ by Lewis base attack on coordinated ligands.

59. (a) Fill in the metal counts for the organometallic compounds included in the Monsanto acetic acid mechanism (Fig. 11.13) and (b) identify the reaction type of each step.

60. Predict the products expected from (a) the ring-opening metathesis polymerization of norbornene (see an organic text for this structure) and (b) the metathesis of ethyne with 2-butyne.

61. From the word descriptions given of the two alternate hypotheses of the mechanism of action of Ziegler–Natta catalysts at the end of the chapter, draw your own two catalytic cycles using $Cp_2Zr(Me)(THF)$ (or the species obtained by dissociation of THF) as catalyst, and do metal atom counts on them. If any counts seem impossible, suggest a modification of the catalyst that would be more suitable.

62. For each of the following synthetic organic industrial processes, choose the name of the process and the organometallic catalyst. Choose among the following names: cyclotrimerization of an alkyne; Ziegler–Natta polymerization of an alkene; hydroformylation of an alkene; alkene metathesis. Choose among the following catalysts: $CpCoL_2$ or $Ni(acac)_2$; $TiCl_4 + AlEt_3$; $WOCl_4 + AlCl_2Et$; $HCo(CO)_4$.

(a) $RCH{=}CH_2 + H_2 + CO \rightarrow RCH_2CH_2CHO$

(b) $3\,RC{\equiv}CR \rightarrow C_6R_6$

(c) $n\,MeCH{=}CH_2 \rightarrow (MeCHCH_2)_n$

(d) $CH_2{=}CHR + CH_2{=}CHR \rightarrow CH_2{=}CH_2 + RCH{=}CHR$.

63. Write a plausible mechanism for the catalytic hydrogenation of $CH_2{=}CH_2$, using as catalyst $[RhH(CO)(PPh_3)_3]$. The first step of your mechanism should be a Lewis base dissociation reaction; the second step should be Lewis base association of $CH_2{=}CH_2$. With this beginning, complete the mechanism of hydrogenation.

Notes

1. T. J. Kealy and P. L. Pauson, *Nature* (*London*), **168**, 1039 (1951).
2. S. A. Miller, T. A. Tebboth, and J. F. Tremaine, *J. Chem. Soc.*, 632 (1952).
3. G. Wilkinson, *J. Organometal. Chem.*, **100**, 273 (1975).
4. The student may wonder why the metal *s* and *p* orbitals have been placed higher in energy in the MO diagrams than the metal *d* electrons, which are occupied after *s* orbitals in the characteristic valence electron configurations of neutral gaseous *d*-block atoms (Chapter 1). The metal *s* and *p* orbitals have much greater overlap with the σ electrons being donated by the ligands than do metal *d* orbitals, which are of smaller $\langle r_{max} \rangle$. Since the ligand σ-donor orbitals are fully occupied, the metal avoids an antibonding repulsion by taking a d^n electron configuration (both in complexes and in organometallic compounds) rather than the characteristic s^2d^{n-2} electron configuration found in free atoms. [C. W. Bauschlicher, Jr., and P. S. Bagus, *J. Chem. Phys.*, **81**, 5998 (1984)].
5. It is likely that MO calculations for some of the geometries in this table have not been made.
6. Ellis has pioneered the synthesis of very air-sensitive and even spontaneously explosive highly reduced carbonylate anions in which the metal ion has oxidation states as low as −4: see, for example, K. M. Chi, S. R. Frerichs, S. B. Philson, and J. E. Ellis, *J. Am. Chem. Soc.*, **110**, 303 (1988).
7. G. Hwang, M. Bodenbinder, H. Willner, and F. Aubke, *Inorg. Chem.*, **32**, 4667 (1993); G. Hwang, C. Wang, F. Aubke, H. Willner, and M. Bodenbinder, *Can. J. Chem.*, **71**, 1532 (1993).
8. L. S. Sunderlin, D. Wang, and R. R. Squires, *J. Am. Chem. Soc.*, **115**, 12060 (1993).
9. From P. S. Braterman, *Metal Carbonyl Spectra*, Academic, New York, 1975, and B. Douglas, D. McDaniel, and J. Alexander, *Concepts and Models of Inorganic Chemistry*, 3rd ed., Wiley, New York, 1993; p. 599.
10. W. Beck, *Angew. Chem. Int. Ed. Engl.*, **30**, 168 (1991).

11. C. A. Tolman, *Chem. Rev.*, **77**, 313 (1975); T. L. Brown and K. J. Lee, *Coord. Chem. Rev.*, **128**, 89 (1993).
12. R. C. Bush and R. J. Angelici, *Inorg. Chem.*, **27**, 681 (1988).
13. D. F. Shriver, P. Atkins, and C. H. Langford, *Inorganic Chemistry*, 2nd ed., Freeman, New York, 1994; p. 668.
14. G. J. Leigh, *Acc. Chem. Res.*, **25**, 177 (1992).
15. M. L. H. Green and D. K. P. Ng, *Chem. Rev.*, **95**, 439 (1995).
16. F. T. Edelmann, *New J. Chem.*, **19**, 535 (1995).
17. F. Calderazzo and G. Pampaloni, *J. Organometal. Chem.*, **500**, 47 (1995).
18. F. G. N. Cloke, *Chem. Soc. Rev.*, 17, 1993; G. B. Deacon and Q. Shen, *J. Organometal. Chem.*, **511**, 1 (1996); W. A. King, S. Di Bella, G. Lanza, K. Khan, D. J. Duncalf, F. G. N. Cloke, I. L. Fragala, and T. J. Marks, *J. Am. Chem. Soc.*, **118**, 627 (1996).
19. The substance C_4H_4, which has no stable form, is treated as the neutral antiaromatic molecule donating 4 π electrons.
20. J. Parry, E. Carmona, S. Coles, and M. Hursthouse, *J. Am. Chem. Soc.*, **117**, 2649 (1995).
21. P. Jutzi, W. Wieland, B. Neumann, and H.-G. Stammler, *J. Organometal. Chem.*, **501**, 369 (1995).
22. H. Köpf and P. Köpf-Maier, *Angew. Chem. Intl. Ed. Engl.*, **18**, 477 (1979); M. L. McLaughlin, J. M. Cronan, Jr., T. R. Schaller, and R. D. Snelling, *J. Am. Chem. Soc.*, **112**, 8949 (1990).
23. On detailed mechanistic investigation, however, many such reactions actually proceed through free radicals generated by a redox reaction between the easily oxidized *s*-block alkyl or aryl and the easily reduced organometallic halide.
24. T. Fehlner, *Chemtracts: Inorg. Chem.*, **4**, 1 (1992).
25. To complicate things further, NO sometimes forms *bent* bonds to metals, in which it acts as if it has an internal double rather than a triple bond: then in the LAB scheme, it is still a 2-electron donor, :N=O:$^-$, but in the RR scheme, it is a 1-electron donor, (·N=O:).
26. C. M. Lukehart, *Fundamental Transiton Metal Organometallic Chemistry*, Brooks/Cole, Monterey, CA, 1985; p. 217.
27. This free radical combination reaction is reversible in solution since the product has chromium with an excessive coordination number of 7.
28. In the LAB counting scheme, an M–M bond is formed [as in Eq. (11.36)] by one metal donating two electrons to the other metal. However, since either M could be the donor while the other acts as the acceptor, on average it is still true that each metal atom donates one electron to the other.
29. T. S. Janik, C. H. Lake, and M. R. Churchill, *Organometallics*, **12**, 1682 (1993).
30. R. H. Crabtree, X. Luo, and D. Michos, *Chemtracts: Inorg. Chem.*, **3**, 245 (1991); D. M. Heinekey and W. J. Oldham, Jr., *Chem. Rev.*, **93**, 913 (1993).
31. D. R. Evans, T. Drovetskaya, R. Bau, C. A. Reed, and P. D. W. Boyd, *J. Am. Chem. Soc.*, **119**, 3633 (1997).
32. C. Hall and R. N. Perutz, *Chem. Rev.*, **96**, 3125 (1996).
33. B. A. Arndtsen and R. G. Bergman, *Science*, **270**, 1970 (1995); *Chem. Eng. News*, Jan. 1, 1996, p. 7.
34. Unsubstituted H–C≡C–H can ionize to give the soft-base C_2^{2-}, which readily forms explosive acetylides with soft-acid metal ions (Section 6.3); this is considered an undesireable outcome by most researchers.
35. B. F. G. Johnson, *J. Chem. Soc. Chem. Comm.*, 211 (1976); B. F. G. Johnson and R. E. Benfield, *J. Chem. Soc. Dalton Trans*, 1554 (1978).
36. R. Hoffman, *Angew. Chem. Intl. Ed. Engl.*, **21**, 711 (1982).
37. As will be mentioned in Chapter 12, the Group 16(VI) elements are also called *chalcogens*, while the Group 15(V) elements are also called *pnicogens*.
38. R. R. Schrock, *Acc. Chem. Res.*, **12**, 98 (1979); L. Li, M. Hung, and Z. Xue, *J. Am. Chem. Soc.*, **117**, 12746 (1995).
39. A. J. Arduengo III, H. V. Rasika Dias, J. C. Calabrese, and F. Davidson, *Organometallics*, **12**, 3405 (1993).
40. F. D. Mango, *Adv. Catal. Relat. Subj.*, **20**, 291 (1969).
41. F. Basolo, *Polyhedron*, **9**, 1503 (1990).
42. J. M. O'Connor and C. P. Casey, *Chem. Rev.*, **87**, 307 (1987).
43. A. E. Stiegman and D. R. Tyler, *Comments Inorg. Chem.*, **5**, 215 (1986).

44. Since the tetracyanoethylene was already coordinated as a η^1 ligand before this reaction begins and becomes a η^2 ligand at the end, the change in the coordination number is only 1 in this case.
45. A variant of this process has been developed that has the advantage of utilizing oxygen and methane itself as reactants in place of the less readily available methanol: M. Lin and A. Sen, *Nature (London)*, **368**, 613 (1994); *Chem. Eng. News*, April 18, 1994, p. 5.
46. J. A. Ewen, *Sci. Am.*, **276**, 86 (May 1997); A. M. Thayer, *Chem. Eng. News*, Sept. 11, 1995, p. 15; H. H. Brintzinger. D. Fischer, R. Mülhaupt, B. Rieger, and R. M. Waymouth, *Angew. Chem. Intl. Ed. Engl.*, **34**, 1143 (1995).
47. J. Corker, F. Lefebvre, C. Lécuyer, V. Dufaud, F. Quignard, A. Choplin, J. Evans, and J.-M. Basset, *Science*, **271**, 966 (1996); *Chem. Eng. News*, Feb. 19, 1996, p. 24.
48. F. D. Mango, J. W. Hightower, and A. T. James, *Nature (London)*, **368**, 536 (1994).
49. See, for example, J. Haggin, *Chem. Eng. News*, Oct. 10, 1994, p. 28; J. A. Cusumano, *J. Chem. Educ.*, **72**, 959 (1995); T. J. Collins, *J. Chem. Educ.*, **72**, 965 (1995).
50. P. Kalck and F. Monteil, *Adv. Organomet. Chem.*, **34**, 219 (1992).
51. S. J. Lippard and J. M. Berg, *Principles of Bioinorganic Chemistry*, University Science Books, Mill Valley, CA, 1994; pp. 13, 70, 133–136.
52. P. J. Craig, *Organometallic Compounds in the Environment*, Wiley, New York, 1986; J. Thayer, *Organometallic Chemistry: An Overview*, VCH, New York, 1988; Chapter 13.
53. M. Kumar, D. Qiu, T. G. Spiro, and S. W. Ragsdale, *Science*, **270**, 628 (1995).
54. B. D. Howes, B. Bennett, R. C. Bray, R. L. Richards, and D. J. Lowe, *J. Am. Chem. Soc.*, **116**, 11624 (1994).

CHAPTER
12

The Elements and Their Physical Properties

With Applications to Materials Science

12.1 Interunit Forces and Physical Properties

In the remainder of this book, we will be concerned with the physical and chemical properties of materials that tend to be of more complex or extended structure than the ions and simple molecules considered so far. In this section, we discuss some general principles that affect the physical properties of inorganic materials; we then illustrate these in the rest of the chapter using the elements themselves as examples. Inorganic materials (i.e., substances valued for their physical properties) are of enormous importance, since the earth we stand on and many of the materials out of which we build the edifices of modern civilization consist of inorganic substances; for many modern purposes (in electronics, spacecraft, and so on) the properties must be such as to withstand extreme conditions, or to show great strength and lightness, or to have various unusual combinations of properties. The principles in this section would hopefully be of use in suggesting the kinds of new materials that you would want to synthesize if you were seeking new materials with certain unusual physical properties.

Fundamentally, many of the physical properties of substances depend on how the **structural units** of the bulk material are **linked** together. For the elements themselves, the structural units are *atoms* of that element, which are linked by shared electron pairs. We will build the bulk element by assembling atoms in much the same manner that we assembled organometallic compounds from isolobal fragments in Section 11.6. (Recall that atoms were included among the isolobal fragments in Table 11.6.) Sharing electron pairs between structural units allows us to build up molecules of varying sizes; if extra electrons are then provided (or taken away) we build up ions of varying sizes.

Structural Types. At one extreme, the bulk material (e.g., a crystal) may consist of numerous, separate, much smaller molecules linked only very loosely together by forces that are not what we normally consider to be chemical bonds. These **monomeric** substances include atomic substances, in which the structural unit is a simple atom such as He, or the more common **molecular** substances, in which the structural unit is a (at least fairly) simple polyatomic molecule.

At the other extreme, the entire piece (e.g., crystal) may consist of structural units that are linked into one giant three-dimensional **network polymeric** molecule or ionic lattice by strong chemical-bonding forces. Ionic lattice substances are like network polymers in that the linkage is in three dimensions, but in these the interunit linking is done by Coulombic attractions of opposite charges, rather than covalent bonding.

The elements themselves encompass both extremes: The structural units of the Group 18(VIII) gases are the same as the "molecules" of these substances; while in Group 14(IV), a diamond crystal is one giant molecule consisting of linked carbon atoms, and a crystal of lead metal similarly consists of linked lead atoms. Although the structural units of an ionic crystalline solid such as CaF_2 are held together by still different chemical forces than those found in diamond or lead, the result is very similar from this point of view: Ionic, covalent, and metallic attachments of structural units to each other are strong, and give strongly linked materials.

Between these extremes there are several intermediate possibilities, as illustrated in Figure 12.1, in which the basic structural unit is represented by a sphere. **Polymeric** substances can be divided into categories. In **linear** or **chain** or **one-dimensional polymers** [often denoted with ${}^{1}_{\infty}$ in front of the formula of the structural or simplest formula unit, for example, ${}^{1}_{\infty}[CH_2CH_2]$ for polyethylene], each structural unit is bonded to two others to give long-chain molecules or ions. In **layer** or **sheet** (or **two-dimensional**) **polymers** (denoted with ${}^{2}_{\infty}$, e.g., ${}^{2}_{\infty}[C]$ for graphite), many or all structural units are bonded to more than two other units each, so that the molecule or ion extends indefinitely in two dimensions. In **network** (or **three-dimensional**) **polymers** (denoted with ${}^{3}_{\infty}$; e.g., ${}^{3}_{\infty}[SiO_2]$ for silicon dioxide), the individual structural units have enough links with orientations allowing the molecule or ion to extend indefinitely in three dimensions.[1]

Between the polymers and the monomers are the **oligomeric** molecules or ions, which contain a medium number of structural units linked together. In **linear** or **chain oligomeric** molecules or ions, each central structural unit is bonded to two others (by bonds represented by lines in Fig. 12.1), but after a definite number of units have been linked, the chain is terminated by a slightly different unit. In **cyclic oligomeric** molecules or ions, all units are bonded to two others and form a ring, so that there are no terminal units; these have the same simplest formulas as the chain polymers. In **cluster oligomeric** molecules or ions, the units are bonded in a small three-dimensional assembly to more than two other units. If all of the units remain on the surface of a polyhedron, as in C_{60}, we have a **polyhedral** molecule or ion; these are related to the layer polymers (e.g., graphite) by the curvature of the sheet and the joining of its dangling edges.

Physical Properties. As postulated in the ideal **kinetic–molecular theory** of gases, molecules capable of existing in the gaseous state exert no attractive forces on each other, move independently of each other, and remain relatively far apart. In practice, however, there are attractive forces between molecules; the weakest of these is the van der Waals force, which occurs between molecules of all substances. The van der Waals force results when a momentary unsymmetrical distribution of electrons in one molecule (a temporary dipole) induces an opposed momentary unsymmetrical distribution in a neighboring molecule; the momentarily oppositely charged ends of the two molecules then attract each other. In small molecules, the van der Waals force is very weak as compared to the average kinetic (thermal) energies that the molecules possess at room temperature. But in very large molecules with numerous highly polarizable

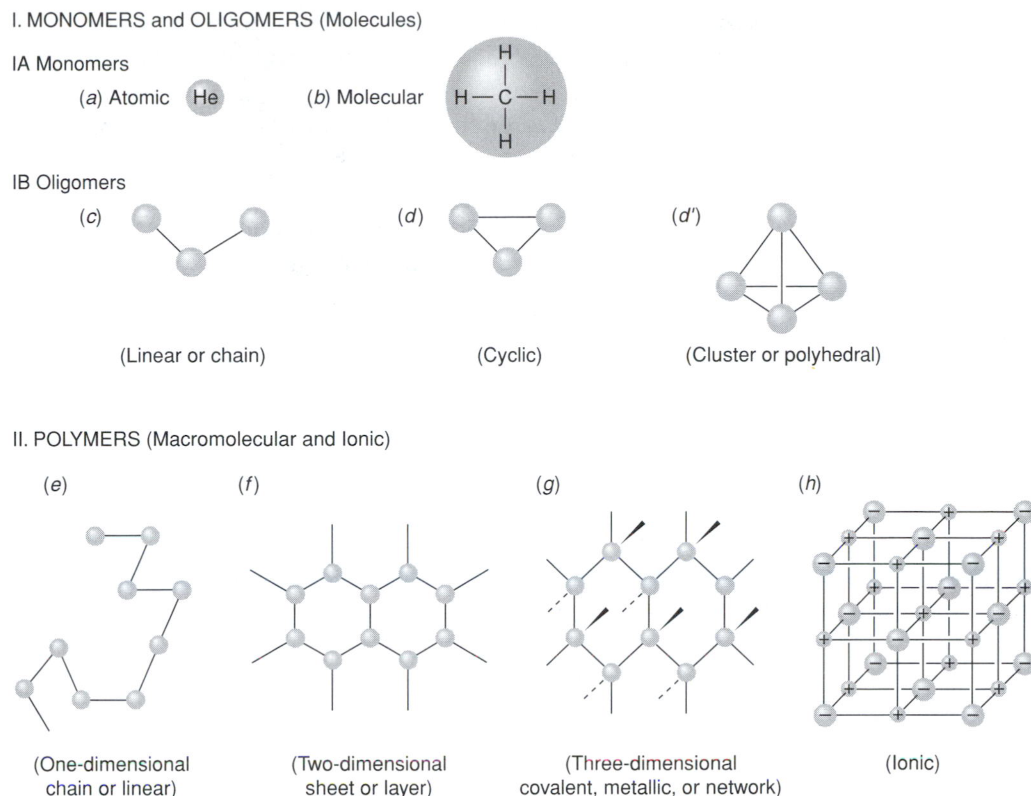

Figure 12.1
Main structural classifications of materials, and their subclassifications. Structural units are represented by spheres and linking forces (covalent bonds or interionic attractions) are represented by lines.

(relatively loosely bound) electrons, the van der Waals forces may be so substantial that the kinetic energy available to molecules at room temperature is not enough to overcome such an attraction; hence, monomeric or oligomeric substances consisting of sufficiently large molecules are not gaseous at room temperature and atmospheric pressure. (Of course, they may become gases at higher temperatures or lower pressures.)

The kinetic–molecular picture of the liquid state postulates that molecules or ions are held close by forces that are strong compared with the thermal energy at that temperature. But the molecules or ions are free to flow around each other, which implies that their intermolecular attractive forces can be stretched. The molecules or ions must also either be relatively small, or if they are chain polymers, the chains must not be rigid; they cannot be layer or network polymers. Medium-sized molecular substances, or smaller molecules linked by somewhat stronger forces such as dipole–dipole forces and hydrogen bonding (i.e., water and alcohols), are most likely to be liquids at room temperature.

In the solid state, molecules or ions are packed closely together and exert attractive forces on one another that are strong enough to hold the units in position in a lattice despite their vibrational energies. Large covalent molecules often have strong enough van der Waals forces to be solid at room temperature, but these forces seldom rival in strength the interunit covalent bonds or Coulombic forces found in the high-melting solid two- or three-dimensional polymeric substances.

Thus *molecular* (and especially monomeric molecular) substances tend to be gases, liquids, or low-melting (say, below 300 °C) solids at room temperature, and to have relatively low heats of fusion and heats of vaporization.[2] Other properties generally follow from the absence of strong intermolecular attractions: molecular substances tend to have higher solubilities and may even be soluble in nonpolar solvents. Like a brick wall with no mortar, these substances are lacking in structural strength: They are deformable (soft in a mechanical, not a hard–soft acid–base (HSAB), sense). The distances between the separate molecules in the solid (or liquid) phase are governed by the relatively long van der Waals radii of whatever atoms are on the outside of the units, so these substances tend to have relatively low densities.

In contrast, the structural units of *macromolecular* or *ionic* substances are held together by strong chemical forces: covalent or ionic bonding, which extends throughout the grain or crystallite of the substance. There is little or no chance for the units to acquire mobility with respect to each other, so these are high-melting solids; these substances have high heats of fusion and vaporization. Since the structural units are strongly bonded together at distances that are the sums, not of long van der Waals radii, but rather of relatively short covalent or ionic radii, the substance tends to have a relatively high density, and may possess greater structural strength, that is, to be hard in a mechanical sense. One important consequence of the higher density of more polymerized materials is that greater degrees of polymerization are favored when the material is under high pressure, since the polymerized material occupies less volume per mole, and hence tends to relieve the stress of the increased pressure.

The subcategories of polymers have some characteristic variations on these typical properties of polymers. The one-dimensional polymers typically differ from other polymers in that the bonding often does not give rigidity to the material. Consequently, chain polymers can be mechanically soft and have low melting points (they may even be liquids at room temperature); but they tend to be viscous (since the long, spaghetti-like molecules can easily become entangled, and hence resistant to flow) and have high boiling points that require prior decomposition of chemical bonds. The two-dimensional polymers tend to be rigid in those two dimensions, but not in the third. Consequently, they often consist of flat crystals that readily slide over each other (some are used as lubricants) and are mechanically soft in that direction. Melting or boiling, however, is not to be expected prior to decomposition. They may even form *intercalation compounds*, in which foreign molecules are sandwiched between successive layers of the layer polymer.

The process of melting network polymeric solids requires stretching and bending strong interunit attractions, therefore it occurs at high temperatures. Melting can happen for ionic solids, since Coulombic attractions are omnidirectional, and another attraction can replace the one lost when an ion becomes mobile. When *covalent* layer or network solids are melted, however, covalent bonds in all but one dimension must be ruptured. In doing so, it may be easiest to rearrange bonds to give the volatile monomeric form of the same compound: Hence, the covalent layer or network solids frequently **sublime** (pass directly from the solid to the gaseous state) at high temperatures. The process of melting or boiling such a solid really is a process of chemical decomposition. For the same reasons, network covalent solids tend to be somewhat harder (in the physical sense) than ionic solids.

12.2 Physical Properties, Allotropes, and Uses of the Nonmetallic Elements

The elements themselves show an enormous diversity in their simplest physical properties, such as their physical states and the ease with which they become gases or individual atoms. At one extreme, helium is a gas (at 1-atm pressure) at any temperature in excess of 4 K. At the other extreme, tungsten metal does not become a gas until a temperature of 5936 K is reached. It is said that tungsten is so nonvolatile at ordinary temperatures that, if the entire universe consisted of tungsten at room temperature, *one atom* of tungsten would be in the vapor phase! These differences are rooted in the structures of the elements. Helium consists of distinct monomeric units (individual atoms), while tungsten, other metals, and many nonmetals contain polymerized structural units (atoms) at all temperatures below their boiling points; chemical bonds must be broken to convert these elements to gaseous atoms. The energy required to do this is the *heat of atomization* or atomization energy of the element (Table 7.1).[3] As pointed out previously in Section 7.3, heats of atomization of elements have an unusual periodicity in that they reach their maximum values in the *middle* of a period.

A second property of the elements showing similar diversity and periodicity is the *density* of the elements, which varies from 0.000090 g cm^{-3} for $H_2(g)$ to 22.5 g cm^{-3} for Os(s) at standard temperature and pressure. Of course, part of this variation is simply due to the fact that the masses of individual atoms increase going down the periodic table; we may divide through by the atomic weights to obtain the number of moles of atoms of each element present in a fixed volume (the atomic density) of the element (Fig. 12.2). These two properties also show the unusual periodic trend of reaching a maximum in the middle of a period: the most strongly polymerized forms of a substance have the highest densities.

We may well wonder why the elements are most effectively polymerized (and thus require the greatest input of energy to be atomized) in the middle of a block of elements. We know, of course, that the bonding between atoms in the nonmetallic elements is pure **nonpolar covalent** bonding. A two-centered nonpolar covalent bond is formed by the overlap of an appropriate orbital on one atom with an orbital on a neighboring atom, with each atom contributing one of the two electrons in the resulting bond. Since each *p*-block atom has four valence orbitals (one *s* and three *p*), the maximum number of two-centered bonds can be achieved when each atom has four valence electrons, one per orbital (Table 12.1). As indicated in this table, the electrons may be shifted among the orbitals so as to achieve one electron per orbital; the resulting electron configuration is called the **prepared-for-bonding electron configuration**.

Table 12.1
Horizontal Periodic Trends in Numbers of Unpaired Electrons Available for Bonding

Lewis structure of atom	Na	Mg	·Al	·Si	·P·	:S·	:Cl·
Valence electron configuration	s^1	s^2	s^2p^1	s^2p^2	s^2p^3	s^2p^4	s^2p^5
Prepared-for-bonding configuration	s^1	s^1p^1	s^1p^2	s^1p^3	s^2p^3	s^2p^4	s^2p^5
Number of unpaired electrons available	1	2	3	4	3	2	1
Lewis structure of atom	Cs	Ba	·Lu	·Hf	·Ta·	·W·	:Re:
Valence electron configuration	s^1	s^2	s^2d^1	s^2d^2	s^2d^3	s^2d^4	s^2d^5
Prepared-for-bonding configuration	s^1	s^1d^1	s^1d^2	s^1d^3	s^1d^4	s^1d^5	s^1d^6
Number of unpaired electrons available	1	2	3	4	5	6	5

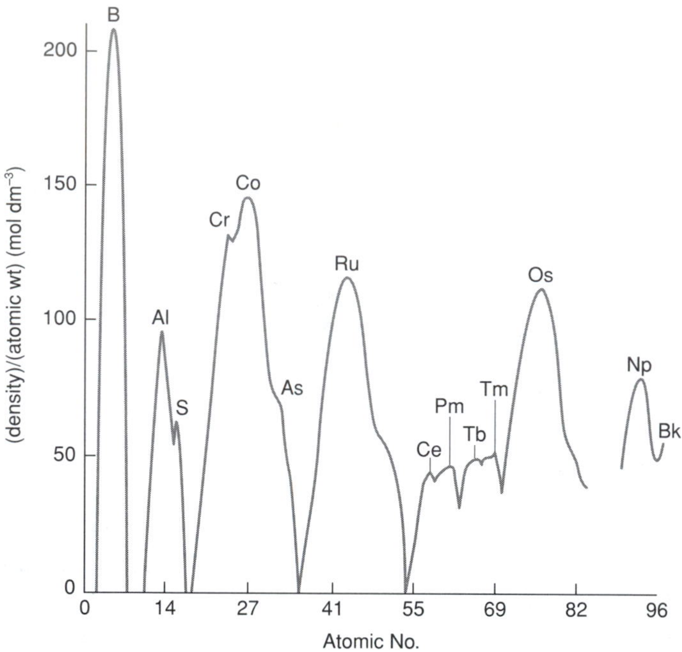

Figure 12.2
The atomic densities (numbers of moles of atoms of the elements present in 1 L or 1 dm^3) of the elements under standard conditions of temperature and pressure.

Adding electrons beyond four per atom reduces the number of covalent bonds that can be formed, since filled orbitals cannot effectively overlap half-filled or filled orbitals on other atoms. Thus Group 15(V) atoms can participate in only three pure covalent bonds; Group 16(VI) atoms can form only two; Group 17(VII) can form only one, and Group 18(VIII) can form no pure covalent bonds.

Consequently, we expect the Group 18(VIII) elements to exist only as monoatomic substances; atoms of Group 17(VII), with one prepared-for-bonding unpaired electron each, are expected to give dimeric molecules, X_2. The Group 16(VI) atoms have two prepared-for-bonding unpaired electrons. Therefore they can link to form either chain polymers or cyclic oligomers, however, the possibility also exists that these two unpaired electrons can be shared between just two atoms to form diatomic double-bonded molecules (X=X). For some of the Group 16(VI) elements, more than one of these possibilities are in fact realized; these alternate structural forms of these elements are known as **allotropes**.

The Group 15(V) elements, with three unpaired electrons in the prepared-for-bonding electron configuration, also have a variety of possibilities. At one extreme, each atom could serve as a link to three other atoms, producing a layer polymeric structure; at the other extreme, two atoms could pair these electrons by forming one triple bond, once again producing diatomic molecules; intermediate possibilities can also be imagined. The Group 14(IV) elements have four unpaired electrons in the prepared-for-bonding electron configuration, but since a quadruple bond is not possible in the *p* block, diatomic molecules are less likely (we saw in Chapter 10 that C_2 molecules have reduced bond orders). Each Group 14(IV) atom could link to four other atoms by single bonds, producing a network polymer; each atom could use two unpaired electrons to form a double bond and the other two electrons to link to other atoms to give layer polymers or oligomeric molecules; each atom could use three unpaired electrons to form a triple bond and one unpaired electron to link to another atom to give a chain polymer.

In summary, one of the main factors that determines the degree of polymerization of the structural units of a material is the number of interunit links that structural unit is able to form. For elements, this is determined by the number of prepared-for-bonding electrons that element can have. Clearly, the choice among possible allotropes for a given element is determined by a second factor: Generally, the formation of the oligomer versus a polymer depends on which type of covalent bond is stronger, σ or π. Hence, we need also to consider not only the number of interunit links, but also the relative strength of each interunit link.

Let us consider, as an example, the limiting possibilities for the atoms of Group 15(VA): three σ bonds to three different atoms in a polymeric structure, or one σ and two π bonds (a triple bond) to one other atom in a diatomic molecule. We can predict the most stable allotrope at low temperatures (when entropy is a negligible factor) by comparing the sums of bond energies present in each allotrope. These bond energies were tabulated in Table 3.1.

Example 12.1

Compute the enthalpy changes for forming (from gaseous atoms) triple-bonded *versus* single-bonded allotropes of nitrogen; phosphorus.

SOLUTION:

Since each atom contributes one-half of its electrons to each of its bonds, we compute that each Group 15(V) atom releases three-halves times its σ bond energy upon forming an allotrope involving only σ bonds. The same atom releases one-half of the σ-bond energy plus two-halves times the π-bond energy upon forming a diatomic allotrope having one triple bond. For nitrogen, the formation of three single bonds releases $\frac{3}{2} \times 167\,\text{kJ} = 250\,\text{kJ mol}^{-1}$ of nitrogen atoms, while the formation of one triple bond releases $\frac{1}{2} \times 167 + 387 = 470\,\text{kJ mol}^{-1}$ of nitrogen atoms. The latter is clearly the preferable alternative for nitrogen. But for phosphorus the single-bonding alternative releases more energy ($301\,\text{kJ mol}^{-1}$ of atoms) than does the triple-bonding alternative ($240\,\text{kJ mol}^{-1}$ of atoms); hence, a single-bonded allotrope is preferred.

This example can be generalized: multiply bonded allotropes (graphite, N_2, O_2, and O_3) are favored for *p*-block elements in the second period; these allotropes have lower degrees of polymerization than singly bonded allotropes, which are favored for *p*-block elements in the third period and below. Figure 12.3 summarizes the degree to which the nonmetallic elements (in their various allotropes) are polymeric or consist of small molecules or individual atoms.

Group 18(VIII). The elements of Group 18(VIII) are monoatomic. The individual atoms are attracted to each other only by the very weak van der Waals forces; hence,

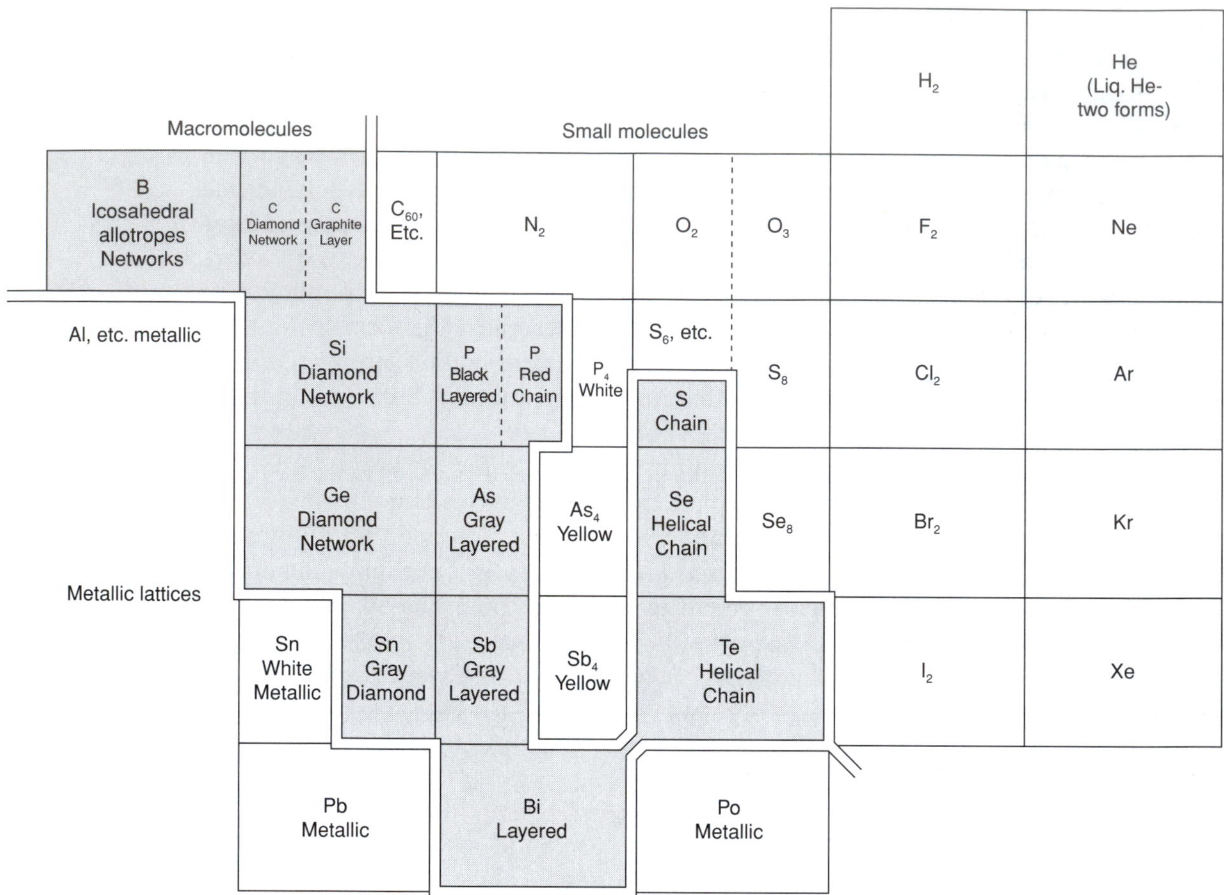

Figure 12.3
The major allotropes of the nonmetals, classified as discrete small molecules, (polymeric) macromolecules, or metallic lattices.

these elements are gases at room temperature, with boiling points that increase as van der Waals forces increase down the periodic table (He, bp −269 °C; Ne, −246 °C; Ar, −186 °C; Kr, −152 °C; Xe, −107 °C; Rn, −62 °C). Although He and Ne are abundant in the universe, their low molecular weights allow them to achieve the velocity necessary to escape the earth's atmosphere readily; only Ar is abundant in our atmosphere (1%). Argon is often used when a very inert atmosphere is needed (as for welding early *d*-block metals). Neon is used in neon lights, which emits colored light upon electrical excitation.

Helium, despite its rarity on earth (it is principally obtained from certain natural gas deposits), is a very valuable material for low-temperature (cryogenic) research due to its low boiling point. Helium as a liquid is also unique in that it has two liquid phases, helium I and helium II. Helium II is called a *superfluid*: It has zero viscosity and wets all surfaces, so that it has the ability to flow spontaneously out of a beaker into which it has been placed. It has many other strange properties that result from the especially strong manifestation of quantum mechanical properties at low temperatures in light particles such as He atoms.

Hydrogen and Group 17(VII). The halogens consist of diatomic molecules; hydrogen also shares this property and is also conveniently discussed here. These diatomic molecules are held together principally by van der Waals forces, so the lighter ones are gases (H_2, colorless, with a bp of −253 °C; F_2, pale yellow, −188 °C; Cl_2, greenish yellow, −34 °C). As the molecular weights increase, however, the van der Waals and other weak covalent bonding interactions build up, so that Br_2 is a dark red liquid (bp 59 °C), whereas I_2 is a purple solid that sublimes readily. In the heavier halogen atoms, the van der Waals forces are supplemented by weak **secondary bonding** forces, which align the molecules in the solid state in directions such that weak electron donation can occur from the halogen HOMOs, the π^* orbitals, to the halogen LUMOs, the σ^* orbital. This weak interaction results in a contact distance between individual I_2 molecules that is longer than the sum of covalent radii but is shorter than the sum of the van der Waals radii.

As we saw in Chapter 6, the halogens (not H_2) are very active oxidizing agents that are quite corrosive and must be handled with some care. Bromine, for example, gives painful burns on contact with the skin; these burns take months to heal. The halogens are used commercially as oxidizing agents to prepare halogenated compounds. Fluorine is principally used to produce UF_6 for nuclear power production (by oxidation of UF_4). Chlorine is used on a very large scale to produce chlorocarbons by the oxidation of hydrocarbons. Bromine is similarly used to produce bromocarbons, and iodine to produce iodocarbons.

Group 16(VI). We first encounter the phenomenon of allotropy among the elements of Group 16(VI). The double-bonded diatomic allotrope X_2 is stable only for oxygen, although S_2 and the like do occur at high temperatures in the vapor phase. As was seen in Chapter 10, although these molecules possess double bonds, they are not diamagnetic, but each has two unpaired electrons in a set of π^* orbitals. Oxygen also has a less stable allotrope **ozone** (O_3), which involves a single and a double bond. The O_2 molecule is now often called **dioxygen** by inorganic chemists to distinguish it from ozone and from the element in general; atoms of oxygen, O, which are the principal form in the atmosphere at altitudes between 180 and 650 km,[4] are called **atomic oxygen**.

The allotrope ozone is produced in low yield by high-energy processes acting upon O_2 (electrical discharges in the laboratory; sunlight in the upper atmosphere). Ozone in the upper atmosphere absorbs high-energy ultraviolet (UV) light that would likely cause skin cancer and increase mutation rates at the earth's surface. There is concern that its concentration may be adversely affected by pollutants such as nitrogen oxides, Freons, or nuclear warfare. Ozone is a much more powerful oxidizing agent than O_2 and is used for this purpose. But it is an undesirable air pollutant in the lower atmosphere, where it is irritating, attacks materials such as rubber, and reacts with other air pollutants such as unburned hydrocarbons to generate irritating pollutants.

The stable allotropes of the heavier Group 16(VI) elements involve two single bonds to each atom [Fig. 12.4(*a,b*)], therefore these may occur as cyclic oligomers or as linear (one-dimensional) polymers. The stable allotrope of sulfur is a yellow solid (mp = 119 °C) that contains the cyclic S_8 molecule. There are a number of additional cyclic allotropes such as S_6, S_7, S_9, S_{10}, S_{11}, S_{12}, S_{18}, and S_{20}. Above 160 °C, liquid S_8 shows some remarkable changes. The S–S bonds in the rings begin to break, giving rise to open chains with unpaired electrons at each end, which join together to give longer and longer one-dimensional polymeric μ-sulfur. The chains may exceed 200,000 sulfur

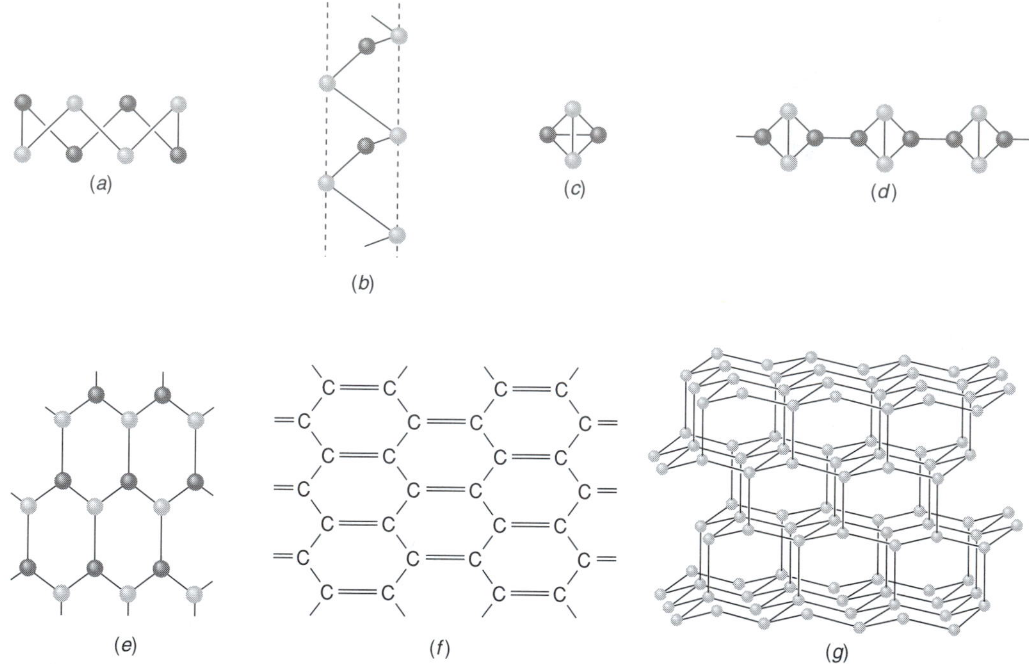

Figure 12.4
The structures of some of the allotropes of the nonmetals. Darker circles represent atoms below the plane of the paper, whereas lighter circles represent those above the plane of the paper. (*a*) S_8; (*b*) the helical structure of gray Se; (*c*) P_4 (white phosphorus); (*d*) red phosphorus (in one possible conformation); (*e*) black phosphorus (rhombohedral form); (*f*) graphite (one resonance structure); (*g*) diamond. [Adapted in part from F. A. Cotton, G. Wilkinson, and P. Gaus, *Basic Inorganic Chemistry*, 3rd ed. Copyright © 1995 by John Wiley & Sons, Inc.]

atoms in length at 180 °C. Consequently, the color darkens and the liquid becomes quite viscous, since the long chains cannot flow effectively. Above 195 °C, entropy effects favor shorter chains, and the liquid decreases in viscosity up to its boiling point of 444 °C. If the hot liquid is rapidly cooled, various polymeric allotropes can be frozen out; these allotropes involve helical chains of sulfur atoms [Fig 12.4(*b*)]. If gaseous sulfur is heated sufficiently, entropy effects prevail and the small-molecule analogues of the allotropes of oxygen finally arise: cherry red S_3 and blue-violet S_2.

The allotropy of selenium is less complex: red selenium contains Se_8 rings like those of S_8; the stable gray "metallic" form of selenium contains one-dimensional polymeric helical chains. Black (commercial) Se contains various sizes of large rings up to about Se_{1000}. Tellurium is still simpler, having only one allotrope, analogous to gray selenium. Polonium adopts a metallic lattice; thus the Group 16(VI) elements show almost the whole range of allotropic forms to be found among all the nonmetals (Fig. 12.3).

Oxygen and sulfur are used industrially far more than the other Group 16(VI) elements. The largest use of pure oxygen is in the basic-oxygen process for purifying iron in producing steel; ozone is used in some organic syntheses and in one process for the purification of drinking water. Most sulfur is used to make sulfuric acid. Selenium is used to decolorize glass; higher concentrations give pink or brilliant red glasses. In accord with its more nearly metallic nature, tellurium is principally used in alloys.

Group 15(V). The triple-bonded diatomic molecule N_2 constitutes 79% of the earth's atmosphere and is readily obtained from it by liquefaction of air followed by fractional distillation (bp of $N_2 = -195\,^\circ C$). In contrast, the diatomic form X_2 for the heavier Group 15(V) elements is found only at high temperatures and low pressures in the vapor phase; allotropes having larger single-bonded molecules are favored under normal conditions.

The simplest phosphorus allotrope is known as white phosphorus, and consists of tetrahedral P_4 molecules [Fig. 12.4(*c*)]. The bond angles in this tetrahedron are abnormally small (60°) and strained, which in part accounts for the fact that this allotrope is thermodynamically unstable and extremely more reactive than the others (e.g., it ignites spontaneously in air). The two other major allotropes are known as red phosphorus and black phosphorus; the former is a one-dimensional polymer in which one of the tetrahedral P–P bonds has opened up and is replaced by a P–P bond that is external to the tetrahedron [Fig. 12.4(*d*)]. This relieves strain and imparts greater stability and resistance to oxidation than is found in white phosphorus. As we may expect, it also increases the melting point (44 °C for white and ~600 °C for red) and lowers its solubility in nonpolar solvents. It also greatly decreases its toxicity: Red phosphorus is essentially nontoxic, while white phosphorus is dangerously toxic, even by absorption through the skin. The two main allotropes called *black phosphorus* contain puckered sheets of phosphorus atoms [Fig. 12.4(*e*)]; consistent with such structures, black phosphorus is flaky, with a graphite-like appearance.

Allotropy is less extensive in As, Sb, and Bi than in P. Tetrahedral As_4 and Sb_4 exist but are not thermodynamically stable at room temperature; the stable modifications of these three elements involve two-dimensional layer structures related to that of black phosphorus. In accord with the polymeric structures, these allotropes have rather high melting points (816 °C for As, which sublimes at 615 °C; 631 °C for Sb; but only 271 °C for Bi.)

The rather inert N_2 is used principally as an inert atmosphere to prevent oxidation during metallurgical and chemical processes. The P_4 molecule finds its principal use in the production of high-purity phosphoric acid. The more metallic As, Sb, and Bi are used in alloys.

Carbon. The stable form of carbon is *graphite*, in which the atoms participate in both single and double bonds, and are linked into a two-dimensional polymeric sheet structure based on six-membered rings [Fig. 12.4(*f*)]. This structure gives graphite its soft, flaky nature, and makes it useful as a lubricant due to the absence of bonding between the layers. Another allotrope, *diamond* [Fig. 12.4(*g*)], is a three-dimensional polymer involving only single bonds, and is only slightly less stable thermodynamically than graphite. Since diamond is bonded in all dimensions, it is the more dense allotrope (3.514 vs. 2.266 $g\,cm^{-3}$ for graphite); hence, although graphite is the more stable allotrope at normal pressures, under very high pressures the more dense form, diamond, is favored and formed. For feasible reaction rates to occur, very high temperatures and the presence of certain transition metals (as catalysts and solvents) are also required. The "melting point" of both forms is very high (<4000 °C) due to the necessity of rupturing so many C–C bonds; at high temperatures depolymerization to small gaseous carbon molecules actually occurs.

Fullerenes. Under certain circumstances, these gaseous small carbon molecules can grow again into a family of new oligomeric allotropes, the *buckminsterfullerenes* (also known as fullerenes or "buckyballs"),[5] the discovery and practical production of which[6] earned Richard Smalley, Robert Curl, and Harold Kroto the 1996 Nobel prize. The C_{60} molecule [Fig. 9.1(*h*); Section 9.5] and the related fullerenes are made by the high-energy evaporation of graphite using a laser or (more practically) an electric arc in a low-pressure atmosphere of helium gas, which controls the rate at which the vaporized small C_n molecules cool. The product of a well-annealed reaction mixture is a sooty material that gives a remarkable mass spectrum [Fig. 12.5(*a*)], with prominent peaks for molecule ions containing 60 and 70 carbon atoms, but with additional peaks present for other fullerenes containing even numbers of carbon atoms, up to 350 or more carbon atoms.

Natural sources of fullerenes have since been found, in deposits that resulted from meteor impacts with the earth, and in regions of lightning strikes.[7] It has also been found that burning some organic materials such as benzene under certain conditions produces soot that contains fullerenes.[8] However, so far they have not been found in the place in which they were first searched for: interstellar space!

If gaseous small C_n molecules were to react to give only six-membered aromatic rings, there would occur a gradual reconstruction of the graphite structure, but as this occurred the edge carbon atoms would be covalently and coordinately unsaturated, that is, they would have "dangling bonds". The emerging layer graphite structure can be made to curve in on itself if some five-membered rings are included in the growing oligomer in place of some six-membered rings. Complete closure to form a polyhedral oligomer of formula C_{20+2m} can occur if the structure includes 12 five-membered rings and m, which must be at least two, six-membered rings; the high-symmetry C_{60} includes 20 six-membered rings. Each atom in the structure is shared between three rings, and each bond is shared between two rings: The bonds shared between two six-membered rings have a bond length of 140.1 pm, while those shared between a six- and a five-membered ring are longer (145.8 pm). Although numerous isomers can be conceived for these chemical formulas, the observed structures all avoid sharing edges between any pentagons, since this would give excessive strain.

Given the complex growth requirement for formation of C_{60} and the other fullerenes, it is surprising that yields can apparently be as high as 20%. Reference has been made earlier (Chapters 1 and 9) to the most abundant and symmetric of these molecules, C_{60} [I_h symmetry, Fig. 9.1(*h*)]. In contrast to the insoluble black graphite and the insoluble colorless diamond, C_{60} dissolves in toluene to give a purple solution. The second most abundant fullerene is C_{70}, which is orange-red: In contrast to the I_h symmetry, soccer ball (European football) shaped C_{60} (with only a single ^{13}C NMR peak), C_{70} has the shape of a rugby ball, with D_{5h} symmetry (Fig. 12.6) and five ^{13}C NMR peaks. The next-higher fullerene, C_{76}, is yellow-green, belongs to the chiral D_2 point group, and has 19 ^{13}C NMR peaks. The C_{78} molecule has been isolated in two isomeric forms, a chestnut-brown allotrope of C_{2v} symmetry with 21 ^{13}C NMR peaks, and a golden-yellow isomer of chiral D_3 symmetry with 13 ^{13}C NMR peaks. The C_{84} molecule has been isolated as a olive-green solution containing at least two isomers.[9]

To obtain pure fullerenes, the sooty product undergoes a complex extraction and separation procedure such as the one illustrated in Figure 12.5(*b*). The complex separation procedures are principally responsible for the current very high cost of purified fullerenes ($945 g^{-1} for C_{60} and $40,000 g^{-1} for C_{76} versus $280 g^{-1} for the "fullerite" mixture), but advances in separation techniques have been made.[10]

Other Allotropes of Carbon. Vaporizing a graphite rod that also includes cobalt and nickel promotes the curling of the growing graphite sheets in one dimension only,

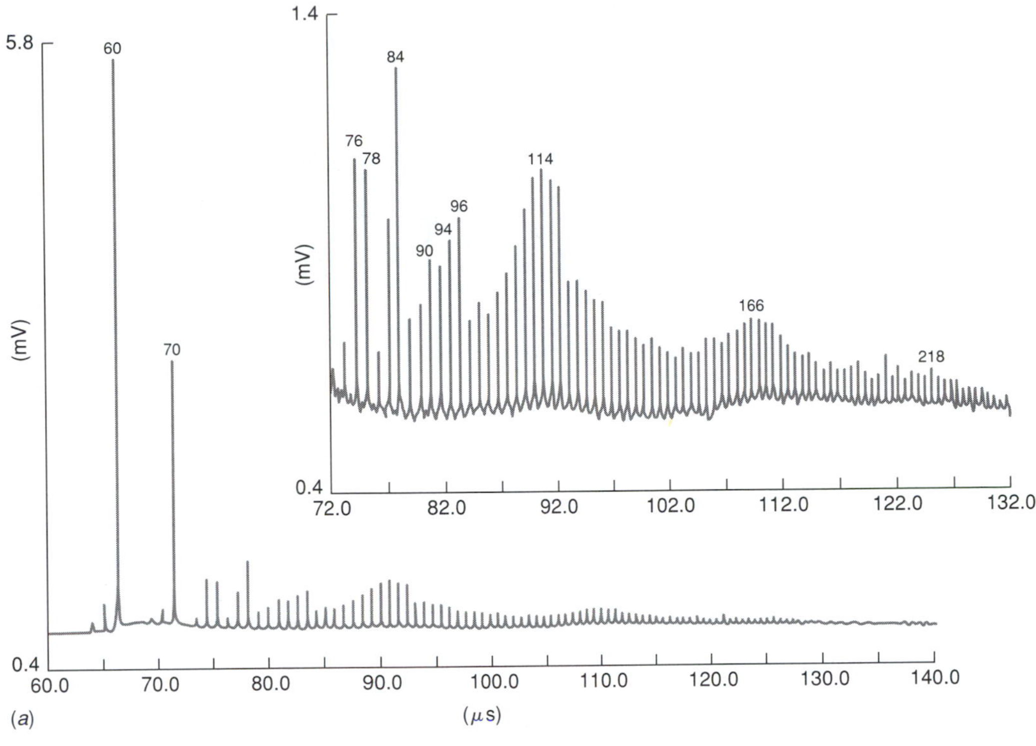

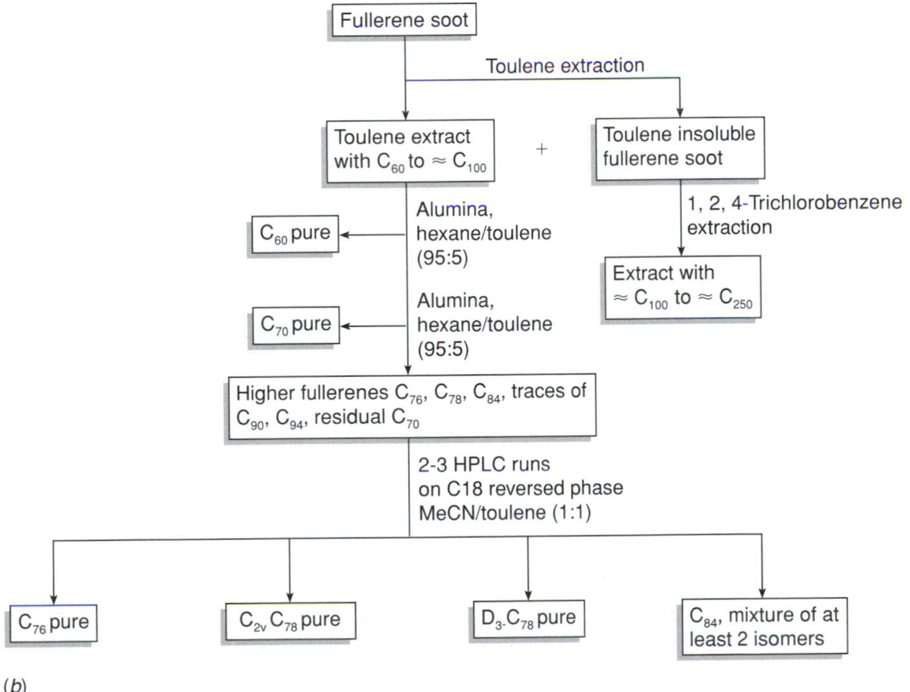

Figure 12.5
(*a*) Mass spectrum of crude fullerene soot, with number of carbons corresponding to each peak indicated; the peaks beyond 76 carbon atoms have been enhanced in sensitivity in the inset. (*b*) A scheme for separating the fullerenes using high-performance liquid chromatography (HPLC). Adapted from F. Diederich and R. L. Whetten, *Acc. Chem. Res.*, **25**, 119 (1992).

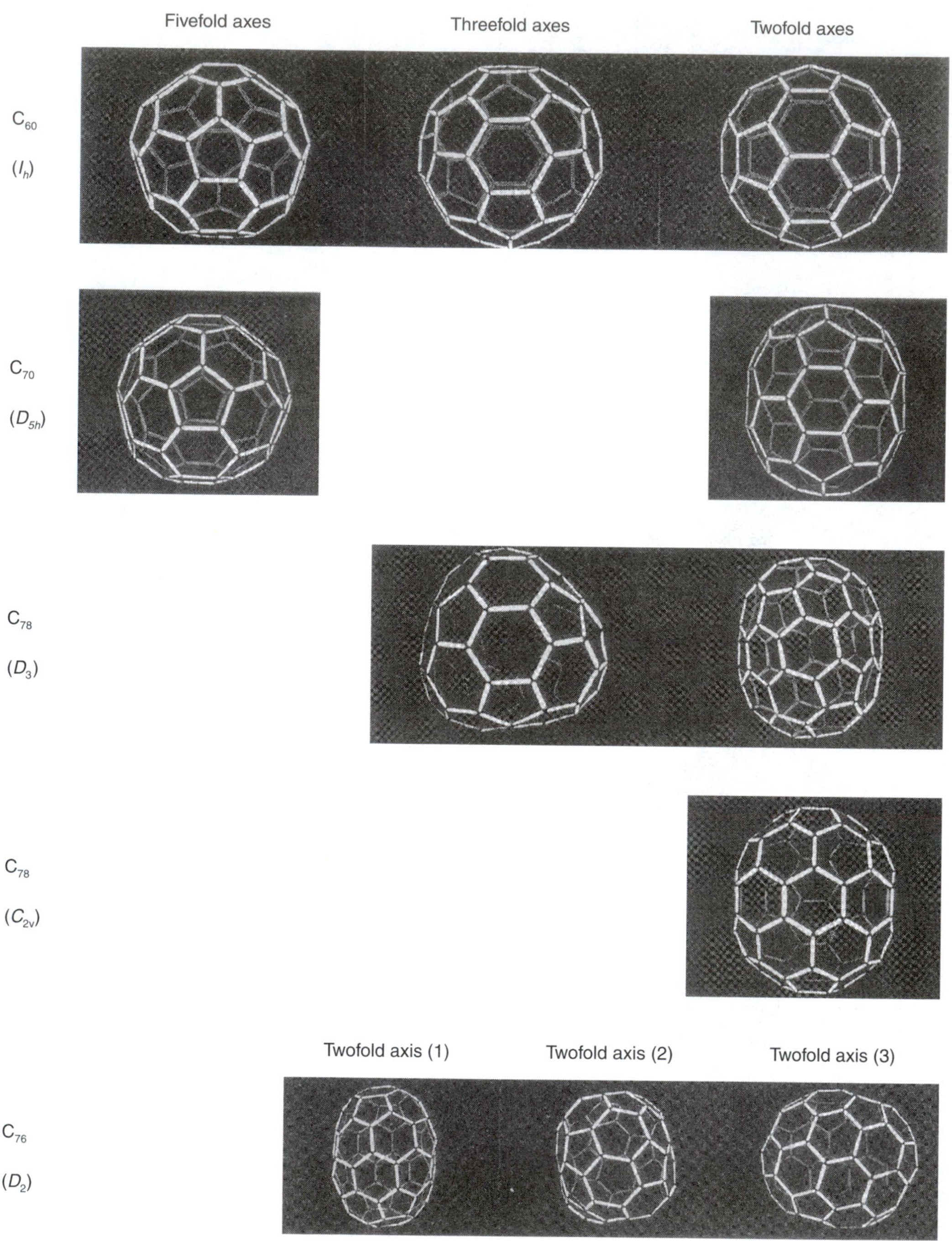

Figure 12.6

Structures of the well-characterized higher fullerenes, viewed along their fivefold, threefold, and/or twofold rotation axes as indicated. [Adapted with permission from W. O. J. Boo, *J. Chem. Educ.*, **69**, 605 (1992); copyright © 1992, Division of Chemical Education, Inc.]

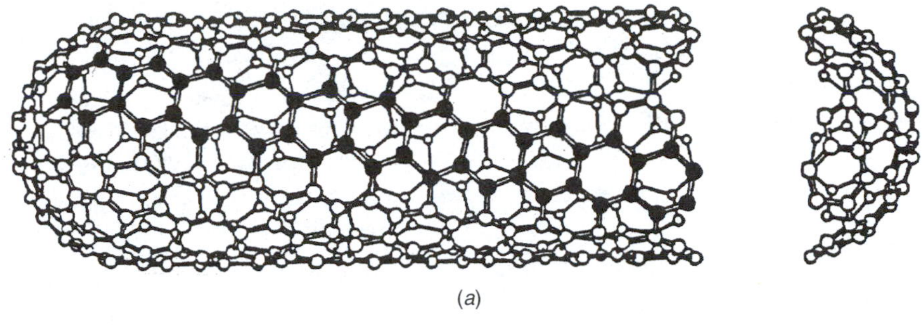

(a)

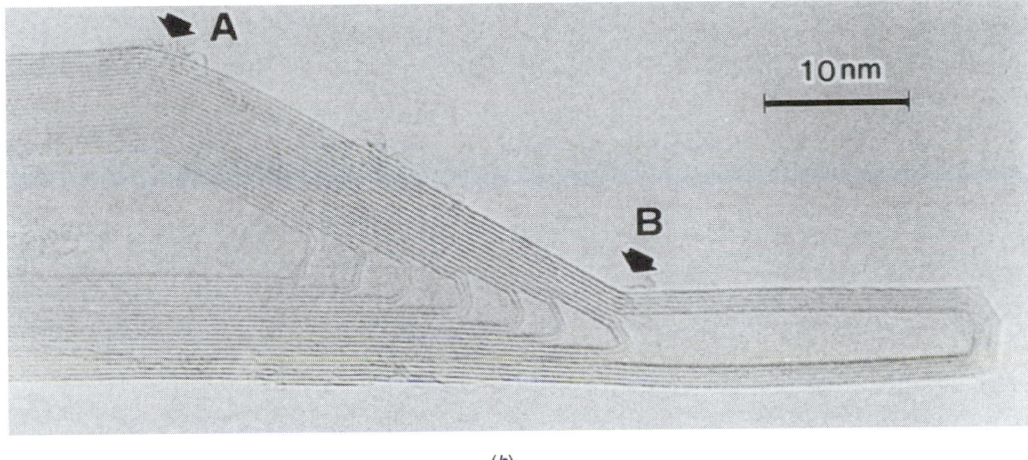

(b)

Figure 12.7
(*a*) Hypothetical structure of a single-layer nanotube with fullerene-like caps. [Adapted with permission from M. S. Dresselhaus, *Nature* (*London*), **358**, 195 (1992). Copyright 1992 Macmillan Magazines Limited]. (*b*) Electron micrograph showing carbon nanotubes contracting via a conical region (between A and B) to give thinner tubes. The separation between the layers of curved carbon sheets seen in the micrograph is 340 pm. [Adapted with permission from S. Iijima, T. Ichihashi, and Y. Ando, *Nature* (*London*), **356**, 776 (1992). Copyright 1992 Macmillan Magazines Limited.]

giving rise to a high yield of another series of carbon allotropes, the *nanotubes* or "buckytubes".[11] These syntheses produce very thin but long helically twisted tubes of carbon hexagons, with diameters of 1380 pm (1.38 nm) but length in the micrometer range [Fig. 12.7(*a*)]. Depending on the conditions under which they are formed and annealed, the hexagonal layers may curl over each other to produce helical tubes two or more layers thick. Electron microscopy [Fig. 12.7(*b*)] shows that larger tubes often contract (via five-membered carbon rings at point A in the figure) to lead to conical regions that then expand (via seven-membered carbon rings at point B) to connect to smaller tubes; the ends of the tubes may be capped with fullerene-like hemispheres. When these hemispheres are oxidized off, the open tubes are capable of absorbing metals such as molten lead: The result may be some of the world's smallest diameter wires, which could be of use in microelectronic devices.[12] Intense electron-beam irradiation of nanotubes and carbon soot results in onion-like nested fullerenes of diameters of the order of 50,000 pm (50 nm), which include about 70 fullerene layers.

Another possible allotrope of carbon could be *carbyne*, $^{1}_{\infty}$[–C≡C–]; evidence for its existence has been reviewed.[13] A number of oligomeric chains related to carbyne, R(–C≡C–)$_n$R (R = CN, CF_3, $SiEt_3$, etc.; *n* up to 250) have been more definitely synthesized and characterized.[14]

Carbon is used industrially in many forms. Graphite is used in making steel; diamond is used for its hardness in cutting and polishing. For many uses, the less pure amorphous forms are more cost effective. Coke, made by heating coal, is used in steel manufacture. Carbon black or soot, made by incomplete burning of hydrocarbons, is used to strengthen rubber in tires. Activated carbon is used to decolorize substances such as sugar. Uses for the fullerenes and nanotubes are speculative at this point in their history: They might prove useful in medicine (a water-soluble C_{60} derivative inhibits the replication of the acquired immunodeficiency syndrome (AIDS) virus in vitro[15]), as very strong fibers (nanotubes), in making diamond coatings, in optical applications, and in electronics (nanotubes).

Other Group 14(IV) Elements. Silicon and germanium essentially have only one allotrope each, with the diamond structure. These have lower melting points than diamond (1420 °C for Si and 945 °C for Ge). Tin, on the other hand, has two allotropes: Its *gray* form, which has the diamond structure, is stable below 13 °C, while its *white* form, stable above that temperature, has a metallic lattice. The much less dense gray allotrope forms and crumbles on prolonged exposure of very pure tin to temperatures below 13 °C; this "tin disease" has been a very troublesome phenomenon in the tin pipes of old organs in European cathedrals. Continuing the type of periodic trend we have seen before (Fig. 12.3), lead shows only a metallic form.

Silicon and germanium are used as semiconductors in the manufacture of transistors, and, of course, in microchips. Tin and lead have the applications of metals: tin in tin plate and alloys (such as bronze); lead in automobile batteries.

Boron. Elemental boron forms numerous very complex allotropes, which are all exceptionally difficult to prepare pure. All of these allotropes feature a structural unit very characteristic of boron chemistry: The icosahedron of 12 boron atoms [Fig. 12.8(*a*)], which we first encountered in Section 10.10 in the cluster anion, $B_{12}H_{12}^{2-}$. In

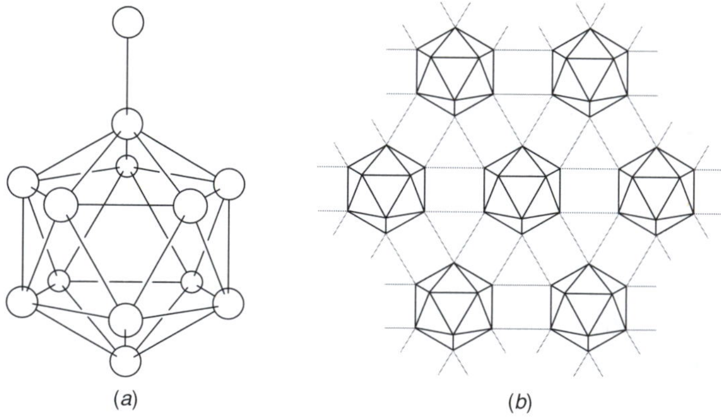

Figure 12.8
(*a*) The structural unit of boron chemistry, the icosahedral B_{12} unit, also showing an additional atom bonded to one of the icosahedral boron atoms. (*b*) The connections (in two dimensions only) of boron icosahedra in the alpha-rhombohedral allotrope of boron. [Adapted from N. N. Greenwood and E. Ernshaw, *Chemistry of the Elements*, Copyright © 1997, 2nd ed., pp. 141 and 142, Butterworth-Heinemann, Oxford.]

the different allotropes of boron the hydrogens of $B_{12}H_{12}^{2-}$ are absent; in their place these icosahedra are linked to each other in different ways via two- or three-centered bonds to form three-dimensional polymers, which have high melting points (e.g., 2300 °C), high atomization energies, and which are very hard and quite resistant to chemical attack. Boron is used in fiber and filament form for certain uses in which high strength and light weight are very important (military aircraft, space shuttles, and bicycle frames). One of the boron allotropes is shown (in two dimensions only) in Figure 12.8(*b*).

Boron, of course, has only three valence electrons and, even in the prepared-for-bonding electron configuration of $2s^1 2p^2$, cannot form ordinary two-centered covalent bonds to each of its five nearest neighbors within the icosahedron and (typically) one nearest neighbor outside the icosahedron. Hence, the best treatment of the bonding in elemental boron is the multicentered approach to covalent bonding embodied in the molecular orbital (MO) theory, although valence bond theory has been modified to cover this type of situation as well.

12.3 Metals and Alloys

Since some metals come from the *p* block of elements but even more come from the *d* and *f* blocks, we need to note whether the same horizontal periodic trend in atomization energies found among the *p*-block nonmetals are also found among the metals of the *d* and *f* blocks. In the *d* block, the maximum atomization energies seem to correspond to (at most) half-filling of the one valence *s* and five valence *d* orbitals. (Since these elements also make some use of the following *p* orbitals, an exact correspondence cannot be made.) This is a consequence of the first rule of Engel and Brewer[16]: *"The bonding energy* (i.e., minus the heat of atomization) *of a metal or alloy depends on the average number of unpaired electrons per atom available for bonding."* This similarity to the principle observed with nonmetals seems to suggest that the bonding in metals, too, is pure covalent bonding—even though metals have quite different properties than nonmetals (such as electrical conductivity) that lead us to call the bonding by a different name, **metallic bonding**.

Note, however, that in the fourth period of the *d* block and especially in both periods of the *f* block, the maximum in atomization energies and bonding occurs well before the valence orbitals can be half-filled. This early maximum is another consequence of the radii of valence *d* and *f* orbitals becoming less than that of the other valence orbitals of these atoms as Z^* increases for the *d* and *f* orbitals on proceeding to the right: Since the valence *d* and *f* orbitals then have poor overlap with orbitals of neighboring atoms, electrons in them are not so readily available for metallic bonding. Since poor overlap is more of a problem for 3*d* orbitals than for 4*d* or 5*d* orbitals, the maximum in atomization energies falls earlier in the fourth period than it does in the fifth or sixth periods. Likewise, 4*f* orbitals are less readily available for bonding than 5*f* orbitals, so *f*-block atomization energies peak earlier in the sixth period than in the seventh. It is interesting to note that these anomalies match the places in Table B1 corresponding to the first elements that commonly fail to achieve the group oxidation number fairly well: Cr in the fourth period of the *d* block, and Ce in the sixth period and Np in the seventh period of the *f* block. This correlation led to the statement that "*electrons and orbitals that are of value in binding atoms to atoms of a different kind are also generally important in binding atoms to their own kind*[17]".

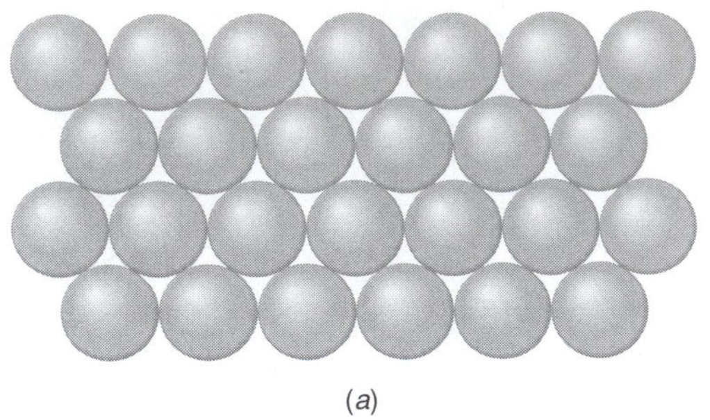

(*a*)

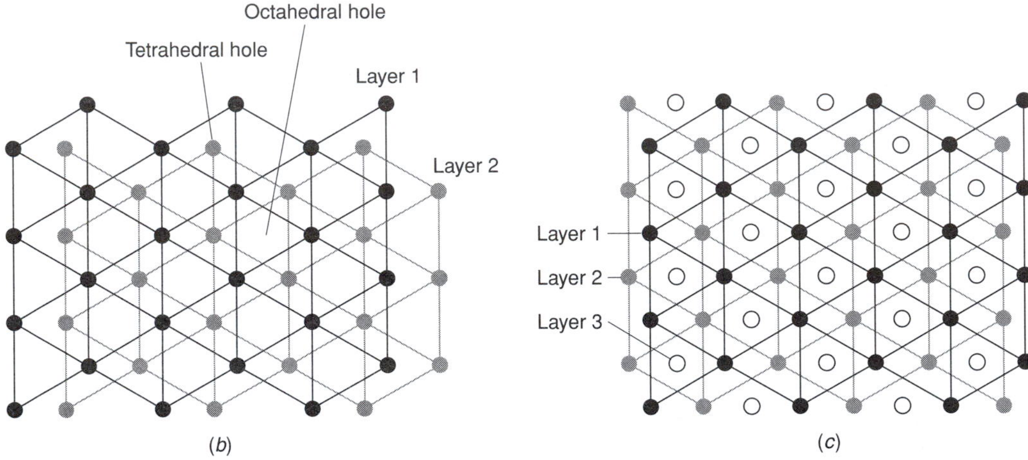

Figure 12.9
Close packing of spheres in (*a*) a single layer; (*b*) two layers, or if the layer of light-colored spheres is repeated not only below but also above the layer of dark spheres, the three-layer pattern of hexagonal close packing; (*c*) the three-layer pattern of cubic close packing. [Adapted with permission from F. A. Cotton, G. Wilkinson, and P. L. Gaus, *Basic Inorganic Chemistry*, 3rd ed, Wiley-Interscience, New York. Copyright © 1995 by John Wiley & Sons, Inc.]

Structures of Metals. Most metals have one of three types of metallic lattice. Two of these are **close-packed** lattices: In a single layer, spheres of equal size can be packed closest in a pattern in which each sphere is surrounded by six other spheres in a hexagonal pattern [Fig. 12.9(*a*)]. In packing another layer below the given layer, we would not superimpose the spheres directly atop each other, but we shift the layers so that each sphere nestles into the cavity formed by three of the spheres in the first layer [Fig. 12.9(*b*)].

The two possible patterns of close packing differ in the way a third layer is placed over the first two. With one possibility, called **hexagonal close packed** (**hcp**), the third (top) layer is matched up exactly with the bottom layer so that the light-colored pattern shown in Figure 12.9(*b*) is repeated. This pattern is then normally continued throughout the whole crystal, with layers of spheres alternately in the A position (light) and B position (dark).

In the alternate pattern of close packing, the third (top) layer does not directly coincide with either the bottom or with the middle layer. Instead, the spheres of the third layer fall directly over the three-sided cavity of both the lower and the middle layer [Fig. 12.9(*c*)]. This pattern is called **cubic close-packed** (**ccp**) or **face-centered cubic** (**fcc**), because, as shown in Figure 12.10, from another perspective this gives rise to a cubic lattice of large spheres. The hcp lattice tends to be preferred by metals in the middle of the *d* block and the ccp/fcc lattice by metals at the right of the *d* block, but many metals show phase transitions between these and other forms (Table 12.2).

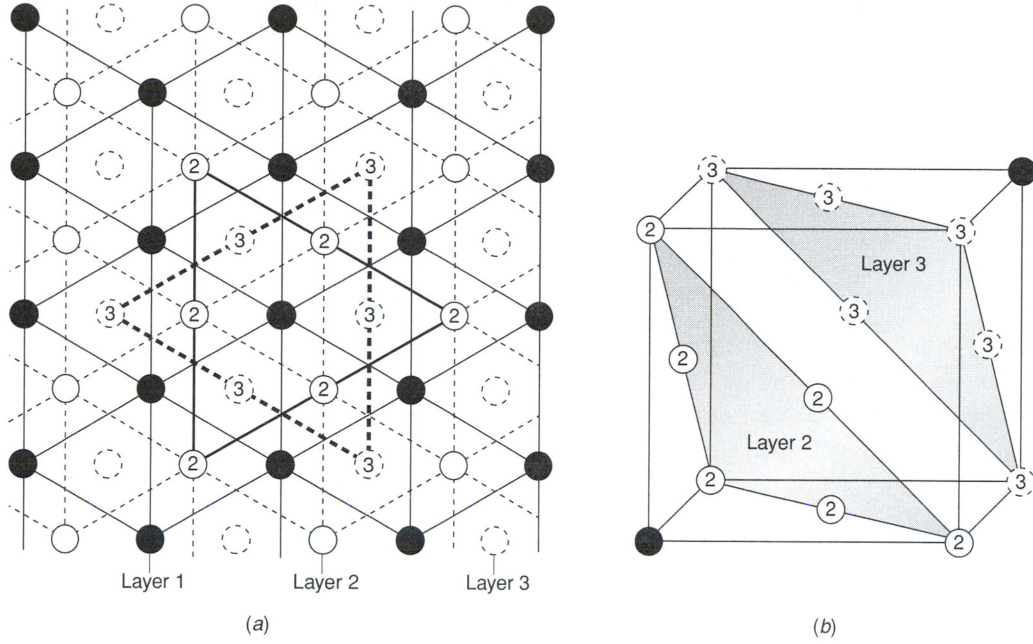

Figure 12.10

Fourteen of the close-packed spheres of Figure 12.9(*c*) shown from another angle to demonstrate their cubic pattern. The numbers (1, 2, 3, 4) on each sphere correspond to the number of the layer in Figure 12.9(*c*) in which each is located; layer 4 is superimposed directly over layer 1. [Adapted from F. A. Cotton, G. Wilkinson, and P. L. Gaus, *Basic Inorganic Chemistry*, 3rd ed. Copyright © 1995 by John Wiley & Sons, Inc.]

Table 12.2

Crystal Structures of the Metals

Li bcc	Be hcp										
Na bcc	Mg hcp	Al ccp									
K bcc	Ca bcc ccp	Sc bcc hcp	Ti bcc hcp	V bcc	Cr bcc	Mn bcc ccp β χ	Fe bcc ccp bcc	Co ccp hcp	Ni ccp	Cu ccp	Zn hcp
Rb bcc	Sr bcc hcp	Y bcc hcp	Zr bcc hcp	Nb bcc	Mo bcc	Tc hcp	Ru hcp	Rh ccp	Pd ccp	Ag ccp	Cd hcp
Cs bcc	Ba bcc	La bcc ccp hcp	Hf bcc hcp	Ta bcc	W bcc	Re hcp	Os hcp	Ir ccp	Pt ccp	Au ccp	

SOURCE: From L. Brewer, *Science*, **161**, 115 (1968).

[a] The room temperature structures are listed lowest for each metal, with other phases in order of temperature stability.

The third common form of metallic lattice is the body-centered cubic (bcc) lattice, in which each metal atom sits at the center of a cube of nearest-neighbor metal atoms, thus having a coordination number of 8. This form of packing is somewhat less dense (32% of the unit cell is space between atoms, as compared to 26% for the close-packed structures), so it tends to be adopted by the metals at the left of the periodic table, which have the fewest prepared-for-bonding electrons to bond atoms closely, and by other metals at higher temperatures, when greater amplitudes of atomic vibration favor a more open structure.

Thus, in contrast with the nonmetals, which have widely varying structures, the metals have very few variations in structure type. Nonetheless, the metals also show substantial variety in their physical properties. This variety is not, therefore, due to differences in the number of interunit links formed to each structural unit (atom), which is 8 or 12 in nearly all cases, but rather to differences in the strengths of those links (covalent bonds). The strengths of these interunit links vary principally because metal atoms vary widely in the number of prepared-for-bonding electrons that they can supply to form the 8 or 12 links. In no case can they supply 8 or 12 electrons; consequently, all linkage involves sharing electron pairs between numerous atoms. Thus the bond order per M–M link is less than 1, but is widely variable, among metals. Just as in the case of boron, MOs can be used to describe the bonding in metals (Section 10.11).

Physical Properties of Metals and Alloys. Some properties of metals are quite variable, and depend markedly on the number of valence electrons available to the metal; among these properties are their heats of atomization and related physical properties such as melting and boiling points. Thus although most metals are high-melting solids, there is one metal that is a liquid at room temperature (mercury, for relativistic reasons[18]) and there are some that melt just above room temperature (cesium or gallium). Except for gallium, these have few valence electrons, and hence low bonding energies.

The strengths of metals are also related to the number of valence electrons available for metallic bonding. Group 1 metals, Group 11 metals (copper, silver, and gold), and some of the subsequent *base metals* such as thallium and lead have few electrons prepared for bonding. These are soft and easily cut, and are also quite malleable (capable of being hammered into thin sheets) and ductile (capable of being drawn into thin wires). On the other hand, metals with many available valence electrons (from the center of the *d* block) have much greater strength and are more useful for structural purposes, although they often tend to break upon hammering (they may be brittle). Of course, one may modify the strength and related properties of either group of metals by preparing alloys, since incorporating other metals alters the average number of valence electrons available for bonding.

The density of metals is a function both of the number of bonding electrons (stronger bonding results in shorter metallic bond distances, hence greater density) and of the atomic weight of the metal. Because of the former factor, it is difficult to get a strong structural metal that is also very light. Metals of atomic weight less than that of iron should generally be lighter, but light metals such as magnesium and aluminum lack many bonding electrons and are rather weak. A very good metal for such purposes (e.g., aircraft manufacture) is titanium, which is very abundant, is lighter than steel, and yet is strong. But its reactivity even with atmospheric nitrogen at high temperatures makes it very difficult to produce and to weld (argon atmospheres are required).

Metals placed under high pressure respond to this stress by adopting a higher density structure, which relieves some of the stress of the high pressure (Le Chatelier's principle). Iron, for example, can adopt the close-packed (dense) hcp structure under high pressures, such as those found in the inner core of the earth. The hcp structure is of relatively low symmetry, which is thought to explain the fact that earthquake waves pass through the earth 3–4% faster when traveling from one pole to the other than when traveling across the equatorial plane. This effect is so large that it has been proposed that the entire inner core may be *one single crystal*, which has grown slowly over the lifetime of the earth.[19] Since a metal crystal is a macromolecule, this would be the largest molecule on (or in) earth, the size of the Earth's moon!

Metals in general are more dense than monomeric or oligomeric nonmetals of similar atomic weight, since their structures involve no long nonbonding (van der Waals) distances between atoms. Hence, under extreme pressure we may expect to convert nonmetals to metals. This phenomenon has been observed in practice with xenon at very high pressures, such as are normally found only in the centers of explosions, and after decades of effort has been claimed with hydrogen at pressures approaching 2 million atmospheres. Data from satellites flying by the planets Jupiter and Saturn suggest that the cores of these very heavy planets may contain metallic hydrogen.[20]

Uses of Metals in the Elemental Form. Only a brief summary, from a periodic table perspective, will be given. The metals of the *s* and *f* blocks are, in general, too chemically reactive and soft to be useful other than as chemical reducing agents: exceptions include NaK, a mixture of Na and K that is liquid at room temperature, as a heat transfer medium in nuclear reactors; and Be and Mg, which are useful as light structural metals, since they do have substantial strength per gram (the toxicity of Be limits its uses to such specialized ones as X-ray transparent windows and in nuclear reactors). Aluminum may also be listed now for its use (when alloyed) in light structural materials; pure Al is soft and melts at a rather low temperature.

The *d*-block metals in Groups 4 and 5 (except V), with their high atomization energies, have high melting points and structural strength and also resist corrosion well, so they are useful for structural purposes. Titanium is also light, so it finds use in jet engines and aircraft and bicycle frames, where light weight is valuable; due to its complicated metallurgy (Chapter 6) it is expensive. Zirconium is used in nuclear reactors since it does not absorb neutrons, but it first must be freed of the omnipresent Hf, which does absorb.

Most of the metals in the middle of the *d* block are used mainly for making alloys of iron (steels), to which they impart various properties of toughness, hardness, and corrosion resistance: Pure iron itself is fairly soft, silvery, and very reactive, therefore it cannot be used for the numerous structural purposes that steel is used for. The metals used this way include Mo, W, and the fourth-period elements from V through Cu.

As shall be amplified in Section 12.6, the six *platinum metals* (Ru to Pd and Os to Pt) are mainly used as heterogeneous catalysts. Platinum and the *coinage metals* Cu, Ag, and Au are prized for their chemical inertness and beauty in jewelry as well as coins; when alloyed, Cu has been very important in human history as a structural material: It was the first metal mined and used by Stone Age humans, and was then succeeded in the Bronze Age by that Cu/Sn alloy, which was followed by brass (a Cu/Zn alloy). Nickel, both pure and as the alloy Monel metal, is also useful for corrosion resistance, especially to F_2. Both Zn and Cd find use as protective coatings for steel (i.e., galvanized steel); mercury has seen much use for its electrical properties as a liquid metal, as well as its ability to form alloys with many metals known as *amalgams*, which are often liquid.

The *p*-block metals Sn, Pb, Sb, and Bi have been used for their alloying properties, not so much with Fe as with Cu or each other; the alloys of Bi in particular have low melting points. Solder is an alloy prized for this purpose, although it contains (and leaches out) toxic lead, as do the lead pipes formerly used in plumbing (Pb is from the Latin *plumbum*).

Alloys. Many alloys are *substitutional,* and can be thought of as solid solutions of one metal in another. For this type of substitution to occur, the two atoms should have metallic radii within 15% of each other, the two metals should show the same type of crystal lattice, and the electronegativities of the two elements should be similar. Thus, Na does not dissolve in K, although both have bcc lattices, since Na is 18% smaller than K. The neighbors of Cu (ccp), Ni and Zn, both have similar radii and electronegativities, but only Ni is also ccp, so only Ni alloys with Cu in any proportion; Zn (hcp) forms solid solutions over only a limited range of composition.

Other alloys form solids with their own distinct lattice types, which are often unrelated to those of their parent elements. These alloys often have approximately constant compositions, and are known as **intermetallic compounds**. Three such intermetallic compound types are known as **Hume–Rothery compounds**. The first of these have the bcc structures known as the *β-brass* structure. *β-Brass* itself is CuZn; other members of this class of Hume–Rothery compounds include AgCd, Cu_3Al, Cu_5Sn, and NiAl. What these compounds have in common is a total of *three s* and *p* prepared-for-bonding electrons for every *two* atoms. (In computing this, Ni is given a s^0d^{10} prepared-for-bonding electron configuration, Cu and Ag are given s^1d^{10} prepared-for-bonding configurations, and Zn and Cd are given $s^1p^1d^{10}$ configurations.) The second class of Hume–Rothery compounds have the *γ-brass* structure, a complex cubic structure formed from metals having 21 *s* and *p* prepared-for-bonding electrons for every 13 atoms. Examples include γ-brass itself, Cu_5Zn_8, Ag_5Hg_8, Cu_9Al_4, $Cu_{31}Sn_8$, and Ni_5Zn_{21}. The third Hume–Rothery type is the ε-brass type, with 7 *s* and *p* prepared-for-bonding electrons for each four atoms: these include $CuZn_3$, $AgCd_3$, and Cu_3Sn.

Zintl phases are intermetallic compounds formed between very electropositive Groups 1 and 2 metals and metals or metalloids of the *p* block or the late *d* block; these often contain the Zintl homopolyatomic anions, which will be discussed in Section 12.5.[21]

Intermetallic compounds between metals from the left side of the *d* block with those of much higher electronegativity from the right side, such as $ZrPt_3$, are also interesting: They may have quite substantial heats of formation (ΔH_f) (up to $\sim 330\,\mathrm{kJ\,mol^{-1}}$). Metals such as Zr from the left side of the *d* block have an insufficient number of valence *s* and *d* electrons to give an optimal prepared-for-bonding electron configuration; those such as Pt from the right side have paired too many of their *s* and *d* electrons to form the maximum number of bonds. But when the two elements are alloyed, electron transfer from Pt to Zr can occur, resulting in a very exothermic ΔH,[22] and allowing each atom to move closer to an optimal s^1d^5 prepared-for-bonding electron configuration. A particularly interesting alloy of this type is Memory Metal[23] or Nitinol (NiTi), which is able to recall its previous shape after it undergoes a phase transition. The sculptor Olivier Deschamps used it to make a statue of a man whose hands point downward in cold weather and upward in warm weather.

According to Miedema's rules,[24] compound formation between two metals is favored not only by a large electronegativity difference, but also by a small difference in charge density of the two metals. Group 1 metals are large because of their use of large valence *s* orbitals, so their charge density is quite different from the charge densities of *d*-block metal atoms, and they do not form compounds with most of them. However, under high pressure the valence *ns* electrons of metals such as potassium move into the smaller $(n-1)d$ orbitals, which allows them to have similar charge densities to, and hence to form intermetallic compounds with, *d*-block metals such as nickel and silver.[25]

Unusual forms of elements or alloys can sometimes be produced by cooling molten elements at a rate (e.g., $1{,}000{,}000°\ s^{-1}$) that is too fast to allow crystallization, thus giving amorphous materials. Amorphous silicon has become of interest for its use in solar energy conversion. Rapid cooling of a mixture of molten aluminum and manganese produces an alloy having grains in the shapes of dodecahedra, that is, with a type of icosahedral symmetry. Although these materials appear crystalline, it is impossible for the unit cell of the lattice of a solid to have fivefold symmetry elements found in the I_h point group. These grains cannot be crystals composed of identical unit cells; they are known as *quasicrystals*, and have unusual properties such as low density and high strength.[26]

12.4 Magnetism, Electrical Conductivity, and Semiconduction

Magnetism. The magnetic properties of metals is another important variable property. In materials in which paramagnetic atoms or ions (those of the *d*- and *f*-block metals) lie close together (not separated by large ligands in discrete complex ions, as in Chapter 8) the magnetic moments can align in cooperation with each other throughout large domains of the solid sample. In **ferromagnets**, all of the individual magnets (metal atoms or ions) in the domain are aligned in the same direction; in **antiferromagnets**, every other metal atom or ion has its magnetic field oriented in the opposite direction (Fig. 12.11).

Ferromagnetism is found in materials in which unpaired electrons in *d* or *f* orbitals interact with unpaired electrons in similar orbitals in neighboring atoms or ions, so as to align their spins. This interaction must be weak, however; otherwise the electrons pair up in the same bonding orbital, and no magnetism results. Thus ferromagnetism is not seen in the fifth-, sixth-, or early fourth-period *d*-block metals, in which the *d* orbitals extend outward far enough for covalent overlap to occur; it is found in the late fourth-period *d*-block metals (Table 12.3) and their oxides, in which overlap is insufficient for pairing, but sufficient for information exchange. Currently, the strongest permanent magnets are made from the alloy $Nd_2Fe_{14}B$.[27]

This weak tendency to order electrons, however, is readily overcome by the entropic tendency to disorder, so that ferromagnetism fades out at higher temperatures and finally disappears at the **Curie temperature** (Fig. 12.12), above which paramagnetic behavior prevails, but below which magnetic susceptibility is larger than predicted by Eq. (8.2).

Antiferromagnetism is found in oxides and other compounds: It results when the two metal ions interact indirectly, through a ligand whose paired electrons couple weakly and in opposite directions with the unpaired electrons of neighboring metal ions [Fig. 12.11(*c*)]. This weakly coupled antiparallel behavior is also opposed by

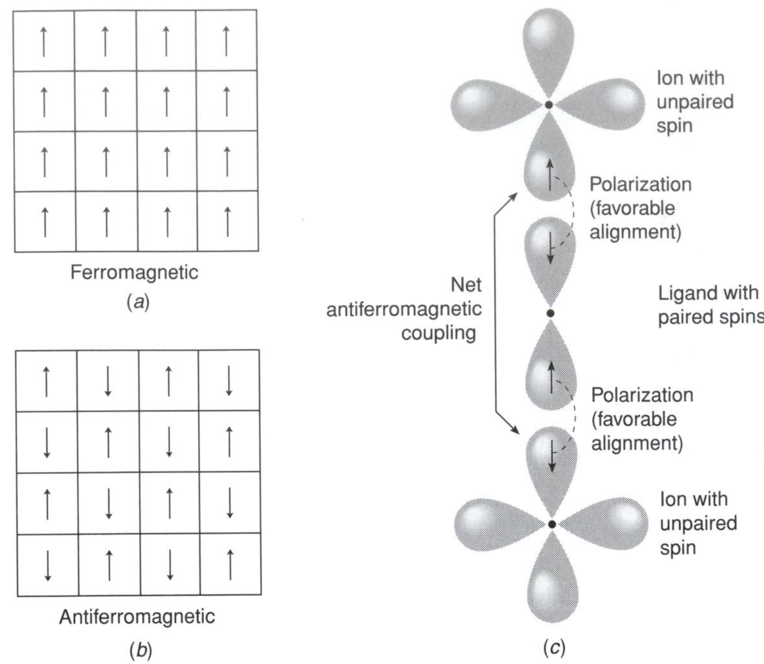

Figure 12.11
Schematic two-dimensional representation of the parallel and antiparallel orientations of the magnetic fields and electron spins of neighboring metal ions in (*a*) ferromagnets and (*b*) antiferromagnets, respectively. (*c*) The antiferromagnetic coupling of electron spins between the paired electrons of a ligand and the unpaired electrons of two neighboring metal ions, resulting in the two metal ions having antiparallel spins. [Adapted from I. S. Butler and J. F. Harrod, *Inorganic Chemistry: Principles and Applications*, Benjamin/Cummings Publishing Co., Redwood City, CA, 1989, p. 434. (*c*) Adapted from D. F. Shriver, P. W. Atkins, and C. H. Langford, *Inorganic Chemistry*, Freeman, New York, 1990, p. 586.]

Table 12.3
Elements Exhibiting Ferromagnetism (in black) and Low-Temperature Superconductivity (in white)

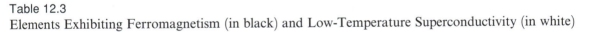

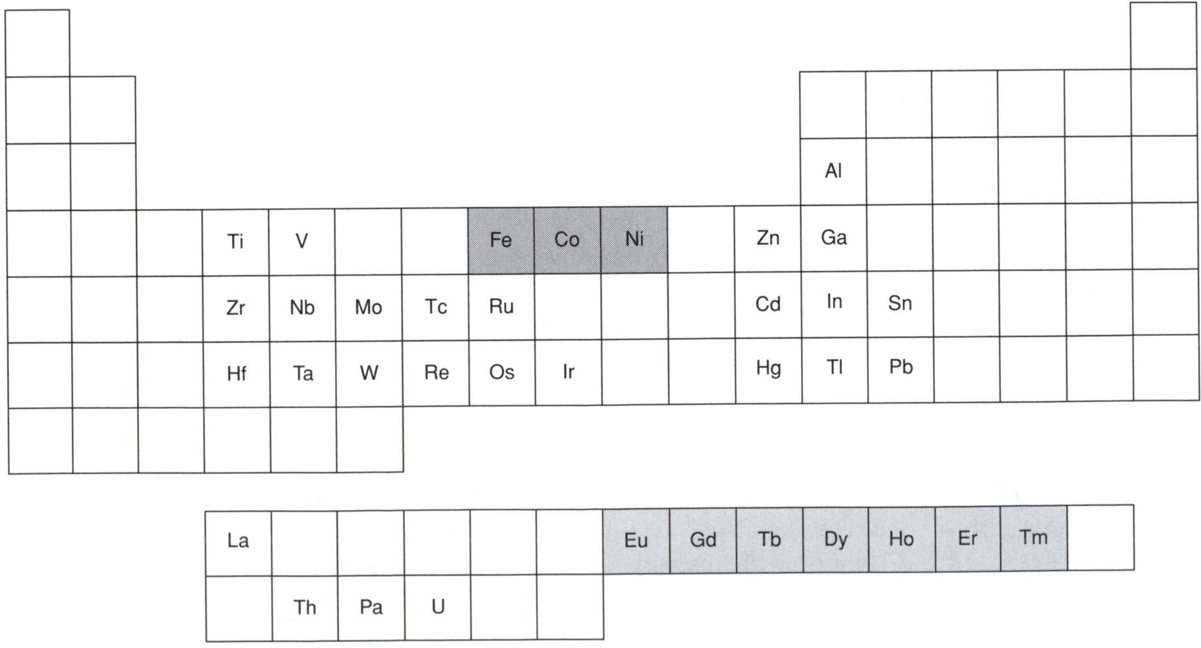

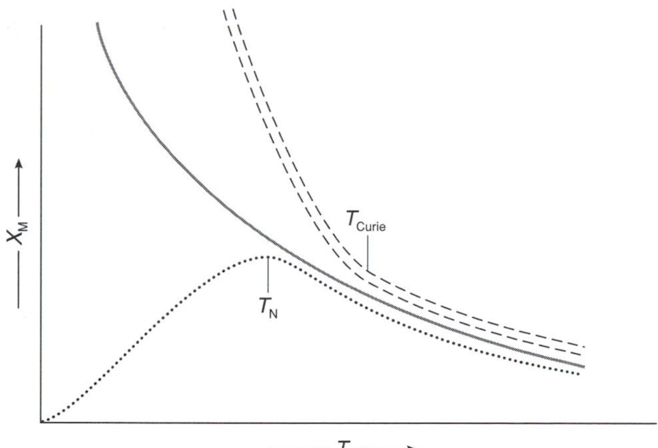

Figure 12.12
Plots of the temperature dependence of magnetic susceptibility of an idealized substance acting (*a*) as a paramagnetic substance (——); (*b*) as a ferromagnetic substance (===) below its Curie temperature; (*c*) as an antiferromagnetic substance (·····) below its Néel temperature T_N.

the entropic tendency toward disorder at higher temperature. Thus, at very low temperatures, antiferromagnetic materials show very low magnetic susceptibility, which increases (Fig. 12.12) with temperature up to the **Néel temperature** T_N, above which normal paramagnetic behavior prevails.

Electrical Conductivity. In Section 10.11, we developed the MO theory for metals (including the concept of the energy band gap, which is important for nonmetallic elements). This theory is important for understanding the properties of electrical and thermal conductivity and metallic luster. In a metallic element (in which there is no band gap, Section 10.11) the band of MOs is only partially filled. The highest occupied molecular orbital (HOMO) in this band at absolute zero is known as the **Fermi level.** However, in an extended solid the MOs are so close together that the lowest unoccupied molecular orbital (LUMO) must lie only infinitesimally higher in energy than the HOMO; as the temperature is raised above absolute zero, thermal energies and entropic tendencies cause some of the higher energy filled MOs to be depopulated in favor of the lower energy (formally) vacant MOs.

Thus for the electrons in the higher energy occupied MO there are available (at only a slightly higher energy) alternative, unoccupied MOs. Any frequency of light will suffice to promote an electron from some occupied to some unoccupied orbital; hence, metals have a characteristic metallic luster resulting from the interaction with light of all frequencies. Some of these accessible orbitals will bring that electron closer to a positive electrode at one edge of a crystal; hence, metals are **conductors** of electricity, even in the solid state where the metal atoms themselves cannot move.[28]

At extremely low temperatures (typically below about 10 K), many metals (Table 12.3) are **superconductors**. Superconductors have two outstanding characteristics below their critical temperature for superconduction. First, they are capable of excluding a magnetic field, a type of perfect diamagnetism known as the Meissner effect that allows a superconductor to levitate a magnet above it. (Note that the metals that can exhibit superconductivity are either composed of diamagnetic atoms or have good overlap of valence orbitals so that electrons can pair well in the MOs; this is in contrast to the requirement for ferromagnetism, so metals that superconduct are not ferromagnetic metals.) Second, and most importantly, the electrical resistance of superconducting metals drops to zero. Superconductivity in metals and alloys is attributed by physicists to the presence of *Cooper pairs* of electrons that are capable of moving together through a metallic lattice in such a way that, when one electron is deflected from its path by collision with a nucleus (i.e., experiences nonzero electrical resistance), it is attracted back onto the path by the other electron. This phenomenon is limited in metals to temperatures below about 20 K, since at higher temperatures vibrations of nuclei disrupt the Cooper pairs. This finding means that superconductivity in metals and alloys can only be achieved practically using the very expensive coolant, liquid helium. The alloys NbTi and Nb_3Sn are used to construct superconducting magnets in the NMR spectrometers used in chemistry research and medically in magnetic resonance imaging (MRI).

The electrical conductivity of metals characteristically decreases at higher temperatures. The mobile conduction electrons also make metals good conductors of heat, since a moving electron can collide with a vibrating atom, and acquire and transfer some of that energy to atoms elsewhere in the crystal. In addition, the malleability and ductility of metals is also related: When a solid is deformed under stress and its atoms change positions, the electrons can readily move to new positions to bond those atoms.

Semiconduction. In the *p* block, we not only find conducting (metallic) elements, at the lower left, but also nonconducting (insulating) elements at the upper right, as well as a diagonal group of seven **semiconducting metalloids**: B, Si, Ge, As, Sb, Se, and Te. The contrast is especially dramatic in Group 14(IV): Lead is a typical metal, with no band gap between its filled **valence band** and empty **conduction band** of MOs. Germanium has a small band gap of 58 kJ mol^{-1} (0.60 eV), while silicon has a larger band gap of 105 kJ mol^{-1} (1.09 eV). In the allotrope of carbon with the same structure, diamond, the band gap is large (580 kJ mol^{-1} or 6.01 eV). The four valence electrons per Group 14(IV) are just sufficient in number to fill the valence band and leave the conduction band empty at absolute zero. Insulators such as diamond and most ionic salts have (when pure and at absolute zero) complete occupation of the valence band and no electron occupation of the conduction band. Thermal energies are insufficient to promote any electrons across the large band gap into the LUMO, where they could conduct an electric current.

Since semiconductors generally have band gaps between 0.5 and 3.5 eV, germanium and silicon are semiconductors. With such modest band gaps, thermal energies at normal temperatures can promote some electrons into the conduction band; the material conducts electricity weakly and is an **intrinsic semiconductor.** The experimentally defining characteristic of semiconductors is the fact that, at higher temperatures, more electrons move into the conduction band, so that (in contrast with metals) the electrical conductivity *increases* with increasing temperature. Their conductivity also increases greatly upon irradiation with light, which promotes electrons to the conduction band.

Although the MOs generated for a metal in Section 10.11 are suitable for explaining the conductivity of lead, they cannot explain why the upper elements in the same group are semiconducting and insulating, since the pattern of these MOs does not result in a band gap. Experimentally, the four valence electrons per Group 14(IV) atom just fill the valence band. Therefore the valence band cannot correspond to the

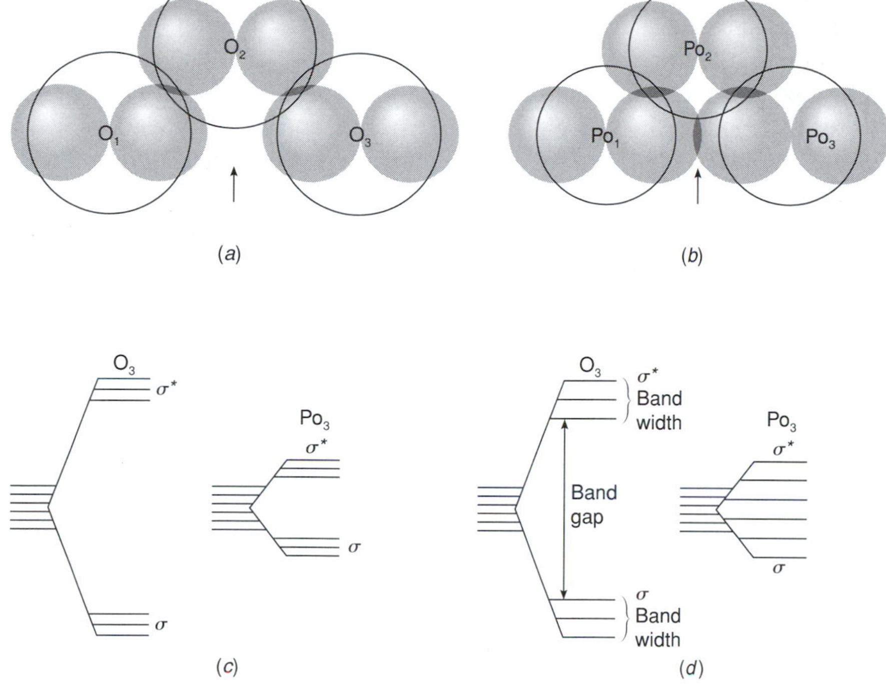

Figure 12.13
Qualitative picture of origin of a bandgap in helical allotropes at the top of Group 16(VI) but not at the bottom, illustrated for three atoms of the extreme (but hypothetical) cases of helical $^{1}_{\infty}$O and $^{1}_{\infty}$Po. (*a*, *b*) The relatively larger size of the valence *p* as compared to valence *s* orbital for Po as compared to O leads to an increased overlap between nonadjacent (first and third) atoms in the case of Po, as indicated above the position of the arrow. (*c*) Better overall overlap for O as compared to Po gives oxygen a larger difference in energies of bonding and antibonding MOs. (*d*) Introducing similar band widths for O and Po leaves a band gap for O, but not for Po.

set of MOs based on valence *s* atomic orbitals, since these would be filled by two electrons per atom. Likewise, the conduction band cannot be explained as due to the set of MOs based on valence *p* orbitals. Instead, as we shall attempt to illustrate, the valence band includes *bonding* MOs and the conduction band includes *antibonding* MOs.[29]

It is easier to illustrate this trend using the Group 16(VI) elements that are one-dimensional (helical) polymers: μ-sulfur, selenium, and tellurium. Unstable μ-sulfur is expected to be a typical insulator; Se has a band gap of 178 kJ mol^{-1}; Te has a gap of 32.2 kJ mol^{-1}. Polonium has a completely different, metallic lattice, but we could extrapolate that it would be a conductor if it adopted the helical geometry, while the hypothetical helical $^{1}_{\infty}$O would be an insulator. The question is why a small band gap develops for Te and Se, a large band gap develops for O and S, and no band gap develops for Po.

For simplicity, let us take a short chain of three Group 16(VI) atoms (Fig. 12.13), each using a singly occupied *s* orbital and one singly occupied *p* orbital; the other two *p* orbitals hold two lone pairs of electrons that we will not need to consider. With a very electronegative second-period atom such as oxygen, the valence *s* and *p* orbitals are compact, and to about the same extent: their "radius ratio" r_p/r_s is 1.15, close enough to 1 that, when two atoms are close enough for their valence *s* orbitals to overlap, their valence *p* orbitals [shaded in Fig. 12.13(*a*)] also can overlap[29]. The valence *p* orbitals do not extend far enough beyond the valence *s* orbtials to reach very well from the first to the third oxygen atom; most of the overlap is between nearest-neighbor oxygen atoms, so we say the bonding is **localized**. Under these conditions, treating the bonding as two-center, two-electron (2c–2e) bonding works well; the valence bond theory does a good job especially in the second period. As another consequence of r_p/r_s being near to 1, the radii are also well-matched for interacting with each other; *s–p* hybridization works well. The organic chemists have sound reasons for using the simpler valence bond model, with hybridization, for their predominantly second-period atoms.

As we go down a group to less electronegative atoms, however, the ratio r_p/r_s increases: it is 1.25 for S, 1.29 for Se, 1.28 for Te, and 1.38 for Po. As shown in Figure 12.13(*b*), this results in a more significant degree of overlap between the first and the third polonium atoms; we say that the bonding is **delocalized**. (It also results in diminished possibilities for *s–p* hybridization.) Under these conditions, which are often faced by inorganic chemists, the MO theory more easily describes properties such as semiconduction.

The widely different band gaps found for Group 16(VI) [or 14(IV) or 15(V)] elements are then a result of two consequences of the above. First, the smaller second-period orbital overlaps its neighbors better than the larger atoms with more diffuse orbitals. This means that, relative to the parent atomic orbitals, bonding MOs fall lowest in energy, and antibonding MOs are raised highest in energy, for the second-period atoms [Fig. 12.13(*c*)]; the energy difference between bonding and antibonding MOs is proportional to one over the bond distance squared.

Second, there are many bonding MOs that together form the valence band, and many antibonding MOs that together form the conduction band; as we saw in Chapter 10, these create different energy levels within each band (Fig. 12.13(*d*)]. The greater total overlap in the second period might normally be expected to create a greater spread of energies among these MOs as well, that is, create a *wider* valence band for oxygen. But this is largely offset by the fact that increased delocalization for polonium increases the importance of positive versus negative overlap of nonadjacent atoms, which spreads energy levels apart. Hence, the widths of the bands are largely unchanged on going from O to Po.

Consequently, the first factor, the degree of total overlap, is dominant when considering elements within these groups. For the second-period element, the distance between the center of the bonding MOs and the center of the antibonding MOs is much greater than the widths of the bands, so there is a large energy gap. For the sixth-period metal, the distance from the center of the band of (more weakly) bonding MOs to the center of the band of (more weakly) antibonding MOs is not sufficent to overcome the band widths of each band, and the two bands merge [Fig. 12.13(*d*)], giving a continuous series of energy levels like those we developed in Section 10.11. In between these elements we find the moderate band gaps characteristic of semiconductors.

Thus we can see that the atomic conditions that determine whether an element is an insulator, semiconductor, or metallic conductor include the electronegativity of the atoms involved, and whether the orbitals involved are of σ or π type. A third factor is the coordination number of the element; high coordination numbers are characteristic of metals. [If we linked each atom in Fig. 12.13(*a*) to additional atoms, the atomic orbitals would overlap more orbitals of nonadjacent atoms.] As we described the allotropes lower in Group 16(VI), we noted the increasing prevalence of secondary bonding interactions, including those between atoms above and below each other along the helix, and between atoms in neighboring helices. These increase the coordination number of each atom, increase the number of non-nearest-neighbor overlaps, increase the delocalization, and also contribute to semiconduction and metallization at the bottom of Groups 16(VI) [and also Group 15(V) but not Group 14(IV)].

Note that even organic chemists need MO theory to describe the bonding situation in π-bonded compounds of carbon. The p orbitals used to form π bonds are more diffuse ($r_p/r_s = 1.23$ for C) than the sp-hybrid orbitals used to form σ bonds, and (in polyenes) they overlap p orbitals on both sides of them: The bonding is both weaker and more delocalized in π-bonded than in σ-bonded systems. Consequently, there is a smaller band gap in extended conjugated π-bonded polyenes than in long σ-bonded alkanes. This difference results in large differences in electrical conductivity among the allotropes of carbon: σ-bonded diamond is an insulator (band gap 6.01 eV); π-bonded but nonpolymeric C_{60} is a semiconductor (band gap ~1.7 eV[30]); the polymeric π-bonded graphite and nanotubes are conductors.

Some of the semiconducting metalloids are extremely important for their semiconducting properties in modern solid-state electronic devices (transistors, etc.). Very high purity is mandatory, since introduction of even small levels of impurities has great effects on semiconduction; Si and Ge must be prepared in an extraordinary state of purity (at least 99.9999999% pure) for the electronics industry. Thus, a small level of As in a Ge semiconductor introduces one extra electron per As atom. Since the occupied MOs of Ge are filled, these electrons must go up into the unoccupied MOs, which are consequently no longer unoccupied, and thus can give rise to more ready conduction. [Such a semiconductor is known as an **n-type** (negative) semiconductor.] Similarly, impurities of Ga in a Ge semiconductor lead to a vacancy or **hole** in the band of occupied MOs, into which, upon the input of a relatively small amount of energy, another electron can jump, also increasing the electrical conductivity and producing a **p-type** (positive) semiconductor.

For many solid-state electronic devices, impurities such as Ga or As are deliberately *doped* at controlled levels (as low as 1 atom in 10^9) into ultrapure Si or Ge to produce n- and p-type **extrinsic semiconductors**. When an n- and a p-type semiconductor are put in contact, for example, the resulting *diode* will conduct an alternating current better when it is flowing from the n to the p semiconductor than in the other direction, hence converting the alternating current (ac) to direct current (dc). This doping is a type of substitutional alloying, so the elements being introduced must have similar sizes and electronegativities to those being replaced.

Although diamond is a nonconductor, it is known that, if it could be suitably doped, it would become a semiconductor; electronic devices made from it would operate many times faster than those made from silicon, and would (because of the high bond energies in diamond) be capable of functioning at temperatures up to 700 °C. However, because of the hardness and high-melting temperature of diamond, it cannot be doped by conventional methods. A process of **chemical vapor deposition (CVD)** has been developed for producing diamond films and crystals from gaseous carbon species, which are allowed to condense in the presence of atomic hydrogen, H, which preferentially attacks the carbon atoms of any graphite that is growing, and which stabilizes carbon in the tetrahedral geometry on the growing surface of the diamond. If gaseous diborane (B_2H_6) is included, the resulting diamond acquires a blue color and becomes a p-type semiconductor. It is more difficult to produce a useful n-type semiconductor: If N is doped in, it is so much more electronegative than C that the excess electrons are localized, and hence are relatively immobile, while P is too much larger than C for effective substitution in the diamond lattice.[31]

12.5 Homopolyatomic Anions and Cations of the Elements: Metal Complexes

In Section 1.8, we predicted that the nonmetallic elements would form monoatomic anions (such as As^{3-}) with charges equal to $8 - n$ (where n is the group number). In fact, the electronegative nonmetallic elements of Groups 14(IV)–16(VI) often fail to form these ions, which would have great negative charge densities on some none-too-electronegative atoms, and which would have to give up some substantial atomization energies on going from the element to the monoatomic ions. Instead, these elements more often form **homopolyatomic ions** (such as $As_7{}^{3-}$) in which lower charge densities per atom are developed, and in which not all of the element–element covalent bond energies need be sacrificed.

These homopolyatomic anions are simultaneously both ions and oligomers or polymers. As we survey the structural chemistry of some of these ions, we may note that the degree of polymerization of these homopolyatomic anions is often less than that of the parent elements, or, if the parent allotrope is a cluster, the homopolyatomic anion is a more open cluster with fewer element–element links. This tendency results from fact that the LUMOs of the parent elements are generally antibonding orbitals, so that addition of extra electrons to them to generate anions reduces bond orders. Often, this process is accompanied by a rearrangement of the geometry found in the corresponding neutral allotrope.

Other homopolyatomic anions of many of the nonmetals have not yet been isolated as pure anions, but may be isolated in the form of metal complexes; a selection of examples of this type is included in this section. If the metal has variable oxidation states, it can be unclear whether the resulting complex is one of a homopolyatomic anion with a metal ion in a high-oxidation state, or of a neutral element as ligand with a metal ion in a low oxidation state, so we also include examples that may in fact belong to the latter category.

Group 17(VII). The only polymerization found in the halogens, of course, is that resulting from secondary bonding, most notably in solid iodine. This polymerization via secondary bonds can persist in the structure of *polyhalide* ions. These ions occur especially for iodine but not at all for fluorine. The best known polyhalide is the *triiodide* ion, $I_3{}^-$ (Fig. 12.14); other polyiodide ions such as $I_5{}^-$ and $I_8{}^{2-}$ are also known as salts of nonacidic cations. Note that, in the formation of these anions, the degree of polymerization has been reduced: from a secondary-bonded polymer in solid elemental iodine to (mainly) one-dimensional oligomers. Such anions are readily formed, either upon partial oxidation of I^-, or on the reaction of I^- with I_2, and are isolated when large nonacidic cations are added to their solutions. The bonding can be thought of as donation of an electron pair from the HOMO of I^- to the LUMO of I_2 (σ^*), which weakens and lengthens the σ bond of the original I_2. Ultimately, either small or no differences develop in the two I–I distances in $I_3{}^-$, which is consistent with treatment of the bonding as three-centered, four-electron (3c–4e) bonding of bond order 0.5 per I–I link.

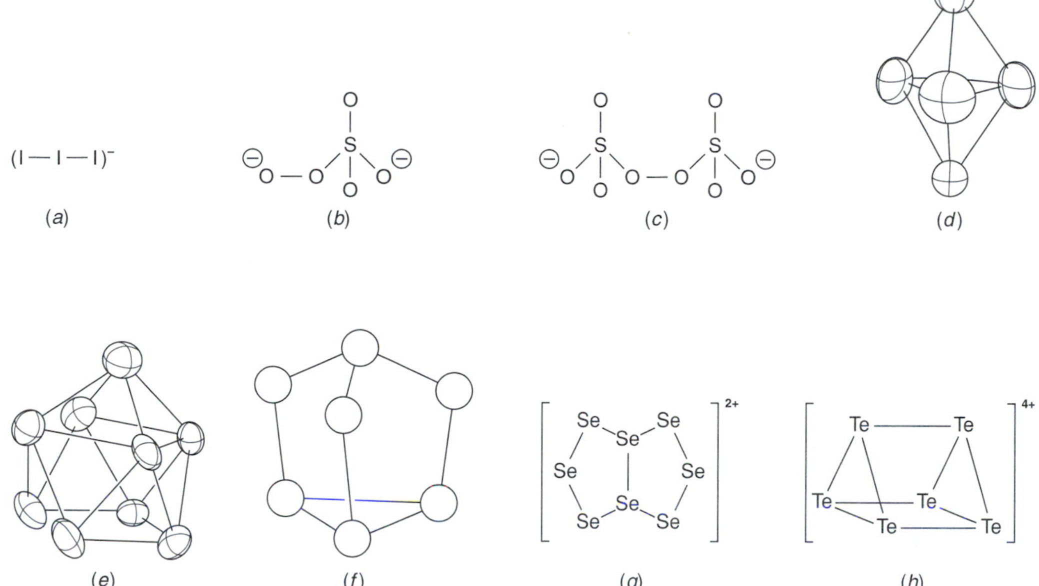

Figure 12.14
Some polyatomic ions of the *p*-block elements: (*a*) triiodide; (*b*) peroxomonosulfate; (*c*) peroxodisulfate; (*d*) the isoelectronic ions Bi_5^{3+} and Pb_5^{2-}; (*e*) Sn_9^{4-}, the isoelectronic ion Bi_9^{5+} is similar in structure; (*f*) As_7^{3-}; (g) Se_8^{2+} cation; (*h*) Te_6^{4+} cation.

Group 16(VI). We mentioned in Section 2.5 that the pK_b values for the S^{2-}, Se^{2-}, and Te^{2-} ions are difficult to measure. In part, this is because of the ease with which these ions are oxidized to ions such as Se_3^{2-}, Te_3^{2-}, and the *polysulfide* ions S_n^{2-}, which are best known for n between 2 and 6. Mixtures of such ions are formed, for example, by dissolving sulfur in a sulfide-ion solution. These ions have chain (but nonlinear) oligomeric structures; S_2^{2-} is perhaps most familiar in the form of the mineral iron pyrites or *fool's gold,* FeS_2. Some minerals contain low concentrations of the radical anions S_3^- and S_2^-; these anions are responsible for the blue color of lapis lazuli and the blue or green colors of ultramarine.

The S_2^{2-} ion is of importance as a bridging ligand in metal complexes; it can result, for example, from the oxidation of two SH^- ligands in two parent complex ions:

$$2\,[(H_2O)_5CrSH]^{2+} + I_2 \rightarrow [(H_2O)_5Cr\text{–}S\text{–}S\text{–}Cr(H_2O)_5]^{4+} + 2\,HI \qquad (12.1)$$

The S_5^{2-} ion is specifically stabilized in some complexes such as $[Pt(S_5)_3]^{2-}$ and $(C_5H_5)_2Ti(S_5)_2$, since it forms six-membered nonplanar chelate rings with the metal ion.

Although the very electronegative nonmetal oxygen normally forms the monoatomic oxide ion (or OH^-) upon reduction, there are some homopolyatomic anions of great importance as well. The biochemical 1-electron reduction of O_2 first produces a radical anion, the *superoxide ion*, O_2^-. This ion is quite reactive and can cause such serious cell damage that an enzyme, *superoxide dismutase,* is needed to catalyze its decomposition:

$$2\,O_2^- + 2\,H^+ \rightarrow H_2O_2 + O_2 \qquad (12.2)$$

The O_2^- ion is stabilized in solids by large nonacidic cations such as K^+ through Cs^+; the superoxide is the normal product of the air oxidation of these metals in preference to M_2O, which gives relatively poor lattice energies due to the size and charge mismatch of these cations with O^{2-}. The compound KO_2 is used in closed atmospheric systems (such as spacecraft) to remove CO_2 from the atmosphere and to replace it with O_2:

$$4\,KO_2(s) + 2\,CO_2(g) \rightarrow 2\,K_2CO_3(s) + 3\,O_2(g) \qquad (12.3)$$

The next step in reduction of O_2 is the *peroxide ion*, O_2^{2-}, which is strongly basic, and is found mainly as *hydrogen peroxide*, H_2O_2. The compound BaO_2 is made by careful oxidation of BaO, while Na_2O_2 (containing some NaO_2) is the normal oxidation product of Na in air. Since the extra electrons in O_2^- and O_2^{2-} occupy antibonding orbitals, their bond lengths exceed those of O_2 (Table 10.2).

There are a number of oxo anions in which one or more oxo groups are replaced by peroxo groups. The casually written formulas and names of these are confusing because they look like impossible oxo anions. For example, H_2SO_5 and $H_2S_2O_8$ (Fig. 12.14), commonly called *Caro's acid* and *persulfuric acid*, respectively, really contain sulfur in its common oxidation number of +6. These are better named peroxomonosulfuric and peroxodisulfuric acid, respectively; they are very strong oxidizing agents. Somewhat similarly, there is a short series of sulfur oxo anions in which two terminal SO_3 groups are bridged by sulfide or polysulfide ions: These are the *polythionate* ions $S_nO_6^{2-}$ ($n = 2$–6), with actual structures $^-O_3S{-}S_{n-2}{-}SO_3^-$.

Group 15(V). We have already mentioned the very powerfully reducing azide ion, N_3^- (Fig. 5.1). The anions formed by the less electronegative elements of this group are usually polyatomic. For example, $^1_\infty[As^-]$ in LiAs adopts a spiral chain structure like that taken by elemental Se and Te, with which it is formally isoelectronic. In some cases such as this, the isolobal analogy may be usefully applied, since the structural unit As^- is isolobal with the S or Se atom. Many of these compounds are also semiconductors. There are also oligomeric cluster anions P_7^{3-}, As_7^{3-}, and Sb_7^{3-} (Fig. 12.14), as well as many more complex types (P_{16}^{2-}, etc.).

Example 12.2

Show how the chain polymeric $^1_\infty[As^-]$ ion might formally be derived by reduction from one of the allotrope types found for Group 15(V) elements.

SOLUTION:

The simplest relationship might be to the chain polymeric allotrope type found in this group, red phosphorus [Fig. 12.4(*d*)]. Addition of four electrons per P_4 structural unit (to give P_4^{4-}) would presumably populate two antibonding σ^* orbitals in the strained P_4 ring. These antibonding interactions would negate two σ bonding interactions, which we will take as two nonadjacent σ bonds in the P_4 unit (Fig. 12.15). Replacing these two bonding

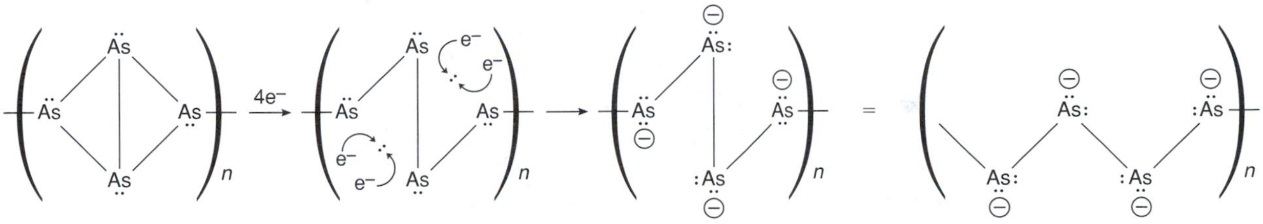

Figure 12.15
Illustration of the reduction of a hypothetical "red arsenic" allotrope (with the structure of red phosphorus) to give the known ${}^{1}_{\infty}[As^-]$ ion.

and two antibonding interactions with four nonbonding interactions (four unshared electron pairs) breaks open the ring to give a chain polymer lacking any incorporated P_4 clusters, as is found for ${}^{1}_{\infty}[As^-]$ and in the isolobal, isoelectronic ${}^{1}_{\infty}[Se]$.

In complex ion chemistry, a different series of polyhomoatomic anions may be stabilized. Of particular note is the *cyclopentaphosphide ion*, P_5^-, which is isoelectronic and isolobal with the $C_5H_5^-$ anion (Chapter 11) and is also aromatic. Simple salts of this anion can be prepared[32] by the reaction of the reactive allotropes of phosphorus (P_4 or red P) with potassium dihydrogen phosphide, $K^+PH_2^-$, in *N,N*-dimethylformamide (DMF); from these, or from the reactive allotropes of the nonmetal, metallocene analogues such as $[Cp^*Fe(As_5)]$[33] and $[(P_5)Mn(CO)_3]$ can be prepared. Related to these are the preparation of some complexes of otherwise-unstable allotropes of Group 15(V) elements in forms that are not known allotropes: triply bonded P_2, As_2, and Bi_2 as 4-, 6-, and 8-electron donating ligands; *cyclo*-P_3, *cyclo*-P_4 and *cyclo*-As_3 ligands; (element)$_4$ tetrahedra; *cyclo*-P_6 and *cyclo*-As_6 (isoelectronic to benzene); and *cyclo*-As_8 (isoelectronic to cyclooctatetraene).[34]

Group 14(IV): Carbides. These are generally prepared either (1) by direct combination of the elements at over 2000 °C, (2) by reduction of the metal oxide with carbon at high temperatures, or (3) by reaction of the metal (or metal ion) with an appropriate hydrocarbon. The C^{4-} ion nominally occurs in Be_2C and Al_4C_3, which react with water to give CH_4, but which in view of the intermediate electronegativities of the cations can scarcely contain pure C^{4-} ions. Two homopolyatomic carbide ions occur more commonly: the triple-bonded $C_2{}^{2-}$ ion (acetylide ion) and the $C_3{}^{4-}$ ion. The less-common $C_3{}^{4-}$ ion occurs in Mg_2C_3, $Ca_3(C_3)Cl_2$, and Li_4C_3, which hydrolyze to give $MeC{\equiv}CH$.

Most salt-like carbides act as salts of the acetylide ion, with short $C{\equiv}C$ bond lengths of 119–124 pm, and hydrolyze to give acetylene, $HC{\equiv}CH$. Many of these can be prepared by reacting the metals with acetylene; the Group 1 metals give M_2C_2, the Groups 2 and 12 metals give MC_2. Calcium carbide (CaC_2) is produced by heating CaO and C at about 2000 °C:

$$CaO + 3\,C \rightarrow CaC_2 + CO \tag{12.4}$$

This compound is important as a source of acetylene and for its reaction with N_2 at 1000 °C to give calcium cyanamide, $CaCN_2$, used ultimately in producing polymers:

$$CaC_2 + N_2 \rightarrow CaCN_2 + C \tag{12.5}$$

Soft-acid "acetylides" can be precipitated from aqueous solutions containing ammonia by the action of acetylene, but these acetylides (Cu_2C_2, Ag_2C_2, Au_2C_2, Hg_2C_2, and HgC_2) are very explosive.

Anions of the electron-poor fullerenes[35] can be obtained by reacting them with Group 1 metals. The low-energy (t_{1u}) LUMO of C_{60} can accommodate up to six electrons; the $E°$ for the first reduction of C_{60} to C_{60}^- is −0.61 V. Here C_{60} behaves as if it were a very large electronegative atom with a 500-pm radius; unlike any true atoms, it forms a −2 charged ion even in the gas phase, without the benefit of any stabilization from lattice energies.[36] The ultimate reduction product, M_6C_{60}, is not as remarkable as the intermediate solid product M_3C_{60}. If the metal ions are of the proper size, M_3C_{60} is superconducting to a higher temperature than any other material other than the cuprate superconductors (Chapter 13): 33 K for Rb_2CsC_{60}. Conduction pathways are provided by the π system of the anions, which are essentially in contact with each other in three dimensions in the solid state, since the "small" cations can fit in holes between the large fullerene anions.

Fullerenes can form another class of metal compounds, the **endohedral** metallofullerenes,[37] in which the metal, often as a +3 cation, fits inside the large ($r = 350$ pm) cavity of the fullerene anion. Their chemical formulas (e.g., $La@C_{60}$) are written using the symbol @ to indicate this unusual intercalation of a metal cation *inside* a nonmetal anion. These are made by high-energy evaporation of graphite rods that were previously impregnated or packed with metal powders, metal oxides, or metal oxides of the Groups 2–4 and the *f*-block elements. These are produced in up to 1% yield, and are difficult to dissolve, so they were not isolated pure until good extraction and separation procedures were developed. The compound $M@C_{60}$ can generally only be extracted with aniline, while the derivatives of C_{82} (which, when metal-free, is not produced in significant quantities) are extracted with toluene under pressure. Some derivatives of C_{80} and C_{84} are also known.

The compound $M@C_{60}$ is known for M = Ca, Sr, Ba, La, Y, Ce, Pr, Nd, and Gd[38]; a fairly similar list of metals is known to form $M@C_{82}$. Crystal structures of $La@C_{82}$ and $Y@C_{82}$ show that the metal ions do not reside at the center of the buckyball, but reside close to the carbon cage.[39] A sample of $Gd@C_{82}$ was irradiated with neutrons in a nuclear reactor, producing radioactive $^{161}Gd@C_{82}$, which emitted a high-energy β particle and decayed to $^{161}Tb@C_{82}$. The Tb isotope remained in the fullerene cage, demonstrating its unusual strength and ability to resist damage from high-energy radiation.[40]

That the fullerenes have large internal cavities is shown by the encapsulation of two or three atoms of smaller metals, in the endohedral metallofullerenes $Y_2@C_{82}$, $Sc_2@C_{84}$, and even $Sc_3@C_{82}$. It has even proved possible to obtain fullerenes in which noble gas atoms, especially helium, have been endohedrally caged inside C_{60} and C_{70}.[41]

Oddly, some very large tetravalent metal atoms (U, Ti, Zr, and Hf) form stable endofullerenes $M@C_{28}$ with a carbon cage that is exceptionally small[42]: In these cases, it is thought that there is direct covalent bonding between these metal atoms and the fullerene.

Both C_{60} and C_{70} have proved capable of bonding *exohedrally* as electron-poor

π-donor ligands to different soft-acid metal ions; these include C_{60} as a 2-electron donor in $[Pt(\eta^2\text{-}C_{60})(PPh_3)_2]$ and C_{60} as a bridging 12-electron donor in $[(Et_3P)_2Pt]_6(\eta^{12}\text{-}C_{60})$.[43] Interestingly, both endohedral and exohedral iron compounds, Fe@C_{60} and (presumably polymerically coordinated) $(Fe^{3+})(C_{60}{}^{3-})$ have been reported, which have strikingly different properties.[44]

The allotrope carbyne is also being stabilized by coordination in complexes such as $\{[CpRe(NO)(PPh_3)]\text{–}C_n\text{–}[CpRe(NO)(PPh_3)]\}$, in which n goes up to 20.[45] These are of interest to see over how long a chain electrons can be transported from one metal to another, a topic to which we shall return at the end of Chapter 16.

Other Group 14(IV) Homopolyatomic Anions. Monoatomic anions are not known for these elements; the simplest homopolyatomic ions are tetrahedral ions such as $Si_4{}^{4-}$ (in Na_4Si_4 and Ba_2Si_4); in calcium silicide, $CaSi_2$, the Si^- structural units are isoelectronic with P or As atoms and form infinite puckered layers as in elemental As or in black phosphorus. But other nonpolymeric anions have been discovered that have finite three-dimensional structures. These metal cluster anions, known as **Zintl anions**[46] after the German chemist who pioneered their study, are found both in Zintl alloys and in more isolated forms with very large nonacidic cations (cryptate-coordinated Group 1 metal ions). The Zintl anions include such species as $Ge_9{}^{4-}$ and $Sn_9{}^{4-}$, and $Sn_5{}^{2-}$ and $Pb_5{}^{2-}$ (Fig. 12.14); the $Sn_9{}^{4-}$, $Pb_9{}^{4-}$, and $Sn_6{}^{2-}$ ions are also known as ligands in chromium carbonyl complexes.[47] Unlike the homopolyatomic anions found in Group 15(V), the formulas of some of these anions are more easily rationalized by considerations of MOs in clusters (Section 10.10) than with conventional two-center Lewis structures.

Borides. Given the low electronegativity of boron, most metal borides[48] are likely to have substantial covalent character. These are often very hard, high-melting substances with high electrical conductivities; they may contain the whole range of degrees of polymerization of the homopolyatomic boron "anion." Some borides with low ratios of B to M (M_4B, etc.) contain isolated boron atoms in unusually shaped holes in lattices or between layers of metal atoms. As the ratio of B to M increases, so does the degree of boron-atom polymerization (Table 12.4). The compound V_3B_2 contains boron atoms in pairs; NiB contains polymeric chains of boron atoms; many M_3B_4 compounds contain double chains. The MB_2 compounds (and M_2B_5) contain polymeric boron layers, with the boron atoms in six-membered rings (as in graphite); these are among the hardest, best conducting, and highest melting borides. The MB_4 compounds contain open networks of boron atoms interlaced with networks of metal atoms; MB_6 compounds contain B_6 octahedra linked in three dimensions to each other with M ions in the cavities; MB_{12} compounds contain B_{12} cuboctahedra (not icosahedra!) closely linked to each other.

The heavier Group 13(III) metals are also beginning to yield a variety of cluster polyatomic anions, for example, $Ga_6{}^{10-}$, $In_4{}^{8-}$, and $Tl_6{}^{6-}$.[49]

Mercurides. Although metal amalgams (alloys with mercury) are common, a few of them contain seemingly discrete polyatomic mercury anions: CsHg, KHg, and Na_3Hg_2 contain square or rectangular Hg_4 units, bonded through electrons shared in the mercury $6p$ orbitals; $Rb_{15}Hg_{16}$ contains these and Hg_8 cubic units.[50] Although these are metallic conductors like alloys, they have ionic features such as a drastic shrinking of the molar volume of the constituent elements.

Table 12.4
Structures of Some Metal Borides

Formula	Examples	B—B bonds	Boron arrangements
M_4B	Mn_4B	Isolated atoms	
M_4B	Ni_3B		
M_4B	Be_2B		
M_3B_2	V_3B_2	Pairs	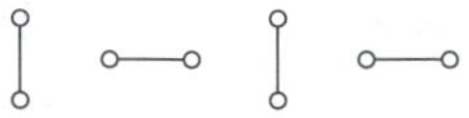
MB	NiB	Zigzag chains	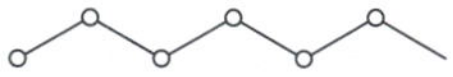
	FeB		
$M_{11}B_8$	$Ru_{11}B_8$	Branched chains	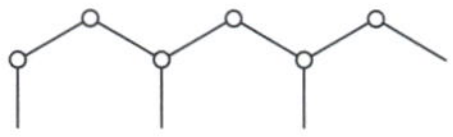
M_3B_4	Ta_3B_4	Double chains	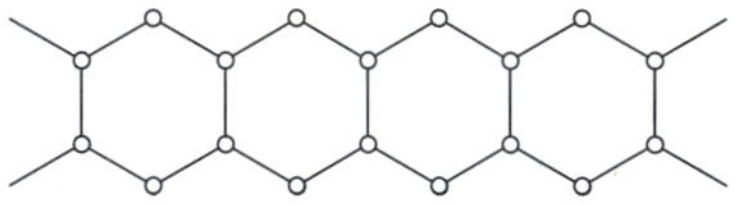
	Cr_3B_4		
MB_2	CrB_2	Two-dimensional layer	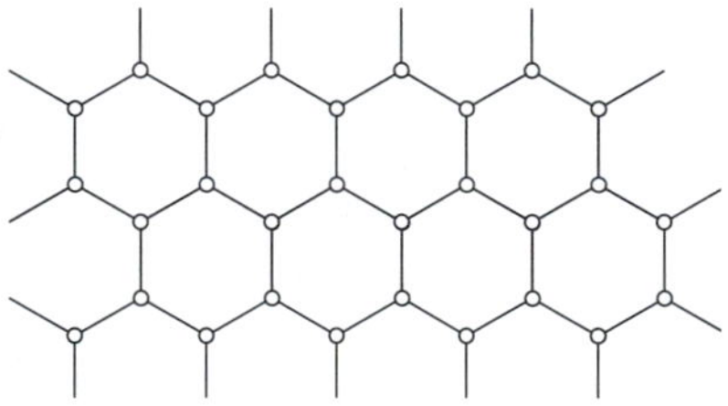
	TiB_2		
	MgB_2		
	ZrB_2		
	GdB_2		
MB_4	LaB_4	Three-dimensional network	Linked B_6 octahedron
MB_6	LaB_6		
MB_{12}	YB_{12}		Linked B_{12} cubooctahedron
MB_{15}	NaB_{15}		Linked B_{12} icosahedron
M_3B_{12}	B_4C		
MB_{66}	YB_{66}		Linked B_{12} $(B_{12})_{12}$ giant icosahedron

SOURCE: F. A. Cotton and G. Wilkinson, *Advanced Inorganic Chemistry*, 5th ed., Wiley, New York, 1988, p. 166.

Polyatomic Cations of Nontransition Elements. Homopolyatomic cations of the *p*-block elements are naturally less abundant than the homopolyatomic anions. One group that has attracted attention is a group of cations obtained by dissolving elemental S, Se, or Te in disulfuric acid containing excess SO_3. These include bright yellow $S_4{}^{2+}$ and bright red $Se_4{}^{2+}$ and $Te_4{}^{2+}$, which have a square arrangement of atoms. Deep blue $S_8{}^{2+}$ and green $Se_8{}^{2+}$ have a cross-linked ring structure [Fig. 12.14(*g*)]; brown $Te_6{}^{4+}$ has a trigonal prismatic structure. Bismuth forms interesting polyhedral cations such as $Bi_5{}^{3+}$ and $Bi_9{}^{5+}$, which are isoelectronic with the $Sn_5{}^{2-}$ and $Sn_9{}^{4-}$ anions. The larger fullerenes have smaller HOMO–LUMO gaps than C_{60}, so both are more easily reduced and oxidized; the cation C_{76}^{+} has been isolated in a salt.[51]

The most familiar homopolyatomic metal cation is the mercury(I) cation Hg_2^{2+}, which occurs commonly in aqueous solution, although it is susceptible to fairly easy oxidation. Longer chains have been prepared: Hg_3^{2+}, Hg_4^{2+}, and an infinite-chain $^1_\infty[Hg_{2.86}^{+}]$, for which a large −1 anion occurs in the lattice alongside each 2.86 Hg atoms of the chain; a triangular Hg_3^{4+} cation, supported by bridging $(Ph)_2PCH_2P(Ph)_2$ ligands, has also been prepared.[52] Bonding in these cations involves mercury 6*s* orbitals, in contrast to the 6*p*-based bonding in the mercuride anions. A diatomic cadmium(I) cation, Cd_2^{2+}, can be isolated with difficulty, but the ease with which mercury forms this cation is not duplicated elsewhere.

Groups 1 and 11 metals form some interesting cluster cations with oxidation numbers less than +1 under protected circumstances [very reducing conditions, in pores of zeolite macromolecules (Chapter 12)]: Au_4^{2+}, Ag_6^{4+}, Rb_7^{5+}, Na_4^{3+}.[53] The bonding in these can be treated approximately as involving just one a_1 or a_{1g} bonding MO, based on *ns* atomic orbitals, occupied by just one or two valence electrons for the entire cluster: This MO resembles the a_g bonding MO of $B_{12}H_{12}^{2-}$ illustrated in Figure 10.29.

12.6 Heterogeneous Catalysis by Metals

Heterogeneous catalysts are widely used in industry for their simplicity of handling (the catalyst does not have to be recovered from solution; gases or liquids can simply be passed through a catalytic bed) and their stability at higher temperatures. A disadvantage of bulk crystalline metallic catalysts[54] is that a low percentage of the metal atoms are on the surface, which is where heterogeneous reactions take place, so often the metals are deposited as finely divided crystallites of high surface areas on porous solids as supports; if the metal crystallite is about 250 pm in diameter, about 40% of its metal atoms are on the surface.

Surface metal atoms have vacant coordination sites and readily form chemical bonds with many reactants used in important industrial processes (Fig. 12.16); the

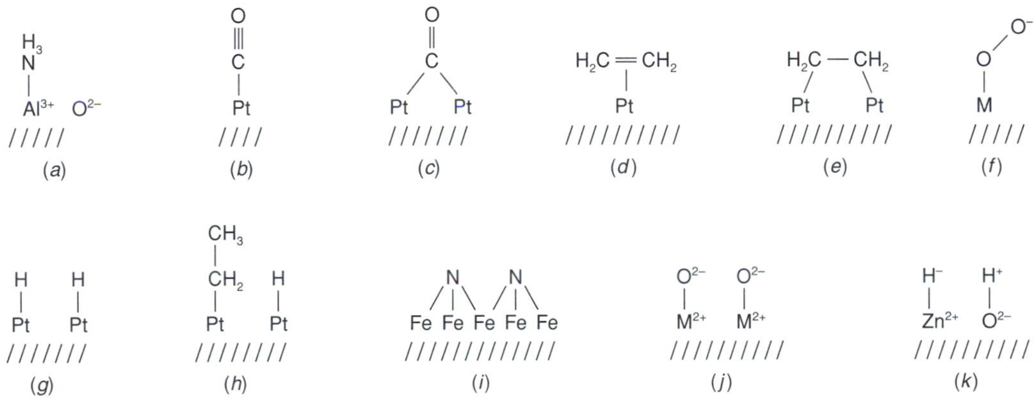

Figure 12.16
Forms of ligands resulting from chemisorption on the indicated metal of (*a*) NH_3; (*b*) and (*c*) CO; (*d*) and (*e*) ethylene; (*f*) and (*j*) O_2; (*g*) and (*k*) H_2; (*h*) C_2H_6; (*i*) N_2. [Adapted from D. F. Shriver, P. W. Atkins, and C. H. Langford, *Inorganic Chemistry*, Freeman, New York, 1990, p. 561.]

process of adsorbing these compounds by formation of covalent (or coordinate covalent) bonds is known as **chemisorption.** Infrared (IR) spectroscopy and a variety of surface techniques can identify these species on surfaces and, for example, distinguish bridging CO ligands [Fig. 12.16(*c*)] from terminal CO ligands [Fig. 12.16(*b*)]. In some cases, the chemisorbed molecule remains intact as a single ligand [Fig. 12.16(*a–d*)]; in homogeneous catalysis (Section 11.7) we called this process Lewis base association. Sometimes the ligand remains intact but suffers a rupturing of its π bonds, as in Figure 12.16(*e–f*); we previously referred to such a process in homogeneous catalysis as oxidative addition. Finally, in some cases **dissociative chemisorption** occurs to break all of the very strong chemical bonds in the ligand [Fig. 12.16(*g–k*)]; except in the last of these cases, this also corresponds to oxidative addition in homogeneous catalysis. The breaking of some of these strong bonds is often the rate-limiting step in reactions of these molecules. Once the molecule or its components is adsorbed, the resulting ligands can *diffuse* from one metal atom to another on the surface (by analogy with fluxional behavior in simple organometallic compounds, Chapters 9 and 11); this allows two different chemisorbed species to come into contact before reacting with each other.

Although adsorption of a reactant molecule is essential for heterogeneous catalysis to proceed, the adsorption must not be too strong, lest a stable species be formed that neither reacts with the other reactant nor desorbs from the surface to allow reuse of the surface site. For example, for the catalytic decomposition of formic acid on metal surfaces

$$HCOOH \rightarrow CO + H_2O \tag{12.6}$$

the reaction proceeds at a reasonable rate only at very high temperatures for metals that form either very stable metal formates or very unstable metal formates; the best catalytic metals are those with formates of intermediate stability (stable enough to form a chemisorbed formate ion, but not so stable that the ion will not further react or desorb), as shown in Figure 12.17.

Hence, not all metals that adsorb a given molecule are effective catalysts; the best catalysts are those with an intermediate tendency to adsorb molecules in general (Table 12.5). The metals at the left of the *d* block are, in general, too strong as Lewis

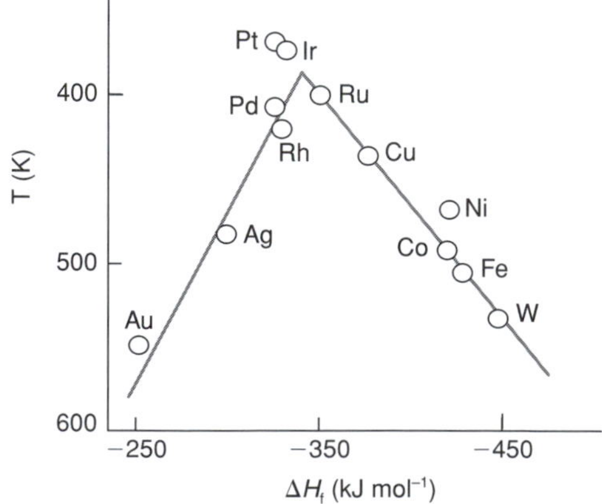

Figure 12.17
Reaction temperature at which a set rate of decomposition of formic acid is achieved (lowest temperatures at the top), versus ΔH_f of the corresponding metal formate. [Adapted from D. F. Shriver, P. W. Atkins, and C. H. Langford, *Inorganic Chemistry*, Freeman, New York, 1990, p. 563.]

Table 12.5
Numbers of Different Common Reactant Gases Chemisorbed by Different Metals[a]

Na 2												Al 4	Si 1				
K 2			Ti 7	V 7	Cr 7	Mn 4	Fe 7	Co 6	Ni 6	Cu 4	Zn 1		Ge 1	As 1			
			Zr 7	Mb 7	Mo 7		Ru 7	Rh 5	Pd 5	Ag 1	Cd 1	In 1	Sn 1	Sb 1			
			Hf 7	Ta 7	W 7		Os 7	Ir 5	Pt 5	Au 4			Pb 1	Bi 1			

[a]These gases are O_2, C_2H_2, C_2H_4, CO, H_2, CO_2, and N_2. Those designated by "7" adsorb all seven of the listed gses; "6" adsorb the first six; "5" the first five, and so on.

acids for good catalysis; the metals at the far right of the *d* block, or in the *p* block, are too weak. Also note from Table 12.5 that the least frequently chemisorbed molecule, N_2, is also among the poorest ligands in simple *d*-block complexes (Chapter 8), while the strong-field ligands such as CO and the π-donor ligands are adsorbed by many metals; the metals that do dissociatively chemisorb nitrogen are those that form stable nitrides (Chapter 14).

Alkenes are catalytically hydrogenated by metals such as Ni; the mechanism, after chemisorption (Lewis base association) of $CH_2{=}CH_2$ and dissociative chemisorption (oxidative addition) of H_2, first involves a reversible *insertion* reaction of one chemisorbed H atom with a chemisorbed $CH_2{=}CH_2$ molecule to give a chemisorbed CH_3CH_2 group:

$$CH_2{=}CH_2(M) + H(M) \rightleftharpoons CH_3CH_2(M) \quad (12.7)$$

This step is followed by reaction with the other chemisorbed H atom to give ethane, which desorbs (*undergoes reductive elimination*):

$$CH_3CH_2(M) + H(M) \rightarrow CH_3CH_3(g) + (M) \quad (12.8)$$

The reaction of $N_2(g)$ with $H_2(g)$ to give the important fertilizer $NH_3(g)$ is extremely slow in the absence of a catalyst, due to the necessity of first breaking the very strong triple bond in N_2. Haber found that an iron catalyst (with small quantities of other promoters) greatly speeds this reaction, since N_2 is dissociatively chemisorbed to give N atoms; H_2 is similarly dissociated. Once the barriers of these strong bonds are overcome, the reaction is fairly straightforward:

$$N(M) + H(M) \rightarrow NH(M)$$

$$NH(M) + H(M) \rightarrow NH_2(M)$$

$$NH_2(H) + H(M) \rightarrow NH_3(M) \rightarrow NH_3(g) \quad (12.9)$$

Desulfurization of fossil fuels is desirable to prevent their contributing to acid rain upon combustion; this desulfurization is accomplished by passing the fuel (containing, e.g., C_2H_5SH) over molybdenum, whereupon dissociative chemisorption occurs to give $C_2H_5S(M)$ and H(M). The C–S bond in the chemisorbed $C_2H_5S(M)$ weakens to leave S(M) behind, while the C_2H_5 group reacts with the chemisorbed H(M) to give C_2H_6, which is desorbed. Although the S remains on the surface, it actually assists the reaction by preventing the catalysis of side reactions to give less valuable products (carbon and hydrogen).

Perhaps the heterogeneous catalysis that is carried out most widely is that of the automobile catalytic converter, which uses a catalyst of Pt and Rh supported on a honeycomb ceramic to catalyze the reduction of NO and the oxidation of CO and unburned hydrocarbons in automobile exhausts; since these soft-acid catalysts strongly bond to sulfur, sulfur in petroleum **poisons** the catalyst and must be removed if present. This process is a major user of the world's platinum, which is normally recovered from junked cars. The discovery of a process that uses the more abundant and less expensive Pd for catalytic conversion caused substantial changes in the market prices of these metals.

One advantage of homogeneous catalysts is that they can be modified by changing functional groups in the attached ligands. Ligands can indeed be attached to the surfaces of metals. Especially in the case of the soft-acid metal gold, organic thiols (RSH) coordinate readily one layer deep, producing a thin-film surface on the outside of which numerous types of R groups can be displayed. For example, in the electronics industry a thin sheet of metal can be coated with an organic film that is photosensitive, then covered with a patterned template and exposed to light, leading to the erosion of the organic film wherever it was exposed to light. Next, a corrosive agent can be applied that will dissolve the thin sheet of metal through the holes in the organic film. The parts of the metal that are still protected by intact organic film then form a printed electrical circuit.

Example 12.3

The Fischer–Tropsch process for the reaction of CO and H_2 over catalysts of iron or cobalt supported on aluminum oxide to give gasoline is complicated, so its mechanism is controversial. This mechanism appears to involve some of the following steps: (a) dissociative adsorption of CO to give C(M) (a carbide) and O(M); (b) reaction of C(M) with two H(M) to give a surface carbene, $CH_2(M)$; (c) reaction of $CH_2(M)$ with H(M) to give $CH_3(M)$; (d) reaction of $CH_3(M)$ with $CH_2(M)$ to give $CH_3CH_2(M)$; (e) continuation of steps analogous to (d) to give longer chain alkyl groups; and (f) finally, reaction of the alkyl groups with H(M) to give the alkane, which desorbs. Classify these steps by the terminology used for homogeneous organometallic catalysts (Section 11.7).

SOLUTION:

(a) Since C(M) has C with an oxidation state of -4 and O(M) has O with an oxidation state of -2, the associated metal atoms must have been oxidized; this is oxidative addition.

(b, c) These involve three steps of the same type: In each step, H is oxidized from the -1 oxidation state when attached to the metal to $+1$ when attached to carbon, so the metal atoms must be reduced; these steps are reductive elimination.

(d, e) The CH_2^{2-} groups are oxidized and joined to alkyl groups in these steps, so again the metal atoms must be reduced: These are also reductive elimination, although final reductive elimination of all bonds of the carbon to the metal is not achieved until (f).

Evidence obtained in 1996 supports a mechanism other than that given in the above example for the Fischer–Tropsch process carried out on two different metals, rhodium and ruthenium. Maitlis and co-workers[55] propagated the process using doubly ^{13}C labeled ethylene derivatives but could not propagate it using doubly labeled ethyl derivatives, and the products were almost entirely doubly labeled. Hence, they propose a mechanism differing from Step (c) onward, in which a surface CH_2 group reacts with a surface vinyl ($CH{=}CH_2$) group instead of a surface CH_3 group, to give a surface allyl ($CH_2{-}CH{=}CH_2$) group. The complete proposed mechanism is shown in Figure 12.18.

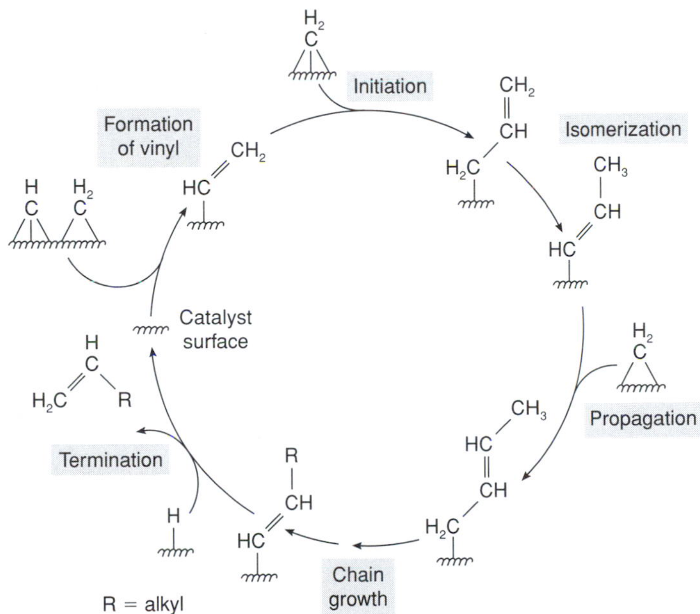

Figure 12.18
Proposed mechanism for the Fischer–Tropsch process on a metallic rhodium surface. [Adapted from *Chem. Eng. News.*, Aug. 12, 1996, p. 32.]

12.7 Metallic Chains, Sheets, and Clusters

Materials scientists and scientists studying catalysts have noted the great contrasts between the properties of bulk metals and those of simple complexes (Chapter 8) and organometallic compounds (Chapter 11) of the same metals. These scientists want to find materials with properties intermediate between the metallic conduction, magnetism, and heterogeneous catalytic abilities of bulk metals, which have overlapping bands of MOs, and the insulating, generally diamagnetic, and homogeneous catalytic properties of simple organometallic compounds, which have discrete well-separated energy levels and MOs. Hence, there has been a great deal of interest in studying the properties of **chains, sheets, and clusters** of metal atoms. Hopefully, these materials might show intermediate properties such as semiconduction due to development of bands, but with band gaps. How many metal atoms must be bonded to each other before a molecule or ion ceases acting like a simple metallic compound and begins acting like a bulk metal?

One-Dimensional Conductors (Chains). Square planar complexes of d^8 metal cations leave the metal ions open to their environment along their z axes, which opens possibilities for metal–metal bonding. Although ions or compounds of this type {i.e., $K_2[Pt(CN)_4]\cdot 3H_2O$} often stack in crystal lattices in columns with metal–metal contacts, the M–M distances are long (i.e., a Pt–Pt distance of 348 pm in the above compound), since the valence d_{z^2} orbitals of the Pt atoms are full, and no bonding can result. However, partial oxidation of compounds such as this (e.g., with Br_2 or Cl_2) result in removal of some electrons, so that a linear chain molecule built on a backbone of Pt–Pt bonding interactions results (the Pt–Pt distance drops to around 288 pm). The resulting products have formulas such as $K_2[Pt(CN)_4]Cl_{0.3}\cdot 3H_2O$ or $K_{1.75}[Pt(CN)_4]\cdot 1.5H_2O$, and have stacks of cations and anions that repeat at different chain lengths, as in $Hg_{2.86}AsF_6$. Because of their polymeric nature, the platinum chains have a band structure of energy levels that is incompletely occupied. Therefore these partially oxidized complexes are metallic conductors and are bronze in color. The original $K_2[Pt(CN)_4]\cdot 3H_2O$ is a white nonconductor.

Two-Dimensional Conductors (Sheets). The loose van der Waals bonding between layers of graphite allows the easy thermal *intercalation* of Group 1 metal atoms between the graphite layers to give *lamellar* compounds LiC_6, NaC_{64}, and MC_8, for M = K, Rb, and Cs. These compounds ignite in air, react explosively with water, and are better conductors than graphite. Some of these compounds are superconductors. Similar results can be obtained by intercalating halogen molecules to partially oxidize the graphite layer to give ${}^{2}_{\infty}[C_8{}^{+}]$ layers alternating with layers of, for example, Br^-; some products of this type have conductivities exceeding that of copper metal in that plane. We shall see additional examples in upcoming chapters where non-metal compounds can be oxidized or otherwise produced to give low-dimensional polymers with incompletely filled bands of MOs, which are also metallic conductors. In the ultimate pursuit of this trend, organic chemists are interested in producing purely organic metals, so that plastics can invade one more market formerly held exclusively by metals!

Clusters. A wide variety of types of metal cluster compounds[56] exist: The variety is in fact too great to be accommodated by a single bonding theory. Among the types of clusters known, we may list the following:

1. The homopolyatomic anions, cations, and allotropes of the *s*-, late *d*-, and *p*-block elements covered in Section 12.5 that have cluster oligomeric structures (e.g., $Au_4{}^{2+}$, As_4, and $Sn_5{}^{2-}$); these are "naked" clusters with no ligands attached to the metal or metalloid atom.

2. Cluster compounds and ions of the *p*-block elements that have ligands or atoms attached to the cluster elements, for example, the polyhedral borane anions, $B_nH_n{}^{2-}$, the bonding of which was discussed in Section 10.10. Details of the chemistry of these species will be given in the chapter appropriate to the type of ligand present.

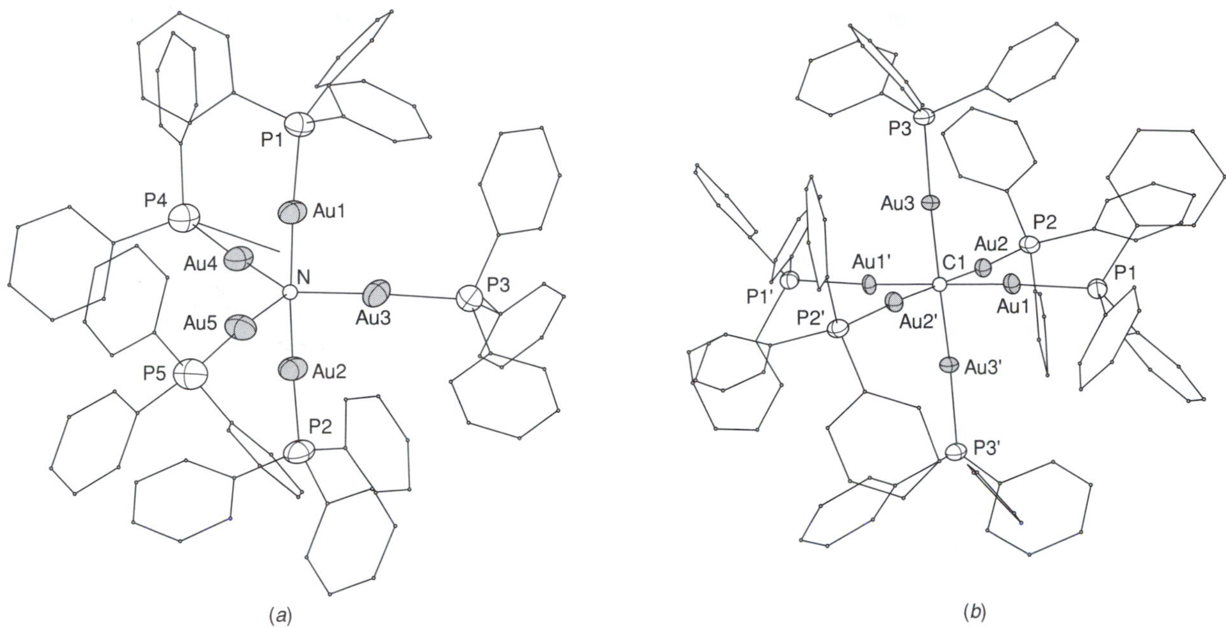

Figure 12.19
Structures of electron-deficient carbon and nitrogen dications $[N(AuPPh_3)_5]^{2+}$ [(*a*), adapted from A. Grohmann, J. Riede, and H. Schmidbaur, *Nature* (*London*), **345**, 140 (1990)] and $[C(AuPPh_3)_6]^{2+}$ [(*b*) adapted from F. Scherbaum, A. Grohmann, B. Huber, C. Krüger, and H. Schmidbaur, *Angew. Chem. Int. Ed. Engl.*, **27**, 1544 (1988)].

3. The polynuclear metal carbonyls and carbonylate anions [i.e., $M_n(CO)_y{}^{z-}$] and their derivatives, which are more complex than the mononuclear and dinuclear carbonyls discussed in Chapter 11.

4. Polynuclear metal carbonyls, carbonylate anions, and derivatives containing a small nonmetal atom (often C) located at the center of the cluster [i.e., $Fe_5C(CO)_{15}$]. Note that the positions of the metal and the carbon atoms in these species are just reversed from those in the endohedral fullerenes; we could perhaps write this formula as $C@Fe_5(CO)_{15}$.

5. Clusters of other metals that also have small nonmetal atoms at the center [e.g., CLi_6; $[C(AuL)_6]^{2+}$; Rb_9O_2]. The compound CLi_6 was reported in the gas phase[57] to have surprisingly strong C–Li bonds (225 kJ mol^{-1}). Computations, however, suggest that most of the bonding involves the Li_6 cluster, with relatively weak bonding of the cluster to the carbon; it can be thought of approximately as $C@Li_6$, with a C^{4-} anion endohedrally located inside a $Li_6{}^{4+}$ cluster. In addition, a number of *aurophilic cations* such as $[C(AuPPh_3)_6]^{2+}$, $[N(AuPPh_3)_5]^{2+}$ (Fig. 12.19), $[O(AuPPh_3)_4]^{2+}$, and $\{P[AuP(i\text{-}Pr)_3]_6\}^{3+}$ have been synthesized,[58] in which relativistic expansion of the gold 5*d* orbitals play a role in facilitating Au–Au bonding interactions, which may reach strengths up to 46 kJ per bonding interaction[59]; these interactions supplement direct Au–nonmetal bonding.

6. Clusters of early *d*-block metal ions in moderately positive oxidation states, with anions such as halides and sulfide incorporated on the outside of the cluster.

These so-called *Keggin phases* will be considered along with other halides and sulfides in Chapter 14.

7. Very large clusters (i.e., $Au_{55}L_n^{x+}$), and *nanoparticles* of thousands of metal atoms, as in colloidal metal particles.

Although the variety of clusters is large, it turns out that the larger clusters of the first four types listed above can be treated by the same set of bonding principles, known as **Wade's rules**[60] or (in view of subsequent contributions) as the **Wade–Mingos–Lauher** principles. In order to introduce this terminology, we first consider small clusters and "subclusters", which can be treated by the 18-electron and 8-electron (octet) rules.

Monomeric noncluster molecules such as CH_4 and $Ni(CO)_4$ obey the octet and 18-electron rules, which means that the valence electron counts (VEC) about the central atom (counting contributions from ligands and any negative or positive overall charges) are 8 and 18, respectively. In related *dimeric* molecules such as ethane, $(CH_3)_2$, and dicobalt octacarbonyl, $[Co(CO)_4]_2$, we classically expect *each* central atom to have 8 or 18 electrons, respectively, but since the shared central atom–central atom (**framework**) bond is counted for each atom, the total VEC must avoid double-counting those electrons: the Lewis structure of ethane contains a total of $2(8) - 2 = 14$ valence electrons; the corresponding structure for $Co_2(CO)_8$ includes $2(18) - 2 = 34$ valence electrons.

Two isolobal trinuclear, triangular molecules are cyclopropane, $(CH_2)_3$, and triosmium dodecacarbonyl, $Os_3(CO)_{12}$; the (Lewis) structures of these molecules show *three* central-atom–central-atom or framework single bonds (links); so the total VEC is $3(8) - 6 = 18$ for cyclopropane and $3(18) - 6 = 48$ for $Os_3(CO)_{12}$.

Some isolobal tetrahedral clusters are P_4, Si_4^{4-}, Si_4R_4 [$R = Si(t\text{-Bu})_4$], and $Rh_4(CO)_{12}$; Lewis structures of any of these include *six* framework bonds, so the total VEC for each of the first three molecules is 20, while that for the *d*-block cluster is 60. Note that the difference between the *p* and *d* block totals is 10 times the number of cluster framework atoms, since each *d*-block atom has 10 more valence electrons than the corresponding *p*-block atom.

Another way of characterizing these clusters and subclusters is by the **number of framework electrons** (**NFE**), which is the number of valence electrons used to bond the framework together. This is, of course, twice the number of framework bonds or central-atom–central-atom links in these classical compounds, so it is (regardless of whether the cluster is composed of *d*- or *p*-block atoms) 2 for the dimers, 6 for the triangular trimers, and 12 for the tetrahedral tetramers. The remaining valence electrons are used to bond the terminal ligands and/or to hold unshared electron pairs (as in P_4, or as with the nonbonding *d* electrons in the *d*-block clusters). Note also that the NFE count is the same for *p*- and *d*- block clusters.

The NFE builds up rapidly with an increase in the number of atoms (n) in a regular closed polyhedron with triangular faces (deltahedron) and classical 2c–2e bonds (i.e., each edge of the polyhedron being a bonding pair of electrons): For n between 3 and 12, it is equal to $6n - 12$. Beyond $n = 4$, framework *p*-block atoms can no longer provide either enough electrons or enough valence orbitals to bond the framework atoms to each other with *2c–2e* bonds and also bond the terminal atoms or to hold lone pairs; yet such clusters exist (consider Sn_5^{2-} and Pb_9^{4-}). The *d*-block atoms, with more valence electrons and orbitals, can hold out a little longer, but discrepancies exist already for $n = 6$ and become quite serious for $n = 8$. Hence, clusters of these dimensions no longer obey the 8- and 18-electron rules for each framework atom.

Before we consider how the Wade–Mingos–Lauher rules treat these larger clusters, let us consider the geometrical and bonding consequences of adding electrons to (chemically reduce) these classical clusters and subclusters. Normally, in such clusters, we expect the LUMO to be a σ^* orbital that we will presume is antibonding with respect to one of the framework atom–framework atom links. So the two extra electrons *negate one bonding interaction*; instead we now add *two nonbonding interactions.* Consequently, *we reduce the number of linkages in the polyhedron, which results in a more open type of structure.*

For example, adding two electrons to $(CH_2)_3$ would give $(CH_2)_3^{2-}$, with a total VEC of 20 electrons. We might look for such a dianion in the Grignard reagent made by reaction of 1,3-dibromopropane with 2 mol of Mg; we would expect a noncyclic (chain) structure, $[:CH_2{-}CH_2{-}CH_2:]^{2-}$. If we were to add two electrons to a tetrahedral cluster, we would expect to rupture one link to give a *butterfly shaped* cluster such as the one found in the structural unit of red phosphorus, or in the cluster anion $[Re_4(CO)_{16}]^{2-}$ [total VEC = 7(4) for Re + 2(16) for the carbonyls + 2 for the charge = 62 valence electrons]. If we were to add yet two further electrons to this more open, less-linked butterfly cluster, we might expect to, and do, destroy the other transannular link to give a square structure, as in S_4^{2+} [VEC = 6(4) − 2 = 22]. Note that we have reduced the total number of links from six in the tetrahedron to five in the butterfly to four in the square.

Wade–Mingos–Lauher Rules. These are based on MO calculations on clusters such as those presented for the closed deltahedral borane anions in Section 10.10. The germane point here is that, for such clusters, only *one* combination of *radial* framework–atom p_z orbital can give a bonding (hence occupied) MO, and only n combinations of the tangential p_x and p_y orbitals can give bonding orbitals, for a total of $n + 1$ bonding framework MOs, which can hold a total of $2n + 2$ framework electrons (NFE). If we also take into account that each boron has a B–H terminal bond in these boranes, or each framework nonmetal atom has an external unshared electron pair in the homopolyatomic anions, cations, and cluster allotropes, we arrive at a total VEC of $4n + 2$ valence electrons for *p*-block clusters, or an VEC of $14n + 2$ valence electrons for *d*-block clusters. These expressions do not grow as rapidly with increasing n as do the values for "classical" structures obeying the 8- and 18-electron rules. They are achievable and in fact usually are those found for the relevant clusters with n between 5 and 12. They do work for the polyhedral borane anions, nearly always work for the homopolyatomic ions of the *p*-block, and usually work for late *d*-block clusters, with discrepancies for $n = 6$ and 7.

Lauher[61] analyzed the bonding in the late *d*-block clusters by carrying out MO calculations on various-sized clusters of Rh atoms, with different geometries as possibilities. He supposed that the stable geometries will be those in which all high-energy antibonding MOs are left unoccupied. Lauher showed that, late in the *d* block, the *d* orbitals of metal atoms are too contracted to have significant overlap with *each other*[62]: Their function in these clusters is to bond with external ligands, and to hold nonbonding (unshared) electron pairs. The valence *s* orbitals (or a hybrid thereof) are generally similarly occupied, leaving the *three valence p orbitals* available to form the framework bonding of the cluster, just as is the case in the polyhedral boranes and

Table 12.6
Electron Counts for Closed Deltahedral Clusters

n	Polyhedron	*p* Block Total VEC	*d* Block Total VEC	*d* Block MOCalc[a]	NFE[b]
5	Trigonal bipyramid	22	72	72	12
6	Octahedron	26	86	86	14
7	Pentagonal bipyramid	30	100	98	16
8	Dodecahedron	34	114	112	18
9	Tricapped trigonal prism	38	128	128	20
10	Bicapped square antiprism	42	142	142	22
11	Octadecahedron	46	156		24
12	Icosahedron	50	170	170	26

[a] The total VEC calculated by Lauher,[61] and generally observed if examples are known; these values should take precedence over those in the previous column, which were calculated more approximately as total VEC = $14n + 2$.

[b] Total NFEs, which equals VEC − $2n$ for *p*-block elements and VEC − $12n$ for *d*-block elements, except when *n* is 7 or 8.

the homopolyatomic ions of the *p* block. One more high-energy antibonding orbital than in the general trend was found for the $n = 7$ and $n = 8$ clusters, leading to the improved total VECs that give regular deltahedral structures for late *d*-block clusters; these results are collected in Table 12.6. The closed deltahedral structures shown in the left column of Figure 12.20 are adopted for total VECs of $4n + 2$ (or total framework electrons = $2n + 2$); boron compounds having these structures are known as closo boranes.

For either *p*- or *d*-block clusters, adding additional electrons results in opening the polyhedral structures to forms that have a smaller total number of links between framework atoms. The details do differ in the two blocks, however. For boron hydrides and their anions the resulting more open structures [Figure 12.20(*b*)] are best envisioned as *fragments* of the deltahedra of the closo boranes having one more vertex; these are adopted for total VECs of $4n + 4$ (or total framework electrons = $2n + 4$). Boron compounds having these structures are known as *nido* boranes. The still more open, basket-like structures in Figure 12.20(*c*) are adopted for total VECs of $4n + 6$ (or total framework electrons = $2n + 6$); boron compounds having these structures are known as arachno boranes. (Not shown are even more open hyphno structures adopted for $4n + 8$ total VECs.) Examples of these compounds and their chemical properties will be detailed in the chapter on hydrides, Chapter 15.

For *d*-block polyhedra, the more open structures are not generally deltahedral fragments but are other less-linked polyhedra. Thus, while the octahedron, with 12 links, is adopted with a total VEC of 86 electrons [as in $Ru_6(CO)_{17}C$]; the pentagonal pyramid, with 10 links, is calculated to be adopted with a total VEC of 88 electrons; and the trigonal prism, with 9 links, is adopted with a total VEC of 90 electrons {as in $[Rh_6(CO)_{15}C]^{2-}$}. Lauher was even able to calculate structures for unusually low VEC totals: for $n = 6$ and VEC = 84, the bicapped tetrahedron (also with 12 links) is adopted [as in $Os_6(CO)_{18}$]. The geometries calculated by Lauher (many of which have not been verified by synthesis of an appropriate example) are shown in Table 12.7.

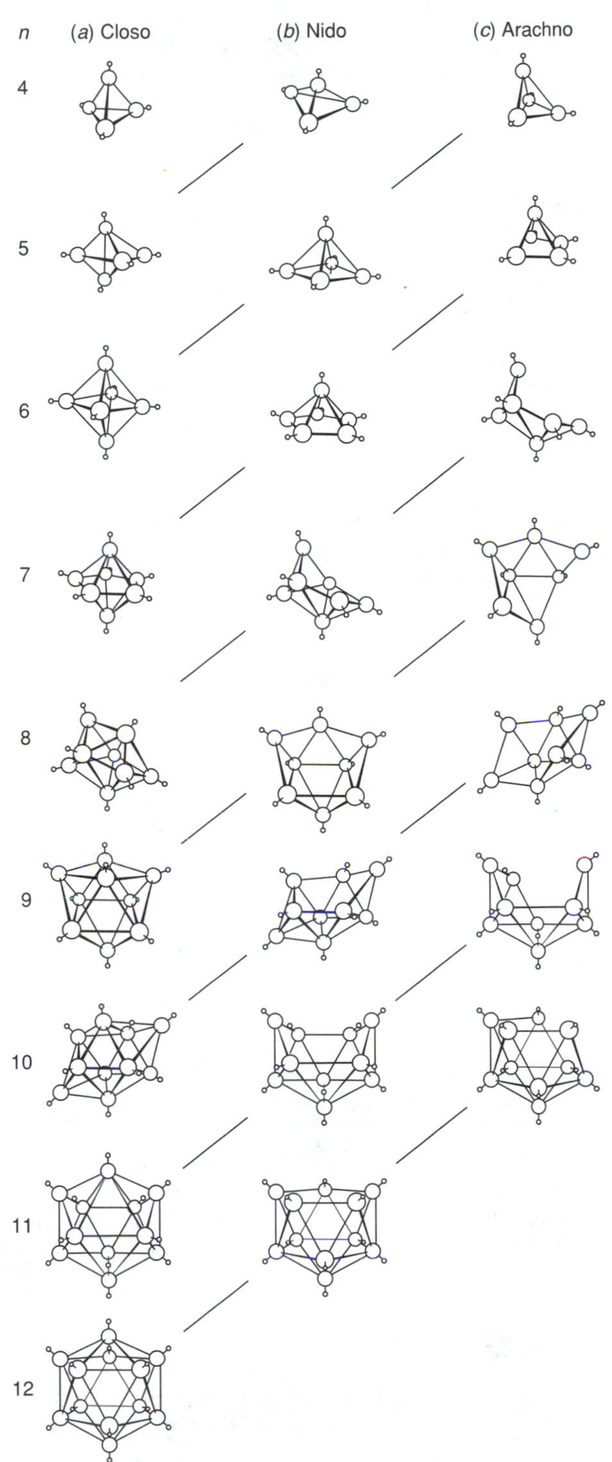

Figure 12.20

Deltahedra and deltahedral fragments for boranes and borane anions. (*a*) The closo boranes containing *n* boron atoms; (*b*), corresponding nido boranes; (*c*), corresponding arachno boranes. (Diagonal lines show the fragmentation patterns leading to the structures of the nido and arachno boranes.) [Adapted with permission from R. W. Rudolph, *Acc. Chem. Res.*, **9**, 446 (1976). Copyright 1976 American Chemical Society.]

Table 12.7
Selected Other *d*-Block Cluster Geometries Predicted by MO Calculations[a]

n	Total VEC	Geometry	Total VEC	Geometry	Total VEC	Geometry
5	74	Square pyramid				
6	88	Pentagonal pyramid	90	Trigonal prism	84	Bicapped tetrahedron
7	102	Capped trigonal prism				
8	114	Square antiprism	120	Cube	110	Bicapped octahedron
9	130	Capped square antiprism	132	Capped cube	126	Tricapped octahedron
14	180	Face centered cube				
15	192	Rhombic dodecahedron				

[a] Calculations by Lauher.[61]

The interstitial carbon atoms located at the center of some of the above clusters apparently arise from a metal-promoted disproportionation of some of the CO ligands:

$$2\,CO(M) \rightarrow C(M) + CO_2(g) \tag{12.10}$$

Since these carbon atoms (or carbide ions) are interstitial, they do not increase the number of vertices of the polyhedron, hence they do not alter n; but they do contribute their valence electrons to the VEC.

Example 12.4
Compute the total VEC and (if possible) select an appropriate geometry for the following clusters: (a) $Pb_9{}^{4-}$; (b) $Te_6{}^{4+}$; (c) $B_{10}H_{10}C_2H_2$; (d) $Os_5(CO)_{16}$; (e) $[Rh_6(CO)_{15}C]^{2-}$.

SOLUTION:

(a) Total VEC $= 9(4) + 4 = 40$ electrons; for $n = 9$; this is $4n + 4$, so a slightly open geometry is predicted (Table 12.7): the capped square antiprism, which is in fact observed (Fig. 12.14).

(b) Total VEC $= 6(6) - 4 = 32$ electrons; for $n = 6$, this is $4n + 8$, which would give a quite open structure. In fact (Fig. 12.14), a trigonal prismatic structure is adopted.

(c) $n = 12$ (10 B and 2 C); total VEC $= 3(10) + 1(10) + 4(2) + 1(2) = 50$ electrons, so icosahedral (Table 12.6).

(d) $n = 5$; total VEC $= 8(5) + 2(16) = 72$ electrons; so trigonal bipyramidal.

(e) This organometallic cluster includes an interstitial carbon atom; its electrons are contributed to the total VEC, but it is not included in the count of n, since it fits in a hole of the polyhedron rather than occupying a corner. Hence, $n = 6$, and total VEC $= 9(6) + 2(15) + 4 + 2 = 90$ valence electrons, so the geometry should be (and is) a trigonal prism.

Although details of the chemistry of late *d*-block metal cluster compounds are beyond the scope of this work, we note that there often is fluxional motion of carbonyl (and other) ligands over the surface of the metal polyhedron: It has been suggested that

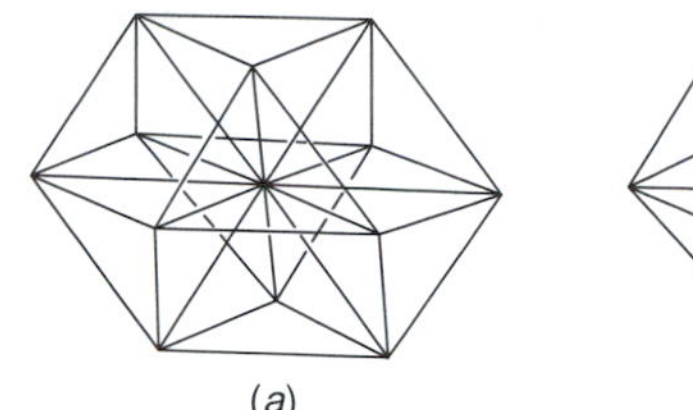
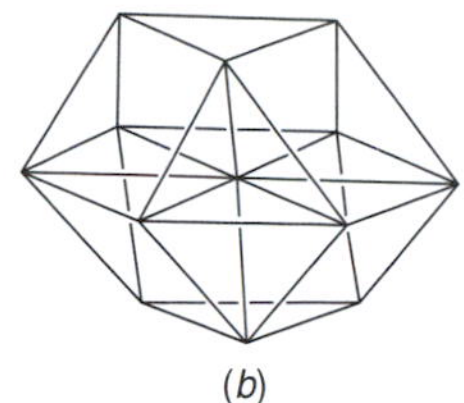

(*a*) (*b*)

Figure 12.21
Close-packed cluster structures. (*a*) The cuboctahedron (O_h point group), which is cubic close packed if an (interstitial) metal atom is included at the center; (*b*) the truncated hexagonal bipyramid (D_{3h}), which under the same conditions is hexagonal close packed. [Adapted from J. W. Lauher, *J. Am. Chem. Soc.*, **100**, 5305 (1978).]

the prevalence of this might be explained as due to bonding of the ligands to the cluster as a whole, rather than to individual atoms. This fluxionality should contribute to catalytic ability of these clusters.

Giant Clusters. Some features of metals and metalloids are not reproduced in the above cluster compounds: They are not close-packed; and they do not have a band structure to their energy levels. That is, the energy levels do not vary by infinitesimal increments: otherwise it would have been impossible for Lauher to specify a definite number of high-energy antibonding orbitals. As alternative geometries for $n = 12$ cluster atoms and VEC = 170 electrons, he did identify two geometries that are fragments of close-packed metallic structures: The D_{3h} truncated hexagonal bipyramid, which is a fragment of a hcp lattice, and the O_h cuboctahedron, which is a fragment of a ccp lattice (Fig. 12.21). Furthermore, it turns out that both of these geometries (and the icosahedron) have central cavities large enough to hold an interstitial metal atom to give 13-atom $M@M_{12}$ fragments in which the central metal atom has the characteristic coordination number of 12 found in these close-packed lattices.

If we were now to add a second layer of close-packed metal atoms around the 13-atom cluster, this $\ell = 2$ layer would contain $10\ell^2 + 2$ (e.g., 42) additional atoms, to give a $M@M_{12}@M_{42}$ cluster containing 55 metal atoms. Such giant clusters[63] (in ligand-coated, partially oxidized forms) as $[M_{55}L_{12}Cl_x]$ can be formed (for M = Rh, Ru, Pt, and Au, and $x = 6$ or 20) by the reduction of appropriate metal salts by B_2H_6 in organic solvents in the presence of the ligands L = R_3P and R_3As. The ^{197}Au Mössbauer spectrum of one such gold complex shows the presence of four types of Au atoms: 12 Au atoms coordinated to PPh_3, 6 Au atoms coordinated to Cl, and 24 "bare" Au atoms, all on the surface, and 13 inner atoms; the Mössbauer parameters of only the inner atoms are very similar to those of gold metal itself. A still larger cluster, $Pt_{309}L_{36}O_{30}$, shows an NMR Knight shift, which is a property very characteristic of metals; the Mössbauer parameters of the 147 inner atoms are similar to those of gold metal.[64]

Still larger clusters can be obtained by producing *colloidal* particles of the metals: colloidal gold has long been known, as the red color of ruby glass produced in the middle ages, and as the pigment "purple of Cassius." Metal clusters can also be produced by molecular beam techniques, and by deposition (say by decomposition of carbonyls) inside the cavities of zeolites (Chapter 13). In these larger clusters, the inner atoms seem to have the properties of bulk metals, while the outer ones behave as if they are in compounds, bonding to ligands, or (if ligands are absent) showing enhanced Lewis acid chemical reactivity and some unusual physical and electronic properties. For these reasons, there is also a great deal of interest in producing nanoscale particles and composites of particles of compounds (particularly semiconductors) as well as of metals.

Very finely divided (and therefore very reactive) metal particles can also be produced chemically (a) by metal-atom reactor techniques (Section 7.3); (b) by the reduction of metal halides with lithium metal and naphthalene, which react to form the strong reducing agent lithium naphthalide, in organic solvents[65]; and (c) by the action of ultrasound on metal carbonyls (or on metals themselves if they have low energies of atomization). Ultrasound produces explosions in a liquid that locally heat the liquid and its contents to around 5000 K, then cause cooling to ambient temperature within a few microseconds. Under these conditions, metal atoms can be cooled into amorphous and even colloidal forms that show enhanced catalytic activity.[66]

Study Objectives

1. Know and explain the relationship between the monomeric, oligomeric, or polymeric structure of a compound, its physical state (solid, liquid, or gas) and its melting point. Exercises 1–4.

2. Compare (qualitatively) the relative heats of atomization or densities of two elements, using periodic trends. Explain these values in terms of the valence or prepared-for-bonding electron configurations of the two elements. Exercises 5–8.

3. Describe the structures and physical properties (allotropes, physical state, color, ease of boiling, and conductivity) of any nonmetallic elements; note periodic trends in degrees of polymerization. Exercises 9–17.

4. Use σ and π bond energies to calculate the relative stabilities of allotropes of elements that differ in their number of σ and π bonds. Exercises 18–27.

5. Give examples of elements that will be most useful (a) as oxidizing agents, (b) as inert gases, (c) as semiconductors, (d) as structural metals, and (e) for applications requiring malleability or ductility. Exercises 28–36.

6. Apply principles from earlier in the text to explain the properties of allotropes such as the fullerenes, or of homopolyatomic ions. Exercises 37–46.

7. Relate the structures of homopolyatomic anions of the *p*-block elements to the structures of (sometimes hypothetical) allotropes of the same elements, or to the structures of allotropes of the elements to the right of them in the periodic table. Exercises 47–53.

8. Classify steps in proposed mechanisms of heterogeneous catalysis. Exercise 54.

9. Identify cluster geometries from calculations of the numbers of cluster atoms and valence electrons, or vice versa; identify giant cluster sizes from the number of shells, or vice versa. Exercises 55–59.

Exercises

1. *Sulfur trioxide exists in three forms, which have the molecular formulas SO_3, S_3O_9, and ${}^{1}_{\infty}[SO_3]$. (a) Classify each of these into one of the structural classifications given in Figure 12.1. (b) Assuming (contrary to actual fact) that the three forms are not readily interconverted, rank these three forms in order of increasing melting point. Which of the three is most likely to be a gas at room temperature?

2. For each of the following physical properties at room temperature, find the structural classification of material likely to be associated with it.

Viscous liquid	Oligomer
Slippery solid	Molecular monomer
Low-boiling gas	Chain polymer
Soft solid	Layer polymer
Hard solid	Network polymer
Low-density solid	Ionic salt
High-density solid	

3. The oxide of a certain element E in its group oxidation state has the following structural forms **1**, **2**, **3**:

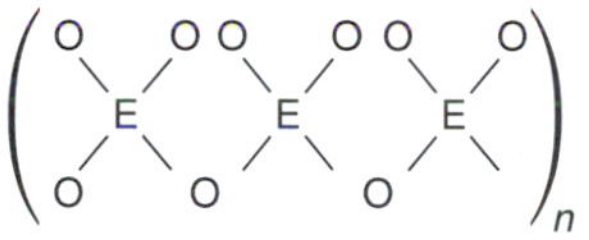

1 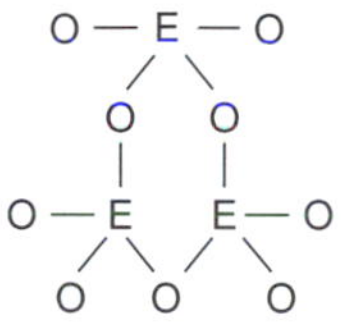 **2** 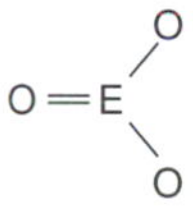 **3**

(a) Give the best description of each structure (e.g., as a monomer, a cluster oligomer, a layer polymer, etc.). (b) Which one of these three structural forms would likely be most volatile? Which would most likely be a high-boiling liquid? (c) Element E is most likely to be which one of these elements? C, Si, Pb, O, Se, Po, Xe.

4. Select one of the following materials as an example of a material likely to exhibit the physical property listed at room temperature. Materials: C_{60}, He, $(Th^{4+})(O^{2-})$, ${}^{3}_{\infty}[C]$, ${}^{2}_{\infty}[C]$, ${}^{1}_{\infty}[S]$. Properties: (a) viscous liquid; (b) slippery solid; (c) low-boiling gas; (d) soft solid; (e) hard solid; (f) low-density solid; and (g) high-density solid.

5. *Without referring to Table 7.1, choose the element from each set that should have the highest atomization energy and the element that should be easiest to atomize. Explain why you made the choice you did: (a) C, Si, Ge, Sn, Pb; (b) V, Nb, Ta; (c) Li, Be, B, C, N, O, F, Ne; and (d) Sr, Zr, Mo, Ru, Pd, Cd, Sn, Te, Xe.

6. Below are listed some important periodic trends in the properties of elements. For each property, a selection of elements from a given group or period are listed. Choose the element from the set at which that property reaches a maximum, or indicate if this property is the same for all elements of the set.

   (a) Atomization energy: Rb, Y, Mo, Rh, Ag, Cd.

   (b) Highest density: Cs, Lu, Os, Hg, At.

(c) Number of prepared-to-bond electrons: Sr, Zr, Mo, Ru, Pd, Cd.

(d) Number of prepared-to-bond electrons: C, Si, Ge, Sn, Pb.

7. (a) Which element of Group 13(III) is the hardest to atomize? Explain why. (b) Which element of Group 7 is the easiest to atomize? Explain why.

8. Suppose that valence *f* orbitals and valence *s* and *p* orbitals all had similar values of Z^* and $\langle r_{max} \rangle$ in a given atom, but that its valence *d* orbitals were spatially and energetically unavailable. At approximately which atom would you expect to find the largest heat of atomization among the *f*-block elements of the seventh period?

9. *Draw the structure of each of the following elements, and characterize each as either a gas, liquid, easily vaporized solid, or hard-to-vaporize solid: (a) boron; (b) white phosphorus; (c) black phosphorus; and (d) molybdenum.

10. In each part, give the symbols of two elements that: (a) occur in the diamond structure; (b) occur in polymeric helical chains; (c) occur in a black, sheet-like allotrope; (d) occur as diatomic molecules; (e) occur as tetrahedral X_4 molecules; (f) occur as X_8 molecules; (g) occur at room temperature in allotropes with multiple bonds; and (h) occur as monoatomic gases.

11. What element occurs as (a) linked icosahedra of atoms and (b) in both a diamond-type lattice and a metallic lattice?

12. *Many allotropes of the nonmetals do not melt and/or boil, but convert to other allotropes on heating (or cooling). Pretending for the moment that this does not happen, classify each allotrope in Figure 12.3 as one of the following: (a) likely to have very high melting and boiling points; (b) likely to be a gas or a liquid at room temperature; or (c) likely to be a solid with a relatively low melting point. Explain, in general, why classes (a), (b), and (c) contain the allotropes they do.

13. (a) Draw the structures of the three major allotropes of phosphorus. (b) Which of the three forms should have the highest vapor pressure at room temperature? (c) Which form is chemically the most reactive? (d) Which has the least stable σ bonds, and why? (e) Which has no counterpart in arsenic or antimony chemistry?

14. *Below are shown the structures (or parts of the structures in the case of polymers) of allotropes of some of the elements.

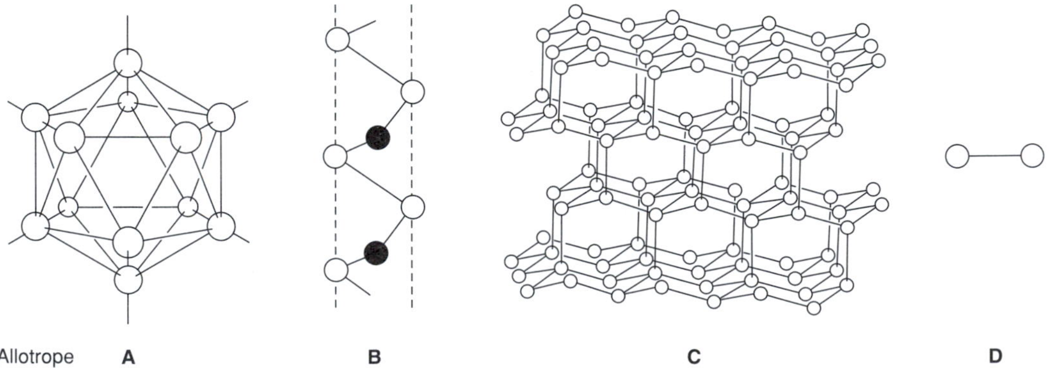

(a) For each structure, list one element that actually has an allotrope (maybe not the stable one) with that structure. (b) List the structures **A**, **B**, **C**, and **D** in order of expected increasing melting points. (Assume the equal strength of all covalent bonds that consist of electron pairs shared between two atoms.)

15. Below are drawn some structures of possible allotropes of elements. For each structure, select the best answer. (a) Describe that type of structure as monomeric molecular, oligomeric, chain (one-dimensional) polymeric, sheet (two-dimensional) polymeric, or network (three-dimensional) polymeric. (b) An allotrope with this structure is likely to be which of the following: a gas; a soft low-melting solid; a slippery high-melting solid; or a hard, dense high-melting solid. (c) An element that actually has a structure similar to the one shown is which of the following: Se, Xe, P, Si.

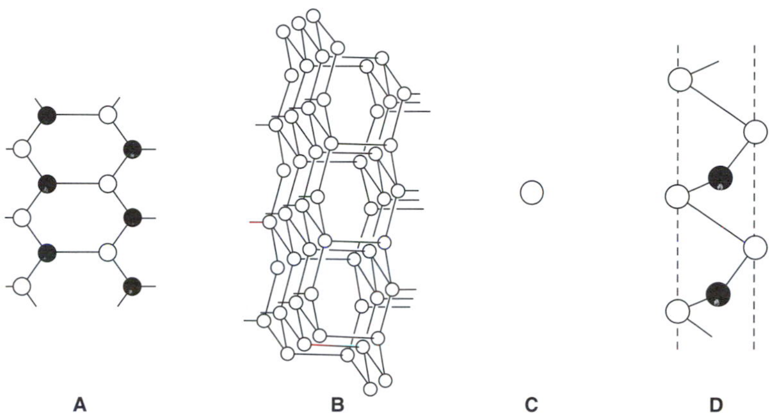

16. Below are shown the structures (or parts of the structures in the case of polymers) of possible allotropes of some of the elements.

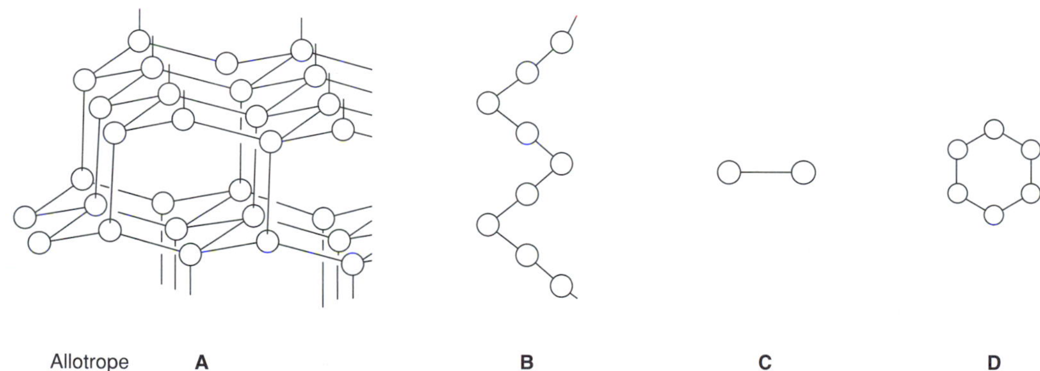

Assume that (somehow) each of these allotropes were possible for one element. Which of the allotropes (**A**, **B**, **C**, or **D**) is most likely to (a) Show the lowest boiling point? (b) Show the greatest mechanical strength (hardness)? (c) Be the most dense? (d) Be a gas at room temperature? (e) List one element that actually has an allotrope (maybe not the stable one) with a structure corresponding to **A**; to **B**; to **C**; to **D**.

17. Below are listed some important periodic trends in the properties of elements. For each property a selection of elements from a given group or period are listed. Choose the element from the set at which that property reaches a maximum, or indicate if this property is the same for all elements of the set.

(a) Predominance of polymeric forms of element: Si, P, S, Cl, Ar.

(b) Tendency to form π bonds: C, Si, Ge, Sn, Pb.

(c) Highest density allotrope of carbon: buckminsterfullerene, diamond, graphite.

(d) Ability to act as an oxidizing agent: Si, P, S, Cl, Ar.

(e) Unreactivity in redox reactions: Si, P, S, Cl, Ar.

18. *From the σ and π O–O bond energies, calculate ΔH for the conversion of 3 mol of O_2 to 2 mol of O_3. Which is more stable?

19. For each of the Group 15(V) and Group 16(VI) elements except Bi and Po, write balanced chemical equations illustrating the process:

$$\text{multiply bonded allotrope} \rightleftharpoons \text{singly bonded allotrope}$$

(One of these allotropes may have to be an imaginary analogue of a known allotrope of another element in the group.) Using the average σ- and π-bond energies from Table 3.1, calculate ΔH for this process in each case. What is the general trend in the stability of multiply bonded allotropes? Can you explain this in terms of overlap of orbitals and bond lengths?

20. Calculate ΔH for the conversion of 6 mol of sulfur *atoms* from the allotropic form S_2 to the allotropic form S_3 (with a structure like that of ozone). Is this reaction favored?

21. Given that the Se–Se σ bond energy is 172 kJ mol^{-1} and the Se–Se π bond energy is 100 kJ mol^{-1}, calculate ΔH for the conversion of 2 mol of Se_3, with a Lewis structure like that of ozone, to 1 mol of the ring compound Se_6. Is this reaction favored by ΔH?

22. *What conditions of temperature and pressure should favor the formation of multiply bonded allotropes for Groups 15(V) and 16(VI), and why?

23. For Group 14(IV) elements, the allotropes either have (1) four single bonds to four different atoms (as in diamond) or (2) two single bonds and one double bond at each atom, as in graphite or buckminsterfullerene. Try to imagine a new allotrope of carbon with two double bonds at each atom. Predict its structure and some physical properties. Estimate ΔH for it decomposing to diamond.

24. *A possible new allotrope of carbon that would be of interest for its electrical properties is carbyne, $^{1}_{\infty}$[C≡C–C≡C–], which contains alternating single and triple bonds. (a) There are how many σ and how many π bonds per carbon atom in this allotrope? (b) What is the stable allotrope of carbon? There are how many σ and how many π bonds per carbon atom in the stable allotrope of carbon? (c) Calculate ΔH for the conversion of 6 mol of C atoms from the new hypothetical allotropic form to the stable allotropic form. Is this reaction favored?

25. For Group 15(V) elements, there are two general types of allotropic structures: (1) those with three single bonds to three different atoms (as in P_4) and (2) those with one triple bond to one other atom (N_2). Try to imagine an allotrope with the intermediate situation, in which each Group 15(V) atom has one single bond to one neighbor and one double bond to another neighbor. (a) Draw a small-molecule allotrope with this structure. Predict some physical properties and compare it with the physical properties of existing allotropes for N; for P. (b) Estimate ΔH for this new allotrope of N decomposing to the more stable allotrope of N. (c) Estimate ΔH for this new allotrope of P decomposing to the more stable allotrope of P.

26. Suppose that a new cyclic allotrope of arsenic, *cyclo*-As_6, were to be discovered. (a) Draw a reasonable Lewis structure (one obeying octet rules) for this allotrope, and determine the number of σ and π bonds present. (b) Estimate ΔH for the conversion of 1 mol of the *cyclo*-As_6 allotrope to the normal allotrope of arsenic.

27. The multiply bonded allotropes of the second period are thermodynamically favored over the singly bonded allotropes. Is this tendency as strong for carbon as for nitrogen and oxygen? Why or why not?

28. *Below are listed some important periodic trends in the properties of elements. For each property, a selection of elements from a given group or period are listed. Choose the element from the set at which that property reaches a maximum, or indicate if this property is the same for all elements of the set.

(a) Highest bond order per atom in element: Sr, Zr, Mo, Ru, Pd, Cd.

(b) Highest melting point: B, Al, Ga, In, Tl.

(c) Highest boiling point: C, Si, Ge, Sn, Pb.

(d) Ferromagnetism of element: Cs, Gd, Os, Pb, Rn.

(e) Alloy with highest average atomization energy: RhAg, RhZr, RhSn, pure Rh.

(f) Ability to semiconduct: N, P, As, Bi.

29. Below are listed some important periodic trends in the properties of elements. For each property, a selection of elements from a given group or period are listed. Choose the element from the set at which that property reaches a maximum, or indicate if this property is the same for all elements of the set.

(a) Ferromagnetism: Gd, Cm, eka-Cm (element below Cm).

(b) Ferromagnetism: Co, Rh, Ir.

(c) Mechanical hardness of element: Rb, Mo, Ag, Sb_4, I_2, Xe.

(d) Highest bond order per atom in element: C, Si, Ge, Sn, Pb.

(e) Highest melting point: Mg, Al, Si, S, Cl.

(f) Highest boiling point: Sr, Zr, Mo, Ru, Pd, Cd, Sn, Te, Xe.

(g) Ductility: Hf, W, Os, Au.

(h) Alloy with highest melting point: RhAg, RhZr, RhSn, pure Rh.

30. Below are listed some important periodic trends in the properties of elements. For each property, a selection of elements from a given group or period are listed. Choose the element from the set at which that property reaches a maximum, or indicate if this property is the same for all elements of the set.

(a) Ability to semiconduct: C, Si, Sn, Pb.

(b) Ability to semiconduct: B, Al, Ba, In, Tl.

(c) Strongest element–element single bond: C, Si, Ge, Sn, Pb.

(d) Largest energy gap between bands of filled and vacant molecular orbitals: C, Si, Ge, Sn, Pb.

(e) Alloy with β-brass structure: Ag_5Cd_8, NiPb, Cu_3Pb, Ni_5Sn_3.

(f) Alloy with γ-brass structure: Ag_5Cd_8, NiPb, Cu_3Pb, Ni_5Sn_3.

(g) Alloy with ε-brass structure: Ag_5Cd_8, NiPb, Cu_3Pb, Ni_5Sn_3.

31. Which element of the second period is the hardest to atomize and to melt? In one sentence, explain why.

32. Consider the following elements: Cs, Ti, Co, Ir, As, Bi. (a) Which one of the metals in this list would be softest? (b) Which of the metals would be most useful for structural purposes, such as building airplanes? (c) Which of the elements would be a semiconductor? (d) Which of the *d*-block metals in the list would give off the most heat on being alloyed with the metal Zr? (e) Which of the metals shows ferromagnetism?

33. *You are the research director for a specialty metallurgical firm, and are looking for ideas for useful new alloys to market. You are considering four possible alloys: TiFe, TiHf, FeOs, and HfOs. (a) Write the "prepared-for-bonding" electron configurations of each of the starting metals (Ti, Fe, Hf, and Os). (b) Assuming that, in the alloys, these electrons would belong half to one metal and half to the other, write the average prepared-for-bonding electron configurations for each metal atom in each of the resulting alloys (TiFe, TiHf, FeOs, and HfOs). (c) What would the average number of unpaired (prepared-for-bonding) electrons be per atom in each of the resulting alloys? (d) Which alloy should have the highest (average) atomization energy?

34. You are the research director for a specialty metallurgical firm, and are looking for ideas for useful new alloys to market. You are considering four possible alloys: TiFe, TiOs, HfFe, and HfOs. (a) Which alloy should have the highest (average) atomization energy? (b) Which two alloys should probably be most similar in hardness and strength? (c) Which alloy would you expect to have the lowest melting point? (d) Which one of the alloys would be the most likely to be show ferromagnetism?

35. Use the computer program "KC-Discoverer" to prepare a periodic plot of the following bulk properties of metals: (a) density; (b) Mohs hardness; (c) melting point; (d) boiling point; (e) ΔH_{fusion}; (f) $\Delta H_{\text{vaporization}}$; (g) $\Delta H_{\text{atomization}}$; (h) thermal conductivity; and (i) electrical conductivity. To what extent do these properties reach their maxima (or minima) in the center of a period?

36. *Which would you expect to be more exothermic and why: the formation of Zintl alloys, or the formation of Hume–Rothery alloys?

37. Given bond energies and neglecting entropy effects, estimate (a) ΔH and (b) $E°$ for the reduction (in acid solution) of atomic oxygen to water. Locate atomic oxygen on the redox predominance diagram of water.

38. *The diatomic molecules C_2 and Zr_2 are known in the gas phase. If you follow the normal rules for drawing Lewis structures, you will come up with a quadruple bond in each of these. For which one of these two is a quadruple bond impossible? For which one is it plausible? Explain why.

39. (a) Match the following species (O_2, O_2^+, O_2^-, and O_2^{2-}) with their stretching frequencies (802, 1097, 1580, 1905 cm^{-1}). (b) Are these stretching frequencies more likely to be observed in the IR or the Raman spectra of these species? Why? [L. Vaska, *Acc. Chem. Res.*, **9**, 175 (1976).]

40. *Consider the following four allotropes of carbon: graphite, diamond, C_{60}, and C_{70}. (a) Assign the structures of each of these to the appropriate structural category: monomer or oligomer, one-dimensional polymer, two-dimensional polymer, and three-dimensional polymer. (b) Only two of these allotropes will dissolve in organic solvents. Which two? Why? (c) One of these was suspected of being present in interstellar space, while another is found in the cones of extinct volcanos. Identify the allotropes, and explain why each is found where it is. (d) Another possible allotrope is the hypothetical diatomic molecule C_2. If this existed, would you expect to find it in interstellar space, in volcanic cones, or on the earth's surface? What does MO theory predict for its bond order? How does this explain its instability?

41. Which fullerene is likely to be a softer base, C_{60} or C_{78}? Why?

42. Would you expect to be able to separate the fullerenes from the other allotropes of carbon by vacuum sublimation? Why or why not?

43. Why do polyhedral borane clusters tend to form −2 charged anions, $B_nH_n^{2-}$? Why do the polyhedral fullerenes not show this tendency?

44. *From lowest to highest energy, the occupied MOs of C_{60} are labeled (according to their irreducible representations) a_g, t_{1u}, h_g, t_{2u}, g_u, g_g, h_g, and h_u. Identify the generator orbitals of a hypothetical central atom in C_{60} that would generate this sequence of MOs.

45. Given that the radii of the nearly spherical C_{60} and its anions are about 500 pm, and with the help of Figure 2.8, predict the pK_b values and the basicity classifications of the anions C_{60} through C_{60}^{6-}. Which of these would you predict could persist in water?

46. Is the reaction of Na(s) with As_n(s) to produce the salt Na_3As(s) (containing a monoatomic anion) expected to be exothermic? Estimate ΔH for this reaction using a thermochemical cycle. [You will need to use the Kaputsinskii equation (4.16) and an extrapolated anionic radius to estimate the lattice energy of the product.]

47. (a) Draw the complete Lewis structure of the As_7^{3-} ion. Can this be done while obeying the octet rule? How many π bonds are present in this ion? (b) Draw a Lewis structure of the hypothetical As_7^{3+} ion that is derived from that of the As_7^{3-} ion without breaking any bonds.

48. *The C–C bond length of graphite is 141.5 pm. Assuming that there is true chemical interaction in the formation of the intercalation compounds of the Group 1 metals with graphite, would you expect the C–C bond lengths in these intercalates (such as RbC_8) to be shorter or longer than 141.5 pm? Why?

49. The cluster P_4 is tetrahedral with T_d symmetry; draw its full Lewis structure. (a) Now suppose that an anion P_4^{2-} is made by reduction of P_4. Assuming the minimal possible change of geometry, draw the full Lewis structure of P_4^{2-} and give its highest possible symmetry point group. (b) Do the same for the further reduced anion P_4^{4-}. (c) Do the same for the still further reduced anion P_4^{6-}. (d) Are any allotropes or polyatomic ions of Group 16(VI) isoelectronic to any of your ions? If so, do they show the same structure?

50. *The CrO_8^{3-} ion is known; clarify this preposterous-looking chemical formula.

51. Show how the structures of each of the following Group 16(VI) cations can be derived by oxidation from the structure of a real or plausible Group 16(VI) allotrope: (a) Se_8^{2+} and (b) Te_6^{4+}.

52. What cations or elemental allotropes in the next group to the right would be isoelectronic and (probably) isostructural with the following: (a) tetrahedral Sn_4^{4-} and (b) square Bi_4^{2-}.

53. Using the isolobal concept, (a) propose some possible *d*-block polymers with structures analogous to those of gray selenium and of black phosphorus. (b) Describe some of the physical properties you would expect these new polymers to show. (c) On the assumption that these new polymers have suitable band gaps and therefore are semiconductors, propose means of doping these two polymers to obtain an *n*-type semiconductor and a *p*-type semiconductor.

54. The Haber process for the production of NH_3 fertilizer from N_2 and H_2 involves, in the first steps, the dissociative chemisorption of N_2 and H_2 on the surface of the iron catalyst. (a) What name was used in Section 11.7 for this type of reaction? (b) The remaining steps combine nitrogen and hydrogen to give ammonia. Write an equation representing any one of these steps. What name was used for this type of reaction?

55. Calculate the total valence electron counts and (if possible) predict the geometries of the following clusters: (a) $B_6H_6^{4-}$; (b) $[Co_8(CO)_{18}C]^{2-}$; (c) $Fe_5(CO)_{15}C$; and (d) $Ni_8(CO)_8(PPh)_6$, assuming that PPh is a 4-electron donor ligand.

56. *Given the formulas and geometries of the following cluster anions, predict their ionic charges (if any): (a) $Rh_7(CO)_{16}$ (capped octahedron, with same VEC as pentagonal bipyramid); (b) bicapped tetrahedral $Os_6(CO)_{18}$; (c) octahedral $Ru_6(CO)_{18}$; (d) octahedral $Co_6(CO)_{15}$; (e) square pyramidal B_5H_5. (f) Which one of these anions would be predicted to be the most strongly basic, and therefore the one most likely to be found only in a fully protonated form? Give the formula of that form.

57. (a) How many shells are present in the close-packed giant cluster $Pt_{309}L_{36}O_{30}$? (b) How many of the Pt atoms in this giant cluster should show metallic properties? (c) How many metal atoms should be present in the giant cluster having one more shell than the above?

58. Both the metallic giant clusters and the onion-like nested fullerenes have multiple concentric shells of atoms. What differences would you expect in the properties of these two types of oligomers?

59. *Several *p*-block *neutral* derivatives of the *closo* polyhedral borane anions can be made by isolobal, isoelectronic substitution of *two* of the BH functionalities. (a) To get a neutral product, what BH functionality is being replaced? It has how many unshared electrons in how many orbitals? (b) Suggest the formulas of some of these neutral borane derivatives (with both substitutions involving the same type of replacement). (c) If the derivative is icosahedral (i.e., $n = 12$), how many isomers can there be of these neutral substituted boranes? (d) In what symmetry point group would each be classified?

Notes

1. This method of denoting polymeric structures has not yet come into universal use; in much common usage the polymeric nature of the material is indicated with an indefinite subscript [e.g., $(CH_2CH_2)_n$] or is not indicated in any manner at all: The simplest formulas are used, for example, C for graphite and SiO_2 for silicon dioxide.
2. It is in principle possible, however, for an ionic compound to be monomeric and to be a gas at room temperature: E. C. Lingafelter, *J. Chem. Educ.*, **70**, 98 (1993).
3. Related to this enthalpy change is the **heat of vaporization** of the element, which is the enthalpy change for the process by which the element in its usual form at room temperature (and 1-atm pressure) is converted to a gas. The two enthalpies are identical if the element in the gaseous state consists only of single atoms and not of small molecules. This identity definitely does not apply to H_2, N_2, O_2, F_2, Cl_2, Br_2, and I_2, of course. For most of the other elements, this is nearly the case, so for most elements there is only a minor difference in the two quantities.
4. L. E. Murr and W. H. Kinard, *Am. Sci.*, **81**, 152 (1993).
5. Many reviews of this topic have been written: R. Taylor and D. R. M. Walton, *Nature* (*London*), **363**, 685 (1993); P. W. Fowler and D. E. Manolopoulos, *An Atlas of Fullerenes*, Clarendon Press, Oxford, UK, 1995; W. O. J. Boo, *J. Chem. Educ.*, **69**, 605 (1992); the entire March 1992 issue of *Accounts of Chemical Research* was devoted to reviews of fullerene chemistry.
6. W. Krätschmer, L. D. Lamb, K. Fostiropoulos, and D. R. Huffman, *Nature* (*London*), **347**, 354 (1990); W. A. Scrivens and J. M. Tour, *J. Org. Chem.*, **57**, 6932 (1992).
7. L. Becker, J. L. Bada, R. E. Winans, J. E. Hunt, T. E. Bunch, and B. M. French, *Science*, **265**, 642 (1994); D. Heymann, L. P. Felipe Chibante, R. R. Brooks, W. S. Wolbach, and R. E. Smalley, *Science*, **265**, 645 (1994); *Chem. Eng. News*, Aug. 1, 1994, p. 4.
8. J. B. Howard, J. T. McKinnon, and M. E. Johnson, *Nature* (*London*), **352**, 139 (1991).
9. F. Diederich and R. L. Whetten, *Acc. Chem. Res.*, **25**, 119 (1992).
10. W. A. Scrivens, P. V. Bedworth, and J. M. Tour, *J. Am. Chem. Soc.*, **114**, 7917 (1992); J. F. Marecek and S. D. Kuduk, *J. Chem. Educ.*, **71**, A141 (1994); J. L. Atwood, G. A. Koutsantonis, and C. L. Raston, *Nature* (*London*), **368**, 229 (1994); B. Nie and V. M. Rotello, *J. Org. Chem.*, **61**, 1870 (1996); for C_{70}, W. A. Scrivens, A. M. Cassell, B. L. North, and J. M. Tour, *J. Am. Chem. Soc.*, **116**, 6939 (1994).
11. S. Iijima and T. Ichihashi, *Nature* (*London*), **363**, 603 (1993); A. Thess, R. Lee, P. Nikolaev, H. Dai, P. Petit, J. Robert, C. Xu, Y. H. Lee, S. G. Kim, A. G. Rinzler, D. T. Colbert, G. E. Scuseria, D. Tománek, J. E. Fischer, and R. E. Smalley, *Science*, **273**, 483 (1996).
12. S. Iijima and P. M. Ayajan, *Nature* (*London*), **363**, 333 (1993); S. C. Tsang, Y. K. Chen, P. J. F. Harris, and M. L. H. Green, *Nature* (*London*), **372**, 159 (1994); M. Freemantle, *Chem. Eng. News*, July 15, 1996, p. 62.
13. Yu. P. Kudryavtsev, S. E. Evsyukov, M. B. Guseva, V. G. Babaev, and V. V. Khvostov, *Russ. Chem. Bull. (Engl. Trans.)*, **42**, 399 (1993).
14. R. J. Lagow, J. J. Kampa, H.-C. Wei, S. L. Battle, J. W. Genge, D. A. Laude, C. J. Harper, R. Bau, R. C. Stevens, J. F. Haw, and E. Munson, *Science*, **267**, 362 (1995).
15. R. Baum, *Chem. Eng. News*, Aug. 2, 1993, p. 3.
16. Brewer, L., in *Electronic Structure and Alloy Chemistry of the Transition Metals*, P. A. Beck, Ed., Interscience, New York, 1963, as cited by W. L. Jolly, *Modern Inorganic Chemistry*, McGraw-Hill, New York, 1984; p. 293.
17. C. S. G. Phillips and R. J. P. Williams, *Inorganic Chemistry*, Oxford University Press, New York, 1966; Vol. II, p. 5.
18. L. J. Norrby, *J. Chem. Educ.*, **68**, 110 (1991).
19. D. Schneider, *Sci. Am.* **275(4)**, 28 (1996); L. Stixrude and R. E. Cohen, *Science*, **267**, 1972 (1995).
20. F. Hensel and P. P. Edwards, *Science*, **271**, 1692 (1996); R. A. Kerr, *Science*, **271**, 1667 (1996).
21. Another classification of alloys is often given as **interstitial alloys**, in which very small non-metallic atoms (e.g, H, N, C, and B) are said to fit into the interstices of close-packed layers of metal atoms. However, there are significant changes of structure and bonding in these metal–nonmetal compounds, and we shall also treat them as compounds in this text (as borides in Section 12.5, as carbides and nitrides in Chapter 14, and as hydrides in Chapter 15).
22. L. Brewer, *Science*, **161**, 115 (1968); L. Brewer, *J. Chem. Educ.*, **61**, 101 (1984).

23. K. R. C. Gisser, M. J. Geselbracht, A. Cappellari, L. Hunsberger, A. B. Ellis, J. Perepezko, and G. C. Lisensky, *J. Chem. Ed.*, **71**, 334 (1994).
24. A. R. Miedema, P. F. de Chatel, and F. R. de Boer, *Physica B*, **100**, 1 (1980).
25. L. J. Parker, T. Atou, and J. V. Badding, *Science*, **273**, 95 (1996); T. Atou, M. Hasegawa, L. J. Parker, and J. V. Badding, *J. Am. Chem. Soc.*, **118**, 12104 (1996).
26. P. W. Stephens and A. I. Goldman, *Sci. Am.* **264(4)**, 44 (1991); A. I. Goldman, J. W. Anderegg, M. F. Besser, S.-L. Chang, D. W. Delaney, C. J. Jenks, M. J. Kramer, T. A. Lograsso, D. W. Lynch, R. W. McCallum, J. E. Shield, D. J. Sordelet, and P. A. Thiel, *Am. Sci.*, **84**, 230, (1996).
27. J. F. Herbst, *Am. Sci.*, **81**, 252 (1993); G. Boebinger, A. Passner, and J. Bevk, *Sci. Am.*, **274(6)**, 59 (1995).
28. The standing waves of MOs can be thought of as the superpositions of two traveling waves moving in opposite directions, each being equally populated. However, if electrodes are attached to the metal, the waves that travel in the direction of the positive electrode will be lower in energy, and will be more populated: Hence, an electric current moves.
29. J. K. Burdett, *Chemical Bonding in Solids*, Oxford, New York, 1995; pp. 148–152; ratios r_p/r_s computed from data on p. 199.
30. R. E. Smalley, *Fullerenes: Synthesis, Properties, and Chemistry*, American Chemical Society, Washington DC, 1992; p. 141.
31. M. W. Geis and J. C. Angus, *Sci. Am.*, **276(4)**, 84 (1992).
32. M. Baudler, D. Düster, and D. Ouzounis, *Z. Anorg. Allgem. Chem*, **544**, 87 (1987); M. Baudler and T. Etzbach, *Chem. Ber.*, **124**, 1159 (1991).
33. O. J. Scherer, C. Blath, and G. Wolmershäuser, *J. Organometal. Chem.*, **387**, C21 (1990).
34. O. J. Scherer, J. Vondung, and G. Wolmershaüser, *Angew. Chem. Intl. Ed. Engl.*, **28**, 1355 (1989); O. J. Scherer, *Angew. Chem. Intl. Ed. Engl.*, **29**, 1104 (1990); O. J. Scherer, R. Winter, G. Heckmann, and G. Wolmershäuser, *Angew. Chem. Intl. Ed. Engl.*, **30**, 850 (1991).
35. R. C. Haddon, *Acc. Chem. Res.*, **25**, 127 (1992).
36. P. D. W. Boyd, P. Bhyrappa, P. Paul, J. Stinchcombe, R. D. Bolskar, Y. Sun, and C. A. Reed, *J. Am. Chem. Soc.*, **117**, 2907 (1995).
37. Reviewed by D. S. Bethune, R. D. Johnson, J. R. Salem, M. S. deVries, and C. S. Yannoni, *Nature* (*London*), **366**, 123 (1993); F. T. Edelmann, *Angew. Chem. Intl. Ed. Engl.*, **34**, 981 (1995).
38. Y. Kubozono, H. Maeda, Y. Takabayashi, K. Hiraoka, T. Nakai, S. Kashino, S. Emura, S. Ukita, and T. Sogabe, *J. Am. Chem. Soc.*, **118**, 6998 (1996).
39. M Takata, B. Umeda, E. Nishibori, M. Sakata, Y. Saito, M. Ohno, and H. Shinohara, *Nature* (*London*), **377**, 46 (1995); H. Suematsu, Y. Murakami, H. Kawata, Y. Fujii, N. Hamaya, O. Shimomura, K. Kikuchi, Y. Achiba, and I. Ikemoto, *Mater. Res. Soc. Symp. Proc.*, **349**, 213 (1994).
40. K. Kikuchi, K. Kobayashi, K. Sueki, S. Suzuki, H. Nakahara, and Y. Achiba, *J. Am. Chem. Soc.*, **116**, 9775 (1994).
41. M. Saunders, H. A. Jimenez-Vásquez, R. J. Cross, S. Mroczkowski, M. L. Gross, D. E. Giblin, and R. J. Poreda, *J. Am. Chem. Soc.*, **116**, 2193 (1994).
42. T. Guo, M. D. Diener, Y. Chai, M. J. Alford, R. E. Haufler, S. M. McClure, T. Ohno, J. H. Weaver, G. E. Scuseria, and R. E. Smalley, *Science*, **257**, 1661 (1992).
43. P. J. Fagan, J. C. Calabrese, and B. Malone, *Acc. Chem. Res.*, **25**, 134 (1992).
44. T. Pradeep, G. U. Kulkarni, K. R. Kannan, T. N. Guru Row, and C. N. R. Rao, *J. Am. Chem. Soc.*, **114**, 2272 (1992).
45. T. Bartik, B. Bartik, M. Brady, R. Dembinski, and J. A. Gladysz, *Angew. Chem. Intl. Ed. Engl.*, **35**, 414 (1996); U. F. H. Bunz, *Angew. Chem. Intl. Ed. Engl.*, **35**, 969 (1996).
46. J. D. Corbett, *Chem. Rev.*, **85**, 383 (1985).
47. B. W. Eichhorn and R. C. Haushalter, *J. Am. Chem. Soc.*, **110**, 8706 (1988); B. W. Eichhorn and R. C. Haushalter, *J. Chem. Soc. Chem. Commun.*, 937 (1990); B. Schiemenz and G. Huttner, *Angew. Chem. Int. Ed. Engl.*, **32**, 297 (1993).
48. B. Aronsson, T. Lundstrom, and S. Rundqvist, *Borides, Silicides, and Phosphides*, Methuen, London, 1965.
49. M. L. Fornasini and M. Pani, *J. Alloys Compounds*, **205**, 179 (1994); Z.-C. Dong and J. D. Corbett, *Inorg. Chem.*, **35**, 2301 (1996).
50. H.-J. Deiseroth and A. Strunck, *Angew. Chem. Intl. Ed. Engl.*, **28**, 1251 (1989); H.-J. Deiseroth, A. Strunck and W. Bauhofer, *Z. Anorg. Allg. Chem.*, **575**, 31 (1989).

51. R. D. Bolskar, R. S. Mathur, and C. A. Reed, *J. Am. Chem. Soc.*, **118**, 13093 (1996).
52. B. D. Cutforth, R. J. Gillespie, and P. K. Ummat, *Rev. Chim. Minér.*, **13**, 119 (1976); B. Hämmerle, E. P. Müller, D. L. Wilkinson, G. Müller, and P. Peringer, *J. Chem. Soc. Chem. Commun.*, 1527 (1989).
53. E. Zeller, H. Beruda, and H. Schmidbaur, *Inorg. Chem.*, **32**, 3203 (1993); C. Linke and M. Jansen, *Inorg. Chem.*, **33**, 2614 (1994); H. G. von Schnering, R. H. C. Gil, W. Hönle, A. Burkhardt, G. Krier, and O. K. Andersen, *Angew. Chem. Intl. Ed. Engl.*, **34**, 103 (1995); P. P. Edwards, P. A. Anderson, and J. M. Thomas, *Acc. Chem. Res.*, **29**, 23 (1996).
54. For example, C. M. Friend, *Sci. Am.* **268(4)**, 74 (1993).
55. P. M. Maitlis, H. C. Long, R. Quyoun, M. L. Turner, and Z.-Q. Wang, *J. Chem. Soc. Chem. Commun.*, 1, 1996; critically discussed by R. L Burwell., Jr., *Chemtracts—Inorg. Chem.*, **8**, 6 (1996).
56. Logically, in order to have a three-dimensional cluster structure as in Figure 12.1(d'), a metal cluster must contain at least four metal atoms, but we will include three-atom triangular systems in this discussion.
57. H. Kudo, *Nature* (*London*), **355**, 432 (1992).
58. H. Schmidbaur, *Chem. Soc. Rev.*, **24**, 391 (1995). This review also discusses a number of linear neutral gold(I) compounds that align in the solid state (due to aurophilic interactions) with Au–Au distances close to those found in Au metal; in one case it even causes two $[R_2Au]^-$ anions to align with each other, despite their mutual electrostatic repulsion! (A. Bauer and H. Schmidbaur, *J. Am. Chem. Soc.*, **118**, 5324 (1996)).
59. N. Rösch, A. Görling, D. E. Ellis, and H. Schidbaur, *Angew. Chem. Intl. Ed. English*, **28**, 1357 (1989); D. E. Harwell, M. D. Mortimer, C. B. Knobler, F. A. L. Anet, and M. F. Hawthorne, *J. Am. Chem. Soc.*, **118**, 2679 (1996).
60. K. Wade, *Adv. Inorg. Chem. Radiochem.*, **18**, 1 (1976); see also R. W. Rudolph, *Acc. Chem. Res.*, **9**, 446 (1976).
61. J. W. Lauher, *J. Am. Chem. Soc.*, **100**, 5305 (1978).
62. In contrast, the early *d*-block metals have better overlap of their *d* orbitals with each other but have less use of the valence (or postvalence) *p* orbitals, so that the clusters of class (f) do not follow these patterns.
63. G. Schmid, *Chem. Rev.*, **92**, 1709 (1992); G. Schmid, *NanoStructured Materials*, **6**, 15 (1995).
64. F. M. Mulder, T. A. Stegink, R. C. Thiel, L. J. de Jongh, and G. Schmid, *Nature* (*London*), **367**, 716 (1994).
65. R. D. Rieke, *Science*, **246**, 1260 (1989); R. D. Rieke, T. P. Burns, R. M. Wehmeyer, and B. E. Kahn, in K. S. Suslick, Ed., *High-Energy Processes in Organometallic Chemistry*, *ACS Symp. Ser. 333*, 223 (1987).
66. K. S. Suslick, S.-B. Choe, A. A. Cichowlas, and M. W. Grinstaff, *Nature (London)*, **353**, 414 (1991).

CHAPTER

13

The Oxides of the Elements

With Applications to Geochemistry, Environmental Chemistry, and Materials Science

13.1 The Physical States and Structures of the Fluorides and Oxides of the Elements

Like the elements themselves, the oxides of the elements include monomers, oligomers, and polymers with a vast variety of physical properties. As in Chapter 12, we will assemble the more complex oxides by linking together their fundamental *structural units*.[1] For most purposes, we will consider the basic structural unit of a *compound* as the unit represented by the simplest or empirical formula of the substance, but with the additional simplification that, if there are two or three atoms of the least abundant element in the simplest formula (e.g., Cl in Cl_2O_7), the structural unit will be reduced to include only one atom of this least abundant atom. Thus, the structural unit of Cl_2O_7 will be considered to be $ClO_{3.5}$.[2] The fractional number of oxygen atoms clearly indicates what we shall see later, that there is an oxygen atom common to two structural units and serving to *link* or *bridge* them. Note that in some applications such as silicate geochemistry (Sections 13.6 and 13.7) the structural unit is called a "nucleus" of the structure.

In contrast to the last two chapters, in this and subsequent chapters we will choose to link our structural units, not by sharing one electron from each unit, but rather by *coordinate covalent bonding*, in which the linking oxide (or other) ion provides the electron pair, while the central atom of the formula unit provides a **vacant coordination site**: a site of potential Lewis acidity. In terms used in Chapter 11, we will use the Lewis acid–base (LAB) approach to assembling our isolobal fragments (structural units), rather than the redox-radical (RR) approach. The LAB approach is justified on two grounds. First, the oxides and halides of the elements involve much more ionic bonds (due to greater electronegativity differences) than are present in the elements themselves or in most organometallic compounds. Second, using the RR approach and Table 11.6, we expect halide ions to form only one bond, and oxide ions to form only two. This will turn out not to be the case: Halide ions can act as bridging or linking ions, and oxide ions can link three central atoms. The "extra" bonds must be treated as coordinate covalent bonds. It is simplest, then, to treat *all* of the bonds to the central atom as coordinate covalent bonds, and use the LAB approach; hence, we return to the approach emphasized in Chapters 2–5.

To a significant extent, the degree of polymerization of the structural units of an oxide or halide (and therefore many of the physical properties of that material) can be understood based on simple predictions of the *number of interunit links that the structural unit is likely to form*, given how close the structural unit actually comes to coordinative saturation, that is, on how close the coordination number of the central atom of the structural unit is to its expected maximum (or penultimate) coordination number. Let us begin by considering certain structural units from which real (or imaginary) oxides of third-period elements might be assembled by polymerization: MgO, SiO_2, SO_3, and "ArO_4." If we simplistically assume a constant penultimate total coordination number of 4 for these central atoms, we see that this is achieved only in the hypothetical ArO_4, which is the only one of these oxides that we would certainly expect to be monomeric. In all other cases, the central atom falls short of its expected (penultimate) coordination number (and falls short of an octet of electrons). Thus the central atoms of these oxide structural units have vacant coordination sites—potential Lewis acid sites—which we might represent by empty parentheses: $(\ )_3MgO$; $(\ )_2SiO_2$; $(\ )SO_3$. The outer oxygen atoms or ions all have unshared electron pairs, and hence represent potential Lewis base sites.

Octets of electrons and coordinative saturation can then be completed for the central atoms in either of two ways: (1) by donation of a second pair of electrons from one oxygen atom to the central atom, forming a π bond; (2) by donation of an unshared electron pair from an oxygen atom or ion in a second structural unit, forming a linking coordinate covalent bond. As discussed earlier, only a few elements form good π bonds, so among the oxides in question only sulfur can utilize this option: SO_3 *sometimes* adopts a double-bonded, monomeric form, $O{=}SO_2$. Since σ bond linkage is normally stronger than π bond linkage, the remaining elements prefer to achieve octets and coordinative saturation by polymerization through bridging oxygen links, as sketched in Figure 13.1.

Even sulfur normally takes this second option; SO_3 is normally found as either an oligomer or a chain polymer, as shown in the figure. In either of these cases, the structure features *one interunit link* per structural unit. More than one structural form is possible for SO_3 (and many other linked materials), because the linking bonds to the two-coordinate oxygen atom may either be turned inward to give cyclic oligomers, or outward to give chain polymers. These different forms are often termed **polymorphs** of SO_3, since they have different crystal lattice types, but the difference between the oligomer and polymer is really more profound than this: Perhaps they should be called **polymerization isomers**.[3] Because links can be formed at different angles, we generally cannot unambiguously predict the exact type of oligomerization or polymerization that will happen to a given coordinately unsaturated structural unit. Instead, we will follow periodic trends in the *degree* of oligomerization or polymerization as we cross periods or groups, and we will watch how physical properties change as a consequence.

Silicon in the SiO_2 structural unit has *two* vacant coordination sites, so we anticipate correctly that it is likely to polymerize to a greater extent than SO_3, forming two links per structural unit, and giving each bridging oxygen atom a coordination number of 2. We could conceive of these two bridging links turning sharply inward to give a double chain polymer, as shown in Figure 13.1(*b*) for the real case of $^{1}_{\infty}[BeCl_2]$, but in fact the resulting bond angles at oxygen are too strained and the partial positively charged Si^{4+} "cations" are brought too close together. Therefore the SiO_2 structural

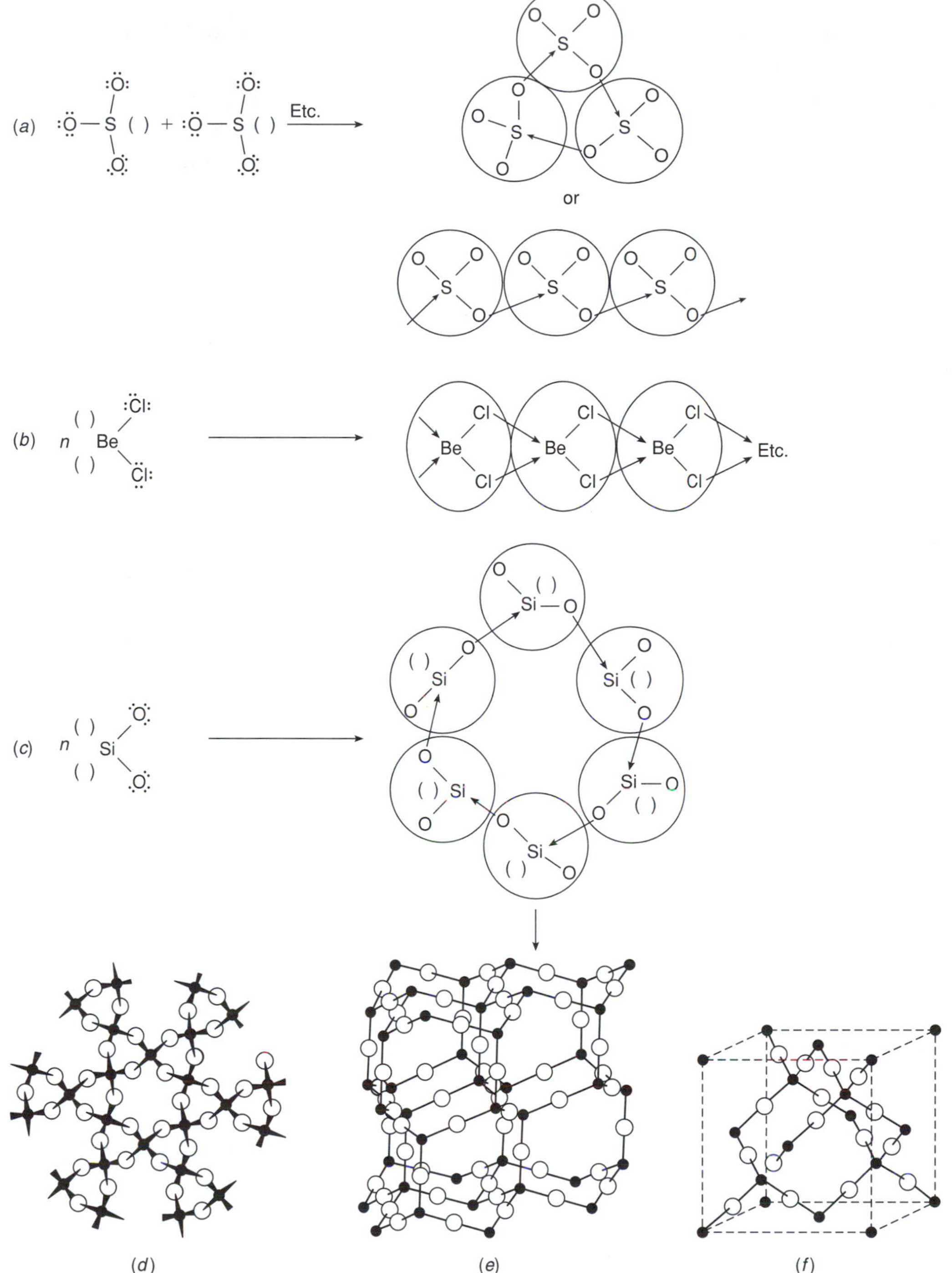

Figure 13.1
Lewis acid–base reactions between structural units (enclosed in circles in products) having vacant coordination sites (represented by parentheses), to generate different polymerization isomers. (Linkage by coordinate covalent bonds in the products is represented by arrows.) (*a*) Oligomerization or polymerization of SO_3 to give S_3O_9 or the one-dimensional polymer of SO_3. (*b*) One-dimensional polymerization of $BeCl_2$. (*c*) Cyclooligomerization of SiO_2 units to give *cyclo*-Si_6O_{12} units, which are still coordinately unsaturated, and hence polymerize further to give various three-dimensional polymerization isomers (Si is represented by closed circles and O by large open circles): (*d*) β-quartz; (*e*) β-tridymite; (*f*) β-cristobalite. [(*d*)–(*f*) Adapted from A. F. Wells, *Structural Inorganic Chemistry*, 5th ed., Oxford, Claredon Press, UK, 1984; pp. 1006–1007.]

unit polymerizes to give one of several network polymerization isomers, such as the three illustrated: β-quartz, β-tridymite, and β-cristobalite. Note that each of these contains large hexameric rings, $(SiO_2)_6$, of varying conformations; formation of the six links to give this ring take care of one vacant coordination site per SiO_2 unit. The other vacant site per unit leads to three-dimensional polymerization, but linkage in different directions result in the different coordination isomers shown, and gives each oxygen atom a coordination number of 2.

The MgO structural unit, with three vacant coordination sites, polymerizes to an ionic lattice; if it indeed had a maximum coordination number of 4, this would be a zinc blende or wurzite lattice (Fig. 4.8), and the oxygen ions would also have a coordination number of 4. In practice, MgO exibits six coordination.

Hence, we see that increasing degrees of polymerization are foreshadowed by increasing numbers of vacant coordination sites at the central atom of the structural unit. The result is an increase in the average coordination number of the oxygen atoms or ions in the polymerized form of the material. If the central atom has no vacant coordination site, it is not a Lewis acid, and the structural unit remains as the monomeric molecular unit: the oxygens do not bridge, so they retain coordination numbers of 1.

The coordination numbers of cations or central atoms are more accurately predicted by carrying out ionic radius ratio calculations (Chapter 4) than by referring to maximum or penultimate total coordination numbers (Chapter 2), which are just the most common outcomes of radius ratio calculations for a given period. Radius ratio calculations, of course, are also not highly reliable, especially since many compounds are not ionic. These calculations do avoid having to prejudge the ionic or covalent nature of the bonding, and should be useful for predicting (and especially for understanding) periodic *trends* in structural types and physical properties such as melting and boiling points, as well as for understanding why dramatic changes occur approximately when they do in a period or a group. The appropriate calculations are outlined below.[4]

1. Calculate the radius ratio r_{cation}/r_{anion}. Referring to Table 4.4, use this ratio to predict the total coordination number (TCN) of the central (normally the metal) atom or ion. If the radius ratio is close to the cutoff points, there is a good deal of uncertainty in the result. Either total coordination number may prove correct: The compound may have one polymerization isomer with one coordination number and one with the other.[5]

2. The number of vacant coordination sites of the central atom in the structural unit is then obtained by subtracting the number of oxygen (or other outer) atoms in the structural unit from the expected total coordination number obtained in Step 1. Also, if the cation or metal is a *p*-block element with less than the maximum oxidation number, space must be reserved for the unshared pair of electrons[6]:

$$\text{(No. of vacant coordination sites around central atom)} = \text{(its expected TCN)} \\ - \text{(No. of its outer atoms)} - \text{(No. of its unshared } sp \text{ electron pairs)} \qquad (13.1)$$

Unless strong π bonding to outer oxygen atoms occurs instead (mainly for central C, N, or S atoms), each vacant coordination site leads to the formation of one bridging link, so Eq. 13.1 gives us the number of links expected per structural unit.

3. If the number of links expected per structural unit is *zero*, the species is predicted to occur as monomeric molecules, with relatively low melting or boiling points.

To predict whether it is a gas, liquid, or low-melting solid at room temperature, we need to look at the size of the molecule and whether or not it could engage in hydrogen bonding.

4. If the number of links expected per structural unit is *fractional* (< 1), we expect a small linear oligomer. As an example, the compound Cl_2O_7 has a structural unit of $ClO_{3.5}$. The radius ratio prediction leads us to expect a coordination number of 4 for Cl, which therefore has 0.5 vacant coordination sites per chlorine. This is remedied by forming *one oxygen link between two structural units*, which produces $O_3Cl–O–ClO_3$ (Cl_2O_7).

5. If the number of links expected per structural unit is *one*, this can be achieved by forming either a cyclic oligomer or a chain polymer.

6. As the number of links increases above one per unit, greater degrees of polymerization are expected, but since the links can be arranged in several ways, various polymerization isomers may be possible. Eventually, network polymerization is generally to be anticipated, especially for more than three links per unit.

7. Occasionally, a *negative* number of vacant coordination sites may be computed: The N in "NF_5" should have a coordination number of 4 from radius ratio calculations, which is exceeded in the structural unit NF_5. In such cases, we expect the structural unit to *expel* an anion: as such, NF_5 is not known, but the ion NF_4^+ is known.

8. If it is desired, calculate the coordination number of the outer (oxygen) atom or ion, using Eq. (4.15); this can be useful in determining an ionic lattice type using Table 4.4.

We begin the illustration of these principles by considering the fluorides of some elements in their group oxidation states. As we discovered earlier (Chapter 3), fluorides serve as ideal illustrations of many phenomena of predominantly ionic compounds. The observed trends we find among many fluorides are simplified by the fact that, while fluoride *ion* readily serves to link units in ionic fluorides, covalently bonded fluorine does not readily serve as a bridging atom: Due to its high electronegativity it does not readily share a second pair of electrons. Hence, we shall see fewer than the expected number of oligomeric structures among fluorides, which allows very sharp contrast between the physical properties associated with monomeric molecular structures and those associated with network (ionic or covalent polymeric) structures (Table 13.1).

Example 13.1

Predict the coordination numbers, the physical states, and the relative melting and boiling points of the fluorides of the *sp*-block fourth-period elements in their group oxidation states. State whether each should have a monomeric, oligomeric, or polymeric or ionic structure.

SOLUTION:

The fluorides in question are KF, CaF_2, GaF_3, GeF_4, AsF_5, SeF_6, and the (unknown) BrF_7. The ionic radius of F^- is 119 pm; the radii of the "cations" of these compounds are also obtained from Table C. The radius ratios are 1.277,[7] 0.958, 0.639, 0.563, 0.504, 0.471, and 0.445, respectively. Hence, the "cationic" coordination numbers are predicted to be 8 for K^+ and Ca^{2+}, and 6 for Ga^{3+} through Br^{7+}; the last two values are close to the 0.414

cutoff radius ratio for a cationic coordination number of 4, so that this possibility remains open for Se^{6+} and Br^{7+}.

The number of vacant coordination sites per central atom are then calculated: this is $(8-1)=7$ in KF, $(8-2)=6$ in CaF_2, $(6-3)=3$ in GaF_3, $(6-4)=2$ in GeF_4, $(6-5)=1$ in AsF_5, $(6-6)=0$ (or negative) in SeF_6, and $(6-7)=-1$ (or more negative) in BrF_7.

We predict that BrF_7 should not exist; only $BrF_6{}^+$ forms for this oxidation state (in combination with large nonbasic fluoro anions). The SeF_6 molecule is coordinately saturated at Se, so it is a monomeric gas (mp −40 °C, Table 13.1). The AsF_5 unit would likely form a cyclic oligomer or chain polymer, except for the reluctance of covalent fluorides to bridge; it is actually a monomeric gas (mp −63 °C). The GeF_4 unit similarly does not form a low-dimensional polymer, but is in fact also a monomeric gas (mp −37 °C).

The remaining three compounds all show enough links per structural unit, and are ionic enough from their electronegativity differences, to make polymeric structures very likely. Hence, we predict that these will be ionic or macromolecular compounds, with high melting and boiling points, and we predict that these will all be solids at room temperature. Our predictions are verified: the melting point of KF is 857 °C, and 1423 °C for CaF_2; the sublimation temperature of GaF_3 is 800 °C.

Example 13.2

Explain the trends in the physical states and boiling points of the Group 14(IVA) tetrafluorides.

SOLUTION:

From Table 13.1, we see that the top three fluorides (CF_4 through GeF_4) are gases with melting points below 0 °C, while the bottom two fluorides are high-melting solids. The calculated radius ratios increase down the group (0.252, 0.454, 0.563, 0.697, and 0.765), as do the predicted central-atom coordination numbers (4, 4 or 6, 6, 6 or 8, 6 or 8) and central-atom vacant coordination sites (0, 0 or 2, 2, 2 or 4, 2 or 4). Our calculations predict that there will be a trend from monomeric gaseous fluorides at the top of the group to (probably) ionic network solids at the bottom, although the calculations are not trustworthy enough to predict exactly where the transitions will occur. The melting points correctly suggest that the top three fluorides are monomeric molecular substances (with increasing melting points due to increasing van der Waals forces); the tin and lead fluorides indeed have polymeric structures, with, in fact, average coordination numbers for Sn and Pb of 6. Two equatorial fluorines bridge adjacent Sn or Pb atoms to give a layer structure; the axial fluorines are terminal (have coordination numbers of 1), and hence are closer to the Sn atoms in SnF_4 (terminal Sn–F distance of 188 pm *vs.* bridging Sn–F distance of 212 pm).[8]

Among the structures of oxides, we find many more bridging covalent oxygen atoms and oligomeric structures, since a covalent oxygen atom following the octet rule readily forms two bonds to two different central atoms, without developing any formal charge. Oxygen, however, also forms π bonds that are relatively strong compared to its σ bonds (Section 3.1), so that if the bonding in the oxide is indeed covalent and the

Table 13.1
Melting Points of the Highest Fluoride of Each Element[a]

HF																	
−83																	
LiF	BeF_2											BF_3	CF_4	NF_3	OF_2	F_2	
848	535											−127	−184	−206	−223	−219	
NaF	MgF_2											AlF_3	SiF_4	PF_5	SF_6	ClF_5	
1012	1263											(1272)	−86	−75	−64	−113	
KF	CaF_2	ScF_3	TiF_4	VF_5	CrF_x[b]	MnF_4	FeF_3	CoF_3	NiF_4	CuF_2	ZnF_2	GaF_3	GeF_4	AsF_5	SeF_6	BrF_5	KrF_2
857	1423	1227	(283)	110		dec.	1102		dec.	755	927	(800)	−37	−63	−40	−61	dec.
RbF	SrF_2	YF_3	ZrF_4	NbF_5	MoF_6	TcF_6	RuF_6	RhF_6	PdF_4	AgF_3	CdF_2	InF_3	SnF_4	SbF_5	TeF_6	IF_7	XeF_6
775	1400	1152	(908)	78	17	33	54			dec.	1100	1150	705	7	−35	5	50
CsF	BaF_2	LuF_3	HfF_4	TaF_5	WF_6	ReF_7	OsF_6	IrF_6	PtF_6	AuF_7	HgF_2	TlF_3	PbF_4	BiF_5			
682	1290	1182		97	2		32	44	57		645	550	500	dec.			
			ThF_4	PaF_5	UF_6	NpF_7	PuF_6										
			900	(500)	64	53	50										

SOURCES: Melting points in degrees Celsius taken from the *Handbook of Chemistry and Physics* and M. C. Ball and A. H. Norbury, *Physical Data for Inorganic Chemists*, Longman, London, 1974; pp. 70–91.

[a] Temperatures enclosed in parentheses represent temperatures of sublimation.

[b] It is currently disputed whether the volatile highest fluoride of chromium is CrF_5 or CrF_6: J. Jacobs, H. S. P. Müller, H. Willner, E. Jacob, and H. Burger, *Inorg. Chem.*, **31**, 5357 (1992).

central atom is relatively good at π bonding, the element may form two bonds to one oxygen in preference to one bond to two oxygen atoms, thus taking a lower total coordination number than our ionic model predicts. The good π bonding of oxygen was the reason why, in Chapter 3, we suggested the use of the penultimate rather than the maximum total coordination number for predicting the formulas of oxo anions; the same reasoning applies to oxides.

Example 13.3

Predict the coordination numbers, the physical states, and the relative melting and boiling points of the oxides of the second-period elements in their maximum oxidation state. State whether each should be monomeric, oligomeric, or polymeric or in an ionic lattice.

SOLUTION:

From Table B2, we see that these oxides are Li_2O, BeO, B_2O_3, CO_2, and N_2O_5. The ionic radius of the oxide ion is 126 pm (Table C); the radius ratios are 0.714, 0.464, 0.325, 0.238, and 0.214, respectively. The least abundant atom in Li_2O is oxygen; in the others it can be taken as the nonoxygen atom. Hence, the coordination numbers predicted are either 8 or 6 for O in Li_2O (hence, 4 or 3 for Li); either 6 or 4 for Be; 4 for B; 4 or 3 for C; and 3 for N.

We now compute the number of vacant coordination sites per structural unit: 3 or 2 for Li in $LiO_{0.5}$; 5 or 3 for Be in BeO; 2.5 for B in $BO_{1.5}$; 2 or 1 for C in CO_2; 0.5 for N in $NO_{2.5}$. The oxides of Li, Be, and B have enough vacant coordination sites for polymeric (ionic) structures to be likely, so we predict that these will be ionic or polymeric covalent compounds, with high melting and boiling points, and we predict that these will all be solids at room temperature. Our predictions are verified: Li_2O has a melting point of 1427 °C, BeO of 2530 °C, and B_2O_3 of 450 °C (but with a boiling point over 1860 °C).[9]

For carbon dioxide, the prediction is uncertain but wrong in either case, since our calculations cannot predict the special stability of oxygen double-bonded to carbon. Carbon dioxide is actually a monomeric molecule and a gas at room temperature, which sublimes at −79 °C at atmospheric pressure. For dinitrogen pentoxide the prediction is correct: In the gas phase, one of the five oxygens bridges or links the two nitrogens, while each of the other oxygen atoms is bonded to only one nitrogen atom; each nitrogen has a coordination number of 3. The solid form of this compound consists of NO_2^+ and NO_3^- ions.

We may also predict lattice types for the ionic compounds, using Table 4.4, and compare them with results given in a more detailed text[10] or by Wells[8]: For Li_2O we predict either an antirutile or antifluorite lattice and find the latter; and for BeO we predict either the NaCl or ZnS lattice and find the latter (wurtzite). The calculations for B_2O_3 do *not* suggest the corundum lattice type; two network polymerization isomers of this compound are known, in which the coordination numbers of boron are 3 and 4, respectively.

When the central *p*-block element is in its group oxidation state −2, we must allow for this in the calculations, as illustrated in Example 13.4.

Example 13.4

Rationalize the structural trends found among the oxides of the Group 16(VI) elements in the +4 oxidation state: Sulfur dioxide (SO_2) is a monomeric covalent molecule; SeO_2 forms a chain polymer; TeO_2 (in one polymerization isomer) forms a layer polymer; and PoO_2 takes the fluorite (CaF_2) ionic lattice structure.

Solution:

The radius ratios for these four dioxides are (unknown), 0.508, 0.881, and 0.857, respectively (the contraction for Po^{4+} is suspect, since few crystal structures have been done of salts of this very radioactive element). From these, we predict *total* coordination numbers of less than 6 for S^{4+}, 6 for Se^{4+}, and 8 for Te^{4+} and Po^{4+}. But at least one position is likely to be occupied by the unshared pair of electrons on each of these ions—since we know from valence shell electron-pair repulsion (VSEPR) theory that unshared electron pairs occupy *more* space than shared electron pairs, it may be that more than one position is needed for each. Given this uncertainty, our predictions of *actual* coordination numbers are necessarily vague: Probably much less than 4 for S^{4+}, less than 4 for Se^{4+}, and less than 6 for Te^{4+} and Po^{4+}. These translate into numbers of vacant coordination sites of much less than 2 for S in SO_2, less than 2 for Se in SeO_2, and less than 4 for Te in TeO_2 and Po in PoO_2.

Such numbers will not predict actual structures, but do illustrate trends: The degree of polymerization should increase on going down this group, as is observed. Actual coordination numbers are 2 for S in SO_2 (which contains a double bond that contributes to the low coordination number), 3 for Se in SeO_2, 4 for Te in TeO_2, and 8 for Po in PoO_2. Sulfur dioxide is a gas of boiling point −10 °C, while SeO_2 and TeO_2 are solids that sublimes at 315 and 450 °C, respectively.

Table 13.2 summarizes similar calculations for the third-period and the early sixth-period elements in their group oxidation states. Figure 13.2 semiqualitatively graphs the trend in melting points as a period is crossed. This trend shows some striking changes of slope.

1. At the right of a period, high oxidation state monomeric oxides of low melting points prevail: Thus the calculations in Table 13.2 successfully predict sharply lower melting points for the compounds P_2O_5, SO_3, Cl_2O_7, Re_2O_7, and OsO_4. As we go to the left, the oxidation number of the central atom decreases and its size increases, both of which result in an increase in the number of vacant coordination sites per central atom. Approximately at the point at which this number rises above 1, the melting points begin to rise sharply as oligomers and chain polymers form. Further left still, very high melting points are exhibited by layer and network polymers and ionic lattice compounds.

2. At the far left, among metal oxides forming ionic lattices, melting and boiling points tend to rise to the right as the charge on the metal ion rises, up to a point. This increase in melting point is a consequence of the increase in Coulombic forces as the ionic charge increases (Chapter 4). Thus melting points increase substantially from KF to CaF_2, from Na_2O to MgO, and from Cs_2O to HfO_2. The peak melting points are reached with ionic lattices of high ionic charge and with network polymeric structures, with no obvious distinction between them.

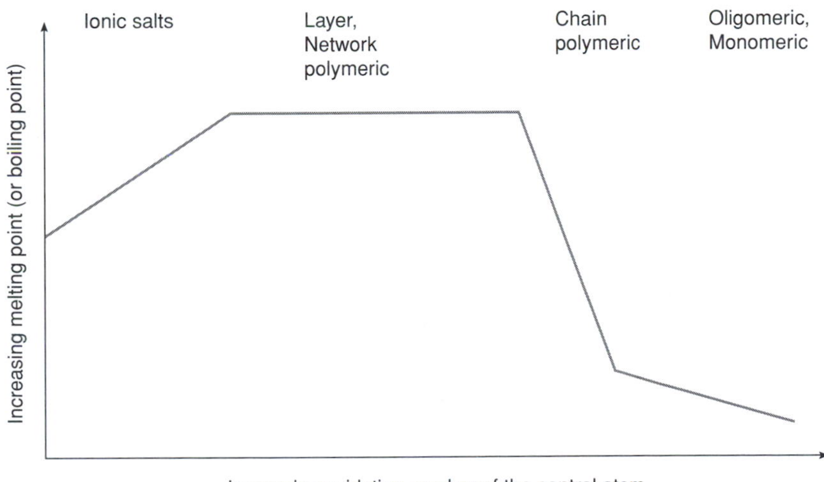

Figure 13.2
Typical trends in melting (or boiling) points of oxides of the elements in their group oxidation states, as the oxidation and group numbers increase.

Table 13.2
Predictions of Physical Properties and Structures of Some Halides and Oxides

Period 3[a]	NaF	MgF_2	AlF_3	SiF_4	PF_5	SF_6	(ClF_7)	
Radius ratio	0.975	0.723	0.563	0.454	0.437	0.361	0.345	
Predicted								
CN of metal	8	8 or 6	6	6 or 4	6 or 4	6 or 4	6 or 4	
Links per unit	7	6 or 4	3	2 or 0	1 or −1	0 or −2	−1 or −3	
Observed								
CN of metal	6	6	6	4	5	6	6	
Links per unit	5	4	3	0	0	0	−1	
Lattice type	NaCl	CaF_2	*	molecular	molecular	molecular	(ClF_6^+)	
Melting point	1012	1263	1272	−86	−75	−64		
Period 3	**NaI**	**MgI_2**	**AlI_3**	**SiI_4**	**PI_5**			
Radius ratio	0.563	0.417	0.325	0.262	0.252			
Predicted								
CN of metal	6	4 or 6	4	4	4			
Links per unit	5	4 or 2	1	0	−1			
Observed								
CN of metal	6	6	4	4	4			
Links per unit	5	4	1	0	−1			
Lattice type	NaCl	*	dimer	molecular	$PI_4^+I^-$			
Melting point	651	>700	191	120				
Period 3	**Na_2O**	**MgO**	**Al_2O_3**	**SiO_2**	**P_2O_5**	**SO_3**	**Cl_2O_7**	
Radius ratio	0.921	0.683	0.532	0.429	0.413	0.341	0.325	
Predicted								
CN of metal	4	6	6	6 or 4	4	4	4	
Links per unit	3	5	4.5	4 or 2	1.5	1	0.5	
Observed								
CN of metal	4	6	6	4	4	4	4	
Links per unit	3	5	4.5	4 or 2	1.5	1	0.5	
Lattice type	anti-fluorite	NaCl	corundum	beta-silica	oligomer	oligomer	Molecular	
Melting point	(1275)	2800	2050	1723	(300)	17	−91	
Period 6	**Cs_2O**	**BaO**	**Lu_2O_3**	**HfO_2**	**Ta_2O_5**	**WO_3**	**Re_2O_7**	**OsO_4**
Radius ratio	1.437	1.183	0.793	0.675	0.619	0.587	0.537	0.476
(Inverse " ")	(0.696)	(0.846)						
Predicted								
CN of metal	3	8	8	6	6	6	6	6
Links per unit	2	7	6.5	4	3.5	3	2.5	2
Observed								
CN of metal	3	6	7	8	6	6	5	4
Links per unit	2	5	5.5	6	3.5	3	1.5	0
Lattice type	[b]	NaCl	[b]	Rutile	[b]	[b]	Layer polymer	Molecular
Melting point	490	1920	2487	2900	1870	1473	296	40

[a] Melting points in degrees Celsius; if in parentheses these are sublimation temperatures. CN = coordination number.
[b] Lattice type is one not described in this text.

13.2 Covalent Oxides: Periodic Trends in Structure and Physical State

As suggested by Figure 13.2, the most important single variable in determining the structure, physical state, and (as we shall see) the acidic or basic properties of an oxide is the oxidation number of the central atom, since this determines the stoichiometry of the structural unit and plays a major role in determining the number of vacant coordination sites in the structural units. Hence, we will organize this discussion by oxidation state/stoichiometry. Next, we follow with the roles of radius ratio, unshared electron pairs in the *p* block, and tendencies to element–oxygen π bonding, all of which play a role in the chemistry of an oxide. Table 13.3 and Figure 13.3 summarize the structures, physical states at room temperature, melting and boiling points, and so on, of most of the important covalent oxides. Note that these oxides are often named using the system of nomenclature for binary covalent substances. Thus although Cl_2O_7 may be perfectly acceptably named chlorine(VII) oxide, it is more commonly named *di*chlorine *hept*oxide. Although this system of nomenclature is most commonly used among compounds of the nonmetals, it is also often used with the more covalent oxides of the *d*-block elements [Mn_2O_7 may also be named either manganese(VII) oxide, or dimanganese heptoxide].

Element Tetroxides. In these compounds, the central atoms exhibit the highest known oxidation state, +8; as indicated in Table 13.3, only four of these are known. The central +8 "ions" are rather small, so we calculate fairly low expected coordination numbers (these are not shown in Table 13.3, but can readily be filled in by the student as an exercise). The observed coordination numbers are tabulated and are 4 in all four tetroxides in these tetrahedral monomeric molecules. This low coordination number is undoubtedly due to the presence of π bonding involving oxygen unshared *p* electron pairs and central-atom *d* orbitals, which are contracted enough in such high oxidation states for significant π bonding even in *p*-block atoms such as Xe. Since the central atoms in these molecules are of relatively high atomic weights, these tetroxides are mainly volatile low-melting *solids*.

Since the +8 oxidation state exceeds the most stable oxidation state of any element, the tetroxides are strongly oxidizing materials. The compound FeO_4 has only been reported by one research group[11]; the compound XeO_4 is very explosive; RuO_4 and OsO_4 are not explosive but are still strong oxidizing agents. The compound OsO_4 is widely used in biology to stain tissues—it oxidizes the organic material in them and is reduced to brown OsO_2—but it is quite hazardous because of its high volatility. It readily oxidizes the organic material in the eye, too.

Dielement Heptoxides (central atom oxidation state of +7). Again, only four of these are known for chlorine and the elements of Group 7. Two $EO_{3.5}$ structural units must at least link through the bridging oxygen atom to give an E_2O_7 molecule; only Re_2O_7 links beyond this minimum (using one-half of its Re atoms, which become six-coordinate and link units into a layer polymer while the other one-half of the rhenium atoms remain four coordinate). This high +7 oxidation state is most stable at the bottom of the *d* block in the colorless Re_2O_7 and the yellow volatile Tc_2O_7; red Mn_2O_7 and Cl_2O_7 are treacherously explosive. The colors in these compounds are not due to the *d*-block ions per se, since their electron configurations are d^0; these are charge-transfer transitions (Section 6.3).

Table 13.3
Higher Oxides of the Elements

A. Tetroxides

Period	4	5	5	6
Oxide	FeO_4	XeO_4	RuO_4	OsO_4
Radius ratio		0.492		
Observed CN(M)	4	4	4	4
Vacant coordination sites	0	0	0	0
Structure type	FeO_4	XeO_4	RuO_4	OsO_4
Physical Properties		explodes gas	mp 25 °C bp 100 °C	mp 40 °C bp 130 °C

B. Dielement Heptoxides

Period	3	4	5	6
Oxide	Cl_2O_7	Mn_2O_7	Tc_2O_7	Re_2O_7
Radius ratio	0.325	0.476	0.555	0.532
Observed CN(M)	4	4	4	4 and 6
Vacant coordination sites	0.5	0.5	0.5	0.5, 1.5
Structure type[a]	Cl_2O_7	Mn_2O_7	Tc_2O_7	Layer
Physical Properties[d]	mp −92 °C bp 82 °C	mp 5 °C exp. 70 °C	mp 19 °C bp 212 °C	mp 220 °C bp 450 °C

C. Trioxides

Period	5	3	4	4	5	5	6	6	7
Oxide[b]	$:XeO_3$	SO_3	SeO_3	CrO_3	TeO_3	MoO_3	WO_3	ReO_3	UO_3
Radius ratio		0.361	0.471	0.487	0.588	0.613	0.622		0.731
Observed CN(M)[c]	3+:	4	4	4	6	6	6	6	6 or 7
Vacant coordination sites	0	1	1	1	3	3	3	3	3 or 4
Structure type[a]	$:XeO_3$	$(SO_3)_3$; Chain	$(SeO_3)_4$; Chain	Chain	Network	Layer	Network	Network	Network
Physical Properties[d]	exp. solid	bp 45 °C	dec 120 °C	mp 196 °C	dec	mp 795 °C	mp 1473 °C	mp 25 °C dec 400 °C	mp 40 °C bp 130 °C

D. Dielement Pentoxides

Period	4	5	2	3	4	4	5	5	6
Oxide[b]	$:Br_2O_5$	$:I_2O_5$	N_2O_5	P_2O_5	As_2O_5	V_2O_5	Sb_2O_5	Nb_2O_5	Ta_2O_5
Radius ratio		0.916	0.227	0.437	0.504	0.571	0.622	0.655	0.655
Observed CN(M)[c]	3+:	3+:+*	2 and 3	4	4 and 6	5	6	6	6 or 7
Vacant coordination sites	0	0*	1 and 0	1.5	1.5 and 3.5	2.5	3	3	3 and 4
Structure type[a]	$:Br_2O_5$	$:I_2O_5$*	$(NO_2)^+$ $(NO_3)^-$	P_4O_{10} Layer	Network	Layer	Network	Network	Network
Properties[d]		dec 300 °C	mp 37 °C	subl 347 °C	dec 315 °C	mp 690 °C	dec 380 °C	mp 1780 °C	dec 1470 °C

E. *p*-Block Dioxides

Period	3	3	4	5	6	2	2	3	4	5
Oxide[b]	$\because ClO_2$	$:SO_2$	$:SeO_2$	$:TeO_2$	$:PoO_2$	$\cdot NO_2$	CO_2	SiO_2	GeO_2	SnO_2
Radius ratio			0.538	0.933	0.908		0.252	0.454	0.563	0.697
Observed CN(M)[c]	2+:·	2+:	3+:	4+:	8	2+·; 3	2	4	4 or 6	6
Vacant coordination sites	0	0	1	2	6	0	0	2	2 or 4	2 or 4
Structure type[a]	$\because ClO_2$	$:SO_2$	Chain	Layer; network	Network	$\cdot NO_2$; N_2O_4	CO_2	Network	Network	Network
Physical Properties[d]	mp −59 °C bp 10 °C	mp −73 °C bp −10 °C	subl 315 °C	subl 450 °C	dec 500 °C	mp −9 °C bp 21 °C	subl −78 °C	mp 1723 °C bp 2230 °C	mp 1116 °C	mp 1127 °C

F. *p*-Block Dielement Trioxides

Period	2	3	4	5	6	2	3	4	5	6
Oxide[b]	$:N_2O_3$	$:P_2O_3$	$:As_2O_3$	$:Sb_2O_3$	$:Bi_2O_3$	B_2O_3	Al_2O_3	Ga_2O_3	In_2O_3	Tl_2O_3
Radius ratio			0.605	0.756	0.983	0.345	0.563	0.639	0.790	0.857
Observed CN(M)[c]	3 and 2+:	3+:	3+:	3+:	≥5	3	6	6	6	6
Vacant coordination sites	0*	1.5	1.5	1.5	≥3.5	1.5	3	3	3	3
Structure type[a]	$:N_2O_3$*	$:P_4O_6$	$:As_4O_6$; Layer	$:Sb_4O_6$; Double chain	Network	Network	Network	Network	Network	Network
Physical Properties[d]	mp −102 °C bp 4 °C	mp 25 °C bp 176 °C	mp 309 °C bp 737 °C	mp 656 °C bp 1425 °C	mp 824 °C bp 1890 °C	mp 450 °C bp 2300 °C	mp 2053 °C	mp 1735 °C	mp 1910 °C	dec 100 °C

[a] "Network" represents either an ionic or a network covalent structure.
[b] The : represents an unshared *sp* electron pair.
[c] An * structure includes additional secondary bonds, which are not counted.
[d] The abbreviation dec is decomposes, subl is sublimes, and exp is explodes.

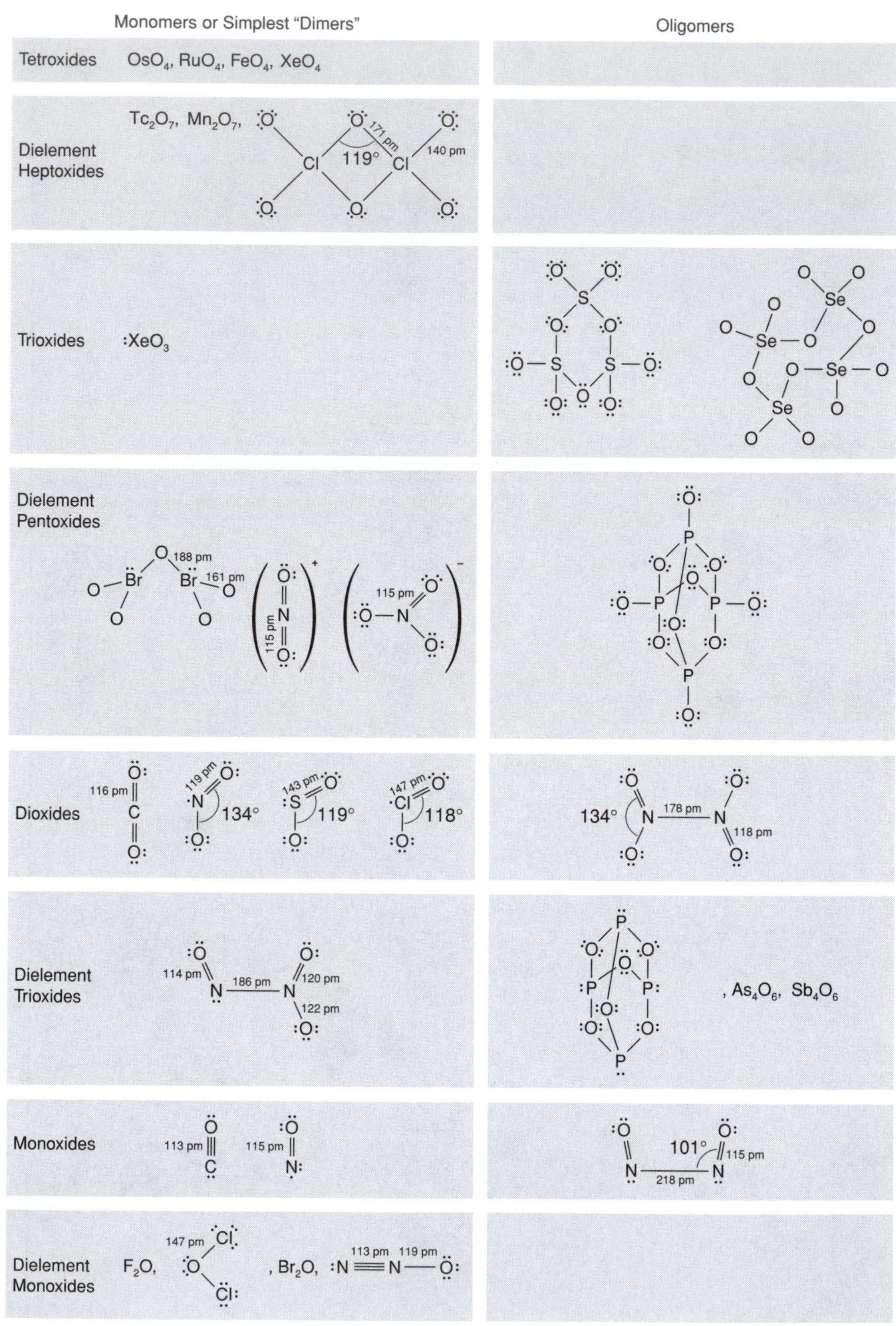

Figure 13.3
Structures of nonpolymeric oxides.

Element Trioxides (central atom oxidation state = +6). This oxidation state is not confined to groups numbered with 6, but can occur in higher groups: The $:XeO_3$ molecule, with a total coordination number of 4, is an explosive, monomeric solid. All other trioxides are found for atoms in Groups 6, 16(VI), and 6*F*, which are expected to have penultimate coordination numbers of 4 or higher, so these EO_3 structural units have at least one vacant coordination site. Both SO_3 and SeO_3 readily form the oligomers $(SO_3)_3$ and $(SeO_3)_4$; these oxides and CrO_3 form chain polymers such as $^1_\infty[CrO_3]$. The largest atoms in these groups take coordination numbers of 6 (or even 7 for one form of UO_3) in layer or network polymeric structures; the best known of these is the ReO_3 structure (Fig. 13.4). Redox stability and color vary considerably here:

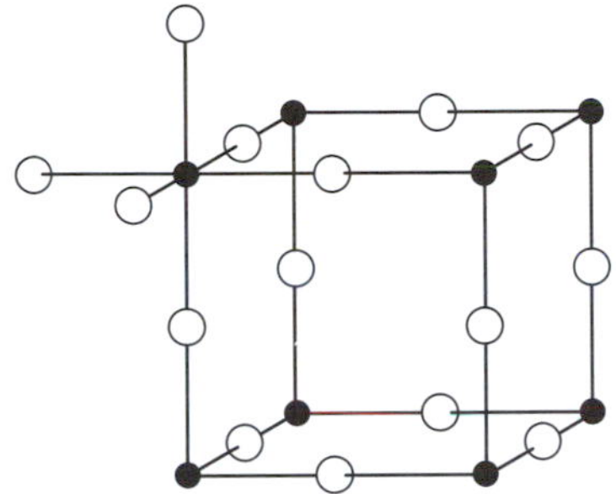

Figure 13.4
Structure of ReO_3, showing octahedral coordination of Re atoms (filled circles) and linear coordination of oxygen atoms (open circles). [Adapted from F. A. Cotton and G. Wilkinson, *Advanced Inorganic Chemistry: A Comprehensive Text*, 5th edition, Wiley-Interscience, New York, 1988, p. 851.]

CrO_3 is red and strongly oxidizing, while MoO_3 and WO_3 are in the most stable oxidation states of these elements and are colorless to yellow; ReO_3 is red, but in this case the central Re^{6+} ion has a d^1 electron configuration.

Dielement Pentoxides. Both $:Br_2O_5$ and $:I_2O_5$ have stereochemically active *sp*-hybrid electron pairs. The approximate structure[12] of $:Br_2O_5$ has a total coordination number of 4 (Fig. 13.3). For $:I_2O_5$, the radius ratio calculation would predict a total coordination number of 8 for iodine. The structure is similar to that of $:Br_2O_5$, with a total coordination number of 4 for iodine; this appears to be the most serious failure of all our calculations in Table 13.3. However, unlike $:Br_2O_5$, the pentoxide $:I_2O_5$ shows several much longer I–O contacts at distances of 223 pm or higher, which are still within the sum of van der Waals radii. This type of weak secondary bonding (Section 12.2) is not simply predicted or counted, but helps bridge the gap between our predictions and the crystal structure.

Among elements in groups numbered with five, dinitrogen pentoxide is also a surprise: In the solid state, its structure consists of an ionic lattice of $[NO_2]^+$ and $[NO_3]^-$ ions (although in other physical states it readily converts to a molecular form, which accounts for its low melting point). Each of these ions involves π bonding between N and O, which is not predictable by radius ratio calculations.

The somewhat larger (and less prone to π bonding) phosphorus atom adopts a coordination number of 4, leaving 1.5 vacant coordination sites in the structural unit, which suffices to form a cluster tetramer, P_4O_{10} (Fig. 13.3). The other, still larger atoms in this series are predicted to have coordination numbers of 6, and approximately do, with some unpredictable deviations to numbers of 5 and 7. These oxides are either layer or network polymers; P_2O_5 also has a layer-structured polymorph.

Element Dioxides. There are too many of these dioxides among the metals to discuss them in detail now; we may note that most of them adopt either the six-coordinate rutile (TiO_2) or eight-coordinate fluorite (CaF_2) lattices (Fig. 4.8). Hence, we concentrate on the dioxides of the *p* block. There are more of these than might have been anticipated.

The Group 16(VI) dioxides $:EO_2$ exhibit a very nice trend showing changes in structure with increasing central-atom radius, as computed in Example 13.4; the simplest of these dioxides, the bent double-bonded SO_2, is shown in Figure 13.3. Surprisingly, there are two fairly stable dioxides with *odd* numbers of valence electrons: the 17-electron brown gas NO_2 and the 19-electron yellow-red gas ClO_2 (the colors being connected with electronic transitions involving their SOMOs). These are examples of a class of molecules known as **free radicals**. These molecules are usually exceedingly reactive, since they tend to pair up with each other (or with other radicals) to share the odd electrons and form additional covalent bonds, releasing the covalent bond energy. These two free radicals can persist, but they are not highly stable: The radical ClO_2 is explosive unless it is kept highly diluted (e.g., when it is used as a bleach and disinfectant); NO_2 is not explosive but dimerizes at higher pressures and lower temperatures to the colorless gas *dinitrogen tetroxide*, N_2O_4.

Clearly, it impossible to draw satisfactory Lewis dot structures for these two substances or to use the octet rule to explain why these two free radicals are so much more stable than most free radicals; MO theory is necessary to adequately account for this phenomenon. It should be noted that (a) ClO_2 has not been reported to dimerize; (b) the dimer of NO_2 has an unusually long N–N single bond of 178 pm (Fig. 13.3), but is diamagnetic and definitely bonded; (c) the isoelectronic CO_2^- anion (17 electrons) is not stable but completely dimerizes to the $C_2O_4^{2-}$ (oxalate) anion, which has an approximately normal C–C bond distance. Explaining trends such as these requires the use of quantitative MO calculations, and has not been without controversy.

The Group 14(IV) dioxides are more regular in their chemistry, but also dramatically show the effects of size and π-bonding ability, which is itself a function of size. The π-bonded carbon dioxide is well known as a gas and as a solid (dry ice) that sublimes at $-78\,^\circ C$; under high pressures it can remain as a liquid at room temperature and is becoming increasingly important as an environmentally friendly solvent.

In SiO_2, the central-atom coordination number normally doubles to 4, giving two vacant coordination sites in the structural unit, which suffices to produce network polymers [Fig. 13.1(*d–f*)]. Silica (SiO_2) is a very important compound that has more than 22 phases (some of them necessarily impure, but at least 12 of which can be pure SiO_2). Many of its forms are familiar to collectors: α-quartz (a major component of granite and sandstone), rose quartz, smoky quartz, flint, heliotrope, jasper, onyx, amethyst, citrine, agate, chalcedony, and others. Industrially useful forms include kieselguhr and diatomaceous earth. Less common polymorphs and polymerization isomers include tridymite and cristobalite [Fig. 13.1(*e*, *f*)], coesite, and the remarkable stishovite, found in Meteor Crater, Arizona, in which silicon is six coordinate in a rutile lattice. It is not surprising that the high pressure presumably produced by the impact of a meteorite is what is needed to force six oxygen atoms around a small silicon atom. The dioxides of the largest Group 14(IV) elements, SnO_2 and PbO_2, adopt a structure that we think of as typically ionic, namely, the rutile lattice. The melting points of these compounds show no discontinuity with the melting points of their neighbors—Coulombic attractions and interunit covalent bonds are both strong interunit forces, about equally capable of resisting melting.

Dielement Trioxides of the *p* Block. A complete set of these is available from Group 15(V) with an unshared pair of *sp* electrons, and a set from Group 13(III) is available without unshared electrons. The Group 15(V) trioxides begin with dinitrogen trioxide, which consists of N_2O_3 molecules not bridged by oxygen but with unusually long N–N bonds of 186 pm length. This structure bears some resemblance to an ionic structure containing a triply bonded NO^+ cation paired with a partially double-bonded NO_2^- ion. The next three dielement trioxides are E_4O_6 cluster oligomers of T_d symmetry (Fig. 13.3); the latter two of these also have polymerization isomers that are layer or double-chain polymers. The Group 13(III) dielement trioxides, with no coordination sites blocked by unshared *sp* electron pairs, all give network covalent or ionic lattices.

Monomeric *p*-Block Monoxides. Only a few examples of the +2 and +1 oxidation states are found among the oxides of the *p* block. There are only two monomeric monoelement monoxides, carbon monoxide, CO, and *nitric oxide*, NO. Both have low polarities and low molecular weights. Therefore they are gases with low solubilities in water and low boiling points (−190 °C and −152 °C, respectively). Nitric oxide has one unpaired electron in the gaseous state, but is diamagnetic in the solid state, since it dimerizes as shown in Figure 13.3. Note that the structure has a very long N–N bond of 218 pm. Related to NO are the triple-bonded NO^+ cation, isoelectronic with CO, and the unstable NO^- anion, isoelectronic with O_2; the latter dimerizes to give the *hyponitrite* ion, $N_2O_2^{2-}$.

In the +1 oxidation state, we may mention four monomeric oxides, the unstable F_2O, Cl_2O, and Br_2O, and the stable *nitrous oxide* or "laughing gas," N_2O, which has a structure unlike the others (Fig. 13.3). Nitrous oxide has an N–N bond order between 2 and 3. These oxides are polar molecules that are considerably more soluble in water than CO or NO. The compound F_2O is explosively unstable and is best named as oxygen difluoride. Finally, a very unstable oxide of nitrogen in the $+\frac{1}{2}$ oxidation state, N_4O, has been prepared by the reaction of nitrosyl chloride, NOCl, with sodium azide, NaN_3; the product is thought of as $N_3^-NO^+$, with a chain of four nitrogen atoms.[13]

13.3 Acidity, Solubility, Practical Uses, and Environmental Chemistry of Volatile Oxides and Oxo Acids

Basic Oxides. To a first approximation, the acid–base properties of oxides of the elements can be thought of as a composite of the very strongly basic properties of the oxide ion and the acidic properties of the element "cation", which can range from nonacidic to very strongly acidic. Recalled from Section 2.7 that many metal oxides are **basic oxides**: The basic properties of the oxide ions, which are very strong, prevail over the lesser acidic properties of most metal cations. We may further subdivide the basic oxides into two categories.

The **soluble basic oxides** are the oxides of the nonacidic and feebly acidic metal cations in the lower left part of the periodic table. These oxides dissolve in water to give hydroxides of these metal ions, so the solution acquires a high pH.

$$BaO(s) + H_2O \rightarrow Ba^{2+}(aq) + 2\,OH^-(aq) \qquad (13.2)$$

The **insoluble basic oxides** are oxides of the weakly acidic and yet more acidic metal cations; by solubility Rule I (Section 4.1) these are insoluble in water, so they do not

Table 13.4
Major Acidic Oxides of the *p*- and *d*-Block Elements[a]

$B_2O_{3(w)}$	$CO_{2(m)}$	$N_2O_{5(s)}$ $N_2O_{3(m)}$			
Al_2O_3	$SiO_{2(w)}$	$P_4O_{10(m)}$	$SO_{3(s)}$	$Cl_2O_{7(vs)}$	
Ga_2O_3	GeO_2	$P_4O_{6(m)}$ $As_2O_{5(m)}$	$SO_{2(m)}$ $SeO_{3(s)}$	$Cl_2O_{(w)}$	
		As_4O_6	$SeO_{2(m)}$	$Br_2O_{(w)}$	
	SnO_2	Sb_2O_5	$TeO_{3(w)}$		$XeO_{4(s)}$
	SnO PbO_2 PbO	Sb_2O_3	TeO_2	$I_2O_{5(s)}$	$XeO_{3(s)}$
		$V_2O_{5(m)}$	$CrO_{3(s)}$	Mn_2O_7	
		Nb_2O_5	$MoO_{3(w)}$	$Tc_2O_{7(vs)}$	$RuO_{4(?)}$
		Ta_2O_5	WO_3	$Re_2O_{7(vs)}$	$OsO_{4(s)}$

[a] Oxides above and to the right of the light and heavy lines are **soluble acidic oxides:** They dissolve in water (at least to some extent) to give oxo acids of the strengths indicated: (vs) = very strong; (s) = strong; (m) = moderate; (w) = weak. Oxides within the light lines are **insoluble acidic oxides:** They are not soluble in water, but do dissolve in (react with) strong bases. Oxides below and to the left of the lines are **amphoteric** (these are not included for the *d*-block elements). The oxides of In and Tl, and Bi_2O_3, are exclusively basic oxides.

give solutions of elevated pH values. Nonetheless, they do dissolve in, and partially neutralize, strong acids, so they show basic properties.

$$Fe_2O_3(s) + 6\,H^+(aq) \rightarrow 2\,Fe^{3+}(aq) + 3\,H_2O \qquad (13.3)$$

Often, however, the acidity of a very strongly acidic "cation" can prevail over the basicity of the oxide ion, so that many oxides of the nonmetals at the right of the periodic table, or of metals in very high oxidation states, are **acidic oxides**. These oxides can again be subdivided into soluble and insoluble categories (Table 13.4).

Soluble Acidic Oxides. The process of dissolving an acidic oxide in water can be envisioned as involving at least two equilibria that influence each other. We may suppose that an oxide first reacts reversibly with water to give an oxo acid:

$$\tfrac{1}{\infty}[SeO_2] + H_2O \rightleftharpoons H_2SeO_3(aq) \qquad (13.4)$$

If the resulting oxo acid is very strongly acidic, strongly acidic, or moderately acidic (Section 3.7), a second equilibrium then proceeds appreciably to the right:

$$H_2SeO_3(aq) + H_2O(aq) \rightleftharpoons H_3O^+(aq) + [HSeO_3]^-(aq) \qquad (13.5)$$

For these three categories of oxo-acid acidity, this equilibrium then shifts the position of the previous equilibrium far enough to the right to allow the original metal oxide to dissolve in neutral water and to lower its pH: So *oxides that can dissolve to give very strongly acidic, strongly acidic, or moderately acidic oxo acids generally are* **soluble acidic oxides**.

The Weakness of Carbonic Acid: An Aside

Solutions of H_2CO_3 are less acidic than we would have expected, since it exists in solution mostly as CO_2 rather than as H_2CO_3 molecules:

$$CO_2(aq) + H_2O(aq) \rightleftharpoons H_2CO_3(aq); \quad K_{eq} = 2 \times 10^{-3} \qquad (13.6)$$

(In the body, a zinc-containing enzyme, carbonic anhydrase, is necessary to catalyze this reaction.) Similar equilibria have been noted for the other monomeric acidic oxides: XeO_4, XeO_3, RuO_4, OsO_4, and SO_2.[14]

To explain this, we note that the dissolution of an *oligomeric* or *polymeric* oxide [as in Eq. (13.4)] generally results in an oxo acid in which the total coordination number of the central atom is unchanged, that is, 4 for Se in $^1_\infty[SeO_2]$ and H_2SeO_3. To add water to a *monomeric* acidic oxide, however, there must be an *increase* in the total coordination number of the central atom (i.e., from 2 for C in CO_2 to 3 in H_2CO_3). There is evidently a significant energy barrier to this increase in coordination number during the process of hydration. In the case of FeO_4, there is apparently no equilibrium concentration of an oxo acid (no H_2FeO_5 or H_4FeO_6, in either of which Fe would exceed its allowable penultimate coordination number), since FeO_4 is reported to be extractable by carbon tetrachloride from a strongly basic solution (i.e., to lack acidic properties).

Insoluble Acidic Oxides. Most other nonmetal oxides, and a significant number of metal oxides, do not undergo sufficient ionization analogous to Eq. (13.5) to allow them to dissolve in neutral water, but they will dissolve in strong bases, partially neutralizing them. These compounds may be termed **insoluble acidic oxides**. The oxo acids that correspond to the hydrated forms of these oxides are (at best) weakly acidic; the anions actually formed on the dissolution of these oxides in strong bases are normally not simple oxo anions, but rather hydroxo anions[15]:

$$Sb_2O_5(s) + 5\,H_2O(s) + 2\,OH^-(aq) \rightleftharpoons 2[Sb(OH)_6]^- \qquad (13.7)$$

Table 13.5 summarizes, in a general way, the locations of the four principal acid–base classes of oxides. Soluble basic oxides and soluble acidic oxides occupy opposite edges of the periodic table, except that metals in very high oxidation states can also give rise to soluble acidic oxides. In between, we find insoluble basic oxides (more to the left) and insoluble acidic oxides (more to the right).

The realms of insoluble acidic oxides and insoluble basic oxides overlap extensively, however. There are many insoluble oxides such as Al_2O_3 that do not dissolve in water, but dissolve in strong acids *and* (in separate experiments) in strong bases:

$$Al_2O_3(s) + 6\,H^+ \rightleftharpoons 2\,Al^{3+}(aq) + 3\,H_2O \qquad (13.8)$$

$$Al_2O_3(s) + 2\,OH^- + 3\,H_2O \rightleftharpoons 2\,[Al(OH)_4]^- \qquad (13.9)$$

Table 13.5
Soluble and Insoluble Acidic and Basic Oxides

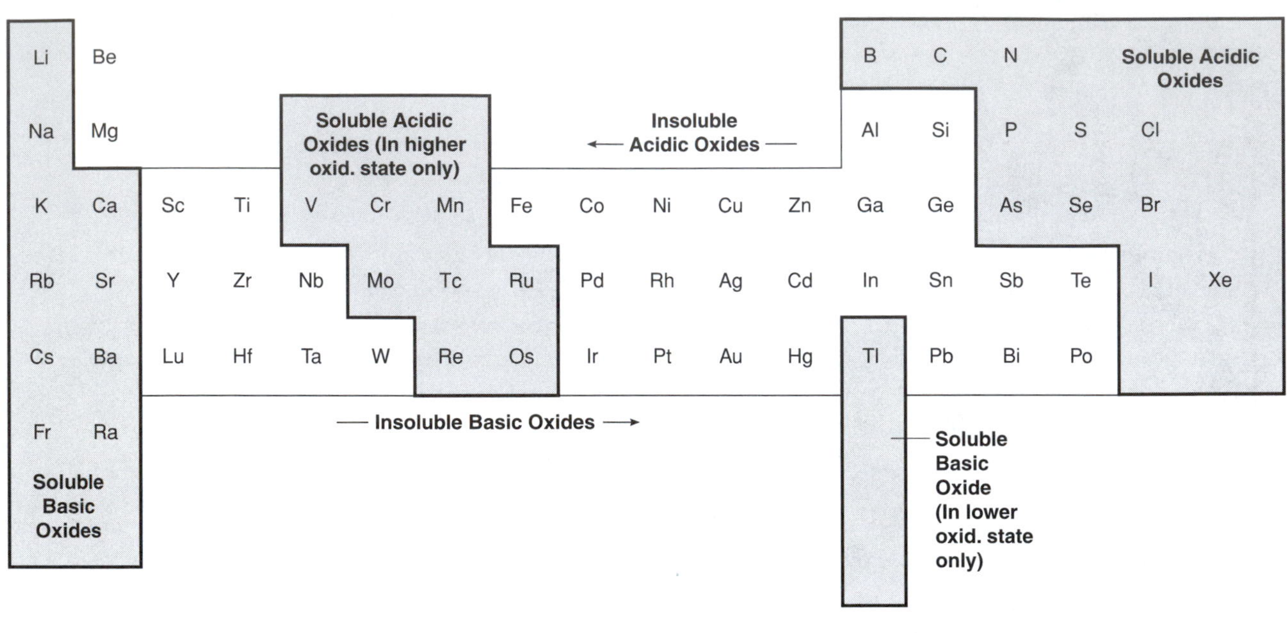

Such oxides are known as **amphoteric oxides.**

Unfortunately, there is no known systematic way of determining which of the many insoluble basic oxides will also show acidic properties (dissolve in strong bases, and hence be classified as amphoteric) and which will not; this problem is a future challenge for some bright student of this textbook! Some insoluble basic oxides derived from even weakly acidic cations (e.g., ZnO) do dissolve in strong bases, and hence are amphoteric; some, derived from moderately or even strongly acidic cations (e.g., Fe_2O_3 and TiO_2), do not. Hence, these compounds are insoluble basic oxides. Judging from the predominance diagrams of Pourbaix, the insoluble basic oxides derived from all weakly, moderately, and strongly acidic cations would probably dissolve in strong bases if the concentrations of base used to dissolve them could be a little higher than is possible in practice.

Finally, three monomeric nonmetal oxides from the upper right portion of the *p* block have such low oxidation numbers for the nonmetal atom that they lack any acidic properties, namely, CO, N_2O, and NO. Since these oxides certainly do not have basic properties, they are neutral oxides. In addition, there are a few oxides, such as the free radicals NO_2 and ClO_2, that do not correspond in oxidation state to a stable or known oxo acid or anion. These oxides can give rise to a mixture of oxo acids or anions by disproportionation (Section 6.9):

$$2\,NO_2 + 2\,OH^- \rightarrow NO_2{}^- + NO_3{}^- + H_2O \qquad (13.10)$$

Because the acidity of a cation rises rapidly with its charge, there are several *d*-block elements possessing several oxidation states (such as Cr) that have one or more oxides that show only basic properties [chromium(II) oxide, CrO], one or more oxides that are amphoteric [chromium(III) oxide, Cr_2O_3], and one or more oxides that possess only acidic properties [chromium(VI) oxide, CrO_3]. Clearly, the higher the oxidation number of a given element, the more acidic the corresponding oxide will be.

Lux–Flood Acid–Base Reactions. Since basic oxides can react with the hydronium ion (a strong aqueous acid) and acidic oxides can react with the hydroxide ion (a strong aqueous base), it is not too surprising that basic oxides such as MgO and acidic oxides such as SiO_2 can react directly with each other:

$$2\,MgO + SiO_2 \rightarrow Mg_2SiO_4 \tag{13.11}$$

In such a reaction, there is a transfer of *oxide ions*; in the Lux–Flood acid–base classification system an oxide-ion donor is a base and an oxide-ion acceptor is an acid. The products of these reactions are salts of oxo acids and, since water is not involved in the reaction, can be salts of oxo anions such as SiO_4^{4-} that are too basic to persist in aqueous solution. As we will discuss later in this chapter, direct reactions of acidic and basic oxides are of enormous practical importance in such areas as control of pollution by gaseous acidic oxides and in the production of materials such as concrete, glass, and ceramics.

Example 13.5

Complete and balance the following chemical equations for reactions of oxides: (a) $N_2O_3 + H_2O \rightarrow$? (b) $BaO + OsO_4 \rightarrow$? (c) $SrO + ZrO_2 \rightarrow$? and (d) $Fe_3O_4 + H^+ \rightarrow$?

SOLUTION:

(a) The compound N_2O_3 is a nonmetal oxide, so it is most likely an acidic oxide. It corresponds in oxidation state to a moderately acidic oxo acid, HNO_2, so it is expected to dissolve in water to give HNO_2:

$$N_2O_3 + H_2O \rightarrow 2\,HNO_2$$

(b) Barium oxide is the oxide of a feebly acidic cation, Ba^{2+}, so the strong basicity of the oxide ion is virtually undiminished: This is a soluble basic oxide. The hypothetical Os^{8+} cation in OsO_4 is so acidic that the oxide itself might also be expected to be acidic. For the relatively large Os, an oxo anion with six oxygens is expected: OsO_6^{4-} (Table 3.3). The product of the reaction of a basic oxide with an acidic oxide is the salt of an oxo anion. By balancing the -4 charge of this anion with two Ba^{2+} cations, we complete and balance this equation:

$$2\,BaO + OsO_4 \rightarrow Ba_2OsO_6$$

(c) The Sr^{2+} ion in SrO is feebly acidic, so the basicity of oxide ion prevails; this is a basic oxide. The Zr^{4+} ion in ZrO_2 is strongly acidic, so its oxide is either basic or amphoteric. Unfortunately, amphotericity is not precisely predictable; if ZrO_2 is also a basic oxide (as it in fact is), there can be no acid–base reaction between these oxides.

(d) The compound Fe_3O_4 is a peculiar oxide (a mixed-metal oxide, to be discussed in Section 13.5), in which the average oxidation state of iron is $+2.67$. This means that two-thirds of the iron ions are moderately acidic Fe^{3+} and one-third are weakly acidic Fe^{2+} ions. Such oxides are insoluble in water, but do dissolve in acids:

$$Fe_3O_4 + 8\,H^+ \rightarrow 2\,Fe^{3+} + Fe^{2+} + 4\,H_2O$$

The reactions of acidic and basic oxides to give salts of oxo anions, like other acid–base neutralization reactions, are generally exothermic. Smith[16] noted that the enthalpy change for the reaction in which 1 mol of oxide ion is transferred from the basic oxide to the acidic oxide, ΔH_{A-B}, can be expressed as a function of acidity parameters a of the acidic oxide (a_A) and the basic oxide (a_B):

$$\Delta H_{A-B}(\text{in kJ mol}^{-1}) = -(a_A - a_B)^2 \tag{13.12}$$

Smith notes the parallel between this equation and the Pauling electronegativity equation (7.16); as in Pauling's equation, one parameter must be arbitrarily set, and he selects $a_{H_2O} = 1.00$. Bratsch has extended this parallel further, and removed the arbitrariness of the water parameter.[17] Figure 13.5 shows that Smith's a parameters are related to the acidity classifications of the element in the oxide. Hence, Eq. (13.12) shows that the greater the difference in the acidities of the two central elements in the two oxides, the more exothermic is the reaction of their oxides to give a salt of an oxo anion.[18]

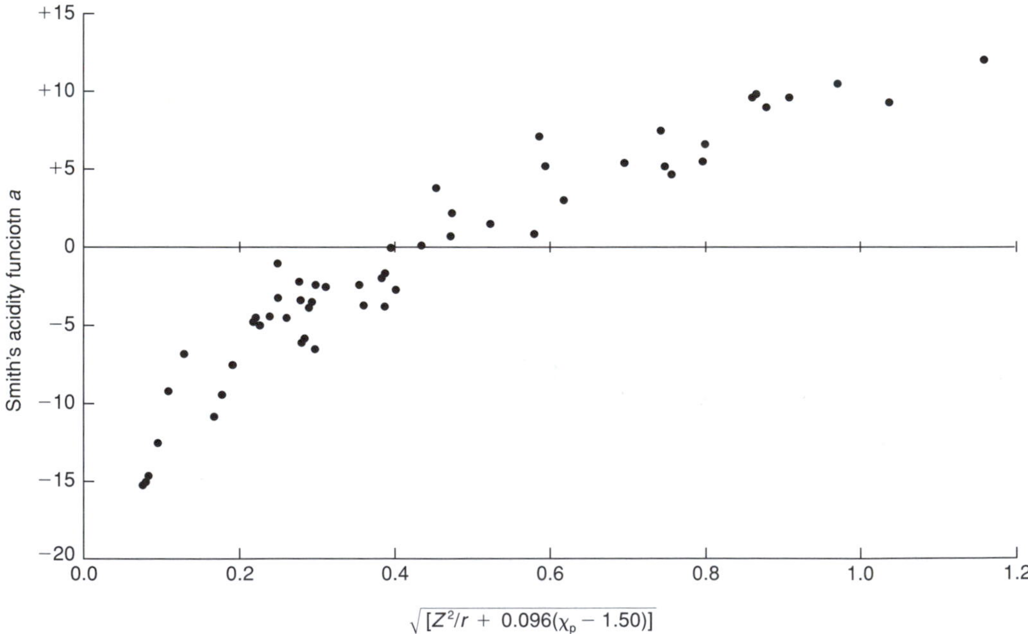

Figure 13.5
Smith's acidity function a for oxides of elements, plotted as a function of the square root of the modified Z^2/r ratio from Eq. (2.12), $\sqrt{[Z^2/r + 0.096(\chi_P - 1.50)]}$. Acidity functions are adapted from D. W. Smith, *J. Chem. Educ.*, **64**, 481 (1987).

Uses and Environmental Chemistry. A number of the volatile (and usually acidic) oxides, and some of the oxo acids formed from them, are of immense significance in our technological civilization. These include some of the chemicals produced in the greatest total tonnage in the world. Any chemicals used in such high tonnage are bound to escape into the environment to some degree. Since these are mainly acidic oxides, we can anticipate that they will not be innocuous components of the environment; since these are volatile oxides we can anticipate that there will be air-pollution problems such as the production of **acid rain** connected with them. In the remainder of this section, we deal with the uses and environmental chemistry of the most important of these oxides; there are many useful sources of this information that should be consulted for more details.[19,20] Given the periodic trends we have seen in the previous sections and chapters, we should also be able to predict some uses and environmental chemistry for the other oxides, should they come into widespread use.

Carbon Oxides. Carbon monoxide is produced industrially on a huge scale as a mixture with H_2, by the reaction of steam with hot coal:

$$C(s) + H_2O(g) \rightarrow CO(g) + H_2(g) \qquad \Delta H^\circ = +131\,\text{kJ mol}^{-1} \tag{13.13}$$

(Since this reaction is quite endothermic, air must periodically be blown through the coal to provide energy via oxidation of the coal.) This mixture of carbon monoxide and hydrogen is known as **water gas**; after adjustment of its hydrogen content it is known as **synthesis gas**, since it is used in the industrial production of a number of important organic chemicals. Intensive research is currently focused on the chemistry of carbon monoxide, since this mixture could be a source of organic chemicals from coal when our present source of many organic chemicals, petroleum, is exhausted. For example, synthesis gas can be converted to hydrocarbons via the Fischer–Tropsch process (Section 12.6):

$$n\,CO + (2n+1)H_2 \rightarrow C_nH_{2n+2} + n\,H_2O \tag{13.14}$$

Environmentally, the main anthropogenic (human-created) source of carbon monoxide is the incomplete combustion of fuel in automobile engines (197 million tons per year). In urban environments, this is a problem due to the high toxicity of carbon monoxide (of which we have no warning, since it is odorless). In terms of the global atmospheric environment, however, carbon monoxide is not considered a serious pollutant, since (a) natural production of CO (i.e., by the oxidation of CH_4 produced by anaerobic decomposition of organic material in swamps and in the tropics) far outweighs human production, and (b) natural processes continuously remove CO from the atmosphere. These processes include microbial degradation in the soil and reactions in the atmosphere with reactive free radicals such as hydroxyl (·OH) and hydroperoxyl (HO_2·). (Free radicals can persist in the atmosphere, especially the thin upper atmosphere, longer than in solution, since the concentration of any species with which they may react is much lower there than in solution.)

Carbon dioxide is produced industrially by a number of reactions, such as the combustion of carbonaceous fuels, and by the **water gas shift reaction**:

$$CO + H_2O = CO_2 + H_2 \tag{13.15}$$

Because of its physical properties, water gas is widely used as a refrigerant: The solid form, dry ice, sublimes at −78 °C, thus cooling its environment without generating any messy liquids or toxic gases. It is also used on a large scale in fire extinguishers and to carbonate beverages.

Although of very low toxicity, carbon dioxide *is* considered to be a potentially serious global atmospheric pollutant. It is, of course, removed from the atmosphere by plant photosynthesis and put into the atmosphere by plant and animal respiration. Presumably, these processes have reached a balance over time. However, the atmospheric concentration of CO_2, as measured in such remote locations as Antarctica and the top of Mauna Loa, Hawaii, has been increasing 0.2% a year since about 1870. This increase is of concern because of the vital role of CO_2 in the atmosphere: It absorbs IR light (heat) emitted by the earth and returns some of it to the surface of the earth, thereby warming it. This process is popularly known as the **greenhouse effect** (Section 9.5). An increase of only 2 or 3 °C would have profound effects on the climate of the earth; between 1880 and 1940 the mean temperature of the earth rose 0.4 °C. This rise might be due to the evident changes in the atmospheric CO_2 concentration, which is thought to be the result of the combustion of fossil fuels and (more importantly) the cutting and burning of many of the tropical forests of the world. A catastrophe in climate is not assured, however; there is an opposing scattering of incoming solar energy by suspended atmospheric particulates, which may be responsible for the cooling of the earth by 0.1 °C since 1940. Current thinking is that more extremes in weather, rather than just warmer weather, may be the main result of increasing concentrations of greenhouse gases.[21]

Nitrogen Oxides and Nitric Acid. Only three nitrogen oxides are stable enough to be of practical use or of environmental importance. One of these oxides, N_2O (commonly called nitrous oxide) is used as an anaesthetic and as a propellant (to provide pressure to expel ingredients) in aerosol cans; it is involved in reactions in the upper atmosphere that could deplete the **ozone layer**.

There are two common sources of NO (commonly called nitric oxide): catalytic oxidation of ammonia, which comes ultimately from petroleum and air, and direct combination of nitrogen and oxygen of the air in an electrical discharge (lightning) or at around 2000 °C (e.g., in a power plant during the burning of coal or in an automobile engine):

$$4\,NH_3 + 5\,O_2 \rightarrow 4\,NO + 6\,H_2O \qquad (13.16)$$

$$N_2 + O_2 \rightleftharpoons 2\,NO \qquad (13.17)$$

After a few days in the atmosphere,[22] NO is oxidized by oxygen to NO_2; hence, in air-pollution work these two are often collectively referred to as NO_x. Nitrogen dioxide and its dimer, dinitrogen tetroxide, are readily interconverted in an equilibrium that is visible due to the brown color of nitrogen dioxide:

$$N_2O_4 \rightleftharpoons 2\,NO_2 \qquad \Delta H = +57\,\text{kJ mol}^{-1} \qquad (13.18)$$

In the solid state, this system is colorless, as it is completely in the form of N_2O_4. At its boiling point (21 °C), the liquid is deep brown due to an 0.1% content of NO_2. The vapor becomes steadily darker with increasing temperature due to the increasing dissociation of the dimer, which is nearly complete at 140 °C.

Nitrogen dioxide is an acidic oxide that reacts with water to produce nitric acid:

$$3\,NO_2 + H_2O \rightarrow 2\,HNO_3 + NO \tag{13.19}$$

This reaction occurs in the atmosphere and is one of the sources of acid rain and of nitrate as a plant nutrient; it also is carried out in industry in the manufacture of nitric acid, which is used on a large scale in the manufacture of ammonium nitrate fertilizer, nylon, steel, and in rockets (as the oxidizer of the rocket fuel).

Nitric acid and NO_x are involved in several environmental problems. The acid rain problem, at least as far as damage to lakes is concerned, apparently results more from the solubilization of toxic metal ions at low pH values, as discussed in Section 3.8, than from the toxicity of nitrate ion or even directly of hydrogen ion.

There has been concern about the injection of NO_x into the upper atmosphere (stratosphere) due to high-temperature combustion in the engines of supersonic transport aircraft (SSTs). Ozone (O_3) is an important component of the upper atmosphere, since it absorbs high-energy UV radiation from the sun, preventing it from reaching the surface of the earth, where it would cause extensive skin cancer and genetic mutations. Nitric oxide is known to catalyze the destructive reaction of ozone with **atomic oxygen** (also present in the upper atmosphere):

$$NO + O_3 \rightarrow NO_2 + O_2 \tag{13.20}$$

$$NO_2 + O \rightarrow NO + O_2 \tag{13.21}$$

The NO produced in Reaction (13.21) is then able to reinitiate Reaction (13.20); thus it is functioning as a catalyst.

In the lower atmosphere, NO_2 is involved in a complex series of **photochemical** reactions in air that is also contaminated with unburned hydrocarbons (from automobile exhaust) and in the presence of bright sunlight (e.g., in Los Angeles). These reactions produce ozone, aldehydes, and organic nitrates such as peroxyacetyl nitrate (PAN) and peroxybenzoyl nitrate (PBN), which are powerful eye irritants and are quite damaging to vegetation.

Physical Properties of Concentrated Oxo Acids. We have not yet had a chance to apply the concepts of this chapter concerning physical properties to the oxo acids discussed in Chapter 3. While we are on the subject of nitric acid, this may be a good time to do so. Unlike oxides, oxo acids have hydrogen atoms directly bonded to oxygen and can thus engage in hydrogen bonding. Hydrogen bonding is substantially stronger than most van der Waals attractions, but is not nearly as strong as Coulombic forces or covalent bonds between units. Hence, pure oxo acids, although smallish molecules, are not gases, but are liquids or solids with rather high boiling points. Pure HNO_3, for example, has a boiling point of 83 °C.

Generally speaking, pure oxo acids are rather difficult to make. Weak oxo acids often decompose to oxides that are polymeric and insoluble in water (Table 13.5) or to gaseous oxides that are readily lost from solution (even in solution, most dissolved CO_2 is not present as carbonic acid, but as hydrated CO_2 molecules). Strong oxo acids such as nitric acid exothermically form hydrated oxo anions and H_3O^+ ions in solution. When one tries to isolate the oxo acids in their anhydrous form, enough stabilization is lost that in many cases the pure oxo acid cannot exist or has greatly reduced stability. The more stable oxo acids are commonly handled as concentrated solutions in water.

In many cases, these are prepared by distilling off water until a mixture of water and acid of constant composition finally distills over (sometimes under reduced pressure to reduce their fairly high boiling points to a temperature low enough that the oxo acid does not decompose). Nitric acid is commonly seen as concentrated HNO_3 (68% HNO_3, 15 *M*). It is corrosive to the skin, reacting with skin protein to produce a yellow material called xanthoprotein.

Sulfur Oxides and Sulfuric Acid. Sulfur dioxide is made commercially by the combustion of sulfur, H_2S, or sulfide ores such as FeS_2. It is a colorless, poisonous gas with a choking odor and a relatively high boiling point (−10 °C), and is useful as a solvent, refrigerant, food preservative, and (mainly) in the manufacture of sulfuric acid. In this process, the SO_2 must first be oxidized by air to SO_3, which is a kinetically slow process, so a catalyst of platinum sponge, V_2O_5, or NO is required. The SO_3 resulting from this oxidation reacts exothermically with H_2O to give H_2SO_4. This reaction is impractical in industry, since a mist of H_2SO_4 would be produced in the air that would pass out into the atmosphere. Consequently, the gas stream is bubbled through concentrated sulfuric acid, with which it reacts to generate a polynuclear sulfuric acid (such as will be discussed later in this chapter):

$$SO_3 + H_2SO_4 \rightarrow H_2S_2O_7 \qquad (13.22)$$

This acid is then *carefully* reacted with the proper amount of water to give concentrated sulfuric acid.

Sulfuric acid is the leading industrial chemical in terms of the number of tons produced per year. Concentrated sulfuric acid is 98% H_2SO_4 by weight, is about 18 *M*, and boils at 338 °C. It has a very strong affinity with water and releases a great deal of heat on absorbing water. Contact with the skin causes dehydration and burns; should a spill occur, the acid should immediately be flushed away with large quantities of water for *at least 15 min*. The dilution of concentrated sulfuric acid should be carried out *cautiously*—the acid should be poured slowly into water with good stirring to dissipate the heat. Adding water to the acid can cause dangerous spattering of concentrated acid. Sulfuric acid even removes the elements of water from many organic molecules, converting carbohydrates to carbon, for example.

The uses of sulfuric acid are so many and varied that the figures for production of sulfuric acid in a given country has been considered a reliable indicator of that country's industrial development. The largest usage is in the production of fertilizer (see the next subsection, Phosphorus Pentoxide and Phosphoric Acid); other major uses are in refining of petroleum, in metallurgy, and in the manufacture of chemicals. For example, organic nitrations to produce explosives typically use a mixture of HNO_3 and H_2SO_4, the latter to remove water. This reaction not only generates toxic nitrogen oxides but produces waste dilute sulfuric acids. Clean H_2SO_4-free nitrating reagents are now being advocated, such as N_2O_5 by itself, or HNO_3 mixed with $Yb(CF_3SO_3)_3$ as an acid catalyst.[23]

Natural sources produce large amounts of SO_2 via decay of organic matter to H_2S, which is rapidly oxidized to SO_2 in the atmosphere. Anthropogenic SO_2 is produced in comparable quantities during the roasting of sulfide ores and the burning of oil and coal, which often contains substantial amounts of FeS_2. In the atmosphere, SO_2 is also oxidized to H_2SO_4. This process is speeded by catalysts such as water droplets and *d*-block metal ions found in atmospheric particles of soot. These conditions, which once prevailed in London fog (as opposed to Los Angeles or photochemical smog), have been found in many cities of the world in which homes were heated by burning soft coal. It may well be imagined that breathing in droplets of sulfuric acid does not have a desirable effect on the lungs and body. The resulting strain on the lungs and heart has shortened the lives of many people during such episodes of smog.

In the vicinity of smelters in which sulfide ores are roasted, the concentration of sulfuric acid and sulfur oxides has been so great that artificial deserts have been created. This problem has been alleviated by the construction of very high smokestacks ("the solution to pollution is dilution"), which unfortunately has resulted in the spread of the sulfur oxides and sulfuric acid over whole continents and has helped cause the current problem of acid rain. Acid rain may have pH values as low as 2.1. (The contributions of coal-fired power plants, though never concentrated enough to give rise to local deserts, are collectively larger than those of smelters.) This acid rain is corroding away many historic monuments and statues made of susceptible salts of basic oxo anions (marble and limestone are largely $CaCO_3$). Lakes that are in contact with limestone deposits are protected by the same reaction, which neutralizes the acid rain, but those that are not so fortunate become quite acidic, with harmful consequences already discussed. Effects on trees and vegetation are now also being discovered.

Due to these problems, a considerable amount of research has been done on methods of control of sulfur dioxide emissions from smelters and power plants. This would seem to be a simple matter, since acidic oxides such as SO_2 and SO_3 would be expected to react readily with inexpensive basic oxides or hydroxides. Thus a solution of $Ca(OH)_2$ can be sprayed down the smokestack of the plant in a **scrubber** to react with the sulfur oxides according to the reaction:

$$Ca(OH)_2 + SO_2 + 0.5\,O_2 \rightarrow CaSO_4 + H_2O \qquad (13.23)$$

(Much of the calcium sulfite expected from this reaction is oxidized by the air to calcium sulfate.)

Finding a suitable chemical reaction is only the first step. A coal-fired power plant emits much more CO_2 than SO_2; CO_2 is also a weakly acidic oxide that would use up much of the $Ca(OH)_2$. Consequently, for each ton of coal burned, up to 0.2 tons of limestone would be required, and an enormous quantity of wet $CaSO_4$ would be generated. We have now converted an air-pollution problem to a water-pollution problem or (if we dry out the wet $CaSO_4$) a solid-waste problem, since there are not enough uses of $CaSO_4$ to be able to market such quantities. A second problem is that the lime solution cools the exhaust gases so much that they no longer rise up out of the smokestack, so the stack gases have to be reheated!

A number of other alternative processes have been studied.[19,20] For example, using $Mg(OH)_2$ instead of $Ca(OH)_2$ has advantages despite the fact that $Mg(OH)_2$, being the hydroxide of a weakly acidic cation, is insoluble in water. After the SO_2 reacts with the slurry of $Mg(OH)_2$, the resulting $MgSO_3$ can be heated (in another location) to regenerate the SO_2:

$$MgSO_3 \rightleftharpoons MgO + SO_2 \qquad (13.24)$$

This reaction is much more feasible than the analogous reaction with $CaSO_3$, because there is a greater mismatch of cation and anion radius with $MgSO_3$, which favors the formation of MgO at a lower temperature than CaO can be formed. The MgO can be recycled to form $Mg(OH)_2$; thus there would be no solid waste problem and no great investment in $Mg(OH)_2$. The SO_2 is formed at a concentration great enough to allow the manufacture of sulfuric acid, which could be sold. However, sulfuric acid is the cheapest acid, and the sale of this acid would not pay for the heat used to decompose the $MgSO_3$.

Although many other alternatives exist, each one has a drawback. It was estimated that equipping the power plants of the United States with devices to remove most of the SO_2 would cost about $32 billion. Unless subsidized by the government, this amount would be added to consumers' electric power bills.

Phosphorus Pentoxide and Phosphoric Acid. The acidic P_4O_{10} (commonly named phosphorus pentoxide from its simplest formula unit, without the di-!) reacts very completely with water, and hence it is used as a drying agent. But it is dangerous to use P_4O_{10} to dry organic liquids, because it releases so much heat that it can set the organic liquid on fire! It is also involved indirectly in the production of high-purity "syrupy" (85% or 15 *M* concentrated) phosphoric acid, made by oxidizing elemental phosphorus in the presence of water:

$$P_4 + 5\,O_2\,(\rightarrow P_4O_{10}) + 6\,H_2O \rightarrow 4\,H_3PO_4 \tag{13.25}$$

This high-purity acid is used in making detergents, toothpaste, and in foods such as colas ($\sim$0.05% H_3PO_4, pH 2.3!), but it is too expensive for the main use of phosphoric acid, in making fertilizers. For this purpose, phosphoric acid, being a moderately weak acid, is produced by the reaction of a strong acid with the moderately basic phosphate ion:

$$Ca_5(PO_4)_3F + 5\,H_2SO_4 \rightarrow 3\,H_3PO_4 + 5\,CaSO_4 + HF \tag{13.26}$$

This reaction mixture is diluted with water and the insoluble calcium sulfate is filtered off; the solution may then be concentrated. The cost of this process is only one-third as much as the preceding process, but as the phosphate rock used, fluoroapatite, contains many impurities, so does the resulting phosphoric acid. This reaction also produces a serious air pollutant, gaseous HF. For use in fertilizers, it is not necessary to protonate the PO_4^{3-} ion completely. Fluoroapatite is too insoluble to be utilized by plants as a nutrient, but partial protonation of the phosphate ion reduces its basicity, so that the salt $Ca(H_2PO_4)_2$ becomes soluble enough to be used as a nutrient ("superphosphate" fertilizer).

The three main nutrient elements provided in fertilizers are K, P, and N; the potassium and phosphorus are provided in the form of K^+ and PO_4^{3-} and the nitrogen either as the NH_4^+ ion or the NO_3^- ion. Interestingly, the compositions of fertilizers are often expressed in terms of acidic and basic oxide contents. The potassium ion is expressed as the weight percent of K_2O. The phosphorus ion content is expressed as the percent of P_4O_{10} or P_2O_5 needed to give the actual content of K and P via acidic oxide–basic oxide reactions. (Since the nitrogen may be present as NH_4^+, which cannot arise from an acidic or basic oxide, its content is given more simply as %N.) A fertilizer labeled 16–48–0 contains 16% N, 48% P_2O_5, and 0% K_2O.

Perchloric Acid. Perchloric acid ($HClO_4$) is available commercially as a 72% solution, with a boiling point of 203 °C. Contact of concentrated $HClO_4$ with organic materials and other easily oxidized materials should be avoided, however, since the products may be treacherous explosives.

13.4 Close Packed Anions, Metal Oxides, and Electrical Conductivity of Solid Ionic Compounds

Close Packed Anions. Before we look at some chemistry of the ionic oxides of metals, we wish to mention an alternative way of looking at crystal lattices that is useful for many ionic compounds. Anions are normally larger than cations and are often in contact with each other in one of two close-packing patterns also found among atoms in metals (Section 12.3; Figs. 12.9 and 12.10); the much smaller cations then fill some of the **holes** or interstices between the anions. Thus if we were packing basketballs and baseballs for shipment in the same large box, we would figure out first how to pack the larger basketballs in the most efficient way possible, with confidence that the baseballs would fit in the spaces between the larger spheres.

There are two types of holes in the hexagonal close-packed (hcp) and cubic close-packed (ccp) lattices: tetrahedral holes and octahedral holes. The locations of these holes may be seen in Figure 12.9(*b*). Between three anions in the layer drawn in black and the anion in neighboring layer drawn in gray there is a small open space, called a **tetrahedral hole**, since it is surrounded by four large spheres; a small cation put in here would have a coordination number of 4.

Figure 12.9(*b*) also shows that this arrangement of the lower layer of ions fills in only one-half of the three-sided cavities of the black layer; the other one-half of the three-sided cavities of the black layer match up with three-sided cavities in the gray layer to produce larger **octahedral holes** surrounded by six larger spheres.

It can be shown that for each sphere in a given layer, there are two tetrahedral holes but only one octahedral hole. The stoichiometry of the salt does not normally allow all of the holes to be filled with cations, but (if close packing is utilized) the structure can be described in terms of the type and fraction of holes occupied.

Although these close-packed descriptions are used more frequently with other types of lattices than those found in Table 4.4 (and illustrated in Fig. 4.8), it is useful to show how they may be applied to these simple lattice types. Thus the two ZnS lattice types differ in the type of close packing: The zinc blende structure (Fig. 4.8) has its sulfide ions cubic close-packed, with zinc ions being located in one-half of the tetrahedral holes. The wurzite structure has an obvious hexagonal pattern (hcp sulfide ions), again with zinc ions in one-half of the tetrahedral holes. Sodium chloride has ccp chloride ions, with sodium ions in all of the octahedral holes. The fluorite structure has cubic close-packed *calcium* ions, with the more abundant fluoride ions filling all of the tetrahedral holes. The rutile structure can only indirectly be derived by distorting a hcp structure; the CsCl structure is not close-packed.

Metal Oxides. The physical properties of the ionic metal oxides contrast strikingly with those of the more volatile oxides we have been discussing. Their very high melting points and low volatility make some of them (such as MgO) useful as **refractories** for providing surfaces capable of withstanding very high temperatures. Magnesium oxide is used not only to line furnaces but to cover the heating elements of electric ranges, since it conducts heat much more readily than it conducts electricity. A related use is that of *thoria* (actually 99% ThO_2 + 1% CeO_2) to provide luminosity to gas flames for lighting purposes: The oxides become white-hot without melting. (Note the common industrial nomenclature by which the most common oxide of a metallic element is named by substituting *-ia* for *-ium* in the name of the metallic element.) The uranium oxide U_3O_8, a waste product of the nuclear power industry, has been found to be an effective catalyst for the incineration of volatile organics, even chlorocarbons, reducing the temperatures at which the latter can be decomposed from 1000 to 400 °C while preventing the formation of dioxins.[24]

Titanium dioxide is used extensively for its intense whiteness when cold (e.g., in white paints, where it has replaced very toxic lead compounds). Naturally occurring TiO_2 (the minerals anatase and rutile) is normally darkened with impurities, however, so that TiO_2 must be manufactured chemically. This process is not easy, since the Ti^{4+} ion is strongly acidic, and consequently TiO_2 is quite insoluble. The titanium ore ilmenite, $FeTiO_3$, dissolves in hot concentrated sulfuric acid, however; after the solution is diluted and the $FeSO_4{\cdot}7H_2O$ is crystallized out, the remaining solution of titanyl sulfate ($TiOSO_4$) is hydrolyzed to give pure TiO_2.

Because of their very high melting points, many ionic and macromolecular oxides and related materials find extensive use as **ceramics**: hard, heat- and chemical-resistant materials that keep their strength even when very hot (as most metals and organic polymers do not). Among the best known of these is **silica**; the related polysilicate ceramics and glasses, which have been known and used for thousands of years, will be discussed later. Nowadays, ceramics are used for making far more than pottery and china. Advanced ceramics are used in abrasives, cutting tools, electrical insulators, heat shields, nuclear fuels, bone implants, and lasing crystals.[25]

The resistance of these materials to heat, chemical attack, and mechanical loads stems from the strong chemical forces linking units, be these covalent bonds or the attractions of oppositely charged ions. Unfortunately, these linking forces are subject to catastrophic failure if they are put under too great a mechanical stress. In ionic oxides, for example, the attraction of the cation for its neighboring anions is very strong. External forces that would result in the anions being pushed away from the cations are resisted mightily—unless the force is strong enough to push the anions far enough from the cations so that their new neighbors are also anions. Suddenly, the attractions are replaced by repulsions, and the material shatters. **Brittleness** is an unfortunate characteristic physical property found in ceramics.

Because ceramics are hard and brittle, they cannot be prepared into useful objects by some of the methods used with organic polymers, such as extrusion through holes or into molds; unlike metals, they are difficult to machine into shape. They can be prepared from molten materials, but these often have impractically high melting points (often as high or higher than the container that would hold them, which might also be a ceramic). Therefore they often must be made by **sintering**: heating finely divided powders in furnaces for long periods of time at very high temperatures (up to 2500 °C), which are nonetheless well below the melting point of the oxide. Under these conditions the ions from different granules gradually diffuse and "weld" the granules together to give a coherent solid, but often voids remain to serve as sources of cracks. Much modern research in ceramics deals with ways to generate fine, homogeneous particles of starting oxides that will sinter together to give dense, strong ceramics with few voids, or will undergo unusual phase transitions when a crack begins that tend to fill the crack (as in zirconia). Or researchers try to overcome some of these problems by creating **composites** of ceramics with either metals or organic or other polymers.

Every solid oxide or solid salt of an oxo anion (in fact, every salt) has a *surface chemistry* that is, in some ways, distinct from the chemistry of the bulk material. Although, for example, the Ca^{2+} and O^{2-} ions in the center of a crystal of CaO do have the expected coordination number of 6, this regular ionic environment must end at the surface: On a clean surface face, the Ca^{2+} must have at least one vacant coordination site, as must each surface O^{2-} ion. As a consequence of these vacant sites, the Lewis acidity of the surface Ca^{2+} ions and the Lewis basicity of the surface O^{2-} ions are not diminished in the normal way. These surface ionic sites have powerful acidic and basic properties that (we shall see later in the chapter) can catalyze chemical reactions, adsorb molecules or ions with the appropriate charges or partial charges, and so on.

Isomorphous Substitution and Mixed-Metal Oxides. There are a number of technologically important oxides, the simplest formulas of which look like oxo salts (e.g., $BaTiO_3$), but that are not formed from acidic oxides and the structures of which do not involve identifiable oxo anions. Such compounds are called **mixed-metal oxides**. The mixed-metal oxides are best regarded as consisting of lattices of oxide ions together with two (or more) different types of metal ions. In many of these compounds, the oxide ions are close-packed, and one kind of metal ion may occupy tetrahedral holes and the other kind, octahedral holes in the close-packed structure. In some of the simplest mixed-metal oxides, a few percent of the metal ions of a second metal (and the corresponding oxide ions) simply enter into the appropriate sites of a metal oxide, leading to **isomorphous substitution** for the first ion and giving a type of solid solution. Dissolving some Cr_2O_3 into the lattice of alumina, Al_2O_3, gives color to the colorless alumina. The gemstone *ruby* is formed, which is used in lasers (Section 17.5). Similarly, dissolving some Ti_2O_3 in Al_2O_3 results in *sapphires.*

The type of cations that can substitute isomorphously for other cations in a metal oxide (or other type of salt) depends on (1) the size of those cations and (2) the charge of those cations. The **first principle of isomorphous substitution** states that *one ion may substitute for another in a lattice if the two ions have identical charges and differ in radii by not more than* 10–20%.[26]

Examination of Table C shows that quite a few sets of ions can be found that have the same charge and very similar ionic radii. Generally speaking, as a salt or mineral is formed by crystallization or the cooling of molten magma, there is little reason for one of these matched types of ions to be preferred over another: Naturally occurring minerals often have a mixture of cations present, which vary depending on the composition of the melt or solution from which the mineral grew. For example, Mg^{2+} and Fe^{2+} not only have identical charges but have very similar radii (86 and 92 pm, respectively). Thus the mineral olivine, with an ideal composition of Mg_2SiO_4, is often "impure" and can contain varying percentages of the Fe^{2+} ion in place of an equal number of Mg^{2+} ions. Thus the formula of olivine is often written $(Mg,Fe)_2SiO_4$ to indicate that there are two magnesium or iron(II) cations present per mole of silicate ion, although there is no definite relationship between the number of magnesium and the number of iron(II) cations.

This principle is often deliberately used in radiochemistry to isolate exceedingly tiny quantities of radioactive elements. Thus, to isolate a tiny amount of radium ion from a large amount of uranium ore, we could (using the solubility principles of Chapter 4) add sulfate ion to precipitate radium sulfate while leaving uranium and most other cations in solution. But this would pose two problems: (1) only a tiny amount of precipitate would be formed, which would be difficult to handle without losing it; and (2) there might be so much solution present that even the low solubility product of radium sulfate would not be exceeded, so no precipitate would form. These problems could be overcome by adding not only sulfate ion but also barium ion. A larger amount of barium sulfate would then precipitate and the radium would substitute isomorphously for the barium. This technique is known as coprecipitation; the barium ion is said to act as a *carrier* for the Ra^{2+}. (Of course, we would then be confronted with the formidable problem of separating the very similar barium and radium ions, but at least we would be working with a much smaller volume of material.)

Another illustration of this process occurs in a familiar experiment of growing crystals of ionic compounds. For example, large, beautiful octahedral crystals of *alum*, $KAl(SO_4)_2 \cdot 12H_2O$, are readily grown from solution. But there exist a whole series of similar compounds that also form large, beautiful octahedral crystals having the same lattice types and the same shapes (these compounds are said to be isomorphous). In the formula for alum, the K^+ ion can be replaced by other +1 cations of similar radius, such as Rb^+ and NH_4^+. The Al^{3+} ion can be replaced by numerous other +3 ions of similar radius, such as Cr^{3+} (giving purple crystals of *chrome alum*) or Fe^{3+} (giving pale violet crystals of *ferric alum*). The sulfate ion can even be replaced by the selenate ion. If nearly any combination of these three ingredients is mixed and crystallized, large crystals of an alum are formed; if a mixture, say, including both Al^{3+} and Cr^{3+} is used, crystals can be grown containing both ions, having whatever shade of light purple you desire! (Such a crystal is sometimes said to be *doped* with a certain percentage of the less abundant ion.)

Perhaps the most extensive case of isomorphous substitution occurs in the minerals monazite and xenotime, MPO_4, and bastnaesite, $MFCO_3$. In 1794, J. Gadolin investigated a mineral obtained from the village of Ytterby, Sweden; from this mineral he extracted a metal oxide that he named *yttria*. But other chemists, in working with this material, kept getting slightly different properties; eventually it was realized that this oxide was a mixture. So the mixture was separated; but the "pure" components also turned out to be mixtures. Eventually all of the *f*-block elements of the sixth period (except Pm), plus La and Y, turned out to be present in these minerals, isomorphously substituted for each other.[27] [Refer to Tables A–C to note the extreme similarities of these elements to each other.] As element after element was discovered, the chemists were harder and harder pressed to come up with new names for them. Thus it came to pass that the humble village of Ytterby has more elements named after it (four) than any of the great cities of the world.

The separation of these elements involves quite a complex process of ion exchange. The most troublesome case of isomorphous substitution, however, is that of the elements Zr (Pauling electronegativity 1.33, ionic radius 86 pm) and Hf (Pauling electronegativity 1.3, ionic radius 85 pm). Hafnium occurs isomorphously substituted in all zirconium compounds to the same extent ($\sim$2%), so there were no chemical discrepancies in the "pure" samples of zirconium prepared in 1825 and thereafter. Consequently, the presence of Hf went undetected for a whole century!

There are often very practical reasons for substituting one metal ion for another in an ionic lattice (or even in compounds such as metalloenzymes). The *s*- and *p*-block metal ions are all colorless. If they can be isomorphously substituted with *d*- and *f*-block ions, the spectra of these ions will give us information on the crystal field environment of the ions in the lattice or enzyme (Section 8.10). Metal ions also have other properties that tell us about their environment: some are fluorescent; some have unpaired electrons with magnetic properties that can be studied; others with appropriate nuclei can be studied by nuclear magnetic resonance (NMR) or Mössbauer spectroscopy. Some of the geologically and biochemically most important metal ions, such as K^+ and Zn^{2+}, however, are "silent metals" that lack most or all of these properties. Silent K^+ (radius = 152 pm) can often usefully be substituted with fluorescent and NMR active Tl^+ (radius = 164 pm); colorless Zn^{2+} (88 pm) is usefully replaced by colored Co^{2+} (88 pm); silent Ca^{2+} (114 pm) can be replaced by Eu^{2+} (131 pm), with seven unpaired electrons.

Nonstoichiometric Oxides. Perhaps the easiest type of mixed-metal-ion oxides to prepare are the **nonstoichiometric oxides** (or compounds, since these phenomena are not restricted to oxides), which are oxides of metals that have cations of more than one oxidation state. Iron(II) oxide as normally prepared gives an actual elemental analysis corresponding to approximately $Fe_{0.95}O$; there is a **defect** consisting of missing iron atoms. The compound, however, must still be electrically neutral; this is accomplished by the replacement of three Fe^{2+} ions by two Fe^{3+} ions, leaving a hole or vacancy but keeping the overall electroneutrality. Since most of the *d*-block metals have cations differing by only one unit of charge, there are many nonstoichiometric *d*-block metal oxides, including those of ideal composition TiO, VO, MnO, FeO, CoO, and NiO. If the +2 and +3 oxidation states are of comparable stability, a wide range of nonstoichiometry can exist: vanadium(II) oxide can range from an actual composition of $VO_{0.79}$ to $VO_{1.29}$, while nickel(II) oxide ranges only from $NiO_{1.000}$ to $Ni_{1.001}$. These nonstoichiometric oxides are often intensely colored due to charge-transfer transitions, which take an electron from one of the M^{2+} to one of the M^{3+} ions. Such transitions are of especially high probability when the two ions are of the same metal.

Electrical Conductivity. An electrical current consists of moving charged particles; in molten or dissolved ionic salts, the ions are free to move and to conduct an electrical current. In most *solid* metal oxides (and other salts), the ions are immobilized by the crystal lattice, so solid ionic compounds are generally not good conductors of electricity. However, some metal oxides have enhanced electrical conductivity due to the easy transfer of *electrons*. In the nonstoichiometric early *d*-block metal oxides of the fourth period, TiO_x and VO_x ($x \approx 1.0$), metal-like conductivity is present due to the overlap of metal t_{2g} orbitals on next-neighboring cations with each other to create extensive MOs through which *d* electrons can travel. This is not possible for the later fourth-period *d*-block metal oxides, in which the *d* orbitals are too contracted to overlap with each other.[28]

Overlap of *d* orbitals also plays a role, not only in making an early *d*-block oxide, TiO_2, a semiconductor with a band gap of 3.23 V, but also a *photocatalyst*. (For other examples of photocatalysts, see Section 17.6.) Titanium dioxide has the ability to harness the energy of sunlight to promote the oxidation in air of almost any organic vapor. In TiO_2 light promotes an electron from the highest occupied molecular orbital (HOMO), an oxide ion lone pair, to the delocalized lowest unoccupied molecular orbital (LUMO), based on Ti 3*d* orbitals, producing a transient excited state resembling $(Ti^{4+})(e^-)(O^{2-})(O^-)$. The excited electrons react with atmospheric O_2 to generate superoxide ions, $O_2{}^-$, which are potent oxidizers of organic vapors. The seven valence electron O^- ion reacts with atmospheric H_2O, removing a hydrogen atom from it to give an HO^- ion in the solid lattice and a hydroxide free radical, ·OH, in the atmosphere; the hydroxide radical is also a potent oxidizer. Titanium dioxide is of great interest to environmental chemists because of its seeming ability, in sunlight, to oxidize almost anything (nitrogen oxides, cigarette smoke, grease, bacteria, and chlorobenzenes).[29]

Nonstoichiometry and improved electrical conductivity can be created by incorporating other metal ions into metal-oxide lattices. It is possible to heat NiO with a source of Li_2O in an oxygen atmosphere; under these conditions a semiconducting nonstoichiometric mixed-metal oxide, Li_xNiO_{1+x}, is produced, in which some of the Li^+ ions are doped into positions normally occupied by Ni^{2+} ions. Since the overall charge balance must be maintained, an equal number of other Ni^{2+} ions must simultaneously be oxidized by the oxygen to Ni^{3+} ions. Thus it becomes feasible for electrical semiconduction to occur by the thermally activated process of electrons hopping between nickel ions of different charges.

In certain types of lattices and under certain conditions *ions* can move in solids, and hence conduct an electric current. There are materials in which the smaller type of ion can move readily, while the larger counterions maintain the rigidity of the solid-state lattice as a whole. Such materials are variously known as solid electrolytes, superionic conductors, and fast-ion conductors. These materials are important for their potential uses in high-energy density batteries and fuel cells and in lasers.[30]

Oxide ions can be the mobile phase in a lattice composed of larger cations, for example, between 500 and 1000 °C in zirconia (ZrO_2) doped with CaO. When a Ca^{2+} ion replaces a Zr^{4+} ion in the zirconia lattice, charge neutrality of the lattice is maintained by deleting an oxide ion as well. The resulting *vacancies* in the lattice act as tetrahedral holes into which other oxide ions can move (and hence carry current) if enough thermal energy is available.

It is more common, of course, for the anions to be larger than the cations. Some solid *iodides* allow superionic conduction of soft-acid cations into tetrahedral holes. For example, the structures of copper(I) and silver(I) tetraiodomercurates, Cu_2HgI_4 and Ag_2HgI_4, feature four-coordinate soft-acid metal ions but only three-coordinate iodide ions; one-fourth of the tetrahedral holes about the large (approximately ccp) iodide ions are vacant. At low temperatures, the vacant holes are ordered and the cations cannot move into them; the compounds have low conductivity and light colors (red for the Cu^+ and yellow for the Ag^+ salt). At a specific temperature (50 °C for the Ag^+ salt) a phase transition occurs, in which the lattice opens up enough for the cations to move between the iodide ions; the cations and vacant holes are then disordered, the motion increases the electrical conductivity dramatically, and the color darkens (to black in the Cu^+ salt and orange in the Ag^+ salt). Similar types of phenomena are found in solid AgI itself, as well as in $RbAg_4I_5$.

Sometimes the holes in which cations can move can be quite large indeed. The β- and β''-aluminas are two classes of mixed-metal oxides built from the oxides of aluminum and sodium; sodium β-alumina has an approximate composition of $Na_{1.2}Al_{11}O_{17.1}$. In these oxides, the Al^{3+} ions are present in dense Al_2O_3 slabs separated from each other by widely spaced oxide-ion "pillars"; there are broad vacancies between these oxide ions. Parked like cars among the pillars of a parking garage are a few of the Na^+ cations, which are quite free to move, and conduct electricity about as well as in aqueous Na^+Cl^-. This property was discovered in 1965 by Yung-Fan Yao and J. T. Kummer of Ford Motor Company, who applied it to devise a revolutionary new battery. This battery operated at 300 °C, using molten sodium as the positive electrode and molten sulfur as the negative electrode; sodium–β-alumina served as a ceramic membrane (analogous to a salt bridge but lighter and less reactive) to separate these reactive liquids while allowing ions (but not electrons) to pass between them. Because of the very open spaces between the alumina slabs, ion exchange is easily carried out on these aluminas: immersion of sodium–β-alumina in a molten salt for a few minutes allows complete exchange of the Na^+ ions for K^+, Rb^+, Cs^+, Cu^+, Ag^+, and Tl^+; divalent and even trivalent cations can exchange for Na^+ ions, then carry current, in the even more open sodium–β''-alumina.

13.5 Spinels, Perovskites, High-Temperature Superconductors, and Cooperative Magnetic Properties in Mixed-Metal Oxides

An important class of mixed-metal oxides are known as the **spinels**, AB_2O_4. (This class is named after the mineral spinel, $MgAl_2O_4$.) In spinels, the oxide ions are ccp. Normally, the A metal ions are +2 charged ions of radius between 80 and 110 pm, which occupy one-eighth of the tetrahedral holes in the oxide-ion lattice. Spinels are known in which the A metal ions are the +2 ions of Mg, Cr, Mn, Fe, Co, Ni, Cu, Zn, Cd, and Sn. Normally, the B metal ions are +3 charged ions of radius between 75 and 90 pm, which occupy one-half of the octahedral holes; these include the +3 ions of Ti, V, Cr, Mn, Fe, Co, Ni, Rh, Al, Ga, and In.

Of particular interest are spinels in which both A and B are the same element. These spinels have stoichiometry $M^{2+}(M^{3+})_2(O^{2-})_4$ or M_3O_4, and seem to share fractional oxidation numbers because of the presence of two different oxidation states in the same compound. The spinels Mn_3O_4, Fe_3O_4, and Co_3O_4 are all intensely black, and much darker than the simple oxides of these metals, due to charge-transfer transitions; Fe_3O_4 has 1 million times the electrical conductivity of Fe_2O_3.

These spinels exhibit the two most common forms of *cooperative* magnetic properties, ferromagnetism and antiferromagnetism (Section 12.4). Unfortunately, it is not so easy to predict in oxides whether the direct ferromagnetic or the indirect antiferromagnetic interaction will prevail. Among the mixed-metal spinels, Mn_3O_4 and Co_3O_4 are antiferromagnetic, while only Fe_3O_4 (the mineral *magnetite* or *lodestone*) shows the much more important property of ferromagnetism. Microscopic crystals of magnetite have been found in a number of different living organisms, including bacteria, pigeons, salmon, and (perhaps) in human brain tissue, which might confer sensitivity to magnetic fields similar to the earth's.[31] Similarly, among the simple *d*-block dioxides, VO_2 and MnO_2 are antiferromagnetic, while CrO_2 is used in magnetic storage devices for its ferromagnetic properties (both magnetite and chromium dioxide have Curie temperatures above room temperature). Spinels in general are very important in the solid-state electronics industry for their electric and magnetic properties.

Another important class of mixed-metal oxides are the **perovskites**, ABO_3, of which the prototype is $CaTiO_3$ (perovskite). It has an unusual ccp lattice of oxide *and calcium* ions (Fig. 13.6), in the octahedral holes of which the much smaller Ti^{4+} ions can "rattle around".

There are many other perovskites, which can be generated from the structure of $CaTiO_3$ through the application of either the first or the **second principle of isomorphous substitution**. This latter principle allows more versatility: although substituting ions must still be about the same size as the ions replaced in order not to change the lattice type, within certain strict limits the *charge* of the entering ion need not be identical to the charge of the departing ion. The basic principle is that the *total* charge of the replacing ions must equal the total charge of the replaced ions. This means that isomorphous substitution can occur even if the new ion C have a charge one greater than the old ion A, *if there is simultaneous substitution by a new ion D with a charge one less than the old ion B*. This conserves the electroneutrality of the salt, since the sum of charges of the new ions C and D equals the sum of charges of the old ions A and B. Table 13.6 shows sets of common ions that frequently substitute for one another.

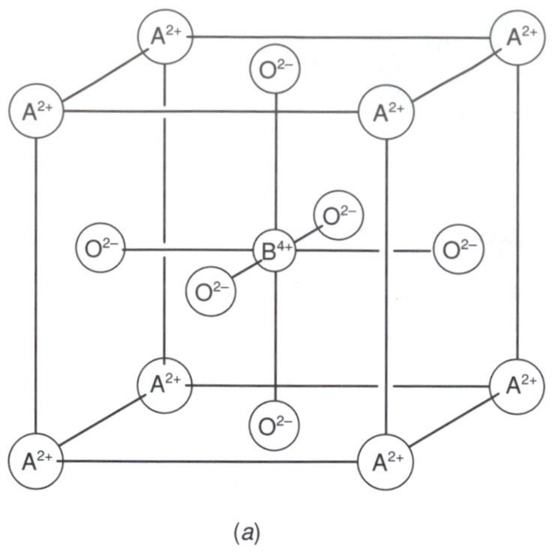

(a)

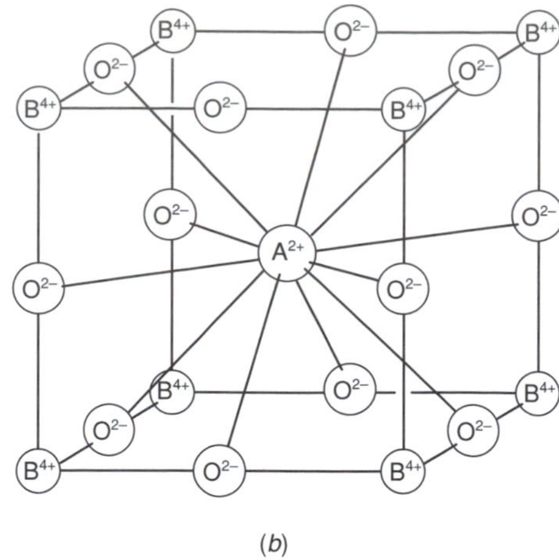

(b)

Figure 13.6
The structure of perovskites, ABO_3 (e.g., A=Ca^{2+}, B=Ti^{4+} in $CaTiO_3$), shown (*a*) with B^{4+} at the center of the unit cell and emphasizing the octahedral coordination of B^{4+}, (*b*) with A^{2+} at the center of the unit cell and emphasizing the dodecahedral (12) coordination of A^{2+}.

Table 13.6
Sets of Common Ions Suitable for Isomorphous Substitution

Range of Radii (pm)	+1 Charge[a]	+2 Charge[a]	+3 Charge[a,b]	+4 Charge[a]
54–67			Al^{3+} (67)	Si^{4+} (54)
74–92	Li^+ (90)	Mg^{2+} (86) Fe^{2+} (92)	Fe^{3+} (78)	Ti^{4+} (74)
100–117	Na^+ (116)	Ca^{2+} (114)	Ln^{3+} (100–117)	
149–152	K^+ (152)	Ba^{2+} (149)		

[a] Radii listed in parentheses are in picometers (pm).
[b] Lanthanide ions = Ln^{3+}.

Example 13.6
Which of the following minerals could arise by isomorphous substitution processes in leucite, $KAlSi_2O_6$? (a) $KYSi_2O_6$; (b) $RbAlSi_2O_6$; (c) $BaBeSi_2O_6$; or (d) $BaAlSi_2O_6$.

SOLUTION:

First, note that *the total charge of all the cations must equal the total charge of all the anions.* This requirement is implicit in the two principles of isomorphous substitution, but it also gives a separate simple test in this example: If the total charge of all cations going into the structure is not the same as the total charge of all cations coming out, the new compound cannot exist, let alone be isomorphous. In (a), Y^{3+} replaces Al^{3+}; in (b), Rb^+ replaces K^+; in (c), Ba^{2+} and Be^{2+} replace K^+ and Al^{3+}; in (d), Ba^{2+} replaces K^+. Substitution product (d) cannot exist.

Also note that *in order for the substitution to be isomorphous, the total number of cations going in must be nearly equal to the total number coming out.* (If the two numbers are slightly unequal, a nonstoichiometric compound is formed.) All four possible substitution products given above obey this principle.

Once these two principles are satisfied, the two principles of isomorphous substitution are satisfied if the cations going into the replacement structures are within 10–20% of the radii of the cations coming out. In (a), Y^{3+} has a radius of 104 pm, which is too much bigger than Al^{3+} (67 pm) for the substitution to be isomorphous. However, (b) and (c) are satisfactory, since in (b) Rb^+ (166 pm) is close in size to K^+ (152 pm), and in (c) Ba^{2+} (149 pm) is close to K^+ (152 pm), and Be^{2+} is close to Al^{3+}.

The perovskite structure type requires that the sum of the charges of the A and B metal ions be +6 and that they be quite different in size from each other. Thus, other perovskites such as $Li^+Nb^{5+}(O^{2-})_3$ are also known, as are fluoride perovskites such as $Na^+Fe^{2+}(F^-)_3$. If the temperature is not too high, the Ti^{4+} ions of perovskite itself tend to be off the center of the lattice unit cell, giving rise to an electric charge separation or dipole. Such materials are known as **ferroelectrics**. Application of mechanical pressure to one side of a perovskite crystal causes the Ti^{4+} ions to migrate, generating an electrical current; application of an electric current causes mechanical motion of the ions. The pressure effect, known as the **piezoelectric effect**,[32] makes perovskites and the form of SiO_2 known as quartz useful in converting mechanical energy to electric energy, as in microphones, sonar, and vibration sensors, or vice versa. The conversion of electrical energy to mechanical energy is useful in sonic and ultrasonic transducers and in headphones and loudspeakers.

Example 13.7

Verify that the unit cell drawings of perovskite shown in Figure 13.6 show the ions in the proper stoichiometry.

SOLUTION:

In the structure of a crystalline solid, unit cells are repeated in all three dimensions. Consequently, any atom or ion appearing on the *surface* of the unit cell also appears in the neighboring unit cell(s). More specifically: (a) each atom or ion found at the *corner* of a unit cell is also shared with seven adjacent unit cells, so it counts as only *one-eighth* of an atom or ion in the given unit cell; (b) each atom or ion found along the *edge* of a unit cell is shared among four unit cells, so it is *one-fourth* in the given unit cell; (c) each atom or ion found inside the *face* of a unit cell is *one-half* in that cell; and (d) each atom or ion found in the *interior* of a unit cell is *completely* in that unit cell.

Hence, in the perovskite unit cell as drawn in Figure 13.6(*a*) we find eight A^{2+} ions on corners of the unit cell, which gives *one* A^{2+} per unit cell (the same result occurs with the unit cell of Figure 13.6(*b*), since there is one A^{2+} in the interior). In Fig. 13.6(*a*), we find one B^{4+} in the interior of the unit cell, which counts as one B^{4+}. The six O^{2-} ions in Fig. 13.6(*a*) are in faces, so they count as three O^{2-} ions, which add up correctly for the stoichiometry ABO_3. (In Fig. 13.6(*b*) there are 12 O^{2-} ions along edges, shared among four touching unit cells, so that this also totals three O^{2-} ions per unit cell.)

Tungsten bronzes, Na_xWO_3, $(0.32 < x < 0.93)$, made by reduction of sodium tungstate, are perovskite-like materials (when $x \geq 0.44$) that are deficient in sodium ions. Their conductivity is based on movement of electrons in delocalized MOs, and ranges from metallic to semiconducting, depending on the value of x—as does their metallic luster and color, which ranges from blue ($x \approx 0.4$) to violet ($x \approx 0.57$) to red, orange, and yellow ($x \approx 0.8$).

High-Temperature Superconductors. In 1987,[33] the scientific world was very much surprised to learn that mixed-metal oxides can show the property of superconductivity (Section 12.4), previously considered characteristic only of metals and alloys; with critical temperatures, T_C, at or below 23 K, these require the use of a very expensive coolant, liquid helium. So it was significant to expand the search to another class of materials, the mixed-metal oxides, and even more significant that the very first of these oxides, $La_{1.85}Sr_{0.15}CuO_4$, showed a higher T_C, 30 K, than any of the metals or alloys. Very soon a mixed-metal oxide system with a T_C of about 90 K was discovered, $YBa_2Cu_3O_7$, familiarly known as "123" for the ratios of the metal ions in the formula. With a T_C exceeding 77 K, the much cheaper and more abundant coolant, liquid nitrogen, could be used. This oxide is prepared by sintering the oxides Y_2O_3, CuO, and the oxide precursor $BaCO_3$ at 950 °C for an extended time, followed by a slow cooling and exposure to O_2 at 500–600 °C, whereupon one-third of the Cu^{2+} ions are evidently oxidized to Cu^{3+} ions. Above T_C these superconductors show the antiferromagnetic properties that are common in metal oxides.

The structure of $YBa_2Cu_3O_7$ is related to that of perovskite; this relationship can be seen if one takes three adjacent A-centered unit cells (Fig. 13.7), and allows Y^{3+} to occupy the A position in the central cell and Ba^{2+} to occupy the A positions in the outer cells. The other alteration is that three perovskite unit cells would contain a total of nine, not seven, oxide ions, and would require an impossibly high oxidation number for at least one of the metals. The actual $YBa_2Cu_3O_7$ structure deletes oxide ions from each of the three cells: along the front and back outer edges in the two Ba centered cells, and along the middle edges in the Y centered cell. This, of course, changes the polyhedra of the copper ions. The d^8 Cu^{3+} ions found on the top and bottom surfaces of the new, larger unit cell of Figure 13.7 (those closer to Ba than to Y) show square planar coordination; bridging oxide ions link these squares into polymeric *chains*, as shown by the arrows in the figure. In contrast, the d^9 Cu^{2+} ions found in the center of the new unit cell (sandwiched between Ba and Y) lose only one oxide neighbor, so show square pyramidal coordination; the square pyramids are also linked by bridging oxide ions, but this time in two dimensions to form *planes* of Cu^{2+} polyhedra. Experimentally, it is found that (a) the diamagnetic Y^{3+} ions can be replaced with most of the lanthanide ions without altering the superconductivity (even though these ions are mostly paramagnetic!); (b) slight replacement (doping) of the Cu^{3+} ions from the chains of CuO_4 square planes with other +3 charged ions such as Ga^{3+} lowers T_C, but not as severely as does (c), slight replacement of the Cu^{2+} ions from the planes of CuO_5 square pyramids with divalent ions such as Zn^{2+}. Reduction of the oxygen content of the superconductor can occur until the composition $YBa_2Cu_3O_6$ is reached. This reduction does not alter the planes of CuO_5 square pyramids, but does reduce the chains of Cu^{3+} containing CuO_4 squares to sticks of linear two-coordinate Cu^+ containing CuO_2 units, and does destroy the superconductivity property.

A variety of related "cuprate" high-temperature superconductors has since been discovered and discussed.[34] Both $YBa_2Cu_4O_8$ and $Y_2Ba_4Cu_7O_{15}$ add additional

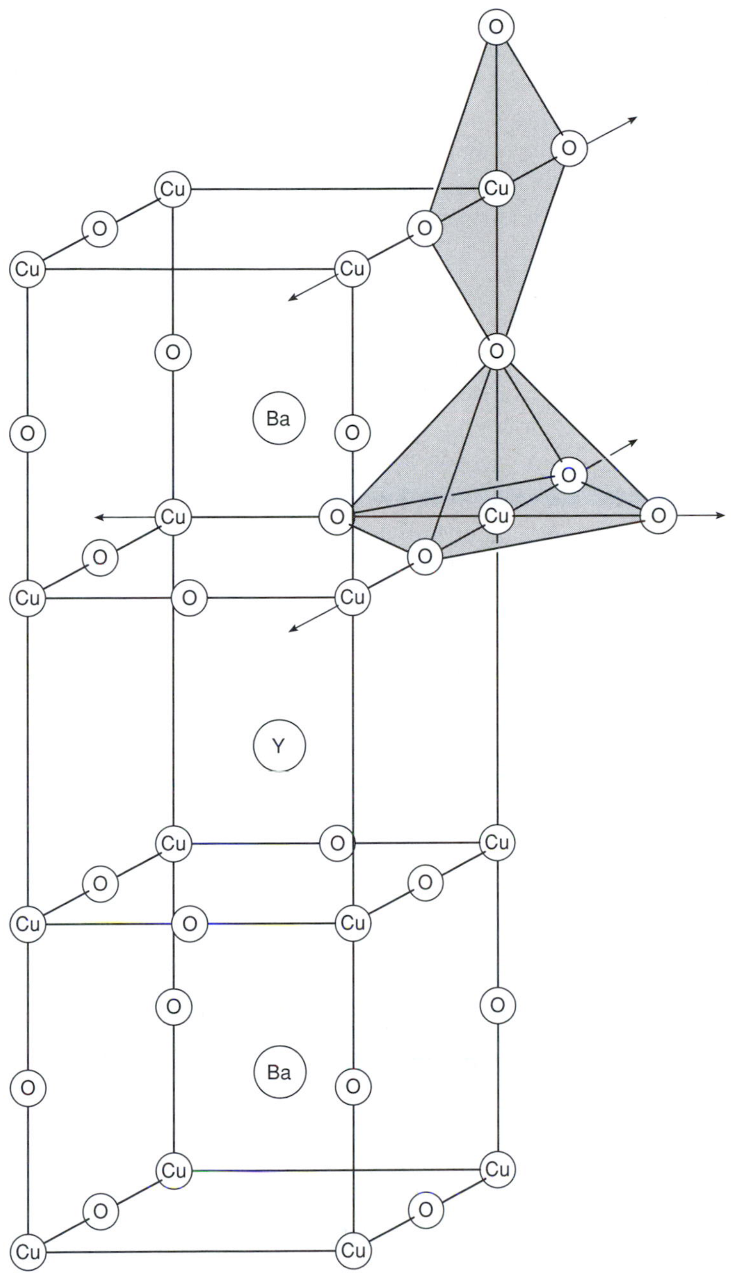

Figure 13.7
The structure of the high-temperature superconductor $YBa_2Cu_3O_7$ as related to three A-centered unit cells of the perovskite structure. At the top the oxide ions from neighboring unit cells have been added in to show the CuO_4 square planar units, which are linked into chains by shared bridging oxide ions in the direction shown by the top arrows. Similarly, nearer to the center the square pyramidal coordination of another copper ion is completed; again sharing of oxide bridges gives rise to sheets extending in the two dimensions indicated by the four arrows in the center.

chains of CuO_4 squares; but the most dramatic improvements in T_C result from incorporating heavy *p*-block metal ions of mercury, thallium, and bismuth, resulting in idealized formulas including $HgBa_2CuO_4$, $Bi_2Sr_2CuO_6$, $Tl_2Ba_2CuO_6$, and $TlBa_2Cu_2O_5$; the highest T_C values yet found, 125 K and 133 K, belong to superconductors having three layers of Cu ions, $Tl_2Ba_2CaCu_3O_{10}$ and $HgBa_2CaCu_3O_8$, respectively. These structures lack chains of CuO_4 squares but still have the planes of cuprate units, confirming their importance in superconductivity. (The first oxide superconductor, $La_{1.85}Sr_{0.15}CuO_4$, also has planes of ordinary Jahn–Teller-distorted CuO_6 octahedra, although the average oxidation state of the Cu^{2+} here is unambiguously greater than +2.)

It is fascinating, of course (and potentially profitable), to speculate about the possible applications of high-temperature superconductors.[35] Some applications involving low magnetic fields and electric current density and using liquid nitrogen cooling are near to realization: microwave devices such as microwave cavities and wave guides and frictionless bearings; superconducting quantum interference devices (SQUIDs) already use high-temperature superconductors.[36]

High magnetic fields cause a breakdown of the superconductivity of these materials at 77 K, so the idea of a magnetically levitated train running at 400 miles per hour using cheap liquid-nitrogen cooling is not just around the corner. But materials scientists continue to be challenged by the characteristics of such oxides: the high temperatures required to get their components to react, and the long times required for so many different metal ions to diffuse over long distances between grains before they can be thoroughly mixed to give homogeneous products; the difficulties of insuring that only one product is produced when so many mixed-metal oxides are possible (with such nonvolatile, insoluble products, normal methods of purification after synthesis are impossible); the brittleness of the products, which are very difficult to draw into wires or to deposit as films (although substantial progress in producing flexible cuprate tapes has occurred).[37] In addition, with the presence of the high Cu^{3+} oxidation state, $YBa_2Cu_3O_7$ is air- and moisture-sensitive, and the production of the thallium- and mercury-containing cuprates is limited by the toxicity and high-temperature volatility of Tl_2O and HgO, which have low lattice energies, and hence relatively high vapor pressures at temperatures at which the other oxides react to give superconductors.

These difficulties have increased the importance of finding alternate routes to high-temperature sintering of oxides for the production of these solids.[38] Some methods will be discussed in Section 14.6. For now, we mention the possibility of obtaining better mixing of the component metal ions by the process of *homogeneously coprecipitating* them at low temperatures in one isomorphously substituted product, or if that is not feasible, of simultaneously precipitating separate but very finely divided, hence well-mixed, metal-ion-containing products. These products need not be oxides if they can be converted to oxides at temperatures below normal sintering temperatures.

Thus, simultaneous precipitation of mixed gelatinous metal *hydroxides* would be good, because these can generally be dehydrated to oxides at comparatively moderate temperatures. Unfortunately, the feeble acidity of Ba^{2+} does not match the weak acidity of Y^{3+} and Cu^{2+}; $Ba(OH)_2$ is a soluble metal hydroxide. Better success may be achieved using the larger carbonate ion, $CO_3{}^{2-}$; its size match with Ba^{2+} allows precipitation of carbonates of all three metals. These metals can be heated to give oxides plus $CO_2(g)$. Success has also been achieved by coprecipitating the *oxalates* of these three metal ions; oxalates (salts of the $C_2O_4{}^{2-}$ ion) are easily oxidized to $CO_2(g)$ and oxides. Because of the charge and size differences among the three metal ions, we may not be expected to be so lucky as to find isomorphous substitution of the three ions in the same salt, however. Other workers have attempted (fruitlessly so far) to find some exotic complex ions that would precipitate containing the three metal ions in the proper ratios, and from which the organic ligands could simultaneously be oxidized. Some success has been achieved by heating a mixture of the three acidic cations, oxalic acid, $H_2C_2O_4$, and urea, NH_2CONH_2. The latter compound slowly hydrolyzes to CO_2 and ammonia, which increases the pH and precipitates the three cations as an intimate mixture of hydroxides, oxalates, and carbonates, all of which are converted to oxides on heating in air at 900 °C.[39]

13.6 Polysilicates: Basic Structural Types, Uses, and Chemistry

Of overwhelming importance in the geochemistry of the earth's crust are the compounds formed by the reaction of the acidic oxide silica, SiO_2, with various basic metal oxides. These are not really mixed-metal oxides such as were discussed in Section 13.5, since definite silicon oxo anions, with covalent Si–O bonds, are present. But most of them do not have the simple silicate ion $SiO_4{}^{4-}$ that we discussed in Chapter 2, but rather have many two-coordinate oxygen atoms covalently linking different silicon atoms into oligomeric or one-, two-, or three-dimensional polysilicate ions. Regardless of the degree of polymerization of the polysilicate ions, these are finally linked into three-dimensional ionic lattices by counterions. Nonetheless, we will see some differences in physical properties caused by the different degrees of polymerization of the covalent bonding in the polysilicate ions themselves. These more complex silicates are the most common of the **polyoxometallates** or **polynuclear** oxo anions (the structural units are often called the "nuclei" of the oxo anions). After examining the polysilicates in some detail, at the end of this chapter we will look briefly at the polyoxometallate anions of other elements.

The simple silicate ion $SiO_4{}^{4-}$ (also called *orthosilicate* ion, to distinguish it from the polysilicate ions) is not found in a wide variety of minerals: It is a very strong base that cannot persist in aqueous solution, but it does occur in nature as insoluble salts of acidic cations. Some of the mineral forms containing the orthosilicate ion are phenacite, Be_2SiO_4; willemite, Zn_2SiO_4; zircon, $ZrSiO_4$; the garnets, $(M^{2+})_3(M^{3+})_2(SiO_4)_3$ (M^{2+} = Ca, Mg, and Fe; M^{3+} = Al, Cr, and Fe); and olivine, $(M^{2+})_2SiO_4$ (M^{2+} = Mg and Fe). Although with these relatively acidic cations there is doubtless some covalent character to the M–O bonds, we will treat these as salts. Isomorphous substitution is very common in minerals such as these, so the formula of, for example, olivine is often written $(Fe,Mg)_2SiO_4$, with commas separating ions that can and do substitute for each other in any proportion.

Chain Oligosilicates. We can suppose that the process of linking simple silicate units is one of acid–base neutralization. In order to make room for a bridging oxygen from one $SiO_4{}^{4-}$, an oxide ion must be removed from another:

$$2\,SiO_4{}^{4-} + 2\,H^+ \rightarrow [O_3Si\text{–}O\text{–}SiO_3]^{6-} + H_2O \quad (13.27)$$

In the resulting disilicate ion, the basic structural unit is now $SiO_{3.5}{}^{3-}$, with one-half vacant coordination site per structural unit and a lower charge density than pres-

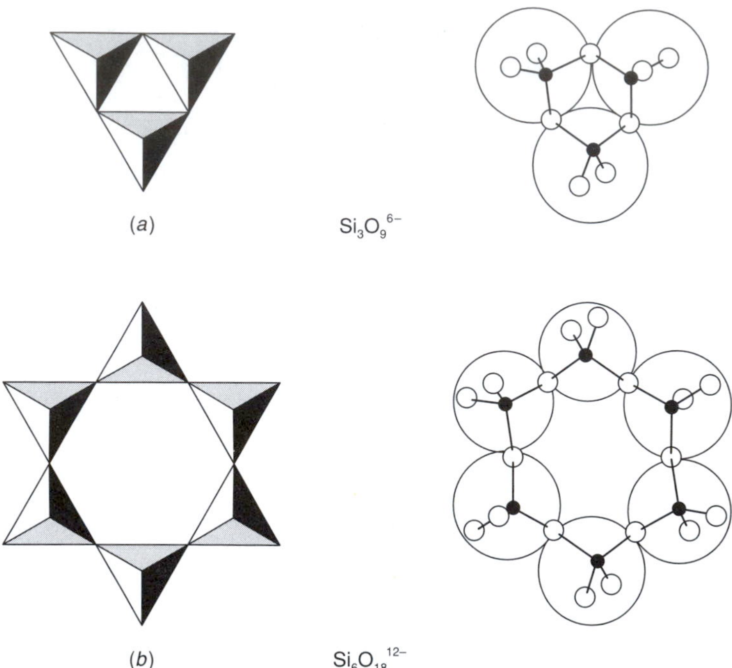

Figure 13.8
Schematic and ball-and-stick models of two cyclic polysilicate ions: (*a*) $Si_3O_9{}^{6-}$ and (*b*) $Si_6O_{18}{}^{12-}$. In the ball-and-stick model, the small filled circles represent Si and the large open circles represent O. In the schematic diagram, each tetrahedron represents the four oxygen atoms coordinated to a silicon atom at the center; bridging oxygens are located at the intersections of neighboring tetrahedra. Structural units, $[SiO_3]^{2-}$, enclosed in circles. [Adapted from N. N. Greenwood and A. Earnshaw, *Chemistry of the Elements*, 2nd ed., p. 349, Copyright © 1997, Butterworth-Heinemann, Oxford.]

ent in the orthosilicate ion; the disilicate ion is consequently less basic than $SiO_4{}^{4-}$. (Note that this polysilicate ion, as well as the others to be discussed, has one negative charge for each nonbridging or **terminal** oxygen atom, as would be suggested by their formal charges.) The disilicate ion is uncommon in nature, being found in the rare mineral thortveitite, $Sc_2Si_2O_7$. Longer chain structures (trisilicates and tetrasilicates) are even rarer.

Cyclic Oligosilicates. More commonly, the ends of these long chains of silicate groups come together to eliminate oxide ions and form *cyclic silicates* (Fig. 13.8):

$$Si_3O_{10}{}^{8-} + 2\,H^+ \rightarrow cyclo\text{-}Si_3O_9{}^{6-} + H_2O \tag{13.28}$$

These cyclic *metasilicates* are oligomers of the $SiO_3{}^{2-}$ structural unit, which has one vacant coordination site per Si nucleus and a still lower charge density of −2 per silicon nucleus. Cyclic trimers, $[SiO_3]_3{}^{6-}$, and hexamers, $[SiO_3]_6{}^{12-}$, are most common and are found in such minerals as benitoite, $BaTi(Si_3O_9)$, and beryl, $Be_3Al_2(Si_6O_{18})$.

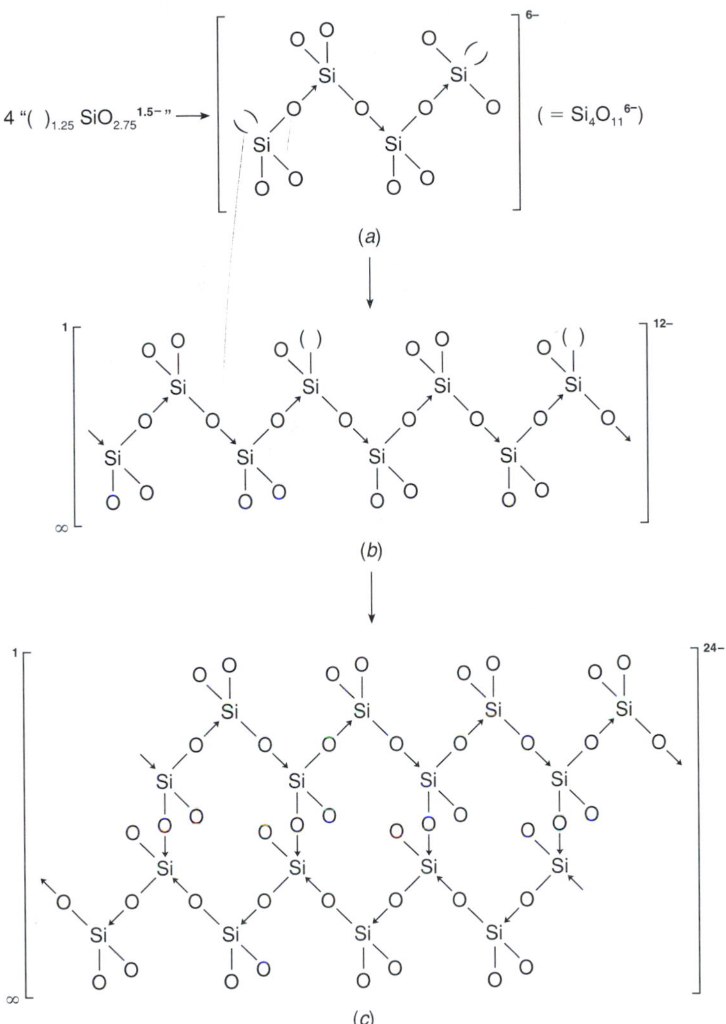

Figure 13.9
Hypothetical steps in the polymerization of an "$( )_{1.25}SiO_{2.75}{}^{1.5-}$" structural unit to give: (*a*) a $[( )_2Si_4O_{11}{}^{6-}]$ simplest formula unit; (*b*) linear polymerization of the simplest formula unit to give chains of ${}^{1}_{\infty}[( )Si_4O_{11}{}^{6-}]$; and (*c*) side-to-side cross-linking of two chains to eliminate the final vacant coordination site and give the double-chain structure, ${}^{1}_{\infty}[Si_4O_{11}{}^{6-}]$.

Chain Polysilicates. Just as we previously saw among the polymerization isomers of oxides such as SO_3, bridging oxygens may alternately be used to form chain polymers, ${}^{1}_{\infty}[SiO_3{}^{2-}]$ or $[SiO_3]_n{}^{2n-}$, rather than oligomers. Chain polymers are more common than cyclic oligosilicates and result in the important class of minerals known as the pyroxenes, which includes enstatite, $MgSiO_3$; diopsite, $CaMgSi_2O_6$; spudomene, $LiAlSi_2O_6$; and pollucite, $CsAlSi_2O_6$. The negative charge still remains at −2 per silicon nucleus.

Further reduction in anion basicity is achieved by a slight expulsion of charge (i.e., oxide ions) from the $SiO_3{}^{2-}$ structural unit found in chains. If this is done once per every four structural units, one obtains a fractional structural unit, $SiO_{2.75}{}^{1.5-}$, with 1.25 vacant coordination sites per nucleus, which is capable of more extensive linking. To envision this, it is better to eliminate decimals from the above structural units and examine the tetranuclear simplest formula unit, $Si_4O_{11}{}^{6-}$ [Fig. 13.9(*a*)]. Three vacant coordination sites are used to link these four nuclei, which leaves two sites still vacant. One of these is used to link the tetrameric units into chains [Fig. 13.9(*b*)]; the last is used to link one chain to its neighbor chain, giving rise to the *double-chain polysilicates*, which contain the polymerized ${}^{1}_{\infty}[Si_4O_{11}{}^{6-}]$ or $[Si_4O_{11}]_n{}^{6n-}$ anion [Fig. 13.9(*c*)].

The double-chain amphibole asbestos minerals[40] such as crocidolite, $Na_2Fe_5(OH)_2[Si_4O_{11}]_2$, and amosite, $(Mg,Fe)_7(OH)_2[Si_4O_{11}]_2$, have long been prized for their fire and heat resistance and for their fibrous nature (undoubtedly rooted in the long-chain structure of the anion), which allows the weaving of insulating, nonflammable garments, as well as the fabrication of more than 3000 other products. Now these minerals are feared, since it has been realized that the inhalation of the tiny fibers of asbestos often leads, after 20 or 30 years, to asbestosis (nonmalignant scarring of the lungs) or rare cancers such as mesothelioma. The Environmental Protection Agency (EPA) has proposed a complete ban on asbestos products. Finding materials to replace asbestos in its 3000 uses poses quite a challenge to the industrial inorganic chemist; a slightly different double-chain silicate, xonotlite, $Ca_6(OH)_2(Si_6O_{17})$, has come into use in construction for this purpose.

Layer Polymeric Silicates. If the side-to-side linking of chains is continued indefinitely, still further oxide ions are eliminated to give a structural unit of $SiO_{2.5}{}^{-}$ of still lower basicity with more vacant coordination sites per Si nucleus (1.5) and a greater ability to link to give a layer or sheet silicate, ${}^{2}_{\infty}[Si_4O_{10}{}^{4-}]$ or $[Si_4O_{10}]_n{}^{4n-}$. The layer polymeric silicates have been extremely important to humans for millennia, due in part to their ready cleavage into thin sheets and other properties that can be related to their layer structures. These minerals include clay minerals such as kaolinite or china clay, $Al_4(OH)_8(Si_4O_{10})$; pyrophyllite, $Al_2(OH)_2(Si_4O_{10})$; talc or soapstone, $Mg_3(OH)_2(Si_4O_{10})$; and serpentine, $Mg_6(OH)_8(Si_4O_{10})$. The contribution of the aluminum or magnesium hydroxide layers to the properties of these minerals will be discussed in Section 13.7; for now we note that the $Mg_6(OH)_8$ layers in serpentine impart a curvature to the sheets of this mineral, which therefore wrap into fibrous tubules. In this form, they are the chrysotile form of asbestos, which is 95% of the asbestos used in the United States and which is not as dangerous as the double-chain forms.

If this process is continued and sheets are linked into three-dimensional polymers, *all* of the oxide ions are ultimately eliminated, *all* of the remaining oxygens are converted into bridging oxygens, and the uncharged oxide silica, ${}^{3}_{\infty}[SiO_2]$, is produced. Having no negative charge, it is no longer basic at all; it is an acidic oxide.

As summarized in Figure 13.10, the successive steps of polymerization of the simple silicate ion have resulted in (1) a successive reduction of the overall ratio of oxygen atoms to silicon atoms, from 4:1 in the orthosilicate ion to 2:1 in silica; (2) an increase in the number of vacant coordination sites, hence linkage possibilities, per silicon nucleus; and (3) a decrease in the charge per silicon nucleus in the anion. The chemical formulas of minerals are often written to set off the polysilicate ions from other anions, such as hydroxide, that may also be present; if this is done it is relatively easy to interpret the formula to tell what kind of polysilicate ion is present.

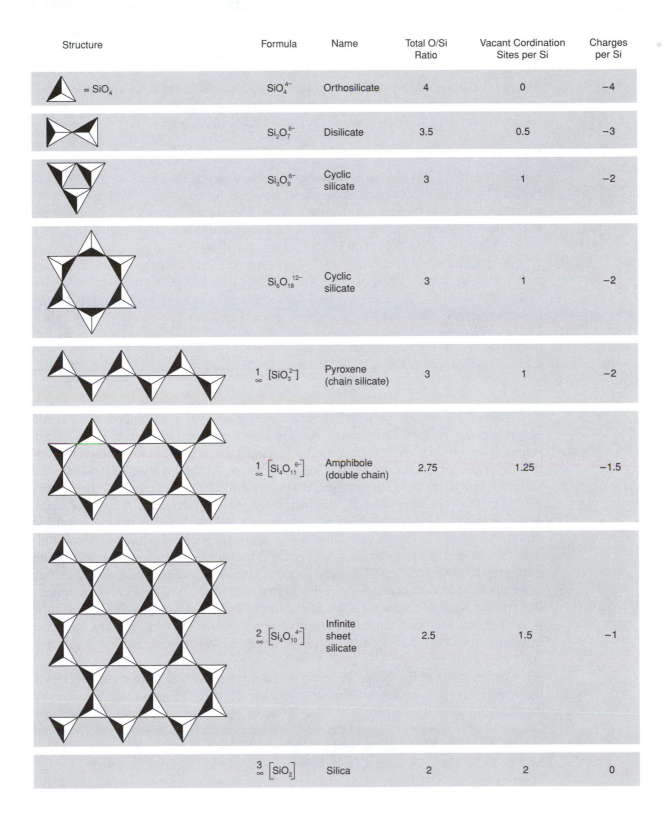

Structure	Formula	Name	Total O/Si Ratio	Vacant Cordination Sites per Si	Charges per Si
 = SiO_4	SiO_4^{4-}	Orthosilicate	4	0	−4
	$Si_2O_7^{6-}$	Disilicate	3.5	0.5	−3
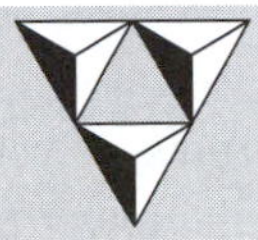	$Si_3O_9^{6-}$	Cyclic silicate	3	1	−2
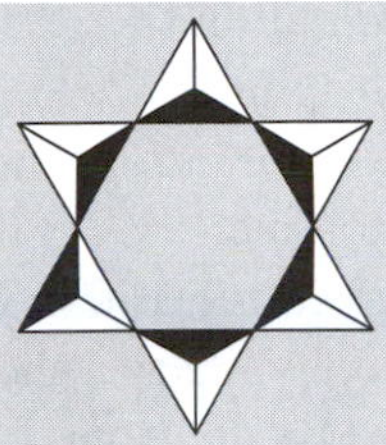	$Si_6O_{18}^{12-}$	Cyclic silicate	3	1	−2
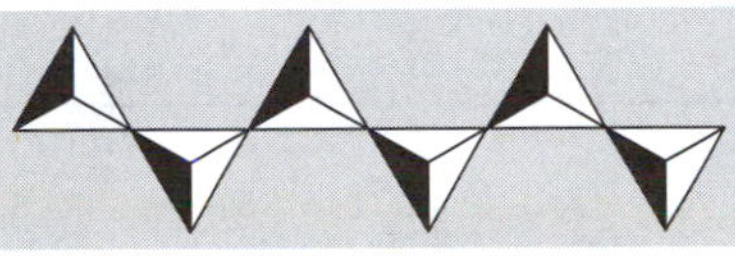	$^1_\infty[SiO_3^{2-}]$	Pyroxene (chain silicate)	3	1	−2
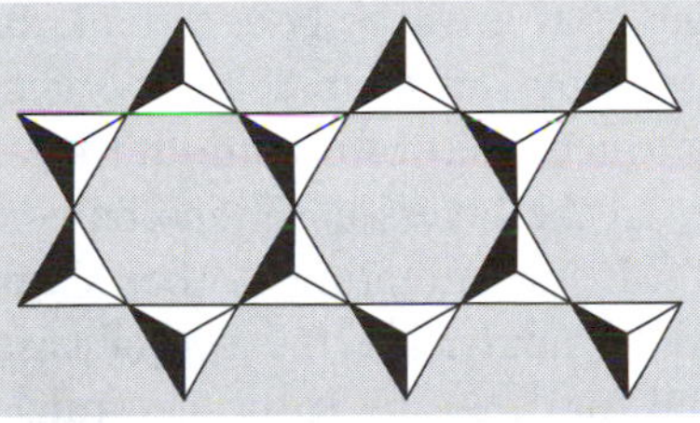	$^1_\infty[Si_4O_{11}^{6-}]$	Amphibole (double chain)	2.75	1.25	−1.5
	$^2_\infty[Si_4O_{10}^{4-}]$	Infinite sheet silicate	2.5	1.5	−1
	$^3_\infty[SiO_2]$	Silica	2	2	0

Figure 13.10

Structures of some polysilicates and resulting ratios of oxygen atoms and charge to the number of silicon atoms present. [Adapted from J. E. Ferguson, *Inorganic Chemistry and the Earth: Chemical Resources: Their Extraction, Use, and Environmental Impact*, Pergamon Press, Oxford, 1982.]

Example 13.8

Identify the degree of polymerization of the polysilicate ions found in the following minerals: (a) pyrophyllite, $Al_2(Si_4O_{10})(OH)_2$; (b) grunerite, $Fe_7(Si_4O_{11})_2(OH)_2$; (c) spessartite, $Mn_3Al_2(SiO_4)_3$; and (d) bustamite, $CaMn(SiO_3)_2$.

SOLUTION:

Calculation of the oxygen/silicon ratio allows a choice among the structures shown in Figure 13.10, with the exception of the choice between chain and cyclic structures. (Even without reference to the figure or the structures of the ions, we can state that the lower the O/Si ratio, the more polymerized the ion is, since the more vacant coordination sites there are per Si nucleus.) The O/Si ratio in these minerals is 10:4 or 2.5 in pyrophillite, 22:8 or 2.75 in grunerite, 4.0 in spessartite, and 3.0 in bustamite. Thus the simple orthosilicate ion is present in spessartite (a type of garnet), and the degree of polymerization increases in the sequence bustamite $<$ grunerite $<$ pyrophillite. Reference to the table (or consideration of the structures) identifies grunerite as a double-chain polysilicate and pyrophyllite as a sheet silicate. Bustamite could be either a ring or a chain silicate; it happens to be the latter.

The various polysilicate ions have negative charges that must be counterbalanced by appropriate cations. The terminal oxygen atoms sticking out from the polysilicate rings, chains, and layers provide negatively charged surfaces that often amount to approximately close-packed surfaces of negative charge. In the layers between chains or silicate layers, there are thus tetrahedral and octahedral (but sometimes other shapes of) holes that are then filled by the cations needed to neutralize the polysilicates' negative charge. In different types of polysilicates, the total charge to be neutralized differs, as does the size of the holes. As a consequence of isomorphous substitution, the cations in most silicate minerals are extensively substituted, and silicates do not usually make economical sources (**ores**) for most of the elements, because they generally have low concentrations of the desired element in hard-to-separate mixtures. But we can predict sources for some of the rarer elements if we are willing to pay the price: Lithium ion (used in treating manic-depressive patients) can generally be found in magnesium minerals; beryllium ion (59 pm) resembles the aluminum ion (67 pm) in a good deal of its chemistry.

The operation of the second principle of isomorphous substitution greatly increases the number of possible substitutions in silicates: Thus the common K^+ ion can be replaced not only by the rare Rb^+ and Tl^+ ions, but also by the common Ba^{2+} ion (Table 13.6). Likewise, the common Ca^{2+} ion can be replaced not only by Sr^{2+}, but also by Na^+ and Y^{3+}, La^{3+}, and the sixth-period *f*-block ions; Mg^{2+} can be replaced by Li^+ or by Fe^{3+}. In many cases, these substitutions are located diagonally to each other in the periodic table. These **diagonal relationships** of elements (especially in the second period) to the elements one group to the right and one period down[41] extend even to the nonmetallic elements, since the similar size of such atoms affects, for example, their maximum coordination numbers in similar ways. [Examination of Table A shows that there is also some similarity in the Pauling electronegativities of second-row elements and those of their diagonal partners in the third row.]

Soil Chemistry. As noted above, increasingly polymerized polysilicate ions have decreasing charges per silicon nucleus, so these ions are less basic. This fact has important consequences in soil chemistry: The more basic the polysilicate anion of a mineral, the more readily it reacts with weak acids, changing the polysilicate structure in the process of **weathering**. Rainwater is acidic even in the absence of sulfur and nitrogen oxides due to dissolved carbon dioxide. This solution of carbonic acid is weakly acidic, and over the ages it reacts with the less polymerized silicate anions to remove oxide ions (as water), thus inducing replacement of the oxide ion with a bridging oxygen to produce a more highly polymerized silicate:

$$Mg_2SiO_4 + 2\,H^+ \rightarrow Mg^{2+}(aq) + H_2O + MgSiO_3 \qquad (13.29)$$

$$MgSiO_3 + 2\,H^+ \rightarrow Mg^{2+}(aq) + H_2O + SiO_2 \qquad (13.30)$$

Thus soils containing large amounts of orthosilicates of weakly acidic cations such as olivine are characterized as "youthful" soils: They may have crystallized from a melt (i.e., magma) recently, or they may be present in a desert region, in which the water necessary to form the acidic rainfall and the hydrated ions is absent. (The "dry valleys" of Antarctica, which seldom see liquid water, also have youthful soils.) The cyclic and chain polysilicates weather somewhat more slowly, as suggested in Eq. (13.30); these compounds are followed by the double-chain silicates.[42]

At the intermediate stage of weathering, such as is found in the temperate regions under a cover of grass or trees, layer silicates such as clays tend to predominate, along with some quartz. As can be seen from Eqs. (13.29) and (13.30), this weathering process is accompanied by a loss of cations, which is especially prominent for nonacidic and feebly acidic cations. Thus the soil is less fertile than it was, due to the loss of the nonacidic plant nutrient potassium ion. (Desert soils, when first irrigated, are often very fertile.) The layer silicates present in the intermediate soils can still hold cations on their negatively charged surfaces, however; these can readily be exchanged for other ions (such as H^+), and are thus released as plants need them; such soils are found in the still-quite-fertile corn and wheat belts of the world.

In the tropics, however, when the trees are cut down and frequent rain and heat speed up the weathering process, the aging process becomes quite advanced. Such soils have high levels of oxides of the most acidic cations present in silicate minerals, such as anatase and rutile, TiO_2; zirconia, ZrO_2; hematite, Fe_2O_3; and gibbsite, $Al(OH)_3$. These soils can no longer hold the less acidic nutrient metal ions, and are quite infertile. When tropical rain forests are removed in "slash and burn" agriculture, the soil can be used for agriculture for only a few years; after that it becomes infertile (and rock-hard).

It is interesting to note that, at the high pressures thought to exist at the boundary between the mantle and the core of the earth, olivine (very abundant in the mantle) is converted first to a spinel, then to a perovskite mineral containing six-coordinate Si^{4+} ions.[43] Reaction of this mineral with iron in the core generates stishovite, the iron silicide FeSi, and FeO. The heat generated by this violent reaction may be involved in the thermal currents that drive continental drift.

Glass. The completely polymerized, acidic silica is often reacted at very high temperatures (1700 °C) with basic oxides to generate polysilicates that are cooled too rapidly to allow any of the orderly polysilicate ions mentioned above to form. The

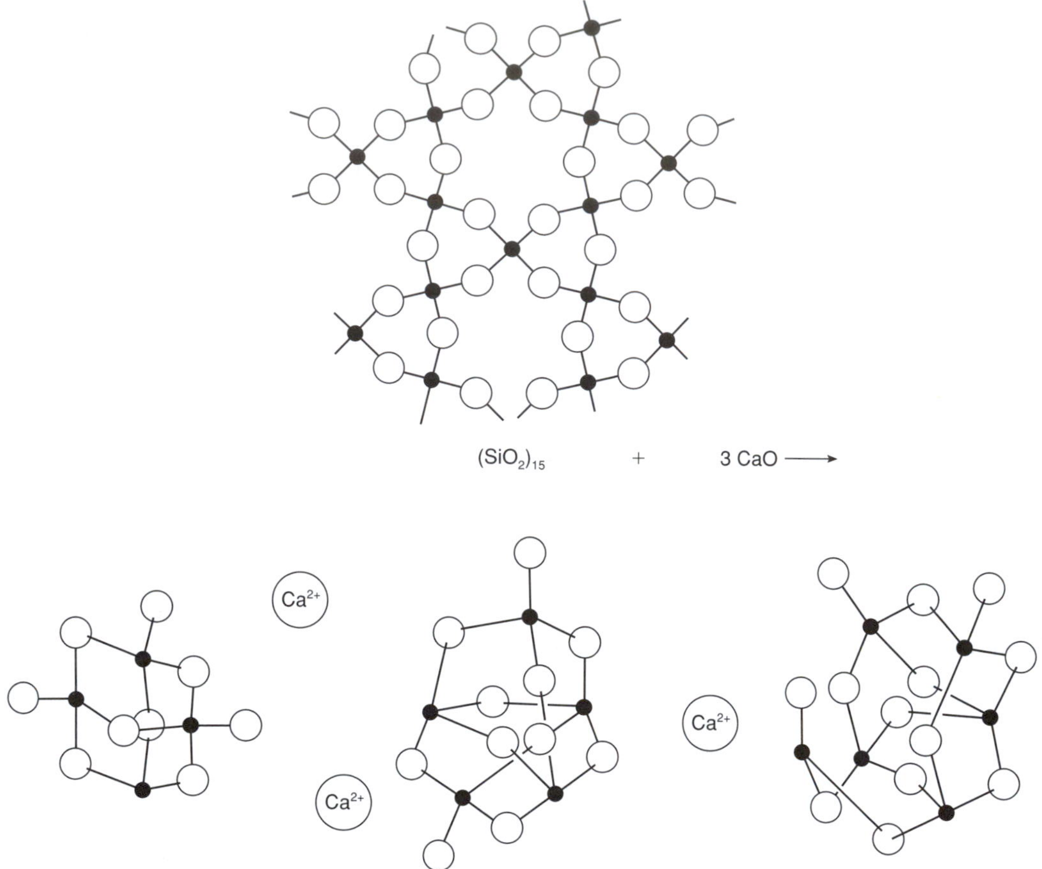

Figure 13.11
Schematic diagram of the breaking of silicon–oxygen bridges in the formation of glass.

resulting material, which has a random mixture of all sorts of polysilicate ions (Fig. 13.11), has no definite freezing point, but instead just thickens on cooling to form an **amorphous** solid or **glass**. Glass is made by fusing (melting) sand and recycled glass with Na_2CO_3 and limestone, which act as sources of the basic oxides Na_2O and CaO.

Specialty glasses are made by altering the composition of acidic and basic oxides in the glass. For example, Pyrex glass, which is unusually resistant to thermal shock, incorporates 10%–25% of the acidic boron oxide B_2O_3. Glass can be colored by incorporating *d*-block metal oxides as part of the basic oxide component. Incorporating 20–60% of the basic PbO gives dense, highly refractive *flint glass,* used in crystal and in optical equipment. Addition of SrO gives a glass that absorbs the X-rays emitted by color television sets; addition of La_2O_3 gives glass with fine optical qualities suitable for camera lenses.

Cement. Portland cement is also generated by high-temperature reactions of acidic and basic oxides. The basic oxide CaO comes from the heating of limestone, while the acidic oxides SiO_2 and Al_2O_3 come from the heating of aluminosilicates such as clay (Section 13.7). After firing, the dry cement principally contains the minerals alite, Ca_3SiO_5; belite, Ca_2SiO_4; calcium aluminate (Section 13.8); and some free CaO and

MgO (from the clay). The formula of alite should look "wrong;" its actual composition is $(Ca^{2+})_3(SiO_4{}^{4-})(O^{2-})$. When water is added to cement to cause it to set, reactions occur such as

$$4\,Ca_3SiO_5 + 12\,H_2O \rightarrow Ca_2(Si_4O_{10}){\cdot}2H_2O + 10\,Ca(OH)_2 \qquad (13.31)$$

Consequently, cement is strongly basic and much more reactive than people commonly realize: It slowly reacts with acidic CO_2 from the air to give $CaCO_3$ (limestone) and SiO_2. The reactivity of the cement used in its construction is what caused the failure of the Biosphere II self-contained, sealed-off ecosystem in Arizona in the early 1990s. Carbon dioxide in its atmosphere was supposed to undergo photosynthesis to generate food for the inhabitants to eat and O_2 for them to breathe, but the CO_2 ended up reacting with the cement instead. Consequently, the O_2 content of the atmosphere of Biosphere II fell from the normal 21% down to 14%, barely enough to sustain human life. Most species in the Biosphere became "extinct", with the few remaining species (cockroaches, ants, katydids, and morning glories) overrunning the environment.[44]

13.7 Aluminosilicates

Isomorphous substitution can involve the silicon in the polysilicate ion itself. The very abundant Al^{3+} (67 pm) replaces Si^{4+} (54 pm), giving rise to the important **aluminosilicates**. This substitution is best indicated by writing the formula of the aluminosilicate ion in brackets, enclosing the structural aluminum ions but not those aluminum ions that serve only to neutralize the negative charge. For instance, the most abundant of all minerals (~60% of the earth's crust) are the **feldspars**, which fall into two categories. One involves simultaneous isomorphous substitution of the 150-pm ions K^+ and Ba^{2+} and of Al^{3+} and Si^{4+}; compositions range from $K[AlSi_3O_8]$ (orthoclase) to $Ba[Al_2Si_2O_8]$ (celsian). The other category, the plagioclase feldspars, involve the smaller 115-pm Na^+ and Ca^{2+} ions; compositions range continuously from $Na[AlSi_3O_8]$ (albite) through $Na_{0.33}Ca_{0.67}[Al_{1.67}Si_{2.33}O_8]$ (labradorite) to $Na_0Ca_1[Al_2Si_2O_8]$ (anorthite).

Thus the mica mineral muscovite may be written as $KAl_2[AlSi_3O_{10}](OH)_2$, showing that only one-third of the Al^{3+} ions enter into the sheet polysilicate ions. (The other Al^{3+} ions, being moderately acidic, serve to bind pairs of polyaluminosilicate sheets together into three-layer structures.) To determine the structure of this material from its formula, one would calculate the ratio of oxygens to silicon *plus nuclear aluminum atoms*: $10/(3+1) = 2.5$, that is, a layer structure, with 1.5 vacant coordination sites in the structural unit. However, the negative charge of this anion is increased by one unit for each framework aluminum atom, since there is one less proton in the nucleus of the aluminum atom than in the nucleus of the silicon atom that it replaced.

Clays. Clays are two-dimensionally polymerized aluminosilicate minerals that are called *secondary minerals*; they are believed to form as a consequence of further reactions of the metal ions and silica released in weathering reactions such as those of Eqs. (13.29) and (13.30). (Note that SiO_2 is substantially more soluble in water than most oxides of very strongly acidic cations. Its total solubility at pH values up to 9 is $\sim 2 \times 10^{-3}\,M$; it dissolves as $H_3SiO_4{}^-$ and as oligomeric silicic acids such as $H_6Si_2O_7$.)

Besides having layer polysilicate or polyaluminosilicate ions, clays (Fig. 13.12) are characterized by having parallel layers of *octahedrally* coordinated Al^{3+} or Mg^{2+} ions

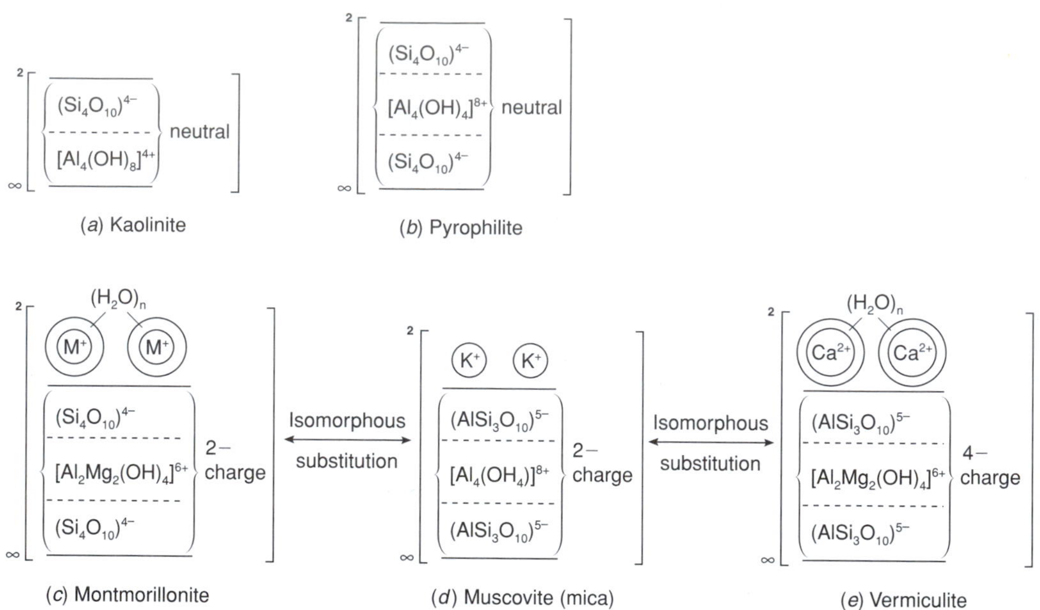

Figure 13.12

Layer structures of selected clays, and their interrelationships via isomorphous or non-isomorphous substitution. (*a*) Kaolinite, ${}^{2}_{\infty}\{[Al_4(OH)_8][Si_4O_{10}]\}$; (*b*) Pyrophillite, ${}^{2}_{\infty}\{[Al_4(OH)_4][Si_4O_{10}]_2\}$; (*c*) Montmorillonite, ${}^{2}_{\infty}\{M_\delta[Mg_\delta Al_{(4-\delta)}(OH)_4][Si_4O_{10}]_2\}$ (illustrated for $\delta = 2$); (*d*) Muscovite, ${}^{2}_{\infty}\{K_2[Al_4(OH)_4][AlSi_3O_{10}]_2\}$ (*e*) Vermiculite ${}^{2}_{\infty}\{(Ca, Mg)_2[Al_2Mg_2(OH)_4][AlSi_3O_{10}]_2\}$.

bridged by OH^- ions; OH^- and/or O^{2-} ions also bridge the adjacent *tetrahedral layers* (planar polysilicate or polyaluminosilicate ions, such as ${}^{2}_{\infty}[Si_4O_{10}{}^{4-}]$) and octahedral layers (such as ${}^{2}_{\infty}[Al_4(OH)_8{}^{4+}]$). These are the only two layers in the simplest clay, kaolinite [Fig. 13.12(*a*)]; the bilayer "sandwiches" are bound to neighboring sandwiches only by comparatively weak hydrogen bonds.

Other important clays involve three layers, two of which are anionic tetrahedral polysilicate or polyaluminosilicate layers sandwiching a central cationic octahedral aluminum hydroxide or magnesium aluminum hydroxide layer. In pyrophyllite [Fig. 13.12(*b*)], the central ${}^{2}_{\infty}[Al_4(OH)_4{}^{8+}]$ layer exactly neutralizes the two outer layers; only hydrogen bonds link one triple layer to another, so that cleavage into sheets is very easy.

In muscovite [mica, Fig. 13.12(*d*)], the two outer layers are each −5 charged aluminosilicate layers, so their total charge exceeds that of the central octahedral layer; charge balance thus requires a fourth layers of cations such as K^+. Two other clays, montmorillonite [Fig. 13.12(*c*)] and vermiculite [the familiar packing material, Fig. 13.12(*e*)] are derived from the structure of muscovite by different isomorphous substitution processes of the second kind.

The cations absorbed on the surfaces of the charged clays (muscovite, vermiculite, and montmorillonite) are of low acidity, and are easily exchanged with ions from solution; the clays perform this function naturally in the soil. (Clays, we recall, are characteristic of moderately weathered soils in the temperate zones of the earth.)

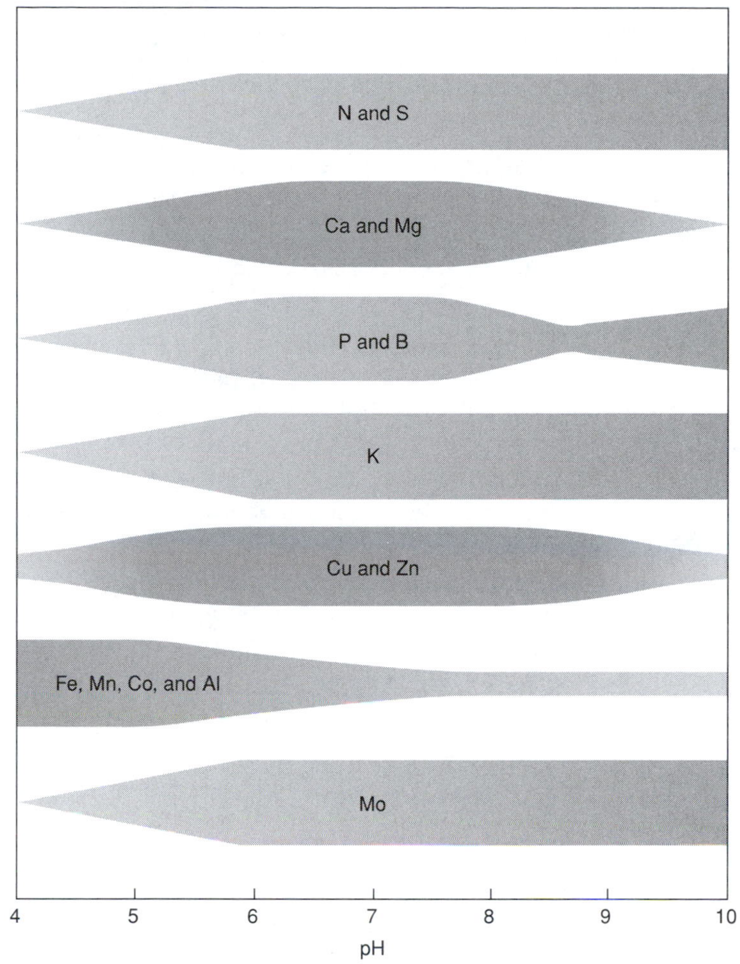

Figure 13.13
Relative availability of nutrient elements in soil as a function of soil pH. (The broader the bar, the more available is the nutrient.) [Adapted from *Environmental Chemistry*, by J. W. Moore and E. A. Moore. Copyright © 1976 by Academic, New York.]

Enormous quantities of hydrated cations can be held to the negatively charged surfaces of clays; these cations include many nutrient ions such as Ca^{2+}, Mg^{2+}, K^+, and Na^+. As slightly acidic rainwater percolates over these clays, these nutrient ions are slowly released by ion exchange with the hydronium ion in the water and made available for plants to use (Fig. 13.13). The largest and the least charged of these ions are held most loosely and are released first (and become depleted first). Recalling that smaller ions have larger secondary hydration spheres, we find that the order of release is Na^+ before K^+ before Mg^{2+} before Ca^{2+}. Since plants make little use of Na^+, the first deficiency that occurs among cationic nutrients is of K^+. Anionic nutrients such as NO_3^- and PO_4^{3-} cannot be bound to clays at all, so they are rapidly lost from soils. Thus we see the need of fertilizers containing (especially) potassium, nitrogen, and phosphorus.

Clays have a variety of uses. When kaolinite is heated in an oven, its hydroxide groups are dehydrated, and it is converted to a mixture of the oxides SiO_2 and Al_2O_3. Kaolinite is a chemical precursor to the ceramic that is the basis of brick, pottery, china, porcelain, and enamels. The cations on the surface of montmorillonite readily adsorb additional water (to form secondary hydration spheres) when montmorillonite is placed in water. These hydration spheres then hydrogen bond to each other, thickening the suspension in water—at least until the suspension is stirred, which breaks up the hydrogen bonds and thins the suspension. Montmorillonite is thus a useful **thixotropic agent** and is used in paints to keep them from flowing while wet, and in drilling muds in the oil industry.

Montmorillonite has, or may have, even greater significance than the above indicates. Its surface acidity is quite substantial, being equivalent to a pH of from 1.5 to −3. If it is washed with acid, so that the cations on the surface are exchanged for H^+, its surface acidity is estimated as being equivalent to a pH of from −5.6 to −8; therefore it is a *very strong acid*. As an insoluble material, however, it is much more easily handled than sulfuric or other strong acids, and it (and related clays such as vermiculite and Fe^{3+} or Cu^{2+} exchanged clays such as bentonite) is finding use in catalyzing typical organic reactions requiring acid catalysis, such as the Friedel–Crafts reaction.[45] Finally, it has been noted that the spacing between charges on the surface of montmorillonite is similar to the spacing between peptide linkages in proteins. This observation and other suggestions have led to the (so far, heretical) view that the first lifemolecules may have originally formed on the surface of montmorillonite clay[46]! In work that perhaps supports this hypothesis, montmorillonite has been used to line up a derivative of the nucleotide adenosine, which then oligomerizes in the manner found in natural RNA and DNA.[47]

Framework Aluminosilicates. To date, we have discussed only one three-dimensional silicate, SiO_2 itself. There are no possibilities for isomorphous substitution in SiO_2, other than by the very rare Ge^{4+} ion, since there is no other ion to be replaced simultaneously. However, a number of structures of aluminosilicates can be imagined as arising from the replacement of Si^{4+} by Al^{3+} *and* a +1 ion. To make room for the extra ion, holes must be opened up in the structure of SiO_2. Due to the resulting open three-dimensional structures, these compounds are known as *framework* aluminosilicates. Since the substitution of two ions for one leads to a change in structure, this change is *not* an example of isomorphous substitution.

Perhaps the best known of the framework aluminosilicates are the **zeolites**,[48] with anions of general formula $[Al_xSi_yO_{2x+2y}]^{x-}$ with $x \leq y$ and an oxygen/(aluminum plus silicon) ratio of 2. Figure 13.14(*a*) shows how the $(Si, Al)O_2{}^{\delta-}$ structural units may be oligomerized into six-membered $[(Si, Al)O_2]_6$ (as in SiO_2 itself) and four-membered $[(Si, Al)O_2]_4$ rings that are interlinked to give large polyhedral $[(Si, Al)_{24}O_{48}]$ clusters known as **sodalite cages**. In Figure 13.14(*b*), the sodalite cage is drawn omitting all oxygen atoms and showing the 24 (Si,Al) nuclei as intersections of links.

In zeolites, sodalite cages are linked to each other to give rise to large, three-dimensional structures that feature open channels. For example, in Figure 13.14(*c*) each square face of each sodalite cage is used twice (shared) between two adjacent sodalite cages to give the smallest possible network polymer, the mineral sodalite, ${}^3_\infty[Na_6(Al_6Si_6O_{24})\cdot 2H_2O]$, which has a central four-sided channel of 220-pm diameter.

In Figure 13.14(*d* and *e*), the sodalite cages are instead linked to adjacent complete cages by converting terminal oxygen atoms of the original sodalite cage [Fig. 13.14(*a*)]

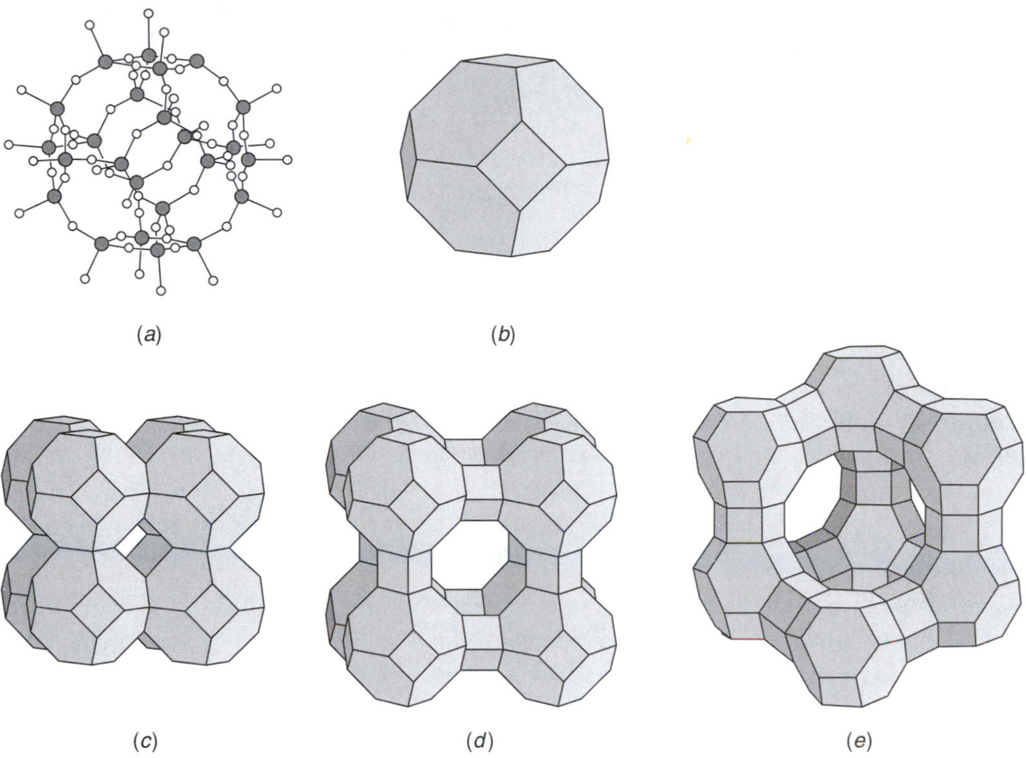

Figure 13.14
Structural representations of the zeolite structure. (*a*) Connection of aluminosilicate nuclei to form an open basket-like cluster known as the sodalite cage $[(Si,Al)_{24}O_{48}]$ (open circles = oxygen; filled circles = silicon or aluminum); (*b*) line representation of the sodalite cage in (*a*), in which vertices represent silicon/aluminum nuclei and bridging oxygens lie along the edges; (*c–e*) connection of sodalite cages to give: (*c*) sodalite, $^3_\infty[Na_6[Al_6Si_6O_{24}]\cdot 2H_2O\}$; (*d*) Linde A, $^3_\infty\{Na_8[Al_8Si_{40}O_{96}]\cdot 24H_2O\}$; (*e*) faujasite, $^3_\infty\{Na_{58}[Al_{58}Si_{134}O_{384}]\cdot 240H_2O\}$. [Adapted from G. T. Kerr, *Sci. Am.*, **261**, 100 (1989).]

to bridging oxygens. In Figure 13.14(*d*), square faces are linked to give an eight-sided channel with a diameter of 410 pm, in the zeolite Linde A, $^3_\infty[Na_8(Al_8Si_{40}O_{96})\cdot 24\,H_2O]$. In Figure 13.14(*e*), hexagonal faces are linked to give a 12-sided channel with a diameter of about 850 pm, in the zeolite faujacite, $^3_\infty[Na_{58}(Al_{58}Si_{134}O_{384})\cdot 240\,H_2O]$. Many other natural and artificial zeolites are known involving other channel sizes, which can have up to 14 sides.[49]

The three-dimensional polyaluminosilicate ion structure must open up as shown to leave room for the cations needed to neutralize the charge of the polyaluminosilicate anion. These cations are normally in the form of hydrated ions; the water of hydration can be driven off by heating to leave a structure with enlarged channels. If the cation of the zeolite is the relatively large K^+ (in Zeolite 3A), a small channel of 300-pm (3 Å) diameter is available; if the cation is Na^+ (in Zeolite 4A), a larger channel of 400-pm diameter is present; if the cation is Ca^{2+} (in Zeolite 5A), a still larger channel of 500-pm diameter is present, since only one-half as many cations are present alongside the channel. The resulting zeolites have cavities of tailored sizes that are used as molecular

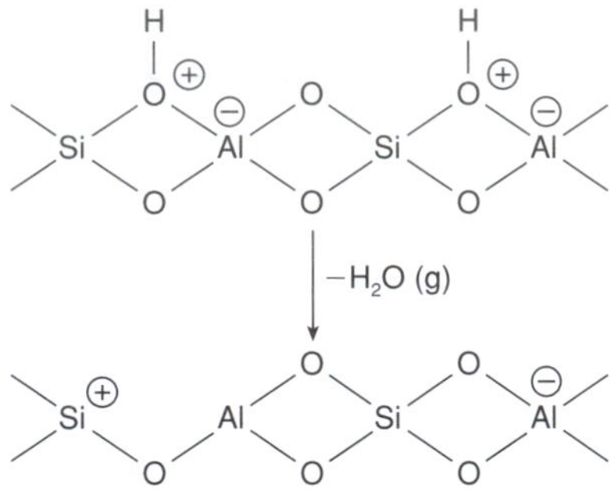

Figure 13.15
Dehydration of the surface of an acid-exchanged zeolite to leave coordinately unsaturated, hence strongly acidic, silicon cationic sites. Formal charges of atoms are shown in circles.

sieves to adsorb molecules of different sizes from liquids. Among the small molecules most readily adsorbed are water molecules, which re-form the hydrated metal ions: Thus these zeolites are very effective drying agents.

Hydrated metal cations in such a structure, being large cations in a lattice with very large insoluble anions, are loosely bound and can readily be exchanged with other cations. Zeolites are used for *water softening*: the removal of ions of +2 charge found in tap water that would precipitate the anions used in detergents. A concentrated solution of NaCl is first percolated through the solid (insoluble) zeolite ion exchanger, replacing whatever hydrated ions are present with hydrated Na^+ ions. Then the tap water is run through the zeolite; Ca^{2+} and other +2 ions become associated with the anion in the solid phase, while Na^+ ions go into solution. When the Na^+ ions in the zeolite are depleted, the solid is again "recharged" by running concentrated NaCl solution through it.

Furthermore, the channels may penetrate through the zeolite in one dimension, or they may intersect other channels that go off in the other two directions. The channels thus may be tailored to be of such a size as to admit only molecules below a certain size, or only straight-chain but not branched-chain hydrocarbons, and so on.

In combination with these possibilities of selectivity, we may note that the dehydrated zeolites have semibare dehydrated metal cations with Lewis acid properties. In addition, the metal ions can be ion exchanged for hydrogen ions, which, if driven off by heat, leave behind positively charged silicon atoms with vacant coordination sites that are powerfully acidic sites[50] (Fig. 13.15). The combination of powerful Lewis acidity with site selectivity makes these of great importance as catalysts. For example, ZSM-5 is used to catalyze the rearrangement of 2 mol of toluene to give 1 mol of benzene and 1 mol of *p*-toluene (*o*-toluene does not fit the cavity); to catalyze the alkylation of benzene with ethylene to give ethylbenzene; to catalyze the conversion of methanol, which can be made from coal and water, into gasoline plus water; and in the **cracking** of petroleum to form gasoline. Analogies have been drawn between the action of zeolites and the actions of enzymes, since both depend on the availability of reactive (e.g., Lewis acidic) sites along with shape selectivity. In addition, the easily recovered and regenerated zeolite and other solid acid catalysts are far more "environmentally friendly" than concentrated sulfuric acid.

The synthesis[51] of zeolites is an art that involves the reaction of sodium silicate, sodium aluminate, and aluminum sulfate solutions in various ratios for extended (e.g., 20–30 h) times at carefully controlled (e.g., $92 \pm 2\,^{\circ}C$) temperatures in the presence of appropriate quaternary ammonium, R_4N^+, or other organic cations (e.g., $[Co(C_5Me_5)_2]^+$ that act as templates to help determine channel sizes. After the synthesis is completed, the product is heated to a high temperature so as to vaporize or burn off the organic template. The size and shape of the organic cation changes the size of the channel that is produced. It is hoped that molecular modeling computer programs can be used to select the proper organic cation to produce a particular cavity size.[52]

It is finally even possible to make aluminum-free, pure-silica zeolite-like **porosils** or **silicalites** by crystallizing SiO_2 from water containing some NaOH and some neutral organic amines, which act as templates. The use of smaller amines such as triethylamine, tripropylamine, and tributylamine gives the ZSM-5 catalysts mentioned above; the use of the larger tetrabutylammonium cation gives another porosil, ZSM-11.

Even larger zeolite-like framework polymers have been made by completing the nonisomorphous substitution of aluminosilicates, by replacing all of the remaining silicon atoms with *phosphorus* atoms. The resulting products, which are polymers of aluminum phosphate, ${}^{3}_{\infty}[AlPO_4]$, have channel diameters as large as 1200 pm (in VPI-5). New titanium silicate molecular sieves have also been developed; one selectively extracts cesium (e.g., radioactive ^{137}Cs) from nuclear wastes.[53]

13.8 *Oligomeric Polyoxometallate Anions of Other *p*- and *d*-block Elements: Selected Structures and Uses

We have seen that elements in combination with oxygen give rise to a rich variety of structures. Henry et al.[54] derived relationships to show that the interaction of the cation of an element with water leads to five main types of products. (a) Slightly charged cations of elements of low electronegativity do not polarize water, so they remain as hydrated ions. (b) Highly charged cations of elements of high electronegativity polarize coordinated water so strongly that they emerge from the process as simple oxo anions. These two trends we saw in Chapter 2; in this chapter we see other possibilities that occur mainly in the middle of the periodic table, for cations of moderate charge and electronegativity. (c) Cations of higher charge and/or electronegativity than those in (a) undergo bridging with hydroxide ions to produce **polymeric or oligomeric hydroxy cations** (such as those of Al^{3+} and Mg^{2+} found in the octahedral layers of clays) that eventually precipitate (as the pH increases) as metal *hydroxides*. (d) Cations of slightly higher charge and/or electronegativity give neutral hydroxides that eliminate water molecules to give neutral *oxides* or *hydrous oxides*. (e) "Cations" of still higher charge and/or electronegativity, such as Si^{4+}, polarize the hydroxide groups to eliminate hydrogen ion, but the oxo group then bridges adjacent structural units to produce mainly oligomeric *polyoxoanions* such as the oligosilicates. In Figure 13.16, the relationship of the charge and electronegativity of the cation to its behavior is plotted (the boundaries of the behavior regions are calculated in a lengthy process based on electronegativity equalization between the cation and the water solvent). It may be seen that, in addition to silicon, polyoxoanions ought also to be formed by Cr^{6+}, Mo^{6+}, W^{6+}, Ta^{5+}, V^{5+}, Sb^{5+}, and B^{3+}.

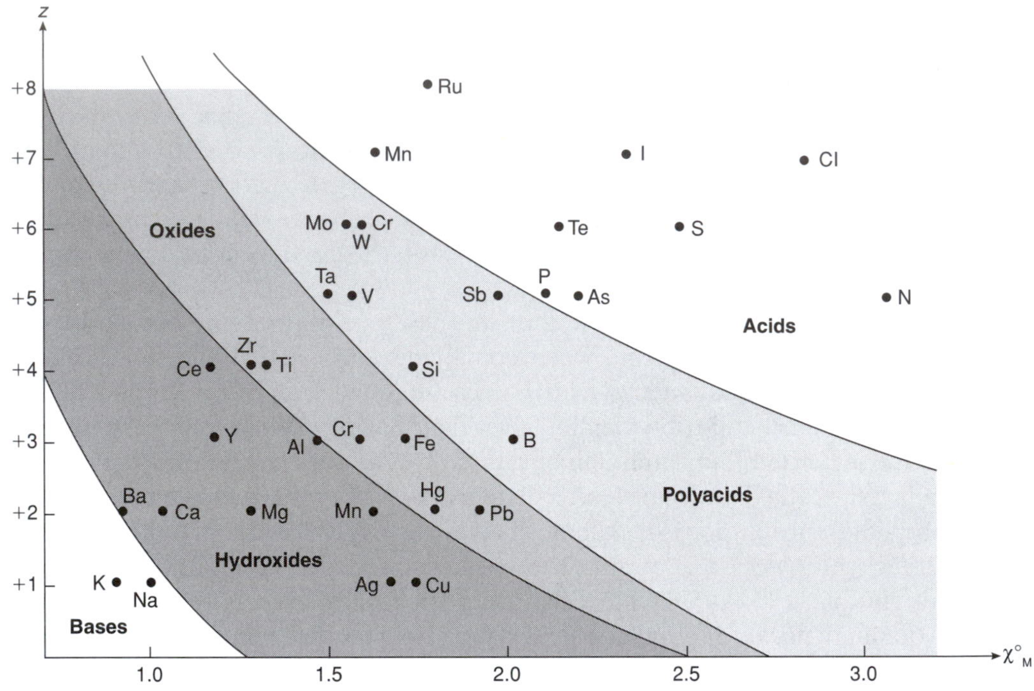

Figure 13.16
Charge-electronegativity diagram for cations of elements in water. [Adapted from M. Henry, J. P. Jolivet, and J. Livage, *Struct. Bonding* (*Berlin*), **77**, 153 (1992).] The electronegativity scale used in this plot is a modified Allred–Rochow scale also used in deriving the boundaries; if Pauling electronegativities are substituted similar results are mainly obtained except in the important cases of Mo and W, the Pauling electronegativities of which are suspect.

Although they are of much less importance than the polysilicates, oligomeric polyoxoanions are indeed formed by most of the predicted *p*- and *d*-block elements. As one might expect from the diagonal relationship of elements, one extensive series of these is the polyborates, in which tetrahedral BO_4 and triangular BO_3 groups are linked together in many ways. The polyborate of the greatest importance is **borax**, $Na_2B_4O_7 \cdot 10\,H_2O$, mined in Death Valley, CA; it actually contains the $[B_4O_5(OH)_4]^{2-}$ ion shown in Figure 13.17(*a*). Although numerous other types of borates are known, in view of the overall scarcity of boron and the limited use of its oxo salts other than borax, we will not discuss these.

Aluminum forms some polyaluminates, of which the most important is the $Al_6O_{18}{}^{18-}$ ion found (as its calcium salt) in Portland cement [Fig. 13.17(*b*)]. In contrast to the less negatively charged polysilicates, this ion reacts rapidly with water to degrade the anion to various dialuminates. This reaction is responsible for the rapid setting of Portland cement.

The polynuclear phosphates are of industrial and especially biochemical importance. The diphosphate ion is also called pyrophosphate because it is formed by heating a dihydrogen phosphate:

$$2\,Na_2HPO_4 + \Delta \rightarrow Na_4P_2O_7 + H_2O \tag{13.32}$$

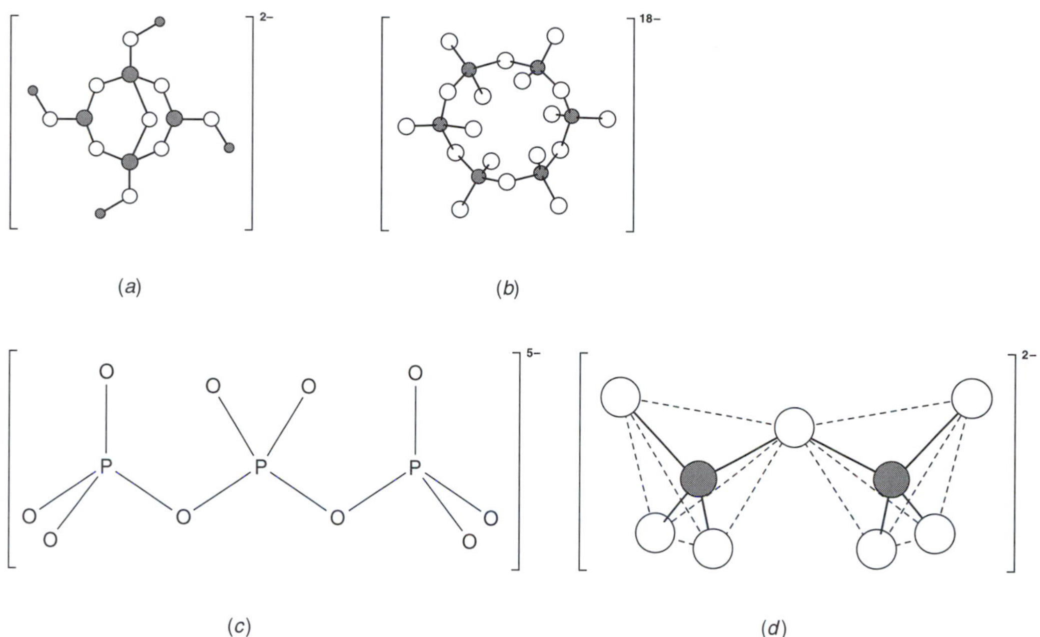

Figure 13.17
Selected oligomeric polynuclear oxo anions of the *p*-block elements: (*a*) the $[B_4O_5(OH)_4]^{2-}$ ion found in borax; (*b*) the $Al_6O_{18}{}^{18-}$ ion found in Portland cement; (*c*) the $P_3O_{10}{}^{5-}$ ion found in the detergent builder sodium tri(poly)phosphate and (covalently bonded to organic groups) in ATP, DNA, and RNA; and (*d*) the structure of the disulfate (and dichromate) ions. [Adapted from N. N. Greenwood and A. Earnshaw, *Chemistry of the Elements*, 2nd ed.; pp. 206, 251, 528, 1009, Copyright © 1997, Butterworth-Heinemann, Oxford.]

This anion has one bridging oxygen atom linking two PO_4 tetrahedral nuclei and occurs (linked through one other oxygen covalently to adenosine) in the important biochemical adenosine diphosphate (ADP).

Of greater importance [Fig. 13.17(*c*)] is the triphosphate or tripolyphosphate ion, $P_3O_{10}{}^{5-}$, synthesized similarly to the diphosphate:

$$2\,Na_2HPO_4 + NaH_2PO_4 + \Delta \rightarrow Na_5P_3O_{10} + 2\,H_2O \qquad (13.33)$$

This compound is used in large quantities as the "phosphate" found in detergent. (It is used as a chelating ligand to tie up the +2 ions in tap water; see Chapter 5.) The environmental problems with using so much of this material in detergents stem from its hydrolysis to give the simple phosphate ions [e.g., the reverse of Reaction (13.33)], which are plant nutrients in lakes. Many lakes that have high inputs of phosphates have been overgrown with algae.

The tripolyphosphate ion, covalently linked to sugar molecules through one of the end oxygen atoms, is found in all nucleic acids (DNA and RNA) and in adenosine triphosphate (ATP). Hydrolysis of one phosphate nucleus off of ATP produces ADP and releases a small, biochemically convenient amount of energy. This hydrolysis is readily reversed in the body, so the system of ADP and ATP functions in energy transfer in the body.

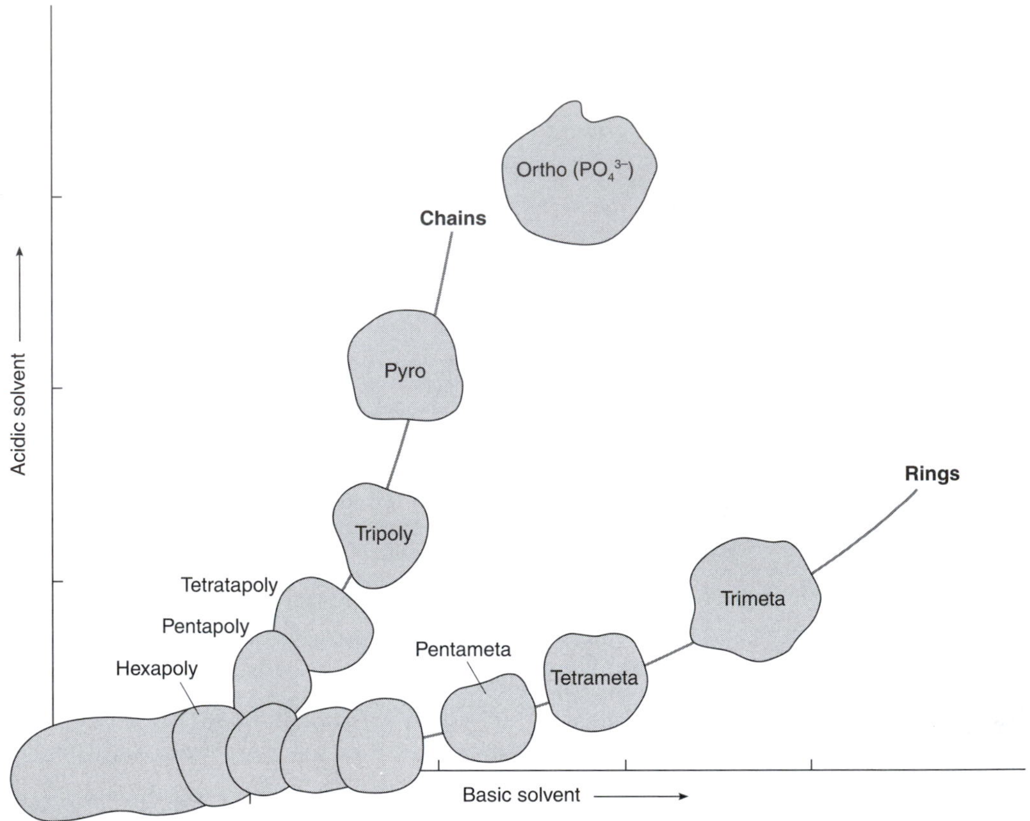

Figure 13.18
Two-dimensional paper chromatogram of a mixture of oligomeric polyphosphates. The sample was spotted at the lower left, then treated with a basic solvent moving to the right, then treated with an acidic solvent to separate chain from cyclic oligophosphates. [Adapted from D. F. Shriver, P. W. Atkins, and C. H. Langford, *Inorganic Chemistry*, Freeman, New York, 1990.]

Additional cyclic and chain oligomeric polyphosphates can be made by the controlled hydrolysis of P_4O_{10} with a limited amount of water; two-dimensional paper chromatographic separation of the products show the variety of oligomeric polyphosphate chains and rings that are produced (Fig. 13.18).

An intermediate in the manufacture of sulfuric acid is the disulfate ion, $S_2O_7^{2-}$ [Fig. 13.17(*d*)]. More familiar is a *d*-block analogue, the dichromate ion, obtained when chromate ion is partially neutralized:

$$2\,HCrO_4^- \rightarrow Cr_2O_7^{2-} + H_2O \tag{13.34}$$

A similar divanadate ion, $V_2O_7^{4-}$, is also formed by vanadium, but generally, vanadium and the heavier elements of Groups 5 and 6 form larger *cluster* polynuclear ions based not on tetrahedra but (since the central atoms are larger) on octahedral structural units (such as $VO_{2.5}$ and MoO_3) with three or more vacant coordination sites linked to each other and to (often central) oxide or other anions (Fig. 13.19).

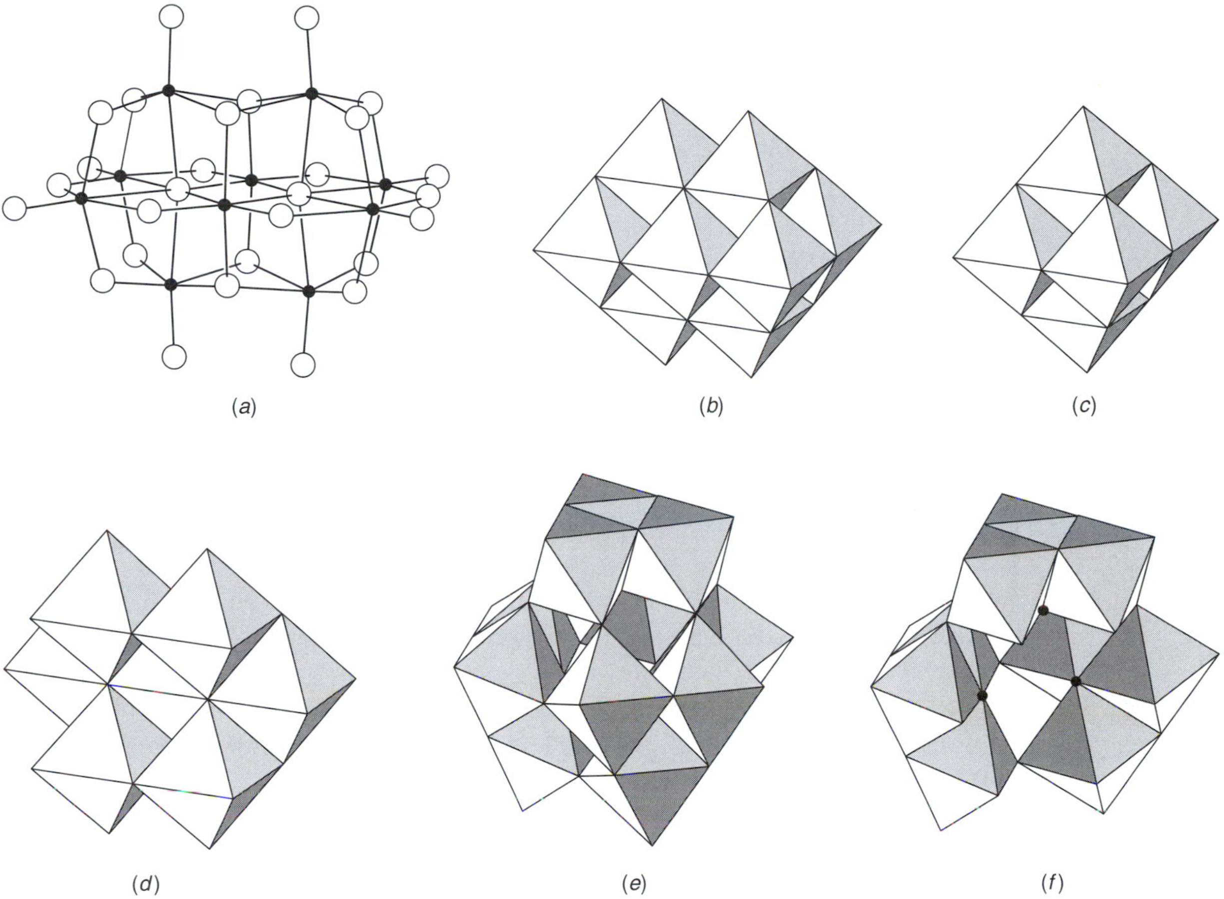

Figure 13.19
Structures of some cluster polynuclear anions of the *d*-block metals: (*a*) the decavanadate ion $V_{10}O_{28}{}^{6-}$; (*b*) the decavanadate ion represented by ten interlinked octahedra (two are obscured); (*c*) the hexaniobate or hexatantalate ions $M_6O_{19}{}^{8-}$ or the hexamolybdate ion $Mo_6O_{19}{}^{2-}$; (*d*) the heptamolybdate ion $Mo_7O_{24}{}^{6-}$; (*e*) the dodecatungstate ion $[H_2W_{12}O_{40}]^{6-}$; and (*f*) the dodecatungstate ion with three octahedral nuclei removed to show the inner cavity; the dark dots represent three of the four oxygen atoms to which the phosphorus or other heteroatom attaches in the phosphotungstate, phosphomolybdate, and related heterpolyatomic anions. [Adapted from N. N. Greenwood and A. Earnshaw, *Chemistry of the Elements*, 2nd ed., pp. 986, 1013, and 1015, Copyright © 1997, Butterworth-Heinemann, Oxford.]

Numerous such ions are known or suspected; we will highlight only a few, such as the decavanadate ion $V_{10}O_{28}{}^{6-}$. Niobium and tantalum do not seem to form monomeric oxo anions (perovskite-type mixed oxides are more common) but do give rise to cluster polyniobates and polytantalates such as $M_6O_{19}{}^{8-}$.

The cluster polymolybdates and polytungstates are especially numerous and elaborate in structure. An interesting one is the hexatungstate ion, $W_6O_{19}{}^{2-}$, with a structure like that of the hexaniobate and hexatantalate ions [Fig. 13.19(*c*)]. This cluster oligomer has six WO_3 structural subunits octahedrally attached to a six-coordinate central oxide ion. Of the three oxygen atoms in the WO_3 structural subunit, one is directed radially outward from the cluster to act as a terminal oxygen, while two are directed tangentially along the surface of the cluster to bridge to adjacent WO_3 subunits. Consequently, the W atom in each octahedron of the structure is surrounded by six oxygen atoms of three types: the central (six-coordinate) oxide ion, at a W–O distance of 233 pm; four bridging (two-coordinate) oxygen atoms with W–O distances of 192 pm; and a terminal oxygen atom coming outward from the cluster, with a W=O distance of 169 pm.

Acidification of solutions of the molybdate ion, MoO_4^{2-}, results in the heptamolybdate ion $Mo_7O_{24}^{6-}$, the octamolybdate ion $Mo_8O_{26}^{4-}$, and even $Mo_{36}O_{112}^{8-}$; mild reduction gives "molybdenum blue", an ion that is apparently so enormous, $[(MoO_3)_{154}(H_2O)_{70}H_x]^{y-}$, that it shows some of the properties of a colloid.[55] The tungstate ion gives rise to a different series of polytungstates, of which the best known are the dodecatungstates $[H_2W_{12}O_{42}]^{10-}$ and $[H_2W_{12}O_{40}]^{6-}$.

Heteropolymetallate Ions. As shown in Figure 13.19(*f*), there is a prominent tetrahedral hole inside the 12 tungstate units present in the $[H_2W_{12}O_{40}]^{6-}$ ion. This tetrahedral coordination site can be filled with any of at least 35 different cations to give a polytungstate ion consisting of a central EO_4^{z-} structural subunit attached to 12 WO_3 subunits that are bridged to each other in a cluster oligomer. Such an ion, containing another element besides tungsten and oxygen, is known as a **heteropolytungstate** or a **Keggin structure**; similar heteropolymolybdates also form.[56] It is interesting to note a structural resemblance of these cluster ions to the endohedral fullerenes (Section 12.5): the Keggin structures amount to wrapping a 12-unit fragment of the planar $^{2}_{\infty}[MoO_3]$ polymer around a central EO_4^{z-} ion to give $(EO_4^{z-})@Mo_{12}O_{36}$.

These ions are particularly important in determining the amount of phosphate contained in unknown solutions or samples; addition of molybdate ion to a solution containing phosphate ion, followed by acidification, produces the yellow phosphomolybdate ion $[PMo_{12}O_{40}]^{3-}$. The intensity of the yellow color can then give a measure of the phosphate ion concentration; greater sensitivity can be achieved by adding a reducing agent that reduces up to six Mo^{6+} ions in the phosphomolybdate ion to Mo^{5+} ions. The resulting *defect structure* is intensely blue due to the same kind of electron-transfer processes as occur in mixed oxides such as Fe_3O_4 (Section 13.5). They also have photochemical properties not unlike those mentioned in Section 13.4 for TiO_2, and can be used to catalyze various redox processes.

Alternatively, the $[PMo_{12}O_{40}]^{3-}$ anion can be precipitated—it is an *extremely* large, *extremely* nonbasic anion: Three negative charges are spread over 13 nuclei; application of Eq. (2.14) to its monomeric structural unit, $[(P,Mo)O_{3.08}]^{0.23-}$, suggests the possibility of a pK_b of 25! This nonbasic anion gives crystalline precipitates with large nonacidic cations such as large organic ammonium ions (this type of anion is used to isolate such ions from natural products). These precipitates have enormous molecular weights of over 1000, and can result in just a few milligrams of phosphate becoming quite an easily weighable amount of a phosphomolybdate precipitate. In addition, the H^+ ion-exchanged forms of these very nonbasic anions are also powerful acids also capable of catalyzing many important industrial organic reactions such as the hydration of alkenes to give alcohols.

Study Objectives

1. Predict trends in melting points among fluorides or oxides (a) across a period of elements, (b) up or down a group of elements, and (c) of a given element in different oxidation states. Exercises 1–5.

2. Know the common nonmetal oxides and tell whether each has a structure based on (a) small molecules, (b) oligomeric molecules, or (c) macromolecules or an ionic lattice. Tell whether each is (a) a solid with high melting and boiling points, or (b) a gas, liquid, or solid with relatively low melting and boiling points. Exercises 6–13.

3. Rank a series of oxides or hydroxides from the same group, period, or element (with different oxidation numbers for the element) in order of increasing basicity. Decide whether a given oxide is likely to be (a) soluble in water, (b) insoluble in water but soluble in strong acids, (c) insoluble in water but soluble in strong bases, or (d) insoluble in water but soluble both in strong acids and in strong bases. Exercises 14–21.

4. Write balanced chemical equations including the products when (a) basic oxides react with water or with strong acids; (b) acidic oxides react with water or with strong bases; (c) amphoteric oxides react with strong acids or with strong bases; (d) acidic and basic oxides react with each other in the absence of water. Rank the degree of exothermicity in different acid–base reactions of oxides. Exercises 22–26.

5. Know the names of, main uses of, and environmental problems (if any) associated with the main oxides of the nonmetals. Exercises 27–30.

6. Know how many tetrahedral and how many octahedral holes occur per anion in close-packed structures, and how these relate to the stoichiometry of the salt. Exercise 31.

7. Identify mixed-metal oxides that could adopt the spinel structure, the perovskite structure, and nonstoichiometric (defect) structures. Exercises 32–38.

8. Given the drawing of a unit cell, determine the simplest formula of the corresponding solid. Determine from the simplest formula whether a given metal ion is present in two oxidation states, and if not describe what kind of substitution would be necessary to achieve this. Exercises 39–42.

9. Given the formula of a silicate mineral, tell whether it contains simple silicate ions, linear oligomeric silicate ions, cyclic or chain polysilicate ions, double chains, layer polysilicate ions, or network structures, and describe the relative basicity of its anion. Exercises 43–50.

10. Given a table of ionic radii, select ions that are likely to be found in the same silicate (or other) minerals by the process of isomorphous substitution; identify some ions that are unlikely to substitute for any other common ions and explain why. Exercises 51–58.

11. Given formulas of several silicate minerals, tell which will be weathered most rapidly. Exercises 59–61.

12. Know which elements are prone to forming polymeric oxo anions; describe the structures and uses of some of these. Exercises 62–67.

Exercises

1. *The "cations" Be^{2+}, Si^{4+}, P^{5+}, Se^{6+}, and Br^{7+} each have a radius of about 50 pm. Assuming this radius: (a) Calculate the radius ratio of these cations to that of the fluoride ion. What two coordination numbers are most likely for each cation? Use the larger of these two numbers in the remainder of the problem. (b) Predict the number of vacant coordination sites in the structural unit of the neutral fluoride of each element. (c) Assuming that these calculations give accurate predictions, classify the fluoride of each of these elements as monomeric, oligomeric, polymeric, or non-existent. (d) With the same assumption, select one of the following melting points as the most likely one for each of the fluorides that does exist: −50 °C; 75 °C; and 1000 °C.

2. The "cations" Al^{3+}, Ge^{4+}, Sb^{5+}, Te^{6+}, and I^{7+} each have a radius of about 70 pm. Assuming this value: (a) Calculate the radius ratio in their fluorides. Each cation would be expected to have what total coordination number? How many vacant coordination sites would be present in each structural unit? (Consider the structural unit to be the neutral fluoride.) (b) Assuming that these calculations give accurate predictions, predict whether the fluoride of each of these cations would be monomeric, oligomeric, polymeric, or nonexistent. (c) With the same assumption, select one of the following melting points as the most likely one for each of the fluorides that does exist: −50 °C; 75 °C; and 1000 °C.

3. Use radius ratio calculations to justify and interpret (in terms of plausible degrees of polymerization of structural units) the melting-point data given in Table 13.1 for the highest fluorides of (a) the fifth-period elements in the *d* block; (b) the seventh-period elements in the *f* block; and (c) the Group 15(V) elements.

4. *Put the following series of oxides in order of increasing melting points, and decide which are gases at room temperature: (a) Na_2O, Cr_2O_3, CO_2, SiO_2, P_4O_{10} and (b) ZrO_2, CO_2, SrO, Rb_2O, Y_2O_3.

5. Assuming that all of the following compounds are stable, predict where melting points will increase and where they will decrease in the following series of oxides. (If radius ratios are unavailable, use periodic trends in total coordination numbers.) (a) Cl_2O_5, Br_2O_5, I_2O_5 and (b) MnO, Mn_2O_3, MnO_2, Mn_2O_7.

6. The following nonmetals show polymerization isomerism of their oxides of the type such that a given oxide can exist as at least two of the three structural types: monomer; oligomer; and polymer. Answer the following questions about one of the following nonmetals: (a) S; (b) Se; (c) N; (d) P; (e) As; and (f) Sb. (1) Read through the Section 13.3 on the nonmetal oxides and identify the oxide or oxides that show polymerization isomerism. (2) If possible, carry out the radius ratio type calculations to predict the expected coordination numbers of the nonmetal and oxygen. Which structural forms correspond to your prediction? Is the radius ratio close to a boundary value, so that you might have predicted isomerism? (3) The melting points, and so on, of these polymerization isomers are complicated by the fact that their bonds can rearrange at some temperature to give rise to another isomer. Assuming that this was not possible, predict whether each polymerization isomer would then be a gas, liquid, low-melting solid, or high-melting solid.

7. *Which of the following oxides are monomeric, which are oligomeric, and which are ionic or macromolecular substances? (a) Na_2O; (b) Cr_2O_3; (c) CO_2; (d) SiO_2; (e) P_4O_{10}; (f) BaO; (g) Tl_2O; (h) SO_2; and (i) Al_2O_3.

8. Selenium trioxide occurs in three forms: (a) as monomeric covalent molecules, (b) as tetrameric covalent molecules, and (c) as chain polymeric molecules. Draw the structure of selenium trioxide in each of these three forms.

9. There is expected to be π bonding from O to the central atom in the tetroxides MO_4. (a) In such π bonding, electrons would be donated from what types of orbitals on oxygen to what types of orbitals on the central atom? (b) Based on periodic trends discussed in Chapters 1 and 3, which two of the following four tetroxides should show the greatest degree of π bonding? Why? Tetroxides: FeO_4; RuO_4; OsO_4; XeO_4.

10. (a) Name the six most important oxides of nitrogen. (b) Draw the Lewis dot structures of any three of these oxides. (c) Which (if any) of these are gases at room temperature?

11. Contrast the structures found in the following solids: (a) CO_2 and SiO_2; (b) BF_3 and AlF_3; (c) BF_3 and BeF_2; and (d) HNO_3 and "HPO_3."

12. *The compound SiO_2 exists in different phases: quartz, cristobalite, coesite, and stishovite. The latter two forms are found only in meteor craters. Carbon dioxide has only one form, which is completely different from any form of SiO_2. The quite different physical properties and conditions of formation of these oxides result from their differing structures, which in turn can be rationalized on the basis of the differing size of C and Si. (a) Do radius ratio calculations for C^{4+}, Si^{4+} and O^{2-}. What are the actual coordination numbers found for C in CO_2 and Si in the common forms of SiO_2 (e.g., quartz)? (b) What are the actual coordination numbers of oxygen in CO_2 and in quartz? Use these numbers to explain why CO_2 is a gas while quartz is a solid with a very high melting point. (c) Stishovite has the less common coordination number for silicon in SiO_2. Identify this number, and predict the lattice type that stishovite is likely to take. (d) Calculate the lattice energies of quartz and of stishovite. (e) Does this calculation necessarily predict the most stable form of SiO_2? Explain why stishovite is only found in meteor craters.

13. (a) If you were to list *all* compounds and ions in which the central atom shows oxidation number +7, which would you list more of and why: fluorides or iodides; oxides or sulfides; and oxo anions or fluoro anions. (b) Would all oxo anions of these elements have equivalent formulas (i.e., would all be EO_4^-)? If not, would the oxo anions of formula EO_4^- have the same, greater, or lesser basicity than the oxo anions of the other formulas? In general, would the salts of oxo anions EO_4^- give precipitates with the same cations as the oxo anions of the other formula? (c) All of the oxides of these elements in the +7 oxidation state would necessarily have the same simplest formula, E_2O_7, but they could have different molecular structures. Discuss how the structure of E_2O_7 might change as one went down a group of elements E, and describe the effects this would have on the its physical state and melting point.

14. *Consider the following oxo acids: (a) H_2SeO_3; (b) H_6TeO_6; (c) $HMnO_4$; (d) H_2PtO_6 (an imaginary acid); (e) H_3PO_4; (f) H_4XeO_6; (g) H_4SnO_3; (h) H_3VO_4; (i) H_2XeO_4; and (j) $HTcO_4$. For each oxo acid give (1) its category of acidity; (2) its name; (3) the

formula of the acidic oxide from which it might be prepared by the addition of water only (i.e., no redox chemistry); and (4) whether it will be soluble or insoluble in water.

15. *Classify each of the following oxides as acidic, basic, or amphoteric or neutral, and decide whether it will be soluble in water: (a) Na_2O; (b) Cr_2O_3; (c) CO_2; (d) SiO_2; (e) P_4O_{10}; (f) BaO; (g) Tl_2O; (h) SO_2; and (i) Al_2O_3.

16. Arrange the following oxides in order of decreasing acidity/increasing basicity: (a) Na_2O, Cr_2O_3, CO_2, SiO_2, P_4O_{10}; (b) ZrO_2, CO_2, SrO, Rb_2O, Y_2O_3; (c) MnO, MnO_2, Mn_2O_3, Mn_2O_7; and (d) TiO_2, TeO_2, SO_2, ThO_2.

17. Suppose that the following five oxides of new elements have been discovered: hoffmon tetroxide, HmO_4 (Hm = fourth period element); dimongium heptoxide, Mi_2O_7 (Mi = sixth period element); yangium dioxide, YaO_2 (Ya = second period element with no unshared *sp* electron pairs); vermillium dioxide, VeO_2 (Ve = fifth period element *with* one unshared *sp* electron pair); and luium dioxide, LmO_2 (Lm = fifth period element with no unshared *sp* electron pair). (a) Which two oxides are most likely to be gases at room temperature? (b) Which oxide is most likely to be strongly acidic? (c) Which oxide is most likely to be basic? Is it likely to be water soluble? (d) Which two oxides are most likely to be strong oxidizing agents? (e) Which oxide is most likely to be useful for making ceramic plates to shield space shuttles from the heat of reentry into the atmosphere?

18. You have produced minute quantities of radioactive element number 109, meitnerium, and are investigating its positive oxidation states. (a) Suppose that you have oxidized it very strongly in a hot acidic solution, and find the radioactivity coming from the vapors above the solution; you conclude that you have a volatile oxide. What oxidation state do you probably have? Give arguments why you chose that oxidation state and not some other. Draw a likely structure of the oxide. (b) Suppose that you subsequently obtain meitnerium in the +6 oxidation state in basic solution, and find that it gives a precipitate not only with Ba^{2+} but also with most acidic cations. What does this lead you to suspect about the formula of the +6 species? (c) You have only a trace of Mt left, in the +7 oxidation state in basic solution. Outline a plan for precipitating it from solution.

19. The research group of Academician V. I. Spitsyn and co-workers[11] reported the first synthesis of the compound iron(VIII) oxide (iron tetroxide): (a) What are the solubility properties and boiling-point-related data given for this compound? Why are these not expected for any other possible oxide of iron (especially Fe_2O_7 or FeO_3)? (b) If these oxides, and the known oxide OsO_4, were to act as acidic oxides upon dissolution in base, predict the formulas of the oxo anions that would be formed; compare with the known oxo anions (Chapter 2). Would FeO_4 be classified as an acidic oxide, a basic oxide, an amphoteric oxide, or a neutral oxide? Contrast this with the classifications of the hypothetical iron oxides Fe_2O_7 and FeO_3, as well as the classification of OsO_4.

20. Consider the following set of oxides: SrO, ZrO_2, TeO_3, XeO_4. (a) Which one of these oxides is a solid, insoluble in water, that reacts with the acidic oxide SiO_2 to give a simple silicate salt? Write an equation for this reaction. (b) Which one of these oxides is a solid that dissolves in water to give a basic solution? Write an equation for this process. (c) Which one of these oxides is a gas that dissolves in water to give an acidic solution? Write an equation for this process. (d) Which oxide would you be most likely to use in a scrubber for removing sulfur oxides from smokestack gases?

21. Consider the following set of oxides: MnO_2, OsO_2, SnO_2, CO_2, OsO_4, XeO_4, FeO_4. (a) Which three of these oxides are high-melting solids? (b) Which four of these oxides are easily vaporized (are gases or at least volatile solids)? (c) Three of the four easily vaporized solids are in the same class of acid–base reactivity. Are these acidic or basic oxides? The fourth of these easily vaporized oxides is, in contrast to the others, a neutral oxide. Which one is this? Briefly explain why.

22. *Complete and balance the following chemical equations (or tell if no reaction will occur):

(a) $Tl_2O + H_2O \rightarrow$
(b) $I_2O_5 + H_2O \rightarrow$
(c) $ClO_2 + OH^- \rightarrow$
(d) $La_2O_3 + H^+ \rightarrow$
(e) $B_2O_3 + OH^- + H_2O \rightarrow$
(f) $FeO + P_4O_{10} \rightarrow$
(g) $MnO + H^+(aq) \rightarrow$

23. Complete and balance the following chemical equations (or tell if no reaction will occur):

(a) $SrO(s) + MoO_3(s) \rightarrow$
(b) $N_2O_5 + H_2O \rightarrow$
(c) $CaO(s) + TeO_3(s) \rightarrow$
(d) $CaO(s) + MnO(s) \rightarrow$
(e) $Cl_2O_7 + H_2O \rightarrow$

24. *Consider the following set of oxides: SrO, ZrO_2, MoO_3, RuO_4. If needed, take the radius of Ru^{8+} to be 52 pm. (a) Which of these oxides will be soluble in water to give a basic solution? Write a chemical equation for this process. (b) Which of these oxides will be soluble in water to give an acidic solution? Write an equation for this process. (c) Which of these oxides (if any) would be monomeric molecular substances? (d) Which of these oxides (if any) would most easily become a gas? (e) Which one of these oxides would most likely be a basic oxide that would be *insoluble* in water? Write an example of an equation for a chemical reaction that would justify classifying it as a basic oxide. (f) Which two of the above oxides will react *most exothermically* with each other? Write an equation for this reaction.

25. Classify each oxide below as acidic, basic, and so on. Then complete and balance the following equations, or note if no reaction is to be expected.

(a) $BaO(s) + P_4O_{10}(s) \rightarrow$
(b) $Na_2O(s) + H_2O \rightarrow$
(c) $Cr_2O_3(s) + H^+(aq) \rightarrow$
(d) $Cr_2O_3(s) + H_2O \rightarrow$
(e) $SiO_2(s) + OH^- \rightarrow$
(f) $TeO_3(s) + I_2O_5(s) \rightarrow$
(g) $Y_2O_3(s) + P_4O_{10}(s) \rightarrow$
(h) $Cl_2O(g) + H_2O \rightarrow$
(i) $Na_2O(s) + V_2O_5(s) \rightarrow$
(j) $Na_2O(s) + BaO(s) \rightarrow$
(k) $Na_2O(s) + P_4O_{10}(s) \rightarrow$
(l) $Fe_2O_3(s) + BaO(s) \rightarrow$

26. Which of the oxide acid–base reactions that you balanced in (a) Exercise 22; (b) Exercise 23; (c) Exercise 25 would be the most exothermic per mole of oxide ion transferred?

27. Describe the air-pollution problems associated with nonmetal oxides specified by your instructor, and devise some possible abatement procedures that might be tried.

28. *(a) Write three balanced chemical equations showing the three steps by which elemental sulfur in coal is converted to sulfuric acid in acid rain. (b) Sulfur dioxide

can be removed from smokestack gases by reaction with ("scrubbing" with) magnesium oxide. Write a chemical equation for this process. (c) Calculate the number of grams of magnesium oxide that would be needed to clean the smokestack gases from burning 1,000,000 g of coal that is 3.2% S by weight.

29. Write three balanced chemical equations showing the three steps by which ammonia is converted industrially to nitric acid.

30. Write an equation showing the low-cost process for making fertilizer-grade phosphoric acid from a common mineral source. Briefly describe one pollution problem that can arise from the manufacture of fertilizer-grade phosphoric acid in this manner.

31. Which of the following structural descriptions is/are inconsistent with the stoichiometry of the salt being described? (a) The compound $CdCl_2$ adapts an hcp lattice of chloride ions in which all of the octahedral holes are occupied by cadmium ions. (b) The compound $CdCl_2$ adapts a ccp lattice of chloride ions in which one-half of the octahedral holes are occupied by cadmium ions. (c) The compound Li_2SO_4 adapts an hcp lattice of sulfate ions in which all of the tetrahedral holes are occupied by lithium ions. (d) The compound $(CH_3)_4NF$ adapts a ccp lattice of tetramethylammonium ions in which all of the octahedral holes are occupied by fluoride ions. (e) The compound Rb_2CsC_{60} adopts a close-packed lattice in which the Rb^+ ions occupy all octahedral holes and the Cs^+ ions occupy all tetrahedral holes.

32. With reference to Tables B and C, describe some low oxidation number metal oxides that might be nonstoichiometric in a manner analogous to $Fe_{0.95}O$. Why would you not expect very high oxidation number oxides such as Mn_2O_7 and OsO_4 to be nonstoichiometric (i.e., $Mn_{2.2}O_7$ or $Os_{1.13}O_4$)?

33. *Which of the following formulas correspond to possible nonstoichiometric oxides? $Ca_{0.95}O$; $Fe_{0.95}O$; $Co_{0.95}O$; $C_{0.95}O$; $Cr_{0.95}O_3$; and $Eu_{0.95}O$.

34. *In the following list, (a) Which oxides are likely to be spinels? (b) Which oxides could be perovskites? $NiFe_2O_4$, $BaFe_2O_4$, $BaTiO_3$, $BeTiO_3$, $BaSO_3$, $TiZn_2O_4$, Ni_3O_4, Pb_3O_4, $NaTaO_3$.

35. In the following list, (a) Which oxides are likely to be spinels? (b) Which oxides could be perovskites? $TiCo_2O_4$, $NaTaO_3$, $SrTiO_3$, Zn_2SiO_4, $ZnSeO_3$, $CoFe_2O_4$, $FeCo_2O_4$.

36. Consider the following series of oxides: MnO, MnO_2, Mn_2O_3, Mn_2O_7, Mn_3O_4. (a) Which of these will be most acidic? (b) Which of these will be most basic? (c) Which is most likely to be soluble in water? (d) Which of these is least likely to show a nonstoichiometric (defect) structure? (e) Which (if any) of these would show a perovskite structure? (f) Which (if any) of these would show a spinel structure? (g) Which (if any) of these would consist of (nonpolymerized) covalent molecules? (h) Which would have the lowest melting point? (i) Which one is a liquid at room temperature? (j) Which would the rest be: gases, or solids? (k) Which is most likely to be amphoteric?

37. Choose the appropriate classification(s)—perovskite, spinel, nonstoichiometric, defect perovskite, superconductor—for each of the following mixed-metal oxides: (a) $BaTiO_3$; (b) $La_{1.85}Sr_{0.15}CuO_4$; (c) $Fe_{0.95}O$; (d) Fe_3O_4; and (e) $YBa_2Cu_3O_{6.9}$.

38. Which of the following formulas correspond to possible nonstoichiometric oxides? $Sr_{0.95}O$, $Fe_{0.95}O$, $Ni_{0.95}O$, $C_{0.95}O$, $Cr_{0.95}O$, and $N_{0.95}O$.

39. *(a) Verify that the large unit cell of $YBa_2Cu_3O_7$ shown in Figure 13.7 indeed shows the correct stoichiometry of ions. (b) If there were no mixing of Cu^{3+} and Cu^{2+} ions between the chains and sheets of copper ions, which type of copper ion would reside in the chains and which in the sheets in order to give the proper stoichiometry?

40. Determine the coordination number and list the nearest neighbors of the oxide ions in (a) perovskite; (b) $YBa_2Cu_3O_7$. (c) Confirm that Eq. (4.15) applies to these mixed-metal oxides if it is modified to read: Σ (Coordination No. of M) (No. of M in formula) = (Coordination No. of anion) (No. of anions in formula).

41. Do crystal field stabilization energies (Chapter 8) favor placing the Cu^{3+} ions in the square planar CuO_4 sites in the chains and the Cu^{2+} ions in the square pyramidal CuO_5 sheets, or do they favor the opposite placement?

42. Assign likely oxidation numbers to each element in the superconductors other than $YBa_2Cu_3O_7$ discussed in Section 13.5. In which of these would it be necessary to substitute other metal ions in order for the superconductor to contain both Cu^{2+} and Cu^{3+} ions? Describe the types of ions you would introduce in order to achieve this.

43. *Select one of the following minerals—(1) wollastonite = $CaSiO_3$; (2) talc = $Mg_3(OH)_2(Si_4O_{10})$; (3) grunerite = $Fe_7(OH)_2(Si_4O_{11})_2$; (4) monticellite = $CaMgSiO_4$; (5) stishovite = SiO_2—as an example of each of the following: (a) contains a monomeric silicate ion; (b) contains a chain polysilicate ion; (c) contains a double-chain polysilicate ion; (d) contains a sheet polysilicate ion; (e) the mineral that would weather the most rapidly; (f) the mineral that would weather the most slowly.

44. Classify each of the following silicates as (1) a sheet polysilicate, (2) a chain or cyclic polysilicate, (3) a simple silicate, or (4) a double-chain polysilicate: (a) bustamite = $CaMn(SiO_3)_2$; (b) spudomene = $LiAl(SiO_3)_2$; (c) tremolite = $Ca_2Mg_5(OH)_2$-$(Si_4O_{11})_2$; (d) coffinite = $U(SiO_4)$; and (e) kaolinite = $Al_2(OH)_4(Si_2O_5)$.

45. Show that you understand the condensed drawings of the fragments of polysilicate structures shown in Figure 13.10 by redrawing them, using closed circles for Si atoms and open circles for O atoms, as in Figure 13.8. Redraw the following: (a) ${}^{1}_{\infty}[SiO_3{}^{2-}]$; (b) ${}^{1}_{\infty}[Si_4O_{11}{}^{6-}]$; and (c) ${}^{2}_{\infty}[Si_4O_{10}{}^{4-}]$.

46. Select the proper structural type—framework aluminosilicate, sheet polysilicate; chain or cyclic polysilicate, simple silicate, double-chain polysilicate, sheet aluminosilicate —for each of the following silicates: (a) ureyite = $NaCr(SiO_3)_2$; (b) natrolite = $Na_2[Al_2Si_3O_{10}]\cdot 2H_2O$; (c) biotite = $K(Mg,Fe)_3[AlSi_3O_{10}](OH)_2$; (d) cummingtonite = $Mg_7(OH)_2(Si_4O_{11})_2$; (e) willemite = $Zn_2(SiO_4)$; (f) celsian = $Ba[Al_2Si_2O_8]$; (g) chabazite = $Ca_6[Al_{12}Si_{24}O_{72}]$; (h) tremolite = $Ca_2(Mg,Fe)_5(OH)_2(Si_4O_{11})_2$; and (i) phenacite = $Be_2(SiO_4)$.

47. *Consider the following silicates and aluminosilicates: tremolite = $Ca_2(Mg,Fe)_5$-$(OH)_2(Si_4O_{11})_2$; ureyite = $NaCr(SiO_3)_2$; chabazite = $Ca_6[Al_{12}Si_{24}O_{72}]$; kaolinite = $Al_4(OH)_8(Si_4O_{10})$; and phenacite = $Be_2(SiO_4)$. (a) Draw the structure of the simplest formula unit of each silicate or aluminosilicate anion except that of chabazite. (b) Write the formula of the (mononuclear) structural unit of each anion, along with its (possibly fractional) charge. (c) List the anions in order of increasing predicted basicity. (d) Briefly explain why aluminum is written inside the brackets in the formula of chabazite but outside the brackets in the formula of kaolinite, $Al_4(OH)_8[Si_4O_{10}]$.

48. The mineral used in construction as an asbestos replacement, xonotlite, $Ca_6(OH)_2$-(Si_6O_{17}), is also a double-chain polysilicate, but it differs slightly in structure: propose a structure for xonotlite.

49. Predict the basicity classification of the anion in each of the following polysilicates or polyaluminosilicates, based on the (untested) hypothesis that these can be estimated from the formula of the mononuclear structural unit obtained by depolymerizing the anion, then counting the number of oxo groups and units of negative charge in the structural unit. (a) $Si_2O_7{}^{6-}$; (b) $Si_3O_9{}^{6-}$; (c) $Si_4O_{11}{}^{6-}$; (d) $Si_4O_{10}{}^{4-}$; (e) anion of orthoclase; (f) anion of celsian; (g) anion of sodalite; (h) anion of ZSM-5; (i) anion of montmorillonite if only one-fourth of the aluminum ions are isomorphously substituted by magnesium ions.

50. Estimate the pK_a values of the corresponding conjugate acids of the silicate and aluminosilicate anions in the previous exercise, assuming that the hypothesis in it has some validity. If so, which of the above polysilicates, when ion exchanged into the H^+ forms, would be very strongly acidic?

51. *You are studying the mineral hornblende, $Ca_2Mg_5(OH)_2(Si_4O_{11})_2$, and find samples in which isomorphous substitution of the magnesium and the calcium has occurred. Which of the following are possible minerals which could result from isomorphous substitution processes in hornblende? (a) $Y_2Mg_5(OH)_2(Si_4O_{11})_2$; (b) $Na_2Mg_5(OH)_2$-$(Si_4O_{11})_2$; (c) $Na_2Mg_3(Fe^{III})_2(OH)_2(Si_4O_{11})_2$; (d) $Y_2Mg_3Li_2(OH)_2(Si_4O_{11})_2$; and (e) $Sr_2Mg_5(OH)_2(Si_4O_{11})_2$.

52. Devise a possible asbestos substitute that would be the aluminosilicate obtained from asbestos by isomorphous substitution of the second kind.

53. The formula of the common form of the mineral garnet, andradite, can be written as $Ca_3Fe_2Si_3O_{12}$; there is a family of garnets including those listed later. For each member of the family, tell which principle (or principles) of isomorphous substitution is/are used to generate that member from common garnet (and confirm that it applies), and identify the oxidation state of any *d*-block metal ions: (a) almandite, $Fe_3Al_2Si_3O_{12}$; (b) grossularite, $Ca_3Al_2Si_3O_{12}$; (c) spessartite, $Mn_3Al_2Si_3O_{12}$; (d) uvarovite, $Ca_3Cr_2Si_3O_{12}$; (e) the synthetic "YAG," used in lasers: $Y_3Al_5O_{12}$; and (f) the synthetic $Y_3Fe_5O_{12}$, of importance for its magnetic properties.

54. *Consider the following mineral pairs. Classify each as related to the other by (1) isomorphous substitution of the first kind; (2) isomorphous substitution of the second kind; (3) non-isomorphous substitution; or (4) a pair that is completely unrelated in structure. (a) $KCr(SO_4)_2{\cdot}12H_2O$ and $RbAl(SO_4)_2{\cdot}12H_2O$; (b) $KCr(SO_4)_2{\cdot}12H_2O$ and $KCr(SO_4)_2$; (c) $Pb_5Cl(PO_4)_3$ and $Pb_5Cl(VO_4)_3$; (d) $Rb[AlSi_3O_8]$ and $Ba[Al_2Si_2O_8]$; and (e) ${}^3_\infty\{[SiO_2]_{24}\}$ and ${}^3_\infty\{Na_{12}[Al_{12}Si_{12}O_{48}]\}$.

55. Explain why Cu^{2+} should be less likely to be found isomorphously substituted in minerals than most other +2 charged *d*-block metal ions of similar size.

56. You are studying the feldspar $K[AlSi_3O_8]$, and find samples in which isomorphous substitution has occurred. Which of the following are possible minerals that could result from isomorphous substitution processes in this feldspar? (a) $Ba[Al_2Si_2O_8]$; (b) $Rb[AlSi_3O_8]$; (c) $Na[AlSi_3O_8]$; and (d) $Ba[AlSi_3O_8]$.

57. The formula of the mineral tremolite is $Ca_2(Mg,Fe)_5(OH)_2(Si_4O_{11})_2$. Interpret the part of the formula which is written $(Fe,Mg)_5$. What is the term for this phenomenon in geochemistry?

58. *You are in charge of disposing of asbestos being removed from schools, where it was used as insulation. Devise a reasonable scheme for chemically destroying the asbestos.

59. The relative availability of nutrient elements in soil as a function of soil pH is shown in Figure 13.13. Using the principles in the book to date, insofar as you can explain why each element is available and unavailable at the pH value given.

60. *Consider the following types of silicates or related soil minerals: sheet aluminosilicates; sheet polysilicates; chain or cyclic polysilicates; simple silicates; three-dimensional polymeric metal oxides (e.g., TiO_2); double-chain polysilicates. (a) Which one would weather most quickly? (b) Which one would be most characteristically found in a desert soil? (c) Which one would be most characteristically found in the soil of a tropical region in which the forest has been cut down?

61. Consider the following minerals and oxides: lime = CaO; ureyite = $NaCr(SiO_3)_2$; chabazite = $Ca_6[Al_{12}Si_{24}O_{72}]$; zirconia = ZrO_2; tremolite = $Ca_2(Mg,Fe)_5(OH)_2(Si_4O_{11})_2$; olivine = $(Fe,Mg)_2(SiO_4)$; kaolinite = $Al_4(OH)_8(Si_4O_{10})$. (a) Which one would weather most quickly? (b) Which one would most characteristically be found on the earth's surface only in the youthful soil of a desert? (c) Which would be the last to weather away? (d) Which is abundant in the mantle, and is converted to a spinel polymerization isomer deep in the mantle?

62. To which type of oligomeric silicate anion is the anion found in Portland cement most closely related?

63. *Give the symbols for six elements that readily form polymeric oxo anions.

64. With reference to Table C, find other "cations," each of a different charge, that could substitute for P^{5+} in the phosphomolybdate ion. Give the resulting formula and charge of each heteromolybdate ion that results; rank these heteromolybdates in order of increasing basicity.

65. Molybdenum forms many heteropolymolybdate ions such as $[PMo_{12}O_{40}]^{3-}$. (a) Are these chain, linked-chain, sheet, and so on polymers like the polysilicates? If so, tell which type of structure they have; if not, tell how their structures differ from the polysilicates. (b) Give two uses for such heteropolymolybdates.

66. Classify the basicity (weakly basic, etc.) of each of the following simple oxo anions. (Also provide the name if that is not given; provide the formula if that is not given.) Then tell whether the polyoxometallic anion listed is a weaker or stronger base than the simple oxo anion. (a) Chromate and dichromate and (b) molybdate and $[PMo_{12}O_{40}]^{3-}$.

67. Complete and balance: $H^+(aq) + [H_2W_{12}O_{40}]^{6-}(aq) + H_3PO_4(aq) \rightarrow$.

68. You are studying the feldspar mineral orthoclase, $K[AlSi_3O_8]$, and find samples in which isomorphous substitution of some of the ions has occurred. (a) State the principles that govern the possible choices of foreign ions that can substitute for given ions in minerals without altering the crystal lattice type (i.e., which can substitute isomorphously). Which of the following formulas represent minerals that could be found in solid solution in orthoclase, as a result of isomorphous substitution (replacement) of ions in orthoclase? (1) $Ba[Al_2Si_2O_8]$, (2) $Rb[AlSi_3O_8]$, (3) $Na[AlSi_3O_8]$, (4) $Ba[AlSi_3O_8]$, (5) $K[YSi_3O_8]$, (6) $K[AlSi_3S_8]$, and (7) $Na[Al_2Si_2O_8]$. (b) You are hoping to find a good mineral source of the element yttrium, Y, for use in high-temperature superconductors. Which of the following minerals would (at least in theory) be likely places in which to find an appreciable concentration of yttrium as the compound indicated? (1) in Bi_2S_3 as Y_2S_3, (2) in orthoclase as $K[YSi_3O_8]$, (3) in $CePO_4$ as YPO_4, (4) in sea water as $Y(H_2O)_n{}^{3+}$, (5) in the core of the earth (largely metallic iron and nickel) as metallic Y, and (6) in deposits of $Ca_2Mg_5(OH)_2(Si_4O_{11})_2$ as $Y_2Mg_5(OH)_2[Al_2Si_2O_{11}]_2$. (c) If you are indeed able to find samples of the minerals $K[YSi_3O_8]$ and $Y_2Mg_5(OH)_2[Al_2Si_2O_{11}]_2$, you would expect them to be examples of which classes of silicate minerals? (1) Framework (three-dimensional) aluminosilicates. (2) Chain or cyclic polysilicates. (3) Simple silicates. (4) Double-chain polysilicates. (5) Sheet aluminosilicates. How many oxygen atoms would you expect to find around each silicon atom in these minerals? Would you anticipate finding a larger or smaller number of oxygen atoms around the yttrium?

Notes

1. Note to the Instructor: If Chapter 12 has not yet been covered, it will be desirable to cover Section 12.1 before beginning this chapter.
2. If we need to avoid fractional numbers of atoms, as in nomenclature, we may write, for example, Re_2O_7 as the *simplest formula unit*.
3. Although isomers of organic compounds have the same molecular formula, which these do not. Among elements, the term *allotropes* covers this distinction; some have advocated using this term among compounds as well.
4. More elaborate but more precise methods for predicting structure are given by I. D. Brown: *Acta Crystallogr., Sect. B*, **44**, 545 (1988); **48**, 533 (1992).
5. Recall that some of the heavier and *d*-block elements show relativistic or crystal-field effects that routinely produce unusually low coordination numbers.
6. In principle, the number of unshared electron pairs should be multiplied by a factor greater than 1, since unshared electron pairs take extra space. However, it is difficult to know in practice exactly what value that factor would have.
7. For radius ratios over 1.0, the inverse radius ratio r_{anion}/r_{cation} should be used in the prediction, as mentioned in Section 4.7.
8. A. F. Wells, *Structural Inorganic Chemistry*, 5th ed., Clarendon Press, Oxford, UK, 1984.
9. Melting and boiling points are taken from M. C. Ball and A. H. Norbury, *Physical Data for Inorganic Chemists*, Longmans, London, 1974, or from the *Handbook of Chemistry and Physics*, 50th ed., Chemical Rubber Publishing Company, Cleveland, OH, 1969. Values were also checked for consistency with data given in the *Alfa Catalog 1983–1984*, Danvers, Morton-Thiokol Inc., MA, 1983.
10. F. A. Cotton and G. Wilkinson, *Advanced Inorganic Chemistry: A Comprehensive Text*, 5th ed., Wiley-Interscience, New York, 1988; N. N. Greenwood and A. Earnshaw, *Chemistry of the Elements*, Pergamon Press, Oxford, 1984.
11. Yu. M. Kiselev, N. S. Kopelev, V. I. Spitsyn, and L. I. Martynenko, *Dokl. Chem.*, **292**, 29 (1987).
12. D. Leopold and K. Seppelt, *Angew. Chem. Intl. Ed. Engl.*, **33**, 975 (1994); K. Seppelt. *Acc. Chem. Res.*, **30**, 111 (1997).
13. A. Schulz, I. C. Tornieporth-Oetting, and T. M. Klapötke, *Angew. Chem. Intl. Ed. Engl.*, **32**, 1610 (1993).
14. Several of these monomeric oxides, being nonpolar covalent molecules, are actually more soluble in carbon tetrachloride than they are in water.

15. Some insoluble acidic oxides, most notably Nb_2O_5 and Ta_2O_5, form polymeric oxo anions such as $Ta_6O_{19}{}^{8-}$ (Section 13.8).
16. D. W. Smith, *J. Chem. Educ.*, **64**, 480 (1987).
17. S. G. Bratsch, *J. Chem. Educ.*, **65**, 877 (1988).
18. The fairly large scatter in this plot confirms that more than just Z^2/r is involved in determining the acidity of oxides.
19. *Kirk–Othmer Encyclopedia of Chemical Technology*, 4th ed., Wiley, New York, 1992.
20. S. E. Manahan, *Environmental Chemistry*, 6th ed., Boca Raton, FL, Lewis Publishers, 1994; J. E. Fergusson, *Inorganic Chemistry and the Earth: Chemical Resources, Their Extraction, Use, and Environmental Impact*, Pergamon, Oxford, UK, 1982.
21. T. R. Karl, N. Nicholls, and J. Gregory, *Sci. Am.*, **276(5)**, 79 (1997).
22. In higher concentrations (as in the lab), NO dimerizes partially to N_2O_2, which is oxidized in seconds by O_2.
23. M. Freemantle, *Chem. Eng. News*, Dec. 23, 1996, p. 7; March 31, 1997, p. 23.
24. G. J. Hutchings, C. S. Heneghan, J. D. Hudson, and S. H. Taylor, *Nature* (*London*), **384**, 341 (1996).
25. H. K. Bowen, *Sci. Am.*, **255(4)**, 169 (1986).
26. Note the similarity to the requirements for forming substitutional alloys (Section 12.3).
27. For geochemical aspects, see G. H. Muecke and P. Möller, *Sci. Am.* **258(1)**, 72 (1988).
28. A. Muñoz-Páez, *J. Chem. Educ.*, **71**, 381 (1994).
29. E. Wilson, *Chem. Eng. News*, Jan. 15, 1996, p. 23; for a lab experiment, see R. F. P. Nogueira and W. F. Jardim, *J. Chem. Educ.*, **70**, 861 (1993).
30. H. Aono, N. Imanaka, and G.-Y. Adachi, *Acc. Chem. Res.*, **27**, 265 (1994); D. F. Shriver and G. C. Farrington, *Chem. Eng. News*, May 20, 1985, p. 42.
31. J. L. Kirschvink, A. Kobayashi-Kirschvink, and B. J. Woodford, *Proc. Natl. Acad. Sci. USA*, **89**, 7683 (1992); *Chem. Eng. News*, May 18, 1992, p. 6.
32. R. B. Seymour and G. B. Kauffman, *J. Chem. Educ.*, **67**, 763 (1990).
33. A. Müller and J. G. Bednorz, *Science*, **217**, 1133 (1987).
34. See, for instance, R. J. Cava, *Sci. Am.*, **263(2)**, 42, (1990); F. J. Adrian and D. O. Cowan, *Chem. Eng. News*, Dec. 21, 1992, p. 24; A. Schilling, M. Cantoni, J. D. Guo, and H. R. Ott, *Nature* (*London*), **363**, 56 (1993); A. W. Sleight, *Acc. Chem. Res.*, **28**, 103 (1995); J. K. Burdett, *Inorg. Chem.*, **32**, 3915 (1993).
35. See, for example, A. M. Thayer, *Chem. Eng. News*, Nov. 27, 1989, p. 9; A. M. Wolsky, R. F. Giese, and E. J. Daniels, *Sci. Am.*, **260(2)**, 61 (1989).
36. J. Clarke, *Sci. Am.*, **271(2)**, 46 (1994).
37. R. Dagani, *Chem. Eng. News*, April 24, 1995, p. 6; R. Dagani, *Chem. Eng. News*, Nov. 11, 1996, p. 38.
38. C. N. R. Rao and J. Gopalakrishnan, *Acc. Chem. Res.*, **20**, 228 (1987).
39. P. I. Djurovich and R. J. Watts, *J. Chem. Educ.*, **70**, 497 (1993).
40. The most common form of asbestos, chrysotile or $Mg_6(OH)_8(Si_4O_{10})$, does not have the double-chain structure, and is not as persistent in the body.
41. T. P. Hanusa [*J. Chem. Educ.*, **64**, 686 (1987)] has argued that the diagonal relationships should involve steeper diagonals, for example, lithium is more closely related to Ca^{2+} through Ba^{2+} than it is to Mg^{2+}; this is certainly the case in terms of our acidity classifications.
42. If the cation is strongly acidic, for example, Zr^{4+} in $ZrSiO_4$ (zircon), the resistance to weathering is also enhanced since the Zr–O bonding is about as strong and chemically resistant as the Si–O bonds, and the structure is therefore network polymeric in covalent bonds. Zircons are so unreactive that, unless subject to intense heat, they can be relied upon to be chemically unchanged since the creation of the earth's crust [D. York, *Sci. Am.*, **268(1)**, 90 (1993)].
43. R. Jeanloz and T. Lay, *Sci. Am.*, **268(5)**, 48, 1993; H. W. Green II, *Sci. Am.*, **271(3)**, 64, 1994.
44. J. E. Cohen and D. Tilman, *Science*, **274**, 1150 (1996); *New York Times*, Nov. 19, 1996, p. B5.
45. P. Laszlo, *Acc. Chem. Res.*, **19**, 121 (1986); S. Okada, K. Tanaka, Y. Nakadaira, and N. Nakagawa, *Bull. Chem. Soc. Jpn.*, **65**, 2833 (1992); E. Angeles, A. Ramírez, I. Martínez, and E. Moreno, *J. Chem. Educ.*, **71**, 533 (1994).
46. A. G. Cairns-Smith, *Sci. Am.*, **252(6)**, 90 (1985); A. G. Cairns-Smith, *Seven Clues to the Origin of Life: A Scientific Detective Story*, Cambridge University Press, Cambridge, UK, 1985; J. P. Ferris, *Chem. Eng. News*, Aug. 27, 1984, p. 22.

47. J. P. Ferris and G. Ertem, *Science*, **257**, 1387 (1992); J. P. Ferris, A. R. Hill Jr., R. Liu, and L. E. Orgel, *Nature* (*London*), **381**, 59 (1996).
48. J. V. Smith, *Chem. Rev.*, **88**, 149 (1980); G. T. Kerr, *Sci. Am.*, **261(1)**, 100 (1989); M. E. Davis, *Acc. Chem. Res.*, **26**, 111 (1993).
49. W. M. Meier, D. H. Olson, and Ch. Baerlocher, "Atlas of Zeolite Structure Types," *Zeolites*, **17**, 1–229 (1996); C. C. Freyhardt, M. Tsapatsis, R. F. Lobo, K. J. Balkus Jr., and M. E. Davis, *Nature* (*London*), **381**, 295 (1996).
50. J. M. Thomas, *Sci. Am.*, **266(4)**, 112 (1992); A. Corma, *Chem. Rev.*, **95**, 559 (1995); W. E. Farneth and R. J. Gorte, *Chem. Rev.*, **95**, 615 (1995); J. A. Cusumano, *J. Chem. Educ.*, **72**, 959 (1995).
51. See, for example, F. Blatter and E. Schumaker, *J. Chem. Educ.*, **67**, 519 (1990); K. J. Balkus, Jr., and K. T. Ly, *J. Chem. Educ.*, **68**, 875 (1991).
52. M. E. Davis, *Chemtech*, Sept. 1994, p. 22; D. W. Lewis, D. J. Willock, C. R. A. Catlow, J. M. Thomas, and G. J. Hutchings, *Nature* (*London*), **382**, 604 (1996).
53. J. Haggin, *Chem. Eng. News*, Sept. 30, 1991, p. 31; R. Baum, *Chem. Eng. News*, July 13, 1992, p. 26.
54. M. Henry, J. P. Jolivet, and J. Livage, *Struct. Bonding* (*Berlin*), **77**, 153 (1992).
55. A. Müller, J. Meyer, E. Krickemeyer, and E. Diemann, *Angew. Chem. Int. Ed. Engl.*, **35**, 1206 (1996).
56. M. T. Pope and A. Müller, *Angew. Chem. Int. Ed. Engl.*, **30**, 34 (1991).

CHAPTER

14

The Halides, Nitrides, and Sulfides of the Elements

With Applications to Materials Science and Biochemistry

In this chapter, we will first examine the halides of the elements, which are widely used in synthesizing other inorganic, organometallic, and organic compounds, including polymers. In the latter one-half of this chapter, we will focus on other classes of inorganic compounds that are typically obtained from the halides by Lewis acid–base reactions: the sulfides, nitrides, and esters and alkoxides of the elements. In Chapter 15, we will look at other classes of compounds that, although often made from the halides, have more pronounced chemistries as reducing agents: the hydrides and simple organometallic compounds of the elements.

14.1 Structural Features of Halides

Oxidation States in Halides. As discussed in Section 6.3, the highest oxidation state of an element is limited by how easy it is to oxidize the anion. Consequently, among halides, a given element exhibits its highest oxidation state in its fluoride (Table 13.1). The highest oxidation state found among the fluorides of all the elements is +7, which is found only for one element at the bottom of the *p* block, iodine, in IF_7, and for two elements at the bottom of the *d* block, rhenium and gold, in ReF_7 and AuF_7. With the more easily oxidized chloride ion, the highest oxidation state achieved is +6, and then only in the *d* and *f* blocks, in WCl_6, $TcCl_6$, $ReCl_6$, and UCl_6. The only known hexabromide is WBr_6, and with the most easily oxidized halide, iodide, only pentaiodides have been achieved: PI_5, NbI_5, TaI_5, and PaI_5.

The highest oxidation state found in halides is, of course, limited to the group oxidation state, but this oxidation state does not persist to the far right in any block of the periodic table even among fluorides. In the *p* block, the other common oxidation state we expect to find is that in which the oxidation number of the central atom equals the group number minus two. Oxidation numbers lower than this are somewhat more common among halides than among oxo anions (Chapter 2) or oxides (Chapter 13), but they are still found, in the *p* block, only in the last three groups. Central atoms having two unshared electron pairs and an oxidation state equal to the group number minus four are found in several halides, such as XeF_4, ClF_3, BrF_3, IF_3, ICl_3, OF_2, SF_2, SCl_2, and $SeCl_2$. Central atoms having three unshared electron pairs and an oxidation state equal to the group number minus six are found in a few halides: XeF_2, KrF_2, and the *interhalogens*, ClF, BrF, BrCl, ICl, and IBr. In the *d* and *f* blocks there is, of course, no such limitations on oxidation states, which can also be equal to the group number minus odd numbers.

Structural Trends among Halides. Structural features among halides have been summavized by Wells.[1] The general procedures for predicting the degree of oligomerization or polymerization of halides were given in Section 13.1, and illustrated there for fluorides. The degrees of polymerization of halides are generally determined by familiar factors: (1) the number of halogen atoms present in the structural unit, which is determined by the group and block of the element; (2) the relative sizes of the "cation" and "anion" in the structural unit; and (3) in the *p* block, the number of unshared electron pairs, which we will show explicitly in their formulas in this section. Due to the importance of the first factor, we organize our studies of geometries according to the number of halogens present. As we go through the halides, we will note the results of the second effect: The iodide of an element may be less polymerized than the fluoride.

By valence shell electron-pair repulsion (VSEPR) theory (Section 3.4) we expect unshared pairs of electrons to affect the shapes of halides in the *p* block; in the *d* block we watch for the square planar preference of the d^8 electron configuration and the existence of Jahn–Teller distortions for some other configurations (Chapter 8).

Oligomeric and polymeric fluorides often have subtle but significant structural differences from the heavier halides, due to the greater electronegativity of fluorine. As mentioned in Section 13.1, fluorine does not readily act as a bridging link in covalent oligomeric structures (consider SiF_4, which does not oligomerize, even though Si can easily accommodate six fluorine atoms about it in SiF_6^{2-}); the less electronegative heavier halides have no such limitation. When an ionic structure is not practical (as for tetrafluorides or still higher fluorides of most nonmetals), a monomeric molecular structure is usually found for a fluoride. In addition, most trifluorides, difluorides, and monofluorides take ionic lattice types similar to those taken by the oxides of the same stoichiometry and radius ratio, while di- and trichlorides, bromides, and iodides often take distinctly nonionic lattice types known as **layer structures**, which will be described later in this section.

When fluorine does bridge in an oligomeric or polymeric covalent fluoride, it acts as a μ_2 ligand, with large bond angles of 130–180°; this angle is too large to contribute to the curvature necessarily found in the surface of a small oligomeric structure. We might suppose that this bridging angle is a consequence of the high polarity of the nonmetal-fluorine bond, which puts a high partial positive charge on the nonmetal atoms, especially in tetrafluorides and higher fluorides. Bridging interactions often would bring these very highly charged nonmetal atoms close together, engendering high degrees of repulsion.

In contrast, the other halogens, as a consequence of their lesser difference in electronegativity from most elements, readily form structures in which the halogens bridge (by normal covalent bonding or weaker secondary bonding). Their bond angles are typically between 70 and 100°, presumably to approximately match the 90° angle found between the two *p* electron pairs being donated to the two bridged atoms. With such a small angle at the halogen, we will see that it is possible to set up four-membered rings involving two bridging halogen atoms and two metal or nonmetal atoms. Even though this brings the two nonhalogen atoms rather close together, this is not such a problem because the nonhalogen atoms are not as positively charged as they are in the corresponding fluorine compound. It also helps that the larger halogens *are* bigger, and thus keep the two nonhalogen atoms further apart than in the case of fluorides.

Heptahalides. As already mentioned, only three heptahalides are known: IF_7, ReF_7, and the recently reported[2] AuF_7. The only known geometry among these is (monomeric) pentagonal bipyramidal.

Hexahalides. A larger number of these compounds are known (Table 14.1); most, as expected, are in Groups 6*F*, 6, or 16(VI), and involve elements that are not very oxidizing in the high +6 oxidation state with "anions" that are difficult to oxidize. With the high coordination number of 6 in the structural units, there is a reluctance to polymerize; hence, most of these compounds consist of monomeric molecules with octahedral geometries. The major exception anticipated was $:XeF_6$, which by VSEPR was expected to be distorted from an octahedron. This structure was controversial for years. Although it is distorted in some manner from an octahedron in the gas phase, in the solid phase it consists mainly of quartets of square pyramidal $:XeF_5^+$ cations bridged by F^- anions.

We ought to mention once again that VSEPR does not always work so well with hexahalide species having an unshared electron pair. Many such hexahalometallate anions (e.g., $:SbX_6^{3-}$ and $:TeX_6^{2-}$) are known, and some of these have O_h symmetry according to crystal structures, while others do not. The unanticipated O_h geometry is favored when both the halogen and the central atom come from lower in the *p* block. Other data suggests that these softer ions may be forced into the octahedral geometry by the packing forces of the lattice, but when this happens, the lone pair then occupies the antibonding (repulsive) *s* orbital and lengthens all M–X bonds.

Pentahalides. Some interesting trends are found in the pentahalides of the Group 15(V) elements. Not unexpectedly for steric reasons, nitrogen does not form pentahalides (but does form the ion NF_4^+); the remainder of the elements do form pentafluorides. The only stable chlorides are PCl_5 and $SbCl_5$; $AsCl_5$ decomposes above $-50\,°C$. Only phosphorus forms a pentabromide and pentaiodide. These trends are in accord with the trend in stability of the +5 oxidation states of these elements in their oxo anions (Section 6.7).

Structurally, pentahalides are more diverse than hexahalides (Fig. 14.1); five basic structural types ((*a–e*) in Fig. 14.1) are common among the pentafluorides and two (*f* and *g*) among the heavier pentahalides. In type (*a*), with (small) fluorine and the smaller central atoms P and As, the VSEPR expected trigonal bipyramidal monomeric geometry results (in which axial and equatorial halogens interchange positions rapidly). But since four and six coordination are somewhat preferred over five coordination, most halides do not adopt this geometry. In types (*b–d*), with fluorine and larger central atoms such as Sb, Bi, and the *d*- and *f*-block elements, fluorines are able to bridge. In some cases such as that of solid SbF_5 [Fig. 14.1(*b*)], oligomerization results in tetramers, $(F_4Sb\text{–}\mu_2\text{-}F)_4$, in which an eight-membered ring is linked together by the two-coordinate bridging halogens. In other cases, such as BiF_5 [Fig. 14.1(*d*)]

Table 14.1
Structures of Hepta-, Hexa-, and Pentahalides of Elements[a]

		Heptafluorides (Pentagonal bipyramidal monomers)				
IF_7						
ReF_7				AuF_7		
		Hexahalides (Octahedral monomers except as noted.)				
SF_6						
SeF_6						
TeF_6		$:XeF_5{}^+F^-$				
MoF_6	TcF_6	RuF_6	RhF_6			
WF_6	ReF_6	OsF_6	IrF_6	PtF_6		
UF_6	NpF_6	PuF_6				
WCl_6	$TcCl_6$					
	$ReCl_6$					
UCl_6						
WBr_6						
			Pentahalides			
PF_5		$:ClF_5$				
AsF_5		$:BrF_5$				
$(SbF_5)_4$		$:IF_5$				
${}^{1}_{\infty}[BiF_5]$						
${}^{1}_{\infty}[VF_5]$	${}^{1}_{\infty}[CrF_5]$					
$(NbF_5)_4$	$(MoF_5)_4$	${}^{1}_{\infty}[TcF_5]$	$(RuF_5)_4$	$(RhF_5)_4$		
$(TaF_5)_4$	$(WF_5)_4$	${}^{1}_{\infty}[ReF_5]$	$(OsF_5)_4$	$(IrF_5)_4$	$(PtF_5)_4$	$(AuF_5)_x$
$(PaF_5)_x$	$(UF_5)_x$					
$PCl_4{}^+PCl_6{}^-$						
$(SbCl_5)_2$						
$(NbCl_5)_2$	$(MoCl_5)_2$					
$(TaCl_5)_2$	$(WCl_5)_2$	$(ReCl_5)_2$	$(OsCl_5)_2$			
$(PaCl_5)_x$	$(UCl_5)_2$					
$PBr_4{}^+Br^-$						
$(NbBr_5)_2$						
$(TaBr_5)_2$	$(WBr_5)_x$	$(ReBr_5)_x$				
$(PaBr_5)_x$	$(UBr_5)_x$					
$PI_4{}^+I^-$						
$(NbI_5)_2$						
$(TaI_5)_2$						
$(PaI_5)_x$						

[a] The first column is for n-halides from the groups numbered n in the periodic table, that is, under heptahalides, for Group 7, 17(VII), or 7*F* halides; each subsequent column to the right is for halides from the next group to the right. Formulas are written to indicate the degree of oligomerization (indicated with numerical subscripts after the parentheses) or polymerization (indicated with ${}^{1}_{\infty}$ for chain polymers, with ${}^{2}_{\infty}$ for layer polymers, and with ${}^{3}_{\infty}$ to indicate network covalent polymers), or with an ionic formulation to indicate an ionic-type lattice. A subscript of x indicates that the structure is unknown, but that the physical properties suggest that the material is not monomeric.

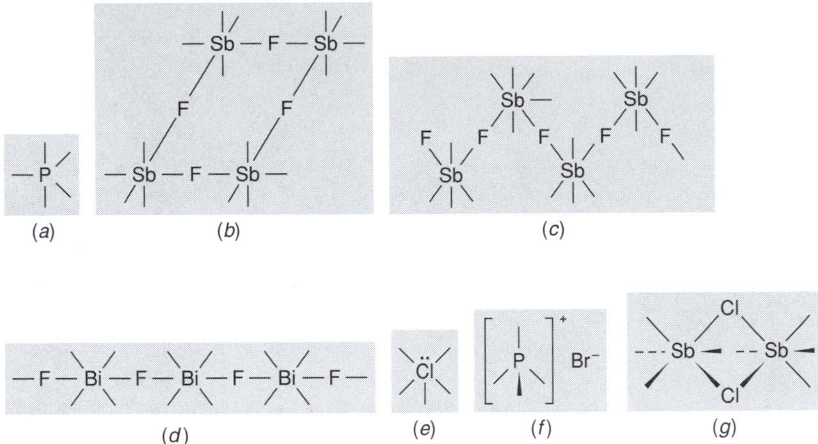

(d) (e) (f) (g)

Figure 14.1
Examples of the main structural types of pentahalides; terminal halogens shown as lines only. (*a*) Monomeric trigonal bipyramidal PF_5. (*b*) Tetrameric structure found for $(SbF_5)_4$. (*c*) Possible one-dimensional (chain) polymeric structure for $^1_\infty[SbF_5]$ in the liquid state. (*d*) Chain polymeric structure of $^1_\infty[BiF_5]$. (*e*) Monomeric square pyramidal $:ClF_5$. (*f*) Ionic $[PBr]^+Br^-$. (*g*) Dimeric structure found for $(NbCl_5)_2$, $(SbCl_5)_2$, and so on.

or liquid SbF_5 [Fig. 14.1(*c*)], chain polymerization occurs instead. In type (*e*), with a space-demanding unshared electron pair on a *p*-block central atom, we expect a pentafluoride to adopt a monomeric square pyramidal geometry; this is in fact done by $:ClF_5$, $:BrF_5$, and $:IF_5$. For type (*f*) with the small central P atom and larger halogen atoms (Br and I), there is not enough room for coordination of all five halogen atoms, so *ionic* structures are adopted in the solid state: $[PBr_4]^+Br^-$; $[PI_4]^+I^-$; and $[PCl_4]^+$-$[PCl_6]^-$. In type (*g*), the larger antimony atom prefers six coordination in $SbCl_5$, which is achieved by forming a halogen-bridged dimer, $Cl_4Sb(\mu_2\text{-}Cl)_2SbCl_4$; this structure is also adopted by a number of *d*- and *f*-block pentachlorides and pentabromides.

Tetrahalides. Over 70 tetrahalides are known, with crystal structures determined for one-half of them: These are too numerous to tabulate in Table 14.2, so we emphasize *p*-block structures which we show in Figure 14.2(*a–e*). (a) With small central atoms in combination with large halogen outer atoms, we expect monomeric tetrahedral molecules to be formed. All the tetrahalides of the Group 14(IV) elements except SnF_4 and PbF_4 in fact have this geometry, as do some *d*-block halides such as $TiCl_4$, $TiBr_4$, TiI_4, $ZrBr_4$, and ZrI_4. (b) In Figure 14.2(*b*), SnF_4 and PbF_4, with larger central atoms and smaller outer atoms, have two-dimensional polymeric structures featuring bridging by one-half of the fluorine atoms. With still larger *f*-block elements in their tetrafluorides and tetrachlorides, each halogen can bridge while the central atoms achieves eight coordination to give three-dimensional polymeric or ionic structures; these are not shown in Figure 14.2.

(c) Two Group 16(VI) atoms with one unshared electron pair, :S and :Se, also form tetrafluorides that are, as expected, seesaw-shaped molecules (Fig. 3.11), but chlorine (in combination with $:S^{IV}$) is too large for this: Solid $:SCl_4$ (unstable $>-30\,°C$) seems to exist as $:SCl_3^+$ and Cl^- ions [Fig. 14.2(*c*)]. (d) The somewhat larger $:Se^{IV}$

Table 14.2
Tetra- and Trihalides of the *p*-Block Elements[a]

Tetrahalides			Trihalides		
CF_4					
SiF_4	$:SF_4$		BF_3	$:NF_3$	
GeF_4	$:SeF_4$		$Al^{3+}(F^-)_3$	$:PF_3$	$::ClF_3$
${}^{2}_{\infty}[SnF_4]$	${}^{1}_{\infty}[:TeF_4]$	$::XeF_4$	$Ga^{3+}(F^-)_3$	$:AsF_3$	$::BrF_3$
${}^{2}_{\infty}[PbF_4]$			$In^{3+}(F^-)_3$	${}^{2}_{\infty}[:SbF_3]$	$::IF_3$
CCl_4			$Tl^{3+}(F^-)_3$	$:Bi^{3+}(F^-)_3$	
$SiCl_4$	$:SCl_3{}^+Cl^-$		BCl_3	$:NCl_3$	
$GeCl_4$	$(:SeCl_4)_4$		${}^{2}_{\infty}[AlCl_3]$	$:PCl_3$	
$SnCl_4$	$(:TeCl_4)_4$		$(GaCl_3)_2$	$:AsCl_3$	
$PbCl_4$	$(:PoCl_4)_x$		${}^{2}_{\infty}[InCl_3]$	${}^{3}_{\infty}[:SbCl_3]$	$(::ICl_3)_2$
CBr_4			${}^{2}_{\infty}[TlCl_3]$	${}^{3}_{\infty}[:BiCl_3]$	
$SiBr_4$			BBr_3		
$GeBr_4$	$(:SeBr_4)_4$		$(AlBr_3)_2$	$:PBr_3$	
$SnBr_4$	$(:SeBr_4)_4$		$(GaBr_3)_2$	$:AsBr_3$	
	$(:PoBr_4)_x$		$(InBr_3)_2$	${}^{3}_{\infty}[:SbBr_3]$	
CI_4			$(TlBr_3)_2$	${}^{3}_{\infty}[:BiBr_3]$	
SiI_4			BI_3		
GeI_4			${}^{1}_{\infty}[AlI_3]$	$:PI_3$	
SnI_4	$(:TeI_4)_4$		$(GaI_3)_2$	${}^{3}_{\infty}[:AsI_3]$	
	$(:PoI_4)_x$		$(InI_3)_2$	${}^{3}_{\infty}[:SbI_3]$	
				${}^{2}_{\infty}[:BiI_3]$	

[a] The first column is for *n*-halides from the groups numbered *n* in the periodic table, that is, under tetrahalides, for Group 4, 14(IV), or 4*F* halides; each subsequent column to the right is for halides from the next group to the right. Formulas are written to indicate the degree of oligomerization (indicated with numerical subscripts after the parentheses) or polymerization (indicated with ${}^{1}_{\infty}$ for chain polymers, with ${}^{2}_{\infty}$ for layer polymers, and with ${}^{3}_{\infty}$ to indicate network covalent polymers), or with an ionic formulation to indicate an ionic-type lattice. A subscript of *x* indicates that the structure is unknown, but that the physical properties suggest that the material is not monomeric.

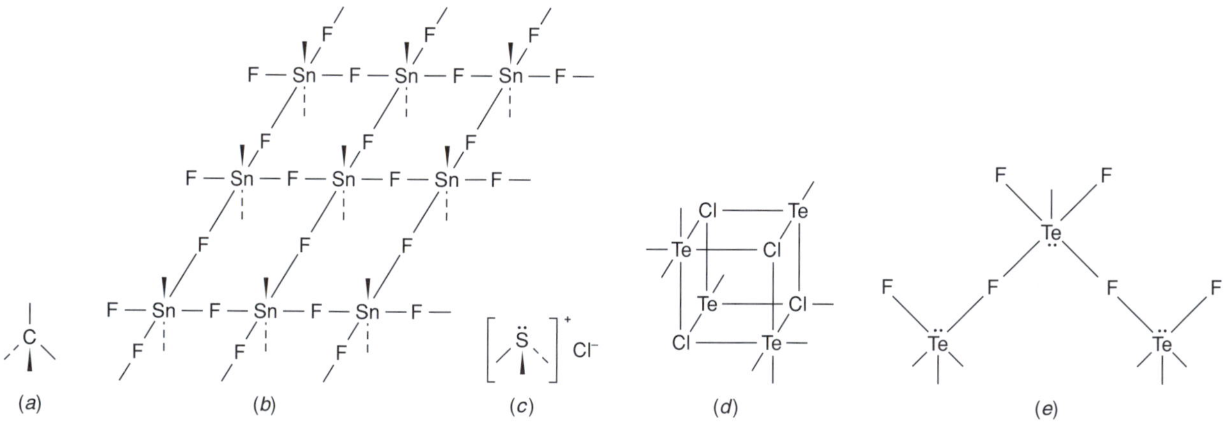

Figure 14.2
Examples of the main structural types of tetrahalides; terminal halogens shown as lines only. (*a*) Monomeric tetrahedral CF_4. (*b*) The layer polymeric structure of ${}^{2}_{\infty}[SnF_4]$. (*c*) Ionic $[:SCl_3]^+Cl^-$. (*d*) The "cubane" structure of Te_4Cl_{16} and related compounds. (*e*) The chain polymeric structure of solid ${}^{1}_{\infty}[:TeF_4]$.

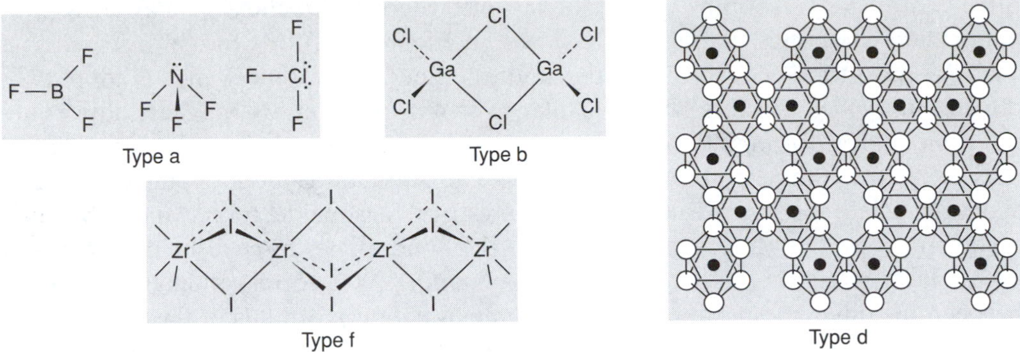

Figure 14.3
Selected examples of main structural types of trihalides. Type a: Monomeric BF_3, NF_3, and ClF_3. Type b: Structure of the dimer Ga_2Cl_6. Type c: Diverse MF_3 ionic structures are not shown. Type d: The $^{2}_{\infty}[AlCl_3]$ layer structure (coordination number of Al = 6; of Cl = 2): The two layers of halogens are represented by open circles and the middle layer of metal atoms by filled circles. Type e: Secondary bonded $:MX_3$ structures are not shown. Type f: Structure of the chain polymer $^{1}_{\infty}[ZrI_3]$. [Adapted from A. F. Wells, "Structural Inorganic Chemistry," 3rd ed., Oxford University Press, London, 1962, and G. J. Miller, "Halides: Solid State Chemistry," in *Encyclopedia of Inorganic Chemistry*, R. B. King, Ed., Wiley, Chichester, UK, 1994; p. 1343.]

and $:Te^{IV}$ atoms not only can accommodate four chlorine or bromine atoms, but even increase their coordination numbers to 6 by forming $(:MX_4)_4$ with the "cubane" structure [Fig. 14.2(*d*)]. The octahedral coordination about the Group 16(VI) atom is distorted by the unshared electron pair on the central atoms, however. In the case of $:TeF_4$ [Fig. 14.2(*e*)], however, chain polymerization is chosen over oligomerization. (f) The tetrahalide XeF_4 is unique, with two *sp*-hybrid unshared electron pairs on the central atom. As expected, it is a square planar monomeric molecule (Fig. 3.11).

Trihalides (Fig. 14.3). (a) In Figure 14.3(*a*), monomeric trihalides of necessity have low total coordination numbers [three for EX_3, four for $:EX_3$ from Group 15(V), and five for EX_3 from Group 17(VII)], so they are not at all abundant; most trihalides are oligomerized or polymerized. The boron trihalides are volatile monomeric molecular compounds. The simple Lewis structure of these species gives less than an octet of electrons to boron unless π bonding is included. The observed short B–X bond distances indicate that there is some degree of π bonding in these halides. From Group 15(V),[3] $:NF_3$, the phosphorus trihalides, $:AsF_3$, $:AsCl_3$, and $:AsBr_3$ are all volatile monomeric compounds having pyramidal structures as expected from VSEPR. All three trifluorides from Group 17(VII) (ClF_3, BrF_3, and IF_3) have the expected T-shaped structure.

(b) In Figure 14.3(*b*), some larger *p*-block trihalides with larger halogen atoms become four coordinate with dimeric structures, as in Ga_2Cl_6 and most of the Group 13(III) tribromides and triiodides. From Group 17(VII), $(ICl_3)_2$ forms a dimer that is similar except that it is square planar at each iodine due to the geometric influence of the two unshared electron pairs.

(c) The trifluorides of most metals are ionic and adopt a three-dimensional ionic lattice suitable to their size: These compounds range from those having six-coordinate metal ions and two-coordinate F^- to those having nine-coordinate metal ions and three-coordinate F^-. Some chlorides of very electropositive metals (such as UCl_3) also adopt these structures (these are not shown in Figure 14.3). (d) In contrast [see Fig. 14.3(*d*)], most trichlorides, tribromides, and triiodides of the *d*-block metals adopt two-dimensional *layer-structure* lattices similar to that shown for $AlCl_3$. These lattices are not quite logical for purely ionic compounds, since between alternate layers of halide ions there is a complete absence of cations. The alternate layers, if completely ionic, should repel each other. But such structures are reasonable for covalent bonding involving μ_2-bridging halogen atoms, since the bond angles at each halogen atom in such a layer structure are good for covalent bonding (90°, corresponding to the angle between the different unshared electron pairs in different *p* orbitals of the halogen).

The subtle variation that can result from changes in radius ratios and halogen electronegativities is illustrated in the halides of aluminum: AlF_3 adopts a typical ionic lattice with a coordination number of 6 for Al and 2 for F and has a very high melting point of 1272 °C. With the larger halogens, aluminum is four coordinate. The compound $AlCl_3$ adopts a two-dimensional layer structure but readily converts to an alternate (gaseous) form above 182 °C (at 1 atm) like that of Ga_2Cl_6. Aluminum bromide consists of dimeric Al_2Br_6, while aluminum iodide uses its bridging iodine to create a chain polymer, ${}^{1}_{\infty}[AlI_3]$.

(e) Another variation is illustrated by the heavier trihalides of Group 15(VA), which contain recognizable pyramidal $:MX_3$ molecules, but in which the large central atom can accommodate a larger total coordination number than 4 (one unshared *p*-electron pair plus three halogens). These trihalides adopt two- or three-dimensional polymeric structures (not shown in Fig. 14.3) utilizing bridging halogen atoms, although the bonds to the bridging halogen atoms are much longer and weaker than the bonds to the atom's "own" halogens. This type of secondary bonding[4] is also found in the lower halides of Se, Te, and in the halogens (Section 12.2) and interhalogens. The weakness of the bonding may often be due to the fact that the electron donation is from fairly low-energy orbitals (halogen lone pairs) to rather high-energy orbitals [antibonding Se–X σ^* orbitals, or high-energy $6p_{j=\frac{3}{2}}$ orbitals in Au, Hg, and so on (Section 1.11)].

Dihalides and Monohalides. Two halogen atoms are almost never sufficient to satisfy the expected total coordination number of an element, so monomeric dihalides are quite scarce. The compound $HgCl_2$ is essentially a monomeric covalent molecule for relativistic reasons (Section 1.11), but even $HgBr_2$ supplements the low coordination number of 2 by adding secondary bonding interactions. Nonmetal dihalides can be monomeric provided that there are at least two unshared electron pairs to complete a total coordination number of 4. Examples of this include the mostly unstable Group 16(VI) dihalides OF_2, OCl_2, and OBr_2 (normally thought of as halogen oxides Cl_2O and Br_2O), SF_2, SCl_2, $SeCl_2$, and $SeBr_2$, and the two Group 18(VIII) difluorides KrF_2 and XeF_2. Reasonably stable molecular monohalides are limited to the five interhalogen compounds IBr, ICl, BrF, BrCl, and ClF.

Most dihalides and monohalides are ionic or polymeric (Fig. 14.4). Metal di- and monofluorides, and the metal salts of the other halogens in which the electronegativity difference is large enough (perhaps 1.8–2.0 or so), adopt one of the ionic lattice types discussed earlier in Chapter 4. Metal dihalides of the heavier halogens in which the electronegativity difference is smaller often adopt distinctive layer-structure lattices; especially common are the $CdCl_2$ and CdI_2 lattice types [Fig. 14.4(*a* and *b*)], two varieties of a structure that provide six coordination for the metal and three coordination for each μ_3-apical halogen atom.

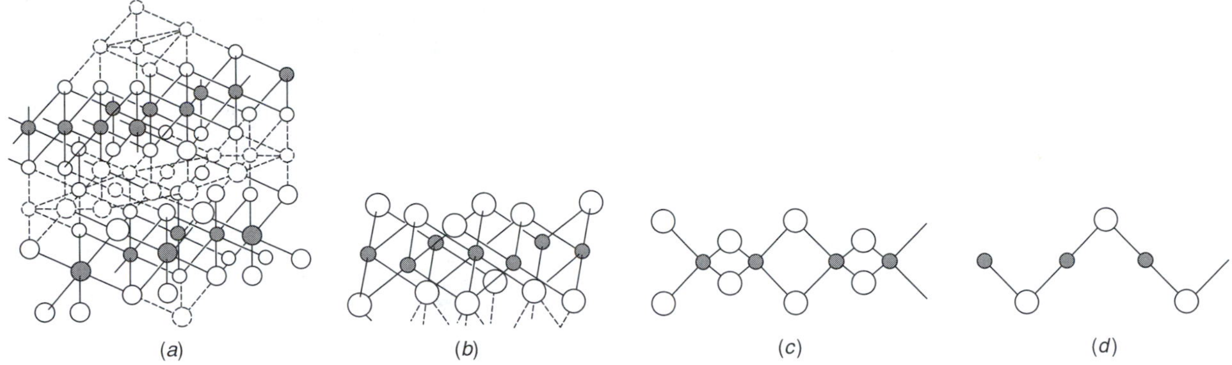

Figure 14.4
Polymeric structures of dihalides and monohalides having some covalent character in their bonding. Darkened circles represent metal atoms; open circles represent halogens. (*a*) The CdI_2 and $CdCl_2$ layer structures; the absent layers of metal atoms are indicated by dotted circles. (*b*) The coordination in each layer of CdX_2. (*c*) The chain structure of ${}^{1}_{\infty}[BeCl_2]$. (*d*) The chain structure of ${}^{1}_{\infty}[AuCl]$. [(*a*) and (*b*) adapted with permission from A. F. Wells, "Structural Inorganic Chemistry," 5th ed., Oxford University Press, London, 1984; p. 170, 259.]

Among metals that need lower coordination numbers than 6 are Be, Au, and Hg (the former for reasons of size, the latter two for relativistic reasons). Both $BeCl_2$ and red HgI_2 have four-coordinate metal atoms and two-coordinate halides with bridging that produces a linear polymer (Fig. 14.4(*c*)]; in contrast, BeF_2 adopts several crystal structures that are analogous to those formed by SiO_2 [Fig. 13.1(*d–f*)]. In AuCl and AuI, the gold atom is two coordinate; the resulting bridging of the halogen also gives a linear polymer [Fig. 14.4(*d*)].

Among the *p*-block elements, fairly stable polymeric dihalides include GeX_2, SnX_2, and PbX_2, and monohalides[5] are known for Ga, In, and Tl. These compounds are more stable than the trihalides for Tl. In these lower halides, the lattices fundamentally involve relatively high total coordination numbers, but there are distortions caused by the unshared electron pairs.

Example 14.1

Select the most plausible of the following structure types (monomer, oligomer, ionic lattice, or layer lattice) for each of the following possible halides: (a) HsF_8; (b) RaF_2; (c) RaI_2; (d) DbF_5; and (e) $AtCl_3$.

SOLUTION:

First, compare the central-atom coordination number present in the structural unit with the maximum total coordination number (or with the number calculated by radius ratios, for better results). If the two are the same, we expect a monomer; if they are nearly the same, we expect an oligomer. Otherwise, we anticipate an ionic lattice or a layer structure, depending on the electronegativity difference of the central atom and the halogen. (a) Eight is a reasonable maximum total coordination number, so we anticipate that HsF_8, if it could be prepared, would be a monomer. (b,c) With the maximum TCN also being about 8, we expect RaF_2 and RaI_2 to be polymers or ionic; the large electronegativity differences suggest ionic lattices, although a layer lattice is not out of the question for RaI_2. (d) The coordination number of 5 in the structural unit is fairly close to the maximum, so an oligomeric structure is possible, although the large eletronegativity difference may favor an ionic lattice. (Note, however, that no ionic lattices are yet known for pentahalides.) (e) The *total* coordination number of 5 in the structural unit is fairly close to the maximum, so an oligomeric structure is possible.

14.2 Physical States and Physical Properties of Halides

The major physical properties of the halides (melting and boiling points; whether each is a solid, liquid, or gas at room temperature) are governed by the principles discussed in Sections 12.1 and 12.2 and applied to the highest fluorides of the elements in Section 13.1. Another important property, their colors, was covered in Section 6.4 for the case of intense coloration and in Chapter 8 for the case of milder coloration due to electronic transitions among *d* or *f* orbitals of the metal ion of the halide. Yet another physical property, solubility, was covered in Chapter 4.

Trends among Halides of a Given Element. There are two significant trends that appear among different halides of the same element. If the halides being compared are *discrete covalent molecules*, the melting and boiling points may be expected to follow the trend fluoride < chloride < bromide < iodide. This trend results from the increase in magnitude of the van der Waals forces holding individual molecules together as the number of electrons in the halogen(s) increases; covalent fluorides have very weak intermolecular van der Waals forces—the very electronegative fluorine atoms do not readily allow their electrons to become unevenly distributed, thus creating a temporary dipole. Consequently, *most monomeric covalent fluorides are gases at room temperature, most such chlorides are liquids, whereas most such iodides are low-melting solids.* For example, CF_4 is a colorless gas of mp $-184\,°C$ and bp $-128\,°C$; CCl_4 is a colorless liquid of mp $-23\,°C$ and bp $77\,°C$; CBr_4 is a colorless solid of mp $48\,°C$ and bp $190\,°C$; and CI_4 is a red solid that melts at $171\,°C$ with decomposition.

On the other hand, if the halides being compared are *ionic compounds*, they can be expected to be solids at room temperature, with melting points that follow the opposite trend, namely, fluoride > chloride > bromide > iodide, since decreasing lattice energies are present for the larger halide ions. For example, the melting points of the sodium halides are 1012 °C for NaF, 801 °C for NaCl, 755 °C for NaBr, and 651 °C for NaI.

Within a group, or within the set of halides of a given element, the contrasting trends in melting points among ionic and covalent materials can produce some reversals. For example, the calcium halides show the following melting points: CaF_2, 1360 °C; $CaCl_2$, 772 °C; $CaBr_2$, 730 °C; and CaI_2, 740 °C. The earliest part of the trend is consistent with decreasing lattice energies in ionic compounds as the halide ion becomes larger, while the latter part of the trend is consistent with increasing van der Waals forces as the halogen atoms in the covalent layer lattices become larger.

Trends Across a Period. Recall that, at the far left of the table, we have colorless metallic fluorides with the high melting points characteristic of substances in ionic lattices in which both metal ion and fluoride ion have coordination numbers higher than 2. These fluorides are insoluble in organic solvents but may dissolve in water. At the far right of the periodic table, we have monomeric covalent nonmetal fluorides with very low melting points—most of these are gases at room temperature. These fluorides are readily soluble even in nonpolar organic solvents, but may react violently with water.

The break between low-melting gaseous monomeric fluorides and high-melting polymeric fluorides is quite sharp: Compare $^{3}_{\infty}[AlF_3]$ (mp 1272 °C]) with its neighbor SiF_4 (mp −86 °C). As a result of the trends discussed above, the sharp break in melting and boiling points found among the fluorides are much less pronounced among the chlorides: $^{3}_{\infty}[MgCl_2]$ melts at 714 °C; $^{2}_{\infty}[AlCl_3]$ melts at 192 °C under pressure; and the liquid $SiCl_4$ melts at −70 °C. The trend is even less pronounced among the iodides: $^{2}_{\infty}[MgI_2]$ 650°; $^{1}_{\infty}[AlI_3]$ 191°; SiI_4 121°; all are solids at room temperature regardless of their structural type.

The fact that the bridging abilities of halide ions are relatively weak (e.g., less than those of oxide ions) has a significant effect on the physical properties of many halides with bridged structures. It often does not take a great deal of energy to break the bridging interactions and produce a less-polymerized structure with quite different properties. As an example, many trihalides with covalent layer structures, upon strong heating, rupture most of the bridges and rearrange to give the dimeric structure of Ga_2Cl_6 [Fig. 14.3(*b*)]. At the temperature of this rearrangement (typically several hundred degrees Celsius) the dimer is a gas. Consequently, the halide passes directly from the solid polymeric form to the gaseous dimeric form without melting; this process is *sublimation*.

The higher halides of some of the elements are often useful because they have a volatility that most derivatives of those elements may lack. Thus they can readily be purified in chemical plants by fractional distillation or sublimation, then converted to other, less volatile forms that would be difficult to purify directly.

One example of this type of purification is the use of the colorless gas, UF_6, in the preparation of uranium metal (or oxide), which is enriched in the fissionable isotope ^{235}U. The gas UF_6 is one of the very few volatile compounds of uranium. Advantage is taken of this property in reprocessing spent nuclear fuel, which contains neptunium, plutonium, and many fission products. This fuel is treated with ClF_3 or BrF_3 to oxidize the uranium to UF_6, which can be swept away from the nonvolatile lower fluorides of most other metals. Enrichment of the isotopes takes advantage of the slightly lower molecular weight of $^{235}UF_6$ as compared with $^{238}UF_6$. The much scarcer but lighter molecules of the former move slightly more rapidly than do the latter. The gas UF_6 is pumped through thousands of porous membranes until the lighter $^{235}UF_6$ molecules are enriched enough for the uranium subsequently produced from it to be fissionable. The process is very expensive and has a very high energy demand, since this process must be carried out so many times on a very reactive gas.

Other examples include the conversion of silicon compounds to the colorless liquid $SiCl_4$, which can be fractionally distilled and freed from boron, arsenic, and so on, before being reduced to the high-purity elemental Si required for making semiconductors. Another example comes in the Kroll process for producing titanium metal from the ore ilmenite, $FeTiO_3$. Heating with carbon and chlorine converts both metals to their chlorides, but the colorless molecular liquid $TiCl_4$ can be fractionally distilled away from the less volatile yellow polymeric solid, $^{2}_{\infty}[FeCl_3]$. The halide is then reduced to the metal with magnesium. A useful procedure for subsequent purification of such metals (e.g., zirconium) is the van Arkel–de Boer process, in which the impure metal is heated in a vacuum with a little iodine vapor to a temperature at which the (deeply colored) iodide of the desired metal vaporizes. The vapor then passes to an electrically heated wire. At the high temperature of the wire, the relatively weak metal–iodine bonds are broken and pure metal is deposited; the iodine diffuses back to continue the process.

Other properties besides unexpected volatility are characteristic of polymeric covalent halides of lower degrees of polymerization. For example, $^{1}_{\infty}[SbF_5]$ is a liquid at room temperature, but its polymeric nature causes it to be an extremely thick, viscous liquid, as the chains become intertwined (recall the case of μ-sulfur from Chapter 13). Although it is a polymer, the Sb–F bonds can be broken and replaced by Sb–solvent coordinate covalent bonds, so that it does dissolve in some organic solvents (it is such a strong Lewis acid that it reacts with most).

The light violet layer polymer $^{2}_{\infty}[CrCl_3]$, in addition to being volatile at 800 °C, resembles the layer allotrope of carbon, graphite, or the layer silicates such as mica in being a soft, slippery solid that readily cleaves into thin shiny flakes. While the bonding within the polymerized layer is fairly strong, that between layers consists only of weak van der Waals attractions. Although this should be a water-soluble chloride by the solubility rules, it is *extremely* slow to dissolve in practice due to its polymeric nature and especially the kinetically inert nature of the d^3 electron configuration (Section 8.8); the polymer can be dissolved, however, if it is treated with a little zinc and acid, which reduces inert Cr^{3+} to labile Cr^{2+}.

Even three-dimensional network polymers involving halides are not strongly bonded enough to be useful for structural purposes: No one nowadays builds pillars (or other buildings) out of salt. If you take infrared (IR) spectra on metal halide plates (NaCl, KBr, CsI, TlBr, etc.), you must be extremely careful not to scratch these very soft materials.

Example 14.2

Compounds **A**, **B**, **C**, and **D** are solid trihalides that all contain the same central atom; none of them forms shiny flakes. The compounds **A**, **B**, and **C** are pale lilac in color, but **D** is black. Compound **A** is insoluble in water and organic solvents, but the other three are soluble in water and alcohol. Compound **A** melts at 1410 °C, **B** at 784 °C, **C** at 684 °C, and **D** at 775 °C. (a) The central element is which one of the following: As, Nd, or Cr? (b) Identify the halogen atom present in each compound.

SOLUTION:

(a) The pale lilac color is a pastel color, not characteristic at all of the *p* block (here if color is present, the color results from charge transfer and is intense) and is more characteristic of the *f*-block elements. Also, AsF_3 would be a gas, and $CrCl_3$, $CrBr_3$, and CrI_3 would be layer compounds that form shiny flakes. The substances $NdCl_3$, $NdBr_3$, and NdI_3 might not form layer compounds, either due to being ionic (there is a larger electronegativity difference for Nd), and/or due to network polymerization resulting from larger coordination numbers (e.g., Nd might be nine coordinate and the halogen three coordinate). Hence, we select Nd, which is confirmed by reference to Table 8.1. Note that *f*-block elements have no significant crystal field effects, so that all halides of Nd could be the same color.

(b) Since Nd^{3+} is a hard weak acid, we expect its fluoride (**A**) to be insoluble, but its other halides to be soluble. Since these compounds are likely ionic, the order of melting points suggests that **A** is the fluoride, **B** is the chloride, **C** is the iodide, and **D** is the bromide. On the other hand, the black color of **D** is likely due to a charge-transfer transition, which would most likely be in the iodide. So probably **C** is the bromide and **D** is the iodide; the mild reversal in melting points may be due to increased covalent character in these two compounds.

14.3 Methods of Synthesis of Halides

Although many of the more important halides of the elements can be purchased commercially, many of them are too reactive to be available, and some of those that are available may have decomposed or become contaminated by the time the purchaser receives them; hence, chemists often have to synthesize halides. A variety of methods are available,[6] depending on the properties of the desired halide. Generally, the simpler ones are satisfactory only for the less reactive halides.

1. Perhaps the simplest method of synthesis is *precipitation* of an insoluble halide by reacting a soluble halide with a soluble salt of the desired cation. This synthesis is generally satisfactory for the chlorides, bromides, and iodides of the soft acids (Section 5.7). The fluoride ion is a hard weak base that gives fairly insoluble fluoride salts with most hard or borderline feebly to strongly acidic metal ions. Many of these, however, are hydrated. Anhydrous fluorides do precipitate with many of the hard feebly acidic cations such as Ca^{2+}.

2. Soluble halides of the nonacidic or feebly acidic cations can be obtained by acid–base neutralization reactions of the hydrohalic acid with the hydroxide, oxide, or carbonate of the cation involved, followed by evaporation of the resulting solution. The chlorides, bromides, and iodides of the nonacidic cations usually crystallize out in anhydrous form (since both cation and anion are large). In the other cases, hydrated salts usually crystallize out, but the water can normally be driven off by applying heat to the crystals in a vacuum.

3. With weakly or moderately acidic cations, hydrated chlorides, bromides, and iodides can be obtained by evaporation of the solution resulting from an acid–base reaction of the hydrohalic acid with an oxide, hydroxide, or carbonate of the cation. But attempts to drive off the water of hydration by heating lead to hydrolysis (Chapter 2):

$$NiCO_3(s) + 2\,HBr(aq) + \Delta \rightarrow [Ni(H_2O)_6]Br_2(s) + CO_2(g) \qquad (14.1)$$

$$[Ni(H_2O)_6]Br_2(s) + \Delta \rightleftharpoons Ni(OH)Br(s) + HBr(g) + 5\,H_2O(g) \qquad (14.2)$$

In some cases, the anhydrous halide can be obtained if the dehydration is carried out in a stream of the anhydrous hydrogen halide, which tends to reverse the above equilibrium.

4. Alternately, after a neutralization reaction the water may be removed from many of the hydrated salts of weakly or moderately acidic cations by chemical reaction with the chemical dehydrating agents 2,2-dimethoxypropane [Eq. (14.3)], triethyl orthoformate [Eq. (14.4)], or thionyl chloride [Eq. (14.5)]. The first two of these are used to produce nonaqueous solutions of metal halides that contain coordinated methanol or ethanol. (Since these are weaker ligands than water, their presence may be unobjectionable.)

$$[Co(H_2O)_6]Cl_2 + 6Me_2C(OMe)_2 \rightarrow [Co(MeOH)_6]Cl_2 + 6\,MeOH + 6\,Me_2C{=}O \tag{14.3}$$

$$[Co(H_2O)_6]Br_2 + 6\,HC(OEt)_3 \rightarrow [Co(EtOH)_6]Br_2 + 6\,EtOH + 6\,HCO_2Et \tag{14.4}$$

If a completely solvent-free and solid halide is needed, thionyl chloride is useful, since it produces gaseous nonbasic byproducts and is itself volatile:

$$[Cr(H_2O)_6]Cl_3(s) + 6\,SOCl_2(\ell) \rightarrow CrCl_3(s) + 6\,SO_2(g) + 12\,HCl(g) \tag{14.5}$$

5. With more acidic cations than the preceding ones, it is necessary to completely avoid the presence of water. (This is sometimes a good idea even with weakly and moderately acidic cations.) Two general strategies are then available, one based on exchange of halide groups for other groups (halogenation) and the other based on redox reactions (some syntheses involve both). Exchange reactions include exchange of one halogen for another:

$$FeCl_3 + (excess)BBr_3 \rightleftharpoons FeBr_3 + BCl_3 \tag{14.6}$$

There are also exchange reactions of halide for oxide:

$$2\,MO + SF_4 \rightarrow 2\,MF_2 + SO_2(g) \tag{14.7}$$

$$Cr_2O_3 + 3\,CCl_4 \xrightarrow{600\,^\circ C} 2\,CrCl_3 + 3\,COCl_2 \tag{14.8}$$

6. Redox reactions can, in principle, be used for most halides, but may be too vigorous for the more active metals. Fluorine is such a powerful oxidizing agent that its oxidations are difficult to control except with the most resistant elements:

$$Xe + 3\,F_2 \xrightarrow{400\,^\circ C} XeF_6 \tag{14.9}$$

Often the fluoride of an element in a high oxidation state (such as CoF_3 or XeF_2 or simply HF) may be used in place of elemental fluorine. This method is particularly convenient for making the lower halide of an element from a more common higher halide:

$$14\,TaI_5 + 16\,Ta \rightarrow 5\,Ta_6I_{14} \tag{14.10}$$

Oxidation with the halogen itself is more convenient with the heavier halogens, but the reaction is still often quite vigorous:

$$2\,Sb + 3\,Cl_2 \rightarrow 2\,SbCl_3 \tag{14.11}$$

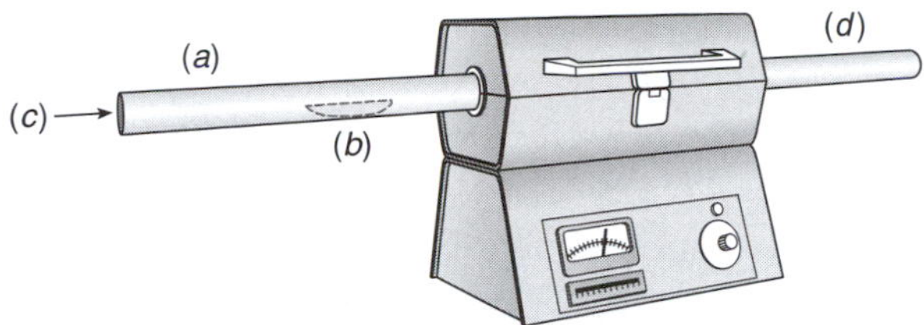

Figure 14.5
Tube furnace used in high-temperature (400–1200 °C) syntheses of volatile metal halides and other materials. (*a*) Tube, which is usually not glass, which would soften or melt in the furnace, but rather Vycor™ or fused silica. (*b*) Ceramic boat (e.g., made of Al_2O_3) to hold the solid reactant; after loading, the boat is pushed into the center of the furnace. (*c*) Inlet for reactant gases or vapors. (*d*) Outlet at which product metal halide vapors (e.g., $(CrCl_3)_2$ at 800 °C) cool and condense (as shiny violet flakes). Sketch of tube furnace from the Lindberg Co.

If a metal with a high heat of atomization (e.g., W) is being oxidized with a halogen, it is often necessary to use temperatures of several hundred degrees Celsius. This reaction can be carried out in a **tube furnace** (Fig. 14.5): vapors of the halogen are passed over the hot metal; the halide formed vaporizes at the high temperature and passes out of the furnace to condense as crystals:

$$2\,Cr(s) + 3\,I_2(g) \xrightarrow{450\,^\circ C} 2\,CrI_3(g) \tag{14.12}$$

7. Finally, there are reactions that combine reduction with halogen exchange at high temperatures. These often involve carbon as a reducing agent to remove the oxygen. Such reactions are economically favorable:

$$Ta_2O_5 + 5\,C + 5\,Br_2 \xrightarrow{460\,^\circ C} 5\,CO + 2\,TaBr_5 \tag{14.13}$$

Example 14.3

Choose reasonable syntheses of the following halides: SF_6, LiF, CoF_2, and PbI_2.

SOLUTION:

The later methods given in the above list tend to be more difficult than the earlier, so if the softness or the low acidity of the cation allows the use of an earlier method, it should be chosen. The hard, feebly acidic cation Li^+ gives an insoluble anhydrous fluoride, so route 1 is the simplest. The Pb^{2+} ion is a soft acid, so its iodide should be (and is) insoluble; route 1 can also be used. (Recrystallization of the yellow PbI_2 from large volumes of hot water gives crystals that sparkle like mosaic gold.)

The weakly acidic cation Co^{2+} would be expected to (and does) give an insoluble hydrated fluoride. Dehydration of the fluoride by heat would be expected to cause some hydrolysis. Direct fluorination of cobalt would likely lead to a higher oxidation state, so an exchange reaction might be preferable. The compound CoF_2 is made by heating $CoCl_2$ in HF.

The fluoride SF_6 is derived from a very acidic "cation" in a very high oxidation state, so direct oxidation with a powerful oxidizing agent is likely to be needed. This gas is prepared by burning sulfur in fluorine.

For experimental details in preparing a given halide (especially if one of the later, more complex methods is to be used), you should consult the original literature or a collected volume.[7]

14.4 Reactions and Uses of Halides

In Chapter 5, we saw that polar covalent bonds tend to be reactive: They readily undergo acid–base reactions that produce ionic salts and products containing nonpolar covalent bonds. Many of the halides of the elements contain polar covalent bonds and are thus quite reactive in acid–base and Lewis acid–base reactions. Thus, while they are too volatile and reactive to be of much interest as materials, they are frequently used as starting materials in syntheses of other compounds. Their physical properties are also sometimes of direct use, and they are sometimes used as oxidizing or reducing agents. In this section, we summarize some of the important types of reactions of the halides of the elements.

Lewis Base Reactions of Halides. Halides of nonacidic and feebly acidic cations often are able to donate their halide ions to Lewis acids, especially if the halide ion and the Lewis acid are matched in hardness or softness. For example, CsF is useful in forming fluoro anions from Lewis acid halides, since the resulting large fluoro anion is stabilized in a lattice by the large Cs^+ cation:

$$CsF(s) + SF_4(g) \rightleftharpoons Cs^+(SF_5)^-(s) \quad (14.14)$$

A few nonmetal halides in lower oxidation states may be able to act as ligands by donating unshared pairs of electrons from the *central atom*, although the high electronegativity of all the halogens tends to discourage such donation. Both $:PF_3$ and $:PCl_3$ can act as soft bases, donating the unshared electron pairs on phosphorus. The complexes of $:PF_3$ resemble those of :CO (Section 11.2); complexes of $:PF_3$ are similarly toxic. The chloro anion $:SnCl_3^-$ is a significant soft base that readily forms complexes with the soft-acid metal ions that feature donation of the unshared electron pair on tin:

$$PtCl_4^{2-} + Cl^- + 5\,SnCl_2 \rightarrow [Pt(SnCl_3)_5]^{3-} \quad (14.15)$$

Lewis Acid Properties of Halides. If the "cation" in a halide is weakly acidic or is in any stronger category than this (Chapter 2), the acidity of the cation is more important than the Lewis base properties of the halide ion, and the metal halide tends to act as a Lewis acid. This is true even for solid di- or trihalides that have achieved their maximum coordination number by the use of bridging halogen atoms. Since electronegative halogen atoms are not good at donating a second or even third pair of electrons, it is not hard to find a stronger Lewis base that will displace the bridging halogen. [Halides of acidic cations that have already reached their maximum total coordination number in a monomeric halide molecule (CCl_4, NF_3, NCl_3, and SF_6) are unlikely to act as Lewis acids, of course.] Five types of Lewis acid reactions of halides are illustrated below.

1. Lewis acidic halides readily react with the free halide ion (as found in the halide of a nonacidic or feebly acidic cation) to give a *halo anion* such as discussed in Section 3.3 [or as shown in Eq. (14.14)]. The relative stability of these tend to follow the hard–soft acid–base (HSAB) principle. Hard acids such as Be^{2+}, B^{3+}, Al^{3+}, Si^{4+}, and P^{5+} tend to form very stable fluoro anions such as $BeF_4{}^{2-}$, $BF_4{}^-$, $AlF_6{}^{3-}$, $SiF_6{}^{2-}$, and $PF_6{}^-$. These anions may be produced even during hydrolysis reactions:

$$3\,SiF_4 + 2\,H_2O \rightarrow SiO_2 + 4\,H^+ + 2\,SiF_6{}^{2-} \tag{14.16}$$

Softer acids, of course, more frequently are observed to give chloro, bromo, or iodo anions.

2. Even a simple property such as *solubility* of a metal halide in an organic solvent is generally a chemical (Lewis acid–base) reaction. Many chlorides, bromides, and especially iodides of metals are soluble in organic solvents that are Lewis bases and form coordinate covalent bonds to the metal ions. Such solvents include alcohols, ketones, and even (for soft-acid metal ions) π-electron donors such as benzene. In general, these solutions would not be expected to contain solvated cations or anions, but rather neutral coordination complexes:

$$^{2}_{\infty}[CrCl_3] + 3\,THF \xrightarrow{Zn} [CrCl_3(THF)_3] \tag{14.17}$$

where THF = tetrahydrofuran. These solutions are then very useful for syntheses of derivatives of the metals that are sensitive to the presence of water, but which will tolerate the presence of organic solvents. Many of these will be found in the latter half of this chapter and throughout Chapter 15.

In contrast, the highly ionic fluorides of the less acidic cations are not usually soluble in organic solvents, since the lattice energies are high in such compounds, and organic solvents do not give nearly as high ionic solvation energies as does water. Thus, for example, NaI is soluble in acetone, while NaCl and NaF are not; so NaI can be used to exchange I for Cl or F in reactive covalent chlorides or fluorides.

$$C_6H_5CH_2Cl + NaI \xrightarrow{\text{acetone}} NaCl(s) + C_6H_5CH_2I \tag{14.18}$$

3. Some strongly Lewis acidic halides such as BF_3 and $AlCl_3$ are widely used for their ability to remove halide ions from reactants. The organic Friedel–Crafts reaction involves reactions of halides (MeCOCl, MeCl, Cl_2, etc.) with benzene (C_6H_6) to give substituted benzenes (MeCOPh, MePh, ClPh) and HCl. In Figure 14.6, we sketch a reasonable mechanism by which such a reaction may occur. (a) We may suppose that the strong Lewis acid first coordinates to the halogen of the organic halide, forming, for example, [$MeCl{:}AlCl_3$]. (b) This intermediate then may enhance the acidity of the carbon atom, allowing it to attract the π-electron pair of the benzene ring, giving an organic cation such as $[CH_3C_6H_6]^+$ and the $AlCl_4{}^-$ anion. (c) Finally, the acidic organic cation transfers its proton to a chlorine of the $AlCl_4{}^-$ ion, giving back the catalytic halide ($AlCl_3$) and producing HCl.

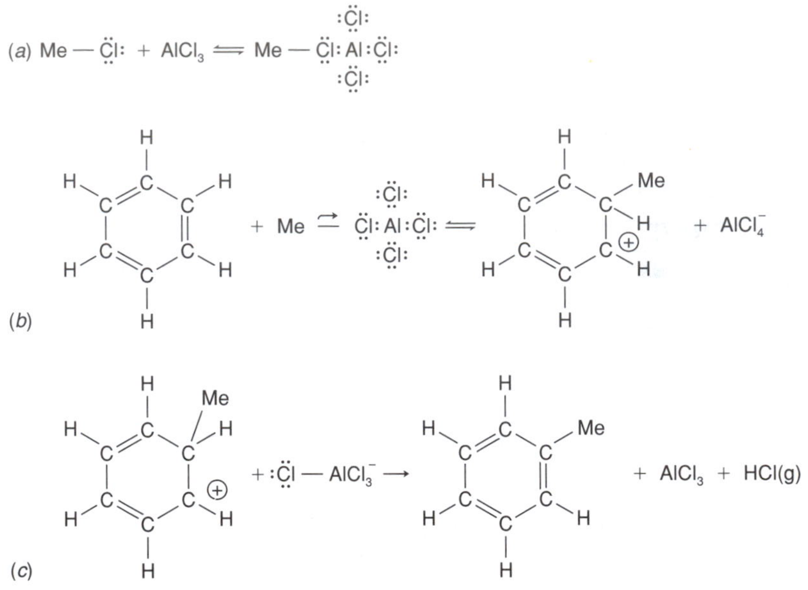

(a) $Me—\ddot{Cl}: + AlCl_3 \rightleftharpoons Me—\ddot{Cl}:Al(:\ddot{Cl}:)_3$

(b) $C_6H_6 + Me \rightleftharpoons Cl:AlCl_3 \rightleftharpoons [C_6H_6Me]^+ + AlCl_4^-$

(c) $[C_6H_6Me]^+ + :\ddot{Cl}—AlCl_3^- \rightarrow C_6H_5Me + AlCl_3 + HCl(g)$

Figure 14.6
Possible mechanism for the Friedel–Crafts reaction of MeCl with benzene, catalyzed by a Lewis acid, to give MePh and HCl(g).

Soft-acid metal ions are often reacted in a similar manner with organic halides as a method of dehalogenation:

$$MeI + Ag^+ \rightleftharpoons [MeI{:}Ag]^+ + OH^- \rightarrow MeOH + AgI(s) \tag{14.19}$$

We may also look at the halogenation reactions in Section 14.3, in which different halide ions are exchanged, as examples of Lewis acid reactions of halides of the elements. Exchange of oxygen for halogen is also of this type:

$$Me_2C{=}O + SF_4 \rightarrow [Me_2C{=}O{:}SF_4] \rightarrow Me_2CF_2 + SOF_2 \tag{14.20}$$

4. Hydrolysis reactions, which results from the Lewis acidity of cations, have already been covered in detail in Chapter 2. They, of course, usually give the hydrogen halide and the oxide or hydroxide of the element. As we also mentioned earlier, partial hydrolysis can produce oxo cations such as UO_2^{2+} or their nonmetal analogues such as $POCl_3$, which we will discuss in Section 14.6. In the case of NCl_3, in which the nitrogen is more electronegative than the chlorine, the H_2O is attracted to the *chlorine* as a center of positive charge and produces ClOH (hypochlorous acid) and NH_3. The prevalence of hydrolysis reactions among halides means that many of these compounds must be handled with rigorous exclusion of air by using glovebags, dryboxes, Schlenk glassware, or vacuum lines (Section 6.2).

5. Solvolysis reactions give rise to some important categories of inorganic compounds, to which we will devote much of the rest of this chapter. These are reactions that are similar to hydrolysis, but involve protic solvents other than water that (like water) can eliminate the hydrogen halides. For example, an alcohol such as MeOH

may be substituted for water:

$$POCl_3 + 3\,MeOH \rightarrow 3\,HCl + PO(OMe)_3 \quad (14.21)$$

The resulting products are known as *inorganic esters* (e.g., trimethyl phosphate) or as *alkoxides* of the elements (Section 14.6). By analogy with the reaction with water, we may suppose that the Lewis base MeOH first is coordinated to the very acidic P^{5+} center. The acidity of the MeOH is then strongly enhanced, and its proton may then be lost (perhaps first to solvent MeOH molecules but ultimately to a chloride group to give HCl).

Similar solvolysis reactions involve organic amines as Lewis bases:

$$POCl_3 + 3\,Me_2NH \rightarrow 3\,HCl + PO(NMe_2)_3 \quad (14.22)$$

The resulting products are known as *amides* of the elements (Section 14.7).

Use of Halides as Oxidizing Agents. As was mentioned earlier in this section on the synthesis of halides, many higher halides are good oxidizing agents, which have the advantage (e.g., as compared to salts of oxidizing oxo anions) of being volatile and of being soluble in organic solvents.

We have already mentioned the use of ClF_3 and BrF_3 as oxidizing agents in the purification of uranium from its fission products:

$$U(s) + 3\,ClF_3(g) \rightarrow UF_6(g) + 3\,ClF(g) \quad (14.23)$$

They can also be used to analyze many compounds for their percent oxygen content, with the excess reactant being easily removed. (Oxygen is probably the most difficult element to determine analytically.)

$$3\,SiO_2 + 4\,BrF_3 \rightarrow 3\,SiF_4 + 2\,Br_2 + 3\,O_2 \quad (14.24)$$

The compound ClF_3 is such a powerful oxidizing agent that even asbestos, wood, and other building materials catch fire in it. Possibly related to its oxidizing properties is the large-scale use of the (safely diluted) gas NCl_3 as a bleach and sterilant for flour.

The compound PtF_6 is such a powerful oxidizing agent that it oxidizes oxygen itself:

$$O_2 + PtF_6 \rightarrow O_2^+PtF_6^- \quad (14.25)$$

Upon discovering this reaction, the chemist Bartlett realized that an oxidizing agent capable of oxidizing O_2 could probably (from consideration of their similar ionization energies) oxidize xenon. The compound PtF_6 did, and the myth of the complete chemical unreactivity of the Group 18(VIII) (inert) gases died. The compound XeF_2 can be used in the very difficult oxidation of bromate to perbromate:

$$BrO_3^- + XeF_2 + H_2O \rightarrow BrO_4^- + Xe + 2\,HF \quad (14.26)$$

Xenon defluoride is a useful oxidizing agent in part because the byproducts are volatile and thus easily removed.

Use of Halides as Reducing Agents. Halides (especially iodides) of elements in low oxidation states can be used as reducing agents, although this is probably a less extensive use. The compound $:SnCl_2$ is widely used as a mild reducing agent; $:PI_3$ is useful for removing oxo groups from some kinds of organic compounds (e.g., converting $R_2S{=}O$ to R_2S). Protection from the oxygen of the air is thus advisable in handling some of the halides, especially the iodides.

Example 14.4

Classify each of the following halides as predominantly a Lewis acid, predominantly a Lewis base, or neither. Then write and balance plausible reactions between appropriate pairs of these halides: (a) $[Me_4N]Cl$; (b) $NiCl_2$; (c) PF_5; and (d) SiI_4.

SOLUTION:

(a) Since the $[Me_4N]^+$ cation is nonacidic, the inherent slight basicity of Cl^- dominates; this halide is basic.

(b) The ion Ni^{2+} is weakly acidic, which is acidic enough to dominate over the basicity of Cl^-; $NiCl_2$ is (mainly) an acidic halide.

(c) With very strongly acidic cations such as P^{5+}, one must look at the structural unit and compare it to reasonable total coordination numbers (if available, radius ratio calculations could be used, but we will be more qualitative): PF_5 does not exhibit the maximum total coordination number expected to be seen in fluorides, so it can add a donor atom; it is an acidic halide.

(d) In SiI_4, the penultimate coordination number, which is more likely for iodides, is already met; this halide would then be neither Lewis acidic nor Lewis basic (although it could react with ligands such as water that have smaller donor atoms).

The appropriate pairs to react would be any acidic halide [(b) or (c)] with any basic halide (a). The equation would be balanced so as to give reasonable total coordination numbers in the halo anion produced. The expected reactions therefore are

(b) + (a): $4\,[Me_4N]Cl + NiCl_2 \rightarrow [Me_4N][NiCl_6]$ (perhaps the tetrahalo complex would be formed for steric reasons);

(c) + (a): $[Me_4N]Cl + PF_5 \rightarrow [Me_4N][PF_5Cl]$ (this may redistribute halide ions to give $(PF_6)^-$ and $(PCl_6)^-$ salts).

14.5 *Catenated and Cluster Halides

A number of halides of the elements in lower oxidation states turn out to be **catenated** compounds, which are compounds containing bonds between the central atoms of the same element[8]; these bonds result in the formation of chain, cyclic, or cluster oligomers or even of polymers. In some cases of catenated or cluster metal halides, the metal–metal bonding is supplemented by metal–halogen bridging interactions.

Chain Catenated Halides. Catenated halides M_mX_n $(m > 1)$ are often synthesized by the reducing action of an element on the normal (noncatenated) halide of the element. A general characteristic of the catenated halides is the relative reactivity of the element–element bond: It is often readily oxidized, for example, by halogen molecules to give back the noncatenated halides; it may be broken apart to give (usually transient) free radicals that can then add to double bonds.

Perhaps the best and longest known of these are from Group 12, the mercury(I) halides, Hg_2X_2; the chloride, bromide, and iodide of the soft Hg_2^{2+} cation are insoluble and more commonly encountered than the fluoride. In these molecules, each mercury atom has linear coordination by one halogen and the other mercury; this is supplemented by secondary bonding to halogens of other molecules.

From Group 13(III), the main catenated halides are those of boron: B_2F_4, B_2Cl_4, and B_2Br_4; these halides are produced from the trihalides by reduction with mercury, and add to double or triple bonds:

$$B_2Cl_4 + H_2C{=}CH_2 \rightarrow Cl_2B{-}CH_2{-}CH_2{-}BCl_2 \tag{14.27}$$

Lower in this group, the compounds with the same MX_2 stoichiometry ($M =$ Ga, In, or Tl) are actually $M^+[MX_4]^-$, although the anion $[Cl_3Ga{-}GaCl_3]^{2-}$ is catenated.

In Group 14(IV), catenated fluorides are more abundant and stable for carbon (e.g., the perfluorocarbons C_nF_{2n+2}) than for silicon (known up to $Si_{16}F_{34}$), but catenated heavier halides are less stable and less abundant for carbon (C_2Cl_6 or C_3Cl_8) than for silicon (known up to Si_6Cl_{14} and Si_4Br_{10}); the combination of small carbon atoms and large halogens gives steric problems that destabilize these few catenated chlorides. The compound Si_2Cl_6 is generated by the high-temperature reaction of $SiCl_4$ and Si, and can be caused to disproportionate to give higher catenated halides and $SiCl_4$; it is a reducing agent specifically useful to remove oxygen atoms from compounds:

$$Si_2Cl_6 + Ph_3PO \rightarrow Ph_3P + Cl_3Si{-}O{-}SiCl_3 \tag{14.28}$$

In Group 15(V), nitrogen forms N_2F_4, a gas that, like N_2O_4, readily dissociates above room temperature to give colored molecules containing unpaired electrons (NF_2, dark blue). There is also FN=NF, in both cis and trans isomers. Phosphorus forms P_2X_4, which is most stable for $X = I$; arsenic and antimony form As_2I_4 and Sb_2I_4.

In Group 16(VI), the catenated fluorides of oxygen, O_2F_2 and O_4F_2, are expectedly rather unstable; O_2F_2 has a very short oxygen–oxygen bond and long oxygen–fluorine bonds. Among the heavier elements in lower oxidation states we find S_2F_2, S_2Cl_2, S_2Br_2, Se_2Cl_2, and Se_2Br_2. The compound S_2F_2 has a more stable isomer, $S{-}SF_2$. The catenating tendency of sulfur is best shown in the dihalosulfanes, S_nX_2, which retain the chain structure found in the allotrope μ-sulfur, terminated by halogen atoms. These compounds are well characterized for $X =$ Cl and Br and $n = 2$–8, but dichlorosulfanes up to about $S_{100}Cl_2$ apparently exist in mixtures, such as those used to dissolve sulfur and vulcanize rubber. Tellurium forms polymeric halides with halogen atoms attached along the chain. In this group, catenation also occurs in relatively high oxidation states: S_2F_{10} is nearly as inert as SF_6, but for some unexplained reason it contrasts with it in being extremely poisonous; Te_2F_{10} is also known.

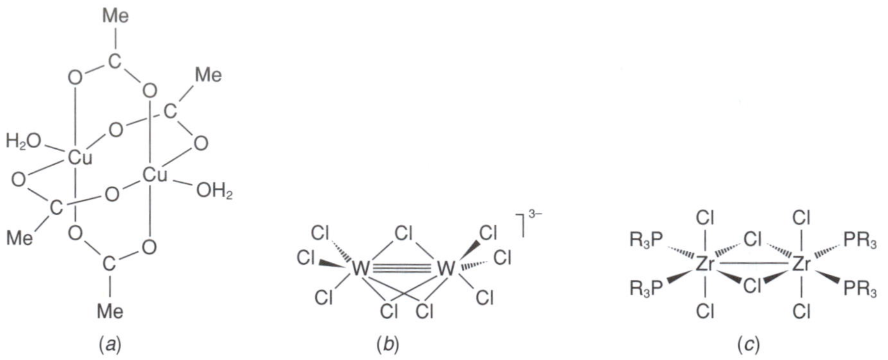

Figure 14.7
Structures of the dimeric: (*a*) hydrated copper(II) acetate; (*b*) $W_2Cl_9^{3-}$; (*c*) $[(R_3P)_2ZrCl]_2$. [Adapted from F. A. Cotton and G. Wilkinson, *Advanced Inorganic Chemistry: A Comprehensive Treatise*, New York, Wiley, 1988, pp. 685, 1085, and 1086.]

***d*-Block Elements.** In the *d* block, catenation is most familiar in the form of the octahalodimetallate ions such as $Re_2Cl_8^{2-}$ and $Mo_2Cl_8^{4-}$ formed by Mo, W, Tc, Re, and Os, and their phosphine-substituted derivatives such as $[Re_2Cl_4(PR_3)_4]^{n+}$. These are examples of the general class of multiply bonded species[9] discussed in Sections 3.1 and 10.2 that also includes metal acetates (Fig. 14.7); in the octahalodimetallate anions there is no bridging ligand such as acetate present, so that the short bond distances are indubitably due to metal–metal bonding. Metal–metal bond orders in these compounds can be as high as 4.

In many such species, a fifth ligand is present on each metal atom along the M–M axis, so that the species are often constructed from two square pyramidal L_5M structural units. The metal orbitals commonly reserved for accepting ligand electrons in such a unit are the following (taking the z direction as the C_4 axis of the square pyramid): s, p_x, p_y, p_z, and $d_{x^2-y^2}$ (or their hybrid equivalents, such as dsp^3 hybrids, although sorting out the hybridization in the *d* block is difficult). This leaves four orbitals available for metal–metal bonding (Fig. 3.3): d_{z^2} (suitable for σ bonding), d_{xz} and d_{yz} (suitable for π bonding), and d_{xy} (suitable for δ bonding). If the metal has a d^4 electronic configuration, so that there is one electron prepared for bonding in each of these four orbitals, the two L_5M units can dimerize to give a metal–metal quadruple bond. Hundreds of such compounds are known for the d^4 ions Cr^{2+}, Mo^{2+}, W^{2+}, Tc^{3+}, and Re^{3+}, in such seemingly simple compounds as the octahalodimetallate ions and chromium(II) and molybdenum(II) acetates, $M_2(O_2CMe)_4{\cdot}2\,H_2O$. If the metal has either a d^3 (e.g., V^{2+}, Mo^{3+}, and W^{3+}) or a d^5 electron configuration (e.g., Tc^{2+}, Re^{2+}, and Os^{3+}), then there can be only three electrons prepared for bonding in these orbitals. Therefore only a triple bond is formed, since the d_{xy} orbital is then either unoccupied or doubly occupied.

These *d*-block metal–metal σ, π, and δ bonds are comparatively weak, due to the contraction, and hence fairly poor overlap of the *nd* orbitals as compared to the valence $(n+1)s$ and $(n+1)p$ orbitals. The most characteristic difference between the quadruply bonding and the triply bonded species is thus not the bond length (δ overlap is so weak that it shortens the bond but little, and may be negated by other factors) but rather the symmetry and conformation. In triply bonded species, the ligands of adjacent ML_5 units repel each other, so they adopt a staggered conformation to minimize repulsion (thus, the complex has, at best, D_{4d} symmetry), while in the quadruply bonded species this conformation would destroy the overlap of adjacent d_{xy} orbitals, so the eclipsed conformation (D_{4h} point group) is adopted. Likewise, there can be doubly or singly bonded species, and there are even mixed-valence species with half-integral bond orders.

The Cr–Cr quadruple bonding is weaker than Mo–Mo or W–W quadruple bonding, since $3d$–$3d$ orbital overlap is poorer than $4d$–$4d$ or $5d$–$5d$ overlap. The quadruple bond in the isolobal and isoelectronic $[Cr_2Me_8]^{4-}$ anion is so weak that it can be ruptured simply upon changing the coordination sphere of a Li^+ counterion:

$$[Li(OEt_2)]_4[Cr_2Me_8] + 4\,Me_2NCH_2CH_2NMe_2 \rightleftharpoons 4\,OEt_2 + 2\,[Li(Me_2NCH_2CH_2NMe_2)]_2[CrMe_4] \qquad (14.29)$$

The stronger Mo–Mo quadruple bond in the molybdenum analogue of this compound is unaffected by this change in the ligand of the counterion.[10]

As the above example indicates, the MX_4 fragments that dimerize to give metal–metal multiple bonds need not be M(halogen)$_4$; a useful series of these are obtained when the four X ligands are the four nitrogen donor atoms of a macrocyclic porphyrin ligand, M^{2+} (porphyrin^{2-}) [as in Fig. 5.4(*a* and *b*)]; M_2(porphyrin)$_2$ dimers have M–M bonds that are quadruple (M = Mo, W), triple (M = Re), double (M = Ru and Os), and single (M = Rh and Ir). The δ bond provides a barrier to rotation of the two halves of the molecule around the M–M quadruple bond; this barrier has been determined and provides a lower estimate of the δ bond strength, 45 kJ mol^{-1} for Mo and 54 kJ mol^{-1} for W.[11]

Because of the weakness of M–M bonds, in the *d* block they are most common in complexes in which there are *bridging* ligands that also help to hold the two metal atoms together. Sometimes simple bridging ligands such as Cl^- suffice to support metal–metal single through triple bonds, as in the complex ion $W_2Cl_9{}^{3-}$ [Fig. 14.7(*b*)], in which two WCl_6 octahedra share a face and its three chloride ions, and $[(R_3P)_2ZrCl]_2$ [Fig. 14.7(*c*)], in which two Zr-centered octahedra share an edge and its pair of chloride ions; one then expects to find the two metal ions to be displaced from the center of the octahedra toward each other.

The most common type of ligand that effectively bridges and helps support a multiple metal–metal bond is not halide, but rather a type in which two donor atoms are separated by one intervening atom, while all three are π conjugated with each other, that is, acetate ion, MeC(=O)O$^-$. Such ligands are not good chelators of a single metal ion, but are effective at simultaneously chelating and bridging a dinuclear $M_2{}^{n+}$ unit, since the ligand and the two metals together form a five-membered ring. This effect is found in the structure of chromium(II) and molybdenum(II) acetate from early in the *d* block, where *d* orbital overlap is still appreciable, but also in the structure of copper(II) acetate, from late in the *d* block, where *d* orbitals are much more contracted relative to valence *s* and *p* orbitals. Copper(II) acetate [Fig. 14.7(*a*)] has a diminished magnetic susceptibility (μ_{eff} = 1.4 BM per Cu atom at room temperature) with pronounced temperature dependence. The bridging–chelating effects of the acetate ligands bring two Cu ions quite close together; the unpaired electrons in the $d_{x^2-y^2}$ orbitals of the two Cu^{2+} ions then interact (either through weak direct overlap of these contracted orbitals, or indirectly via the π orbitals of the ligand) to give incomplete antiferromagnetic coupling of the electrons.

Cluster and Polymeric Halides. In the *p* block, there are two elements that produce cluster or polymeric halides: boron and carbon. Heating B_2Cl_4 and B_2Br_4 produces low yields of a series of B_nCl_n ($n = 4$, 8–12) or B_nBr_n ($n = 7$–10) clusters[12] similar to those found in the boron hydrides (Section 10.10; Chapter 15), but which have $2n$ rather than the expected $2n + 2$ framework electrons; the expected $B_6X_6{}^{2-}$ and $B_7Br_7{}^{2-}$ are also known.[13] A complexed aluminum(I) bromide cluster, $Al_4Br_4(NEt_3)_4$, containing a square of aluminum atoms, has also been characterized.[14]

The fullerene C_{60} has been halogenated to give various cluster halides: Halogenation is not accomplished as readily as might have been expected for a molecule with so many double bonds, and is another sign that C_{60} is electron deficient. Bromination gives $C_{60}Br_6$, $C_{60}Br_8$, and $C_{60}Br_{24}$, but this can be reversed by heating to 200 °C[15]; fluorination for 1 week at 275 °C gives mainly $C_{60}F_{48}$.[16]

The more common allotrope of carbon, graphite, can be fluorinated to various degrees, eventually giving rise to white nonconducting graphite fluoride, ${}^{2}_{\infty}[CF]$. In contrast, some halide molecules such as Mo_2Cl_{10} insert themselves between layers of graphite to form intercalation compounds (Section 12.7). Chlorine and bromine, or Cl_2 plus metal chlorides, or strongly oxidizing halides such as SbF_5, not only intercalate between layers of graphite but also remove electrons from their conduction bands to give oxidized graphite cation layers plus layers of halide or halometallate ions.

Cluster halides of the *d*-block metals[17] typically involve both metal–metal and metal–halogen bridge bonding; many of these are thermally very robust. Perhaps the simplest of these are the triangular clusters, $[Re_3X_9L_3]$ or $[Re_3X_9]_n$, characteristic of rhenium(III) halides and halo anions [Fig. 14.8(*a*)]. In these clusters, the Re–Re distances are quite short and indicate double bonding along each edge of the triangle. In addition, each edge of the triangle is bridged by a μ_2-halogen, while each rhenium atom has two axial terminal halogen atoms. One remaining equatorial position can be occupied by a variety of relatively labile ligands, including halide ions in salts such as $[R_4N]_3[Re_3Cl_{12}]$, neutral ligands, or bridging halides from other Re_3X_9 units in the very stable halides, $[Re_3X_9]_n$.

Two cluster halide structures are based on octahedra of metals, which also involve two kinds of ligands: kinetically inert bridging halogens, and kinetically labile terminal ligands. The $M_6X_8L_6$ structure [Fig. 14.8(*b*)] involves eight μ_3 (apical) halogen atoms, and (not shown) six much more labile ligands, each being located outside a metal atom along its fourfold axis. This structure is mainly found for M = Mo(II) and W(II) and X = Cl, Br, and I, but not F. The simplest examples of this class of cluster halides are, for L = halide ion, the halo anions as found in $[(C_4H_9)_4N]_2[W_6X_{14}]$, and, for L = bridging halogen from another cluster, the "dihalides", $[M_6X_{12}]_n$. In these clusters, each M–M edge or link corresponds to a shared pair of electrons, so that these clusters are not electron deficient. Since the terminal ligands are so much more labile than the bridging halogens, the formulas of these clusters are often written as $[M_6Cl_8]^{4+}$ (X^- or $L)_6$ to emphasize the inner, inert part of the cluster.

The $M_6X_{12}L_6$ structure [Fig. 14.8(*c*)] involves, instead of the eight μ_3 halogen bridges on each face of the octahedron, 12 μ_2 halogen bridges, one along each edge of

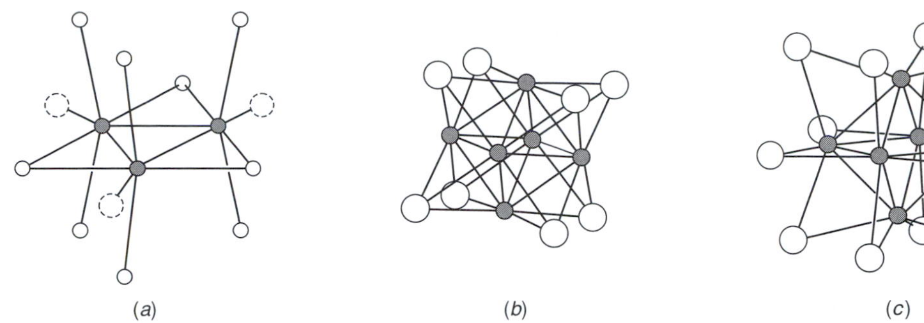

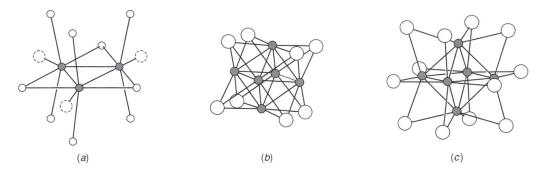

Figure 14.8
Cluster halide units for *d*-block metals; metal atoms shown as filled circles and halogens as open circles. (*a*) The $Re_3Cl_9L_3$ cluster; positions for labile terminal ligands L shown as broken open circles. (*b*) The $M_6X_8L_6$ cluster; not shown are the positions for labile terminal ligands, which lie along fourfold axes outside of each metal atom. [Adapted from F. A. Cotton and G. Wilkinson, *Advanced Inorganic Chemistry*, 5th ed., Wiley, New York, 1988, p. 846.] (*c*) The $M_6X_{12}L_6$ cluster; not shown are the positions for labile terminal ligands, which lie along fourfold axes outside of each metal atom. [Adapted from L. Pauling, *The Nature of the Chemical Bond*, 3rd ed., Cornell University Press, Ithaca, NY, 1960.]

the octahedron. Once again, labile terminal ligands are also located outside the metal atoms along the fourfold axes. A greater variety of metals and oxidation states are possible in the inert cores of these clusters: $[M_6X_{12}]$ can be, for example, $[Pd_6Cl_{12}]^0$, $[Pt_6Cl_{12}]^0$, $[W_6Cl_{12}]^{6+}$, $[Ta_6Cl_{12}]^{2+}$, $[Nb_6Cl_{12}]^{2+}$, $[Ta_6Cl_{12}]^{4+}$, $[Nb_6Cl_{12}]^{4+}$, $[Zr_6I_{12}]^0$, and even $[Nb_6F_{12}]^{3+}$. Again, the labile positions may be filled by halide ions to give halo anions, by neutral ligands to give complexes, or by halogens bridging from other clusters in the cases of the neutral halide clusters $[Pd_6Cl_{12}]$, $[Pt_6Cl_{12}]$, and $[Zr_6I_{12}]$. Obviously, a variety of electron configurations and metal–metal bond orders must be involved in these clusters; generally each M–M link or octahedron edge does not correspond to a bonding electron pair, so these are generally electron-deficient clusters. Many endohedral clusters (Section 12.5) can also be made containing small main group or even *d*-block atoms at the center of the metal octahedron, for instance, $[Be@Zr_6Cl_{12}]$ and $[C@Zr_6Cl_{12}]Cl_2$; the electron counts in these show a fair degree of regularity.[18]

One final class of cluster halides are obtained by the high-temperature reduction of Group 3 or *f*-block trihalides with the corresponding metals, and are typified by Gd_4Cl_6, Y_4Cl_6, and Y_4Br_6 (Fig. 14.9), in which (a) the octahedra of metal atoms are condensed into a chain polymer, ${}^1_\infty[Gd_4{}^{6+}]$, by sharing atoms between opposite edges; (b) μ_3-Cl^- ions bridge the triangular faces of the octahedra; there are only six in the formula unit since some of these chloride ions are also shared between chains of metal atoms to give a more polymerized final structure. Further reduction can result in sheet polymers of metal atoms in "monohalides" such as ${}^2_\infty[Zr_2Cl_2]$, ${}^2_\infty[Sc_2Cl_2]$, and ${}^2_\infty[Gd_2Cl_2]$, which are actually layer structures somewhat like $CdCl_2$ except that double sheets of metals are sandwiched between two sheets of halogen atoms, ${}^2_\infty[\cdots X–M–M–X\cdots]$. Clearly, during this process the boundary between metal halides and metals begins to be blurred.

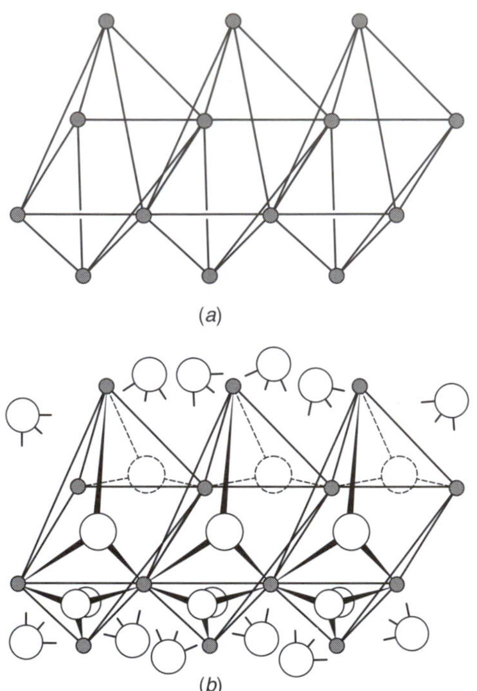

Figure 14.9
The chain cluster structure found in the Group 3 and *f*-block subhalides such as Gd_4Cl_6. (*a*) The chain of Gd octahedra formed by sharing edges, $^1_\infty[Gd_4]$. (*b*) The capping of Gd clusters with μ_3-Cl bridges, some of which also link chains (not shown). Due to their large number, some Gd–Gd and Gd–Cl links have not been drawn in this figure.

14.6 More Complex Oxygen Derivatives: Oxo Halides, Silicones, Inorganic Esters, Alkoxides

A number of important oxygen compounds of the nonmetals can be thought of as reaction products of the halides of the elements with water, inorganic hydroxides, or organic hydroxides (alcohols); the reactions producing these (in practice or sometimes in concept only) are examples of the solvolysis reactions of halides discussed in Section 14.4.

Oxo Halides. Partial hydrolysis of halides, in theory, should produce numerous examples of oxo halides, EO_yX_z. Numerous oxo halides are known, although partial hydrolysis is very difficult to control and is not a practical way to produce these compounds. They can be valuable as a source of oxo functional groups (such as C=O from $COCl_2$) in synthetic reactions that will be discussed later in this section. Oxo halides can be monomeric and soluble in organic solvents when the corresponding oxide would not be, and the corresponding halide might not exist or be suitably stable or reactive [e.g., monomeric CrO_2Cl_2 as compared to $^1_\infty[CrO_3]$ and nonexistent $CrCl_6$]. Only a few oxo halides will be specifically mentioned.

Phosgene, $COCl_2$, was mentioned in Chapter 2 as a very toxic high-temperature hydrolysis product of CCl_4, but it is made industrially by oxidizing CO with Cl_2 in the presence of activated carbon. Combining NO with fluorine, chlorine, or bromine gives the *nitrosyl halides*, NOX, which are reactive gases; the two *nitryl halides* NO_2F and NO_2Cl also exist. The *phosphoryl halides*, POX_3, are gases or volatile liquids having tetrahedral structures with short P–O bonds, suggesting partial double-bond character. The halides are mostly made by the action of O_2 on PX_3. The compound NOF_3 is also known and unexpectedly has a short N–O bond (116 pm), suggesting double-bond character, and long N–F bonds (143 pm), suggesting that the bonding should be accounted for with Lewis resonance structures such as $[O{=}NF_2]^+F^-$.[19] Their nearest relatives from the *d* block are the tetrahedral vanadyl halides, $VOCl_3$ and $VOBr_3$, which are also made by redox reactions:

$$V_2O_5 + 3\,C + 3\,Cl_2 \xrightarrow{300\,^\circ C} 2\,VOCl_3 + 3\,CO \tag{14.30}$$

Sulfur forms the *thionyl halides* SOF_2, $SOCl_2$, and $SOBr_2$ and the *sulfuryl halides* SO_2F_2 and SO_2Cl_2. Thionyl chloride is made by the reaction of SCl_2 and SO_3:

$$SO_3 + SCl_2 \rightarrow SO_2 + O{=}SCl_2 \tag{14.31}$$

We have already mentioned the use of thionyl chloride as a dehydrating agent. Sulfuryl chloride is made by the combination of SO_2 and Cl_2. In concept, further partial hydrolysis of the sulfuryl halides leads to the important acids *chlorosulfuric acid*, $ClSO_2(OH)$, and *fluorosulfuric acid*, $FSO_2(OH)$. In practice, these acids are made by the direct reaction of SO_3 with anhydrous HCl or HF. The former is used in organic syntheses and the latter is a component of "superacids" (Section 4.10). Related to the sulfuryl halides but from the *d* block are the red-brown gas CrO_2F_2 and the deep red liquid CrO_2Cl_2; salts of the halochromate ions also exist.

Silicones and Related Polymers. Most of the familiar polymers in our lives are organic molecules, but polymer chemists leave no stones unturned in their search for new materials having desired but difficult-to-obtain properties. Therefore more and more research is going into the syntheses and characterization of inorganic polymers.[20] The earliest polymers that were (substantially) inorganic were the **silicones**, obtained by hydrolysis of the (organometallic) silicon chlorides such as R_2SiCl_2. This reaction produces relatively unstable hydroxides [the silanediols, $R_2Si(OH)_2$] that readily lose water to give cyclosiloxanes, $[R_2SiO]_n$ ($n = 3$–6), which have rings with a backbone of –Si–O–Si–O– linkages. Ring opening and polymerization can be induced by KOH or H_2SO_4 to produce polymers, $^1_\infty[R_2SiO]$, containing on the order of 100,000 $[R_2SiO]$ units. The success of the ring-opening polymerization is very sensitive to the nature of the R group; this reaction is ready for R = Me but often does not succeed with large alkyl and aryl groups.

Chain polymers, of course, are not truly infinite in length but must eventually be terminated by some chemically distinct group (perhaps created by a side reaction or an impurity). Inclusion of some of the siloxane $[R_3Si]_2O$ in the polymerization mixture gives rise to comparatively short chains, $R_3Si–O–[R_2Si–O–]_nSiR_3$, the silicone oils. On the other hand, it may be desired to create a solid polymer, which is more likely with at least some degree of two- or three-dimensional polymerization. Chemical cross-linking of the chains every 100 to 1000 silicon atoms produces silicone elastomers or rubbers that are superior to other rubbers in their inertness, elasticity, strength, and flexibility over a wide temperature range. Inclusion of $RSiCl_3$ and/or $SiCl_4$ in the original hydrolysis reaction gives rise to polymers still having OH groups attached to the Si atoms in the polymer. Heat treatment in the presence of a catalyst eliminates water between neighboring chains, cross-linking them to give three-dimensional polymeric silicone resins useful in electronics and in high-temperature paints. Addition of $B(OH)_3$ to the cyclosiloxane undergoing ring-opening polymerization produces the bizarre polymer known as "bouncing putty" or Silly Putty™.

The remarkable properties of these silicone polymers[21] contrast very much with those of Nature's silicon polymers, the polysilicates such as the chain-polymeric asbestos minerals, particularly in their flexibility. The main reason for this lies in the fact that the silicones are uncharged species with only weak van der Waals attractions between chains, while polysilicate ions sit in lattices and are attracted to counterions. The properties of natural polysilicates can be modified (a) by exchanging their cations for tetraalkylammonium cations, which, being large and nonacidic, reduce the lattice energies, or (b) more profoundly, by using the OH or O^- groups on the surfaces of polysilicates as the "solvolytic" reactant, treating them with the silicon halide chlorotrimethylsilane, Me_3SiCl. This treatment introduces nonpolar, hydrophobic Me_3SiO groups on the polysilicate surface in place of ionic O^- or hydrogen-bonding OH groups. Such treatments are often used, for example, in modifying the surfaces of silica chromatographic columns to make them more retentive for nonpolar substances. If an oligomeric silicate ion is so treated, a nonpolar, organic-soluble oligomeric siloxane-silicate can be produced: For example, trimethylsilylation of the quasicubic spherosilicate ion, $[Si_8O_{20}]^{8-} = [(^-OSi)_8(\mu_2\text{-}O)_{12}]$, produces[22] the acetonitrile-soluble $[(Me_3SiOSi)_8(\mu_2\text{-}O)_{12}]$.

Another reason for the unusual flexibility of silicone polymers lies in the bonding of the Si–O–Si bridge unit. This unit typically shows a bond angle of about 143° in siloxanes, which is much larger than expected from VSEPR theory; the Si–O bond is also particularly strong. It is believed that oxygen atoms can act as π-donor ligands to silicon in siloxanes; the empty orbitals accepting these electrons are probably the relatively low-energy C–Si σ^* orbitals. Although the double-bond character of the Si–O bond is only partial, it may be readily variable, allowing the Si–O–Si backbone of these chains to bend easily to different angles by fractionally changing the π bond order.

Similarly, the $SiCl_2$, $GeCl_2$, or $SnCl_2$ groups chelated in the center of phthalocyanine macrocyclic ligands (Fig. 14.10) undergo hydrolysis to diols, which dehydrate to give siloxanes, germoxanes, or stannoxanes. However, the rigid, flat, bulky phthalocyanine rings located around the six-coordinate silicon atoms do not allow any possibility of bending nor of cross-linking. These polymers form orderly stacked fibers, not flexible, randomly twisted chains. This is advantageous for one use in particular, since partial oxidation of these chains with a dopant of I_2 partially depopulates the band of MOs running along the chain, creating holes in it that allow the polymer to become a one-dimensional electrical conductor similar in properties to those made by partial oxidation of stacked square planar d^8 metal complexes (Section 12.7).

Alkoxides/Inorganic Esters. Alkoxide ions, RO^-, mentioned in Section 2.6, are broadly comparable, not only to the hydroxide ion, but also to the fluoride ion, F^-, in being hard bases that can donate many unshared electron pairs to metal atoms. Hence, there are some similarities in structures and properties between the alkoxides of the

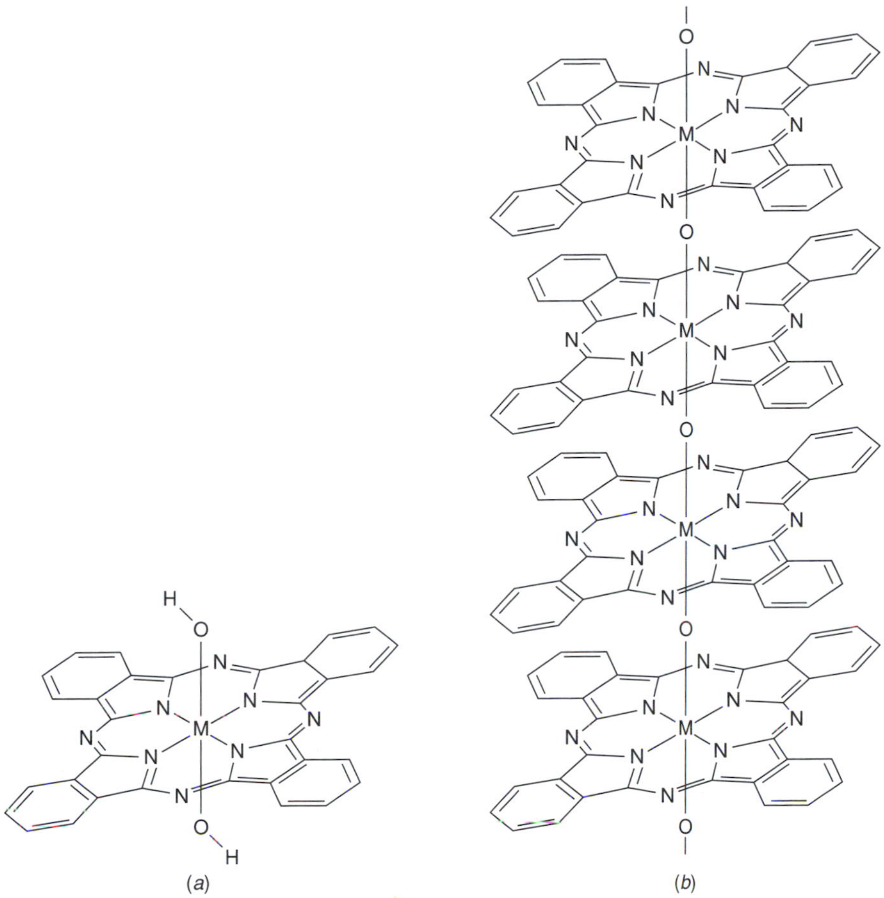

Figure 14.10
(*a*) A metallophthalocyanine diol monomer and (*b*) the corresponding metallophthalocyanine chain polymer. [Adapted from J. E. Mark, H. R. Allcock, and R. West, *Inorganic Polymers*, Prentice-Hall, Englewood Cliffs, NJ, 1992, p. 260.]

elements and the hydroxides and fluorides. But alkoxide ions differs in some important respects. (1) In contrast to the fluoride ion, they are generally very strong bases that are hydrolyzed completely in water. (2) Due to this basicity, they tend to be better bridging groups than fluoride ions; however, one side of them is blocked by the R group, which limits the bridging ability in a controllable way as the bulk of the R group is varied. (3) In contrast to most derivatives of either ion, their derivatives are potentially quite soluble in nonpolar organic solvents due to the presence of the hydrophobic R groups —provided that they are not polymerized by the presence of excessive numbers of bridging interactions.

Some of the most important varieties of alkoxides are shown in Figure 14.11. By appropriate organic substitution, the basic alkoxide ion [Fig. 14.11(*a*)] can be converted to a chelating anion (*b*; a "glycolate"). The aromatic equivalents, the aryloxide ions [Fig. 14.11(*c*)] are less strongly basic and can exist in water; appropriate substitution converts them to chelating ligands (*d*; the "catecholates"). Finally, introducing conjugation with a β-keto group (*e*, a β-diketonate anion, e.g., with $R_1 = R_2 = Me$,

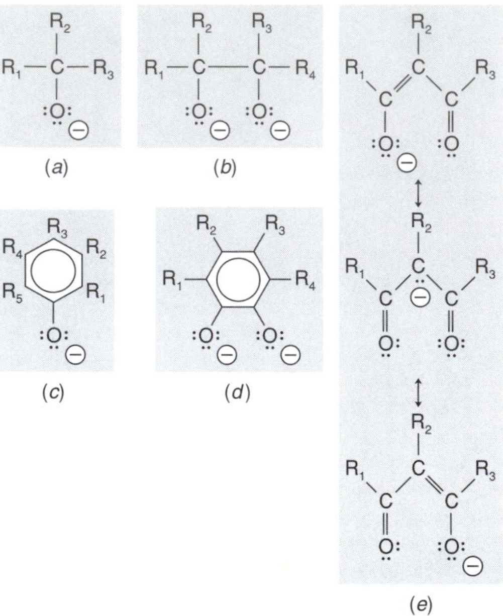

Figure 14.11

Varieties of alkoxide anions. (*a*) The basic alkoxide ion; (*b*) a chelating alkoxide ion ("glycolate" or "diolate" ion); (*c*) the basic aryloxide (phenolate) ion; (*d*) a chelating aryloxide (catecholate) ion; and (*e*) the resonance structures that contribute to the lessened basicity of β-diketonate anions.

the acetylacetonate ion or 2,4-pentanedionate anion) lowers the basicity still further while introducing the possibility of chelation (as well as ambidentate behavior as either a soft or a hard base).

Alkoxides of the more acidic cations can be generated by solvolysis reactions using the alcohol, phenol, or diketone in appropriate solvents, such as the alcohol itself. Alcohols are Lewis bases that coordinate to strongly Lewis acidic chlorides via their oxygen atoms. Then both the C–O and H–O bonds are polarized toward the Lewis acid. In some cases, both the C–O and the H–O bonds are broken and the organic group is transferred to the halogen:

$$PCl_5 + ROH \rightarrow RCl + HCl + POCl_3 \tag{14.32}$$

Usually, only the H–O bond is broken, and (especially in the presence of a basic amine to remove the proton from the coordinated alcohol) an **inorganic ester** is produced:

$$PCl_3 + 3\,ROH + 3\,R_3N \rightarrow P(OR)_3 + 3\,[R_3NH]^+Cl^- \tag{14.33}$$

This product is known as a trialkyl phosphite and is a useful soft base due to the unshared pair of electrons on the phosphorus. The naming of such derivatives as inorganic esters is based on the possible (but usually impractical) synthesis of such compounds by the reaction of an alcohol and an inorganic acid:

$$B(OH)_3 + 3\,ROH \xrightarrow{H_2SO_4} B(OR)_3 + 3\,H_2O \tag{14.34}$$

The corresponding trialkyl phosphates can be made by related routes:

$$6\,Et_2O + P_4O_{10} \rightarrow 4\,(EtO)_3PO \tag{14.35}$$

Tributyl phosphate is an important solvent for separating lanthanides. These phosphates, and especially the corresponding sulfates such as dimethyl sulfate, Me_2SO_4, still have polarizable C–O bonds, and hence are powerful *alkylating agents*. As such compounds can alkylate DNA, they may be powerful mutagens and/or carcinogens and can be quite toxic.

A series of organophosphates (and thiophosphates) very important in modern agriculture is the *organophosphate* pesticides, which interfere with nerve transmission in insects (and mammals) by inhibiting cholinesterase. An especially toxic ester is tetraethyl pyrophosphate, $(EtO)_2PO{-}O{-}PO(OEt)_2$. Some common pesticides contain a thio (S=) group in place of the oxo group: for example, parathion, $S{=}P(OEt)_2(OC_6H_4NO_2)$. Also related are some chemical warfare agents, the nerve gases, such as Sarin or GB, $Me_3PF({=}O)(OCHMe_2)$, the lethal dose of which may be 1 mg. Although very toxic, these pesticides have the advantage that they undergo further hydrolysis to the starting alcohols and phosphoric acid, so they do not persist in the environment. Of course, not all organophosphates are toxic: The biochemical compounds DNA, RNA, ATP, and ADP are all organophosphates. Other relatively common inorganic esters include nitroglycerin (the trinitrate of the organic alcohol glycerin) and amyl nitrite, $C_5H_{11}ONO$ (an antidote for cyanide poisoning).

The distinction between inorganic esters and alkoxides is fairly small: Esters generally include very acidic nonmetal atoms, while alkoxides tend to be derivatives of metals; the latter also do not generally have free oxo groups in addition to RO groups. But due to their similarities, $(RO)_4Si$ and $(RO)_4Ti$ can be named either way, as tetraalkyl silicates and titanates, or silicon and titanium alkoxides. These are useful for their hydrolysis to SiO_2 and TiO_2 (the latter in heat-resistant paints; the former, being produced in uniform microparticles, can be sintered to give quartz glass at a temperature of only 1150 °C rather than the 2000 °C needed with normal SiO_2).

Alkoxides of active metals such as Al, Na, and K cannot be made by Reaction (14.33) but instead involve the redox reaction of the metal with the alcohol:

$$2\,Na + EtOH \rightarrow 2\,EtO^-Na^+ + H_2 \tag{14.36}$$

This reaction is analogous to that of sodium with water in the activity series of metals (Section 6.5) but is somewhat less violent. Sodium and potassium alkoxides are useful very strong bases in nonaqueous solvents; an aluminum alkoxide is a useful reducing agent in organic chemistry.

The recent spectacular growth in importance of the solid-state properties of metal and mixed-metal oxides (Chapter 13) has contributed to an increase in interest in the properties of metal and mixed-metal alkoxides.[23] Although the electronic and magnetic properties of metal oxides can be very desirable, their physical properties often make them difficult to synthesize or to fabricate in the form desired. For example, the standard "shake-and-bake" synthesis of the 123 superconductor $YBa_2Cu_3O_7$ seldom gives rise to pure crystalline products, because it is so difficult to mix the precursor oxides of the three metals. These are hard, high-melting solids that, despite extensive grinding, do not break down to fine enough powders to mix well with each other. The brittle oxide powder then cannot readily be drawn into wires or applied as fine films in modern electronic devices.

Alkoxides of the metals are readily converted, either by hydrolysis, combustion, or thermal decomposition, to the corresponding oxides and only volatile, easily removed byproducts. Provided that their polymerization can be controlled, the alkoxides of the elements, like the fluorides, can be made volatile and soluble in nonpolar organic solvents; in either the gaseous or solution state, they can be intimately mixed with each other prior to conversion to the oxides.

The basic requirement for either volatility or solubility of an alkoxide in an organic solvent are similar: the alkoxide is preferably monomeric or if necessary oligomeric, but should not be polymeric. Alkoxides of metals with oxidation number 6, like the corresponding fluorides, are monomeric [e.g., $W(OPh)_6$ and $U(OEt)_6$], and although they are not as volatile as the corresponding fluorides due to their higher molecular weights, they can be sublimed in vacuum. As the metal oxidation number drops, the degree of oligomerization increases: $Nb_2(OMe)_{10}$; $Ti_4(OEt)_{16}$; and $Al_4(OEt)_{12}$. With oxidation numbers of 2 or 1, however, polymeric alkoxides are formed that are generally nonvolatile and often insoluble in organic solvents. [An exception is $Tl_4(OEt)_4$, which has a cubane structure and is (due to its content of toxic thallium) dangerously volatile.]

The problem of nonvolatility can be overcome by varying the R group. Very bulky R groups such as the *tert*-butyl group [$R_1 = R_2 = R_3 =$ Me in Fig. 14.11(*a*)] reduce the degree of oligomerization or polymerization by steric effects: $Ti(OBu)_4$ is monomeric, $Al_2(OBu)_6$ is a dimer, $Be_3(OBu)_6$ is a trimer, and lithium *tert*-butoxide is probably a hexamer. Very low coordination numbers can sometimes be achieved in metal 2,4,6-tris(*tert*-butyl)phenoxides [barium 2,4,6-tris(*tert*-butyl)phenoxide is a dimer].[24] By replacing CH_3 substituents in the R group with CF_3 substituents, we increase volatility by reducing van der Waals forces, since fluorine-containing groups are very nonpolarizable. Finally, including additional neutral donor atoms in the R groups may make it possible for a small number of R groups to satisfy the maximum coordination number of the metal atom. A simple alkoxide group can be converted into a polydentate chelating group[25] by replacing every third CH_2 group with an –O– group; $Ba(OCH_2CH_2OCH_2CH_2OCH_2CH_2OCH_3)_2$ is a monomer in which two alkoxides presumably give barium a reasonable coordination number of 8 by tetradentate chelation.

Three major techniques have emerged for the synthesis and fabrication of mixed-metal oxides: **homogeneous coprecipitation** (Section 13.5), **sol–gel syntheses**, and **metal–organic chemical vapor deposition (MOCVD).** The first two of these methods attempt to simultaneously precipitate some compound of the different metals desired in the oxide in a manner such that the three compounds are intimately mixed; the mixture is then heated or burned to convert the compounds to oxides. In sol–gel syntheses[26] metal alkoxides may be dissolved and mixed in an organic solvent such as the parent alcohol. This solution is then carefully treated with a limited amount of water, to produce first of all a **sol** (colloidal dispersion of a solid polynuclear hydroxide in the liquid), and finally condensations of the polynuclear hydroxide to a more firmly cross-linked solid **gel** (colloidal dispersion of the liquid in a network of the solid oxyhydroxide). When this is filtered off and partially dried, it is still soft and can be formed into the desired shapes readily, before it is fired in a tube furnace to finally dehydrate it and produce the mixed-metal oxide. Differing cation acidities may cause difficulties in both homogeneous coprecipitation and sol–gel synthesis.

In metal–organic chemical vapor deposition, the vapors of the different metal alkoxides can be decomposed thermally or with a laser or burned to give volatile organics or CO_2 and H_2O and metal oxides, which can be deposited as thin films on an appropriate substrate such as a semiconductor. Also useful for MOCVD are metal β-diketonates; many +3 charged metal ions form six-coordinate monomeric tris(β-diketonates), which are soluble in organic solvents, sublimable, and even show aromatic properties. Similar properties can be imparted to +2 charged metal bis(β-diketonates) or to higher coordinated tris(β-diketonates) of larger +3 charged metals by also incorporating polyether ligands.[27] Again there are challenges, since it is unlikely that a set of alkoxides or β-diketonates of yttrium, barium, and copper can be found that have vapor pressures in a 1:2:3 ratio, so that pure $YBa_2Cu_3O_7$ is deposited on the substrate. It is possible to mix the components in this ratio in an organic solvent that is rapidly evaporated, however.

In order to assure the proper composition of the mixed-metal oxide final product, much effort has gone into attempting to devise mixed-metal alkoxides that are single compounds already containing the metal ions in the proper proportions. Bimetallic alkoxides can be prepared by Lewis acid–base reactions between alkoxides of non-acidic or feebly acidic metal ions and those of more acidic metal ions:

$$2\,Na^+OEt^- + Sn(OEt)_4 \rightarrow Na_2[Sn(OEt)_6] \qquad (14.37)$$

Such a product is analogous to a salt of a fluoro anion ($Na_2[SnF_6]$) or a hydroxo anion ($Na_2[Sn(OH)_6]$). More unexpected was the reaction of other pairs of alkoxides such as the isopropoxides of Zn and Al to give volatile bimetallic products:

$$Zn(OPr)_2 + 2\,Al(OPr)_3 \rightarrow Zn[(\mu_2\text{-}OPr)_2Al(OPr)_2]_2 \qquad (14.38)$$

In these compounds, advantage is taken of the bridging capability of the alkoxide group; the tetraalkoxyaluminate anion acts as a chelating ligand to bind the zinc ion. Since then trimetallic alkoxides have been synthesized as well; but no one has (to the authors' knowledge) yet synthesized a trimetallic alkoxide, β-diketonate, or other complex that incorporates in one compound the metals Y, Ba, and Cu in a 1:2:3 ratio.

Although they are not of use in MOCVD processes, some mention should be made of chelating alkoxide and aryloxide ligands [Fig. 14.11(*b* and *d*)], which are useful in solubilizing some very insoluble hard-acid metal oxides such as SiO_2. The latter dissolves readily in a strongly alkaline solution of ethylene glycol, $HOCH_2CH_2OH$, or of catechol, to give monomeric anions such as $[Si(O_2C_6H_4)_3]^{2-}$ and $[Si(OCH_2CH_2O)_2 \cdot (OCH_2CH_2OH)]^-$.[28] These glycolate anions can be converted to other chemical forms of silicon much more readily than can SiO_2 itself, and prove useful in sol–gel syntheses of zeolites such as sodalite and the aluminum-free silicalite.[29]

14.7 Nitrides and Carbides of the Elements

Ammonia, NH_3, has three hydrogen atoms that can undergo solvolysis with metal halides to produce three general types of derivatives, the amides (MNH_2), the imides (M_2NH), and the nitrides (M_3N). In many of the amides or imides to be discussed, the remaining hydrogen atoms have first been replaced with organic groups, which can help facilitate the synthesis by preventing further solvolysis.

Amides. As an example of a metal amide we mention "sodamide," $Na^+NH_2^-$, which is prepared by the reaction of sodium metal with liquid ammonia in the presence of a catalytic amount of Fe^{3+}. (Recall that without the catalyst, this reaction gives gold or deep blue solutions containing Na^- ions or solvated electrons, Section 6.5.) This compound is used in synthesis and as a very strong base. At the other extreme is the reaction of a strongly acidic nonmetal oxide SO_3 with NH_3 to give *sulfamic acid*, commonly written as NH_2SO_2OH but actually having a solid-state structure closer to H_3NSO_3. This strong acid is useful in analytical chemistry since it is a readily purified nondeliquescent solid that can be weighed out accurately, in contrast to strong oxo acids, which are deliquescent liquids.

Two substituted amides of phosphorus are of importance: *hexamethylphoshoramide*, $(Me_2N)_3P{=}O$, and tris(dimethylamino)phosphine, $(Me_2N)_3P$. These are made from dimethylamine and $POCl_3$ or PCl_3. The former is a useful aprotic solvent that is a good Lewis base; the latter is used as a ligand and in syntheses.

Imides. Two steps of solvolysis of ammonia or of a secondary organic amine, RNH_2, can give rise to an element–nitrogen double bond if this is favorable: For example, imido analogues of the tungstate and perrhenate ions, $[W(N\text{-}t\text{-}Bu)_4]^{2-}$ and $[Re(N\text{-}t\text{-}Bu)_4]^-$ have been synthesized.[30] More commonly, the nitrogen atoms of imido groups bridge between pairs of atoms of the element to give ring oligomers or chain polymers. One well-known example is the ring compound *borazine* (borazole), $B_3H_3N_3H_3$ [Fig. 14.12(*a*)]:

$$3\,(NH_4)_2SO_4 + 6\,NaBH_4 \rightarrow 2\,B_3H_3N_3H_3 + 3\,Na_2SO_4 + 18\,H_2 \qquad (14.39)$$

This reaction may proceed through the intermediate formation of H_3BNH_3.[31] Borazine has been called "inorganic benzene", because its physical properties are very similar to those of C_6H_6. Although a B and an N atom are isoelectronic to two C atoms, the (partial) B=N double bond in borazole is not equally shared as in benzene but is clearly polarized toward the more electronegative nitrogen atom, resulting in much greater chemical reactivity for borazine than for benzene.

Solvolysis of sulfur halides (e.g., S_2Cl_2) in polar solvents leads to a variety of imides of sulfur, the best known of which is heptasulfur imide, S_7NH. This compound has an eight-membered ring structure very much like that of elemental sulfur as S_8. The hydrogen atom is weakly acidic, so salts of the S_7N^- anion are readily made.

Nitrides. The nitrides of the elements are commonly grouped into four categories: *covalent*, *salt-like or ionic*, *metallic* or interstitial, and *polymeric*, although there are generally not clear-cut differences in properties and bonding types between end members of each classification. We will also discuss a series of *nitrido anions*.

The *covalent* organosilicon compounds $[R_3Si]_3N$ (R = H and Me) are of interest because (in contrast to the predictions of VSEPR theory) they are not pyramidal but are planar or nearly planar at the nitrogen atom: As in the case of the siloxanes discussed in Section 14.6, this planarity is attributed to π donation of the lone pair on the N or O to lowest unoccupied molecular orbitals (LUMOs) on the silicon atoms, which are C–Si or H–Si σ^* orbitals.

There is a very important series of molecular covalent nitrides of phosphorus, the **cyclopolyphosphazenes**, $[-N{=}PX_2-]_n$ ($n = 3, 4$, and 5; X = F, Cl, and Br), which are readily made by the reaction:

$$n\,PCl_5 + n\,NH_4Cl + \Delta \rightarrow [NPCl_2]_n + 4n\,HCl \qquad (14.40)$$

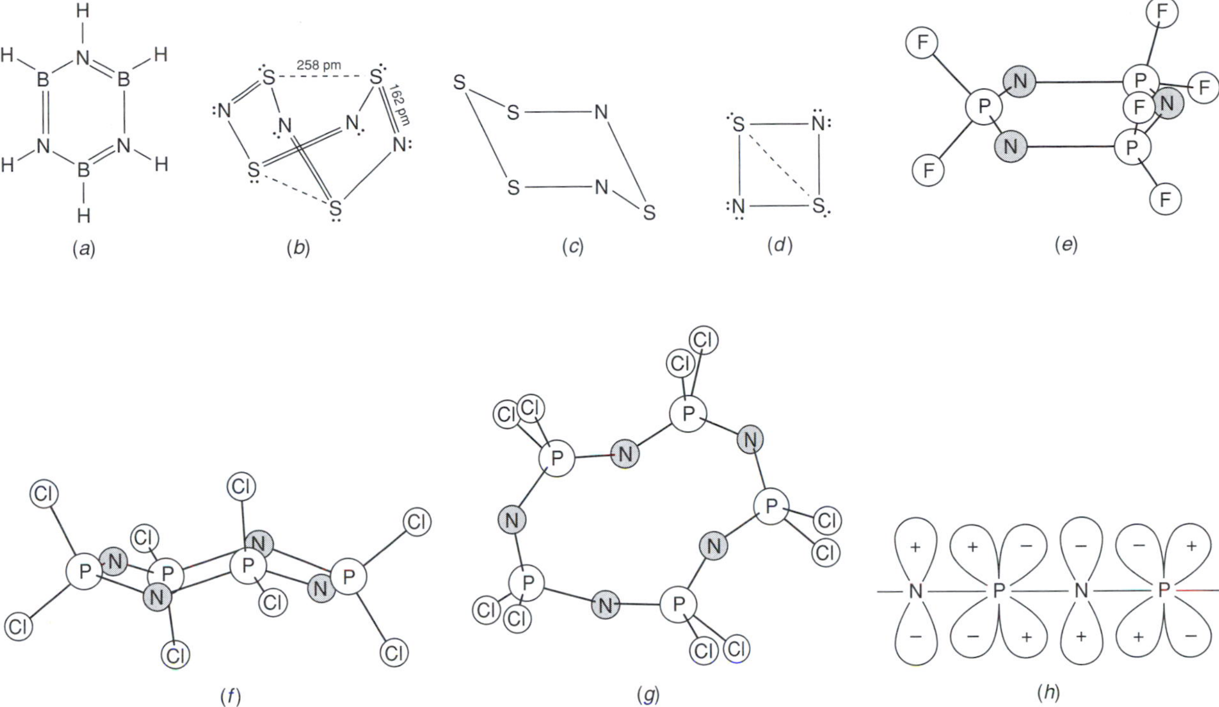

Figure 14.12
The structure of (*a*) borazine (one resonance structure); (*b*) tetrasulfur tetranitride (one resonance structure); (*c*) tetrasulfur dinitride; (*d*) disulfur dinitride; (*e*) the trimer of NPF_2; (*f*) the tetramer of $NPCl_2$; (*g*) the pentamer of $NPCl_2$; and (*h*) a sketch of one type of π bonding involving nitrogen *p* and phosphorus *d* orbitals that may occur in cyclophosphazenes.

Structures of the trimer, tetramer, and pentamer are shown in Figure 14.12(*e*–*g*). Note that the $-N{=}PX_2-$ structural unit is isoelectronic and isolobal to the $-O{-}SiR_2-$ unit of cyclosiloxanes and silicones; we shall shortly see that cyclophosphazenes also undergo ring-opening polymerization reactions.

Although there is double bonding between the P and N atoms in these compounds, the nature of this bonding has been a subject of controversy. In the common models for π bonding in the cyclophosphazenes, participation of *d* orbitals on the phosphorus atom *is* invoked, so that π-symmetry *p* electron pairs on N can be shared with lobes of appropriate sign of *d* orbitals of the adjacent P atom on either side [Fig. 14.12(*h*)]. Since the *d* orbitals contain nodal planes, the delocalization around the ring is not as extensive as it is in benzene rings. This means, for example, that there is no more aromaticity for $4n + 2$ π electrons than there is for $4n$ π electrons, and that there is no serious disruption of this interaction in nonplanar conformations of these compounds, which are less rigid than aromatic organic molecules.

There are several significant molecular nitrides of sulfur, the most important of which is *tetrasulfur tetranitride*, S_4N_4. Like S_7NH, this can be made by the reaction of S_2Cl_2 and NH_3, but this time it is in a nonpolar solvent (CCl_4 or C_6H_6):

$$6\,S_2Cl_2 + 16\,NH_3 \rightarrow S_4N_4 + S_8 + 12\,NH_4Cl \qquad (14.41)$$

The orange-yellow crystals of S_4N_4 are stable to air but may explode on being heated or struck. The structure of S_4N_4 [Fig. 14.12(*b*)] apparently involves secondary sulfur–sulfur bonding and partial double-bond character in each S–N bond (since all are of nearly equal length), but many resonance structures or a MO treatment are needed to account for its properties. Reaction of S_4N_4 with sulfur gives a second nitride, S_4N_2 [Fig. 14.12(*c*)], whereas heating S_4N_4 vapor over silver wool depolymerizes it to give the simplest (but also explosive) nitride, S_2N_2 [Fig. 14.12(*d*)]. This compound also undergoes an important ring-opening polymerization reaction that will be discussed shortly.

The *salt-like nitrides* behave approximately as salts of the N^{3-} (nitride) ion and include Li_3N and M_3N_2 (M = Group 2 or 12 metal). The most active metals (K, etc.), however, do not form nitrides. The nitrides of Li and Mg result when these metals are ignited in a nitrogen or ammonia atmosphere or even in air; nitrides can also be made by thermal decomposition of amides.

Nitrido anions are the *d*-block complexes of the nitride ion, in which N^{3-} may form a triple bond to the *d*-block metal (as in the anions $[OsO_3N]^-$, $[VCl_3N]^-$, and $[MCl_4N]^-$ (M = Mo, Re, Ru, and Os) and $[WCl_5N]^{2-}$. Complete substitution of nitrogen for oxygen results in some extremely basic anions such as $MN_4{}^{7-}$ for Group 5 or 15 (V) (M = P, Ta, and Nb), and $MN_4{}^{6-}$ for Group 6 (M = Cr, Mo, and W), so lower coordination numbers than those found in oxo anions may be adopted (e.g., three-coordinate Ga^{3+}, Fe^{3+}, or Mn^{3+} in $MN_3{}^{6-}$; note the decreasing coordination numbers in the series $GaF_6{}^{3-}$, $GaO_4{}^{5-}$, and $GaN_3{}^{6-}$).[32] Because of the low coordination numbers in the structural unit, many nitride anions form polymeric structures with bridging nitrido groups. One such series of compounds are the "nitrido zeolites," which include $(P_{12}N_{24})^{12-}$ cages clustered in the manner of the $(Al_6Si_6O_{24})^{6-}$ units of sodalite [Fig. 13.14(*c*)].[33]

Metallic nitrides occur for the *f*- and early *d*-block metals and have stoichiometries such as MN, M_2N, and M_4N. Nitrogen atoms occupy some or all of the interstices in cubic or hexagonal close-packed (hcp) metal lattices (Section 12.3) in these compounds; they can be prepared by heating the metals in N_2 or NH_3 at 1200 °C, by electrolysis of the metal in liquid NH_3, or by solid-state reactions of the metal halide with Li_3N or NaN_3.[34] These materials are usually very hard, chemically inert materials with metallic luster and conductivity, and they find use in crucibles, thermocouple sheaths, and the like.

A *polymeric nitride* of historical interest is *Millon's base*, hydrated $[Hg_{2n}N_n]^{n+}$-$(OH^-)_n$, which has a polymeric cation in which each nitrogen is attached to four mercury atoms and each mercury to two nitrogen atoms; the external anion undergoes ready ion exchange.

The polymeric covalent nitrides of formula MN [M = Group 13(III) element except Tl] are isoelectronic with and structurally related to diamond and graphite. The reaction of BCl_3 with excess NH_3 at 750 °C produces the slippery white hexagonal boron nitride, BN, with a layer structure much like that of graphite [Fig. 12.4(*f*)]. At high temperature and pressure this is converted to cubic BN with the diamond structure [Fig. 12.4(*g*)], a material so hard that it will scratch diamond. The polarity of the B–N bonds makes these materials less inert than carbon, however: BN is slowly hydrolyzed by water. The nitrides of Al, Ga, and In are similar to cubic BN.

With the discovery of the conversion of graphite to C_{60}, there has also been interest in the possibility of converting hexagonal BN to a fullerene analogue such as $B_{30}N_{30}$, but the presence of five-membered rings in the buckyball structure as applied to $B_{30}N_{30}$ requires not only many B–N bonds, which are favorable to π bonding, but also some

B–B and N–N bonds, which are not; a MO study suggests that this can be avoided by a less symmetric structure in which six- and *four*-membered rings alternate.[35] To date, a boron–nitrogen buckyball has not been synthesized, but a type of tubular "bucky-tube" structure has been reported.[36]

Other polymeric nitrides, although not isoelectronic with or structurally similar to diamond, are of a great deal of interest as materials. *Silicon nitride*, Si_3N_4, has properties that are of great interest to materials scientists and ceramics chemists. Silicon nitride (Si_3N_4) is made by heating SiO_2 and coke (C) in N_2 and H_2 at 1500 °C. It is almost completely inert chemically, is very hard and dense, has high resistance to thermal shock, and retains its high strength, shape, and resistance to corrosion even above 1000 °C. Unfortunately, it is extremely difficult to fabricate and sinter components in the shape desired, so research continues on making composites of it with other materials such as Al_2O_3.

Theoretical calculations have suggested that the replacement of Si with C in the above nitride to give carbon nitride, C_3N_4, would result in a material that would be significantly harder than diamond. Synthesis of this material poses formidable difficulties, however, since the atomization energies of both carbon and N_2 are so high. Thin films of C_3N_4 have been made, however, by using lasers to vaporize graphite targets while a radiofrequency discharge is used to split N_2 molecules; a complex synthetic route has also been reported.[37]

Polymeric nitrides of phosphorus and sulfur are also important. Of great practical use are the polyphosphazene polymers, synthesized by heating carefully purified hexachlorocyclotriphosphazene to 250 °C, whereupon ring-opening polymerization gives a chain polymer $[-N{=}PCl_2-]_n$ in which n is about 10,000. Subsequent reaction of the reactive P–Cl bonds of this polymer with various reactants gives a variety of useful polymers.[38] Most of those in current use are alkoxides, $[-N{=}P(OR)_2-]_n$, resulting from reactions of the chloro polymer with sodium alkoxides. Elastomers result if the side group is methoxide or ethoxide; fluoroalkoxides impart fire resistance and water repellence. Polyphosphazene polymers tend to resist burning or oxidative breakdown better than linear organic polymers. They have many uses, such as in O-rings, gaskets, shock absorbers, and nonburning insulating materials. Solvolysis of the chloro polymers with methylamine, amino acids, or drugs containing amino groups produces water-soluble polymers with biomedical applications.

A polymer with quite different properties results when disulfur dinitride is simply allowed to sit at 0 °C for several days. The resulting polymer, *polythiazyl*, ${}^{1}_{\infty}[SN]$, is a bronze-colored solid with metallic luster; it conducts electricity about as well as mercury, and at very low temperature it is a superconductor (it conducts electricity with no resistance). In short, it appears to be a metal! Polythiazyl is a one-dimensional polymer (Fig. 14.13), for each chain of which it is impossible to draw satisfactory Lewis structures. The best structures have nine electrons on each sulfur; one must be in a π^* orbital. This orbital apparently overlaps with corresponding antibonding orbitals on adjacent S–N units along the chain to generate a half-filled conduction band. Consistent with this, the metallic conductivity of ${}^{1}_{\infty}[SN]$ is in one direction only; it is a "one-dimensional metal."

Carbides. Carbides are classified in the same categories as are the nitrides. We discussed the salt-like or ionic carbides back in Section 12.5, pointing out that they are built up, not only from the C^{4-} ion, but more commonly from the $C_2{}^{2-}$ and $C_3{}^{4-}$ ions.

The *metallic* carbides of the *d*-block metals do not contain C_2 units and have typical stoichiometries of MC, M_2C, and M_3C; the small carbon atoms are considered to

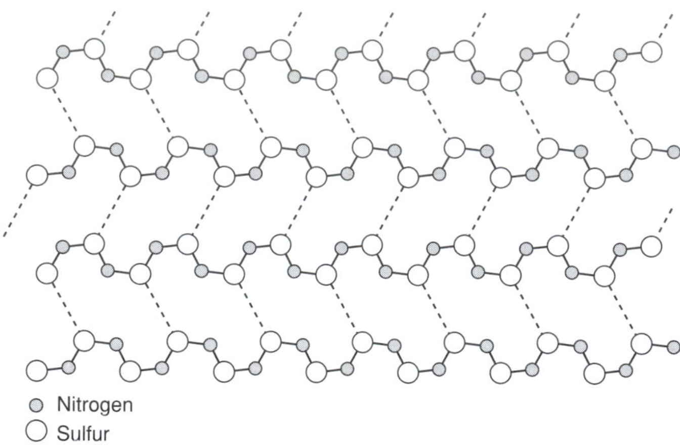

Figure 14.13
The $^1_\infty$[SN] chains (with interchain linkage indicated by dashes) in polythiazyl. [Adapted from A. G. McDiarmid, C. M. Mikulski, M. S. Saran, P. J. Russo, M. J. Cohen, A. A. Bricht, A. F. Garito, and A. J. Heeger, *Inorganic Compounds with Unusual Properties*, R. B. King, Ed., Advances in Chemistry Series No. 150, 1976, p. 63. Copyright © 1976, American Chemical Society.]

occupy holes or interstices in the close-packed lattices of metal atoms. These conducting compounds are stable to water and are very high melting and hard; TaC and WC are widely used in the production of high-speed cutting tools. Cementite, Fe_3C, is an important constituent of steel, being formed during the reduction of iron ores by carbon. Superconductivity at temperatures up to 12 or even 23 K has been found in the carbides TaC, NbC, and in boride/carbide and boride/nitride compounds such as YNi_2B_2C and $La_3Ni_2B_2N_3$.[39]

The *f*-block metals form carbides of formula MC_2, which contain C_2 groups but also M^{3+} ions instead of M^{2+} ions. The C_2 groups have longer C≡C bonds (128–134 pm), which suggests the presence of $C_2{}^{3-}$ radical anions, but additional complex interactions are present that cause these carbides to be metallic conductors. They hydrolyze to give a mixture of hydrocarbon products. The compound $Lu_2Cl_2C_{2-x}$, which contains $C_2{}^{4-}$ ions, is superconducting to 8 K.[40]

The most important *polymeric carbides* include the semiconductor silicon carbide or carborundum, SiC, with a structure related to that of diamond; it is used as a refractory and abrasive (it is extremely hard) that is chemically quite unreactive. Boron carbide, B_4C, contains B_{12} icosahedra linked directly to each other and also linked via C_3 bridges; it has similar properties to SiC. Both B_4C and SiC are prepared by heating B_2O_3 or SiO_2 with carbon in an electric furnace.

A class of *cluster* carbides of early *d*-block metals, known as **metallocarbohedranes** or **met-cars** have also been produced since 1992; the simplest and most thoroughly studied of these are M_8C_{12} (M = Ti, Zr, Hf, V, Cr, Mo, W, and Fe). These compounds originally appeared during the study of reactions of metal atoms with hydrocarbons in mass spectrometers, then were synthesized by using metal–graphite composite electrodes in the fullerene generation apparatus, and by laser vaporization of TiC and ZrC.[41]

Although the met-cars are stable, the separation problems have not yet been solved, so their crystal structures are not known. The geometries of these cluster mol-

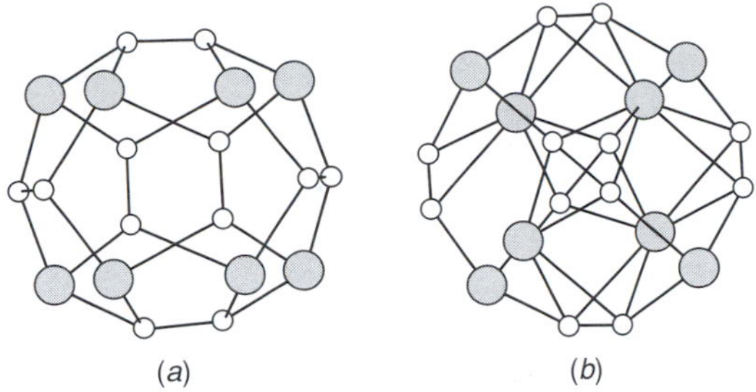

Figure 14.14
Proposed structures for the metallocarbohedranes M_8C_{12}; metal atoms are shaded. (*a*) The dodecahedral (T_h) structure based on a cube of M atoms; (*b*) The T_d structure (M–M bonds omitted for clarity), based on a tetrahedron of inner metal atoms each capped with an outer metal atom. [Adapted from Z. Lin and M. B. Hall, *J. Am. Chem. Soc.*, **115**, 11165 (1993). Copyright © American Chemical Society.]

ecules are controversial. Some evidence points to the originally proposed pentagonal dodecahedral structure (in the unusual T_h point group), which has 8 equivalent Ti atoms in 12 pentagonal faces; each face contains two titanium and three carbon atoms from two C_2 units [Fig. 14.14(*a*)]. Computations[42] clearly favor another structure of T_d symmetry [Fig. 14.14(*b*)], which has two different kind of Ti atoms and all C_2 units being similar to C_2^{2-} ions. The outer Ti atoms of either geometry have vacant, Lewis acid coordination sites, so the gas-phase coordination chemistry of these clusters has been studied: They form adducts such as $[Ti_8C_{12}(H_2O)_8]^+$, $[Ti_8C_{12}(NH_3)_8]^+$, and $Ti_8C_{12}(\pi\text{-bonding molecule})_4^+$ in mass spectrometers .[43]

14.8 Sulfides, Selenides, Tellurides, and Arsenides of the Elements

Since the sulfide ion is the archetype of soft bases, while the oxide ion is the archetype of hard bases, there is little if any resemblance in structure, chemistry, or uses of corresponding element sulfides and oxides. As an illustration among the nonmetals, a very large variety of molecular sulfides and selenides of phosphorus and arsenic are known (Fig. 14.15). With one exception, these structures are different than those of the corresponding oxides (Section 13.2). Catenation is often a feature in the structures of sulfides, arsenides, phosphides, and so on. Note the presence of P_3 units in the structures of P_4S_3, β-P_4S_4, and α-P_4S_5, while P_2 units are present in the structures of α-P_4S_4 (and As_2 units in the corresponding structure of the important arsenic mineral *realgar*, As_4S_4), β-P_4S_5, P_4S_6, and P_4S_7. Although the larger Sb and Bi atoms expand from oligomeric to polymeric structures in both the dielement trioxides and trisulfides, the details again differ: while the oxides form two-dimensional polymers, the sulfides give one-dimensional polymers (supplemented with secondary bonding, which is seldom seen in oxides). The sulfides P_4S_3, P_4S_{10}, and Sb_2S_3 are used in matches as a mixture with an oxidant (e.g., $KClO_3$) that is ignited by friction.

The best-known and most important molecular sulfide may be carbon disulfide, CS_2, which has double bonds like those of carbon dioxide. This foul-smelling liquid, however, is very volatile and extremely flammable. It may catch fire even in the

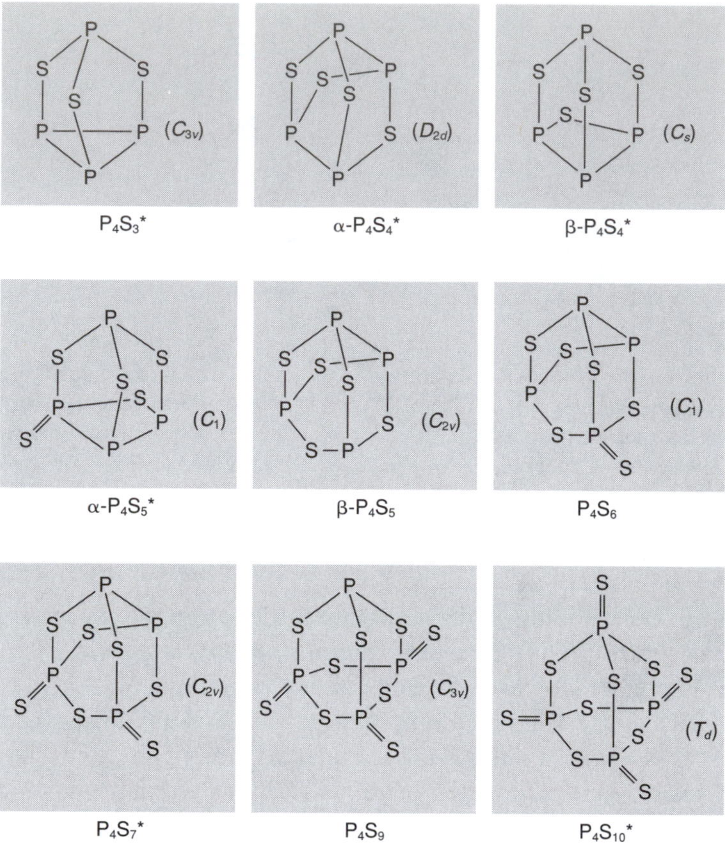

Figure 14.15
Molecular structures of P_4S_n molecules; for the starred (∗) cases many or all of the corresponding P_4Se_n, As_4S_n, and As_4Se_n molecules also are known. [Adapted from F. A. Cotton and G. Wilkinson, *Advanced Inorganic Chemistry*, 5th ed., Wiley, New York, 1988; p. 403.] Some corresponding oxide structures are shown in Figure 13.3.

absence of a flame or spark if heated to 100 °C. Some important soft-base ligands are produced by the reaction of CS_2 with bases: amines, R_2NH, give *dithiocarbamate* ions, $R_2N–C(=S)S^-$, while alkoxides give *xanthate* ions, $RO–C(=S)S^-$. Silicon disulfide is, not surprisingly, completely unlike CS_2, CO_2, or SiO_2: ${}^{1}_{\infty}[SiS_2]$ adopts the chain structure found in ${}^{1}_{\infty}[BeCl_2]$ [Fig. 14.4(*c*)]. Note that the oxide shows behavior parallel to that of fluorides, in using bridging to produce large rings so as to avoid close approaches of positively charged atoms, while the sulfide shows behavior parallel to that of chlorides, bromides, and iodides, in using bridging to produce small (four-membered) rings.

The closer resemblance of sulfides, selenides, and tellurides (and even arsenides) to chlorides, bromides, and iodides than to oxides (or fluorides) is especially pronounced among the metal chalcogenides, which tend to be much more covalent than oxides and thus more frequently adopt the same types of layer lattices that chlorides, bromides, and iodides of the same stoichiometry adopt. Since the sulfide ion is a much better reducing agent than the oxide ion, metal sulfides are far more often intensely colored—even black—due to charge-transfer absorptions.

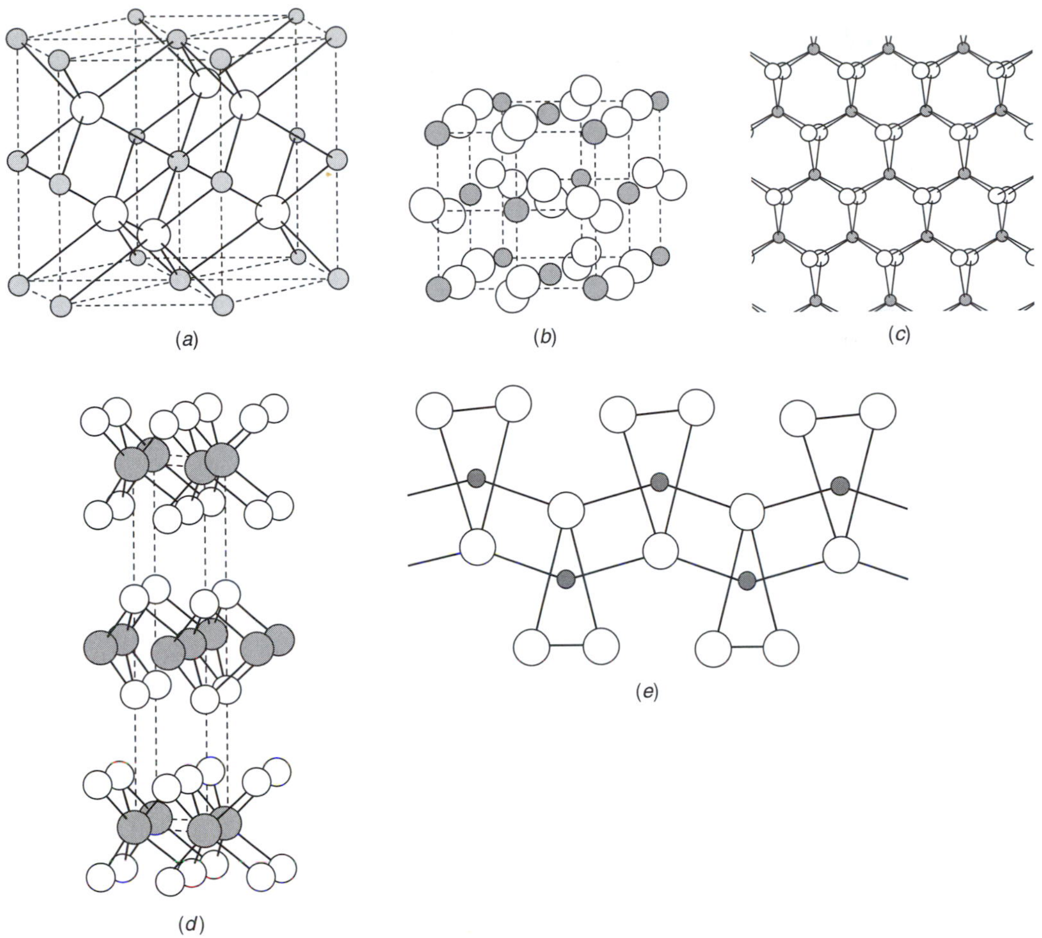

Figure 14.16
Metal sulfide structures: Metal atoms are represented by shaded circles. (*a*) The nickel arsenide (NiAs) structure; (*b*) the pyrites (FeS_2) structure; (*c, d*) the molybdenum disulfide (MoS_2) layer structure from different perspectives; (*e*) schematic of one fiber cut from the NbS_3 layer structure. [(*a*)–(*c*) Adapted from A. F. Wells, "Structural Inorganic Chemistry," 3rd ed. Copyright © 1962 by the Clarendon Press, Oxford, UK. (*d*) Adapted from C. E. Stanton, SB. T. Nguyen, J. M. Kesselman, P. E. Laibinis, and N. S. Lewis, Semiconductors, R. B. King, Ed., in *Encyclopedia of Inorganic Chemistry*, Wiley, Chichester, 1994, p. 3727. [(*e*) Adapted from E. Sandre, A. LeBlanc, and M. Danot, *J. Chem. Educ.*, **68**, 809 (1991).]

Monosulfides. Layer lattices are generally not possible in monosulfides, since layer structures sandwich one layer of metal atoms with two layers of "anions," as in the $CdCl_2$ and $CrCl_3$ structures seen earlier. Hence, monosulfides can adopt familiar lattice types, such as the zinc blende and wurzite (ZnS) lattices (Fig. 4.8) when radius ratios are appropriate for four coordination, and the NaCl type when six coordination is called for; the sulfides of Groups 1 and 2 in general behave chemically as ionic sulfides. However, many monosulfides of electropositive *d*-block metals adopt a peculiar structure, the nickel arsenide (NiAs) structure [Fig. 14.16(*a*)]. In this structure, each nonmetal atom is coordinated to six metal atoms, while each metal atom is surrounded by six sulfur and two other metal atoms close enough to each other [along the vertical dashed lines in Fig. 14.16(*a*)] to allow metal–metal bonding. These sulfides and arsenides consequently show metallic luster, and some metal sulfides show superconducting properties at very low temperatures. The NiAs structure is very unlikely for an ionic compound, as it would require close approach of cations to each other.

Disulfides. These compounds show two major structural types, which are completely different from each other and from the structures of dioxides. One type may have perplexed you when you encountered it in Chapter 5 (and especially Table 5.5) among the mineral sources of borderline and soft-acid metals, since the metal oxidation numbers in such minerals as pyrites (FeS_2), arsenopyrite (FeAsS), smaltite ($CoAs_2$), and sperrylite ($PtAs_2$) appear to be improbably high. In fact, this group of disulfides (diarsenides, etc.) contain catenated anions such as $S_2{}^{2-}$ (Section 12.5) along with rather normal Fe^{2+}, Co^{2+}, and Pt^{2+} ions. The most common lattice type for this type of disulfide, and so on, is known as the **pyrites** structure [Fig. 14.16(*b*)], after the mineral FeS_2 (pyrites, also called "fool's gold" for its golden metallic luster). This structure is really just the NaCl lattice type, in which each Na^+ ion is replaced by a M^{2+} ion, and each Cl^- ion is replace by a $S_2{}^{2-}$ ion. Perhaps we might think of these as "mono(disulfides)".

By contrast, the other main class of disulfides, are "di(monosulfide)s"; these typically adopt one of two layer structures. Many *d*-block metal di(monosulfide)s adopt the CdI_2 layer structure. A slightly different layer structure [Fig. 14.16(*c*)] is found for the important sulfide MoS_2, which is widely used as a solid lubricant and as a catalyst. The Mo atom in MoS_2 is six coordinate, but its geometry is that of a trigonal prism rather than an octahedron. Such a geometry, not infrequently found with sulfur ligands, is unusual by VSEPR theory because it brings sulfur atoms much closer together than does the octahedron. This implies an unexpected lack of repulsion of negative ions; the S–S distance in such compounds is often short enough to suggest secondary bonding.

In common with the layer allotrope of carbon, ${}^{2}_{\infty}[MoS_2]$ and ${}^{2}_{\infty}[WS_2]$ sheets can be wrapped around to produce oligomeric nested fullerene structures, which have smaller buckyballs inside of larger ones. Milligram-scale syntheses of these are based on reduction of MO_3 and reaction with H_2S.[44]

Trisulfides. Compounds such as MoS_3 and WS_3 are readily prepared by precipitation, but not in crystalline form, so it is not known if these are, as seems likely, tri(monosulfides). Another class of MS_3 materials is exemplified by NbS_3, which clearly cannot be a tri(monosulfide). It has a layer structure [Fig. 14.16(*e*)] in which one anion is a disulfide ion and the other is a monosulfide.[45] Since these ions are of unequal size, the layers are puckered to accommodate disulfides on alternate sides of the layer.

Physical and Chemical Properties. As with other layer allotropes and compounds, it is sometimes possible to insert other atoms or flat molecules between the layers of some of these disulfides to give intercalation complexes. This insertion is often accompanied by a partial redox reaction, giving rise to formulas such as Li_xTiS_2. This reaction can be carried out in an electrochemical cell with electrodes of Li metal, TiS_2, and an electrolyte consisting of a lithium salt dissolved in an inert polar organic solvent. Once run, the cell becomes a "high-energy cell" capable of producing over 2 V and storing twice the energy per unit volume as does a Ni–Cd rechargeable battery; these batteries are rapidly rechargeable because of the simplicity of the intercalation process.

Monosulfides such as ZnS, CdSe, and CdTe are isoelectronic and isostructural with the semiconductors silicon and germanium and are known as "II–VI" compounds (from their former group numbers). Zinc sulfide, ZnS, is white when pure, and is used in white paint in a mixture with $BaSO_4$ known as *lithopone*, which is obtained by the reaction:

$$ZnSO_4(aq) + BaS(aq) \rightarrow ZnS(s) + BaSO_4(s) \qquad (14.42)$$

Zinc sulfide is also widely used in cathode-ray tubes and radar screens, since it fluoresces upon being struck by X-rays, cathode rays, or radiation. The color of light emitted may be varied by partially substituting such ingredients as Cd in place of Zn and Se in place of S. Cadmium sulfide is an important yellow pigment and a useful phosphor, and CdTe may have potential as a semiconductor.

Similar in properties and structure are the "III–V" monoarsenides such as gallium arsenide, GaAs, and the other arsenides, phosphides, and antimonides of the Group 13(III) metals. The III–V compounds GaAs and GaP are now widely used as light-emitting diodes (LEDs) in pocket calculators, wrist watches, scientific instrument displays, and the like, in which electrical energy is converted to light.[46] The color of light emitted depends on the energy gap; GaAs itself, with an energy gap of 138 kJ, emits IR light, while GaP (218 kJ) emits green light; a mixture with 40% P and 60% As emits red light. Doping Zn in place of Ga creates a *p*-type semiconductor; doping Te in place of As gives a *n*-type semiconductor. Recently, semiconducting lasers have been designed based on these compounds.

These compounds are also useful in the reverse process of converting light energy into electrical energy; InSb is used as a photoconductive detector of IR light. Several of these compounds are being investigated for use in the direct conversion of sunlight to electricity; at the present time, Cu_2S–CdS and silicon are commercially preferred. Semiconducting properties are also found in the sulfides, selenides, and arsenides, M_2S_3, of arsenic, antimony, and bismuth.

If we go beyond trisulfides we could finally obtain stoichiometries rich enough in sulfur to break up polymerization and give molecular sulfides for metals. The nearest thing to this occurs among **thio anions** $[MS_4]^{n-}$, which are readily prepared for several metals; $[MoS_4]^{2-}$ and $[WS_4]^{2-}$ are well known and seem to resemble the molybdate and tungstate anions (Chapter 2). However, these are soft bases, and sulfur has a strong ability to bridge, so these anions readily form insoluble polymeric "salts" with borderline and soft acids. Similar to these are some anions found in Zintl compounds formed from a Group 1 or 2 metal, a Group 13(III) or 14(IV) metal or metalloid, and phosphorus or arsenic: In addition to nonacidic or feebly acidic cations, many of these contain anions such as $SiAs_4{}^{8-}$ and $SiP_4{}^{8-}$, which are isoelectronic to the corresponding thio or oxo anions, but which clearly are too strongly basic to persist in water or in any but very basic solvents such as 1,2-diaminoethane (ethylenediamine).

Chevrel Phases. A final parallel between the behavior of halides and sulfides lies in the formation of molybdenum sulfide, selenide, and telluride clusters, $[Mo_6(\mu_3\text{-}S)_8]^{n-}$, which have the same structure as the molybdenum halide inner-cluster ion $[Mo_6X_8]^{4+}$ [Fig. 14.8(*b*)], but in which the redox chemistry is more prominent. The neutral chalcogenides $[Mo_6(\mu_3\text{-}S)_8]$, which are electron deficient as compared to $[Mo_6X_8]^{4+}$, are prepared by reacting molybdenum and S, Se, or Te at 1000 °C. The French chemist Chevrel (after whom these compounds are named) discovered that various metals

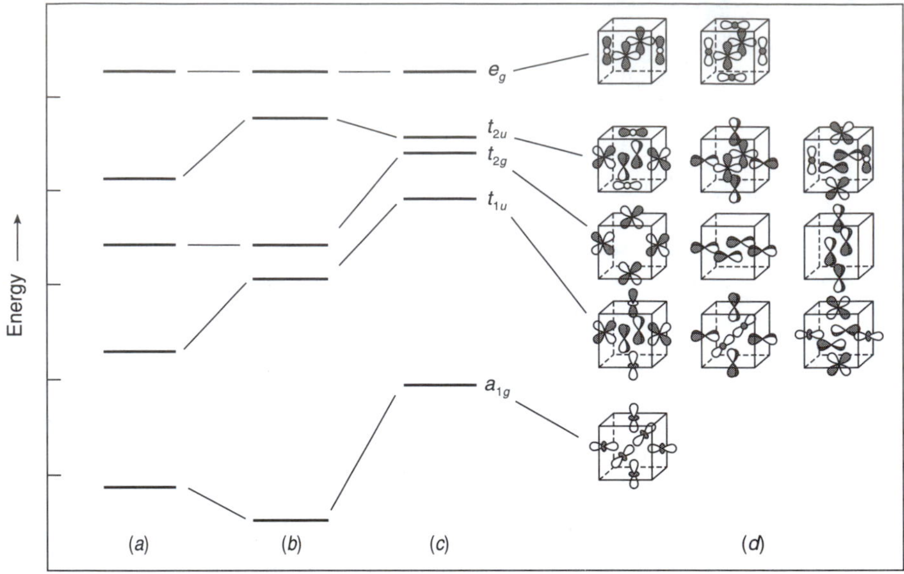

Figure 14.17
Metal atom-based bonding MOs and their energies for Mo_6 clusters as calculated by various methods. (*a*) The SCR-SW-Xα calculations for $[Mo_6Cl_{14}]^{2-}$; (*b*) Fenske–Hall calculations for $[Mo_6Cl_{14}]^{2-}$; (*c*) extended Hückel calculations for $[Mo_6S_8]^{4-}$; and (*d*) diagrams for the MOs. [Adapted from F. A. Cotton and G. Wilkinson, *Advanced Inorganic Chemistry*, 5th ed.; Wiley, New York, 1988; p. 1080.]

could be reacted with these compounds, reducing them to give compounds such as $Li_{0.8}Mo_6S_8$ and $PbMo_6S_8$. These Chevrel phases display remarkable superconductivity, especially in the presence of strong magnetic fields, which is a property not shared by the copper oxide superconductors (Chapter 13) but which is obviously very important in the construction of powerful superconducting magnets. In Figure 14.17, the similarity of $[Mo_6X_8]^{4+}$ and $[Mo_6S_8]^{n-}$ shows up in the results of MO calculations, which also explain the varying degrees to which the clusters can be reduced, since the highest energy levels, e_g, are nonbonding and therefore readily accommodate varying numbers of electrons.

Biological Metal Sulfide Clusters. Sulfide cluster (as opposed to polymer) building is facilitated if we block one of the sulfur lone pairs by alkylating it to give the *thiolate* (alkyl sulfide or *mercaptide*) ions, RS^-, which are analogous to the alkoxide ions. Interest in complexes of these ligands stems from the discovery by biochemists of a class of iron-containing proteins in which the RS^- anion is derived from the amino acid cysteine. One series of these proteins, the *rubredoxins*, involves one iron atom tetrahedrally coordinated to four cysteine RS^- groups ($CysS^-$); these complexes are involved in redox reactions in anaerobic bacteria. Both the iron(II) and iron(III) forms are high-spin tetrahedral, with but a small change in bond lengths, so that the redox reactions interconverting them proceed rapidly.

The other series of proteins are the *ferredoxins*, which include both two-iron and four-iron varieties; in both of these proteins similar redox reactions can readily occur. The two-iron ferredoxins contain four-membered rings with bridging sulfides, $[Fe(\mu_2\text{-}S)_2Fe]^{n+}$, with tetrahedral coordination completed at each iron by adding two $CysS^-$ ligands to each iron. In the normal redox state the iron is iron(III), so that the inner ring is $[Fe(\mu_2\text{-}S)_2Fe]^{2+}$ with each iron being high spin, but the two irons are brought so close (272 pm) by the bridging sulfide ions that complete antiferromagnetic coupling (Section 12.4) occurs, giving a diamagnetic compound. This compound is readily reduced at about 0.0 V to give a mixed iron(II)–iron(III) species, $\{(CysS)_2[Fe(\mu_2\text{-}S)_2Fe](SCys)_2\}^{3-}$, with one unpaired electron.

The four-iron ferredoxins involve a central cubane core, $[Fe_4(\mu_3\text{-}S)_4]^{n+}$, akin to that seen earlier in Te_4Cl_{16} [Fig. 14.2(*d*)], but with only tetrahedral coordination about each iron being completed by one additional terminal $CysS^-$ group coordinated to each iron. Model compounds for these ferredoxins are easily synthesized by reactions in air-free methanol:

$$4\,FeCl_3 + 4\,HS^- + 6\,RS^- + 4\,MeO^- \rightarrow [Fe_4(\mu_3\text{-}S)_4(SR)_4]^{2-} + RS\text{–}SR + 12\,Cl^- + 4\,MeOH \qquad (14.43)$$

As synthesized, this cluster contains two Fe(II) and two Fe(III) irons, but is diamagnetic; in different proteins such clusters are either easily and reversibly reduced by one electron at about 0.0 V, or easily oxidized by one electron at about +0.75 V.

Another extremely important redox enzyme involving iron, sulfur, and molybdenum in clusters is the enzyme *nitrogenase*, utilized by *Rhizobium* organisms that live in the nodules of legumes; these reduce otherwise-inert N_2 molecules to bioactive NH_3 at room temperature and pressure. After many years of effort, two metal-containing cofactors of this enzyme were finally crystallized, and their structures determined[47]: each cofactor has two cubane units linked to each other by sulfur bridges; one dicubane cofactor has two linked $[Fe_4(\mu_3\text{-}S)_4]$ units and the other has one $[Fe_4(\mu_3\text{-}S)_4]$ unit linked to a $[Fe_3(\mu_3\text{-}S)_4Mo]$ unit. The chemistry of nitrogen fixation is discussed more fully in Section 17.9.

It is also possible to prepare more open clusters without inorganic sulfide ligands, using only thiolate ligands, for example, $[(RSCd)_4(\mu_2\text{-}SR)_6]^{2-}$, which has an analogous structure to that of P_4O_6 (Fig. 13.3). Clusters of this sort (as well as those of clusters similar to the rubredoxins and ferredoxins) not only serve as models for metalloenzymes and proteins, but also provide sources of soft acid metal ions in a soft base environment and which (if charged) are soluble in water and organic solvents, in contrast to the extremely insoluble simple and layer sulfides.

Example 14.5

(a) Given the statement that P_4O_6 and $[(RSCd)_4(\mu_2\text{-}SR)_6]^{2-}$ have analogous structures, describe the structure of the latter ion. (b) What pure hydrocarbon has an analogous structure to these compounds? (c) Describe the most likely bromo analogue to $[(RSCd)_4(\mu_2\text{-}SR)_6]$ that can be generated by replacing only one type of thiolate ligand.

SOLUTION:

(a) Oxide ion, O^{2-} in P_4O_6 in μ_2-bridging, as is $\mu_2\text{-}SR^-$; $:P^{3+}$ in P_4O_6 has a total coordination number of 4, with three positions being filled by bridging oxide ions; a similar statement can be made about tetrahedral $RSCd^+$. Hence, the structure of the cadmium cluster thiolate anion can be generated from that of P_4O_6 by replacing each P with RSCd and each O with SR. With the Lewis acid–base counting scheme (Chapter 11) these two compounds could be described as isolobal, since both $:P^{3+}$ and tetrahedral $RSCd^+$ have three empty lobes, and both of the anions have two filled lobes that are used in the bridging interaction.

(b) By referring to Table 11.6 and using the redox-radical (RR) approach, the carbon species that is isolobal with P is CR; the one that is isolobal with μ_2-O (or O^{2-}) is CR_2. Letting R be hydrogen, the hydrocarbon is $[(CH)_4(\mu_2\text{-}CH_2)_6]$ or $C_{10}H_{16}$, the high-symmetry (T_d) hydrocarbon commonly known as adamantane. (We have not used the Lewis acid–base approach since Table 11.6 lists only Lewis bases; the carbon species isolobal to P^{3+} would of course be CH^{3+}).

(c) The complex $[(RSCd)_4(\mu_2\text{-}SR)_6]$ contains both bridging and terminal thiolate groups. Bromide is more likely to be a terminal ligand, since the more basic sulfide ion makes a better bridging ligand. Hence, we are more likely to synthesize $[(BrCd)_4(\mu_2\text{-}SR)_6]^{2-}$ than $[(RSCd)_4(\mu_2\text{-}Br)_6]^{2-}$.

A number of unusual metal–sulfur cluster-containing enzymes have recently been isolated from *hyperthermophile* microorganisms, which thrive at extreme conditions (over 100 °C) and in bizarre environments such as the H_2S- and metal sulfide-rich "black smokers" at the crests of the mid-oceanic ridges (Section 5.9). Some of these enzymes utilize very unusual metals such as tungsten.[48] Microorganisms such as the ones containing these enzymes seem to constitute a separate kingdom of species—some biologists are beginning to feel, the oldest kingdom of living species on earth. A German patent lawyer (with a Ph.D. in organic chemistry) has been advocating the controversial hypothesis that the mid-oceanic ridges are the site of the origin of life, and hypothesized that the first carbon–carbon bond forming reactions could have occurred on the positively charged surfaces of layered metal sulfides such as FeS_2. He and a scientist with a laboratory have produced experimental verification[49] that carbon–carbon bonds can be formed from reactants thought to be present in the black smokers, MeSH and CO, under the influence of a mixture of iron and nickel sulfides (also present) at 100 °C:

$$2\,\text{MeSH} + \text{CO} \rightarrow \text{MeC(=O)SMe} + H_2S \qquad (14.44)$$

The methyl thioacetate produced by this reaction then hydrolyzes to give acetic acid. If the iron/nickel sulfide contains traces of selenide (also present), acetic acid can be produced directly from CO and H_2S. This reaction bears a remarkable similarity to the vital biosynthesis of carbon chains in more advanced organisms using vitamin B_{12} (Section 11.9) and acetyl-coenzyme A (CoA):

$$\text{Me–cobalt} + \text{CO} + \text{HS–CoA} \rightarrow \text{MeC(=O)S–CoA} \qquad (14.45)$$

If this particular hypothesis proves out, then the metal–sulfur clusters found in our bodies may be our inheritance from ancient times in the depths of the ocean, with some necessary adjustments such as the replacement of metals such as tungsten (rare and unavailable at the earth's surface) with more common lighter metals such as molybdenum. Note that both this hypothesis and the other controversial hypothesis that life began on montmorillonite clay (Section 13.7) invoke the catalytic powers of the acidic charged surfaces of metal layer compounds to concentrate reactants that must have been present in low concentration, and to arrange them (via the template effect) so as to enhance the probability of the entropy-defying synthesis of large complex self-replicating organic molecules from small molecules in very dilute solution.

14.9 Relationships among Ionic, Covalent, and Metallic Bonding

In introductory chemistry courses, and earlier in this text, chemical bonding has been traditionally treated as being of two types: ionic and covalent. The division of compounds into two types raises the question as to how one tells which type of bonding is present in a given compound. Structurally, this can sometimes but not always be answered by looking at the crystal structures: Layer structures point toward covalent bonding, as do acute bond angles in bridging halogen atoms in oligomeric halide molecules. On a more fundamental level, it is common to try to cite a particular electronegativity difference, such as 1.7 or 1.8, between two elements in a binary compound; but there are a significant number of uncomfortable results of applying this simple procedure [e.g., classifying $SiF_4(g)$ as an ionic compound, even though it might be possible to rationalize this as suggested by Lingafelter[2] in Chapter 12]. In addition, although this leads to the classification of metals and alloys as pure covalent compounds, these have enough distinctive properties (including conductivity, of course, but also very high coordination numbers and low bond orders) that many would add a third category of bonding, *metallic bonding*, which we have now seen (in polythiazyl, graphite halides, etc.) does not necessarily require the presence of metal atoms!

One of the earliest suggestions that compounds could be divided into *three* fundamental classes was made by van Arkel[50] over 50 years ago, when he drew a triangle representing bonding types, with metallic, covalent, and ionic bonding at the vertices, as exemplified by Cs, F_2, and CsF as extreme examples. More recently, Sproul[51] and Jensen[52] rendered this tripartite division quantitative by using both electronegativity *differences* and *average electronegativities* to classify binary compounds [Fig. 14.18(*a*)].

They suggest that (1) *covalent compounds* [right corner of triangle in Figure 14.18(*a*)] are characterized, not only by low electronegativity differences as previously recognized, but also by *high average* electronegativities, in contrast to (2) *metals* and *metallic compounds* (e.g., alloys, left corner), which have low electronegativity differences and *low average electronegativities.* (3) *Ionic compounds* are characterized by *high* electronegativity *differences* (top corner of the triangle), which (by its nature) requires intermediate average electronegativities.

An oversimplified view of the effects of electronegativity differences can be obtained by examining *diatomic* "molecules" of average electronegativity 2.5, but with electronegativity differences ranging from 0.0 (in C_2) to 1.0 (BN) to 1.9 (BeO) to 3.0 (LiF). This corresponds to a movement from the bottom center of the van Arkel diagram almost to the apex. Figure 10.13 shows how the atomic orbital and MO energies change during this transit. With LiF but not at all with C_2, the energy levels are closely connected to those of the more electronegative (F) atom: We say the electrons are *localized on the more electronegative atom*, and that the bonding is consequently ionic.

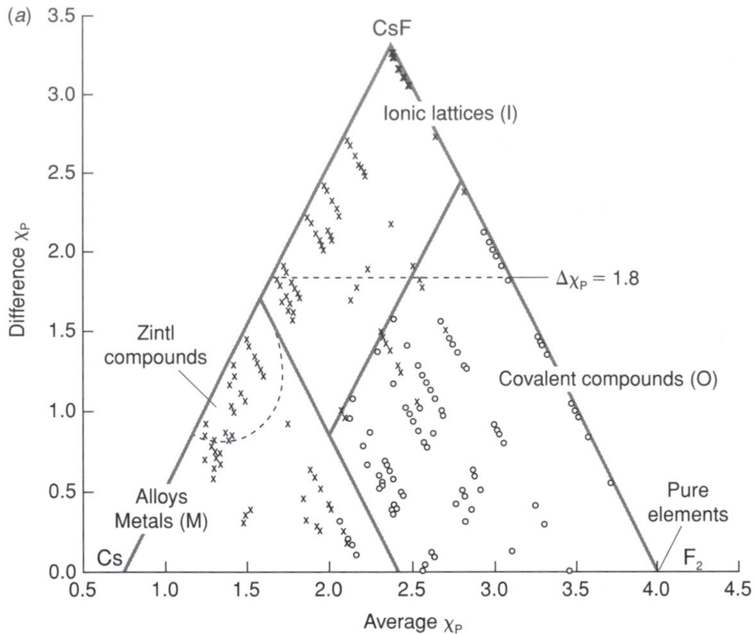

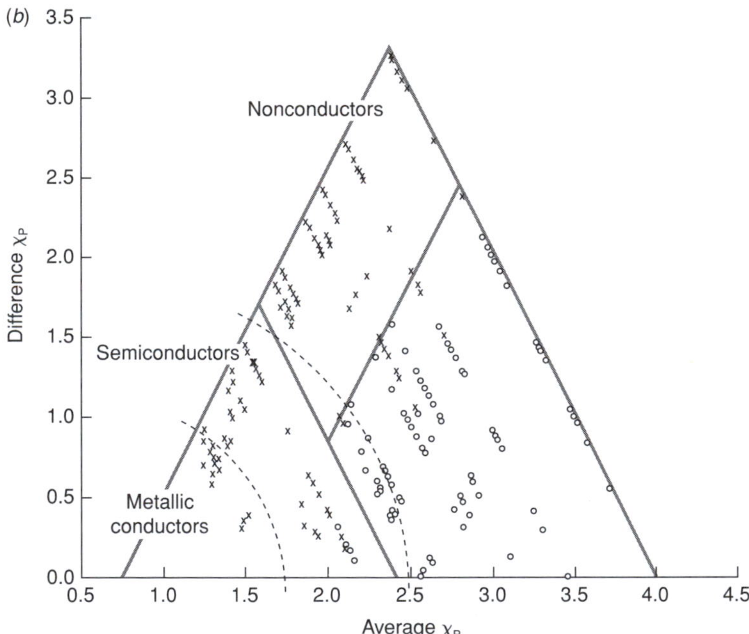

Figure 14.18
(*a*) Triangular graph of differences in Pauling electronegativities (χ_p) versus average Pauling electronegativities, showing Sproul's best division of the binary compounds involved into three regions. Compounds are symbolized by M if classified by Wells as metallic, O if classified as covalent, and I if ionic. [Adapted from G. D. Sproul, *J. Phys. Chem.*, **98**, 6699 (1994).] The dashed line represents the traditional bipartite division of ionic and covalent compounds based on a Pauling electronegativity difference of 1.8; the approximate locations of Zintl alloys are also indicated.
(*b*) The author's qualitative suggestion of the locations of metallic conductors, semiconductors, and nonconductors (insulators) on the Sproul–van Arkel triangle.

Note, however, that this picture for diatomics can only apply in the gas phase: All of these materials polymerize in the solid state.

Next, we consider a horizontal transit across the base of the van Arkel triangle, invoking pure covalent bonding of the second-period elements as examples, first in a hypothetical chain polymer resembling our hypothetical H_{10} molecule (Fig. 10.15) or our helical Group 16(VI) allotropes (Fig. 12.13). As the constituent atoms get more electronegative from Li to F, the orbital radius ratio r_p/r_s (Section 12.4) decreases from about 2.0 in Li to about 1.0 in F; the overlaps become more localized, not on a single atom, but in the overlap region between two neighboring atoms. As this happens, it becomes better to pull pairs of atoms together and concentrate the bonding between pairs of atoms. Thus, although it is computed that $^{1}_{\infty}$[Li] would prefer the delocalized structure corresponding to that illustrated in Figure 10.15 for H_{10}, by the time we reach F_{10}, the structure breaks up into five F_2 molecules (this is known as a *Peierls distortion* of the chain), each with a large band gap between the 2p HOMO and the σ^* LUMO.

Going back to the left from the bonding-localized F_2 molecules, we encounter progressively increasing diffuseness in the valence *p* orbitals (Section 12.4), which not only enhances delocalized bonding in the Li_{10} or $^{1}_{\infty}$[Li] chain, but also allows chains to overlap each other, increasing the coordination number of Li from 2 in the interior of the chain to 8 in the body-center cubic (bcc) structure; this further adds to the delocalization and the consequent disappearance of the band gap, so that $^{3}_{\infty}$[Li] becomes metallic. Thus, although the left corner of the van Arkel triangle includes metallic solids, no separate bonding theory is needed to describe it, so that some authors argue against having a third (metallic) category of bonding.[53] (The bonding is pure covalent in both lithium metal and in F_2, but a new set of metallic *properties* result from the way that bonding works out at the left side of the triangle.)

To the right of the region of metallic properties, such as high electrical conductivity, (lower left of the triangle) and to the left of the regions of ionic properties (top of the triangle) and the region of localized covalent bonding (lower right), there should be a central region in which intermediate electronegativities combine with intermediate coordination numbers to give the property of semiconduction. The relationship of the band gap in binary 1:1 compounds to the electronegativities of the elements has long been noted: The band gap falls as the electronegativities of the two elements approach 1.75 (Fig. 14.19).

The quantitative van Arkel diagram of Sproul,[51] which uses Pauling electronegativities, may be modifiable to show a domain of semiconductors [Fig. 14.18(*b*)]: roughly speaking, those that lie between a radius of 1.0 and 1.8 electronegativity units out from the metallic apex of the diagram (Cs, $\chi_P = 0.70$). This slice of the triangle includes compounds with the requisite energy gap, including binary 1:1 compounds such as those included in Figure 14.19, other semiconductors such as MoS_2, $MoSe_2$, $MoTe_2$, WS_2, and WSe_2 (average $\chi_P \sim 2.3$, very small electronegativity differences, and band gaps of $\sim$1 eV[54]); and (at the top) Zintl compounds, which are characteristically semiconductors.[55]

This figure should not be taken too quantitatively, however, since elements or binary compounds can have different allotropes or polymerization isomers with different coordination numbers (and degrees of secondary bonding), and consequently, quite different band gaps. Recall that diamond is an insulator with a band gap of 5.48 eV, C_{60} is a semiconductor with a band gap of 1.7 eV, and graphite is a conductor.

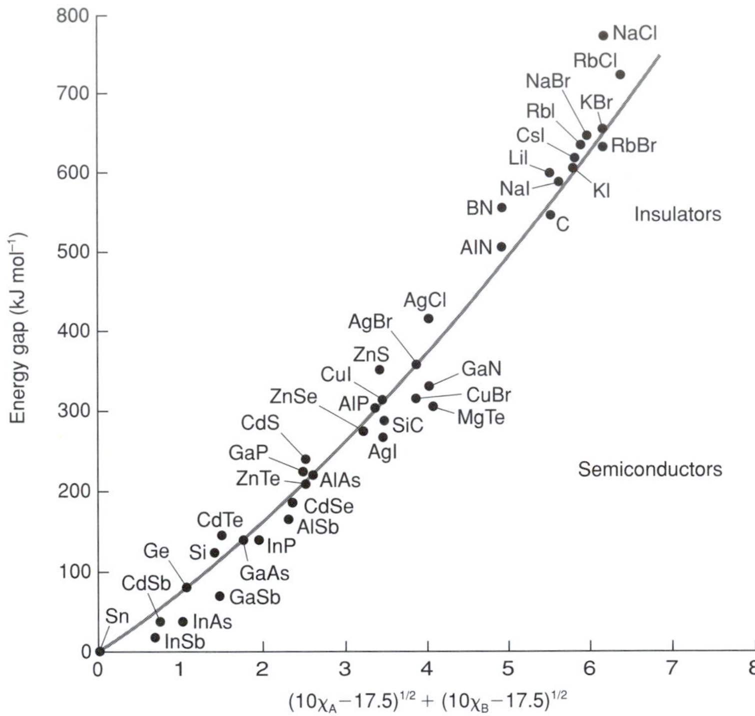

Figure 14.19
Empirical relationship of the energy gap important in semiconduction (Section 12.4) to the electronegativities of the elements present in a I–VII, II–VI, or III–V compound or a Group 14(IV) element. [Adapted from N. B. Hannay, "Solid-state Chemistry." Copyright © 1967 by Prentice-Hall Englewood Cliffs, NJ.] Unfortunately, the electronegativity scale used to derive this graph is not specified; it is not the current Pauling scale.

Also, compare the conductivity of gold ($\chi_P = 2.54$, coordination number = 12) with that of diamond ($\chi_P = 2.55$, coordination number = 4).

For too long, chemists have concentrated only on ionic and covalent bonding, letting the physicists deal with metallic bonding, which they have explained in a language quite foreign to chemists. Now there is a renewed interest in understanding, in a chemical language, important materials such as metallic conductors and semiconductors, which can come from a surprisingly large number of classes of compounds; as calculations on these materials are interpreted in chemical terms, provisional and empirical explanations such as those offered above will surely be modified and clarified.

Study Objectives

1. Given the formulas of a halide or a related group of halides, select the most likely structure type: (a) among the large categories of monomers, oligomers, or polymers; and (b) within a category, of the type of detailed structure most likely for the halogen atom involved. Exercises 1–7.

2. Tell whether a given halide is likely to be a gas, a liquid or low-melting solid, or a high-melting solid at room temperature. Predict some of its other physical properties, such as color and solubility. Exercises 8–17.

3. Choose an appropriate type of synthesis for a given halide. Exercises 18–24.

4. Classify a given halide as predominantly a Lewis acid or predominantly a Lewis base. Pick appropriate halides with which it might react. Exercises 25–30.

5. Give examples of halides that are likely to be good oxidizing agents; good reducing agents. Write balanced chemical equations showing such functions. Exercises 31, 32.

6. Describe some characteristic methods of synthesis, reactivity types, or bonding characteristics of catenated and cluster halides. Exercises 33–42.

7. Give or classify reasonable methods of preparation of examples of the oxygen derivatives of the elements discussed in Section 14.6. Exercises 43–51.

8. Give reasonable methods of preparation of the following types of compounds of the elements: (a) amides; (b) imides; and (c) nitrides. Exercises 52–55.

9. Know the practical uses of the types of polymers or macromolecules discussed in this chapter. Exercises 56–59.

10. Compare and contrast the structures and physical properties of the sulfides of the elements with the oxides and halides of the same elements. Exercises 60–67.

11. Identify, with the aid of a van Arkel diagram, the differences between ionic, covalent, and metallic materials. Exercises 68–70.

Exercises

1. In the following polymeric structures of metal halides, the smaller darkened circle represents the metal ion. Identify the degree of polymerization of each structure, and the coordination number of the metal ion and each kind of halide ion in (a) α-$PdCl_2$; (b) NbI_4; (c) $ReCl_4$; and (d) $PaCl_5$.

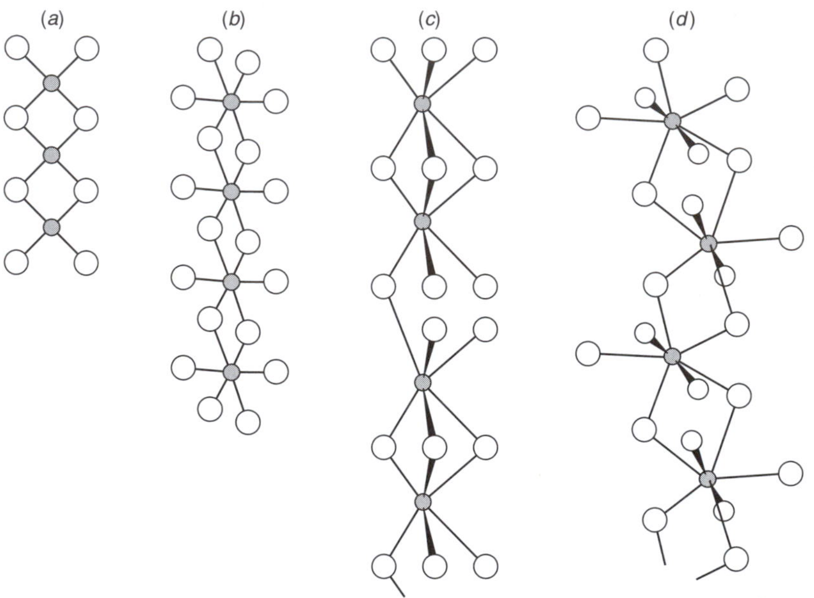

Figure 14.20
Chain structures for analysis in Exercise 1. (*a*) α-$PdCl_2$; (*b*) NbI_4; (*c*) $ReCl_4$; and (*d*) $PaCl_5$.

2. *Give the structural types of the following halides, in all of which the central atom has the group oxidation number −2: InI; $SnCl_2$; $SbBr_3$; $TeCl_4$; and IF_5. Discuss the reasons for the trends in this series.

3. Examine a copy of Table A, the Pauling electronegativity table. Assuming that metal–halogen bonding will be ionic for Pauling electronegativity differences in excess of 1.8 but polar covalent for differences less than 1.8, identify all metals that you would expect to have ionic fluorides but polar covalent chlorides. For each stoichiometry of halide, illustrate with one set of examples how the fluorides and chlorides differ in structure.

4. *Choose one of the following two structural classifications (monomeric or dimeric molecular; macromolecular or ionic) for each of the following halides: (a) CF_4; (b) AgI; (c) KF; (d) $SiCl_4$; (e) $BiCl_3$; and (f) SiI_4.

5. Choose one of the following structural classifications (monomeric, oligomeric, covalent polymeric, or ionic) for each of the following halides: (a) AuI; (b) RbBr; (c) $NiCl_2$; (d) $BiBr_3$; (e) LaF_3; (f) $SnCl_2$; (g) $PrCl_3$; and (h) WF_6.

6. Choose one of the following structural classifications (monomeric, oligomeric, covalent polymeric, or ionic) for each of the following halides: (a) ICl; (b) CsCl; (c) GeF_2; (d) GeF_4; (e) PbF_4; (f) ClF; (g) AlF_3; and (h) $AlBr_3$.

7. Predict the geometry of the monomeric structural unit of each of the following halides or derivatives using VSEPR theory, and compare it to the actual structure of the (perhaps polymerized) material. Discuss the reasons for any discrepancies. (a) SF_6; (b) XeF_6; (c) SF_4; (d) TeF_4; (e) $TeCl_4$; (f) PCl_3; (g) $SbCl_3$; (h) IF_7; (i) PI_5.

8. *Choose one of the following three categories of physical state at room temperature (gas; liquid or solid with a low melting point; solid with a high melting point) for each of the following halides: (a) CF_4; (b) AgI; (c) KF; (d) $SiCl_4$; (e) $BiCl_3$; and (f) SiI_4.

9. Choose one of the following three categories of physical state at room temperature (gas; liquid or solid with a low melting point; solid with a high melting point) for each of the following halides: (a) AuI; (b) RbBr; (c) $NiCl_2$; (d) $BiBr_3$; (e) LaF_3; (f) $SnCl_2$; (g) $PrCl_3$; and (h) WF_6.

10. Choose one of the following three categories of physical state at room temperature (gas, liquid, or solid with a low melting point; solid with a high melting point) for each of the following halides: (a) ICl; (b) CsCl; (c) GeF_2; (d) GeF_4; (e) PbF_4; (f) ClF; (g) AlF_3; and (h) $AlBr_3$.

11. (a) Which one of the following chromium(III) halides would be least likely to crystallize in a layer structure: CrF_3; $CrCl_3$; $CrBr_3$; CrI_3. (b) Briefly describe a difference in physical properties that you would likely notice between that halide and the ones that crystallized in layer structures. (c) Which of these halides is black? (d) Which, if any, of these halides are colorless?

12. From each of the following series of halides, pick the one that should have the lowest melting point, and explain why: (a) IF, BrF, ClF, F_2; (b) MgF_2, $MgCl_2$, $MgBr_2$, MgI_2; (c) InI, $SnCl_2$, $SbBr_3$, $TeCl_4$, IF_5; and (d) SiF_4, $SiCl_4$, $SiBr_4$, SiI_4.

13. *The compound $GeCl_4$ is a liquid with a melting point of 223 K; $GeCl_2$ is a solid with a melting point of about 400 K. Is this the normal trend to be expected for two monomeric covalent compounds of these formulas? Is this the normal trend to be expected for two ionic compounds of these stoichiometries? If neither, explain why the melting points fall in this order.

14. The highest bromides of the third-period elements are listed, together with their physical states and melting points:

Bromide	NaBr	$MgBr_2$	$AlBr_3$	$SiBr_4$	PBr_5	S_2Br_2
Physical state	Solid	Solid	Solid	Liquid	Solid	Liquid
mp (°C)	755	700	97	5	250	−40

(a) Some of these bromides show ionic structures, some are monomeric covalent molecules, and some are dimeric covalent molecules. Given the physical properties of the bromides and their periodic trends, list the most logical structural classification for each bromide. If you choose ionic, tell what you think the ions are. (b) List the most likely total coordination number for the central atom in each structure. (c) Four of the bromides shown above are white, one is yellow, and one is red. Assign logical colors to each one.

15. *On the bench in a professor's laboratory are six bottles containing the products of the syntheses of the six halides $CrCl_3$, HgI_2, NF_3, NI_3, $SnCl_2$, and $SnCl_4$. Unfortunately, the labels have come loose. Examine the bottles, and reattach the labels correctly. Bottle A contains a white solid; bottle B contains an intensely red solid; bottle C contains a colorless gas; bottle D contains light violet flakes; bottle E contains a colorless liquid; and bottle F consists of broken glass, being all that remains after an explosive attempted preparation of its would-be contents.

16. On the counter are five bottles. Bottle A contains a colorless gas; bottle B contains a layered gold-colored crystalline solid; bottle C contains a colorless solid; bottle D is broken, being all that is left after the product explodes; bottle E contains a colorless liquid. Tell which of the following five halides each bottle contains: PbI_2; SF_6; SI_4; PbF_2; and $SiCl_4$.

17. The boiling points of the fluorides of the third-period elements drop off after the fluoride of which element? Why?

18. *For each of the halides listed, choose one of the following four general synthetic routes as the most appropriate (workable yet simple) method of synthesis of the anhydrous halide: Method 1, anhydrous salt results from evaporating the reaction mixture of the metal hydroxide plus the hydrohalic acid; method 2, the salt precipitates in anhydrous form upon mixing solutions containing the cation and the anion; method 3, the hydrated halide can be dehydrated by an appropriate chemical dehydrating agent; and method 4, a strictly anhydrous method of synthesis (halogen exchange or redox) is required. Halides: (a) CF_4; (b) AgI; (c) KF; (d) $SiCl_4$; (e) $BiCl_3$; and (f) SiI_4.

19. For each of the halides listed, choose one of the four general synthetic routes listed in the previous exercise as the most appropriate (workable yet simple) method of synthesis of the anhydrous halide: (a) AuI; (b) RbBr; (c) $NiCl_2$; (d) $BiBr_3$; (e) LaF_3; (f) $SnCl_2$; (g) $PrCl_3$; and (h) WF_6.

20. *For any of the halides assigned from one of the two previous exercises, complete and balance a reaction showing a plausible synthesis of that halide in anhydrous form.

21. Complete and balance chemical equations showing the best way to make each of the following halides in anhydrous form. (a) TlI_3; (b) $CrCl_3$; (c) $EuCl_2$; (d) PtF_6; (e) TlI; and (f) IF_7.

22. Which of the following are good ways to synthesize the product shown in boldface? Which are inefficient ways to synthesize it, or would not work?

(a) $FeCl_3 \cdot 6H_2O + \Delta \rightarrow \mathbf{FeCl_3} + 6\,H_2O$

(b) $Ca^{2+}(aq) + 2\,F^-(aq) \rightarrow \mathbf{CaF_2}(s)$

(c) $2\,Na + F_2 \rightarrow 2\,\mathbf{NaF}$

(d) $Ag_2O + SOCl_2 \rightarrow 2\,\mathbf{AgCl} + SO_2$

23. Briefly explain why nickel iodide is green when it crystallizes out of cold water, but turns black after pumping with the vacuum pump.

24. Consider the synthesis of one of the halides listed at the end of this question. (1) Discuss the applicability of each of the seven synthetic methods listed in Section 14.3 to this halide. (2) Look up the synthesis of the halide in an advanced text, such as Cotton and Wilkinson, *Comprehensive Inorganic Chemistry*, 5th ed., Wiley-Interscience, New York, 1988. What methods are given there? Do they agree with your prediction? (3) Look up the actual details of the synthesis—in the reference cited in the advanced text, in an inorganic lab manual, in *Inorganic Syntheses*, in Georg Brauer's *Handbook of Preparative Inorganic Chemistry*, or in *Chemical Abstracts*. Are there important details (such as carrying out the synthesis in a dry or nonoxidizing atmosphere) that you would have expected from the principles of this book? Are there important details that you could not have anticipated? HALIDES: (a) $MoCl_5$; (b) $[Mo_6Cl_8]Cl_4$; (c) VCl_4; (d) $TiBr_4$; (e) $CrBr_3$; and (f) CrI_3.

25. *Classify each of the following halides as predominantly a Lewis acid, predominantly a Lewis base, or neither (or both): (a) CF_4; (b) AgI; (c) KF; (d) $SiCl_4$; (e) $BiCl_3$; and (f) SiF_4.

26. Classify each of the following halides as predominantly a Lewis acid, predominantly a Lewis base, or neither (or both): (a) HgI_2; (b) RbBr; (c) $BaCl_2$; (d) $BiBr_3$; (e) CeF_4; (f) PF_3; (g) $FeCl_3$; (h) SF_6; (i) AsF_5; and (j) IF_7.

27. Classify each of the following halides as predominantly a Lewis acid, predominantly a Lewis base, or neither (or both): (a) XeF_6; (b) $AuCl_3$; (c) LiI; (d) CCl_4; and (e) PCl_3.

28. Complete and balance plausible Lewis acid–base reactions between appropriate pairs of halides in (a) Exercise 26; and (b) Exercise 27.

29. *Complete and balance equations showing the products from the following pairs of reactants, or state that no reaction will occur: (a) $PF_5 + CsF$; (b) $XeF_6 + CF_4$; (c) $RbBr + BaBr_2$; (d) $XeF_6 + BF_3$; and (e) $NaF + AlF_3$.

30. Complete and balance equations showing the products from the following pairs of reactants, or state that no reaction will occur: (a) $SnCl_3^- + IrCl_6^{3-}$; (b) $ClF_3 + UO_3$; (c) $PI_5 + ZnI_2$; and (d) MeI + NaCl + acetone solvent.

31. Complete each of the following Lewis acid–base reactions, or state that no reaction will occur. If a reaction does occur, classify the function of each reactant in the given reaction as that of a Lewis acid, a Lewis base, an oxidizing agent, or a reducing agent.

(a) $RbI + AlI_3 \rightarrow$

(b) $CCl_4 + SbCl_5 \rightarrow$

(c) $IF_7 + BF_3 \rightarrow$

(d) $CsF + PF_5 \rightarrow$

(e) $SnCl_2 + PbCl_4 \rightarrow$

(f) $PbO + ClF_3 \rightarrow$

32. List four halides that would likely be strong oxidizing agents; four that would likely be useful reducing agents. Complete and balance plausible redox reactions between appropriate pairs of these halides.

33. Give actual examples of catenated halides from each of the following groups: (a) 16(VI); (b) 15(V); (c) 14(IV); and (d) 13(III).

34. Draw resonance structures with bond orders consistent with the information that the catenated halide O_2F_2 has an extraordinarily short O–O bond and a rather long O–F bond.

35. *Predict plausible products from the following possible reactions of catenated halides: (a) $Si_2Cl_6 + SO_3$; (b) $B_2Br_4 + Ph_3PO$; (c) $Si_2Cl_6 + HC{\equiv}CH$; and (d) $Hg_2Br_2 + Br_2$.

36. *Determine the metal–metal (or element–element) bond orders in the following compounds: (a) $Tc_2Cl_8^{2-}$; (b) $Tc_2Cl_8^{3-}$; (c) $[Ru_2(O_2CMe)_4(acetone)_2]$; and (d) $[Ru_2(O_2CMe)_4Cl_2]^-$.

37. A diplatinum(III) tetraacetate cation (dihydrated) has recently been prepared [T. G. Appleton et al., *J. Am. Chem. Soc.*, **114**, 7305 (1992)]. Predict the Pt–Pt bond order in this compound.

38. Is the monomeric unit of copper(II) acetate hydrate isolobal and isoelectronic with other units that would be expected to form single M–M bonds? Does the copper in this dimer obey the 18-electron rule? Why or why not—or is this not to be expected, and if so, why not?

39. (a) A recent report of the discovery of an unsupported Co=Co bond (i.e., with no bridging ligands) was subsequently refuted. Consult the original literature [*Angew. Chem. Intl. Ed. Engl.*, **30**, 1120 (1991); **31**, 1341 (1992)] to determine what experimental difficulties allowed this disagreement to arise, and what the product was finally identified as. (b) A similar report, followed by a revised proposal, dealt with a proposed quintuple bond in a Tc(II) complex. Consult the original literature [*Dokl. Chem.*, **288**, 147 (1986); *Inorg. Chem. Acta*, **189**, 53 (1991)] to determine what experimental difficulties allowed this disagreement to arise, and what bond order was finally proposed for the compound.

40. Calculate the average metal oxidation number in (a) $Mo_2Cl_8{}^{4-}$; (b) $Os_2Cl_8{}^{2-}$; (c) Re_3Cl_9; (d) $Mo_6Cl_8{}^{4+}$; (e) $W_6Cl_{12}{}^{6+}$; (f) $Ta_6Cl_{12}{}^{2+}$; and (g) Gd_4Cl_6.

41. *Calculate whether the following types of clusters obey the 18-electron rules (Chapter 11): (a) $Mo_2Cl_8{}^{4-}$; (b) $Os_2Cl_8{}^{2-}$; (c) the Re_3Cl_9 inert inner cluster; (d) the $Re_3Cl_9L_3$ overall cluster; (e) the $Mo_6Cl_8{}^{4+}$ inner cluster; (f) the $Mo_6Cl_{14}{}^{2-}$ overall cluster; (g) the $W_6Cl_{12}{}^{6+}$ inner cluster; and (h) the W_6Cl_{18} overall cluster.

42. Using L to represent a generic neutral ligand, show equations for substitution reactions replacing only the right number of labile halide ligands to give neutral products, starting with the following cluster halo anions: (a) $Nb_6Cl_{18}{}^{4-}$; (b) $Re_3Cl_{12}{}^{3-}$; and (c) $Mo_6Cl_{14}{}^{2-}$. (Write the formulas of the products to show the inert inner-cluster halogens inside brackets but any remaining labile halogens, plus substituted ligands, outside the brackets.)

43. Suggest common names and geometries for the compounds CrO_2F_2 and CrO_2Cl_2. Explain their colors.

44. Would you expect the ring-opening polymerization of $[Me_2SiO]_4$ to give $^{1}_{\infty}[Me_2SiO]$ to be accompanied mainly by a favorable enthalpy or a favorable entropy change? Explain why.

45. *Write balanced chemical equations illustrating reasonable methods of preparation of the following compounds: (a) ZnS; (b) CaS; (c) KOC_2H_5; (d) $As(OMe)_3$; and (e) $(Me_3Si)_2O$

46. Briefly tell what is wrong with the following proposed synthesis of barium isopropopoxide (as a precursor to the superconductor); suggest instead a synthesis that would work. $[BaI_2(HO\textit{i}\text{-}Pr)_2] + 2\,{:}NMe_3 \rightarrow Ba(O\textit{i}\text{-}Pr)_2 + 2\,[Me_3NH]I$

47. Which of the following are good ways to synthesize the product shown in boldface? Which are inefficient ways to synthesize it, or would not work?

(a) $2\,Li + 2\,MeOH \rightarrow 2\,\mathbf{LiOMe} + H_2$

(b) $Cu + 2\,MeOCH_2CH_2OH \rightarrow \mathbf{Cu(MeOCH_2CH_2O)_2} + H_2$

(c) $KCl + HOEt \rightarrow \mathbf{KOEt} + HCl$

48. Complete and balance each of the following reactions, or state that no reaction will occur. (a) $Li + MeOH \rightarrow$ and (b) $BCl_3 + MeNH_2$ (from its hydrochloride) $+ \Delta \rightarrow$.

49. Which are the most suitable starting materials for synthesizing trimethyl phosphite: $P_4 + MeOH$; $PCl_3 + MeOH$; or $PCl_3 + MeOH + R_3N$. Why is your choice the best?

50. *The oxides from which the 1:2:3 high-temperature superconductor, $YBa_2Cu_3O_7$, are usually made are high melting and mostly insoluble in any solvent, so it is quite difficult to mix them well enough by grinding to obtain good-quality single crystals of the superconductor. Consequently, many researchers have attempted to find compounds of Y, Ba, and Cu that are soluble in the same solvent, so that the 1:2:3 superconductor can be precipitated in a finely divided well-mixed form, for example, by hydrolysis of the dissolved compounds on adding water. Which of the following compounds of yttrium would likely both be soluble in some solvent and would react with water by hydrolysis to precipitate yttrium oxide or hydroxide in high yield? (a) Y_2S_3; (b) YF_3; (c) $Y(OMe)_3$; (d) $Y(OCH_2CH_2OCH_2CH_2OMe)_3$; (e) YPO_4; (f) $K[YSi_3O_8]$; and (g) $Y_2Mg_5(OH)_2[Al_2Si_2O_{11}]_2$.

51. For each of the following oxo halides, write an equation showing how it is synthesized, and (if possible) select one method of synthesis of halides (Section 14.3) that describes the synthesis: (a) $COCl_2$; (b) SO_2Cl_2; (c) NOCl; (d) $VOCl_3$; and (e) $POCl_3$.

52. *Write balanced chemical equations illustrating reasonable methods of preparation of the following compounds: (a) $VO(NMe_2)_3$; (b) $B_3Cl_3N_3Me_3$; (c) S_4N_4; and (d) $[N{=}P(OMe)_2]_3$

53. Are the following reasonable methods of preparation of the compounds listed in boldface? (a) $3\,LiCl + 4\,NH_3 \rightarrow \mathbf{Li_3N} + 3\,NH_4Cl$ and (b) $Cr(NH_2)_3 + \Delta \rightarrow \mathbf{CrN} + 2\,NH_3$.

54. Determine the metal–metal (or element–element) bond orders in the following compounds: (a) minimum average Ga–N bond order in $GaN_3{}^{6-}$; (b) average P–N bond order in $[Ph_3PNPPh_3]^+$; and (c) average B–N bond order in borazole.

55. For which one of the two metals listed would the following syntheses be more likely to succeed: (a) $2\,M + 2Me_2NH \rightarrow 2\,MNMe_2 + H_2$, **M = Na or Cu?** (b) $3\,M + N_2 \rightarrow M_3N_2$, **M = Mg or Pb?** (c) $3\,MCl_3 + 3\,NH_3 \rightarrow$ *cyclo*-(Cl–M–N–H)$_3$, **M = Sc or B**?

56. *Suggest an inorganic polymer or macromolecule that might be suitable if you need the following properties: (a) a semiconductor that would convert electrical energy to UV light; (b) a polymer that would exhibit metallic conduction in one dimension only; (c) a macromolecule that would exhibit metallic conduction in two dimensions only; (d) a polymer that might slowly hydrolyze to release the anticancer drug cisplatin in the blood; (e) a ceramic that would maintain its strength to 1000 °C; (f) a lubricating oil of low viscosity; and (g) a solid lubricant.

57. Write a chemical formula (showing the structural unit and degree of polymerization) for and list a suitable use or significant property for the following: (a) a silicone; (b) a polyphosphazene; (c) polythiazyl; (d) gallium arsenide; and (e) borazole.

58. Which would be the most suitable material for use as a ceramic that would retain its strength to 1000 °C: Si_3N_4; MoS_2; $H_3B_3N_3H_3$. Briefly explain why the other two are unsuitable.

59. For each of the following pairs of compounds briefly describe two similarities between the two (in structure, uses, and/or physical properties) and one difference in chemical properties. (a) Gallium arsenide and silicon; (b) borazine (borazole) and benzene; (c) boron nitride and diamond; (d) molybdenum(IV) sulfide and graphite (two similarities only); and (e) polythiazyl and mercury (one similarity only).

60. *Which of the following lattice types are layer structures? (a) Fluorite; (b) rock salt; (c) cadmium iodide; (d) chromium(III) chloride; (e) rutile; (f) nickel arsenide; (g) graphite; and (h) diamond.

61. Tell for each of the following types of halide or chalcogenide which of the lattice types listed in Exercise 60 is most likely, assuming that M is a *d*-block metal with an appropriate radius ratio: (a) MO; (b) MS; (c) MO_2; (d) $(M^{4+})(S^{2-})_2$; (e) MTe; (f) MF; (g) MF_2; and (h) MI_2.

62. Which sulfide of phosphorus *does* correspond in structure to an oxide of phosphorus?

63. Some phosphorus sulfides occur in isomeric forms: (a) α and β-P_4S_4; (b) α- and β-P_4S_5. Estimate ΔH for interconverting the α isomeric form to the β from bond-dissociation energies. The P=S bond dissociation energy is $+335\ \mathrm{kJ\,mol^{-1}}$; the P–S energy is not known, but can be estimated from the Pauling electronegativity equation.

64. *What is the most logical oxidation state to assign to (a) S in each of the phosphorus sulfides; (b) P in each of the phosphorus sulfides; (c) Nb in NbS_3; (d) Mo in $[Mo_6Cl_8]^{4+}$; (e) Mo in $[Mo_6S_8]$; and (f) Mo in $PbMo_6S_8$?

65. Figure 14.17(*c*) illustrates the results of MO calculations done on $Mo_6S_8{}^{4-}$. With this cluster charge, what is the oxidation number and the d^n electron configuration of Mo? How many metal valence electrons would be present in this Mo_6 cluster? Write its valence electron configuration, using the symmetry labels given in Figure 14.17.

66. *(a) Write the formulas of the two amines that can be generated from adamantane, $C_{10}H_{16}$, by replacing one type of carbon group completely with isolobal nitrogen groups. (b) Write the formula of the hydrocarbon that would be structurally most similar to the ferredoxin cluster. (The common name of this hydrocarbon is cubane.) (c) Which hydrocarbon, adamantane or cubane, would be more stable and why?

67. The interesting, intensely colored biologically significant complexes of sulfide and thiolate ions with iron(III) and copper(II) (the latter are known as "copper blue" proteins) have proved somewhat difficult to model (build simpler analogues of) because of a common side reaction. Explain the intense colors, what this side reaction is, and why it occurs.

68. Draw a van Arkel triangle with the usual orientation. (a) Show the approximate location of ionic compounds, covalent compounds, and metals on this triangle by inserting in the triangle the letters "I", "C", and "M" respectively. (b) What atomic property is graphed along the vertical axis of a van Arkel triangle? (c) What atomic property is graphed along the horizontal axis?

69. (a) At which corner of the van Arkel triangle (top, lower left, lower right) is the bonding most delocalized? (b) At which corner is it most localized on the more electronegative of the two atoms? (c) At which corner is it most localized in a two-center, two-electron (2c–2e) bond between the two atoms? (d) Where in the triangle is semiconduction located?

70. Consider these three materials: Cs, a metallic conductor; Cs^+Au^-, a semiconductor; Cs^+F^-, an insulator. (a) Along which side of the van Arkel diagram would these three substances fall? (b) How much does the coordination number (say, of Cs) change along this side as compared to moving along the other sides of the triangle?

(c) Does the bonding become more localized or more delocalized from Cs to CsF? In what sense of the word "localized" or "delocalized"? (d) Which material (if any) would have, on the average, the lowest energy occupied energy levels? (e) Which material, if any, would have no energy gap between its occupied and unoccupied MOs? (f) Which material would likely have the narrowest band of occupied energy levels?

Notes

1. A. F. Wells, *Structural Inorganic Chemistry*, Clarendon Press, Oxford, UK, 1984; pp. 407–429.
2. A. A. Timakov, V. N. Prusakov, and Yu. V. Drobyshevskii, *Dokl. Chem.*, **291**, 442 (1986).
3. The Compound $:NI_3$ is a famous explosive (it is normally found as $NI_3 \cdot NH_3$), $:NBr_3$ explodes above $-100\,^\circ C$, and $:NCl_3$ is explosive unless diluted with other gases.
4. N. W. Alcock, *Adv. Inorg. Chem. Radiochem.*, **15**, 1 (1972).
5. P. Paetzold, *Angew. Chem. Intl. Ed. Engl.*, **30**, 544 (1991).
6. A. J. Downs and C. J. Adams, in J. C. Bailar, H. J. Emeleus, R. Nyholm, and A. F. Trotman-Dickenson, Eds., *Comprehensive Inorganic Chemistry*, Pergamon Press, Oxford, UK, 1973; Vol. 2, pp. 1253–1257.
7. *Inorganic Syntheses*, McGraw-Hill, New York, 1939–date; G. Brauer, *Handbook of Preparative Inorganic Chemistry*, 2 vols., 2nd ed., Academic, New York, 1963.
8. We have already seen catenation illustrated in the structures of the homopolyatomic anions and cations of the elements (Section 12.5).
9. F. A. Cotton and R. A. Walton, *Multiple Bonds between Metal Atoms*, 2nd ed., Clarendon Press, Oxford, UK, 1993.
10. S. Hao, S. Gambarotta, and C. Bensimon, *J. Am. Chem. Soc.*, **114**, 3556 (1992).
11. J. P. Collman and H. J. Arnold, *Acc. Chem. Res.*, **26**, 586 (1993).
12. J. A. Morrison, *Chem. Rev.*, **91**, 35 (1991).
13. A. Franken, H. Thomsen, and W. Preetz, *Z. Naturforsch.*, **51b**, 744 (1996); W. Preetz and J. Fritze, *Z. Naturforsch.*, **39b**, 1472 (1984).
14. C. Dohmeier, D. Loos, and H. Schnöckel, *Angew. Chem. Int. Ed. Engl.*, **35**, 129 (1996).
15. P. R. Birkett, P. B. Hitchcock, H. W. Kroto, R. Taylor, and D. R. M. Walton, *Nature (London)*, **357**, 479 (1992); F. N. Tebbe, R. L. Harlow, D. B. Chase, D. L. Thorn, G. C. Campbell, J. C. Calabrese, N. Herron, R. J. Young, Jr., and E. Wasserman, *Science*, **256**, 822 (1992).
16. A. A. Gakh, A. A. Tuinman, J. L. Adcock, R. A. Sachleben, and R. N. Compton, *J. Am. Chem. Soc.*, **116**, 819 (1994).
17. A. Simon, *Angew. Chem. Int. Ed. Engl.*, **27**, 159 (1988).
18. J. D. Corbett, *Pure Appl. Chem.*, **64**, 1395 (1992).
19. V. Plato, W. D. Hartford, and K. Hedberg, *J. Chem. Phys.*, **53**, 3488 (1970).
20. H. R. Allcock, *Chem. Eng. News*, March 18, 1985, p. 22; J. E. Mark, H. R. Allcock, and R. West, *Inorganic Polymers*, Prentice-Hall, Englewood Cliffs, NJ, 1992.
21. The name "silicone" derives from an early, extremely faulty analogy between the simplest unit of these polymers and ketones such as acetone, $Me_2C{=}O$.
22. P. A. Agaskar, *Inorg. Chem.*, **29**, 1603 (1990).
23. D. C. Bradley, R. C. Mehrotra, and D. P. Gaur, *Metal Alkoxides*, London, Academic, New York, 1978; D. C. Bradley, *Philos. Trans. R. Soc. London*, **A330**, 167 (1990); L. G. Hubert-Pfalzgraf, *Appl. Organomet. Chem.*, **6**, 627 (1992); R. C. Mehrotra, *Chemtracts: Inorg. Chem.*, **2**, 389 (1990).
24. S. R. Drake, D. J. Otway, M. B. Hursthouse, and K. M. Abdul Malik, *Polyhedron*, **11**, 1995 (1992).
25. W. A. Hermann, N. W. Huber, and O. Runte, *Angew. Chem. Int. Ed. Engl.*, **34**, 2187 (1995); W. S. Rees, Jr., and D. A. Moreno, *J. Chem. Soc. Chem. Commun.*, 1759 (1991).
26. C. Sanchez and J. Livage, *New J. Chem.*, **14**, 513 (1990); L. L. Hench and J. K. West, *Chem. Rev.*, **90**, 33 (1990); J. D. Mackenzie, *J. Non-Crystalline Solids*, **100**, 162 (1988); D. R. Ulrich, *J. Non-Crystalline Solids*, **121**, 465 (1990).
27. R. E. Sievers, S. B. Turnipseed, L. Huang, and A. F. Lagalante, *Coord. Chem. Rev.*, **128**, 285 (1993); G. Malandrino, R. Licata, F. Castelli, and I. L. Fragalà, *Inorg. Chem.*, **34**, 6233 (1995).

28. R. M. Laine, K. Y. Blohowiak, T. R. Robinson, M. L. Hoppe, P. Nardi, J. Kampf, and J. Uhm, *Nature* (*London*), **353**, 642 (1991).
29. B. Herreros and J. Klinowski, *J. Phys. Chem.*, **99**, 1025 (1995).
30. A. A. Danoupoulos, G. Wilkinson, B. Hussain, and M. B. Hursthouse, *J. Chem. Soc., Chem. Commun.*, **896** (1989).
31. T. Wideman and L. G. Sneddon, *Inorg. Chem.*, **34**, 1002 (1995).
32. R. Niewa and H. Jacobs, *Chem. Rev.*, **96**, 2053 (1996); D. H. Gregory, M. G. Barker, P. P. Edwards, and D. J. Siddons, *Inorg. Chem.*, **34**, 5195 (1995).
33. W. Schnick, *Stud. Surf. Sci. Catal*, **84**, 2221 (1994).
34. E. G. Gillan and R. B. Kaner, *Inorg. Chem.*, **33**, 5693 (1994); T. Wade, C. B. Ross, and R. M. Crooks, *Chem. Mater.*, **9**, 248 (1997); L. E. Toth, *Refractory Material, Vol. 7: Transition Metal Carbides and Nitrides*, Academic, New York, 1971.
35. I. Silaghi-Dumitrescu, I. Haiduc, and D. B. Sowerby, *Inorg. Chem.*, **32**, 3755 (1993).
36. E. J. M. Hamilton, S. E. Dolan, C. M. Mann, H. O. Colijn, and S. G. Shore, *Chem. Mater.*, **7**, 111 (1995); N. G. Chopra, R. J. Luyken, K. Cherrey, V. H. Crespi, M. L. Cohen, S. G. Louie, and A. Zettl, *Science*, **269**, 966 (1995).
37. C. Niu, Y. Z. Lu, and C. M. Lieber, *Science*, **261**, 334 (1993); M. Todd, J. Kouvetakis, T. L. Groy, D. Chandrasekhar, D. J. Smith, and P. W. Deal, *Chem. Mater.*, **7**, 1422 (1995).
38. J. E. Mark, H. R. Allcock, and R. West, *Inorganic Polymers*, Prentice-Hall, Englewood Cliffs, NJ, 1992; Chapter 3.
39. R. J. Cava, H. Takagi, H. W. Zandbergen, J. J. Krajewski, W. F. Peck Jr, T. Siegrist, B. Batlogg, R. B. van Dover, R. J. Felder, K. Mizuhashi, J. O. Lee, H. Eisaki, and S. Uchida, *Nature* (*London*), **367**, 252 (1994); R. J. Cava, H. W. Zandbergen, B. Batlogg, H. Eisaki, H. Takagi, J. J. Krajewski, W. F. Peck Jr, E. M. Gyorgy, and S. Uchida, *Nature* (*London*), **372**, 245 (1994).
40. A. Simon, H. Mattausch, R. Eger, and R. K. Kremer, *Angew. Chem. Int. Ed. Engl.*, **30**, 1188 (1991).
41. B. C. Guo, K. P. Kerns, and A. W. Castleman, Jr., *Science*, **255**, 1411 (1992); S. F. Cartier, Z. Y. Chen, G. J. Walder, C. R. Sleppy, and A. W. Castleman, Jr., *Science*, **260**, 195 (1993); Z. Y. Cho, G. J. Waldmer, and A. W. Castleman, Jr., *J. Phys. Chem.*, **96**, 9581 (1992); S. F. Cartier, B. D. May, B. J. Toleno, J. Purnell, S. Wei, and A. W. Castleman, Jr., *Chem. Phys. Lett.*, **220**, 23 (1994); J. S. Pilgrim and M. A. Duncan, *J. Am. Chem. Soc.*, **115**, 6958 (1993).
42. Z. Lin and M. B. Hall, *J. Am. Chem. Soc.*, **115**, 11165 (1993); M.-M. Rohmer, M. Bénard, C. Bo, and J.-M. Poblet, *J. Am. Chem. Soc.*, **117**, 508 (1995).
43. H. T. Deng, K. P. Kerns, and A. W. Castleman, Jr, *J. Am. Chem. Soc.*, **118**, 446 (1996); Y. G. Byun and B. S. Freiser, *J. Am. Chem. Soc.*, **118**, 3681 (1996).
44. R. Tenne, L. Margulis, M. Genut, and G. Hodes, *Nature* (*London*), **360**, 444 (1992); Y. Feldman, E. Wasserman, D. J. Srolovitz, and R. Tenne, *Science*, **267**, 222 (1995).
45. E. Sandre, A. LeBlanc, and M. Danot, *J. Chem. Educ.*, **68**, 809 (1991); J. Rouxel, *Acc. Chem. Res.*, **25**, 328 (1992).
46. For use of LEDs in teaching see G. C. Lisensky, R. Penn, M. J. Geselbracht, and A. B. Ellis, *J. Chem. Educ.*, **69**, 151 (1992).
47. M. K. Chan, J. Kin, and D. C. Rees, *Science*, **260**, 792 (1993).
48. M. K. Johnson, D. C. Rees and M. W. W. Adams, *Chem. Rev.*, **96**, 2817 (1996).
49. C. Huber and G. Wächtershäuser, *Science*, **276**, 245 (1997).
50. A. E. van Arkel, *Moleculen en Kristallen*, van Stockum, s'Gravenhage, 1941; A. E. van Arkel, *Molecules and Crystals*, Butterworths, London, 1949; W. B. Jensen, *Bull. Hist. Chem.*, **13–14**, 47 (1992–1993).
51. G. D. Sproul, *J. Chem. Educ.*, **70**, 531 (1993); G. D. Sproul, *J. Phys. Chem.*, **98**, 6699 (1994).
52. W. B. Jensen, *Educ. Chem.*, **31**, 10 (1994); W. B. Jensen, *J. Chem. Educ.*, **72**, 395 (1995).
53. W. P. Anderson, J. K. Burdett, and P. T. Czech, *J. Am. Chem. Soc.*, **116**, 8808 (1994); L. C. Allen and J. F. Capitani, *J. Am. Chem. Soc.*, **116**, 8810 (1994).
54. C. E. Stanton, SB. T. Nguyen, J. M. Kesselman, P. E. Laibinis, and N. S. Lewis, Semiconductors, in R. B. King, Ed., *Encyclopedia of Inorganic Chemistry*, Wiley, Chichester, UK, 1994; p. 3725.
55. S. M. Kanzlarich, Zintl Compounds, in R. B. King, Ed., *Encyclopedia of Inorganic Chemistry*, Wiley, Chichester, UK, 1994; p. 4467.

CHAPTER

15

Hydrides, Alkyls, and Aryls of the Elements

With Applications to Materials Science and Organic Chemistry

In this last chapter of descriptive and materials science chemistry, we will study some of the chemistry of the hydrides[1] and the σ-bonded organometallic compounds of the elements (with an emphasis on the *s*- and *p*-block elements). Hydrides have grown greatly in importance since the beginning of the Atomic Age, due to the importance of hydrogen in nuclear fusion and energy storage. Despite the familiarity of hydrogen and its compounds, these compounds have still given rise to some recent major scientific controversies. Sigma-bonded organometallic compounds have been of importance since the discovery of the Grignard reagents, RMgX, by Victor Grignard early in the twentieth century. These reagents are no doubt familiar to you from your organic chemistry course, since they are of great importance in organic synthesis.

15.1 Classification of Hydrides: Hydrides of Metals

The element hydrogen is uniquely difficult to locate in the periodic table. Although it is conventionally located in Group 1 and, like the other Group 1 members, predominantly exhibits a +1 oxidation state, it differs in lacking core electrons. It is considerably smaller and more electronegative than the other members of the group and consequently does not form a simple cation H^+ unless strongly hydrogen bonded to a Lewis base such as water. Also, under normal conditions it is not a metal.

An alternate location for hydrogen in the periodic table is in Group 17(VII); it is a nonmetal that forms a diatomic molecule like the other members of this group. Although it does not use *p* electrons, hydrogen is one electron shy of a noble gas electron configuration and sometimes forms the hydride ion, H^-, which resembles the halide ions. This resemblance, however, is limited: For example, the electron affinity of H (Table 7.5) is much less than that of the halogen atoms. Consequently, H^- is a much stronger reducing agent ($E^\circ = -2.25\,V$; Fig. 6.2) than the halide ions.

A third location that has been proposed is above carbon. Although H, of course, does not have four valence electrons, its Pauling electronegativity is reasonably close to that of carbon, and there is a fair degree of resemblance between the chemical properties of the hydrides of many of the *s*- and *p*-block elements and the σ-bonded organometallic compounds of the same elements. Without necessarily advocating this periodic table position, we consider the hydrides and the organometallic derivatives of the elements together in this chapter.

Because the electronegativity of hydrogen (2.2) is about in the middle of the Pauling scale, its compounds include polar covalent compounds in which the hydrogen is partially positively charged, ionic and polar covalent compounds in which it is partially negatively charged, and some in which there is little or no polarity. This variation in polarity results in a considerable range of NMR chemical shifts being found in hydrides of the elements, from δ +16 ppm in some strongly acidic compounds (protonated ethers) in which there is little electron density around H, to δ 0 or below in some hydrides in which the H atom is partially negatively charged. (In some organometallic hydride complexes of *d*-block metals, the chemical shifts can be −10 to −20 ppm or more; factors beyond high electron density around H are involved here.)

There are no fast dividing lines between bonds of hydrogen to elements in which it is partially positively charged, neutral, and partially negatively charged, so we will divide the hydrides into classifications based more on structural properties than on bond types; these are shown in Table 15.1.

Table 15.1
Hydrides of the Elements

A. Molecular (Covalent) Hydrides (gases at room temperature unless otherwise noted)							
B_2H_6	CH_4	NH_3	$H_2O(\ell)$	$HF(\ell)$			
	SiH_4	PH_3	H_2S	HCl			
$Ga_2H_6(\ell)$	GeH_4	AsH_3	H_2Se	HBr			
	SnH_4	SbH_3	H_2Te	HI			
Catenated Molecular Hydrides (liquids unless otherwise noted)							
B_nH_{n+6} ($n = 4$–10)		C_nH_{2n+2} (no limit to n)		N_2H_4	H_2O_2		
B_nH_{n+4} ($n = 5$–20)		C_nH_{2n} (no limit to n)					
(solid for $n > 9$)		(gas for $n < 5$)					
		(solid for large n)					
		Si_nH_{2n+2} (n up to 8)		P_nH_{n+2}	H_2S_n (n up to 8)		
		(gas for $n = 2$)					
		Ge_nH_{2n+2} (n up to 5)					
B. Salt-Like (Saline) and *Polymeric* (*Borderline*) Hydrides (Solids at room temperature unless otherwise noted; polymeric hydrides are in *italics*)							
LiH	*BeH_2*						
NaH	*MgH_2*				*AlH_3*		
KH	CaH_2		*CuH*	*ZnH_2*			
RbH	SrH_2						
CsH	BaH_2	EuH_2, YbH_2					
C. Metallic (Interstitial) Hydrides (solids at room temperature)							
ScH_2	TiH_2	VH	CrH		NiH		
		VH_2					
YH_2	ZrH_2	NbH			$PdH_{<1}$		
YH_3		NbH_2					
LuH_2	HfH_2	TaH					
LuH_3							
The *f*-block elements of the sixth period form both MH_2 and MH_3 (except for M = Eu and Yb).							
AcH_2	ThH_2			NpH_2	PuH_2	AmH_2	CmH_2
	Th_4H_{15}	PaH_3	UH_3	NpH_3		AmH_3	

Molecular Hydrides. The **molecular** (or covalent) hydrides are found in Groups 13(III)–17(VII). Their structures are those of discrete molecules. Although they are fundamentally members of the same class, for convenience we have listed the molecular hydrides that involve catenation of the other element separately. The chemistry of these molecular hydrides is diverse enough to justify two separate sections (Sections 15.2 and 15.3).

Saline and Polymeric Covalent Hydrides. At the left of the periodic table, we find solid salt-like or **saline** hydrides, which are seen to be salts of the H^- ion with non-acidic and feebly acidic cations of the *s* block. These compounds show typical lattice structures: The NaCl type for the Group 1 hydrides, and a distorted $PbCl_2$ type of structure for the feebly acidic Group 2 hydrides. (The compound MgH_2, which may also be considered a saline hydride, adopts the rutile structure.) In these compounds, the H^- ion seems to show roughly the same size as the F^- ion; but the apparent radius of H^- varies from 126 to 154 pm in these compounds, since in this anion, unlike any other, the (one) proton in the nucleus cannot fully control the radius of twice its number of electrons. Hence, this anion is quite compressible, that is, is soft (a soft base). Since this anion readily loses its extra electron (is a strong reducing agent), these salts react with many solvents in which they might have dissolved. They can be dissolved in molten Group 1 halides; electrolysis of such a solution liberates $H_2(g)$ at the anode, giving evidence for the existence of H^- in this solution.

Some *s*- and *p*-block cations that are a little more acidic give somewhat salt-like hydrides in which the hydrogen seems to serve as a bridging hydride ligand, giving *polymeric covalent* or borderline hydrides. Some of these solid hydrides [$(MgH_2)_x$, $(BeH_2)_x$, and $(ZnH_2)_x$] have not had their crystal structures determined, but ${}^3_\infty$[CuH] adopts a ZnS lattice, while ${}^3_\infty$[AlH_3] adopts an AlF_3 type structure.

Metallic Hydrides. Finally, there is a category of **metallic** hydrides of many of the *d*- and *f*-block elements. These are also solids, but they differ from the salt-like and polymeric hydrides in that they retain metallic conductivity. They have perhaps provoked the most controversy, in part because they are often nonstoichiometric (Section 13.4). (They generally can be made stoichiometric under sufficiently forcing conditions, but there is a significant entropy effect favoring nonstoichiometry resulting from a deficiency of hydrogen atoms randomly distributed among excess sites.) This nonstoichiometry led to an original classification of these compounds as *interstitial* hydrides, in which tiny hydrogen atoms fit in holes in metallic lattices without altering the lattice types, but it is now clear that the lattice types are altered. For example, UH_3 has a lattice in which each hydrogen is surrounded by four uranium atoms and each uranium atom has a coordination number of 12!

There is controversy about the nature of the hydrogen in these species; some models suppose that the hydrogen is present as H^+, while other models postulate the presence of H^- ions. It is known that the metallic conduction band (Sections 10.11 and 12.3) among *f*-block metals should accommodate six valence electrons per atom and that the *f*-block trihydrides are substantially poorer conductors than the *f*-block dihydrides. The reduced conductivity of MH_3, however, can be explained either by the presence of both M^{3+} and $3\,H^+$ ions filling the conduction band with six electrons, or by M^{3+} and $3\,H^-$ ions leaving no electrons to go into the conduction band. A film has been produced that switches between being a metallic mirror (of YH_2 or LaH_2) and being a transparent semiconducting window (of YH_3 or LaH_3) as the H_2 pressure is changed.[2]

The hydride of palladium appears to be unique, however: The lattice type is not altered, and it is never stoichiometric, with no more than 0.8 hydrogen atoms incorporated per palladium atom. Dihydrogen can be purified from other gases by its high rate of diffusion through palladium alloys (it diffuses in the form of H atoms!) An extremely controversial claim in the realm of palladium–deuterium chemistry was that, under certain conditions, some of the nuclei of the chemisorbed deuterium atoms may undergo "cold fusion" to give nuclei of helium; this claim seems to have been refuted.[3]

Other than in the unique case of Pd, none of the Group 7–10 metals form stable hydrides under normal pressure; this is the so-called *hydride gap* in the periodic table. How significant this gap is may be questioned, however, since hydrides can be formed from alloys of these metals with metals to the left in the periodic table, or by reacting Group 1 or 2 metal hydrides with a metal from Groups 7–10 and H_2 under pressure.[4] Some of these ternary hydrides have cations from Groups 1 and 2 with complex anions such as octahedral IrH_6^{3-} and MH_6^{4-} (M = Fe, Ru, and Os), five-coordinate CoH_5^{4-}, tetrahedral MnH_4^{2-}, square planar MH_4^{2-} (M = Pd and Pt), and tricapped trigonal prismatic TcH_9^{2-} and ReH_9^{2-}. These latter two complex anions can be obtained as discrete, stable anions in appropriate solvents. These and other hydrides such as $LaNi_5H_6$ (and the less expensive $FeTiH_x$, $x < 1.95$) are of potential importance as energy-storage solids, since they may contain more hydrogen, a very clean-burning fuel, per unit volume than the difficult-to-handle liquid hydrogen. This hydrogen can be released as H_2 by moderate heating.

15.2 Electron-Precise and Electron-Rich Molecular Hydrides

The molecular hydrides found in Groups 13(III)–17(VII) are a diverse group of compounds; some have partially positive hydrogen and some partially negative hydrogen. Since each hydrogen atom contributes only one electron to these molecules, these substances have weak van der Waals forces between them and are mostly gases at room temperature. Recall that the nomenclature of the molecular hydrides was covered in Section 2.8. Now is a good time to review this material, while noting that some alternate nomenclature recommended by International Union of Pure and Applied Chemistry (IUPAC) is also beginning to come into use among inorganic chemists. The replacement of the *-ine* ending in Group 15(V) by the same *-ane* ending found in Groups 13(III) and 14(IV) (e.g., PH_3 as "phosphane"; H_2S or more frequently its derivatives are now sometimes seen named as "sulfane"). In cases of catenation, the common method of indicating numbers of each type of atom (e.g., H_2S_4 as "dihydrogen tetrasulfide") has more recently been supplanted by the use of names such as "tetrasulfane"; for elements such as boron and phosphorus that form hydrides with a common number of boron or phosphorus atoms but differing numbers of hydrogen atoms, the number of hydrogen atoms in the given compound is then enclosed in parentheses [e.g., B_5H_9 is pentaborane(9) and B_5H_{11} is pentaborane(11)].

The hydrides of the heavier *p*-block elements are not thermodynamically stable, as suggested by their positive standard enthalpies of formation (Table 15.2); neither are they kinetically stable to decomposition. In addition, they are neither kinetically nor thermodynamically stable to oxidation by oxygen: They are **spontaneously flammable** in air. Other hydrides such as methane have kinetic barriers to reaction with atmospheric oxygen: Sparks or flames are needed to ignite them. Among the molecular

Table 15.2
Standard Enthalpies of Formation[a] ($kJ\,mol^{-1}$) and Stability to Oxidation of Some Hydrides[b,c]

LiH −91	BeH_2	B_2H_6 +32	CH_4 −75	NH_3 −46	H_2O −286	HF −269
wat	wat	flam	stab	stab	stab	stab
NaH −56	MgH_2 −76	AlH_3 −11	SiH_4 +31	PH_3 +5	H_2S −20	HCl −92
wat, flam	wat		flam	flam	stab	stab
KH −58	CaH_2 −174		GeH_4 +90	AsH_3 +67	H_2Se +86	HBr −36
wat, flam	wat		flam	unst	stab	stab
RbH −54	SrH_2 −177		SnH_4 +163	SbH_3 +145	H_2Te +154	HI +26
wat, flam	wat		unst	unst	unst	stab
CsH −56	BaH_2 −171		PbH_4 +250	BiH_3 +278		
wat, flam	wat		unst	unst		
		Catenated, Miscellaneous Hydrides (*ΔH_f only*)				
	TiH_2 −124	UH_3 −127	C_2H_6 −85	N_2H_4 +50	H_2O_2 −188	
		B_5H_9 +63	Si_2H_6 +72	P_2H_4 +21	H_2S_2 −23	
		$B_{10}H_{14}$ +33	Ge_2H_6 +162	As_2H_4 +147		
			Sn_2H_6 +274	Sb_2H_4 +239		
				HN_3 +294	NH_2OH −106	

SOURCES: Standard states assumed; Data from J. C. Bailar, Jr., H. J. Emeleus, R. Nyholm, and A. F. Trotman-Dickinson, *Comprehensive Inorganic Chemistry*, Pergamon, Oxford, UK, 1973, Chapter 2; and M. C. Ball and A. N. Norbury, *Physical Data for Inorganic Chemists*, Longman, London, 1974; p. 92.

[a] Units are in kilojoules per mole ($kJ\,mol^{-1}$).

[b] Abbreviations: stab = (kinetically) stable to air oxidation; flam = spontaneously inflammable in air; unst = thermally unstable at or near room temperature; wat = oxidized by water with release of H_2.

[c] The free energies of formation of most of these compounds (except in the last two groups) are on the order of 30 $kJ\,mol^{-1}$ more positive than these enthalpies (since gaseous H_2 must be ordered in order to form these hydrides).

hydrides, only H_2O_2, H_2O, and HF are completely resistant to air oxidation; even HCl can be oxidized to Cl_2 and H_2O in the presence of a catalyst.

One explanation for the relative dearth of thermodynamically stable hydrides of the elements may be found in the Pauling electronegativity equation (7.16). By this equation, element–hydrogen bonds should be stronger (Table 15.3) than the average of the element–element and hydrogen–hydrogen bond energies; therefore, to the degree that the electronegativities of hydrogen and the element differ, hydrides should be more stable than their component elements. At the far left and the far right of the periodic table the electronegativities do differ greatly, and the hydrides are thermodynamically stable; but in the center of the table, the elements have Pauling electronegativities close to that of hydrogen.

By Eq. (7.19), such hydrides should have ΔH_f values approximately equal to zero; in fact, many of the hydrides have very endothermic heats of formation. The Pauling equation does not take size explicitly into account, and the very small size of the hydrogen atom results in the H–H bond being the strongest of any single homopolar (element–element) bond (Table 3.1), while the orbital size mismatch of hydrogen with many other atoms may contribute to their comparatively poor orbital overlap, and hence bond strengths. Hence, the decomposition of many hydrides to the element and H_2 is thermodynamically favorable, and such reactions are often relatively rapid as well.

Table 15.3
Covalent- and Hydrogen-Bond Energies

A. Element–Hydrogen Covalent Bond Dissociation Energies ($kJ\,mol^{-1}$)

H–H 432						
Li–H 243		B–H 389	C–H 411	N–H 386	O–H 459	F–H 565
Na–H 197			Si–H 318	P–H 322	S–H 363	Cl–H 428
K–H 180	Cu–H 276			As–H 247	Se–H 276	Br–H 362
Rb–H 163	Ag–H 230				Te–H 238	O–H 295
Cs–H 176	Au–H 285					

SOURCE: Data from J. E. Huheey, E. A. Keiter, and R. L. Keiter, *Inorganic Chemistry: Principles of Structure and Reactivity*, 4th ed., Harper-Collins, New York, 1993, p. A-25.

B. Hydrogen Bond Energies ($kJ\,mol^{-1}$)

$H_2NH\cdots NH_3$	$HOH\cdots OH_2$	$FH\cdots FH$
17	22	29
	$HSH\cdots SH_2$	
	7	

SOURCE: Data from D. F. Shriver, P. Atkins, and C. H. Langford, *Inorganic Chemistry*, 2nd ed., Freeman, New York, 1994, p. 407.

Bonds of elements to hydrogen are often easily identified in infrared (IR) spectrum because the bond involves such a light atom, and infrared (IR) stretching frequencies depend on the (reduced) mass of the atoms in the bond, as well as the electronegativity differences of the atoms involved (hence, strength of the bond). If both atoms are light, the IR stretching frequency may be at quite high frequencies ($4100\,cm^{-1}$ for H–F; $2900\,cm^{-1}$ for C–H); even if the atom is heavy, the frequency generally exceeds $2000\,cm^{-1}$. However, substituting the heavy hydrogen isotope *deuterium* (*D* or ^2H) causes the vibration to be shifted to considerably lower frequencies: in the case of a bond to a heavy atom, to approximately 0.7 times the frequency of the undeuterated bond.

Classification of Molecular Hydrides. The molecular hydrides can be divided into three categories based on the relative numbers of electron pairs and bonds in their Lewis structures. An **electron-precise hydride** is one in which the number of electron pairs matches the total number of element–element and element–hydrogen links, so that (a) there are no electron pairs left over to serve as unshared electron pairs (Lewis base sites); and (b) there is no shortage of electron pairs as compared to links, so that there is no need for one electron pair to link three or more atoms in electron-deficient bonding. An **electron-rich hydride** is one in which there are unshared (lone) electron pairs on the central atom. An **electron-deficient hydride** is one in which there are too few electron pairs to allow a Lewis structure to be drawn with an octet about the central atom.

Example 15.1
Classify and name each of the following hydrides: (a) HF; (b) NH_3; (c) H_2S_3; (d) RaH_2; (e) TiH_2; (f) $Li[AlH_4]$; (g) P_5H_7; and (h) P_5H_5.

SOLUTION:

Nearly all *p*-block hydrides are molecular hydrides; these can be subdivided into electron-rich, electron-precise, and electron-deficient hydrides based on their Lewis structures (or, with care, on their groups and formulas). (a, b) Both HF and NH_3 are electron-rich molecular hydrides, named hydrogen fluoride and ammonia (common name); (c) H_2S_3 is also an electron-rich (catenated) hydride, named dihydrogen trisulfide or trisulfane; (d) since Ra^{2+} is feebly acidic, RaH_2 is a saline hydride, named radium hydride; (e) since Ti^{2+} is a *d*-block metal, its hydride will be metallic, and would probably be named titanium(II) hydride; (f) the hydride $Li[AlH_4]$ contains a complex anion, and would be named (Section 8.11) lithium tetrahydridoaluminate(III). This anion (not molecule) is electron-precise since a good Lewis structure can be drawn for it. (g, h) These electron-rich molecular hydrides are called pentaphosphane(7) or pentaphosphine(7), and pentaphosphane(5) or pentaphosphine(5), respectively.

Electron-Precise Hydrides. For a hydride to be both electron precise and to obey the octet "rule" (i.e., not to be hypervalent), the central atom must be from Group 14(IV). The electron-precise hydride of carbon, methane (CH_4), is of course familiar from organic chemistry; it is thermodynamically stable (in the absence of oxygen).

As we know, organic chemistry covers far more than the chemistry of methane and its derivatives, because the hydrides of carbon can also include **catenated** compounds: Compounds that involve element–element (in this case, carbon–carbon) bonds. When catenation occurs for a given element, it greatly expands the number of possible derivatives of a given class that exist for that element.

The element showing the greatest tendency to catenation is carbon. The C–C single bond is the second-strongest single bond (Table 3.1), but this is actually a hindrance to the catenating ability of carbon![5] Kinetic factors are probably most important: Even ethane is *thermodynamically* unstable with respect to decomposition to methane, hydrogen, and carbon. Steric barriers are present in hydrocarbons that inhibit the approach of reactive molecules such as O_2 to the small carbon atoms; other mechanisms of decomposition may require as a difficult first step the breaking of a strong C–H bond.

Inorganic chemists have long hoped that the element just below carbon, silicon, could also be induced to show extensive catenation. Early attempts focused on the hydrogen compounds of silicon. A number of catenated silicon hydrides, Si_nH_{2n+2}, can be made, but these are spontaneously flammable in air. In order to study them, Alfred Stock invented vacuum-line techniques for handling volatile materials in the complete absence of air (Section 6.2). The corresponding series of hydrides of germanium are even less extensive, due to their greater tendency to decompose thermally and/or catch fire. Stannane and silane are toxic (the latter probably due to the finely divided silica formed in the lungs upon oxidation). Although germane is not known to be toxic, it is germane to note the precedent of its neighboring hydrides.

The problem seems to be that the larger central Si atom is not adequately shielded from attacking O_2 by the small H atoms, and the Si–H bond is significantly weaker than the C–H bond. As we shall see in the second part of this chapter, if the small H atoms are replaced by larger CH_3 groups, kinetic thermal stability and the absence of spontaneous flammability of catenated silanes and germanes is then achieved.

The non-catenated hydrides of Group 14(IV) (and most of the hydrides of the other groups of the *p* block) contain relatively few electrons, hence they have weak intermolecular van der Waals forces and are gases at room temperature. There are enough additional electrons in the catenated molecular hydrides for most of these to be liquids at room temperature.

Finally, it is in principle possible to have electron-precise (although hypervalent) hydrides in Group 15(V) such as PH_5, and in Group 16(VI) such as SH_6. Halides of these types of course exist, and some organic derivatives of these types will be encountered at the end of this chapter, but no such hydrides exist.

Electron-Rich Hydrides. The structures of the electron-rich molecular hydrides of the nonmetals are essentially as predicted by valence shell electron-pair repulsion (VSEPR) (Section 3.4), but we may note (Table 3.2) that the bond angles diminish down a given group, ultimately reaching the 90° bond angles expected if the bonding to hydrogen involves only the *p* orbitals (not the *s* orbital) of the central atom.

As a consequence of their unshared electron pairs, these hydrides have Lewis base properties. The hydrides that have central atoms sufficiently more electronegative than hydrogen also have Lewis acid properties. These will be discussed in more detail in Section 15.5. One consequence for those that have both properties to a great degree is that they can react with other molecules of the same type, forming **hydrogen bonds**.

The hydrides of the three most electronegative small nonmetal atoms (fluorine, oxygen, and nitrogen) show elevated melting and boiling points due to the presence of hydrogen bonding between molecules (Fig. 15.1). Consequently, the hydrides HF and H_2O are liquids at room temperature despite their low numbers of (polarizable) electrons. The association of water molecules has already been discussed (Section 4.4); crystalline HF contains zigzag chains of hydrogen-bonded HF molecules (Fig. 15.2); the hydrogen bond between H and F is so strong that it persists partially in the vapor state below 80 °C in a ring hexamer, $(HF)_6$.

Hydrogen bonds between two electron-rich hydride molecules have strengths roughly one-twentieth as strong as the normal covalent bond between the electron-rich central atom and its hydrogen (Table 15.3). Nonetheless, the hydrogen bond has important consequences, such as those mentioned in the previous paragraph; it is detected by the broadening and shift to lower frequency of the element–hydrogen stretching frequency in the IR spectrum.

When the hydrogen bond is between a neutral molecule and an *anion,* the hydrogen-bond strength can be considerably stronger: for the hydrogen bond between water and a chloride ion it is 55 kJ mol^{-1} and for the hydrogen bond between HF and a fluoride ion it is 165 kJ mol^{-1}! Solid salts of the latter ion, such as $K^+[HF_2]^-$, can readily be isolated. In many of these latter salts, the hydrogen is found located midway between the two electron-rich atoms; in the normal weaker hydrogen bonds between two molecules, the hydrogen is located considerably closer to the atom to which it is covalently bonded than to the atom to which it is hydrogen bonded.

Scientists and science fiction writers[6] have long speculated about the consequences if neutral water molecules would suddenly begin forming strong rather than weak hydrogen bonds between each other. The boiling point of water would then be elevated, its IR spectrum would be altered, and much more. In the late 1960s, scientific reports began appearing reporting the growth of such a new form of water, dubbed "polywater," which grew only inside of very tiny capillaries in minute quantities.[7] When working on such surfaces, contaminants can be a severe problem, and the

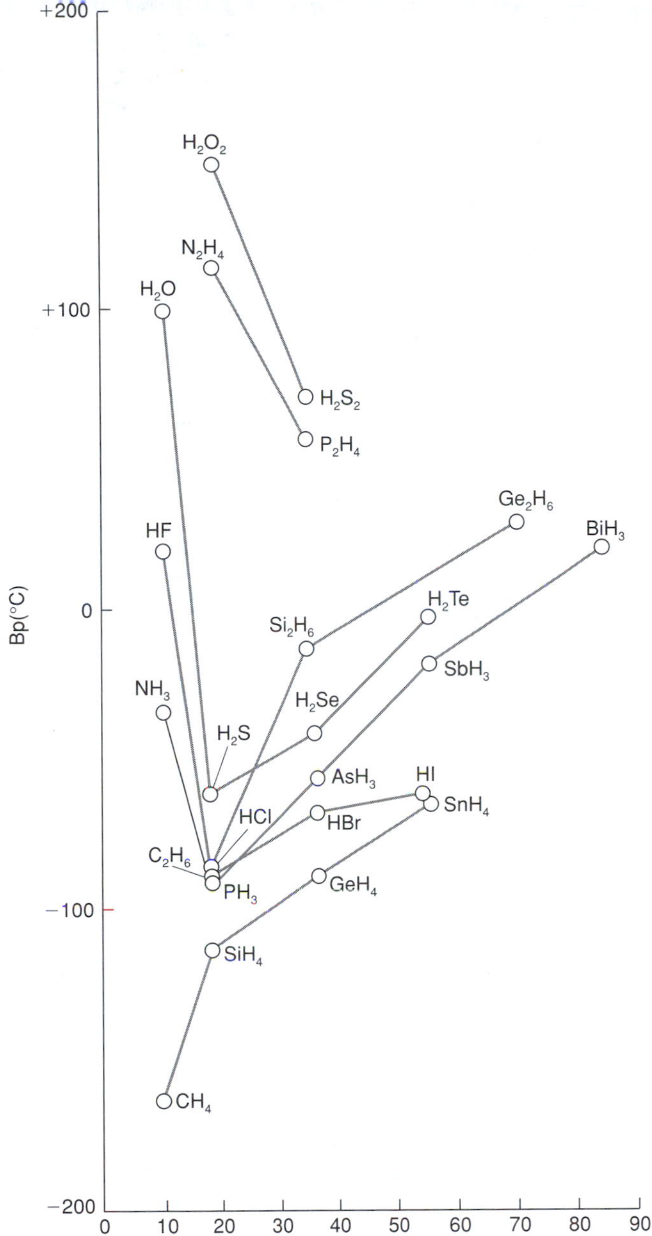

Figure 15.1

Normal boiling points of *p*-block molecular hydrides (including catenated hydrides), as a function of the number of electrons in each molecule.

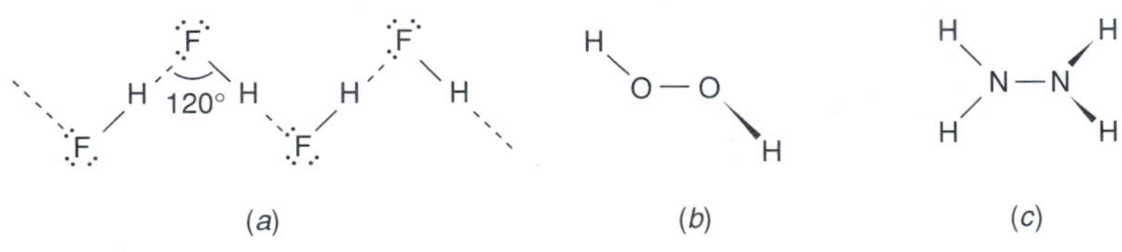

Figure 15.2

The structures of (*a*) crystalline HF, showing hydrogen bonding; (*b*) hydrogen peroxide, H_2O_2; and (*c*) hydrazine, N_2H_4.

enhanced boiling point and unusual IR spectrum eventually turned out to be due to contamination.

The catenated electron-rich hydrides have additional structural features of interest. In order to avoid repulsions of the electrons on neighboring nitrogen or oxygen atoms, the hydrides **hydrogen peroxide**, H_2O_2, and **hydrazine**, N_2H_4, adopt the less symmetrical C_2 conformations shown in Figure 15.2. Related to these two compounds is **hydroxylamine**, NH_2OH; due to hydrogen bonding, H_2O_2 and N_2H_4 are liquids at room temperature; NH_2OH is a solid. Due to the low N–N, N–O, and O–O single-bond energies these compounds are explosive or otherwise unstable, for example, because formation of N_2 is thermodynamically so advantageous. Nearly 100 (mostly unstable) catenated hydrides of phosphorus are also known[8]; the linear P_nH_{n+2} structures prevail for $n = 2$–4; cyclic P_nH_n prevails for $n = 5$ and 6; polycyclic P_nH_{n-2} through P_nH_{n-18} prevail for $n = 6$–22. The hydride P_2H_4 is actually responsible for the spontaneous flammability observed for PH_3 produced in swamps (Section 6.9); other linear oligophosphines are prevalent.

Adding a proton to the unshared electron pair of an electron-rich hydride results in the formation of a cation:

$$:NH_3 + H^+ \rightleftharpoons NH_4^+ \tag{15.1}$$

The **ammonium** ion, NH_4^+, resembles the potassium ion in size-related chemistry. The hydroxylammonium cation (NH_3OH^+), the hydrazinium cations $N_2H_5^+$ and $N_2H_6^{2+}$, and H_3O^+ and H_2F^+ are all stable but more acidic than the ammonium ion.

The cations formed by the heavier hydrides in the periodic table (e.g., H_2S and HCl) are so acidic that they transfer their protons completely to water, but they sometimes can be formed in solids or in acidic solvents. For example, the PH_4^+ (**phosphonium**) ion is found in some salts, but in water it completely transfers its protons:

$$PH_4^+ + H_2O \rightleftharpoons PH_3 + H_3O^+ \qquad K = 2.5 \times 10^{13} \tag{15.2}$$

The less acidic ammonium ion also undergoes the corresponding reaction to an extent sufficient to make its solution slightly acidic, but not to result in complete decomposition:

$$NH_4^+ + H_2O \rightleftharpoons NH_3 + H_3O^+ \qquad K = 5.5 \times 10^{-10} \tag{15.3}$$

Finally, recall from Section 2.7 that these molecular hydrides (such as $:NH_3$) are not only Lewis bases, but they are also conjugate acids of nonmetal anions (such as NH^{2-} and N^{3-}). Thus they can, under the right circumstances, act as either acids or bases. The hydrides of the good hydrogen-bonding elements N, H, and F act as both acids and bases and autoionize appreciably as pure liquids:

$$2\,HF \rightleftharpoons H_2F^+ + F^- \qquad K = 10^{-10} \tag{15.4}$$

$$F^- + HF \rightleftharpoons FHF^- \tag{15.5}$$

$$2\,H_2O \rightleftharpoons H_3O^+ + OH^- \qquad K = 10^{-14} \tag{15.6}$$

$$2\,NH_3 \rightleftharpoons NH_4^+ + NH_2^- \qquad K = 10^{-30} \tag{15.7}$$

Example 15.2

Predict whether each of the following hydrides is a gas, liquid, or a solid at room temperature; and acidic or basic or autoionizing. (a) FrH; (b) AtH; (c) HN_3 (hydrazoic acid); and (d) P_7H_5.

SOLUTION:

(a) The compound FrH is a saline hydride: An ionic salt that is solid. It contains H^-, which is very strongly basic.

(b) The compound AtH is an electron-rich molecular hydride of Group 17(VII), which should be acidic. Since it is not catenated, it is probably a gas, although the high molecular weight might lead to its being a liquid.

(c) The hydride HN_3 is an electron-rich molecular hydride that is catenated. It is acidic (ionizes to give azide ion), and is likely to be good at hydrogen bonding, since it contains N and H bonded to N. Consequently, it is likely to be a liquid that could autoionize.

(d) The compound P_7H_5 is also an electron-rich molecular hydride that is catenated. It has unshared electron pairs and should be Lewis basic; at least some acidity would be anticipated. It would not be good at hydrogen bonding, and would not be expected to autoionize significantly. Due to catenation it would be a liquid or even a solid.

15.3 Electron-Deficient Molecular Hydrides

The simple boron hydride BH_3, although it forms stable adducts with Lewis bases such as $H_3B{:}CO$, rapidly dimerizes to give *diborane*, B_2H_6, which has two bridging hydrogen atoms [Fig. 15.3(*a*)]. Since hydrogen, even as the hydride ion, does not have two pairs of electrons to donate, this is an example of a *three-center, two-electron* (3c–2e) bonding MO in which the one electron pair of H^- is shared with two boron atoms (Section 10.5). All of the boranes, and nearly all of the borane anions, to be discussed in this section involve such multicentered electron bonding of H to two B atoms, and/or of different B atoms with each other.

The hydrides of the other Group 13(III) elements are much less well known than the boron hydrides. Digallane, Ga_2H_6, was finally prepared in 1989 in 5% yield[9]; it readily oligomerizes in the solid state to give a material resembling the polymeric $^3_\infty[AlH_3]$.

An attractive alternative to multicentered bonding for these oligomers and polymers of MH_3 is addition of a hydride ion to form *hydrido anions*, which are more stable than the corresponding hydrides. Aluminum forms both the AlH_4^- and the AlH_6^{3-} ions; the former is much more important. The compound $LiGaH_4$ is stable but unimportant; $LiInH_4$ and $LiTlH_4$ exist below 0 °C. Boron forms the fairly water-stable BH_4^- ion (common name borohydride, more formal name tetrahydroborate). The BH_4^- ion itself can act as a ligand, using bridging hydrogen atoms as 3c–2e donor atoms in forming complexes such as $Be(BH_4)_2$, $Al(BH_4)_3$, and $U(BH_4)_4$. The latter complex is polymeric, involving bridging chelating borohydride groups around uranium atoms with coordination numbers of 14!

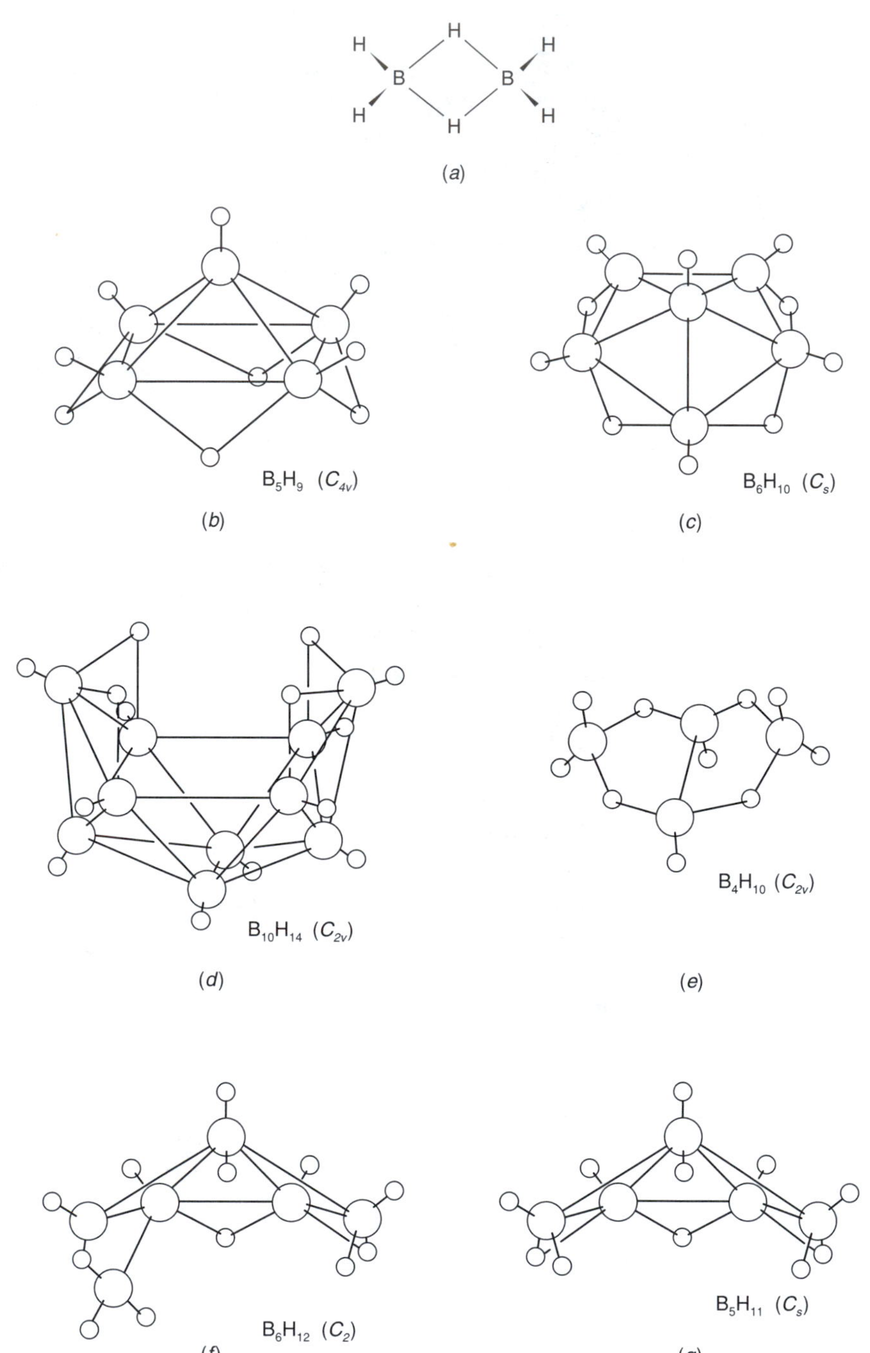

Figure 15.3

The structures of boron hydrides. (*a*) Diborane, B_2H_6; (*b*)–(*d*) the nido-boranes B_5H_9, B_6H_{10}, and $B_{10}H_{14}$; (*e*)–(*g*) the arachno-boranes B_4H_{10}, B_5H_{11}, and B_6H_{12}. In (*b*)–(*g*) the larger circles represent boron atoms and the smaller circles hydrogen atoms. [Adapted from N. N. Greenwood and A. Earnshaw, *Chemistry of the Elements*, 2nd ed., p. 154, Copyright © 1997, Butterworth-Heinemann, Oxford.]

There are about 50 catenated (cluster) hydrides of boron. Although many or most of these catch fire in air, they have sufficient thermal stability that a very extensive chemistry of them is known, which can only be hinted at in this chapter. There are two major categories, the *nido-boranes*, B_nH_{n+4}, represented in Figure 15.3 by the structures of B_5H_9, B_6H_{10}, and $B_{10}H_{14}$, and the *arachno-boranes*, B_nH_{n+6}, represented by the structures of B_4H_{10}, B_5H_{11}, and B_6H_{12} [more examples are partially illustrated in Figure 12.20(*b*,*c*)]. The arachno-boranes are characterized by more "open" structures than the nido-boranes. These boranes are liquids unless they contain more than nine boron atoms. The two most important and stable of these catenated boranes are pentaborane(9) and decaborane(14).

Boron also forms about 50 catenated or cluster boron hydride anions, most of which fall in three series. The $B_nH_n^{2-}$ ions ($n = 6$–12) have regular polyhedral *closo-borane* structures [Fig. 12.20(*a*)] and are (especially for $n = 10$ and 12) very stable and nonbasic (as seen for monoatomic and oxo anions, the basicity decreases as the size increases). The *nido*-$B_nH_{n+3}^-$ ions have structures derived from those of the nido-boranes, B_nH_{n+4}, by removal of one bridging proton. The *arachno*-$B_nH_{n+5}^-$ ions ($n = 2$, 3, 5, 9, and 10) have structures similarly derived from those of the arachno-boranes.

The structures of closo-, nido-, and arachno-boranes and borane anions may also be correlated with the results of Wade's rule calculations (Section 12.7). Recall that the closo-borane anions $B_nH_n^{2-}$ possess a total of $4n + 2$ valence electrons, of which $2n$ are used to bond the n terminal H atoms to the n cluster boron atoms; thus, the number of framework electrons is $2n + 2$. This number can be used, with the help of Table 12.6, to select the geometry of the cluster closo anion.

Figure 12.20 shows diagonal lines connecting structures of *closo*-$B_nH_n^{2-}$ borane anions (left column) to structures of *nido*-B_nH_{n+4} boranes (center column) and *arachno*-B_nH_{n+6} boranes (right column). Boranes connected by these diagonal lines have the same number of framework electrons (NFE). To see this, let us begin with octahedral $B_6H_6^{2-}$, which has $2n + 2 = 14$ framework electrons. To move along the diagonal line to a *nido*-B_5 borane, we *remove one BH group while leaving its two framework electrons behind,* that is, we cut out a framework-electron-free BH^{2+} group. Subtracting BH^{2+} from $B_6H_6^{2-}$ leaves $B_5H_5^{4-}$, and changes the cluster boron geometry from octahedral to square pyramidal, but the cluster still has 14 framework electrons. This anion is very basic, and picks up four H^+ ions to give the final *nido*-B_5H_9. The four incoming protons attach to the B–B framework electrons, forming 3c–2e bonds without withdrawing those electrons from the framework, so that the total of 14 framework electrons is preserved. Now, however, n is only 5, so that $14 = 2n + 4$ framework electrons, as is characteristic of nido-boranes.

In the next step, one of the BH^{2+} groups from *nido*-B_5H_9 square base is also removed, leaving a $B_4H_8^{2-}$ with 14 [$= 2(4) + 6$] framework electrons. The ion $B_4H_8^{2-}$ is also protonated to give the final *arachno*-B_4H_{10} with a "butterfly" geometry.[10]

Carboranes. Since the closo-borane anions are nonbasic, there are no neutral closo-boron hydrides. However, we can generate neutral closo species if we carry out two isolobal substitutions of BH^- groups by neutral CH groups. The resulting $B_{n-2}H_{n-2}C_2H_2$ species are the very stable and important (*closo*-)**carboranes.** Thus, the 6 cluster atom pseudooctahedral neutral closo-carborane is $B_4C_2H_6$, which has two possible isomers: one in which the two carbons are adjacent (1,2- or *cis*-$B_4C_2H_6$), and one in which they are across the octahedron (1,6- or *trans*-$B_4C_2H_6$). Likewise, the 12 cluster atom pseudoicosahedral carborane is $B_{10}C_2H_{12}$, which can have three isomers: 1,2- or *ortho*-$B_{10}C_2H_{12}$, 1,7- or *meta*-$B_{10}C_2H_{12}$, and 1,12- or *para*-$B_{10}C_2H_{12}$ [Fig. 15.4(*b*–*d*)].

It is also possible to abstract a BH^{2+} group from *ortho*-$B_{10}C_2H_{12}$ to generate a nido dianion, *ortho*-$[B_9C_2H_{11}]^{2-}$, which has the type of structure suggested in Figure 12.20 for the *nido*-B_{11} carborane, but with two adjacent carbon atoms on the open five-membered ring at the top (flat) surface [Fig. 15.4(*e*)]. This open ring resembles the cyclopentadienide ion (Chapter 11), and, like it, readily forms metallocenes:

$$Fe^{2+} + 2\,[B_9C_2H_{11}]^{2-} \rightarrow [Fe(B_9C_2H_{11})_2]^{2-} \tag{15.8}$$

Another carborane anion that resembles the cyclopentadienide ion (if we replace three CH groups in $C_5H_5^-$ with isoelectronic, isolobal BH^- groups) is the very highly basic *meta*-$[C_2B_3H_5]^{4-}$ ion. This ion is so basic that it can readily coordinate *d*-block metal ions on *both faces*; with appropriate sequences of reactions it is then possible to construct *multidecker sandwich* compounds such as $[CpCo(C_2B_3H_5)CoCp]$ or even the hexadecker $[Cp^*Co(C_2B_3H_5)Co(C_2B_3H_5)Co(C_2B_3H_5)Co(C_2B_3H_5)CoCp^*]$ where $Cp = C_5H_5$ and $Cp^* = C_5\,Me_5$.[11]

Example 15.3

Assign structural types and (if appropriate) compute numbers of framework electrons for the following phosphanes, boranes, and carboranes. If possible, support your conclusion with a plausible Lewis structure: (a) $B_7C_2H_9$; (b) B_7H_{11}; (c) P_7H_9; and (d) P_7H_5.

SOLUTION:

Since phosphanes are electron-rich and boranes are electron-poor hydrides, different principles and classifications apply: Numbers of framework electrons are more useful for boranes, while Lewis structures can be drawn only for phosphanes.

(a) Seven BH groups contribute 14 framework electrons, while two CH groups contribute 6 framework electrons, for a total of 20. Since the number of framework atoms is 9, $20 = 2n + 2$, so this is a closo-borane as shown for $n = 9$ in Figure 12.20(*a*); isomers are possible.

(b) Seven BH groups contribute 14 framework electrons; the extra four H atoms contribute one each for a total of 18, which equals $2n + 4$, so this is a nido-borane as shown in Figure 12.20(*b*).

(c) Linear (chain) phosphines (PH_3, P_2H_4, P_3H_5, ...) fit the formula P_nH_{n+2}; P_7H_9 fits this formula, so it could be a linear molecule (PH_2–PH–PH–PH–PH–PH–PH_2) or have branches in the chain similar to alkanes [:P–$(PH–PH_2)_3$].

(d) Cyclic phosphines fit the formula P_nH_n, but P_7H_5 fits the formula P_nH_{n-2}, which would apply to bicyclic phosphines. The structure could be that of *bicyclo*[2,2,1]heptaphosphine(5) or some isomer thereof:

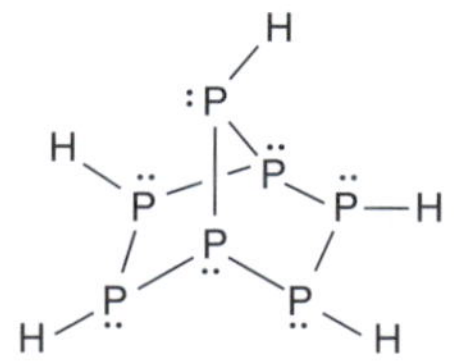

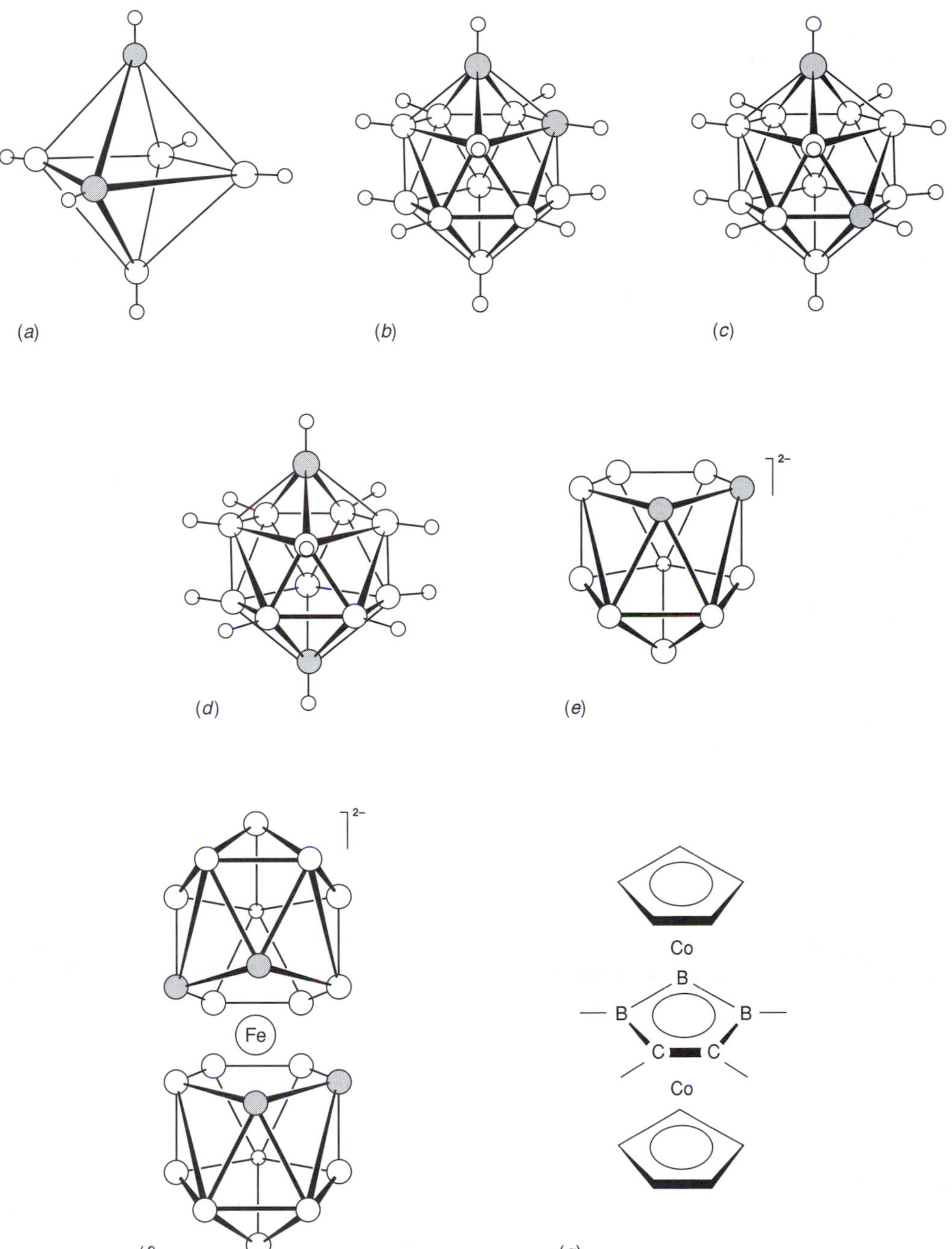

Figure 15.4
Structures of carboranes and derivatives. (*a*) *cis*- or 1,2-$B_4C_2H_6$; (*b*) *ortho*- or 1,2-$B_{10}C_2H_{12}$; (*c*) *meta*- or 1,7-$B_{10}C_2H_{12}$; (*d*) *para*- or 1,12-$B_{10}C_2H_{12}$; (*e*) $B_9C_2H_{11}{}^{2-}$; (*f*) the metallocene $[Fe(B_9C_2H_{11})_2]^{2-}$; (*g*) a multidecker sandwich, $[CpCo(B_3C_2H_5)CoCp]$. [Adapted from D. F. Shriver, P. W. Atkins, and C. H. Langford, *Inorganic Chemistry*, 2nd ed., Freeman, New York, 1994; pp. 477, 478.]

15.4 Syntheses of the Hydrides and Hydride Anions of the Elements

There are three common synthetic routes to hydrides: **direct combination of the elements**, **protonation of an anion**, and **displacement of halide by hydride**. Catenated hydrides are often made by **pyrolysis** and by **acid–base condensation reactions**. Hydride anions are often made by **deprotonation of a hydride**.

Direct Combination. The *salt-like* and *metallic* hydrides are prepared by direct combination of the metal and H_2. Due to the strong H–H bond and the low electron affinity of the hydrogen atom, this reaction is not nearly so easy as the reaction of these active metals with halogens: high temperatures (e.g., 300 °C) are required. This reaction is also used in the syntheses of lithium aluminum hydride ($LiAlH_4$), Li_3AlH_6 (under different conditions), and $NaAlH_4$:

$$M + Al + 2\,H_2 \rightarrow MAlH_4 \tag{15.9}$$

where M = Li or Na. As mentioned earlier, the metallic hydrides normally are produced in a nonstoichiometric form; excess pressure is needed to get fairly stoichiometric or higher hydrides.

For *molecular hydrides,* direct combination of hydrogen and a nonmetal is seldom the method of choice: with fluorine the reaction is explosive, while with many of the nonmetals the reaction is strongly endothermic and does not go. This is, however, one route for the industrial preparation of HCl and HBr and is the basis for the extremely important *Haber–Bosch* process for the production of ammonia. In the industrial synthesis of NH_3, which on a mole scale is produced in larger quantities than any other industrial chemical, hydrogen is first produced by the re-forming of methane from natural gas with steam over a nickel catalyst at 750 °C:

$$CH_4 + 2\,H_2O \rightleftharpoons CO_2 + 4\,H_2 \tag{15.10}$$

Air is then injected; some of the hydrogen burns in the oxygen. The resulting water and the CO_2, CO, and so on, are removed in a scrubber, leaving behind a *synthesis gas* containing the N_2 from the air in a 1:3 mixture with H_2. The mixture is then passed over a promoted iron catalyst at 400 °C and 200-atm pressure to give about a 15% yield of ammonia:

$$N_2 + 3\,H_2 \rightleftharpoons 2\,NH_3 \tag{15.11}$$

The ammonia is condensed by refrigeration (bp −34 °C), and the synthesis gas is recycled.

Protonation of an Anion. A fairly general approach to many of the molecular hydrides is the treatment of a salt of the anion of the element with strong acid (if the anion is not strongly basic, e.g., F^-, Cl^-, S^{2-}, Se^{2-}, and Te^{2-}) or with water if the anion is very strongly basic.

$$CaF_2 + H_2SO_4 \rightarrow CaSO_4 + 2\,HF \tag{15.12}$$

$$FeS + 2\,HCl(aq) \rightarrow Fe^{2+}(aq) + 2\,Cl^-(aq) + H_2S(g) \quad (15.13)$$

$$Ca_3P_2 + 6\,H_2O \rightarrow 3\,Ca(OH)_2 + 2\,PH_3 \quad (15.14)$$

This route was originally used by Stock to prepare silane and diborane mixed, however, with catenated silanes and boranes. Because the separation of these very flammable compounds is quite tedious, this method is no longer preferred for these compounds.

Displacement of Halide by Hydride. A third general route, suitable especially for the elements toward the center of the periodic table, is the reaction of NaH, $LiAlH_4$, or $NaBH_4$ with a halide of the element. This reaction is run typically in ether solution, although $NaBH_4$ can be used in aqueous solution. The reactions can be regarded, according to one's preference, either as reductions of the halide of the element or as acid–base reactions in which the $:H^-$ ion displaces halide ions.

$$2\,BF_3(g) + 6\,NaH(s) \rightarrow B_2H_6(g) + 6\,NaF(s) \quad (15.15)$$

$$SbCl_3 + NaBH_4 \rightarrow SbH_3(g) + NaBH_xCl_{4-x} \quad (15.16)$$

$$Si_3Cl_8 + 2\,LiAlH_4 \rightarrow Si_3H_8 + 2\,LiCl + 2\,AlCl_3 \quad (15.17)$$

$$Cu^{2+}(aq) + NaBH_4 \rightarrow CuH(s) \quad (15.18)$$

$$3\,LiAlH_4 + AlCl_3 + \text{ether} \rightarrow 3\,LiCl + 4\,AlH_3\text{:ether} \quad (15.19)$$

Pyrolysis. Catenated covalent hydrides are often made from the noncatenated hydrides by pyrolysis, silent electrical discharge, or mercury-photosensitized irradiation. These processes eliminate hydrogen:

$$2\,GeH_4 \rightarrow Ge_2H_6 + H_2 \quad \text{(electrical discharge)} \quad (15.20)$$

Diborane undergoes pyrolysis at very moderate temperatures:

$$2\,B_2H_6 \rightarrow B_4H_{10} + H_2 \quad (25\,^\circ C, 10\text{ days}) \quad (15.21)$$

Higher boranes are produced with slight variations in the conditions: pyrolysis at 250 °C in the presence of H_2 gives B_5H_9; pyrolysis in the presence of dimethyl ether at 150 °C gives $B_{10}H_{14}$.

Some borane anions are produced by pyrolysis of other anions: pyrolysis of NaB_3H_8 gives salts of $B_{10}H_{10}{}^{2-}$ and $B_{12}H_{12}{}^{2-}$ along with $BH_4{}^-$.

Acid-Base Condensation Reactions. Boron hydrides, being electron deficient, have some Lewis acid properties, and will react on heating with Lewis bases such as boron hydride anions to give larger boron clusters and H_2 (which we can envision as the elimination product of H^+ from the Lewis acid and H^- from the anion):

$$2\,B_5H_9 + 5\,B_9H_{14}{}^- \rightarrow 5\,B_{11}H_{14}{}^- + 9\,H_2 \quad (15.22)$$

$$B_{10}H_{14} + 2\,BH_4{}^- \rightarrow B_{12}H_{12}{}^{2-} + 5\,H_2 \quad (15.23)$$

$$4\,BF_3 + 5\,BH_4{}^- \xrightarrow{Et_2O} 2\,B_3H_8{}^- + 3\,BF_4{}^- + 2\,H_2 \quad (15.24)$$

Syntheses of carboranes from boron hydrides also loosely follow this pattern. The compound 1,2-$B_{10}C_2H_{12}$ is synthesized from the boron hydride $B_{10}H_{14}$ in two steps. The first step involves attack by the Lewis base SEt_2 and the second involves the Lewis base HC≡CH (acetylene):

$$B_{10}H_{14} + 2\,SEt_2 \rightarrow B_{10}H_{12}(SEt_2)_2 + H_2 \qquad (15.25)$$

$$B_{10}H_{12}(SEt_2)_2 + HC{\equiv}CH \rightarrow 1{,}2\text{-}B_{10}C_2H_{12} + H_2 + 2\,SEt_2 \qquad (15.26)$$

Conversion of this (ortho) 1,2-$B_{10}C_2H_{12}$ to the meta isomer is a matter of pyrolysis at 500 °C; conversion of the meta to the para isomer (1,12-$B_{10}C_2H_{12}$) requires pyrolysis at 700 °C. Note, by comparison with the pyrolysis temperatures for diborane, how much thermal stability has been gained in these icosahedral carboranes!

The $[B_9C_2H_{11}]^{2-}$ anion used to generate metallocenes is obtained, not by a condensation, but by a *degradation* of 1,2-$B_{10}C_2H_{12}$. The main step here is attack by the strong hard base OEt^-, which removes the hard acid B^{3+} as triethyl borate:

$$B_{10}C_2H_{12} + OEt^- + 2\,EtOH \rightarrow [B_9C_2H_{12}]^- + B(OEt)_3 + H_2 \qquad (15.27)$$

$$[B_9C_2H_{12}]^- + H^- \rightarrow [B_9C_2H_{11}]^{2-} + H_2 \qquad (15.28)$$

Deprotonation of a Hydride. The *nido*- and *arachno*-borane anions are often made by straightforward deprotonation of the corresponding *nido*- and *arachno*-borane, respectively; a framework hydrogen (one involved in 3c–2e bonding) rather than a terminal hydrogen is removed. For the smaller boranes, a very strong base, Na^+H^-, is used; for the larger boranes, a weaker base such as $:NMe_3$ may suffice, since the resulting larger anion is not as basic.

A few special syntheses of hydrides of other elements also merit mention. Arsenic may be detected at very low levels via the synthesis of arsine in the Marsh test, in which an arsenic-containing sample is reduced with zinc and acid to arsine, which is then allowed to contact a hot glass plate, where it decomposes to give a silvery mirror of arsenic. Phosphine is made industrially by the disproportionation of white phosphorus in base:

$$P_4 + 3\,KOH + 3\,H_2O \rightarrow PH_3 + 3\,KH_2PO_2 \qquad (15.29)$$

Hydrogen peroxide is made industrially by reducing the oxygen of the air with a hydrogen-rich organic compound, 2-ethylanthraquinol, which is then regenerated with H_2. Polysulfanes, H_2S_n, can be made by the condensation of the appropriate sulfur chloride with hydrogen sulfide:

$$S_nCl_2 + 2\,H_2S \rightarrow H_2S_{n+2} + 2\,HCl \qquad (15.30)$$

Hydrazine is made by the *Raschig process*, in which ammonia and basic sodium hypochlorite solutions are combined in the presence of glue or gelatin:

$$NH_3 + OCl^- \rightarrow NH_2Cl + OH^- \qquad (15.31)$$

$$NH_2Cl + NH_3 \rightarrow [NH_2NH_3]^+ + Cl^- \qquad (15.32)$$

$$NH_2NH_3{}^+ + OH^- \rightarrow H_2O + N_2H_4 \qquad (15.33)$$

15.5 Reactions and Uses of the Hydrides

The uses of the hydrides of the elements depend on some characteristic types of reactivity found to varying degrees in hydrides across the periodic table.

Hydrides as (Lewis) Acids. Hydrides in which the element–hydrogen bond is partially positively charged can act as *Lewis acids*, for example, as *proton sources*. This property is characteristic in electron-rich hydrides when the element is more electronegative than hydrogen, but is also found with elements such as Te.

Many of the electron-rich molecular hydrides are extremely toxic materials. Hydrogen fluoride burns the skin and precipitates calcium ions in the tissues, producing excruciating pain. It is a particularly dangerous substance in part because at first it fails to produce pain, so that one may not notice that contact of skin and acid has occurred. The greatest caution should be exercised to prevent fluorides from contacting acids or HF from contacting the skin. In the event that contact does occur, not only should the burn be washed with copious amounts of water, but prompt medical attention must be obtained—otherwise surgical removal of the affected area may be needed! The other hydrogen halides are "merely" corrosive substances whose fumes are choking and should not be inhaled.

Hydrogen fluoride is used in the manufacture of freons (chlorofluorocarbons) for aerosols and in the manufacture of synthetic cryolite (Na_3AlF_6) used in making aluminum metal (Section 6.8). It is also a useful solvent for many biochemicals. It attacks glass, producing H_2SiF_6, and hence must be used in plastic containers. Anhydrous HCl is used in the manufacture of anhydrous metal halides and some metals (Mg, Al, Ti, etc.); aqueous hydrochloric acid has many uses, the largest of which is the pickling of steel (removal of surface oxides).

The compounds H_2S, H_2Se, and H_2Te are sources of their soft-base anions, and hence are very poisonous. Their revolting odors usually alert us to the hazard, but our nasal receptors that detect H_2S are numbed, creating a dangerous situation. Ammonia, hydrazine, and hydroxylamine are all toxic; phosphine, arsine, and stibine are among the most toxic gases known. These substances also have revolting smells. You may also recall from Section 6.9 that some of these compounds can arise in natural waters under conditions of very low E°, and can cause untoward environmental consequences. Arsine and stibine react with metals to give arsenides and antimonides, which are useful dopants in semiconductors.

The electron-precise hydrocarbons, of course, cannot be hydrolyzed by water, or even by strong bases. The compound SiH_4 and the higher silanes are stable to water, but are rapidly hydrolyzed by dilute bases:

$$SiH_4 + OH^- \rightarrow \{H_4Si\text{–}OH\}^- \rightarrow H_2 + [H_3SiO]^-$$

$$[H_3SiO]^- + 3\,OH^- \rightarrow SiO_4{}^{4-} + 3\,H_2 \tag{15.34}$$

In the presumed intermediate, $H_4Si\text{–}OH^-$, the hydrogens attached to oxygen have a partial positive charge, while those attached to silicon have a slight negative charge, which no doubt facilitates their elimination as H_2 molecules. (Note that this reaction is also an oxidation of silane.) With a larger central atom, GeH_4 is not hydrolyzed even by 30% NaOH. This failure to hydrolyze has been cited as an effect of the opposite polarizations of the $Si^{\delta+}\text{–}H^{\delta-}$ and $Ge^{\delta-}\text{–}H^{\delta+}$ bonds, which would be due to the higher electronegativity of Ge than Si, which is in turn due to the scandide contraction, Chapter 1.

Many electron-deficient hydrides also show Lewis acid properties in reactions with Lewis bases such as water and amines. The lighter boranes are hydrolyzed by water:

$$B_2H_6(g) + 6\,H_2O \rightarrow 2\,\{H_3B{-}OH_2\} \rightarrow 2\,B(OH)_3(aq) + 6\,H_2(g) \qquad (15.35)$$

The corresponding reaction of the BH_4^- anion is much slower, due both to the negative charge and the better coordinative saturation of the boron atom; AlH_4^- and GaH_4^- are rapidly hydrolyzed (oxidized) by water.

Diborane also acts as a Lewis acid with amines, sulfides, ethers, and so on. If the Lewis base is strong and sterically uncrowded, two molecules of it may end up on the same boron atom of the diborane, resulting in *unsymmetrical cleavage:*

$$B_2H_6 + 2\,{:}NH_3 \rightarrow [BH_2(NH_3)_2][BH_4] \qquad (15.36)$$

Symmetric cleavage is more typical:

$$B_2H_6 + 2\,{:}NMe_3 \rightarrow 2\,H_3BNMe_3 \qquad (15.37)$$

A special case of acidity is that of the larger *closo*-borane clusters: The neutral forms of these, *closo*-B_nH_{n+2}, are too strongly acidic to be isolated, since the aromatic delocalization in the anions (see Section 10.10) is destroyed on protonation. Salts of these large nonbasic anions are easily extracted from water into organic solvents, which is useful in separating metal ions.[12]

Hydrides as (Lewis) Bases. Hydrides in which the hydrogen is largely or fully negatively charged act as *hydride sources*, and hence are strong bases. This includes the saline and intermediate hydrides (Table 15.1A) as well as hydride anions. The uses of these hydrides in the synthesis of other hydrides has already been discussed in Section 15.4; H^- in general transfers from the less electronegative element to the more electronegative one:

$$AlH_4^- + GeCl_4 \rightarrow GeH_4 + AlCl_4^- \qquad (15.38)$$

This finding is in line with expectations from the Pauling electronegativity analyses of hard–soft acid–base (HSAB) trends, because the product $AlCl_4^-$ contains the greater number of polar (and therefore, relatively the strongest) bonds.

A characteristic property of the electron-rich hydrides is the Lewis basicity of the unshared electron pair. Many of these hydrides (NH_3, N_2H_4, NH_2OH, PH_3, and H_2O) form complexes with metals; those with NH_3 are called ammine complexes and those with H_2O are called hydrated ions, of course. Other Group 15(V), 16(VI), and 17(VII) hydrides also have unshared pairs of electrons and presumably can coordinate to metals, but they would then normally be deprotonated, since (1) these hydrides are more acidic than the ones above them (Section 2.7), and (2) coordination enhances the acidity of a ligand (Section 5.10). Thus bubbling H_2S through a solution of a soft metal ion does not give $[M(H_2S)_n]^{x+}$; instead, deprotonation occurs to give the sulfide of the metal.

Addition Reactions of Hydrides. Hydrides that have relatively weak M–H bonds (such as those of heavier *p*-block elements such as Sn) can readily be cleaved to generate radicals ($R_nM\cdot$ and $H\cdot$). This reaction may occur on photolysis, pyrolysis, or on reaction with a *radical initiator* such as O_2 or *tert*-butyl hydroperoxide, which readily breaks apart into radicals that in turn can react with R_nH to generate one of the above radicals. These radicals can then undergo reaction with reactive substrates such as alkenes.

The most important addition reactions are those of diborane and of silane (or trichlorosilane, $HSiCl_3$) to alkenes and alkynes, which generate organoboron or organosilicon compounds that are of great use in organic synthesis of other hydrocarbons, alcohols, aldehydes, ketones, or silicones. (The main driving force of this reaction is the fact that the C–H bond produced is much stronger than the Si–H or B–H bond that is broken.) This addition is cis- and anti-Markovnikov (the silyl or boryl group goes on the less substituted end of an alkene chain):

$$6\,RCH{=}CH_2 + B_2H_6 \rightarrow 2\,(RCH_2CH_2)_3B \tag{15.39}$$

These reactions are often catalyzed by 16-electron *d*-block complexes, which suggests that oxidative addition may be involved, but they are also sometimes initiated with a radical initiator. Because of the versatility of reaction types of hydrides, we can envision or hypothesize several types of mechanisms (radical, oxidative, reductive, Lewis acid, and Lewis base). Experiments in kinetics (Chapter 16) are necessary to determine the mechanism in a given case, which may not apply to a seemingly similar situation in which the conditions are slightly different.

Hydrazine, hydroxylamine, and phosphine add to certain classes (e.g., ketones and aldehydes) of unsaturated organic compounds (often followed by elimination of H_2O) to give useful derivatives: the hydrazones, $R_2C{=}N{-}NH_2$, the oximes, $R_2C{=}N{-}OH$, and the compound $[P(CH_2OH)_4]Cl$, used in flameproofing cotton cloth.

Hydrides as Reducing Agents. The reducing ability of hydrides is very important in many applications, including as fuels. As mentioned earlier, the metallic hydrides have been greatly investigated for their possible use as an energy source in fuel cells because they are a much more convenient form in which to store hydrogen than as gaseous H_2. They have also been tested as catalysts and as reactive starting materials for synthesis. For example, UH_3 forms as a powder that is more conveniently reactive than uranium metal, and hence it is used in the synthesis of some uranium compounds. Two saline hydrides find important uses: NaH as a powerful reducing agent and strong base in nonaqueous solvents, and CaH_2, which is substantially less reactive, as a drying agent for organic solvents, since it reacts with water to give $Ca(OH)_2$ and H_2.

$$CaH_2 + 2\,H_2O \rightarrow Ca(OH)_2 + 2\,H_2 \tag{15.40}$$

Many of the hydrides of the less electronegative nonmetals, or the catenated hydrides, are important reducing agents and even rocket fuels; the use of hydrocarbons as fuels is of course common knowledge. The boron hydrides burn with green flames and release enormous amounts of energy per gram upon combustion. These hydrides were produced in quantity and were tested as rocket fuels during the 1960s.[13] This project foundered because of the expense of the boranes (boron is a rare element, and *ortho*-carborane is as expensive as gold), and because of the corrosive nature of the acidic B_2O_3 that is ejected. Hydrazine, N_2H_4, is also an important reducing agent with a high fuel value. Organic derivatives of hydrazine are currently used for liquid-fueled rockets; they also have the advantage of being spontaneously inflammable in the oxidant used, so that no ignition system, which could malfunction, is required:.

$$Me_2NNH_2(\ell) + 2\,N_2O_4(\ell) \rightarrow 3\,N_2(g) + 4\,H_2O(g) + 2\,CO_2(g) \tag{15.41}$$

Ammonia as a Source of Nitrogen Compounds. Ammonia is used in enormous quantities in fertilizers as anhydrous (liquid) ammonia or as derivatives (ammonium salts and urea). It is also used in several major chemical processes: the manufacture of plastics based on amines, refrigeration, and the manufacture of nitrogen compounds in general. Some ammonia is oxidized over a Pt catalyst at 800 °C to give nitric oxide:

$$4\,NH_3 + 5\,O_2 \rightarrow 4\,NO + 6\,H_2O \tag{15.42}$$

The NO is subsequently oxidized by air to NO_2, which is converted to HNO_3, some of which is used to make nitrogen-based explosives.

Example 15.4

Tell whether each of the following reactions involves transfer of hydrogen as H^+, H·, or H^-. Tell whether reactants or products are favored in each equilibrium: (a) $NH_2^- + PH_3 \rightleftharpoons NH_3 + PH_2^-$; (b) $CH_3Te \rightleftharpoons CH_2TeH$; and (c) $2\,CaH_2 + BF_4^- \rightleftharpoons 2\,CaF_2 + BH_4^-$.

SOLUTION:

(a) The H^+ ion is transferred; products are favored because the less basic anion is formed, and PH_2^- is less basic because it is larger. (b) Completion of these Lewis structures would show an unpaired electron on Te in $CH_3Te\cdot$ and on carbon in $\cdot CH_2TeH$, so the reaction is one in which H· is transferred. The reactant is favored because the extra C–H bond is stronger than the Te–H bond in the product. (c) The H^- ion is transferred; the products are favored because electronegativity differences are maximized in the CaF_2 product.

15.6 Alkyls and Aryls of the Elements: Classification and Structures

There are broad similarities between the σ-bonded alkyls and aryls of the metals and metalloids, and the corresponding hydrides. These similarities are rooted in the similar electronegativities of carbon and of hydrogen. This extends even to nomenclature (Section 8.11): alkyls (generically symbolized in formulas as R) and aryls (Ar) are named by modifying the names of the corresponding hydrides (PH_3 is phosphine, PPh_3 is triphenylphosphine).

But there are also subtle differences between alkyls and aryls on one hand, and hydrides on the other. Foremost of these is the fact that an alkyl or aryl group is substantially larger than a hydride group, and its size (i.e., cone angle, Section 11.2) can be altered by varying the size of the groups attached to carbon. For simplicity, we shall begin by emphasizing the methyl and phenyl derivatives of the elements, in which steric effects are minimized, but later we shall see some very different chemistry that becomes possible when steric effects are maximized.

Characteristic noncatenated methyl compounds (or phenyl compounds if the methyl compound is unstable) are listed in Table 15.4. (Catenated organometallic compounds involving elements in low oxidation states will be discussed later in this chapter.)

These methyl compounds can be put into categories similar to those of the hydrides. There is, however, no category of metallic methyls, and the realm of **ionic** methyls is diminished as compared to that of ionic or saline hydrides. The pyramidal $:CH_3^-$ ion is known in methylpotassium, $K^+:CH_3^-$, but this adopts the nickel arsenide crystal lattice type, which is not expected for ionic compounds (Section 14.8). The alkyls and aryls of calcium, strontium, and barium are traditionally assumed to be ionic, but no structural data exists to confirm or refute this. We may note that the larger, more asymmetric $:CH_3^-$ and $:C_6H_5^-$ ions cannot be expected to give lattice energies as substantial as those found in corresponding ionic hydrides, nor can they allow overlap in all directions among neighboring metal atoms, which would contribute to metallic conduction.

Table 15.4
Noncatenated Methyl Derivatives of the Elements[a]

A. Molecular (Covalent) Methyls[b]							
Electron Deficient		*Electron Precise*			*Electron Rich*		
	Me_3B(g)	Me_4C(g)			Me_3N(g)	Me_2O(g)	MeF(g)
		Me_4Si	Ph_5P(s)		Me_3P	Me_2S	MeCl(g)
Me_2Zn	Me_3Ga	Me_4Ge	Me_5As		Me_3As	Me_2Se	MeBr
Me_2Cd	Me_3In(s)	Me_4Sn	Me_5Sb		Me_3Sb	Me_2Te	MeI
Me_2Hg	Me_3Tl(s)	Me_4Pb	Ph_5Bi(s)[c]	Me_6Te	Ph_3Bi	Me_4Te	Ar_FXe^+
			Me_5Ta(s)	Me_6W(s)			
B. Ionic and Oligomeric and Polymeric Covalent Methyls[d]							
$(LiMe)_4$	${}^{1}_{\infty}[BeMe_2]$						
$(NaMe)_4$	${}^{1}_{\infty}[MgMe_2]$				$(AlMe_3)_2$(l)		
K^+Me^-	$Ca^{2+}(Me^-)_2$		$(CuMes)_5$				
Rb^+Me^-	$Sr^{2+}(Me^-)_2$		$(AgMes)_4$				
Cs^+Me^-	$Ba^{2+}(Me^-)_2$		$(AuMes)_5$				

[a] The symbol Me is used for the CH_3 group; Ph is used for the C_6H_5 group; Mes is the mesityl or 2,4,6-$Me_3C_6H_2$ group; Ar_F represents some fluorinated aryl groups. Phenyl derivatives are shown if the corresponding methyl is unknown.

[b] All are liquids at room temperature unless otherwise noted.

[c] Methyl derivative is known but is much less stable.

[d] All are solids at room temperature unless otherwise noted.

Oligomeric or Polymeric Covalent Alkyls and Aryls. Corresponding roughly to the polymeric hydrides are the **oligomeric or polymeric electron-deficient alkyls and aryls**. Polymeric chains are formed via 3c–2e bridging methyl groups in $^{1}_{\infty}[BeMe_2]$ and $^{1}_{\infty}[MgMe_2]$. Structural data are mostly available for the alkyls and aryls of lithium and aluminum. These compounds are oligomeric in contrast to the hydrides, which are ionic or polymeric—again the influence of the larger size of the alkyl or aryl group can be seen. In the solid state, methyllithium and methylsodium involve tetramers [Fig. 15.5(*a*)]; each methyl anion is strongly bonded by its one electron pair in one (sp^3 hybridized) orbital to the three nearest Li ions, each having a vacant 2*s* valence orbital and also using its normally postvalence 2*p* orbital. Thus the bonding can be described as four-centered, two-electron (4c–2e), although a more complete molecular orbital (MO) picture can be developed in which the four electron pairs contributed by the four methyl groups fill the four bonding MOs of the tetrahedral cluster molecule.[14]

Methylcopper(I) is an explosive polymeric solid, but greater stability is present in the cyclic aryl oligomers Cu_5Mes_5 [Fig. 15.5(*b*)], Ag_4Mes_4, and Au_5Mes_5 (Mes = 2,4,6-trimethylphenyl or mesityl).[15] (This structure illustrates the tendency of a bridging aryl group to orient itself at right angles to the two bridged metal atoms, so that the π electrons of the aryl group can also become involved in the bonding.) The cluster or ring sizes adopted by the oligomeric electron-deficient alkyls and aryls vary with the size of the alkyl or aryl group, and with the physical state: For example, the Group 11 aryls become mainly dimers in solution.

Methyl groups bridge two aluminum atoms in the structure of hexamethyldialane (or trimethylaluminum dimer), giving a structure [Fig. 15.5(*c*)] resembling that of diborane. Compared to a bridging halide ion, which can contribute four or six electrons to the bonding scheme, a bridging alkyl group, which can contribute only two electrons, is an electron-deficient bridging group. Consequently, organometallic compounds of the above elements are more often seen incorporating bridging halides in place of the bridging alkyls. Far more common than the dialkylmagnesium compounds are the oligomeric **Grignard reagents**, $(RMg\text{–}\mu\text{-}X)_n$, where X = halide; more common than the dimeric trialkylaluminums are dimers such as $Al_2Me_4(\mu\text{-}Cl)_2$ [Fig. 15.5(*d*)]; organolithium compounds are often contaminated by incorporation of lithium halides in their clusters.

The general electron-deficiency and the failure to reach the maximum coordination number of the (at least feebly acidic) metal ions associated with the ionic and oligomeric and polymeric covalent alkyls results in them having, not only the strongly Lewis basic properties of the $:CH_3^-$ carbanion, but also some Lewis acid properties. They are more often found as adducts incorporating Lewis bases such as ethers or chelating amines than as pure alkyls or aryls.

Although these are laboratory curiosities rather than compounds of practical importance, we may now recall the existence of *perlithiocarbons* such as CLi_6, which were mentioned in Section 12.7. Dilithiomethane, CH_2Li_2, is stable enough to have had its crystal structure determined[16]; volatile hexalithiobenzene, C_6Li_6, and solutions of perlithioruthenocene, $(C_5Li_5)_2Ru$, have been prepared but not structurally characterized,[17] leaving theoreticians free to predict exotic structures for these materials.[18] A permagnesioruthenocene, $[C_5(MgCl)_5]_2Ru$, has also been reported.[19]

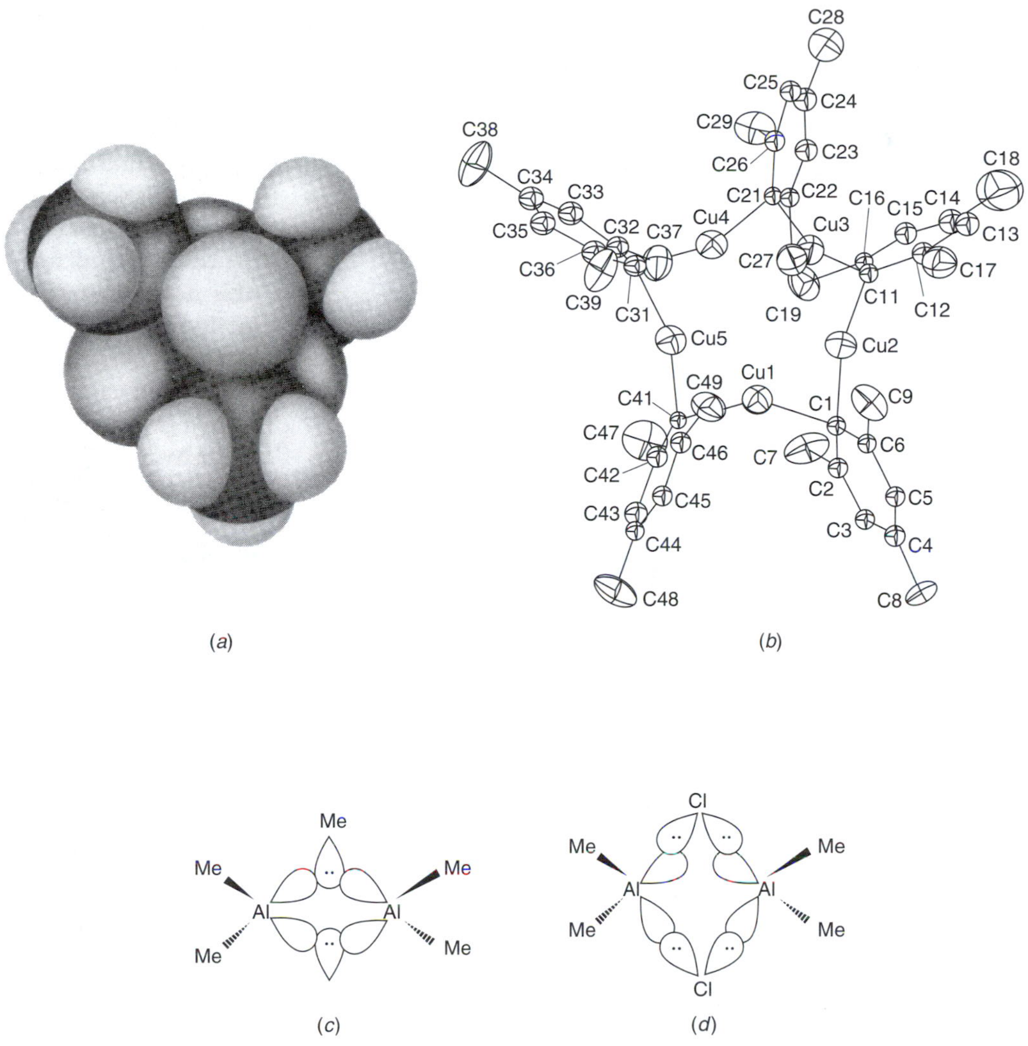

Figure 15.5
(a) The structure of the methyllithium tetramer: each methyl group (with dark-colored carbon atoms) bridges three (light-colored) lithium atoms and each lithium atom bridges three methyl groups. (One methyl group is obscured from view.) [Reprinted from E. Weiss and E. A. C. Lucken, *J. Organometal. Chem.*, **2**, 197 (1964) with permission from Elsevier Science.] (b) The structure of the mesitylcopper(I) puckered pentamer. [Adapted from E. M. Meyer, S. Gambarotta, C. Floriani, A. Chiesi-Villa, and C. Guastini, *Organometallics*, **8**, 1067 (1989).] (c) Structure of hexamethyldialane, Al_2Me_6. (d) Structure of di-μ-chlorotetra-methyldialane.

Electron-Deficient (Monomeric) Molecular Alkyls and Aryls. Although these compounds are electron deficient and generally both Lewis acidic and Lewis basic, the alkyls and aryls of the Group 12 metals and the Group 13(III) elements other than aluminum are monomeric. Due to their consequent moderate molecular weights (numbers of electrons), they are generally liquids; the corresponding phenyl compounds are generally solids.

An Aside

The reasons for the failure of these compounds, except of aluminum, to dimerize or oligomerize may be diverse. Aluminum has the lowest electronegativity of the Group 12 or 13(III) elements: Thus the Al–C bond is the most polar, so that we can anticipate a greater degree of additional attraction between $Al^{\delta+}$ and $CH_3{}^{\delta-}$ groups in different monomeric structural units. The Group 12 methyl compounds show the same very low coordination number of 2 that is found in the Group 11 mesityl compounds. In these groups, the empty *np* orbitals are more or less postvalence orbitals, but most substituents on the Group 11 or 12 element *other than carbon* are electronegative enough to lower these enough in energy to make them available as acceptor orbitals; *bridging* methyl groups are also poor donors, since they must use 3c–2e bonding. In the case of mercury (and thallium) relativistic effects, which also favor two coordination, play a major role: Dialkylmercury and diarylmercury compounds are completely lacking in Lewis acid properties, and trialkyl- or triaryl-thallium compounds are less stable than two-coordinate dialkyl- or diaryl-thallium cations, $TlR_2{}^+$ and $TlAr_2{}^+$.

Electron-Precise (Monomeric) Compounds. Since the electronegativity of carbon is reasonably close to those of the other Group 14(IV) elements, the bonds of alkyl and aryl groups to these elements basically behave as nonpolar bonds, so the Group 14(IV) alkyls and aryls have neither Lewis acidic nor Lewis basic properties. Since the decomposition of alkyls and aryls does not readily produce the very stable H–H bond, the alkyls and aryls are often more stable thermally than the corresponding hydrides. In fact, the alkyls of lead(IV) are strikingly more stable than any lead(II) organics, while the halides and oxides of lead(IV) are much less stable than the organics of lead(IV) or the lead(II) halides. The reasons for this are rooted in the diffuse size of the 6*p* as compared to the 6*s* orbitals: r_p/r_s for Pb is 1.45 as compared to r_p/r_s for C = 1.23 (see Section 12.4 and recall that relativistic effects contract the 6*s* orbital of Pb). This makes sp^3 hybridization for lead difficult and especially sensitive to the electronegativity of the substituents on lead.[20]

The stability of the Group 14(IV) alkyls and aryls is such that more reactive substituted derivatives are more important than the alkyls and aryls themselves. Halides such as Me_3SiCl and Me_2SiCl_2 are important for their hydrolysis and solvolysis reactions, which lead to polymers such as the silicones (Section 14.6), and which can replace reactive, water-attracting OH groups on the surfaces of clays, silica, and so on, with water-repellent, relatively inert Me_3SiO (trimethylsiloxy) groups. In compounds containing R_3SiO groups (such as silicones and $R_3SiOSiR_3$), the Si–O–Si bond angle is found to be unusually large and easily deformed; this variable bond angle contributes to the flexibility of silicone polymers, and is attributed to the presence of some π donation of oxygen unshared electron pairs back to empty C–Si σ^* orbitals. Organotin compounds and their derivatives are widely used as fungicides, antifouling additives for paints, and stabilizers for plastics, and often turn up in environmental analyses, which attests to their relative stability.

The increase in stability of electron-precise alkyls and aryls (in contrast to the corresponding hydrides) allows the extension of this family beyond Group 14(IV) into the two following groups, even though this calls for oxidation states that are formally higher than +4, in compounds with alkyl or aryl groups that are formally quite reducing. Pentaphenyl derivatives of all of the Group 15(V) elements except N are known, while pentamethyl derivatives are known except for N and P but are generally less stable than the pentaphenyls. In Group 16(VI), the series of electron-precise alkyls has been extended to include the octahedral tellurium compound, Me_6Te.[21]

The VSEPR theory predicts trigonal bipyramidal geometries for the Group 15(V) alkyls and aryls, and this geometry generally is found. But Ph_5Sb and Ph_5Bi adopt square pyramidal geometries. These peculiar exceptions are assumed to be due to crystal packing forces acting on these fluxional molecules since, for example, Me_5Sb is trigonal bipyramidal, and Ph_5Sb crystallizes in a trigonal bipyramidal geometry from cyclohexane solution.

The bonding in these pentacoordinate compounds is best described using the valence s, p_x, and p_y orbitals of the Group 15(V) element to bond to the three equatorial methyl groups with conventional (two-centered, two-electron) bonds, while the bonds to the two axial methyl groups involve four electrons from these two anions but only the p_z orbital from the central atom. Hence, this gives rise to a 3c–4e bonding interaction consisting of an occupied bonding σ molecular orbital and an occupied σ^{nb} nonbonding orbital; the antibonding σ^* orbital remains vacant. Consequently, the bond order per element–methyl link in the axial direction is only 0.5, while it is 1.0 in the equatorial plane. Unsurprisingly, the element–methyl bond distances are over 7 pm longer for axial methyls than for equatorial methyls in PMe_5 and $SbMe_5$. Surprisingly, they are about the same in trigonal bipyramidal $BiMe_5$.[22]

The nominally high oxidation states in these hypervalent alkyls and aryls are more stable if at least some electronegative substituents are present. Thus one more commonly encounters organometallics in these oxidation states if halogens or oxygen are also present, as in the tetrahedral oxides R_3EO or the trigonal bipyramidal R_3EX_2, in which the more electronegative halogen atoms occupy the axial positions that have the lower bond order (less covalent overlap). These R_3EX_2 derivatives are well known for R = aryl or alkyl and X = the common halogens, although the structure of Ph_3PI_2 proved a surprise. One of the two iodine atoms is not bonded to phosphorus, but rather is bonded to the other iodine atom, so that the compound is apparently an adduct of I_2 as a Lewis acid coordinated to triphenylphosphine as a Lewis base![23]

Some simple alkyls and aryls of *d*-block metals with formulas similar to the *p*-block compounds exist, but these are unstable and even explosive at room temperature: yellow $TiMe_4$, red $ZrMe_4$, square pyramidal $TaMe_5$, and trigonal prismatic red WMe_6.[24] Since *d*-block metals have additional valence orbitals, these are not really electron-precise molecules, and they lack any strong-field ligands normally needed to stabilize organometallic compounds. They also do not obey the 18-electron rule. In compounds obeying the 18-electron rule and having π-acceptor ligands, there is no problem with also having metal–methyl or metal–phenyl σ bonds (Chapter 11). One derivative that is stable, even though it lacks π-acceptor ligands, is *methylrhenium trioxide*, $MeReO_3$. This compound is a catalyst for olefin (alkene) metathesis, alkene oxidation, and conversion of aldehydes to alkenes.[25]

Electron-rich Alkyls and Aryls. These are well-known compounds: The Group 17(VII) compounds (*halocarbons*) and the Group 16(VI) compounds (*ethers* and *organic sulfides, selenides, and tellurides*) are among the classical types of organic compounds. In group 15(V), the *organic amines* are similarly familiar and need not be mentioned further. The later Group 15(V) derivatives, although somewhat less familiar to organic chemistry students, are still of importance. Triphenylphosphine, -arsine, and -stibine are stable white solids that are important soft-base ligands and are, from the organic chemists' point of view, excellent nucleophiles. Trimethylphosphine, -arsine, -stibine, and -bismuthine are more volatile liquids; the combination of volatility and a point of vulnerability—their unshared pairs of electrons—means that they can catch fire or oxidize readily in air to give species such as trimethylphosphine oxide, $Me_3P{=}O$. Spontaneous flammability is not a problem with the very odiferous sulfides, and so on, of Group 16(VI), but they can be chemically oxidized to give compounds such as *dimethyl sulfoxide*, $Me_2S{=}O$, and *dimethyl sulfone*, $Me_2S({=}O)_2$.

The organic chemistry of the elements has even been extended to Group 18(VII): Arylxenon cations Ar_FXe^+, where Ar_F is a fluorinated phenyl group such as C_6F_5 or 2,6-$F_2C_6H_3$, have been isolated as salts of nonbasic fluoro anions.[26] We should also mention one electron-rich but hypervalent methyl, *tetramethyltellurium*(IV),[27] the synthesis of which soon led to the preparation of the electron-precise but hypervalent Me_6Te.

Example 15.5

Tell whether each of the following organometallics is likely to be a gas, liquid, or a solid, and explain this fact in terms of the molecular, ionic, oligomeric, or polymeric structure of the organometallic compound: (a) RbPh; (b) $PbEt_4$; (c) $SnPh_4$; (d)$BePh_2$; and (e) BuLi.

SOLUTION:

(a) RbPh is expected to be ionic Ph^-Rb^+, hence it is a solid.

(b) The compound $PbEt_4$ is molecular, and of high molecular weight, so it should be a liquid or low-boiling solid (in fact it is a liquid).

(c) The compound $SnPh_4$ is molecular, and of even higher molecular weight, so it should be (and is) a solid.

(d) The compound $BePh_2$ is expected to be a chain polymer, hence it is a liquid or solid.

(e) The compound BuLi is likely to be oligomeric, hence it is of moderate molecular weight, and is probably a liquid or a low-melting solid.

15.7 Methods of Synthesis of Alkyls and Aryls

A number of methods exist for the synthesis of σ-bonded organometallic compounds. The first three methods apply to a limited set of metals, but include the original sources of the organometallics used in the other two, more general methods. (1) The *direct reaction of an alkyl or aryl halide with the element*. (2) *Metalation* of a hydrocarbon. (3) The *reaction of unsaturated organic compounds with a metal hydride or with a metal plus hydrogen*. The two methods of the most general applicability are (4) *the reaction of organometallic compounds with metal halides*, and (5) the *reaction of the more active metal with the organometallic derivative of a less active metal*.

1. The method of most commercial importance for selected elements is the *direct reaction of an alkyl or aryl halide with the element.*

$$4\,\mathrm{MeCl} + 8\,\mathrm{Li} \rightarrow [\mathrm{MeLi}]_4 + 4\,\mathrm{LiCl} \quad (15.43)$$

This reaction is an oxidation–reduction reaction (related to the activity series of metals, Section 6.5) that the most active metals undergo most readily, and that the least active metals often do not undergo. With sodium the direct reaction is *too* exothermic: halocarbons and sodium metal may react explosively. In general, iodocarbons are the most reactive halocarbons for this type of reaction, while fluorocarbons are not generally useful. The mechanism of the reaction may often involve electron transfer to the carbon–halogen σ^* lowest unoccupied molecular orbital (LUMO) of the halocarbon, which is lowest in energy for iodocarbons. Some metals, under conditions of direct reaction, instead promote the Wurtz coupling reaction of two intermediate alkyl radicals to give the R_2 alkane.

With a divalent metal such as magnesium, the product of direct reaction is not normally the pure dimethyl compound plus $MgCl_2$ but is instead the mixed complex commonly known as a **Grignard reagent**:

$$\mathrm{MeBr} + \mathrm{Mg} \rightarrow \mathrm{MeMgBr} \quad (15.44)$$

This synthesis is generally done in a Lewis base solvent such as an ether, which coordinates tenaciously to the magnesium atom. The reaction requires the exclusion of air and the presence of a dry solvent, and often a little iodine is needed to clean off the surface coating of MgO that impedes this exothermic reaction. Organozinc compounds are prepared in a similar manner.

The organosilicon halides used to make silicone polymers are prepared by passing gaseous MeCl over heated silicon mixed with copper as a catalyst:

$$2\,\mathrm{MeCl} + \mathrm{Si}\,(300\,^{\circ}\mathrm{C}) \rightarrow \mathrm{Me_2SiCl_2} \quad (15.45)$$

The reaction also produces a few percent of byproducts such as Me_3SiCl, $MeSiCl_3$, $MeHSiCl_2$, $SiCl_4$, and disilanes.

The least active (electronegative) metals may have to be alloyed with an active metal such as Na to promote the reaction, as in the production of tetraethyllead for use in leaded gasoline:

$$4\,\mathrm{EtCl} + 4\,\mathrm{PbNa} \rightarrow \mathrm{Et_4Pb} + 3\,\mathrm{Pb} + 4\,\mathrm{NaCl} \quad (15.46)$$

Organomercury compounds can also be prepared in this manner, using the sodium–mercury alloy (sodium amalgam). For other inactive metals, one may use either gaseous metal atoms generated by the metal atom reactor (Section 7.3) or very finely divided metal powders, freshly generated from a metal halide and a very active metal such as potassium. These are not, however, commercially viable methods.

2. The *metalation* reaction (metal–hydrogen exchange reaction) can be considered to be an HSAB reaction producing the hard acid H^+. Hard-base salts of the Hg^{2+}, Tl^{3+}, and Pd^{2+} ions often metalate unsaturated organic compounds under quite mild conditions, readily giving products containing the soft-acid–soft-base Hg–C, Tl–C,

and Pd–C bonds:

$$Hg^{2+}(aq) + C_6H_6 \rightarrow [C_6H_5Hg]^+ + H^+(aq) \quad (15.47)$$

Such reactions are commonly known as *mercuration*, *thallation*, and *palladation* reactions. Permercuration of electron-rich alkene systems is even possible, leading to compounds such as the permercurated cyclopentadiene $C_5(HgOOCCH_3)_6$ and the permercurated ferrocene $Fe[C_5(HgOOCCF_3)_5]_2$.[28] This reaction, in contrast to direct reaction, is useful for the electronegative (least active) metals.

3. The *reaction of unsaturated organic compounds with a metal hydride or with a metal plus hydrogen* (Section 15.5) is important for the elements boron, aluminum, and silicon. The *hydroboration* reaction is extremely important in organic chemistry, since the organoboranes produced can be converted to many important organic chemicals:

$$B_2H_6 + 6\,H_2C{=}CH_2 \rightarrow 2\,Et_3B \quad (15.48)$$

The resulting organoboranes can be converted to hydrocarbons by treatment with H^+, to alcohols by treatment with H_2O_2, and to aldehydes, ketones, or tertiary alcohols by treatment with CO followed by H_2O_2. *Hydrosilylation* using silicon hydrides is important for producing silanes for use in making silicones. The driving force of these reactions is the formation of C–H bonds that are stronger than the element–H bonds being broken.

Organoaluminum compounds, which are very important (along with titanium compounds) in the Ziegler–Natta catalysts for producing organic polymers such as polyethylene, are produced industrially by the direct reaction of aluminum metal, the organic compound, and hydrogen:

$$2\,Al + 3\,H_2 + 6\,CH_2{=}CH_2 \rightarrow 2\,Et_3Al \quad (15.49)$$

This reaction is economical on the industrial scale but is impractical to carry out on the laboratory scale, since H_2 under pressure must be heated.

4. The *reaction of organometallic compounds with metal halides* provides (at least formally) a Lewis acid–base exchange of the methyl anion for a halide ion: in accord with the HSAB principle, the soft organic anion ends up coordinated with the more electronegative (soft) metal, while the driving force of the reaction is the formation of ionic salts of very electropositive elements:

$$2\,MeMgX + CdCl_2 \rightarrow Me_2Cd + MgX_2 + MgCl_2 \quad (15.50)$$

This reaction normally begins with a Grignard reagent (as above), an organolithium compound, or an organoaluminum compound, but can involve two compounds of the same element, redistributing alkyl or aryl and halo groups:

$$Ph_2Hg + HgCl_2 \rightleftharpoons 2\,PhHgCl \quad (15.51)$$

The redistribution reaction may be driven by entropy considerations, or when the mixed-ligand product is more insoluble (as in the above example); in other cases, the reverse reaction is favored if an adduct or complex formed by the inorganic halide is insoluble or very stable. Related to these reactions is the synthesis of methylrhenium trioxide by an exchange reaction between Me_4Sn and Re_2O_7.

5. Alkyl and aryl derivatives of a more active metal can often be prepared by the *reaction of the more active metal with the organometallic derivative of a less active metal*, such as an organomercury compound. In accord with the activity series and in contrast to Method 4, the organic group ends up associated with the "cation" of the less electronegative metal. This reaction is useful for preparing organometallic compounds uncontaminated with halide and with no coordinated ether solvent molecules:

$$Me_2Hg + \text{excess Na} \rightarrow 2\,MeNa + \text{Na amalgam} \qquad (15.52)$$

6. There is a second type of acid–base *metalation* reaction, in which an organolithium compound, especially in the presence of a chelating Lewis base ligand such as $Me_2N{-}CH_2CH_2{-}NMe_2$ to enhance the anionic character of the organic group, reacts with a more acidic organic compound to produce a less acidic organic compound and another organolithium compound:

$$BuLi(Me_2N{-}CH_2CH_2{-}NMe_2) + C_6H_6 \rightarrow C_4H_{10} + PhLi(Me_2N{-}CH_2CH_2{-}NMe_2) \qquad (15.53)$$

This reaction is based on acid strength, rather than softness.

Example 15.6

Which of the two syntheses listed under each letter is the more likely to work? What type of organometallic synthesis is being attempted in each case?

(a) $MeI + Mg \rightarrow MeMgI$ or $MeF + Mg \rightarrow MeMgF$

(b) $C_4H_9Li + C_6H_6 \rightarrow C_6H_5Li + C_4H_{10}$ or $C_5H_5Li + C_6H_6 \rightarrow C_6H_5Li + C_5H_6$

(c) $MeBr + Mg \rightarrow MeMgBr$ or $MeBr + Cd \rightarrow MeCdBr$

(d) $Me_2Be + BaCl_2 \rightarrow Me_2Ba + BeCl_2$ or $Me_2Ca + CdCl_2 \rightarrow Me_2Cd + CaCl_2$

SOLUTION:

(a) Direct reaction of a alkyl halide with a metal is being attempted; iodides react most rapidly and fluorides most poorly, so $MeI + Mg \rightarrow MeMgI$.

(b) Metalation of a hydrocarbon is being attempted. The more acidic hydrocarbon will react with the lithium derivative of the less acidic hydrocarbon to liberate the less acidic hydrocarbon. Since cyclopentadiene is the most acidic of the three hydrocarbons and butane is the least, $C_4H_9Li + C_6H_6 \rightarrow C_6H_5Li + C_4H_{10}$.

(c) This is the direct reaction of an alkyl halide with a metal; the least active of the two metals is Cd, so the reaction most likely to go is with Mg: $MeBr + Mg \rightarrow MeMgBr$.

(d) This reaction is of an organometallic compound with a metal halide; the organic group ends up bonded to the most electronegative metal, so $Me_2Ca + CdCl_2 \rightarrow Me_2Cd + CaCl_2$.

15.8 Reactions and Uses of the Alkyls and Aryls

Ionic and Polymeric Alkyls and Aryls. Characteristically, the metal atom of the ionic and polymeric molecular alkyls is of relatively low acidity and is much less electronegative than the carbon atom, so these alkyls and aryls (and those of zinc and cadmium) either contain (for example) the methide ion, CH_3^-, or may act as sources of that soft base, and their dominant chemistry is as Lewis bases. However, CH_3^- is also a powerful reducing agent that can transfer electrons to an appropriate reactant. Sometimes it is difficult to predict in advance which function (Lewis base or reducing agent) will predominate in a given situation.

Organolithium compounds are extensively used in organic chemistry as a source of carbon anions (carbanions[29]) for the construction of more extensive carbon chains:

$$Li^+Me^- + Me_2C{=}O \rightarrow Me_3CO^-Li^+ \tag{15.54}$$

Even more well known for this use in organic chemistry are the Grignard reagents such as MeMgBr.

Electron-Deficient Alkyls and Aryls. In Groups 12 and 13(III), the metal ion is often at least moderately acidic, so not only the basic properties of the alkyl or aryl group, but also the acidic properties of the electron-deficient metal or metalloid, are important.

There are many Lewis base reactions of these compounds. Organoaluminum compounds are used for alkylations and as strong bases in industry. The diorganozinc and diorganocadmium compounds, which contain more electronegative metals, and hence less partially negatively charged organo groups, are also used for this purpose when a milder reagent is needed. The vapors of diethylzinc can be used as a mild dry base capable of neutralizing the acids that cause the paper of old books to deteriorate.[30]

The electron-deficient alkyls and aryls of Group 13(III) can become electron precise by acting as Lewis acids. They readily form adducts with Lewis bases:

$$Me_3B + {:}NMe_3 \rightarrow Me_3B{:}NMe_3 \tag{15.55}$$

Adducts of Group 13(III) methyls with Group 15(V) hydrides are important in **metal–organic chemical vapor deposition** (MOCVD) of thin, very pure films of "III(V)" semiconductors (Sections 14.6, 14.8). The component methyls and hydrides, being volatile, are readily freed from impurities by distillation. The adduct, on heating, readily vaporizes, then decomposes smoothly (by combination of the acidic H^+ groups from the Group 15(V) hydride with the basic CH_3^- groups of the Group 13(III) methyl, or by decomposition to free radicals) to deposit very pure semiconductor films on a suitable surface:

$$(CH_3)_3Ga{:}AsH_3 \xrightarrow{600\,^\circ C} GaAs(s) + 3\,CH_4(g) \tag{15.56}$$

Group 13(III) alkyls and aryls can also react with R^- or Ar^- Lewis bases to give complex electron-precise anions:

$$Ph_3B + Na^+Ph^- \rightarrow Na^+ + BPh_4^- \tag{15.57}$$

This compound (sodium tetraphenylborate) is an important source of a large nonbasic anion and is used to precipitate large nonacidic cations. This reaction is even found to a limited extent with the less acidic metal cations of Groups 12, 11, or even earlier groups. Organic copper(I) is usually encountered, not as its polymeric methyl compound, but as lithium dimethylcuprate or related organocuprate anions, which find extensive use in organic synthesis of new carbon–carbon bonds upon reaction of dialkylcuprate with an alkyl halide. Syntheses have been reported of trialkyl- and triarylcadmate ions, magnesiate ions, and even a linear dialkylsodate (R_2Na^-) ion![31] Some transition metal ions that do not form stable neutral alkyls do give relatively stable anions (in which more coordination sites are blocked), such as $ReMe_8{}^{2-}$, $PtMe_6{}^{2-}$, and $MnMe_6{}^{2-}$.[32]

A characteristic reaction of many electron-deficient alkyls with hydrogen atoms on the β-carbon (second carbon atom) is **β-hydrogen elimination**:

$$R_nM{-}CH_2{-}CH_2{-}D \rightleftharpoons R_nM{-}D + CH_2{=}CH_2 \qquad (15.58)$$

This reaction is the reverse of hydrometalation, and is favored when M is electron deficient, or is a catalytically active *d*-block metal (it is known as an elimination reaction in *d*-block chemistry, Table 11.8). The reaction is thought to involve the formation of a four-membered ring in which the end hydrogen atom (here shown as a deuterium atom) is bonded in an electron-deficient manner both to the metal and to the carbon. As a consequence of the ease of this elimination, ethyl derivatives of the electron-deficient and *d*-block metals are scarcer than methyls; for synthesizing higher alkyls it is useful to go to groups lacking a β-hydrogen, such as neopentyl groups, $-CH_2{-}CMe_3$, trimethylsilylmethyl groups, $-CH_2{-}SiMe_3$, or benzyl groups, $CH_2{-}Ph$. Thus there are many neopentyl derivatives of *d*-block metals, such as $Cr(CH_2CMe_3)_4$; in $[Co(norbornyl)_4]^+$ the apparent computed cobalt oxidation number, +5, is implausibly high.[33]

Electron-Precise Alkyls and Aryls. Addition of aryl or alkyl anions to electron-precise aryls or alkyls is not at all common, but this can happen. For example, pentaphenylbismuthane can add another phenyl anion to give the six-coordinate $BiPh_6{}^-$ anion, and one case of a pentaorganosilicate(IV) has been reported.[34]

Electron-Rich Alkyls and Aryls. The electron-rich alkyls and aryls of the elements of Groups 15(V)–17(VII) can achieve electron precision, by donating unshared electron pairs from the central atom. This function is most significant in Group 15(V) and only occasionally manifests itself in Group 17(VII), where the very electronegative halogen atom is reluctant to donate any additional electrons. Thus trimethyl- and triphenyl-phosphine and -arsine are important soft-base ligands, forming numerous complexes with soft or borderline-acid metal ions. Fewer complexes of trimethyl- or triphenyl-stibine and -bismuthine are known, perhaps due to the lower covalent bond energies. Similarly, organic sulfides and selenides are soft bases, forming numerous complexes. Organic ethers, R_2O, and organic amines, R_3N, act as hard and borderline bases in forming complexes; the latter can experience steric problems in complex formation.

These alkyls and aryls also achieve electron precision when they react with methyl halides to displace the halide ion to produce *quaternary -onium ions*:

$$Me_3P{:} + MeI \rightarrow Me_4P^+ + I^- \qquad (15.59)$$

The cation produced is analogous to the phosphonium ion mentioned previously but is more stable since it cannot readily transfer a proton and lose its positive charge. Such ions are named using the suffix *-onium*: Me_4P^+ is the tetramethylphosphonium ion, Ph_4As^+ is the tetraphenylarsonium ion, Me_3S^+ is the trimethylsulfonium ion, and Ph_2I^+ is the diphenyliodonium ion. These are very large nonacidic ions, and are often used to precipitate large nonbasic anions. Those from Groups 16(VI) and 17(VII) readily transfer their methyl groups (as methyl "cations," not as CH_3^-) to more basic substances, however, and thus are alkylating agents that may be mutagenic.

Other Reactions. Of particular importance in organic syntheses are the results of attempts to synthesize the as-yet-unknown pentamethylphosphorane by the reaction of methyllithium with tetramethylphosphonium chloride. A molecule of methane is eliminated and the product is $Me_3P{=}CH_2 \leftrightarrow Me_3P^{\oplus}{-}CH_2^{\ominus}$. Using Ph_3P gives the well-known *Wittig* reagent of organic chemistry, which readily reacts with carbonyl groups to transfer the $=CH_2$ functional group:

$$Ph_3P{=}CH_2 + Me_2C{=}O \rightarrow Ph_3P{=}O + Me_2C{=}CH_2 \tag{15.60}$$

Like the Si–O bond in siloxanes and silicones, the P–O bond in the product, triphenylphosphine oxide, has some π character to it, which gives it added bond energy; formation of this strong bond drives this reaction to the right.

The electron-precise and electron-rich organo halo derivatives are subject to hydrolysis and to related solvolyses, such as reaction with alcohols to give alkoxides. Such reactions are not as vigorous as those discussed in Chapter 2 for the pure halides of the same element in the same oxidation state, since the alkyl–element bond is relatively nonpolar and does not produce as much positive charge on the central-atom "cation" as do polar halogen–element bonds. Thus, while both $SiCl_4$ and Me_2SiCl_2 hydrolyze vigorously on contact with water, the reaction of Me_2SiCl_2 to give (ultimately) silicones is less violent:

$$Me_2SiCl_2 + 2\,H_2O \rightarrow Me_2Si(OH)_2 \rightarrow H_2O + \tfrac{1}{\infty}[Me_2SiO] \tag{15.61}$$

The unsubstituted electron-precise and electron-rich alkyls and aryls, having basically neutral or even partially negatively charged central atoms protected by large alkyl or aryl groups, are not subject to hydrolysis or oxidation by O_2, in contrast to the corresponding hydrides. While silane and germane are spontaneously flammable in air, and stannane and plumbane decompose even in the absence of air at room temperature, the corresponding tetramethyls are thermally stable at room temperature and do not catch fire in air (in the absence of a flame or spark). Since organic groups such as Me and Ph are larger than the hydride group, they more effectively hinder the approach of reactive substances such as H_2O and O_2.

However, even the electron-precise alkyls and aryls of the heaviest elements are subject to fairly ready thermal decomposition, since the bonds of large atoms of heavy elements to small carbon atoms are relatively weak. Tetraalkyllead compounds generally decompose above 100 °C, producing elemental lead and alkyl radicals. The function of the gasoline additive tetraethyllead was to produce radicals such as $C_2H_5^{\cdot}$, which terminate the chain reactions that occur during the combustion of gasoline, thus reducing the possibility of explosive combustion (knocking). In order to prevent the engine from becoming coated with lead salts such as PbO (from oxidation of Pb), ethylene bromide was also added, which reacted with the Pb to produce $PbBr_2$. This compound is volatile at engine temperatures. This product, however, was a serious air pollutant, so the use of lead additives in gasoline has been drastically curtailed.

The reaction of the electron-deficient alkyls and aryls with water or alcohols proceeds differently than the reactions of the electron-rich organo halo derivatives, since the most reactive sites in these compounds are the very strongly basic alkyl or aryl carbanions. Hydrolysis of these produce an alkane or an arene, often with explosive violence, so that fires involving these materials (or electron-deficient hydrides) must not be fought with water!

$$(CH_3Li)_4 + 4\,H_2O \rightarrow 4\,CH_4(g) + 4\,Li^+(aq) + 4\,OH^-(aq) \qquad (15.62)$$

Attack by O_2 is also very easy: These compounds are spontaneously inflammable (**pyrophoric**), therefore they must be handled with rigid exclusion of air. There is an interesting gradation in stability among the Group 12 methyls. Dimethylzinc is decomposed by water and is spontaneously flammable; dimethylcadmium is decomposed by water but is less quickly oxidized by air, whereas dimethylmercury is unaffected either by water or by air. This inertness, of course, means that dimethylmercury is among the most persistent of the methyls in the environment.

Example 15.7

Describe how each of the following might react with either an alkyllithium compound or an alkyl halide to achieve a more nearly electron-precise product, and predict the formula of the product: (a) PPh_3; (b) BPr_3; (c) $ZnPh_2$; and (d) Cu_5Mes_5.

SOLUTION:

(a) $:PPh_3$ is electron rich, so it is a Lewis base that can displace halide from an alkyl halide to give an electron-precise phosphonium ion: $:PPh_3 + MeI \rightarrow [PPh_3Me]^+I^-$.

(b) The compound BPr_3 is electron deficient, so it is a Lewis acid that can coordinate the carbanion of an alkyllithium to give an electron-precise borate ion: $4\,BPr_3 + (PrLi)_4 \rightarrow 4\,Li^+[BPr_4]^-$.

(c) The compound $ZnPh_2$ is also electron deficient, and needs two carbanions to become electron precise: $2\,ZnPh_2 + (MeLi)_4 \rightarrow 2\,(Li^+)_2[ZnPh_2Me_2]^{2-}$.

(d) The compound Cu_5Mes_5 is electron deficient and might be predicted to react with three alkyllithium units to give a -3 charged electron-precise cuprate ion. In practice, such a highly charged ion is not formed, and Cu can only approach electron precision by achieving a coordination number of two: $4\,Cu_5Mes_5 + 5(MeLi)_4 \rightarrow 20\,Li^+[Cu(Mes)(Me)]^-$.

15.9 *Catenated Alkyls and Aryls of the *p*-Block

When the central atom of a *p*-block organometallic is in a low oxidation state [e.g., +1 in Group 13(III), +2 in Group 14(IV), +1 in Group 15(V)], the central atom is not coordinately saturated and does not have an octet of electrons, so that the organo-

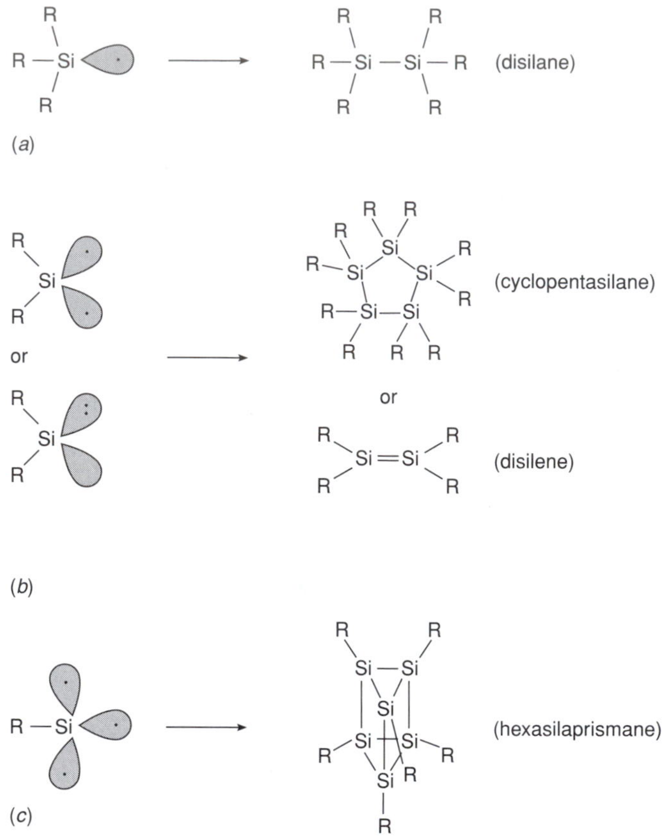

Figure 15.6
Oligomerization of divalent silicon alkyl and aryl structural units (monomeric fragments, shown on left) to give catenated alkyl and aryls (on right). (*a*) Dimerization of $R_3Si\cdot$. (*b*) Cyclooligomerization or dimerization of R_2Si: (a silylene). (*c*) Cluster formation by oligomerization of RSi:·; cyclooligomerization (to give a silicon-based aromatic) and dimerization (to give a disilaalkyne) are not known to occur.

metallic structural unit (monomeric fragment) tends to dimerize, oligomerize, or polymerize, as sketched in Figure 15.6 using silicon as an example.

The steric bulk of the organic group on the *p*-block element, relative to the size of the atom of the element, plays an important role in determining the degree of oligomerization or polymerization of the monomeric unit. Even the moderate bulk of the three phenyl groups in a Ph_3[Group 14(IV) atom]· structural unit is a disadvantage if the central atom of the molecule is very small: The compound hexaphenylethane, Ph_6C_2, is so sterically strained that it tends to break apart in solution to give the triphenylmethyl free radical, $Ph_3C\cdot$, but there is no such tendency when the central atom is larger, as in Ph_6Si_2 (hexaphenyldisilane).

The greatest tendency for R_2Si structural units to oligomerize or polymerize is found when R is small, but R should be large enough compared to the central atom to protect it from hydrolysis or oxidation. If the central atom is also small, then hydride is bulky enough: Organic hydrocarbons are protected kinetically from attack by water or

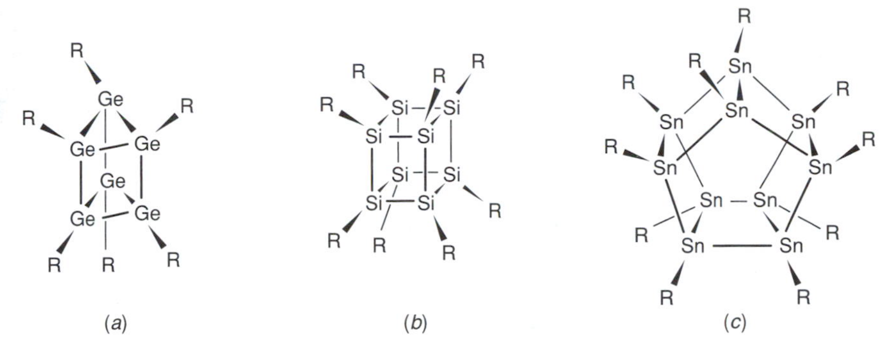

(a) (b) (c)

Figure 15.7
Cluster oligomers of heavier Group 14(IV) elements (symbolized by E). (*a*) Prismanes $(RE)_6$, known, for example, for E = Si with R = 2,6-diisopropylphenyl and for E = Ge with R = $CH(SiMe_3)_2$. (*b*) Cubanes $(RE)_8$, known, for example, for E = Si, Ge, or Sn with R = 2,6-diethylphenyl. (*c*) Pentagonal prismane known for E = Sn with R = 2,6-diethylphenyl. [Adapted from D. F. Shriver, P. W. Atkins, and C. H. Langford, *Inorganic Chemistry*, 2nd ed., Freeman, New York, 1994; p. 448.]

oxygen. Recalled from the beginning of this chapter, however, that the high flammability and other forms of reactivity of the polysilanes, Si_nH_{2n+2}, has limited the value of n to date to about 8. But with the larger methyl group, there seems to be no limit at all to the extent of catenation in the fully methylated polysilanes $Me(SiMe_2)_nMe$: The methyl group is bulky enough to protect the larger silicon, germanium, and tin atoms from attack by oxygen and water (the nonpolar nature of the bonds of the element to carbon also contribute to this stability). In addition to the linear methylated polysilanes, germanes, and stannanes,[35] there are also several methylated cyclic silanes,[36] germanes, and stannanes $(Me_2Si)_n$, $(Me_2Ge)_n$, and $(Me_2Sn)_n$ (with n ranging from 3 to 6 and even upward), and even a variety of methylated bicyclic polysilanes such as $MeSi(SiMe_2SiMe_2)_3SiMe$. As we shall see in Section 15.10, if the alkyl or aryl group is too bulky, extensive oligomerization and polymerization are prevented, and instead dimeric disilenes or even monomeric silylenes may result instead.

Monomeric RSi, RGe, and RSn fragments can be assembled into clusters often named informally based on their central-atom geometry: 4-membered "tetrahedranes," 6-membered "prismanes," 8-membered "cubanes", and 10-membered "pentagonal–prismatic" species (Fig. 15.7).[37]

The actual assembly of these structural units into the dimeric chain, cyclic, or cluster silanes, germanes, or stannanes is generally achieved by the reaction of the appropriate alkyl- or aryl-element halide with a metallic reducing agent such as Li, Na/K alloy, or Mg with $MgBr_2$, although thermal cracking of the strained three-membered cyclic ring compounds can also be used to generate bicyclic or cluster compounds.[38]

Many of these compounds are strikingly colored: $(ArSn)_8$ is deep red and $(ArSn)_{10}$ is orange. Even the oligosilanes $Me(SiMe_2)_nMe$ show intense ultraviolet (UV) absorptions that gain intensity and move toward the visible region of the spectrum as n increases (up to about $n = 30$). The UV spectral properties of oligosilanes are similar to those of oligoalkenes such as R–(CH=CH)$_n$–R, but contrast with those of the alkanes, R–$(CH_2)_n$–R, which do not absorb in the normal UV range of the spectrum. The UV spectra of oligoalkenes are the result of electronic transitions from filled π to empty π^* molecular orbitals, the energies of which become closer as the chain lengths become longer (Section 10.5). The oligosilanes, germanes, and stannanes have no π or π^* orbitals. These transitions are considered to be from bonding MOs arising from delocalized overlap of σ-bonding orbitals of the chain or cycle Si or Ge atoms with each other to antibonding MOs arising from delocalized overlap of σ^* antibonding orbitals of the chain or cycle Si or Ge atoms with each other; recall that such MOs for linear H_n systems were discussed in Section 10.5. Due to the poorer overlap of the larger Si and Ge orbitals with each other, these bonding and antibonding MOs are closer to each other for Si than for C in alkanes; consequently, the electronic transitions appear in the UV or even the visible spectra of these compounds. One practical consequence of this is that polymeric silanes are sensitive to UV light and are cleaved by them. This photosensitivity makes these polymers of low interest as structural materials but very useful in the production of integrated circuits: A coating of polysilane can be exposed to a pattern of UV light and degraded only where exposed to the light; the degraded polymer is then washed away and the underlying layer of another material is then chemically modified only in the areas that had been exposed to UV light.[39]

The low-energy σ^* antibonding MOs of oligo-, cyclo-, and polysilanes and germanes such as $(Me_2Si)_6$ readily accept electrons from Group 1 metals to form radical anions such as $[(Me_2Si)_6]^-$; the electron spin resonance (ESR) spectra of such anions shows that the unpaired electron is located equally on all six ring silicon atoms, as expected for an electron in a fully delocalized MO.

Finally, ${}^1_\infty[SiMe_2]$ [or, alternately, $(SiMe_2)_6$] has proved useful as a precursor to the ceramic silicon carbide, SiC. When either of these precursors are heated to 450 °C, pyrolysis causes one of the methyl groups to cleave, then insert CH_2 groups in the Si–Si bonds to give ${}^1_\infty[HMeSi{-}CH_2{-}]$ (polycarbosilane), which can be melted and spun into fibers. Polycarbosilane pyrolyzes further at 1300 °C to eliminate CH_4 and H_2, giving strong fibers of silicon carbide.[40]

Group 15(V) Organic Polymers. Phosphorus, arsenic, and antimony form catenated compounds of two major types. Dimers of the type Me_4P_2 and Me_4As_2 can be made by reduction of dialkylelement halides with metals; the stinking toxic liquid Me_4As_2, known as "dicacodyl," was one of the first organometallic compounds made. Many R_4Sb_2 and R_4Bi_2 compounds show the interesting property of **thermochromism**: their colors change with temperature. Thus tetramethyldistibine is pale yellow at −180 °C but darkens as it warms, becoming bright red at −17 °C; it then melts to give a pale yellow liquid. Tetramethyldibismuthine is a violet-blue solid that melts to give a red-yellow liquid. Structural studies have shown that the bright colors are associated with the presence of secondary bonds weakly linking antimony or bismuth atoms of different molecules into a type of chain polymer in the solid state only (Fig. 15.8); these secondary bonds have about 20% of the strength of the primary Bi–Bi bonds within the R_4Bi_2 dibismuthine molecules.[41]

Reduction of alkyl- and aryl-element dihalides with metals (even some as inactive as Hg) gives three-, four-, five-, and six-membered cyclic oligomers such as $(PhAs)_6$; alkyl and phenyl substituents favor the formation of six-membered rings, while bulkier groups favor the formation of five- or four-membered rings; as suggested in Figure 15.9, the substituents are further apart when attached to a smaller sized ring (note that the separation is greatest for the "two-membered" ring, e.g., a double-bonded dimeric species, which will be discussed in Section 15.10). These cyclic oligophosphines and

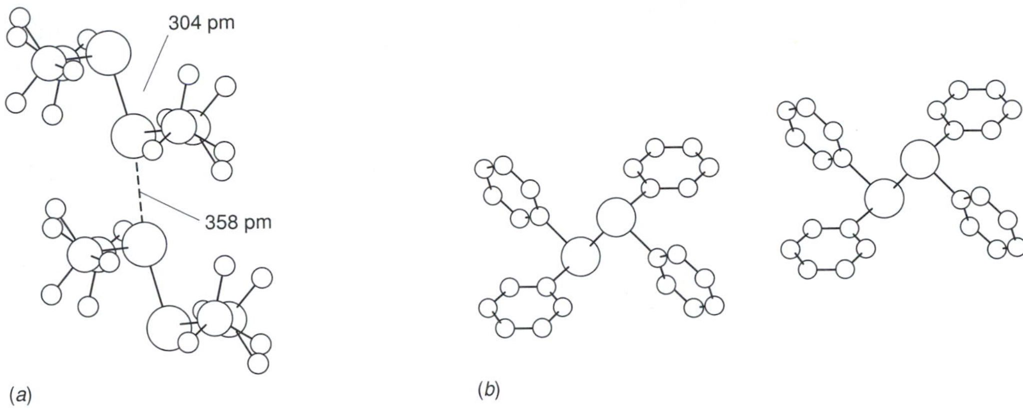

Figure 15.8
(*a*) Crystal structure of the thermochromic dibismuthine $(Me_3Si)_4Bi_2$, in which the covalent Bi–Bi bond length is 304 pm and the secondary Bi···Bi bond length is 358 pm; the corresponding numbers for $(Me_3Si)_4Sb_2$ are 287 and 399 pm. (*b*) Crystal structure of the nonthermochromic Ph_4Sb_2, in which the closest intermolecular Sb–Sb distances are 429 pm. [Adapted from A. A. Ashe III, *Adv. Organometal. Chem.*, **30**, 76 (1990).]

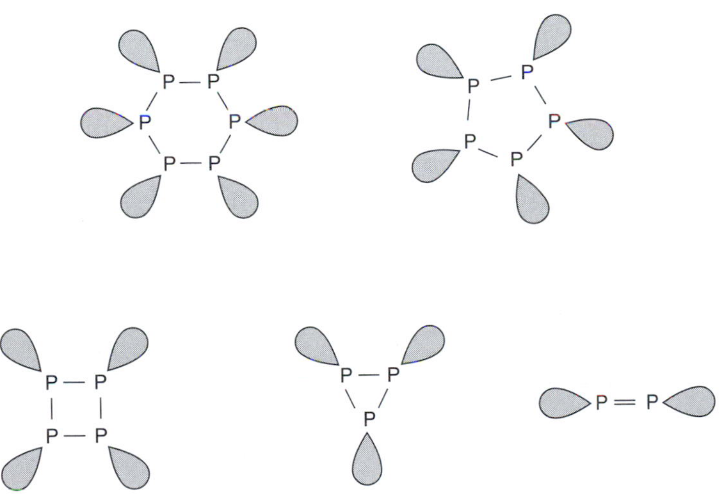

Figure 15.9
Representation of rings of cyclophosphines and -arsines (actual rings are often puckered), showing that bulky substituents (represented by large shaded ellipses) favor small ring sizes; the bulkiest favor the "two-membered" ring of a dimeric double-bonded species or ultimately a monomer.

oligoarsines are capable of acting as monodentate, bidentate, or tridentate ligands with *d*-block metal carbonyls; often coordination is accompanied by a change in the size of the ring. The six-membered arsenic ring $(t\text{-BuAs})_6$ is large enough that it can act as a macrocyclic ligand, complexing a neutral Ni or Pd atom in its center: $Ni(t\text{-BuAs})_6$.[42]

Group 13(III) Cluster Oligomers. Exact organic analogues of the numerous boron hydrides have not been made, but nonisoelectronic organometallic dimers and clusters have been made for most of the elements of this group. Perhaps least surprising are the dimeric $(R_2E)_2$ alkyls synthesized for E = Al and Ga and R = $CH(SiMe_3)_2$.[43] Tetrahedrane clusters $(RE)_4$ have been obtained for E = B, Al, and Ga, and In and R = CMe_3, CH_2CMe_3, and $C(SiMe_3)_3$[44]; and icosahedral $[CB_{11}Me_{12}]^-$ carboranyl and $[Al_{12}(CHMe_2)_{12}]^{2-}$ aluminate anions have been reported.[45]

15.10 *Multiply Bonded and Divalent Alkyls and Aryls of the Heavier *p*-Block Elements

Although multiply bonded allotropes of the *p*-block elements from the third period and below are known in high-temperature vapors (e.g., P_2, As_2, Sb_2, S_2, and S_3), compounds of these elements containing multiple bonds were unknown until comparatively recently,[46] although this was not for want of trying to duplicate, especially in silicon chemistry, the interesting and important chemistry of carbon-based alkenes, alkynes, and aromatic compounds. We note by comparing parts (A) and (B) of Table 3.1 (after omitting consideration of the anomalously weak σ N–N and O–O bonds) that π bonds are generally weaker than homoatomic σ bonds, so that addition, oligomerization, or polymerization reactions (analogous to the polymerization of $CH_2{=}CH_2$ to give polyethylene) are normally favored. But alkenes can be readily isolated and handled because a large activation energy is required to break the strong C–C π bond in preparation for its (noncatalyzed) polymerization to give σ bonded polymers.

Pi bonds are proportionately still weaker than σ bonds among the third-period elements and below than for the second-period element carbon, and require relatively small activation energies to undergo polymerization. Success in synthesis was achieved only when it was realized that, if the central atom of the element is small and/or the outer atoms of the compound are large, there can be sufficient steric hindrance to polymerization and to the approach of oxygen or other reactive molecules to allow some of these molecules to be synthesized.[47] However, if a silicon species is to contain a protected Si=Si bond, it can have at most only two other substituents (rather than three as in $R_3Si{-}SiR_3$), so to achieve the same degree of steric protection as is found in the polysilanes even larger organic groups must be used. To prepare a compound RSi≡SiR with a protected triple bond, the cone angle (steric bulk) of the one organic group on each Si would have to be close to 180°; to date no such triply bonded Group 14(IV) compounds have been synthesized.

The first bona fide **disilene**[48] (compound containing an Si=Si bond) was an $R_2Si{=}SiR_2$ in which R was the *mesityl* group, 2,6-$Me_2C_6H_3$–[49]; it and related disilenes have Si=Si bond lengths of about 215 pm that are 10% shorter than that of an Si–Si bond, which is similar to the shortening of the C=C bond relative to the C–C bond, and, at least in some of these compounds, the two Si atoms and the four nearest carbon atoms of the R groups are all coplanar, as in alkenes. Many such disilenes (or silenes containing Si=C bonds, which for brevity we are not considering here) use similar protective R groups such as the 2,4,6-(*t*-Bu)$_3C_6H_2$, the $(Me_3Si)_3C$, and the $(Me_3Si)_2CH$ groups; the **digermene** ($R_2Ge{=}GeR_2$) derivative of the latter R group has a Ge=Ge bond length of 235 pm, 4% less than a Ge–Ge bond length, while the **distannene** ($R_2Sn{=}SnR_2$) derivative of the same R group has an Sn=Sn bond length of 276 pm, only 2% shorter than the corresponding single bond; but in contrast to the case in ethylene, the central six atoms are not entirely coplanar.

Although these disilenes, digermenes, and distannenes are protected by their bulky R groups against oligomerization or polymerization, smaller reagents can reach the double bonds and add to them, as happens with alkenes: halogens and hydrogen halides add across the double bonds to give $R_2XSi–SiXR_2$ and $R_2HSi–SiXR_2$; alcohols also add to disilenes (but not to alkenes) to give $R_2HSi–SiR_2(OR')$. Further evidence of the intrinsic reactivity of the double bonds in disilenes, digermenes, and distannenes come from their intense yellow or orange colors, which are due to π–π^* electronic transitions. In simple alkenes, these transitions occur in the UV ($\sim$200 nm), so the alkenes are colorless, but in disilenes, they may occur at 400 nm, thus indicating that the energy separation of π and π^* molecular orbitals is only one-half as much in the disilenes as in the alkenes.

With still larger R groups (e.g., one mesityl and one 2,4,6-$(Me_3SiCHSiMe_3)_3C_6H_2$ group on each Si),[50] steric strain may develop between the R groups of one Si atom and those of another, leading to reversible dissociation of $R_2Si{=}SiR_2$ into two R_2Si: monomers, known as **silylenes.** Similarly, a digermene has been prepared with four R groups, each of which contains two Me_3Si groups {e.g., $[(Me_3Si)_2CH]_2Ge{=}Ge[CH(SiMe_3)_2]$}; when another Me_3Si group is added to one-half of the R groups, the Ge=Ge bond is irreversibly broken to give the **germylene** $[(Me_3Si)_2CH]\cdot[(Me_3Si)_3C]Ge$:.[51] The tendency to dissociate is perhaps even more apparent with Sn and even Pb, for which a monomeric diarylstannylene $[2,4,6\text{-}(CF_3)_3C_6H_2]_2Sn$: and dialkylplubylene $[(Me_3Si)_2CH]_2Pb$: and a diarylplumbylene $[2,4,6\text{-}(CF_3)_3C_6H_2]_2Pb$: have been prepared.[52] Here we must keep in mind the general periodic trend (Table B2) for the stability of the (group number minus two) oxidation state to increase down the *p*-block.

Similarly, **diphosphenes** RP=PR and **diarsenes** RAs=AsR have been prepared with the sterically protecting R groups; these have P=P and As=As bond lengths that are 8–10% shorter than single-bond distances, and may be compared with azobenzene, PhN=NPh. Phosphorus is, however, smaller than silicon, so it is less reluctant to engage in π bonding, as seen in the $R_3P{=}CH_2$ representation of the structure of the Wittig reagents. Indeed, relatively unprotected triple bonds of phosphorus and arsenic to carbon are known in the **phospha-alkynes** $Me_3C–C{\equiv}P$[53]: and 2,4,6-$(Me_3C)_3C_6H_2$–C≡P: and the arsa-alkyne 2,4,6-$(Me_3C)_3C_6H_2$–C≡As:.

It is possible to incorporate the heavier Group 15(V) atoms into aromatic rings (in place of one isolobal CH of benzene or one nitrogen atom of pyridine) to give the aromatically stabilized compounds **phosphabenzene** or **phosphinine** (*cyclo*-C_5H_5P:) and arsabenzene; the heavier analogues incorporating Sb and Bi can also be made but are of low stability.[54] From Group 13(III) **borabenzene** derivatives such as *cyclo*-$C_5H_5B{:}NEt_3$ and $[cyclo\text{-}C_5H_5BH]^-$ are known.[55]

There are a number of five-membered ring compounds containing a heteroatom analogous to the cyclopentadienide ion, *cyclo*-$C_4H_4CH^-$. From Group 14(IV) there are silolyl anions (silacyclopentadienides, *cyclo*-$C_4R_4SiR^-$) and germolyl anions; from Group 15(V) there are, in addition to the organic pyrrole, *cyclo*-C_4H_4NR, the phospholes, *cyclo*-$C_{5-n}H_{5-n}P_nR_n$ ($n = 1$–5) and their anions *cyclo*-$C_{5-n}H_{5-n}P_n^-$.[56] Many of these (and the corresponding phosphabenzene, etc.) are computed to have fairly substantial degrees of aromatic character,[57] although is not always enough to make the ring planar (like cyclopentadienide) rather than pyramidal at the heteroatom. Complexes of the phosphabenzenes, phospholes, and phospholide anions such as $[Cr(PC_5H_5)_2]$ and $[Cr(PC_5H_5)_6]$ are also known.[58]

Study Objectives

1. Based on periodic table position, classify the type of hydride (if any) an element is likely to form, and suggest a likely formula. Classify molecular hydrides as electron precise, electron rich, or electron deficient. (Review) Tell the name of a hydride given its formula, or vice versa. Assign oxidation numbers in a given hydride. Exercises 1–4.
2. Based on the Pauling electronegativity of an element, predict whether its hydride is likely to have an endothermic or an exothermic heat of formation or whether it is nonexistent; predict the polarity of the element–hydrogen bond. Exercises 5–8.
3. Use its classification and (if applicable) molecular size and hydrogen-bonding ability to predict whether a given hydride is likely to be a gas, a liquid, or a solid at room temperature, and whether it is likely to be flammable, toxic, strongly reducing, and acidic or basic. Exercises 9–14.
4. Assign numbers of framework electrons and classify the structure of a given borane or carborane as closo-, nido-, or arachno-; assign the structure of a given polyphosphane as chain (linear), cyclic, or polycyclic. Exercises 15–18.
5. Give a reasonable synthesis of a given hydride or hydride anion. Exercises 19–21.
6. Classify reactions of hydrides and predict relative tendencies of different hydrides to undergo those reactions. Exercises 22–26.
7. Based on periodic table position, classify the type of alkyl or aryl (if any) an element is likely to form, and suggest a likely formula; predict its most likely physical state. (Review) Tell the name of an alkyl or aryl given its formula, or vice versa; assign oxidation numbers in an alkyl or aryl. Exercises 27–32.
8. Give a reasonable synthesis of a given alkyl or aryl compound. Exercises 33–34.
9. Predict or classify reactions likely to occur for given alkyl or aryl compounds. Exercises 35–41.
10. Compare and explain the relative stability of (especially catenated) methyls and phenyls versus hydrides the same element; predict sensitivity of methyls and phenyls to heat, oxygen, and moisture. Explain any intense colors among these catenated organometallics. Exercises 42–48.
11. Describe the steric properties of alkyl and aryl substituents that, during oligomerization, favor formation of large ring oligomers, small ring oligomers, double-bonded dimers, or low-valent monomers. Describe some properties of the resulting products. Exercises 49–50.
12. Discuss the bonding properties and aromatic characteristics of aromatic ligands in which a carbon atom has been replaced with another nonmetal atom or with a *d*-block metal atom. Exercises 51–52.

Exercises

1. *Name and classify each of the following hydrides. If the hydride is molecular, also classify it as electron rich, and so on: (a) H_2S_5; (b) Si_2H_6; (c) B_2H_6; (d) $NaBH_4$; (e) AsH_3; (f) SnH_4; (g) PH_3; (h) B_4H_{10}; and (i) UH_3.

2. Name and classify each of the following hydrides. If the hydride is molecular, also classify it as electron rich, and so on: (a) HF; (b) N_2H_4; (c) CH_4; (d) TiH_2; (e) NH_2OH; (f) MgH_2; and (g) H_2O_2.

3. Write the formulas of, and classify, the following hydrides: (a) bismuthine; (b) stibine; (c) plumbane; (d) hydrazine; (e) hydroxylamine; (f) nonaborane(15); (g) pentagermane; (h) diborane; and (i) hexaphosphane(6).

4. For each of the hydrido anions listed for Groups 7–10, give its name, its central-atom oxidation number, the central-atom *d* electron count, and determine whether the complex ion obeys the 18(16)-electron rule. (a) octahedral $IrH_6{}^{3-}$; (b) octahedral $FeH_6{}^{4-}$; (c) five-coordinate $CoH_5{}^{4-}$, (d) square planar $PdH_4{}^{2-}$; (e) tricapped trigonal prismatic $ReH_9{}^{2-}$; and (f) linear $PtH_2{}^{2-}$.

5. *Predict which of the following elements do not form hydrides; which form thermodynamically stable hydride(s); which form thermodynamically unstable hydride(s). (a) C; (b) As; (c) O; (d) Br; (e) Xe; (f) Sr; and (g) Nd.

6. Give a likely formula of one hydride for each of the elements from Exercise 5 that forms a hydride, and indicate whether the hydrogen is likely to be present as an anion, a cation, or, if in a covalent bond, as a partially negatively charged atom, a partially positively charged atom, or as a neutral atom in a nonpolar covalent bond.

7. Predict which of the following elements do not form hydrides; which form thermodynamically stable hydride(s); which form thermodynamically unstable hydride(s). (a) Hf; (b) Fe; (c) Os; (d) Ge; (e) B; and (f) Tl.

8. Give a likely formula of one hydride for each of the elements from Exercise 7 that forms a hydride, and indicate whether the hydrogen is likely to be present as an anion, a cation, or, if in a covalent bond, as a partially negatively charged atom, a partially positively charged atom, or as a neutral atom in a nonpolar covalent bond.

9. *Predict whether each of the following hydrides is a gas, liquid, or a solid: (a) NaH; (b) HF; (c) HCl; (d) H_2S_5; (e) $Na^+B_3H_8{}^-$; and (f) SnH_4.

10. Predict whether each of the following hydrides is a gas, liquid, or a solid: (a) N_2H_4; (b) CuH; (c) PuH_2; (d) SiH_4; (e) Si_8H_{18}; (f) B_2H_6; (g) $NaBH_4$; and (h) UH_3.

11. Which of the following catenated hydrides are likely to be strongly hydrogen bonded, and consequently what physical state would you expect for each at room temperature: (a) H_2O_2; (b) H_2S_2; (c) N_2H_4; (d) P_2H_4; (e) HN_3 (hydrogen azide); (f) P_7H_3; and (g) B_2H_6.

12. Which one (if any) of each of the following pairs of compounds (as pure liquids) is more likely to show significant autoionization: (a) NH_3 or PH_3; (b) HF or HCl; (c) H_2O or H_2S; (d) N_2H_4 or P_2H_4; and (e) CH_4 or SiH_4.

13. *Classify each of the following boranes or boron hydride anions as either closo-, nido-, or arachno-, and tell what that implies about their structures: (a) $B_7H_7{}^{2-}$; (b) B_9H_{15}; (c) $B_3H_8{}^-$; (d) B_2H_6; (e) B_6H_{10}; (f) B_6H_{12}; and (g) $B_5H_8{}^-$.

14. Decide whether each of the following phosphines has a linear, cyclic, or polycyclic structure; draw a plausible Lewis structure to support your conclusion. (a) P_3H_5; (b) P_5H_5; (c) P_7H_3.

15. Below are shown some drawings of polyhedral borane structures containing 10 boron atoms. Classify each as nido, arachno, or closo. Give the formula of the boron hydride or boron hydride anion having that structure.

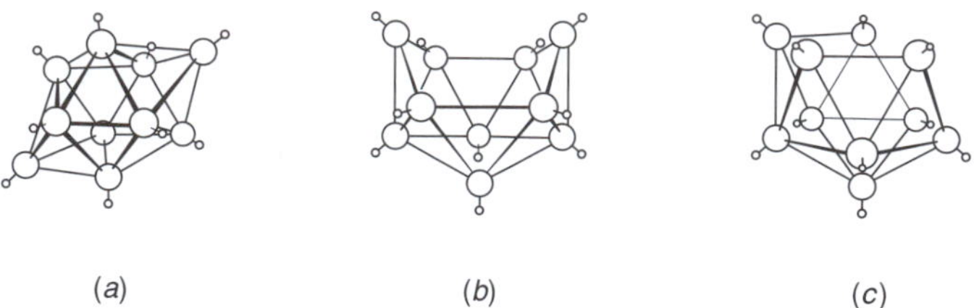

16. Below are shown some drawings of polyhedral borane structures containing six boron atoms. Classify each as nido, arachno, or closo. Give the formula of the boron hydride or boron hydride anion having that structure; tell how many framework electrons that structure has.

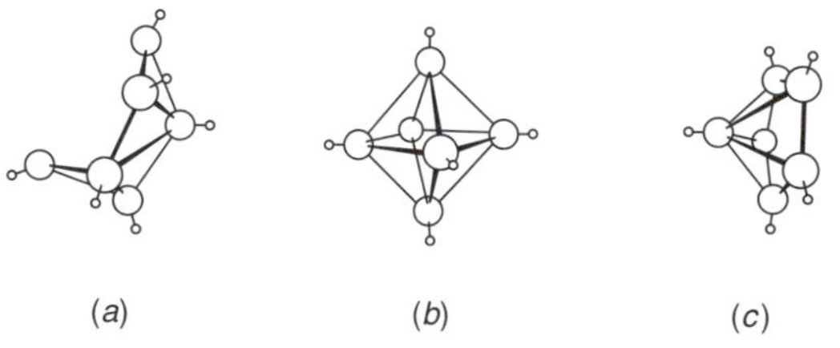

17. *Complete and balance an equation showing a reasonable synthesis of each of the following: (a) $LiAlH_4$; (b) HF; (c) NH_3; (d) N_2H_4; (e) H_2S_3; (f) H_2S_5; and (g) Si_2H_6.

18. Complete and balance an equation showing a reasonable synthesis of each of the following: (a) B_2H_6; (b) $NaBH_4$; (c) AsH_3; (d) SnH_4; (e) PH_3; (f) B_4H_{10}; (g) UH_3; and (h) TiH_2.

19. For each of the following hydrides or anions, select a reasonable method of synthesis from those mentioned at the beginning of Section 15.4 and write an equation showing a plausible reaction that might produce that hydride or anion: (a) $B_{10}H_{14}$; (b) H_2S; (c) $B_5H_8^-$; (d) Si_2H_6; (e) *ortho*-$B_{10}C_2H_{12}$; (f) $B_9C_2H_{11}^{2-}$; and (g) P_7H_3.

20. Rank each of the following series in order of increasing acidity: (a) HF, HCl, HBr, HI; (b) SiH_4, PH_3, SH_2, ClH; (c) $[Cr(SH_2)_6]^{3+}$, $[Cr(NH_3)_6]^{3+}$, $[Cr(OH_2)_6]^{3+}$; (d) $N_2H_6^{2+}$, N_2H_4, $N_2H_5^+$; (e) H_2O, H_2S, H_2Se, H_2Te; (f) NH_4^+, PH_4^+, AsH_4^+, SbH_4^+.

21. *Rank each of the following series in order of increasing basicity: (a) CH_4, NH_3, AsH_3, PH_3; and (b) NH_3, NH_2OH, NH_2NH_2.

22. Only one of the hydrides discussed in this chapter shows any marked properties as an oxidizing agent. Which one is this likely to be?

23. *In each of the following equilibria, tell whether the hydrogen is being transferred as H^+, $H\cdot$, or H^-; tell whether the equilibrium should favor products or reactants.

(a) $HF(aq) + I^-(aq) \rightleftharpoons HI(aq) + F^-(aq)$

(b) ${}^3_\infty[AlH_3](s) + BF_3(\ell) \rightleftharpoons {}^3_\infty[AlF_3](s) + 0.5\,B_2H_6(g)$

(c) $(CH_3)_3Sn\cdot + CH_4 \rightleftharpoons (CH_3)_3SnH + CH_3\cdot$

(d) $CH_3^- + HF \rightleftharpoons CH_4 + F^-$

24. Write the formula of the more likely type of product from the cleavage of 1 mol of diborane by 2 mol of the reactant specified: (a) NH_3; (b) $N(CHMe_2)_3$; and (c) PPh_3.

25. Complete the following addition reactions of hydrides by filling in the missing organic reactants.

(a) $B_2H_6 + ??? \rightarrow 2\,(Me_2CH\text{–}CH_2)_3B$

(b) $H_2N\text{–}NH_2 + ??? \rightarrow Me_2C\text{=}N\text{–}NH_2 + H_2O$

(c) $H_2N\text{–}OH + ??? \rightarrow$ *cyclo*-$C_6H_{10}\text{=}N\text{–}OH + H_2O$

(d) $PH_3 + ??? + HCl \rightarrow [P(CH_2OH)_4]Cl$

26. You are considering building a small rocket with sufficient O_2 as oxidizer and the following as fuels: (a) H_2; (b) C_2H_6; (c) B_2H_6; (d) PH_3; (e) N_2H_4; and (f) HN_3. You may use a total of 1.00 kg of fuel and oxidizer together. Which fuel will give you the most exothermic reaction per kilogram of load on the rocket? Take the products of combustion to be H_2O and (b) CO_2 ($\Delta H_f = -393\,kJ\,mol^{-1}$); (c) B_2O_3 ($\Delta H_f = -1272\,kJ\,mol^-$); (d) P_4O_{10} ($\Delta H_f = -2942\,kJ\,mol^{-1}$); and (e,f) N_2. In addition to doing the calculations, discuss any practical drawbacks of each fuel.

27. Name each of the following organometallic compounds: (a) $BiPh_3$; (b) $BiPh_5$; (c) $[Ph_4P]Cl$; (d) $[Et_4N]_3[CrCl_6]$; (e) $[PMe_4]Cl$; (f) $[Ni(PMe_3)_2Br_2]$; and (g) Al_2Ph_6.

28. *Write the formulas of the following compounds: (a) triphenylgermane; (b) dimethyldiphenylgermane; (c) hexamethyldialane; (d) dodecamethylcyclohexasilane; and (e) triphenylstibine.

29. Assign oxidation numbers to the central or metal atom in the following species: (a) $[PMe_4]Cl$; (b) $[Ni(PMe_3)_2Br_2]$ (assign Ni and P); and (c) Al_2Ph_6.

30. *Give the chemical formula of (a) a gaseous Group 13(IIIA) methyl compound; (b) a solid Group 16(VIA) methyl or phenyl compound; (c) a covalent methyl compound that has less than eight valence electrons about the central atom; (d) a covalent methyl compound that has more than eight valence electrons about the central atom; (e) any polymeric covalent methyl compound; and (f) any ionic methyl compound.

31. Tell whether each of the following organometallic derivatives is likely to be a gas, liquid, or a solid, and explain this fact in terms of the molecular, ionic, oligomeric, or polymeric structure of the organometallic compound: (a) Me_4Pb; (b) MeK; (c) $(MeLi)_4$; (d) $SiMe_4$; (e) PMe_3; and (f) BMe_3.

32. Classify each of the following organometallic compounds as electron deficient, electron precise, or electron rich: (a) $(MeLi)_4$; (b) $SiMe_4$; (c) PMe_3; and (d) BMe_3.

33. *Give reasonable syntheses of the following organometallic compounds: (a) MeMgI; (b) MeK; (c) ether-free Me_2Mg; (d) Me_2SiCl_2; (e) Et_3B; (f) Me_6Al_2; (g) $(PhCH_2)_3SnCl$; and (h) Me_2Hg.

34. Classify each of the following attempted syntheses, and predict whether the reactions will favor reactants or products:

(a) $3\,MeK + TlBr_3 \rightarrow Me_3Tl + 3\,KBr$

(b) $3\,MeK + Tl \rightarrow Me_3Tl + 3\,K$

(c) $C_6H_6 + Hg(C_2H_3O_2)_2 \rightarrow PhHg(C_2H_3O_2) + HC_2H_3O_2$

(d) $C_6H_6 + Ba(C_2H_3O_2)_2 \rightarrow PhBa(C_2H_3O_2) + HC_2H_3O_2$

(e) $5\,Me_3C_6H_2F + 10\,Au \rightarrow (Me_3C_6H_2Au)_5 + 5\,AuF$

(f) $Me_3C_6H_2I + Mg \rightarrow Me_3C_6H_2MgI$

(g) $Cl_3SiH + CH_2{=}CH_2 \rightarrow Cl_3Si{-}CH_2CH_3$

(h) $KH + CH_2{=}CH_2 \rightarrow KCH_2Me$

35. *Complete the following equations to show the products expected (if any) from (Lewis) acid–base reactions of the following pairs of substances: (a) $Me_3Tl + Me_3As$; (b) $Me_3Tl + Me_4Sn$; (c) $Me_3Tl + Me_2Se$; (d) $Me_6Al_2 + H_2O$; (e) $Me_6Al_2 + Me_2O$; (f) $Me_3As + MeI$; (g) $MeLi + Me_4Pb$; (h) $MeLi + Me_3In$; (i) $Me_5As + Me_3Ga$; and (j) $MeLi + HgCl_2$.

36. (Review) Give an example of a water-soluble and a water-insoluble salt for each of the following cations and anions; explain your choices: (a) Ph_4P^+ and (b) Ph_4B^-.

37. Organometallic halides are halides that can undergo hydrolysis in much the same manner as inorganic halides. In each pair, select the organometallic or organo-elemental halide that is more likely to undergo fast hydrolysis, and explain why: (a) Me_4Si or Me_3SiCl; (b) Me_3GeCl or $GeCl_4$; (c) Me_3CCl or Me_3SiCl; (d) Ph_4PCl or Ph_3PCl_2; (e) Me_2Zn or Me_2Hg; and (f) Me_2TlCl or Me_3PbCl.

38. Write an equation showing how each of the following might react with either an alkyllithium compound or an alkyl halide to achieve a more nearly electron-precise compound or ion: (a) $AsMe_3$; (b) BMe_3; (c) $PtMe_4$; (d) $(AuMes)_5$; (e) Me_6Al_2; and (f) ${}^{1}_{\infty}[BeMe_2]$.

39. *Suggest suitable organometallic/hydride adducts that could be pyrolyzed to yield thin films of the following semiconductors, writing equations for the pyrolyses: (a) InP; (b) CdS; (c) BP; and (d) CuI.

40. The compounds Me_3P, Me_3As, and Me_3Sb are very flammable; would you also expect them to be easily hydrolyzed? Why or why not?

41. Predict products of the following reactions. (a) $Me_2I^+ + OH^-$; (b) $Me_6Al_2 + 2\,PH_3$; and (c) pyrolysis of the product from (b).

42. A diversity of structures are possible for the Group 15(V) organoelement dihalides, R_3EX_2, especially given that the penultimate total coordination number of six may be reached for the larger E elements. List as many plausible structures as you can for this series of compounds.

43. *From each pair of compounds given, pick the one that will be most stable (thermally and/or to oxidation) and explain why: (a) Si_3H_8 or Si_3Me_8; (b) $AsMe_3$ or $AsPh_3$; (c) $GeMe_4$ or $AsMe_3$; (d) $CrEt_4$ or $Cr(CH_2CMe_3)_4$; (e) C_2Me_6 or C_2Ph_6; (f) PbH_4 or $PbMe_4$; (g) CH_4 or SiH_4; and (h) Si_4H_{10} or Si_4Me_{10}.

44. Briefly explain why, among the three compounds Ph_6C_2, Ph_6Si_2, and Ph_6Ge_2, you may expect the silicon compound to be the most stable.

45. Give a plausible explanation of the failure of all attempts to prepare fully alkylated or phenylated analogues of the closo-, nido-, and arachno-boron hydrides.

46. Of each pair of compounds, which one should absorb the longest wavelength light? (a) $Me(CH_2CH_2)_{20}Me$ or $Me(CH{=}CH)_{20}Me$; (b) C_8H_{18} or Sn_8Me_{18}; (c) Si_8Me_{18} or

Sn_8Me_{18}; (d) Me_4N_2 or Me_4Sb_2; (e) Me_4Bi_2(s) or Me_4Bi_2(l); and (f) Ar_6Si_6 or $Ar_{10}Sn_{10}$.

47. Name each of the following catenated organoelemental compounds: (a) Me_2S_3; (b) Me_4N_2; (c) $(MeP)_5$; (d) Me_4Bi_2; (e) Ge_5Me_{12}; and (f) $(PhSn)_{10}$.

48. Draw Lewis structures (showing also any empty valence orbitals) of the simplest possible fragments (structural units) from which the following catenated organometallics are assembled by oligomerization. How many prepared-for-bonding electrons does each of these structural units have? (a) R_4In_4; (b) R_4Ga_2; (c) *cyclo*-$R_{12}Si_6$; (d) R_6Ge_2; (e) $Ar_{10}Sn_{10}$; (f) R_6Ge_6; (g) R_5Sb_5; (h) $R_4Sb_2(\ell)$; and (i) R_4Sb_2(s).

49. If you carry out four reactions that assemble each of the four structural units $[2,4,6\text{-}Me_3C_6H_2]_2Sn$, $[2,4,6\text{-}(Me_2CH)_3C_6H_2]_2Sn$, $[2,4,6\text{-}(Me_3C)_3C_6H_2]_2Sn$, and $[2,4,6\text{-}(adamantyl)_3C_6H_2]_2Sn$ [adamantyl = tricyclic $(CH)_3(CH_2)_6C$] to different stages of oligomerization, which structural unit probably produces: (a) $R_{12}Sn_6$, a cyclohexastannane; (b) R_8Sn_4, a cyclotetrastannane; (c) $R_2Sn{=}SnR_2$, a distannene; (d) R_2Sn:, a stannylene?

50. Consider the three compounds $Ar_2C{=}CAr_2$, $Ar_2Si{=}SiAr_2$, and $Ar_2Sn{=}SnAr_2$. (a) Which should have its highest occupied molecular orbital (HOMO) at the highest energy of all? (b) Which should be the best electron donor? (c) Which is most likely to be colorless? (d) Which is most likely to be colored? (e) Which would be the softest base, based on classifications using HOMOs and LUMOs?

51. Predict the hapticity and nature of the metal–ligand bonding in the two phosphabenzene complexes mentioned in the text, $Cr(C_5H_5P)_2$ and $Cr(C_5H_5P)_6$, such that each complex can obey the 18-electron rule. Describe the likely approximate positions in the spectrochemical series of the phosphabenzene ligands in these two bonding modes.

52. By isoelectronic analogy with the known aromatic properties of benzene, pyridine, and arsabenzene, it was proposed that *metallabenzenes* of structure *cyclo*-$(L_4M)C_5H_5$, *cyclo*-$(L_2X_2M)C_5H_5$, and *cyclo*-$(L_3M)C_5H_5$ could be prepared. (a) M could be which *d*-block metals in each class of metallabenzene? (b) Consult the literature [J. R. Bleeke, *Acc. Chem. Res.*, **24**, 271 (1991), and references cited therein] to determine what members of these classes can in fact be made, how they are made, and to what extent they show aromatic properties.

Notes

1. K. M. Mackay, "Hydrides," in *Comprehensive Inorganic Chemistry*, Vol. 1, J. C. Bailar H. J. Emeléus, R. Nyholm, and A. F. Trotman-Dickenson, Eds., Pergamon Press, Oxford, UK, 1973.
2. J. N. Huiberts, R. Griessen, J. H. Rector, R. J. Wijngaarden, J. P. Dekker, D. G. de Groot, and N. J. Koeman, *Nature* (*London*), **380**, 231 (1996).
3. R. Dagani, *Chem. Eng. News*, 34, June 5, 1995, and papers cited therein; J. R. Huizenga, *Cold Fusion: The Scientific Fiasco of the Century*, Oxford University Press, Oxford, UK, 1993.
4. G. J. Miller, H. Deng, and R. Hoffman, *Inorg. Chem.*, **33**, 1330 (1994); M. Kritikos and D. Noréus, *J. Solid State Chem.*, **93**, 256 (1991); K. Kadir and D. Noréus, *J. Alloys Compounds*, **209**, 213 (1994); K. Kadir, M. Kritikos, D. Noréus, and A. F. Andresen, *J. Less-Common Metals*, **172–4**, 36 (1991); W. Bronger and G. Auffermann, *J. Alloys Compounds*, **228**, 119 (1995); W. Bronger, S. Hasenberg, and G. Auffermann, *Z. Anorg. Allg. Chem.*, **622**, 1145 (1996).
5. D. W. Smith, *J. Chem. Educ.*, **65**, 414 (1988); K. F. Purcell and J. C. Kotz, *Introduction to Inorganic Chemistry*, Saunders: Philadelphia, 1980; p. 188.
6. K. Vonnegut, *Cat's Cradle*, Gollancz, London, 1962.
7. F. Franks, *Polywater*, MIT Press, Cambridge, MA, 1981.
8. M. Baudler, *Angew. Chem. Intl. Ed. Engl.*, **26**, 419 (1987); M. Baudler and K. Glinka, *Chem. Rev.*, **94**, 1273 (1994).
9. A. J. Downs, M. J. Goode, and C. R. Pulham, *J. Am. Chem. Soc.*, **111**, 1936 (1989).
10. In practice in this case, the two added protons become terminal protons. To keep the framework electron count constant at $14e^- = 2n + 6$, we must pretend that these two hydrogens also stay with the framework.
11. X. Wang, M. Sabat, and R. N. Grimes, *J. Am. Chem. Soc.*, **116**, 2687 (1994).
12. J. Plesek, *Chem. Rev.*, **92**, 269 (1992).
13. A. Dequasie, *The Green Flame: Surviving Government Secrecy*, American Chemical Society, Washington DC, 1991; reviewed in *Chem. Eng. News*, Feb. 3, 1992, p. 52.
14. F. M. Bickelhaupt, N. J. R. van Eikema Hommes, C. F. Guerra, and E. J. Baerends, *Organometallics*, **15**, 2923 (1996).
15. E. M. Meyer, S. Gambarotta, C. Floriani, A. Chiesi-Villa, and C. Guastini, *Organometallics*, **8**, 1067 (1989).
16. G. D. Stucky, M. M. Eddy, W. H. Harrison, R. Lagow, H. Kawa, and D. E. Cox, *J. Am. Chem. Soc.*, **112**, 2425 (1990).
17. J. R. Baran, Jr., C. Hendrickson, D. A. Laude, Jr., and R. J. Lagow, *J. Org. Chem.*, **57**, 3759 (1992); A. Bretschneider-Hurley and C. H. Winter, *J. Am. Chem. Soc.*, **116**, 6468 (1994).
18. Y. Xie and H. F. Schaefer III, *Chem. Phys. Lett.*, **179**, 563 (1991); B. J. Smith, *Chem. Phys. Lett.*, **207**, 403 (1993).
19. K. N. Seneviratne, A. Bretschneider-Hurley, and C. H. Winter, *J. Am. Chem. Soc.*, **118**, 5506 (1996).
20. M. Kaupp and P. v. R. Schleyer, *J. Am. Chem. Soc.*, **115**, 1061 (1993); J. W. Kollis, *Chemtracts: Inorg. Chem.*, **5**, 62 (1993).
21. L. Ahmed and J. A. Morrison, *J. Am. Chem. Soc.*, **112**, 7411 (1990).
22. S. Wallenhauer and K. Seppelt, *Inorg. Chem.*, **34**, 116 (1995).
23. S. M. Godfrey, D. G. Kelly, C. A. McAuliffe, A. G. Mackie, R. G. Pritchard, and S. M. Watson, *J. Chem. Soc., Chem. Commun.*, 1163 (1991). In solution, this compound is $[IPPh_3]^+I^-$.
24. C. Pulham, A. Haaland, A. Hammel, K. Rypdal, H. P. Verne, and H. V. Volden, *Angew. Chem. Int. Ed. Engl.*, **31**, 1464 (1992); M. Kaupp, *J. Am. Chem. Soc.*, **118**, 3018 (1996); V. Pfennig and K. Seppelt, *Science*, **271**, 626 (1996).
25. W. A. Hermann, J. G. Kuchler, J. F. Felixberger, E. Herdtweck, and W. Wagner, *Angew. Chem. Int. Ed. Engl.*, **27**, 394 (1988); W. A. Hermann, W. Wagner, U. N. Weissner, U. Volkhardt, and H. Komber, *Angew. Chem. Int. Ed. Engl.*, **30**, 1636 (1991).
26. D. Naumann and W. Tyrra, *J. Chem. Soc. Chem. Commun.*, 47 (1989); H. J. Frohn and S. Jakobs, *J. Chem. Soc. Chem. Commun.*, 625 (1989); H. J. Frohn, S. Jakobs, and G. Henkel, *Angew. Chem. Intl. Ed. Engl.*, **28**, 1506 (1989).
27. R. W. Gedridge, Jr., D. C. Harris, K. T. Higa, and R. A. Nissan, *Organometallics*, **8**, 2817 (1989).

28. Y.-H. San, M. J. Heeg, and C. H. Winter, *Organometallics*, **13**, 3009 (1994).
29. C. Lambert and P. von R. Schleyer, *Angew. Chem. Int. Ed. Engl.*, **33**, 1129 (1994).
30. J. Krieger, *Chem. Eng. News*, Aug. 7, 1989, p. 6.
31. H. Tang, M. Parvez, and H. G. Richey, Jr., *Organometallics*, **15**, 5281 (1996); S. S. Al-Juaid, C. Eaborn, P. B. Hitchcock, K. Izod, M. Mallien, and J. D. Smith, *Angew. Chem. Intl. Ed. Engl.*, **33**, 1268 (1994).
32. R. J. Morris and G. S. Girolami, *J. Am. Chem. Soc.*, **110**, 6245 (1988).
33. P. J. Davidson, M. F. Lappert, and R. Pearce, *Acc. Chem. Res.*, **7**, 209 (1974); P. J. Davidson, M. F. Lappert, and R. Pearce, *Chem. Rev.*, **76**, 219 (1976); E. K. Byrne and K. H. Theopold, *J. Am. Chem. Soc.*, **111**, 3887 (1989).
34. S. Wallenhauer, D. Leopold, and K. Seppelt, *Inorg. Chem.*, **32**, 3948 (1993); A. H. J. F. de Keijzer, F. J. J. de Kanter, M. Schakel, R. F. Schmitz, and G. W. Klumpp, *Angew. Chem. Int. Ed. Engl.*, **35**, 1127 (1996).
35. L. R. Sita, K. W. Terry, and K. Shibata, *J. Am. Chem. Soc.*, **117**, 8049 (1995).
36. E. Hengge and R. Janoschek, *Chem. Rev.*, **95**, 1495 (1995).
37. N. Wiberg, C. M. M. Finger, and K. Polborn, *Angew. Chem. Int. Ed. Engl.*, **32**, 1054 (1993); S. Nagase, *Acc. Chem. Res.*, **28**, 469 (1995).
38. L. R. Sita, *Acc. Chem. Res.*, **27**, 191 (1994).
39. R. West, in *Encyclopedia of Inorganic Chemistry*, R. B. King, Ed., Wiley, Chichester, UK, 1994, p. 3389.
40. M. Birot, J.-P. Pillot, and J. Dunoguès, *Chem. Rev.*, **95**, 1443 (1995).
41. A. J. Ashe III, *Organometallics*, **1**, 1408 (1982); A. A. Ashe III, *Adv. Organometal. Chem.*, **30**, 77 (1990); L. L. Lohr and A. J. Ashe III, *Organometallics*, **12**, 343 (1993).
42. E. Hey-Hawkins, M. Pink, H. Oesen, and D. Fenske, *Z. Anorg. Allgem. Chem.*, **622**, 689 (1996).
43. W. Uhl, M. Layh, and T. Hildenbrand, *J. Organometal. Chem.*, **364**, 289 (1989).
44. T. Mennekes, P. Paetzold, R. Boese, and D. Bläser, *Angew. Chem. Intl. Ed. Engl.*, **30**, 173 (1991); E. P. Schram and N. Sudha, *Inorg. Chim. Acta*, **183**, 213 (1991); W. Uhl, W. Hiller, M. Layh, and W. Schwarz, *Angew. Chem. Intl. Ed. Engl.*, **31**, 1364 (1992); R. D. Schluter, A. H. Cowley, D. A. Atwood, R. A. Jones, and J. L. Atwood, *J. Coord. Chem.*, **30**, 25 (1993).
45. B. T. King, Z. Janousek, B. Grüner, M. Trammel, B. C. Noll, and J. Michl, *J. Am. Chem. Soc.*, **118**, 3313 (1996); W. Hiller, K.-W. Klinkhammer, W. Uhl, and J. Wagner, *Angew. Chem. Intl. Ed. Engl.*, **30**, 179 (1991).
46. N. C. Norman, *Polyhedron*, **12**, 2431 (1993).
47. See, for example, A. H. Cowley, *Acc. Chem. Res.*, **17**, 386 (1984).
48. R. West, *Angew. Chem. Intl. Ed. Engl.*, **26**, 1201 (1987).
49. R. West, M. J. Fink, and J. Michl, *Science*, **214**, 1343 (1981).
50. N. Tokitoh, H. Suzuki, R. Okazaki, and K. Ogawa, *J. Am. Chem. Soc.*, **115**, 10428 (1993).
51. P. Jutzi, A. Becker, H. G. Stammler, and B. Neumann, *Organometallics*, **10**, 1647 (1991).
52. H. Grützmacher, H. Pritzkow, and F. T. Edelmann, *Organometallics*, **10**, 23 (1991); S. Brooker, J.-K. Büijink, and F. T. Edelmann, *Organometallics*, **10**, 25 (1991).
53. J. F. Nixon, *Chem. Soc. Rev.*, **24**, 319 (1995).
54. A. J. Ashe, *Acc. Chem. Res.*, **11**, 153 (1978).
55. D. A. Hoic, J. R. Wolf, W. M. Davis, and G. C. Fu, *Organometallics*, **15**, 1315 (1996); D. A. Hoic, W. M. Davis, and G. C. Fu, *J. Am. Chem. Soc.*, **117**, 8480 (1995).
56. W. P. Freeman, T. D. Tilley, L. M. Liable-Sands, and A. L. Rheingold, *J. Am. Chem. Soc.*, **118**, 10457 (1996); F. Mathey, *Coord. Chem. Rev.*, **137**, 1 (1994).
57. J. A. Chamizo, J. Morgado, and P. Sosa, *Organometallics*, **12**, 5005 (1993); B. Goldfuss and P. v. R. Schleyer, *Organometallics*, **14**, 1554 (1995).
58. C. Elschenbroich, M. Nowotny, J. Kroker, A. Behrendt, W. Massa, and S. Wocadlo, *J. Organometal. Chem.*, **459**, 157 (1993).

CHAPTER
16

Inorganic Reaction Mechanisms

With Applications to Materials Science, Organic Chemistry, Environmental Chemistry, and Biochemistry

The field of study of the rate (kinetics) and mechanisms of inorganic reactions is a complex and challenging area that can only be surveyed in a book of this scope. First, it is based on experiments in which the concentrations of various reactants, products, and intermediates are determined using a variety of techniques. The rates of inorganic chemical reactions vary enormously with some *labile* species (Section 8.8) reacting in very short periods measured in nanoseconds (10^{-9} s), while some *inert* species may require years. Consequently, a large variety of physical methods must be employed in kinetics measurements, such as stopped-flow or other methods (Table 16.1), which are beyond the scope of this book. These topics are generally covered in textbooks of physical chemistry, or more specialized works on kinetics.

Table 16.1
Methods for Studying Kinetics of Reactions, with Their Time Scales in Seconds[a]

Method	Time scale
Laser flash photolysis	$\geq 10^{-14}$
Pulse radiolysis	$\sim 10^{-9}$
Mössbauer spectroscopy (^{57}Fe)	10^{-9}–10^{-6}
EPR (transition metals)	10^{-9}–10^{-8}
Temperature-jump spectrometry	$\geq 10^{-8}$
NMR (^{1}H)	$\sim 10^{-5}$
Chemical mixing	$\geq 10^{-3}$

[a]SOURCE: I. Bertini, H. B. Gray, S. J. Lippard, and J. S. Valentine, *Bioinorganic Chemistry*, University Science Books, Sausalito, CA, 1994; p. 335.

Second, once the measurements of concentrations have been obtained, there follows the process of interpreting them consistently and logically to try to develop a **rate law** for the reaction and to determine the associated rate constants that quantitatively characterize how fast a reaction goes, then to identify a **mechanism** for the reaction, and to classify the mechanism according to standard types. This process basically involves hypothesizing a number of possible mechanisms or sequences of **steps** or **ele-**

mentary reactions involved in a reaction, then showing that all but one are inconsistent logically with the measurements. As shall be shown, the process is not simple, and there is always the possibility that another hypothetical mechanism not yet considered may fit the data as well. So this is a demanding part of the work as well, the procedures of which we can only sketch incompletely. A number of mechanisms have been presented already in this text, with the more thoroughly developed examples including the study of exchange rates of coordinated water molecules and other ligands in Section 8.8, fluxionality in Section 9.3, organometallic catalysis in Sections 11.8 and 11.9, and catalysis by metals in Section 12.6. We will add a number of additional results in this chapter.

Third, there is the investigation of the rates of a number of different reactions that proceed with the same mechanism, with the goal of determining what happens electronically during the course of the reaction, or at least during its key step or steps. Most studies of mechanisms involve specific classes of reactions among just a few elements in one block of the periodic table, and do not consider separate blocks. The blocks of the periodic table have been segregated in mechanistic studies even in time: most kinetic and mechanistic work involving main group elements appears to have been done before the 1970s, while most work now being done involves *d*-block elements in complexes or especially in organometallics or metalloenzymes or their models. Since one goal of this textbook is to generalize, obtaining the broad view of periodicity in chemical reactivity across the whole periodic table, we will attempt to compare some results for reaction types that occur in different blocks across the periodic table. We begin with mechanisms of **ligand-substitution reactions**, which organic chemists refer to as **nucleophilic substitution reactions**, then follow with mechanisms for oxidation–reduction reactions.

16.1 Rate Laws and Stoichiometric Reaction Mechanisms for Ligand Substitution Reactions

Recall from general or physical chemistry courses that the rate law for a chemical reaction such as

$$aA + bB \rightarrow cC + dD \tag{16.1}$$

tells how the rate of appearance of a product such as C, $d[C]/dt$, or the rate of disappearance of a reactant such as B, $-d[B]/dt$, is related to the concentrations of the reactants A and B. Only on the occasion of a one-step reaction or the rate-limiting (slow) step of a multistep reaction can the rate law be written for this reaction by inspection:

$$\text{Rate} = (1/c)d[C]/dt = -(1/b)d[B]/dt = k[A]^a[B]^b \tag{16.2}$$

Since one does not know a priori whether the mechanism one is investigating will turn out to have just one step, the rate law of a reaction under investigation *cannot* be written by inspecting the balanced equation; it must be deduced by analyzing the measurements of the concentrations of a product or reactant as a function of time, and as one varies the concentrations of the different reactants.

Sometimes one finds that doubling the concentration of a reactant A doubles the rate at which the concentration of a product C increases, or doubles the rate at which another reactant B disappears. In this case, we say that the rate law is *first order* with respect to reactant A:

$$\text{Rate} = k[\text{A}]^1[\text{B}]^n \tag{16.3}$$

For other reactions, one finds that doubling the concentration of A quadruples the rate of the reaction; the rate law is then *second order* with respect to A:

$$\text{Rate} = k[\text{A}]^2[\text{B}]^n \tag{16.4}$$

In yet other occasions, one finds that doubling the concentration of A has no effect on the rate of the reaction; the rate law is then *zero order* with respect to A. Further experiments would then be undertaken to determine what effect, if any, doubling, tripling, and so on, the concentration of B has on the rate of the reaction, which would determine the value of n in the above rate laws. Then, the *overall order* of the reaction with respect to all reactants can be obtained by adding the orders with respect to the individual reactants. Finally, experimental concentrations can be substituted in the rate law of the correct form to obtain the value of the ***rate constant k***.

Most reactions, however, consist of more than one step. Let us consider a two-step possibility, using as an example a ligand-exchange reaction of an octahedral complex L_5MX, in which the first step is the dissociation of the ligand X:

$$L_5MX \xrightarrow{k_1} L_5M + X \tag{16.5}$$

This step is then followed by a second step in which a new ligand Y enters

$$L_5M + Y \xrightarrow{k_2} L_5MY \tag{16.6}$$

There are now two possibilities, the first of which is that Step (16.5) is the slow or rate-determining step. In this case, we can write the first-order rate law by inspection:

$$\text{Rate} = k_1[L_5MX]; \qquad k_1 = k \tag{16.7}$$

The second possibility is that Step (16.5) is part of a rapid equilibrium reaction that has had time for its reverse to be established

$$\text{Reverse rate} = k_{-1}[L_5M][X] \tag{16.8}$$

This expression is inconvenient, since it is expressed in terms of the concentration of a *reaction intermediate*, L_5M, that we did not add in known concentration, and that is probably either too short lived or present in too low a concentration to be measured. Hence, we must eliminate $[L_5M]$ by substitution.

To do this, we assume that the fast first reaction and its reverse justify the *steady-state approximation* that, by the time the second step is established, the concentration of the intermediate L_5M is neither increasing nor decreasing with time, so that $d[L_5M]/dt = 0$. All three reactions affect $[L_5M]$—(16.5) by increasing it and (16.6) and

(16.8) by diminishing $[L_5M]$; combining those effects, we can then write

$$d[L_5M]/dt = 0 = k_1[L_5MX] - k_{-1}[L_5M][X] - k_2[L_5M][Y] \quad (16.9)$$

By solving this for $[L_5M]$, we obtain

$$[L_5M] = \frac{k_1[L_5MX]}{k_{-1}[X] + k_2[Y]} \quad (16.10)$$

All concentrations on the right side of this expression are measurable. Hence, if we now assume that Step (16.6) is the slow step, and write a rate law based on it,

$$\text{Rate} = k_2[L_5M][Y] \quad (16.11)$$

we can convert this rate law (16.11) to a form with only measurable concentrations:

$$\text{Rate} = \frac{k_1k_2[L_5MX][Y]}{k_{-1}[X] + k_2[Y]} \quad (16.12)$$

The denominator of this equation contains a sum of two expressions, either of which may be much smaller than the other, and hence could be neglected. If $k_2[Y]$ is the smaller of the two terms (i.e., the second step is slower than the reverse of the first), the expression simplifies to

$$\text{Rate} = k_1k_2[L_5MX][Y]/k_{-1}[X] \quad (16.12a)$$

If $k_{-1}[X] \ll k_2[Y]$, then the above expression simplifies to

$$\text{Rate} = k_1[L_5MX] \quad (16.7)$$

which is the rate law we expected for the other mechanism, the one in which the first step is rate limiting but does not reach equilibrium. So precautions must be taken to eliminate false conclusions about the mechanism such as this.

A frequently occurring mechanism that can be deceptive is one in which, in a slow first step, the departing ligand is replaced, not by the incoming ligand, but by a solvent molecule, which is subsequently rapidly replaced by the incoming ligand. Since the concentration of solvent cannot be varied, the term in concentration of solvent ends up being absorbed into an apparent first-order rate constant.

Of course, three-step or even more complex mechanisms can be written and may indeed apply, which explains why very complicated rate laws may apply. And, as we saw in Section 11.8 for catalytic cycles, many reactions may proceed simultaneously through two or more mechanisms, in which case the rates again add up to give complex rate laws.

Classifications of Stoichiometric Mechanisms. Assuming that all of these difficulties can be overcome, a mechanism is then identified (by eliminating all alternatives), which consists of a series of steps, one of which is identified as rate limiting. This series of steps is called the **stoichiometric mechanism** of the reaction. Study of rate laws of ligand substitutions allows one to group most of the resulting stoichiometric mechanisms into one of three classifications: a dissociative, associative, or interchange mechanism.[1]

A **dissociative mechanism** (symbolized with a capital ***D***) involves a step in which an (at least somewhat stable and long-lasting) **intermediate** with a reduced coordination number is formed. That is, the leaving group (commonly symbolized as X in a reaction) has departed in Step (16.5) before the entering group (conventionally Y) attaches to the metal atom in Step (16.6).

An **associative mechanism** (symbolized with a capital ***A***) involves a step in which a somewhat persistent intermediate is formed, where the metal or central atom has a higher coordination number. A typical *A* stoichiometric mechanism has steps such as

$$[\mathrm{L_3M{-}X}] + \mathrm{Y} \underset{k_{-1}}{\overset{k_1}{\rightleftharpoons}} [\mathrm{Y{-}L_3M{-}X}] \tag{16.13}$$

$$[\mathrm{Y{-}L_3M{-}X}] \xrightarrow{k_2} [\mathrm{L_3M{-}Y}] + X \tag{16.14}$$

The rate law for such a mechanism works out to have the form:

$$\mathrm{Rate} = \frac{k_1[\mathrm{L_3MX}][\mathrm{Y}]}{k_{-1} + k_2} \tag{16.15}$$

An **interchange mechanism** (symbolized with a capital ***I***) takes place in one step without forming a fairly stable, durable intermediate: Instead an **activated complex** forms containing both the entering and leaving ligands, but this complex has very little or no stabilization energy, and almost immediately passes on to products or reverts to reactants. The activation complex is the most difficultly achieved point of highest energy during the passage from reactants to products. One might draw a structure of the activated complex connecting the two ligands X and Y to the metal M with broken rather than solid lines, to indicate that they are connected by less than a full covalent bond, and add a superscript such as ‡ and use a different type of bracket, to make it clear that this species is extremely ephemeral. So a typical *I* stoichiometric mechanism would have steps such as

$$[\mathrm{L_5M{-}X}] + \mathrm{Y} \rightarrow \{\mathrm{Y}\cdots\mathrm{L_5M}\cdots\mathrm{X}\}^{\ddagger} \rightarrow [\mathrm{L_5M{-}Y}] + \mathrm{X} \tag{16.16}$$

In practice, *I* stoichiometric mechanisms are by far the most commonly seen.

Although the transient nature of the activated complex precludes its observation, we have previously seen plausible models when we discussed the structure of hydrated ions (Section 4.3), which have two spheres of hydration: A more strongly held primary hydration sphere, and a secondary hydration sphere that very rapidly (in $\sim$1 ms) exchanges water molecules with the solvent water. It could be the case, in interchange mechanisms, that one ligand is present in the primary sphere, while the other is in the secondary sphere.[2] The *Eigen–Wilkins mechanism* envisions the very fast formation and breakup of such an "encounter complex" as reaching equilibrium (with an equilibrium constant *K*) before any other steps occur:

$$\mathrm{L_5MX} + \mathrm{Y} \rightleftharpoons (\mathrm{L_5MX, Y}) \tag{16.17}$$

This step is then followed by the slow, difficult interchange of the X and Y ligands between the first and second coordination spheres:

$$(L_5MX, Y) \xrightarrow{k} (L_5MY, X) \tag{16.18}$$

During this rate-determining step, we might imagine the entering ligand squeezing into the (probably already crowded) primary hydration sphere while the leaving group has to overcome its attraction to the metal ion in order to get out of the primary and into the secondary hydration sphere. Finally, the leaving ligand rapidly diffuses out of the second coordination sphere to give the product:

$$(L_5MY, X) \xrightarrow{\text{fast}} L_5MY + X \tag{16.19}$$

The rate law for this works out to be

$$\text{Rate} = \frac{kK[L_5MX]_0[Y]}{1 + K[Y]} \tag{16.20}$$

It is helpful that the equilibrium constant K for formation of the encounter complex (L_5MX,Y) can be estimated using the *Fuoss–Eigen equation:*

$$K = 4/3\pi a^3 N_A e^{-V/RT} \tag{16.21}$$

where N_A is Avogadro's number and V is the Coulombic potential energy of attraction of the complex cation and the Y anion at their closest distance of approach a.[3]

Although the rate laws for the three types of stoichiometric mechanisms, *D*, *A*, and *I*, are different, in practice it can be difficult to distinguish them if one of the terms in the denominator is undetectably small. Table 16.2 indicates the limiting forms that the rate laws take depending on the relative magnitude of the denominator term in [Y]. It can be seen that, if one has been unable to explore both very high and very low concentrations of the entering Y ligand, observation of an overall first-order rate law could mean either a *D* or an *I* stoichiometric mechanism, and observation of an overall second-order rate law could mean either an *A* or an *I* stoichiometric mechanism.

Table 16.2
Limiting Forms of Rate Laws Depending on Magnitude of Denominator Term in [Y][a,b]

Stoichiometric Mechanism	Very Large [Y] Term	Very Small [Y] Term
D	$k_1[L_5MX]$	$\dfrac{k_1k_2[L_5MX][Y]}{k_{-1}[X]}$
A	$\dfrac{k_1[L_5MX][Y]}{k_{-1}+k_2}$	$\dfrac{k_1[L_5MX][Y]}{k_{-1}+k_2}$
I	$k[L_5MX]$	$kK[L_5MX][Y]$

[a] For example, $k_2[Y]$, $K[Y]$, [Y].
[b] Adapted from B. E. Douglas, D. H. McDaniel, and J. J. Alexander, *Concepts and Models of Inorganic Chemistry*, 3rd ed., Wiley, New York, 1994; p. 498.

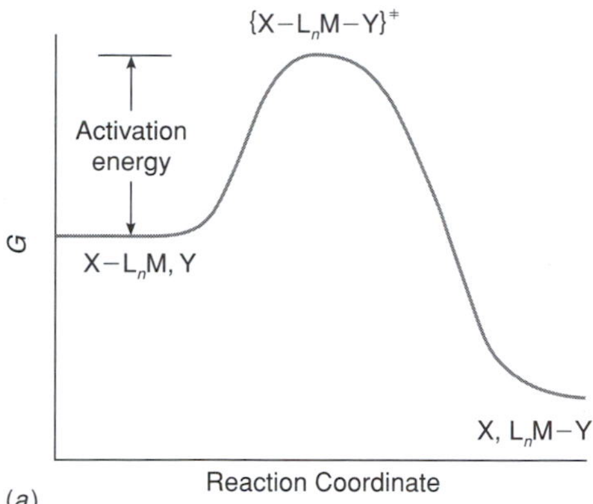

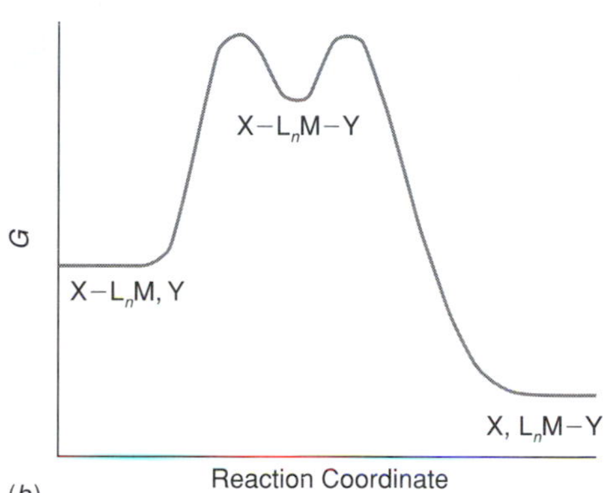

Figure 16.1
Reaction profiles for a substitution reaction of $[ML_nX]$ with Y to give $[ML_nY]$ and X (*a*) by an interchange (*I*) mechanism and (*b*) by an associative (*A*) or dissociative (*D*) mechanism.

Sometimes sensitive or fast physical methods can detect intermediates such as $[L_5M]$ or $[L_3MXY]$, which provides positive evidence for either the *D* or *A* mechanism that involves such an intermediate; the *I* alternative mechanism involves only an undetectable, rapidly exchanging activated complex. Or perhaps the proposed reaction intermediate can be isolated or detected under other, more favorable conditions: The isolation of the large, five-coordinate anion $[Ni(CN)_5]^{3-}$ from solution using a correspondingly large +3 charged cation (Section 4.8) rendered more plausible the suggestion that this was the intermediate in the exchange reaction between free CN^- and the $[Ni(CN)_4]^{2-}$ ion.

We can represent the free energy differences between *I* and *A* or *D* mechanisms as they proceed by drawing **reaction profiles** (Fig. 16.1). In either type of reaction, the energy of the reactant or reactants must increase in order to distort the existing stable bonds (by increasing their vibrations), until finally enough energy, the **activation energy,** is supplied to reach the "top of the hill", an activated complex in which existing bonds are stretched enough that the new bonds, or new bond angles, can form. In an *I* mechanism, this activated complex either reverts back to the original reactants, or it proceeds onward to give the final products [Fig. 16.1(*a*)]. In an *A* or *D* mechanism there is a stopping point along the way to products: the intermediate, which is in a high "valley" in the energy profile [Fig. 16.1(*b*)]. How long the intermediate lasts before enough energy is available to drive it over the "hill" to the left back to reactants or over the "hill" to the right on to products will vary depending on how high the energy walls around the intermediate are.

16.2 Intimate Mechanisms for Ligand Substitution Reactions

Other kinetic studies of ligand substitution reactions have a different focus: Instead of focusing on changing the concentrations of reactants in one specific reaction, they study one type of reaction, and vary the specific ligands, or metal ions, involved. The purpose of this type of study is not to develop the sequence of steps involved, hence not to determine the stoichiometric mechanism, but rather to reveal the **intimate mechanism** of the reaction: to develop a more detailed picture of the electronic nature of the reaction during the key rate-limiting step. The kinetic data for such a study can involve controlled changes in the electronic natures of the leaving ligands X, the entering ligands Y, the "spectator ligands" L, and the central metal atom M. These studies allow an independent classification of the associative or dissociative nature of a reaction.

Thus, if we use a wide variety of leaving groups X with a common entering group, we may find that the nature of the leaving group makes a large difference in the values of the rate constants for this type of reaction, as shown in Table 16.3 for the substitution of water for the X ligand in the cobalt(III) complexes $[Co(NH_3)_5X]^{n+}$. This finding would lead us to infer that a key factor in determining the rate of this type of reaction is the strength of the M–X bond being broken.

Table 16.3
Rate Constants for Acid Aquation of Some Octahedral Complexes of Co(III) at 25 °C[a]

Complex	k (s^{-1})	Complex	k (s^{-1})
$\{Co(NH_3)_5[OP(OMe)_3]\}^{3+}$	2.5×10^{-4}	$[Co(NH_3)_5Cl]^{2+}$	1.8×10^{-6}
$[Co(NH_3)_5(NO_3)]^{2+}$	2.4×10^{-5}	$[Co(NH_3)_5(SO_4)]^{+}$	8.9×10^{-7}
$[Co(NH_3)_5I]^{2+}$	8.3×10^{-6}	$[Co(NH_3)_5F]^{2+}$	8.6×10^{-8}
$[Co(NH_3)_5(H_2O)]^{3+}$	5.8×10^{-6}	$[Co(NH_3)_5N_3]^{2+}$	2.1×10^{-9}
		$[Co(NH_3)_5(NCS)]^{2+}$	3.7×10^{-10}

SOURCE: Data from M. L. Tobe, *Adv. Inorg. Bioinorg. Mech.* **2**, 1 (1984).

Such evidence would be supported by another study of the effect of the entering group Y if that study found little effect of varying Y on the values of the rate constants, as, for example, shown in Table 16.4 for the substitution of a Y anion for water as the leaving group in the complex ion $[Co(NH_3)_5(H_2O)]^{3+}$. This would lead us to infer

Table 16.4
Rate Constants for Substitution Reactions of $[Co(NH_3)_5(H_2O)]^{3+}$ with Y^{n-} at 45 °C

Y^{n-}	k (s^{-1})
H_2O	10.0×10^{-5}
N_3^-	10.0×10^{-5}
SO_4^{2-}	2.4×10^{-5}
Cl^-	2.1×10^{-5}
NCS^-	1.6×10^{-5}

SOURCE: Data from R. G. Wilkins, *The Study of Kinetics and Mechanisms of Reactions of Transition Metal Complexes*, Allyn & Bacon, Boston, 1974.

that the strength of the bond being formed between the incoming Y ligand and M is immaterial to the rate, because the bond is formed after the rate-determining step. Both of these pieces of evidence finally lead us to classifying the rate-determining step of this kind of ligand substitution reaction as **dissociative**. Classifications of intimate mechanisms are symbolized using lower case letters; thus this would be a ***d*** intimate reaction mechanism.

In other cases, we study the effect of the entering Y ligands and find that the rate of substitution is strongly influenced by the nature of the entering Y ligand, as seen in Table 16.5 for the replacement of one water molecule in $[Ti(H_2O)_6]^{3+}$ by diverse ligands Y. In such a case, we deduce that the nature of the M–Y bond being formed is affecting the rate-determining step, so that this step is **associative (*a*)**.

Table 16.5
Rates of Substitution of $[Ti(H_2O)_6]^{3+}$ by Y^{n-} at 13 °C

Y^{n-} ($n = 0$)	k ($M^{-1}s^{-1}$)	Y^{n-} ($n = 1$)	k ($M^{-1}s^{-1}$)
$ClCH_2CO_2H$	6.7×10^2	NCS^- (8–9 °C)	8.0×10^3
$MeCO_2H$	9.7×10^2	$ClCH_2CO_2^-$	2.1×10^5
H_2O	8.6×10^3	$MeCO_2^-$	1.8×10^6

SOURCE: Data from H. Diebler, *Z. Phys. Chem.* **68**, 64 (1969).

Ideally, we would like to have a reaction classified both stoichiometrically and intimately. When we do this, we find that, whereas an associative reaction in the stoichiometric sense involves an associative rate-determining step in the intimate sense (hence, the label A is kept), and likewise for a dissociative reaction (which keeps the label D), it is possible to subdivide the category of interchange mechanisms. Although these involve both X and Y in the activated complex, the degree of association of the incoming ligand Y in the activated complex can either be quite significant or rather insignificant. If the association of Y with M in (the primary sphere of) the activated complex is significant, so that the identity of Y strongly affects the rate, then the intimate classification is a, and overall the mechanism is labeled I_a. If the association of Y with M (in the primary sphere) has not proceeded very far in the activated complex, so that the rate is insensitive to the identity of Y, but is in fact more sensitive to the identity of X, which has begun to leave the primary sphere, then the intimate classification is d, and the mechanism is labeled I_d.

Thus, if both stoichiometric and intimate mechanisms have been determined, we can select the proper choice from the four overall classes of mechanisms: A; D; I_a; and I_d.

Periodic Trends in Intimate Mechanism Type. Associative (I_a or especially A) mechanisms increase the total coordination number of the central atom in the intermediate or activated complex, whereas dissociative (I_d or especially D) mechanisms decrease the total coordination number. If the central atom in the starting reactant already has its maximum total coordination number, we would be most surprised if it adopted an A mechanism, which would not only give it an extraordinarily high total coordination number, but require it to keep that high number for a period of time in a fairly stable intermediate. Thus, the characteristic substitution reactions of tetracoordinate carbon in organic compounds are either I_d or I_a, and certainly not A; I_d

or D mechanisms have been observed for tetrahedral boron (BH_4^-) and beryllium (BeF_4^{2-}). In the third period, where a higher maximum total coordination number prevails, A as well as I_a mechanisms can be found for tetrahedral Si and P (tetrahedral phosphorus also shows some D mechanisms).[4]

In the fourth period, associative (especially I_a) mechanisms are, as expected, common for square planar complexes, which can add a ligand (with initial bonding being electron donation to the empty metal p_z orbital) to give a five-coordinate intermediate or activated complex without exceeding the maximum coordination number of 6 for that period. These mechanisms would be expected to be, and are, less common for octahedral complexes in this period. However, associative mechanisms are known for some of the early, large fourth-period d-block metals, and throughout the fifth and sixth periods, where the resulting seven-coordinate intermediates or activated complexes do not exceed the maximum coordination number.

Dissociative (D or more commonly I_d) mechanisms would be expected to be, and are, more common for octahedral complexes of fourth-period d-block metals (the resulting intermediate is five-coordinate) than for square planar complexes. For dissociating d^8 square planar complexes, the intermediates would be three-coordinate 14-electron species; nonetheless some such mechanisms are known, particularly for bulky ligands.

A summary for many metal ions (including a few from the p block) is shown in Table 16.6; you should be able to rationalize some of the exceptions.

Table 16.6
Main Intimate Mechanisms for Substitutions of Octahedral Complexes[a]

Mg^{2+}										Al^{3+}
	TiO^{2+}, $\mathbf{Ti}^{3+}$	VO^{2+}, $\mathbf{V}^{2+}$, $\mathbf{V}^{3+}$	$\mathbf{Cr}^{3+}$	Mn^{2+}	Fe^{2+}, Fe^{3+}	Co^{3+}, Co^{2+}	Ni^{3+}, Ni^{2+}	Cu^{2+}	Zn^{2+}	Ga^{3+}
			$\mathbf{Mo}^{3+}$		Ru^{2+}, Ru^{3+}	Rh^{3+}			$\mathbf{Cd}^{2+}$	$\mathbf{In}^{3+}$
						Ir^{3+}				

SOURCE: Data from B. E. Douglas, D. H. McDaniel, and J. J. Alexander, *Concepts and Models of Inorganic Chemistry*, 3rd ed., Wiley, New York, 1994; p. 503.

[a]Ions showing mainly **associative** mechanisms are indicated in **boldface**; those showing both types of intimate mechanisms are in normal type; those showing mainly *dissociative* mechanisms are indicated in *italics*.

Distinguishing Mechanism Types. There can be difficulties in distinguishing the A class of mechanism from the I_a, and in distinguishing the D class of mechanism from the I_d. The distinction can be made when the intermediates characteristic of the A and D classes last long enough to be detected in one way or another. The methods that can be used include (a) some sensitive, fast spectroscopic method; (b) the isolation or detection of the intermediate under some other (more favorable) reaction conditions; (c) discovering the products of side reactions of the intermediate with other materials in the solution; and (d) discovery that the intermediate undergoes a *cis–trans* or other isomerization.

As an example of the latter, it is characteristic for organic S_N2 reactions at tetrahedral carbon to proceed with inversion of configuration [Fig. 16.2(a)]: the pentacoordinated activated complex $\{Y\cdots R_3C\cdots X\}^{\ddagger}$ does not persist long enough to change

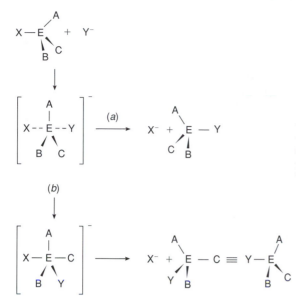

Figure 16.2
Alternate pathways for associative substitution reactions at tetrahedral central atoms. (*a*) Pathway for inversion of configuration. (*b*) Pathway for retention of configuration following a Berry pseudorotation of the trigonal bipyramidal intermediate.

its stereochemistry. On the other hand, if the central atom is the larger Si, this substitution reaction often proceeds partly with retention of configuration, since the five-coordinate intermediate is stable for long enough to undergo the Berry pseudorotation type of fluxional behavior (Section 9.3), which interchanges axial and equatorial ligands, and hence may interchange the entering group and one of the spectator ligands. Hence, we would call this an *A* mechanism; this conclusion is supported by the actual isolation (with large nonacidic counterions) of a number of five-coordinate silicon anions such as SiF_5^-, $[ArSiF_4]^-$, and $[Ar_2SiF_3]$.[5]

Similarly, the association of a ligand Y with a *d*-block square planar complex of stoichiometry $[PtTC_2X]$ (T = the ligand trans to the leaving group X; C = one of the two ligands cis to it, Figure 16.3) generates an intermediate or activated complex with a trigonal bipyramidal structure in which the two C ligands now occupy axial positions. In an I_a mechanism, this is an activated complex from which X will depart very

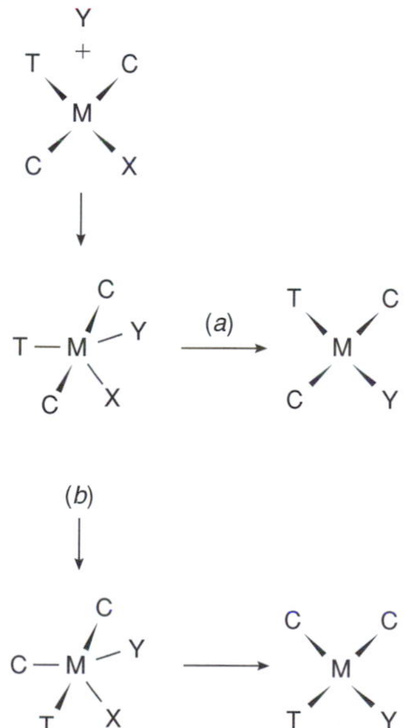

Figure 16.3
Alternate pathways for associative substitution reactions at a square planar metal ion. (*a*) Pathway for retention of original (trans) geometry. (*b*) Pathway for isomerization to other (cis) geometry following a Berry pseudorotation of the trigonal bipyramidal intermediate.

rapidly, leading to the same isomeric form of $[PtTC_2Y]$ (viz., cis or trans) as was found in the starting material. But in an *A* mechanism, this intermediate may last long enough to undergo Berry pseudorotation, which interchanges C and T ligands. In this case, we will see some cis products from trans starting materials, and vice versa.

Other methods have been devised and applied recently to assist in the identification of the intimate mechanism of reactions. One of these is based on the study of rates as a function of temperature, and leads to the *entropy of activation* $\Delta S^\ddagger$ of the reaction. The theory is that an associative mechanism reduces the number of particles in solution, hence it reduces disorder and gives a strongly negative $\Delta S^\ddagger$, characteristically below $-10\,J\,mol^{-1}\,K^{-1}$. A strongly positive $\Delta S^\ddagger$, characteristically above $+10\,J\,mol^{-1}\,K^{-1}$, is expected for a dissociative mechanism: For example, $+24.4\,J\,mol^{-1}\,K^{-1}$ for the reaction of $[Cr(CO)_4PPh_3(P(OPh)_3)]$ with CO. Intermediate values must be treated with caution, especially for polar solvents and charged metal complexes that can have secondary hydration or solvation spheres that are disturbed during the reaction.[6]

The other method is based on the study of rates as a function of pressure, and leads to the *volume of activation* $\Delta V^\ddagger$ of the reaction; an associative reaction should lead to a decrease in total volume in the intermediate or activated complex, as the incoming ligand sheds its hydration layer. As an example, the volumes of activation for exchange of coordinated water among fourth-period +2 charged *d*-block metal ions vary from negative values at the left $\{-4.1\,cm^3\,mol^{-1}$ for $[V(H_2O)_6]^{2+}$ and $-5.4\,cm^3\,mol^{-1}$ for $[Mn(H_2O)_6]^{2+}\}$ to positive values at the right $\{+3.8\,cm^3\,mol^{-1}$ for $[Fe(H_2O)_6]^{2+}$, $+6.1\,cm^3\,mol^{-1}$ for $[Co(H_2O)_6]^{2+}$, $+7.2\,cm^3\,mol^{-1}$ for $[Ni(H_2O)_6]^{2+}\}$, suggesting a gradual change from associative to dissociative mechanisms as the metal ions become smaller.

The difficulty with interpreting these values lies in the aqueous solvent effects that we saw in Sections 4.3 and 4.4: The two ligands are also breakers or makers of the structure that exists in the water itself, which can result in extraneous volume changes. So the debate[7] is still going on concerning the reliability of the conclusions drawn on the basis of volumes of activation. For example, a recent computational study of water exchange mechanisms for fourth-period *d*-block ions has supported a dissociative mechanism for ions with +2 charge, in contradiction to the conclusions based on $\Delta V^\ddagger$ values, but has supported associative mechanisms for ions with +3 charge.[8]

Example 16.1

Ligand exchange reactions of dialkyl chalcogenides ER_2 (E = S, Se, and Te; R = Me and Et) with $[TaCl_5(ER_2)]$ in inert solvents are very much slowed down when R is Et rather than Me, and show very negative entropies of activation ($\Delta S^\ddagger$ from −96 to $-108\,J\,mol^{-1}\,K^{-1}$). Is this reaction dissociative or associative?

SOLUTION:

They very negative $\Delta S^\ddagger$ values argue for increasing order in the activated complex, hence an associative mechanism. (Since an inert solvent and uncharged reactants were used, there should be little possibility of complications from multiple solvation spheres). The dependence of the rate on the identity of the incoming ligand also argues for an associative mechanism: The lower rate with R = Et could be a consequence of steric hindrance in the activated complex.

16.3 Entering Group, Leaving Group, and Metal Effects on Ligand Substitution Rates

Entering Groups and Nucleophilicity. The rate constants of many associative (I_a or A) reactions have been determined for a variety of incoming ligands at inert square planar Pt(II) centers, and it proves roughly possible to rate the entering ligands into an order of increasing rates of nucleophilic substitution at a given central atom:

$$Y : H_2O < Cl^- < I^- < H^- < PR_3 < CO, CN^- \tag{16.22}$$

If the rate constants for substitution reactions of other Y ligands on a specific Pt complex are compared with the rate constant for substitution by MeOH on the same Pt complex, it is possible to quantify this series, determining a list of **nucleophilicity parameters** n_{Pt} (Table 16.7). Note that the above series and these nucleophilicity parameters are in order of increasing softness: soft bases tend to be good nucleophiles, that is, undergo rapid A or I_a substitution reactions. Note also what a profound effect the nature of the entering group can have on the rate constants in this table: They vary by a factor of almost 1 billion (10^9)!

Leaving Groups. Likewise, the rate constants of many dissociative (I_d or D) reactions have been determined for a variety of leaving groups, and it proves roughly possible to rate these X groups into an order of increasing rates of nucleophilic substitutability at a given central atom:

$$X : H_2O > Cl^- > I^- > H^- > PR_3 > CO, CN^- \tag{16.23}$$

Table 16.7
Rate Constants for Ligand Displacement in Some Square Planar Pt(II) Complexes[a]
trans-[Pt(pyridine)$_2$Cl$_2$] + Y → *trans*-[PtCl(pyridine)$_2$Y] + Cl$^-$

Y	k_2 (M^{-1}s^{-1})	n_{Pt}
MeOH (25 °C)	2.7×10^{-7}	0.00
MeO$^-$ (25 °C)	Very slow	<2.4
$N(CN)_2^-$	3.03×10^{-4}	2.87
Cl^-	4.5×10^{-4}	3.04
NH_3	4.7×10^{-4}	3.07
N_3^-	1.55×10^{-3}	3.58
I^-	1.07×10^{-1}	5.46
$(Me_2N)_2CS$	0.30	5.87
$(MeNH)_2CS$	2.5	6.79
$(PhNH)_2CS$	4.13	7.01
CN^- (25 °C)	4.00	7.14
PPh_3 (25 °C)	249	8.93

SOURCE: From U. Belluco, L. Cattalini, F. Basolo, and A. Turco, *J. Am. Chem. Soc.* **87**, 241 (1965); R. G. Pearson, H. Sobel, and J. Songstad, *J. Am. Chem. Soc.* **90**, 319 (1968); M. Becker and H. Elias, *Inorg. Chim. Acta* **116**, 47 (1986).

[a] In MeOH, at 30 °C except as noted.

Clearly, this list was obtained by reversing the above Y series: Good entering groups (good nucleophiles, soft bases) make poor leaving groups. Again, the nature of the leaving group can have a profound effect on the rate of a substitution reaction; chemists are beginning to appreciate the value of hard nonbasic leaving ligands such as $CF_3SO_3^-$ (trifluoromethanesulfonate) in starting materials, since these groups can be quickly replaced by other ligands in substitution reactions.[9]

The leaving ability of a ligand bearing unshared pairs of electrons can also be enhanced by adding the appropriate hard or soft acid. It has been found that basic, hard ligands such as F^- and CO_3^{2-} have their leaving rates enhanced if hard acids (H^+, Be^{2+}, and Al^{3+}) are present. These hard acids attach to the lone pairs and begin to convert the ligands to less charged forms such as HF and HCO_3^-, which are less strongly bound to the metal atom. Likewise, borderline and soft ligands such as Cl^- or I^- can have their leaving rates greatly enhanced when they bridge to added soft acid ions such as Ag^+ or Hg^{2+}:

$$Y^- + [L_5MI]^+ + Ag^+ \rightarrow \{Y\cdots L_5M\cdots I\cdots Ag\}^+ \rightarrow [YML_5]^+ + AgI(s) \qquad (16.24)$$

Rates of hydrolysis of the tetrahedral phosphorus atom in phosphate esters and thioesters such as DNA, pesticides, and nerve gases are of obvious practical importance; many important enzymes use two metal ions to assist this hydrolysis. A recent model compound study[10] has achieved a spectacular 10^{13}-fold enhancement of this rate by attaching two La^{3+} ions in a complex designed so that one hard La^{3+} ion is coordinated to the hard oxygen base of the poor alkoxide leaving group, enhancing its leaving ability, while the other La^{3+} group is coordinated both to a terminal phosphonate oxo group and to a water molecule, using the template effect (Section 5.10) to locate the water molecule near to the phosphorus center, and hence enhancing its rate of attack as an entering group. All known ribozymes (catalytic RNA) similarly use two metal ions to catalyze the hydrolysis of phosphate esters.

Intimate substitution mechanisms of *d*-block tetrahedral complexes have been studied for dichromate, vanadate, ferrate, and perrhenate, and were found to be associative. Dissociation of a very strongly basic group such as O^{2-} from the primary coordination sphere of the very strongly acidic Cr(VI), V(V), Fe(VI), and Re(VII) is unlikely (O^{2-} is a poor leaving group). In contrast, substitution mechanisms in $Ni(CO)_4$ are found to be dissociative. The Ni–CO bond is known to be readily broken, as, for example, in the Mond process for purifying nickel (Section 11.2).

It is interesting to note that leaving-group effects cause a different stereochemical result for four-coordination associative substitution in the *p* block than in the *d* block, which of course involves the tetrahedral rather than square planar geometry. When association of an incoming ligand with a chiral silane generates a trigonal bipyramidal intermediate, the stereochemistry of the final products indicates that the leaving group, which is (as expected) the most electronegative group, departs from the axial position.[4] In contrast, it departs from the equatorial position in the *d*-block intermediate or activated complex in Figure 16.3.

The difference in the *d*- and *p*-block trigonal bipyramidal intermediates or activated complexes is that the *p*-block intermediate is hypervalent, while the *d* block intermediate is not. In Section 10.5, we discussed hypervalent bonding in terms of three-center, four-electron (3c–4e) bonding, which involves the axial substituents in the trigonal bipyramid. The bond order per axial substituent is only 0.5, so that these substituents are more likely to leave than equatorial ones; the 3c–4e bonding scheme works best with electronegative atoms as terminal (i.e., axial) atoms.[11]

Metal–Atom Effects. It might be anticipated that the central (metal) atom has some say in rates, and recall from Section 8.8 that both the acidity of a cation and (if a *d*-block metal ion) its octahedral site stabilization energy have very strong effects on the rates of coordinated water exchange; this would apply to exchanges of other ligands as well.

Among d^8 square planar complexes, the rates depend strongly on the period of the periodic table: Rates for comparable reactions are typically 10^7 times faster for Ni(II) than Pt(II) and 10^5 times faster for Pd(II) than Pt(II). The inertness of square planar Pt(II) complexes makes their kinetics much easier to study, so that the large majority of studies on square planar complexes are for this metal ion. We recall (Section 8.8) that large crystal field splittings are required to stabilize the square planar geometry over alternate geometries such as (octahedral or) the trigonal bipyramidal geometry found in the intermediate or activated complex. Crystal field splittings are largest in the sixth period and smallest in the fourth period; the more the starting square planar geometry is stabilized, the larger the activation energy will be, and the slower the rate of substitution.

Example 16.2

The room temperature rate of exchange of metal carbonyls with free CO decreases dramatically from $Ni(CO)_4$ ($1 \times 10^{-2}\ s^{-1}$) to $Fe(CO)_5$ ($5 \times 10^{-10}\ s^{-1}$) to $Cr(CO)_6$ (estimated at $9 \times 10^{-14}\ s^{-1}$). Offer an explanation.

SOLUTION:

Such an enormous variation in rates speaks to large differences in activation energies, such as could be due to differences in crystal field stabilization energies: $Ni(CO)_4$ has a d^{10} central metal atom, with no crystal field stabilization energy, while $Cr(CO)_6$ is an octahedral d^6 complex with a very strong-field ligand, which would have a very large octahedral site stabilization energy as a barrier to achieving a dissociative square pyramidal activated complex (Section 8.8). The d^8 trigonal bipyramidal $Fe(CO)_5$ would have smaller crystal field effects, and hence it would show the intermediate rate.

16.4 Spectator Ligand Effects on Ligand Substitution Rates

Spectator ligand effects are formally equivalent to substituent effects on nucleophilic substitution reactions in organic chemistry. First, we will discuss steric effects that favor dissociative mechanisms and hinder associative mechanisms; these have their counterpart in steric hindrance effects in organic chemistry. Second, we will discuss electron-donating or withdrawing effects, which are also well known in organic chemistry. Third, we will discuss base-assisted hydrolysis reactions, which are inorganic counterparts of various types of organic neighboring-group participation. The author does not know of an organic counterpart to the fourth type, the trans effect in square planar complexes.

Table 16.8
Effect of Nonleaving Ligands on Acid Hydrolysis Rates of Some Co(III) Complexes at 25 °C

A. *trans*-$[Co(N-N)_2Cl_2]^+ + H_2O \longrightarrow [CoCl(N-N)_2(H_2O)]^{2+} + Cl^-$ [a]

N—N	*k* (s^{-1})
$NH_2CH_2CH_2NH_2$	3.2×10^{-5}
$NH_2CH_2CHMeNH_2$	6.2×10^{-5}
d, l-$NH_2CHMeCHMeNH_2$	1.5×10^{-4}
meso-$NH_2CHMeCHMeNH_2$	4.2×10^{-4}
$NH_2CH_2CMe_2NH_2$	2.2×10^{-4}
$NH_2CMe_2CMe_2NH_2$	Instantaneous
$NH_2CH_2CH_2NHMe$	1.7×10^{-5}

B. *trans*-$[Co(N_4)LCl]^{n+} + H_2O \longrightarrow$ *trans*-$[Co(N_4)L(H_2O)]^{(n+1)+} + Cl^-$

Complex[b,c]	*k* (s^{-1})
trans-$[Co(cyclam)Cl_2]^+$	1.1×10^{-6}
trans-$[Co(cyclam)(NCS)Cl]^+$	1.1×10^{-9}
trans-$[Co(cyclam)(CN)Cl]^+$	4.8×10^{-7}
trans-$[Co(tet\text{-}b)Cl_2]^+$	9.3×10^{-4}
trans-$[Co(tet\text{-}b)(NCS)Cl]^+$	7.0×10^{-7}
trans-$[Co(tet\text{-}b)(CN)Cl]^+$	3.4×10^{-4}

[a] R. G. Pearson, C. R. Boston, and F. Basolo, *J. Am. Chem. Soc.* **75**, 3089 (1953).
[b] Cyclam = 1,4,8,11-Tetraazacyclotetradecane.
[c] Tet-*b* = *d,l*-1,4,8,11-Tetraaza-5,5,7,12,12,14-hexamethylcyclotetradecane.

Steric Effects. Table 16.8 lists the dissociative (first-order) rate constants of two series of octahedral cobalt(III) complexes involving (a) two chelating nitrogen-donor spectator ligands and (b) one macrocyclic nitrogen-donor spectator ligand. In each case, it will be noted that increasing the number of methyl substituents on the spectator ligands increases the rate constant by a moderate but significant factor. Steric effects are also seen in some associative reactions of square planar complexes such as the substitution of H_2O for Cl^- in *cis*-$[Pt(PEt_3)_2LCl]$. Here increasing the steric effects of the spectator ligand L decreases the rate constant moderately: for L = pyridine, $k = 8 \times 10^{-2}$; for L = 2-methylpyridine, $k = 2.0 \times 10^{-4}$ (the ortho methyl group blocks approach of H_2O above the square plane); for L = 2,6-dimethylpyridine, $k = 1.0 \times 10^{-6}$ (the two ortho methyl groups block approach of H_2O both above and below the square plane).

Electron-Donor Effects. Better electron-donating substituents (spectator ligands) can be expected to enhance the rates of dissociative reactions, since they donate more total electron density to compensate for the reduction in the total number of ligands. Thus, replacement of X^- by H_2O is much faster in $[Ni(NH_3)_5X]^+$ than in $[Ni(H_2O)_5X]^+$. The effect is enhanced if the cis ligands are good π donors.

Base-Assisted Hydrolyses. The rate of hydrolysis of $[Co(NH_3)_5Cl]^{2+}$ is a million times faster in basic solution than in acidic solution; the same is true of complexes of

RNH_2, but there is no comparable effect in the complexes $[Co(pyridine)_4Cl_2]^+$ or $[Co(CN)_5Cl]^{3-}$. A number of studies support the notion that the enhanced acidity of coordinated NH_3 and RNH_2 (Section 5.10) is involved, and that one of the coordinated NH_3 or RNH_2 molecules is first deprotonated to become a coordinated amide ion (NH_2^- or RNH^-). Such a ligand is not only a powerful electron donor, but it is also a very good π-donor ligand, which can readily replace the electrons lost when the Cl^- ligand leaves. The mechanism of this reaction is considered to involve the following steps:

$$[Co(NH_3)_5Cl]^{2+} + OH^- \rightleftharpoons [Co(NH_3)_4(NH_2)Cl]^+ + H_2O \qquad (16.25)$$

$$[Co(NH_3)_4(NH_2)Cl]^+ \xrightarrow{\text{slow}} [Co(NH_3)_4(NH_2)]^{2+} + Cl^- \qquad (16.26)$$

$$[Co(NH_3)_4(NH_2)]^{2+} + H_2O \xrightarrow{\text{fast}} [Co(NH_3)_5(OH)]^{2+} \qquad (16.27)$$

Example 16.3

You want to enhance the rate of replacement of the anionic ligand by solvent water in each of the following complexes, and can choose to do this by adding either a strong acid, a strong base, or Ag^+. Which would be effective in each case, and why? (a) $[CO(NH_2Me)_5\text{-}ONO_2]^{2+}$; (b) $[Co(pyridine)_5Br]^{2+}$; (c) $[Co(pyridine)_5OMe]^{2+}$; and (d) $[Co(NH_3)_5(CN)]^{2+}$.

SOLUTION:

(a) The leaving nitrate group is hard, so Ag^+ will not help by coordinating to it. It is nonbasic, so H^+ will not attach to it. But OH^- can deprotonate one spectator NH_2Me group to an amide group ($NHMe^-$), which will greatly accelerate the rate of substitution.

(b) Pyridine has no acidic protons, so it cannot be deprotonated by OH^-. The leaving bromide group is nonbasic, so H^+ will not attach to it, but it is soft, so Ag^+ would coordinate and thereby enhance the rate of substitution.

(c) In this case, the leaving methoxy group is hard and basic, thus H^+ should enhance its rate of leaving.

(d) The best choice would probably be OH^-, which would have a profound rate-enhancing effect by deprotonating a spectator NH_3 ligand to give an amide ligand. The CN^- ion is both soft and basic, and coordinated CN^- has an unshared pair of electrons on nitrogen, therefore both Ag^+ and H^+ could also enhance the rate by attaching to nitrogen; in the case of H^+, very toxic HCN gas would, however, be generated!

Trans Effect. Perhaps the most spectacular examples of spectator ligand effects occur in the associative reactions of square planar complexes. In such complexes, the ligand trans to the leaving group has a very strong effect on the rate of substitution, while the cis ligands have a very mild effect. Furthermore, it is noted that, in studying the reactions of cis and trans isomers in substitution reactions, that the products always retain the same geometry: substitution of *cis*-$[PtL_2X_2]$ by Y gives *cis*-$[PtL_2XY]$; substitution of *trans*-$[PtL_2X_2]$ by Y gives *trans*-$[PtL_2XY]$.

The value of this type of evidence in distinguishing *I* from *A* mechanisms was mentioned earlier (Section 16.2). The specificity of the reaction in fact supports the I_a mechanism, in which the activated [MC_2TXY] complex shown in the middle of Figure 16.3 rapidly proceeds to the same geometric isomer along pathway (*a*). And the evidence strongly supports the specific transition state shown, in which the relatively uninfluential cis C spectator ligands occupy the axial positions of the trigonal bipyramidal activated complex, while the important entering Y, leaving X, and trans-ligand T share the equatorial positions, and the same molecular orbitals (MOs).

Even leaving rates aside, fairly dramatic influences of the trans ligand can be seen in bond distances in square planar Pt(II) complexes (because of their inertness, these are the most commonly studied type of square planar complex.) The Pt–Cl bond distances vary by several picometers in the cis- and trans isomers of the compound [$Pt(PEt_3)_2Cl_2$]: This distance is much shorter (229.4 pm) in the trans isomer, in which two Cl atoms are trans to each other and share overlap with a Pt $6p_x$ orbital. It is much longer (average 237.6 pm) in the *cis* isomer, in which a Cl atom and a PEt_3 ligand share overlap with a common Pt $6p$ orbital. Evidently the soft, good σ donor PEt_3 overlaps very well with Pt, making the bonding to Cl noticeably more ionic, hence longer. We might then anticipate that this square planar *thermodynamic trans influence* would carry over to the trigonal bipyramidal activated complex (as a *kinetic trans effect*), in which the three equatorial ligands only (X, Y, and T) share bonding through overlap with the equatorial Pt $6p_x$, $6p_y$ and $5d_{x^2-y^2}, 5d_{xy}$ orbitals. The Cl might then be a more ready leaving group when it has a good soft σ donor such as PEt_3 sharing equatorial-plane MOs with it as compared to when it has a borderline donor such as Cl^- in the equatorial plane.

The kinetic trans effect is confirmed by kinetic measurements in the compounds *trans*-[$PtCl(PEt_3)_2L$]: The second-order rate constant is 10,000 times larger when $L = PEt_3$ as compared to when $L = Cl^-$. By expanding the database of ligands studied, one can get an approximate order of trans ligands T in increasing magnitude of their trans-labilizing effects in substitution reactions:

$$\text{T:}\quad OH^- \ll H_2O < NH_3 < \text{pyridine} < Cl^- < Br^- < SCN^- < I^-, NO_2^-, Ph^- \ll Me^- < H^-, PR_3 < CO, CN^-, Bu \tag{16.28}$$

This series is not only in the order of increasing softness, increasing position in the spectrochemical series, and increasing nucleophilicity [cf. Eq. (16.22)], but, for the most part, is in the order of increasing σ-donor ability as a soft base.

The top three members of the series, however, are not good σ donors, but rather are good π acceptor ligands. They have little influence in the square planar starting complex on Pt–Cl bond distances, and are considered to have their main effect by stabilizing the trigonal bipyramidal transition state, which also lowers the activation energy for the substitution, Figure 16.1(*a*), by draining off some of the extra electron density that the fifth ligand brings to the Pt atom.

Synthetic Applications of the Trans Effect. Judicious use of the trans-effect series (16.28) along with a clever choice of starting materials and the order of addition of different ligands can allow one to synthesize almost any geometrical isomer one may want of a square planar Pt(II) complex.

Example 16.4

Which isomer of $[Pt(NH_3)_2Cl_2]$ is obtained by reacting: (a) 2 mol of HCl with $[Pt(NH_3)_4]^{2+}$ and (b) 2 mol of NH_3 with $[PtCl_4]^{2-}$.

SOLUTION:

Ammonia (NH_3) falls below Cl^- in the trans-directing series (16.31). Hence, for (a): After the first Cl^- has produced $[Pt(NH_3)_3Cl]^+$ [Fig. 16.4(*a*)], the second Cl^- will be directed to replace the (boldfaced) NH_3 trans to Cl^-, giving *trans*-$[Pt(NH_3)_2Cl_2]$. For (b): After the first NH_3 has produced $[Pt(NH_3)Cl_3]^-$ [Fig. 16.4(*b*)], the second NH_3 will be directed to replace the Cl^- trans to Cl^-, giving *cis*-$[Pt(NH_3)_2Cl_2]$.

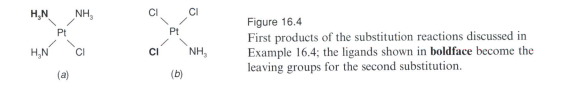

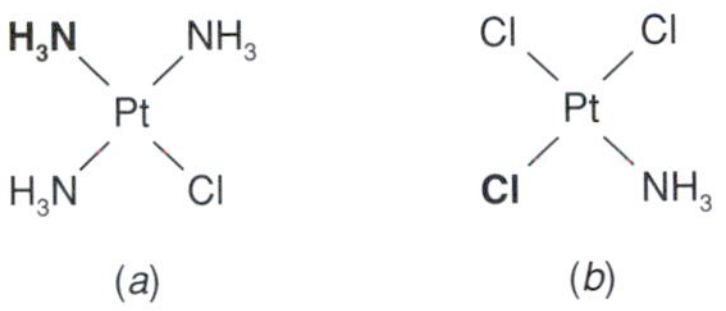

Figure 16.4
First products of the substitution reactions discussed in Example 16.4; the ligands shown in **boldface** become the leaving groups for the second substitution.

16.5 Classification of Redox Reaction Mechanisms

Redox reactions are nearly always second-order reactions, since one species must oxidize another (the rare exceptions are intramolecular redox reactions). In Chapter 6, we treated redox reactions as electron-transfer reactions, but mechanistic studies have shown that some redox reactions involve the transfer of atoms or ions from one species to another. The fundamental classification scheme for the stoichiometric mechanisms of redox reactions was established in the 1950s by the Nobel prize winner Henry Taube.[12] He recognized that an atom-transfer reaction would require that the atom being transferred from one metal to another would generally first have to enter the coordination sphere of the second metal ion as a bridging ligand via ligand substitution. On the other hand, since electrons are capable of passing through barriers such as ligands via the quantum mechanical process of tunneling, some electron transfers could proceed without any change in the primary coordination sphere of either reactant:

$$[\mathrm{Fe(bipyridyl)_3}]^{3+} + [\mathrm{Co(terpyridyl)_2}]^{2+} \rightarrow [\mathrm{Fe(bipyridyl)_3}]^{2+} + [\mathrm{Co(terpyridyl)_2}]^{3+} \quad (16.29)$$

Reactions such as this during which the ligands remain coordinated in the primary spheres of their respective central atoms (i.e., no ligand exchange or substitution occurs) are classified as **outer-sphere electron transfer reactions.** Reactions in which an intermediate or activated complex is formed in which a ligand of one metal atom bridges[13] to the other metal atom are classified as **inner-sphere electron-transfer reactions.**

If the electron transfer results in a change of ± 1 in the oxidation state of each central atom or metal atom, the reaction is called a **one-electron** or **one-equivalent process**; Reaction (16.29) is an example of a one-electron process (nearly all outer-sphere electron transfers are one-electron processes). Oxidants that characteristically engage in one-electron processes are expected to oxidize sulfite ion to dithionate ion by a two-step outer-sphere process:

$$SO_3^{2-} - e^- \rightarrow SO_3^{\cdot -}; \qquad 2\,SO_3^{\cdot -} \rightarrow S_2O_6^{2-} \tag{16.30}$$

If the electron transfer results in changes of ± 2 in the central-atom oxidation states, it is called a **two-equivalent** or **two-electron process**, and is not infrequently found in inner-sphere electron transfer, especially involving *p*-block elements. Oxidants that characteristically engage in two-electron processes are expected to oxidize sulfite ion to sulfate ion.

$$SO_3^{2-} - 2e^- + H_2O \rightarrow SO_4^{2-} + 2\,H^+ \tag{16.31}$$

16.6 Outer-Sphere Electron-Transfer Processes

Self-Exchange Reactions. Among outer-sphere mechanisms, it is useful first to consider *self-exchange redox reactions*, in which electrons are exchanged between ions of the same element that differ in charge. To follow these reactions, one of the ions is made radioactive, as indicated by an asterisk ($*$):

$$[Fe(H_2O)_6]^{3+} + [Fe^*(H_2O)_6]^{2+} \rightarrow [Fe(H_2O)_6]^{2+} + [Fe^*(H_2O)_6]^{3+} \tag{16.32}$$

Except for the position of the radioactive label, products and reactants in this reaction are the same, but nonetheless the reaction has a significant activation energy of $32\,kJ\,mol^{-1}$ and a modest second-order self-exchange rate constant, k_{11}, of $4\,L\,mol^{-1}\,s^{-1}$.

As shown in Table 16.9, self-exchange rate constants can vary over a wide range; the **Frank–Condon principle** is useful for understanding this great variability. Two concepts are involved here. The first states that electron rearrangements occur much more rapidly than rearrangements of nuclei, so that bond distances cannot alter (via slow vibrations) during the very short time during which an electron is transferred.

The second is that electron transfer is fastest when the energy levels of the two orbitals are the same, so that no input of energy is required. This concept can be illustrated by using Reaction (16.32): A problem arises because the radius of Fe^{3+} is 14 pm smaller than that of Fe^{2+}. Since the radius cannot change during the electron transfer, this spontaneous transfer of electrons from the original ions with these radii would produce $[Fe(H_2O)_6]^{3+}$ with Fe–O bond lengths 14 pm too long, and $[Fe(H_2O)_6]^{2+}$ ions with bond lengths 14 pm too short. These are both vibrationally excited, high-energy forms of the respective ions: Both would lose energy as heat by vibrating to reach the equilibrium bond lengths, and energy would have been created out of nothing, in violation of the first law of thermodynamics.

Consequently, the electron transfer is most probable when enough vibrational activation energy has been put into each ion to bring them to the same energy levels, which in the case of self-exchanging ions means that the radii have reached the same

Table 16.9
Self-Exchange Outer-Sphere Redox Rate Constants and Standard Reduction Potentials

Complex Ions	Charges	k_{11} (L mol^{-1} s^{-1})	$E°$ (V)
$Ce(H_2O)_n$	4+, 3+	4	+1.72[a]
$Cr(H_2O)_6$	3+, 2+	2×10^{-5}	−0.42
$Mo(CN)_8$	3−, 4−	3×10^4	+0.72
$W(CN)_8$	3−, 4−	7×10^4	+0.46
MnO_4	1−, 2−	3000	+0.56
$Fe(H_2O)_6$	3+, 2+	4	+0.77
Fe(phenanthroline)$_3$	3+, 2+	3×10^7	+1.00
$Fe(CN)_6$	3−, 4−	740	+0.36
$Ru(H_2O)_6$	3+, 2+	50	+0.24
$Ru(NH_3)_6$	3+, 2+	2800	+0.10
Ru(bipyridyl)$_3$	3+, 2+	4.2×10^8	+1.53
$Co(NH_3)_6$	3+, 2+	8×10^{-6}	+0.06
$Co(NH_2CH_2CH_2NH_2)_3$	3+, 2+	7.7×10^{-5}	−0.24
Co(bipyridyl)$_3$	3+, 2+	18	
Co(terpyridyl)$_2$	3+, 2+	48	
Co(phenanthroline)$_3$	3+, 2+	40	+0.33
$IrCl_6$	2−, 3−	2×10^5	+0.87

SOURCE: Data from B. E. Douglas, D. H. McDaniel, and J. J. Alexander, *Concepts and Models of Inorganic Chemistry*, 3rd ed., Wiley, New York, 1994; p. 533; D. Katakis and G. Gordon, *Mechanisms of Inorganic Reactions*, Wiley, New York, 1987; p. 279; and D. F. Shriver, P. Atkins, and C. H. Langford, *Inorganic Chemistry*, 2nd ed., Freeman, New York, 1994; Appendix 2.

[a] Strongly acidic Ce^{4+} is complexed even by feebly basic anions such as SO_4^{2-}, so its reduction potential depends on the acid used to keep it in solution.

intermediate value. (For reactions between ions of different elements, the radii would not reach the same values, but the orbital energies would.) The fifth- and sixth-period ions of a given metal in Table 16.9 differ in radius by only 4–6 pm. Since fourth-period ions of a given metal differ by 18–20 pm, higher activation energies are needed to bring them to equal radii.

Some of the slowest self-exchanging systems are the cobalt(II,III) systems, because electrons are being exchanged between antibonding $e_g{}^*$ orbitals, resulting in large changes in radii. A compounding factor is that the exchange also generally requires each ion to switch between high-spin d^7 and low-spin d^6 electron configurations, which serves further to decrease the rate of self-exchange.

It may also be noted in Table 16.9 that self-exchange rates are much faster for ligands having π bonding systems (CN^-, bipyridyl, terpyridyl, phenanthroline) than for complexes of the same metal with purely σ donor ligands (H_2O, NH_3, and $NH_2CH_2CH_2NH_2$). The former group of ligands have empty π^* orbitals that can accept electrons being transferred, then pass them on to the receiving metal ions, whereas the purely σ-donor atoms do not have this capacity. Thus outer-sphere electron transfer can be characterized as *direct* (from one metal to another) in the case of the purely σ-donor ligands but *indirect* (from one metal to ligands to the other metal) in the case of the ligands with accessible π bond systems.

Cross Reactions. R. A. Marcus won the 1992 Nobel prize for deriving the **Marcus equations**, including the **Marcus cross-relation** relating the rate constants k of reactions between diverse oxidizing and reducing ions to the rates of self-exchange of the oxidizing species and reducing species (k_{11} and k_{22}), and to the equilibrium constant, K, for the redox reaction:

$$k_{12}{}^2 = f k_{11} k_{22} K \tag{16.33}$$

In this equation, f is a complex parameter that includes k_{11}, k_{22}, K, and the encounter rate; for our approximate calculations, the value of f can be taken as 1. The value of the equilibrium constant K can be calculated in the standard way from the cell potential $E°$ for the reaction:

$$\log K = nFE°/2.303RT = nE°/0.0592 \quad (\text{at } 25°\text{C}) \tag{16.34}$$

However, if the energy connected with a very high cell potential greatly exceeds the energy required to adjust the metal–ligand bond lengths, the rate actually *decreases* with increasing $E°$; this may happen in practice with $E°$ values substantially over 1 V.

Example 16.5

Calculate the rate constant for the reaction of Ce^{4+}(aq) with $[Fe(CN)_6]^{4-}$ to give Ce^{3+}(aq) and $[Fe(CN)_6]^{3-}$.

SOLUTION:

All of the half-reactions in Table 16.9 are one-electron reductions, so $n = 1$. For this reaction, $E° = E_{\text{reduction}} - E_{\text{oxidation}} = 1.72\text{ V} - 0.36\text{ V} = 1.36\text{ V}$. By using Eq. (16.34), $\log K = 1.36/0.0592 = 23.0$, so $K = 1.0 \times 10^{23}$. The values of k_1 and k_2 are also taken from Table 16.9 and substituted into Eq. 16.33: $k_{12}{}^2 = 1 \times 4 \times 740 \times 1 \times 10^{23} = 3 \times 10^{26}$, so $k_{12} = 1.7 \times 10^{13}$. Since $E°$ is quite high, the actual rate is likely less than this.

A somewhat unusual factor of the Marcus equation is that the rate of an outer-sphere redox reaction depends on the position of the equilibrium. At the beginning of this chapter, we pointed out that rates of chemical reactions and positions of thermodynamic equilibrium are separate concepts, and in the extreme case of potentially explosive reactions (very exothermic reactions with very slow rates, Section 6.2) they are unconnected. But here they are related; another case is in the classic experiment of the activity series of metals (Section 6.5 and Experiment A.6), in which thermodynamic favorability of a reaction (activity) is qualitatively judged by the kinetic rate of evolution of H_2 bubbles.

All of the self-exchange half-reactions in Table 16.9 are one-electron reactions, so that the outer-sphere redox reactions we can calculate are also one-electron half-reactions. Two-electron outer-sphere transfers are much rarer: They are of very low probability by quantum mechanics; but a few cases are known, such as $Hg + Tl^{3+} \rightarrow Hg^{2+} + Tl^{+}$. Most balanced redox reactions involving two or more electrons that proceed by outer-sphere electron transfer therefore have the problem of *noncomplementarity* previously discussed in Section 6.2. Therefore, unless *d*-block ions

Table 16.10
Second-Order Rate Constants for Selected Inner-Sphere Reactions with Variable Bridging Ligands

Oxidant	Reductant	Bridging Ligand	k (L mol^{-1} s^{-1})
$[Co(NH_3)_6]^{3+}$	$[Cr(OH_2)_6]^{2+}$		8×10^{-5}
$[CoF(NH_3)_5]^{2+}$	$[Cr(OH_2)_6]^{2+}$	F^-	2.5×10^5
$[CoCl(NH_3)_5]^{2+}$	$[Cr(OH_2)_6]^{2+}$	Cl^-	6.0×10^5
$[CoI(NH_3)_5]^{2+}$	$[Cr(OH_2)_6]^{2+}$	I^-	3.0×10^6
$[Co(NCS)(NH_3)_5]^{2+}$	$[Cr(OH_2)_6]^{2+}$	NCS^-	1.9×10
$[Co(SCN)(NH_3)_5]^{2+}$	$[Cr(OH_2)_6]^{2+}$	SCN^-	1.9×10^5
$[Co(NH_3)_5(OH_2)]^{3+}$	$[Cr(OH_2)_6]^{2+}$	H_2O	0.1

SOURCE: From F. Basolo and R. Pearson, *Mechanisms of Inorganic Reactions*, 2nd ed., Wiley, New York; 1967.

with many stable oxidation states are involved, these reactions must generally proceed through multistep mechanisms involving unfavorable free-radical intermediates such as $\cdot Tl^{2+}$. However, we shall see many cases of (formal) two-electron redox changes among inner-sphere redox reactions.

16.7 One-Electron Inner-Sphere and Long-Range Electron-Transfer Processes

For an inner-sphere mechanism to be plausible, one complex must have a good bridging ligand, which has more than one pair of sufficiently basic unshared electron pairs. Ammonia and pyridine cannot fill this role, and apparently neither can water, as illustrated by the rate constants in Table 16.10.

The bridging species can be a saturated organic or inorganic ligand, or it can be an organic or inorganic ligand involving a conjugated π system. Examples of each type are shown in Figure 16.5.

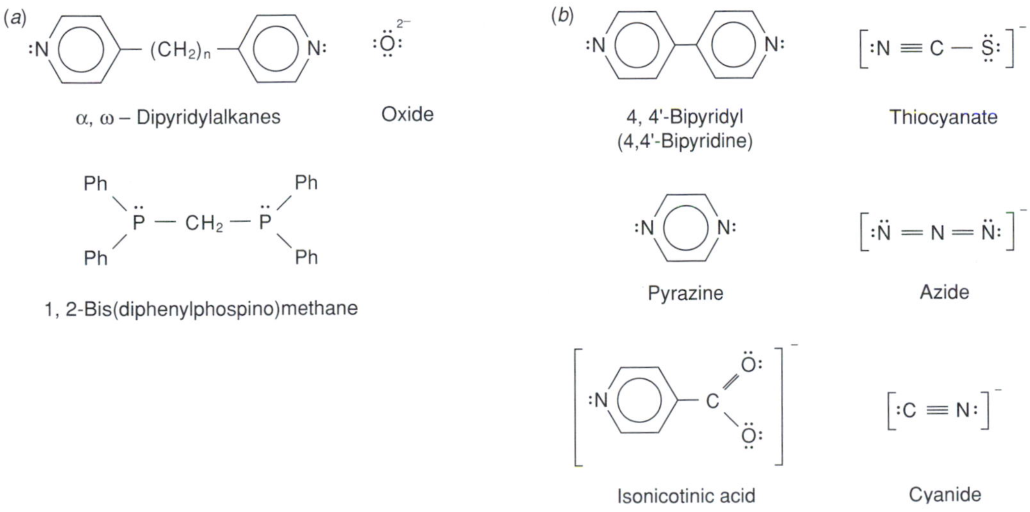

(a)

:N⟨⟩—(CH₂)ₙ—⟨⟩N: α, ω – Dipyridylalkanes

$:\ddot{O}:^{2-}$ Oxide

$Ph_2\ddot{P}—CH_2—\ddot{P}Ph_2$ 1, 2-Bis(diphenylphospino)methane

(b)

4, 4'-Bipyridyl (4,4'-Bipyridine)

$[:N\equiv C—\ddot{S}:]^-$ Thiocyanate

Pyrazine

$[:\ddot{N}=N=\ddot{N}:]^-$ Azide

Isonicotinic acid

$[:C\equiv N:]^-$ Cyanide

Figure 16.5
Organic and inorganic examples of (*a*) saturated bridging ligands; (*b*) conjugated bridging ligands.

Inner-sphere electron-transfer processes must proceed through more than one step, beginning with the formation of a bridged μ complex by a (dissociative or associative) ligand substitution:

$$[Co^{III}(NH_3)_5Cl]^{2+} + [Cr^{II}(H_2O)_6]^{2+} \rightarrow [(H_3N)_5Co^{III}\text{–}Cl\text{–}Cr^{II}(OH_2)_5]^{4+} + H_2O \tag{16.35}$$

This reaction is then followed by the redox step:

$$[(H_3N)_5Co^{III}\text{–}Cl\text{–}Cr^{II}(OH_2)_5]^{4+} \rightarrow [(H_3N)_5Co^{II}\text{–}Cl\text{–}Cr^{III}(OH_2)_5]^{4+} \tag{16.36}$$

Finally, the new bridged complex breaks apart into the final[14] products:

$$[(H_3N)_5Co^{II}\text{–}Cl\text{–}Cr^{III}(OH_2)_5]^{4+} + H_2O \rightarrow [(H_3N)_5Co^{II}(OH_2)]^{2+} + [ClCr^{III}(OH_2)_5]^{2+} \tag{16.37}$$

In the preceding one-electron process, cobalt is reduced by one oxidation number and chromium is oxidized by one oxidation number. In effect, a chlorine *atom* is transferred from cobalt to chromium. Two-electron processes will be discussed in the following section.

Rate-Limiting Step. Inner-sphere electron-transfer processes all involve at least the three steps exemplified by Reactions (16.35)–(16.37); any of these three steps can be rate limiting.

Reactions in which the *second* (the redox) step [e.g., Step (16.36)] is rate limiting are quite common and quite diverse in rates as the metal ions and bridging ligands are varied. The Frank–Condon principle applies to the electron-transfer process in this step, so the Marcus equation has relevance, and the process of the second (electron-transfer) step itself bears a greater or lesser resemblance to the outer-sphere process, depending on whether the bridging ligand is saturated or conjugated; these two cases will be discussed in greater detail at the end of this section.

Least common is for the *final* step of type (16.37) to be rate limiting; this happens only when the final binuclear complex is inert at both ends, as in the cases of $[(aq)V^{III}(\text{–}\mu\text{–}OH\text{–})_2V^{III}(aq)]^{4+}$ and $[Ru^{II}(NH_3)_5\text{–}\mu\text{–}Cl\text{–}Cr^{III}(H_2O)_5]^{4+}$. Note that, if the final binuclear complex is labile or otherwise unstable at only one end, the bridging group may not in fact be transferred to the second metal center, as in the following reaction:

$$\begin{aligned}[(CN)_6Fe^{III}]^{3-} + [Co^{II}(CN)_5]^{3-} &\rightarrow [(CN)_5Fe^{III}\text{–}CN\text{–}Co^{II}(CN)_5]^{6-} \\ &\rightarrow [(CN)_5Fe^{II}\text{–}CN\text{–}Co^{III}(CN)_5]^{6-} \\ &\rightarrow [(CN)_6Fe^{II}]^{4-} + [Co^{III}(CN)_5]^{2-}\end{aligned} \tag{16.38}$$

In this case, the bridge probably breaks apart in this manner because CN^- is a much stronger field ligand when attached at the C end than when attached at the N end.

Although this inner-sphere reaction results in electron transfer but not in transfer of an atom or group, the bridge was nonetheless crucial in the mechanism.

The *first* step of type (16.35), substitution of the bridging atom into the inner sphere of the other complex, can be rate determining. If, as is often the case, this reaction is dissociative in nature, then the rate constants for the inner-sphere reaction will be very similar for a variety of incoming bridging ligands, as, for example, in the data for the oxidation of $[V(H_2O)_6]^{2+}$ by a number of cobalt(III) complexes (Table 16.11). These rate constants are furthermore similar to other dissociative ligand-substitution rate constants for $[V(H_2O)_6]^{2+}$.

Table 16.11
Rate Parameters for Some Reductions by V^{2+} at 25 °C

Oxidant[a]	k ($M^{-1}\,s^{-1}$)	$\Delta S^{\ddagger}$ ($J\,mol\,K^{-1}$)
$[Co(NH_3)_5C_2O_4H]^{2+}$	12.5	−54
$[Co(NH_3)_5C_2O_4]^{+}$	45.3	
cis-$[Co(NH_3)(en)_2(N_3)]^{2+}$	10.3	−50
cis-$[Co(H_2O)(en)_2(N_3)]^{2+}$	16.6	−50
trans-$[Co(en)_2(N_3)_2]^{+}$	26.6	−46
trans-$[Co(H_2O)(en)_2(N_3)]^{2+}$	18.1	−67
Cu^{2+}	26.6	−57.7

[a] Ethylenediamine = en = $NH_2CH_2CH_2NH_2$.

Assignment of Redox Processes as Inner or Outer Sphere. Since the first and last steps mentioned above are ligand-substitution reactions that may independently be known to be fast (involving a labile complex), or slow (involving an inert complex), it is sometimes possible to use logic to assign the mechanism of a redox process.

- Outer-sphere electron transfer is required when the rate of the redox reaction exceeds the rates of substitution reactions of either of the two reactants, as, for example, when two *inert* complexes undergo a *rapid* redox reaction with each other.

- If inert complexes are found to *transfer atoms* or *ligands rapidly* in a redox reaction, the mechanism can be unambiguously assigned as *inner sphere*. More specifically, if the reactant metal bearing the atom or ligand being transferred is in an oxidation state that is inert in substitution reactions [e.g., Co(III) in Step (16.35), and the product metal, after acquiring the ligand, is in an inert oxidation state (e.g., Cr(III)] in Step (16.37), but the redox reaction is fast, then alternative mechanisms involving rapid outer-sphere electron transfer either before or after rapid ligand exchange of the bridging ligand can be excluded.

- In some cases, if the bridging ligand is oxidizable (a soft base), the bridged intermediate may have an intense charge-transfer absorption (Section 6.4), so that a fleeting color may appear during the reaction.

- Inner-sphere reaction rates would be expected to be more sensitive to the nature of the second donor atom of the bridging ligand. The rates of reduction by Cr^{2+}(aq) of thiocyanato complexes of cobalt(III) are quite sensitive to the linkage isomer present: $[Co(NH_3)_5(SCN)]^{2+}$ is reduced 10^5 times as fast as $[Co(NH_3)_5(NCS)]^{2+}$. This result is attributed to the preference of the hard acid Cr^{2+} for bonding to the terminal N atom of $[Co(NH_3)_5(SCN)]^{2+}$ rather than to the terminal S atom of $[Co(NH_3)_5(NCS)]^{2+}$. Since the azido ligand (N_3^-) also presents a terminal N atom to the incoming metal ion, its complexes are also reduced about 10^4 times faster than thiocyanato complexes with terminal S atoms. Contrarily, if it is found that an azido and a S-terminal thiocyanato complex are reduced by some reducing agent at about the same rate, the reaction is probably outer sphere.

- In additional support of an inner-sphere mechanism, it may be observed that, when running a chlorine atom transfer reaction in a solution containing radioactive $^{36}Cl^-$, very little of the radioactive chloride ion is incorporated into the product.
- If both reacting complexes also exchange ligands rapidly, then the classification of the reaction as outer sphere or inner sphere is not obvious and is more difficult to make.

Example 16.6

(a) Write a hypothetical outer-sphere electron-transfer mechanism for the reduction of $[Cr(H_2O)_6]^{2+}$ with $[Co(NH_3)_5I]^{2+}(aq)$ in acid solution to give (ultimately) $[Cr(H_2O)_5I]^{2+}$, $Co^{2+}(aq)$, I^-, and NH_4^+. What evidence could rule out this mechanism and why?
(b) Suppose you wanted to consider a mechanism for this reaction that was neither inner sphere nor outer sphere, but was **free radical**, involving an intermediate of free iodine atoms, I·(aq). What kind of evidence might support such a mechanism?

SOLUTION:

(a) Since Co(III) would be slow to exchange its I^- ligand, let us begin with the electron transfer:

$$[Co(NH_3)_5I]^{2+} + [Cr^{II}(H_2O)_6]^{2+} \rightarrow [Co(NH_3)_5I]^{+} + [Cr(H_2O)_6]^{3+}$$

$$\text{labile } [Co(NH_3)_5I]^{+} + 5\,H^+ \rightarrow Co^{2+}(aq) + I^-(aq) + 5\,NH_4^+$$

$$[Cr(H_2O)_6]^{3+} + I^- \rightarrow [ICr^{III}(OH_2)_5]^{2+} + H_2O$$

If the reaction were fast, it could not proceed by this mechanism: The last of these three steps would necessarily be slow, since $[Cr(H_2O)_6]^{3+}$ is inert in substitution reactions. Also, including radioactive I^- in the reaction mixture might show that none of it is incorporated in the final product; radioactive I^- should be incorporated in this mechanism.

(b) The first step of a free radical mechanism might be $[Co(NH_3)_5I]^{2+} \rightarrow [Co(NH_3)_5]^{2+} + I\cdot(aq)$. The free radical I·(aq), is then an intermediate that could perhaps be observed by its color, or trapped in a reaction with some additional species included in the solution, such as thiosulfate ion or an alkene. The next steps would be

$$I\cdot + [Cr(H_2O)_6]^{2+} \rightarrow I^- + [Cr(H_2O)_6]^{3+}$$

$$[Cr(H_2O)_6]^{3+} + I^- \rightarrow [ICr^{III}(OH_2)_5]^{2+} + H_2O$$

The last step would again necessarily be slow, so this reaction must be found to be slow.

Saturated Bridging Ligands and Long-Distance Outer-Sphere Electron Transport. The use of saturated organic ligands such as the α,ω-dipyridylmethanes results in an inability to transmit the electron via the ligand, so that the electron-transfer step in such cases is essentially outer sphere, even though a bridging ligand is present. Inert complexes of saturated organic ligands are useful to study because the distance between the two metal centers is fixed and knowable, which is not the case in solution studies of bimolecular outer-sphere electron transport reactions, and the distances between the oxidizing and reducing centers can be systematically varied.

The most important examples of such inert bridged complexes are various metalloproteins involved in redox reactions that have two or more metal centers bridged by long protein connections, at metal–metal distances from 1000 to 2500 pm. Despite the distances involved, the rates of electron transfer must be, and are, substantial (e.g., 1.1×10^4/s for the yeast cytochrome *c*/cytochrome *c* peroxidase pair, separated by a distance of ~1700 pm).[15]

These reactions are more readily studied if additional metal redox centers, in particular $[Ru(NH_3)_5]^{n+}$ units, are chemically bonded to histidine units along the protein chain. If, for example, one Ru unit involves Ru(III) and another Ru(II), then the outer-sphere self-exchange process can be studied, for which the equilibrium constant is 1, which simplifies analysis. Alternately, Ru(III) units can be attached to histidines at various distances from an iron(II) heme group of a cytochrome *c*, and the rate of the electron transport can be studied. In one case, Harry Gray and co-workers inserted five ruthenium groups at distances of 450, 840, 1110, 1230, and 1480 pm from the heme group of cytochrome *c*.[16]

A number of such studies have shown an approximately exponential decay in the rate of electron transport as a function of distance, as shown in Figure 16.6. Substan-

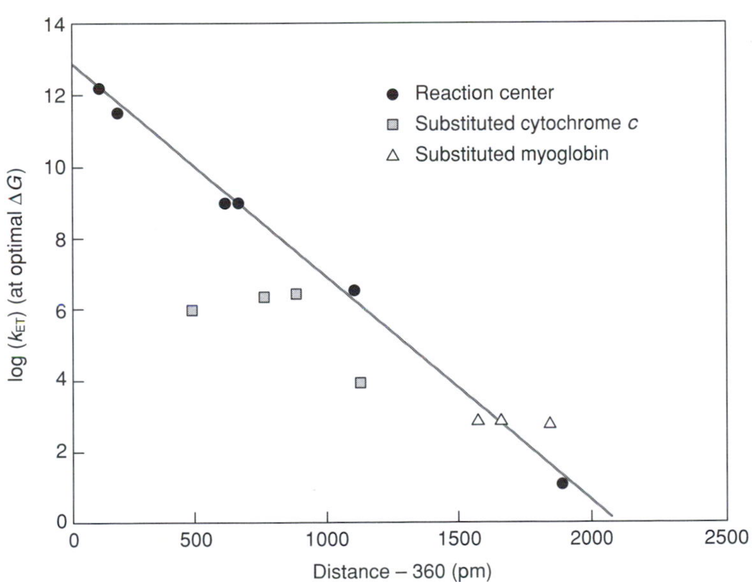

Figure 16.6
Plot showing the log of the electron-transfer rate constant, $\log k_{ET}$, as a function of metal–metal distance for several natural and ruthenium-modified proteins. [Adapted with permission from S. J. Lippard and J. M. Berg, *Principles of Bioinorganic Chemistry*, University Science Books, Sausalito, CA, 1994; p. 249.]

tial deviations are noted, however, and the rates of long-distance electron transport are no longer considered to be just a function of distance alone. In the above pentaruthenated cytochrome *c*, the rate of electron transfer was faster to the Ru(III) located 1230 pm from the heme Fe(II) than to the Ru(III) located 840 pm away. It has been proposed[17] that the nature of the pathway between the two sites is vital: Covalent σ bonds transmit electrons about twice as well as (the weaker) hydrogen bonds, which transmit the electrons about five times as well as the equivalent distance through empty space. Transmission is easiest of all, however, if the transmission can occur via conjugated or overlapping π-electron systems.

Conjugated Bridges and Intervalence Charge Transfer. When the bridging ligand is conjugated, electron transfer can occur from one metal to the other via the ligand (indirect electron transfer); the rate can then be quite sensitive to the nature of the ligand. For example, a pyrazine ligand and a 1,2-bis(diphenylphosphino)methane ligand give approximately the same length of bridge, but the rate of electron transport over the bridge is 30 times as fast for the pyrazine bridge,[18] since the process can involve the bridge, rather than just tunnel around (or under) it.

The base pairs stacked along the axis of the double helix of DNA offer a different type of π-overlapping bridge through which electrons may be transferred. Metal complexes of flat aromatic nitrogen donor ligands can be inserted (intercalated) between adjacent stacked sets of base pairs in DNA. Electron transport can then be measured between two such metal complexes at known distances apart, and was found to be remarkably fast: on the order of $10^{10}\ s^{-1}$ when a Ru(II) or Os(II) complex transfers an electron to a Rh(III) complex over 4000-nm away from the first metal complex.[19] When just a Rh(III) complex is intercalated and the DNA is irradiated with ultraviolet (UV) light, the oxidation-sensitive guanine transfers electrons up to 3700 nm to the metal complex, causing oxidative damage to the guanine bases.[20]

Interesting bridged redox-active complexes result if two ruthenium atoms with identical ligands but differing in oxidation number by one unit are attached at the two identical ends of conjugated organic ligands such as pyrazine or 4,4′-bipyridyl: $[(H_3N)_5Ru(II){-}NC_nH_mN{-}Ru(III)(NH_3)_5]^{5+}$. The 4,4′-bipyridyl *mixed-valence complex* shows an intense, very low energy absorption band in the near-infrared (IR), at 1050 nm, due to a *metal–metal* or **intervalence**[21] **charge-transfer transition**. The photon of light promotes the electron from the Ru(II) atom to the Ru(III) atom, producing a Ru(III) atom and a Ru(II) atom, respectively. Since this photochemical process is not spontaneous, the bond lengths do not have to be preequalized, but the process is related to the spontaneous process that is governed by the Frank–Condon principle, and rate information can be derived from the energy of the charge-transfer band.[22]

The energy of this intervalence charge-transfer transition can vary widely, and three limiting cases may be discussed, according to a classification scheme proposed by Robin and Day.[23] Complexes showing a distinct intervalence charge-transfer band are Robin and Day **Class II** intervalence compounds. In these compounds, the extra electron is localized on an identifiable Ru(II) (or other metal) ion, but there is electronic interaction with the other metal center.

If the two metal centers are connected by a saturated bridge (or not connected), then this band is not observed in accessible regions of the spectrum, and a variety of tests allow one to tell which ruthenium atom is Ru(II) and which is Ru(III). The extra electron is fully localized on one of the metal atoms, and this complex is a member of Robin and Day's **Class I** of intervalence compounds.

Table 16.12
Robin and Day Classification of Mixed-Valence Compounds

Class I	Class II	Class III
Two metals in very different environments	Two metals in similar environments	Two metals in identical environments
Extra electron fully localized on one metal	Extra electron not fully localized on one metal	Extra electron fully delocalized on both metals
d–d Electronic spectra of both ions seen	*d–d* Electronic spectra of both ions modified	New electronic spectra seen
Intervalence charge-transfer bands at high energy	Intervalence charge-transfer bands in visible or near IR	Intervalence charge transfer bands not seen
Insulator	Semiconductor (if polymeric)	Metallic conductor (if polymeric)
Diamagnetic or paramagnetic	Ferromagnetic or antiferromagnetic at low temperatures (if polymeric)	May be ferromagnetic at high temperatures (if polymeric)

At the other extreme are **Class III** compounds, in which the extra electron is fully delocalized, so that, despite the presence of a fractional average oxidation number, it is impossible to assign separate integral oxidation numbers, and the separate centers give rise neither to separate electronic transitions nor to intervalence charge-transfer absorptions. Instead they participate fully in electronic transitions among distinctive MOs (Chapter 10). An example would be the multiply bonded $Tc_2Cl_8{}^{3-}$ ion, in which each Tc must be assigned an oxidation number of +2.5. In this case, the "bridging" is the result of direct metal–metal bonding, but it is possible to involve bridging ligands, as in the cluster cation $Ta_6Cl_{12}{}^{3+}$ (Section 14.5). Important characteristics distinguishing the three classes are summarized in Table 16.12.

The value of the Robin and Day classification scheme is enhanced by the fact that it can be extended to cover, not only discrete complex ions, but polymeric materials, in which case the intervalence interactions extend in as many as three dimensions. The classic example of an intervalence charge-transfer transition in a polymeric Class II compound is in the pigment known as Prussian blue, $KFeFe(CN)_6$. This three-dimensional ionic lattice compound is built from "ferrocyanide" ions, $[Fe^{II}(CN)_6]^{4-}$, with each terminal N donor atom of the bridging cyanide ions being coordinated to an Fe^{3+} ion. The Fe^{3+} ions then become six coordinate with N donor atoms, while K^+ ions fill holes in the network. Polymeric Class I compounds would include many mixed-metal oxides such as $(Pb^{2+})_2(Pb^{4+})(O^{2-})_4$ or Pb_3O_4. Polymeric Class III compounds include the metals themselves, or the bronze-colored metallic conductor Ag_2F, which contains charged sheets of silver atoms with one-half positive charge per Ag atom.

Polymeric Robin and Day Class II and Class III (sometimes called Class IIIB) compounds have some additional physical properties not found in the discrete binuclear complexes: Class IIIB compounds are metallic conductors with metallic luster and bulk magnetic properties such as ferromagnetism; Class II polymeric compounds are semiconductors that can show ferromagnetic or antiferromagnetic coupling at low temperatures (Section 12.4).

Two network polymers related to Prussian blue but prepared from $[Cr(CN)_6]^{3-}$ are room temperature "organic" magnets: $(Cr^{2+})_{0.36}(Cr^{3+})_{1.76}(CN^-)_6$ (with a critical temperature $T_C = 270\,K$ and containing both $[Cr(CN)_6]^{3-}$ and $[Cr(CN)_6]^{4-}$ in its network lattice) and $(V^{2+})_{0.47}(V^{3+})_{0.67}[Cr^{3+}(CN^-)_6]$ (with $T_C = 315\,K$).[24] An even higher T_C of 348 K has been obtained from a difficult-to-characterize material, $V[C_2(CN)_4]_x \cdot y(CH_2Cl_2)$, obtained by reacting the oxidizing, potentially bridging ligand tetracyanoethylene, $(NC)_2C{=}C(CN)_2$, with the reducing vanadium hexacarbonyl, $V(CO)_6$. These "organic" magnets could have significant advantages over metallic magnets, since they would be more flexible and could be produced as thin films at much lower temperatures.[25]

16.8 Two-Electron Inner-Sphere Electron-Transfer Processes

Although two-electron inner-sphere electron-transfer processes are probably more common in the *p* block, where stable oxidation states generally differ by ±2, an example from the *d* block is the reaction involving Pt(II) and labeled Pt(IV):

$$\begin{aligned} Cl^- + [Cl_4Pt]^{2-} + [{}^*PtCl_6]^{2-} &\rightarrow [Cl_5Pt^{II}{-}Cl{-}{}^*Pt^{IV}Cl_5]^{5-} \\ &\rightarrow [Cl_5Pt^{IV}{-}Cl{-}{}^*Pt^{II}Cl_5]^{5-} \\ &\rightarrow [Cl_6Pt]^{2-} + [{}^*PtCl_4]^{2-} + Cl^- \end{aligned} \qquad (16.39)$$

This reaction can be regarded as the transfer of a six valence electron chlorine cation, Cl^+, from the platinum(IV) atom to the unshared electron pair in the d_{z^2} orbital of the platinum(II) atom.

Conversely, it can also be regarded as a type of ligand-substitution reaction: An electron-pair is donated from the platinum(II) d_{z^2} orbital of the attacking $PtCl_5{}^{3-}$ "ligand" to the Lewis acid Cl^+, displacing its other ligand, another $:PtCl_5{}^{3-}$.

$$[Cl{-}Pt{:}\,(Cl)_4]^{3-} + [Cl{-}Pt(Cl)_4{-}Cl]^{2-} \rightarrow [Cl{-}Pt(Cl)_4{-}Cl]^{2-} + [{:}Pt(Cl)_4{-}Cl]^{3-} \qquad (16.40)$$

Electronically, this may not be at all unreasonable: the d_{z^2} orbital of the Pt(II) is likely the highest occupied molecular orbital (HOMO) of the $PtCl_4{}^{2-}$, while the Cl–Pt σ^* orbital may be, if not the lowest unoccupied molecular orbital (LUMO), one of the lowest lying empty orbitals of the $PtCl_6{}^{2-}$. Since the energies of halogen–metal σ^* orbitals depend strongly on the halogen (being lowest for iodine and highest for fluorine), it is not surprising that the rates of these reactions are very dependent on the identity of the halogen. For self-exchange in the following ruthenocene–ruthenocenium system:

$$Cp_2RuX^+ + {}^*RuCp_2 \rightarrow Cp_2Ru + [X^*RuCp_2]^+ \quad (16.41)$$

the rate constant k_{11} is found to be 16.5 for X = Cl, 1600 for X = Br, and 2.1×10^6 for X = I.[26]

Similar reactions are known with nonmetal ligands displacing others around Cl^+:

$$R_3P + Cl\text{–}C_5Cl_5 \rightarrow R_3PCl^+ + C_5Cl_5^- \quad (16.42)$$

$$R_3P + I\text{–}I \rightarrow [R_3P\cdots I\cdots I] \rightarrow [R_3PI]^+I^- \rightarrow R_3PI_2 \quad (16.43)$$

In examples of Reaction (16.43), using different R groups and bromine as well as iodine, intermediates such as $[Ph_3P\cdots Br\cdots Br]$ can sometimes be isolated,[27,28] in addition to the expected organophosphoranes R_3PX_2 and phosphonium halides $[R_3PX]^+X^-$.

Bridging halogen atoms play a key role in the properties of some semiconducting mixed-valence Robin and Day Class II linear polymeric complexes, $^1_\infty[L_4M^{II}\cdots X\text{–}L_4M^{IV}\text{–}X\cdots]^{n\pm}$, in which X is Cl, Br, or I; M^{II} is square planar Pt^{2+}, Pd^{2+}, or Ni^{2+}, M^{IV} is octahedral Pt^{4+}, Pd^{4+}, or Ni^{4+}, and L_4 is four amine ligands, four halide ligands, two bidentate amine ligands, and so on.[29] The classic example of this type of compound is Wolffram's red salt, $^1_\infty[(\text{ethylamine})_4Pt^{II}\cdots Cl\text{–}(\text{ethylamine})_4Pt^{IV}\text{–}Cl\cdots]Cl_4\cdot 4\,H_2O$, in which the Pt^{IV}–Cl bond length is 226 pm while the secondary $Pt^{II}\cdots Cl$ bond length is 313 pm. In such a chain, the oxidation states of the two types of platinum can be interchanged by a halogen–cation transfer process of a very simple type: a concerted in-phase vibrational movement of the axial halogen atoms. As the axial ligand is changed from Cl to Br to I, the two bond lengths become more nearly equal, while the electrical conductivity improves and the intervalence charge-transfer absorption band moves to lower energy. The same consequences occur as the metal atom is changed from Pt to Pd to Ni. This trend reaches the ultimate conclusion in the compound $^1_\infty[(1,2\text{-cyclohexanediamine})_2Ni^{III}\text{–}Br\text{–}(1,2\text{-cyclohexanediamine})_2Ni^{III}\text{–}Br\text{–}]Br_4$, in which the two Ni–Br bond lengths are equal. Both metal atoms may therefore be assigned the oxidation number +3; they are strongly antiferromagnetically coupled to each other. The material therefore is at least close to being a Robin and Day Class III polymer; it has a very small band gap of 1.28 eV.

Two-Electron Processes Not Involving Halogen Bridges. Other two-electron processes involve the transfer of the oxygen atom, for example, in the reduction of $V^{IV}O^{2+}(aq)$ by $[V(H_2O)_6]^{2+}$, which produces a long-lasting bridged complex, $[(aq)V^{III}(\text{–}\mu\text{–OH–})_2V^{III}(aq)]^{4+}$, presumably by oxygen atom transfer and proton redistribution.

The numerous oxidation–reduction reactions involving *p*-block oxo anions[30] with each other or with monoatomic anions would seem to offer a rich source of oxygen atom transfers. Proof that this can happen has been obtained using hypochlorous acid labeled with the oxygen-18 isotope:

$$Cl^{18}OH + {:}NO_2^- \rightarrow Cl^- + {}^{18}ONO_2^- + H^+ \quad (16.44)$$

In fact, many reactions of this type are found to be quite slow unless they are catalyzed by the addition of hydrogen ions; the rate laws are often first or second order in $[H_3O]^+$. For example, the oxidation of halide ions by chlorate:

$$6\,X^- + ClO_3^- + 6\,H^+ \rightarrow 3\,X_2 + Cl^- + 3\,H_2O \qquad (16.45)$$

follows the rate law:

$$\text{Rate} = k[ClO_3^-][X^-][H^+]^2 \qquad (16.46)$$

The first three steps of the mechanism (followed by other fast ones) are interpreted as follows:

$$2\,H^+(aq) + ClO_3^- \rightleftharpoons H_2OClO_2^+ \; [= ClO_2^+(aq)] \qquad \text{(Fast)}$$

$$X^- + H_2OClO_2^+ \rightarrow X\text{–}ClO_2 + H_2O \qquad \text{(Slow)}$$

$$XClO_2 + X^- \rightarrow X_2 + ClO_2^- \qquad \text{(Fast)} \qquad (16.47)$$

It would thus appear that, in many cases, oxygen does not function as a bridging ligand in the mechanism: The bridge is instead the lone pair on the central atom of the anion being oxidized. For this bridge to form [as in the slow step of Reaction (16.47)], an oxide ion must be dissociated. However, oxide ions are very poor leaving groups, due to their double negative charge. The function of the hydrogen ion or ions is to convert them to water molecules or hydroxide ions, which are much better leaving groups, and which leave behind Lewis acidic (hydrated) oxo cations or oxo molecules such as ClO_2^+, Cl^+, NO^+, NO_2^+, CO_2, SO_2, and SO_3. In the slow step, these electron-deficient oxo cations or molecules transfer from the lone pair of water to the lone pair of the reducing agent, to form (as above) an oxo halide, which then undergoes rapid hydrolysis (Section 14.6).

In general, the higher the oxidation state (nominal charge) and the smaller the size of the central atom of the oxo anion, the slower its water-exchange and redox reactions of this type. Thus, the rates of reactions of the oxo anions of chlorine fall in the order $ClO_4^- < ClO_3^- < ClO_2^- < ClO^-$, and among halate ions, $ClO_3^- < BrO_3^- < IO_3^-$. This is understandable in terms of the relative acidities of the central atoms (i.e., $Cl^{7+} > Cl^{5+} > Cl^{3+} > Cl^+$), with the more acidic central atom being more reluctant to let the water molecule be a leaving group. It also explains why, although perchloric acid is a very potent oxidizing agent with a high $E°$, hypochlorite or hypochlorous acid generally acts much more quickly. This kinetic slowness but thermodynamic potency also contributes to the formidable explosive potential of perchlorates and nitrates (Section 6.2).

Study Objectives

1. Review the use of rate data to establish orders of reactions. Use information on the detection of intermediates to classify the stoichiometric mechanism as associative, interchange, or dissociative. Exercises 1–4.
2. Predict the periodicity of the effects of changing central (metal) atoms on (a) the associative or dissociative nature of a reaction, or (b) the rate of a series of reactions of common mechanistic type. Exercises 5–10.
3. Interpret rate data involving a series of ligands to classify the intimate mechanism of a series of substitution reactions as dissociative or associative. Exercises 11–15.
4. Predict or interpret the effects of changing the entering or leaving group on the rate of a type of substitution reaction. Exercises 16–20.
5. Predict or interpret the effects of changing spectator ligands on the rate of a type of substitution reaction. Exercises 21–23.
6. Predict or explain trans effects in square planar complexes; use them to design syntheses of specific geometric isomers of square planar complexes. Exercises 24–26.
7. Use the Marcus equation to calculate rates of outer-sphere cross-reactions, or given those rates, calculate other factors in the Marcus equation. Exercises 27–29.
8. Write mechanisms for inner-sphere or outer-sphere electron-transfer reactions; deduce from the evidence which mechanism is possible in a given case. Exercises 30–34.
9. Describe how the nature of a bridging ligand affects intervalence charge transfer and electron transport in bridging metal complexes and materials. Exercises 35–39.

Exercises

1. *In the study of a ligand substitution reaction $[MX_4]^{2-} + Y^- \rightarrow [MX_3Y]^{2-} + X^-$, a species $[MX_4Y]^-$ is identified via its characteristic visible spectrum. Is this reaction, likely to be associative (A), dissociative (D), interchange (I), or can no conclusion be drawn?
2. In the study of a ligand substitution reaction $[MX_6]^{3-} + Y^- \rightarrow [MX_5Y]^{3-} + X^-$, it proves impossible to detect any intermediate by the spectroscopic or other methods available in your laboratory. Is this reaction associative (A), dissociative (D), interchange (I), or can no conclusion be drawn? Explain your reasoning.
3. In Section 9.3, we discussed the rates of fluxional exchange in species such as metal η^1-cyclopentadienyls:

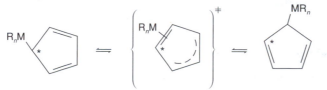

(a) Justify classifying this mechanism as one of the following: associative (A), dissociative (D), or interchange (I). (b) Suppose that, on doubling the concentration of $R_nM(\eta^1\text{-Cp})$, you found that the rate of this exchange quadrupled. Would this be consistent or inconsistent with the above proposed mechanism? Why? If it is inconsistent, suggest another mechanism that should be considered.

4. In Section 8.8, the rate of exchange of coordinated water between primary and secondary hydration spheres was discussed. When such an exchange occurs as described, would we be likely to classify the mechanism as associative (A), dissociative (D), or interchange (I)?

5. You are about to undertake kinetic comparisons of intimate mechanisms of ligand substitutions of some fourth-period metal ions such as $[Cr(H_2O)_6]^{3+}$ with their fifth-period counterparts such as $[Mo(H_2O)_6]^{3+}$. The ions of which period are more likely to follow dissociative mechanisms? The ions of which period are more likely to follow associative mechanisms?

6. *You are about to undertake kinetic comparisons of rates and intimate mechanisms of fluoride-ion exchanges of SF_6 and free F^- versus SeF_6 and free F^-, in a suitable inert solvent. Which hexafluoride is more likely to follow a dissociative mechanism? Which hexafluoride is more likely to follow an associative mechanism? Which hexafluoride is more likely to react rapidly?

7. Water exchange mechanisms for $[M(H_2O)_6]^{3+}$ are more likely to occur by dissociation rather than association as one goes to the right across a block of the periodic table. Explain. Would this be more noticeable in the d or the f block? Why?

8. Which late fourth-period d-block +2 charged hydrated ion, $[M(H_2O)_6]^{2+}$ ($M = Mn$ through Zn), should exchange water least rapidly, and why?

9. *Which late fifth-period d-block +3 charged hydrated ion, $[M(H_2O)_6]^{3+}$ ($M = Tc$ through Ag), should dissociatively exchange water least rapidly, and why?

10. Arrange the following in order of increasing rates of exchange (by a dissociative mechanism) of coordinated and solvent water: $[Os(H_2O)_6]^{2+}$, $[Fe(H_2O)_6]^{2+}$, $[Zn(H_2O)_6]^{2+}$, and $[Ru(H_2O)_6]^{2+}$.

11. Substitution reactions of $Pt(PF_3)_4$ with a series of ligands are found to proceed at a rate that is basically independent of the identity of the entering ligand. Is this reaction dissociative or associative?

12. Ligand exchange reactions of dialkyl ethers OR_2 ($R = Me$ or Et) with $[MCl_5(OR_2)]$ ($M = Ta$ or Sb) in inert solvents are about the same whether R is Me or Et, and show very positive entropies of activation. (a) Is this reaction dissociative or associative? (b) Fill in the unidentified exponents in the rate law for this reaction: Rate = $k[MCl_5(OR_2)]^x[OR_2]^y$.

13. *Suggest a method of determining whether the rate of formation of sixth-period f-block complex ions such as $[La(H_2O)_7Y]^{2+}$ from the octahydrated ions and Y^- is associative or dissociative.

14. Complex ions such as $[SiF_5]^-$ can be isolated with large nonacidic counterions, but CF_5^- cannot. Of what relevance, if any, is this to the question of whether the associative substitution reactions of tetrahedral carbon and silicon compounds proceed by A or by I_a mechanisms?

15. The volumes of activation for exchange of coordinated and free water among fourth-period *d*-block +3 charged metal ions are all negative, and vary from $-12.2\ cm^3\ mol^{-1}$ at the left for $[Ti(H_2O)_6]^{3+}$ to $-5.4\ cm^3\ mol^{-1}$ at the right for $[Fe(H_2O)_6]^{3+}$. (a) Describe the horizontal periodic trend in associative versus dissociative character of water exchange among these ions. (b) Compare these values with those given in the text for +2 charged ions. What do these values suggest concerning the effects of additional metal-ion positive charge on the associative versus dissociative character of water exchange? Is this reasonable?

16. The ClO_4^- ligand is actually a slightly faster leaving group than $CF_3SO_3^-$, but it is nonetheless preferred to use $[M(NH_3)_5(O_3SCF_3)]^{2+}$ as a starting material to make substituted complexes $[M(NH_3)_5Y]^{2+}$. Why?

17. The substitution reaction of solvent water to replace F^- in $[Cr(H_2O)_5F]^{2+}$ has a rate constant of $6.2 \times 10^{-10}\ s^{-1}$ in neutral solution, but $1.4 \times 10^{-8}\ s^{-1}$ in acid solution. (a) Explain the reason for the rate enhancement in acid solution. (b) Would you expect a similar rate enhancement if you added Ag^+ instead of H^+? (c) Would you expect a similar rate enhancement with $[Cr(H_2O)_5Br]^{2+}$ in acid solution? (d) Would you expect a rate enhancement with $[Cr(H_2O)_5Br]^{2+}$ if you added Ag^+ instead of H^+? (e) Would you expect a rate enhancement with $[Co(NH_3)_5(CO_3)]^+$ in acid solution?

18. The order of ability for X substituents to act as leaving groups from chiral silicon or phosphorus(V) centers is

$$X : Cl^-, Br^-, {}^-O_2CMe > F^-, {}^-SR > {}^-OMe, H^-$$

Compare this with the corresponding series given in the text for groups leaving Pt complexes [Eq. (16.23)]. (a) Are there any specific reversals between the two series? (b) Does the silicon–phosphorus series show the same kind of general periodic trend as the platinum series? Why might this be?

19. *Table 16.7 gives the nucleophilicity parameters n_{Pt} for several nucleophiles; other values are given by M. L. Tobe, in *Comprehensive Coordination Chemistry*, Vol. 1, G. Wilkinson, Ed., Pergamon, Oxford, 1987, p. 281. (a) Values of n_{Pt} of <2.2 and 4.18 are found for the two halide ions not given in Table 16.7. Complete the set of values for the halide ions. (b) Values of 2.90, 3.22, and 7.17 are found for the following sulfur donor ligands: PhS^-; Ph_2S; and PhMeSO. Associate each value with one of the ligands, and explain your choices.

20. (a) The three n_{Pt} values of 3.48, 5.75, and 7.11 are found for the three pseudohalide ions selenocyanate, azide, and thiocyanate. Associate each value with one of the ligands, and explain your choices. (b) The organometallic ligands of Group 15(V) have the following values: 8.93 for Ph_3P; 6.89 for Ph_3As; and 6.79 for Ph_3Sb. Explain these values, and make some rough predictions for the values for Ph_3N and Ph_3Bi.

21. *The reaction of $CrCl_3$ with liquid NH_3 to give yellow $[Cr(NH_3)_6]Cl_3$ often has a low yield, since the reaction tends to stop with the production of pink $[Cr(NH_3)_5Cl]Cl_2$. However, adding a small piece of Na metal to the liquid NH_3 results in complete conversion of the $CrCl_3$ to $[Cr(NH_3)_6]Cl_3$. Offer two plausible explanations of this rate enhancement for the final step of the substitution process.

22. (a) Would you expect the reaction of $CrCl_3$ with liquid NMe_3 to give $[Cr(NMe_3)_{6-x}Cl_x]^{(3-x)+}$ to proceed as readily and as completely as the corresponding reaction with liquid NH_3 mentioned in Exercise 21? Why or why not? (b) Would you expect the completeness of this reaction to be enhanced by adding a cube of Na? Why or why not?

23. You are comparing the effect of ethyl substituents next to the nitrogen donor atoms in the bidentate ligands 1,10-phenanthroline and 2,9-diethyl-1,10-phenanthroline (see Table 8.5 for structures) and want to see how they affect ligand substitution rates in the following complexes. In each case, first tell whether you expect the substitution process to be associative or dissociative, and second, tell whether you expect the 2,9-diethyl-1,10-phenanthroline to show an enhanced rate of substitution, a reduced rate of substitution, or neither. (a) $[Pt(phenanthroline)Cl_2]$; (b) $[Si(phenanthroline)Cl_2]^{2+}$; (c) $[Si(phenanthroline)_2Cl_2]^{2+}$; and (d) $[Be(phenanthroline)F_2]$.

24. Predict the geometries of the complexes produced by the following reactions, taking Cl^- as the leaving group:

(a) $[PtCl_3(CH_2{=}CH_2)]^- + NH_3 \rightarrow$

(b) *cis*-$[PtCl_2(SCN)(PR_3)]^- + SCN^- \rightarrow$

(c) $[PtCl_3(pyridine)]^- + NO_2^- \rightarrow$

25. *Design syntheses of the (a) cis and (b) trans isomers of $[PtCl_2(CN)(NH_3)]^-$ using $PtCl_4{}^{2-}$, NH_3, and CN^-.

26. Design syntheses of the (a) cis and (b) trans isomers of $[PtCl_2(NO_2)(pyridine)]^-$ using $[PtCl_4]^{2-}$, pyridine, and $NO_2{}^-$.

27. Using data in Table 16.9, identify the products and calculate the approximate rate constants expected for the following outer-sphere one-electron transfer reactions: (a) $[Fe(CN)_6]^{4-} + IrCl_6^{2-}$; (b) $[Fe(CN)_6]^{4-} + [Fe(phenanthroline)_3]^{3+}$; (c) $[Fe(CN)_6]^{4-} + MnO_4{}^-$; and (d) $[Fe(phenanthroline)_3]^{3+} + MnO_4{}^{2-}$.

28. Using data in Table 16.9, identify the products and calculate the approximate rate constants expected for the following outer-sphere one-electron transfer reactions: (a) $Fe^{2+}(aq) + [Fe(phenanthroline)_3]^{3+}$; (b) $Fe^{2+}(aq) + IrCl_6^{2-}$; and (c) $Cr^{2+}(aq) + Fe^{3+}(aq)$.

29. The observed rate constant for the outer-sphere one-electron-transfer reaction of $[Co(bipyridyl)_3]^{3+}$ with $[Co(terpyridyl)_2]^{2+}$ to give $[Co(bipyridyl)_3]^{2+}$ and $[Co(terpyridyl)_2]^{3+}$ is 64 L mol s^{-1}. Use this data and the rate constants for self-exchange from Table 16.9 to calculate the approximate value of $E°$ for this reaction.

30. *(a) Write a hypothetical inner-sphere redox mechanism for the reaction of $[Cr(H_2O)_6]^{2+}$ with *cis*-$[Co(NH_2CH_2CH_2NH_2)_2I_2]^+(aq)$ in acid solution to give (ultimately) *cis*-$[Cr(H_2O)_4I_2]^+$, $Co^{2+}(aq)$, and $[NH_3CH_2CH_2NH_3]^{2+}$. (b) What evidence could support this mechanism and why?

31. For which reducing agent, $Cr^{2+}(aq)$ or $Eu^{2+}(aq)$, would it be easier to prove that a reduction involved an inner-sphere mechanism? Why?

32. Suppose that you observe a transient bright-red color during the reaction in acidic solution of $[Co(NH_3)_5(N_3)]^{2+}$ with $Fe^{2+}(aq)$, which then fades away as the reaction ultimately gives $NH_4{}^+$, the azide ion $N_3{}^-(aq)$, $Fe^{3+}(aq)$, and $Co^{2+}(aq)$. Does the transient color suggest that this reaction has an *inner-* or *outer-sphere* mechanism? Write a plausible formula for the transient bright-red species.

33. If the rate of the reaction of $[Rh(CN)_6]^{4-}$ with $[Co(C_2O_4)_3]^{3-}$ to give $[Rh(CN)_6]^{3-}$ and $[Co(C_2O_4)_3]^{4-}$ were found to be rapid, what would this suggest about the mechanism of the reaction and why?

34. Suppose that you are studying the rate of reduction of $[Co(NH_3)_5(SR)]^{2+}$ (R = alkyl) with Cr^{2+}(aq) to give an unidentified chromium(III) product, and Co^{2+}(aq). What would you conclude about the mechanism of the reaction (and why) if you found (a) that the rate of the reaction was insensitive to the size of the R group; or (b) that the rate was strongly affected by the bulk of the R group, and a transient intense color were observed when R was not bulky? In each case, also predict the identity of the chromium(III) containing product.

35. *Using labile $[Cr(H_2O)_6]^{2+}$ as reducing agent, the rates of reduction of the following para-substituted pyridine complexes of inert cobalt(III) are to be studied: (**I**) $[Co(NH_3)_5(NC_5H_5)]^{3+}$; (**II**) $[Co(NH_3)_5(NC_5H_4\text{-}p\text{-}CH_2NH_2)]^{3+}$; and (**III**) $[Co(NH_3)_5(NC_5H_4\text{-}p\text{-}CONH_2)]^{3+}$. (a) Which cobalt complex is least likely to be reduced by an inner-sphere mechanism? (b) The reduction of (**III**) proceeds with a rate constant $k = 17.4$, while the reduction of (**I**) proceeds with a rate constant $k = 4.0 \times 10^{-3}$. What is a likely explanation of the difference? (c) What is the likely chromium-containing product from the reduction of (**III**)? From the reduction of (**I**)? (d) Would you expect the reduction of (**II**) to proceed more rapidly than that of (**III**)? Why or why not?

36. Consider the mixed-valence complex ion $[(H_3N)_5Os\text{–}N{\equiv}N\text{–}Os(NH_3)_5]^{5+}$. Assume that the conformation of this complex is such that the four equatorial ammine groups on each Os are directly in line with each other. (a) Assign the average oxidation number of Os. (b) Would this ion have an inversion center (Chapter 9) if it were a Robin and Day Class I intervalence compound? Class II? Class III? For each possible classification, assign the symmetry point group of this ion. (c) How could the appearance or nonappearance of the symmetric N≡N stretching frequency in the IR spectrum of this ion help assign the Robin and Day classification? (d) How would the presence or absence of intense color (an intense visible absorption) of this ion help assign the Robin and Day classification?

37. The iron atoms in the following mixed-valence biferrocenylene complex have been found to be equivalent:

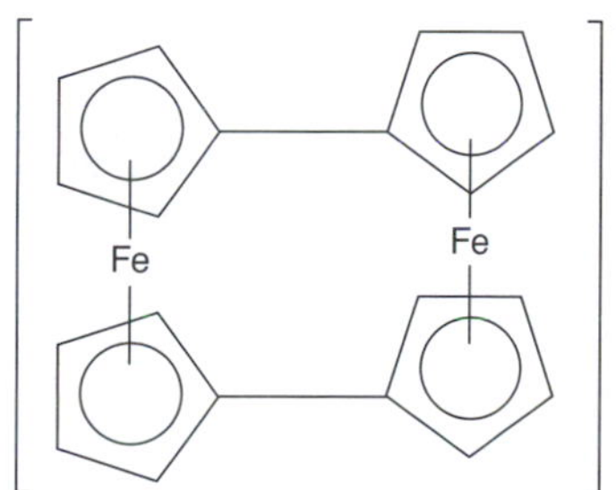

(a) Classify this in the Robin and Day scheme. Would this compound be expected to show an intervalence charge–transfer band? Would its spectrum be expected to resemble a superposition of the spectra of ferrocene and the ferrocenium ion?
(b) Suppose that you devised a synthetic scheme by which you could link units like this into a chain polymer, $^1_\infty\{[Fe_2(C_{10}H_6)_2]^+[X^-]\}$ (X = a large nonbasic anion). Would you predict this new material to be an insulator, a semiconductor, or a metallic conductor? Might it be ferromagnetic at moderately high temperatures?

(c) Speculate on how your answers to (a) and (b) might be changed if, instead, you started with a methylene-bridged analogue of the above ion, $[Fe_2(C_5H_4{-}CH_2{-}C_5H_4)_2]^+$.

38. Consider an extensive series of mixed-valence ruthenium(II,III) complexes $[(H_3N)_5Ru^{II}{-}(A{-}A)_n{-}Ru^{III}(NH_3)_5]^{5+}$, in which both the nature of A and the number of repeating units of the bridging unit, *n*, are being varied. (a) For –A–A– = $-CH_2{-}CH_2-$, would you expect to find the rate of intervalence electron transfer to increase, decrease, or not change as *n* increases? (b) For a common value of *n*, would you expect to find the intervalence electron transfer to be faster for –A–A– = $-CH_2{-}CH_2-$ or for –A–A– = –C≡C–? Which bridging group would be more likely to result in a Robin and Day Class I material? Which in a Class II material? (c) Some bridging –A–A– groups, such as –S–S– or a segment of the DNA molecule, tend to curl themselves into helices. How might this affect the relationship you found between *n* and the rate of intervalence electron transfer in these complexes? (d) Suppose that you could attach an $(H_3N)_5Ru^{II}$ group to one end and a $Ru^{III}(NH_3)_5$ group to the other end of a polythiazyl chain, $^1_\infty[SN]$ (Section 14.8). What Robin and Day classification might you expect for a solid sample of this material?

39. In each set of materials, assign one as likely to belong to to Robin and Day Class I, one to Class II, and one to Class III, and one as not an intervalence compound: (a) $YBa_2Cu_3O_7$ at low temperatures, black Co_3O_4, $[Co(NH_3)_6]_2[Co(EDTA)]_3$ where $EDTA^{4-}$ = ethylenediamentetraacetate ion, $Co_2(CO)_8$ and (b) $[Re_2Cl_8]^{3-}$, $[BrO]^+[BrO_3^-]$, $[NO_2]^+[NO_3]^-$, $[ClFe(\mu\text{-}Cl)_3Fe(\mu\text{-}Cl)_3FeCl]$.

Notes

1. The classifications introduced in this and Section 16.2 were introduced in C. H. Langford and H. B. Gray, *Ligand Substitution Processes*, Benjamin, New York, 1996.
2. Of course, multiple hydration spheres are especially characteristic of aqueous solutions. However, most kinetic data is for aqueous solutions, and even if the primary sphere is composed of ligands such as NH_3 or Cl^-, we would expect hydrogen bonding of water to this hydrogen sphere. In addition, even if the solvent is not water, when the metal complex is positively charged and the entering ligand is negatively charged, it might still be able to associate with the complex in a second coordination sphere by the type of ion pairing that leads to reduced activity coefficients in analytical chemistry (Section 3.2).
3. The parameter K depends strongly on *a*. The entering ligand is more likely to enter the secondary coordination sphere if both the ligand and the complex are large, so that *a* is large. This likelihood is justified on statistical grounds: small ions are less likely to bump into each other.
4. D. Katakis and G. Gordon, *Mechanisms of Inorganic Reactions*, Wiley-Interscience, New York, 1987; pp. 181–187.
5. R. R. Holmes, *Chem. Rev.*, **90**, 17 (1990).
6. J. D. Atwood, *Inorganic and Organometallic Reaction Mechanisms*, 2nd ed., VCH, New York, 1997; p. 14.
7. C. H. Langford, *Inorg. Chem.*, **18**, 3288 (1979); T. W. Swaddle, *Inorg Chem.*, **19**, 3203 (1980).
8. R. Åkesson, L. G. M. Pettersson, M. Sandström, and U. Wahlgren, *J. Am. Chem. Soc.*, **116**, 8705 (1994).
9. N. E. Dixon, G. A. Lawrance, P. A. Lay, and A. M. Sargeson, *Inorg. Chem.*, **22**, 846 (1983).
10. A. Tsubouchi and T. C. Bruice, *J. Am. Chem. Soc.*, **116**, 11614 (1994); *Chem. Eng. News*, Dec. 19, 1994, p. 7.
11. We saw earlier (Section 3.4) that the VSEPR principle includes the concept that the most electronegative substituent prefers the axial position in a trigonal bipyramid.
12. His Nobel lecture has been reprinted: H. Taube, *Science*, **226**, 1028 (1984).

13. Recall (Section 5.1) that good bridging ligands have more than one pair of unshared electrons available to donate to two metal atoms, and include such species as halide ions, hydroxide ion, and pseudohalide ions.
14. Since the first product, $[(H_3N)_5Co(OH_2)]^{2+}$, is labile and unstable in the presence of excess water, it subsequently undergoes further ligand exchange to give $[Co(H_2O)_6]^{2+}$.

15. S. J. Lippard and J. M. Berg, *Principles of Bioinorganic Chemistry*, University Science Books, Mill Valley, CA, 1994; pp. 244–253.
16. D. S. Wuttke, M. J. Bjerrum, T. Jy Chang, J. R. Winkler, and H. B. Gray, *Biochim. Biophys. Acta*, **1101**, 168 (1992); R. M. Baum, *Chem. Eng. News*, Feb. 22, 1993, p. 20.
17. D. N. Beratan, J. N. Betts, and J. N. Onuchic, *Science*, **252**, 1285 (1991).
18. C. Creutz, *Prog. Inorg. Chem.*, **30**, 1 (1983).
19. M. R. Arkin, E. D. A. Stemp, R. E. Holmlin, J. K. Barton, A. Hörmann, E. J. C. Olson, and P. F. Barbara, *Science*, **273**, 475 (1996).
20. D. B. Hall, R. E. Holmlin, and J. K. Barton, *Nature* (*London*), **382**, 731 (1996).

21. It would be more modern to think of this term as "interoxidation state charge transfer."

22. G. M. Brown, H. J. Krentzien, M. Abe, and H. Taube, *Inorg. Chem.*, **18**, 3374 (1979).
23. M. B. Robin and P. Day, *Adv. Inorg. Chem. Radiochem.*, **10**, 248 (1967).
24. O. Sato, T. Iyoda, A. Fujishima, and K. Hashimoto, *Science*, **271**, 49 (1996); S. Ferlay, T. Mallah, R. Ouahès, P. Veillet, and M. Verdaguer, *Nature* (*London*), **378**, 701 (1995).
25. J. S. Miller and A. J. Epstein, *Chem. & Eng. News*, Oct. 2, 1995, p. 30.
26. T. M. Shea, S. P. Deraniyagala, D. B. Studebaker, and T. D. Westmoreland, *Inorg. Chem.*, **35**, 7699 (1996).
27. N. Bricklebank, S. M. Godfrey, A. G. Mackie, C. A. McAuliffe, R. G. Pritchard, and P. J. Kobryn, *J. Chem. Soc. Chem. Commun.*, 355 (1992); N. Bricklebank, S. M. Godfrey, A. G. Mackie, C. A. McAuliffe, and R. G. Pritchard, *J. Chem. Soc. Dalton Trans.*, 2261 (1993).
28. A dissenting interpretation is given by M. A. H. A. Al-Juboori, P. N. Gates, and A. S. Muir, *J. Chem. Soc. Dalton Trans.* 1441 (1994).
29. R. J. H. Clark, *Chem. Soc. Rev.*, **13**, 219 (1984); R. J. H. Clark, *Chem. Soc. Rev.*, **19**, 107 (1990).
30. J. O. Edwards, *Inorganic Reaction Mechanisms*, Benjamin, New York, 1964, Chapter 8.

CHAPTER

17

Advanced Topics: Excited Electronic States, Photochemistry, and Activated Molecules

With Applications to Materials Science, Atmospheric Chemistry, and Biochemistry

17.1 Electronic States and Term Symbols

For many atoms and ions and some molecules the electronic configurations that we have derived [e.g., $1s^2 2s^2 2p^2$ for C, $1s^2 2s^2 2p^6 3s^2 3p^6 3d^2$ for V^{3+}, and $\sigma_g(2s)^2 \sigma_u$*-$(2s)^2 \sigma_g(2p_z)^2 \pi_u{}^4 \pi_g{}^{*2}$ for O_2] are inadequate to describe all of their chemical and spectroscopic properties fully. In the case of O_2, this does not explain the existence of different forms of O_2, singlet oxygen and triplet oxygen, with greatly different reactivity and even magnetic properties, even though each form has the above electronic configuration. In the case of *d*-block metal ions, the electron configuration by itself does not explain why the spectrum of a d^1 ion such as Ti^{3+} ion has only one absorption band, but the spectrum of a d^2 ion such as Ti^{2+} or V^{3+} has more than one ultraviolet (UV), visible, or near-infrared (IR) absorption band (Section 8.4; Fig. 8.5). In that section, it was suggested that the spectra of d^2 through d^8 ions are influenced by the different ways that the *d* electrons repel each other, depending on which of the five *d* orbitals are occupied.

Let us begin to analyze such phenomena by seeing how many ways the last two valence electrons can be arranged in the available orbitals in each of the above cases (C, V^{3+}, and O_2). Each distinct arrangement of electrons is known as a **microstate** of that overall electron configuration. To begin, let us write all possible microstates of the $\pi_g{}^{*2}$ electron configuration of O_2. There are two $\pi_g{}^*$ orbitals; let us designate the empty orbitals as $\pi_x{}^*(\ )$ and $\pi_y{}^*(\ )$. The two electrons may enter either with spin up ($m_s = +\frac{1}{2}$) or spin down ($m_s = -\frac{1}{2}$), which we may designate by inserting ↑ and ↓, respectively, in the parentheses. There are six allowed ways of doing this, that is, six microstates: $\pi_x{}^*(\uparrow\downarrow)\pi_y{}^*(\ )$; $\pi_x{}^*(\ )\pi_y{}^*(\uparrow\downarrow)$; $\pi_x{}^*(\uparrow)\pi_y{}^*(\uparrow)$; $\pi_x{}^*(\downarrow)\pi_y{}^*(\downarrow)$; $\pi_x{}^*(\uparrow)\pi_y{}^*(\downarrow)$; $\pi_x{}^*(\downarrow)\pi_y{}^*(\uparrow)$. [Two other arrangements that we can conceive of, $\pi_x{}^*(\uparrow\uparrow)\pi_y{}^*(\ )$ and $\pi_x{}^*(\ )\pi_y{}^*(\downarrow\downarrow)$, are impossible because then the two electrons are identical in all respects, which violates the Pauli exclusion principle.]

In general, the total number of ways N that x electrons can be placed in y orbitals of equivalent energy with either of two different spins is given by

$$N = \frac{(2y)!}{x!(2y-x)!} \tag{17.1}$$

When we write the possible microstates for atoms and monoatomic ions, it will be most useful to label the orbitals, not as (p_x, p_y, and p_z), but rather using the corresponding values of the magnetic quantum number $m_\ell(1, 0, -1)$ (Section 1.4). This allows us to compute two useful parameters: (1) M_S, the total spin component along the z axis of all of the electrons, each of which has a spin component m_s of either $+\frac{1}{2}$ or $-\frac{1}{2}$:

$$M_S = \sum m_s \tag{17.2}$$

(2) M_L, the total orbital angular momentum component along the z axis of all the orbitals occupied by electrons:

$$M_L = \sum m_\ell \tag{17.3}$$

Example 17.1

Write all possible microstates for the p^2 electron configuration of the C atom. Compute M_S for each; compute M_L for each.

SOLUTION:

There are three p orbitals, which we may designate (in empty form) as 1( )0( )−1( ). For two electrons being placed with either of two spins in three equivalent orbitals, the number of microstates is $6!/2!(4!) = 6 \times \frac{5}{2} = 15$. Among these, there are six microstates in which the two electrons have the same spin. Three have all spins up, and hence have $M_S = +1$: 1(↑)0(↑)−1( ); 1(↑)0( )−1(↑); 1( )0(↑)−1(↑). Three have all spins down, and hence have $M_S = -1$: 1(↓)0(↓)−1( ); 1(↓)0( )−1(↓); 1( )0(↓)−1(↓). There remain nine ways of locating the electrons with opposing spins, and hence $M_s = 0$. Six ways are analogous to the first six: 1(↑)0(↓)−1( ), 1(↑)0( )−1(↓), 1( )0(↑)−1(↓); 1(↓)0(↑)−1( ), 1(↓)0( )−1(↑), 1( )0(↓)−1(↑). The last three have two electrons in the same orbital, so they cannot have analogues among the first six identical-spin possibilities: 1(↑↓)0( )−1( ), 1( )0(↑↓)−1( ), 1( )0( )−1(↑↓).

Only one microstate has $M_L = 2$, which requires two electrons in $m_\ell = 1$ orbitals: 1(↑↓)0( )−1( ); one has $M_L = -2$, 1( )0( )−1(↑↓). (*Note*: When written this way, the microstates with high orbital angular momentum even *look* lopsided.) $M_L = 1$ requires one electron in the $m_\ell = 1$ orbital and the other in the $m_\ell = 0$ orbital: 1(↑)0(↑)−1( ); 1(↓)0(↓)−1( ); 1(↑)0(↓)−1( ); 1(↓)0(↑)−1( ). Four microstates have $M_L = -1$: 1( )0(↑)−1(↑); 1( )0(↓)−1(↓); 1( )0(↑)−1(↓); 1( )0(↓)−1(↑). To have $M_L = 0$, either both electrons must have $m_\ell = 0$, or one must have $m_\ell = +1$ and the other $m_\ell = -1$; the remaining five microstates fill this condition.

Fortunately, each microstate does not result in a separate energy level, or the results of crystal field theory would be hopelessly oversimplified, and spectra would consist of an enormous number of overlapping absorption bands. [For the d^2 V^{3+} ion, there are $10!/2!(8!) = 10 \times \frac{9}{2} = 45$ microstates!] Instead each distinct energy level or **electronic state** (energy level that is available to a collection of electrons) involves (usually) contributions from more than one of these microstates, so that a given electron configuration gives rise to a relatively small number of electronic states.

The most important of these states is the energetically favored (lowest energy) **ground** electronic state. Hund's rule tells us that in this state the electrons are not paired with each other in a given orbital until all orbitals in the set have been occupied once by electrons of the same spin. In our $\pi_g{}^{*2}$ example, this is clearly the case for two microstates: $\pi_x{}^*(\uparrow)\pi_y{}^*(\uparrow)$ and $\pi_x{}^*(\downarrow)\pi_y{}^*(\downarrow)$. These must be of equal energy, and participate in the same electronic state. The same conclusion can be drawn for the first six microstates we identified for p^2 carbon in Example 17.1.

Term Symbols. Each electronic state can be designated by a **term symbol**. For example, the ground electronic states are designated by the term symbols ^3P for C, ^3F for V^{3+}, and $^3\Sigma_g{}^-$ for O_2. The superscripts in these term symbols indicate the **spin multiplicity** of the electronic state: the number of ways in which that number of unpaired electrons can be achieved, which is equal to one more than the total number of unpaired electrons in the molecule or ion [Eq. (17.4)]. Thus, there is only one way to arrange the spins of the electrons in a set of equivalent orbitals so as to produce no unpaired electrons: by putting equal numbers with spin up and with spin down. There are two ways to achieve one net unpaired electron: by having one extra with spin up, or by having one extra with spin down.

$$\text{Multiplicity of state} = (\text{No. unpaired e}^-) + 1 = 2S + 1 \tag{17.4}$$

As indicated by this equation, the multiplicity can also be computed from the total **spin quantum number S** of the electronic state; the total spin is simply the maximum value of M_S found among the microstates participating in that electronic state.

When a term symbol is pronounced out loud, the superscript "1" is read as "singlet", "2" as "doublet", "3" as "triplet", and so on. The ground states of C, V^{3+}, and O_2 that we have been discussing are therefore called *triplet* electronic states, since each has two unpaired electrons, and there are three ways in which the two unpaired electrons can be achieved: both electrons may have spin up; both may have spin down; and there is a more complex third (hybrid) way that need not concern us here. Atoms or ions with singlet electronic states have no unpaired electrons, so they are diamagnetic; atoms or ions with doublet or higher states are paramagnetic.

The letter part of the term symbol has a different origin for molecules or polyatomic ions than it does for atoms and monoatomic ions. Later we shall see that for molecules or polyatomic ions it is an irreducible-representation label; if there is only one electron involved, it is simply the capitalized version of the irreducible representation of the orbital of that electron. The letters used for term symbols in spherical atoms and monoatomic ions have a spectroscopic[1] origin, although they do correspond to irreducible representations in the point group of a sphere.

The letter stands for the **total orbital angular momentum quantum number L** of the atom or ion (Table 17.1), which is equal to the maximum value (among the different microstates) of the sums M_L of the angular momentum quantum numbers m_ℓ (recall values of m_ℓ from Table 1.2B) of the individual orbitals being occupied

$$L = \text{maximum value of } M_L \tag{17.5}$$

Total orbital angular momentum results from unequal occupation of the equivalent atomic orbitals oriented in different directions, which usually causes the atom as a whole to lose its spherical symmetry—only in an S electronic state does it keep spherical symmetry.

For higher values of L, there is degeneracy as indicated in Table 17.1. Thus, for $L = 1$, there are three possible values of M_L $(1, 0, -1)$, corresponding to three possible orientations of the atom; this is analogous to the degeneracy in orbital orientations for a p atomic orbital (for which quantum number $\ell = 1$). The orbital degeneracy is simply equal to $2L + 1$.

Table 17.1
Letters Used in Atomic Term Symbols

Letter	S	P	D	F	G	H	I	K
Value of L	0	1	2	3	4	5	6	7
Degeneracy of L	1	3	5	7	9	11	13	15

Let us consider some simple examples of atomic term symbols. The hydrogen atom with a $1s^1$ electron configuration has two degenerate microstates: $1s(\uparrow)$ and $1s(\downarrow)$; its term symbol is ^2S. The superscript 2 indicates that there are two spin possibilities; the letter S indicates the absence of orbital angular momentum (since an s orbital has none). The helium atom has a $1s^2$ electron configuration that can only be achieved one way, $1s(\uparrow\downarrow)$, so its term symbol shows a singlet electronic state, ^1S. *Any filled set of orbitals by itself gives a nondegenerate* ^1S *electronic state*, which contributes neither spin nor orbital angular momentum to the atom, so filled sets of orbitals (such as all core orbitals) can be ignored in assigning electronic states.

In the electron configuration of boron, the only possibilities of degeneracy are with the last valence electron, $2p^1$, which can be arranged in six ways (six microstates): $2p_x(\uparrow)$, $2p_x(\downarrow)$, $2p_y(\uparrow)$, $2p_y(\downarrow)$, $2p_z(\uparrow)$, $2p_z(\downarrow)$. These are all equal in energy, and participate in the ^2P electronic state, which has a doublet spin multiplicity times a threefold orbital degeneracy. By very similar reasoning, the $3d^1$ Ti^{3+} ion has one electronic state, represented by the term symbol ^2D, which has doublet spin multiplicity times a fivefold orbital degeneracy: 10 equivalent microstates, or ways of arranging the one electron in the five $3d$ orbitals.

For the fluorine atom, with five p electrons, the number of possible microstates is $6!/5!(1)! = 6$, the same number as for boron with one p electron. These amount to one *hole* being present in the filled set of p orbitals: This hole has the same possibilities as boron's one electron, and gives rise to the same term symbol, ^2P. Note that the term symbol for one-electron or one-hole atoms or ions uses the capitalized version of the letter designation of the orbital involved.

Carbon, with two p electrons, is a more complex example. As indicated earlier, not all microstates are energetically equivalent; there are distinct electronic states, including the ground state and one or more (in this case, two) **excited states.** These have been

Table 17.2A
Multiple Terms of Various Electronic Configurations

s^2, p^6, and d^{10}	^1S
p and p^5	^2P
p^2 and p^4	$^3P, {}^1D, {}^1S$
p^3	$^4S, {}^2D, {}^2P$
d and d^9	^2D
d^2 and d^8	$^3F, {}^3P, {}^1G, {}^1D, {}^1S$
d^3 and d^7	$^4F, {}^4P, {}^2H, {}^2G, {}^2F, {}^2D, {}^2D, {}^2P$
d^4 and d^6	$^5D, {}^3H, {}^3G, {}^3F, {}^3F, {}^3D, {}^3P, {}^3P, {}^1I, {}^1G, {}^1G, {}^1F, {}^1D, {}^1D, {}^1S, {}^1S$
d^5	$^6S, {}^4G, {}^4F, {}^4D, {}^4P, {}^2I, {}^2H, {}^2G, {}^2G, {}^2F, {}^2F, {}^2D, {}^2D, {}^2D, {}^2P, {}^2S$

SOURCE: From J. C. Davis, Jr., *Advanced Physical Chemistry: Molecules, Structure, and Spectra*, 1965, The Ronald Press Co., New York.

Table 17.2B
The Splitting of Atomic Electronic States in Complexes of O_h Symmetry

Atomic term	Number of states	Terms in O_h symmetry
S	1	A_{1g}
P	3	T_{1g}
D	5	$T_{2g} + E_g$
F	7	$T_{1g} + T_{2g} + A_{2g}$
G	9	$A_{1g} + E_g + T_{1g} + T_{2g}$
H	11	$E_g + T_{1g} + T_{1g} + T_{2g}$
I	13	$A_{1g} + A_{2g} + E_g + T_{1g} + T_{2g} + T_{2g}$

determined and are listed in Table 17.2A; for carbon these are ^3P, ^1D, and ^1S. We can see that there is just one paramagnetic electronic state for carbon, ^3P, which is the state with the greatest spin multiplicity; by Hund's rule this must be the ground state. We can see that the ^1D electronic state has the greatest orbital degeneracy, fivefold. This $L = 2$ state includes one microstate each with $M_L = 2$ (e.g., $1(\uparrow\downarrow)0(\ )-1(\ )$, $M_L = 1$, $M_L = 0$, $M_L = -1$, and $M_L = -2$. There are $3 \times 3 = 9$ microstates in ^3P, $1 \times 5 = 5$ in ^1D, and one in ^1S, for a total of 15 microstates. Note that this total agrees with the total calculated for p^2 from Eq. (17.1).

It would be nice to be able to assign each microstate to one of the corresponding electronic states. This assignment is often not possible, but it is possible and useful to see that, for example, the 15 microstates we have drawn for a p^2 electron configuration (Example 17.1) match the characteristics that we expect for the combined microstates of the ^3P, ^1D, and ^1S electronic states of p^2. To do this, we should organize our different microstates in a table (Table 17.3), with total spin M_s, the sum of the spins ($+\frac{1}{2}$ or $-\frac{1}{2}$) of the individual electrons, across the top, and the M_L totals for the occupied orbitals along the vertical axis.

The ^3P electronic state must involve the majority of these microstates, nine, in a centered 3×3 block. It is easy to see six of these: The six that have nonzero total spins, in the left and right columns of Table 17.3. Three more microstates must be involved from the center as well, with M_L values of 1, 0, and -1. There are two or three possi-

Table 17.3
Arrangement of Microstates for p^2 Electron Configuration According to Their Total Spins and Angular Momenta

Total spin M_S[a]	1	0	−1
$M_L = 2$		1(↑↓)0( )−1( )	
$M_L = 1$	1(↑)0(↑)−1( )	1(↑)0(↓)−1( ) 1(↓)0(↑)−1( )	1(↓)0(↓)−1( )
$M_L = 0$	1(↑)0( )−1(↑)	1( )0(↑↓)−1( ) 1(↑)0( )−1(↓) 1(↓)0( )−1(↑)	1(↓)0( )−1(↓)
$M_L = -1$	1( )0(↑)−1(↑)	1( )0(↑)−1(↓) 1( )0(↓)−1(↑)	1( )0(↓)−1(↓)
$M_L = -2$		1( )0( )−1(↑↓)	

[a] Total spin M_S = sum of spins ($+\frac{1}{2}$ or $-\frac{1}{2}$) of individual electrons.

bilities for each of these; we have no way of telling which three these are—in fact, they are some linear combination of the different microstates listed.

The ^1D electronic state must involve five microstates, all from the center $M_S = 0$ column. The top $M_L = 2$ and bottom $M_L = -2$ microstates are easily identified, but again this cannot be done clearly in the middle of the column. Since the ^1S microstate must come from the very center of the table ($M_L = 0$, $M_S = 0$), it also cannot be identified. Nonetheless we have found the proper number of microstates in the proper (M_L, M_S) blocks of the table, so we have confirmed that ^3P, ^1D, and ^1S are the proper three electronic states for the p^2 electronic configuration.

For our other two-electron example, $3d^2$ V^{3+}, the states are similar to those of carbon in spin degeneracy (triplets and singlets), but show greater orbital degeneracy [there is a $M_L = 4$ microstate, 2(↑↓)1( )0( )−1( )−2( ), which is part of the $L = 4$ or G state]. There are also a larger number of electronic states, five: ^3F, ^3P, ^1G, ^1D, and ^1S. The total number of microstates involved in the d^2 electron configuration is $3 \times 7 + 3 \times 3 + 1 \times 9 + 1 \times 5 + 1 \times 1 = 45$, as expected from Eq. (17.1).

Example 17.2

Select the "extreme" four microstates for the ^{3}F electronic state for the d^2 electron configuration; select the "extreme" two microstates for the ^{1}G electronic state.

SOLUTION:

By "extreme" we mean those with the highest and lowest possible values of M_L and M_S; the high values match the L and S values of the electronic states. Hence, for ^3F ($L = 3$, $S = 1$), the four we want are 2(↑)1(↑)0( )−1( )−2( ), 2(↓)1(↓)0( )−1( )−2( ), 2( )1( )0( )−1(↑)−2(↑), and 2( )1( )0( )−1(↓)−2(↓). For ^1G ($L = 4$, $S = 0$), the two that we want are 2(↑↓)1( )0( )−1( )−2( ), and 2( )1( )0( )−1( )−2(↑↓).

Among the term symbols representing electronic configurations for a given electronic configuration, the **ground-state** term is readily picked out by application of Hund's rules; since these indicate that electrons should occupy orbitals *with parallel spins* insofar as possible, it follows that (1) *the term with the greatest spin multiplicity lies lowest in energy*. Among those of maximum spin multiplicity, (2) *the ground-state term will be the one with the greatest total orbital angular momentum.*[2] From among the three choices for carbon, we select the ^3P as the ground state, while the ground state for the d^2 V^{3+} ion is ^3F.

17.2 Electronic States for Diatomic Molecules and Their Consequences

The term symbol letterings of Tables 17.1 and 17.2 apply only to free atoms and ions; in complexes of lower than spherical symmetry the labels appropriate to the symmetry point group of the species are used instead. For diatomic molecules in the $D_{\infty h}$ or $C_{\infty v}$ point groups, the letter designation of total orbital angular momentum is replaced by a Greek letter that is the irreducible representation from the point group of the molecule that indicates the symmetry of the molecule as a whole when having that particular molecular orbital (MO) occupation. In the $D_{\infty h}$ or $C_{\infty v}$ point groups, orbital degeneracy can only be twofold, except for the nonorbitally degenerate Σ states. The ground state for the *p*-orbital-based π^{*2} O_2 molecule is $^3\Sigma_g{}^-$ and its singlet excited states are $^1\Delta_g$ [including the microstates $\pi_x{}^*(\downarrow\uparrow)\pi_y(\ )$ and $\pi_x(\ )\pi_y{}^*(\downarrow\uparrow)$] and $^1\Sigma_g{}^+$ [including the microstate $\pi_x{}^*(\uparrow)\pi_y{}^*(\downarrow)$].[3] In the $^3\Sigma_g{}^-$ and the $^1\Sigma_g{}^+$ states, the two oxygen π^* orbitals are equally occupied, so the molecule as a whole retains its cylindrical symmetry, and the Σ (capital σ) designation is appropriate. In the $^1\Delta_g$ excited state, one π^* orbital is doubly occupied while the other is unoccupied, so the molecule has nonzero orbital angular momentum, and the less-symmetric Δ irreducible representation is appropriate.

As would be expected from Hund's rule, the ground state of O_2 is $^3\Sigma_g{}^-$; it is O_2 in this electronic state whose electron configuration was discussed in Section 10.1. The other two forms are known as *singlet oxygen*. The more commonly encountered, longer lasting, and less high-energy form of singlet oxygen is $O_2(^1\Delta_g)$, which is 92 kJ higher in energy than O_2; $O_2(^1\Sigma_g{}^+)$ is 155 kJ higher in energy than triplet O_2. Both forms of singlet oxygen have different chemical properties than $O_2(^3\Sigma_g{}^-)$, not only because of their higher energy content but because their reactivity is not hindered by the presence of two unpaired electrons. Electrons do not readily change spin, so reactions of O_2 with other diamagnetic species are inhibited because they would characteristically give first products with two unpaired electrons, which are generally high-energy excited states of those products. So triplet O_2 often must react by one-electron free-radical mechanisms, a limitation that does not apply to singlet oxygen. Ordinary triplet O_2 is really a fairly sluggish oxidant as compared to other very electronegative nonmetals of similar electronegativity such as Cl_2. By contrast, singlet oxygen is more reactive: For example, it is able to engage in concerted reactions (Section 10.7) with dienes to generate cyclic peroxides:

$$CH_2{=}CH{-}CH{=}CH_2 + O_2(^1\Delta_g) \rightarrow H_2C\langle\begin{matrix} HC{=}CH \\ O{-}O \end{matrix}\rangle CH_2 \qquad (17.6)$$

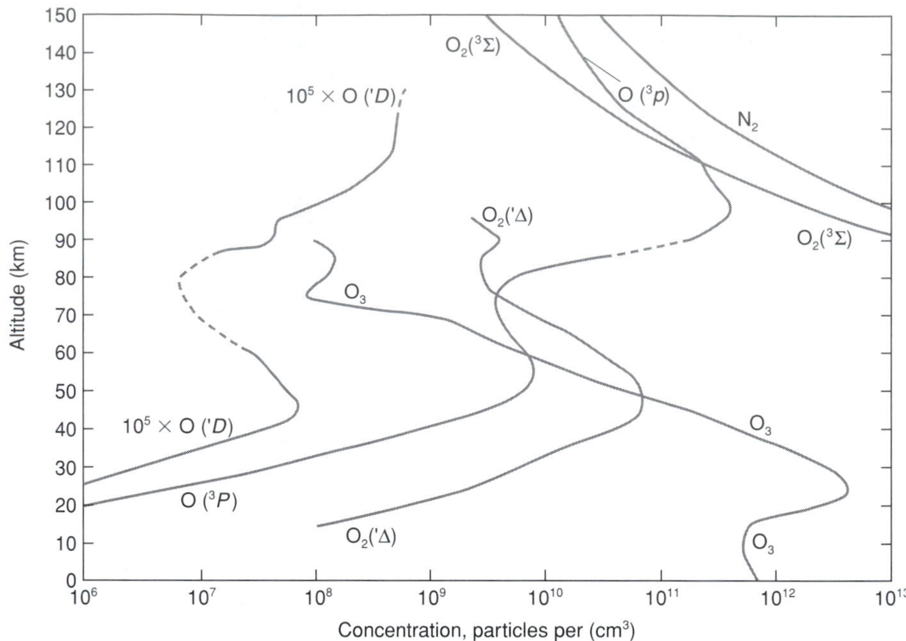

Figure 17.1
Variation in typical concentrations of different forms of elemental O_2 versus altitude for an overhead sun at low latitudes. [Adapted from J. Heicklen, *Atmospheric Chemistry*, Academic, New York, 1976; p. 9.]

Singlet oxygen is formed in certain chemical reactions in which the mechanisms do not allow the production of unpaired electrons:

$$H_2O_2 + Cl_2 \rightarrow 2\,Cl^- + 2\,H^+ + O_2(^1\Delta_g) \tag{17.7}$$

$$H_2O_2 + ClO^- \rightarrow Cl^- + H_2O + O_2(^1\Delta_g) \tag{17.8}$$

The production of the O_2 in these reactions is accompanied by the production of a red chemiluminescent glow. It can also be formed from triplet O_2 by photochemical irradiation in the presence of a sensitizer (a dye or a complex ion such as $[Ru(bipyridyl)_3]^{2+}$, which can first be converted from its ground singlet to an excited triplet state).

In the upper atmosphere, which is subject to high-energy irradiation, much triplet O_2 is converted to singlet O_2, and even dissociated into *atomic oxygen*, in both of its corresponding forms: $O(^3P)$ and $O(^1D)$. These very reactive forms are quite corrosive to spacecraft as they re-enter the atmosphere at high temperatures; Figure 17.1 illustrates the variation in the concentration of these different forms of the element oxygen (along with ozone, O_3) at different altitudes.

Isolobal (Section 11.6) with atomic oxygen are species such as the carbenes, R_2C:, which are important transient reactants in organic chemistry and can also exist in singlet and triplet electronic states of differing reactivity. Although Hund's rule predicts that the triplet state will always be the ground state if the two orbitals are of the same type, a violation has been detected in the planar, antiaromatic $D_{\infty h}$ form of cyclooctatetraene, which arises as a transition state during inversions of the stable nonplanar form: the singlet form is lower in energy than the triplet state.[4]

Another molecule with a complex electron configuration to begin with (as a result of *sp* hybridization, Section 10.2) is C_2, which has numerous electronic states; 13 of these have been observed experimentally along with their bond distances, which range from 123 pm ($^3\Sigma_u{}^+$ state) and 124 pm ($^1\Sigma_g{}^+$ ground state) through 136 pm ($^3\Delta_g$) to 154 pm ($^3\Pi_g$).[5] Finally, we may mention the case of what is probably the highest spin diatomic molecule of all, Gd_2, which has a $^{19}\Sigma$ ground state.[6] The Gd atom is one of the 21 or so *d*- and *f*-block gaseous atoms that has an anomalous (as opposed to characteristic, Section 1.7) electron configuration, $4f^7 5d^1 6s^2$, with a ^9D ground state; when two such atoms bond to each other, most of the valence orbitals, which have relatively small values of $\langle r_{max} \rangle$, do not overlap effectively with each other, so that nearly all of the valence electrons remain unpaired, but do, however, couple ferromagnetically (Section 12.4) with each other.

17.3 Electronic States for Octahedral Complexes

In octahedral complexes, the electronic states are also labeled with irreducible representations, rather than with $S, P, D, F, G \ldots$. But in addition to the change in labeling style, there is a more fundamental difference: there cannot be higher than threefold orbital degeneracy in the O_h point group. So, although we relabel an S electronic state using A_{1g} and a P electronic state using T_{1g} (Table 17.2B), D and higher electronic states must be split into new electronic states, none of which has more than threefold degeneracy in orbital angular momentum. For example, the ^3F state of a d^2 electronic configuration has sevenfold orbital degeneracy that cannot persist in an octahedral complex: it splits into three new electronic states, $^3T_{2g}$, $^3T_{1g}$, and $^3A_{2g}$, which have threefold, threefold, and onefold orbital degeneracy, respectively. Each of these states also has threefold spin degeneracy, so together these electronic states derived from ^3F account for $3 \times 7 = 21$ of the total of 45 microstates of the d^2 ion such as V^{3+}.

In a similar manner, we can analyze and if necessary split the four excited energy states. The ^3P state is simply relabeled $^3T_{1g}$, and ^1S is relabeled $^1A_{1g}$; together these account for 10 microstates. The ^1D energy state is split into ^1E_g and $^1T_{2g}$ energy states, just as the d set of individual orbitals is split into e_g and t_{2g} sets (Chapter 8): 5 microstates. The ^1G state of the free V^{3+} ion is split into $^1A_{1g}$, $^1T_{2g}$, $^1T_{1g}$, and ^1E_g states: 9 final microstates, for a grand total of 45.

The five free-ion electronic states that we just relabeled and split are those found in a ligand-free gaseous ion, for example, in the presence of the ultimate in weak-field ligands, none at all. If we go to a complex with real but weak-field ligands, each of the split electronic states should branch out in energy from the corresponding original free-ion energy state, as shown at the left side of Figure 17.2.

On the other hand, let us consider the situation if we have the ultimate in strong-field complexes, a "dominant-field" complex: one in which interelectronic repulsions are negligible compared to the completely dominant effects of the ligand field, which splits the d orbitals into two types, e_g and t_{2g}, differing in energy by Δ_o, the crystal field splitting. In such a dominant-field complex (far right side of Fig. 17.2), a d^2 ion would have three energetically distinct electron configurations: $t_{2g}{}^2$ (lowest in energy), $t_{2g}{}^1 e_g{}^1$ (higher in energy by Δ_o), and $e_g{}^2$ (highest in energy, by another Δ_o). But if we lower the ligand field strength from dominant-field to merely strong field, we expect interelec-

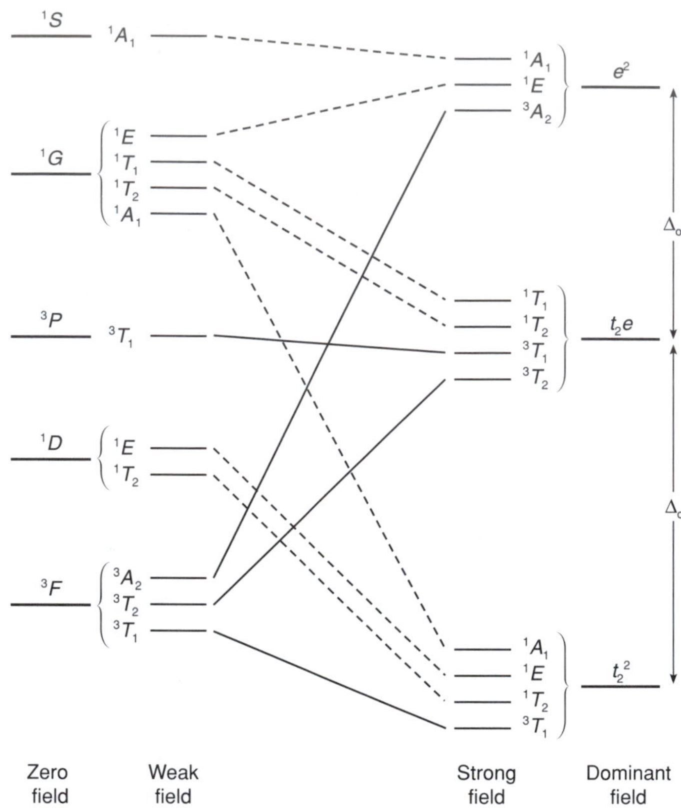

Figure 17.2
Energy levels of a d^2 ion in an octahedral complex as the strength of the ligand field is varied. (Since all energy levels are *gerade*, the "g" has been omitted.) [Adapted from F. A. Cotton, G. Wilkinson, and P. L. Gaus, *Basic Inorganic Chemistry*, 3rd ed., John Wiley & Sons, Inc., New York, 1995; p. 527.]

tronic repulsions to become significant, and to split these three energy levels with a total of 45 microstates into the same series of new energy states, none of which has more than threefold orbital degeneracy. Let us analyze how each of the three dominant-field energy levels splits.

First, the ground ${t_{2g}}^2$ state involves $6!/2!(4!) = 15$ microstates, and is related to the case of the p^2 carbon atom, which gives ^3P, ^1D, and ^1S energy states. As can be seen from Table 17.2B, ^3P is relabeled $^3T_{1g}$, which (since it is the only spin triplet) is the lowest in energy. Likewise ^1S is relabeled $^1A_{1g}$, but the fivefold orbital degeneracy of ^1D cannot persist—it is split into $^1T_{2g}$ and ^1E_g. At the lower right of Figure 17.2 we see these strong-field energy levels radiating out in energy to the left from the dominant-field ${t_{2g}}^2$ energy level.

In the top right of the diagram, the dominant-field ${e_g}^2$ case involves $4!/2!(2!) = 6$ microstates (like the O_2 molecule), and splits as a result of interelectronic repulsions into three similar electronic states, including one triplet, which goes lowest in energy. The spin triplet is orbitally singly degenerate: In the $D_{\infty h}$ point group it was labeled $^3\Sigma_g{}^-$, but in O_h the $^3A_{2g}$ label is appropriate. One of the two singlets is orbitally doubly degenerate, and in the $D_{\infty h}$ point group was labeled $^1\Delta_g$, but in O_h it is labeled ^1E_g; the other singlet is orbitally nondegenerate and has the highest irreducible representation label—in the $D_{\infty h}$ point it is ${}^1\Sigma_g{}^+$, but in O_h it is ${}^1A_{1g}$.

The middle dominant-field electron configuration of $t_{2g}{}^1e_g{}^1$ differs from the others in the computation of its microstates, since the two electrons must be in two distinct types of orbitals, so that Eq. 17.1 does not apply. No combination is impossible by the Pauli exclusion principle, so the first electron can choose any of three t_{2g} orbitals with either spin (6 possibilities), while the second independently chooses either e_g orbital with either spin, for a total of 24 microstates. Since this case is not covered in Table 17.2, we will accept the result that these fall into the four electronic states ${}^3T_{1g} + {}^3T_{2g} + {}^1T_{1g} + {}^1T_{2g}$; again the triplet states will fall below the singlet states in energy.

Finally, it is necessary to draw lines to connect the weak-field left side of Figure 17.2 with its strong-field right side. The result is known as a *correlation diagram*. By no coincidence, we have used the same 11 term symbols on each side of the diagram, so in some cases we can simply connect a term symbol on the left to the same term symbol on the right. The only ambiguity comes when a given term symbol is used twice (or more often) on each side; this actually happens rather often in this case. For example, ${}^3T_{1g}$ occurs twice on the left side of the figure, once associated with the ${}^3F$ free-ion electronic state [this is sometimes indicated by writing ${}^3T_{1g}(F)$], and once associated with the ${}^3P$ free-ion electronic state [this is then written as ${}^3T_{1g}(P)$]. Similarly, two ${}^3T_{1g}$ terms are found on the right. In connecting ${}^3T_{1g}$ terms on the left and the right, *our lines (energy levels) should never cross*. Thus, the *lower* ${}^3T_{1g}$ state on the left is connected to the *lower* one on the right.[7]

17.4 Tanabe–Sugano Diagrams and Spectra of *d*-Block Complexes

At last, we are ready to attempt to explain more fully the observation noted at the beginning of the chapter that d^2 complexes, unlike their d^1 counterparts, have more than one electronic absorption band in their spectra. Now the question is why are the spectra not more complex, because a full electronic spectrum of an octahedral d^2 complex would include separate transitions from the ground ${}^3T_{1g}(F)$ to *each* of the 10 excited states lying at higher energies!

Fortunately for us, there is the principle that electrons do not readily change their spins, either during chemical reactions (Section 17.2) or during electronic transitions. So the transitions from the ${}^3T_{1g}$ ground state to the seven singlet excited states are **spin-forbidden**: They are very improbable, and have very low molar absorptivities ($\varepsilon < 1$) in the spectrum, so they are not normally observed. We may expect to see at most only the **spin-allowed** transitions from the triplet ground state to the three triplet excited states, which have been highlighted in Figure 17.2 by being drawn with solid (rather than with dashed) lines.

If we consult the spectrum of the d^2 hydrated ion in Figure 8.5, that of $[V(H_2O)_6]^{3+}$, we in fact notice only two bands. The third is weakened as a result of another principle: The transition from an electronic state that correlates to $t_{2g}{}^2$ on the right to one that correlates to $e_g{}^2$ on the right involves the simultaneous excitation of *two* electrons from the t_{2g} level in the complex to the e_g level. This is also improbable, although not as improbable as the electron changing its spin, so that sometimes one does observe this third band, although not in case of $[V(H_2O)_6]^{3+}$.

Racah Parameters. Motion to the right in Figure 17.2 corresponds to changing ligands so as to move them up in the spectrochemical series, thus increasing the magnitude of the crystal field splitting Δ_o. Although, as we pointed out in Section 8.4, the value of Δ_o can often be obtained simply as the energy of the lowest energy electronic transition in the near-IR (or visible, or perhaps UV) spectrum, we have two or three electronic transitions that can be measured; use of more than one of these to evaluate Δ_o should lead to a more accurate value. In addition, a second parameter, known as a **Racah parameter**, can also be obtained, which contains useful information in its own right about the covalency in the metal–ligand bonds. This parameter is important in modifying the simple crystal field theory to become the more subtle ligand field theory.

To do this, we need to make our correlation diagram (Fig. 17.2) quantitative; we need to do more than connect the term symbols. Equations can be derived for the energies of the free-ion electronic states in terms of the interelectronic repulsions. A number of complicated integrals are involved in this quantum mechanical calculation, which is usually treated empirically rather than computationally. For the empirical treatment, it is first noted that all of the integrals can be grouped into three specific types of combinations, which are called the Racah parameters A, B, and C, which always turn out to have positive numerical values. The energies of the free-ion electronic states of a d^2 configuration can then be expressed in terms of these three collected types of integrals:

$$\begin{aligned} \mathrm{E}(^1S) &= A + 14\,B + 7\,C \\ \mathrm{E}(^1G) &= A + 4\,B + 2\,C \\ \mathrm{E}(^1D) &= A - 3\,B + 2\,C \end{aligned} \qquad (17.9)$$

$$\begin{aligned} \mathrm{E}(^3P) &= A + 7\,B \\ \mathrm{E}(^3F) &= A - 8\,B \end{aligned} \qquad (17.10)$$

These equations are used to locate the relative positions of the five free-ion electronic states on the left side of Figure 17.2.[8]

When we evaluate electronic spectra, we are evaluating the *differences* between the two energy levels involved in that electronic transition. Under these circumstances, the A term always cancels out, so we do not need to know it. Also, normally we are only observing transitions between triplet states, which do not involve the C term, so we also do not need to know it. Hence, for the spectrum of the free ion, the Racah parameter B is the only one involved, and its value is readily obtained: The energy separation of the two triplet states equals $15B$.

As we move from a free ion to a weak-field octahedral complex, the Racah equations (17.9 and 17.10) do not lose their validity, but the value of the Racah B parameter does change from that found in the gaseous free ion to one with a new value, B', which is less than that of the B parameter of the gaseous free ion due to the effects of covalency. With two or three observed electronic transitions, our goal will be to extract the values of Δ_o and of the Racah parameter B' in the complex ion.

Also, as we move from a free ion to a weak-field octahedral complex, we know that many of the free-ion electronic states are split by the action of the crystal field, so that the crystal field splitting Δ_o also becomes involved in the energy level diagram. States that connect on the right to the dominant-field $t_{2g}{}^2$ configuration [i.e., $^3T_{1g}(F)$] have crystal

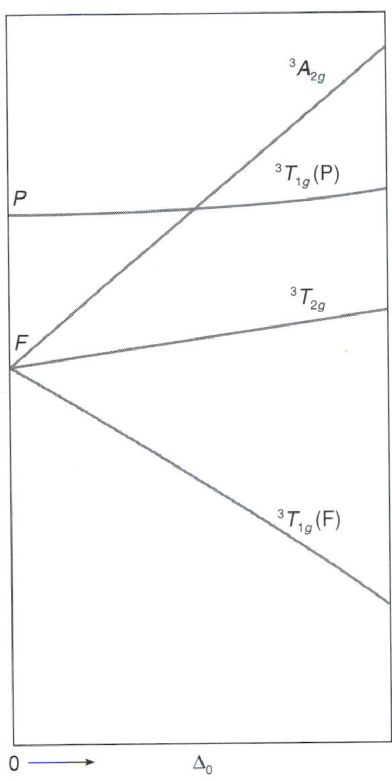

Figure 17.3
Orgel diagram for d^2 and weak-field d^7 ions in an octahedral complex, showing only triplet states. [Adapted from A. B. P. Lever, *Inorganic Electronic Spectroscopy*; 2nd ed., Elsevier, Amsterdam, The Netherlands, 1986; p. 85.]

field stabilization energies (Section 8.6) of ($-0.4\,\Delta_o$ per electron)(2 electrons) $= -0.8\,\Delta_o$, so their energy levels have a slope of -0.8 on the graph. States that connect on the right to $e_g{}^2$ have a slope of (+0.6 per electron)(2 electrons) $= +1.2$; states that connect on the right to $t_{2g}{}^1e_g{}^1$ have the intermediate slope of $-0.4 + 0.6 = +0.2$.

The result is known as an *Orgel diagram*, and is shown for d^2 in Figure 17.3. It will be noted that, as Δ_o increases, the ^3F electronic state is split, with the $^3T_{1g}(F)$ state being progressively stabilized, while the $^3T_{2g}$ and especially the $^3A_{2g}$ states are destabilized (this is in agreement with our qualitative results in Fig. 17.2). Note also that, whereas the energy levels of the $^3T_{2g}$ and $^3A_{2g}$ states depend linearly on Δ_o, there is curvature in the energies of the two $^3T_{1g}$ levels, resulting from their configuration interaction with each other.[7]

Tanabe–Sugano Diagrams. More commonly used for actual evaluation of spectra are the Tanabe–Sugano diagrams, which incorporate two modifications beyond the simple Orgel diagram of Figure 17.3. First, we know that the energies of the electronic transitions depend, not only on the arrangements of electronic states characteristic of the electron configuration (e.g., d^2), but also on what the specific metal ion is. To make a more generally useful diagram, it is necessary to factor out the metal ion; this is done by dividing both the vertical energy scale and the horizontal Δ_o scale by the value of B' characteristic of the metal ion in the octahedral complex. Second, since the energies of spectral transitions are due to *differences* in energies of certain excited states from that of the ground state, it is easier to use the diagram if the baseline *is* the energy of the ground state. Then the vertical scale measures the *difference* in energy of the excited state in question from that of the ground state, divided by B'. The result is the

Tanabe–Sugano diagram for any d^2 ion, which is shown in Figure 17.4 along with the corresponding diagrams for d^1 through d^9 ions.

The Tanabe–Sugano diagrams for d^2, d^3, and d^8 have been simplified by showing only the energy levels of the states of maximum multiplicity, since it is only these that will be involved in observable spectral bands. The diagrams for d^1 and d^9 are complete, since all electronic states are doublets (with one unpaired electron each). But the full detail is shown for d^4, d^5, d^6, and d^7, since in these cases the ground electronic state changes as the crystal field strength changes: In weak-field complexes, the high-spin state (state of higher multiplicity) is the ground state (states of higher multiplicity are shown with boldface lines in the diagram); in strong-field complexes, the low-spin state (state of lower multiplicity) is the ground state (states of lower multiplicity are shown with dashed lines in the diagram). At the position of the vertical line in each of these diagrams, spin crossover occurs, and a different state becomes the ground state. This tilts all energy levels in a Tanabe–Sugano diagram, since the ground state, whatever it is, is taken as level across the bottom of the diagram.

Let us now analyze some spectra, starting with the simplest case, that of a d^1 ion such as $[Ti(H_2O)_6]^{3+}$. In this case, there is only one possible electronic transition, ν_1 ($^2T_{2g} \rightarrow {}^2E_g$), and its energy is Δ_o. The line for the unique excited state in the d^1 Tanabe–Sugano diagram has a slope of +1, since E, the energy of the transition and (when divided by B') the y ordinate, equals Δ_o (which, when divided by B', is the x ordinate). Note that we cannot determine what B' is, since we have only one measurable parameter.

The next simplest type of Tanabe–Sugano diagram is not that of d^2, but rather that found for d^3 and d^8 (these two differ only in the term symbols used), for example, $[Cr(H_2O)_6]^{3+}$ and $[Ni(H_2O)_6]^{3+}$. In these cases, the energy level for the first excited state also has a slope of +1, and the value of Δ_o can also be obtained simply by measuring the lowest energy absorption, ν_1. But now we also have two other absorptions, both of which are in fact seen (Fig. 8.5). Therefore we have the possibility, not only of measuring the Racah parameter B, but of reducing any errors in our measurement of Δ_o.

However, we note that the energy level of the second excited state (a T_{1g} state) shows curvature, because of configuration interaction or "hybridization" with the nearby T_{1g} excited state.[7] This complicates the expressions for the energies of the transitions to both of these states, which become

$$\nu_1(A_{2g} \rightarrow T_{2g}) = \Delta_o \tag{17.11}$$

$$\nu_2[A_{2g} \rightarrow T_{1g}(F)] = 7.5\,B' + 1.5\,\Delta_o - 0.5\sqrt{[225\,B'^2 + \Delta_o{}^2 - 18\,B'\Delta_o]} \tag{17.12}$$

$$\nu_3[A_{2g} \rightarrow T_{1g}(P)] = 7.5\,B' + 1.5\,\Delta_o + 0.5\sqrt{[225\,B'^2 + \Delta_o{}^2 - 18B'\Delta_o]} \tag{17.13}$$

From these equations it follows that B' is easily found[9]:

$$15\,B' = \nu_3 + \nu_2 - 3\nu_1 \tag{17.14}$$

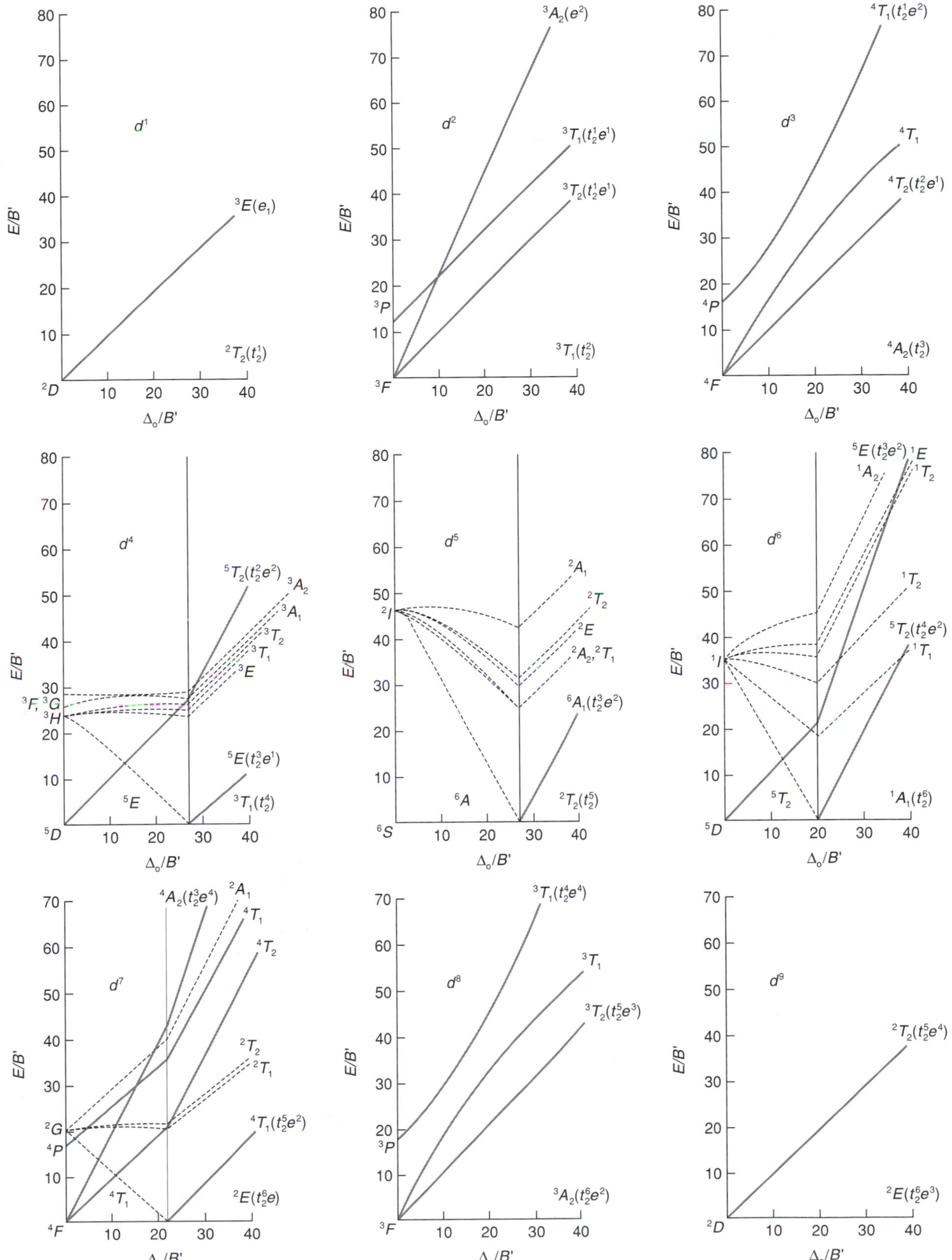

Figure 17.4

Tanabe–Sugano diagrams for d^1 through d^9 metals in octahedral ligand fields. (The g subscripts have been omitted for clarity.) [Adapted from K. F. Purcell and J. C. Kotz, *Inorganic Chemistry*, Saunders, Philadelphia, 1977; pp. 584–585.]

Example 17.3

The octahedrally hydrated Ni^{2+} ion has a near-IR absorption at 1176 nm and visible absorptions at 649 and 385 nm. Calculate Δ_o and B' for this complex.

SOLUTION:

These wavelengths should first be converted to frequencies in reciprocal centimeters by dividing them into 10^7 (Section 8.4): $10^7/1176 = 8500\,\text{cm}^{-1}$; similarly, $\nu_2 = 15{,}400\,\text{cm}^{-1}$ and $\nu_3 = 26{,}000\,\text{cm}^{-1}$. The value of ν_1, $8500\,\text{cm}^{-1}$, is the value of Δ_o. Applying Eq. (17.14), we find $15\,B' = 26{,}000 + 15{,}400 - 25{,}500 = 15{,}900\,\text{cm}^{-1}$, so $B' = 1{,}060\,\text{cm}^{-1}$.

The case of the d^2 spectrum (and the similar weak-field d^7) is discussed next. It appears that one can also obtain Δ_o directly by measuring ν_1 in this case as well, but it turns out that this is only approximately so. The reason is that the ground state is now one of the T_{1g} states engaging in curvature, so its energy level also involves quadratic terms. (This is disguised by the Tanabe–Sugano diagram, which artificially flattens the ground energy level, but it can still be seen in the energy level of the other T_{1g} state.) Consequently, the slope of the energy level of the first excited (T_{2g}) level is no longer exactly +1, but is a little less, and it includes some curvature. In this case, the *ratio* of ν_2 to ν_1 along with the Tanabe–Sugano diagram can be used to find the values of Δ_o and B', as shown in Example 17.4.

Example 17.4

(a) By using the Tanabe–Sugano diagram for a d^2 ion, graphically estimate the ratio ν_2/ν_1 that should be found for each of the following values of Δ_o/B': 20; 30; and 40. (b) The first two absorption bands of $[V(H_2O)_6]^{3+}$ are found at 17,800 and 25,700 cm^{-1}. Determine the values of Δ_o and B'.

SOLUTION:

(a) On a Tanabe–Sugano diagram for a d^2 ion, draw vertical lines from the ${}^3T_{1g}$ baseline to the lowest excited state, ${}^3T_{2g}$, and measure its length; divide this measurement into the length of the vertical line from the baseline to the ${}^3T_{1g}(P)$ excited state. Results from several such measurements are shown in Figure 17.5.

(b) The ratio of the two observed frequencies is 25,700/17,800 = 1.44. As shown on Figure 17.5, this ratio is extrapolated to the curve and then downward to give a Δ_o/B ratio of about 31. Now we turn to the Tanabe–Sugano diagram for d^2 itself, and draw a vertical line up from the x ordinate of $\Delta_o/B = 31$ to the line for the first excited state, then directly across to the y ordinate, which has a value $E/B' = 29$. But we know that E for this ν_2 transition is 17,800, so $B' = 17{,}800/29 = 610\,\text{cm}^{-1}$. We also know that $\Delta_o/B = 31$, so $\Delta_o = 31 \times 610 = 19{,}000\,\text{cm}^{-1}$. (Note that this is significantly different than the value of ν_1, which is only roughly equal to Δ_o.)

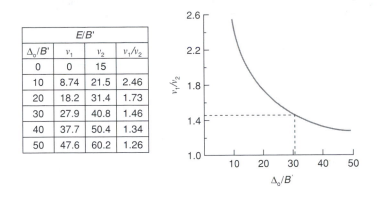

E/B'			
Δ_o/B'	ν_1	ν_2	ν_1/ν_2
0	0	15	
10	8.74	21.5	2.46
20	18.2	31.4	1.73
30	27.9	40.8	1.46
40	37.7	50.4	1.34
50	47.6	60.2	1.26

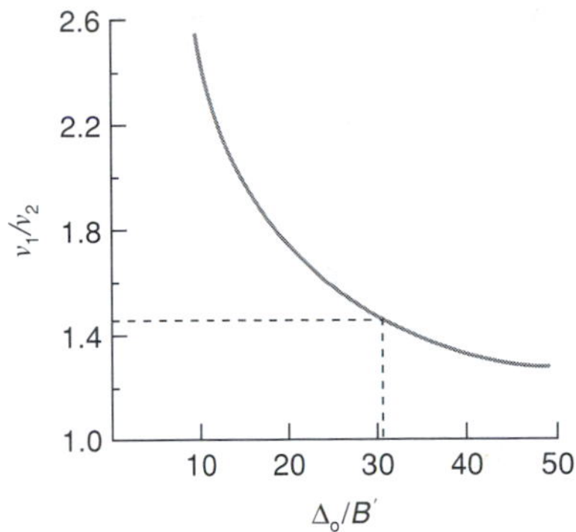

Figure 17.5
Values of the ν_2/ν_1 ratio for d^2 or weak-field d^7 octahedral complex ions. [Adapted from G. L. Miessler and D. A. Tarr, *Inorganic Chemistry*, Prentice-Hall, Englewood Cliffs, NJ, 1991; p. 333.]

Nephelauxetic Series of Ligands. Values of the Racah B and C parameters have been obtained for a number of free gaseous d-block metal ions from their spectra; these are tabulated in Table 17.4. However, the values of B' that we obtain spectroscopically from complexes of these metal ions are always found to be lower than those obtained for the corresponding free ions. The ratio of the B' Racah parameter found in a complex to the B parameter of the same metal ion in ligand-free form is known as the **nephelauxetic parameter** β of the ligand:

$$\beta = B'/B \tag{17.15}$$

For example, β for the H_2O ligand can be obtained from B' for the $[V(H_2O)_6]^{3+}$ ion (Example 17.4) and from B for the free V^{3+} ion (Table 17.4), and is computed as 0.71 in this case.

Table 17.4
Racah B and C Parameters for Free Gaseous d-Block Metal Ions

Configuration	Ion	B	C	Configuration	Ion	B	C
$3d^2$	Ti^{2+}	718	2629	$3d^9$	Cu^{2+}	1240	
	V^{3+}	861	4165	$3d^{10}$	Cu^+	1220	
	Cr^{4+}	1039	4238	$4d^3$	Mo^{3+}	610	
$3d^3$	Sc^+	480		$4d^6$	Rh^{3+}	720	
	V^{2+}	766	2855	$4d^7$	Rh^{2+}	620	4002
	Cr^{3+}	918	3850	$4d^8$	Pd^{2+}	683	2620
	Mn^{4+}	1064		$5d^2$	Os^{6+}	780	
$3d^4$	Cr^{2+}	830	3430	$5d^3$	Re^{4+}	650	
	Mn^{3+}	1140	3675		Ir^{6+}	810	
$3d^5$	Mn^{2+}	960	3325	$5d^4$	Os^{4+}	700	
$3d^6$	Fe^{2+}	1058	3901	$5d^6$	Ir^{3+}	660	
	Co^{3+}	1100			Pt^{4+}	720	
$3d^7$	Co^{2+}	971	4366	$5d^8$	Pt^{2+}	600	
$3d^8$	Ni^{2+}	1041	4831				

SOURCE: From A. B. P. Lever, *Inorganic Electronic Spectroscopy*, 2nd ed., Elsevier, New York, 1986; p. 115.

Since the Racah parameter is a measure of interelectronic repulsions of metal *d* electrons with each other, this suggests that these repulsions are reduced upon forming a complex ion. Repulsions are reduced because the metal *d* electrons in fact are involved in MOs that, to greater or lesser extents, delocalize them from the metal atom out onto the ligands. (Nephelauxetic comes from Greek words meaning cloud-expanding.) This delocalization causes different *d* electrons to be farther apart, hence it decreases their interelectronic repulsions.

Just as ligands can be arranged in order of their increasing effect on Δ_o (the spectrochemical series, Section 8.5), so they can be arranged in order of the tendency for the nephelauxetic parameter β to fall below the ideal value of 1.00, which is most nearly achieved with F^- as a ligand. Such a series is known as a **nephelauxetic series** of ligands:

$$F^- > H_2O > NH_3 > C_2O_4{}^{2-}, NH_2CH_2CH_2NH_2 > Cl^- > CN^- > Br^- > I^- \quad (17.16)$$

In Section 8.5, we saw that the crystal field splitting Δ_o in a complex can be expressed approximately as a product of an *f* factor of the ligands and a *g* factor of the metal ion [Eq. (8.3)]. A similar approximate factoring of the nephelauxetic parameter into an *h* factor for the ligand and a *k* factor of the metal has been suggested[10]:

$$\beta = 1 - hk \quad (17.17)$$

Some values for these two parameters are given in Table 17.5. Although it does not always give highly accurate results,[10] this equation could be used, along with *f* and *g* factors and either Tanabe–Sugano diagrams or appropriate equations to predict spectral positions of absorption bands. For example, Eq. (17.18) and (17.19) apply to the two lowest energy absorption bands of low-spin (strong-field) d^6 complexes:

$$\nu_1 \text{ (strong-field } d^6) = \Delta_o - 4\,B' + 86\,B'^2/\Delta_o \quad (17.18)$$

$$\nu_2 \text{ (strong-field } d^6) = \Delta_o + 12\,B' + 2B'^2/\Delta_o \quad (17.19)$$

Example 17.5

Using appropriate *f*, *g*, *h*, and *k* factors, estimate the positions of the first two absorption bands of the following low-spin Co^{3+} complexes: $[Co(NH_2CH_2CH_2NH_2)_6]^{3+}$ and $CoI_6{}^{3-}$. (a) Estimate these positions using Tanabe–Sugano diagrams. (b) Estimate these positions using the equations given. (c) The latter of these complexes has not been prepared, and if it were prepared, there would be difficulty in observing these two bands. Explain.

SOLUTION:

The *g* factor of Co^{3+} is 18,200 cm^{-1} (Table 8.4); its *k* factor is 0.33 (Table 17.5); its Racah *B* factor is 1100^{-1} (Table 17.4). The *f* factor of the $NH_2CH_2CH_2NH_2$ ligand is 1.28 (Table 8.5); the *f* factor of the I^- ligand is not given, but can be extrapolated from those of F^-, Cl^-, and Br^- down to perhaps 0.66. The *h* factor of $NH_2CH_2CH_2NH_2$ is 1.5, while that of I^- is 2.7.

(a) $\Delta_o = f \times g = 1.28 \times 18{,}200 = 23{,}300\ \text{cm}^{-1}$ for $[Co(NH_2CH_2CH_2NH_2)_6]^{3+}$ and $0.66 \times 18{,}200\ \text{cm}^{-1} = 12{,}000\ \text{cm}^{-1}$ for $[CoI_6]^{3-}$, respectively. $B' = \beta B = B(1 - hk) =$

Table 17.5
Ligand h and Metal k Nephelauxetic Parameters for Selected Ligands

Ligand[a]	h	Metal	k
$\mathbf{F}^-$	0.8	Mn(II)	0.07
$H_2\mathbf{O}$	1.0	V(II)	0.1
$Me_2NCH\mathbf{O}$	1.2	Ni(II)	0.12
$(H_2N_2)C\mathbf{O}$	1.2	Mo(III)	0.15
$\mathbf{N}H_3$	1.4	Cr(III)	0.20
$H_2\mathbf{N}CH_2CH_2\mathbf{N}H_2$	1.5	Fe(III)	0.24
$C_2\mathbf{O}_4{}^{2-}$	1.5	Rh(III)	0.28
$\mathbf{Cl}^-$	2.0	Ir(III)	0.28
$\mathbf{C}N^-$	2.1	Tc(IV)	0.3
$\mathbf{Br}^-$	2.3	Co(III)	0.33
$\mathbf{N}_3{}^-$	2.4	Mn(IV)	0.5
$\mathbf{I}^-$	2.7	Pt(IV)	0.6
$(EtO)_2P\mathbf{S}_2{}^-$	2.8	Pd(IV)	0.7
$(EtO)_2P\mathbf{Se}_2{}^-$	3.0	Ni(IV)	0.8

SOURCE: From C. K. Jørgensen, *Oxidation Numbers and Oxidation States*, Springer, New York, 1969; p. 106.
[a] Donor atoms of ligands shown in boldface.

$1100(1 - 1.5 \times 0.33) = 556\,\text{cm}^{-1}$ and $1100(1 - 2.7 \times 0.33) = 120\,\text{cm}^{-1}$, respectively (this product of large h and k values should be treated as suspect[10]). Then the Δ_o/B' value is 42 for $[Co(NH_2CH_2CH_2NH_2)_6]^{3+}$. Consulting the d^6 Tanable–Sugano diagram, we estimate E/B' is about 37 for ν_1 (the transition from the ${}^1A_{1g}$ ground state to ${}^1T_{1g}$) and is about 52 for ν_2 (the transition to ${}^1T_{2g}$). Hence the predicted energy E of $\nu_1 = (E/B') \times B' = 37 \times 556 = 20{,}600\,\text{cm}^{-1}$ and that of $\nu_2 = 52 \times 556 = 28{,}900\,\text{cm}^{-1}$; the experimental values are $21{,}550\,\text{cm}^{-1}$ and $29{,}600\,\text{cm}^{-1}$. The Δ_o/B' value of 100 for $[CoI_6]^{3-}$ is far off the scale of the Tanabe–Sugano diagram, so no predictions can be made.

(b) By applying Eq. (17.18) and (17.19), we find, for $[Co(NH_2CH_2CH_2NH_2)_6]^{3+}$, $\nu_1 = 23{,}300 - 4(556) + 86(13.27) = 22{,}200\,\text{cm}^{-1}$ and $\nu_2 = 23{,}300 + 12(556) + 2(13.27) = 30{,}000\,\text{cm}^{-1}$, in reasonable agreement with the above-mentioned observed values. For $CoI_6{}^{3-}$, $\nu_1 = 12{,}000 - 4(120) + 86(1.2) = 11{,}600\,\text{cm}^{-1}$ and $\nu_2 = 12{,}000 + 12(120) + 2(1.2) = 13{,}400\,\text{cm}^{-1}$.

(c) The reason is that $CoI_6{}^{3-}$ is the complex of an oxidizing borderline acid with a reducing soft base; since the softnesses are not well matched, the Co^{3+} could well oxidize the six I^- to give $CoI_4{}^{2-} + 0.5\,I_2 + I^-$ (Section 6.6). If the complex is stable, it will have intense charge-transfer absorptions (Section 6.4) that may completely hide the weaker *d–d* electronic absorptions.

In Section 8.5, we expressed surprise that the spectrochemical series of ligands showed neither the periodic trend of increasing with ligand charge, nor that of depending on the hardness or softness of the ligand. The nephelauxetic series of ligands (as shown by their h values) can be seen, by way of contrast, to be connected to the softness of the ligand donor atom. Since softer donor atoms have lower electronegativities, which are therefore closer to those of *d*-block metals, they can participate in more effective covalent bonding with the metal ions (have more covalent overlap within the MOs). We can also see, somewhat contrarily, that the *k* values of metals increase with increasing oxidation states, although this is not normally true of their softnesses.

17.5 Luminescence, Lasers, and Spin–Orbit Coupling

Luminescence: Fluorescence and Phosphorescence. Often after light has been absorbed by an atom, molecule, or ion, and an electron has been promoted to produce an excited electronic state, the excess energy is rapidly emitted as another photon and the system returns to the ground state. This process is known as *spontaneous deexcitation*, and occurs after random but short time intervals. But sometimes the system passes from the excited electronic state to a lower excited electronic state that is still above the ground state. Then light may be emitted, but in this case the light is of a lower energy (longer wavelength) than was involved in the excitation, and the atom, molecule, or ion **luminesces**. [If the energy is generated, not by light absorption, but by a chemical reaction such as that of Eq. (17.7), the term **chemiluminescence is** used.]

There are two subcategories of luminescence. The first is known as **fluorescence**; it occurs when light is emitted from an excited state of the same spin multiplicity as the ground state. Since this process is spin-allowed, it characteristically happens more rapidly (in nanoseconds) than the other type of property, **phosphorescence**.

Phosphorescence requires that the lower energy excited state be of a different spin multiplicity than the ground state; the relative slowness of the photon emission is because the process is spin-forbidden. Phosphorescence is found in a number of d^3 chromium(III) complexes in which the ground state is $^4A_{2g}$ but the lowest energy excited state is ^2E_g (see the d^3 Tanabe–Sugano diagram including doublet states, Fig. 17.6). Several Cr^{3+} complexes show phosphorescent emission of the same wavelength of red light (627 nm, 16,000 cm^{-1}). Since the energy level of the ^2E excited state is practically parallel to that of the ground state, the energy of the transition is independent of Δ_o.

When phosphorescence occurs, the lower energy excited state then must also be of a different spin multiplicity than the original, higher energy excited state generated by the original absorption ($^4T_{2g}$ in the case of Cr^{3+}, or $t_{2g}{}^2e_g{}^1$ in the dominant-field description). The process by which the original $^4T_{2g}$ state is converted to ^2E_g ($t_{2g}{}^3$ in the dominant-field description) is known as *intersystem crossing*. It requires that the difference in energy of the two excited states ($^4T_{2g}$ and ^2E_g) be virtually equal to the energy of one of the vibrational modes of the lattice or surrounding medium. Energy is then transferred to the lattice (and degraded to heat), while an electron simultaneously changes its spin; this intersystem crossing process is not forbidden by spin considerations.

The most important case of Cr^{3+} phosphorescence occurs in the gem *ruby*. Ruby is basically corundum (Al_2O_3) in which Cr^{3+} ions have been isomorphously substituted (Section 13.4). Since Cr^{3+} is a larger ion than Al^{3+} (Table C), when it is substituted into the Al_2O_3 lattice it has a shorter than normal Cr–O distance, resulting in a larger than normal crystal field splitting, giving it its unusual red color [when Ti^{3+} is similarly substituted in corundum, the blue gemstone *sapphire* results; when Cr^{3+} is substituted, not into corundum, but into beryl (Section 13.6), an elongated Cr–O bond and weaker crystal field splitting produces the green color of *emeralds*].

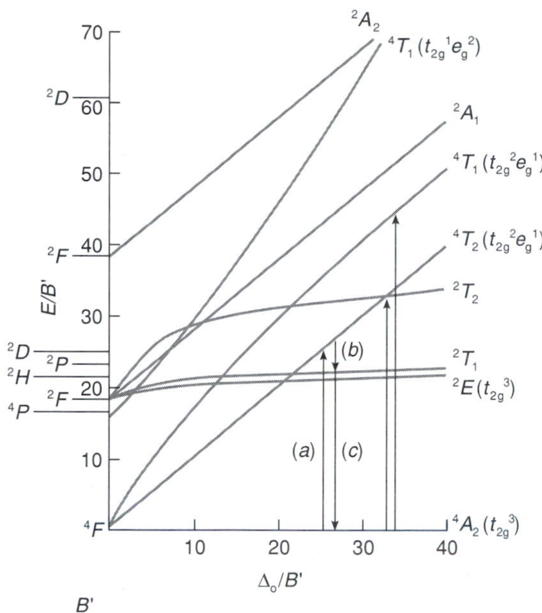

Figure 17.6
The Tanabe–Sugano diagram for the d^3 electron configuration, including lines of lower spin multiplicity. The arrows represent: (*a*) excitation of a Cr^{3+} ion in ruby; (*b*) the intersystem crossing in ruby; and (*c*) phosphorescent light emission in ruby. [Adapted from D. F. Shriver, P. Atkins, and C. H. Langford, *Inorganic Chemistry*, 2nd ed., W. H. Freeman and Company, New York, 1994; p. 594.]

In the case of ruby, the difference in energy of the two excited states ($^4T_{2g}$ and ^2E_g) matches the energy of one of the lattice vibrational modes very well, so intersystem crossing proceeds very efficiently. If intense light flashes are used to excite the Cr^{3+} ions to the $^4T_{2g}$ excited state, they can degrade to the ^2E_g state very well, to the point that more Cr^{3+} ions are in the ^2E_g excited state than are in the $^4A_{2g}$ ground state: a *population inversion*. As photons of red light are very slowly emitted by phosphorescence, the population inversion allows them to excite other photons of the same wavelength to be emitted, a process known as *stimulated emission*. If these photons are bounced back and forth between mirrors at the ends of the ruby, the effect cascades, producing an intense beam of monochromatic light in which the light waves are perfectly in phase with each other. This is the basis of the ruby *laser* (an acronym for light amplification by stimulated emission of radiation), invented by Theodore Maiman in 1960.

Phosphorescence of *f*-block Ions. Ruby lasers and phosphors have largely been supplanted in commercial applications by lasers and phosphors based on sixth-period *f*-block metal (especially Eu and also Y ions); such phosphors have made possible the ubiquitous phenomenon of color television. The *f*-block metal ions are also usefully substituted isomorphously (Section 13.7) for Ca^{2+} ions in enzymes, where their fluorescence allows their sensitive detection.

There are several advantages of *f*-block metal ions in these applications. First, the number of electronic states that can arise from partially occupied *f* orbitals is very

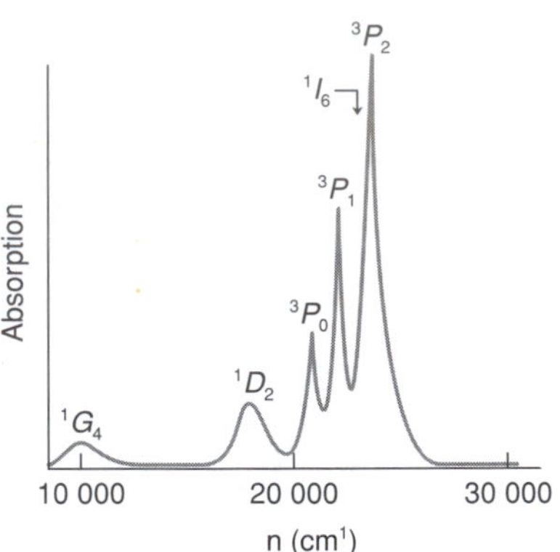

Figure 17.7
Spectrum in the visible region of the green f^2 Pr^{3+}(aq) ion. [Adapted from D. F. Shriver, P. Atkins, and C. H. Langford, *Inorganic Chemistry*, 2nd ed., W. H. Freeman and Company, New York, 1994; p. 604.]

large. A listing of just the ground electronic states of the *f*-block +3 charged ions hints at this diversity. Starting from the ends of the *f* block and working to the middle, we find: ^1S for colorless La^{3+} and Lu^{3+}; ^2F for colorless Ce^{3+} and Yb^{3+}; ^3H for green Pr^{3+} and Tm^{3+}; ^4I for lilac Nd^{3+} and Er^{3+}; ^5I for pink Pm^{3+} and yellow Ho^{3+}; ^6H for yellow Sm^{3+} and Dy^{3+}; ^7F for pale pink Eu^{3+} and Tb^{3+}; and ^8S for colorless Gd^{3+}. Above these (in most cases) there are numerous excited electronic states of diverse spin multiplicities. Since these are not populated, it becomes easy to achieve population inversion between different excited states, so that stimulated emission can easily be achieved, involving an electronic transition from a higher energy excited state to a lower energy, unpopulated excited state of different multiplicity.

Second, recall that the valence $4f$ orbitals of these metal ions have $\langle r_{max} \rangle$ radii that are small in comparison to the $\langle r_{max} \rangle$ values of their core $5s$ and $5p$ and valence $6s$ orbitals, so that they interact very little with their ligands; crystal field effects are virtually nonexistent for the $4f$-block metal ions. This absence of crystal field effects has several important consequences. (a) It makes it difficult for the ions in their excited electronic states to lose their excess energy by transferring vibrational energy to the very distant ligands, so the excited states of some of these ions have relatively long lives. (b) Coupling to ligand vibrations is what makes the absorption bands of *d*-block metal ions so broad: Since this coupling is absent with $4f$ ions, their spectra are notably sharper (Fig. 17.7). (c) Since this coupling also improves the intensity of the symmetry-forbidden *d–d* transitions in octahedral complexes, the spectra of *f*-block metals are even less intense (but more complex, given the number of electronic states), so that these ions have pale pastel colors. (d) Since crystal field splitting essentially does not occur, the electronic states can be labeled using free-ion labels in almost any complex environment.

An interesting application of these properties of the *f*-block ions has been proposed: as taggants to identify sources of explosives.[11] Tiny quantities of particles containing two different *f*-block ions can be added to the explosive; one *f*-block ion absorbs photons of IR light from an IR diode laser, and transfers the energy to the other *f*-block ion, which emits photons of visible light that can be detected. Enough distinctive pairs of sixth-period *f*-block ions are possible for each manufacturer to add its own "signature" to its explosive.

Spin–Orbit Coupling. Close examination of Figure 17.7 shows the presence of more than one sharp absorption with the same term symbol, differing only in a subscript at the right of the term symbol. These separate absorption bands result from the fact that the ^3P electronic state has been split by **spin–orbit coupling**. This splitting is due to the interaction of two types of magnetic moments in these atoms, molecules, or ions: the magnetic moment associated with the spin(s) of the unpaired electron(s) (i.e., with the spin quantum number S), and the magnetic moment associated with the orbital angular momentum (and its quantum number L) and the rotation of the electron cloud in these asymmetric and unevenly occupied orbitals. These two magnetic moments interact in that the spin system can either align in such a way as to add to the magnetic moment associated with motion in the orbitals, or the spin system can align so as to oppose the orbital system, or there may be intermediate degrees of partial alignment.

As was mentioned in Section 1.11, this interaction can be characterized by a new *total angular momentum* quantum number J, which is the vector sum of L and S resulting from the different possible alignments. For example, a single unpaired electron ($S = \frac{1}{2}$) in a d orbital ($L = 2$) can either give a total angular momentum $J = 2 + \frac{1}{2} = \frac{5}{2}$, or $J = 2 - \frac{1}{2} = \frac{3}{2}$. In other cases, J can also take intermediate values between $J = L + S$ and $J = L - S$, which differ by 1: $L + S - 1, \ldots, L - S + 1$. The subscript found to the right of the term symbols in Figure 17.7 is the J value for that state. Each J state includes $2J + 1$ microstates.

It is interesting that we can add to Hund's rules for selecting maximum spin multiplicity, then maximum orbital angular momentum, the following for 4*f*-block ions: the ground electronic state on the left side of this block of elements has the minimum total angular momentum ($J = L - S$), the ground electronic state on the right side of the block has the maximum total angular momentum ($J = L + S$).

The energy magnitude of this interaction is very much a function of the size of the atom: in H it is a completely insignificant $0.4\,\text{cm}^{-1}$; in K it is a minor $58\,\text{cm}^{-1}$. Its effects become noticeable in atoms of atomic number over 40: it splits electronic spectra (as in Fig. 17.7) and photoelectron spectra (Chapter 10). In the heaviest atoms, it can be a very dominant energy term: Not only would the electronic peaks of a 5*f*-block ion be much further split than shown in Figure 17.7, but the splitting could exceed those of the different electronic states, or even of the parent valence orbitals. Hence, spin–orbit coupling even splits subshells of orbitals (i.e., the 6*p* orbitals, Section 1.11) into energy levels that fill sequentially ($6p_{1/2}$ before $6p_{3/2}$). Thus, the heaviest atoms are better characterized using the four quantum numbers n, ℓ, j, m_j instead of the usual n, ℓ, m_ℓ, m_s. For most atoms of atomic number over 40 but not near to 100, this is not necessary or even advisable, but there are still perceptible splittings of the energy states based on M_L and M_S, which can be indicated using the J value as a subscript.

17.6 Photochemically Assisted Substitution and Redox Processes

Absorption of a photon by a metal complex increases the energy content of a complex, typically by a few hundred kilojoules per mol. In some cases, this very substantial quantity of energy can be harvested to carry out important chemical reactions,[12] provided this can be done rapidly enough to avoid rapid loss of the excess energy by spontaneous deexcitation or luminescence (Section 17.5). As one example, we think of elemental mercury as being very inert, but when it is photoexcited to the $^3P_1$ state, it is capable of η^2 coordination of aromatic hydrocarbons, and can then transfer its excess energy to the ligands, cleaving C–H, C–C, C–O, or C–N bonds.[13] Normally unreactive TiO_2, when photoexcited, initiates the oxidation of air pollutants (Section 13.4).

Prompt Reactions. If the chemical reaction occurs very rapidly (in <50 fs) this is called a *prompt reaction*. Practical examples include substitution reactions of metal carbonyls that are initiated by UV irradiation such as Reactions (11.41), (11.51), and the following[14]:

$$Cr(CO)_6 + h\nu \rightarrow Cr(CO)_5 + CO$$
$$Cr(CO)_5 + Xe \rightarrow [Cr(CO)_5Xe] \qquad (17.20)$$

In the first step of this reaction sequence, the UV irradiation promotes a $\pi \rightarrow \sigma^*$ electronic transition, giving a photoexcited state with a weakened Cr–CO bond, so that CO dissociates. It is possible to study the kinetics of the prompt reactions of the photochemically generated 16-electron $M(CO)_5$ with solvents as seemingly inert as heptane or liquid or supercritical Xe, then study the relatively slower reactions of the complexes of these ligands with other ligands included in the solution to produce more stable organometallic products.[15]

Besides metal carbonyls, metal–metal bonded compounds are often photochemically active. These compounds are often intensely colored due to metal $\sigma \rightarrow \sigma^*$ transitions [$Mn_2(CO)_{10}$ is bright yellow, $Fe_2(CO)_9$ is gold, and $Co_2(CO)_8$ is purple-black]; excitation of this transition leads to cleavage of the M–M bond and formation of free radicals, which can then undergo free radical reaction mechanisms with haloalkanes such as CCl_4 or $CHBr_3$:

$$Mn_2(CO)_{10} + h\nu \xrightarrow{\text{fast}} 2{\cdot}Mn(CO)_5$$
$${\cdot}Mn(CO)_5 + CCl_4 \xrightarrow{\text{slow}} {\cdot}Mn(CO)_5(ClCCl_3)$$
$${\cdot}Mn(CO)_5(ClCCl_3) \xrightarrow{\text{fast}} Mn(CO)_5Cl + {\cdot}CCl_3 \qquad (17.21)$$

(Note that this reaction mechanism is passing through 17- and 19-electron species that are unstable but react very rapidly.)

Photosubstitution reactions of complexes such as $[Cr(NH_3)_5X]^{2+}$, $[Cr(NH_3)_4X_2]^+$, and $[Cr(NH_3)_3X_3]$ have been studied. These reactions are interesting in that they often (but not always) proceed differently than thermal substitutions, in that NH_3 may be replaced rather than halide. With reference to Figure 17.6, we can see that the photoactive excited state could either be the $^4T_{1g}$, which with one electron in an $e_g{}^*$ orbital would be expected to follow a dissociative mechanism more readily, or it may be the ^2E_g state, which has one vacant t_{2g} orbital [i.e., can be represented with microstates such as $(\uparrow\downarrow)(\uparrow)(\ )$], so it might be expected to follow an associative mechanism more readily.[16]

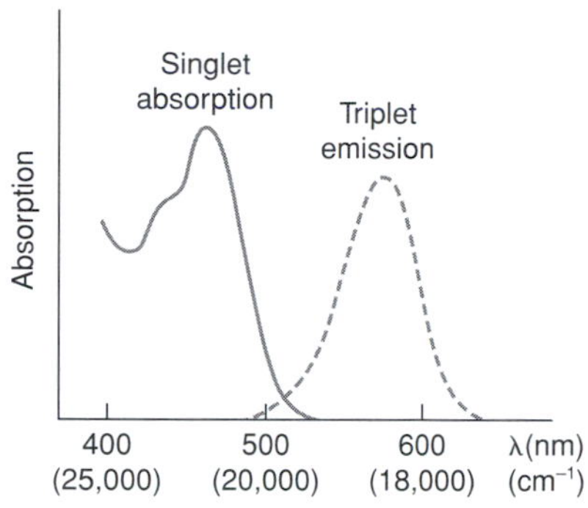

Figure 17.8
The singlet absorption and triplet phosphorescence spectra of $[Ru(bpy)_3]^{2+}$. [Adapted from D. F. Shriver, P. Atkins, and C. H. Langford, *Inorganic Chemistry*, 2nd ed., W. H. Freeman and Company, New York, 1994; p. 603.]

Delayed Reactions. The other condition that can allow success for a photochemical reaction is that the photochemically excited state of the complex has a long lifetime (e.g., in microseconds), so that it can participate in *delayed reactions.* An important example of this is $[Ru(bipyridyl)_3]^{2+}$, the excited state of which has the potential of carrying out a very important redox reaction, the splitting of water into H_2 and O_2. If this can be practically achieved, this would be an important method of capturing the energy of the sun, storing it briefly in the excited state of $[Ru(bipyridyl)_3]^{2+}$, then transferring the energy to give a very clean-burning fuel, H_2.

The complex $[Ru(bipyridyl)_3]^{2+}$ has an intense charge-transfer absorption band at 452 nm; absorption of a photon of this wavelength causes a **metal–ligand charge transfer** of an electron from the highest occupied molecular orbital (HOMO) $t_{2g}{}^6$ set of electrons of Ru(II) to an empty π^* orbital [the lowest unoccupied molecular orbital (LUMO)] of one of the bipyridyl ligands (we will abbreviate these as "bpy"), to give a singlet state complex with electron configuration $t_{2g}{}^5\pi^{*1}$. This can be represented with orbital diagrams as $t_{2g}{}^5(\uparrow\downarrow)(\uparrow\downarrow)(\uparrow)\pi^{*1}(\downarrow)$: the absorption is spin-allowed, and the resulting complex can be described as $^1[Ru^{III}(bpy)_2(bpy^{\cdot -})]^{2+}$. This complex undergoes rapid intersystem crossing to give the lower energy and longer lasting triplet state complex $^3[Ru^{III}(bpy)_2(bpy^{\cdot -})]^{2+}$ or just $^3[Ru(bpy)_3]^{2+}$ for short.

This triplet excited state may either undergo a delayed reaction or, if this does not happen quickly enough (within 1 μs), it may show phosphorescence back to the ground state, emitting bright orange light. The shorter wavelength of the absorption and the longer wavelength of the phosphorescent emission are contrasted in the spectrum illustrated in Figure 17.8.

Or a delayed reaction of $^3[Ru(bpy)_3]^{2+}$ may occur. The net effect of the intersystem crossing is to put one electron in a relatively high energy, previously unoccupied LUMO, π^*, so that this photoexcited complex is now a very much stronger reducing agent (by 2.12 V) than the original $[Ru(bpy)_3]^{2+}$. In addition, it has removed one electron from the relatively low energy, previously filled HOMO, t_{2g}, so that the photoexcited complex is also a very much stronger oxidizing agent (by 2.12 V) than the

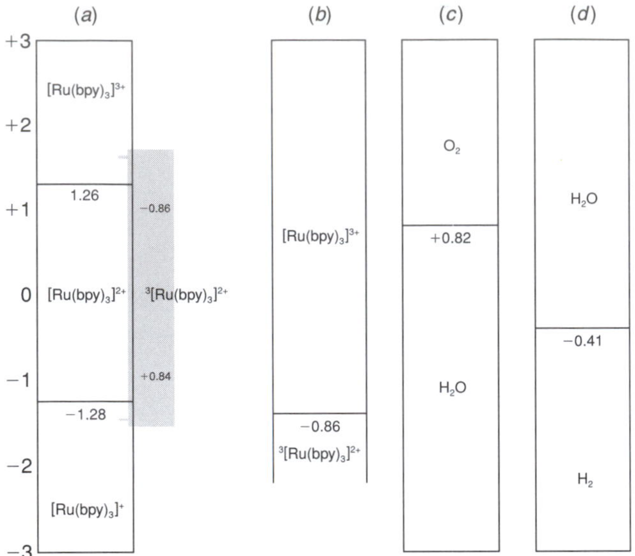

Figure 17.9
(a) The redox predominance diagram for $[Ru(bpy)_3]^{n+}$; the thermodynamically unstable triplet excited state is shown at the right. (b) Representation of the reducing powers of the triplet excited state, showing its $E°(-0.86\,V)$ when being oxidized to $[Ru(bpy)_3]^{3+}$. (c) The redox predominance diagram of H_2 at pH = 7. (d) The redox predominance diagram of O_2 at pH = 7.

original $[Ru(bpy)_3]^{2+}$. So absorption of light can produce a species that is simultaneously a strong oxidizing and a strong reducing agent!

In Figure 17.9(a), we show the redox predominance diagram (Section 6.1) of the three stable oxidation states of the $[Ru(bpy)_3]^{n+}$ complex; the triplet photoexcited state of the +2 diagram must be indicated at the right of the diagram, since it is thermodynamically unstable (it will oxidize and reduce itself when it phosphoresces back to the ground state). However, we can show one of the boundaries of this unstable species in a separate diagram; we have done so for the excited state acting as a reducing agent in Figure 17.9(b).

In 1975, Creutz and Sutin[17] sparked great interest in the photoexcited state of $[Ru(bpy)_3]^{2+}$ when they pointed out that $^3[Ru(bpy)_3]^{2+}$ is a strong enough reducing agent to transfer its π^* electron to water, reducing it to H_2 at pH 7 [compare parts (b) and (d) of Fig. 17.9 to verify this]:

$$\begin{aligned} &2\,^3[Ru(bpy)_3]^{2+} + 2\,H_2O \\ &\quad \rightarrow H_2 + 2\,OH^- + 2\,[Ru(bpy)_3]^{3+} \qquad E = +0.45\,V \end{aligned} \tag{17.22}$$

In addition, the oxidized ruthenium complex resulting from this reaction is a strong enough oxidizing agent to oxidize water to O_2 at pH 7 [compare parts (a) and (c) of Fig. 17.9 to verify this]:

$$\begin{aligned} &4\,[Ru(bpy)_3]^{3+} + 2\,H_2O \\ &\quad \rightarrow O_2 + 4\,H^+ + 4\,[Ru(bpy)_3]^{2+} \qquad E = +0.44\,V \end{aligned} \tag{17.23}$$

The adding of these two reactions, the original photoexcitation, and the neutralization of the OH^- and H^+ byproducts leaves simply the splitting of water into H_2 and O_2; the Ru complex serves as a catalyst.

As almost always happens when an idea for producing a clean, efficient fuel (or, it seems, anything else wonderful) comes up, the difficulty is in the details. Unfortunately, both of these reactions are noncomplementary: the ruthenium reactions involve one electron, while the H_2 reaction is a two-electron reduction, and the O_2 reaction is a four-electron oxidation. Thus multistep mechanisms are required, but the photoexcited ruthenium complex does not oblige us by hanging around long enough to carry out several steps. So many experiments have been carried out using intermediate *quenching agents* that can be more rapidly reduced and yet will last long enough to carry out the multistep mechanism of water splitting. Some of these have succeeded in photochemically splitting water,[18] but none of these has yet overcome the other difficulties that must be overcome before a practical water-splitting process can been achieved.

17.7 Chlorophyll and Photosynthesis

There have been many places throughout the text where we have introduced examples of the functions, not only of inorganic principles, but of specific biologically utilized metal complexes in biochemistry: ion-transporting antibiotics and metalloenzymes with Lewis acid functions (e.g., carboxypeptidase) in Section 5.10; oxygen-transporting heme, hemoglobin, and myoglobin in Section 8.10; cobalamin and Vitamin B_{12} in Section 11.9; rubredoxins and ferredoxins in Section 14.8. We wish to close out this chapter with some additional examples of biochemically important metal complexes.[19]

We begin with the chemistry of photosynthesis, which is by no means fully understood and involves a number of steps beyond the scope of this text. It is known that photosynthesis involves two photochemical reaction centers, called *photosystem I* (PS I) and *photosystem II* (PS II). Photosystem I has the final goal of reducing carbon dioxide by four electrons to a carbohydrate:

$$6\,CO_2 + 24\,e^- + 24\,H^+ \rightarrow C_6H_{12}O_6 + 6\,H_2O \tag{17.24}$$

Photosystem II has the primary responsibility of using the energy of a photon to produce energy in the form of 2 mol of adenosinetriphosphate (ATP), with O_2 being a byproduct (albeit one of considerable interest to us) that comes from the oxidation of water:

$$2\,H_2O \rightarrow 4\,H^+ + 4\,e^- + O_2(^3\Sigma_g{}^-) \tag{17.25}$$

Photosystem I is based on one form of chlorophyll, a Mg^{2+} porphyrin complex [Fig. 5.4(*a*)], which as a result of its many conjugated double bonds has strong absorption bands in the red and blue regions of the visible spectrum (allowing the characteristic green color of light to be transmitted). Both absorption bands involve promoting an electron from the porphyrin π HOMO to the π^* LUMO, and produce a singlet excited state. Fluorescence or degradation of the photochemical energy through molecular vibrations would be expected to follow rapidly, but this is undesirable in the case of chlorophyll: The excited electron and its chemical energy need to be transferred (ultimately) to CO_2, not dissipated in fluorescence or vibrations. The rigidity of the conjugated macrocyclic ring helps by reducing the ease of vibration; the ring is further stiffened by the Mg^{2+} ion at the center, which also favorably adjusts the rate of intersystem crossing to the triplet state.

The loss of energy to vibrations or to fluorescence is also minimized by having an electron acceptor close to the "special pair" of chlorophyll molecules to quickly accept the excited electron. It is also necessary to prevent the reversibility of this electron transfer by quickly transferring the electron to other redox systems further and further from the original special pair, until it is used to reduce a quinone molecule to a radical anion or (with two electrons) a hydroquinone molecule. This molecule can diffuse away to continue the sequence of reactions, which has many further steps where the electron is passed to other redox systems (one of which involves iron–sulfur clusters). This rapid electron transfer is made possible by having the initial electron acceptors built into the same structure in close proximity to each other (Fig. 17.10).

The task of photosystem II in producing oxygen from water is complicated by the fact that this is a four-electron oxidation that can have a problem of noncomplementarity (Section 6.2) and that requires a strong oxidizing agent. The strong oxidizing agent is known as the manganese-containing oxygen-evolving complex (Mn–OEC),[20] and is a (probably) tetranuclear manganese cluster oxide also containing one Ca^{2+} and one Cl^- ion. This cluster oxide can be oxidized in four steps, one electron and (presumably) one manganese at a time, so as to be able to accept four electrons from water. It is not known what oxidation states of manganese are involved in this cluster oxide at any of the four steps: Mn(II), Mn(III), and/or Mn(IV) may participate. Again, the four electrons are then shuttled through several other redox systems [including another iron–sulfur cluster and a copper(I)–copper(II) system] while the solar energy is used to produce ATP, and the electrons are ultimately transferred to chlorophyll a_1 in photosystem I, to restore it to its original ground state and prepare it to interact with another photon.

Related to these large, complex light-harvesting systems, and also to complexes showing long-range electron transfer (Section 16.7), are 9000-pm long π-conjugated molecules (connected by three zinc porphyrin links), which are terminated on one end by a functional group that absorbs blue-green light. The energy of the resulting excited state is transmitted through the zinc–porphyrin links to the distinct functional group at the other end, where a photon of red light is emitted, with 76% success at transmission.[21] In photosynthesis, however, the efficiency of transmission is nearly 100%.

17.8 Biochemical Oxygen Transport and Activation

Earlier in this chapter, we suggested that the fact that O_2 has a triplet ground state ($^3\Sigma_g^-$) inhibits its reactivity under many conditions, since in reacting with diamagnetic molecules an initial high-energy triplet state must be produced. In addition, a rather strong double bond in O_2 must be cleaved to form the ultimate oxidation products that result from biochemical oxidations. However, *d*-block metal ions have unpaired electrons with which these O_2 electrons can be paired, and they are readily oxidized by one or two (or sometimes more) electrons, so they offer rich possibilities for coordinating and activating O_2 for biochemical purposes.

The reduction of O_2 is a four-electron process, and the product from three of the stages are known, well-characterized species: first the superoxide ion, O_2^-, of bond order 1.5; then the peroxide ion, O_2^{2-}, of bond order 1.0; ultimately, of course, one

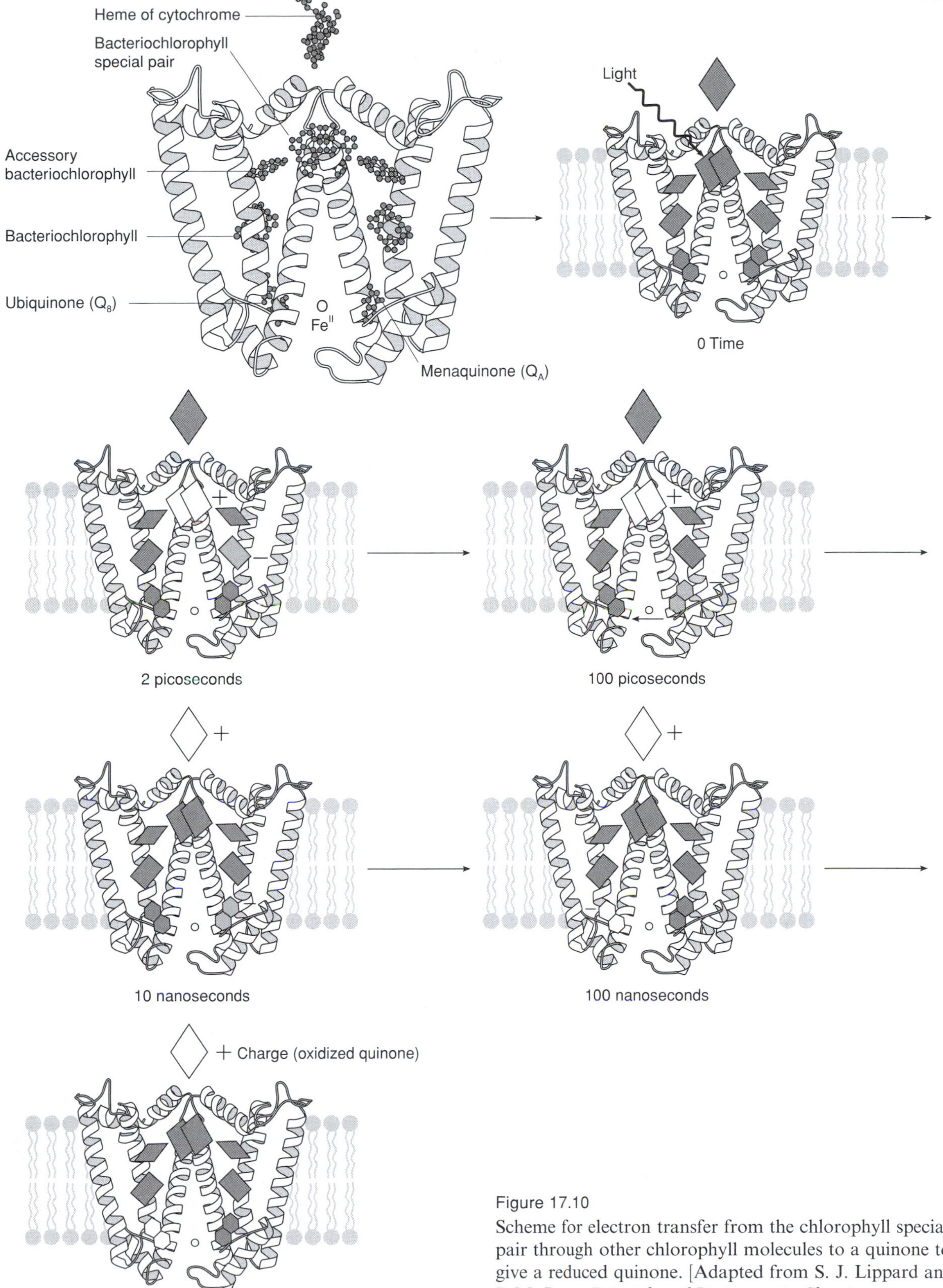

Figure 17.10

Scheme for electron transfer from the chlorophyll special pair through other chlorophyll molecules to a quinone to give a reduced quinone. [Adapted from S. J. Lippard and J. M. Berg, *Principles of Bioinorganic Chemistry*, University Science Books, Sausalito, CA, 1994; pp. 374–375.]

Table 17.6
Variation of O–O Stretching Frequencies and Bond Distances with Bond Order

Species	ν_{O-O} (cm^{-1})	d_{O-O} (pm)	Bond Order
O_2^+	1,905	112	2.5
O_2	1,580	121	2.0
O_2^-	1,097	133	1.5
O_2^{2-}	802	149	1.0

SOURCE: Adapted from S. J. Lippard and J. M. Berg, *Principles of Bioinorganic Chemistry*, University Science Books, Sausalito, CA, 1994; p. 287.

obtains two oxide ions, hydroxide ions, or water molecules, with zero O–O bond order. Detailed crystal structures are notoriously difficult to obtain for large biochemical molecules such as metalloenzymes and metalloproteins, so unraveling just what happens during the coordination and activation of O_2 requires not only evidence concerning bond distances, which vary considerably with bond order as shown in Table 17.6, but also evidence from other spectroscopic methods such as vibrational (IR and Raman) spectroscopy. The O–O stretching frequency varies considerably with bond order (Table 17.6). Superoxide ions usually bridge two metal ions in a nonlinear bridge with either a cis or a trans conformation (M/O–O\M or M/O\O/M, respectively); the O–O bond lengths are in the range of 110–130 pm, and the O–O stretching frequency is in the range of 1075–1195 cm^{-1}. Peroxide ions may either form three-membered chelate rings with one metal ion, or may also bridge two metal ions in the trans conformation; the O–O bond length is in the range of 140–150 pm, and the O–O stretching frequency is in the range of 790–930 cm^{-1}.

In Section 8.10, we discussed heme, an iron–porphyrin complex that functions as part of two O_2 carrying metalloproteins in the body: hemoglobin and myoglobin. The heme molecule in these metalloproteins bonds one O_2 molecule per iron atom. The O_2 acts as a terminal ligand for iron; when coordinated it shows an O–O stretching band at about 1105 cm^{-1}, suggesting a coordinated superoxide ion. The Fe–O–O unit is bent at the central oxygen atom, which brings the outer oxygen atom within hydrogen-bonding distance of a distant histidine proton. This additional interaction helps hemoglobin and myoglobin discriminate in favor of O_2 as a ligand and against CO, which coordinates linearly at the iron, and hence is not optimally located for this additional hydrogen bond.

If the oxygen is coordinated as a O_2^- ion, then the iron must be oxidized (reversibly) to iron(III). Oxyhemoglobin and oxymyoglobin are observed to be diamagnetic; O_2^- has one unpaired electron that can couple antiferromagnetically with the one unpaired electron of a low-spin d^5 Fe^{3+} ion. This coupling is not the only way in which this diamagnetism could be achieved: For example, we could have had bonding between two zero-spin species, low-spin d^6 Fe^{2+} and singlet O_2. But its reasonableness is enhanced by the study of a d^3 chromium(III) porphyrin complex that has two unpaired electrons rather than four: Again antiferromagnetic coupling of the oxygen with one metal unpaired electron is indicated.

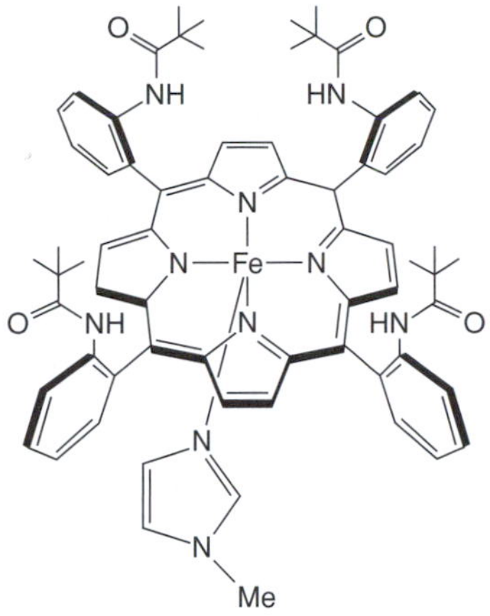

Figure 17.11
Structure of a picket-fence porphyrin. [Adapted from S. J. Lippard and J. M. Berg, *Principles of Bioinorganic Chemistry*, University Science Books, Sausalito, CA, 1994; p. 290.]

The superoxide ligand can be displaced by other ligands such as chloride ion, or released after protonation as the radical HO_2[22]; in either case the hemoglobin is left as the iron(III) derivative *methemoglobin*, which will not carry O_2; each day about 3% of our total hemoglobin is inactivated in this way. Protein-free heme is even more rapidly deactivated by being irreversibly oxidized by O_2 to methemoglobin. The mechanism has been worked out, and involves the following steps (Porph represents the porphyrin ring):

$$\mathrm{PorphFe^{II}} + \mathrm{O_2} \rightleftharpoons \mathrm{PorphFe^{III}} - (\mathrm{O_2}^-) \tag{17.26}$$

$$\mathrm{PorphFe^{III}} - (\mathrm{O_2}^-) + \mathrm{PorphFe^{II}} \rightleftharpoons \mathrm{PorphFe^{III}} - (\mathrm{O_2}^{2-}) - \mathrm{Fe^{III}Porph} \tag{17.27}$$

$$\mathrm{PorphFe^{III}} - (\mathrm{O_2}^{2-}) - \mathrm{Fe^{III}Porph} \rightarrow 2\,\mathrm{PorphFe^{IV}}{=}(\mathrm{O^{2-}}) \tag{17.28}$$

$$\mathrm{PorphFe^{IV}}{=}(\mathrm{O^{2-}}) + \mathrm{Fe^{II}Porph} \rightarrow \mathrm{PorphFe^{III}} - (\mathrm{O^{2-}}) - \mathrm{Fe^{III}Porph} \tag{17.29}$$

This sequence of reactions does not happen with hemoglobin or myoglobin, however, since the globin portion of the protein prevents two heme molecules from approaching each other in Step (17.27). This has been confirmed by preparing sterically hindered "picket-fence" heme complexes (Fig. 17.11), which likewise cannot undergo the second step above, even though no globin protein is present.

Two other O_2 carriers, hemerythrin and the copper-based hemocyanins, are also of importance; in contrast to heme, each of these carriers uses two metal atoms per dioxygen molecule bound.

Hemerythrin is a nonporphyrin based dinuclear iron complex that involves, in the oxygen-free deoxyhemerthyrin form, two antiferromagnetically coupled high-spin iron(II) atoms (one six coordinate and one five coordinate) that are bridged by two carboxylate groups from glutamate and aspartate amino acids, and one hydroxide ion [Fig. 17.12(*a*)]. The situation would seem to be set up for bonding the $O_2(^3\Sigma_g^-)$ as another bridging ligand,[23] pairing the two electrons with unpaired electrons on two iron atoms to give a complex involving two six-coordinate Fe(III) atoms and a

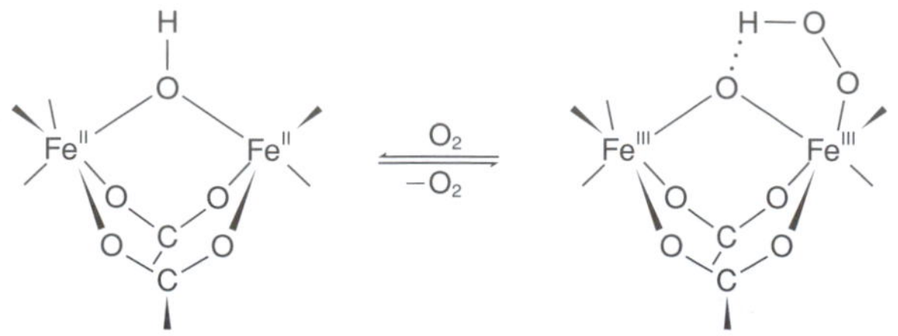

(a) Deoxyhemerythrin (b) Oxyhemerythrin

Figure 17.12
Scheme showing the environment of the diiron core in hemerythrin (*a*) before and (*b*) after reversible O_2 binding. (The proton location could not be determined, but is postulated.) [Adapted from S. J. Lippard and J. M. Berg, *Principles of Bioinorganic Chemistry*, University Science Books, Sausalito, CA, 1994; p. 287.]

bridging diamagnetic peroxide ion, μ-O_2^{2-}. This expectation is partially met: In oxyhemerythrin the iron atoms are present as high-spin iron(III) atoms, and the oxygen seems to function as a peroxide ion, as suggested by its O–O stretching vibration at $844\,cm^{-1}$. But the iron-57 Mössbauer spectrum shows that the two iron atoms are different in oxyhemerythrin, and the Raman spectra show that the two oxygen atoms of the O_2 are not equivalent. The crystal structure of oxyhemerythrin has been refined[24] to show that the O_2 molecule is not functioning as a bridging ligand, but is terminally bound (probably as a hydroperoxide ligand, $^{-}$OOH) to one of the two irons [Fig. 17.12(*b*)].

Despite their name, the hemocyanins contain neither heme nor cyanide; the term means "blue blood," in reference to the bright blue color of their oxidized forms (the deoxy forms are colorless). These oxygen carriers are used by many species of arthropods and mollusks, and do not contain iron at all, but contain d^{10} Cu(I) in the deoxy form and d^9 Cu(II) in the oxy form[25]; each copper atom is coordinated by N donor atoms of three histidine amino acids. The O–O stretching vibration of oxyhemocyanin is found at $744\,cm^{-1}$, which also suggests a peroxide ion. But other evidence capped by that of a crystal structure determination[26] shows that the peroxide ion is symmetrically bridging in this case—not in the

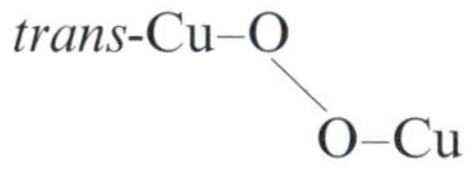

trans-Cu–O
O–Cu

manner that one might expect, but rather in an unusual form:

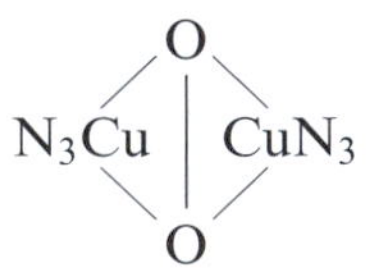

O
N₃Cu | CuN₃
O
A

The possibility of this unusual bridging mode for the peroxide ion had been anticipated, since it had earlier been observed in a model inorganic O_2 complex of copper with a tridentate chelating nitrogen donor ligand; in this model complex the

Table 17.7
Major Types of Reactions Catalyzed by Cytochrome P-450

Reaction Type	Simplified Example	Typical Substrate
Aliphatic hydroxylation	Cyclohexane → cyclohexanol	Pentobarbital
Aromatic hydroxylation	Benzene → phenol	Phenobarbital
Alkene epoxidation	Cyclohexene → cyclohexene oxide	Aldrin
N-Dealkylation	$MeN(H)Me \rightarrow MeNH_2 + H_2C{=}O$	Methadone
O-Dealkylation	$PhOMe \rightarrow PhOH + H_2C{=}O$	Codeine
Oxidative deamination	$Me_2CHNH_2 \rightarrow Me_2C{=}O + NH_3$	Amphetamine
S-Oxidation	$MeSMe \rightarrow Me_2S{=}O$	Chlorpromazine
Reductive dehalogenation	$PhCH_2Br \rightarrow PhMe$	Halothane

SOURCE: From S. J. Lippard and J. M. Berg, *Principles of Bioinorganic Chemistry*, University Science Books, Sausalito, CA, 1994; p. 304.

O–O distance was found to be 141 pm.[27] A similar complex has been prepared in which the N_3 donor atoms come from a macrocyclic amine ligand, 1,4,7-triisopropyl-1,4,7-triazacyclononane, *cyclo*-(*i*-$PrNCH_2CH_2)_3$.[28] This complex, when exposed in CH_2Cl_2 solution to O_2, forms a peroxide-bridged adduct with structure (**A**), but when the same oxygenation is carried out in tetrahydrofuran (THF) solution, a complex of the same formula results, but this complex has a much larger O–O distance of 229 pm, indicating that the O–O bond in the peroxide bridge has been cleaved. Even more remarkably, the reaction is reversible by adding excess of one solvent to the species in the other solvent; in acetone, an equilibrium mixture is produced. For the first time, the process of O–O bond cleavage has been observed reversibly, which is of great importance: In the action of photosystem II, the tetranuclear manganese complex must form an oxygen–oxygen bond from two water molecules; in the action of various oxidizing enzymes such as the cytochrome P-450 to be discussed next, the O–O bond must be cleaved.

Cytochrome P-450 catalyzes the oxidation of organic substances by O_2, performing essential roles in metabolism, biosyntheses, and detoxification (Table 17.7); this enzyme is found, for example, in the liver, where the oxidation normally generates hydroxylated species that have enhanced water solubility, so they can better be excreted in the urine. This important process rids the body of toxins; unfortunately, some of the oxidized forms (such as the epoxides) are carcinogenic. It is named as a *P*igment that absorbs at *450* nm; it characteristically includes one heme unit in which one axial coordination site is occupied by a sulfur donor atom from cysteine. Its mechanism of action is likely to include some fairly unprecedented steps, since room temperature oxidation of alkanes in aqueous solution is quite remarkable.

The currently accepted catalytic cycle for the action of P-450 is shown in Figure 17.13, although the Steps from 4–6 become increasingly speculative. The mechanism is thought to involve high-spin five-coordinate and low-spin six-coordinate species including both iron(II) and iron(III), and may even involve some iron(IV) or iron(V) chemistry.

The resting state of the enzyme (shown at the left of the figure) involves water as the second axial ligand, in the sixth coordination site. In Step 1, a hydrocarbon enters a hydrophobic pocket of the protein, near the heme, expelling the water from the pocket and the iron axial coordination site. In Step 2, an electron is added to reduce the iron

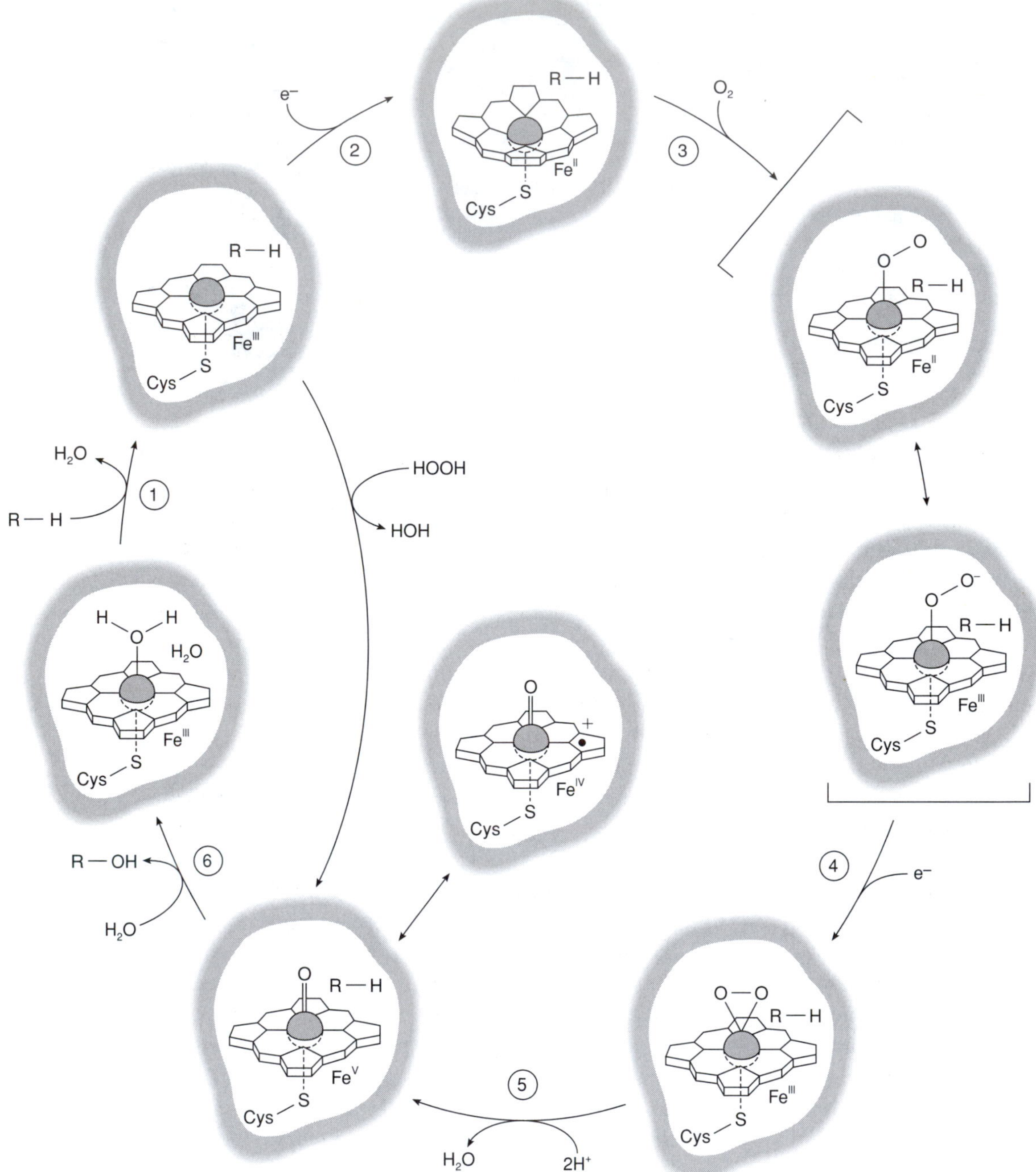

Figure 17.13

Catalytic cycle (beginning at the left center or 9:00 position) proposed for cytochrome P-450. [Reprinted with permission from S. J. Lippard and J. M. Berg, *Principles of Bioinorganic Chemistry*, University Science Books, Sausalito, CA, 1994; p. 308.]

to iron(II), so that it can (in Step 3) coordinate an O_2 molecule: just as with heme in hemoglobin, the O_2 is a terminal ligand, and may be envisioned as a coordinated O_2^- ion (O–O stretching frequency is at $1140\,cm^{-1}$) attached to iron(III), or as neutral O_2 attached to iron(II). From here on the evidence becomes less unequivocal: in Step 4, a second electron may be added, producing a coordinated peroxide complex.

In Step 5, the important cleavage of the O–O bond is postulated to occur: two protons come in, removing one oxide ion as water and leaving an *oxenoid* species, which could be described as an oxygen atom coordinated to iron(III), or as an oxide ion coordinated to iron(V), or equivalently as Fe(V)=O, or even as oxygen double bonded to iron(IV) with one electron having been oxidized from the π HOMO of the porphyrin ring, leaving it as a radical cation.

Step 6 involves the most unprecedented step of all: The hydrocarbon RH, waiting in its hydrophobic pocket, is finally oxidized to an alcohol. Concerted addition of the RH to the Fe(V)=O bond, or an acid–base reaction, cannot be ruled out, but the best thinking seems to be that a free radical sequence of reactions occurs here:

$$R{-}H + Fe^{V}{=}O \rightarrow [R\cdot + Fe^{IV}{-}OH] \rightarrow ROH + Fe^{III} \qquad (17.30)$$

The difficulty here is that radical reactions of this type normally are very unselective: the high-energy R· radical can rearrange or react with all sorts of bonds. But P-450 is characterized by good selectivity in its reactions. So it is postulated that the reactions of Eq. (17.30) must occur within 10^{-9} s, so as to preclude the normal competing reactions of the radical; confinement within the pocket no doubt helps. Finally, the ROH product of the reaction dissociates, and water reenters to complete the catalytic cycle.

In support of some of the uncertain steps, it is observed that addition of hydrogen peroxide to the resting form of P-450 leads immediately to the oxenoid intermediate and to enhanced rates of alkane oxidation. In addition, model compounds support a similar concept of, in effect, transferring an oxygen atom to the iron: a type of picket-fence heme has been found to be an effective catalyst for alkane oxidation, with oxygen atoms being provided by a peroxycarboxylic acid, or by iodosylbenzene (PhI=O), a known oxygen transfer reactant, with an unstable green intermediate being isolated at low temperatures, which has been characterized as having an Fe=O bond and radical cation character in the porphyrin ring.

Even more potent than cytochrome P-450 is the bacterial enzyme methane monooxygenase (MMO), the only enzyme capable of oxidizing CH_4 (to CH_3OH). This is a nonheme iron enzyme; in its reactive form two Fe(IV) iron atoms are bridged in a diamond-shaped core by two oxide ions, while each iron(IV) is also coordinated to a separate oxygen donor atom of a bridging carboxylate group. Isolation of this very reactive species was obtained by reacting MMO in solution with O_2, then quickly freezing the solution. How the reactive intermediate oxidizes the very inert CH_4 molecule is not yet known.[29]

17.9 Biochemical Nitrogen Activation

Nitrogen, in chemically bound forms such as the amino acids and porphyrins, is one of the most important elements of life, but on Earth the main reservoir of this element is as the elemental N_2, which nearly all forms of life are unable to utilize; lack of availability of nitrogen is the limiting factor for growth of organisms in many environments.

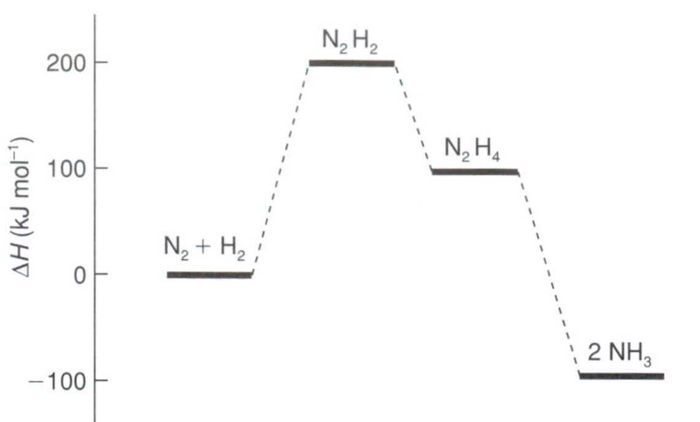

Figure 17.14
Energetics of N_2 reduction through the potential intermediate stages of diazene and hydrazine. [Adapted with permission from I. Bertini, H. B. Gray, S. J. Lippard, and J. S. Valentine, *Bioinorganic Chemistry*, University Science Books, Sausalito, CA, 1994; p. 416.]

The triple bond (bond energy of $942\,\text{kJ mol}^{-1}$) in N≡N is enormously strong, and must be cleaved during the reactions needed to get it into the reduced forms (amines, amides, etc.) in which it is found in cells. The barrier to the simplest and most reduced form, NH_3, is not thermodynamic (ΔH_f for $NH_3 = -50\,\text{kJ mol}^{-1}$), but kinetic: The two compounds showing the intermediate stages of reduction that one would expect before reaching complete N–N bond cleavage, HN=NH (diazene) and N_2H_4 (hydrazine) are indeed disfavored thermodynamically (Fig. 17.14).

The industrial process of nitrogen fixation (the Haber process, Section 15.4) overcomes the kinetic barrier by the use of high temperatures (400–500 °C) and pressures (200 atm) and a metallic iron catalyst that breaks the N≡N bond by dissociative chemisorption (Section 12.6). The process is quick and efficient, but the conditions required are *much* more drastic than those used by blue-green algae and by bacteria found in the root nodules of legumes, which carry out this reaction at ambient temperature and pressure, using the **nitrogenase** enzymes. These enzymes involve two proteins, the more interesting of which has two joined redox-active ferredoxin-type Fe_4S_4 clusters (Section 14.8) in close proximity to a cluster of similar stoichiometry but also involving a molybdenum atom (seven iron atoms per molybdenum atom).[30] The action of nitrogenase does not release free diazene, which is unstable, or hydrazine, which is toxic; ammonia is the first product released by the enzyme. Dihydrogen is also produced, according to the stoichiometry:

$$N_2 + 10\,H^+ + 8\,e^- \rightarrow 2\,NH_4{}^+ + H_2 \qquad (17.31)$$

The enzyme is capable of reducing several other π-bonded substrates such as C_2H_2 and HCN to C_2H_4 and $MeNH_2$; these reductions commonly involve $2\,e^- + 2\,H^+$, $4\,e^- + 4\,H^+$, $6\,e^- + 6\,H^+$, or $8\,e^- + 8\,H^+$.

Because the enzyme nitrogenase was very difficult to crystallize, the work on understanding its action had to rely a good deal on work on model inorganic complexes that fix and reduce dinitrogen. The first complexes of N_2 were ruthenium complexes synthesized in 1965,[31] which used hydrazine as a reducing agent and as the source of N_2, but then it was discovered that N_2 itself could be used with reduced forms of ruthenium:

$$2\,[\mathrm{Ru(NH_3)_5(H_2O)}]^{2+} + \mathrm{N_2} \rightarrow [\mathrm{Ru(NH_3)_5N_2Ru(NH_3)_5}]^{4+} + 2\,\mathrm{H_2O} \qquad (17.32)$$

Soon emphasis shifted to earlier *d*-block metals such as Mo, V, and Ti. In 1975, it was found[32] that acidification of a Mo(0) complex produced NH_3:

$$[\mathrm{Mo(Ph_2PCH_2CH_2PPh_2)_2(N_2)_2}] + 6\,\mathrm{H^+} \rightarrow 2\,\mathrm{NH_3} + \mathrm{N_2} + \mathrm{Mo^{VI}} \qquad (17.33)$$

Of course, it is known that *d*-block metals around Mo readily engage in multiple bonding to atoms such as nitrogen. This evidence began to suggest the possibility that the reduction might proceed so as to compensate the loss of the very strong (942 kJ mol^{-1}) nitrogen–nitrogen bond by the formation of two fairly strong (calculated 680 kJ mol^{-1}) nitrogen–molybdenum triple bond(s). Evidence from model compounds that such a reaction could occur finally came in 1995 when a three-coordinate Mo(III) complex of bulky amine ligands, $Mo(NRAr)_3$, where Ar = aryl, was exposed to an atmosphere of N_2: the red-orange solution turned purple. Upon warming to room temperature, this solution turned gold, and the nitrido complex $N{\equiv}Mo(NRAr)_3$ was isolated.[33]

$$\mathrm{N_2} + 2\,\mathrm{Mo(NRAr)_3} \rightarrow \mathrm{(ArRN)_3Mo{-}N{\equiv}N{-}Mo(NRAr)_3}$$
$$\rightarrow 2\,\mathrm{N{\equiv}Mo(NRAr)_3} \qquad (17.34)$$

Calculations[34] have suggested that the mechanism of this reaction proceeds through a bent (zigzag) transition state $[(ArRN)_3Mo{=}\dot{N}{-}\underset{\cdot\cdot}{N}{=}Mo(NRAr)_3]$; cleavage of the NN bond while maintaining linearity is forbidden by symmetry. The activation energy of 97 kJ mol[35] is significantly lowered by (a) relativistic stabilization by about 30 kJ mol^{-1} of each Mo≡N bond produced (relativistic effects are only a few percent with such fifth-period metals, but can prove crucial), and (b) the π-donating properties of the three NRAr ligands, since calculations indicated that three H, Me, or Cl ligands on Mo would not result in cleavage of the triple bond.

Although Reaction (17.34) was progress, the system still failed to model nitrogenase adequately in that it was not catalytic; the molybdenum complex is used up in the reaction. The starting $Mo(NRAr)_3$ has been regenerated[36] by transferring the nitrido group to another molybdenum compound, but results were still unsatisfactory if only catalytic quantities of $Mo(NRAr)_3$ were used. On another front, catalysis was achieved[37] (using $TiCl_4$ as catalyst) of the reaction of N_2 with Li metal and Me_3SiCl to give $(Me_3Si)_3N$ and titanium trimethylsilylamides, although it must be admitted that the other reactants are much more reactive than those used by bacteria!

Thus the models seemed to point to a key role for the molybdenum atom in coordinating, then perhaps multiply bonding to, the N_2 molecule. However, these hypotheses suffered a severe blow when the crystal structure of the iron–molybdenum protein was finally resolved.[38] There was no particular surprise in the structure of the all-iron double Fe_4S_4 cluster: The two cubanes are joined by two bridging atoms from two cysteine amino acids [Fig. 17.15(*a*)]. More astonishing was the incompletely resolved structure of the iron–molybdenum double cluster [Fig. 17.15(*b*)]. There is a hole in the center of this cluster where a N_2 molecule could fit, and there is an unidentified Y bridging ligand; light atoms such as N atoms are difficult to identify clearly in crystal

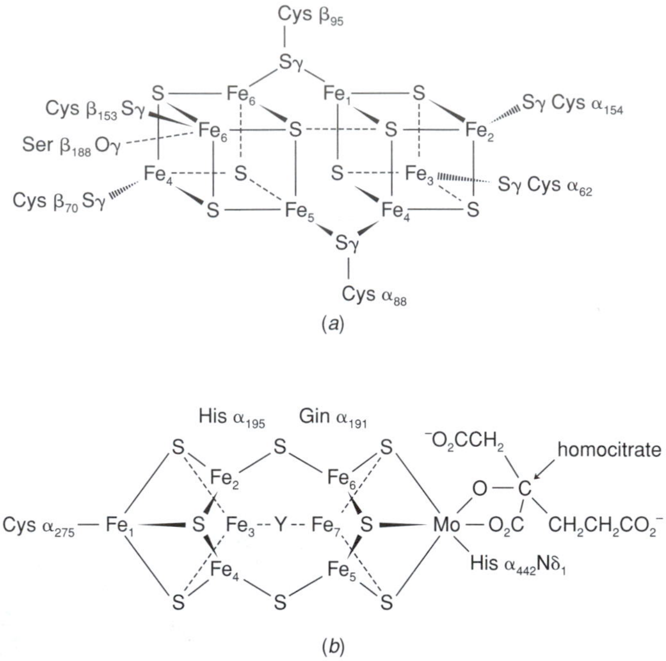

Figure 17.15

Details of the crystal structure of the Fe–Mo protein of nitrogenase from *Azotobacter vinelandii*: (*a*) the joined pair of Fe_4S_4 clusters and (*b*) the Fe_7Mo double cluster, with an unidentified bridging ligand indicated by "Y". [Adapted from I. Bertini, H. B. Gray, S. J. Lippard, and J. S. Valentine, *Bioinorganic Chemistry*, University Science Books, Sausalito, CA, 1994; pp. 443, 444.]

structures of large molecules, when in the vicinity of heavy atoms such as Fe and S. But the metal environment of the Y ligand and of the hole is completely composed of iron atoms! The molybdenum sits on the outside of the cluster, and appears to be coordinately saturated, with a coordination number of 6: There seems to be no room for nitrogen! Finally, two of the coordination sites on Mo are occupied by an unusual ligand, the homocitrate ion, which no one had thought to attempt to include in a model compound.

So the mechanism of action of nitrogenase, upon which all life on Earth depends, is perhaps a deeper mystery than ever. In addition, it must be admitted that the principles of inorganic chemistry, as embodied in the model compound work, and which we have spent much time trying to convey in this book, seem at this point to have led us up a blind alley and not to a true picture of how this aspect of life works. But, in defense of these principles, it could be claimed that these principles did allow researchers to generate *reasonable hypotheses*, which when investigated gave illuminating discoveries. The chemistry of life is a composite of many kinds of chemistry, and even in the case of metalloenzymes cannot be said to be dominated by or determined by inorganic principles alone. Biochemists have found inorganic principles to be useful, and have adopted much of the language of inorganic chemistry to describe them, but they have not given up the study of organic chemistry, or physical chemistry, or of course of the principles of biochemistry! In addition, knowledge can be transferred in more than one direction: industrial chemists seeking to improve catalytic processes are hoping to get new ideas by seeing how Nature does the job.

Study Objectives

1. Write microstates for a given electron configuration; compute M_L and M_S values for these microstates. Exercises 1–3.
2. Interpret term symbols of electronic states of atoms, molecules, or ions; write some "extreme" microstates that would be part of these electronic states. Given Table 17.2, select possible term symbols for a specified electron configuration, and select the ground electronic state from among them. Exercises 4–10.
3. Write term symbols appropriate for use with linear or octahedral molecules or complex ions. Exercises 11–15.
4. Know the functions of correlation diagrams and of Tanabe–Sugano diagrams; use the latter to interpret spectra of octahedral *d*-block metal complexes. Exercises 16–26.
5. Explain why B' Racah parameters for complexes differ from B parameters for free ions, and tell how these relate to the nephelauxetic series of ligands. Exercises 27–29.
6. Use appropriate *f*, *g*, *h*, *k*, and Racah B factors to predict the positions of absorption bands, either graphically using Tanabe–Sugano diagrams, or with the appropriate equations. Exercises 30–32.
7. Know what spin–orbit coupling, as quantified by J, is. Exercises 33–35.
8. Describe the functions of the metal ions, ligands, and proteins in the bioinorganic systems described in this chapter; illustrate by explaining why specific alterations would be deleterious. Exercises 36–40.

Exercises

1. *(a) How many microstates would be expected for a p^3 electronic configuration? (b) Write all microstates among these for which $M_S = \frac{3}{2}$; $M_S = -\frac{3}{2}$. (c) Write all microstates among these for which $M_L = 2$; $M_L = -2$. (d) Write all microstates among these for which $M_L = 0$.

2. (a) How many microstates would be expected for a d^2 electronic configuration? (b) Write all microstates among these for which $M_S = 1$; $M_S = -1$. (c) Write all microstates among these for which $M_L = 4$; $M_L = -4$.

3. (a) How many microstates would be expected for an f^2 electronic configuration? (b) Write all microstates among these for which $M_S = 1$. (c) Write all microstates among these for which $M_L = 6$; $M_L = -6$.

4. *Write term symbols for the following valence electron configurations, none of which give rise to more than one electronic state: (a) $1s^2$; (b) $3s^2 3p^6$; (c) $3d^9$; (d) $4f^{13}$; (e) ${\sigma_g}^2({\sigma_u}^*)^1$; (f) ${\sigma_g}^2{\pi_u}^1$; and (g) ${\sigma_g}^2(\delta_g)^1$.

5. Write the term symbols for the electronic states having the following quantum numbers (S, L): (a) $(\frac{1}{2}, 0)$; (b) $(3, 0)$; (c) $(3, 4)$; and (d) $(\frac{3}{2}, 2)$.

6. In your own words, describe what each of the following term symbols signifies, and tell how many microstates will be associated with each, with what values of M_L and M_S: (a) ^1S; (b) $^1{\Sigma_g}^+$; (c) ^2S; (d) $^2\Pi_u$; and (e) ^4F.

7. Identify the ground state within the sets of electronic states given in Table 17.2A for the following electronic configurations: (a) d^3; (b) d^4; (c) d^5; and (d) p^3.

8. *A microstate $1(\uparrow)0(\ )-1(\uparrow)$ cannot be part of which of the following electronic states: ^3S, ^1F, ^1S, and ^2D.

9. A microstate $1(\uparrow\downarrow)0(\uparrow)-1(\ )$ cannot be part of which of the following electronic states: ^3S, ^1F, ^1S, and ^2D.

10. The d^2 electronic states include triplet states, and include a G state, but do not include a ^3G state. Why not? To answer this, try to draw the "extreme" microstates that would be included in this state.

11. The nitrogen in our atmosphere does not give nearly as rich a chemistry as the oxygen does. (a) What term symbol(s) would apply to electronic states of N_2 molecules having the normal (σ, π) electron configuration? Are there as many electronic states as for O_2? (b) What term symbol(s) would apply to atomic N in the upper atmosphere? Why is atomic N likely to be less abundant than atomic O?

12. What electronic states should exist for the CH fragment (Fig. 11.8)? In which electronic state would CH most be most likely to undergo a concerted cycloaddition reaction with 1,3-butadiene? What would the product be, and how many unpaired electrons (if any) would it have?

13. If Gd_2 has a $^{19}\Sigma$ ground state, how many *paired* valence electrons are there in this ground state of this molecule?

14. *Using Table 17.2, determine the possible term symbols to use for ions having the following valence electron configurations in octahedral complexes: (a) d^{10}; (b) d^1; and (c) d^2.

15. Using Table 17.2, determine the possible term symbols to use for ions having the following valence electron configurations in octahedral complexes: (a) p^4; (b) d^5; and (c) f^1.

16. If you were constructing a correlation diagram like that of Figure 17.2 for the following electronic configurations, how many free-ion energy levels would be found on the left, and how many dominant-field energy levels would be found on the right, for the following electron configurations? (a) d^1; (b) d^3; and (c) d^4.

17. The ^1D state of free gaseous V^{3+} lies 10,640 cm^{-1} above its ground state; the ^3P state lies 12,920 cm^{-1} above the ground state. Calculate B for the free V^{3+} ion.

18. Compute the energy separations (in terms of the Racah parameters A, B, and C) that should be found on the left side of the correlation diagram for d^2 (Fig. 17.2). With what relative values of B and C, which are positive in value, would ^3P be the lowest energy excited state rather than ^1D?

19. *Consult the Tanabe–Sugano diagrams to determine how many reasonably intense (i.e., spin-allowed) d–d electronic transitions are expected in the spectra of *weak-field* octahedral complexes of ions having the following electron configurations, and show how they could be labeled [e.g., $\nu_1(^2T_{2g} \rightarrow {}^2E_g)$]: (a) d^4; (b) d^5; (c) d^6, and (d) d^7.

20. *Consult the Tanabe–Sugano diagrams to determine how many reasonably intense (i.e., spin-allowed) d–d electronic transitions are expected in the spectra of *strong-field* octahedral complexes of ions having the following electron configurations, and suggest how many are reasonably distinct enough in energy level to have a chance of giving separate and resolvable absorption bands: (a) d^4; (b) d^5; (c) d^6, and (d) d^7.

21. Students reported the absorption spectra for NiL_6 complexes involving the following ligands (wavelengths in nanometers were reported). In each case, compute Δ_o and B' for the complex. (a) MeOH: 993, 586, 318 nm; (b) $C_2O_4^{2-}$: 1182, 705, and 382 nm; and (c) pyridine: 957, 600, and 370 nm.

22. Students reported the absorption spectra for octahedral nickel(II) complexes involving the following ligands (wavelengths in nanometers were reported). In each case, compute Δ_o and B' for the complex. (a) $EDTA^{4-}$: 878, 588, and 379 nm; (b) NH_3: 945, 575, and 355 nm; and (c) $NH_2CH_2CH_2NH_2$: 870, 575, and 355 nm.

23. Use the spectrum of $[Co(H_2O)_6]^{2+}$ given in Figure 8.5 to determine graphically the values of Δ_o and B' for this weak-field complex. (There is a weak shoulder in this spectrum at $16{,}000^{-1}$, which should be assigned to the electronic transition among these three that is expected to be weakest.)

24. The VF_6^{3-} ion has its two lowest energy absorptions at 14,800 and 23,350 cm^{-1}. (a) Determine values of Δ_o and B' for this ion. (b) Are these absorptions in the near-IR, visible, or UV spectra? What is the likely color of this ion?

25. *Compute Δ_o and B' for the following octahedral complexes of chromium(III), given the listed absorption energies. (a) CrF_6^{3-}: 34,400, 22,700, and 14,900 cm^{-1}; (b) $[Cr(C_2O_4)_3]^{3-}$: 38,100 (theoretically), 23,900, and 17,500 cm^{-1}; and (c) $[Cr(NH_2CH_2CH_2NH_2)_3]^{3+}$: 46,500 (theoretically), 28,500, and 21,800 cm^{-1}.

26. Suppose that, by the judicious choice of our ligands, we could make a series of complexes of various metals in each of which $\Delta_o = 20{,}000\ cm^{-1}$ and $B' = 1000\ cm^{-1}$. Predict the positions of the spin-allowed electronic transitions using the appropriate Tanabe–Sugano diagrams for metal ions having the following electron configurations: (a) d^1; (b) d^2; (c) d^3; and (d) d^4.

27. Arrange the ligands used by students in Exercise 21 into (a) a spectrochemical series; (b) a nephelauxetic series. (c) Compute the nephelauxetic parameters of these ligands.

28. *Arrange the ligands involved in Exercise 25 into (a) a spectrochemical series; (b) a nephelauxetic series. (c) Compute the nephelauxetic parameters of these ligands.

29. The Racah B parameters of metal ions in Table 17.4, like the pairing energies P in Table 8.2, are presumably functions of some or all of these characteristics of the metal ions: (a) their charges; (b) their d-orbital electron configuration; and (c) the principal quantum number of the d orbitals. To the extent possible, select sets of these parameters that are controlled for the other two variables that show how the values of B and/or P depend on each of these three characteristics of the ions.

30. Calculate likely wavenumbers at which to look for the first two absorption bands of the following low-spin Co^{3+} complexes: (a) $[Co(NH_3)_6]^{3+}$; (b) $[Co(C_2O_4)_3]^{3-}$; and (c) $[Co(CN)_6]^{3-}$.

31. Calculate likely wavenumbers at which to look for the three absorption bands of the following Ni^{2+} complexes: (a) $[Ni(NH_3)_6]^{2+}$; (b) $[Ni(C_2O_4)_3]^{4-}$; and (c) (the unknown) $[Ni(CN)_6]^{4-}$.

32. Make the necessary extrapolations to predict (using Tanabe–Sugano diagrams) the positions of the two prominent absorption bands of the following V^{3+} complexes: (a) $[V(NH_3)_6]^{3+}$; (b) $[V(C_2O_4)_3]^{3-}$; (c) $[V(CN)_6]^{3-}$; and (d) $[V(SMe)_6]^{3-}$.

33. *For the sixth-period f-block ions with the valence electron configurations given, compute the possible J values that can be associated with the ground electronic state, and then tell which J value is preferred for each ion. (a) f^1 and f^{13}; (b) f^2 and f^{12}; and (c) f^3 and f^{11}.

34. For the sixth-period f-block ions with the valence electron configurations given, compute the possible J values that can be associated with the ground electronic state, and then tell which J value is preferred for each ion. (a) f^4 and f^{10}; (b) f^5 and f^9; (c) f^6 and f^8; and (d) f^7.

35. For the sixth-period f-block ions with the valence electron configurations given, compute the number of microstates associated with the ground (L, S) electronic state, and then compute the number of microstates associated with each J value of that electronic state; finally, show that the two numbers are compatible. (a) f^1 and f^{13}; (b) f^2 and f^{12}; (c) f^3 and f^{11}; (d) f^4 and f^{10}; (e) f^5 and f^9; (f) f^6 and f^8; and (g) f^7.

36. *Suggest two reasons why it would hinder the function of chlorophyll if Y^{3+} or Nd^{3+} (and an anion) were substituted for the Mg^{2+}.

37. Why would it probably hinder the function of chlorophyll to hydrogenate the porphyrin ring every one-fourth of the way around the macrocyclic ring?

38. Contrast the ways in which the following metalloporphyrins function when embedded in their proteins as compared to when they are free molecules in solution: (a) heme and (b) chlorophyll.

39. Why would the substitution of Mg^{2+} or Zn^{2+} for the Fe^{2+} probably hinder the function of (a) heme and (b) cytochrome P-450?

40. Use the following standard reduction potentials to add HO_2 (hydrogen superoxide) and H_2O_2 (hydrogen peroxide) to the redox predominance diagram for oxygen. Will these species have to be to the side of the diagram? If so, what does that indicate about them? $O_2 \rightarrow HO_2$, -0.125 V; $HO_2 \rightarrow H_2O_2$, 1.51 V; $O_2 \rightarrow H_2O_2$, 0.695 V; and $H_2O_2 \rightarrow H_2O$ 1.763 V.

41. Work the following crossword puzzle (Fig. 17.16).

ACROSS

2. Principal ionic constituent of igneous rocks
9. Color of liquid bromine
10. Symbol of the element for which a coin is named
11. Symbol for an element prepared by the Hall process
12. Principal source of magnesium
14. A chalcogen
15. A halide that gives a red liquid on oxidation
19. A synthetic element named after a woman
20. A pseudohalide used to prepare detonators
21. Forms 15 across
23. Where most rare gases are found
24. Main body of the hydrosphere
25. A rare gas that forms compounds
27. The most abundant alkali
29. The solid phase of an important liquid
31. Viscous, black liquid hydrocarbons
32. See 19 across
33. An alkaline earth that gives a crimson flame test
34. Prefix for eight
37. See 32 down
38. Mineral form of Mg_2SiO_4
41. A heavy alkali
42. An important constant
43. An unreactive rare gas
44. Condensation from the atmosphere
45. First $5f$ series element
46. A radioactive halogen
47. The commercially most important halogen
48. Group 17/VIIA elements
49. A multidentate complex

DOWN

1. Prefix for four
3. It forms a purple solution in CCl_4 on oxidation
4. A scarce Group 13/IIIA element
5. Geometrical designation of two groups on the same side
6. A silicate mineral that has lubricant properties
7. The basic kind of matter
8. The element preceding 13 down
12. Generic name for a silicon hydride
13. A rare earth used to prepare a red phosphor
16. O_3
17. A mineral that has a sheetlike structure
18. To remove activity
22. Symbol for prescription (pharm)
26. A metallic Group 15/VA element
28. The most abundant *metal*
30. Symbol for element with $3d^4 4s^2$ configuration
32. A Group 9 element that forms many inert d^6 coordination complexes
35. Obsolete name for Ce(IV)
36. RNH_2
37. Its mineral source is chromite
38. A semiprecious gem
39. Atomic number 3
40. A platinum metal
46. Forms photosensitive halides
47. Metallic element in limestone

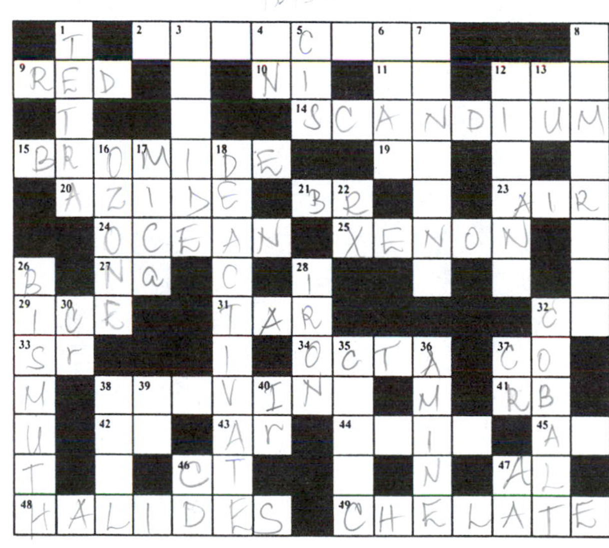

Figure 17.16
An inorganic chemistry crossword puzzle by Leonard F. Druding. [Reprinted with permission from Leonard F. Druding. *J. Chem. Educ.*, **50**, 773 (1973). Copyright © 1973, Division of Chemical Education, Inc.]

Notes

1. The historical origin of the letters to represent electronic states is based on characteristics of the spectra that result when electronic transitions occur involving these states: spectral bands involving *S* states are sharp, *P* states give the principal absorptions, *D* states give diffuse bands, and *F* states give fundamental absorptions. From these letters, the letters that represent the individual atomic orbitals later evolved.
2. These rules only work to select the ground state—they cannot be applied to rank the various excited states.
3. For methods of deriving these labels in homonuclear diatomic molecules, using methods similar to those given in Section 17.1 for atoms and ions, see M. L. Campbell, *J. Chem. Educ.*, **73**, 749 (1996).
4. W. C. Lineberger, P. G. Wenthold, W. T. Borden, and D. A. Hrovat, *Science*, **272**, 1456 (1996).
5. R. Hoffman, *Am. Sci.*, **83**, 309 (1995).
6. R. J. Van Zee, S. Li, and W. Weltner, Jr., *J. Chem. Phys.*, **100**, 4010 (1994).
7. The reason for this is that, if the two lines, which represent energy levels, were to approach on the way to crossing each other, the electronic states would undergo configuration interaction or "hybridization" in a manner analogous to *sp* hybridization (Section 10.2). This interaction results in the lower level being stabilized, hence driven back down in the diagram, while the upper level becomes even more unstable, going higher in the diagram; it also introduces quadratic terms in the energy level expression, producing curved rather than straight lines in energy level diagrams.
8. To locate the singlet states accurately with respect to the triplet states, it is necessary to have some information on the relative magnitudes of the *C* and *B* Racah parameters. Experimentally, the *C* parameter is approximately four times as large as the *B* parameter, and for a given ion a more accurate ratio can be obtained, and is used for subsequent diagrams.
9. If, as is often not the case, the highest energy ν_3 is not obscured by a much more intense charge-transfer absorption (Section 6.4). If ν_3 is obscured, then other methods, such as the graphical ones outlined below for the d^3 case, must be used.
10. C. K. Jørgensen, *Oxidation Numbers and Oxidation States*, Springer, New York, 1969, p. 106. This equation fails for very large values of both *h* and *k*.
11. "Unique Excitation, Emission Forms Basis of New Taggants," *Chem. Eng. News*, Jan. 27, 1997, p. 24.
12. Useful collections of articles on inorganic photochemistry and its applications to biochemistry and materials are found in two issues of *J. Chem. Educ.*, **60** (October 1983); **74** (June 1997).
13. L. A. Fowley, J. C. Lee, Jr., R. H. Crabtree, and P. E. M. Siegbahn, *J. Organometal. Chem.*, **504**, 57 (1995); L. A. Fowley, J. C. Lee, Jr., R. H. Crabtree, and P. E. M. Siegbahn, *Organometallics*, **15**, 1157 (1996).
14. M. B. Simpson, M. Poliakoff, J. J. Turner, W. B. Maier, III, and J. G. McLaughlin, *J. Chem. Soc. Chem. Commun.*, 1355 (1983)
15. X.-Z. Sun, M. W. George, S. G. Kazarian, S. M. Nikiforov, and M. Poliakoff, *J. Am. Chem. Soc.*, **118**, 10525 (1996); M. Jyo-o, H. Takeda, K. Omiya, Y. Ishikawa, and S. Arai, *Bull. Chem. Soc. Jpn.*, **66**, 3618 (1993).
16. R. B. Jordan, *Reaction Mechanisms of Inorganic and Organometallic Systems*, Oxford University Press, New York, 1991, Chapter 7.
17. C. Creutz and N. Sutin, *Proc. Natl. Acad. Sci. USA*, **72**, 2858 (1975).
18. M. Grätzel, Ed., *Energy Sources through Photochemistry and Catalysis*, Academic, New York, 1983.
19. More detailed texts in bioinorganic chemistry are available: S. J. Lippard and J. M. Berg, *Principles of Bioinorganic Chemistry*, University Science Books, Sausalito, CA, 1994; I. Bertini, H. B. Gray, S. J. Lippard, and J. S. Valentine, *Bioinorganic Chemistry*, University Science Books, Sausalito, CA, 1994.
20. V. K. Yachandra, K. Sauer, and M. P. Klein, *Chem. Rev.*, **96**, 2927 (1996).
21. R. W. Wagner and J. S. Lindsey, *J. Am. Chem. Soc.*, **116**, 9759 (1994).
22. An enzyme, superoxide dismutase, is employed to clear the body of the possibly hazardous O_2^- by causing it to disproportionate into O_2 and H_2O_2.

23. Since octahedrally coordinated metal ions have only three ligands on a face, they can utilize only three small bridging ligands, so that if O_2 were to be added as a bridging ligand, something else would have to cease bridging.

24. R. E. Steinkemp, L. C. Sieker, L. H. Jensen, J. D. McCallum, and J. Sanders-Loehr, *Proc. Natl. Acad. Sci. USA*, **82**, 713 (1985).
25. K. D. Karlin, S. Kaderli, and A. D. Zuberbühler, *Acc. Chem. Res.*, **30**, 139 (1997).
26. K. C. Magnus and H. Tong-That, *Inorg. Biochem.*, **27**, 20 (1992).
27. N. Kitajima, K. Fujisawa, C. Fujimoto, Y. Moro-oka, S. Hashimoto, T. Kitagawa, K. Toriumi, K. Tatsumi, and A. Nakamura, *J. Am. Chem. Soc.*, **114**, 1277 (1992).
28. J. A. Halfen, S. Mahapatra, E. C. Wilkinson, S. Kaderli, V. G. Young, Jr., L. Que, Jr., A. D. Zuberbühler, and W. B. Tolman, *Science*, **271**, 1397 (1996); *Chem. Eng. News*, March 11, 1996, p. 6.
29. L. Que, Jr., and Y. Dong, *Acc. Chem. Res.*, **29**, 190 (1996); L. Shu, J. C. Nesheim, K. Kauffmann, E. Münck, J. D. Lipscomb, and L. Que, Jr., *Science*, **275**, 515 (1997).

30. *Azotobacter vinelandii* and *A. chroococcum* can activate an alternate form of nitrogenase containing vanadium when they are starved for molybdenum.

31. A. D. Allen and C. V. Senoff, *J. Chem. Soc. Chem. Commun.*, 621 (1965).
32. J. Chatt, A. J. Pearman, and R. L. Richards, *Nature* (*London*), **253**, 39 (1975).
33. C. E. Laplaza and C. C. Cummins, *Science*, **268**, 861 (1995).
34. Q. Cui, D. G. Musaev, M. Svensson, S. Sieber, and K. Morokuma, *J. Am. Chem. Soc.*, **117**, 12366 (1995); K. M. Neyman, V. A. Nasluzov, J. Hahn, C. R. Landis, and N. Rösch, *Organometallics*, **16**, 995 (1997).
35. C. E. Laplaza, M. J. A. Johnson, J. C. Peters, A. L. Odom, E. Kim, C. C. Cummins, G. N. George, and I. J. Pickering. *J. Am. Chem. Soc.*, **118**, 8623 (1996).
36. C. E. Laplaza, A. R. Johnson, and C. C. Cummins, *J. Am. Chem. Soc.*, **118**, 709 (1996).
37. M. Hori and M. Mori, *J. Org. Chem.*, **60**, 1480 (1995).
38. J. Kim and D. C. Rees, *Science*, **257**, 653 (1992); J. Kim and D. C. Rees, *Science*, **257**, 1677 (1992); M. K. Chan, J. Kim and D. C. Rees, *Science*, **260**, 792 (1993); J. Kim and D. C. Rees, *Nature* (*London*), **360**, 553 (1992).

APPENDICES

APPENDIX

A

Discovery Laboratory Experiments for Part I

The following laboratory experiments are designed for use in Part I of the text to allow students to practice the scientific method while observing reactions of ions and simple molecules in the laboratory (or in a classroom demonstration–discussion). Opportunities are given for students to draw up their own hypotheses about inorganic reactivity, and attempt to verify or disprove them, before the textbook presents the relevant theory. It is suggested that each experiment be performed before the chapter and section indicated below.

Experiment No.	Experiment Name	Best Done before Section
1.	Some Reactions of Cations	2.1
2.	Nonaqueous Reactions of Metal Ions and Compounds	3.2
3.	Some Reactions of Oxo Anions	3.6
4.	Reaction of Anions with Cations	4.1
5.	Competitive Precipitation and Complexation Reactions	5.3
6.	Periodicity in the Activity (Electromotive) Series of Metals	6.5
7.	The Widely Varying Colors of *d*-Block Metal Complexes	8.5

These experiments are by no means intended to provide a complete laboratory experience. Introductory-level experiments dealing with inorganic materials (Part II) have been produced by the Division of Inorganic Chemistry (Solid-State Subdivision) of the American Chemical Society;[1] there is also a manual of advanced-level experiments appropriate to Chapter 8 and Part II.[2] We suggest using a few of these experiments to give the students experience in discovering scientific principles themselves and in designing their own experiments; once they have mastered these, we go on to other types of experiments dealing with syntheses and characterization.

For instructors not having a laboratory component to accompany Part I, we suggest doing the experiments as demonstration–discussions or as discovery videos[3] with discussion, in a game format we call "Publish or Perish"; see the Instructor's Manual (available to adopters of the text upon request to the Publisher) for details.

A.1 Experiment 1: Some Reactions of Cations

We will begin our study of the chemistry of the elements with a laboratory investigation (or classroom demonstration and discussion) of their chemistry under a very common set of circumstances: when the elements have positive oxidation numbers and are in aqueous solution. We will start with the presumption that an element in a positive oxidation state is present as a cation. Of course, we cannot obtain cations by themselves, so in this experiment we will use the chlorides of these elements—the chloride anion does not substantially affect the results of most of these experiments. We will observe a most elementary reaction of these chlorides—dissolving them in water—and will find that something may indeed happen chemically to the cations when this is done. Finally, we will try to find periodic trends in the degree to which this chemical reaction occurs, by relating the tendency to react to the basic periodic properties of atoms discussed in Chapter 1.

1.a. Take five test tubes; to each add 2–3 mL of distilled water. Add *nothing* to the first test tube. Measure the pH of the distilled water using first long-range and then short-range pH papers. Record the pH in the data table below. (Do not presume that the pH of the distilled water is 7.0!) Also, feel the test tube to note qualitatively the temperature of the distilled water.

1.b. Add LiCl (enough to cover the larger end of a standard nickel spatula) to the second test tube. Stir until it is dissolved; then measure and record its pH as before. Also note whether there is any detectable change in temperature.

1.c. To the third test tube, add a similar amount of dry $ZnCl_2$, stir to dissolve, measure its pH, and record any temperature change. To the fourth test tube, add a similar amount of *fresh, anhydrous* $AlCl_3$, stir to dissolve, measure the pH, and record any temperature change.

1.d. GO TO THE HOOD to obtain a micro test tube (closed with a tight stopper or a septum) containing 0.5 mL of $TiCl_4$. **Cautiously** open the test tube and pour the contents into the (larger) fifth test tube of distilled water. Stir. Measure the pH of the solution and (using moistened pH paper) of the gas being evolved from the test tube. Cautiously feel the bottom of the test tube to note any temperature change. Record your pH values and observations of temperature changes and any other visible or audible changes in the table below.

Solute	Cation Radius	χ_P	Charge	pH	Observations
Distilled water					
LiCl	90	0.98			
$ZnCl_2$	88	1.65			
$AlCl_3$	67	1.61			
$TiCl_4$	74	1.54			

2. What do you think happened in the test tubes in which reactions occurred? On a separate report sheet, write plausible chemical equations that would account for your observations.

3. Fill in the oxidation numbers (cation charges) of the (non-chlorine) elements in the above table, then look at the three periodic properties listed there (radius of the cation, Pauling electronegativity χ_P of the cation, and cation charge). Which one of these varies most significantly in this series of four compounds? Finally, decide how the tendency of a cation to undergo the reaction you described in Step 2 depends on this periodic property. Write your conclusion on the report sheet.

4. "Sky-writing" involves spraying $TiCl_4$ from an airplane into the air. Explain the chemistry of sky-writing. How would you have to handle compounds like $TiCl_4$ to prevent this reaction from happening?

5. In the hood, carry out the same sort of experiment, adding spatula tipfuls (or a few drops taken using a *dry* eyedropper) of the following compounds to 2–3 mL of distilled water. What is the significant periodic variable in this series of compounds? Can you write a conclusion relating the reaction tendency in this series of compounds to this variable?

Solute	Cation Radius	χ_P	Charge	pH	Observations
$BiCl_3$	117	2.02			
$SbCl_3$	90	2.05			
PCl_3	(~66)	2.19			

6. Design and carry out an experiment to determine whether the Pauling electronegativity of the cation has any effect on this reaction tendency. Use some, but not all, of the chlorides from the following list: $CaCl_2$; $SrCl_2$; $MnCl_2$; $FeCl_2$; $ZnCl_2$; and $SnCl_2$ (estimated cation radius of 126 pm); $Pb(NO_3)_2$ and $Hg(NO_3)_2$ (chlorides not suitable here); and $PrCl_3$ (or other *f*-block trichloride) and $BiCl_3$. *Hint*: Rather than starting by testing all of these chlorides, look at the periodic properties of the cations first, and pick out only the set or sets of compounds to test that will give you the comparison that you want.

7. Double-check your conclusion in Step 5 by checking the results with these two Group 14/(IV) chlorides: $SnCl_4$ and CCl_4. First, use your principles to predict what will happen, then do the test. Can you explain any discrepancy between theory and observation?

8. A far-too-frequent experience of people who have to make up solutions of metal salts such as $SnCl_2$, $Hg(NO_3)_2$, $BiCl_3$, and so on, and get a cloudy solution is to assume that their compound or water was contaminated; hence, they throw out the solution and try again, only to get the same result. Looking at the equations you wrote in Step 2, suggest what must be done to get clear solutions of these metal ions. Test your answer by trying it with one of the above three salts.

A.2 Experiment 2: Nonaqueous Reactions of Metal Ions and Compounds

In this experiment, we will see whether the reactivity of cations observed in Experiment 1 is still present in the absence of water, in nonaqueous solutions. We will also try to identify products other than hydroxides or oxides that may form when cations react.

1. In a test tube, take 1 mL of 0.5 *M* HCl. Using first long, then short-range pH papers, measure the pH of this solution. Then **IN THE HOOD** add an eyedropper-full of pyridine

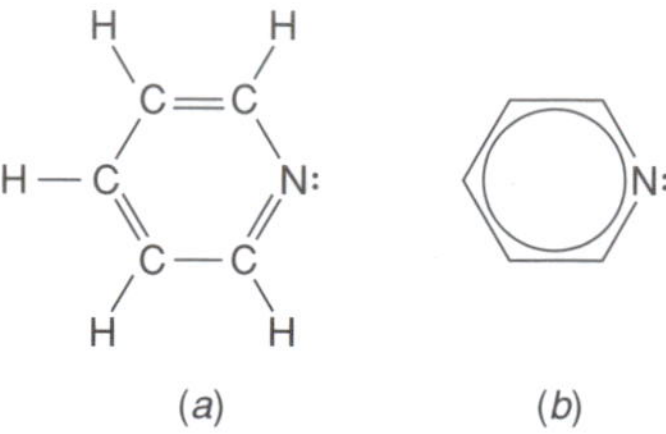

Figure A.1
The Lewis dot structure of pyridine, (*a*) completely drawn out, and (*b*) in the type of shorthand organic chemists use, implying the presence of carbon and hydrogen atoms at the corners of the hexagon.

(py), C_5H_5N (Fig. A.1), to the test tube. Stir with a glass rod, then measure the pH of the solution again. What term do we use to describe what the pyridine has done to the HCl solution? What class of compound reacts with acids such as HCl in a manner similar to pyridine? Does it seem to be necessary for that class of compound to contain hydroxide or oxide ions?

2. Try to write an equation for the above reaction, using full Lewis dot structures, to show how the pyridine reacts with the HCl to give the observed result. **IN THE HOOD**, try holding the stopper from a bottle of concentrated HCl over a watch glass in which you have put an eyedropper-full of pyridine. Does it appear that this reaction can proceed even when water is completely absent? (*Note*: The water in the air and in the concentrated HCl are not involved in this reaction.)

3. Looking back at your results in Experiment 1, select the chlorides of three of the more acidic cations from that experiment (*Suggestions*: $TiCl_4$, $BiCl_3$, $SnCl_4$, and $SbCl_3$). If you select a liquid chloride, dissolve one-half of an eyedropperful in 5 mL of dry toluene in a test tube. If you select a solid chloride, dissolve two large spatula-tipfuls of it in 2–3 mL of acetone. To each test tube of a chloride that you prepare in this manner, add one-half of an eyedropperful of pyridine **(IN THE HOOD).** Describe the results.

4. By analogy with the reaction in Step 1 of this experiment, what is probably happening in this reaction? Draw Lewis dot structures of the products you think might have been produced in this reaction. The function of the pyridine in this reaction is to act as what class of compound? The function of the metal cation (or metal chloride) is to act as what kind of compound?

5. In each of three test tubes, take 3 mL of a 1 *M* aqueous solution of cesium chloride, CsCl. To each add two good spatula-scoopfuls or a half-eyedropperful of one of your chlorides from the previous part of this experiment. Stir. Let sit awhile if no reaction is immediately apparent. Explain your results. Draw possible structures of the products. What is the function of the cesium chloride, and which ion (Cs^+ or Cl^-) is performing this function? (*Note*: Although water is present, it is not participating in this reaction either. If you are skeptical of that statement, as any good chemist ought to be, try the experiment with the three chlorides and 3 mL of water instead of the CsCl solution, and see if the results are quite the same.)

6. Add a half-eyedropperful of CCl_4 to 3 mL of the 1 *M* CsCl solution. Explain the results (or lack of results) with CCl_4.

7. Repeat Step 5 using 3 mL of a saturated aqueous solution of tetramethylammonium chloride, $(CH_3)_4N^+Cl^-$, in place of the CsCl in Step 5. Are your results similar? Write a Lewis dot structure of a plausible product for one of the reactions. Which ion (chloride or tetramethylammonium) is directly involved? Compare or contrast the behavior of the two nitrogen-containing species, pyridine and the tetramethylammonium ion, using their Lewis dot structures to explain their similarity or dissimilarity. In summary, what structural feature seems to be essential for a species to act as a base?

A.3 Experiment 3: Some Reactions of Oxo Anions

The principles we have developed to predict the extent of hydrolysis of cations are most useful with the metallic elements when they are in not too high an oxidation state. When we are dealing with nonmetals or metals in high oxidation states, the notion of a "cation" of the nonmetal or metal is too far divorced from reality to enable us to make reasonably accurate predictions—for one thing, only limited numbers of "cationic" radii for such species have been tabulated. Our best approach to systematizing the chemistry of nonmetals in aqueous solution with positive oxidation numbers is to start with them in their "final" form—as **oxo anions**. In this experiment, we will try to find simple but useful principles for predicting some of the properties of oxo anions.

1. Using long- and short-range pH test papers, check the pH of the distilled water. Then measure and compare the pH values of solutions of the following two salts of oxo anions: $NaClO$ and $NaClO_4$. Likewise measure and compare the pH values of solutions of Na_2SO_3 and Na_2SO_4. All of these salts contain oxo anions having general formulas $MO_x{}^{y-}$. In each pair of compounds, what component of the general formula are we varying? What effect does this variation have on the pH of the solutions? Predict which of each of the following pairs of salts will have the highest pH: $NaNO_2$ or $NaNO_3$; Na_3AsO_3, or Na_3AsO_4. Test your hypothesis by measuring and comparing the pH values within each pair of salts.

2. What type of reaction is occurring here to produce the pH values observed? Write an equation to illustrate this type of reaction, using the most reactive of the above oxo anions as an example.

3. Can you suggest any physical reason for the relationship of this reaction tendency to the structural variables you investigated in Step 1?

4. Design an experiment to determine the effect of the *charge* -*y* of an oxo anion on its basicity, and carry out the experiment. You may use the data from Step 1; in addition solutions of the following salts are available: Na_3PO_4; Na_4SiO_4; and $NaIO_3$. (You may decide that you have enough compounds to set up two experiments; if so, set up and do both of them.) What relationship of charge and basicity do you observe? How would you explain this relationship?

5. Predict the trend in pH values of the following series of solutions, two of which are too intensely colored to be tested by pH paper: K_3VO_4, K_2CrO_4, and $KMnO_4$. (If a pH meter is available, you may want to check the pH values of these solutions.)

6. Below are listed the oxo anions of the later *p*-block elements in their highest oxidation states. Note that the number of oxygen atoms (**oxo groups**) changes down a group. How would you explain the fact that the number does not remain constant?

$CO_3{}^{2-}$	$NO_3{}^{-}$			
$SiO_4{}^{4-}$	$PO_4{}^{3-}$	$SO_4{}^{2-}$	$ClO_4{}^{-}$	
$GeO_4{}^{4-}$	$AsO_4{}^{3-}$	$SeO_4{}^{2-}$	$BrO_4{}^{-}$	
$SnO_6{}^{8-}$	$SbO_6{}^{7-}$	$TeO_6{}^{6-}$	$IO_6{}^{5-}$	$XeO_6{}^{4-}$

7. Note that down a group, not only the number of oxo groups but also the charge on the oxo anion changes. Looking at your previous conclusions, would you expect these simultaneous changes in structure of oxo groups to have the same effects on basicity (i.e., to reinforce each other's effects)? If so, describe the basicity trend for oxo anions down a group of the periodic table. Or would you expect the two changes to have opposite effects on the basicity of the oxo anions? If the latter is the case, you need to determine which of the two effects is dominant. To do this, compile the pH values of as many of the oxo anions listed above as you have data for. If possible, test some additional solutions to add to your data (*Suggestion*: Na_2CO_3). What is the observed basicity trend for oxo anions down a group of the periodic table?

8. Putting together all of your trends, identify the oxo anion listed in Step 6 that would most strongly undergo the type of chemical reaction you wrote in Step 2.

A.4 Experiment 4: Reactions of Anions with Cations

In this experiment, we will investigate what happens when cations of varying acidity are combined with oxo anions of varying basicity.

1. We will study the reactions of the following eight cations: Cs^{+}, K^{+}, Ag^{+}, Mg^{2+}, Sr^{2+}, Hg^{2+}, Zn^{2+}, and Al^{3+}. Recalling the principles of Chapter 2, give the category of acidity of each cation. Then try to arrange the ions within each category in order of increasing acidity, so that, when you list all eight ions, they will all be in order of increasing acidity. Finally, check yourself by listing the pK_a values for the ions (Table 2.2). (You probably will not be precisely right, since the rules of thumb in Chapter 2 are not exact, but the ordering you predict should put the pK_a values reasonably close to the correct order.)

2. We will now study the reactions of the above eight cations with the following four oxo anions: $SiO_4{}^{4-}$, $SO_4{}^{2-}$, $PO_4{}^{3-}$, and $ClO_4{}^{-}$. List these four anions in order of increasing basicity, and give the category of basicity and the approximate pK_b for each.

3. Test the reactions of each of the eight cations in Step 1 with each of the four anions in Step 2. For each test, mix equal volumes (say, one eyedropperful) of the two solutions, mix well, and allow a minute (if necessary) for the reaction to occur. Note whether the test tubes get hot or cold. Describe the reactions and list your observations in tabular form, listing the eight cations down one side of the table in a logical order, and listing the four anions across the top of the table in a logical order.

4. What do we call the kind of reaction that occurs in some of the test tubes? Write the formulas of some of the products.

5. How does the tendency for this kind of reaction appear to relate to the acidity and basicity of the cations and anions involved?

6. Predict which categories of metal cations would give insoluble salts with each of the following anions: selenate, permanganate, chromate, carbonate, and nitrate. If a solution of one of these ions is available, test your predictions using the eight available cations. Check your answers for one of the anions that is not available by looking up the solubilities of the salts of that ion in the *Handbook of Chemistry and Physics.*

7. The perbromate ion was first synthesized in the 1960s in minute quantities by the decay of radioactive $^{*}SeO_4^{2-}$. How would you go about separating the BrO_4^- ion from the selenate ion?

8. In Section 2.3, we studied principles that could be used to predict the solubilities of metal hydroxides. Classify the basicity of the hydroxide ion. We could consider the hydroxide ion as a type of oxo anion. Are the principles in Section 2.3 consistent with the principles you derived in Step 5 of this experiment?

A.5 Experiment 5: Competitive Precipitation and Complexation Reactions

In nature (in our physical environment or in cells), we find mixtures of several metal ions with countless numbers of ligands. Although a given metal ion might tend to combine (give a complex or a precipitate) with a given ligand, this may not happen in nature because some other ligand may form a more stable complex or a more insoluble precipitate. In this experiment, we look for some generalizations to help us anticipate, in a complicated mixture, which metal ions will combine with which ligands.

l.a. Prepare a mixture of 10 mL of 2.0 *M* KI solution with 10 mL of 2.0 *M* KF solution (**AVOID CONTACT WITH ACID! HF IS A VERY DANGEROUS POISON!**). Stir well, then put 1 mL of this mixture in each of six test tubes. To each of the six test tubes, add 1 mL of a different one of the following six solutions: 1 *M* LiCl, 1 *M* $AgNO_3$, 0.5 *M* $SrCl_2$, 0.5 *M* $HgCl_2$, 0.5 *M* $MgCl_2$, 0.5 *M* $CuSO_4$. Record your results, noting the colors of the products.

1.b. Design and carry out some experiments that will enable you to determine what the precipitate is in each case. Identify each precipitate in Step 1.a.

1.c. Which metal ions have fluorides that are less soluble than their iodides? Which metals ions have iodides that are less soluble than their fluorides? Which of the fundamental atomic properties we have used as predictors (Pauling electronegativity, ionic charge, ionic radius, etc.) seems to be most useful to predict the relative solubility of iodides versus fluorides? Use it to predict the identities of four more insoluble fluorides; four more insoluble iodides. Verify your predictions with your instructor before proceeding.

2.a. Take 10 mL of an 0.5 *M* Na_2S solution. Test its pH. What other anions (besides S^{2-}) are present? Put 1 mL in each of eight test tubes. To each of the eight test tubes add 1 mL of a different one of these eight solutions: 1 *M* LiCl, 1 *M* $AgNO_3$, 0.5 *M* $MgCl_2$, 0.5 *M* $CuSO_4$, 0.5 *M* $SrCl_2$, 0.5 *M* $HgCl_2$, 0.33 *M* $PrCl_3$ (or another *f*-block +3 ion), and 0.33 *M* $BiCl_3$. Record the results.

2.b. Carry out experiments that will enable you to determine the identity of the precipitate in each case.

2.c. Which metals form sulfides that are less soluble than their hydroxides or oxides? Which metals prefer to remain as hydrated ions, or to form hydroxides or oxides, rather than to precipitate sulfides? Which fundamental atomic property seems to work best to predict the relative tendency to form sulfides versus hydroxides, oxides, or hydrated ions? Predict four more metals that will have sulfides more insoluble than hydroxides; four more that will have hydroxides more insoluble than sulfides. Verify with your instructor.

3.a. Mix 10 mL of a saturated urea $[(NH_2)_2C{=}O]$ solution with 10 mL of a saturated thiourea $[(NH_2)_2C{=}S]$ solution. Stir well, then put 2 mL in each of eight test tubes. To each of these test tubes add a different one of the same eight metal-ion solutions used in Step 2.a. Record your results.

3.b. Identify your products and write plausible formulas for them.

3.c. Which metals react with thiourea in preference to urea? Which prefer to react with urea or stay in hydrated form?

4. Do you see any relationship between the results of the experiment with the F^-/I^- mixture, that with the S^{2-}/OH^- mixture, and that with the urea/thiourea mixture? If so, give the relationship.

5.a. Prepare a mixture of 10 mL of 0.25 *M* Na_4SiO_4 and 5 mL of 0.5 *M* Na_2S. (The mixture is now 0.167 *M* in each.) Put 1.5 mL of this mixture in each of eight test tubes. To each of the eight test tubes add 0.5 mL of one of the eight metal-ion solutions you have been using. Record the results and identify the products.

5.b. The largest number of minerals are silicates or sulfides, which presumably arise as the result of competitive precipitation experiments. Draw a periodic table in which you indicate the metals you would expect to find in nature as silicates. (Geochemists refer to these as the *lithophile* elements.) Indicate the metals that you would expect to find in nature as sulfides. (Geochemists refer to these as the *chalcophile* elements.) Also indicate any area of the tables for which you are not yet able to make predictions.

Alternate Version of This Experiment

This version is for students who have done the qualitative analysis scheme (qual scheme) in general chemistry.

a. List all of the metals (in the general known sample) that precipitate as chlorides in group I (note that these would precipitate as sulfides in group II if you neglected to add the chloride ion). List all of the metals that precipitate as sulfides in group II *or as sulfides* in group III. List all of the metals that precipitate as *hydroxides or oxides* in group III. List all of the metals that precipitate as carbonates in group IV. List all of the metals that remain as hydrated ions in group V.

b. Identify the donor atom of the carbonate ion (group IV) and the water molecule (group V). Consolidate your list from the previous Step (a) into two larger lists: one of metal ions that prefer to precipitate with the sulfide ion (sulfur donor atom); the other of metal ions that prefer to coordinate with oxygen donor atoms.

c. Which of the fundamental atomic properties we have used as predictors (Pauling electronegativity, ionic charge, ionic radius, etc.) would be most useful to predict, which metal ions prefer to coordinate with sulfur donors, and which prefer to coordinate with oxygen donors?

d. Use your relationship to predict six more metal ions that should belong to each list. Verify these with your instructor. Can you also decide which of the six additional oxygen-preferring metal ions would fall in qual scheme group III, which in qual scheme group IV, and which in qual scheme group V? (This solution requires principles explained earlier in this text.)

e. Go back and answer the question in Step 5b, predicting the results of Step 5a.

A.6 Experiment 6: Periodicity in the Activity (Electromotive) Series of Metals

A. Introduction

In this experiment, we will study the stability of low oxidation states of the metallic elements (i.e., in the zero oxidation state, the metals themselves). We will examine the relative reactivity (*activity*) of different metals with the hydrogen ion and will list the different elements in order of decreasing reactivity with the hydrogen ion (such a list is called an activity or electromotive series of elements). But in contrast with the usual general chemistry experiment, we will also attempt to discover the periodicity in such an activity series so that we can also predict the activities of other metals not yet tested and also (in Chapters 7 and 12) gain some insight into the bonding that occurs in metals.

The reaction of some metals with standard 1 M hydrogen ion is dangerously exothermic, so we will begin our tests by studying the reactivity of metals with pure cold water, in which the concentration of the hydrogen ion is only 10^{-7} M. In terms of predominant species such a reaction can usually be summarized as

$$M(s) + n\ H_2O \rightarrow M(OH)_n(s \text{ or } aq) + n/2\ H_2(g) \tag{A.1}$$

Only very reactive metals undergo this reaction with cold water; their relative activity will be judged by the relative rate of evolution of H_2. If this reaction is not perceptible, we will try to speed the rate of the reaction by using hot water. Metals that show no activity to hot water will then be reacted with cold (roughly 1 M) hydrochloric acid:

$$M(s) + n\ H^+ \rightarrow M^{n+}(aq) + n/2\ H_2 \tag{A.2}$$

If cold hydrochloric acid produces no reaction, then hot hydrochloric acid will be tested. A certain number of metals fail to react with the hydrogen ion at all; these are said to be less active than hydrogen and are listed below it in an activity series. To rate their relative activity, stronger oxidizing agents than H^+ must be used; we will use oxidizing metal ions.

$$n\ M(s) + m\ N^{n+}(aq) \rightarrow m\ N(s) + n\ M^{m+} \tag{A.3}$$

B. Forming Hypotheses

There are more than 80 metals in the periodic table; testing all these would be very time consuming and very expensive too. If we can find some form of periodicity in the tendency of metals to react with oxidizing agents, we will not need to test all 80 metals or memorize the results. Let us first form hypotheses as to how a metal's activity might relate to each of these fundamental properties of a metal: its position in the periodic table, its ionization potential, and its electronegativity.

1. Recalling general chemistry, describe how metallic properties of an element (such as their activities) relate to their position in the periodic table. Can you use this principle to predict the order of decreasing activity of the elements Ag, Al, Ca, Cu, Hg, Mg, Mn, Na, Ni, Pb, and Zn? Why or why not?

2. The *first ionization energy* (or first ionization potential) is the energy required to remove one electron from a gaseous atom of an element to produce a +1 ion:

$$M(g) + \text{energy} \rightarrow M^+(g) + e^- \qquad (A.4)$$

How do you think the activity of an element ought to be related to its first ionization potential? Table 7.2 presents the first ionization energies of the elements. Predict a decreasing order of reactivity of the above elements based on their first ionization energies.

3. How do you think the Pauling electronegativity of an element and its activity might be related? Predict a decreasing order of reactivity of the above elements based on their Pauling electronegativities. [Lead goes to Pb^{2+} in this experiment, so the Pauling electronegativity for Pb(II) should be used to predict the activity of lead.]

4. The concept of electronegativity is complicated by the fact that there is more than one way of defining and measuring electronegativity. Allred and Rochow's scale of electronegativity is presented in Table 1.7. Predict a decreasing order of reactivity of the above elements based on their Allred–Rochow electronegativities.

C. Experimental

1. Heat a 400-mL beaker of water on a ringstand to near boiling. This will be used from time to time throughout the experiment.

2. The following metals are to be tested: Ag, Al, Ca, Cu, Hg, Mg, Mn, Na, Ni, Pb, and Zn. *Certain metals require special precautions:*

- *Sodium*—Use only a very tiny cube of metal. Do not point the test tube of water at anyone; do this reaction in the hood behind the glass shield.
- *Aluminum* is coated with thin films of tightly adhering oxide that must be cleaned off before its reactions can be observed. To do this, put a *few* granules of the metal in a large test tube with 2 mL H_2O and 2 mL 6 *M* HCl. Heat in the hot water bath until vigorous reaction just begins (a few minutes are required), then quickly remove the test tube, dilute the acid with cold distilled water, pour off the diluted acid, half-fill the test tube with distilled water, pour this off, and again half-fill the test tube and pour off nearly all of the water (leave enough to keep the metal out of contact with the air). Observe whether the metal is reacting with this cold water—if not, go immediately to Step 4.

3. Put a few granules of each of your metals in large test tubes that contain about 5 mL of distilled water. Observe whether bubbling occurs, and if so, which metal gives the most rapid reaction.

4. If you observe no reaction with cold water (or if you observe only a very faint reaction), put the test tube in your hot water bath (which should be *very hot* but *not boiling*). Observe whether bubbles of H_2 now form.

5. From any unreactive metals pour off all but about 2 mL of the water, then add 2 mL of 6 *M* HCl. Observe over a period of a few minutes; note the relative bubbling rates.

6. For any metals that are still unreactive, put the test tube in the beaker of almost-boiling water and heat for a few minutes. Note relative bubbling rates.

7. Arrange the above 11 metals, insofar as possible, in order of decreasing reactivity (an *activity series*).

8. In each of six test tubes, place 2 mL of 0.5 *M* $Pb(C_2H_3O_2)_2$ solution. To each test tube, add a few granules (or one drop) of the following six metals: Cu, Hg, Ag, Mg, Mn, and Zn. Observe whether reaction occurs (wait 10 mins before deciding "no reaction"). The Pb^{2+} ion [in $Pb(C_2H_3O_2)_2$] is capable of oxidizing which of these metals? What is the observed product? Write balanced equations for the reactions that occur.

9. As a generalization, how does the activity of metals which react with the Pb^{2+} ion compare with the activity of Pb itself?

10. Devise a series of experiments by which you can determine the relative positions of the metals Ag, Cu, and Hg in the activity series, using any of the following reactants: Ag(s), Cu(s), Hg(l), 1 *M* $AgNO_3$, 0.5 *M* $Hg(NO_3)_2$, 0.5 *M* $CuSO_4$. Carry out and describe the experiments and your results, and complete the activity series in Step 7.

D. Conclusions

Decide which periodic property of a metallic element (periodic table position, ionization potential, Pauling electronegativity, Allred–Rochow electronegativity) correlates most strongly with the activity of metals. Justify your choice.

E. Applications

1. Based on your conclusions, predict the products of the following reactions (if they go), and describe the vigor of the expected reaction: (a) $Cu + AuCl$; (b) $La + H_2O$; (c) $Au + HBr$; (d) $Ti + HCl$; (e) $Be + H_2O$; (f) $U + HCl$; and (g) $Pt + Hg(NO_3)_2$.

2. Many of the metals can be conveniently prepared by the *thermite* reaction:

$$2\ Al + (e.g.)M_2O_3 \rightarrow 2\ M + Al_2O_3 \qquad (A.5)$$

Which of the following metals could be produced in this way: (a) Sc; (b) La; (c) U; (d) Cr; (e) Fe; (f) Ca; and (g) Bi.

3. Many of the least-active metals have been prized by humans for ages for their durability (and scarcity); these are sometimes referred to as the noble metals. Which of the metals would most likely fail to react with oxygen (under neutral conditions) and hence might be called noble metals?

F. Alternate Version of the Experiment

This version is for students who carried out an activity series experiment in general chemistry. Go through this experiment omitting Part C, the experimental part; substitute your data sheet (or a textbook activity series) for this experiment. Draw up your hypotheses in Part B for the elements you tested or for the elements in your textbook activity series.

A.7 Experiment 7: The Widely Varying Colors of *d*-Block Metal Complexes

Not only are the *d*-block hydrated metal ions of widely different colors, but the colors of the metal ions can be changed by replacing the coordinated water with other ligands. As you do this experiment you will be looking for regularities in the color changes induced by the ligands. But also enjoy the beautiful colors produced: The georgeous colors are what persuaded the author of this textbook to become an inorganic chemist!

A. Preparation of Complexes and Their Ordering by Visual Inspection

1. Different students will be assigned different metal ions, and will prepare separate complexes of that ion with seven different ligands. Some students will be assigned the Ni^{2+} ion, some the Co^{2+} ion, and some will be assigned the Cu^{2+} ion.

2. Take a series of seven test tubes in a rack. To each test tube add 2 mL of an 0.1 *M* solution of your metal ion. In turn, add 2 mL of one of the following solutions to each of the test tubes: (1) H_2O (pure); (2) 1.0 *M* sodium glycinate, $Na^+NH_2CH_2CO_2^-$; (3) $K_2C_2O_4$ (saturated); (4) 1.0 *M* NH_3, buffered; (5) 1.0 *M* pyridine (Fig. A.1); (6) 1.0 M sodium ethylenediaminetetraacetate, $(Na^+)_4[(^-O_2CCH_2)_2NCH_2CH_2N(CH_2CO_2^-)_2]$ or just *EDTA* for short; and (7) 1.0 *M* buffered ethylenediamine, $H_2NCH_2CH_2NH_2$. Stopper each test tube and mix its contents. *Note*: The Co^{2+} complexes should be run as soon as possible after they are prepared, since many of them undergo air oxidation.

3. Does the color of a transition metal complex depend on the ligand present? Arrange the seven test tubes in "rainbow" order, such that the colors lead naturally from one test tube to the next. List the ligands in order, starting from the end that is closest to the color of the pure water test tube.

4. Is the ligand ordering for Ni^{2+} complexes reasonably consistent with the ordering of the ligands for complexes of the Co^{2+} ion, and of the Cu^{2+} ion? As a group, check the possibility that making slight rearrangements would make a more nearly metal-independent listing of ligands in order. Your final group list is a qualitative *spectrochemical series of ligands.*

5. Such a listing has uncertainties due to the limited ability of our eyes to discriminate similar colors. Hence, we will check our ordering by measuring the spectra of these complexes. The spectra should be measured in the visible and near-IR region on spectrophotometers available to you; the detailed instructions depend on the type of instrument, so will be provided by your instructor.

B. Analysis of the Spectra and Contents of the Lab Report

1. Attach all spectra, with arrows drawn to indicate any peaks. Tabulate the wavelengths of those peaks.

2. Convert your wavelength measurements to energy units of reciprocal centimeters using the equation:

$$\text{Wavenumber in cm}^{-1} = 10^7/\text{wavelength in nanometers} \qquad (A.6)$$

3. In the Cu^{2+} complexes, you will observe only one (possibly asymmetrical) *d*-to-*d* transition. The energy of this transition in reciprocal centimeters is the octahedral ligand field splitting, Δ_o. In the other octahedral complexes you will observe more than one *d*-to-*d* transition; the longest wavelength (lowest energy) absorbance[4] in these complexes equals Δ_o. Record Δ_o for each of your complexes.

4. The energy of Δ_o can be approximately factored into two components, an *f* factor due to the ligand and a *g* factor due to the metal ion:

$$\Delta_o = f_{\text{ligand}} \times g_{\text{ion}} \tag{A.7}$$

By convention we assign an *f* factor of 1.000 to the water molecule as a ligand. From the Δ_o value for your hydrated ion, compute the *g* factor for your metal ion, and list it.

5. Your experimental *g* factor can then be used to calculate the *f* factors for the other ligands in your complexes. To obtain the *f* factor, we note that

$$\Delta_{\text{hydrated ion}} = f_{\text{water}} \times g_{\text{ion}} = 1.00 \times g_{\text{ion}} \tag{A.8}$$

Dividing Eq. (A.7) by Eq. (A.8) and canceling out the g_{ion} factor, we obtain

$$\Delta_{\text{complex ion}}/\Delta_{\text{hydrated ion}} = f_{\text{ligand}} \tag{A.9}$$

Calculate *f* factors for each of your ligands.

6. The *f* factors finally allow us to make the spectrochemical series semiquantitative. List your ligands in increasing order of their *f* values, indicating small and large differences between corresponding ligands with “$<$” and “$\ll$”, respectively. Compare this spectrochemical series with the qualitative spectrochemical series you obtained by eyeball in Part A.

7. Analyze the periodicity of your spectrochemical series: can you note any grouping of nitrogen- and oxygen-donor ligands in the spectrochemical series?

Notes

1. A. B. Ellis, M. J. Geselbracht, B. J. Johnson, G. C. Lisensky, and W. R. Robinson, *Teaching General Chemistry: A Materials Science Companion*, American Chemical Society, Washington DC, 1993.
2. G. S. Girolami, T. B. Rauchfuss, and R. J. Angelici, *Synthesis and Technique in Inorganic Chemistry, A Laboratory Manual*, University Science Books, Sausalito, CA, 1999.
3. Lyubov V. Hoffman, “Design and Evoluation of Effectiveness of Audio-Visual Discovery Lab Experiments in Teaching Descriptive Inorganic Chemistry,” M. S. Thesis, Middle Tennessee State University, 1996. Faculty who adopt this text may obtain copies of the video forms of Experiments A.1 and A.6 from the author by writing the author on letterhead stationery (Box 405, Chemistry Department, Middle Tennessee State University, Murfreesboro, TN 37132) and sending him two blank videocassette tapes.
4. The other one or two low-intensity (*d*-to-*d*) absorbances in the complexes of Ni^{2+} and Co^{2+} arise from the different ways in which *d* electrons in t_{2g} and e_g orbitals can repel each other; this will be discussed much later (in Chapter 17) and the energies of the other absorbances will be analyzed using the ligand field theory to give a nephelauxetic series of ligands in order of covalent π-bonding ability. In addition, at very low wavelengths (high energies) there may be some absorbances that are offscale in intensity. These very intense absorbances are charge-transfer absorbances and are not due to *d*-to-*d* transitions, so they need not be noted.

APPENDIX

B

Character Tables

A. The Nonaxial Groups

C_1	E
A	1

C_s	E	σ_h			
A'	1	1	x, y, R_z	x^2, y^2, z^2, xy	$xz^2, yz^2, x(x^2-3y^2), y(3x^2-y^2)$
A''	1	−1	z, R_x, R_y	yz, xz	$z^3, xyz, z(x^2-y^2)$

C_i	E	i			
A_g	1	1	R_x, R_y, R_z	$x^2, y^2, z^2, xy, xz, yz$	
A_u	1	−1	x, y, z		All cubic functions

B. The C_n Groups

C_2	E	C_2			
A	1	1	z, R_z	x^2, y^2, z^2, xy	$z^3, xyz, z(x^2-y^2)$
B	1	-1	x, y, R_x, R_y	yz, xz	$xz^2, yz^2, x(x^2-3y^2),\ y(3x^2-y^2)$

C_3	E	C_3	C_3^2	$\varepsilon=\exp(2\pi i/3)$		
A	1	1	1	z, R_z	x^2+y^2, z^2	$z^3, x(x^2-3y^2),\ y(3x^2-y^2)$
E	1	ε	ε^*	$(x, y), (R_x, R_y)$	$(x^2-y^2, xy), (yz, xz)$	$(xz^2, yz^2), [xyz, z(x^2-y^2)]$
	1	ε^*	ε			

C_4	E	C_4	C_2	C_4^3			
A	1	1	1	1	z, R_z	x^2+y^2, z^2	z^3
B	1	-1	1	-1		x^2-y^2, xy	$xyz, z(x^2-y^2)$
E	1	i	-1	$-i$	$(x, y), (R_x, R_y)$	(xz, yz)	$(xz^2, yz^2), [x(x^2-3y^2), y(3x^2-y^2)]$
	1	$-i$	-1	i			

C_5	E	C_5	C_5^2	C_5^3	C_5^4	$\varepsilon=\exp(2\pi i/5)$		
A	1	1	1	1	1	z, R_z	x^2+y^2, z^2	z^3
E_1	1	ε	ε^2	ε^{2*}	ε^*	$(x, y), (R_x, R_y)$	(yz, xz)	(xz^2, yz^2)
	1	ε^*	ε^{2*}	ε^2	ε			
E_2	1	ε^2	ε^*	ε	ε^{2*}		(x^2-y^2, xy)	$[xyz, z(x^2-y^2)], [x(x^2-3y^2), y(3x^2-y^2)]$
	1	ε^{2*}	ε	ε^*	ε^2			

C_6	E	C_6	C_3	C_2	C_3^2	C_6^5	$C_6 = C_3 \times C_2$	$\varepsilon=\exp(2\pi i/6)$	
A	1	1	1	1	1	1	z, R_z	x^2+y^2, z^2	z^3
B	1	-1	1	-1	1	-1			$x(x^2-3y^2),\ y(3x^2-y^2)$
E_1	1	ε	$-\varepsilon^*$	-1	$-\varepsilon$	ε^*	(x, y), (R_x, R_y)	(xz, yz)	(xz^2, yz^2)
	1	ε^*	$-\varepsilon$	-1	$-\varepsilon^*$	ε			
E_2	1	$-\varepsilon^*$	$-\varepsilon$	1	$-\varepsilon^*$	$-\varepsilon$		(x^2-y^2, xy)	$[xyz, z(x^2-y^2)]$
	1	$-\varepsilon$	$-\varepsilon^*$	1	$-\varepsilon$	$-\varepsilon^*$			

C_7	E	C_7	C_7^2	C_7^3	C_7^4	C_7^5	C_7^6	$\varepsilon=\exp(2\pi i/7)$		
A	1	1	1	1	1	1	1	z, R_z	x^2+y^2, z^2	z^3
E_1	1	ε	ε^2	ε^3	ε^{3*}	ε^{2*}	ε^*	(x, y), (R_x, R_y)	(xz, yz)	(xz^2, yz^2)
	1	ε^*	ε^{2*}	ε^{3*}	ε^3	ε^2	ε			
E_2	1	ε^2	ε^{3*}	ε^*	ε	ε^3	ε^{2*}		(x^2-y^2, xy)	$[xyz, z(x^2-y^2)]$
	1	ε^{2*}	ε^3	ε	ε^*	ε^{3*}	ε^2			
E_3	1	ε^3	ε^*	ε^2	ε^{2*}	ε	ε^{3*}			$[x(x^2-3y^2), y(3x^2-y^2)]$
	1	ε^{3*}	ε	ε^{2*}	ε^2	ε^*	ε^3			

C_8	E	C_8	C_4	C_2	C_4^3	C_8^3	C_8^5	C_8^7	$C_8 = C_4 \times C_2$	$\varepsilon=\exp(2\pi i/8)$	
A	1	1	1	1	1	1	1	1	z, R_z	x^2+y^2, z^2	z^3
B	1	-1	1	1	1	-1	-1	-1			
E_1	1	ε	i	-1	$-i$	$-\varepsilon^*$	$-\varepsilon$	ε^*	(x, y), (R_x, R_y)	(xz, yz)	(xz^2, yz^2)
	1	ε^*	$-i$	-1	i	$-\varepsilon$	$-\varepsilon^*$	ε			
E_2	1	i	-1	1	-1	$-i$	i	$-i$		(x^2-y^2, xy)	$[xyz, z(x^2-y^2)]$
	1	$-i$	-1	1	-1	i	$-i$	i			
E_3	1	$-\varepsilon$	i	-1	$-i$	ε^*	ε	$-\varepsilon^*$			$[x(x^2-3y^2), y(3x^2-y^2)]$
	1	$-\varepsilon^*$	$-i$	-1	i	ε	ε^*	$-\varepsilon$			

C. The C_{nv} Groups

C_{2v}	E	C_2	$\sigma_v(xz)$	$\sigma_v(yz)$			
A_1	1	1	1	1	z	x^2, y^2, z^2	$z^3, z(x^2-y^2)$
A_2	1	1	−1	−1	R_z	xy	xyz
B_1	1	−1	1	−1	x, R_y	xz	$xz^2, x(x^2-3y^2)$
B_2	1	−1	−1	1	y, R_x	yz	$yz^2, y(3x^2-y^2)$

C_{3v}	E	$2C_3$	$3\sigma_v$			
A_1	1	1	1	z	x^2+y^2, z^2	$z^3, x(x^2-3y^2)$
A_2	1	1	−1	R_z		$y(3x^2-y^2)$
E	2	−1	0	$(x, y), (R_x, R_y)$	$(x^2-y^2, xy), (xz, yz)$	$(xz^2, yz^2), [xyz, z(x^2-y^2)]$

C_{4v}	E	$2C_4$	C_2	$2\sigma_v$	$2\sigma_d$			
A_1	1	1	1	1	1	z	x^2+y^2, z^2	z^3
A_2	1	1	1	−1	−1	R_z		
B_1	1	−1	1	1	−1		x^2-y^2	$z(x^2-y^2)$
B_2	1	−1	1	−1	1		xy	xyz
E	2	0	−2	0	0	$(x, y), (R_x, R_y)$	(xz, yz)	$(xz^2, yz^2), [x(x^2-3y^2), y(3x^2-y^2)]$

C_{5v}	E	$2C_5$	$2C_5^2$	$5\sigma_v$			
A_1	1	1	1	1	z	x^2+y^2, z^2	z^3
A_2	1	1	1	−1	R_z		
E_1	2	2 cos 72°	2 cos 144°	0	$(x, y), (R_x, R_y)$	(xz, yz)	(xz^2, yz^2)
E_2	2	2 cos 144°	2 cos 72°	0		(x^2-y^2, xy)	$[xyz, z(x^2-y^2)], [x(x^2-3y^2), y(3x^2-y^2)]$

C_{6v}	E	$2C_6$	$2C_3$	C_2	$3\sigma_v$	$3\sigma_d$			
A_1	1	1	1	1	1	1	z	x^2+y^2, z^2	z^3
A_2	1	1	1	1	−1	−1	R_z		
B_1	1	−1	1	−1	1	−1			$x(x^2-3y^2)$
B_2	1	−1	1	−1	−1	1			$y(3x^2-y^2)$
E_1	2	1	−1	−2	0	0	$(x, y), (R_x, R_y)$	(xz, yz)	(xz^2, yz^2)
E_2	2	−1	−1	2	0	0		(x^2-y^2, xy)	$[xyz, z(x^2-y^2)]$

D. The C_{nh} Groups

C_{2h}	E	C_2	i	σ_h			
A_g	1	1	1	1	R_z	x^2, y^2, z^2, xy	
B_g	1	−1	1	−1	R_x, R_y	xz, yz	
A_u	1	1	−1	−1	z		$z^3, xyz, z(x^2 - y^2)$
B_u	1	−1	−1	1	x, y		$xz^2, yz^2, x(x^2 - 3y^2), y(3x^2 - y^2)$

C_{3h}	E	C_3	C_3^2	σ_h	S_3	S_3^5	$\varepsilon = \exp(2\pi i/3)$		
A'	1	1	1	1	1	1	R_z	$x^2 + y^2, z^2$	$x(x^2 - 3y^2), y(3x^2 - y^2)$
E'	1	ε	ε^*	1	ε	ε^*	(x, y)	$(x^2 - y^2, xy)$	(xz^2, yz^2)
	1	ε^*	ε	1	ε^*	ε			
A''	1	1	1	−1	−1	−1	z		z^3
E''	1	ε	ε^*	−1	$-\varepsilon$	$-\varepsilon^*$	(R_x, R_y)	(xz, yz)	$[xyz, z(x^2 - y^2)]$
	1	ε^*	ε	−1	$-\varepsilon^*$	$-\varepsilon$			

C_{4h}	E	C_4	C_2	C_4^3	i	S_4^3	σ_h	S_4			
A_g	1	1	1	1	1	1	1	1	R_z	$x^2 + y^2, z^2$	
B_g	1	−1	1	−1	1	−1	1	−1		$x^2 - y^2, xy$	
E_g	1	i	−1	$-i$	1	i	−1	$-i$	(R_x, R_y)	(xz, yz)	
	1	$-i$	−1	i	1	$-i$	−1	i			
A_u	1	1	1	1	−1	−1	−1	−1	z		z^3
B_u	1	−1	1	−1	−1	1	−1	1			$xyz, z(x^2 - y^2)$
E_u	1	i	−1	$-i$	−1	$-i$	1	i	(x, y)		$(xz^2, yz^2), [x(x^2 - 3y^2), y(3x^2 - y^2)]$
	1	$-i$	−1	i	−1	i	1	$-i$			

C_{5h}	E	C_5	C_5^2	C_5^3	C_5^4	σ_h	S_5	S_5^7	S_5^3	S_5^9	$\varepsilon = \exp(2\pi i/5)$		
A'	1	1	1	1	1	1	1	1	1	1	R_z	$x^2 + y^2, z^2$	
E_1'	1	ε	ε^2	ε^{2*}	ε^*	1	ε	ε^2	ε^{2*}	ε^*	(x, y)		(xz^2, yz^2)
	1	ε^*	ε^{2*}	ε^2	ε	1	ε^*	ε^{2*}	ε^2	ε			
E_2'	1	ε^2	ε^*	ε	ε^{2*}	1	ε^2	ε^*	ε	ε^{2*}		$(x^2 - y^2, xy)$	$[x(x^2 - 3y^2), y(3x^2 - y^2)]$
	1	ε^{2*}	ε	ε^*	ε^2	1	ε^{2*}	ε	ε^*	ε^2			
A''	1	1	1	1	1	−1	−1	−1	−1	−1	z		z^3
E_1''	1	ε	ε^2	ε^{2*}	ε^*	−1	$-\varepsilon$	$-\varepsilon^2$	$-\varepsilon^{2*}$	$-\varepsilon^*$	(R_x, R_y)	(xz, yz)	
	1	ε^*	ε^{2*}	ε^2	ε	−1	$-\varepsilon^*$	$-\varepsilon^{2*}$	$-\varepsilon^2$	$-\varepsilon$			
E_2''	1	ε^2	ε^*	ε	ε^{2*}	−1	$-\varepsilon^2$	$-\varepsilon^*$	$-\varepsilon$	$-\varepsilon^{2*}$			$[xyz, z(x^2 - y^2)]$
	1	ε^{2*}	ε	ε^*	ε^2	−1	$-\varepsilon^{2*}$	$-\varepsilon$	$-\varepsilon^*$	$-\varepsilon^2$			

C_{6h}	E	C_6	C_3	C_2	C_3^2	C_6^5	i	S_3^5	S_6^5	σ_h	S_6	S_3	$\varepsilon = \exp(2\pi i/6)$		
A_g	1	1	1	1	1	1	1	1	1	1	1	1	R_z	$x^2 + y^2, z^2$	
B_g	1	−1	1	−1	1	−1	1	−1	1	−1	1	−1			
E_{1g}	1	ε	$-\varepsilon^*$	−1	$-\varepsilon$	ε^*	1	ε	$-\varepsilon^*$	−1	$-\varepsilon$	ε^*	(R_x, R_y)	(xz, yz)	
	1	ε^*	$-\varepsilon$	−1	$-\varepsilon^*$	ε	1	ε^*	$-\varepsilon$	−1	$-\varepsilon^*$	ε			
E_{2g}	1	$-\varepsilon^*$	$-\varepsilon$	1	$-\varepsilon^*$	$-\varepsilon$	1	$-\varepsilon^*$	$-\varepsilon$	1	$-\varepsilon^*$	$-\varepsilon$		$(x^2 - y^2, xy)$	
	1	$-\varepsilon$	$-\varepsilon^*$	1	$-\varepsilon$	$-\varepsilon^*$	1	$-\varepsilon$	$-\varepsilon^*$	1	$-\varepsilon$	$-\varepsilon^*$			
A_u	1	1	1	1	1	1	−1	−1	−1	−1	−1	−1	z		z^3
B_u	1	−1	1	−1	1	−1	−1	1	−1	1	−1	1			$x(x^2 - 3y^2), y(3x^2 - y^2)$
E_{1u}	1	ε	$-\varepsilon^*$	−1	$-\varepsilon$	ε^*	−1	$-\varepsilon$	ε^*	1	ε	$-\varepsilon^*$	(x, y)		(xz^2, yz^2)
	1	ε^*	$-\varepsilon$	−1	$-\varepsilon^*$	ε	−1	$-\varepsilon^*$	ε	1	ε^*	$-\varepsilon$			
E_{2u}	1	$-\varepsilon^*$	$-\varepsilon$	1	$-\varepsilon^*$	$-\varepsilon$	−1	ε^*	ε	−1	ε^*	ε			$(xyz, z(x^2 - y^2))$
	1	$-\varepsilon$	$-\varepsilon^*$	1	$-\varepsilon$	$-\varepsilon^*$	−1	ε	ε^*	−1	ε	ε^*			

E. The D_n Groups

D_2	E	$C_2(z)$	$C_2(y)$	$C_2(x)$			
A	1	1	1	1		x^2, y^2, z^2	xyz
B_1	1	1	−1	−1	z, R_z	xy	$z^3, z(x^2 - y^2)$
B_2	1	−1	1	−1	y, R_y	xz	$yz^2, y(3x^2 - y^2)$
B_3	1	−1	−1	1	x, R_x	yz	$xz^2, x(x^2 - 3y^2)$

D_3	E	$2C_3$	$3C_2$	(x axis coincident with C_2)		
A_1	1	1	1		$x^2 + y^2, z^2$	$x(x^2 - 3y^2)$
A_2	1	1	−1	z, R_z		$z^3, y(3x^2 - y^2)$
E	2	−1	0	$(x, y), (R_x, R_y)$	$(x^2 - y^2, xy), (xz, yz)$	$(xz^2, yz^2), [xyz, z(x^2 - y^2)]$

D_4	E	$2C_4$	$C_2(= C_4^2)$	$2C_2'$	$2C_2''$	(x axis coincident with C_2')		
A_1	1	1	1	1	1		$x^2 + y^2, z^2$	
A_2	1	1	1	−1	−1	z, R_z		z^3
B_1	1	−1	1	1	−1		$x^2 - y^2$	xyz
B_2	1	−1	1	−1	1		xy	$z(x^2 - y^2)$
E	2	0	−2	0	0	$(x, y), (R_x, R_y)$	(xz, yz)	$(xz^2, yz^2), [x(x^2 - 3y^2), y(3x^2 - y^2)]$

D_5	E	$2C_5$	$2C_5^2$	$5C_2$	(x axis coincident with C_2)		
A_1	1	1	1	1		$x^2 + y^2, z^2$	
A_2	1	1	1	−1	z, R_z		z^3
E_1	2	2 cos 72°	2 cos 144°	0	$(x, y), (R_x, R_y)$	(xz, yz)	(xz^2, yz^2)
E_2	2	2 cos 144°	2 cos 72°	0		$(x^2 - y^2, xy)$	$[xyz, z(x^2 - y^2)], [x(x^2 - 3y^2), y(3x^2 - y^2)]$

D_6	E	$2C_6$	$2C_3$	C_2	$3C_2'$	$3C_2''$	(x axis coincident with C_2')		
A_1	1	1	1	1	1	1		$x^2 + y^2, z^2$	
A_2	1	1	1	1	−1	−1	z, R_z		z^3
B_1	1	−1	1	−1	1	−1			$x(x^2 - 3y^2)$
B_2	1	−1	1	−1	−1	1			$y(3x^2 - y^2)$
E_1	2	1	−1	−2	0	0	$(x, y), (R_x, R_y)$	(xz, yz)	(xz^2, yz^2)
E_2	2	−1	−1	2	0	0		$(x^2 - y^2, xy)$	$[xyz, z(x^2 - y^2)]$

F. The D_{nd} Groups

D_{2d}	E	$2S_4$	C_2	$2C_2'$	$2\sigma_d$	(x axis coincident with C_2')		
A_1	1	1	1	1	1		x^2+y^2, z^2	xyz
A_2	1	1	1	−1	−1	R_z		$z(x^2-y^2)$
B_1	1	−1	1	1	−1		x^2-y^2	
B_2	1	−1	1	−1	1	z	xy	z^3
E	2	0	−2	0	0	$(x, y), (R_x, R_y)$	(xz, yz)	$(xz^2, yz^2), [x(x^2-3y^2), y(3x^2-y^2)]$

D_{3d}	E	$2C_3$	$3C_2$	i	$2S_6$	$3\sigma_d$	(x axis coincident with C_2)		
A_{1g}	1	1	1	1	1	1		x^2+y^2, z^2	
A_{2g}	1	1	−1	1	1	−1	R_z		
E_g	2	−1	0	2	−1	0	(R_x, R_y)	$(x^2-y^2, xy); (xz, yz)$	
A_{1u}	1	1	1	−1	−1	−1			$x(x^2-3y^2)$
A_{2u}	1	1	−1	−1	−1	1	z		$y(3x^2-y^2), z^3$
E_u	2	−1	0	−2	1	0	(x, y)		$(xz^2, yz^2), [xyz, z(x^2-y^2)]$

D_{4d}	E	$2S_8$	$2C_4$	$2S_8^3$	C_2	$4C_2'$	$4\sigma_d$	(x axis coincident with C_2')		
A_1	1	1	1	1	1	1	1		x^2+y^2, z^2	
A_2	1	1	1	1	1	−1	−1	R_z		
B_1	1	−1	1	−1	1	1	−1			
B_2	1	−1	1	−1	1	−1	1	z		z^3
E_1	2	$\sqrt{2}$	0	$-\sqrt{2}$	−2	0	0	(x, y)		(xz^2, yz^2)
E_2	2	0	−2	0	2	0	0		(x^2-y^2, xy)	$[xyz, z(x^2-y^2)]$
E_3	2	$-\sqrt{2}$	0	$\sqrt{2}$	−2	0	0	(R_x, R_y)	(xz, yz)	$[x(x^2-3y^2), y(3x^2-y^2)]$

D_{5d}	E	$2C_5$	$2C_5^2$	$5C_2$	i	$2S_{10}^3$	$2S_{10}$	$5\sigma_d$	(x axis coincident with C_2)		
A_{1g}	1	1	1	1	1	1	1	1		x^2+y^2, z^2	
A_{2g}	1	1	1	−1	1	1	1	−1	R_z		
E_{1g}	2	2 cos 72°	2 cos 144°	0	2	2 cos 72°	2 cos 144°	0	(R_x, R_y)	(xz, yz)	
E_{2g}	2	2 cos 144°	2 cos 72°	0	2	2 cos 144°	2 cos 72°	0		(x^2-y^2, xy)	
A_{1u}	1	1	1	1	−1	−1	−1	−1			
A_{2u}	1	1	1	−1	−1	−1	−1	1	z		z^3
E_{1u}	2	2 cos 72°	2 cos 144°	0	−2	−2 cos 72°	−2 cos 144°	0	(x, y)		(xz^2, yz^2)
E_{2u}	2	2 cos 144°	2 cos 72°	0	−2	−2 cos 144°	−2 cos 72°	0			$[xyz, z(x^2-y^2)]$
											$[x(x^2-3y^2), y(3x^2-y^2)]$

D_{6d}	E	$2S_{12}$	$2C_6$	$2S_4$	$2C_3$	$2S_{12}^5$	C_2	$6C_2'$	$6\sigma_d$	(x axis coincident with C_2')		
A_1	1	1	1	1	1	1	1	1	1		x^2+y^2, z^2	
A_2	1	1	1	1	1	1	1	−1	−1	R_z		
B_1	1	−1	1	−1	1	−1	1	1	−1			
B_2	1	−1	1	−1	1	−1	1	−1	1	z		z^3
E_1	2	$\sqrt{3}$	1	0	−1	$-\sqrt{3}$	−2	0	0	(x, y)		(xz^2, yz^2)
E_2	2	1	−1	−2	−1	1	2	0	0		(x^2-y^2, xy)	
E_3	2	0	−2	0	2	0	−2	0	0			$[x(x^2-3y^2), y(3x^2-y^2)]$
E_4	2	−1	−1	2	−1	−1	2	0	0			$[xyz, z(x^2-y^2)]$
E_5	2	$-\sqrt{3}$	1	0	−1	$\sqrt{3}$	−2	0	0	(R_x, R_y)	(xz, yz)	

G. The D_{nh} Groups

D_{2h}	E	$C_2(z)$	$C_2(y)$	$C_2(x)$	i	$\sigma(xy)$	$\sigma(xz)$	$\sigma(yz)$			
A_g	1	1	1	1	1	1	1	1		x^2, y^2, z^2	
B_{1g}	1	1	−1	−1	1	1	−1	−1	R_z	xy	
B_{2g}	1	−1	1	−1	1	−1	1	−1	R_y	xz	
B_{3g}	1	−1	−1	1	1	−1	−1	1	R_x	yz	
A_u	1	1	1	1	−1	−1	−1	−1			xyz
B_{1u}	1	1	−1	−1	−1	−1	1	1	z		$z^3, z(x^2-y^2)$
B_{2u}	1	−1	1	−1	−1	1	−1	1	y		$yz^2, y(3x^2-y^2)$
B_{3u}	1	−1	−1	1	−1	1	1	−1	x		$xz^2, x(x^2-3y^2)$

D_{3h}	E	$2C_3$	$3C_2$	σ_h	$2S_3$	$3\sigma_v$	(x axis coincident with C_2)		
A_1'	1	1	1	1	1	1		x^2+y^2, z^2	$x(x^2-3y^2)$
A_2'	1	1	−1	1	1	−1	R_z		$y(3x^2-y^2)$
E'	2	−1	0	2	−1	0	(x, y)	(x^2-y^2, xy)	(xz^2, yz^2)
A_1''	1	1	1	−1	−1	−1			
A_2''	1	1	−1	−1	−1	1	z		z^3
E''	2	−1	0	−2	1	0	(R_x, R_y)	(xz, yz)	$[xyz, z(x^2-y^2)]$

D_{4h}	E	$2C_4$	C_2	$2C_2'$	$2C_2''$	i	$2S_4$	σ_h	$2\sigma_v$	$2\sigma_d$	(x axis coincident with C_2')		
A_{1g}	1	1	1	1	1	1	1	1	1	1		x^2+y^2, z^2	
A_{2g}	1	1	1	−1	−1	1	1	1	−1	−1	R_z		
B_{1g}	1	−1	1	1	−1	1	−1	1	1	−1		x^2-y^2	
B_{2g}	1	−1	1	−1	1	1	−1	1	−1	1		xy	
E_g	2	0	−2	0	0	2	0	−2	0	0	(R_x, R_y)	(xz, yz)	
A_{1u}	1	1	1	1	1	−1	−1	−1	−1	−1			
A_{2u}	1	1	1	−1	−1	−1	−1	−1	1	1	z		z^3
B_{1u}	1	−1	1	1	−1	−1	1	−1	−1	1			xyz
B_{2u}	1	−1	1	−1	1	−1	1	−1	1	−1			$z(x^2-y^2)$
E_u	2	0	−2	0	0	−2	0	2	0	0	(x, y)		$(xz^2, yz^2), [x(x^2-3y^2), y(3x^2-y^2)]$

D_{5h}	E	$2C_5$	$2C_5^2$	$5C_2$	σ_h	$2S_5$	$2S_5^3$	$5\sigma_v$	(x axis coincident with C_2)		
A_1'	1	1	1	1	1	1	1	1		x^2+y^2, z^2	
A_2'	1	1	1	−1	1	1	1	−1	R_z		
E_1'	2	2 cos 72°	2 cos 144°	0	2	2 cos 72°	2 cos 144°	0	(x, y)		(xz^2, yz^2)
E_2'	2	2 cos 144°	2 cos 72°	0	2	2 cos 144°	2 cos 72°	0		(x^2-y^2, xy)	$[x(x^2-3y^2), y(3x^2-y^2)]$
A_1''	1	1	1	1	−1	−1	−1	−1			
A_2''	1	1	1	−1	−1	−1	−1	1	z		z^3
E_1''	2	2 cos 72°	2 cos 144°	0	−2	−2 cos 72°	−2 cos 144°	0	(R_x, R_y)	(xz, yz)	
E_2''	2	2 cos 144°	2 cos 72°	0	−2	−2 cos 144°	−2 cos 72°	0			$[xyz, z(x^2-y^2)]$

D_{6h}	E	$2C_6$	$2C_3$	C_2	$3C_2'$	$3C_2''$	i	$2S_3$	$2S_6$	σ_h	$3\sigma_d$	$3\sigma_v$	(x axis coincident with C_2')		
A_{1g}	1	1	1	1	1	1	1	1	1	1	1	1		x^2+y^2, z^2	
A_{2g}	1	1	1	1	−1	−1	1	1	1	1	−1	−1	R_z		
B_{1g}	1	−1	1	−1	1	−1	1	−1	1	−1	1	−1			
B_{2g}	1	−1	1	−1	−1	1	1	−1	1	−1	−1	1			
E_{1g}	2	1	−1	−2	0	0	2	1	−1	−2	0	0	(R_x, R_y)	(xz, yz)	
E_{2g}	2	−1	−1	2	0	0	2	−1	−1	2	0	0		(x^2-y^2, xy)	
A_{1u}	1	1	1	1	1	1	−1	−1	−1	−1	−1	−1			
A_{2u}	1	1	1	1	−1	−1	−1	−1	−1	−1	1	1	z		z^3
B_{1u}	1	−1	1	−1	1	−1	−1	1	−1	1	−1	1			$x(x^2-3y^2)$
B_{2u}	1	−1	1	−1	−1	1	−1	1	−1	1	1	−1			$y(3x^2-y^2)$
E_{1u}	2	1	−1	−2	0	0	−2	−1	1	2	0	0	(x, y)		(xz^2, yz^2)
E_{2u}	2	−1	−1	2	0	0	−2	1	1	−2	0	0			$[xyz, z(x^2-y^2)]$

The D_{nh} Groups (*continued*)

D_{8h}	E	$2C_8$	$2C_8^3$	$2C_4$	C_2	$4C_2'$	$4C_2''$	i	$2S_8^3$	$2S_8$	$2S_4$	σ_h	$4\sigma_v$	$4\sigma_d$	(x axis coincident with C_2')		
A_{1g}	1	1	1	1	1	1	1	1	1	1	1	1	1	1		x^2+y^2, z^2	
A_{2g}	1	1	1	1	1	−1	−1	1	1	1	1	1	−1	−1	R_z		
B_{1g}	1	−1	−1	1	1	1	−1	1	−1	−1	1	1	1	−1			
B_{2g}	1	−1	−1	1	1	−1	1	1	−1	−1	1	1	−1	1			
E_{1g}	2	$\sqrt{2}$	$-\sqrt{2}$	0	−2	0	0	2	$\sqrt{2}$	$-\sqrt{2}$	0	−2	0	0	(R_x, R_y)	(xz, yz)	
E_{2g}	2	0	0	−2	2	0	0	2	0	0	−2	2	0	0		(x^2-y^2, xy)	
E_{3g}	2	$-\sqrt{2}$	$\sqrt{2}$	0	−2	0	0	2	$-\sqrt{2}$	$\sqrt{2}$	0	−2	0	0			
A_{1u}	1	1	1	1	1	1	1	−1	−1	−1	−1	−1	−1	−1			
A_{2u}	1	1	1	1	1	−1	−1	−1	−1	−1	−1	−1	1	1	z	z^3	
B_{1u}	1	−1	−1	1	1	1	−1	−1	1	1	−1	−1	−1	1			
B_{2u}	1	−1	−1	1	1	−1	1	−1	1	1	−1	−1	1	−1			
E_{1u}	2	$\sqrt{2}$	$-\sqrt{2}$	0	−2	0	0	−2	$-\sqrt{2}$	$\sqrt{2}$	0	2	0	0	(x, y)		(xz^2, yz^2)
E_{2u}	2	0	0	−2	2	0	0	−2	0	0	2	−2	0	0			$[xyz, z(x^2-y^2)]$
E_{3u}	2	$-\sqrt{2}$	$\sqrt{2}$	0	−2	0	0	−2	$\sqrt{2}$	$-\sqrt{2}$	0	2	0	0			$[x(x^2-3y^2), y(3x^2-y^2)]$

H. The S_n Groups

S_4	E	S_4	C_2	S_4^3			
A	1	1	1	1	R_z	x^2+y^2, z^2	$xyz, z(x^2-y^2)$
B	1	−1	1	−1	z	x^2-y^2, xy	z^3
E	1	i	−1	$-i$	$(x, y), (R_x, R_y)$	(xz, yz)	$(xz^2, yz^2), [x(x^2-3y^2), y(3x^2-y^2)]$
	1	$-i$	−1	i			

S_6	E	C_3	C_3^2	i	S_6^5	S_6	$S_6 = C_3 \times C_i$		$\varepsilon = \exp(2\pi i/3)$
A_g	1	1	1	1	1	1	R_z	x^2+y^2, z^2	
E_g	1	ε	ε^*	1	ε	ε^*	(R_x, R_y)	(x^2-y^2, xy), (xz, yz)	
	1	ε^*	ε	1	ε^*	ε			
A_u	1	1	1	−1	−1	−1	z		$z^3, x(x^2-3y^2), y(3x^2-y^2)$
E_u	1	ε	ε^*	−1	$-\varepsilon$	$-\varepsilon^*$	(x, y)		$(xz^2, yz^2), [xyz, z(x^2-y^2)]$
	1	ε^*	ε	−1	$-\varepsilon^*$	$-\varepsilon$			

S_8	E	C_8	C_4	S_8^3	C_2	S_8^5	C_4^3	S_8^7		$\varepsilon = \exp(2\pi i/8)$	
A	1	1	1	1	1	1	1	1	R_z	x^2+y^2, z^2	
B	1	−1	1	−1	1	−1	1	−1	z		z^3
E_1	1	ε	i	$-\varepsilon^*$	−1	$-\varepsilon$	$-i$	ε^*	(x, y), (R_x, R_y)		(xz^2, yz^2)
	1	ε^*	$-i$	$-\varepsilon$	−1	$-\varepsilon^*$	i	ε			
E_2	1	i	−1	$-i$	1	i	−1	$-i$		(x^2-y^2, xy)	$[xyz, z(x^2-y^2)]$
	1	$-i$	−1	i	1	$-i$	−1	i			
E_3	1	$-\varepsilon^*$	$-i$	ε	−1	ε^*	i	$-\varepsilon$		(xz, yz)	$[x(x^2-3y^2), y(3x^2-y^2)]$
	1	$-\varepsilon$	i	ε^*	−1	ε	$-i$	$-\varepsilon^*$			

I. The Groups $C_{\infty v}$ and $D_{\infty h}$ for Linear Molecules

$C_{\infty v}$	E	$2C_\infty^\phi$	$\cdots$	$\infty\sigma_v$	$C_{\infty v} = C_\infty \wedge C_s$		
$A_1 \equiv \Sigma^+$	1	1	$\cdots$	1	z	x^2+y^2, z^2	z^3
$A_2 \equiv \Sigma^-$	1	1	$\cdots$	-1	R_z		
$E_1 \equiv \Pi$	2	$2\cos\phi$	$\cdots$	0	$(x, y); (R_x, R_y)$	(xz, yz)	(xz^2, yz^2)
$E_2 \equiv \Delta$	2	$2\cos 2\phi$	$\cdots$	0		(x^2-y^2, xy)	$[xyz, z(x^2-y^2)]$
$E_3 \equiv \Phi$	2	$2\cos 3\phi$	$\cdots$	0			$[x(x^2-3y^2), y(3x^2-y^2)]$
$\vdots$	$\vdots$	$\vdots$	$\vdots$	$\vdots$			

$D_{\infty h}$	E	$2C_\infty^\phi$	$\cdots$	$\infty\sigma_v$	i	$2S_\infty^\phi$	$\cdots$	∞C_2	$D_{\infty h} = D_\infty \times C_i$		
$A_{1g} \equiv \Sigma_g^+$	1	1	$\cdots$	1	1	1	$\cdots$	1		x^2+y^2, z^2	
$A_{2g} \equiv \Sigma_g^-$	1	1	$\cdots$	-1	1	1	$\cdots$	-1	R_z		
$E_{1g} \equiv \Pi_g$	2	$2\cos\phi$	$\cdots$	0	2	$-2\cos\phi$	$\cdots$	0	(R_x, R_y)	(xz, yz)	
$E_{2g} \equiv \Delta_g$	2	$2\cos 2\phi$	$\cdots$	0	2	$2\cos 2\phi$	$\cdots$	0		(x^2-y^2, xy)	
$\cdots$	$\cdots$	$\cdots$	$\cdots$	$\cdots$	$\cdots$	$\cdots$	$\cdots$	$\cdots$			
$A_{1u} \equiv \Sigma_u^+$	1	1	$\cdots$	1	-1	-1	$\cdots$	-1	z		z^3
$A_{2u} \equiv \Sigma_u^-$	1	1	$\cdots$	-1	-1	-1	$\cdots$	1			
$E_{1u} \equiv \Pi_u$	2	$2\cos\phi$	$\cdots$	0	-2	$2\cos\phi$	$\cdots$	0	(x, y)		(xz^2, yz^2)
$E_{2u} \equiv \Delta_u$	2	$2\cos 2\phi$	$\cdots$	0	-2	$-2\cos 2\phi$	$\cdots$	0			$[xyz, z(x^2-y^2)]$
$E_{3u} \equiv \Phi_u$	2	$2\cos 3\phi$	$\cdots$	0	-2	$2\cos 3\phi$	$\cdots$	0			$[x(x^2-3y^2), y(3x^2-y^2)]$
$\vdots$	$\vdots$	$\vdots$	$\vdots$	$\vdots$	$\vdots$	$\vdots$	$\vdots$	$\vdots$			

J. The Higher Order Groups

T_d	E	$8C_3$	$3C_2$	$6S_4$	$6\sigma_d$			
A_1	1	1	1	1	1		$x^2+y^2+z^2$	xyz
A_2	1	1	1	-1	-1			
E	2	-1	2	0	0		$(2z^2-x^2-y^2, x^2-y^2)$	
T_1	3	0	-1	1	-1	(R_x, R_y, R_z)		$[x(z^2-y^2), y(z^2-x^2), z(x^2-y^2)]$
T_2	3	0	-1	-1	1	(x, y, z)	(xy, xz, yz)	(x^3, y^3, z^3)

O_h	E	$8C_3$	$6C_2$	$6C_4$	$3C_2(=C_4^2)$	i	$6S_4$	$8S_6$	$3\sigma_h$	$6\sigma_d$			
A_{1g}	1	1	1	1	1	1	1	1	1	1		$x^2+y^2+z^2$	
A_{2g}	1	1	-1	-1	1	1	-1	1	1	-1			
E_g	2	-1	0	0	2	2	0	-1	2	0		$(2z^2-x^2-y^2, x^2-y^2)$	
T_{1g}	3	0	-1	1	-1	3	1	0	-1	-1	(R_x, R_y, R_z)		
T_{2g}	3	0	1	-1	-1	3	-1	0	-1	1		(xz, yz, xy)	
A_{1u}	1	1	1	1	1	-1	-1	-1	-1	-1			
A_{2u}	1	1	-1	-1	1	-1	1	-1	-1	1			xyz
E_u	2	-1	0	0	2	-2	0	1	-2	0			
T_{1u}	3	0	-1	1	-1	-3	-1	0	1	1	(x, y, z)		(x^3, y^3, z^3)
T_{2u}	3	0	1	-1	-1	-3	1	0	1	-1			$[x(z^2-y^2), y(z^2-x^2), z(x^2-y^2)]$

I_h	E	$12C_5$	$12C_5^2$	$20C_3$	$15C_2$	i	$12S_{10}$	$12S_{10}^3$	$20S_6$	15σ			
A_g	1	1	1	1	1	1	1	1	1	1		$x^2+y^2+z^2$	
T_{1g}	3	$\frac{1}{2}(1+\sqrt{5})$	$\frac{1}{2}(1-\sqrt{5})$	0	-1	3	$\frac{1}{2}(1-\sqrt{5})$	$\frac{1}{2}(1+\sqrt{5})$	0	-1	(R_x, R_y, R_z)		
T_{2g}	3	$\frac{1}{2}(1-\sqrt{5})$	$\frac{1}{2}(1+\sqrt{5})$	0	-1	3	$\frac{1}{2}(1+\sqrt{5})$	$\frac{1}{2}(1-\sqrt{5})$	0	-1			
G_g	4	-1	-1	1	0	4	-1	-1	1	0			
H_g	5	0	0	-1	1	5	0	0	-1	1		$(2z^2-x^2-y^2, x^2-y^2, xy, yz, zx)$	
A_u	1	1	1	1	1	-1	-1	-1	-1	-1			
T_{1u}	3	$\frac{1}{2}(1+\sqrt{5})$	$\frac{1}{2}(1-\sqrt{5})$	0	-1	-3	$-\frac{1}{2}(1-\sqrt{5})$	$-\frac{1}{2}(1+\sqrt{5})$	0	1	(x, y, z)		
T_{2u}	3	$\frac{1}{2}(1-\sqrt{5})$	$\frac{1}{2}(1+\sqrt{5})$	0	-1	-3	$-\frac{1}{2}(1+\sqrt{5})$	$-\frac{1}{2}(1-\sqrt{5})$	0	1			(x^3, y^3, z^3)
G_u	4	-1	-1	1	0	-4	1	1	-1	0			$[x(z^2-y^2), y(z^2-x^2), z(x^2-y^2), xyz]$
H_u	5	0	0	-1	1	-5	0	0	1	-1			

Source: B. E. Douglas and C. A. Hollingsworth, "Symmetry in Bonding and Spectra: An Introduction," 1985, Orlando, FL; Academic Press, pp. 391–408.

Answers to Selected Exercises

Chapter 1

1. a. No planes, 12 spheres, yes; **b.** 2 planes, 8 spheres, yes; **c.** 4 planes, 0 spheres, no; **d.** 5 planes, 0 spheres, no; **e.** 4 planes, 4 spheres, yes; **f.** 117 planes, 114 spheres, yes.

6. a. $5g < 5f < 5d < 5p < 5s$. **b.** $5g < 5f < 5d < 5p < 5s$. **c.** $5s < 5p < 5d < 5f < 5g$.

10. For $6s$, $Z^* = 2.85$; for $4f$, $Z^* = 11.65$.

15. a. $7s^1$; **b.** $4s^2 4p^3$; **c.** $6s^2 5d^8$; **d.** $6s^2 4f^{10}$; **e.** $4s^2 4p^2$; **f.** $5s^2$; **g.** $6s^2 6p^5$; **h.** $6s^2 5d^4$; **i.** $7s^2 5f^9$; **j.** $4s^2 3d^5$; **k.** $7s^2 5f^3$.

19. a. Xe and Te^{2-}; **b.** Mn; **c.** Pr^{3+}; **d.** Hg and Au^-.

22. a. C +4, O −2; **b.** N +3, O −2; **c.** N +3, O −2; **d.** H +1, O −1; **e.** H +1, C −4; **f.** I +5, F −1; **g.** Os +8, O −2.

24. a. The Fe^{2+} electron configuration is $3d^6$; Lewis symbols are not normally used for *d*-block atoms or ions. **b.** Lewis symbol Sn^{4+}, electron configuration $5s^0 5p^0$. **c.** Lewis symbol $:P^{3+}$, electron configuration $3s^2$. **d.** Lewis symbol $:Sb^{3+}$, electron configuration $5s^2$.

28. Six valence electrons: O, S, Se, Te, Po, Cr, Mo, W, Nd, U, Sg. Commonly give up or share all six: S, Se, Mo, W, U. Atomic numbers: 116, 124, 142, 156, 166.

29. a. +4; **b.** −1 or +5; **c.** +2 (or +3); **d.** +6; **e.** +2; **f.** +4.

34. a. 157 pm; **b.** 97 pm; **c.** 217 pm; **d.** 217 pm.

40. a. Ne < F < C < Li; **b.** Be < Ca < Ba < Ra; **c.** B < Al ≈ Ga < In ≈ Tl; **d.** V < Nb = Ta ≈ Db.

44. a. For $4s$, +3.75; for $3d$, +6.25; c. for $3p$, +14.75. **b.** For $4s$, 193 pm; for $3d$, 76 pm; for $3p$, 32 pm.

46. a. Increase. **b.** Decrease. **c.** Less rapidly. **d.** Stay the same; decrease; less rapidly. **e.** Increase; increase; less rapidly; after first filling of *d* and *f* orbitals; scandide contraction, lanthanide contraction, and relativistic effects.

52. a. Rule 1: $1s^2(2s, p)^8(3s, p)^8 3d^{10} 4s^2$. Rule 3: $1 \times 0.35 = 0.35$. Rule 4a: $18 \times 0.85 = 15.3$. Rule 4b: $10 \times 1 = 10$. Rule 6: $Z^* = 30 - 10 - 15.3 - 0.35 = 4.35$.
b. Rule 1 same as above. Rule 2: Ignore $4s^2$. Rule 3: $9 \times 0.35 = 3.15$. Rule 5: $18 \times 1 = 18$. Rule 6; $Z^* = 30 - 3.15 - 18 = 8.85$.
c. For $4s$: $\langle r_{max} \rangle = 52.9 \times (3.7)^2/4.35 = 166$ pm. For $3d$: $\langle r_{max} \rangle = 52.9 \times 3^2/8.85 = 53.8$ pm.
d. $\chi_{A-R} = 0.74 + (3590 \times 4.00/121^2) = 1.72$. This should be (and is) close to the Pauling value.

57. a. Z^* for the valence (s) orbitals, using the characteristic electron configurations: Na 2.20, Cu 4.20; $\langle r_{max} \rangle$: Na, 216 pm, Cu 172 pm; χ_{AR}: Na 1.02, Cu 1.50.
b. Z^* for the valence (s) orbitals: W 3.45, Nd 2.85, U 2.85; $\langle r_{max} \rangle$: W 270 pm; Nd 327 pm; U 359 pm; χ_{AR}: W 1.40, Nd 1.07, U 1.18. U appears to be more closely related to Nd.

60. The $6d$ orbitals will be expanded and raised in energy; spin–orbit coupling will split them into a set of three and a set of two. The best three $6d$ orbitals and the $7s$ orbital can hold eight easily-ionized valence electrons, hence allow Lr^{3+} through Hs^{8+}. But just possibly the relativistic shielding of the other two *d* orbitals might suffice to raise them in energy and spatial extent enough that electrons in them could also be shared with very electronegative atoms such as F or O, allowing a +9 oxidation state. (This is sheer fancy on my part, not supported by any calculations whatsoever.) Presumably, this would be most likely in the first element to have nine valence electrons, Mt.

62. a. $8s^2 5g^6$, $8s^2 6f^6$, $8s^2 7d^{10}$. **b.** > 1.38, > 2.54, < 2.2. **c.** +6, +3, +5 or higher, +2, +4 or higher, +2. **d.** Rn^{6+} > 62 pm, No. 121 difficult to extrapolate, eka–Np^{5+} > 89 pm, eka–No^{2+} > 124 pm, eka–Os^{4+} < 77 pm, eka–Pb^{2+} > 133 pm. (These are nonrelativistic predictions.)

64. a. $1s^2 2s^2 2p^6\ 3s^2 3p^6\ 4s^2 3d^{10} 4p^6\ 5s^2 4d^{10} 5p^6\ 6s^2 4f^{14} 5d^{10} 6p^6\ 7s^2 5f^{14} 6d^{10} 7p^6\ 8s^2 5g^{18} 6f^{14} 7d^{10}$.
b. $8s^2 7d^{10}$; $7d^{10}$. **c.** 4 nodal planes, 0 nodal spheres, poor shielders, poor penetrators. **d.** $Z^* = 4.35$.
e. $\chi_{A-R} = 0.74 + [3590 \times (4.35 - 0.35)/150^2] = 0.74 + 0.64 = 1.38$. The Pauling value should be much higher. Khalidium falls two periods below mercury, which strongly experiences relativistic and lanthanide contraction effects not reflected in the Allred–Rochow calculations. The relativistic effects should be much stronger in khalidium than in mercury.

Chapter 2

8. a. U^{3+}, $Z^2/r = 0.077$ so weakly acidic; Ag^+, $Z^2/r = 0.008$ and $\chi_P > 1.8$ so feebly acidic; Pa^{5+}, $Z^2/r = 0.272$ so very strongly acidic; C^{4+}, $Z^2/r = 0.533$ and $\chi_P > 1.8$ so very strongly acidic; As^{3+}, $Z^2/r = 0.125$ and $\chi_P > 1.8$ so strongly acidic; Tl^+, $Z^2/r = 0.006$ so nonacidic; Th^{4+}, $Z^2/r = 0.148$ so moderately acidic.
b. Precipitation of oxide or hydroxide with Pa^{5+}, As^{3+}, possibly with Th^{4+}. This could be cleared up by adding excess acid for As^{3+} and Th^{4+}, but not for Pa^{5+}.
c. U^{3+}(aq), Ag^+, Pa_2O_5 or $Pa(OH)_5$, in original form such as CCl_4, $As(OH)_3$ or As_2O_3, Tl^+(aq), ThO_2 or $Th(OH)_4$.

12. a. For +1 ions with $\chi_P < 1.8$: To be nonacidic, r must be $>1/.01 = 100$ pm. Qualifying ions: Na^+, K^+, Rb^+, Cs^+, Fr^+, Tl^+, from Groups 1 and the bottom of Group 13 (III). To be feebly acidic: r must be <100 pm but $>1/.04 = 25$ pm. Qualifying ion: Li^+, from the top of Group 1.
For +1 ions with $\chi_P > 1.8$: To be feebly acidic, $r > 100$ pm, that is, Cu^+, Ag^+, Au^+, from Group 11.
b. For +2 ions with $\chi_P < 1.8$: To be nonacidic $r > 400$ pm—none. To be feebly acidic $r > 100$ pm: Ba^{2+}, (Ra^{2+}), Sr^{2+}, Ca^{2+}, Eu^{2+}, Dy^{2+}, Tm^{2+}, Yb^{2+}, No^{2+}, Cd^{2+}: from the lower part of Group 2, and from the *f* (and one from the *d*) block. To be weakly acidic, r should be between 40 and 100 pm: Mg^{2+} and Be^{2+} from the *s* block and Ti^{2+}, V^{2+}, Cr^{2+}, Mn^{2+}, and Zn^{2+} from the *d* block qualify.
For +2 ions with $\chi_P > 1.8$: To be weakly acidic r should be >100 pm; Pb^{2+} from the *p* block and Hg^{2+}, Ag^{2+}, and Pd^{2+} from the *d* block. To be moderately acidic, r should be between 40 and 100 pm: Fe^{2+}, Co^{2+}, Ni^{2+}, and Cu^{2+} from the *d* block, and Ge^{2+} from the *p* block.

14. a. 0.219, 0.027, 0.008, 0.148, and 0.417, respectively; **b.** strongly acidic, feebly acidic, nonacidic, moderately acidic, and very strongly acidic; **c.** very strongly acidic, feebly acidic, feebly acidic, moderately acidic, and very strongly acidic; **d.** Most nearly neutral: Ba^{2+} and Ag^+; do not exist: B^{3+} and As^{5+}.

16. a. For $K^+ Z^2/r = 0.007$ and $\chi_P < 1.8$, so nonacidic. Potassium chloride will dissolve smoothly in water; it will not fume in air. **b.** For $Nb^{5+} Z^2/r = 0.32$, so very strongly acidic. $NbCl_5$ will fume in air and will react violently and irreversibly with water to give HCl and $Nb(OH)_5$ or Nb_2O_5. **c.** $Z^2/r = 0.13$ and $\chi_P < 1.8$ for Al^{3+}, so moderately acidic. $AlBr_3$ might fume in air some; it will react with moderate heat with water. **d.** $Z^2/r = 0.533$, so very strongly acidic, but steric hindrance is expected. CBr_4 will not fume, nor will it react with or dissolve in water. **e.** $Z^2/r = 0.027$ and $\chi_P < 1.8$, so feebly acidic. BaI_2 will not fume; it will dissolve smoothly in water. **f.** $Z^2/r = 0.71$, so very strongly acidic. IF_7 will fume in air and react violently and irreversibly with water.

20. a. UF_6; **b.** $COCl_2$; **c.** SO_2F_2. Maximum coordination number. The number of halogen atoms around the central atom add up to the maximum coordination number in CCl_4 and SF_6, so there is no room for a water molecule to enter and begin the hydrolysis. In the others the number of groups attached is less than the maximum coordination number, so there is room for an H_2O molecule to attack.

25. a. I^- (lowest charge, largest size); **b.** C^{4-} (highest charge).

26. Bb^- (Z^2/r ratio 0.003) < Dd^- (0.006) < Cc^{2-} (0.018) < Ee^{3-} (0.025) < Aa^{3-} (0.064).

32. The predominance regions of the oxides of the nonacidic and feebly acidic cations are either off scale to the right (high pH) or nearly are so, and so do not overlap the predominance region of Mg^{2+}, which according to Figure 2.4(*c*) extends up to pH 8.5. Hence, the oxides of the nonacidic and feebly acidic cations should react with Mg^{2+}(aq) to precipitate $Mg(OH)_2$.

36. a. Na^+(aq) is nonacidic, so overlaps the predominance range of Fe_2O_3—no reaction. (The strongest acid, Fe^{3+}, remains combined with the strongest base, O^{2-}.) **b.** Fe^{3+} is moderately acidic, so its predominance range goes up to 1–6. Since Li^+ is feebly acidic, the predominance range of LiOH comes down to 11.5–14. The two ranges do not overlap, so reaction will occur to give Li^+ and $Fe(OH)_3$. (Fe^{3+} and OH^- are the strongest acid and base.) **c.** From Figure 2.10, the predominance ranges of H_2S and Te^{2-} do not overlap, so the reaction will produce products with overlapping ranges: HS^- and HTe^-. (Of the two bases in competition for the acid H^+, Te^{2-} is stronger than HS^-, so gets the H^+.) **d.** Since Cl^- is nonbasic, HCl is very strongly acidic. The predominance ranges of HCl and the strongly basic S^{2-} do not overlap, so these will react to give Cl^- and either H_2S or HS^-, depending on the final pH. (S^{2-} is a stronger base than Cl^-, so gets the H^+.) **e.** From Figure 2.10, the predominance ranges of HTe^- and Se^{2-} do not overlap, so these will react to give HSe^- and Te^{2-}. (Se^{2-} is a stronger base than Te^{2-}.) **f.** From Figure 2.4, the predominance ranges of Ca^{2+} and Al_2O_3 [similar to $Al(OH)_3$] overlap, so there will be no reaction. (Al^{3+} is the stronger acid, so gets the strong base O^{2-}.) **g.** From Figure 2.4, the ranges of $Ca(OH)_2$ and Al^{3+} do not overlap, so they will react to give Ca^{2+} and $Al(OH)_3$.

39. The cations in order of increasing acidic strength are Cr^{2+} (in CrO) < Cr^{3+} (in Cr_2O_3) < Cr^{4+} (in CrO_2) < Cr^{6+} (in CrO_3). The order of base strength of the oxides is the converse: $CrO_3 < CrO_2 < Cr_2O_3 < CrO$.

46. a. Fe(II) set: Fe^{2+} = iron(II) ion; FeS = iron(II) sulfide. Fe(III) set: Fe^{3+} = iron(III) ion; $FeCl_3$ = iron(III) chloride; FeN = iron(III) nitride.
b. Sn(II) set: Sn^{2+} = tin(II) ion; $SnCl_2$ = tin(II) chloride; Sn_3N_2 = tin(II) nitride. Sn(IV) set: Sn^{4+} = tin(IV) ion; SnO_2 = tin(IV) oxide; Sn_3N_4 = tin(IV) nitride.
c. Tl(I) set: Tl^+ = thallium(I) ion; TlCl = thallium(I) chloride; Tl_2O = thallium(I) oxide. Tl(III) set: Tl^{3+} = thallium(III) ion; TlN = thallium(III) nitride; Tl_2O_3 = thallium(III) oxide.

47. a. Cu^+, Cu^{2+}, CuO, CuCl, Cu_3N_2; **b.** Cr^{3+}, Cr_2O_3, CrF_3, Cr^{6+}, CrO_3, CrF_6.

Chapter 3

2. From left to right of figure: **a.** A = $4p_x$, B = $2s$; A = $4d_{xz}$, B = $2p_x$; A = $1s$, B = $3d_{z^2}$. **b.** zero, positive, positive. **c.** (does not apply), π, σ.

6. a. $Z^* = 4.95$ for $3d$ and 3.45 for $4s$; $\langle r_{max} \rangle = 96.2$ pm for $3d$ and 210 pm for $4s$. **b.** $Z^* = 4.95$ for $4d$ and 3.45 for $5s$; $\langle r_{max} \rangle = 146$ pm for $4d$ and 245 pm for $5s$. **c.** $Z^* = 4.95$ for $5d$ and 3.45 for $6s$; $\langle r_{max} \rangle = 171$ pm for $5d$ and 270 pm for $6s$. Ratio of $\langle r_{max} \rangle$ values of d/s orbitals is 0.458 for Cr, 0.596 for Mo, and 0.632 for W. These calculations slightly show the effect. The quantum defect is involved: due to the small increase in n^*, the $(n+1)s$ orbital gains less in size relative to the nd orbital for higher values of n.

8. a. Pb–Pb < Sn–Sn < Ge–Ge < Si–Si < C–C. **b.** Li–Li < Be–Be (but Be_2 molecule itself is unstable) < B–B < C–C. **c.** $\delta < \pi < \sigma$.

10. In the order shown: δ, π, σ. Bond dissociation energies: $\delta < \pi < \sigma$ (In general: A different order might result if widely varying sizes of orbitals were compared). For bonding interactions, signs of wavefunctions should match in regions of overlap (while changing on crossing nodal planes).

13.

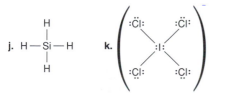

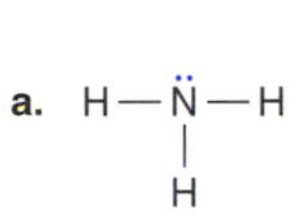

a. N donor atom; b. N and two O donor atoms; c. four O donor atoms ($2-$); d. S and three O donor atoms ($2-$)

18. a. $Cs[AlCl_4]$; **b.** $[Si(C_5H_5N)_4Cl_2]Cl_2$; **c.** $[Pb(C_5H_5N)_2Cl_4]$.

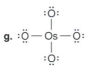

22.

a. :Ö=C=Ö: b. $[$:Ö—N̈=Ö:$]^-$ c. $[$:N≡O:$]^+$ d. CH_4 e. SiH_2Br_2 f. IF_5 g. OsO_4 h. $[$:Ö=N=Ö:$]^+$ i.

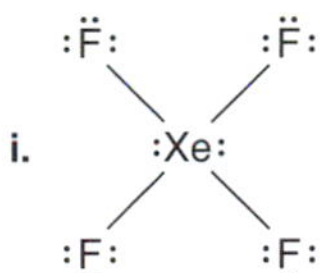

j. SiH_4 k. $[ICl_4]^-$

24.

a. $[$:Ö=N=Ö:$]^+$ 180°, linear

b. XeF_4 90°, planar

c. $Te(CH_3)_2Cl_2$ <90° and <120°, seesaw

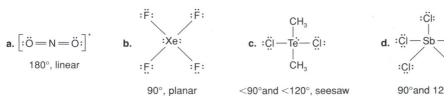

d. $SbCl_5$ 90° and 120°, trigonal bipyramidal

e. 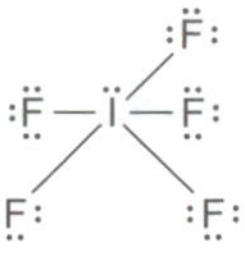 IF_5 — <90°, square pyramidal
f. 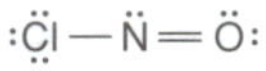 :Cl—N=O: — <120°, bent
g. 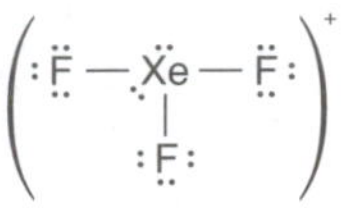 $(XeF_3)^+$ — <90°, T-shaped
h. 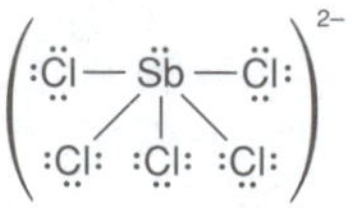 $(SbCl_5)^{2-}$ — <90°, square pyramidal

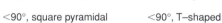

i. 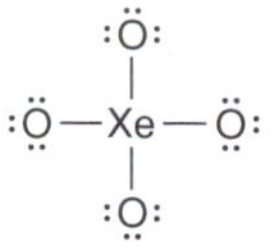 XeO_4 — 109.5°, tetrahedral
j. 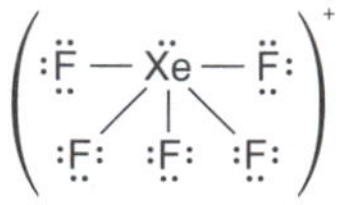 $(XeF_5)^+$ — <90°, square pyramidal
k. 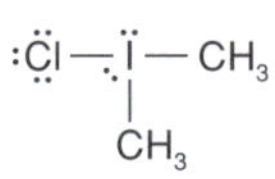 $ICl(CH_3)_2$ — <90°, T-shaped
l. 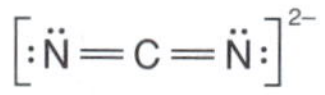 $[:N=C=N:]^{2-}$ — 180°, linear

29. **a.** $B^{3+} + 3\ O^{2-}$ gives $BO_3{}^{3-}$. **b.** $:Br:^{3+}$, with 2 unshared electron pairs, can accommodate two O^{2-} to give $BrO_2{}^{-}$. **c.** As^{5+}, with no unshared electron pairs, can accommodate four O^{2-} to give $AsO_4{}^{3-}$. **d.** $:S^{4+}$, with one unshared electron pair, can accommodate three O^{2-} to give $SO_3{}^{2-}$.

31. **a.** $BO_4{}^{5-}$; **b.** $BO_3{}^{3-}$; **c.** $AsO_4{}^{5-}$; **d.** $RuO_5{}^{2-}$; **e.** $IO_4{}^{3-}$.

35. (One way of reasoning) Increasing negative charge increases the attraction for the H of water, hence increases basicity. More oxo groups allow the dispersal of the negative charge, hence reduce basicity. In the pyrophosphate ion four negative charges are dispersed over six oxo groups; deleting two oxo groups per negative charge leaves M^-, which would be moderately basic. In the tripolyphosphate ion, there are five negative charges dispersed over eight oxo groups; this also works out to moderately basic M^-.

39. **a.** $ReO_4{}^-$ (or perhaps $ReO_6{}^{5-}$), $AsO_3{}^{3-}$, $SO_2{}^{2-}$, $ClO_3{}^-$, $SbO_6{}^{7-}$, $PO_4{}^{3-}$. **b.** 22.6 (or perhaps −6.8), −3.5, 1.0, 16.9, −27.2, 2.2. **c.** Nonbasic (or very strongly basic), (very) strongly basic, moderately or strongly basic, nonbasic, very strongly basic, moderately basic.

41. Canceling one negative charge per each two oxo groups, we obtain: **a.** MO, nonbasic; **b.** $M^{2.5-}$, very strongly basic; **c.** $M^{2.5-}$; very strongly basic. $MO_5{}^{2-} < MO_5{}^{5-} \approx MO_7{}^{6-}$.

46. $TeF_7{}^- < SbF_6{}^- = WF_8{}^{2-} < TaF_7{}^{2-} < TaF_8{}^{3-} = PbF_6{}^{2-} = IF_4{}^- < ZrF_7{}^{3-} < ClF_2{}^- = BeF_4{}^{2-} = GaF_6{}^{3-}$; the last three may be feebly basic.

48. **a.**

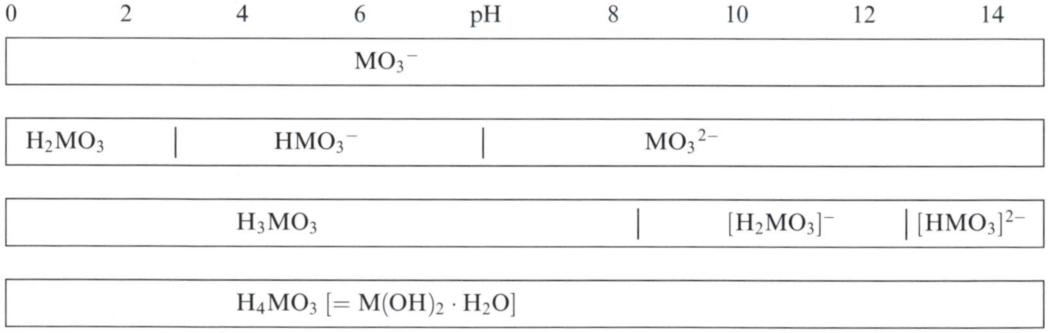

pH: 0 2 4 6 8 10 12 14

pH 0–14
$MO_3{}^-$

H_2MO_3	$HMO_3{}^-$	$MO_3{}^{2-}$

H_3MO_3	$[H_2MO_3]^-$	$[HMO_3]^{2-}$

pH 0–14
$H_4MO_3\ [= M(OH)_2 \cdot H_2O]$

b. None from pH 0–14, but the last gives $[H_3MO_3]^- = [M(OH)_3]^-$ at pH values in excess of 15.2. **c.** $MO_3{}^-$ and $MO_3{}^{2-}$ are not protonated; $[HMO_3]^{2-}$; none for $MO_3{}^{4-}$. **d.** HMO_3 is strongly acidic; H_2MO_3 is moderately acidic; H_3MO_3 is weakly acidic. (H_4MO_3 is weakly acidic or even nonacidic.)

53. **a.** Strongly acidic; **b.** weakly acidic; **c.** H_6TeO_5 weakly acidic; **d.** assuming a penultimate total coordination number of 6, H_3MtO_6 would be very strongly acidic.

56. **a.** +1, Li^+(aq); **b.** +3, $Al(OH)_3$ or Al_2O_3; **c.** +6, since W is a good π bonder the oxo anion is probably $WO_4{}^{2-}$, which should be feebly basic and could persist. **d.** +3, 121^{3+}(aq) or $121(OH)_3$; **e.** +6, $[HRnO_6]^{3-}$ or $[H_2RnO_6]^{2-}$; **f.** +5, eka-Np_2O_5; **g.** +2, eka-No^{2+}(aq); **h.** +4 (or higher), if +4 eka-OsO_2 or eka-$Os(OH)_4$; **i.** +2, eka-Pb^{2+}(aq).

59. **a.** Po^{4+} is strongly acidic, so would *not* be present as a cation. **b.** With penultimate total coordination number 6 the anion would be $PoO_5{}^{6-}$, which would be very strongly basic and would *not* be present;

with the number 4 the anion would be PoO_3^{2-}, which would be moderately basic and *might* be present (more likely as the partially protonated form $[HPoO_3]^-$).

62. **a.** N 3, Br 4, P 4, Sb 6 (or 4), Np 6 (or 4). **b.** NO_3^-, BrO_3^-, PO_4^{3-}, SbO_6^{7-}, NpO_6^{7-}. **c.** nitrate, bromate, phosphate, antimonate, neptunate. **d.** nonbasic, nonbasic, moderately basic, very strongly basic, very strongly basic. **e.** dissolved, dissolved, dissolved, sediments, sediments. **f.** phosphate ion, as HPO_4^{2-} and $H_2PO_4^-$.

65. **a.** SiO_4^{4-}, $pK_b = -8.0$ so very strongly basic; **b.** TeO_6^{6-}, $pK_b = -17.0$ so very strongly basic; **c.** BrO_4^-, $pK_b = 22.6$ so nonbasic; **d.** SO_3^{2-}, $pK_b = 6.7$ so moderately basic; **e.** OCl^-, $pK_b = 5.5$ so moderately basic; **f.** NpO_6^{5-}, $pK_b = -6.8$, so very strongly basic; **g.** NO_2^-, $pK_b = 11.2$ so (feebly) basic; **h.** FeO_4^{2-}, $pK_b = 12.4$, so feebly basic.

67. **a.** Uranyl sulfate; **b.** titanium(III) chloride; **c.** sulfuryl chloride; **d.** PO_4^{3-} has group V oxidation number, is phosphate; Fe is Fe^{3+}; iron(III) phosphate. **e.** IO_6^{5-} has group VII oxidation number, is periodate; Ag is Ag^+; silver(I) periodate. **f.** TeO_6^{6-} has group VI oxidation number, is tellurate; Hg is Hg^{2+}; mercury(II) tellurate.

68. **a.** $BiONO_3$. **b.** $Sr(ClO_4)_2$. **c.** $EuSO_3$. **d.** $Fe_3(PO_4)_2$. **e.** $CrCO_3$. **f.** $Cs^+ + PO_4^{3-}$ gives Cs_3PO_4. **g.** $Zn^{2+} + BrO_4^-$ gives $Zn(BrO_4)_2$. **h.** $K^+ + BrO_4^-$ gives $KBrO_4$. **i.** $Ca^{2+} + BrO_2^-$ gives $Ca(BrO_2)_2$. **j.** $Ca^{2+} + BO_3^{3-}$ gives $Ca_3(BO_3)_2$.

75.

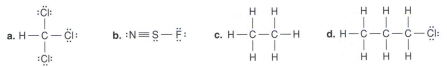

```
      :Cl:                                    H   H              H   H   H
       |                                      |   |              |   |   |
a. H — C — Cl:        b. :N≡S—F:       c. H — C — C — H     d. H — C — C — C — Cl:
       |                                      |   |              |   |   |
      :Cl:                                    H   H              H   H   H
```

e. Five (–C≡O:) and four unshared electron pairs around Fe.

```
         H   H    H    H
         |   |    |    |
f. :Cl — C — Si — Si — C — Cl:
         |   |    |    |
         H   H    H    H
```

Chapter 4

1. $CsBrO_4$; $CePO_4$; $BaSeO_4$; $Hg_5(IO_6)_2$; TiO_2.

4. **a.** All chromates are soluble except those of Ca^{2+}, Sr^{2+}, Ba^{2+}, Ra^{2+}, Tl^+, Eu^{2+}, Tm^{2+}, Yb^{2+}. (This prediction is not highly accurate. We may note that the solubilities of salts of feebly acidic cations and feebly basic anions are difficult to predict accurately. Also, our classification of CrO_4^{2-} as a feebly basic anion is not realistic, since its pK_b is 7.5, which makes it a weakly basic anion with quite different solubility rules. We might have predicted this if we had taken into account the effect of the comparatively low electronegativity of Cr on the basicity of its oxo anion.) **b.** Ferrates would be predicted in the same manner as chromates. (Little is known of their solubilities.) **c.** All pertechnetates are soluble except those of (Na^+?), K^+, Rb^+, Cs^+, and Fr^+. **d.** All silicates are insoluble except those of Na^+, K^+, Rb^+, Cs^+, Fr^+, (Li^+, Ca^{2+}, Sr^{2+}, and Ba^{2+}?). (Actually, these last three silicates are insoluble.) **e.** Since TeO_6^{6-} is very strongly basic, we make the same prediction as in **d.**

10. **a.** Since tropylium ion would be large and of low charge, we would (ignoring any possible reactions due to its organic chemistry) expect it to be a nonacidic cation that would best be precipitated by large nonbasic anions such as PF_6^-, nitrate, and perchlorate. **CAUTION**: Before attempting such a reaction read Chapter 6!! After reading Chapter 6, you may alter your choice!!! **b.** $C_4O_4^{2-}$ is a large anion that is likely either feebly basic or nonbasic, so large, feebly acidic Ba^{2+} (or perhaps nonacidic Cs^+) should precipitate and stabilize it. **c.** $C_5H_5^-$ is large (since it is not an oxo or fluoro anion, we cannot readily assess its basicity); a large +1-charged cation such as Cs^+ or Tl^+ may precipitate and stabilize it.

14. **a.** Favorable, positive; **b.** Unfavorable, negative; **c.** Favorable, positive; **d.** favorable, positive; **e.** unfavorable, negative; **f.** favorable, positive.

15. **a.** Zero; **b.** zero; **c.** unfavorable, positive; **d.** unfavorable, positive; **e.** favorable, negative; **f.** zero.

19. **a.** $[(C_4H_9)_4N]^+$ **b.** Al^{3+} **c.** $[(C_4H_9)_4N]^+$ **d.** Al^{3+} or Li^+ **e.** Rb^+

21. **a.** The smallest and most highly charged cation, Al^{3+}. **b.** Small, acidic cations: Zn^{2+} and Al^{3+}. **c.** Large K^+. **d.** Large −1 charged anion ClO_4^-. **e.** Nonacidic K^+. **f.** Acidic Zn^{2+} and Al^{3+}. **g.** Combinations of acidic cations and basic anions: $Al_4(SiO_4)_3$ and Zn_2SiO_4. **h.** For $Zn(ClO_4)_2$, $Al(ClO_4)_3$, $ZnSeO_4$, or $Al_2(SeO_4)_3$, the structure-making cation and structure-breaking anion cancel each other's entropy effects. The lattice energies are low due to the great difference in size of the cation and anion, so hydration energies predominate and the salt is soluble. For K_4SiO_4 or K_2SeO_4: The structure-breaking cation and structure-making anion cancel each other's entropy effects. The lattice energies are relatively low since the stoichiometries are not 1:1, so these are soluble.

24. LiF, -1033 kJ mol^{-1}; NaCl, -779 kJ mol^{-1}; AgCl, -910 kJ mol^{-1}; $CaCl_2$, -2249 kJ mol^{-1}.

32. **a.** Radius ratio = 0.802; Coordination number of Ce = 8, of O = 4; fluorite (CaF_2) lattice type. **b.** $-11{,}035$ kJ mol^{-1}.

34. The Z^2/r ratios and expected lattice types are **a.** NaAt 116/226 = 0.513 NaCl; KAt 152/226 = 0.672 NaCl; RbAt 0.735 CsCl; CsAt 0.800 CsCl; **b.** $MgAt_2$ 0.380 SiO_2; $CaAt_2$ 0.504 TiO_2; $SrAt_2$ 0.584 TiO_2; $BaAt_2$ 0.659 TiO_2; **c.** MgPo 0.380 ZnS; CaPo 0.504 NaCl; SrPo 0.584 NaCl; BaPo 0.659 NaCl.

38. **a.** 8 for La, 8 for N, CsCl; **b.** 6 for Sc, 6 for N; NaCl; **c.** -7918 kJ mol^{-1}.

42. For example, $FePO_4$. Insoluble. Both Fe^{3+} and $PO_4{}^{3-}$ form very large hydrated ions; on forming the solid salt many moles of water molecules are liberated. Entropy effects.

44. **a.** Acidic Fe^{3+} is an electrostatic structure maker but nonbasic $ClO_4{}^-$ is an electrostatic structure breaker; thus the entropy effects tend to cancel out. Instead the size mismatch results in a lattice type with a poor Madelung constant and a lattice energy that is less favorable than the hydration energies. **b.** Basic $PO_4{}^{3-}$ is also an electrostatic structure maker, so the release of numerous water molecules from their large hydrated ions causes precipitation. **c.** Although both ions are electrostatic structure breakers, the reason for the insolubility is that both ions are large and form a better lattice than they do hydrated ions. **d.** Both ions are hydrophobic structure makers. Again a good lattice is formed; also, precipitation of these ions allows water molecules to reenter the iceberg cavities and be disordered again.

46. **a.** Sr^{2+} feebly acidic, Ga^{3+} acidic, $[(CH_3)_4N]^+$ nonacidic. **b.** Ga^{3+} **c.** Ga^{3+}, Sr^{2+} **d.** $[(CH_3)_4N]^+$; the large cation and large anion are matched in size and radius, and form a more stable lattice than they do hydrated ions. **e.** The smaller and/or more highly charged cations are mismatched in size (acidity/basicity) with $ClO_4{}^-$.

52. **a.** $Pu(PF_6)_4$; **b.** $Pu(PF_6)_4$. [In each case $Pu(SO_4)_2$ is probably also an acceptable answer.]

53. **a.** $Ba^{2+}(aq) + SO_4{}^{2-}(aq) \rightarrow BaSO_4(s)$; **b.** $2\ H^+\ (aq) + SeO_3{}^{2-}(aq) \rightarrow H_2SeO_3(aq)$; **c.** $4\ H^+(aq) + SiO_4{}^{4-}(aq) \rightarrow Si(OH)_4(s)$; **d.** $3\ Th^{4+}(aq) + 4\ PO_4{}^{3-}(aq) \rightarrow Th_3(PO_4)_4(s)$; **e.** $HSO_3{}^-(aq) + H^+(aq) \rightarrow H_2O(l) + SO_2(g)$.

55. **a.** Since Co^{2+} is a weakly acidic cation, the soluble salt needed would be one of a nonbasic (or perhaps weakly basic) anion; $Co(NO_3)_2$ would be suitable. *The Handbook of Chemistry and Physics* shows that cobalt(II) nitrate is found as a hexahydrate, formula mass = 291.04. To make 0.2 mol of cobalt(II) vanadate, 0.6 mol or 174.6 g of $Co(NO_3)_2 \cdot 6H_2O$ should be dissolved in water. Since $VO_4{}^{3-}$ is a moderately basic anion, its salt with a nonacidic cation such as Na^+ should be soluble. *The Handbook of Chemistry and Physics* shows that sodium vanadate occurs either anhydrous (formula mass = 183.94), as a decahydrate (formula mass = 364.06), or as a hexadecahydrate (formula mass = 472.15). 0.4 mol of sodium vanadate should be dissolved in water, and the solution then mixed with the cobalt(II) nitrate solution. The amount of sodium vanadate needed is 73.6 g of the anhydrous salt, or 145.6 g of the 10-hydrate, or 188.8 g of the 16-hydrate.

b. Among the possible answers: React a solution of 67.9 g (0.40 mol) of $AgNO_3$ with a solution of 64.4 g (0.20 mol) of $Na_2SO_4 \cdot 10H_2O$.

57. **a.** HgS(s); **b.** $HgCl_2(aq)$; **c.** $Hg(CH_3)_2(\ell)$; **d.** $Hg(CH_3)_2(\ell)$.

62. In general, the very weak acids or very strongly basic anions should be titrated in a strongly basic solvent covering effective pH values over 14 (i.e. NH_3 or $(CH_3)_2SO$). The very strong acids or nonbasic anions should be titrated in a strongly acidic solvent covering effective pH values below 0 (i.e., HCOOH and above in Fig. 4.10). The acids and bases of intermediate strength could be titrated in water or a similar solvent such as ethanol. In practice, one would also need to consider the solubility of the material, and whether it reacted with the solvent. (**a**) Use the base strength listed in Table 2.5 and apply the above general principles. (**b**) These are strong to very strong acids with pK_a values around 0; the solvent should include this effective pH in the middle of its range (i.e., C_2H_5OH or CH_3COOH would be the best). (**c**) A weak acid with pK_1 = 8.5, so H_2O, C_2H_5OH, or $(CH_3)_2SO$. (**d**) A strong acid with pK_1 = -2.9, so perhaps C_2H_5OH or HF. (**e**) For this ion, pK_3 = 17.5, so use $(CH_3)_2SO$ or NH_3. (**f**) These are very strong acids with unspecified pK_a values, so one would select the most strongly acidic solvent available: HF or HSO_3F from Figure 4.10, or (as indicated by the text) the superacid $HSO_3F + SbF_5$.

Chapter 5

2. **a.** NaN_3; **b.** $Cr(CN)_2$; **c.** $Ag_2S_2O_3$; **d.** $Hg(N_3)_2$; **e.** $Ba[N(CH_3)_2]_2$.

6. **a.** No—forms four-membered ring with metal ion; **b.** Yes—forms five-membered rings with metal ion; **c.** No—forms four-membered ring; **d.** No—only one donor atom; **e.** Yes—forms five-membered rings; **f.** No—has no donor atoms. (It could be noted that four-membered chelate rings often do form in the absence of better possibilities.)

11. **a.** CH_3–O–CH_2–CH_2–O–CH_2–CH_2–O–CH_3 (let us call this "A"); CH_3^-; N_3^-; Cl^-; H_2O; NH_3; C_6H_5–CH(NH_2)–C(=O)–O^- (let us call this "B"); $S_2O_3^{2-}$. **b.** "A" is tridentate; "B" is bidentate. **c.** N_3^-, Cl^-, $S_2O_3^{2-}$. **d.** Each Mo has a coordination number of 7.

15. **a.** C; **b.** E; **c.** A, C, D, E; **d.** B; **e.** B.

19. **a.** B (monodentate) < A (chelate) < C (macrocyclic); **b.** chelate: A, C; macrocyclic, C; bridging, none, although CN^- is potentially a bridging ligand.

23. **a.** N donor atom is borderline; **b.** N donor atom is borderline and O donor atoms are hard; **c.** O donor atoms are hard; **d.** O donor atoms are hard and S donor atom is soft.

25. **a.** $K^+ < Cu^+ < Ag^+ < Au^+$; **b.** $F^- < Cl^- < Br^- < I^-$; **c.** $Mo^{6+} < Mo^{4+} < Mo^{2+}$; **d.** $BF_3 \le B(OCH_3)_3 < B(CH_3)_3$; **e.** $Fe^{3+} < Fe^{2+} < FHg^+ < CH_3Hg^+$; **f.** $(CH_3)_2O < (CH_3)_2S < (CH_3)_2Se$.

29. **a.** Products; **b.** products; **c.** reactants; **d.** products; **e.** products.

36. Very small size, and high charge in the case of B^{3+}.

39. The HSAB principle favors the reactants, since the softest acid–softest base combination is HgI_2. But strength considerations favor products, since the strongest acid–strongest base combination is $Hg(OH)_2$. Hence, the two effects approximately cancel each other out.

40. CdTe, AgI, TiO_2, $PtAs_2$. (EuSe and $TiTe_2$ should undergo hydrolysis.)

44. **a.** Group II, SnS; **b.** group III, $Pr(OH)_3$ or Pr_2O_3; **c.** forms complex $CuCl_2^-$ ion in group I so carries over to group II as Cu_2S; **d.** group IV, $SrCO_3$; **e.** group II, Sb_2S_3; **f.** group III, Eu_2O_3 or $Eu(OH)_3$; **g.** Forms $AuCl_2^-$ in group I, carries over to group II and precipitates as Au_2S; **h.** group II, $RaCO_3$; **i.** group V, Rb^+(aq).

45. Large nonbasic anions: ClO_4^-, ReO_4^-, PF_6^-, probably ICl_4^-.

48. **a.** Zr soluble, Ag insoluble, Sb soluble; **b.** Zr insoluble, but hydrolyzed to H_2S and insoluble ZrO_2, Ag insoluble, Sb insoluble; **c.** Zr insoluble, Ag soluble (answer depends on method used to classify acidity of Ag^+), Sb insoluble; **d.** Zr Group III, Ag Group I, Sb Group II; **e.** Zr yes, Ag no, Sb no; **f.** Zr silicate, Ag sulfide, Sb sulfide.

49. **a.** A, $LaPO_4$; **b.** B, $PtAs_2$; **c.** A, $ZrSiO_4$; **d.** B, Sb_2S_3; **e.** B, PbTe.

55. Soft-acid metal ions; cysteine groups.

59. **a.** Ac^{3+}; **b.** Mt^{2+}; **c.** Po^{2-}; **d.** Mt^{2+}; **e.** Fr^+.

63. The last ligand.

71. Coordination of a ligand enhances its acidity. Both NH_3 and H_2O are not very acidic to start with, so even upon coordination they remain of low acidity. But PH_3 is more acidic than NH_3, and H_2S is more acidic than H_2O, so upon coordination they may well become acidic enough to ionize. (The ionized forms are then good soft-base bridging ligands; other factors to be seen later, such as the ease of oxidation of P–H and S–H bonds, undoubtedly also contribute.)

Chapter 6

3. **a.** CoO_2; **b.** H_2Se; **c.** Eu^{2+}; **d.** IrO_4^{2-}; **e.** AmO_2^{2+}; **f.** Group 1, 2, 3, and 18(VIII) elements; Zr, Hf, Ta, Au, Zn, B, Al, Ga, Si, F; those *f*-block elements not shown in Figure 6.24 plus Sm and Yb. (These answers allow for approximate 0.5-V overvoltages.)

4. Br

5. **a.** Ta; **b.** Pr has no +5 oxidation state—otherwise Bi.

14. **a.** Cl_2; **b.** Zn, Bi, Sc, Tm, Os; **c.** None.

17. **a.** NO_3^-, N_2, (N_2O_4, NO_2, NH_3OH^+, $N_2H_5^+$); **b.** for example, $2\ NO_3^- + 5\ Mn + 12\ H^+ \rightarrow N_2 + 6\ H_2O + 5\ Mn^{2+}$; **c.** MnO_4^-, MnO_2, (MnO_4^{2-}); **d.** for example, $2\ MnO_4^- + 2\ NH_4^+ \rightarrow N_2 + 2\ MnO_2 + 4\ H_2O$; $3\ MnO_2 + 2\ NH_4^+ + 4\ H^+ \rightarrow N_2 + 3\ Mn^{2+} + 6\ H_2O$.

18. **a.** Unstable **b.** stable **c.** unstable **d.** unstable **e.** E^0 (C_2^{2-}) is not given, but as the anion of an electronegative nonmetal it is likely to be a good reducing agent. However, Ca^{2+}, as the cation of a very electropositive metal, is *very* difficult to reduce. So this salt is probably stable. **f.** Since Ag^+ is an oxidizing cation, it is likely that this salt is unstable. **g.** $E^0(BrO_3^-) - E^0(Cr^{2+}) > 1$ V, so unstable. **h.** $E(Tl^{3+}) - E(N_3^-) > 1$ V, so unstable.

22. **a.** Since Ag^+ is only found in the upper part of the Pourbaix diagram, it will be a good oxidizing agent. **b.** F_2 is the strongest oxidizing agent. **c.** No. $E(BrO_4^-)$ minus $E(Ag^+/Ag^{2+})$ is negative. **d.** Yes. $E(Ag^+)$ minus $E(H^-)$ exceeds 1 V.

25. **a.** Species capable of oxidizing Mn should have nonoverlapping predominance areas lying above Mn. In the N diagram, these are NO_3^- and N_2; among metastable species, N_2O_4, NO_2, NH_3OH^+, $N_2H_5^+$.

b. MnO_4^-, MnO_4^{2-}, MnO_2, Mn_2O_3, and Mn_3O_4.

c. As the salt of a nonacidic cation and a nonbasic anion, ammonium permanganate should be insoluble. Since the predominance areas of NH_4^+ and MnO_4^- are separated by more than 1 V, it is thermodynamically unstable and is likely to be explosive.

d. The predominance areas of water, MnO_2, and N_2 all overlap each other. Hence, they could coexist as final products.

e. As the salt of a nonacidic cation and the feebly basic MnO_4^{2-} anion, ammonium manganate should be soluble but also thermodynamically unstable and explosive if it can be prepared.

f. The predominance area of Mn falls entirely below that of water and N_2, so we would expect the Mn to be further oxidized and either the water or the N_2 (in practice the water) to be further reduced.

(g)1. The predominance areas of nitrate and Mn metal fail to overlap by more than 1 V, so this could be an explosive combination. **(g)2.** The predominance areas of these two oxidizing agents overlap, so it should be possible to grind them together safely, as long as no dusts or organics get in!

30. **a.** Nk above +2.50 V and Nk^- below. H_6SsO_6 above +1.65 V, Ss from +1.65 V down to +0.10 V, and Ss^{2-} below +0.10 V. Hm^{2+} above +0.85 V and Hm below. Pz^{2+} above −2.30 V and Pz below.

b. $H_6SsO_6 + 6\,H^+ + 6\,e^- \rightleftharpoons Ss + 6\,H_2O$.

c. Pz **d.** Nk **e.** none **f.** Yes.

31. **a.** BiI_3 and BiF_3; **b.** SbI_5 and InI_3; **c.** PAt_5 and PF_5; **d.** FO_4^- and ReO_4^-. Charge-transfer; see Sec. 6.4.

33. See Table D for list of elements. **a.** (Te is not a metal) Tb $\geq$ Tm > Th > Ta $\geq$ Ti > Tl > Sn > Tc > W; **b.** K > Pr $\geq$ Pm > Pu > Pa > Pd > Pt.

35. **a.** Reacts with all; **b.** unreactive with 1, 2, perhaps 3; **c.** reacts with 2 (and 3?—may be passivated); **d.** reacts with all; **e.** unreactive with 1, 2, perhaps 3; **f.** reacts with all; **g.** reacts with 2, [and 3?—might be (actually is) passivated by 3]. **h.** Reacts with 2, 3; **i.** unreactive with all; **j.** reactive with all; **k.** reacts with 3; **l.** reacts with 2, 3; **m.** reacts with 2, might be passivated by 3 (in practice does dissolve in dilute HNO_3); **n.** reacts with all.

42. CO.

45. The H^+ of a nonoxidizing acid will not oxidize gold, so the anion must do the oxidation. Selenate is a stronger oxidizing agent than sulfate.

49. F^- and O^{2-}.

52. **a.** Since E^0 for SCN^- is +0.77V, NO_3^-, $Cr_2O_7^{2-}$, ClO_4^-, Br_2; **b.** since E^0 for N_3^- is −3.09 V, all of the reagents can oxidize it.

55. **a.** Carbon **b.** none **c.** electrolysis **d.** none **e.** carbon **f.** electrolysis **g.** more active metal.

59. **a.** Chemical conversion; **b.** reduction; **c.** beneficiation. **d.** purification; **e.** purification.

61. **a.** Reduction $FeO + C \rightarrow Fe + CO$, and so on; purification $(P + Si + Mn) + CaO + O_2 \rightarrow Ca_3(PO_4)_2 + CaSiO_3 + MnO$, and so on.

b. Beneficiation $(Al_2O_3 + TiO_2 + Fe_2O_3) + OH^- \rightarrow [Al(OH)_4]^-$ only; conversion $Al(OH)_3$ to Al_2O_3; reduction $2\,Al_2O_3 + 3\,C +$ electrical energy $\rightarrow 4\,Al + 3\,CO_2$.

c. Beneficiation with detergents is a physical separation; conversion and reduction are combined as $Cu_2S + O_2 \rightarrow 2\,Cu + SO_2$; for purification see Exercise 59e.

d. Conversion $TiO_2 + 2\,Cl_2 + 2\,C \rightarrow TiCl_4 + 2\,CO$; reduction $TiCl_4 + 2\,Mg \rightarrow Ti + 2\,MgCl_2$

68. **a.** PuO_2, in sludge.

b. React: probable overall reaction is $Pu + 2\,H_2O \rightarrow PuO_2 + 2\,H_2$, although first step could be $2\,Pu + 6\,H^+ \rightarrow 2\,Pu^{3+} + 3\,H_2$.

c. Acidic; may be protected in basic solution by forming adhering coating of insoluble Pu_2O_3.

d. Pu^{3+}, in solution.

70. **a.** Disproportionate at some pH: $Am(OH)_4$ or $Am_2O_5 \cdot H_2O$; CrO_2; Mn_2O_3, Mn_3O_4, or MnO_4^{2-}; Bi_4O_7 or BiO_2; I_3^- or I_2.

b. Oxidize water: AmO_3; no Cr species; MnO_4^- in acid solution; Bi_2O_5(?); no I species(?).

c. Am metal; Cr metal; Mn metal (not shown); BiH_3; none for N or I (assuming 0.5-V overvoltage).

73. 14.

76. a. Does not exist. **b.** Strongly basic solution. **c.** Strongly basic solution. **d.** Action of base on element. **e.** Acidic solution with HNO_3.

80. a. $FeO_4{}^{2-}$; **b.** $BrO_4{}^{-}$; **c.** $PoO_6{}^{6-}$.

Chapter 7

1. a. -2630 kJ mol^{-1}; **b.** -3401 kJ mol^{-1}; **c.** CaF_2 theoretical lattice energy -2628 kJ mol^{-1}. CaO: predict CsCl lattice type, for which lattice energy is -3571 kJ mol^{-1}; actually has NaCl lattice type, for which latttice energy is -3540 kJ mol^{-1}.

4. a. $+535$ kJ mol^{-1}; **b.** No, -2.77 V.

9. a. For Pb part of cycle, $Pb(g) \rightarrow Pb^{2+}(g) \rightarrow Pb^{2+}(aq)$, and $\Delta H = IE(1) + IE(2) + \Delta H_{hyd} = +716 + 1450 - 1480$ kJ.
For Mg part of cycle, $Mg^{2+}(aq) \rightarrow Mg^{2+}(g) \rightarrow Mg(g) \rightarrow Mg(s)$, and $\Delta H = -\Delta H_{hyd} - IE(1) - IE(2) - \Delta H_{atom} = +1922 - 738 - 1451 - 146$ kJ.
b. The overall $\Delta H = +273$ kJ. **c.** No. -1.42 V.

13. a. C and Pb, since covalent bonds are strongest at the top of the *p* block; **b.** Ta and V, since covalent bonds are strongest at the bottom of the *d* block; **c.** C and Ne, since C has the maximum number of unpaired prepared-for-bonding electrons, 4; **d.** Mo and Xe, since Mo has the maximum number of unpaired prepared-for-bonding electrons, 6.

16. Mo; Cl; Ar; Na.

18. a. -38 kJ mol^{-1}; **b.** $+328$ kJ mol^{-1}.

23. 319 kJ mol^{-1} in CCl_4; 366 kJ mol^{-1} in $(CH_3)_3CCl$.

26. $\Delta H = 4\,\Delta H_{BE}(C{-}Cl) - 4\,EA(Cl) + 4\,\Delta H_{atom}(Na) + 4\,IE(1)(Na) + 4\,U(NaCl) - \Delta H_{atom}(C) = 4(327) - 4(349) + 4\,(107) + 4\,(496) + 4\,(-787) - 717 = -1541$ kJ. $E^0 = -3.99$ V (neglecting entropy), so the reaction could be explosive.

33. χ_P for Ge(II) $= 1.83$. Yes; the electronegativity of an element in a lower oxidation state should be lower.

35. Considering only the fluorides in Table 7.6, the data suffice for Cl, Br, Se, As, Ge, I, Te, Sb, In, and Xe if we assume that the Xe–Xe bond energy is zero. For Xe, $E(XeF_n) = (0 + 155)/2 + 96.5(\chi_P - 3.98)^2$: for Xe(VI), $\chi_{Xe} = 3.27$; for Xe(IV), $\chi_{Xe} = 3.24$. For Cl, we end up with square roots of negative numbers except in the case of Cl(V), for which $\chi_{Cl} = 3.25$. (Most other elements in the above list give more reasonable results than these two.)

38. a. χ_P for Cu(I) is 1.92 while χ_P for Cu(II) is 2.11, a normal trend; **b.** χ_P for Au(I) $= 2.33$ while χ_P for Au(III) $= 2.42$, also a normal trend.

40. a. 320 kJ mol^{-1}; **b.** Gallicium norrellide—there is no electronegativity difference between these elements, hence no ionic stabilization energy for this bond. **c.** +3, +1.

44. a. For *dissociation*, $\Delta H = 4.184(5.54 \times 10.00 + 0.30 \times 4.28 + 30.67 \times 2.03) = 497$ kJ; for *ionization*, $\Delta H = 4.184(3.78 \times 9.73 + 0.10 \times 4.28 + 20.79 \times 37.40) = 3409$ kJ. **b.** For dissociation, 323 kJ; for ionization, 632 kJ. Relatively speaking, KI is easier to *ionize*, since its ionization involves separating larger ions that are already further apart.

Chapter 8

3. a. f_{xyz} (intercepts corners of cube); $(f_{x^3}, f_{y^3}, f_{z^3})$ (intercept centers of faces of cube); $f_{y(x^2-z^2)}$, $f_{x(z^2-y^2)}$, $f_{z(y^2-x^2)}$ (intercept centers of edges of cube). **b.** Greatest repulsion (pointed directly at octahedral ligands: $(f_{x^3}, f_{y^3}, f_{z^3})$. Greatest stabilization (furthest from ligands): f_{xyz}. **c.** These orbitals (e.g., 4*f*) do not actually overlap ligand orbitals well, since the ligand orbitals are outside the larger (e.g., 5*s* and 5*p*) core orbitals of the metal ion.

6. a. $t_{2g}{}^6$, $-2.4\Delta_o$ (+2P), 0 unpaired e$^-$, $[Co(CN)_6]^{3-}$. **b.** $t_{2g}{}^3e_g{}^2$, $0\Delta_o$, five unpaired e$^-$, $[Fe(H_2O)_6]^{3+}$. **c.** $t_{2g}{}^4e_g{}^2$, $-0.4\Delta_o$, four unpaired e$^-$, $[Co(H_2O)_6]^{3+}$. **d.** $t_{2g}{}^5$, $-2.0\Delta_o$ (+2P), one unpaired e$^-$, $[Fe(CN)_6]^{3-}$.

11. a. 5.92 BM and 0.0159 cgs units; **b.** 0 BM and 0 cgs units; **c.** 4.90 BM and 0.0109 cgs units; **d.** 1.73 BM and 0.00136 cgs units; **e.** 4.90 BM and 0.0109 cgs units.

13. a. Electron–electron repulsions differ among different pairs of *d* orbitals if more than one *d* electron is present. **b.** A; 19,000 cm^{-1}. **c.** The chromium complex transmits visible light below about 17,000 cm^{-1}—red light—and at about 23,000 cm^{-1}—blue light—so it would be violet. The titanium complex transmits visible light below about 21,000 cm^{-1}, which is centered approximately in the yellow range.

18. After taking 78% of the tabulated pairing energies and dividing by the g values of the metal ions, one obtains the following minimum f values. **a.** 0.90 [F^- or $(C_2H_5)_2NCS_2^-$]; **b.** 1.95 (maybe CO); **c.** 1.37 (phenanthroline); **d.** 2.65 (probably none; maybe CO); **e.** 1.04 (NCS^-); **f.** 1.31 [$NH(CH_2CH_2NH_2)_2$ or bipyridyl].

20. $f = \Delta_o/g$; for A = 0.94 (CH_3COOH); for B = 1.18 ($NH_2CH_2COO^-$); for C = 1.41 (phenanthroline).

23. **a.** Upper part, since the donor C atom has an empty p orbital in the Lewis structure, so that the ligand can be a π-acceptor; **b.** lower part, since the donor N atom has two unshared electron pairs, so that the ligand can be a π-donor; **c.** middle, since the donor O atom has only one unshared electron pair, and the ligand is thus only a sigma donor.

26. $N{\equiv}C{-}\underline{O}^- = C{\equiv}N{-}\underline{O}^-$ (O donors) < $\underline{N}{\equiv}C{-}O^-$ (N donor) < $\underline{C}{\equiv}N{-}O^-$ (C donor)

27. ΔH_{hyd} values come from Table 2.1, and are available beginning with V^{2+}. V^{2+}: CFSE $= -1.2\ \Delta_o = (f_{ligand} \times g_{ion}) \times 12.0\ J/1000\ cm^{-1} = -170$ kJ, which is 8.9% of the hydration energy. $Cr^{2+} = -101$ kJ mol^{-1} = 5.6%. $Fe^{2+} = -48$ kJ = 2.5%. $Co^{2+} = -86$ kJ = 4.2%. $Ni^{2+} = -124$ kJ = 5.9%. $Cu^{2+} = -94$ kJ = 4.5%.

36. d_{z^2} will be repelled strongly; d_{xz} and d_{yz} will feel little effect; d_{xy} and $d_{x^2-y^2}$ will be strongly stabilized in a linear complex (ligands at ends of z axis).

41. **a.** 6700 cm^{-1} expected for octahedral; since observed value is only about one-half of this, complex may be tetrahedral $NiCl_4^{2-}$. **b.** 6570 cm^{-1} expected for tetrahedral; since observed ν_1 is much larger than this, the complex is likely square planar, not tetrahedral; **c.** Square planar, not tetrahedral, since the latter would have absorptions in the near-IR; **d.** Square planar, since only these are diamagnetic (tetrahedral and octahedral are paramagnetic).

45. **a.** $t_{2g}{}^6e_g{}^1$, $t_{2g}{}^3e_g{}^1$, $t_{2g}{}^6e_g{}^3$. **b.** The ligands along the z direction move away from the metal ion (while the others move in), giving a distorted octahedron.

49. Since Pt is in the sixth period, its complexes are likely to be strong field. Pt^{4+} prefers octahedral, since it is d^6, which has a large CFSE ($-2.4\Delta_o$). Pt^{2+} prefers square planar, since it is d^8.

54. Twist one square (say, the top one) by 45° with respect to the bottom square of ligands. This will result in less repulsion between the two squares of ligands. If we take the z axis as the one connecting the two squares, this twisting will reduce the repulsions between the ligands and the d_{xy} orbital, raise the repulsions involving the $d_{x^2-y^2}$ ligand, and have no effect on the repulsions involving the d_{z^2} ligand. (The twisting alters the choice of location of the x and y axes, which makes the problem difficult for the other two d orbitals, but these answers can be confirmed in Table 8.6.)

56. Complex (a), since Co^{3+} oxidizes water and would be moderately acidic even if it did not, so that its hydroxide would likely precipitate; (b), since Fe^{3+} is moderately acidic, so that its hydroxide would likely precipitate; and (d) since Tc^{4+} is strongly acidic, so that its hydroxide or oxide would precipitate.

59. **a.** 1 (substitution reaction in aqueous solution); **b.** 4 (substitution of a volatile ligand); **c.** 3 (substitution on a metal ion in a more labile oxidation state [Cr(VI), d^0]); **d.** 3 (substitution on a metal ion in a more labile oxidation state [Co(II), d^7])

64. **a.** Br^-, CH_3^-, $C_6H_5^-$, and $H_2N\text{-}CH_2\text{-}CH_2\text{-}NH_2$ in A; CN^-, CO, $(CH_3)_2O$, and $SnCl_3^-$ in B. **b.** Pb^{4+} in A; Ni^{2+} in B.

67. **a.** C_6CrO_{12}; **b.** $C_6H_{16}N_6Pt$; **c.** $C_2H_{12}N_6Pt$; **d.** $C_4H_{16}Br_2N_4Pt$; **e.** $CH_{12}N_2O_8U$.

74. **a.** Diamminedichloroplatinum(II); **b.** potassium carbonylpentacyanoferrate(III) or (2-); **c.** tetraphenylarsonium tetracarbonylferrate(-II) or (2-); **d.** benzenebis(triphenylphosphine)iron; **e.** tetraethylammonium hexachlorochromate(III) or (3-); **f.** ammineaquabromochloro-fluoroiodoplatinum(IV); **g.** bis(benzene)chromium(I) or (1+) iodide.

77. **a.** Potassium hexaacyanoferrate(II), potassium hexacyanoferrate(III), iron(III) hexacyanoferrate(III), iron(II) hexacyanoferrate(II), and potassium iron(II) hexacyanoferrate(III) *or* potassium iron(III) hexacyanoferrate(II); **b.** The cyano groups are acting as bridging ligands, using their other donor atom, nitrogen. **c.** Prussian blue contains both Fe^{3+} and Fe^{2+} ions, between which electrons (charge) can be transferred readily upon by light absorption; charge-transfer absorptions are especially intense (Sec. 6.4).

Chapter 9

2. Object **a.** C_4, no i, yes σ, no S_n; **b.** C_5, no i, yes σ, no S_n; **c.** C_3, no, no, no; **d.** C_4, no, no, no S_4; **e.** C_3, yes, yes, yes; **f.** C_5, yes, yes, yes (S_{10}); **g.** C_5, no, no, no; **h.** C_5, yes, yes, yes (S_{10}).

6. **a.** C_{4v}; **b.** C_{5v}; **c.** C_3; **d.** C_4; **e.** D_{3d}; **f.** I_h; **g.** D_5; **h.** I_h.

9. **a.** O_h; **b.** C_{4h}; **c.** C_{2v}; **d.** D_3.

11. **a.** D_{6h}; **b.** D_{3d}.

17. **a.** No for chirality, yes for polarity; **b.** no, yes; **c.** yes, yes; **d.** yes, yes; **e.** no, no; **f.** no, no; **g.** yes, no; **h.** no, no.

23. **a.** 2 π electrons = C_s = 4 NMR peaks = *dihapto*-cyclooctatetraene; 4 π electrons = C_s = 4 NMR peaks = *tetrahapto*-cyclooctatetraene; 6 π electrons = C_s = 4 NMR peaks = *hexahapto*-cyclooctatetraene; 8 π electrons = C_{8v} = 1 NMR peak = *octahapto*-cyclooctatetraene; **b.** one peak at $\delta = 5.55$ ppm; **c.** η^4.

28. **a.** (Previously answered as Exercise 8.3a) f_{xyz}; $(f_{x^3}, f_{y^3}, f_{z^3})$; $f_{y(x^2-z^2)}, f_{x(z^2-y^2)}, f_{z(y^2-x^2)}$; **b.** f_{xyz}: A or B, since singly degenerate (consult character table operations to choose between these); others, T, since in triply degenerate sets; **c.** u.

33. **a.** $(p_x, p_y, p_z) = T_2$; $(d_{z^2}, d_{x^2-y^2}) = E$; $(d_{xz}, d_{yz}, d_{xy}) = T_2$; **b.** A_1.

37. $p_z = A_{2u}$; $(p_x, p_y) = E_u$; $d_{z^2} = A_{1g}$; $(d_{x^2-y^2}, d_{xy}) = E_g$; $(d_{xz}, d_{yz}) = E_g$.

39. **a.** 15; **b.** T_{1u}; **c.** A_{1g}, E_g and T_{2g}; **d.** C_{3v}; Yes; C_{3v} is a much lower symmetry point group, so that fewer of the vibrational bands fall in IR- and Raman-inactive irreducible representations.

44. SO_3^{2-}, since it is pyramidal and in a lower symmetry point group (C_{3v}) than planar SO_3 (D_{3h}).

Chapter 10

4. **a.** $\sigma_g{}^2\sigma_u{}^{*2}\sigma_g{}^2\pi_u{}^2$, bond order 2, 2 unpaired e$^-$; **b.** $\sigma_g{}^2\sigma_u{}^{*2}\sigma_g{}^2\pi_u{}^3$, bond order 2.5, 1 unpaired e$^-$; **c.** $\sigma_g{}^2\sigma_u{}^{*2}\sigma_g{}^2\pi_u{}^4\pi_g{}^{*3}$, bond order 1.5, 1 unpaired e$^-$; **d.** $\sigma_g{}^2\sigma_u{}^{*2}\sigma_g{}^2\pi_u{}^4\pi_g{}^{*4}$, bond order 1, 0 unpaired e$^-$; **e.** $\sigma_g{}^2$, bond order 1, 0 unpaired e$^-$.

6. **a.** $\sigma_g{}^2\pi_u{}^4\delta_g{}^2$, bond order 4, 2 unpaired e$^-$, SOMOs = δ_g; **b.** $\sigma_g{}^2\pi_u{}^4\delta_g{}^4$, bond order 5, 0 unpaired e$^-$, HOMO = δ_g, LUMO = $\sigma_g(5s)$; **c.** $\sigma_g{}^2\pi_u{}^4\delta_g{}^4\sigma_g(5s)^2\delta_u{}^{*4}\sigma_u{}^*(5s)^2\pi_g{}^{*2}$, bound order 2, 2 unpaired e$^-$, SOMOs = $\pi_g{}^*$; **d.** $\sigma_g{}^2\pi_u{}^4\delta_g{}^4\phi_u{}^4\phi_g{}^{*4}$, bound order 5, 0 unpaired e$^-$, HOMO = $\phi_g{}^*$, LUMO = $\delta_u{}^*$; **e.** $\sigma_g{}^2\pi_u{}^4\delta_g{}^2$, bound order 4, 2 unpaired e$^-$, SOMOs = δ_g; **f.** $\sigma_g{}^2\pi_u{}^4\delta_g{}^4\phi_u{}^4\phi_g{}^{*4}\delta_u{}^{*4}$, bound order 3, 0 unpaired e$^-$, HOMO = $\delta_u{}^*$, LUMO = $\pi_g{}^*$.

10. **a.** N_2 has a bond order of 3 while N_2^+ has a bond order of only 2.5, so has the longer and weaker bond; Cl_2 has a bond order of 1 while Cl_2^+ loses an antibonding electron to give a bond order of 1.5, hence has the shorter and stronger bond. **b.** s and p orbitals decrease their overlap going down a group in the p block, while d orbitals increase their overlap going down a group in the d block (this trend is helped by relativistic effects). **c.** The Group 14(IV) diatomic molecules involve π bonding, which decreases more rapidly down a group than does the σ bonding in the Group 17(VII) diatomic molecules.

17. **a.** Orbitals of the same irreducible representation: $\sigma_g(d_{z^2})$ with $\sigma_g(5s)$, also $\sigma_u{}^*(d_{z^2})$ with $\sigma_u{}^*(5s)$, if the parent d_{z^2} and $5s$ orbitals are of comparable energy, which is more likely at the left side of the d block, where the Z^* values are similar. **b.** $\sigma_g(f_{z^3})$ with $\sigma_g(d_{z^2})$ and $\sigma_g(s)$; $\pi_u(f_{xz^2}, f_{yz^2})$ with $\pi_u(d_{xz}, d_{yz})$; $\delta_g(f_{xyz}, f_{z(x^2-y^2)})$ with $\delta_g(d_{xy}, d_{x^2-y^2})$.

22. **a.** $\sigma^2\sigma_{nb}{}^2\pi^4\sigma_{nb}{}^2\pi^{*1}$, with bond order 2.5; **b.** weaker; **c.** longer; **d.** the photoelectron spectrum of NO should have an additional band with fine structure due to ionization of the π^* electron.

25. **a.** $\eta = \frac{1}{2}(\text{IE} - \text{EA}) = \frac{1}{2}(15.58 + 2.2) = 8.89$ eV, so CO (7.9 eV) is softer. **b.** CO; **c.** CO; **d.** CO, CO.

29. During the attachment of another electron (responsible for the electron affinity) C_2 gains an electron in its relatively low-energy LUMO, $\sigma_g{}^{nb}$, while O_2 and N_2 each gain an electron in a relatively high-energy orbital, $\pi_g{}^*$.

32.

Orbital	+ Overlaps	– Overlaps	Net Overlaps	
① │ ❷ │ ③ │ ❹ │ ⑤	0	4	−4	$\sigma_g{}^*$
① │ ❷ │ ④ │ ❺	0	2	−2	$\sigma_u{}^*$
① │ ❸ │ ⑤	0	0	0	$\sigma_g{}^{nb}$
① ② │ ❹ ❺	2	0	+2	σ_u
① ② ③ ④ ⑤	4	0	+4	σ_g

The energy level diagram has five equally spaced levels.

37.

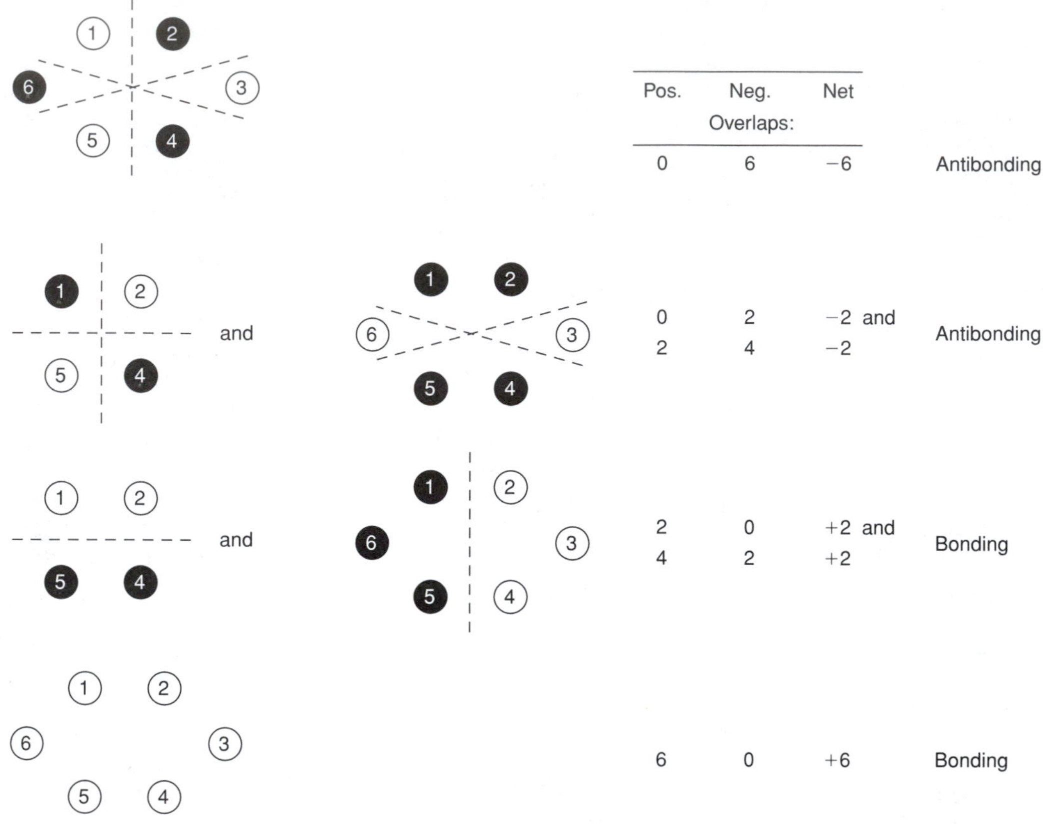

Pos.	Neg.	Net	
Overlaps:			
0	6	−6	Antibonding
0 2	2 4	−2 and −2	Antibonding
2 4	0 2	+2 and +2	Bonding
6	0	+6	Bonding

The energy level diagram has four equally spaced levels; the middle two levels are each doubly degenerate.

42. a_2'' orbital overlaps with metal s, p_z, d_{z^2}, and f_{z^3} orbitals, and $C_5H_5^-$ is acting as a π-donor ligand; e_1'' orbitals overlap with metal (p_x, p_y), (d_{xz}, d_{yz}), and (f_{xz^2}, f_{yz^2}) orbitals, and $C_5H_5^-$ is acting as a π donor ligand; e_2'' orbitals overlap with metal $(d_{xy}, d_{x^2-y^2})$ and $[f_{xyz}, f_{z(x^2-y^2)}]$ orbitals, and $C_5H_5^-$ can act as a π-acceptor ligand.

48.

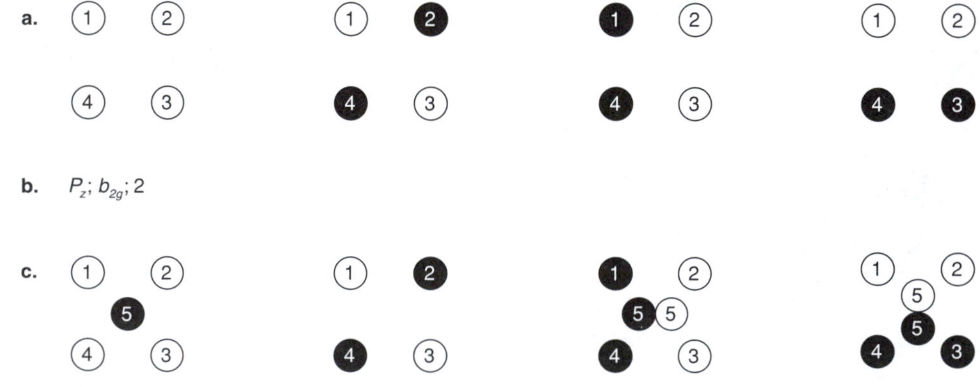

d. (Reading from the bottom of the figure) $a_{1g}(\uparrow\downarrow)e_u(\uparrow\downarrow\uparrow\downarrow)a_{2u}{}^{nb}(\uparrow\downarrow)b_{2g}{}^{nb}(\uparrow\downarrow)e_u{}^*(\uparrow\uparrow)a_{1g}{}^*$

e. 2; $\frac{1}{2}$.

56. (a) +10, (b) +2, (c) +2, (d) −6, (e) +6, (f) −2, (g) −2, (h) −10 net overlaps. The energy level diagram has six equally-spaced levels, with the two center ones each being doubly degenerate.

Chapter 11

1. **a.** Four ligands are needed for a square planar complex. These eight electrons leave eight valence electrons (d^8) to be supplied by the metal. **b.** Neutral metal atoms: Fe, Ru, Os; +1 ions: Co^+, Rh^+, Ir^+; +2 ions: Ni^{2+}, Pd^{2+}, Pt^{2+}; +3 ions: Cu^{3+}, Ag^{3+}, Au^{3+}.

4. **a.** 18, yes; **b.** 18, yes; **c.** 16, no; **d.** $8 + 8 + 2 = 18$, yes; **e.** $6 + 8 + 4 = 18$, yes; **f.** $6 + 10 + 2 = 18$, yes.

8. **a.** +1, pentacarbonylcobalt(I) ion, trigonal bipyramid; **b.** −1, pentacarbonylcobaltate(1−) ion, tetrahedral; **c.** +1, pentakis(methyl isonitrile)cobalt(1+) ion, trigonal bipyramid; **d.** 0, tetracarbonyl(triphenylphosphine)iron, trigonal bipyramidal with *equatorial* bulky phosphine.

14. $V(CO)_6$ has 17 valence electrons and is d^5; the maximum coordination number of 6 prevents its dimerization. $[Pt(CO)_4]^{2+}$ has 16 valence electrons and is d^8, for which square planar is preferred; $[Cu(CO)_n]^+$ and $[Hg(CO)_2]^{2+}$ have 16 or 14 valence electrons and are d^{10}; relativistic effects may make some *p* orbitals unavailable for bonding.

15. For example:

a. $NbCl_5 + 6\ Na + 6\ CO + 2$ diglyme $\rightarrow$ $[Na(diglyme)_2][Nb(CO)_6] + 5\ NaCl$

b. $Fe + 5\ CO \rightarrow Fe(CO)_5$

c. $Fe(CO)_5 + PPh_3 \xrightarrow{h\nu} Fe(CO)_4(PPh_3) + CO$

d. $CoCl_2 + 5\ P(OPh)_3 + \frac{1}{2} H_2 + AgPF_6 \rightarrow \{Co[P(OPh)_3]_5\}[PF_6] + HCl + AgCl$

e. $PtCl_4{}^{2-} + H_2 + 2\ P(t\text{-}Bu)_3 \rightarrow Pt[P(t\text{-}Bu)_3]_2 + 2\ HCl + 2\ Cl^-$

18. Since CO is high in the spectrochemical series, Δ_o is large, and UV rather than visible light is absorbed. Since visible light is not absorbed by the complex, it cannot initiate the dissociation of the ligand.

22. More than the usual M–C≡O, we would emphasize the following: **a.** Ti=C=O; **b.** Cr≡C–O⁻; **c.** Pt–C≡O.

24. **a.** 20 electrons; **b.** 20 electrons; **c.** 23 electrons; **d.** 18 electrons.

28. **a.** 2 C_6H_6 have 12 electrons, so M should have 6: Cr, Mo, W; **b.** 2 Cp^- have 12 electrons (AB method), so M^{2+} should have 6: Fe, Ru, Os; **c.** C_6H_6 and Cp^- should total 12 electrons, so M^+ should have 6: Mn, Tc, Re; **d.** 2 C_4H_4 have 8 electrons, so M should have 10: Ni, Pd, Pt. **e.** $C_7H_7{}^+$ and C_6H_6 have 6 electrons each, so M^- should have 6: V, Nb, Ta; **f.** $C_3H_3{}^+$ has 2 electrons, and Cp^- has 6, so M should have 10 electrons: Ni, Pd, Pt.

32. $[(Ti,Hf)(\eta^5\text{-}Cp)_2(\eta^1\text{-}Cp)_2]$ = 8-coordinate with 16 valence electrons; $[Zr(\eta^5\text{-}Cp)_3(\eta^1\text{-}Cp)]$ = 10-coordinate with 20 valence electrons; $[(Th,U,Np)(\eta^5\text{-}Cp)_4]$ = 12-coordinate with 24 valence electrons (Th), 26 (U), 27 (Np).

37. **a.** Six-coordinate Mn^I with 18 valence electrons (d^6) in pentacarbonylchloromanganese(I) (or pentacarbonylmanganese chloride, common name); **b.** seven-coordinate Fe(III) (AB counting; Fe^0 by RR counting) with 18 valence electrons (d^5 by AB counting, d^8 by RR counting) in diiron nonacarbonyl (common name); **c.** five-coordinate Co^0 with 18 valence electrons (d^9) in dicobalt octacarbonyl (common name); **d.** four-coordinate Ni(II) with 16 valence electrons (d^8) in dichlorobis(triphenylphosphine)nickel(II); **e.** four-coordinate Pd(II) with 16 valence electrons (d^8) in dicarbonyldichloropalladium(II) or dicarbonylpalladium dichloride (common name).

42. **a.** Taking NO as NO^+ (LAB method): Cr(-IV); Mn(-III); Fe(-II); Co(-I); Ni(0) (by the RR method, all have oxidation number 0); **b.** Taking NO as NO^+ and CN as CN^- (LAB method): Cr^0; Mn(I); Fe(II).

47. **a.** $[Na(diglyme)_2][Mn(CO)_5]$; **b.** $[Na(diglyme)_2]_2[Fe(CO)_4]$; **c.** $[Na(diglyme)_2][Co(CO)_4]$; **d.** $[Na(diglyme)_2][CpFe(CO)_2]$.

49. The only −1 anion X^- from Table 11.1 (or 11.6) not forming an oxidized dimer X_2 is $[V(CO)_6]^-$, which does not form $V_2(CO)_{12}$ since the latter would have seven-coordinate V.

58. **a.** $Pt[P(t\text{-}Bu)_3]_4$; **b.** $[IrCl(PPh_3)_3] + Cl_2$; **c.** $[MeC(=O)Fe(CO)_4]^-$; **d.** $[Fe(CO)_4]^{2-} + [(CH_3)_3O]^+$; **e.** $[Fe(CO)_4]^{2-} + Hg^{2+}$ (a cyclic compound would probably result in order to give the coordination numbers of 2 to Hg^{2+} and 6 to Fe); **f.** *cyclo*-$C_4H_4Cl_2 + Fe(CO)_5 \rightarrow [(\eta^4\text{-}C_4H_4)FeCl_2(CO)_2] + 3\ CO$; **g.** $[Cp_2TiMe_2] + Me^-Li^+ \rightarrow CH_4$ + product.

Chapter 12

1. **a.** SO_3 is monomeric (molecular); S_3O_9 is oligomeric; $\frac{1}{\infty}[SO_3]$ is (chain or linear) polymeric. **b.** $SO_3 < S_3O_9 < \frac{1}{\infty}[SO_3]$; SO_3 is most likely a gas.

5. **a.** Highest is C, lowest is Pb. Smaller atoms give the strongest covalent bonds, which must be broken in order to atomize an element. **b.** Highest is Ta, lowest is V. The 5*f* orbitals overlap each other better than 3*f* orbitals do. **c.** Highest is C, lowest is Ne. Carbon can participate in four covalent bonds in its prepared-for-bonding electron configuration $4s^1 4p^3$; Ne can form none since all its valence orbitals are full. **d.** Highest is Mo, lowest is Xe (for same reason as Ne, above). Molybdenum can participate in as many as six covalent bonds in its prepared-for-bonding electron configuration $5s^1 5d^5$.

9. For structures see Figures 12.8, 12.4, and 12.9 (for a metal such as Mo). **a.** hard-to-vaporize solid; **b.** easily vaporized solid; **c.** hard-to-vaporize solid; **d.** hard-to-vaporize solid.

12. **a.** All macromolecular forms (B, diamond, graphite, Si, Ge, black and red P, gray As, gray Sb, Bi, polymeric S, helical Se, Te). Reason: to boil, these must be converted to small molecules by rupturing covalent bonds. **b.** The smallest of the small molecules: He, Ne, Ar, Kr, Xe, H_2, F_2, Cl_2, Br_2, O_2, O_3, N_2. Reason: These molecules are held together in the liquid or solid state only by very weak van der Waals forces. **c.** The larger of the small molecules: I_2, S_8, S_6 and so on, Se_8, P_4, As_4, Sb_4. Reason: Having more electrons, these molecules are held together in the liquid or solid state by stronger van der Waals forces.

14. **a.** A: boron; B: S or Se or Te; C: C or Si or Ge or Sn; D: N or O or H or halogens or P or As. **b.** D < B < A < C.

18. +210 kJ mol^{-1}. O_2.

22. High temperature and low pressure. High temperature increases the importance of the $T\Delta S$ term in $\Delta G = \Delta H - T\Delta S$. The multiply bonded allotropes involve larger numbers of smaller molecules than do the singly bonded allotropes, so their formation increases the disorder. Low pressures will favor the gas state, which (at low temperatures) the smaller multiply bonded molecules will adopt before the larger singly bonded ones. This also gives a favorable entropy effect.

24. **a.** One σ and one π; **b.** graphite: 1.5 σ and 0.5 π. **c.** This reaction converts 0.5 π bonds per atom to 0.5 σ bonds per atom; for six atoms $\Delta H = 3(256) - 3(346) = -270$ kJ. Yes.

28. **a.** Mo; **b.** B; **c.** C; **d.** Gd; **e.** RhZr; **f.** As.

33. **a.** Ti $4s^1 3d^3$, Fe $4s^1 3d^7$ or $4s^2 3d^6$, Hf $6s^1 5d^3$, Os $6s^1 5d^7$ or $6s^2 5d^6$; **b.** (ignoring principal quantum numbers) TiFe $s^1 d^5$, TiHf $s^1 d^3$, FeOs $s^1 d^7$ or $s^2 d^6$, HfOs $s^1 d^5$; **c.** TiFe 6, TiHf 4, FeOs 4, HfOs 6. **d.** HfOs.

36. Zintl alloys, since the electronegativity differences are greater between their component metals.

38. Impossible for C_2; plausible for Zr_2. A quadruple bond must include a δ bond, which requires overlap of d orbitals.

40. **a.** Graphite two-dimensional polymer; diamond three-dimensional polymer; C_{60} and C_{70} oligomers; **b.** C_{60} and C_{70}, since no strong forces (chemical bonds) need be broken to separate the clusters. **c.** Interstellar space: C_{60} (low pressure should favor the smaller molecules, which are volatile); volcano cones: diamond (high pressures favor the most dense, most polymerized form). **d.** Interstellar space; bond order 2, or one per C atom, which is lower than all other allotropes, so less bond energy is released in forming this allotrope from the atoms of C.

44. $a_g = s$; $t_{1u} = p$; $h_g = d$; $t_{2u} + g_u = f$; $g_g + h_g = g$; h_u = some h orbitals (presumably).

48. Longer, since the LUMOs of graphite would be antibonding (π*) orbitals.

50. It is likely a complex containing four peroxide (O_2^{2-}) ions coordinated to a Cr^{5+} central ion.

56. **a.** This $n = 7$ cluster, as a neutral species, has 95 valence electrons; to give the proper number for a pentagonal bipyramid it needs three more, so would be a -3 anion (Note: by the $14n + 2$ rule, -5 anion would be expected). **b.** This neutral cluster has 84 valence electrons, which is OK for a $n = 6$ bicapped tetrahedron: no charge. **c.** The neutral cluster has 84 valence electrons, but needs 86, so would have a -2 charge. **d.** The neutral cluster has 84 valence electrons, but needs 86, so would have a -2 charge. **e.** Square pyramidal is a nido structure, which for $n = 5$ should have $4n - 4 = 24$ valence electrons. The neutral cluster has 20, so should acquire a -4 charge. **f.** $B_5H_5^{4-}$ is the smallest and most highly charged, so would be most basic. The fully protonated form would be B_5H_9.

59. **a.** BH^-; three electrons in three orbitals. **b.** Replacing two BH^- in $B_{12}H_{12}^{2-}$ with two CH gives, for example, $B_{10}H_{10}C_2H_2$. **c.** Three. **d.** "ortho" is C_{2v}; "meta" is C_{2v}; "para" is D_{5d}.

Chapter 13

1. **a.** 0.420; near the border between four and six-coordination. (Use 6 for remainder of problem.) **b.** Be 4, Si 2, P 1, Se 0, Br -1. **c.** Be polymeric, Si oligomeric or polymeric, P oligomeric, Se monomeric, Br nonexistent. **d.** Be 1000°, Si 1000° or 75°, P 75°, Se -50°.

4. **a.** $CO_2(g) < P_4O_{10} < Na_2O < Cr_2O_3$, SiO_2; **b.** CO_2 (g) $\ll Rb_2O < SrO < Y_2O_3 < ZrO_2$.

7. Small molecules: CO_2, SO_2. Oligomeric molecules: P_4O_{10}. Ionic or macromolecular: the rest.

12. **a.** The radius ratio $C^{4+}/O^{2-} = 30/126 = 0.238$, which is close to the 0.225 value predicted to divide species of carbon coordination number 3 and coordination number 4. For Si^{4+}, the radius ratio is $54/126 = 0.429$, which is close to the 0.414 boundary value between coordination numbers 4 and 6.

Carbon actually has the coordination number 2 in the doubly bonded CO_2, while Si has the coordination number 4 in quartz.

b. The coordination number of oxygen is 1 in CO_2; hence CO_2 is a monomeric gas. The coordination number of oxygen is 2 in quartz, which means that it bonds one silicon to another in a hard-to-disrupt, high-melting polymer.

c. The other likely coordination number of Si, from the radius ratio calculations in **(a)** is 6. With this coordination number, stishovite would be expected to adopt the rutile (TiO_2) lattice type.

d. For quartz $U = (138{,}900)(2.201)(+4)(-2)(1 - 1/7)(1/180) = -11{,}650$ kJ mol^{-1}; for stishovite $U = 138{,}900(2.408)(+4)(-2)(1 - 1/7)(1/180) = -12{,}740$ kJ mol^{-1}.

e. These calculations appear to suggest that stishovite is the thermodynamically favored form, which it is not. The rarity of coordination number 6 in oxygen compounds of silicon suggests that it represents too close a crowding of oxygens around silicon. Hence the true Si–O distance is likely larger than 180 pm due to oxygen–oxygen repulsions. This would, in turn, reduce the lattice energy of stishovite. (To the extent that the bonding in SiO_2 is covalent, these calculations would not be highly accurate anyway.) Forcing six oxygens close together around small oxygen would require high pressure, such as would be found when a meteor hit the surface of the earth.

14. **a.** moderate, selenous acid, SeO_2, soluble; **b.** weak, telluric acid, TeO_3, insoluble; **c.** very strong, permanganic acid, Mn_2O_7, soluble; **d.** very very strong, perplatinic acid, PtO_5, soluble; **e.** moderate, phosphoric, P_4O_{10}, soluble; **f.** strong, perxenic acid, XeO_4, soluble; **g.** weak, stannous acid, SnO, insoluble; **h.** moderate, vanadic acid, V_2O_5, soluble; **i.** strong, xenic acid, XeO_3, soluble; **j.** very strong, pertechnetic acid, Tc_2O_7, soluble.

15. **a.** basic and soluble; **b.** amphoteric or neutral and insoluble; **c.** acidic and soluble; **d.** acidic and insoluble; **e.** acidic and soluble; **f.** basic and soluble; **g.** basic and soluble; **h.** acidic and soluble; **i.** basic and insoluble—moderately acidic cations often (but not always) can also form hydroxo anions [e.g. $Al(OH)_4^-$] in strongly basic solutions; hence, this oxide may also be amphoteric.

22. **a.** $Tl_2O + H_2O \rightarrow 2\,Tl^+(aq) + 2\,OH^-(aq)$; **b.** $I_2O_5 + H_2O \rightarrow 2\,H^+(aq) + 2\,IO_3^-(aq)$; **c.** $ClO_2 + 2\,OH^- \rightarrow ClO_3^-(aq) + ClO_2^-(aq) + H_2O$; **d.** $La_2O_3 + 6\,H^+(aq) \rightarrow 2\,La^{3+}(aq) + 3\,H_2O$; **e.** $B_2O_3 + 2\,OH^-(aq) + 3\,H_2O \rightarrow 2\,[B(OH)_4]^-(aq)$; **f.** $6\,FeO + P_4O_{10} \rightarrow 2\,Fe_3(PO_4)_2$; **g.** $MnO + 2\,H^+(aq) \rightarrow Mn^{2+}(aq) + H_2O$.

24. **a.** $SrO + H_2O \rightarrow 2\,Sr^{2+} + 2\,OH^-$; **b.** $MoO_3 + H_2O \rightarrow 2\,H^+(aq) + MoO_4^{2-}(aq)$; **c.** RuO_4; **d.** RuO_4; **e.** ZrO_2, $ZrO_2(s) + 2\,SO_3 \rightarrow Zr(SO_4)_2(s)$; (f) $SrO + MoO_3 \rightarrow SrMoO_4$.

28. **a.** $S + O_2 \rightarrow SO_2$; $SO_2 + \frac{1}{2}O_2 \rightarrow SO_3$; $SO_3 + H_2O \rightarrow H_2SO_4$. **b.** $MgO(s) + SO_2(g) \rightarrow MgSO_3(s)$.
c. 1,000,000 g coal × 3.2 g S/100 g coal × 1 mol S/32.0 g S × 1 mol SO_2/1 mol S × 1 mol MgO/1 mol SO_2 × 40.3 g MgO/1 mol MgO = 40,300 g MgO.

33. $Fe_{0.95}O$, $Co_{0.95}O$, $Eu_{0.95}O$.

34. **a.** $NiFe_2O_4$, Ni_3O_4; **b.** $BaTiO_3$, $NaTaO_3$

39. **a.** Clearly shown are one Y and two Ba ions. There are eight Cu ions shown on top and bottom corners, each shared with eight other cells, so this gives one Cu ions (in the chains). There are also eight Cu ions shown along the edges, which are each shared with four other cells, so this gives two more Cu ions (in the planes). There are a total of six O ions along edges shown in the top cube and six more in the bottom cube; these are shared among four cells to give a total of three O ions. There are a total of eight O ions within faces, each shared by two cells, for four more O ions, for a grand total of seven O ions; hence $YBa_2Cu_3O_7$. **b.** The stoichiometry works out properly for the composition $Y^{3+}(Ba^{2+})_2(Cu^{3+})(Cu^{2+})_2(O^{2-})_7$, for example, with one-third of the Cu ions being Cu^{3+}. Hence, if these were ordered, these would be the ones in the chains.

43. **a.** Monticellite; **b.** wollastonite; **c.** grunerite; **d.** talc; **e.** monticellite; **f.** stishovite.

47. **a.** ${}^{1}_{\infty}[Si_4O_{11}]^{6-} =$

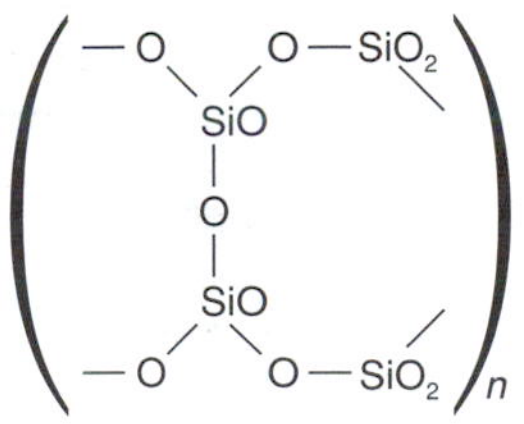

—O O—SiO₂
SiO
O
SiO
—O O—SiO₂ n

${}^{2}_{\infty}[Si_4O_{10}]^{4-} =$

—O O—SiO
SiO
O
SiO
—O O—SiO
O
n

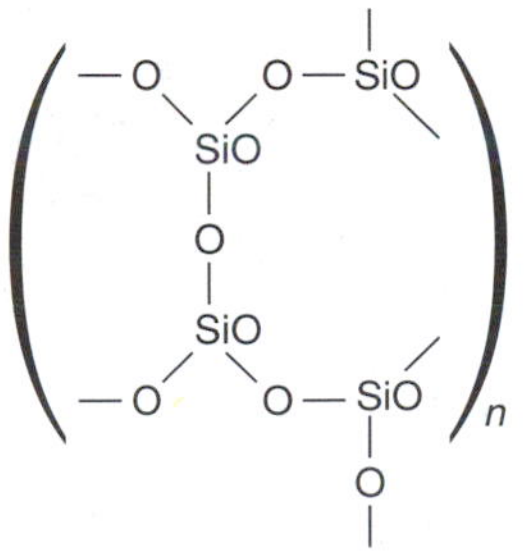

${}^{1}_{\infty}[SiO_3]^{2-} = (\text{–O–}SiO_2\text{–})_n$; $(SiO_4)^{4-}$ = as written;

b. Tremolite $(SiO_{2.75})^{1.5-}$, ureyite $(SiO_3)^{2-}$, chabazite $(EO_2)^{0.33-}$, kaolinite $(SiO_{2.5})^-$, phenacite $(SiO_4)^{4-}$; **c.** chabazite anion $[Al_{12}Si_{24}O_{72}]^{12-}$ (−1/3 charge per nucleus) < kaolinite anion $(Si_4O_{10})^{4-}$ (−1 charge per nucleus) < tremolite anion $(Si_4O_{11})^{6-}$ (−1.5 charge per nucleus) < ureyite anion $(SiO_3)^{2-}$ (−2 charge per nucleus) < phenacite anion $SiO_4{}^{4-}$ (−4 charge per nucleus). **d.** Al is part of the aluminosilicate anion in chabazite but is an external counterion (in a cationic layer) in kaolinite.

51. **a.** No—cannot substitute 2 Y^{3+} for 2 Ca^{2+}. **b.** No—cannot substitute 2 Na^+ for 2 Ca^{2+}. **c.** Yes. You can substitute the similarly sized 2 Na^+ for 2 Ca^{2+} if you simultaneously substitute the similarly sized 2 Fe^{3+} for 2 Mg^{2+}. **d.** Yes. You can substitute the similarly sized 2 Y^{3+} for 2 Ca^{2+} if you simultaneously substitute the similarly-sized 2 Li^+ for 2 Mg^{2+}. **e.** Yes. You can substitute the similarly sized 2 Sr^{2+} for 2 Ca^{2+}.

54. **a.** (1); **b.** (4); **c.** (1); **d.** (2); **e.** (3).

58. To "destroy" the asbestos you would want to convert it to a nondangerous form, one without the double-chain structure. This could be done, for example, by: (a) heating it with a basic oxide to make a less-polymerized structure: $Mg_3(Si_4O_{11}) + 5\ MgO \rightarrow 4\ Mg_2SiO_4$; (b) heating it with an acidic oxide to make a more-polymerized structure: $Mg_3(Si_4O_{11}) + SiO_2 \rightarrow Mg_2(Si_4O_{10}) + MgSiO_3$.

60. **a.** Simple silicates; **b.** simple silicates; **c.** three-dimensional polymeric metal oxides.

63. B, (Al,) Si, P, (S), V, Nb, Ta, (Cr,) Mo, W. (Many other elements can be added if one includes heteropolymolybdates and -tungstates.)

Chapter 14

2. As one proceeds to the right in the series, the total coordination number of the formula unit (i.e., the number of halides plus one for the lone pair of electrons) gets closer to the maximum coordination number (six or perhaps higher). Hence, the degree of polymerization drops off. InI and $SnCl_2$ are macromolecular; $SbBr_3$ is macromolecular only via weak secondary bonds; $TeCl_4$ forms an oligomer Te_4Cl_{16}; and IF_5 is a monomeric molecule.

4. **a.** Molecular; **b.** macromolecular (or ionic); **c.** ionic; **d.** molecular; **e.** macromolecular or ionic; **f.** molecular.

8. **a.** Gas; **b.** solid with high mp; **c.** solid with high mp; **d.** liquid or solid with low mp; **e.** liquid or solid with low mp (low mp due to polymerization being by weak secondary bonds only); **f.** liquid or solid with low mp.

13. Neither. The normal trend for ionic compounds would be for $GeCl_4$ to have the higher melting point due to the higher lattice energy with a 4+ ion. The normal trend for monomeric covalent compounds would be for $GeCl_4$ to have the higher melting point due to stronger van der Waals forces when more electrons are present. Actually, $GeCl_4$ is monomeric covalent but $GeCl_2$ does not meet a reasonable coordination number for Ge, so is macromolecular. Hence, $GeCl_2$ has the higher melting point.

15. A. $SnCl_2$; B. HgI_2; C. NF_3; D. $CrCl_3$; E. $SnCl_4$; F. NI_3.

18. **a.** Method 4; **b.** Method 2; **c.** Method 1; **d.** Method 4; **e.** in principle Method 3, but since $BiCl_3$ can be obtained from solution in anhydrous form, Method 1 can be used *with a large excess of HCl*; **f.** Method 4.

20. Among the possible answers: **18a.** $CO_2 + SF_4 \rightarrow CF_4 + SO_2$; **18b.** $Ag^+(aq) + I^-(aq) \rightarrow AgI(s)$; **18c.** $KOH(aq) + HF(aq) \rightarrow$ (after evaporation) KF(s); **18d.** $Si + 2\ Cl_2 \rightarrow SiCl_4$; **18e.** Bi_2O_3 + large excess of HCl $\rightarrow$ (after evaporation) $BiCl_3(s)$; **18f.** $Si + 2\ I_2 \rightarrow SiI_4$; **19a.** Although $Au^+(aq) + I^-(aq)$ should work, Au^+ oxidizes water, so this is impractical; instead $2\ Au + I_2 \rightarrow 2\ AuI$; **19b.** $RbOH(aq) + HBr(aq) \rightarrow$ (after evaporation) RbBr(s); **19c.** $NiCl_2 \cdot 6H_2O + 6\ SOCl_2 \rightarrow NiCl_2 + 6\,SO_2 + 12\ HCl$; **19d.** $BiBr_3$ analogous to $BiCl_3$; **19e.** $La^{3+}(aq) + 3\ F^-(aq) \rightarrow LaF_3(s)$ (could precipitate as a hydrate that would have to be dehydrated; evidently does not); **19f.** $SnCl_2 \cdot 2H_2O + 2(MeCO)_2O \rightarrow SnCl_2 + 4\ MeCO_2H$; **19g.** $PrCl_3 \cdot 7\,H_2O + 7\ SOCl_2 \rightarrow PrCl_3 + 7\,SO_2 + 14\ HCl$; **19h.** $2\ WO_3 + 3\ SF_4 \rightarrow 2\ WF_6 + 3\ SO_2$.

25. **a.** neither; **b.** Lewis acid (if Ag^+ is first classified as a weak acid; if first classified as feebly acidic, neither); **c.** Lewis base; **d.** Lewis acid; **e.** Lewis acid; **f.** Lewis acid.

29. **a.** $PF_5 + CsF \rightarrow Cs[PF_6]$; **b.** $XeF_6 + CF_4 \rightarrow$ NR; **c.** $RbBr + BaBr_2 \rightarrow$ NR; **d.** $XeF_6 + BF_3 \rightarrow [XeF_5][BF_4]$; **e.** $3\ NaF + AlF_3 \rightarrow Na_3[AlF_6]$.

35. **a.** $SO_2 + Cl_3Si{-}O{-}SiCl_3$; **b.** $Ph_3P + Br_2B{-}O{-}BBr_2$; **c.** $Cl_3Si{-}CH{=}CH{-}SiCl_3$; **d.** 2 $HgBr_2$.

36. **a.** 4; **b.** 3.5; **c.** 2; **d.** 1.5.

41. For purposes of these calculations, we expect μ_2- and μ_3- halide ions to have enough electrons to donate a pair to each metal atom bridged. In the multiply bonded dimers, the s, p_x, p_y, and d_{xy} orbitals are used to bond the halide ligands, so the metal-metal bonding can involve the other four d orbitals, hence give a maximum of one δ bond. **a.** The individual metal Lewis acid is Mo^{2+}, with a d^4 electron configuration, so this contributes 4 valence electrons. The metal–metal electron configuration is $\sigma^2\pi^4\delta^2$, for a quadruple Mo–Mo bond, so the four bonds to the neighboring Mo atom contribute 4 valence electrons. Each Mo^{2+} has four terminal Cl^- ligands, which contribute 8 valence electrons, for a total of **16** valence electrons. **No. b.** Os^{3+} is d^5, so contributes 5 valence electrons. The metal–metal electron configuration is $\sigma^2\pi^4\delta^2\delta^{*2}$, for a triple Os–Os bond, so the three bonds to the neighboring Os atom contribute 3 valence electrons. Each Os^{3+} has four terminal Cl^- ligands, which contribute 8 valence electrons, for a total of **16** valence electrons. **No. c.** [See Fig. 14.8(*a*)]. The ion Re^{3+} is d^4, so it contributes 4 electrons. Each Re is double bonded to two neighbors, so it gets 4 more electrons from these four metal–metal bonds. Two terminal Cl^- ions contribute 4 electrons; two bridging Cl^- contribute 4 electrons. The total is **16** electrons; No. **d.** Adding an outer ligand brings the total to **18** electrons; Yes. **e.** Mo^{2+} is d^4, so contributes 4 electrons. Four single bonds to other Mo atoms contribute 4 electrons. Four apical Cl^- ions about each Mo contribute 8 electrons, for a total of **16** valence electrons. No. **f.** Adding the terminal Cl^- gives a total of **18** valence electrons; **Yes. g.** W^{3+} has 3 valence electrons. If we assume (contrary to the text) that each of its four W–W links is a single bond, each W thereby acquires 4 more valence electrons. The four bridging Cl^- ions per W contribute 8 electrons, for a total of **15** valence electrons. **No. h.** Adding a terminal ligand brings the total to **17** valence electrons; **No.**

45. a. ZnS is insoluble, so $Zn^{2+}(aq) + S^{2-}(aq) \rightarrow ZnS(s)$. **b.** CaS is soluble but easily hydrolyzed, so redox might be best: $CaSO_4 + 4\,C + \Delta \rightarrow CaS + 4\,CO$. **c.** K^+ is nonacidic, so $2\,K + 2\,EtOH \rightarrow H_2 + 2\,KOEt$. **d.** As^{3+} is strongly acidic, so $AsCl_3 + 3\,MeOH + 3\,Me_3N \rightarrow As(OMe)_3 + 3\,[Me_3NH]Cl$. **e.** Me_3Si^+ is very strongly acidic, so $2\,Me_3SiCl + H_2O \rightarrow$ (through a silanol intermediate) $(Me_3Si)_2O + 2\,HCl$.

50. (d) $Y(OCH_2CH_2OCH_2CH_2OMe)_3$.

52. a. Two steps. First make $VOCl_3$ by careful hydrolysis of VCl_5 or partial halogen exchange on V_2O_5 or by oxidation of VCl_3. Then $VOCl_3 + 3\,[Me_2NH_2]Cl + \Delta \rightarrow VO[NMe_2]_3 + 6\,HCl$. **b.** $3\,MeNH_2 + 3\,BCl_3 +$ HCl acceptor $\rightarrow B_3Cl_3N_3Me_3 + 6\,HCl$. **c.** $6\,S_2Cl_2 + 16\,NH_3 \rightarrow S_4N_4 + S_8 + 12\,NH_4Cl$. **d.** $[N{=}PCl_2]_3 + 6\,MeONa \rightarrow [N{=}P(OMe)_2]_3 + 6\,NaCl$.

56. a. A light-emitting diode of energy gap > GaP (say, ZnSe or AlP or ZnS); **b.** polythiazyl, $^1_\infty$[SN]; **c.** an intercalate of a layer-structure compound such as graphite.AsF_5 or NbS_2.intercalate or BN.intercalate; **d.** cisplatin bonded to polyphosphazine, $^1_\infty\{N{=}P(NH_2)[NH_2Pt(NH_3)Cl_2]\}$; **e.** Si_3N_4; **f.** a silicone $^1_\infty[Me_2SiO]$; **g.** MoS_2.

60. c, CdI_2; d, $CrCl_3$; g, graphite.

64. a. (See Fig. 14.15.) The bridging –S– and terminal =S units would logically show −2 oxidation states [S(-II)]. **b.** P_4S_3 contains three S(-II) units, a P atom forming three bonds to more electronegative S so acting as P(III), and a P_3 ring forming three such bonds, so acting as $P_3{}^{3+}$, that is, with three P atoms of +1 oxidation state; the average P oxidation state is +1.5. α-P_4S_4 has two pairs of P_2 units forming four bonds each to S atoms, hence acting as $P_2{}^{4+}$ with each P atom having a +2 oxidation state. β-P_4S_4 has a single P(III) unit and a bent P_3 unit, in which the central P atom acts as P^I and the outer P atoms as P(II); the average oxidation state again is +2. α-P_4S_5 contains a P(III) atom and a chain of three P atoms, one being P(IV), one P(I), and one P(II), for an average P oxidation state of +2.5. β-P_4S_5 contains two P(III) atoms and a dimeric pair of P(II) atoms, for an average P oxidation state of +2.5. P_4S_6 contains a pair of P(II) atoms, a P(III) atom, and a P(V) atom, for an average P oxidation state of +3. P_4S_7 contains two P(V) atoms and a pair of P(II) atoms for an average P oxidation state of +3.5. P_4S_9 contains three P(V) atoms and a P(III) atom, for an average P oxidation state of +4.5. P_4S_{10} contains P(V) atoms. **c.** NbS_3 is a polymer built from S^{2-} and $S_2{}^{2-}$ units, so the logical oxidation number for Nb is IV. **d.** II. **e.** +2.67. **f.** Assuming Pb is Pb^{2+}, +2.33.

66. a. Adamantane is $(\mu_2\text{-}CH_2)_6(\mu_3\text{-}CH)_4$, so the two isolobal amines are $(\mu_2\text{-}CH_2)_6(\mu_3\text{-}N)_4$ (hexamethylenetetramine) and $(\mu_2\text{-}NH)_6(\mu_3\text{-}CH)_4$; **b.** $(\mu_3\text{-}CH)_8$ or C_8H_8; **c.** Adamantane, which can have normal bond angles, while cubane is forced to have 90° bond angles, so has ring strain.

Chapter 15

1. a. hydrogen pentasulfide (or pentasulfane), molecular electron rich; **b.** disilane, molecular electron precise; **c.** diborane, molecular electron deficient; **d.** sodium tetrahydridoborate (common = borohydride), (electron precise) hydride anion; **e.** arsine, molecular electron rich; **f.** stannane, molecular electron precise; **g.** phosphine, molecular electron rich; **h.** tetraborane-10; molecular electron deficient; **i.** uranium(III) hydride; metallic.

5. No hydride: (e) Xe. Thermodynamically stable hydrides: (a) C, (c) O, (d) Br, (f) Sr, (g) Nd. Thermodynamically unstable hydrides: (b) As.

9. **a.** Solid since salt-like; **b.** Liquid since molecular with hydrogen bonding; **c.** Gaseous since molecular; **d.** Liquid since catenated molecular; **e.** Solid since it is a salt; **f.** Gaseous since molecular.

13. **a.** Closo-, a regular polyhedron; **b.** arachno-, very open; **c.** derived from *arachno*-B_3H_9; **d.** nido-, semiopen; **e.** nido; **f.** arachno; **g.** derived from *nido*-B_5H_9.

17. **a.** $Li + Al + H_2 \rightarrow LiAlH_4$ (or other answer); **b.** $CaF_2 + H_2SO_4 \rightarrow CaSO_4 + 2\ HF$; **c.** $N_2 + 3\ H_2 \rightarrow 2\ NH_3$; **d.** $NaOCl + 2\ NH_3 \rightarrow NaCl + H_2O + N_2H_4$; **e.** $SCl_2 + 2\ H_2S \rightarrow H_2S_3 + 2\ HCl$ (or analogous to 8f); **f.** $2\ Na_2S + S_8 \rightarrow 2\ Na_2S_5$; $Na_2S_5 + 2\ HCl \rightarrow 2\ NaCl + H_2S_5$ (or analogous to 8e); **g.** $2\ Si_2Cl_6 + 3\ LiAlH_4 \rightarrow 2\ Si_2H_6 + 3\ LiAlCl_4$ (or other reactions)

21. **a.** $CH_4 < AsH_3 < PH_3 < NH_3$; **b.** $NH_2OH < NH_2NH_2 < NH_3$

23. **a.** H^+, reactants; **b.** H^-, products; **c.** $H\cdot$, reactants; **d.** H^+, products.

28. **a.** Ph_3GeH; **b.** Ph_2GeMe_2; **c.** Al_2Me_6; **d.** Si_6Me_{12}; **e.** $SbPh_3$.

30. (For example) **a.** BMe_3; **b.** Ph_2Te_2; **c.** BMe_3; **d.** Me_5As; **e.** $BeMe_2$; **f.** K^+Me^-.

33. **a.** Direct reaction $MeI + Mg \rightarrow MeMgI$; **b.** Direct reaction $MeI + 2\ K \rightarrow MeK + KI$ may be violent; alternately $Me_2Hg + 2\ K \rightarrow 2\ MeK + Hg$; **c.** $Me_2Hg + Mg \rightarrow Me_2Mg + Hg$; **d.** Direct reaction (with Cu catalyst): $2\ MeCl + Si \rightarrow Me_2SiCl_2$; **e.** Hydroboration: $6\ CH_2{=}CH_2 + B_2H_6 \rightarrow 2\ Et_3B$; **f.** $3\ Me_2Hg + 2\ Al \rightarrow Me_6Al_2 + 3\ Hg$; **g.** $3\ PhCH_2Li + SnCl_4 \rightarrow (PhCH_2)_3SnCl$; **h.** $HgCl_2 + 2\ MeMgBr \rightarrow Me_2Hg + MgBr_2 + MgCl_2$.

35. **a.** Me_3Tl (Lewis acid) + $AsMe_3$ (Lewis base) $\rightarrow Me_3Tl{:}AsMe_3$; **b.** $SnMe_4$ is neither Lewis acid nor Lewis base; no reaction; **c.** $Me_3Tl + SeMe_2$ (Lewis base) $\rightarrow Me_3Tl{:}SeMe_2$; **d.** Me_6Al_2 (contains very strong base, Me^-) $+ 6\ H_2O \rightarrow 2\ Al(OH)_3 + 6\ CH_4$; **e.** Me_6Al_2 (Lewis acid) + $2\ Me_2O$ (Lewis base) $\rightarrow 2\ Me_3Al{:}OMe_2$; **f.** Me_3As (Lewis base) + $MeI \rightarrow [Me_4As]^+I^-$; **g.** MeLi (Lewis base) + Me_4Pb (neither) $\rightarrow$ no reaction; **h.** $MeLi + Me_3In$ (Lewis acid) $\rightarrow Li^+[InMe_4]^-$; **i.** Me_5As (weak Lewis acid) + Me_3Ga (Lewis acid) $\rightarrow$ no reaction; but possibly due to the relatively high coordination number of As it would donate a methyl group, acting as a Lewis base: $Me_5As + Me_3Ga \rightarrow [AsMe_4][GaMe_4]$; **j.** 2 MeLi (contains soft Lewis base Me^-) + $HgCl_2$ (contains borderline Lewis base Cl^-) $\rightarrow HgMe_2 + 2\ LiCl$.

39. **a.** $Me_3In{:}PH_3 \xrightarrow{\Delta} InP(s) + 3\ CH_4(g)$; **b.** $(Me_2)Cd{:}SH_2 \xrightarrow{\Delta} CdS(s) + 2\ CH_4(g)$; **c.** $Me_3B{:}PH_3 \xrightarrow{\Delta} BP(s) + 3\ CH_4(g)$; **d.** Perhaps $MeCu{:}IH \xrightarrow{\Delta} CuI(s) + CH_4(g)$.

43. **a.** Si_3Me_8, steric inhibition of oxygen attack on Si; **b.** $AsPh_3$, greater steric inhibition of O_2 attack, less volatile, (and phenyl more electron withdrawing); **c.** $GeMe_4$, no unshared electrons on central atom to be attacked; **d.** $Cr(CH_2CMe_3)_4$, no hydrogen substituent on second carbon atom; **e.** C_2Me_6, since phenyl derivative has excessive steric interference of bulky phenyl groups; **f.** $PbMe_4$, steric inhibition of oxygen attack on Pb; **g.** CH_4, maximum coordination number reached so attack by O_2 difficult; **h.** Si_4Me_{10}, steric inhibition of oxygen attack on Si.

Chapter 16

1. Associative.

6. SF_6 is more likely to react by a dissociative mechanism, since S^{6+} is smaller and has reached its maximum total coordination number and is therefore unlikely to associate another F^-. SeF_6 is more likely to react by an associative mechanism. SeF_6 is more likely to react rapidly: Dissociation of F^- from very strongly acidic S^{6+} or Se^{6+} is likely to have a high activation energy.

9. These fifth-period +3 charged ions are almost certainly strong-field, hence they would have octahedral site stabilization energies reaching a maximum at the d^6 ion, $[Rh(H_2O)_6]^{3+}$, which should show the slowest rate of dissociative exchange.

13. Study the rate as a function of the variation of the ligand Y^-; if the rate depends substantially on the identity of Y^-, an associative mechanism is indicated. Alternatively, measure the entropy of activation by studying the temperature dependence of the rate; if this is strongly negative, an associative mechanism is indicated. (The volume of activation could also be studied.)

19. **a.** Softer ligands are more nucleophilic toward Pt, so the logical sequence is $F^- < 2.2$; Cl^- 3.04; Br^- 4.18; I^- 5.46. **b.** Sulfur in oxidized form (in a sulfoxide) is least soft, so it is least nucleophilic: PhMeSO 2.90. The remaining ligands are both soft, but PhS^- is a stronger base, so it would likely be more nucleophilic: Ph_2S 3.22; PhS^- 7.17.

21. (a) If the Na and liquid NH_3 react to give solvated electrons (Section 6.6) that persist to react with $CrCl_3$, they would reduce it to give the labile d^4 ion Cr^{2+}, which would react far more rapidly with NH_3 than the inert d^3 Cr^{3+} ion (Section 8.8). (b) The Na and liquid NH_3 may react to give the amide ion NH_2^-, which is very strongly basic (even more so than OH^-), so would readily deprotonate a coordinated NH_3 group in $[Cr(NH_3)_5Cl]^{2+}$ to give a rate-enhancing coordinated NH_2^- group.

25. **a.** $PtCl_4^{2-} + NH_3 \rightarrow [PtCl_3(NH_3)]^-$, then $+ CN^- \rightarrow$ *cis*-$[PtCl_2(CN)(NH_3)]^-$, since Cl^- is the better trans-director; **b.** $PtCl_4^{2-} + CN^- \rightarrow [PtCl_3(CN)]^-$, then $+ NH_3 \rightarrow$ *trans*-$[PtCl_2(CN)(NH_3)]^-$, since CN^- is the better trans-director.

30. **a.** $[Co(NH_2CH_2CH_2NH_2)_2I_2]^+ + [Cr^{II}(H_2O)_6]^{2+} \rightarrow$
$[(NH_2CH_2CH_2NH_2)_2Co^{III}(-\mu I-)_2Cr^{II}(OH_2)_4]^{3+} + 2\ H_2O$;
$[(NH_2CH_2CH_2NH_2)_2Co^{III}(-\mu I-)_2Cr^{II}(OH_2)_4]^{3+} \rightarrow$
$[(NH_2CH_2CH_2NH_2)_2Co^{II}(-\mu I-)_2Cr^{III}(OH_2)_4]^{3+}$;
$[(NH_2CH_2CH_2NH_2)_2Co^{II}(-\mu I-)_2Cr^{III}(OH_2)_4]^{3+} + 2\ H_2O \rightarrow$
$[(NH_2CH_2CH_2NH_2)_2Co^{II}(OH_2)_2]^{2+} +$ *cis*-$[I_2Cr(OH_2)_4]^+$;
labile $[(NH_2CH_2CH_2NH_2)_2Co^{II}(OH_2)_2]^{2+} + 4\ H^+ \rightarrow Co^{2+}(aq) + 2\ NH_3CH_2CH_2NH_3{}^{2+}$.
b. If this reaction is fast, then an outer-sphere mechanism also involving ligand exchange can be ruled out, since both the iodine-bearing reactant and the iodine-bearing products are inert in substitution reactions. Radioactive iodide ions added to the solution would not be incorporated in the product by this mechanism. The bridged intermediate, having reducing iodine atoms and (somewhat) oxidizing metal atoms, might have a characteristic intense, fleeting color due to charge-transfer absorptions; recall from Chapter 14 that CrI_3 is black. The same reaction, if tried with *trans*-$[Co(NH_2CH_2CH_2NH_2)_2I_2)]^+$, should give neither *cis*- nor *trans*-$[I_2Cr(OH_2)_4]^+$.

35. **a. I**: It does not have the second unshared electron pair it would need to function as a bridging ligand. **b. III** proceeds via an inner-sphere mechanism while **I** proceeds via an outer-sphere mechanism. **c.** $[(NC_5H_4C(NH_2){=}O{:}Cr(H_2O)_5]^{3+}$ from **III** (with Cr bonded to the hard O donor atom); $[Cr(H_2O)_6]^{3+}$ from **I. d. II**, with a saturated CH_2 group breaking up the conjugation, should be reduced more slowly than the fully conjugated **III.**

Chapter 17

1. **a.** $N = 6!/3!(6-3)! = 20$. **b.** $M_S = +\frac{3}{2}$ requires all three electrons to have spin $+\frac{1}{2}$, which can only be achieved one way: $1(\uparrow)0(\uparrow){-}1(\uparrow)$; similarly for $M_S = -\frac{3}{2}$, $1(\downarrow)0(\downarrow){-}1(\downarrow)$. **c.** $M_L = 2$ requires two of the three electrons to have $m_l = +1$ while the third has $m_l = 0$, which can be accomplished in two ways: $1(\uparrow\downarrow)0(\uparrow){-}1(\ )$ and $1(\uparrow\downarrow)0(\downarrow){-}1(\ )$. Similarly for $M_L = -2$, $1(\ )0(\uparrow){-}1(\uparrow\downarrow)$ and $1(\ )0(\downarrow){-}1(\uparrow\downarrow)$. **d.** $M_L = 0$ requires one electron to have $m_l = 1$, one to have $m_l = 0$, and one to have $m_l = -1$, which can be achieved eight ways: $1(\uparrow)0(\uparrow){-}1(\uparrow)$, $1(\uparrow)0(\uparrow){-}1(\downarrow)$, $1(\downarrow)0(\uparrow){-}1(\uparrow)$, $1(\uparrow)0(\downarrow){-}1(\uparrow)$, $1(\uparrow)0(\downarrow){-}1(\downarrow)$, $1(\downarrow)0(\uparrow){-}1(\downarrow)$, $1(\downarrow)0(\downarrow){-}1(\uparrow)$, $1(\downarrow)0(\downarrow){-}1(\downarrow)$.

4. **a.** ${}^1S$; **b.** ${}^1S$; **c.** ${}^2D$; **d.** ${}^2F$; **e.** ${}^2\Sigma_u{}^+$; **f.** ${}^2\Pi_u$; **g.** ${}^2\Delta_g$.

8. The microstate $1(\uparrow)0(\ ){-}1(\uparrow)$ has $M_s = +1$, so quantum number $S \geq 1$, so $2S + 1 \geq 3$, so this is part of a spin triplet or higher, and cannot be part of ${}^1F$ or ${}^1S$. The microstate has $M_L = 0$, so $L \geq 0$, which does not rule out any of the term symbols.

14. **a.** ${}^1A_{1g}$; **b.** ${}^2T_{2g}$ and ${}^2E_g$; **c.** (from ${}^3F$) ${}^3T_{1g} + {}^3T_{2g} + {}^3A_{2g}$, (from ${}^3P$) ${}^3T_{1g}$, (from ${}^1G$) ${}^1A_{1g} + {}^1E_g + {}^1T_{1g} + {}^1T_{2g}$, (from ${}^1D$) ${}^1T_{2g} + {}^1E_g$, (from ${}^1S$) ${}^1A_{1g}$.

19. **a.** One, $\nu_1({}^5E_g \rightarrow {}^5T_{2g})$; **b.** None; **c.** One, $\nu_1({}^5T_{2g} \rightarrow {}^5E_g)$; **d.** Three, $\nu_1({}^4T_1g \rightarrow {}^4T_{2g})$, $\nu_2({}^4T_{1g} \rightarrow {}^4T_{1g})$, $\nu_3({}^4T_{1g} \rightarrow {}^4A_{2g})$ (at some values of Δ_o/B', the identifications of ν_2 and ν_3 would be reversed).

20. **a.** Five transitions (from${}^3T_{1g}$ ground state to each of five triplet excited states), but all are close in energy to each other, so no resolvable bands are expected, only one composite band. **b.** Five transitions. One resolvable band and two composite bands likely. **c.** Five transitions; at least the two lowest energy bands should be resolvable. **d.** Three transitions. One resolvable and one composite band.

25. **a.** $\Delta_o = 14{,}900\ cm^{-1}$; $15B' = 22{,}700 + 34{,}400 - 3(14{,}900) = 12{,}400\ cm^{-1}$, so $B' = 827\ cm^{-1}$;
b. $\Delta_o = 17{,}500\ cm^{-1}$; $15B' = 38{,}100 + 23{,}900 - 3(17{,}500) = 9500\ cm^{-1}$, so $B' = 633\ cm^{-1}$;
c. $\Delta_o = 21{,}800\ cm^{-1}$; $15B' = 46{,}500 + 28{,}500 - 3(21{,}800) = 9600\ cm^{-1}$, so $B' = 640\ cm^{-1}$.

28. **a.** $F^- < C_2O_4{}^{2-} < NH_2CH_2CH_2NH_2$; **b.** $F^- > NH_2CH_2CH_2NH_2 > C_2O_4{}^{2-}$; **c.** $\beta = B'/B = B'/918 = 0.90$ for F^-, 0.69 for $C_2O_4{}^{2-}$, and 0.70 for $NH_2CH_2CH_2NH_2$.

33. **a.** The ground ${}^2F$ state has $L = 3$ and $S = \frac{1}{2}$, so $J_{max} = L + S = \frac{7}{2}$; this is the ground J value for the heavier f^{13} Yb^{3+} ion. $J_{min} = L - S = \frac{5}{2}$; this is the ground J value for the f^1 Ce^{3+} ion. **b.** The ground ${}^3H$ state has $L = 5$ and $S = 1$. The ground J value for $Tm^{3+} = 5 + 1 = 6$; the ground J value for $Pr^{3+} = 5 - 1 = 4$; there is also an intermediate J value of 5. **c.** The ground ${}^4I$ state has $L = 6$ and $S = \frac{3}{2}$. The ground J value for $Er^{3+} = 6 + \frac{3}{2} = \frac{15}{2}$; the ground J value for $Nd^{3+} = 6 - \frac{3}{2} = \frac{9}{2}$; there are also intermediate J values of $\frac{11}{2}$ and $\frac{13}{2}$.

36. Several possible answers suggest themselves. (1) Since Y^{3+} and Nd^{3+} are larger than Mg^{2+}, they would not fit in the center of the porphyrin ring, and therefore might not keep it as rigid and free from large-amplitude vibrations. (2) Especially in the case of lilac Nd^{3+} with its multiple electronic states, the HOMO would probably no longer be the porphyrin π MO, and the LUMO would probably no longer be the π^*; the energy of an excited electron on the metal ion might not be sufficient for transferring electrons ultimately to CO_2. (3) These ions fluoresce or phosphoresce readily, which is undesirable. (4) These ions, with their multiple electronic states, could greatly and unfavorably alter the rates of intersystem crossing in the π system of the porphyrin ring.

Index

Page entries followed by the letter "t" indicate location of tables; see also the List of Tables on pages xii–xiv.

A

Absorption spectroscopy, *see also* Spectrochemical series of ligands; Color
  catenated aryls and alkyls, 825–826, 829
  complex ions, 365–369, 384–386
  molecular orbital theory, 470–471
Abundance, elements, 284–287, 285t
Acceptors, 99
Acetic acid, synthesis, 577
Acetylide, 257
Acid–base predominance diagram
  anions, 73–76
  cations, 60–64
  definition, 60
  oxo anions, 121–123
  reactivity prediction, 77–78
Acidic hydrides, 74–76
Acidic oxides, 674–678
  insoluble, 675–676
  soluble, 674–675
Acidity, *see* Cation; pK_a
Acid mine drainage, 128
Acid rain, 128, 679, 681, 683
Activated complex, 843
Activation energy, 254, 845
Activity coefficient, 101
Activity series, 935–937 (Experiment 6)
  thermochemical analysis, 323–325
  redox analysis, 266–270
Adducts, 99
Adenosine triphosphate (ATP), 713
Agostic interaction, 556
Alkane activation, 556
Alkene
  hydroformylation, 574–576
  hydrogenation with Wilkinson's catalyst, 576
  insertion reaction, 570, 633
  metathesis, 578–579
  Ziegler–Natta polymerization, 579–580
Alkoxides, 756–761
Alkylating agents, 759, 822
Alkylidene complex, 561–562
Alkyls and aryls, 811t
  catenated compounds, 823–828
  electron-deficient compounds, 813–814, 820–821
  electron-precise compounds, 814–815, 821
  electron-rich compounds, 815–816, 821–822
  hydride similarities, 810–811
  multiply bonded and divalent compounds of heavier *p*-block elements, 828–829
  nomenclature, 398–402, 810
  oligomeric or polymeric covalent compounds, 812
  reactions and uses, 820–823
  synthesis
    halide reaction with element, 817
    metal substitution, 819
    metalation, 817–818
    organometallic compound reaction with metal halides, 818–819
    unsaturated organic compound reactions, 818
Alkyne, bridging ligands, 556
Allotropes, 600–606, 609–610
Alloys, 616–617
Allred–Rochow electronegativities, 37–38, 37t, 340
β-Alumina, 690
Aluminosilicates
  clays, 705–708
  feldspars, 705
  framework aluminosilicates, 708–711
Alums, 688
Ambidentate ligands, 192
Amides, 762
Ammonia, 179, 271–2, 804, 810, 914
Ammonium ion, 101, 256, 798
Ammonium ion, tetraalkyl, 101, 152, 162, 177
Amphoteric compounds, 124, 178–179, 676
Angular part, electronic wave function, 10–13
Anion
  acid–base predominance diagrams, 73–74
  basicity, 70–71
  charge and radius effects on basicity, 71
  hydration and hydrolysis, 68–73
  nomenclature, 79
  oxo anion, *see* Oxo anion
  partially protonated, 73–74
Antibiotics, cation migration in cells and, 160, 217
Antibonding, 372, 460–461, 486
Antiferromagnet, 617, 619, 867, 908
Aprotic solvent
  amphoteric, 178–179
  molten ionic salts, 179
  nonpolar, weakly solvating, 176–178
  polar, strongly solvating, 178
Aqua regia, 270–271
Arsenic, toxicity, 224
Aryls, *see* Alkyls and Aryls
Asbestos, 700
Associative mechanism, 843, 847–848, 852
Atmophile, 212
Atomic radius, 6; *see also* Ionic radius
  bond strength effects, 92–93, 95
  calculated radii, 35–36
  comparison of types, 30, 31, 31t
  covalent radius, 30–31, 33, 32t
  metallic radius, 30, 32–33, 32t
  periodic table trends, 33–36
  van der Waals radius, 30, 32
Atomization energy, 316–317, 317t, 325–326, 599, 611
ATP, 713
Aurophilic cation, 637
Autotrophic bacteria, 273
Axial position, 107
Azrdes, 192, 257

B

Back-bonding, 484
BAL, 225
Basicity, *see* Anion; Oxo anion; pK_b
Basic oxides, 76–77, 673–674
Bending mode, 444
Beneficiation, 276–277
Bent's rule, 110

Benzene, molecular orbital theory, 497–498, 503–505
Berry pseudorotation, 434
Bioamplification, toxins, 176
Biosphere II, 705
Body-centered cubic lattice, 614
Bohr magneton, 364
Boiling point prediction
  fluorides, 661–663
  halides, 738–740
  oxides, 663–664
Bond angle, 108–110, 194–195
Bond energy, 92–93, 93t, 95–97, 331–336, 332t, 333t, 468–470, 794t
Bond length, 468–470, 470t
Bond order, 470t
  molecular orbital theory, 461–462, 468, 487
  valence bond theory, 95–96, 461
Borabenzene, 829
Boranes, 501–502, 799–801, 805, 808
Borax, 712
Borazine, 762
Borides, 629, 630t
Born exponent, 164–165, 165t, 323
Born–Haber cycle, 315–316, 322
Borohydride, 799
Boron, allotropes, 610–611
Boron monofluoride, molecular orbital theory, 480–481
Boron trifluoride, molecular orbital theory, 501–502
Bridging ligand, 192
British Anti-Lewisite (BAL), 225
Brittleness, 686
Buckminsterfullerene, 605–606, 628–629
  discovery, 4
  infrared spectroscopy, 444–446
  symmetry, 419
Bucky tubes, 609, 765

C

Cadmium, toxicity, 221
Capped octahedral geometry, 106–107
Capped trigonal prismatic geometry, 106–107
Carbamates, dithio, 768
Carbene complex, 561–562
Carbides, 627–629, 765–767
Carbon, allotropes, 603–606, 609–610
Carbon dioxide, 679–680
Carbon disulfide, 767–768
Carbonic acid, 675
Carbon monoxide, 278, 280, 398, 679
  alkylation, 584
  molecular orbital theory, 478–480, 484
Carbon tetrachloride, 67–68
Carbonyl
  apical carbonyls, 558
  bridging, 556–558
  charge, 534
  dinuclear carbonyls, 553–554
  infrared spectroscopy of metal carbonyls, 447–448, 536–538, 556, 558
  insertion reaction, 569–570
  ligand substitution, 537–538
  π-donor ligands, 537–548
  polyene carbonyl derivatives, 543–545
  polynuclear carbonyls, 554
  reductive ligation in synthesis, 536
  toxicity of metal carbonyls, 536
Carboranes, 801–803, 805
Carborundum, 766
Carboxypeptidase A, 218, 386–387
Carbyne, 610, 629
Carbyne complex, 562
Caro's acid, 626
Catalysts, 177, 254, 573–584, 631–635, 689, 708, 818, 905
Catenated alkyls, 823–828
Catenated aryls, 823–828
Catenated halides
  chains, 749
  d-block elements, 750–752
Catenated hydrides, 795, 797–798, 801
Cation
  acidity, 56, 58, 928–929 (Experiment 1)
  categories of acidity, 62
    charge and radius effects, 65–66
    electronegativity effects, 66–67
    rationale for classification, 61
  hydration, 55–56
  hydrolysis constants, 58–60
  ionic radius, *see* Ionic radius
  nomenclature, 79
  nonacidic cations, 62, 101
  oxo cation, 68, 79
  water exchange rates, 389–390
Cement, 704–705, 712
Ceramics, 686
CFSE, *see* Crystal field stabilization energy
Chain, metallic, 635–636
Chalcogende ion, 79
Chalcophile, 212
Character tables, 438–448, 439t, 940t–948t
Charge-transfer absorption, 265–266, 270
Chelate effect, 197
Chelating agents, medicine, 225–227
Chelating ligand, 192–195
Chemical Abstracts, searching, 401–402
Chemical vapor deposition (CVD), 623
Chemiluminescence, 886, 898
Chemisorption, 632
Chevrel phases, 771–772
Chiral molecules, 431–432
Chlorophyll, 195–196, 395, 905–906
Chromium ion, hydrolysis, 58–61
Cisplatin, 221, 224, 391
Clay, 703–708
Close-packed anions, 685
Close-packed lattice, metals, 612, 614
Clusters, 715–716
  halides, 752–753
  metallic
    electron counts, 638–640, 640t
    giant clusters, 643
    molecular orbital theory, 510–513
    types, 635–638
    Wade–Mingos–Lauher rules, 638–640, 642
Cobalamin, *see* Vitamin B_{12}
Color
  highest halides, 265–266
  hydrated metal ions, 357–358, 358t, 369
  thermochromism, 368, 826
Complex ion, definition, 99
Concerted reaction, orbital symmetry rules, 498–501
Conductivity, *see* Electrical conductivity
Cone angle, 538
Constructive interference, 8
Cooperative binding, 397
Cooper pairs, 620
Coordinate covalent bond, 99
Coordination complexes, definition, 99
  nomenclature of, 398–401
Coordination number, 100, 102–104, 107, 168–170, 660
Coprecipitation, 687–688, 696–697, 760
Core electron, 21
Core of earth, 272, 615, 703
Core orbital, 24–25
Corrosion, 300–301
Coulomb's law, 164
Counterion, 99
Covalent bonding, 89, 775–778
Covalent radius, 30–31, 33
Cracking, 710
Crown ethers, 177, 218, 227
Cryptands, 177
Crystal field effects
  atomic radii, 377–378, 384
  Pauling electronegativity, 378–379
  thermodynamic consequences, 374–376
Crystal field splitting, 361, 890–891, 898, 900
  absorption spectra, 384–386

cubic coordination geometry, 382–383
distorted octahedral geometry, 380
d-orbital energy levels, 382–383
periodic trends, 367–368
spectroscopy, 365–367, 380–381, 386–387
square planar geometry, 381, 391
tetrahedral complexes, 383
Crystal field stabilization energy (CFSE), 360, 362–363, 374–375, 379, 383–384, 388, 390
Crystal field theory, introduction, 357–361
Crystallization, 688
Cubic close-packing, 612, 685
Curie's law, 364
Curie temperature, 617, 691
CVD, 623
Cyanate, 192
Cyclobutadiene, molecular orbital theory, 494–497
Cyclometalation reaction, 567
Cyclopentadienide ion, ligand properties of, 436–438, 542–543
Cyclopentaphosphide ion, 627
Cyclophosphazenes, 762–763
Cytochrome P-450, 911–913

D

d-Block complex
synthesis, 391–5
substitution reactions
aqueous solution, 392
metal ions in labile oxidation states, 393
nonaqueous solvents, 392–393
weakly bound or volatile ligands, 393–394
d-Block elements, 6–7, 22–23, 96, 249, 357–359, 368, 615, 750–752
d-Block organometallic chemistry
bridging ligands
alkynes, 556
carbonyls, 556–558
catalysts, 581, 583–584
catalytic cycles, 573–580
dinuclear and polynuclear organometallics, 548–556
18-electron rule, 532–534
mononuclear metal carbonyls, 534–539
multibonding ligands, 559–562
π-donor ligands, 539–548
reaction types
analysis of, 570–572
elimination reaction, 569
insertion reactions, 569–570
Lewis acid association or dissociation, 565
Lewis acid–base attack on coordinated ligands, 565–566
Lewis base association or dissociation, 563–565
oxidative addition, 567–568
reductive elimination, 567–568
de Broglie relationship, 8, 10
Defect structure, 716
Degeneracy, 441, 443, 445–446, 882, 884
Deliquescence, 76
δ Bond, 91, 96–97, 467–468, 750–751
Desferrioxamine, 225, 228
Desiccant, 173
Destructive interference, 8
Diagonal relationships of elements, 701
Diamagnetism, 363
Diamond, 605, 610
Diarsenes, 829
Diborane, 487, 799, 805, 808, 809
Differentiation, elements, 212–213
Diffraction, 8
Digermene, 828
Dimethyl sulfoxide (DMSO), 180, 816
Dinosaurs, extinction of, 272
Diodes, 623, 771
Diphosphate ion, 712–713
Diphosphenes, 829
Disilene, 828
Displacement reaction, 269
Disproportionation reaction, 262, 292, 326
Dissociative chemisorption, 632
Dissociative mechanism, 843, 847–848, 851
Distannene, 828
Disulfate ion, 714
Divanadate ion, 714
DNA, 215, 221, 759, 866
Dodecahedral geometry, 106–107
Donors, 99
d Orbital, 13, 15, 359–360, 382–383, 467
Double bond, 95
Drago equation, 341
covalent bond cleavage enthalpy, 343–345
hard–soft acid–base principle comparison, 345
parameters, 341–342, 342t, 344t
Drierite, 386
Dry box, 258–259

E

EDTA, 225
Effective nuclear charge, 18, 20–24, 23t, 330
Effective pH, 179–180
Effective principal quantum number, 19
Eigen–Wilkins mechanism, 843
18-Electron rule, 532–534, 540, 638, 815
Einstein's theory of relativity, 39
Electrical conductivity, 619–620, 689
Electrochemical cell, 243
Electromotive force (emf), 244–245
Electron affinity, 321, 321t, 340, 483
Electron configuration, 6, 879, 887
Electron diffraction, 9
Electron spin resonance (ESR), 435, 826, 839
Electronegativity, 6, 36–38
Allred–Rochow electronegativities, 37–38, 340
cation acidity effects, 66–67
Mulliken electronegativity, 340, 483
oxo anion basicity effects, 119
Pauling electronegativity, *see* Pauling electronegativity
spectroscopic electronegativity, 340
Electronic chemical potential, 340
Electronic spectroscopy, *see* Absorption spectroscopy
Electronic state, 881–885
diatomic molecules, 885–887
octahedral complexes, 887–889
Electrophile, 99
Electrostatic structure breaker, 161–162
Electrostatic structure maker, 157, 160
Ellingham diagram, 278, 280
emf, *see* Electromotive force
Enantiomer, 431–432
Endohedral metallofullerene, 628
Energy level diagram, 361, 383t, 462, 466, 473, 477–478, 482, 488–489, 494, 533
Enthalpy change
allotrope bonding, 601
Drago equation, 341, 343
Hess's law, 315, 324
polar covalent bonding, 205, 334–336
precipitation reactions, 154–157, 156t, 162–163
Enthalpy of formation, metal halides, 322
hydrides, 793t
Entropy of activation, 850
Entropy change
polar covalent bonding, 205
precipitation reactions, 154–157, 160, 162
Enzymes, metal functions, 215, 386, 911, 913, 914–916
Equatorial position, 107
Equilateral triangular geometry, 106

ESR, *see* Electron spin resonance
Esters, inorganic, 758–759
Ethylenediaminetetraacetic acid (EDTA), 225
Ewens–Bassett nomenclature, 399
Excited state, 882–883
Explosion, redox reactions, 254–257
Extrinsic semiconductor, 623

F

f-Block elements, 6–7, 23, 96, 250, 368, 615, 899–901
f Factors, 371t
Feldspar, 705
Fermi level, 619
Ferredoxin, 772–773
Ferrocene, 531–2, 540
Ferroelectrics, 693
Ferromagnets, 617, 867
Fertilizer, 684
Fischer–Tropsch process, 634–635, 679
Flash photolysis, 839
Fluorescence, 898
Fluorides, prediction of physical properties, 661–663
Fluoro anion, 129, 149–154
Fluxional molecules, 433–438, 435t
Fool's gold, 625, 770
f Orbital, 13, 15, 361, 468
Forbidden transition, 368
Formal charge, 115
Formic acid, 632
Formation constants of complexes, 376–7
Framework aluminosilicates, 708–711
Frank–Condon principle, 858
Free energy change, metal oxidation, 324–325
Free radical, 363, 467, 672, 809, 864
Friedel–Crafts acylation, 545
Frontier orbital, *see* Highest occupied molecular orbital; Lowest unoccupied molecular orbital; Singly occupied molecular orbital
Fullerene; *see* Buckminsterfullerene; Endohedral metallofullerene
Fuoss–Eigen equation, 844

G

Generator orbital, 492–497
Geochemical classification of elements, 212t, 212
Geometric isomer, octahedral complexes, 431
Germylenes, 829
Gibbs free energy change
  Ellingham diagram, 278, 280
  precipitation reactions, 154–157
  redox reactions, 244
Glovebag, 258
g Factors, 367t
g Orbital, 15
Graphite, 605, 610, 636
Graphite fluoride, 752
Greenhouse effect, 446, 680
Grignard reagent, 531, 789, 812, 817, 818, 820
Ground electronic state, 881, 885
Group 13 (III) elements, cluster oligomers, 828
Group 14 (IV) elements
  physical properties, 610–611
  polymerization, 627–629
Group 15 (V) elements
  organic polymers, 826–827
  physical properties, 605–606, 609–610
  polymerization, 626–627
Group 16 (VI) elements
  physical properties, 603–604, 664–665
  polymerization, 625–626
Group 17 (VII) elements, polymerization, 624
Group 18 (VIII) elements, physical properties, 601–603
Group theory, 438

H

H_3, molecular orbital theory, 486–488
Haber–Bosch process, 804, 914
Halides
  alkoxides, 756–761
  bridging, 731, 751
  catenated halides
    chains, 749
    d-block elements, 750–752
  cluster halides, 752–753
  colors, 265–266, 266t
  dihalides, 736–737
  heptahalides, 731
  hexahalides, 731
  Lewis acid properties, 744–746
  Lewis base reactions, 744
  monohalides, 736
  ore extraction, 281
  organometallic substitution reactions, 548–550
  oxidation states, 729–730
  oxidizing agent applications, 747
  oxo halides, 754–755
  pentahalides, 731, 733
  physical property trends, 661–663, 663t, 666t, 738–741
  reducing agent applications, 748
  silicone polymers and, 755–756
  structural trends, 730–731
  synthesis, 741–744
  tetrahalides, 733, 735
  trihalides, 735–736
Hapticity, 546–548
Hard acid, 200, 220
Hard base, 200
Hard–soft acid–base (HSAB) principle, 933–5 (Experiment 5)
  applications
    biological functions and toxicology of elements, 220–224
    geochemical classification and differentiation of elements, 212–214
    medicinal chemistry, 225–228
    qualitative analysis scheme for metal ions, 209–211
    solubility estimation for halides and chalcogenides, 207–209
  borderline acids and bases, 198, 203, 220
  classifications, 199t, 199–200
  Drago parameter comparison, 345
  electronegativity effects, 335–336
  equation, 198
  hard acids, 200, 220
  hard bases, 200
  molecular orbital theory, 483–485
  polar covalent bonding, 204–206, 335–336
  redox chemistry, 270–273
  relative hardness and softness, 201–203, 208
  soft acids, 199, 220–221, 224
  soft bases, 199–200, 224
  softness modification, 203–204
  strength of acid or base, effects, 206–207, 209
  substitution effects, 338–339
Heat of atomization, *see* Atomization energy
Hemerythrin, 909–910
Hemocyanin, 909–911
Hemoglobin
  coordination, 396–397
  heme structure, 395, 908
  methemoglobin, 909
  oxidation to hematin, 397
  oxygen binding, 397
  toxicity of soft bases, 398
Hess's law, 315, 324
Heteropolytungstate, 716
Heterotrophic bacteria, 273
Hexagonal close-packing, 612, 685
High-spin complex, 364, 369, 397
High-temperature superconductor, 694–696
Highest occupied molecular orbital (HOMO), 466, 470–471, 475–476, 479–480, 483–485, 498–501, 532, 619, 903
HOMO, *see* Highest occupied molecular orbital
Homopolyatomic ions
  group 14 (IV), 627
  group 15 (V), 626

group 16 (VI), 625–626
group 17 (VII), 624
HSAB principle, *see* Hard–soft acid–base principle
Hume–Rothery compounds, 616
Hund's rules, 885–886, 901
Hybridization, 91, 472, 505, 622
Hydrated radius, 159t
Hydration energy, 56, 57t
Hydration enthalpy
anions, 69, 69t
cations, 56, 57t
Hydration number, 159
Hydrazine, 798, 806, 810
Hydrides, 552, 556
acidic, 74–76
addition reactions, 809
classification, 790–792, 790t, 794–795
infrared spectroscopy, 794
Lewis acids, 807–808
Lewis bases, 808
metallic hydrides, 791–792
molecular hydrides, 791
electron-deficient hydrides, 794, 799–802
electron-precise hydrides, 794–796
electron-rich hydrides, 794, 796–799
nomenclature, 79, 792
nuclear magnetic resonance, 790
polymeric covalent hydrides, 791
reducing agents, 809–810
saline hydrides, 791
stability, 792–793, 793t, 795
synthesis
acid–base condensation reactions, 805–806
anion protonation, 804–805
deprotonation of hydride, 806
direct combination of elements, 804
halide displacement by hydride, 805
pyrolysis, 805
β-Hydride elimination reaction, 570, 821
Hydride gap, 792
Hydrido anion, 799
Hydroboration, 818
Hydrogen
periodic table placement, 789–790
physical properties, 603
Hydrogen bond, 794t, 796, 798
Hydrogen fluoride
molecular orbital theory, 477–478
toxicity, 807
Hydrogen peroxide, 798, 806
Hydrolysis, acid production by metal ions, 58
Hydrometallurgy, 277
Hydrophobic structure maker, 162
Hydrosilyation, 818
Hydrothermal vents, 213
Hydroxylamine, 798
Hyperaccumulator plants, 227–228
Hypervalent molecule, 97, 491, 508, 816, 852
Hypophosphorous acid, 299

I
Ice, structure of, 161
Icebergs in liquid water, 161
Identity operation, 421, 424
Imides, 762
Inductive effects, 338
Inert-atmosphere drybox, 258, 260
Inert pair effect, 40
Infrared spectroscopy
fluxional molecules and, 435
hydrides, 794
ligand complexation analysis, 446–448
metal carbonyls, 447–448, 536–538, 537t, 556, 558
molecular symmetry and absorptions, 443–446
signal origin, 444–445
Inner sphere, 157
Inner-sphere electron transfer reactions
assignment, 863–864
bridging ligands, 861–862
rate-limiting step, 862–863
two-electron transfer processes, 868–870
Inorganic chemistry, overview, 3–5
Intercalation compound, 598, 636
Interchange mechanism, 843
Interhalogens, 730
Intermediates, 841, 843
Intermetallic compounds, 616
Intersystem crossing, 898
Intervalence charge transfer, 866–867
Intimate mechanism, 846–850
Intrinsic semiconductor, 620
Inversion operation, 422–423
Ion channels, 217–218
Ionic bonding, 164–167, 775–778
Ionic charge, 6
Ionic equation, 174
Ionic radius, 30, 33–34, Table (inside back cover)
anion basicity effects, 71
cation acidity effects, 65–66
crystal field effects, 377–378, 384
geometry effects, 384–385
hydrated, 159t
in complexes as a function of geometry, 384, 385t
radius ratio, 168–170, 660, 664
Ionization energy, 483
calculation, 330
correlation with element activity, 323–324
first ionization energy, 316, 318t
periodic table trends
horizontal trends, 328–329
isoelectronic trends, 331
vertical trends, 329
second ionization energy, 316, 319t
successive energies, 327–328
third ionization energy, 320t
Ion pair, 33, 164–165
Iron
deficiency in Antarctic ocean, 128
toxicity, 225
Irreducible representation, 464, 471, 485, 885
point groups, 438, 440–443
Irving–Williams series, 376, 381, 395
Isoelectronic ions, 33
Isoelectronic fragment, 560
Isolobal fragment, 560–561, 626, 638, 802, 886
Isomorphous substitution, 687–689, 691–693, 692t, 701, 705

J
Jahn–Teller distortion, 380–381, 385–386, 730

K
Kaolinite, 708
Kapustinskii equation, 171, 173
Keggin phases, 638
Keggin structure, 716
Kinetic energy, particles, 10
Kinetic–molecular theory of gases, 596–597
Koopman's theorem, 476, 483
Kroll process, 280

L
LAB approach, *see* Lewis acid–base approach
Laser, 899
Lanthanide contraction, 36, 38
Latimer equation, 69, 172
Lattice energy, 163–167, 171, 322
Lattice type, 167–170, 169t
Layer structures, 605, 662, 700, 705, 730, 770
Lead, toxicity, 221
Leaving groups, 851–852
LEDs, 771
Lewis acid–base (LAB) approach, organometallic synthesis, 540–541, 544, 548–551, 559, 657–658
Lewis acid–base concept, 98–101, 178–179, 206, 271, 929–931 (Experiment 2)
Lewis structures, drawing, 101–105, 132–133
Lewis symbols of atoms, 25–26
LFSE, *see* Ligand field stabilization energy

Ligand
- chelate effect, 197
- macrocyclic effect, 197
- spectrochemical series of ligands, 369–374
- types
  - bridging, 192
  - chelating, 192–195
  - macrocyclic, 195–196
  - monodentate, 192
  - π donor, 197

Ligand field stabilization energy (LFSE), 360, 374
Ligand field theory, 358, 387–391
Ligand substitution reactions
- distinguishing mechanism types, 848–850
- entering groups and nucleophilicity, 851
- intimate mechanisms, 846–850
- leaving groups, 851–852
- metal–atom effects, 853
- rate law, 840–844
- spectator ligand effects, 853–857
- types of stoichiometric mechanisms, 842–843

Light-emitting diodes (LEDs), 771
Linear geometry, 106
Lithiocarbons, 637, 812
Lithophile, 212
Lithopone, 771
Low-spin complex, 364, 369, 397
Lowest unoccupied molecular orbital (LUMO), 466–467, 470–471, 475, 478, 480, 483–485, 498–501, 532–533, 619, 903
Luminescence, 898
LUMO, *see* Lowest unoccupied molecular orbital
Lux–Flood acid–base reactions, 677–678

M

Macrocyclic effect, 197
Macrocyclic ligand, 195–196
Madelung constant, 166, 169t, 171, 173
Magnetic moment, 364
Magnetic quantum number, 12–13
Magnetic susceptibility, 364
Magnetism, 617, 619, 867–868
Mannich condensation, 545
Marcus equations, 860
Maximum coordination number, 67, 103–104
Maximum total coordination number, 104
Medicinal chemistry, 225–228
Melting point prediction
- fluorides, 661–663
- halides, 738–740
- oxides, 663–664

Mercuration, 818
Mercurides, 629
Mercury
- polymers, 631
- toxicity, 221, 273

Metal carbonyl, *see* Carbonyl
Metal crystal
- close-packed lattices, 612, 614
- heterogeneous catalysts, 631–635
- magnetic properties, 618t, 619–620
- molecular orbital theory, 513–515
- structures, 613t, 612–614
- physical properties, 614–615

Metalation, 817–819
Metal–ligand charge transfer, 903
Metallacyclopropane, 546
Metallic bonding, 611, 775–778, 513–515
Metallic radius, 30, 32–33
Metallocarbohedranes, 766
Metallocene, 532–533, 540, 542–543, 545, 580
Metallocenium ion, 543
Metalloenzyme, 215, 218, 386–387
Metallothionein, 227
Metal–organic chemical vapor deposition (MOCVD), 760–761, 820
Metasilicate, 698
Metastable species, 262–263
Methane, molecular orbital theory, 505–506
Methane monooxygenase (MMO), 913
Methyl compounds, *see* Alkyls
Microstate, electron configuration, 879–884, 888–889
Millon's base, 764
Minerals, sources for elements, 213, 214t
Mirror plane, 421–422, 425
Mixed-metal oxides, 687
MMO, 913
MOCVD, *see* Metal–organic chemical vapor deposition
Molecular orbital theory
- cluster compounds, 510–513
- cyclic molecules, 492–498
- D_{nh} complexes, 501–505
- frontier orbitals, 466–467
- heteroatomic diatomic molecules, 476–483
- homonuclear diatomic molecules, 461–476
- linear molecules, 485–491
- metals, 513–515
- molecules with high order point groups, 505–509
- overview, 358, 459–461
- photoelectron spectroscopy, 475–476
- semiconductors, 620–623
- valence bond theory comparison and translation, 459–460, 474–475

Molecular point group
- assignment, 427–430
- character tables, 438–441
- definition, 425–426
- irreducible representation, 438, 440–443
- types, 426–427

Momentum, 8
Mond process, 852
Monodentate ligand, 192
Monomeric substances, 595
Montmorillonite, 708, 775
Mössbauer spectroscopy, 435, 839
Mulliken electronegativity, 340, 483
Muscovite, 705–706
Myoglobin, 397

N

Nanotubes, 609, 765
Natural waters, elements in, 127t, 125–128, 221–224, 222–223t, 296–297
Néel temperature, 619
Nephelauxetic parameter, 895–897
Nephelauxetic series of ligands, 895–897
Nernst equation, 264, 288, 290
Nerve impulse transmission, 217
Net ionic equation
- interpretation, 174–175
- writing, 173–174

NFE, 638
n-fold improper axis, 423–424
n-fold proper axis, 421, 424
Nitric acid, 680–681
- synthesis, 810
- toxicity, 398

Nitric oxide, biological functions, 4, 398
Nitrides, 762–765
Nitrido anion, 764
Nitrogen, biochemical activation, 913–916
Nitrogenase, 773, 914–916
Nitrogen dioxide, 680–681
Nitrosonium ion, 550–551
NMR, *see* Nuclear magnetic resonance
Nodal spheres, planes and surfaces 13–17
Node, 9–10
Nomenclature, *see* class of compound involved
Nonacidic cations, 62, 101, *see also* Cation categories; Ammonium, tetraalkyl
Nonaqueous solvents, 175–180
Nonpolar covalent bond, 599
Nonstoichiometric oxides, 689
N-type semiconductor, 623

Nuclear magnetic resonance (NMR), 386
  active nuclei, 688–689
  bridging carbonyls, 559
  enantiomer detection, 433
  fluxional molecules and, 433–438
  hydrides, 552, 790
Nuclear quadrupole resonance (NQR), 339, 435t, 438
Nucleophile, 99, 851
Nucleophilic substitution reaction, *see* Ligand substitution reaction
Nucleus, stability in superheavy elements, 42–44
Number of framework electrons (NFE), 638

O

Octahedral geometry, 106–107
Octahedral hole, 685
Octahedral site stabilization energy (OSSE), 388, 390
Octet rule, 67
Oligomeric structure, 596
Orbital, 7, 12t
  antibonding, 460–461, 486
  cross sections, 15–18
  degeneracy, 441, 443
  electron energies, 18–19
  hybrid orbitals, 91, 472, 505, 622
  order of energies, 24–25
  overlap, 91–97, 460–461, 464
  relativistic effects, 39–41
  types, 13, 15
  valence orbital potential energy, 462–463, 471
Orbital angular momentum quantum number, 12
Ore extraction
  abundance of elements, 284–287
  aluminum, 280–281
  beneficiation, 276–277
  copper, 282–283
  halides, 281
  hydrometallurgy, 277
  mineral sources of elements, 213–214
  oxidation, 277–279
  pollution control, 283
  reduction, 277–279
  refinement, 281–283
  roasting, 277
  titanium, 280
Organometallic compound nomenclature
  Chemical abstracts, searching, 401–402
  Ewens–Bassett nomenclature, 399
  ligand naming, 398
  metal naming, 399
  numbering prefixes, 400
  order of naming, 399
  parentheses, 399
Organophosphates, 759
Orgel diagram, 891
Origins of life, 708, 774–775
OSSE, *see* Octahedral site stabilization energy
Outer sphere, 158–159
Outer-sphere electron transfer reactions
  assignment, 863–864
  cross reactions, 860–861
  direct versus indirect reactions, 859
  long-distance electron transport, 865–866
  self-exchange reactions, 858–859
Overpotential, 254
Oxidation half-reaction, 243, 251
Oxidation numbers, Table (inside front cover)
  calculation, 29–30, 134–136
  effects on electronegativity, 337–338
  periodic trends, 28–29
Oxidation–reduction, *see* Redox
Oxides, *see also specific oxides*
  acidic oxides, 674–678, 674t, 676t
  applications, 679–684
  basic oxides, 673–674, 676t
  isomorphous substitution, 687–689
  Lux–Flood acid–base reactions, 677–678
  metal oxides, 685–687
  mixed-metal oxides, 687–691
  nonstoichiometric oxides, 689
  periodic trends in structure and physical state
    dielement heptoxides, 667–668
    dielement pentoxides, 669, 671
    dielement trioxides, 672–673
    element dioxides, 669, 671–672
    element tetroxides, 667–678
    element trioxides, 671
    monoxides, 673
  physical property prediction, 663–664, 668t–669t
Oxidizing agent, 244, 298–299, 747
Oxo acid
  acidity of, 124–125
  applications, 682–684
  physical properties, 681–682
Oxo anion
  basicity, 117t, 931–932 (Experiment 3)
    charge effects, 117–118
    classification, 116–117
    electronegativity effects, 119
    number of oxo groups, effects, 118–119
    prediction, 117, 119–120
  fifth and sixth periods, 114–115
  formal charge, 115–116
  formulas, 111–116, 113t, 131
  hydroxo anions, 114, 120–121, 123–124
  Lewis structures, 132–133
  nomenclature, 129–131
  oxidation number assignment, 134–136
  penultimate total coordination number, 111
  resonance structures, 115
  solubility rules, 149–154
  synthesis, 297–300
Oxo cation, 68, 79
Oxo group, 117–118
Oxo halides, 754–755
Oxygen
  allotropes, 603
  biochemical transport and activation, 397, 906, 908–913
Ozone, 603, 680–681

P

Pairing energy, 362, 362t
Palladation, 818
Paramagnetism, 357, 363–364, 459
Passivation, 276, 300
Pauling electronegativity, Table (inside front cover)
  bond energy effects, 331, 334–335
  crystal field effects, 378–379
  differences versus averages in bonding classification, 775, 777
  electronegative metals, 268–269
  electropositive metals, 268
  hardness and softness of acids and bases, 204–205, 335
  nonmetals, 273
  oxidation numbers, effects of changing, 337–338
  periodic trends, 36–38, 331
  reactivity trend analysis, 335–336
  substitution effects, 338–339
p-Block elements, 6–7, 26, 93, 105, 247–248, 616
  multiply bonded and divalent alkyls and aryls, 828–829
  polymerization, 630–631
Peierls distortion, 777
Pentagonal bipyramidal geometry, 106–107
Pentagonal planar geometry, 109–110
Perchloric acid, 684
Periodic table
  acidity–basicity trends, 71–73
  atomic and ionic radii trends, 33–36, inside back cover
  bond strength trends, 92–93, 96
  crystal field splitting trends, 367–368
  density trends, 599
  diagonal relationships, 701

Periodic table (*continued*)
  electronegativity trends, 36–38, 331, inside front cover
  halides, physical property trends, 738–741
  hydride gap, 792
  intimate mechanism type trends, 847–848
  ionization energy trends
    horizontal trends, 328–329
    isoelectronic trends, 331
    vertical trends, 329
  long form, 7t
  octet rule, 67
  organization, 5–7
  oxidation number trends, 28–29, inside front cover
  oxidation state trends of nonmetals
    negative oxidation states, 275–276
    positive oxidation states, 273–275
  physical property trends
    fluorides, 661–663
    oxides, 663–673
  spectrochemical series of ligands, 370–372
  unpaired electron trends, 599–600
  valence electron configurations, 26
Perovskites, 691, 693–694
Peroxide, 626
Persulfuric acid, 626
Phase-transfer catalyst, 177
Phosgene, 68, 755
Phospha-alkynes, 829
Phosphabenzene, 829
Phosphazenes, cyclopoly, 762–763
Phosphinine, 829
Phosphonium ion, 798
Phosphorescence, 898–901
Phosphoric acid, 684
Phosphorus, allotropes, 605
Phosphorus pentafluoride, molecular orbital theory, 502–503
Phosphorus pentoxide, 684
Photochemical reactions
  delayed reactions, 903–905
  overview, 901–902
  prompt reactions, 902
Photoelectron spectroscopy, 475–476, 483
Photosensitive compounds, 265
Photosynthesis, 905–906
Photosystem I, 905–906
Photosystem II, 905–906
π-Acceptor ligands, 372–373, 398
π Bond, 91, 95–96
π-Bonding system, 488–490, 508–509
π-Donor ligands, 197, 372, 537–548
Piezoelectric effect, 693
Pig iron, 282
pK_a, 59t, 216t
  cation classification by acidity, 61–62, 64–67
  definition, 58–59
  precipitation dependence of metal hydroxides–oxides, 61
pK_b, anion classification by basicity, 70–71, 70t, 117t, 193t
Pnicogenide ion, 79
Polar covalent bond, 204–206, 331, 334–336, 445
Polarity, molecular symmetry, 430, 432
Polymeric hydroxy cations, 711
Polymeric structures, 596
Polymerization isomer, 658, 660
Polymorph, 658
Polyoxoanions, 711–712
Polyoxometallates, 697
Polysilicates
  chain oligosilicates, 697–698
  chain polysilicates, 699–700
  cyclic oligosilicates, 698
  glass, 703–704
  layer polymeric silicates, 700, 702
  soil chemistry, 703
Polythiazyl, 765
Polythionates, 626
Population inversion, 899
p Orbital, 13, 15, 89, 91–92, 95
Pore model, 160, 217
Porosils, 711
Porphyrins, 395, 908–909
Portland cement, 712
Postvalence orbitals, 24, 97
Pourbaix diagram
  concentration effects, 292, 296
  d-block elements, 294
  definition, 245, 287–288
  f-block elements, 295
  interpretation, 290–291
  iron, 288, 290–292, 296
  p-block elements, 293
  plutonium, 291–292
  s-block elements, 293
  window of stability of water, 290
Precipitation, thermodynamics, 154–157, 156t, 162–163
Predominance diagrams, *see* Acid-base predominance diagrams; Redox predominance diagrams
Prepared-for bonding electron configuration, 599
Primary hydration sphere, 157
Principal quantum number, 12, 15, 18–19, 27
Protic solvents, 179–180
Prussian blue, 867–868
Pseudohalides, 192–193
Pseudochalcogen, 561
P-Type semiconductor, 623
Pulse radiolysis, 839
Pyrites, *see* Fool's gold

Q

Quadruple bonds, *see* δ Bond
Qualitative analysis scheme, metal ions, 209–211, 210t
Quantum mechanics, 7–10
Quantum numbers, 9, 12–13, 15
Quasicrystal, 617

R

Racah parameters, 890–891, 895–896
Radial probability function, 15
Radius ratio, 168–170, 660, 664
Raman spectroscopy, 448, 539
Rate constant, 841
Rate law, 839
  ligand substitution reactions, 840–844
  order of reaction, 841
Reaction mechanism
  determination, 839–840
  fluxional exchange, 433–438
  homogeneous organometallic catalysis, 573–580
  heterogeneous catalysis by metals, 631–635
  ligand substitution reactions
    distinguishing mechanism types, 848–850
    exchange of coordinated water, 387–391
    entering groups and nucleophilicity, 851
    intimate mechanisms, 846–850
    leaving groups, 851–852
    metal–atom effects, 387–391, 853
    rate law, 840–844
    spectator ligand effects, 853–857
    types of stoichiometric mechanisms, 842–843
  redox reaction mechanisms
    assignment, 863–864
    inner-sphere electron transfer reactions, 857, 861–870
    intervalence charge-transfer, 866–868
    one-electron process, 858, 861–868
    outer-sphere electron transfer reactions, 857–861
    two-electron process, 858, 868–870
Reaction profile, 845
Redox half-reactions, 243–244, 251, 254
Redox predominance diagram
  concentration effects, 264
  construction, 261
  d-block elements, 249
  electronegative nonmetals, 273–275
  f-block elements, 250

iron, 246–247
metastable species, 262–264
overview, 245–246
p-block elements, 247–248
photochemistry, 904
prediction of reactivity, 251
s-block elements, 247–248
silver, 261–263
water at pH 0, 253–254
Redox radical (RR) approach, organometallic synthesis, 540–541, 544, 548–551, 559, 657
Redox reaction balancing, 251–253
Redox reaction mechanisms
assignment, 863–864
inner-sphere electron transfer reactions, 857, 861–870
intervalence charge-transfer, 866–868
one-electron process, 858, 861–868
outer-sphere electron transfer reactions, 857–861
two-electron process, 858, 868–870
Redox reaction safety, 254–257
Reducing agent
definition, 244
halides, 748
safety in handling, 257–258, 260
Reduction half-reaction, 243, 246–247, 251
Reduction potential, hydrated metal ions, 267
Refractory, 685
Relativistic effects, 39, 68, 96, 275, 323, 468, 534, 614, 637, 814, 915
Repulsion energy, 164
Resonance structures, 115
Rotational mode, 444
Robin and Day classification, 866–868
RR approach, *see* Redox radical approach
Rubredoxin, 772
Ruby, 898–899

S

Safety, 254–260, 304
Salts
nomenclature, 79
formula writing, 80
Sandwich molecules, 503, 540
s-Block elements, 6–7, 247–248
Scandide contraction, 35
Schlenkware, 260
Schrödinger wave equation, 10–11, 460
Scrubber, 683
Secondary bonding, 603, 623, 624, 730, 736, 770
Secondary hydration sphere, 158–159, 390, 707, 843, 850
Secondary quantum number, 12
Selenium, toxicity, 224
Semiconductors, 610, 620–624, 689–690, 694, 766, 771, 776–777, 820, 867
Sepulchrate ligand, 395
Sheet, metallic, 635–636
Shielding, 18, 21
Siderophile, 212, 272, 277
Siderophore, 228
σ Bond, 91–92, 548–556
σ-Donor ligands, 372
Silanes, 807
Silica, 672, 686
Silicalites, 711
Silicate
aluminosilicates, *see* Aluminosilicates
minerals, 697
polysilicates, *see* Polysilicates
Silicon nitride, 765
Silicone polymers, 755–756, 817
Siloxanes, 755
Silylenes, 829
Singlet oxygen, 885–886
Singly occupied molecular orbital (SOMO), 467
Sintering, 686
Slag, 281–282
Slaking, 77
Slater's rules, 19–20, 38
Smog, 682–683
Smokestack scrubber, 683
Sodalite cages, 708
Sodamide, 762
Sodium hypochlorite, 299–300
Soft acid, 199, 220–221, 224
Soft base, 199–200, 224
Soil polysilicates, 703
Sol–gel synthesis, 760
Solubility rules, 932–3 (Experiment 4)
applications, 149
hard–soft acid–base principle for soft bases, 207–209
rule I, 150
rule II, 151, 173
solvents, 149–150
tendency III, 151–152, 162, 172–173, 175
tendency IV, 153–154
Solvents, *see* Aprotic solvents; Protic solvents
Solvolysis, 179
SOMO, 467
s Orbital, 13
Speciation, 191
Spectator ion, 174
Spectator ligand effects
base-assisted hydrolyses, 854–855
electron-donor effects, 854
steric effects, 854
trans effect, 855–856
Spectrochemical series of ligands, 938–939 (Experiment 7)
definition, 369–370
donor and acceptor ligands, 372–373
f factors, 370–371, 371t, 373
g factors, 367t, 370, 373
periodic trends, 370–372, 897
Spectroscopic electronegativity, 340
s–p Hybridization, 471–474, 478–479
Spin-allowed transition, 889
Spin angular quantum number, 13
Spin crossover, 365
Spinels, 691
Spin-forbidden transition, 889
Spin multiplicity, 881
Spin–orbit couping, 40, 901
Spin pairing, 362
Spin quantum number, 881
Square antiprismatic geometry, 106–107, 110
Square planar geometry, 109, 381, 391, 534, 536
SQUIDs, 696
Standard reduction potential, 193t, 245, 247t–250t, 267t, 298t
Steady-state approximation, 841
Steel, 282
Stimulated emission, 899
Stoichiometric mechanism, 842–844
Stokes radius, 159
Stretching mode, 444, 537, 908
Strong-field complex, 362
Structural units, 657–658, 661
Sublimation, 598, 739
Sulfamic acid, 762
Sulfides
biological metal sulfide clusters, 772–775
Chevrel phases, 771–772
disulfides, 770
monosulfides, 769–770
overview, 767–768
physical and chemical properties, 770–771
trisulfides, 770
Sulfur
allotropes, 603–604
desulfurization of fossil fuels, 634
poisoning of catalysts, 634
Sulfur dioxide, 570, 682–684
Sulfur hexafluoride, 68, 506–508
Sulfuric acid, 682–683
Sulfur trioxide, 658
Superacid, 180
Superconducting quantum interference devices (SQUIDs), 696
Superconductor, 619–620, 694–696
Superfluid, 602
Superoxide ion, 625–626
Swamp gas, 296
Symmetry element, 419, 421–424, 421t
Symmetry operation, 419, 421–424, 421t

Symmetry point group, *see* Molecular point group
Synthesis gas, 679

T
Taggant, explosives, 900–901
Tanabe–Sugano diagram, 891–894
TASO, 501–511
Temperature-jump, 839
Template effect, 215, 218, 395, 852
Term symbols, 881–885
Terminal atom symmetry adapted orbital (TASO), 501–511
Tetraethyllead, 817, 822
Tetrahedral geometry, 106–107
Tetrahedral hole, 685, 690
Tetrasulfur tetranitride, 763–764
Thallation, 818
Thermite reaction, 280
Thermochemical radius, 171–172, 172t
Thermochromism, 368, 826
Thio anions, 771
Thiocyanate, 192–193
Thiosulfate, 192–193
Titanium dioxide, 686, 689
Total coordination number, 104–105, 107, 111
Total orbital angular momentum, 881–882, 885, 901
Trans effect, 855–856
Translational mode, 444
Transuranium elements, 42t
  stability, 42–44
  synthesis, 44
Tremolite, 725
Tricapped trigonal prismatic geometry, 106–107
Trigonal bipyramidal geometry, 106–107, 109, 533
Triphosphate ion, 713
Triple bond, 95
Tube furnace, 743
Tungsten bronzes, 694

U
Uranium enrichment, 739
Uranocene, 540
Usanovich base, 271

V
Vacant coordination site, 657–658, 660–661
Vacuum line, 258
Valence bond theory, 358, 459–460, 474
Valence electron, 21–22
  anions, 27–28
  anomalous configurations, 27
  cations, 28
  configuration, 24, 27–28, 464, 466
  number of an atom, 25–26, 100
  superheavy elements, 41–42
Valence electron count (VEC), 534, 540–541, 559, 638, 640, 642–643
Valence orbital potential energies, 463t
Valence shell electron-pair repulsion model (VSEPR), 105, 108, 111, 195, 411–412, 427, 535, 730, 796, 815
van Arkel diagram, 775–777
van der Waals force, 596–598, 603
van der Waals radius, 30, 32, 49, 598
VEC, *see* Valence electron count
Vibrational mode, 444, 447–448
Virtual orbital, 24
Visible spectroscopy—*see* Absorption spectroscopy
Vitamin B_{12}, 215, 218, 581–584
Volume of activation, 850
VSEPR, *see* Valence shell electron-pair repulsion model

W
Wade–Mingos–Lauher rules, 638–640, 642
Water
  acid–base strength effects, 100
  element in and pollution, *see* Natural waters
  photochemical spliting of, 905
  polywater, 796
  redox predominance diagram at pH 0, 253–254
  softening of, 710
  structure
    ice, 161
    liquid, 161–162
Water gas, 679
Water gas shift reaction, 679–680
Wave functions, 9–13, 15–18
Weak-field complex, 362
Weathering, 703
Wilkinson's catalyst, 576
Wittig reagent, 822

X
Xanthates, 768
Xanthine oxidase, 584
XeF_2, molecular orbital theory, 490–491
X-ray diffraction, 8–9, 435

Z
Zeolites, 708–711
Ziegler–Natta polymerization, 579–580, 818
Zintl anions, 629
Zintl phases, 616

Table C

Shannon–Prewitt Crystal Ionic Radii[a]

Anions			Cations																	
15 (V)	16 (VI)	17 (VII)	1	2	3	4	5	6	7	8	9	10	11	12	13 (III)	14 (IV)	15 (V)	16 (VI)	17 (VII)	18 (VIII)
			H																	He
			Li +1 90	Be +2 59											B +3 41	C +4 30	N +5 27	O	F	Ne
N −3 132	O −2 126	F −1 119	Na +1 116	Mg +2 86											Al +3 67	Si +4 54	P +5 52	S +6 43	Cl +7 41	Ar
	S −2 170	Cl −1 167	K +1 152	Ca +2 114	Sc +3 88	Ti +2 100 +3 81 +4 74	V +2 93 +3 78 +4 72 +5 68	Cr +2 94 +3 75 +4 69 +6 58	Mn +2 97 +3 78 +4 67 +7 60	Fe +2 92 +3 78 +4 72	Co +2 88 +3 75 +4 67	Ni +2 83 +3 74	Cu +1 91 +2 87	Zn +2 88	Ga +3 76	Ge +2 87 +4 67	As +3 72 +5 60	Se +4 64 +6 56	Br +7 53	Kr
	Se −2 184	Br −1 182	Rb +1 166	Sr +2 132	Y +3 104	Zr +4 86	Nb +3 86 +4 82 +5 78	Mo +4 79 +5 75 +6 73	Tc +4 78 +5 74 +7 70	Ru +3 82 +4 76 +5 70	Rh +3 80 +4 74 +5 69	Pd +2 100 +4 75	Ag +1 129 +2 108 +3 89	Cd +2 109	In +3 94	Sn +4 83	Sb +3 90 +5 74	Te +4 111 +6 70	I +5 109 +7 69	Xe
	Te −2 207	I −1 206	Cs +1 181	Ba +2 149	Lu +3 100	Hf +4 85	Ta +3 86 +4 82 +5 78	W +4 80 +5 76 +6 74	Re +4 77 +5 72 +7 67	Os +4 77 +5 71 +7 66	Ir +3 82 +4 76 +5 71	Pt +2 94 +4 76	Au +1 151 +3 99	Hg +2 116	Tl +1 164 +3 102	Pb +2 133 +4 91	Bi +3 117 +5 90	Po +4 108 +6 81	At +7 67	Rn
			Fr +1 194	Ra	Lr	Rf	Db	Sg	Bh	Hs	Mt									
					3*F*	4*F*	5*F*	6*F*	7*F*	8*F*	9*F*	10*F*	11*F*	12*F*	13*F*	14*F*	15*F*	16*F*		
				6	La +3 117	Ce +3 115 +4 101	Pr +3 113 +4 99	Nd +3 112	Pm +3 111	Sm +3 110	Eu +2 131 +3 109	Gd +3 108	Tb +3 106 +4 90	Dy +2 121 +3 105	Ho +3 104	Er +3 103	Tm +2 117 +3 102	Yb +2 116 +3 101		
				7	Ac +3 126	Th +4 108	Pa +3 118 +4 104 +5 92	U +3 116 +4 103 +5 90 +6 87	Np +3 114 +4 101 +5 89 +6 86 +7 85	Pu +3 114 +4 100 +5 88 +6 85	Am +3 111 +4 99	Cm +3 111 +4 99	Bk +3 110 +4 97	Cf +3 109 +4 96	Es	Fm	Md	No +2 124		

SOURCES: Data are for cations with coordination number of six and are from R. D. Shannon and C. T. Prewitt, *Acta Cryst.*, **B25**, 925 (1969), and R. D. Shannon, *Acta Cryst.*, **A32**, 751 (1976).

[a] Units are picometers (pm).